University of St. Francis Library
S0-ACI-326

McGRAW-HILL ENCYCLOPEDIA OF

Science & Technology

13

PAR-PLAN

McGRAW-HILL ENCYCLOPEDIA OF Science & Technology

13

PAR-PLAN

9th Edition

An international reference work in twenty volumes including an index

McGraw-Hill
New York Chicago San Francisco Lisbon London Madrid Mexico City
Milan New Delhi San Juan Seoul Singapore Sydney Toronto

Library of Congress Cataloging-in-Publication Data

McGraw-Hill encyclopedia of science & technology—9th ed.
p. cm.
Includes bibliographical references and index.
ISBN 0-07-913665-6 (alk. paper)
1. Science—Encyclopedias. 2. Technology—Encyclopedias.
I. Title: Encyclopedia of science & technology.
II. Title: McGraw-Hill encyclopedia of science and technology.
III. Title: Encyclopedia of science and technology.
Q121.M3 2002
503—dc21 2001057910

ISBN 0-07-913665-6 (set)

McGraw-Hill

A Division of The **McGraw-Hill** *Companies*

1 2 3 4 5 6 7 8 9 0 DOW/DOW 0 8 7 6 5 4 3 2

This book set was printed on acid-free paper.

It was set in Garamond Book and Neue Helvetica Black Condensed by TechBooks, Fairfax, Virginia. The art was prepared by TechBooks. The book was printed and bound by R. R. Donnelley & Sons Company, The Lakeside Press.

Organization of the Encyclopedia

The *McGraw-Hill Encyclopedia of Science & Technology* presents pertinent information in every field of modern science and technology. The 7100 articles are arranged alphabetically in the first 19 volumes (the 20th volume contains the indexes and ancillary materials). The range of article titles included in each volume is indicated on the spine and front cover (for example, volume 1 contains articles with titles starting with "Aar" up to "Ano"). Thus the reader may quickly locate an article by its title.

Broad survey articles are available for each of the disciplines covered; even readers with little prior knowledge of that discipline will find the basic concepts covered in these articles. From the survey article, the reader may proceed to more specialized articles using the cross-referencing system. These cross references are set in small capitals for emphasis and are inserted at the relevant points in the text. This edition contains some 62,000 cross references. For example, in a survey article such as **Digital computer**, the reader is directed to numerous other articles such as **Computer peripheral devices, Computer storage technology, Microprocessor**, and **Programming languages**. The references may lead to subjects that have not occurred to the reader. The article **Solvent** has such diverse cross references as **Coordination chemistry, Halogenated hydrocarbon, Industrial health and safety**, and **Water pollution**. The cross references not only lead to articles of greater specialization but also help illuminate the context of the article and the broader connections among topics.

The pattern of proceeding from the general to the specific has been employed not only in the plan of the *Encyclopedia* but within the body of the articles. Each article begins with a definition of the subject, followed by sufficient background material to give a frame of reference and permit the reader to move into the detailed text of the article. Within the text are centered heads and two levels of sideheads that outline the article; they are intended to enhance understanding, and can guide the user that prefers to read selectively the sections of a long article.

Alphabetization of article titles is by word, not by letter, with a comma providing a stop in occasional inverted article titles (so that subject matter can be grouped). Two examples of sequence are:

Air
Air-cushion vehicle
Air mass
Air-traffic control
Aircraft fuel

Earth, age of
Earth, heat flow in
Earth crust
Earth tides
Earthquake

Numerous illustrations, both line drawings and images, contribute to the utility, clarity, and interest of the text. Each illustration (as well as each table) is called out in boldface at its first mention in the text. This emphasis enables the reader to move from an illustration to the point in the text where the illustration is often discussed in detail.

To meet the needs of the *Encyclopedia*'s broad readership, measurements are given in dual systems of units: The U.S. Customary System is used throughout the text along with equivalent measurements in the International System of Units. In particular cases, such as references to measurements in some illustrations or tables, conversion factors may be given for simplicity.

The contributor's full name appears at the end of an article section or an entire article. Each author is identified in an alphabetical Contributors list in volume 20, which cites the university, laboratory, business, or other organization with which the author is affiliated and the titles of the articles written by that contributor.

Most of the articles contain bibliographies citing useful sources. The bibliographies are placed at the ends of articles or occasionally at the ends of major sections in long articles. For additional bibliographies, the reader should refer to related articles as indicated by cross references.

Thus, the alphabetical arrangement of article titles, the text headings, the cross references, and the bibliographies permit the reader to pursue a particular interest by simply taking a volume from the shelf. However, the reader can also find information in the *Encyclopedia* by using the Analytical Index and the Topical Index in volume 20. The Analytical Index contains each important term, concept, and person—170,000 entries in all—mentioned throughout the 19 text volumes. It guides the reader to the volume numbers and page numbers concerned with a specific point. The reader wishing to consult everything in the *Encyclopedia* on a particular aspect of a subject will find that the Analytical Index is the best approach. A broader survey may be made through the Topical Index, which groups all article titles of the *Encyclopedia* under 87 general headings. For example, under "Geophysics" 60 articles are listed, and under "Organic chemistry," 193. The Topical Index thus enables the reader to identify quickly all articles in the *Encyclopedia* in a particular subject area.

The Study Guides in volume 20 provide highly structured outlines of 15 major scientific disciplines and relate groups of *Encyclopedia* articles to each outline heading. By following a guide, the reader is led through pertinent *Encyclopedia* articles in a sequence that provides an overall grasp of the discipline.

A useful feature is the section "Scientific Notation" in volume 20. It clarifies usage of symbols, abbreviations, and nomenclature, and is especially valuable in making conversions between the International System, U.S. Customary, and metric measurements.

McGRAW-HILL ENCYCLOPEDIA OF

Science & Technology

13

PAR-PLAN

P

Parablastoidea

A small class of primitive blastozoan echinoderms containing three genera found in the early Middle Ordovician in eastern Canada; northeastern, eastern, south-central, and western United States; and near Leningrad, Russia. Parablastoids have a bud-shaped theca or body with well-developed pentameral symmetry. Thecal plates include basals, radials, and sometimes other small plates in the lower theca; large, distinctive, triangular-to-parabolic deltoids between the ambulacra in the upper theca; and several sets of small plates surrounding and covering the central mouth on the summit (see **illus.**). Parablastoids have single or multiple slits through the lower deltoids that are connected by internal folds (cataspires) to pores that open between the single ambulacral plates at the edge of each ambulacrum; apparently they were respiratory organs. Short-to-long biserial brachioles were attached to the edges of the five ambulacra and served as the main food-gathering structures. A stem with one-piece columnals attached the theca to the sea floor, suggesting that parablastoids were attached, medium- to high-level suspension feeders. Although they converged on blastoids in thecal design and way of life, parablastoids had differences in their plating, ambulacra, and respiratory structures to indicate a separate origin and history. That is the justification for assigning parablastoids and blastoids to different classes. *See* CRINOZOA; ECHINODERMATA.

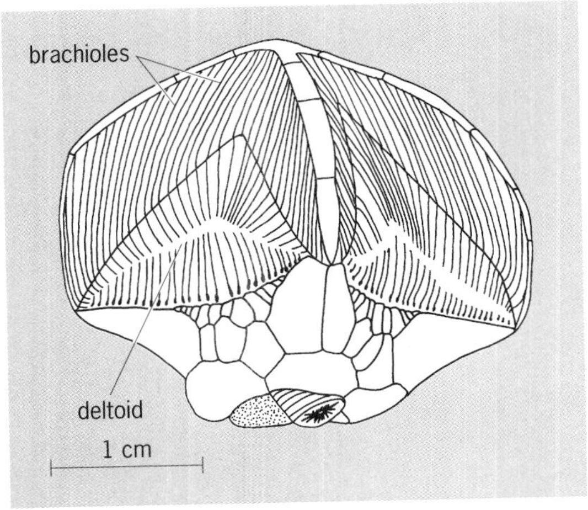

Side view of *Blastoidocrinus carchariaedens* from the Middle Ordovician of New York, showing the thecal plating, large triangular deltoids, and numerous short brachioles attached to the ambulacra. (*After R. C. Moore, ed., Treatise on Invertebrate Paleontology, pt. S, pp. S293–S296, Geological Society of America and University of Kansas Press, 1968*)

James Sprinkle

Bibliography. R. C. Moore (ed.), *Treatise on Invertebrate Paleontology*, pt. S, S293–S296, 1968; J. Sprinkle, *Morphology and Evolution of Blastozoan Echinoderms*, Mus. Compar. Zool. Harv. Univ. Spec. Publ., 1973.

Parabola

A member of the class of curves that are intersections of a plane with a cone of revolution. It is obtained (see **illus.**) when the cutting plane is parallel to an element of the cone. *See* CONIC SECTION.

In analytic geometry the parabola is defined as the locus of points (in a plane) equally distant from a fixed point F (focus) and a fixed line (directrix) not through the point. It is symmetric about the line through F perpendicular to the directrix. To construct a parabola, pin one end of a piece of string to a point F and fasten the other end to one end of a ruler whose length equals that of the string. If the other end of the ruler slides along a line, and the string is kept taut by a pencil, the point of the pencil will trace an arc of a parabola. Hippocrates of

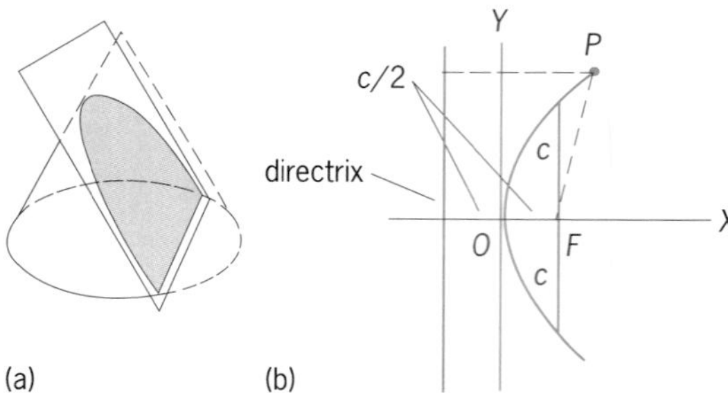

Parabola as (*a*) conic section and (*b*) locus of points.

Chios (about 430 B.C.) showed that one of the three famous problems of antiquity, duplication of the cube, can be solved by use of parabolas. The problem is to construct the edge of a cube whose volume is twice that of a given cube. If c denotes the edge of the given cube, then the desired edge is obtained by considering the two parabolas whose equations in rectangular cartesian coordinates are $x^2 = cy$, $y^2 = 2cx$. They intersect at the origin and a point $P(x_0, y_0)$, with $x_0^3 = 2c^3$.

All parabolas are similar; they differ only in scale. For a discussion of the optical property of parabolas *see* ANALYTIC GEOMETRY.

The curve has numerous other properties of interest in both pure and applied mathematics. (For example, the trajectory of an artillery shell, assumed to be acted upon only by the force of gravity, is a parabola; and the circle that circumscribes the triangle formed by any three tangents of a parabola goes through the focus.) Archimedes found the area bounded by an arc of a parabola and its chord; for example, the area bounded by the parabola $y^2 = 2cx$ and its latus rectum (the chord through F perpendicular to the axis) is $^2/_3c^2$. The length of the parabolic arc that is cut off by the latus rectum is $[\sqrt{2} + \ln(\sqrt{2} + 1)]c$. The volume obtained by revolving this about the axis of the parabola is $\pi c^3/4$, and the surface is $^2/_3(2\sqrt{2} - 1)\pi c^2$. Leonard M. Blumenthal

Parachute

A flexible, lightweight structure, generally intended to retard the passage of an object within or through the atmosphere by materially increasing the resistive surface. A parachute is a decelerator or air-braking device in the general form of an oblate hemisphere. It comprises a canopy and cords, which form the suspension and attachment between canopy and object (see **illus.**). The theory of parachute devices can be traced back to Leonardo da Vinci; practical application dates from late in the eighteenth century. The parachute, first employed for exhibition purposes, was used as a life-saving vehicle starting with the first years of World War I, and by 1916 was adopted for widespread military use.

Characteristics. A parachute canopy is a membrane which relies upon pressure differential across it to maintain its inflated shape. The differential is created by entrapment of an air mass on the inside and movement of the air on the outside. Inasmuch as its foremost purpose is to resist the force propelling any body that is to be decelerated, and because that force is most often gravity, the decelerator itself should be extremely light. Rules for determining stresses in thin-walled pressure vessels apply to the parachute canopy, which, by the nature of its shape and loading, experiences tension forces only. A parachute is usually composed of one basic element—fiber. When fiber is converted to thread, thence to cloth, cord, webbing, and tape, parachute construction becomes a matter of structural assembly. Important characteristics of a parachute include porosity (either through the fabric or numerous vents), strength of materials, aerodynamic behavior, dynamic behavior, weight, and ability to deploy freely.

Application. The classical application of the parachute is as a human-carrying apparatus. It is the only suitably demonstrated device for emergency descent from aircraft. In various forms and arrangements, parachutes have been used to (1) deploy paratroops in assault operations; (2) distribute supplies from aircraft; (3) restrict rate of descent of bombs, mines, and flares; (4) drop land and water vehicles; (5) decelerate airplanes, drones, and missiles during landing; and (6) recover space capsules, warheads,

Parachute descending. (*U. S. Army Soldier Systems Command*)

drones, and weather recorders. Sport parachuting, or skydiving, which usually includes a period of free fall before deployment of the parachute, has gained in popularity. Although applications are limited by available materials (usually nylon, silk, cotton, rayon, or plastic film) and design or fabrication practices, usage expands with the development of new techniques, fabrics, and plastics. *See* AERODYNAMIC FORCE. S. E. Weaver

Bibliography. T. W. Knacke, *Parachute Recovery Systems Design Manual*, 1992; D. Poynter, *The Parachute Manual*, vol. 1, 3d ed., 1984, vol. 2, 4th ed., 1991; D. Poynter, *Parachuting: The Skydivers' Handbook*, 6th ed., 1992.

Paraffin

A term used variously to describe either a waxlike substance or a group of compounds. The former use pertains to the high-boiling residue obtained from certain petroleum crudes. It is recovered by freezing out on a cold drum and is purified by crystallization from methyl ethyl ketone. Paraffin wax is a mixture of 26- to 30-carbon alkane hydrocarbons; it melts at 52–57°C (126–135°F). Microcrystalline wax contains compounds of higher molecular weight and has a melting point as high as 90°C (194°F). The name paraffin is also used to designate a group of hydrocarbons—open-chain compounds of carbon and hydrogen with only single bonds, of the formula C_nH_{2n+2}, where n is any integer. This usage is obsolete. *See* ALKANE; WAX, PETROLEUM. Allen L. Hanson

Parainfluenza virus

A member of the genus *Paramyxovirus* of the family Paramyxoviridae which is associated with a variety of respiratory illnesses. The virus particles range in size from 90 to 200 nanometers, agglutinate red blood cells, and (like the influenza viruses) contain a receptor-destroying enzyme. They differ from the influenza viruses in their large size, their possession of the larger ribonucleoprotein helix characteristic of the paramyxoviruses, their tendency to lyse as well as agglutinate erythrocytes, and their generally poor growth in eggs. In culture these viruses grow well in primary monkey or human epithelial cells; although they produce little cytopathic effect in the culture cells, they are easily recognized by the hemadsorption method. They are relatively unstable upon storage, even at freezing temperatures. Laboratory diagnosis may be made by hemagglutination-inhibition, complement fixation, and neutralization tests of the patient's blood. *See* COMPLEMENT-FIXATION TEST; EMBRYONATED EGG CULTURE; PARAMYXOVIRUS; TISSUE CULTURE.

Four subgroups are known, designated parainfluenza 1, 2, 3, and 4. Types 1, 2, and 3 are distributed throughout the world, but thus far type 4 has been found only in the United States. Parainfluenza 1 and 3 are ubiquitous endemic agents producing infections all through the year. Types 2 and 4 occur more sporadically. With all of the parainfluenza viruses, most primary infections take place early in life. About half of the first infections with parainfluenza 1, about two-thirds of those with parainfluenza 2, and three-fourths of those with parainfluenza 3 produce febrile illnesses. The target organ of type 3 is the lower respiratory tract, with first infections frequently resulting in bronchial pneumonia, bronchiolitis, or bronchitis. Type 1 is the chief cause of croup, but the other types have also been incriminated, to the extent that one-half of all cases of croup can be shown to be caused by parainfluenza viruses. Antibody surveys of normal populations in the United States indicate that type 3 infection generally occurs very early, and most infants have type 3 neutralizing antibody before they are 2 years old. By the time they enter school, a large proportion of children have acquired antibodies to types 1 and 2 also. Almost all adults possess parainfluenza antibodies, but reinfection is common, particularly with type 3. Although preexisting antibody does not prevent reinfection, it appears to reduce the severity of the illness.

Type 1. This group includes the Sendai virus, also known as the hemagglutinating virus of Japan (HVJ) or influenza D, and the hemadsorption virus type 2 (HA-2). The widespread HA-2 virus appears to be one of the chief agents producing croup in children. In adults it produces symptoms like those of the common cold. Immunization with experimental inactivated parainfluenza type 1 vaccine stimulates the appearance of antibody in the blood but not in nasal secretions. Recipients of such a vaccine remain susceptible to reinfection. In contrast, infection stimulates antibody appearance in the nasal secretions and concomitant resistance to reinfection; however, reinfections may occur with time.

Type 2. This group includes the croup-associated (CA) virus, or acute laryngotracheobronchitis virus of children. Simian viruses 5 and 41 appear to be antigenically related to parainfluenza 2.

Type 3. This virus is also known as hemadsorption virus type 1 (HA-1). Most infants have antibody to type 3 by the time they are 2 years old—usually before they have experienced infection with the other parainfluenza types. Upon subsequent infection with types 1 or 2, a "booster" response is seen in type 3 antibodies, indicating a sharing of antigens.

Type 4. The prototype of this virus is the M-25 strain. It has been associated only with malaise in young children or mild illnesses limited to the upper respiratory tract. It has failed to produce any cytopathic effects in cultures and can be recognized only by the hemadsorption technique. *See* ANIMAL VIRUS; INFLUENZA; VIRUS CLASSIFICATION.

Joseph L. Melnick; M. E. Reichmann

Bibliography. R. M. Chanock, M. A. Mufson, and K. M. Johnson (eds.), Comparative biology and ecology of human virus and mycoplasma respiratory pathogens, *Progr. Med. Virol.*, 7:208–272, 1965.

Parainsecta

A class of hexapod (six-legged) arthropods consisting of the orders Collembola and Protura. Class Parainsecta (also known as Ellipura) is one of several classifications of those hexapods, which, while similar to insects, are sufficiently different that most researchers believe they deserve separate class status.

Collembola and Protura have frequently been lumped with the Diplura (Entotrophi) into one class (Entognatha) on the basis of their mouthparts, which are inside the head capsule rather than exposed as in the Insecta. However, some recent morphological and paleontological studies support a closer link between insects and Diplura. In addition, there are a number of features found in Protura and Collembola that are not found in Diplura. These include the developmental process: Both Collembola and Protura have epimorphic development (that is, development of the young is completed in the egg). In the Parainsecta nervous system the abdominal ganglia are reduced to one ganglion or fused to the last thoracic ganglion, whereas Diplura and primitive insects have at least seven abdominal ganglia. The internal cephalic supportive structures of Parainsecta are different from those seen in Diplura. The postantennal organ present in many Collembola may be homologous with the similar organ of Tömösváry of Protura; no such organ occurs in Diplura or insects. The clypolabium and labium (upper and lower lips) are immobile, unlike those of other adult hexapods. Two unique features of the Parainsecta are the presence of a ventral midline or groove and coxal vesicles (fluid-filled sacs, generally associated with leg bases) that are limited to the first abdominal segment. In the Collembola these vesicles are fused into a single ventral tube. The digestive tract lacks anterior or posterior enlargements, which are often present in Diplura and primitive insects. Other distinguishing features of the Parainsecta are the absence of styli and unpaired pretarsal claws.

Both Collembola and Protura contain very small animals, generally less than 4 mm long. Protura are exclusively, and Collembola primarily, soil and litter inhabitants. Genetic evidence concerning relationships is at present confusing, with some studies showing a closer link between Diplura and Collembola than between either and Protura, and others showing a closer link between Collembola and Protura. In addition, differences of interpretation concerning the phylogenetic importance of various morphological features leave the status of the Parainsecta ambiguous. *See* ARTHROPODA; COLLEMBOLA; DIPLURA; INSECT PHYSIOLOGY; INSECTA; PROTURA.

Kenneth Christiansen

Parallax (astronomy)

The apparent angular displacement of a celestial object due to a change in the position of the observer. With a baseline of known length between two observations, the distance to the object can be determined directly.

Geocentric parallax. The rotation of the Earth or the linear separation of two points on its surface can be used to establish distances within the solar system. The parallax determined is scaled to the equatorial radius of the Earth, which is equal to 6378 km (3963 mi). At the mean distance of the Moon, this baseline subtends an angle of 57′02″61, and at the mean distance to the Sun it amounts to 8″794148. This latter distance is defined as the astronomical unit and serves as a measure of distances within the solar system. One astronomical unit is 149,597,870.66 ± 0.02 km (92,955,807.25 ± 0.01 mi) in length, and its high precision results from tracking interplanetary space probes. *See* ASTRONOMICAL UNIT; EARTH ROTATION AND ORBITAL MOTION.

Heliocentric or trigonometric parallax. The astronomical unit is the baseline for the measure of stellar parallaxes or distances and ultimately every other distance in the universe outside the solar system. Observations made from the Earth in its orbit on opposite sides of the Sun are scaled to the astronomical unit, although as much as twice that distance may be available. The stellar parallax is given in units of arc-seconds and is by definition the reciprocal of the distance in parsecs. One parsec is the distance at which one astronomical unit subtends an angle of one second of arc and equals 206,264.8 astronomical units or 3.2616 light-years, one of which is 63,240 astronomical units. *See* LIGHT-YEAR; PARSEC.

Attempts to measure stellar parallax were unsuccessful before 1838, when the parallaxes of three different stars were measured independently. One was the triple star Alpha and Proxima Centauri, whose parallax of about 0″76 is still the largest known. Stellar parallaxes accumulated slowly until about 1900, when long-focus refracting telescopes had become available and the photographic process was introduced into astrometry. *See* ALPHA CENTAURI; ASTRONOMICAL PHOTOGRAPHY.

The third edition of the *Yale Parallax Catalogue*, published with a supplement in 1962, contained 10,194 parallaxes determined for 6399 different stars. The fourth edition of the catalog, published in 1995, contains 15,993 parallaxes of 8112 stars. Observations included in the earlier volume were almost entirely obtained by using long-focus refractors with apertures between 0.5 and 1.0 (20 and 40 in.), and measurements and their reductions were made by hand. With few exceptions, measurements were made only along the direction of right ascension, since most parallactic motion occurs in that direction.

Improvements in observations. Since 1962, a number of improvements have been made which result in a great reduction in parallax error. These include measurement in both coordinates, a large increase in the number of comparison stars constituting the reference frame, and the introduction of computers, which allowed more rigorous solutions to be made. Precision was also improved through a more careful balance between observations made before and after midnight, when the Earth is on alternate sides of its orbit.

The external mean error of a single parallax determination made before 1962 averages about ±0″016. By making use of the improved techniques mentioned above, this error has been reduced by about half. The automatic measuring machines on which most work has been done since that date, along with the fine-grain emulsions which were developed a few years later, cut the error in half again. The charge-coupled device (CCD) has all but replaced the photographic plate, and reduces the error still further. Most parallaxes now being determined have errors between ±0″001 and ±0″003. *See* CHARGE-COUPLED DEVICES.

The size of the error divided by the parallax itself determines the uncertainties in derived parameters of a star such as its intrinsic luminosity. The criterion for a good parallax is entirely arbitrary, but some reasons exist for setting the ratio not far from 0.15; that is, any parallax with an error less than 15% its size is useful for calibrating the physical properties of stars. A reduction of the error by about 10, as has been done since 1962, raises the distance to which a good parallax can be found by the same amount, and raises the volume of space (and hence the number of stellar candidates) by a factor of 1000. The 1962 and 1995 editions of the *Yale Parallax Catalogue* show the superiority of the work done since the earlier date. Only 376, or about 4%, of the parallaxes listed in the earlier edition qualify as good data under this constraint, whereas in the later edition the number is 1223.

Ground-based programs. The principal reason for this threefold increase is the inauguration of the parallax program of the U.S. Naval Observatory, for which the low-coma 1.5-m (60-in.) reflecting telescope was designed and built. Improvements made on other long-focus refractors have also expanded the list of worthy parallaxes. Unlike many of their predecessors, these programs emphasize the types of stars that are likely to lie at small distances from the solar system, and thus have low-percentage errors. *See* ABERRATION (OPTICS).

Hipparcos. In 1989 the European Space Agency launched the first space satellite entirely devoted to astrometry, named *HIPPARCOS* (from *High Precision Parallax Collecting Satellite*, and also honoring the ancient astronomer Hipparchus). In 1993 it completed its mission by obtaining positions, parallaxes, and proper motions for 118,322 stars listed in its final catalog. Its precision averages about ±0″0015 for stars brighter than its completeness limit of about the 8th magnitude, and rises to perhaps twice that amount at its ultimate limiting magnitude of 12. This precision is about equal to that of the best of current ground-based work for the brighter stars and less for faint stars. The advantage of *HIPPARCOS* lies in the much greater numbers of stars, including all the bright ones, with quality parallaxes. Among the common stars, its greatest contribution is in the calibration of the luminosities of the upper-middle main sequence and the giant and subgiant stars. The contributions to stellar knowledge from this satellite are enormous. *See* BINARY STAR; HERTZSPRUNG-RUSSELL DIAGRAM; SATELLITE (ASTRONOMY).

Statistical and secular parallax. Distance estimates are possible based on the assumption that the apparent mean space motion of a large number of stars with similar characteristics is a reflection of the peculiar motion of the Sun, derived from previous observations. Two related methods apply; the first equates the mean drift of the stars to the Sun's motion vector. The second requires measurements of radial velocities as well as proper motions, and assumes that the radial velocity distribution at the solar apex and antapex (the directions toward and away from which the Sun is moving, respectively) matches that in the transverse velocities along the direction normal to that of the assumed solar motion. These methods are referred to as secular and statistical parallax, respectively, although the terms are sometimes confused.

Indirect parallaxes. Several methods for the establishment of distance are frequently and incorrectly labeled parallax. These methods, such as spectroscopic or photometric parallax, have in common the reverse of the procedure typical of all true parallax methods. In any true parallax determination, a direct comparison of the same distance in linear and angular measure leads directly to a distance. By contrast, such methods as spectroscopic and photometric parallax involve a comparison of an object to another of known distance and like luminosity, and scaling the distance as necessary. *See* ASTROMETRY.

Arthur R. Upgren

Bibliography. W. F. van Altena, J. T. Lee, and E. D. Hoffleit, *Catalogue of Trigonometric Stellar Parallaxes*,1995; S. Debarbat et al. (eds.), *Mapping the Sky: Past Heritage and Future Directions*,1988; H. K. Eichhorn and R. J. Leacock (eds.), *Astrometric Techniques*,1994; European Space Agency, *The Hipparcos and Tycho Catalogues*, SP-1200,1997; European Space Agency, *Proceedings of the ESA Symposium: Hipparcos—Venice '97*,1997.

Parallel circuit

An electric circuit in which the elements, branches (elements in series), or components are connected between two points with one of the two ends of each component connected to each point. The **illustration** shows a simple parallel circuit. In more complicated electric networks one or more branches of the network may be made up of combinations of

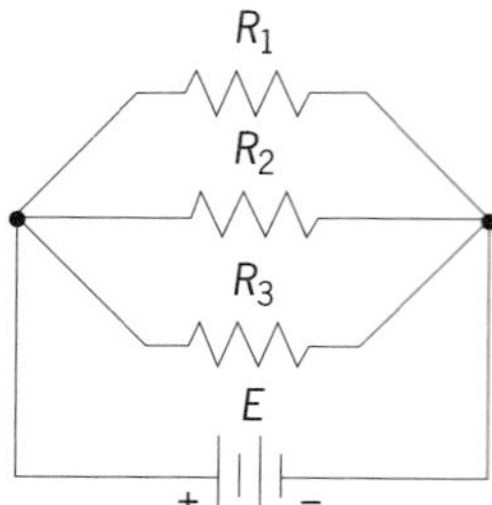

Schematic of a parallel circuit.

series or series-parallel elements. *See* CIRCUIT (ELECTRICITY).

In a parallel circuit the potential difference (voltage) across each component is the same. However, the current through each branch of the parallel circuit may be different. For example, the lights and outlets in a house are connected in parallel so that each load will have the same voltage (120 V), but each load may draw a different current (0.50 A in a 60-W lamp and 10 A in a toaster).

For a discussion of parallel circuits *see* ALTERNATING-CURRENT CIRCUIT THEORY; DIRECT-CURRENT CIRCUIT THEORY. Clarence F. Goodheart

Paramagnetism

A property exhibited by substances which, when placed in a magnetic field, are magnetized parallel to the field to an extent proportional to the field (except at very low temperatures or in extremely large magnetic fields). Paramagnetic materials always have permeabilities greater than 1, but the values are in general not nearly so great as those of ferromagnetic materials. Paramagnetism is of two types, electronic and nuclear.

Paramagnetic substances. The following types of substances are paramagnetic:

1. All atoms and molecules which have an odd number of electrons. According to quantum mechanics, such a system cannot have a total spin equal to zero; therefore, each atom or molecule has a net magnetic moment which arises from the electron spin angular momentum. Examples are organic free radicals and gaseous nitric oxide.

2. All free atoms and ions with unfilled inner electron shells and many of these ions when in solids or in solution. Examples are transition, rare-earth, and actinide elements and many of their salts. This includes ferromagnetic and antiferromagnetic materials above their transition temperatures. For a discussion of these materials.

3. Several miscellaneous compounds including molecular oxygen and organic biradicals.

4. Metals. In this case, the paramagnetism arises from the magnetic moments associated with the spins of the conduction electrons and is called Pauli paramagnetism. *See* ANTIFERROMAGNETISM; FERRIMAGNETISM; FERROMAGNETISM.

Relatively few substances are paramagnetic. Aside from the Pauli paramagnetism found in metals, the most important paramagnetic effects are found in the compounds of the transition and rare-earth elements which have partially filled $3d$ and $4f$ electron shells respectively.

Electronic paramagnetism. This arises in a substance if its atoms or molecules possess a net electronic magnetic moment. The magnetization arises because of the tendency of a magnetic field to orient the electronic magnetic moments parallel to itself. The magnitudes of electronic magnetic moments are of the order of a Bohr magneton, which is equal to 9.27×10^{-21} electromagnetic unit, or emu (erg/gauss). *See* ELECTRON SPIN.

Nuclear paramagnetism. This arises when there is a net magnetic moment due to the magnetic moments of the nuclei in a substance. An example is solid sodium, in which each sodium atom has a nuclear magnetic moment of 2.217 nuclear magnetons. One nuclear magneton is equal to 5.05×10^{-24} emu. Nuclear magnetic moments are about 10^3 times smaller than electron magnetic moments. As a result, nuclear paramagnetism produces effects 10^6 times smaller than electron paramagnetic or diamagnetic effects. Therefore, it is usually impossible to detect nuclear paramagnetism by static methods since it will be masked by electronic effects. (An exception is the case of nuclear paramagnetism arising from the protons in solid hydrogen.) However, paramagnetic effects of nuclei are directly observable in resonance experiments. *See* DIAMAGNETISM; MAGNETIC RESONANCE; NUCLEAR MOMENTS.

Langevin theory. The Langevin theory of paramagnetism treats the paramagnetic substance as a classical (non-quantum-mechanical) collection of permanent magnetic dipoles with no interactions between them. The dipoles are the magnetic moments of the paramagnetic atoms or ions in the substance. The first task of a theory of paramagnetism is to account for the experimentally observed susceptibility (ratio of magnetization to applied field). *See* MAGNETIC SUSCEPTIBILITY.

If an external magnetic field is applied to the paramagnet, each magnetic dipole experiences a torque. Associated with the force which produces this torque is a potential energy given by Eq. (1), where μ is the

$$V = -\mu H \cos\theta \qquad (1)$$

magnetic moment of the dipole, H is the applied magnetic field intensity, and θ is the angle between the dipole and the direction of H. Now, in the absence of thermal agitation each permanent magnetic dipole will become oriented in such a way that this potential energy is minimized, that is, oriented parallel to the magnetic field. With all the dipoles lined up, the magnetization (magnetic moment per unit volume), if there are N dipoles per unit volume, would be given by Eq. (2), where the direction of the magnetization

$$M = N\mu \qquad (2)$$

would be that of the applied field. Note that in this case an arbitrarily small magnetic field causes all the dipoles to line up so that the susceptibility $\chi = M/H$ would be infinite. In the actual case there is thermal agitation which in part offsets the aligning tendency of the magnetic field. The Langevin theory takes this into account and predicts the paramagnetic susceptibility as a function of temperature.

In the presence of thermal agitation, the magnetic dipoles are not all lined up in the direction of the magnetic field, but there is some distribution of angles made with the field. In this case the magnetization is given by Eq. (3), where cos θ is the aver-

$$M = N\mu\,\overline{\cos\theta} \tag{3}$$

age of the cosine of the angle between dipole and field. The average is taken over the distribution of dipoles in thermal equilibrium. According to statistical mechanics, this average is given by Eq. (4), where

$$\overline{\cos\theta} = \frac{\int e^{(-V/kT)} \cos\theta \, d\Omega}{\int e^{(-V/kT)} \, d\Omega} \tag{4}$$

$d\Omega$ is the element of solid angle and $e^{(-V/kT)}$ is the Boltzmann distribution in energy $V = -\mu H \cos\theta$ [Eq. (1)] of a dipole at angle θ with respect to the applied field at absolute temperature T. The integrations may be performed, and the result is $L(a)$, the Langevin function of $a = \mu H/kT$. The result may be combined with Eq. (3) to give Eq. (5). If $a \ll 1$, then $L(a) \cong b\ a/3$ so that Eq. (6) holds. This is a

$$M = N\mu L(a) \tag{5}$$

$$M \cong \frac{N\mu^2 H}{3kT} \tag{6}$$

good approximation except at low temperatures or extremely high fields. The susceptibility is given by Eq. (7). The $1/T$ dependence of the susceptibility is

$$\chi = \frac{M}{H} = \frac{N\mu^2}{3kT} = \frac{C}{T} \tag{7}$$

known as the Curie law. The Curie law was established empirically by P. Curie in 1895 and is obeyed by many gases, liquids, and solids. There are some paramagnetic solids which obey the Curie-Weiss law $\chi = C/(T - \Theta)$ in a certain temperature range. Here Θ is the Curie temperature. The modification often arises because of effective interactions between the dipoles which are neglected in the preceding development. It may also be due to distortion effects. *See* CURIE-WEISS LAW; LANGEVIN FUNCTION.

Experimental data for the paramagnetic susceptibility are often expressed in terms of the effective magnetic moment which must be used for μ in the Curie law [Eq. (7)] in order to give the observed slope of the curve of χ plotted against $1/T$.

Quantum theory. The quantum-mechanical theory of paramagnetism was worked out in detail by J. H. Van Vleck in 1928. This theory is based on the fact that the magnetic moment of the permanent magnetic dipole arises from the total angular momentum of the electrons in the paramagnetic atom, ion, or molecule. Thus an atom with total angular momentum quantum number J has $(2J + 1)$ energy levels in a magnetic field. A collection of such atoms will be distributed among these levels according to a Boltzmann distribution. The magnetization of such a system may be computed by finding the average component of angular momentum parallel to the field. The result is Eq. (8), where g is the spectroscopic

$$M = NgJ\mu_B B_J(a^*) \tag{8}$$

splitting factor (the measure of the energy level splittings of the system), μ_B is the Bohr magneton, $a^* = gJ\mu_B H/kT$, and $B_J(a^*)$ is the Brillouin function of a^* expressed in Eq. (9). The Brillouin function also

$$B_J(a^*) = \frac{2J+1}{2J} \coth \frac{(2J+1)a^*}{2J} - \frac{1}{2J} \coth \frac{a^*}{2J} \tag{9}$$

enters the theory of ferromagnetism. If a^* is much less than unity, which is a good approximation except at very low temperatures or in large fields, then Eq. (10) holds. In this case a Curie law again prevails as in Eq. (11). The effective magneton number

$$B_J(a^*) \cong \frac{g(J+1)\mu_B H}{3kT} \tag{10}$$

$$\chi = \frac{M}{H} = \frac{NJ(J+1)g^2\mu_B^2}{3kT} \tag{11}$$

is defined by $g\sqrt{J(J+1)}$ and is the quantity usually given in experimental results. *See* NONRELATIVISTIC QUANTUM THEORY.

If only the electron spin contributes to the total angular momentum, $J = {}^1/_2$ and $B_{1/2}(a^*) = \tanh(a^*)$ so that, except at low temperatures or high fields, Eq. (12) holds, which agrees with the classical result.

$$\chi = \frac{N\mu_B^2}{3kT} \tag{12}$$

This case is referred to as the spin-only case.

Rare-earth ions. The paramagnetism of rare-earth ions at room temperature is summarized by some representative examples in **Table 1**.

The calculated effective magneton numbers in Table 1 are the theoretical values for isolated ions. The experimental values are derived from Eq. (11), using experimental values of the paramagnetic susceptibility χ. There is good agreement for all rare-earth ions with the exception of europium and samarium. The experimental results of Table 1 refer to the paramagnetic behavior of rare-earth ions in crystals; different salts of the same ion give the same results.

The experimental result is therefore that at room temperature a crystal containing a number of

TABLE 1. Paramagnetism of some trivalent rare-earth ions

Ion	Electron configuration	Effective magneton number: Calculated	Effective magneton number: Experimental
Ce^{3+}	$4f^1 5s^2 p^6$	2.54	2.4
Nd^{3+}	$4f^3 5s^2 p^6$	3.62	3.5
Sm^{3+}	$4f^5 5s^2 p^6$	0.84	1.5
Eu^{3+}	$4f^6 5s^2 p^6$	0.00	3.4
Gd^{3+}	$4f^7 5s^2 p^6$	7.94	8.0
Yb^{3+}	$4f^{13} 5s^2 p^6$	4.54	4.5

TABLE 2. Paramagnetism of iron-group ions

Ion	Electron configuration	Effective magneton number: Calculated with J	Calculated with S only	Experimental
Ti^{3+}, V^{4+}	$3d^1$	1.55	1.73	1.8
V^{3+}	$3d^2$	1.63	2.83	2.8
Cr^{3+}, V^{2+}	$3d^3$	0.77	3.87	3.8
Mn^{3+}, Cr^{2+}	$3d^4$	0.00	4.90	4.9
Fe^{3+}, Mn^{2+}	$3d^5$	5.92	5.92	5.9
Fe^{2+}	$3d^6$	6.70	4.90	5.4
Co^{2+}	$3d^7$	6.54	3.87	4.8
Ni^{2+}	$3d^8$	5.59	2.83	3.2
Cu^{2+}	$3d^9$	3.55	1.73	1.9

trivalent rare-earth ions has the paramagnetic susceptibility of that number of free trivalent ions. The reason that there is little influence of the crystalline electric fields on the magnetic behavior is that the electrons responsible for the magnetic moments are in the $4f$ state and therefore occupy an electronic shell lying well inside the ion, a shell that is shielded from outside influence by the $5s$ and $5p$ electrons. This is in contrast to the behavior of iron-group ions discussed later.

At lower temperatures the influence of the crystalline electric fields on the electrons becomes more important and the behavior of the susceptibility can become quite complex. In this case, the susceptibility depends upon the orientation of the magnetic field with respect to the crystal axes.

The behavior of europium and samarium at room temperature is still explainable on the basis of a theory of free ions if the effect of Van Vleck paramagnetism is included.

Van Vleck paramagnetism. This arises when the energy states of an atom or ion divide into two groups, those within an energy kT of the ground (lowest energy) state and those which are separated from the ground state by an energy greater than kT. Here k is Boltzmann's constant and T is the absolute temperature. The situation is shown in the **illustration**. The low-lying states give rise to a susceptibility which follows a Curie law. If these low-lying states arise from a single value of the total angular momentum J, as in the illustration, then the quantum-mechanical derivation applies [Eq. (11)]. The high-lying states give rise to a small temperature-independent susceptibility, an effect which is known as Van Vleck paramagnetism. In intermediate cases, such as in the trivalent europium and samarium ions, the upper states are only a little more than kT away from the ground state so that the temperature dependence is still more complicated.

Iron-group ions. The paramagnetism of iron-group ions in crystals is summarized in **Table 2**.

Quenching of orbital angular momentum is exhibited in crystal containing ions of the iron group. The last three columns of Table 2 indicate that the orbital angular momentum makes no contribution to the magnetic moment but the iron-group ions behave in crystals as free ions with only the spin S contributing to the magnetic effects. This is evidenced by the fact that the spin-only values of the effective magneton numbers agree well with the experimental results. The orbital angular momentum is quenched because the $3d$ electronic shell, which gives rise to the paramagnetism, is outermost for the iron group; it is therefore exposed to the strong crystalline electric fields arising from neighboring ions. These asymmetric electric fields decouple the orbital angular momentum from the spin angular momentum. This means that the energy levels are no longer specified by the total angular momentum quantrum number J; S alone may determine the levels. More precisely, the $(2L + 1)$ degenerate orbital angular momentum states of orbital angular momentum quantum number L may be split by the crystal fields so that the lowest orbital state is nondegenerate (singlet). Then there is no possibility of orienting the orbital angular momentum by a magnetic field so that only the spin contributes to the magnetic moment. It is often said that the crystal field "locks" the orbital angular momentum so that it cannot be oriented by a

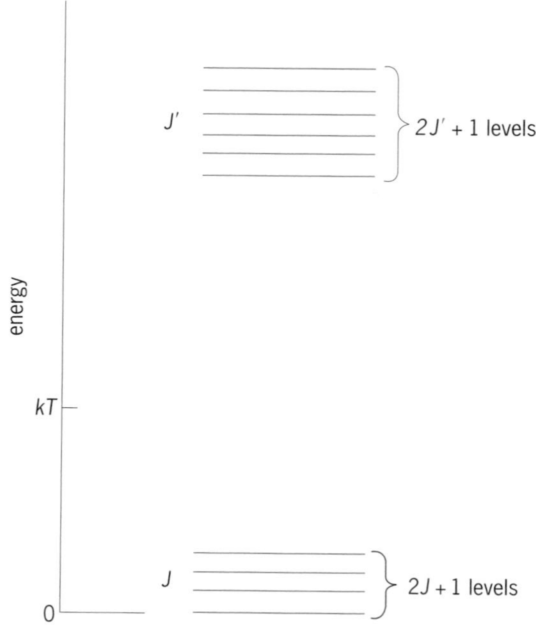

Energy levels in Van Vleck paramagnetism.

magnetic field. Partial quenching occurs when the orbital degeneracy is only partially removed by the crystal field. Partial quenching and anisotropic effects can also be caused by spin-orbit coupling. This influence may be observed in spin-resonance and specific-heat experiments. *See* ADIABATIC DEMAGNETIZATION; MAGNETIC RESONANCE.

Pauli paramagnetism. This is the paramagnetism associated with the conduction electrons of a metal. A metal is usually described in terms of a collection of positive ions with closed shells which are arranged on a crystal lattice plus electrons which are essentially free to move about the crystal. Each electron has an intrinsic spin angular momentum, and these momenta give rise to a paramagnetic magnetic moment. At first sight it would seem to be correct to apply the Langevin formula to this "gas" of electrons, but the experimental facts are that the paramagnetic susceptibility of conduction electrons is about one-hundredth of that predicted by the Langevin formula [Eq. (7)]. Furthermore, the susceptibility is temperature independent rather than varying as $1/T$ (Curie law). The explanation was given by W. Pauli in 1927: The electrons obey the quantum statistics of E. Fermi and P. A. M. Dirac rather than the classical statistics which are used in the derivation of the Langevin formula. This means that a given energy state can be occupied at most by two electrons, and their spin angular momenta must be in opposite directions. As a result the net angular momentum is zero, even on application of a magnetic field. Thus most of the electrons in a metal contribute in sum no magnetic moment. That is to say, an electron's spin angular momentum may not orient parallel to an applied magnetic field because there is already an electron in that energy state with its spin parallel to the field. There are, however, a few electrons which are not "paired off," and the spins of these can be oriented by the field. These electrons contribute to the susceptibility according to a Curie law, but the number of them is proportional to the temperature. The combination of the two temperature dependences leads to a temperature-independent susceptibility, smaller than the prediction of the Langevin formula for N electrons per unit volume because only a fraction of these may contribute. The Pauli susceptibility may be written (for a free electron gas) as Eq. (13), where N

$$\chi = \frac{3N\mu_B^2}{2kT_F} \qquad (13)$$

is the number of electrons per unit volume and kT_F is the Fermi energy characteristic of the metal. The fraction of electrons contributing to the susceptibility at temperature T is of the order T/T_F. For sodium, for example, $kT_F = 3.12$ eV, $T_F = 37{,}000$ K. The Pauli paramagnetism of metals has been observed in spin-resonance experiments. The total susceptibility arises from Pauli paramagnetism and diamagnetic contributions from conduction electrons and ion cores. *See* EXCLUSION PRINCIPLE; FREE-ELECTRON THEORY OF METALS.

Elihu Abrahams; Frederic Keffer

Bibliography. D. Craik, *Magnetism*, 1994; C. Kittel, *Introduction to Solid State Physics*, 7th ed., 1996; A. H. Morrish, *The Physical Principles of Magnetism*, 1965, reprint 1980; J. H. Van Vleck, *The Theory of Electric and Magnetic Susceptibilities*, 1932.

Parameter

An auxiliary variable, functions of which give the coordinates of a curve or surface. The coordinates of a curve are functions of one parameter. A curve in 3-space has parametric equations (1).

$$x = f(t) \qquad y = g(t) \qquad z = h(t) \qquad (1)$$

The coordinates of a surface are functions of two parameters, shown in Eqs. (2). When constant values

$$x = f(u, v) \qquad y = g(u, v) \qquad z = h(u, v) \qquad (2)$$

are assigned to the parameters u or v, these equations represent the parametric curves on the surface. General surface curves are obtained by replacing u and v by functions of a new parameter t.

An arbitrary constant in an equation is also called a parameter. Variations in the values of the parameter generate a system of equations which may represent a family of curves or surfaces. Such families are called one-parameter, two-parameter, and so on, according to the number of independent parameters. *See* PARAMETRIC EQUATION.

Louis Brand

Parametric amplifier

A highly sensitive low-noise amplifier for ultrahigh-frequency and microwave radio signals, utilizing as the active element an inductor or capacitor whose reactance is varied periodically at another microwave or ultrahigh frequency. A varactor diode is most commonly used as the variable reactor. The varactor is a semiconductor *pn* junction diode, and its junction capacitance is varied by the application of a steady signal from a local microwave oscillator, called the pump. Amplification of weak signal waves occurs through a nonlinear modulation or signal-mixing process which produces additional signal waves at other frequencies. This process may provide negative-resistance amplification for the applied signal wave and increased power in one or more of the new frequencies which are generated. *See* VARACTOR.

The parametric amplifier operates through a periodic variation of a circuit parameter, in this case, the capacitance of the varactor. Similar amplification is possible by using a periodically variable inductor which may utilize a saturable magnetic material such as a ferrite. The inductive form of this device has not found application because it requires much larger amounts of pumping power to vary the inductance significantly at microwave frequencies.

Effects. Parametric effects are also found with variable resistive elements such as the varistor or rectifying diode. These elements do not commonly provide amplification of signals but are widely used as frequency converters in radio receivers to translate signals from high frequencies to lower or intermediate frequencies for more convenient signal amplification. *See* VARISTOR.

The varactor diode is a simple *pn* junction semiconductor device designed to maximize the available reactance variation, to minimize signal losses in electrical resistance, and to require low microwave power from the pump. The junction is biased with a low dc voltage in the nonconducting direction and the applied voltage of the microwave pump wave causes the total bias to vary periodically at the microwave pumping frequency. The capacitance of a reverse-biased semiconductor diode varies with voltage, decreasing as the bias voltage is increased. Common diodes with an abrupt junction between the *p* and *n* regions have a capacitance variation according to the equation shown below, where C is the capacitance in farads, k is a constant depending upon the area of the junction and the doping concentrations of impurities in the two semiconductor regions, V_i is an internal or "built-in" voltage (a few tenths of a volt in most diodes), and V_b is the applied bias voltage (in the nonconducting direction). When used in a parametric amplifier, V_b includes a steady dc component plus a microwave ac value, varying periodically with time at a very high rate.

$$C = \frac{k}{\sqrt{V_i + V_b}}$$

There are several possible circuit arrangements for obtaining useful parametric amplification. The two most common are the up-converter and the negative-resistance amplifier. In both types, the pump frequency is normally much higher than the input-signal frequency. In the up-converter, a new signal wave is generated at a higher power than the input wave. In the negative-resistance device, negative resistance is obtained for the input-signal frequency, causing an enhancement of signal power at the same frequency. *See* NEGATIVE-RESISTANCE CIRCUITS.

When a varactor is energized with two simultaneous voltage waves of different frequencies, such as a pump wave at a frequency f_p and a signal wave at a frequency f_s, the current induced may have a complex waveform which contains other frequencies. Mathematical analysis shows that the first-order and most significant of these new frequencies are the sum ($f_p + f_s$) and the difference ($f_p - f_s$). Depending on the varactor characteristics and the strength of the applied waves, still other frequencies can be produced with frequencies such as $nf_p \pm mf_s$, where m and n are integers of any value, including zero. Normally only the sum or the difference frequency is used.

Analysis also shows that, if the varactor and its surrounding network are so tuned that significant power is generated by the varactor and dissipated at the new frequency $f_p - f_s$ and not at $f_p + f_s$, the varactor will present a negative-resistive input impedance to an input wave at the frequency f_s. Instead of being absorbed by the amplifier network, the input wave will be reflected back down the input transmission line with an increased power. The wave at the frequency $f_p - f_s$ is not utilized outside the amplifier and is called, therefore, the idler frequency. This is the mechanism of the negative-resistance form of the parametric amplifier, the most common type. A functional block diagram of this type of amplifier is shown in the **illustration**.

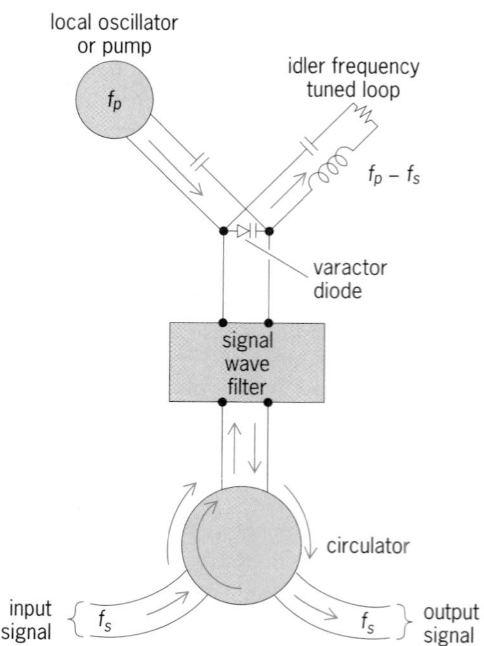

Negative-resistance-type parametric amplifier.

Alternatively, if the network is so designed that the sum frequency $f_p + f_s$ is generated and transmitted out by a second transmission line, and if significant energy is not generated or dissipated at the difference frequency $f_p - f_s$, the power output at $f_p + f_s$ can be greater than the power input at f_s by a factor which may approach the ratio of these two frequencies $(f_p + f_s)/f_s$. This is the mechanism of the up-converter form of the parametric amplifier, so named because the signal wave is converted to a new higher frequency band.

Advantages. The most important advantage of the parametric amplifier is its low level of noise generation. In most other amplifiers, current flows through the active device by discretely charged electrons passing from one electrode to another. Random fluctuations in the rate of such flow normally occur, providing a finite background level of electrical noise (called shot noise) in the output which may override very weak signals, destroying their usefulness. Similar weak fluctuation noise is generated by resistive elements at normal temperatures in amplifier networks (called thermal noise). In the parametric amplifier, relatively few electrons pass from one electrode to another in the varactor, reducing the shot-noise effects. Resistance and thermal noise generation effects are also minimized by proper diode and network design.

The parametric amplifier finds its greatest use as the first stage at the input of microwave receivers where the utmost sensitivity is required. Its noise performance has been exceeded only by the maser. Maser amplifiers are normally operated under extreme refrigeration using liquid helium at about 4° above absolute zero. The parametric amplifier does not require such refrigeration but in some cases cooling to very low temperatures has been used to give improved noise performance that is only slightly poorer than the maser. *See* MASER. M. E. Hines

Bibliography. J. C. Decroly et al., *Parametric Amplifiers*, 1973; K. H. Loecherer and C. D. Brandt, *Parametric Electronics: An Introduction*, 1981.

Parametric arrays

Arrays of sources (or receivers) of sound formed by variation of appropriate parameters of the propagation medium. Normally, these parameters are the local sound speed c and the particle velocity u, which vary because of the presence of large-amplitude pump, or primary, sound waves.

Parametric acoustic sources. The usual parametric source configuration simply consists of a directional transducer (often a plane piston or planar array) driven at two frequencies near the transducer resonance, forming a dual-frequency sound beam called the primary beam. Because sound-wave propagation is not a completely linear process, signals at new frequencies are formed effectively through the interaction of sound with sound as the beam progresses and are generated along the length of the primary beam. The lowest of these new frequencies is the difference of the two primary frequencies, and so the primary beam acts as an end-fire array of sources at the difference frequency. The effective length of the array will be determined by the attenuation of the primary beam, which occurs either as a result of small-signal absorption or, for sufficiently high primary amplitudes, as a result of nonlinear losses due to the generation of harmonics of the primary frequencies and other intermodulation components, such as the sum-frequency component.

Applications. Most applications of parametric sources have been to underwater acoustics. Because the effective length of a parametric source can be made quite long in practice, it is possible to generate highly directional difference-frequency beams, and because the primary amplitude is shaded very gradually along the length of the array, these beams can be made practically side-lobe-free. **Figure 1** illustrates

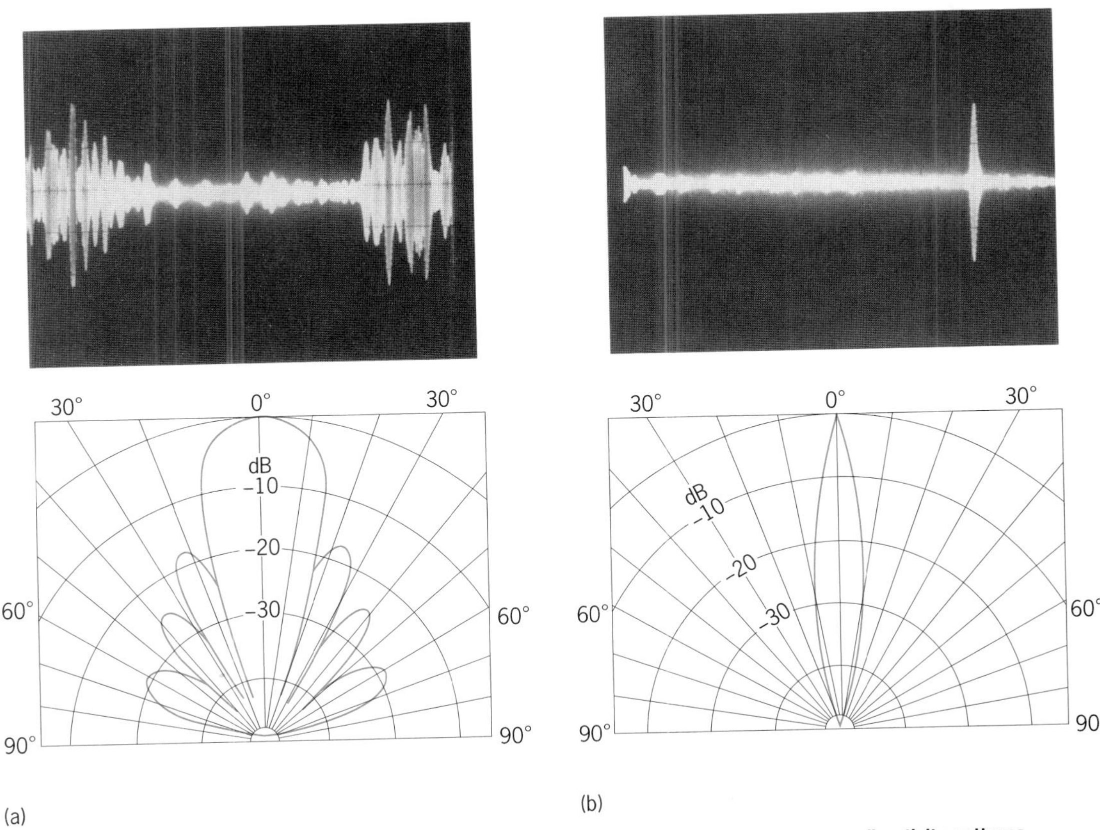

Fig. 1. Oscillograms of echoes received from a cylindrical target (top), and the respective source directivity patterns (bottom). (*a*) Conventional acoustic source. (*b*) Parametric acoustic source. (*Naval Underwater Systems Center*)

the difference between a parametric source and a conventional source of the same size and frequency. In each case, the projector was a plane piston transducer 9.8 in. (0.25 m) in diameter. The conventional beam (Fig. 1*a*) was obtained by simply driving the projector directly at 25 kHz. For parametric operation the projector was driven at primary frequencies of 250 ± 12.5 kHz in order to generate the 25-kHz beam depicted in Fig. 1*b*. Figure 1 also illustrates oscillograms of echoes received from a cylindrical target 0.5 in. (1.3 cm) in diameter and about 3.0 in. (7.6 cm) in length at a range of 118 ft (36 m). The conventional echo is obscured by reverberation from the surface and bottom of the quarry in which the experiment was performed, whereas the parametric source exhibits no reverberation. Thus, the parametric source may be expected to be useful in reverberation-limited situations where one desires a narrow beam from a small projector. Such applications include precision fathometry, subbottom profiling, echo ranging, communications, and Doppler navigation logs. In order to obtain the advantages of a parametric source, however, one must be willing to tolerate low efficiency and low search rate. *See* DIRECTIVITY; ECHO SOUNDER; SONAR.

Theory. The basic theoretical approach is that outlined by P. J. Westervelt, in which the secondary pressure wave is generated by a volume distribution of acoustic sources whose strength depends on the primary pressure. The acoustic source density $q(\mathbf{r}',t)$, due to a finite-amplitude (primary) acoustic pressure field $p_0(\mathbf{r}',t)$, is given by Eq. (1), where ρ is the static

$$q = \beta\rho^{-2}c^{-4}\frac{\partial p_0{}^2}{\partial t} \qquad (1)$$

density, c is the ambient sound speed, and β involves the parameter of nonlinearity of the fluid ($\beta \approx 1.2$ for air and 3.5 for water). Then the secondary pressure field is given by Eq. (2), where $\mathbf{r}$ is the position vector

$$p(\mathbf{r},t) = \frac{\rho}{4\pi}\int\int\int \frac{d^3\mathbf{r}'}{|\mathbf{r}-\mathbf{r}'|} \times \frac{\partial}{\partial t}\left[q\left(\mathbf{r}', t - \frac{|\mathbf{r}-\mathbf{r}'|}{c}\right)\right] \qquad (2)$$

of the observation point and $\mathbf{r}'$ is that of the source point. The primary pressure p_0 must be evaluated at the retarded time, $t - |\mathbf{r}-\mathbf{r}'|/c$, for each source point and then integrated over all regions having appreciable primary-wave intensity.

In the most common parametric-source configuration, the primary waveform consists of two discrete frequency components of equal amplitude which generate a difference-frequency wave, that is, the secondary signal of interest. The primary beam is usually produced by a directional piston transducer. One can express the (averaged) acoustic pressure at the face of the transducer by Eq. (3), where f_0 and f are

$$P_0\left[\cos 2\pi(f_0 - {}^1\!/_2 f)t - \cos 2\pi(f_0 + {}^1\!/_2 f)t\right] = 2P_0 \sin(2\pi f_0 t)\sin(\pi f t) \qquad (3)$$

the average and difference of the primary frequencies and P_0 is a constant. In the absence of linear and nonlinear absorption, the amplitude of each primary component in the far field would be approximately $P_0R_0D_0(\theta',\varphi')/r'$, where r' is the radial distance from the center of the transducer, θ' and φ' are angular coordinates, $D_0(\theta',\varphi')$ is the directivity function for the primary beam (assumed to be similar at the two primary frequencies), and R_0 is the Rayleigh length given by Eq. (4), that is, the ratio of transducer area A_0 to the primary wavelength λ_0.

$$R_0 = \frac{A_0}{\lambda_0} = \frac{A_0 f_0}{c} \qquad (4)$$

The root-mean-square source level SL_0 for each primary component is then given by Eq. (5), where SL_0

$$SL_0 = 20\log\frac{P_0R_0}{\sqrt{2}} \qquad (5)$$

is given in decibels above 1 micropascal-meter, P_0 is given in micropascals, and R_0 is given in meters. The difference-frequency source level SL may be defined in terms of the parametric gain G by Eq. (6).

$$G = SL - SL_0 \qquad (6)$$

(G is always a negative quantity.) **Figure 2** shows computed values of the parametric gain for $f_0/f = 10$ in water. The parameter for the curves is αR_0, the amount of primary absorption loss occurring within the Rayleigh length. The abscissa is a scaled

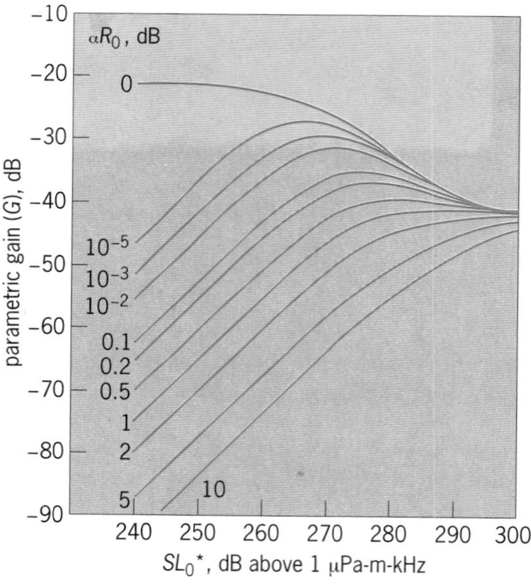

Fig. 2. **Parametric-gain curves for $f_0/f = 10$ in water. (*After M. B. Moffett and R. H. Mellen, Model for parametric acoustic sources, J. Acoust. Soc. Amer., 61:325–337, 1977*)**

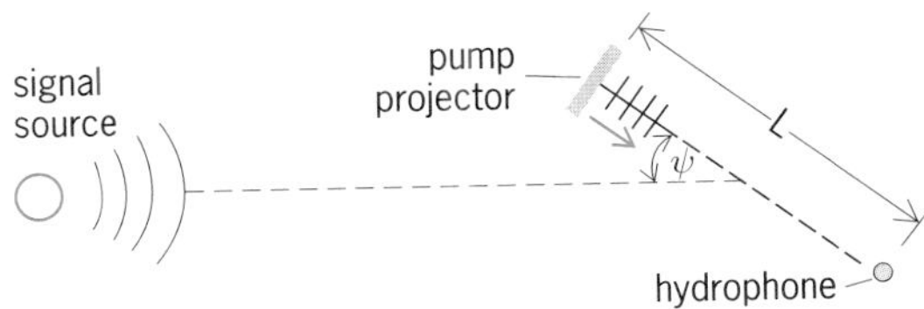

Fig. 3. Parametric acoustic receiver. (*Naval Underwater Systems Center*)

root-mean-square primary source level given by Eq. (7), where f_0 is given in kilohertz. At low lev-

$$SL_0^* = SL_0 + 20 \log f_0 \qquad (7)$$

els the gain is proportional to SL_2^*, but at high levels, when the primary waveform undergoes nonlinear distortion and consequent loss of energy to harmonics, saturation, illustrated by the leveling-off of the gain curves at the right of Fig. 2, occurs.

Parametric acoustic receivers. The parametric acoustic receiving array consists of a large-amplitude pump source directed at a receiving hydrophone, such as in **Fig. 3**. When an acoustic signal arrives at the array, it interacts with the pump wave to form sum- and difference-frequency components which are received at the hydrophone. After filtering, these side bands can be detected, and their level will be proportional to the signal amplitude. The beam pattern (for rotation about the angle ψ) is identical to that of a continuous endfire line array whose length is equal to the pump-hydrophone separation distance L.

The potential advantages of parametric receivers over their conventional counterparts are not as obvious as for parametric sources. Since the beam pattern is identical to that of a continuous end-fire line array, one would expect that the simplicity of the latter would render it more suitable than a parametric receiver, which requires a power source and sophisticated circuitry to filter the side bands from the strong pump component. Furthermore, the parametric receiver cannot be electrically steered away from the end-fire direction, whereas electrical beam steering is standard practice with conventional line arrays. Still, the fact that no hardware is required between the pump source and the receiving hydrophone may eventually prove to be a useful feature. *See* SOUND; UNDERWATER SOUND.

Mark B. Moffett

Parametric equation

A type of mathematical equation used, typically, to represent curves in a plane or in space of three dimensions. In principle, however, there is no limitation to any particular number of dimensions. A parameter is actually an independent variable. In elementary analytic geometry a curve in the xy plane is often studied, in the first instance, as the locus of an equation $y = F(x)$ or $G(x,y) = 0$. The form $y = F(x)$ is not adequate for the complete representation of certain curves, whereas the form $G(x,y) = 0$ may be adequate. The circle $x^2 + y^2 - 16 = 0$ affords an example. But the form $G(x,y) = 0$ is not always convenient. The parametric form $x = f(t)$, $y = g(t)$ is often the most convenient; moreover, it is often the naturally occurring form of representation of the curve. For the circle $x^2 + y^2 - 16 = 0$, one possible parametric representation is $x = 4 \cos t$ and $y = 4 \sin t$.

Parametric curves. A pair of equations $x = f(t)$, $y = g(t)$, where f and g are continuous functions defined for some interval of values of t, for example, $a \leqq t \leqq b$, is said to define a parametric curve. The locus of points (x,y) obtained in this way is not always what would be regarded as a curve by a layman. If one thinks of t as time, the equations define the motion of the point (x,y) as t increases from a to b. Clearly the path can cross itself, double back on itself, or the point may even remain motionless. Even more surprising, the point may pass through every position within an entire square unit of area.

In calculus the customary restrictions placed on f and g in the case of a parametric curve are that f and g have derivatives f', g' which are continuous and such that $f'(t)$ and $g'(t)$ are not zero at the same time. Exceptions for isolated values of t are sometimes permitted. The result of these conditions is that a part of the curve near any one point on it can be represented either in the form $y = F(x)$ or in the form $x = F(y)$, where F is a differentiable function with a continuous derivative (a result of implicit-function theory). This implies that the parametric curve really looks like a curve in the intuitive sense and has a continuously turning tangent. The curve may cross itself, but the part generated by small variations of t from any fixed value t_0 is what is called a smooth arc.

The arc length L of a smooth arc from t_0 to t_1 is Eq. (1).

$$L = \int_{t_0}^{t_1} \left[\left(\frac{dx}{dt}\right)^2 + \left(\frac{dy}{dt}\right)^2 \right]^{1/2} dt \qquad (1)$$

The corresponding differential formula is Eq. (2),

$$ds^2 = dx^2 + dy^2 \qquad (2)$$

where s is arc length measured along the curve.

In polar coordinates the formula is Eq. (3).

$$ds^2 = dr^2 + r^2 d\theta^2 \qquad (3)$$

The curvature of a curve is defined as $K = d\phi/ds$, where s is arc length and ϕ is the angle from a chosen fixed direction to the tangent drawn in the direction of increasing s. The parametric formula for K is Eq. (4), where the plus or minus sign is to be chosen

$$K = \pm \frac{x'y'' - y'x''}{(x'^2 + y'^2)^{3/2}} \qquad (4)$$

according to whether ds/dt is positive or negative.

Here Eqs. (5) hold.

$$x' = \frac{dx}{dt} \qquad x'' = \frac{d^2x}{dt^2} \qquad y' = \frac{dy}{dt} \qquad y'' = \frac{d^2y}{dt^2} \tag{5}$$

If $y = F(x)$ is obtained from $x = f(t)$, $y = g(t)$, it follows by the technique of differentials that $dy = F'(x)\,dx$ and $dx = f'(t)\,dt$, $dy = g'(t)\,dt$, whence Eq. (6) holds, and likewise Eq. (7). Higher derivatives

$$F'(x) = \frac{g'(t)}{f'(t)} \tag{6}$$

$$\frac{d^2y}{dx^2} = \frac{dF'(x)}{dx} = \frac{f'(t)\,g''(t) - g'(t)f''(t)}{[f'(t)]^3} \tag{7}$$

of F can also be found in terms of derivatives of f and g.

In some cases it is necessary to include what occurs as $t \rightarrow \pm\infty$ to complete a parametric representation in a natural way. For instance, Eqs. (8)

$$x = \frac{2t}{1+t^2} \qquad y = \frac{1-t^2}{1+t^2} \tag{8}$$

represent the circle $x^2 + y^2 = 1$ except for the point $(0, -1)$. This point is obtained in the limit as either $t \rightarrow -\infty$ or $t \rightarrow +\infty$.

The cycloid $x = a(t - \sin t)$, $y = a(1 - \cos t)$ is an example of a curve that is easily and naturally represented in parametric form but is represented only with great awkwardness by a single equation in x and y.

Curves in space of three dimensions are represented parametrically in the form $x = f(t)$, $y = g(t)$, $z = h(t)$. The conditions for a smooth arc are similar to the conditions in the plane case.

Parametric surface. A parametric surface in space of three dimensions is defined by $x = f(u, v)$, $y = g(u, v)$, $z = h(u, v)$ where f, g, h are continuous functions of the two parameters u, v. In order to have a surface which conforms to the intuitive ideas of a smooth surface, it is sufficient to impose the condition that, for (u, v) in a certain square (or other region) in the $u\,v$ plane, the functions f, g, h possess continuous first partial derivatives such that the three jacobian determinants (9) are never zero at the same point

$$j_1 = \begin{vmatrix} \dfrac{\partial g}{\partial u} & \dfrac{\partial g}{\partial v} \\ \dfrac{\partial h}{\partial u} & \dfrac{\partial h}{\partial v} \end{vmatrix} \qquad j_2 = \begin{vmatrix} \dfrac{\partial h}{\partial u} & \dfrac{\partial h}{\partial v} \\ \dfrac{\partial f}{\partial u} & \dfrac{\partial f}{\partial v} \end{vmatrix}$$

$$j_3 = \begin{vmatrix} \dfrac{\partial f}{\partial u} & \dfrac{\partial f}{\partial v} \\ \dfrac{\partial g}{\partial u} & \dfrac{\partial g}{\partial v} \end{vmatrix} \tag{9}$$

(u, v). Under these conditions a sufficiently small square containing a given point in the uv plane is "mapped" onto a smooth piece of surface, and (j_1, j_2, j_3) are components of a vector which is perpendicular to the plane tangent to the surface. The area A of this piece of surface is given by Eq. (10), where the

$$A = \int\int \sqrt{j_1^2 + j_2^2 + j_3^2}\,du\,dv \tag{10}$$

double integral is extended over the square in the uv plane. For a discussion of implicit functions and jacobian determinants *see* PARTIAL DIFFERENTIATION. *See also* ANALYTIC GEOMETRY; CALCULUS.

Angus E. Taylor

Paramo

A biological community, essentially a grassland, covering extensive high areas in equatorial mountains of the Western Hemisphere. Paramos occur in alpine regions above timberline and are controlled by a complex of climatic and soil factors peculiar to mountains near the Equator. The richly diverse flora and the fauna of the paramos are adapted to severely cold, mostly wet conditions. Humans have found some paramos suitable for living and use. Since paramos were defined originally in Spanish-speaking countries, no English equivalent has been given to the term. *See* GRASSLAND ECOSYSTEM.

Geographic and altitudinal zones. Geographically, paramos are limited to the Northern Andes and adjacent mountains, where the climate is generally wetter at high altitudes than in the Central Andes, where related communities are called punas. The only paramo reported outside Colombia, Ecuador, and Venezuela is in Costa Rica. *See* PUNA.

In a broad sense paramos extend from the highest Andean forests to the upper altitudinal limit of vascular plants, the latter corresponding generally to permanent snowline (**Fig. 1**). Areas above this have been termed eolian since only airborne nutrients are available for nonvascular plants. In equatorial alpine regions, the timberline is often lower than the 13,000 ft (4000 m) of Fig. 1, with grassy paramos beginning several hundred feet below this. The lower portions form a transition zone with the forest, termed subparamo or paramillo. Thickets of shrubs and some trees, many in the composite and heath families, become fewer with increasing altitude.

Species composition. Typical paramo is dominated by large bunch grasses with narrow leaves, often appearing like a lush meadow, always green (**Fig. 2**), with the genera *Calamagrostis, Festuca*, and *Agrostis* having the most species. Perennial herbs with brilliant flowers, or growing in compact cushions, are interspersed between the grasses (**Fig. 3**). In Colombia, Venezuela, and northern Ecuador, the paramos are conspicuously dotted with many species of the genus *Espeletia*, family Compositae, reported also in Los Llanganati, a remote eastern area of Ecuador. These plants are known as frailejon (monk's cowl) because of top-heavy rosettes of large leaves which terminate thick stems and bear blooming heads resembling sunflowers, in some species near the ground, in others elevated to 20 ft (6 m) or more (**Fig. 4**). The occurrence in the same habitat of such composites, bunch grasses, and cushion growths gives many paramos a unique appearance.

Above the typical paramo there exists a zone sometimes termed superparamo, consisting of scattered, low-growing vegetation on rocks and gravel areas. Bunch grass and frailejon are lacking; some perennials not seen readily below include low cushions such as *Draba*, many as dwarf shrubs such as *Hinterhubera*. The species and appearance of the vegetation make the higher levels near the snow seem more like tundra than true paramo. The limit of vascular plants in equatorial regions (paramos) is shown in Fig. 1 at lower level than in the low-latitude alpine regions farther from the Equator (punas). *See* PLANTS, LIFE FORMS OF; TUNDRA.

However, striking differences of paramo and puna, as compared with polar and temperate alpine tundras, include absence of those rhythms and organisms with seasonal adaptation to temporarily favorable growing conditions of summer, which partly offset some of the very severe winter conditions of high latitude and altitude. *See* ALTITUDINAL VEGETATION ZONES.

Climate and soil factors. On tropical mountains, variation in day- and night-length (which serve as signals for plants' seasonal changes) are much smaller than in other cold climates. Depending on altitude and local cloudiness, the incoming solar energy may warm plants and animals above the temperature of the surrounding air and soil. As a result of the outgoing thermal radiation, people and other organisms are chilled every night. It is not surprising therefore that paramos are considered inhospitable areas with penetrating cold, strong winds and heavy fog or mist which saturates air and soil. Early morning sun is usually lost when clouds sweep up from lower altitudes. Nights are always cold, often producing needle ice in the soil and snow in higher parts. Temperatures vary little throughout the year, but considerably diurnally. A range from 32°F (0°C), or slightly above, to around 68°F (20°C) is frequent in a 24-h period. Differences among paramos are more in degree of wetness, depending upon their exposure and position in the Andean ranges, with the eastern slope catching most of the moisture-laden winds. Soils in typical paramo are very black, deep, rich in organic matter, often peaty, frequently saturated even to being swampy, and quite acid.

Plant forms. Many paramo plants present xeromorphic features, as hairy or coriaceous leaves, densely compacted or overlapping in rosettes. The latter growth form occurs in paramo plants other than frailejon. It is also prominent in large *Lobelia* and *Senecio* species of East African mountains, where it is attributed to diurnal temperature changes similar to those mentioned above for the paramos.

Fauna and human habitation. A common paramo animal, evidenced by abundant scat, is the rabbit (*Lepus*). Dantas or tapirs are hunted in some areas.

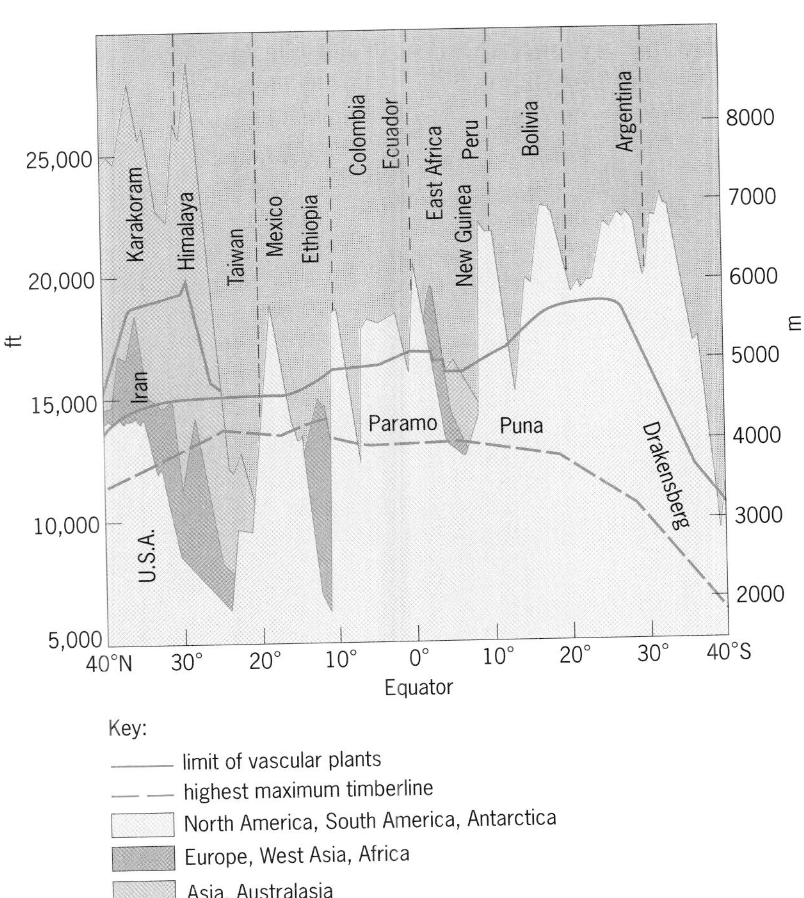

Fig. 1. Alpine and eolian regions of the tropics and subtropics, showing the broad extent of paramo areas. (*After H. E. Wright, Jr., and W. H. Osburn, eds., Arctic and Alpine Environments, Indiana University Press, 1968*)

Wolves, bear, elk, deer, bats, butterflies, bees, many beetles and birds, and one myriapod have been reported in Ecuadorean paramos. Small black and green frogs are abundant in swampy areas.

In Colombia and Venezuela, paramos are rather sparsely inhabited by humans, but they are used as

Fig. 2. Paramo de Chisaca, Cordillera Oriental, Colombia (altitude approximately 12,950 ft or 3950 m), showing large bunch grasses in foreground.

Fig. 3. Paramo del Ruiz, Cordillera Central, Colombia (altitude approximately 13,000 ft or 4000 m), showing compact cushions of *Plantago rigida* between grasses.

Fig. 4. Paramo del Cocuy, Cordillera Oriental, Colombia (altitude approximately 14,700 ft or 4500 m), showing dense stand of *Espeletia lopezii* (known as frailejon).

pasture and the grasses are cut for roof thatching. In these countries the potato is the chief crop, but in Ecuador barley and wheat are grown by Quechua Indians up to nearly 13,200 ft (4000 m). The abundant frailejon is used in several ways, and is often injured by humans' repetitious burning of the paramos.

Harriet G. Barclay

Bibliography. H. G. Barclay, Human ecology of the paramos and the punas of the high Andes, *Proc. Okla. Acad. Sci.*, 43:13–34, 1963; J. H. Brown and A. C. Gibson, *Biogeography*, 2d ed., 1998; M. J. Coe, *Ecology of the Alpine Zone of Mount Kenya*, 1967; G. W. Prescott, Ridgepole of the Western Hemisphere, *Ward's Bull.*, 7(48):1–5, 1967; H. E. Wright, Jr., and W. H. Osburn (eds.), *Arctic and Alpine Environments*, 1968.

Paramyxovirus

A group of viruses that belong to the genus *Paramyxovirus* of the family Paramyxoviridae. The family includes two other genera, *Morbillivirus* and *Pneumovirus*. The genus *Parainfluenza* includes viruses such as mumps and Newcastle disease (in fowl and humans). Measles virus is a member of the genus *Morbillivirus*, and respiratory syncytial virus belongs to the genus *Pneumovirus*. Related members exist in nonhuman species; human measles virus is related to canine distemper and bovine rinderpest virus. Simian and bovine parainfluenza viruses also are known. *See* PARAINFLUENZA VIRUS; RESPIRATORY SYNCYTIAL VIRUS.

Like influenza viruses, the paramyxoviruses are ribonucleic acid (RNA)–containing viruses and possess an ether-sensitive lipoprotein envelope. They range in size from 100 to 300 nanometers, and sometimes larger pleomorphic forms are present. The ribonucleoprotein helix of the paramyxoviruses is 18 nm in diameter, twice as thick as that of influenza. *See* ANIMAL VIRUS; RIBONUCLEIC ACID (RNA).

The diseases caused by mumps and measles viruses are well known. The importance of the parainfluenza and respiratory syncytial viruses in causing severe respiratory disease in infants and young children was recognized chiefly after 1960. The most important respiratory tract pathogen of young infants is respiratory syncytial virus, especially in cases of bronchiolitis and pneumonia, and it is the only respiratory virus which has its greatest frequency of human infection in the first 6 months of life. Parainfluenza viruses are next in importance, particularly in cases of croup, but they more commonly infect older children rather than infants. The parainfluenza viruses also frequently cause common colds and other mild upper respiratory illnesses in children and adults. *See* MEASLES; MUMPS; NEWCASTLE DISEASE; PARAINFLUENZA VIRUS; VIRUS CLASSIFICATION.

Joseph L. Melnick; M. E. Reichmann

Bibliography. D. W. Kingsbury (ed.), *The Paramyxoviruses*, 1990; D. O. White and F. J. Fenner, *Medical Virology*, 4th ed., 1994.

Paranoia

A mode of thought, feeling, and behavior characterized centrally by false persecutory beliefs, more specifically referred to as paranoidness. The paranoid individual falsely believes he or she is being intentionally harassed, plotted against, threatened, demeaned, interfered with, wronged, exploited, or otherwise persecuted by malevolent others—either a specific person or a group. Commonly associated with these core persecutory beliefs are properties of suspiciousness, fearfulness, hostility, hypersensitivity, rigidity of conviction, and an exaggerated sense of self-reference. These properties are evident with varying degrees of intensity and duration.

Paranoidness represents a specific mode of processing semantic information that has triggering and structuring causes. Triggering causes determine when the paranoid mode is activated; structuring causes are those specific semantic contents that produce the distinctive characteristics of the paranoid class. The historical factors that lead to the formation of the paranoid mode are not well understood.

The paranoid mode can be triggered at either biological or psychological levels. Common precipitating biological causes are brain trauma or tumor, thyroid disorder, cerebral arteriosclerosis, and intoxication with certain drugs, including alcohol, amphetamines, cocaine, other psychostimulants, and hallucinogens such as mescaline or lysergic acid diethylamide (LSD). They can produce disordered activity of central dopaminergic and noradrenergic pathways. At the psychological level, triggering causes include false arrest, birth of a deformed child, social isolation, deafness, and intensely humiliating experiences. *See* NORADRENERGIC SYSTEM.

The paranoid mode is resistant to modification by psychotherapeutic or pharmacological methods. Acute psychotic states of paranoidness accompanied by high levels of anxiety are usually responsive to neuroleptic medication. *See* PSYCHOPHARMACOLOGY.

Paranoidness can occur by itself, or it can be associated with other mental conditions such as depression or schizophrenia, and its course can be quite variable. In some cases it may persist or fluctuate for years, whereas in others it subsides inexplicably. *See* SCHIZOPHRENIA. Kenneth Mark Colby

Bibliography. K. M. Colby, Modeling a paranoid mind, *Behav. Brain Sci.*, 4:515–560, 1981; D. W. Swanson, P. J. Bohnert, and J. A. Smith, *The Paranoid*, 1970.

Parasexual cycle

A series of events, discovered in filamentous fungi, which lead to genetic recombination outside the standard sexual cycle.

Until the 1940s the standard sexual cycle (an alternation of nuclear fusion and meiosis) was the only route to formal genetic analysis and planned breeding. Since that time diverse alternative processes leading to recombination have been discovered in microorganisms. The processes occurring in certain filamentous fungi were termed the parasexual cycle by Guido Pontecorvo. There are three essential steps in the cycle: heterokaryosis; fusion of unlike haploid nuclei in the heterokaryon to yield heterozygous diploid nuclei; and recombination and segregation at mitosis by two independent processes, mitotic crossing-over and haploidization.

The features and operation of the cycle can be outlined in the homothallic ascomycete *Aspergillus nidulans*, the species in which the cycle was first demonstrated. Wild-type *A. nidulans* grows on a minimal medium containing a sugar and inorganic salts, producing green, haploid, uninucleate, vegetative spores (conidia). It is possible to produce mutant strains differing from wild type in various characteristics such as conidial color or growth requirement for an exogenous supply of one or other essential nutrient. Heterokaryons are synthesized by growing together pairs of mutant strains which differ from wild type and each other in nutritional requirements and, preferably, conidial color. Hyphal fusions yield some heterokaryotic cells which carry genetically different nuclei in a common cytoplasm. The heterokaryon, somewhat comparable with a heterozygote, is nutritionally wild type; it can be grown in balanced form on minimal medium which does not permit growth of either parent homokaryon. Most of the conidia formed by a heterokaryon are of one or the other parental haploid type. However, infrequent fusions of unlike nuclei in the hyphae give rare conidia with heterozygous diploid nuclei. Diploid strains are isolated by plating conidia from a heterokaryon on minimal medium which selects only the heterozygotes. The resulting strains are fairly stable at mitosis. They are recognized mainly by their phenotypes, as compared with the parent haploids; their larger conidial diameter; and their ability to show mitotic segregation.

Mitotic crossing-over was first discovered by C. Stern in *Drosophila* and is a rare but regular occurrence in a small percentage of all mitotic divisions at the four-strand stage of mitosis in diploid nuclei. When such crossing-over occurs, there is usually only one exchange, reciprocal between nonsister chromatids, in the whole genome. The subsequent chromatid segregation is mitotic and may yield a nucleus, still diploid, homozygous for all alleles linked in coupling and distal to the exchange. All other loci which are heterozygous remain heterozygous. The modalities of this process permit the ordering of genes on a linkage group arm relative to each other and to their centromere. Mitotic haploidization is independent of mitotic crossing-over and its results are, in effect, like those of meiosis without crossing-over. Haploidization is usually, perhaps always, a stepwise process in which chromosomes are lost successively until a stable haploid state is reached. Haploidization permits the ready assigning of a gene to its linkage group and permits association of linkage group arms mapped by mitotic crossing-over. Detection of mitotic segregants depended initially on visual markers (such as conidial color) revealed in homozygous mutant or haploid condition. Many devices are now available for the automatic selection of mitotic segregants.

Some or all of the features of the parasexual cycle have been demonstrated in many species of fungi. Diploid strains find use in formal genetic analysis and in biochemical research. Mitotic segregation is used for genetic analysis both in species which have a sexual cycle and in imperfect species which have no known sexual cycle. In the former, mitotic analysis is a valuable additional analytical tool; in the latter, it provides the only approach to formal genetic analysis and planned breeding.

Processes similar to the elements of the parasexual cycle have been exploited in the somatic cell genetics of higher plants and animals where, as in fungi, they may overcome limitations of sexual processes. Amphiploids of *Nicotiana* have been produced by selection of the somatic hybrids formed in a mixture of protoplasts from two species; this technique may open the way to the combination of very divergent plant genomes. Techniques are also available for the fusion of cultured animal cells, within and between species. Intraspecific hybrids have a potential in formal genetic analysis and in, for example, tests of allelism of phenotypically indistinguishable conditions determined by recessive mutations. Interspecific hybrids, such as those of mouse and humans, show a sequential loss of whole chromosomes, usually preferentially those of one particular component species; cytological analysis and phenotypic tests of resulting somatic variants provide a means, somewhat parallel to that of haploidization in fungi, for allocation of a gene to its chromosome. *See* BREEDING (PLANT); FUNGI; GENETICS; RECOMBINATION (GENETICS); SOMATIC CELL GENETICS. J. Alan Roper

Bibliography. G. C. Ainsworth and A. S. Sussman (eds.), *The Fungi*, vol. 2, 1966; M. J. Carlile and S. Watkinson, *The Fungi*, 1994; E. Moore-Landecker, *Fundamentals of the Fungi*, 4th ed., 1996; G. Pontecorvo, Alternatives to sex: Genetics by means of somatic cells, in *2d John Innes Symposium*, Oxford, 1975.

Parasitology

The scientific study of parasites and of parasitism. Parasitism is a subdivision of symbiosis and is defined as an intimate association between an organism (parasite) and another, larger species of organism (host) upon which the parasite is metabolically dependent. Implicit in this definition is the concept that the host is harmed, while the parasite benefits from the association. Since well-adapted parasites may be nonpathogenic, parasitism merges into another subdivision of symbiosis in which both partners benefit, that is, mutualism. Parasitism also blends with another ecological relationship, predation. For example, ichneumon fly larvae, which ultimately kill their caterpillar hosts by internal consumption, are indeed predators but are called parasitoids, reflecting the similarity of their life-style to true parasitism. By the definition, above, most species of organisms may properly be called parasites. Although technically parasites, pathogenic bacteria and viruses and nematode, fungal, and insect parasites of plants are traditionally outside the field of parasitology.

Parasites often cause important diseases of humans, wildlife, livestock, and pets. For this reason, parasitology is an active field of study; advances in biotechnology have raised expectations for the development of new drugs, vaccines, and other control measures. However, these expectations are dampened by the inherent complexity of parasites and host–parasite relationships, the entrenchment of parasites and vectors in their environments, and the vast socioeconomic problems in the geographical areas where parasites are most prevalent. Therefore, most parasitologists believe that parasite control or eradication will be a slow and expensive process.

Parasite–host relationships. The ecological and physiological relationships between parasites and their hosts are multifarious and constitute some of the most impressive examples of biological adaptation known. Much of classical parasitology has been devoted to the elucidation of one of the most important aspects of host–parasite ecological relationships, namely, the dispersion and the transmission of parasites to new hosts.

Parasite life cycles range from simple to highly complex. Simple life cycles (transmission from animal to animal) are direct and horizontal with adaptations that include high reproduction rates, and the production of relatively inactive stages (cysts or eggs) that are resistant to environmental factors such as desiccation, ultraviolet radiation, and extreme temperatures. The infective stages are passively consumed when food or water is contaminated with feces that contain cysts. The cysts are then activated in the gut by cues such as acidity to continue their development. Other direct-transmission parasites, such as hookworms, actively invade new hosts by penetrating the skin. Physiologically more complicated are those life cycles that are direct and vertical, with transmission being from mother to offspring. The main adaptation of the parasite for this type of life cycle is the ability to gain access to the fetus or young animal through the ovaries, placenta, or mammary glands of the mother.

Many parasites have taken advantage of the food chain of free-living animals for transmission to new hosts. During their life cycle, these parasites have intermediate hosts that are the normal prey of their final hosts. Parasites may ascend the food chain by utilizing a succession of progressively larger hosts. This process, called paratenesis, may have been helped in its evolution by the common habit among carnivores of opportunistic scavenging. Another adaptation to food-chain transmission is for parasites to physically impede or alter the behavior of intermediate hosts, thus raising their chances of being caught by the final host. Parasites achieve this by growing in the host's central nervous system or by producing substances that affect host behavior or physiology. This group includes serious pathogens such as the neurotropic larval tapeworms that cause cysticercosis, gid, and hydatid disease in humans as well as animals. *See* FOOD WEB.

Vectors are intermediate hosts that are not eaten by the final host, but rather serve as factories for the production of more parasites and may even carry them to new hosts or to new environments frequented by potential hosts. Blood-sucking athropods such as mosquitoes and tsetse flies are well-known examples. After acquiring the parasite from an infected host, they move to another host, which they bite and infect. Their efficiency as vectors is enhanced if they permit parasite multiplication internally. In aquatic environments, the role of vector is often taken by snails or blood-sucking leeches. Snails are important vectors for two-host trematodes (flukes), which increase their numbers greatly in the snail by asexual reproduction. The stages that leave the snail may either infect second intermediate hosts that are eaten by carnivorous final hosts, may encyst on vegetation that is eaten by herbivorous hosts, or in the case of the blood flukes (schistosomes) may swim to and directly penetrate the final host.

Parasite–parasite relationships. Besides their ecological relationship with their hosts, parasites interact with other parasites of their own or other species. Many parasites are monoecious (hermaphroditic), allowing self-fertilization, which is an obvious advantage when only one parasite has colonized a host. In dioecious parasites, the opposite sexes must find each other in or on the body of the host. This is facilitated by site specificity in the host and a high density of parasites, but much evidence implicates more refined mechanisms such as pheromones. If crowding may sometimes facilitate sexual contact, it may also be inimical when parasites must share limited resources. Some parasites have evolved mechanisms to limit intraspecific competition. The human pork tapeworm (*Taenia solium*) appears to have a preemptive mechanism against colonization by others of its kind, either through direct chemical action of one worm on another or through the mediation of the host's immune system. Many worm infections exhibit concomitant immunity, whereby the primary infection sensitizes the host to resist subsequent reinfection by the same species while the primary infection worm is spared. *See* PHEROMONE.

Parasites must also compete interspecifically: seldom is a host infected with just one species of parasite. A variety of parasites can be found in certain niches in the host, such as the intestine. Two coexisting species of intestinal worms may adjust to each other's presence by shifting their locations to partially overlapping segments of the bowel, even though in single infections each would have inhabited the identical site. The several species of intestinal coccidia of chickens have become specialized to occupy relatively discrete intestinal regions.

Parasites with free-living transmission stages interact ecologically with other free-living organisms. They are subject to predation, against which they compensate with very high reproductive rates.

Physiologic interactions. The physiologic interactions of parasites with their hosts are as complex and varied as the ecological interactions. Metabolic dependency is the key to parasitism, and parasites employ many ways to feed off their hosts. The simplest is exhibited by the common intestinal roundworm, *Ascaris*, which consumes the host's intestinal contents. Parasites require from their hosts not only energy-yielding molecules but also basic monomers for macromolecular synthesis and essential cofactors for these synthetic processes. Many examples of the specific absence of key parts of energy-yielding or biosynthetic pathways in parasites are known, and these missing enzymes, cofactors, or intermediates are supplied by the host. Theoretically, differences in metabolism between parasite and host are the targets for rational chemotherapy, but these differences are largely useless for this purpose, since the parasite basically parasitizes an already existent host pathway. Other differences must and are being identified for targeted chemotherapy.

Tapeworms are more complex than *Ascaris* in nutritional requirements from the host. They lack a gut, but their surface actively takes up, by facilitated diffusion or active transport, small molecules such as amino acids and simple sugars. They may actually outcompete the host's intestine, as is dramatically illustrated by the broad tapeworm of humans, *Diphyllobothrium latum*, which absorbs vitamin B_{12} so avidly that the host may develop pernicious anemia. The vertebrate intestine is an organ that regulates the concentrations of small molecules at its mucosal surface. But tapeworms do not regulate the uptake of these small organic molecules very well, relying on the host to provide the optimal mix, to which the tapeworm becomes adapted. This has led to the theory that parasites may gain more than just nutrients from the host by also parasitizing host regulatory mechanisms.

Genetics. Because of the obvious absence of organs relating to sensory perception, motility, and digestion, and also because of reductions in biosynthetic and bioregulatory capabilities, parasites have been described as degenerate organisms. In studies comparing genome size of certain parasites with their nonparasitic relatives, the genome size of the parasites was not reduced, suggesting that parasites are not degenerate but instead are highly adapted organisms. The apparent degeneracy of parasites is matched by special requirements for their success. For example, consider a parasite that alternates between a poikilothermic invertebrate host and a homeothermic vertebrate host, which is a common pattern. Different life cycle stages are genetically programmed to respond at the proper times to quite different developmental cues supplied by disparate host physiologies. For example, many parasites are now known to have heat-shock protein genes. The heat-shock response to elevated temperatures could be the signal for a parasite to switch to an alternate metabolism upon infecting a warm-blooded animal. If free-living stages are involved in the parasite's life cycle, then complex host-finding behavior and invasive mechanisms may be required for successful infection, all of which must be encoded in the deoxyribonucleic acid (DNA).

Immunology. The special problems of parasitic existence are well illustrated by the fact that parasites survive and thrive, as foreign invaders, in the bodies of animals with highly evolved immune defense systems. This system's efficiency, which normally protects animals from bacteria and viruses, is underscored by the dire consequences of its failure in congenital or acquired immunodeficiency syndromes. Parasites, by coevolving with their hosts, have the ability to evade the immune response. The best-known evasive tactic is antigenic variation, as found in African trypanosomes, which have a complicated genetic mechanism for producing alternative forms of a glycoprotein that virtually cover the entire parasite. Each type of variant surface glycoprotein is antigenic, stimulating the synthesis of host blood proteins called antibodies that specifically bind to it and no other variant surface glycoprotein. Antibody binding is followed by the trypanosome's death and clearance from the blood. By going through a genetically programmed (but largely unpredictable) sequence of variant surface glycoproteins, the trypanosome population in a host stays one step ahead of immunity and is not eliminated. Other possible immune escape mechanisms in parasites have been discovered and probably cooperate to prolong parasite survival. In the first place, parasites may avoid stimulation of the immune response by poorly expressing antigens, by localizing in sites not well serviced by the immune system, or by synthesizing antigens that resemble those of the host (molecular mimicry), taking advantage of the host's built-in safeguards against autoimmunity. Second, if the immune system is specifically stimulated by parasite antigens, the parasites may avoid its consequences by locally degrading antibodies and other effector molecules. Third, they may redirect the immune response into nonprotective channels by interfering with its normal control processes, which could account for the fact that many of the antibodies and effector cells produced in response to parasitic infection are ineffective and even irrelevant against the parasites.

This is not to say that parasites are altogether exempt from the effects of immunity. Rather than completely eliminating parasites, the immune system more often functions to control their populations in the host. Thus a balance is achieved between hosts and parasites that have lived in long evolutionary association, with both surviving through compromise. Enhancing these particular antiparasite mechanisms and neutralizing the parasite's evasion mechanisms would tip the balance in favor of the host.

The adaptation of parasites to the host immune response has gone even beyond its evasion. Some, perhaps many, parasites may exploit it to their own advantage. Immune exploitation may take several forms: the formation of protective capsules, the movement of parasites or their products through tissues, and the use of host immune molecules as nutrient sources or growth factors. Thus, to label parasites as degenerate organisms is to ignore their intricate adaptations to their way of life. *See* MEDICAL PARASITOLOGY; POPULATION ECOLOGY.

Raymond T. Damian

Bibliography. R. T. Damian, The exploitation of host immune responses by parasites, *J. Parasit.*, 73:3–13, 1987; P. T. Englund and A. Sher (eds.), *The Biology of Parasitism: A Molecular and Cellular Approach*, 1988; E. R. Noble et al., *Parasitology: The Biology of Animal Parasites*, 6th ed., 1989; D. Rollinson and R. M. Anderson (eds.), *Ecology and Genetics of Host-Parasite Interactions*, 1985; G. D. Schmidt and L. S. Roberts, *Foundations of Parasitology*, 5th ed., 1995.

Parasympathetic nervous system

A portion of the autonomic system. It consists of two neuron chains, but differs from the sympathetic nervous system in that the first neuron has a long axon and synapses with the second neuron near or in the organ innervated. In general, its action is in opposition to that of the sympathetic nervous system, which is the other part of the autonomic system. It cannot be said that one system, the sympathetic, always has an excitatory role and the other, the parasympathetic, an inhibitory role; the situation depends on the organ in question. However, it may be said that the sympathetic system, by altering the level at which various organs function, enables the body to rise to emergency demands encountered in flight, combat, pursuit, and pain. The parasympathetic system appears to be in control during such pleasant periods as digestion and rest. The alkaloid pilocarpine excites parasympathetic activity while atropine inhibits it. *See* SYMPATHETIC NERVOUS SYSTEM.

Because the parasympathetic transmission of efferent impulses is restricted to cranial nerves III (oculomotor), VII (facial), IX (glossopharyngeal), and X (vagus), and to sacral segments 1, 2, 3, and 4, it is also known as the craniosacral system.

Results of stimulation of the parasympathetic components are listed for the following nerves:

1. Oculomotor (III) cranial nerve: eye accommodation in near vision, constriction of the pupil.

2. Facial (VII) cranial nerve: secretion, vasodilation, and constriction of ducts in the submaxillary and sublingual salivary glands and in the lacrimal, nasal, and buccal glands.

3. Glossopharyngeal (IX) cranial nerve: secretion, vasodilation, and constriction of ducts in parotid glands.

4. Vagus (X) cranial nerve: decrease in heart rate, coronary constriction, tracheobronchial constriction, increased smooth-muscle mobility in esophagus, gastrointestinal secretion, internal and external pancreas reduction, increase in gastrointestinal motility, inhibition of ileocolic sphincter, and constriction of pancreatic and biliary ducts.

5. Sacral: evacuation of rectum and bladder, pelvic vasodilation, erection, and secretion of genital tract glands (prostate and Cowper's).

In birds and reptiles the parasympathetic system is very similar to that in mammals. In amphibians and jawed fishes, only the parasympathetic elements of the oculomotor (III) and vagus (X) nerves are distinct. In agnathans (lampreys and hagfish) there is no direct parasympathetic system, but the vagus (X) nerve does innervate the alimentary canal. *See* AUTONOMIC NERVOUS SYSTEM; SPINAL CORD.

Douglas B. Webster

Parathyroid gland

An endocrine organ usually associated with the thyroid gland. Embryologically, the glandular primordia are formed in the endoderm of the third and fourth pharyngeal pouches and are associated with the similarly derived primordia of the thymus. There may be from one to three pairs of the small glands present in individuals of the various vertebrate classes, although two pairs appear most frequently. They are characteristically within, on, or near the thyroid gland. In response to lowered serum calcium concentration, a hormone is produced which promotes bone destruction and inhibits the phosphorus-conserving activity of the kidneys. *See* THYROID GLAND.

Anatomy. Parathyroid glands occur in all vertebrate classes with the exception of the fishes, although cells that appear to be homologs of parathyroid cells are found in cyclostomes at the dorsal and ventral ends of all pharyngeal pouches. Whether they function as parathyroid tissue, however, is unknown. Data relating to the details of the comparative anatomy of these glands are summarized in **Table 1**.

In humans, there are typically four glands situated as shown in **Fig. 1**; however, the number varies between three and six, with four appearing about 80% of the time. Variations in the positioning of the glands (or more properly, glandules) along the craniocaudal axis occur but, excepting parathyroid III which may occasionally be found upon the anterior surface of the trachea, the relation to the posterior surface of the thyroid is rarely lost. The extent of variation in primates generally is greater than that observed in all other mammals.

Macroscopically, the glandules appear as oval, round, or irregularly shaped objects, generally having a smooth, highly vascularized surface, although the latter may be slightly roughened (horse) or finely serrated (cattle, swine). Both size and weight are quite variable and appear to have no apparent correlation with total body weight. The color varies from reddish brown through yellowish pink to grayish red or white, depending upon the proportion of fat cells to secreting cells and the amount of lipid in the secreting cells.

Histology. The histological structure of the parathyroids is based upon nonsecretory supportive or stromal elements and secreting or parenchymal elements.

The glandules are enclosed in a smooth connective tissue capsule from which trabeculae, bearing lymphatics, blood vessels, and nerves, emerge to penetrate the gland and to divide its substance into a roughly lobular pattern. The blood vessels are connected to a rich capillary network, characteristic of the endocrine glands generally, that is entwined in reticular fibers found among the parenchymal cells. In addition, fat cells also occur and increase in number with advancing age.

Parenchymal cells are arranged in interconnecting cords and clumps of cells as well as in occasional follicular masses enclosing a gelatinous protein material called colloid; the latter resembles thyroid

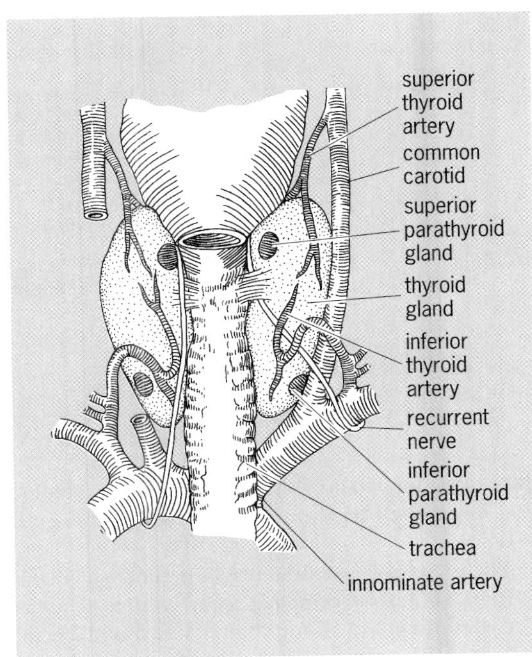

Fig. 1. Common positions of human parathyroid glands on the posterior aspect of the thyroid. (*After W. H. Hollinshead, Anatomy of the endocrine glands, Surg. Clin. N. Amer., 32(4):1115–1140, 1952*)

TABLE 1. Comparative anatomy of parathyroid glands

Vertebrate class, order, and examples	Number of glandules, location, and relations*
Amphibia	
Urodeles	
Salamanders	2–3 pairs of glandules separate from thyroid, ventral to thymus, lateral to arterial arches
Anura	
Frogs, toads	2–3 pairs of glandules lie laterally next to external jugular vein; gill remnant is immediately dorsal in frog, just ventral in toad
Reptilia	
Saurians (suborder)	
Lizards	1 pair of glandules, III only; usually attached to thymus
Serpentes (suborder)	
Snakes	2 pairs of glandules which may be II and III or III and IV (probably latter); if so, III is near temporomaxillary joint next to bifurcation of CC; IV lies between thymus III and IV
Chelonia	
Turtles	2 pairs of glandules, III at cranial end, IV at caudal end of thymus
Crocodylia	
Crocodiles	1 pair of glandules, III only; next to thymus, near origin of collateral cervical artery
Aves	
Galliformes	
Fowl (chicken)	2 pairs of glandules, III at inferior pole of thyroid, IV next to III; when fused, connection may be more or less tenuous
Others	In birds other than fowl, there is wide variation in number and position; it is usually associated with thyroid or CC in some way
Mammalia	
Monotremes	
Platypus	3 pairs of glandules, II at bifurcation of CC, III at dorsal end of thymus, IV on lateral surface of thyroid
Marsupialia	
Opossum	2 pairs of glandules, III medial of branching of CC, IV connected with or embedded in thymus IV; also found between the thyroid and CC
Insectivores	
Moles	1 pair of glandules, III near middle CC
Chiroptera	
Bats	2 pairs of glandules, III (larger) on dorsal outer surface of thyroid, IV (smaller) on inner (tracheal) surface of thyroid
Primates	
Monkeys, apes, humans	2 pairs of glandules; in monkeys and apes, extent of variation in number and position exceeds that observed in other mammals; generally, an external pair (III) related to thymus; an internal pair buried in the thyroid
Lagomorpha (rabbits)	2 pairs of glandules, III associated with thyroid, either at superior or inferior pole of lateral lobes, free or sunken into a depression on thyroid; often more closely related to CC; IV at middle of lateral lobe of thyroid; exceptionally at inferior pole
Rodentia	
Muridae (rats, mice)	1 pair of glandules, III only; usually in cavity on lateral surface of thyroid; cavity usually more superficial in mouse than rat
Caviidae (guinea pig)	2 pairs of glandules, III usually removed from thyroid, sometimes vestigial; IV is embedded in thyroid
Carnivores	
Canidae (dogs)	2 pairs of glandules, III on or in front of lateral cranial surface of thyroid; sometimes more or less embedded in surface; IV is small, deeply buried in thyroid; position variable
Felidae (cats)	2 pairs of glandules, III similar to dog; IV near middle of thyroid nearer tracheal than outer surface
Perissodactyla (horse)	2 pairs of glandules, III near (within 1 cm or 0.4 in.), on, or partially embedded in upper medial border of thyroid; occasionally on tracheal surface; IV entirely embedded in thyroid
Artiodactyla	
Suidae (swine)	1 pair of glandules, III only; usually well removed from thyroid, either in thymus or with terminal branches of CC
Bovidae	
Sheep, goats	2 pairs of glandules, III near bifurcation of CC; IV may be either free on surface of thyroid or deeply embedded
Cattle	2 pairs of glandules, III near bifurcation of CC; IV usually lies on inner tracheal surface of thyroid

*Positions given are subject to extensive variation. III or IV is used to designate glandules according to pharyngeal site of origin. In mammals, III is external to the thyroid; IV, to a greater or lesser extent, within its substance. CC is the common carotid artery.

colloid in appearance but does not contain iodine and appears to have no comparable physiological activity.

There are two general types of parenchymal cells (**Fig. 2**): chief, or principal, cells, which occur in the parathyroids of all vertebrates; and oxyphil cells, which are found in humans only after the fourth to seventh year, and in older macaque monkeys and cattle. These observations support the belief that the parathyroid hormone is produced by the chief cells.

Chief cells. These occur in so-called light or dark forms, according to their appearance under the light-microscope; chief cells are generally smaller than oxyphils, are polygonal in outline, and contain a weakly to moderately acidophilic cytoplasm.

Dark chief cells are thought to be active in hormone production and are distinguished by numerous 100–400-nanometer cytoplasmic granules which are distinctively stained by iron and chrome hematoxylin, Bodian silver stain, and aldehyde fuchsin; nuclei are larger and somewhat more vesicular than their light counterparts. Electron microscopy reveals that the granules, which are membrane-bound aggregates of particles about 10–20 nm, initially appear next to a prominent Golgi apparatus and then pass through the cytoplasmic matrix to be released through the plasma membrane; the particles eventually enter the circulation after transiting the cytoplasm of capillary endothelial cells. The cytoplasm contains sparse amounts of glycogen, and profiles of the cisternae of the granular endoplasmic reticulum occur in parallel arrays.

Light chief cells, which predominate, are thought to be an inactive or nonsecreting phase of a secretory cycle in which the dark cell is the active participant. The cytoplasm of these cells is less chromophilic and less granular than the dark variety, and the nucleus is smaller and denser. Fine structure is characterized by a sharply reduced content of secretory granules, a reduction in the extent of the Golgi apparatus and of the granular reticulum, and a marked increase in glycogen content. Significantly, there are no granules near the Golgi apparatus, and those remaining are at the periphery of the cell.

Oxyphil cells. These are larger than chief cells and have a smooth, rather rounded contour, a strongly acidophilic granular cytoplasm, and a small dark-staining nucleus which may be pycnotic; light and dark forms are said to exist. Oxyphils form only a small part of the cell population and occur singly or in small clusters. Their fine structure is remarkable in that the cytoplasm is literally packed with mitochondria to the virtual exclusion of other organelles; secretory granules occur rarely. The cell has a demonstrable oxidative enzymatic activity, but no definitive function has been discovered; a small body of evidence suggests that some cells may produce hormones. Cells intermediate in structure between chief and oxyphil cells have been found, lending support to the belief that they might be participants in the secretory cycle of the gland.

Embryology. Developmentally, the parathyroids of all vertebrates originate in the endodermal lining of certain pharyngeal pouches. However, differences in details of the site of origin occur among the vertebrate classes (**Table 2**).

The broader aspects of parathyroid development in the various classes are comparable and differ only in detail; the formation in man is given as a representative example. At 35–37 days of development, the endodermal cells that are destined to become the secreting cells of the gland enlarge, become less acidophilic, and aggregate into clumps of cells that constitute the glandular primordia. One primordium is established on the dorsolateral surface of each of the paired third and fourth pharyngeal pouches. Unlike the primordia of the fourth pouches, those of the third are intimately associated with thymic primordia (thymus III). Parathyroid III accompanies thymus III as the latter migrates caudally. When the inferior border of the developing thyroid is reached, the parathyroid primordium is released, while thymus III continues its migration.

TABLE 2. Comparative embryology of parathyroid glands

Vertebrate class, order, and examples	Location of parathyroid primordia: Pharyngeal pouch*	Portion of pouch
Pisces		
Cyclostomes		
Lamprey	?	
Chondrichthyes		
Sharks, skates, and rays	Not known to occur	
Osteichthyes		
Perch, trout	Not known to occur	
Amphibia		
Urodeles		
Salamanders	II(1),III,IV	Ventral
Anura		
Frogs, toads	II(3),III,IV	Ventral
Reptilia		
Sauria (suborder)		
Lizards	II†,III,IV†	Ventral
Serpentes (suborder)		
Snakes	II(?),III,IV,V†	Ventral
Chelonia		
Turtles	III,IV	Ventral
Crocodilia		
Crocodiles	III	Ventral
Aves		
Galliformes		
Fowl (chicken)	III,IV,V(2)	Ventral
Others	III,IV	Ventral
Mammalia		
Monotremes		
Platypus	II,III,IV	Ventral
Marsupialia		
Opossum	III,IV	Dorsal
Insectivores		
Moles	III,IV(1)	Dorsal
Chiroptera		
Bats	III,IV	Dorsal
Primates		
Monkeys, apes, humans	III,IV	Dorsal
Lagomorpha (rabbits)	III,IV	Dorsal
Rodentia		
Muridae (rats, mice)	III,IV†	Dorsal
Caviidae (guinea pig)	III,IV	Dorsal
Carnivores		
Canidae (dogs)	III,IV	Dorsal
Felidae (cats)	III,IV	Dorsal
Perissodactyla (horse)	III,IV	Dorsal
Artiodactyla		
Suidae (swine)	III,IV†	Dorsal
Bovidae		
Sheep, goats	III,IV	Dorsal
Cattle	III,IV	Dorsal

*Numbers in parentheses indicate frequency of occurrence: 1, frequent; 2, occasional; 3, seldom.

†Parathyroid primordium is either not formed or undergoes involution shortly after formation.

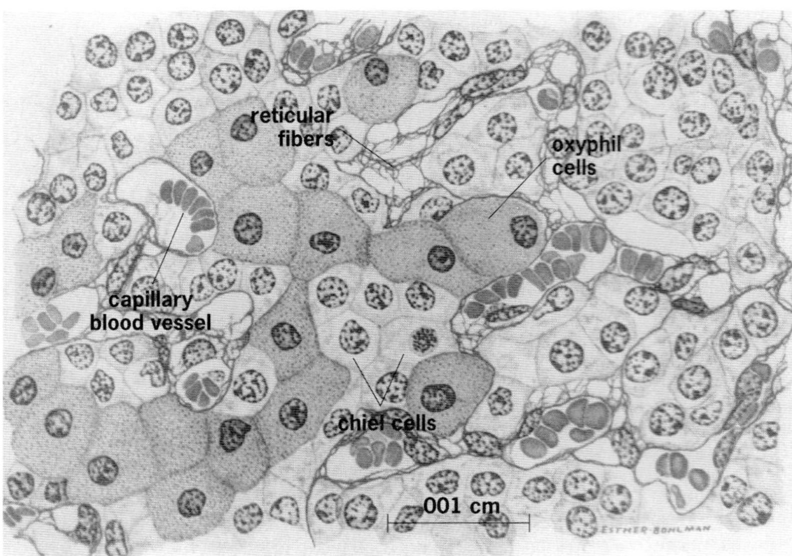

Fig. 2. Section of human parathyroid gland, demonstrating the two main parenchymal cell types. Zenker-formal fixation, Mallory-azan stain. (*After D. Bloom and D. W. Fawcett, A Textbook of Histology, 9th ed., Saunders, 1968*)

In consequence of this movement, parathyroid III comes to lie inferior to parathyroid IV. Occasionally fragmentation of parathyroid III occurs during migration, the fragments later developing into isolated masses of functional tissue, accessory parathyroids. For this reason it is often difficult or may even be impossible to render an animal entirely free of parathyroid tissue by surgical means.

During the fourth to fifth months, chief cells appear and a sinusoidal circulatory pattern is established among the interconnecting cords of cells. The onset of functional activity probably coincides with the appearance of this histological pattern. Variations in size, number, and position of matured glands are known to occur, but complete absence is quite rare and is usually associated with other abnormalities of such severity as to ensure fetal or neonatal death. William E. Dossel

Physiology. The parathyroid glands are essential for the regulation of calcium and phosphate concentrations in the extracellular fluids of amphibians and higher vertebrates, but are not present in fishes. Calcium values must be maintained within a small range in order for cells to function normally. Removal or decreased function of the parathyroid glands results in hypocalcemia, increased irritability of nerves, uncontrolled muscle contractions (tetany), and sometimes death. Alternatively, excessive amounts of parathyroid hormone cause increased removal of calcium from bone, resulting in hypercalcemia and abnormal deposition of calcium and phosphate in soft tissues. *See* PARATHYROID HORMONE.

In the normal animal, calcium in the blood and extracellular fluids is regulated at a concentration ranging from 9 to 11 mg/100 ml (mg %). The secretion of parathyroid hormone is controlled directly by the calcium content of the blood. Parathyroid hormone output is stimulated when circulating calcium concentrations decrease below normal; in response to the hormone, blood calcium values will increase. Conversely, parathyroid hormone secretion decreases when calcium values increase above the normal range. This will result in a gradual fall of calcium concentrations back to normal. *See* CALCIUM METABOLISM.

In addition to raising blood calcium values, parathyroid hormone increases the excretion of phosphate in the urine. This urinary loss of phosphate results indirectly in decreased values of blood phosphate. Although phosphate does not appear to be regulated as precisely as calcium, this renal mechanism for elimination of phosphate serves to maintain a balance between phosphate and calcium in the extracellular fluids. Another important effect of parathyroid hormone on the kidney is to increase the renal reabsorption of calcium, thus reducing the loss of calcium in the urine and thereby conserving calcium in the body.

In fishes, some of the calcium-regulating function of parathyroid function appears to be carried out in fishes by calcitonin, a hormone secreted by the ultimobranchial glands.

Mechanism of action. Parathyroid hormone has two major target organs, bone and kidney. It acts on bone in several ways with both short-term and long-term effects. Short-term changes include a rapid uptake of bone fluid calcium into osteoblast cells, which in turn pump the calcium into the extracellular fluids. Long-term effects of parathyroid hormone include stimulating the activity and increasing the number of osteoclasts, bone cells which act to break down bone matrix and release calcium from bone. All of these effects will result in increased blood calcium values. *See* BONE.

Parathyroid hormone also inhibits the renal reabsorption of phosphate, thus increasing the urinary output of phosphate. Phosphate reabsorption across the renal tubule is dependent upon sodium transport, and parathyroid hormone interferes with this sodium-dependent phosphate transport in the proximal tubule. *See* KIDNEY.

There are reports that parathyroid hormone stimulates calcium uptake into the body across the intestine. However, this is not a direct effect. Parathyroid hormone stimulates the production of the most active metabolite of vitamin D, 1,25-dihydroxycholecalciferol, during vitamin D synthesis. This metabolite of vitamin D directly stimulates the intestinal absorption of calcium. *See* ENDOCRINE SYSTEM (VERTEBRATE); PARATHYROID GLAND DISORDERS; VITAMIN D. Nancy B. Clark

Bibliography. S. A. Binkley, *Endocrinology*, 1994; P. Felig, J. D. Baxter, and L. A. Frohman, *Endocrinology and Metabolism*, 3d ed., 1994; A. Gorbman et al., *Comparative Endocrinology*, 1983; M. E. Hadley, *Endocrinology*, 5th ed., 1999; P. K. Pang and A. Epple, *Evolution of Vertebrate Endocrine Systems*, 1980; J. A. Parsons (ed.), *Endocrinology of Calcium Metabolism*, 1982.

Parathyroid gland disorders

Disorders involving excessive or deficient blood levels of parathyroid hormone caused by abnormal functioning of the parathyroid gland. Parathyroid hormone is responsible for keeping the concentration of calcium in blood within a narrow normal range. If the blood calcium concentration falls, the parathyroid glands respond by secreting hormone which tends to increase the concentration of calcium. Conversely, an increase in blood calcium concentration above the normal range normally suppresses parathyroid hormone secretion. Parathyroid hormone acts directly on bone and kidney, and indirectly on the intestine, to increase the concentration of calcium in blood. It also acts on the kidney to increase excretion of phosphate in the urine, causing a lowering of the concentration of phosphorus in blood.

Hyperparathyroidism. This group of disorders is characterized by excessive parathyroid hormone secretion. Primary hyperparathyroidism is defined as increased parathyroid hormone secretion despite elevated blood calcium. The disorder is readily diagnosed by demonstration of increased blood calcium and parathyroid hormone, and reduction in blood phosphorus. Symptoms and signs range from overt loss of calcium from bone and formation of calcium stones in the kidney to nonspecific manifestations such as weakness and fatigue. With the advent of routine automated measurement of blood calcium, primary hyperparathyroidism is often diagnosed in individuals with no symptoms. The disorder occurs most frequently in women after menopause. There are also inherited forms of the disease. The nonhereditary form of the disease is commonly due to a single benign parathyroid tumor, and very rarely is caused by a parathyroid gland cancer. The hereditary forms are caused by benign enlargement of several or all four of the parathyroid glands. Treatment of primary hyperparathyroidism involves surgical removal of the abnormal gland(s). Secondary hyperparathyroidism is defined as excessive secretion of parathyroid hormone in response to reduction in blood calcium caused by kidney or intestinal disorders or vitamin D deficiency. Treatment is directed at the underlying disorder (for example, kidney transplantation for kidney failure). *See* CALCIUM METABOLISM; VITAMIN D.

Hypoparathyroidism. This group of disorders involves a deficiency in parathyroid hormone secretion or metabolism. Manifestations include involuntary muscle contractions, or generalized seizures in the most extreme cases. Diagnosis is made by demonstrating decreased blood calcium and increased blood phosphorus despite normal kidney function. In most forms of hypoparathyroidism, parathyroid hormone is low or undetectable in blood despite reduced blood calcium because the parathyroid glands are unable to respond appropriately by secreting parathyroid hormone. Hypoparathyroidism can be caused by parathyroid gland destruction, for example, by inadvertent surgical removal. Certain rare forms of hypoparathyroidism are due to resistance to parathyroid hormone action rather than hormone deficiency, and are termed pseudohypoparathyroidism. Blood parathyroid hormone is actually elevated in this form of hypoparathyroidism, because the normal parathyroid glands try to compensate for the inability of the kidney to respond to the hormone. Treatment of all forms of hypoparathyroidism involves administration of calcium and large amounts of vitamin D to restore a normal blood calcium and phosphorus. *See* PARATHYROID GLAND.

Allen M. Speigel

Bibliography. J. P. Bilezikian, R. Marcus, and M. A. Levine (eds.), *The Parathyroids: Basic and Clinical Concepts*, 1994; J. D. Wilson and D. W. Foster (eds.), *Williams Textbook of Endocrinology*, 9th ed., 1998.

Parathyroid hormone

The secretory product of the parathyroid glands. Parathyroid hormone (PTH) is a single-chain polypeptide composed of 84 amino acids. The peptide has no intrachain disulfide linkages and has a molecular weight of 9500. The sequences of human, bovine, and porcine parathyroid hormone are known, and the gene for human parathyroid hormone has been cloned and sequenced (see **illus.**).

Chemistry and biosynthesis. Parathyroid hormone is produced from a larger polypeptide precursor by means of two enzymic cleavages. The gene product is prepro-PTH, a polypeptide of 115 amino acids. The leader sequence is removed in the microsomes, generating pro-PTH, a peptide of 90 amino acids. The 6-amino acid pro-sequence is removed in the Golgi apparatus, generating the final 84-amino acid storage and secretory form of the hormone.

Biological activity resides in the amino terminal portion of the molecule; synthetic parathyroid hormone containing only amino acid residues 1–34 has been shown to contain complete biological activity. The native 84-amino acid secretory form of the peptide is rapidly cleaved into amino-terminal and carboxy-terminal fragments in the body of the organism, but whether amino-terminal fragments constitute an important component of total circulating biological activity is unknown. The multiple forms and fragments of parathyroid hormone in the circulation have complicated the establishment of valid immunoassays for the hormone. Knowledge of the primary structure of the hormone has, however, permitted synthesis of active competitive inhibitors.

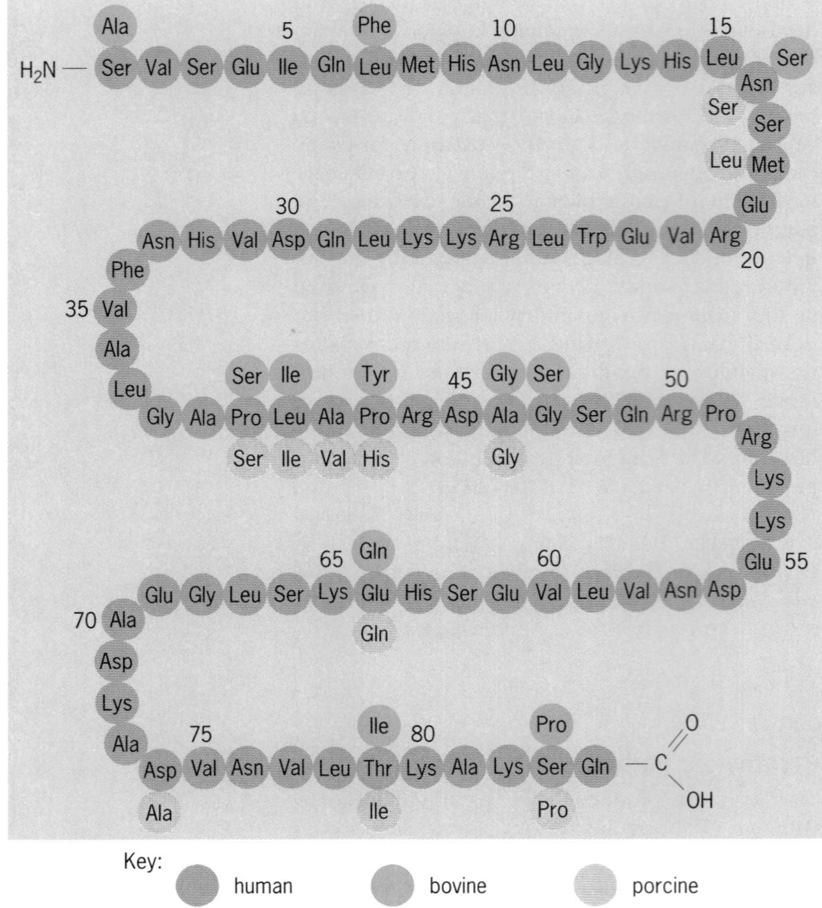

Amino acid sequences of human, bovine, and porcine parathyroid hormone. The backbone in the figure is the human sequence, with substitutions in the bovine and porcine sequences as indicated. (*After H. T. Keutmann et al., The complete amino acid sequence of human parathyroid hormone, Biochemistry, 17:5723–5729, 1978*)

Regulation of secretion. The major regulator of parathyroid hormone secretion is the serum concentration of calcium ions, to which the parathyroid cells are exquisitely sensitive. Only a limited amount of parathyroid hormone is stored in secretory granules, so that a hypocalcemic stimulus must ultimately influence biosynthesis as well as secretion of the hormone. Parathyroid secretory protein is a large, acidic glycoprotein which is stored and cosecreted with parathyroid hormone in roughly equimolar amounts; the biological function of parathyroid secretory protein is unknown.

Target tissues and actions. The principal target tissues of parathyroid hormone are bone and the kidney; intestinal effects previously attributed to the hormone are now known to be indirectly mediated by 1,25-dihydroxyvitamin D. *See* BONE; KIDNEY.

The actions of parathyroid hormone at the level of bone are complex. Parathyroid hormone stimulates preformed osteoclasts (the large, multinucleate cells principally responsible for bone resorption) to resorb bone. This process results in the removal of both bone mineral (largely calcium and phosphate ions) and bone matrix (largely collagen). Chronic stimulation with parathyroid hormone results in the recruitment of a larger population of osteoclasts, with an increase in the surface area of bone participating in the resorption process. Chronic stimulation by parathyroid hormone also results in an increase in bone formation (mediated by osteoblasts) and mineralization, the net result therefore being an increase in all aspects of bone metabolism and turnover.

Parathyroid hormone has three important actions in the kidney: (1) it stimulates calcium reabsorption, thereby reducing the quantity of calcium lost into the urine; (2) it inhibits phosphate reabsorption, thereby increasing the quantity of phosphate excreted into the urine; and (3) it stimulates the synthesis of 1,25-dihydroxyvitamin D, the final hormonal product of vitamin D metabolism. The stimulation of 1,25-dihydroxyvitamin D synthesis results in an increase in 1,25-dihydroxyvitamin D–mediated calcium absorption in the small intestine, enhancing the flow of exogenous calcium into the organism.

In all of its target cells, the interaction of parathyroid hormone with its receptors results in a stimulation of adenylate cyclase activity and cyclic adenylic acid (AMP) formation. In many cells that are responsive to parathyroid, the net regulation of cellular physiology appears to result from a complex interaction of the hormone's effects on the cytosolic concentration of cyclic AMP and free calcium ions.

Physiology. Parathyroid hormone is responsible for the fine regulation of the serum calcium concentration on a minute-to-minute basis. This is achieved by the acute effects of the hormone on calcium resorption in bone and calcium reabsorption in the kidney. The phosphate mobilized from bone is excreted into the urine by means of the hormone's influence on renal phosphate handling. Parathyroid hormone also stimulates calcium absorption in the intestine, this being mediated indirectly by 1,25-dihydroxyvitamin D. Thus, a hypocalcemic stimulus of parathyroid hormone secretion results in an increased influx of calcium from three sources (bone, kidney, and intestine), resulting in a normalization of the serum calcium concentration without change in the serum phosphate concentration. *See* CALCIUM METABOLISM.

Disorders. The syndromes of overproduction and underproduction of parathyroid hormone are known as hyperparathyroidism and hypoparathyroidism. Hyperparathyroidism is a common disorder and usually results from a benign tumor (adenoma) involving one of the parathyroid glands. The hallmarks of hyperparathyroidism are an increase in the serum calcium concentration and a decrease in the serum phosphate concentration, and it may be complicated by bone loss and the formation of calcium-containing kidney stones. The hallmarks of hypoparathyroidism are a reduction in the serum calcium concentration and an increase in the serum phosphate concentration, with neuromuscular irritability and occasionally frank seizures being the principal complications. Hypoparathyroidism is uncommon. Pseudohypoparathyroidism is a rare form of the disorder which results from target tissue resistance to the effects of the hormone. *See* PARATHYROID GLAND; PARATHYROID GLAND DISORDERS; THYROCALCITONIN; VITAMIN D. Arthur E. Broadus

Bibliography. S. A. Binkley, *Endocrinology*, 1994; P. Felig, J. D. Baxter, and L. A. Frohman, *Endocrinology and Metabolism*, 3d ed., 1994.

Parazoa

A name proposed for a subkingdom of animals which includes the sponges. Erection of a separate subkingdom for the sponges implies that they originated from protozoan ancestors independently of all other Metazoa. This theory is supported by the uniqueness of the sponge body plan and by peculiarities of fertilization and development. Much importance is given to the fact that during the development of sponges with parenchymella larvae, the flagellated external cells of the larva take up an internal position as choanocytes after metamorphosis, whereas the epidermal and mesenchymal cells arise from what was an internal mass of cells in the larva. These facts suggest that either the germ layers of sponges are reversed in comparison with those of other Metazoa or the choanocytes cannot be homologized with the endoderm of other animals. Either interpretation supports the wide separation of sponges from all other Metazoa to form the subkingdom Parazoa or Enantiozoa. *See* CALCAREA; DEMOSPONGIAE; METAZOA; PORIFERA.

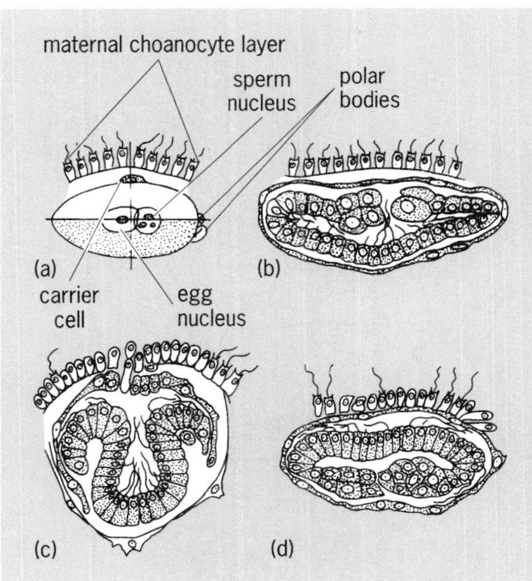

Fig. 1. Development of *Sycon*. (*a*) Fertilized egg beneath the maternal choanocyte layer. (*b*) Stomoblastula. (*c*) Blastula in process of eversion. (*d*) Inverted blastula.

Development. On the other hand, there are cogent arguments in favor of the basic similarity of the development of sponges and other Metazoa. Detailed studies of the embryology of Calcarea with amphiblastula larvae suggest an explanation for the reversal of the germ layers seen so strikingly in the development of a parenchymella. The egg cell of *Sycon* (**Fig. 1**), for example, always lies beneath the layer of choanocytes of a flagellated chamber with its long axis parallel to that layer. The sperm enters from a carrier cell at the pole adjacent to the maternal choanocyte layer, and this pole is determined as ectoblastic. Three meridional cleavages are followed by an equatorial cleavage which separates two tiers of eight cells each. The tier in contact with the choanocyte layer gives rise to ectomesenchyme; the other eight cells furnish the choanocytes. Further divisions of the flagellated cells lead to the formation of a hollow blastula with an opening to the outside between the ectomesenchymal cells. The flagella at this stage are located on the internal surface of the embryo. The polarity of the embryo at this stage is identical to that of a hydrozoan planula, where the pole next to the blastostyle and hence near the source of food is ectoblastic. In *Sycon*, however, the blastula is inverted at this stage, so that the flagella of the flagellated cells are on the external surface, and this region of the larva is anterior as the amphiblastula swims. The ectomesenchymal cells are now at the posterior pole. *See* BLASTULATION; CLEAVAGE (EMBRYOLOGY); HYDROZOA.

Gastrulation and larvae. The amphiblastula of *Sycon*, upon settling on its anterior end, gastrulates by invagination of the flagellated hemisphere. After the blastopore closes, an osculum breaks through at the opposite, free end. Were it not for the fact that pores also develop in the body wall, such a simple attached olynthus would bear a striking resemblance to a planula developing into a polyp.

The parenchymella larva may be interpreted as an amphiblastula with an accelerated development. No trace of the inversion of the surfaces is apparent, except for the external position of the flagella. The ectomesenchymal cells develop precociously at the posterior pole and eventually fill the interior of the larva, displacing the blastocoele. Gastrulation by invagination is no longer possible; instead the peripheral flagellated cells migrate internally to take up their places in the flagellated chambers. The most accelerated development in sponges is seen in the parenchymellae of spongillids, in which choanocytes begin to differentiate from the internal mass of blastomeres while the larvae are still free-swimming. The external flagellated cells are phagocytized by amebocytes after fixation of the larva to the substrate, and thus take no part in choanocyte formation. *See* GASTRULATION; INVERTEBRATE EMBRYOLOGY; PHAGOCYTOSIS.

Thus, although it cannot be denied that a reversal of the germ layers is a fact in many sponge parenchymellae, this process can be explained as

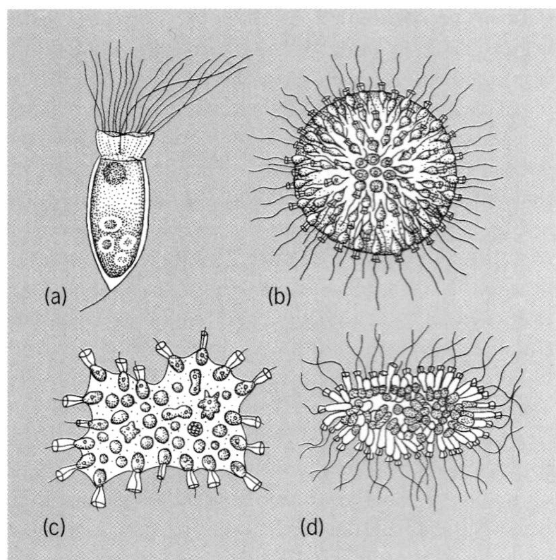

Fig. 2. Choanoflagellate, or craspedomonadine, protozoa. (*a*) *Salpingoeca infusionum*, showing cytoplasmic filaments arising from the collar. (*b*) *Sphaeroeca volvox*, a colonial planktonic form. (c) *Proterospongia haeckeli*. (*d*) Free-swimming aggregate of cells of *Grantia compressa* which formed after dissociation of adult cells.

a consequence of the inversion of the surfaces of the larva which brings the flagellated cells into an anterior position with their flagella directed outward. Clear evidence of this process is seen only in the prolonged development of the amphiblastula larvae of certain Calcarea among existing sponges.

Phylogeny. The striking similarity in structural details between the sponge choanocyte and zooflagellates of the group called choanoflagellates (**Fig. 2**), or craspedomonadines, has led most zoologists to look to this group of protozoa as being ancestral to the sponges. Choanoflagellates are holozoic and capture food by means of the collar. *Proterospongia*, found in both fresh and marine waters, has choanocyte-like cells embedded in a gelatinous mass in which ameboid cells wander. It shows a striking similarity to free-swimming aggregates of cells sometimes found in cultures of dissociated sponges. Choanocytes not only resemble choanoflagellates but also certain cells among other Metazoa. Some corals have flagellated cells with collars of cytoplasmic filaments, and mollusks and vertebrates have microvilli on the cells making up ciliated epithelia. *See* KINETOPLASTIDA.

The phenomenon of inversion of the surfaces which occurs in the blastulae of certain calcareous sponges bears a striking resemblance to a similar process in the development of daughter colonies as well as sexually produced young in *Volvox*. However, a volvocine ancestry for the sponges is difficult to support on any other grounds. *Volvox* cells lack a collar, have two flagella, are photosynthetic, and have postzygotic reduction divisions. Furthermore, the Volvocales are exclusively fresh-water organisms, whereas the earliest fossil sponges are found in marine strata. *See* ANIMAL KINGDOM.

Willard D. Hartman

Parenchyma

A ground tissue chiefly concerned with the manufacture and storage of food. The primary functions of plants, such as photosynthesis, assimilation, respiration, storage, secretion, and excretion—those associated with living protoplasm—proceed mainly in parenchymal cells. Parenchyma is frequently found as a homogeneous tissue in stems, roots, leaves, and flower parts. Other tissues, such as sclerenchyma, xylem, and phloem, seem to be embedded in a matrix of parenchyma; hence the use of the term ground tissue with regard to parenchyma is derived. The parenchymal cell is one of the most frequently occurring cell types in the plant kingdom. *See* PLANT ANATOMY; PLANT PHYSIOLOGY.

Origin. Parenchymal cells are differentiated in the primary plant body (shoot, root) from apical growing zones or meristems. The ground meristem that initates pith, cortex, and leaf mesophyll is the seat of development of parenchyma. In the secondary plant body parenchymal cells are derived from the vascular cambium and cork cambium in the form of ray tissues, xylem and phloem parenchyma, and phelloderm. *See* APICAL MERISTEM; LATERAL MERISTEM.

Variations. Typical parenchyma occurs in pith and cortex of roots and stems as a relatively undifferentiated tissue composed of polyhedral cells that may be more or less compactly arranged and show little variation in size or shape (**Fig. 1**). The mesophyll, that is, the tissue located between the

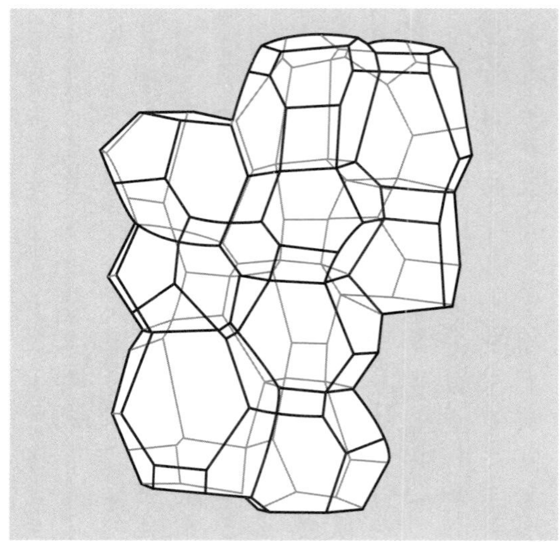

Fig. 1. Schematic representation of a group of cells from parenchyma of *Asparagus* root, illustrating regular arrangement of cells in a compact tissue.

upper and lower epidermis of leaves, is a specially differentiated parenchyma called chlorenchyma because its cells contain chlorophyll in distinct chloroplastids (**Figs. 2** and **3**).

This chlorenchymatous tissue is the major locus of photosynthetic activity and consequently is one of the more important variants of parenchyma. Specialized secretory parenchymal cells are found lining resin ducts and other secretory structures. *See* PHOTOSYNTHESIS; SECRETORY STRUCTURES (PLANT).

Cell contents. Parenchymal cells include the protoplast, common to all living cells, usually with a large vacuole located in the center and the cytoplasm and nucleus next to the cell wall. The kinds of inclusions are partly related to the position of the cell in the plant. Thus cells in leaf mesophyll always have chloroplasts. Cells in root cortex commonly contain grains of storage starch. Cells in pith and cortex of stems contain starch grains and crystals, usually of calcium oxalate. Substances in crystal form are usually interpreted as waste products and are lost with leaf fall, shedding of bark, or decay of pith. Parenchyma in flower parts or fruits may have yellow or red plastids in the cytoplasm or purple or blue anthocyanin pigments in the vacuole.

Cell walls. Parenchymal cells are typically thin and contain cellulose, hemicelluloses, and pectins. They may become thick, hard, and lignified, for example, in wood parenchyma. *See* CELL WALLS (PLANT); CELLULOSE; HEMICELLULOSE; PECTIN.

Cell shape. The three-dimensional cell shape of parenchyma is frequently a function of cell arrangement. In a compact tissue, parenchymal cells are almost isodiametric and polyhedral with an average of approximately 14 sides in contact with 14 other cells. In more loosely arranged tissues with many air spaces, the average number of contacts with adjoining cells is reduced. Parenchymal cells may deviate from the isodiametric shape. Those in the vascular tissues are often conspicuously elongated. Parenchymal cells in the mesophyll and in other tissues with large intercellular spaces may be variously lobed. *See* CORTEX (PLANT); EPIDERMIS (PLANT); FLOWER; FRUIT; LEAF; PHLOEM; PITH; ROOT (BOTANY); SCLERENCHYMA; STEM; XYLEM. Robert L. Hulbary

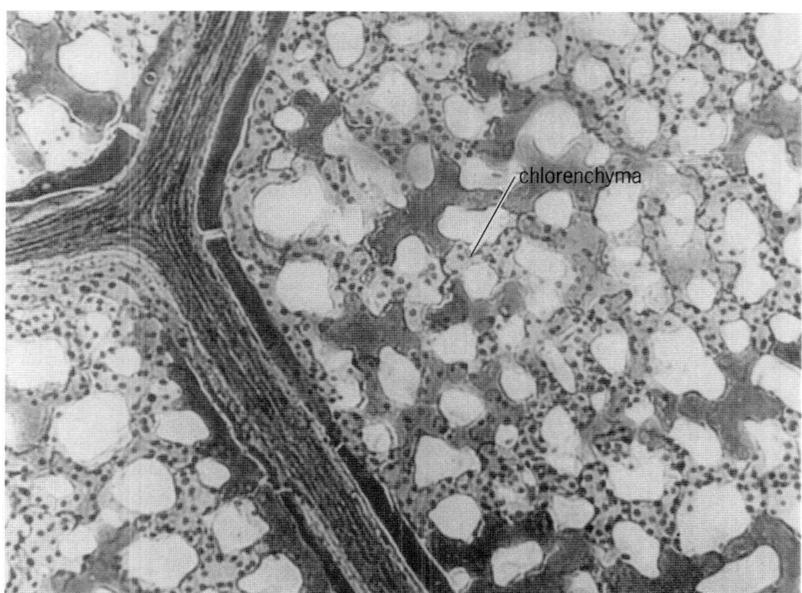

Fig. 3. Section of *Olearia rami* leaf cut parallel to surface, showing mesophyll or chlorenchyma with prominent intercellular spaces. (*Photograph by R. B. Wylie*)

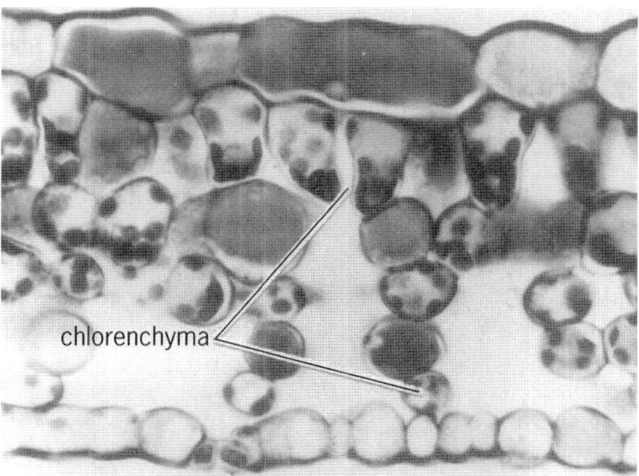

Fig. 2. Transverse section of a leaf of *Tolmiea menziesii*, showing mesophyll or chlorenchyma with prominent intercellular spaces. (*Photograph by R. B. Wylie*)

Pareto's law

A concept developed by the Italian economist Vilfredo Pareto (sometimes called the 20-80 rule) describing the frequency distribution of an empirical relationship fitting the skewed concentration of a variate-values pattern. The phenomenon wherein a small percentage of a population accounts for a large percentage of a particular characteristic of that population is an example of Pareto's law. When the data are plotted graphically, the result is called a maldistribution curve. To take a specific case, an analysis of a manufacturer's inventory might reveal that less than 15% of the component part items account for over 90% of the total annual usage value. Other analyses may show that a small percentage of suppliers account for most of the late or substandard deliveries; a few engineers get most of the patents; a few employees are responsible for most of the quality rejects, errors, accidents, absences, thefts, and so on. *See* INVENTORY CONTROL.

The key to benefiting from Pareto's law is to recognize its existence in a given situation and then to handle the case accordingly. For example, if there is no Pareto's law effect in a population—for example, the number of assembly errors in a factory is equally spread among all the employees—then one course of action should be taken by the engineer; but if Pareto's law is present and 5–10% of the employees are making 85–90% of the errors, then a very different course of action should be taken. If this important distinction is not made, not only will the engineer's action be the wrong one for the problem, but more importantly the engineer's entire approach to engineering management will be wrong.

Good management requires the devotion of a major portion of the manager's time and energy to the solution of the major problems—with the less important problems delegated to subordinates, handled by an automatic system, dealt with through management by exception, or in some cases simply ignored. A manager who devotes too much time to unimportant things is in peril of failure. All things should not be considered equally when in fact they are not of equal impact on the total problem.

A Pareto's law analysis distinguishes the significant or vital few from the insignificant or trivial many. If it reveals that, for example, 2 out of 20 items (10%) account for 80% of the total problem, then by focusing attention on only that 10% of the population, up to 80% of the problem can be resolved.

The mathematics required to calculate and graph the curve of Pareto's law is simple arithmetic. The **illustration** shows a typical curve, where the cumulative percentage of the total annual usage value (Σ) is plotted against the cumulative percentage of the total number of parts (N) of the different components used in manufacture of a company's products.

It should be noted, however, that the calculations need not be done in all cases. It may suffice to merely make a rough approximation of a situation in order to determine whether or not Pareto's law is present and whether benefits may subsequently accrue. For example, a firm with 100 employees can easily review its personnel records and pick out the 5 employees having the worst chronic absenteeism. Suppose those 5 employees were absent for 120 of the 160 total days of work lost by all 100 employees. Pareto's law would apply, as 5% of the population accounts for 75% of the problem, while the remaining 95% of the employees account for only 25% of the total absences. The value of such a discovery would be the realization that absenteeism could be reduced by 75% by dealing with only 5 out of 100 employees. Vincent M. Altamuro

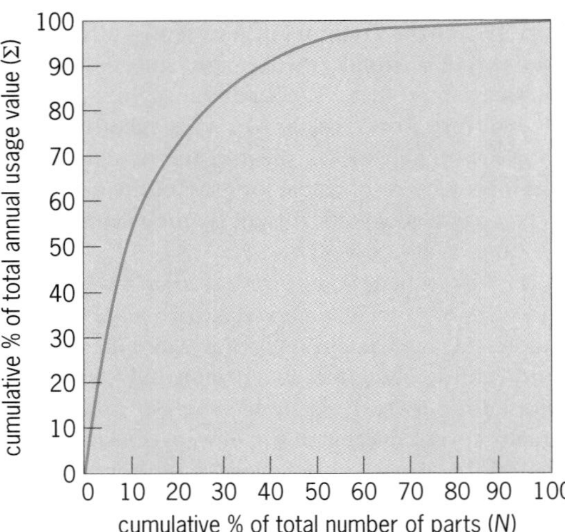

Typical curve representing Pareto's law.

Bibliography. J. M. Juran, *Quality Control Handbook*, 4th ed., 1988; M. G. Kendall, *Dictionary of Statistical Terms*, 4th ed., 1986; M. O. Lorenz, *Methods of Measuring the Concentration of Wealth*, Amer. Stat. Ass. Pub., vol. 9, pp. 209–219, 1904–1905.

Parity (quantum mechanics)

A physical property of a wave function which specifies its behavior under simultaneous reflection of all spatial coordinates through the origin, that is, when x is replaced by $-x$, y by $-y$, and z by $-z$. If the single-particle wave function ψ satisfies Eq. (1),

$$\psi(x,y,z) = \psi(-x,-y,-z) \qquad (1)$$

it is said to have even parity. If, on the other hand, Eq. (2) holds, the wave function is said to have odd

$$\psi(x,y,z) = -\psi(-x,-y,-z) \qquad (2)$$

parity. These two expressions can be combined in Eq. (3), where $P = \pm 1$ is a quantum number, parity,

$$\psi(x,y,z) = P\psi(-x,-y,-z) \qquad (3)$$

having only the two values +1 (designated as even parity) and −1 (odd parity). More precisely, parity is defined as the eigenvalue of the operation of space inversion. Parity is a concept that has meaning only for fields or waves and therefore has meaning only in classical field theory or in quantum mechanics. *See* QUANTUM MECHANICS.

Conservation. The conservation of parity follows from the inversion symmetry of space, that is, the invariance of the Schrödinger equation $H\psi = E\psi$ (the wave equation satisfied by the wave function ψ) to the inversion of space coordinates, $\mathbf{r} \to -\mathbf{r}$. As usual, this is shown formally by defining a transformation operator: Let $\mathcal{P}$ be the parity operator which inverts space; that is, $\mathcal{P}$ acting on a wave function yields the wave function at the inverse point of space, $\mathcal{P}\psi(\mathbf{r}) = \psi(-\mathbf{r})$. Similarly, for an operator A, $\mathcal{P}A(\mathbf{r})\mathcal{P}^{-1} = A(-\mathbf{r})$. [Here, $\mathbf{r}$ stands for the cartesian coordinates of all the particles of the system.] The statement that the world is symmetrical to inversion means that the hamiltonian H after inversion is the same as before, that is, $\mathcal{P}H\mathcal{P}^{-1} = H$; and thus $\mathcal{P}H - H\mathcal{P} = [\mathcal{P},H] = 0$. Since $\mathcal{P}$ commutes with H, it is a constant of the motion. Further, H and $\mathcal{P}$ can be simultaneously diagonal; that is, the eigenfunctions of H can be simultaneously eigenfunctions of $\mathcal{P}$. In fact, if for an eigenvalue E of H there is only a single eigenfunction (nondegenerate level), this eigenfunction must be an eigenfunction of $\mathcal{P}$. As for the eigenvalues of $\mathcal{P}$, since $\mathcal{P}^2 = 1$, it follows that the possible eigenvalues of $\mathcal{P}$ are +1 or −1. That is, an eigenfunction of $\mathcal{P}$ satisfies $\mathcal{P}\psi_\pm(\mathbf{r}) = \psi_\pm(-\mathbf{r}) = \pm\psi_\pm(\mathbf{r})$, where the upper (lower) sign indicates an eigenfunction of positive (negative) parity, also known as even (odd) parity, as stated above. *See* EIGENFUNCTION; EIGENVALUE (QUANTUM MECHANICS); NONRELATIVISTIC QUANTUM THEORY.

The parity (or inversion) operator, which changes **r** to −**r**, has the alternative interpretation that the coordinate values remain unchanged but the coordinate axes are inverted; that is, the positive x axis of the new frame points along the old negative x axis, and similarly for y and z. If the original frame was right-handed, then the new frame is left-handed. [A cartesian coordinate system (frame, for short) is called right-handed if it is possible to place the right hand at the origin and point the thumb and first and second fingers along the positive x, y, and z axes, respectively.] Thus, parity would be conserved if the statement of physical laws were independent of the handedness of the coordinate system that was being used. Of course, the fact that most people are right-handed is not a physical law but an accident of evolution; there is nothing in the relevant laws of physics which favors a right-handed over a left-handed human. The same holds for optically active organic compounds, such as the amino acids. However, the statement that the neutrino is left-handed is a physical law. *See* NEUTRINO.

All the strong interactions between hadrons (for example, nuclear forces) and the electromagnetic interactions are symmetrical to inversion, so that parity is conserved by these interactions. As far as is known, only the weak interactions fail to conserve parity. Thus parity is not conserved in the weak decays of elementary particles (including beta decay of nuclei); in all other processes the weak interactions play a small role, and parity is very nearly conserved. Likewise, in energy eigenstates, weak interactions can be neglected to a very good approximation, and parity is very nearly a good quantum number, so that each atomic, nuclear, or hadronic state is characterized by a definite value of parity, and its conservation in reactions is an important principle. *See* FUNDAMENTAL INTERACTIONS; WEAK NUCLEAR INTERACTIONS.

Orbital parity. The parity of a one-particle state of orbital angular momentum l is given by $P = (-)^l$, that is, even (+1) for $s, d, \ldots,$ states, and odd (−1) for $p, f, \ldots,$ states. Thus the deuteron, whose state is a linear combination of $^3S_1$ and $^3D_1$, has even parity; there cannot be any admixture of the odd parity state $^3P_1$. The orbital parity of an n-particle system is the product of the parities of the $n - 1$ relative orbital angular momentum states: $P_{\text{orb}} = (-)^{l_1 + \cdots + l_{n-1}}$. Thus the parity of an atom is the product of the parities of the one-electron wave functions; any configurations which mix must have the same parity. The Laporte rule of atomic spectroscopy, which states that an electric dipole transition can occur only between states of opposite parity, depends on the fact that the electric dipole radiation field has odd parity. *See* ANGULAR MOMENTUM; SELECTION RULES (PHYSICS).

Intrinsic parity. The total parity of a system of particles is the product of the (total) orbital parity and the intrinsic parities of the particles. ("Particles" here can mean any subsystems.) But the intrinsic parity of a conserved particle is irrelevant and can be omitted, as was done implicitly in the examples discussed above. For if the particle is conserved in a reaction, so is the contribution of its intrinsic parity to the total parity, so that its intrinsic parity is irrelevant to the balance of parity in the reaction. In fact, if a particle is conserved in all reactions, its intrinsic parity can never be determined. The photon is an unconserved particle; its intrinsic parity is odd. The parity of the π^0 meson (a pseudoscalar) is odd, so that to conserve parity it must be emitted by a nucleon into a P state. By charge independence, the charged π meson must also be emitted in a P state; it is natural to call the parity of the charged π meson odd also, which amounts to defining the parity of the neutron and proton (or the u and d quarks) to be the same. An electron by itself is conserved, but an electron plus a positron can annihilate. Thus the product of the parities of an electron and a positron is well defined. According to the Dirac equation of relativistic quantum theory, the product of their parities is −1. The same result holds for any fermion particle-antiparticle pair. Thus the parity of positronium is −1 times its orbital parity, that is, $-(-)^l$. *See* ELEMENTARY PARTICLE; QUARKS; RELATIVISTIC QUANTUM THEORY.

Spin and momentum correlations. The symmetry of the strong and electromagnetic interactions with respect to inversion implies statements about possible correlations of momenta and spins of the particles emitted as a result of such reactions. The principle is that the probability of a configuration of momenta and spins must be a scalar; that is, it must not change under inversion of the coordinate system. A momentum **p** is a vector like **r** (it changes sign under inversion: $\mathbf{p} \to -\mathbf{p}$). On the other hand, orbital angular momentum $\mathbf{L} = \mathbf{r} \times \mathbf{p}$ is a pseudovector, since under inversion $\mathbf{L} \to +\mathbf{L}$; the same must hold for spin angular momentum **S**. Thus $\mathbf{S} \cdot \mathbf{p}$ is a pseudoscalar (it changes sign under inversion), and so such a term cannot occur in the angular distribution of a parity conserving process. This term, $\mathbf{S} \cdot \mathbf{p}$, in an angular distribution, would correlate a particle's spin with its momentum; that is, it would imply a polarization in the momentum direction, or longitudinal polarization, which is accordingly not produced in strong and electromagnetic reactions.

Nonconservation. One of the selection rules which follows from parity conservation is the following: A spin zero boson cannot decay sometimes into two π mesons and sometimes into three π mesons, because these final states have different parities, even and odd respectively. But the positive K meson is observed to have both these decay modes, originally called the θ and the τ mesons, respectively, but later shown by the identity of masses and lifetimes to be decay modes of the same particle. This τ-θ puzzle was the first observation of parity nonconservation. In 1956, T. D. Lee and C. N. Yang made the bold hypothesis that parity also is not conserved in beta decay. They reasoned that the magnitude of the beta-decay coupling is about the same as the coupling which

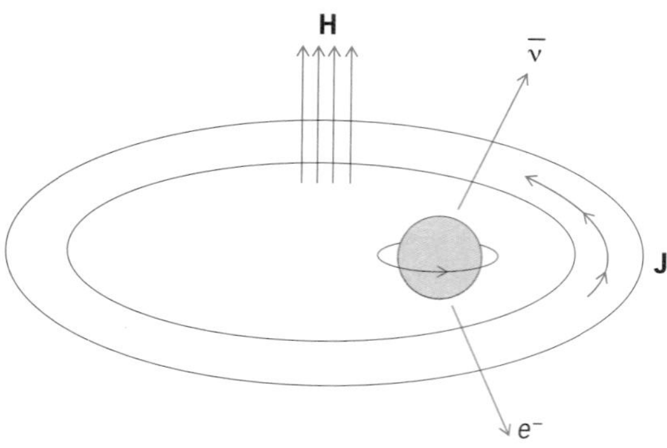

Beta decay from polarized cobalt-60 nuclei. When the spin axes of the cobalt nuclei are not polarized, the preferential emission of the electrons and antineutrinos in the directions shown does not occur.

leads to decay of the *K* meson, and so these decay processes may be manifestations of a single kind of coupling. Also, there is a very natural way to introduce parity nonconservation in beta decay, namely, by assuming a restriction on the possible states of the neutrino (two-component theory). They pointed out that no beta-decay experiment had ever looked for the spin-momentum correlations that would indicate parity nonconservation; they urged that these correlations be sought.

In the first experiment to show parity nonconservation in beta decay, the spins $\mathbf{S}_{Co}$ of the beta-active nuclei cobalt-60 were polarized with a magnetic field **H** at low temperature; the decay electrons were observed to be emitted preferentially in directions opposite to the direction of the ^{60}Co spin (see **illus.**). Thus a $\mathbf{S}_{Co} \cdot \mathbf{p}_e$ correlation was found or, in terms of macroscopic quantities, an $\mathbf{H} \cdot \mathbf{p}_e$ correlation. The magnitude of this correlation shows that the parity-nonconserving and parity-conserving parts of the beta interaction are of equal size, substantiating the two-component neutrino theory.

It was at first somewhat disconcerting to find parity not conserved, for that seemed to imply a handedness of space which would then not be the empty thing which (since the demise of the ether hypothesis) many physicists think it to be. That is, an ether would be needed to provide a standard of handedness at each point of space, to tell ^{60}Co which direction to decay into. But this is not really the situation; the saving thing is that anti-^{60}Co decays in the opposite direction. Thus, after all, there is nothing intrinsically left-handed about the world, just as there is nothing intrinsically positively charged about nuclei. What really exists here is a correlation between handedness and sign of charge.

The present theory of weak interactions is the so-called standard model of the electroweak gauge theory, in which weak and electromagnetic interactions are the consequence of the exchange of the gauge bosons W^+, W^-, Z^0, and γ (the photon). The exchange of $W^\pm$ bosons produces what are called charged-current weak interactions, in which each fundamental fermion that emits or absorbs a $W^\pm$ loses or gains a unit of charge (more precisely, makes a transition between up-type and down-type). This transformation of fundamental fermions is responsible for the weak decays of particles. It is maximally parity-violating in the sense that the $W^\pm$ couples only to the left-hand chiral part of the fermions (chiral coupling); this means that if the mass of the fermion is negligible compared to its energy (which is always true for neutrinos), the fermion is 100% longitudinally polarized. At energies small compared to the mass of the $W^\pm$, this part of the standard model is identical to the older weak interaction theory. *See* ELECTROWEAK INTERACTION; INTERMEDIATE VECTOR BOSON; STANDARD MODEL.

Neutral-current weak interactions. The exchange of Z^0 bosons produces what are called neutral-current weak interactions, in which the fundamental fermions that emit and absorb the Z^0 are unchanged (except in energy and momentum). Since these interactions do not result in decays, they are not as obvious as the charged-current weak interactions, and in fact were not discovered until after their prediction by the standard model. The coupling of the Z^0 to fundamental fermions has two terms: One is a chiral coupling (like the coupling of $W^\pm$) and has opposite sign for up-type and down-type fermions; the other is a vector coupling (like the coupling of the photon γ) and is proportional to the charge of the fermion. The ratio of strengths of these two terms is proportional to $\sin^2 \theta_W$, where the weak angle θ_W is the one new parameter of the standard model. A strong test of the standard model is that when neutral-current weak interactions are described by this form of coupling of the Z^0, the same value of θ_W is found, for all reactions involving any of the fundamental fermions.

At energies not negligibly small compared to the mass of the Z^0, the neutral-current weak interactions are not negligible compared to electromagnetic interactions and can be observed directly in lepton-lepton or lepton-hadron scattering. But at lower energies these interactions can be observed only through the parity nonconservation that they cause. *See* NEUTRAL CURRENTS.

Parity nonconservation in atoms. Parity nonconservation has been observed in several atoms. An example is the work by Bouchiat and colleagues (1986) on cesium. The principle used here is that the M1 radiative decay $7S_{1/2} \rightarrow 6S_{1/2}$ (ground state) is highly forbidden (because the valence electron wave functions in the two states are nearly orthogonal), with a decay rate of only $10^{-6}\ \mathrm{s}^{-1}$; compared to this the parity-nonconserving E1 transition amplitude that results from the mixing of $P_{1/2}$ states into the $S_{1/2}$ states by the parity-violating part of the weak interaction (Z^0 exchange) between the valence electron and the nucleus is not small. This parity-nonconserving E1 transition amplitude, proportional to $\mathbf{S}_e$ (the spin of the valence electron), is detected by observing its interference with a parity-conserving transition amplitude, proportional to **E** (and also $\mathbf{S}_e \times \mathbf{E}$), where **E** is a static electric field applied to the atom. The interference results in circular polarization of light (longitudinal polarization of photons) emitted in the fluorescent transition $7S \rightarrow 6P$ in a direction perpendicular both to the electric field and to the direction of incident circularly polarized light which resonantly excites the $6S \rightarrow 7S$ transition; that is, the observed parity nonconservation signal is $\sim \mathbf{S}_{\text{fluor}} \times \mathbf{S}_{\text{excit}} \cdot \mathbf{E}$, where $\mathbf{S}_{\text{fluor}}$ and $\mathbf{S}_{\text{excit}}$ are the spins of the photons emitted and absorbed in the fluorescent and excitation transitions, respectively. The result shows to high precision that the ratio of coupling of the Z^0 to the u and d quarks is as predicted by the standard model with the value of θ_W as determined by other experiments, and does so at an energy (more precisely, the momentum transferred by the Z^0) much lower than that of any nonatomic experiment. *See* ATOMIC STRUCTURE AND SPECTRA.

Parity nonconservation in nuclei. Parity nonconservation has been observed in many nuclei. Nuclear parity nonconservation effects depend on nuclear and hadronic structure (in contrast to atomic parity nonconservation, where the Z^0 exchanged between the nucleus and an electron has low momentum and so couples to the total weak charge of the nucleus, which is merely the sum of the weak charges of the quarks). Hence, nuclear parity nonconservation is more a test of the understanding of this structure than of the fundamental neutral-current weak interaction. (However, it might detect any additional Z^0-like bosons that are not coupled to leptons and hence unobservable in other experiments.) On the whole, there is a trade-off between the size of the parity violation and size and complexity of the nucleus. At one extreme is a 7% longitudinal polarization of neutrons resonantly scattered from lanthanum-139 (^{139}La) at 0.73 eV, due to the mixing of a great many *s*-wave states into the narrow *p*-wave resonance; it seems impossible, in view of the lack of knowledge of the structure of such a nucleus, to quantitatively relate this result to the nucleon-nucleon parity-nonconserving interaction. At the other extreme is a 2×10^{-7}% longitudinal polarization (equivalently, a 2×10^{-7}% dependence of the cross section on longitudinal polarization) observed in low-energy (less than 50 MeV), proton-proton scattering. (In order to gather enough statistics to observe such a small effect, individual events are not observed, but rather detector currents, that is, macroscopic time averages.) The results agree with prediction. *See* NUCLEAR STRUCTURE.

As a result of parity mixing of nuclear energy levels, radiative transitions between them are mixtures of electric and magnetic multipoles, and so nuclear gamma rays have a circular polarization. Unfortunately, the efficiency of detecting their polarization (by scattering or transmission through magnetized iron) is only a few percent. But offsetting this, some nuclei have parity doublets, pairs of levels with nearly the same energy, of the same spin but opposite parity; the near degeneracy means that the parity-nonconserving nuclear force mixes relatively strongly the two states of the pair (and only these, which simplifies the theoretical calculation). Further enhancement results if the radiative decay rates of the two levels to a common state are very different: Even a small admixture of the rapidly decaying state into the other results in a relatively large admixture of the opposite parity multipole into its decay state, giving a relatively large circular polarization, typically of the order of 0.1%. There are such doublets in four light nuclei, neon-14 (^{14}N), fluorine-18 (^{18}F), fluorine-19 (^{19}F), and neon-21 (^{21}Ne); measurement of the circular polarization of all will yield information on the isobaric-spin dependence of the parity-nonconserving nuclear forces. *See* I-SPIN.

Charles J. Goebel

Bibliography. E. G. Adelberger and W. C. Haxton, Parity violation in the nucleon-nucleon interaction, *Annu. Rev. Nucl. Part. Sci.*, 35:501–558, 1985; M. Bouchiat and L. Pottier, An atomic preference between left and right, *Sci. Amer.*, 250(6):100–111, June 1984; M. Gardner, *The New Ambidextrous Universe*, 3d ed., 1991; W. M. Gibson and B. R. Pollard, *Symmetry Principles in Elementary Particle Science*, 1980; G. E. Mitchell, J. D. Bowman, and H. A. Weidenmueller, Parity violation in the compound nucleus, *Rev. Mod. Phys.*, 71:445-457, 1999; R. Novick (ed.), *Thirty Years of Parity Nonconservation: A Symposium Honoring T. D. Lee*, 1987.

Parkinson's disease

A progressive disorder of the nervous system that mainly affects elderly people, with peak onset in the 60s and 70s. Males and females are equally affected. The basic mechanism of the disease is not known. The disease usually occurs sporadically. However, 10–15% of the time it runs in families.

Characteristics. Parkinson's disease is characterized by abnormalities of motor function, several of which predominate, but all do not necessarily occur in all individuals. Slowness of movement and an inability to start a movement are hallmarks of the disease. For example, when a patient wants to get up from a chair, he or she may have considerable difficulty initiating the movements. The motor disturbance also results in diminished facial expression and a decreased rate of blinking. The second important manifestation is stiffness and rigidity so that the person encounters increased resistance when attempting to move a limb and a joint. The third manifestation, in some individuals, is a tremor that may be quite asymmetrical, occurring in just one hand, or may involve both hands and the trunk. The tremor is slow, with a frequency of two or three per second, and is much worse when the arm is at rest than when it is involved in a purposeful movement.

As the disease progresses, problems with balance become quite limiting, and falls may occur frequently. Falls tend to happen when the person first stands up and tries to walk, because there is difficulty in moving the feet. Walking must, therefore, be initiated by leaning forward. This leaning persists and the person is unable to stop, which results in either a fall or running into objects. Alternatively, with disease progression, episodes of "freezing" may occur, during which voluntary movement becomes impossible. Falls can also occur during such episodes. Finally, some individuals have an associated dementia, which appears to be an integral part of the Parkinson's disease process, although in others it may be a manifestation of Alzheimer's disease. *See* ALZHEIMER'S DISEASE.

Mechanisms. The basic pathologic change is degeneration of a group of nerve cells deep within the center of the brain in an area called the substantia nigra. These cells use dopamine as their neurotransmitter to signal other nerve cells. As these cells degenerate and stop functioning, dopamine fails to reach the areas of the brain that affect motor functions. The possible role of toxins in the disease process has aroused considerable interest. The toxin MPTP, an illicit narcotic analog of demerol, has been found to produce clinical symptoms and pathological changes that are very similar to those observed in Parkinson's disease both in younger people and in experimental animals.

Treatment. Therapy for Parkinson's disease is aimed at replacing dopamine. Since the blood–brain barrier prevents dopamine from entering the brain from the bloodstream, a precursor of dopamine (L-dopa) that will enter the brain is given. L-Dopa is usually administered as part of a compound that inhibits the enzymes that break down L-dopa in the liver, thus making a greater part of it available to the brain. The therapy can be spectacularly successful, allowing those who would otherwise be disabled to function at near-normal levels for an additional 10–20 years. However, the efficacy of L-dopa therapy can decline after about 5–7 years. Active research is ongoing to find other approaches to therapy. One approach is the transplantation to the brain of cells that make dopamine. *See* MOTOR SYSTEMS; NERVOUS SYSTEM (VERTEBRATE); SENSATION.

Guy Mead McKhann

Bibliography. S. Fahn et al. (eds.), *Recent Developments in Parkinson's Disease*, 1986; A. Fisher et al. (eds.), *Alzheimer's and Parkinson's Disease: Strategies for Research and Development*, 1986; J. Jankovic and E. Tolosa, *Parkinson's Disease and Movement Disorders*, 3d. ed., 1998.

Parsec

A unit of measure of astronomical distances. One parsec is equivalent to 3.084×10^{13} km or 1.916×10^{13} mi. There are 3.26 light-years in 1 parsec. The parsec is defined as the distance at which the semimajor axis of Earth's orbit around the Sun (1 astronomical unit) subtends 1 second of arc. Thus, because the angle is small, the equation below holds.

$$\frac{1 \text{ astronomical unit}}{1 \text{ parsec}} = 1 \text{ second} = \frac{1}{206{,}265}$$

A parsec is then 206,265 astronomical units; its accuracy depends on the precision with which the distance from Earth to Sun is measured. At a distance of 1 parsec, the parallax is 1 second of arc. The nearest star is approximately 1.3 parsecs distant; the farthest known galaxy is several billion parsecs. *See* PARALLAX (ASTRONOMY).

Jesse L. Greenstein

Parsley

A biennial, *Petroselinum crispum*, of European origin belonging to the plant order Umbellales. Parsley is grown for its foliage and is used to garnish and flavor foods. It contains large quantities of vitamins A and C and has been grown for 2000 years or more. Two types, plain-leafed and curled, are grown for their foliage; Hamburg parsley (*P. crispum* var. *tuberosum*), also called turnip-rooted parsley, is grown for its edible parsniplike root. Propagation is by seed. Harvesting begins 70–80 days after planting for foliage varieties, and 90 days after planting for Hamburg parsley. *See* APIALES.

H. John Carew

Parsnip

A hardy biennial, *Pastinaca sativa*, of Mediterranean origin belonging to the plant order Umbellales. The parsnip is grown for its thickened taproot and is used primarily as a cooked vegetable. Propagation is by seed; cultural practices are similar to those used for carrot, except that a longer growing season is required. Parsnip seed retains its viability only 1–2 years. Harvesting begins in late fall or early winter, usually 100–125 days after planting. Exposure of mature roots to low temperatures, not necessarily freezing, improves the quality of the root by favoring the conversion of starch to sugar. *See* APIALES; CARROT.

H. John Carew

Partial differentiation

A mathematical operation performed on functions of more than one variable. In this article only two or three variables are considered; however, the principles apply to functions of n variables, for any positive integer $n > 1$. If $z = f(x,y)$, the partial derivative $\partial z/\partial x$ is defined as the derivative of $f(x,y)$ with respect to x, with y being regarded as fixed;that is,

$$\frac{\partial z}{\partial x} = \lim_{h \to 0} \frac{f(x+h, y) - f(x, y)}{h}$$

Another notation for $\partial z/\partial x$ is $f_1(x,y)$. The other first partial derivative is $\partial z/\partial y$, also written $f_2(x,y)$. For values at particular points the notation is

$$\left(\frac{\partial z}{\partial x}\right)_{(a,b)} = f_1(a, b)$$

In the case of a function of three variables, $f(x,y,z)$, the expression is

$$\frac{\partial f}{\partial z} = f_3(x, y, z)$$

The second derivatives of $f(x,y)$ are given by Eqs. (1). It can happen that $f_{12}(x,y) \neq f_{21}(x,y)$, but

$$f_{11}(x,y) = \frac{\partial}{\partial x}\left(\frac{\partial f}{\partial x}\right) \quad f_{12}(x,y) = \frac{\partial}{\partial y}\left(\frac{\partial f}{\partial x}\right)$$
$$f_{21}(x,y) = \frac{\partial}{\partial y}\left(\frac{\partial f}{\partial y}\right) \quad f_{22}(x,y) = \frac{\partial}{\partial y}\left(\frac{\partial f}{\partial y}\right) \qquad (1)$$

this will not happen in common practice, especially with elementary functions. If f_1, f_2, f_{12}, f_{21} are defined in a neighborhood of (a,b), and if f_{12}, f_{21} are continuous at (a,b), then $f_{12}(a,b) = f_{21}(a,b)$. In addition, there are more delicate theorems relating to this matter.

Differentials. The notions of a differential, and of the differentiability of a function, are fundamental in the theory of partial differentiation. The requirement that $f(x,y)$ be differentiable is not the same as the requirement that $f_1(x,y)$ and $f_2(x,y)$ both exist; it is a more inclusive requirement. The geometric meaning of f being differentiable at (a,b) is that the surface defined by $z = f(x,y)$ has a tangent plane not parallel to the z axis when $x = a$, $y = b$. In analytic terms the condition is that if

$$\epsilon = f(a+h, b+k) - f(a, b) - f_1(a,b)h - f_2(a,b)k$$

then
$$\lim_{(h,k)\to(0,0)} \frac{\epsilon}{|h| + |k|} = 0$$

When f is differentiable at (a,b), the expression $f_1(a,b)\,dx + f_2(a,b)\,dy$ is called the differential of f at (a,b) with independent increments dx and dy. It is a linear function of dx and dy, and among all linear functions $A\,dx + B\,dy$, it is the best approximation (in a definite sense of the word) to the expression

$$f(a+dx, b+dy) - f(a, b)$$

In the usual notation $z = f(x,y)$, the differential, evaluated at (x,y), is written as Eq. (2). Here dx and dy

$$dz = \frac{\partial z}{\partial x}\,dx + \frac{\partial}{\partial y}\,dy \qquad (2)$$

are independent variables and dz is a dependent variable.

A sufficient condition that f be differentiable at (a,b) is that the partial derivatives f_1, f_2 be defined at all points near (a,b), and continuous at (a,b).

Chain rule. The prime importance of the differentiability concept is that the differentiability property is needed in proving the chain rule for functions of several variables. This rule asserts that a differentiable function of a differentiable function is differentiable, and the rule tells how to compute partial derivatives of the composite function. For example, if $x = f(s,t)$, $y = g(s,t)$, where f and g are differentiable, and if $z = F(x,y)$, where F is differentiable, then the composite function is $G(s,t) = F[f(s,t), g(s,t)]$, and its differential is

$$\frac{\partial F}{\partial x}\,dx + \frac{\partial F}{\partial y}\,dy$$

where dx and dy, instead of being independent, are given by Eqs. (3), where ds and dt are independent.

$$dx = \frac{\partial f}{\partial s}\,ds + \frac{\partial f}{\partial t}\,dt \qquad dy = \frac{\partial g}{\partial s}\,ds + \frac{\partial g}{\partial t}\,dt \qquad (3)$$

Then $z = G(s,t)$ is differentiable as a function of s and t, and

$$\frac{\partial G}{\partial s} = \frac{\partial F}{\partial x}\frac{\partial f}{\partial s} + \frac{\partial F}{\partial y}\frac{\partial g}{\partial s} \qquad \frac{\partial G}{\partial t} = \frac{\partial F}{\partial x}\frac{\partial f}{\partial t} + \frac{\partial F}{\partial y}\frac{\partial g}{\partial t}$$

These equations, expressing the formal part of the chain rule, are often written in the form

$$\frac{\partial z}{\partial s} = \frac{\partial z}{\partial x}\frac{\partial x}{\partial s} + \frac{\partial z}{\partial y}\frac{\partial y}{\partial s} \qquad \frac{\partial z}{\partial t} = \frac{\partial z}{\partial x}\frac{\partial x}{\partial t} + \frac{\partial}{\partial y}\frac{\partial y}{\partial t}$$

At one occurrence, the status of x and y is that of independent variables, as in $z = F(x,y)$, where they are called variables of the first class. But x, y also occur as dependent variables, depending on the independent variables s, t, which are called variables of the second class.

The chain rule is valid for situations in which there are any number of variables of the first class and, quite independently, any number (the same or different) of the second class.

A typical use of the chain rule occurs when transformations are made on the variables in a problem. For example, one may switch from rectangular to polar coordinates. Then derivatives with respect to x, y, or both, must be converted into expressions involving derivatives with respect to r and θ. Transformations of variables are quite extensively used in studying partial differential equations.

Another interesting instance of the chain rule occurs in the so-called particle differentiation in the flow of fluids. If $\rho = F(x,y,z,t)$ is the density at (x,y,z) in the fluid at time t, and if in a given motion one follows a certain selected particle, denoting the density of the fluid at this particle by $\rho = G(t)$, then

$$G'(t) = \frac{d\rho}{dt} = \frac{\partial\rho}{\partial x}\frac{dx}{dt} + \frac{\partial\rho}{\partial y}\frac{dy}{dt} + \frac{\partial\rho}{\partial z}\frac{dz}{dt} + \frac{\partial\rho}{\partial t}$$

Here ρ has two different meanings: On the left $\rho = G(t)$, and on the right $\rho = F(x,y,z,t)$. In dx/dt, dy/dt, dz/dt, the point (x,y,z) is the position of the particle being followed. Here the variables of the first class are x, y, z, t, and there is just one variable of the second class, namely, t. The role of t also is different on the left and on the right in the equation.

Taylor developments. There is a Taylor's formula with remainder and a Taylor's series of functions of several variables. The easiest way to deal with these things is to think of them as being reduced back to the case of one variable by a device. If one wants to express $f(a+h, b+k)$ by a formula proceeding by terms of various degrees in h and k, consider $g(t) = f(a+th, b+tk)$, develop $g(t)$ in powers of t, and then set $t = 1$. The chain rule is needed to compute the derivatives of g. The general formula is Eq. (4).

$$g^n(t) = \left[\left(h\frac{\partial}{\partial x} + k\frac{\partial}{\partial y}\right)^n f(x,y)\right]_{\substack{x=a+th,\\ y=b+tk}} \qquad (4)$$

Here a symbolic notation with rather evident meaning is used on the right.

Implicit functions. Suppose $F(x,y,z)$ is a function of three variables whose domain of definition is a certain collection of points (x,y,z) in space of three dimensions. As a general rule, it will not be the case that the locus of points for which $F(x,y,z) = 0$ is the graph of an equation $z = f(x,y)$, where f is a single-valued function of two variables. But it may happen that, if (x_0,y_0,z_0) is a point of the locus $F(x,y,z) = 0$, there is a neighborhood of (x_0,y_0,z_0), consisting of all points inside a certain rectangular box centered at (x_0,y_0,z_0), such that the part of the locus $F(x,y,z) = 0$ inside this box is the graph of a function $z = f(x,y)$. There is a standard "implicit function theorem" which covers this situation. It states: Suppose F and its first partial derivatives F_1, F_2, F_3 are continuous throughout some specified neighborhood N of (x_0,y_0,z_0). Suppose also that $F(x_0,y_0,z_0) = 0$ and $F_3(x_0, y_0z_0) \neq 0$. Then there exist certain positive constants a, b, c and a function f of x and y meeting all the following conditions. Let B denote the boxlike region composed of all (x,y,z) such that $|x - x_0| < a$, $|y - y_0| < b$, $|z - z_0| < c$, and let R denote the rectangle in the xy plane composed of all (x,y_0) such that $|x - x_0| < a$, $|y - y_0| < b$. The region B is contained in N; the function f is defined in R, and the graph of $z = f(x,y) = 0$; f is continuous and has continuous first partial derivatives in R, given by

$$f_1(x, y) = -\frac{F_1(x, y, z)}{F_3(x, y, z)} \qquad f_2(x, y) = -\frac{F_2(x, y, z)}{F_3(x, y, z)}$$

where $z = f(x,y)$.

This theorem has two kinds of generalizations: One of the type in which F is a function of n variables and f is a function of $n - 1$ variables, and the other of the type in which the locus $F(x,y,z) = 0$ is replaced by a locus defined by k equations in n variables ($n > k$), while the equation $z = (x,y)$ is replaced by k equations involving k functions of $n - k$ variables. Sample: $F(x,y,z,u,v) = 0$, $G(x,y,z,u,v) = 0$, $u = f(x,y,z)$, $v = g(x,y,z)$. Implicit function theorems of this second type are proved by mathematical induction with respect to k. The conditions in these theorems involve jacobian determinants.

Jacobians. If $F_1, \ldots, F_k$ are k functions of $z_1, \ldots, z_k$, determinant (5)

$$J = \begin{vmatrix} \frac{\partial F_1}{\partial z_1} & \frac{\partial F_1}{\partial z_2} & \cdots & \frac{\partial F_1}{\partial z_k} \\ \frac{\partial F_2}{\partial z_1} & \cdots & & \frac{\partial F_2}{\partial z_k} \\ \cdots & \cdots & \cdots & \cdots \\ \frac{\partial F_k}{\partial z_1} & \cdots & & \frac{\partial F_k}{\partial z_k} \end{vmatrix} \qquad (5)$$

is called the jacobian of $F_1, \ldots, F_k$ with respect to $z_1, \ldots, z_k$, and is denoted by

$$J = \frac{\partial(F_1, \ldots, F_i)}{\partial(z_1, \ldots, z_k)}$$

Notice that the subscripts on the F's are for distinguishing different functions, and do not indicate partial derivatives.

The general implicit function theorem for a system of equations

$$F_1(x_1, \ldots, x_r, z_1, \ldots, z_k)$$
$$= 0, \ldots, F_k(x_1, \ldots, x_r, z_1, \ldots, z_k) = 0\text{"}$$

guarantees a local solution of the form

$$z_1 = f_1(x_1, \ldots, x_r) \cdots z_k = f_k(x_1, \ldots, x_r)$$

near a set of values $x_1 = a_i$, $z_j = b_j$ for which $F_1 = \cdots = F_k = 0$ and $J \neq 0$. This is on the assumption that the F's have continuous first partial derivatives. The derivatives of the f's are given by the formulas

$$\frac{\partial f_i}{\partial x_p} = -\frac{J_{ip}}{J}$$

where J_{ip} is what J becomes when its ith column is replaced by $\partial F_1/\partial x_p, \ldots, \partial F_k/\partial x_p$.

If $u = f(x,y)$, $v = g(x,y)$ defines a one-to-one mapping of a region R_1 of the xy plane onto a region R_2 of the uv plane, and if it is known that f and g have continuous partial derivatives and the jacobian $j = [\partial(f, g)/\partial(x, y)]$ is never zero in R_1, then a simple closed curve C_1 and R_1 maps onto a simple closed curve C_2 in R_2 and, as a point P_1 goes counterclockwise around C_1, its image P_2 goes counterclockwise or clockwise around C_2 according to whether $J > 0$ or $J < 0$. Also, if A_1, and A_2 are the areas enclosed by C_1 and C_2 respectively, there is some point inside C_1 such that A_2/A_1 is the value of $|J|$ at that point. If a double integral with respect to u and v, over the region R_2, is converted into a double integral with respect to x and y, over the region R_1, $du\,dv$ is replaced by $|J|dx\,dy$. These results generalize to the case of mappings in space of more than two dimensions. For example, in the passage from rectangular coordinates x, y, z to spherical polar coordinates r, θ, ϕ, by the equations

$$x = r \sin\phi \cos\theta$$
$$y = r \sin\phi \sin\theta$$
$$z = r \cos\phi$$

the jacobian $\partial(x,y,z)/\partial(r,\theta,\phi)$ has the value $-r^2 \sin\phi$, and $dx\,dy\,dz$ is replaced by $r^2 \sin\phi\; dr\,d\theta\,d\phi$ in triple integrals.

If the equations $u = f(x,y)$, $v = g(x,y)$ define a one-to-one mapping from the xy plane to the uv plane (in restricted regions), then

$$\frac{\partial(u, v)}{\partial(x, y)} = \left[\frac{\partial(x, y)}{\partial(u, v)}\right]^{-1}$$

Functional dependence. If $f(x,y)$ and $g(x,y)$ are functionally dependent in a region R of the xy plane, then $[\partial(f,g)/\partial(x,y)] = 0$ in that region. An example of functional dependence would be: $g(x,y) = [f(x,y)]^2 + \sin[f(x,y)]$. In general, f and g are called functionally dependent in R if there is some function F of u and v such that $F[f(x,y), g(x,y)] = 0$ at all points of R, and yet $F(u,v)$ is not zero throughout any two-dimensional portion of the uv plane. Conversely, if $[\partial(f,g)/\partial(x,y)] = 0$ at all points (x,y) in a neighborhood of (x_0,y_0), then usually f and g are functionally dependent in some (perhaps smaller) neighborhood of the point.

Homogeneous functions. One calls a function $F(x_1,\ldots,x_k)$ positively homogeneous of degree n if $F(tx_1,\ldots,tx_k) = t^nF(x_1,\ldots,x_k)$ for all $t > 0$ and for all $(x_1,\ldots,x_k)$ in the domain of definition of F. The index n need not be an integer. If F is differentiable and positively homogeneous of degree n, the Euler relation

$$x_1\frac{\partial F}{\partial x_1} + \cdots + x_k\frac{\partial F}{\partial x_k} = nF(x_1,\ldots,x_k)$$

holds. Conversely, if F is differentiable in an open region which contains $(tx_1,\ldots,tx_k)$ for all $t > 0$, provided it contains $(x_1,\ldots,x_k)$, then the validity of Euler's relation in the region implies that F is positively homogeneous in degree n.

Lagrange's method in extremal problems. If $F(x,y,z)$ is a differentiable function of three independent variables in an open region R of (x,y,z) space, and if F reaches a relative maximum or minimum value at a point of R, then necessarily $(\partial F/\partial x) = (\partial F/\partial y) = (\partial F/\partial z) = 0$ there. Sufficient conditions, and tests for discrimination between maximum and minimum values, are sometimes stated in terms of second partial derivatives. Lagrange's method is concerned with the situation in which x, y, z are not independent, but are restricted by a side condition $G(x,y,z) = 0$, where G is a specified function. Example: What is the maximum value of $x^2y^2z^2$ subject to the restriction $(x^2/25) + (y^2/16) + (z^2/9) - 1 = 0$? On the assumption that F and G have continuous first partial derivatives and that one never has $G = G_1 = G_2 = G_3 = 0$ at one point, the Lagrange procedure is to set $u = F + \lambda G$, where λ is a parameter. Then, among all the values of F attained for (x,y,z) such that $G(x,y,z) = 0$, if there is a maximum or minimum value, it will occur for an (x,y,z) point which satisfies the equations $F_i + \lambda G_i = 0$ $(i = 1, 2, 3)$ and $G = 0$, for a certain value of λ. These four equations can in theory be solved for x, y, z, λ, and the extreme value can be located. The Lagrange method extends to other numbers of variables and to more than one side condition. *See* CALCULUS; DETERMINANT; DIFFERENTIATION; OPERATOR THEORY; PARAMETRIC EQUATION. Angus E. Taylor

Bibliography. T. M. Apostol, *Mathematical Analysis,* 2d ed., 1974; R. Courant, *Differential and Integral Calculus,* 2 vols., 1936–1937, reprint 1992; W. Fulks, *Advanced Calculus,* 4th ed., 1988; W. Kaplan, *Advanced Calculus*, 4th ed., 1991; A. E. Taylor and R. W. Mann, *Advanced Calculus,* 3d ed., 1983.

Particle accelerator

An electrical device which accelerates charged atomic or subatomic particles to high energies. The particles may be charged either positively or negatively. If subatomic, the particles are usually electrons or protons and, if atomic, they are charged ions of various elements and their isotopes throughout the entire periodic table of the elements. Before the advent of accelerators, the only sources of energetic particles for research were cosmic rays, or the naturally occurring radioactive atoms that emit subatomic particles such as electrons and alpha particles at various energies ranging from kilovolts to over 8 MeV. *See* RADIOACTIVITY.

The energy of an accelerated atomic or subatomic particle is usually expressed in units of electronvolts (eV). An electronvolt is the amount of energy that a particle with unit charge, such as an electron or proton, receives when accelerated through a potential difference of 1 V. Therefore, when a proton is accelerated with an electrostatic accelerator operating at 10^6 V, its energy will be 10^6 eV; equivalently, if the particle has q units of charge, its resultant energy in this example will be q MeV. Commonly used multiple units are kiloelectronvolts (keV), megaelectronvolts (MeV), gigaelectronvolts (GeV), and teraelectronvolts (TeV). *See* ELECTRONVOLT.

Very high energy atomic and subatomic particles are accelerated in outer space, and constantly bombard the Earth in the form of cosmic rays ranging in energy from hundreds of kiloelectronvolts to hundreds of thousands of megaelectronvolts. These natural sources of particles are uncontrollable and must be used for research in whatever energy, intensity, and kind of natural radiation is available. Particle accelerator devices, on the other hand, allow these same particles to be accelerated to precise energies with complete control of the intensity and energy over wide ranges. The particles may also be directed to collide with specific target materials in whatever way desired so as to greatly expand knowledge of the fundamental interactions of charged particles with each other and with other materials. Moreover, with accelerators it is possible to produce a wide variety of secondary beams of exotic particles with such short half-lives that they would otherwise be unavailable for controlled experimentation. *See* COSMIC RAYS; ELEMENTARY PARTICLE.

Accelerators that produce various subatomic particles at high intensity have many practical applications in industry and medicine as well as in basic research. Electrostatic generators, pulse transformer sets, cyclotrons, and electron linear accelerators are used to produce high levels of various kinds of radiation that in turn can be used to polymerize plastics, provide bacterial sterilization without heating, and manufacture radioisotopes which are utilized in industry and medicine for direct treatment of some illnesses as well as research. They can also be used to provide high-intensity beams of protons, neutrons, heavy ions, pi mesons, or x-rays that are used for cancer therapy and research. The x-rays used in industry are usually produced by arranging for accelerated electrons to strike a solid target. However, with the advent of electron synchrotron storage rings that produce x-rays in the form of synchrotron radiation, many new industrial applications of these x-rays have been realized, especially in the field of solid-state microchip fabrication and medical diagnostics. *See* ISOTOPIC IRRADIATION; RADIATION BIOLOGY; RADIATION CHEMISTRY; RADIOACTIVITY AND RADIATION APPLICATIONS; RADIOGRAPHY; RADIOISOTOPE; RADIOLOGY; SYNCHROTRON RADIATION.

Particle accelerators fall into two general classes—electrostatic accelerators that provide a steady dc potential, and varieties of accelerators that employ various combinations of time-varying electric and magnetic fields.

Electrostatic accelerators. Electrostatic accelerators in the simplest form accelerate the charged particle either from the source of high voltage to ground potential or from ground potential to the source of high voltage. The high-voltage dc potential may be either positive or negative, and consequently positive or negative particles will be accelerated by being either attracted to or repelled from the high voltage.

All particle accelerations are carried out inside an evacuated tube so that the accelerated particles do not collide with air molecules or atoms and may follow trajectories characterized specifically by the electric fields utilized for the acceleration. Usually the evacuated tube is provided with a series of electrodes arranged to have gradually increasing potentials from ground to the maximum high-voltage potential. In this way, the high voltage is distributed uniformly along the acceleration tube, and the acceleration process is thereby simplified and better control permitted. The maximum energy available from this kind of accelerator is limited by the ability of the evacuated tube to withstand some maximum high voltage. This limitation in energy was a severe problem in the early days of nuclear and atomic research, when only electrostatic accelerators were available, and led to the development of the second kind of accelerator, which uses time-varying electric or magnetic fields, or both, and is not restricted by any particular limit of potential that can be maintained.

Time-varying field accelerators. In contrast to the high-voltage-type accelerator which accelerates particles in a continuous stream through a continuously maintained increasing potential, the time-varying accelerators must necessarily accelerate particles in small discrete groups or bunches. Since the voltage on any given electrode is varying in time, at certain times the voltage will be suitable for acceleration, while at other times it would actually decelerate the particles. For this reason the electrodes must be arranged so that the particle bunches appear in their vicinity only when the voltage is correct for acceleration.

Linear accelerators. An accelerator that varies only in electric field and does not use any magnetic guide or turning field is customarily referred to as a linear accelerator or linac. In the simplest version of this kind of accelerator, the electrodes that are used to attract and accelerate the particles are connected to a radio-frequency (rf) power supply or oscillator so that alternate electrodes are of opposite polarity. In this way, each successive gap between adjacent electrodes is alternately accelerating and decelerating. If these acceleration gaps are appropriately spaced to accommodate the increasing velocity of the accelerated particle, the frequency can be adjusted so that the particle bunches are always experiencing an accelerating electric field as they cross each successive gap. In this way, modest voltages can be used to accelerate bunches of particles indefinitely, limited only by the physical length of the accelerator construction.

All conventional (but not superconducting) research linacs usually are operated in a pulsed mode because of the extremely high rf power necessary for their operation. The pulsed operation can then be adjusted so that the duty cycle or amount of time actually on at full power averages to a value that is reasonable in cost and practical for cooling. This necessarily limited duty cycle in turn limits the kinds of research that are possible with linacs; however, they are extremely useful (and universally used) as pulsed high-current injectors for all electron and proton synchrotron ring accelerators. Superconducting linear accelerators have been constructed that are used to accelerate electrons and also to boost the energy of heavy ions injected from electrostatic machines. These linacs can easily operate in the continuous-wave (cw) rather than pulsed mode, because the rf power losses are only a few watts.

The Continuous Electron Beam Accelerator Facility (CEBAF) uses two 400-MeV superconducting linacs to repeatedly accelerate electrons around a racetracklike arrangement where the two linacs are on the opposite straight sides of the racetrack and the circular ends are a series of recirculation bending magnets, a different set for each of five passes through the two linacs in succession. The continuous electron beam then receives a 400-MeV acceleration on each straight side or 0.8 GeV per turn, and is accelerated to a final energy of 4 GeV in five turns and extracted for use in experiments. The superconducting linacs allow for continuous acceleration and hence a continuous beam rather than a pulsed beam. This makes possible many fundamental nuclear and quark structure measurements that are impossible with the pulsed electron beams from conventional electron linacs. *See* SUPERCONDUCTING DEVICES.

Circular accelerators. As accelerators are carried to higher energy, a linac eventually reaches some practical construction limit because of length. This problem of extreme length can be circumvented conveniently by accelerating the particles in a circular path maintained by either static or time-varying magnetic fields. Accelerators utilizing steady magnetic fields as guide paths are usually referred to as cyclotrons or synchrocyclotrons, and are arranged to provide a steady magnetic field over relatively large areas that allow the particles to travel in an increasing spiral orbit of gradually increasing size as they increase in energy. After many accelerations through various electrode configurations, the particles eventually achieve an orbit as large as the maximum size of the magnetic field and are extracted for use in research and other kinds of applications.

A special kind of circular accelerator for electrons using a static magnetic field is called a microtron. A constant frequency in a small acceleration gap or cavity near the edge of a circular-shaped uniform magnetic field is arranged so that it accelerates the electrons tangentially. Electrons are provided at an energy of one or a few megaelectronvolts so that they are moving close to the velocity of light. This means that when the electrons are accelerated their velocity can increase only slightly because the velocity of light cannot be exceeded according to the laws of special relativity. The acceleration and increase in energy are consequently accomplished by increasing their mass instead of their velocity. With constant velocity, the acceleration frequency can also be constant and arranged in such a way that the circular orbits increase in circumference by one or more beam bunches or wavelengths on each revolution of larger diameter. Much larger microtrons in a racetracklike configuration of two opposed halfmoon-shaped magnets have been built with an electron linac between them and aligned with the outermost racetrack orbit.

Practical limitations of magnet construction and cost have kept the size of circular proton accelerators with static magnetic fields to the vicinity of 100 to 1000 MeV. Most circular accelerators can easily operate over a range of 10 to 1 and some as high as 30–50 to 1. However, increasing the size and coupling of smaller to larger machines is necessary for larger ranges in energy. For even higher energies, up to 400 GeV per nucleon in the largest conventional (not superconducting) proton synchrotron in operation, it is necessary to vary the magnetic field as well as the electric field in time. In this way the magnetic field can be of a minimal practical size, which is still quite extensive for a 980-GeV accelerator (6500 ft or 2000 m in diameter). This circular magnetic containment region, or "racetrack," is injected with relatively low-energy particles that can coast around the magnetic ring when it is at minimum field strength. The magnetic field is then gradually increased to stay in step with the higher magnetic rigidity of the particles as they are gradually accelerated with a time-varying electric field. Again, when the particles achieve an energy corresponding to the maximum magnetic field possible in the circular guide field, they are extracted for utilization in research programs, and the magnetic field is cycled back down to its low or near-zero value and the acceleration process is repeated.

Although electron synchrotrons were formerly used in high-energy physics research and as synchrotron radiation sources, they are now used exclusively as injectors for storage rings that provide the research capabililty.

Focusing. One of the chief problems in any accelerator system is that of maintaining spatial control over the beam; this requires some form of focusing. In addition, the presence of focusing elements makes it possible to trade off beam cross-sectional area and angular divergence as the specific utilization may require.

A wide variety of focusing principles and devices are used in different accelerator types. The natural bowing-out of magnetic field lines in a cyclotron magnet results in a net focusing action as will be discussed below; so also does the electrostatic field line configuration between drift tubes of many simple linear accelerators. But these systems are classed generally as weak-focusing approaches. The particle beams under their control can make relatively large excursions from the desired equilibrium orbits with the consequence that large vacuum envelopes are essential if the beams are not to be lost through collision with the envelope walls.

Fortunately, an alternate strong-focusing approach, first described in 1952, makes possible very substantial improvements in this area. It uses electrostatic and magnetic lens systems arranged alternately positive and negative (in the sense of convex and concave optical elements) so that, as in the optical analog, they have a net- and strong-focusing action on charged particle beams. This alternation of positive and negative elements has been implemented in many ways, with many devices and for a variety of geometries. Perhaps its most frequent appearance is in quadrupole (four-pole) or hexapole (six-pole) magnets used as variable-focal-length, variable-astigmatism elements in beam transport both within large accelerators themselves and in extensive beam-transport systems external to the accelerator.

Superconducting magnets. The study of the fundamental structure of nature and all associated basic research require an ever increasing energy in order to allow finer and finer measurements on the basic structure of matter. Since the voltage-varying and magnetic-field-varying accelerators also have limits to their maximum size in terms of cost and practical construction problems, the only way to increase particle energies even further is to provide higher-varying magnetic fields through superconducting magnet technology, which can extend electromagnetic capability by a factor of 4 to 5. Large superconducting cyclotrons and superconducting synchrotrons are in operation. *See* MAGNET.

Storage rings. Beyond the limit just described, the only other possibility is to accelerate particles in opposite directions and arrange for them to collide at certain selected intersection regions around the accelerator. The main technical problem is to provide adequate numbers of particles in the two colliding beams so that the probability of a collision is moderately high. This intensity is usually expressed as luminosity and is in units of number of particles per square centimeter per second. Such storage ring facilities are in operation for both electrons and protons. Besides storing the particles in circular orbits, the rings can operate initially as synchrotrons and accelerate lower-energy injected particles to much higher energies and then store them for interaction studies at the beam interaction points.

TABLE 1. Operating characteristics of particle accelerators

Accelerator type	Particle accelerated	Energy range*	Beam current (average; peak) or beam intensity (luminosity)	Duty cycle	Energy spectrum†	Beam geometry	Development status (2000)
ELECTROSTATIC ACCELERATORS							
Tandetron	p, d, α, heavy ions	To 3 MV	1μA; 10 μA	Continuous	~0.01%	Small focal spot	1–3 MV operating
Cockcroft-Walton	p, d, α, e, heavy ions	To 4 MV	1 mA; 10 mA	Continuous	~0.01%	Small focal spot	4 MeV operating
Dynamitron	p, d, α, e, heavy ions	To 4 MV	1 μA; 50 mA	Continuous	~0.01%	Small focal spot	4 MV operating
Tandem Van de Graaff	p, d, α, e, heavy ions	To 20 MV	1 μA; 50 μA	Continuous	~0.01%	Small focal spot	20 MV operating
Tandem pelletron	p, d, α, e, heavy ions	To 26 MV	1 μA; 50 μA	Continuous	~0.01%	Small focal spot	26 MV operating
Vivitron	p, d, α, e, heavy ions	To 18 MV	1 μA; 50 μA	Continuous	~0.01%	Small focal spot	18 MV operating
TIME-VARYING FIELD ACCELERATORS							
Circular magnetic types (radio-frequency resonance accelerators)							
Microtron	e^-	To 855 MeV	10 μA; 100 mA	0.1%	0.1%	Small focal spot	30–855 MeV operating
Sector or isochronous cyclotron	p, d, α, heavy ions	To 590 MeV (p); 90 MeV/u	20 μA; 2 mA	Continuous	~0.01%	Internal target or external beam with small focal spot	590 MeV (p), 90 MeV/u (heavy ions) operating Small focal spot 200 MeV/u operating
Superconducting cyclotron	Heavy ions	200 MeV/u	1 μA; 100μA	Continuous	~0.01%	Internal or external beam with small focal spot	200 MeV/u operating
Synchrotron (weak focusing)	p, e, heavy ions	1–6 GeV (p); 2 GeV/u	0.1 μA; depends on duty cycle	30%	0.1%	Internal targets; external beam of fair collimation at lower intensity	Mostly closed down
Alternating-gradient synchrotron	p, e^+, e^+, heavy ions: mass 12–197 mass 12–208	10–980 GeV (p) 11.4 GeV/u 160 GeV/u	1.0 μA (p) 10^8–10^{10} ions/s 10^8–10^{10} ions/s	30% 30% 30%	0.1% 0.1% 0.1%	(p) internal targets; external beam of fair collimation	980 GeV (p) operating (superconducting); heavy-ion machines operating; gold (^{197}Au) and lead (^{208}Pb) beams operating
Linear accelerators							
Heavy-ion linear accelerator	p, d, α, heavy ions	To 30 MeV/u	10 μA; 130 mA	~10%	0.5%	Well-collimated and well-focused external beam	30 MeV/u (heavy ion) operating (normal and heavy superconducting)
Linear accelerator	p	50–800 MeV	1 mA; 100 mA	~10%	0.1%	Well-collimated and well-focused external beam	800 MeV (p) operating
TJNAF recirculating superconducting linear accelerator	e^-	0.5–4 GeV	200 μA	Continuous	0.003%	Well-collimated and well-focused external beam	4 GeV operating
Electron linear accelerator	e^+, e^-	6 MeV to 50 GeV; 100 GeV (CM) in e^+, e^- collider mode	60 μA; 400 μA 10^{29}/cm^{2}s	~6%	~ 0.2% 0.06%	Well-collimated well-focused external beam	50 GeV operating (SLED mode)
Colliding-beam storage rings							
Electron-positron storage ring	e^+, e^-	0.3–208 GeV (CM)	~ 10^{31}/cm^2 s	Continuous	0.1%	Small-diameter internal beam	208 GeV (CM) closed down
Proton-proton collider	pp	14 TeV (CM)	~ 10^{34}/cm^2 s	Continuous	0.1%	Small-diameter internal beam	14 TeV (CM) under construction
Proton-antiproton collider	$p\bar{p}$	1.96 TeV (CM)	3×10^{31}/cm^2 s	Continuous	0.1%	Small-diameter internal beam	1.96 TeV (CM) operating
Electron-proton collider	e-p	27 GeV (e), 820 GeV (p)	2×10^{31}/cm^2 s	Continuous	0.1%	Small-diameter internal beam	27 GeV (e), 820 GeV (p) operating
Heavy-ion collider	p, heavy ions	100 GeV/u (^{197}Au), 250 GeV (p)	10^{27}/cm^2 s (^{197}Au), 1.4×10^{31}/cm^2 s (p)	Continuous	0.1%	Small-diameter internal beam	100 GeV/u (^{197}Au) operating

*Voltage range is given for electrostatic accelerators. u = atomic mass unit. CM = center of mass.
†Spread in energy of beam expressed as a percentage of total energy of beam; that is, 1% for the cyclotron means for a 1-MeV beam a spread in energy of 0.01 MeV.

Large proton synchrotrons have been used as storage-ring colliders by accelerating and storing protons in one direction around the ring while accelerating and storing antiprotons (negative charge) in the opposite direction. The proton and antiproton beams are carefully programmed to be in different orbits as they circulate in opposite directions and to collide only when their orbits cross at selected points around the ring where experiments are located. The antiprotons are produced by high-energy proton collisions with a target, collected, stored, cooled, and eventually injected back into the synchrotron as an antiproton beam. It takes 24–36 h to collect and prepare an antiproton beam for injection and use in the collider mode of operation, and the intensity is limited ($^1/_{1000}$ or less than the proton beam) because of the difficulty in producing an antiproton beam.

Electron-positron synchrotron accelerator storage rings have been in operation for many years in the basic study of particle physics, with energies ranging from 2 GeV + 2 GeV to 104 GeV + 104 GeV. The by-product synchrotron radiation from many of these machines is used in numerous applications. In fact, the production of synchrotron radiation is proportional to the rf power necessary to keep the circulating stored beam on a stable orbit. This synchrotron radiation loss forces the machine design to larger and larger diameters, characterized by the Large Electron Positron Storage Ring (LEP) at CERN, near Geneva, Switzerland (closed down in 2000), which was 17 mi (27 km) in circumference. Conventional rf cavities enable electron-positron acceleration only up to 50–70 GeV (limited by synchrotron radiation loss) while higher energies of 100–150 GeV require superconducting cavities.

Advanced linacs. Although circular machines with varying magnetic fields have been developed because linacs of comparable performance would be too long (many miles), developments in linac design and utilization of powerful laser properties may result in a return to linacs that will outperform present ring machines at much lower cost. As a first example, the 20-GeV electron linac at Stanford University, Palo Alto, California, has been modified to provide simultaneous acceleration of positrons and electrons to energies as high as 50 GeV, while operating in what is called the SLED mode. After acceleration the electrons and positrons are separated by a magnet, and the two beams are magnetically directed around the opposite sides of a circle so that they collide at one intersection point approximately along a diameter extending from the end of the linac across the circle. This collider arrangement is much less expensive than the 17-mi (27-km) ring at CERN and provides electron-positron collisions of comparable energies but at lower intensities.

Performance characteristics. This introduction has shown that there are many kinds of accelerators, varying in size and performance characteristics and capable of accelerating many kinds of particles at varying intensities and energies, in the form of bunched or pulsed beams of particles or as a steady dc current (**Table 1**).

Electrostatic Accelerators

Electrostatic accelerators are used to accelerate atomic or subatomic particles to high energies by means of a high voltage potential that is maintained through some electrical or mechanical means of transporting charge from ground to the high voltage potential. Modern machines of larger size all use mechanical charge transport, but electrical transport systems are widely used in small Cockcroft-Walton or dynamitron-type electrostatic accelerator units. The most commonly used electrostatic generator, invented by R. J. Van de Graaff, utilizes a high-speed rubberized-fabric belt to transport the charge. Many machines utilizing this principle have been home-made or built by various research groups throughout the world.

An improved charging system, the Pelletron® electrostatic accelerator, manufactured by National Electrostatics Corporation, utilizes a chain-charging system instead of the rubberized-fabric belt. The continuous-looped charging chain is fabricated of alternate metal and insulated links and travels at high speed on special pulleys from ground to high potential, an arrangement mechanically similar to the charging belt system. Instead of the charge being sprayed onto the charging belt system from sharp corona points, each metal link is charged and discharged by induction methods so that no electrical sparking or erosion is involved.

Most electrostatic accelerators are housed inside a large high-pressure vessel that is filled to a pressure as high as 15 atm (220 lb/in.^{2} or 1.5×10^6 pascals) with very dry insulating gas. The insulating gas may be pure sulfur hexafluoride (SF_6) or a mixture with carbon dioxide (CO_2) and nitrogen (N_2). This high-pressure insulating gas allows the machine to be housed in a much smaller space than would be possible in air at atmospheric pressure. *See* ELECTRICAL INSULATION.

Basic design. The basic high-voltage electrostatic generator may be considered in the form of a van de Graaff high-voltage generator (**Fig. 1**) in which electric charge is carried on an insulated belt from ground to an insulated high-voltage terminal. The high-voltage terminal usually contains an ion source that produces protons, other types of positive ions, or electrons that can then be accelerated by the high voltage of the generator through an evacuated tube—the acceleration tube—back to ground potential. *See* ION SOURCES.

A motor at the base of the machine with a grounded pulley drives either a fabric or rubberized fabric belt over a second pulley that is mounted inside the high-voltage terminal and connected or grounded to it. A comblike set of points is arranged at the base of the machine to spray charge onto the belt, and a similar device removes the charge when it arrives at the high-voltage terminal.

The acceleration tube is fabricated of alternate rings of insulating glass or ceramic and metal electrodes glued or bonded together to provide a hollow tubelike structure, extending from the high-voltage

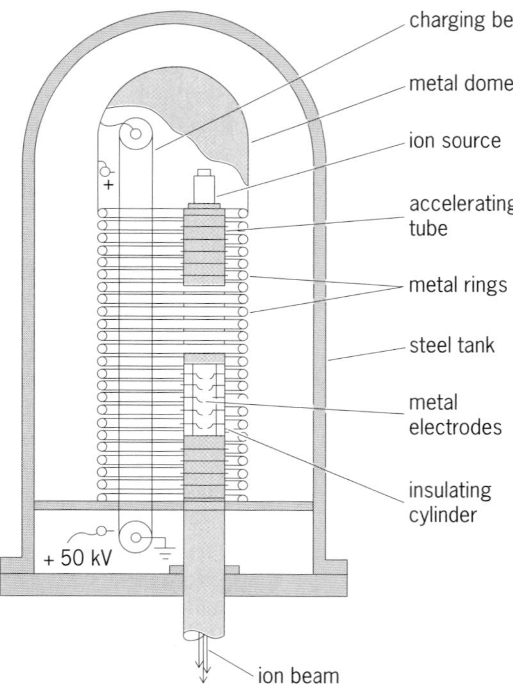

Fig. 1. Van de Graaff generator for operation in high-pressure gas.

terminal to the base of the machine. Power for the operation of the ion source, which often requires a hot filament and an electric arc power supply, is usually supplied by a generator driven by a belt from the charging pulley. In some designs the generator is inside the charging pulley itself. The acceleration tube is evacuated by vacuum pumps at the base of the machine.

The electrical voltage from the high-voltage terminal to ground is graded in a series of steps corresponding to each of the metal electrodes of the acceleration tube and the basic insulating support column itself. This gradient is provided by resistors strung together in series from the high-voltage terminal to ground, each connected to the metal electrodes and the metal rings. The resistor string then provides a stepped high voltage from the metal dome or terminal of the machine uniformly downward from terminal voltage to ground with each of the metal rings and corresponding electrodes in the tube. In this way, the injected ion from the source is accelerated by a continually accelerating electric field as it moves toward ground potential. After accelerating through the entire gradient of a machine operating at 1 MV, a charged particle has 1 MeV of energy for each unit charge. A similar machine but with a Pelletron charging system instead of a charging belt is called a Pelletron generator.

This basic design is used only for accelerators with capabilities up to a few megavolts, with a 6-MV unit being the largest commercially available. The main operational difficulty with this design is that the ion source, which is inside the high-voltage terminal, requires regular maintenance, and this in turn requires depressurization of the pressure vessel for maintenance and repressurization for operation. This procedure can take a day or more with the larger machines.

Electrical charging generators. The basic accelerator design with mechanical charging systems (Van de Graaff or Pelletron) is limited to continuous particle beam currents of a few hundred microamperes because of the limited amount of charge that can be transported mechanically to the high-voltage terminal. This limitation prompted the design of electrical charging systems that are capable of transporting a thousand times as much charge or hundreds of milliamperes.

Cockcroft-Walton accelerator. The original Cockcroft-Walton accelerator transports charge to a high-voltage terminal electrically by using voltage multiplying circuitry. In its most rudimentary form, it contains a high-voltage transformer and a voltage-multiplying circuit (**Fig. 2**). The multiplier circuit consists of two stacks of series-connected capacitors, cross-linked by voltage rectifiers. One set of capacitors is connected to the transformer. During the negative portion of the transformer cycle, the first capacitor in this stack, C_1, becomes charged. During the positive portion, some of this charge is transferred to the first capacitor in the second set, C_2. This process continues, with the transformer effectively pumping charge through the rectifier chain, until all capacitors are charged. Ultimately, the potential difference across each capacitor is $2V$, where V is the peak transformer output voltage, and the voltage across each complete set of N capacitors is $2NV$. Since the second set is referenced to ground, the voltage across it is constant except for ripple caused by power drain. To reduce ripple, alternating voltages at frequencies of several kilohertz are usually supplied to the transformer. Direct-current voltages of either sign can be obtained depending on the orientation of the rectifiers. Modern supplies, which use solid-state rectifiers, have many stages of multiplication

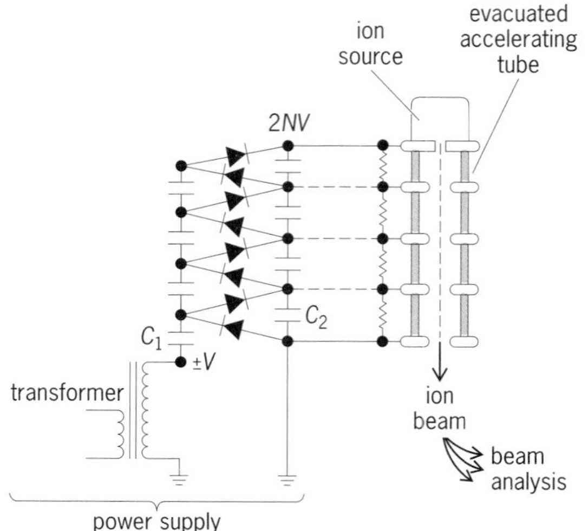

Fig. 2. Cockcroft-Walton power supply and accelerator. The multiplier circuit shown has four stages ($N = 4$).

with relatively modest transformer voltages. Modern Cockcroft-Walton machines are used mainly as ion injectors for large synchrotron facilities. They are open-air insulated (no pressure vessel) for convenient and rapid ion-source maintenance, and limited to voltages of 0.5–1.0 MV. All of these large dc machines are being replaced by radio-frequency quadrupole accelerators, which are a type of linear accelerator capable of accelerating directly from an ion source to energies suitable for injection into conventional linear accelerators. *See* SEMICONDUCTOR RECTIFIER; VOLTAGE-MULTIPLIER CIRCUIT.

Dynamitron accelerator. Although the Cockcroft-Walton voltage multiplication system provides moderately high voltages (~1 MV) at high currents, it is difficult to go to much higher voltages because the multiplier arrangement is essentially a series-driven system from ground up to the high voltage. An alternative system, known as the dynamitron accelerator, that operates on a parallel rather than series arrangement can easily drive to much higher voltages (4–5 MV). Most other cascade-type systems operate at low frequency (60–500 Hz) and rely on large energy storage in inductance or capacitance or both. In contrast, the dynamitron uses high-frequency (100 kHz) and low-capacitive coupling with consequently low energy storage, which helps minimize the consequences of component failure.

Two large, semicylindrical rf electrodes operate at approximately 150 kV at 100 kHz, with a resonant inductor and oscillator system (**Fig. 3**). Each split corona shield on the high-voltage column of the accelerator acts as an rf pickup electrode, which in turn drives an individual rectifier. As a pair, the split corona shields also help provide a smooth dc potential distribution, just as do the circular equipotential rings in an electrostatic accelerator.

Modern units use a special solid-state rectifier for rf rectification, which simplifies the design and increases the maximum current capacity. Power is coupled into each rectifier through the shield-to-electrode capacitance.

Low-energy machines. The Cockcroft-Walton series-fed voltage multiplier and the dynamitron parallel-fed voltage multiplier characterize the two main methods of voltage production among a myriad of compact machines. Many machines are small compact accelerators connected to a separate solid-state, high-voltage power supply with a high-voltage cable rated up to 1 MV capacity. Modern production methods require small accelerators for crucial diagnostic capabilities and often in the production process itself.

Neutron tube. One interesting example of a tiny accelerator system is the neutron tube accelerator, devised for logging boreholes drilled for oil and mineral prospecting. These accelerators operate at 0.1 MV or less, but this is sufficient for deuterium ions to produce 14-MeV neutrons when they strike a tritium target. The ion source, accelerator, power supplies, and radiation detectors are assembled into a long probe a few inches in diameter which is lowered into the borehole. The accelerator is pulsed on for a few microseconds, during which large numbers of neutrons are produced. These neutrons slow down and are absorbed. The time distribution and types of radiation which result are related to the local nature of the stratum surrounding the borehole.

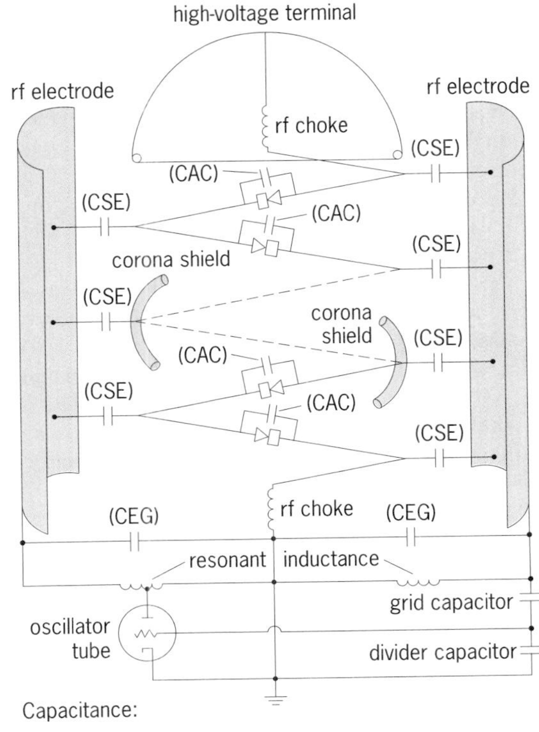

Fig. 3. Radio-frequency coupling and direct-current rectification system utilized in the dynamitron accelerator. (*After R. F. Shea, ed., 1st National Particle Accelerator Conference, IEEE Trans. Nucl. Sci., NS-12(3):227–234, 1965*)

Tandem electrostatic generator. The tandem electrostatic generator utilizes the basic Van de Graaff generator principle (Fig. 1) in a special form. The pressure vessel is in the form of a long cylinder, and the high voltage terminal is supported in the middle of the cylinder by an insulating column structure extending from both ends of the cylinder. The charging belt operates from the high-voltage terminal to either or both ends of the cylinder, and an acceleration tube passes through from one end of the cylinder to the other so that accelerated particles can pass through the machine. The support column, high-voltage terminal, metal rings, and so forth, are all arranged concentrically about the axis of the cylindrical pressure vessel. This structure is actually two basic electrostatic generators, on the same axis and sharing a common high-voltage terminal.

Negative ions from a negative-ion source outside one end of the cylindrical structure are injected into the acceleration tube at one end of the machine, and then accelerated to the high-voltage terminal. There they pass through a thin foil usually made of carbon approximately 10^{-6} in. (0.02 μm) thick (or a gas stripper), and electrons are stripped away from them, leaving them as positive ions. The particles are then accelerated out of the other end of the acceleration tube just as though they were produced by an ion source in the high-voltage terminal. In this way, a 1-MV machine can provide 2-MeV protons, giving them 1-MeV acceleration as a negative ion and 1 MeV as a positive ion.

Even more advantage is gained when heavy ions of various atomic nuclei are accelerated. The negative ion again achieves 1 MeV acceleration with a 1-MV machine; however, the foil will probably strip the heavy ion to a charge state of two or three units of positive charge, which means it then achieves an additional 2 or 3 MeV of energy, reaching a total energy of 3 or 4 MeV.

The tandem electrostatic generator has been widely used as a research instrument because of the convenience of having the ion source and complicated ion-producing devices of various kinds outside the accelerator at ground potential rather than inside the high-voltage terminal as in the basic electrostatic accelerator.

Folded tandem design. If a shorter tandem electrostatic accelerator is desired, it is possible to combine the two end-to-end machines, making a tandem by folding the two machines together 180° at the high-voltage terminal into a common column structure that now contains two acceleration tubes rather than one. In this way, one tube accelerates the negative ions to the high-voltage terminal, where they are stripped and reversed in direction by a 180° bending magnet and then accelerated back to ground potential by the second acceleration tube.

Large machines. Very large tandem Van de Graaff generators and tandem Pelletron generators have been manufactured commercially, with the largest tandem Van de Graff generator rated at 20 MV, and the largest tandem Pelletron generator rated at 25 MV.

The largest tandem electrostatic accelerator, with a design voltage of 35 MV, is the Vivitron in Strasbourg, France. This accelerator has a charging belt running from one end to the other, and features seven concentric partial electrostatic shields called porticos. Each of the seven concentric shells operate in successive 5-MV steps starting next to the high-voltage terminal. The first shell is common and concentric to each end of the high-voltage column (containing the acceleration tubes and charging belt) at the 30-MV point. The next portico is then common to the 25-MV point and so on, building the concentric shells out to the pressure vessel walls. Each portico is mechanically supported from the next by radially mounted insulators which ultimately support the high-voltage terminal and two high-voltage columns. The accelerator turned out to have many design flaws. It now operates at up to 18 MV.

Tandem electrical charging generators. The tandem or two-stage acceleration principle has also been used with the electrical charging generators, specifically the dynamitron and a smaller lower-powered version of the machine called a tandetron. Two 4-MV tandem dynamitrons were built for research use; however, ion accelerators are no longer produced, and only high-current electron accelerators are manufactured for industrial processing. On the other hand, the smaller, lower-powered tandetron is used exclusively for ion acceleration and is widely used in ion implanting, carbon dating, and many different kinds of atomic and solid-state basic research programs. The tandetrons are made in 1-, 2-, and 3-MV sizes and are small enough to fit into most available laboratories. A special 0.7-MV folded version is made for helium backscattering studies and utilizes a specially designed 180° permanent-magnet bending arrangement inside the high-voltage terminal. *See* ION IMPLANTATION; MASS SPECTROSCOPE; RADIOCARBON DATING.

Use as injectors. The ease of energy adjustment, ease of change of ion species, and the high-quality ion beam, which can be easily pulsed to match the input requirements of linacs, cyclotrons, or synchrotrons, make the tandem an ideal injector for other kinds of accelerators, which then may be used to boost the limited energy of the tandem accelerator. Many tandem research laboratories have booster accelerators in the form of either normal or superconducting linacs and cyclotrons.

Applications. Although the original driving need for larger machines in the study of nuclear physics has now largely moved on to energies beyond the practical capability of electrostatic machines, they will continue to support the field as injectors into other machines and be used as powerful tools in the study of atomic, molecular, and solid-state physics. Many application studies and commercial processing operations are carried out with heavy ions from tandem accelerators. The smaller electrostatic accelerators are in constant demand in all fields of research, and industrial applications are expanding on a steady basis. *See* NUCLEAR PHYSICS. Harvey E. Wegner

Circular Accelerators

Circular accelerators utilize a magnetic field to bend charged-particle orbits and confine the extent of particle motion.

Cyclotrons. The cyclotron is a circular accelerator in which the particle orbits start at the center and spiral outward in a guide magnetic field which is constant in time. The basic cyclotron concept, put forward in 1930, is given by Eq. (1), the Lawrence

$$\omega = \frac{qB}{m} \tag{1}$$

equation for the angular frequency of rotation of a particle in a plane perpendicular to a magnetic field. Here, q and m are the charge and mass of the particle and B is the magnetic induction of the perpendicular magnetic field. (SI or mksa units are used throughout this section of the article.) This equation can be

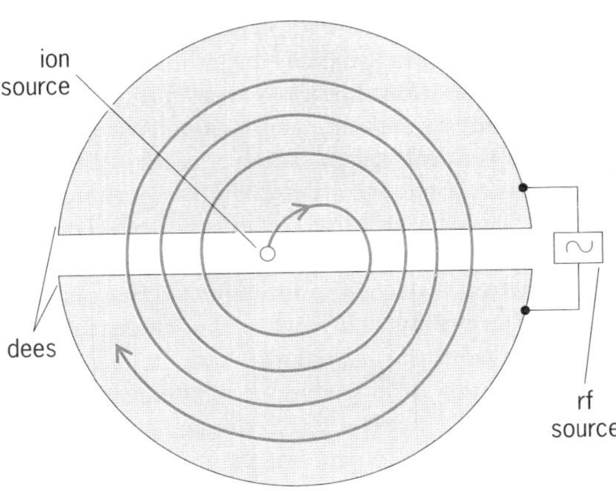

Fig. 4. Principle of the cyclotron. An ion in the accelerator follows the arrowed path. The magnetic field is perpendicular to the page.

derived by noting that (1) $\omega = v/r$, where v is the linear velocity and r is the orbital radius; (2) the magnetic force is perpendicular to the velocity and is given by qvB; (3) this perpendicular force produces a centripetal acceleration and, from Newton's second law, gives $qvB = mv^2/r$. Solving the last equation for v/r leads directly to Eq. (1). The striking statement of Eq. (1) is that if field, charge, and mass are all constant, then angular velocity is likewise constant and, in particular, does not depend on the linear velocity. Faster-moving particles travel on circles of larger radius—the increase in the length of a revolution due to the larger radius just matches the increase in speed, and fast or slow particles thus require the same amount of time to make a 360° revolution. Thus, an accelerating electrode operating at constant frequency can be used. This is the key idea of the Lawrence cyclotron. *See* MAGNETISM; NEWTON'S LAWS OF MOTION.

In such a cyclotron (**Fig. 4**), a pair of D-shaped accelerating electrodes, called dees, is electrically attached to an rf voltage source whose frequency is matched to the rotation frequency of the particle in a surrounding magnetic field. Ions are introduced at the center, either from an ion source placed there or from an external ion source, and immediately encounter the high voltage on the facing dee. Those which enter the gap between the two dees at a time when the rf voltage produces an accelerating electric field are accepted for acceleration; the remainder are lost. Beyond the accelerating gap the particle passes inside the dee, which is an electrically shielded region; at the same time it is pulled into a circular path by the perpendicular magnetic field, and after 180° it arrives again at the gap between the dees. Since the frequency of the rf source which drives the dees is selected to match the rotational frequency of the particle, the voltage between the dees will have reversed while the particle is making its 180° turn, and therefore the particle is again accelerated as it crosses from dee to dee. This process can clearly be repeated as often as desired. The particle will thus be repetitively accelerated and, as it speeds up, will gradually move on circles of larger and larger radius. Finally the particle will come to the limit of the magnetic field. At this point an additional electrode—referred to as a deflector—can be inserted to direct all particles out along a single path.

Limitations. This theoretical picture of exactly synchronized orbits and voltage is an oversimplification, and more detailed investigation reveals two fundamental problems with the cyclotron concept. As any particle speeds up, its total energy E will rise, and therefore, according to the equation $E = mc^2$ of relativistic mechanics, its mass m will increase, the speed of light c being a constant. In a constant magnetic field the gradual increase in mass causes the rotation frequency to steadily decrease as the particle speeds up, and the particle then gradually lags behind a constant rf frequency. A second limitation comes from the magnetic field factor in Eq. (1). To prevent particles from drifting away from the plane because of their initial velocities there must be a restoring force which pushes particles back toward the plane. A way to obtain the needed restoring force is to make the magnetic field lines bend toward the axis of rotation of the particle. However, according to Maxwell's equations, the field strength perpendicular to the plane of motion must then decrease as the radius becomes larger, producing a slowing-down according to Eq. (1). This effect adds to that of the mass increase, and the two effects limit the energy of the Lawrence cyclotron. The highest energy achieved by such a cyclotron was 23 MeV for a beam of protons. *See* MAXWELL'S EQUATIONS.

Synchrocyclotron. A way to avoid the energy limit on the Lawrence cyclotron is to slow down the frequency of the rf power source at the same rate as the slowing down of the rotational frequency of the particle. But this condition can be maintained only for a reference particle at a particular "synchronous phase" (**Fig. 5**). The usefulness of the process

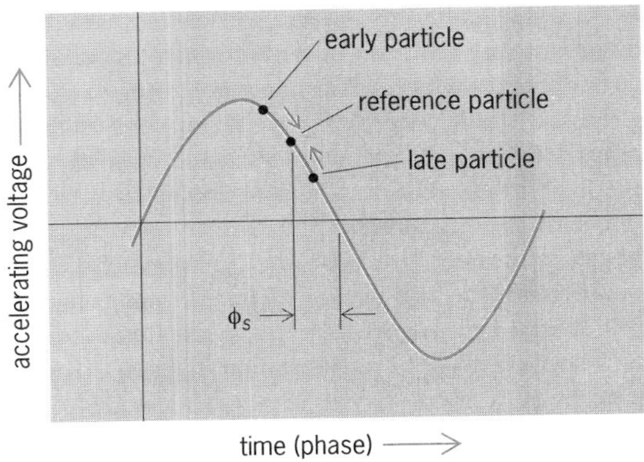

Fig. 5. Principle of phase stability in the synchrocyclotron. The reference particle has resonance energy and synchronous phase, ϕ_s. The early particle accelerates faster to larger radius, where it has a lower rotation frequency which moves it later in time toward the central particle. The late particle is correspondingly moved earlier in time toward the central particle.

Fig. 6. Partially assembled magnet for the 500-MeV TRIUMF cyclotron, showing the spiral iron sectors of the lower pole. (*TRIUMF, Vancourer, BC*)

depends on the stability of the motion of neighboring nonsynchronous particles. Will particles which are not exactly at the design values of energy and phase remain close to those values in synchronism with the rf voltage (stability) and accelerate, or diverge from the design values (instability) and not accelerate?

Particles which arrive at the accelerating gap before the synchronous reference particle experience a larger accelerating voltage and energy gain than the reference particle; this causes their mass to increase faster than that of the reference particle, resulting in their revolving more slowly according to Eq. (1). The larger energy gain also causes these early particles to reach a larger radius and thus a lower magnetic field than the reference particle, slowing their revolution even more. They thus gradually lose their early lead and are forced toward the synchronous phase. Similarly, particles which reach the accelerating gap late receive a smaller energy increase, but consequently revolve faster and gradually catch up with the reference particle. This restoring action relative to the phase of the synchronous particle is referred to as phase focusing. It occurs only if the synchronous phase is on the falling side of the rf voltage wave; if it is on the rising side, the phase motion is unstable. The principle of phase stability was discovered toward the end of World War II, and made possible the frequency-modulated cyclotrons or "synchrocyclotrons" that provided the leading advances in ion beam energy in the postwar years. The largest synchrocyclotrons have achieved energies of about 1000 MeV with proton beams; their size is limited by cost and construction difficulty.

In changing from the Lawrence cyclotron to the synchrocyclotron, however, a very valuable beam property is given up, namely, beam intensity. In the Lawrence cyclotron, particles can start their acceleration journey from the ion source on every rf cycle, because every rf cycle is exactly identical. The synchrocyclotron, in contrast, is a batch device. A group of particles leave the source, and the frequency of the accelerating system is then steadily lowered to stay matched with the group while it is being accelerated. During this relatively long period, particles emerging from the ion source cannot be accelerated, since the radio frequency of the accelerating system does not match their frequency of revolution in the magnetic field nearby; particles can be "captured" and accelerated again only after the first group has reached full energy and the radio frequency has been returned to its initial value. This results in a large intensity loss. A synchrocyclotron typically accelerates only 1/100 as many particles per second as a Lawrence cyclotron under comparable conditions. Largely because of this low beam intensity, very few synchrocyclotrons remain in use.

Isochronous cyclotron. Responding to the need for higher beam intensity at energies above 20 MeV, the isochronous (that is, equal-time) cyclotron (sometimes referred to as azimuthally varying-field or sector-focused cyclotron) introduced in the 1950s made constant-frequency operation possible at higher energies. Although the concept had been published by L. H. Thomas in 1938, its practical realization was delayed by the complexity of both the magnet design and the associated orbit calculations. The basic idea can be understood by starting with Eq. (1). As the particle speeds up, its mass increases, as discussed earlier, and this will tend to lower its angular frequency ω; however, if the magnetic field B is increased in a compensating way, the frequency can be kept constant. Since faster (heavier) particles move in circles of larger radius, a magnetic field is required whose strength increases with the radius—the opposite situation to that found in Lawrence cyclotrons. Such a field can readily be built, but Maxwell's equations then require the field lines to curve away from the axis and hence produce force components pushing particles above or below the central plane further away from it. In such an axially defocusing field, very few particles would survive the acceleration process, and so an extra source of axial focusing is required.

The discussion so far has assumed that the magnetic field is axially symmetrical. If azimuthal variations are allowed, additional terms in Maxwell's equations come into play and can provide additional focusing forces. In a simple example, the magnetic field comes from three sector (wedge-shaped) magnets which meet at a point in the center. In the regions between the magnets, the orbits are almost straight, since the magnetic field is very weak there. They then bend sharply through 120° within each of the magnets. As a result, orbits cross the edge of each magnet at a nonperpendicular angle, and the ion velocity at the field edge has a component in the radial direction. But the magnetic field lines at the edge of any magnet are bowed outward, so the field there has an azimuthal component, which points away from the magnet above the median plane and toward it below. Taking the vector product of the radial component of velocity times the azimuthal component of the field gives an axial force toward the median plane, which pushes particles that are out of the plane back toward it. This force is axially focusing whether the particles are above or below the plane, and whether they are entering or leaving the magnet. This "edge" focusing is then an additional force introduced by the sector structure; it can be used to override the defocusing that arises from increasing the average field with radius in order to maintain a constant angular frequency in Eq. (1).

Isochronous cyclotrons have been built in a broad spectrum of sizes and types. The sectors of many such cyclotrons have a spiral shape (**Fig. 6**) to increase the edge axial focusing for higher-energy ions; this introduces a strong-focusing effect (discussed above). In the ring or separated-sector cyclotron (**Fig. 7**), a magnet coil is wound around each sector magnet, leaving open space between sectors for rf accelerating cavities. The beam from a preaccelerator (sometimes a smaller cyclotron) is injected horizontally between the sectors where the magnetic field is very low. The largest spiral-sector and separated-sector cyclotrons (Figs. 6 and 7) produce over 200 microamperes of protons, and are known as meson factories, since large quantities of mesons are produced when the protons hit a target. The record for proton energy and current is held by the cyclotron at PSI (Paul Scherrer Institut), near Zurich, Switzerland, which delivers 2 milliamperes at 590 MeV to a spallation neutron source (Fig. 7). *See* MESON.

Several laboratories have multistage cyclotron systems. Some of these systems are designed for heavy-ion beams, and electrons are stripped off between cyclotrons to produce higher charge states and higher energies. Other laboratories inject ions into a cyclotron from an electrostatic or linear accelerator, again with the help of electron stripping. Cyclotrons are also used to inject ions into synchrotron and storage rings, which can further accelerate, store, and cool the ions for precision experiments.

An important development in nuclear science research is the acceleration of radioactive ion beams to enable the interactions of unstable isotopes to be studied. Cyclotrons are actively used in this area in several ways. One is in the production of radioactive species by bombardment of a target with a cyclotron beam. If the beam consists of heavy ions, a thin target is used, allowing high-energy radioactive ions to emerge for capture and transport directly to the experimental apparatus. If the beam consists of protons, a thick target is used, and low-energy radioactive ions are collected for subsequent acceleration in a cyclotron or linear accelerator. The heavy-ion process is referred to as nuclear fragmentation, the proton process as isotope production on line (ISOL).

Superconducting cyclotron. A powerful extension of cyclotron capabilities has been provided by the use of superconducting magnets—which, in contrast to previous advances, is a technological rather than a conceptual change. In practice, the only superconducting element in the magnet is usually the main coil, which is typically housed in an annular cryostat. Conventional room-temperature components, including pole tips, the rf accelerating system, the vacuum system, and the ion source, are inserted in the warm bore of the cryostat from top and bottom. The superconducting coil allows the strength of the magnetic field to be greatly increased up to a level of approximately 5 teslas (50,000 gauss), or three times higher than was previously typical of cyclotrons. This increase in field reduces the linear size of the cyclotron to approximately one-third, areas to approximately one-ninth, and so on, compared to a normal cyclotron of the same energy. The result is a large reduction in the cost of many cyclotron components. The largest operating superconducting cyclotron, at Michigan State University, is only 9.5 ft (2.9 m) in overall diameter (**Fig. 8**), but can accelerate heavy ions to energies as high as 200 MeV per nucleon. The first cyclotron with separate superconducting sector magnets is being built at RIKEN, near Tokyo, to reach 400 MeV per nucleon.

Separated-orbit cyclotrons (SOCs), in which the bending and focusing fields of the sector magnets are specially tailored for each orbit, were first conceived in the early 1960s for the acceleration of

Fig. 7. Separated-sector ring cyclotron at PSI in Switzerland, which accelerates protons to 590 MeV. The four rf cavities are visible between sector magnets. (*PSI*)

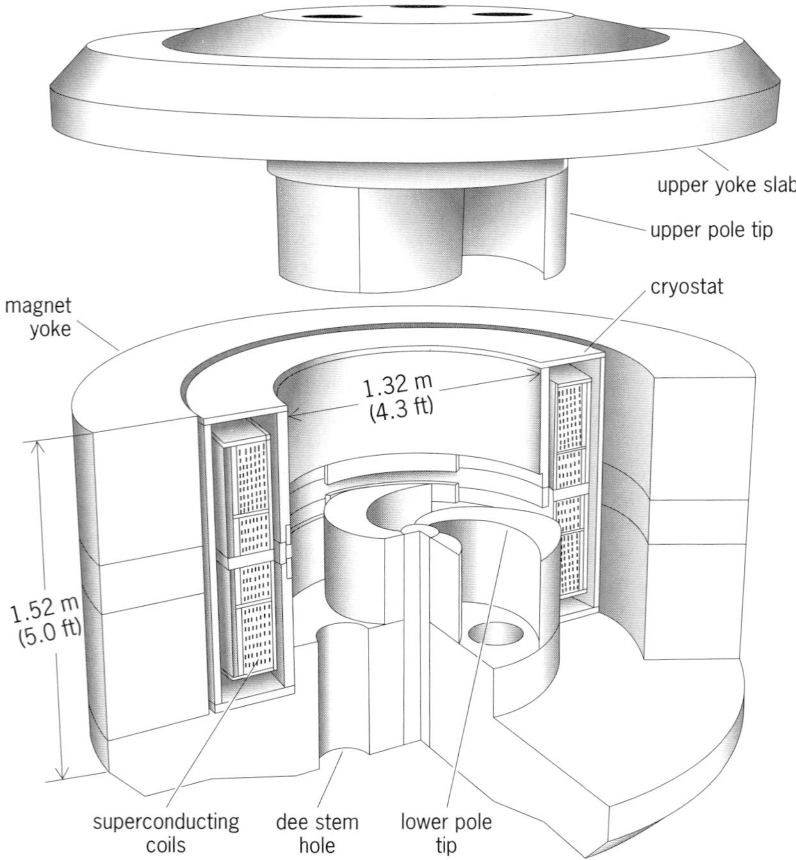

Fig. 8. Conceptual drawing of superconducting cyclotron at Michigan State University. The upper pole cap shown in the raised position is used for maintenance. (*After H. G. Blosser, The Michigan State University superconducting cyclotron program, IEEE Trans. Nucl. Sci., NS-26(2), pt. 1:2040–2047, 1979*)

high-current (milliampere) proton beams to gigaelectronvolt energies. Their complexity and cost, using normal magnets, deterred potential builders. However, a prototype superconducting SOC has been built in Munich, Germany, and has accelerated a S^{14+} ion beam through six turns to 70 MeV. This is the ultimate in superconducting cyclotron design, since the rf accelerating cavities as well as the magnet are superconducting, and the vacuum is maintained by cryogenic pumping.

FFAG. Another early concept that continues to attract interest is the fixed-field alternating-gradient (FFAG) accelerator, essentially a sector-focused ring synchrocyclotron. This idea was developed in the late 1950s, and electron models were successfully built and tested, but a full-scale version for protons seemed prohibitively expensive at that time. New technology, however, in the form of metallic-alloy rf tuners, has created a resurgence in interest, as it has allowed development of rf cavities producing very high effective accelerating fields (~100 kV/m) at low (megahertz) frequencies and low Q values (~1–15). The low Q allows simple broadband operations, while the high fields permit higher repetition rates (kilohertz) to be used than in synchrotrons, so that the charge per pulse can be much lower, and beam and rf stability more easily maintained. FFAGs are therefore attractive as high-intensity proton machines in the 0.1–10 GeV range. A prototype proton FFAG was commissioned in Tsukuba, Japan, in 2000, and a 250-MeV FFAG for proton therapy has received funding approval. Even more ambitious FFAGs for acceleration of protons and muons are being considered for neutrino factories and muon colliders. As muons are produced with a wide angular spread, the very large transverse and momentum acceptances of FFAGs are very advantageous. *See* Q (ELECTRICITY).

Ion sources. An important property of an accelerated ion is its charge state q. The maximum energy of an ion in a cyclotron is proportional to the square of its charge (and inversely proportional to its mass). Thus, for heavier ions, there is a strong incentive to produce ions in high charge states.

Originally the standard ion source was based on the principle of the Penning ion gauge (PIG), in which electrons oscillating in a magnetic field ionize the source gas. The highest-charge-state ions produced by a PIG source in useful intensities were, for example, $q = 5$ for nitrogen (N^{5+}). The electron cyclotron resonance (ECR) source uses microwave-heated electrons in a magnetically confined plasma to produce much higher charge states. For example, the ECR source will produce fully stripped N^{7+} beams. Since maximum ion energy is proportional to the square of ion charge, this doubles the nitrogen energy from that available from the same cyclotron with a PIG source. For heavier ions, fully stripped argon A^{30+} and uranium up to U^{60+} have been accelerated from an ECR source. This source, which is too large to fit in the center of a cyclotron, is placed outside and connected to an injection system which brings the beam axially into the cyclotron center region. Many cyclotrons have upgraded their capabilities by installing ECR sources, and over 30 are in service. These sources are also employed in linacs and synchrotrons.

Polarized protons and deuterons (that is, with their spins aligned) can be produced in special ion sources. Like the ECR source, polarized ion sources are too large to fit into the center of a cyclotron, so ions are injected from an external source. Polarized ions are used for nuclear science experiments which probe the spin-dependent characteristics of nuclear reactions or structure.

Cyclotron applications. There are over 200 cyclotrons in operation. More than 70 are used principally for basic research in nuclear physics and chemistry, aimed at understanding the structure of nuclei, nuclear forces, and nuclear reactions. Another 130 are installed in laboratories and hospitals for the pursuit of practical applications, including isotope production for medical and other uses, cancer therapy using proton and secondary neutron beams, and the study and irradiation of materials. *See* NUCLEAR REACTION; NUCLEAR STRUCTURE; RADIOLOGY.

Michael Craddock

Electron synchrotron. An electron synchrotron is a circular accelerator optimized to accelerate electrons or positrons to high energies. The acceleration is achieved as the particles pass through electric fields created in a resonating rf cavity. A string of bending magnets along a circular path forces the particles to travel along a closed loop leading through the accelerating cavity at each revolution. In this way, each particle is accelerated many times during the accelerating cycle by the same cavity, and even a modest accelerating field will eventually lead to a high particle energy. *See* CAVITY RESONATOR.

To make this acceleration mechanism function properly, certain conditions have to be met. The oscillating field in the cavity has to be in the accelerating phase every time a bunch of particles arrives. Thus the time it takes the particles to orbit once around the ring must be an integral multiple of the rf period (the synchronicity condition). In an electron synchrotron the particles are injected at a kinetic energy of some tens of megaelectronvolts and therefore always have a velocity close to the speed of light. As a consequence the revolution frequency of the particles is constant, and it is possible to use a fixed-frequency accelerating cavity. This is the primary difference between electron synchrotrons and proton synchrotrons or cyclotrons, where the frequency of the accelerating rf fields must be adjusted during acceleration by factors of 10 or more because the much heavier protons travel, at typical injection energies, at speeds much slower than the speed of light. A linear accelerator or a microtron is usually used as a preaccelerator to inject the particles into the synchrotron at an energy of tens or hundreds of megaelectronvolts.

In the past, electron synchrotrons have been used extensively for research in high-energy physics and as sources of synchrotron radiation. However, they are now constructed and used exclusively as injectors into storage rings. Electron synchrotrons have been built in sizes from 6.5 ft (2 m) in diameter for an energy of 100 MeV, up to 650 ft (200 m) in diameter (**Table 2**) for a maximum particle energy of 12 GeV.

Acceleration and beam control. The operation of the electron synchrotron is determined by its accelerating cycle, which may range from a few milliseconds up to seconds. At the beginning of a cycle the preaccelerator is triggered to produce a particle beam that is injected into the synchrotron. The beam from the injector will have certain characteristics as required for optimum acceptance by the synchrotron. The synchronicity condition is fullfilled only for particles which arrive at the right time for acceleration in the synchrotron cavity. The ratio of the radio frequency to the revolution frequency is called the harmonic number and is equal to the maximum number of places (called buckets) around the ring where particles can be placed for stable acceleration. While these buckets rotate around the ring at nearly the speed of light, the injector has to deliver a beam structured and timed so that the particles arrive at the injection point at the same time as the buckets arrive there. Either one or all of the buckets can be filled with particles.

While the synchronicity condition is very important, it has to be fulfilled only approximately. The principle of phase stability provides a restoring force for all particles which do not exactly meet the synchronicity condition, tying together the characteristics of the magnetic guide field and the rf fields in such a way that the particles stay trapped close to and oscillating about the ideal equilibrium position. This is important during the accelerating process. As the magnetic guide field is increased, the principle of phase stability adjusts the particle energy to be proportional to the rising magnetic field. After executing many turns during which the particle energy is increased by a few megaelectronvolts per turn, the particles reach the maximum design energy.

At this point the particle beam is extracted and guided to an experimental station or to a storage ring. The extraction is commonly achieved by the use of a pulsed high-field kicker magnet that deflects the beam out of the synchrotron. After extraction the magnetic guide field is reduced again to the injection condition, and a new accelerating cycle can begin.

The maximum energy that can be achieved in a synchrotron is limited by either the maximum magnetic field necessary to keep the beam on its orbit or by the effect of synchrotron radiation. As the particles are made to follow curved trajectories forming closed orbits, they lose energy by way of this

TABLE 2. Some parameters of electron synchrotrons

Parameters	Cornell University	DESY II*	National Synchrotron Light Source (NSLS)	Stanford University
Maximum beam energy, GeV	12	9.2	0.75	3.5
Radius, ft (m)	328 (100)	153 (46.6)	14.8 (4.51)	69.6 (21.2)
Cycling rate, Hz	60	12.5	1	10
Maximum bending magnet field, teslas	0.4	1.13	1.31	1.3
Injection energy, MeV	150	200	70	120
Particles per pulse	3×10^{10}	4×10^{10}	10^{10}	2×10^{10}

*Deutsches Elektronen Synchrotron, located in Hamburg, Germany.

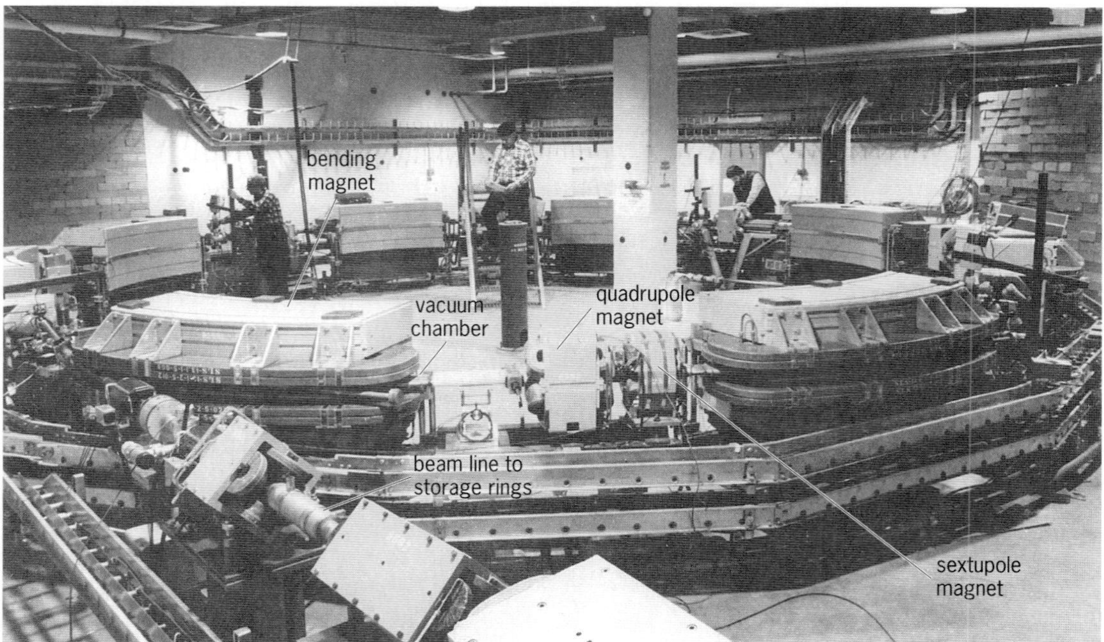

Fig. 9. The 750-MeV electron synchrotron at the national synchrotron Light Source (NSLS) at Brookhaven, National Laboratory.

radiation. The energy loss per turn increases as the fourth power of the energy, and the maximum energy is reached when this energy loss is as large as the maximum energy gain per turn in the accelerating cavities.

While a stronger magnet and rf system could increase the maximum energy somewhat, there is another fundamental limit on this energy. During the first phase of an accelerating cycle the beam cross section shrinks from its value at injection (typically a few square millimeters), inversely proportional to its energy. However, the statistical emission of synchrotron radiation photons causes a rapid increase in the beam cross section as the energy is raised. The limit is reached when the beam fills the available aperture in the vacuum pipe.

The maximum intensity in a synchrotron is determined by the performance of the preinjector as well as by beam instabilities. Individual particles can be lost because of collisions with other particles or because of the electromagnetic interaction of all particles in one bunch with the surrounding vacuum chamber or with other bunches. In all cases the limitation on the beam current is more severe at low energies than at high energies, which is why synchrotrons are used as a step-up injector between the preaccelerator and a storage ring. In a storage ring it would not be possible to accumulate a high beam current at the low energy of the preinjector. In a booster synchrotron a small beam current is accelerated repeatedly to higher energies where instabilities in the storage ring are less severe and high beam currents can be accumulated.

Magnet system. The magnetic guide field in a synchrotron consists of dipole fields to keep the particles on a circular path and a focusing field generated by quadrupoles to keep the beam from diverging. The arrangement of bending and focusing magnets is called the magnet lattice. In the older synchrotrons, both bending and focusing were performed by one type of magnet, and the resulting magnet lattice was called a combined-function lattice. As magnet technology progressed and more flexibility was desired, electron synchrotrons began to employ the so-called separated-function lattice, where bending and focusing is done by different magnets with separate power supplies (**Fig. 9**). To enhance the beam current capability, the newer synchrotrons also use sextupole magnets to correct for effects caused by the energy spread in the beam. In analogy to light optics, these effects are called chromatic aberrations of the particle beam focusing.

Vacuum system. To avoid loss of particles during acceleration, the particles circulate in a vacuum pipe embedded in the lattice magnets. In older synchrotrons this vacuum pipe had to be made from an insulating material to allow the rapid cycling of synchrotrons at 50 or 60 Hz without creating problems due to eddy currents caused by the magnetic fields in the vacuum chamber. In the newer synchrotrons used exclusively as injectors for storage rings, the cycling rate is only about 10–15 Hz or less, and eddy currents are reduced to a negligible level at this lower cycling rate. This makes it possible to use thin stainless steel vacuum pipes, thus reducing cost and improving reliability *See* EDDY CURRENT.

Helmut Wiedemann

Proton synchrotron. The highest-energy particle accelerators are proton synchrotrons (**Table 3**). While many new uses for this type of particle accelerator

TABLE 3. Large proton synchrotrons currently in operation

Synchrotron	Date of commissioning	Approximate radius, ft (m)	Beam energy, GeV
Proton synchrotron (PS), CERN, Switzerland*	1959	330 (100)	28
Alternating-gradient synchrotron (AGS), Brookhaven, Upton, New York	1961	410 (125)	33
IHEP, Serpukhov, Russia	1967	770 (235)	76
Super proton synchrotron (SPS), CERN	1976	3600 (1100)	400
Tevatron, Fermilab	1983	3300 (1000)	980
Main Injector, Fermilab	1999	1700 (520)	150

*The CERN PS and the Main Injector at Fermilab are used principally as injectors for the SPS and the Tevatron, respectively.

are being realized, such as medical applications, the primary use of the proton synchrotron remains in the exploration of nuclear and high-energy particle physics. Beams of protons are accelerated to high energy and either are extracted toward fixed targets of material or are directed into collision with other beams of particles (commonly protons, antiprotons, or electrons) to study the most elementary forces of nature. In the case of fixed-target experiments, the proton beams are often used to produce secondary beams of other elementary particles, such as mesons and neutrinos, which are in turn used as probes for physics experiments.

Like the cyclotron and the synchrocyclotron, the proton synchrotron makes repetitive use of accelerating fields, in this case by guiding the protons in a nearly circular path by using a time-varying magnetic field which changes in synchronism with the increasing momentum of the particles. The acceleration is performed by radio-frequency cavities, whose oscillating electric fields point back and forth along the design trajectory of the accelerator. The electric field in the cavities is arranged such that particles are accelerated as they pass through, and then as the field reverses direction, a gap is left in the particle beam. In this way the proton beam is not uniformly continuous about the circumference of the accelerator, but is bunched. The acceleration takes place over a typically short portion of the circumference (**Fig. 10**), while most of the device's circumference is made up of bending magnets which serve to bring the particles back to the accelerating cavities. As the momentum (and hence, energy) of the particles increases, the fields of the guiding magnets must also increase to maintain the same central trajectory. In addition, as the speed of the particles increases, the frequency of the rf cavities must be adjusted to maintain synchronism. Most modern proton synchrotrons use separate types of magnets to perform the bending (dipole magnets) and the necessary focusing (quadrupole magnets) of the proton beams. A certain number of straight sections are also necessary in the accelerator to perform other functions such as injection and extraction of the particles to and from the accelerator, and various beam diagnostics and control.

Accelerator cascade. As proton beams are accelerated, their transverse dimensions shrink, and hence smaller apertures may be used for the guiding magnets and accelerator hardware. Thus, to generate beams of very high energy protons, it is most economical to use a cascaded series of accelerators, each optimized to accelerate the particles over a particular range of energy. For example, in the Fermilab accelerator complex at Batavia, Illinois (**Fig. 11**), H^- ions (hydrogen atoms each with an extra electron) are generated in a Cockcroft-Walton electrostatic generator and accelerated across a potential of 750 kV. The beam is next transported to a linear accelerator which increases the total kinetic energy of each particle to 400 MeV. The ions are then stripped of their electrons, leaving a pure proton beam, upon injection into a booster synchrotron. This synchrotron accelerates the particles to an energy of 8 GeV and transfers the bunches of protons into the Main Injector. At this point, the protons are moving at over 99% the speed of light. The Main Injector accelerates the beam to an energy of 150 GeV, and then the particles are transported, in a single revolution, into the Tevatron accelerator (**Fig. 12**). Two such injections are required to fill the teratron. Finally, the Tevatron, a synchrotron made up of superconducting

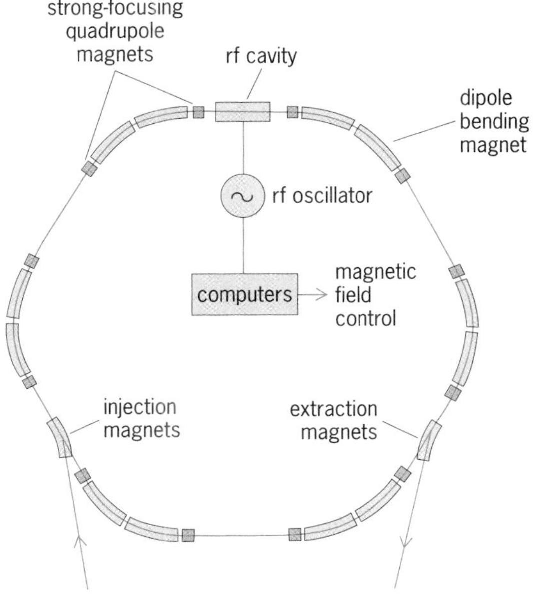

Fig. 10. Idealized proton synchrotron.

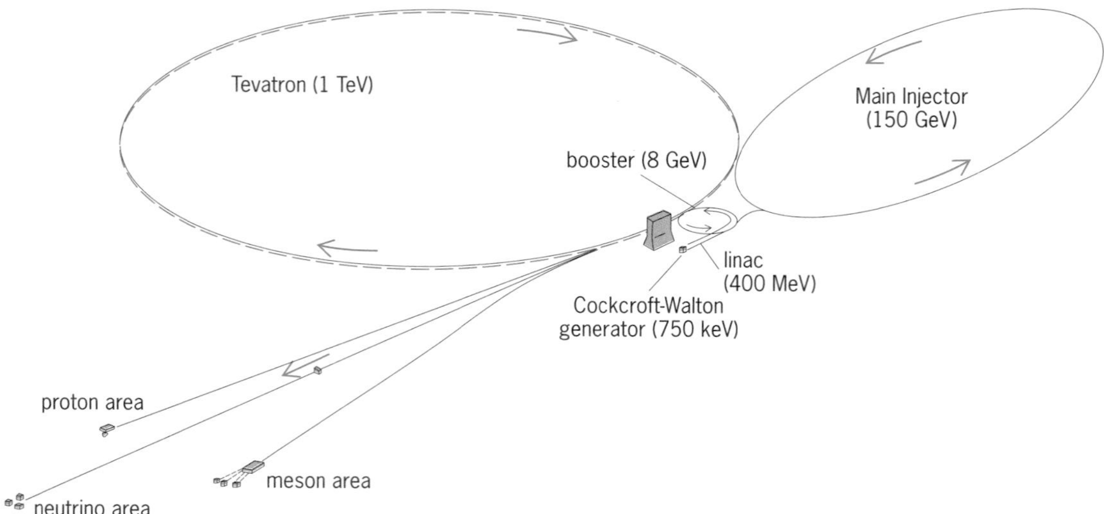

Fig. 11. Fermilab facility, showing sequence of accelerators. Extracted protons are routed to three experimental areas by a "proton switchyard." Part of the extracted beam is used to produce beams of secondary particles.

magnets, accelerates the particles to an energy of nearly 1 TeV. At that point, the particles can be extracted and transported to the fixed-target experiments. The entire acceleration process takes roughly 20 s to complete.

Particle stability. Implied in this description of a synchrotron is an ideal particle, which at each moment has exactly the right energy and the right time of passage through the accelerating structure so that it receives exactly the right increment of energy to stay in accord with the plan of the accelerator system. But in fact, most (if not all) particles do not meet these very special requirements. At relatively low energies, particles which have a lower momentum than the ideal tend to lag behind because of their lower velocities, and higher-momentum particles move ahead of the ideal particle. At higher energies, as the velocities of the particles approach the velocity of light, increments in energy do not increase their speed significantly. But their momentum continues to increase, and hence the bending magnets do not bend the particles' trajectories as much. The particles with slightly excessive momentum then move in longer orbits around the circumference and hence actually take longer to circulate.

Fig. 12. Tunnel of the main accelerator at Fermilab. The superconducting magnets of the Tevatron sit near the floor. The magnets above it are remnants of the original Fermilab 400-GeV synchrotron, which has been decommissioned. (*Fermi National Accelerator Laboratory*)

Since the fields in the rf cavities are changing with time, they can be tuned to give different increments of energy to the particles that arrive at the cavities too soon or too late. At lower energies, the particles arriving early receive less than the ideal increment in energy, and those arriving late receive more. A particular particle which has too little energy will keep receiving larger-then-ideal increments until it eventually receives too much. It will then begin arriving at the rf cavities late and hence begin receiving less energy than required. In this way, particles will oscillate about the ideal energy. These energy oscillations are called synchrotron oscillations. If the ideal energy of the particles is high enough, the slightly higher energy particles will arrive at the cavities later than desired, and the phase of the rf oscillating fields must be adjusted so as to give these late particles less energy rather than more. In many proton synchrotrons, there is a transition energy at which the two effects determining the revolution period—the particle's speed and the particle's energy—have offsetting effects. At this energy, the protons all circulate with the same orbital period, and phase stability is momentarily lost. If this energy is crossed quickly enough, and the phase of the rf system switched quickly enough, then no harm is done to the quality of the proton beam.

In addition to having a small spread in energy, the individual protons in the beam enter the accelerator with a variety of trajectories, close to the ideal but slightly off in either displacement or direction. The synchrotron thus contains magnetic elements which can focus the particles, redirecting them toward the ideal trajectory. Early, weak focusing proton synchrotrons used so-called gradient magnets to generate focusing fields that weakened with increasing radius of the particle trajectory. These magnets had the disadvantage that as the design energy, and hence the circumference of the orbit, is increased, aperture also must be increased to allow for particle oscillations. This circumstance led to the invention, in 1952, of alternating-gradient focusing or strong focusing, which allows the size of the beam tube and magnets to be decoupled from the circumference of the accelerator. Most modern proton synchrotrons use quadrupole magnets to perform this focusing. A quadrupole magnet focuses the beam in one plane (horizontal, say) and defocuses it in the other plane (vertical). But a pair of quadrupoles, of opposite polarities, can be arranged to have a net focusing effect in both planes simultaneously.

An individual particle oscillates transverse to the ideal trajectory with a frequency determined by the focusing magnets. This motion is called betatron motion, and the number of transverse oscillations that a particle undergoes during one revolution about the accelerator is termed the "tune" of the accelerator. A particle may undergo many betatron oscillations during one revolution, while it may take hundreds of revolutions to perform one synchrotron oscillation.

Extraction. In a fixed-target synchrotron facility, once the beam is accelerated to its final energy the particles are slowly extracted from the accelerator and sent toward external experimental areas. There, beams of secondary particles, such as mesons and neutrinos, can be created if required, and the final particle beams are directed toward the experimental apparatus to study particle interactions with matter. Frequently, the experimental equipment can handle only a certain rate of events, and so a smooth, continuous delivery of particles to the apparatus is preferred to a single shot of an intense particle beam. (Of course, many other experiments do prefer the latter.) The method of slow resonant extraction is generally used to meet this request. By putting the particles in the synchrotron into resonance with error fields in the accelerator (often introduced purposefully), in a well-controlled fashion, the transverse amplitudes of the particles are made to grow until some of the particles can find themselves on the opposite side of a septum made up of very thin, electrically charged wires. The electric field on one side of the wires deflects the particles into a separate beam pipe and beam-transport channel which guides them toward the experimental areas. The rate of growth of the particle amplitudes is controlled by a feedback system, which ensures that the particles do not all leave at once but are gradually spilled out of the accelerator.

The extraction process can take tens of seconds if desired. Ideally, it may be desirable for experiments to receive the beam continuously from the accelerator, but the actual duty factor—the slow spill time divided by the total cycle time—will be less than one because of the injection, acceleration, and reset time. A fast resonant extraction can also be performed, whereby the entire beam is brought into resonance very quickly, on the order of milliseconds. The extraction rate can also be modulated, changing as the external beam is switched quickly among users with differing requirements.

In some instances, a very fast extraction process is required. Notably in superconducting accelerators, the beam may have to be taken out of the synchrotron in order to avoid damage to the superconducting magnets. In this case, a manually or electronically generated signal will cause kicker magnets with very fast rise times, on the order of microseconds, to energize and deflect the beam through an extraction channel toward a special beam absorber. The kicker magnets are fast enough to extract the entire beam in a single revolution.

Superconducting synchrotrons. In any synchrotron, the maximum energy of the protons is dictated by attainable field strength of the magnets used to bend the particle trajectory along its course. In conventional (warm) accelerators, the magnetic field is produced by using electromagnets whose field shape and uniformity are governed by the shape of the gap in the iron, through which the particles circulate (**Fig. 13*a***). But in iron, magnetic fields saturate at

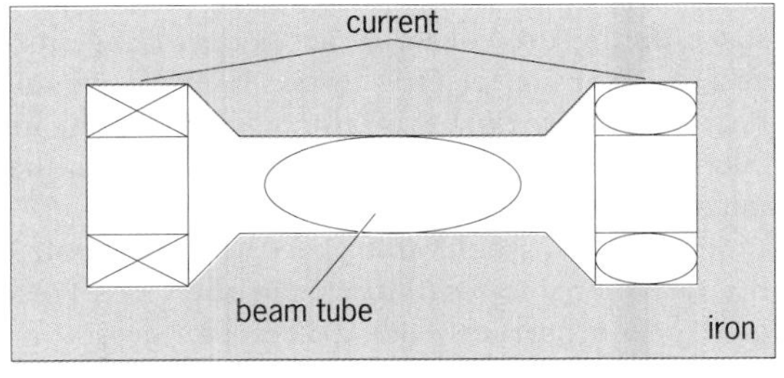

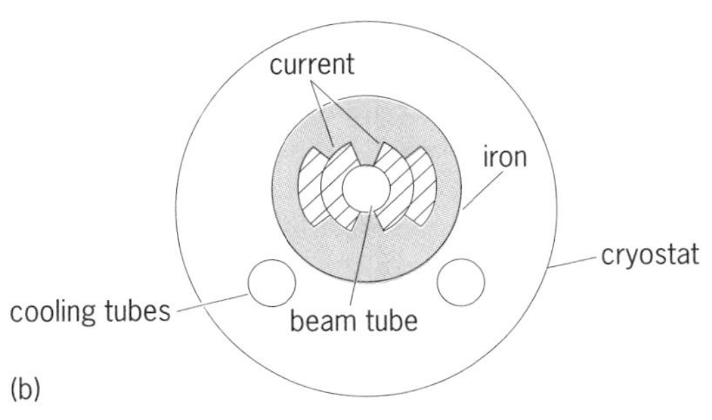

Fig. 13. Cross sections of dipole magnets used in particle accelerators. (*a*) Conventional iron magnet. (*b*) Superconducting magnent.

strengths of approximately 2 T. A significant advance in accelerator technology occurred with the invention of superconducting accelerator magnets. These magnets use superconducting materials for the cable used to carry the current which generates the magnetic field. When cooled to temperatures near absolute zero, these materials lose all electrical resistivity, and very large current densities can be obtained with no heat lost in the cable due to resistance. The fields of these magnets are determined primarily by the placement of the strands of superconducting cable; the iron serves primarily to provide structural support (Fig. 13*b*). Accelerator magnets of this type have been produced in the laboratory which reach fields of roughly 10 T. Niobium-titanium (Nb-Ti) alloys have been the primary superconductor employed. In the presence of the strong magnetic fields involved, these alloys must be cooled to about 4.5 K (−451°F) by using liquid helium as coolant.

In addition to providing higher fields, accelerators with superconducting magnets allow beams to be stored for long periods of time (as in a colliding-beams accelerator, or during slow spill in a fixed-target accelerator) at high field with relatively low electrical power consumption. The most power consumption comes from running the refrigeration system to keep the magnets at their superconducting temperature. (The bus work carrying electricity to the magnets also consumes some power.) Energy deposition in the magnets, due to radiation, heat leaks, or perhaps most importantly the beam itself, must be carefully controlled in these systems.

Michael J. Syphers

Heavy-ion synchrotron. There are close similarities between heavy-ion synchrotrons and colliders, and their respective proton counterparts. Special considerations of accelerator design for heavy ions arise mostly from their tendency to change their charge states during the course of acceleration and storage, and their high electric charge. In discussing the heavy-ion accelerator, it is customary to express its energy in terms of energy per each nucleon in the ion (GeV/u) instead of the total energy.

Many proton synchrotrons have been augmented by a heavy-ion injector to accelerate heavy ions. For a given peak magnetic (bending) field of a synchrotron, the maximum energy attainable is proportional to the ratio of the charge state to the atomic mass number of the ion species being accelerated. For protons this ratio is unique and equals 1, whereas for heavy ions it varies according to ion species accelerated and the degree of stripping (how many electrons are removed from an atom). The stripping can be accomplished by passing ion beams through thin metal or carbon foils. In this process, a significant attenuation of beam intensity occurs by the ion's interactions, mostly scattering with foil. Light ions can readily be stripped to full charge states. Heavier ions, however, require sequences of acceleration and stripping stages to reach fully stripped states. Fully stripped ions as heavy as gold and lead have been accelerated to energies of 11.4 GeV/u and 160 GeV/u, respectively.

TABLE 4. Major heavy-ion synchrotrons

Facility and location	Beam energy, GeV/u	Range ion types
Saturne II	1.15	Lithium (Li)
(Saclay, France)†	0.69	Krypton (Kr)
GSI	2.0	Deuterium (D)
(Darmstadt, Germany)	1.0	Uranium (U)
Bevalac	2.1	Carbon (C)
(Berkeley, California)	0.96	Uranium (U)
Synchrophasotron-Nuclotron	6	Deuterium (D)
(JINR, Dubna, Russia)	4.3	Uranium (U)*
AGS, Brookhaven,	14.5	Oxygen (O)
(Upton, New York)	11.4	Gold (Au)
SPS, CERN	200	Sulfur (S)
(Geneva, Switzerland)	160	Lead (Pb)

*Beam intensity less than 10^5 ions per second.
†Closed down in 1997.

Most of the major heavy-ion synchrotrons (**Table 4**) are utilized for nuclear physics studies. An 800 MeV/u heavy-ion synchrotron in Japan was constructed with medical applications such as cancer therapy as its primary purpose.

Heavy-ion accelerators require a significantly higher vacuum than their proton counterparts. This is to prevent beam loss from the change of the charge state by electron pick-up or stripping processes through collisions with residual vacuum-chamber gas. The rf power must be boosted for heavy ions to overcome the increased inertia caused by the lower charge-to-mass ratio, and to keep the beam energy synchronized with the rising magnetic fields during acceleration.

Satoshi Ozaki

Linear Accelerator

A linear accelerator accelerates particles in a straight line by means of electric fields developed across a series of accelerating gaps in sequence. Each gap is transversed once and the electric fields must be produced along the entire orbit. Extremely high field strengths and hence high rf power levels are required for room-temperature structures and high beam currents. Superconducting structures requiring low power have been developed for certain applications.

Principles of operation. The synchronism between particles and accelerating field may be achieved by either of two methods: traveling-wave acceleration, whereby a wave with an accelerating field component whose phase velocity is equal to the particle velocity is utilized; or standing-wave acceleration, whereby a standing-wave pattern is produced by the superposition of forward and backward waves in the structure. In the latter case, either the forward-wave phase velocity is made equal to the particle velocity or, more frequently, a set of drift tubes is introduced into the structure to shield the particles from the fields when they are passing through regions where the fields would otherwise be decelerating.

The energy to which a particle can be accelerated is proportional to the particle's charge, the square root of both the length of the accelerating structure and the total power of the rf power sources feeding it, and the $-{}^1/_4$ power of the wavelength. Thus, length and power are equally instrumental in attaining high energies and short wavelengths are also advantageous in terms of power economy. Countering this advantage of short wavelengths are the following disadvantages: (1) higher power levels are generally available from power tubes operating at longer wavelengths and the accelerator guide is capable of high power handling; (2) the absolute tolerances of the mechanical structure are larger at larger wavelengths; (3) longer wavelengths allow for higher beam currents. In practice, electron accelerators use wavelengths of 3–20 cm; proton accelerators, 30–200 cm; and heavy-ion accelerators, 300–1000 cm. In general, the high power levels required for room temperature operation result in pulsed rather than continuously operating machine design, but there is no fundamental reason for pulsed operation and some continuously operating machines are under design.

Successful acceleration of particles requires that acceleration be phase-stable and that the beam of particles remain focused along the orbit. There will be a net defocusing action in a field region where the accelerating field increases in time as the particle crosses the region (**Fig. 14***a*). This is because the defocusing transverse momentum imparted as the particle leaves the region is greater than the focusing momentum imparted as the particle enters it. However, the action is phase-stable, since a particle arriving late will experience a larger acceleration than a faster-traveling particle which arrives early. If the field decreased in time during particle passage, there would be transverse focusing but longitudinal phase instability. Thus focusing and phase stability are incompatible, and this appeared to be a serious obstacle to the design of long linear accelerators. It has been circumvented in various ways as follows: (1) Charge can be included in the beam to terminate field lines inside the beam and cause a convergent action even at phase-stable transit phases. This charge has typically been induced by grids placed in the field (Fig. 14*b*). In this case, there is focusing momentum in all parts of the region, and in an increasing field the acceleration is also phase-stable. Unfortunately, the grids also intercept some of the beam and therefore prohibit high-current operation. (2) As the particle velocity approaches the speed of light, the incompatibility becomes irrelevant because, for relativistic velocities, the action of the radial electric time-varying field is almost canceled by the accompanying time-varying magnetic field; also, because the velocity is almost invariable, the particles are in neutral equilibrium longitudinally. (3) External magnetic fields (solenoidal or strong-focusing) or electrostatic lenses have been used. (4) By alternating the phase difference, or by suitable field shaping, a region of limited phase stability is attainable.

Electron accelerators. Virtually all electron linear accelerators are of the traveling-wave type. The wave is produced in a loaded waveguide excited in a transverse magnetic (TM) mode which has a longitudinal electric field component. The loading is typically achieved by means of disks which are sized to produce the correct phase velocity to match the particles. The field pattern may be visualized as translating along the axis with a velocity equal to that of the electron beam. Typically solenoid focusing is used over the first few feet of the accelerator, after which the electrons approach the velocity of light whereby all radial forces may be neglected. The momentum component transverse to the axis remains constant, and since the longitudinal component increases continuously, the angle of divergence of the beam decreases continuously. For uniform energy gain per

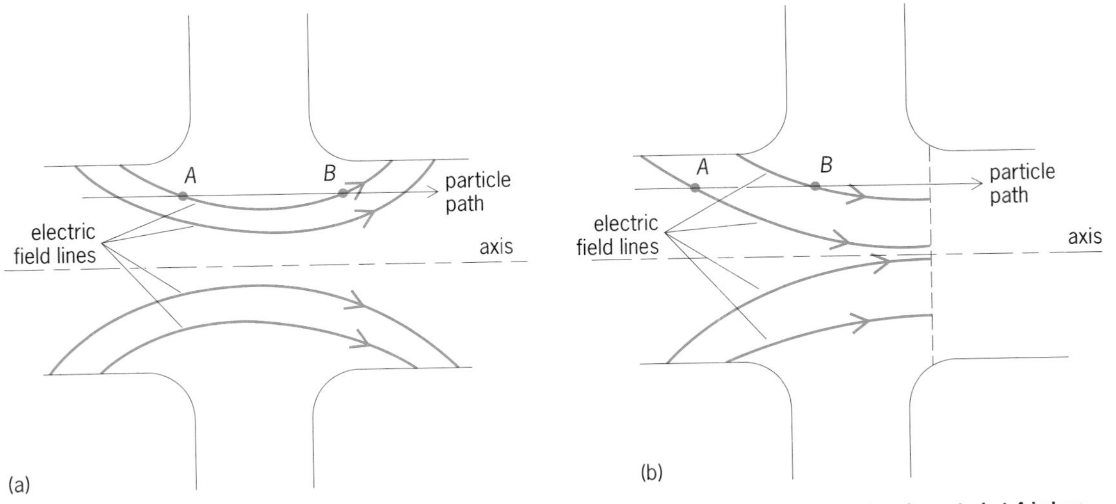

Fig. 14. Phase focusing in a linear accelerator. (*a*) Gap without grids. Focusing transverse momentum imparted at *A* is less than defocusing momentum imparted at *B*. (*b*) Gap with grids. Focusing momentum is imparted at both *A* and *B*.

unit length, this corresponds to a beam angle varying as the inverse of distance along the machine and to a beam radius increasing logarithmically. A set of weak magnetic lenses can be used to produce a decrease in beam diameter. *See* MAGNETIC LENS; PHASE VELOCITY; WAVEGUIDE.

Accelerator guides must be manufactured to very close tolerances in order to control the phase velocity to the required accuracy, since there is no phase stability once the energy has grown to several times the rest energy of the electron. However, in practice the accelerator is divided into many separate accelerator guides, each with its own power source. These separate sections may be individually tuned and phased during operation, thus somewhat relieving the tolerance requirements.

Electrons are injected into the accelerator with an energy of typically 80–120 keV, corresponding to a velocity of approximately half the speed of light, and the first accelerator section is often a special bunching device to concentrate the electrons near the crest of the traveling wave in the succeeding sections.

Performance. The total energy capability from a linear accelerator has no fundamental limit since electrons accelerated in a straight line do not lose any appreciable energy because of radiation. The performance of a given accelerator is closely tied to the available power source and the voltage breakdown capability of the accelerating guide. Most machines operate at microwave frequencies and utilize klystron amplifiers driven from a common master oscillator as power sources. *See* KLYSTRON.

The acceleration of electrons at room temperature requires peak radio-frequency power levels of tens of megawatts per source. Costs of power and available power sources limit the beam duty cycle to a few tenths of a percent. Two 400–500-MeV linacs of higher duty factor have been constructed with long pulses and fast repetition rates achieved at the expense of low acceleration rates. A superconducting electron linac which can operate continuously with dissipation of about 1 W/m has been constructed, utilizing standing-wave structures of niobium metal and operated at 1 to 4 K (−458 to −452°F). However, metallurgical and electron loading problems limited the accelerating gradients to about 3 MeV/m. At this level these accelerating guides are useful with recirculating beams as in the microtron (discussed below).

The largest electron linac is 10,000 ft (3050 m) long (**Fig. 15**). Besides producing electrons directly for basic research, it injects two different synchrotron storage rings with electrons and positrons. The Stanford Linear Accelerator Center (SLAC) has increased the output energy of this 20-GeV machine to 50 GeV.

The advent of free-electron lasers as radiation sources and the development of very high power, millimeter-wave radio-frequency sources, brought renewed interest in electron linear accelerators. A key factor in the design of these machines is the brightness of the electron gun, and this has led to the development of rf electron guns. By utilizing a photocathode excited by a pulsed laser, very short electron bunches (~5 ps) may be produced at output energies of 2–5 MeV. Because of the high energy

Fig. 15. Stanford Linear Accelerator Center. (*a*) Aerial view showing the 2-mi (3.2-km) accelerator and associated buildings and storage rings. (*b*) The accelerator installed in its underground tunnel.

and short bunch length, very bright beams may be produced with peak currents in excess of 100 A. *See* LASER.

Laser acceleration. Since about 1990 there has been a significant amount of theoretical, and some practical study of the acceleration of electrons at very high acceleration rates of the order of 1 GeV/m with the use of high-power lasers. A number of the experimental studies have produced acceleration. These include inverse Cerenkov and inverse free-electron-laser acceleration.

Electromagnetic radiation is emitted whenever a charged particle passes through any medium in which the phase velocity of light is less than the particle velocity. This is known as Cerenkov radiation, and is emitted in a conical surface of half angle ϕ, whose value is given by Eq. (2), where n is the

$$\cos\phi = \frac{1}{\beta n} \qquad (2)$$

index of refraction in the medium and β is the particle velocity divided by the velocity of light. Conversely a particle can remain synchronously in phase with an electromagnetic wave in a medium if the particle and wave vector cross at this Cerenkov angle. In this situation the particle may be accelerated, and this is termed inverse Cerenkov acceleration. A practical experiment carried out at the Accelerator Test Facility at Brookhaven National Laboratory in Upton, New York, utilized an axicon lens to focus a radially polarized laser beam onto the same axis as a relativistic electron beam passing through a gas cell. When the electron and light beams were synchronized, acceleration of the electron beam was observed.

In another experiment at the Accelerator Test Facility, inverse free-electron-laser acceleration was achieved. In this scheme the electron beam is made to wiggle back and forth though a light beam from a powerful carbon dioxide (CO_2) laser by applying a series of alternating dipole fields. With proper synchronization of the laser field and the electron beam, accelerating gradients of several hundred megaelectronvolts per meter should be attainable.

An experiment using the inverse free-electron-laser accelerator and the inverse Cerenkov accelerator in tandem has been proposed. Laser wake-field acceleration has also been achieved at Argonne National Laboratory and at JAERI in Tokyo.

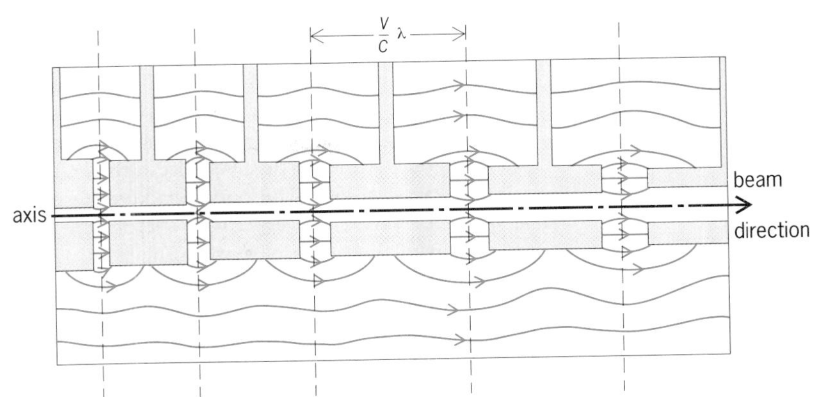

Fig. 16. Electric field configuration in a standing-wave proton accelerator employing a cavity utilizing an Alvarez structure operating in the TM_{010} mode. Drift tubes supported on stems shield the particles when the phase is decelerating. Particles cross gaps at approximately the same phase.

Proton accelerators. Proton linear accelerators are of the standing-wave type since protons do not reach relativistic velocities at typical injection voltages which are in the range 500–2000 keV. The most usual structure is that due to L. W. Alvarez (**Fig. 16**),

TABLE 5. Heavy-ion linear accelerators

Injector linacs

Name	Location	Resonator type*	Output energy†	Highest-mass ion accelerated	Operating frequency	Injector type†	Injector energy
Super HILAC	Berkeley, California	Alvarez	8.5 MeV/amu	^{40}Ar	200 MHz	RFQ	200 keV/amu
LITL	Tokyo University, Japan	Vane-type RFQ	800 keV/amu	^{84}Kr	100 MHz	—	—
Hyperion II	Saclay, France	Vane-type RFQ	187.5 keV/amu	^{40}Ar	200 MHz	—	—
RILAC	Saitama, Japan	Quarter-wave resonator	0.5–4 MeV/amu	^{197}Au	17–45 MHz	—	—

Posttandem machines

Location	Resonator type	Resonator material	Operating frequency	Velocity range	Total voltage, MV	Number of resonators
Argonne, Illinois	Split-ring	Niobium	145.5 MHz	0.06–0.16c	> 40	42
Stony Brook, New York	Split-ring	Lead	151.7 MHz	0.06–0.21c	20	40
Seattle, Washington	Quarter-wave	Lead	150 MHz	0.06–0.21c	26	32
Bucharest, Romania	Spiral	Copper	—	0.06–0.10c	6	20
Heidelberg, Germany	Spiral and split-ring	Copper	108.5 MHz	0.06–0.10c	12.5	40
Karlsruhe, Germany	Helix	Niobium	108 MHz	0.04–0.08c	~3.5	3
Saclay, France	Helix	Niobium	135 MHz	0.04–0.08c	~25	50

*RFQ = radio-frequency quadrupole.
† amu = atomic mass unit.

whereby the particles pass inside copper drift tubes where no electric field is present, and are accelerated in the region between the drift tubes. The distance between the center of each drift tube, or cell length, is equal to $v\lambda/c$, where λ is the free-space exciting wavelength, v is the particle velocity at the cell, and c is the speed of light The particles cross each gap at the same phase, usually 20–30° before the time maximum of the electric accelerating field. Focusing is achieved by means of magnetic quadrupoles situated inside the drift tubes. A series of cells is a resonant cavity if each cell resonates at the same frequency. Resonators are sized to match available power sources at the typical operating frequency (around 200 MHz). Special resonant posts or stems are used to improve the energy transfer along the resonator and ease the mechanical and beam loading tolerances. Resonators are placed in tandem to achieve energies up to 200 MeV with the Alvarez structure. Above 200 MeV, the efficiency of the Alvarez structure falls off, and special coupled cavity structures are necessary. The largest proton linac in operation is the 800-MeV accelerator at the Los Alamos Meson Physics Facility (LAMPF).

In some accelerators a structure called the radio-frequency quadrupole (RFQ) is used as an injector instead of the dc injectors previously favored. This device utilizes a special vane-type accelerating structure which also provides quadrupole focusing by electric fields near the axis. Injection energies of 20–50 keV and output energies of 1–2 MeV are typical for such structures.

Heavy-ion accelerators. Modern heavy-ion accelerators are divided into two classes: high-current pulsed machines used mainly for injection into circular accelerators, and low-current continuous-wave machines, which are built as postaccelerators for tandem accelerators (**Table 5**).

In general, the pulsed machines utilize an electron cyclotron resonance ion source or electron-beam ion source, operating at output voltages of 10–100 kV, followed by either a rf quadrupole or a Wideröe structure to increase the output velocity to about 5% of the speed of light.

The posttandem machines employ independently phased resonant accelerating cavities with only a few accelerating gaps and are either copper room-temperature devices or cryogenic cavities made from lead or titanium operating at liquid-helium temperatures. Structures used are either helical, split-ring, or quarter-wave resonant. Kenneth Batchelor

Microtron

The microtron is an electron accelerator in which the beam energy is raised in steps by recirculating short electron bunches through the same rf accelerating sections. A magnetic guide field provides stable orbits which the electrons follow as they are accelerated. Compared with linear accelerators, only moderate rf power is required because the same accelerating structure is used repeatedly.

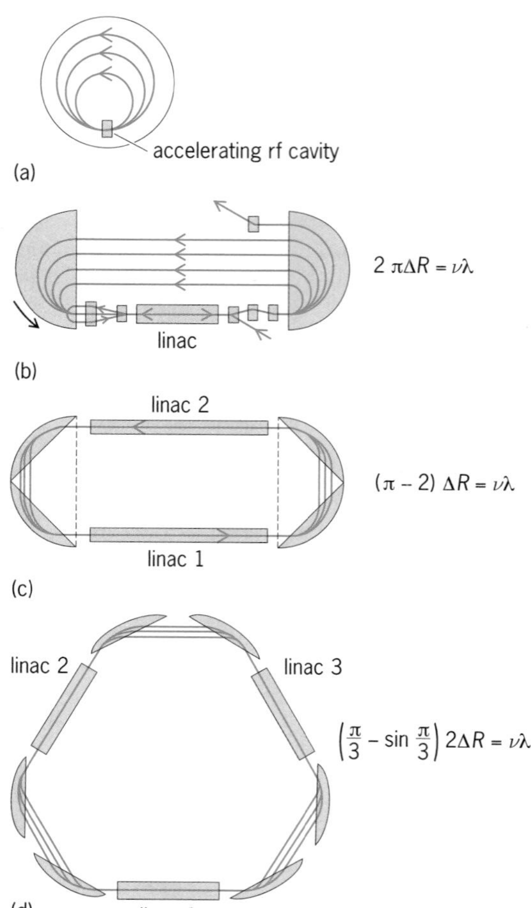

Fig. 17. Various designs for microtrons. (*a*) Circular microtron. (*b*) Race-track microtron. (*c*) Double-sided microtron. (*d*) Hexatron. Also shown for the types in which the particles travel close to the speed of light is the restriction on the change in orbit radius, ΔR, in terms of the rf wavelength, λ, and the number of such wavelengths, ν, by which the orbit length increases between successive passages through the linac.

Principle of operation. The basic principle of all microtrons containing one or more accelerating sections is embodied in the coherence condition. This condition requires that the geometry of all orbits be adjusted to ensure that the revolution time back to each accelerating section equals an integral multiple of the rf period.

For the circular microtron (**Fig. 17***a*), where the electrons start with zero kinetic energy in front of a short microwave resonator, this condition is met by Eq. (3). Here, ΔW is the energy gain per turn;

$$\Delta W = \frac{\nu}{\nu_1 - \nu} \cdot E \qquad (3)$$

$$E_0 = 0.511 \text{ MeV}$$

ν_1, is the revolution time for the first turn in units of the rf period; and ν is the increase in the revolution time between successive revolutions, also in units of the rf period. The optimum values are $\nu_1 = 2$ and

$\nu = 1$ leading to $\Delta W = 0.511$ MeV. Due to thermal heating by rf losses in the cavity walls, the necessary field strength can be achieved only in pulsed operation.

Much more space for the acceleration section is obtained by splitting the 360° magnet into two separated 180° magnets. However, the velocity of the injected electrons has to be close to the speed of light in this race-track microtron. In modern pulsed race-track microtrons, this is achieved by a hairpin-shaped orbit (Fig. 17*b*) which first guides the beam twice in opposite directions through the accelerating section (linac) before the actual microtron starts. The maximum energy of electrons emerging from the race-track microtron is limited to about 1 GeV because the weight of the magnets increases rapidly with energy. By replacing the 180° magnets with two segment-shaped 90° dipoles, the electron energy can be approximately doubled with the same amount of iron, and a second common axis for another linac is obtained (Fig. 17*c*). To achieve even higher energies, higher-order variants of the microtron geometry have been proposed (Fig. 17*d*).

For particles traveling close to the speed of light, c, successive orbits between two accelerating sections increase in length by an integral number of rf wavelengths (Fig. 17). The resulting restriction on the change in orbit radius, ΔR, is met by adjusting ΔW and the magnetic field in the bending magnets, B, to give the prescribed value of ΔR, according to Eq. (4), where q is the electron charge.

$$\Delta R = \frac{\Delta W}{qBc} \tag{4}$$

Focusing. Due to the fact that the revolution time in all microtrons is proportional to the particle energy, the central phase must be placed on the decaying slope of the rf wave. For instance, electrons starting with higher energy than the central particle will be delayed after one revolution and hence will meet the wave at lower rf field strength so that the energy deviation is reduced. Electrons that have energies lower than the central particle are corrected correspondingly. This effect of phase focusing leads to a very small energy spread of the electrons.

In the circular microtron, transverse focusing takes place only in the horizontal plane due to the geometrical effect that all electrons return to the starting point. Magnetic lenses on the accelerator axes and special field configurations in the bending magnets can be used to focus the beam both in the race-track microtron and in higher-order microtrons. In the latter, additional effort is required to compensate for the vertical defocusing in the fringe field of the inclined entrance and exit edges.

Applications. The circular microtron finds applications in medicine and industry, furnishing a beam in the energy range of 10–50 MeV. Racetrack microtrons are used in nuclear physics and other

Fig. 18. Third stage of the MAMI continuous-wave microtron cascade. Injection beam energy is 180 MeV, and maximum energy is 855 MeV. One of the two 500-ton (450-metric-ton) bending magnets is in the foreground, and the 7.5-MeV linac and 90 return pipes are visible.

research fields, and as injectors for synchrotrons and storage rings.

MAMI microtron cascade. In 1990, MAMI (for Mainz Microtron), a three-stage continuous-wave racetrack microtron cascade with a maximum energy of 855 MeV, was set into operation at the University of Mainz, Germany (**Fig. 18**). It delivers a high-quality beam with an intensity of up to 100 microamperes for experiments in nuclear physics and for the production of coherent x-rays.

A double-sided microtron is being added to increase the final energy to 1.5 GeV. The edge-defocusing in this machine is compensated by a decrease in field perpendicular to the magnet edge. The mean bending field thereby drops during acceleration. Therefore, the energy gain per turn must also decrease, in order to fulfill the coherence condition. The energy gain reduction is realized by a corresponding phase slide down on the accelerating wave. Due to phase focusing, this happens automatically if the beam is injected with the correct energy and phase. Simulations demonstrated that the stability against mistunings and drifts can be improved significantly by operating one of the two linacs at the first subharmonic frequency. Karl-Heinz Kaiser

Thomas Jefferson National Accelerator Facility

The Thomas Jefferson National Accelerator Facility, or Jefferson Lab (JLab), in Newport News, Virginia, reached design energy in 1995. (It was formerly known as the Continuous Electron Beam Accelerator Facility, or CEBAF.) The scientific purpose of the 4-GeV continuous-wave superconducting recirculating electron accelerator is study of the structure of the nuclear many-body system, its quark substructure, and the strong and electroweak interactions governing nuclear matter. Such research requires electron beams with a unique combination of technical characteristics. The beams must have sufficient energy to provide the kinematic flexibility required to study the transition region between the hadron-meson description of nuclear matter and the quark-gluon description; high current, to allow precise measurement of relatively small electromagnetic cross sections; high duty factor, to allow coinciding observation of both the scattered electron and the nuclear fragments from a given interaction in an experiment; and high beam quality, to allow high-resolution experiments. *See* ELECTROWEAK INTERACTION; FUNDAMENTAL INTERACTIONS; GLUONS; QUARK-GLUON PLASMA; QUARKS; STRONG NUCLEAR INTERACTIONS.

To attain this combination of characteristics, the accelerator's beam performance objectives are an energy ranging from 0.5 to 4.0 GeV, a maximum current of 200 μA, a duty factor of 100%, a transverse emittance (radius times angular divergence) of $\sim 2 \times 10^{-9}$ meter-radian, and an energy spread of 2.5×10^{-5} times the beam energy. Beams are extracted from the accelerator for simultaneous use in three experimental halls.

A continuous-wave device is the approach of choice to produce a high-quality continuous beam. Low peak current for a given average current lowers emittance, and continuously operating rf systems can be controlled more precisely in phase and amplitude, thereby leading to smaller energy spread and smaller variations of average energy. Conceptually, the most straightforward continuous-wave accelerator would be a single linac in which the beam could attain the requisite energy in a single traversal. However, to reach an energy of 4 GeV in one traversal of a linac made up of the superconducting accelerating cavities used in Jefferson Lab would require 2600 ft (800 m) of active accelerating length. To avoid the high capital cost of such a linac, and to minimize total accelerator length, the beam is passed five times through a parallel pair of much shorter linacs connected by recirculation arcs (**Fig. 19**).

This approach is made possible by the fact that electrons move at very nearly the speed of light c at quite modest energies. At only 50 MeV, for instance, the electron velocity is $0.99995c$. Once fully relativistic, an electron's velocity is essentially independent of energy. Thus beams at different energies can pass together through the linacs, all maintaining the proper phase relative to the rf field.

In this five-pass system, five electron beams at five different energies are simultaneously present on the same trajectory in the linacs. Each beam requires a separate recirculation path tuned to accommodate its energy. The beam transport lines in the recirculation arcs connecting the two linacs are achromatic and isochronous, provide matching in all phase-space coordinates, and are designed with adequate bend radii and strong focusing to minimize quantum excitation to preserve beam quality.

Accelerating cavities. The accelerator uses 1497-MHz, five-cell (20-in. or 0.5-m) superconducting niobium cavities. A waveguide at one end acts as the fundamental rf power input coupler and as a coupler for extracting some of the higher-order modes that are generated by the beam current; at the other end of the cavity, two waveguides serve as couplers to extract higher-order modes. In longitudinal section, the inner surfaces of the cells comprise elliptical segments; this elliptical shape reduces multipacting, a process that degrades the achievable electric field by leading to excessive heat loss in the cavity wall.

Cavities are paired for operation at 2 K (-456°F) within liquid-helium cryostats called cryounits. Every four such cryounits are integrally linked to form the accelerator's basic operating unit, the cryomodule. Each linac has 20 cryomodules for a total of 338 cavities in the accelerator, including 18 cavities in the injector. Cryomodules are interconnected by warm sections of beam line containing vacuum equipment, beam monitors, and magnets to focus and guide the beam.

Injector and distribution system. The 45-MeV injector provides a high-quality electron beam that is sufficiently relativistic to stay in phase with the rf system and the higher-energy, recirculated electron beams present in the first linac. The beam originates in an

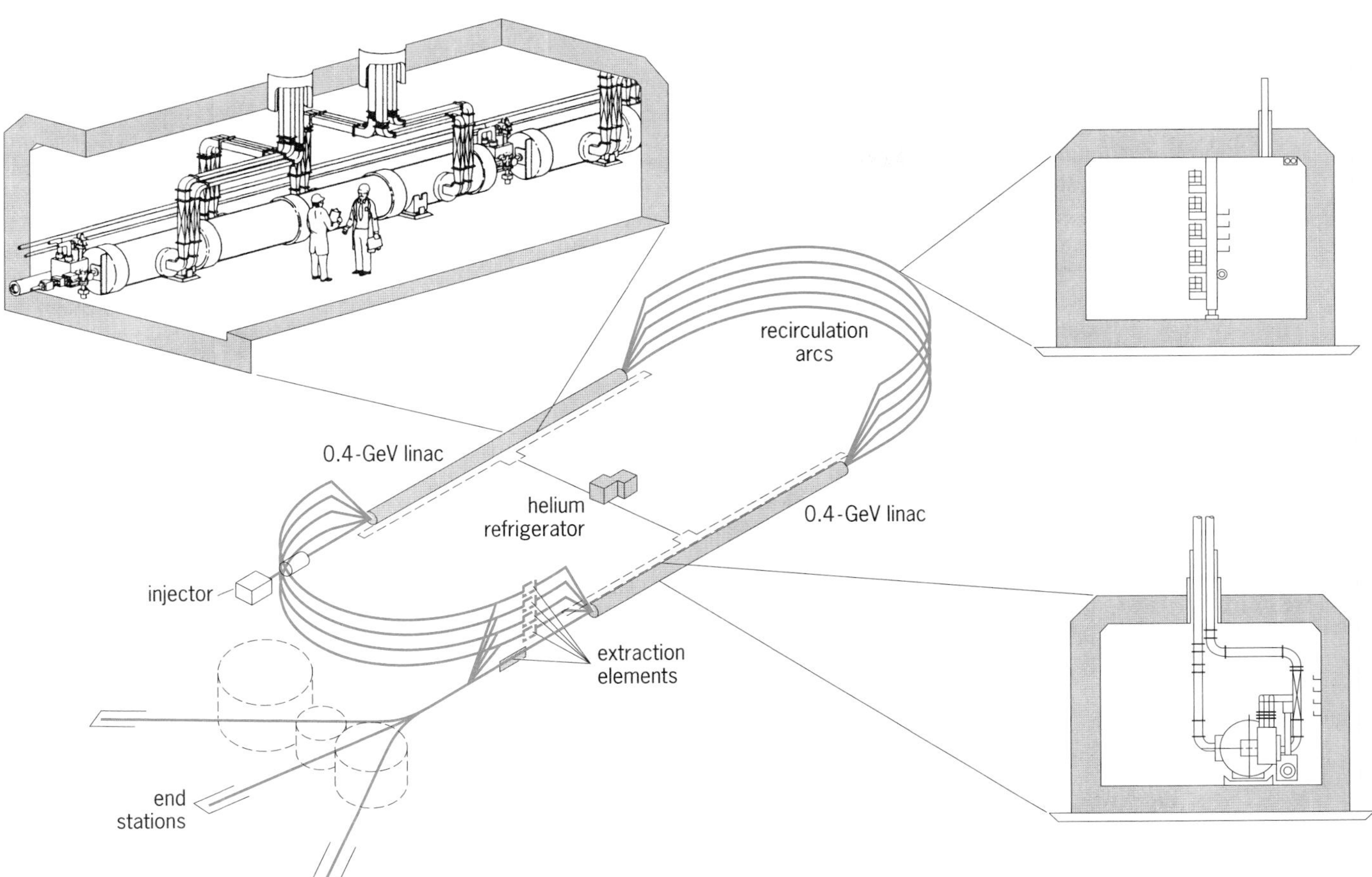

Fig. 19. Configuration of the Thomas Jefferson National Accelerator Facility (Jefferson Lab).

electron gun, passes through bunching, capture, and initial acceleration (up to 0.5 MeV) regions that operate at room temperature, is further bunched and accelerated to just over 5 MeV in a two-cavity cryounit, and then is accelerated to injection energy in two full-sized cryomodules.

The multiuser beam distribution system has two key elements: the injector and the rf separator (deflecting cavities) in the extraction line. The injector creates three interspersed bunch trains that can have different bunch charges. The rf separator deflects the beam, the optics amplify the initial deflection, and septum magnets extract the beam for simultaneous delivery to all three end stations. Hermann A. Grunder

Storage Rings and Colliders

A storage ring consists of an annular vacuum chamber embedded in a ring of bending and focusing magnets, in which a beam of long-lived, charged particles [e^- (electron), e^+ (positron), p (proton), or $\overline{p}$ (antiproton)], or two counterrotating charged particle–antiparticle beams (e^- and e^+, or p and $\overline{p}$) can be stored for many hours. Storage rings are the major components of circular particle colliders in which two particle beams are made to collide head-on. Storage rings also serve as particle accumulators in which successive particle bunches from an accelerator are added to produce a stored beam of relatively high intensity and with desired space, time, and momentum characteristics suitable for acceleration to higher energies for use in a collider.

Principles of colliding-beam systems. The motivation for using colliding-beam systems is found in the following kinematic considerations. Unlike fixed-target systems (cyclotrons, synchrotrons, linear accelerators) in which a beam of accelerated particles traverses a fixed target, colliding-beam systems use the full energy of each particle to produce reactions. When two particles interact, the center-of-mass energy measures the energy available for reactions. For a fixed-target accelerator which produces a beam of high-energy relativistic particles hitting stationary target particles, this useful energy is only a fraction of the available beam energy and increases slowly as its square root. By contrast, when two particles of the same mass and energy collide head-on in a storage ring, all of the available energy is useful. For instance, the CERN storage ring LEP [Large Electron Positron Storage Ring, which was closed down in 2000 to make way for the Large Hadron Collider (LHC)] operated with e^+ and e^- beams of up to 104 GeV yielding a center-of-mass energy of 208 GeV. With a fixed e^- target, an e^+ beam of 42,000,000 GeV would be required to reach the same center-of-mass energy.

Low reaction rate. The high center-of-mass energy of colliding-beam systems is obtained at the expense of reaction rate, which is proportional to the density of target particles and to the beam intensity. It is helpful to think of one stored beam as the target for the other, and to consider the collision of a "beam bunch" with a "target bunch." In a typical e^+e^- storage ring the beam density is about 10^{14} particles/cm^3; a fixed target contains about 10^{24} electrons/cm^3. This huge difference is partially compensated by (1) the fact that at the interaction point each beam bunch collides with each target bunch f times per second (where f is the number of stored particle revolutions per second) and (2) the large intensity of the stored beams compared to accelerator beams which is the result of the accumulation of particles injected over a period of time. Because of the importance of these two factors for obtaining useful reaction rates, beam storage was used in all high-energy colliding beam systems placed in operation before 1987. However, there are limits to the density and intensity of beams that can be stored and made to collide, and thus colliding-beam reaction rates are always much lower than fixed-target reaction rates. Storage ring experiments are designed to take full advantage of the high center-of-mass energy within the limitations imposed by the relatively low reaction rates.

Luminosity. The performance of storage-ring colliding-beam systems is measured by the luminosity. By definition, this parameter is equal to the reaction rate, or number of interactions per second, divided by the interaction cross section.

The LEP storage ring had a circumference of 17 mi (27 km). To reach much higher e^+ and e^- energies than were achieved at LEP, the size of the storage ring would have to become unreasonably large. Instead, it appears possible to obtain useful luminosities in direct head-on collisions between two beams from two very high energy linear accelerators by focusing each beam to a very small cross-sectional area. The SLAC-Linac-Collider (SLC) project is a prototype which started taking data at a center-of-mass energy of 92 GeV in 1989.

Circular positron-electron colliders. Positron-electron collisions occur in a storage ring composed of bending and focusing magnets enclosing a doughnut-shaped vacuum chamber in which counterrotating beams of e^+ and e^- are stored for periods of several hours. The two beams are made to collide with each other about 10^6 times per second in straight interaction sections of the vacuum chamber (**Fig. 20**). These are surrounded with detectors for the observation of collision products.

Collisions between e^+ and e^- particles are of particular interest since most of the resulting processes proceed through annihilation of both initial-state particles resulting in a particularly simple final state with the quantum numbers of the photon. This is one of the main reasons why e^+e^- collisions have proved to be particularly fruitful in deepening understanding of the fundamental structure of matter.

Beam instabilities. When particles from beam 1 (e^+ or e^-) pass through a more intense beam 2 (e^- or e^+) in the interaction region, they feel a strong force due to the electromagnetic field set up by beam 2. Thus, beam 2 acts as a very nonlinear lens on beam 1 and tends to cause a diffusive growth in the size of the latter. The force increases rapidly with the density of particles in beam 2 until it becomes so strong that the beam 1 area becomes suddenly very large and the luminosity decreases drastically; this is beam-beam instability. There also exist single-beam instabilities which are caused by electromagnetic forces between particles in the same beam or by fields set up by induced currents in the walls of the vacuum chamber. Maximum luminosity is achieved by operating with two beams of equal intensity just below the beam-beam instability limit, which in a well-designed storage ring is reached before single-beam instabilities become a problem.

Synchrotron radiation. The limitations discussed so far apply equally well in principle to e^+e^- as to p-p or $\bar{p}$-p colliding-beam systems. There are important differences, however. Because they have a small mass, e^+ and e^- in circular orbits radiate a substantial amount of energy in the form of synchrotron radiation, just as they do in electron synchrotrons. The power radiated is proportional to the total number of orbiting particles (positions and electrons) and to the fourth power of the energy, and it is inversly proportional to the square of the orbit radius. This power, which has to be supplied by the rf system to keep particles on the design orbit, is appreciable and costly; at the peak operating energy it amounts typically to several megawatts in multi-gigaelectronvolt storage rings. In the design of a storage ring for a given energy and luminosity, the most economic solution requires a careful balance between the cost of magnets and buildings (which increases with increasing orbit radius) and the cost of the rf system (which decreases with increasing orbit radius since less power

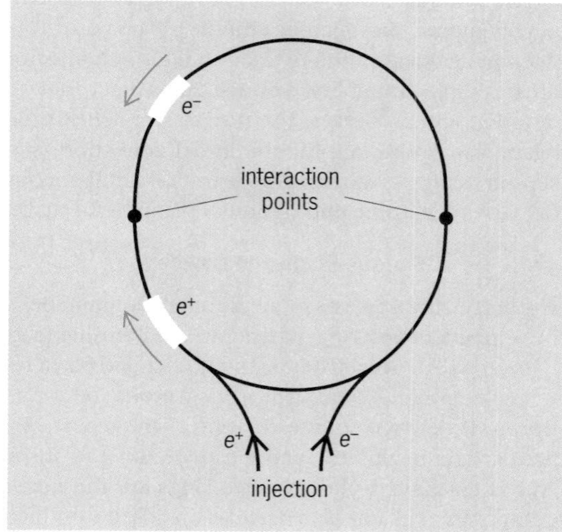

Fig. 20. Schematic diagram of e^+e^- storage ring. This ring has two interaction regions and one bunch of particles circulating in each direction

is needed). At the highest operating energies the amount of available rf power limits the luminosity, since for fixed values of radiated power and orbit radius only a steep decrease in the number of orbiting particles can compensate for an increase in energy.

There are also very beneficial consequences from the emission of synchrotron radiation. This emission dampens the amplitude of e^- and e^+ oscillations around the equilibrium orbit, and produces a gradual transverse polarization of the stored beams. As in electron synchrotrons, the radiation itself constitutes a very intense source of x-rays which has proved very useful in various branches of biology, chemistry, and physics. Not only are all operating e^+e^- colliders used as sources of synchrotron radiation, but many e^- storage rings have been built and are used specifically for investigations with this radiation.

Ring. The ring consists of circular sectors of bending and focusing magnets and of straight sections for rf cavities and interaction regions. The bending magnets guide the beam particles around the ring, and the focusing magnets drive them toward the equilibrium orbit around which they execute radial and vertical oscillations. The use of separate magnets for guiding and for focusing permits tight control of beam size to maximize the luminosity as the energy of operation is changed.

Injector. To achieve high luminosity, it is desirable to inject and accumulate e^- and e^+ with energies close to the desired collision energy. Linear accelerators and synchrotrons are used as injectors. The e^- are accelerated directly; e^+ are collected from a target bombarded with high-energy e^- and then accelerated. Accumulation occurs through the addition of particles first to the circulating e^+ beam, then to the circulating e^- beam until desired intensities have been reached. Typical accumulation times are seconds for e^- and minutes for e^+.

Vacuum chamber and system. Very low pressures of 10^{-9} and 10^{-10} torr (10^{-7} and 10^{-8} Pa) are required to prevent the outscattering of beam particles by collisions with gas molecules, which reduces storage times below useful values. Special techniques are used to achieve such high vacuum in the presence of high-power synchrotron radiation hitting the chamber walls.

The rf power system. The power from the rf system replaces the energy emitted as synchrotron radiation and accelerates the stored beams to the desired energy. In a multigigaelectronvolt storage ring, several megawatts of power at 400–500 MHz are typically delivered to several rf cavities located in the ring.

Beam structure. The stored beams consist of bunches (typically 2 in. or 5 cm long), since rf power is being supplied continuously. For a given number of stored particles, maximum luminosity is achieved with the smallest number of bunches in each beam. This number equals half the number of interaction regions.

Interaction region. Magnet-free straight sections at the center of which e^+e^- collisions occur are provided for the installation of detectors. Special magnets are usually installed on each side of the interaction region to focus the beams to a very small area at the interaction point so as to maximize the luminosity; this is called low beta insertion. Karl Strauch

Proton-proton and $\bar{p}p$ colliders. Proton-proton and proton-antiproton storage rings or circular colliders provide the highest-energy particle beams used to search for and measure the fundamental constituents of matter and their associated forces. Hadron colliders, like circular electron-positron colliders, are synchrotrons and are constructed of a large number of long magnets and other equipment distributed in an approximately circular or racetrack configuration. The magnets have a small aperture or evacuated beam pipe through which the beam can pass repeatedly for many millions of revolutions while being deflected by the magnetic field into the necessary curved trajectory. The so-called bend or dipole magnets cause charged particles, such as protons or antiprotons, traveling in a vacuum chamber to follow the circular layout of the magnets. Additional magnetic elements (quadrupoles), interspersed in a regular way with the bend magnets, focus the beam of particles in a manner analogous to that in which an optical lens focuses light. The beams are injected into the storage rings in special regions or straight sections. Radio-frequency (rf) cavities used to accelerate the beams are also located in these regions, which are free from standard magnetic components.

In proton-proton colliders (**Fig. 21*a***), the two counterrotating beams are of the same charge, but moving in opposite directions. Each beam must be provided with a separate aperture with opposite magnetic field direction in order to follow the same curved trajectory. This can be achieved with two rings of magnets (above or below one another, or side by side), or the magnets can be built to contain the two apertures of opposite magnetic field direction. The beams must then be brought together at specific azimuthal locations around the rings to intersect or pass through one another in order that collisions between individual protons may take place. These regions, called interaction regions, are where high-energy physics detectors are assembled to measure the events from the proton-proton interactions.

Proton-antiproton colliders (Fig. 21*b*) have the advantage that only one ring of magnets with one aperature is required. The protons and antiprotons are identical except that they have opposite charge. Because of this, the counterrotating beams can be guided in opposite directions in one ring of magnets with one sign of magnetic field. In this case, collisions are possible at many places around the ring's circumference. As the experiments are located at only a few locations, it is desirable to separate the beams from one another so that they collide only at the experimental areas. Separation into (typically) helical orbits within the single magnetic aperture

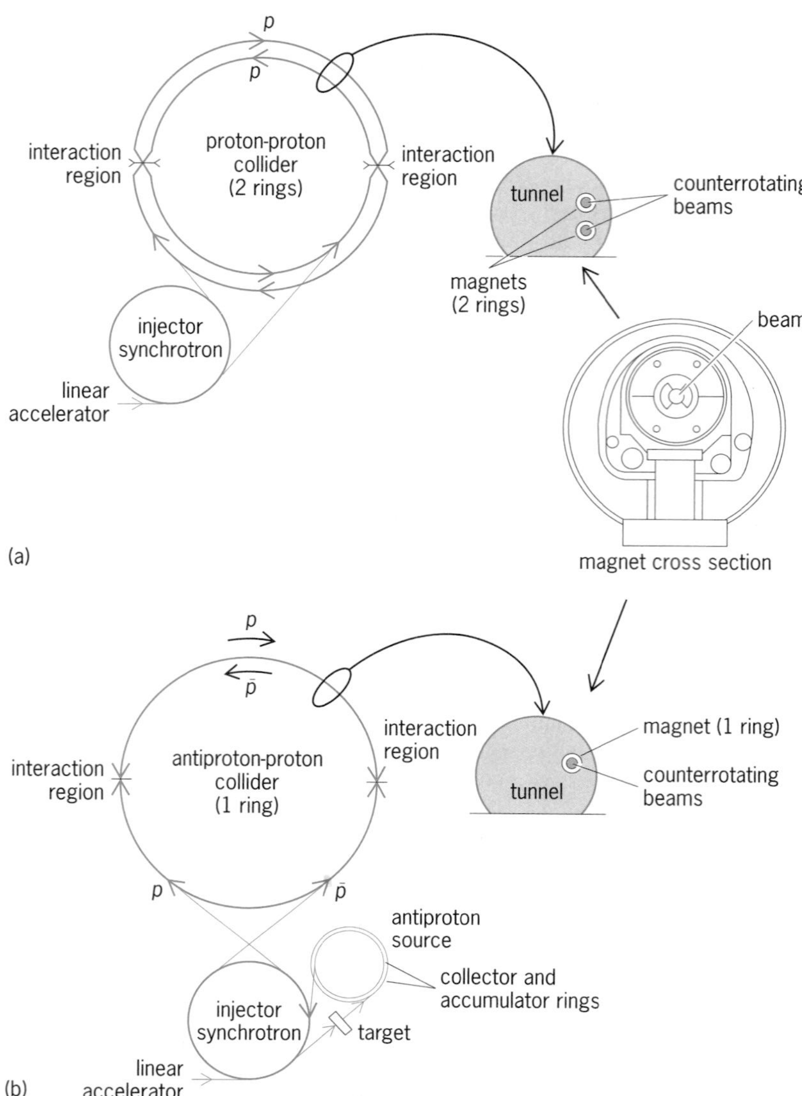

Fig. 21. Layout of colliders. (*a*) Proton-proton collider (two rings). (*b*) Proton-antiproton collider (one ring). The magnet cross section, common to both systems, is shown.

can be provided by electrostatic deflectors. Though proton-antiproton colliders have a cost advantage in that only one magnet ring is required, they have the disadvantage that the antiprotons must be produced and the antiproton beam intensity is usually limited.

Collision rate. At the experimental interaction regions, the beams in a proton-proton or proton-antiproton collider are tightly focused to make their spot size as small as possible. This is to enhance the collision interaction rate needed for high-energy physics experiments. As in positron-electron colliders, luminosity and the cross section together determine the rate at which collisions take place. The total cross section for all types of interactions in proton-proton collisions is of the order of 100 millibarns (10^{-25} cm^2). Interesting events on the threshold of understanding have cross sections 10^{-8} of this, in the nanobarn to picobarn region (10^{-34} to 10^{-36} cm^2). Luminosities of 10^{30} to 10^{34} cm^{-2} s^{-1} are achievable in antiproton-proton and proton-proton colliders. The higher range of luminosity is realistic only in proton-proton colliders. Antiproton-proton colliders are limited by the number of antiprotons that can feasibly be produced and collected for use in the collider.

Maximum energy. The primary limitation to the energy of proton-proton and proton-antiproton colliders is the strength of the magnetic guide field required to bend the protons or antiprotons in the circular orbit. Magnetic fields of the order of 4 to 9 T are produced by superconducting magnets. This field strength then sets the scale of the required circumference of the magnet ring. A continuous array of bending magnets with a $3^1/_3$-T field arranged on a 0.6-mi-radius (1-km) circle would provide the appropriate bending strength for 1-TeV-momentum particles. In practice, space required for magnetic focusing, other necessary beam manipulations such as acceleration from the injection energy (with rf cavities), the injection and extraction processes, and the interaction regions increase circumference of the racetrack to about $1^1/_3$ times that needed just for bending. Thus, the Fermilab Tevatron (**Table 6**), a 0.98-TeV collider with 4.0-T magnetic field, is 3.9 mi (6.3 km) in circumference.

The development of magnets using wire carrying superconducting current has made it possible to operate large rings of magnets at high magnetic excitation for long periods of time with reasonable power usage. If conventional copper magnets were used, power demands would severely limit the maximum beam energy that could be practically realized.

Whereas the energy of proton colliders is limited by the strength of the magnetic field and the circumference that can be realized at reasonable expense, electron storage rings are limited by the energy loss from synchrotron radiation. Both electron and proton rings are forced to larger ring circumferences as the particle energy increases. As discussed above, the Tevatron collider is 3.9 mi (6.3 km) in circumference for 0.9-TeV beam energy, whereas the largest electron collider (LEP at CERN) had a circumference of 17 mi (27 km) and beam energy of 104 GeV. LEP has been decommissioned, and the LEP tunnel is being used for installation of the Large Hadron Collider (LHC), a proton-proton collider which will have an energy of 7 TeV per beam, or 70 times of energy of the electron ring.

Thus, proton-proton and proton-antiproton storage rings would seem to have a large energy advantage over electron rings. This apparent advantage is, however, offset by the fact that the electron is a fundamental particle whereas the proton is made up of quarks and other constituents, only one of which is primarily involved in the collision process. Effectively the average useful energy of the protons is reduced by a factor of 3–6. Also, precision interaction energy measurements are more difficult with protons than with electrons. Conversely, proton colliders have a broader exploration potential because

TABLE 6. Proton-proton and proton-antiproton collider storage rings*

Name	Laboratory	Location	Beam energy	$\bar{p}$ stack rate	Luminosity (L), cm^{-2} s^{-1}	Circumference	Type	Status
Intersecting Storage Rings (ISR)	CERN	Geneva, Switzerland	32 GeV		10^{32}	0.6 mi (1 km)	*pp*, 2 rings, conventional magnets	First *pp* ring, decommissioned
Super Proton-Antiproton Synchrotron ($Sp\bar{p}S$)	CERN	Geneva, Switzerland	315 GeV	4.8×10^{10}/h	2.8×10^{30}	4.3 mi (6.9 km)	$\bar{p}p$, 1 ring, conventional magnets	First collisions 1981
Tevatron	Fermilab	Batavia, Illinois	0.98 TeV	6×10^{10}/h	3×10^{31}	3.9 mi (6.3 km)	$\bar{p}p$, 1 ring, superconducting magnets	First collisions 1985
Large Hadron Collider (LHC)	CERN	Geneva, Switzerland	7 TeV		10^{34}	17 mi (27 km)	*pp*, 2-beam aperture in one magnet, superconducting magnets	Under construction
Relativistic Heavy Ion Collider (RHIC)	Brookhaven	Upton, New York	100 GeV/u 250-GeV protons		1.4×10^{31} (protons) 1×10^{27} (gold)	2.4 mi (3.8 km)	Heavy ions, 2 rings, superconducting magnets	First collisions 2000
HERA (Hadron Electron Ring Anlage)	DESY	Hamburg, Germany	26–820 GeV		2.0×10^{31}	3.9 mi (6.3 km)	*e-p*, 2 rings, superconducting magnets	First collisions 1991

* Including heavy-ion and electron-proton (*e-p*) colliders.

very high constituent interaction energies are possible, though at much lower effective luminosities because of the statistics of the energy distribution of the quarks within the proton.

Acceleration and beam control. As in proton synchrotrons, a cascaded chain of accelerators, each providing a factor of 10–20 in acceleration energy gain, is used. The final acceleration is performed in the storage ring itself. In order that colliders can be efficiently used, beams must be stored and collided for the order of a day before a fresh set of particles is injected, accelerated, and stored.

A beam of particles can be characterized by its density distribution in six-dimensional phase space, that is, by the number of particles in a volume element $dx_x\, dydp_y dzdp_z$. Here, p_x is the momentum conjugate to the coordinate x, and so on. In turn, the emittance is a characterization of the volume in phase space occupied by the beam. Typically in accelerators, emittance may be regarded as the product of two loosely coupled components, transverse emittance and longitudinal emittance. Transverse emittance is related to the distribution of position and angle of the particle trajectories and consequently is related to the beam spot size in the focusing lattice of the accelerator. Longitudinal emittance is related to the momentum distribution of the particles and their distribution along the major direction of motion around the ring; it includes the momentum spread of the beam and is important in determining the amount of rf accelerating voltage required and the bunch length of the beam.

Protons in synchrotrons and storage rings differ from their electron counterparts in that great care must be taken that the beam emittance is not enlarged during the acceleration process, during the transfer from one ring to another, or during the storage time when collisions are under way. Such phase space dilution decreases the luminosity and performance of the collider. Many sources can cause dilution and even very small nonlinear magnetic fields, variations in power-supply current (ripple), or rf noise from the accelerating cavities can play a major role. An important problem in proton collider design is to calculate accurately how this dilution is expected to take place over hours or days of stored-beam operation. Electron storage rings do not have this problem because the synchrotron radiation which limits the energy also provides a damping mechanism for any enlargement of transverse spot size. *See* RIPPLE VOLTAGE.

Antiproton sources. The antiprotons in proton-antiproton storage rings are not available in nature and must be produced in order to be injected and stored in the collider each time the beams are replenished. The storage rings and associated equipment which provide for the antiproton acceleration is called the antiproton source (Fig. 19*b*). Typically, 20–150 GeV protons extracted from a stage of the proton injector chain are used to strike a target. Secondary particles produced from this target, of as large a momentum and angular acceptance as possible, are then gathered into a collector or debuncher synchrotron. Some of the particles captured in this collector ring will be antiprotons. Their momentum spread can be reduced by debunching the beam or converting very short bunches of particles with large momentum spread into a continuous uniform-density current with small momentum spread. This beam is then passed to the accumulator synchrotron, which stores and accumulates each successive cycle of antiproton beam as it is produced.

Fig. 22. One of the Fermilab colliding beam detectors used to observe collision events. The detector is made up of a central tracking region surrounding the beam pipe in which the colliding particles counterrotate. Outside the central tracking region is a superconducting magnetic solenoid coil that provides a magnetic field to help analyze the momentum of the produced particles. External to the solenoid are energy calorimeter modules or towers, which absorb, detect, and measure the energy of the events. As shown, the calorimeter modules are pulled away from the central region and coil for maintenance and checkout.

Once a sufficient quantity of antiprotons has been accumulated and it is desirable to refill the collider, antiprotons are extracted from the accumulator, accelerated, and stored along with protons in the collider ring.

Antiprotons produced in the production target have a large spread in angle and velocity. Even after the manipulations in the collector-debuncher, they are still hot in the sense that they are spread over a large region of phase space (or emittance) that must be reduced to achieve the small beam size necessary for high-luminosity collisions. This cooling process is done in the accumulator. It is accomplished by a stochastic process that relies on monitoring for a very short time a sample of the beam and kicking it back to beam center with high-frequency kickers. This is done repeatedly over long periods of time as the different particles mix with different neighbors, making for different beam sample populations. This combination of damping and mixing in order to create new samples results in cooling or reducing the beam's emittance.

Existing storage rings. The first proton storage ring, the Intersecting Storage Rings (ISR) at CERN, is now decommissioned (Table 6). The next hadron collider to operate, the Super Proton Antiproton Synchroton (Sp̄pS), also at CERN, made use of the Super Proton Synchrotron (SPS), which is no longer operated for proton-antiproton collider physics.

The Tevatron is the highest-energy accelerator-synchrotron-storage ring in operation (800–900 GeV, with engineering runs at 980 GeV and plans for physics data taking there). It is also used as a fixed target proton synchrotron; the acceleration of protons in either mode of operation is discussed above. The antiproton source has two rings, the debuncher and the accumulator, operating at 8 GeV. Antiprotons are produced by extracting and targeting a 120-GeV beam from the main ring. The resultant 8-GeV antiprotons are collected first in the debuncher source ring at an approximately 2.5-s-cycle period and then are transferred to the accumulator source ring, where they are allowed to build up for a number of hours and the phase space of the antiproton beam is reduced or cooled. Six equally spaced bunches of 10^{11} protons and 3×10^{10} antiprotons are injected into the Tevatron and accelerated to full field, where they are stored and collided for typically 15–20 h. Collisions take place at the two straight sections where the large experimental detectors are located (**Fig. 22**). The beams are separated elsewhere by electrostatic separators.

A second collider using superconducting magnets to circulate the proton beam is the HERA (Hadron Electron Ring Anlage) in Germany. This collider has two very different energy beams of electrons and protons contained in two separate rings, and collides 820-GeV protons against 27-GeV electrons in 170 bunches. It is the only collider of this type.

The Large Hadron Collider, under construction at CERN, is a proton-proton collider which will have a beam energy of 7 TeV and a luminosity of 10^{34}. Thus, when it comes into operation, it will be the foremost collider for high-energy physics research. There will be over 1200 main dipole magnets, each 30 ft (9 m) long, operating at a magnetic field of 9 T. The magnets are of a unique two-in-one design, with both beam apertures in one cold iron mass and cryostat. The magnets will operate at 2 K (−456°F).

Helen T. Edwards

Heavy-ion collider. The theory of strong interactions predicts that a phase transition from so-called normal nuclear structure to plasma of quarks and gluons will occur in the extreme states of high temperature and high density created by relativistic heavy-ion collisions. Such states resemble the conditions of the early universe, shortly after the big bang. To investigate this new phase of nuclear matter requires a machine capable of providing collisions of heavy ions at energies of the order of 200 GeV/u.The Relativistic Heavy-Ion Collider (RHIC), a colliding-beam accelerator facility at Brookhaven National Laboratory in Upton, New York, is dedicated to the investigation of this new phase of nuclear matter, and experiments with its four detectors began in 2000. RHIC can accelerate, store, and collide ions as heavy as gold with a maximum beam energy of 100 GeV/u. *See* QUARK-GLUON PLASMA; RELATIVISTIC HEAVY-ION COLLISIONS.

A heavy-ion synchrotron is used as an injector to a heavy-ion collider. Since heavy ions are available only in positive charge states, heavy-ion colliders, like proton-proton colliders, must have two separate storage rings intersecting at several collision points. In addition, for collisions of unequal species of ions, such as proton on gold, the two rings must fulfill either of two conditions in order to maintain the synchronized bunch crossings: (1) the two rings are operable at different magnetic fields to accommodate the difference in the magnetic rigidities of unequal species that have the same rotation frequency (or velocity); or (2) the energies of both beams are relativistic enough that the difference in the rotation frequencies is small and can be accommodated by an adjustment of the average radius of the closed orbits within the dynamic aperture of the rings.

The intrabeam scattering (Coulomb interaction among ions in the beam bunch), which scales as the fourth power of the ion charge divided by the square of the ion mass, and is thus enhanced for heavier ions, results in the rapid growth of transverse and longitudinal dimensions of the beam bunches. The transverse growth of the beam affects the lattice design (favoring short lattice cells and strong focusing) and magnet aperture. The longitudinal growth influences the rf system requirements. Since the bunch growth is particularly rapid at low energies, the time allowed for the injection beam stacking and for acceleration to the top energy must be kept very short (a few minutes). Transverse beam growth beyond the dynamic aperture (the so-called good-field region of the magnetic fields) of the accelerator lattice and longitudinal growth beyond the rf bucket (or capture phase space) result in beam loss, which reduces the beam lifetime.

In addition to intrabeam scattering, there are a number of unavoidable beam depletion mechanisms, which are dictated by the physics of heavy-ion collisions. The highly Lorentz-contracted Coulomb field of heavy ions such as gold is very intense and produces copious electron pairs as the bunches pass through each other. Ions passing through the cloud of electrons thus produced have a high probability of capturing an electron. This change of charge state causes ions to be lost from the stable orbit, and is a significant beam-loss mechanism. The breakup of the nucleus into several lighter fragments (Coulomb dissociation) by the intense Coulomb field is also significant, but is much smaller than electron capture from the pair production. The beam loss from ordinary nuclear interactions is insignificant. Satoshi Ozaki

Bibliography. D. A. Bromley (ed.), *Large Electrostatic Accelerators*, 1974; P. J. Bryant and K. Johnsen, *The Principles of Circular Accelerators and Storage Rings*, Cambridge University Press, 1993; A. W. Chao (ed.), *Handbook of Accelerator Physics and Engineering*, 1999; D. A. Edwards and M. J. Syphers, *An Introduction to the Physics of High Energy Accelerators*, 1993; H. T. Edwards, The Tevatron energy doubler: A superconducting particle accelerator, *Annu. Rev. Nucl. Part. Sci.*, 35:605–660, 1985; H. A. Grunder and F. B. Selph, Heavy-ion accelerators, *Annu. Rev. Nucl. Sci.*, 27:353–392, 1977; S. Humphries, Jr., *Charged Particle Beams*, Wiley, 1990; S. Humphries, Jr., *Principles of Charged Particle Acceleration*, Wiley, 1986; Institute of Electrical and Electronics Engineers, *Proceedings of the 1999 Particle Accelerator Conference*, and similar conferences every 2 years; J. D. Lawson, *Physics of Charged-Particle Beams*, 2d ed., 1988; S.-Y. Lee, *Accelerator Physics*, World Scientific, 1999; S. Myers and E. Picasso, The LEP collider, *Sci. Amer.*, 263(1): 54, July 1990; *1998 Linear Accelerator Conference Proceedings*, and similar conferences every 2 years; *Proceedings of the 2000 European Particle Accelerator Conference*, and other conferences in this series, Vienna; *Proceedings of the 15th International Conference on Cyclotrons and Their Applications*, Caen, France, 1998, and other conferences in this series; R. E. Rand, *Recirculating Electron Accelerators*, 1984; M. Reiser, *Theory and Design of Charged Particle Beams*, 1994; W. Scharf, *Particle Accelerators and Their Uses*, Harwood, 1986; H. Schopper (ed.), *Advances in Accelerator Physics and Technology*, 1993; T. Wangler, *Principles of RF Linear Accelerators*, 1998; H. Wiedemann, *Particle Accelerator Physics*, 2d ed., vol. 1, 1999, vol. 2, 1998.

Particle detector

A device used to detect and measure radiation characteristically emitted in nuclear processes, including gamma rays or x-rays, lightweight charged particles (electrons or positrons), nuclear constituents (neutrons, protons, and heavier ions), and subnuclear constituents such as mesons. The device is also known as a radiation detector. Since human senses do not respond to these types of radiation, detectors are essential tools for the discovery of radioactive minerals, for all studies of the structure of matter at the atomic, nuclear, and subnuclear levels, and for protection from the effects of radiation. They have also become important practical tools in the analysis of materials using the techniques of neutron activation and x-ray fluorescence analysis. *See* ACTIVATION ANALYSIS; ELEMENTARY PARTICLE; NUCLEAR REACTION; NUCLEAR SPECTRA; PARTICLE ACCELERATOR; PROSPECTING; RADIOACTIVITY; X-RAY FLUORESCENCE ANALYSIS.

Classification by use. A convenient way to classify radiation detectors is according to their mode of use: (1) For detailed observation of individual photons or particles, a pulse detector is used to convert each such event (that is, photon or particle) into an electrical signal. (2) To measure the average rate of events, a mean-current detector, such as an ion chamber, is often used. Radiation monitoring and neutron flux measurements in reactors generally fall in this category. Sometimes, when the total number of events in a known time is to be determined, an integrating version of this detector is used. (3) Position-sensitive detectors are used to provide information on the location of particles or photons in the plane of the detector. (4) Track-imaging detectors image the whole three-dimensional structure of a particle's track. The output may be recorded by immediate electrical readout or by photographing tracks as in the bubble chamber. (5) The time when a particle passes through a detector or a photon interacts in it is measured by a timing detector. Such information is used to determine the velocity of particles and when observing the time relationship between events in more than one detector. *See* TIME-OF-FLIGHT SPECTROMETERS.

Ionization detectors. Any radiation-induced effect in a solid, liquid, or gas can be used in a detector. To be useful, however, the effect must be directly or indirectly interpretable in terms of either the quantity or quality (that is, type, energy, and so on) of the incident radiation or both. The ionization produced by a charged particle is the effect commonly employed.

Gas ionization detectors. In the basic type of gas ionization detector, an electric field applied between two electrodes separates and collects the electrons and positive ions produced in the gas by the radiation to be measured. Depending on the intensity of the electric field, the charge signal in the external circuit may be equal to the charge produced by the radiation, or it may be much larger. In a proportional counter, the output charge is larger than the initial charge by a factor called the gas amplification. A Geiger-Müller counter provides still larger signals, but each signal is independent of the original amount of ionization. All three types of gas ionization detectors can be used as pulse detectors, mean current detectors, or, with an indicator such as a quartz-fiber electroscope, as integrating detectors. *See* GEIGER-MÜLLER COUNTER; IONIZATION CHAMBER.

Position-sensitive and track-imaging detectors. Position-sensitive detectors and track-imaging detectors are nearly all based on the ionization process. Multiwire proportional chambers and spark chambers are position-sensitive adaptations of gas detectors. The signal division or time delay that occurs between the ends of an electrode made of resistive material is sometimes used to provide position sensitivity in gas and semiconductor detectors. Track-imaging detectors rely on a secondary effect of the ionization along a particle's track to reveal its structure. In Wilson cloud chambers, ionization triggers condensation along particle tracks in a supercooled vapor; bubble formation in liquids is the basis for operation of bubble chambers. A secondary effect of ionization is also employed in photographic emulsions used as radiation detectors where ionization triggers the formation of an image.

Semiconductor detectors. In this type of detector, a solid replaces the gas of the previous example. The "insulating" region (depletion layer) of a reverse-biased *pn* junction in a semiconductor is employed. Choice of materials is very limited; very pure single crystals of silicon or germanium are presently the only fully suitable materials, although other semiconductors can be used in noncritical applications. Collection of the primary ionization is normally used, but an avalanche mode is sometimes employed in which an intense electric field causes charge multiplication. Although semiconductor devices are mostly used as pulse detectors, mean-current and integrating modes are possible when the significant leakage currents of *pn* junctions can be tolerated.

Since solids are approximately 1000 times denser than gases, absorption of radiation can be accomplished in relatively small volumes. A less obvious but fundamental advantage of semiconductor detectors is the fact that much less energy is required (~3 eV) to produce a hole-electron pair than that required (~30 eV) to produce an ion electron pair in gases. This results in better statistical accuracy in determining radiation energies. For this reason, semiconductor detectors have become the main tools for nuclear spectroscopy, and they have also made neutron activation analysis and x-ray fluorescence analysis of materials practical tools of great value. *See* CRYSTAL COUNTER; JUNCTION DETECTOR.

Scintillation detector. In addition to producing free electrons and ions, the passage of a charged particle through matter temporarily raises electrons in the material into excited states. When these electrons fall back into their normal state, light may be emitted and detected as in the scintillation detector. The early scintillation detectors consisted of a layer of powder (zinc sulfide, for example) that, when struck by charged particles, produced light flashes, which were observed by eye and counted. The meaning of the term "scintillation detector" has changed with time to refer to the combination of a scintillator and a photomultiplier tube that converts light scintillations into signals that can be processed electronically. Various organic and inorganic crystals, plastics, liquids, and glasses are used as scintillators, each having particular virtues in regard to radiation absorption, speed, and light output. *See* SCINTILLATION COUNTER.

Neutral particles. Neutral particles, such as neutrons, cannot be detected directly by ionization. Consequently, they must be converted into charged particles by a suitable process and then observed by detecting the ionization caused by these particles. For example, high-energy neutrons produce "knock-on" protons in collisions with light nuclei, and the protons can be detected. Slow neutrons are usually detected by using a nuclear reaction in which the neutron is captured and a charged particle is emitted. For example, in boron trifluoride (BF_3) detectors, neutrons react with boron to produce alpha particles which are detected.

Other detector types. Although ionization detectors dominate the field, a number of detector types based on other radiation-induced effects are used. Notable examples are (1) transition radiation detectors, which depend on the x-rays and light emitted when a particle passes through the interface between two media of different refractive indices; (2) track detectors, in which the damage caused by charged particles in plastic films and in minerals is revealed by etching procedures; (3) thermo- and radiophotoluminescent detectors, which rely on the latent effects of radiation in creating traps in a material or in creating trapped charge; and (4) Cerenkov detectors, which depend on measurement of the light produced by passage of a particle whose velocity is greater than the velocity of light in the detector medium. *See* CERENKOV RADIATION; PARTICLE TRACK ETCHING; TRANSITION RADIATION DETECTORS.

Large detector systems. The very large detector systems used in relativistic heavy-ion experiments and in the detection of the products of collisions of charged particles at very high energies, typically at the intersection region of storage rings, deserve special consideration. These detectors are frequently composites of several of the basic types of detectors discussed above and are designed to provide a detailed picture of the multiple products of collisions at high energies. The complete detector system may occupy a space tens of feet in extent and involve tens or hundreds of thousands of individual signal processing channels, together with large computer recording and analysis facilities.

The time projection chamber (TPC) typifies this class of detector. This chamber, in its initial configuration, is a cylinder 6.5 ft (2 m) long and 6.5 ft (2 m) in diameter, containing a gas, with a thin-plane central electrode that splits the chamber longitudinally into two drift spaces, each 3 ft (1 m) in length. A high potential (approximately 150 kV) is applied to this electrode to cause the electrons produced by

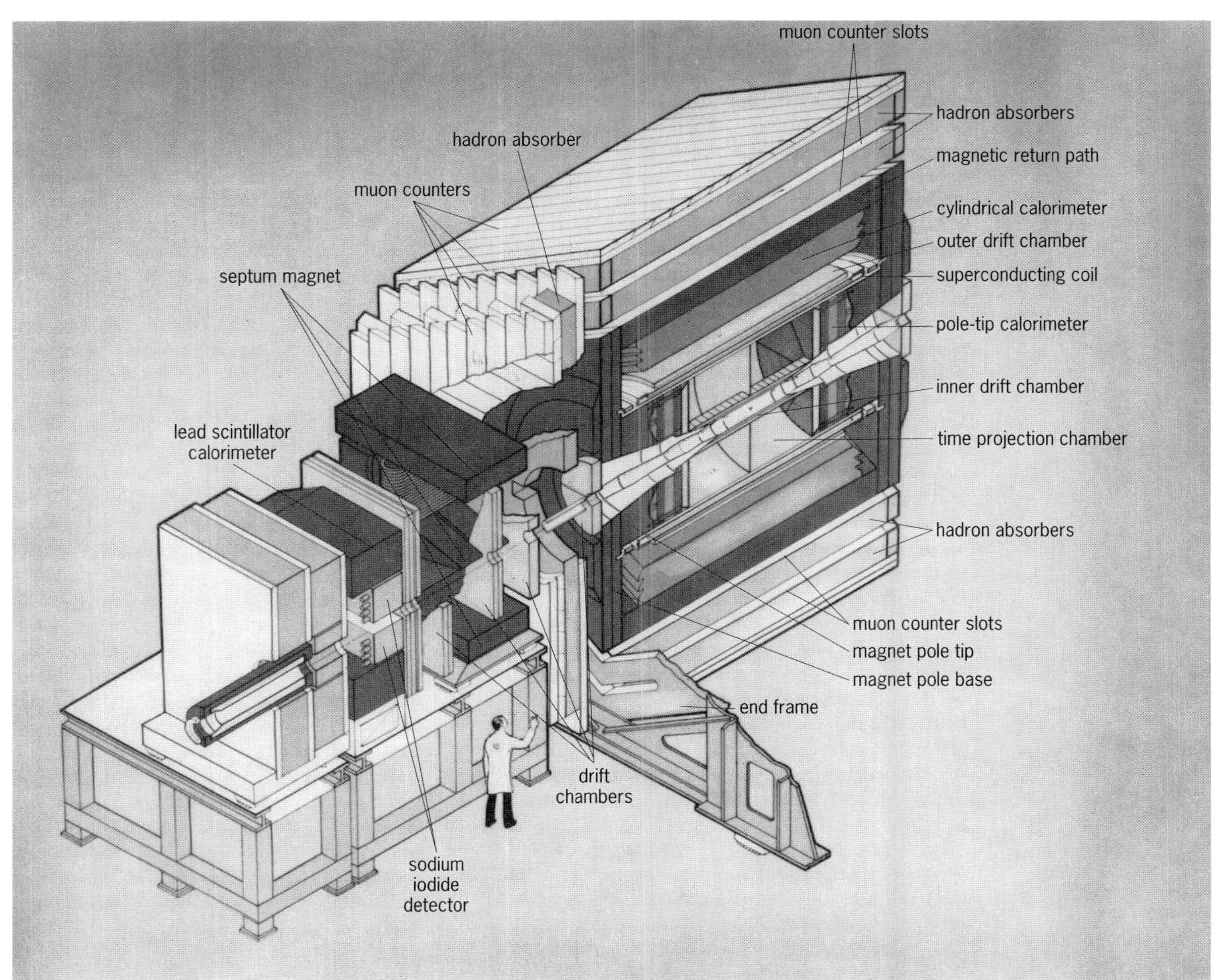

Time projection chamber designed for the PEP (Positron-Electron Projection) storage ring at Stanford University, with surrounding detection equipment and magnets.

ionization along tracks to drift to the end-cap regions of the cylinder. The end caps contain arrays of a few thousand gas proportional wire chambers designed to locate the position of signals with an accuracy of a small fraction of an inch in both radius and azimuth. The signal amplitude is maintained proportional to the original ionization in the track segment. When this information is combined with the drift time information, the effect is to provide the capability to measure the amount of ionization in individual cells a small fraction of an inch in linear dimensions throughout the whole volume of the chamber. The chamber is immersed in an axial magnetic field so that the bending of the tracks can be used to provide particle momentum information. Finally, the central time projection chamber is surrounded by other detector chambers (see **illus.**) to provide information on particles leaving the time projection chamber and also to perform ray calorimetry. The time projection chamber may be thought of as a modern version of the cloud or bubble chamber with prompt electrical readout of track information.

Very complex large-area and -volume chambers are also used in heavy-ion physics experiments. Here identification of the type of ion is usually important, and this is achieved partly by measuring the ionization or energy loss at several points along the track and by combining this information with time-of-flight or magnetic bending information. Complex combinations of gas, semiconductor, and scintillation detectors are frequently used in these detector systems. Fred S. Goulding

Bibliography. G. Charpak, *Research on Particle Detectors*, 1995; C. F. Delaney and E. C. Finch, *Radiation Detectors: Physical Principles and Applications*, 1992; February issues of *IEEE Trans. Nucl. Sci.*, 1970 to present; K. Klein Knecht, *Detectors for Particle Radiation*, 1988; G. F. Knoll, *Radiation Detection and Measurement*, 3d ed., 1999.

Particle flow

Particle flow is important to many industrial processes, including the pneumatic conveying of solids, transport of solids in liquids (slurries), removal of particulates from gas streams for pollution control, combustion of pulverized coal, and drying of particulates in the food and pharmaceutical industries. The vast majority of flow problems in industrial design involve the flow of gas or liquids with suspended solids. *See* DRYING; FLUIDIZED-BED COMBUSTION; PIPELINE.

Properties. The basic concepts of particle flows can be illustrated by referring to a specific example: the motion of a particle-laden fluid through a converging duct similar to a venturi entrance (**illus.** *a*). A key parameter in fluid-particle flows is the Stokes number, which is the ratio of the response time of a particle to a time characteristic of a flow system. Particle response time is the time a particle takes to respond to a change in carrier flow velocity. For example, if a particle is released from rest into a flow, the particle response time is the time required for the particle to reach 63% of the flow velocity.

If the Stokes number is small (say, less than 0.1), the particles have sufficient time to respond to the change in fluid velocity, so the particle velocity approaches the fluid velocity (illus. *b*). However, if the Stokes number is large (say, greater than 10), the particles have little time to respond to the varying fluid velocity and the particle velocity shows little change.

Loading. The relative concentration of the particles in the fluid is referred to as loading. The loading may be defined in several ways, but one definition frequently used is the ratio of particle mass flow to fluid mass flow. Many industrial applications involve highly loaded particle flows.

Coupling. Another term encountered frequently in the particle flow terminology is coupling. If the particle loading is small, the fluid will affect the particle properties (velocity, temperature, and so forth),

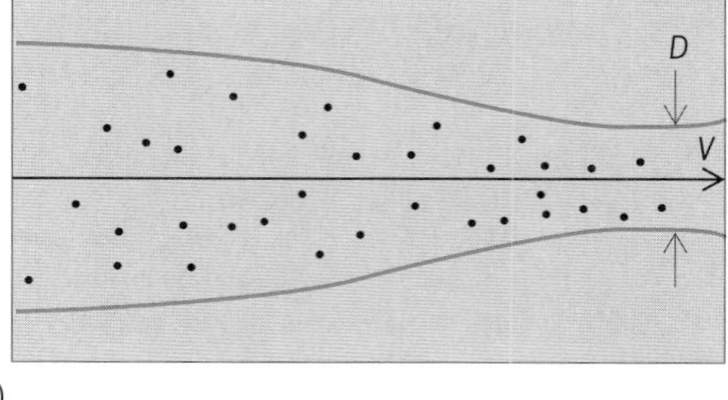

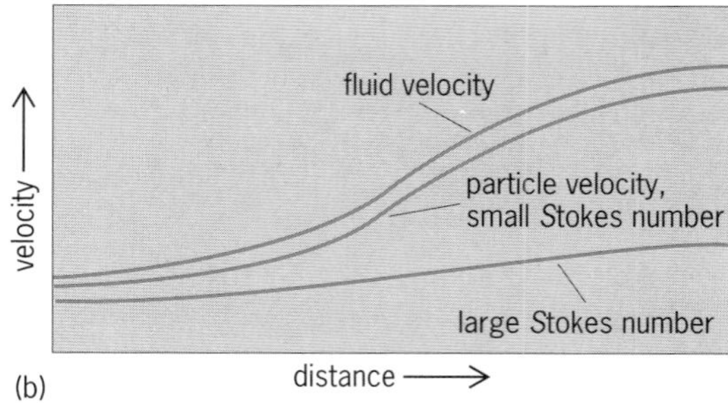

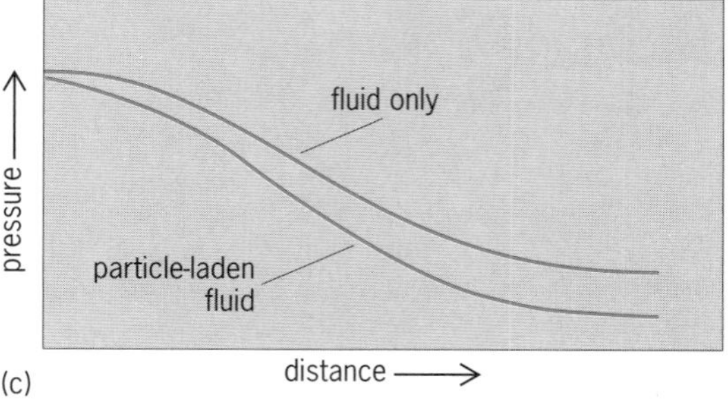

Fluid-particle flow in a converging duct. (*a*) Flow geometry. (*b*) Fluid and particle velocities along the duct in the cases of small and large Stokes numbers. (*c*) Effect of loading on pressure distribution along the duct.

but the particles will not influence the fluid properties. This is referred to as one-way coupling. If the conditions are such that there is a mutual interaction between the particles and fluid, the flow is two-way coupled.

For the above flow configuration, low particle loading will not affect the pressure distribution, so the flow will be one-way coupled. At high loadings, the pressure drop in the fluid is larger (illus. *c*) because of the momentum needed by the fluid to accelerate the particles. This interaction is defined as two-way coupling. Two-way coupling effects are reduced with increasing Stokes number because the particles undergo less acceleration.

Dilute and dense flows. If the particle motion is controlled by the action of the fluid on the particle, the flow is termed dilute. But, if the particle concentration is sufficiently high, the particles will collide with each other and their motion will be dependent on particle-particle collisions; the flow is then regarded as dense.

Pneumatic transport. An interesting example of a gas-particle flow is pneumatic transport in a horizontal duct. At high velocities the particles are kept in suspension by the fluid motion (homogeneous flow). The flow is essentially dilute. In this case, there is no acceleration of the gas, but the particles are continually accelerating as they collide with the wall and lose their velocity before reentrainment into the gas stream. The force required to accelerate the particles toward the gas flow velocity leads to a higher pressure gradient (larger pressure loss per unit length of tube) or a two-way coupling effect.

As the velocity of the airflow in the horizontal pipe is reduced, the particles tend to settle out on the tube wall. The velocity at which this occurs is called the saltation velocity. The accumulation of particles on the wall leads to a reduced cross-sectional area for the flow, a larger pressure loss for a given flow rate, and the onset of unsteady flow. A series of particle-gas flow patterns are observed as the particles tend to accumulate on the wall. First, dunelike formations occur. A further reduction in velocity leads to the dense-flow condition, where the particles nearly fill the pipe and move through it in a sequence of slugs. For even lower velocities, the particles will plug the pipe, and the air flows through a stationary packed bed or the bed may slide through the pipe. In this case, the particle-particle interaction controls the motion. *See* FLOW OF SOLIDS.

Slurry transport. Slurry transport is similar to pneumatic transport but, because of buoyancy, the liquid is more capable than a gas of supporting the particles. A slurry is often treated as a single-phase non-newtonian fluid with empirical relationships for shear stress. As with pneumatic transport, there is a critical velocity at which the solid phase begins to settle on the pipe wall, which is called the deposition velocity. It is very important to design a slurry transport system with adequate velocities to avoid the immense problems associated with plugging.

Measurements. Measurements of particle flows are generally difficult to carry out. Instrumentation such as phase-Doppler anemometry, flow visualization with digital image analysis, and other nonintrusive techniques have provided important information. Measurements in dense particle systems are particularly difficult. *See* FLOW MEASUREMENT.

Numerical models. Significant advances have been made in numerical models for particle flows. These models require high-speed computational and extensive storage capability. Also, many of the models are based on empirical input that must be provided by experiment. Still, the models are sufficiently viable to be used to complement the design of particle flow systems. *See* COMPUTATIONAL FLUID DYNAMICS; FLUID FLOW; PARTICULATES. Clayton T. Crowe

Bibliography. G. Hetsroni (ed.), *Handbook of Multiphase Systems*, 1982; G. E. Klinzing, *Gas-Solid Transport*, 1981; G. Matsui et al., *Proceedings of the International Conference on Multiphase Flows—'91 Tsukuba*, vol. 3, 1991; M. C. Roco (ed.), *Particulate Two-Phase Flow*, 1993; C. A. Shook and M. C. Roco, *Principles and Practice of Slurry Flow*, 1993.

Particle track etching

A technique of selective chemical etching to reveal tracks of heavy nuclear particles in a wide variety of solid substances. Developed in order to see fossil particle tracks in extraterrestrial materials, the technique finds application in many fields of science and technology.

Identification of nuclear particles. An etchable track is produced if the charged particle has a sufficiently high radiation-damage rate and if the damaged region in the solid is permanently localized. Thus only highly ionizing particles are detectable; only nonconductors record tracks; and radiation-sensitive plastics can detect lighter particles than can radiation-insensitive minerals and glasses. The conical shape of the etched track depends on the ratio of the rate of etching along the track to the bulk etching rate of the solid. Careful measurements with accelerated ions of known atomic number Z and velocity $v = \beta c$ (where c = velocity of light) have shown that this ratio is an increasing function of Z/β. Measurements of the shapes of etched tracks thus serve to identify particles (**Fig. 1**). The most sensitive detector, a plastic known as CR-39, detects particles with Z as low as 1 (**Fig. 2**) provided $Z/\beta \gtrsim 10$. The clarity and contrast of the images and its high sensitivity make CR-39 a very attractive track-etch detector for nuclear physics, cosmic-ray research, element mapping, personnel neutron dosimetry, and many other applications. Minerals, on the other hand, are insensitive to particles with $Z/\beta \lesssim 150$ and are therefore useful in recording rare, very heavily ionizing, relatively low-energy nuclei with $Z \gtrsim 25$. *See* CHARGED PARTICLE BEAMS.

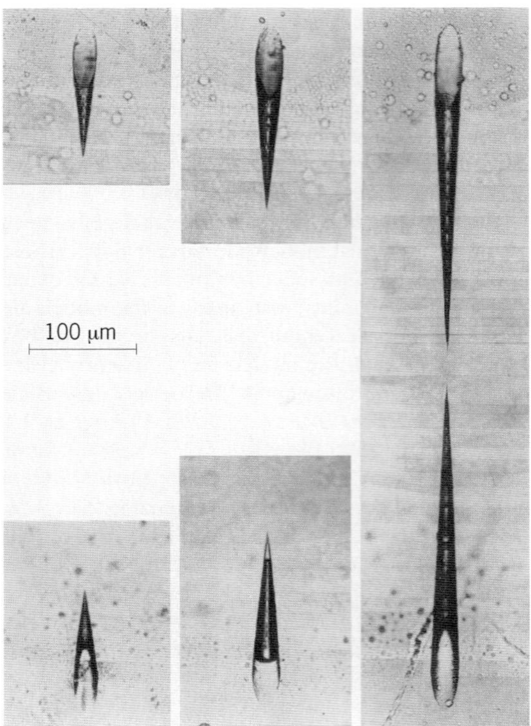

Fig. 1. Upper images show increasing radiation damage produced by a high-energy uranium nucleus in the cosmic radiation as it penetrates a thick stack of Lexan plastic sheets and slows down. The lengths of the etched cones in these three sheets, taken at intervals of several millimeters from within the stack, permit the atomic number and energy of the nucleus to be determined. Lower images show the exit points of the tracks.

Solar and galactic irradiation history. The lunar surface, meteorites, and other objects exposed in space have been irradiated by charged particles from a variety of sources in the Sun and the Galaxy. The particles of lowest energy are produced by the expanding corona of the Sun—the solar wind. Arriving in prodigious numbers, but penetrating only some millionths of a centimeter, the solar wind particles quickly produce an amorphous radiation-damaged layer on crystalline grains. At depths from about 0.00004 to 0.04 in. or 1 micrometer to 1 mm, individual tracks produced in solar flares (sporadic, energetic outbursts on the Sun) can be resolved by electron microscopy or sometimes by optical microscopy. At depths greater than 1 mm, most of the particle tracks are produced by heavy nuclei in the galactic cosmic radiation. *See* SOLAR WIND.

Comparison of fossil particle tracks in lunar rocks and meteorites with spacecraft measurements of present-day radiations has established that solar flares and galactic cosmic rays have not changed over the last 2×10^7 years—the typical time a lunar rock exists before being shattered by impacting interplanetary debris. Observations of grains in stratified lunar cores and lunar and meteorite breccias (**Fig. 3**) enable the particle track record to be extended back more than 4×10^9 years in time. Breccias, which are complex grain assemblages, often contain grains that have high solar flare track densities on their edges, indicating exposure to free space prior to breccia formation. Dating work, using spontaneous-fission tracks, indicates that some of the breccias were assembled soon after the beginning of the solar system some 4.6×10^9 years ago. The study of the time history of energetic radiations in space, using lunar samples and meteorites, elucidate various dynamic processes such as rock survival lifetimes, microerosion of rocks, and the formation and turnover rates of planetary regoliths. *See* METEORITE; MOON.

Nucleosynthesis. In terrestrial crystals, which are well shielded from external radiations, the dominant source of tracks is the spontaneous fission of ^{238}U. Certain meteorites and lunar rocks (Fig. 3*a*) contain additional fission tracks due to the presence of ^{244}Pu (half-life, 8×10^7 years) when the crystal was formed. After corrections are made for chemical fractionation of Pu and U, the data indicate that a large spike of newly synthesized elements was produced at the time of formation of the Galaxy, followed by a period of continuous synthesis and relatively rapid mixing. *See* MILKY WAY GALAXY; NUCLEOSYNTHESIS.

Current solar and galactic irradiation. Studies of tracks in a piece of glass from the *Surveyor 3* spacecraft after a 2.6-year exposure on the lunar surface, and of tracks in plastic detectors exposed briefly above the Earth's atmosphere in rockets, led to the surprising discovery that the Sun preferentially ejects heavy elements in its flares rather than an unbiased sample of its atmosphere. *See* SUN.

The existence of galactic cosmic rays with atomic number greater than 30 was discovered in 1966 when fossil particle tracks were first studied in meteorites. Since then many stacks of various types of plastics and nuclear emulsions up to 220 ft^2 (20 m^2) in area have been exposed in high-altitude balloons and in Skylab in order to map out the composition of the heaviest, rarest cosmic rays. Several particles heavier than uranium have been detected, indicating that cosmic rays originate in sources where synthesis

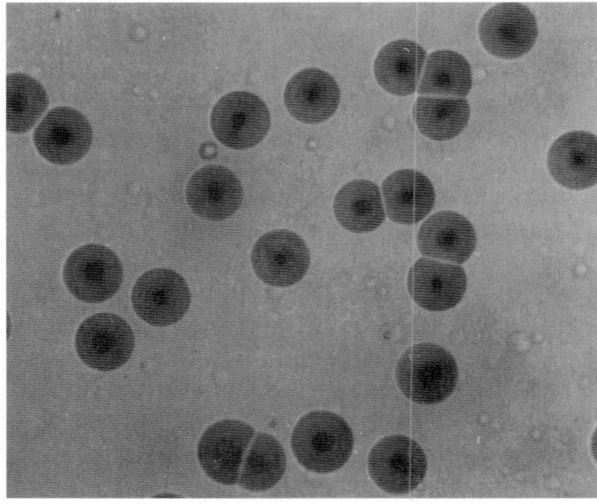

Fig. 2. Etched conical pits in a sheet of allyl diglycol carbonate (CR-39 plastic) irradiated at normal incidence with 60-MeV alpha particles. (*From P. B. Price et al., Do energetic heavy nuclei penetrate deeply into the Earth's atmosphere?, Proc. Nat. Acad. Sci., 77:44–48, 1980*)

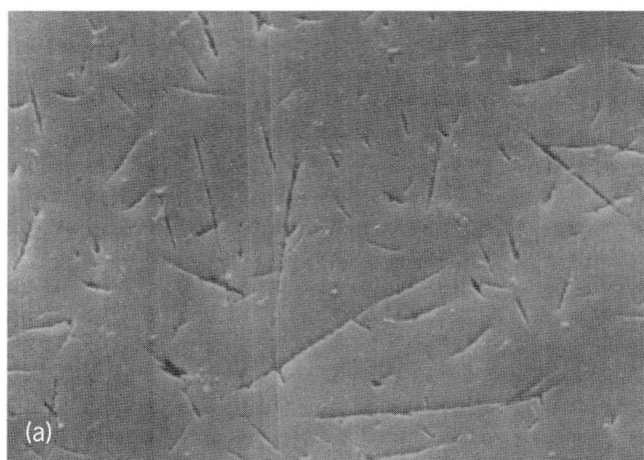

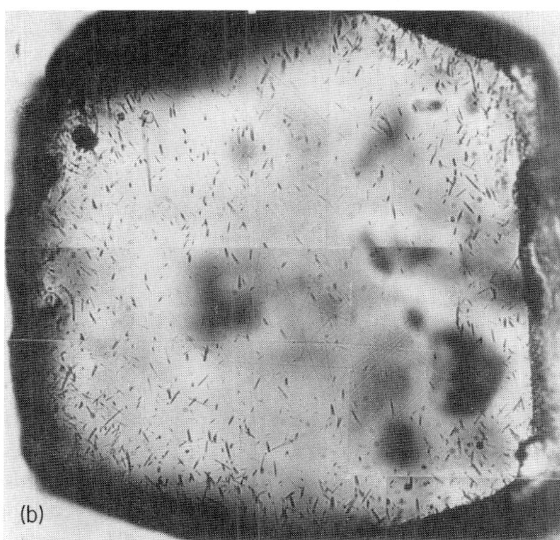

Fig. 3. Etched fossil particle tracks. (*a*) Tracks from spontaneous fission of ^{238}U and ^{244}Pu in a zircon crystal from a lunar breccia, seen by scanning electron microscopy (*from D. Braddy et al., Crystal chemistry of Pu and U and concordant fission track ages of lunar zircons and whitlockites, Proceedings of the 6th Lunar Science conference, pp. 3587–3600, 1975*). (*b*) Tracks of energetic iron nuclei from solar flares, showing a decreasing concentration from edges to center of a 150-μm-diameter olivine crystal from a carbonaceous chondritic meteorite. The irradiation occurred about 4×10^9 years ago, before the individual grains were compacted into a meteorite (*from P. B. Price et al., Track studies bearing on solar-system regoliths, Proceedings of the 6th Lunar Science Conference, pp. 3449–3469, 1975*).

has proceeded explosively beyond uranium. Exposures of giant detectors of about 1100 ft^2 (100 m^2) for a year in space are planned. Hybrid detectors using stacks of track-recording plastics to measure range, and photomultiplier tubes to measure light from Cerenkov radiation and scintillation detectors, have made it possible to determine relative abundances of the isotopes of very heavy elements in the cosmic rays. These data have a bearing both on the history of cosmic rays and on nucleosynthesis in their sources. *See* COSMIC RAYS.

Nuclear and elementary particle physics. Unique advantages of etched-track detectors are their ability to distinguish heavy-particle events in a large background of lightly ionizing radiation and their ability to detect individual rare events by a specialized technique such as electric-spark scanning or ammonia penetration through etched holes. These advantages have permitted such advances as the measurement of very long fission half-lives; the discovery of ternary fission; the determination of fission barriers; the production of numerous isotopes of several far-transuranic elements; the discovery of several light, neutron-rich nuclides such as ^{20}C at the limit of particle stability; and highly sensitive searches for magnetic monopoles, superheavy elements, and anomalously dense nuclear matter in nature and in accelerators. *See* MAGNETIC MONOPOLES; NUCLEAR FISSION; TRANSURANIUM ELEMENTS.

Geochronology. The spontaneous fission of ^{238}U, present as a trace-element purity, gives tracks that can be used to date terrestrial samples ranging from rocks to human artifacts. Because fission tracks are erased in a particular mineral at a well-defined temperature (for example, 212°F or 100°C in apatite), one can use the apparent fission-track ages as a function of distance from the heat source to measure the thermal (tectonic) history of regions. Examples are the rate of sea-floor spreading (about 1 cm per year) and the surprisingly rapid rate of uplift of the Alps (1.2×10^{-2} to 5.6×10^{-2} in. or 0.3 to 1.4 mm per year) and of the Wasatch range in Utah (4×10^{-3} to 1.6×10^{-2} in. or 0.01 to 0.4 mm per year). Because of its very high uranium concentration (typically a few hundred parts per million), a few tiny zircon crystals less than 4×10^{-3} in. (0.1 mm) in diameter suffice to give a fission track age of sedimentary volcanic ash and thus to determine absolute ages of stratigraphic boundaries. Occasional, anomalously low fission-track ages can sometimes be used to locate valuable ore bodies and even petroleum. *See* FISSION TRACK DATING.

Geophysics and element mapping. In these applications a track detector placed next to the material being studied records the spatial distribution of certain nuclides that either spontaneously decay by charged particle emission or are induced to emit a charged particle by a suitable bombardment. A resolution of a few micrometers is easily attained. Such micromaps make it possible to identify radionuclides in atmospheric aerosols, to measure the distribution and transport of radon, thorium, and uranium, and to measure sedimentation rates at ocean floors. Thermal neutrons, readily available in a reactor, can be used to map ^{235}U (via fission), ^{10}B (via alpha particles), ^{6}Li (via tritons), ^{14}N (via protons), and several other nuclides. Deuterium, a tracer in biological studies, and lead and bismuth can also be mapped by using different irradiations. *See* RADIOACTIVE TRACER.

Practical applications. Filters are produced by irradiating thin plastic sheets with fission fragments and then etching holes to the desired size. Uses include biological research, wine filtration, and virus sizing. In virus sizing a single hole is formed that separates two halves of a conducting solution. When a virus or other tiny object passes through the hole, the resistance increases drastically, and the size, shape, and speed of the object can be determined by analyzing the electric signal.

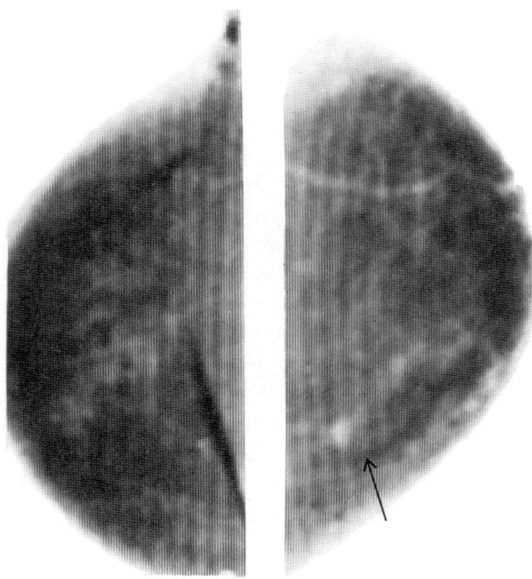

Fig. 4. Reconstructed images of right and left breasts of a patient taken by using a 3600-MeV carbon-ion beam at the Lawrence Berkeley Laboratory Bevalac accelerator. Arrow indicates 1-cm (0.4-in.) carcinoma. (*From C. A. Tobias et al., Lawrence Berkeley Laboratory*)

There are numerous other uses. A uranium exploration method relies on a survey of radon emanation, as measured by alpha-particle tracks in plastic detectors, to locate promising locations in which to drill. CR-39 plastic detectors are used in a Fresnel zone-plate imaging technique to make high-resolution images of the thermonuclear burn region in laser fusion experiments. Plastic detectors are also used in conjunction with a beam of high-energy heavy ions to take radiographs of cancer patients that reveal details not detectable in x-rays. **Figure 4** shows reconstructed images of right and left breasts of a patient taken by using a 3600-MeV carbon-ion beam at the Lawrence Berkeley Laboratory Bevalac accelerator. A single beam pulse is passed through each breast as it is immersed in water, and the stopping points of the ions are recorded in a stack of 30 cellulose nitrate sheets, each 0.01 in. (0.025 cm) thick, which are located behind the water bath. After the sheets are etched, the information on each sheet is converted to digital data by a scanning system, and two images capable of revealing slight differences in density are displayed on a television screen. The patient dose for such images is about 20 to 100 millirads (200 to 1000 micrograys), which is considerably lower than the usual dose from x-ray mammographic imaging. *See* RADIOGRAPHY. P. Buford Price

Bibliography. B. G. Cartwright, E. K. Shirk, and P. B. Price, A nuclear-track-recording polymer of unique sensitivity and resolution, *Nucl. Instrum. Meth.*, 153: 457–460, 1978; R. L. Fleischer, Where do nuclear tracks lead?, *Amer. Sci.*, 67:194–203, 1979; R. L. Fleischer, P. B. Price, and R. M. Walker, *Nuclear Tracks in Solids*, 1975; R. M. Walker, Interaction of energetic nuclear particles in space with the lunar surface, *Annu. Rev. Earth Planet. Sci.*, 3:99–128, 1975.

Particle trap

A device used to confine charged or neutral particles where their interaction with the wall of a container must be avoided. Electrons or protons accelerated to energies as high as 1 teraelectronvolt (10^{12} electronvolts) are trapped in magnetic storage rings in high-energy collision studies. Other forms of magnetic bottles are designed to hold dense hot plasmas of hydrogen isotopes for nuclear fusion. At the other end of the energy spectrum, ion and atom traps can store isolated atomic systems at temperatures below 1 millikelvin. Other applications of particle traps include the storage of antimatter such as antiprotons and positrons (antielectrons) for high-energy collision studies or low-energy experiments. *See* ANTIMATTER; NUCLEAR FUSION; PARTICLE ACCELERATOR; PLASMA (PHYSICS); POSITRON.

Charged-particle traps. Charged particles can be trapped in a variety of ways. An electrostatic (Kingdon) trap is formed from a thin charged wire. The ion is attracted to the wire, but its angular momentum causes it to spiral around the wire in a path with a low probability of hitting the wire.

A magnetostatic trap (magnetic bottle) is based on the fact that a charged particle with velocity perpendicular to the magnetic field lines travels in a circle, whereas a particle moving parallel to the field is unaffected by it. In general, the particle has velocity components both parallel and perpendicular to the field lines and moves in a helical spiral. In high-energy physics, accelerators and storage rings also use magnetic forces to guide and confine charged particles. The earliest form of such a trap is the cyclotron, used to accelerate protons. A tokamak has magnetic field lines configured in the shape of a torus, confining particles in spiral orbits (**Fig. 1***a*). This type of bottle is used to contain hot plasmas in nuclear fusion studies. Another type of bottle uses a magnetic mirror. As the particle approaches the ends of the bottle (Fig. 1*b*), the pinched magnetic field causes it to spiral into ever-tightening helixes. The time-averaged circular motion of the particle is effectively a current loop with an associated magnetic moment that is repelled by the gradient of the magnetic bottle field. Ions are trapped in the Earth's magnetic field (the van Allen belts) by this effect. This type of trap has been used to accumulate positrons for simulations of intergalactic space and to help produce positronium for laser spectroscopy. *See* ELECTRON MOTION IN VACUUM; POSITRONIUM; VAN ALLEN RADIATION.

The radio-frequency Paul trap uses inhomogeneous radio-frequency electric fields to confine particles, forcing them to oscillate rapidly in the alternating field (**Fig. 2**). If the amplitude of oscillation (micromotion) is small compared to the trap dimensions, the trap may be thought of as increasing the (kinetic) energy of the particle in a manner that is a function of the particle position. The particle moves to the position of minimal energy and is therefore attracted to the center of the trap where the oscillating

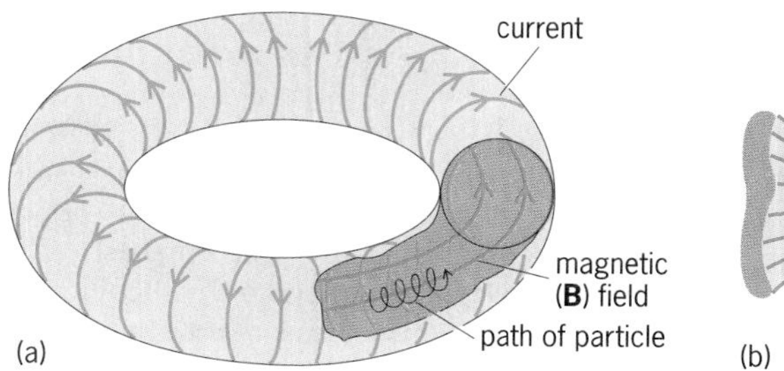

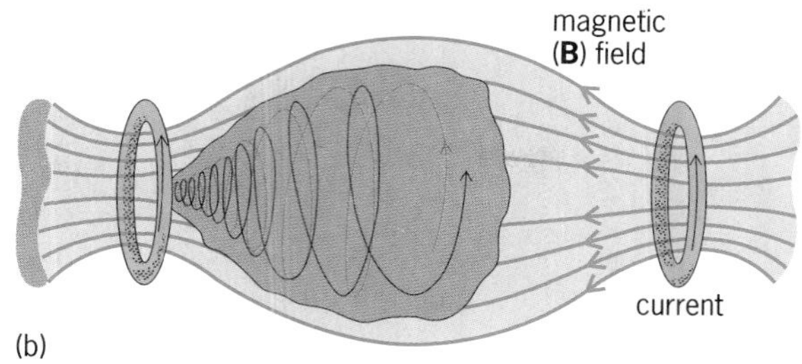

Fig. 1. Magnetic bottles. (*a*) Cutaway view of a tokamak magnetic bottle used to confine hot plasmas. (*b*) Orbit of a charged particle in a magnetic mirror trap as the particle is reflected from the pinched field ends of the trap.

electric fields are weakest. At the center of the trap, the fields are exactly zero, and a single, cold ion or electron trapped there is essentially at rest with almost no micromotion. This trap has stability limits. If the frequency of oscillation is too low, the amplitude of the micromotion approaches the trap dimensions and the oscillations become unstable. On the other hand, if the frequency is too high, the energy associated with the oscillation approaches zero together with the depth of the pseudopotential created by the position dependence of the kinetic energy. The great advantage of this trap is that a single particle is driven to the region of space where the trapping fields are zero. Single ions or a small, countable number of ions have been stored, laser-cooled, and visualized in these traps. When a large number of ions are stored in this type of trap, the mutually repulsive Coulomb forces tend to push the ions into regions where the micromotion is large. This effect is referred to as rf heating.

The Penning trap, with the same electrode configuration as the Paul trap, uses a combination of static electric (**E**) and magnetic (**B**) fields instead of oscillating electric fields. The electric fields create a harmonic potential along the z axis but are antitrapping in the radial direction. The repulsive radial forces are overcome by a static magnetic field directed along the z axis. This additional field causes the particles to move in the x-y plane with a combination of a circular cyclotron motion due to the magnetic field and an **E** × **B** drift magnetron motion about the trap axis. In this trap, tens of thousands of ions can be stored and cooled to millikelvin temperatures. *See* MAGNETRON.

Neutral-particle traps. Uncharged particles such as neutrons or atoms are manipulated by higher-order moments of the charge distribution such as the magnetic or electric dipole moments.

Magnetic traps of neutral particles use the fact that atoms usually have a magnetic dipole moment on which the gradient of a magnetic field exerts a force. The atom can be in a state whose magnetic energy increases or decreases with the field strength, depending on whether the moment is antiparallel or parallel to the field. A magnetic field cannot be constructed with a local maximum in a current-free region, but a local minimum is possible, allowing particles seeking a weak field to be trapped. In the magnetic field configuration first used to trap an atom (**Fig. 3**), a particle remains in a trapping state as long as its magnetic moment can follow the changing external magnetic field, that is, if the precession frequency of its magnetic dipole moment $\nu = \mu B/h$ (where h is Planck's constant) is high compared to its instantaneous orbital frequency.

Laser traps use the strong electric fields of the laser beam to induce an electric dipole moment on the atom. A laser field tuned below the atomic resonance polarizes the atom in phase with the driving field; the instantaneous dipole moment points in the same direction as the field. Thus the energy of the atom is lowered if it is in a region of high laser intensity. The high-intensity trapping region is formed simply by focusing the beam of a laser. The attraction of the atom to the laser field tuned below resonance resembles the electrostatic attraction of a dust particle to a charged rod. If the laser is tuned above the resonance of the atom, its instantaneous polarization is out of phase with the driving field. In this situation, the atom seeks regions of space with the weakest driving field. For the case in which the atom is surrounded by a laser field with a hole in the middle, this type of trap is analogous to a high-frequency version of the Paul trap. Transparent particles of 0.02 to 10 micrometers suspended in water can also be trapped with a focused laser beam, using the same principles as the neutral atom trap. With these larger particles the trapping forces are helpfully analyzed in terms of the forces on a lens in an inhomogeneous light

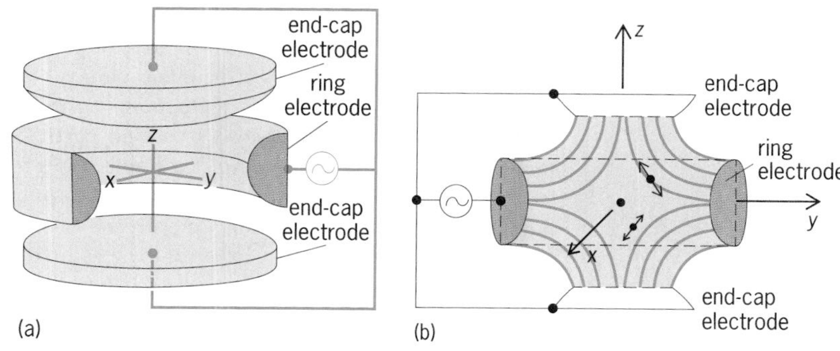

Fig. 2. Radio-frequency Paul trap consisting of two end caps and a ring electrode. (*a*) Cutaway view (*after G. Kamas, ed., Time and Frequency Users's Manual, National Bureau of Standards Technical Note 695, 1977*). (*b*) Cross section, showing the amplitude of the instantaneous oscillations for several locations in the trap.

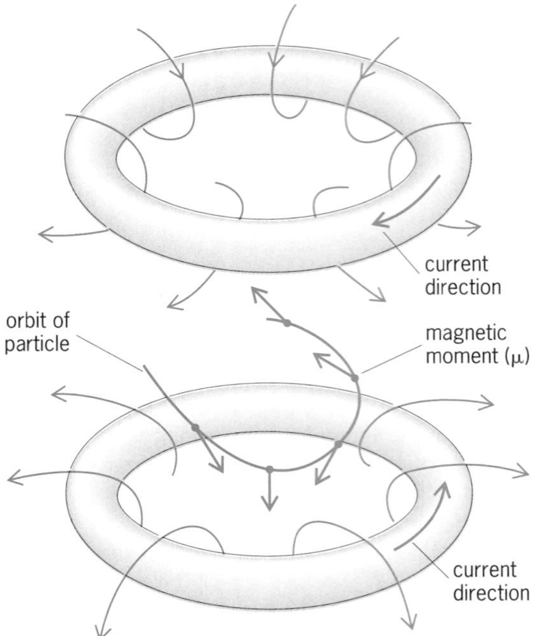

Fig. 3. **Spherical quadrupole magnetic trap for slowly moving atoms, showing how the orientation of the atom's magnetic moment remains constant with respect to the local orientation of the magnetic field.**

field. By considering the change in momentum between the incident and refracted-light fields, it can be shown that a particle acts as a converging lens and is drawn to regions of high laser intensity. *See* LASER.

Magnetooptic hybrid traps use, instead of the dipole forces induced by the laser field, the scattering force that arises when an atom absorbs photons with momentum h/λ, where λ is the wavelength. An inhomogeneous magnetic field separates the magnetic substates of an atom in a position-dependent manner. These states interact differently with circularly polarized light. It is possible to arrange a combination of laser beams with proper polarizations to create net scattering forces that drive the atom into the region of zero magnetic field. Such a trap requires much lower laser intensities and weaker magnetic fields. *See* LASER COOLING.

Applications. Particle traps are typically used to hold samples that are too hot or too cold to be permitted to interact with the walls of a containment vessel, or that would react violently with them. Charged-particle traps have been used to confine particles for a very long time. For example, a single positron (named Priscilla by the experimenters) was kept at liquid helium temperatures for 3 months in a radio-frequency trap. Such traps were used to measure the *g*-factor of the electron and positron to a precision of 4 parts in 10^{15}. This experimental value can be compared with the value calculated by quantum electrodynamics with similar precision. Antiprotons created with high-energy accelerators have been slowed down, trapped, and cooled for over 65 h in a form of Penning trap to be used in precision measurements of the mass of the antiproton and for the creation of antihydrogen. *See* ELECTRON SPIN; QUANTUM ELECTRODYNAMICS.

Long trapping times translate into long measurement times, and potentially more accurate measurement because of the Heisenberg uncertainty principle. Trapped mercury ions form the basis of a microwave clock with a precision of better than 10^{-14}. Laser-cooled ions in traps may make possible atomic clocks with a frequency uncertainty of less than 10^{-18}.

Laser cooling of ions in these traps to temperatures over a million times colder than that of the vacuum chamber walls has resulted in the Coulomb crystallization of a controlled number of ions. Studies of the quantum interactions of few-body systems under controlled conditions are now possible. *See* ATOMIC CLOCK; UNCERTAINTY PRINCIPLE.

Neutral-particle traps are usually weaker than charged-particle traps. Consequently, only extremely cold atoms can be trapped, and the storage time is limited by collisions between high-temperature background atoms in the vacuum chamber and the trapped atoms. In well-designed traps at liquid-helium temperatures, the vacuum is below 10^{-12} torr (10^{-10} pascal), storage times of several hours have been achieved. Despite the difficulty of trapping neutral particles, there are many instances where their storage and manipulation is feasible. For example, the storage of atoms cooled to a few tens of microkelvins is now possible. Work has been undertaken to cool high-density gases to microkelvin temperatures in order to study their quantum-statistical behavior, with the formation of a Bose condensate or a degenerate Fermi gas as the ultimate goal. Another example of neutral-atom manipulation would be the formation of very slow, intense beams of atoms (including atomic fountains) for use in atomic clocks, atomic-beam interferometers, or sensitive probes in the search for the breakdown of fundamental symmetries such as time reversal invariance. *See* BOSE-EINSTEIN STATISTICS; FERMI-DIRAC STATISTICS; MOLECULAR BEAMS; NEUTRON; SYMMETRY LAWS (PHYSICS); TIME REVERSAL INVARIANCE.

The trapping of particles in the micrometer and submicrometer range has enhanced the possibility of manipulating them. For example, the single-focused-beam laser trap can serve as "laser tweezers" to hold and move microscopic particles such as bacteria and viruses under an optical microscope without apparent damage to the organism. Yeast cells held in this way have reproduced in the trap through 2.5 life cycles. These laser tweezers possess the unique ability to reach inside a living cell and manipulate the contents without puncturing the cell membrane. For example, organelles inside a protozoan cell have been moved about, making it possible to measure the viscosity and elastic properties of the cytoplasm in the region of the trapped organelles. In another application, submicrometer spheres have been attached to the ends of a single deoxyribonucleic acid (DNA) molecule. By using a laser tweezer to grab the sphere, the DNA strand can be stretched out into a

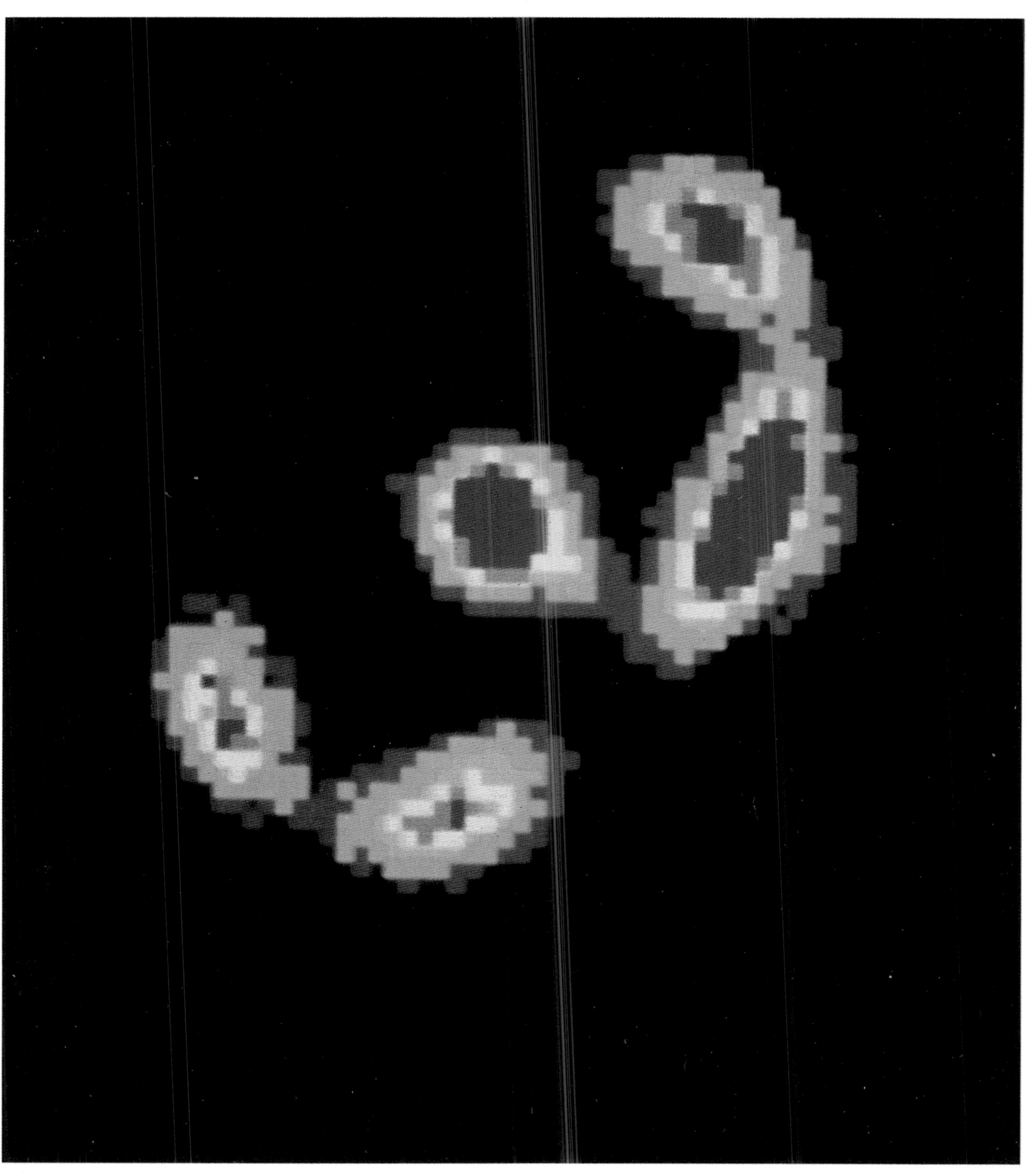

False-color image of individual ions in a particle trap. The ions have been cooled to the point where they form an ordered array called an ion crystal. This array has six ions, one of which is not visible in the image. Five ions lie in a plane at the vertices of a regular pentagon; a sixth ion lies at the center of the pentagon. The five visible ions are strongly fluorescing mercury ions. The nonfluorescing ion at one corner of the pentagon may be a heavier isotope of mercury or a molecular ion such as $HgOH^+$.

straight-line filament for examination. *See* CELL ORGANIZATION; DEOXYRIBONUCLEIC ACID (DNA).

Steven Chu

Bibliography. E. Arimondo, W. D. Phillips, and F. Strumia (eds.), *Laser Manipulation of Atoms and Ions*, 1992; A. Bárány et al. (eds), Proceedings of Workshop and Symposium on the Physics of Low-Energy Stored and Trapped Particles, *Phys. Scripta*, vol. T22, 1988; S. Chu, Laser manipulation of atoms and particles, *Science*, 263:861, 1991; S. Chu, Laser trapping of neutral particles, *Sci. Amer.*, vol. 266, no. 2, February 1992; S. Chu and C. Wieman (eds.), Special issue on laser cooling, *J. Opt. Soc. Amer.*, B6:2058, 1989; W. Itano, J. Bergquist, and D. Wineland, Laser spectroscopy of trapped atomic ions, *Science*, 237:612-617, 1987.

Particulates

Solids or liquids in a subdivided state. Because of this subdivision, particulates exhibit special characteristics which are negligible in the bulk material. Normally, particulates will exist only in the presence of another continuous phase, which may influence the properties of the particulates. A particulate may comprise several phases. **Table 1** categorizes particulate systems and relates them to commonly recognized designations. Fine-particle technology deals with particulate systems in which the particulate phase is subject to change or motion. Particulate dispersions in solids have limited and specialized properties and are conventionally treated in disciplines other than fine-particle technology. *See* ALLOY; GEL.

The universe is made up of particles, ranging in size from the huge masses in outer space—such as galaxies, stars, and planets—to the known minute building blocks of matter—molecules, atoms, protons, neutrons, electrons, neutrinos, and so on. Fine particle technology is concerned with those particles which are tangble to human senses, yet small compared to the human environment—particles that are larger than moecules but smaller than gravel. Fine particles are in abundance in nature (as in rain, soil, sand, minerals, dust, pollen, bacteria, and viruses) and in industry (as in paint pigments, insecticides, powdered milk, soap, powder, cosmetics, and inks). Particulates are involved in such undesirable forms as fumes, fly ash, dust, and smog and in military strategy in the form of signal flares, biological and chemical warfare, explosives, and rocket fuels.

TABLE 1. Types of particulate systems

System			Hydrosol	Aerosol	Powder
Continuous phase		Solid	Liquid	Gas	None (or gas)
Dispersed or particulate phase	Gas	Sponge	Foam	—	—
	Liquid	Gel	Emulsion	Mist Spray Fog Rain	—
	Solid	Alloy	Slurry Suspension	Fume Dust Snow Hail	Single phase Multi-phase (ores, flour)

Functional attributes. The special properties that particulates possess or accentuate over bulk matter include subdivision, surface area, critical size, curvature effects, interaction with radiation, and mobility.

Bulk materials are subdivided also to increase surface area for purposes of achieving chemical reactions and increased mass or heat transfer, or of altering apparent characteristics (as in coating or encapsulation).

Critical sizes exist or must be met for such phenomena as nucleation, separation, and liberation or beneficiation (including ore dressing and protein shift, a change in the protein content of a food ingredient as the result of mechanical fractionation). Special properties are also exhibited by particles when their size approaches that of a single domain. *See* FOOD ENGINEERING; NUCLEATION; ORE DRESSING.

The curvature of a particle surface affects some of the properties of the material composing the particle, such as surface tension and vapor pressure. Curvature also influences the amount of charge that a particle may acquire in an ionized field. *See* ELECTRIC CHARGE; SURFACE TENSION; VAPOR PRESSURE.

Interactions of particles with electromagnetic radiation (such as light and infrared radiation) include absorption, transmission, reflection, and scattering. *See* ABSORPTION OF ELECTROMAGNETIC RADIATION; REFLECTION OF ELECTROMAGNETIC RADIATION; SCATTERING OF ELECTROMAGNETIC RADIATION.

Mobility of particulates, which is important in pneumatic and hydraulic conveying, involves such considerations as particle inertia, diffusion, and drag.

Processing and uses. There are many cases where a particulate is either a necessary or an inadvertent intermediary in an operation. Areas that may be involved in processing particulates may be classified as follows:

- Size reduction
  - Mechanical (starting with bulk material)
    - Grinding
    - Atomization
    - Emulsification
    - Physicochemical (conversion to molecular dispersion)
      - Phase change (spray drying, condensation)
      - Chemical reaction
- Size enlargement (agglomeration, compaction)
  - Pelletizing
  - Briqueting
  - Nodulizing
  - Sintering

Separation or classification
  Ore beneficiation
  Protein shift
Deposition (collection, removal)
Coating or encapsulation
Handling
  Powders
  Gas suspensions (pneumatic conveying)
  Liquid suspensions (non-newtonian fluids)

See ATOMIZATION; CONVECTION (HEAT); DRYING; SINTERING.

End products in which particulate properties themselves are utilized include the following:

Mass or heat transfer agents (used in fluidized beds)
Recording (memory) agents
  Electrostatic printing powders and toners (for optical images)
  Magnetic recording media (for electronic images)
Coating agents (paints)
Nucleating agents
Control agents (for servomechanisms)
  Electric fluids
  Magnetic fluids
Charge carriers (used in propulsion, magnetohydrodynamics)
Chemical reagents
  Pesticides, fertilizers
  Fuels (coal, oil)
  Soap powders
  Drugs
  Explosives
Food products

See EXPLOSIVE; FERTILIZER; FLUIDIZED-BED COMBUSTION; FOOD; INK; ION PROPULSION; MAGNETIC RECORDING; MAGNETOHYDRODYNAMIC POWER GENERATOR; PAINT; PESTICIDE; PETROLEUM; SOAP.

Characterization. The processing or use of particulates will usually involve one or more of the following characteristics:

Physical
  Size (and size distribution)
  Shape
  Density
  Packing or concentration
Chemical
  Composition
  Surface character
Physiochemical (including adhesive, cohesive)
Mechanical or dynamic
  Inertial
  Diffusional
  Fluid drag
    Dilute suspensions (Stokes' law, Cunningham factor, drag coefficient)
    Concentrated suspensions (hindered settling, rheology, fluidization)
Optical (scattering, transmission, absorption)
  Refractive index (including absorption)
  Reflectivity
Electrical
  Conductivity
  Charge
Magnetic
Thermal
  Insulation (conductivity, absorptivity)
  Thermophoresis

See CONDUCTION (ELECTRICITY); CONDUCTION (HEAT); REFRACTION OF WAVES.

Many of the characteristics of particulates are influenced to a major extent by the particle size. For this reason, particle size has been accepted as a primary basis for characterizing particulates. However, with anything but homogeneous spherical particles, the measured "particle size" is not necessarily a unique property of the particulate but may be influenced by the technique used. Consequently, it is important that the techniques used for size analysis be closely allied to the utilization phenomenon for which the analysis is desired.

Particle size. Size is generally expressed in terms of some representative, average, or effective dimension of the particle. The most widely used unit of particle size is the micrometer (μm), equal to 0.001 mm (1/25,400 in.). Another common method is to designate the screen mesh that has an aperture corresponding to the particle size. The screen mesh normally refers to the number of screen openings per unit length or area; several screen standards are in general use, the two most common in the United States being the U.S. Standard and the Tyler Standard Screen Scales. The nominal size range of various common materials is shown in **Fig. 1**.

Particulate systems are often complex (**Fig. 2**). Primary particulates may exist as loosely adhering (as by van der Waals forces) particles called flocs or as strongly adhering (as by chemical bonds) particulates called agglomerates. Primary particles are those whose size can only be reduced by the forceful shearing of crystalline or molecular bonds. *See* CHEMICAL BONDING; INTERMOLECULAR FORCES.

Mechanical dispersoids are formed by comminution, decrepitation, or disintegration of larger masses of material, as by grinding of solids or spraying of liquids, and usually involve a wide distribution of particle sizes. Condensed dispersoids are formed by condensation of the vapor phase (or crystallization of a solution) or as the product of a liquid- or vapor-phase reaction; these are usually very fine and often relatively uniform in size. Condensed dispersoids and very fine mechanical dispersoids generally tend to flocculate or agglomerate to form loose clusters of larger particle size.

Because of the difficulty of distinguishing between the various stages of flocculation, size analysis methods usually aim at measuring discrete particles, because this state is more readily defined and reproduced. Some methods, because of their nature, will

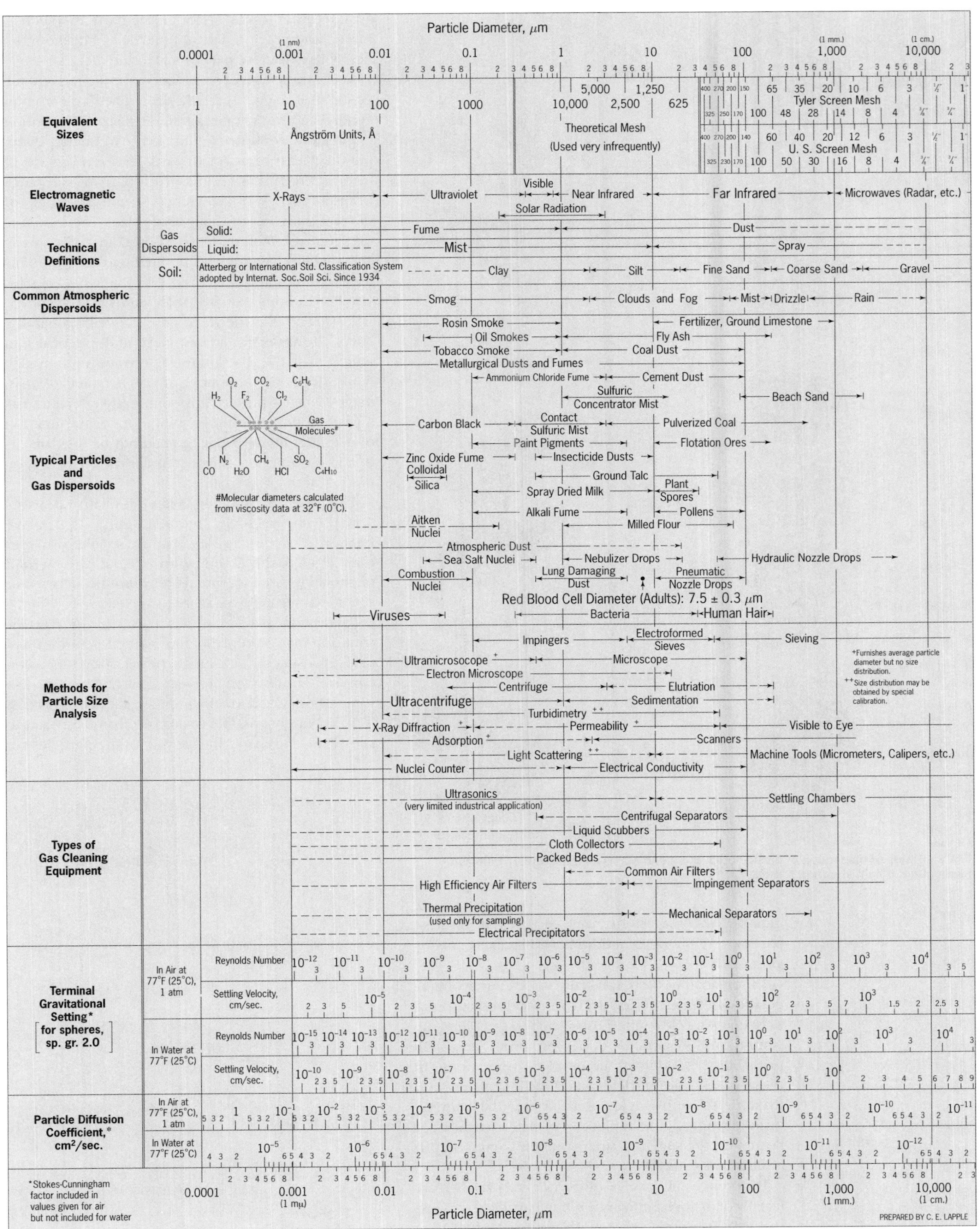

Fig. 1. Characteristics of particles and particle dispersoids. 1 cm/s = 3.281×10^{-2} ft/s; 1 cm²/s = 1.0764×10^{-3} ft²/s; 1 atm = 10^2 kPa. *(After C. E. Lapple, Particle technology: The little things in life, Stanford Res. Inst. J., 5:95–102, 1961)*

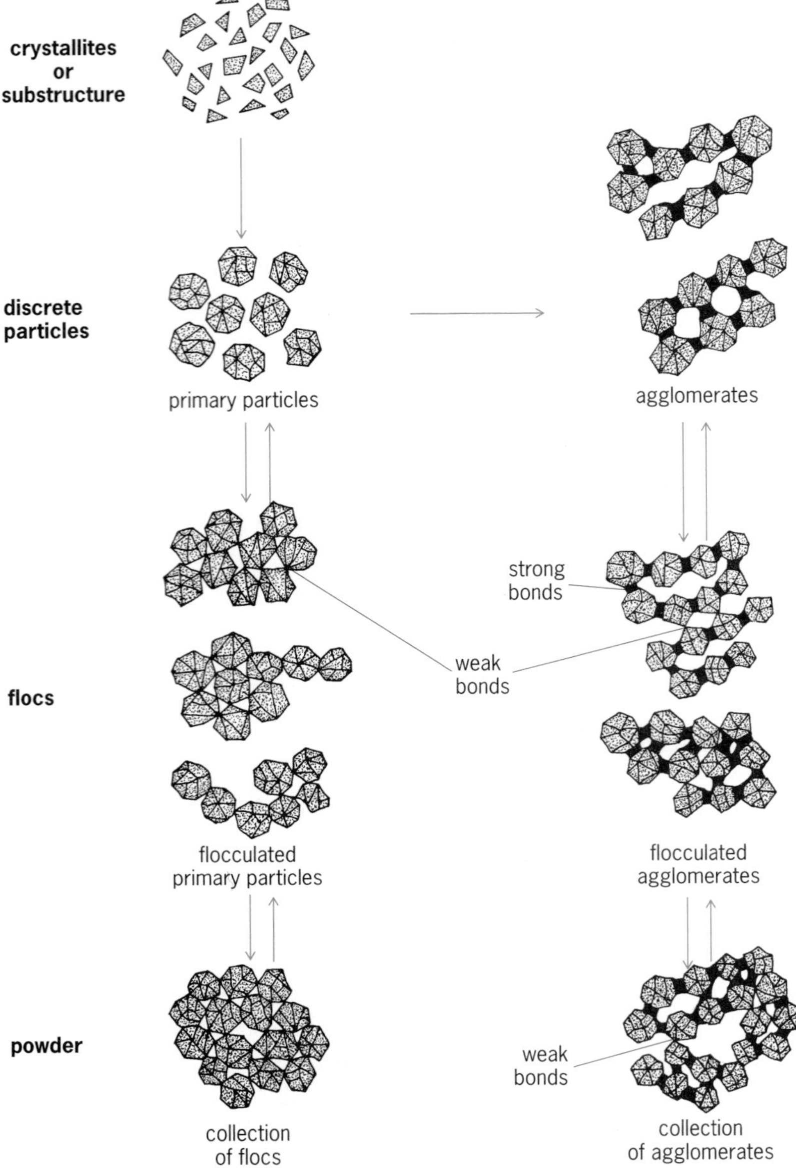

Fig. 2. States of dispersion of particulates. The double arrows imply reversibility with application of light shearing forces.

measure the size of primary particles as discrete particles, while other methods will measure the agglomerate.

Unfortunately, there is usually some degree of flocculation despite the relatively small adhesive forces between particles. Often this is due to the high concentrations of particles (close proximity and high probability of contact) in practical operations. In those cases, the particles may be flocculated in a dynamic sense. Particles in a floc may be continuously dislodged and replaced by other particles or flocs with the state of dispersion being fixed only in a statistical sense.

Differences in state of dispersion of particles are usually of major significance when particle size is less than 10 μm in diameter. With particles larger than 10 μm the magnitude of the surface forces holding particles in flocculated clusters decreases rapidly relative to other forces acting on the particles (such as inertia and gravity), and the particles behave more nearly like discrete particles in a statistical sense.

Methods of representing size. Most real systems are composed of a range of particle sizes. The two common general methods for representing size distribution graphically are shown in **Fig. 3**. The frequency distribution gives the fraction of particles $d\phi$ (on whatever basis desired) that lie in a given narrow size range dD as a function of the average size of the range (or of some function of the average size). It is essential that the size increment be chosen in a systematic and specified manner (as implied by the denominator of the ordinate of Fig. 3*a*). The shape of the curve will depend on how the increment is chosen, and it will usually take the form of a skewed probability curve. However, it may also have multiple peaks, in which case the distribution is termed multimodal. If an arithmetic increment of size is taken—that is, $d\psi(D) = dD$—the result is commonly referred to as a percent-per-micrometer plot. While the function of diameter in the ordinate need not be the same as that in the abscissa, the area under the curve will be unity if both are the same.

The lower curve of Fig. 3 is a cumulative distribution, the integral of the frequency curve. It gives the fraction ϕ of the particles that are smaller or larger than a given size D. Being an integral, the method of choosing size increments need not be selected or specified. *See* INTEGRATION; STATISTICS.

Bases for representation. As discussed above, size distribution may be represented by the fraction of particles in a specific size range, or over or under a certain size. The fraction, however, must be specified on some basis, which may be number, surface or mass. It is theoretically possible to convert from one basis to another. If the shape and density of the particles are

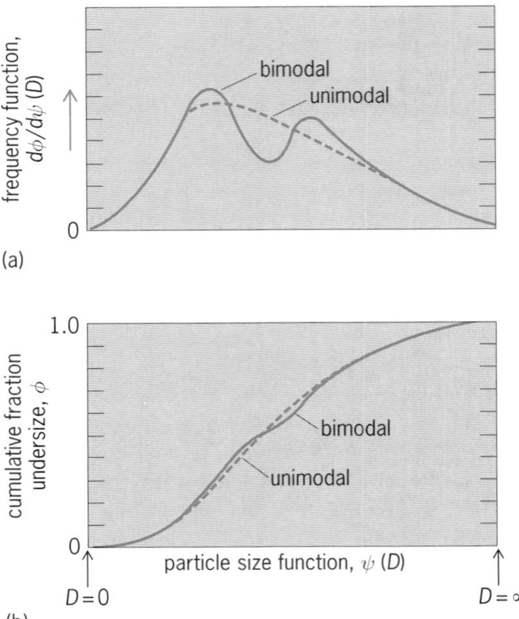

Fig. 3. Methods for representing size distribution. (*a*) Frequency distribution. (*b*) Cumulative distribution.

TABLE 2. Bases for expressing size distribution

Type of distribution basis	Value of *x* or *y*
Number or count	0
Linear	1
Surface	2
Volume (or mass)	3

statistically uniform (do not vary with size), Eq. (1)

$$\phi_x = \frac{\sum_0^D D^{x-y}\Delta\phi_y}{\sum_0^\infty D^{x-y}\Delta\phi_y} = \frac{\int_0^D D^{x-y}d\phi_y}{\int_0^\infty D^{x-y}d\phi_y} \qquad (1)$$

may be used to convert a distribution from one basis (ϕ_y) to another (ϕ_x). Here ϕ_x is the cumulative fraction smaller than size D on a basis corresponding to x; ϕ_y is the cumulative fraction smaller than size D on a basis corresponding to y; D is particle diameter; and x and y are the characteristic subscripts or exponents used to define a specific basis. **Table 2** gives the value of x or y corresponding to various bases for expressing size distribution.

The measurement basis used may markedly influence the precision of the desired answer. A few large particles overlooked in a count analysis will not significantly affect the count distribution, but may result in serious errors in the corresponding calculated mass or volume distribution. Similarly, the count distribution calculated from a mass measurement may be grossly in error at the fines end.

If shape varies with size, conversion between bases requires a knowledge of the shape factor and how it varies with size. This knowledge is usually not available. Variation in particle composition introduces additional complications.

Analytical relationships. Various analytical relationships have been proposed for representing size distribution, notably the Rosin-Rammler, the Roller, the Nukiyama-Tanasawa, and the log-probability distributions. Advantages have been claimed for each, usually on the basis that one or the other gives a better fit of data in a particular instance. In addition, claims based on various simplifying assumptions have been made for theoretical justification, and have been made especially in the case of the log-probability distribution.

Actually, it is easier to find exceptions to than agreements with each. There are times, however, when analytical convenience may justify one. The log-probability relationship is particularly useful in this respect. The log-probability frequency distribution is given by Eq. (2), or, in the normalized form by

$$\frac{d\phi_x}{d\ \ln D} = \left[\frac{1}{(\sqrt{2\pi})\ \ln\sigma}\right] \times \exp\left\{-\left[\frac{\ln D - \ln D_{mx}}{(\sqrt{2\pi})\ln\sigma}\right]^2\right\} \qquad (2)$$

Eq. (3), where ϕ_x is the cumulative fraction smaller

$$\frac{d\phi_x}{d\ \ln(D/D_{mx})} = \left[\frac{1}{(\sqrt{2\pi})\ \ln\sigma}\right] \times \exp\left\{-\left[\frac{\ln(D/D_{mx})}{(\sqrt{2\pi})\ln\sigma}\right]^2\right\} \qquad (3)$$

than size D on a basis corresponding to x; D is the particle diameter; σ is the standard geometric deviation; and D_{mx} is median diameter on the basis of a property corresponding to x.

The values of σ and D_{mx} are characteristic constants for a given size distribution. If a material follows a log-probability distribution on one basis (x), it also does on any other basis (y)—with the same value of the standard geometric deviation (σ) but a different value of median size (D_{my}) corresponding to the new basis (y). This is a unique property of the log-probability distribution. Probability graph papers are commercially available (**Fig. 4**) and are convenient for graphical representation of size distribution data even if the distribution does not follow a probability relationship. In addition, the assumption of a log-probability distribution as an approximation permits simple conversion from one basis of representing size distribution, mean size, or median size to another (as described below).

The solid line of Fig. 4 shows a typical size distribution. If the size distribution follows a log-probability relationship, it will be a straight line on this type of plot, as shown by the two broken lines, and can

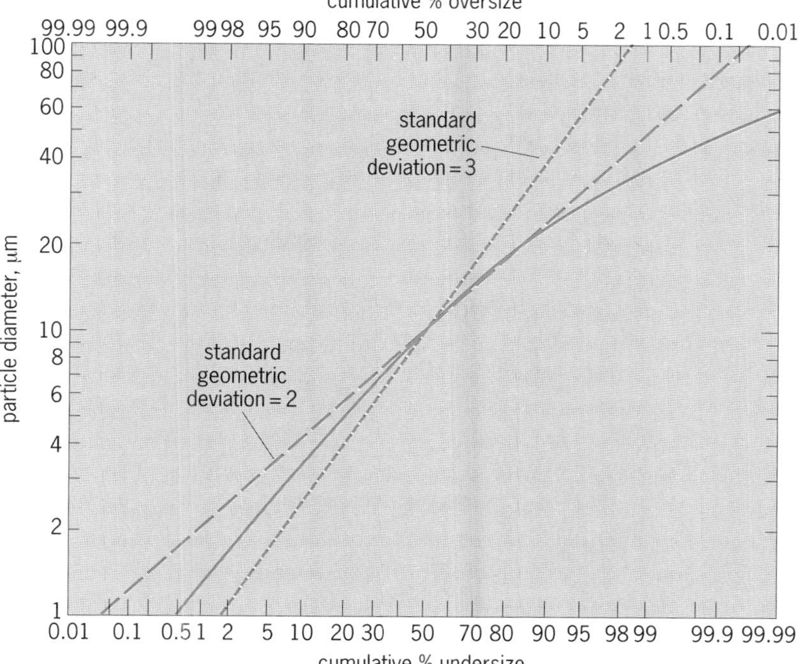

Fig. 4. Probability graph for showing typical size distributions.

be completely characterized by two numbers, (1) a median diameter, corresponding to the 50% cumulative size, and (2) a standard geometric deviation, a number equal to or greater than unity, given by the ratio of the 84.13% to the 50% size (or of the 50% to the 15.87% size).

The median diameter is a measure of the general size level, whereas the standard geometric deviation is a measure of the degree of uniformity. A completely uniform material (all particles the same size) would show up as a horizontal line in Fig. 4 and have a standard geometric deviation of 1.0. A completely heterogeneous material would be represented by a vertical line, which would have a standard geometric deviation of infinity.

A more realistic but more complex distribution is the upper-limit log-probability distribution proposed by R. A. Mugele and H. D. Evans.

Average particle diameter. For many purposes, it is adequate to represent a particulate material by an average size. The type of average size depends on the nature of the application and can often be established quantitatively by a consideration of the physics involved.

The general size level represented by a particulate material can be expressed in terms of the median size given by Eq. (4), where D_{mx} is the median

$$\int_0^{D_{mx}} d\phi_x = \int_{D_{mx}}^1 d\phi_x = {}^1/_2 \qquad (4)$$

diameter based on a property corresponding to x, and ϕ_x is the cumulative fraction of all particles (based on a property corresponding to x) smaller than size D.

Thus, in terms of a specific property x, half the material is finer than the median size D_{mx}, the other half coarser. In terms of mass, a mass median diameter means that half the mass of a powder consists of particles coarser than the mass median diameter, and half finer. Therefore, the median diameter is given by the 50% point on the cumulative distribution curve obtained in terms of the property desired.

If a material obeys a log-probability relationship, the various median diameters are related to each other by the relationship in Eq. (5), where D_{mx} is

$$D_{mx} = D_{my} \exp \{(x - y) \ln^2 \sigma\} \qquad (5)$$

the median diameter on the basis of a property corresponding to x; D_{my} is the median diameter on the basis of property corresponding to y; x and y are the characteristic subscripts or exponents used to define

TABLE 3. Types of, and interrelationships between, mean diameters*

Definition of mean diameter			Diameter ratios for log-probability distribution		
$\bar{D}_{qp}$	q	p	Standard geometric deviation, σ	(Mass median / Mean) $D_{m3}/\bar{D}_{qp}$	(Mean / Number median) $\bar{D}_{qp}/D_{m0}$
Geometric mean	0	0	2	4.23	1.000
			3	37.37	1.000
			4	319.1	1.000
Linear mean (number mean)	1	0	2	3.32	1.272
			3	20.4	1.829
			4	122.1	2.614
Surface mean (surface-to-number mean)	2	0	2	2.61	1.617
			3	11.18	3.343
			4	46.69	6.833
Volume mean (volume-to-number mean)	3	0	2	2.06	2.06
			3	6.11	6.11
			4	17.86	17.86
Surface-to-diameter	2	1	2	2.06	2.06
			3	6.11	6.11
			4	17.86	17.86
Volume-to-diameter	3	1	2	1.62	2.61
			3	3.34	11.18
			4	6.83	46.69
Sauter (volume-to-surface mean)	3	2	2	1.27	3.32
			3	1.83	20.44
			4	2.61	122.1
DeBrouckere (mass mean)	4	3	2	0.786	5.38
			3	0.547	68.2
			4	0.382	835

*After C. E. Lapple, Particle-size analysis and analyzers, *Chem. Eng.*, 75:149–156, 1968.

TABLE 4. Types of size analysis methods

Size-discriminating property: Type	Size-discriminating property: Mechanism	Measurement technique
Geometric	Physical barrier	Sieving (wet or dry) Ultrafiltration
Mechanical or dynamic (in fluids)	Inertia	Impaction on surface Pressure pulse (sonic)
	Terminal settling velocity in force field (gravity, centrifugal, electrical)	Elutriation (liquid or gas; single or series fractionation) Sedimentation (liquid or gas; layer or suspension; differential or integral) Decantation Photosedimentation
	Diffusion	Particle displacement Particle deposition
Optical	Imaging	Light microscopy Ultramicroscopy (mean size only) Electron microscopy
	Spectral transmission	Extinction
	Scattering	Single particle count (right angle, angular, forward, polarization) Macroscopic (mean size only)
	Diffraction	Light, x-ray, laser
Electrical	Resistance Capacitance Charge	Current flow pulses Voltage pulses Triboelectric Induction Corona
Thermal	Particle deposition	Migration in thermal gradient
Physico-chemical	Condensation	Nuclei growth with controlled supersaturation

a specific basis as given in Table 2; and σ is the standard geometric deviation. Thus, D_{m0} and D_{m3} may be defined, respectively, as number and volume (or mass) median diameters.

However, while median diameters are readily obtained from size distribution measurements, the effective role of particles in some specific phenomenon can usually be expressed in terms of a mean diameter, defined by Eq. (6), where $\overline{D}_{qp}$ is the mean

$$\overline{D}_{qp} = \left[\frac{\int_0^1 D^q d\phi_n}{\int_0^1 D^p d\phi_n} \right]^{[1/(q-p)]} \qquad (6)$$

diameter; D is any particle diameter; ϕ_n is the fraction of all particles smaller than size D on a number basis; and q and p are the characteristic subscripts and exponents used to define mean diameter.

If the material follows a log-probability distribution, Eq. (6) reduces to Eq. (7), where q, p, and x

$$\overline{D}_{qp} = D_{mx} \exp \{[(q+p-2x)/2] \ln^2 \sigma\} \qquad (7)$$

are the same characteristic subscripts or exponents previously defined.

Table 3 lists some typical mean diameters defined in terms of their p and q values. These are arranged in order of increasing size. Values of x corresponding to the various bases are given in Table 2.

Table 3 also gives the ratio of each mean to median size (on both a mass and count basis) for various degrees of uniformity (expressed in terms of standard geometric deviation), assuming that the material follows a log-probability distribution, to show the vast difference between the various mean and median sizes.

With a uniform material ($\sigma = 1$), all the ratios would be unity, and all mean and median diameters would be the same. As the material becomes less uniform (as σ increases), the differences between the various means and medians become greater. For a log-probability distribution, the geometric mean and number median diameter are identical. This would not be true for another type of distribution.

Precision and significance of analysis. The precision and significance of size analyses are governed by (1) sampling of particulate source, (2) preparation of sample analysis, (3) type size-discrimination property employed, (4) type of measurement property employed, (5) precision of various analytical procedures, and (6) analyses and interpretation of data.

Sampling involves at least two steps: sampling in the field or plant, and selection of small aliquots for individual analysis. Sampling becomes more difficult as the particle size becomes larger and the size distribution broader because of particle segregation tendencies. Sample preparation can be a determining factor in the significance of the size analysis, especially in the degree of particle dispersion achieved. With liquids, dispersion is improved by the use of dilute suspensions, the addition of dispersing agents, and the use of viscous liquids. *See* CHEMOMETRICS.

Analysis methods. There are two classes of size analysis equipment. In one, the class of counters, each particle is observed individually and its discrimination property measured directly. In the other, particles are measured in bulk (in terms of total number, surface area, or mass) after being subjected to a discriminating action. There are hundreds of size analysis methods, each employing some combination of size discrimination and measurement technique. **Table** 4 summarizes the general types of discriminating properties available together with associated variations in measurement techniques. Most techniques do not measure size but some other property that is markedly, but not usually solely, influenced by size. It has become conventional to report the results in terms of a size by some assumed or empirical relationship even though the size does not represent any real dimension.

In some operations (such as chemical reactions, catalysis, comminution, and certain optical effects), a knowledge of the total surface area of the particulate matter may be sufficient. The more common methods for measuring total surface rather than size distribution are the permeability methods (as exemplified by the Blaine and the Fischer Sub-Sieve Sizer instruments) in which the resistance to flow of air through a mass of powder of known porosity is measured. The specific surface of the powder (or the equivalent of Sauter diameter in Table 3) is determined from a semi-empirical equation. In adsorption methods, exemplified by the Brunauer-Emmett-Teller (BET) method, the amount of gas (or liquid, dye, or radioactive materia) adsorbed on particles is used as a measure of surface. While permeability methods tend to measure the outer exposed surface of primary particles, an adsorption method may include surface associated with internal fissures. Turbidimetric methods and methods involving measurement of energy changes (such as heat of solution, heat of wetting, and coercive force) are also used to measure total surface. *See* ADSORPTION.

Particle concentration. The concentration of particles in a dispersoid is often expressed in terms of grains per cubic foot of gas, where 7000 grains = 1 lb. In air-pollution work, concentrations are sometimes expressed as grams per cubic meter (1 g/m^3 = 0.437 grain/ft^3). Process dust concentrations normally range from 0.01 to 100 grain/ft^3 (0.02 to 200 g/m^3) with 1–10 grains/ft^3 (2–20 g/m^3) being common in process ventilation work. In air conditioning applications concentrations of particulate matter generally range from 0.1 to 10 grains/1000 ft^3 (0.2 to 20 g/m^3). In operations such as pneumatic conveying, the concentrations range from 0.1 to 50 lb solid/lb gas or 0.1 to 50 kg solid/kg gas (about 50–20,000 grains/ft^3 or 100–45,000 g/m^3). In fluidized-solid systems concentrations frequently exceed 1000 lb solid/lb gas or 1000 kg solid/kg gas.

Figure 1 shows a listing of particle sizes of common materials and related items, as well as methods of size analysis. **Figure 5** shows a summary of concentrations of particulates suspended in air.

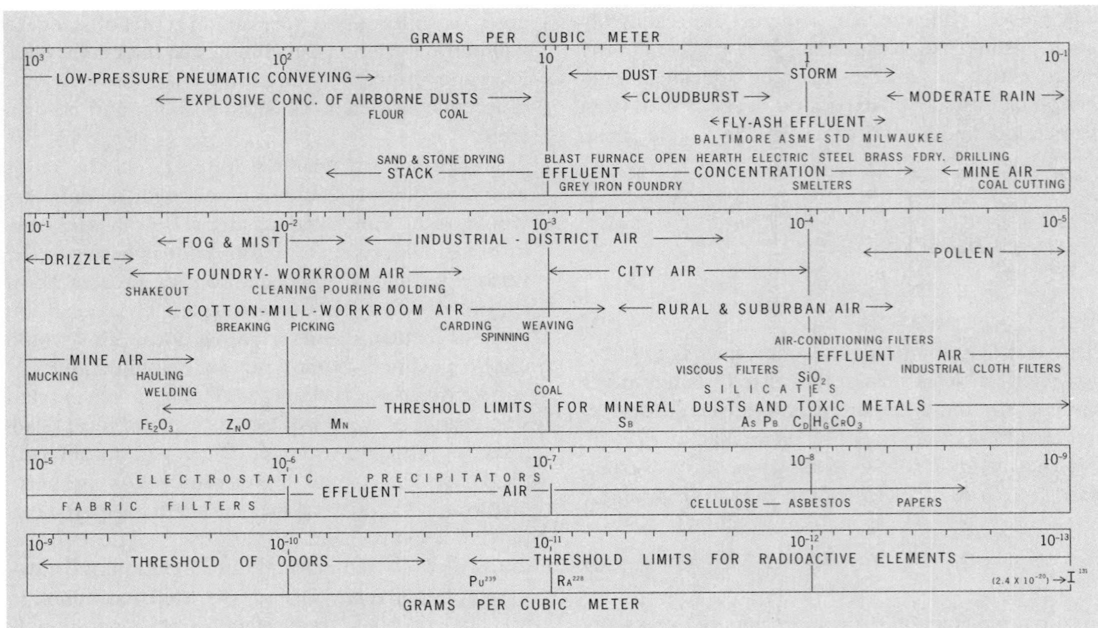

Fig. 5. Concentration of particulates suspended in air under various conditions. 1 g/m^3 = 0.437 grain/ft^3. (*From M. W. First and P. Drinker, Arch. Ind. Hyg. Occup. Med., 5:387–388, 1952*)

Particulate mechanics. A particle moving in a fluid encounters a frictional resistance given by Eq. (8),

$$F_r = C_D A_p \rho \frac{u_r^2}{2} \quad (8)$$

where F_r is the resisting force in newtons; C_p is the dimensionless drag coefficient; A_p is the area of the particle projected in a plane normal to the direction of motion in m²; ρ is the fluid density in kg/m³; and u_r is the velocity of the particle relative to the fluid in m/s. The drag coefficient C_D is a function of the particle shape, the proximity of bounding surfaces, and the Reynolds number N_{Re} (= $D_p u_r \rho/\mu$, where D_p is the particle diameter or other representative dimension in m; and μ is the fluid viscosity in Pa·s). Drag coefficient data are available for a wide variety of particle shapes. *See* REYNOLDS NUMBER.

If a particle suspended in a fluid is acted upon by a force, it will accelerate to a terminal velocity at which the resisting force due to fluid friction just balances the applied force. If a particle falls under the action of gravity, this velocity is known as the terminal gravitational settling velocity and is given by Eq. (9), where u_t is the settling velocity in m/s; g_L

$$u_t = \sqrt{\frac{2 g_L M_p (\rho_p - \rho)}{\rho \rho_p A_p C_D}} \quad (9)$$

is the acceleration due to gravity, 9.807 m/s²; M_p is the particle mass in kilograms; ρ_p is particle density in kg/m³; and other terms are as previously defined.

For spherical particles present in dilute concentration, C_D is approximately constant at 0.44 for $N_{\mathrm{Re}} > 500$. For $N_{\mathrm{Re}} < 1$, C_D is equal to $24/N_{\mathrm{Re}}$, and the resisting force is given by Eq. (10), which is known

$$F_r = 3\pi \mu u_r D_p \quad (10)$$

as Stokes' law. Corresponding to this the terminal gravitational settling velocity is given by Eq. (11),

$$u_t = \frac{g_L D_p^2 (\rho_p - \rho)}{18\mu} \quad (11)$$

commonly known as Stokes' law of settling, which is usually applicable for particles smaller than 50 μm in diameter.

When the particle size approaches the magnitude of the mean free path of the molecules of the suspending fluid, the frictional resistance to motion becomes less than (and the settling velocity becomes greater than) that indicated by the above equations. To correct for this molecular slip flow effect, the resistance to motion calculated from the above equations must be divided by (or the settling factor multiplied by) a factor k_m, commonly known as the Stokes-Cunningham correction factor, given by Eq. (12),

$$k_m = 1 + k_{me} \frac{\lambda}{D_p} \quad (12)$$

where k_{me} is a dimensionless constant and λ is the mean free path of the fluid molecules in m. The value k_{me} has been shown experimentally to lie between 1.3 and 2.3 for different gases, particle sizes, and materials. Based on the data of R. A. Millikan, Eq. (13)

$$k_{me} = 1.644 + 0.552 e^{-(0.656 D_p/\lambda)} \quad (13)$$

is obtained, where λ is based on simple kinetic theory ($\lambda = 3\mu\rho\,\bar{v}$), and the mean molecular speed $\bar{v}$ is given in ft/s or m/s by Eq. (14), where R is the gas con-

$$\bar{v} = \sqrt{\frac{8RT}{\pi M}} \quad (14)$$

stant, 8.31×10^3 J/(°C) (kg-mole); T is the absolute temperature in °R or K; and M is the molecular weight of the gas in kg/kg-mole. For particles in gases at atmospheric pressure, the Stokes-Cunningham factor becomes significant for particles smaller than 1 μm. At high altitude or low pressure, this factor can become extremely large.

Particles suspended in a fluid partake of the molecular motion of the suspending fluid and hence acquire diffusional characteristics analogous to those of the fluid molecules. This random zigzag motion of the particles, commonly known as brownian motion, is obvious under the microscope for particles smaller than 1 μm. The average displacement of a particle in time t is given by Eq. (15), where Δs is

$$\Delta s = \sqrt{\frac{4 D_v t}{\pi}} \quad (15)$$

the average linear displacement along a given axis, regardless of sign, in m; D_v is the diffusion coefficient for the particle in m²/s; and t is the time in s. The Einstein equation for the diffusion coefficient D_v for spherical particles is Eq. (16), where N_A is Avogadro's

$$D_v = \frac{k_m RT}{3\pi \mu N_A D_p} \quad (16)$$

number, 6.023×10^{26} molecules/kg-mole, and other terms are as previously defined. This diffusion coefficient may also be used to estimate the migration rate of particles when a concentration gradient exists. *See* BROWNIAN MOVEMENT; DIFFUSION.

Figure 1 shows settling velocities and diffusion coefficients with particles of various sizes in air and water. *See* AIR FILTER; COLLOID; DUST AND MIST COLLECTION; EMULSION; FOAM. Charles L. Lapple

Bibliography. T. Ariman and T. N. Veziroglu (eds.), *Particulate and Multiphase Processes*, 3 vols., 1987; J. K. Beddow (ed.), *Particulate Systems: Technology and Fundamentals*, 1983; R. Clift, *Particulate Technology*, 1991; N. Fuchs, *Mechanics of Aerosols*, 1989; H. E. Hesketh, *Fine Particles in Gaseous Media*, 2d ed., 1986; R. H. Perry and D. Green (eds.), *Perry's Chemical Engineers' Handbook*, 7th ed., 1997.

Pascal's law

A law of physics which states that a confined fluid transmits externally applied pressure uniformly in all directions. Blaise Pascal, using the mercury-column barometer of Evangelista Torricelli, demonstrated the decrease in atmospheric pressure with increasing height and determined that atmospheric force at a point exerted equal pressure in all directions. More exactly, in a static fluid, force is transmitted at the velocity of sound throughout the fluid. The force acts normal to any surface. This natural phenomenon is the basis of the pneumatic tire, balloon, hydraulic jack, and related devices. *See* HYDROSTATICS; TORRICELLI'S THEOREM. Karl Arnstein; Robert S. Ross

Paschen-Back effect

An effect on spectral lines obtained when the light source is placed in a very strong magnetic field, first explained by F. Paschen and E. Back in 1921. In such a field the anomalous Zeeman effect, which is obtained with weaker fields, changes over to what is, in a first approximation, the normal Zeeman effect. The term "very strong field" is a relative one, since the field strength required depends on the particular lines being investigated. It must be strong enough to produce a magnetic splitting that is large compared to the separation of the components of the spin-orbit multiplet. *See* ATOMIC STRUCTURE AND SPECTRA; ZEEMAN EFFECT.

The **illustration** shows, as an example, the Paschen-Back effect of the red line $2s^2S$–$2p^2P$ of lithium. The natural separation of this doublet is very small, only 0.0175 nanometer, so that in the field of 44,200 oersteds (3.52×10^6 amperes/m) for which the diagram is drawn the normal Zeeman splitting of 0.0929 nm greatly exceeds it. The Paschen-Back effect is therefore practically complete, and the resulting pattern is nearly a normal triplet. There is still, however, a residual splitting, which theory gives as two-thirds of the field-free separation. The weak field patterns shown in the figure require a field of only about 1800 oersteds (1.4×10^5 A/m) and correspond to the anomalous patterns obtained in the Zeeman effect.

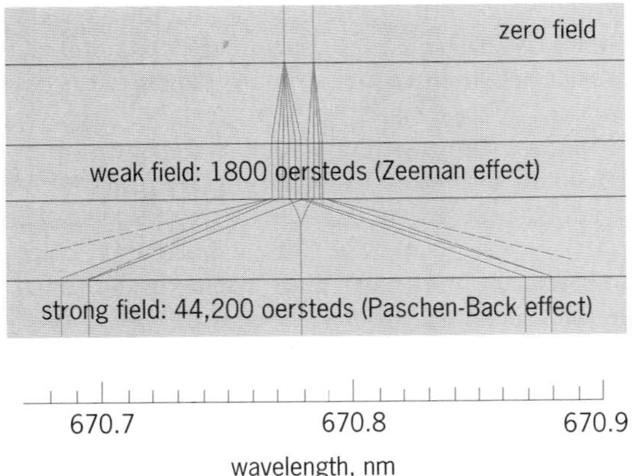

Zeeman and Paschen-Back effects of red lithium doublet, whose natural separation is 0.0175 nm.1 oersted = 79.5 A/m.

Theory explains the transformation from the Zeeman to the Paschen-Back effect as due to the uncoupling of the orbital and spin vectors L and S by the magnetic field. Whereas in a weak field these vectors are coupled magnetically to form a resultant J, a sufficiently strong field causes them to precess independently about the field direction. The correlation of the energy levels between weak and strong fields follows from the principle that the magnetic field is incapable of changing the angular momentum component along the field; that is, the magnetic quantum number M of a level remains constant for all field strengths. A further rule is that two levels having the same value of M (which in the strong field equals $M_L + M_S$) do not cross each other. These rules lead to the correlation of lines shown in the figure. Certain lines fade out during the transition, as is indicated by the change from a solid to a dotted line, and these are the ones for which the direction of polarization would be altered. F. A. Jenkins; W. W. Watson

Passeriformes

The largest and most diverse order of birds, which is found worldwide, including most oceanic islands but excluding Antarctica, in all terrestrial habitats. The most closely related species may be other land birds such as the Coraciiformes and the Piciformes. *See* CORACIIFORMES; PICIFORMES.

Classification. The Passeriformes is divided into three major suborders, as outlined below. The affinities of most families within those suborders is still much disputed.

Order Passeriformes
  Suborder Eurylaimi
    Family: Eurylaimidae (broadbills; 14 species)
      Philepittidae (false sunbirds; 4 species)
      Pittidae (pittas; 24 species)
      Acanthisittidae [Xenicidae] (New Zealand wrens; 4 species)
  Suborder Furnarii
    Family: Dendrocolaptidae (woodcreepers; 52 species)
      Furnariidae (ovenbirds; 218 species)
      Formicariidae (antbirds; 239 species)
      Rhinocryptidae (tapaculos; 30 species)
  Suborder Tyranni
    Family: Tyrannidae (tyrant flycatchers; 377 species)
      Pipridae (manakins; 52 species)
      Cotingidae (cotingas; 79 species)
      Oxyruncidae (sharpbill; 1 species)
      Phytotomidae (plant cutters; 3 species)

Suborder Oscines
Family: Menuridae (lyrebirds; 2 species)
Atrichornithidae (scrubbirds; 2 species)
Palaeospizidae (fossil)
Alaudidae (larks; 78 species)
Hirundinidae (swallows)
Motacillidae (pipits, wagtails; 54 species)
Campephagidae (cuckoo shrikes; 70 species)
Palaeoscinidae (fossil)
Pycnonotidae (bulbuls; 123 species)
Irenidae (fairy bluebirds, leafbirds; 14 species)
Laniidae (shrikes; 65 species)
Prionopidae (helmet shrikes; 9 species)
Vangidae (vanga shrikes; 13 species)
Bombycillidae (waxwings; 8 species)
Dulidae (palm chat; 1 species)
Prunellidae (hedge sparrows; 12 species)
Mimidae (mockingbirds, thrashers; 31 species)
Cinclidae (dippers; 5 species)
Turdidae (thrushes; 309 species)
Timaliidae (babblers; 276 species)
Troglodytidae (wrens; 60 species)
Sylviidae (Old World warblers; 349 species)
Muscicapidae (Old World flycatchers; 113 species)
Monarchidae (monarch flycatchers; 91 species)
Maluridae (Australian wrens; 26 species)
Acanthizidae (thornbills; 67 species)
Ephthianuridae (Australian chats; 5 species)
Orthonychidae (rail babblers; 19 species)
Rhipiduridae (fantails; 40 species)
Eopsaltriidae (Australian robins; 40 species)
Pachycephalidae (whistlers; 46 species)
Aegithalidae (long-tailed tits; 7 species)
Remizidae (bearded tits; 10 species)
Paridae (chickadees; 47 species)
Sittidae (nuthatches; 23 species)
Neosittidae (tree runners; 3 species)
Certhiidae (creepers; 6 species)
Rhabdornithidae (Philippine creeper; 1 species)
Climacteridae (Australian creepers; 6 species)
Dicaeidae (flowerpeckers; 58 species)
Nectariniidae (sunbirds; 117 species)
Zosteropidae (white-eyes; 83 species)
Meliphagidae (honeyeaters; 172 species)
Vireonidae (vireos; 43 species)
Emberizidae (buntings; 318 species)
Catamblyrhynchidae (plush-capped finch; 1 species)
Thraupidae (tanagers; 240 species)
Tersinidae (swallow tanager; 1 species)
Parulidae (wood warblers; 125 species)
Icteridae (orioles, blackbirds, troupials; 95 species)
Fringillidae (finches; 122 species)
Drepanididae (Hawaiian honey creepers; 24 species)
Estrildidae (waxbills, grass finches; 127 species)
Ploceidae (weaver finches; 117 species)
Passeridae (sparrows; 27 species)
Sturnidae (starlings; 111 species)
Oriolidae (orioles; 25 species)
Dicruridae (drongos; 20 species)
Callaeidae (wattlebirds; 3 species)
Grallinidae (mud-nest builders; 4 species)
Artamidae (wood swallows; 13 species)
Cracticidae (butcher-birds; 10 species)
Pityriasididae (bristlehead; 1 species)
Ptilonorhynchidae (bower birds; 18 species)
Paradisaeidae (birds of paradise; 42 species)
Corvidae (crows, jays; 106 species)

Several groups within the suborder Oscines are apparent. The largest and most definite are the New World oscines with nine primary feathers on each wing, which include the families from the Vireonidae to the Fringillidae. Another is the Australian radiation, which includes the Dicruridae to the Paradisaeidae. Perhaps yet another diverse Australian group would include the Meliphagidae, Zosteropidae, Dicaeidae, Climacteridae, Pachycephalidae, Eopsaltridae, Rhipiduridae, Orthonychidae, Ephthianuridae, Acanthizidae, Maluridae, and Monarchidae. The Old World insect eaters include the families Prunellidae, Mimidae, Cinclidae, Turdidae, Timaliidae, Troglodytidae, Sylviidae, Muscicapidae. Several smaller groups could be listed, but knowledge of the evolutionary history and relationships within the perching birds is still poor.

Fossil record. Because they are small land birds, the passeriforms have a poor fossil record. It is not even certain whether *Palaeospiza bella* is correctly assigned to the Passeriformes. No definite passeriform fossils have been identified in the Northern Hemisphere that predate the beginning of the Miocene in spite of a number of other small avian fossils found in earlier deposits. The lack of passeriform fossils from the well-worked Northern Hemisphere has led some paleontologists to postulate that the birds originated in the Southern Hemisphere and invaded the northern regions in the mid-Tertiary. Passeriform fossils are found rarely in the Miocene, become somewhat more common in the Pliocene, and are most abundant in the Quaternary. However, the available fossils provide no useful information on the evolutionary history of the passeriforms.

Characteristics. The perching birds are small to medium-sized birds, ravens (*Corvus corax*) being the largest. The wings are short to medium in length and vary from rounded to pointed. A few species, including lyrebirds, scrubbirds, and New Zealand wrens, are almost flightless. The tail varies from nearly absent to long. The bill is widely variable in shape. It may be hooked or straight to strongly decurved, depending largely on food and feeding habits, and it can be small and weak or large and strong. Passeriforms have legs of short to medium length that are usually strong. The four toes show the usual avian arrangement, three in front and a well-developed hallux behind. Most forms can walk or climb well, or both. Plumage varies widely, from all black to mostly white and from bright colors and bold patterns to cryptic coloration.

Feeding habits and food choices show wide variation. Most species eat insects or small animals; however, some groups have become specialized on a diet of nectar, seeds, fruits, small vertebrates, or even leaves and the waxy covering of berries for part of the year. Methods of obtaining food varies as well. Insect eaters include flycatchers and swallows, which catch insects in the air; creepers and nuthatches, which probe crevices in bark; thrashers, which dig into ground litter and soft earth; and dippers feeding under water. Moreover, groups of passerine birds that feed on seeds, on nectar, or by flycatching have evolved independently several times.

Song is important to most perching birds for species recognition and courtship. The pair bond is usually strong, with both sexes incubating and caring for the young, which remain in the nest until they are able to fly.

Many arctic and cold temperature passerine species migrate to warmer areas for the cold months. Some of the migratory flights measure several thousand miles.

Economic significance. The value of the perching birds to humans is difficult to state in simple terms. However, many species are important in literature and song. In addition, the value of birds, and largely perching birds, to human enjoyment from bird watching and bird feeding during winter months is large. Such activity is perhaps of greater total economic value today than hunting and other commercial activities that are associated with wild birds. *See* AVES.

Walter J. Bock

Passive radar

A receive-only radar used for search, tracking, surveillance, identification, guidance, and mapping. The techniques used are similar to those in radiometry; passive radar could be said to be a branch of radiometry. Military applications are quite different, however, from other uses of radiometry in that the source is usually not known to exist until detected by the radar, and the operator has no control over the position of the source but must determine it and sometimes track it. There are often other sources in the vicinity, so that discriminating between competing sources is important in military applications. The operation of passive radars depends upon the detection of microwave or infrared radiation from warm bodies. *See* INFRARED RADIATION; MICROWAVE; RADIOMETRY.

All bodies, solid, liquid, and gaseous, emit electromagnetic radiation in the form of noise, the amount of noise depending upon the absolute temperature of the body. Energy is radiated at all frequencies, including microwaves, with a maximum in the infrared range. In many applications, very high-gain receivers are required, and the accuracy of the measurements is severely limited by small time variations in the gain of the amplifiers. R. H. Dicke solved this problem in 1946 with a radiometer in which the receiver input is switched between the antenna and a known reference noise source at a rate faster than the gain variations of the amplifiers. The detected output of the receiver is switched synchronously at the same rate, resulting in two output signals, one due to the noise received by the antenna and the other to the reference noise source. These two signals are subtracted to form an output signal proportional to the difference in the input noise temperatures. The reference noise power can be adjusted until the output signal is zero, indicating that the received noise power is equal to the known reference noise power. Radiometers of this type can be used to measure the absolute power received from a source, and this is required in some applications. In military applications, however, it is usually only necessary to separate a noise source from its surroundings, so an adjustment of the known reference noise source is not necessary. *See* HEAT RADIATION.

Many potential military targets radiate high noise power: ships at sea, exhaust from trucks, tanks, missiles, and airplanes, and factory chimneys, to name a few. Unlike an active radar, a passive radar cannot determine the range to a target. However, using the high antenna directivity obtainable at microwave and infrared wavelengths, a passive radar can locate a source of radiation accurately in direction and discriminate between nearby targets. The angular resolution is given approximately by the equation $A = 1.22\lambda/D$, where A is the angular resolution in radians, λ is the wavelength, and D is the diameter of the antenna used. An antenna with a diameter of 39 in. (100 cm) has an angular resolution of about 0.0037 radian at a frequency of 100 GHz; an angular resolution of 1 milliradian can be achieved at infrared with a 12-in. (30-cm) antenna diameter.

A passive radar can track a target closely and be used to direct weapon fire toward it. A passive radar, mounted on a missile, can be used to home the missile in on a target by using just the pointing information provided by the radar. The power required to operate such a radar is quite small because there is no transmitter. Ground surveillance and mapping can be accomplished with an airborne ground scanner. This type of radar provides an infrared picture of the terrain and any targets which may be present. Radars of this type can often see through visual camouflage.

The absence of transmitted power makes the location, and even the existence, of a passive radar difficult to determine. Even if the position of a passive radar is known, its frequency cannot be determined; for this reason and because of the high angular resolution, it is difficult to jam. *See* RADAR. Clyde L. Ruthroff

Bibliography. J. T. Miller, *Principles of Infrared Technology*, 1994; M. Schlessinger, *Infrared Technology Fundamentals*, 1994; F. T. Ulaby, R. K. Moore, and A. K. Fung, *Microwave Remote Sensing: Active and Passive*, 3 vols., 1981, 1982, 1986.

Pasteurella

A genus of gram-negative, nonmotile, nonsporulating, facultatively anaerobic coccobacillary to rod-shaped bacteria which are parasitic and often pathogens in many species of mammals, birds, and reptiles. It was named to honor Louis Pasteur in 1887. The organisms grow well on enriched complex media, such as blood agar, and preferably at increased atmospheric carbon dioxide and reduced oxygen pressures. They require organic nitrogen sources and various vitamins, and are capable of both aerobic and anaerobic respiration and carbohydrate fermentation. Nitrates are reduced to nitrites. The base composition of the genome deoxyribonucleic acids range from 38 to 45 mol % guanine plus cytosine. The group is closely interrelated with the genera *Haemophilus* and *Actinobacillus*. *See* ACTINOBACILLUS; HAEMOPHILUS.

Genetic studies have shown that the three groups constitute a family, Pasteurellaceae. The genetically redefined genus *Pasteurella* contains organisms that exhibit the following phenotypic features: positive reactions for alkaline phosphatase, indophenol oxidase, catalase, and fermentation of D-glucose, D-galactose, D-fructose, D-mannose, and sucrose; negative reactions for hemolysis, arginine dihydrolase, production of acid from L-sorbose, L-rhamnose, *m*-inositol, adonitol, and salicin, and splitting of esculin.

The genus contains at least 10 species. *Pasteurella multocida* causes hemorrhagic septicemia in various mammals and fowl cholera, and is occasionally transmitted to humans, mainly in rural areas. Human pasteurellosis may include inflammation in bite and scratch lesions, infections of the lower respiratory tract and of the small intestine, and generalized infections with septicemia and meningitis. *Pasteurella dagmatis* (*P.* "gas," *P.* "new species," or *P. pneumotropica*-type Henriksen), *P. canis*, and *P. stomatis* may cause similar, though generally less severe, infections in humans after contact with domestic or wild animals. Other *Pasteurella* species appear to be confined to particular hosts, for example, *P. gallinarum* and *P. avium* (*H. avium*); and several unnamed species, to birds. The control of pasteurellosis in domestic animals is an economically important task, but the problems of immunoprophylaxis are widely unresolved. *See* PASTEURELLOSIS.

The bovine pathogen *P. haemolytica*, the human parasite *P. ureae*, and *P. pneumotropica* (types Jawetz and Heyl), which are common in rodents and some other mammals, do not belong strictly to the genus *Pasteurella* but are closely related to the *Actinobacillus* group. *Pasteurella aerogenes* (frequently occurring in swine) and the *Pasteurella*-like "SP" group (which may affect guinea pigs and humans) are not true pasteurellas, and their precise taxonomic positions at the genus level are presently unknown. The etiologic agent of plague is no longer classified as a *Pasteurella* species but has been transferred to the genus *Yersinia*. *See* YERSINIA.

Although drug-resistant *Pasteurella* strains have been encountered, human *Pasteurella* infections are as a rule readily sensitive to the penicillins and a variety of other chemotherapeutic agents. *See* ANTIBIOTIC; DRUG RESISTANCE; MEDICAL BACTERIOLOGY; PLAGUE. Walter Mannheim

Bibliography. C. Adlam and J. M. Rutter (eds.), *Pasteurella and Pasteurellosis*, 1989; M. Kilian, W. Frederiksen, and E. L. Biberstein (eds.), *Haemophilus, Pasteurella and Actinobacillus*, 1981; N. R. Krieg and J. G. Holt (eds.), *Bergey's Manual of Systematic Bacteriology*, vol. 1, 2001.

Pasteurellosis

A variety of infectious diseases caused by the coccobacilli *Pasteurella multocida* and *P. haemolytica*. However, the term pasteurellosis also applies to diseases caused by any *Pasteurella* species. For example, *P. caballi* is considered to have a causal role in upper respiratory infections, pneumonia, peritonitis, and mesenteric abscesses of horses; and *P. granulomatis* has a causal role in a chronic disease of cattle in Brazil characterized by a progressive fibrogranulomatous process. All *Pasteurella* species occur as commensals in the upper respiratory and alimentary tracts of their various hosts. Although varieties of some species cause primary disease, many of the infections are secondary to other infections or result from various environmental stresses. *Pasteurella* species are generally extracellular parasites that elicit mainly a humoral immune response. Several virulence factors have been identified. *See* VIRULENCE.

Pasteurella multocida. This heterogeneous species consists of capsular (polysaccharide) and somatic (lipopolysaccharide) types. The former antigens are designated by capital letters and the latter by numbers. For example, an important serotype causing fowl cholera is designated A:3; E:2 strains cause hemorrhagic septicemia in cattle in Africa (see **table**). Serotyping is useful in studying the epidemiology of *P. multocida* diseases. *Pasteurella multocida* is the

Principal pasteurelloses and their hosts

Species	Serogroup	Biotype	Diseases	Hosts
Pasteurella multocida	A		Fowl cholera	Chickens turkeys, ducks, other avian species
			Wide variety of infections	Many animal species
			Variety of secondary infections	Many animal species and humans
	B		Epidemic hemorrhagic septicemia	Mainly cattle and water buffaloes
			Hemorrhagic septicemia	Elk, deer, other wild ruminants
	D		Infectious atrophic rhinitis	Swine
			Secondary invader in pneumonia	Swine
			Various infections	Various animal species and humans
	E		Epidemic hemorrhagic septicemia	Cattle and water buffaloes in Africa
	F		Infrequent infections	Avian species, various animals
Pasteurella haemolytica		A	Primary and secondary pneumonia	Cattle, sheep, goats
			Bovine pneumonic pasteurellosis	Cattle
			Septicemia	Nursing lambs
		T	Septicemia	Feeder lambs

most prevalent species of the genus causing a wide variety of infections in many domestic and wild animals, and humans. It is a primary or, more frequently, a secondary pathogen of cattle, swine, sheep, goats, and other animals. As a secondary invader, it is often involved in pneumonic pasteurellosis of cattle (shipping fever) and in enzootic or mycoplasmal pneumonia of swine. Particular serotypes are the primary cause of the important diseases fowl cholera and epidemic hemorrhagic septicemia of cattle and water buffaloes. The latter disease occurs in many tropical and subtropical countries, particularly in southeast Asia and Africa; however, it is rare in the United States and South America. *Pasteurella multocida* is one of the causes of the pleuropneumonia form of snuffles in rabbits. It is responsible for a variety of sporadic infections in many animals, including abortion, encephalitis, and meningitis. It produces severe mastitis in cattle and sheep, and toxin-producing strains are involved in atrophic rhinitis, an economically important disease of swine. Hemorrhagic septicemia, caused by capsular type B strains, has been reported in elk and deer in the United States.

Pasteurella haemolytica. Two different biotypes, A and T, of *P. haemolytica* are recognized based on pathogenicity, antigenic nature, and biochemical activity. Serotypes based upon differences in capsular antigens are designated by numbers. Specific serotypes are associated with certain animal hosts and diseases; for example, serotype 1 of biotype A is the principal cause of bovine pneumonic pasteurellosis (shipping fever). All strains of *P. haemolytica* produce a soluble cytotoxin (leukotoxin) that kills various leukocytes of ruminants, thus lowering the primary pulmonary defense. The importance of the cytotoxin in the pathogenesis of pneumonic pasteurellosis has been clearly demonstrated experimentally. This species has a primary or secondary role in pneumonia of cattle, sheep, and goats. As mentioned above, it is the principal cause of the widespread pneumonic pasteurellosis of cattle. Shipping fever occurs frequently in cattle that have been stressed by transport. Various viruses or mycoplasmas may be the primary cause of a secondary pneumonic pasteurellosis, although *P. haemolytica* is considered the primary agent in most outbreaks. Stresses other than those related to transport may also predispose cattle to the disease. Other important diseases caused by certain serotypes of *P. haemolytica* are mastitis of ewes and septicemia of lambs.

Diagnosis and treatment. All of the *Pasteurella* species can be isolated by culturing appropriate clinical specimens on blood agar. They are identified generically by their basic characteristics and identified to species by a number of conventional biochemical tests. *Pasteurella* isolates are routinely subjected to antimicrobial susceptibility tests. Multiple drug resistance is frequently encountered, particularly in fowl cholera and bovine pneumonic pasteurellosis. Treatment is effective if initiated early. Among the drugs used are penicillin and streptomycin, tetracyclines, chloramphenicol, sulphonamides, and some cephalosporins.

Prevention. Sound sanitary practices and segregation of affected animals may help limit the spread of the major pasteurelloses. However, the endemic character of pasteurellosis makes control difficult. Live vaccines and bacterins (killed bacteria) are used for the prevention of bovine and ovine pneumonic pasteurellosis, fowl cholera, epidemic hemorrhagic septicemia, and atrophic rhinitis of swine. *See* PASTEURELLA.

G. R. Carter

Bibliography. C. Adlam and J. M. Rutter (eds.), *Pasteurella and Pasteurellosis*, 1989; B. E. Patten (ed.), *Pasteurellosis in Production Animals*, 1993.

Pasteurization

The treatment of foods or beverages with mild heat, irradiation, or chemical agents to improve keeping quality or to inactivate disease-causing microorganisms. Originally, Louis Pasteur observed that spoilage of wine and beer could be prevented by heating them a few minutes at 122–140°F (50–60°C). Today pasteurization as a thermal treatment is applied to many foods, including liquid eggs, crab, fruit juices, pickles, sauerkraut, smoked fish, beer, wine, and dairy products. In foods consumed directly, destruction of pathogens to protect consumer health is paramount, while in products without public health hazards, control of spoilage microorganisms is primary. In fermentation processes, the raw material may be pasteurized to eliminate microorganisms that produce abnormal end products, or the final product may be heated to stop the fermentation at the desired level. In many cases, such as with some dairy products, mild heat treatments may be utilized for all of these purposes.

Milk and dairy products probably represent the most widespread use of pasteurization. Several time-temperature combinations have been approved as equivalent: 145°F (63°C) for 30 min; 161°F (72°C) for 15 s; 191°F (89°C) for 1 s; 194°F (90°C) for 0.5 s; 201°F (94°C) for 0.1 s; 204°F (96°C) for 0.05 s; or 212°F (100°C) for 0.01 s. These precise heat treatments are based on the destruction of the rickettsia *Coxiella burnetii*, which is considered the most heat-resistant nonsporeforming pathogen found in milk. Absolute control of the thermal treatment is essential for safety. Pasteurization of milk has successfully eliminated the spread of diseases such as diphtheria, tuberculosis, and brucellosis through contaminated milk. *See* DAIRY MACHINERY; FOOD MANUFACTURING; MALT BEVERAGE; MILK; WINE.

Francis F. Busta

Patent

Common designation for letters patent, which is a certificate of grant by a government of an exclusive right with respect to an invention for a limited period of time. A United States patent confers the right to exclude others from making, using, or selling the patented subject matter in the United States and its territories. A United States patent covering a process also, under certain conditions, prohibits the unlicensed sale in the United States of articles made by that process anywhere in the world. Portions of those rights deriving naturally from it may be licensed separately, as the rights to sell, to use, to make, to have made, and to lease. Any violation of this right is an infringement.

An essential substantive condition which must be satisfied before a patent will be granted is the presence of patentable invention or discovery. To be patentable, an invention or discovery must relate to a prescribed category of contribution, such as process, machine, manufacture, composition of matter, plant, or design. In the United States there are different classes of patents for different members of these categories.

Utility patents. Utility patents, which include electrical, mechanical, and chemical patents, are the most familiar; they each have a term beginning upon issue and ending 17 years later. This 17-year class of patent is granted (or issued) for a process, manufacture, machine, or composition of matter which meets the statutory criteria. A process may, for example, be a method of inducing or promoting a chemical reaction, of controlling a computer, or of producing a desired physical result (such as differential specific gravity ore separation by flotation). "Manufacture" means any article of manufacture, and includes such diverse items as waveguides, transistors, fishing reels, hammers, buttons, and corks. "Machines" has its broadest conventional meaning, and "composition of matter" includes, for example, drugs, genetically altered living matter, herbicides, and alloys; patentability is not precluded by the fact that the subject matter is alive. This class of patent affords protection for the claimed invention and, under the appropriate circumstances, can also afford protection for equivalents of the subject matter claimed in the patent which do substantially the same thing as that claimed subject matter in substantially the same way to achieve the same result. The extent of protection for such equivalents is subject to the limitations provided by the prior art and the history of the prosecution of the patent application. Thus mere changes in form, material, inversion, or rearrangement will not avoid infringement of the utility patent. A straightforward substitution of one active device for another, as a transistor for an electronic tube, is not normally such a change as to avoid infringement of a utility patent.

Design patent. This class of patent is granted for any new, original, and ornamental design for an article of manufacture. To the extent that shape is determined by functional, rather than ornamental, considerations, it is not the proper subject matter for a design patent. Unlike the utility patent, the design patent may be avoided by a change in appearance, although the essential function may be retained. The design patent is issued for 14 years.

Plant patents. This class of patent is granted to one who discovers and asexually reproduces any distinct and new variety of plant other than a tuber-propagated plant or a plant found in an uncultivated state. The right of exclusion extends only to the asexual propagation of such a plant. The term of the plant patent is 17 years from the date of issue by the Patent and Trademark Office.

Procedure. The discussion in the balance of this article is limited to utility patents. In the United States, letters patent are granted upon the making of written application to the Patent and Trademark Office. The usual elements of this application are (1) the title; (2) a brief summary of the invention; (3) a brief description of the drawings, which are required where necessary for the understanding of the subject matter to be patented; (4) a detailed description of the invention and the manner and process of making and using it in such full, clear, concise, and exact terms as to enable any person skilled in the art to which it pertains, or with which it is most nearly connected, to make and use it, and setting forth the best mode contemplated by the inventor for carrying out the invention; (5) the claims, which define the scope of the requested exclusionary right and which particularly point out and distinctly claim that right; (6) a brief abstract of the technical disclosure; (7) an oath or declaration stating, among other things, that the applicant is the original and first inventor, that the invention was not known or used by others in the United States or patented or described in a printed publication in this or a foreign country before the applicant's invention, or printed or described in a printed publication anywhere in the world or in public use or on sale in this country more than 1 year prior to the date of filing of the United States application.

Owing to the complexities of obtaining proper patent protection, it is customary to retain an attorney who has been registered to practice before the Patent and Trademark Office for the purpose of preparing the application and pursuing it through that office.

When the application is received, it is examined at the Patent and Trademark Office by an examiner. The examiner searches the prior art, consisting of prior patents and other publications, to determine whether the application and claims comply with the rigorous standards required by the statute for patentability. If it does not, the examiner points this out in a letter of rejection, specifying the particulars of the basis for rejection. If the application, on the other hand, is allowable, the examiner sends the applicant a notice of allowance. Following the date when the Patent Office mails a rejection and within the time period specified by the Patent Office, ordinarily 3 months but never more than 6 months, the applicant or the attorney must respond. Normally the application would be amended to make it comply with the statutes. If not, the examiner would issue a final rejection, from which the applicant could appeal to the Board of Appeals and Interferences. If the amended application was found to be in accord with the statute, the examiner would issue a notice of allowance. The patent is issued upon the payment of the issue fee. The entire process requires about $1\frac{1}{2}$ years on the average.

As early in the pendency of the application as convenient, it is imperative that the patent examiner be informed of the prior art closest to the invention being claimed known to the applicant or to the applicant's attorney. If it is desired to obtain corresponding patent coverage in foreign countries, it is essential that there be no public disclosure of the invention before the United States application is filed.

Invention. Patents are granted for new inventions or discoveries which are within the statutory classes of patentable subject matter, which were not known in the United States before the invention date and not printed anywhere in the world before that date, which are the subject of applications timely filed, and in which the subject matter is not obvious to one skilled in the art to which the invention relates.

It is generally considered that there are two discernible steps in invention. The first is thinking of the invention; the second is constructing it. The first is termed conception. The second is called reduction to practice.

Conception. The formation in the mind of the inventor of a definite and permanent idea of the complete and operative invention as it is thereafter to be applied in practice is conception. It is not merely a perception of what is done or considered desirable to do, but, going beyond this, how it is to be done in terms of a currently realizable instrumentality or group of instrumentalities. Because conception is a mental act, there must be some external, verifying manifestation in the form of impartation to another if the act is to be established in later controversy over priority of invention. Although oral transmission is adequate in theory, the perishability of the human memory favors the unchanging written word, dated, signed by the inventor, and witnessed by the corroborating party. The corroborating party should be a person who fully understands the invention.

It is this which makes the keeping of written records by the inventor in the course of his or her work so important, because a failure to have this verified external manifestation of the conception may cost the inventor the patent if the date becomes important and is challenged. Because the record is not a proof but only a document capable of proof, it is important that the recording be in some permanent form in which undetectable alteration is practically impossible; otherwise the weight of the proof will be diminished. Predating of the document is damaging, although a record, made a few days later, of a previously observed event is valuable, if appearing on its face as such. The parties involved in a record of conception will be tested for verity by cross examination should the dates ever be challenged.

Frequently the moment of conception is clear only in retrospect, when there is a long sequence of experimental effort directed to attaining the desired result. There is no dramatic thunderclap ushering in the birth of most inventions; hence, it is desirable to keep current dated and witnessed records of all work. Such records tend further to corroborate the conception by providing a clue to the entire thought train and revealing the completeness of understanding by the skill with which its principles are later applied.

Conception alone does not give the inventor any vested right in the invention.

Reduction to practice. Two forms of reduction to practice are recognized, actual and constructive. The filing of a patent application which does not become abandoned is a constructive reduction to practice.

Actual reduction to practice requires that the invention be carried out in a physical embodiment demonstrating its practicability. In a process or method, this is sufficiently done when the steps are actually carried out to produce the desired result. In a machine or article of manufacture, it is required that there be a construction showing every essential feature of the claimed invention. Practicality is demonstrated by operating the apparatus under the conditions which it is anticipated will be encountered in actual service. For example, the testing of an automobile hot-water heater by using water from the hot-water tap has been held not to be an actual reduction to practice because some conditions might occur in a motor vehicle which would not be observed in a stationary installation with a heat source of relatively unlimited capacity. Materials are reduced to practice when they are produced, unless utility is not self-evident, for example, in the case of drug compounds, where it is required to establish that the drug is useful for the purpose stated, not merely harmless. To be effective, the reduction to practice must be by the inventor, by one acting as his or her agent, or by one who has acquired rights from the inventor.

A reduction to practice which results from diligent efforts following conception of the invention gives a vested right in the invention, unless followed by an abandonment. If there is a gap in such diligence, the effective date of the right to assert ownership of the invention is only that date which can be connected by a continuous train of diligence with the reduction to practice. Like conception, reduction to practice must be corroborated by a third-party witness, and it is advisable to have a contemporary written record of what was done and what was observed, accompanied by the date of the observations, to refresh the recollection of the witnesses. The witness must have sufficiently acquainted himself or herself with the internal details of any apparatus to be able later to establish the identity between what was demonstrated and what is sought later to be patented. The diligence required is that which is reasonable under all the circumstances. The safest course is to make every effort promptly to reduce to practice consistent with the inventor's physical, intellectual, and financial capacity. For example, if reduction to practice would be clearly within the inventor's financial means, alternative attempts to secure financing could not constitute diligence.

Interferences. When two or more persons are claiming substantially the same invention, a contest to determine priority between the two is instituted by the Patent and Trademark Office. This proceeding is termed an interference. That party who made the invention first will prevail in the interference and be awarded the patent. Making the invention is proved by corroborated acts of reduction to practice and a continuous bridge of diligence, if possible, to the act of conception, provided that those acts have been in the United States. If any or all of the acts were performed outside the United States, the inventor is limited to the date of his or her first efforts in the United States. Exception is made for foreign inventors when they have first filed a corresponding application for patent in a foreign country which is a signatory to the Paris Convention. In such instance the inventor will be credited with a date corresponding to his or her first foreign filing date if the filing in the United States occurs within a year of that date and there has been no sale or publication of the invention in the United States more than 1 year before the actual filing of the application there.

An applicant may provoke an interference with an issued patent which claims an invention that the applicant believes should rightly belong to him or her. This is done within 12 months of issue of the patent by filing in the Patent and Trademark Office, as part of a pending application for patent, all the claims which it is sought to contest, applying them to his or her disclosure, and requesting the declaration of an interference.

Following declaration of an interference, normally the parties are called upon to file preliminary statements under oath setting forth pertinent data surrounding the genesis of the invention and its disclosure to others. When these preliminary statements have been received and approved by the Patent and Trademark Office, the applicant is notified of the setting of a period of time during which motions may be filed. During this time, access to the adversary's application is permitted, marking one of the few times that the secrecy with which the Patent and Trademark Office surrounds each patent application is penetrated. During the motion period, various motions for modification or termination of the interference may be presented and set for hearing. When an issued patent is involved, there may be no motion for dissolution of the interference upon the ground that the claim in issue is unpatentable. After the motions are disposed of, times for taking the testimony are set, during which the parties may take sworn statements from their witnesses before duly qualified officers, which are then filed with the Patent and Trademark Office, accompanied by any proper exhibits. The burden of proof in the interference is on the party who is entitled to the later filing date and is therefore known as the junior party. If the junior party does not take testimony during the time allotted, the interference is terminated in favor of the senior party without any testimony being taken by him or her.

The conduct of an interference proceeding is frequently an arduous and complex matter, being fraught with many technicalities.

Inventor. In the United States only a natural person may be an inventor, as distinguished from a corporation. Inventors may be either sole or joint.

Assignment. Patents and applications for patents have the general attributes of personal property, and interests in them are assignable by instrument in writing. Such assignments will be recorded by the Patent and Trademark Office upon filing of a request accompanied by a copy of the assignment and payment of the proper fee. To be good against subsequent purchasers without notice, the assignment must be recorded within 3 months from its date, or before the date when such transfer of rights was made.

Witnesses. No joint inventor can serve as a corroborating witness for another joint inventor in connection with the invention which they have jointly made. Beyond this, the rules normally governing witnesses apply. A corroborating witness should fully understand the invention.

Enforcement. Enforcement is available against the manufacturer, the user, and the seller in the United States who perform acts within the scope of the patent. Those who induce infringement by another and those who contribute to the infringement by another by, for example, knowingly providing special parts with no use except in the infringement, are also infringers. Enforcement of patents is through the federal judicial system, action being initiated by the patent owner in the federal judicial district where the defendant resides, or where the defendant has committed the alleged act of infringement and has a regular and established place of business. Action may also be started by one accused of infringement in any place where the patent owner resides or is licensed to do business. Damages, no less than a reasonable royalty, may be awarded, and an injunction granted, prohibiting further infringement by the defendant. Interest on the award prior to the judgment and subsequent to infringement is normally awarded. If damages are awarded, there can be recovery for a period not longer than 6 years preceding the filing of the complaint.

Patents may also be enforced by the International Trade Commission (ITC), which can prevent imports into the United States of goods that infringe a valid patent. No damages are awarded in an ITC proceeding of this type. All appeals from district courts and from the ITC in such cases are heard by the Court of Appeals for the Federal Circuit.

When the infringer is the United States government, or a supplier of the government operating with the authorization and consent of the government, the suit must be filed against the government in the U.S. Court of Claims.

A patent may become unenforceable through improper use, for example, use as a part of an act in violation of the antitrust law.

Under the 6-year statute of limitations, an action may be maintained on a patent up to 6 years after its expiration, the accounting being limited in such instance to damages for infringing activities within the life of the patent.

Licenses. Licenses to operate under a patent may be granted, either nonexclusive or exclusive, and may be in writing or may arise as a necessary implication of other actions of the patentee. Except for an exclusive license, licenses are not ordinarily recorded by the Patent and Trademark Office.

Foreign filing. A United States patent is void if the United States inventor should file an application for the same invention in any foreign country before 6 months from the date when he or she filed in the United States, unless license to do so is obtained from the Commissioner of Patents and Trademarks. Because the United States is a party to the Paris Convention, applications filed in foreign countries that are also members of the Paris Convention within 12 months of the date when the parent case is filed in the United States are accorded an effective filing date in the foreign countries which is the same as the date of filing in the United States. The procedural details of foreign filing vary from country to country and from time to time. The Patent Cooperation Treaty, to which the United States, Japan, most European countries, and other countries are party, has greatly facilitated filing of patent applications in foreign countries. In particular it has established a standardized application and claim format acceptable in the patent offices of all member countries.

Nuclear and atomic energy. Patents may not be issued for inventions or discoveries useful solely in the utilization of special nuclear material or atomic energy in an atomic weapon, nor does any patent confer rights upon the patentee with respect to such uses. As a substitute for the patent incentive for disclosure in this field, there is a mandatory provision in the law that requires anyone making an invention or discovery useful in the production of special nuclear material or atomic energy, in the utilization of such special nuclear materials in the production of an atomic weapon, or in the utilization of atomic energy weapons, to report such invention promptly to the Department of Energy, unless it is earlier described in an application for patent. Awards may then be requested from the Patent Compensation Board of the Department of Energy.

Aeronautics and astronautics. No patent may be issued for an invention having significant utility in the field of aeronautics or space unless there be filed with the Commissioner of Patents and Trademarks a sworn statement of the facts surrounding the making of the invention and establishing that the invention was made without relation to any contract with the National Aeronautics and Space Administration. Such a contract right is subject to waiver by the administrator, but in the event of waiver the administrator is required by law to retain a license for the United States and foreign governments.

Marking. A patented product may carry a notice of this fact, including the patent number. The affixation of such notice is of advantage in establishing constructive notice and fixing the period for which damages may be collected. In many cases damages may not be collected, in the absence of such marking, without actual notice to the infringer. By statute, false marking is a criminal offense. The marking "Patent pending" or "Patent applied for" gives no substantive rights, but may give rise to sanction under the above-mentioned statute if without foundation in fact.

Foreign patents. The principles guiding most foreign patent systems are essentially the same as those underlying the United States system: the granting of a carefully defined exclusionary right for a limited term of years in return for a laying open of the invention through letters patent. In the socialist countries the grant is sometimes made for innovation as well as invention and establishes a right to compensation. There are some differences in the classes and terms of patent, and in the nature of subject matter which may be patented. The most significant departures are the requirement in other countries that there be no public divulgation of the invention before effective date of the patent application, the measuring of the term, in most instances, from the filing date, and compulsory licensing.

In some examination countries the mounting magnitude of material to be searched so slowed the rate of disposition of patent applications that the delays and the accumulation of pending patent applications created intolerable uncertainties. For example, in

Japan the practice of deferred examination has been introduced, according to which the application is given cursory formal examination, and published after 18 months, but examination on the merits is delayed. Examination on the merits is made upon request, but must be made before enforcement of the patent or the end of 7 years, whichever first occurs. The theory was that only a few of the patents are ever brought to litigation, and many are without value after 7 years, so that the load on the examining staff would be reduced to manageable proportions. In fact, however, the system created additional uncertainties for the business community.

Furthermore, in Japan after the patent application has been allowed but before it is issued as a patent, it is published for opposition. Interested parties can then further delay issuance of the patent by asserting prior art not previously considered by the Patent Office alleging it prevents the issuance of a valid patent. Opposition proceedings can last for years.

To eliminate duplicate searching of the same invention by the different patent offices and to standardize the requirements for patent applications, as well as the form of patent claims, the Patent Cooperation Treaty, as noted above, has been entered into by many countries. Under this treaty the application is examined by a competent searching authority, the criteria for which are set by the treaty. The application, together with the results of that search, is then published. Any country to which the applicant subsequently applies has the benefit of that search, which may be utilized by that country in determining whether the invention is in fact patentable in that country.

In somewhat similar fashion the countries of Europe have by treaty established a European Patent Office in Munich to which application can be made. The European Patent Office conducts the search on behalf of the other countries of Europe who are members of that treaty. As in the case of the Patent Cooperation Treaty, further searching may be done by each country, but need not be.

Copyrights and trademarks. Authors of literary, musical, artistic, dramatic, and certain other types of intellectual works such as computer programs protect their creations by copyright, which is provided under the federal law. Copyright protection exists in such original works of authorship when they are fixed in any tangible medium of expression. The exclusive right protected by a copyright is, in general, the right to copy or print the work, to sell or distribute such copies, and to dramatize, perform, or translate the work. The protection provided by copyright extends only to the particular form or expression of the author's idea—as distinguished from the idea itself, which may be freely used by others. The term of copyright protection extends for the author's life plus 50 years. Copyrights are registered in the Copyright Office, which is part of the Library of Congress. Registration is a requirement for enforcement of the copyright. Printed forms are provided by the Copyright Office to facilitate the registration of copyrights; registration of the copyright is effected by completing the forms and sending them to the Copyright Office together with copies of the work copyrighted. Infringers of the copyright are liable for the actual damages of the copyright owner, as well as the infringer's profits—addition to an injunction. The United States has become a signatory to the Berne Convention. As a result, notice of copyright on the copyrighted work is no longer essential; it is, however, highly desirable because monetary recovery for infringement by another is affected by such marking.

A trademark is a word, symbol (that is, a design, a letter, or a number), or any combination of these elements that is used by a merchant or manufacturer to identify the origin or sponsorship of his or her goods and services from competing goods or services. Technically, a service mark identifies a service, but for purposes of brevity the term trademark should be read to include service mark here. A trade name identifies a company and should not be confused with a trademark, which identifies a product.

A trademark serves three separate functions: it can indicate the ownership and origin of products or services; it can serve as an advertising device; and it provides a guarantee of quality or degree of product or service consistency. Not every word, name, symbol, or design can function as a trademark. The common descriptive name of a product (generic name) cannot be a trademark.

A trademark derives its essential strength in the marketplace by consumers who recognize the differentiating character of such word, symbol, or other device. Trademarks may be registered, after being used on goods described in the registration, in each of the states, and may also be registered federally in the Patent and Trademark Office. Federal registration may be obtained after the mark is used in "commerce" as defined by the Constitution; furthermore, after November 1989 applications to obtain federal registrations could be filed on the basis of good-faith "intent to use" the specified trademark on the goods or services set out in the application. The application does not mature into a registration until a declaration of actual use is filed within the prescribed period of time. In addition, of course, the trademark must be of such nature as to distinguish the goods of the manufacturer or merchant from the goods of others. When registered by the Patent and Trademark Office, the owner of the mark may show such registration by using the symbol ®. Rights in trademarks may last indefinitely; certificates of registration must, however, be renewed periodically. Federal registration of a trademark provides nationwide constructive notice of the registrant's claim of ownership. Any person who uses on his or her goods (or services) a mark likely to cause confusion in the marketplace with goods bearing a registered mark is liable in a civil action (with certain exceptions) to an injunction preventing further use of the confusingly similar mark on similar goods and to pay damages sustained by the plaintiff as well as defendant's profits. The district courts of the United States have jurisdiction of actions arising out of infringement of federally registered trademarks, with appeals being taken to the regional circuit courts of appeal. Federally registered trademarks, under certain circumstances, may also be protected from infringement through exclusion by the Custom Service of goods with infringing marks; they may also be enforced by the International Trade Commission, which can order the borders closed to goods of foreign origin carrying marks that infringe the registered mark.

Donald W. Banner

Bibliography. H. L. Barber, *Copyrights, Patents and Trademarks: Protect Your Rights Worldwide*, 2d ed., 1994; Matthew Bender and Co., *Patent Law Annual*, annually; Matthew Bender and Co., *Patent Law Developments: Enforceability of Rights*, 1965; Matthew Bender and Co., *Patent Law Developments: Proprietary Intellectual Rights*, 1964; A. W. Deller, *Deller's Walker on Patents*, 1964; P. Goldstein, *Copyright, Patent Trademark and Related State Doctrines: Cases and Materials on the Law of Intellectual Property*, 4th ed., 1999; T. A. Penn and R. D. Foltz, *Understanding Patents and Other Forms of Legal Protection*, 1990; P. D. Rosenberg, *Patent Law Basics*, 1992; P. D. Rosenberg, *Patent Law Fundamentals*, 2d ed., 3 vols., 1980; U. S. Bureau of National Affairs Legal Services Staff, *United States Patents Quarterly*, quarterly; U.S. Department of Commerce, Patent Office, *Manual of Patent Examining Procedures*; H. C. Wegner, *Patent Law in Biotechnology, Chemicals, and Pharmaceuticals*, 1992.

Paterinida

A small extinct order of inarticulate brachiopods that ranges in age from Early Cambrian to Middle Ordovician. The shell is chitinophosphatic in composition, and its outline is circular or elliptical. The ventral (pedicle) valve is more convex than the dorsal (brachial) valve. The pedicle was absent or emerged between the valves. The ventral valve has a flattened posterior area (pseudointerarea) that is divided by a triangular notch which is closed to a variable extent by a plate (homeodeltidium). Comparable, but not as extensive, structures occur at the posterior end of the dorsal (brachial) valve. The muscle scars are unusual compared to other inarticulates, and form narrow triangular tracks radiating from the posterior extremity of each valve (see **illus.**). Except for their shell composition and lack of articulation, the Paterinida resemble articulate brachiopods. Members were presumably epifaunal and sessile. *See* BRACHIOPODA; INARTICULATA.

Merrill W. Foster

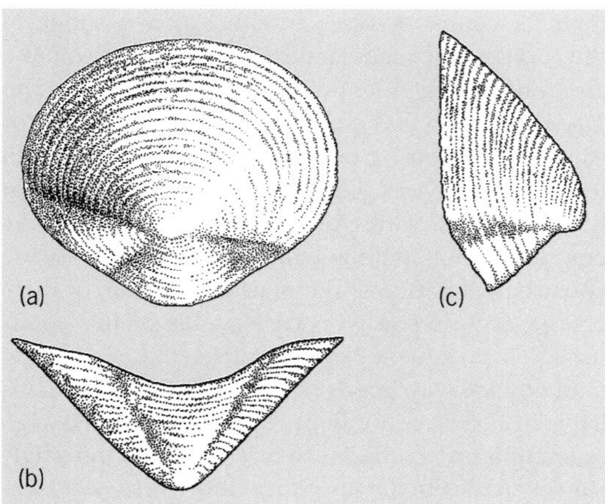

***Paterina*: (*a*) exterior, (*b*) posterior, and (*c*) lateral view of pedicle (ventral) valve. (*After C. D. Walcott, Cambrian Brachiopoda, USGS Monogr., vol. 51, 1912*)**

Pathogen

Any agent capable of causing disease. The term pathogen is usually restricted to living agents, which include viruses, rickettsia, bacteria, fungi, yeasts, protozoa, helminths, and certain insect larval stages.

Pathogenicity is the ability of an organism to enter a host and cause disease. The degree of pathogenicity, that is, the comparative ability to cause disease, is known as virulence. The terms pathogenic and nonpathogenic refer to the relative virulence of the organism or its ability to cause disease under certain conditions. This ability depends not only upon the properties of the organism but also upon the ability of the host to defend itself (its immunity) and prevent injury. The concept of pathogenicity and virulence has no meaning without reference to a specific host. For example, gonococcus is capable of causing gonorrhea in humans but not in lower animals. *See* BACTERIA; FUNGI; IMMUNITY; MEDICAL MYCOLOGY; MEDICAL PARASITOLOGY; MYIASIS; PLANT PATHOLOGY; PLANT VIRUSES AND VIROIDS; PROTOZOA; RICKETTSIOSES; VIRULENCE; VIRUS.

Daniel N. Lapedes

Pathology

The study of the etiologies, mechanisms, and manifestations of disease. Techniques and knowledge gained from other disciplines, including anatomy, physiology, microbiology, biochemistry, and histology, are utilized. The information obtained from the study of pathology is necessary prior to developing methods with which to control and prevent disease.

Etiology. The concepts concerning the etiologies of disease have undergone tremendous evolution. For much of history, etiologies were associated with mystical forces. Skeletal remains from prehistoric times had a surprising number of trepanned skulls, that is, in which holes had been cut. It is thought that a belief in evil spirits, trapped within the body and manifesting their presence by producing illness, promoted this surgery. A major shift in thought occurred in the fourth century B.C. when Hippocrates taught that disease is a natural occurrence of the biological world. Unfortunately, during the Middle Ages there was a return to mysticism. The enlightenment of these Dark Ages began in the sixteenth century. Vesalius, an anatomist, broke with tradition and published what he saw rather than the accepted dogma. An understanding of the normal anatomy had to be gained before disease responses could be appreciated. In the seventeenth century modern physiology was born as a result of the work of William Harvey on the circulatory system. The rate of development of scientific knowledge became explosive in the nineteenth century and continues. An entirely new perspective in biology developed and provided a basis for the modern study of disease. A key event was the development of the light microscope. With the aid of this instrument, J. Schleiden (1804–1881) and T. Schwann (1810–1882) understood that all life is composed of cells. R. Virchow (1821–1902) applied the cell theory to the pathogenesis of disease. He recognized that diseases were the product of the alteration of cell structure and function and that these changes were often reflected in gross anatomic alterations.

Thus, with the light microscope it became possible to correlate the observed signs and symptoms in an individual with cellular changes. In its early stages pathology was very descriptive. Diseases were understood and categorized, in part, by how gross and microscopic anatomy was altered. In the last half of the nineteenth century, by using this approach to pathology, coupled with microbiological techniques, it was learned that the major causes of human death were biotic agents: protozoa, bacteria, viruses, and fungi. Infectious diseases took a heavy toll in human lives. Better sanitation and public health measures were instrumental in controlling these diseases, and the production of antibiotics and immunization procedures further reduced their importance.

Mechanisms. Understanding the mechanisms underlying disease responses is a significant task. Whereas the light microscope enabled scientists to look at cells and observe their behavior, the development of the electron microscope by B. von Borries and E. Ruska in 1938 in Germany enabled scientists to look into cells with greater resolution than before. Recognition of subcellular structures became relatively easy. The term molecular biology was introduced in 1938. It is a field of biology which is concerned with the analysis of cell structure and function. Smaller and smaller components of the cells are dissected out and manipulated, in order to understand their normal function and to see what occurs when systems are perturbed in various ways. An offshoot of this science is molecular pathology, which approaches the study of disease mechanism in the same way. *See* MOLECULAR BIOLOGY; MOLECULAR PATHOLOGY.

It is now apparent that all diseases reflect changes at the molecular level. Scientists are beginning to understand what these biochemical alterations are in some diseases. An example of a disease whose underlying mechanism is understood is xeroderma pigmentosum, the result of an inherited disorder. Affected individuals are very sensitive to the ultraviolet rays of sunlight. Their skin ages rapidly, and they develop skin cancer, often as early as their late teens. It is now known that this disease is due to a defect in the enzyme which repairs the damage done to deoxyribonucleic acid (DNA) by ultraviolet light. Unfortunately there is no known cure. Individuals with xeroderma pigmentosum must avoid sunlight. This inhibits the progress of the disease but does not prevent it, since apparently other agents produce the same type of damage as ultraviolet light.

Another rare disease, also attributable to an inborn genetic error, has a more optimistic prognosis now that the biochemical basis of its manifestations has been worked out. This is phenylketonuria, which is characterized by mental retardation, too little pigment in the skin, and sometimes eczema. These responses have been traced to the absence of an enzyme which converts one of the amino acids, phenylalanine, obtained in the diet, to another product. Thus, phenylalanine accumulates in the blood and spinal fluid. If the disorder is recognized shortly after birth, the effects can be minimized by giving the child a diet which is very low in phenylalanine. Phenylalanine cannot be completely eliminated because it is necessary for the synthesis of proteins. In summary, the goal of molecular pathology is the explanation of the pathogenesis of disease on the basis of active or reactive portions of molecules, energy states, or alterations in molecular constituents. *See* PHENYLKETONURIA.

Manifestations. A diagnosis of a disease is made by observation of manifestations of the process. Alterations within cells are reflected in changes in cellular function and often in structure. When enough cells are involved in a disease response, tissues, organs, and the whole organism may exhibit various lesions. Historically, alterations in gross or microscopic anatomy were very important when identifying a disease process. In human pathology, these observations were used in conjunction with the subjective manifestations, or symptoms, reported by the affected person. This kind of information is still very important. However, added to this is a whole realm of biochemical and physical assays which, in many cases, are very sensitive and give definitive answers to the type of disease process which is occurring.

The interaction between an etiologic agent and an individual may evoke one or more responses. Cells can respond to injury in a limited number of ways. Cellular responses may result in degeneration, death, disturbances of growth and specialization, teratogenesis, inflammation, neoplasia, or aging. The nature of the response is modified by the agent, dose, portal of entry, or duration of exposure, as well as by host factors such as age, sex, nutrition, species, and individual susceptibility. For instance, an important cause of death of humans is a specific type of cancer of the lung, bronchogenic carcinoma. A cell or cells are stimulated to undergo an abnormal proliferation, and a malignant process occurs. It is now known that this disease is produced in response to some agent in cigarette smoke. Since all smokers do not eventually have this cancer, there is obviously individual variation in the cellular response.

Branches. There are many branches of pathology. Divisions are made depending upon focus of interest. Clinical pathology is concerned with diagnosis of disease. As medicine has expanded, subspecialties such as surgical pathology and neuropathology have developed. Experimental pathology attempts to study disease mechanisms under controlled conditions. General pathology covers all areas, but in less detail, and serves in medical education. *See* CLINICAL PATHOLOGY.

Another area of pathology is environmental pathology, which deals with disease processes resulting from physical and chemical agents. At present, the leading causes of death have environmental agents as the known or suspected major etiologic factors; these diseases include heart disease, atherosclerosis, and cancer. It is believed that with understanding, many such diseases, like those produce in response to biotic agents, can be brought under control.

In summary, pathology considers the factors which are known or suspected to be related to alterations of form or function. It attempts to produce or implicate an etiologic agent and seeks to delineate the mechanisms involved in producing the sequential changes in the body. *See* DISEASE.

N. Karle Mottet; Carol Quaife

Pathotoxin

A chemical of biological origin, other than an enzyme, that plays an important causal role in a plant disease. Most pathotoxins are produced by plant pathogenic fungi or bacteria, but some are produced by higher plants, and one has been reported to be the product of an interaction between a plant and a bacterial pathogen. Some pathogen-produced pathotoxins are highly selective in that they cause severe damage and typical disease symptoms only on plants susceptible to the pathogens that produce them. Others are nonselective and are equally toxic to plants susceptible or resistant to the pathogen involved. A few pathotoxins are species-selective, and are damaging to many but not all plant species. In these instances, some plants resistant to the pathogen are sensitive to its toxic product.

Selectivity and disease specificity. Selective pathotoxins, sometimes called host-specific toxins, provided the first mechanisms to account for the remarkable specificity of several important plant diseases. Often, only a single group of plant varieties, which possess a single specific genetic factor, is susceptible to a particular disease. Pathogen-produced selective pathotoxins are equally specific; they are up to 400,000 times more toxic to plants susceptible to the pathogen involved than to those that are resistant. From a scientific viewpoint, selective pathotoxins have been most valuable as model systems in investigations of the physiology of disease and the nature of disease resistance in plants. Substituted for the pathogens that produce them, they have made it possible to follow metabolic changes in diseased plants without complications introduced by metabolic activities of a living pathogen. Moreover, pathological events that occur over a period of days or weeks in infected plants can be compressed to a few hours with toxin treatments.

Selective pathotoxins have had useful practical applications in relation to two destructive epidemics, Victoria blight of oats in the late 1940s and southern corn leaf blight in 1970. Each of these epidemics followed the introduction of a single agronomically desirable genetic character into nearly all commercial cultivars of a crop, and in each case this desirable character rendered the entire population susceptible to a previously unimportant fungal pathogen.

In Victoria blight of oats, the genetic factor, Pc-2, introduced from the cultivar Victoria, conferred a high degree of resistance to the crown rust pathogen and, at the same time, extreme susceptibility to a new fungal pathogen, *Helminthosporium victoriae*. In artificial culture *H. victoriae* produced a metabolite, victorin, which was highly toxic only to oat cultivars that carried the Pc-2 factor and that hence were susceptible to Victoria blight. After all attempts to combine resistance to rust and Victoria blight by conventional breeding techniques had failed, victorin was used to screen 80,000,000 oat seedlings for resistance to this disease. A few of the survivors (**Fig. 1**) were resistant to both crown rust and Victoria blight. Pathotoxins produced by other plant pathogens have been used in a similar way to screen for resistance to diseases of sorghum, citrus, and sugarcane. Pathotoxins are being used as screening agents for disease resistance in conjunction with regeneration of plants from protoplasts and cell-suspension cultures.

In the southern corn leaf blight epidemic, the introduced character was a cytoplasmic factor for male

Fig. 1. Oat seedling resistant to Victoria blight obtained by screening a susceptible population with the pathotoxin victorin, a product of the pathogen *Helminthosporium victoriae*.

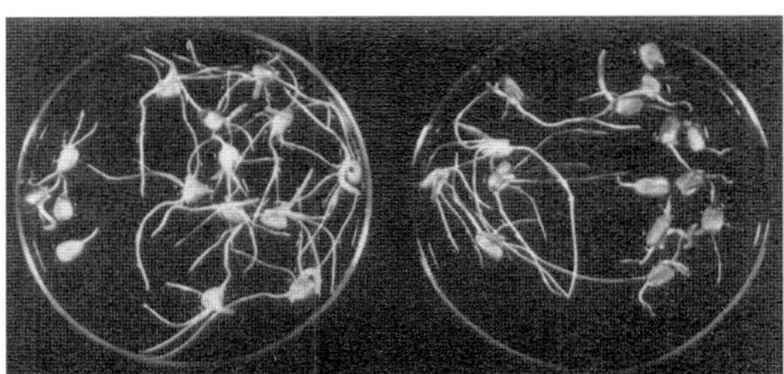

Fig. 2. Separation of corn seedlings resistant (N) and susceptible (T) to southern corn leaf blight by treatment with a pathotoxin produced by *Helminthosporium maydis* race T.

sterility, designated Tms for Texas male sterile, which made possible the production of hybrid corn seed without the tedious and costly process of mechanical detasseling. Southern corn leaf blight, caused by *H. maydis*, had long been known as a minor disease in the southern regions of the United States. In 1970, however, a new race of the fungus, *H. maydis* race T, highly virulent to plants with Tms cytoplasm, swept north into the Corn Belt and destroyed more than 1,000,000 acres (400,000 hectares) of the crop. Although producers of hybrid corn seed immediately reverted to mechanical detasseling, inadequate supplies of seed with resistant, normal cytoplasm made it necessary to plant much of the corn acreage in 1971 with varying mixtures of seeds with normal and Tms cytoplasm. The selective pathotoxin produced by *H. maydis* race T provided an efficient agent for testing thousands of lots of seeds to verify the proportions of the two cytoplasms in mixtures (**Fig. 2**).

Chemistry and mode of action. Most chemically characterized pathotoxins are nonselective, and some (the gibberellins, for example) function as plant growth regulators. The gibberellins were discovered in Japan during studies of the bakanae or "foolish seedling" disease of rice, caused by the fungus *Gibberella fujikuroi*. In culture this fungus produces gibberellins, and plants infected by the fungus have excessively long internodes, which typify the effects of this group of plant hormones. Fusicoccin (**Fig. 3***a*), produced by *Fusicoccum amygdali*, which causes a wilt disease of peach and almond trees, is another nonselective pathotoxin with growth-regulator properties. This chemical also causes stomates to open. Tentoxin (Fig. 3*b*), a product of *Alternaria alternata*, is species-selective; it causes a striking variegated chlorosis in cucumber, cotton, lettuce, and many other sensitive plants, but has no effect on tobacco, radish, willow, or other insensitive species. Tentoxin binds specifically to coupling factor 1 in chloroplasts of sensitive species, and presumably acts by disrupting energy generation in the photophosphorylative electron transport system. *See* GIBBERELLIN.

The only selective pathotoxins fully characterized chemically are the AM-toxins (Fig. 3*c*), produced by *A. mali*, which attacks apples. AM-toxin I has been reported to disrupt the cellular permeability of susceptible apple tissues at a concentration of $10^{-9}M$, whereas concentrations 10,000 times higher were required to produce similar effects on resistant tissues. HMT-toxin (Fig. 3*d*), produced by *H. maydis* race T, causes rapid swelling and uncoupling of electron transport in mitochondria from Tms corn plants, but has no such effects on mitochondria from disease-resistant plants with normal cytoplasm. These results suggest that HMT-toxin disrupts energy generation in the mitochondrial system, and such a site of action would account for the cytoplasmic inheritance of resistance to southern corn leaf blight.

Extensive attempts to purify other pathotoxins, especially victorin, have been frustrated by instability

(a)

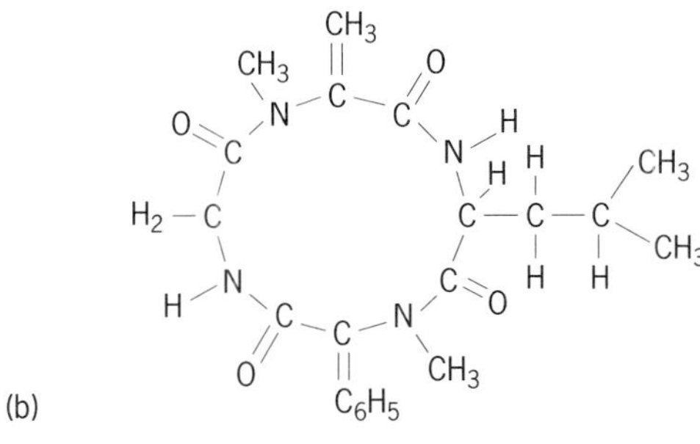

(b)

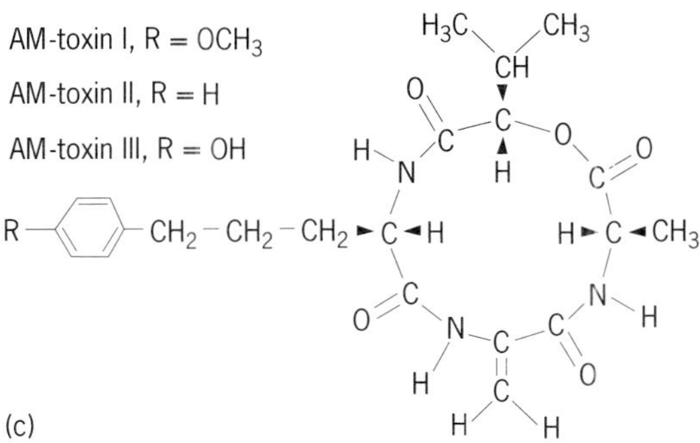

(c)

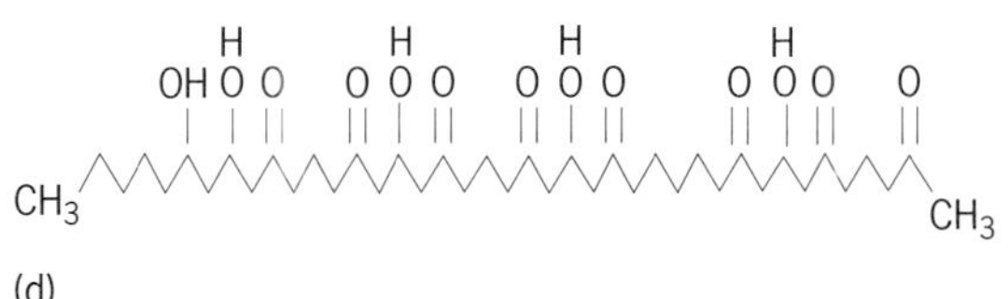

(d)

Fig. 3. Chemical structures of four pathotoxins. (*a*) Fusicoccin. (*b*) Tentoxin. (*c*) AM-toxin. (*d*) HMT-toxin.

of the toxin, but these, like the AM-toxins, rapidly disrupt cell permeability. Such disruption would result in the mixing of cell contents, and this in turn would block energy generation through electron transport systems. It thus appears that pathotoxins interfere with the energy generation required for cell maintenance and repair, either by effects on specific energy-generating sites or by disruption of compartmentalizing membranes. *See* PLANT PATHOLOGY; TOXIN.

Harry Wheeler

Bibliography. R. D. Durbin (ed.), *Toxins in Plant Disease*, 1981.

Pattern formation (biology)

The mechanisms that ensure that particular cell types differentiate in the correct location within the embryo and that the layers of cells bend and grow in the correct relative positions. Pattern formation is one of four processes that underlie development, the others being growth, cell diversification, and morphogenesis. Pattern formation can more easily be understood by first briefly clarifying the importance of the other three processes.

Growth refers to an increase in size of a tissue, organ, or body. It may occur through any of a number of events, such as increase in cell number, increase in cell size, or production of extracellular material by a cell. Cell diversification occurs throughout development. At the outset there is only one cell, the fertilized egg, but as development proceeds new cell types with different characteristics emerge. In the human body, for example, there are over 200 identified cell types, each with a precise function and location. The challenge in the study of cell diversification in an embryo is to establish how these cells become different even though they all have the same deoxyribonucleic acid (DNA) complement. *See* ANIMAL GROWTH; CELL DIFFERENTIATION; PLANT GROWTH.

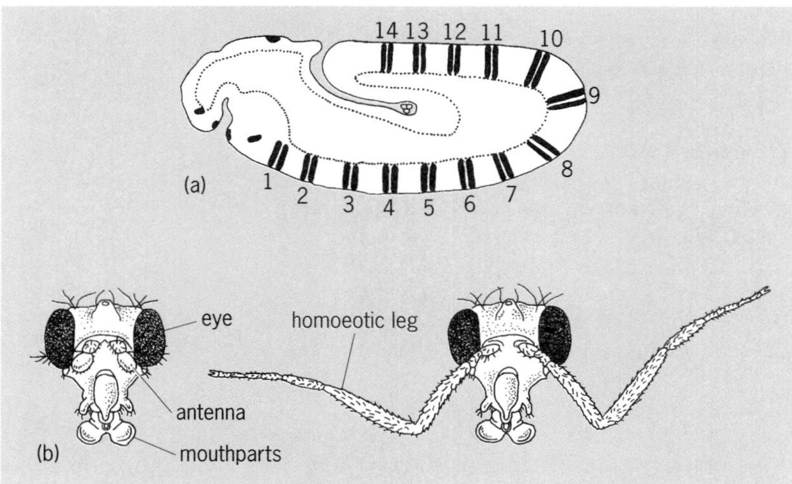

Segmentation and mutagenesis in *Drosophila*. (*a*) Tissue section of the developing *Drosophila* larva showing 14 segments as revealed with antibodies to proteins produced by segment polarity genes. Two such gene products are shown, engrailed (spots) and wingless (stripes). (*b*) Once segment number has been established, homeotic genes act to give each segment a character. Mutations in such genes alter the character of given segments, turning one body part into another. In this example, a mutant (antennapaedia) has converted the antennae into legs.

Morphogenesis is the creation of form, that is, the molding of a tissue into a recognizable and precise shape. During development, many cells migrate from one part of the embryo to another, and sheets or layers of cells bend to form pockets, bulges, thickenings, and grooves. Morphogenesis involves cells sticking together or breaking away, changing shape, dividing, and making extracellular material. *See* ANIMAL MORPHOGENESIS; PLANT MORPHOGENESIS.

Cell arrangement. Pattern formation is the creation of a predictable arrangement of cell types in space during embryonic development. The central importance of the study of pattern formation to the understanding of how embryonic development proceeds can be understood by comparing the cell types that are present in a human arm and a human leg. Although the two limbs are made up of identical sets of cell types, such as skin cells, bone cells, muscle cells, nerve cells, and blood cells, anyone can recognize an arm as distinct from a leg. The difference is not what cell types are present but how they are arranged in space relative to one another. Thus, arms and legs have five digits—but each finger is a different shape and length, and fingers are easily distinguished from toes.

Types of patterns. The types of patterns of cell types found in animals and plants can be conveniently described as simple or complex.

Simple patterns. Simple patterns involve the spatial arrangement of identical or equivalent structures such as bristles on the leg of a fly, hairs on a person's head, or leaves on a plant. Such equivalent patterns are thought to be produced by mechanisms that are the same or very similar in the fly and the plant. The glands in the growing skin of a salamander, for example, are found in a regular hexagonal array, and new glands emerge in the pattern only when the older glands move farther apart than a certain threshold distance in the growing animal. This suggests that the glands themselves produce an inhibitor of new gland development that diffuses into the skin all around the gland. Only when the glands become separated by a prescribed distance in the growing skin does the level of the inhibitor fall low enough to allow new gland formation. The same lateral inhibition type of mechanism could easily explain how leaf patterns develop at the growing apices of plants. The exact nature, timing, and distribution of an inhibitor would determine the exact pattern that leaves have one to another. Such arrangements, called phyllotactic patterns, may be highly varied, from spiral to alternate.

Complex patterns. Complex patterns are those that are made up of parts that are not equivalent to one another. In the vertebrate limb, for example, the structure of the arm is different at each level, with one bone (humerus) in the upper arm, two bones (radius and ulna) in the lower arm, and a complex set of bones making up the wrist and the hand. The theoretical framework that allows a basis for understanding how nonequivalent parts are patterned during development is called positional information.

Two stages exist in the positional information framework. First, a cell must become aware of its position within a developing group, or field, of cells. This specification of cellular position requires a mechanism by which each cell within a field can obtain a unique value or address. The second component is the interpretation of the positional address by a cell to manifest a particular cell type by the expression of a particular set of genes. Positional specification mechanisms have been examined in many different systems ranging from vertebrate limbs to insect larvae and coelenterate polyps such as *Hydra*. In each of these diverse animal systems, common principles of positional specification emerge.

Segmentation pattern in Drosophila. A similar principle is seen in the larva of *Drosophila*. The advantage of this system is that genetic knowledge of the fruit fly *Drosophila* is more comprehensive than for any other animal, and mutants have been used in analyzing the positional specification mechanisms. Genetic analysis, both in terms of looking for mutants that show interesting pattern alterations and in terms of molecular genetics and the power of techniques for manipulating deoxyribonucleic acid (DNA), have led to the unraveling of the early stages of pattern formation in this animal. The *Drosophila* larva is segmented and has three regions: head, thorax, and abdomen. The head is derived from a fused set of segments, the thorax has three segments, and the abdomen has eight. Each segment is different, and so it is of interest to determine how this pattern is brought about during development. The explanation is complex but can be summarized in five stages, each involving a particular set of genes. The first four stages involve positional specification and the creation of 14 segments in the larva. The final stage involves the interpretation of positional signals created at earlier stages and is the stage at which the segments differentiate themselves from each other.

The first stage involves the establishment of gradients of morphogens in the newly laid insect egg. In *Drosophila*, there are two gradients of proteins in the egg, one with its high point at the anterior end, and one at the posterior end. Once established, the gradients must be interpreted to establish a segmental pattern. Initially, seven segments, which appear to be equivalent at the outset, form within the embryo. This development involves two sets of genes, called gap genes and pair rule genes. There are a number of gap genes, each one turned on in response to maternally derived gradients. If these genes are mutated, they create large gaps in the larval body. The pair rule genes, of which there are also a number, express in response to the gap gene products. The products of the pair rule genes appear as seven stripes down the length of the embryo from the head to the tip of the abdomen. The exact position of the stripes for each pair rule gene is different, and between them they cover the full length of the larva. The 14 stripes eventually become 14 segments with the expression of the fourth group, the segment polarity genes. The stripes can be visualized by using antibodies to the protein product of the genes (see **illus.**). Finally, with the 14 segments established, the function of the fifth group of genes becomes obvious. These are called homeotic genes, and they are concerned with giving each segment an anatomical character of its own. Mutants in the homeotic genes, therefore, may be very dramatic, turning one region of a larva into another. Because the larva will metamorphose into a fly, these mutants may effect the final anatomy of the fly. Many examples of these homeotic mutations exist. *See* EMBRYONIC DIFFERENTIATION; GENE ACTION.

Pattern formation genes. It seems that the two-step positional information framework is correct in principle, but the interpretation steps following creation of initial spatially regulated patterns of gene expression are very complex and involve a series of different gene classes. Perhaps the most surprising realization is that the patterning mechanisms are largely universal in animals. *See* DEVELOPMENTAL BIOLOGY.

Nigel Holder

Bibliography. M. Fougerean and R. Stora (eds.), *Cellular and Molecular Aspects of Developmental Biology*, 1986; A. F. Hopper and N. H. Hart, *Foundations of Animal Development*, 2d ed., 1985; G. Malacinski, *Pattern Formation*, 1984; V. Walbot and N. Holder, *Developmental Biology*, 1987; C. C. Wylie (ed.), *Determinative Mechanisms in Early Development*, 1986.

Pauropoda

A class, and perhaps the most obscure group, of the Myriapoda. They are pale creatures, no more than 0.04–0.08 in. (1–2 mm) in length, inhabiting damp situations in leaf litter, under bark, stones and debris, and in humus and similar detritus. Their size and retiring habits suggest rarity; this, however, has been refuted by later studies that show them to occur in huge numbers in suitable habitats. Apparently very widely distributed as a class, they have been undiscovered only in deserts and in the arctic and antarctic regions. Like millipedes, they are progoneate and have one pair of maxillae, and their trunk segments display a certain degree of amalgamation manifest in the tendency to form diplotergites, but not diplosegments. Their peculiar bifurcate antennae and adult complement of 12 trunk segments with nine pairs of functional legs are distinctive within the myriapod complex (see **illus.**). All pauropods lack eyes, spiracles, tracheae, and a circulatory system.

What they eat and how they live is virtually unknown. They probably subsist largely or entirely upon decaying plant and animal matter; some may consume microscopic animals. Studies based upon a few forms reveal an anamorphic ontogeny similar to that of diplopods. The young pauropod hatches with three pairs of legs, then molts successively until the adult complement of nine is attained.

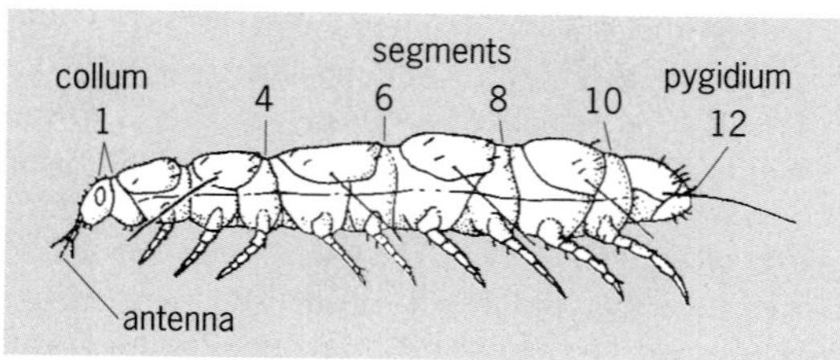

***Pauropus silvaticus*, fully extended adult, enlarged. (*After R. E. Snodgrass, A Textbook of Arthropod Anatomy, Cornell University Press, 1952*)**

The class currently consists of two families with less than 10 genera; there are probably fewer than 60 species known. *See* MANDIBULATA.

Ralph E. Crabill

Pavement

An artificial surface laid over the ground to facilitate travel. In this article only road pavements are discussed. The engineering involved is closely similar to that for airport surfacing, another major type of pavement. A pavement's ability to support loads depends primarily upon the magnitude of the load, how often it is applied, the supporting power of the soil underneath, and the type and thickness of the pavement structure. Before the necessary thickness of a pavement can be calculated, the volume, type, and weight of the traffic (the traffic load) and the physical characteristics of the underlying soil must be determined. *See* AIRPORT ENGINEERING.

Traffic load. Traffic data can readily be obtained. Traffic is counted to learn the total volume and the proportion of heavy vehicles. Loadometer scales are placed next to the road and trucks are weighed, front and rear wheels separately. Traffic trends are studied. Such data provide a basis for estimating the total volume of traffic to be carried by a pavement during its service life. They also permit an estimate of the magnitude and frequency of the expected load.

Base and surface courses of pavements are designed to withstand many applications of load over a prolonged period, in some cases up to 30 years. In structural design it is also necessary to give consideration to the direction of traffic. For example, a pavement from a mine to a nearby railroad siding may carry a high percentage of heavily loaded trucks on the inbound lanes and a low percentage on the return lanes.

In general, the larger the volume of heavy vehicles on a highway, the greater the structural capacity required in the pavement. But equally as important as the volume of heavy vehicles are the magnitude of the applied loads and the conditions that will influence the effect of those loads on the pavement. Under the action of vehicular traffic, the surface of a pavement is subjected to a series of highly concentrated forces applied through the wheels of the vehicle. These forces exert an influence throughout the depth of the pavement.

The applied loads vary considerably depending upon the number and spacing of the wheels of each vehicle, the gross weight of the vehicle, and the distribution of that weight among the axles. Two vehicles of the same gross weight may differ widely in the wheel loads applied to the highway. A relatively small truck may cause a load of higher unit stress than a larger vehicle with larger tires and more axles. Consideration of the actual wheel loads has become increasingly important with the large increase in the volume of heavy vehicles on most highways.

All highways, regardless of their design, have some limit to their ability to support the frequent application of a heavy load. A large number of load applications can be supported by a given pavement if the load does not exceed a particular magnitude, but once this magnitude is exceeded, distress and failure of the pavement becomes increasingly evident. Weather also influences the ability of the pavement to support a load. Definite load limitations are imposed by law on many highways. Further limitations are often imposed on specific highways of lighter design, especially during such conditions as spring freezing and thawing.

In the preceding discussion wheel loads were treated as static loads. For adequate design analysis the effects of dynamic loads must be evaluated. The vertical force exerted on pavement by a moving wheel may be considered to be the sum of the static weight of the wheel and the impact or dynamic force from the wheel's movement over irregularities in the pavement surface. The static load is a constant factor except as the movement of the vehicle along the highway sways or oscillates the load.

An exceedingly variable factor, the dynamic force generated by a moving wheel depends upon (1) the magnitude of the static wheel load; (2) the operating speed of the vehicle; (3) the type, size, and cushioning properties of the tire equipment; and (4) the smoothness of the pavement. An increase in static wheel load or pavement roughness, a decrease in the cushioning qualities of the tires, and, within limits, an increase in vehicle speed all result in increased dynamic forces.

The importance of the foregoing variables and factors has long been recognized by design engineers. The difficulty has been to express the data in terms that could be rationally applied to design formula and then correlated with the performance of foundation soils and materials used to construct the pavement. The problem has not been given a rigorous mathematical solution but rather has depended largely upon field observations under actual operating conditions.

Several methods of load evaluation have been developed and used by various highway agencies and organizations. All of the methods are more or less empirical in approach and thus are subject to revision and adjustment when field observations indicate changing conditions of traffic, climate, or soil performance. No universal method of load evaluation has been developed so far and indeed cannot be until a method is devised that can be readily adjusted when other factors influencing the performance of a pavement structure change.

Methods of evaluation in use include (1) numerical count method, in which the number of vehicles using a particular highway is actually counted and the weight of various vehicles listed as light, medium, heavy, and extra heavy; (2) wheel load method, in which factors based upon the actual weight of the wheel load are used; (3) load frequency method, in which the wheel load weights are combined with the volumetric count of the commercial vehicles; and (4) equivalent wheel load method, in which destructive effects of the actual wheel loads being applied to a pavement are expressed in terms of a standard wheel load. *See* LOADS, DYNAMIC.

Evaluation of subgrade. Factors that must be considered in evaluating the ability of the underlying soil or subgrade to support the pavement include (1) type of soil, such as loam or clay; (2) gradation and variation in particle size; (3) strength or bearing value; (4) modulus of deformation; and (5) swell or volume change characteristics and related properties. Measurement of the supporting power of the subgrade presents numerous difficulties. Tests sometimes used include the plate bearing test, the direct shear method, the triaxial compression test, and the bearing ratio procedure. *See* SOIL MECHANICS.

Some soil types are unsuitable for supporting pavements because they have low bearing values or undergo changes in volume with variation in moisture content. For example, it is desirable to excavate peat and muck along the road and replace it with soil of higher bearing value.

Rigid and flexible pavements. Once the grading operation has been completed and the subgrade compacted, construction of the pavement can begin. Pavements are either flexible or rigid. Flexible pavements have less resistance to bending than do rigid pavements. Both types can be designed to withstand heavy traffic. Selection of the type of pavement depends among other things, upon (1) estimated construction costs; (2) experience of the highway agency doing the work with each of the two types; (3) availability of contractors experienced in building each type; (4) anticipated yearly maintenance costs; and (5) experience of the owner in maintenance of each type.

Flexible pavements. Flexible paving mixtures are composed of aggregate (sand, gravel, or crushed stone) and bituminous material. The latter consists of asphalt products, which are obtained from natural asphalt products or are produced from petroleum; and tar products, which are secured in the manufacture of gas or coke from bituminous coal or in the manufacture of carbureted water-gas from petroleum distillates. Structural strength of a bituminous pavement is almost wholly dependent upon the aggregate, which constitutes a high percentage of the volume of the mixture and forms the structure that carries the wheel load stresses to the base layers. The bituminous material cements the aggregate particles into a compact mass with enough plasticity to absorb shock and jar; it also fills the voids in the aggregate, waterproofing the pavement.

Among the many types of bituminous surface that are used are surface treatments, penetration macadams, road mixes, and plant mixes, as well as variations of these. With surface treatment a liquid bituminous material is applied over a previously prepared aggregate base.

In building penetration-macadam pavement, a base, usually of crushed stone or gravel, is constructed in layers and firmly compacted by rolling. Often during the rolling, water is applied to make what is termed a water-bound base, or macadam. Then, the keyed and wedged fragments in the upper portion of the base are bonded in place by working alternate applications of bituminous material and choke stone into the surface voids.

With road-mix designs a base is constructed, and a layer of aggregate and bituminous material, mixed on the road with a motor grader, is then uniformly spread and compacted with rollers.

When a plant-mix design is used, construction of a base is also necessary, but the aggregate and bituminous materials are mixed at a central plant, trucked to the job, and then spread or placed with a paving machine and compacted. With the plant-mix procedure more accurate mixing is possible.

For flexible pavements designed to support heavy loads a subbase built of materials similar to those of the base but of poorer quality may also be used. Thickness of the wearing surface, the base, and the subbase depends upon the design load. A typical flexible pavement design for light loads or a 5-ton (4.5 metric ton) axle loading is shown in **Fig. 1**. Where a flexible pavement is designed to carry a large traffic volume, for example, 2000–5000 vehicles per day, including 150–300 heavy commercial vehicles, the gravel subbase would be increased

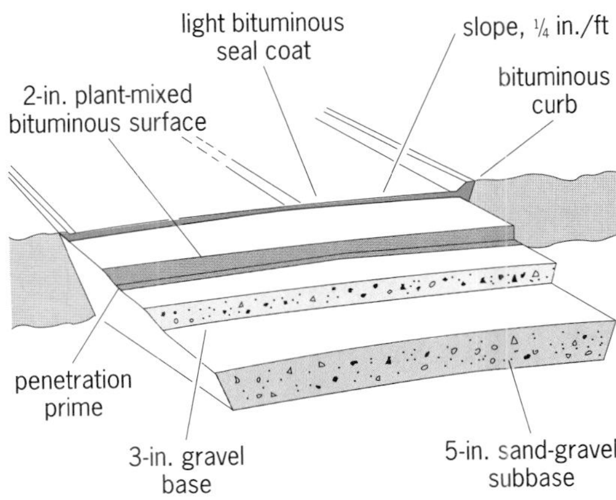

Fig. 1. Flexible pavement design for a city collector street with maximum traffic load of 5 tons (4.5 metric tons) per axle. Right-of-way is 60 ft (18 m) wide and the pavement width is 38 ft (12 m). Berms or boulevards at the sides are sloped in order to drain toward the street. 1 in. = 2.5 cm; 1/4 in./ft = 2 cm/m.

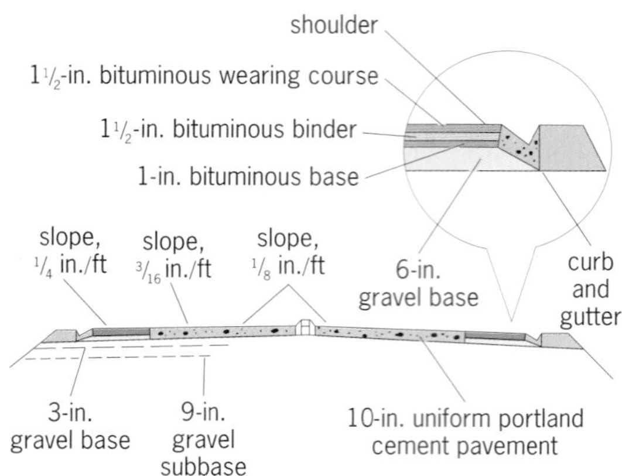

Fig. 2. Rigid main roadway with flexible shoulders. The design is for maximum loads of 9 tons (8 metric tons) per axle. 1 in. = 2.5 cm. 1 in./ft = 8 cm/m.

to 10 in. (2.5 cm), the gravel base to 5 in. (13 cm), and the plant-mixed bituminous surfacing to 4 in. (10 cm). It is assumed that the maximum wheel load would be restricted to 9 tons (8 metric tons) per axle. Some flexible pavement designs have called for mixing asphalt with the gravel base material to provide greater pavement strength. Bases treated in this manner are termed bituminous stabilized bases.

Rigid pavements. Coarse aggregate, fine aggregate, and portland cement are mixed with clean acid-free water to produce the concrete used for rigid pavements. The coarse aggregate may consist of coarse gravel or crushed stone and the fine aggregate of sand or crushed-stone screenings. The thickness of the pavement slab may vary from 6 in. (15 cm) for light traffic to 18 in. (45 cm) or more for airport pavements accommodating heavy aircraft. A layer of granular material such as sand, sandy gravel, or slag is generally used as a subbase under the concrete slab to prevent frost heave and to increase the supporting power of the underlying soil. A rigid pavement is shown in cross section in **Fig. 2**.

No steel reinforcement is used with bituminous or flexible pavements. With rigid pavements, especially those designed for heavy loads, reinforcement is often used to strengthen the pavement and to prevent cracking. The reinforcement may consist of welded wire fabric or bar mats assembled by tying transverse and longitudinal steel rods together at their point of intersection (**Fig. 3**). The reinforcement is usually placed about 2 in. (5 cm) below the upper surface of the concrete slab. *See* CONCRETE.

In constructing rigid pavement a longitudinal

Fig. 3. A rigid pavement reinforced by welded steel wire mesh. Underneath the mesh are 6 in. (15 cm) of concrete. A 3-in. (8-cm) layer of concrete is being laid on top.

control joint is used between adjacent lanes. Transverse joints, such as expansion and contraction joints to prevent cracking of the pavement when the temperature changes, may also be included. With flexible pavement no joints are used. *See* HIGHWAY ENGINEERING. Archie N. Carter

Bibliography. E. R. Brown, *New Pavement Materials*, 1988; S. Brown (ed.), *Pavement Management Systems*, 1993; R. Haas, W. R. Hudson, and J. P. Zaniewski, *Modern Pavement Management*, 2d ed., 1994; Y. H. Huang, *Pavement Analysis and Design*, 1992; Society of Automotive Engineers, *Vehicle-Pavement Interaction: Where the Truck Meets the Road*, 1988; A. F. Stock (ed.), *Concrete Pavements*, 1988; P. Ulliditz, *Pavement Analysis*, 1988; E. J. Yoder and M. W. Witczak, *Principles of Pavement Design*, 2d ed., 1975.

Pawl

The driving link or holding link of a ratchet mechanism, also called a click or detent. In **Fig. 1** the driving pawl at *A*, forced upward by lever *B*, engages the teeth of the ratchet wheel and rotates it counterclockwise. Holding pawl *C* prevents clockwise rotation of the wheel when the pawl at *A* is making its return stroke. Pawl and ratchet are an open, upper pair.

Driving and holding pawls likewise engage rack teeth on the plunger of a ratchet lifting jack, such as those supplied with automobiles. A ratchet wheel with a holding pawl only, acting as a safety brake, is fastened to the drum of a capstan, winch, or other powered hoisting device.

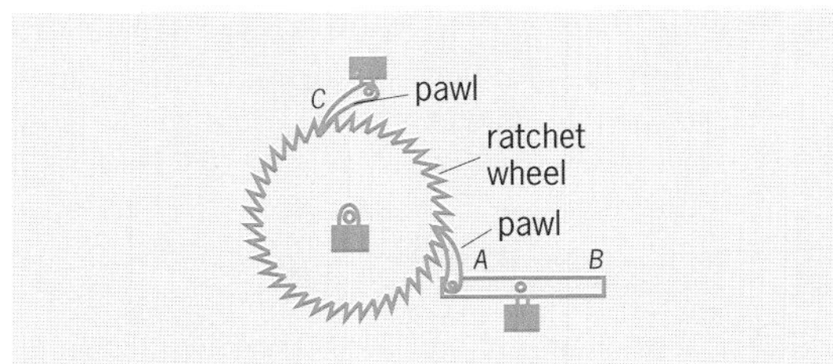

Fig. 1. Holding and driving pawls with a ratchet wheel.

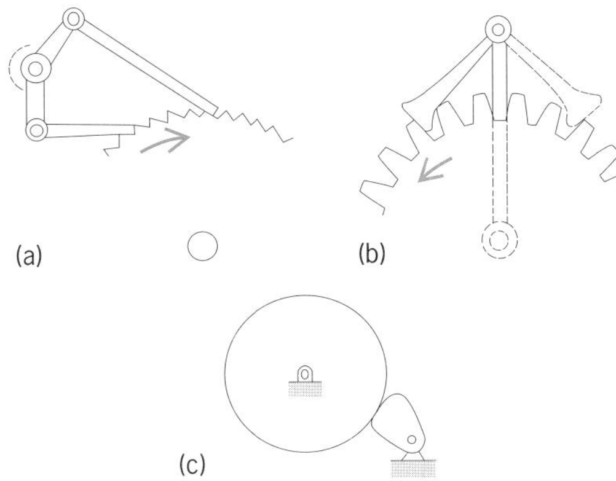

Fig. 2. Some pawl applications. (*a*) Double-acting pawl. (*b*) Reversible pawl. (*c*) Cam pawl.

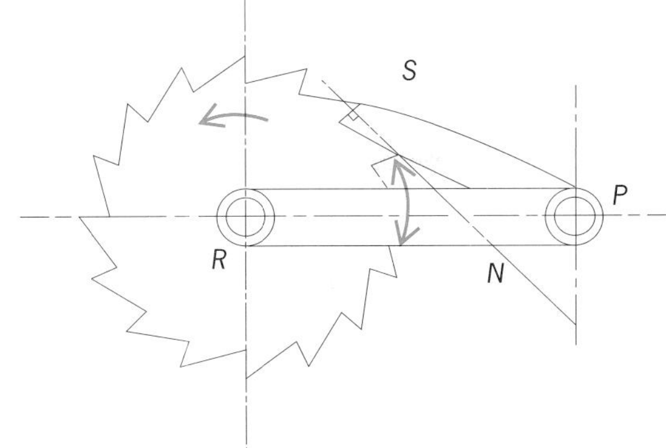

Fig. 3. Essential geometry of pawl design.

A double pawl can drive in either direction (**Fig. 2***a*) or be easily reversed in holding (Fig. 2*b*). A cam pawl (Fig. 2*c*) prevents the wheel from turning clockwise by a wedging action while permitting free counterclockwise rotation. This technique is used in the automobile hill-holder to prevent the vehicle's rolling backward.

The spacing mechanism of a typewriter, although frequently termed an escapement, is properly a ratchet device in which a holding pawl is withdrawn from a spring-loaded rack to allow movement of the carriage, while an arresting pawl is introduced momentarily to permit the holding pawl to engage the next tooth of the rack.

In designing a pawl, care should be taken that the line of contact has a normal *N* passing between centers *R* and *P* (**Fig. 3**); otherwise the pawl will ride out of ratchet step *S*. The pallet of an escapement mechanism is closely related to the pawl. *See* ESCAPEMENT; RATCHET. Douglas P. Adams

Bibliography. H. H. Mabie and F. W. Ocvirk, *Mechanisms and Dynamics of Machinery*, 4th ed., 1987; W. J. Patton, *Kinematics*, 1979; C. E. Wilson et al., *Kinematics and Dynamics of Machines*, 2d ed., 1992.

Paxillosida

An order of sea stars and members of the class Asteroidea. The name is derived from the club-shaped plates, or paxillae, that form the sea star's upper skeletal surface and that have tiny spinelets or granules covering their tips (see **illus.**). Paxillosida encompasses six families, the largest being the Astropectinidae, Luidiidae, and Porcellanasteridae. The Ctenodiscidae, Goniopectinidae, and Radiasteridae are represented by comparatively few members. Astropectinids and luidiids are primarily predators of mollusks and other echinoderms. The former are found over a wide range of depths, whereas the latter live in relatively shallow water. Porcellanasterids are deep-water asteroids that swallow sediment in

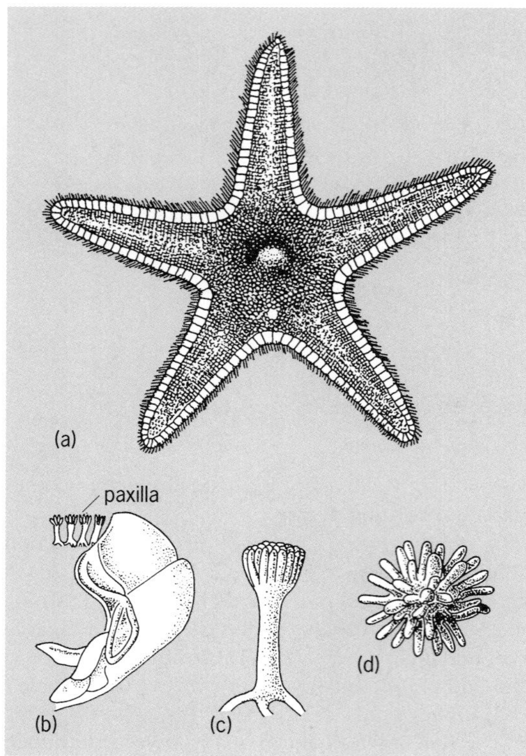

Paxillosida. (*a*) Complete astropecten, a typical paxillosid (*after R. Koehler, Echinodermes: Resultats des Campagnes Scientifiques accomplies sur son yacht par Albert Ier Prince Souverain de Monaco, Imprimerie de Monaco, 1909*). (*b*) Arm arrangement, enlarged to show paxillae; (*c*) paxilla in side view and (*d*) top view (*after W. K. Fisher, Starfishes of the Philippine Seas and Adjacent Waters, Bull. U.S. Nat. Mus. 110, vol. 3, 1919*).

bulk as they bury themselves to a level just below the surface.

The geologic range of astropectinids extends from Jurassic to Recent; luidiids are known from Miocene to Recent. The Ctenodiscidae are primarily Recent, but a single Cretaceous specimen is known. The other families have not yet been recognized from the fossil record. Asteroids are not easily preserved, and so paxillosidans almost certainly were more common in the past than their fossil record suggests.

In addition to the presence of paxillae, paxillosidans are characterized by a number of unusual features that have been thought to indicate a primitive phylogenetic position among living asteroids. Although the tube feet of most asteroids have suckered disks, those of most paxillosidans are pointed. The digestive system of most asteroids is relatively complex and complete, terminating in an anus, whereas that of the paxillosidans is simple, saclike, and lacking an anus in some members. Most asteroids can extrude their stomach during feeding, but that ability is limited in paxillosidans. A brachiolarian larval stage has been recognized in the development of most asteroids, yet that stage is thought to be absent from paxillosidans.

Research has demonstrated that both suckered tube feet and an anus are present initially but are lost during the ontogeny of certain paxillosidans. Some researchers believe a reduced brachiolarian stage can be recognized as well. Loss of structures during development implies evolutionary reduction rather than primitive organization, and such changes probably reflect an evolutionary adaptation to habit and habitat. Paxillosidans live on unconsolidated sandy or muddy substrates, and they feed by ingesting sediment or shells. Because the large sac-like gut holds shells that are too large to be expelled through an anus and that must be ejected through the mouth, the anus was lost from some. Most paxillosidans bury themselves, a unique ability among the asteroids, so that the paxillae protect the aboral surface from sediment. The brachiolaria is an attached stage that is not suited to the soft, shifting substrates typical of paxillosidans, as is also true of suckered, clinging tube feet. Certain predatory paxillosidans feed at dawn and dusk, then bury themselves in the sediment during digestion. The threat of larger predators, such as fish, may have stimulated evolution of the relatively rapid feeding and hiding behavior. *See* ASTEROIDEA; ECHINODERMATA.

Daniel B. Blake

Bibliography. D. B. Blake, A classification and phylogeny of post-Paleozoic sea stars (Asteroidea: Echinodermata), *J. Nat. Hist.*, 21:481–528, 1987; D. B. Blake, Somasteroidea, Asteroidea, and the affinities of *Luidia (Platasterias) latiradiata*, *Palaeontology*, 25:167–191, 1982; R. D. Burke et al. (eds.), *Echinoderm Biology*, 1988.

Pea

The pea is one of the oldest cultivated crops. It is a native to western Asia from the Mediterranean Sea to the Himalaya Mountains. It appears to have been carried to Europe as early as the time of the lake dwellers of prehistoric times. Peas were introduced into China from Persia about A.D. 400; they were introduced into the United States in very early Colonial days.

Description. Garden peas (*Pisum sativum*; **Fig. 1**) have wrinkled seed coats at maturity when dry; field peas (*P. arvense*) have a smooth seed coat. Both types are annual leafy plants with stems 1–5 ft (0.3–1.5 m) long. Each leaf bears three pairs of leaflets and ends in a slender tendril. The blossoms are reddish-purple, pink, or white, with two or three on each flower stalk. Five to nine round seeds are enclosed in a pod about 3 in. (8 cm) long. Seed color varies from white to cream, green, yellow, or brown. Smooth-seeded varieties may be harvested fresh for freezing or canning, or harvested dry as edible peas. Dry peas may be split or ground and prepared in various ways, such as for split-pea soup. *See* LEGUME.

Culture. Peas require a cool growing season. They should be sown in early spring as soon as a fine, firm seedbed can be prepared. Early-seeded peas develop before the heat of early summer can harm them and usually produce higher yields. Peas are seeded 2–3 in. (5–8 cm) deep. Then the fields are usually rolled

Fig. 1. Typical wrinkled-seed type of pea, showing fruit (pod) containing seeds (peas). (*Dumas Seed Co., Inc.*)

immediately to firm the seedbed and aid germination. In the Palouse area of Washington and Idaho, peas are usually planted in early April, while elsewhere irrigated fields are often seeded in March. Garden peas may be seeded as early as February in home gardens in the southern half of the United States.

Applications of lime, sulfur, molybdenum, phosphorus, and potassium are commonly made for pea production.

Harvesting. The peas are mowed and windrowed with a swather. Self-propelled viners, which have replaced stationary viners, pick up the windrowed crop and separate the fresh green peas from the vines and pods. To maintain high quality, this must be done within 2 h after the peas have been mowed. *See* AGRICULTURAL SOIL AND CROP PRACTICES.

A tenderometer is used to determine when green peas for processing should be harvested. Dry peas are harvested with a regular grain combine when the moisture content in the seed drops below 14%. Adjustments are made to minimize damage to the pea seed.

The Grain Division of the Agricultural Marketing Service, U.S. Department of Agriculture, has set standards and issues a grade certificate to define the relative market value of different-quality dry peas.

Kenneth J. Morrison

Production. Wisconsin, Washington, Minnesota, Oregon, Illinois, New York, Pennsylvania, Utah, and Idaho are important in the production of peas harvested green. In dry- and seed-pea production the leading states are Washington, Idaho, Oregon, California, Colorado, Montana, North Dakota, Wyoming, and Minnesota. Seed for all garden varieties is grown primarily in the same areas that grow the dry commercial peas. Washington and Idaho produce most of the total seed and dry commercial pea crop. The two main export markets for dry peas are Europe and Latin America. The European market is somewhat sporadic, depending on the total production of Europe's usually large acreage of peas.

William A. Beachell

Diseases. Diseases are often major limiting factors to acceptable yields in many pea production areas. Causes of major pea diseases include bacteria, fungi, and viruses; nonparasitic problems and nematode-induced diseases are generally of less consequence.

Bacterial diseases. Bacterial blight of pea is caused primarily by *Pseudomonas pisi*, occasionally by *P. syringae*. Symptoms produced by these seed- and rain-borne pathogens are very similar, consisting first of small, often angular, water-soaked spots on leaves (**Fig. 2**) and stipules. When these lesions are numerous, they often coalesce into large, brown, dead areas covering major portions of the foliage, which then dries prematurely to give a blighted appearance. Pods develop water-soaked, slightly sunken lesions that turn brown in 6–10 days. Severely diseased plants have peas which are greatly reduced in yield and quality.

Fungal diseases. The main fungal diseases of peas are root rots, wilts, foliage blights, and mildews. Root rots are the major problem, especially in the United States' most important pea-producing area—the Midwest. They may be caused by *Aphanomyces euteiches*, *Fusarium solani* f. sp. *pisi*, *Pythium* spp., *Ascochyta pinodella*, and *Rhizoctonia solani*. *Aphanomyces euteiches* causes greater economic loss than all of the others combined. This is because no effective control has been found, even though the Aphanomyces root rot problem has existed since 1925. Warm and wet soils favor the disease. Highly infested fields can be identified by laboratory techniques and can be eliminated from production. These root rot pathogens live naturally in the soil and in the roots of crops other than peas; thus rotation is of limited benefit, and root rots are a recurrent

Fig. 2. Bacterial blight lesions on pea leaf.

problem. Generally roots show various amounts of brownish discoloration and decay. When severe, the root system may be largely destroyed (**Fig. 3**) and the stunted plants turn yellow and die prematurely.

Two widespread wilt diseases of pea are caused by *Fusarium oxysporum* f. sp. *pisi* race 1 and race 2. Race 5 of the same fungus causes a locally important wilt in Washington. These soil organisms become established in bothersome numbers due to too many pea crops. They attack the roots and grow through the water-conducting vessels up into the stem, interfering with the passage of water to the leaves. Affected plants wilt, sometimes unilaterally (**Fig. 4**), the lower leaves turn yellow, and the plants die in an early stage of development or soon after flowering.

Fungus-induced foliage blights include four diseases of variable occurrence and severity, most of them being encouraged by contaminated seed, too many consecutive pea crops, and rainy weather. Ascochyta leaf and pod spot is caused by *Ascochyta pisi*. A similar disease, Mycosphaerella blight, is caused by *Mycosphaerella pinodes*. *Colletotrichum pisi* causes pea anthracnose. All three pathogens cause reddish-brown spots of various sizes on the leaves, stipules, and pods. When leaf lesions are numerous, leaves shrivel and die and the plant appears blighted. Septoria blotch is caused by *Septoria pisi*, and is found primarily on lower, somewhat senescent leaves and stipules. It is characterized by large, yellow to straw-colored areas with indefinite margins.

Powdery mildew is widespread and often quite troublesome on late-maturing plants. It is caused by *Erysiphe polygoni*, and appears as a superficial whitish "powder" on foliage and pods. Dry weather favors its development. Downy mildew, caused by *Peronospora pisi*, produces tan spots of irregular size and shape on the upper leaf surfaces. A mouse-gray, cottony growth of the fungus develops on the undersides of the foliage immediately beneath the tan spots.

Fig. 4. Near-wilt of pea showing unilateral wilt.

Fig. 3. Aphanomyces root rot of pea (right); healthy root (left).

Viral disease. Viral diseases of peas are widespread and common, but are generally of minor consequence. Pea enation mosaic and pea streak are the most important, but pea stunt and seed-borne mosaic also occur. The causal viruses are aphid-borne, often from nearby perennial legumes. *See* PLANT VIRUSES AND VIROIDS.

Disease control. Many of the pea diseases mentioned can be controlled through the use of resistant cultivars. For example, there are cultivars resistant to one or more of the three pea wilt diseases, Ascochyta leaf and pod spot, powdery mildew, downy mildew, and both pea mosaics. Use of disease-free seed, crop rotation, field sanitation, and strict aphid control are helpful disease control measures. Chemical seed protectants are important, and fungicides may sometimes be needed for powdery mildew control. *See* PLANT PATHOLOGY.

D. J. Hagedorn

Bibliography. D. J. Hagedorn, *Compendium of Pea Diseases*, 1984.

Peach

A deciduous fruit tree species (*Prunus persica*) that originated and was first cultivated in western China. It is adapted to relatively moderate climates in the temperate zone. Although most peach cultivars require a substantial amount of winter chilling (temperatures between 32 and 45°F, or 0–7°C) to ensure adequate breaking of winter dormancy and uniform budbreak, peach wood is susceptible to winter injury at temperatures below −15°F (−25°C) and dormant fruit buds are injured by temperatures below 0°F (−18°C). Consequently, commercial cultivation is limited to lower latitudes in the temperate zone or to higher latitudes where large bodies of water have a moderating influence on climate. The principal peach-growing regions in North America, ranked in order of commercial production, are central California, Georgia and the Carolinas, the mid-Atlantic region, the Great Lakes region, and the Pacific northwestern region. Other important peach-growing regions in the world include Italy, southern France, Spain, Japan, China, Argentina, southern Brazil, Chile, South Africa, and southeastern Australia. *See* ROSALES.

Cultivars. Peach cultivars can vary greatly and are usually distinguished by their fruit types. Peach fruits are covered with short epidermal trichomes called fuzz (smooth-skinned peaches are called nectarines) and at maturity are usually yellow or white with a red blush. The internal flesh is also yellow or white. Clingstone cultivars have a relatively firm flesh that adheres to the pit at maturity, and are primarily used for canning. Some principal North American clingstone canning cultivars are Halford, Loadel, Andross, and Carolyn.

Freestones usually have a softer flesh that separates from the pit at fruit maturity, and are primarily used for the fresh market, freezing, and drying. There are more than 150 freestone peach cultivars grown commercially in the United States. The more important western varieties are Springcrest, O'Henry, Fay Elberta, and June Lady. Redhaven, Blake, and Redglobe are important eastern varieties, and June Gold is an important low-chilling variety grown in warm southern locations.

Genetic dwarf peach trees occur naturally; though there have been attempts to develop dwarf peach tree cultivars, none is commercially acceptable.

Propagation and cultivation. Peach cultivars are clonally propagated primarily by budding onto selected rootstocks. The most common rootstock in most of the United States is the Lovell cultivar; however, Nemaguard is commonly used in central California where soil nematodes are a problem, and Halford is often used in the Great Lakes region. Marianna and Myrobolan plum rootstocks are also used commercially. Peach trees grow best on well-drained sandy soils or gravelly loam soils but can also tolerate heavier, clay loams. Traditionally, trees are planted 16–22 ft (5–7 m) apart and trained to an open vase-shaped form with three to five main scaffold branches originating from the trunk at approximately 18–24 in. (0.5–0.8 m) from the ground. There has been a trend to plant trees at higher densities and in hedgerows, but these planting systems have been hampered by the excessive vigor of peach tree growth. Fruit is borne almost exclusively on 1-year-old extension shoots. Trees require extensive pruning to prevent overloading, maintain tree size and shape, and promote regrowth of new fruiting wood in lower areas of the tree. Most peach cultivars are self-pollinating, often set large numbers of fruit, and require fruit thinning by hand or with poles.

Harvesting. Fruit is harvested when its dark-green external coloring changes to a pale yellow or white. In some regions commercial fruit harvest extends from late spring to early fall. Mechanical harvesting is used on some processed fruit, but fruit for fresh market is hand-picked. Theodore M. De Jong

Diseases. Brown rot, caused by *Monilinia fructicola* and *M. laxa*, is a destructive fungus disease of the peach throughout the world. Blossoms, twigs, and fruit are infected. Major loss is from the decay of the fruit, which is converted into a soggy mass unfit for human consumption. Control is achieved by spraying or dusting the trees at regular intervals with powdered sulfur or with captan, an organic fungicide. For best results it is essential also to control the plum curculio, a common fruit insect, whose punctures facilitate entrance of the fungus into the fruit. *See* COLEOPTERA; FUNGISTAT AND FUNGICIDE.

In contrast to brown rot, which requires strenuous efforts for its control, peach scab (*Cladosporium carpophilum*), although universally present, is readily controlled by one or two sprays of sulfur shortly after the blossom petals drop. Leaf curl caused by *Taphrina deformans*, a leaf-distorting fungus, can be prevented by one spray of lime sulfur, bordeaux mixture, or ferbam applied before the buds begin to swell.

Bacterial spot, caused by *Xanthomonas pruni*, is a serious disease of peaches in the United States, China, Japan, and New Zealand. The bacteria kill small groups of cells in the leaves, twigs, and fruit. Infected leaves drop prematurely and this devitalizes the trees mainly through reduced photosynthesis, the most serious long-time effect of the disease. Diseased fruit is edible, but its appearance is marred and its market value reduced. The bacteria survive the winter in small cankers formed on the twigs. Control is difficult because of the prolonged infection period. A mixture of zinc sulfate and hydrated lime, applied as a spray at intervals of 10–14 days, reduces the severity of the infection in most years.

Other fungus diseases affecting peaches are rust caused by *Tranzschelia discolor*, peach blight caused by *Coryneum beijerinckii*, and mildew caused by *Sphaerotheca pannosa*. These diseases cause serious losses in various parts of the world. *Rhizopus nigricans*, occurring throughout the world, causes rapid decay of harvested and stored peaches.

Among the more than 20 virus diseases known to

affect peaches in the United States, yellows, rosette, phony peach, and mosaic are the most serious. The first two kill the trees in a few years, whereas phony peach and mosaic reduce both quality and quantity of fruit, and the trees eventually become worthless. *See* PLANT VIRUSES AND VIROIDS.

Peaches are also injuriously affected by nutrient deficiencies in the soil, the symptoms of which often resemble those resulting from disease organisms. Fluctuations in temperature during the winter, particularly in the southern states, frequently kill many peach trees. *See* FRUIT, TREE; PLANT PATHOLOGY. John C. Dunegan

Bibliography. W. H. Chandler, *Deciduous Orchards*, 1942; N. F. Childers (ed.), *The Peach*, 1975; J. Janick and J. N. Moore (eds.), *Advances in Fruit Breeding*, 1975; M. N. Westwood, *Temperate Zone Pomology*, 3d ed., 1993.

Peanut

A self-pollinated, one- to six-seeded legume. It is cultivated throughout the tropical and temperate climates of the world. The oil, expressed from the seed, is of high quality, and a large percentage of the annual world production is used for this purpose. In the United States some 65% goes into the cleaned and shelled trade, the end products of which are roasted or salted peanuts, peanut butter, and confections. *See* LEGUME; ROSALES.

Origin and Description

Peanuts originated in Bolivia and northeastern Argentina where a large number of wild forms are found. The cultivated species, *Arachis hypogaea*, was grown extensively by Indians in pre-Columbian times. Merchant ships carried seed to many continents during the early part of the sixteenth century. Although grown in Mexico before the discovery of America, the peanut was introduced in the United States from Africa.

Botanically, peanuts may be divided into three main types, Virginia, Spanish, and Valencia, based on branching order and pattern and the number of seeds per pod. Marketing standards in the United States include grades for Virginia (large-seeded, grown mostly in North Carolina and Virginia), runner (small-seeded Virginia, grown mostly in Georgia, Florida, and Alabama, with some in Texas and Oklahoma), Spanish (grown mostly in Oklahoma and Texas), and Valencia (grown mostly in New Mexico). *See* SEED.

The peanut's most distinguishing characteristic is the yellow flower, which resembles a butterfly (papilionaceous) and is borne above ground. Following fertilization the flower wilts and, after 5–7 days, a positively geotropic (curving earthward) peg or ovary emerges. Penetrating the soil 0.8–2.8 in. (2–7 cm), the peg assumes a horizontal position and the pod begins to form (**Fig. 1**). *See* FLOWER; PLANT MOVEMENTS.

Fig. 1. A typical Virginia-type peanut.

The pod, a one-loculed legume, splits under pressure along a longitudinal ventral suture. Pod size varies from 0.4 by 0.2 in. (1 by 0.5 cm) to 0.4 by 3.2 in. (2 by 8 cm), and seed weight varies from 0.08 to 0.2 oz (0.2 to 5 g). The number of seeds per pod usually is two in the Virginia and runner type, two or three in the Spanish, and three to six in the Valencia.

The plant may be upright, prostrate, or intermediate between these forms. The main stem is usually upright and may be very short in some cultivars. The leaves are even-pinnate with four obovate to elliptic leaflets. Leaves occur alternately and have a 2:5 phyllotaxy (arrangement on stems). *See* LEAF.

Peanuts are harvested by running a special wing-type plow under the plants. After wilting they are either stacked or allowed to cure in windrows before picking. The main production areas in the United States extend southward from Virginia and westward to Oklahoma and Texas. Peanuts are under strict acreage controls. *See* AGRICULTURAL MACHINERY; AGRICULTURAL SCIENCE (PLANT); AGRICULTURAL SOIL AND CROP PRACTICES. Astor Perry

Diseases

The peanut plant is susceptible to 46 described diseases that reduce yield of fruit (pods) at least 30%. Causal agents include fungi, nematodes, viruses, bacteria, and insects. Yield of pods is reduced primarily by premature loss of leaves and decay of stems, pods, and pegs. For economical production, disease control measures used by growers involve pesticides, crop rotation, deep burial of surface organic matter, and disease-resistant cultivars.

Fungal diseases. Fungi cause the greatest number of diseases and are of greatest concern to growers. Peanut seed harbors fungi that cause rots, poor germination, and seedling diseases. The most common are species of *Rhizopus* and *Aspergillus*, and consequently remedial procedures must be used. Seed is relatively free of fungi when pods are harvested, but becomes infected because of mechanical injury incurred during the shelling operation and improper handling. It is a common and necessary practice to treat seed prior to planting, and if seed is not coated

Fig. 2. Blue damage of seed coat and cotyledon (*D. C. Norton*)

with a broad-spectrum fungicide serious reduction in stand will result. Leafspot diseases caused by *Cercospora arachidicola* and *Cercosporidium personatum* are the most serious diseases that cause premature leaf loss wherever peanuts are grown (**Fig. 2**). Annual crop losses range from 15 to 50% in yield of pods in many areas. Infection begins on lower leaves near the soil, and movement of fungal units (inoculum) to leaves and subsequent secondary spread of the fungi (conidia) is by wind, rain, insects, or machinery. Foliage is sprayed with fungicides before infection and at 7–14-day intervals through the season for optimum suppression of diseases. The most intensive use of foliar fungicides is in the southeastern United States, where six or seven applications are made each season. Use of crop rotation where peanuts are grown only one out of three years significantly reduces severity of leaf spots. Fungicides used for suppression of leaf spots are also effective against rust caused by *Puccinia arachidis*, a very serious threat to production. The rust fungus is endemic in the West Indies, and has become established throughout Asia and Oceania in Australia and Africa. Web blotch caused by *Phoma arachicola* is widely distributed and may cause significant damage. *See* FUNGISTAT AND FUNGICIDE.

Diseases of subterranean plant parts are the most destructive and difficult to control, and growers must depend primarily on cultural practices for disease suppression. Southern blight, caused by *Sclerotium rolfsii*, is the most devastating and occurs in all peanut-growing areas. The fungus attacks succulent tissues at the ground line and either kills plants or causes pod and root decay. Because the fungus depends on organic matter for growth and subsequent invasion, growers remove organic litter from soil surface by deep plowing and plant on flat or raised beds. During cultivation soil is kept away from planting rows so that sclerotia (survival propagules) and organic litter will not be moved to plants and initiate infection. Sometimes a soil fungicide and a nematicide are applied over the row 45–50 days after planting to control southern blight. Other fungi that are widely distributed and cause decay by acting singly or in combination include *Rhizoctonia* spp., *Pythium* spp., and *Fusarium* spp. *Cylindrocladium* spp. and *Sclerotinia* spp. cause sporadic losses in the eastern United States. A pod rot initiated by calcium deficiency is prevented by application of gypsum after flowering.

Nematodes. Plant parasitic nematodes that feed on roots of plants occur in all peanut-growing areas of the world. Most damaging are the northern root-knot nematode (*Meloidogyne hapla*) and the peanut root-knot nematode (*M. arenaria*) which feed on the inside of root, peg, and fruit and cause galls on these parts. Sting nematodes (*Belonolaimus gracilis*) are very destructive and are known to occur only in the lighter soils of the United States. The sting nematodes feed on the outside of the peanut root, peg, and fruit and cause a restricted root system and smaller fruit. The lesion nematode (*Pratylenchus brachyurus*) feeds on the inside of the peanut root, peg, and fruit and causes necrotic lesions on these parts. Nematodes cause damage directly by feeding on roots and other underground plant parts and indirectly by interacting with soil-borne fungi to increase decay. If populations are sufficiently high, nematicides are applied to soil to reduce damage. *See* NEMATA.

Viral diseases. There are five described virus diseases of peanuts—rosette, mosaic, ringspot, stunt, and mottle. Of these, rosette and stunt are the most significant to growers.

Bacterial diseases. *Pseudomonas solanacearum*, the cause of bacterial wilt, occurs in all peanut-growing areas but causes yield losses only occasionally. Most of the peanut cultivars grown currently in the United States are resistant to bacterial wilt.

Insects. The Southern corn root worm (*Diabrotica undecimpunctata howardii*) interacts with soil-borne fungi and bacteria to cause fruit rot. The potato leaf hopper (*Empoasca fabae*) secretes a toxic substance on the leaves and causes a disorder known as hopper burn. The tobacco thrip (*Frankliniella fusca*) causes puckered leaflets and retards the growth of seedlings; however, reduction in yield of pods is minimal. *See* PLANT PATHOLOGY. R. H. Littrell

Processing

Most peanuts are shelled and processed into peanut butter, salted peanuts, and confections. A few are boiled in 6% salt brine, roasted in shell, salted in shell, exported, or crushed for oil. Outside the United States, peanut oil is the most common end product. It is well suited for deep fat frying and stir frying because it has a high smoke point (446°F or 230°C) and a lack of flavor carry-over. *See* FAT AND OIL (FOOD); SOLVENT EXTRACTION.

Factors affecting flavor and quality. Peanuts must be carefully handled throughout growing, harvesting, curing, storage, shelling, subsequent storage, transportation, processing, packaging, and storage of finished products. Improper handling will produce an unacceptable consumer product with poor flavors. Good roasted peanuts have a slightly sweet taste with a roasted flavor and an appropriate aftertaste. Peanuts are high in oil content (45–50%), and they are easily oxidized to produce off flavors, which cover up good roasted flavor. Some off flavors are bland, stale, fruity, musty, rancid, and beany.

Three major problems face the peanut manufacturers: flavor, foreign material, and aflatoxin. Flavor defects can occur at any stage and can adversely affect product flavor and consumer acceptance. Drought-stressed peanuts have spotted surfaces which are associated with flavor defects. At harvest, not all peanuts are of the same maturity so it is important that they be harvested at optimum maturity to give best flavor, greatest storage stability, and longest product shelf life. Immature peanuts have flavor problems because of high levels of free arginine, threonine, tyrosine, and lysine. Excessive drying also produces off flavors that are associated with acetaldehyde, ethanol, and ethyl acetate. Foreign material includes sticks, stones, small rocks, and glass. A series of screens and blowers will remove much of this material; electronic sorters are also used. Unfortunately, some foreign material may still get into the final product and can cause costly consumer complaints such as broken teeth. Sorting is also used to remove faulty nuts that are discolored and may have off flavors and contain aflatoxins. Aflatoxins are produced by molds such as *Aspergillus flavus* and are associated with peanuts that have become wet or have not been dried properly. United States government regulations minimize the amount of aflatoxins in peanut products. After shelling, cleaning, and aflatoxin testing, peanuts—now ready for processing—are stored in refrigerated warehouses at 32–37°F (0–3°C) with 65% relative humidity to ensure protection against insects, rancidity, and mold growth. *See* AFLATOXIN.

Roasting and oil cooking. Roasting (referred to as dry roasting in the peanut industry) uses hot air (320–446°F or 160–230°C) to develop desirable color, texture, and flavor. An internal temperature of 290°F (143°C) must be reached to develop the roasted peanut flavor and peanutty aftertaste. These hot peanuts must be properly cooled with ambient or refrigerated air to prevent overroasting, which gives burnt flavors. Most of the roasters are of the flat-bed type, but a few gas-fired rotary roasters are used by some smaller processors.

Oil cooking (referred to as oil roasting in the peanut industry) uses hot cooking oils (320–347°F or 160–175°C) to develop the desirable characteristics. Generally, these peanuts have a higher flavor intensity.

Peanut butter. Major peanut butter manufacturers further clean the raw shelled peanuts to remove additional foreign material. Peanuts are roasted, cooled, blanched (to remove the skins which give specks, and to remove the hearts, which impart a grayish color), sorted to remove unblanched and off-colored nuts, ground in a primary mill (usually a plate mill), combined with various ingredients, ground in the secondary mill (usually a cutting mill) to give the desirable particle size, deaerated, cooled, packaged, and then tempered before shipping. Ingredients may include sweeteners, salt, and hydrogenated vegetable oil to improve smoothness, spreadability, and flavor. In the United States, the Food and Drug Administration standards require that peanut butter contain at least 90% peanuts.

There are four basic types of peanut butters. The most common type is molasses flavored (molasses is a sweetener). The second type has roasted peanut flavor with the peanutty aftertaste. The third type has a dark roast and contains some of the burnt flavor with very little roasted and peanutty flavors. The first three types use sweeteners, salt, and stabilizers to prevent oil separation. The fourth is known as the old-fashioned type; it does not contain stabilizers and may not contain sweeteners or salt. The peanut oil separates to the top of the jar and must be remixed before using.

Roasted peanuts. The product known as cocktail peanuts is usually cooked in oil, cooled, and salted. Cocktail peanuts usually have a higher intensity of roasted peanut flavors. Dry roasted peanuts are often coated with gums, sweeteners, and salt and then roasted to make products such as honey roasted peanuts. There are a large number of regional products in this category.

Peanut candies and confections. Wide variations exist in formulations, with peanuts being used as a topping, a component in nut rolls and clusters, coated nuts, peanut brittles, and peanut butter cups (peanut-butter-filled chocolate confections). Peanuts used for candy bars are usually overroasted (a dark roast) so that the chocolate does not cover up the peanut flavor.

In-shell peanuts. These are of two types. There is the so-called roasted-in-shell, which is roasted, cooled, and packaged. The other type, known as salted-in-shell, is placed in a sodium chloride brine solution under vacuum, dried slowly, roasted, and packaged. *See* FOOD ENGINEERING; FOOD MANUFACTURING. Clyde T. Young

Bibliography. American Peanut Research and Education Association Inc., *Peanut: Culture and Uses*, 1973; C. R. Jackson and D. K. Bell, *Diseases of Peanut (Groundnut) Caused by Fungi*, Univ. Georgia Coll. Agr. Exp. Sta. Res. Bull. 56, 1969; E. W. Lusas, D. R. Erickson, and W.-K. Nip (eds.), *Food Uses of Whole Oil and Protein Seed*, 1989; H. E. Pattee and C. T. Young, *Peanut Science and Technology*, 1982; D. M. Porter, D. H. Smith, and R. Rodriquez-Kabana (eds.), *Compendium of Peanut Diseases*, 2d ed., 1997; J. G. Woodroof, *Peanuts: Production, Processing, Products*, 3d ed., 1983.

Pear

Any of approximately 20 species of deciduous tree fruits in the genus *Pyrus*. About half of the species are native to Europe, North Africa, and the Middle

East around the Mediterrean Sea; the others are native to Asia. Pear culture is documented to have started as early as 1100 B.C. Pears are best adapted to temperate climates with warm, dry summers and cold winters. They require winter cold to break the dormant period but are injured by temperatures below −10 to −15°F (−23 to −26°C). Commercial pear production in the United States is concentrated in the interior valleys of California, Oregon, and Washington.

Commercial types and varieties. Nearly all United States pear production is of the European pear, *P. communis*. The Bartlett variety comprises over 75% of the United States pear crop. Other European pear varieties include d'Anjou, Bosc, Comice, Seckel, and Winter Nelis. The European pear is noted for its soft, juicy flesh. The skin color is medium green to yellow, depending on fruit maturity and the variety. Skin texture can be smooth or rough. Fruit shape ranges from the classic pear shape (round base with narrow neck) to a rounded oblong shape with no clearly defined neck area.

The crisp-fleshed Asian pear, *P. pyrifolia*, is the second most popular type of pear grown worldwide. In the United States, Asian pear production is found primarily in California, with limited commercial production in Oregon and Washington. The Asian pear is characterized by a crisp, juicy flesh that has a gritty texture, and has been referred to as the sand pear. The fruit shape is round, the skin color is generally yellow to amber at maturity, and the skin texture is smooth or corky. Asian pears are sometimes marketed commercially as pear apples because of their flavor and fruit texture; however, they are not a hybrid of apple and pear. The most popular variety of Asian pear grown worldwide is 20th Century. Other commercial varieties include Shinseiki, Hosui, Ya Li, and Kosui. *Pyrus communis* and *P. pyrifolia* hybrids, such as Kieffer and Leconte, have been developed for limited commercial use in the southeastern United States. *Pyrus ussuriensis* has also been selected for Asian pear varieties. The snow pear, *P. nivalis*, is produced in Europe for cider and perry (a fermented liquor).

Propagation and cultivation. The European pear is usually propagated onto seedlings grown from Bartlett or Winter Nelis. Vegatatively propagated *P. communis* rootstocks known as the Old Home × Farmingdale series have been developed for size control and precocity. European pears are also grown on quince (*Cydonia oblonga*) rootstocks, primarily for size control or dwarfing. *Pyrus calleryana* has also been used for European pear rootstock. Soil type and winter hardiness affect the rootstock selection. The Asian pear is not used as a rootstock for European pears because of pear decline, which results from incompatability between the Asian and European pear species. Quince is also incompatible with Bartlett, and interstems of Old Home × Farmingdale are used to alleviate incompatibility symptoms. Pear orchard care and culture are similar to apple. *See* APPLE; QUINCE.

Disease and insect pests. Most commercial pear varieties are susceptible to fireblight, a bacterial disease caused by *Erwinia amylovora*. Fireblight is vectored by insects, such as honeybees, and is spread from the blossoms to the shoot tips and branches. Properly timed sprays of copper or streptomycin provide partial control, but removal of infected branches is always recommended. Resistant varieties are another option, but they generally lack commercial appeal. Fireblight-resistant pear varieties include Harrow Delight, Harvest Queen, and Moonglow. Warm, humid climates favor the development of fireblight, commercial pear production is limited to the western United States. Pear psylla (*Psylla pyricola*) is the major insect pest of pears. To date, chemical control is the preferred method, as there are limited biological control methods. Psylla-resistant varieties have not been developed. *See* PLANT PATHOLOGY.

Harvesting and storage. For optimum eating quality, European pears are picked immature and ripened off the tree. Unlike apples, pears do not continue to ripen in cold storage. Pears are best stored at a temperature of 30 to 32°F (−1 to 0°C) and about 90% humidity. Winter pears, such as d'Anjou, require cold storage for several months to ensure proper ripening after removal from storage. For best dessert quality, pears should be allowed to ripen at 60–70°F (15–21°C) at about 80–90% humidity. Asian pears are picked at the firm-ripe stage, and can be held in cold storage for several months. Asian pears should not soften for optimum eating quality. *See* FRUIT, TREE.

Kathleen M. Williams

Bibliography. M. N. Westwood, *Temperature Zone Pomology*, 1978; K. M. Williams, *Growing Pears in North Carolina*, 1987.

Pearl

The term pearl is applied properly to any mollusk-formed calcareous concretion that displays an orient and is lustrous. There are two major groups of bivalved mollusks in which gem pearls may form: the salt-water pearl oyster (*Pinctada*), of which there are several species; and a number of genera of fresh-water clams. Usually, jewelers refer to salt-water pearls as Oriental pearls, regardless of their place of discovery, and to those from fresh-water bivalves as fresh-water pearls. *See* GEM.

Formation. Between the body mass and the valves of the mollusk extends a curtainlike tissue called the mantle. Epithelial cells on the side of the mantle toward the shell perform the several stages of the shell-secreting process during the life of the mollusk. One of the stages of shell building is the secretion of nacre, the colorful, lustrous, mother-of-pearl material. In order for a pearl to form, a tiny object such as a parasite or a grain of sand must work through the mantle, carrying with it epithelial cells. When this happens, secretion of nacre around the invading object builds a pearl within the body of the mollusk. Whole pearls form within the body mass of the mollusk, in contrast to blister pearls, which form as protrusions on the inner surface of the shell. Edible oysters produce lusterless concretions, but never pearls.

Pearls occur in a great variety of shapes. The term baroque is used for the common, irregularly shaped forms. The most common and most desirable shape is the spherical or nearly spherical; this is the shape usually chosen for necklaces. Other desirable shapes include those called button, pear, egg, and drop. The colors, more or less in the order of desirability, are rosé, white rosé, light-cream rosé, slightly greenish rosé, white, black, cream, gold, and blue-gray.

Pearls are composed of many tiny overlapping plates of nacreous material. Nacre consists of prismatic pseudohexagonal aragonite crystals (oriented so that the long crystallographic axis is at right angles to the plane of the platelet) held together by conchiolin, a hornlike organic secretion. Chemical analysis of pearls shows calcium carbonate ($CaCO_3$), organic material, and water; the relative quantities vary with the species of the mollusk, the position of the pearl within the shell, and other factors. $CaCO_3$ content is usually 90–92%, but may be somewhat lower. Organic matter usually makes up 4–6% and water 2–4%. *See* ARAGONITE.

Producing areas. The major pearl-producing region in the world is the Persian Gulf. From its pearl fisheries come most of the natural pearls used for gem purposes, but output is now only a fraction of that of a long period during the nineteenth century through the 1920s. Since World War II, the sale of Oriental pearls has never regained earlier peaks. In the Persian Gulf, the important species *Pinctada margaritifera* is found at depths of from about 4 to 8 fathoms (7–14 m) on broad banks that extend for many miles into the gulf from both shores. Since pearl-producing mollusks are only one of many types of mollusk, fishing is a great gamble, and the waste of nonbearing mollusks is enormous. In times of demand, depletion of the mollusk supply is a grave problem. The supply has apparently never fully recovered from the depletion caused by the heavy fishing of the 1920s. Less important pearl sources include the coast of northern Venezuela, to the north of Australia, between Sri Lanka and India, in the Red Sea, and in the South Pacific.

Many of the rivers of the central portion of the United States have pearl-bearing mussels. In the nineteenth century, several rich finds were made that led to rapid exploitation and virtual exhaustion in those regions of the supplies of pearl mollusks; these were principally of the genus *Unio*. Fresh-water pearls vary in color from almost pure white to many that are more strongly tinted than the majority of the salt-water variety.

Cultured pearl. The substitute for natural pearls, to which the name cultured pearl has been given, is usually made by inserting a large bead into a mollusk to be coated with nacre. When the pearl-bearing mollusk (*Pinctada martensii*) reaches maturity at $3^1/_2$ years, the mollusks are gathered and prepared for pearl cultivation using larger nuclei. Smaller beads may be put in younger mollusks.

Workers trained for the task cut a channel into the foot mass or body cavity and insert a large sphere (bead), plus a small section of mantle tissure, with the epithelial cells next to the bead. Ten or more small beads (0.08–0.12 in. or 2–3 mm) may be placed in one mollusk. The number decreases with size. Perhaps two 0.24-in. (6-mm) beads are inserted, but any larger nucleus is placed one to a mollusk. Beads are prepared from large American fresh-water shells. The mollusks are placed in cages suspended from rafts in sheltered bays and usually left for 10 months to $3^1/_2$ years, except for periodic cleaning and inspection. The rate of nacre accretion in Honshu waters is only about 0.0059 in. (0.15 mm) annually, so that the diameter increases about 0.012 in. (0.30) mm annually; thus a 0.28-in. (7.2-mm) cultured pearl usually has a bead center of 0.24 in. (6.0 mm). The rate of accumulation in Kyushu is about double. More rapid nacre accumulation is encountered in South Seas culture stations, which are now producing larger species of *Pinctada*. Baroque fresh-water pearls have been produced without bead nuclei in Japan. *See* BIVALVIA; MOLLUSCA.

Richard T. Liddicoat, Jr.

Peat

A dark-brown or black residuum produced by the partial decomposition and disintegration of mosses, sedges, trees, and other plants that grow in marshes and other wet places. Forest-type peat, when buried and subjected to geological influences of pressure and heat, is the natural forerunner of most coal.

Peat may accumulate in depressions such as the coastal and tidal swamps in the Atlantic and Gulf Coast states, in abandoned oxbow lakes where sediments transported from a distance are deposited, and in depressions of glacial origin. Moor peat is formed in relatively elevated, poorly drained moss-covered areas, as in parts of Northern Europe. *See* COAL; HUMUS.

In the United States, where the principal use of peat is for soil improvement, the estimated reserve on an air-dried basis is 13,827,000 short tons (12,544,000 metric tons). In Ireland and Sweden peat is used for domestic and even industrial fuel. In Germany peat is the source of low-grade montan wax.

Gilbert H. Cady

Pebble mill

A tumbling mill that grinds or pulverizes materials without contaminating them with iron. Because the pebbles have lower specific gravity than steel balls, the capacity of a given size shell with pebbles is considerably lower than with steel balls. The lower capacity results in lower power consumption. The shell has a nonmetallic lining to further prevent iron contamination, as in pulverizing ceramics or pigments

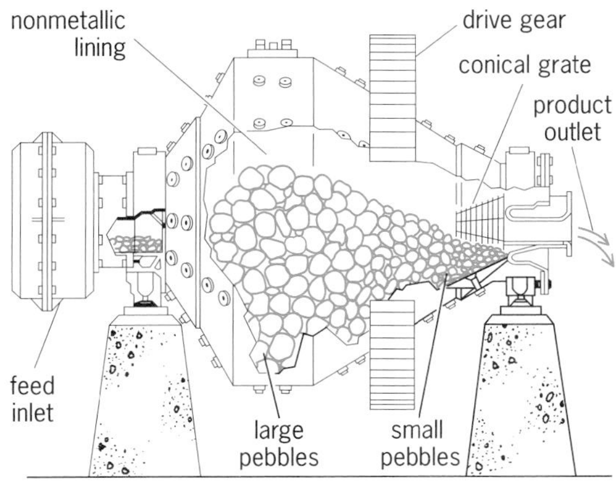

Diagrammatic sketch of a conical pebble mill.

(see **illus.**). Selected hard pieces of the material being ground can be used as pebbles to further prevent contamination. *See* CRUSHING AND PULVERIZING; TUMBLING MILL. Ralph M. Hardgrove

Pecan

A large tree (*Carya illinoensis*) of the family Juglandaceae, and the nut from this tree. Native to valleys of the Mississippi River and tributaries as far north as Iowa, to other streams of Texas, Oklahoma, and northern and central Mexico, this nut tree has become commercially important throughout the southern and southwestern United States and northern Mexico. Limited plantings have been made in other regions, being reported of interest in such diverse areas as Arizona and California, South Africa, Israel, Brazil, and Peru. The tree's major importance is in areas with a long growing season of over 200 days and midsummer average temperatures of 79°F (26°C) or higher, though a few early-ripening cultivars are grown in slightly cooler regions. There are different cultivars for arid, irrigated regions and humid summer regions. The United States cultivars have an indistinct winter chilling requirement but fruit best where the coldest winter month averages less than 61°F (16°C). *See* JUGLANDALES.

Nuts and kernels of the Cherokee pecan cultivar developed by the U.S. Department of Agriculture. (*From Pecan Quarterly*)

Botany. The tree is deciduous with compound, pinnate leaves. Flowers are unisexual with staminate flowers borne on year-old wood and pistillate flowers formed terminally on new spring growth. The period of pollen shedding of a cultivar usually does not overlap the time of stigma receptivity, so a choice of cultivars is necessary in large plantings to ensure cross pollination. Pollen is windblown for distances up to 0.6 mi (1 km), and usually there is enough diversity of bloom habits to ensure fruit set in smaller orchards and home plantings. Because of the bloom habits, pecans are heterozygous and seedlings are highly variable.

Production. Nuts from native seedlings and from named grafted cultivars are marketed in roughly equal amounts. The former are more important to Texas, Oklahoma, and Louisiana while the latter are more important in Georgia, Alabama, other southeastern states and in new producing areas of western Texas, New Mexico, and Arizona. The tree requires good soil fertility and culture, and usually some spraying to give satisfactory production. Total annual production varies as much as 20% from year to year because of strong tendencies toward biennial bearing. Heavy production in one season, combined with premature leaf drop from disease, insects, or poor nutrition will result in light cropping the next year.

Breeding. As in all agricultural endeavors, the cultivar is the base for success. Old selections like Stuart and Western, dating from the late nineteenth century, are still grown extensively. Intensive breeding work produced cultivars (see **illus.**) that come into production early and have 0.25–0.28 oz (7–8 g) nuts with over 60% kernel content. Combined with tree spacings as close as 33 ft (10 m), these can make orchards profitable sooner and double or triple yields. *See* NUT CROP CULTURE. Ralph H. Sharpe

Processing. Harvesting of pecans is done either by mechanical means or by collecting the nuts that have fallen to the ground. Contact of pecan nuts with the ground, particularly with wet ground, may contaminate the meat with pathogenic microorganisms, such as *Salmonella*, which can enter intact pecan shells. Processing, therefore, involves not only stabilizing the nut meat against spoilage microorganisms such as mold but also separation of the nut meat from the shell, removal of immature or damaged meats, and separation of shell fragments and pith from the nut meat.

Pecans are usually purchased from individual growers by wholesalers, who may process the pecans or sell them to pecan shellers. Mechanically harvested pecans must be separated from sections of tree limbs and other debris. Prior to storage in large lots in collection centers, inshell pecans are dried to 4.5–6% moisture by using low-temperature dry flowing air in trailers or bins. Moisture content below 6% is needed to prevent mold growth. Inshell pecans may be processed immediately or may be stored at 34°F (1°C) and 65% relative humidity prior to processing.

Approximately 94% of the total pecan crop is marketed as shelled pecans and 6% as inshell pecans. Inshell pecans are processed by first passing the nuts on a perforated conveyor belt under a vacuum hood to remove light nuts. This step is followed by a light spray of coloring agent, drying, application of a thin coat of wax, and packaging.

The processing of shelled pecans is more involved. Since pecan nut meat is a high-value item, shellers must recover most of the nut meats. The highest-valued product is the pecan halves, and the lowest-valued is the pecan meal. Thus, pecan shelling requires the utmost care in maintaining pecan pieces as large as possible. The pecan nut consists of a hard outer shell, and soft pithy material between the two halves of the meat and between the shell and the meat. The pith has a bitter flavor, and the hard shell can cause tooth damage to consumers, so the nut meats must be free of these materials. In addition, larvae of the pecan weevil in the nut mix with the meat fragments and must be removed. Pecan shelling operations involve several types of separation processes.

The first step in pecan shelling is grading the nuts into several size categories. Cracking machines are set to handle specific pecan sizes so that the nuts receive just the right amount of impact to crack the shell without damaging the nut meat. The nuts, separated into the proper size categories, are conditioned to add moisture to the meat, preventing them from shattering in the cracking machine. Conditioning is done by soaking the nuts in chlorinated warm water, followed by draining, and then equilibrating at room temperature. Use of chlorinated water in all operations where water contacts the inshell pecans and the nut meats minimizes microbial counts in the final product.

Cracking is accomplished by positioning individual nuts on a carrier that advances them at high speed between an anvil and a spring-operated impact tool. The cracked nuts are conveyed to a rotating cylinder that loosens the meat from the shell. The mixture of nut meat, pith, and shell fragments enters a primary water flotation tank, where the large shell fragments and most of the pith settles while the floating nut meats are removed by a conveyor. This mixture of nut meats of various sizes, including a few shell fragments and pith, is dried with hot air and separated by size by means of vibrating screens. The halves retained on the top screen are virtually free of shell and pith. These are conveyed to color sorters, where dark-colored ones are ejected electronically.

The sound nut meats are either packaged raw or fried in oil and packaged as roasted pecans. The smaller fragments may contain shell fragments and pith. They go through electronic sorting machines where off-colored ones and some shell fragments are ejected. These small nut meats contain shell fragments and pith which are separated by pressure flotation. They are placed in a tank full of water and pressurized to force water into the shell fragments and pith, which settle. The floating nut meats are skimmed from the top of the tank and air dried. Some nut meat fragments are of the same size as the pecan weevil larvae. These fragments are first electronically sorted to remove those with off-color. The larvae may be separated by flotation in dilute ethanol solution, where the larvae float while the nut meats settle. Following alcohol flotation, the nut meats are floated under pressure and air dried. As an alternative to alcohol flotation, larvae are removed by inspectors, who pick out the fluorescent larvae as the nut meats are passed on a conveyor under ultraviolet light. *See* FLOTATION.

Shelled pecans rapidly become rancid, so they are stored at 14°F (−10°C) or below in odor-free rooms. For applications such as ice cream, which requires very low microbial counts in the product, the shelled pecans may be decontaminated by exposure to ethylene oxide gas. *See* ICE CREAM. Romeo T. Toledo

Bibliography. J. Ogawa and H. English, ***Diseases of Temperate Zone Tree Fruit and Nut Crops***, 1991; C. R. Santerre (ed.), *Pecan Technology*, 1994.

Pectin

A group of polysaccharides occurring in the cell walls and intercellular layers of all land plants. They are extractable with hot water, dilute acid, or ammonium oxalate solutions. Pectins are precipitated from aqueous solution by alcohol and are commercially used for their excellent gel-forming ability.

Source and isolation. Commercially, the primary source of pectin is the peel of citrus fruits such as lemon and lime, although orange and grapefruit may be used. A secondary source is apple pomace and sunflower heads.

Plant tissue is treated with sulfurous acid to inactivate pectic enzymes and washed with water to remove bitter glycosides and free sugars. The tissue is then suspended in boiling water at pH 1.5–3.0, obtained by addition of hydrochloric, sulfuric, or nitric acid. Acid extraction is continued for 30–60 min, and the mixture is filtered with filter aid and precipitated with isopropanol. The precipitate is pressed, dried, and milled to approximately 60-mesh.

Structure and properties. Native pectin is a mixture of polysaccharides, with the major component a polymer of α-D-galacturonic acid, mainly as the methyl ester (see **illus.**) and often with some acetyl

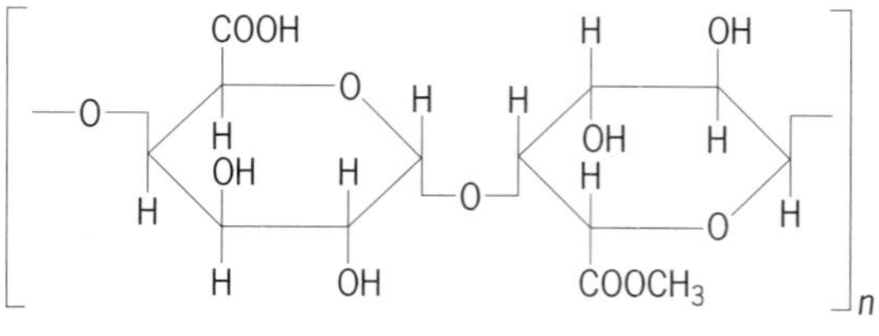

Partial methyl ester of pectic acid.

groups on the hydroxyls at C-2, although some may be on the C-3 hydroxyls. The range of acetyl content may vary from pectin to pectin over the range 0.2–4%, with apple pectin being 0.45%. A minor component is a polymer of α-L-arabinofuranosyl units (an arabinan), joined by α-L-(1→5) linkages. Sugar units are substituted occasionally by α-L-arabinofuranosyl groups linked α-D-(1→3). Another polysaccharide component is a linear chain of β-D-galactopyranosyl units (a galactan) linked (1→4). Arabinan and galactan are largely broken down and removed during the extraction and purification steps. Commercial pectin can be regarded as mainly an α-D-galacturonan with varying degrees of methyl esterification.

Gel-forming properties of pectins are largely dependent on their content of methyl ester groups. Pectins are classified according to their degree of methylation. Those with methyl content 70% or greater are called high-methoxyl (or rapid-set) pectins, while those of methyl content less than 50% are called low-methoxyl pectins. Intermediate values of 60–65% are characteristic of slow-set pectins. Amidated pectins are those that have been treated with ammonium hydroxide to partially saponify the methyl ester groups and to replace some methyl ester groups with amide groups. Such pectins have reduced sensitivity to a calcium ions in gel formation.

Immature plant tissues contain an insoluble pectin precursor, protopectin, which, due to the action of various pectinases, becomes more soluble as tissue maturity and ripening proceed.

The molecular weight range is 30,000–300,000, and the linear nature of pectin is illustrated by its film-forming ability, by x-ray diffraction, viscosity, and other physical measurements.

Methyl groups are removed slowly by acid hydrolysis to leave at intermediate stages a partially methylated polysaccharide, pectinic acids, wherein the ester groups are randomly distributed. With the enzyme pectinesterase, obtained from such sources as roots, leaves, and fruits of all higher plants and also from a number of microorganisms, the ester groups are quickly removed. However, at intermediate stages the molecule contains both segments that have the full complement of ester groups and segments that are completely devoid of ester groups. Pectic acid has a lower water solubility than pectinic acid and is more affected by multivalent ions, such as calcium, that cause gelatin by forming structural cross-links.

Aqueous solutions of commercial pectin of 2–3% concentration may be easily prepared in warm or hot water. These solutions are viscous, with the viscosity depending on the polymer molecular weight range, the pH, the degree of methylation, and the solution composition, such as the presence of polyvalent ions and sugar. The free acid groups of pectin can be titrated in the normal manner to give titration curves that resemble those of simple, monobasic acids. Pectins are usually characterized by their anhydrouronic acid content, their solution viscosity (non-newtonian), and their degree of methylation.

Uses. Pectin is widely used in the food industry, principally in the preparation of gels. Food gels based on pectin have been known for over 500 years. Pectins are standardized on the basis of their jelly grade, that is, the number of weight units of sucrose that one weight unit of pectin will support in a standard 65%-sugar jelly containing a specified amount of acid. Good commercial-grade pectins have jelly grades in the 150–300 range. Pectin has been used in the preparation of jams, jellies, preserves, canned fruits, fruit juices, and confectionery products. It is also used as a stabilizer in some dairy products and frozen desserts, such as sherbet. Low-methoxyl pectins are useful as stabilizers for pudding mixes and also as edible protective coatings for sausages, almonds, candied dried fruit, and soft dates.

Pectin has also been used for the treatment of diarrhea, and several commercial products have a pectin-based formula. Intravenously administered pectin shortens the blood coagulation time and is thus useful in controlling hemorrhage. Work has shown that orally administered pectin lowers blood cholesterol and is an aid in improving carbohydrate tolerance in diabetics.

Pectins, in combination with plant hemicelluloses and lignin, may be useful dietary constituents in preventing coronary heart disease, diverticular disease, ulcerative colitis, and a variety of other Western diseases. D-Galacturonic acid prepared from pectin can be used to synthesize vitamin C. Pectin has numerous other medical and pharmaceutical uses. *See* COLLOID; GEL; GUM; HEMICELLULOSE; POLYSACCHARIDE.

Roy L. Whistler; Jams R. Daniel

Bibliography. M. Fishman, L. Fisheman, and J. Joseph (eds.), *Chemistry and Function of Pectins*, 1986; R. H. Walter (ed.), *The Chemistry and Technology of Pectin*, 1997.

Pectolite

A mineral inosilicate with composition $Ca_2NaSi_3O_8(OH)$, crystallizing in the triclinic system. Crystals are usually acicular in radiating aggregates (see **illus.**).

Globular masses of needlelike pectolite crystals with calcite at left, found in Paterson, New Jersey. (*American Museum of Natural History specimen*)

There is perfect cleavage parallel to the front and basal pinacoids yielding splintery fragments elongated on the *b* crystal axis. The hardness is 5 on Mohs scale, and the specific gravity is 2.75. The mineral is colorless, white, or gray with a vitreous to silky luster. Pectolite, a secondary mineral occurring in cavities in basalt and associated with zeolites, prehnite, apophyllite, and calcite, is found in the United States at Paterson, Bergen Hill, and Great Notch, New Jersey. *See* SILICATE MINERALS.

Cornelius S. Hurlbut, Jr.

Pediculosis

Human infestation with lice. There are two biological varieties of the human louse, *Pediculus humanus*, var. *capitis* and var. *corporis*, each showing a strong preference for a specific location on the human body. *Pediculus humanus capitis* colonizes the head and *P. h. corporis* lives in the body-trunk region. *See* ANOPLURA.

These lice are wingless insects which are ectoparasites. Their mouthparts are modified for piercing skin and sucking blood. The terminal segments of their legs are modified into clawlike structures which are utilized to grasp hairs and clothing fibers.

After a blood meal, the male and female lice copulate, and the female deposits (cements) eggs to hair shafts (*P. h. capitis*) or clothing fibers (*P. h. corporis*); these eggs are called nits and are visible to the unaided eye. After emerging from the egg, a sexually immature louse feeds on blood and goes through three stages (instars) of development before becoming a sexually mature adult louse. This life cycle requires approximately 3 weeks.

Lice are important vectors of human diseases. Their habit of sucking blood and their ability to crawl rapidly from one human to another transmit relapsing fever (spirochetal). The body fluids and feces of infected lice transmit these diseases. Epidemic typhus (*Rickettsia prowazeki*), European epidemic relapsing fever (*Borrelia recurrentis*), and trench fever (*Rochalimaea quintana*) are transmitted by the body louse. The role of head lice and pubic lice (*Pthirus pubis*) in disease transmission has not been demonstrated, and their involvement in the transmission of hepatitis and acquired immune deficiency syndrome (AIDS) remains to be determined. However, mild-to-severe dermatitis, which appears to be an allergic response to the secretions and excretions of the lice, may occur during louse infestation. *See* ACQUIRED IMMUNE DEFICIENCY SYNDROME (AIDS); HEPATITIS.

Historically, evidence of head lice infestations of humans has been indicated by the presence of *Pediculus h.c.* eggs (nits) on hair combs dating back to the first century B.C. in the Middle East. Egyptian mummies have also been found with evidence of head lice. With the advent of regular bathing, the human louse populations were reduced, and the number of cases of louse-borne diseases diminished. Disruption of human populations during disasters caused by nature or humans, such as crowding into refugee camps, tends to increase louse populations owing to unsanitary conditions. Infestation of school children by head lice continues to be a public health problem; these school-centered infestations are cosmopolitan.

Pediculosis can be controlled by public education, good personal hygiene, and the use of insecticides, such as lindane or pyrethrin with piperonyl butoxide.

Richard Sudds

Bibliography. L. R. Barker, J. R. Burton, and P. D. Zieve, *Principles of Ambulatory Medicine*, 5th ed., 1998; W. K. Joklik et al., *Zinsser Microbiology*, 20th ed., 1995; T. M. Peters, *Insects and Human Society*, 1988; S. A. Schroeder, M. A. Krupp, and L. M. Tierney, *Current Medical Diagnosis and Treatment*, 1988, 1987; H. Zinsser, *Rats, Lice, and History*, 1935.

Pedinoida

An order of Diadematacea, making up those genera which possess solid spines and a rigid test. The ambulacra show typical diadematoid structure, and the tubercles are noncrenulate. The single known family, Pedinidae, includes 15 genera, ranging from the Late Triassic onward, though only one genus, *Caenopedina*, survives today; the latter inhabits the Indo-Pacific and Caribbean seas, ranging from the continental shelf down to 660 ft (2000 m). *See* DIADEMATACEA; ECHINODERMATA; ECHINOIDEA.

Howard B. Fell

Bibliography. R. E. Moore (ed.), *Treatise on Invertebrate Paleontology*, U-3(1):340–366, 1966.

Pedology

Defined narrowly, a science that is concerned with the nature and arrangement of horizons in soil profiles; the physical constitution and chemical composition of soils; the occurrence of soils in relation to one another and to other elements of the environment such as climate, natural vegetation, topography, and rocks; and the modes of origin of soils. Pedology so defined does not include soil technology, which is concerned with uses of soils.

Broadly, pedology is the science of the nature, properties, formation, distribution and function of soils, and of their response to use, management, and manipulation. The first definition is widely used in the United States and less so in other countries. The second definition is worldwide. *See* SOIL; SOIL CONSERVATION; SOIL MECHANICS.

Roy W. Simonson

Pegasiformes

The sea moths or sea dragons, a small order of peculiar actinopterygian fishes also known as the Hypostomides. The precise relationships are unknown. The body is encased in a broad, bony framework anteriorly and has bony rings posteriorly, simulating the seahorses and pipefishes (see **illus.**). Unlike those fishes, which have the mouth at the tip of the produced snout, sea moths have enlarged nasal bones which form a rostrum that projects well forward of the small, toothless mouth. The preopercle, pterosphenoid, intercalar, entopterygoid, and metapterygoid bones are absent, and the gill cover has a single bone. The greatly expanded horizontal pectoral fin belies its appearance and does not function in aerial gliding. The pelvic fin is abdominal and consists of a slender spine and one or two long rays. The short, opposed dorsal and anal fins have no spines. There is no swim bladder. *See* GASTEROSTEIFORMES.

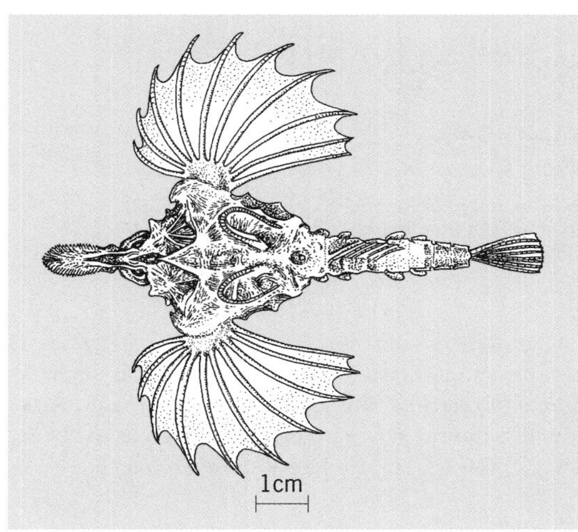

Sea moth (*Pegasus draconis*). (*After D. S. Jordan and J. O. Snyder, vol. 24, Leland Stanford University Contributions to Biology, 1901*)

There is a single family, Pegasidae, with one genus, *Pegasus*, and four or five species. They live amidst vegetation on Indo-Pacific shores from East Africa to Japan, Australia, and Hawaii. They rarely exceed 4 in. (10 cm) in length. There is no fossil record. *See* ACTINOPTERYGII. Reeve M. Bailey

Pegasus

The Winged Horse, in astronomy, an autumnal constellation. Pegasus is usually identified by the four bright stars α, β, γ, and α situated on the corners of a large square known as the Great Square in Pegasus (see **illus.**) The constellation is represented by a winged horse. Markab (the Saddle), a navigational star, occupies the southwestern corner of the square. The star Alpheratz at the opposite corner is really in the constellation Andromeda. The star at the northwestern corner is a red star, known as Scheat, a giant irregular variable. Diagonally opposite on the southeastern corner of the square is Algenib. Enif, another navigational star, lies in the nose of the horse. *See* CONSTELLATION. Ching-Sung Yu

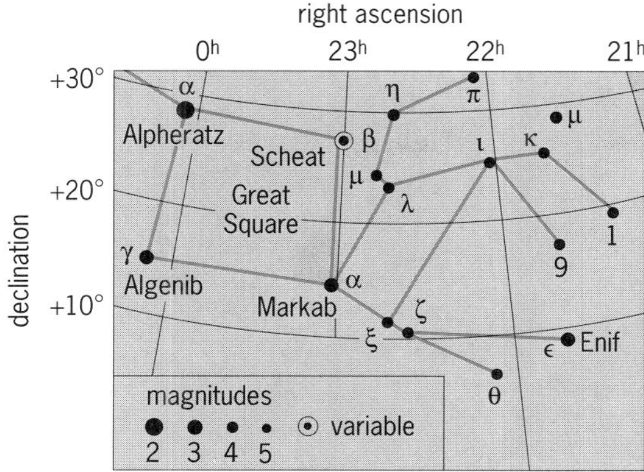

Line pattern of the constellation Pegasus. The grid lines represent the coordinates of the sky. The apparent brightness, or magnitude, of the stars is shown by the sizes of the dots, graded by appropriate numbers.

Pegmatite

Exceptionally coarse-grained and relatively light-colored crystalline rock composed chiefly of minerals found in ordinary igneous rocks. Extreme variations in grain size also are characteristic (**Fig. 1**), and close associations with dominantly fine-grained aplites are common. Pegmatites are widespread and very abundant where they occur, especially in host rocks of Precambrian age, but their aggregate volume in the Earth's crust is small. Individual bodies, representing wide ranges in shape, size, and bulk composition, typically occur in much larger bodies of intrusive igneous rocks or in terranes of metamorphic rocks, from most of which they are readily distinguished by their unusual textures (**Fig. 2**) and often by concentrations of relatively rare minerals. Many pegmatities have been economically valuable as sources of clays, feldspars, gem materials, industrial crystals, micas, silica, and special fluxes, as well as beryllium, bismuth, lithium, molybdenum,

Fig. 1. Contrasting textures in pegmatite of quartz monzonite composition, New Hampshire. Upper part is relatively fine-grained and granitoid, and remainder comprises large, roughly faced crystals of alkali feldspar (light) set in slightly smoky quartz (dark).

Fig. 2. Sill of faintly layered quartz diorite pegmatite, 6 to 12 in. (15 to 30 cm) thick, in fine-grained granitic gneiss, North Carolina. Sodic feldspar (light), quartz (darker), and muscovite (very dark) are the principal pegmatite minerals. (*From R. H. Jahns, The study of pegmatites, Econ. Geol., 50th Anniv. Vol., pp. 1025–1030, 1955*)

rare-earth, tantalum-niobium, thorium, tin, tungsten, and uranium minerals. *See* APLITE; CLAY MINERALS; FELDSPAR; GEM; MICA.

Compositional types. Most abundant by far are pegmatite bodies of granitic composition, corresponding in their major minerals to granite, quartz monzonite, granodiorite, or quartz diorite. In some regions or districts they contain notable quantities of the less common or rare elements such as As, B, Be, Bi, Ce, Cs, La, Li, Mo, Nb, Rb, Sb, Sn, Ta, Th, U, W,

Fig. 3. Four contrasting zones in upper part of thick, gently dipping dike of very coarse-grained granite pegmatite overlain by dark-colored metamorphic rocks, New Mexico. From top down, the zones consist of: quartz with alkali feldspars, apatite, beryl, and muscovite (light band); quartz (dark band); quartz with giant lathlike crystals of spodumene (thick band with comblike appearance); and quartz, alkali feldspars, spodumene, lepidolite, and microlite (mined-out recesses). (*From E. N. Cameron et al., Internal structure of granitic pegmatites, Econ. Geol., Monogr. 2, 1949*)

Y, and Zr. Diorite and gabbro pegmatites occur in many bodies of basic igneous rocks, but their total volume is very small. Syenite and nepheline syenite pegmatites are abundant in a few regions characterized by alkaline igneous rocks. Some of them are noted for their contents of As, Ce, La, Nb, Sb, Th, U, Zr, and other rare elements.

Shape and size. Individual bodies of pegmatite range in maximum dimension from a few inches to more than a mile, and in shape from simple sheets, lenses, pods, and pipes to highly complex masses with many bulges, branches, and other irregularities. Tabular to podlike masses of relatively small size commonly occur as segregations or local injections within bodies of cogenetic igneous rocks, where they represent almost the full spectrum of known pegmatite compositions. Granitic and syenitic pegmatites also occur in metamorphic rocks, some of them distributed satellitically about igneous plutons. Their contacts with the enclosing rocks may be sharp or gradational. Many bodies formed in strongly foliated or layered rocks are conformable with the host-rock structure (Fig. 2); others are markedly discordant, as if their emplacement were controlled by cross-cutting fractures or faults.

Mineral composition. Essential minerals (1) in granitic pegmatities are quartz, potash feldspar, and sodic plagioclase; (2) in syenitic pegmatites, alkali feldspars with or without feldspathoids; and (3) in diorite and gabbro pegmatites, soda-lime or lime-soda plagioclase. Varietal minerals such as micas, amphiboles, pyroxenes, black tourmaline, fluorite, and calcite further characterize the pegmatites of specific districts. Accessory minerals include allanite, apatite, beryl, garnet, magnetite, monazite, tantalite-columbite, lithium tourmaline, zircon, and a host of rarer species.

Most pegmatites are mineralogically simple, with only one or two varietal constituents and a few sparse accessories. More complex pegmatites typically contain much albite (commonly the variety cleavelandite) and groups of minerals that reflect unusual concentrations of rare elements such as beryllium (beryl, chrysoberyl, gadolinite, phenakite), boron (axinite, tourmaline), and lithium (amblygonite, lepidolite, petalite, spodumene, triphylite-lithiophilite, zinnwaldite). Cavities and pockets in some pegmatite bodies contain beautifully formed crystals, and such occurrences have yielded transparent gemstones of beryl, garnet, quartz, spodumene, topaz, tourmaline, and other minerals.

Texture. The most distinctive features of pegmatites are coarseness of grain and great variations in grain size over short distances (Fig. 1 and **Fig. 3**). The average grain size for all occurrences is on the order of 4 in. (10 cm), but mica and quartz crystals 10 ft (3 m) across, beryl and tourmaline crystals 10 to 20 ft (3 to 6 m) long, feldspar crystals 30 ft (9 m) long, and spodumene crystals nearly 50 ft (15 m) long have been found, as have individual crystals of allanite, columbite, monazite, and topaz weighing hundreds of pounds.

On much smaller scales, sodic plagioclase is intergrown with host potash feldspar to form perthite, and quartz with feldspars in cuneiform fashion to form graphic granite. More or less regular intergrowths of hematite and magnetite in muscovite, and of muscovite, garnet, or tourmaline in quartz also are common. Evidence of corrosion and replacement of some minerals by others typically is present over a wide range of scales. *See* IGNEOUS ROCKS.

Internal structure. Many pegmatite bodies are grossly homogeneous, whereas others, including nearly all of those with economic mineral concentrations, comprise internal units of contrasting composition or texture. Such internal zoning is systematically expressed as successive layers in nearly flat-lying bodies with tabular form (Figs. 2 and 3), and elsewhere as layers or shells that also reflect the general form of the respective bodies (**Fig. 4**). In horizontal section the zones tend to be arranged concentrically about a single or segmented core, but in vertical section they are typically asymmetric and less continuous (Fig. 4). Some pegmatite bodies are distinguished by relatively fine-grained outer zones that are more regular and continuous in three dimensions. Associations of major minerals within successive zones consistently follow a definite sequence of worldwide application.

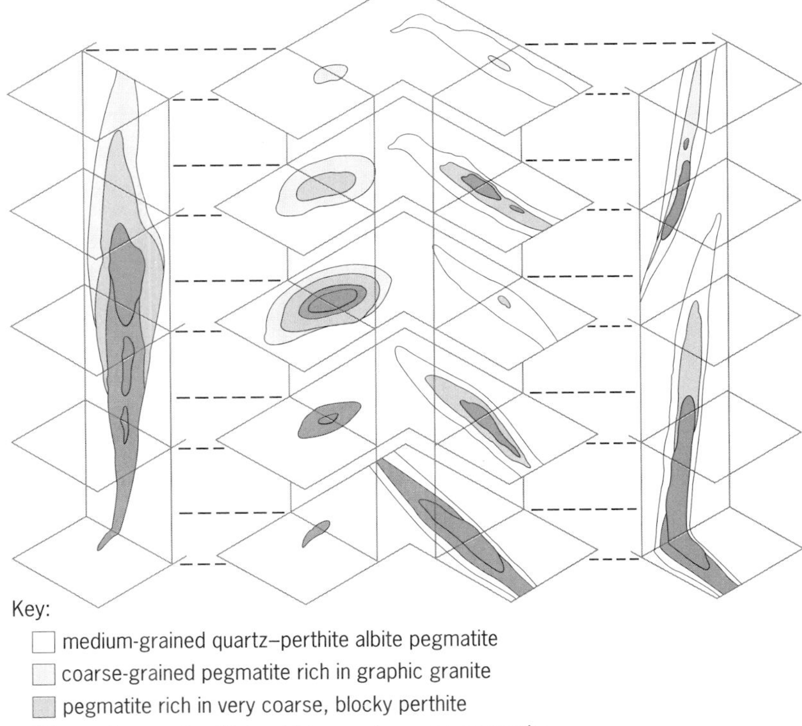

Fig. 4. Isometric diagram comprising vertical sections of a cucumber-shaped pegmatite body (left) and parts of two pegmatite dikes (right), together with successive horizontal sections of all three bodies (center). Internal zoning is shown in idealized form without a specific scale, and its configuration within relatively horizontal bodies can be visualized through a figurative vertical compression of the bodies shown here.

Fracture-filling masses of tabular form transect parts of some pegmatite bodies. They range in thickness from less than an inch to nearly 10 ft (3 m), and most are composed of quartz or of quartz, feldspars, muscovite, and other common minerals in various combinations. Less regular masses of lenticular, branching, or network form, and a few inches to several hundred feet in maximum dimension, appear to have developed wholly or in part at the expense of earlier pegmatite. Typical examples are quartz, potash feldspar, and muscovite replaced by aggregates of cleavelandite or sugary albite, alkali feldspar and quartz by muscovite, and potash feldspar and spodumene by lepidolite. Some replacement masses also contain notable concentrations of rare minerals.

Origin. Many pegmatites must have formed from the residual fractions of crystallizing rock melts (magmas) or from the initial fractions of deeply buried crustal materials undergoing partial melting. Such fractions would be silicate liquids relatively rich in water, halogens, and other volatile constituents, and also in many of the less common elements. They could remain within the crystallized or partly melted parent rocks to form bodies of segregation pegmatite with irregular and gradational boundaries, or they could be forced away to form homogeneous or zoned bodies of pegmatite elsewhere in the parent rocks or in rocks beyond. *See* MAGMA.

Other pegmatites, generally regarded as metamorphic in origin, consist of materials probably derived from the enclosing rocks via aqueous pore fluids and concentrated at favorable sites in the form of secretionary or replacement bodies. Such transfer of materials could account for numerous veins, pods, and irregular bodies of pegmatite with relatively simple composition and internal structure, and for the development of pegmatite on a grand scale in Precambrian terranes of metamorphism and granitization in regions such as Fennoscandia, India, central Africa, and eastern Canada. *See* METAMORPHIC ROCKS; METAMORPHISM.

Long-time studies of pegmatites and experimental investigations of pertinent silicate systems indicate that the unifying factor in pegmatite genesis is the presence of a high-temperature aqueous fluid. When such a fluid exsolves from a crystallizing water-rich magma, it serves as a low-viscosity medium into which other constituents can be preferentially transferred from the magma, and through which they can diffuse rapidly but selectively in response to temperature-induced concentration gradients. Thus it is an effective vehicle for the nourishment of large crystals and development of coarse textures, and for the gross segregation of constituents to form zonal structure in many pegmatite bodies. Upward gravitational rise of low-density aqueous fluid also can account for observed vertical asymmetry of zonal structure (Fig. 4).

The aqueous fluid promotes exsolution of alkali feldspars, mobilization of the released albitic constituents, and extensive reactions among earlier-formed crystals both before and after all magma has been used up. Thus pegmatites of igneous origin reflect metasomatic activities as well. Moreover, the transfer of materials and the mineral replacements implied in the origin of metamorphic pegmatites can be expected to occur if an interstitial aqueous fluid is available in the hot host rocks and a temperature or pressure gradient is present, whether or not a magma is at hand. *See* METASOMATISM. Richard H. Jahns

Bibliography. D. Hyndman, *Petrology of Igneous and Metamorphic Rocks*, 2d ed., 1985; A. Miyashiro, *Metamorphic Petrology*, 1994; B. B. Rao, *Metamorphic Petrology*, 1986.

Pelecaniformes

A small order of diverse aquatic, mainly marine, fish-eating birds that includes the pelicans, boobies, and cormorants. The members of the order, which is found worldwide, are very different, and some researchers believe that it is an artificial group; however, all members are characterized by several unique features. Although a few scientists place the whale-billed stork (Balaenicipitidae) in the Pelecaniformes, that is a controversial inclusion. The five suborders and nine families of the order Pelecaniformes are listed below.

Order Pelecaniformes
  Suborder Phaethontes
    Family: Prophaethontidae (fossil)
      Phaethontidae (tropic birds; 3 species)
  Suborder Odontopterygia
    Family Pelagornithidae (fossil)
  Suborder Pelecani
    Family Pelecanidae (pelicans; 8 species)
  Suborder Sulae
    Family: Sulidae (boobies, gannets; 9 species)
      Plotopteridae (fossil)
      Phalacrocoracidae (cormorants; 33 species)
      Anhingidae (anhingas; 4 species)
  Suborder Fregatae
    Family Fregatidae (frigate birds; 5 species)

Fossil record. The pelecaniforms have an extensive, and indeed exciting, fossil record. *Prophaethon* is an Eocene primitive tropical bird, and *Limnofregata* is an early Eocene primitive frigate bird. Fossil pelicans are known from the lower Miocene, boobies from the Oligocene, and anhingas from the late Miocene. The Eocene *Protoplotus* may possibly belong to a distinct family but be related to the Plotopteridae and to cormorants from the Eo-Oligocene phosphorite beds of France. Most families have a relatively complete record continuing to the present. Of greater interest are the two purely fossil families, the Pelagornithidae and the Plotopteridae.

The Pelagornithidae, or pseudodontorns, are a remarkable family of extinct gigantic marine birds with long bills that bore numerous bony projections up to 2 in. (5 cm) long, the pseudoteeth found in deposits dating from the early Eocene to the late Miocene or early Pliocene from many locations throughout the world. They were flying birds with wingspans as great as 18–20 ft (5.5–6 m), longer than any known albatross and rivaling that of the largest Teratornithidae. The pseudodontorns appear to have had behavior patterns similar to those of albatrosses. *See* CICONIIFORMES.

A second remarkable group is the Plotopteridae, flightless, wing-propelled diving relatives of the cormorants that have been found in a number of sites around the North Pacific from California to Japan. They are known from the late Oligocene to their disappearance in the middle Miocene. The plotopterids range in size from medium-sized cormorants to birds larger than emperor penguins. A nondescript femur suggests a form that would be the largest known diving bird. The plotopterids are closely convergent to the penguins, but evolved from a different ancestral stock and in a different region of the world. The reasons for the extinction of the plotopterids is as mysterious as for the Pelagornithidae. *See* SPHENISCIFORMES.

Characteristics. The pelecaniforms are medium-sized to large marine birds that were characterized by a foot with four toes united in a common web (totipalmate). The legs are short and stout or weak, and they are used for swimming and perching. The birds are poor walkers, but all living forms—with the exception of the flightless Galápagos cormorant, *Nannopterum harrisi*—are excellent fliers, particularly the frigate birds. The bill varies widely in size and shape, from the short, stout bill of tropical birds to the long, hooked bill of frigate birds and the long, flat, flexible bill of pelicans. All species have a bare throat pouch; the largest such pouch is found in the pelicans, which use it as a fishing net. The plumage is black, gray, or white. All pelecaniforms feed on fish, crustaceans, and squid that are caught by diving from the air or the water surface, taken from the water surface, or stolen from other birds. The clavicle is fused to the sternum for additional protection against injury from the impact with water during dives.

Pelecaniforms are usually gregarious, breeding in colonies. The pairs are monogamous, and both sexes incubate and care for the young, which remain in the nest.

Economic significance. The Chinese and Japanese once trained cormorants for fishing, but that method is now uneconomical. Several species of cormorants are valuable as guano producers, and so in some areas large artificial platforms are built to encourage nesting of cormorants. The guanay (*Phalacrocorax bougainvilli*) is the chief contributor to the guano deposits of Peruvian islands and is considered to be the most valuable wild bird. Several species, especially the brown pelican of North America (*Pelecanus occidentalis*), have been severely reduced in numbers because of concentrations of DDT in their food. *See* AVES.

Walter J. Bock

Pelmatozoa

A division of the Echinodermata made up of those forms which are anchored to the substrate during at least a part of the life history. Formerly treated as a formal unit of classification with the rank of subphylum, pelmatozoans are now realized to be a heterogeneous assemblage of forms with similar habits but dissimilar ancestry, their common features having arisen by convergent evolution. Most pelmatozoan echinoderms are members of the subphylum Crinozoa, but some echinozoans also exhibit a sedentary, anchored life, with modifications for such existence. *See* CRINOZOA; ECHINODERMATA; ECHINOZOA; ELEUTHEROZOA.

Howard B. Fell

Peltier effect

A phenomenon discovered in 1834 by J. C. A. Peltier, who found that at the junction of two dissimilar metals carrying a small current the temperature rises or falls, depending upon the direction of the current. Many different pairs of metals were investigated; bismuth and copper were among the first. The temperature rises at a junction where the flow of positive charge is from Cu to Bi and falls where the flow is from Bi to Cu. A reversible output of heat occurs at the first-named junction and a reversible intake at the second. In view of the experiments of Quintus Icilius (1853), which established that the rate of intake or output of heat is proportional to the magnitude of the current, it can be shown that an electromotive force resides at a Cu-Bi junction, directed from Bi to Cu. Electromotive forces of this type are called Peltier emf's. *See* SEEBECK EFFECT; THERMOELECTRICITY; THOMSON EFFECT.

John W. Stewart

Pelycosauria

An extinct order of reptiles of the subclass Synapsida. They are characterized by a temporal fossa that lies low on the side of the skull. The group is known from rocks of the upper Carboniferous and lower and middle Permian. Three suborders are included: Ophiacodonta, primitive, partially aquatic carnivores; Sphenacodontia, advanced active carnivores of the early Permian; and Edaphosauria, including early Permian herbivores (possibly molluscivores) and derived Caseidae. Caseids became the dominant herbivores during the middle Permian in the United States, and are known as well from western Europe and Russia. Pelycosaurs ranged from about 2 to 15 ft (0.6 to 5 m) in total length, the largest being late members of the Caseidae. The majority inhabited lowland, deltaic environments. They are best known from Permian deposits in northern Texas and Oklahoma. Late in the early Permian the sphenacodonts gave rise to more advanced mammallike reptiles, the therapsids. *See* SYNAPSIDA; THERAPSIDA.

Everett C. Olson

Bibliography. M. J. Benton, *Vertebrate Paleontology*, 1991; R. L. Carroll, *Vertebrate Paleontology and Evolution*, 1988.

Pendulum

A rigid body mounted on a fixed horizontal axis, about which it is free to rotate under the influence of gravity. The period of the motion of a pendulum is virtually independent of its amplitude and depends primarily on the geometry of the pendulum and on the local value of g, the acceleration of gravity. Pendulums have therefore been used as the control elements in clocks, or inversely as instruments to measure g.

Pendulum motion. In the schematic representation of a pendulum shown in the **illustration**, the line OC makes an instantaneous angle θ with the vertical. In rotary motion of any rigid body about a fixed axis, the angular acceleration is equal to the torque about the axis divided by the moment of inertia I about the axis. If m represents the mass of the pendulum, the force of gravity can be considered as the weight mg acting at the center of mass C. Therefore, the angular acceleration α is determined by relation (1)

$$-mgh \sin\theta = I\alpha = I\frac{d^2\theta}{dt^2} \tag{1}$$

where h is the distance OC, and t represents time.

If the amplitude of motion is small, $\sin\theta \approx \theta$ in radian measure. In this approximation the motion is simple harmonic. Equation (2) has for its solution Eq. (3) where the amplitude A and the phase δ are

$$-mgh\theta = I\frac{d^2\theta}{dt^2} \tag{2}$$

$$\theta = A \sin(\omega t - \delta) \tag{3}$$

arbitrary constants. The angular frequency ω is given by Eq. (4). The period T, time for a complete vibra-

$$\omega^2 = \frac{mgh}{I} \tag{4}$$

tion (for example, from the extreme displacement right to the next extreme displacement right), and frequency f, number of vibrations per unit time, are given by Eq. (5).

$$T = \frac{1}{f} = \frac{2\pi}{\omega} = 2\pi\sqrt{\frac{I}{mgh}} \tag{5}$$

See HARMONIC MOTION.

The actual form of a pendulum often consists of a long, light bar or a cord that serves as a support for a small, massive bob. The idealization of this form into a point mass on the end of a weightless rod of length L is known as a simple pendulum. An actual pendulum is sometimes called a physical or compound pendulum. In a simple pendulum, the lengths h and L become identical, and the moment of inertia I equals mL^2. Equation (5) for the period becomes Eq. (6).

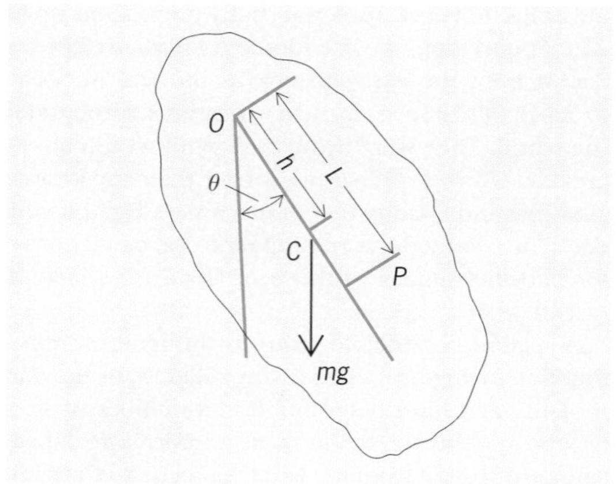

Schematic diagram of a pendulum. *O* represents the axis, *C* is the center of mass, and *P* the center of oscillation.

$$T = 2\pi\sqrt{\frac{L}{g}} \tag{6}$$

Because the value of g in metric units (about 9.8 m/s^2) is very nearly equal to π^2, a simple pendulum 1 m (3.28 ft) in length has a period very close to 2 s; the time for a single swing from right to left or left to right is approximately 1 s.

Center of oscillation. Equation (6) can be used to define the equivalent length of a physical pendulum. Comparison with Eq. (5) shows that Eq. (7) holds.

$$L = \frac{I}{mh} \tag{7}$$

The point P on line OC of the figure, whose distance from the axis O equals L, is called the center of oscillation. Points O and P are reciprocally related to each other in the sense that if the pendulum were suspended at P, O would be the center of oscillation.

The proof of this relation follows from the parallel axis property of moments of inertia. If the moment of inertia of the pendulum about its center of mass is defined by Eq. (8), then Eq. (9) holds; and if defined by Eq. (7), then Eq. (10) holds.

$$I = mb^2 \tag{8}$$

$$I = m(h^2 + b^2) \tag{9}$$

$$hL = h^2 + b^2 \tag{10}$$

For a given value of b (the radius of gyration about the center of mass) and L, h is determined by Eq. (10) to be either of the quantities given in Eq. (11). The

$$h = \frac{L}{2} \pm \sqrt{\frac{L^2}{4} - b^2} \tag{11}$$

sum of these two values is equal to L: therefore, if one value of h is the distance OC, then CP must represent the other value of h that will give the same equivalent length.

If some particular body with a definite value for b is to be mounted about an arbitrary axis to make a pendulum, Eq. (10) shows that L can never be less than $2b$, and that L will have this minimum value (and the period T will be a maximum) if h is made equal to b.

Center of percussion. The points O and P share another reciprocal property. If the body is free to move in the plane of the figure, instead of fixed on an axis, and an impulsive force is applied to the body at O, the initial motion of the body will be a rotation about P. For this reason P is sometimes called the center of percussion about O.

If the motion of a pendulum is not limited to small amplitudes, Eq. (2) is not an adequate substitute for the correct Eq. (1). The angular velocity, $d\theta/dt$, can be derived as a function of displacement by multiplying both sides of Eq. (1) by $d\theta/dt$ and then integrating. The result, which can also be obtained directly from the principle of conservation of energy, is given in Eq. (12), where θ_0 is the maximum displacement

$$\left(\frac{d\theta}{dt}\right)^2 = 2\omega^2(\cos\theta - \cos\theta_0) \qquad (12)$$

or amplitude of the motion. Here ω is still the characteristic constant of the pendulum defined by $\omega^2 = mgh/I = g/L$, although the relation between ω and frequency is no longer as simple as in the approximate Eq. (5).

To obtain θ as a function of time, introduce an angle ψ by relation (13), where $k \equiv \sin(\theta_0/2)$.

$$\sin\psi \equiv \frac{1}{k}\sin\frac{\theta}{2} \qquad (13)$$

Equation (12) becomes Eq. (14). If time is chosen

$$\begin{aligned}\omega\, dt &= \pm\frac{d\theta}{\sqrt{2(\cos\theta - \cos\theta_0)}}\\ &= \pm\frac{d\theta}{2\sqrt{\sin^2(\theta_0/2) - \sin^2(\theta/2)}}\\ &= \pm\frac{d\psi}{\sqrt{1 - k^2\sin^2\psi}}\end{aligned} \qquad (14)$$

zero when θ is zero, then Eq. (15) holds, where

$$\omega t = F(k, \psi) \qquad (15)$$

$F(k,\psi)$ is the standard elliptic integral of the first kind, as defined by Eq. (16). Conversely, the angle

$$F(k, \psi) = \int_0^{\psi} \frac{dz}{\sqrt{1 - k^2\sin^2 z}} \qquad (16)$$

θ can be expressed as an elliptic function of time, as in Eq. (17).

$$\sin\frac{\theta}{2} = k\sin(\omega t) \qquad (17)$$

The accurate expression for the period T, obtained from Eq. (15), can be written in terms of the complete elliptic integral of the first kind, $K(k)$, as Eq. (18). Numerical values for the ratio of the pe-

$$\omega T = 4F\left(k, \frac{\pi}{2}\right) = 4K(k) \qquad (18)$$

riod for amplitude θ_0 to the period for infinitesimal amplitude are listed in the **table**.

Ratio of the period for amplitude θ_0 to the period for infinitesimal amplitude

θ_0	$T(\theta_0)/T(O)$
0°	1.0000
20°	1.0077
40°	1.0313
60°	1.0732
80°	1.1375
100°	1.2322
120°	1.3729
140°	1.5944
160	2.0075
180°	∞

Pendulum types. The following paragraphs describe important types of gravity pendulums.

Kater's reversible pendulum. This type is designed to measure g, the acceleration of gravity. It consists of a body with two knife-edge supports on opposite sides of the center of mass as at O and P (and with at least one adjustable knife-edge). If the pendulum has the same period when suspended from either knife-edge, then each is located at the center of oscillation of the other, and the distance between them must be L, the length of the equivalent simple pendulum. The value for g follows from Eq. (6) or Eq. (18).

Ballistic pendulum. This is a device to measure the momentum of a bullet. The pendulum bob is a block of wood into which the bullet is fired. The bullet is stopped within the block and its momentum transferred to the pendulum. This momentum is determined from the amplitude of the pendulum swing. *See* BALLISTICS.

Spherical pendulum. This is a simple pendulum mounted on a pivot so that its motion is not confined to a plane. The bob then moves over a spherical surface. A Foucault pendulum is a spherical pendulum suspended so that its plane of oscillation is free to rotate. Its purpose is to demonstrate the rotation of the Earth. If such a pendulum were mounted at the North Pole, the rotation of the Earth under the pendulum would make it appear to a terrestrial observer that the plane of the pendulum's motion rotated 360° once every day. The plane of motion of a Foucault pendulum set up at a lower latitude rotates at a reduced rate, proportional to the sine of the latitude. *See* FOUCAULT PENDULUM.

Torsional pendulum. Despite its name, a torsional pendulum is not a pendulum. It is an example of a torsional harmonic oscillator, consisting of a disk or other body of large moment of inertia mounted on one end of a torsionally flexible rod. The other end of the rod is held fixed. If the disk is twisted and released, the torsional pendulum oscillates harmonically. Gravitation plays no part in its motion. *See* CLOCK; DIMENSIONAL ANALYSIS; SCHULER PENDULUM.

Joseph M. Keller

Bibliography. A. P. Arya, *Introduction to Classical Mechanics*, 2d ed., 1997; V. D. Barger and M. G. Olsson, *Classical Mechanics: A Modern Perspective*, 2d ed., 1995; J. L. Meriam and L. G. Kraige, *Engineering Mechanics*, 4th ed., 1998; L. Shepley and R. Matzner *Classical Mechanics*, 1991.

Penicillin

One of a series of beta-lactam antibiotics, all of which possess a four-ring beta-lactam structure fused with a five-membered thiazolidine ring. These antibiotics are nontoxic and kill sensitive bacteria during their growth stage by the inhibition of biosynthesis of their cell wall mucopeptide. *See* PLANT CELL.

The antibiotic properties of penicillin were first recognized by A. Fleming in 1928 from the serendipitous observation of a mold, *Penicillium notatum*, growing on a petri dish agar plate of a staphylococcal culture. The mold produced a diffuse zone which lysed the bacterial cells. Commercial production of penicillin came from the pioneer work of E. Chain and H. W. Florey in 1938, first in England and then in the United States, where it was developed from an academic project into a collaborative war effort between industry and government research. Penicillin (as penicillin G) was made available to the allied troops in Europe in the latter part of the World War II.

Penicillin is produced from the fungal culture *P. chrysogenum* that was isolated from a moldy cantaloupe. The pharmaceutical industry uses highly mutated strains cultured in large, highly aerated, stirred tank fermentors controlled to optimize antibiotic production and to efficiently use raw materials such as corn syrup, corn steep liquor, and cottonseed flour. Temperature, pH, dissolved oxygen, soluble nitrogen and ammonia levels, and sugar feed rates are important control factors. The biosynthesis of penicillin is known in detail, and all the enzymes involved in the formation of this secondary metabolite have been isolated and purified.

Penicillin G, the most commonly fermented penicillin, is produced by the addition of a precursor phenylacetic acid to the growing culture. Use of phenoxyacetic acid as a precursor produces penicillin V. Both penicillins are recovered by extraction into organic solvents at acid pH and precipitation as their potassium or sodium salt.

Penicillin G is generally given by injection against penicillin-sensitive streptococci such as pneumococci (meningitis), and in treatment of endocarditis and gonorrhea. Penicillin G procaine salt is used for intramuscular injection to provide quick distribution of the antibiotic. Penicillin G benzathine salt is used for slower release. Penicillin G can be given in combination with probenecid, a compound which delays urinary excretion. Penicillin V is acid stable and is usually given orally. It is effective in the treatment of upper respiratory infections and periodontal work. *See* GONORRHEA; MENINGITIS.

More effective semisynthetic penicillins are produced by coupling different side chains to the active penicillin nucleus 6-aminopenicillanic acid. This nucleus is produced from either penicillin G or penicillin V by using specific immobilized enzymes. The various side chains confer different antibiotic properties on the penicillin.

The fermented penicillin G and penicillin V are susceptible to destruction by an enzyme (beta-lactamase) produced by certain bacteria which makes them resistant. The penicillins methicillin, oxacillin, nafcillin, cloxacillin and dicloxacillin all are resistant to hydrolysis by beta-lactamases and are used to treat staphylococcal infections. Cloxacillin and dicloxacillin are used orally. Ampicillin and amoxicillin are penicillins with extended spectra as they are effective against many gram-negative bacteria. They are used mainly orally against streptococci and other respiratory-tract pathogens, including *Haemophilus influenzae*, in the treatment of sinusitis, bronchitis, and pneumonia. They are used extensively in pediatrics and against *Listeria monocytogenes* and *Salmonella* spp. *See* HAEMOPHILUS; SALMONELLOSES; STREPTOCOCCUS.

Hetacillin, pivampicillin, and bacampicillin are effective prodrug forms of ampicillin. Amoxicillin is formulated with a beta-lactamase inhibitor clavulanic acid as Augmentin for the treatment of ampicillin-resistant infections. Carbenicillin, ticarcillin, azlocillin, mezlocillin, and piperacillin were developed to combat pseudomonad, enterobacter, and serratia infections resistant to ampicillin. They are used mainly to treat urinary *Pseudomonas aeruginosa* infections, sepsis from burns, and chronic infections of the respiratory tract. *See* ANTIBIOTIC; DRUG RESISTANCE; PSEUDOMONAS.

D. A. Lowe

Bibliography. I. Kawamoto, *Antibiotics I: Lactams and Other Microbial Agents*, 1992; S. F. Queener, J. A. Webber, and S. W. Queener (eds.), *Beta-Lactam Antibiotics for Clinical Use*, 1986.

Penis

The male organ of copulation, or phallus. In mammals the penis consists basically of three elongated masses of erectile tissue. The central corpus spongiosum (corpus urethrae) lies ventral to the paired corpora cavernosa. The urethra runs along the underside of the spongiosum and then normally rises to open at the expanded, cone-shaped tip, the glans penis, which fits like a cap over the end of the penis. Loose skin encloses the penis and also forms the retractable foreskin, or prepuce. The organ is held firmly in place by fibrous tissue and ligaments that bind it to the posterior surface of the pubic arch. In some mammals, although not in man, there may be spines on the glans penis, or there may be a bone (the baculum) within the penis; both of these characters are useful in classifying certain types of mammals.

Erection of the penis is caused by nervous stimulation resulting in engorgement of the spiral helicine arteries and the plentiful venous sinuses of the organ. In most mammals other than Primates the penis is retracted into a sheath when not in use.

In submammalian forms the penis is not as well developed. Crocodilians, turtles, and some birds have a penis basically like that of mammals, although not as well developed, lying in the floor of the cloaca. When erected, it protrudes from the cloaca and functions in copulation. In most birds the penis is rudimentary. Snakes and lizards have paired hemipenes, which are quite different organs serving the same function. They are sacs in the posterior wall of the cloaca which can be everted and used as intromittent organs. They are retracted by muscles and lack erectile tissue. The surface of the hemipenis is frequently covered with spines and ornamentation important in classifying these animals.

Other vertebrates lack a penis, although various functionally comparable organs may be developed such as the claspers on the pelvic fins of sharks and the gonopodia on the anal fins of certain teleost fishes. *See* COPULATORY ORGAN. Thomas S. Parsons

Pennatulacea

An order of the subclass Alcyonaria, commonly called the sea pens. These animals lack stolons and live with their bases embedded in the soft substratum of the sea. The colony consists of a distal rachis bearing many polyps and a polypless proximal peduncle, whose terminal end sometimes expands to form a bladder. The colony of *Pennatula* looks like a feather (**Fig. 1***a*), being formed of numerous secondary polyps which arise from leaf-shaped lateral expansions of the very elongated primary axial or terminal polyp. In the other form of colony the polyps arise directly from the primary one, as in *Veretillum* and *Renilla* (Fig. 1*b* and *c*) and *Cavernularia* (**Fig. 2**). The colony has a horny unbranched axial skeleton composed of pennatulin and some calcium

Fig. 2. *Cavernularia habereri* (preserved state).

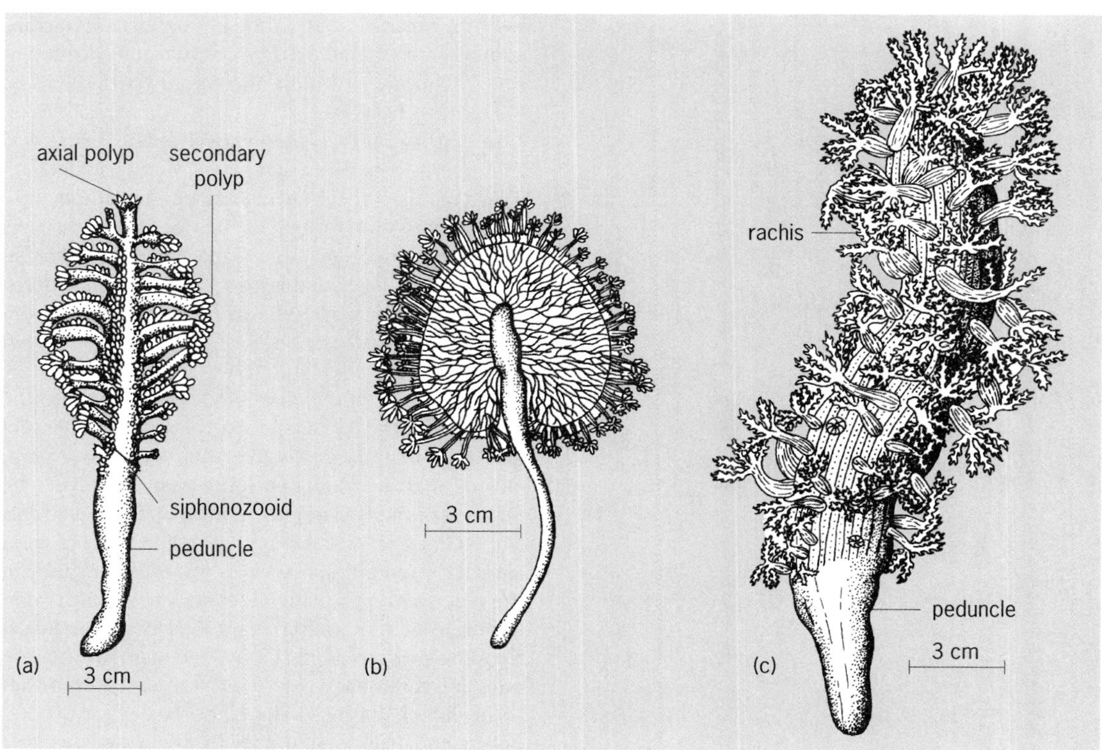

Fig. 1. Pennatulacea colonies. (*a*) Young colony of *Pennatula phosphorea*. (*b*) *Renilla amethystina*. (*c*) *Veretillium cynomorium*.

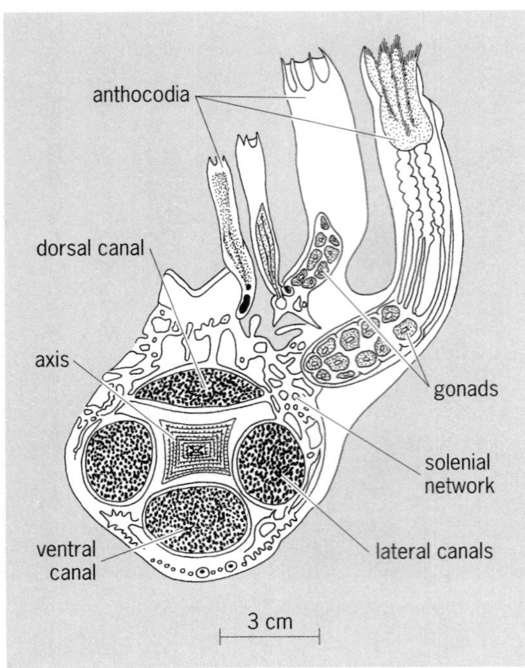

Fig. 3. Cross section of *Funiculina quadrangularis*.

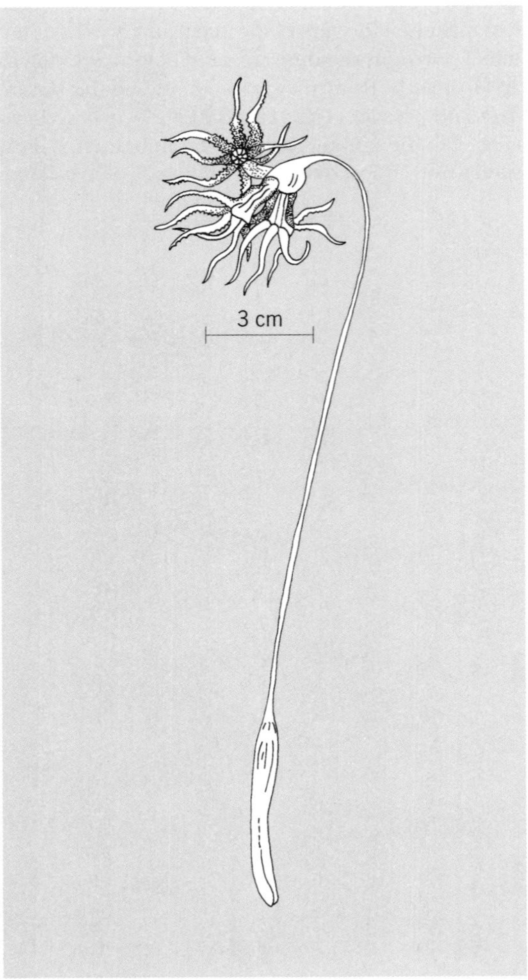

Fig. 4. *Umbellula encrinus*.

carbonate and phosphate, but *Cavernularia*, including the luminous species, has a rudimentary one and *Renilla* lacks a skeleton. Smooth, calcareous spicules in oval, rod, spindle, and needle shapes are found in the coenenchyme.

The gastrovascular cavity of the primary polyp is divided into four principal canals which run longitudinally around the axial skeleton through a spongy tissue containing solenial networks (**Fig. 3**). The cavity of each secondary polyp is interconnected by such canal systems. The musculatures are strongly developed, especially in the peduncle. Longitudinal muscles contract the body, and circular ones serve to force the water out of the main canals. In some littoral species the polyp contracts or expands according to the tidal or diurnal and nocturnal rhythms.

All Pennatulacea are dimorphic. The two kinds of polyps resemble those of the other orders. In addition to the genera indicated above, the order also includes *Virgularia*, *Umbellula* (**Fig. 4**), *Pteroeides*, and others. *See* ALCYONARIA. Kenji Atoda

Pennsylvanian

A major division of late Paleozoic time, considered either as an independent period or as the younger subperiod of the Carboniferous. In North America, the Pennsylvanian has been widely recognized as a geologic period and derives its name from a thick succession of mostly nonmarine, coal-bearing strata in Pennsylvania. Radiometric ages place the beginning of the period at approximately 320 million years ago and its end at about 290 million years ago. In northwestern Europe, strata of nearly equivalent age are commonly designated as Upper Carboniferous and in eastern Europe as Middle and Upper Carboniferous. *See* CARBONIFEROUS.

In North America, the Pennsylvanian Period was characterized by the progressive growth and enlargement of the Alleghenian-Ouachita-Marathon orogenic belt, which formed as the northwestern parts of the large continent Gondwana (mainly northwestern Africa, the area that is now Florida, and northern South America) collided against and deformed the eastern and southern parts of the North American continent.

During the Pennsylvanian and continuing into the early part of the Permian, the orogeny generally progressed with a large jawlike rotational movement, closing from northeast to southwest. However, the outlines of the two cratons on opposite sides of this jaw collided in such a way that the continents were irregularly forced against each other. This resulted in the reactivation of many older structures and, more importantly, in forming numerous new northwest trending shear folds and associated tear faults. These new intracratonic structures comprise important large graben basins, half-graben basins, and horst uplifts and half-horst uplifts that were particularly well developed as intermontane troughs and

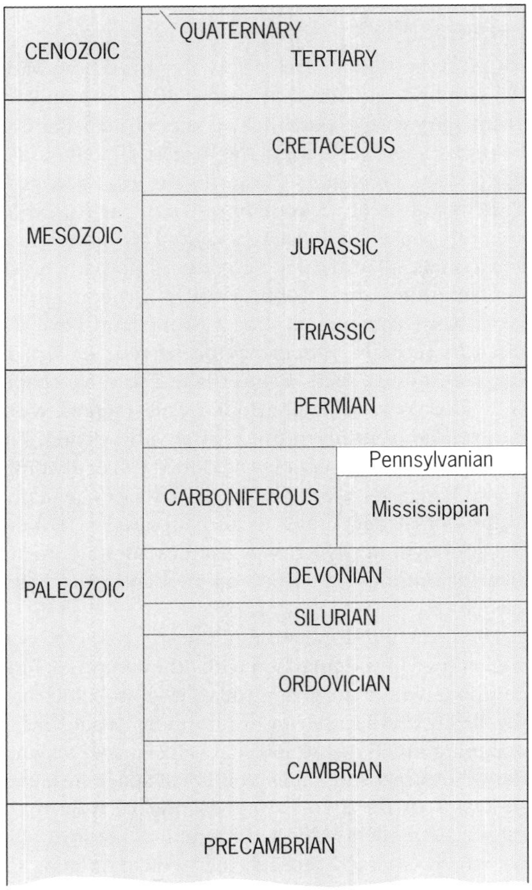

Era	Period	
CENOZOIC	QUATERNARY	
	TERTIARY	
MESOZOIC	CRETACEOUS	
	JURASSIC	
	TRIASSIC	
PALEOZOIC	PERMIAN	
	CARBONIFEROUS	Pennsylvanian
		Mississippian
	DEVONIAN	
	SILURIAN	
	ORDOVICIAN	
	CAMBRIAN	
PRECAMBRIAN		

block-faulted mountains in Oklahoma, Texas, New Mexico, Colorado, Arizona, Utah, Nevada, Wyoming, and Montana. *See* CRATON; FAULT AND FAULT STRUCTURES; GRABEN; HORST; OROGENY.

The uplifted orogenic belt contributed great amounts of clastic sediments onto the edge of North America either as terrestrial clastic deposits where the adjacent colliding blocks were uplifted, or as dark marine turbidites and other marine clastics where one side of the colliding blocks was tectonically depressed beneath the other side. This contrast in the type of deposits resulted because of the irregularies along the colliding margins and the wrinkling and tearing of the North America craton. This is well shown by the mainly nonmarine fluvial clastics that form the Pennsylvanian deposits in the Appalachians sector as compared to the deep-water marine turbidites in the Ouachita and Marathon sectors which formed in fore-deep basins just ahead of the advancing orogenic belt. The intracratonic horsts and half-horsts were in many places sources of large amounts of sediments for adjacent graben basins and half-graben basins. Some of these basins had marine deposits, others had evaporite deposits, and still others were essentially nonmarine. Yet others, such as the Permian Basin of western Texas and southern New Mexico, were more complicated as two rapidly subsiding subbasins, the Midland and Delaware basins, were separated by a much less rapidly subsiding horstlike structure, the Central Basin Platform, and surrounded on the east, north, and west by relatively stable cratonic blocks. *See* BASIN; TURBIDITE.

Much of North America remained a stable, low-lying cratonic platform during the Pennsylvanian and was covered by a relatively thin veneer of shallow-water marine carbonates and marine and nonmarine clastic sediments. These were deposited as sea level repeatedly rose and fell as the polar glaciers of southern Gondwana contracted and expanded. *See* CARBONATE MINERALS; MARINE SEDIMENTS.

In Pennsylvania and along the western part of the present Allegheny Plateau, Pennsylvanian strata are predominantly nonmarine deposits made up of channel sandstones, floodplain shales, siltstones, sandstones, and coals. Farther to the west in the Illinois-Kentucky basin, the nonmarine clastics intertongue through a complex series of distributary deltas with shallow marine beds. In western Missouri and eastern Kansas, marine beds form the greater proportion of the sediments.

During the Pennsylvanian, the western margin of cratonic North America was much farther east of the present coast and extended north-northeastward from southeastern California across eastern Nevada, eastern Idaho, and western Montana into easternmost British Columbia and Alberta. This margin was tectonically unstable and was being pushed westward into an oceanic basin during the Pennsylvanian Period. Later Permian, Mesozoic, and Cenozoic orogenic movements and their structures obscure much of this Pennsylvanian history; however, it is possible to identify several basins and uplifts along the old cratonic margin. Locally, basins on this margin include complete successions of marine sediments across the Mississippian and Pennsylvanian boundary, as at Arrow Canyon near Las Vegas, Nevada, where a stratigraphic reference section for the lower boundary of the Pennsylvanian was established in 1996.

During Pennsylvanian time, the paleoequator extended across North America from southern California to Newfoundland, through northwestern Europe into the Ukrainian region of eastern Europe, and across parts of northern China. This was a time of extensive coal deposition in a tropical belt that appears to have included areas from 15 to 20° north and south of the paleoequator. Coal of this age is abundant and relatively widespread and has great economic importance. Its geographic distribution strongly influenced the location and development of the Industrial Revolution in the mid-1800s. These coals supplied the energy for the rapid increase in the standard of living in western Europe and North America during the late 1800s, were one of the major prizes of the Franco-Prussian war, and influenced many military objectives of World Wars I and II. As a source of readily available combustible fuel, coals of Pennsylvanian age remain of major importance. *See* COAL.

Petroleum is commonly trapped in nearshore

marine deposits of Pennsylvanian age, particularly in carbonate banks near the edge of shelves, in longshore bars and beaches, in reefs and mounds, and at unconformities associated with transgressive-regressive shore lines. Many of these traps contribute significantly to petroleum production and include the Horseshoe atoll of the Texas Panhandle and carbonate banks around the Midland basin in Texas. *See* PETROLEUM.

Pennsylvanian paleogeography changed significantly during the period as the supercontinent Pangaea gradually was formed by the joining together of Gondwana and Laurasia. North America and northern Europe, which had been combined into the continent Laurasia since the late Silurian, and South America and northwestern Africa, which formed the northern part of the continent of Gondwana, came together along the Ouachita–Southern Appalachian–Hercynian geosyncline. The result was an extensive orogeny, or mountain-building episode, which supplied the vast amounts of sediments that make up most of the Pennsylvanian strata in the eastern and midwestern parts of the United States.

Evidence in the form of well-developed tree rings, less diverse fossil floras and faunas, and glacial deposits indicates that temperate and glacial conditions were common in nonequatorial climatic belts during Pennsylvanian time. Climatic fluctuations during the period caused significant increases and decreases in the amount of water that was temporarily stored in the glaciers in Gondwana and contributed to eustatic changes of sea level. The Early Pennsylvanian (Morrowan Stage) was a time of generally low sea levels, and marine sediments are preseved only along lower parts of the cratonic margins and marginal basins. Sea levels rose high enough by the middle of the Atokan Stage (early part of the Middle Pennsylvanian) to extend onto, and across, many of the cratonic platforms. At least 60 of these major cycles of sea-level fluctuations repeatedly exposed much of the cratonic areas during Pennsylvanian time so that on the cratons these strata have many unconformities. *See* CYCLOTHEM; PLATE TECTONICS; UNCONFORMITY.

Charles A. Ross; June R. P. Ross

Bibliography. E. D. McKee et al., Paleotectonic Investigations of the Pennsylvanian System in the United States, *USGS Prof. Pap.*, no. 853 (3 pts.), 1975; C. A. Ross and J. R. P. Ross, Biostratigraphic zonation of late Paleozoic depositional sequences, in C. A. Ross and D. Haman, Timing and Depositional History of Eustatic Sequence: Constraints on Seismic Stratigraphy, *Cushman Found. Foram. Res. Spec. Publ.*, no. 24, pp. 151–168, 1987; C. A. Ross and J. R. P. Ross, Late Paleozoic sea levels and depositional sequences, in C. A. Ross and D. Haman (eds.), Timing and Depositional History of Eustatic Sequence: Constraints on Seismic Stratigraphy, *Cushman Found. Foram. Res. Spec. Publ.*, no. 24, pp. 137–149, 1987; J. W. Skehan et al., The Mississippian and Pennsylvanian (Carboniferous) Systems in the United States, *USGS Prof. Pap.*, no. 1110-A–DD, pp. A1–DD16, 1979.

Pentamerida

An extinct order of brachiopods that lived from Middle Cambrian to Late Devonian (520 to 390 million years ago) with a peak during the Silurian Period. Pentamerides are among the largest brachiopods, with adults ranging 1–15 cm (0.4–6 in.) in length. The two valves, always composed of calcite ($CaCO_3$), are both convex, with a larger ventral (pedicle) valve and a smaller dorsal (brachial) valve. Externally, most pentamerides are smooth except for concentric growth-line increments, but a proportion have radial ribs (costae) which may be sinuous or sharp-crested in cross section. Internally, the valve space is divided by variably developed walls (septa), with one median septum in the ventral valve dividing it into two, and two septa in the dorsal valve, dividing it into three, for a total of five divisions. The septa in the dorsal valve unite to form a platform termed the spondylium, which was used as an attachment area for the muscles which opened and closed the valves.

The form and disposition of the soft parts are not well known. By comparison with other groups of brachiopods which are living today, it is probable that the fleshy, food-gathering lophophore would have occupied much of the internal valve space, leaving the other organs in a relatively small space near the posterior of the two valves. During early growth stages, all pentamerides had a functional fleshy pedicle which emerged through the pedicle opening in the ventral valve to enable the animal to attach to the substrate. However, in the majority of pentameride genera and families, this function did not persist into adult life; the pedicle atrophied and the pedicle opening became closed with secondary calcite deposits. Most pentamerides lived vertically as adults, with the posterior parts of both valves settled downward into the substrate. These posterior parts of the valves were much thicker and heavier than the anterior parts, and this arrangement was the primary cause of the stability of the animal in this vertical life position. Many pentamerides, including *Pentamerus*, which lived during the Early Silurian, occurred in enormous quantities, dominating the midshelf benthic community, which is named after them. In some cases the pentameride shells were packed as tightly as modern oyster beds, and thus occur in rock-forming quantities after death.

Within the order Pentamerida there are currently 215 genera, which have been described from Paleozoic rocks in all continents. The order is subdivided into the rather smaller suborder Syntrophidina, with 90 genera ranging in age from the Middle Cambrian to the Early Devonian, and the suborder Pentameridina, which ranged from the Middle Ordovician to the Late Devonian (see **illus.**). The acme of the syntrophidines was in the Early Ordovician; for example, in the paleocontinent of Baltica, which included Scandinavia eastward to the Urals, the endemic syntrophidine *Lycophoria* occurred in great abundance, dominating its shallow to midshelf community and

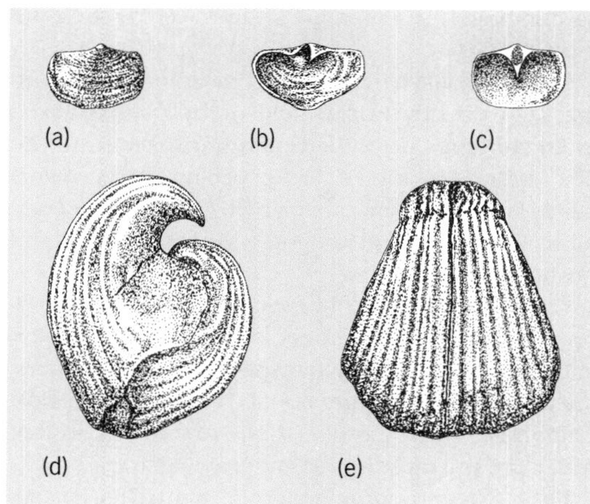

Pentamerida. (*a–c*) *Imbricatia* (Syntrophiidina): (*a*) pedicle valve exterior and (*b*) interior; (*c*) brachial valve interior. (*d–e*) *Conchidium* (Pentameridina): (*d*) lateral view and (*e*) ventral view. (*After R. C. Moore, ed., Treatise on Invertebrate Paleontology, pt. H, Geological Society of America and University of Kansas Press, 1965*)

also providing a useful indicator to the margins of the old continent. *See* BRACHIOPODA. L. R. M. Cocks

Bibliography. *Treatise on Invertebrate Paleontology*, vol. H: *Brachiopoda*, University of Kansas Press and Geological Society of America, 1965, 2d ed., 2001.

Pentastomida

A class of bloodsucking arthropods, parasitic in the respiratory organs of vertebrates, that frequently are referred to as the Linguatulida or tongue worms. The adults are vermiform, with a short cephalothorax and an elongate, annulate abdomen that may be cylindrical or flattened. On the ventral surface the cephalothorax bears two pairs of hooklike, retractile claws. Respiratory, circulatory, and excretory organs are absent. The digestive tract is a straight tube with cuticle-lined anterior and posterior ends. The nervous system consists of subesophageal ganglia and a circumesophageal ring. Paired nerves extend from the ganglia to cephalic structures, and a midventral pair extends posteriad the length of the body. The sexes are separate; the females are two to five times the size of the males. The male genital pore is medial and ventral, at the anterior border of the abdomen. The female genital pore is anterior in the Cephalobaenida but located at or near the posterior end of the body in the Porocephalida.

The class is divided into two orders: the Cephalobaenida, a more primitive group and the Porocephalida, a more specialized one. The first has six-legged larvae and the other four-legged larvae. The mitelike form of the larvae, with short stumpy legs, demonstrates relationship to the arthropods. Characteristic arthropod features include the presence of (1) jointed appendages in the larvae; (2) stigmata or breathing pores in the body wall; (3) specialized reproductive organs, especially those of the male; and (4) ecdysis or molting of larvae and nymphs. More than 50 species have been described.

These animals are parasitic, both as larvae and as adults, in a wide variety of vertebrates, chiefly in tropical regions. With the exception of two genera—*Reighardia*, whose only species occurs in birds, and *Linguatula*, with species only in mammals—all other pentastomes are parasites of reptiles. Adults occur in snakes, lizards, turtles, and crocodiles. Larval stages are encysted in fishes, amphibians, reptiles, birds, and mammals. Life cycles are not well known; usually an intermediate or transfer host harbors the larval stage, but development may be completed in a single host. Late larval stages are sometimes described as nymphs. Fertilization is internal, and the eggs contain fully formed larvae when passed from the uterus. The eggs are ingested by a vertebrate, and the larva emerges in its digestive tract. The larva bores through the tissues and soon encapsulates. In the capsule it undergoes metamorphosis and attains the adult form, except that it has a circle of spines around each annulus and is not sexually mature. Some authors believe that the nymph leaves the cyst and migrates to another location, where it encysts again. When liberated in the intestine of the final host, the larvae migrate to the lungs or air passages, where they become sexually mature. Human infection occurs frequently in Africa and Asia, where humans are accidental intermediate hosts of the nymphal form. The liver is a common site of infection, and large numbers of larvae may produce serious and even fatal effects. *See* ARTHROPODA; CEPHALOBAENIDA; POROCEPHALIDA.

Horace W. Stunkard

Bibliography. H. R. Hill, Annotated bibliography of the Linguatulidae, *Bull. S. Calif. Acad. Sci.*, 47:56–73, 1948; S. P. Parker (ed.), *Synopsis and Classification of Living Organisms*, 2 vols., 1982; L. W. Sambon, A synopsis of the family Linguatulidae, *J. Trop. Med. Hyg.*, 25:188–208, 391–428, 1922; C. W. Stiles and A. Hassal, Key-catalogue of the Crustacea and arachnoids of importance in public health, *Hyg. Lab. Bull.*, no. 148, pp. 197–289, 1927.

Pentlandite

A mineral having composition $(Fe,Ni)_9S_8$. Pentlandite is the major ore of nickel. It crystallizes in the isometric system, but crystals are rare. It is usually massive, showing a well-defined octahedral parting. The hardness is 3.5–4 (Mohs scale) and the specific gravity varies from 4.6 to 5.0, depending on the ratio of iron to nickel; greater amounts of iron cause an increase in the specific gravity. The luster is metallic and the color yellowish bronze. Pentlandite is usually associated with pyrrhotite, which it closely resembles in appearance, but the two can be distinguished by the octahedral parting and lack of magnetism of pentlandite. It is found at many localities in small amounts, but its chief occurrence is at Sudbury, Ontario, where it is mined on a large scale as a nickel ore. *See* NICKEL. Cornelius S. Hurlbut, Jr.

Pepper

The garden pepper, *Capsicum annuum* (family Solanaceae), is a warm-season crop originally domesticated in Mexico. It is usually grown as an annual, although in warm climates it may be perennial. This species includes all peppers grown in the United States except for the "Tabasco" pepper (*C. frutescens*), grown in Louisiana. Other cultivated species, *C. chinense*, *C. baccatum*, and *C. pubescens*, are grown primarily in South America. Some 10–12 strictly wild species also occur in South America. Peppers are grown worldwide, especially in the more tropical areas, where the pepper is an important condiment. *Piper nigrum*, the black pepper, a tropical climbing vine, is botanically unrelated to the common pepper.

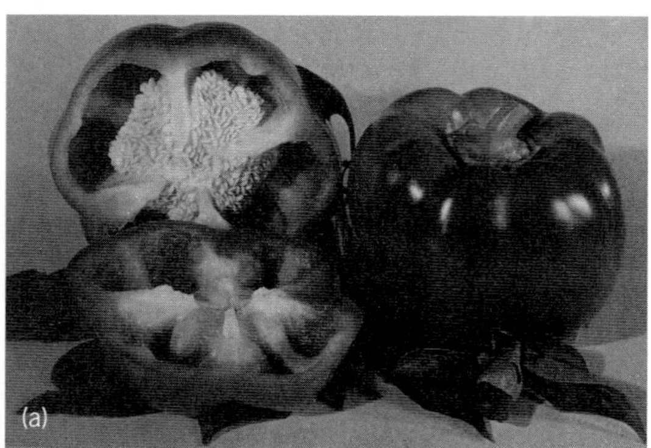

Pepper. (*a*) Sweet pepper. (*b*) Long Red cayenne pepper plant. (*W. Atlee Burpee Co.*)

Cultivation and harvesting. Propagation is by planting seed directly in the field or by plants started in greenhouses or hotbeds and transplanted to the field after 6–10 weeks. Field spacing varies: plants 12–24 in. (30–60 cm) apart in 30–36 rows are common. Long warm, but not hot, seasons favor yield and quality.

Green peppers are harvested when fully developed but before appearance of red or yellow color, about 60–80 days after transplanting. Hot peppers, picked red ripe, require about 70–90 days. Florida, California, North Carolina, and New Jersey, in that order, are the most important growing areas.

Uses. Sweet (nonpungent) peppers (**illus.** *a*), harvested fully developed but still green, are widely used in salads or cooked with other foods. Popular varieties (cultivars) are California Wonder and Yolo Wonder. Perfection pimento, harvested red ripe, is used for canning. Paprika is made from ripe red pods of several distinct varieties; the pods are dried and ground. *See* PAPRIKA; PIMENTO.

Hot (pungent) peppers (illus. *b*) may be picked before ripening for fresh use, pickles, or canning. "Wax" peppers are yellow instead of green before ripening. Ripe pods dried and ground are used to make red pepper powder or mixed with spices to make chili powder. Cayenne and Anaheim Chili, respectively, are the major varieties for grinding. Hot sauce is made from the fresh fruit of the Tabasco pepper ground with vinegar. Pungency is due to the compound capsaicin, located in fragile glands on the internal partitions or "ribs" of the fruit. Seed and wall tissue are not pungent.

The ripe color of most varieties is red, a few varieties are orange-yellow, and in Latin America brown-fruited varieties are common. Nutritionally, the mature pepper fruit has three to four times the vitamin C content of an orange, and is an excellent source of vitamin A. *See* ASCORBIC ACID; VITAMIN A.

Paul G. Smith

Diseases. Virus diseases are the most common field problem of peppers throughout the world. Four potyviruses—pepper mottle, pepper veinal mottle, potato Y, and tobacco etch—are particularly troublesome. Potato Y virus is more widely distributed and known to be present on pepper in Africa, Asia, Australia, the Americas, and Europe, while pepper veinal mottle is restricted to Africa, and pepper mottle to Central and North America. Tobacco etch is known to occur on pepper in North America, but its presence elsewhere has not been established. Early infection with each of these viruses causes stunted plants with vein-banding mottle leaf symptoms and fruits reduced in size and severely disfigured in shape, with streaked ovulary walls.

Tobacco mosaic virus causes only a transient, chlorotic mottle and some reduction in yield, but early infection with cucumber mosaic may result in multiple symptoms that may include severe stunting, systemic granular mottling to shoestring leaf malformation, and extreme reduction in size and number of fruits set. Oak-leaf, concentric line patterns in leaves are caused by alfalfa mosaic infection. Tomato spotted wilt virus causes conspicuous chlorotic and necrotic ring and line patterns on leaves and fruits, with progressive necrosis of one or more growing tips.

Delay in field incidence of potato Y virus has been reported from repeated application of oil sprays. Cultivars with resistance to three of the potyviruses and tobacco mosaic, namely, Yolo Y, Florida VR-2, and Delray Bell, have been developed at the Florida Agricultural Experiment Station.

Bacterial leaf spot (caused by *Xanthomonas vesicatoria*) results in dark brown, translucent lesions that coalesce and cause defoliation under warm, rainy, and windy conditions. Pathotypes of the bacterium have been distinguished. Copper fungicides provide adequate control, except under conditions favorable for the disease.

Leaf spot caused by *Cercospora capsici* [mostly circular lesions to 0.4 in. (1 cm) in diameter with gray-white centers and dark margins] may cause severe defoliation if lesions are abundant. Carbamate fungicides have been recommended for control.

Phytophthora capsici causes a rapid, wet, systemic rot of plants of all ages. The disease spreads quickly in wet weather, when control is practically impossible, but the disease usually disappears when dry weather returns.

Bacterial canker (caused by *Corynebacterium michiganense*) has been found to infect pepper naturally in Australia, California, and Israel. *See* PLANT PATHOLOGY.

A. A. Cook

Peppermint

The mint species *Mentha piperita* (family Lamiaceae), a sterile interspecific hybrid believed to have occurred in nature from the hybridization of fertile *M. aquatica* with fertile *M. spicata*. Peppermint grows in wet, marshy soil along streams in Europe at latitudes of 40–55°. Overwintering stolons, the leafless underground stems, produce new plants, giving the species a perennial habit.

Strains. Numerous clonal strains brought by pioneers for medicinal use in colds and stomach disorders or as flavoring herbs have persisted in North America in wet areas for more than 100 years. These and a few strains cultivated prior to 1890 are frequently called American peppermint to distinguish them from the single clonal strain called Black Mitcham peppermint, obtained from Mitcham, England, near London. Due to its excellent flavor and aroma, Black Mitcham, with a red anthocyanin stem color, became the exclusive cultivar source of United States peppermint oil from 1910 to 1972. A similar, green-stemmed strain called White Mitcham proved less winterhardy and is no longer cultivated. Irradiation breeding during 1955–1972 produced two *Verticillium*-resistant varieties. These were vegetatively propagated and distributed to farmers by certified seed-producing agencies at the state universities of the mint-producing states. Mitcham peppermint is also grown in Bulgaria, Italy, and Russia and to a limited extent in other countries but is restricted by its long-day–short-night photoperiod to an area north of 40° latitude that is not subject to severe mid-June freezes.

Cultivation and harvesting. In mint farming the dormant stolons are scattered by hand in 10–12-in.-deep (25–30-cm) furrows and covered with soil in rows 36–38 in. (90–95 cm) apart that can be cultivated for weed control with a maize cultivator. In the United States machine planting and preemergence application of terbicil herbicides (such as Sinbar) for weed control are standard practice to minimize labor costs. Peppermint has rather high water requirements; hence supplementary overhead or sprinkler irrigation may be used on the organic soils of the Midwest area (southern Michigan, northern Indiana, Wisconsin) and must be used on alluvial soils of the Willamette Valley of western Oregon and the Madras area of central Oregon. In all these areas the first-year crop is row mint, and all subsequent crops are meadow mint in a solid stand, whereas rill or ditch irrigation necessitating row culture has generally been used on the volcanic soil of eastern Washington and in part on alluvial soils of the Snake River Valley of Idaho.

The crop is mown and windrowed, as hay, about the time that peak vegetative growth is attained or flower buds are beginning to form (July 15–September 10). The windrowed, partially dried hay is picked up by a loader that chops the hay and blows it into a mobile metal distillation tank mounted on four wheels or on a dump truck chassis.

Peppermint oil. The oil of commerce is obtained by steam distillation from the partially dried hay. Steam under pressure traverses the chopped hay in the mobile tank, removing the oil, and is then cooled in a condenser. The oil floats on the water in a receiving receptacle and is removed to metal drums for storage and sale. Stills are expensive since automatic valves and stainless steel construction are needed. The oil is complex, having 3 major constituents and over 80 minor ones that are necessary to the flavor and aroma. Oil yields in pounds per acre (kilograms per hectare) are generally 40–65 (45–73) in the Midwest, 65–85 (73–96) in the Willamette Valley and Idaho, and 85–110 (96–125) in eastern Washington. The United States produces 50,000–80,000 acres (202–324 km^2) of peppermint. The main uses of peppermint oil are to flavor chewing gum, confectionary products, toothpaste, mouthwashes, and medicines, and as a carminative in certain medical preparations for the alleviation of digestive disturbances. *See* SPICE AND FLAVORING.

Diseases. Principal diseases are Verticillium wilt, which occurs everywhere, and Puccinia rust in the western United States. In the Willamette Valley, where rust is severe, fields are flamed with a propane burner to interrupt the rust cycle in early spring as plants are emerging. Genetic control of Verticillium wilt is through use of the new varieties.

Merritt J. Murray

Pepsin

A proteolytic enzyme found in the gastric juice of mammals, birds, reptiles, and fish. It is formed from a precursor, pepsinogen, which is found in the stomach mucosa. Pepsinogen is converted to pepsin either by hydrochloric acid, naturally present in the stomach, or by pepsin itself. A pepsin inhibitor, peptide, remains attached to the pepsin molecule at pH values above 5.0, preventing activation of the enzyme. The approximate molecular weight of pepsin is 34,500 and that of pepsinogen is 42,600. Maximum activity occurs at pH 1.8, and broad specificity for peptide bonds is demonstrated. *See* ENZYME; PH.

Pepsin is prepared commercially from the glandular layer of fresh hog stomachs. Official National Formulary (NF) pepsin has an activity of 1:3000 and digests 3000–3500 times its weight of egg albumin in 2.5 h at 126°F (52°C). Activities up to 1:30,000 are available commercially.

Pepsin is a part of the crude preparation known as rennet, used to curdle milk in preparation for cheese manufacture. Pepsin is also used to modify soy protein and gelatin (providing whipping qualities), to modify vegetable proteins for use in nondairy snack items, to make precooked cereals into instant hot cereals, to prepare animal and vegetable protein hydrolysates for use in flavoring foods and beverages, and to prevent the formation of cloudiness in beer during refrigeration. *See* CHEESE; FOOD ENGINEERING; MILK.

Myron Solberg

Bibliography. H. Holzer and J. Tschesche (eds.), *Biological Functions of Proteinases*, 1980; T. Nagodawithana and G. Reed (eds.), *Enzymes in Food Processing*, 3d ed., 1997; L. Stryer, *Biochemistry*, 4th ed., 1995.

Peptide

A compound that is made up of two or more amino acids joined by covalent bonds which are formed by the elimination of a molecule of H_2O from the amino group of one amino acid and the carboxyl group of the next amino acid. Peptides larger than about 50 amino acid residues are usually classified as proteins.

In the reaction shown, the α linkage between the α-amino group of alanine and the α-carboxyl group of glutamic acid is by far the most common type of linkage found in peptides. However, there are a few cases of γ linkage in which the side-chain carboxyl group of one amino acid is linked to an α-amino group of the next amino acid, as in glutathione and in part of the structure of the bacterial cell wall. Linear peptides are named by starting with the amino acid residue that bears the free α-amino group and proceeding toward the other end of the molecule.

```
    COOH
     |
    CH2
     |                                  -H2O
    CH2           +          CH3       ----->
     |                        |
 H — C — NH2           H2N — C — H
     |                        |
    COOH                     COOH
L-Glutamic acid          D-Alanine

                 COOH
                  |
                 CH2
                  |
                 CH2
                  |
              H — C — NH2        CH3
                  |               |
                  C ——————— N — C — H
                  ||        H     |
                  O              COOH
            L-Glutamyl-D-alanine
```

Sometimes the configuration of the asymmetric carbon atom of each residue is included in the name, as shown in the reaction. A three-letter notation is convenient to use, for example, L-Glu-D-Ala.

Occurrence. Peptides of varying composition and length are abundant in nature. Glutathione, whose structure is γ-L-glutamyl-L-cysteinyl-glycine, is the most abundant peptide in mammalian tissue. Hormones such as oxytocin (8), vasopressin (8), glucagon (29), and adrenocorticotropic hormone (39) are peptides whose structures have been deduced; in parentheses are the numbers of amino acid residues for each peptide. Some of the hormone regulatory factors which are secreted by the hypothalamus are peptides that govern the release of hormones by other endocrine glands. In the peptides indigenous to mammalian tissues all of the amino acid residues are of the L-configuration.

Many of the antimicrobial agents produced by microorganisms are peptides that can contain both D- and L-amino acid residues. Penicillin contains a cyclic peptide as part of its structure. Other peptide antibiotics include the gramicidins, the tyrocidines, the polymyxins, the subtilins, and the bacitracins. *See* ANTIBIOTIC.

Synthesis. For each step in the biological synthesis of a peptide or protein there is a specific enzyme or enzyme complex that catalyzes each reaction in an ordered fashion along the biosynthetic route. However, although the biological synthesis of proteins is directed by messenger RNA on cellular structures called ribosomes, the biological synthesis of peptides does not require either messenger RNA or ribosomes. *See* RIBONUCLEIC ACID (RNA); RIBOSOMES.

In the methods most commonly used in the laboratory for the chemical synthesis of peptides, the α-carboxyl group of the amino acid that is to be added to the free α-amino group of another amino acid or peptide is usually activated as an anhydride, an azide, an acyl halide, or an ester, or with a carbodiimide. To prevent addition of the activated amino acids to one another, it is essential that the α-amino group of the carboxyl-activated amino acid be blocked by some chemical group (benzyl oxycarbonyl-, *t*-butyloxycarbonyl-, trifluoroacetyl-) that is stable to the conditions of the coupling reaction; such blocking groups can be removed easily under other conditions to regenerate the free α-amino group of the newly added residues for the next coupling step. In order to ensure that all the new peptide bonds possess the α linkage, the reactive side chains of amino acid residues are usually blocked during the entire synthesis by fairly stable chemical groups that can be removed after the synthesis. For a successful synthesis of a peptide or protein, all of the coupling steps should be complete, and none of the treatments during the progress of the synthesis should lead either to racemization or to alteration of any of the side chains of the amino acids. In the earlier techniques for the chemical synthesis of peptides the reactions were carried out in the appropriate solvents, and the products at each step were purified if necessary by crystallization ("solution method"). An innovation devised by R. B. Merrifield uses a "solid phase" of polystyrene beads for the synthesis. The peptide, which is attached to the resin beads, grows by the sequential addition of each amino acid. The product can be washed by simple filtration at each step of the synthesis. *See* AMINO ACIDS; PROTEIN.

James M. Manning

Peracarida

A superorder of the superclass Crustacea, subclass Eumalacostraca. The Peracarida includes the orders Amphipoda, Cumacea, Isopoda, Mictacea, Mysidacea, Spelaeogriphacea, Thermosbaenacea, and Tanaidacea. In these orders the young develop within the mother's ventral thoracic marsupium, which they leave at an advanced stage of development. The marsupium is formed by from one to seven pairs of membranous oostegites that extend inward from the coxae of the thoracic legs. The eggs or developing young lie free in the space between the ventral surface of the thorax and the overlapping oostegites.

The other basic feature of the Peracarida is an accessory incisor process (lacinia mobilis) in the adult that is known elsewhere only in the primitive class Remipedia. Some larval Decapoda and Euphausiacea have a presumed lacinia mobilis, but adults lack it. The lacinia mobilis, on the left mandible or both mandibles, is formed from modification of the anterior spine of the spine-row. Besides aiding in cutting, the lacinia mobilis helps align the incisor processes for occlusion.

Morphology. As with other Malacostraca, the thorax consists of eight somites, the first of which is fused with the head. A carapace is present except in the Amphipoda, Isopoda, and Mictacea; in the last, small carapace folds cover the bases of the mouthparts posterior to the mandibles. When present, the carapace does not coalesce with more than four thoracic somites, as follows: one segment in the Spelaeogriphacea and the Thermosbaenacea, one or two in the Tanaidacea, one to four in the Mysidacea, and three or four in the Cumacea.

Epipodites are present on the maxillipeds of the orders with a carapace, and their movements maintain water flow along the inner surface of the carapace. In the Mysidacea, suborder Mysida, this inner surface is highly vascular, and gaseous exchange takes place through it. Members of the Spelaeogriphacea and the mysid suborder Lophogastrida, however, have epipodial gills on some of the thoracic legs, and in the Cumacea the epipodite of the maxilliped forms a complex gill. Isopoda have flattened epipodites on the maxillipeds; the epipodites are not branchial but form a lateral wall over the other mouthparts. Respiration in isopods takes place through the surface of the pleopods. In Amphipoda, gaseous exchange occurs in the epipodial gills attached to the inner surfaces of the coxae of the thoracic legs. There are no epipodites on the amphipod maxillipeds.

The thorax is followed by an abdomen of six segments bearing up to five pairs of biramous swimming legs or pleopods. The sixth segment has a pair of uropods, which, together with the telson, forms a tail fan. Pleopods are absent in all female and some male Cumacea, may be reduced or absent in female Tanaidacea, and may be rudimentary in female Mysidacea. The abdominal appendages are sharply divided into two groups in Amphipoda; the three anterior pairs are turned forward and are natatory, while the last three, all called uropods, are turned backward. In the Isopoda, some or all of the pleopods function as gills, and the surfaces may have folds to increase the respiratory area. One of the anterior pairs of pleopods—or in the suborder Valvifera, the uropods—may be modified into an operculum protecting the delicate pleopods. One or two pairs of pleopods of some Mysidacea and Isopoda are modified in the males to assist in the transfer of the sperm to the female.

The embryo in the Amphipoda is ventrally concave and lies with the dorsal side toward the outside of the egg. The position is reversed in the other orders. Mysidacea and Amphipoda leave the egg with a full complement of appendages; hatchlings in the other orders lack the eighth thoracic leg. Lophogastrid Mysidacea have both antennal and maxillary nephridia, the latter being small. Amphipoda and nonlophogastrid Mysidacea have antennal nephridia, while Cumacea, Tanaidacea, and Isopoda have maxillary nephridia.

Relationships among Peracarida. There is no consensus as to relationships within the Peracarida, and a number of diverse "family trees" have been proposed, resulting from differences in whether particular characters are considered to be primitive or specialized. For example, the presence of stalked eyes or carapace is considered primitive by some, advanced by others. The **illustration** shows one of a number of possible patterns of relationships. The taxon Peracarida itself is considered invalid or doubtful by some authorities. With so much of the family tree lost

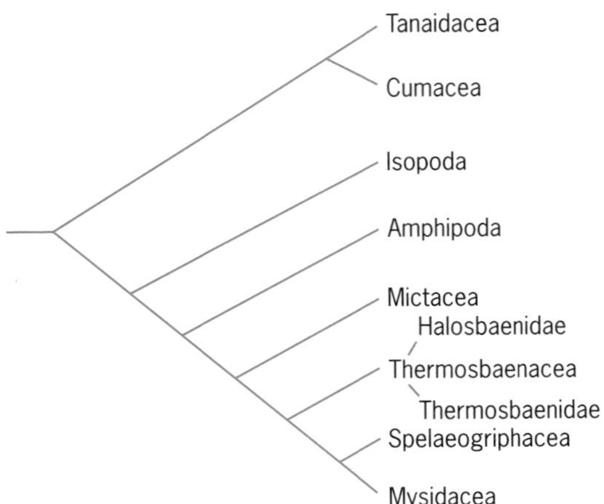

Phylogenetic relationships within the Peracarida.

forever, attempts to restore it will always be somewhat subjective. *See* CRUSTACEA; EUMALACOSTRACA.

Thomas E. Bowman

Bibliography. L. G. Abele (ed.), *The Biology of Crustacea*, vol. 1: *Systematics, The Fossil Record, and Biogeography*, 1982; R. Lankester (ed.), *A Treatise on Zoology*, pt. 7, fasc. 3, 1909; F. R. Schram, *Crustacea*, 1986; F. R. Schram (ed.), *Crustacean Issues 1: Crustacean Phylogeny*, 1983.

Percent

A ratio comparison of two quantities expressed by using 100 equal parts, or hundredths; symbolized %. There are three major uses of percent: part of a whole, rate, and comparison of any two quantities.

Part of a whole. The basic idea of percent is as a ratio that shows a part of a whole. The technical name for the whole is base.

Ratios as percents. If 89 out of 100 problems are correct, the part/whole comparison shows 89/100 or 89% correct. If the whole is not already divided into equal parts, an equivalent ratio to 100 is found.

Certain ratios are easy to express as hundredths. When 3 baskets are made in basketball out of 10 attempts, an equivalent ratio using 100 is found from which the percent is obvious: 3/10 = 30/100 = 30%.

For 35 hits out of 126 times at bat, the ratio 35/126 is not easily expressed as a ratio using 100. Hundredths will be more obvious by dividing 35 by 126, and then reading the number of hundredths to find the percent: $35 \div 126 \approx 0.278$ or 27.8%. The percent is obtained by moving the decimal point two places to the right.

Another method is to set up a part-to-whole proportion with n representing the part out of 100, as in Eqs. (1). The ratio is 27.8%.

$$\frac{\text{part}}{\text{whole}} \quad \frac{35}{126} = \frac{n}{100} \quad \frac{\text{part}}{\text{whole}}$$

$$126n = 3500 \text{ (multiplying each side by } 126 \times 100) \quad (1)$$

$$n = 27.8$$

The usage for the word percentage has changed over time from its original meaning of the "part." In sports statistics, pct. for percentage really means average, not percent. If a batting average of .278 were a percent, it would be reported as 27.8%.

Discounts and savings from bargain sales are examples of a part of a whole. A discount of $35 (the part of the whole) on a coat with an original price of $105 (the whole or base is 105) is a savings of 35/105 or 33.3% off the original price, as in Eq. (2).

$$\frac{35}{105} = \frac{1}{3} = \frac{33^1/_3}{100} = 33^1/_3\% \text{ or } 33.3\% \quad (2)$$

Percent decrease shows a relative amount of decrease. The amount of decrease, the part, is compared to the original amount, the whole or base. If the value of a house drops $9000 from an original price of $80,000, the percent decrease is 9000/80,000, 0.1125, or 11.25%.

Percent of a number. To find a part when the percent and the whole (base) are known, a percent of a number is found.

There are three ways to find the part saved in buying a chair marked at $300 (the whole or base) with a 25% discount: using a fraction and division, as in notation (3); using multiplication by a fraction or decimal, as in notation (4); or using a proportion, as in Eqs. (5).

$$\begin{aligned} 25\% \text{ of } \$300 &= {}^1/_4 \text{ of } \$ \\ {}^1/_4 \text{ of } \$300 &= \$300 \div 4 \text{ or } \$75 \end{aligned} \quad (3)$$

$$25\% \text{ of } \$300 \text{ is } 0.25 \times \$300 \text{ or } {}^1/_4 \times 300 = \$75 \quad (4)$$

$$\begin{aligned} \frac{25}{100} &= \frac{n}{300} \\ 25 \times 300 &= 100n \\ n &= 75 \end{aligned} \quad (5)$$

The advantage of the first method is that it can be done by easy computation, often mentally. The second method is easier on a calculator. The third method illustrates the proportional connection for percent. In this example, the sale price is $300 − $75 or $225. The sale price can also be found by taking 75%, that is, 100% − 25%, of $300.

Income tax is a part (the tax) of the whole or base (income). For example, a tax of $12,500 on income of $65,000 is 12,500/65,000 or 19.2%.

Calculator use. On calculators, percents can be expressed as decimals and calculations can be done with the decimals. The amount saved in a 35% sale on a $467 desk can be keystroked as .35 × 467 =. If a percent key, %, is used, conversions are not made to decimals and the % key is pressed last: 467 × 35%. (On some calculators, = is pressed after the %.) After seeing $163.45 as the amount saved, pressing the minus sign (−) and = keys shows the balance after the discount, $467 − $163.45 = $303.55.

If $83 is saved on a printer originally priced at $425, keystroking 83 ÷ 425% shows 19.5, a 19.5% saving.

Rate. While percent always means a comparison to 100, percent can show a rate of so many per 100, not so many out of 100. A sales tax of 6% means a rate of 6% for each dollar, and this 6% is in addition to the dollar. The tax amount can often be calculated mentally by multiplying the 6% rate by the number of dollars. The tax on a $30 book is 6 × 30% or 180%, $1.80, and the total cost is $31.80. The total is 106% of the original price.

Methods of calculation are the same. On a chair priced at $275 with a sales tax of 8.5%, the tax is 8.5% of $275. The tax can be found by solving either 0.085 × 275 = *n* or Eq. (6). Note in the proportion that the

$$\frac{8.5}{100} = \frac{n}{275} \qquad (6)$$

$$n = \$23.38$$

bases of 100 and 275 appear as the second number for each ratio. The keystroking on a calculator is 275 × 8.5%, and perhaps = also. After finding the tax, $23.38, it must be added to the price to find the total cost, which is $298.38.

Mental computation is needed, for example, to calculate the amount of tip on a restaurant bill of $62.50, as in notation (7).

10% tip:	15% tip:
$\frac{1}{10}$ of $62.50 = $6.25	10% is $6.25
Move the decimal point	5% more is $3.13
one place to the left	15% is $9.38

20% tip:
10% is $6.25
20% is $12.50
Or divide by 5,
since 20% = $^1/_5$ (7)

Interest paid or interest received is done by using rates. Simple interest at a rate of 6% means $6 per $100 for a full year. If interest is calculated monthly on the unpaid balance and the yearly rate is 18%, the monthly rate is approximately 18% ÷ 12 or about 1.5%. At this rate, a balance of $650 for one month results in interest of 1.5% of $650, or $9.75.

With compound interest, the amount of interest is added each compounding period, and this total amount is subject to compounding for the next period. For example, $1.00 invested at 8% will be worth 108% of $1.00 at the end of one year, or $1.08. The worth at the end of the second year is 108% of $1.08, or $(1.08)^2$. At this same rate, after 10 years the compounded value is $(1.08)^{10} = 2.1589247$, or $2.15, using 1.08 as a factor for 10 times.

Similarly, if the birth rate, above deaths, of a country is 4% per year, evaluating $(1.04)^n$ when $n = 18$ shows that the population will double in about 18 years.

It is through compound interest that an annuity builds up value over time. Income from an annuity depends on the rate on investment as well as the number of years that payment is to be made before the annuity matures.

Comparing any two quantities. Percent is used to compare any two quantities, but special care must be given to the base for the comparison. Comparison of city A with a population of 42,000 people and city B with 67,000 people will depend on the base. A compared to B is 42,000/67,000, 62.7%; A is 62.7% of B. B compared to A is 67,000/42,000, 1.595, or 159.5%; B is 159.5% of A. Total sales are often compared year to year. For example, if sales this year are $240,000 and last year they were $175,000, then this year is 240,000/175,000 or 137.1% of last year, an increase of 37.1%. *See* ARITHMETIC.

Joseph N. Payne

Bibliography. G. D. Allinger and J. N. Payne, *Percent Level G*, 1991; E. F. Krause, *Mathematics for Elementary Teachers*, 1991; H. L. Schoen (ed.), *Estimation and Mental Computation*, 1986.

Perception

Those subjective experiences of objects or events that ordinarily result from stimulation of the receptor organs of the body. This stimulation is transformed or encoded into neural activity (by specialized receptor mechanisms) and is relayed to more central regions of the nervous system, where further neural processing occurs. Most likely, it is the final neural processing in the brain that underlies or causes perceptual experience, and so perceptionlike experiences can sometimes occur without external stimulation of the receptor organs, as in dreams.

In contemporary psychology, interest generally focuses on perception or the apprehension of objects or events, rather than simply on sensation or sensory processes (such as vision or hearing). While no sharp line of demarcation between these topics exists, it is fair to say that sensory qualities are generally explicable on the basis of mechanisms within the receptor organ, whereas object and event perception entails higher-level activity of the brain. For example, color vision appears to be based upon the activity of cone cells in the retina of the eye. Accordingly, an adequate explanation of why a particular hue is seen is provided by an analysis of the activation of those cells that are primarily sensitive to a particular range of wavelengths of light in the visible spectrum. However, the perception of an object's size, shape, orientation, or the like cannot be accounted for solely by events within the retina of the eye or by the direct neural transmission of these events into the brain. *See* COLOR VISION; HEARING (HUMAN); SENSATION; VISION.

Since objects or events are not experienced only through vision, the term perception obviously applies to other sense modalities as well. Certainly things and their movement may be experienced through the sense of touch. Such experiences derive from receptors in the skin (tactile perception), but more importantly, from the positioning of the fingers with respect to one another when an object is

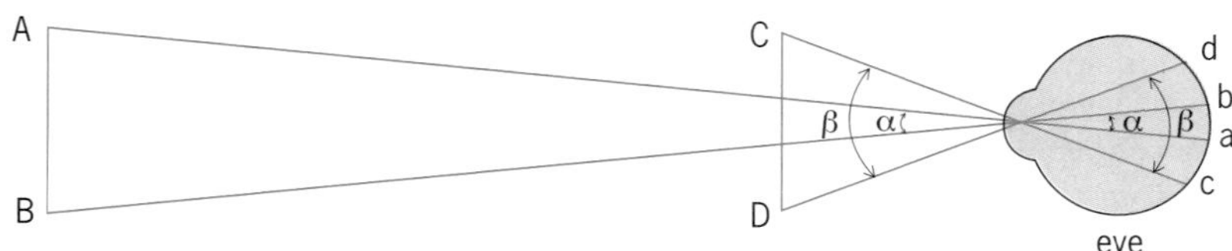

Fig. 1. Objects of equal size A-B and C-D at different distances from the observer produce rentinal images of unequal size a-b and c-d. The size of these retinal images can also be described in terms of the visual angles subtended by the objects at the eye for a-b and c-d.

grasped, the latter information arising from receptors in the muscles and joints (haptic or tactual perception). The position of the parts of the body are also perceived with respect to one another whether they are stationary (proprioception) or in motion (kinesthesis), and the position of the body is experienced with respect to the environment through receptors sensitive to gravity such as those in the vestibular apparatus in the inner ear. Auditory perception yields recognition of the location of sound sources and of structures such as melodies and speech. Other sense modalities such as taste (gustation), smell (olfaction), pain, and temperature provide sensory qualities but not perceptual structures as do vision, audition, and touch, and thus are usually dealt with as sensory processes. In this article the emphasis will be almost exclusively on visual perception. *See* OLFACTION; PAIN; PROPRIOCEPTION.

In many respects, the eye is analogous to a camera. Thus, both eye and camera admit light through a small opening and focus that light by means of a lens upon a light-sensitive surface such that any given point in the environment is represented by a single point on the retina or film. The sum of all these points can be thought of as an image of the external scene. Moreover, the principles of optics apply equally to both eye and camera. For example, distant objects yield smaller images than nearer objects of the same size, and all objects produce images that are inverted and left-right reversed. But perception cannot be explained simply on the basis of the "picture" on the retina, as if nothing more were required but the neural transmission of this picture back into the visual area of the brain. First of all, the eye is in a constant state of tremor, which has important consequences for vision, and eye movements are essential to bring various regions in the scene into the center of the retina so that they can be seen clearly. But more important for the present discussion, if perception were nothing more than the registration in the brain of the retinal image, the perceived size of objects would vary as a function of their distance, the perceived shape of objects would vary as a function of their slant, the perceived orientation of objects would change as the head was tilted, the perceived direction of objects would vary as a function of the position of the eyes within the head, objects would be seen to move when the observer moved, and so forth.

Constancy. By and large, these perceptual properties of objects remain remarkably constant despite variations in distance, slant, and retinal locus caused by movements of the observer. This fact, referred to as perceptual constancy, is perhaps the hallmark of perception and more than any other, serves to characterize the field of perception.

Size. Consider first the problem of size perception. Except at very great distances, objects appear to be about the same size whether seen nearby or farther away. For example, an object at one end of a room will not appear to be any smaller than one of the same size next to the observer despite the fact that the size of the retinal image (or visual angle) subtended by the more distant object will be much smaller than that subtended by the near object (**Fig. 1**). Size perception is constant and thereby correct (or veridical). This would not be true if perception depended exclusively on the size of the object in the retinal "picture." Also, size constancy is perceptual and not simply an act of judgment or knowledge. In other words, objects appear to be the same size at varying distances. It is not the case that they appear to be small at a distance, but that it is realized or reasoned that they are not. This is supported by the fact that animals and young children behave as if their perception of object size, despite variation of object distance, is constant. It is unlikely that they would do so if constancy were a matter of reasoning rather than of perception.

Shape. A similar problem arises in the perception of shape. Object shape also remains perceptually constant. A circle seen from the side projects an elliptical rather than a circular image on the retina. Therefore, if perceived shape were simply a function of retinal-image shape, the circle would appear to be elliptical from most vantage points. The fact that it is seen as circular and not elliptical is referred to as shape constancy.

Orientation and position. The orientation of an object's image on the retina changes when the head is tilted, and the position and stability of an image changes when the observer moves the entire body, head, or eyes. Nonetheless, objects maintain their perceived orientation, location, and stability with such tilts or motions. These properties are referred to as orientation constancy and position constancy, respectively.

Lightness. Whether a surface reflects most or little of the light it receives is a function of its physical characteristics. There is little variation in the perceived lightness of such surfaces—whether they appear black, gray, or white—despite enormous difference in the amount of light falling upon them and in the amount of light reaching the eye from them (luminance). This property is referred to as lightness constancy. *See* LUMINANCE.

Experiments. A typical experiment on constancy, for example, size constancy, may involve a subject viewing an object such as a triangle at a particular distance. The subject is instructed to compare the size of that triangle (called the standard) with a series of other triangles of graduated size shown one at a time (called the comparison series) at a much closer distance; the subject then selects the triangle that appears to be the same size as the standard. The comparison series can be presented alternately in ascending and descending order for several trials and the average match computed. The same procedure is then followed with the standard at an entirely different distance. By using many subjects and averaging their average judgments for each distance, the degree of constancy achieved can be determined for the population sampled for a range of distances. There are alternatives to this manner of presenting the comparison objects, referred to as the psychophysical methods. Experiments of this kind on constancy usually yield average matches that reflect a high degree of constancy except when very great distances are utilized, when the adequacy of distance perception is much reduced, or when young children are tested. *See* PSYCHOPHYSICAL METHODS.

Basis for constancy. While there is as yet no accepted explanation of perceptual constancy, there is general agreement that it is based on taking account of certain additional information available to the perceptual system. For instance, in the above example, the far object is seen to be far away; however, the perceptual system could compute by a process analogous to multiplication that the small retinal image yielded by a distant object represents an object equal in size to that represented by a larger retinal image projected by a nearer object. In other words, the computation would entail multiplying image size by detected distance, and since the size of the image is inversely proportional to distance, perceived size should remain constant as long as the object's distance is correctly perceived. Where distance is underestimated or not adequately registered, the object's size should be computed as small. Therefore at very great distances, size perception no longer remains correct or constant—things then do look diminutive. There is evidence that constancy requires focused attention, and that without it, perception regresses to conform to the retinal image alone.

Veridicality and illusion. Constancy implies veridicality, that is, agreement with the real properties of the object. Veridicality is vital for survival. An animal must be able to perceive that an approaching but still distant object is an adult lion and not a cub, and it must be able to distinguish whether a moving retinal image signifies a moving external thing or is caused by its own motion while viewing a stationary thing. But the process that has evolved—or possibly has been learned—and that leads to veridicality and constancy can often lead to nonconstancy or illusion. That is why investigators take seriously and investigate illusions of all kinds. It is presupposed that such illusions, once understood, will lead to an improved understanding of perceptual processing in general.

Motion perception. Perception is not only of objects but of events as well. The kind of event that has been investigated in great detail is the perception of motion. As already noted, perceived movement cannot simply be explained by the motion of an object's retinal image since image motion caused by observer or eye movement does not lead to perceived object movement. Moreover, an object tracked by smooth-pursuit eye movements will appear to move, although in that case there is essentially no motion of the object's image over the retina. Similarly, an afterimage will appear to move during eye movement even in a completely darkened room. Where ordinarily the movement of the retinal image caused by the moving eye is computed to signify "no object motion," thus yielding position constancy (since the image motion and eye motion are equal in magnitude), the same computational rule must signify "object motion" in the case of the afterimage.

Mechanisms. In any event, motion will be perceived when an object or afterimage changes its angular direction with respect to the observer at a rate above some threshold value, and it does not matter whether the information concerning such change of angular direction is based upon eye movement, with retinal image stationary, or retinal-image movement, with eyes stationary. Clearly it is the combined information from retina and eye that matters, and thus perception is not explicable simply in terms of events in the receptor mechanism. There are cells in the visual nervous system that are responsive to motion of contours over the retina. However, the discharging of these so-called motion-detector cells should not be thought of as the explanation or cause of perceived motion, at least in animals high on the evolutionary scale where the eyes move. Rather these mechanisms should be thought of as providing the information that an image is displacing over the retina. What that information then leads to as far as motion perception is concerned depends upon other information, such as what the eyes are doing at that moment.

Induced movement. Other phenomena also attest to the operation of more complex processes underlying motion perception. For example, when clouds pass in front of the Moon, the Moon appears to move in the opposite direction. This kind of illusion is referred to as induced movement. It is studied in a dark room by moving a large object such as a luminous rectangular perimeter back and forth while the observer views a stationary luminous spot inside the rectangle. Typically the spot appears to move in a direction opposite to that of the rectangle. While induced movement is not yet fully understood, it seems evident that it is based upon the change in position of spot and rectangle relative to one another. The rectangle is taken as the frame of reference, and the relative motion between spot and rectangle is thus interpreted by the perceptual system as motion of the spot. If change of position of one object relative to others is information that can produce an illusion, then it is probable that it is important in daily life. Therefore when an object moves in the environment, there are two independent determinants that account for the perception of it as moving: it changes its angular direction with respect to the observer at an above-threshold rate, and it changes its position relative to neighboring objects. Of the two, the latter may be the stronger factor.

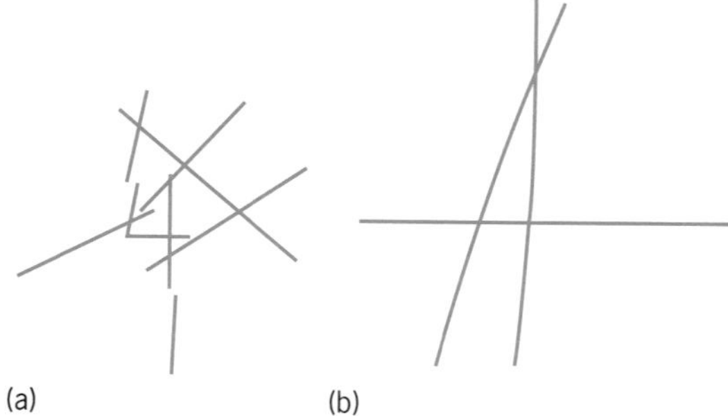

Fig. 2. Perceptual organization. (*a*) The figure of a four is immediately and spontaneously perceived despite the presence of other overlapping and adjacent lines. (*b*) The four, although physically present, is not spontaneously perceived and is even difficult to see when one knows it is there.

Since one's body is itself an object in the field, it would seem to follow that if the body is surrounded by a structure that could serve as a frame of reference for it, that is, for one's own self, motion of that structure would induce movement of the self in the opposite direction. In fact, this effect occurs often in daily life, as when a nearby automobile moves and the observer in another automobile is stationary. This effect is studied in the perception laboratory by surrounding a stationary observer with a moving drum or artificial room. When observers experience themselves as moving, the moving, surrounding room appears to be stationary.

Stroboscopic motion. Still another illusion of motion occurs when a stationary object is first seen briefly in one location, and, following a short interval, is seen in another location. This phenomenon, known as stroboscopic motion (or alternatively, as apparent motion, beta motion, or the phi phenomenon), underlies the perception of motion in movies or television. A moving picture is based on the successive projection of stationary frames in the film strip. Thus there is no motion itself on the screen, but motion is nonetheless perceived. Many erroneously believe that this effect is based on the neural persistence of the image of a frame during the interval between frames, but clearly this only explains the absence of a flickering effect. The motion perceived must be based upon the tendency of the perceptual system to interpret the successive appearance of an object in different locations as signifying motion of the object. The effect occurs only if the temporal and spatial separations are within a particular range of values. It is worth emphasizing that in both induced and stroboscopic motion, there is no motion over the retina of the object which is seen to move.

Causality. Another type of event besides motion that has been investigated concerns the factors that lead to the impression that one object has caused a certain behavior in another object. For example, when one billiard ball hits another, the movement of the second ball is experienced as having been caused by the movement of the first. This kind of effect is studied in the laboratory by a sequence of visual transformations (as in an animated cartoon) in which the behavior of the visual objects A and B can be experimentally manipulated. It has been found that the impression of causality occurs if and only if certain stimulus conditions prevail: the movement of B must occur immediately upon impact, the direction of B must be the same as that of A, and so forth. For these reasons it is now claimed that causality can be directly perceived, whereas in the past it would have been held that it is simply a judgment based on knowledge and that all that is directly perceived is the sequence of motions.

Form perception. One question requiring study is why a triangular object appears triangular when a triangular image of the object is present on the retina. Before attempting to answer this question, it is important first to make certain distinctions. Form perception means the experience of a shaped region in the field. Recognition means the experience that the shape is familiar. Identification means that the function or meaning or category of the shape is known. For those who have never seen the shape before, it will be perceived but not recognized or identified. For those who have, it will be perceived as a certain familiar shape and also identified. Recognition and identification obviously must be based on past experience, which means that through certain unknown processes, memory contributes to the immediate experience that one has of the triangle, giving the qualities of familiarity and meaning.

Perceptual organization. But even if the discussion is restricted to form or shape perception per se, there are still many aspects of such perception that require explanation. For example, there is perceptual organization. The figure of a 4 in **Fig. 2*a*** is seen as one unit, separate from other units in the field, even if these units overlap. This means that the parts of the figure are grouped together by the perceptual system into a whole, and these parts are not grouped with the parts of other objects. Moreover, a region will usually be seen as a figure in front of a background (ground) rather than as a hole in a surface. In general, the contour tends to give shape to the region on only one side at any one time, the one organized as a figure. A change in such organization can thus lead to a very different perceived shape (**Fig. 3**).

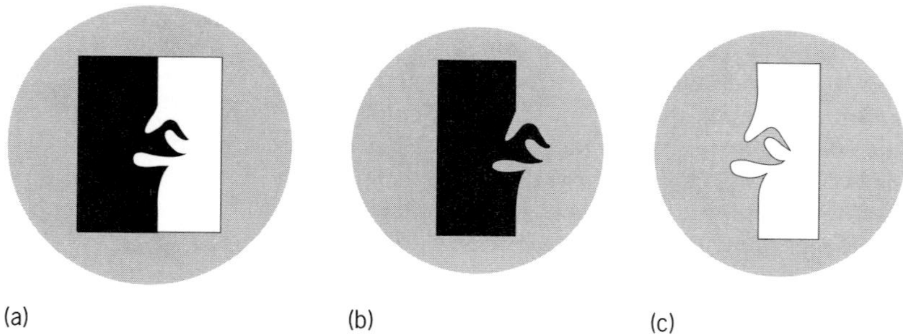

Fig. 3. Figure-ground organization. The pattern in *a* tends to be organized as a black figure on a white ground or a white figure on a black ground. The shapes of these two figures are very different despite the fact that the central contour is the same in the two cases, as is clearly indicated in *b* and *c*.

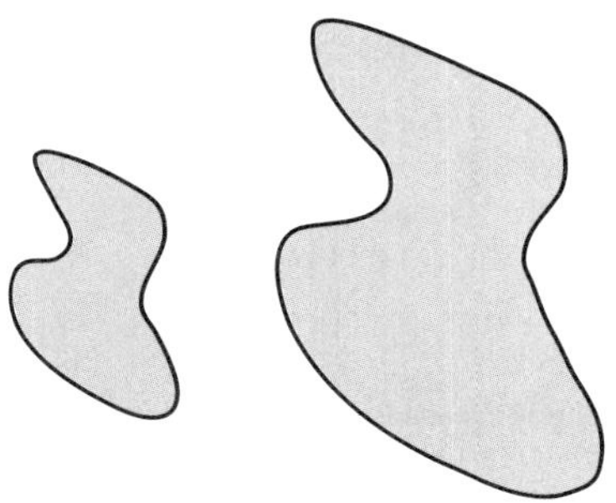

Fig. 4. Transposition of form; the two shapes look the same despite the difference in size.

Perceptual organization is an achievement of perceptual systems without which the perception of the world would presumably be an incoherent mosaic of unrelated points and patches of different lightnesses and hues. It is held that this organization is based on certain principles, many of which are known. For example, in Fig. 2*b* the figure of a 4 that is physically present is not spontaneously perceived, as it is in Fig. 2*a*, because each of its sides is organized as part of the longer lines by virtue of the principle known as good continuation. The necessity for such organization is one reason why form perception cannot be regarded as given directly by the presence of a retinal image of a particular shape.

Geometrical relationships. There are other problems about form perception that remain to be unraveled. For example, the size of a figure can vary, as can its locus on the retina or even its color or type of contour, without affecting its perceived shape (**Fig. 4**). This means that the perception of a particular shape does not depend upon the stimulation of particular cells of the retina or on the action of particular cells in the visual area of the brain, the occipital cortex, to which the retinal pattern is projected. What matters for form perception is the patterning or geometrical relationships between the parts of a figure. In other words, just as a melody can be transposed in key or octave, so a form can be transposed in various ways. Common to both cases is the fact that while the parts are no longer the same, the experience of the whole remains unchanged. This implies that the perceptual system is focusing on what does not vary with transposition, namely, the relational information about the melody or form. In the former it is the rhythm and pitch relationships, and in the latter the geometrical relationship of contour components to one another.

Physical contours. Ordinarily, form perception depends upon the detection of the contours which delineate a figure, and the contours are represented on the retina by an abrupt change of light intensity (or luminance) across each edge. Since sensory physiologists have discovered cells in the brain that "detect" the presence of contours or edges in particular orientations on the retina, some might be inclined to believe that form perception simply reduces to the detection of all the contours of a figure. Besides the problems of organization and transposition already mentioned, there are further difficulties with such a reductionistic explanation. Physical contours need not be present on the retina for "subjective" or illusory contours to be perceived (**Fig. 5**). Moreover, a figure can be perceived when it is moved behind an opaque surface containing a narrow slit so that only a thin slice of it is visible at any moment, a phenomenon referred to as anorthoscopic perception. These facts suggest that physical contours are ordinarily the markers of the locations of the boundaries of a figure but are not essential for form perception.

Orientation. A further fact about form perception is that it is dependent upon orientation (**Fig. 6**). It is a commonplace observation that printed or written words are difficult to read when inverted, and faces look very odd or become unrecognizable when upside down. Simple figures also look different when their orientation is changed: a square looks like a diamond when tilted by 45°. It is not easy to explain these effects on the basis of the altered orientation of the figure's contours on the retina since its perceived shape will not be changed simply by tilting

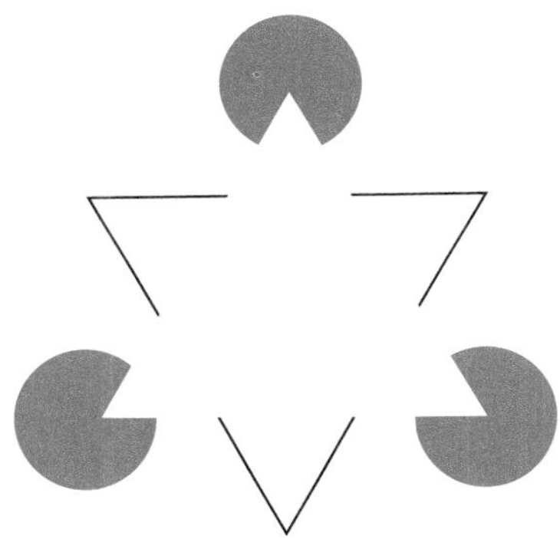

Fig. 5. Illusory contours. Although most of the contours of the upright triangle are not physically present, the triangle is perceived as a solid white figure with contours.

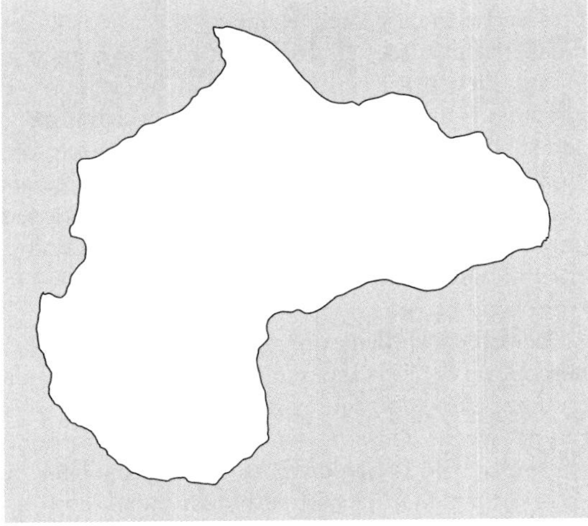

Fig. 6. This figure is not recognizable. If, however, it is tilted 90° clockwise, it will be recognized as an outline map of Africa.

the head (and therefore the eyes). In fact, the square tilted in the environment will look like a diamond regardless of head position. What seems to matter here are changes in the regions of the figure that are regarded as "top," "bottom," and "sides." The assignment of these locations to a figure by the perceptual system plays an important role in form perception.

Attention. Then there is the question of the role of attention in perception. Is form perception the automatic result of the neural transmission of an image from retina to brain, or is attention necessary? Suppose, as in a well-known experiment, the observer is shown an array of many letters of the alphabet in a very brief presentation, less than a second. Are they all perceived? There is disagreement among psychologists on this question, but the fact is that unless attention is focused on certain letters during or shortly after their presentation, they will not be remembered, even directly afterward. Therefore these unattended letters are either not perceived or, if they are, they do not lead to the establishment of even fleeting memories. In another famous experiment on hearing, observers are presented with two voices simultaneously reading different messages, and they are required to attend only to one of them by repeating each word in it. Afterward, memory tests indicate that very little if any of the unattended message is recalled. This experiment brings into the laboratory the well-known situation in everyday life where a person tries to listen only to one other person when several people are speaking at once, the so-called cocktail party effect. The impression is of perceiving only that person's conversation which is being attended to, and the rest is experienced as mere background "noise." Similar effects of inattention have been found for visual perception. What is not yet clear from these experiments, however, is whether the unattended figures or messages are not identified or are not even perceived as specific visual forms or organized auditory units.

One theory states that attention is not necessary to process certain elementary features in the visual array, such as colors, edges, or motions. Indeed, the presence of many such elements seems to be detected "in parallel." On the other hand, when combining such features in perception, they must be integrated—for example, perceiving a red triangle—and such feature integration does require attention. Support for this theory derives from experiments suggesting that integrated features can be processed only serially. Further support derives from discoveries that different object features are processed in entirely different regions of the brain.

Geometrical illusions. Related to the topic of form perception is the misperception of the size or direction of parts of figures that constitutes many of the geometric illusions. In an illusion figure, one particular part is perceived to be either longer or shorter than another part, although they are objectively equal (**Fig.** 7*a*); or the direction of a contour is perceived to be different from that of another contour although they are the same (Fig. 7*b*). For reasons still not understood despite a century of investigation of these illusions, the background or context of the rest of the figure affects these parts. One theory is that the illusion pattern is interpreted by the perceptual system as three-dimensional—whether the observer is consciously aware of this or not—and thus constancy operations are inappropriately applied. For example, in the Ponzo illusion illustrated in Fig. 7*a*, the converging lines function as the perspective representation of parallel lines receding into the distance. Consequently the horizontal lines are taken to be at different distances from the observer and, if so, would have to appear different in size. However, there are difficulties with this theory, and several other theories have been advanced and investigated.

Fig. 7. Geometrical illusions. (*a*) The Ponzo illusion in which the two horizontal lines of equal length appear unequal. (*b*) The Poggendorff illusion in which the two oblique line segments are aligned with one another (that is, are colliner) but appear to be misaligned.

Innate or learned? A central problem is whether the perception of properties such as form and depth or the achievement of veridical perception as in the constancies is innately determined or is based on past experience. By "innate" it is meant that the perception is the result of evolutionary adaptation and thus is present at birth or when the necessary neural maturation has occurred. By "past experience" it is meant that the perception in question is the end result of prior exposure to certain relevant patterns or conditions, a kind of learning process. Despite centuries of discussion of this problem by philosophers, and considerable experimental work on it during the century since psychology became an empirical science, there is still no final answer to the question. However, one method of studying vision in infants which has made it possible to provide some answers makes use of the fact that infants tend to look at presented objects but within a matter of seconds will cease doing so (habituation). The direction of their eyes can be observed and recorded. If another object is then presented, they will look at it provided that it appears sufficiently different from the first object (dishabituation). In this way it is possible to ascertain what objects appear similar to the infant, which in turn presupposes adequate vision of such objects. Using this and other methods it has been determined that infants, even shortly after birth, perceive shape and achieve figure-ground organization and perceptual constancy such as the size and shape. It seems clear that certain kinds of perception are innate, but past experience also is a determining factor.

Depth perception. The perception of the distance of objects is present in various species of animals early enough to rule out past experience as a causal factor. One experiment demonstrating this consists of placing a very young animal in the center of a large horizontal sheet of glass. To one side an opaque cloth is directly under the glass but, to the other side, the floor is several feet below. When placed in the center of this "visual cliff," the young of virtually all species tested will avoid moving to the side of the glass that appears to entail a steep drop. The experiment makes use of an innate fear of height, but of course only if that height can be perceived should the animals be expected to avoid one side of the glass. What is not yet known with certainty, however, is which of the many "cues" to distance is relevant in this experiment. Since a similar effect occurs when vision is permitted in only one eye, some investigators believe the relevant innate cue here is one governed by the displacement of the retinal image resulting from movement of the observer, known as movement parallax. In fact, a similar, vivid depth effect can be achieved in adult observers by motion of an array of points or contours with the observer stationary, currently referred to as depth (or structure) through motion, the stereokinetic effect, or the kinetic depth effect. By using the method of habituation with infants, it has been found that very young infants perceive distance and depth within objects, particularly when motion of the infant or object is allowed; binocular depth perception based on steroposis seems to take several months to develop.

Of course, one of the important cues to depth derives from the slightly different images the two eyes receive by virtue of their slightly different position in the head, an effect known as retinal disparity. Such depth perception will occur even if nothing recognizable (or even no visible contours) is presented to each eye, but instead a random array of small elements or black and white dots is presented to each eye (the so-called random-dot stereogram). These stereograms are constructed by selecting some region in one picture and displacing a corresponding region in the other picture with respect to the background. In viewing monocularly, one perceives only a random array of dots. In viewing binocularly, however, the observer perceives the region as floating in front of or behind the remainder of the pattern.

By way of contrast, cues to depth can be based on the content of the scene itself, such as perspective, shadow, or the interposition of one object in front of another. Since these cues are effective in creating in impression of depth even in pictures, they are called pictorial cues. There is some experimental evidence that past experience plays a role in the utilization of this kind of stimulus information in the perception of depth, both in real-life scenes and in pictures. Moreover, it seems plausible that this kind of experience could and would be acquired. Patterns of this kind are encountered virtually all the time in daily experience under conditions where depth perception is given by other information that has an innate basis. Thus the pictorial cues could be acquired by a process of associative learning.

Picture perception. The topic of picture perception has been receiving increasing attention. It is a complex problem because observers simultaneously perceive what the picture represents, for example, depth, and that it is something that is two-dimensional on a surface. Of particular interest is the question of whether or not past experience with pictures is necessary to perceive them appropriately or whether exposure to certain kinds of environments, such as the carpentered world of technologically advanced societies, plays a crucial role in picture perception. These questions have led to the study of picture perception in animals, in young children, and in "primitive" societies. If object perception is examined separately from depth perception in such pictures, it seems clear that the objects themselves are universally perceived and recognized in all societies and by very young children and some animals. This is particularly true if one removes the sometimes obfuscating effect of the background of a picture.

Basis of constancy. As to the question of whether or not perceptual constancy is learned or innate, the results of the habituation method with infants indicate the very early presence of constancy. In the case of lightness constancy, there is wide agreement that it is based on stimulus relationships, specifically on the ratios of luminance of neighboring regions. Indeed, there is evidence that the crucial information about such luminance ratios is given only at the edges between differing regions. If ratios govern perceived lightness, it would explain why constancy prevails, that is, why changes in overall illumination do not affect perceived lightness. It is unlikely that such a mechanism for constancy would have to be learned. However, a further problem here is that the perceptual system must be able to distinguish edges caused by lightness differences from those caused by illumination differences, such as shadows or three-dimensional corners. The perceptual system seems to be able to do so, but the mechanism is unknown.

Optical distortions. There has, however, been a good deal of research concerned with the capacity of animals and humans to adapt perceptually to optical distortions created by viewing the world through prisms and lenses of various kinds. For example, if an observer views the world through lenses that invert the retinal image—that is, turn it around 180°F from its customary orientation and cause the world to appear upside down—will the observer, after wearing the lenses for days or weeks, ultimately perceive the world as upright again? Within certain limits, there is evidence that adaptation does occur to prismatically created optical displacements, tilts, changes of curvature, and the like, although the precise nature, basis, and mechanism of such perceptual learning are not yet fully understood. However, the evidence is not convincing that adaptation to a reinverted retinal image occurs.

Some of this research bears on the question of the learned basis of constancy. For example, experiments have been conducted in which certain prisms or lenses worn by the subject create a new rate or direction of motion of the retinal image of a stationary scene resulting from the observer's movement. At the outset, position constancy is abolished, and the scene appears to move whenever the observer moves. Following continual viewing through these devices combined with movement of the observer, motion of the scene begins to lessen and eventually it ceases. Therefore position constancy is known to be subject to relearning even if, as seems to be the case, it did not have to be learned in the first place by the infant.

Experience and form perception. It has never been entirely clear why past experience has been considered essential for the emergence of various kinds of perception, particularly in the case of form. One might ask what is perceived prior to the occurrence of the necessary past experience. Nor has it been clear how experience would play a determining role such that, following it, the appearance of objects and events would be different than before it. One difficulty is the fact that if past experience is "carried" in the brain in the form of specific memories, these memories (rather than other irrelevant ones) must be contacted or evoked by the stimulus now impinging on the sense organ if they are to play a role in shaping the percept. Ordinarily, some psychologists believe, memories are evoked on the basis of similarity to what is now perceived, yielding recognition and identification. For past experience to determine form perception per se, however, the memories would have to be evoked before the form is perceived, and that seems unlikely, or at least it poses difficulties. This analysis suggests that the role of experience in form perception may be limited. It cannot entirely determine the form perceived, but might be able to enrich it in certain respects once it is perceived, and a bridge to the relevant memories is thus established. The impression of depth in the case of a picture of a three-dimensional object might well be an example of this kind. *See* MEMORY.

Characteristics. One very important principle is that, with few exceptions, perception is not determined by knowledge about the nature of the objects or events in a given situation. Thus, illusions are not abolished simply by knowing they are illusions. Perception follows its own laws, but this does not necessarily deny that past experience can automatically and unconsicously influence perception in certain cases.

Ambiguity. Another importance fact is that the stimulus impinging on a sense organ, such as the retinal image of an object, is ambiguous as concerns certain properties of the object it represents. An image subtending a given visual angle logically can represent an object in the world of any size, depending upon its distance. Conversely, different visual angles can represent an object of the same size. Therefore, the same stimulus can represent varying objects, and different

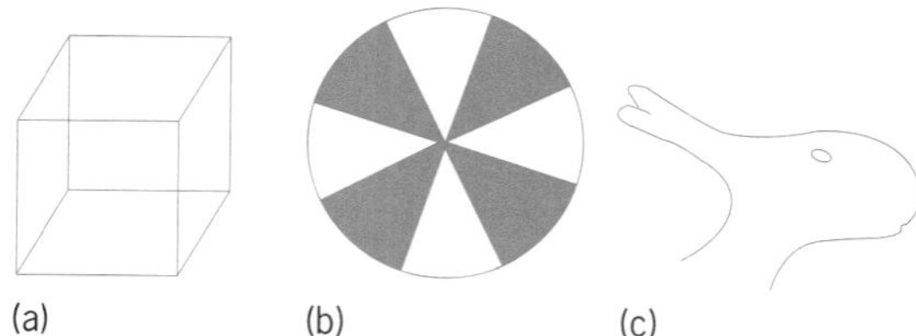

Fig. 8. Reversible figures. (*a*) The Necker cube, an example of reversible perspective. (*b*) A variation of the Maltese cross in which either the colored or white cross can be seen as the figure. (*c*) The duck or rabbit figure.

stimulus can represent the same object. Ambiguity also refers to the fact that the same stimulus can, over time, give rise to alternating perceptions. Such stimulus patterns are known as reversible figures (**Fig. 8**). Reversal has been much studied in the perception laboratory. However, research has shown that if the observer is not familiar with the figure and is not told it is reversible, then such reversal often will not occur. In the case of other patterns that can represent two (or more) objects or events, they often tend to be perceived in only one of these ways, suggesting that the perceptual system has some basis for a preference for one over the other. For example, when no other depth information is available, an expanding or contracting image of an object will generally be seen as an object approaching or receding rather than as one changing its size.

The Gestalt psychologists suggested that there is a preference for the simplest of all possible perceptions, which they referred to as the tendency toward *Prägnanz*, and they based it on their corresponding notion that brain processes tend toward a minimum of energy distribution. Thus Fig. 8*a* was held to be seen as three-dimensional rather than two-dimensional because as such it is regular and symmetrical. There is now renewed interest in this concept, since it is possible to formulate it in terms of information theory. *See* INFORMATION THEORY (BIOLOGY).

Relationships. Still another general characteristic of perception is that what is perceived is often determined by relationships or context as mentioned above. Thus, percepts are often relationally determined. There are various examples of this kind, such as the fact that a vertical rod seen in a dark room within a tilted rectangle will appear tilted from the upright in the opposite direction, the so-called rod-and-frame effect. Clearly it is the angular relationship of rod to rectangular frame that matters here. One explanation of the effect is that the rectangle is taken as the frame of reference, as if it were a surrogate of the vertical and horizontal axes of space. That being so, a rod not aligned with either axis of the rectangle must appear tilted. Induced motion and lightness perception are other examples.

Visual capture. One other general fact about perception concerns the dominance of vision over other sense modalities. That is, when a conflict is created such that vision indicates one value and touch or hearing another value, not only is vision dominant in determining what is perceived, but what is felt or heard conforms to what is seen. This latter fact is referred to as visual capture. Some examples have already been mentioned. Induced movement of the self implies that, despite sensory information that one is stationary, visual motion of the surroundings leads to the experience that one is in motion in the opposite direction. Another example is a straight rod that appears curved when viewed through a wedge-shaped prism. If one then runs one's hand along the rod while viewing it (which of course, with eyes closed, would lead to the impression of a straight rod), it will feel as if it is curved. Thus vision captures touch. Sound localization is also dominated by vision. While a speaker inside one's automobile in a drive-in theater will produce sounds of voices localized inside the car before the movie begins, after it does commence, the voices are perceived as coming from the actors seen on the distant screen. Here vision captures hearing.

Theories. Three different theories of perception have dominated this field over the last century.

Helmholtz theory. The first, associated with H. von Helmholtz, is that perception is the result of a process of unconscious inference concerning what the stimulus reaching the sense organ most probably represents in the world. Helmholtz believed that the inference process is based on past experience and is unconscious because the individual obviously is not aware of making it. For example, a white surface will appear white whether seen in sunlight or in dim indoor illumination because allowance is made for the differing illumination on the basis of experience. The achievement of veridical perception despite the ambiguity of the stimulus is thus accounted for in terms of a process of inference.

Gestalt theory. The second theory, represented by Gestalt psychology and founded by M. Wertheimer, W. K. Köhler, and K. Koffka, arose in part as a reaction against the empiricist view as it developed in the hands of psychologists such as Wundt and Titchener. These latter theorists had argued that individuals do in fact perceive on the basis of the stimulus impinging on the sense organ (which the empiricists called sensations) but that they interpret them in accord with what they have learned. To avoid confusion interpretations with true sensations, Titchener advocated training observers to become expert at attending to their sensory experiences (analytical introspection). The Gestaltists mounted a powerful critique of this approach, noted many logical and empirical difficulties with it, and emphasized the danger of distorting the description of the way the world appears in order to fit a preconceived theory. They also objected to the assumption for which there was no solid evidence, that perception is a function of prior experience. Instead they postulated innate organizing or interactive processes in the brain as the basis of perception. For example, stroboscopic motion perception was held to result from a neural interaction of the regions in the brain where the stimulus from each flashing object was transmitted. In general, the achievement of veridical perception despite the ambiguity of the stimulus was held to be explicable on the basis of relational determination and perceptual organization. Stimuli were said to interact and that is why relationships matter.

Psychophysical theory. The third theory represented and attempt to get around the difficulties concerning the ambiguity of the stimulus and those that arise because perception does not seem to correspond in any direct way with the stimulus. It was those same difficulties, for example, that size perception is not simply a direct function of visual angle, that led to the first two theories. According to this third theory, represented chiefly by James J. Gibson and his associates, there are features of the stimulus that do correspond to, or are correleated with, the perception, but they depend upon a more abstract or higher-order analysis of the stimulus. This approach has been characterized as a psychophysical or direct theory since it maintains that perception can be completely explained on the basis of stimulation, and therefore it is not necessary to posit such nonobservable hypothethical events as unconscious inference, interpretation of sensation, or neural interaction. As example of a higher-order feature of the stimulus is the gradient of density of texture that is present when an image of a uniform or random distribution of elements on a plane is projectd onto the retina. For example, the blades of grass and space between them from the nearby region of a field will yield relatively large retinal images and separations among them, but the farther away the region of grass, the smaller these images and separations will be. Thus a gradient of texture density could well be the stimulus for the apparent depth of a plane such as the ground. With this redefinition, proponents of this theory argue that the stimulus information is in fact not ambiguous.

Difficulties. There are contemporary psychologists who espouse one or another of these three theories, with or without certain modifications. However, each kind of theory encounters certain difficulties in accounting for all the known facts. Therefore, most investigators in perception today extract those aspects of these theories that seem to have been verified or that seem most plausible. Thus there is general agreement with the Gestalt view that a process of perceptual organization must be assumed and that relationships among stimulus components are indeed the basis of many perceptions. There is substantial agreement with the helmholtzian view that constancy often entails a process very much like unconscious inference, of computing object properties on an basis of available information, but it is not yet clear that such a process is necessarily based on past experience. There are many other examples in perception which suggest that what is perceived is the result of an intelligent, thoughtlike mental operation or inference. If so, such an operation must be very rapid (since perception occurs within a fraction of a second), not verbal (since animals and young children seem to perceive much as adults do), and unconscious. Finally, many researchers agree with the direct theory that there are certain higher-order attributes of the stimulus that can be abstracted (which were previously undetected or overlooked) and that can account for certain perceptions.

Trends. Independent of these theories are certain trends in the field. One approach that has flourished is to view perception as a stage in the temporal sequence of processing of information (called information processing). Perception follows earlier stages such as the neural encoding of sensory stimulation, but it precedes later stages such as the establishment of memories. Processing at each state requires a certain amount of time and, in principle, can be isolated by certain experimental techniques. *See* INFORMATION PROCESSING (PSYCHOLOGY).

Another development arises out of the interest in simulating the perceptual and other intelligent achievements of animals and humans by computers. Thus investigators in the field of artificial intelligence have begun to develop computer programs that would make it possible to arrive at a correct representation of the outer scene, given the same kind of information the eye receives in the form of an image (referred to as machine vision). Connectionism, or parallel distributed processing, challenges the assumptions underlying the "symbolic paradigm" of the artificial intelligence approach to perception, namely, that mental processes can be modeled as programs running on a digital computer. Instead, connectionism proposes that mental processes can be thought of as large-scale dynamic networks of simple neuronlike processing units. Perception (as well as other cognitive processes) can then be understood as the flux of global patterns of activation over the entire network. Since the pattern of activation tends toward a state of minimum energy, the theory is in some respects similar to Gestalt theory except that it is compatible with contemporary knowledge of neutral networks. *See* COGNITION; NEURAL NETWORK.

There is also renewed interest in the relationship between perception and imagery. Research has suggested that imagery has characteristics similar to those of perception and that perception itself can sometimes be affected by imagining. *See* ARTIFICIAL INTELLIGENCE; CONSCIOUSNESS; INTELLIGENCE.

Irvin Rock

Bibliography. K. R. Boff, L. Kaufman, and J. P. Thomas, *Handbook of Perception and Human Performance*, 2 vols., 1986; I. Rock, *Perception*, 1984; R. Sekuler and R. Blake, *Perception*, 3d ed., 1994.

Perciformes

The typical spiny-rayed fishes, also known by the ordinal names Acanthopteri and Percomorphi. This is the largest order of vertebrates; the approximately 7500 species include 41% of all fishes. Perciformes include a diversity of structural types and sizes, from a length of less than $^1/_2$ in. (1.3 cm) to a weight of nearly 1 ton (0.9 metric ton). The characters of the Perciformes include fin spines, usually present; a pelvic fin which, if present, is usually thoracic or jugular in position; the pelvic girdle usually attached to the cleithra, sometimes connected by ligaments; the pelvic fin usually with a spine and 5 soft rays, the latter occasionally reduced; the pectoral fin base more or less vertical, usually placed well up on the side; an upper jaw bordered largely or entirely by premaxillae; orbitosphenoid and mesocoracoid absent; a swim bladder without a duct; usually a posttemporal which is forked, articulating to the skull; scales usually ctenoid, sometimes secondarily cycloid, absent, or variously modified; caudal fin with 17 principal rays (15 branched) or fewer; hyoid arch with 4 branchiostegal rays attached to the outer face of the epihyal and ceratohyal above the prominent angle of the ceratohyal, plus 1–4 rays, usually 2–3, attached to the edge of the ceratohyal below the angle, the number rarely increased or further reduced.

Adaptive radiation. Perciform fishes dominate the modern vertebrate life of the oceans and have done so throughout the Cenozoic. The group first appeared in the Upper Cretaceous, after which it underwent a rapid adaptive radiation; many of the basic structural types, as well as most major perciform derivatives such as the Pleuronectiformes and Tetraodontiformes, were present in the Eocene. A few families of perciforms have been notably successful in fresh water; the Cichlidae in Africa and South America, the Centrarchidae in North America, the Percidae in North America and Eurasia, the Anabantidae in Africa and southeastern Asia, and many other families have achieved limited success in invading fresh waters. Other families, including the Chiasmodontidae and Cyclopteridae, have effectively adapted to life in the deep seas, and still others, such as the Scombridae, Stromateidae, and Coryphaenidae, have become specialized for pelagic existence. It is in the shore areas, the offshore banks, the coral reefs, the coastal beaches and lagoons, and the intertidal zone, however, that the perciforms have attained their ultimate achievement. Here the enormous variety attests the adaptive effectiveness of the group.

Classification. Ichthyologists are rapidly extending knowledge of phyletic relationships in this great order of fishes so it is inevitable that classification is changing and controversial. For example, groups ranked as separate orders by some are included as subgroups of the Perciformes by others, and as a consequence of new discoveries, families are transferred from one order to another. Fifteen suborders are herein recognized.

Mugiloidei. Three important families of shore fishes are included: the Mugilidae (mullets), Sphyraenidae (barracudas), and Polynemidae (threadfins). Some authors rank each as a separate suborder, while others treat the group as a distinct order. The Atherinidae (silversides) were formerly included here but that family is now placed in the order Atheriniformes. Mugiloids are rather elongate, terete fishes with a short spinous dorsal fin that is well separated from the soft dorsal. The suborder includes 22 genera and about 150 species. *See* ATHERINIFORMES.

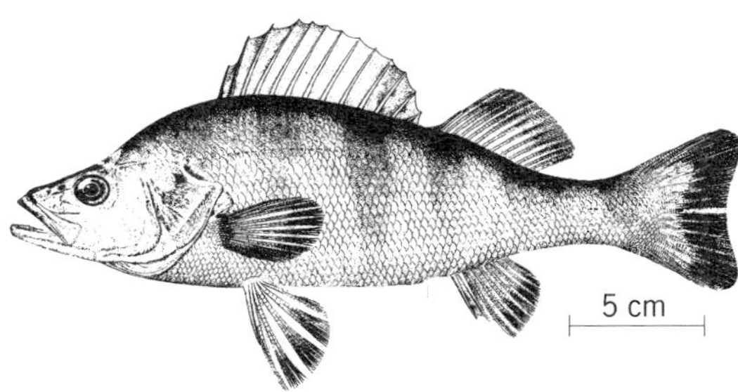

Fig. 1. Yellow perch (*Perca flavescens*), member of the family Percidae. (*After G. B. Goode, Fishery Industries of the United States, sect. 1, 1884*)

Anabantoidei. The labyrinth fishes include 3 families of fresh-water fishes from southeastern Asia and Africa: Channidae (snakeheads), sometimes ranked as a separate order; Luciocephalidae; and Anabantidae (climbing perches, gouramies, and relatives), a group that has been split into 4 families by some recent authors. Anabantoids have a specialized labyrinthine mechanism in the gill region that permits air breathing. Snakeheads are swamp-living fishes that are valued as food in southeastern Asia; gouramies and their relatives are important food fishes and many anabantids, such as the paradisefish and Siamese fighting fish, are aquarium favorites. The suborder includes 18 genera and about 80 species.

Percoidei. This suborder includes the typical perciforms (**Fig. 1**), most of which lack highly modified features such as are found in most other suborders. Approximately 63 families, 650 genera, and nearly 4000 species are included, that is, about 53% of the species in the Perciformes or almost 22% of all Recent fishes. Some of the larger or more important families are the Serranidae (sea basses and groupers), Centrarchidae (sunfishes), Percidae (perches; Fig. 1), Apogonidae (cardinalfishes), Echeneidae (remoras; **Fig. 2**), Carangidae (jacks, scads, and pompanos), Lutjanidae (snappers), Pomadasyidae (sweetlips, grunts), Sparidae (porgies), Sciaenidae (drums), Chaetodontidae (butterflyfishes), Embiotocidae (surfperches), Cichlidae (cichlids), Pomacentridae (damselfishes), Labridae (wrasses), and Scaridae (parrotfishes). Their importance in human life covers the entire range of human-fish interactions. About three-fourths are marine shore fishes, a few inhabit the deep sea, and the rest live in fresh water.

Stromateoidei. This suborder consists of 2 or more families, the Stromateidae (butterfishes), which by some are split into as many as 5 smaller families, and the Icosteidae (ragfishes). These are mostly high-seas fishes, some of which live in association with or feed on coelenterates. Except for the ragfish all have teeth located in pharyngeal pockets in the throat. Some are excellent food fishes. There are 15 genera and between 65 and 70 species.

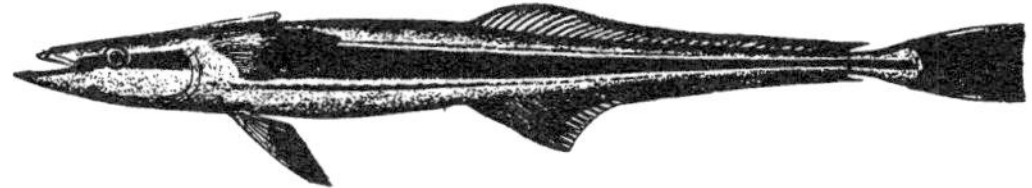

Fig. 2. Sharksucker (*Echeneis naucrates*), member of the family Echeneidae; length to 38 in. (96 cm). (*After D. S. Jordan and B. W. Evermann, The Fishes of North and Middle America, U.S. Nat. Mus. Bull. 47, 1900*)

Cottoidei. These, the mail-cheeked fishes, are placed by some in a separate order (Cottiformes or Scorpaeniformes). They are characterized by the expanded third infraorbital bone which crosses the cheek as a brace against the preopercle. Some families, such as the Triglidae (searobins) and Platycephalidae (flatheads), are predominantly shore or bottom fishes of the tropics and subtropics; others, including the Agonidae (poachers) and Cyclopteridae (lumpfishes and snailfishes), inhabit cold northern seas, either in shallow or deep water. The Scorpaenidae (scorpionfishes) include a large number of tropical shore fishes, some venomous, and such valuable food fishes as the live-bearing rockfishes and redfish of northern oceans. The Cottidae (sculpins) abound on coasts of northern continents and many live in cool fresh-water lakes and trout streams. The Cottoidei include about 16 families, roughly 227 genera, and more than 900 species.

Dactylopteroidei. The single family, Dactylopteridae (flying gurnards), consists of 2 genera and 4 species of marine shore fishes characterized by tremendously expansive pectoral fins. Their relationships have been disputed; some workers rank them as a separate order, while others place them in the Cottoidei.

Blennioidei. This, one of the 4 large suborders of the Perciformes, contains chiefly small fishes that live in coral and rock reefs or in the tidal zone. Such groups as the large family Blenniidae (combtooth blennies) and Dactyloscopidae (sand stargazers) are mostly tropical or subtropical; others, for example, Stichaeidae (pricklebacks), live on cool or cold coasts or in deep water. Most of the fishes of the permanently frigid waters surrounding Antarctica, the Nototheniidae and their relatives, belong here. In blennioids pelvic fins are either lacking or are located on the throat, and there is an exact numerical correspondence between dorsal and anal soft rays and the vertebrae between them. Many blennioids are slender or eellike in form. There are about 30 families, roughly 230 genera, and slightly over 1000 species.

Schindlerioidei. The single family Schindleriidae consists of the genus *Schindleria* with 2 species. These tiny oceanic fishes are transparent and neotenic.

Ammodytoidei. The sand lances are marine, cool- or cold-water shore fishes with slender, eel-shaped bodies. The relationships are not well understood; they are sometimes placed in the Percoidei. The 2 families include 4 genera and about 8 species.

Callionymoidei. This suborder consists of 2 closely related families known as dragonets, colorful little bottom fishes of marine waters of the tropics and subtropics. The body is elongate, the head is flattened, and one or more bones in the opercular area are spinous. The gill aperture is commonly a small pore located high on the head. There are 12 genera and about 44 species.

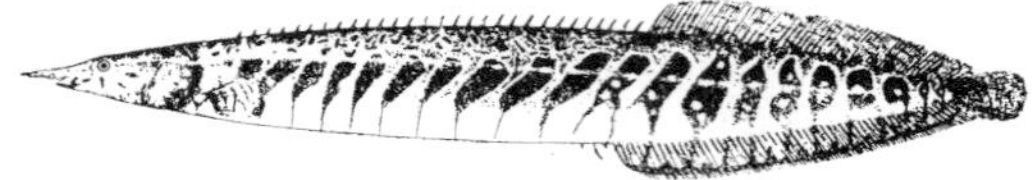

Fig. 3. Spiny eel (*Mastacembelus circumcinctus*); length to 7 in. (18 cm). (*After H. M. Smith, The Fresh-Water Fishes of Siam or Thailand, U.S. Nat. Mus. Bull. 188, 1945*)

Gobioidei. The gobies make up one of the predominant groups of fishes in shallow shore waters, reefs, and estuaries of the tropics and subtropics; a few live in temperate waters. Some enter fresh waters near the coasts, and many skip nimbly over mudflats or about tide pools many feet from water. They vary from the smallest of fishes to nearly 2 ft (60 cm) in length, from somber hues to the brilliant and gaudy, and from short and stocky to a sinuous, eel-shaped build. All lack a lateral line. Most belong to the family Gobiidae, in which the pelvic fins are more or less united to form a sucking disk on the breast. The number of known kinds is being increased rapidly, in part because the development of scuba gear has made the habitat of secretive reef species available to scientific collectors. The 4–6 families usually recognized include over 200 genera and perhaps nearly 1000 species, but these figures are tentative because of the incomplete state of knowledge of the group.

Acanthuroidei. The Acanthuridae (surgeonfishes), Zanclidae (moorish idols), and Siganidae or Teuthidae (rabbitfishes) are mostly inhabitants of the tropical and subtropical Indo-Pacific shores, though a few enter streams and some surgeonfishes live in the Atlantic. Many are highly colorful and showy in aquariums or on their native reefs. They are chiefly herbivorous. The 9 genera include nearly 100 species.

Kurtoidei. This suborder consists of a single genus, *Kurtus*, with 2 species of small Indo-Pacific fishes. They have a unique ossification that encloses the upper part of the swim bladder, and a conspicuous occipital hook in the male that is used to hold the eggs during brooding.

Scombroidei. Scombroids are moderate-sized to large shore and oceanic fishes, all of which have the premaxillae fixed. The Trichiuridae (cutlassfishes) and Gempylidae (snake mackerels) are compressed, elongate or eel-shaped fishes with enlarged coniniform teeth; the Scombridae (mackerels and tunas) are streamlined, fast-swimming denizens of the high seas; and the Istiophoridae (billfishes) and Xiphiidae (swordfishes) have tremendously produced bills. The suborder contains 6 families, which are divided into 40 genera and about 80 species.

Mastacembeloidei. The spiny eels (**Fig. 3**) are sometimes classified as a separate order. Though eellike in shape and in having the pectoral girdle suspended from the vertebral column instead of the skull as in most fishes, these are not true eels but modified perciform fishes. They inhabit swamps, marshes, and brackish water in Africa and southeastern Asia. There are 2 families, 4 genera, and about 50 species.

Economics. From an economic standpoint the Scombridae, including the oceanic tunas and mackerels, rank first among the perciforms. The Sciaenidae or drums, the Serranidae or sea basses, the Scorpaenidae or rockfishes, the Carangidae or jacks, the Cichlidae of tropical fresh waters, the Percidae of temperate fresh waters, and other groups also support important commercial fisheries. Some of these and other families, such as the Centrarchidae or sunfishes, the Istiophoridae or sailfishes and marlins, and the Pomatomidae or bluefishes are valued in sport fisheries, and hence are of great recreational and indirect economic importance. *See* ACTINOPTERYGII; FISHERIES ECOLOGY; OSTEICHTHYES. Reeve M. Bailey

Percopsiformes

A small order of actinopterygian fishes that is also known as the Salmopercae and Amblyopsiformes. The order is thought to be remotely related to the codfishes (Gadiformes) and toadfishes (Batrachoidiformes). The characters (see **illus.**) include single, ray-supported dorsal and anal fins, each usually with one to four anterior spines; pelvic fin, if present, subabdominal in position, with three to eight soft rays; pelvic girdle, if present, attached to the postcleithra; upper jaw bordered by premaxillae; no orbitosphenoid bone; swim bladder without a duct; and body covered with cycloid or ctenoid scales. *See* BATRACHOIDIFORMES; GADIFORMES; SCALE (ZOOLOGY); SWIM BLADDER.

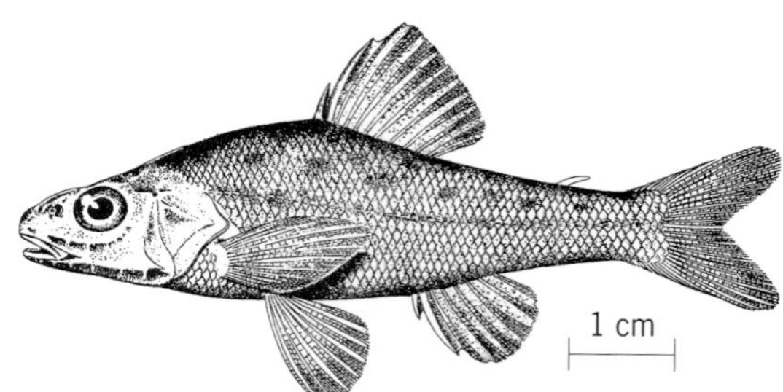

Sand roller (*Percopsis transmontana*). (*After D. S. Jordan and B. W. Evermann, The Fishes of North and Middle America, U.S. Nat. Mus. Bull. 47, 1900*)

The order comprises three families, five genera, and eight Recent species of North American freshwater fishes. Eocene and Miocene fossil genera from North America are assigned to the group. The species attain a maximum length of 6 in. (15 cm) and inhabit sluggish or standing surface waters or subterranean cavern pools and streams. The cavefishes, Amblyopsidae, long regarded as relatives of the killifishes (Cyprinodontidae), have been shown to belong to this order; their specialized characters—including reduced pelvic fins, pigmentation, and eyes; cycloid scales; and highly developed tactile senses—are correlated with their life in underground waters. *See* ACTINOPTERYGII; TELEOSTEI. Reeve M. Bailey

Performance rating

A procedure for determining the value for a factor which will adjust the measured time for an observed task performance to a task time that one would expect of a trained operator performing the task, utilizing the approved method and performing at normal pace under specified workplace conditions. Since the introduction of stopwatch time study more than a hundred years ago, there has been a need to adjust an observed operator time to a "normal time." Normal time is the time that a trained worker requires to perform the specified task under defined workplace conditions, employing the assumed philosophy of "a fair day's work for a fair day's pay."

Normal performance must ultimately be subjectively based on the observer's opinion of what represents a fair or equitable pace of activity for a worker with a specified task under specified conditions. It is assumed under the "fair day's work for a fair day's pay" philosophy that skilled operators of average physique for the tasks they are performing, performing widely differing tasks under widely varying conditions, will be equally fatigued at the end of an average day's work. During a normal workday, a specified amount of time is set aside as nonwork time and is called allowances. The performance rating process is concerned with determining normal pace during the work portion of an average day and must, therefore, consider the fatigue recovery aspects of allowance times occurring during the day.

The following two equations relate factors in determining how much time a worker will be allowed per unit of output:

$$\text{Standard time} = \text{normal time} + \text{allowances}$$

$$\text{Normal time} = \text{observed time} \times \text{rating factors}$$

If the observed time for a task is adjusted by the performance rating factor to determine normal time, and allowance time is added for nonwork time, the standard time will represent the allowed time per unit of production. For example, if the observed time is 60 min, and the rating factor is estimated to be 80%, and the allowances are 15% of normal time, then the normal time is $60 \times 0.8 = 48$ min, and the standard time is $48 + 0.15(48) = 55.2$ min. In a typical 8-h shift, the expected production from this worker would be $8(60)/55.2 = 8.7$ units of production.

Two frequently used benchmarks of normal pace are walking and dealing playing cards. These benchmarks are walking at a pace of 3 mi/h (1.33 m/s), and dealing a deck of 52 cards into four piles 1 ft (30 cm) apart in 0.5 min. Films of these activities being performed at different speeds are frequently viewed by performance raters so as to have their ratings be more consistent and more standardized.

What has been observed in rating has been variously called speed, effort, tempo, pace, or some other word connoting rate of activity or exhibited effort. In 1927 a technique referred to as leveling, which required the evaluation of four factors—skill, effort, conditions, and consistency—was developed by the Westinghouse Corporation. Evaluation scales for the first two factors, effort and skill (see **table**) were employed in determining values for these factors. The factors of skill, conditions, and consistency were employed to adjust what was observed to what was intended as standard for the specified task. However, if an average trained operator is observed performing the specified task under specified conditions, there is no underlying theoretical justification for considering these factors. Conditions and consistency, as adjustment factors, are not in common use today in employing this technique.

Leveling factors for performance rating

Effort			Skill		
Category	Code	Value	Category	Code	Value
Excessive	A1	+0.13	Superskill	A1	+0.15
	A2	+0.12		A2	+0.13
Excellent	B1	+0.10	Excellent	B1	+0.11
	B2	+0.08		B2	+0.08
Good	C1	+0.05	Good	C1	+0.05
	C2	+0.02		C2	+0.03
Average	D	+0.00	Average	D	0.00
Fair	E1	−0.04	Fair	E1	−0.05
	E2	−0.08		E2	−0.10
Poor	F1	−0.12	Poor	F1	−0.16
	F2	−0.17		F2	−0.22

The word "effort" has been defined in the past in such unfortunate terms as "demonstration of the will to work effectively," which is possibly unobservable and likely to be unmeasurable and irrelevant. Regardless of an employee's will, it is the pace of a worker that the employer has a right to expect (that is, employers do not pay for trying, they pay for accomplishment).

In 1949 the original Westinghouse method was revised extensively, and the new plan, called the performance rating plan, included three major classifications of factors—dexterity, effectiveness, and physical demand. Each classification contains from two to four factors. As indicated above, there is no theoretical justification for any factor other than pace, if the appropriate employee utilizes the specified method under specified conditions.

If two tasks of widely varying job difficulty are rated without consideration of the inherent relative job difficulty, and the two employees are paid on an incentive basis, the employee with the more difficult task is penalized in terms of potential incentive earnings, because although it may be an equally demanding task to perform at day work pace, the more difficult task is more difficult to increase in pace in comparison to the simpler task. The objective rating technique, first published in 1946, considers job difficulty and, in so doing, adjusts for this problem. Such subfactors as amount of body movement, use of foot pedals, bimanualness, eye-hand coordination, handling requirements, and weight are considered in evaluating the primary factor—job difficulty. Pace is the other primary factor considered.

The most commonly employed rating technique throughout the history of stopwatch time study, including the present, is referred to today as pace rating. A properly trained employee of average skill is time-studied while performing the approved task method under specified work conditions. Rating consists only of determining the relative pace (speed) of the operator in relation to the observer's concept of what normal pace should be for the observed task, including consideration of expected allowances to be applied to the standard. *See* HUMAN-FACTORS ENGINEERING; METHODS ENGINEERING; WORK MEASUREMENT. Philip E. Hicks

Bibliography. R. M. Barnes, *Motion and Time Study*, 7th ed., 1980; M. E. Mundel, *Measuring Total Productivity in Manufacturing Organization: Algorithms and PC Programs*, 1987; M. E. Mundel and D. L. Danner, *Motion and Time Study: Improving Productivity*, 7th ed., 1994; B. W. Niebel, *Motion and Time Study*, 9th ed., 1992.

Performing arts medicine

A subspecialty in medicine that deals with problems specific to the activities of dancers, musicians, and vocalists. It is an outgrowth of the fields of sports medicine and occupational medicine.

Performing artists have medical problems similar to those of many athletes in that their activities involve certain parts of the body repetitively performing a specific motion. Athletes and performance artists often practice at least 3–5 h each day on their own and then may have team practice or rehearsal and finally actual competition or a performance. Unlike athletics, however, the performing arts are typically not seasonal, and so time for rest and rehabilitation is limited.

The vast majority of injuries in performers can be attributed to overuse. Unlike athletes, who may lift weights, run, or swim as a part of their training regimen, performing artists train by executing the same motions as when they perform. This adds to the overuse incurred during their work schedule. Many performing artists, particularly dancers, avoid lifting weights because they fear development of defined musculature and as a result may not develop proper muscle tone. Poor technique often plays an important role in overuse injuries: poor body positioning and inefficient use of abdominal musculature adversely affect dancers, singers, and musicians. Inappropriate choice of dance style, musical instrument, or vocal repertoire also may lead to overuse injuries. A musician should select an instrument that can be held and played comfortably, a singer should select pieces that fit his or her register, and a dancer should choose a dance form that complements his or her lower extremity anatomical structure. Another problem stems from the fact that professional performance artists do not have trainers, and they may be unaware of proper training techniques including warm-up, cool-down, stretching and strengthening exercises, and even proper nutrition.

Types of injuries. Each field of the performing arts tends to have its own characteristic injuries, which afflict the parts of the musculoskeletal system subjected to the greatest stress. Tendinitis in dancers occurs most commonly in the rectus femoris tendon in the thigh and the Achilles tendon in the leg. Stress fractures most often affect the metatarsal bones in the foot, the fibula in the leg, and the spine. Many dancers also suffer from pain related to mechanical stress on the kneecap, a consequence of jumping and deeply bending the knees. Kneecap pain is particularly prevalent among dancers whose lower extremity alignment is not ideally suited to their dance form and dancers who use poor technique.

Eating disorders are also common in dancers, particularly female ballet dancers, who strive to remain thin. For dancers, weight control is a major issue, and malnutrition is common; some dancers even have anorexia nervosa. In some cases, dietary disorders, low body weight, and intense training routines combine to produce amenorrhea. This lack of normal menstrual cycling can have disturbing secondary effects such as osteoporosis, which increases the likelihood of stress fracture. Most musicians, singers, and dancers perform on empty stomachs, and eat a large meal shortly before bedtime; doing so tends to result in the escape of stomach acids into the esophagus and even laryngitis and tends to add body fat. *See* ANOREXIA NERVOSA; MALNUTRITION; OSTEOPOROSIS.

Tendinitis in musicians most commonly occurs in the extensor tendons of the wrist and fingers, the flexor tendons of the fingers, and the rotator cuff muscles in the shoulder, producing pain with use of the upper extremities. Neck pain is also common, as is compression of a nerve (nerve entrapment) in the upper extremities. Nerve entrapments often manifest themselves with pain, weakness, and numbness and tingling in the fingers and arm.

Musicians playing wind instruments suffer from problems related to use of the embouchure, which is the opening from the upper airway into a wind instrument. Coordinated action of the lips and tongue in relation to the palate, teeth, and mouthpiece controls the flow of air into the instrument to create tone, pitch, and volume. Overuse produces neck muscle fatigue, pain at the articulation between the upper and lower jaws, and lip sores.

Singers are prone to overuse injuries of the vocal cords. These injuries take the form of swelling (laryngitis), nodules, and even hemorrhage. Excessive muscle tension in the neck, inadequate abdominal support, and singing beyond one's range may contribute to vocal cord injury. Vocal cord abuse may be aggravated by allergy, upper respiratory tract infection, and breathing smoke or dry air, such as during airplane travel.

Evaluation and treatment. Determining the cause of a performing artist's medical problem is most important for treatment. Data on the performing artist's schedule, equipment, environment, and technique—as well as general stamina, specific muscular strength, and flexibility—are important in evaluating an injured performer. Identifying and changing specific factors that contributed to the injury not only help treat the problem but also prevent recurrence. Watching the artist perform is very helpful in detecting generalized tension and, particularly, harmful nuances of body positioning and muscle use.

Most performing artists want to remain active despite injury, and so the treatment should be designed to permit such activity whenever possible. Some injuries heal with relative rest, which may involve continuing to perform and rehearse but decreasing the practice schedule and eliminating certain parts of the repertoire that are particularly painful or that stress use of a certain body part. For the vocalist, relative rest may involve staying in midregister and at moderate volume intensity. Physical therapy modalities such as ultrasound, ice, and whirlpool are frequently used for spine and extremity problems, and hydration and steam inhalation are recommended for inflamed vocal cords. In addition, nonsteroidal anti-inflammatory medications are often prescribed. If poor technique, inadequate warm-up, poor nutrition, or psychological stress is contributing to the problem, it should be addressed. Relaxation techniques are especially helpful. During the rehabilitative phase of treatment, exercises to gain or maintain endurance, flexibility, and strength are critical. *See* SPORTS MEDICINE.

Carol C. Teitz

Bibliography. E. S. Gorman and C. A. Warfield, Pain syndromes in musicians, *Hosp. Practice*, 22:68, 71, 1987; W. G. Hamilton, Foot and ankle injuries in dancers, *Clin. Sports Med.*, 7:143–173, 1988; J. A. Howard and A. T. Lovrovich, Wind instruments: Their interplay with orofacial structures, *Med. Prob. Perform. Art.*, 4:59–79, 1989; A. J. Ryan and R. E. Stephens, *Dance Medicine*, 1987; R. T. Sataloff, Common diagnoses and treatments in professional voice users, *Med. Prob. Perform. Art.*, 2:15–20, 1987; L. M. Vincent, *Competing with the Sylph: Dancers and the Pursuit of the Ideal Body Form*, 1979.

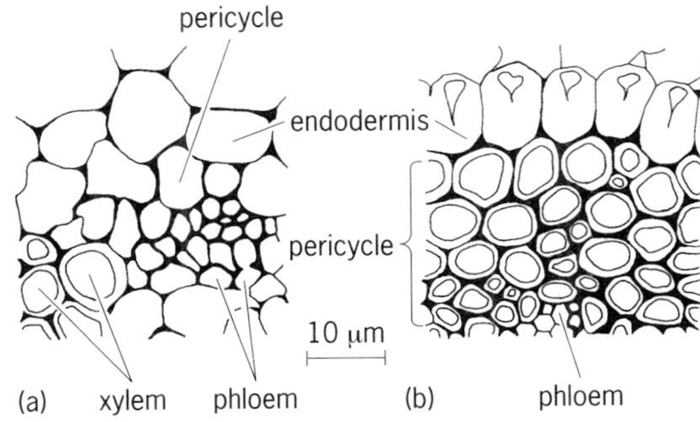

Fig. 1. Pericycles. (*a*) Part of transection of root *Actaea alba*, including xylem and phloem. Pericycle thin-walled and one cell in radial dimension. (*b*) Part of transection of root of *Smilax herbacea*, including phloem. Pericycle thick-walled and four or five cells in radial dimension (brace).

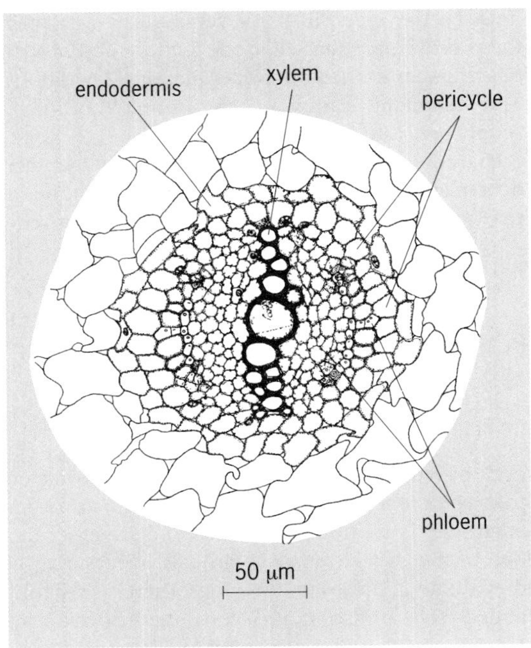

Fig. 2. Transection of central part of sugarbeet. (*After K. Esau, Hilgardia, 9(8), 1935*)

Pericycle

As commonly defined, the outer boundary of the stele of plants. Originally it was interpreted as a band of cells between the phloem and the innermost layer (endodermis) of the cortex. Such pericycle is commonly found in roots and, in lower vascular plants, also in stems. In higher vascular plants, however, a distinct layer of cells may not be present between the phloem and the cortex. The homogeneous groups of fibers located on the outer boundary of the primary phloem in many stems are not pericyclic in origin but develop in the earliest part of the phloem, whose remaining cells are obliterated. The pericycle, if present, may be composed of parenchyma or sclerenchyma cells with relatively thin or heavily thickened walls. It may be one to several layers in radial dimensions (**Fig. 1**).

Primordia of branch roots commonly arise in the pericycle in seed plants, most frequently outside the xylem ridges (**Fig. 2**). The first cork cambium may also arise in the pericycle of those roots that have secondary vascular tissues. In roots, a part of the vascular cambium itself (that outside the primary xylem ridges) originates from pericycle cells. *See* CORTEX (PLANT); ENDODERMIS; LATERAL MERISTEM; PARENCHYMA; PHLOEM; PLANT TISSUE SYSTEMS; ROOT (BOTANY); SCLERENCHYMA; STEM; XYLEM.

Vernon I. Cheadle

Pericyclic reaction

Concerted (single-step) processes in which bond making and bond breaking occur simultaneously (but not necessarily synchronously) via a cyclic (closed-curve) transition state. Although a given reaction may appear formally to be pericyclic, it cannot be assumed to be a concerted process. In each case, the detailed mechanism of the reaction must be established experimentally. Pericyclic reactions can be promoted either by heat or by light; the stereochemistry of the reaction is determined by the mode of activation employed and the number of electrons that are delocalized in the transition state. *See* CHEMICAL BONDING; STEREOCHEMISTRY.

Classification. Four types of pericyclic reactions that are frequently encountered in organic chemistry are electrocyclic processes, cycloadditions, sigmatropic shifts, and cheletropic reactions.

Electrocyclic processes. These are reactions that involve either cyclization across the termini of a conjugated π-system with concomitant formation of a new σ-bond or the microscopic reverse. The sequence of steps involved in the forward reaction must be the same, in the reverse order, as that in the reverse direction when the forward and reverse reactions are carried out under identical conditions. This statement is known as the principle of microscopic reversibility.

The effect of the mode of activation upon the stereochemistry of an electrocyclic process is shown in reaction (1), where Me = methyl, for the hexatriene-

light conrotatory — heat disrotatory (1)

Me H H Me — Me H H Me — Me H Me H

(3) [trans] — (1) conrotatory, disrotatory — (2) [cis]

cyclohexadiene interconversion (a six-electron electrocyclic process). Thus, when *trans,cis,trans*-2,4,6-octatriene [structure (**1**)] is heated, disrotatory motion of the two terminal 2*p* orbitals occurs; that is, they rotate in opposite directions, thereby resulting in exclusive formation of *cis*-5,6-dimethylcyclohexa-1,3-diene (**2**). The corresponding photochemical process results in conrotatory motion of the termini in structure (**1**); that is, the two terminal 2*p* orbitals rotate in the same direction, thereby yielding *trans*-5,6-dimethylcyclohexa-1,3-diene (**3**) exclusively.

The stereochemistries of the corresponding thermal and photochemical four-electron processes (for example, butadiene-cyclobutene interconversion) are just the opposite of the results cited above for the hexatriene-cyclohexadiene interconversion. The thermal four-electron process occurs in conrotatory fashion, whereas the corresponding photochemical reaction is disrotatory.

Cycloadditions. These occur when two (or more) π-electron systems react under the influence of heat or light to form a cyclic compound with concomitant formation of two new σ-bonds that join the termini of the original π-systems. The stereochemistry of this reaction is classified with respect to the two molecular planes of the reactants. Thus, if σ-bond formation occurs from the same face of the molecular plane across the termini of one of the component π-systems, the reaction is said to be suprafacial on that component. If instead σ-bond formation occurs from opposite faces of the molecular plane, the reaction is said to be antarafacial on that component. This distinction is illustrated in reaction (2) for two thermal processes, where symbol ≠ indicates the structure of the transition state. Reaction (2*a*) shows the Diels-Alder [4+2] cycloaddition of butadiene (**4**) to ethylene (**5**), a six-electron pericyclic reaction in which additions across the termini of the diene (four-electron component) and dienophile (two-electron component) both occur suprafacially. Reaction (2*b*) shows a [14+2] cycloaddition in which σ-bond formation occurs suprafacially on the two-electron component tetracyanoethylene (**6**)] and antarafacially on the fourteen-electron component [heptafulvalene (**7**)]. [Note the trans relationship between the two circled hydrogen atoms in the product (**8**)]. In the orbital representations of the transition states for these reactions, the π-lobes which interact constructively to form new σ-bonds are shaded and are connected by dotted lines. *See* DIELS-ALDER REACTION.

A different mode of activation, that is, light rather

diene — dienophile — heat — ≠ (2*a*)

(4) + (5)

heat — NC NC H CN H CN — ≠ (2*b*)

(7) NC CN NC CN (6)

H H NC NC CN CN (8)

than heat, is required to promote cycloaddition between *trans*-stilbene (**9**) and tetramethylethylene (**10**) [reaction (3), where Me = methyl, Ph = phenyl].

(9) light (10) Ph = phenyl (3)

This photochemical [2+2] cycloaddition proceeds through the lowest-energy electronically excited state (rather than through the ground state). Two new σ-bonds are formed; the overall process is suprafacial on both components. *See* PHOTOCHEMISTRY.

Sigmatropic shifts. These reactions involve migration of a σ-bond that is flanked at either (or both) ends by conjugated π-systems. Either one or both ends of the σ-bond may migrate to a new location within the one or more flanking π-systems. The shift is said to be of order $[i,j]$ when the σ-bond migrates from the [1,1] position to some distant position, $[i,j]$, along the one or more conjugated chains. Thermal rearrangement of 3-methyl-1,5-hexadiene (**11**) to 1,5-heptadiene (**12**) is an example of a sigmatropic shift [reaction (4), where Me = methyl.]. This reaction, the Cope

old σ-bond heat [3,3] new σ-bond (4) (11) (12)

rearrangement, is a sigmatropic shift of order [3,3].

The stereochemistry of a $[i,j]$ sigmatropic shift is classified with respect to the movement of the migrating group along the π-system with respect to the plane that contains the π-electron system. If the migrating group moves across one face of the π-system, the reaction is suprafacial. If instead the migrating group crosses the molecular plane during this process, the reaction is antarafacial. In both instances, migration can occur with either retention or inversion of configuration of the migrating group. Reaction (5) shows the thermal rearrangement of structure (**13**) to structure (**14**); this involves a suprafacial sigmatropic shift of order [1,3] that proceeds with inversion of the configuration of the migrating group. [Note that the substituents OAc and D are mutually

(13) heat D = deuterium OAc = $OCCH_3$ (with =O) (14) (5)

trans in structure (**13**), but in the product, structure (**14**), they are cis].

Cheletropic reactions. These are reactions that involve extrusion of a fragment via concerted cleavage of two σ-bonds that terminate at a single atom or the reverse process. Cheletropic fragmentations may be either linear or nonlinear [reaction (6*a*)]. An example is reaction (6*b*), in which 7-norbornadienone (**15**)

nonlinear π-system linear (6*a*)

(15) temperature < 173 K (–149°F) → CO + (6*b*)

decomposes at very low temperature via cheletropic extrusion of carbon monoxide (CO) with resulting aromatization. (In this example, it is not possible to demonstrate experimentally whether carbon monoxide is extruded in linear or nonlinear fashion.)

Theory. R. B. Woodward and R. Hoffmann introduced an application of molecular orbital theory that permits prediction of rates and products of pericyclic reactions. They utilized symmetry properties of molecular orbitals to estimate relative energies of diastereoisomeric transition states for structurally similar pericyclic reactions. In their approach, the energy changes that accompany orbital interactions as reactants are converted to products along the reaction coordinate are followed via construction and analysis of orbital correlation diagrams. An element of symmetry is chosen that is preserved throughout the reaction and that bisects bonds that are either being broken or formed during the pericyclic reaction. Central to their argument is the principle of conservation of orbital symmetry; that is, the symmetry of individual molecular orbitals with respect to the chosen symmetry element will be preserved as the reaction coordinate is traversed. If the variation in total energy of the system with reaction geometry along the reaction coordinate is small, little resistance to reaction will occur, and the reaction is predicted to occur readily; that is, the reaction is said to be symmetry-allowed. However, if any of the occupied molecular orbitals experience considerable destabilization while traversing the reaction coordinate, hindrance to reaction will result; that is, the reaction is symmetry-forbidden. It should be noted that these arguments apply to the rates of pericyclic reactions and not to the position of thermodynamic equilibria (which is determined by the relative stability of reactants versus products).

In an alternative theoretical approach to understanding pericyclic reactions, the transition state is examined directly, and attempts to estimate the degree of electronic stabilization (allowedness) or destabilization (forbiddenness) inherent in that transition state are made. One such approach emphasizes the importance of frontier orbitals (highest-occupied–lowest-unoccupied molecular orbitals) in determining the course of a pericyclic reaction. Thus, if donation of electrons from occupied to unoccupied frontier orbitals results in a stabilizing interaction, the reaction should be allowed; if instead this results in a destabilizing interaction, the reaction will be forbidden.

A second approach directly estimates the energies of relative transition states by noting whether delocalization of electrons results in an aromatic (symmetry-allowed) or an antiaromatic (symmetry-forbidden) transition state. As an example, ground state (thermal) [4+2] cycloadditions proceed via a six-electron cyclic transition state that is Hückel-aromatic and thus is symmetry-allowed. Thermal [2 + 2] cycloadditions, however, proceed via a four-electron cyclic transition state that is isoconjugate with cyclobutadiene and thus is Hückel-antiaromatic and symmetry-forbidden.

Another example is provided by the butadiene-cyclobutene interconversion. The disrotatory electrocyclic process is isoconjugate with cyclobutadiene and hence antiaromatic (a symmetry-forbidden ground-state reaction). In contrast, the corresponding conrotatory process results in an array of π-orbitals that resembles a Möbius strip (that is, the top of one π-terminus overlaps with the bottom of the other). This latter array is isoconjugate with so-called Möbius cyclobutadiene and is aromatic (a symmetry-allowed ground-state reaction). *See* CONJUGATION AND HYPERCONJUGATION; DELOCALIZATION; MOLECULAR ORBITAL THEORY; ORGANIC REACTION MECHANISM; PHYSICAL ORGANIC CHEMISTRY; WOODWARD-HOFFMANN RULE. Alan P. Marchand

Bibliography. G. Desimoni et al. (eds.), *Natural Products Synthesis through Pericyclic Reactions*, 1983; S. N. Ege, *Organic Chemistry*, 4th ed., 1998; R. J. Fessenden and J. S. Fessenden, *Organic Chemistry*, 6th ed., 1998; I. Fleming, *Frontier Orbitals and Organic Chemical Reactions*, 1976; E. A. Halevi, *Orbital Symmetry and Reaction Mechanisms: The OCAMS View*, 1992; G. M. Loudon, *Organic Chemistry*, 3d ed., 1995; T. H. Lowry and K. S. Richardson, *Mechanism and Theory in Organic Chemistry*, 3d ed., 1990; A. P. Marchand and Roland E. Lehr, *Pericyclic Reactions*, vols. 1 and 2, 1977; R. B. Woodward and R. Hoffmann, *The Conservation of Orbital Symmetry*, 1970.

Periderm

A group of tissues which replaces the epidermis in the plant body. Its main function is to protect the underlying tissues from desiccation, freezing, heat injury, mechanical destruction, and disease. Although periderm may develop in leaves and fruits, its main function is to protect stems and roots. The fundamental tissues which compose the periderm are the phellogen, phelloderm, and phellem (**Fig. 1**). *See* EPIDERMIS (PLANT).

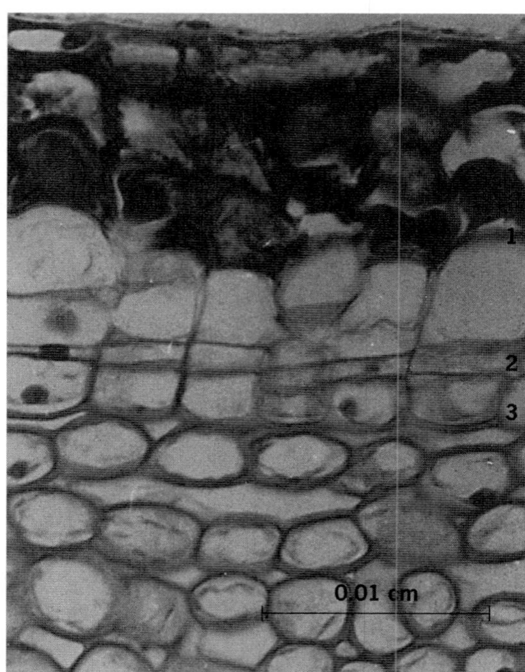

Fig. 1. Cross section of the stem periderm of *Sambucus nigra*: (1) phellem; (2) phellogen; (3) phelloderm.

Phellogen. The phellogen is the meristematic portion of the periderm and consists of one layer of initials. These exhibit little variation in form, appearing rectangular and somewhat flat in cross and radial sections, and polygonal in tangential sections (**Fig. 2**).

It is commonly accepted that the first phellogen of the shoot is derived from mature parenchymatous or collenchymatous cells of the cortex or from epidermal cells. Subsequent phellogens may originate from parenchymatous cells of the cortex or phloem. Phellogen initation occurs a short time before or concurrently with the rupture of the epidermis. Such a rupture may result from an internal increase in diameter or from external damage. The cells that give rise to the phellogen usually cannot be distinguished from other cells of the cortex by their shape. However, shortly before their first division, those cells lose their central vacuoles, their cytoplasmic volumes increase, and prominent quantities of tannins, leucoanthyocyanins, and chlorogenic acid are accumulated. The time at which the first phellogen is

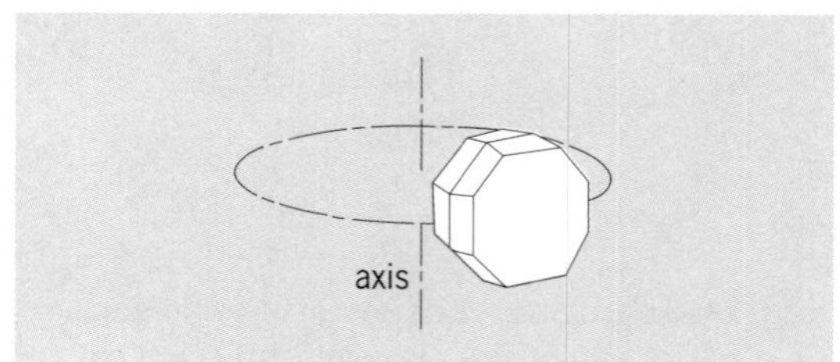

Fig. 2. Schematic representation of a phellogen cell, shown in its natural orientation.

initiated varies with species, organs, and environmental conditions. In most species, phellogen is initiated within a month or two after germination, or at least within the first year. However, in a few temperate and subtropical tree species (for example, *Ceratonia siliqua*) the initiation of the phellogen and the subsequent formation of the periderm take place only after several years.

The cell divisions that follow phellogen initiation are mostly periclinal (that is, walls are formed parallel with the circumference). However, occasional anticlinal divisions occur which increase the circumference of the phellogen initial layer. *See* CORTEX (PLANT); PARENCHYMA; STEM.

Periods of phellogen activity alternate with periods of inactivity, even during the same year. The activity of the phellogen within a growth period shows variations in rate and can be modified by environmental conditions. Short-day conditions stimulate phellogen activity, while long-day conditions hinder it, as with *Robinia pseudacacia*. Extreme temperatures—both high and low—accelerate activity, as with *Acacia* sp. Hyperhydric conditions seem also to cause both early initiation and intensive activity of the phellogen, as with *Eucalyptus* sp. In many cases, conditions that inhibit activity of the vascular cambium stimulate the activity of the phellogen, and vice versa. The first phellogen may enter dormancy and resume activity several times, but in most cases it finally loses its meristematic activity and undergoes a complete differentiation. In such a case, a new layer of phellogen, or subsequent phellogen, is initiated in the tissues beneath. *See* PHOTOPERIODISM; PLANT-WATER RELATIONS.

The phellem produced by subsequent phellogens will usually enclose layers of dead cells of the cortex and phloem. Such a dead outer periderm is named rhytidome, or shell bark. *See* PHLOEM.

The manner in which the subsequent periderms are formed determines to a large extent the form of rhytidome exfoliation. Usually when the first periderm is shallow beneath the epidermis, the subsequent periderms are formed in overlapping patches, and the rhytidome is exfoliated in the form of scales (scaly bark). When the first periderm is initiated deep in the cortex, subsequent periderms will also encircle the axis and the rhytidome will be exfoliated in the form of a hollow cylinder (ring bark).

The phellogen of the roots is usually initiated in the pericycle. *See* PERICYCLE; ROOT (BOTANY).

Phelloderm. The phelloderm cells are phellogen derivatives formed inward. The number of phelloderm layers varies with species, season, and age of the periderm. In some species, the periderm lacks the phelloderm altogether. The phelloderm consists of living cells with photosynthesizing chloroplasts and cellulosic walls. These cells may show a less organized tier arrangement than the phellogen or phellem cells and are hard to distinguish from the parenchymatous cells of the cortex and phloem. Sclerified phelloderm cells are found in various species of *Eucalyptus* and *Acacia*.

Phellem. The phellem, or cork, cells are phellogen derivatives formed outward. These cells are arranged in tiers with almost no intercellular spaces except in the lenticel regions. After completion of their differentiation, the phellem cells die and their protoplasts disintegrate. The cell lumens remain empty, excluding a few species in which various crystals can be found. Distinguishable rings of thin-walled and wide-lumened cells and thick-walled and narrow-lumened cells develop seasonally or intraseasonally. The walls of the cork cells consist of a thin cellulosic primary wall and a thick suberized secondary wall (suberin lamella). In some species an inner cellulosic lamella can also be found. The remarkable impermeability of the suberized cell walls is largely due to their impregnation with waxes, tannins, cerin, friedelin, and phellonic and phellogenic acids. In various species, non-suberized cells named phelloids may also be found in the phellem.

Except for a very few cork-producing species, the production of phellem cell layers may be as low as one or two cell layers per month. This rate can be accelerated under various environmental conditions. *See* CELL WALLS (PLANT).

Lenticels. Lenticels are loose-structured openings that develop usually beneath the stomata and that facilitate gas transport through the otherwise impermeable layers of phellem (**Fig. 3**).

The lenticels have a round or oblong shape in surface view. In their center, a mass of loose, thin-walled cells (complementary tissue) is formed. In a few cases these cells are released outward while in other cases they are held in place by bands of normal dense phellem cells (closing layers).

The lenticels develop prior to or concurrently with

Fig. 3. A lenticel as seen in a cross section of the stem periderm of *Sambucus nigra*.

the initiation of the first phellogen and may persist for many years. The lenticels of subsequent periderms are formed continuous with the lenticels of the first periderm. More lenticels may be formed in subsequent periderms, usually inside cracked periderm layers.

Cork. Cork is the most important commercial product of the periderm. Commercial cork is produced mainly by two species, *Quercus suber* and *Q. occidentalis*, both natives of the west Mediterranean region. This cork consists mostly of layers of phellem cells with perennial lenticels and is a product of a subsequent phellogen which is formed only after the first phellogen has been exfoliated (**Fig. 4**). The phellogen of the cork oaks may function for scores of years.

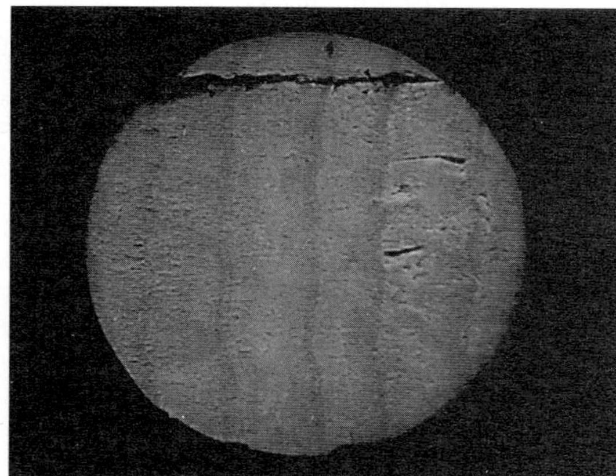

Fig. 4. Surface view of a bottle cork. A few annual rings and a perennial lenticel can be distinguished.

Wound periderm. A special type of periderm which is worth notice is the wound periderm. This is also a protective tissue that develops within injured plant organs beneath wound surfaces. Its formation is rapid and is accelerated under high temperature and high humidity conditions. The development of a similar periderm is also associated with any kind of normal plant abscission. Prior to abscission, a separation layer is formed between the abscising and the persisting tissues, and below this layer a phellogen is initiated. *See* ABSCISSION; BARK; SCLERENCHYMA.

Y. Waisel; H. Wilcox

Bibliography. G. B. Cooke, *Cork and the Cork Trees*, 1961; K. Esau, *Anatomy of Seed Plants*, 4th ed., 1990; A. Fahn, *Plant Anatomy*, 3d ed., 1982; N. Harris and K. J. Oparka (eds.), *Plant Cell Biology: A Practical Approach*, 1994.

Peridotite

A rock consisting of more than 90% of millimeter-to-centimeter-sized crystals of olivine, pyroxene, and hornblende, with more than 40% olivine. Other minerals are mainly plagioclase, chromite, and garnet. Much of the volume of the Earth's mantle probably is peridotite.

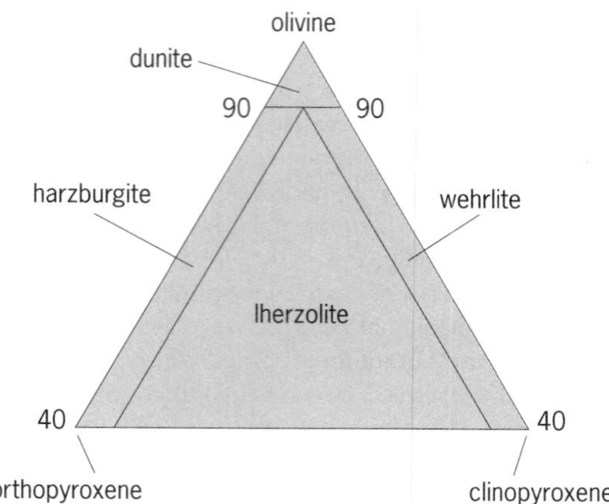

Fig. 1. Diagram of various peridotites which depend on principal mineral content. The numbers indicate 90 vol % olivine; and 40 vol % olivine and 60 vol % pyroxene (ortho or clino or both).

Mineralogy. Peridotites have various names depending upon the subordinate minerals (**Fig. 1**). Generally the aluminous mineral is mentioned as a modifier, as in feldspathic lherzolite. Other minerals, if significant, are similarly used as modifiers—hornblende peridotite, for example.

Structure and texture. Some peridotites are layered or are themselves layers; others are massive. Many layered peridotites occur near the base of bodies of stratified gabbroic complexes. Other layered peridotites occur isolated, but possibly once composed part of major gabbroic complexes. *See* GABBRO.

Both layered and massive peridotites can have any of three principal textures: (1) rather well-formed crystals of olivine separated by other minerals (**Fig. 2**); (2) equigranular crystals with straight grain boundaries intersecting at about 120°; and (3) long crystals with ragged curvilinear boundaries. The first

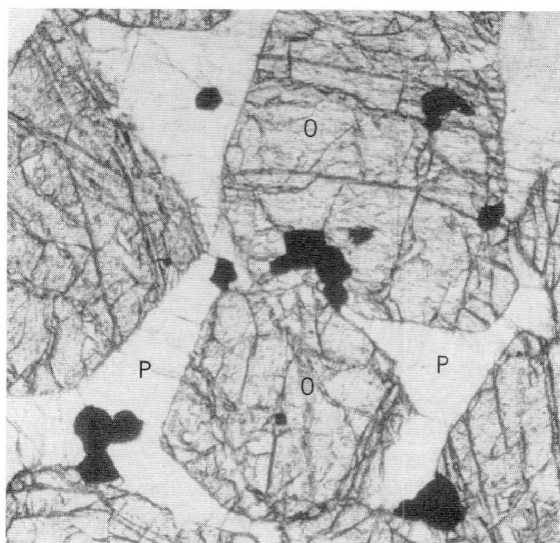

Fig. 2. Millimeter-sized crystals of olivine (O) with good crystal form are outlined by plagioclase (P) which fills the space between the accumulated crystals of olivine. Small crystals of black chromite occur in the olivine and plagioclase.

texture probably reflects the original deposition of an olivine sediment (or cumulate) from magma. The second texture may result from slow cooling whereby recrystallization leads to a minimization of surface energy. The third texture probably results from internal deformation.

Physical properties. Peridotites are comparatively dense rocks (3.3–3.5 g/cm^3 or 1.9–2.0 oz/in.^{3}) with rather high velocities ($V_p \approx 5$ mi/s or 8 km/s) for body-wave propagation. The electrical conductivity is sensitive to oxidation state, lattice defects, and temperature. Remanent magnetization is variable in peridotites, being particularly high in altered varieties rich in small grains of magnetite.

Occurrence. Peridotites have three principal modes of occurrence corresponding approximately to their textures: (1) Peridotites with well-formed olivine crystals occur mainly as layers in gabbroic complexes; so-called Alpine peridotites generally have irregular crystals and occur as more or less serpentinized lenses bounded by faults in belts of folded mountains such as the Alps, the Pacific coast ranges, and the Appalachian piedmont. (2) Peridotite nodules in alkaline basalts and diamond pipes generally have equigranular textures, but some have irregular grains. (3) Peridotite also occurs on the walls of rifts in the deep sea floor and as hills on the sea floor, some of which reach the surface. Many Alpine peridotites occur in the ophiolite association: peridotite, gabbro, diabase sill-and-dike complex, pillow basalt, and red chert. Some peridotites rich in amphibole have a concentric layered structure and form parts of plutons called Alaskan-type zoned ultramafic complexes. Certain volcanic rocks, sometimes called komatiites, are sufficiently rich in olivine and pyroxene to be termed peridotite. Small pieces of peridotite have been found in lunar breccias.

Composition. Peridotites are rich in magnesium, reflecting the high proportions of magnesium-rich olivine. The compositions of peridotites from layered igneous complexes vary widely, reflecting the relative proportions of pyroxenes, chromite, plagioclase, and amphibole.

Special interest focuses on the compositions of nodules of peridotites found in certain basalts and diamond pipes (kimberlites) because they may provide clues to the composition of the Earth's mantle. There are serious difficulties in evaluating the analytical data, particularly for minor and trace elements because of the complex history which such rocks have undergone. At least two principal varieties of peridotites exist in the Earth's mantle: (1) lherzolites, which have relatively high proportions of basaltic ingredients (in the minerals garnet and clinopyroxene); and (2) harzburgites, which have relatively low proportions of basaltic ingredients (because garnet and clinopyroxene are lacking or minor; see **table**). *See* AMPHIBOLE; BASALT; CHROMITE; GARNET; PYROXENE.

Origin. Layered peridotites are igneous sediments and form by mechanical accumulation of dense olivine crystals (Fig. 2). Peridotite nodules are pieces

Composition of two principal varieties of peridotite*

Ingredients	A	B
SiO_2	47.52	43.70
TiO_2	<0.02	0.25
Al_2O_3	0.63	2.75
Cr_2O_3	0.27	0.28
Fe_2O_3	2.37	1.38
FeO	3.50	8.81
MnO	0.10	0.13
MgO	43.21	37.22
CaO	0.49	3.26
Na_2O	0.06	0.33
K_2O	0.02	0.14
H_2O^{+}†	1.67	1.94
H_2O^{-}‡	0.07	0.05
Total	99.93	100.24

*A is garnet harzburgite, sample number PHN 1569, Thaba Putsoa kimberlite, Lesotho, South Africa. B is garnet lherzolite, sample number PHN 1611, Thaba Putsoa kimberlite, Lesotho, South Africa.
†Water released above 110° C (230° F).
‡Water released at or below 110° C (230° F).

of mantle rock more or less modified by partial melting. Mantle lherzolites may be the principal source rock for basaltic magmas, whereas mantle harzburgites probably form both from the crystalline residue left after basaltic magma migrates out of lherzolite and from a crystalline accumulation of early solidification products of some basaltic magmas within the mantle. Alpine peridotites of the ophiolite association probably formed in the oceanic crust and uppermost mantle by transfer of partial melt from the mantle to the crust, followed by tectonic emplacement of both crust and mantle along thrust faults in mountain belts. Peridotites associated with Alaskan-type ultramafic complexes probably formed in the root zones of volcanoes. *See* PLATE TECTONICS.

Use. Peridotite is an important rock economically. Where granites have intruded peridotite, asbestos and talc are common. Pure olivine rock (dunite) is quarried for use as refractory foundry sand and refractory bricks used in steelmaking. Serpentinized peridotite is locally quarried for ornamental stone. Tropical soils developed on peridotite are locally ores of nickel. The sulfides associated with peridotites are common ores of nickel and platinoid metals. The chromite bands commonly associated with peridotites are the world's major ores of chromium. *See* IGNEOUS ROCKS. Alfred T. Anderson, Jr.

Bibliography. F. R. Boyd and H. O. A. Meyer (eds.), *The Mantle Sample: Inclusions in Kimberlites and Other Volcanics*, American Geophysical Union, 1979; A. R. McBirney, *Igneous Petrology*, 2d ed., 1992; L. R. Wager and A. Hall, *Igneous Petrology*, 1987; P. J. Wyllie (ed.), *Ultramafic and Related Rocks*, 1976, reprint 1979.

Period doubling

A scenario for the transition of a natural process from regular motion to chaos. Various natural processes develop in time in a way that depends upon prevailing environmental details. A quantity that specifies the particular state of the environment of a process is called a parameter, and is taken as a fixed value over the course of development of the process. It is a frequent natural occurrence for a process to have a regular and easily describable motion for some range of parameters, but to have complex, irregular, and difficult-to-describe motions for other ranges of parameters. In the context of fluid flow, the latter circumstance is termed turbulence. In a more general context it is called chaos (which includes fluid turbulence but presages an underlying generality). *See* FLUID FLOW; TURBULENT FLOW.

Sometimes, as the environmental parameters are varied, a process may systematically exhibit more irregular motions, turning over into chaotic motion beyond some parameter value. In analogy to the phenomenology of phase transitions, this circumstance is termed a transition to chaos. There are a variety of qualitatively different transitions to chaos, each termed a scenario. Some of them are now well understood.

Period doubling is one frequently encountered scenario leading to chaos for which a full theoretical account exists. Since it occurs in a wide variety of processes of significantly divergent physical characters (for example, fluid flow, chemical reactions, and electronic devices), it is sensible to consider it as a phenomenon in its own right. Indeed, this panphysical character is a motivation to pose the study of the nature of complex motions as a subject in itself, and hence justifies the embracing generalization from turbulence to chaos. The affirmation of this vantage point is afforded by the fact that period doubling, wherever it appears, appears in a way that is quantitatively universal. This means that accurate numerical relations that govern this scenario hold for every such physical process, be it a fluid or a Josephson junction device.

Qualitative behavior. In order to observe this scenario, it is sufficient that all but one parameter is held fixed. Over some range of this varied parameter (it shall be defined to increase over the range of investigation) the motion is observed to be periodic. Above a certain value of the parameter the motion grows more complicated (a bifurcation has occurred): after the amount of time T for which the motion exactly repeated itself just prior to the bifurcation, the motion now slightly fails to do so, exactly repeating, however, after another T seconds. That is, the period has doubled from T to $2T$. As the parameter is further increased, the error to repeat after the first half of the new period systematically increases. A still further increase of parameter produces another bifurcation resulting in a new doubling of the period: the motion slightly fails to repeat after two roughly periodic cycles, exactly doing so after four. As the parameter is further increased, there are successive period-doubling bifurcations, more and more closely spaced in parameter value until at a critical value the doubling has occurred an infinite number of times, so that the motion is now no longer periodic and hence of a more complex character than had yet been encountered. Unpredictably complex motions occur for values of the parameter above its critical value, although ranges of parameter still exist for which the system exhibits new periodic motions. Indeed, any period-doubling system exhibits the same sequence of truly chaotic motion and interspersed periodicities as its parameter increases. Thus there is a strong degree of qualitatively universal behavior for all systems experiencing this scenario.

Quantitative behavior. However, there is also a precise quantitative universality. That is, without knowing the system (or its equations) essentially all measurable quantities can be predicted: By looking at the data alone, it would not be possible to guess the physical system responsible for that data. In order to describe these facts, some formal definitions are required. Denoting the parameter by p, let p_n be the value of p at the nth bifurcation, so that, for large n, T_n, the new periodicity is asymptotically given by Eq. (1) for an appropriate value of T_0. The theory of this scenario then predicts that Eq. (2) is satisfied, where δ is a universal mathematical constant of nature with value given by Eq. (3). The content of

$$T_n \sim 2^n T_0 \tag{1}$$

$$\frac{p_{n+1} - p_n}{p_{n+2} - p_{n+1}} \sim \delta \tag{2}$$

$$\delta = 4.6692016\ldots \tag{3}$$

Eq. (2) is that the p_n's accumulate to the critical value p_∞ geometrically for large n at rate δ, as given by Eq. (4).

$$p_\infty - p_n \sim \delta^{-n} \tag{4}$$

Furthermore, the errors to reproduce after T_n seconds when the period is $2T_n$ at each value of time over the entire motion are related to the errors at the same time when the periodicity was T_n in a precise way. To specify this, define $x_n(t)$ as the value of any quantity in the system when the periodicity is T_n. The error to reproduce after half the period is then given by Eq. (5). The theory of period doubling then determines that Eq. (6) is satisfied, where σ is a

$$d_n(t) = x_n(t) - x_n\left(t + \frac{T_n}{2}\right) \tag{5}$$

$$d_{n+1}(t) \sim \sigma\left(\frac{t}{T_{n+1}}\right) d_n(t) \tag{6}$$

universal function which in first approximation has two distinct values (and their negatives) given by α^{-1} and α^{-2} where α is given by Eq. (7).

$$\alpha = 2.50290787\ldots \tag{7}$$

The meaning of σ is that the quantitatively universal aspects of this scenario are scaling relationships for the motion: in first approximation, after the overall variation of $x(t)$ is measured to set a "yardstick" for the given system, all the details of $x(t)$ can be constructed from Eq. (6) by successive scalings. At p_∞, the motion $x_\infty(t)$ is self-similar at all scales. That is, the "noise" in x_∞ is not statistical in nature, but rather the complex outcome of repeated elaboration of details at all scales. It is a technical exercise to compute the power spectrum of $x_n(t)$ once Eq. (6) is known. *See* FRACTALS.

Thus the manner in which the parameter axis is laid out, given by Eq. (4), and the manner in which the system modifies its response at each parameter value, given by Eq. (6), are precisely determined apart from some overall scales in a way completely independent of the nature of the given system. That is, δ and $\sigma(t)$ are measurable quantities (and have been measured in a variety of real physical systems) and basically exhaust all the information that could be measured, or could be computed from the specific equations governing each specific system investigated. Thus, reminiscent of thermodynamics, questions can be posed and answered in a general manner that bypasses the specific mechanisms governing any particular system.

Universality of the scenario. The results discussed above hold over some parameter range for almost all nonlinear dissipative systems with enough "parts" (degrees of freedom), where enough means three or more (or even infinitely more). For some such systems the relevant range of parameters is sufficiently large to make period doubling a prominent part of the system's repertoire. For others it may be unmeasurably small, and so of no physical consequence. Indeed there do exist alternative scenarios, some of which satisfy theories highly reminiscent of that of period doubling, that can prominently mark a system's transition to chaos.

All of the features described are the consequences of a rigorous renormalization group theory, so that this scenario is similar in spirit to second-order phase transitions. In particular, δ is a critical exponent, α is a rescaling parameter, and the function σ is constructed from the fixed point of an appropriate transformation. The other scenarios akin to period doubling differ in that the renormalization group transformations differ. *See* PHASE TRANSITIONS; RENORMALIZATION. Mitchell J. Feigenbaum

Bibliography. J. Argyris, G. Faust, and M. Haase, *An Exploration of Chaos*, 1994; R. L. Devaney, *A First Course in Dynamical Systems: Theory and Experiment*, 1992; J. Gleick, *Chaos: Making a New Science*, 1992; E. Lorenz, *The Essence of Chaos*, 1993.

Periodic motion

Any motion that repeats itself identically at regular intervals. If $x(t)$ represents the displacement of any coordinate of the system at time t, a periodic motion has the property defined by Eq. (1) for every value of

$$x(t + T) = x(t) \quad (1)$$

the variable time t. The fixed time interval T between repetitions, or the duration of a cycle, is known as the period of the motion. Frequency is the number of repeating cycles per unit time, and is numerically equal to the reciprocal of the period T.

The motion of the escapement mechanism of a watch, the motion of the Earth about the Sun, and the more complicated motion of the crankshaft, piston rods, and pistons in an engine running at uniform speed are all examples of periodic motion.

The vibration of a piano string after it is struck is a damped periodic motion, not strictly periodic according to the definition. Although the motion very nearly repeats itself, and with a fixed repetition time, each successive cycle has a slightly smaller amplitude. *See* DAMPING.

Any periodic motion can be expressed as a Fourier series—a sum of sine and cosine terms whose frequencies of integral multiples of the frequency f of the periodic motion. Thus Eq. (2) holds, where the

$$x(t) = A_0 + \Sigma_n A_n \cos(2\pi nft) + \Sigma_n B_n \sin(2\pi nft) \quad (2)$$

A's and B's are constant coefficients, and the sums may be taken over all positive integer values of n. For the special case in which the coefficients all vanish for $n > 1$ *see* HARMONIC MOTION; FOURIER SERIES AND TRANSFORMS.

Many systems with more than one degree of freedom, whose motion is not simply periodic, are multiply periodic. The motion may be resolved into parts (for example, horizontal and vertical components, radial and tangential components), each of which is periodic, but with periods that are not commensurate. One example is the vibration of a bell, whose overtone frequencies are not simply related to the fundamental frequency. The motion of the solar system is multiply periodic because it never exactly repeats itself, even though each planet moves periodically. *See* VIBRATION; WAVE MOTION. Joseph M. Keller

Periodic table

A list of elements (atoms) ordered along horizontal rows according to atomic number (the number of electrons in an atom and also the charge on its nucleus). In the periodic table (**Fig. 1**) the rows are arranged so that elements with nearly the same chemical properties occur in the same column (group) and each row ends with a noble gas (closed-shell element that is generally inert). For chemists, the position of atoms in the periodic table provides the most powerful guide for classifying the expected properties of molecules and solids made from these particular atoms. *See* INERT GASES.

The origin of the periodic table was explained in the 1920s in terms of the basic physical laws (quantum mechanics) obeyed by the electrons of an atom. Thus, the rows in the periodic table correspond to the shell number, n, and groups correspond to a

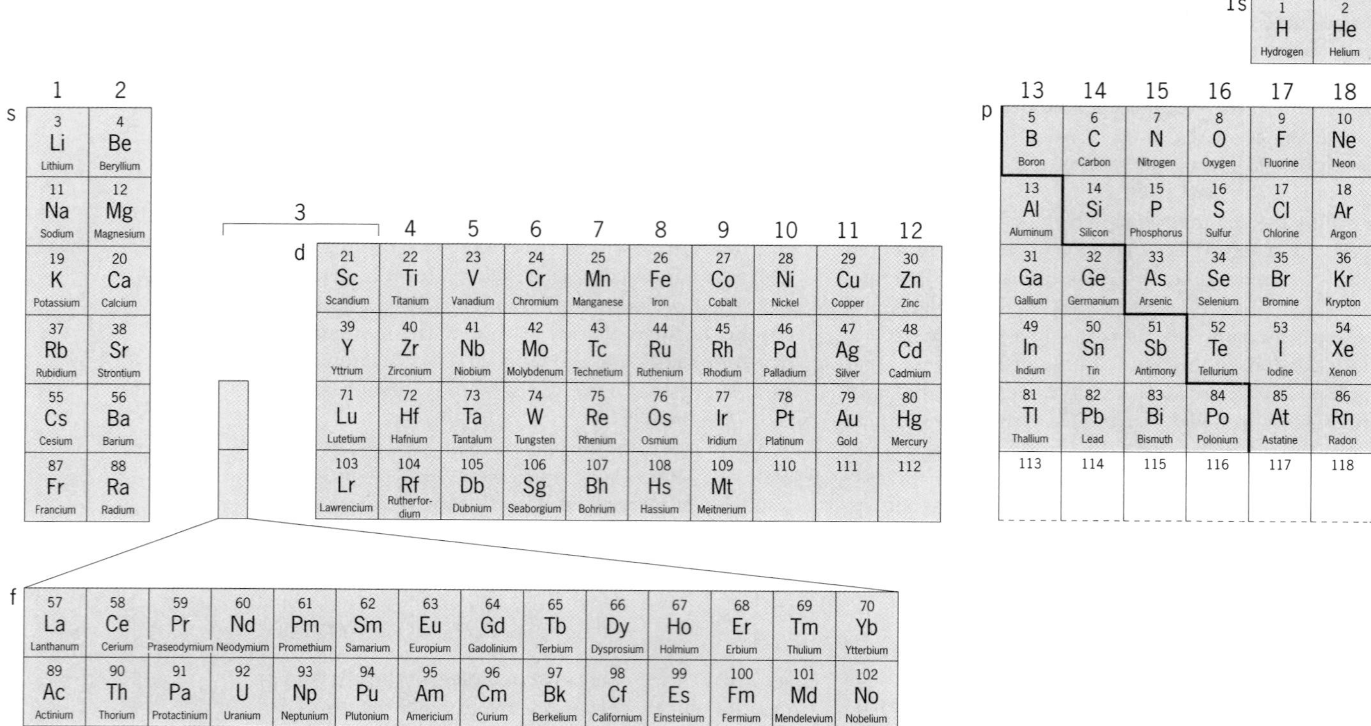

Fig. 1. Two-dimensional periodic table. The atomic numbers are listed above the symbols identifying the elements. The heavy line separates metals from nonmetals.

particular electronic configuration designated by the number and type of electrons in the outermost shell. These electrons govern chemical properties and are known as valence electrons. For example, the electronic configuration for nitrogen is $2s^2 2p^3$, where $n = 2$ is the row number also the electronic shell number; *s* and *p* are quantum-mechanical indices that tell the shape of an electron's orbit around an atom, and the superscripts give the number of *s*-type and *p*-type electrons present in the valence shell. The indices, *s*, *p*, *d*, and *f*, also designate blocks in the table (Fig. 1). When a group in the table is descended, *n* increases in steps, but the type and number of valence electrons remain the same; for example, in group 15, N (nitrogen) = $2s^2 2p^3$, P (phosphorus) = $3s^2 3p^3$, As (arsenic) = $4s^2 4p^3$, Sb (antimony) = $5s^2 5p^3$, and Bi (bismuth) = $6s^2 6p^3$. *See* ELECTRON CONFIGURATION; QUANTUM MECHANICS; VALENCE.

Configuration energy. Additional information from the physical laws of atoms can be incorporated into the periodic table, and greatly enhance its organizing capability. For example, configuration energy adds a third dimension to the periodic table (**Fig. 2**). The configuration energy is defined in terms of the ionization energy (*I*), the energy required to remove an electron from an atom. Configuration energy is given by the formula below, where *a* and *b* are the

$$\mathrm{CE} = \frac{aI_{ns} + bI_{np}}{a + b}$$

number of *s* and *p* valence electrons. The formula gives the energy of an average valence electron for the atom in question (see **table**). For nitrogen, the configuration energy is $(2I_{2s} + 3I_{2p})/5$, where I_{2s} is the energy required to remove a 2*s* electron from nitrogen, $\mathrm{N}(2s^2 2p^3) \rightarrow \mathrm{N}^+(2s2p^3)$, with electronvolts as units.

Besides enhancing the organizing capability of the periodic table, the concept of configuration energy explains many long-standing puzzles about the table itself. It explains the existence of the metalloid band of elements (configuration energy is nearly constant in this band) and why these elements divide the metals from the nonmetals. Elements possessing configuration energies with magnitudes greater than those of the metalloids are nonmetals; those with lower configuration energies are metals. The topography of a three-dimensional form of the periodic table graphically brings out this division (Fig. 2). Likewise, the decreasing magnitudes of configuration energies down a group in the three-dimensional table (Fig. 2) demonstrate and quantify the long-known fact that the elements in a group do not have exactly the same chemistry. For example, the chemistry of carbon is quite different from that of tin, even though both are in group 14. This differentiation is called metallization. *See* METAL; METALLOID; NONMETAL.

Other properties. Another property of the periodic table is resolved by configuration energy: the limitation of the fluoride oxidation states in the elements nitrogen (N), oxygen (O), fluorine (F), chlorine (Cl), bromine (Br), helium (He), neon (Ne), argon (Ar), and krypton (Kr). These nine atoms cannot reach a fluoride oxidation state equal to the number of

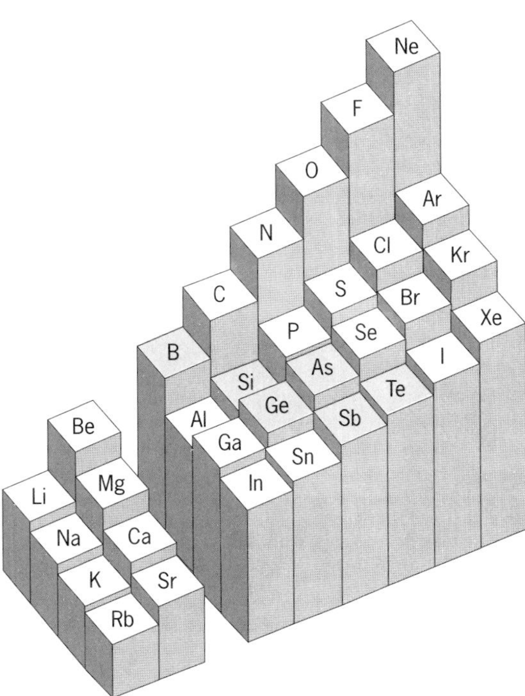

Fig. 2. Three-dimensional periodic table. Only *s* and *p* blocks are displayed. The tinted blocks are the metalloid elements. (*After L. C. Allen, Electronegativity is the average one-electron energy of the valence-shell electrons in ground-state free atoms, J. Amer. Chem. Soc., 111:9003–9014, 1989*)

their valence electrons, as can all other representative atoms (atoms whose valence shell is made up of *s* and *p* electrons only). They have the highest nine configuration energies in the periodic table; some fractions of their valence electrons have ionization energies too high to engage in chemical bonding. For example, iodine heptafluoride (IF_7) exists but not chlorine heptafluoride (ClF_7).

Similarly, it has previously been believed that atomic radius was an independent property, but again (for representative atoms) radius can be expressed as a simple inverse function of configuration energy. Likewise, the multiple bonds found for representative atoms of the $n = 2$ row, but much less frequently for $n = 3$, 4, 5, or 6, can be shown to follow directly from the high values of configuration energy for $n = 2$.

Chemical bonding. Configuration energy also quantifies the chemical bonding designations, covalent (C), ionic (I), and metallic (M), which have always been associated with the periodic table but have not been adequately defined. The concept of configuration energy can be used to organize types of chemical bonding by the periodic table (**Fig. 3**). In such an organization, the elements of a given row and their binary combinations are displayed around a triangle (Fig. 3*a*). This concept can be extended to construct a schematic triangle for all the binary solids that can be made from the stable (nonradioactive) and noninert (He, Ne, and Ar) elements in the periodic table (Fig. 3*b*). Atoms along the horizontal axis in such a construction are positioned according to the magnitude of their configuration energy values (instead of fixed positions as in Fig. 3*a*), and this triangle represents 78(79)/2 = 3081 possible binary solids. It proves possible to divide this type of arrangement into four subregions: metallic, ionic, covalent, and metalloid bonding (metalloid bonding made from the atoms B, Si, Ge, As, Sb, and Te); and the pair numbers specifying a binary compound place it in one of the four subregions, thereby defining what type of bonding characterizes the compound. In this organization of chemical bonding by the periodic table (Fig. 3), the compounds at the points of the triangle are well known, but near the center of the legs of these diagrams novel features become apparent. Near the middle along the horizontal axis are the metalloids, and going into the triangle there is the 3-5 semiconductor aluminum phosphide (AlP). Close to the center of the ionic-covalent leg are the polymeric materials aluminum fluoride (AlF_3) and aluminum chloride ($AlCl_3$). Close to the center of the metallic-ionic leg are Zintl phases, that is, semiconductors with a metallic sheen and ceramic-type brittleness. These latter have not found a commercial application, but constructing such triangles helps focus on new combinations of elements. *See* SEMICONDUCTOR.

Significance. The periodic table organizes the properties of the nearly 20 million known molecules and solids. This works well first, because molecules and solids are made up of atoms that are only very slightly perturbed when they link together; therefore, they largely retain their identity; and second, because atomic electronic structure is intrinsically simple. Even though the motion of electrons around a nucleus is a highly complex, many-particle swirl

Configuration energies (in electronvolts) of selected elements

$n = 1$	$n = 4$	$n = 5$	$n = 6$
H 13.61	K 4.34	Rb 4.18	Cs 3.89
He 24.59	Ca 6.11	Sr 5.70	Ba 5.21
$n = 2$	Sc 6.8	Y 5.9	
	Ti 7.4	Zr 6.6	
Li 5.39	V 8.1	Nb 7.4	
Be 9.32	Cr 8.6	Mo 8.2	
B 12.13	Mn 9.2	Tc 9.0	
C 15.05	Fe 9.9	Ru 9.8	
N 18.13	Co 10.4	Rh 10.6	
O 21.36	Ni 11.0	Pd 11.3	
F 24.80	Cu 10.8	Ag 11.7	
Ne 28.31	Zn 9.39	Cd 8.99	Hg 10.44
$n = 3$			
Na 5.14			
Mg 7.65			
Al 9.54	Ga 10.39	In 9.79	
Si 11.33	Ge 11.80	Sn 10.79	
P 13.33	As 13.08	Sb 11.74	
S 15.31	Se 14.34	Te 12.76	
Cl 16.97	Br 15.88	I 13.95	
Ar 19.17	Kr 17.54	Xe 15.27	

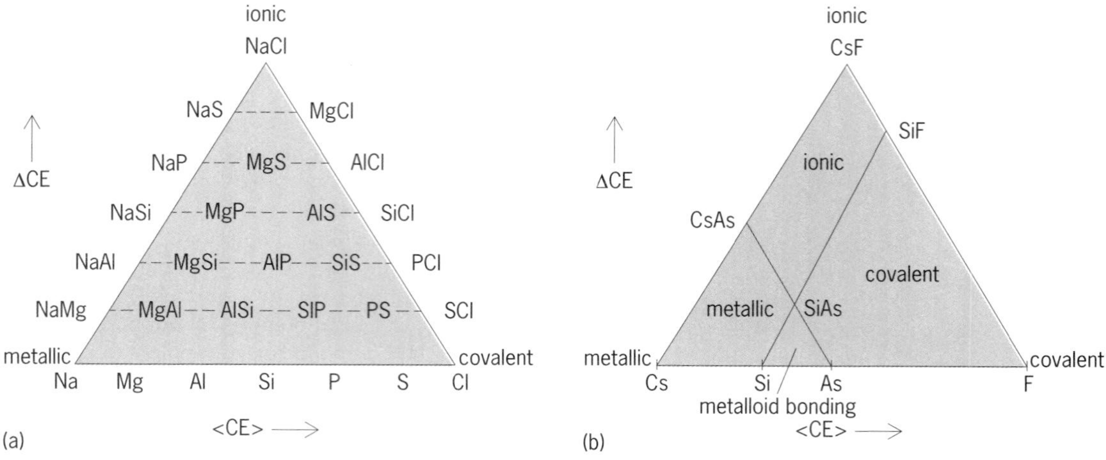

Fig. 3. Types of chemical bonding organized by the periodic table showing difference in configuration energy (ΔCE) between pairs of atoms, and average of configuration energy (<CE>) for pairs of atoms. (*a*) Bonding for $n = 3$ elements. The base line of the triangle is the $n = 3$ row of the periodic table. (*b*) Diagram for all binary solids made from the 78 stable, bond-forming elements in the periodic table. Each solid in the triangle is specified by a pair of numbers, [<CE>, ΔCE], using configuration energy values from the table. For example, SiAs corresponds to the number pair [12.21, 1.75]. Dividing lines in the triangle designate the bonding type for a solid whose number pair falls into one of the four subregions metallic, ionic, covalent, and metalloid bonding, which is intermediate between metallic and covalent (these solids are semiconductors). (*After L. C. Allen et al., Van Arkel-Ketelaar triangles, J. Mol. Struct., 300:647–655, 1993*)

requiring the largest digital computers to determine accurately, to a very good approximation each electron can be individually assigned a type, *s*, *p*, *d*, or *f*, and a shell number, *n*. The ability to label electrons individually by these indices is the other reason why the periodic table works so well. On a grander scale, the claim can be made that it would not be possible to have the science of chemistry—or any insight at all into the matter out of which everything on Earth is made—were it not for these two fundamental simplicities. *See* ATOMIC STRUCTURE AND SPECTRA.

Electronegativity. Traditionally, the best auxiliary aid in the use of the periodic table has been the long-recognized chemical concept of electronegativity. But this quantity has not previously been applicable to metallic bonding. Configuration energy has been shown to be fully compatible with the traditional electronegativity scales of L. Pauling and of A. L. Allred and E. G. Rochow, and also to encompass metallic bonding (because of its strong correlation with energy level spacings). Thus, configuration energy can be identified as a generalization of traditional electronegativity scales. *See* ELECTRONEGATIVITY.

Applications. The lack of numerical or analytic connection between the traditional two-dimensional periodic table (Fig. 1) and methods used to predict the structure and reactivity of molecules and solids has long reduced the table's usefulness. However, configuration energy, introduced as a new dimension of the periodic table, is just the average atomic energy level, and simultaneously the average density of states, for the atoms out of which the molecular-orbit-energy-level diagrams and energy bands in solids are constructed, thereby typing the periodic table directly to present-day research techniques. *See* BAND THEORY OF SOLIDS; ENERGY LEVEL (QUANTUM MECHANICS); MOLECULAR ORBITAL THEORY; MOLECULAR STRUCTURE AND SPECTRA.

Leland C. Allen

Bibliography. L. C. Allen, Extension and completion of the periodic table, *J. Amer. Chem. Soc.*, 114:1510–1511, 1992; J. E. Huheey, E. A. Keiter, and R. L. Keiter, *Inorganic Chemistry*, 5th ed., 2002; R. J. Puddephatt and P. K. Monaghan, *The Periodic Table of the Elements*, 2d ed., 1986; D. F. Shriver, P. W. Atkins, and C. H. Langford, *Inorganic Chemistry*, 3d ed., 1999.

Periodontal disease

An inflammatory lesion caused by bacteria affecting the tissues housing the roots of the teeth. The disease, sometimes called pyorrhea, increases in prevalence and severity with increasing age, and is a principal cause of tooth loss in adult humans throughout the world. When only the gum tissue or gingiva is affected the disease is called gingivitis, but when the process extends into the deeper structures it is known as periodontitis. The diseased tissues appear abnormally red and slightly swollen, and they tend to bleed, sometimes profusely, when the teeth are brushed. In some cases the gums may become thickened and scarred, and they may recede, exposing the root surfaces. As the disease advances, the attachment of the gum to the tooth is lost, creating a periodontal pocket; a large portion of the gum tissue is destroyed; and the bone surrounding the roots is resorbed (see **illus.**). The teeth become loose, abscesses form, and extraction is required.

Etiology. Both gingivitis and periodontitis are caused by bacteria that form plaques on the surfaces of the teeth at the gingival sulcus or pocket.

These plaques may contain 250 or more separate microbial species. Plaques of any microbial composition can cause gingivitis, but specific bacteria appear to be necessary for induction of periodontitis. Among the bacteria involved in periodontitis are various species of *Porphyromonas*, *Bacteroides*, *Actinobacillus*, *Eikenella*, *Fusobacterium*, *Wolinella*, and other less well-characterized species. Spirochetes are present in active lesions, but their role remains unclear. The bacteria extend apically along the interface between the tooth root and the gingival tissue and cause periodontal pockets to form.

Pathogenesis. The principal features of the pathogenesis of periodontitis have been described. The lesions begin as an acute inflammatory response, followed by a dense accumulation of lymphoid cells. There is a net loss of collagen in the areas nearest the junctional epithelium and periodontal pocket, with scarring and fibrosis of the connective tissues at more distant sites. The junctional epithelium is converted into an ulcerated pocket epithelium, the alveolar bone housing the tooth roots resorbed, and the periodontal ligament is destroyed. Products released by infiltrating leukocytes, including prostaglandins, interleukins and collagenase, and other hydrolytic enzymes, are involved in tissue destruction.

Bacterial colonization and extension activate several host defense mechanisms. The most effective of these is the accumulation of functional neutrophilic granulocytes between the surface of the plaque and the gingival tissue. These cells tend to counter and limit microbial extension. The bacteria appear to invade the periodontal connective tissues, where they induce immunopathologic and other destructive inflammatory reactions in the host, and these lead, in major part, to the observed tissue destruction. Periodontal destruction is episodic, with periods of exacerbation characterized by highly acute inflammation, followed by periods of quiescence.

Predisposing factors. Although bacteria are essential for induction of the disease, predisposing factors are also important, though their elucidation is not complete. Individuals who manifest functionally abnormal neutrophilic granulocytes or monocytes are unusually susceptible to the severe early-onset forms of periodontitis. The leukocyte abnormality appears to be genetically transmitted. The early-onset forms have been designated as prepubertal, juvenile, and rapidly progressive periodontitis; adult periodontitis has a later onset and does not seem to be related to leukocyte abnormalities. Some persons with acquired immune deficiency syndrome (AIDS) manifest a highly destructive, unique form of periodontitis. Other predisposing conditions include unusually stressful situations, and periods of hormone imbalance occurring at puberty, during pregnancy, and in some women taking birth control drugs.

Prevention and treatment. Good daily oral hygiene practices, including vigorous brushing of the exposed surfaces of the teeth and use of dental floss, interproximal brushes, and other devices to clean between the teeth, constitute the most effective measures to prevent periodontal disease. Basic ingredients of treatment of existing disease include bringing the infection under control and establishing conditions which preclude reinfection. All of the microbial deposits must be removed from the crown and root surfaces. In individuals with severe forms of periodontitis, these procedures may be supplemented by use of antibiotics either systematically or directly into the pocket. These procedures usually lead to reduction of the inflammation and to some shrinkage in the gums, but the periodontal pockets remain. Based on the traditional view that treated pockets may become reinfected and the disease may continue to spread, surgical treatment may be performed with the aim of

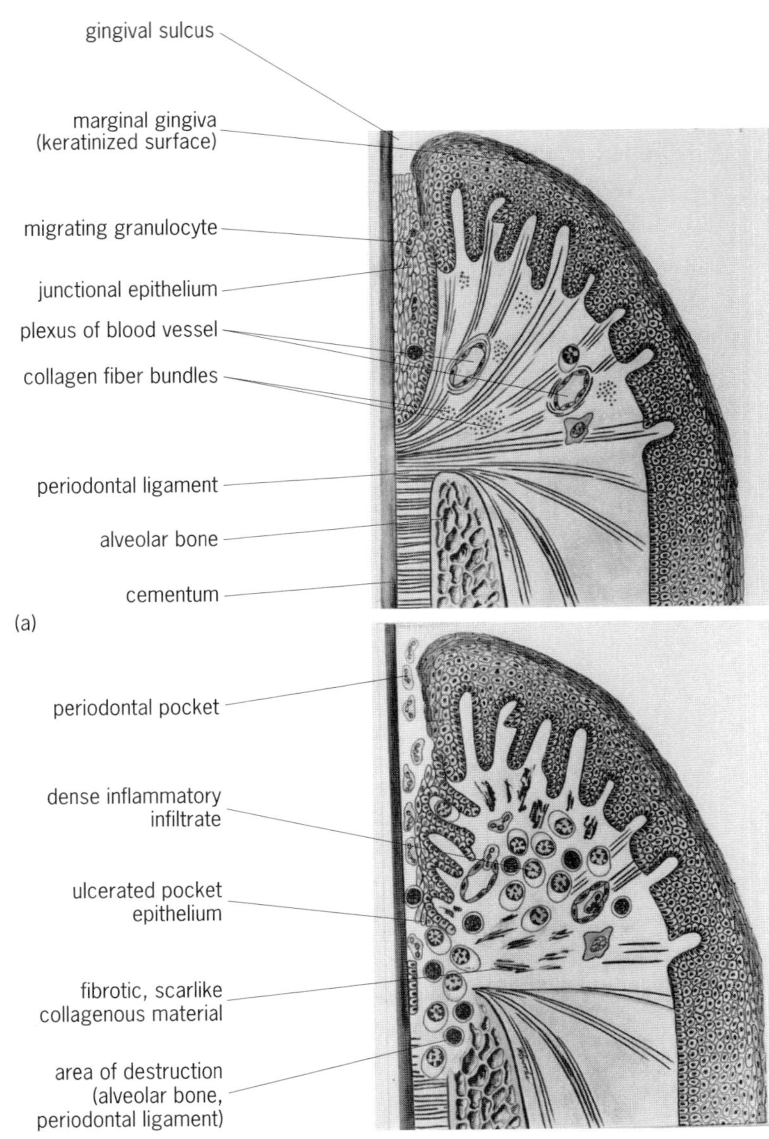

Structural features of (*a*) the normal periodontium and (*b*) the advanced stage of periodontal disease. (*After S. Schluger, R. C. Page, and R. Yuodalis, Periodontal Disease: Basic Phenomena, Clinical Management and Restorative Interrelationships, Lea and Feibiger, 1990*)

reducing pocket depth and restoring normal tissue contours. Alternatively, regenerative procedures use various grafting materials, including freeze-dried decalcified bone or bone substitutes and guided tissue regeneration. To perform guided tissue regeneration flaps are opened in the gingival tissue and the root surfaces are thoroughly cleaned; a porous membrane is placed around the tooth covering the bone defect with or without placing grafts, and flaps covering the membranes are sutured into place. The membrane permits the wound site to become populated with cells having the capacity to generate new bone, cementum, and periodontal attachment. *See* TOOTH DISORDERS.

Roy C. Page

Bibliography. R. C. Page, Periodontal therapy: Prospects for the future, *J. Periodontal*, 64:744–753, 1993; S. Schluger et al., *Periodontal Disease*, 2d ed., 1990.

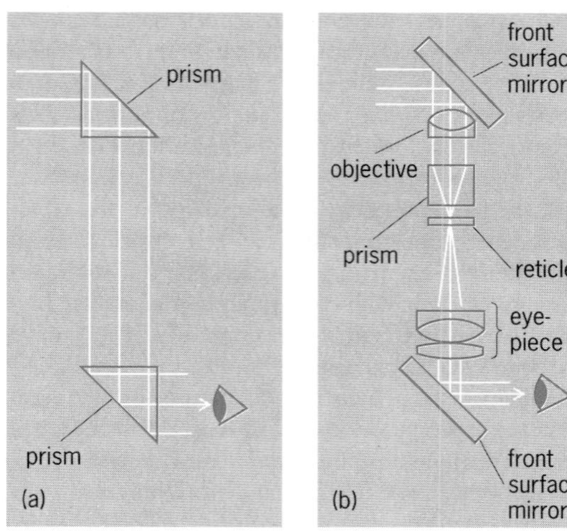

Fig. 1. Tank periscope. (*a*) Simple, with parallel reflecting surfaces. (*b*) With terrestrial telescope.

Perischoechinoidea

A subclass of Echinoidea lacking stability in the number of columns of plates that make up the ambulacra and interambulacra. The ambulacral columns vary from 2 to 20, the interambulacral from 1 to 14. There are two orders, Bothriocidaroida and Echinocystitoida. *See* ECHINODERMATA; ECHINOIDEA.

Howard B. Fell

Periscope

An optical instrument that permits viewing along a displaced or deflected axis, providing an observer with the view from a position which may be inaccessible or dangerous. Periscopes range in complexity from the simple unit-power tank periscope to the complex multielement submarine periscope.

Tank periscope. This device, intended to protect the user from bullets, employs a pair of plane, parallel, reflecting surfaces (either mirrors or prisms), so arranged in a mount that the path of light through the instrument forms a crude letter Z (**Fig. 1***a*). If powers greater than unity are desired or if the periscope is to be used for sighting, a terrestrial telescope can be added to the periscope, either as a simple, internally contained system (Fig. 1*b*) or entirely in front of or behind the periscope itself, as desired. The reflecting elements of the system which are responsible for deflecting the optical axis are independent of the refracting (telescope) elements which provide the optical power. *See* TELESCOPE.

It is also possible to arrange the two mirrors of a periscope at right angles to each other, in which case the observer views an inverted image while facing away from the direction from which light enters the instrument. By adding an inverting (astronomical type) telescope to this system, the image is reinverted for the observer.

Periscopes of this type cannot be used for scanning the horizon by rotating the upper mirror because of the image rotation which accompanies such movement. In the panoramic sight (**Fig. 2**), this difficulty is overcome by providing the system with a dove prism which rotates at half the angular speed of the right-angle prism through the action of a differential gear linkage. The combined inversions of the dove prism and the amici prism at the bottom completely compensate for the inversions of the telescope system, while the relative motions of the right-angle prism and the dove prism maintain the image erect during scanning. *See* MIRROR OPTICS; OPTICAL PRISM.

Submarine periscope. In this device, it is necessary to employ a telescope system having a wide field of view and uniform illumination across a field which can be fitted into a long, narrow tube whose

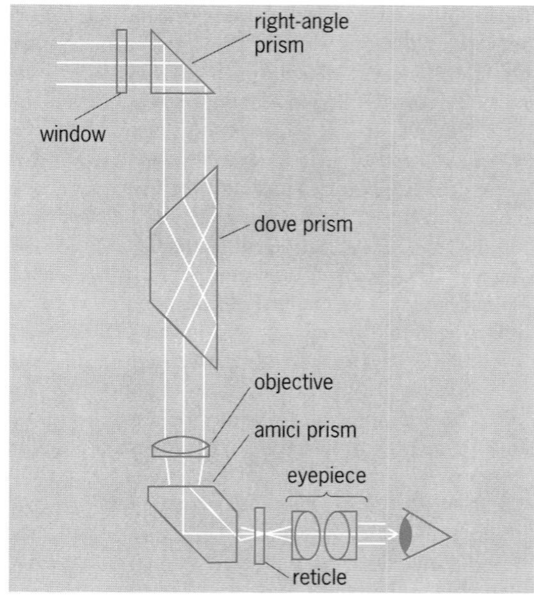

Fig. 2. Panoramic sight, with erect image.

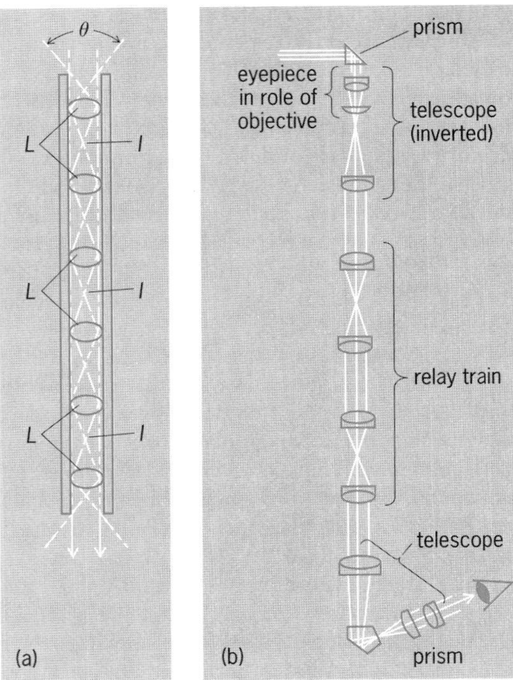

Fig. 3. Periscope relay train. (*a*) Showing lenses *L*, inversions *I*, and angle of view *θ*. (*b*) Between a pair of facing telescopes in a submarine periscope.

length-to-diameter ratio may be 50 or greater. This is achieved by utilizing a plurality of lenses so spaced along the length of the tube as to cause the incoming principal rays from the edge of the field to be deviated from side to side within the tube. In general, the greater the number of lenses, the wider the field of view. One example of the periscopic relay train is shown in **Fig. 3*a*** and employs six lenses with three inversions. The typical submarine periscope (Fig. 3*b*) may be considered to be a pair of telescopes facing each other, with such a relay train between them. The usual magnification of submarine periscopes is 6, although some U.S. Navy periscopes have dual magnifications of 6 and 1.5, the latter being achieved by inserting an inverted galilean telescope into the optical path before the top objective. *See* LENS (OPTICS); MAGNIFICATION.

The submarine periscope can be provided with a built-in rangefinder for fire-control purposes. A conventional coincidence or split-field type of rangefinder may be attached either vertically or horizontally to the upper end of the periscope, the objective of which receives an image from each of the entrance windows of the rangefinder. *See* RANGEFINDER (OPTICS); SUBMARINE.

Other types. Various modifications of the basic optical systems described here are employed as viewing periscopes in military aircraft and as viewing devices in particle accelerators and nuclear reactors. The cystoscope and endoscope are slender, sometimes mechanically flexible periscopes used for visual examination and photography of body cavities inaccessible to direct observation; an entirely different basis for the design of such instruments is in the use of bundles of optical fibers. *See* OPTICAL FIBERS.

Edward K. Kaprelian

Bibliography. Optical Society of America, *Handbook of Optics*, 2d ed., 1996; A. G. Smith, *Easy to Make Periscope*, 1990; B. H. Walker, *Optical Engineering Fundamentals*, 1995.

Perissodactyla

An order of herbivorous, odd-toed, hoofed mammals, including the living horses, zebras, asses, tapirs, rhinoceroses, and their extinct relatives. They are defined by a number of unique specializations, but the most diagnostic feature is their feet. Most perissodactyls have either one or three toes on each foot, and the axis of symmetry of the foot runs through the middle digit. The woodchucklike hyraxes, or conies, are apparently closely related to perissodactyls, although there is still some question about the relationships.

The perissodactyls (other than hyraxes) are divided into three groups: the Hippomorpha (horses and their extinct relatives); the Titanotheriomorpha (the extinct brontotheres); and the Moropomorpha (tapirs, rhinoceroses, and their extinct relatives). *See* HORSE PRODUCTION; RHINOCEROS; TAPIR.

Origins. Perissodactyls were once thought to have evolved in Central America from the phenacodontids, an extinct group of archaic hoofed mammals placed in the invalid taxon Condylarthra. However, in 1989 a specimen recovered from deposits in China about 57 million years old was described and named *Radinskya*. This specimen shows that perissodactyls originated in Asia some time before 57 million years ago (Ma), and were unrelated to North American phenacodonts. *Radinskya* is very similar to the earliest relatives of the tethytheres (elephants, manatees, and their kin). This agrees with other evidence that perissodactyls are more closely related to tethytheres than they are to any other group of mammals.

By 55 Ma, the major groups of perissodactyls had differentiated, and migrated to Europe and North America. Before 34 Ma, the brontotheres and the archaic tapirs were the largest and most abundant hoofed mammals in Eurasia and North America. After these groups became extinct, horses and rhinoceroses were the most common perissodactyls, with a great diversity of species and body forms. Both groups were decimated during another mass extinction about 5 Ma, and today only five species of rhinoceros, four species of tapir, and a few species of horses, zebras, and asses cling to survival in the wild. The niches of large hoofed herbivores have been taken over by the ruminant artiodactyls, such as cattle, antelopes, deer, and their relatives.

Horses. From their Asian origin, the hippomorphs spread all over the northern continents. North America became the center of evolution of true horses, which occasionally migrated to other continents. In Europe, the horselike palaeotheres substituted for true horses. From *Protorohippus* (once called *Hyracotherium* or *Eohippus*), a dog-sized ancestor with four toes on the front feet that lived 55–50 Ma, horses evolved into many different lineages living side by side. The three-toed horses *Mesohippus* and *Miohippus* (from beds dated about 30–37 Ma) were once believed to be sequential segments on the unbranched trunk of the horse evolutionary tree. However, they coexisted for millions of years, with five different species of the two genera living at the same time and place. From *Miohippus*-like ancestors, horses diversified into many different ecological niches, with some species adapted for eating soft, leafy vegetation, and others with high-crowned teeth for eating tough, gritty grasses. About 15 Ma, there were as many as 12 different lineages of three-toed horses in North America, each with slightly different ecological specializations; and they were analogous to the diversity of modern antelopes in East Africa. On two different occasions (*Pliohippus* and *Dinohippus*), three-toed horses evolved into lineages with a single toe on each foot. About 5 Ma, most of these three-toed and one-toed horse lineages became extinct, leaving only *Dinohippus* to evolve into modern *Equus*.

On a few occasions in the last 20 million years, true horses migrated back from North America to Eurasia, and eventually to Africa (evolving into zebras and wild asses) and South America. At the end of the last ice age (about 10,000 years ago), horses became extinct in the New World. However, they were reintroduced to their ancestral homeland by Columbus in 1493. Wild horses that have escaped from domesticated stock are known as mustangs.

Most extinct horses were browsers and ate soft, leafy vegetation, but all living horses are grazers, using their sharp incisors and mobile lips to crop low-growing grasses. The only common wild horse, the plains zebra, lives in large herds (up to 100 individuals) and migrates over large areas of grasslands in search of food. However, desert-dwelling asses and Grevy's zebra live in small herds, with a stallion guarding a small harem of mares. Most species of wild horses, including the Grevy's and mountain zebras, all species of onagers and asses, and Przewalski's horse (an ancestor of domesticated horses), are nearly extinct in the wild.

Brontotheres. These beasts began as pig-sized, hornless animals about 53 Ma, and quickly evolved into multiple lineages of cow-sized animals with long skulls and no horns. Between 37 and 34 Ma, their evolution culminated with huge, elephant-sized beasts bearing paired blunt horns on their noses. The extinction of brontotheres about 34 Ma was due to a global climatic change (triggered by the first Antarctic glaciers) that eliminated most of the soft, leafy vegetation on which they fed. *See* EXTINCTION (BIOLOGY).

Moropomorphs. The earliest moropomorphs, such as *Homogalax*, from strata about 55 million years old, are virtually indistiguishable from the earliest horses. From this unspecialized ancestry, a variety of archaic tapirlike animals diverged. Most retained the simple leaf-cutting teeth characteristic of tapirs and, like brontotheres, died out about Ma when their forest habitats shrank. Only the modern tapirs, with their distinctive long proboscis, survive in the jungles of Central and South America (three species), and southeast Asia (one species). All are stocky, piglike beasts with short stout legs, oval hooves, and a short tail. They have no natural defenses against large predators (such as jaguars or tigers), so they are expert at fleeing through dense brush and swimming to make their escape.

The horselike clawed chalicotheres are closely related to some of these archaic tapirs. They apparently used their claws to haul down limbs and branches to eat leaves (much as ground sloths did), rather than for digging. *Chalicotherium* had such long forelimbs and short hindlimbs that it apparently knuckle-walked like a gorilla, with its claws curled inward.

Rhinoceroses have been highly diverse and successful throughout the past 50 million years. They have occupied nearly every niche available to a large herbivore, from dog-sized running animals, to several hippolike forms, to the largest land mammal that ever lived—the 18-ft-tall (6-m), 44,000-lb (20,000-kg) *Paraceratherium* (once called *Baluchitherium* or *Indricotherium*). Most rhinoceroses were hornless. Rhinos with horns first appeared about 28 Ma; two different lineages independently evolved paired horns on the tip of the nose. Other groups evolved a single nasal horn, or huge horn on the forehead, or a pair of horns in tandem on nose and forehead. Unlike the horns of cattle, sheep, and goats, rhino horns are made of cemented hair fibers, and have no bony core.

Between 20 and 5 Ma, rhinos diversified into several browsing (leaf-eating) lineages, and hippolike grazing lineages, and browser-grazer pairs of rhinos were found all over the grasslands of Eurasia, Africa, and North America. The mass extinction event that occurred about 5 Ma wiped out North American rhinos and decimated most of the archaic rhino lineages in the Old World. During the ice ages, woolly rhinos and their relatives were common all over Eurasia. Their only surviving descendant is the endangered Sumatran rhinoceros. Only a few hundred individuals still live in the mountainous jungles of Sumatra.

Four other species of rhino survive in Asia and Africa, but all are on the brink of extinction because of heavy poaching for their horns. The Javan rhinoceros is rarely seen in the dense jungles of Java and Vietnam; fewer than 50 individuals may still be living in the wild. The Indian rhinoceros and the African rhinos (the browsing black rhinoceros and the grazing white rhinoceros) inhabit grasslands, although they prefer to hide in dense vegetation. Rhinos have very poor eyesight but excellent senses of smell and hearing, so they can detect danger long before others can see them. Most live in small family groups of females with their calves; however, bull rhinos tend to be solitary. *See* MAMMALIA.

Donald R. Prothero

Bibliography. R. L. Carroll, *Vertebrate Paleontology and Evolution*, 1988; D. R. Prothero, The rise and fall of the American rhino, *Nat. Hist.*, 96:26–33, 1987; D. R. Prothero and R. M. Schoch (eds.), *The Evolution of Perissodactyls*, 1989; R. J. G. Savage and M. R. Long, *Mammal Evolution: An Illustrated Guide*, 1986.

Peritoneum

The membranous lining of the coelomic, especially the abdominal, cavity, composed mainly of flattened epithelial cells that produce a small amount of watery, or serous, fluid. In the embryo the coelomic wall is lined by this membrane, which continues over the developing viscera so that they are suspended and supported by the reflected peritoneum, principally from the dorsal body wall, but also from the ventral body wall in the region of the liver.

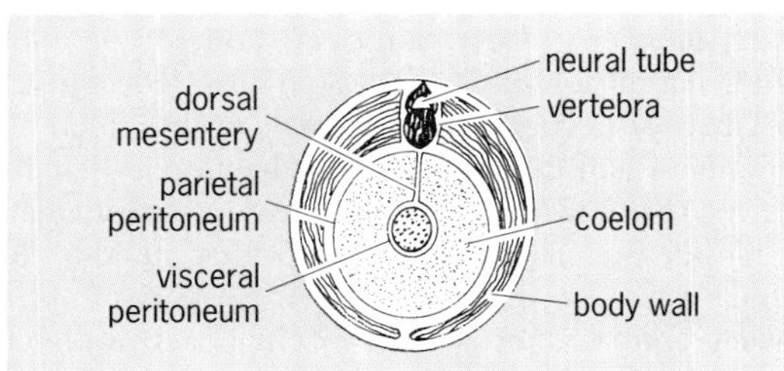

Diagrammatic cross section of a vertebrate, showing the relations of the layers of peritoneum and the coelom. (*After H. M. Smith, Evolution of Chordate Structure, Holt, Rinehart and Winston, 1960*)

As the organs develop, enlarge, and assume their adult form and arrangement, the supporting peritoneum becomes modified, some parts being lost and others becoming thickened or twisted, or otherwise adapting to normal growth (see **illus.**). After development is completed, those portions which line the interior of the body wall are called parietal peritoneum, and the supporting sheets are known collectively as mesenteries, many areas of which have received specific names. The remaining peritoneum, which covers most of the organs, is called visceral peritoneum and forms the outer layer, or serosa, of the walls of most of the gastrointestinal tract. The fluid-filled space between the serosa and the parietal peritoneum is the remnant of the coelom, or body cavity. *See* EPITHELIUM; FETAL MEMBRANE.

Thomas S. Parsons

Peritonitis

Inflammation of the peritoneum, the serous membrane which lines the abdominal cavity and surrounds most of the abdominal organs. The condition may be caused by infectious organisms or foreign substances introduced into the abdominal cavity. The small amount of serous fluid normally present as a lubricant acts as an excellent culture medium for bacterial growth and also as a means of spreading invading materials. The source of such substances or organisms is commonly a gastrointestinal inflammation, especially if perforation has occurred. Appendicitis, peptic ulcer, cancer of the bowel, gallbladder disease, and dysentery are common sources of infection that may produce peritonitis, as well as blood-borne forms of tuberculosis and pneumonia. *See* PERITONEUM.

Infection may also stem from spread of bacterial organisms from the female organs, the kidneys, and the pancreas. Each form of peritonitis may show both common and specific features.

Contamination of the abdominal cavity may occur from penetration of the abdominal wall during an accident or following a wound. Bile, blood, or fluid from a ruptured abdominal cyst or ectopic pregnancy may also induce a peritoneal inflammation.

Where peritonitis occurs, the normally glistening peritoneum becomes dull, the blood vessels engorge, and a fibrin-containing exudate is produced which may later lead to adhesions. A tendency for localization is apparent, with loops of intestine or other organs forming pockets of inflammation.

The clinical course is quite variable, depending on the agent involved, the type and severity of reaction, and concomitant disease. *See* APPENDICITIS; BACILLARY DYSENTERY; GALLBLADDER DISORDERS; PNEUMONIA; PREGNANCY DISORDERS; TUBERCULOSIS.

Edward G. Stuart; N. Karle Mottet

Peritrichia

A specialized subclass of the class Ciliatea composed of a large group of unusual-looking ciliate protozoa. It has excited the curiosity of microscopists for nearly 300 years. Many are sessile and stalked, while some form colonies which may reach a large size. A number are attached as ectocommensals to a variety of animals and plants. A free-swimming stage in the life cycle, indispensable for distribution, is known as the telotroch. It is a small, mouthless form equipped with a single girdle of posteriorly located locomotor cilia. This is quite unlike the morphology of the mature, sedentary form, which is an inverted bell form atop a long stalk. The body is naked of ciliature except for conspicuous wreaths of buccal ciliary organelles at the oral end. Much is made by many protozoologists of the fact that the adoral zone of membranelles, in this instance, winds counterclockwise toward the mouth. Actually, this may be of little real importance, although convenient in taxonomic keys. *Vorticella* (**illus.** *a*) and *Epistylis* are probably the best-known stalked forms. The former is a solitary ciliate, the latter a colony builder. *Trichodina* (illus. *b*) belongs to the group of mobile peritrichs. Its species are associated with a wide variety of invertebrate and vertebrate hosts on which its actions range from those of a harmless commensal to a pathogenic parasite. The

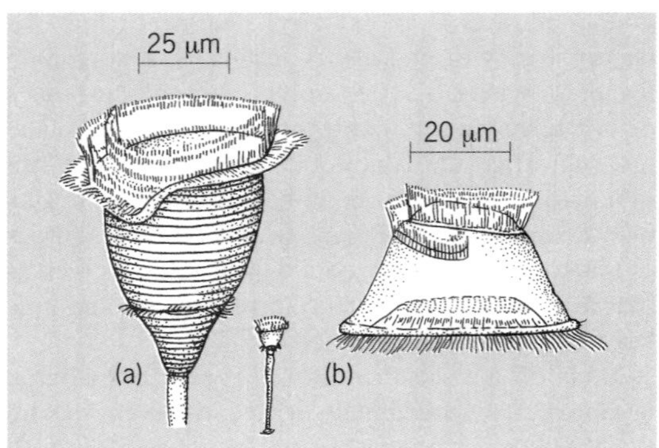

Peritrichida. (*a*) *Vorticella*, a stalked peritrich. (*b*) *Trichodina*, a mobile peritrich.

subclass Peritrichia contains a single order, the Peritrichida, whose species may have evolved from such holotrichian orders as the Hymenostomatida or the Thigmotrichida. *See* CILIATEA; CILIOPHORA; HYMENOSTOMATIDA; PROTOZOA; THIGMOTRICHIDA.

John O. Corliss

Permafrost

Perennially frozen ground. Two classes of frozen ground are generally distinguished: seasonally frozen ground, which freezes and thaws on an annual basis; and permafrost, which is formed when subsurface earth materials remain below 0°C (32°F) for more than a year, without regard to composition, phase, or cementation by ice. Permafrost occurs at most locations where the mean annual temperature at the ground surface is below freezing. Because substantial differences often exist between the temperature at the surface and that measured in the air, climate statistics do not provide a reliable guide to the existence or distribution of permafrost. Although permafrost is ultimately a climatically determined phenomenon, its presence or absence is strongly influenced by local factors, including microclimatic variations, circulation of ground water, type of vegetation cover, and thermal properties of subsurface materials.

The range of temperatures experienced annually at the ground surface in permafrost terrain decreases progressively with depth, down to a level at which only minute annual temperature variation occurs, termed the level of zero annual amplitude (**Fig. 1**). A zone of near-surface, seasonally frozen ground above permafrost called the active layer often experiences complex heat-transfer processes. Below the permafrost table, the upper limit of material that experiences a maximum annual temperature of 0°C (32°F), heat transfer occurs largely by conduction. In situations where the bottom of the active layer is not in direct contact with the top of the permafrost, the permafrost is a relic of a past colder interval and is not in equilibrium with the present-day climate.

Distribution. Approximately 25% of the Earth's land surface is underlain by permafrost. Substantial areas of subsea permafrost also occur around the land margins of the Arctic Ocean. Permafrost distribution is often classified on the basis of its lateral continuity (**Fig. 2**). In the Northern Hemisphere the various classes form a series of concentric zones that roughly follow the parallels of latitude. Southward deviations in the extent of permafrost in Siberia and central Canada are a response to lower mean annual temperatures in the continental interiors. In the continuous permafrost zone, perennially frozen ground underlies most locations, the principal exception being beneath large bodies of water that do not freeze to the bottom annually. In the discontinuous zone, permafrost may be widespread, but a substantial proportion of the land surface can be underlain by seasonally frozen ground because of variations in local factors. Some scientists also refer to a zone in which permafrost is sporadic, occurring largely as isolated patches that reflect combinations of local factors favorable to its formation or maintenance. Because the thermal properties of peat are conducive to its formation and maintenance, subarctic bogs are the primary locations of permafrost in southerly parts of the subarctic lowlands. In the Southern Hemisphere, permafrost occurs in ice-free areas of the Antarctic continent and in some of the subantarctic islands. Permafrost also occurs extensively in such midlatitude mountain ranges as the Rockies, Andes, and Himalayas, as a response to progressively colder mean annual temperatures at higher elevations. Given sufficient altitude and favorable local conditions, patches of permafrost can exist even in the

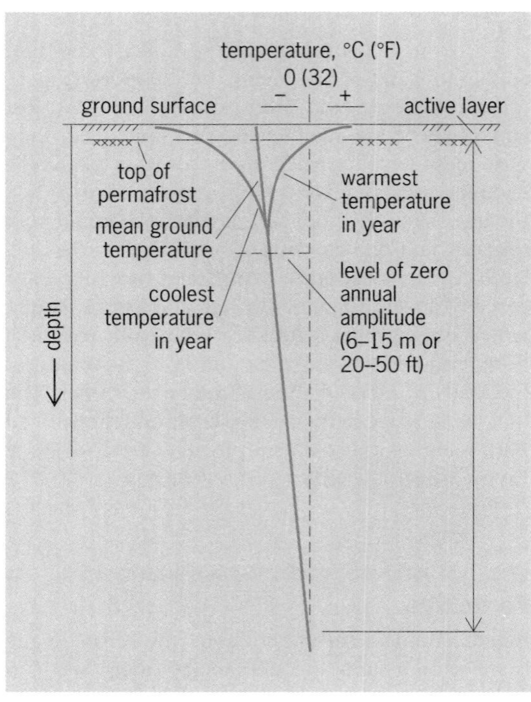

Fig. 1. Typical ground thermal regime in permafrost (area to right of broken line). (*After P. J. Williams, Pipelines and Permafrost: Science in a Cold Climate, Carleton University Press, 1986*)

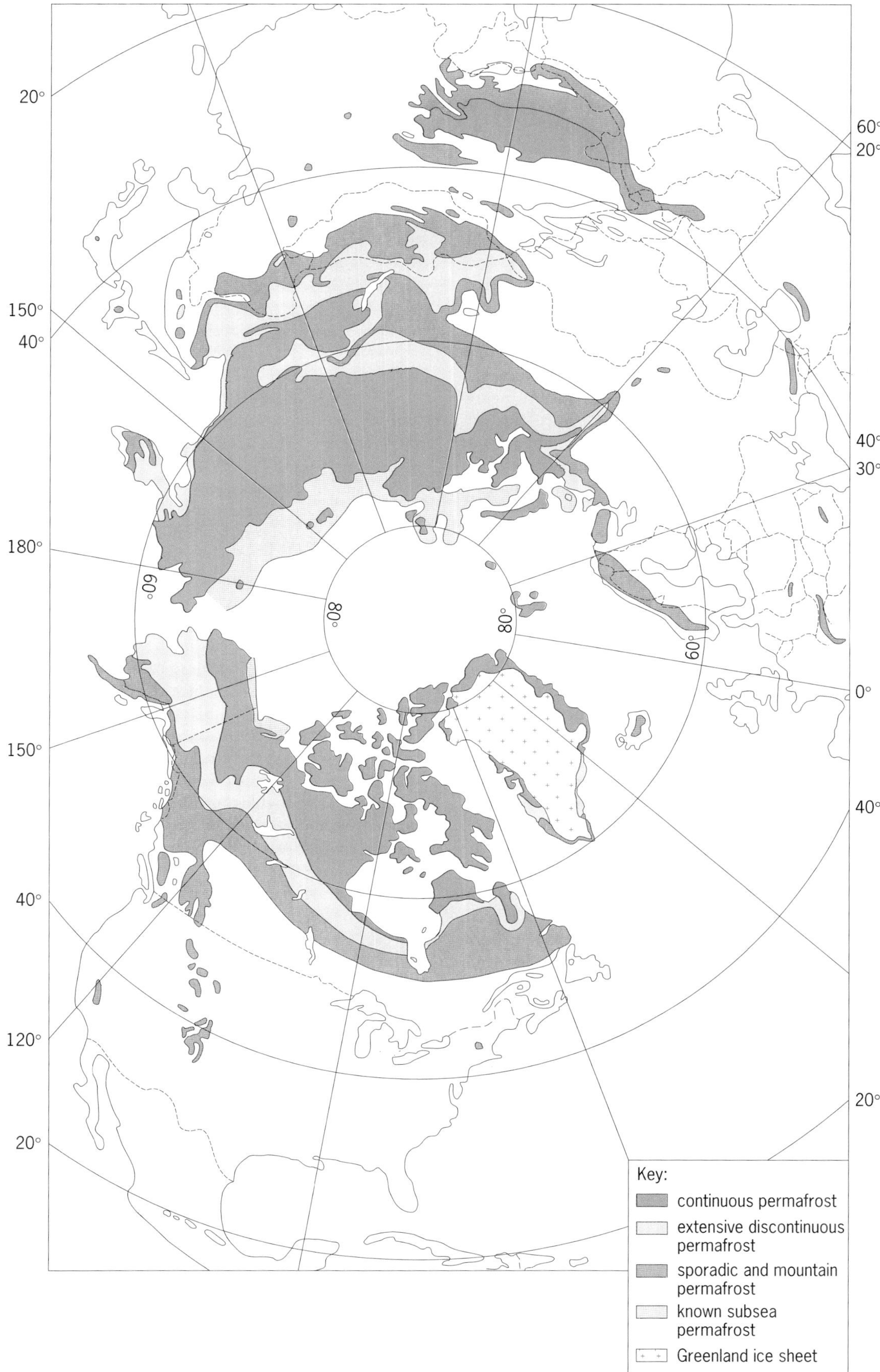

Fig. 2. Map of permafrost distribution in the Northern Hemisphere. (After *J. A. Heginbottom, Geological Survey of Canada*)

subtropics, such as in the crater of Mauna Kea at an elevation of 4140 m (13,580 ft) on the island of Hawaii. *See* ANTARCTICA; ARCTIC OCEAN; BOG; PEAT.

The thickness of permafrost at a specific location is determined by the mean annual temperature at the ground surface, the thermal properties of the substrate, and the amount of heat flowing from the Earth's interior. Permafrost thickness ranges from very thin layers to about 1500 m (4920 ft) in some unglaciated parts of Siberia. In general, permafrost thickness increases toward the poles. *See* EARTH, HEAT FLOW IN.

Ice content and surface features. Although the presence of ice is not a criterion in the definition of permafrost, ground ice is responsible for many of the distinctive features and problems in permafrost regions. Ice can occur within permafrost as small individual crystals in soil pores, as lenses or layers of nearly pure ice parallel to the ground surface, and as variously shaped intrusive masses formed when water is injected into soil or rock and subsequently frozen. Many remarkable landforms exist in permafrost regions, although only some are unambiguous indicators of the presence of permafrost. Surface features formed under cold, nonglacial conditions are known collectively as periglacial landforms. Periglacial features that serve as good indicators of the presence of permafrost include ice wedges, vertically oriented ice inclusions created when water produced from melting snow seeps into cracks formed in fine-grained sediments during severe cold-weather events earlier in winter. Repeated many times over centuries or millennia, this process produces tapered wedges of foliated ice more than a meter wide near the surface and extending several meters or more into the ground. Viewed from the air, networks of ice wedges form striking polygonal patterns over extensive areas of the Arctic. Other landforms diagnostic of the presence of permafrost are pingos, ice-cored hills frequently over 10 m (33 ft) in height that can form when freezing fronts encroach from several directions on saturated sediments remaining after drainage of a deep lake. Palsas are smaller mound-shaped forms often found in subarctic peatlands, and can form in a variety of ways. Thawing of ice-rich permafrost, whether through climatic changes or local disturbance, can result in irregular subsidence of the ground surface and can form a series of pits, mounds, and water-filled depressions known collectively as thermokarst terrain.

Other periglacial landforms occur frequently in association with permafrost, but are not necessarily diagnostic of its presence. When ice-rich permafrost or another impermeable layer prevents infiltration of water, the soil on a hillside may become vulnerable to a slow, flowlike process known as solifluction, giving rise to a network of lobes and terraces that impart a crenulated or festooned appearance to the slope. Other small landforms frequently encountered in subpolar and polar regions, collectively referred to as patterned ground, include small hummocks and networks of striking geometric forms arranged into circles, polygons, and stripes of alternating coarse and fine-grained sediment. The bold appearance of these forms is often accentuated by vegetation patterns. *See* GLACIAL GEOLOGY.

Fig. 3. Linear thermokarst feature in ice-rich permafrost terrain near Prudhoe Bay, Alaska. (*Frederick E. Nelson*)

Engineering considerations. Construction in permafrost regions requires special techniques at locations where the terrain contains ice in excess of that contained within soil pores. Prior to about 1970, many projects in northern Alaska and elsewhere disturbed the surface significantly, triggering thermokarst processes and resulting in severe subsidence of the ground surface, disruption of local drainage patterns, and in some cases destruction of the engineered works themselves. The route of a winter road constructed in 1968–1969 by bulldozing the tundra vegetation and a thin layer of soil can be seen as a scar on the ground surface (**Fig. 3**). This disturbance altered the energy regime at the ground surface, leading to thawing of the underlying ice-rich permafrost and subsidence of up to 2 m (6.6 ft) along the road, which became unusable several years after construction.

Environmental restrictions in North America, based on scientific knowledge about permafrost, have been established to regulate construction activities and minimize their impact on terrain containing excess ice. The Trans-Alaska Pipeline, which traverses 1300 km (812 mi) from Prudhoe Bay on the Arctic coastal plain to Valdez on Prince William Sound near the Gulf of Alaska, carries oil at temperatures above 60°C (140°F). To prevent development of thermokarst and severe damage to the pipe, the line is elevated in sections of the route where surveys indicated the presence of excess ice. To counteract conduction of heat into the ground, many of the pipeline's vertical supports are equipped with heat pipes that cool the permafrost in winter, lowering the mean annual ground temperature and preventing thawing during summer. In several short sections of ice-rich terrain where local aboveground conditions required burial of the line, the pipe is enclosed in thick insulation and refrigerated. *See* PIPELINE.

Other unusual engineering techniques devised for use in ice-rich permafrost include constructing heated buildings on piles, which allows circulation of air beneath the structures and prevents conduction of heat to the subsurface. Roads, airfields, and building complexes are frequently situated atop thick gravel pads or other insulating materials. In relatively large settlements, water and sewage are transported in insulated, elevated pipes known as utilidors.

Effects of climatic change. Despite implications contained in the term permafrost, perennially frozen ground is not permanent. Permafrost extended farther into the midlatitudes during past episodes of continental glaciation. Evidence for the former existence of permafrost, including ice-wedge casts and pingo scars, has been found in many European countries now far removed from the permafrost regions. In North America, evidence that permafrost existed during colder intervals is scattered along the glacial border from New Jersey to Washington state.

Much of the world's permafrost is within only a few degrees of the melting point of ice and is potentially unstable. Many scenarios of global climatic change suggest that warming may be most severe in the polar regions. Pronounced warming in those portions of the Arctic containing large areas of thaw-sensitive permafrost could lead to development of thermokarst terrain at regional scales, damage many existing engineered works, and cause the boundaries of the permafrost zones (Fig. 2) to be displaced poleward. Moreover, a large amount of carbon is stored in the permafrost regions. An increase in the active-layer thickness or widespread disappearance of permafrost could lead to the release of much of this carbon, intensifying the effects of global warming. The relationship between permafrost and future changes in climate involves much more than temperature fluctuations, and constitutes an important area for research. *See* CLIMATE HISTORY; CLIMATE MODIFICATION. Frederick E. Nelson

Bibliography. G. H. Johnston, *Permafrost: Engineering Design and Construction*, 1981; F. E. Nelson and S. I. Outcalt, A computational method for prediction and regionalization of permafrost, *Arctic Alpine Res.*, 19(3):279–288, 1987; A. L. Washburn, *Geocryology: A Survey of Periglacial Processes and Environments*, 1980; P. J. Williams, *Pipelines and Permafrost: Science in a Cold Climate*, 1986; P. J. Williams and M. W. Smith, *The Frozen Earth: Fundamentals of Geocryology*, 1989.

Permeance

The reciprocal of reluctance in a magnetic circuit. It is the analog of conductance (the reciprocal of resistance) in an electric circuit, and is given by Eq. (1), where **B** is the magnetic flux density, **H** is the magnetic field strength, and the integrals are respectively over a cross section of the circuit and around a path within it. *See* CONDUCTANCE.

$$P_m = \frac{\text{magnetic flux}}{\text{magnetomotive force}} = \frac{\int\int \mathbf{B}\cdot d\mathbf{S}}{\oint \mathbf{H}\cdot d\mathbf{l}} \qquad (1)$$

From Eq. (1), it can be shown that Eq. (2) is valid,

$$P_m = \frac{\mu A}{l} \qquad (2)$$

where A is the cross-sectional area of the magnetic circuit, l its length, and μ the permeability. If the material is ferromagnetic, as is often the case, then μ is not constant but varies with the flux density, and the complete magnetization curve of B against H may have to be used to determine the permeance.

If there are parallel magnetic circuits with the total flux divided between them, the permeances simply add, so that the combined permeance is given by Eq. (3), where $P_1, P_2, \ldots$, are the permeances of the individual branches.

$$P_m = P_1 + P_2 + \cdots \qquad (3)$$

For permeances in series, on the other hand, as for an iron circuit with an air gap, the reciprocals add, as given by Eq. (4).

$$\frac{1}{P_m} = \frac{1}{P_1} + \frac{1}{P_2} + \cdots \qquad (4)$$

See MAGNETISM; RELUCTANCE. A. Earle Bailey

Permian

The name applied to the last period of geologic time in the Paleozoic Era and to the corresponding system of rock formations that originated during that period. The Permian Period commenced approximately 290 million years ago and ceased about 250 million years ago. The system of rocks that originated during this interval of time is widely distributed on all the continents of the world.

The Permian System is presently divided into three series: the lower Cisuralian Series with type sections in the western Ural region of Russia; the middle Guadalupian Series with type sections in western Texas and southern New Mexico; and the upper Lopingian Series with type sections in southern China.

Permian rocks contain evidence for a paleogeography that was greatly different from present geography. The Permian Period was a time of variable and changing climates, and during much of this time latitudinal climatic belts were well developed. During the later half of Permian time, many long-established lineages of marine invertebrates became extinct and were not immediately replaced by new fossil-forming lineages. Rocks of Permian age contain many resources, including petroleum, coal, salts, and metallic ores. *See* LIVING FOSSILS.

Eastern European Permian. The Permian System was proposed in 1841 by R. Murchison for a succession of marine, brackish, evaporitic, and nonmarine deposits exposed in the former province of Perm on the western flank of the Ural Mountains in Russia (**Fig. 1**). During the Permian Period, this

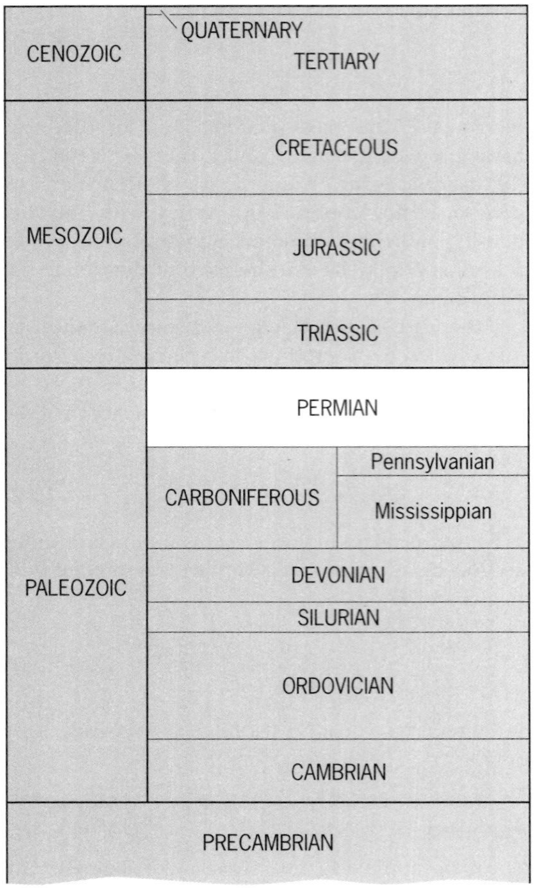

Era	Period	
CENOZOIC	QUATERNARY	
	TERTIARY	
MESOZOIC	CRETACEOUS	
	JURASSIC	
	TRIASSIC	
PALEOZOIC	PERMIAN	
	CARBONIFEROUS	Pennsylvanian
		Mississippian
	DEVONIAN	
	SILURIAN	
	ORDOVICIAN	
	CAMBRIAN	
PRECAMBRIAN		

region was near the boundary between the stable Russian Platform on the west and the tectonically active Ural Geosyncline to the east (**Fig. 2**). The ancestral Ural Mountains were gradually formed by the uplift and compressive collapse of the Ural Geosyncline during this time. The geosynclinal deposits are mostly clastic sediments that reach a thickness of several thousand meters. During the early part of the Permian, the edge of the Russian Platform was marked by the development of extensive carbonate banks and reefs. These carbonate deposits contrast markedly with the thick clastic beds of the geosynclinal facies and the thin shales and limestones of the platform facies.

Continental deposition. By the middle of Early Permian time, the marine connection from the Russian Platform southward into the Tethyan ocean was closed, apparently by uplifts along the southern and western margins of the platform. During middle Permian time, the Russian Platform was covered by a broad, shallow, brackish gulf which had limited water circulation with an ocean to the north. Brackish conditions fluctuated with evaporites, redbeds, and other continental deposits. During later Permian time, detrital material eroded from the uplifted ancestral Ural Mountains, filled the remnants of the deformed geosynclinal depression, and were progressively transported westward onto the Russian Platform. Continental deposition continued into Triassic time.

Boundaries. Both the lower and upper boundary of the Permian System have been subject to considerable discussion. As Murchison originally defined the system, the lower boundary was at the base of the evaporitic facies of the Kungurian beds and the upper boundary at the top of the continental facies of the Tatarian beds. Based on similarities in lithology, Murchison correlated the thick Artinskian clastics that lie beneath the Kungurian within the Ural Geosyncline with the lower Carboniferous of western Europe. He believed the limestones and shales that lie beneath the Kungurian beds on the Russian Platform were part of the upper Carboniferous. Later studies showed that these original correlations were not correct. A. P. Karpinsky in 1889 described Artinskian ammonoids that were closely related to faunas considered Permian in age in other parts of the world. The suggestion to lower the Permian boundary in its type region to include these Artinskian faunas gained general acceptance.

Cisuralian Series. The limestones of the Ufa Plateau continued to be classified as upper Carboniferous until after 1930. During the 1930s V. E. Ruzhentsev recognized prolific ammonoid faunal assemblages that were present in both the carbonates and shales of the Ufa Plateau and the Artinskian clastics. Based on these studies, Ruzhentsev subdivided the original Artinskian into a lower Asselian Stage, a middle Sakmarian Stage, and an upper (revised) Artinskian Stage (Fig. 1). In the 1990s, these stages were combined with the Kungurian Stage to form the lower Permian Cisuralian Series (Fig. 1).

The Asselian, Sakmarian, and Artinskian stages include normal marine faunas with abundant fusulines, conodonts, and ammonites, which are extremely useful in interregional correlation. Large, massive reef mounds and pinnacle reefs, 1800 ft (600 m) or more in thickness, are common on the edge of the Russian Platform (Fig. 2), and these pass eastward into thick shales and sandstones in the Ural Geosyncline and westward into thin dolomitic limestone, dolostone, and evaporites.

On the Russian Platform and on the margin of the Ural Geosyncline, the upper part of the Artinskian beds intertongues with Kungurian red shales and evaporites. The Kungurian extends stratigraphically higher and locally is the source of major salt, anhydrite, and potash salts, as near Solikamsk. Fossils are rare and not diverse, and show affinities at some localities to Artinskian species.

Middle Permian deposits. On the Russian Platform, the Ufimian beds lie at the base of the middle Permian and are less continuously distributed and are red shales and sandstones, variegated clays, and marly shales less than 450 ft (150 m) thick. They primarily contain a meager fresh-water fauna of bivalves and ostracodes. Eastward near the Ural Mountains, the Ufimian reaches 4500 ft (1500 m) and is coarser grained.

The Kazanian Stage unconformably overlies the Ufimian and consists of about 300 ft (100 m) of

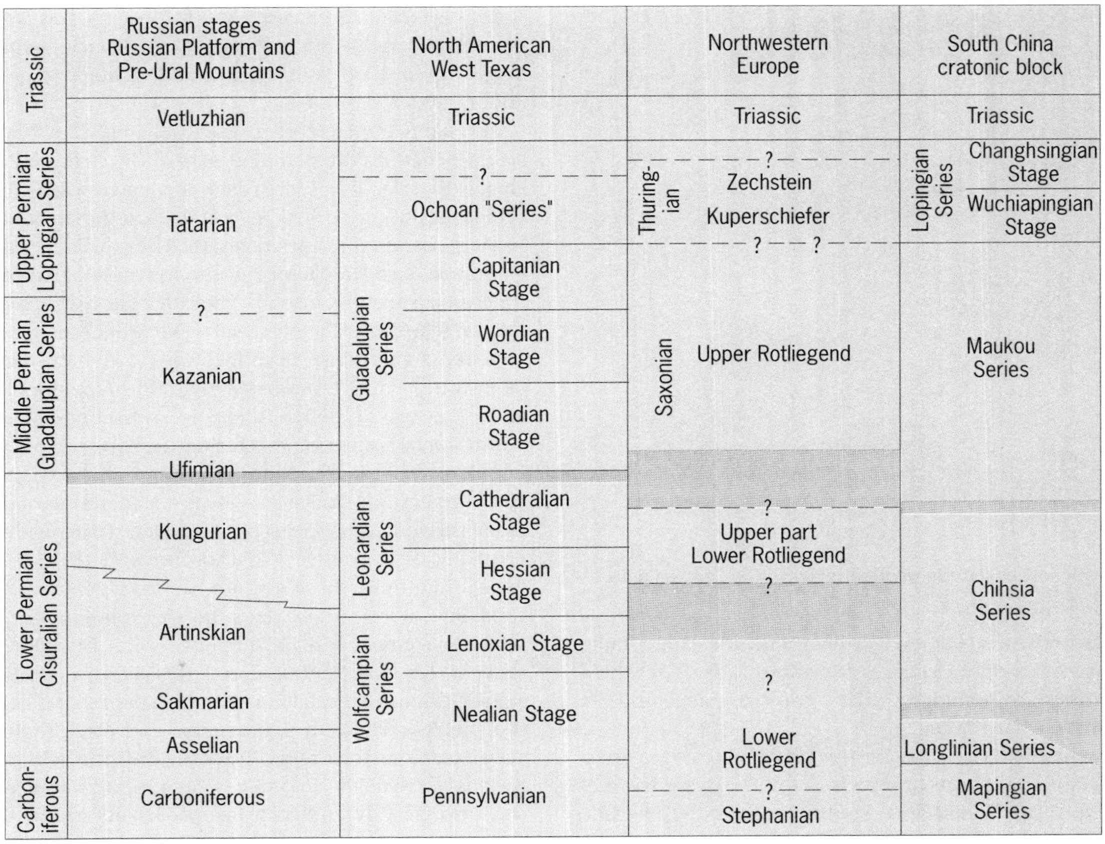

	Russian stages Russian Platform and Pre-Ural Mountains	North American West Texas	Northwestern Europe	South China cratonic block
Triassic	Vetluzhian	Triassic	Triassic	Triassic
Upper Permian Lopingian Series	Tatarian	?	Thuring-ian: ?	Lopingian Series: Changhsingian Stage
		Ochoan "Series"	Thuring-ian: Zechstein, Kuperschiefer	Lopingian Series: Wuchiapingian Stage
	?	Guadalupian Series: Capitanian Stage	? ?	
Middle Permian Guadalupian Series	Kazanian	Guadalupian Series: Wordian Stage	Saxonian: Upper Rotliegend	Maukou Series
		Guadalupian Series: Roadian Stage		
	Ufimian			
Lower Permian Cisuralian Series	Kungurian	Leonardian Series: Cathedralian Stage	? Upper part Lower Rotliegend	Chihsia Series
		Leonardian Series: Hessian Stage	?	
	Artinskian	Wolfcampian Series: Lenoxian Stage		
	Sakmarian	Wolfcampian Series: Nealian Stage	?	
	Asselian		Lower Rotliegend	Longlinian Series
Carbon-iferous	Carboniferous	Pennsylvanian	? Stephanian	Mapingian Series

Fig. 1. Correlation of major Permian rock units.

greenish-gray impure limestone that has brackish-water or restricted marine faunas that are mostly bryozoans and brachiopods. The Kazanian deposits were laid down in an elongate shallow basin that paralleled the western flank of the Ural Mountains. This basin was closed to the south, where it locally includes evaporites, and open to the north, where a few additional marine invertebrates, including rare ammonoids, are locally present.

Upper Permian deposits. The youngest stage of the Russian Permian is the Tatarian, which is formed of brightly colored, variegated sandstone, conglomerate, and other clastic beds that were deposited by rivers, streams, and lakes, and includes some dune sands. Plants and terrestrial animals, including insects and vertebrates, are well known from these sediments. The original Tatarian of Murchison had five main faunal zones; however, only the lower two

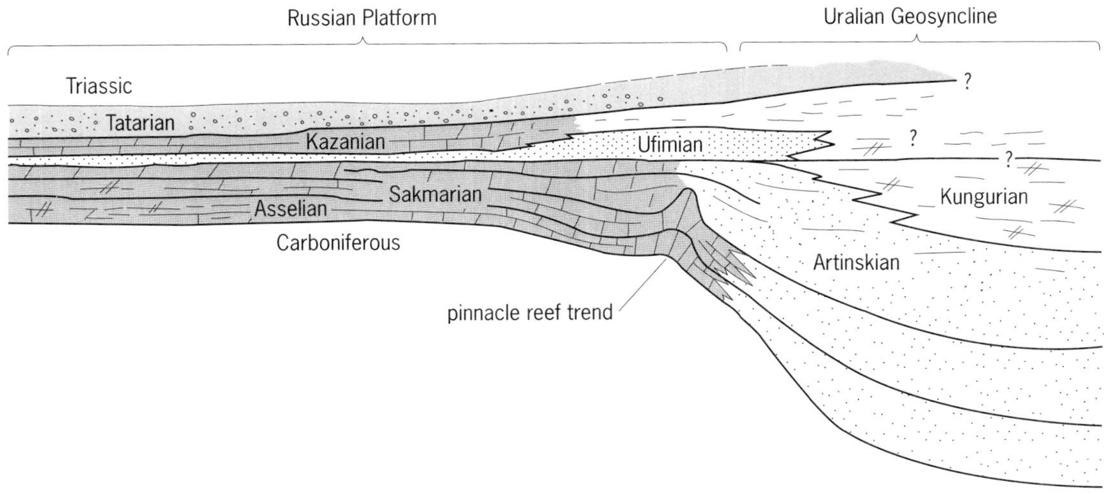

Fig. 2. East-west section of the Permian System in its type area.

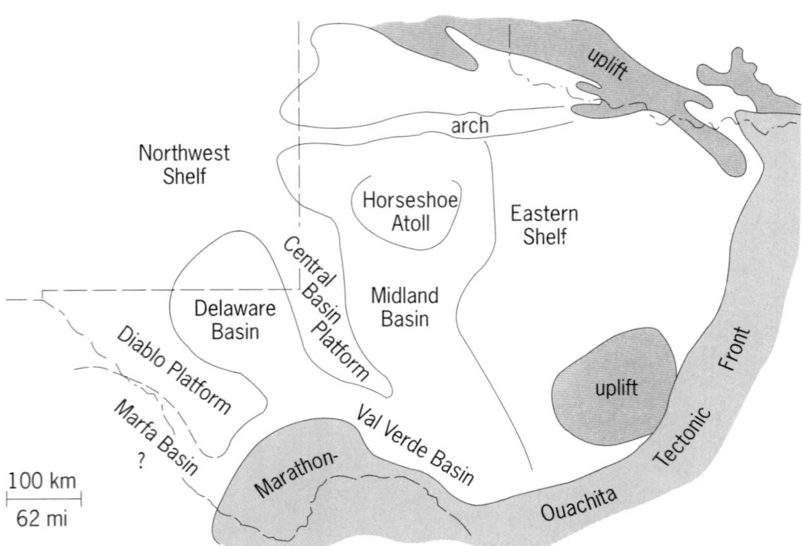

Fig. 3. Depositional and structural relationships in western Texas during Permian time.

are now considered Permian, and the name Tatarian is restricted to these two zones. The upper three zones are included in the Vetlushian Stage and are Triassic. *See* REDBEDS.

Western Texas and southern New Mexico. One of the finest Permian sections known is in the Delaware Basin of western Texas and southeastern New Mexico (**Fig. 3**), where the Permian sequence reaches a thickness of more than 12,000 ft (4000 m). The rocks of the lower 6600 ft (2200 m) are of marine origin and are highly fossiliferous. Although a few fossils from this region were described in 1858, this succession of strata remained virtually unknown until 1909. Intensive study followed the discovery of oil in the region about 1920. In 1939 a committee of Permian specialists proposed to subdivide the American Permian section into four series, and this usage was widely accepted. It has become the standard section for North America and, to a considerable degree, for the world. It is divided into four stages in ascending order: Wolfcampian Stage, 1500 ft (500 m); Leonardian Stage, about 3000 ft (1000 m); Guadalupian Stage, about 3000 ft (1000 m); and Ochoan Stage, about 4500 ft (1500 m).

The Wolfcampian correlates closely with the Russian Asselian, Sakmarian, and part of the Artinskian stages (Fig. 1). The Leonardian correlates with the upper part of the Artinskian and the Kungurian, and has larger and more varied faunas. The Guadalupian cannot be correlated in detail with the middle part of the Russian section, because the latter has restricted faunas. The Ochoan is virtually unfossiliferous, except for a thin zone near the top (in the Rustler Dolostone) which contains productid brachiopods and a few other types of Paleozoic invertebrates.

Wolfcampian and Leonardian. During the Wolfcampian Epoch, most of the region was a broad, shallow marine intracratonic basin in which a rich and varied fauna thrived. A fore-deep basin bordered the northern flank of the active Marathon orogenic belt. By Leonardian time, three distinct basins (Fig. 3) were subsiding more rapidly than the surrounding area, which became a broad shelf occupied by wide, shallow lagoons. Light-colored, fossiliferous limestones (Victorio Peak Limestone) accumulated on the shelves, while black limestone and black shale accumulated in the basins. Evidently the threshold to the basins, which was somewhere in Mexico, was so shallow that water in the basins was density-stratified and the bottom was stagnant and foul, so that almost no benthic organisms could survive. The black Bone Spring Limestone is generally barren of fossils.

Guadalupian. During Guadalupian time (middle Permian), the basins continued to deepen, and as the climate became markedly arid, surface water flowed radially out of the basins onto the adjacent platforms to replace the water lost by evaporation in the shallow lagoons. Narrow limy banks grew along the margins of the basin to form the great Capitan Reef. **Figure 4** shows the complex relations within the Leonardian and Guadalupian deposits along the face of the Guadalupe Mountains. Probably no other great

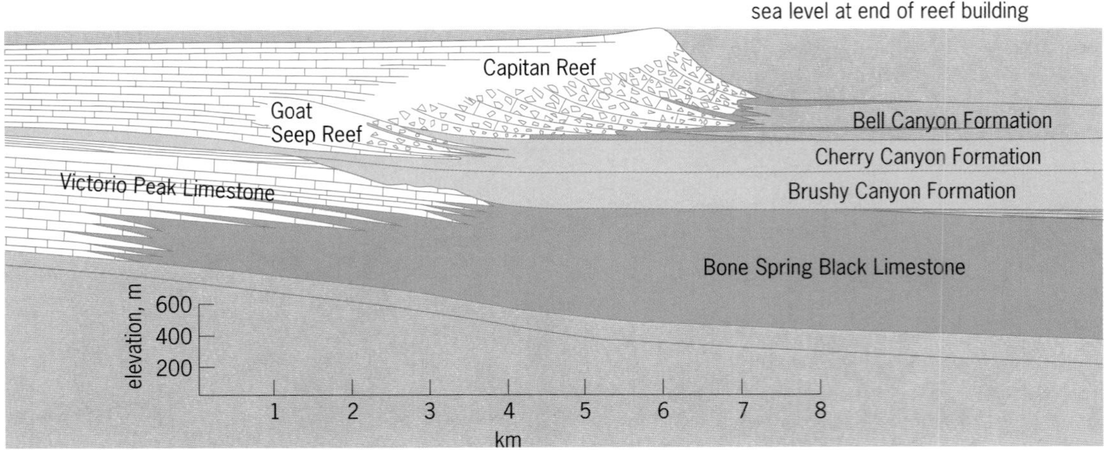

Fig. 4. Section across the Capitan Reef Complex. 1 m = 3.3 ft; 1 km = 0.6 mi.

reef complex is so well exposed or has been so intensively studied as that of the Capitan Reef. Massive deposits of reef talus were derived from the growing front of the reef. These dip steeply into the basin, become finer down dip, and grade out into thin tongues of calcarenite. The back-reef deposits are calcareous for distances of 1.5–6 mi (2–10 km) and then grade rapidly into gypsiferous shales and anhydrite. Gray, beach, and offshore-bar sands intertongue from the landward margins of the lagoon. Farther back the sands pass into redbeds.

Late Permian. During Ochoan time, the basins became evaporitic under intensely arid conditions. Enormous deposits of anhydrite and, later, of halite were precipitated. Interbedded in the salt in the center of the Delaware Basin are several lenses of potash salts—sylvite, carnallite, and polyhalite.

Other areas. In central Texas, Oklahoma, and Kansas, the Wolfcampian equivalents are largely of marine origin. Marine conditions persisted well into Leonardian time in central Texas, but in Kansas a very large salt deposit (in the Wellington Shale) is followed by a thick redbed sequence. By Guadalupian time, most of the deposits in central Texas were nonmarine redbeds in which several thin marine dolostones intertongued from the west. In Oklahoma, all but the lower part of the Wolfcampian is in redbed facies because of the local influence of the Oklahoma mountains.

In other areas of the central and southern Rocky Mountain states, block faulting during Wolfcampian time resulted in a series of horsts, which were local sources of sediments, and adjacent grabens, which became local basins of deposition. Although most of these basins had initial deposits of marine origin, they were mainly filled by the end of Wolfcampian time. Subsequently, Leonardian and early Guadalupian sediments formed thin, blanketlike continental deposits over much of the region. Dunes, redbeds, and local evaporites are common.

The western margin of the North American craton during the Permian was located far to the east of the present west coast. The Permian margin extended generally northward from southern California through southern and eastern Nevada, southeastern Idaho, eastern British Columbia, and just into the eastern edge of Alaska. The southern part of this margin had a thick succession of Wolfcampian, Leonardian, and early Guadalupian limestones deposited on it. In northern Utah and Idaho, sandstone becomes more prevalent, and in the Guadalupian, phosphorite and chert become important deposits. Farther north, sandstone and chert continue to increase in abundance and form most of the Permian sediments on the old cratonic margin.

West of the Permian cratonic margin, Permian strata are found in four or five tectonically disturbed belts. Although these belts include rocks of the same age, each structural belt contains different fossil faunal assemblages that are quite distinct, and in each the lithologic succession and depositional history appears to be independent. These different assemblages of faunas and sediments were apparently deposited at considerable distances from one another, and their present-day geographic proximity indicates that each belt has been added by tectonic accretion to the western margin of the North American craton in post-Permian time—that is, during the Mesozoic or Cenozoic. Of all these structural belts, the best known is the Cache Creek (British Columbia) belt, which extends from southern British Columbia north-northwestward into southern Yukon. The Cache Creek belt is composed of oceanic ribbon cherts, dark shales and sandstones, basic igneous flows, and massive limestone reefs having a tropical fossil fauna and flora which show affinities to the Tethyan faunal realm rather than to the Midcontinent-Andean realm of North American and South American cratons.

Northwestern Europe. Permian beds are widespread in northwestern Europe and the North Sea and produce large amounts of gas and oil, which have had an important influence on the economies of northwestern European countries since the late 1960s. These Permian beds are divisible into two parts. The lower part (Rotliegend beds) is mainly red sandstone. The upper part (Thuringian Series) consists of conglomerate, chalcopyritic shale, dolomitic limestone, evaporites, and shale. The Rotliegend beds are subdivided into the Autunian Stage (or Lower Rotliegend) in the lower part and Saxonian Stage (or Upper Rotliegend) in the upper part. Saxonian strata are younger than the last major movements of the Hercynian orogeny, and an unconformity separates them from the overlying Thuringian.

The Thuringian contains a basal conglomerate, a thin copper-bearing shale (Kupferschiefer), dolomitic limestones (Zechstein), and seven important evaporitic units that near Stassfurt include potassium salts. The Thuringian sea extended east across Poland, and possibly connected with the Kazanian sea of the Russian Platform. The Magnesian Limestone of England is a thin western tongue of the Thuringian Series.

The Atlantic Ocean did not start to open until middle Mesozoic time, so that the Permian deposits of northwestern Europe are very similar to, and are thought to have been formerly continuous with, Permian beds found along the central east coast of Greenland, on Spitsbergen, and on the Canadian Arctic islands. They also connected, by way of the Barents shelf, to the northern entrance to the Ural Geosyncline and Russian Platform.

Tethyan regions. South of the Hercynian orogenic belt, the sediments and faunas and floras of the Permian change markedly in Mediterranean Europe, northern Africa, and southern and central Asia (**Fig. 5**). Limestone, dark shales and sandstones, cherts, and basaltic volcanics are common. The species and generic diversity is great in contrast to that of northwestern Europe and the Russian Platform.

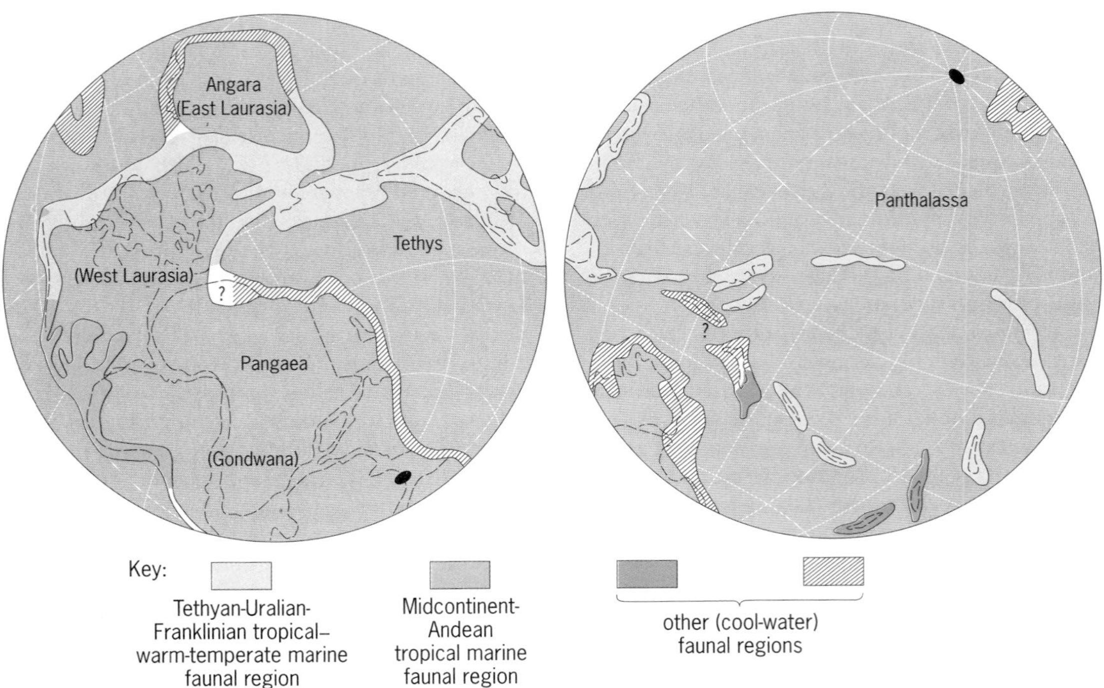

Fig. 5. Paleogeography during the earliest part of the Permian Period.

The marine Tethyan invertebrate fauna evolved rapidly during the middle part of the Permian and formed a distinctive biogeographic realm, called the Tethyan realm. This diverse fauna is characterized by the verbeekinid fusulinaceans (foraminiferal Protozoa) and colonial and solitary waagenophyllid corals (Rugosa). Later orogenic movements during the Mesozoic and Cenozoic have greatly complicated the interpretation of these Permian deposits; however, several linear belts of Permian rocks appear to be present. Each may have included shallow reef deposits and shallow- and deeper-water clastic and carbonate deposits, and some belts may have fringed a number of small fragments of cratonic crustal blocks. In post-Permian time, sea-floor spreading and crustal subduction apparently displaced these several depositional belts and cratonic fragments and lodged them as accreted terranes in orogenic belts against larger cratons, such as Europe, central and eastern Asia, and western North America.

The Permian Tethyan faunas were apparently tropical, and mostly shallow-water because of the abundance of calcareous algae, bryozoans, brachiopods, and echinoderms, and the reeflike nature of most of the limestones. The close association of these limestones with basic igneous rocks (including pillow lavas), ribbon cherts, and dark sandstones and shales suggests that much of the Tethyan region was made up of island arcs and oceanic carbonate plateaus similar to those of the present tropical Pacific Ocean.

Southern China. The lower part of the Permian in southern China includes the upper part of the Maping Stage and the Longlinian (Changshan) Stage which are widespread, but locally deeply eroded, beneath a middle lower Permian unconformity. Overlapping this unconformity are extensive limestones of the Chihsia Series which comprise the remainder of the lower Permian in the area. An extensive unconformity at the top of the Chihsia represents the mid-Permian sea-level event. Higher, widely distributed limestones of the Maukou Series are present. They show major lateral variations in lithologic facies.

Lopingian Series. The type area for the upper series of the Permian, the Lopingian Series, overlies the Maukou in continuous succession in much of southern China. This series is divided into the the Wuchiapingian Stage (below) and the Changhsingian Stage (above) and the top of the succession passes conformably into the Lower Triassic.

Gondwana continents. The Permian successions are remarkably similar in all of the parts of southern Africa, Australia, South America, Pakistan and India, and Antarctica that formed the large supercontinent of Gondwana during the late Paleozoic. In southern Africa these deposits form the lower part of the Karoo System and commence with a tillite (Dwyka Tillite), which is followed by dark shales (Ecca Series) overlain by sandstones and red shales (Beaufort Series). The Ecca includes the late Paleozoic coals of southern Africa, and the Beaufort has a distinctive and extensive mammallike reptilian fauna.

The Permian deposits of much of South America have close similarities to those of southern Africa. In the Paraná Basin the Guata Group with glacial sediments, coal, and marginal marine beds, and, in its upper part, *Eurydesma* (Bivalvia) and *Glossopteris* (plant) is of probable Early Permian age. Above, the Irati Formation has the same distinctive reptiles found in the Beaufort of southern Africa and also has an extensive plant fossil succession.

Antarctica also has a generally similar late Paleozoic succession to those found in southern Africa and southern South America, including coal and plant and reptile fossils.

Australia, Pakistan, and India have late Paleozoic successions preserved in a number of fault-bounded basins. The adjacent parts of the Gondwanan craton were apparently emergent. Many of these basins include tillites and glacial marine deposits in their Carboniferous and earliest Permian deposits. These are commonly followed by marine deposits having some early and middle Permian limestones. Several important New South Wales coalfields, such as the Newcastle Coal Measures, are of late Permian age and have nontropical Tatarian insect faunas. Coal formation continued into the Mesozoic in many of these basins.

Paleogeography. During the Permian Period, several important changes took place in the paleogeography of the world. The joining of Gondwana to western Laurasia (Fig. 5), which had started during the Carboniferous, was completed during Wolfcampian time (earliest Permian). The addition of eastern Laurasia (Angara) to the eastern edge of western Laurasia finished during Artinskian time (middle to latest early Permian) and completed the assembly of the supercontinent Pangaea (**Fig. 6**). The climatic effects of these changes were dramatic. Instead of having a circumequatorial tropical ocean, such as during the middle Paleozoic, a large landmass with several high chains of mountains extended from the South Pole across the southern temperate, the tropical, and into the north temperate climatic belts. One very large world ocean, Panthalassa and its western tropical branch, the Tethys, occupied the remaining 75% of the Earth's surface, with a few much smaller cratonic blocks, island arcs, and atolls. *See* CONTINENTAL DRIFT; CONTINENTS, EVOLUTION OF.

Early Permian glacial deposits of the Gondwanan continents lie within an area about 40° from the south paleopole. Recently reported Permian glacial beds in eastern Angara would have been within 30 to 40° of the north paleopole. Middle and late Permian sediments, such as sand dunes, marine beds having a few invertebrate fossils with warmer-water affinities, and coals, suggest warmer conditions but not tropical or subtropical. Considerable evidence also suggests the world climate became progressively milder through a series of fluctuating warming and cooling steps during the later part of the early Permian and late Permian. Desert conditions became widespread in many parts of tropical and subtropical Pangaea with dune sands, evaporites, redbeds, and calcic soil zones.

Vast deposits of salt and anhydrite accumulated in Kansas, New Mexico, and the Permian Basin in western Texas. On the eastern part of the Colorado Plateau, extensive dune sands were deposited, such as in the Canyon DeChelly area. Elsewhere similar conditions are shown by the evaporites of the Kungurian on the Russian Platform and the dune sands and salt deposits of the Thuringian of northwestern Europe. *See* PALEOGEOGRAPHY; SALINE EVAPORITES.

Life. Most marine invertebrates of the Early Permian were continuations of well-established phylogenetic lines of middle and late Carboniferous ancestry. During early Permian time, these faunas, dominated by brachiopods, bryozoans, conodonts, corals, fusulinaceans, and ammonoids, gradually evolved into a number of specialized lineages. The

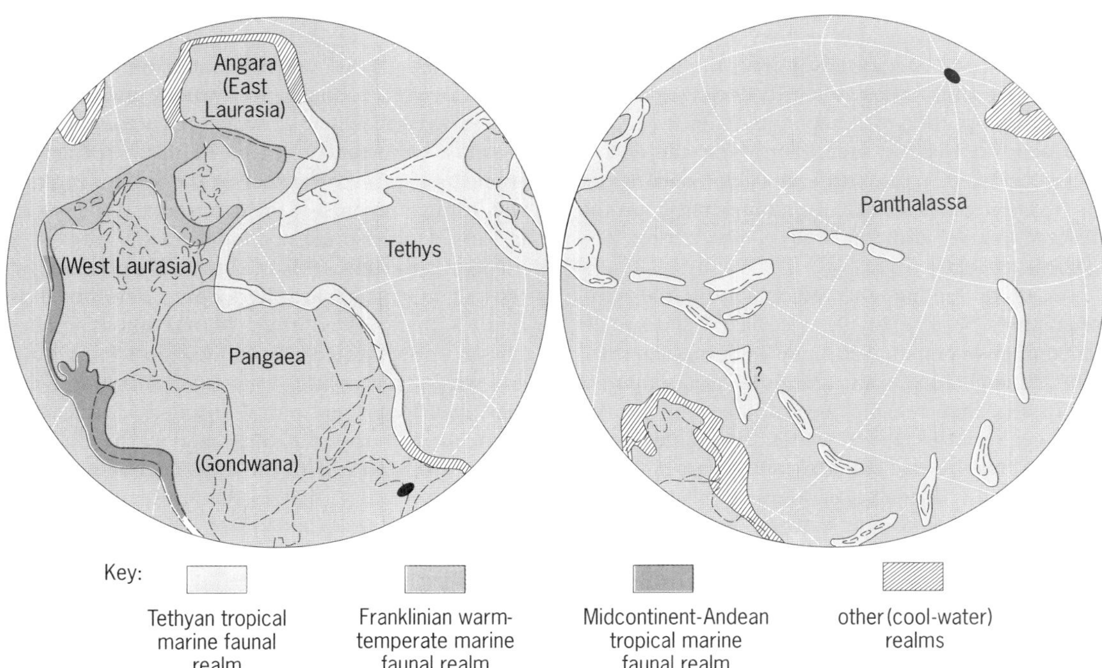

Fig. 6. Paleogeography near the middle of the Permian Period.

tropical shallow-water faunas of southwestern North America evolved almost in isolation from those of the Tethyan region, because faunal exchanges had to cross either the cooler waters of a temperate shelf or deep waters of Panthalassa. Fluctuating climates permitted rare dispersals of some faunas between these two tropical faunal realms during the early part of early Permian time. The closure of the Uralian seaway during Artinskian time, however, extended that dispersal path north around Angara and into cold boreal waters through which the tropical species could not disperse. With the extension of this dispersal path, the faunas of the two tropical realms evolved independently, with only extremely rare dispersals between them.

The Siberian traps, an extensive outflow of very late Permian basalts and other basic igneous rocks (dated at about 250 million years ago), are considered by many geologists as contributing to climatic stress that resulted in major extinctions of many animal groups, particularly the shallow-water marine invertebtares. The end of the Permian is also associated with unusually sharp excursions in values of the carbon-12 isotope (^{12}C) in organic material trapped in marine sediments, suggesting major disruption of the ocean chemistry system.

Foraminiferans. The warm to tropical shallow-water foraminiferal faunas during the Permian were dominated by the fusulinaceans. One group, the verbeekinids, which evolved in the Tethyan realm, formed the distinctive foraminiferal fauna of reefs and atolls of the Tethyan realm. Schwagerinids were less abundant and occurred in sandy sediments adjacent to the reefs. Many of the schubertellids appeared to be adapted to lagoonal environments. In the tropical Midcontinent-Andean realm along the western coast of Pangaea, the schwagerinids filled more of the shallow water niches than in the Tethyan realm, but they were not as diverse in number of new genera and species. Away from the paleotropics in both realms, species diversity decreases rapidly, nearly in proportion to the amount of limestone that was deposited in the succession. Near the end of Guadalupian time, fusulinaceans declined markedly. In the Lopingian specialized genera persisted in the Tethyan region until the end of the period. *See* FORAMINIFERIDA.

Brachiopods. Among the brachiopods, the strophomenids became highly specialized and important framework builders in many bioherms and small reefs. Rhynchonellids and spiriferids are of interest because of their relict pattern of distribution after their evolutionary diversification during the Devonian. Locally they are abundant and important. Near the end of the Guadalupian, brachiopods also were greatly reduced, and about half of those that survived the Lopingian became extinct before the Triassic. Only a few genera and species in one-fifth of the families that were present in the early Permian survived into the Triassic. *See* BRACHIOPODA.

Bryozoans. These have much the same history as other marine invertebrates during the Permian. Several families of cryptostomes, such as the polyporids, hyphasmoporids, and nikiforovellids, and several families of trepostomes, such as the eridotrypellids and araxoporids, became increasingly diverse during the middle part of the Permian. Their greatest geographical diversity was in the Tethys. During Guadalupian time, many genera within the bryozoan families became extinct and, by the end of that epoch, more than 10 families were extinct. Six more families became extinct during the Lopingian (latest Permian). At least five families range into the Triassic. *See* BRYOZOA.

Ammonoids. This important Permian fossil group also had a rapid middle Permian diversification, followed by a rapid reduction in genera late in Guadalupian time. Of about 55 Guadalupian genera, 15 survived into the earliest Lopingian (earliest late Permian), where they evolved into about 50 genera. Only 6 of these survived into the late Lopingian (late late Permian), however; they evolved into nearly 30 genera. Only one ammonoid genus survived the end of the Permian and ranged into the Early Triassic.

Insects. Terrestrial faunas included insects which showed great advances over those of the Carboniferous Coal Measures. Several modern orders emerged, among them the Mecoptera, Odonata, Hemiptera, Trichoptera, Hymenoptera, and Coleoptera. Extensive insect faunas are known from the lower Permian rocks of Kansas and Oklahoma, the Permian of Russia, and the upper Permian of Australia. *See* INSECTA.

Land plants. During the Permian, plants, including lepidodendrons and cordaites, were well adapted to the moist conditions of low-lying coal swamps. Several lineages also adapted to the drier, well-drained conditions of mountains and alluvial plains, particularly conifers. In glaciated areas of Gondwana, the tongue ferns *Glossopteris* and *Gangamopteris* are common and were apparently adapted to cold climates.

Vertebrates. Of the vertebrates, labyrinthodont amphibians were common and varied; however, reptiles showed the greatest evolutionary radiation and the most significant advances. Reptiles are found in abundance in the lower half of the system in Texas and throughout most of the upper part of the system in Russia and also are common in Gondwana sediments. Of the several Permian reptilian orders, the most significant was the Theriodonta, or mammallike reptiles, that evolved in the Triassic into mammals. These reptiles carried their bodies off the ground and walked or ran like mammals. Unlike most reptiles, their teeth were varied—incisors, canines, and jaw teeth as in the mammals—and all the elements of the lower jaw except the mandibles showed progressive reduction. Most of the known theriodonts are from South Africa and Russia. *See* PALEOZOIC; REPTILIA; THERAPSIDA. Charles A. Ross; June R. P. Ross

Bibliography. C. O. Dunbar and K. M. Waage, *Historical Geology*, 3d ed., 1969; A. L. DuToit, *Geology of South Africa*, 3d ed., 1954; H. Falke (ed.), *The Continental Permian in West Central and South Europe*, 1976; C. R. Handford et al., *Regional Cross*

Sections of the Texas Panhandle: Precambrian to Mid-Permian, 1982; R. T. Magginett, C. E. Stevens, and P. Stone (eds.), *Early Permian Fusulinids from the Owens Valley Group, East-Central California*, 1988; N. D. Newell et al., *The Permian Reef Complex of the Guadalupe Mountains Region, Texas and New Mexico*, 1953; C. A. Ross and J. R. P. Ross, Permian, in R. A. Robinson and C. Teichert (eds.), *Treatise on Invertebrate Paleontology*, pt. A, pp. 291–350, 1979; S. M. Stanley, *Earth and Life Through Time*, 1985; D. H. Tarling, *Paleomagnetism: Principles and Applications in Geology, Geophysics, and Archaeology*, 1983.

Permittivity

A property of a dielectric medium that determines the forces that electric charges placed in the medium exert on each other. If two charges of q_1 and q_2 coulombs in free space are separated by a distance r meters, the electrostatic force F newtons acting upon each of them is proportional to the product of the charges and inversely proportional to the square of the distance between them. Thus, F is given by Eq. (1), where $1/(4\pi\epsilon_0)$ is the constant of proportionality, having the magnitude and dimensions necessary to satisfy Eq. (1). This condition leads to a value for ϵ_0, termed the permittivity of free space, given by Eq. (2), where c is the velocity of light in vacuum.

$$F = \frac{q_1 q_2}{4\pi\epsilon_0 r^2} \quad \text{newtons} \tag{1}$$

$$\epsilon_0 = \frac{1}{4\pi 10^{-7} c^2} \simeq 8.8542 \times 10^{-12} \quad \text{farads/meter} \tag{2}$$

In all that follows, dielectric media are taken to be homogeneous and isotropic. If now the charges are placed in such a medium, the force on each of them is reduced by a factor ϵ_r, where ϵ_r is greater than 1. This dimensionless scalar quantity is termed the relative permittivity of the medium, and the product $\epsilon_0\epsilon_r$ is termed the absolute permittivity ϵ of the medium.

A consequence is that if two equal charges of opposite sign are placed on two separate conductors, then the potential difference between the conductors will be reduced by a factor ϵ_r when the conductors are immersed in a dielectric medium compared to the potential difference when they are in vacuum. Hence a capacitor filled with a dielectric material has a capacitance ϵ_r times greater than a capacitor with the same electrodes in vacuum would have. Except for exceedingly high applied fields, unlikely normally to be reached, ϵ_r is independent of the magnitude of the applied electric field for all dielectric materials used in practice, excluding ferroelectrics. *See* CAPACITANCE; CAPACITOR; FERROELECTRICS.

Polarization. Since, for a given fixed distribution of charges, the introduction of a dielectric medium reduces the electric field everywhere by a factor ϵ_r, the medium must produce a flux P coulombs/meter^2, termed the polarization, which opposes the flux D produced by the charges. The reduction in electric field E can be expressed in terms of ϵ_r [Eq. (3)] or in terms of P [Eq. (4)]. The dimensionless ratio of P to $\epsilon_0 E$ is known as the electric susceptibility χ_e of the dielectric. Eliminating D between Eqs. (3) and (4) yields Eq. (5). Since the value of ϵ_r for vacuum is unity, χ_e represents the component of the relative permittivity which is due to the electrical response of the medium itself to the applied field. *See* ELECTRIC SUSCEPTIBILITY; POLARIZATION OF DIELECTRICS.

$$E = \frac{D}{\epsilon_0\epsilon_r} \quad \text{volts/meter} \tag{3}$$

$$E = \frac{D - P}{\epsilon_0} \quad \text{volts/meter} \tag{4}$$

$$\chi_e = \frac{P}{\epsilon_0 E} = \epsilon_r - 1 \tag{5}$$

Dipole moment. When two charges of equal magnitude q and opposite signs are separated by a fixed distance d, they constitute a dipole. If the line joining the charges were to make a right angle with an electric field E, the dipole would experience a torque qEd. The torque in unit field is termed the dipole moment μ. Thus Eq. (6) holds. In a dielectric material in which a polarization P exists, an elementary cuboid may be imagined of volume δv with two faces perpendicular to P of area δA, separated by a distance δx. Its dipole moment is $P\delta A\delta x$, or $P\delta v$, and so the polarization P is the dipole moment per unit volume of dielectric. This dipole moment is determined by the polarizability of the submicroscopic entities of the material. *See* DIPOLE; DIPOLE MOMENT.

$$\mu = qd \quad \text{coulombs} \times \text{meter} \tag{6}$$

Mechanisms of polarization. Polarization may arise due to both nonpolar and polar molecules.

Nonpolar molecules. Even if the distribution of charges within a molecule is such that it has no dipole moment, the application of an electric field E can cause a change in this distribution and so cause each molecule to have an induced dipole moment. The dipole moment μ_e caused by the displacement of electrons within the atom, and the dipole moment μ_a caused by the displacement of atoms within the molecule, give rise to electronic and atomic polarization respectively. Normally electronic polarization is larger than atomic polarization by a factor of, say, 5–10, but for some materials, for example, for alkali halide crystals, atomic polarization is considerable.

The electric field E_l that acts at the molecular site and causes charge displacements is termed the local field. Although E_l is proportional to the applied field E, E_l is in general not equal to E, since E_l contains an additional component arising from neighboring induced dipoles. The induced dipole moments and E_l are continuously fluctuating, and a rigorous treatment demands the use of sophisticated statistical mechanics. Time-averaged values are implied here throughout. *See* STATISTICAL MECHANICS.

Since μ_e and μ_a are proportional to E_l, electronic polarizability α_e and atomic polarizability α_a may be defined by Eqs. (7) and (8). The electronic and

$$\alpha_e = \frac{\mu_e}{E_l} \quad \text{farads} \times \text{meter}^2 \tag{7}$$

$$\alpha_a = \frac{\mu_a}{E_l} \quad \text{farads} \times \text{meter}^2 \tag{8}$$

atomic polarizations P_e and P_a are obtained from the induced dipole moments μ_e and μ_a by multiplying by the number density of molecules N. Thus the induced polarization P_i is given by Eq. (9).

$$P_i = N(\alpha_e + \alpha_a)E_l \quad \text{coulombs/meter}^2 \tag{9}$$

Polar molecules. Some molecules have a charge distribution such that, even when no electric field is applied, the molecule possesses a dipole moment. Such molecules are termed polar molecules, and their dipole moment μ_p is termed permanent, to distinguish it from the induced dipole moment which, in common with nonpolar molecules, they also possess. In general, depending on the structure of the material, the molecules will have some freedom of rotational movement. In the absence of an applied electric field, the permanent dipole moments are randomly directed, and so the sum of their components in any given direction is zero. However, when a steady field is applied, the summation of the dipole moments in the field direction yields a net positive value. Thus the ordering effect of the field is in competition with the thermal disordering effect of the environment, and the resulting orientational polarization P_0 is inversely proportional to absolute temperature T and, except for exceedingly high fields unlikely normally to be reached, directly proportional to E_l, where E_l now includes a component arising from neighboring permanent dipoles. It is easily shown, following a calculation carried out by P. Langevin in 1905 relating to magnetic dipoles, that for all realistic fields P_0 is given by Eq. (10), where k is Boltzmann's constant. Thus

$$P_0 = \frac{N\mu_p^2 E_l}{3kT} \tag{10}$$

the effective orientational polarizability α_p is given by Eq. (11). The total polarizability α of a polar molecule is thus given by Eq. (12), and the total polarization P is given by Eq. (13). Similarly, the total electric

$$\alpha_p = \frac{\mu_p^2}{3kT} \quad \text{farads} \times \text{meters}^2 \tag{11}$$

$$\alpha = \alpha_e + \alpha_a + \alpha_p \quad \text{farads} \times \text{meters}^2 \tag{12}$$

$$P = N\alpha E_l \quad \text{coulombs/meters}^2 \tag{13}$$

susceptibility χ_e is the sum of three components, χ_{el}, χ_a, and χ_p. *See* POLAR MOLECULE.

Local field. Substituting the expression for P of Eq. (13) into Eq. (5) gives Eq. (14). In order to eval-

$$\chi_e = \epsilon_r - 1 = \frac{N\alpha}{\epsilon_0} \times \frac{E_l}{E} \tag{14}$$

uate this equation, it is necessary to know E_l/E. It can be shown that E_l is given by Eq. (15), where the

$$E_l = E + \frac{\gamma P}{\epsilon_0} \quad \text{volts/meter} \tag{15}$$

dimensionless constant γ is dependent on the internal arrangement of molecules in the material. For the special cases of either solids with cubic atomic structure, or media in which the molecules are in complete disorder, $\gamma = 1/3$; then E_l is given by Eq. (16).

$$E_l = E + \frac{P}{3\epsilon_0} \quad \text{volts/meter} \tag{16}$$

A classical calculation carried out by H. A. Lorentz gives this expression for E_l, which is therefore known as the Lorentz field.

Substituting the expression for P of Eq. (5) into Eq. (16) gives Eq. (17). Substituting this value for E_l/E into Eq. (14) gives Eq. (18). This is the Clausius-

$$\frac{E_l}{E} = \frac{\epsilon_r + 2}{3} \tag{17}$$

$$\frac{\epsilon_r - 1}{\epsilon_r + 2} = \frac{N\alpha}{3\epsilon_0} \tag{18}$$

Mossotti equation, relating the macroscopic quantity ϵ_r with the submicroscopic quantity α, for materials for which $\gamma = 1/3$. It takes account of dipole-dipole interaction, whereas failure to do so would result in the expression for ϵ_r given in Eq. (14) with the term E_l/E put equal to unity.

Complex permittivity. When an ac electric field is applied to a dielectric, each polarization mechanism can only operate effectively up to a limiting frequency. Then, further increase in frequency will cause the mechanism to become gradually less effective until it virtually ceases to operate. The frequency range over which the polarization is diminishing is known as the dispersion region. Because of the springlike nature of the forces involved in electronic and atomic polarization, their disappearance is accompanied by resonant behavior of ϵ_r. The dropping out of the orientational polarization is, however, a relaxation process, possibly involving a broad distribution of relaxation times, and therefore it occurs over a much wider frequency range. Typically, the midfrequencies f_0 involved would be about 10^{15} Hz for electronic polarization and 10^{12} Hz for atomic polarization, while for orientational polarization f_0 might be anywhere in a wide range, say 10^4 to 10^{10} Hz, depending on the material and its temperature. *See* RESONANCE (ALTERNATING-CURRENT CIRCUITS).

Each dispersion region is associated with an energy loss, or absorption, and the energy loss per period per unit volume, J, exhibits a peak at f_0. The existence of energy loss means that, on a phasor diagram, D, instead of being in phase with E, lags E

by a loss angle δ. Thus it is evident from Eq. (3) that ϵ_r is a complex quantity, given by Eq. (19), where Eq. (20) holds. Similarly χ_e is also complex, given by

$$\epsilon_r = \epsilon_r' - j\epsilon_r'' \quad (19)$$

$$\frac{\epsilon_r''}{\epsilon_r'} = \tan\delta \quad (20)$$

Eq. (21), where the real and imaginary parts of χ_e are related to those of ϵ_r by Eqs. (22) and (23). From this it follows that J is given by Eq. (24) or (25), where E

$$\chi_e = \chi_e' - j\chi_e'' \quad (21)$$

$$\chi_e' = \epsilon_r' - 1 \quad (22)$$

$$\chi_e'' = \epsilon_r'' \quad (23)$$

$$J = 2\pi\epsilon_0\epsilon_r''E^2 \quad \text{joules/meter}^3 \quad (24)$$

$$J = 2\pi\epsilon_0\epsilon_r'E^2\tan\delta \quad \text{joules/meter}^3 \quad (25)$$

is the root-mean-square value of the applied field (in V/m), and ϵ_r'' and $\tan\delta$ are known as the loss factor and loss tangent respectively.

At optical frequencies it is usual to characterize the material by its complex refractive index $\hat{n}$, given by Eq. (26), where n is the real part of the refractive

$$\hat{n} = n - jk \quad (26)$$

index and k is the absorption coefficient. Refractive index $\hat{n}$ is related to ϵ_r through the relevant expressions for propagation velocity, so that $\epsilon_r = \hat{n}^2$, with Eqs. (27) and (28) resulting. The use of these rela-

$$\epsilon_r' = n^2 - k^2 \quad (27)$$

$$\epsilon_r'' = 2nk \quad (28)$$

tionships allows values of ϵ_r at high frequencies to be derived from optical measurements. For low k, $\epsilon_r' \simeq n^2$. *See* ABSORPTION OF ELECTROMAGNETIC RADIATION; REFRACTION OF WAVES.

The frequency variations of ϵ_r' and ϵ_r'' are not independent (**Fig. 1**). The drop in ϵ_r' between the effective beginning and end of a dispersion region, namely $\Delta\epsilon_r'$, is called the strength of the dispersion, and is given by the Kramers-Kronig relationship, Eq. (29).

$$\Delta\epsilon_r' = \frac{2}{\pi}\int_{-\infty}^{+\infty} \epsilon_r''\, d(\ln f) \quad (29)$$

If ϵ_r' and ϵ_r'' are plotted against $\log_{10} f$, the reduction in ϵ_r' for the dropping out of a given polarization mechanism is therefore about 1.466 times the area beneath the ϵ_r'' curve for that mechanism. In practice, the integral will be truncated to avoid the inclusion of area attributable to other mechanisms. *See* DISPERSION RELATIONS.

In addition to the submicroscopic polarization mechanisms, the existence of macroscopic discontinuities at internal structure defects or at blocking electrodes, or the inclusion of impurities causing conducting regions, result in behavior classified as

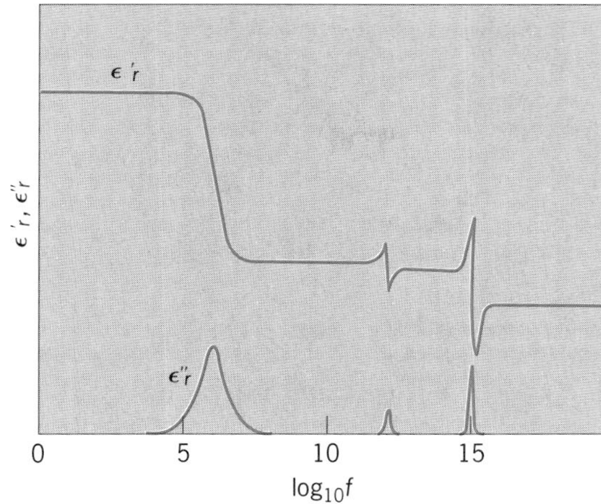

Fig. 1. **Variation of ϵ_r' and ϵ_r'' with frequency *f*. Abscissa is $\log_{10} f$, where *f* is expressed in hertz.**

Maxwell-Wagner effects. These give rise to an effective polarization and associated loss, the frequency behavior (though not the temperature behavior) of which is similar to that of orientational polarization. The dispersion may lie in the region of 1 Hz or lower. *See* CRYSTAL DEFECTS.

Some dielectrics may exhibit an ohmic conductivity σ because they contain free charge carriers. This causes an energy loss $\sigma E^2/f$ per period. Comparison with Eq. (24) indicates that the associated apparent ϵ_r'' is given by Eq. (30). Thus away from dispersion

$$\epsilon_r'' = \frac{\sigma}{2\pi\epsilon_0 f} \quad (30)$$

regions, a curve of $\log \epsilon_r''$ versus $\log f$ is a straight line of negative unit slope, with an intercept yielding σ. With this information, a correction can be made to measurements in the dispersion regions to give the value of ϵ_r'' attributable to polarization processes alone.

Orientational polarization. When orientational polarization is operative, it is usually the dominant polarization present, and it normally gives rise to a stronger dispersion than do other mechanisms.

Debye theory. The classical theory of this mechanism is due to P. Debye, and for a single relaxation time τ, the variation of ϵ_r with angular frequency ω is given by Eq. (31) where ϵ_∞ and ϵ_s are the relative permit-

$$\frac{\epsilon_r - \epsilon_\infty}{\epsilon_s - \epsilon_\infty} = \frac{1}{1 + j\omega\tau} \quad (31)$$

tivities at frequencies much higher and much lower respectively than those of the dispersion region arising from the relaxation of permanent dipoles, but not so far removed that other dispersion mechanisms are operative between these frequency limits. Equating real and imaginary parts gives Eqs. (32) and (33).

$$\frac{\epsilon_r' - \epsilon_\infty}{\epsilon_s - \epsilon_\infty} = \frac{1}{1 + \omega^2\tau^2} \quad (32)$$

$$\frac{\epsilon_r''}{\epsilon_s - \epsilon_\infty} = \frac{\omega\tau}{1 + \omega^2\tau^2} \quad (33)$$

See COMPLEX NUMBERS AND COMPLEX VARIABLES.

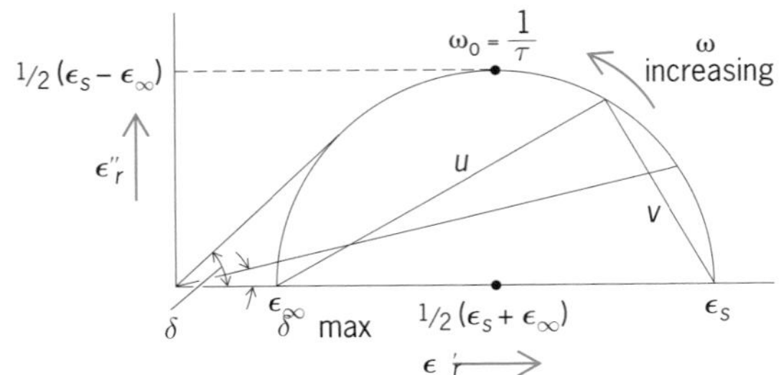

Fig. 2. Plot of ϵ''_r versus ϵ'_r for Debye behavior.

This shows that if ϵ''_r is plotted against ϵ'_r (**Fig. 2**), a semicircle of radius $^1/_2(\epsilon_s - \epsilon_\infty)$ results, with its center on the ϵ'_r axis at $^1/_2(\epsilon_s + \epsilon_\infty)$ making intercepts of ϵ_s and ϵ_∞ on that axis. The loss factor ϵ''_r has its maximum value when $\omega = \omega_0 = 1/\tau$, and according to Eq. (24) this corresponds to a maximum value of J given by Eq. (34). The maximum value of

$$J_{\max} = \pi\epsilon_0(\epsilon_s - \epsilon_\infty)E^2 \qquad \text{joules/meter}^3 \tag{34}$$

tan δ occurs when $\omega\tau = (\epsilon_s/\epsilon_\infty)^{1/2}$ and is given by Eq. (35). For any point on the semicircle the frequency is given by Eq. (36), where u and v are the

$$\tan\delta_{\max} = {}^1/_2(\epsilon_s - \epsilon_\infty)(\epsilon_s\epsilon_\infty)^{-1/2} \tag{35}$$

$$\frac{\omega}{\omega_0} = \frac{v}{u} \tag{36}$$

distances of the point from the intercepts of the semicircle with the ϵ'_r axis, as shown in Fig. 2.

The power loss W associated with orientational polarization rises monotonically with increase in ω, and if plotted to a suitable scale against log $\omega\tau$ has the same shape as the χ'_p curve (where χ_p, as introduced above, is the component of the electric susceptibility that is due to orientational polarization), but reversed (**Fig. 3**). As $\omega \to \infty$, W rises asymptotically to a theoretical value given by Eq. (37), which

$$W = \frac{\epsilon_0(\epsilon_s - \epsilon_\infty)E^2}{\tau} \qquad \text{watts/m}^3 \tag{37}$$

is twice the power loss at $\omega\tau = 1$. However, at very high frequencies, the Debye expression for ϵ''_r, and so Eq. (37) for W, may not hold. Various modifications have been suggested to make the calculation more satisfactory at the higher frequencies.

Cole-Cole plot. Frequently, experimental results yield a circular arc, rather than a semicircle, with its center below the abscissa. This modified plot (**Fig. 4**) is given by the Cole-Cole equation (38), where

$$\frac{\epsilon_r - \epsilon_\infty}{\epsilon_s - \epsilon_\infty} = \frac{1}{1 + (j\omega\tau)^{1-\alpha}} \tag{38}$$

$0 < \alpha < 1$. The maximum value of ϵ''_r still occurs at $\omega = \omega_0 = 1/\tau$, but is reduced to $\epsilon''_{\omega 0}$ (see **table**). For any point on the arc the frequency is given by Eq. (39). Such behavior can be predicted if a suitable

$$\frac{\omega}{\omega_0} = \left(\frac{v}{u}\right)^{1/(1-\alpha)} \tag{39}$$

distribution of relaxation times is postulated, as is also the case for the skewed-arc plot.

Cole-Davidson plot. A variety of other shapes are observed, such as the skewed arc which approximates to a straight line intersecting the ϵ'_r axis at an acute angle ξ at the high frequency end, curving round into a circular arc intersecting the ϵ'_r axis at right angles at the low frequency end (**Fig. 5**). Such a curve

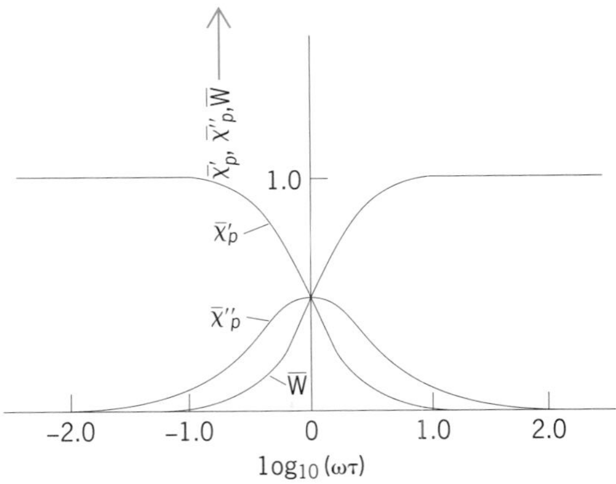

Fig. 3. Plot of $\bar{\chi}'_p = \chi'/\chi_{ps}$, $\bar{\chi}''_p = \chi''/\chi_{ps}$, and $\bar{W} = W\tau/\epsilon_0\chi_{ps}E^2$ for Debye behavior (χ_{ps} is the static orientational susceptibility).

Values of permittivity in the dispersion region for orientational polarization

Plot	$\frac{\epsilon_r - \epsilon_\infty}{\epsilon_s - \epsilon_\infty}$	Frequency ω	ϵ'_r	ϵ''_r
Debye	$\frac{1}{1 + j\omega\tau}$	$\omega_0 = \frac{1}{\tau}$	$\epsilon'_{\omega 0} = \frac{1}{2}(\epsilon_s + \epsilon_\infty)$	$\epsilon''_{\omega 0} = \frac{1}{2}(\epsilon_s - \epsilon_\infty)$
Cole-Cole	$\frac{1}{1 + (j\omega\tau)^{1-\alpha}}$	$\omega_0 = \frac{1}{\tau}$	$\epsilon'_{\omega 0} = \frac{1}{2}(\epsilon_s + \epsilon_\infty)$	$\epsilon''_{\omega 0} = \frac{1}{2}(\epsilon_s - \epsilon_\infty)\tan\frac{(1-\alpha)\pi}{4}$
Cole-Davidson	$\frac{1}{1 + (j\omega\tau)^2}$	$\omega_0 = \frac{1}{\tau}$	$\epsilon'_{\omega 0} = \epsilon_\infty + (\epsilon_s - \epsilon_\infty)2^{-\beta/2}\cos\frac{\pi\beta}{4}$	$\epsilon''_{\omega 0} = (\epsilon_s - \epsilon_\infty)2^{-\beta/2}\sin\frac{\pi\beta}{4}$
		$\omega_m = \frac{1}{\tau}\tan\frac{\pi}{2(1+\beta)}$	$\epsilon'_{\omega m} = \epsilon_\infty + (\epsilon_s - \epsilon_\infty)\left[\cos\frac{\pi}{2(1+\beta)}\right]^\beta \cos\frac{\pi\beta}{2(1+\beta)}$	$\epsilon''_{\omega 0} = (\epsilon_s - \epsilon_\infty)\left[\cos\frac{\pi}{2(1+\beta)}\right]^\beta \sin\frac{\pi\beta}{2(1+\beta)}$

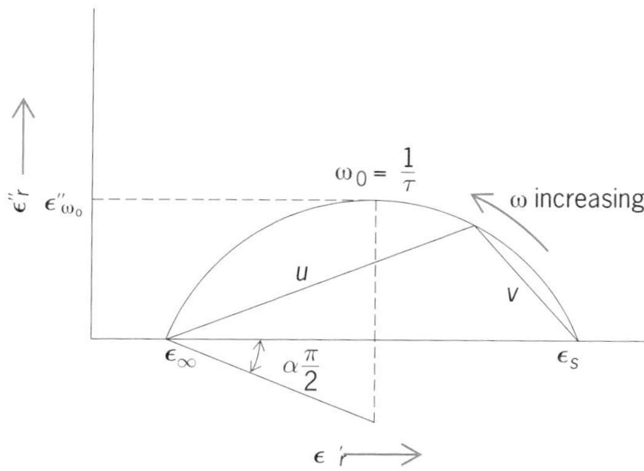

Fig. 4. Cole-Cole plot. Value of $\epsilon''_{\omega 0}$ is given in the table.

is represented by the Cole-Davidson equation (40),

$$\frac{\epsilon_r - \epsilon_\infty}{\epsilon_s - \epsilon_\infty} = \frac{1}{(1 + j\omega\tau)^\beta} \tag{40}$$

where $0 < \beta < 1$. The maximum value of ϵ'_r occurs not at ω_0 but at angular frequency ω_m given in the table, which also gives the values of ϵ'_r and ϵ''_r at ω_0 and ω_m.

Time domain. It follows from the foregoing that if a steady dc voltage were to be applied to a specimen exhibiting Debye behavior, or if the specimen were to be short-circuited after the polarization was fully established, then the circuit current would decay exponentially with a time constant τ. However, for many materials, particularly polymers, over a range of time the current decays as t^{-n}, a relationship known as the Curie-von Schweidler law. In the frequency domain, this corresponds to the type of linear relationship between ϵ'_r and ϵ''_r that is approached at the high-frequency end of the Cole-Davidson plot. The modifications to the Debye equation, devised to yield plots which fit such experimental observations, are sometimes criticized on the ground that they are not based on physical models. It has been suggested particularly with regard to materials that obey the Curie-von Schweidler law, that explanations should be sought not in terms of modifications to the Debye equation based on postulated distributions of relaxation times but in terms of quite different concepts arising directly from the linear relationships between ϵ'_r and ϵ''_r observed over quite wide ranges of frequency. However, a physical model based on protonic conduction has been put forward which yields a distribution of relaxation times from which such behavior can be predicted.

James H. Calderwood

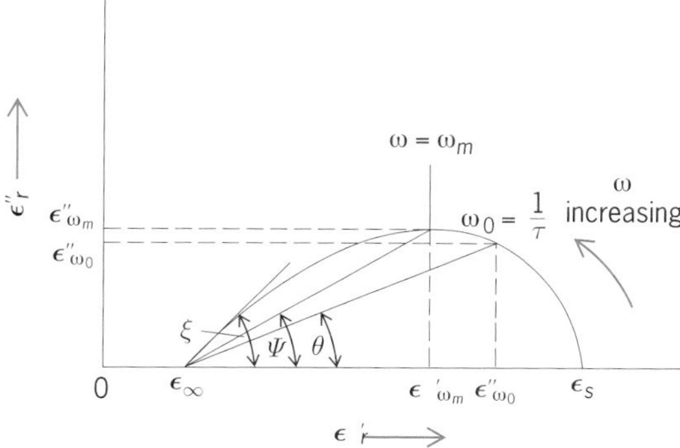

Fig. 5. Cole-Davidson plot. $\xi = (\pi/2)\beta$; $\psi = (\pi/2) \times [\beta/(1 + \beta)]$; $\theta = (\pi/4)\beta$. Values of ω_m, $\epsilon'_{\omega 0}$, $\epsilon''_{\omega 0}$, $\epsilon'_{\omega m}$, and $\epsilon''_{\omega m}$ are given in the table.

Bibliography. C. J. F. Böttcher et al., *Theory of Electric Polarization*, 2d ed., vol. 1: *Dielectrics and Static Fields*, 1973, vol. 2: *Dielectrics in Time Dependent Fields*, 1978; W. T. Coffey et al., *Molecular Diffusion and Spectra*, 1984; P. Debye, *Polar Molecules*, 1929; V. I. Gaiduk and J. R. McConnell, *Dielectric Relaxation and Dynamics of Polar Molecules*, 1995; A. K. Jonscher, *Dielectric Relaxation in Solids*, 1983; H. Kliem, *Dielctric small signal response by protons in amorphous insulators*, IEE Trans. Electrical Insulation, 24, 185-197, 1989; B. K. P. Scaife, *Principles of Dielectrics*, 1989; A. von Hippel, *Dielectrics and Waves*, 2 vols., 1954.

Perovskite

A minor accessory mineral, formula $CaTiO_3$, occurring in basic rocks. Perovskite has given its name to a large family of materials, synthetic and natural, crystallizing in similar structures. The crystal structure is ideally cubic, with a framework of corner-sharing octahedra, containing titanium (Ti) or other relatively small cations surrounded by six oxygen (O) or fluorine (F) anions (see **illus.**). Within this framework are placed calcium (Ca) or other large cations, surrounded by twelve anions. Tilting of the octahedra and other distortions often lower the symmetry from cubic, giving the materials important ferroelectric properties and decreasing the coordination of the central cation. This flexibility gives the structure the ability to incorporate ions of different sizes and

Ideal cubic perovskite structure: anions are at the corners of each octahedron, small cations (gray circles) are at the center of each octahedron surrounded by six anions, while the large cation (colored circle) is in the central cavity surrounded by twelve anions. (*After A. Navrotsky and D. J. Weidner, eds., Perovskite: A Structure of Great Interest to Geophysics and Materials Science, American Geophysical Union, 1989*)

charges. Substitution of niobium (Nb), cerium (Ce), and other rare-earth elements in natural calcium titanate ($CaTiO_3$) is common and can make perovskite an ore for these elements. *See* COORDINATION CHEMISTRY; CRYSTAL STRUCTURE; RARE-EARTH ELEMENTS.

At pressures over 200,000 atm (20 gigapascals), reached in the Earth's lower mantle at depths below 700 km (420 mi), silicate minerals transform to a perovskite structure with a composition close to magnesium silicate ($MgSiO_3$). Synthesized in tiny quantities at great expense in the laboratory, this silicate perovskite may in fact be the most abundant single mineral on Earth. Its properties may relate to plate tectonics, seismic discontinuities, and the origin of deep earthquakes. *See* EARTHQUAKE; PLATE TECTONICS.

A number of synthetic perovskites are of major technological importance. Barium titanate ($BaTiO_3$) and lead zirconate-titanate (PZT) ceramics form the basis of a sizable industry in ferroelectric and piezoelectric materials crucial to transducers, capacitors, and electronics. Lanthanum chromate ($LaCrO_3$) and related materials find applications in fuel cells and high-temperature electric heaters. *See* FUEL CELL.

It is anticipated that the high-temperature oxide superconductors, with critical temperatures above the boiling point of liquid nitrogen, will open up a variety of possible applications; they are largely based on the perovskite structure. In these, another fundamental feature of this structure plays a crucial role, namely its ability to incorporate large deviations from the ideal composition by cation or anion vacancies and by intergrowth, on a molecular scale, of other structural elements. One important family of oxide superconductors, the strontium-lanthanum-copper oxide ($Sr_xLa_{2-x}CuO_{4-y}$) type, is based on the potassium-nickel fluoride (K_2NiF_4) structure, which can be described as an intergrowth of perovskite and rock salt layers. Another group, the so-called 1:2:3 compounds like yttrium-barium-copper oxide ($YBa_2Cu_3O_{7-y}$), has complex structures derived from perovskite by the ordered removal of oxygen atoms, leaving both planes and chains of copper atoms. Superconductivity in both these types depends strongly on the oxygen content, which is controlled by preparation conditions and determines the formal oxidation state of copper. *See* CERAMICS; OXIDE; SOLID-STATE CHEMISTRY; SUPERCONDUCTIVITY.

Alexandra Navrotsky

Peroxide

A chemical compound that contains the peroxy (—O—O—) group. The simplest member of the series is hydrogen peroxide (HOOH), and the higher members of the series result from substituting a group for one or both of the hydrogen atoms in HOOH. Peroxidic compounds occur in nature, and they have commercial applications.

Peroxidic compounds in nature. Both plants and animal cells reduce oxygen by one electron to form superoxide as a by-product of the reduction of oxygen to water, the process in which the cell burns foodstuff to make energy. Two superoxide radical ions react to give a dismutation reaction, that is, simultaneous oxidation and reduction, in which one ion is reduced to hydrogen peroxide and one is oxidized back to oxygen [O_2; reaction (1)]. In 1968

$$2O_2^{\bullet -} + 2H^+ \rightarrow H_2O_2 + O_2 \qquad (1)$$

J. McCord and I. Fridovich discovered the enzyme that catalyzes this reaction in cells, superoxide dismutase. Virtually all aerobically living cells contain some form of superoxide dismutase. Cells also contain enzymes called lipoxygenases, which catalyze the substitution of oxygen into the position next to a double bond in unsaturated fatty acids. This position is called the allylic position, and the allylic hydrogen atom is particularly easily removed in free-radical reactions [reaction (2)].

$$\underset{\text{Lipid}}{LH} + O_2 \rightarrow \underset{\substack{\text{Lipid}\\ \text{hydroperoxide}}}{LOOH} \qquad (2)$$

See FREE RADICAL; EICOSANOIDS; ENZYME; SUPEROXIDE CHEMISTRY.

Lipid hydroperoxides can also be produced by the nonenzymatic oxidation of lipids, a process called autoxidation or lipid peroxidation. This is the process that causes butter and other fats to turn rancid when they are exposed to air. Lipid peroxidation involves an initiation step in which radicals are formed, followed by two propagation steps [reactions (3) and (4)] in which the number of radicals is conserved.

$$L^{\bullet} + O_2 \rightarrow LOO^{\bullet} \qquad (3)$$

$$LH + LOO^{\bullet} \rightarrow L^{\bullet} + LOOH \qquad (4)$$

See AUTOXIDATION.

A termination reaction then occurs in which two radicals combine and nonradical products are formed.

The reactions that cause fats such as butter to become rancid also occur in cells, causing the gradual deterioration of the fats. However, lipoxygenase enzymes also catalyze these processes, producing complex lipid hydroperoxides that play a role in many biological processes. For example, all of the prostaglandin and leukotriene hormones are produced from oxidation of the 20-carbon fatty acid that contains four double bonds, arachidonic acid, to the compound prostaglandin G, which contains both an endoperoxide bridge and a hydroperoxide bond [structure (**1**)].

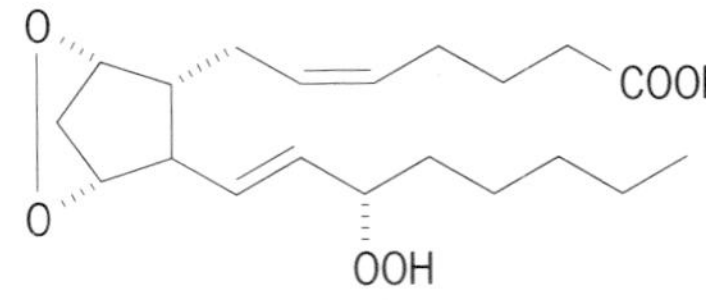

O O COOH OOH

(**1**)

Nitric oxide (NO) is a hormone produced by a number of types of cells, including those in the cells of the artery lining (called endothelial cells), where its function is to reduce muscle tone and thus blood pressure. Compounds such as nitroglycerin release nitric oxide. Since nitric oxide is a free radical, it combines with superoxide [reaction (5)], which is

$$\bullet NO + O_2^{\bullet -} \rightarrow {}^{-}O{-}O{-}N{=}O \qquad (5)$$

(2)

produced by many of the same types of cells that release nitric oxide. This gives the peroxidic compound peroxynitrite (2). The related acid, peroxynitrous acid (3), a potent oxidant in the cell, is formed by the reaction of peroxynitrite with the hydrogen ion [reaction (6)].

$${}^{-}O{-}O{-}N{=}O + H^+ \rightarrow HO{-}O{-}N{=}O \qquad (6)$$

(3)

Peroxides in commerce. In commerce, peroxides are used in oxidations, purifications, syntheses, polymerizations, and oxygen generation. Inorganic peroxides include persulfates, sodium peroxide, and bivalent metal peroxides. Organic peroxides include peroxyacetic acid, benzoyl peroxide, cumene peroxide, and a number of addition products of hydrogen peroxide and aldehydic compounds.

Inorganic peroxides. Peroxydisulfates, familiarly called persulfates, are produced by electrolytic oxidation of aqueous sulfuric acid or ammonium bisulfate. The potassium salt of persulfuric acid (KHS_2O_8) is used in the manufacture of styrene-butadiene synthetic rubber, and smaller quantities are used in hair bleaches. Peroxymonosulfates, also called monopersulfates, are salts of peroxymonosulfuric acid (Caro's acid, $HO{-}SO_2{-}OOH$); they are powerful oxidants. *See* OXIDIZING AGENT.

Hydrogen peroxide. The most widely used peroxy compound is hydrogen peroxide, a waterlike liquid manufactured as aqueous solutions. Hydrogen peroxide is not combustible; water is a safe diluent and coolant. With organic compounds, hydrogen peroxide can form detonable mixtures. Directions for safe handling and storage of hydrogen peroxide are available.

Hydrogen peroxide is manufactured by electrolytic and organic oxidation processes. The former involves electrolytic production of the peroxydisulfate intermediate, followed by steam hydrolysis to hydrogen peroxide. One organic process uses an anthraquinone that is dissolved in organic solvents. The quinone, similar to that shown in (4), is catalytically hydrogenated to the hydroquinone (5), which undergoes autoxidation. Isopropyl alcohol also can be

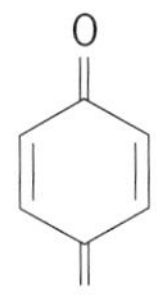

(4)  (5)

oxidized at moderate temperatures and pressures to give hydrogen peroxide and acetone coproducts. *See* ELECTROCHEMICAL PROCESS.

Hydrogen peroxide applications include commercial bleaching, which consumes more than half of that produced. Applications include bleaching of practically all wool and cellulosic fibers, as well as of major quantities of synthetics, and paper and pulp mill bleaching of ground wood and chemical pulps.

Hydrogen peroxide also is used to synthesize epoxides and glycols from unsaturated petroleum hydrocarbons, terpenes, and natural fatty oils. The resultant products are plasticizers, stabilizers, diluents, and solvents for vinyl plastics and protective coating formulations. Power generation applications include use in specialized propulsion units for aircraft, missiles, torpedoes, and submarines. The hot oxygen-system mixture from catalytically decomposed hydrogen peroxide powers the feed pumps of many large liquid-propellant rockets. *See* BLEACHING; HYDROGEN PEROXIDE; JET PROPULSION; ROCKET PROPULSION.

Organic peroxides. As a group, organic peroxides are more hazardous than the inorganic peroxides. Many of the organic compounds are flammable or detonable, thus restricting their availability. Manufacture is chiefly through reaction of the organic substrate with hydrogen peroxide. However, air oxidation also can sometimes be used.

Peracetic acid [$CH_3(C{=}O)OOH$] is prepared from acetic acid and hydrogen peroxide. Peracetic acid, as well as performic or perpropionic, may also be generated from hydrogen peroxide, organic acid, and catalyst. Major applications of peracetic acid are in the synthesis of epoxidized and hydroxylated compounds and as a bactericide, fungicide, and sterilizing agent for processing equipment.

Cumene hydroperoxide [$C_6H_5{-}C(CH_3)_2{-}OOH$] is another peroxide used in commerce. It is a colorless to pale-yellow liquid, produced by air oxidation of isopropyl benzene. It cleaves in acid solution to give phenol and acetone, and this reaction is one of the commercial syntheses of phenol.

Benzoyl peroxide. Benzoyl (or dibenzoyl) peroxide ($C_6H_5{-}CO{-}O{-}O{-}CO{-}C_6H_5$) is the most important aromatic acyl peroxide. It is a colorless, crystalline solid, melting point 106–108°C (223–226°F), that is virtually insoluble in water but very soluble in most organic solvents. It is stable at room temperature but explosible with heating or scratching when in the pure state. For this reason, it is shipped in paper barrels and plastic bags, rather than glass bottles, since crystals caught in the threads could explode when the cap is unscrewed. It is manufactured by reaction of benzoyl chloride and alkaline hydrogen peroxide. Major uses include polymer manufacture and flour bleaching. *See* POLYMERIZATION.

Benzoyl peroxide is synthesized from the reaction of benzoyl chloride with hydrogen peroxide in alkaline aqueous solution or with sodium peroxide in aqueous acetone. Recrystallization from hot solvents is hazardous; peroxides should be heated only in dilute solution. Benzoyl peroxide is an example of a peroxidic initiator, a material that decomposes at a controlled rate at moderate temperatures to give free radicals. Benzoyl peroxide has a half-life (10 h at 73°C or 163°F in benzene) for decomposition that is convenient for many laboratory and commercial processes. This, plus its relative stability among peroxides, makes it one of the most frequently used initiators, for example in vinyl polymerization. It also is the preferred bleaching agent for flour, and is used to bleach many commercial waxes and oils. Benzoyl peroxide is the active ingredient in commercial acne preparations.

Hazards. Benzoyl peroxide itself is neither a carcinogen nor a mutagen. However, when coapplied with carcinogens to mouse skin it is a potent promoter of tumor development. Many peroxidic compounds also are tumor promoters. All peroxides should be treated as potentially explosive, but benzoyl peroxide is one of the least prone to detonate. *See* MUTAGENS AND CARCINOGENS.

Reactions. Like all peroxides, benzoyl peroxide decomposes by rupture of its oxygen-oxygen peroxide bond to give two benzoyloxyl radicals. Thus, benzoyl peroxide provides a source of both phenyl and benzoyloxyl radicals that can perform chemical reactions. For example, the phenyl radicals can add to the double bond of vinyl chloride ($H_2C{=}CHCl$) to form a new radical and thus initiate the polymerization of the vinyl monomer.

Accelerated and induced decomposition. Many materials cause benzoyl peroxide to decompose at a rate accelerated over that observed in pure benzene. For example, solvents with high reactivity toward free radicals (such as dimethyl ether) cause a greatly accelerated decomposition. Compounds that can act as nucleophiles, reducing agents, or electron donors also can greatly accelerate the decomposition; examples are amines and sulfides. Transition-metal ions such as iron or copper also cause an increased rate of decomposition; this can be utilized in commercial practice to achieve a lower temperature or more rapid decomposition. *See* OXIDATION-REDUCTION.

W. A. Pryor

Bibliography. F. A. Cotton and G. Wilkinson, *Advanced Inorganic Chemistry: A Comprehensive Text*, 6th ed., 1999; D. C. Nonhebel and J. C. Walton, *The Chemistry of Peroxides*, 1984; W. A. Pryor, *Free Radicals*, 1966; W. A. Pryor (ed.) *Free Radicals in Biology*, vols. 1-6, 1976-1983; D. Swern (ed.), *Organic Peroxides*, vol. 2, 1971; A. Wataru (ed.), *Organic Peroxides*, 1992.

Peroxisome

Peroxisomes are intracellular organelles that are found in all eukaryotes except the archezoa (original lifeforms). They are bounded by a single membrane and contain a granular matrix and sometimes a paracrystalline core. In electron micrographs, peroxisomes appear round with a diameter of 0.1–1.0 micrometer (see **illus.**), although there is evidence that in some mammalian tissues peroxisomes form an extensive reticulum (network). The name peroxisome derives from the fact that many enzymes, particularly the oxidases, that generate hydrogen peroxide, and catalase, the enzyme that catalyzes the dismutation of hydrogen peroxide to water and molecular oxygen, reside in the organelle. However, peroxisomes contain more than 50 characterized enzymes and perform many biochemical functions, including detoxification. *See* CELL ORGANIZATION.

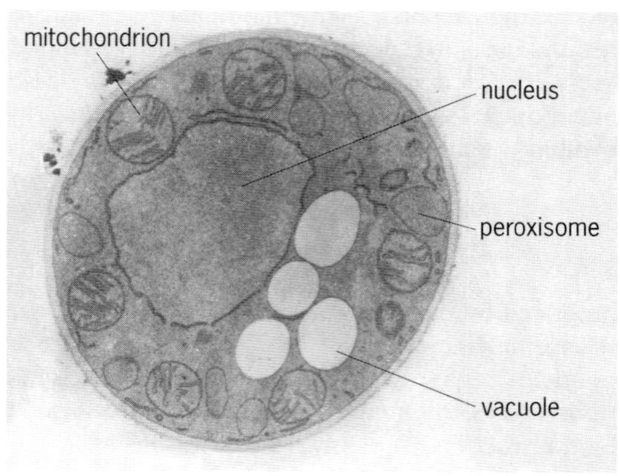

Electron micrograph of a cell of the yeast *Yarrowia lipolytica* grown in fatty acid–containing medium.

Function. Peroxisomes are important for lipid metabolism. In humans, the β-oxidation of fatty acids greater than 18 carbons in length occurs in peroxisomes, while fatty acids of shorter length undergo β-oxidation primarily in mitochondria. In yeast, all fatty acid β-oxidation occurs in peroxisomes. Peroxisomes contain the first two enzymes—alkyl dihydroxyacetonephosphate (DHAP) synthase and DHAP acyl transferase—required for the synthesis of plasmalogens (a group of phospholipids in which a fatty acid group is replaced by a fatty aldehyde). Peroxisomes also play important roles in cholesterol and bile acid synthesis, purine and polyamine catabolism, and prostaglandin metabolism. In plants, peroxisomes are required for photorespiration. *See* LIPID METABOLISM; PHOTORESPIRATION.

Peroxisomes are related to two other organelles, the glyoxysomes of plants and the glycosomes of certain parasites, particularly the trypanosomatids. Together these organelles constitute the microbody family of organelles. All three organelles contain enzymes involved in the β-oxidation of fatty acids. In addition, glyoxysomes harbor the enzymes of the glyoxylate pathway, while glycosomes contain a number of glycolytic enzymes. *See* ENZYME.

Peroxisomes show dramatic changes in number, volume, and protein composition in response to changes in environmental or physiological conditions. Various yeasts show one or few peroxisomes when grown on glucose, but peroxisomes readily proliferate and there is an increased overall peroxisomal protein content when yeasts are grown on carbon sources such as oleic acid and methanol (the metabolism of these sources requires peroxisomal activity). Similar peroxisomal multiplication effects have been observed in some mammals, particularly rodents, treated with peroxisome proliferative agents such as the fibrate family of hypolipidemic drugs (substances that produce a decrease in the level of lipids in the blood).

Human disorders. A number of recessively inherited peroxisomal disorders have been described and grouped into three categories. Group I is the most severe and is characterized by a general loss of peroxisomal function. Many of the enzymes normally localized to the peroxisome are instead found in the cytosol. Mature peroxisomes are absent in the cells of these patients, and instead peroxisomal ghosts are found. These ghosts do not contain matrix enzymes, but most do have a normal complement of peroxisomal membrane proteins. Among the diseases found in group I are Zellweger syndrome, neonatal adrenoleukodystrophy, and infantile Refsum disease. Patients with these disorders usually die within the first years after birth and exhibit neurological and hepatic (liver) dysfunction, along with craniofacial dysmorphism (malformation of the cranium and the face).

Group II peroxisomal disorders are characterized by a loss of peroxisomal function less severe than in group I. Examples are rhizomelic chondrodysplasia punctata, in which DHAP acyl transferase, alkyl DHAP synthase, phytanic acid oxidase, and thiolase are absent from peroxisomes, and Zellweger-like syndrome, in which DHAP acyl transferase and the enzymes of peroxisomal β-oxidation are absent.

Group III comprises those genetic diseases in which one peroxisomal enzyme is disfunctional. Examples are acatalasemia, X-linked adrenoleukodystrophy, and deficiencies of the individual enzymes of peroxisomal β-oxidation.

Peroxisome biogenesis. Peroxisomes do not contain deoxyribonucleic acid (DNA). They do not have independent protein synthesis machinery, as do mitochondria and chloroplasts. Accordingly, all peroxisomal proteins are encoded in the nucleus. Most peroxisomal proteins, both soluble and membrane-associated, are posttranslationally imported into peroxisomes. Import of proteins into peroxisomes requires energy, but it is uncertain whether there is a requirement for an electrochemical gradient across the peroxisomal membrane. Interestingly, peroxisomes are capable of oligomeric protein import. Peroxisomes can arise by growth and fission of preexisting peroxisomes; however, there is accumulating evidence for the de novo synthesis of peroxisomes from vesicular precursors.

Most peroxisomal proteins are synthesized without any precursor extensions or posttranslational modifications. Many matrix proteins are targeted to peroxisomes by a conserved tripeptide peroxisomal targeting signal, called PTS1, located at their extreme carboxyl termini. A small subset of matrix proteins, the best known being the β-oxidation enzyme 3-ketoacyl-CoA thiolase, are targeted by an aminoterminal PTS2. PTS2 motifs may or may not be cleaved from the newly synthesized protein within the peroxisomal matrix. Little is known about the signals that target proteins to the peroxisomal membrane.

Mutants of peroxisome assembly (*pex* mutants) have been made in a number of yeast species and in Chinese hamster ovary cells. These *pex* mutants mimic human peroxisomal biogenesis disorders, particularly Zellweger syndrome. Complementation of *pex* mutants has led to the identification of more than 20 genes whose proteins are required for peroxisome assembly. Pex proteins are localized both to the cytosol and to peroxisomes. Pex proteins that act as PTS receptors, as docking proteins for PTS receptors, and as regulators of peroxisome division have been identified. Nevertheless, the functions of many Pex proteins remain unknown. *PEX* genes isolated from yeast and rodents have been used to identify the corresponding human genes through sequence homology. Many of these human *PEX* genes have been shown to be mutated in the various complementation groups of Zellweger syndrome. The gene encoding Pex7p, the PTS2 receptor, has been shown to be mutated in patients of rhizomelic chondrodysplasia punctata. The identification of human *PEX* genes by homology to *PEX* genes of yeast has been one of the triumphs of the use of simple model systems to identify human genes. *See* GENE; PROTEIN.

Richard A. Rachubinski

Bibliography. B. Alberta et al., *Molecular Biology of the Cell*, 3d ed., Garland Publishing, 1994; G. M. Cooper, *The Cell: A Molecular Approach*, ASM Press, 1997.

Peroxynitrite

A nitrogen oxyanion containing an O—O peroxo bond that is a structural isomer of the nitrate ion. These species are generally distinguished as $ONOO^-$ and NO_3^-, respectively. Other names for peroxynitrite include pernitrite and peroxonitrite; the systematic name recommended by the International Union of Pure and Applied Chemistry (IUPAC) is oxoperoxonitrate (1−). Energy calculations indicate that there are two stable conformations of $ONOO^-$, for which all of the atoms lie in a plane with the peroxo O—O and N=O bonds forming dihedral angles of approximately 0° (cis isomer) or 180° (trans isomer):

$$\mathrm{O{=}N{-}O{-}O^-}\ \text{(cis)} \rightleftharpoons \mathrm{O{=}N{-}O{-}O^-}\ \text{(trans)}$$

See CHEMICAL BONDING.

Characteristics. Although they are nearly isoenergetic, the cis isomer appears to be slightly more stable than the trans isomer. Peroxynitrite is a strong base and is protonated in weakly acidic solutions, forming peroxynitrous acid (ONOOH; IUPAC name: hydrogen oxoperoxynitrate). Calculations indicate that the gas-phase molecule also possesses stable cis and trans conformations; in this case, the energy of the cis isomer is further stabilized relative to the trans by an internal hydrogen bond between the bound hydrogen atom and the N═O terminal oxygen atom. In the trans isomer, the peroxo O—H bond is almost perpendicular to the plane of the other atoms. As expected from the high stability of the nitrogen dioxide ($^{\bullet}NO_2$) radical relative to the hydroxyl radical ($^{\bullet}OH$), the peroxo O—O bond in ONOOH is very weak. Its bond dissociation energy is estimated at ~22 kcal/mol compared to ~51 kcal/mol for HOOH. The acid dissociation constant of ONOOH, defined by the equilibrium $ONOOH \rightleftharpoons ONOO^- + H^+$, is ~2 × 10^{-7} *M*. This value is unusually high for a peroxo acid; the relatively weak O—H bond is attributed to the strongly electron-withdrawing character of the N═O substituent group. One consequence of the relatively high acidity of ONOOH is that, unlike hydrogen peroxide and organic peroxides, the anion is a major form in neutral solutions. *See* ACID AND BASE; HYDROGEN BOND; HYDROXYL.

Formation. Peroxynitrite is formed in nitrate salts or nitrate-containing solutions when exposed to ionizing radiation or ultraviolet light. Solutions can also be prepared by a variety of chemical reactions, including the reaction of hydrogen peroxide with nitrous acid (1); reaction of the hydroperoxide anion

$$HOOH + HNO_2 \rightarrow ONOOH + H_2O \qquad (1)$$

with organic and inorganic nitrosating agents (2); reaction of ozone with the azide ion (3); or, apparently, reaction of O_2 with compounds capable of generating the nitroxyl anion (NO^-) [4]. These prepa-

$$HOO^- + RONO \rightarrow ROH + ONOO^- \qquad (2)$$

$$2O_3 + N_3^- \rightarrow ONOO^- + N_2O + O_2 \qquad (3)$$

$$2O_2 + NH_2OH + 2OH^- \rightarrow ONOO^- + HO_2^- + 2H_2O \qquad (4)$$

rations invariably contain unreacted materials or decomposition products, particularly nitrite ion, which can significantly modulate the peroxynitrite chemical reactivity. Peroxynitrite is also formed in radical-radical coupling reactions, notably superoxide ($^{\bullet}O_2^-$) with nitric oxide ($^{\bullet}NO$) [reaction (5)], and hydroxyl radical with nitrogen dioxide [reaction (6)].

$$^{\bullet}NO + {}^{\bullet}O_2^- \rightarrow ONOO^- \qquad (5)$$

$$^{\bullet}OH + {}^{\bullet}NO_2 \rightarrow ONOOH \qquad (6)$$

See SUPEROXIDE CHEMISTRY.

Peroxynitrite has been isolated as the tetramethylammonium salt by carrying out reaction (5) in liquid ammonia. Formation of peroxynitrite in both solids and solutions is indicated by the appearance of yellow coloration, which is due to tailing of intense near-ultraviolet absorption bands into the visible region.

Reactions. The peroxynitrite anion is relatively stable in both solid matrices and aqueous solution; it has, for example, been found in soil samples recovered from the Martian landscape. Peroxynitrous acid, however, undergoes spontaneous decomposition to give nitric acid and/or nitrous acid plus O_2. Both the decomposition rate and product distributions depend upon the medium conditions. Below pH 5, decomposition is first-order and only nitric acid is formed; the ONOOH lifetime is ~1 s at room temperature. Although, in principle, the reaction could occur by a concerted intramolecular rearrangement, theoretical calculations suggest that the lowest-energy pathways involves homolytic cleavage of the O—O bond to form a hydrogen-bonded {$^{\bullet}NO_2$, $^{\bullet}OH$} radical pair that subsequently recombines with rearrangement to nitric acid. This conclusion is supported by kinetic studies made in alkaline solutions, where the decomposition rate slows progressively with increasing solution alkalinity and is accompanied by formation of increasing amounts of O_2 and NO_2^- in 1 : 2 proportions. In this reaction domain, the O_2 yield exhibits an exceptionally complex dependence upon added radical scavengers that is quantitatively predicted by a dynamical model based upon known radical reactions. Crucial steps include dissociation of ONOOH to $^{\bullet}NO_2$ and $^{\bullet}OH$ radicals, dissociation of $ONOO^-$ to $^{\bullet}NO$ and $^{\bullet}O_2^-$ radicals, oxidation of $ONOO^-$ by $^{\bullet}OH$, and a radical chain cycle involving dinitrogen trioxide (N_2O_3). *See* CHEMICAL DYNAMICS.

Decomposition. Peroxynitrite decomposition is catalyzed by other Lewis acids, including metal ions and carbon dioxide. Higher-valent oxo ions have been detected as intermediates in decompositions catalyzed by metalloporphyrins, indicating that these reactions proceed by oxidation-reduction cycling of the metal ion. The reactant species in the carbon dioxide–catalyzed decomposition are $ONOO^-$ and CO_2, which apparently combine to form an adduct called nitrosoperoxycarbonate ($ONOOCO_2^-$; IUPAC name: 1-carboxylato-2-nitrosodioxidane). Because association is the rate-limiting step in this reaction, the intermediate does not accumulate, and its properties must be inferred by indirect means. Kinetic studies have established that the intermediate lifetime is shorter than ~1 ms and that CO_2 acts as a true catalyst; that is, it is the immediate product of the decomposition. However, when NO_2^- or other reductants are present, alternate decomposition pathways are expressed that lead to conversion of carbon dioxide to bicarbonate ion (HCO_3^-). It is therefore possible under some reaction conditions for dehydration of HCO_3^- to CO_2 to become rate-limiting in the catalytic decomposition cycle. In the lowest-energy configuration calculated for nitrosoperoxycarbonate, the —NO and —CO_2 substituents are cis about the peroxy O—O bond and oriented perpendicular to each other. The peroxy bond is even weaker than in ONOOH, having a calculated bond dissociation energy of ~9 kcal/mol. This difference can be attributed to the greater stability of the carbonate radical ($^•CO_3^-$) formed upon O—O bond homolysis of $ONOOCO_2^-$ than the hydroxyl radical formed upon O—O bond homolysis of ONOOH. Calculations suggest that, as with ONOOH, the decomposition of $ONOOCO_2^-$ to NO_3^- and CO_2 involves homolytic O—O bond cleavage to form an intermediary {$^•NO_2$, $^•CO_3^-$} radical pair. *See* COMPUTATIONAL CHEMISTRY.

Oxidation. Peroxynitrite is a powerful oxidant that has been shown to react with a wide variety of inorganic and organic reductants, as well as hydroxylate and nitrate aromatic compounds, including benzene. Its pH-dependent one-electron standard reduction potential is estimated to be ~2.1 V (pH 0). Nitrosoperoxycarbonate is also a strong oxidant that is capable of oxidizing compounds whose reduction potentials are as high as ~1.3 V. It nitrates phenolic compounds more effectively than ONOOH, but appears unreactive toward benzene and other simple aromatics. Kinetic studies have demonstrated that, unlike most nitration reactions, conversion of tyrosine to 3-nitrotyrosine does not involve electrophilic attack of the aromatic ring by an intermediary nitronium (NO_2^+) ion, but probably proceeds via one-electron oxidation of the phenol to the corresponding phenoxy radical, followed by radical coupling to the $^•NO_2$ generated in the initial redox step. Metal-catalyzed phenol nitrations by ONOOH are thought to occur by similar redox mechanisms.

Peroxynitrite and nitrosoperoxycarbonate can engage in both one-electron and two-electron oxidations of reacting partners. Two-electron oxidations are generally limited to good nucleophiles that have highly polarizable atoms or functional units; for ONOOH, these reactions occur by direct bimolecular interaction of the reactants. In contrast, most (if not all) one-electron oxidations occur by rate-limiting unimolecular activation of ONOOH with rate parameters that are identical to those for its decomposition. Furthermore, the maximal product yields obtained for reactions of both ONOOH and $ONOOCO_2^-$ are always substantially less than limits based upon stoichiometric consumption of the oxidant. This behavior indicates the existence of at least two intermediates along the reaction pathway, only one of which is oxidizing. Two fundamentally different hypotheses concerning the nature of these intermediates have been advanced. One is that the unreactive and reactive species are different configurational isomers similar to the cis-trans isomers of ONOOH; the other is that the unreactive species are the {$^•NO_2$,$^•OH$} and {$^•NO_2$,$^•CO_3^-$} radical pairs that either rapidly recombine to form NO_3^- or escape the cage to give as reactive species discrete $^•NO_2$ and $^•OH$ (or $^•CO_3^-$) radicals. Although this issue is unresolved, the dynamical behavior of the reactive species indicates that they have reactivities that are comparable to free $^•OH$ and $^•CO_3^-$ radicals. *See* OXIDATION-REDUCTION; REACTIVE INTERMEDIATES.

Physiological aspects. Interest in these reactions has been greatly stimulated by recognition that $^•NO$ and $^•O_2^-$ radicals are generated in the bloodstream, neuronal tissues, and phagocytic cells of animals in sufficient quantities to form peroxynitrite [reaction (5)]. Correspondingly, major roles for this powerful oxidant have been proposed both in diseases and tissue damage associated with oxidative stress and in natural cellular defense mechanisms against microbial infection. Indirect evidence that endogenously generated peroxynitrite might be involved in cellular damage includes the observations that tyrosyl nitration levels in proteins from affected tissues of patients suffering from oxidative diseases are exceptionally high, that tissues can be protected from oxidative damage by treatments that inhibit $^•NO$ formation, and that ONOOH is highly toxic to cells and can cause substantial damage to proteins, carbohydrates, nucleic acids, membrane lipids, and cellular antioxidant systems. Physiological buffer systems contain high concentrations of CO_2; consequently, a major fraction of the peroxynitrite formed by living systems is expected to be converted to nitrosoperoxycarbonate before reacting with biological components. Because $ONOOCO_2^-$ has a considerably shorter lifetime than ONOOH, it is unclear whether its formation will act to protect cells from oxidative damage by promoting peroxide decomposition or will potentiate the damage by facilitating reactions with biological components.

Several additional factors that make it difficult to assess the contributions of these peroxy compounds to cellular oxidative processes are that ONOOH reacts rapidly with oxyhemoglobin and peroxidases (which could thereby constitute major sinks for its oxidizing equivalents), that ONOOH is freely diffusible across biological membranes (and therefore not confined to a particular intracellular environment), and that peroxidases also catalyze formation of other reactive nitrogen species (that might account, at least in part, for the elevated levels of tyrosyl group nitration detected at sites of oxidative damage). *See* BIOINORGANIC CHEMISTRY.

James K. Hurst

Bibliography. J. O. Edwards and R. C. Plumb, The chemistry of peroxonitrites, *Prog. Inorg. Chem.*, 46:599–635, 1994; Forum presenting various viewpoints about reaction mechanisms, *Chem. Res. Toxicol.*, 11:710–721, 1988; W. A. Pryor and G. L. Squadrito, The chemistry of peroxynitrite: A product from the reaction of nitric oxide with superoxide, *Amer. J. Physiol. (Lung Cell. Mol. Physiol. 12)*, 268:L699–L722, 1995.

Perpetual motion

The expression perpetual motion, or perpetuum mobile, arose historically in connection with the quest for a mechanism which, once set in motion, would continue to do useful work without an external source of energy or which would produce more energy than it absorbed in a cycle of operation. This type of motion, now called perpetual motion of the first kind, involves only one of the three distinct

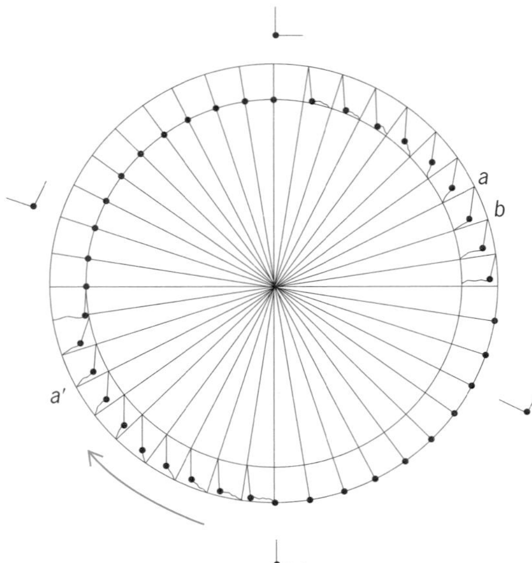

Fig. 1. Wheel of the Marquis of Worcester.

concepts presently associated with the idea of perpetual motion.

First kind. Perpetual motion of this type refers to a mechanism whose efficiency exceeds 100%. Clearly such a mechanism violates the now firmly established principle of conservation of energy, in particular that statement of the principle of conservation of energy embodied in the first law of thermodynamics. (Indeed, the first law of thermodynamics is sometimes stated as "A perpetuum mobile of the first kind cannot exist.") Thus, with the establishment of the energy conservation principle in the middle of the nineteenth century, the possibility of obtaining perpetual motion of the first kind could be denied. However, prior to that time, some ingenious machines, generally dependent on gravity for their "operation," had been devised. One of the most famous of these is the wheel of the Second Marquis of Worcester (**Fig. 1**). *See* CONSERVATION OF ENERGY.

The original model is described as having an outer rim 14 ft (4.3 m) in diameter, a concentric inner rim 12 ft (3.7 m) in diameter, and 40 spokes. Forty 50-lb (23-kg) weights are fastened at the center of ropes which are 2 ft (0.6 m) long. The weights are then attached to the wheel in the following fashion. One end of a given rope is secured to spoke *a*, say at the outer rim, and the other end of this rope is fastened to the adjacent spoke in a clockwise direction, spoke *b*, at the inner rim. All weights are attached in this way. The actual orientation of the weights is indicated both on the wheel in Fig. 1 and in the small sketches appearing outside the periphery of the wheel. To assess the motional tendency of the wheel, consider a pair of spokes making up a wheel diameter, say the pair *a-a′*. The weight supported by spoke *a* is supported at the outer rim, whereas the weight supported by spoke *a′* is supported at the inner rim. Thus there is a clockwise torque associated with the spoke pair *a-a′*. Since, from Fig. 1, this appears to be the case for all spoke pairs that make up wheel diameters, the wheel presumably rotates clockwise perpetually. After describing the wheel and commenting upon the torque imbalance, the Marquis enigmatically remarks, "Be pleased to judge the consequence."

Also in the category of perpetual motion of the first kind were the proposals for hydrodynamical devices, essentially waterwheels which could resupply their own millstreams. These devices, many of extreme complexity, usually involved the hydrostatic paradox. In the example in **Fig. 2**, the water would supposedly flow continuously because its weight in the large vessel exceeds that in the tube.

Second kind. Perpetual motion of the second kind refers to a device that extracts heat from a source and then converts this heat completely into other forms of energy, a process which satisfies the principle of conservation of energy. A dramatic scheme of this type would be an ocean liner, which extracts heat from the nearly limitless oceanic source and then uses this heat for propulsion. This type of perpetual motion is, however, precluded by the second law of thermodynamics, also established in the nineteenth century, which is sometimes stated as "A perpetuum mobile of the second kind cannot exist."

Third kind. The third type of perpetual motion is, in contrast to the two types described above wherein useful output was the goal, merely a device which can continue moving forever. It could result in actual systems if all mechanisms by which energy is dissipated could be eliminated. For example, if all bearing friction could be removed, a wheel spinning in a vacuum would continue to spin indefinitely. Since experience indicates that dissipative effects in mechanical systems can be reduced, by lubrication in the case of friction, for example, but not eliminated, mechanical perpetual motion of the third kind can be approximated but never achieved.

An example of a genuine case of this kind occurs in a superconductor. A superconductor is a metal which, if cooled to a very low temperature, loses all of its resistance to the passage of a direct current; that is, a dissipative effect has been truly eliminated rather than reduced. If a direct current is caused to

Fig. 2. Perpetual-motion device of the first kind based on the hydrostatic paradox.

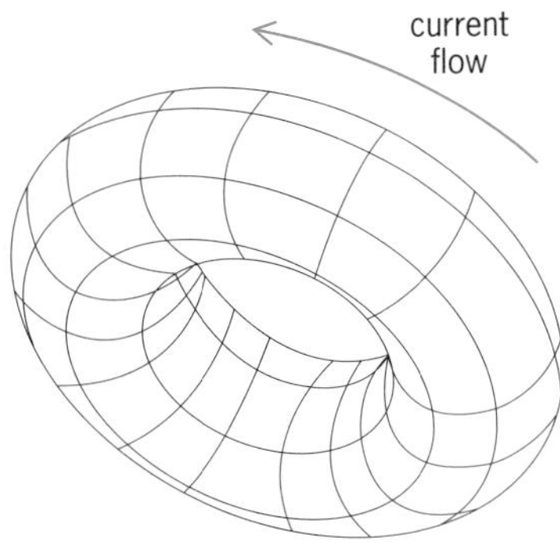

Fig. 3. Perpetual motion of the third kind.

flow in a superconducting ring (**Fig. 3**), this current will continue to flow undiminished in time without application of any external force. *See* SUPERCONDUCTIVITY.

This last example leads perilously close to the microscopic realm where, of course, perpetual motion is the rule. Electrons in atoms and atoms themselves are in constant motion, this latter motion manifesting itself, for example, in brownian motion. The existence of this microscopic perpetual motion does not, in any event, force a reevaluation of the conclusions above, for the traditional ideas of perpetual motion deal with the macroscopic world of machines and devices. Since the laws of thermodynamics are themselves applicable to matter in macroscopic quantities and, as such, contain implicitly the consequences of the microscopic behavior, the impossibility of the useful types of perpetual motion, that is, of the first or second kinds, is at present deemed an unequivocal conclusion. *See* BROWNIAN MOVEMENT; THERMODYNAMIC PRINCIPLES.

Modern examples. Designs and working models of devices promising perpetual motion, in particular of the first kind, continue to be described frequently. Since the impossibility of perpetual motion of the first or second kind is now commonly accepted, the inventors of these devices are usually quick to dissociate their inventions from perpetual motion and to attribute device operation to "new (and not understood) principles." Though mechanical systems continue to be popular in this realm, the new breed of device is more likely to feature an operating system based on electromagnetic concepts. Particularly popular are "motors" incorporating complex and often ingenious configurations of permanent magnets, which are claimed to operate indefinitely with no power supplied. That this type of "motor" can operate, even for extended periods, need not be a contradiction of any of the known laws of physics. However, the operation is limited by the energy available, namely, the energy contained in the magnetic field established by the configuration of magnets. Thus, the dream of perpetual motion continues unrealized and, in the view of essentially all scientists, unrealizable.

K. L. Kliewer

Bibliography. A. W. Ord-Hume, *Perpetual Motion: The History of an Obsession*, 1977; K. S. Pitzer and L. Brewer, *Thermodynamics*, 3d ed., 1995; M. Tinkham, *Introduction to Superconductivity*, 1975, reprint 1980.

Perseus

A compact circumpolar constellation of the northern sky, like its neighbor, Cassiopeia, on the east. Both constellations lie in a brilliant part of the Milky Way. The prominent stars in Perseus form the letter A (*see* **illus.**). This group is represented by the figure of the

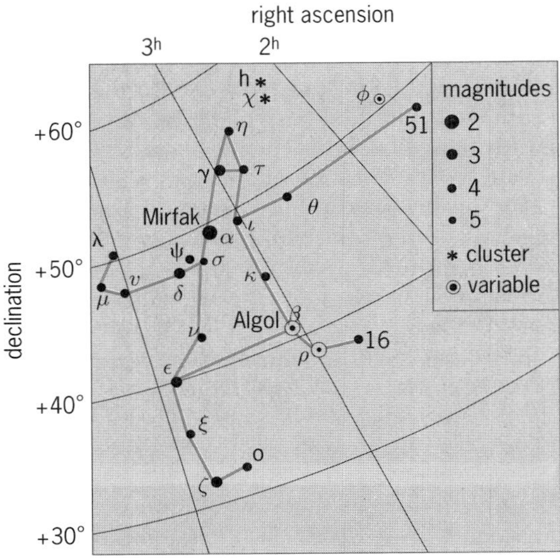

Line pattern of the constellation Perseus. The grid lines in the chart represent the coordinates of the sky. The apparent brightness, or magnitude, of the stars is shown by the size of the dots, which are graded by appropriate numbers as indicated.

hero Perseus. The conspicuous curved arc of stars, bright and easy to identify, is commonly known as the Segment of Perseus. Mirfak, a navigational star, lies in the right shoulder. The constellation is noted for its clusters of stars. Just above the head are the famous double clusters *h* and χ in Perseus. Algol, the Demon Star, which is an eclipsing variable, is located in this constellation. *See* CASSIOPEIA; CONSTELLATION.

Ching-Sung Yu

Persian melon

A long-season cultivar of muskmelon, *Cucumis melo*, of the gourd family, Cucurbitaceae. The fruit is 7–10 in. (18–25 cm) in diameter, weighs 6–8 lb (2.7–3.6 kg), and is round and without sutures; it has dark-green skin and thin, abundant netting. The flesh is deep orange, very thick and firm, and distinctly sweet in flavor. The fruit separates from the stem at full maturity and soon becomes overripe. Plants are large and vigorous with large leaves. The vines usually bear andromonoecious flowers, and pollination is performed by bees. Persian melons require a long frost-free growing season of about 115 days. Average growing temperatures of 70–75°F (21–24°C) are preferred. Fruits are about 90% water and contain 7.5 g of carbohydrate per 100 g of tissue. The flesh is very rich in potassium, vitamin A, and vitamin C. *See* VIOLALES.

In the United States the Persian melon is a very minor crop and is grown chiefly in the Central Valley of California, where low humidity and absence of summer rain tend to prevent the fungus diseases that often defoliate plants in humid areas. It is a luxury product that is very susceptible to damage during transit and storage.

Important foliar diseases are powdery and downy mildews, which are intensified by high humidity. Sulfur dusts are effective for powdery mildew control. Fusarium and Verticillium wilts inhabit the soil and cause wilting of the foliage and early death. Watermelon mosaic virus, cucumber mosaic virus, and squash mosaic virus are transmitted by insects and cause severe losses. Aphids, cucumber beetles, leaf miners, and red spider mites are the major insect pests. Nematodes attack the roots of the plants grown on infested soils. *See* MUSKMELON; PLANT PATHOLOGY. Oscar A. Lorenz

Bibliography. S. P. Doolittle et al., *Muskmelon Culture*, USDA Agr. Handb. 216, 1961.

Persimmon

A deciduous fruit tree species, *Diospyros kaki*. Persimmons originated in the subtropical regions of China but were cultivated more extensively in Japan where, until 1900, it was the most important fruit. Now, although it is only the fifth most important crop there, Japan leads in world production. Persimmons have been introduced to a number of temperate zone countries; however, they have not attained substantial popularity, and only small commercial plantings exist in the United States (primarily in California), Italy, Brazil, and Israel.

Persimmon cultivars adapt to a wide climatic range. Although they have a low chilling requirement, they can tolerate temperatures as low as 5°F (−15°C) if they are dormant. However, areas with late spring and early fall temperatures below 27°F (−3°C) should be avoided, as young growth and maturing fruit will be damaged. Mean annual temperatures averaging 57–59°F (14–15°C) are required for good growth and quality.

Cultivars. For horticultural purposes, persimmons are classified as astringent or nonastringent. Astringent persimmons have water-soluble tannins in the flesh that decrease as the fruit softens to ripeness. They are conical in shape. Nonastringent persimmons, are firm when ripe, as their soluble tannins decrease with pollination. They have an oblate shape. Astringent cultivars produce best in cool climates, whereas nonastringent types require hot, dry climates for good quality. Both astringent and nonastringent cultivars are subdivided into pollination-constant and pollination-variant types. Pollination-variant persimmon flesh darkens in response to the presence of seeds, produced by pollination, but pollination-constant persimmon flesh does not. Among major persimmon cultivars grown in Japan, California, and New Zealand, wide variation exists among the astringent and nonastringent, pollination-constant and variant cultivars, suggesting centuries of chance and deliberate cross breeding. Nonastringent cultivar production has increased at the expense of astringent cultivars.

Propagation and cultivation. Persimmons are clonally propagated by budding onto seedling rootstocks. Seed extracted from mature fruits of *D. lotus*, *D. kaki*, and *D. virgiana* are germinated in the greenhouse in the fall. Seedlings are transplanted in outdoor nursery rows when temperatures are over 55°F (13°C). After a growing season, seedlings are of sufficient size to chip-bud in September. The dormant, budded whip is maintained in the nursery row until the danger of frost subsides. During the following spring, it is removed from the nursery rows and held in moist sawdust as a bare root whip until planting. Trees are planted at 12–15 ft (3.5–4.5 m) in the row, with 16–18 ft (.5–5.5 m) between rows. Whips are topped at 20 in. (50 cm) when planted and are trained to a modified central leader tree through five dormant seasons. Persimmons bear laterally on current season's growth. As persimmons are vigorous and tend to alternate crops, mature-tree dormant pruning is used to maintain an upright tree and thin crop load. Persimmons can be monoecious, dioecious, and hermaphroditic. The major cultivars are dioecious, with female flowers capable of producing fruit without fertilization. Male pollinating trees may be planted at a ratio of one male to eight females to reduce fruit drop and increase fruit quality. This is a common practice among nonastringent cultivars. However, pollination produces seeds, an undesirable quality. *See* REPRODUCTION (PLANT).

Diseases. Persimmons are relatively disease free in California. The most troublesome disease is crown gall, caused by *Bacterium tumifaciens*. The major symptom is large galls at the juncture of large roots and the trunk on young trees. These eventually die and slough off, inviting secondary infections that girdle the roots and crown. The disease is easily avoided by soaking rootstock seeds in sodium hypochlorite prior to germination and stratifying the seeds in nursery beds rather than refrigerating them. If the nursery stock is not clean, the young trees can become susceptible to disease any time the trunk or crown roots are wounded during digging and planting. Another common disease in California is mushroom root rot, caused by *Armillaria mellea*. This fungal disease can exist in the soil for years on the roots of previously infected hosts. It attacks new root growth in spring and early summer, when soil temperatures range between 62 and 75°F (17 and 24°C), and results in a slow decline of the tree, manifesting as poor shoot growth, leaf drop, and eventual death. During the host summer months, trees can suddenly collapse.

The best method of preventing disease is by developing a resistant rootstock. The most damaging postharvest disease is caused by the omnipresent fungus *Penicillium*. It is controlled by avoiding injury during picking and transport. *See* PLANT PATHOLOGY.

Harvesting. Multiple, selective harvests are done in late fall to ensure that the fruits achieve the desirable deep orange-red color. Fruits are harvested by closely clipping the stem above the calyx with clippers.

Astringent cultivars are marketed fresh or dry, made into jams and jellies, or frozen as a pulp for cooking. Nonastringent persimmons, which can be maintained in cold storage for as long as 2 months, are primarily marketed as fresh fruit. *See* FRUIT, TREE.

Louise Ferguson

Bibliography. J. Janik, *Advances in New Crops*, 1990.

Personality theory

A branch of psychology concerned with developing a scientifically defensible model or view of human nature—the modern parlance, a general theory of behavior.

Personality theory is an outgrowth of nineteenth-century French and German psychiatry. Nineteenth-century psychiatry was a practical rather than an academic discipline, concerned with the origins and treatment of neurotic and even psychotic disorders. It dealt with abnormal rather than normal thought processes, focusing on the unconscious rather than the conscious mind. Perhaps most distinctively, it considered a single person as the unit of study, rather than a single perception or behavioral response as in academic psychology. These themes continue today. Thus, with only a few exceptions, the prominent personality theorists have been psychiatrists or clinical psychologists, and personality psychology has developed largely outside the mainstream of academic psychology. The major exceptions are those personality theorists with a behaviorist orientation who stress the role of learning in personality development—John Dollard, Neal Miller, and Albert Bandura.

Personality theory as a portion of academic psychology came into existence in the late 1930s with the publications of Gordon Allport and Ross Stagner. The field has grown steadily, until today, in conjunction with social psychology, personality is one of the most active areas of research in all psychology.

There are two primary usages for the word "personality" in ordinary language. In the first sense, personality refers to the distinctive impression a person makes on others. In thinking of someone (a parent, a friend, a favorite or hated teacher), that person's face is rarely evoked; rather, the thought concerns the unique way that person acted on a particular occasion, the kind of atmosphere that person created by his or her presence. This is known as the surface definition of personality. In the second sense, which is usually preferred by philosophers, theologians, playwrights, and novelists, personality refers to the inner core of a person, his or her soul or "true" nature. This inner nature may be expressed in terms of the unique impression a person makes on others, but personality is defined in terms of its core rather than surface aspects.

Basis for Theories

A personality theory is a set of assumptions and expectations about oneself and other people that guides one's perceptions of, and actions with regard to, oneself and others. Personality theories answer three kinds of questions: How or in what ways are people all alike—for example, are humans naturally aggressive? How or in what ways are people different—for example, what were the sources of Beethoven's creativity? What is the meaning of a particular, inexplicable action by someone directly or indirectly known—for example, why did Lee Harvey Oswald assassinate John F. Kennedy? However they may differ, all personality theories eventually respond to these three questions, but of course they provide unique answers to them.

Differences. Personality theories differ with regard to certain features which are often used as a means of evaluating them.

Scope. Theories differ in terms of their scope. Cognitive dissonance theory and attribution theory, for example, each offer a single psychological principle to explain the manner in which people think about social behavior. On the other hand, psychoanalysis has great scope and complexity, and attempts to account for all important aspects of social and psychological experience. Academic psychologists tend to prefer theories with limited scope; clinical psychologists and psychiatrists tend to prefer theories with larger scope.

Explicitness. Theories also differ in terms of their explicitness—the degree to which it is clear what a theory would predict in a particular case. Those personality theories inspired by behaviorism and learning theory (Dollard, Miller, Bandura, B. F. Skinner) tend to be admirably explicit; those inspired by psychoanalysis (Sigmund Freud, Carl Jung, Alfred Adler, Erik Erikson) tend to be inexplicit—one never quite knows what these theories will predict in a given case.

Theory versus fact. Theories differ in terms of the degree to which they conform to empirical facts. This, however, is a subtle judgment because what counts as a fact depends on the theory, and all theories point to certain facts that support them, and ignore facts that contradict them. That is, all theorists ignore the facts that embarrass their theories, but this tendency is a deeply human characteristic.

Supporters. Finally, theories differ in the degree to which they attract advocates and supporters. Far and away the two most influential perspectives on personality are the behaviorist models of Skinner, Bandura, and other social learning theorists, and the psychoanalytic model of Freud and his later followers. Although behaviorism and psychoanalysis represent widely divergent perspectives on human nature, the work of Dollard and Miller was largely aimed at synthesizing these apparently antithetical traditions. Curiously, the popularity of various theories often seems as closely related to the politics of science and the charisma of the theorist as it does to the intellectual merit of the ideas involved.

Classification. Most personality theories can be classified in terms of two broad categories, depending on their underlying assumptions about human nature. On the one hand, there are a group of theories that see human nature as fixed, unchanging, deeply perverse, and self-defeating. These theories emphasize self-understanding and resignation; in the cases of Freudian psychoanalysis and existentialism, they also reflect a distinctly tragic view of life—the sources of human misery are so various that the best that can be hoped for is to control some of the causes of suffering. On the other hand, there are a group of theories that see human nature as plastic, flexible, and always capable of growth, change, and development. Human nature is basically benevolent; therefore bad societies are the source of personal misery. Social reform will produce human happiness if not actual perfection. These theories emphasize self-expression and self-actualization—the cases of Carl Rogers and Abraham Maslow, they reflect a distinctly optimistic and romantic view of life.

Core Ideas

There are six core ideas with which any competent theory of personality must deal. Theories differ in terms of how well they treat these ideas, which provide another means for evaluating them.

Motivational theory. Motivational theory is concerned with what makes people go, what provides the energy or reasons for their actions. Motives are the major, if not primary, explanatory variables in personality theory; people's actions are normally explained by appealing to motivational concepts. Motivational terms vary widely across personality theories; nonetheless, each theory tends to specialize in one of two broad types of motivational concepts.

Biological theories. The first type of motivational concept is oriented toward biology or physiology. This category includes terms such as drive, need, instinct, passion, and urge, so that when a theorist uses these terms, a biological form of motivational theory is being endorsed. Biological theories (such as psychoanalysis) describe human motivation in terms of a small number of fixed and unchanging motives, and everyone is assumed to share this same set of motives. These motives are usually described as unconscious, as beyond rational control, and as becoming harmful or dangerous if frustrated or ignored for a long time. *See* INSTINCTIVE BEHAVIOR; PSYCHOANALYSIS.

Psychological theories. The second category of motivational concepts is psychological or mental in origin, and includes terms such as values, expectations, attitudes, preferences, and goals; and so, when a theorist uses these terms, a psychological form of motivation is being considered. Theorists of this type (Allport, G. A. Kelly, D. C. McClelland) tend to think of human motivation in terms of a relatively large number of personalized or idiosyncratic motives, so that each person's complement of motives is unique. These motives are usually regarded as conscious, as rationally chosen, and as capable of being modified through education or self-reflection. In contrast with biological motives, which are seen as driving behavior, psychological motives are seen as guiding and directing it. Consequently, psychological motives—the desire for fame, power, or revenge—are more subtle in their operation than the biological motives. *See* MOTIVATION.

Personality development. Again, personality theories differ considerably in the degree to which they deal with personality development. For some (Erikson), their theory of development *is* their theory of personality. Others, especially the behaviorists, rarely consider the subject.

Classic view. Among those theories concerned with personality development, two views prevail. In the first, classic developmental theory, personality development is seen as proceeding through a series of levels or stages which are usually correlated with a person's age. Personality is thought to change markedly as a person moves from one stage to the next; that is, the movement from one stage to the next is accompanied by a qualitative transformation in the structure of personality. Classic developmental theory also assumes that earlier experiences have a lasting impact on later development; those experiences that occur earliest are the most significant. Thus the events of the first year of life are often crucial for, and are reflected in, adult personality. The sources of adult neurosis, therefore, are located in infancy and early childhood according to this classic view. Europeans such as Freud, Jung, and Erikson by and large adopted this view of personality development.

Learning. In contrast to the classic view, many British and American theorists adopt what might be called the "development as learning" view. For them, most notably the factor analysts (R. B. Cattell, H. J. Eysenck) and behaviorists (Skinner, Bandura), development is linear rather than stagelike. No structural transformations take place during development; an adult differs from a child primarily in being more experienced, in having learned more ways of dealing with the world. Children are inexperienced adults, with essentially the same needs, capacities, and interests. In this case, development is not tied to or programmed by age; development is a function of sheer experience—the more experience, the more development. These theorists also tend to deny that the timing of experiences is important; early experiences are no more crucial than later experiences. The effects of early childhood trauma are readily overcome in later development. That which makes an experience important is its frequency, not its timing. The sources of adult neurosis, therefore, are not located in a person's childhood but in the current circumstances of that person's life. Since development is not stagelike, the trend of development can be altered at any time by changing a person's living circumstances.

Moral development. One aspect of personality development is so important that it has become a separate field of study in itself. The topic used to be called socialization, but now many psychologists focus on the problem of values and their transmission and on the moral implications of conformity, rebellion, and delinquency. Despite the seeming trendiness of this interest in moral development, it is actually a very old preoccupation among personality theorists—for example, it is the central problem in classic psychoanalytic theory. There are four problems to be distinguished with regard to moral development.

Effects of moralization. The first contrast concerns the effects of moralization on personality development. Psychoanalytic theory, for example, describes moral development as absolutely necessary for the survival of civilization but as devastating for the individual who undergoes it. The moralization process transforms an individual from an antisocial child to a civilized member of society, but at the cost of that child's spontaneous, natural tendencies, and ultimately at the cost of that person's happiness. To live in society, the natural instincts of a child must be ruthlessly suppressed. According to psychoanalysis, most pleasure comes from the gratification of instincts. By definition, then, living in society is unpleasurable and contrary to human instinctual nature. In contrast to psychoanalysis, classic sociological theory maintains that children become truly human only when they internalize the morality of their culture. Far from crushing or stultifying one's personality, the process of internalizing or incorporating the morals of one's family and culture, according to classic sociology, deepens and enriches one's personality and gives meaning and purpose to one's life. From this viewpoint, children of nature, happy savages, and persons who have somehow escaped from the "bondage" of civilization are shallow, empty, and less than human. It is ties to others, concern for their welfare, and a feeling of respect for the rules and traditions of one's culture that makes someone a whole person and a well-rounded personality.

Causes. The second problem concerns the causes of moral development—what instigates, triggers, or sets off this process? Some feel that moral development is somehow preprogrammed in the psyche, that it is an unfolding process that occurs naturally and spontaneously in every human infant. According to J. Piaget, for example, the child-training practices of well-meaning adults in reality only interfere with the moralization process. To the degree that adults intervene in children's spontaneous growth, they impede the normal process of moral development as it unfolds in each child. The alternative viewpoint is that without clear, direct, and even forceful adult intervention, children will remain forever childlike, antisocial, narcissistic, and impulsive—unfit as companions for themselves or other people. On this second view, best exemplified by Freud, there is no natural unfolding, no spontaneous evolution of personality. Parents and society are responsible for training children, and adult personality reflects the quality of this training.

Socialization. The third question concerns what it means to be socialized—what distinguishes a person whose moral development is normal from someone who has not developed normally? Traditionally there have been two answers to this. On the one hand, psychoanalysts and behaviorists think of moral development in terms of the number of rules and values that a person has internalized, and in terms of that person's capacity to feel guilt. A person with a well-developed conscience has internalized many rules and tends to feel guilty if breaking one of the rules. Conversely, criminals have internalized few rules and feel little guilt about breaking them. An alternative way of stating this first definition of moral development is in terms of a person's attitudes toward authority—to be well socialized means having positive attitudes toward authority and therefore being willing to comply with the requests and orders of authority figures. Conversely, persons who are poorly socialized are hostile to authority, and much of their behavior can be seen as defiance or rebellion against authority.

A second way of thinking about socialization has to do with social sensitivity, especially sensitivity to other persons' needs and expectations. Here, the more responsive one is to the expectations of one's family and other members of society, the more socialized one is and the more closely will one's actions conform to the rules and values of society. A person may feel guilty when violating these expectations. From this perspective, criminals are insensitive to the needs and expectations of others. Consequently, the ability to respond to others' expectations is seen as a fundamental part of what it means to be human.

Autonomy. The final question in theories of moral development concerns the topic of autonomy. For many theorists the defining feature of moral development is a person's willingness to conform to the norms and laws of family and culture. Conformity is essentially equated with moral development; psychoanalytic and behaviorist theories are good examples of this approach. The alternative, exemplified in the works of W. McDougall and the existentialists, emphasizes autonomy as the essential feature of moral behavior. These writers argue that when one conforms to the rules of society and to others' expectations, one is not acting in a truly moral way. That which is necessary for true moral behavior is personal, autonomous choice. Morally mature people may act contrary to the laws of their society and expectations of their families; in these cases they act autonomously.

Self concept. Personality theories differ greatly in terms of how they deal with the self concept. Behaviorists, psychoanalysts, and factor theorists tend to give the topic a skimpy treatment at best. Among those who deal with the self concept, it is usually seen as a central aspect of personality. Some, notably Jung, regard the self as a state of being—attaining selfhood is the major goal of life; others see the self as equivalent to personality (Allport, Carl Rogers); and a third group see the self concept as a view or image that one holds of oneself, which image serves to filter and screen action. But in this third case the self concept is just one aspect of personality. Thus, there are four ways of dealing with the self concept: minimize its importance; postulate it as a state of being; equate it with personality; and regard it as one component of personality.

Assuming that people can spell out their self-images, how do these images compare with how the individual would like to be viewed? The correspondence between real and ideal self-images is often regarded as an index of mental health. Phenomenologically oriented theorists (such as Rogers) use this correspondence as a way of evaluating the effects of psychotherapy—as one's adjustment improves, one's real self-image will become more like one's ideal self-image.

Unconscious processes. The idea that an unconscious but dynamically active mind operates outside the control of the individual and causes a person to do things he or she would not consciously do, is one of the oldest and most interesting notions in psychiatric theory and popular psychology. Although traditional psychiatric theory accepts the notion of a dynamic unconscious as an article of faith, other psychologists are suspicious of the idea. Not surprisingly, then, personality theories differ considerably in terms of the importance they assign to unconscious processes. Generally speaking the older the theory and the more it is concerned with explaining the origins of neurotic disorders, the more heavily will it appeal to unconscious processes. Conversely, the more recent the theory and the more it is concerned with problems other than neurotic disorders—creativity, self-actualization, personal soundness, status—the less will that theory be concerned with unconscious processes.

Among those theories that are concerned with unconscious mental processes, many issues are still disputed. Perhaps the most important issue concerns how thoughts, wishes, and desires become unconscious in the first place. There are probably four answers to this problem. Jung argues that the nature of the mind is such that a portion of it is and always will be unconscious, that the most that can be done is be alert for signs of unconscious thoughts and wishes leaking through into everyday conduct. Freud maintains that thoughts become unconscious through a direct action of the mind; one portion of the mind, an unconscious censor, decides that certain thoughts are unacceptable and blocks them out of consciousness by directing mental energy against them.

A third viewpoint, strongly rejected by Freud, is called dissociationism. This view, widely popular in the nineteenth century, gained prominence through the work of Ernest Hilgard. Here the notion is that the mind operates along separate channels, and the contents of consciousness depend on the channel one has tuned in; the contents of any channel will be unconscious so long as one is tuned in elsewhere. The final viewpoint comes from existentialism. Existentialists argue that people often lie to themselves about certain unpleasant facts (about being too fat, too lazy, and such). These lies include the lie that certain memories have been "forgotten," or that the person did not realize what he or she was doing or saying on particular occasions. The unconscious is composed of these lies.

Emotional dynamics. The topic of emotional dynamics concerns the various psychological factors that regulate one's life, how they are interrelated, and most importantly, how these factors can be put out of kilter. Ultimately, then, emotional dynamics concerns how and why people break down psychologically, how they become neurotic or even psychotic, and what may be done to help them. The study of emotional dynamics overlaps all the earlier central ideas—personality development, motivation, the self concept, and unconscious processes are involved to some degree in every explanation of why and how people break down.

Five approaches to the origin of neurosis can be found in the various theories of personality.

Classic psychodynamics. The first of these is the classic psychodynamic view that certain powerful instinctual impulses or potent emotions and memories somehow get sealed off in the unconscious mind where they then begin to fester like a psychic boil. Eventually these buried unconscious impulses and memories begin to affect everyday behavior, but always in peculiar ways that are self-defeating for the person involved. Such problems are relieved or cured by a careful, elaborate, and often time-consuming exploration of the person's unconscious mind and personality development. The goal is to bring the forgotten memories to consciousness where they may be analyzed and overcome.

Phenomenology. The second approach is the phenomenological perspective of such theorists as Jung, Rogers, and the existentialists. Here the assumption is that people are motivated by a need for personal growth and development, for self-actualization and transcendence. Unfortunately, according to these theories, the path of life is strewn with difficulties, and certain traumatic events in one's life can block the process of personal growth. Over time these frustrated strivings for wholeness and completeness turn into neurotic symptoms. The cure for neurosis, in these cases, is to remove the psychological forces that block the process of self-actualization.

Learning theory. The third approach comes from learning theory. Here the primary assumption is that the unconscious is irrelevant for understanding neurotic symptoms, because the origins of neurotic behavior are available to public inspection; they are learned patterns of behavior that turn out to be inappropriate and self-defeating. But because they are learned, neurotic symptoms can, in principle, be unlearned. Two kinds of neurotic behavior patterns can be learned. On the one hand, a person may learn to fear certain objects, people, or situations, although fear is inappropriate and may become very disruptive in later life. On the other hand, people can learn certain habits, certain ways of acting that may work in one special situation but be terribly inappropriate in other situations. So, for example, a child may learn that it can get its way with the parents by throwing tantrums. In later life, when that person's wishes are opposed by others, the person may react with anger, screams, and hostile emotional displays. In this second case, the maladaptive behavior pattern is like a bad habit that needs to be unlearned. The essential feature of any learning theory approach to the origins of neurotic behavior is that the source of the problem is the behavior itself—the problem is not buried in the unconscious mind. If the behavior can be changed, the problem will be eliminated. Learning approaches to the study of neurosis tend to challenge the validity of psychodynamic and phenomenological approaches.

Sociocultural theory. A fourth way of thinking about neurosis may be called the sociocultural theory, a viewpoint that comes from anthropology and sociology. Here mental illness is seen as a function of changes in society and the social environment. Wars, economic depressions, famines, earthquakes—any anthropogenic or natural disaster that uproots people, disturbs their standard of expected life style—will produce stress. When this stress reaches a certain level, disordered behavior breaks out in the population—alcoholism, drug addiction, violent crime, and mental illness are all symptoms of a society under pressure. Environmentally produced stress may be confined to a remote mountain village—a mudslide or flood may affect only a localized and very specific population. On the other hand, a major economic depression may affect the entire population of a country.

In sociocultural theory, emotional disorders, because they are caused by environmental conditions, are treated by modifying environmental factors. Abused children are placed in foster homes where presumably they will receive better care; underprivileged children are provided federally financed preschool education in order to stimulate their intellectual development. Within clinical psychology and psychiatry, group therapy and family therapy are based on sociocultural theory.

Biophysical model. A final way of thinking about emotional dynamics is the biophysical model. This model assumes that biological or neurological defects, not environmental stress or frustrated desires, produce mental disorders. This viewpoint is consistent with medical theory and practice, but not very popular with psychologists. It received its strongest support with the discovery around the turn of the century that a distinctive group of mental patients, classified dementia paralytica, were in fact suffering from an advanced state of syphilis. This finding led to the hope that some kind of neurological defect underlies all other forms of mental disorder, from anxiety attacks to schizophrenia.

Biophysical theory assumes biological causes for emotional disorders. Consequently, this theory recommends biophysical treatment for such disorders. Whatever the emotional problem, it will be treated with biochemical or physical means.

These six topic areas—motivation, personality development, moral development, the self concept, the unconscious, and emotional dynamics—form the substance of personality theory. All theories treat these six topics and differ largely in terms of how they handle them. Taken together, the various personality theories reflect the best modern judgment regarding the nature of human nature. *See* NEUROTIC DISORDERS; PSYCHOSIS. Robert Hogan

Bibliography. R. B. Ewen, *Introduction to Theories of Personality*, 4th ed., 1992; P. Fonagy and A. Higgitt, *Personality Theory and Clinical Practice*, 1985.

PERT

An acronym for program evaluation and review technique; a planning, scheduling, and control procedure based upon the use of time-oriented networks which reflect the interrelationships and dependencies among the project tasks (activities). The major objectives of PERT are to give management improved ability to develop a project plan and to properly allocate resources within overall program time and cost limitations; and to control the time and cost performance of the project, and to replan when significant departures from budget occur.

Background. In 1958 the U.S. Navy Special Projects Office, concerned with performance trends on large military development programs, introduced PERT on the Polaris weapons system. Since that time the use of PERT has spread widely throughout the United States and the rest of the industrialized countries. At about the same time that the Navy was developing PERT, the DuPont Company, concerned with the increasing cost and time required to bring new products from research to production, and to overhaul existing plants, introduced a similar technique called the critical path method (CPM).

Requirements. The basic requirements of PERT, in its time or schedule form of application, are the following:

1. All individual tasks required to complete a given program must be visualized in a clear enough manner to be put down in a network composed of events and activities. An event denotes a specified program accomplishment at a particular instant in time; in effect, it represents a state of the project system. An activity represents the time and resources that are necessary to progress from one event to the next. Emphasis is placed on defining events and activities with sufficient precision so that there is no difficulty in monitoring actual accomplishment as the program proceeds. **Figure 1** shows a typical operating-level PERT network from the electronics industry. Events are shown by squares, and activities are designated by arrows leading from predecessor to successor events.

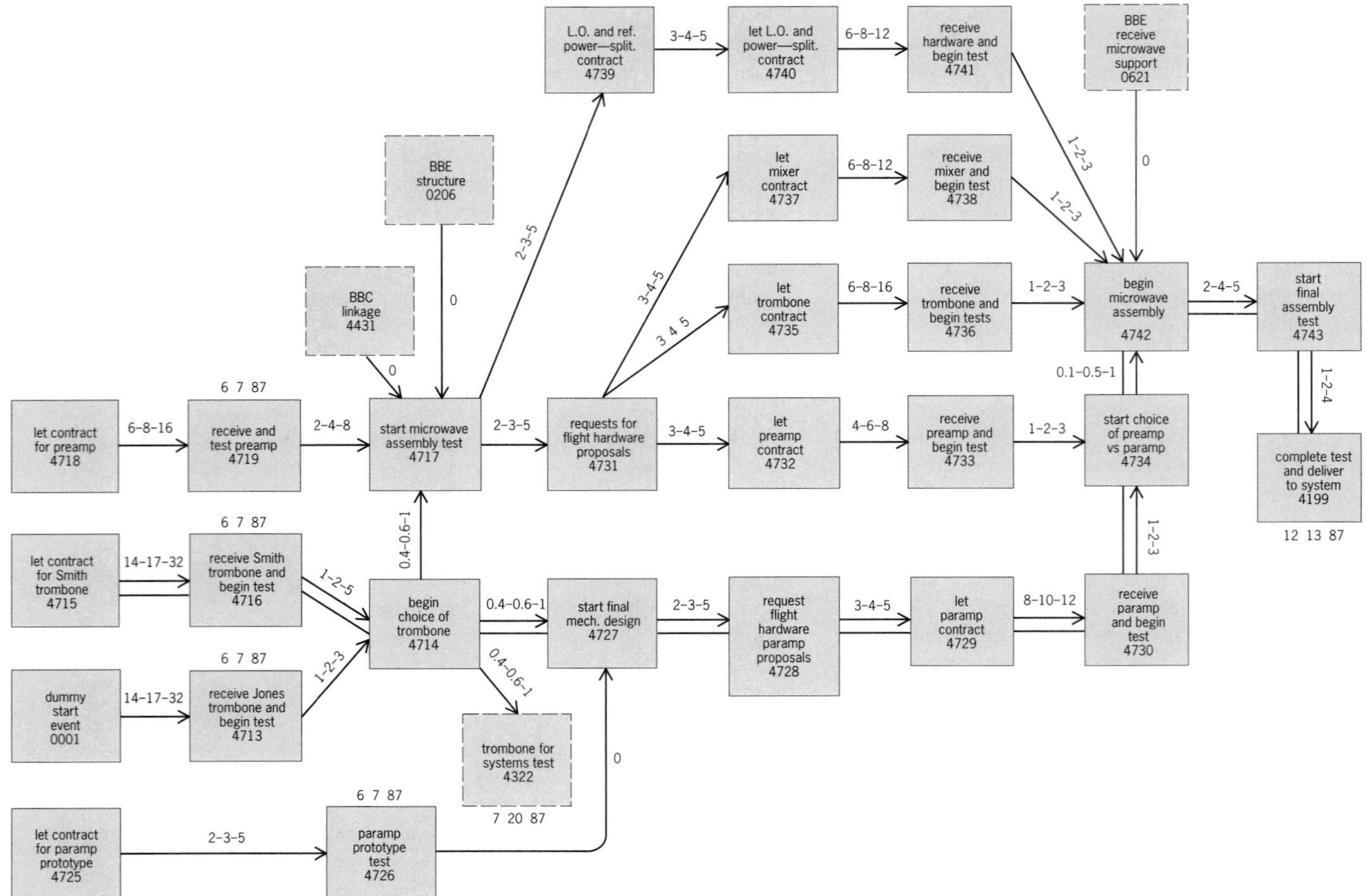

Fig. 1. Typical PERT network of an electronic module development project. Arrows and lines denote the critical path. (*Applied Physics Laboratory*, *Johns Hopkins University*)

2. Events and activities must be sequenced on the network under a logical set of ground rules. The activity sequencing is not arbitrary, but rather it is based on technological constraints; a foundation must be dug before the concrete can be poured. The network logic is merely the requirement that an event is said to occur when all predecessor activities are completed, and only then can the successor activities begin. The initial event, without predecessors, is self-actuated when the project begins, and the occurrence of the final event (without successor activities) denotes completion of the project. This logic requires that all activities in a network must be completed before the project is complete, and no "looping" of activities in the network is allowed. Another technique, called GERT, relaxes these logic constraints.

3. Time estimates can be made for each activity of the network on a three-way basis (the three numbers shown along the arrows in Fig. 1). Optimistic (minimum), most likely (modal), and pessimistic (maximum) performance time figures are estimated by the person or persons most familiar with the activity involved. The three-time estimates are used as a measure of uncertainty of the eventual activity duration; they represent the approach used in PERT to express the probabilistic nature of many of the tasks in development-oriented and nonrepetitive programs. It is important to note, however, that for the purposes of critical path computation and reporting, the three-time estimates are reduced to a single expected time T_E, and it is used in the same way that CPM employs a single (deterministic) time estimate of activity duration time.

4. Finally, critical path and slack times are computed. The critical path is that sequence of activities and events on the network that will require the greatest expected time to accomplish. Slack time is the difference between the earliest time that an activity may start (or finish) and its latest allowable start (or finish) time, as required to complete the project on schedule. Thus, for any event, it is a measure of the spare time that exists within the total network plan. If total expected activity time along the critical path is greater than the time available to complete the project, the program is said to have negative slack time. This figure is a measure of how much acceleration is required to meet the scheduled program completion date.

5. The difference between the pessimistic (b) and optimistic (a) activity performance times is used to compute the standard deviation (σ) of the

RUN 1 ENDING EVENT
BY PATHS OF CRITICALITY
CHART AJ LR SN 9 ELECTRONIC MODULE (ILLUSTRATIVE NETWORK) SYSTEM W 034 DATE 06-07-87

EVENT PREDECESSOR	SUCCESSOR	NOMENCLATURE	DEP	DATE EXPECTED	ALLOWED	DATE SCHD/ACT	PROB	SLACK	EXP TIME	EXP VAR
4004-715	4004-716	REV DATE (SMITH TROMBONE RECD-BEG TEST)	98		05-26-87	A06-07-87		-1.6	+	
4004-716	4004-714	SMITH TROMBONE TESTED	0146	06-23-87	06-12-87			-1.6	+ 2.3	.4
4004-714	4004-727	TROMBONE CHOSEN-BEGIN MECH DESIGN	0146	06-28-87	06-16-87			-1.6	+ 3.0	.5
4004-727	4004-728	RFP PARAMP FLIGHT HARDWARE	0146	07-20-87	07-08-87			-1.6	+ 6.1	.7
4004-728	4004-729	PARAMP CONTRACT LET	0146	08-17-87	08-05-87			-1.6	+10.1	.8
4004-729	4004-730	PARAMP RECEIVED		10-26-87	10-14-87			-1.6	+20.1	1.3
4004-730	4004-734	PARAMP TESTED	0146	11-09-87	10-28-87			-1.6	+22.1	1.4
4004-734	4004-742	CHOICE BETWEEN PREAMP-PARAMP	0146	11-13-87	11-01-87			-1.6	+22.6	1.4
4004-742	4004-743	COMPL MICROWAVE ASSY	0146	12-09-87	11-28-87			-1.6	+26.5	1.6
4004-743	4004-199	COMPL FINAL TEST MICWAVE ASSY-DELIVERED	0146	12-25-87	12-13-87	12-13-87	.12	-1.6	+28.6	1.9
4000-001	4004-713	REV DATE (JONES TROMBONE RECD-BEG TEST)	99		05-29-87	A06-07-87		-1.3	+	
4004-713	4004-714	JONES TROMBONE TESTED	0146	06-21-87	06-12-87			-1.3	+ 2.0	.1
4004-714	4004-717	TROMBONE CHOSEN-BEGIN MICWAVE ASSY TEST	0146	06-28-87	07-02-87			+ .5	+ 3.0	.5
4004-717	4004-731	RFP FOR FLIGHT HDW-MIXER-TROMB-PREAMP	0146	07-20-87	07-24-87			+ .5	+ 6.1	.7
4004-717	4004-739	COMPL MICWAVE ASSY TEST-RFP LOC OSCIL	0146	07-20-87	07-24-87			+ .5	+ 6.1	.7
4004-731	4004-735	TROMBONE CONTRACT LET	0146	08-17-87	08-21-87			+ .5	+10.1	.8
4004-731	4004-737	MIXER CONTRACT LET	0146	08-17-87	08-21-87			+ .5	+10.1	.8
4004-739	4004-740	CONTRACT LET FOR LOC OSCIL AND PWR SPLT	0146	08-17-87	08-21-87			+ .5	+10.1	.8
4004-735	4004-736	TROMBONE RECEIVED		10-14-87	10-18-87			+ .5	+18.5	1.8
4004-737	4004-738	MIXER RECEIVED		10-14-87	10-18-87			+ .5	+18.5	1.8
4004-740	4004-741	LOC OSC-PWR SPLITTER RECEIVED		10-14-87	10-18-87			+ .5	+18.5	1.8
4004-736	4004-742	TROMBONE TESTED	0146	10-28-87	11-01-87			+ .5	+20.5	1.9
4004-738	4004-742	MIXER TESTED	0146	10-28-87	11-01-87			+ .5	+20.5	1:9
4004-741	4004-742	LOC OSC-PWR SPLITTER TESTED	0146	10-28-87	11-01-87			+ .5	+20.5	1.9

Fig. 2. Typical PERT computer output. First three paths of Fig. 1 are shown here. (*After J. J. Moder and C. R. Phillips, Project Management with CPM and PERT, 2d ed., Van Nostrand–Reinhold, 1970*)

hypothetical distribution of activity performance times [$\hat{\sigma} = (b - a)/6$]. The PERT procedure employs these expected times and standard deviations (σ^2 is called variance) to compute the probability that an event will be on schedule, that is, will occur on or before its scheduled occurrence time. The procedure merely adds the expected time (T_E) and variances (σ^2) of the activities on the critical path to get the mean and variance of the hypothetical distribution of project duration times. [See columns headed EXP TIME and EXP VAR in **Fig. 2**, and the total values of 28.6 and 1.9, respectively, for the last activity (4004-743 to 4004-199) on the critical path.] The normal distribution is then used to approximate the probability of meeting the project schedule, as the area under the normal distribution curve to the left of (earlier than) the project scheduled completion date. *See* GERT.

A computer-prepared analysis of the illustrative network contains data on the critical path (first group of activities in the table) and slack times for the other, shorter network paths. Note that the events (points in time) are labeled in the network, but the computer output is by activities (identified by event numbers) which also have descriptive labels. Under the column heading PROB(ability), note the figure 0.12 for the final activity on the critical path (4743-4199). This analysis indicates that the expected completion time of 12/25/87 results in a low probability of meeting the scheduled time of 12/13/87. This computer output (slack order report) is the most important of a number of outputs provided by most PERT computerized systems. Other reports may give greater or lesser details for different levels of management; they may deal with estimated and actual costs by activities (system called PERT COST); and so forth.

In the actual utilization of PERT, review and action by responsible managers is required, generally on a biweekly basis, concentrating on important critical path activities. Where necessary, effective means of shortening critical path time must be found by applying new resources or additional funds, often obtained from those activities that can afford them because of their slack condition. Alternatively, sequencing of activities along the critical path may be compromised to reduce overall duration. A final alternative may be a change in the scope of the work along the critical path to meet a given program schedule. Utilization of PERT requires constant updating and reanalysis, since the outlook for completion of activities in a complex program is constantly changing. Systematized methods of handling this aspect have been developed.

Advantage. A major advantage of PERT is the kind of planning required to create an initial network. Network development and critical path analysis reveal interdependencies and problem areas before the program begins that are often not obvious or well defined by conventional planning methods. Another advantage, especially where there is a significant amount of uncertainty, is the three-way estimate. If there is a minimum of uncertainty, the single-time approach may be used while retaining the advantages of network analysis.

In summary, it should be stated that while the developments of PERT and CPM were independent, they are both based on the same network logic to represent the project plan. PERT emphasizes the time performance of a project, including a probabilistic treatment of the uncertainty in the activity performance times and scheduled completion dates, while CPM treats time deterministically and addresses the problem of minimizing total (direct plus indirect) project cost as a function of scheduled project duration. The acronym PERT has become a generic term for network-based project management schemes that have evolved over the years. These schemes are hybrids of both PERT and CPM, but they are most often referred to as PERT. Finally, mention should be made of GERT, which denotes a generalization of the PERT/CPM network logic to complex situations where branching at events and closed loops of activitites are required to adequately portray a complex project plan in the form of a network. *See* ACTIVITY-BASED COSTING. Joseph J. Moder

Bibliography. D. G. Malcolm et al., Applications of a technique for R and D program evaluation, *Operations Res.*, 7(5):646–669, 1959; J. J. Moder and S. E. Elmaghraby, *Handbook of Operations Research, Models and Applications*, vol. 2, 1978; J. J. Moder and C. R. Phillips, *Project Management with CPM and PERT*, 3d ed., 1995.

Perthite

Any of the oriented intergrowths of potassium- and sodium-rich feldspars, $(K,Na)AlSi_3O_8$, whose proportions are determined in part by the initial composition of the alkali feldspar from which they exsolved and whose physical properties are thus somewhat variable. The early stages of perthite formation from homogeneous, usually monoclinic $(K,Na)AlSi_3O_8$ may be observed experimentally by high-magnification electron microscopy.

Spinodal decomposition and homogeneous nucleation are the dominant mechanisms of exsolution, producing compositional modulations that initially are lamellar in nature and coarsen with increasing time and temperature of annealing from 10 to 50 nanometers or more. The coherent K- and Na-rich lamellae are subparallel and have interfaces oriented within a few degrees of the $(\bar{8}01)$ or $(\bar{6}01)$ crystallographic planes, a consequence of minimizing the elastic strain energy between them. Details of the exsolution process vary from one bulk composition and rock annealing history to another, and the anorthite $(CaAl_2Si_2O_8)$ component is more important in some perthites than others. Water has a critical effect on cation diffusion rates and the kinetics affecting the final textures. One suggested path for the coarsening of a cryptoperthite from the earliest stages of exsolution is shown in the **illustration**. If the final K- and Na-rich lamellae or particles are submicroscopic, the composite feldspar is called cryptoperthite. If the particles are small enough and the feldspar relatively clear, Rayleigh-type scattering of light may occur, giving rise to the beautiful blue-to-whitish luster of the semiprecious gem called moonstone. If coarsening has progressed to the micrometer scale and can be seen on a polarizing microscope, the composite is called microperthite; and if the two feldspars are visible to the eye in hand specimen, it

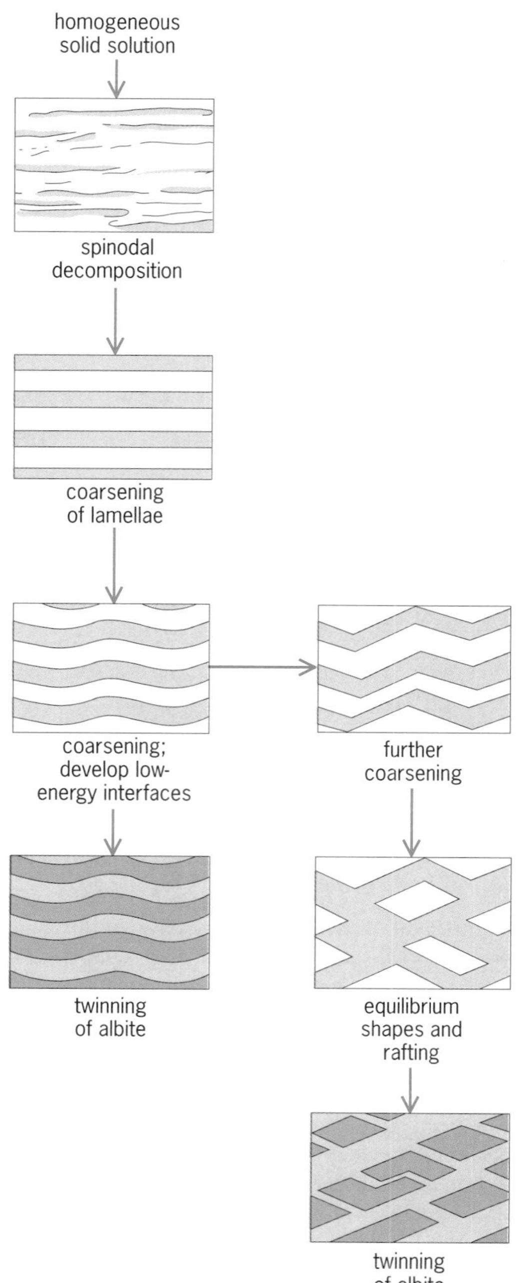

Schematic showing the development of two perthitic intergrowth textures. (*After G. W. Lorimer and P. E. Champness, The origin of the phase distribution in two perthitic alkali feldspars, Philos. Mag., 28:1401, 1973*)

is called perthite or macroperthite. Often the albite phase will appear as white veins or blotchy patches against a colored K-rich phase, which may be green to blue microcline (amazonite) or dull pink to orange-brown orthoclase. *See* ALBITE; ANORTHITE; ANORTHOCLASE; CRYSTAL STRUCTURE; FELDSPAR; IGNEOUS ROCKS; MICROCLINE; ORTHOCLASE.

Paul H. Ribbe

Perturbation (astronomy)

Departure of a celestial body from the trajectory it would follow if moving only under the action of a single central force. Perturbations may be caused by either gravitational or nongravitational forces.

Corrections to elliptic orbits. In the solar system, orbits of planets may be adequately represented by mean elliptical elements to which are added small corrections due to the mutual planetary attractions. Although such motion is referred to as disturbed, it is as much a consequence of the law of gravitation as is undisturbed elliptic motion.

Another method of representing perturbed motion is to augment the position derived from the mean ellipse by the actual displacements in the coordinates due to the disturbing forces. These perturbations of the elements, and perturbations of the coordinates, are represented by infinite series; usually many terms are required to represent the disturbed motion accurately. These analytical expressions are referred to as general perturbations and, with their associated mean elements, form a general theory of the motion. In some instances, such as the orbit of an outer satellite of Jupiter or the motion of a comet moving with nearly parabolic velocity, the analytical expressions representing the perturbing function become so involved (mainly because of lack of convergence of the Fourier series) that general perturbations are not attempted. Instead, the perturbed positions are computed from a step-by-step numerical integration of the equations of motion; this is known as the method of special perturbations.

Long- and short-term disturbances. Planetary orbits are subject to two classes of disturbances: secular, or long-term, perturbations; and periodic, or relatively short-term, perturbations. Secular perturbations, so called because they are either progressive or have excessively long periods, arise because of the relative orientation of the orbits in space. They cause slow oscillatory changes of eccentricities and inclinations about their mean values with accompanying changes in the motions of the nodes and perihelia. The periods of time involved in these oscillations may extend from 50,000 to 2,000,000 years. Periods and major axes of orbits are not affected by secular change. For the orbit of the Earth, the present inclination to the invariable plane is 1°35′. This will diminish to a minimum of 47′ in approximately 20,000 years. The eccentricity, presently 0.017, is diminishing also and will reach a minimum of 0.003 in about 24,000 years.

Periodic perturbations arise from the relative positions of the planets in their orbits. When the disturbed and disturbing planets are aligned on the same side of the Sun, the perturbation reaches a maximum, and reduces to minimum when alignment is reached on opposite sides of the Sun. The size of a periodic perturbation is a function of the mass of the disturbing body and of the length of time the two planets remain near the point of closest approach. Periodic perturbations continually shift a planet away from the position it would occupy in undisturbed motion, moving it above or below the orbital plane, nearer to or farther from the Sun, and forward or backward in the orbit.

Commensurable motions. If the mean motion of the disturbed planet were exactly a submultiple, say $^1/_2$, of the mean motion of the disturbing planet, the maximum perturbation produced by their close approach would always occur in the same part of the disturbed orbit. The displacement in position of the disturbed planet would increase with each coincidence until the character of the orbit became modified to the point where exact commensurability of the mean motions would cease to exist.

Because the solar system is middle-aged, cosmically speaking, few examples of commensurability of mean motions exist today. None is found in the motions of the major planets. Cases of near commensurability exist which give rise to long-period periodic terms of large amplitude. As an example, the periods of Jupiter and Saturn are nearly in the ratio of 2:5. Thus, after nearly five revolutions of Jupiter, the two planets return to approximately the same juxtaposition. Their line of coincidence, however, sweeps slowly around Jupiter's orbit, completing a circuit in about 850 years and thus producing a perturbation of this period.

Among the four inner planets, the periodic perturbations are small, amounting in orbital longitude at most of 0.25′ for Mercury, 0.5′ for Venus, 1′ for Earth, and 2′ for Mars. Periodic perturbations of the outer planets are larger, reaching in the case of the long-period terms to 30′ for Jupiter, 70′ for Saturn, 60′ for Uranus, and 35′ for Neptune.

Because the amplitude of a periodic perturbation depends on the mass of the disturbing planet, observational measurement of this amplitude affords a method of determining the disturbing mass. For the planets Mercury, Venus, and Pluto, which do not have satellites, this is the only method of determining the mass. As a consequence of the mutual perturbations of the planets, the distance of a planet from the Sun is, on the average, decreased by the action of planets closer to the Sun, and increased by planets farther from the Sun; this mean effect represents a perturbation of the radius vector with a constant value.

The orbits of the minor planets are affected in varying degree by the attractions of the major planets. Those orbits passing close to Jupiter suffer large perturbations which, if the mean motions were commensurable with that of Jupiter, would be augmented at each close approach until the trajectories were sufficiently altered to reduce the commensurability. In the overall distribution of mean motions of the minor planets there are noticeable gaps near the points where the period would be an exact submultiple ($^1/_2$, $^1/_3$ $^2/_5$, . . .) of the period of Jupiter. In cases of near commensurability, observational determination of the amplitude of the long-period perturbation affords a method for measuring the mass of Jupiter. A small group of minor planets, called the Trojan asteroids, has been so completely captured by Jupiter that they oscillate about the 60° points which form equilateral triangles with Jupiter and the Sun. *See* TROJAN ASTEROIDS.

Effect on comets. Planetary perturbations also affect the orbits of comets. Studies of the motion of Halley's comet indicate that the time from one perihelion passage to the next has varied by almost 5 years because of perturbations. Most comets approach the Sun at nearly parabolic speeds in randomly oriented orbits, but if a comet approaches close to one of the more massive major planets, the planet may so alter the trajectory that the comet pursues an elliptical orbit thereafter. A number of short-period comets whose orbits agree only in that they all pass close to Jupiter illustrate the perturbing effect of this planet on cometary orbits.

Nongravitational causes. Material forming the tails of comets is subject to a nongravitational type of perturbation. This rarefied matter which is given off by the head of the comet is forced into a trajectory away from the Sun by the pressure of solar radiation.

Associated with many of the periodic comets are swarms of smaller particles which appear as meteors upon collision with the upper atmosphere of the Earth. The density of these swarms is so tenuous that they cannot hold themselves together by their own gravitation, and planetary perturbations of speed and direction soon spread the components completely around the orbit. The annual meteor showers, such as the Perseids, reflect this dispersal of particles along the orbit. The effect of the Earth's attraction on a meteor trajectory depends on the relative velocity, that is, whether the Earth is overtaking the meteor or meeting it head on. Once the meteor enters the upper reaches of the Earth's atmosphere its motion is subject to a nongravitational perturbation caused by atmospheric drag. This resistance to the passage of the particle is evidenced by the trail of incandescent gas and vapor which forms until the particle is consumed or continues in its trajectory greatly decelerated. *See* METEOR.

Perturbations of satellite orbits. The motions of planetary satellites, natural and artificial, reflect both gravitational and nongravitational perturbations. The centrifugal force arising from the rotation of a planet causes a deformation or oblateness of figure. In such a case the central mass does not attract as if it were concentrated at its center. For a close satellite the principal perturbation arises from the attraction of this equatorial bulge. The effect of this attraction on an otherwise undisturbed satellite orbit is a gradual regression of the line of nodes on the equatorial plane and a rotation of the line of apsides. Both rotations vary with the inclination of the satellite orbit. Nearer to the primary, the tidal forces may become so great that a satellite would be literally torn to pieces. For a fluid satellite of the same density as the planet, the limit within which this disruptive perturbation occurs is about $2^1/_2$ times the radius of the planet. *See* SATURN.

Satellite motions are also disturbed by the direct attraction of other satellites, the Sun, and, to a lesser amount, by other planets. Observation of the orbital displacements caused by the mutual perturbations of satellites in the systems of Jupiter and Saturn makes possible the determination of the masses of these satellites. The solar attraction is significant in the orbits of the outer satellites of Jupiter and Saturn, reaching to one-ninth the planet's attraction for the eighth satellite of Jupiter. So greatly disturbed is this satellite that it is not possible to derive a general theory for its motion.

The orbit of the Moon is disturbed mainly by the Sun, with some changes in motion due to the oblateness of the Earth, the figure of the moon, and smaller perturbations caused by the planets. The attraction of the Sun on the Moon is more than twice the Earth's attraction, but because both the Earth and Moon are free to move it is only their relative acceleration with respect to the Sun which determines the motion. This relative acceleration toward the Sun is always less than $^1/_{80}$ of the acceleration of the Moon toward the Earth. The eccentricity and inclination of the Moon's orbit oscillate slowly about their mean values, while the line of apsides advances with an average period of almost 9 years and the nodes regress through one revolution in 18.6 years.

The observed motion of the lunar node and perigee affords one means of measuring the oblateness of the Earth. The present lunar theory incorporates the value 1/294. The *International Astronomical Union System of Astronomical Constants* (1976) contains a reference ellipsoid of revolution for the Earth having a flattening of 1/298.257. This value has been derived mainly from measures of the motions of the nodes and apsides of artificial Earth satellites. Lunar and solar perturbations of artificial Earth satellite orbits are minor for orbits 500 mi (800 km) above the surface but grow with increasing distance from the Earth. Atmospheric drag perturbations are significant at this altitude, but decrease with increasing altitude. *See* CELESTIAL MECHANICS.

Raynor L. Duncombe

Bibliography. J. M. A. Danby, *Fundamentals of Celestial Mechanics*, 2d ed., 1988; J. E. Prussing and B. A. Conway, *Orbital Mechanics*, 1993; A. E. Roy, *Orbital Motion*, 3d ed., 1988; P. K. Seidelmann (ed.), *Explanatory Supplement to the Astronomical Almanac*, 1992; V. Szebehely, *Adventures in Celestial Mechanics: A First Course in the Theory of Orbits*, 1989.

Perturbation (mathematics)

A modification in the mathematical structure of a problem changing the problem from one that can be solved exactly, the unperturbed problem, to one, the perturbed problem, for which it is usually possible to obtain only an approximate solution. The methods employed for this purpose form perturbation theory. These methods attempt to express the solution of the perturbed problem in terms of the properties of the solutions of the unperturbed problem.

Examples. Examples of perturbation problems can be found in nearly every branch of mathematics and physics, and in astronomy. The simplest case occurs in ordinary algebra. Suppose that the roots of the equation $f(x) = 0$ are known (the unperturbed problem), and that the roots of the equation $f(x) + \epsilon g(x) = 0$ are to be found (the perturbed problem). The parameter ϵ measures the size of the perturbation. Another set of examples occurs in linear differential equations and in particle dynamics. Possible perturbations include changes in the forces considered to be acting on the particle as well as changes in initial conditions. *See* PERTURBATION (ASTRONOMY).

Several examples occur in partial differential equations. One physical realization occurs in the theory of wave propagation where the perturbations can be changes in the index of refraction, changes in initial conditions, or changes in the nature or shape of the surfaces encountered by the waves. All of these changes can occur separately or concurrently. The first of these changes is called a volume perturbation, the second a perturbation of initial conditions, and the third a perturbation of boundary conditions. Similar examples can be taken from quantum mechanics, where the volume perturbation corresponds to a change in the hamiltonian, and perturbation of initial conditions to quantum-mechanical time-dependent perturbation theory. Other partial differential equations of physics, such as the Laplace equation, the diffusion equation, and the equations of hydrodynamics, furnish further examples. *See* PERTURBATION (QUANTUM MECHANICS).

As a final illustration of these various types of perturbation, consider possible modifications in an equation (as well as boundary and initial conditions) describing the motion of particles such as neutrons or electrons moving through a medium which can scatter and absorb them. The equation is known as the Lorentz-Boltzmann equation and changes in it occur as a consequence of modifications of the laws of scattering and absorption, that is, because of changes in the medium.

All of these problems are linear and can therefore be cast into an equation of the form $A\psi = \lambda\psi$, where ψ is the unknown quantity, λ is a constant, and A is an operator involving among other possibilities differentiation and integration. The quantity ψ may be a scalar, a vector, or more generally a matrix quantity. When solutions can be obtained for only special values of λ, the eigenvalues, the equation is called the eigenvalue equation, and the associated problem is called the eigenvalue problem. The operator A contains the perturbation; that is, A equals $A_0 + \epsilon A_1$, where A_0 is the unperturbed operator and ϵA_1, the perturbing term.

Iteration method. The method generally employed to obtain an approximate solution is called the iteration method. Rewrite the equation $A\psi = \lambda\psi$ as $(A_0 - \lambda)\psi = -\epsilon A_1\psi$. Let the unperturbed solution be ϕ_0, where $(A_0 - \lambda_0)\phi_0 = 0$. Then ψ_1, a first approximation to ψ, is obtained as a solution of $(A_0 - \lambda)\psi_1 = -\epsilon A_1\phi_0$. A second approximation is the solution of $(A_0 - \lambda)\psi_2 = -\epsilon A_1\psi_1$. The nth approximation is obtained in terms of the $(n-1)$ approximation, ψ_{n-1}, from the equation $(A_0 - \lambda)\psi_n = -\epsilon A_1\psi_{n-1}$. It is assumed that the properties of the unperturbed operator, A_0, are completely known so that the solution of these equations can be obtained. If the sequence $\phi_0, \psi_1, \ldots, \psi_n \ldots$, converges, it will converge to a solution of the problem. For an eigenvalue problem, the procedure must be modified. The first approximation to λ is λ_0; the nth approximation is λ_n. Then the equation determining ψ_n in terms of ψ_{n-1} is $(A_0 - \lambda_{n-1})\psi_n = \epsilon A_1\psi_n$. It is important for the practicality of this procedure that the approximation λ_n can be expressed in terms of the approximation ψ_{n-1} and the operators A_0 and ϵA_1.

In a related and more familiar formulation both ψ and λ are expanded in a power series in ϵ; that is, $\psi = \phi_0 + \epsilon\phi_1 + \epsilon_2\phi_2 + \cdots$ and $\lambda = \lambda_0 + \epsilon\lambda_1 + \epsilon_2\lambda_2 + \cdots$. Then the equation $A\psi = \lambda\psi$ reduces to a set of equations for ϕ_n. For example, the equation for ϕ_1 is $(A_0 - \lambda_0)\phi_1 = -(A_1 - \lambda_1)\phi_0$ and the equation for ϕ_2 is $(A_0 - \lambda_0)\phi_2 = -(A_1 - \lambda_1)\phi_1 + \lambda_2\phi_0$, and so on. This formulation is more complex, and often yields slower rates of convergence than the method just outlined.

The iteration method can be generalized in two respects. First, it is not necessary to use ϕ_0 as the zeroth approximation to ϕ. If by reason of other information a better approximation, say ψ_0, is known, the iteration sequence starts with the equation $(A_0 - \lambda)\psi_1 = -\epsilon A_1\phi_0$. Second, the iteration method can be employed in the treatment of nonlinear as well as the linear problems discussed in detail here.

For the iteration method to be at all possible, it is necessary for the sequence $\psi_0, \psi_1, \ldots$ to exist and to converge. The usefulness of the method increases with increasing rate of convergence. The sequence exists only if the singularities of the perturbation are not too strong, or if the initial zero approximation is properly chosen, or both. When the sequence exists, it will converge for a range in values of the parameter ϵ. The largest value of ϵ is the radius of convergence. This is found to be that value of ϵ for which the equation $A\psi = \lambda\psi$ has at least two degenerate solutions, that is, solutions with identical values of λ. There are various methods of increasing the radius of convergence. For example, the general techniques of analytic continuation, such as the Euler transformation, can often be employed. A clever choice of ψ_0, the zeroth approximation, will often produce the desired effect. The variational method can generate the appropriate choice for ψ_0. A more general method was developed by I. Fredholm in which the solution of $A\psi = \lambda\psi$ is given as the ratio of two functions, each of which can be expressed as a series of ϵ. For a wide class of operators A, each of these series will have an infinite radius of convergence.

The eigenvalue problem can be reduced to the problem of the solution of a set of homogeneous linear simultaneous equations which is generally but not always infinite. A nontrivial solution of these equations is possible only if the determinant of the coefficients is zero. Because the coefficients involve the eigenvalue λ, this condition yields an equation, the secular equation, which determines the possible values of λ. The determinant is known as the secular determinant. An example follows. Let the solutions of the unperturbed problem be $\phi^{(p)}$ with eigenvalues $\lambda^{(p)}$; that is, $A_0\phi^{(p)} = \lambda^{(p)}\phi^{(p)}$. Moreover, suppose that the set $\phi^{(p)}$ is complete, which roughly means that an arbitrary function can be represented as a linear combination of $\phi^{(p)}$. Therefore let ψ, the solution of the perturbed problem, be $C_0\phi^{(0)} + C_1\phi^{(1)} + C_2\phi^{(2)} + \cdots$, where C_p are constants. By substituting this expression for ψ in the equation $A\psi = \lambda\psi$ and employing the properties of the set $\phi^{(p)}$ which follow from the nature of the operator A_0, it is possible to obtain a set of equations for C_p. In a typical case these equations have the form: and so on. The

$$C_0(\lambda - \lambda^{(0)}) + C_1(\epsilon A_1)_{10} + C_2(\epsilon A_2)_{20} + \cdots = 0$$

$$C_0(\epsilon A_1)_{01} + C_1(\lambda - \lambda^{(1)}) + C_2(\epsilon A_1)_{21} + \cdots = 0$$

$$C_0(\epsilon A_1)_{02} + C_1(\epsilon A_1)_{12} + C_2(\lambda - \lambda^{(2)}) = 0$$

elements $(\epsilon A_1)_{pq}$ are numbers which depend upon $\phi^{(p)}$, $\phi^{(q)}$, and the operator ϵA_1. The consequent secular equation is obtained by setting the determinant of the coefficients of C_p in this sequence of equations equal to zero. The solution of these simultaneous equations for the coefficients C_p can be obtained by the iteration method, which yields a particular representation of each of the approximations ψ_n. If there are only a finite number of C_p, the secular determinant is of finite order and reduces to a finite polynomial in λ so that in the finite case solutions of the secular equation can always be obtained without using perturbation methods. Once allowed values of λ are known, corresponding values of C_p can be determined. *See* EIGENVALUE (QUANTUM MECHANICS).

Degenerate perturbation theory. A special technique is required when the unperturbed problem is degenerate, that is, when there are several solutions of the equation $A_0\phi = \lambda\phi$ for a single value of the eigenvalue λ. The number of such independent solutions is the order of the degeneracy. The corresponding method is designated degenerate perturbation theory. The objective of the special method adopted for this case is the determination of the appropriate linear combinations of these degenerate solutions for use as the initial approximation, ψ_0, in the iterative method. To this end, all terms of the equations for C_p are dropped when p refers to an unperturbed solution which is not one of the degenerate solutions under consideration, and only those C_p which do refer to these degenerate solutions are retained. The resulting secular equation for λ has a number of roots equal to the order of the degeneracy. For each root there is a corresponding set of values for C_p which determine a particular linear combination of the degenerate unperturbed solutions. Each of these combinations can be employed as the initial approximation, ψ_0, in the iterative method. It is often the case that the determination of the possible ψ_0 is sufficient for the evaluation of the major effects of the perturbation. Herman Feshbach

Bibliography. H. Baumgartel, *Analytic Perturbation Theory for Matrices and Operators*, 1985; E. J. Hinch, *Perturbation Methods*, 1991; J. Kevorkian and J. D. Cole, *Perturbation Methods in Applied Mathematics*, 1991; P. M. Morse and H. Feshbach, *Methods of Theoretical Physics*, 1953; A. H. Nayfeh, *Introduction to Perturbation Techniques*, 1981, reprint 1993.

Perturbation (quantum mechanics)

An expansion technique useful for solving complicated quantum-mechanical problems in terms of solutions for simple problems. Perturbation theory in quantum mechanics provides an approximation scheme whereby the physical properties of a system, modeled mathematically by a quantum-mechanical description, can be estimated to a required degree of accuracy. Such a scheme is useful because very few problems occurring in quantum mechanics can be solved analytically. Consequently an approximation technique must be employed in order to give an approximate analytic solution or to provide suitable algorithms for a numerical solution. Even for problems which admit an exact analytic solution, the exact solution may be of such mathematical complexity that its physical interpretation is not apparent. For these situations, perturbation techniques are also desirable. Here the discussion of the application of perturbation techniques to quantum mechanics will be limited to the domain of nonrelativistic quantum theory. Applications of a similar but mathematically more intricate nature have also been made in quantum electrodynamics and quantum field theory. *See* NONRELATIVISTIC QUANTUM THEORY; QUANTUM ELECTRODYNAMICS; QUANTUM FIELD THEORY; QUANTUM MECHANICS.

Perturbation theory is applied to the Schrödinger equation, $H\Psi = (H_0 + \lambda V)\Psi = i\hbar(\partial/\partial t)\Psi$ [where $\hbar$ is Planck's constant h divided by 2π, and $(\partial/\partial t)$ represents partial differentiation with respect to the time variable t], for which the exact hamiltonian H is split into two parts: the approximate (unperturbed) time-independent hamiltonian H_0 whose solutions of the corresponding Schrödinger equation are known analytically, and the perturbing potential λV. The basic idea is to expand the exact solution Ψ in terms of the solution set of the unperturbed hamiltonian H_0

by means of a power series in the coupling constant λ. Such a procedure is expected to be successful if the system characterized by the unperturbed hamiltonian closely resembles that characterized by the exact hamiltonian. Supposedly the differences are not singular in character, but change as a continuous function of the parameter λ.

Perturbation theory is used in two contexts to provide information about the state of the system, which in quantum mechanics is determined by the wave function Ψ. If λV is time-independent, an objective may be to find the stationary states of the system Ψ_n whose time dependence is given by $\exp(-iE_n t/\hbar)$, where $i = \sqrt{-1}$ and E_n represents the energy of the stationary state labeled by n. If λV is either time-independent or time-dependent, an objective may be to find the time evolution of a state which at some specified time was a stationary state of the unperturbed hamiltonian. The perturbing potential is then considered as causing transitions from the original state to other states of the unperturbed hamiltonian, and application of time-dependent perturbation theory provides the probability of such transitions.

Time-independent perturbation theory. In time-independent perturbation theory, the stationary state $\Psi_n = \psi_n \exp(-iE_n t/\hbar)$, which satisfies the time-independent Schrödinger equation (1), is solved

$$(H_0 + \lambda V)\psi_n = E_n \psi_n \tag{1}$$

by a perturbation technique employing the normalized set of eigenfunctions of H_0, Eq. (2). The ap-

$$H_0 \phi_m = \epsilon_m \phi_m \tag{2}$$

proach used in Rayleigh-Schrödinger perturbation theory, which was first utilized by Lord Rayleigh in his study of acoustic modes of vibration, is to expand both the eigenfunctions ψ_n and the eigenvalues E_n in a power series in λ, as in Eqs. (3) and (4),

$$\psi_n = \sum_{m=0}^{\infty} \lambda^m \psi_n^{(m)} = \sum_{m=0}^{\infty} \lambda^m \sum_s a_{ns}^{(m)} \phi_s \tag{3}$$

$$E_n = \sum_{m=0}^{\infty} \lambda^m E_n^{(m)} \tag{4}$$

where s is a label that ranges over the complete set of the unperturbed set ϕ_s [the symbolic sum over s representing a sum over discrete (bound-state) unperturbed eigenfunctions, as well as an integral over continuum unperturbed eigenfunctions], and $a^{(m)}{}_{ns}$ are numerical coefficients to be determined. The labels s are arranged so that in the limit $\lambda \to 0$, $E_n \to \epsilon_n$, and $\psi_n \to \phi_n$. If only one eigenfunction ϕ_n has the value ϵ_n as its associated eigenvalue (that is, the eigenvalue is nondegenerate), then equating the coefficients of like powers of λ in the time-independent Schrödinger equation leads to Eqs. (5)–(7), where

$$a_{ns}^{(0)} = \delta_{ns} \qquad E_n^{(0)} = \epsilon_n \tag{5}$$

$$a_{nn}^{(1)} = 0 \; a_{ns}^{(1)} = \frac{-V_{sn}}{\epsilon_s - \epsilon_n} \text{ for } n \neq s \tag{6}$$

$$E_n^{(1)} = V_{nn} \qquad E_n^{(2)} = -\sum_m{}' \frac{V_{nm} V_{mn}}{\epsilon_m - \epsilon_n} \tag{7}$$

generically $V_{mn} = \int \phi_m^* V \phi_n$ (integration taking place over all configuration space), and the prime indicates that the $m = n$ term is excluded. The expansion also provides recursion relations from which higher-order contributions may be obtained. The criteria for validity of the expansion is usually taken to be $|a_{ns}^{(1)}| \ll 1$ for $n \neq s$, although formal proofs of convergence exist only in a limited number of cases.

Degenerate perturbation theory. If $N > 1$ zero-order (unperturbed) eigenfunctions have the same value of the eigenvalue ϵ_n, the zeroes in the denominators invalidate the preceding expressions, and one must employ degenerate perturbation theory. Essentially this consists of taking appropriate linear combinations of the zero-order degenerate wave functions for the unperturbed basis. The degenerate set $\{\phi_s\}$ is replaced by the j linear combinations of Eq. (8), and

$${}^j\phi_s = \sum_{s'=1}^{N} {}^j b_{ss'} \phi_{s'} \qquad j = 1, 2, \ldots, N \tag{8}$$

matching like powers of λ then leads to the condition of Eq. (9). The first-order energy corrections for the

$$\sum_{s'=1}^{N} {}^j b_{ss'} \left[V_{ks'} - \delta_{ks'} {}^j E_s^{(1)} \right] = 0 \tag{9}$$

jth combination, ${}^j E_s^{(1)}$, are found by setting the determinant of the coefficients of ${}^j b_{ss'}$ equal to zero. This technique, which may be considered as a first-order diagonalization process, together with the normalization condition on the coefficients, gives the coefficients ${}^j b_{ss'}$ for the jth combination of unperturbed wave functions. In general, the first-order shifts ${}^j E_s^{(1)}$ split the original unperturbed energy levels ϵ_s, and higher-order correction formulas contain denominators involving differences of the first-order shifts. If the shifts are the same for two or more different values of j, a linear combination of the corresponding ${}^j\phi_s$ must then be taken for the unperturbed starting basis for the second-order treatment. The appropriate combination is determined by a second-order diagonalization process. Higher-order degeneracies are removed in a similar fashion.

Brillouin-Wigner expansion. An alternative approach to time-independent perturbation theory is provided by the Brillouin-Wigner expansion. In this approach the exact time-independent wave function ψ_n is expanded in terms of unperturbed eigenfunctions

$$\psi_n = \sum_s a_{ns} \phi_s$$

which leads to relation (10). The basic idea here is

$$a_{nm}(E_n - \epsilon_m) = \lambda \sum_s a_{ns} V_{ms} \tag{10}$$

to use this expression to generate an iterative expansion for the off-diagonal coefficients a_{ns} (that is, for $n \neq s$). It follows that Eq. (11) holds, where the

$$E_m - \epsilon_m = \lambda V_{mm} + \lambda(a_{mm})^{-1} \sum_s{}' a_{ms} V_{ms} \quad (11)$$

prime indicates only a sum over off-diagonal a's, and the expansion is provided by repeated iteration of Eq. (12). The convergence of the Brillouin-Wigner

$$\frac{a_{ms}}{a_{mm}} = \frac{\lambda V_{sm}}{E_m - \epsilon_s} + \sum_k{}' \frac{\lambda a_{mk} V_{sk}}{E_m - \epsilon_s} \quad (12)$$

series is usually much faster than the convergence of the corresponding Rayleigh-Schrödinger series, but since the energy levels E_m are given only implicitly (to a given order in terms of a solution of a polynomial), the Brillouin-Wigner series expansion may be less convenient, although formally the expansion is much simpler.

Applications. Among the problems for which time-independent perturbation theory is employed are estimation of level splitting in the Zeeman effect, calculation of electric and magnetic susceptibilities, and a host of other problems concerning energy-level determination in atomic and molecular physics. Special techniques are employed in many-body perturbation theory in which a product basis of single-particle unperturbed eigenfunctions is utilized.

WKB method. An approximation technique for solving the time-independent Schrödinger equation by exploiting the classical limit of quantum mechanics is the WKB (Wentzel-Kramers-Brillouin) method. Here one substitutes $\psi = A \exp\,[iW/\hbar]$ into Schrödinger's equation, and expands the amplitude A and phase W in powers of h. In the zeroth approximation as $h \rightarrow 0$, the phase W satisfies the classical Hamilton-Jacobi equation; higher-order contributions give the peculiarly quantum-mechanical effects. The applicability of the WKB approximation is usually confined to one-dimensional potential well or barrier problems, or three-dimensional problems for which the well or barrier depends only on the radial coordinate. The WKB approximation is expected to give reliable results when $d|\lambda(x)|/dx \ll 1$, $\lambda(x)$ being the local de Broglie wavelength $h/p = h\,[2m(E - \lambda V)]^{-1/2}$. The WKB method has been used to give an approximate treatment of alpha-particle emission from a nucleus, as well as of a number of other problems where penetration of a barrier is involved. *See* HAMILTON-JACOBI THEORY; WENTZEL-KRAMERS-BRILLOUIN METHOD.

Time-dependent perturbation theory. In time-dependent perturbation theory, where the perturbing potential may or may not be time-dependent, the time-dependent wave function Ψ is expanded in terms of the set of time-dependent unperturbed eigenfunctions by means of expansion coefficients which are themselves time-dependent. This method is sometimes known as the method of variation of constants. Specifically one expresses a particular state, labeled by n, by means of Eq. (13). The coefficients c then obey the rate relation, Eq. (14), where

$$\Psi_n = \sum_s c_{ns}(t)\phi_s \exp\frac{-i\epsilon_s t}{\hbar} \times \exp\left[\frac{-i\lambda}{\hbar}\int_{t_0}^{t} V_{ss'}\,dt'\right] \quad (13)$$

$$i\hbar\frac{\partial c_{nm}}{\partial t} = \lambda \sum_{s\neq m} c_{ns} V_{ms} \exp\left[\frac{i}{\hbar}\int_{t_0}^{t} \gamma_{ms}\,dt'\right] \quad (14)$$

$\gamma_{ms} = (\epsilon_m + \lambda V_{mm}) - (\epsilon_s + \lambda V_{ss})$ is the difference of the perturbed energies for the states m and s. The coefficient c_{ns} is defined in Eq. (13) so as not to include the phase involving the integral over V_{ss}, in order to prevent so-called secular terms (those which have a nonoscillatory behavior at large times) from appearing in an iterative expansion of the coefficients. The appearance of such terms would inhibit convergence of the resulting expansion. Usually, in time-dependent problems, one assumes that at time $t = t_0$ the system is in state n so that $c_{ns}(t_0) = \delta_{ns}$. The problem then is to determine the probability of transition to other states at a later time.

Transient perturbations. For transient perturbations, that is, those for which $V \rightarrow 0$ for $t \rightarrow \pm\infty$, the initial time t_0 may be taken as $-\infty$, and it is clear that as $t \rightarrow \infty$, the expansion coefficients of Eq. (15) become constant since the ϕ's are eigenfunctions of the

$$a_{ns}(t) = c_{ns}(t) \exp\left[\frac{-i\lambda}{\hbar}\int_{t_0}^{t} V_{ss}\,dt'\right] \quad (15)$$

final hamiltonian operator. Calculations of $a_{ns}(\infty)$ to successive orders in λ, leading to what is called successive Born approximations, give the long-term excitation probability $P_{nm}(\infty) = |a_{nm}(\infty)|^2$ for excitation from initial state n to final state m. Problems such as scattering of particles in a continuum state from atomic systems, and inelastic collisions of atomic systems, are amenable to this type of treatment. *See* SCATTERING EXPERIMENTS (ATOMS AND MOLECULES).

Persistent perturbations. When a persistent perturbation is present, the perturbing potential which was originally zero ultimately reaches a nonvanishing time-independent limit. A first-order solution for the coefficients c gives Eq. (16), where the notation of Eqs. (17) and (18) is used. The coefficients $\beta_{nm}(t)$

$$c_{nn}(t) = 1 \qquad c_{nm}(t) = \alpha_{nm}(t) + \beta_{nm}(t) \quad n \neq m \quad (16)$$

$$\alpha_{nm}(t) = \int_{t_0}^{t}\left[\exp\frac{i}{\hbar}\int_{t_0}^{t'} \gamma_{mn}\,dt''\right] \times \frac{d}{dt'}\left[\frac{V_{mn}(t')}{\gamma_{mn}(t')}\right] dt' \quad (17)$$

$$\beta_{nm}(t) = \frac{-\lambda V_{mn}(t)}{\gamma_{mn}(t)} \exp\left[\frac{i}{\hbar}\int_{t_0}^{t} \gamma_{mn}\,dt'\right] \quad (18)$$

merely provide for the first-order propagation of the initial state in time, whereas the coefficients $\alpha_{nm}(t)$ give the excitation probability $P_{nm}(t) = |\alpha_{nm}(t)|^2$. If the perturbation increases from zero to its asymptotic value in such a short time that the exponential in the definition of α_{nm} is essentially constant, that is, in times shorter than $\hbar/(\epsilon_m - \epsilon_n)$, then $|\alpha_{nm}(\infty)|$ may be approximated by $|\lambda V_{mn}/\gamma_{mn}|$ where the relevant quantities are evaluated at their asymptotic values. This approximation, known as the sudden approximation, leads to useful results for transition probabilities arising from short-time perturbations, such as occur in the radioactive decay of ^{3}H atom into a doubly ionized ^{3}He ion.

Adiabatic approximation. Yet another approach, the adiabatic approximation, is useful when the perturbing potential changes very slowly. This technique employs a set of basis wave functions χ_s (with a time-independent phase) and associated eigenvalues $E_s(t)$ satisfying Eq. (19), which would follow for the exact

$$[H_0 + \lambda V(t)]\chi_s = E_s(t)\chi_s \tag{19}$$

solution of the time-dependent Schrödinger equation if the time variation of the potential were neglected. The basic idea is to expand the exact time-dependent wavefunction Ψ_n in terms of this basis, as in Eq. (20), obtain the rate equation giving the

$$\Psi_n = \sum_s c_{ns}(t)\,\chi_s \exp\left[\frac{-i}{\hbar}\int_{t_0}^{t} E_s(t')\,dt'\right] \tag{20}$$

time-development of the coefficients, and solve for these coefficients to the desired order in λ by an iterative technique, assuming that the coefficients c_{ns} differ little from their initial values. The use of the basis χ rather than an unperturbed basis hopefully provides for a more satisfactory lower-order treatment. This approach has been successfully used in atomic-collision problems for which the interaction distance changes slowly with time. *See* PERTURBATION (ASTRONOMY); PERTURBATION (MATHEMATICS).

David M. Fradkin

Pesticide

A material useful for the mitigation, control, or elimination of plants or animals detrimental to human health or economy. Algicides, defoliants, desiccants, herbicides, plant growth regulators, and fungicides are used to regulate populations of undesirable plants which compete with or parasitize crop or ornamental plants. Attractants, insecticides, miticides, acaricides, molluscicides, nematocides, repellants, and rodenticides are used principally to reduce parasitism and disease transmission in domestic animals, the loss of crop plants, the destruction of processed food, textile, and wood products, and parasitism and disease transmission in humans. These ravages frequently stem from the feeding activities of the pests. Birds, mice, rabbits, rats, insects, mites, ticks, eel worms, slugs, and snails are recognized as pests.

Materials and use. Materials used to control or alleviate disease conditions produced in humans and animals by plants or by animal pests are usually designated as drugs. For example, herbicides are used to control the ragweed plant, while drugs are used to alleviate the symptoms of hay fever produced in humans by ragweed pollen. Similarly, insecticides are used to control malaria mosquitoes, while drugs are used to control the malaria parasites—single-celled animals of the genus *Plasmodium*—transmitted to humans by the mosquito.

Sources. Some pesticides are obtained from plants and minerals. Examples include the insecticides cryolite, a mineral, and nicotine, rotenone, and the pyrethrins which are extracted from plants. A few pesticides are obtained by the mass culture of microorganisms. Two examples are the toxin produced by *Bacillus thuringiensis*, which is active against moth and butterfly larvae, and the so-called milky disease of the Japanese beetle produced by the spores of *B. popilliae*. Most pesticides, however, are products which are chemically manufactured. Two outstanding examples are the insecticide DDT and the herbicide 2,4-D.

Evaluation. The development of new pesticides is time-consuming. The period between initial discovery and introduction is frequently cited as being about 5 years. Numerous scientific skills and disciplines are required to obtain the data necessary to establish the utility of a new pesticide. Effectiveness under a wide variety of climatic and other environmental conditions must be determined, and minimum rates of application established.

Insight must be gained as to the possible side effects on other animals and plants in the environment. Toxicity to laboratory animals must be measured and be related to the hazard which might possibly exist for users and to consumers. Persistence of residues in the environment must be determined. Legal tolerances in processed commodities must be set and directions for use clearly stated. Methods for analysis and detection must be devised. Economical methods of manufacture must be developed. Manufacturing facilities must be built. Sales and education programs must be prepared.

Regulation and restriction. By the mid-1960s, it became apparent that a number of the new pesticides, particularly the insecticide DDT, could be two-edged swords. The benefits stemming from the unmatched ability of DDT to control insect pests could be counterbalanced by adverse effects on other elements of the environment. Detailed reviews of the properties, stability, persistence, and impact upon all facets of the environment were carried out not only with DDT, but with other chlorinated, organic insecticides as well. Concern over the undesirable effects of pesticides culminated in the amendment of the Federal Insecticide, Fungicide, and Rodenticide Act (FIFRA) by Public Law 92-516, the Federal Environmental Pesticide Control Act (FEPCA) of 1972. The purpose behind this strengthening of earlier laws was to prevent exposure of either humans or the environment to unreasonable hazard from pesticides through rigorous registration procedures, to classify pesticides for general or restricted use as a function of acute toxicity, to certify the qualifications of users of restricted pesticides, to identify accurately and label pesticide products, and to ensure proper and safe use of pesticides through enforcement of FIFRA.

Selection and use. Pesticides must be selected and applied with care. Recommendations as to the product and method of choice for control of any pest problem—weed, insect, or varmint—are best obtained from county or state agricultural extension specialists. Recommendations for pest control and pesticide use can be obtained from each state agricultural experiment station office. In addition, it is necessary to follow explicitly the directions, restrictions, and cautions for use on the label of the product container. Insecticides are a boon to the production of food, feed, and fiber, and their use must not be abused in the home, garden, farm, field, forest, or stream. George F. Ludvik

Persistence. Considerable scientific evidence has been accumulated that documents the deleterious effects on the environment of several of the organic insecticides. Over 800 compounds are used as pesticides and, with increasing technological sophistication, the list can be expected to grow. Until World War II pesticides consisted of inorganic materials containing sulfur, lead, arsenic, or copper, all of which have biocidal properties, or of organic materials extracted from plants such as pyrethrum, nicotine, and rotenone. During World War II a technological revolution began with the introduction of the synthetic organic biocides, particularly DDT.

There is an increasing number of organic biocides. They vary widely in toxicity, in specificity, in persistence, and in the production of undesirable derivatives. The factors that determine how, when, and where a particular pesticide should be used, or whether it should be used at all, are therefore necessarily different for each pesticide. The use of chemicals to achieve a measure of control over the environment has become an integral part of technology, and there are no valid arguments to support the proposition that the use of all of these chemicals should cease.

Some take the view that pests include insects, fungi, weeds, and rodents. Many insects, however, particularly the predatory species that prey upon the insects that damage crops, are clearly not pests. Fungi are essential in the processes that convert dead plant and animal material to the primary substances that can be recycled through living systems. In many ecosystems rodents are also an important link in the recycling process. On the other hand, an increasing number of species traditionally considered desirable are becoming "pests" in local contexts. Thus chemicals are used to destroy otherwise useful vegetation along roadsides or in pine plantations, or vegetation that might provide food and shelter to species undesirable to human populations. The control of any component of the global environment through the use of biocidal chemicals is therefore a problem much more complex than is initially suggested by the connotation of the word pest. The highly toxic chemicals that kill many different species are clearly undesirable in situations in which control over only one or two is sought.

Other pest-control measures. Sophisticated methods of pest control are continually being developed. One technique involves the raising of a large number of insect pests in captivity, sterilizing them with radioactivity, and then releasing them into the environment, where they mate with the wild forms. Very few offspring are subsequently produced. Highly specific synthetic insect hormones are being developed. In an increasing number of pest situations, a natural predator of an insect has been introduced, or conditions are maintained that favor the propagation of the predator. The numbers of the potential pest species are thereby maintained below a critical threshold. An insect control program in which use of insecticides is only one aspect of a strategy based on ecologically sound measures is known as integrated pest management. *See* AGRICULTURAL CHEMISTRY; CHEMICAL ECOLOGY; FUNGISTAT AND FUNGICIDE; HERBICIDE; INSECT CONTROL, BIOLOGICAL; INSECTICIDE.

Robert W. Risebrough

Bibliography. A. J. Burn, T. H. Coaker, and P. C. Jepson (eds.), *Integrated Pest Management*, 1988; R. Carson, *Silent Spring*, 25th anniversary ed., 1987; G. P. Georghiou and T. Saito (eds.), *Pest Resistance to Pesticides*, 1983; R. Honeycutt, *Advances in Understanding the Mechanics of Pesticide Movement*, 1994; P. Hurst, N. Dudley, and A. Hay, *The Pesticide Handbook*, 1989; R. L. Metcalf and W. Luckmann (eds.), *Introduction to Insect Pest Management*, 3d ed., 1994; H. F. Van Emden, *Pest Control: New Studies in Biology*, 2d ed., 1989; G. W. Ware, *The Pesticide Book*, 5th ed., 1999.

Petalite

A rare pegmatitic mineral with composition $LiAlSi_4O_{10}$. Its economic significance is markedly disproportionate to the number of its occurrences. It is the only basic raw material suitable for production of a group of materials known as crystallized glass ceramics (melt-formed ceramics). These extremely fine-grained substances, formed by controlled nucleation and crystal growth, are based on a keatite-type structure (stuffed silica derivative). Among the desirable properties of such submicroscopic aggregates are their exceedingly low thermal expansion and high strength, making them suitable for use in cooking utensils and telescopic mirror blanks.

Petalite is mined from a large lithium-rich pegmatite at Bikita, Rhodesia, the only major world source. Other significant occurrences are at Varuträsk, Sweden (largely exhausted); Karibib district, South-West Africa; and Londonderry, Western Australia. About 30 other localities are reported, chiefly for specimens only.

Petalite is characterized as follows: white, pink, pale green, or gray to black; monoclinic (Pa); hardness $6^1/_2$; optically (+); α = 1.504–1.507,

$\beta = 1.510\text{–}1.513$, $\gamma = 1.516\text{–}1.523$, $\gamma - \alpha = 0.011\text{–}0.017$, $2V = 82\text{–}84^\circ$ optic plane {001}; $\beta \wedge Z = 24\text{–}30^\circ$; cleavages, {001} perfect, {201} good, lamellar twins {001} common.

In pegmatites it is associated with other lithium minerals such as spodumene, lepidolite, eucryptite, and amblygonite, and common also with the rare cesium silicate, pollucite. It is an early-formed pegmatitic species and may be converted isochemically to a spodumene-quartz intergrowth via the reaction

$$\underset{\text{Petalite}}{LiAl_4O_{10}} \rightarrow \underset{\text{Spodumene}}{LiAlSi_2O_8} + \underset{\text{Quartz}}{2SiO_2}$$

See LITHIUM; PEGMATITE. E. William Heinrich

Petrifaction

One of the most remarkable mechanisms by which the remains of extinct organisms are preserved in the fossil record is the process of petrifaction. In petrifactions (though chiefly in the case of plants rather than animals) the original shape and topography of the tissues, and occasionally even minute cytological details, are retained relatively underformed.

Formation. Petrifaction was adopted as a scientific term before knowledge existed of the geochemical mechanism or processes involved. It was formerly widely believed that in the formation of a petrifaction the organic matter of the organism or tissue was replaced molecule by molecule with mineral material entering in solution in percolating ground water. It is now evident that what actually happens is that the mineral fills cell lumena and the intermicellar interstices of cell walls with insoluble salts depositing from solution. Petrifaction is hence a form of mineral emplacement or embedding, by which the organic residues are filled with solid substance which infiltrates in solution. The most common substances involved in petrifactions are silica, SiO_2, and calcium carbonate, $CaCO_3$ (calcite). Occasionally phosphate minerals, pyrite, hematite, and other less common minerals make up all or part of the petrifaction matrix. The most perfect preservation of original structure is found in siliceous petrifactions. The clear, transparent, microcrystalline silica renders excellent optical properties to thin sections of such fossils and makes possible the use of transmitted light in microscopic study, in much the same manner as with recent biological material.

Calcified types. An unusual type of calcareous petrifaction known as coal balls occurs in Carboniferous coal seams of parts of Europe and North America. They comprise nodular, usually spheroidal or ovoidal masses of relatively uncompressed plant tissues completely permeated with calcite or dolomite. They represent irregularly spaced and localized areas of mineral precipitation with resulting petrifaction of the coal-forming plant debris. Mineralization occurred while the coal was still in the peat stage. After mineralization the infiltrated plant parts failed to compress so that their structure and cellular organization are preserved. Much of what is known of the internal organization and anatomy of ancient plants and the evolution of their organs, both vegetative and reproductive, is derived from petrifactions, which occur throughout the geologic record from Precambrian to Recent. Coal balls, though limited to the Carboniferous, have provided an unusually comprehensive body of knowledge on the morphology and anatomy of the rich and varied vegetation of that geologic period. *See* COAL BALLS.

Silicified types. Despite the widespread abundance of petrifactions (chiefly of plants but also of the hard parts of animals) throughout the geologic column, there is very little known of the geochemical mechanisms which induce their formation. The problem is particularly baffling in the case of silicification; in many examples entire trunks of trees are completely mineralized with no visible evidence of the sequence of changes following sedimentation. The fact that silicified plant parts often show virtually no physical distortion or compression indicates that the process occurs relatively rapidly. On the other hand, the amount of silica in ground water (±70–100 parts per million) is so small that the process must proceed with extreme efficiency. It is probable that decay processes proceeded to at least the early stages, since the histology of silicified (also calcified) plant tissues exhibits varying stages of degradative alteration, to nearly total loss of structure. The percentage by weight of organic matter retained in silicified wood may range from a few hundredths of 1% to more than 15%. Ordinarily it is only a few percent. *See* FOSSIL; PALEOBOTANY; PETRIFIED FORESTS.

Elso S. Barghoorn

Petrified forests

Exposures containing appreciable numbers of petrified tree trunks, either standing upright or lying prostrate in the enclosing sedimentary rocks; sometimes called fossil forests. The best-known examples are the Petrified Forest of Arizona, and the fossil forests near Cairo, Egypt; near Calistoga, California; near Vantage Bridge, Washington; and in Yellowstone National Park, Wyoming (see **illus.**).

The Petrified Forest of Arizona is made up of hundreds of prostrate silicified tree trunks, logs, and stumps lying scattered at random like an ancient log drive. These are of Triassic age, roughly 175 million years old, and occur in a portion of the Painted Desert, which owes its brilliant hues to the varicolored layers of the Chinle formation. The wood is mainly agatized or changed to chalcedony and shows an unusually varied and beautiful coloration of reds, browns, yellows, and purples. The majority of the petrified trees are conifers (*Araucarioxylon*), distantly related to the araucarian pines of South America and Australia. One huge log over 100 ft (30 m) long has been left by erosion across a ravine about 40 ft (12 m) wide, forming a natural span known

Upright petrified trees located on Specimen Ridge in Yellowstone National Park, Wyoming.

as Agate Bridge. Here and there are exposures of shale beds containing many impressions of leaves and seeds representing a humid, subtropical forest bordering the streams of a lowland savanna.

An even more extensive fossil forest lies in the northeastern portion of Yellowstone National Park, Wyoming. Here the majority of the petrified tree trunks are found still standing upright in positions of original growth in the enclosing medium of volcanic tuffs and breccia. Even more unusual is the occurrence here of not merely a single fossil forest, but a vertical succession of 27 buried forests—one above the other—in a thickness of over 2000 ft (600 m) of volcanic debris. The fossilized trees and the impressions of leaves, twigs, needles, and cones in associated ash layers indicate a forest of over 100 species of warm-temperate to subtropical trees typical of a humid, lowland environment. These petrified forests are roughly 50 million years old, belonging to the Eocene Epoch. *See* PALEOBOTANY.

Erling Dorf

Petrochemical

Any of the chemicals derived from petroleum or natural gas. The definition of petrochemicals has been broadened to include the whole range of aliphatic, aromatic, and naphthenic organic chemicals, as well as carbon black and such inorganic materials as sulfur and ammonia.

Petrochemicals are made or recovered from the entire range of petroleum fractions, but the bulk of petrochemical products are formed from the lighter (C_1-C_4) hydrocarbon gases as raw materials. These materials generally occur in natural gas, but they are also recovered from the gas streams produced during refinery operations, especially cracking. Refinery gases are particularly valuable because they contain substantial amounts of olefins that, because of their double bonds, are much more reactive then the saturated (paraffin) hydrocarbons. Also important as raw materials are the aromatic hydrocarbons (benzene, toluene, and xylene) that are obtained from various refinery product streams. For example, catalytic reforming processes convert nonaromatic hydrocarbons to aromatic hydrocarbons by dehydrogenation and cyclization. *See* PETROLEUM; PETROLEUM PRODUCTS.

Thermal cracking processes (such as coking) are focused primarily on increasing the quantity and quality of gasoline and other liquid fuels, but also produce gases, including lower-molecular-weight olefins such as ethylene ($CH_2{=}CH_2$), propylene ($CH_3CH{=}CH_2$), and butylenes (butenes, $CH_3CH{=}CHCH_3$ and $CH_3CH_2CH{=}CH_2$). Catalytic cracking is a valuable source of propylene and butylene, but it is not a major source of ethylene, the most important of the petrochemical building blocks. *See* CRACKING; ETHYLENE.

The starting materials for the petrochemical industry (**Table 1**) are obtained from crude petroleum in one of two ways. They may be present in the raw crude oil and are isolated by physical methods, such as distillation or solvent extraction; or they are synthesized during the refining operations. Unsaturated (olefin) hydrocarbons, which are not usually present in natural petroleum, are nearly always manufactured as intermediates during the refining sequences (**Table 2**). *See* DISTILLATION; PETROLEUM PROCESSING AND REFINING; SOLVENT EXTRACTION.

The main objective in producing chemicals from petroleum is the formation of a variety of

TABLE 1. Hydrocarbon intermediates used as sources of petrochemical products

Carbon number	Hydrocarbon type: Saturated	Unsaturated	Aromatic
1	Methane		
2	Ethane	Ethylene Acetylene	
3	Propane	Propylene	
4	Butanes	*n*-Butenes Isobutene Butadiene	
5	Pentanes	Isopentenes (isoamylenes) Isoprene	
6	Hexanes Cyclohexane	Methylpentenes	Benzene
7		Mixed heptenes	Toluene
8		Di-isobutylene	Xylenes Ethylbenzene Styrene
9			Cumene
12		Propylene tetramer Tri-isobutylene	
18			Dodecylbenzene
6–18		*n*-Olefins	
11–18	*n*-Paraffins		

TABLE 2. Major sources of petrochemical intermediates

Hydrocarbon	Source
Methane	Natural gas
Ethane	Natural gas
Ethylene	Cracking processes
Propane	Natural gas, catalytic reforming, cracking processes
Propylene	Cracking processes
Butane	Natural gas, reforming and cracking processes
Butene(s)	Cracking processes
Cyclohexane	Distillation
Benzene, Toluene, Xylene(s), Ethylbenzene	Catalytic reforming
Alkylbenzenes	Alkylation
$>C_9$	Polymerization

TABLE 3. Chemicals produced from methane (natural gas)

Basic derivatives and sources	Uses
Carbon black	Rubber compounding Printing ink, paint
Methanol	Formaldehyde (mainly for resins) Methyl esters (polyester fibers), amines, and other chemicals Solvents
Chloromethanes	Chlorofluorocarbons for refrigerants, aerosols, solvents, cleaners, grain fumigant

well-defined chemical compounds, including (1) chemicals from aliphatic compounds; (2) chemicals from olefins; (3) chemicals from aromatic compounds; (4) chemicals from natural gas; (5) chemicals from synthesis gas (carbon monoxide and hydrogen); and (6) inorganic petrochemicals.

An aliphatic petrochemical is an organic compound that has a chain of carbon atoms, be it normal (straight), such as *n*-pentane ($CH_3CH_2CH_2CH_2CH_3$); or branched, such as, *iso*-pentane (2-methylbutane, $CH_3CHCH_3CH_2CH_3$); or unsaturated. Unsaturated compounds, called olefins, include important starting materials such as ethylene ($CH_2{=}CH_2$), propylene ($CH_3CH{=}CH_2$), 1-butene ($CH_3CH_2CH_2{=}CH_2$), *iso*-butylene [$(CH_3)_2C{=}CH_2$], and butadiene ($CH_2{=}CHCH{=}CH_2$).

An aromatic petrochemical is an organic compound that contains, or is derived from, a benzene ring.

A significant proportion of the basic petrochemicals are converted into plastics, synthetic rubbers, and synthetic fibers. These materials, known as polymers, are high-molecular-weight compounds made up of repeated structural units. The major polymer products are polyethylene, polyvinyl chloride, and polystyrene, all derived from ethylene, and polypropylene, derived from propylene. Major raw material sources for synthetic rubbers include butadiene, ethylene, benzene, and propylene. Among synthetic fibers the polyesters, which are a combination of ethylene glycol and terephthalic acid (made from xylene), are the most widely used. They account for about one-half of all synthetic fibers. The second major synthetic fiber is nylon, its most important raw material being benzene. Acrylic fibers, in which the major raw material is the propylene derivative acrylonitrile, make up most of the remainder of the synthetic fibers. *See* MANUFACTURED FIBER; POLYACRYLATE RESIN; POLYAMIDE RESINS; POLYESTER RESINS; POLYMER; POLYMERIZATION; POLYOLEFIN RESINS; POLYURETHANE RESINS; POLYVINYL RESINS; RUBBER.

Aliphatics. Methane, obtainable from petroleum and natural gas or as a product of various conversion (cracking processes), is an important raw material source for aliphatic petrochemicals (**Table 3**; **Fig. 1**). Ethane, propane, and butane are also important sources of petrochemicals because of their relative ease of conversion to ethylene, which provides more valuable routes to petrochemicals (**Table 4**; **Fig. 2**). The higher-boiling members of the paraffin series are less easy to fractionate from petroleum in pure form; moreover, the number of compounds formed in each particular chemical treatment makes the separation of individual products quite difficult. *See* PARAFFIN.

Halogenation. The ease with which chlorine and bromine can be introduced into the molecules of all the hydrocarbon types present in petroleum has

TABLE 4. Chemicals from ethylene

Basic derivatives and sources	Uses
Ethylene oxide	Ethylene glycol (polyester fiber and resins, antifreeze) Di- and triethylene glycols Ethanolamines Nonionic detergents Glycol esters
Ethyl alcohol	Acetaldehyde Solvents Ethyl acetate
Polyethylene	
Low-density	Film, injection molding
High-density	Blow molding, injection molding
Styrene (from ethylene and benzene)	Polystyrene and copolymer resins Styrene-butadiene rubber and latex Polyesters
Ethylene dichloride	Vinyl chloride Scavenger in antiknock fluid Ethyleneamines
Ethyl chloride	Tetraethyllead Minor amounts for ethylations (such as ethyl cellulose)
Ethylene dibromide	Scavenger in antiknock fluid
Acetyls	Plastics and chemical intermediates
Linear alcohols and α-olefins	Detergents, plasticizers

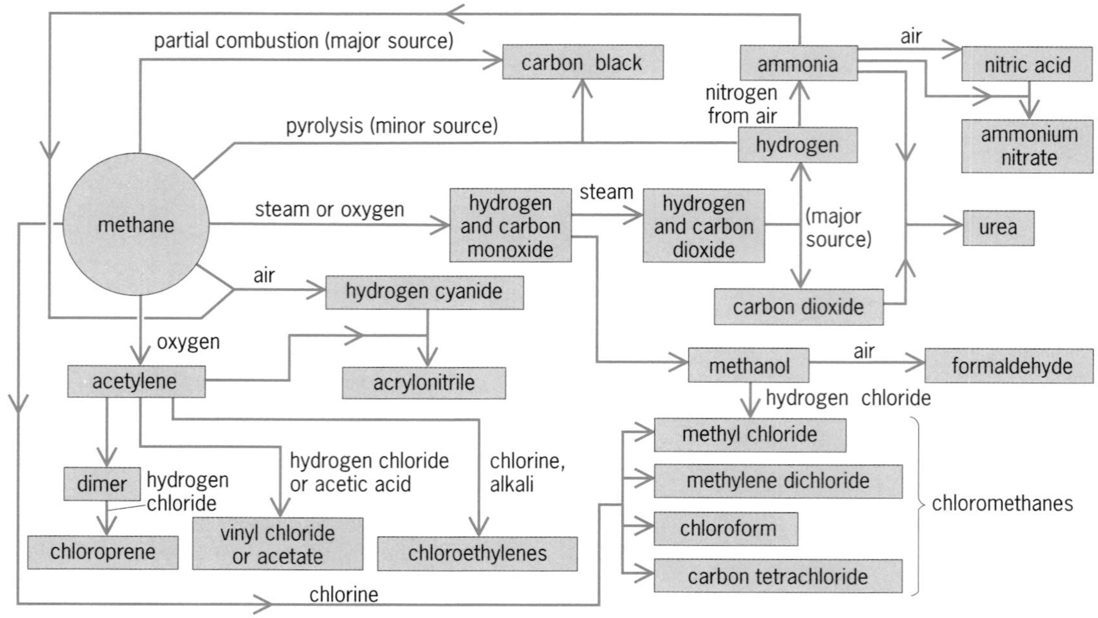

Fig. 1. Chemicals from methane.

resulted in the commercial production of a number of widely used compounds. With saturated hydrocarbons the reactions are predominantly substitution of hydrogen by chlorine or by bromine, and are exothermic, difficult to control, and inclined to be explosive [reaction (1)].

$$RH + Cl_2 \longrightarrow RCl + HCl \quad (1)$$

Moderately high temperatures are used, about 250–300°C (480–570°F), for the thermal chlorination of methane; but as the molecular weight of the paraffin increases, the temperature may generally be lowered. A mixture of chlorinated derivatives is always obtained; and many variables, such as choice of catalyst, dilution of inert gases, and presence of other chlorinating agents (antimony pentachloride, sulfuryl chloride, and phosgene), have been tried in an effort to direct the path of the reaction.

An example of the chlorination reaction is the formation of ethyl chloride by the chlorination of ethane [reaction (2)]. Ethyl chloride (CH_3CH_2Cl) is

$$CH_3CH_3 + Cl_2 \longrightarrow CH_3CH_2Cl + HCl \quad (2)$$

also prepared by the direct addition of hydrogen chloride (HCl) to ethylene ($CH_2{=}CH_2$) or by reacting ethyl ether ($CH_3CH_2OCH_2CH_3$) or ethyl alcohol (CH_3CH_2OH) with hydrogen chloride. The chlorination of *n*-pentane and *iso*-pentane does not take place in the liquid or vapor phase below 100°C (212°F) in the absence of light or a catalyst, but above 200°C (390°F) it proceeds smoothly by thermal action alone. The hydrolysis of the mixed chlorides obtained yields all the isomeric amyl (C_5) alcohols except *iso*-amyl alcohol. Reaction with acetic acid produces the corresponding amyl acetates, which find wide use as solvents.

The alkyl chloride obtained on substituting an equivalent of one hydrogen atom by a chloride atom in kerosine is used to alkylate benzene or naphthalene for the preparation of a sulfonation stock for use in the manufacture of detergents and antirust agents. Similarly, paraffin wax can be converted to a hydrocarbon monochloride mixture, which can be used to alkylate benzene, naphthalene, or anthracene. The product finds use as a pour-point depressor for retarding wax crystal growth and deposition in cold lubricating oils. *See* HALOGENATED HYDROCARBON; HALOGENATION.

Nitration. Hydrocarbons that are usually gaseous (including normal and *iso*-pentane) react smoothly in the vapor phase with nitric acid to give a mixture of nitro compounds, but there are side reactions, mainly oxidation. Only mono nitro derivatives are obtained with the lower paraffins at high temperatures, and they correspond to those expected if scission of a C-C and C-H bond occurs. For example, ethane yields nitromethane and nitroethane [reaction (3)]. Propane yields nitromethane, nitroethane, 1-nitropropane, and 2-nitropropane [reaction (4)].

$$2CH_3CH_3 + 2HNO_3 \longrightarrow CH_3NO_2 + CH_3CH_2NO_2 + 2H_2O \quad (3)$$

$$4CH_3CH_2CH_3 + 4HNO_3 \longrightarrow CH_3NO_2 + CH_3CH_2NO_2 + CH_3CH_2CH_2NO_2 + CH_3CH(NO_2)CH_3 + 4H_2O \quad (4)$$

The nitro derivatives of the lower paraffins are colorless and noncorrosive and are used as solvents or as starting materials in a variety of syntheses. For example, treatment with inorganic acids and water yields fatty acid (RCO_2H) and hydroxylamine (NH_2OH) salts, and condensation with aldehydes ($RCH{=}O$) yields nitroalcohols [$RCH(NO_2)OH$]. *See* NITRATION; NITRO AND NITROSO COMPOUNDS; NITROPARAFFIN.

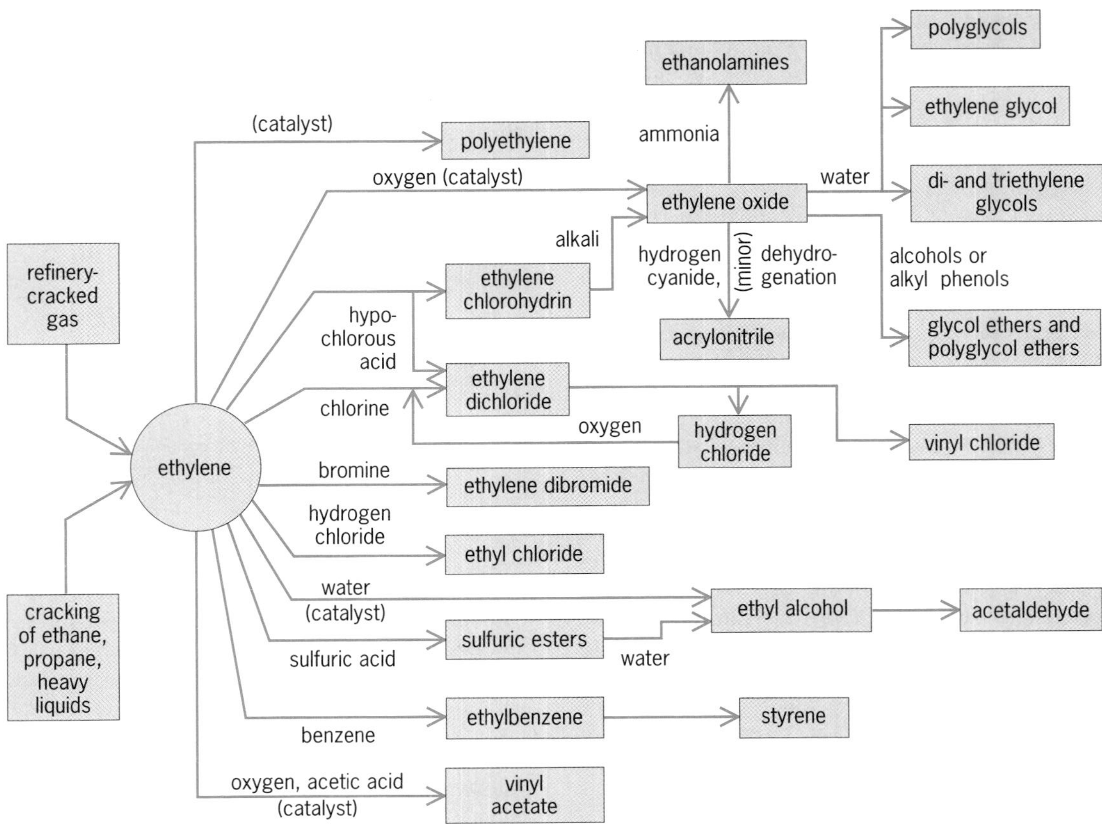

Fig. 2. Chemicals from ethylene.

Oxidation. The oxidation of hydrocarbons and hydrocarbon mixtures is generally uncontrollable, and the complex mixtures of products have made resolution of the reaction sequences extremely difficult. Therefore, except for the preparation of mixed products having specific properties, such as fatty acids, hydrocarbons higher than pentanes are not used for oxidation because of the difficulty of isolating individual compounds. For example, when propane and butane are oxidized in the vapor phase, without a catalyst, at 270–350°C (520–660°F) and at 50–3000 psi (345–20,685 kPa), a wide variety of products are obtained, including C_1–C_4 acids, C_2–C_7 ketones, ethylene oxide, esters, formals, and acetals. *See* OXIDATION PROCESS; OXIDATION-REDUCTION.

In contrast to the oxidation of paraffins, cyclohexane is somewhat selective in its reaction with air at 150–250°C (300–480°F) in the liquid phase in the presence of a catalyst, such as cobalt acetate. Cyclohexanol is the initial product, but prolonged oxidation produces adipic acid. Oxidation of cyclohexane and methylcyclohexane over vanadium pentoxide at 450–500°C (840–930°F) affords maleic and glutaric acids.

The preparation of carboxylic acids from petroleum, particularly from paraffin wax, for esterification to fats or neutralization to form soaps is comparatively slow with very little reaction taking place at 110°C (230°F), with a wax melting at 55°C (130°F) after 280 hours. At higher temperatures the oxidation proceeds more readily; maximum yields of mixed alcohol and high-molecular-weight acids are formed at 110–140°C (230–285°F) at 60–150 psi (414–1034 kPa); higher temperatures (140–160°C; 285–320°F) result in more acid formation [reaction (5)]. Acids from formic (HCO_2H) to that with a

$$\underset{}{\text{Paraffin wax}} \longrightarrow \underset{\text{Alcohol}}{ROH} + \underset{\text{Acid}}{RCO_2H} \qquad (5)$$

10-carbon atom chain [$CH_3(CH_2)_9CO_2H$] have been identified as products of the oxidation of paraffin wax. *See* CARBOXYLIC ACID; ESTER; SOAP.

Alkylation. Alkylation chemistry contributes to the efficient use of C_4 olefins generated in the cracking operations. *Iso*-butane has been added to butenes (and other low-boiling olefins) to give a mixture of highly branched octanes (such as heptanes) by a process called alkylation. Typically, sulfuric acid (85–100%), anhydrous hydrogen fluoride, or a solid sulfonic acid is employed as the catalyst in these processes. The first step in the process is the formation of a carbocation by combination of an olefin with an acid proton [reaction (6)], followed by the

$$(CH_3)_2C{=}CH_2 + H^+ \longrightarrow (CH_3)_3C^+ \qquad (6)$$

addition of the carbocation to a second molecule of olefin to form a dimer carbocation. The extensive branching of the saturated hydrocarbon results in high octane. In practice, mixed butenes are employed (*iso*-butylene, 1-butene, and 2-butene), and the product is a mixture of isomeric octanes that has an octane number of 92–94. With the phase-out of leaded additives in gasoline, octane improvement is a major challenge for the refining industry, and alkylation provides a route to high-octane products. *See* ALKYLATION (PETROLEUM); GASOLINE; OCTANE NUMBER.

Thermolysis. Thermal cracking is the major process for generating ethylene and the other olefins that are the reactive building blocks of the petrochemical industry. In addition to thermal cracking, other important processes that generate sources of hydrocarbon raw materials for the petrochemical industry include catalytic reforming, alkylation, dealkylation, isomerization, and polymerization. *See* CATALYSIS; DEHYDROGENATION; HETEROGENEOUS CATALYSIS; ISOMERIZATION; REFORMING PROCESSES.

At the high temperatures of refinery processes (usually approximately 500°C, or 950°F, and higher), cracking is favored. Unfortunately, in the cracking process certain products interact with one another to produce products of increased molecular weight relative to the molecular weight of the original feedstock. Thus some products are taken from the cracking units as useful products (such as olefin gases, gasoline, and other distillates), but other products include cracked residuum and coke.

Olefins. Olefins (C_nH_{2n}) are the basic building blocks for a host of chemical syntheses (**Tables 4, 5,** and **6**). These unsaturated materials enter into polymers and rubbers, and with other reagents react to form a wide variety of useful compounds, including alcohols, epoxides, amines, and halides. In addition, ethane (also available from natural gas and cracking processes) is an important source of ethylene that, in turn, provides more valuable routes to petrochemical products (Fig. 2). *See* ALKENE.

Ethylene is usually made by cracking gases—ethane, propane, butane, or a mixture of these as might exist in a refinery's off-gases. Ethylene and coproducts are produced from steam cracking of hydrocarbons such as ethane, propane, butane, naphtha, and gas oil in tubular reactor coils installed in externally fired heaters. Alternative processes include naphtha cracking, whole crude oil cracking, and heavy oil cracking processes; in such cases significant quantities of higher-molecular-weight olefins and aromatics are also produced. Other processes include ethanol dehydrogenation, syngas-based processes, dehydrogenation of paraffins, oxidative coupling of methane, and olefin disproportionation.

TABLE 5. Chemicals from propane and propylene and their uses

Basic derivatives and sources	Uses
Isopropyl alcohol	Acetone
	Solvents, drugs, and chemicals
Cumene	Phenol and acetone
Acrylonitrile	Acrylic fibers
	Nitrile elastomers and acrylonitrile-butadiene-styrene resins
Polypropylene	Molding, fiber, and film
Propylene oxide	Propylene glycol
	Dipropylene glycol and polypropylene glycol
	Polyurethane
Oxochemicals	
Isooctyl alcohol	Phthalate esters
Butyraldehydes (propylene only)	Intermediate for butanols, 2-ethylhexanol, *n*-butyric acid
Dodecene (tetramer)	Dodecyl benzene
	Dodecyl phenol
Nonene (trimer)	Decyl alcohol and nonylphenol
Epichlorohydrin	Glycerol and epoxy resins
Polyisoprene	Elastomers

TABLE 6. Chemicals from butane and butylene and their uses

Basic derivatives and sources	Uses
Butadiene	Styrene-butadiene rubber and resins
	Polybutadiene
	Adiponitrile
	Nitrile rubber
	Acrylonitrile-butadiene-styrene plastics
sec-Butyl alcohol	Methyl ethyl ketone
Butyl rubber (from isobutylenes)	Tire products
Polybutenes (from isobutylene)	Lube oil additives
	Caulking and sealing compounds
	Adhesives
	Rubber compounding

Ethylene manufacture via the steam cracking process is widely practiced throughout the world. The operating facilities are similar to gas oil cracking units, operating at temperatures of 840°C (1550°F) and at low pressures (24 psi; 165 kPa). Steam is added to the vaporized feed to achieve a 50-50 mixture, and furnace residence times are only 0.2–0.5 second. Ethane extracted from natural gas is the predominant feedstock for ethylene cracking units. Propylene and butylene are largely derived from catalytic cracking units and from cracking a naphtha or light gas oil fraction to produce a full range of olefin products.

The olefinic materials (RCH=CHR′) present in gaseous products of cracking processes offer promising source materials. Cracking paraffin hydrocarbons and heavy oils also produces olefins. For example, cracking ethane (CH_3CH_3), propane ($CH_3CH_2CH_3$), butane ($CH_3CH_2CH_2CH_3$), and other feedstocks such as naphtha and gas oil produces ethylene. Propylene ($CH_3CH{=}CH_2$) is produced from thermal and catalytic cracking of naphtha and gas oils, as well as propane and butane.

Hydroxylation. The earliest method for converting olefins into alcohols involved their absorption in sulfuric acid to form esters, followed by dilution and hydrolysis, generally with the aid of steam. In the case of ethyl alcohol, the direct catalytic hydration of ethylene can be used. Ethylene is readily absorbed in 98–100% sulfuric acid at 75–80°C (165–175°F), and both the mono- and diethyl sulfates are formed; hydrolysis takes place readily on dilution with water and heating.

The direct hydration of ethylene to ethyl alcohol is practiced over phosphoric acid on diatomaceous earth or promoted tungstic oxide under 100 psi (690 kPa) pressure and at 300°C (570°F). Purer ethylene is required in direct hydration than in the acid absorption process and the conversion per pass is low, but high yields are possible by recycling. Propylene and the normal butenes can also be hydrated directly.

One of the first alcohol syntheses practiced commercially was that of isopropyl alcohol from propylene. Sulfuric acid absorbs propylene more readily than it does ethylene, but care must be taken to avoid polymer formation by keeping the mixture relatively cool and using acid of about 85% strength at 300–400 psi (2068–2758 kPa) pressure; dilution with an inert oil may also be necessary. Acetone is readily made from isopropyl alcohol, either by catalytic oxidation or by dehydrogenation over metal (usually copper) catalysts. Secondary butyl alcohol is formed on absorption of 1- or 2-butene by 78–80% sulfuric acid, followed by dilution and hydrolysis. Secondary butyl alcohol is converted into methyl ethyl ketone by catalytic oxidation or dehydrogenation.

There are several methods for preparing higher alcohols. One method, the Oxo reaction, involves the direct addition of carbon monoxide (CO) and a hydrogen (H) atom across the double bond of an olefin to form an aldehyde (RCH=O) that, in turn, is reduced to the alcohol (RCH_2OH). Hydroformylation (the Oxo reaction) is brought about by contacting the olefin with synthesis gas (1:1 carbon monoxide–hydrogen) at 75–200°C (165–390°F) and 1500–4500 psi (10,342–31,027 kPa) over a metal catalyst, usually cobalt. The active catalyst is held to be cobalt hydrocarbonyl $HCO(CO)_4$, formed by the action of the hydrogen on dicobalt ocatcarbonyl.

A wide variety of olefins participate in this reaction, and the olefins that contain terminal unsaturation ($—CH=CH_2$) are the most active. The hydroformylation reaction is not specific, and the hydrogen and carbon monoxide add across each side of the double bond. Propylene gives a mixture of 60% *n*-butyraldehyde and 40% *iso*-butyraldehyde. Terminal ($RCH=CH_2$) and nonterminal ($RCH=CHR'$) olefins, such as 1- and 2-pentene ($CH_3CH_2CH_2CH=CH_2$ and $CH_3CH_2CH=CHCH_3$), give the same distribution of straight- and branched-chain C_6 aldehydes because of isomerization. Simple branched structures add mainly at the terminal carbon; *iso*-butylene forms 95% *iso*-valeraldehyde and 5% trimethylacetaldehyde. Commercial application of the synthesis has been most successful in the manufacture of *iso*-octyl alcohol from a refinery C_3C_4 dimer, decyl alcohol from propylene trimer, and tridecyl alcohol from propylene tetramer. *See* ALCOHOL.

The hydrolysis of ethylene chlorohydrin ($HOCH_2CH_2Cl$) or the cyclic ethylene oxide produces ethylene glycol ($HOCH_2CH_2OH$). The main use for this chemical is for antifreeze mixtures in automobile radiators and for cooling aviation engines; considerable amounts are used as ethylene glycol dinitrate in low-freezing dynamites. Propylene glycol is also made by the hydrolysis of its chlorohydrin or oxide.

Glycerin [$HOCH_2CH(OH)CH_2OH$] can be derived from propylene by high-temperature chlorination to produce alkyl chloride, followed by hydrolysis to allyl alcohol and then conversion with aqueous chloride to glycerol chlorohydrin, a product that can be easily hydrolyzed to glycerol (glycerin).

Halogenation. At ordinary temperatures, chlorine and bromine react with olefins by addition. For example, ethylene dichloride ($ClCH_2CH_2Cl$) is made in this way from ethylene. At slightly higher temperatures, substitution occurs. In the chlorination of propylene, a rise of 50°C (90°F) changes the product from propylene dichloride [$CH_3CH(Cl)CH_2Cl$] to allyl chloride ($CH_2=CHCH_2Cl$).

Polymerization. The polymerization of ethylene under pressure (1500–3000 psi; 10,342–20,684 kPa) at 110–120°C (230–250°F) in the presence of a catalyst or initiator, such as a 1% solution of benzoyl peroxide in methanol, produces a polymer in the 2000–3000 molecular-weight range. Polymerization at 15,000–30,000 psi (103,422–206,843 kPa) and 180–200°C (355–390°F) produces a wax melting at 100°C (212°F) and 15,000–20,000 molecular weight, but the reaction is not straightforward since there are branches in the chain. Considerably lower pressures can be used over catalysts composed of aluminum alkyls (R_3Al) in the presence of titanium tetrachloride ($TiCl_4$), supported chromic oxide (CrO_3), nickel (NiO), or cobalt (CoO) on charcoal, and promoted molybdena-alumina (MoO_2-Al_2O_3), which at the same time give products more linear in structure. Polypropylenes can be made in similar ways; and mixed monomers, such as ethylene-propylene and ethylene-butene mixtures, can be treated to give high-molecular-weight copolymers of good elasticity. Lower-molecular-weight polymers, such as the dimers, trimers, and tetramers, are used as constituents of gasoline. The materials are normally prepared over an acid catalyst. Propylene trimer (dimethylheptenes) and tetramer (trimethylnonenes) are applied in the alkylation of aromatic hydrocarbons for the production of alkylaryl sulfonate detergents, and as olefinic feedstocks in the manufacture of C_{10} and C_{13} oxo-alcohols. Phenol is alkylated by the trimer to make nonylphenol, a chemical intermediate for the manufacture of lubricating oil detergents and other products. *See* PHENOL.

Iso-butylene also forms several series of valuable products; the di- and tri-*iso*-butylenes make excellent motor and aviation fuel additives; they can also be used as alkylating agents for aromatic hydrocarbons and phenols and as reactants in the oxo-alcohol synthesis. Poly-*iso*-butylenes in the viscosity range of 10,000 m^2/s or 55,000 SUS (Saybolt Universal Seconds) [at 38°C; 100°F] have been used as viscosity index improvers in lubricating oils. 1-Butene ($CH_3CH_2CH=CH_2$) and 2-butene ($CH_3CH=CHCH_3$) participate in polymerization reactions by the way of butadiene ($CH_2=CHCH=CH_2$), the dehydrogenation product, which is copolymerized with styrene (23.5%) to form GR-S (styrene-butadiene) rubber, and with acrylonitrile (25%) to form GR-N (nitrite) rubber. Derivatives of acrylic acid (butyl acrylate, ethyl acrylate, 2-ethylhexyl acrylate, and methyl acrylate) can be homopolymerized using peroxide initiators, or copolymerized with other monomers to generate acrylic resins.

Oxidation. The most striking industrial olefin oxidation process involves the oxidation of ethylene by air using a silver catalyst at 225–325°C (435–615°F) to give pure ethylene oxide in yields ranging from 55 to 70%. Analogous higher-olefin oxides can be prepared from propylene, butadiene, octene, dodecene, and styrene via the chlorohydrin route or by reaction with peracetic acid. Acrolein ($CH_2{=}CHCHO$) is formed by air oxidation or propylene over a supported cuprous oxide catalyst or by condensing acetaldehyde and formaldehyde. When acrolein and air are passed over a catalyst, such as cobalt molybdate, acrylic acid is produced; or if acrolein is reacted with ammonia and oxygen over molybdic oxide, the product is acrylonitrile. Similarly, propylene may be converted to acrylonitrile.

Oxidation of the higher olefins by air is difficult to control, but at temperatures between 350 and 500°C (660 and 930°F) maleic acid is obtained from amylene (2-methyl-2-butene) and a vanadium pentoxide catalyst; higher yields of the acid are obtained from hexene, heptene, and octene. *See* OXIDATION PROCESS; OXIDATION-REDUCTION.

Aromatics. Aromatic compounds are valuable starting materials for a variety of chemical products. Reforming processes have made benzene, toluene, xylene, and ethylbenzene economically available from petroleum sources. A further source of supply is the aromatic-rich liquid fraction produced in the cracking of naphtha or light gas oils during the manufacture of ethylene and other olefins. The aromatics are generally recovered by extractive or azeotropic distillation, by solvent extraction (with water-glycol mixtures or liquid sulfur dioxide), or by adsorption. Naphthalene and methylnaphthalenes are present in catalytically cracked distillates. *See* AROMATIC HYDROCARBON; BENZENE; POLYNUCLEAR HYDROCARBON.

The major aromatic feedstocks are benzene (C_6H_6), toluene ($C_6H_5CH_3$), xylene ($CH_3C_6H_4CH_3$), and the two-ring condensed aromatic compound naphthalene ($C_{10}H_8$) [**Table 7**; **Fig. 3**]. Benzene is used to make styrene ($C_6H_5CH{=}CH_2$), the basic ingredient of polystyrene plastics. It is also used to make paints, epoxy resins, glues, and other adhesives. Toluene is used primarily to make solvents, gasoline additives, and explosives. Xylene is used in the manufacture of plastics and synthetic fibers and in the refining of gasoline. Naphthalene is used to manufacture insecticides. Toluene is used as a source of trinitrotoluene (TNT); and alkylation with propylene, followed by dehydrogenation, yields methylstyrene, which can be used for polymerization. Alkylation of toluene with propylene tetramer yields a product suitable for sulfonation to a detergent-grade surfactant. *See* NITROAROMATIC COMPOUND; STYRENE.

TABLE 7. Chemicals from cycloaliphatics and aromatics and their uses

Chemical	Uses
Benzene	Styrene
	Cyclohexane
	Phenol
	Detergent alkylate
	Maleic anhydride
	Aniline
	DDT
Toluene	Dealkylation to benzene
	Solvents
	Toluene diisocyanate
	Motor and aviation gasoline
	TNT
Xylenes	*p*-Xylene
	o-Xylene
	m-Xylene
	Solvents and chemicals
	Gasolines
Ethyl benzene	Styrene
Cyclohexane	Nylon intermediates
	Nonnylon uses (cyclohexanone and adipic acid)
Naphthalene	Phthalic anhydride
	Insecticides
	β-Naphthol and mothballs

Ortho-xylene is oxidized by nitric acid to phthalic anhydride, *m*-xylene to *iso*-phthalic acid, and *p*-xylene with nitric acid to terephthalic acid. These acid products are used in the manufacture of fibers, plastics, and plasticizers. Phthalic anhydride is also produced by the aerial oxidation of naphthalene at 400–450°C (750–840°F) in the vapor phase at about 25 psi (172 kPa) pressure over a fixed-bed vanadium pentoxide (V_2O_5) catalyst. Terephthalic acid is produced in a similar manner from *p*-xylene.

Chemicals from natural gas. Natural gas can be used as a source of hydrocarbons (such as ethane and propane) that are higher-molecular-weight than methane and that are important chemical intermediates. The preparation of chemicals and chemical intermediates from natural gas should not be restricted to those described here, but should be regarded as some of the building blocks of the petrochemical industry. In summary (Table 3; Fig. 1), natural gas can be a very important source of petrochemical intermediates and solvents. *See* NATURAL GAS.

The availability of hydrogen from catalytic reforming operations has made its application economically feasible in a number of petroleum-refining operations. Previously, the chief sources of large-scale hydrogen (used mainly for ammonia manufacture) were the cracking of methane (or natural gas) and the reaction between methane and steam. In the latter, at 900–1000°C (1650–1830°F) conversion into carbon monoxide and hydrogen results [reaction (7)].

$$CH_4 + H_2O \longrightarrow CO + 3H_2 \quad (7)$$

If this mixture is treated further with steam at 500°C (930°F) over a catalyst, the carbon monoxide present is converted into carbon dioxide and more

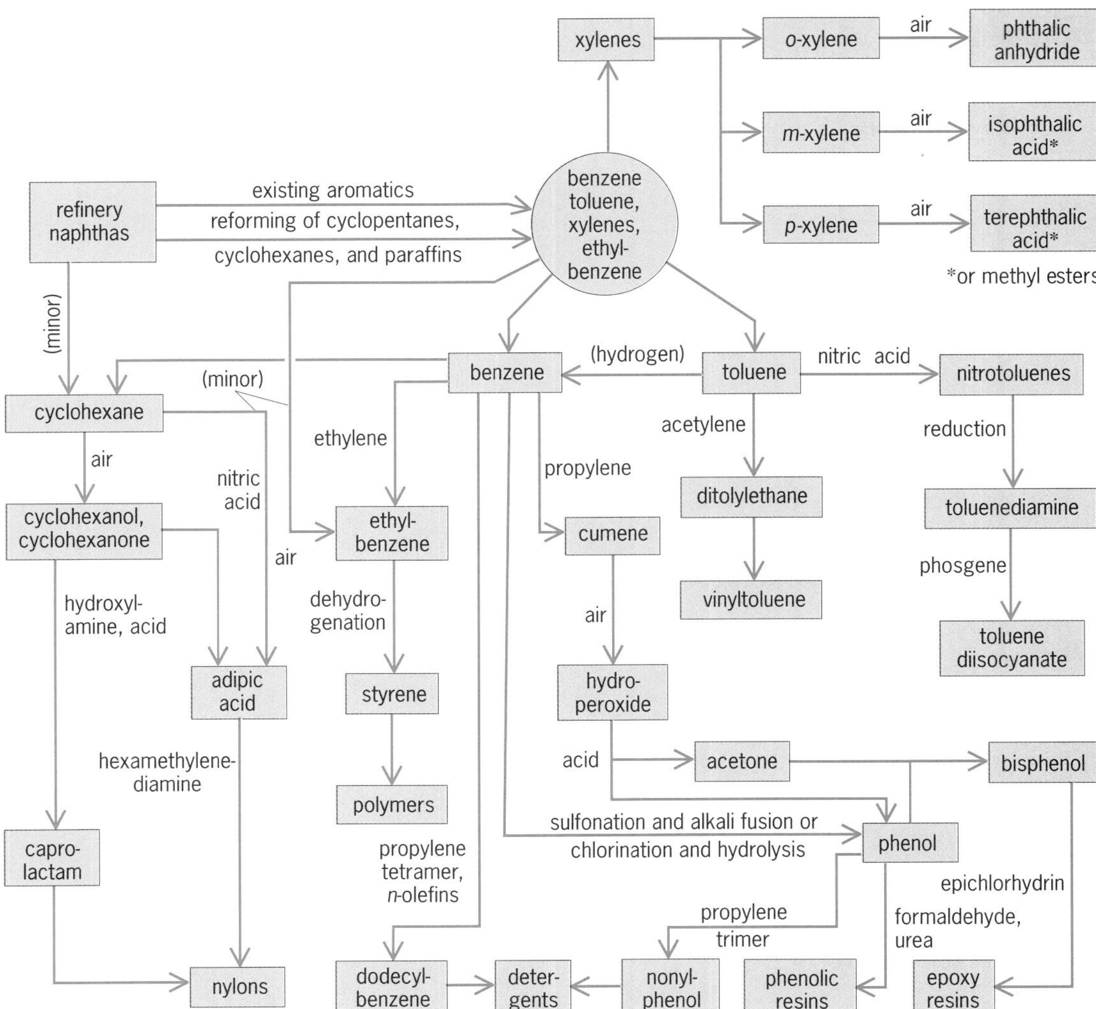

Fig. 3. Chemicals from cycloaliphatic compounds and from aromatic compounds.

hydrogen is produced [reaction (8)]. The reduction

$$CO + H_2O \longrightarrow H_2 + CO_2 \qquad (8)$$

of carbon monoxide by hydrogen is the basis of several syntheses, including the manufacture of methanol and higher-boiling alcohols. In fact, the synthesis of hydrocarbons by the Fischer-Tropsch reaction (9)

$$nCO + 2nH_2 \longrightarrow (CH_2)_n + nH_2O \qquad (9)$$

occurs in the temperature range 200–350°C (390–660°F), which is sufficiently high for the water-gas shift to take place in presence of the catalyst [reaction (10)]. The major products are olefins and

$$CO + H_2O \longrightarrow CO_2 + H_2 \qquad (10)$$

paraffins, together with some oxygen-containing organic compounds that may be varied by changing the catalyst or the temperature, pressure, and carbon monoxide–hydrogen ratio. *See* FISCHER-TROPSCH PROCESS.

The hydrocarbons formed are mainly aliphatic, and on a molar basis methane is the most abundant; the amount of higher hydrocarbons usually decreases gradually with increased molecular weight. *Iso*-paraffin formation is more extensive over zinc oxide (ZnO) or thoria (ThO_2) at 400–500°C (750–930°F) and at higher pressure. Paraffin waxes are formed over ruthenium catalysts at relatively low temperatures (170–200°C; 340–390°F), high pressures (1500 psi; 10,342 kPa), and with a carbon monoxide–hydrogen ratio. The more highly branched product made over the iron catalyst is an important factor in a choice for the manufacture of automotive fuels. On the other hand, a high-quality diesel fuel (paraffinic character) can be prepared over cobalt.

The small amount of aromatic hydrocarbons found in the product covers a wide range of possibilities. In the C_6–C_9 range, benzene, toluene, ethylbenzene, xylene, *n*-propyl- and *iso*-propylbenzene, methylethylbenzenes, and trimethylbenzenes have been identified; naphthalene derivatives and anthracene derivatives are also present.

Synthesis gas. Synthesis gas is a mixture of carbon monoxide (CO) and hydrogen (H_2) that is the beginning of a wide range of chemicals (**Fig. 4**). Many

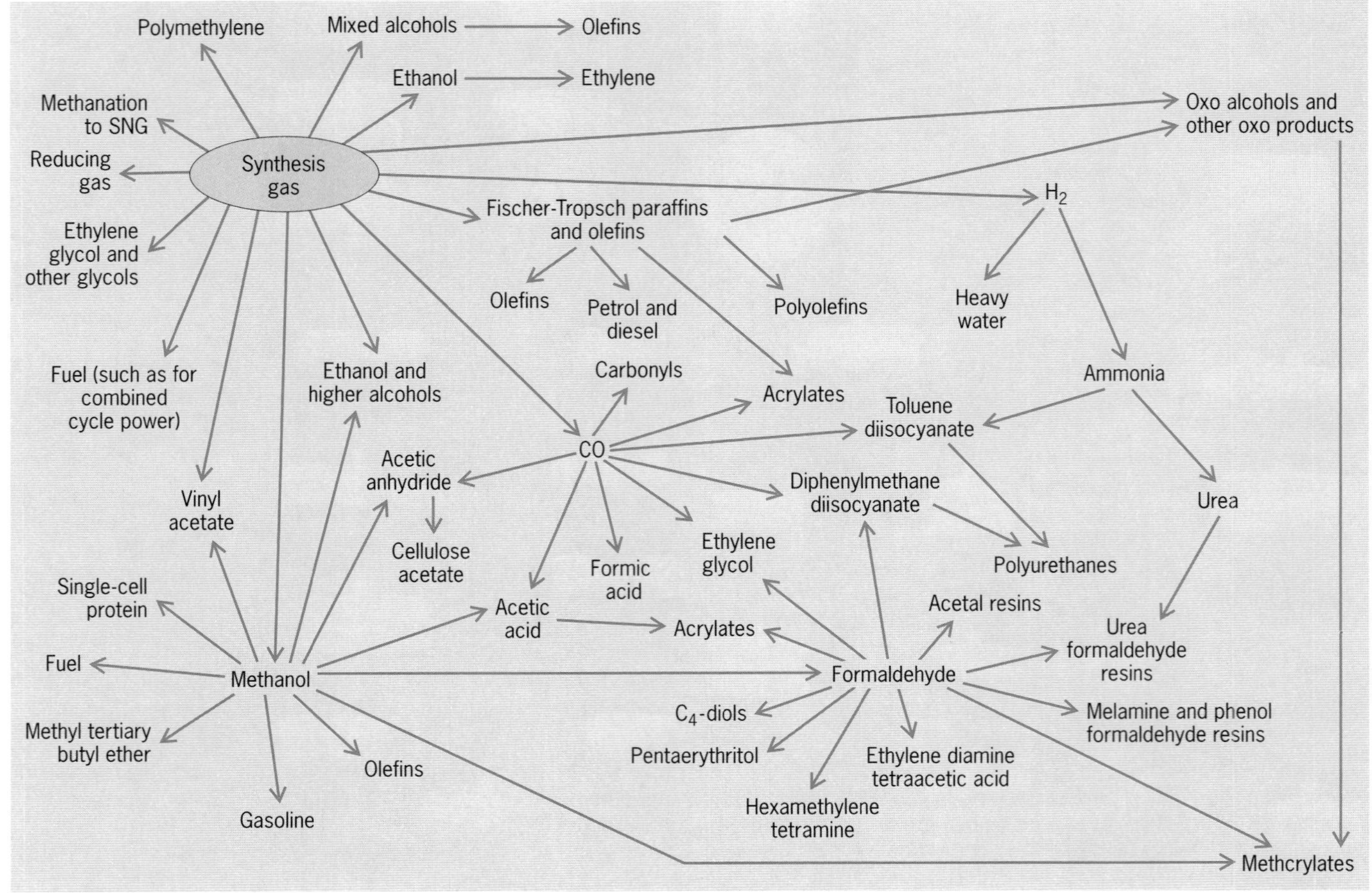

Fig. 4. Chemicals from synthesis gas. *(After J. G. Speight, The Chemistry and Technology of Petroleum, 3d ed., p. 852, Marcel Dekker, New York, 1999)*

refining catalysts promote the steam-reforming reaction (11) that leads to a product gas containing more

$$CH_4 + H_2O \longrightarrow CO + 3H_2 \qquad (11)$$

hydrogen and carbon monoxide and fewer unsaturated hydrocarbon products than the gas product from a noncatalytic process. The catalyst also influences the reactions rates in the thermal cracking reactions, which can lead to higher gas yields and lower tar and carbon yields.

Synthesis gas can be produced from heavy oil by a partial oxidizing via the oxidation process that leads to the formation of carbon monoxide and hydrogen [reaction (12)]. The initial partial oxidation step con-

$$[2CH]_{petroleum} + O_2 \longrightarrow 2CO + H_2 \qquad (12)$$

sists of the reaction of the feedstock with a quantity of oxygen insufficient to burn it completely, making a mixture consisting of carbon monoxide, carbon dioxide, hydrogen, and steam. *See* OXYGEN.

Inorganic petrochemicals. An inorganic petrochemical is one that does not contain carbon atoms; typical examples are sulfur (S), ammonium sulfate [$(NH_4)_2SO_4$)], ammonium nitrate (NH_4NO_3), and nitric acid (HNO_3). Of the inorganic petrochemicals, ammonia is by far the most common. Ammonia is produced by the direct reaction of hydrogen with nitrogen, with air being the source of nitrogen [reaction (13)].

$$N_2 + 3H_2 \longrightarrow 2NH_3 \qquad (13)$$

Ammonia production requires hydrogen from a hydrocarbon source. Traditionally, the hydrogen was produced from a coke and steam reaction, but refinery gases, steam reforming of natural gas (methane) and naphtha streams, and partial oxidation of hydrocarbons or higher-molecular-weight refinery residual materials (residua, asphalt) are the sources of hydrogen. The ammonia is used predominantly for the production of ammonium nitrate (NH_4NO_3) as well as other ammonium salts and urea (H_2HCONH_2) that are major constituents of fertilizers. *See* AMMONIA; AMMONIUM SALT; FERTILIZER; UREA.

Carbon black (also classed as an inorganic petrochemical) is made predominantly by the partial combustion of carbonaceous (organic) material in a limited supply of air. The carbonaceous sources vary from methane, to aromatic petroleum oils, to coal tar by-products. The carbon black is used primarily in the production of synthetic rubber. *See* CARBON BLACK.

Sulfur, another inorganic petrochemical, is obtained by the oxidation of hydrogen sulfide [reaction (14)]. Hydrogen sulfide is a constituent of

$$H_2S + O_2 \longrightarrow H_2O + S \qquad (14)$$

natural gas and also of the majority of refinery gas streams, especially those off-gases from hydrodesulfurization processes. Most of the sulfur is converted to sulfuric acid for the manufacturer of fertilizers and other chemicals. Other uses for sulfur include the production of carbon disulfide, refined sulfur, and pulp and paper industry chemicals. *See* PAPER; SULFUR; SULFURIC ACID. James G. Speight

Bibliography. R. A. Meyers (ed.), *Handbook of Petroleum Refining Process*, McGraw-Hill, New York, 1997; J. G. Speight, *The Chemistry and Technology of Coal*, 2d ed., Marcel Dekker, New York; J. G. Speight, *The Chemistry and Technology of Petroleum*, 3d ed., Marcel Dekker, New York, 1999; J. G. Speight, *The Desulfurization of Heavy Oils and Residua*, 2d ed., Marcel Dekker, New York, 2000; J. G. Speight (ed.), *Fuel Science and Technology Handbook*, Marcel Dekker, New York, 1990; J. G. Speight (ed.), *Petroleum Chemistry and Refining*, Taylor & Francis, New York, 1998.

Petrofabric analysis

The systematic study of the fabrics of rocks, generally involving statistical study of the orientations and distributions of large numbers of fabric elements. The term fabric denotes collectively all the structural or spatial characteristics of a rock mass. The fabric elements are classified into two groups: (1) megascopic features, including bedding, schistosity, foliation, cleavage, faults, joints, folds, and mineral lineations; and (2) microscopic features, including the shapes, orientations, and mutual arrangement of the constituent mineral crystals (texture) and of internal structures (twin lamellae, deformation bands, and so on) inside the crystals. The aim of fabric analysis is to obtain as complete and accurate a description as possible of the structural makeup of the rock mass with a view to elucidating its kinematic history. The fabric of a sedimentary rock, for example, may retain evidence of the mode of transport, deposition, and compaction of the sediment in the size, shape, and disposition of the particles; similarly, that of an igneous rock may reflect the nature of the flow or of gravitational segregation of crystals and melt during crystallization. The fabrics of deformed metamorphic rocks (tectonites) have been most extensively studied by petrofabric techniques with the objective of determining the details of the history of deformation and recrystallization. *See* FAULT AND FAULT STRUCTURES; FOLD AND FOLD SYSTEMS; STRUCTURAL GEOLOGY; STRUCTURAL PETROLOGY.

Techniques. The methods of investigation of fabrics include mapping the distribution and recording the orientations of megascopic structures in the field, and collecting selected specimens of recorded orientation for laboratory study. Measurements of the orientations of microscopic structures are made on cut, polished, or etched surfaces of oriented

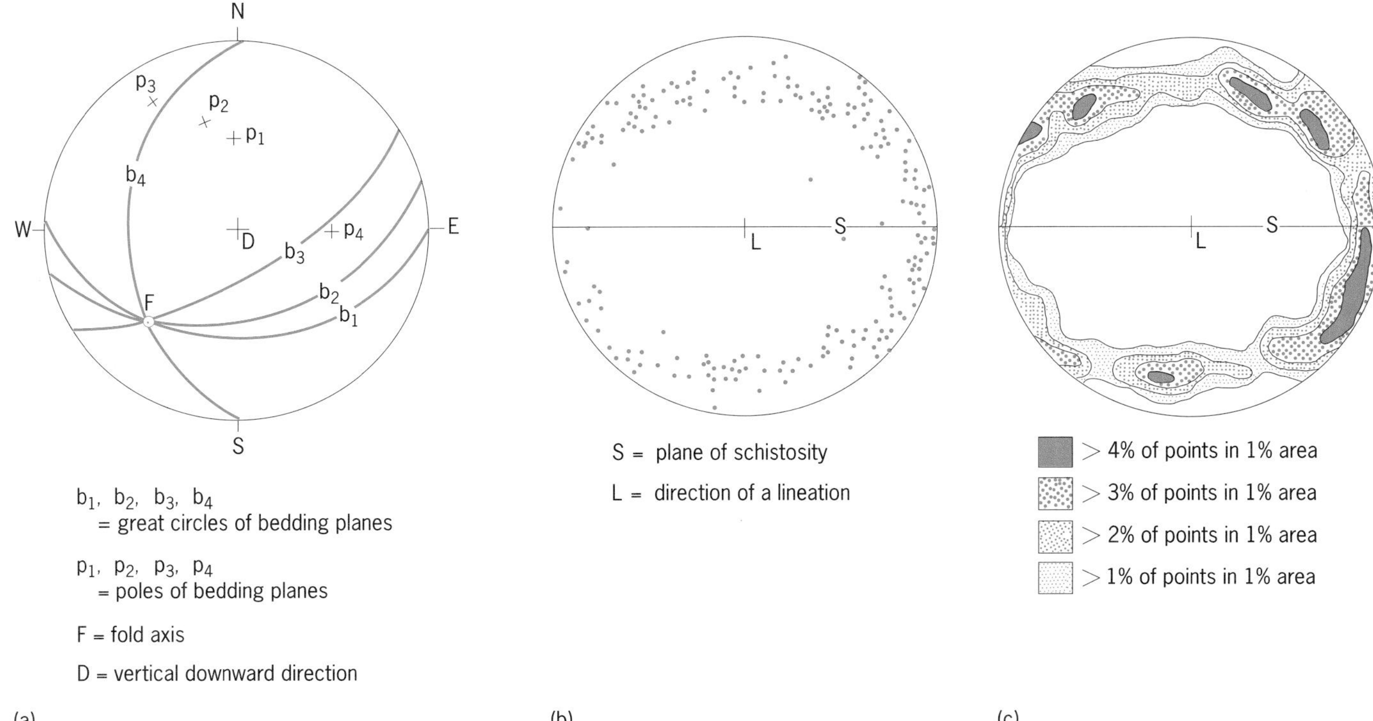

Fig. 1. Representation of the orientations of structures in spherical projections. (*a*) Stereographic projection of the orientations of four bedding planes in a cylindrical fold. The poles of the planes are distributed on a great circle perpendicular to the fold axis. (*b*) Equal-area projection of 200 *c* axes of quartz crystals in a metamorphic quartzite. The crystal axes show a strong preferred orientation. (c) Same data as *b*, contoured to show the distribution of the crystal axes.

specimens with a binocular microscope or in thin sections of known orientation with a polarizing microscope and universal stage. The orientations of planar and linear structures are analyzed by means of spherical (stereographic and equal-area) projections; each plane or line is considered to pass through the center of the sphere, so that the orientations of planes or lines are represented by a great circle or a point, respectively, in the projections (**Fig. 1***a*). If the equal-area projection is employed, the distribution of large numbers of points on the projection (representing directions such as poles of bedding, mineral lineations, or crystallographic axes of mineral grains) may be contoured to give a statistical representation of the preferred orientation of the fabric elements (Fig. 1*b* and *c*). The spatial distribution of groups of elements may be analyzed on maps or thin sections by statistical techniques. The fields of homogeneity of subfabrics of different elements must be defined. *See* PETROGRAPHY.

Applications. Significant information may be derived directly from a synthesis of the fabric data. The geometry of a rock mass is described in quantitative terms. In areas that have suffered repeated deformation it is generally possible by association to identify groups of structures assignable to separate phases of deformation. It is also possible in some instances to establish the sequence of the deformations from simple geometric considerations.

Although an exhaustive study of tectonic fabrics on all scales is seldom made, analysis of some elements of the microfabric is commonly done to supplement the information obtained from megascopic structures. Tectonites usually display textures that clearly originated by recrystallization, and many common minerals (such as quartz, micas, calcite, and dolomite) show preferred orientations of their crystal directions (Fig. 1*b* and *c*) analogous to those in deformed metals. Microscopic textures are especially informative in determining the sequence of recrystallization, and deformation in a rock, as recrystallization obliterates preexisting textures and internal structures in the mineral grains. Thus microstructures induced after metamorphism are preserved, whereas those produced before or during metamorphism are removed in the process of recrystallization and growth of new mineral phases. Fortunately, a record of the deformation during metamorphism is also retained in the textures and preferred orientations of crystals, though the exact interpretation

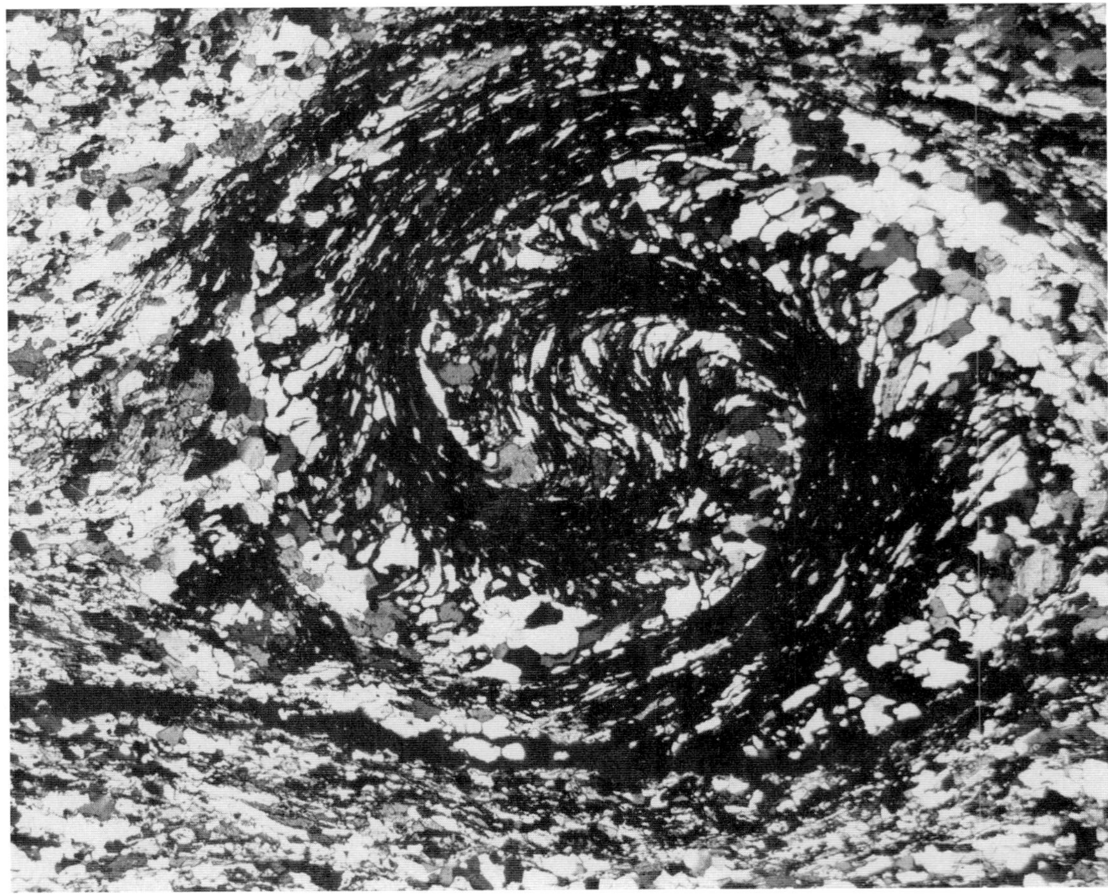

Fig. 2. Photomicrograph of a thin section of schist from Proctorsville, Vermont, showing a "snowball" garnet (crossed polarizers). The garnet (circular black area) contains many inclusions of quartz, feldspar, and mica arranged in a helical distribution as a result of rotation of the garnet during its growth. Garnet is 6 mm (0.2 in.) in diameter. (*Courtesy of J. L. Rosenfeld*)

of these records is still not fully understood. One spectacular example of such a record is seen in the trains of mineral inclusions in "rolled," or "snowball," garnets, which were rotated because of strain in the enclosing rock during their growth (**Fig. 2**).

Traditionally the strain history has been inferred from fabrics, largely by the use of symmetry arguments of the type introduced by B. Sander in his classic studies of rock fabrics: The overall symmetry of a fabric is considered to be the same as that of the deformation that produced the fabric. The general validity of such arguments is borne out by experiments, if the influence of the predeformational fabric and the sequence of incremental contributions to the strain (that is, the history of strain) are taken into account.

This approach does not yield a unique picture of the strain history, however, and more precise interpretation of fabrics is based on results of experimental deformation of rocks and mineral crystals at high temperatures and pressures. Certain types of folding, microscopic textures, and preferred orientations of minerals have been reproduced in the laboratory under known conditions closely analogous to those existing at depth in the Earth's crust. The results of these experiments provide a reliable basis for specifying the mode of origin of many features of rock fabrics. Theoretical analysis and experiments with dimensionally matched models are also contributing appreciably to the understanding of rock structures and the development of rock fabrics. *See* HIGH-PRESSURE MINERAL SYNTHESIS. John M. Christie

Bibliography. K. Bucher and M. Frey, *Petrogenesis of Metamorphic Rocks*, 6th ed., 1994; R. Mason, *Petrology of the Metamorphic Rocks*, 2d ed., 1990; A. Miyashiro, *Metamorphic Petrology*, 1994; B. B. Rao, *Metamorphic Petrology*, 1986; F. J. Turner, *Metamorphic Petrology*, 2d ed., 1981.

Petrography

The description of rocks with goals of classification and interpretation of origin. Most schemes for the classification of rocks are based on the size of grains and the proportions of various minerals. Interpretations of origin rely on field relations, structure, texture, and chemical composition as well as sizes and proportions of different kinds of grains. The names of rocks are based on the sizes and relative proportions of different minerals; boundaries between the names are arbitrary. The conditions of formation of a rock can be estimated from the types and textures of its constituent minerals. The **illustration** shows examples of petrographic features.

Rock description. The description of rocks begins in the field with observation of the shape and structure of bodies of rock at the scale of centimeters to kilometers. The geometrical relations between and structures within mappable rock units are generally the domain of field geology, but are simply rock descriptions at a reduced scale.

A petrographer can correctly name most rocks in which most crystals are larger than about 0.04 in. (1 mm) simply by examining the rock with a 10-power magnifying lens. Rocks with smaller grains require either microscopical examination or chemical analysis for proper classification.

Sizes, shapes, and orientations of grains and voids

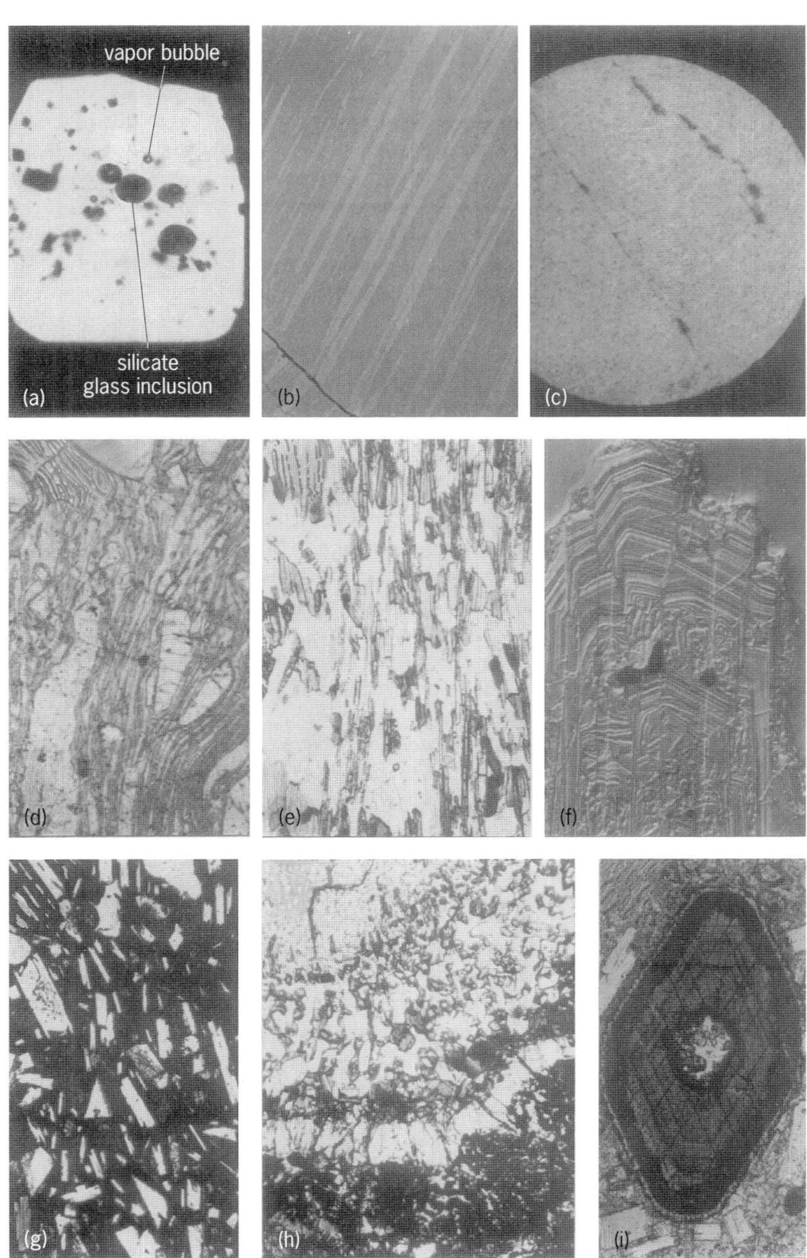

Examples of petrographic features. Images *b*, *c*, and *f* are reflected light images; the crystal in *f* has been etched and coated with a reflective coating to reveal contrasts. Images *d*, *e*, *g*, *h*, and *i* are transmitted light images, all in plane polarized light. Image *a* shows an anisotropic crystal of olivine under crossed polarized light. (*a*) Faceted crystal of olivine (Mg_2SiO_4) erupted from Kilauea volcano in 1959. (*b*) Lamellar intergrowth of hematite (Fe_2O_3, in light gray) and ilmenite ($FeTiO_3$). (*c*) Globule of sulfide minerals surrounded by basaltic glass from a submarine lava near Kilauea volcano. (*d*) Welded ash flow tuff from the Jemez Mountains, New Mexico. (*e*) Schist with aligned crystals of mica and sillimanite from southeastern Quebec, Canada. (*f*) Crystal of plagioclase ($CaAl_2Si_2O_8$–$NaAlSi_3O_8$) erupted from Fuego volcano, Guatemala, 1974. (*g*) Crystals in an andesitic lava flow from Mount Shasta, California. (*h*) Zones of minerals in a metamorphic rock. (*i*) Crystal of amphibole in an andesite from Mount Shasta.

are the most important features of a rock relevant to its origin. The same features also affect density, porosity, permeability, strength, and magnetic behavior. It is also essential to know the identity, abundance, and compositions of minerals constituting the grains in order to name a rock and infer its conditions of formation.

Rock sections. Sections of rock are used for measurement of grain size and other purposes. There are two kinds of rock sections: flat surfaces (commonly polished) which are looked at, and thin slices which are looked through. To make rock sections, the rock is first impregnated with a plastic to hold it together. The rock is then sawed into a wafer a few millimeters thick using a cutting edge embedded with diamonds. The wafer is smoothed with abrasives, glued to a glass slide, and ground to a thinness of about 0.001 in. (0.03 mm). The wafer may be polished but not thinned, or it may be both thinned and polished.

The polarizing microscope is used by petrographers to routinely examine sections of rock. Polarized light is useful in identifying minerals because most minerals are optically anisotropic and this property affects the transmission and reflection of light in characteristic ways. Rock slices are mounted in standard glue, and the glue helps reveal the relative refractive index of ground mineral grains when they are viewed under the microscope. Opaque materials are polished and then examined under the petrographic microscope using reflected light (sometimes called a metallographic microscope). Identification of minerals under reflected light depends on variable reflectivity, color, hardness, and reflection anisotropy (see illus.).

Separation of minerals. Separation of grains of different minerals is necessary for some chemical and physical studies. Magnetic and electrostatic devices and liquids with densities up to about 4.2 g/cm^3 (2.4 oz/in.^{3}) are commonly used. Similar methods are used industrially to concentrate ore minerals.

Grains. In sedimentary and some igneous rocks, grains commonly consist of more than one crystal. To determine grain size in sedimentary rocks, the petrographer must be able to discern individual particles of deposition and determine whether they are obvious pebbles in a conglomerate or fecal pellets in a mud. In igneous rocks, aggregates or intergrowths which formed individual entities during crystallization must be recognized. Discernment of grains and determination of grain size are partially subjective.

Grain size is measured by sorting of loose grains and by examination of two-dimensional sections. Grains can be loosened from certain sediments and sedimentary rocks by agitation or by the dissolution of cementing material in an appropriate solvent. Loose grains may be freed from igneous rocks such as some tuffs and pumices by dissolving the glass in an appropriate acid. The grains may then be sorted mechanically either by using sieves or by measuring the dimensions of individual grains. Alternatively the disaggregated grains may be sorted dynamically according to their rate of fall through air or water. For certain purposes the dynamical size is more important than the geometrical size.

The measurement of grain size using sections is complex because grains have three dimensions. The grain size evident in a section is smaller than that revealed by measurement of loose grains. Shape and orientation effects render interpretation of measurements in sections difficult. Crystals and grains may also be imaged using x-rays.

Grains in a rock vary in size. The size distribution characterizes the variation and has important implications for origin: for example, well-sorted sands form in environments where currents deliver and deposit grains of similar settling velocity. Even-grained, igneous rocks may suggest a particular range of cooling rates. Size grading of grains (large grains concentrated near the base of a particular unit) suggests that the grains settled through a liquid or gas. Examples exist in both sedimentary (water-laid) and igneous (melt-laid) rocks.

Grain shapes are measured by two parameters: sphericity (equidimensionality) and roundness (smoothness). Most crystals develop flat faces and sharp, angular edges and corners as they grow from liquids and gases, but are rounded by abrasion and solution. Needles and wafers may have the same sphericity and roundness, but they differ in shape. Some crystals in igneous rocks may have the same shape, but they differ in habit if their bounding faces differ crystallographically. The surface texture of particles may be revealed at high magnification with the aid of a scanning electron microscope. The surface features may indicate impacts between grains, etching of defects, overgrowths, and so on. Shape is a complex, varied, and revealing feature of grains.

Voids. These have size and shape also. Round voids (vesicles) surrounded by many grains indicate shaping due to surface tension and therefore the presence or former presence of a liquid, because most solids are too strong to be shaped by the weak forces of surface tension. Cavities with angular, faceted crystals projecting into them suggest growth of crystals from a gas or liquid.

Minerals in grains and rocks. Minerals, and sometimes their composition, are identified microscopically according to their optical properties. Compositional zoning is common and can reveal the progressive crystallization of igneous rocks. More complete and quantitative chemical analyses are possible at the scale of 0.0002 in. (0.005 mm) with the electron probe x-ray microanalyzer. Ion-beam microanalyzers can determine ages of crystallization of some grains or parts of grains in a rock. The compositions of certain associated minerals in rocks can help establish the temperature and pressure of crystallization or recrystallization, if the minerals were in equilibrium (zoned minerals cannot be in equilibrium).

The occurrence of certain garnets and diamonds in some rocks indicates that the rocks are derived from depths of 60 mi (100 km) or more. The identity and composition of certain minerals in sedimentary rocks can suggest certain kinds of source rocks, the weathering, and tectonic environments. For example, because uraninite (U_3O_8) is readily oxidized and dissolved, round grains of it in certain ancient conglomerates suggest transport in an environment with much less atmospheric oxygen than at present.

The proportions of various minerals in a rock are measured under the microscope. Commonly a mechanical stage moves a thin section by fixed increments, and the minerals encountered at the grid points are added up. The result is called the mode. It must be corrected for a thickness effect if some of the minerals occur mostly as grains as small as, or smaller than, the thickness of the section. Other methods are used to find the mode of loose grains or of polished surfaces. The mode together with the grain size generally suffices to name the rock. *See* CHEMICAL MICROSCOPY; REFLECTING MICROSCOPE; SECONDARY ION MASS SPECTROMETRY (SIMS); X-RAY FLUORESCENCE ANALYSIS; X-RAY SPECTROMETRY.

Structural modification within minerals. Structural characterization of certain minerals can help elucidate the cooling history of igneous rocks. The degree of order in certain mineral structures can be revealed optically or by x-ray crystallography or Mössbauer spectral analyses. Such properties may be inherited by the sand grains in sedimentary rocks; consequently, it may be possible to recognize sandstones derived from eroded volcanoes. *See* MÖSSBAUER EFFECT; X-RAY CRYSTALLOGRAPHY.

Intergrowths of crystals within grains. Mineral intergrowths provide a vast array of textures useful in interpreting how rocks form, but are difficult to generalize. Intergrowths within grains reveal twinning, alteration, and unmixing. Twinning may reveal certain conditions of growth or recrystallization in the presence of nonuniform stress. Particles of clay dispersed throughout minerals indicate partial alteration at temperatures below a few hundred degrees Celsius. Regular intergrowths of feldspars, pyroxenes, amphiboles, and oxide and sulfide minerals commonly are interpreted in terms of unmixing of an initially uniform (homogeneous) solid during cooling. The widths and compositions of the separate portions help evaluate the cooling history. For example, the lamellae of nickel-rich alloy in the metal of meteorites have been used to infer the cooling rate and size of the parent body. Certain regular intergrowths between unlike minerals such as quartz and feldspar are interpreted to result from competing, simultaneous growth. Sieve textures where many small grains are included as islands in a large crystal of another mineral may indicate successive growth. Reaction and solution of the smaller grains are suggested if they are widely separated and if it can be inferred that initially the small grains formed an interlocking or touching aggregate. The various textures are named in an elaborate terminology that is falling into disuse in favor of simple words, pictures, and drawings. *See* CRYSTALLOGRAPHY.

Orientations of grains. Fabric and grain distribution in rocks suggest various processes of rock formation. The alignment of large crystals suggests laminar flow or deformation. If aligned parallel to the boundary of a rock body, such fabric may be the principal evidence of igneous origin. Parallel arrangements of platy or leaflike crystals yield a foliation (which means something else to foresters). Foliation may reveal the deformational history of the rock. For example, slates are metasedimentary rocks which recrystallized in a stress field such that micas grew parallel to each other. The shiny flat surfaces of slates are formed by the cleavage surfaces of aligned micas.

The crystallographical orientations are determined by studying thin sections with a device called a universal stage. It consists of two hemispheres of glass between which the rock section is sandwiched. Three or more mutually perpendicular axes of rotation allow the section to be rotated in three dimensions. The rotations are necessary to determine the orientations of the grains. The orientations of grains in rocks help reveal how the grains were deposited or deformed. The petrographer can determine from the orientations whether a particular rock has rotated loosely relative to bedrock, and can thereby guide engineers in their search for firm foundations for bridges and dams. *See* PETROFABRIC ANALYSIS.

Liquids and gases in rocks. Many crystals in ores contain inclusions of gas and liquid. The freezing temperatures of the inclusions help reveal the salinities of ore-depositing solutions. The temperatures of homogenization of liquid and vapor can yield information about the temperature and pressure at the time of ore formation.

Applications. Petrographers study organic as well as inorganic objects, and petrographic analyses are useful to both paleontologists and petroleum geologists. The quality of coal is revealed with polarizing and reflecting light microscopes. Fossil wood was one of the first objects studied with a polarizing microscope, in 1827. The daily and seasonal tree-ring-like increments of growth of certain corals and mollusks are revealed microscopically, and have permitted paleontologists to estimate the greater number of days in the year hundreds of millions of years ago. Tiny fossils in cherty rocks about 2 billion years old reveal the existence of algae at a time when the atmosphere probably began to contain free oxygen. The families of shell fragments in limestones are evident under the microscope, and are routinely used for purposes of correlation and characterization of the environment of deposition. Inclusions of petroleum and brine in crystals of silicates and salt in rocks help scientists infer how petroleum formation is connected with cementation and other modifications of buried sediments. *See* PALEONTOLOGY; PETROLEUM GEOLOGY.

Petrographers also study synthetic objects. The textures of metals and alloys are scrutinized by petrographers in order to understand what makes these materials strong and resistant to corrosion. Flaws in glasses and ceramics are revealed by microscopical and polarizing techniques. Rims of hydrated glass on obsidian artifacts increase in thickness with time after burial and help anthropologists date them. The microscopic roughness of the cutting edges of stone tools can reveal their principal use. Fragments of minerals and rocks in some pottery can help point to its source and help trace prehistoric routes of trade. The industrial, agricultural, and natural sources of particles in the air and water may be established from petrographic study. *See* MINERALOGY; PETROLOGY.

Alfred T. Anderson, Jr

Bibliography. H. Blatt and R. J. Tracy, *Petrology: Igneous, Sedimentary, and Metamorphic*, 2d ed., 1996; R. V. Dietrich and B. J. Skinner, *Rocks and Rock Minerals*, 1979; D. W. Hyndman, *Petrology of Igneous and Sedimentary Rocks*, 2d ed., 1985; A. Philpotts, *Principles of Igneous and Metamorphic Petrology*, 1990; L. A. Raymond, *Petrology: The Study of Igneous, Sedimentary, and Metamorphic Rocks*, 1995.

Petrolatum

A smooth, semisolid blend of mineral oil with waxes crystallized from the residual type of petroleum lubricating oil. The wax molecules contain 30–70 carbon atoms and are straight chains with a few branches or naphthene rings. They are microneedles and hold a large amount of oil in a gel. Petrolatums are useful because they cling, lubricate, and resist both moisture and oxidation. They serve as lubricants in baking and candymaking; as carriers in polishes, cosmetics, and ointments; as rust preventives; as waterproofing agents for paper; and in other uses calling for an inert greaselike material. *See* LUBRICANT; PETROLEUM PRODUCTS; WAX, PETROLEUM.

J. K. Roberts

Petroleum

Unrefined, or crude, oil is found underground and under the sea floor, in the interstices between grains of sandstone and limestone or dolomite (not in caves). Petroleum is a mixture of liquids varying in color from nearly colorless to jet black, in viscosity from thinner than water to thicker than molasses, and in density from light gases to asphalts heavier than water. It can be separated by distillation into fractions that range from light color, low density, and low viscosity to the opposite extreme. In places where it has oozed from the ground, its volatile fractions have vaporized, leaving the dense, black parts of the oil as a pool of tar or asphalt (such as the Brea Tar Pits in California). Egyptians used such tar for embalming the dead, and Mesopotamians used it for adhering bricks together. Much of the world's crude oil is today produced from drilled wells. *See* PETROLEUM ENGINEERING.

Composition. Petroleum consists mostly of hydrocarbon molecules. The four main classes of hydrocarbons are paraffins (also called alkanes), olefins (alkenes), cycloparaffins (cycloalkanes), and aromatics. Olefins are absent in crude oil but can be formed in certain refining processes.

Alkanes consist of straight or branched chains of carbon atoms surrounded by hydrogen atoms. Each carbon atom is capable of forming four chemical bonds with other atoms, while each hydrogen atom can from only one bond. The simplest hydrocarbon is one carbon atom bonded to four hydrogen atoms (chemical formula CH_4), and is called methane (structure **1**). The proportions of hydrogen and carbon

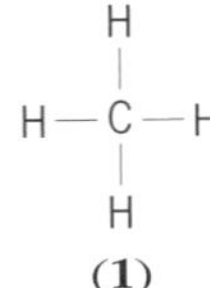

(1)

atoms in the alkane series is expressed as C_nH_{2n+2}, where n is the number of carbon atoms in each molecule. *See* ALKANE; PARAFFIN.

Branched-chain alkanes are called isoalkanes. An example is isooctane (2,2,4-trimethylpentane, C_8H_{18}). Sometimes the carbon atoms are joined as a closed ring, called a cycloalkane. This can occur only for a ring of three to seven carbon atoms, although only five- and six-membered-ring compounds occur naturally. *See* ALICYCLIC HYDROCARBON.

When carbon atoms are joined by a double bond in a straight or branched chain, they are called olefins (alkenes). If double bonds occur in a ring, only every other carbon-carbon bond can be double. The simplest six-membered (aromatic) ring is called benzene (C_6H_6; structure **2**). Napthalene ($C_{10}H_8$; structure **3**)

(2) (3)

is formed when a four-carbon chain containing two double bonds is attached at each end to adjacent carbons in a benzene ring. It is also possible to have aromatics with multiple fused rings. *See* AROMATIC HYDROCARBON.

Petroleum usually contains all of the possible hydrocarbon structures except alkenes, with the number of carbon atoms per molecule going up to a hundred or more (see **table**). These fractions include compounds that contain sulfur, nitrogen, oxygen, and metal atoms. Sulfur and oxygen may be covalently bonded to nonring hydrocarbon compounds or may occur as one atom of a carbon ring. Nitrogen is found only as one atom of a ring. Metal atoms occur in the lubricating oil range of some crude oils as salts of organic acids (similar to acetic acid but much larger). The proportion of compounds containing these atoms increases with increasing size of the molecule.

Asphaltic molecules contain many cyclic compounds in which the rings contain sulfur, nitrogen, or oxygen atoms; these are called heterocyclic compounds. An example is pyridine (structure **4**). There

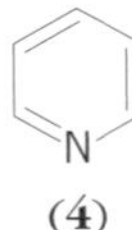

(4)

Range of carbon numbers per molecule for the fractions of petroleum commonly produced by distillation

Carbon numbers	Fraction
1–2	Natural gas
3–4	Liquefied petroleum gas (LPG)
5–11	Gasoline
12–14	Kerosene or jet fuel
15–19	Fuel oil or diesel fuel
20–40	Heavy gas oil, light to medium lubricating oils
40+	Asphalt (residue left by distillation)

can also be porphyrin structures having four nitrogen-containing rings with the nitrogens facing each other and linked weakly to a metal atom between the four. Some refineries use processes to remove the metals from heavy gas oil fractions before catalytic cracking (a process used to reduce the molecular weight of hydrocarbons by breaking molecular bonds). This is because the metals may deposit on the cracking catalyst and make it less active. *See* CRACKING; HETEROCYCLIC COMPOUNDS; PORPHYRIN.

Asphalt contains a high proportion of heavy, viscous oil in the lubricating oil range, plus a number of high-molecular-weight aromatic-heterocyclic solids (asphaltenes). The asphaltenes are dispersed in the oil in a very fine colloidal form by similar molecules of about half the molecular weight, called resins. Typical road asphalt contains about 25% asphaltenes, 5–15% resins, and the rest viscous oil, as does the asphalt used in roofing tar and shingles. *See* ASPHALT AND ASPHALTITE; PAVEMENT; ROOF CONSTRUCTION.

Composition changes in refining. In refining petroleum at the present time, it is normally necessary to change the hydrocarbon structures to improve the amount and performance of the desired products. Crude oil usually has too many heavy fractions. This is remedied by catalytically cracking the heavy gas oil or even thermally cracking the asphalt and making gas, gasoline, and fuel oil from it. The gasoline obtained directly from the crude oil has too low an octane number for use in modern automobiles, so it is catalytically reformed or isomerized to raise the octane number. Catalytic refining processes use catalysts which speed the reactions and preferentially make the desired products. Sometimes the small olefins resulting from catalytic and thermal cracking can be linked together by another catalyst to make a gasoline-range isoalkane. This was how isooctane was first made. Alkylate (a gasoline component) made this way is currently used in premium gasoline. *See* GASOLINE; OCTANE NUMBER; PETROCHEMICAL; PETROLEUM PROCESSING AND REFINING; PETROLEUM PRODUCTS.

Lubricating oils, which have had wax and aromatics removed, are often treated catalytically with hydrogen to remove undesired trace components. The asphalt is air-blown to make it more suitable for manufacturing roof shingles or for paving roads. In air-blowing, part of the resins are converted into asphaltenes and the product becomes stiffer, better able to stand hot weather.

Some of the intermediate refining products go to petrochemical plants where they are converted catalytically into rubber, plastics, and textile fibers.

The yield of valuable products from modern refineries is usually slightly greater than the amount of crude oil from which they are made. Chemical additives are used in most refinery products to improve their end-use performance.

Origin. It is generally agreed that petroleum formed by processes similar to those which yielded coal, but was derived from small animals rather than from plants. Geophysical knowledge of plate tectonics has revealed that moving plates of the Earth's surface layers have piled up mountain ranges where the plates collided. The erosion of these mountains by water and ice, wind, and gravity led to the grinding up of enormous amounts of rock into sand, silt, and mud which settled to the bottom of lakes, rivers, and seas. The lakes and seas teemed with mostly small organisms. Some of these organisms extracted calcium from the salts in the water and reacted the calcium with carbon dioxide to form calcium carbonate shells that protected them from predators.

Dead organisms have been buried in mud over millions of years. Further layers deposited over these mud layers have in some cases reached a thickness of thousands of feet, and compacted the layers beneath them, until the mud has become shale rock. When a layer was particularly rich in broken sea shells, it was compacted into limestone. Sandy layers became sandstone. The deep interior of the Earth is as hot as the Sun's surface, and heat moving up toward the Earth's surface makes the temperature in a given layer hotter the deeper it is buried. Thus, the mud layers were heated and compressed by the layers above. The bodies of the organisms in the mud were decomposed and converted into fatty liquids and solids. Heating these fatty materials over a very long time caused their molecules to break into smaller fragments and combine into larger ones, so the original range of molecular size was spread greatly into the range found in crude oil. Bacteria were usually present, and helped remove oxygen from the molecules (which began mostly as carbohydrates, comprising carbon, oxygen, and hydrogen) and turned them into hydrocarbon compounds. The great pressure of the overlying rock layers helped to force the oil out of the compacted mud (shale) layers into less compacted limestone, dolomite, or sandstone layers next to the shale layers. *See* DOLOMITE; LIMESTONE; ORGANIC GEOCHEMISTRY; PETROLEUM GEOLOGY; SANDSTONE; SEDIMENTOLOGY; SHALE.

Water with dissolved salts is present everywhere underground. Oil is less dense than water, so it moves up through the permeable rock wherever it can go. If it can reach the surface, it does. But if the rock layer which it enters has been bent by tectonic processes into a ridge or a dome structure, the oil is trapped there. A liquid's pressure in underground porous rock is equal to the pressure of an equal depth of water, and can even be higher. Trapped oil is under enough pressure that it can keep in solution a considerable amount of the hydrocarbon gas made by the oil generation processes.

At depths greater than about 25,000 ft (7620 m), the temperature is so high that the oil conversion processes go all the way to natural gas and soot. Crude oil is found generally at depths of a few hundred feet to about 20,000 ft (6100 m). Layers have often had overlying layers eroded away, so strongly converted oil can be found at shallower depths. Natural gas formed by the conversion processes is now also found over a variety of depths which do not indicate the depth and temperature of their origin. *See* NATURAL GAS.

Nature of oil reservoirs. The oil formed by the natural thermal and bacterial processes was squeezed out of the compacting mud layers into sandstone or limestone layers and migrated upward in tilted layers. Tectonic processes caused such uptilting and bulging of layers to form ridges and domes. When the ridges and domes were covered by shale already formed, the pores of the shale were too tiny to let the oil through, so the shale acted as a sealing cap. In some cases, the migrating oil came to a place where the tilted sandy or limy layer through which they were oozing had been eroded away and a new layer of shale formed as a horizontal layer above, cutting off the tilted layer. When the oil could not rise farther, it was trapped. Often, a number of layers were bent or folded at the same time, and the oil migrated simultaneously into a number of stacked layers in a given ridge or domal structure, underlain and surrounded by water. Porous rock in such a structure that contains oil or gas is called an oil or gas reservoir.

Over geologic time, sandy deserts have been formed on land. The sand was later submerged and covered with mud, which became shale. Oil has in some cases migrated into such buried desert sands and been trapped.

Shallow seas and vast lakes have existed in past times and were later buried, then uplifted. Rivers with sand bars and muddy bottoms and deltas have also contributed source rocks and reservoir rocks.

Certain mixtures of sand, silt, and mud retained their animal content while the conversion processes generated hydrocarbons, and in some cases tarry oils, which probably remained where they were generated. The Athabasca Tar Sands and the Orinoco Tar Belt may have formed this way. The heavy oil from these deposits requires preprocessing, that is, thermal cracking or hydrocracking, prior to refinery processing. It is therefore less economical to process, relative to lighter crude oils. *See* OIL SAND.

In other cases, the organic fatty acids, which could have been thermally converted to hydrocarbon liquids, reacted with alkaline minerals present in their source rock, as did the oil shale deposits in Colorado and elsewhere. Organic acids are characterized by the formula RCOOH, where R is the large hydrocarbon end of the molecule, and H is a positively charged hydrogen ion which can react with an alkali (in this case, sodium bicarbonate, HOCOONa) to form water, HOH, and RCOOCOONa. Upon heating in a pyrolysis reactor, part of the compound is converted to CO_2 (carbon dioxide), the alkali is regenerated, and the rest becomes a hydrocarbon vapor, which can be condensed to produce shale oil. This oil is usually high in sulfur and nitrogen and metal compounds, and before entering a normal refining processes, it must be hydrotreated to reduce the amount of these deleterious compounds. The hydrogen necessary for the hydrotreating can be made from part of the hydrocarbons produced, but this consumes a significant part of the shale oil. Disposal of the spent alkaline rock is also a potential environmental problem. Thus, the recovery of oil suitable for further refining from shale oil deposits is generally not economical, compared to normal petroleum. *See* OIL SHALE.

Even with light crude oils, the recovery from typical reservoirs is not as high as might be thought. Multiple-layer reservoirs will typically contain oil-bearing layers with a wide range of permeability. When recovery from the highest-permeability layers is as complete as it can be, the low-permeability layers will usually have been only slightly depleted, despite all efforts to improve the recovery. The approximate success of various oil recovery processes is as follows (varying from case to case): primary depletion (by solution gas liberation and pumping), ~10–20%; secondary water flooding, ~15–35%; tertiary processes (propane slug, carbon dioxide slug), ~10–15%; steam flooding (including steam soak and backflow), ~5–40%. The recovery from primary, secondary, and tertiary processes is additive, from steam flooding is not. Despite recovery efforts, half or more of the oil originally present in United States oil reservoirs is still in them. *See* PETROLEUM ENHANCED RECOVERY; PETROLEUM RESERVES; PETROLEUM RESERVOIR ENGINEERING.

Oil is still being formed by natural processes—perhaps as much as hundreds or a few thousands of barrels per year. Of course, formation cannot keep up with consumption, so known oil and gas deposits are being depleted at such a rate that most of these will be consumed by some time in the twenty-first century. Elmond L. Claridge

Heavy oil. Heavy oil and tar sand oil (bitumen) are petroleum hydrocarbons found in sedimentary rocks. They are formed by the oxidation and biodegradation of crude oil, and occur in the liquid or semiliquid state in limestones, sandstones, or sands. *See* BITUMEN.

These oils are characterized by their viscosity; however, density (or API gravity) is also used when viscosity measurements are not available. Heavy oils have gas-free viscosities between 100 and 10,000 millipascal seconds (centipoise) at reservoir conditions, and their densities range from 1000 kg/m^3 (10°API) to 943 kg/m^3 (20°API) measured at 15.6°C and atmospheric pressure. Extraheavy oils or tar sand oil (bitumen) has gas-free viscosities greater than 10,000 millipascal seconds (centipoise) and densities greater than 1000 kg/m^3 (less than 10°API). Heavy oils contain 3 wt % or more sulfur and as much as 200 ppm vanadium. Titanium, zinc, zirconium, magnesium, manganese, copper, iron, and aluminum are other trace elements that can be found in these deposits. Their high naphthenic acid content makes refinery processing equipment vulnerable to corrosion. *See* OIL AND GAS FIELD EXPLOITATION.

Worldwide sources of heavy oils are estimated at over 950 billion cubic meters (6000 billion barrels) compared to conventional oil reserves of 258 billion cubic meters (1620 billion barrels). However, the true resource base is believed to be much higher since heavy oil discoveries are not well defined or documented. Presently, major heavy oil deposits are found in Canada, Venezuela, and Russia. Canada has the world's largest heavy oil reserves at the Athabasca Tar Sands, Cold Lake, and Peace River. Deposits in Venezuela are located in the Orinoco Belt. Russia has reserves in the Lena-Anabar basin (eastern Siberia) and Tataristan. There are other countries where such deposits are known to exist, including the United States (Utah, Texas, and California), Turkey (Bati Raman), China, and Indonesia. These resources are generally found within 1500 m (4500 ft) of the Earth's surface. Production, transportation, and refining these hydrocarbons require technologies that are expensive and environmentally sensitive.

Heavy oils that are in the liquid state in the reservoir as well as at surface conditions can be recovered using techniques known as steam injection, in situ combustion, or chemical flooding with conventionally drilled oil wells. These processes, known as in situ recovery methods, all aim at decreasing the viscosity of the oil so that it will flow easily through the porous reservoir rock to the production wells. *See* PETROLEUM ENHANCED RECOVERY.

In steam injection, steam generated at the surface is injected down through the wells into the reservoir. Steam injection temperatures up to 250°C have been used. The viscosity of the oil decreases as the hot fluids contact the oil in the reservoir rock, increasing the flow toward the production wells.

For in situ combustion, air is injected down the well, and when it contacts the oil reservoir combustion starts with some initial downhole heating. The continuous injection of air or air-water mixtures helps drive the combustion front through the reservoir, displacing more oil toward the production wells. Temperatures are as high as 500°C in this application. Both of these heat injection methods require specialized production equipment to withstand the high temperatures.

There are several different types of chemical flooding; however, the method known as carbon dioxide flooding has found application in heavy oil fields. Carbon dioxide is injected into the reservoir through the wells and dissolves in the oil, thus reducing its viscosity as it moves through the porous rock. Oil is recovered by the continuous injection of carbon dioxide or injection and production from the same wellbore, known as the huff and puff method.

The application of these methods may increase heavy oil recovery from 1–3% to 10–30% of the oil in place.

Extraheavy oils or tar sand bitumens—semisolid or solid hydrocarbons found close to the Earth's surface—are mined. Following the mining operation, the tar sands are crushed into smaller-sized particles, and then the bitumen is extracted. There are three main processes for extracting bitumen from tar sands. The hot water process uses sprayed hot water, often with added steam and caustic to separate the bitumen from the sand. The resulting mixture is sent to a flotation cell to separate the sand from the bitumen. The bitumen is treated to reclaim the solvents, and clean bitumen is sent for various end uses. In the solvent extraction process, crushed sand is mixed with an organic solvent, and then the dissolved bitumen and sand are separated. Following processing, the solvent is treated for recycling. The thermal processes involve the thermal decomposition of the bitumen. Crushed sand is sent to a reactor where it is heated. The product is a thermally cracked liquid, and the by-products are coke and sand. Another type of thermal process, known as the Taciuk process, has been developed and tested in Canada. The oil produced by the above methods is called synthetic crude oil or syncrude. With today's technology it is possible to produce 0.159 m^3 (1 barrel) of syncrude from 2 tons of tar sand. *See* SYNTHETIC FUEL.

Ender Okandan

Bibliography. G. D. Hobson (ed.), *Modern Petroleum Technology*, 5th ed., 1984; G. D. Hobson and E. N. Tiratsoo, *Introduction to Petroleum Geology*, 2d ed., 1985; R. R. Kinghorn, *An Introduction to the Physics and Chemistry of Petroleum*, 1983; E. Okandan (ed.), *Heavy Crude Oil Recovery*, Martinus Nijhof, 1984; J. H. Tatsch, *Petroleum Deposits; Origin, Evolution, and Present Characteristics*, 1974; B. P. Tissot, *Petroleum Formation and Occurence*, Springer-Verlag, 1984; *Unitar 7th International Conference on Heavy Crude and Tar Sands*, Beijing, China, October 27–30, 1998; A. J. Walker, *The Oil and Gas Book*, 1985; R. R. Wheeler and M. Whited, *Oil from Prospect to Pipeline*, 5th ed., 1985.

Petroleum engineering

An eclectic discipline comprising the technologies used for the exploitation of crude oil and natural gas reservoirs. It is usually subdivided into the branches of petrophysical, geological, reservoir drilling, production, and construction engineering. After an oil or gas accumulation is discovered, technical supervision of the reservoir is transferred to the petroleum engineering group, although in the exploration phase the drilling and petrophysical engineers have played a role in the completion and evaluation of the discovery.

Petrophysical engineering. The petrophysical engineer is perhaps the first of the petroleum engineering group to become involved in the exploitation of the new discovery. By the use of down-hole logging tools, hydrocarbon analysis from the circulating mud and drill cuttings, and visual or laboratory analysis of cores or drill cuttings, the petrophysicist estimates the porosity, permeability, and fluid content of the reservoir rock. In addition, the petrophysical engineer is called upon to assist the geologist, geophysicist, and drilling engineer with subsurface information. *See* WELL LOGGING.

Geological engineering. The geological engineer, using petrophysical and petrographic data, seismic surveys conducted during the exploration operations, and an analysis of the regional and environmental geology, develops inferences concerning the lateral continuity and extent of the reservoir. However, this assessment usually cannot be verified until additional wells are drilled and the geological and petrophysical analyses are combined to produce a firm diagnostic concept of the size of the reservoir, the distribution of fluids therein, and the nature of the natural producing mechanism. Additional geophysical data, particularly three-dimensional seismic, may be useful to help delineate a complex field. As the understanding of the reservoir develops with continued drilling and production, the geological engineer, working with the reservoir engineer, selects additional drill sites to further develop and optimize the economic production of oil and gas. *See* PETROLEUM GEOLOGY; SEISMIC EXPLORATION FOR OIL AND GAS.

Reservoir engineering. The reservoir engineer, using the initial studies of the petrophysicist and geological engineers together with the early performance of the wells drilled into the reservoir, attempts to assess the producing rates (barrels of oil or millions of cubic feet of gas per day) that individual wells and the entire reservoir are capable of sustaining. One of the major assignments of the reservoir engineer is to estimate the ultimate production that can be anticipated from both primary and enhanced recovery from the reservoir. The ultimate production is the total amount of oil and gas that can be secured from the reservoir until the economic limit is reached. The economic limit represents that production rate which is just capable of generating sufficient revenue to offset the cost of operating the reservoir. The proved reserves of a reservoir are calculated by subtracting from the ultimate recovery of the reservoir (which can be anticipated using available technology and current economics) the amount of oil or gas that has already been produced. *See* PETROLEUM RESERVES.

Primary recovery operations are those which produce oil and gas without the use of external energy except for that required to drill and complete the wells and lift the fluids to the surface (pumping). Enhanced oil recovery, or supplemental recovery, is the amount of oil that can be recovered over and above that producible by primary operation by the implementation of schemes that require the input of significant quantities of energy. In modern times waterflooding has been almost exclusively the supplementary method used to recover additional quantities of crude oil. However, with the realization that the discovery of new petroleum resources will become an increasingly difficult achievement in the future, the reservoir engineer has been concerned with other enhanced oil recovery processes that promise to increase the recovery efficiency above the average 33% experienced by the United States (which is somewhat above that achieved in the rest of the world). The restrictive factor on such processes is the economic cost of their implementation. *See* PETROLEUM ENHANCED RECOVERY.

In the past the reservoir engineer was confined to making predictions of ultimate recovery by using analytical equations for fluid flow in the framework of an overall definition of the geological and lithological description of the reservoir. Extensive references to reservoirs that are considered to have analogous features, and history matching, such as curve fitting and extrapolation of the declining production, are important tools of the reservoir engineer. The analytical techniques are limited in that reservoir heterogeneity and the competition between various producing mechanisms (solution gas drive, water influx, gravity drainage) cannot be accounted for, nor can one reservoir be matched exactly by another. The reservoir engineer has also been able to use mathematical modeling to simulate the performance of the reservoir. The reservoir is divided mathematically into segments (grid blocks), and the appropriate flow equations and material balances are repeatedly applied to contiguous blocks. Extensive history matching is still required to achieve a reliable predictive model, but this can usually be achieved more quickly and reliably than with the use of analytical solutions and analogy. Numerical simulation is almost routine with the power of desktop computers; complex simulation of enhanced recovery processes such as thermal recovery or miscible flooding may be most effectively run on a powerful mainframe computer. *See* COMPUTER; MICROCOMPUTER; PETROLEUM RESERVOIR ENGINEERING.

Drilling engineering. The drilling engineer has the responsibility for the efficient penetration of the earth by a well bore, and for cementing of the steel casing from the surface to a depth usually just above the target reservoir. The drilling engineer or another specialist, the mud engineer, is in charge of the fluid that is continuously circulated through the drill pipe and back up to surface in the annulus between the drill pipe and the borehole. This mud must be formulated so that it can do the following: cool the bit; carry the drill cuttings to the surface, where they are separated on vibrating screens; gel and hold cuttings in suspension if circulation stops; form a filter cake over porous low-pressure intervals of the earth, thus preventing undue fluid loss; and exert sufficient pressure on any gas- or oil-bearing formation so that the fluids do not flow into the well bore prematurely, blowing out at the surface. In special situations, air may be used instead of mud. As drilling has gone deeper and deeper into the earth in the search of additional supplies of oil and gas, higher and higher pressure formations have been encountered. This has required the use of positive-acting blow-out preventers that can firmly and quickly shut off uncontrolled flow due to inadvertent unbalances in the mud system. Important new technology being used routinely by the drilling engineer includes downhole motors to rotate the bit, novel bit designs, measurement of borehole and formation characteristics while drilling, and horizontal drilling. *See* OIL AND GAS WELL DRILLING.

Production engineering. The production engineer, upon consultation with the geological, petrophysical, reservoir, or completion engineers, plans the completion procedure for the well. This involves a choice of setting a liner across the formation or perforating a casing that has been extended and cemented across the reservoir, selecting appropriate pumping techniques, and choosing the surface collection, dehydration, and storage facilities. The production engineer also compares the productivity index of the well (barrels per day per pounds per square inch of drawdown around the well bore) with that anticipated from the measured and inferred values of permeability, porosity, and reservoir pressure to determine whether the well has been damaged by the completion procedure. Such comparisons can be supplemented by a knowledge of the rate at which the pressure builds up at the well bore when the well is abruptly shut-in. Using the principles of unsteady-state flow, the reservoir engineer can evaluate such a buildup to assess quantitatively the nature and extent of well bore damage. Damaged wells, or wells completed in low-permeability formations, can be stimulated by acidization, hydraulic fracturing, additional perforations, or washing with selective solvents and aqueous fluids. *See* OIL AND GAS WELL COMPLETION.

Construction engineering. Major construction projects, such as the design and erection of offshore platforms, require the addition of civil engineers to the staff of petroleum engineering departments, and the design and implementation of natural gasoline and gas processing plants require the addition of chemical engineers. *See* CIVIL ENGINEERING; OIL AND GAS, OFFSHORE; PETROLEUM; PETROLEUM PROCESSING AND REFINING.

The technology has become increasingly sophisticated and demanding with the implementation of new recovery techniques, deeper drilling, and the expanding frontiers of the industry into the hostile territories of arctic regions and deep oceans.

Todd M. Doscher; R. E. Wyman

Data analysis and mapping. In petroleum exploration, relational databases and advanced computer graphics have radically altered the work environment of computers from processing applications to interpretive usages. There is a heavy emphasis on facile gathering of data and extraction of selected items to provide effective displays and interpretations.

The analysis of subsurface petroleum information poses a dual problem: there is either too much or insufficient data for any given application. In exploration, the volumes of data from seismic processing and interpretation can strain the limits of most mass-storage devices, and sophisticated processing routines can require supercomputers. On the other hand, data from wells can be minimal to nonexistent and may require extreme extrapolation and model-based solutions. Production applications can also be computer-intensive. As an example, reservoir oil simulation is feasible only in advanced modules on supercomputers. In most cases, there is no happy medium. Because of this dichotomy, computing environments in the petroleum industry run the full range from laptop field applications to main-frame installations. The computing requirements have become visually oriented, requiring a large dependency on interactive, graphic-display, and visualization routines.

In general, petroleum computing can be viewed on three levels: geological computing, geophysical computing, and engineering applications. Geological computing trends have focused on database and spatial system configurations, with specialty applications such as cross-section balancing or geochemical modeling. Geophysical computing tends to be computer-intensive; interpretive installations are, like all interactive workstation environments, driven by graphics. Engineering applications are also computer-intensive; they are generally classified as either simulation or process types.

Geological computing. Proliferation of high-quality workstation environments has given an individual geoscientist the tools to create maps and perform analyses on a large scale and to incorporate data to a much larger extent than was previously possible. The demand for data has led to the development of major products for exploration applications. Databases, language structures, and hardware that facilitate access and management of data have become a necessary part of computing environments in companies of all sizes.

In addition to database software, geographic information systems (GIS), initially developed for resource planning and cultural analysis, have been adapted to geological computing. The internal structure of geographic information systems differs from conventional computer-aided design (CAD) technology in the referencing of digital elements in a geographic (spatial) relation to all other elements in the database, not to layers (two-dimensional drawings) as symbols independent of their locations. *See* DATABASE MANAGEMENT SYSTEM; GEOGRAPHIC INFORMATION SYSTEMS.

Geophysical computing. Geophysical computing is based on an interpretive workstation dedicated to tasks of seismic interpretation. The display, manipulation, and interpretation of both two- and three-dimensional seismic data is done interactively using both stand-alone workstations and integrated main-frame systems. Seismic lines and traces are displayed in color, and the system permits interactive identification of horizons from line to line. Most of these programs have automated facilities for enhancement of attributes as well as for correction of errors in interpretation between seismic lines (mis-ties). The computer-aided interpretation frees the explorationist or production geophysicist from the extremely large volume of paper-record sections; at the same time it allows for consistent and realistic mapping data.

Engineering computing. The demands of engineering computing in development of petroleum resources are intense, and hardly any area remains untouched. The giant reservoir simulation routines known as black-oil simulators have given way to even more sophisticated segmented-reservoir simulators. These modeling packages generally require supercomputing speeds to adequately produce reasonable representations of reservoirs. The addition of geostatistics (kriging) and fractal geometries in the simulation procedure permits analysis of reservoir heterogeneities. Brian Robert Shaw

Bibliography. American Association of Petroleum Geologists, *Geobyte*, bimonthly; *Computer and Geosciences*, eight times per year; P. A. Dickey, *Petroleum Development Geology*, 3d ed., 1986; J. H. Doveton, *Log Analysis of Subsurface Geology: Concepts and Computer Methods*, 1986; K. K. Landes, *Petroleum Geology of the U.S.*, 2d ed., 1975; L. W. LeRoy (ed.), *Subsurface Geology: Petroleum, Mining, Construction*, 5th ed., 1987; A. I. Levorsen, *Geology of Petroleum*, 2d ed., 1967; F. K. North, *Petroleum Geology*, 1985; *Petroleum Engineer International*, monthly; B. P. Tissot and D. H. Welte, *Petroleum Formation and Occurrence*, rev. ed., 1992.

Petroleum enhanced recovery

Technology to increase oil recovery from a porous formation beyond that obtained by conventional means. Conventional oil recovery technologies produce an average of about one-third of the original oil in place in a formation. Conventional technologies are primary or secondary. Primary technologies rely on native energy, in the form of fluid and rock compressibility and natural aquifers, to produce oil from the formation to wells. Secondary technologies supplement the native energy to drive oil to producing wells by injecting water or low-pressure gas at injection wells. The target of enhanced recovery technologies is that large portion of oil that is not recovered by primary and secondary means. *See* PETROLEUM ENGINEERING.

Process fundamentals. The boundary between secondary and enhanced recovery technologies is blurry. Many of the challenges encountered by secondary technologies are identical to those encountered by enhanced recovery technologies. Those challenges include reducing residual oil saturation, improving sweep efficiency, fitting the technology to the reservoir heterogeneities, and minimizing up-front and operating costs.

Residual oil remains trapped in a porous rock after the rock has been swept with water, gas, or any other recovery fluid. The residual oil saturation is the percentage of the pore space occupied by the residual oil. The residual oil saturation depends on the pore size distribution and connectivity, the interfacial tension between a recovery agent and the oil, the relative wettability of the rock surfaces with respect to the recovery agent and the oil, the viscosity of the fluids, and the rate at which the fluids are moving through the rock. The qualitative relationship between residual oil saturation and these various factors is shown in **Fig. 1**.

The sweep efficiency specifies that portion of a reservoir that is contacted by a recovery fluid. Sweep efficiency increases with volume of injected fluid. It also depends on the pattern of injection and production wells in a formation, on the mobility of the oil and the recovery fluid, and on heterogeneities in the formation (**Fig. 2**).

Oil reservoirs, which consist of porous rock formations, have complex structures created by the whims of nature in deposition, diagenetic, and tectonic processes. The combined effect of these processes yields formations that are heterogeneous at

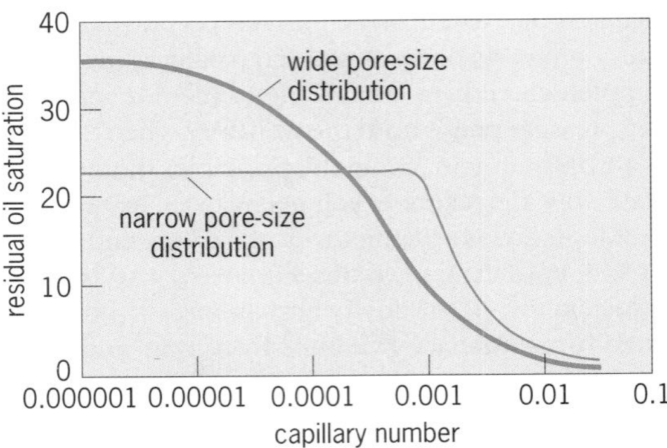

Fig. 1. The capillary number is the ratio of viscous forces to capillary forces in the rock, or simply the viscosity of the recovery agent times its velocity through the rock divided by the interfacial tension between the recovery agent and the oil. Capillary numbers for secondary technologies fall to the left of the graph; thus residual oil saturations for these processes lie between 20 and 40%.

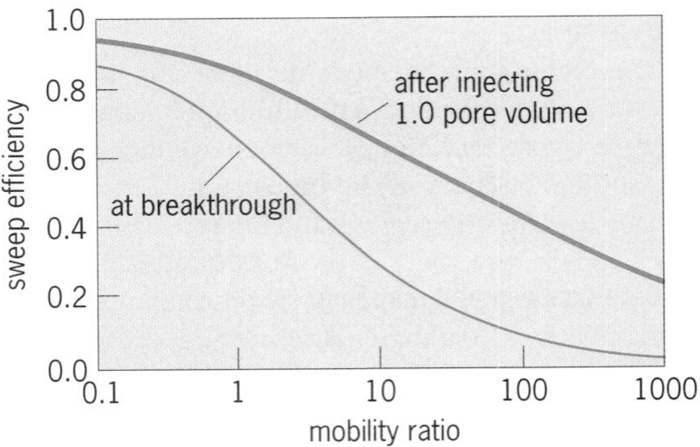

Fig. 2. The sweep efficiency of a five-spot pattern (a producer at each corner of a square area with an injector in the center) is shown as a function of the ratio of the mobility of the recovery agent to the mobility of the oil. Breakthrough sweep efficiency is that fractional volume swept when the recovery agent first arrives at the producing wells.

microscopic and macroscopic scales. The size distribution and connectivity of pores, the presence of clay particles in the pores, and the mineralogy of the surfaces of the pores all contribute to heterogeneity at the pore level. Layering or bedding of the formation produces heterogeneity at a slightly larger scale. At a larger scale, winding of river channels, anomalies in reef structures, and fractures caused by tectonic process contribute to additional heterogeneity. These various levels of heterogeneity affect the ease with which oil can be recovered. At the microscopic scale, heterogeneities affect residual oil saturation, as shown in Fig. 1. At larger scales, heterogeneities decrease sweep efficiency. Enhanced oil recovery technology must be selected or adapted to cope with the particular heterogeneities of each reservoir. *See* PETROLEUM GEOLOGY.

For economic reasons, access to reservoirs is extremely limited. The distance between wells that penetrate a formation is often 1000 ft or more. Limited access to the formation compounds the challenge of adapting recovery technology to cope with heterogeneity. Another consequence of limited access is that months of operation may be required to see the benefit of enhanced recovery technology because the rate at which fluids move through formation is a few feet per day. So, after starting injection of a recovery agent, one might wait as much as one year before any increased oil productivity results. The delayed benefit of a technology translates to increased economic risk. An enhanced recovery technology with high up-front and operating costs will likely fail to yield acceptable oil recovery performance. *See* OIL AND GAS WELL DRILLING.

Process technologies. A wide variety of processes have been considered for enhancing oil recovery: thermal processes, high-pressure gas processes, and chemical processes. The fundamental motivations for these processes are found in the relationships of Figs. 1 and 2. Specifically, low residual oil saturation can be obtained by selecting a recovery fluid that provides a very low interfacial tension between the oil and the fluid. With very low interfacial tension, the capillary number is large. And high sweep efficiency can be obtained by selecting a recovery agent with low mobility or by increasing the mobility of the oil.

In thermal processes, the mobility of the oil is increased by heating the oil-bearing formation, thus reducing the oil's viscosity. Thermal processes consist of cyclic steam injection, continuous steam drive, and in situ combustion. Typically, thermal processes have been applied to formations with high oil viscosity, usually greater than 100 millipascal seconds (centipoise). Such formations are rarely candidates for secondary recovery technologies, so the oil saturations are relatively high. Cyclic and continuous steam injection processes are widely used for recovery of high-viscosity oils in southern California, Venezuela, and western Canada. In the cyclic steam injection process, the same well serves as an injector and producer. First, steam is injected for a period of one month. After a short soak period of one week, the same well is put on production for several months to a year. In continuous steam injection, the injected steam propagates through the reservoir as it moves from the injector toward the producer. Steam injection processes are limited to shallow reservoirs, less than 610 m (2000 ft) below the surface. For deeper reservoirs, or for reservoirs in formations less than 15 m (50 ft) thick, heat losses in the injection well and to surrounding formations can be too large for profitable operation with steam injection. For such reservoirs, in situ combustion processes may be more suitable. In these processes, air is injected to produce combustion of a portion of the oil in the formation. The remainder of the oil is driven ahead of the combustion front by steam and gases produced by combustion. *See* PETROLEUM.

Carbon dioxide and natural gas (mostly methane) dominate applications of high-pressure gas processes, although nitrogen is injected into some reservoirs. Gas injection processes are particularly suited to reservoirs with low permeability. With sufficiently high operating pressure, the interfacial tension between the injected gas and the displaced oil reduces to zero; that is, the gas becomes miscible with the oil. The pressure required to develop miscibility depends on the compositions of the gas and the oil, and on the reservoir temperature. Most miscible carbon dioxide applications operate between 10,342 and 17,237 kPa (1500 and 2500 psi). For miscible natural gas and nitrogen processes, the pressures must be greater than 27,580 kPa (4000 psi). If a high-pressure gas process is to be applied to a reservoir, the reservoir must be deep enough to contain the pressure without fracturing and leaking to other formations or the surface. Usually, a reservoir must be at least 610 m (2000 ft) deep for every 6895 kPa (1000 psi) of operating pressure. High-pressure carbon dioxide processes have been applied to many oil reservoirs in western Texas and southeastern New Mexico. The carbon dioxide for these reservoirs is transported in pipelines from northern New Mexico and southern Colorado, where nearly pure carbon dioxide is found in deep formations.

Chemical processes include polymer, micellar-polymer, caustic or alkaline, and microbial, as well as combinations of these processes. The mobility of water injected in a secondary process can be reduced by adding a suitable polymer, usually polyacrylamide or a xanthan gum. With decreased mobility, the sweep efficiency of a waterflood improves. Polymer injection, however, does not increase the ultimate recovery (does not reduce the residual oil saturation) but increases the rate of recovery. With surfactants, very low interfacial tension between oil and water can be produced. A wide variety of surfactants have been used, including surfactants produced by sulfonation of crude oils. In caustic processes, surfactants are produced in situ by injecting alkaline agents, dissolved in water, into a reservoir. The alkaline agents react with portions of the oil in the reservoir to produce surfactants. A major obstacle to success of polymer and surfactant processes is loss of the chemicals by precipitation and adsorption onto mineral surfaces. Precipitation is caused by reaction of the chemicals with divalent ions, such as magnesium and calcium. High-surface-area clays are the main actor in adsorption. Chemical processes are also limited to formations with moderate to high permeabilities. While many chemical processes were tested in the United States during the early 1980s, there are few in operation today. *See* CLAY MINERALS; PETROLEUM MICROBIOLOGY; POLYMER; SURFACE AND INTERFACIAL CHEMISTRY; SURFACTANT.

Future developments. With the decline in oil prices in 1985 and 1986, interest and research efforts in enhanced oil recovery technologies waned. Many enhanced recovery operations were terminated as prices remained low. The surviving operations in the United States are dominated by steam injection (in California) followed by carbon dioxide injection in western Texas.

While continued instability in oil prices in the 1990s suppressed technological development, there was limited effort on technologies that are less costly to implement. For example, combinations of alkaline, surfactant, and polymer processes have been implemented in a few reservoirs in recent years. These processes rely on the capacity of some oils to produce useful surfactants when reacted with alkaline agents. The cost of alkaline agents is low. Microbial processes provide another example of continuing field testing. Microbial processes may potentially generate useful low-cost chemicals (surfactants and polymers) and gases (such as carbon dioxide) from bacterial colonies growing in a reservoir formation.

Recent developments in horizontal and multilateral drilling technologies have provided greater access to reservoirs, and they may stimulate some renewed interest in enhanced recovery technology. Increased access to reservoirs by large-scale tunneling into the formations has been proposed and was tested sporadically over the last 30 years. Very high well densities are possible by short-reach drilling from tunnels. Whether such technology will ever gain wider application is hard to project. An extreme example of reservoir access is found in the Athabasca Tar Sands near Fort McMurray in northern Alberta, Canada. There, the near-surface tar sands are transported by large shovels and trucks to a central processing facility which separates the tar from the sand with a thermal process and converts the tar to coke, sulfur, and a high-quality crude oil. Large increases in oil production from these tar sands are planned. *See* OIL SAND.

Richard L. Christiansen

Bibliography. *Enhanced Oil Recovery*, National Petroleum Council, U.S. Department of Energy, 1984; D. W. Green and G. P. Willhite, *Enhanced Oil Recovery*, Textbook Series, Society of Petroleum Engineers, 1998; *Improved Oil Recovery*, Interstate Oil Compact Commission, Oklahoma City, 1983; L. W. Lake, *Enhanced Oil Recovery*, Prentice Hall, Englewood Cliffs, NJ, 1989.

Petroleum geology

The practice of utilizing geological principles and applying geological concepts to the discovery and recovery of petroleum. Related fields in petroleum discovery include geochemistry and geophysics. The related areas in petroleum recovery are petroleum and chemical engineering. *See* CHEMICAL ENGINEERING; GEOCHEMISTRY; GEOPHYSICS.

Occurrence of petroleum. Petroleum occurs in a liquid phase as crude oil and condensate, and in a gaseous phase as natural gas. The phase is dependent on the kind of source rock from which the petroleum was formed and the physical and thermal environment in which it exists. As a liquid, oil may be readily transported from producing fields to points of consumption, thus moving in a global market. Natural gas is moved chiefly by pipelines and is therefore tied to land-locked markets. *See* NATURAL GAS; PETROLEUM.

Most petroleum occurs at varying depths below the ground surface, but generally petroleum existing as a liquid (crude oil) is found at depths of less than 20,000 ft (6100 m) while natural gas is found both at shallow depths and at depths exceeding 30,000 ft (9200 m). In some cases, oil may seep to the surface, forming massive deposits of oil or tar sands, such as the Athabasca oil sands of Alberta, Canada, and the tar sands of the Faja de Orinoco in Venezuela. Natural gas also seeps to the surface but escapes into the atmosphere, leaving little or no surface trace. *See* OIL SAND.

Most petroleum is found in sedimentary basins in sedimentary rocks, although many of the 700 or so sedimentary basins of the world contain no known significant accumulations. Although oil or gas accumulations are commonly known as pools, oil and gas occur in relatively small voids in rocks, such as pores and fractures, and not as underground pools or streams. *See* BASIN; SEDIMENTARY ROCKS.

Several conditions must exist for the accumulation of petroleum: (1) There must be a source rock, usually high in organic matter, from which petroleum can be generated. (2) There must be a mechanism for the petroleum to move, or migrate. (3) A reservoir rock with voids to hold petroleum fluids must exist. (4) The reservoir must be in a configuration to constitute a trap and be covered by a seal—any kind of low-permeability or dense rock formation that prevents further migration. If any of these conditions do not exist, petroleum either will not form or will not accumulate in commercially extractable form.

Petroleum source. Source rocks for most petroleum generation are those containing high concentrations of animal and plant organic matter. To yield high concentrations of total organic carbon, accumulation of the animal or plant remains should be in a chemical reducing environment so that the carbon is not oxidized. Accumulated organic matter eventually must be buried at depths sufficient to have temperatures of at least 140°F (60°C). On average, temperature increases with depth at a rate of about 18°F/1000 ft (10°C/300 m), but such a geothermal gradient exhibits wide variations. Organic matter exposed to sufficiently high temperatures, over time, changes or matures through stages to a liquid or gaseous hydrocarbon. The process of converting organic matter to petroleum is essentially a cooking process. *See* ORGANIC GEOCHEMISTRY.

Petroleum migration. Rocks that form source beds are commonly fine-grained and low in permeability, and thus make poor reservoirs. In some cases, however, the generated petroleum hydrocarbons do not leave the source rock or place of formation. Excellent examples are the so-called oil shales of the western United States, where vast quantities of petroleum hydrocarbons occur but extraction and recovery are expensive. Generally, migration from the source rock to a porous and permeable reservoir is necessary for commercial extraction. Such migration is initiated as buried source rocks are compacted and fluids are expelled and moved laterally and vertically to reservoirs, but the precise mechanisms for primary migration from source rocks to reservoirs are subjects of continuing debate. *See* OIL SHALE.

Petroleum trapping and reservoir. Migration ends once petroleum liquids and gases are trapped in a reservoir, or if they are not trapped, when they escape to the surface. The reservoir must have sufficient porosity and permeability, either with pores or fractures, to accumulate fluids. Those voids are generally filled initially with water, so that emplacement of petroleum must involve the displacement of some or all of the water. Most reservoir rocks are either sandstone or carbonate; these rocks show wide ranges of porosity and permeability, and they vary significantly in reservoir quality, as demonstrated by the efficiency in extraction. In fact, actual recovery of oil from a reservoir varies from as little as 5% of total original volume to as much as 95%, depending largely on the quality of the reservoir.

A trap is any arrangement of strata that allows the accumulation of oil and gas and precludes further migration. A variety of geologic mechanisms exist. Strata may be arranged by folding to give an anticline or a convex-upward trap. Areal changes in permeability of a reservoir from porous to impermeable will stop migration and cause a trapping of oil and gas. A reservoir may terminate in any direction due to subsequent erosion or to the original process by which it formed, such as a coral reef; such terminations are pinch-out traps. A fault or a vertical displacement of strata may result in permeable strata abutting impermeable strata, creating a fault trap. Or traps may be formed by vertical movement of salt or shale to form diapir traps.

Reservoirs must be covered by a seal of impervious strata, such as salt or shale, to prevent further migration and to effect petroleum accumulation. With changing geologic conditions over time, the reservoir may be breached or modified, causing the trapped oil and gas to escape. This movement, known as secondary migration, may be to the ground surface or to another reservoir and trap. *See* DIAPIR; LIMESTONE; SALT DOME; SANDSTONE; SHALE.

Prospecting. The aim of petroleum geologists is to find traps or accumulations of petroleum. The trap not only must be defined but must exist where other conditions such as source and reservoir rocks occur. Most structures or rock configurations that could contain and hold petroleum in fact do not. Thus, even after applying the best scientific information, a well that is drilled to an accurately defined trap may not encounter hydrocarbons.

A variety of techniques are used by the petroleum geologist to reconstruct geologic events and define an area or a prospect to be drilled. Geologic structures shown at the surface, such as folds or unconformities, can be projected into the subsurface. Accumulation of hydrocarbons at the ground surface may occur as seeps. In the early days of oil and gas exploration, reliance was placed on such surface manifestations; and in remote, unexplored areas of the world, surface conditions are yet a clue to the existence of petroleum at depth.

Most of the world's oil and gas accumulations show no surface manifestation. To locate these traps, the geologist must rely on subsurface information and data gathered by drilling exploratory wells and data obtained by geophysical surveying. These data, once interpreted, are used to construct maps, cross sections, and models that are used to infer or to actually depict subsurface configurations that might contain petroleum. Such depictions are prospects for drilling. If, on drilling, the trap is found to be as reconstructed by the petroleum geologist and if all other conditions for oil and gas accumulation exist, a discovery is made. If the prospect does not exist as envisaged, the well encounters no oil and is recorded as a dry hole. On average, less than one exploratory well in ten contains commercial quantities of oil or gas. Exploration efficiency has improved remarkably in recent years with advances in the technology of seismic reflection acquisition and processing. Running seismic surveys in a closely spaced grid allows three-dimensional imaging of the Earth's subsurface. In certain areas, especially offshore where seismic is shot through a uniform medium of seawater, character of reflections commonly allows direct detection of hydrocarbon accumulations. Accordingly, in areas amenable to advanced seismic technology, success rates for oil and gas discovery are 35% or higher. *See* GEOPHYSICAL EXPLORATION; OIL AND GAS WELL DRILLING.

Economic deposits. Oil and gas must be trapped in an individual reservoir in sufficient quantities to be commercially producible. That quantity is determined by the price or value of oil or gas and the cost to find it. In the United States, extensive drilling since around 1890 has led to the discovery of nearly 30,000 oil and gas fields. These fields range in size from some with less than 100,000 barrels (16,000 m^3) of oil to the largest field so far discovered in the United States, Prudhoe Bay in Alaska, which contained more than 10^{10} bbl (1.6×10^8 m^3) of recoverable oil. Worldwide, 25% of all oil discovered so far is contained in only ten fields, seven of which are in the Middle East. Fifty percent of all oil discovered to date is found in only 50 fields.

Worldwide, about 4×10^6 oil and gas wells have been drilled. About 85% percent of the wells have been drilled in the United States. As a result, most of the large and fairly obvious fields in the United States have been discovered, except those possibly existing in frontier or lightly explored areas such as Alaska and the deep waters offshore. Few areas of the world remain entirely untested, but many areas outside the United States are only partly explored, and advanced techniques have yet to be deployed in the recovery of oil and gas found so far. *See* PETROLEUM RESERVES.

Although in the United States the most obvious and most readily detectable prospective traps have been tested, many traps remain that are subtle and not easily definable. It is known that petroleum is accessible in the subtle traps because many have been found, commonly by accident. This occurs as a particular prospect is being tested and drilling encounters an unexpected accumulation. The challenge to the petroleum geologist is to develop models and to use ever-advancing technology to improve the ability to detect these subtle structures.

In addition to exploration for difficult-to-detect traps, greater efforts in petroleum geology along with petroleum engineering are being made to increase recovery from existing fields. Of all oil discovered so far, it is estimated that there will be recovery of only 35% on the average. In the United States alone, the amount of oil known in existing reservoirs and classed as unrecoverable is more than 3.25×10^{11} bbl (5.2×10^{10} m^3), twice the volume produced to date and 50 times the amount of oil the United States uses every year. Recovering some part of this huge oil resource will require geological reconstruction of reservoirs, a kind of very detailed and small-scale exploration. These reconstructions and models have allowed additional recovery of oil that is naturally movable in the reservoir. If the remaining oil is immobile because it is too viscous or because it is locked in very small pores or is held by capillary forces, techniques must be used by the petroleum geologist and the petroleum engineer to render the oil movable. If oil is too viscous or heavy to flow, steam can be injected into the reservoir to raise the temperature and thus lower the oil viscosity. If oil is locked in small pores, gas can be injected to expand the fluid and cause it to escape and move. William L. Fisher

Bibliography. P. A. Dickey, *Petroleum Development Geology*, 3d ed., 1986; K. K. Landes, *Petroleum Geology of the U. S.*, 2d ed., 1975; L. W. LeRoy et al. (eds.), *Subsurface Geology: Petroleum, Mining, Construction*, 5th ed., 1987; A. I. Levorsen, *Geology of Petroleum*, 2d ed., 1967; L. B. Magoon and W. G. Dow (eds.), *The Petroleum System—from Source to Trap*, 1994; F. K. North, *Petroleum Geology*, 1985; B. P. Tissot and D. H. Welte, *Petroleum Formation and Occurrence*, rev. ed., 1992.

Petroleum microbiology

Those aspects of microbiology in which crude oil, refined petroleum products, or pure hydrocarbons serve as nutrients for the growth of microorganisms or are altered as a result of their activities. Applications of petroleum microbiology include oil pollution control, enhanced oil recovery, microbial contamination of petroleum fuels and oil emulsions, and conversion of petroleum hydrocarbons into microbial products.

Biodegradation of petroleum hydrocarbons. Many species of bacteria, fungi, and algae have the enzymatic capability to use petroleum hydrocarbons as food. Biodegradation of petroleum requires an appropriate mixture of microorganisms, contact with oxygen gas, and large quantities of utilizable nitrogen and phosphorus compounds and smaller amounts of other elements essential for the growth of all microorganisms. Part of the hydrocarbons are converted into carbon dioxide and water and part into cellular materials, such as proteins and nucleic acids. The requirement for a mixture of different microorganisms arises from the fact that petroleum is composed of a wide variety of different groups of hydrocarbons, whereas any specific microorganism is highly specialized with regard to the type of hydrocarbon it can digest. For example, different bacteria are involved in the degradation of aliphatic cyclic and aromatic hydrocarbons. The bacterial genera that contain the most frequently isolated hydrocarbon degraders are *Pseudomonas*, *Acinetobacter*, *Flavobacterium*, *Brevibacterium*, *Corynebacterium*, *Arthrobacter*, *Mycobacterium*, and *Nocardia*. The fungal genera that contain oil utilizers include *Candida*, *Cladosporium*, *Rhodotorula*, *Torulopsis*, and *Trichosporium*.

By the use of genetic engineering techniques, a *Pseudomonas* species has been constructed which has the enzymatic capability of degrading several different groups of hydrocarbons. Although practical applications have not yet been found for this particular strain, it has provided a precedent for patenting genetically constructed microorganisms. *See* BIODEGRADATION; GENETIC ENGINEERING.

Clean-up of oil spills. Oil pollution results from natural hydrocarbon seeps, accidental spills, and intentional discharge of oily materials into the environment. Once the oil is released and comes into contact with water, air, and the necessary salts, microorganisms present in the environment begin the natural process of petroleum biodegradation. If this process did not occur, the world's oceans would soon become completely covered with a layer of oil. The reason that oil spills become a pollution problem is that the natural microbial systems for degrading the oil become temporarily overwhelmed.

Various methods have been tested for increasing the rates of petroleum biodegradation and thereby stimulating microbial clean-up of oil spills, the most successful of which involve modification of environmental parameters, such as addition of nitrogen and phosphorus fertilizers. Since most environments have an indigenous population of oil-degrading microorganisms that can multiply rapidly under appropriate conditions, additions of "seed" bacteria have little value in stimulating clean-up of oil spills. *See* WATER POLLUTION.

Enhanced oil recovery. The largest potential application of petroleum microbiology is in the field of enhanced oil recovery. Microbial products, as well as viable microorganisms, have been used as stimulation agents to enhance oil recovery from petroleum reservoirs. Xanthan, a polysaccharide produced by *Xanthomonas campestris*, is used as a waterflood thickening agent in oil recovery. Emulsan, a lipopolysaccharide produced by a strain of *Acinetobacter calcoaceticus*, stabilizes oil-in-water emulsions. A number of other microbial products are being tested for potential application in enhanced oil recovery processes, such as polymer flooding, micellular flooding, wetting of reservoir rock, reduction of oil-water interfacial tensions, and release of hydrocarbons from tar sands and oil shale. Field tests have indicated that injection of viable microorganisms with their nutrients into petroleum reservoirs can lead to enhanced oil recovery, presumably due to production of carbon dioxide gas, acids, and surfactants. *See* PETROLEUM ENHANCED RECOVERY.

Contamination of hydrocarbons. The ability of microorganisms to utilize petroleum also has its detrimental aspects, particularly with respect to the deterioration of petroleum fuels, asphalt coatings, and oil emulsions used with cutting machinery. All hydrocarbons become contaminated if they come into contact with water during storage. The problem is most serious with kerosine-based jet aircraft fuels which are consumed at very high rates by jet engines; thus, even a small amount of microbial sludge in the fuel becomes a hazard.

A significant economic problem is microbial degradation of asphalt pipeline coatings leading to metal corrosion. Petroleum-in-water emulsions are widely used as coolants and lubricants for high-speed metal-cutting machines. These emulsions rapidly become contaminated with microorganisms, resulting in breakage of the emulsion and potentially causing dermatitis to the workers. Antimicrobials that concentrate at oil-water interfaces are most effective at reducing the rate of microbial contamination of hydrocarbons. *See* CORROSION.

Conversion of petroleum hydrocarbons. A variety of valuable materials, such as amino acids, carbohydrates, nucleotides, vitamins, enzymes, antibiotics, citric acid, long-chain dicarboxylic acids, and biomass can be produced by microbial processes using petroleum hydrocarbons as substrates. Most of these materials also can be produced from more traditional microbial feedstocks, such as molasses. The main advantage of using hydrocarbons as substrates is their lower cost. Also, certain products, such as tetradecane-1,14-dicarboxylic acid, a raw material for preparing perfumes, are synthesized in higher yields on hydrocarbon than on carbohydrate substrates.

The most active area of research and development in petroleum microbiology since the mid-1960s has been in the large-scale production and concentration of microorganisms for animal feed and human food. Dried microbial cells are collectively referred to as single-cell protein. The advantages of single-cell protein over the more traditional sources of protein include the following: (1) Microorganisms grow rapidly. For example, yeast growing on hydrocarbons will double their mass every 4 h, leading to more than a hundredfold increase in yeast protein in 30 h. (2) Microorganisms have a high protein content. Depending on the particular type of microorganism and the growth conditions, the protein content of microbes varies from 30 to 70%. By comparison, legumes, wheat, and rice contain 25, 12, and 8% protein, respectively. (3) Microorganisms can be easily modified genetically to produce favorable traits. It takes many years to alter animals and plants genetically. Most strains that are used today for food have undergone continuous strain improvement for centuries. It should be possible in a relatively short time to improve microbial food strains by mutation and selection as well as by genetic engineering. (4) Microorganisms can be cultivated on a large scale anywhere in the world, independent of soil or climatic conditions. A medium-sized single-cell protein plant, occupying an area of 3 acres (1.2 hectares), produces the same amount of protein as 300,000 acres (120,000 hectares) devoted to soybeans.

In spite of the above advantages, single-cell protein has not yet played a significant role in providing protein for animal feed or human consumption. Manufacturing costs are still quite high when compared with the major competing protein source, soy, which is produced with little fertilizer and requires only a limited amount of processing to produce the soy meal. There is no way to predict when all of the conditions necessary to make single-cell protein a major force in the food and feed markets will occur, but many scientists are optimistic about its potential. *See* ANIMAL FEEDS; BACTERIAL PHYSIOLOGY AND METABOLISM; INDUSTRIAL MICROBIOLOGY.

Eugene Rosenberg

Bibliography. A. L. Demain, J.E. Davis, and R. M. Atlas (eds.), *Manual of Industrial Microbiology and Biotechnology*, 2d ed., 1999; I. Goldberg, *Single Cell Protein*, 1986; D. L. Gutnick and E. Rosenberg, Oil tankers and pollution: A microbiological approach, *Annu. Rev. Microbiol.*, 31:379–396, 1977; N. Kosaric (ed.), *Biosurfactants*, 1993; J. G. Leahy and R. R. Colwell, Microbial degradation of hydrocarbons in the environment, *Microbiol. Rev.*, 54:305–315, 1990; J. H. Litchfield, Single-cell protein, *Science*, 219:740–746, 1983; E. Rosenberg, Exploiting microbial growth on hydrocarbons: New markets, *Trends Biotech.*, 11:419–423, 1993.

Petroleum processing and refining

The separation of petroleum into fractions and the treating of these fractions to yield marketable products. Petroleum is a mixture of gaseous, liquid, and solid hydrocarbon compounds that occurs in sedimentary rock deposits throughout the world. In the crude state, petroleum has little value but, when refined, it provides liquid fuels (gasoline, diesel fuel, aviation fuel), solvents, heating oil, lubricants, and the distillation residuum asphalt, which is used for highway surfaces and roofing materials. *See* PETROLEUM; PETROLEUM PRODUCTS.

Crude petroleum (oil) is a mixture of compounds with different boiling temperatures that can be separated into a variety of fractions (**Table 1**). Since there is a wide variation in the composition of crude petroleum, the proportions in which the different fractions occur vary with origin. Some crude oils have higher proportions of lower-boiling components, while others have higher proportions of residuum (asphaltic components).

Petroleum processing and refining involves a series of steps by which the original crude oil is converted into products with desired qualities in the amounts dictated by the market. In fact, a refinery is essentially a group of manufacturing plants that vary in number with the variety of products in the mix. Refinery processes must be selected and products manufactured to give a balanced operation; that is, crude oil must be converted into products according to the demand for each. For example, the manufacture of products from the lower-boiling portion of petroleum automatically produces a certain amount of higher-boiling components. If the latter cannot be sold as, say, heavy fuel oil, these products will accumulate until refinery storage facilities are full. To prevent such a situation, the refinery must be flexible and able to change operations as needed. This usually means more processes, such as thermal processes to change excess heavy fuel oil into gasoline with coke as the residual product, or vacuum distillation processes to separate heavy oil into lubricating oil stocks and asphalt.

Early refineries (70 years ago) were predominantly distillation units with, perhaps, ancillary units to remove objectionable odors from the various product streams. The refinery of today (**Fig. 1**), the result of major evolutionary trends, is a highly complex operation. Most of the evolutionary adjustments to refineries have occurred since 1940. In the petroleum industry, as in many other industries, supply and demand are key factors in efficient and economic operation. Innovation is also key.

Distillation. Distillation, the first method by which petroleum was refined, remains a major process. The original technique involved a batch operation in which the still, a cast-iron vessel, was mounted on brickwork over a fire and the volatile materials passed through a pipe or gooseneck at the top of the still to a condenser, a coil of pipe (worm) immersed in a tank of running water.

In a modern refinery, after dewatering and desalting operations (**Table 2**), distillation remains the prime means by which petroleum is refined. The distillation section of a modern refinery (**Fig. 2**) is the most flexible unit in the refinery since conditions can be adjusted to process a wide range of feedstocks from the lighter crude oils to the heavier, more viscous crude oils. The maximum preferential temperature (in the vaporizing furnace or heater) to which the feedstock can be subjected is 350°C (660°F), unless the goal is to produce a thermally decomposed residuum. Thermal decomposition (cracking) occurs above this temperature, which can lead to coke deposition in the heater pipes or in the tower itself, resulting in failure of the distillation unit. Higher temperatures (up to 393°C; 740°F) are used in distillation units, but the petroleum residence time must be adjusted to prevent cracking.

TABLE 1. Petroleum fractions and their uses*

Fraction	Boiling range °C	Boiling range °F	Uses
Fuel gas	−160 to −40	−260 to −40	Refinery fuel
Propane	−40	−40	Liquefied petroleum gas (LPG)
Butane(s)	−12 to −1	11–30	Increases volatility of gasoline, advantageous in cold climates
Light naphtha	−1 to 150	30–300	Gasoline components, may be (with heavy naphtha) reformer feedstock
Heavy naphtha	150–205	300–400	Reformer feedstock; with light gas oil, jet fuels
Gasoline	−1 to 180	30–355	Motor fuel
Kerosine	205–260	400–500	Fuel oil
Stove oil	205–290	400–550	Fuel oil
Light gas oil	260–315	500–600	Furnace and diesel fuel components
Heavy gas oil	315–425	600–800	Feedstock for catalytic cracker
Lubricating oil	>400	>750	Lubrication
Vacuum gas oil	425–600	800–1100	Feedstock for catalytic cracker
Residuum	>600	>1100	Heavy fuel oil, asphalts

*From J. G. Speight (ed.), *The Chemistry and Technology of Petroleum*, 3d ed., Marcel Dekker, New York, 1999.

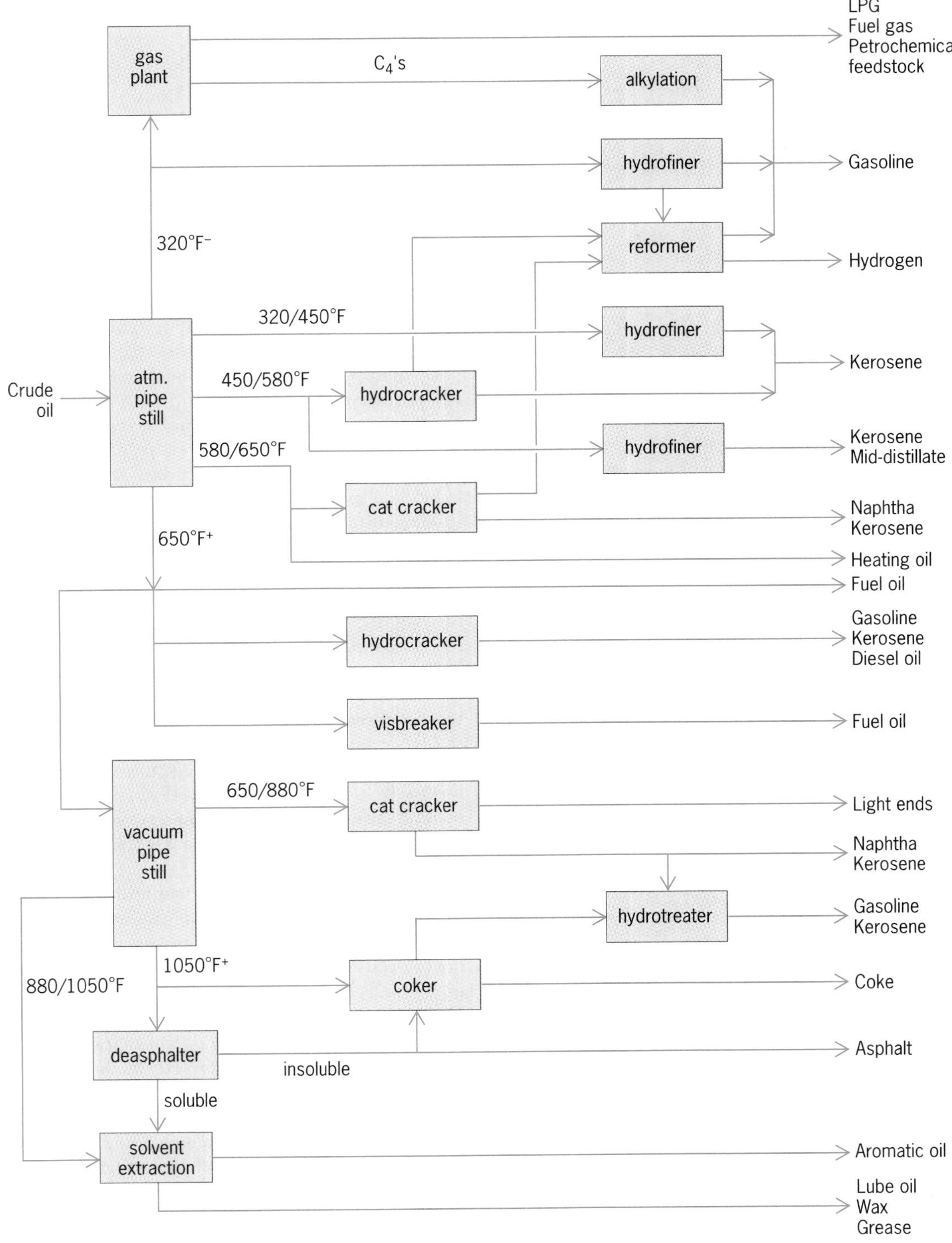

Fig. 1. Schematic of a petroleum refinery showing the various units. (*After J. G. Speight, ed., The Chemistry and Technology of Petroleum, 3d ed., Marcel Dekker, New York, 1999*)

Atmospheric distillation. In a present-day petroleum distillation unit (actually a collection of distillation units), a tower is used for fractionation (separation). The feedstock to the tower (pipe still) flows through one or more pipes arranged within a large furnace (pipe-still heater or pipe-still furnace) where the feed is heated to a temperature at which a predetermined portion of the feed changes into vapor. The heated feed is introduced into a fractional distillation tower where the nonvolatiles or liquid portions pass downward to the bottom of the tower and are pumped away, while the vapors pass upward through the tower and are fractionated into gas oils, kerosine, and naphthas.

TABLE 2. Desalting and dewatering operations

Separation method	Temperature, °F (°C)	Type of treatment
Chemical	140–210 (60–99)	0.05–4% solution of soap in water 0.5–5% solution of soda ash in water
Electrical	150–200 (66–93)	10,000–20,000 V
Gravity	180–200 (82–93)	Up to 40% water added
Centrifugal	180–200 (82–93)	Up to 20% water added (sometimes no water added)

Pipe-still furnaces vary greatly and can accommodate 50,000 (US) barrels (bbl; 2,100,000 US gallons) or more of crude petroleum per day. The walls and ceiling are insulated with firebrick, and the interior of the furnace is divided into two sections: a smaller convection section where the oil first enters the furnace, and a larger section (fitted with heaters) where the oil reaches its highest temperature.

Heat exchangers, used to preheat the feed, are made of bundles of tubes arranged within a shell so that the unheated feedstock passes through a series of tubes opposite the direction of the heated feedstock also passing through the shell. In this way, hot products from the distillation tower are cooled, passing the heat to the incoming cold crude oil. *See* HEAT EXCHANGER.

The primary fractions from a distillation unit contain some proportion of the lighter constituents characteristic of a lower-boiling fraction. These lower-boiling materials are removed by stripping (stabilization) before storage or further processing.

Vacuum distillation. Vacuum distillation is used in petroleum refining to separate the less volatile products, such as lubricating oils, from petroleum without subjecting the high-boiling products to cracking conditions. When the feedstock (reduced crude, atmospheric residuum) is required for the manufacture of labricating oils, further fractionation without cracking is desirable, which can be achieved by distillation under vacuum conditions.

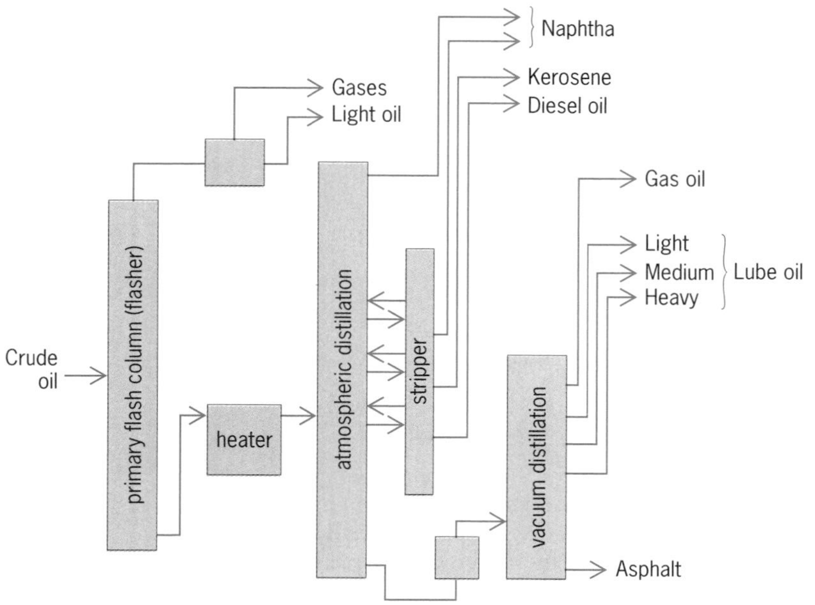

Operating pressure for vacuum distillation is usually 50–100 mm of mercury (6.7–13.3 kilopascals) [atmospheric pressure = 760 mm of mercury] and, in order to minimize large fluctuations in pressure in the vacuum tower, the units are necessarily larger in diameter than the atmospheric units. Some vacuum distillation units have diameters on the order of 45 ft (14 m). By this means, a heavy gas oil that has a boiling range in excess of 315°C (600°F) at atmospheric pressure may be obtained at temperatures of around 150°C (300°F), and lubricating oil, having a boiling range in excess of 370°C (700°F) at atmospheric pressure may be obtained at temperatures of 250–350°C (480–660°F). The feedstock and product temperatures are kept below the temperature and residence time where cracking will occur. The partial pressure of the feedstock constituents is reduced further by steam injection. The steam added to the column, principally for the stripping of asphalt in the base of the column, is superheated in the convection section of the heater.

The fractions obtained by vacuum distillation of the reduced crude (atmospheric residuum) from an atmospheric distillation unit depend on whether or not the unit is designed to produce lubricating or vacuum gas oils. In the former case, the fractions include (1) heavy gas oil, an overhead (volatile) product used as catalytic cracking stock or a light lubricating oil; (2) lubricating oil (light, intermediate, and heavy); and (3) asphalt (or residuum), the bottom product, which is used directly as, or to produce, asphalt or is blended with gas oil to produce a heavy fuel oil. Atmospheric and vacuum distillation are major parts of refinery operations, and no doubt will continue to be used as the primary refining operation.

Other distillation processes. As distillation techniques in refineries became more sophisticated, they were able to process a wider variety of crude oils to produce marketable products or feedstocks for other refinery units. However, it became apparent that the distillation units in the refineries were incapable of producing specific product fractions. In order to accommodate this type of product demand, refineries, beginning about midcentury, incorporated azeotropic distillation and extractive distillation in their operations.

All compounds have definite boiling temperatures,

Fig. 2. Distillation section of a refinery. (*After J. G. Speight, ed., The Chemistry and Technology of Petroleum, 3d ed., Marcel Dekker, New York, 1999*)

but a mixture of chemically dissimilar compounds will sometimes cause one or both of the components to boil at a temperature other than that expected. A mixture that boils at a temperature lower than the boiling point of any of the components is called an azeotropic mixture. When it is desired to separate close-boiling components, the addition of a nonindigenous component will form an azeotropic mixture with one of the components of the mixture, thereby lowering the boiling point and facilitating separation by distillation.

The separation of components of similar volatility may become economic if an azeotropic entrainer can be found that effectively changes their relative volatility. It is also desirable that the entrainer be reasonably cheap, stable, nontoxic, and readily recoverable from the components. In practice it is probably recoverability that limits the application of extractive and azeotropic distillation. The majority of successful processes are those in which the entrainer and one of the components separate into two liquid phases on cooling if direct recovery by distillation is not feasible. A further restriction in the selection of an azeotropic entrainer is that the boiling point of the entrainer be in the range of 10–40°C (18–72°F) below that of the components. *See* AZEOTROPIC DISTILLATION; AZEOTROPIC MIXTURE; DISTILLATION.

Thermal processes. Cracking is used to thermally decompose higher-boiling petroleum constituents and products to lighter oils. The first units were installed in refineries in 1913 as a means of producing more gasoline. In the early (pre-1940) processes used for gasoline manufacture, the major variables involved to achieve the cracking of the feedstock to lighter products with minimal coke formation were feedstock type, time, temperature, and pressure.

As refining technology evolved, the residuum or heavy distillate from a distillation unit became the feedstock for the cracking process. The residua produced as the end products of distillation processes, and even some of the heavier virgin oils, often contain substantial amounts of asphaltic materials, which preclude their use as fuel oils or lubricating stocks. However, subjecting these residua directly to thermal processes has become economically advantageous, since the end result is the production of lower-boiling materials. The asphaltic materials in the residua are regarded as the unwanted coke-forming constituents.

As the thermal processes evolved and catalysts were used with more frequency, poisoning of the catalyst (and this reduction in the lifetime of the catalyst) become a major issue for refiners. To avoid catalyst poisoning, it became necessary to remove from the feedstock as much of the nitrogen and metals (such as vanadium and nickel) as possible. Most of the nitrogen, oxygen, and sulfur as well as the metals are contained in, or associated with, the asphaltic fraction (residuum).

A number of thermal processes, such as residuum separation (flash distillation), vacuum flashing, visbreaking, and coking, became widely used by refiners to upgrade feedstocks by removing the asphaltic fraction.

Thermal cracking. One of the earliest conversion processes used in the petroleum industry was the thermal decomposition of higher-boiling materials into lower-boiling products. This process is known as thermal cracking.

The majority of the thermal cracking processes use temperatures of 455–540°C (850–1005°F) and pressures of 100–1000 psi (690–6895 kPa). For example, the feedstock (reduced crude) is preheated by direct exchange with the cracking products in the fractionating columns. Cracked gasoline and heating oil are removed from the upper section of the column. Light and heavy distillate fractions are removed from the lower section and are pumped to separate heaters. Higher temperatures are used to crack the more stable light distillate fraction. The streams from the heaters are combined and sent to a soaking chamber where additional time is provided to complete the cracking reactions. The cracked products are then separated in a low-pressure flash chamber where a heavy fuel oil is removed as bottoms. The remaining cracked products are sent to fractionating columns. *See* DISTILLATION COLUMN.

The thermal cracking of higher-boiling petroleum fractions to produce gasoline is now virtually obsolete. The antiknock requirements of modern automobile engines together with the different nature of crude oils (compared to those of 50 years ago) has reduced the ability of the thermal cracking process to produce gasoline on an economic basis. Very few new units have been installed since the 1960s, and some refineries may still operate the older thermal cracking units.

Visbreaking. Visbreaking (viscosity breaking) is a mild thermal cracking operation that can be used to reduce the viscosity of residua to allow the products to meet fuel oil specifications. Alternatively, the visbroken residua can be blended with lighter product oils to produce fuel oils of acceptable viscosity. By reducing the viscosity of the residuum, visbreaking reduces the amount of light heating oil that is

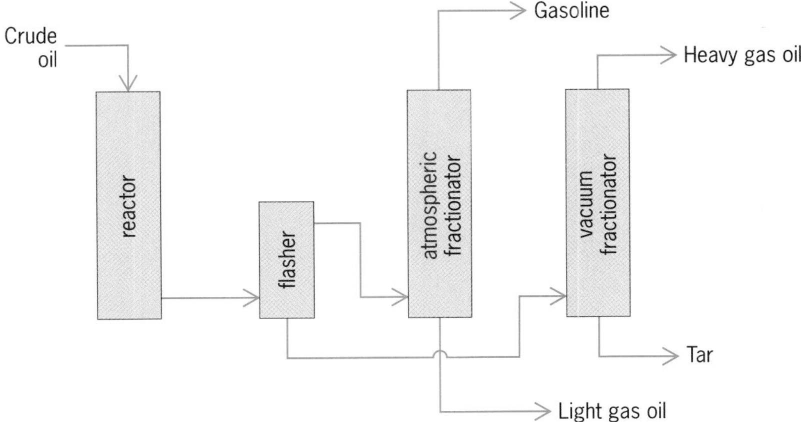

Fig. 3. Visbreaking process. (*After J. G. Speight, ed., The Chemistry and Technology of Petroleum, 3d ed., Marcel Dekker, New York, 1999*)

required for blending to meet fuel oil specifications. In addition to the major product, fuel oil, material in the gas oil and gasoline boiling range is produced. The gas oil may be used as additional feed for catalytic cracking units or as heating oil.

In a typical visbreaking operation (**Fig. 3**), a crude oil residuum is passed through a furnace where it is heated to a temperature of 480°C (895°F) under an outlet pressure of about 100 psi (689 kPa). The heating coils in the furnace are arranged to provide a soaking section of low heat density, where the crude oil residuum remains until the visbreaking reactions are completed and the cracked products are then passed into a flash-distillation chamber. The overhead material from this chamber is then fractionated to produce a low-quality gasoline as an overhead product and light gas oil as bottom. The liquid products from the flash chamber are cooled with a gas oil flux and then sent to a vacuum fractionator. This yields a heavy gas oil distillate and a residual tar of reduced viscosity. Visbreakers come as soaker type (Fig. 3) or coil type. In the coil type, the heating and thermal changes occur in the heating coil.

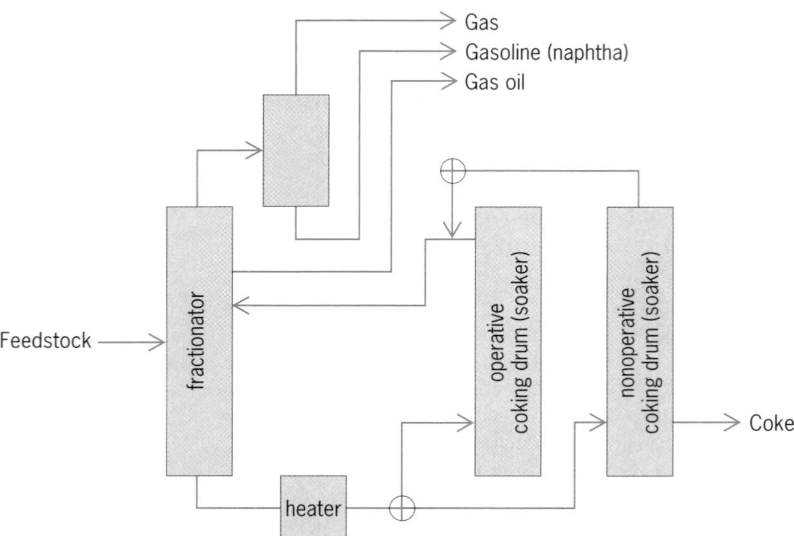

Fig. 4. Delayed coking process. (*After J. G. Speight, ed., The Chemistry and Technology of Petroleum, 3d ed., Marcel Dekker, New York, 1999*)

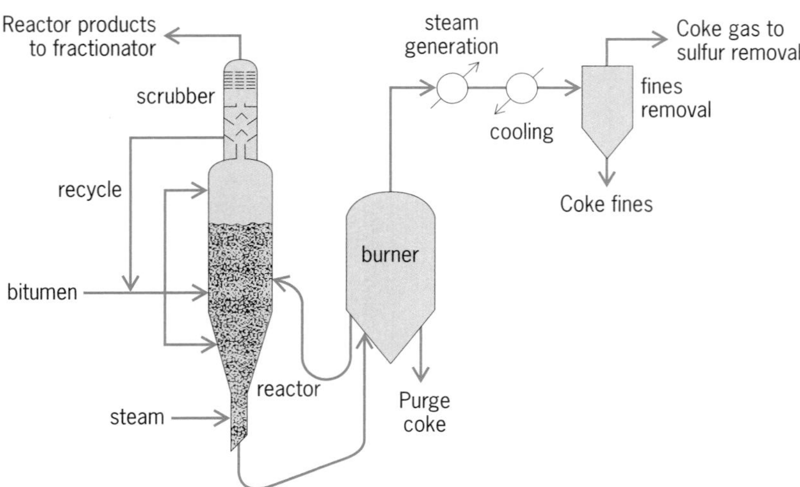

Fig. 5. Fluid coking pocess. (*After J. G. Speight, ed., The Chemistry and Technology of Petroleum, 3d ed., Marcel Dekker, New York, 1999*)

Delayed coking. Delayed coking is a thermal process for continuously converting (atmospheric or vacuum) residua into lower-boiling products, such as gases, naphtha, fuel oil, gas oil, and coke. Gas oil, often the major product of a coking operation, serves primarily as a feedstock for catalytic cracking units. The coke obtained is typically used as fuel; but specialty uses, such as electrode manufacture, and production of chemicals and metallurgical coke are also possible, increasing the value of the coke. For these uses, the coke may require treatment to remove sulfur and metal impurities. *See* COKE; FUEL OIL; NAPHTHA.

Delayed coking is a semicontinuous process (**Fig. 4**) in which the heated charge is transferred to large soaking (or coking) drums, which provide the long residence time needed to allow the cracking reactions to proceed to completion. The feed to these units is normally an atmospheric residuum, although cracked residua are also used. The feedstock is introduced into a product fractionator where it is heated and the lighter fractions are removed as a side streams. The fractionator residuum, including a recycle stream of heavy product, is then heated in a furnace whose outlet temperature varies from 480 to 515°C (895 to 960°F), and enters one of a pair of coking drums where the cracking reactions continue. The cracked products leave as distillate, and coke deposits form on the inner surface of the drum. The temperature in the coke drum ranges from 415 to 450°C (780 to 840°F), and pressures are from 15 to 90 psi (101 to 620 kPa). Overhead products go to the fractionator, where the naphtha and heating oil fractions are recovered. The nonvolatile material is combined with preheated fresh feed and returned to the coking drum. *See* CRACKING.

For continuous operation, two drums are used; while one is on stream, the other is being cleaned. A coke drum is usually on stream for about 24 hours before becoming filled with porous coke, which is removed hydraulically. Normally, 24 hours are required to complete the cleaning operation and prepare the coke drum for use.

Fluid coking. Fluid coking (**Fig. 5**) is a continuous process that uses a fluidized bed to convert atmospheric and vacuum residua into more valuable products. The residuum is sprayed into a fluidized bed of hot, fine coke particles, which permits the coking reactions to be conducted at higher temperatures with shorter contact times than are possible in delayed coking. These conditions result in decreased coke yields and the recovery of greater quantities of more valuable liquid product.

Fluid coking uses two vessels, a reactor and a burner. Coke particles are circulated between them to transfer heat (generated by burning a portion of the coke) to the reactor. Steam is introduced at the bottom of the reactor to fluidize the coke particles. *See* FLUIDIZATION.

Flexicoking. Flexicoking is also a continuous process (**Fig. 6**) that is related to fluid coking. The unit uses the same configuration as the fluid coker but has a gasification section in which excess coke can be gasified to produce refinery fuel gas. *See* COKING (PETROLEUM).

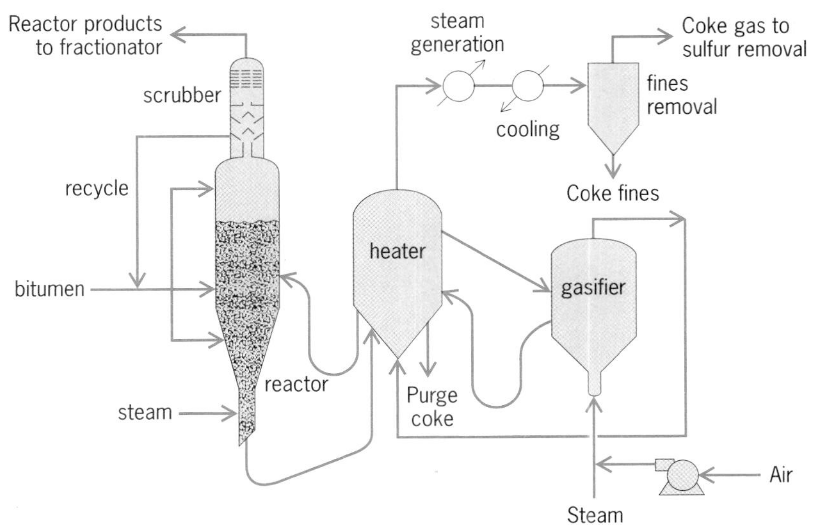

Fig. 6. Flexicoking process. (*After J. G. Speight, ed., The Chemistry and Technology of Petroleum, 3d ed., Marcel Dekker, New York, 1999*)

Catalytic cracking. There are many processes in a refinery that use a catalyst to improve process efficiency (**Table 3**). The original incentive arose from the need to increase gasoline supplies in the 1930s and 1940s. Since cracking could virtually double the volume of gasoline from a barrel of crude oil, cracking was justifiable on this basis alone.

Catalytic cracking (**Fig. 7**) was introduced to refineries in 1936 as a method for producing higher-octane gasoline. This process is basically the same as thermal cracking, but differs by the use of a catalyst, which directs the course of the cracking reactions to produce more of the desired higher-octane hydrocarbon products. Catalytic cracking has a number of advantages over thermal cracking.

1. The gasoline produced has a higher octane number because of the presence of isoparaffins and aromatics, which also have greater chemical stability than monoolefins and diolefins.

2. Olefinic gases are suitable for polymer gasoline manufacture.

3. Smaller quantities of methane, ethane, and ethylene are produced by catalytic cracking.

4. Sulfur compounds are changed in such a way that the sulfur content in catalytically cracked gasoline is lower than in thermally cracked gasoline.

Catalytic cracking also produces less coke and more of the useful gas oils than does thermal cracking. *See* ALKENE; GASOLINE; OCTANE NUMBER.

Catalytic cracking is regarded as the modern method for converting high-boiling petroleum fractions, such as gas oil, into gasoline and other low-boiling fractions. The usual commercial process involves contacting a gas oil faction with an active catalyst at a suitable temperature, pressure, and residence time so that a substantial part (>50%) of the gas oil is converted into gasoline and lower-boiling products, usually in a single-pass operation. However, during the cracking reaction, carbonaceous material is deposited on the catalyst, which markedly reduces its activity. Removal of the deposit is usually accomplished by burning the catalyst in air.

The several processes currently used in catalytic cracking differ mainly in the method of catalyst handling, although there is overlap with regard to catalyst type and the nature of the products. The catalyst, which may be an activated natural or synthetic material, is employed in bead, pellet, or microspherical form and can be used as a fixed bed, moving bed, or fluid bed. Fixed-bed was the first process to be used commercially and involves a static bed of catalyst in several reactors, which allows a continuous flow of feedstock to be maintained. Thus, the cycle of operations consists of (1) flowing feedstock through the catalyst bed, (2) discontinuing the feedstock flow

TABLE 3. Catalytic processes

Process	Materials charged	Products recovered	Temperature of reaction	Type of reaction
Cracking	Gas oil, fuel oil, heavy feedstocks	Gasoline, gas, and fuel oil	875–975°F 470–525°C	Dissociation or splitting of molecules
Hydrogenation	Gasoline to heavy feedstocks	Low-boiling products	400–850°F 205–455°C	Mild hydrogenation; cracking; removal of sulfur, nitrogen, oxygen, and metallic compounds
Reforming	Gasolines, naphthas	High-octane gasolines, aromatics	850–1000°F 455–535°C	Dehydrogenation, dehydroisomerization, isomerization, hydrocracking, dehydrocyclization
Isomerization	Butane, C_4H_{10}	Isobutane, C_4H_{10}		Rearrangement
Alkylation	Butylene and isobutane, C_4H_8 and C_4H_{10}	Alkylate, C_8H_{18}	32–50°F 0–10°C	Combination
Polymerization	Butylene, C_4H_8	Octene, C_8H_{16}	300–350°F 150–175°C	Combination

* From J. G. Speight (ed.), *The Chemistry and Technology of Petroleum*, 3d ed., Marcel Dekker, New York, 1999.

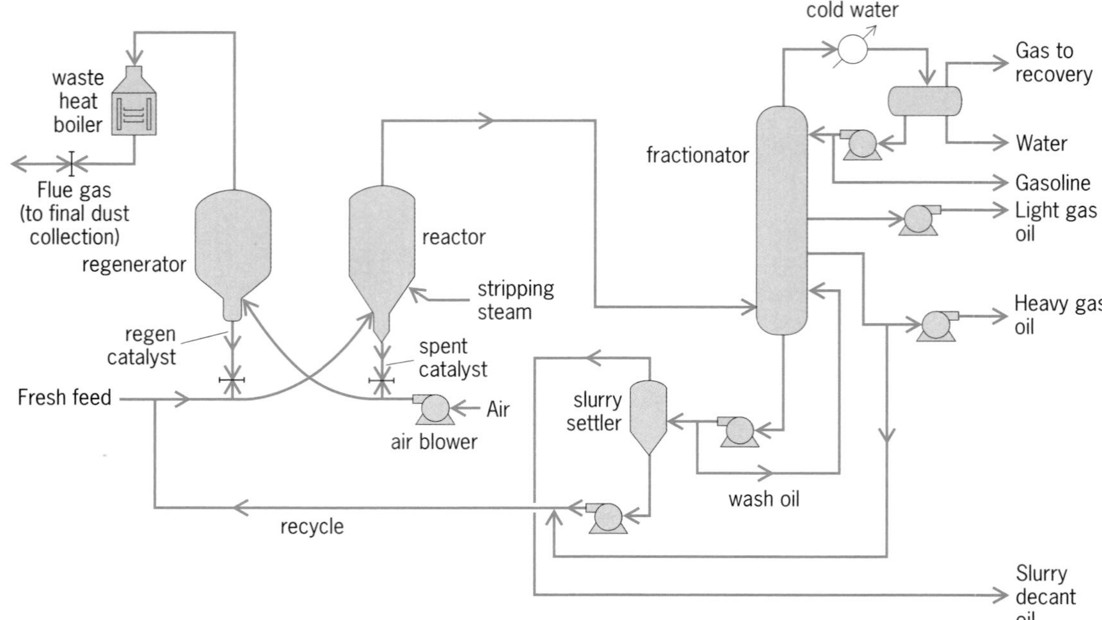

Fig. 7. Catalytic cracking process. (*After J. G. Speight, ed., The Chemistry and Technology of Petroleum, 3d ed., Marcel Dekker, New York, 1999*)

and removing the coke from the catalyst by burning, and (3) Putting the reactor back on stream. *See* CHEMICAL REACTOR.

The moving-bed process uses a reaction vessel (in which cracking takes place) and a kiln (in which the spent catalyst is regenerated), and catalyst movement between the two is provided by various means.

The fluid-bed process differs from the fixed-bed and moving-bed processes insofar as a powdered catalyst is circulated essentially as a fluid with the feedstock. The several fluid catalytic cracking processes in use differ primarily in mechanical design. Side-by-side reactor-regenerator construction and unitary vessel construction (the reactor either above or below the regenerator) are the two main mechanical variations.

Cracking of crude oil fractions occurs over many types of catalytic materials, but high yields of desirable products are obtained with hydrated aluminum silicates. These may be either activated (acid-treated) natural clays of the bentonite type of synthesized silica-alumina (SiO_2-Al_2O_3), or silica-magnesia (SiO_2-MgO) preparations. Their activity to yield essentially the same products may be enhanced to some extent by the incorporation of small amounts of other materials, such as the oxides of zirconium, boron (which has a tendency to volatilize away on use), and thorium. Natural and synthetic catalysts can be used as pellets, beads, or powder; all of the forms require replacement because of attrition and gradual efficiency loss. It is essential that they are able to withstand the physical impact of loading and thermal shocks, and the chemical action of carbon dioxide, air, sulfur and nitrogen compounds, and steam. Synthetic catalysts, or certain selected clays, appear to be better in this regard than average untreated natural catalysts.

Hydroprocesses. The use of hydrogen in thermal processes was perhaps the single most significant advance in refining technology during the twentieth century. The process uses the principle that the presence of hydrogen during a thermal reaction of a petroleum feedstock will terminate many of the coke-forming reactions and enhance the yields of the lower-boiling components, such as gasoline, kerosine, and jet fuel. Hydrogenation processes for the conversion of petroleum fractions and petroleum products may be classified as destructive and nondestructive. *See* HYDROGENATION.

Destructive hydrogenation (hydrogenolysis or hydrocracking) is characterized by the conversion of the higher-molecular-weight constituents in a feedstock to lower-boiling products. Such treatment requires severe processing conditions and the use of high hydrogen pressures to minimize the polymerization and condensation reactions that lead to coke formation. *See* HYDROCRACKING.

Nondestructive hydrogenation is used for improving product quality without appreciable alteration of the boiling range. Nitrogen, sulfur, and oxygen compounds undergo reaction with the hydrogen, forming ammonia, hydrogen sulfide, and water, respectively. Unstable compounds that might lead to the formation of gums or insoluble materials are converted to more stable compounds.

Hydrotreating. Hydrotreating (nondestructive hydrogenation) is carried out by charging the feed to the reactor, together with hydrogen in the presence of catalysts, such as tungsten-nickel sulfide (W-NiS), cobalt-molybdenum-alumina (Co-Mo-Al_2O_3), nickel oxide-silica-alumina (NiO-SiO_2-Al_2O_3), and platinum-alumina (Pt-Al_2O_3). Most processes use cobalt-molybdena catalysts that generally contain about 10% molybdenum oxide and less than 1% cobalt oxide supported on alumina (Al_2O_3). The temperatures used are in the range of 300–345°C (570–655°F), while the hydrogen pressures are about 500–1000 psi (3447–6895 kPa).

The reaction generally takes place in the vapor phase but, depending on the application, may be a mixed-phase reaction. Generally it is more economical to hydrotreat high-sulfur feedstocks prior to catalytic cracking than to hydrotreat the cracking products. The advantages of hydrotreating are that (1) sulfur is removed from the catalytic cracking feedstock, and corrosion is reduced in the cracking unit; (2) carbon formation during cracking is reduced, resulting in higher conversions; and (3) the cracking quality of the gas oil fraction is improved.

Hydrocracking. Hydrocracking (destructive hydrogenation) is similar to catalytic cracking with hydrogenation superimposed and with the reactions taking place either simultaneously or sequentially. Hydrocracking was initially used to upgrade low-value distillate feedstocks, such as cycle oils (high aromatic products from a catalytic cracker which usually are not recycled to extinction for economic reasons), thermal and coker gas oils, and heavy-cracked and straight-run naphthas. Catalyst improvements and modifications have made it possible to yield products from gases and naphtha to furnace oils and catalytic cracking feedstocks. The hydrotreating catalysts are usually cobalt plus molybdenum or nickel plus molybdenum (in the sulfide) form impregnated on an alumina (Al_2O_3) base. The hydrotreated operating conditions are such that appreciable hydrogenation of aromatics will not occur [1000–2000 psi (6895–13,790 kPa) hydrogen and about 370°C (700°F)]. The desulfurization reactions are usually accompanied by small amounts of hydrogenation and hydrocracking.

In a hydrotreating process, the feedstock is heated and passed with hydrogen gas through a tower or reactor filled with catalyst pellets. The reactor is maintained at a temperature of 260–425°C (500–800°F) at pressures from 100 to 1000 psi (690 to 6895 kPa), depending on the particular process, the nature of the feedstock, and the degree of hydrogenation required. After leaving the reactor, excess hydrogen is separated from the treated product and recycled through the reactor after the removal of hydrogen sulfide. The liquid product is passed into a stripping tower where steam removes the dissolved hydrogen and hydrogen sulfide; after cooling, the product is stored or, in the case of feedstock preparation, pumped to the next processing unit.

The most common hydrocracking process is a two-stage operation (**Fig. 8**) that maximizes the yield of transportation fuels and has the flexibility to produce gasoline, naphtha, jet fuel, or diesel fuel to meet seasonal demand. The processes also operate at higher temperatures than the hydrotreating processes and use a more temperature-resistant catalyst.

Reforming processes. When the demand for higher-octane gasoline developed during the early 1930s, attention was directed to improving the octane number of fractions within the boiling range of gasoline. Straight-run (distilled) gasoline frequently had very low octane numbers, and any process that would improve the octane numbers would aid in meeting the demand for higher-octane-number gasoline. Such a process (thermal reforming) was developed and used widely, but to a much lesser extent than thermal cracking. Thermal reforming converts (reforms) gasoline into higher-octane gasoline. The equipment for thermal reforming is essentially the same as for thermal cracking, but uses higher temperatures.

Thermal reforming. In a thermal reforming process, the feedstock, such as 205°C (400°F) end-point naphtha or a straight-run gasoline, is heated to 510–595°C (950–1100°F) in a furnace, much the same as a cracking furnace, with pressures from 400 to 1000 psi (2758 to 6895 kPa). As the heated naphtha leaves the furnace, it is cooled or quenched by the addition of cold naphtha and then enters a fractional distillation tower where the heavy products are separated. The remainder of the reformed material leaves the top of the tower and is separated into gases and reformate. The reformate has a high octane number due to cracking of the longer-chain paraffins into higher-octane olefins.

The products of thermal reforming are gases, gasoline, and residual oil or tar, the latter being formed in very small amounts (about 1%). The amount and quality of gasoline, known as reformate, is very dependent on the temperature. As a rule, the higher the reforming temperature, the higher the octane number, but the lower the yield of reformate.

Catalytic reforming. Like thermal reforming, catalytic reforming converts low-octane gasoline into high-octane gasoline (reformate). While thermal reforming could produce reformate with octane numbers of 65–80 depending on the yield, catalytic reforming produces reformate with octane numbers on the order of 90–95.

Catalytic reforming usually is carried out by feeding a naphtha (after pretreating with hydrogen if necessary) and hydrogen mixture to a furnace for

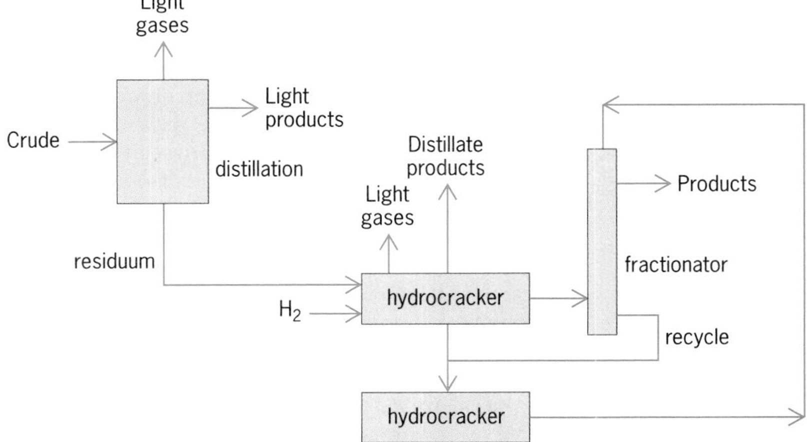

Fig. 8. Two-stage hydrocracking process. (*After J. G. Speight, ed., The Chemistry and Technology of Petroleum, 3d ed., Marcel Dekker, New York, 1999*)

heating to the desired temperature, 450–520°C (840–965°F), and then passed through fixed-bed catalytic reactors at hydrogen pressures of 100–1000 psi (689–6895 kPa). Normally pairs of reactors are used in series, with heaters located between adjoining reactors to compensate for the endothermic reactions taking place. Sometimes as many as five reactors are kept on stream in series while one or more is being regenerated. The on-stream cycle of any one reactor may vary from several hours to many days, depending on the feedstock and reaction conditions.

The commercial processes available for use are broadly classified as the fixed-bed, moving-bed, and fluid-bed types. Fixed-bed processes use predominantly platinum-containing catalysts in units equipped for cycle, occasional, or no regeneration, whereas the fluid- and moving-bed processes use mixed nonprecious metal oxide catalysts in units equipped with separate regeneration facilities.

Dehydrogenation is a major chemical reaction in catalytic reforming, producing large quantities of hydrogen gas. The hydrogen is recycled though the reforming reactors, providing the atmosphere necessary for the chemical reactions and preventing carbon deposition on the catalyst, thus extending its operating life. An excess of hydrogen, above whatever is consumed in the process, is produced. As a result, catalytic reforming processes are unique in that they are the only petroleum refinery processes to produce hydrogen as a by-product. *See* DEHYDROGENATION.

The reforming catalyst composition is dictated by the composition of the feedstock and the desired reformate. The catalysts used are principally molybdena-alumina (Mo_2O_3-Al_2O_3), chromia-alumina (Cr_2O_3-Al_2O_3), or platinum on a silica-alumina (Pt-SiO_2-Al_2O_3) or alumina (Pt-Al_2O_3) base. Nonplatinum catalysts are widely used in regenerative process for feeds containing sulfur, which poisons platinum catalysts, although pretreatment processes (hydrodesulfurization) may permit platinum catalysts to be used.

The purpose of platinum on the catalyst is to promote dehydrogenation and hydrogenation reactions, that is, the production of aromatics, participation in hydrocracking, and rapid hydrogenation of carbon-forming precursors. For the catalyst to have an activity for isomerization of both paraffins and naphthenes—the initial cracking step of hydrocracking—and to participate in paraffin dehydrocyclization, it must have an acid activity. The balance between these two activities is most important in a reforming catalyst. In fact, in the production of aromatics from cyclic saturated materials (naphthenes), it is important that hydrocracking be minimized to avoid loss of the desired product; and, thus, the catalytic activity must be moderated relative to the case of gasoline production from a paraffinic feed, where dehydrocyclization and hydrocracking play an important role. *See* REFORMING PROCESSES.

Isomerization processes. During World War II, isomerization processes found commercial applications for making high-octane aviation gasoline components and additional feed for alkylation units. The lowered alkylate demands in the postwar period led to the majority of the butane isomerization units being shut down. In recent years the greater demand for high-octane motor fuel has resulted in new butane isomerization units being installed.

The earliest process of note was the production of isobutane, which is required as an alkylation feed. The isomerization may take place in the vapor phase, with the activated catalyst supported on a solid phase, or in the liquid phase with a dissolved catalyst. In the process, pure butane is mixed with hydrogen (to inhibit olefin formation) and passed to the reactor, at 110–170°C (230–340°F) and 200–300 psi (1379–2068 kPa). The product is cooled, the hydrogen is separated, and the cracked gases are then removed in a stabilizer column. The stabilizer bottom product is passed to a superfractionator where the normal butane is separated from the isobutane.

During World War II, aluminum chloride was the catalyst used to isomerize butane, pentane, and hexane. Since then, supported metal catalysts have been developed for use in high-temperature processes, which operate in the range 370–480°C (700–900°F) and 300–750 psi (2068–5171 kPa), while aluminum chloride plus hydrogen chloride is universally used for the low-temperature processes. *See* ALKANE.

A nonregenerable aluminum chloride catalyst is used with various carriers in a fixed-bed or liquid contactor. Platinum or other metal catalyst processes are used in fixed-bed operation, and can be regenerable or nonregenerable. The reaction conditions vary widely, depending on the particular process and feedstock [40–480°C (100–900°F) and 150–1000 psi (1034–6895 kPa)]. *See* HETEROGENEOUS CATALYSIS; HOMOGENEOUS CATALYSIS; ISOMERIZATION.

Alkylation processes. The combination of olefins with paraffins to form higher isoparaffins is called alkylation. Since olefins are reactive (unstable) and are responsible for exhaust pollutants, their conversion to high-octane isoparaffins is desirable when possible. In refinery practice, only isobutane is alkylated, by reaction with isobutene or normal butene, and isooctane is the product. Although alkylation is possible without catalysts, commercial processes use aluminum chloride, sulfuric acid, or hydrogen fluoride as catalysts, allowing the reactions to take place at low temperatures and minimizing undesirable side reactions, such as polymerization of olefins.

Alkylate is composed of a mixture of isoparaffins having octane numbers that vary with the olefins from which they were made. Butylenes produce the highest octane numbers, propylene the lowest, and pentylenes the intermediate numbers. All alkylates, however, have high octane numbers (>87), making them particularly valuable.

The alkylation reaction as now practiced in petroleum refining is the reaction through catalysis, of an olefin (ethylene, propylene, butylene, and amylene) with isobutane to yield high-octane branched-chain hydrocarbons in the gasoline boiling range. Olefin feedstock is derived from the gas produced in a catalytic cracker, while isobutane is recovered from refinery gases or produced by catalytic butane isomerization. To accomplish this, either ethylene or propylene is combined with isobutane at 50–280°C (125–450°F) and 300–1000 psi (2068–6895 kPa) in the presence of a metal halide catalyst, such as aluminum chloride. Conditions are less stringent in catalytic alkylation; olefins (propylene, butylenes, or pentylenes) are combined with isobutane in the presence of an acid catalyst (sulfuric acid or hydrofluoric acid) at low temperatures and pressures (1–40°C, 34–104°F; 14.8–150 psi, 6.9–1034 kPa).

Sulfuric acid, hydrogen fluoride, and aluminum chloride are the general catalysts used commercially. Sulfuric acid is used with propylene and higher-boiling feeds, but not with ethylene, because it reacts to form ethyl hydrogen sulfate. The acid is pumped through the reactor and forms an air emulsion with the reactants, and the emulsion is maintained at 50% acid. The rate of catalyst deactivation varies with the feed and isobutane charge rate. Butene feeds cause less acid consumption than the propylene feeds.

Aluminum chloride is not widely used as an alkylation catalyst, but when it is, hydrogen chloride is used as a promoter and water is injected to activate the catalyst as an aluminum chloride/hydrocarbon complex. Hydrogen fluoride is used for alkylation of higher-boiling olefins bacause it is more readily separated and recovered from the products. *See* ALKYLATION (PETROLEUM).

Polymerization processes. In the petroleum industry, polymerization is the process by which olefin gases are converted to liquid products that may be suitable for gasoline (polymer gasoline) or other liquid fuels. The feedstock usually consists of propylene and butylenes from cracking processes or may even be selective olefins for dimer, timer, or tetramer production.

Polymerization may be accomplished thermally or in the presence of a catalyst at lower temperatures. Thermal polymerization is regarded as not being as effective as catalytic polymerization, but it can be used to polymerize saturated materials that cannot be induced to react by catalysts. The process consists of vapor-phase cracking of, for example, propane and butane, followed by a prolonged period at high temperature (510–595°C; 950–1100°F) for the reactions to proceed to near completion.

Olefins can also be conveniently polymerized by means of an acid catalyst in which an olefin-rich feed stream is contacted with a catalyst (sulfuric acid, copper pyrophosphate, phosphoric acid) at 150–220°C (300–425°F) and 150–1200 psi (1034–8274 kPa), depending on feedstock and product requirement.

Phosphates are the principal catalysts used in polymerization units; the commercially used catalysts are liquid phosphoric acid, phosphoric acid on kieselguhr (diatomite), copper pyrophosphate pellets, and phosphoric acid film on quartz. The last is the least active, but the most often used and the easiest to regenerate by washing and recoating, the serious disadvantage is that tar must occasionally be burned off the support. The process using liquid phosphoric acid catalyst is the one most often associated with attempts to raise production by increasing temperature. *See* POLYMERIZATION.

Solvent processes. Solvent processes, such as deasphalting, occur after the distillation units and are used chiefly to produce asphalt, deasphalted oil, and aromatic oils (Fig. 1). *See* SOLVENT EXTRACTION.

James G. Speight

Bibliography. R. A. Myers (ed.), *Handbook of Petroleum Refining Process*, McGraw-Hill, New York, 1997; J. G. Speight, *The Chemistry and Technology of Petroleum*, 3d ed., Marcel Dekker, New York, 1999; J. G. Speight, *The Desulfurization of Heavy Oils and Residua*, 2d ed., Marcel Dekker, New York, 2000; J. G. Speight (ed.), *Fuel Science and Technology Handbook*, Marcel Dekker, New York, 1990; J. G. Speight (ed.), *Petroleum Chemistry and Refining*, Taylor & Francis, New York, 1998.

Petroleum products

Petroleum products are those fractions derived from petroleum that have commercial value as a bulk product (**Table 1**). Petrochemicals, in contrast, are individual chemicals, derived from bulk fractions, that are used as the basic building blocks of the chemical industry. Gases and liquid fuels are currently the main products of the petroleum industry (**Table 2**). However, other products, such as lubricating oils, waxes, and asphalt, have also added to the value of petroleum resources. *See* PETROCHEMICAL; PETROLEUM.

Petroleum products are hydrocarbon compounds, containing combinations of hydrogen and carbon with various molecular forms. Many compounds occur naturally. Other compounds are created by commercial processes for altering one combination to form another. Each combination has its unique set of chemical and physical properties. *See* PETROLEUM PROCESSING AND REFINING.

Many products replace similar products historically derived from nonpetroleum sources—for example: illuminating oil to replace sperm oil from whales; synthetic rubber to replace natural rubber from trees; and manufactured fibers to replace textiles from animals and vegetation. Each new use for a product imposes additional specifications regarding application and manufacturing methods. Specifications for petroleum products are based on properties such as density and boiling range to assure that a petroleum product can perform its intended task.

Gaseous products. Fuels with four or fewer carbons in the hydrogen-carbon combination have boiling points lower than room temperature. Therefore, these products are normally gases. Common classifications for these products are as follows.

Natural gas. This is predominantly methane (CH_4), which has the lowest boiling point and least complex structure of all hydrocarbons. Natural gas from an underground reservoir, when brought to the surface, may contain other, higher-boiling-point hydrocarbons, and is often referred to as wet gas. Wet gas is processed to remove the entrained hydrocarbons that are higher-boiling than methane. The high-boiling hydrocarbons that are isolated and liquefied are called natural gas condensates. *See* METHANE; NATURAL GAS.

Still gas. This is a broad term applied to low-boiling hydrocarbon mixtures, and is the lowest-boiling fraction isolated from a distillation (still) unit in the refinery. If the distillation unit is separating only low-boiling hydrocarbon fractions, the still gas will be almost entirely methane with only trace amounts of ethane (CH_3CH_3) and ethylene ($CH_2{=}CH_2$). If the distillation unit is handling higher-boiling fractions, the still gas may also contain propane ($CH_3CH_2CH_3$), butane ($CH_3CH_2CH_2CH_3$), and their respective isomers. Fuel gas and still gas are terms that are often used interchangeably. However, fuel gas is intended to denote the product's future use as a fuel for boilers, furnaces, or heaters. *See* ETHYLENE.

TABLE 1. Petroleum products

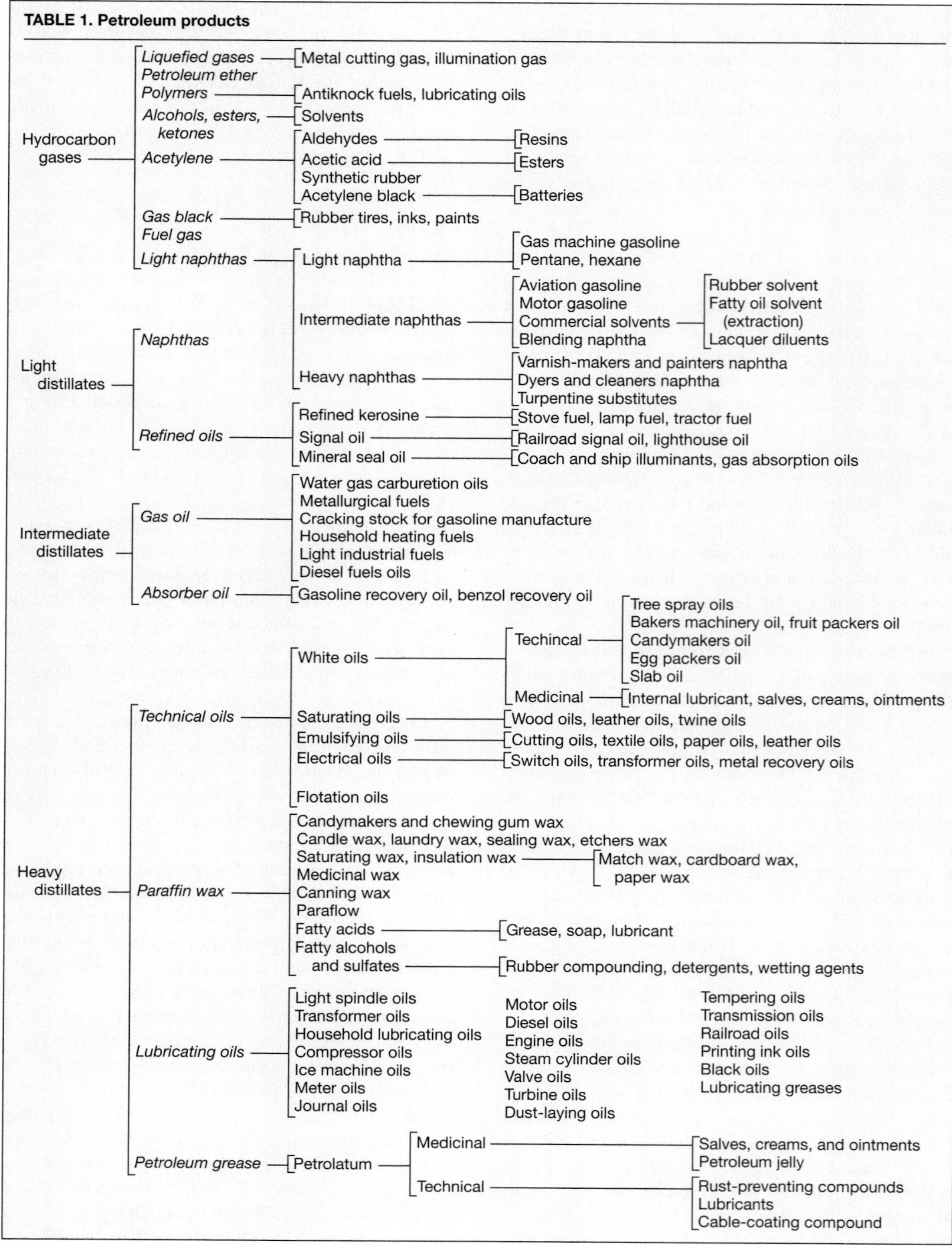

- Hydrocarbon gases
  - *Liquefied gases* — Metal cutting gas, illumination gas
  - *Petroleum ether*
  - *Polymers* — Antiknock fuels, lubricating oils
  - *Alcohols, esters, ketones* — Solvents
  - *Acetylene*
    - Aldehydes — Resins
    - Acetic acid — Esters
    - Synthetic rubber
    - Acetylene black — Batteries
  - *Gas black* — Rubber tires, inks, paints
  - *Fuel gas*
  - *Light naphthas* — Light naphtha
- Light distillates
  - *Naphthas*
    - Light naphtha — Gas machine gasoline; Pentane, hexane
    - Intermediate naphthas — Aviation gasoline; Motor gasoline; Commercial solvents (— Rubber solvent; Fatty oil solvent (extraction); Lacquer diluents); Blending naphtha
    - Heavy naphthas — Varnish-makers and painters naphtha; Dyers and cleaners naphtha; Turpentine substitutes
  - *Refined oils*
    - Refined kerosine — Stove fuel, lamp fuel, tractor fuel
    - Signal oil — Railroad signal oil, lighthouse oil
    - Mineral seal oil — Coach and ship illuminants, gas absorption oils
- Intermediate distillates
  - *Gas oil* — Water gas carburetion oils; Metallurgical fuels; Cracking stock for gasoline manufacture; Household heating fuels; Light industrial fuels; Diesel fuels oils
  - *Absorber oil* — Gasoline recovery oil, benzol recovery oil
- Heavy distillates
  - *Technical oils*
    - White oils
      - Techincal — Tree spray oils; Bakers machinery oil, fruit packers oil; Candymakers oil; Egg packers oil; Slab oil
      - Medicinal — Internal lubricant, salves, creams, ointments
    - Saturating oils — Wood oils, leather oils, twine oils
    - Emulsifying oils — Cutting oils, textile oils, paper oils, leather oils
    - Electrical oils — Switch oils, transformer oils, metal recovery oils
    - Flotation oils
  - *Paraffin wax*
    - Candymakers and chewing gum wax
    - Candle wax, laundry wax, sealing wax, etchers wax
    - Saturating wax, insulation wax — Match wax, cardboard wax, paper wax
    - Medicinal wax
    - Canning wax
    - Paraflow
    - Fatty acids — Grease, soap, lubricant
    - Fatty alcohols and sulfates — Rubber compounding, detergents, wetting agents
  - *Lubricating oils*
    - Light spindle oils; Transformer oils; Household lubricating oils; Compressor oils; Ice machine oils; Meter oils; Journal oils
    - Motor oils; Diesel oils; Engine oils; Steam cylinder oils; Valve oils; Turbine oils; Dust-laying oils
    - Tempering oils; Transmission oils; Railroad oils; Printing ink oils; Black oils; Lubricating greases
  - *Petroleum grease* — Petrolatum
    - Medicinal — Salves, creams, and ointments; Petroleum jelly
    - Technical — Rust-preventing compounds; Lubricants; Cable-coating compound

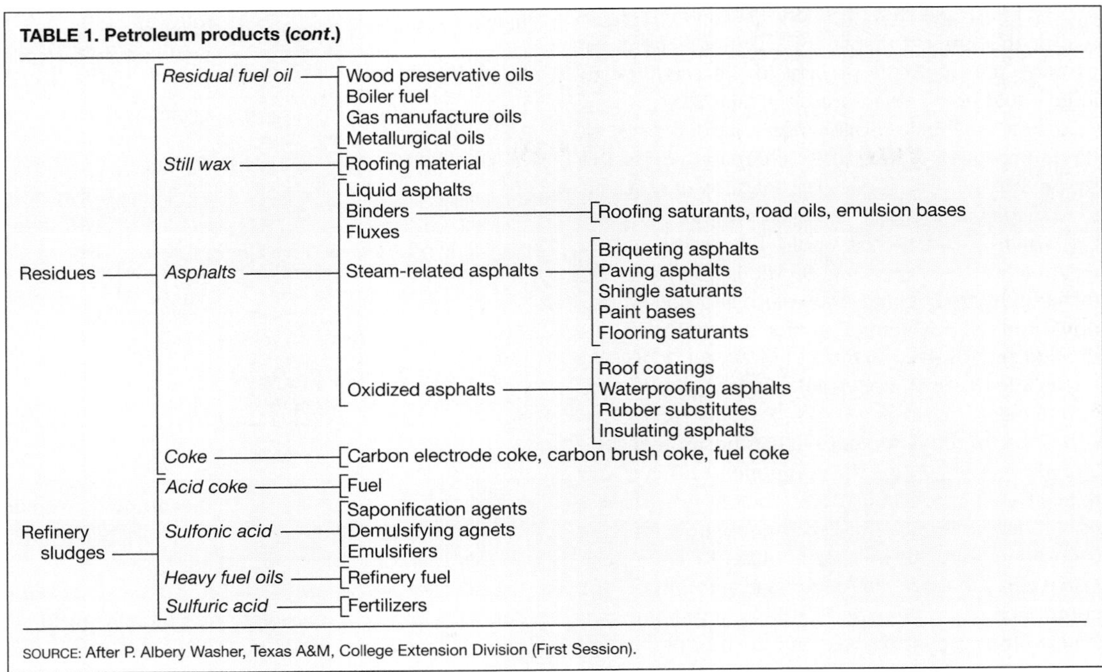

TABLE 1. Petroleum products (*cont.*)

Residues	*Residual fuel oil*	Wood preservative oils	
		Boiler fuel	
		Gas manufacture oils	
		Metallurgical oils	
	Still wax	Roofing material	
	Asphalts	Liquid asphalts	
		Binders	Roofing saturants, road oils, emulsion bases
		Fluxes	
		Steam-related asphalts	Briqueting asphalts
			Paving asphalts
			Shingle saturants
			Paint bases
			Flooring saturants
		Oxidized asphalts	Roof coatings
			Waterproofing asphalts
			Rubber substitutes
			Insulating asphalts
	Coke	Carbon electrode coke, carbon brush coke, fuel coke	
Refinery sludges	*Acid coke*	Fuel	
	Sulfonic acid	Saponification agents	
		Demulsifying agnets	
		Emulsifiers	
	Heavy fuel oils	Refinery fuel	
	Sulfuric acid	Fertilizers	

SOURCE: After P. Albery Washer, Texas A&M, College Extension Division (First Session).

Liquefied petroleum gas (LPG). This is composed of propane and butane and is stored under pressure in order to keep these hydrocarbons liquefied at normal atmospheric temperatures. Before liquefied petroleum gas is burned, it passes through a pressure relief valve that causes a reduction in pressure and the liquid vaporizes (gasifies). Winter-grade liquefied petroleum gas is mostly propane, the lower-boiling of the two gases, that is easier to vaporize at lower temperatures. Summer-grade liquefied petroleum gas is mostly butane. *See* LIQUEFIED PETROLEUM GAS (LPG).

Gasoline. Gasoline (motor fuel) is a complex mixture of hydrocarbons that boils below 200°C (390°F) and is intended for most spark-ignition engines (such as those used in passenger cars, light-duty trucks, motorcycles, and motorboats). The properties of gasoline are intended to satisfy the requirements of smooth and clean burning, easy ignition in cold weather, minimal evaporation in hot weather, and stability during long storage periods. *See* GASOLINE; INTERNAL COMBUSTION ENGINE.

The hydrocarbon constituents in this boiling range are those that have 4 to 12 carbon atoms in their molecular structure. Gasolines can vary widely in composition; even those with the same octane number may be quite different. For example, low-boiling distillates with high aromatic content (above 20%) can be obtained from some crude oils. The variation in the content of aromatics, as well as normal paraffins, branched paraffins, cyclopentanes, and cyclohexanes, involve the characteristics of any individual crude oil and influence the octane number of the gasoline. *See* AROMATIC HYDROCARBON.

During the first decade of the twentieth century, the gasoline produced was originally present in crude oil or condensed from natural gas. However, it was soon discovered that if the heavier portions of petroleum (for example, the fraction which boiled higher than kerosine, such as gas oil) were heated to more severe temperatures, thermal degradation

TABLE 2. Commercial names and uses for major petroleum products

Crude oil cuts	Refinery blends	Consumer products
Gases	Still gases	Fuel gas
	Propane/butane	Liquefied petroleum gas (LPG)
Light/heavy naphtha	Motor fuel	Gasoline
	Aviation turbine, Jet-B	Jet fuel (naphtha type)
Kerosine	Aviation turbine, Jet-A	Jet fuel (kerosine type)
	No. 1 fuel oil	Kerosine (range oil)
Light gas oil	Diesel	Auto and tractor diesel
	No. 2 fuel oil	Home heating oil
Heavy gas oil	No. 4 fuel oil	Commercial heating oil
	No. 5 fuel oil	Industrial heating oil
	Bright stock	Lubricants
Residuals	No. 6 fuel oil	Bunker C oil
	Heavy residual	Asphalt
	Coke	Coke

(or cracking) occurred to produce smaller molecules within the range suitable for gasoline. Therefore, gasoline not originally present in the crude petroleum could be manufactured. *See* CRACKING.

At first, cracked gasoline was regarded as an inferior product because of its comparative instability on storage; but as more gasoline was required, the petroleum industry revolved around processes by which this material could be produced (for example, catalytic cracking, thermal and catalytic reforming, hydrocracking, alkylation, and polymerization), and the problem of storage instability was addressed and resolved. *See* ALKYLATION (PETROLEUM); HYDROCRACKING; POLYMERIZATION; REFORMING PROCESSES.

Because of the differences in composition of gasoline fractions, blending the components is necessary to meet gasoline specifications. The physical process of blending is simple, but determining how much of each component to include is difficult. The operation is carried out by simultaneously pumping all the components of a gasoline blend into a pipeline that leads to the gasoline storage, but the pumps must be set to deliver automatically the proper proportion of each component. Baffles in the pipeline are often used to mix the components as they travel to the storage tank.

Regular and premium grades. These are the relative classifications for the octane numbers of gasoline. An octane number is a measure of a gasoline's ability to resist spontaneous detonation. It is critical that detonation be at a precise time for a gasoline-air mixture in a spark-ignition engine. That time is determined by the electrical spark system. After ignition, the course of the detonation should progress smoothly, with a flame front moving across the combustion chamber.

If the fuel has a low octane number, the temperature and pressure wave generated by the spark-timed flame front can allow the remaining fuel-air mixture to spontaneously ignite, causing an extra pressure pulse (knock), more fuel energy to be lost as heat, and the engine to deliver less power. *See* SPARK KNOCK.

The knocking tendencies of two pure hydrocarbons are used as references to relate final engine conditions to an octane number. One of these hydrocarbons, isooctane (2,2,4-trimethylpentane), is assigned an octane number of 100, and the other hydrocarbon (*n*-heptane) is assigned an octane number of 0. Mixtures of these reference hydrocarbons are assigned octane numbers equal to the volume percent of isooctane in the mixture. To determine the octane rating of a sample fuel, its knock intensity in a test engine is compared to those from these reference mixtures. *See* OCTANE NUMBER.

Research (RON or R) and motor octane (MON or M) grades. These grades identify other test engine variables (such as engine speed and intake air temperature), with the motor octane number being representative of the performance of the engine at higher revolutions per minute and higher air intake temperature. Both ratings are important to a multicylinder automobile engine because it usually operates over a wide range of conditions. The arithmetic average of the research and motor octane numbers [(RON + MON)/2] is used for marketing purposes.

TABLE 3. Trends in gasoline specifications

Time frame	Specifications
Pre-1994	Phase-out of lead in gasoline Lower Reid vapor pressure
1995–2000	Reformulated gasoline (Phase I) Lower aromatics content: 25 vol % maximum Benzene: 1 vol % maximum Oxygen: 2 wt % minimum Reid vapor pressure: 8.1 psi maximum
Post-2000	Reformulated gasoline (Phase II) Sulfur: 30 ppm maximum Aromatics content: 22 vol % maximum Benzene: 0.8 vol % maximum Oxygen: >2 wt % Olefins: 4 vol % Reid vapor pressure: 7.0 psi maximum

Leaded and unleaded grades. These grades of gasoline denote whether the gasoline mix includes lead additives. Lead compounds, such as tetraethyllead and tetramethyllead, are additives to improve the octane rating of a gasoline mix. However, the introduction of the catalytic muffler to automobile engines (to reduce exhaust emissions) led to the phase-out of leaded gasoline (**Table 3**). *See* LEAD.

Volatility. This is also an important property of gasoline, since it is related to ease of startup, length of warmup, extent of carburetor icing, speed of acceleration, occurrence of vapor lock, quality of manifold distribution, and extent of crankcase dilution. Volatility comprises two extreme properties: (1) enough low-boiling hydrocarbons to vaporize easily in cold weather; and (2) enough high-boiling hydrocarbons to remain a liquid in an engine's fuel supply system during hot periods.

The reduction of the lead content of gasoline and the introduction of reformulated gasoline have been very successful in reducing automobile emissions due to changes in gasoline composition (Table 3). Further improvements in fuel quality have been proposed for the early decades of the twenty-first century. These projections are accompanied by a noticeable and measurable decrease in crude oil quality. The reformulated gasoline will help meet environmental regulations for liquid fuel emissions, but will be subject to continuous review because of the potential for environmental impact. *See* AUTOMOBILE.

Aviation fuel. This comes in two types: gasoline and jet fuel. Aviation gasoline, now usually found in use in light aircraft and older civil aircraft, has a narrower boiling range (38–170°C or 100–340°F) than conventional (automobile) gasoline (−1 to 200°C or 30–390°F). The narrower boiling range ensures better distribution of the vaporized fuel through the more complicated induction systems of aircraft engines. Since aircraft operate at high altitudes (5.3 km or 3.3 mi) where the prevailing pressure (0.5 atm or 7.4 psi) is less than the pressure at the surface of the Earth (1.0 atm or 14.7 psi), the vapor pressure of aviation gasoline must be limited to reduce boiling in the tanks, fuel lines, and carburetors. Jet fuel is classified as aviation turbine fuel, and the specifications and ratings relative to octane number are replaced with properties connected with the ability of the fuel to burn cleanly. *See* AIRCRAFT FUEL; JET FUEL.

Solvents. Petroleum solvents (naphthas) have been available since the early days of the petroleum industry. They are valuable as solvents because of their nonpoisonous character and good dissolving power. The wide range of naphthas available and the varying degree of volatility possible offer products suitable for many uses. *See* NAPHTHA.

Petroleum naphtha is a generic term applied to refined, partly refined, or unrefined petroleum products. Naphthas are prepared by several methods, including (1) fractionation of distillates or crude petroleum, (2) solvent extraction, (3) hydrogenation of distillates, (4) polymerization of unsaturated (olefinic) compounds, and (5) alkylation processes. The naphtha may also be a combination of product streams from more than one process.

The main uses of petroleum naphthas fall into the general areas of (1) solvents (diluents) for paints, (2) dry-cleaning solvents, (3) solvents for cutback asphalts, (4) solvents in rubber industry, and (5) solvents for industrial extraction processes. Turpentine, the traditional solvent for paints, has been almost completely replaced by the cheaper and more abundant petroleum naphthas. *See* SOLVENT.

Kerosine. Kerosine was the major refinery product before the onset of the automobile age and the ensuing emphasis on gasoline. Kerosine originated as a straight-run (distilled) petroleum fraction that boiled between approximately 205 and 260°C (400 and 500°F). In the early days of petroleum refining, some crude oils contained kerosine fractions of very high quality; but other crude oils, such as those having a high proportion of asphaltic materials, must be thoroughly refined to remove aromatics and sulfur compounds before a satisfactory kerosine fraction can be obtained.

The kerosine fraction is essentially a distillation fraction of petroleum. The quantity and quality of the kerosine vary with the type of crude oil; some crude oils yield excellent kerosine, while others produce kerosine that requires substantial refining. Kerosine is a very stable product, and additives are not required to improve the quality. Apart from the removal of excessive quantities of aromatics, kerosine fractions may need only a lye (alkali) wash if hydrogen sulfide is present. Kerosine is used as a fuel for heating and cooking, jet engines, and lamps, for weed burning, and as a base for insecticides. *See* KEROSINE.

Diesel fuel. Diesel fuel is a distillate product that has a higher boiling point than gasoline (or naphtha) but that also must self-ignite easily. This is determined through the cetane rating, derived from the reference fuel *n*-cetane. Cetane number is a measure of the tendency of a diesel fuel to knock in a diesel engine. The scale is based upon the ignition characteristics of two hydrocarbons, *n*-hexadecane (cetane) and 2,3,4,5,6,7,8-heptamethylnonane. *See* CETANE NUMBER; DIESEL ENGINE; DIESEL FUEL.

Cetane has a short delay period during ignition and is assigned a cetane number of 100; heptamethylnonane has a long delay period and is assigned a cetane number of 15. Just as the octane number is meaningful for automobile fuels, the cetane number is a means of determining the ignition quality of diesel fuels, and is equivalent to the percentage by volume of cetane in the blend with heptamethylnonane, which matches the ignition quality of the test fuel.

Sulfur content is an important specification for diesel fuels, as a predictor of sulfur emissions into the atmosphere. While most sulfur compounds are removed from gasoline prior to use, the sulfur content of diesel fuel (and fuel oil) requires consideration. *See* SULFUR.

Diesel fuel oil is essentially the same as furnace fuel oil, but the proportion of cracked gas oil is usually less since the high aromatic content of the cracked gas oil reduces the cetane value of the diesel fuel.

Fuel oil. Fuel oil is classified as two main types: distillate and residual. It is also divided into subtypes based on use (**Table 4**). *See* FUEL OIL.

Distillate fuel oil is vaporized and condensed during a distillation process, has a definite boiling range, and does not contain high-boiling constituents such as asphaltic components. A fuel oil that contains any amount of the residue from crude distillation or thermal cracking is a residual fuel oil. Diesel fuel oil is generally a distillate fuel oil, but residual oil has been successfully used to power marine diesel engines, and mixtures of distillate and residua have been used on locomotive diesel engines. *See* LOCOMOTIVE.

The terms "distillate fuel oil" and "residual fuel oil" are losing their significance, since fuel oil is now made for specific uses and may be distillate, residuum, or a mixture of the two. The terms "domestic fuel oil" and "heavy fuel oil" are more indicative of the uses.

TABLE 4. Properties of types of fuel oils

Types	Properties
No. 1 fuel oil	Similar to kerosine or range oil (fuel used in stoves for cooking) Defined as a distillate intended for vaporizing in pot-type burners and other burners where a clean flame is required
No. 2 fuel oil	Often called domestic heating oil Similar to diesel and higher-boiling jet fuels Defined as a distillate for general-purpose heating in which the burners do not require the fuel to be completely vaporized before burning
No. 4 fuel oil	A light industrial heating oil that is intended where preheating is not required for handling or burning Two grades differing primarily in safety (flash) and flow (viscosity) properties
No. 5 fuel oil	A heavy industrial oil that often requires preheating for burning and, in cold climates, for handling
No. 6 fuel oil	A heavy residuum oil Commonly referred to as Bunker C oil when it is used to fuel oceangoing vessels Preheating required for both handling and burning

Domestic fuel oil. This term is applied to fuel oil used primarily in the home, and includes kerosine, stove oil, and furnace fuel oil. Stove oil is a straight-run (distilled) fraction from crude oil, whereas other fuel oils are usually blends of two or more fractions. The straight-run fractions available for blending into fuel oils are heavy naphtha, light and heavy gas oil, and residua. Cracked fractions such as light and heavy gas oil from catalytic cracking, cracking coal tar, and fractionator bottoms from catalytic cracking may also be used as blends to meet the specifications of the different fuel oils.

Heavy fuel oil. This term includes a variety of oils, ranging from distillates to residual oils that must be heated to 260°C (500°F) or higher before they can be used. In general, heavy fuel oil consists of residual oil blended with distillate oil to suit specific needs and to meet designed specifications. Included among the category of heavy fuel oil are various industrial oils; when used to fuel ships, heavy fuel oil is called bunker oil.

Heavy fuel oil usually contains residuum that is mixed (cut back) to a specified viscosity with gas oils and fractionator bottoms. For some industrial purposes where flames or flue gases contact the product (ceramics, glass, heat treating, open hearth furnaces), the fuel oil must be blended to have a minimum specified sulfur content.

Lubricating oil. After kerosine, the early petroleum refiners wanted paraffin wax for the manufacture of candles. Demand for the lubricating oils that were the by-products of paraffin wax manufacture did not grow until heavy industry increased, after the 1890s; at this time, petroleum largely replaced animal oils (such as lard oil, sperm whale oil, and tallow) and vegetable oils as the source of lubricants. *See* LUBRICANT.

Lubricating oils are distinguished from other fractions of crude oil by a high boiling point (>400°C or 750°F), as well as high viscosity. Materials suitable for the production of lubricating oils are composed principally of hydrocarbons containing 25–35 carbon atoms per molecule, whereas residual stocks may contain hydrocarbons with 50–80 carbon atoms per molecule. *See* VISCOSITY.

Lubricating oil manufacture was well established by 1880, the method used depending on whether the crude petroleum was processed primarily for kerosine or for lubricating oils. Usually the crude oil was processed for kerosine, and primary distillation separated the crude into three fractions—naphtha, kerosine, and a residuum. To increase the production of kerosine, the cracking distillation technique was used, and this converted a large part of the gas oils and lubricating oils into kerosine.

The development of vacuum distillation provided the means of separating more suitable lubricating oil fractions with predetermined viscosity ranges and removed the limit on the maximum viscosity that might be obtained in a distillate oil. Vacuum distillation prevented residual asphaltic material from contaminating lubricating oils, but did not remove other undesirable materials such as acidic components or components that caused the oil to thicken excessively when cold and become very thin when hot. *See* DISTILLATION.

Lubricating oils may be divided into many categories according to the types of service they are intended to perform. However, there are two main groups: oil used in intermittent service, such as motor and aviation oil; and oil designed for continuous service, such as turbine oil.

Oils used in intermittent service must show the least possible change in viscosity with temperature, and these oils must be changed at frequent intervals to remove the foreign matter that collects during service. The stability of such oils is, therefore, of less importance than the stability of oils used in continuous service for prolonged periods without renewal. Oils used in continuous service must be extremely stable because such engines operate at fairly constant temperature without frequent shutdown.

White oil, insulating oil, and insecticides. The term "white distillate" is applied to all the refinery streams with a distillation range between approximately 80 and 360°C (175 and 680°F) at atmospheric pressure and with properties similar to the corresponding straight-run distillate from atmospheric crude distillation. Light distillate products, such as naphthas, kerosine, jet fuels, diesel fuels, and heating oils, are manufactured by appropriate blending of white distillate streams.

For many years, most white distillate products originated from naphthenic stocks, but now they are prepared from paraffinic, mixed-base, or naphthenic fractions, depending on the final use of the oil. Naphthenic crude oils give products of high specific gravity and viscosity, desirable in pharmaceutical use; whereas paraffinic stocks produce oils of lower specific gravity and lower viscosity suitable for lubrication purposes.

White oils. This category of petroleum products falls into two classes: those often referred to as technical white oils, which are employed for cosmetics, textile lubrication, insecticide vehicles, paper impregnation, and so on; and pharmaceutical white oils, which may be employed as laxatives or for the lubrication of food-handling machinery. The colorless character of these oils is important in some cases, as it may indicate the chemically inert nature of the hydrocarbon constituents. Textile lubricants should be colorless to prevent the staining of light-colored threads and fabrics. Insecticide oils should be free of reactive (easily oxidized) constituents so as not to injure plant tissues when applied as sprays. Laxative oils should be free of odor, taste, and also hydrocarbons, which may react during storage and produce unwanted by-products. These properties are attained by the removal of oxygen-, nitrogen-, and sulfur-containing compounds, as well as reactive hydrocarbons by, say, sulfuric acid.

The medicinal oils require a test showing minimal color change, but depending on its intended use, a technical oil showing rather marked color change may be satisfactory. The only further distinction between pharmaceutical and technical oils is that the high-quality medicinal oils are made as viscous as possible, whereas the technical oils are likely to be made of the less viscous fractions.

Insulating oils. These fall into two classes: those used in transformers, circuit breakers, and oil-filled cables; and those employed for impregnating the paper covering of wrapped cables. The first are highly refined fractions of low viscosity and a comparatively high boiling temperature range that resemble heavy burning oils, such as mineral seal oil, or the very light lubricating fractions known as nonviscous neutral oils. The second are usually highly viscous products, often naphthenic distillates, and are not usually highly refined.

The insulating value of fresh transformer oils seems to vary little with chemical constitution, but physical purity, including freedom from water, is highly significant. A water content of 0.1% lowers an original dry insulating value by a factor of about 10; higher water content causes little additional change.

Insecticides. These are petroleum oils that are usually applied in water-emulsion form and have marked killing power for certain species of insects. For many applications in which their own effectiveness is too slight, the oils serve as carriers for active poisons, as in household and livestock sprays.

The physical properties of petroleum oils, such as their solvent power for waxy coatings on leaf surfaces and insect bodies, make them suitable as carriers for more active fungicides and insecticides. The additive substance may vary from fatty acids and soaps, the latter intended chiefly to affect favorably the spreading properties of the oil, to physiologically active compounds such as pyrethrum, nicotine, rotenone, thiocyanates, methoxychlor, and lindane. *See* INSECTICIDE.

Grease. Grease is a lubricating oil in which a thickening agent has been added for the purpose of holding the oil to surfaces that must be lubricated. The most widely used thickening agents are soaps of various kinds, and grease manufacture is essentially the mixing of soaps with lubricating oils.

Soap is made by chemically combining a metal hydroxide with a fat or fatty acid [reaction (1)]. The

$$\underset{\text{Fatty acid}}{RCO_2H} + \underset{\text{Sodium hydroxide}}{NaOH} \longrightarrow \underset{\text{Soap}}{RCO_2^-Na^+} + \underset{\text{Water}}{H_2O} \qquad (1)$$

most common metal hydroxides used for this purpose are calcium hydroxide, lye, lithium hydroxide, and barium hydroxide. Fats are chemical combinations of fatty acids and glycerin. If a metal hydroxide is reacted with a fat, a soap containing glycerin is formed. Frequently a fat is separated into its fatty acid and glycerin components, and only the fatty acid portion is used to make soap. Commonly used fats for grease-making soaps are cottonseed oil, tallow, and lard. Among the fatty acids used are oleic acid (from cottonseed oil), stearic acid (from tallow), and animal fatty acids (from lard). *See* FAT AND OIL; SOAP.

Grease may contain 30–50% soap, and although the fatty acid influences the properties of a grease, the metal in the soap has the most important effect. For example, calcium soaps form smooth buttery greases that are resistant to water but are limited in use to temperatures under about 95°C (200°F). Soda (sodium) salts form fibrous greases that disperse in water and can be used at temperatures well over 95°C (200°F). Barium and lithium soaps form greases similar to those from calcium soaps, and can be used at both high temperatures and very low temperatures; hence barium and lithium soap greases are known as multipurpose greases. In addition to soap and oil, grease may contain additives that are used to improve the ability of the grease to stand up under extreme bearing pressures, to act as a rust preventive, and to reduce the tendency of oil to seep or bleed from a grease. Graphite, mica, talc, or fibrous material may be added to grease that is used to lubricate rough machinery, and other chemicals can make grease more resistant to oxidation.

Wax. Petroleum wax can be categorized as two types: paraffin wax in petroleum distillates, and microcrystalline wax in petroleum residua. The melting point of wax is not directly related to its boiling point, because waxes contain hydrocarbons of different chemical structure. Nevertheless, waxes are graded according to their melting point and oil content. *See* WAX, PETROLEUM.

Paraffin wax is a solid crystalline mixture of straight-chain (normal) hydrocarbons ranging from C20 to C30 and higher. Wax constituents are solid at ordinary temperatures (25°C or 77°F), whereas petrolatum (petroleum jelly) does contain both solid and liquid hydrocarbons. *See* PARAFFIN; PETROLATUM.

Wax production by a process called wax sweating was originally used in Scotland to separate wax fractions with various melting points from the wax obtained from shale oils. Wax sweating is still used to some extent but is being replaced by the more convenient wax recrystallization process. In wax sweating, a cake of slack wax is slowly warmed to a temperature at which the oil in the wax and the lower-melting waxes become fluid and drip (or sweat) from the bottom of the cake, leaving a residue of higher-melting wax. *See* DEWAXING OF PETROLEUM.

The amount of oil separated by sweating is now much smaller than it used to be, due to the development of highly efficient solvent dewaxing techniques. Wax sweating is now more concerned with the separation of slack wax into fractions with different melting points. The three main methods used in modern refinery technology are: (1) solvent dewaxing, in which the feedstock is mixed with one or more solvents; then the mixture is cooled down to allow the formation of wax crystals, and the solid phase is separated from the liquid phase by filtration; (2) urea dewaxing, in which urea forms adducts with straight-chain paraffins that are separated by filtration from the dewaxed oil; (3) catalytic dewaxing, in which straight-chain paraffin hydrocarbons are selectively cracked on zeolite-type catalysts and the lower-boiling reaction products are separated from the dewaxed lubricating oil by fractionation.

Wax recrystallization, like wax sweating, separates wax into fractions; but instead of relying upon differences in melting points, the process makes use of the different solubilities of the wax fractions in a solvent such as a ketone. When a mixture of ketone and wax is heated, the wax usually dissolves completely; and if the solution is cooled slowly, a temperature is reached at which a crop of wax crystals is formed. These crystals will be of the same melting point, and if they are removed by filtration, a wax fraction with a specific melting point is obtained. If the clear filtrate is cooled further, a second batch of wax crystals with a lower melting point is obtained. Thus, alternate cooling and filtration can subdivide the wax into a large number of wax fractions, each with different melting points.

Recrystallization can also be applied to the microcrystalline waxes obtained from intermediate and heavy paraffin distillates, which cannot be sweated. Indeed, the microcrystalline waxes have higher melting points and differ in their properties from the paraffin waxes obtained from light paraffin distillates; and thus, wax recrystallization has made new kinds of waxes available.

Asphalt. Asphalt is a residuum that cannot be distilled even under the highest vacuum since the temperatures required to volatilize the residuum promote the formation of coke. Asphalts have complex chemical and physical compositions that usually vary with the source of the crude oil. *See* ASPHALT AND ASPHALTITE.

Asphalt manufacture commences with distilling everything possible from crude petroleum until a residuum with the desired properties is obtained. This is usually done in two stages: (1) Distillation at atmospheric pressure removes the lower-boiling fractions and yields a reduced crude that may contain higher-boiling (lubricating) oils, asphalt, and even wax. (2) Distillation of the reduced crude under vacuum removes the oils (and wax) as volatile overhead products, and the asphalt remains as a bottom (or residual) product. If the residuum obtained by distillation does not meet the desired specification, blowing the residuum with air can then be applied. Propane deasphalting also produces asphalt.

There are wide variations in refinery operations and in the types of crude oils and different asphalts that will be produced. Blending with higher- and lower-softening-point asphalts may make asphalts of intermediate softening points. If lubricating oils are not required, the reduced crude may be distilled in a flash drum that is similar to a distillation tower but has few, if any, trays. Asphalt descends to the base of the flasher as the volatile components pass out of the top. Asphalt can be made softer by blending the hard asphalt with the extract obtained in the solvent treatment of lubricating oils. Soft asphalts can be converted into harder asphalts by oxidation (air blowing).

Road oil is liquid asphalt that is intended for easy application to earth roads to provide a strong base or a hard surface and will maintain a satisfactory passage for light traffic. Liquid road oils, cutback asphalt, and emulsion asphalt are of recent date, but use of asphaltic solids for paving goes back to the European practices of the early 1800s. Cutback asphalt is a mixture in which hard asphalt has been diluted with a lighter oil to permit application as a liquid without drastic heating. Cutback asphalt may be described as rapid-, medium-, and slow-curing, depending on the volatility of the diluent that governs the rate of evaporation and consequent hardening.

An asphaltic material may be emulsified with water to permit application without heating. Such emulsions are normally of the oil-in-water type. They reverse or break on application to a stone or earth surface, so that the oil clings to the stone and the water disappears. In addition to their usefulness in road and soil stabilization, they are useful for paper impregnation and waterproofing. The emulsions are chiefly the soap or alkaline type and the neutral or clay type. The former break readily on contact, but the latter are more stable and probably lose water mainly by evaporation. Good emulsions must be stable during storage or freezing, suitably fluid, and amenable to control for speed of breaking.

In the post-1980 period, a shortage of good-quality asphalts developed in the United States and in other parts of the world. Due to the tendency of refineries to produce as much liquid fuel (such as gasoline) as possible to meet market demand, residua that would have once been used for asphalt manufacture was being used to produce liquid fuels (and coke). Now asphalt, once the garbage product of a refinery, has again become a valuable refinery product.

Coke. Petroleum coke is the residue left by the noncatalytic destructive distillation (thermal decomposition with simultaneous removal of distillate) of petroleum residua. The coke formed in catalytic cracking operations is usually nonrecoverable because it adheres to the catalyst employed as fuel for the process. *See* COKE.

The composition of the coke varies with the source of the crude oil, but in general, large amounts of high-molecular-weight complex hydrocarbons (rich in carbon but correspondingly poor in hydrogen) make up a high proportion.

Petroleum coke is employed for a number of purposes, but the major use is in the manufacture of carbon electrodes for aluminum refining, which requires a high-purity carbon (that is, low in ash and sulfur-free). In addition, petroleum coke is employed in the manufacture of carbon brushes, silicon carbide abrasives, and structural carbon (such as pipes and Rashig rings), as well as in the manufacture of calcium carbide (CaC_2) from which acetylene is produced [reaction (2)].

$$\underset{\text{Calcium carbide}}{CaC_2} + \underset{\text{Water}}{H_2O} \longrightarrow \underset{\text{Acetylene}}{HC\equiv CH} \qquad (2)$$

In the refining before the 1920s, coke was an unwanted by-product and was usually discarded. As more coking operations became integral parts of refinery operations in the post-1920 era, coke was produced in significant amounts, and the demand for coke as an industrial fuel and for graphite/carbon electrode manufacture increased after World War II.

The use of coke as a fuel must proceed with some caution. With the acceptance by refiners of the heavier crude oils as refinery feedstocks, the higher content of sulfur and nitrogen in these oils yields a product coke containing substantial amounts of both elements. Both will produce unacceptable pollutants—sulfur oxides (SO_x) and nitrogen oxides (NO_x)—during combustion. They must also be regarded with caution in any coke that is scheduled for electrode manufacture, and removal procedures for these elements are continually being developed.

Sulfonic acids and sulfuric acid sludge. Petroleum distillates are generally treated with sulfuric acid to dissolve unstable or colored substances and sulfur compounds, as well as to precipitate asphaltic materials. When drastic conditions are employed, as in the treatment of lubricating fractions with large amounts of concentrated acid or when fuming acid is used in the manufacture of white oils, considerable quantities of petroleum sulfonic acids are formed [reaction (3)]. Two general methods are applied for the

$$\underset{\text{Paraffin}}{RH} + \underset{\text{Sulfuric acid}}{H_2SO_4} \longrightarrow \underset{\text{Sulfonic acid}}{RSO_3H} + \underset{\text{Water}}{H_2O} \qquad (3)$$

recovery of sulfonic acids from sulfonated oils and their sludge. In one case the acids are selectively removed by adsorbents or by solvents (generally low-molecular-weight alcohols); and in the other case the acids are obtained by salting-out with organic salts or bases.

Petroleum sulfonic acids may be roughly divided into those soluble in hydrocarbons and those soluble in water. Because of their color, hydrocarbon-soluble acids are referred to as mahogany acids, and the water-soluble acids are referred to as green acids. The composition of each type varies with the nature of the oil sulfonated and the concentration of the acids produced. In general, those formed during light acid treatment are water-soluble; oil-soluble acids result from more drastic sulfonation.

Sulfonic acids are used as detergents made by the sulfonation of alkylated benzene. The number, size, and structure of the alkyl side chains are important in determining the performance of the finished detergent. The salts of mixed petroleum sulfonic acids have many other commercial applications, as anticorrosion agents, leather softeners, and flotation agents; they have been used in place of red oil (sulfonated castor oil) in the textile industry. Lead salts of the acids have been employed in greases as extreme-pressure agents, and alkyl esters have been used as alkylating agents. The alkaline earth metal (magnesium, calcium, and barium) salts are used in detergent compositions for motor oils, and the alkali metal (potassium and sodium) salts are used as detergents in aqueous systems.

The sulfuric acid sludge from sulfuric acid treatment is frequently used as a source (through thermal decomposition) to produce sulfur dioxide (SO_2) that is returned to the sulfuric acid plant, and sludge acid coke. The coke, in the form of small pellets, is used as a substitute for charcoal in the manufacture of carbon disulfide. Sulfuric acid coke is different from other petroleum coke in that it is pyrophoric in air and also reacts directly with sulfur vapor to form carbon disulfide. *See* SULFURIC ACID. James G. Speight

Bibliography. American Society for Testing and Materials, *Annual Book of Standards*, 2000; G. W. Mushrush and J. G. Speight, *Petroleum Products: Instability and Incompatibility*, Taylor & Francis, New York, 1995; J. G. Speight, *The Chemistry and Technology of Petroleum*, 3d ed., Marcel Dekker, New York, 1999; J. G. Speight, *Fuel Science and Technology Handbook*, Marcel Dekker, New York, 1990; J. G. Speight (ed.), *Petroleum Chemistry and Refining*, Taylor & Francis, New York, 1998.

Petroleum reserves

Proved reserves are the estimated quantities of crude oil liquids which with reasonable certainty can be recovered in future years from delineated reservoirs under existing economic and operating conditions. Thus, estimates of crude oil reserves do not include synthetic liquids which at some time in the future may be produced by converting coal or oil shale, nor do reserves include fluids which may be recovered following the future implementation of a supplementary or enhanced recovery scheme.

Indicated reserves are those quantities of petroleum which are believed to be recoverable by already implemented but unproved enhanced oil recovery processes or by the application of enhanced recovery processes to reservoirs similar to those in which such recovery processes have been proved to increase recovery.

Thus, crude oil reserves can be called upon in the future with a high degree of certainty, subject of course to the limitations placed on production rate by fluid flow within the reservoir and the capacity of the individual producing wells and surface facilities to handle the produced fluids. It is important to bear in the mind the distinction between resources and reserves. The former term refers to the total amount of oil that has been discovered in the subsurface, whereas the latter refers to the amount of oil that can be economically recovered in the future. The ratio of the ultimate recovery (the sum of currently proved reserves and past production) to the resource or original oil in place is the anticipated recovery efficiency. *See* PETROLEUM ENHANCED RECOVERY.

Levels. In earlier years crude oil reserves were estimated by first defining the volume of the resources from drilling data, the nature of the natural producing mechanism from the performance of the reservoir, particularly the rate of decline in productivity, and then applying a recovery factor based on analogy with similar reservoirs. Although more sophisticated technology is in use today, earlier rule-of-thumb estimates have proved to be surprisingly valid.

Reserves are increased by the discovery of new reservoirs, by additions to already discovered reservoirs by continued drilling, and by revisions due to a better-than-established anticipated performance or implementation of an enhanced recovery project. Reserves are decreased by production, and by negative revisions due to poorer-than-anticipated performance or less-than-projected reservoir volumes.

In the 15 years between 1954 and 1969, reserves of crude oil in the United States remained relatively stable. In 1969 the reserves jumped to a value of 39 $\times 10^9$ bbl (6.2×10^9 m^3) as a result of the discovery of the gigantic Prudhoe Bay oil field on the North Slope of the Brooks Range in Alaska. From 1969 to the early 1990s, the crude oil reserves declined to less than 25×10^9 bbl (40×10^8 m^3) in the United States, which is the lowest level since 1950.

Discovery and production. In retrospect, it can be seen that reserves in the United States were sustained during the 1950s and early 1960s as a result of extensions and revisions, the latter due primarily to the installation of waterfloods, rather than as a result of significant new discoveries. The discovery at Prudhoe Bay indicated the necessity of exploration in new frontiers if new reserves were to be added to the United States total. Subsequent exploration failures in other frontier areas, such as the eastern Gulf of Mexico, the Gulf of Alaska, and the Baltimore Canyon, however, again emphasized the limited occurrence of crude oil in the Earth's crust.

The existence of an accumulation of petroleum fluids in the Earth's crust represents the fortuitous sequence of several natural events: the concentration of a large amount of organic material in ancient sediments, sufficient burial and thermal history to cause conversion to mobile fluids, and the migration and eventual confinement of the fluids in a subsurface geological trap. Although some oil and gas have been produced in 33 of the 50 states, 4 states account for 85% of the proved reserves: Texas and Alaska account for over 60%, with California and Louisiana ranking third and fourth. Further evidence for the spotty concentration of oil in the crust is gleaned from the fact that although there are some 10,000 producing oil fields in the United States, some 60 of them account for over 40% of the productive capacity.

In this context, it was to be anticipated that oil discovery and production would reach a peak followed by a monotonic decline in reserves and productive capacity. Only the date of peaking might be in doubt, but even this was predicted with a high degree of accuracy at least a decade before it occurred in the late 1960s in the United States. Additional oil discoveries will be made in some of America's remaining new frontiers, such as the deep ocean and in the vicinity of the Arctic Circle; however, most estimates of the amount of oil remaining to be found have been steadily decreasing, and it is highly unlikely that the decline will be substantially affected by future events.

Natural gas. Natural gas reserves are estimated in a manner similar to that used for the estimation of crude oil reserves. Some natural gas is found in association with crude oil in the same reservoir, and some is found in reservoirs which do not have a gas-oil contact, that is, nonassociated gas. Gas represents a higher maturation level than that of crude oil, and therefore some geographical areas are more gas-prone than others. The history of reserves of natural gas in the United States parallels that of crude oil, with a peak having been reached in the late 1960s. The ratio of gas reserves to oil reserves in the United States is approximately twice the ratio for the world. *See* NATURAL GAS; PETROLEUM. Todd M. Doscher

Bibliography. Worldwide report, *Oil Gas J.*, published in last issue of each year.

Petroleum reservoir engineering

The technology concerned with the prediction of the optimum economic recovery of oil or gas from hydrocarbon-bearing reservoirs. It is an eclectic technology requiring coordinated application of many disciplines: physics, chemistry, mathematics, geology, and chemical engineering. Originally, the role of reservoir engineering was exclusively that of counting oil and natural gas reserves. The reserves—the amount of oil or gas that can be economically recovered from the reservoir—are a measure of the wealth available to the owner and operator. It is also necessary to know the reserves in order to make proper decisions concerning the viability of downstream pipeline, refining, and marketing facilities that will rely on the production as feedstocks.

The scope of reservoir engineering has broadened to include the analysis of optimum ways for recovering oil and natural gas, and the study and implementation of enhanced recovery techniques for increasing the recovery above that which can be expected from the use of conventional technology.

Original oil in place. The amount of oil in a reservoir can be estimated volumetrically or by material balance techniques. A reservoir is sampled only at the points at which wells penetrate it. By using logging techniques and core analysis, the porosity and net feet of pay (oil-saturated interval) and the average oil saturation for the interval can be estimated in the immediate vicinity of the well. The oil-saturated interval observed at one location is not identical to that at another because of the inherent heterogeneity of a sedimentary layer. It is therefore necessary to use statistical averaging techniques in order to define the average oil content of the reservoir (usually expressed in barrels per net acre-foot) and the average net pay. The areal extent of the reservoir is inferred from the extrapolation of geology and fluid content as well as the drilling of dry holes beyond the productive limits of the reservoir. The definition of reservoir boundaries can be heightened by study of seismic surveys, particularly 3-D surveys, and analysis of pressure buildups in wells after they have been brought on production. *See* PETROLEUM GEOLOGY; WELL LOGGING.

If the only mobile fluid in the reservoir is crude oil, and the pressure during the production of the crude remains above the bubble point value (at which point the gas begins to come out of solution), the production of crude oil will be due merely to fluid expansion, and the appropriate equation (1). Here c is

$$\text{Production} = N_P = (c \cdot V \cdot \Delta P)B_0 \qquad (1)$$

the compressibility of the fluid, V is the volume of crude oil in the reservoir, and ΔP is the decrease in average reservoir pressure associated with the production of N_P barrels of crude oil measured at the surface. The term in parentheses is the expansion measured in reservoir barrels, and B_0 is the formation volume factor (inverse of shrinkage) which corrects subsurface measurements to surface units. Thus, if N_P is plotted as a function of $(c \cdot \Delta P)B_0$, the slope of the resulting straight line will be V, the volume of the original oil in place in the reservoir.

For most reservoirs, the material balance equation is considerably more complicated. Few reservoirs would produce more than 1 or 2% of the oil in place if fluid expansion alone were relied upon. The reasons for the increased complexity of a material balance equation when the pressure falls below the bubble point are as follows: (1) The pore volume occupied by fluids in the reservoir changes with reservoir pressure as does the volume of the immobile (connate) water in the reservoir. (2) As the pressure declines below the bubble point, gas is released from the oil. (3) The volume of gas liberated is greater than the corresponding shrinkage in the oil, and therefore the liberated gas becomes mobile upon establishing a critical gas saturation, and the gas will flow and be produced in parallel with oil production. (4) There may be an initial gas cap above the oil column (above a gas-oil contact), and some of this gas, depending upon the nature of the well completion, may flow immediately upon opening the well to production. (5) Below the oil column, there may be an accumulation of water (below a water-oil contact), and if sufficiently large, the water in this aquifer will expand under the influence of the pressure drop and replace the oil that is produced from the oil column.

Thus the simple material balance equation (1) for fluid expansion must be expanded to take into account all the changes in fluid and spatial volumes and the production of oil and gas. By the suitable manipulation of the various terms in the resulting equation, the material balance can be reduced to the equation of a straight line with the slope and intercept of the line providing information about the magnitude of the original oil in place and the significance of the water encroachment and the magnitude of the gas cap. However, the accuracy of the material balance equation depends strongly on having obtained accurate knowledge about the *PVT* (pressure-volume-temperature) behavior of the reservoir crude, and knowing the average reservoir pressure corresponding to successive levels of reservoir depletion. There are significant limits to acquiring such information.

The material balance equation is not time-dependent; it relates only average reservoir pressure to production. Although it can be extrapolated to provide information on the amount of oil and gas that may be produced in the future when the pressure falls to a given level, it does not provide any information as to when the pressure will fall to such a level. It can be used to estimate the ultimate production from the reservoir if an economically limiting pressure can be specified. This usually can be done since pressure and rate of production are of course explicitly interrelated by the equations for fluid flow, and the economic limit of production is that rate at which the economic income is insufficient to pay for the cost of operation, royalties, taxes, overheads, and other facets of maintaining production.

Fluid flow in crude oil reservoirs. In order to develop some understanding of the future performance of a reservoir on a real-time basis, the reservoir engineer has two options. One is to predict the flow rate history of the reservoir by using the equations for fluid flow, and the second is to extrapolate the already known production history of the reservoir into the future by using empiricism and know-how generated by experience with analogous reservoirs.

The flow of fluids in the reservoir obeys the differential form of the Darcy equation, which for radial flow is Eq. (2), where q is the volumetric flow rate

$$q = \frac{k \cdot h \cdot 2\pi r \left(\dfrac{\Delta P}{\Delta r}\right)}{\mu} \qquad (2)$$

measured in subsurface volumes, r is the radial distance from the well bore to any point in the reservoir, $(\Delta P/\Delta r)$ is the pressure gradient at the corresponding value of the radius, μ is the viscosity of the mobile fluid, and $\boldsymbol{k}$ is the permeability of the reservoir to the mobile fluid.

This equation could be integrated directly if $\Delta P/\Delta r$ and the pressure at the outer limits of the reservoir were constant. However, this condition is true only where there exists a strong natural water drive resulting from a contiguous aquifer. Steady-state conditions then prevail and the flow equation is simply Eq. (3), where q is in barrels per day, k

$$q = \frac{7.08kh(P_e - P_w)}{\mu \ln\left(\dfrac{r_e}{r_w}\right)} \qquad (3)$$

in darcies, P_e (pressure at the outer boundary) and P_w (pressure at the well bore) in pounds per square inch, μ in centipoises, and r in feet.

When steady-state conditions do not exist, the Darcy equation must be combined with the general radial diffusivity equation (4), where ϕ, the porosity,

$$\frac{\partial^2 P}{\partial r^2} + \frac{1}{r}\frac{\partial P}{\partial r} = \frac{\phi \mu c}{k}\frac{\partial P}{\partial t} \qquad (4)$$

is the only term not elsewhere defined. Appropriate boundary conditions must be specified in order to obtain an integrated analytical equation for fluid flow in the reservoir. Solutions have been given for the important two sets of boundary conditions: constant production rate at the well bore, and constant production (well bore) pressure. Solutions are available for both the bounded reservoir in which the pressure declines from its initial value at the physical reservoir boundary, or in the case of multiple wells in a reservoir at the equivalent no-flow boundary between wells, and the infinite reservoir (usually only a transient condition in real reservoirs) in which the pressure has not declined below initial value at the boundary.

However, the use of the solutions to these equations still does not permit prediction of reservoir performance. The solutions are usually presented in terms of dimensionless time and dimensionless pressure or dimensionless cumulative production. The permeability to the mobile fluid and its compressibility enter into these dimensionless parameters, and neither of these functions is constant, nor can they be explicitly calculated. The reasons are readily traced.

It has already been noted that as the pressure in the reservoir is reduced, the gas saturation builds up. Accompanying this increase, there is an increase in the permeability to the mobile gas phase, and a decrease in the permeability to oil. The changes are not linear (see **Fig. 1**). Since the pressure is not uniform throughout the reservoir, neither will the gas saturation be uniform, nor will the corresponding permeabilities to the gas and oil. The compressibility of the reservoir fluids will obviously vary directly as the saturation of the compressible gas. Thus, the integrated forms of the radial diffusivity equation cannot be used with any high degree of accuracy over the entire reservoir. The modern high-speed digital computer has made it possible, however, to represent the entire reservoir by a three-dimensional grid system in which the reservoir is mathematically subdivided, and the fluid flow equations and material balance across each block are iteratively and compatibly solved for adjacent blocks. The use of mathematical simulation techniques requires a detailed specification of the reservoir geometry and lithology, and the relative permeabilities to the mobile fluids under conditions of both diffuse and stratified flow. Again, such knowledge is limited, and as a result it is necessary to calibrate the mathematical simulator by history-matching available production and pressure history for the reservoir. Unique predictions are difficult to achieve and as a result such simulations must be constantly updated. Pressure-transient analyses of individual wells are analyzed by superimposing the flow regimes for the flowing period and the shut-in period to yield data on effective permeability to the flowing fluids and the average reservoir pressure at the time of shutting in the well. Such data are important in calibrating the reservoir simulator for a given reservoir.

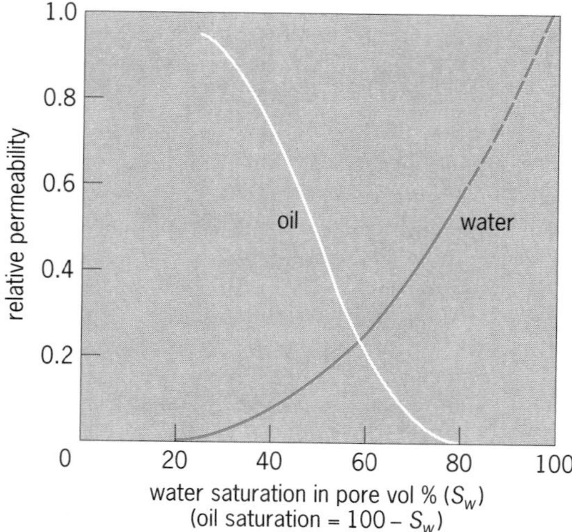

Fig. 1. Typical relative permeability relationship for oil and water in a porous medium.

Before the high-speed digital computer was available to the reservoir engineer, and today when the data available for numerical simulation are not available or when the cost of such simulation studies cannot be justified, the reserves may be predicted by decline curve analysis. Depending on the production mechanism, the production rate will decline in keeping with exponential, hyperbolic, or constant-percentage decline rates. The appropriate coefficients and exponents for a particular decline rate are estimated by history matching, and then the decline curves are extrapolated into the future. A variety of nomographs are available for facilitating the history matching and extrapolation.

Reservoir recovery efficiency. The overall recovery of crude oil from a reservoir is a function of the production mechanism, the reservoir and fluid parameters, and the implementation of supplementary recovery techniques. In the United States, recovery shows a geographical pattern because of the differences in subsurface geology in the various producing basins. Many reservoirs in southern Louisiana and eastern Texas, where strong natural water drives, low-viscosity crudes, and highly permeable, uniform reservoirs occur, show a recovery efficiency well over 50% of the original oil in place. Pressure maintenance, the injection of water to maintain the reservoir pressure, is widely used where the natural water influx is not sufficient to maintain the original pressure. In California, on the other hand, strong water drives are virtually absent, the crude oil is generally more viscous than in Louisiana, and the reservoirs, many of which are turbidites, show significant heterogeneity. Both the viscous crude oil and the heterogeneity permit the bypassing of the oil by less-viscous water and gas. Once a volume of oil is bypassed by water in water-wet rocks (the usual wettability), the oil becomes trapped by capillary forces, which then require the imposition of unfeasible pressure gradients or the attainment of very low interfacial tensions for the oil to be released and rendered mobile. As a result, the overall displacement in California reservoirs is relatively poor, and even after the implementation of water flooding, the recovery efficiency in California will probably be only 25–30%.

In general, recovery efficiency is not dependent upon the rate of production except for those reservoirs where gravity segregation is sufficient to permit segregation of the gas, oil, and water. Where gravity drainage is the producing mechanism, which occurs when the oil column in the reservoir is quite thick and the vertical permeability is high and a gas cap is initially present or is developed on producing, the reservoir will also show a significant effect of rate on the production efficiency.

The overall recovery efficiency in the United States is anticipated to be only 33% unless new technology can be developed to effectively increase this recovery. Reservoir engineering expertise, together with geological and petrophysical engineering expertise, is being used to make very detailed studies of the production performance of crude oil reservoirs in an effort to delineate the distribution of residual oil and gas in the reservoir, and to develop the necessary technology to enhance the recovery. Significant strides have been made in California, where the introduction of thermal recovery techniques has increased the recovery from many viscous oil reservoirs from values as low as a few percent to well over 50%. *See* PETROLEUM ENHANCED RECOVERY.

Todd M. Doscher

Well Testing

Well testing broadly refers to the diagnostic tests run on wells in petroleum reservoirs to determine well and reservoir properties. The most important well tests are called pressure transient tests and are conducted by changing the rate of a well in a prescribed way and recording the resulting change in pressure with time.

The information obtained from pressure transient tests includes estimates of (1) unaltered formation permeability to the fluid(s) produced in the well; (2) altered (usually reduced) permeability near the well caused by drilling and completion practices; (3) altered (increased) permeability near the well created by deliberately stimulating the well by injecting either an acid that dissolves some of the formation or a high-pressure fluid that creates fractures in the formation; (4) distances to flow barriers located in the area drained by the well; and (5) average pressure in the area drained by the well. In addition, some testing programs may confirm hypothesized models of the reservoir, including important variations of formation properties with distance or location of gas/oil, oil/water, or other fluid/fluid contacts.

Tests. Pressure transient tests may be broadly categorized as either single-well tests or multiwell tests. Common single-well tests include pressure buildup tests, pressure drawdown tests, injection tests, and injection falloff tests. In pressure buildup tests, a well is produced at a fixed or stabilized rate for a given period of time, following which the well is shut-in (well production is stopped), and the change in the downhole pressure is measured with time.

In drawdown or flow tests, a well is produced at specified rates, while the pressure downhole is measured as it declines with time. For injection wells, the preferred tests are injection tests and injection falloff tests. Injection tests are run by injecting into a well using a specified rate schedule (usually constant), and measuring the bottomhole pressure as it increases with time. Injection falloff tests are run by injecting a fluid into a well at fixed rate for a given period of time, then shutting-in the well and measuring the bottomhole pressure as it declines with time.

The major multiwell tests are interferences tests and pulse tests. An interference test involves producing (or injecting into) one well (called the active well) and measuring the pressure response in one or more offset wells, called observation wells. The size of the pressure change and the time at which it is large enough to measure are related to the permeability and porosity/compressibility product in the area between the active and observation wells.

Interference tests may indicate pressure changes at an observation well that may have been caused by reservoir activities at wells other than the active well. For that reason, pulse tests may be a more practical alternative for sending a pressure signal from an active well to observation wells. In a pulse test, the active well is produced (or injected into) for a few hours, shut-in, and returned to production at the same rate for several cycles. In this way, a signal is sent from active wells to observation wells. This signal is usually clearly distinguishable from pressure responses to other activities in the reservoir.

Interpretation. Pressure transient tests are usually interpreted by comparing the observed pressure-time response to the predicted response by a mathematical model of the well/reservoir system. The method is an example of an inverse problem, in which an observed response from a system that is modeled mathematically, but with a model having parameters that are unknown—in this case, well and formation properties. The solution to this inverse problem is the unknown properties themselves. Often, however, (1) the mathematical model is usually highly idealized and (2) the solution is not unique—more than one set of formation properties may appear to be accurate.

Graphical techniques are used to calculate permeability. More sophisticated graphical techniques involve matching changes in pressure to preplotted analytical solutions (type-curve matching).

Regression analysis is used to match observed pressure-time data to mathematical models. Although analytical solutions are being found for more and more complex reservoir models each year, many reservoirs are still so complex that their behavior cannot be described accurately by analytical solutions. In such cases, finite-difference approximations to the governing flow equations can be used in commercial reservoir simulators, the reservoir properties treated as unknowns, and properties found that fit the observed data well.

W. John Lee

Special Core Analysis Using Computerized Tomography

Special core analysis is used to determine the spatial distribution of solids, voids, and fluids in a rock

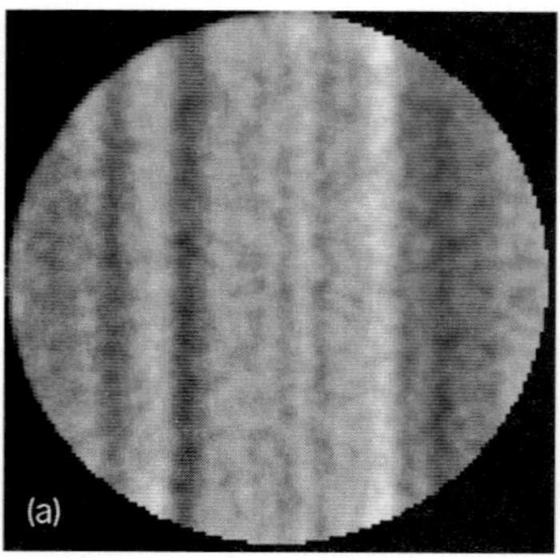

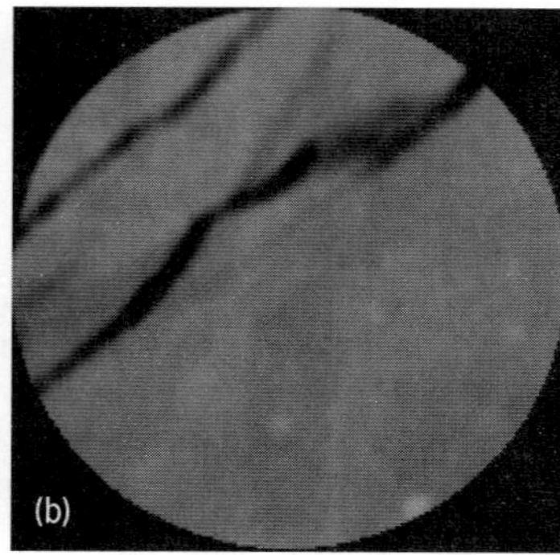

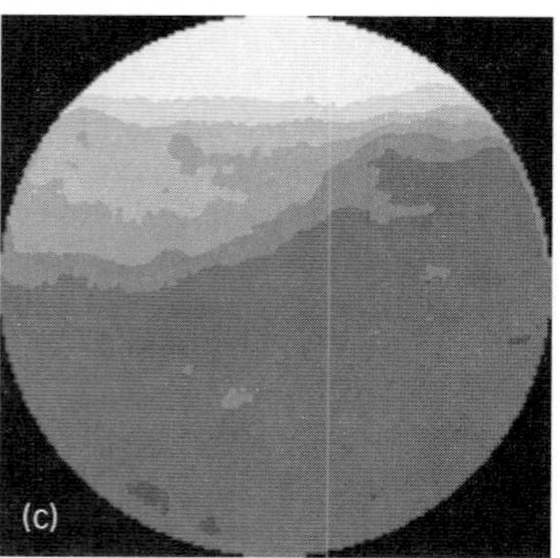

Fig. 2. Core analysis examples using x-ray computerized tomography for rock: (*a*) layers, (*b*) fractures, and (*c*) fluid phases.

sample and to quantify the sample rock's ability to transport oil, gas, and water. The rock solid is classified by its lithology, composition, and density. The void space is characterized by its porosity, spatial distribution, and connectivity. The fluid saturation is described by the fraction of the void volume that it occupies. Multiphase fluid transport properties of a rock are described by absolute and relative permeabilities. Special core analysis results are used in reservoir prediction models to design and optimize the recovery processes.

X-rays, a form of short-wavelength radiation, are able to penetrate many materials. The attenuation of an x-ray beam as it passes through matter is controlled by a combination of Compton scattering and photoelectric effects, and is sensitive to the density and the apparent atomic number of the material. The ability of x-rays to pass through matter and yet be affected by it led to the development of computerized tomographic (CT) mathematical algorithms and instruments known today as CT scanners or imagers. CT technology has had a significant impact on special core analysis procedures. It provides a nondestructive technique for obtaining the spatial distribution of solids and fluids in a rock sample under static or dynamic conditions. *See* ATOMIC NUMBER; COMPTON EFFECT; COMPUTERIZED TOMOGRAPHY; X-RAYS.

Special core analysis uses CT to determine the distributions of densities, atomic numbers, lithologies, porosities, and fluids in reservoir rock samples. Density and apparent atomic number maps are obtained through dual-energy scanning of dry extracted cores. The same maps can be generated for saturated cores if some assumptions are made about porosity and fluid saturations. The lithology of the rock is obtained by density–atomic number cross-plots, similar to wellbore logging analysis. The structure of the rock is mapped by a nondestructive method using x-ray CT. A layered sample with a diameter of 5 cm is identified by a gray-scale display (**Fig.** 2*a*). In this display, light shades denote high density and low porosity; dark shades denote low density and high porosity. The layers are parallel to each other, and the two dark layers at the left are about 3 mm thick. Fractures are also detected by x-ray CT (Fig. 2*b*). Open fractures without solid deposits appear as dark, low-density elongated features. The figure shows an oil-filled shale with a diameter of 10 cm. The fracture width is less than the resolution of the x-ray scanner, and hence the fracture appears wider than it actually is.

There are two primary methods for determining porosity. In the first method, dual-energy scanning is used to determine the overall density of the sample. The porosity distribution is determined using estimates of the densities of the rock matrix and the fluids. In the second method, two high-energy scanning sequences, dry and saturated, are acquired. Subtracting the dry images from the saturated ones yields a map of the net fluid distribution in the sample. Quantitative porosity distribution is obtained either by combining the CT responses of the rock and the saturating fluid, or by calibrating the net fluid maps to external average porosity values.

X-ray CT provides the distribution of residual saturations in a rock sample and helps explain fluid displacement processes. It is also used to track fluid distributions in a sample during flow tests. In two-phase flow cases, one energy level, typically greater than 120 kV, is used. The fluids must have a strong x-ray attenuation contrast to be detected. The contrast between the fluids must be increased as the porosity of the sample is decreased. Elements such as iodine are used to artificially tag one of the fluids. Fluid saturations in a rock are tracked by careful calibrations and by computing CT attenuation changes. The independent determination of three-phase saturations using CT requires fluid tagging and dual-energy scanning at every saturation stage. It is a difficult process to perform without altering the intensive properties of the fluids. A net map of three fluid phases in a sample with a diameter of 5 cm is shown in Fig. 2*c*. The light-shaded region at the top denotes a gravity-driven low-density phase. This phase is tagged with iodine and has a high x-ray attenuation. The dark-shaded region in the lower half of the figure denotes an oil phase bordered at the top with a water-saturated region. This image was obtained during a dynamic displacement experiment with three nonmiscible phases. Abraham S. Grader

Petroleum Reservoir Models

Reservoir behavior can be simulated using models that have been constructed to have properties similar either to an ideal geometric shape of constant properties or to the shape and varying properties of a real (nonideal) oil or gas reservoir. *See* MODEL THEORY; SIMULATION.

In early reservoir modeling, analog electrical physical models were tested against the mathematical solutions that could be obtained for a single fluid by the method of conformal transformation for such ideal geometries. *See* CONFORMAL MAPPING.

For application to petroleum reservoirs, it is necessary to predict the simultaneous flow behavior of more than one fluid phase having different properties (water, gas, and crude oil). The permeability, the relative permeability, and the density and viscosity of each phase constitute its transport properties for calculating its flow. The relative permeability is a factor for each phase (oil, water, gas) which, when multiplied by the permeability for a single phase such as water, will give the permeability for the given phase. It varies with the volume fraction of the pore space occupied by the phase, called the saturation of the given phase. Generally, the relative permeability of the water phase depends only on its own saturation, and likewise for the gas phase. The relative permeability of the oil phase is a function of the saturations of both gas and water phases. *See* FLUID FLOW; FLUID-FLOW PRINCIPLES; FLUID MECHANICS.

For mathematical predictions, the flow is equal to a coefficient times a driving force divided by a resistance. For flow through porous, permeable rocks, the relationship is called Darcy's law [Eq. (2)], in which the coefficient is called the permeability, the driving force is the cross-sectional area times a combination of the gravity force and the pressure gradient per unit area, and the resistance is the fluid viscosity. The equation is written for each phase present. For use in three-dimensional reservoir models, the equations are converted into partial differential equations for each phase (with the three space dimensions and time as independent variables), and then into numerical finite difference equations to allow for finite steps in each dimension and in time. The reservoir is subdivided into rectilinear blocks for the spatial steps, and the porosity, permeability, permeability for each phase, saturation of each phase, temperature, and pressure are specified initially for each grid block of the reservoir model. It is then possible to compute flows under the driving forces between blocks over the entire volume (including blocks specified to contain either injection wells or production wells, where fluids are injected or withdrawn), over a time step small enough not to incur mass-balance errors larger than a specified maximum. The computations must include changes of given components, such as natural gas between liquid phase and gas phase, and the changes in transport properties of the phases with temperature, pressure, and saturation. By repeating such calculations using computers, the process can be carried from an initial state to a desired end point, usually dictated by an economic minimum oil rate or maximum water-oil or gas-oil ratio at producing wells. *See* FINITE ELEMENT METHOD.

Using three-dimensional grids, the number of grid blocks may easily become quite large, approaching a million in some cases, and thus computer reservoir simulation can tax the abilities of the highest capability of computer available. For mega-cell simulation studies, parallel processing is used rather than conventional sequential processing. If the reservoir architecture is modeled using geostatistical variances, it will be necessary to scale the multi-million-cell model to a size with, at most, a few hundred thousand cells. If the crude-oil, gas- and water-phase chemical compositions are initially given and allowed to vary by specified processes as the computation continues, further demands are placed on computer capabilities, and certain compromises are usually required.

For short-term predictions of oil production rates where the rate is declining with time, decline curve analysis issued to give a suitable exponent between rate and time for the available data, so that further prediction to a given end point can be made.

An experimental prediction can also be obtained by constructing a laboratory model in which the geometric shape is similar to that of a given oil or gas reservoir. The injection rate and fluid properties are modified from those of the actual reservoir so that given combinations of the fluid properties, pressure differences, and flow rates can be made by expressing ratios of forces or flow resistances, and the dimensions of each combination cancel out (giving dimensionless numbers). The design method of such models is called dimensionless scaling. Dimensionless numbers are generally found by examining the groups which occur when the partial differential equations for the system involved are cast into dimensionless form. One of the earliest users of dimensionless modeling was Osborne Reynolds, whose experimental work on fluid flow in pipes led to the use of the Reynolds number [Re; Eq. (5)]. He observed

$$\mathrm{Re} = \frac{\rho V L}{\mu} \tag{5}$$

that streamline or turbulent regions of fluid flow in pipes could be determined simply by the value of the dimensionless number formed by the pipe diameter (L) times the flow velocity (V) times the fluid density (ρ), divided by the fluid viscosity (μ). It has been found that laboratory models will exhibit appropriate flow behavior when given properties are changed from those in an oil reservoir but certain important dimensionless groups are made the same for both model and reservoir. The important set of dimensionless groups vary for different reservoir processes. *See* DIMENSIONAL ANALYSIS; DIMENSIONLESS GROUPS; REYNOLDS NUMBER.

The compilation of the reservoir data for either computer modeling or experimental models is a major task. Expert geophysicists, geologists, petrophysicists, logging experts, and reservoir engineers are required. In order to ameliorate the need for experts, computer-based expert systems have been developed for processing data. One method, called neural networks, involves computing methods in which weighting factors are applied to given parts of the data used to compute a given result. A set of data, which should give a certain result, is fed to the system. The weighting factors are varied until the computation gives the known answer. This is done for several different learning sets, and adjusted as needed to give the best set of answers. Then sets are tried for which the answers are withheld until the computations are made. If the neural network still gives the correct answers, it is concluded that the network has learned the proper weighting and may be used when the correct answer is unknown. *See* NEURAL NETWORK.

Elmond L. Claridge

Bibliography. J. K. Ali, Neural Networks: A New Tool for the Petroleum Industry?, *Soc. Petrol. Eng. Pap.*, no. 27561, presented at the European Petroleum Computer Conference, Aberdeen, March 1994; J. J. Arps, Analysis of decline curves, *Trans. AIME*, 160: 219–227, 1945; L. E. Baker, Three-Phase Relative Permeability Correlations, *Soc. Petrol. Eng. Pap.*, no. 17639, presented at the SPE/DOE Enhanced Oil Recovery Symposium, Tulsa, April 1988; H. B. Crichlow, *Modern Reservoir Engineering: A Simulation Approach*, Prentice Hall, Englewood Cliffs, NJ, 1977; R. C. Earlougher, Jr., Advances in Well Test Analysis, *SPE Monog. Ser.*, no. 5, 1977; A. S. Emanuel et al., Reservoir Performance Prediction Methods Based on Fractal Statistics, *Soc. Petrol. Eng. Res. Eng.*, p. 311, August 1989; R. N. Horne, *Modern Well Test Analysis*, 2d ed., Petroway, Palo Alto, CA, 1995; M. K. Hubbert, The Physical Basis of Darcy's Law and Its Importance in the Exploration and Production of Petroleum, *Soc. Petrol. Eng. Pap.*, no. 639, 1963; P. K. Hunt, P. Engler, and C. Bajsarowicz, Computed tomography as a core analysis tool: Applications, instrument evaluation, and image improvement techniques, *SPE Formation Eval.*, pp. 1203–1210, September 1988; R. E. Johnstone and M. W. Thring, *Pilot Plants, Models, and Scale-Up Methods in Chemical Engineering*, McGraw-Hill, New York, 1957; W. J. Lee, Well Testing, *SPE Textbook Ser.*, no. 1, 1981; C. S. Matthews and D. G. Russell, Pressure Buildup and Flow Tests in Wells, *SPE Monog. Ser.*, no. 1, 1967; M. Muskat, *Physical Principles of Oil Production*, McGraw-Hill, New York, 1949; S. Siddiqui, P. J. Hicks, Jr., and A. S. Grader, Verification of Buckley-Leverett three-phase theory using computerized tomography, *J. Petrol. Sci. Eng.*, 15:1–21, 1996; H. J. Vinegar and S. L. Wellington, Tomographic imaging of three-phase flow experiments, *Rev. Sci. Instrum.*, pp. 96–107, January 1987; S. Y. Wang, S. Ayral, and C. C. Gryte, Computer-assisted tomography for the observation of oil displacement in porous media, *Soc. Petrol. Eng. J.*, pp. 53–55, February 1984.

Petrology

The study of rocks, their occurrence, composition, and origin. Petrography is concerned primarily with the detailed description and classification of rocks, whereas petrology deals primarily with rock formation, or petrogenesis. Experimental petrology reproduces in the laboratory the conditions of high pressure and temperature which existed at various depths in the Earth where minerals and rocks were formed. A petrological description includes definition of the unit in which the rock occurs, its attitude and structure, its mineralogy and chemical composition, and conclusions regarding its origin. For a discussion of mineral identification, petrographic analysis, and the classification of rocks *see* MINERALOGY; PETROGRAPHY; ROCK.

Igneous rocks. Volcanic rocks are igneous rocks that reach the Earth's surface before solidification. They occur as lavas or as pyroclastic (fragmental) rocks.

Volcanic activity varies greatly in intensity, duration, periods between eruptions, and quantities of gases, liquid rock, and solidified fragments expelled. The important factors influencing these differences are chemical composition of the magma; amount of gas dissolved in it; extent of crystallization or cooling before eruption; and configuration of the conduit and depth to the magma chamber. *See* MAGMA.

Volcanic structures are of a variety of types (**Table 1**), influenced by the style of activity of the volcano. Volcanic rocks are presently forming mostly at plate boundaries: at the midocean ridges where voluminous basalts are produced and at convergent plate boundaries (island arcs and continental margins) where most explosive and more silicic volcanism occurs. *See* VOLCANO.

Intrusive igneous rocks occur in many different types of units or intrusive masses, which are classified chiefly by their shape and structural relations to their wall rocks (**Table 2**). Bodies that crystallize at great depths (such as batholiths) are referred to as plutonic; those consolidated under shallow cover are designated as hypabyssal. *See* PLUTON.

The crystallization of the larger intrusives may result in profound alterations in the adjacent wall rocks (exomorphism). Where stocks and batholiths have invaded sedimentary rocks, an aureole of contact metamorphism is developed. This results from recrystallization under increased temperature and may be accompanied by chemical transformations (pyrometasomatism) produced by hydrothermal solutions generated during the latter stages of magmatic differentiation. Where batholiths have been intruded into rocks which are already regionally metamorphosed, the contact rocks formed are injection gneisses or migmatites. *See* CONTACT AUREOLE.

Igneous rocks make room for themselves by forceful injection (dilatance), by engulfing wall rock blocks (magmatic stoping), or by subsidence of overlying rocks. The hypothesis of granitization maintains that granites result from the wholesale transformation of sedimentary or metamorphic rock layers by solutions operating through mineral replacement or by ionic emanations acting through solid diffusion.

Blocks of wall rock included in an intrusive mass are xenoliths; their partial destruction by reaction may produce irregular clumps of mafic minerals called schlieren. In some instances such endomorphic effects are sufficiently intensive to result in

TABLE 1. Types of volcanic structure

Name	Characteristics
Shield	Low height, broad area; formed by successive fluid flows accumulating around a single, central vent
Cinder cone	Cone of moderate size with apex truncated; circular in plan, gently sloping slides; composed of pyroclastic particles, usually poorly consolidated
Spatter cone	Small steep-sided cone with well-defined crater composed of pyroclastic particles, well consolidated (agglomerate)
Composite cone	Composed of interlayered flows and pyroclastics; flows from sides (flank flows) common, as are radial dike swarms; slightly concave in profile, with central crater
Caldera	Basins of great size but relatively shallow; formed by explosive decapitation of stratocones, collapse into underlying magma chamber, or both
Plug dome	Domal piles of viscous (usually rhyolitic) lava, growing by subsurface accretion and accompanied by outer fragmentation
Cryptovolcanic structures	Circular areas of highly fractured rocks in regions generally free of other structural disturbances; believed to have formed either by subsurface explosions or sinking of cylindrical rock masses over magma chambers

modification of the composition of the magma (syntexis). *See* XENOLITH.

Crystallizing under equilibrium conditions, early magmatic minerals react with remaining fluid to yield new species (**Fig. 1**). Interruption of the sequence will yield liquid fractions richer in silicon dioxide, alkalis, iron, and water than the original magma, and crystalline fractions richer in calcium and magnesium than the parent magma (magmatic differentiation).

Igneous rocks occur in distinct associations (**Table 3**) which can be put in a specific tectonic context. The sources of magma are the mantle and lower crust of the Earth. The diversity of magmas is caused by variations in the source rock, variations in the conditions and depth of origin, differentiation at various depths, assimilation of other rocks by the magma, mixing of two magmas which originated separately, and unmixing (immiscibility) of the melt. A scheme which shows the possible complexity of the origin of common igneous rock associations is given in **Table 4** . *See* IGNEOUS ROCKS.

Sedimentary rocks. Sedimentary rocks are broadly divided into two classes: clastic sediments, such as sandstones and conglomerates, which are composed largely of fragments of preexisting rocks and minerals and chemical sediments, such as evaporites and many limestones, which form as chemical precipitates from oceans or lakes. With the exception of material deposited by glaciers (till or the consolidated form tillite), sedimentary rocks show bedding or stratification. This separation into generally parallel layers (beds, strata) results from sorting according to grain size during deposition, from differences in composition or texture, or from variations in the rate of deposition. The development of most clastic sediments proceeds in the following stages: There is a source rock, any older rock or, for organic sediments, a supply of organically originated material. By weathering, the older rock is mechanically comminuted, chemically altered, or both, to form unconsolidated surficial rock debris called mantle. Particles are transported by streams, ocean and lake currents, wind, glaciers, or by the direct action of gravity which causes particles to slide and roll down slopes. Material moved by rolling, suspension, or solution is deposited. Deposits usually are consolidated by the processes of cementation (sandstones), compaction (shales), and recrystallization (limestones).

Chemical changes accompanying consolidation are termed diagenetic. Weathered material not transported may become a residual sedimentary rock

TABLE 2. Characteristics of intrusive igneous rock masses

Name	Shape	Structural relations to wall rocks	Size and other features
Dikes	Tabular, lensoid	Discordant	Few feet to hundreds of miles long
Sills	Tabular, lensoid	Concordant	Up to several hundred feet thick
Laccoliths	Plano-convex or doubly convex lenses	Generally concordant	1–4 mi (1.5–6.5 km) in diameter; several thousand feet thick
Volcanic necks	Pipelike	Discordant	Few hundred feet to a mile in diameter; cores of eroded volcanoes
Stocks	Irregular, with steep walls	Crosscutting	A small batholith or its upward projection; outcrop area less than 40 mi^2 (104 km^2)
Batholiths	Irregular, contacts dip steeply or outward; no bottoms known	(1) Discordant (2) Concordant in general, may be crosscutting in detail	Some cover 16,000 mi^2 (41,000 km^2); some composite intrusives of varied petrology
Plutons	Irregular	Usually crosscutting	Usually large; used as general name for intrusive masses that do not fit other definitions

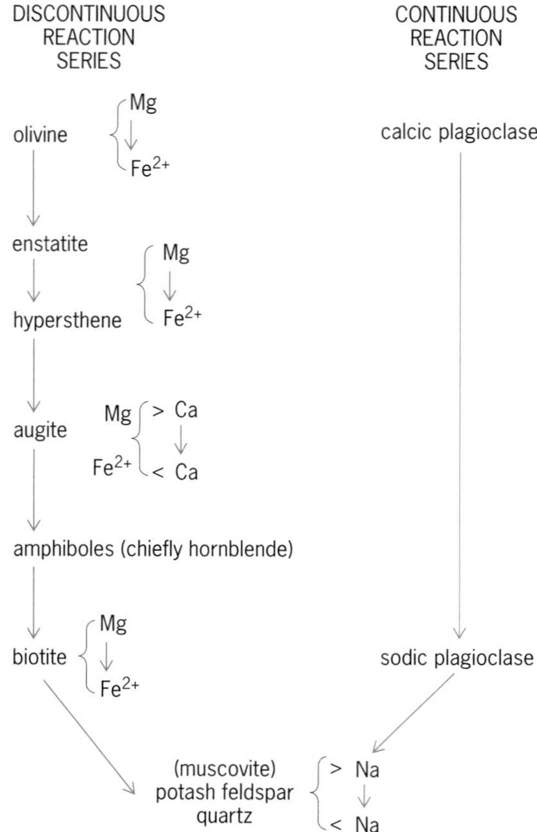

Fig. 1. Reaction series of Bowen (modified).

(bauxite). Sedimentary rocks are deposited either on land areas (continental) or in ocean waters (marine). Most marine sedimentation takes place on the submarine extensions of the continents called continental shelves. Examples of types of sedimentary deposits are listed in **Table 5**. Features characteristically found in sedimentary rocks, in addition to stratification, are cross-bedding, concretions, ripple marks, mud cracks, and fossils. *See* DIAGENESIS; SEDIMENTOLOGY; WEATHERING PROCESSES.

A formation, which is the basic unit of stratigraphy, is a series of rocks deposited during a specific unit of geologic time and consisting either of a particular rock type or of several types deposited in a sedimentary cycle. Such a cycle is the changing sequence of deposits reflecting, for example, advance or retreat of marine waters in a particular area.

However, while sandstone may be deposited at one time in one place in the sedimentary basin, limestone may be formed simultaneously elsewhere. Such lateral variation in a formation is referred to as facies. *See* CYCLOTHEM; FACIES (GEOLOGY); STRATIGRAPHY.

By means of detailed studies of the fossils of a formation and its lithology, composition, structure, and distribution, the paleoecology of the area may be reconstructed. Correlation of formations is attempted chiefly on the basis of fossils, with supplementary data from the lithology, stratigraphic position, insoluble residues (in acid-soluble rocks), heavy detrital minerals (in clastic rocks) and in drill holes by electrical conductivity, radioactivity, and seismic-wave velocities. *See* SEDIMENTARY ROCKS.

Metamorphic rocks. Metamorphism transforms rocks through combinations of the factors of heat, hydrostatic pressure (load), stress (directed pressure), and solutions. Most of the changes are in texture or mineral composition; major changes in chemical composition are called metasomatism. The major types of metamorphism are presented in **Table 6**. Rocks that can serve as parent material for metamorphic derivatives include igneous, sedimentary, and older metamorphic rocks as well. The complexity of the possible metamorphic mineral assemblages stems not only from the variety of possible parent rocks and from the imposition of the several kinds of metamorphism but also from variation in the intensity of particular types of metamorphism (grade), and from the difficulty of readily achieving chemical equilibrium through solid-state reactions. Various features characteristic of metamorphic rocks include foliation (slaty cleavage, schistosity, and

TABLE 3. Igneous rock associations

Name	Main rock types	Environment
Oceanic basalts	Tholeiitic basalt	Mid-ocean ridges, much of the ocean floor
Spilites	Spilite, keratophyre, pillow basalt	Edges of island arcs; obducted ocean floor
Alkali basalts	Olivine basalt, trachyte, phonolite	Oceanic islands, atectonic continental sites
K-rich basalts	Leucite basalt, trachyte	Isolated atectonic continental sites
Tholeiitic flood basalts	Basalt (generally olivine-free), diabase	Continental flood basalt regions, associated with rifting
Layered mafic intrusives	Gabbro, norite, anorthosite, peridotite	Continental rifting sites
Calc-alkalic volcanic rocks	Basalt, andesite, rhyolite	Convergent plate margins, island arc and continental margin
Alpine peridotites	Peridotite, serpentinite	Convergent plate boundaries
Precambrian anorthosite	Andesine or labradorite anorthosite, norite, syenite, monzonite	Domed pluton of massifs in Precambrian terrains
Granite batholith	(a) Simple: granite, granodiorite (b) Complex: gabbro, tonalite, granodiorite, minor granite	Precambrian shields; cores of mountain ranges
Minor granitic intrusive	Granite (some alkalic), quartz syenite, syenite, diorite	Hypabyssal, in mountain ranges and as their outliers
Nepheline syenite	Feldspathoidal rocks, carbonatites	(a) Simple plutons (b) Ring complexes
Lamprophyre	Minette, kersantite, camptonite	Dike swarms

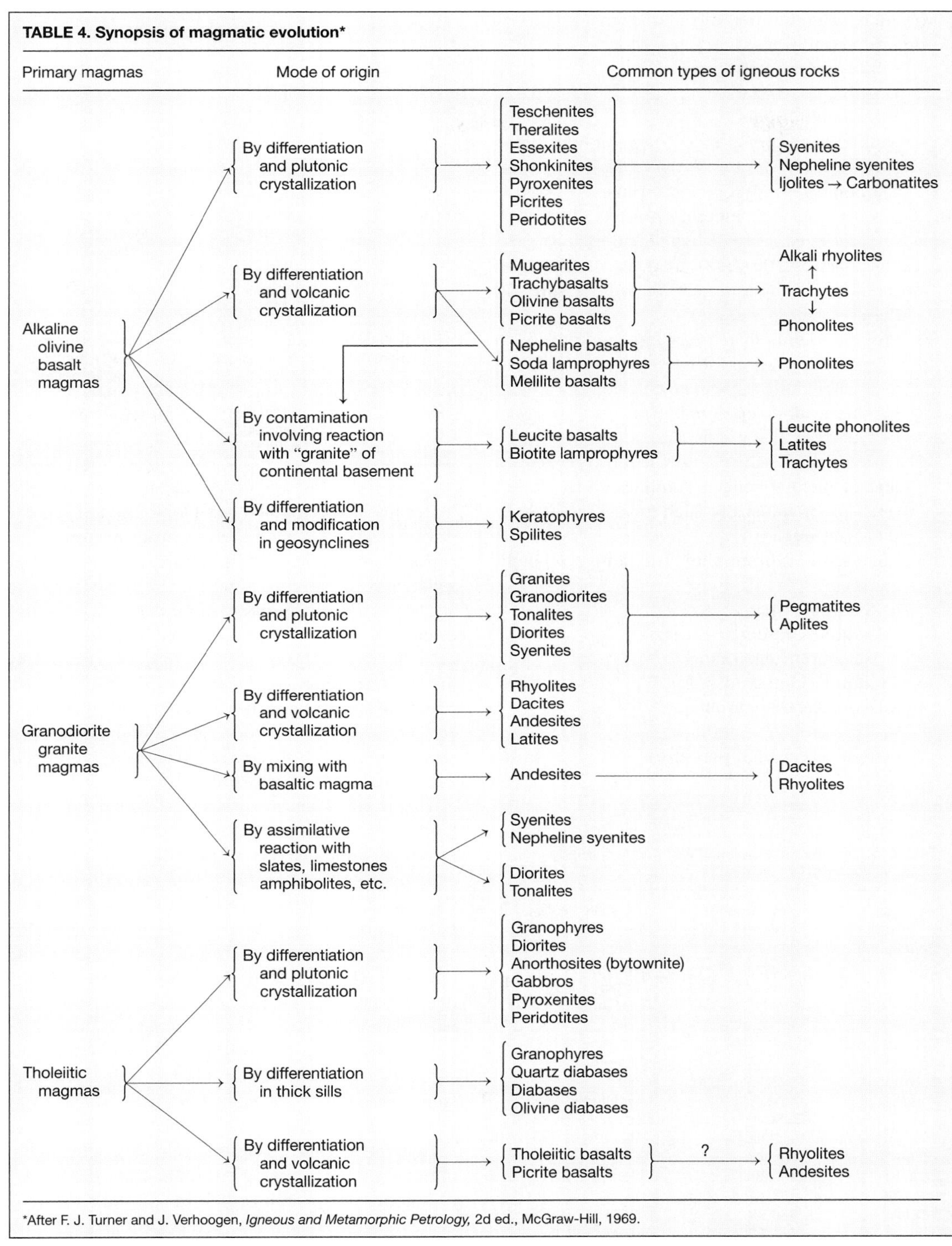

TABLE 4. Synopsis of magmatic evolution*

Primary magmas	Mode of origin	Common types of igneous rocks	
Alkaline olivine basalt magmas	By differentiation and plutonic crystallization	Teschenites, Theralites, Essexites, Shonkinites, Pyroxenites, Picrites, Peridotites	→ Syenites, Nepheline syenites, Ijolites → Carbonatites
	By differentiation and volcanic crystallization	Mugearites, Trachybasalts, Olivine basalts, Picrite basalts	→ Alkali rhyolites ↑ Trachytes ↓ Phonolites
		Nepheline basalts, Soda lamprophyres, Melilite basalts	→ Phonolites
	By contamination involving reaction with "granite" of continental basement	Leucite basalts, Biotite lamprophyres	→ Leucite phonolites, Latites, Trachytes
	By differentiation and modification in geosynclines	Keratophyres, Spilites	
Granodiorite granite magmas	By differentiation and plutonic crystallization	Granites, Granodiorites, Tonalites, Diorites, Syenites	→ Pegmatites, Aplites
	By differentiation and volcanic crystallization	Rhyolites, Dacites, Andesites, Latites	
	By mixing with basaltic magma	Andesites	→ Dacites, Rhyolites
	By assimilative reaction with slates, limestones amphibolites, etc.	Syenites, Nepheline syenites; Diorites, Tonalites	
Tholeiitic magmas	By differentiation and plutonic crystallization	Granophyres, Diorites, Anorthosites (bytownite), Gabbros, Pyroxenites, Peridotites	
	By differentiation in thick sills	Granophyres, Quartz diabases, Diabases, Olivine diabases	
	By differentiation and volcanic crystallization	Tholeiitic basalts, Picrite basalts	? → Rhyolites, Andesites

*After F. J. Turner and J. Verhoogen, *Igneous and Metamorphic Petrology,* 2d ed., McGraw-Hill, 1969.

gneissic structure), lineation, banding, and relict structures. *See* METAMORPHISM; METASOMATISM.

The facies principle is employed in attempting to reconstruct the environment under which a metamorphic rock was developed. A metamorphic facies consists of all rocks, without respect to chemical composition, that have been recrystallized under equilibrium, within a particular environment of stress, temperature, load, and solutions. The first two factors are considered critical. The facies are named after metamorphic rocks deemed diagnostic of such restricted conditions. In practice, a group of related rocks of different compositions is assigned to a particular facies upon presence of such a key assemblage. Facies and their type descriptions are as follows:

A. Facies of contact metamorphism. Load pressure low, generally 100–3000 bars (10–300 megapascals). Water pressure highly variable, in some cases possibly exceeding load pressure, in a few cases very low. Facies listed in order of increasing temperature for given range of pressure conditions.

1. Albite-epidote hornfels.
2. Hornblende hornfels.
3. Pyroxene hornfels.
4. Sanidinite—corresponds to minimum pressures (load, $P_{H_2}O$, P_{CO_2}) and maximum temperatures—pyrometamorphism.

B. Facies of regional metamorphism. Load and water pressures generally equal and high (3000–12,000 bars or 300–1200 MPa). Facies listed in order of increasing temperature and pressure.

1. Zeolite (very low grade).
2. Prehnite-pumpellyite (very low grade).
3. Greenschist (low grade).
   a. Quartz-albite-muscovite-chlorite.
   b. Quartz-albite-epidote-biotite.
   c. Quartz-albite-epidote-almandine.
4. Glaucophane schist (represents a divergent line of metamorphism conditioned by development of unusually high pressures at low temperatures).
5. Almandine amphibolite (medium to high grade).
   a. Staurolite-quartz.
   b. Kyanite-muscovite-quartz.
   c. Sillimanite-almandine.
6. Granulite (high grade).
   a. Hornblende granulite.
   b. Pyroxene granulite.
7. Eclogite (very high pressure).

TABLE 5. Selected examples of sedimentary deposits under various environments

Agent	Deposit	Resulting rock
Continental		
Streams	Valley fill	Sandstone
	Alluvial fan	Conglomerate
	Delta	Siltstone
Lakes	Varved clay	Shale
Springs		Travertine
		Siliceous sinter
Swamps	Peat	Coal
Wind	Dune	Sandstone
	Dust	Loess
	Volcanic ash	Tuff
Glaciers	Moraine	Tillite
Ground water	Stalactite	Dripstone
Gravity	Talus	Breccia
	Avalanche	Conglomerate
	Landslide	
Marine		
Breakers and alongshore currents	Beach	Sandstone
		Conglomerate
Longshore currents		Sandstone
		Shale
Marine organisms	Reefs and other shell deposits	Shell limestone
		Coquina
		Diatomite
Marine water	Evaporites	Rock salt
		Rock anhydrite
Marine water	Colloidal precipitates	Phosphorite
		Manganese oxide concretions
		Chert

TABLE 6. Types of metamorphism and their factors

Type	Factors	Changes in rock
Cataclastic	Stress, low hydrostatic pressure	Fragmentation, granulation
Contact (thermal)	Heat, low to moderate hydrostatic pressure	Recrystallization to new minerals or coarser grains; rarely melting
Pyrometa-somatism	Heat, additive hydrothermal solutions, low to moderate hydrostatic pressure	Reconstitution to new minerals; change in rock composition
Regional (dynamic)	Heat, weak to strong stress, moderate to high hydrostatic pressure, ±nonadditive solutions	Recrystallization to new minerals or coarser grains; parallel orientation of minerals to produce foliation

Because experimental petrology has allowed the development of a "petrogenetic grid" of mineral reactions, most facies may be subdivided if mineral assemblages are carefully detailed. This subdivision can allow inference of pressure-temperature conditions of metamorphism.

Regional variations in grade may be mapped by means of isograds, lines formed by the intersection of planes of isometamorphic intensity with the Earth's surface. These are defined on the appearance of a specific mineral known to reflect a major increase in the intensity of metamorphism.

The primary cause of stresses acting during regional metamorphism is diastrophism of the mountain-building type. These stresses may precede, accompany, or follow the heating, and the order of thermal and structural events can be unraveled by petrographic study. Most regional metamorphic rocks record several heating and deformation episodes. The higher temperatures may result from deep burial, owing to the geothermal gradient of the Earth, in part to concentrations of radiogenic heat, or in part to heat supplied by cooling masses of magma. In contact metamorphism this last is the sole heat source. *See* GEOLOGIC THERMOMETRY.

William Ingersoll Rose, Jr.

Experimental mineralogy and petrology. One aim of mineralogy and petrology is to decipher the history of igneous and metamorphic rocks. Detailed study of the field geology, the structures, the petrography, the mineralogy, and the geochemistry of the rocks is used as a basis for hypotheses of origin. The conditions at depth within the Earth's crust and mantle, the processes occurring at depth, and the whole history of rocks once deeply buried are deduced from the study of rocks now exposed at the Earth's surface. One approach used to test hypotheses so developed is experimental petrology; the term experimental minerals refers to similar studies involving minerals rather than rocks (mineral aggregates).

The experimental petrologist reproduces in the laboratory the conditions of high pressure and high temperature encountered at various depths within

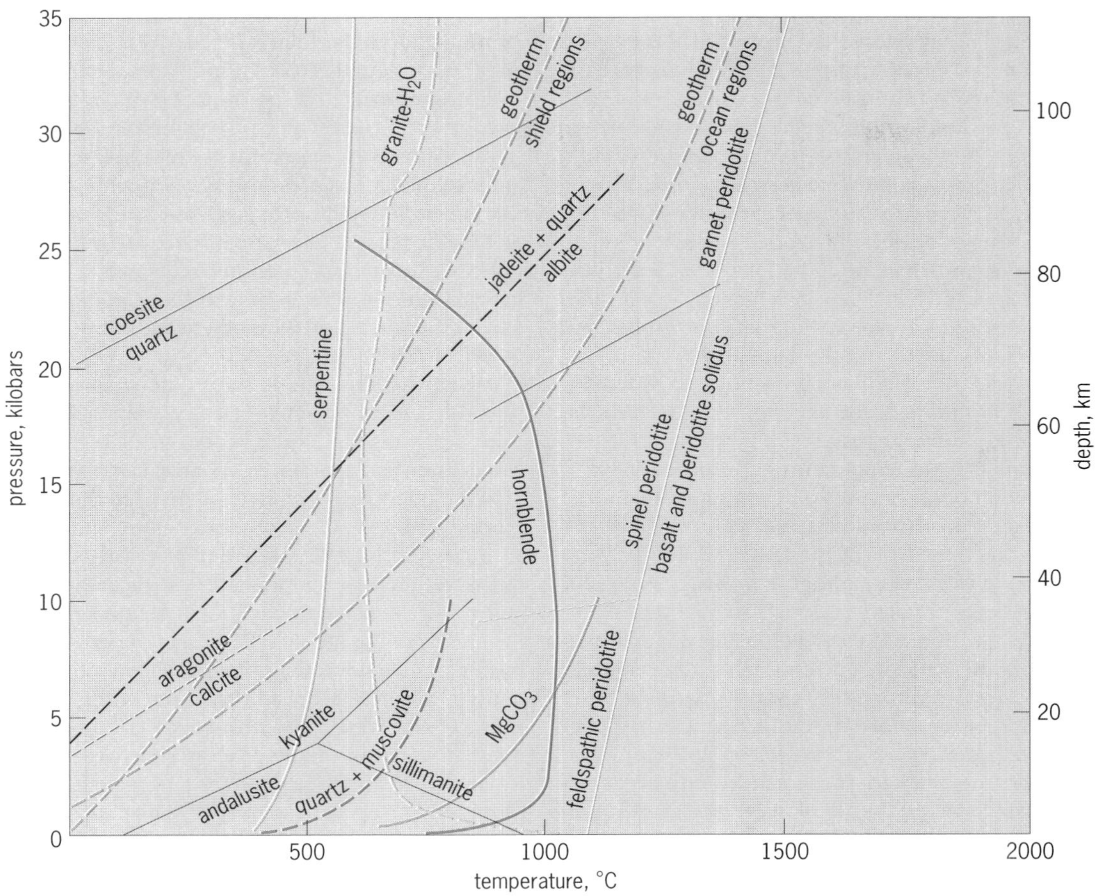

Fig. 2. **Selected results in experimental mineralogy and petrology, averaged from many sources. The two geotherms show the temperature distribution with depth in the Earth. The solidus for dry basalt and peridotite provides the upper temperature limit for magma generation to occur in the mantle. The mineral facies of a peridotite upper mantle are shown below the solidus. The lower limit for magma generation is given by the curve granite-H_2O. Solid-solid mineral transitions and reactions plotted are quartz-coesite, the breakdown of albite, calcite-aragonite, and the polymorphic transitions among kyanite-andalusite-sillimanite. A decarbonation reaction is shown by the curve for $MgCO_3$ (magnesite), and dehydration reactions are shown for serpentine, for the assemblage muscovite + quartz, and for hornblende in mafic or ultramafic rocks. Note the effect of pressure on the hornblende reaction in the garnet-peridotite facies. 1 kilobar = 10^2 megapascals; °F = (°C × 1.8) + 32.**

the Earth's crust and mantle where the minerals and rocks were formed. By suitable selection of materials the petrologist studies the chemical reactions that actually occur under these conditions and attempts to relate these to the processes involved in petrogenesis.

The experiments may deal with the stability range of minerals and rocks in terms of pressure and temperature; with the conditions of melting of minerals and rocks; or with the physical properties or physical chemistry of minerals, rocks, and rock melts (magmas), or of the vapors, gases, and solutions coexisting with the solid or molten materials. These experiments may thus be related to major geological processes involved in the evolution of the Earth: the conditions of formation of magmas in the mantle and crust and their subsequent crystallization either as intrusions or lava flows; the evolvement of gases by the magmas during their crystallization and the precipitation of ore deposits from some of them; the processes leading to the development of volcanic arcs and mountain ranges; and the metamorphism and deformation of the rocks in the mountain chains, and thus the origin and development of the continents. Representative experimental results are shown in **Fig. 2**.

Experimental probes into the Earth. The pioneer work of James Hall of Edinburgh on rocks and minerals sealed within gun barrels at high pressures and temperatures earned him the title "father of experimental petrology." His experiments, published early in the nineteenth century, heralded an era of laboratory synthesis in mineralogy and petrology. Experimental petrology gained tremendous impetus in 1904, when the Geophysical Laboratory was founded in Washington, D.C., and new techniques were developed for the study of silicate melts at carefully controlled temperatures. The results obtained by N. L. Bowen, J. F. Schairer, and others have persuaded most petrologists that physicochemical principles can be successfully applied to processes as complex as those occurring within the Earth, and they have formed the basis for much of the work performed at high temperatures and high pressures. The design of equipment capable of maintaining high temperatures simultaneously with high pressures has been successfully achieved only since about 1950, with a few notable exceptions. The extent of the experimental probe into the Earth varies with the type of apparatus used. The Tuttle cold-seal pressure vessel, or high-pressure test tube, reproduces conditions at the base of the average continental crust. Internally heated, hydrostatic pressure vessels reproduce conditions similar to those just within the upper mantle. Opposed-anvil devices, or simple squeezers, readily provide pressures of 200 kilobars or more, but temperatures attainable are considerably lower than those existing at depths in the Earth corresponding to these pressures.

The deepest experimental probe is provided by a variety of internally heated opposed-anvil devices which can produce pressures of 200 kbar (20 GPa) with simultaneous temperatures of 2000°C (3630°F) or more, although the pressure and temperature measurements become less accurate with the more extreme conditions. This deepest probe produces conditions equivalent to depths of 180–240 mi (300–400 km) in the upper mantle, which are not very deep compared with the 1740-mi (2900-km) depth at the core, but are well within the Earth's outer 300 mi (500 km) where many petrological processes have their origins. *See* HIGH-PRESSURE MINERAL SYNTHESIS.

Studies related to Earth structure. Geophysical studies and inferences about the compositions of deep-seated rocks based on petrological studies of rocks now exposed at the surface, together with laboratory measurements of physical properties of selected materials, provide the basis for theories concerning the composition of the Earth and those parts of the Earth which may be treated as units. *See* EARTH INTERIOR.

At high pressures, most silicate minerals undergo phase transitions to denser polymorphs, and it is believed that polymorphic transitions within the mantle contribute to the high gradients of seismic-wave velocities occurring at certain depths. Experimental studies of the olivine-spinel transition have been successfully correlated with the beginning of the transition zone of the upper mantle at a depth of about 210 mi (350 km). It has been widely held that the Mohorovičić (M) discontinuity marks a change in chemical composition from basalt of the lower crust to peridotite of the upper mantle, but recent experimental confirmation that basalt undergoes a phase transition to eclogite, its dense chemical equivalent, suggests that the M discontinuity could be a phase transition. This now appears to be unlikely for several reasons, but more detailed experimental studies are required because there remains a strong possibility that this phase transition is involved in tectonically active regions of high heat flow where the M discontinuity is poorly defined. *See* MOHO (MOHOROVIČIĆ DISCONTINUITY).

Igneous petrology. The Earth's mantle transmits shear waves and is therefore considered to be crystalline rather than liquid; but the eruption of volcanoes at the Earth's surface confirms that melting temperatures are reached at depth in the Earth from time to time and from place to place. One group of experiments is therefore concerned with the determination of fusion curves for minerals believed to exist in the Earth's mantle and of melting intervals for mineral aggregates, or rocks. The position of the solidus curve provides an upper temperature limit for the normal geotherm, and the experiments dealing with melting intervals at various pressures have direct bearing on the generation of basaltic magmas at various depths within the mantle. They also provide insight into the processes of crystallization and fractionation of these magmas during their ascent to the surface. This is one of the major problems of igneous petrology. *See* IGNEOUS ROCKS; MAGMA.

Many reactions among silicate minerals are very sluggish, and the presence of water vapor under pressure facilitates the experiments. Water is the most abundant volatile component of the Earth, with carbon dioxide a poor second, and the water in the experiments often plays a role as a component as well as a catalyst. In the presence of water vapor under pressure, melting temperatures of silicate minerals and rocks are markedly depressed, and curves for the beginning of the melting of rocks in the presence of water provide lower limits for the generation of magmas. It is widely believed that many of the granitic rocks constituting batholiths, the cores of mountain ranges, were derived by processes of partial fusion of crustal rocks in the presence of water. The hypothesis appears to be consistent with experimental results on feldspars and quartz in the presence of water at pressures up to 10 kbar (1 GPa) and with similar experiments on plutonic igneous rock series ranging in composition from gabbro to granite. However, evidence suggests that some magmas forming batholiths, and andesite lavas of equivalent composition in tectonically active regions, originated directly from mantle material. Experimental petrologists are therefore extending their studies of the minerals and rocks to granitic and andesitic compositions to pressures greater than 10 kbar (1 GPa), corresponding to upper mantle conditions, with and without water present. *See* SILICATE PHASE EQUILIBRIA.

The path of crystallization of a magma and the mineral reaction series thus produced are apparently quite sensitive to many variables, including pressure, pressure or fugacity of water, and oxygen fugacity. Techniques have been developed for studying the effect of oxygen fugacity on crystallization paths and also for controlling oxygen fugacity at very low values while the water pressure is simultaneously maintained at very high values.

Metamorphic petrology. When rocks are metamorphosed, they recrystallize in response to changes in pressure, temperature, stress, and the passage of solutions through the rock. The facies classification is an attempt to group together rocks that have been subjected to similar pressures and temperatures on the basis of their mineral parageneses. The metamorphic facies are arranged in relative positions with respect to pressure (depth) and temperature scales, and the experimental petrologist attempts to calibrate these scales by delineating the stability fields of minerals and mineral assemblages under known conditions in the laboratory. Potentially the most useful reactions for this purpose are solid-solid reactions involving no gaseous phase, such as the reactions among the polymorphs of Al_2SiO_5, kyanite, sillimanite, and andalusite.

Progressive metamorphism of sedimentary rocks produces a series of dehydration and decarbonation reactions with progressive elimination of water and carbon dioxide from the original clay minerals and carbonates. Experimental determination of these reactions, along with the solid-solid reactions, provides a petrogenetic pressure-temperature grid, in which mineral assemblages occupy pigeonholes bounded by specific mineral reactions. If mineral assemblages in metamorphic rocks are matched with the grid which has been calibrated in the laboratory, this provides estimates of the depth and temperature of the rock during metamorphism. Unfortunately, the temperatures of dissociation reactions are very sensitive to partial pressures of the volatile component involved in the reaction, and therefore there are complications introduced in application of the grid to metamorphic conditions. However, continued experimental studies of the stability of minerals and mineral stabilities under a wide range of laboratory conditions (in the presence of H_2O and CO_2 gas mixtures, for example, and with oxygen fugacity varied as well) may eventually provide a guide to the composition of the pore fluid that was present during metamorphism. If this composition is known, then the experimental data can be applied with greater confidence. *See* METAMORPHIC ROCKS. Peter J. Wyllie

Bibliography. T. F. W. Barth, *Theoretical Petrology*, 2d ed., 1962; M. G. Best, *Igneous and Metamorphic Petrology*, 1982; K. Bucher and M. Frey, *Petrogenesis of Metamorphic Rocks*, 6th ed., 1994; G. V. Chilinger, H. J. Bissel, and R. W. Fairbridge (eds.), *Carbonate Rocks*, 2 vols., 1967; Y. Guegen and V. Palciauskas, *Introduction to Rock Physics*, 1994; H. H. Hess and A. Poldervaart (eds.), *Basalts*, 2 vols., 1967; A. Miyashiro, *Metamorphic Petrology*, 1994; A. Philpotts, *Principles of Igneous and Metamorphic Petrology*, 1990; P. C. Ragland, *Basic Analytical Petrology*, 1989; F. J. Turner, *Metamorphic Petrology*, 2d ed., 1981; M. Wilson, *Igneous Petrogenesis*, 1988.

Petromyzontida

The order comprising the lampreys, sometimes called Petromyzontiformes, which are eellike, jawless vertebrates (class Agnatha). Lampreys differ from Myxiniformes as follows: the single nasal opening is on the dorsal side of the head and ends internally as a blind sac; the mouth is surrounded by a circular oral disk and provided with a rasping tongue (both disk and tongue are set with horny teeth); seven pairs of gill pouches open separately to the exterior; two pairs of semicircular canals are present; adults have well-developed eyes; dorsal and caudal fins are separate, and both are supported by fin rays; there are separate sexes and a distinct larval stage; and they either live in fresh water or enter fresh water to breed if they live in the sea. *See* MYXINIFORMES.

Lampreys usually spawn on the gravelly bottoms of streams. The larval stage, known as ammocoetes, differs from the adult in being wormlike and toothless, and in having a broad, hoodlike upper lip and rudimentary eyes. This stage may last for 3–5 years during which the larvae burrow under stones or in the muddy bottom, leaving only their heads to protrude, and feed on microscopic organisms strained from the water. After they have metamorphosed to adults, most lampreys aggressively attack other fish to which they attach themselves by suction, using their oral disks; then they rasp through skin and scales with their tongues and suck the blood and flesh of the host. This parasitic habit has resulted in serious damage to commercial fisheries. Lampreys are rarely used as food.

The Petromyzontida have a worldwide distribution and are divided into eight Recent genera, of which *Petromyzon* and *Lampetra* are well known. Two genera of fossil lampreys are currently known. A single specimen, *Hardistiella*, has been found in the lower Carboniferous (Mississippian) of Montana, and 14 specimens of *Mayomyzona* have been recovered from the upper Carboniferous (Pennsylvanian) near Mazon Creek, Illinois. Details of the structure of the four genera indicate a relationship to the Osteostraci and Anaspida, extinct ostracoderms of the Silurian and Devonian periods, from which they may be descended. *See* ANASPIDA; JAWLESS VERTEBRATES; OSTEOSTRACI. Robert H. Denison; Everett C. Olson

Bibliography. M. W. Hardisty and I. C. Potter (eds.), *The Biology of Lampreys*, 1971; P. Janvier and R. Lund, *Hardistiella montanaensis*, n. gen. and sp. from the Lower Carboniferous of Montana, *J. Vert. Paleontol.*, 2:407–413, 1983.

Pewter

A tarnish-resistant alloy of lead and tin always containing appreciably more than 63% tin. Other metals are sometimes used with or in place of the lead; among them are copper, antimony, and zinc. Pewter is commonly worked by spinning and it polishes to a characteristic luster. Because pewter work-hardens only slightly, pewter products can be finished without intermediate annealing.

Early pewter, with high lead content, darkened with age. With less than 35% lead, pewter was used for decanters, mugs, tankards, bowls, dishes, candlesticks, and canisters. The lead remained in solid solution with the tin so that the alloy was resistant to the weak acids in foods.

Addition of copper increases ductility; addition of antimony increases hardness. Pewter high in tin (91% tin and 9% antimony or antimony and copper, for example) has been used for ceremonial objects, such as religious communion plates and chalices, and for cruets, civic symbolic cups, and flagons. *See* ALLOY STRUCTURES; LEAD ALLOYS; TIN ALLOYS. Frank H. Rockett

pH

An expression for the effective concentration of hydrogen ions in solution. The activity of hydrogen ions or, more correctly, hydronium ions, that is, hydrated hydrogen ions $H(H_2O)_n^+$, affects the equilibria and kinetics of a wide variety of chemical and biochemical reactions. Because these effects are activity dependent, it is extremely important to distinguish between the hydrogen-ion concentration and activity. The concentration, or total acidity, is obtained by titration and corresponds to the total concentration of hydrogen ions available in a solution, that is, free, unbound hydrogen ions as well as hydrogen ions associated with weak acids. The hydrogen-ion activity refers to the effective concentration of unassociated hydrogen ions, the form that directly affects physicochemical reaction rates and equilibria. This activity is therefore of fundamental importance in many areas of science and technology. The relationship between hydrogen-ion activity (a_{H^+}) and concentration (C) is given by Eq. (1), where the activity coefficient γ is a

$$a_{H^+} = \gamma C \qquad (1)$$

function of the total ionic strength (concentration) of the solution and approaches unity as the ionic strength approaches zero; that is, the difference between the activity and the concentration of hydrogen ion diminishes as the solution becomes more dilute. *See* ACTIVITY (THERMODYNAMICS); CHEMICAL EQUILIBRIUM; HYDROGEN ION.

The effective concentration of hydrogen ions in solution is expressed in terms of pH, which is the negative logarithm of the hydrogen-ion activity [Eq. (2)].

$$\text{pH} = -\log_{10} a_{H^+} \qquad (2)$$

Because of the negative logarithmic (exponential) relationship, the more acidic a solution, the smaller the pH value (see **table**). The pH of a solution may have little relationship to the titratable acidity of a solution that contains weak acids or buffering substances; the pH of a solution indicates only the free hydrogen-ion activity. If total acid concentration is to be determined, an acid-base titration must be performed. *See* ACID AND BASE; BUFFERS (CHEMISTRY); TITRATION.

Two methods, electrometric and chemical indicator (optical), are used for measuring pH. The more commonly used electrometric method is based on the glass (or polymeric) pH electrode used along with a reference electrode and pH meter (an electronic voltmeter that provides a direct conversion of potential differences to values of pH). This procedure is based on measurement of the difference between the pH of a test solution and that of a standard solution. The pH scale is defined by a series of reference buffer solutions that are used to calibrate the pH measurement system. The instrument measures the potential difference developed between the pH electrode and a reference electrode of constant potential. The difference in potential obtained when the electrode pair is removed from the standard solution and placed in the test solution is converted to the pH value. In the indicator method, the pH value is obtained by simple visual comparison of the color of pH-sensitive dyes to standards (for example, color charts) or by use of calibrated optical readout devices (photometers), often in combination with fiber-optic sensors. *See* ACID-BASE INDICATOR; ELECTRODE; REFERENCE ELECTRODE.

The determination of pH is one of the most widely performed analytical measurements, ranging in application from basic research and clinical analyses to industrial process control and environmental monitoring. Although electrometric pH measurements are among the most specific with respect to possible interferences, a number of potential sources of error can affect the measurements and care must be exercised to ensure the accuracy of the measurements.

Richard A. Durst

Bibliography. R. G. Bates, *Determination of pH: Theory and Practice*, 2d ed., 1973; G. Eisenman (ed.), *Glass Electrodes for Hydrogen and Other Cations*, 1967; Y. C. Wu, W. F. Koch, and R. A. Durst, *Standardization of pH Measurements*, NBS Spec. Publ. 260-53, 1988.

Approximate pH of some common solutions

Solution	pH	a_{H^+}
Battery acid	0	1
Cola soft drink	2	10^{-2}
Vinegar	3	10^{-3}
Beer	5	10^{-5}
Pure water	7	10^{-7}
Milk of magnesia	10	10^{-10}
Household ammonia	12	10^{-12}
Caustic drain cleaner	14	10^{-14}

pH regulation (biology)

The processes operating in living organisms to preserve a viable acid-base state. In higher animals, much of the body substance (60–70%) consists of complex solutions of inorganic and organic solutes. For convenience, these body fluids can be subdivided into the cellular fluid (some two-thirds of the total) and the extracellular fluid. The latter includes blood plasma and interstitial fluid, the film of fluid that bathes all the cells of the body. For normal function, the distinctive compositions of these various fluids are maintained within narrow limits by a process called homeostasis. A crucial characteristic of these solutions is pH, an expression representing the concentration (or preferably the activity) of hydrogen ions, $[H^+]$, in solution. The pH is defined as $-\log [H^+]$, so that in the usual physiological pH range of 7 to 8, $[H^+]$ is exceedingly low, between 10^{-7} and 10^{-8} M. Organisms use a variety of means to keep pH under careful control, because even small deviations from normal pH can disrupt living processes. *See* HOMEOSTASIS; PH.

Normal values. The most accessible and commonly studied body fluid is blood, and it, therefore, provides the most information on pH regulation. Blood pH in humans, and in mammals generally, is about 7.4. This value indicates that blood is slightly alkaline, because neutrality, the condition in which the concentration of hydrogen ions ($[H^+]$) equals the concentration of hydroxyl ions ($[OH^-]$), is pH 6.8 at mammalian body temperature of 98°F (37°C). The pH within cells, including the red blood cells, is typically lower by 0.2–0.6 unit, and is thus close to neutrality. The cell fluid is more complex and far less accessible than blood, but is of obvious importance and has received considerable research attention with techniques for measuring cell pH such as micro pH electrodes, the distribution characteristics of weak acids and bases, nuclear magnetic resonance, and fluorescent dyes.

In most animals other than warm-blooded mammals, blood pH deviates from the familiar value of 7.4. The major reason is that body temperature has an important influence on pH regulation, and as a rule, the blood pH of invertebrates and lower vertebrates varies inversely with body temperature (see **illus.**). For example, the blood pH of a reptile at 86°F (30°C) may be 7.5, but when the same reptile is tested at 50°F (10°C) its blood pH may be 7.8. Values of blood pH from a great many animals fall on a broad continuum at body temperatures between 32 and 104°F (0 and 40°C; see illus.), and humans and other mammals occupy a stationary position near the high-temperature end. The significance of this regulatory pattern is controversial, but one hypothesis is that the net electric charge state of proteins remains constant when pH varies in this manner. Preservation of charge state will stabilize the three-dimensional structure of proteins that may be essential for their normal roles as enzymes or membrane receptor molecules. Consequently, animals that experience significant changes in body temperature have no single normal pH at which they regulate,

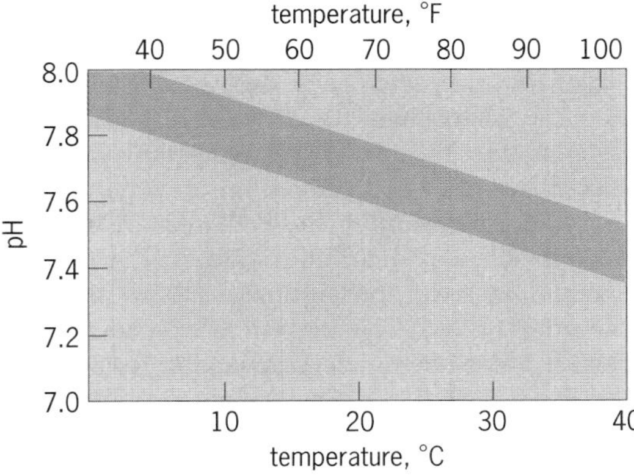

The pH of blood samples from animals at various body temperatures varies inversely with temperature. (*After D. C. Jackson, Blood acid-base control in ectotherms, in C. R. Taylor, K. Johansen, and L. Bolis, eds., A Companion to Animal Physiology, Cambridge University Press, 1982*)

but rather a series of values depending on body temperature.

Blood. Blood pH regulation is necessary because metabolic and ingestive processes add acidic or basic substances to the body and can displace pH from its proper value. In addition, organisms excrete waste products and exchange substances with their surroundings, and these processes can affect acid-base balance. The immediate impact on pH of any of these events, however, is moderated by the chemical buffers of the blood and other body fluids. Buffers are weak acids and bases that are partially dissociated at physiological pH. Their dissociation properties help set the fluid pH at its normal value and resist displacements of pH away from that value by combining with most H^+ or OH^- added to the system. Buffering is a passive, physicochemical process, though, that cannot precisely regulate the pH of a living organism. For true regulation, active physiological mechanisms are required that can alter the acid-base composition of the blood in a controlled fashion. *See* BUFFERS (CHEMISTRY).

It is conventional to identify these control mechanisms with their effects on the principal buffer of the extracellular fluid, carbon dioxide (CO_2). Carbon dioxide is produced by cellular metabolism and distributes readily throughout the body because of its high solubility and rapid diffusion. In solution, CO_2 is hydrated to carbonic acid (H_2CO_3) which dissociates almost completely to H^+ and bicarbonate ions (HCO_3^-). Each of these reactions is reversible, as shown below. Further dissociation of HCO_3^- to

$$CO_2 + H_2O \rightleftharpoons H_2CO_3 \rightleftharpoons H^+ + HCO_3^-$$

carbonate and H^+ is negligible at normal blood pH. Dissolved CO_2 can equilibrate and exchange with CO_2 gas, so that the dissolved CO_2 corresponds to a particular partial pressure of CO_2 (P_{CO_2}). The ready loss of CO_2 from the body, by diffusion into either the surrounding water or air, is a key property of this molecule and accounts for its importance in pH regulation. To regulate blood pH, organisms have mechanisms to independently control P_{CO_2} and $[HCO_3^-]$.

P_{CO_2} control. To control P_{CO_2}, animals utilize gas exchangers, such as lungs or gills. By appropriately adjusting the rate at which air ventilates the lungs or water flows through the gills, the respiratory loss of CO_2 is matched to the metabolic production of CO_2, and body fluid P_{CO_2} is controlled. The value of blood P_{CO_2} can be quite different depending on the habitat of an organism. Fishes and other aquatic animals have low blood P_{CO_2}, on the order of 2–3 mmHg (0.3 kilopascal), because of the large volumes of water that must be breathed to obtain enough O_2 from it. Due to the close link between fish respiration and O_2 supply, the gills are not commonly employed for pH control. In contrast, air breathers generally have higher values of P_{CO_2} [human arterial blood P_{CO_2} is close to 40 mmHg (5.3 kPa)], and their breathing is less tied to O_2 uptake in the more O_2-rich aerial environment. Consequently, air breathers

can use their lungs to rapidly and accurately control P_{CO_2} and pH. Animals such as lungless salamanders that rely on passive diffusion of CO_2 and O_2 through the body surface and have no active breathing mechanism may lack the accurate and rapid control of P_{CO_2}. *See* RESPIRATION.

Bicarbonate control. To control $[HCO_3^-]$, animals use various ion exchange or excretion mechanisms that differentially transfer strong cations (such as Na^+) and strong anions (such as Cl^-) and cause a shift in the equilibrium state of CO_2 and other weak acids. Assuming that P_{CO_2} is set by the respiratory system, these ion exchanges will exert their effects on pH and $[HCO_3^-]$. In most terrestrial vertebrates, the kidneys serve this function and can produce and excrete urine having a highly acid or alkaline composition, depending on the needs of the organism. In fishes, the gills are the primary site for ionic pH regulation. Unlike the kidneys, which are strictly excretory organs, the gills exchange ions with the water passing through; for example, in a fresh-water fish, ammonium ion (NH_4^+) or H^+ ions may be pumped into the water in exchange for Na^+, or HCO_3^- may be exchanged for ambient Cl^-; these processes respectively raise or lower the blood $[HCO_3]$. Other tissues, such as the amphibian skin, may also excrete or exchange ions for acid-base regulation, but each of these mechanisms, whether kidneys, gills, or skin, shares the common effect of controlling the ionic balance of the organism and thereby controlling $[HCO_3^-]$ and pH. Ionic mechanisms are slower than respiratory P_{CO_2} changes, but are probably the principal pH regulators in fishes and in animals with passive gas exchanges. *See* KIDNEY; OSMOREGULATORY MECHANISMS; RESPIRATORY SYSTEM.

Cells. The respiratory and ionic mechanisms already discussed serve the interests of all the body fluids, although their immediate effects may be on the blood. The cells, while benefitting from the stability afforded by these whole body mechanisms, also have local means for their own pH regulation. An acute acid load on a cell, whether from its intrinsic metabolism or from an external source, is dealt with first by the cell's chemical buffering capacity, a capacity that exceeds that of the blood by severalfold. Other cellular mechanisms include the conversion of organic acids to neutral compounds through metabolic transformations, and the transfer of acid equivalents from the primary cell fluid, the cytoplasm, into cellular organelles. Ultimately, however, to restore its normal acid-base state, the cell must expel the excess acid to the exterior; cell membrane ion exchangers have been described in several cell types that can do this. For example, in mammalian cells, a coupled Na^+/H^+ exchange exists. In more easily studied giant cells of various invertebrates, other coupled ion transfers, such as $H^+ + Cl^-$ out in exchange for $Na^+ + HCO_3^-$ in, have been described.

Presumably, all cells possess similar transmembrane transport systems to independently regulate the pH of their contained fluid and to act as a second line of defense behind the overall body pH control mechanisms. Cell pH regulation is complicated by the heterogeneous nature of cell contents. The cytoplasm is the bulk of the cell fluid, but is itself the fluid bathing the cell organelles, and the fluid within these organelles differs in pH from the cytoplasm. For example, mitochondrial fluid is more alkaline than the cytoplasm, so that a H^+ gradient exists between these two compartments that is created and maintained by H^+ pumps on the inner mitochondrial membrane. Oxidative phosphorylation, the transfer of metabolic energy to adenosine triphosphate (ATP), is thought to depend on this H^+ gradient. *See* CHEMIOSMOSIS.

Significance. The integrity of cellular function requires a normal pH in the extracellular fluid bathing the cells and, above all, in the fluids within the cells. Examples of vital functions dependent on normal pH include membrane permeability to charged molecules and catalyzed metabolic reactions. The permeability function underlies the excitable properties of nerve and muscle that give rise to action potentials and nerve transmission. A seemingly small reduction in blood pH, from 7.4 to 7.0, can severely depress nervous function in a human. The basis for this effect is probably a change in the distribution of fixed charges within membrane ion channels. For metabolic reactions, enzymes have pH optima where their catalysis of substrate reactions is maximal. Disturbances in pH can alter surface charges on these protein molecules, leading to changes in their three-dimensional structure, and loss of their catalyzing properties. *See* BIOPOTENTIALS AND IONIC CURRENTS; CELL PERMEABILITY; ENZYME. Donald C. Jackson

Bibliography. N. L. Jones, *Blood Gases and Acid-Base Physiology*, 2d ed., 1987; H. A. Neidig and J. N. Spencer, *pH: Acids and Bases*, 1992.

Phaeophyceae

A class of plants, commonly called brown algae, in the chlorophyll *a-c* phyletic line (Chromophycota). Brown algae occur almost exclusively in marine or brackish water, where they are attached to rocks, wood, sea grasses, or other algae. Approximately 265 genera and 1500 species are recognized, arranged in about 15 orders. *See* ALGAE.

Characteristics. Phaeophyceae are characterized primarily by biochemical and ultrastructural features. The cells are typically uninucleate and contain one or more chloroplasts with or without pyrenoids. Photosynthetic pigments include chlorophyll *a* and *c*, β-carotene, and several xanthophylls, principally fucoxanthin. Varying proportions of these pigments cause the thalli to range in color from yellow-brown to olive green. Food reserves are chiefly laminaran (a polysaccharide in which the glucose units have β-1,3-linkages) and mannitol (a soluble sugar alcohol). The cell walls are composed of an inner layer of cellulose and an outer layer of alginate (a linear polymer of mannuronic and glucuronic acids) and fucans (highly branched sulfated polysaccharides). The mucilage that is often secreted copiously by the larger brown algae (kelps and rockweeds) is composed of various polysaccharides, especially fucans. Motile reproductive cells are usually pear-shaped and have two laterally inserted flagella, an anteriorly directed tinsel flagellum, and a posteriorly directed smooth flagellum. *See* ALGINATE.

Structure. The simplest thallus exhibited by Phaeophyceae is generally considered to be an erect, unbranched or branched, uniseriate filament arising from a prostrate filamentous base. Some phycologists, however, believe that the unicellular algae assigned to the order Sarcinochrysidales of the class Chrysophyceae may be considered primitive brown algae. Many Phaeophyceae are crustose or bladelike. Complex thalli differentiated into macroscopic organs are produced by kelps and rockweeds. Growth is diffuse or, more often, localized at the apex of the thallus (apical growth), at the base of uniseriate filaments (trichothallic growth; see **illus.**), or in special intercalary tissues (transition meristem of kelps, meristoderm of kelps and rockweeds). Colorless uniseriate hairs with a basal meristem (phaeophycean hairs) are common.

Life histories and reproduction. Most Phaeophyceae have a life history involving an alternation of both somatic and cytological phases. The diploid thallus (sporophyte) bears unilocular sporangia in which the initial nucleus undergoes meiosis. The resulting haploid nuclei undergo several mitotic divisions so that 32 or 64 zoospores are commonly produced within a sporangium. A zoospore develops directly into a haploid thallus (gametophyte), which produces isogametes, anisogametes, or eggs and spermatozoids. The two somatic phases may be identical in size and form (isomorphic) or as different (heteromorphic) as the microscopic gametophyte and 100-ft-long (30-m) sporophyte of the giant kelp, *Macrocystis*. The more primitive brown algae are markedly plastic in their life histories, with one or both somatic phases, and sometimes an interpolated juvenile phase, recycling by means of zoospores. Rockweeds, such as *Fucus*, have a life history with only one somatic phase, a diploid thallus that produces eggs and sperm. The gametes, however, do not incorporate the nuclei that result directly from meiosis (as in animals), but instead the derivatives of those nuclei. The intervening mitotic divisions are interpreted by some phycologists as a manifestation of a vestigial or incipient haploid somatic phase.

Embryonic sporophyte of *Desmarestia anceps*, showing trichothallic growth resulting in an embryonic stipe and several pairs of embryonic branches.

Classification. The order Ectocarpales comprises simple filamentous forms with diffuse growth and isomorphic somatic phases. In the Sphacelariales, the filaments have longitudinal cell divisions (pluriseriate) and terminate in conspicuous apical cells that initiate growth. The somatic phases are isomorphic, and sexual reproduction is anisogamous or oogamous. In the Chordariales, the filaments are often compacted into a firm or mucilaginous branched thallus, and the two somatic phases are heteromorphic. The thallus of Dictyotales is a blade in which growth is effected by apical cells, either singly at the tip of each branch or in marginal rows. Sexual reproduction is oogamous, with the sperm unique in lacking a posteriorly directed flagellum. The meiospores formed on the isomorphic soporophyte are also distinctive within the Phaeophyceae in that they are nonmotile. Thalli of Dictyosiphonales are parenchymatous and in the form of a solid or hollow cylinder or a blade. The orders Laminariales (kelps) and Fucales (rockweeds), which exhibit the most specialized anatomy in the algae (including food-conducting tissues), are very interesting biologically and economically important. Large kelplike thalli may be formed in the Desmarestiales, but growth is pseudoparenchymatous, being effected primarily by divisions of a row of cells at the base of filaments that terminate the branches, and secondarily by the production of external corticating filaments as well as internal thickening filaments (see illus.). *See* FUCALES; LAMINARIALES.

Distribution. The geographic distribution of Phaeophyceae is bimodal. Kelps are most abundant and diverse on surf-swept rocky shores of the North Pacific, but they form an ecologically important vegetation belt in the lower intertidal and upper subtidal zones on all cold-water shores except Antarctica, where they are replaced by members of the Desmarestiales. Rockweeds are similarly abundant on cold-water shores, forming conspicuous belts both in the upper intertidal zone, but they reach their peak of diversity in Australia and New Zealand.

Rockweeds also form extensive stands in salt marshes in the Northern Hemisphere. Tropical waters, on the other hand, support a diverse array of Dictyotales and members of the fucalean family Sargassaceae. Five genera of brown algae have been recorded from fresh-water habitats, especially rapidly flowing streams, but so infrequently that it is impossible to deduce a meaningful pattern of geographic distribution. Paul C. Silva; Richard L. Moe

Bibliography. H. C. Bold and M. J. Wynne, *Introduction to the Algae: Structure and Reproduction*, 2d ed., 1985; M. N. Clayton, Evolution of the Phaeophyta with particular reference to the Fucales, *Prog. Phycol. Res.*, 3:11–46, 1984; S. P. Parker (ed.), *Synopsis and Classification of Living Organisms*, 2 vols., 1982.

Phagocytosis

A mechanism by which single cells of the animal kingdom, such as smaller protozoa, engulf and carry particles into the cytoplasm. It differs from endocytosis primarily in the size of the particle rather than in the mechanism; as particles approach the dimensions and solubility of macromolecules, cells take them up by the process of endocytosis.

Feeding mechanism. Cells such as the free-living amebas or the wandering cells of the metazoa move about their environment by ameboid movement. They often can "sense" the direction of a potential food source and move toward it (chemotaxis). If, when the cell contacts the particle, the particle has the appropriate chemical composition, or surface charge, it adheres to the cell. The cell responds by forming a hollow, conelike cytoplasmic process around the particle, eventually surrounding it completely. Although the particle is internalized by this sequence of events, it is still enclosed in a portion of the cell's surface membrane and thus isolated from the cell's cytoplasm. The combined particle and membrane package is referred to as a food or phagocytic vacuole, one of the many normal components of cytoplasm in cells capable of phagocytosis. *See* FEEDING MECHANISMS (INVERTEBRATE); VACUOLE.

Not all cell contacts with particles result in phagocytosis, primarily because the particles fail to stick to the cell surface. Most amebas respond to moving objects such as fine glass needles or swimming paramecia or other food organisms which repeatedly contact them by attempting phagocytosis. This seems to be a nonspecific response to mechanical stimulation rather than a specific response to the chemical properties of the object. There is another type of nonspecific phagocytosis (surface phagocytosis) in which the cell engulfs any small object which it traps in a surface depression during normal locomotion.

Defense mechanism. At the end of the nineteenth century E. Metchnikoff first proposed that the ameboid cells of the metazoa can selectively remove foreign particles, bacteria, and other pathogens by phagocytosis. He demonstrated his proposition by introducing a thorn into a transparent starfish larva and noting the large numbers of wandering cells that accumulated around the thorn. Subsequent studies by Metchnikoff and others established that the macrophages and some white blood cells of humans can remove bacteria and other foreign particles, such as splinters, which invade the body. Of particular significance in this defense mechanism is the highly specific interaction between the surface of the bacteria and the surface of the macrophage. The same mechanism governs the transplant of organs from one individual to another: If the normal activities of the macrophages are not suppressed by drugs, most transplants fail. *See* TRANSPLANTATION BIOLOGY.

Immunity factors. The surface of the macrophage may adhere spontaneously to the surface of the bacteria or the foreign cell with no intervening factors. However, substances in the blood and tissue fluids usually act to facilitate sticking, thereby accelerating phagocytosis. These substances (opsonins) are usually antibodies and blood factors which give the body immunity to foreign cells and viruses. The concentration of specific opsonins in the host, the rate at which the macrophages move by chemotaxis toward the invading bacteria, the ability of the macrophage to stick to the bacteria, and the completeness with which the macrophage can digest or destroy the microbes, balanced against the rate of microbe entry and growth, determine the course of the disease caused by the microorganism. *See* DISEASE; IMMUNITY; OPSONIN.

Digestion and destruction. After the foreign particle or microorganism is trapped in a vacuole inside the macrophage, it is usually digested. To accomplish this, small packets (lysosomes) of lytic proenzymes are introduced into the phagocytic vacuole, where the enzymes are then dissolved and activated. However, not all ingested microbes are killed and digested; some (for example, the tubercle bacillus) survive and multiply. In such instances the body must rely on other methods to combat the infection just as it must when the microbe initially fails to stick to the surface of the macrophage. The destructive abilities of the enzymes secreted into the phagocytic vacuole are supplemented by the ionic conditions found in the cytoplasm. The vacuole fluid is rapidly equilibrated with the cytoplasm and becomes therefore rich in potassium ions and deficient in calcium ions. This condition, which is necessary and normal within cells, is intolerable to most forms of cellular life when it exists both inside and outside the cell membrane. Thus the microbe loses the life-supporting juxtaposition of very differently composed salines if it becomes trapped in the cytoplasm of a wandering cell by the process of phagocytosis. *See* ENDOCYTOSIS; LYSOSOME. Philip W. Brandt

Pharetronidia

A subclass of sponges of the class Calcarea in which the skeleton is formed of quadriradite spicules cemented together into a compact network, or includes

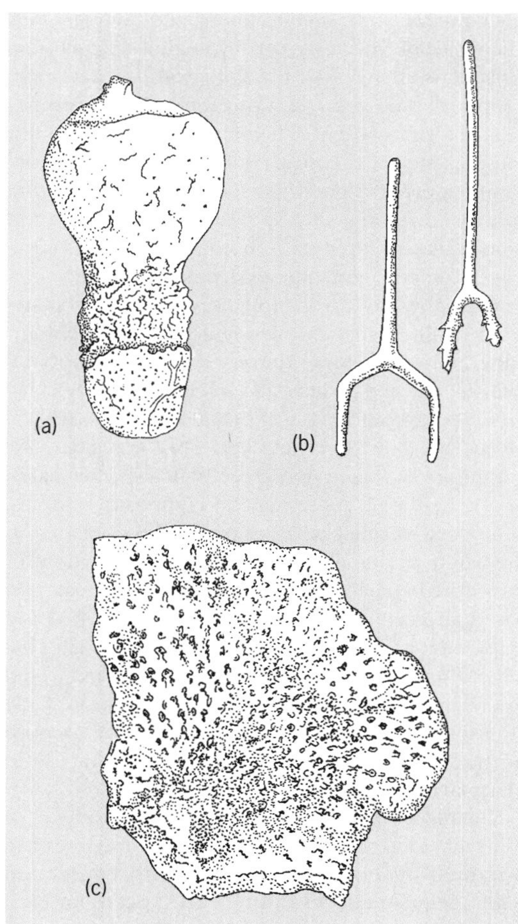

Representative pharetronidians. (*a*) *Petrobiona massiliana*. (*b*) Tuning-fork spicules of *Minchinella*; (*c*) *M. lamellosa*.

an aspiculous massive basal skeleton formed of irregular calcitic spherulites (see **illus.**). Spicules shaped like tuning forks are characteristically present, and free triradiates, quadriradiates, and diactinal spicules may occur. *Murrayona* has a surface layer of scale-like spicules.

The characteristics of the pharetronidians fail to support their inclusion in either the calcinean or calcaronean line of the Calcarea, and they are best regarded as a separate subclass. They have left a rich fossil record that begins in the Permian and is best developed in the Mesozoic Era. Recent forms are typical inhabitants of cryptic habitats of Indo-Pacific coral reefs. *Petrobiona* occurs in caverns in the Mediterranean. *See* CALCAREA. Willard D. Hartman

Pharmaceutical chemistry

The chemistry of drugs and of medicinal and pharmaceutical products. The important aspects of pharmaceutical chemistry are as follows:

1. Isolation, purification, and characterization of medicinally active agents and materials from natural sources (mineral, vegetable, microbiological, or animal) used in treatment of disease and in compounding prescriptions.

2. Synthesis of medicinal agents not known from natural sources, or the synthetic duplication, for reasons of economy, purity, or adequate supply, of substances first known from natural sources.

3. Semisynthesis of drugs, whereby natural substances are transformed by means of comparatively simple steps into products which possess more favorable therapeutic or pharmaceutical properties.

4. Determination of the derivative or form of a medicinal agent which exhibits optimum medicinal activity and at the same time lends itself to stable formulation and elegant dispensing.

5. Determination of incompatibilities, chemical and biological, between the various ingredients of a prescription.

6. Establishment of safe and practical standards, with respect to both dosage and quality, to assure uniform and therapeutically reliable forms for all medication.

7. Improvement and promotion of the use of chemical agents for prevention of illness, alleviation of pain, cure of disease, and search for new therapeutic agents, particularly where no satisfactory remedy now exists.

Chemistry in its various facets, along with cognate disciplines from the physical and biological sciences, is essential for pharmaceutical chemistry. It is difficult to characterize any individual procedure or reaction as solely or preponderantly pharmaceutical. For example, the procedure for converting alcohol to ether, an excellent general anesthetic, may with slight modification be adjusted for the production of ethylene, an important industrial chemical; and the step by which adiponitrile is hydrogenated to diaminohexane, a synthetic fiber intermediate, is useful for the conversion of streptomycin to dihydrostreptomycin. Drug standardization uses many techniques and methods of analytical chemistry.

In the nineteenth century, beginning with the isolation of morphine from opium by F. W. A. Sertüurner (1805), an apothecary from Einbeck, Germany, effort was primarily directed toward the isolation of the active constituents of plants; this resulted in the discovery and description of many alkaloids, glycosides, carbohydrates, volatile oils, and fixed oils. The terms plant chemistry and pharmaceutical chemistry were used interchangeably, and this field of investigation remains active today and promises to continue so for years come, because it is estimated that less than 3% of the known flora have been examined.

The food and drug legislation of 1906 led to the broadening of research activities to include the development of analytical procedures for the maintenance of the quality and potency of drugs. These endeavors are continuing and expanding, as can be seen in the quinquennial revisions of the *National Formulary* and the *U.S. Pharmacopeia*, both official standards, as supplemented by state and federal regulations and tentative methods proposed by agencies charged with the enforcement of these official standards. *See* PHARMACOGNOSY; PHARMACOLOGY; PHARMACY.

Walter H. Hartung

Bibliography. J. N. Delgado (ed.), *Wilson and Gisvold's Textbook of Organic Medicinal and Pharmaceutical Chemistry*, 10th ed., 1998; W. O. Foye, *Principles of Medicinal Chemistry*, 4th ed., 1995; United States Pharmacopeial Convention, Inc., *U.S. Pharmacopeia and the National Formulary*, USP XXII–NF XVII, 1990.

Pharmaceuticals testing

Techniques used to determine that pharmaceuticals conform to specified standards of identity, strength, quality, and purity. Conformance to these standards assures pharmaceuticals which are safe and efficacious, of uniform potency and purity, and of acceptable color, flavor, and physical appearance. Pharmaceuticals are medicinal products which are prescribed by medical doctors and dispensed through pharmacies and hospitals. They are usually taken orally or by parenteral injection.

Standards or specifications and their attendant procedures are designed to provide desired characteristics and acceptable tolerances for all raw materials, intermediates, and finished products. These standards thus provide an objective determination of whether pharmaceuticals are properly constituted. Two components are vital to the makeup of such standards: appropriate analytical procedures to permit a comprehensive examination, and a list of specifications to define the acceptable limits for each property tested. Standards are established by the pharmaceutical manufacturer, official compendiums (*U.S. Pharmacopeia* and the *National Formulary*), and regulations promulgated by the U.S. Food and Drug Administration (FDA).

The steps in the production cycle for pharmaceuticals must be rigidly and uniformly controlled so that each phase is completely accurate. The four control phases are generally designated as raw materials, manufacturing procedures, finished product testing, and control of identity.

Raw materials. These are usually referred to as components and are purchased on specifications. If the raw materials are officially recognized in the *Pharmacopeia* or the *National Formulary*, the specifications are provided in monographs in these compendiums. For raw materials which are not covered by these compendiums, specifications are prepared by the pharmaceutical manufacturer on the basis of requirements for the finished product. In the case of antibiotics there are federal regulations which specify acceptable characteristics of identity, strength, quality, and purity. These criteria are correlated with physical, chemical, biological, and microbiological examinations consisting of identity tests, safety tests, tests for limiting contaminants, and assay procedures. *See* ANTIBIOTIC.

Physical specifications include such characteristics as bulk density, mesh size, color, odor, extraneous contamination such as fibers, and homogeneity. Chemical specifications usually include such characteristics as chemical or physiological potency, melting point, boiling range, optical rotation, moisture, heavy-metals content, chemical identity, solubility, and presence of chemical contaminants. In addition, complex drugs cannot be adequately evaluated and controlled without such special instrumentation and skills as ultraviolet spectrophotometry, infrared spectrophotometry, nonaqueous titrimetry, column chromatography, gas chromatography, paper chromatography, thin-layer chromatography, polarography, x-ray diffraction, x-ray fluorescence, spectrophotofluorimetry, and radioactive tracer techniques. *See* CHROMATOGRAPHY; POLAROGRAPHIC ANALYSIS; RADIOACTIVE TRACER; TITRATION; X-RAY DIFFRACTION.

Samples are taken upon receipt of a specific batch of raw material. These may be composite samples, composed of small portions from each container in the batch, or one or more random small portions of the entire batch. In some cases the sampling may be based on a statistical, or square-root, plan (except in shipments of five or less containers, when all containers must be sampled). The samples are tested to ensure conformance to each specification. Only after the raw material has been checked against each of the specifications can it be approved for use in pharmaceuticals.

Manufacturing procedures. To ensure products of the highest quality, pharmaceuticals must be manufactured under strictly regulated procedures and with adequate checks during each operation. Each dosage unit, though produced by different people from various raw materials, must furnish the patient with the exact kind and amount of drug specified on the label, in a safe and suitable formulation.

Pharmaceutical manufacturers must make products in conformance with the Good Manufacturing Practices as prescribed by the FDA. These regulations provide criteria on the following: buildings, equipment, personnel, components, master-formula and batch-production records, production and control procedures, product containers, packaging and labeling, laboratory controls, distribution records, stability, and complaint files. Proper production of drug products requires highly qualified and trained personnel, adequate manufacturing facilities and equipment, and competent testing laboratories. The quality-control system must provide regular and continuous use of all reasonable procedures, methods, and operations that are necessary to ensure uniform safety and effectiveness of the pharmaceutical.

Batch-production records must describe each manufacturing step in detail. The completion of each step or operation is initialed by the operator. Exact processing temperatures, specific mixing times, designated equipment, and precise details of operations, such as mixing sequence, filtration, or compression, are carefully specified on the batch-production record. All raw materials are double-checked for identity and quality before being incorporated in the process.

In-process assays are used to ensure homogeneity of mixing or completeness of a reaction in the manufacturing process. Such assays can range from simple pH measurements to complex chromatographic determinations. In accord with the FDA's Good Manufacturing Practices regulations, there must be adequate in-process controls, such as checking the weights and disintegration time of tablets, the fill of liquids, the adequacy of mixing, the homogeneity of suspensions, and the clarity of solutions. Upon completion of the manufacturing operation, batch-production records are checked by competent and responsible personnel for actual yield against theoretical yield of a batch of drug and to ensure that each step has been performed and signed for.

Representative samples of the pharmaceutical are taken by inspectors and submitted to the chemical or biological testing laboratory for final assays. Although the Good Manufacturing Practices regulations provide for testing adequately representative samples, the antibiotic regulations provide for a sampling ratio. Samples of the finished antibiotic dosage form must be collected by taking single tablets at intervals throughout the entire processing of the batch so that the quantities tableted during the sampling intervals are approximately equal. In no case may more than 5000 tablets be compressed during each interval of sampling. The nature of the production of a given batch of a pharmaceutical determines the method used to obtain an adequately representative sample. The regulations also provide for a reserve sample, of at least twice the quantity of drug required to conduct all the tests performed on the batch of drug, to be retained at least 2 years after distribution has been completed.

Finished-product testing. Finished pharmaceuticals must conform to appropriate standards of identity, strength, quality, and purity. Accordingly each batch of a pharmaceutical must satisfy five requirements: conformance with (1) the label claim for potency, (2) homogeneity standards, (3) standards of pharmaceutical elegance, (4) identity specifications, and (5) regulatory standards if they are applicable to the specific pharmaceutical.

Potency standards require that the drug meet label claims within specified limits. Monographs of the *Pharmacopeia* or the *National Formulary* usually specify maximum and minimum acceptance limits for official products. Limits for unofficial products are established by the manufacturer and usually parallel those for official products. New drugs and antibiotics are the subject of formal applications which must be approved by the FDA. A part of such application provides for specific acceptance limits on the potency of the product. In the case of antibiotics these standards are specified in regulations which are issued for every antibiotic preparation.

Potency assays vary in complexity from simple tests on pharmaceuticals containing only one active ingredient (ascorbic acid tablets) to very complex chemical and biological tests on pharmaceuticals containing two or more active ingredients (antibiotic and steroid combination tablets). Similarly, biological products must meet analogous standards for potency.

Some special types of pharmaceuticals require additional complex tests. All parenteral products, intended for injection, must meet sterility requirements. Tests are frequently required on parenteral products for absence of pyrogens and for safety (toxicity). In some cases ophthalmic solutions require sterility tests. Products such as parenterals and ophthalmic preparations should be examined for conformance to appropriate standards on foreign particles. These additional tests are mandatory to obviate undesirable physiological reactions.

Statistical quality-control trend charts on certain characteristics of a pharmaceutical, such as tablet weights, ampul-fill volume, or random assay values, assure conformance to the appropriate standards. Stability tests, which are usually more complex and specific than the production potency tests, ensure the effectiveness and safety of the pharmaceutical during normal shelf life. At the same time such tests confirm the absence of any harmful degradation substances. As a result of these tests, many pharmaceuticals have a full-effectiveness expiration date on the label.

The most perfect pharmaceutical is no better than the container in which it is held. In accordance with the Good Manufacturing Practices regulations, suitable specifications, test methods, and cleaning and sterilization procedures must be used to assure that containers, closures, and other component parts of drug packages are suitable for their intended use. They must not be reactive, additive, or absorptive to any extent that significantly affects the identity, strength, quality, or purity of the drug and must furnish adequate protection against its deterioration or contamination.

Pharmaceutical elegance refers to the physical appearance of the dosage units. These standards include inspection to ensure that solutions are sparkling clear, tablets are not capped or chipped, parenteral ampuls are free of floaters, and colored products are of the right hue or shade. These standards govern physical quality.

Identity. Identity is the final requirement in pharmaceutical testing. Identity techniques guarantee proper labeling, that is, that the right product is in the right bottle with the right label. Serious consequences would result from a bottle of strychnine tablets carelessly labeled as saccharin tablets. To maintain the identity of the product, extensive checks are made throughout the manufacturing operation, including the use of duplicate label tags on all bulk goods, and very rigid controls are applied to printing, storage, and application of labels on finished pharmaceuticals.

Only when all operations in the production of pharmaceuticals, from securing raw materials to labeling the final container, are rigidly controlled through testing and checking procedures can it be assured that the pharmaceutical is pure, safe, and efficacious. *See* BIOASSAY; QUALITY CONTROL.

W. Brooks Fortune

Bibliography. American Pharmaceutical Association, *Evaluations of Drug Interactions*, 4th ed., 1988; H. C. Ansel, *Introduction to Pharmaceutical Dosage Forms and Drug Delivery Systems*, 5th ed., 1989; K. A. Connors et al., *Chemical Stability of Pharmaceuticals: A Handbook for Pharmacists,* 2d ed., 1986; *U.S. Pharmacopeia*, 22d ed., 1990; *U.S. Pharmacopeia*, *National Formulary*, XVII, 1990.

Pharmacognosy

The general biology, biochemistry, and economics of nonfood natural products of value in medicine, pharmacy, and other health professions. The products studied are of biologic origin, either plant or animal. They may consist of entire organs, mixtures obtained by exudation or extraction, or chemicals obtained by extraction and subsequent purification.

Pharmacognosy literally means knowledge of drugs, as do pharmacology and pharmacy. The center of interest in pharmacology, however, is on the mode of action of all drugs on the animal body, particularly on humans. In pharmacy major attention is directed toward provision of suitable dosage forms, their production and distribution. Pharmacognosy is restricted to natural products with attention centered on sources of drugs, plant and animal, and on the biosynthesis and identity of their pharmacodynamic constituents.

Sources of materials. Organs, or occasionally entire plants or animals, are dried or frozen for preservation and are termed crude drugs. They may be used medicinally in essentially this form, as in the case of the cardiac drug, digitalis, or the endocrine drug, thyroid, or as sources of mixtures or of chemicals obtained by processes of extraction.

Mixtures obtained by exudation from living plants include such drugs as opium, turpentine, and acacia. Processes of extraction are required to obtain such mixtures as peppermint oil (steam distillation), podophyllum resin (percolation), and parathyroid extract (solution). For a discussion of classes of natural products with medically significant members of this type *see* ESSENTIAL OILS; FAT AND OIL (FOOD); GUM; TERPENE; WAX, ANIMAL AND VEGETABLE.

Pure chemicals may be extracted from a crude drug (for example, the glycoside digitoxin from digitalis or the hormone insulin from pancreas), from a mixture obtained by exudation (for example, the alkaloid morphine from opium), or from an extracted mixture (for example, the terpene menthol from peppermint oil). For a discussion of natural products of this type *see* ALKALOID; GLYCOSIDE; HORMONE.

Vitamins as a class of natural products are within the scope of pharmacognosy, although many are obtained commercially by laboratory synthesis. Included also are antibiotics and biologicals (serums, vaccines, and diagnostic biological products). *See* ANTIBIOTIC; VITAMIN.

The general biology of pharmacognosy is largely descriptive. It includes the taxonomic position of the natural source of the product, the part of the plant or animal yielding the drug, the scientific and common names of the biologic source, the gross and histologic anatomic characterization of the part used, and the principal uses of the product in the health professions.

The biochemistry is both descriptive and experimental. It includes the chemical nature and percentage of the medically significant constituent, the mechanisms of biosynthesis of the constituent, and the role of the constituent in the economy of the plant. Attention is also given to mechanisms of biosynthesis by the use of radioactive precursors of medically active constituents to follow biosynthesis step by step. The isolation and chemical identification of new, potentially useful plant and animal constituents are an important aspect of biochemical research in pharmacognosy.

The economic aspects include the discovery and study of natural sources of crude drugs and their derivatives, development of cultivated sources where feasible, improvement of the yield of useful constituents, and protection of medically useful crop plants from their natural enemies. Methods of harvesting, drying, curing or other processing treatment, storing, packaging, and shipping enter into the commerce of drugs.

For a single drug, for example, menthol from peppermint oil, these several aspects are frequently inseparable. The commercial grower must know the species of *Mentha* yielding commercially profitable quantities of the essential oil and of menthol, percentage yields from various species, conditions suitable for growth, natural enemies of the growing plant such as specific viruses, methods of extraction of the oil and of isolation of the menthol, and proper conditions for packaging, storing, and shipping the purified drug.

Uses of materials. Medical uses are chiefly as therapeutic, prophylactic, or diagnostic agents. Prior to the twentieth century the materia medica of all countries was preeminently of natural products; it still is in some underdeveloped countries. Research in the twentieth century has contributed many synthetic and semisynthetic drugs to the modern materia medica, but a significant number of the crude drugs are still the drugs of choice in therapy or serve as the source of widely used purified mixtures or chemicals. Digitalis and its glycosides, the alkaloids of opium and belladonna, penicillin, thyroid, insulin, and poliomyelitis vaccine are examples.

Pharmaceutical uses are chiefly in the production of palatable and stable dosage forms: gums and mucilages in emulsions and suspensions, starch and lactose in tablets, sugar and essential oils in elixirs, oils and waxes in ointments. Many natural products of insignificant or questionable therapeutic value continue to be used in home remedies.

Uses in other health professions include antiseptics, protectives, and local anesthetics used by dentists; rodenticides, insecticides, and other pesticides used in the protection of the public health; and a variety of prophylactic and therapeutic agents used by veterinarians.

A large number of natural products of value in the health professions are used also in cosmetology (essential oils, gums, fats, and waxes), in the culinary arts (spices, essential oils, and condiments), and in industry (naval stores, mucilages, fats, and waxes).

The role of a medically active chemical produced and used by the animal organism is usually well understood. Physiologic function within the organism and therapeutic use by humans are usually closely related, as in the case of pepsin, thyroid, or the sex hormones.

Corresponding knowledge of medically active plant constituents is almost nonexistent. The role of menthol in the economy of *Mentha piperita*, of digitoxin in *Digitalis purpurea*, of morphine in *Papaver somniferum*, or of reserpine in *Rauwolfia serpentina* is unknown. Through the centuries, however, it has been discovered that certain plants relieve the symptoms of or cure certain diseases. With the discovery of alkaloids in the early nineteenth century and of the other major classes of medically active plant constituents, the chemicals responsible for therapeutic actions have been identified one by one, but not the function of these chemicals in their respective plant sources.

Types of materials. Classes of therapeutic agents have frequently been discovered by study of biosynthesized medicinal chemicals. Most such classes, in fact, have been developed from chemicals orginally known from crude drugs or from their exudates or extractives. The first uses of opium as a narcotic and analgesic drug are lost in antiquity, but its position in the medical practice of the day has been primary for over 2000 years. Morphine was among the first alkaloids to be isolated and has been widely used for more than 150 years; from its study has developed a class of analgesic drugs of wide application in medical practice. *See* ANALGESIC; MORPHINE ALKALOIDS.

An analogous pattern has given the modern classes of hypotensive drugs and tranquilizers. The Indian drug rauwolfia after centuries of use in folk medicine eventually found its way into scientific medical practice in the Orient. Study of its chemical derivatives during the 1930s and 1940s revealed the presence of many alkaloids, one of which, reserpine, was shown to be an effective antihypertensive agent. Subsequent therapeutic use of reserpine and other rauwolfia products demonstrated the tranquilizing action. A large class of drugs having hypotensive action, tranquilizing effect, or both developed rapidly. Various species of the genus have been characterized morphologically, and there has been intensive study of practical methods of culture. *See* TRANQUILIZER.

Another major class of modern drugs of natural origin, the antibiotics, has been developed largely since the beginning of World War II. The prototype, penicillin, was discovered in part as a result of fortuitous accident, but the many other commercially available antibiotics have been developed as a result of carefully planned systematic search.

Not infrequently the clue that has led to collection and scientific investigation of a crude drug as a possible source of medically significant constitutents has been use of the drug by natives for a nonmedical, but to them desirable, purpose such as narcosis or as a poison against wild animals or humans. The use of opium as a narcotic by the laity undoubtedly preceded its medical use. Coca was chewed or sucked by the Indians of South America to increase endurance; condemned by the Spanish conquerors, it nevertheless was introduced into medical practice in Europe. Discovery of the local anesthetic action of the alkaloid cocaine led to the development of a new and important class of therapeutic agents, the local anesthetics. *See* COCA; NARCOTIC OPIATES.

A second native drug from South America, curare, was first used by the Indians as an arrow poison to kill wild animals for food. The neuromuscular paralysis caused in game by the drug suggested therapeutic use as a muscular relaxant. Studies of the plants yielding crude curares revealed several species of two principal genera. *Strychnos* and *Chondodendron*, as the main sources. A number of crystalline alkaloids were isolated from crude curares; eventually the alkaloid tubocurarine was identified as having the therapeutic potentialities suggested by the paralyzing action of the native drug. The botanical source was established as *C. tomentosum*.

Comparable studies of the African arrow poisons inee and kombe added ouabain and strophanthin to the class of cardioactive glycosides, of which digitoxin is the most widely used. The arrow poisons are prepared from African species of *Strophanthus* and are used by natives of both the eastern and western coasts.

The alkaloid physostigmine, also from an African poison and useful in the treatment of glaucoma, was discovered as a result of use of its plant source as a human poison in the trial by ordeal of those accused of offenses. The alkaloid is the toxic constituent of the seeds of *Physostigma venenosum*, which were fed to the accused. Toxic symptoms were taken as evidence of guilt; those who vomited the material were considered guiltless.

Synthetic materials. Development of synthetic drugs related chemically to the active constituent of a natural product has frequently followed investigation of native use of the natural product as drug or poison. The objective of such development is usually to produce a drug having fewer undesirable side effects while retaining the useful therapeutic action. Substitutes for morphine, reserpine, cocaine, tubocurarine, and physostigmine are among a host of synthetic drugs which accomplish the objective to a greater or lesser degree and whose discovery depended on study of natural products.

Intermediates useful in the laboratory synthesis of drugs often exist as therapeutically inactive chemicals in natural products. Plants and animals biosynthesize many such compounds with chemical structures similar to, but not identical with, medicinally useful substances. A slight change in molecular configuration may yield a potent therapeutic agent. A simple example is pinene, a chemical abundant in turpentine oil and convertible by laboratory procedures into camphor. The resulting "synthetic" camphor is actually semisynthetic and possesses the therapeutic and most other properties of natural camphor.

An important class of natural intermediates are the steroids, widely distributed in both plants and animals. Some chemical variations are active physiologically and as drugs, for example, sex and adrenal cortical hormones. Natural sources, glands of domesticated animals used as food by humans, are not available in quantities adequate to fulfill the drug needs for these products. Many plants contain steroids suitable as intermediates for sex hormones. Natural intermediates readily converted into adrenal cortical hormones are uncommon, and extensive search for such steroids has been made. Field studies involve collection and identification of plants judged to be potentially good sources of steroids, preliminary extraction and determination of the presence or absence of these intermediates, and further collection, drying, and preserving of larger quantities of promising species. *See* STEROID.

Systematic screening of plants of reputed therapeutic value and indigenous to a country or other restricted geographic area is a costly and time-consuming procedure—a major reason it has been done for relatively few regions. Notwithstanding, such surveys give promise of uncovering new sources for known classes of drugs, adding to knowledge of such little-known classes of hallucinogens and anticarcinogenic drugs, developing entirely new classes of therapeutic agents and providing profitable natural sources of intermediates useful in drug synthesis. *See* ANTIMICROBIAL AGENTS; BIOCHEMISTRY; BIOSYNTHESIS; PATHOLOGY; PHARMACOLOGY; PHARMACEUTICAL CHEMISTRY; PHARMACY; PLANT PHYSIOLOGY. Richard A. Deno

Bibliography. W. A. Creasy, *Drug Disposition in Humans: The Basis of Clinical Pharmacology*, 1979; G. E. Trease and W. C. Evans, *Pharmacognosy*, 12th ed., 1983; V. E. Tyler et al., *Pharmacognosy*, 9th ed., 1988.

Pharmacology

The science of detection and measurement of the effects of drugs or other chemicals on biological systems. The effect of chemicals may be beneficial (therapeutic) or harmful (toxic) when administered to humans, other mammals, or other living systems. The pure chemicals or mixtures may be of natural origin (plant, animal, or mineral) or may be synthetic compounds.

The broad area covered may be conveniently divided into a number of categories: chemotherapy, the use of chemicals to destroy invading organisms such as bacteria and molds in or on the host; pharmacotherapy, the use of drugs to restore or replace normal function in various tissue cells, organs, or integrated units; pharmacodynamics, studies on the mechanism of action of drugs which may utilize physiological, biochemical, or electrical techniques; toxicology, the study of the poisonous effects of chemicals; psychopharmacology, the study of the effects of chemicals on the behavior of humans or animals; biochemical pharmacology, the effects of chemicals on biochemical reactions in living systems, and the effects of these systems on the chemicals, that is, their metabolism; structure-activity relationship, relationship of biological activity to chemical structure and molecular properties; and clinical pharmacology, the study and evaluation of the effects of drugs in humans. *See* CHEMOTHERAPY; PATHOLOGY; TOXICOLOGY.

The chemicals which have a beneficial effect in humans (such as restoring function or behavior toward the normal state, relieving pain, destroying harmful invading organisms, or aiding in the diagnosis of disease) are called drugs; those chemicals which produce only harmful effects are called poisons. All drugs may be poisonous, however, if administered in large enough amounts. *See* POISON.

It is a fundamental property of drugs that their effects increase in intensity as the dose is increased. Determination of this relationship delineates the dose-response curve. This relationship can be usefully summarized by the dose required to produce a standardized effect and also by the slope of the curve. In many situations the numerical expression of the dose required to produce 50% of the maximal response permits comparison of drugs having the same type of action. This ED_{50} (dose effective at the 50% level) is usually the standard-effect dose used to compare the potency of drugs having the same action, since it can be determined with greater precision than other doses producing a smaller or larger percentage of the maximal effect. Measurement of toxicity as manifested by lethal effects is expressed similarly as the LD_{50} (lethal dose 50). The relative safety of drugs is estimated by means of the therapeutic index, which is the ratio of the LD_{50} to the ED_{50}. The higher the ratio, the safer the drug. *See* EFFECTIVE DOSE 50; LETHAL DOSE 50.

The necessity to use such terms as ED_{50} and LD_{50}, which are averages, implies that individuals have been found to differ in the amount of drug that each will require to produce the desired effect. It is common knowledge that various individuals tolerate widely varying quantities of ethyl alcohol. Individual differences occur in the response to all drugs, and what may be an effective dose in one person may not be effective in another. Individuals may also differ qualitatively in their response to a drug. In some cases these differences have been shown to be due to genetic differences. This has led to expanded research of genetic influences on drug actions, which is generally referred to as pharmacogenetic research.

Depending upon the effect under study, placebos (pills or other preparations containing starch, sugar, and coloring matter but no pharmacologically active agent) may also produce a favorable response in some individuals. The administration of a placebo to a person with hypertension may result in a temporary fall in blood pressure, for example; and in several studies of agents used to relieve pain, as many as 35% of the individuals tested obtained relief of pain from the administration of the placebo. Observations such as these have pointed out the necessity of using a placebo control or a positive control (a drug of established activity) in experiments designed to evaluate a new drug. Because the attitude of the physician who administers a drug may also influence the way patients respond, this variable is controlled by use of the double-blind technique; that is, neither the individual nor the doctor knows which sample is drug and which is placebo.

The use of animals has been and is absolutely essential to progress in pharmacology and other medical sciences; however, different species may differ markedly in their response to drugs both quantitatively and qualitatively. Drugs which produce sedation in one species may produce excitement in another. Fortunately, most chemicals produce similar effects in many species, including humans. Unfortunately, not all drugs found effective in animals are effective in humans, presumably because of differences in metabolism or excretion or unrecognized differences in the nature of the disease process or the physiological mechanism of maintenance of the same function in different species. Study of these differences has yielded much basic information on mechanism of actions of drugs and on physiological and biochemical differences among species. *See* BIOASSAY; PHARMACEUTICAL CHEMISTRY; PHARMACY.

Charles J. Kensler

Bibliography. A. Gilman et al. (eds.), *Goodman and Gilman's The Pharmacological Basis of Therapeutics*, 9th ed., 1996; B. G. Katzung, *Basic and Clinical Pharmacology*, 8th ed., 2000; D. H. Robertson and C. R. Smith, *Manual of Clinical Pharmacology*, 1981.

Pharmacy

The health profession concerned with the discovery, development, production, and distribution of drugs. Drugs are substances (other than devices) used to diagnose, prevent, cure, or relieve the symptoms of disease. For relations to closely allied fields *see* MEDICINE; PHARMACEUTICAL CHEMISTRY; PHARMACOLOGY.

General practice. This part of the profession is carried on in exclusive prescription pharmacies, semiprofessional pharmacies, and drug stores. It consists of compounding and dispensing drugs on order of the physician, dentist, or veterinarian; serving as consultant on drugs to the health professions and to the public; and selling other health supplies such as antiseptics, bandages, and home remedies. Combination of nonprofessional with professional activities is customary in the United States, however, and is commonly termed retail pharmacy.

Hospitals. In addition to qualities characteristic of general pharmacy practice, a hospital pharmacy includes special administrative features, provision of drugs for nursing stations, manufacturing of pharmaceutical preparations, teaching of nurses and medical and pharmacy interns, service to the hospital committee on pharmacy and therapeutics, preparation and revision of a hospital formulary, and monitoring the drug regimen of the individual patient (clinical pharmacy). The pharmacist may have charge of investigational drugs, radioactive pharmaceuticals, medical and surgical sterile supplies, and gaseous drugs for inhalation therapy.

Research. One type of research is in pharmaceutical chemistry, synthetic if the objective is to produce new and improved drugs by laboratory procedures, and analytic if the objective is to provide improved methods of assay for quality control of pharmaceutical production. Research may be in product development, aimed at provision of more palatable, stable, economical dosage forms. It may be on drugs from natural sources (pharmacognosy), or on the mode of action of drugs from any source (pharmacology). It draws heavily on investigations in organic chemistry, biochemistry, and microbiology.

Manufacturing. The pharmaceutical industry is the mass-production medium of drugs and suitable dosage forms. It produces some of its own raw materials, but depends on the closely allied chemical industry for most of them. It sponsors and conducts extensive pharmaceutical research.

Pharmaceutical distribution and promotion are specialized activities within the pharmaceutical industry which are now commonly called pharmacy administration; this term also includes the commercial phases of retail pharmacy.

Jurisprudence. This is a highly specialized area in pharmacy. General pharmacy practitioners and hospital pharmacists are subject to federal and state laws on drugs and pharmacy practice. A board of pharmacy serves as the law enforcement agency and the examining and licensing body in each state; various federal agencies administer the provisions of the several federal laws governing pharmacy.

Other special fields of pharmacy include journalism, administration of national and state associations, and provisions of modern drug standards.

Education. From 1932 to 1960 the education of pharmacists consisted of collegiate studies for at least 4 years in basic science, mathematics, general education, and professional areas. In 1960 the minimal educational program was extended to a total of 5 years of collegiate study. A flood of new and complex drugs, released at the rate of hundreds per year since the end of World War II, has made necessary for the pharmacist a strong background in the physical and biological sciences, highly specialized training in drugs, and the sociologic and humanistic understanding desirable in all professional people. Richard A. Deno

Bibliography. *Remington's Pharmaceutical Sciences*, 17th ed., 1985; A. Wertheimer, *Pharmacy Practice: Social and Behavioral Aspects*, 3d ed., 1989.

Pharynx

A chamber at the oral end of the vertebrate alimentary canal, leading to the esophagus. Because of divergent specializations in the various classes of vertebrates, it cannot be described in general terms except at embryonic stages. In adult humans it is divided anteriorly by the soft palate into a nasopharynx and an oropharynx, lying behind the tongue but anterior to the epiglottis; there is also a retropharyngeal compartment, posterior to both epiglottis and soft palate. The nasopharynx receives the nasal passages and communicates with the two middle ears through auditory tubes. The retropharynx leads to the esophagus and to the larynx, and the paths of breathing and swallowing cross within it. In adult fishes, the pharynx is not segregated from the mouth cavity but is pierced by a number of paired gill slits. *See* ESOPHAGUS; LARYNX; PALATE.

Embryology. Shortly after the germ layers of the embryo are in place, the pharyngeal cavity appears as a simple enlargement of the anterior end of the endoderm tube. Its lateral or lateroventral walls promptly become thickened as a series of paired pillars, the pharyngeal segments (branchial arches or visceral arches), separated from their fellow by paired pouches which push outward from the endodermal lining of the cavity. These pharyngeal pouches are approached on each side by corresponding branchial grooves which push inward from the head ectoderm. The matching grooves and pouches may meet. Their touching surfaces may then become thinned as closing plates, and they may actually break through as pharyngeal clefts. The mouth perforates into the pharyngeal cavity in a similar way anteroventrally. This embryonic appearance of the pharynx as an enlarged anterior gut chamber whose walls are marked by pouches, grooves, clefts, and intervening pillarlike segments, is one of the few characteristics found in all members of the phylum.

The more primitive vertebrate embryos, such as those of lampreys and sharks, have up to seven or even more pairs of large open pharyngeal clefts of fairly uniform size (**Fig. 1**). Embryos of reptiles, birds, and mammals show fewer segments and sharp reduction of the more posterior ones (**Fig. 2**). Also, several of the last pharyngeal pouches fail to break through as clefts. Whereas in most fishes several posterior clefts become enlarged and the intervening segments become equipped with gills and valves during development, few vestiges of either pouches or clefts remain at adult stages of land animals. Nevertheless, in all vertebrates the solid tissues of the pharyngeal segments give rise to numerous structures of the head and neck.

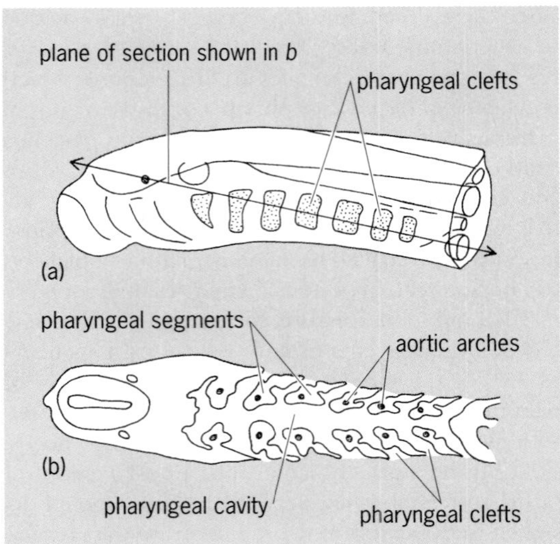

Fig. 1. Lamprey embryo. (*a*) Side view and (*b*) horizontal section of pharynx region.

Each pharyngeal segment of the embryo is lined by ectoderm (the future epidermis) on the outside, and by endoderm (the future mucous membrane) on the inside. The mesoderm enclosed between these layers is of hypomeric or lateral-plate origin, and differentiates chiefly into muscles and arteries. In addition to these constituents, tongues of neural crest cells migrate down into the pharyngeal segments and differentiate there into skeletal elements, except in

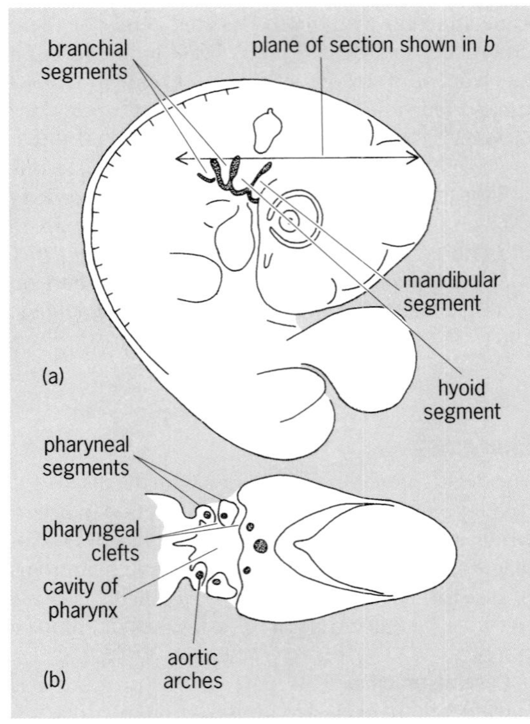

Fig. 2. Chick embryo. (*a*) Side view and (*b*) horizontal section of pharynx region.

cyclostome fishes. *See* NEURAL CREST; RESPIRATORY SYSTEM.

Pharyngeal derivatives. The first pair of pharyngeal segments, called the mandibular segments, enclose the mouth between them, and their skeleton-forming cells produce the mandibular cartilages, which are the earliest rudiments of the lower jaw. Their muscle-forming cells differentiate into a dorsal group of muscles whose principal function in all gnathostome vertebrates is to snap the jaws shut, and a ventral group which supports the tongue by forming a sheet between the halves of the lower jaw. All these muscles are supplied by the motor division of the trigeminal nerve.

The second pair is called the hyoid segments. They arise just posterior to the mandibular segments, usually separated from them temporarily by hyomandibular clefts. They lie ventral to the ear vesicles on the sides of the embryonic head. Their skeleton-forming cells form important parts of the tongue support, and their muscles are used in some vertebrates for opening the mouth, for tongue manipulation, for facial expression, and other functions. The motor division of the facial nerve innervates the hyoid group of muscles. In higher vertebrates the hyomandibular pouch is involved in the formation of the auditory tube and the middle ear space. *See* EAR (VERTEBRATE).

The third and all more posterior pairs are called branchial segments since in fishes they usually form gill-bearing arches; they tend to be repressed or dispersed in the development of the higher vertebrates. Their skeleton-forming cells either form a succession of jointed rodlike bones or cartilages for the support of gills, as in fishes, or join the hyoid skeleton in tongue support and concentrate ventrally in the larynx cartilages, as in terrestrial vertebrates. In fishes the muscles derived from the first branchial segment (the third in the pharyngeal series) are strictly innervated by the glossopharyngeal nerve, and those of all the rest of the segments by individual branches of the vagus nerve. They function for manipulation of the gills, and for grinding and swallowing food. In higher vertebrates, the branchial segments become less and less distinct during development and the nerve supply, while still derived from the glossopharyngeal and vagus nerves, is not so clearly segmental, since the striated muscles themselves become arranged in a more or less continuous pharynx-constrictor sheet or congregate in the laryngeal cartilages, serving nerve mechanisms of swallowing and sound production. *See* SPEECH.

Histology. The pharynx is in general lined by simple mucous membrane and backed by fibrous connective tissue and a double layer of striated muscle. In bony fishes its skeletal arches may be thickly studded with simple teeth, or may even be developed into grinding or crushing plates. Terrestrial vertebrates show simple tubular glands emptying, usually in great numbers, through the mucous membrane and tonsillar collections of lymphoid tissue in the submucosa. *See* TONSIL.

Gross derivatives. Elaborate gill pouches are developed in all aquatic groups and subject to many special adaptations. They differ sharply in design in lampreys, hagfishes, cartilaginous fishes, and the bony fishes. A median ventral evagination from the posterior end of the pharynx gives rise to the entire respiratory system of the land vertebrates, including lungs, larynx, and trachea. A similar but often dorsal evagination, usually from the pharynx-esophagus boundary, gives rise to the air bladder in bony fishes. *See* SWIM BLADDER.

The epithelium of the pharyngeal pouches and of the pharynx floor produces a constellation of endocrine glands and other structures. *See* PARATHYROID GLAND; THYMUS GLAND; THYROID GLAND; ULTIMOBRANCHIAL BODIES. William W. Ballard

Defects and disorders. Congenital defects of the human pharynx that are commonly seen are malformed or split uvulae, or soft palates, and extension of a cleft palate backward to the pharyngeal region. *See* CLEFT LIP AND CLEFT PALATE.

Inflammations may be local or part of a systemic involvement. Acute pharyngitis may be caused by almost any irritant and typically does not involve an infection by a microorganism. Acute follicular pharyngitis is caused by infectious bacteria, usually streptococci; it is also called septic sore throat. Acute tonsillitis involves the masses of lymphoid tissue found in the back of the pharynx. The tonsils may also be the seat of peritonsillar abscesses. All of these inflammations may persist as subacute or chronic diseases, but usually the more prolonged forms occur as a result of repeated attacks or low-grade, persistent infections. *See* STREPTOCOCCUS.

The most common benign tumor of this region is the papilloma; hemangiomas, fibromas, and other less common nonmalignant growths are also found here. The malignant tumors of the pharynx include several varieties of carcinoma and sarcoma, particularly lymphosarcomas in children. *See* ONCOLOGY.

Nervous disorders seen not infrequently are paresthesias (abnormal sensation) and hyperesthesia (increased sensation). Neuralgia of the glossopharyngeal nerve is marked by severe pain in the neck-ear-jaw region. Motor disorders which affect swallowing may originate from local irritations or from central nervous system disease, as in the case of rabies. *See* PHARYNX; SOMESTHESIS.

Edward G. Stuart/N. Karle Mottet

Bibliography. W. Becker et al., *Ear, Nose, and Throat Diseases: A Pocket Reference*, 1994; K. Haskins (ed.), *Nose and Paranasal Sinuses: Larynology and Bronchoesophagology*, 1994; E. Jarvik, *Basic Structure and Evolution of Vertebrates*, 2 vols., 1981.

Phase (astronomy)

The changing fraction of the disk of an astronomical object that is illuminated, as seen from some particular location. The monthly phases of the Moon are a familiar example (see **illus.**). When the Sun is approximately on the far side of the Moon as seen from Earth (conjunction), the dark side of the Moon

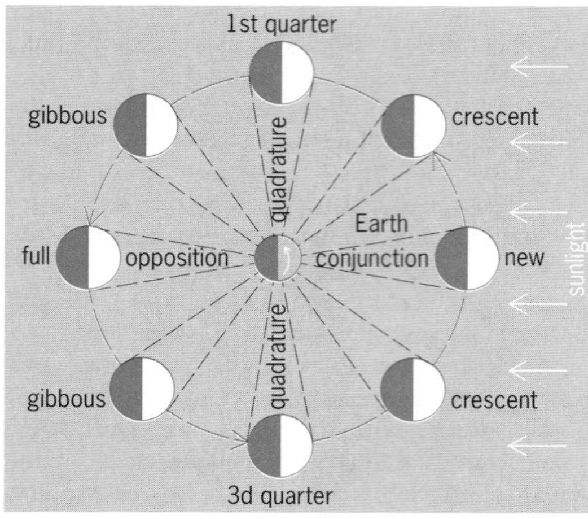

From Earth, different fractions of the illuminated half of the Moon are seen at different times as the Moon goes through a 29.53-day cycle of phases.

faces the Earth and there is a new moon. The phase waxes, beginning with crescent phases, as an increasing fraction of the illuminated face of the Moon is seen. At quadrature, when half the visible face of the Moon is illuminated, the phase is called the first-quarter moon, since the Moon is now one-quarter of the way through its cycle of phases. The waxing moon continues through its gibbous phases until it is in opposition; the entire visible face of the Moon is illuminated, the full moon. During the full moon, the Moon and the Sun are on opposite sides of the Earth, a configuration known as a syzygy. Then the Moon wanes, going through waning gibbous, third-quarter, and waning crescent phases until it is new again. The cycle of moon phases takes approximately 29.53 days and explains the origin of the word month.

The Earth, Moon, and Sun are not directly in line at the times of new moon and full moon, because the Moon's orbit around the Earth is inclined 5° to the plane of the Earth's orbit around the Sun. When they are directly in line, a lunar eclipse occurs when the Earth's shadow falls on the Moon, and a solar eclipse when the Moon's shadow falls on the Earth. *See* ECLIPSE; MOON.

Galileo discovered the phases of the planet Venus when he observed the sky with his telescope in 1610. Giovanni Zupus discovered the phases of the planet Mercury in 1639. Because of the angle at which the outer planets are seen from Earth, and because of their great distance, they do not appear to go through phases as seen from Earth. Mars, Jupiter, Saturn, Uranus, and Neptune have now all been seen by spacecraft as crescents. Jay M. Pasachoff

Phase (periodic phenomena)

The fractional part of a period through which the time variable of a periodic quantity (alternating electric current, vibration) has moved, as measured at any point in time from an arbitrary time origin. In the case of a sinusoidally varying quantity, the time origin is usually assumed to be the last point at which the quantity passed through a zero position from a negative to a positive direction. It is customary to choose the origin so that the fractional part of the period is less than unity.

In comparing the phase relationships at a given instant between two time-varying quantities, the phase of one is usually assumed to be zero, and the phase of the other is described, with respect to the first, as the fractional part of a period through which the second quantity must vary to achieve a zero of its own (see **illus.**). In this case, the fractional part of

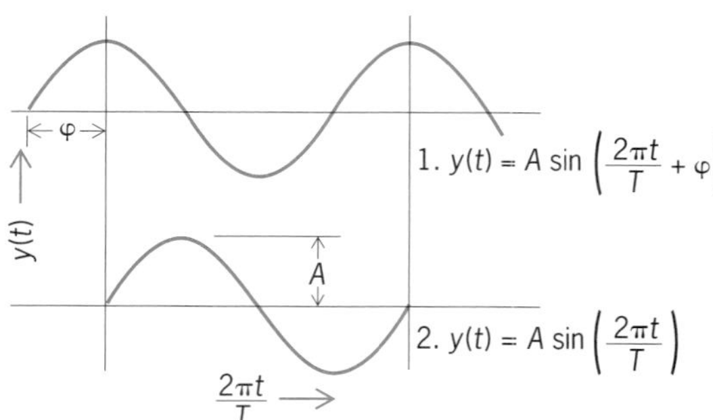

Illustration of the meaning of phase for a sinusoidal wave. The difference in phase between waves 1 and 2 is φ and is called the phase angle. For each wave, *A* is the amplitude and *T* is the period.

the period is usually expressed in terms of angular measure, with one period being equal to 360° or 2π radians. Thus two sine waves of a given frequency are said to be 90°, or $\pi/2$, out of phase when the second must be displaced in time, with respect to the first, by one-fourth period in order for it to achieve a zero value. *See* PHASE-ANGLE MEASUREMENT; SINE WAVE. William J. Galloway

Phase-angle measurement

Measurement of the time delay between two periodic signals. The phase difference between two sinusoidal waveforms that have the same frequency and are free of a dc component can be conveniently described as shown in **Fig. 1**. It can be seen that the phase angle can be considered as a measure of the time delay between two periodic signals expressed as a fraction of the wave period. This fraction is normally expressed in units of angle, with a full cycle corresponding to 360°. For example, in Fig. 1, where the voltage v_1 passes through zero $^1/_8$ cycle before a second voltage v_2, it leads by 360°/8 or 45°. Phase angle is usually defined from the fundamental component of each waveform; therefore distortion of either or both signals can give rise to errors, the extent of which depends on the nature of the distortion and the method of measurement. *See* DISTORTION (ELECTRONIC CIRCUITS).

Many techniques have been developed to measure

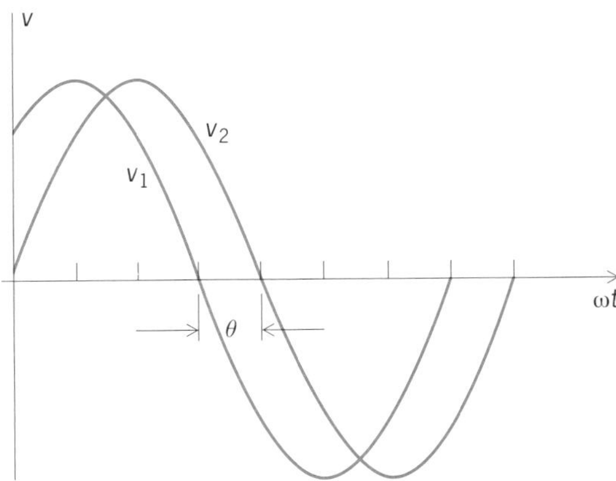

Fig. 1. Phase angle θ between voltages v_1 and v_2.

phase differences or to generate signals with known phase differences.

Three-voltmeter method. This method, also known as the law-of-cosines method, can be used when the voltages involve a common point. **Figure 2*a*** shows three terminals *a*, *b*, and *c*. If a high-impedance voltmeter is used to measure the amplitudes of the signal voltages v_{ab} and v_{bc} and the differential voltage between them v_{ca}, they can be plotted to give a triangle, as in Fig. 2*b*. The phase angle θ can then be determined from the law of cosines in trigonometry, as shown by Eq. (1).

$$\theta = \cos^{-1}\left(\frac{v_{ab}^2 + v_{bc}^2 - v_{ca}^2}{2v_{ab}v_{bc}}\right) \qquad (1)$$

See TRIGONOMETRY.

Zero-crossing-detector phase meter. The majority of modern phase-measuring devices are based on the use of zero-crossing detectors. The time at which each signal crosses the zero-voltage axis is determined, usually by means of a squaring-up circuit

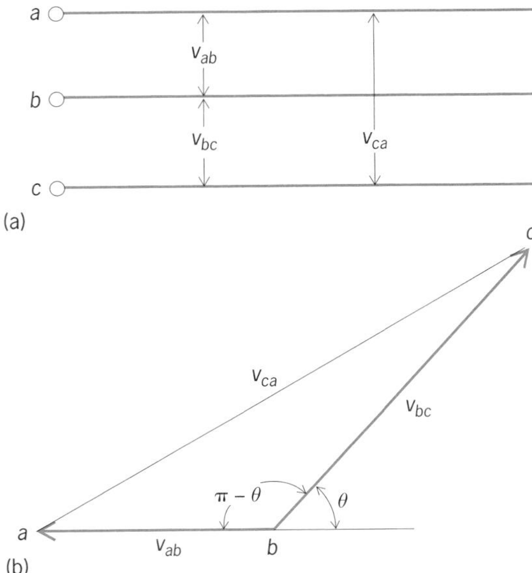

Fig. 2. Voltages employed in three-voltmeter method. (*a*) Circuit diagram. (*b*) Vector diagram.

(for example, an overdriven amplifier) followed by a high-speed comparator. This produces, in each channel, a trigger pulse that is used to drive a bistable flip-flop. The output from the bistable is a rectangular wave, the duty cycle of which is proportional to the phase difference between the input signals. If this signal is integrated by means of a suitable filter, a dc voltage is produced that is an analog representation of the phase angle. This voltage is then displayed on a panel meter (analog or digital) suitably scaled in degrees or radians. Instrumentation using this principle is capable of measuring phase differences to approximately $\pm 0.05^\circ$ over a wide range of amplitudes and frequencies. **Figure 3** is a timing diagram illustrating the principles involved. *See* AMPLIFIER; COMPARATOR; ELECTRIC FILTER; MULTIVIBRATOR; SWITCHING CIRCUIT; WAVE-SHAPING CIRCUITS.

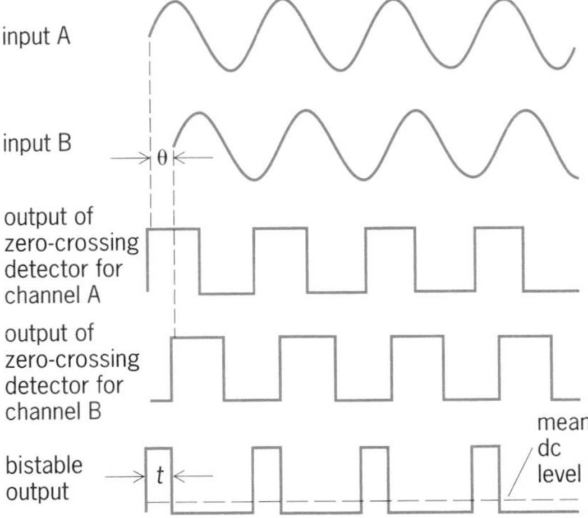

Fig. 3. Zero-crossing-detector timing diagram.

FFT phase meter. If the waveforms to be measured are sampled at a number of equally spaced points within each cycle and these samples are digitized, a fast Fourier transform (FFT) algorithm can be used to compute, among other parameters, the phase difference between the signals. This technique has the advantage that only the phase of the fundamental need be evaluated, so that the effects of distortion are virtually eliminated. The accuracy from commercial instrumentation of this kind is still modest when compared with the technique's theoretical limits.

Crossed-coil phase meter. Also known as the Tuma phase meter, this is the basic element of power-factor meters and of synchroscopes. When used for power-factor indication, it contains two movable coils A and B on a common shaft (**Fig. 4**). The two coils move as a unit, the angle β between them remaining fixed. There is no restraining spring acting between them (as there is in a conventional electrodynamic meter movement), and the system will revolve as long as the currents produce an average nonzero torque. *See* POWER-FACTOR METER.

The meter also contains a fixed coil C, which carries the load current. The fixed coil is made in two

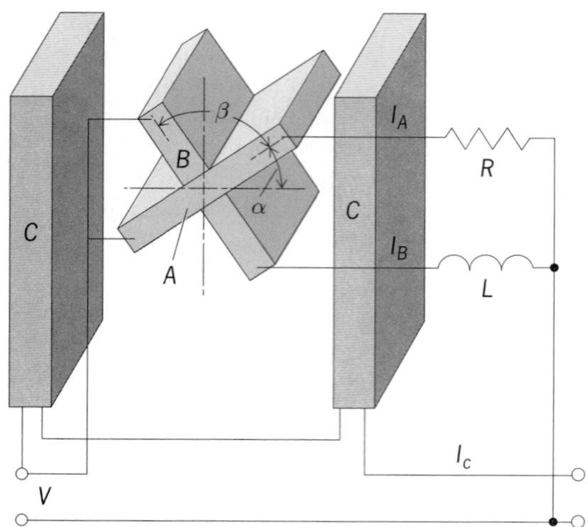

Fig. 4. Crossed-coil phase meter.

sections so that its magnetic field will be nearly uniform in the vicinity of the movable-coil assembly.

The instantaneous force induced on coil A is proportional to the product of the instantaneous currents in coils A and C and therefore to the sine of the angle between the planes of A and C. If the currents in A and C are sinusoidal, with phase displacement θ, the average torque T is proportional to expression (2), and this can also be expressed as relation (3).

$$(I_A \sin \omega t) I_C \sin (\omega t + \theta) \cos \propto \tag{2}$$

$$T \propto I_A I_C \cos\theta \cos \propto \tag{3}$$

The movable coil A is placed in a resistive circuit so that its current I_A is in phase with the applied voltage V. The other movable coil B is placed in an inductive circuit so that its current lags the voltage by approximately 90°.

In practice, the circuit of coil A cannot be made completely noninductive and, equally, the circuit of coil B cannot be nonresistive. The phase angle between the currents in coils A and B is therefore less than 90°. The scaling of the phase meter is constructed to take this into account.

If the current in coil B leads the current in fixed coil C by the angle θ, and the current in coil B lags the current in coil A by the angle ϕ, then the current in coil A will lead the current in the fixed coil by the angle $\theta + \phi$, and the average torque on A is proportional to $\cos(\theta + \phi) \cos \alpha$ whereas the average torque on B is proportional to $\cos \theta \cos (\alpha + \beta)$ and in the opposite direction. If these torques do not cancel, the shaft will rotate ($\propto$ will vary) until the two become equal, or when Eq. (4) is satisfied.

$$\cos(\theta + \phi) \cos \propto = \cos\theta \cos(\alpha + \beta) \tag{4}$$

From this equation it follows that if β equals ϕ, then $\propto$ equals θ, and the phase meter indicates directly the phase angle between the sources supplying coil C and the crossed coils. A change of 5° in phase angle produces a deflection of 5° of the phase-meter pointer.

Oscilloscope methods. There are many methods available that use an oscilloscope to measure phase difference. The easiest and most convenient is to use a dual-channel oscilloscope, one signal being applied to each channel. The phase difference may readily be calculated from the relative positions of the two waveforms on the oscilloscope screen. *See* OSCILLOSCOPE.

Phase-order indicators. These devices are used to indicate which phase voltage of a polyphase circuit leads or lags another. If the voltage vectors of a polyphase circuit are as indicated in **Fig. 5*a***, the voltage from neutral to line 2 reaches a maximum $^1/_3$ of the wave period after the voltage of line 1 and $^1/_3$ of the wave period before the voltage of line 3. The phase order (or sequence) is then said to be 1-2-3.

Relative motion between the armature conductors and the magnetic field induces the voltages in an alternator. Therefore, if the alternator were to be rotated in the opposite direction, the order in which the phase voltages reached maximum would be reversed. The phase sequence would then be 1-3-2, as shown in Fig. 5*b*.

A miniature three-phase motor designed to rotate clockwise when connected to a three-phase system possessing a phase sequence of 1-3-2 is used as a phase-sequence indicator. Counterclockwise rotation indicates a 1-2-3 phase sequence.

A common type of phase-order indicator consists of an inductance and two lamps connected in a Y configuration to the three-phase lines as in **Fig. 6**. In the absence of the inductor, equal currents would flow through the lamps; however, the inductor produces a 90° phase shift in the current injected at point *n*. This produces an imbalance in the current through the lamps, the overall effect being to increase the current in lamp 2 when the phase sequence is 1-2-3,

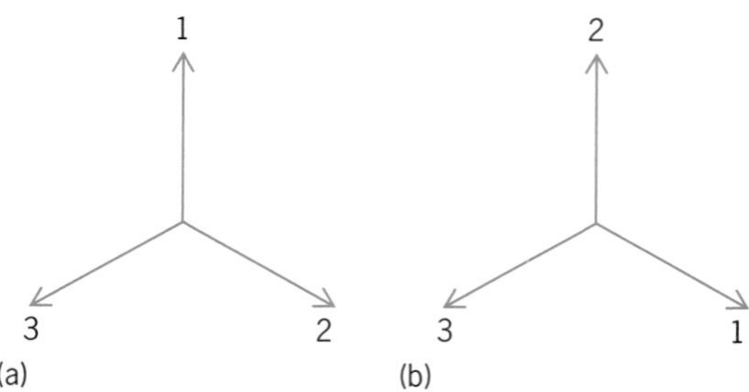

Fig. 5. Phase sequence. (*a*) Sequence 1-2-3. (*b*) Sequence 1-3-2.

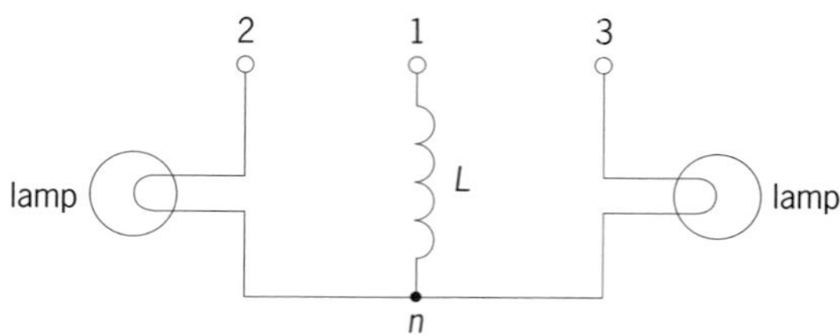

Fig. 6. Circuit diagram of phase-sequence indicator.

and to increase the current in lamp 3 when the phase sequence is 1-3-2.

Phase-relation indicators. Frequently employed in the electrical utility industry, these devices are used to indicate when two generators or sources of alternating voltage are in phase with one another. If two voltages reach maximum at the same time, they are in phase. If two generators are to be connected in parallel, they should have the same voltage, frequency, and phase. A voltmeter and a tachometer (or frequency meter) can be used to indicate equality of the voltage and frequency. The phase relation between the two sources is shown by means of phasing lamps or by means of a synchroscope or synchronizer.

Phasing lamps. Phasing lamps placed across the open switch used to parallel two generators will often suffice to indicate an in-phase condition. Depending on the relative phase of the two generators, the lamp voltage varies from the sum to the difference of the generator voltages. As the frequencies of the two generators approach one another, the lamps flicker, changing from full brightness to dim at a decreasing rate. If the two frequencies are equal, the lamp will maintain a fixed brilliance. Usually the oncoming generator is set with a slightly higher frequency so that it will provide additional power rather than be an additional load on the system. As the lamps slowly go through the dim phase, the switch is closed, connecting the incoming generator to the system.

Synchronizer or synchroscope. This is based on the Tuma phase meter. The current in the fixed coil is supplied by one generator; the current in the movable crossed-coils is supplied by the other generator. If the two generators are in synchronism, their frequencies are equal and the movable coils will take up a position depending on their relative phase angle. If the frequencies are different, the phase will continue to vary and the movable coils will rotate, the direction of rotation indicating which generator is running faster. As in the case of phasing lamps, the incoming generator is generally running at a slightly higher speed; the two are connected together when the synchroscope pointer drifts past the zero mark. *See* SYNCHROSCOPE.

High-frequency measurement. Conventional phase meters, such as the zero-crossing-detector type, have an upper frequency limit of a few hundred kilohertz. This limit is imposed mainly by the ability of the arrangement consisting of a comparator and a flip-flop to maintain a clean and precise rectangular waveform under conditions of high-speed operation. In order to measure phase angle at frequencies between about 100 kHz and several gigahertz, it is necessary to down-convert the radio-frequency signals to a frequency that can be handled correctly by the phase meter. A superheterodyne system that can fulfill this function is shown in **Fig.** 7. *See* HETERODYNE PRINCIPLE.

Sampling systems are sometimes used to achieve the same result. Any phase difference caused by the mixers can be measured by applying zero phase angle (for example, via a power splitter) and adjusting a compensating trimmer in one channel for a zero reading. In order to ensure that phase errors due to radio-frequency mismatch are minimized, these instruments normally terminate the radio-frequency signals in a 50-ohm load. At radio frequencies, a major source of error can arise due to differences in cable length, and users of such equipment must ensure that such effects are minimized.

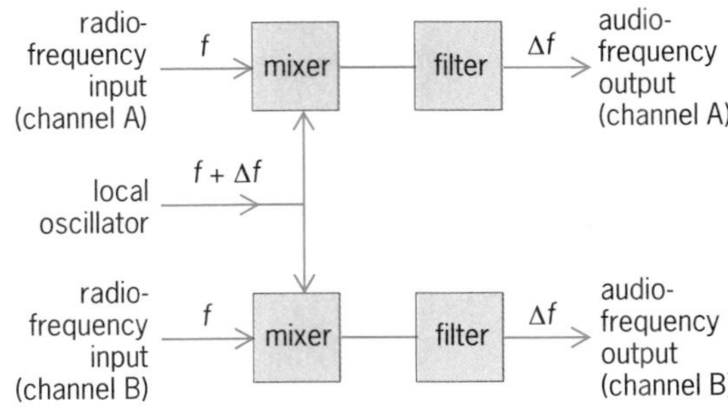

Fig. 7. High-frequency phase measurement using down-conversion technique.

At microwave frequencies, instruments such as slotted lines, air lines, and vector network analyzers are also used for phase-angle measurements. *See* MICROWAVE MEASUREMENTS.

Phase-angle generation techniques. Many techniques have been developed for generating signals of known phase difference. A classic method involves establishing two signals in quadrature and mixing them together via calibrated voltage dividers to produce the required phase angle. Such methods have been superseded by modern digital techniques, which offer greater accuracy, versatility, and ease of use.

Although the various digital techniques differ in detail, the principles of operation are all similar. A clock and address counter are used to sequentially address the memory locations, thus producing bit patterns that, when applied to a digital-to-analog converter, will produce a stepped approximation to the stored waveform. Each step is maintained for one clock period, at the end of which the next set of digital data is applied to the digital-to-analog converter. A low-pass filter is used to remove the steps. If two memories are used, then a dual-channel generator is available. A phase displacement between the two signals can then be obtained by either deliberately offsetting the relative memory locations or recalculating the data contained in each memory location in accordance with the required phase angle. *See* DIGITAL-TO-ANALOG CONVERTER.

Such techniques are used in national metrology laboratories to provide phase-angle generating capabilities with uncertainties as good as $\pm 0.01^\circ$. Commercial instrumentation using similar techniques is also available. *See* ELECTRICAL MEASUREMENTS.

J. Hurll

Bibliography. M. M. Berlin and F. C. Getz, Jr., *Principles of Electronic Instrumentation and Measurement*, 1988; C. G. Coombs, *Electronic Instrument*

Handbook, 3d ed., 1999; Dranetz Engineering Laboratories Inc., *Application Handbook for Precision Phase Measurement*, 1975; A. D. Helfrick and W. D. Cooper, *Modern Electronic Instrumentation and Measurement Techniques*, 1990; L. Schnell (ed.), *Technology of Electrical Measurements*, 1993; L. M. Thompson, *Electrical Measurements and Calibration: Fundamentals and Applications*, 2d ed., 1994.

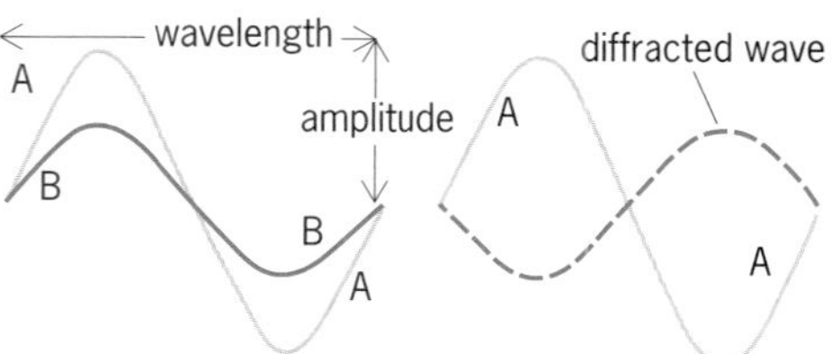

Fig. 1. Effect of a partially absorbing object on a light wave. The diffracted wave is half a wavelength out of phase with the incident wave. A = incident wave; B = transmitted wave.

Phase-contrast microscope

A microscope used for making visible differences in phase or optical path in transparent or reflecting specimens. It is an important instrument for studying living cells and is used in biological and medical research.

Microscopy of transparent objects. When a light wave passes through an absorbing object, it is reduced in amplitude and intensity. Since the human eye and the photographic plate are sensitive to variations in intensity, such an object will give a visible image when viewed through an ordinary microscope. A perfectly transparent object does not absorb light, so that the intensity remains unaltered, and such an object is essentially invisible in an ordinary microscope. However, the light that has passed through the transparent object is slowed down by it and arrives at the eye a minute fraction of a second later than it would otherwise have done. Such delays, or in technical language, phase changes or differences in optical path, are not detected by the eye, and the fundamental problem in the microscopy of transparent objects, of which living cells are important examples, is to convert them into visible intensity changes. For further discussions of basic principles *See* ABSORPTION OF ELECTROMAGNETIC RADIATION; DIFFRACTION; INTERFERENCE OF WAVES; LIGHT.

Simple theory. A consideration follows of what happens when a light wave passes through a partially absorbing object, such as a stained biological specimen. In **Fig. 1**, let A be the incident wave. When the wave passes through the object, energy is absorbed so that the transmitted wave B is reduced in height or amplitude. The intensity is proportional to the square of the amplitude. Wave B can also be represented as the sum of the incident wave A and another wave shown by a dotted line. This wave has a real physical existence and represents the light scattered or diffracted by the object. The German physicist E. Abbe was the first to stress the importance of diffraction in image formation and showed that the final image was formed by interference or addition of the incident and diffracted light. If the incident wave and the diffracted wave are added together algebraically, that is, upward displacements being treated as positive and downward ones as negative, the resultant is wave B. In the case of a partially absorbing object, therefore, the trough of the diffracted wave coincides with the crest of the incident wave, that is, the two waves differ in phase by half a wavelength.

Consideration of a perfectly transparent refractile object follows. Since no light is absorbed, the height of the wave is unchanged, but if the object has a higher refractive index than its surroundings, the wave will be delayed and can be represented by wave B in **Fig. 2**. Once again, wave B can be represented as the sum of the incident wave A and a diffracted wave (Fig. 2*b*). This diffracted wave is no longer half a wavelength out of phase with wave A. If wave B is only slightly delayed, the diffracted wave is approximately one-quarter of a wavelength out of phase with A. Suppose now that the incident wave A and the diffracted wave could somehow be separated and their phase relationships altered; in particular, suppose the phase difference between them could be increased from about one-quarter of a wavelength to half a wavelength, as in Fig. 2*c*. The crest of one

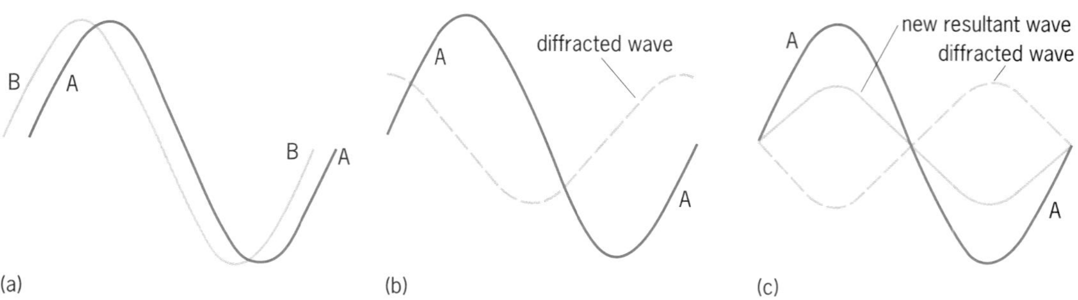

Fig. 2. Effect of a transparent object on a light wave. (*a*) Incident wave A is unaltered in amplitude but delayed in phase and is transmitted as wave B. (*b*) B can again be represented as the sum of the incident wave A and a diffracted wave. The diffracted wave is no longer half a wavelength out of phase with A, but if the phase difference can be changed to half a wavelength (c) the new resultant wave will resemble that produced by a partially absorbing object.

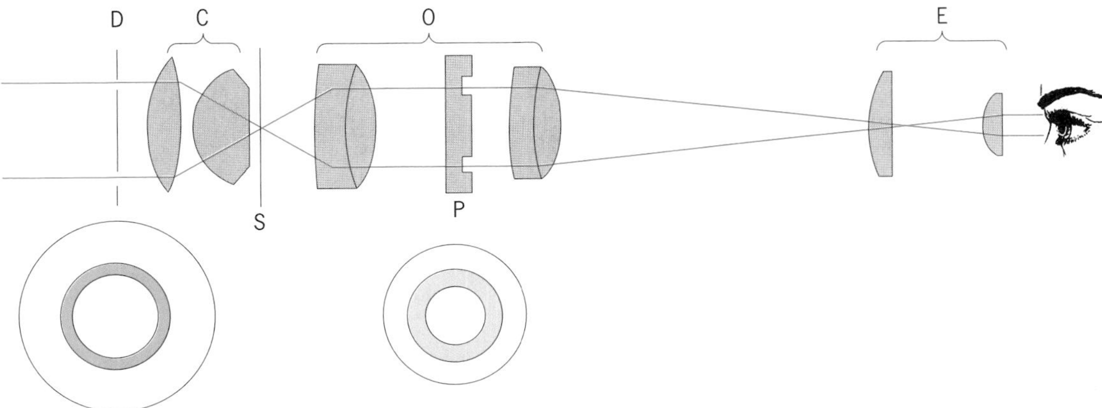

Fig. 3. Diagram of a phase-contrast microscope. D = annular diaphragm; C = substage condenser; P = phase plate at rear focal plane of objective; S = specimen; E = eyepiece.

wave would now coincide with the trough of the other, and the new resultant wave would be indistinguishable from wave B in Fig. 1. In other words, the invisible transparent object would be indistinguishable from a partially absorbing object. This was first realized by Frits Zernike, who was awarded the 1953 Nobel Prize in physics for his invention of phase contrast.

Practical realization. The essential features of a phase-contrast microscope are shown in **Fig. 3**. The practical problem is to find some way of separating the incident or direct light from that diffracted by the object. This is done by placing a diaphragm D of easily recognizable shape, such as an annulus, at the front focal plane of the substage condenser C. Light from each point of the focal plane passes as a parallel pencil of rays through the specimen S and is brought to a focus at the rear focal plane P of the objective O. Thus, on removing the eyepiece, an image of the annulus will be seen at the back of the objective lens. This image corresponds to the incident light. In addition, when a specimen is present, some light is diffracted by it and spreads out to fill the whole of the back lens of the objective. Thus, apart from the small area of overlap over the image of the annulus, the direct and diffracted waves are essentially separated at the plane P. A phase plate is now inserted at this level. This can be a transparent disk with an annular groove of such dimensions that it coincides exactly with the image of the diaphragm D. All the direct light now passes through the groove in the phase plate, whereas the diffracted light passes mainly outside the groove. Since the diffracted light has to pass through a greater thickness of transparent material than the direct light, a phase difference, depending on the refractive index of the phase-plate material and on the thickness of the groove, is introduced between them. If this phase difference is about one-quarter of a wavelength, the basic conditions for phase contrast will have been achieved (Fig. 2). If the phase plate is made to retard the incident wave by a quarter of a wavelength, the crests and troughs of the two waves will coincide, giving a resultant of greater amplitude. Refractile details will appear bright (negative contrast) instead of dark (positive contrast). The commercial Anoptral system is a type of negative phase contrast.

In practice, a partially absorbing layer of metal or dye is usually deposited over the phase-plate annulus. Figure 2*c* shows that the amplitude of the direct wave is usually greater than that of the diffracted wave so that, although the resultant wave is reduced in amplitude, it is still far from zero, and a transparent object will not appear perfectly black, but gray. If the amplitude of the direct wave is reduced after it has passed through the object, it can be made equal to that of the diffracted wave so that the resultant is zero, and the object appears perfectly black. Unfortunately, in order to match the amplitude of all possible diffracted waves, it would be necessary to make a variable-absorption phase plate. Such a device is expensive, and in practice most manufacturers use fixed phase plates with absorptions of between 50 and 80%. Each objective must have a corresponding substage diaphragm. The commonest arrangement is a rotatable wheel in the substage condenser containing three or four diaphragms of different sizes and also a clear circular aperture which enables the instrument to be used with conventional illumination.

Image interpretation. The phase change or optical path difference ϕ introduced by an object is defined by the relationship $\phi = (n_o - n_m)t$, where n_o is the refractive index of the object, n_m that of the surrounding medium, and t the object thickness; ϕ is usually expressed either in terms of the wavelength, in which case t must also be expressed in the same unit, or in angular measure. In the latter notation, one wavelength is equal to 360° or 2π radians. Although the phase-contrast microscope converts variations in ϕ into variations in intensity, the relationship between these quantities is not a simple one. In general, as ϕ increases, the image becomes darker; but beyond a certain value of ϕ, it becomes lighter again, and finally, for very large values of ϕ, the contrast actually becomes reversed, so that the image becomes brighter than the background. Thus, it cannot be assumed that because one part of the image appears darker than another, it will necessarily correspond to a region of greater optical path. Even these basic relationships are disturbed because the theoretical requirement of complete separation between the direct and diffracted light can never be achieved in practice. This results in the appearance of a bright halo around every dark detail and an accentuation of edges and sharp discontinuities in the object. Although these effects do not allow the instrument to be used for quantitative measurements of optical path, they are not really disadvantageous for purely observational work.

Biological applications. The phase-contrast microscope is a routine instrument for the examination of living cells. It made it possible to study the structure of living cells under excellent optical conditions and with no loss in resolving power. Accurate observations thus became much easier to make, and in particular, the use of phase-contrast cinemicrography made it possible to study changes in cell structure during the movement and division of cells with great clarity. The method is also useful for the study of unstained tissue sections and for the comparison of material in the electron and optical microscopes.

A quantitative application is microrefractometry. It follows from the basic definition of phase change, namely $\phi = (n_o - n_m)t$, that if the refractive index n_m of the mounting medium is made equal to that of the object n_o, ϕ becomes zero irrespective of the object thickness t, and since the intensity is a function of ϕ, the object will become invisible. The phase-contrast microscope can thus be used as a very sensitive null indicator for measuring refractive indices. The principle is to immerse the object in a series of media of graded refractive index until one is found that makes the object invisible. This is one of the most sensitive methods of microrefractometry available, and it can be extended to the quantitative refractometry of living cells (**Fig. 4**). In this case, a suitable nontoxic medium must be used. The most suitable medium

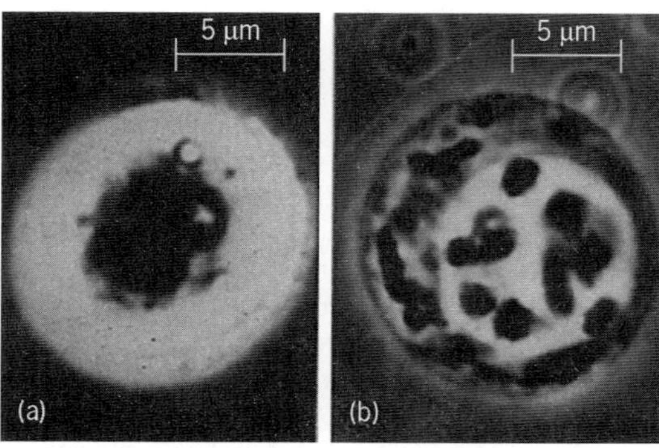

Fig. 4. Phase-contrast photograph of living locust spermatocyte. (*a*) In a saline medium. The refractive index of the cell is so much higher than the index of the medium that all internal detail is obscured. (*b*) Similar cell immersed in 9% bovine plasma albumin solution. The refractive index difference is much less, and the internal nucleoplasm and chromosomes can be clearly seen.

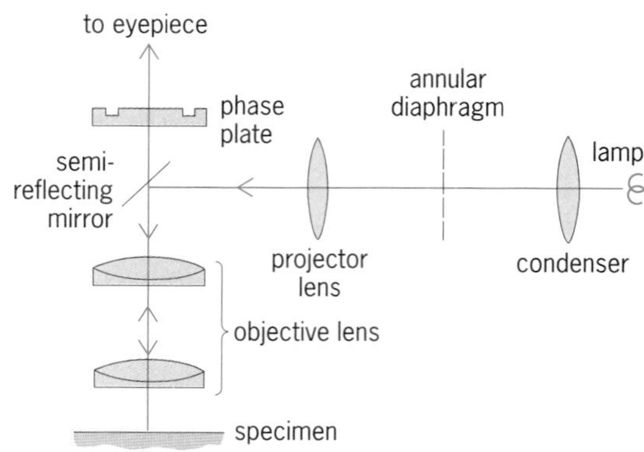

Fig. 5. Phase-contrast microscope with vertical illumination for reflecting specimens.

is a concentrated protein solution such as Armour's bovine plasma albumin fraction V, with added salts. The importance of cell refractometry rests on the fact that there is a linear relationship between the refractive index n of a solution and the concentration C of the dissolved substance. Thus, $n = n_o + \alpha C$ where n_o is the refractive index of the solvent, for example, water, and α is a characteristic constant known as the specific refraction increment. The mean value of α for most cellular constituents may be taken as 0.0018, and the formula can be used to calculate C, the total solid concentration in the cell. The water concentration can also be deduced, and if the cell volume is known, the dry and wet mass of the cell is also known. *See* INTERFERENCE MICROSCOPE.

Nonbiological applications. The study of transparent specimens such as crystals and fibers calls for little comment except to point out that, if the specimen is birefringent, a single polarizer must be used in order to avoid a confused image.

Industrial applications include the examination of chemicals, oils, waxes, soaps, paints, foods, plastics, rubber, resins, emulsions, and textiles.

Surface structure can often be studied by phase contrast with vertical illumination. Either the specimen must be naturally reflecting, or a thin film of reflecting material must be deposited on it or on a surface replica. A typical arrangement is shown in **Fig. 5**. A condenser lens forms an image of a lamp on an annular diaphragm. A semireflecting cover slip is placed just above the objective. A projector lens and the objective combine to form an image of the annulus at a plane just above the semireflector, and a phase plate is inserted here (compare Fig. 3). Normally, it is desirable to place the phase plate at the rear focal plane of the objective; but if this were done here, the incident light would have to pass through the metalized part of the phase plate and would lose intensity. Stray light and glare would also occur because of reflections at the surface of the phase plate. Apart from these modifications, the system is similar to that for phase contrast with transmitted light. Differences in height in the reflecting specimen will produce differences in optical path length, and these will be converted into intensity differences. Robert Barer

Bibliography. A. Briggs, *Acoustic Microscopy*, 1992; T. G. Rochow and P. Tucker, *Introduction to Microscopy by Means of Light, Electrons, X-Rays, or Acoustics*, 2d ed., 1994; G. Wade (ed.), *Acoustic Imaging: Cameras, Microscopes, Phased Arrays, and Holographic Systems*, 1976.

Phase equilibrium

A general field of physical chemistry dealing with the various situations in which two or more phases (or states of aggregation) can coexist in thermodynamic equilibrium with each other, with the nature of the transitions between phases, and with the effects of temperature and pressure upon these equilibria. Many superficial aspects of the subject are largely qualitative, for example, the empirical classification of types of phase diagrams; but the basic problems always are susceptible to quantitative thermodynamic treatment, and in many cases, statistical thermodynamic methods can be applied to simple molecular models.

Thermodynamics requires that when two phases, α and β, are free to exchange heat, mechanical work, and matter (chemical species), the temperature T, the pressure P, and the chemical potential (partial molar free energy) μ_i of each particular component i must be equal in both phases at equilibrium. Algebraically, equilibrium exists when $T_\alpha = T_\beta$, $P_\alpha = P_\beta$, $\mu_{i,\alpha} = \mu_{i,\beta}$.

These conditions of thermal, mechanical, and material equilibrium need not all be present if the equilibrium between phases is subject to inhibiting restrictions. Thus, for a solution of a nonvolatile solute in equilibrium with the solvent vapor, the condition of equality of solute chemical potentials $\mu_{2,\alpha} = \mu_{2,\beta}$ need not apply, since there can be no solute molecules in the vapor phase. Similarly, in osmotic equilibria, in which solvent molecules can pass through a semipermeable membrane, whereas solute molecules cannot, $\mu_{1,\alpha} = \mu_{1,\beta}$ and $T_{1,\alpha} = T_{2,\beta}$, but the solute chemical potentials μ_2 are unequal, as are the pressures on opposite sides of the membrane. *See* OSMOSIS; SOLUTION.

If a system consists of P phases and C distinguishable components, there are $C + 2$ thermodynamic variables (C chemical potentials μ_i, plus the temperature and pressure) which are interrelated by an equation for each phase. Since there are P independent equations relating the $C + 2$ variables, one needs to fix only $F = C + 2 - P$ variables to define completely the state of the system at equilibrium; the other variables are then beyond control. This relation for the number of degrees of freedom F, or variance, is called the phase rule and was first derived by Willard Gibbs in 1873. It has proved to be a powerful tool in interpreting and classifying types of phase equilibria.

When chemical changes may occur in the system, the number of components C is the number of independent components, that is, the number of components whose amounts can be varied by the experimenter; this is equal to the total number of chemical species present less the number of independent chemical equilibria between them.

An invariant system has no degrees of freedom ($F = 0$), for which the number of phases $P = C + 2$. For a one-component system, such an invariant point is a triple point at which three phases coexist at a single temperature and pressure only; for a two-component system, a quadruple point (four phases) would be invariant. *See* TRIPLE POINT.

In a univariant system ($F = 1$), $P = C + 1$. With a one-component system, one can fix the temperature at which two phases (liquid and gas, for instance) can coexist in equilibrium; then the pressure (here the vapor pressure) is determined and not subject to external control. A univariant system is described by a line in a phase diagram, for example, a plot of vapor pressure versus temperature. The differential equation (1) for such a univariant line

$$\frac{dP}{dT} = \frac{\Delta H}{T\,\Delta V} \qquad (1)$$

in a one-component system was first deduced by B. P. E. Clapeyron in 1834. In the equation ΔH and ΔV are the enthalpy change (heat absorbed) and volume change, respectively, for the transition from one phase to another. *See* VAPOR PRESSURE.

In systems of two or more components, more complicated equations for the univariant line replace the Clapeyron equation, but in special cases they may reduce to the simpler form. For example, in the chemical decomposition of a solid calcium carbonate to solid calcium oxide and gaseous carbon dioxide, reaction (2), the three-phase system (two solids and

$$CaCO_3(s) \rightleftharpoons CaO(s) + CO_2(g) \qquad (2)$$

one gas) is univariant (since the number of independent components C is two), the equilibrium pressure is the decomposition pressure of calcium carbonate ($CaCO_3$), and the ΔH and ΔV of the Clapeyron equation are the enthalpy and volume changes associated with the chemical reaction.

Binary Systems

Phase diagrams and binary systems containing two components are easily classified. Typical examples of the important classes (liquid-gas, liquid-liquid, solid-liquid, solid-solid) have been selected for description.

Liquid-gas equilibrium. In a one-component system, liquid and vapor are in equilibrium at the boiling point. For two-component systems, the two-phase situation is bivariant and more complex. A complete temperature-pressure-composition diagram would be three-dimensional, so most phase diagrams are made for either constant pressure or constant temperature. **Figure 1** shows the simplest type of binary liquid-vapor temperature-composition diagram, exemplified by the system carbon tetrachloride-stannic chloride, which forms essentially ideal solutions. The regions labeled G and L are one-phase

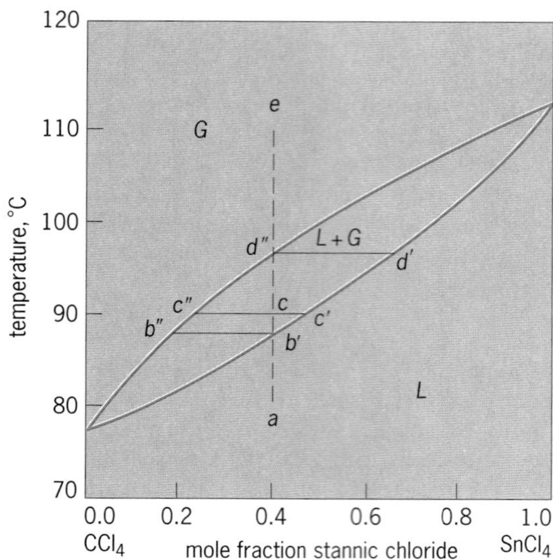

Fig. 1. Binary liquid-vapor temperature-composition diagram for the system carbon tetrachloride + stannic chloride. *G* = gas (vapor) phase; *L* = liquid phase; *L* + *G* = two coexisting phases. Pressure = 1 atm (10^2 kPa).

regions, gas (vapor) and liquid, respectively; the region labeled $L + G$ is a two-phase region in which liquid and vapor coexist. If the temperature of a liquid mixture of 40 mole percent $SnCl_4$ (mole fraction = 0.40) is increased at a constant pressure of 1 atm, the change in the system can be traced along the straight line $ab'cd''e$. At low temperatures, only one phase, the liquid, is present, but at 87.5°C (190°F), point b', a vapor phase appears. The composition of this vapor phase is given by point b'' (mole fraction $SnCl_4 = 0.18$), and the two conjugate phases are connected on the diagram by the tie line $b''b'$. As the temperature is increased further, more vapor is formed; since the vapor is rich in CCl_4, this component becomes relatively depleted in the liquid phase, and the liquid composition moves along the line $b'c'd'$, while the vapor composition moves along the line $b''c''d''$.

At 90°C (194°F) the overall composition of the two-phase system is represented by point c, but the compositions of vapor and liquid separately are given by the two ends of the tie line, points c'' and c', respectively (mole fractions of 0.22 and 0.47, respectively). The relative amounts of the two phases are given by the lever arm principle of physics. The ratio of the moles of vapor to moles of liquid is given by the ratio of the length cc' to the length $c''c$, here 0.07:0.18, or 28%, in the vapor phase. Further increase in temperature produces more and more vapor until, at 97°C (207°F), the liquid phase (point d', mole fraction 0.62) has become vanishingly small; at higher temperatures, it disappears, and only the vapor phase (point d'', mole fraction = 0. 40) remains. Further increase in temperature (along the line $d''e$) is uneventful.

In this simple system, there are no maxima or minima in the liquid and vapor curves; consequently, such liquid mixtures can be separated completely into the two pure components by fractional distillation. Systems with maximum boiling points (acetone + chloroform; **Fig. 2**) or minimum boiling points (ethanol + benzene; **Fig. 3**) cannot be so separated into the pure substances. At the maximum or minimum, the composition of the liquid is identical with that of the vapor with which it is in equilibrium; continued boiling will not alter these compositions, so that these solutions are called constant-boiling mixtures or azeotropes. It should be noted that the solid and liquid lines are smooth curves tangent to each other at such a point; any phase diagram which shows a sharp corner is thermodynamically incorrect. *See* AZEOTROPIC MIXTURE; DISTILLATION.

Maximum boiling mixtures are associated with

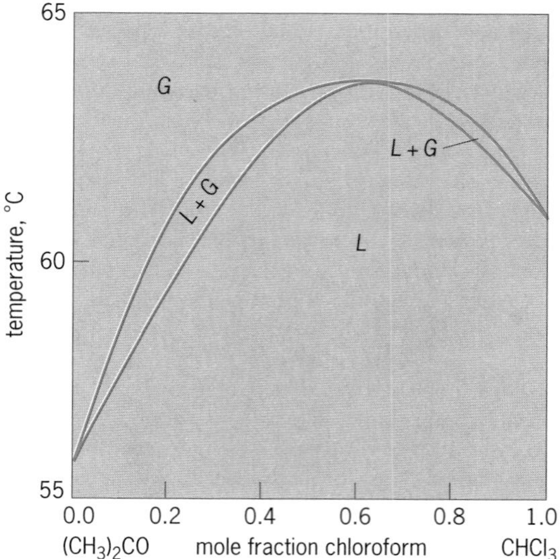

Fig. 2. Temperature-composition diagram for acetone + chloroform, showing maximum boiling point. Pressure = 1 atm (10^2 kPa).

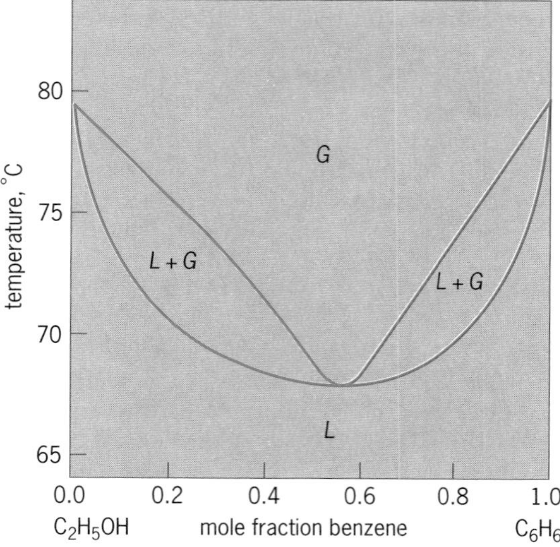

Fig. 3. Temperature-composition diagram for ethanol + benzene, showing minimum boiling point.

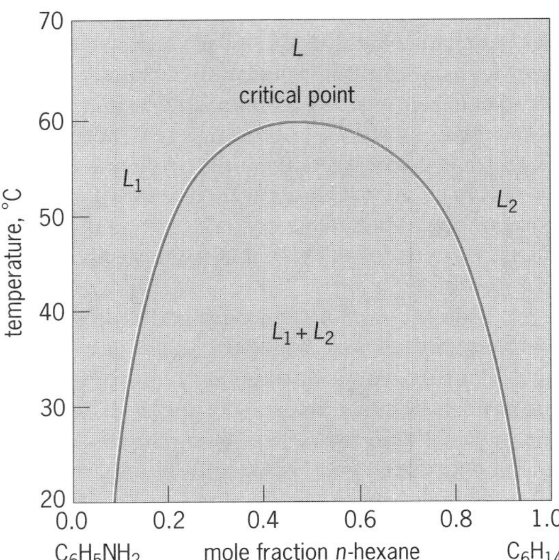

Fig. 4. Liquid-liquid equilibrium for aniline + *n*-hexane.

negative deviations from ideal behavior; this type of deviation usually arises from strong attractions between molecules of the different species (sometimes called compound formation). Minimum boiling mixtures are associated with positive deviations from ideal behavior; this usually arises when the attraction between two unlike molecules (1-2) is weaker than the average of two like pairs (1-1 and 2-2); extreme examples of this arise when one component may be described as associated. The simpler type of phase diagram (Fig. 1) occurs when the two components mix nearly ideally or when the boiling points are very different.

Liquid-liquid equilibrium. When two liquids are sufficiently different in their intermolecular forces, they may not mix in all proportions but instead be only partially miscible, part of the phase diagram being occupied by a two-phase region of two immiscible liquid phases. Most liquids (but not all) become more miscible as the temperature increases and are completely miscible at the critical solution temperature (also called the consolute temperature). The system aniline + *n*-hexane (**Fig. 4**) is a typical example of such liquid-liquid immiscibility and critical phenomena. The liquid-liquid phase boundary and the critical solution temperature are only slightly dependent upon pressure. The size of the two-phase region increases with pressure if (as is usual) the two liquids expand on mixing at constant pressure.

Solid-liquid equilibrium. When two substances are completely miscible with each other and form a complete series of solid solutions, the solid-liquid phase diagrams are entirely analogous to the liquid-vapor diagrams illustrated above. Those with no maximum or minimum, usually associated with nearly ideal liquid and solid solutions, such as methane + krypton, are called type I, according to the Bakhuis Roozeboom classification; those with a maximum melting point, an exceedingly rare type exemplified by *d*-carvoxime + *l*-carvoxime, are type II; and those with a minimum melting point (bromobenzene + iodobenzene), type III. *See* SOLID SOLUTION.

In most binary systems, however, extensive solid solutions are impossible because of the incompatibility of the size, shape, and crystal lattices of the two components. In the absence of solid solution formation, the addition of a solute to a liquid solvent invariably depresses the freezing point, that is, the temperature at which, upon cooling, the first trace of solid solvent appears. This depression of the freezing point of a dilute solution is a convenient method (the so-called cryoscopic method) for determining the molecular weight of a solid solute. *See* MOLECULAR WEIGHT.

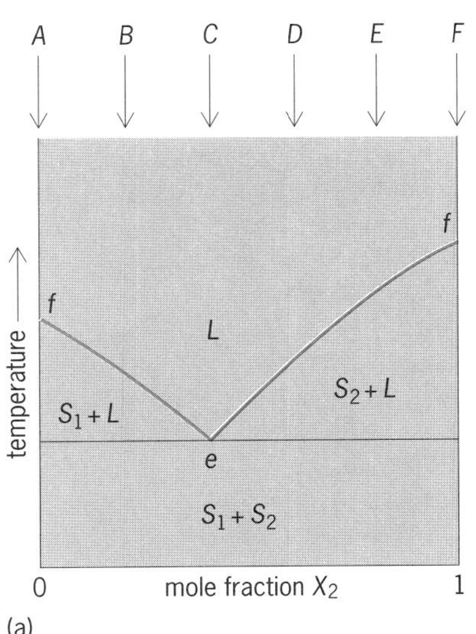

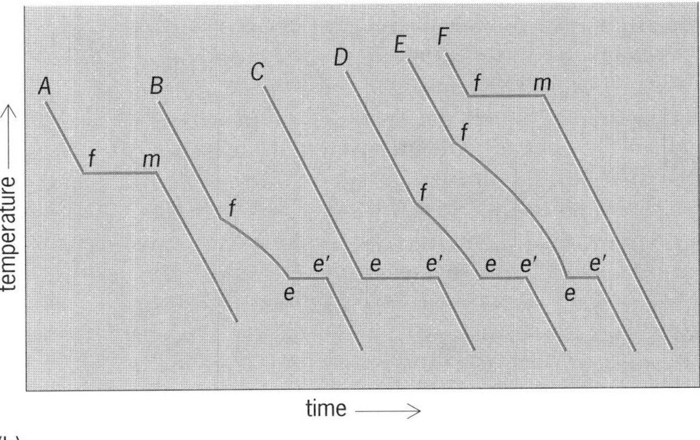

Fig. 5. Solid-liquid equilibrium. (*a*) Schematic diagram when there are no compounds of solid solutions. Points *f* are freezing points of pure substance; *e* is the eutectic point. *A, B, C, D, E*, and *F* are the compositions corresponding to the cooling curves of part *b*. (*b*) Schematic cooling curves for system shown in part *a*; *f* = freezing point, *m* = melting point, *e* = beginning of eutectic freezing, *e′* = end of eutectic freezing.

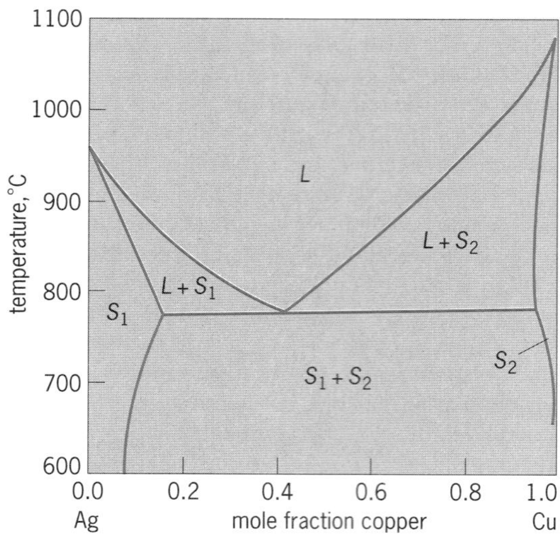

Fig. 6. Solid-liquid equilibrium in the system silver + copper (type-V solid solutions).

As the freezing-point curve of a liquid continues to lower temperatures with higher concentrations of the second component, the conventional roles of solute and solvent become reversed, and one speaks of the solubility of a solute rather than of the depression of the freezing point of the solvent; no point of demarcation exists, and the two situations are in fact two aspects of the same phenomenon.

The second component also has a freezing-point-solubility curve marking the temperature at which the solution is in equilibrium with pure solid 2. The point of intersection of the two phase boundaries is the eutectic point which defines a eutectic temperature and a eutectic composition. Below this temperature, the system consists of two solid phases ($S_1 + S_2$ in **Fig. 5*a***).

Solid-liquid phase diagrams are conveniently determined by thermal analysis of cooling curves. An initially homogeneous liquid is cooled gradually, and the temperature plotted against time. Figure 5*a* shows a simple eutectic-type phase diagram, and Fig. 5*b* sketches typical cooling curves obtained for various compositions (those marked *A* to *F* in Fig. 5*a*). Curves *A* and *F* are for the pure components; each shows a single temperature arrest, a horizontal section of the curve for the melting point (between *f* and *m*), at which the temperature remains constant from the time the first bit of solid is formed until the last bit of liquid disappears. *See* THERMAL ANALYSIS.

Curves *B* and *E* are for solutions rich in components 1 and 2, respectively. At a temperature below the melting point of the pure substance, the first bit of solvent begins to freeze out; this is indicated by a change in slope (point *f*). The freezing section of the cooling curve (between points *f* and *e*) is not horizontal; as solvent freezes out, the liquid phase becomes richer in solute, and the freezing point is further depressed. When the composition of the liquid phase reaches the eutectic composition, the second component begins to freeze out as well as the first. No further change in liquid composition occurs, so the temperature remains constant until no liquid remains (between points *e* and *e′*). The solid which freezes out at the eutectic (the eutectic mixture) appears superficially very different from either pure solid; it is a mixture of very small crystals of each of the two components which have crystallized together. This microcrystalline two-phase mixture is in no sense a compound.

Curve *D* is a cooling curve for still another composition, and curve *C* is for the eutectic composition. Alone, curve *C* is indistinguishable from that for the freezing of a pure substance; only by combining the information from a series of cooling curves can one construct the whole diagram.

When the components form an incomplete series of solid solutions (partial miscibility), the phase diagram can be of the eutectic type illustrated by the system silver + copper shown in **Fig. 6**. Indeed, since the mutual solubilities are never exactly zero, eutectic diagrams, such as that in Fig. 5*a*, are in fact merely extreme examples of this more general case.

When a solid phase upon melting transforms into a liquid phase and a solid phase of different composition, one speaks of incongruent melting and a peritectic-type phase diagram. A simple example is the type-IV solid-solution diagram illustrated by the system silver chloride + lithium chloride (**Fig. 7**); similar phenomena occur in systems without any appreciable solid-solution formation.

Many solid-liquid phase diagrams are complicated by the existence of intermediate crystalline phases of different crystal structure. Usually, these intermediate phases are at compositions close to simple mole ratios of the components and have the two kinds of molecules distributed in a regular arrangement; consequently, it is convenient to call these compounds,

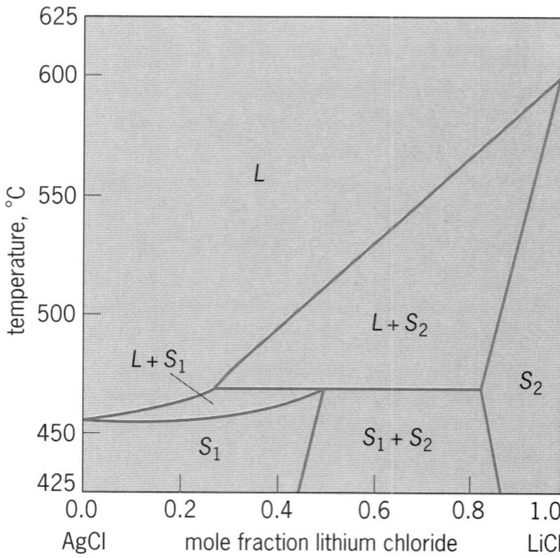

Fig. 7. Solid-liquid equilibrium in the system silver chloride + lithium chloride (type-IV solid solutions).

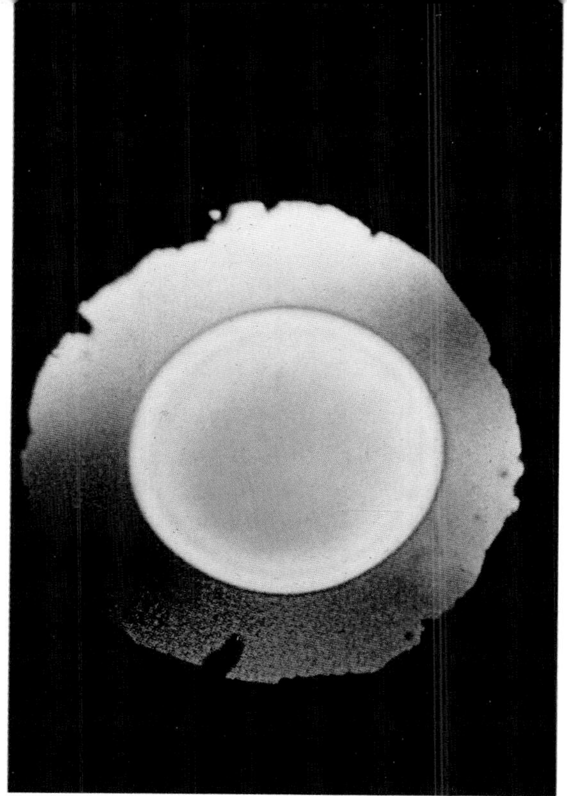

(a)

Phase equilibrium is the coexistence of two or more phases in thermodynamic equilibrium with each other; such equilibria are affected by temperature and pressure. Photomicrographs magnified approximately 500X, using polarized light (crossed Nicols), of (*a*) single crystal and (*b*) 11 crystals of benzene I coexisting with liquid benzene, at about 680 atmospheres and at room temperature; (*c*) potassium nitrate under pressure, where the central portion is potassium nitrate IV (~3000 atmospheres) and surrounding it is potassium nitrate II (~2500 atmospheres). Benzene I and potassium nitrate II and IV are dense forms of the solids favored at these high pressures. Photomicrographs are of specimens contained in a diamond-anvil high-pressure cell. (*National Institute of Standards and Technology*)

(b)

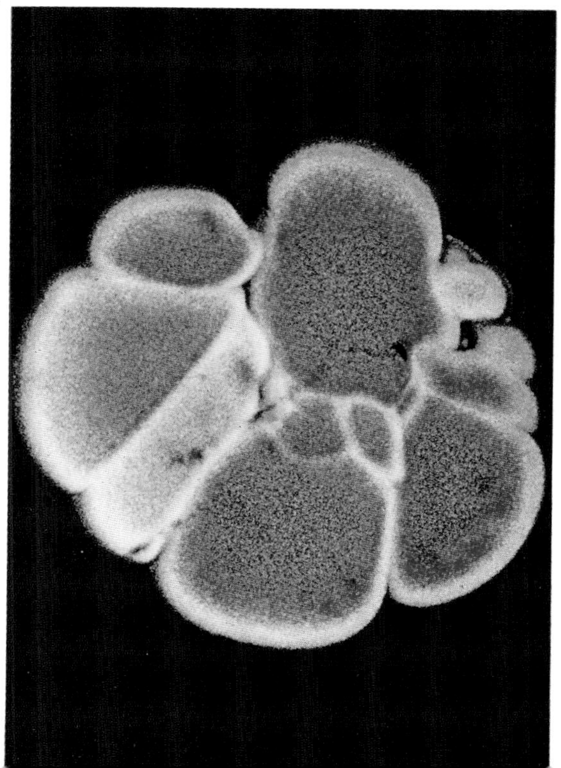

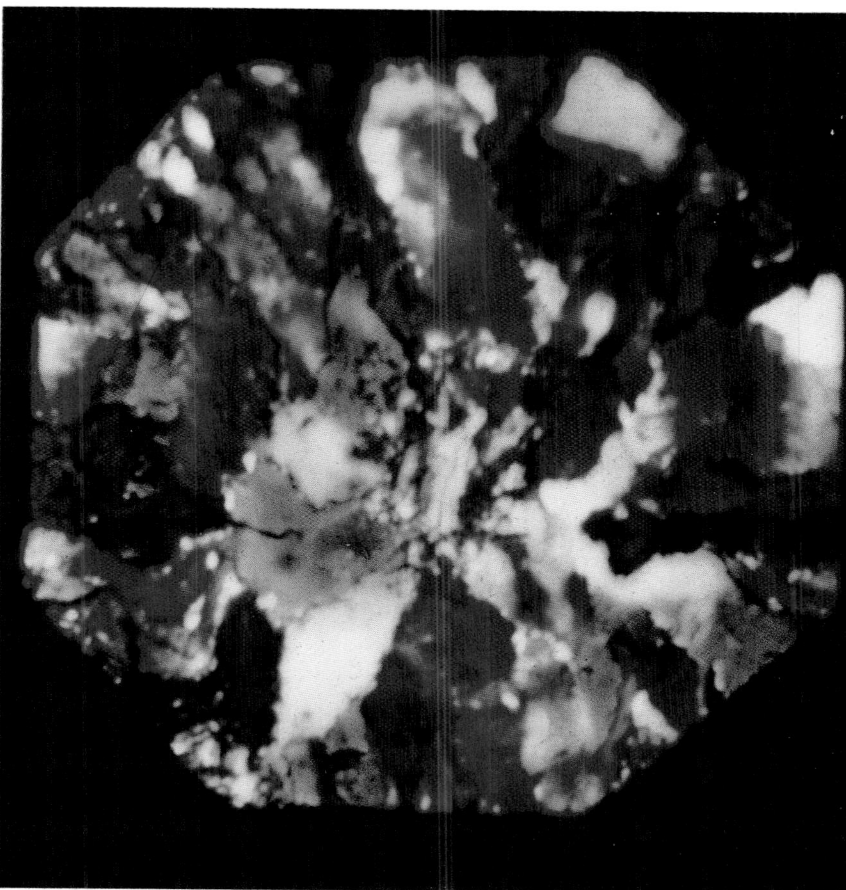

(c)

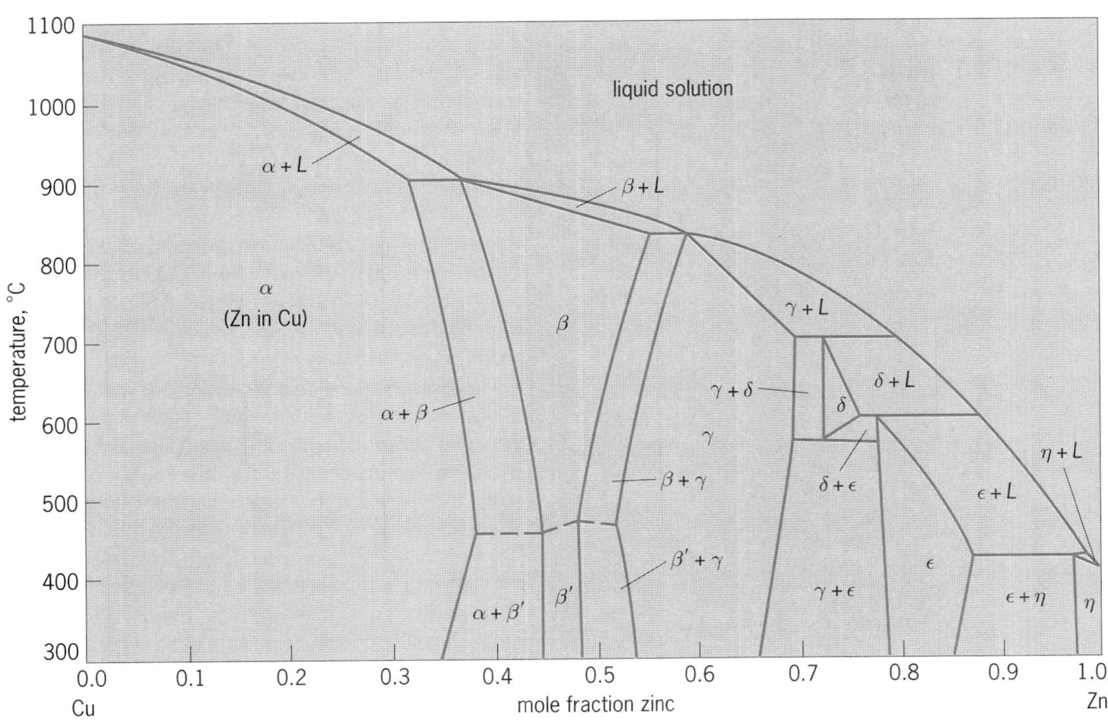

Fig. 8. Temperature-composition diagram for the system copper + zinc (brass). Note the six different solid phases. The broken line separating β from β' denotes a second-order, order-disorder transition.

even if no specific chemical interactions can be demonstrated unequivocally. There is a rich variety of such complex phase diagrams, especially among binary systems of ionic salts ($CaCl_2$-KCl) and among binary systems of metals (alloys). **Figure 8** shows the Cu + Zn system (brass), in which a whole series of crystal structures appears: α-brass has the crystal structure of pure copper (face-centered cubic) with an occasional Zn atom in the lattice; β-brass has a body-centered cubic structure with a Cu-Zn ratio of approximately 1:1; γ-brass has a very complex structure related to the formula Cu_5Zn_9; η-brass has the crystal structure of pure Zn (hexagonal, close-packed); the δ and ϵ phases have still different structures. Note the five peritectic transitions [for example, at 600°C (1112°F) where the ϵ-solid melts incongruently to the δ-solid and a liquid solution rich in zinc].

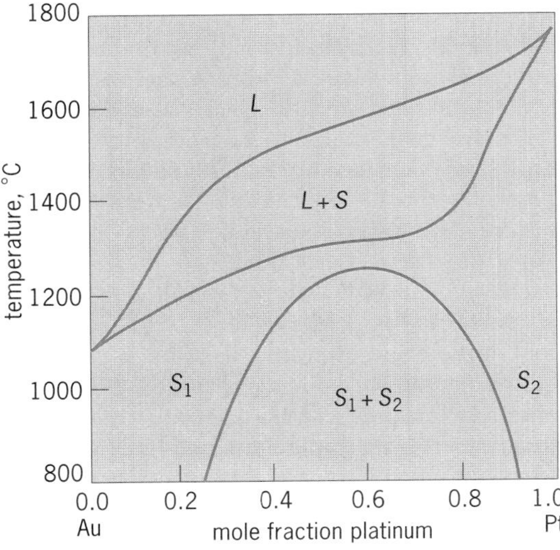

Fig. 9. Solid-liquid and solid-solid equilibria in the system gold + platinum. Critical region of the solid- solid phase boundary nearly touches the solidus curve of the solid-liquid phase boundary. If these touched and coalesced, a type-IV diagram (Fig. 7) would result.

Solid-state equilibrium. These are of several types. One can have solid critical solution temperatures (with solid solutions of the same crystal structure) which are analogous to the more familiar liquid-liquid case (the system gold + platinum shown in **Fig. 9**). More common are the transitions between one crystalline form and another, which can occur even in pure substances. One such occurs in the system nitrogen + carbon monoxide (**Fig. 10** shows solid-solid, solid-liquid, and liquid-vapor transitions for this system). *See* TRANSITION POINT.

Multicomponent Systems

As one proceeds from binary systems to systems with three or more components, the phase diagrams become more complex. Each component adds another dimension to the representation of the phase equilibria. Thus, for three components, two dimensions are required to represent the phase diagrams for a single temperature and pressure; these are conveniently depicted by a triangular diagram in which each vertex represents a pure component. **Figure 11** shows a diagram of a ternary system in which three liquid phases can coexist. Even when there are three

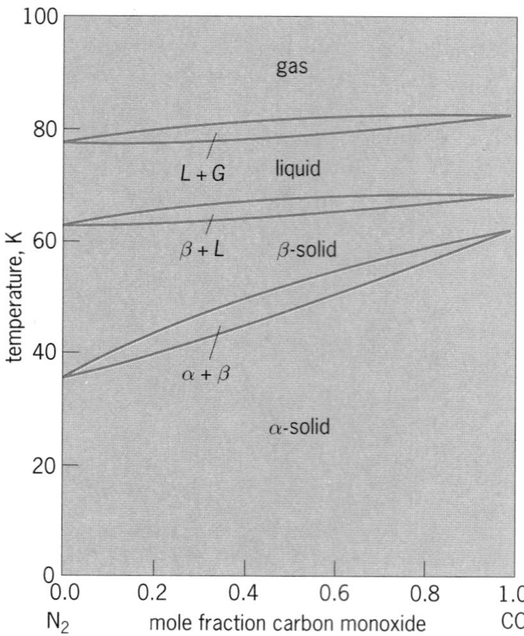

Fig. 10. Phase equilibria for nitrogen + carbon monoxide, being solid-solid, solid-liquid, and liquid-vapor.

phases, the system is still bivariant. (An example of such a system is water + succinonitrile + diethyl ether.)

A special case of a three-component system is that in which there are two immiscible solvents and a third component, soluble in both, distributed between the two phases. The ratio of the concentrations of the solute in the two solvents is the distribution coefficient; in dilute solutions, this is independent of the concentration, but at higher concentrations, nonideal behavior of the solute can produce systematic variations of the distribution coefficient which ends in the most concentrated solutions with the ratio of the solubilities in the saturated solutions.

Distribution effects are important in separating similar materials. A small difference in distribution coefficients is amplified by multistage equilibria, such as those used in countercurrent extraction and partition chromatography.

The examples used to illustrate the various types of phase equilibria are not supposed to suggest that these types are restricted to the particular kind of chemical substances shown. In general, examples of each could have been selected from many kinds of substance such as metals, nonmetallic elements, inorganic salts, and organic nonelectrolytes. *See* ALLOY STRUCTURES; CHEMICAL EQUILIBRIUM; CHEMICAL THERMODYNAMICS; COUNTERCURRENT TRANSFER OPERATIONS; CRYSTAL STRUCTURE; EXTRACTION; INTERFACE OF PHASES; SILICATE PHASE EQUILIBRIA; SULFIDE PHASE EQUILIBRIA. Robert L. Scott

Bibliography. G. M. Barrow, *Physical Chemistry*, 6th ed., 1996; R. Ginell, *Association Theory: The Phases of Matter and Their Transformations*, 1979; I. N. Levine, *Physical Chemistry*, 4th ed., 1995; Research and Education Association Staff, *The Essentials of Physical Chemistry*, vols. 1 and 2, 1994; F. E. Wetmore and D. J. LeRoy, *Principles of Phase Equilibria*, 1951.

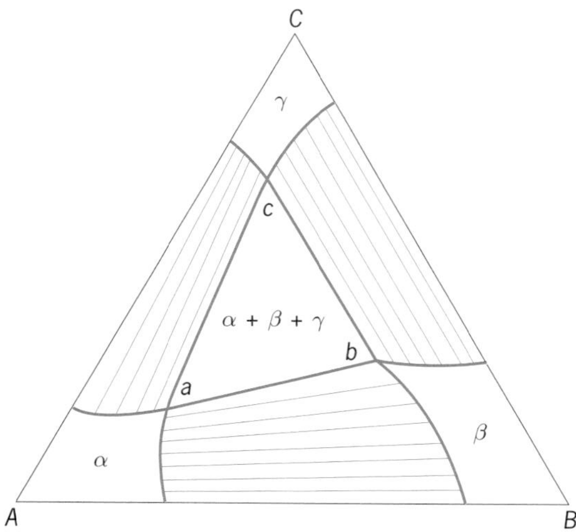

Fig. 11. Schematic diagram of a three-component system at a fixed temperature and pressure. Points *A*, *B*, and *C* represent the pure liquids. The composition corresponding to a point in the diagram is determined by the positions along a line from each vertex to the opposite side; thus point *b* is 20% *A*, 50% *B*, 30% *C*. Regions α, β, and γ correspond to single phases rich in *A*, *B*, and *C*, respectively; $\alpha + \beta + \gamma$ is a three-phase region; the three saturated solutions have the compositions given by the points *a*, *b*, and *c*. Three two-phase regions, $\alpha + \beta$, $\alpha + \gamma$, and $\beta + \gamma$, are indicated by drawing in the tie lines.

Phase inverter

A circuit having the primary function of changing the phase of a signal by 180°. The phase inverter is most commonly employed as the input stage for a push-pull amplifier. Therefore, the phase inverter must supply two voltages of equal magnitude and 180° phase difference. A variety of circuits are available for the phase inversion. The circuit used in any given case depends upon such factors as the overall gain of the phase inverter and push-pull amplifier, the possibility that the input to the push-pull amplifier may require power, space requirements, and cost. *See* PUSH-PULL AMPLIFIER.

Overall fidelity of a phase inverter and push-pull amplifier can be adversely affected by improper design of the phase inverter. The principal design requirement is that frequency response of one input channel to the push-pull amplifier be identical to the frequency response of the other channel. In this respect, popular phase-inversion circuits are capable of providing precisely 180° phase difference only after careful selection of components. Some phase-inverter circuits can perform inversion at only one frequency; at other frequencies, distortion is introduced because of unequal frequency-response characteristics.

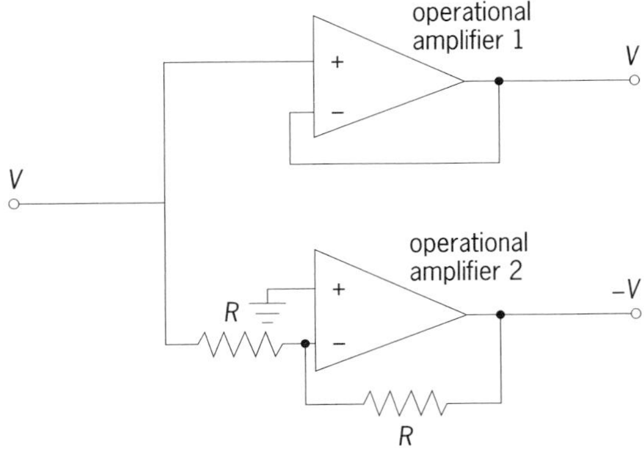

Fig. 1. Phase inverter using operational amplifiers.

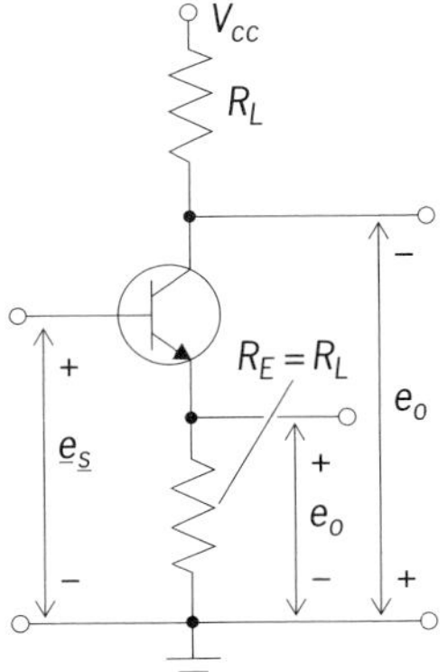

Fig. 2. Single-transistor inverter.

Operational-amplifier inverter. A model phase inverter using two operational amplifiers is shown in **Fig. 1**. Operational amplifer 1 is a voltage follower; the output voltage is equal to the input voltage. Operational amplifier 2 is a unity-gain inverter, and so the output is the negative of the input. The two operational amplifiers can be fabricated in one integrated circuit chip. The circuit is shown in simplified form in that each operational amplifier needs a dual voltage supply, perhaps plus and minus 15 V. Furthermore, the circuit needs some extra resistors to eliminate the effects of the dc offset voltages and currents. The frequency responses of operational amplifiers 1 and 2 must be identical over the frequency range of operation so that the outputs will truly be V and $-V$ as shown. As noted above, this is a requirement for all phase-inverter circuits. The output resistance of each operational-amplifier path in Fig. 1 is very low because of the feedback. This is a desirable property, not found in the circuits discussed below. *See* AMPLIFIER; OPERATIONAL AMPLIFIER.

Paraphase amplifiers. An amplifier that provides two equal output signals 180° out of phase is called a paraphase amplifier. If coupling capacitors can be omitted, the simplest paraphase amplifier is illustrated in **Fig. 2**. Approximately the same current flows through R_L and R_E, and therefore if R_L and R_E are equal, the ac output voltages from the collector and from the emitter are equal in magnitude and 180° out of phase. The gain of the circuit is less than unity, which is one factor that limits its applicability. A second important factor is that the addition of coupling capacitors and biasing resistors, necessary when the circuit is coupled to the push-pull stage, causes the phase inversion to be other than 180° over the frequency range of expected operation. One of the most stable and important paraphase amplifiers is the emitter-coupled phase inverter (**Fig. 3**). If the emitter resistance R_E is large compared to the impedance seen looking into the emitter of each transistor, current i_1 will equal i_2. Under this condition the voltage at one collector is exactly the negative of that at the other collector, and push-pull operation is achieved. The collector-to-collector gain is equal to the gain which would be provided by a single-transistor grounded-emitter amplifier with collector load R_L. *See* TRANSISTOR.

If the phase-inverter circuit is to produce two voltages 180° out of phase, the equivalent circuits governing the behavior of the two output voltages must be identical. The midfrequency gain of each must be identical, and the phase-shift functions must be identical. The phase-shift requirements are often compromised in the interests of simplicity of the final circuit and freedom from critical adjustments of key circuit parameters.

Transformer inverter. The simplest form of phase-inverter circuit is a transformer with a center-tapped secondary (**Fig. 4**). Careful design of the transformer assures that the secondary voltages are equal. The transformer forms a good inverter when the inverter must supply power to the input of the push-pull

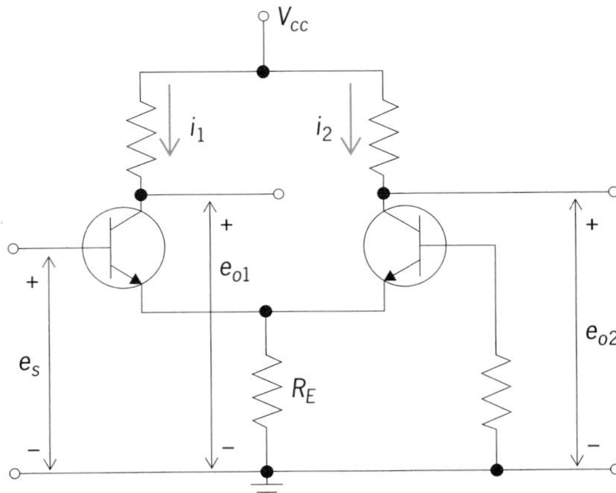

Fig. 3. Emitter-coupled phase inverter.

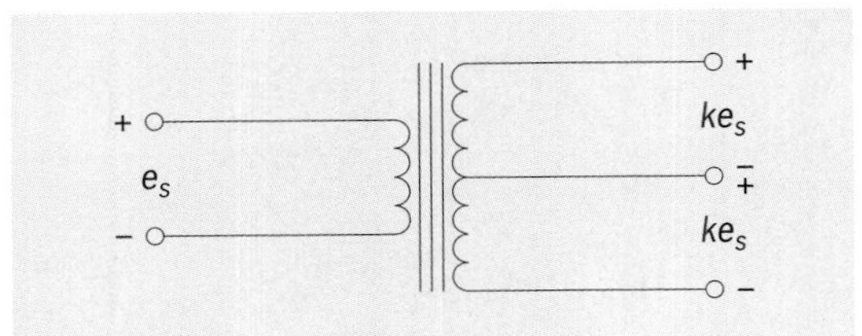

Fig. 4. Transformer, simplest form of phase-inverter circuit. The turns ratio is represented by *k*.

amplifier. The turns ratio can be adjusted for maximum power transfer. *See* TRANSFORMER.

The transformer inverter has several disadvantages. It usually costs more, occupies more space, and weighs more than a transistor circuit. Furthermore, some means must be found to compensate for the frequency response of the transformer, which may not be as uniform as that which can be obtained from solid-state circuits. *See* PHASE (PERIODIC PHENOMENA); PHASE-ANGLE MEASUREMENT.

Harold F. Klock

Bibliography. P. Chirlian, *Analysis and Design of Integrated Electronic Circuits*, 2d ed., 1986; M. Ghausi, *Electronic Devices and Circuits: Discrete and Integrated*, 1995; S. Gibilisco (ed.), *Encyclopedia of Electronics*, 2d ed., 1990.

Phase-locked loops

Electronic circuits for locking an oscillator in phase with an arbitrary input signal. A phase-locked loop (PLL) is used in three fundamentally different ways: (1) as a demodulator, where it is employed to follow (and demodulate) frequency or phase modulation; (2) as a tracker of a carrier or synchronizing signal which may vary in frequency with time; and (3) as a frequency synthesizer, where an oscillator is locked to a multiple of an accurate reference frequency. When operating as a demodulator, the phase-locked loop may be thought of as a matched filter operating as a coherent detector. When used to track a carrier, it may be thought of as a narrow-band filter for removing noise from the signal and regenerating a clean replica of the signal. When used as a frequency synthesizer, a voltage-controlled oscillator (VCO) is divided down to a reference frequency that is locked to a frequency derived from an accurate source such as a crystal oscillator. *See* DEMODULATOR; ELECTRIC FILTER; FREQUENCY-MODULATION DETECTOR; FREQUENCY MULTIPLIER; OSCILLATOR; PHASE-MODULATION DETECTOR.

Basic operation. The basic components of a phase-locked loop are shown in **Fig. 1**. The input signal is a sine or square wave of arbitrary frequency. The voltage-controlled-oscillator output signal is a sine or square wave of the same frequency as the input, but the phase angle between the two is arbitrary. The output of the phase detector consists of a direct-current (dc) term, and components of the input frequency and its harmonics. The low-pass loop filter removes all alternating-current (ac) components, leaving the dc component, the magnitude of which is a function of the phase angle between the voltage-controlled-oscillator signal and the input signal. If the frequency of the input signal changes, a change in phase angle between these signals will produce a change in the dc control voltage in such a manner as to vary the frequency of the voltage-controlled oscillator to track the frequency of the input signal.

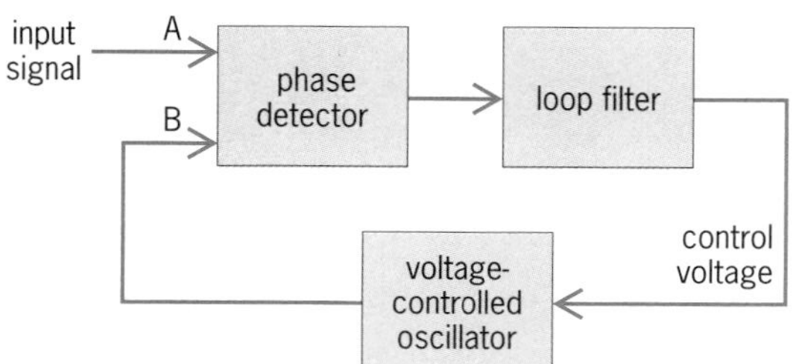

Fig. 1. Basic components of a phase-locked loop.

Two qualities of the loop specify its performance: the lock range and the capture range. The lock range is the maximum change in input frequency for which the loop will remain locked. It is governed by the dc gain of the loop. As the input frequency is changed, the change in phase of the two signals to the phase detector will produce a dc control voltage that will change the frequency of the voltage-controlled oscillator. As the input frequency is further changed, the phase angle will continue to increase until it reaches 0 or 180°, when the loop will unlock. If an amplifier is added to the loop, a greater control voltage will be generated which will decrease the phase error in the phase detector, and hence further detuning can occur before unlocking takes place.

The capture range is that range of frequencies that the loop will lock to if it is initially unlocked. Suppose the loop is unlocked and the voltage-controlled oscillator is running at frequency f_1. If the input signal f_2 is applied but is out of the pull-in range, a beat note $f_1 - f_2$ will appear at the output of the phase detector. The filter components will govern the amplitude of this beat note at the input to the voltage-controlled oscillator. If the frequency difference is reduced, the frequency of the beat note will decrease and the amplitude at the voltage-controlled oscillator input will increase. At some point, the amplitude will drive the voltage-controlled oscillator far enough over in frequency to match the input frequency and locking will occur. The lower the roll-off frequency of the loop filter, above which its attenuation begins to increase, the less will be the capture range.

Phase-locked loops are generally designed to have narrower capture range than lock range. This is the advantage of a phase-locked loop over more conventional types of filters. The capture range is analogous to filter bandwidth and may be made as narrow as desired by suitable choice of the resistance-capacitance (*RC*) filter, while the center frequency may be any desired value.

Voltage-controlled oscillators. Voltage-controlled oscillators may take many forms, two common types being the reactance modulator and the varactor modulator. These circuits are commonly used in narrow-band phase-locked loops where narrow lock and capture ranges are desired. *See* VARACTOR.

Another common form of voltage-controlled oscillator utilizes a multivibrator type of circuit where the timing capacitor charging current may be varied by a dc control voltage. This circuit relies on the fact

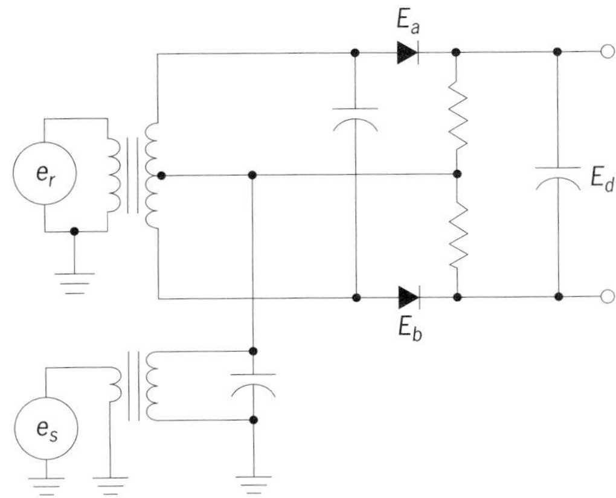

Fig. 2. Diode phase detector.

that the time required to charge the timing capacitance is inversely proportioned to the frequency of oscillation, hence the control voltage will vary the frequency of the oscillator. *See* MULTIVIBRATOR.

Reactance- and varactor-type modulators are used with crystal-controlled oscillators where tuning ranges of 0.25–0.5% are desired, or with inductance-capacitance (*LC*) oscillators where tuning ranges of 5% or more can be achieved. Multivibrator types of circuits are used where wide tuning ranges of up to 1000 to 1 have been achieved. Conversely, phase noise and stability are best with a crystal-controlled oscillator and least with multivibrator-type circuits.

Phase detectors. Two commonly used phase detectors are the diode phase detector and the double-balanced phase detector. The diode phase detector has been used historically because of lower cost and better performance at high frequencies (above 50 MHz). With the advent of monolithic integrated circuits, the double-balanced phase detector has become useful at lower frequencies. *See* INTEGRATED CIRCUITS.

Diode detector. In the diode detector (**Fig. 2**) the voltages applied to the two peak rectifiers the sum and difference of the two input signals, e_r and e_s. Rectified output voltages are equal to the amplitudes of each of the sums E_a and E_b. The phase detector output voltage E_d is equal to the difference of the two rectified voltages, $E_a - E_b$. Considering that e_r and e_s are sinusoidal waves of the same frequency but varying phase angle, the dc output E_d will be zero when the two input signals are in quadrature (90°), a maximum positive value when in phase, and a maximum negative value when 180° out of phase.

Double-balanced detector. The double-balanced phase detector (**Fig. 3*a***) works well at low frequencies. The upper four transistors operate as a double-pole double-throw switch driven by the reference signal e_r. The lower transistors operate as an amplifier, their collector currents being proportional to the input signal e_s. Figure 3*b* shows the output waveform for different phase angles between the two input signals. The output signal shown is that which would be obtained if the filter capacitor *C* were removed.

Digital phase/frequency detector. A third type of phase detector, used in frequency synthesizers, is shown in **Fig. 4**. This digital three-state phase/frequency comparator with a charge-pump (switched current source) output circuit has the advantage of producing a maximum dc component at its output when not in lock, so that the loop is driven to lock even when a large error exists between the two frequencies at the detector input. The output of this detector has a three-state capability; depending on which input signal is higher or lower in frequency, its

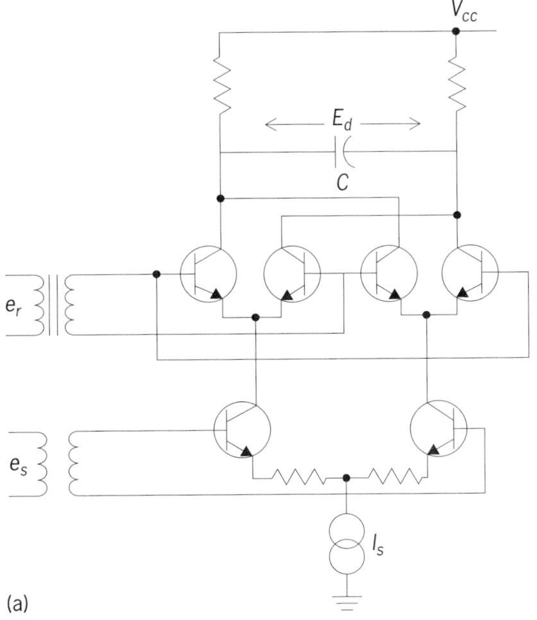

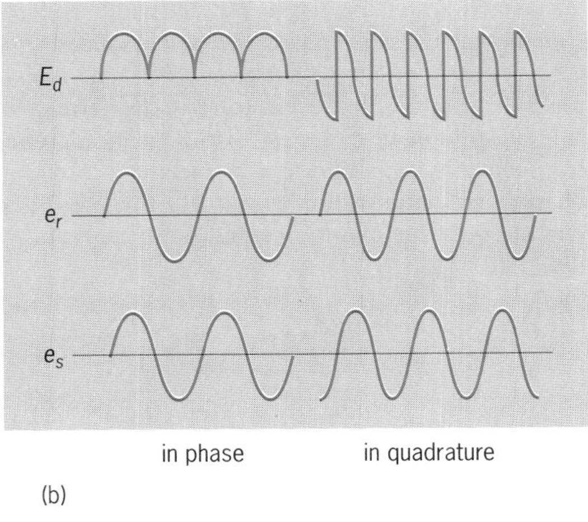

Fig. 3. Double-balanced phase detector. (*a*) Circuit. (*b*) Waveforms.

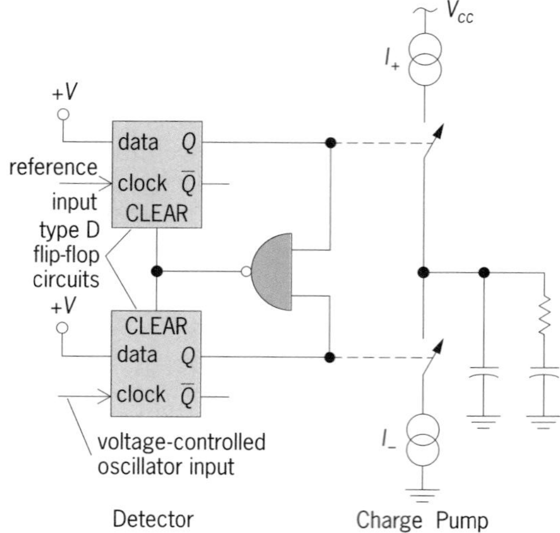

Fig. 4. Digital phase/frequency detector. Q = output; $\overline{Q}$ = inverting output.

output is a positive or negative current source or an open circuit. When the two input frequencies are the same, the detector's output will drive high or low depending on the phase error between the input signals. When a very small phase error exists between the triggering edges of the two input signals A and B, either the pull-up or the pull-down signal will generate a small output error signal to maintain lock. Unlike the linear phase detectors described above, this circuit operates at zero phase error between its inputs, and under these conditions the output is a high impedance, with essentially no output ripple voltage. This is an ideal type of phase detector for loops operating with very quiet input signals; however, any noise will drive the output high or low until the next cycle edge occurs, causing a major disturbance to the loop. This property of the digital-logic-gate type of phase detector makes it a poor choice for demodulator functions operating with noisy input signals.

Loop filter. The loop filter determines the dynamic characteristics of the loop: its bandwidth and response time. This filter is generally composed of a lag-lead network (**Fig. 5**). The transfer function of the voltage-controlled oscillator has the characteristic of an integrator: a step change in the dc control voltage will produce a ramp change in output phase, increasing without bound. The low-pass filter characteristic begins attenuating frequencies above

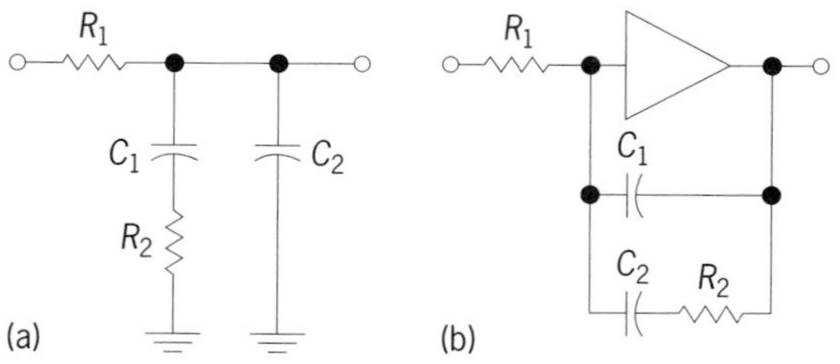

Fig. 5. Loop filter configurations. (*a*) Passive. (*b*) Active.

its break point determined by R_1C_1; at a higher frequency determined by R_2C_1, no further attenuation occurs. This is important because the phase shift associated with the voltage-controlled oscillator is 90°, and that associated with the roll off due to R_1C_1 is also approaching 90°. With a phase shift of 180°, the loop will oscillate and become unstable. The leading phase shift associated with R_2C_1 reduces the phase shift around the loop and enhances stability.

Uses of phase-locked loops. The most widespread use of phase-locked loops is undoubtedly in television receivers. Synchronization of the horizontal oscillator to the transmitted sync pulses is universally accomplished with a phase-locked loop; here it is desirable to have a stable, noise-free reference source to generate the scanning line since the eye is intolerant of any jitter (or phase noise). The color reference oscillator is often synchronized with a phase-locked loop; here it is necessary to maintain less than 5° of phase error between the transmitted color reference signal and the locally generated reference in the receiver. In both of these applications, the voltage-controlled oscillator signal is the useful output, and the design of the loop filter is such as to produce a narrow-band characteristic so that noise is reduced and the filter "remembers" the proper phase during short periods of noise and loss of signal. *See* TELEVISION RECEIVER.

Phase-locked loops are also used as frequency demodulators. Since the voltage-controlled oscillator is locked to the incoming carrier frequency, and the control voltage is proportional to the voltage-controlled oscillator frequency, as long as the loop remains locked, the control voltage will be a replica of the modulating signal.

Stereo decoding. Phase-locked loops have been applied to stereo decoders made on silicon monolithic integrated circuits. This design technique eliminates the coils used in previous decoder designs. The circuit consists of a 76-kHz voltage-controlled oscillator, which, after being divided by 4, is locked to the transmitted 19-kHz pilot carrier. The phase-locked loop therefore acts as a carrier regenerator (or narrow-band filter). The 76 kHz is also divided by 2 to obtain 38 kHz that is applied to a double-balanced demodulator, which decodes the stereo signal into left and right channels. *See* STEREOPHONIC RADIO TRANSMISSION.

Amplitude demodulation. High-performance amplitude demodulators may be built by using phase-lock techniques. If, instead of a diode detector, a double-balanced demodulator is used, the lower port of which is driven with the signal to be demodulated and the upper port is driven with a noise-free replica of the carrier, a very linear (low-distortion) demodulator results. The phase-locked loop locks to the carrier frequency; the loop filter should be narrow enough that modulation is not present in the phase-locked loop. The phase detector is driven in quadrature with the input signal so that very little amplitude information is demodulated in the loop. The voltage-controlled oscillator signal is also phase-shifted 90°

and applied to a second phase detector, which is also driven with the signal to be demodulated. Since the carrier and reference are in phase in this detector, the output will be proportional to the amplitude of the incoming carrier. Demodulators using this technique have been built with linearity better than 0.5% distortion for 90% modulated signals. *See* AMPLITUDE-MODULATION DETECTOR.

Frequency synthesis. A frequency-synthesizer application of a phase-locked loop is used in the majority of AM-FM radio and television receivers and most modern two-way communications equipment. A crystal oscillator generates a reference signal F_x, which is divided down by a number M to a reference frequency F_r. A voltage-controlled oscillator running at the frequency to be synthesized is also divided down by a number N to the reference frequency to which it is phase-locked. In this way, the voltage-controlled oscillator can be tuned to exact mutiples of the reference frequency. In an example of a local oscillator for an FM broadcast receiver, the crystal oscillator may operate at 4 MHz, and be divided by 160 in the M counter to produce a reference frequency of 25 kHz. It is desired to tune the local oscillator from 98.6 to 118.6 MHz in 25-kHz steps. In this case, the divide ratio in the N counter would vary between 3944 and 4744. In radio and communications applications, fast switching speed is not a primary requirement, and loops are designed to have a relatively narrow closed-loop bandwidth in order to reduce noise associated with the reference frequency. A microprocessor is used to load N in the programmable divider and provide for frequency display, keyboard reading, and so forth in a radio application. A variety of single- and multiple-chip synthesizers are available from manufacturers of integrated circuits. Most of these integrated circuits contain the crystal oscillator, both M and N counters, and the phase detector. Varactor-tuned inductance-capacitance (LC) oscillators are usually used in these applications. *See* RADIO RECEIVER.

Data-disk synchronization. With the requirements for large amounts of mass storage, the computer industry has turned to magnetic recording techniques using both hard and floppy magnetic disks. Optical storage techniques have become available for extremely high-density data storage. In all of these systems, it is necessary to lock a clock signal to the data being read from the disk or an internal clock signal. This is accomplished with a phase-locked loop using a digital-logic-gate type of phase detector described above. The voltage-controlled oscillator is locked to either the system clock or the encoded data signal from the disk depending on the state of the read gate. In this way, the voltage-controlled oscillator signal is available to the coding and decoding logic circuits in the drive and is coherent in phase with the data being processed at all times. Generally, response speed is important in phase-locked loops for disk applications, and loop bandwidth may approach 15% of the data rate or hundreds of kilohertz. Many loops are designed for both high- and low-speed operation by switching the gain of the phase detector. Single-chip integrated circuits are available from many manufacturers to provide the phase-lock functions for both fixed and floppy disk drives. Resistance-capacitance (RC)–coupled multivibrator voltage-controlled oscillators are often employed in these applications. *See* COMPUTER STORAGE TECHNOLOGY; OPTICAL RECORDING.

Thomas B. Mills

Bibliography. R. E. Best, *Phase-Locked Loops: Theory, Design, and Applications*, 4th ed., 1999; J. B. Encinas, *Phase Locked Loops*, 1993; F. M. Gardner, *Phaselock Techniques*, 2d ed., 1979; W. C. Lindsey and C. M. Chie (eds.), *Phase-Locked Loops*, 1985; D. H. Wolaver, *Phase-Locked Loop Circuit Design*, 1991.

Phase modulation

A technique used in telecommunications transmission systems whereby the phase of a periodic carrier signal is changed in accordance with the characteristics of an information signal, called the modulating signal. Phase modulation (PM) is a form of angle modulation. For systems in which the modulating signal is digital, the term "phase-shift keying" (PSK) is usually employed. *See* ANGLE MODULATION.

In typical applications of phase modulation or phase-shift keying, the carrier signal is a pure sine wave of constant amplitude, represented mathematically as Eq. (1), where the constant A is its amplitude,

$$c(t) = A \sin \theta(t) \tag{1}$$

$\theta(t) = \omega t$ is its phase, which increases linearly with time, and $\omega = 2\pi f$ and f are constants that represent the carrier signal's radian and linear frequency, respectively.

Phase modulation varies the phase of the carrier signal in direct relation to the modulating signal $m(t)$, resulting in Eq. (2), where k is a constant of pro-

$$\theta(t) = \omega t + km(t) \tag{2}$$

portionality. The resulting transmitted signal $s(t)$ is therefore given by Eq. (3). At the receiver, $m(t)$ is re-

$$s(t) = A \sin \{\omega t + km(t)\} \tag{3}$$

constructed by measuring the variations in the phase of the received modulated carrier.

Phase modulation is intimately related to frequency modulation (FM) in that changing the phase of $c(t)$ in accordance with $m(t)$ is equivalent to changing the instantaneous frequency of $c(t)$ in accordance with the time derivative of $m(t)$. *See* FREQUENCY MODULATION.

Properties. Since only the phase of the carrier signal $c(t)$ is affected by the modulation process, the amplitude of the modulated signal $s(t)$ is constant over time. Consequently the peak power and average power in $s(t)$ are constant as well, independent of the form of the modulating signal $m(t)$ and the value of k, and completely determined by the amplitude of $c(t)$. Because of this feature, phase modulation has several advantages over other forms of modulation, such as amplitude modulation (AM). Among the advantages are superior noise and interference rejection, enhanced immunity to signal fading, and reduced susceptibility to nonlinearities in the transmission and receiving systems. Phase modulation also exhibits the capture effect, whereby the stronger of two signals with nearly equal amplitudes suppresses the other. *See* DISTORTION (ELECTRONIC CIRCUITS); ELECTRICAL INTERFERENCE; ELECTRICAL NOISE.

Although theoretically the bandwidth of $s(t)$ is infinite and independent of $m(t)$, in practical terms it is strongly affected by the bandwidth of $m(t)$. Thus, under certain conditions, if the power of $m(t)$ is distributed over the range of frequencies from f_1 hertz at the lower end of the spectrum to f_2 hertz at the upper end, the amount of power in $s(t)$ contained outside the frequency range from $f - f_1$ hertz to $f - f_2$ hertz and from $f + f_1$ hertz to $f + f_2$ hertz is of minor consequence. The bandwidth of $s(t)$ is then taken to be $2f_2$ hertz, centered at f hertz, the linear frequency of $c(t)$.

Phase modulators and demodulators. For modulating signals that produce only small variations in phase, a simple form of phase modulation can be achieved by generating a carrier signal in a local oscillator, shifting its phase by 90°, and multiplying it by the modulating signal. Adding this product to the original unmodulated carrier signal yields a good approximation of the desired result. For larger phase variations, switching-circuit modulators are frequently employed. *See* PHASE MODULATOR.

Some demodulators, known as zero-crossing detectors, rely on the fact that the points in time when the modulated carrier signal assumes a value of 0 are representative of the original modulating signal. Other demodulators employ phase-locked loops, which track the phase of the modulated carrier signal by adjusting the output of the demodulator through a feedback mechanism. Phase-locked loops find application in low signal-to-noise ratio environments. *See* PHASE-LOCKED LOOPS; PHASE-MODULATION DETECTOR.

Phase-shift keying. When the modulating signal $m(t)$ is digital, so that its amplitude assumes a discrete set of values, the phase of the carrier signal is "shifted" by $m(t)$ at the points in time where $m(t)$ changes its amplitude. The amount of the shift in phase is usually determined by the number of different possible amplitudes of $m(t)$. In binary phase-shift keying (BPSK), where $m(t)$ assumes only two amplitudes, the phase of the carrier differs by 180°. An example of a higher-order system is quadrature phase-shift keying (QPSK), in which four amplitudes of $m(t)$ are represented by four different phases of the carrier signal, usually at 90° intervals.

The bandwidth of a PSK signal, defined as the range of frequencies over which the power in the modulated carrier signal is distributed, is determined by the rate R at which the phase of the carrier signal is shifted, and is independent of the number of amplitudes of $m(t)$. To a good approximation, it can be taken to equal $2R$ hertz, extending over the frequency range from $f - R$ to $f + R$ hertz. Thus the higher-order systems such as QPSK require less bandwidth per unit information than BPSK. *See* MODULATION.

Hermann J. Helgert

Bibliography. B. P. Lathi, *Modern Digital and Analog Communication Systems*, 3d ed., Oxford University Press, 1998; M. Schwartz, *Information Transmission, Modulation, and Noise*, 4th ed., McGraw-Hill, 1990.

Phase-modulation detector

A device which recovers or detects the modulating signal from a phase-modulated carrier. Any frequency-modulation (FM) detector with minor modifications will detect phase-modulated waves. *See* FREQUENCY-MODULATION DETECTOR; PHASE MODULATION.

Modification of FM detector. In standard FM broadcast, the only difference between FM and phase modulation (PM) is the manner in which the modulation index varies with the modulating frequency. The modulation index is independent of the modulating frequency in PM but is inversely proportional to the modulating frequency in FM. Therefore an FM detector, when used to detect a phase-modulated wave, produces an output voltage which is proportional to the modulating frequency, assuming the original modulating signal to be of constant amplitude. Consequently, a low-pass filter with a single reactive element, such as an RC (resistance-capacitance) filter, is needed in the output of the FM detector which is used to detect a phase-modulated wave.

A simple RC filter which might be used to convert an FM detector into a PM detector is shown in **Fig. 1**. The RC time constant of the filter should equal the reciprocal of the lowest modulating frequency component in radians per second. However, the resistance R involved in the time constant is the total resistance in parallel with the capacitance C, not just the resistance R_F. Commercial FM broadcasting ordinarily utilizes the characteristics of FM for modulating frequencies below about 2100 Hz and the characteristics of PM for frequencies above 2100 Hz for the purpose of maintaining improved signal-to-noise ratio at high modulating frequencies, as compared with pure FM. This technique is usually known as preemphasis. Then the FM detector must include a filter such as the RC filter shown in Fig. 1, but the

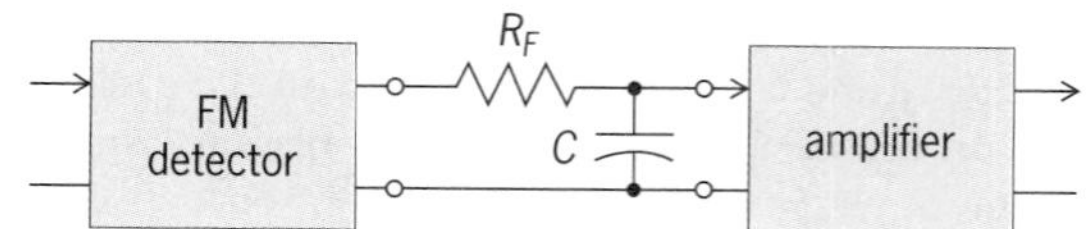

Fig. 1. An *RC* filter which is used to convert a frequency-modulation detector into a phase-modulation detector.

crossover frequency must be approximately 2100 Hz instead of the lowest modulating frequency. This filter is usually known as a deemphasis filter and has a time constant of approximately 75 microseconds ($RC = 1/6.28 \times 2100$). This filter must be added to an FM detector to convert it into a PM detector. *See* ELECTRIC FILTER.

Multiplier circuit. A multiplier circuit or chip may be used as a phase detector, provided its frequency capability, or passband, includes the carrier frequency. **Figure 2** shows how the multiplier serves as a phase detector. The phase-modulated carrier is applied to one multiplier input while the unmodulated carrier frequency is applied to the other input. These signals are shown as square waves because PM and FM are usually severely clipped to remove amplitude modulation with its associated noise. Also Fig. 2 is easier to understand with square waves. The output is maximum positive when the inputs V_1 and V_2 are in phase (Fig. 2*a*), but the average value of the output decreases to zero as the relative phase of the two inputs shifts to 90° (Fig. 2*b*). Then as the relative input phase increases to 180°, the output becomes maximum negative (Fig. 2*c*). After the carrier and other high frequencies are filtered from the output, the output voltage is a linear function of the phase difference between the two inputs (Fig. 2*d*). The unmodulated carrier may be generated in the receiver by the use of a phase-locked loop (PLL). *See* AMPLITUDE MODULATOR; PHASE-LOCKED LOOPS.

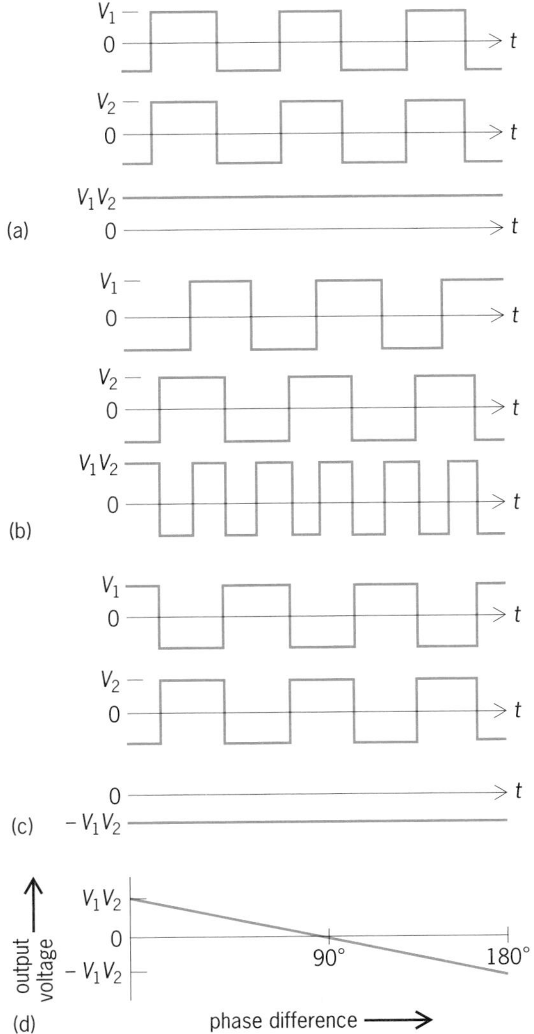

Fig. 2. Phase detection by multiplication. (*a*) Inputs V_1 and V_2 are in phase. (*b*) Relative phase of inputs is 90°. (*c*) Relative phase of inputs is 180°. (*d*) Output voltage as a function of phase difference between the inputs.

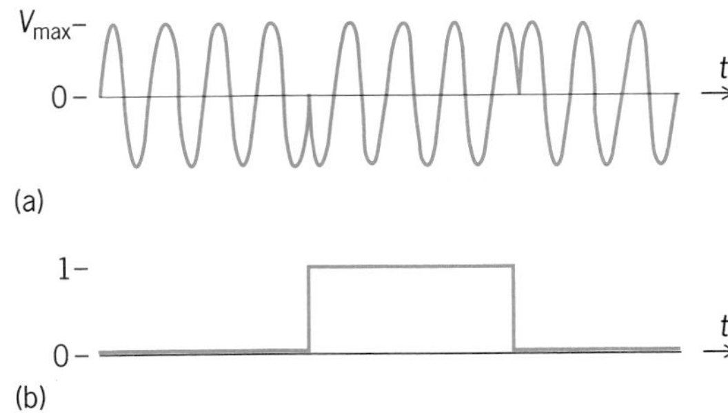

Fig. 3. Phase-shift keying modulation. (*a*) Modulated signal. (*b*) Pulse-code modulating signal.

Demodulation of PSK. Phase-modulation detectors are often called upon to demodulate a type of pulse-code modulation known as phase-shift keying (PSK). In PSK modulation the carrier phase shifts abruptly by 180° whenever the pulse-code modulating signal changes from the zero level to the one level, or vice versa (**Fig. 3**). When the multiplier-type demodulator is used for PSK, the output filter must allow the output voltage to change from maximum positive to maximum negative during a time that is short compared with the time allotted to either a one or a zero in the pulse-code modulation. *See* PULSE MODULATION.

Since a PLL has both a phase detector and a voltage-controlled oscillator (VCO), the PLL may be used as a complete phase demodulator, including the carrier regeneration. Only a long-time-constant filter is needed between the phase detector output and the VCO to provide a quasi-dc voltage to the VCO in order to maintain its lock to the carrier frequency.

Demodulation of FSK. Another commonly used type of pulse-code modulation is frequency-shift keying (FSK), in which the carrier frequency changes abruptly by an amount Δf as the modulating signal changes from the zero level to the one level, or vice versa. Therefore, FSK is basically a type of FM, and any type of FM demodulator may be used to recover the pulse-code modulating signal. However, phase-locked loops are commonly used. Thus a phase-locked loop may be used to detect either FSK or PSK, but an additional long-time-constant filter is required between the output and the VCO in the PSK detector in order to hold the VCO at the carrier frequency without abrupt changes in phase.

Although the fidelity of a PLL in demodulating an FM signal is as good as the linearity of the VCO frequency to its control voltage, this linearity is not important in the reception of FSK because the only requirement is to faithfully identify the two different voltage levels in the output of the PLL. The same is true for the reception of PSK, except that the linearity of the phase detector instead of the VCO would be involved in this case. *See* MODULATION; MODULATOR.

Charles L. Alley

Bibliography. S. Haykin, *Communication Systems*, 4th ed., 2000; M. S. Roden, *Analog and Digital Communication Systems*, 4th ed., 2000; W. Tomasi, *Fundamentals of Electronic Communications Systems*, 1994.

Phase modulator

An electronic circuit that causes the phase angle of the modulated wave to vary (with respect to the unmodulated carrier) in accordance with the modulating signal. Since frequency is the rate of change of phase, a phase modulator will produce the characteristics of frequency modulation (FM) if the frequency characteristics of the modulating signal are so altered that the modulating voltage is inversely proportional to frequency. Commercial FM transmitters normally employ a phase modulator because a crystal-controlled oscillator can then be used to meet the strict carrier-frequency control requirements of the Federal Communications Commission. The chief disadvantage of phase modulators is that they generally produce insufficient frequency-deviation ratios, or modulation index, for satisfactory noise suppression. Frequency multiplication can be used, however, to increase the modulation index to the desired value, since the frequency deviation is multiplied along with the carrier frequency. Many different types of phase modulators have been devised. A few of the typical and more commonly used modulators are described in this article. *See* FREQUENCY MODULATION; PHASE MODULATION; PHASE-MODULATION DETECTOR.

Types. A simple modulator is shown in **Fig. 1**. In this circuit the modulating voltage changes the capacitance of the varactor diode. The phase shift depends upon the relative magnitudes of the capacitive reactance of the varactor diode and the load resistance R. Therefore the phase shift varies with the modulating voltage and phase modulation (PM) is accomplished. However, the phase shift is not linearly related to the modulating voltage if the PM exceeds a few degrees, because the phase shift is not linearly related to the capacitance and the capacitance of the varactor diode is not linearly related to the modulating voltage. *See* VARACTOR.

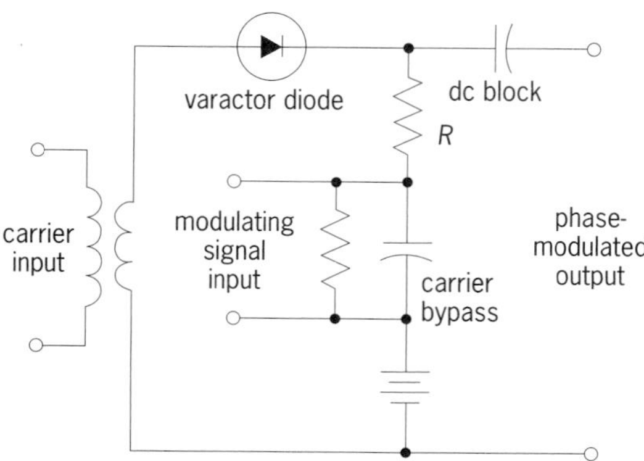

Fig. 1. A simple phase modulator.

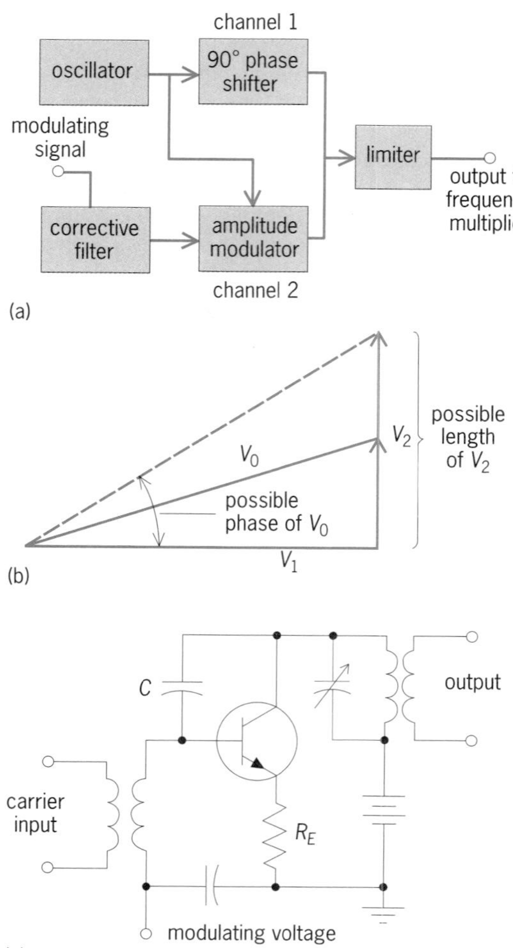

Fig. 2. A method for the production of phase modulation. (*a*) Block diagram. (*b*) Phasor diagram showing the combination of voltages. (*c*) A circuit which performs the necessary functions, except corrective filtering and limiting, in a single transistor circuit.

Figure 2 illustrates the principle of operation of a phase modulator which provides greater phase shift for a given distortion than the simple phase modulator of Fig. 1. The block diagram of Fig. 2*a* shows that the oscillator signal is passed through two channels, 1 and 2. In channel 1 the phase of the signal is shifted 90°, and in channel 2 the signal is amplitude-modulated. The outputs of channel 1 and channel 2 are then recombined and passed through a limiter to remove the residual amplitude modulation (AM). The phasor diagram shows how PM is produced when the amplitude-modulated carrier voltage V_2 is added to the 90° phase-shifted carrier voltage V_1. The phasor sum V_0 is the hypotenuse of the phasor triangle and varies in phase with respect to the channel 1 voltage as the channel 2 voltage varies in amplitude. The phasor sum V_0 also varies somewhat in amplitude, but a limiter can be used to remove this AM, as previously mentioned and shown in Fig. 2. This phase modulator is linear, or distortionless, providing first, that the AM is distortionless and second, that the change in phase is proportional to the change in amplitude of the amplitude-modulated signal from channel 2.

But elementary trigonometry and Fig. 2*b* show that the phase of V_0 is the angle whose tangent is V_2/V_1. Therefore linear modulation is attained only for small phase angles where the tangent of the angle is approximately equal to the angle itself. This requirement limits the total phase variation to about 20° if the distortion is limited to 54. This limit would permit only 10° phase deviation on each side of the unmodulated, or carrier, position.

A simple circuit employing a single transistor can perform the basic functions of a phase modulator, as shown in Fig. 2*c*. The transistor is a base-modulated amplifier which has low gain because of the unbypassed resistor R_E in the emitter circuit. The current through the small capacitance C leads the input voltage by approximately 90° and adds to the amplitude-modulated collector current in the collector load. The small voltage gain, perhaps less than 1, is required in the transistor amplifier for two reasons. First, the current through capacitor C is then essentially proportional to the carrier input, which is essentially constant, as assumed in the phasor diagram of Fig. 2*b*. Second, the capacitance C can cause the amplifier to be unstable unless the voltage gain is low.

The phase deviation can be doubled for a given distortion level if the base-modulated amplifier of Fig. 2 is replaced by a balanced modulator. The reason for this improvement is that the carrier is suppressed in the balanced modulator output and only the sideband frequencies are present in channel 2 output to add to channel 1 output. Thus the carrier is not present to produce the unmodulated phase offset shown by the solid line labeled V_0 in Fig. 2*b*. The improvement gained by suppressing the AM carrier can be visualized if the modulating signal is a single frequency so there are only two side frequencies. Each of these side frequencies can be viewed as a rotating phasor which adds vectorially to the carrier phasor. The upper side frequency is higher than the carrier. Therefore when the carrier phasor is used as a reference, the upper side frequency appears to rotate counterclockwise. Similarly, the lower side-frequency phasor appears to rotate clockwise because its frequency is lower than the carrier frequency. The vector sum of these three phasors produces AM as shown in **Fig. 3*a***. However, PM is produced when the side frequencies are shifted 90° in phase with respect to the carrier (or vice versa) as shown in Fig. 3*b*. Then the phase is seen to deviate symmetrically on either side of the phase-shifted carrier and the peak-to-peak deviation can be at least twice as great for a given distortion as when the phase deviates only in one direction. The peak-phase deviation may be about 30° if 8% harmonic distortion is tolerable. A simple balanced modulator circuit is shown in Fig. 3*c*. *See* AMPLITUDE MODULATION.

When the balanced modulator is used in the block diagram of Fig. 2*a* and the frequency multiplier and a power amplifier follow the limiter, a complete FM transmitter results and is known as the Armstrong system of phase (or frequency) modulation.

Phase-shift keying (PSK). A comparatively simple circuit may be used to phase-modulate a pulse-coded signal. In this type of signal, only two different voltage levels are used to represent ones and zeros in binary coded signals. Therefore, a simple phase inverter and switching circuit may be used to obtain 180° phase-shift modulation as the signal changes from a zero to a one, or vice versa. Such a circuit is shown in **Fig. 4**, where the carrier is applied through a transformer to the single-pole double-throw transistor switch. When the modulating signal is at "zero level," transistor T_1 is forward-biased to saturation so the lower end of the input transformer is connected through the low-saturation resistance of T_1 through the dc blocking capacitor C_5 to the output terminal. As the modulating signal changes to "one" level, transistor T_1 is cut off and transistor T_2 becomes saturated, thus connecting the top end of the input transformer through T_2 to the output. Since the signal at the upper end of the transformer is 180° out of phase with the signal at the lower end, the center tap being the reference, the output phase shifts 180° at each transition between zero and one, or vice versa. Capacitors C_1 and C_2, like C_3, block dc so proper bias levels may be maintained on the

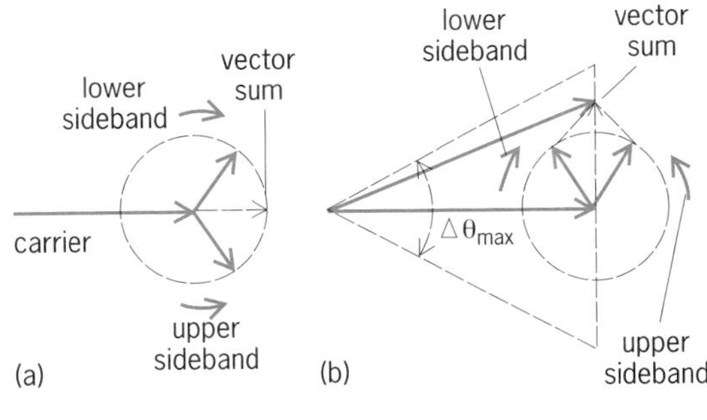

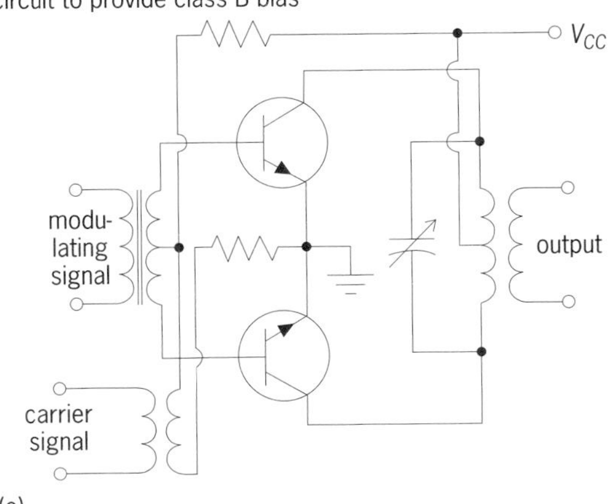

Fig. 3. Illustration of the principle of phase modulation using a balanced modulator. (*a*) Amplitude modulation. (*b*) Phase modulation. (*c*) A balanced modulator.

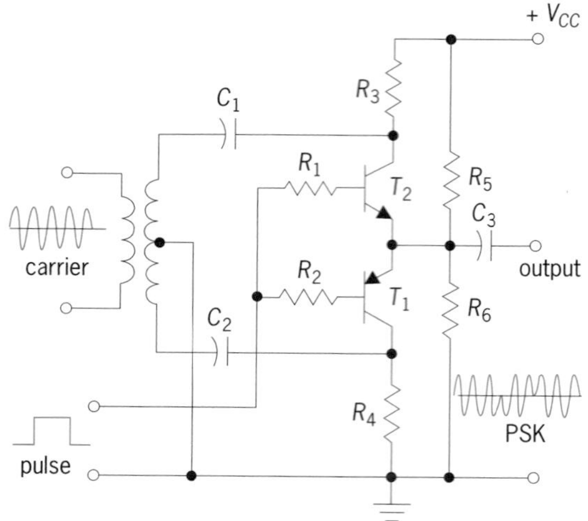

Fig. 4. Phase modulators for phase-shift keying.

transistors. Thus their reactances need to be negligible at the carrier frequency. Resistors R_1, R_2, R_3, and R_4 control the transistor bias currents, and resistors R_5 and R_6 maintain the average voltage, or dc component, at the emitters of the two transistors at about $V_{CC}/2$ volts. The carrier frequency is limited by the switching characteristics of the transistors, but may be above 100 MHz with careful transistor selection. At high frequencies, the input transformer should be tuned or bifilar-wound. *See* MODULATION; PULSE MODULATION.

The examples given above are typical of the many different phase-modulating schemes that may be used. Charles L. Alley

Bibliography. S. Haykin, *Communication Systems*, 4th ed., 2000; M. S. Rodin, *Analog and Digital Communication Systems*, 4th ed., 2000; M. Schwartz, *Information Transmission, Modulation and Noise*, 4th ed., 1990.

Phase rule

A relationship used to determine the number of state variables F, usually chosen from among temperature, pressure, and species compositions in each phase, which must be specified to fix the thermodynamic state of a system in equilibrium. It was derived by J. Willard Gibbs between 1875 and 1878. The phase rule (in the absence of electric, magnetic, and gravitational phenomena) is given by Eq. (1), where C is the

$$F = C - P - M + 2 \quad (1)$$

number of chemical species present at equilibrium, P is the number of phases, and M is the number of independent chemical reactions. Here phase is used to indicate a homogeneous, mechanically separable portion of the system, and the term independent reactions refers to the smallest number of chemical reactions which, upon forming various linear combinations, includes all reactions which occur among the species present. The number of independent state variables F is referred to as the degrees of freedom or variance of the system.

Examples. The system constituted by liquid water and water vapor contains two phases ($P = 2$) and one component ($C = 1$), and there are no chemical reactions ($M = 0$). Therefore this system has $F = 1 - 2 - 0 + 2 = 1$ degree of freedom. This is in accord with the observation that the vapor and liquid forms of water exist in equilibrium only for values of temperature and pressure along the coexistence curve, so that specifying either of these variables fixes the other.

At high temperatures the three reactions shown in (2) occur between sulfur and oxygen in the gas

$$S + O_2 \longrightarrow SO_3 \quad (2a)$$

$$SO_2 + {}^1\!/_2O_2 \longrightarrow SO_3 \quad (2b)$$

$$S + {}^3\!/_2O_2 \longrightarrow SO_3 \quad (2c)$$

phase. There are only two independent chemical reactions in this single-phase, four-component system, since the last reaction is the sum of the first two. Therefore this system has $F = 4 - 1 - 2 + 2 = 3$ degrees of freedom.

Derivation. The derivation of the phase rule starts with the experimental observation that a single phase of C nonreacting components has $C + 1$ degrees of freedom (for example, once temperature, pressure, and $C - 1$ mole fractions are specified, no other intensive variables of the system can vary). Now consider chemical and phase equilibrium in a more general system in which there are C components and P phases, and in which M independent chemical reactions occur. It might appear, based on the experimental observation above, that such a system should have $P(C + 1)$ degrees of freedom. However, since the system is in equilibrium, the degrees of freedom are reduced as follows:

1. At equilibrium the temperature of each phase must be the same. Thus it is not possible to set the temperature of each of the P phases separately; once the temperature of one phase is specified, the temperature of all phases is fixed. Consequently the requirement that the temperature must be the same in all phases at equilibrium eliminates $P - 1$ degrees of freedom.

2. At equilibrium the pressure must be the same in each phase. This eliminates another $P - 1$ degrees of freedom.

3. At equilibrium the chemical potential (partial molar Gibbs free energy) of each species must be the same in each phase. This restriction, which holds for each of the C components, eliminates an additional $C(P - 1)$ degrees of freedom.

4. For each chemical reaction (3) to be in equi-

$$aA + bB + \cdots \longrightarrow rR + sS + \cdots \quad (3)$$

librium, Eq. (4) must be satisfied, where μ_i is the

$$a\mu_A + b\mu_B + \cdots \longrightarrow r\mu_R + s\mu_S + \cdots \quad (4)$$

chemical potential of species i. (If this equation is

satisfied in one phase, it is, by the equality of the chemical potentials of a given species in its various phases, satisfied in all phases.) This requirement eliminates another M degrees of freedom.

Therefore, the actual number of degrees of freedom in a multicomponent, multiphase, chemically reacting system is given by Eq. (5), which is the

$$F = P(C+1) - 2(P-1) - C(P-1) - M$$

$$= C - P - M + 2 \qquad (5)$$

same as Eq. (1). *See* CHEMICAL EQUILIBRIUM; CHEMICAL THERMODYNAMICS; PHASE EQUILIBRIUM; THERMODYNAMIC PROCESSES. Stanley I. Sandler

Phase-transfer catalysis

A process in which the rate of an organic reaction in a heterogeneous two-phase system is enhanced by the addition of a compound that transfers one of the reactants across the interface between the two phases.

An important factor, which contributes to the slowness of many organic reactions, is the lack of homogeneity of the reaction mixture. This is particularly the case with nucleophilic substitution reactions (1), where RX is an organic reagent and Nu^- is

$$RX + Nu^- \rightarrow RNu + X^- \qquad (1)$$

the nucleophilic reagant. The nucleophilic reagent is frequently an inorganic anion, which is soluble in water in which the organic substrate is insoluble, but is insoluble in the organic phase. The encounter rate between Nu^- and RX is consequently low as they can only meet at the interface of the heterogeneous system. The water-soluble anion is also frequently highly solvated by water molecules, which stabilize the anion and thus reduce its nucleophilic reactivity. These problems have been overcome in the past by the use of polar aprotic solvents, which will dissolve both the organic and inorganic reagents, or by the use of homogeneous mixed-solvent systems, such as water:ethanol or water:dioxan. Homogeneous reaction systems can also be established by the use of surfactants, and the interfacial area can be increased by rapid agitation or mechanical emulsification. Although these procedures may increase the rate of reaction, they have disadvantages, such as difficulties in the isolation and purification of the products and, on an industrial scale, they can be costly in terms and solvents and energy.

Phase-transfer catalysis involves the transportation of the inorganic anion, Nu^-, from the aqueous phase into the organic phase by the formation of a nonsolvated ion pair with a cationic phase-transfer catalyst, Q^+. It was originally believed that this process involved the formation of the ion pairs in the aqueous phase, followed by their partition between the aqueous and organic phases. A more current explanation is that with highly lipophilic catalyst, the reactive ion pair $[Q^+Nu^-]$ is formed at the interface between the aqueous and organic phases, followed by rapid transportation into the bulk of the organic phase (**Fig. 1**). The rate of the reaction is enhanced, because the encounter rate of the nucleophile, Nu^-, with the organic reagant, RX, in the single phase will be significantly higher than at the interface. Moreover, as the anion is transferred virtually without any water of solvation, its nucleophilic reactivity can be considerably higher in the organic phase than in the aqueous phase. Rate enhancements of greater than 10^7 have thus been observed.

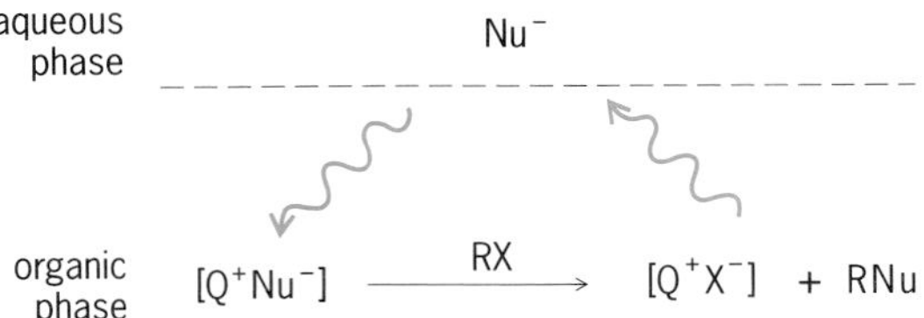

Fig. 1. Formation and transportation of the reactive ion pair in phase-transfer catalysis.

It has been noted, however, that several reactions in which the nucleophile is generated by abstraction of a proton from an organic substrate under basic two-phase conditions are ineffective when the reaction is conducted using the aqueous-organic two-phase conditions, described above, and are better accomplished using solid powdered potassium hydroxide [KOH; or what is even better, an intimate mixture of powdered potassium hydroxide and potassium carbonate (K_2CO_3)] instead of aqueous sodium hydroxide. It has been proposed that under these conditions the pH of the medium can be greater than 30. It has been further observed that the efficiency of many phase-transfer catalyzed liquid:liquid two-phase reactions can be improved by the omission of the aqueous solvent and the use of finely divided solid inorganic salts in solid:liquid two-phase conditions.

Although the process does not yet appear to have widespread application, phase-transfer catalyzed gas:liquid and gas:solid two-phase reaction systems have also been reported.

The efficiency of the catalyst in both liquid:liquid and solid:liquid two-phase systems depends upon the ability of the cation, Q^+, to transfer the anion across the interface of the heterogeneous system; that is, in the case of the aqueous:organic two-phase systems, the rate should be proportional to the partition coefficient of the ion pair $[Q^+Nu^-]$ and the water solubility of $[Q^+X^-]$ should be higher than that of $[Q^+Nu^-]$. It has been found that bulky quaternary ammonium cations, R_4N^+, such as tetra-*n*-butylammonium and methyltrioctylammonium ions, and phosphonium cations, R_4P^+, such as tetra-*n*-butylphosphonium cations, have a high propensity to form "strong" ion pairs, which have a high solubility in organic solvents, and are excellent phase-transfer catalysts.

As the partition coefficient of the ion pair is a significant factor in the catalytic effect, the choice of the organic phase is important. The most commonly used organic phases are the semipolar solvents, dichloromethane and 1,2-dichlorobenzene, and the less polar solvent, toluene.

An alternative phase-transfer catalytic procedure utilizes polyethers, which are capable of complexing alkali metal cations. Ion pairs of the complexed cation and the nucleophilic anion are formed and are transported across the water:organic phase interface in the same manner as the quaternary ammonium and phosphonium salts. Polyether catalysts as shown in **Fig. 2***a* may be acyclic [glymes; (**1**)], or monocyclic [crown ethers; (**2**)], or polycyclic (cryptands; (**3**)]. The most commonly used polyether catalysts are the crowns, such as dicyclohexanono-18-crown-6, which complex readily with potassium and sodium ions to form complexes of type (**4**) [Fig. 2*b*].

Advances in phase-transfer catalyst design has led to increased use of tridents [tris(polyalkoxyalkyl)-amines], of which TDA-1 [tris(3,6-dioxaheptyl)-amine; (**8**) Fig. 2*d*] is an example. These compounds are acyclic forms of the cryptands and have the capability to form complexes with a wide range of cationic species. The trident catalyst therefore has a potential use in oxidation:reduction reactions through the transport of transition-metal cations, as well as in the standard nucleophilic substitution reactions. This type of compound has particularly good application in the catalysis of solid:liquid two-phase reactions.

There are also descriptions of the use of bis-quaternary ammonium salts as phase-transfer catalysts. Compounds such as BBDE [bis(2-(benzyldiethylammonio)ethyl)ether dichloride; (**9**); Fig. 2*e*] not only have the capability of transporting two equivalents of a mononucleophilic species across the interface of two phases, but also efficiently transport dinucleophilic species, for example, the phthalate dianion. *See* COORDINATION COMPLEXES; MACROCYCLIC COMPOUND.

Applications. Phase-transfer catalysts can be used to enhance reaction rates in nucleophilic substitution reactions, oxidation-reduction reactions, generation of reactive intermediates, asymmetric induction, and carbonylation reactions.

Nucleophilic substitution reactions. By using phase-transfer catalysis, it is possible to enhance the rate of most nucleophilic substitution reactions, as in reaction (1), with a wide range of nucleophilic anions.

Kinetic studies have shown that the nucleophilicity of the nonhydrated halide anions, produced upon dissolution of the quaternary-onium halides in organic solvents, follow the order fluorine > chlorine > bromine > iodine. Under these conditions the fluoride ion is a strong base and can be used in preference to the hydroxide ion in base-catalyzed reactions. *See* STERIC EFFECT (CHEMISTRY).

Oxidation and reduction reactions. Purple benzene (or purple dichloromethane) and yellow benzene (or yellow dichloromethane) are readily obtained by the dissolution of permanganate or dichromate ions, respectively, in the otherwise colorless benzene (or dichloromethane) by using phase-transfer catalysts, and the solutions have been used for a wide range of oxidation reactions in high yields with minimal side

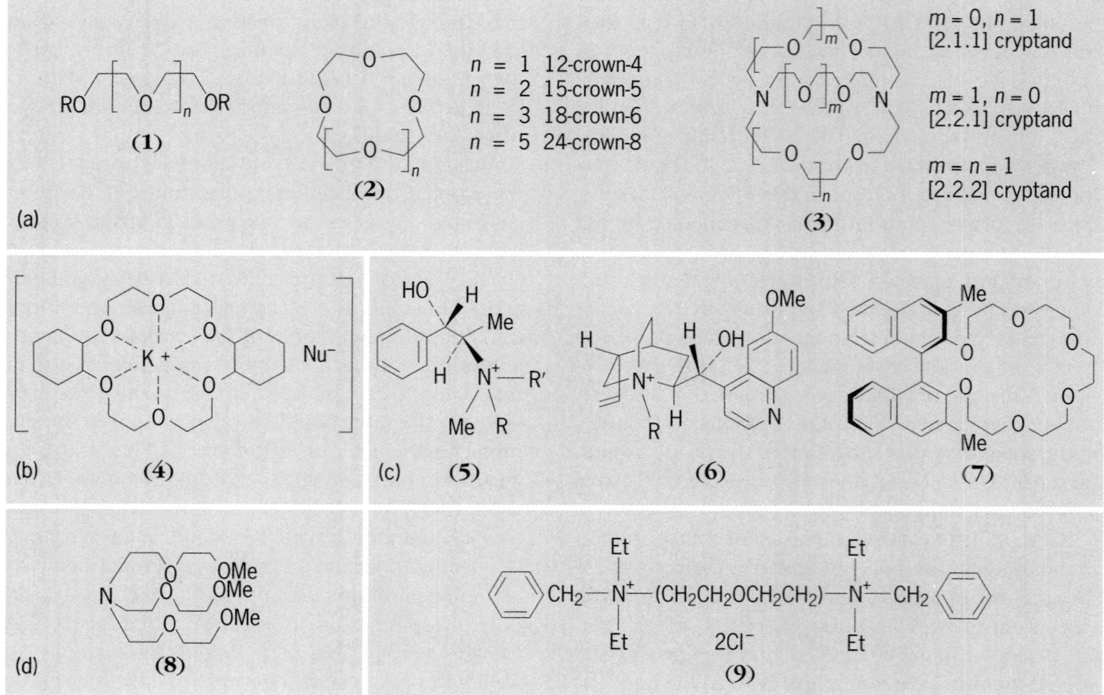

Fig. 2. Structures of some involved in phase-transfer catalysis. (*a*) Typical polyether catalysts used in phase-transfer catalyst procedure. (*b*) Typical complex formed with a polyether catalyst. (*c*) Chiral catalysts used in asymmetric induction. (*d*) Typical tris(polyoxaalkyl)amine (trident) catalyst. (*e*) Example of bis-quaternary ammonium salt catalyst. Me = CH_3 Et = C_2H_5.

reactions. Solid tetraalkylammonium permanganates and dichromates are potentially explosive.

Hydride reductions have been conducted in organic solvents by using tetraalkylammonium borohydrides.

Diborane, B_2H_6, produced from the reaction of tetraalkylammonium borohydride in dichloromethane with simple alkyl halides [reaction (2)] has an

$$2[Q^+BH_4^-] + 2CH_3I \longrightarrow 2CH_4 + 2[Q^+I^-] + B_2H_6 \qquad (2)$$

activity similar to that of diborane in ether and can be used for all the usual reduction and hydroboration reactions. *See* HYDROBORATION; OXIDATION-REDUCTION.

Generation of reactive intermediates. Reactive carbanions, which normally require strictly anhydrous reaction conditions for their preparation, can be generated by using the phase-transfer catalytic technique [reaction (3)], where Z = acyl, alkoxycarbonyl, cyano, ano, or nitro, and R = H, alkyl, or aryl.

$$\underset{(org)}{RCHZ_2} + \underset{(aq)}{HO^-} \xrightarrow{[R'_4N^+X^-]} \underset{(org)}{[RCZ_2^-R'_4N^+]} \xrightarrow{R''X} \underset{(org)}{RR''CZ_2} \qquad (3)$$

Similarly, other systems, which are labile in the presence of water such as RCOCN, can be synthesized by the phase-transfer catalytic procedure.

Tetraalkylammonium hydroxides react with chloroform to yield, initially, the trichloromethyl carbanion, which loses a chloride ion to generate the highly reactive dichlorocarbene [reaction (4)]. It has

$$\underset{(org)}{CHCl_3} \xrightarrow{HO^-(aq)} \underset{(interface)}{CCl_3^- + H_2O} \xrightarrow{[Q^+Cl^-]} \underset{(org)}{[Q^+CCl_3^-]} \longrightarrow \underset{(org)}{[Q^+Cl^-]} + \underset{(org)}{\ddot{C}Cl_2} \qquad (4)$$

been proposed that the abstraction of the proton from the chloroform occurs at the interface and that the role of the catalyst is the transfer of the trichloromethyl carbanion into the bulk of the organic solvent. The dichlorocarbene reacts rapidly and in high product yield with a wide range of organic substrates.

Other dihalogenocarbenes, and vinylidene and vinyl carbenes have been obtained by phase-transfer catalyzed procedures. *See* REACTIVE INTERMEDIATES.

Carbonylation reactions. Reactions of organic substrates with metal carbonyl compounds, which frequently require high pressures of carbon monoxide (CO), are feasible under relatively mild conditions when conducted under phase-transfer catalytic conditions. For example, the reaction of iodoalkanes with dicobalt octacarbonyl to give acylcobalt tetracarbonyls [reaction (5)] is catalyzed by quaternary

$$\underset{\text{Iodoalkane}}{RCH_2I} + \underset{\text{Dicobalt octacarbonyl}}{Co_2(CO)_8} \xrightarrow[Q^+Cl^-]{\text{aqueous NaOH}} \underset{\text{Acylcobalt tetracarbonyl}}{RCH_2COCo(CO)_4} \qquad (5)$$

ammonium salts. Subsequent reaction of the acylcobalt derivative with alkenes or alkyl halides leads to formation of ketones, while further reaction with a base leads to carboxylic acids. Reduction and oxidation reactions, which are initiated by metal carbonyl derivatives, are also catalyzed by the addition of quaternary ammonium salts. *See* METAL CARBONYL.

Asymmetric induction. Chiral ammonium catalysts having both a chiral quaternary nitrogen atom and chirality within the carbon skeleton, such as structures (5) and (6) [Fig. 2*c*], have been used successfully in the stereochemical control of reactions involving carbonyl-containing compounds. A critical feature of the chiral catalyst is the presence of a β-hydroxyethyl substituent on the chiral nitrogen atom, which can hydrogen-bond with the carbonyl group of the organic substrate, thereby preferentially presenting one "face" of the substrate for reaction. Reactions in which an enantiometric excess is >60% have been observed.

Stereochemical control of organic reactions has also been accomplished by using chiral crown ethers, such as structure (7). *See* ASYMMETRIC SYNTHESIS.

Solid-supported phase-transfer catalysis. As an elaboration of liquid:liquid catalysis, the quaternary ammonium group or crown ether is attached to an organic polymeric support, such as polystyrene. Quaternary ammonium salts have also been immobilized on zeolites. Although the efficiency of such catalysts differs little from the conventional soluble phase-transfer catalysts, there are advantages in the use of triphase (that is liquid:liquid:catalyst) systems. The catalyst is readily removed by filtration from the reaction system after use and, on an industrial scale, there is a potential for continuous phase-transfer catalyzed processes. *See* CATALYSIS; HETEROGENEOUS CATALYSIS; HOMOGENEOUS CATALYSIS; QUATERNARY AMMONIUM SALTS; STEREOCHEMISTRY; ZEOLITE.

R. Alan Jones

Bibliography. H. des Abbayes, Hydroxide ion initiated reactions under phase-transfer catalysis conditions, *Israel J. Chem.*, 26:249–262, 1985; E. V. Dehmlow and S. S. Dehmlow, *Phase-Transfer Catalysis*, 3d ed., 1993; W. E. Keller (ed.), *Compendium of Phase-Transfer Reactions and Related Synthetic Methods*, 2 vols., 1979, 1987; C. Liotta, C. M. Starks, and M. Halpern, *Phase-Transfer Catalysis: Fundamentals, Applications, and Industrial Perspectives*, 1994; M. Rabinovitz, Y. Cohen, and M. Halpern, Metal carbonyls in phase-transfer catalysis, *Angew. Chem. Int. Ed. Engl.*, 25:960–970, 1986; G. Soula, Tris(polyoxaalkyl) amines (Trident), a new class of solid-liquid phase-transfer catalysts, *J. Org. Chem.*, 50:3717–3721, 1985; C. M. Starks (ed.), *Phase-Transfer Catalysis: New Chemistry, Catalysts, and Applications*, American Chemical Society Symposium Series, no. 326, 1987; W. P. Weber and G. W. Gokel, *Phase-Transfer Catalysis in Organic Synthesis*, 1977.

Phase transitions

Changes of state brought about by a change in an intensive variable (for example, temperature or pressure) of a system. Some familiar examples of phase transitions are the gas-liquid transition (condensation), the liquid-solid transition (freezing), the normal-to-superconducting transition in electrical conductors, the paramagnet-to-ferromagnet transition in magnetic materials, and the superfluid transition in liquid helium.

Characteristics

Typically the phase transition is brought about by a change in the temperature of the system. The temperature at which the change of phase occurs is called the transition temperature (usually denoted T_t). As examples: the solid-liquid transition occurs at the melting point (denoted T_m); the liquid-gas transition occurs at the boiling point (denoted T_b). The condition of temperature and pressure, unique for each substance, at which the distinction between liquid and gaseous fluids disappears is called the critical point (denoted T_c). T_c is also used to denote the critical temperature below which ferromagnetism begins to grow in some materials.

Order and entropy. The two phases above and below the phase transition can be distinguished from each other in terms of some ordering that takes place in the phase below the transition temperature. For example, in the liquid-solid transition the molecules of the liquid get "ordered" in space when they form the solid phase. In a paramagnet the magnet moments on the individual atoms can point in any direction (in the absence of an external magnetic field), but in the ferromagnetic phase the moments are lined up along a particular direction, which is then the direction of ordering. Thus in the phase above the transition the degree of ordering is smaller than in the phase below the transition. One measure of the amount of disorder in a system is its entropy, which is the negative of the first derivative of the thermodynamic free energy with respect to temperature. When a system possesses more order, the entropy is lower. Thus at the transition temperature the entropy of the system changes from a higher value above the transition to some lower value below the transition. *See* ENTROPY; FERROMAGNETISM; PARAMAGNETISM.

Continuous and discontinuous transitions. This change in entropy can be continuous or discontinuous at the transition temperature. In other words, the development of order in the system at the transition temperature can be gradual or abrupt. This leads to a convenient classification of phase transitions into two types, namely, discontinuous and continuous.

Discontinuous transitions involve a discontinuous change in the entropy at the transition temperature. A familiar example of this type of transition is the freezing of water into ice. As water reaches the freezing point, order develops without any change in temperature. Thus there is a discontinuous decrease in the entropy at the freezing point. This is characterized by the amount of latent heat that must be extracted from the water for it to be "ordered" into the solid phase (ice). Discontinuous transitions are also called first-order transitions.

In a continuous transition, entropy changes continuously, and hence the growth of order below T_c is also continuous. There is no latent heat involved in a continuous transition. Continuous transitions are also called second-order transitions. The paramagnet-to-ferromagnet transition in magnetic materials is an example of such a transition.

Order parameter. The degree of ordering in a system undergoing a phase transition can be made quantitative in terms of an order parameter. At temperatures above the transition temperature the order parameter has a zero value, and below the transition it acquires some nonzero value. For example, in a ferromagnet the order parameter is the magnetic moment per unit volume (in the absence of an externally applied magnetic field). It is zero in the paramagnetic state since the individual magnetic moments in the solid may point in any random direction. Below the critical transition temperature T_c, there exists a preferred direction of magnetic ordering. As the temperature is decreased below T_c, more and more individual magnetic moments start to align along the preferred direction of ordering, leading to a continuous growth of the magnetization or the macroscopic magnetic moment per unit volume in the ferromagnetic state. Thus the order parameter changes continuously from zero above to some nonzero value below the transition temperature. In a first-order transition the order parameter would change discontinuously at the transition temperature.

The behavior of systems very near the onset of phase transitions is an important field of study in itself. For a detailed description of this behavior *see* CRITICAL PHENOMENA

Examples

Several modern examples of phase transitions or quasi-phase transitions will be discussed. By quasi-phase transition is meant an apparent, diffuse transition which may have an underlying thermodynamic basis but for which kinetic factors also play a role in determining the apparent transition temperature. A number of well-known transitions are mentioned only briefly or excluded from the discussion here; for these transitions *see* FERROMAGNETISM; SUPERCONDUCTIVITY.

Glass transitions. When certain substances are taken very rapidly from the liquid state to the solid state, or condensed from the vapor phase to the

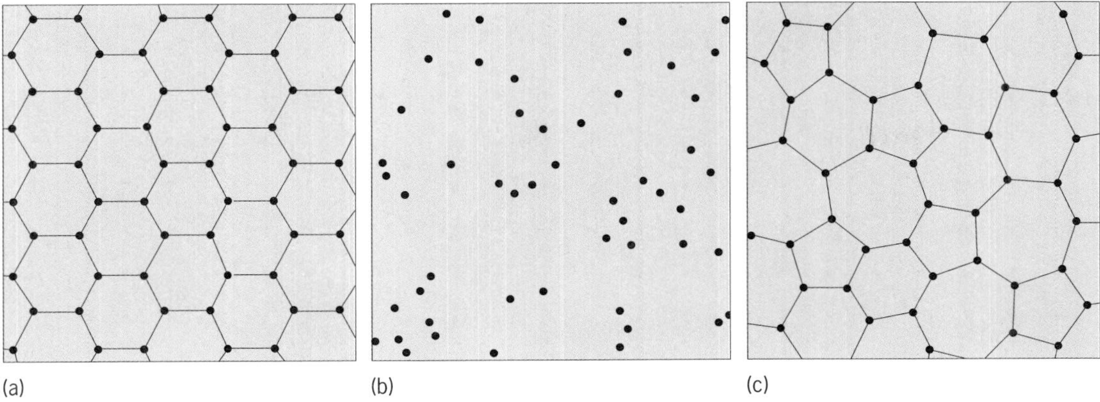

Fig. 1. Two-dimensional representation of atomic arrangements in three phases: (*a*) Crystal. (*b*) Gas. (*c*) Glass. Each dot represents position of an atom at a given instant of time. (*After E. W. Montroll and J. L. Leibowitz, Fluctuation Phenomena, North-Holland Publishing, 1979*)

solid phase, they are found to have an amorphous or glassy structure whose atomic arrangement differs from those of both crystals and gases. Similar features are observed in substances composed of chainlike molecules in random arrangement. In the crystal, the atoms are placed on a regular, periodic lattice, and only discrete interatomic spacings between the atoms are possible (**Fig. 1***a*); this describes a state of long-range order. In the gas, the atoms are constantly colliding with each other and the walls of the container (Fig. 1*b*)—there is no long-range order. The glass phase has a structure intermediate between that of the crystal and gas (Fig. 1*c*). The local environment or short-range order of each atom in the glass shown is similar to that of the crystal, but there is no long-range order. Some glasses have atomic arrangements, at any instant of time, very similar to those of a liquid of the same composition. The structure of certain amorphous metallic alloys is similar to that which would be obtained by packing hard spheres into a container to obtain the largest density.

The glass transition temperature T_g is defined to be the temperature below which the atoms are frozen into relatively stable positions, even though the glassy phase is a metastable one whose energy is higher than the crystalline phase. When a glass is heated above T_g, the material can transform into the crystalline state by the formation of nucleation centers in the solid state. In metallic glasses, this crystallization process can take place with only a little heating above T_g. In other glasses, such as fused silica, crystallization does not take place above T_g, no matter how long one waits; the next transition in such a material as the temperature is increased is to the liquid state, that is, melting. The amorphous solid-liquid transition is an example of a quasi-phase transition as defined above. The glass transition appears to possess an underlying equilibrium thermodynamic character but is modified by kinetic effects. Some examples of insulating or semiconducting glasses are silicon dioxide (SiO_2), arsenic selenide (As_2Se_3), silicon, (Si), and germanium (Ge). Examples of metallic glasses are $Pd_{80}Si_{20}$, $Fe_{80}B_{20}$, $Gd_{65}Co_{35}$, $Zr_{40}Cu_{60}$, and $Mg_{70}Zn_{30}$. Both classes of materials have been intensively studied with a view toward significant technological applications. *See* AMORPHOUS SOLID; GLASS; METALLIC GLASSES.

Ferromagnetism and antiferromagnetism. As a preliminary to discussing several more complicated magnetic phase transitions, the fundamentals of simple ferromagnetic and antiferromagnetic transitions will be outlined. The fundamental interaction energy E_{ij} between two spins S_i and S_j can be represented by Eq. (1), where $\mathcal{J}_{ij}$ is called the exchange interaction

$$E_{ij} = -\mathcal{J}_{ij}\vec{S}_i \cdot \vec{S}_j \qquad (1)$$

because it originates in the quantum-mechanical requirement that the wave function of a system of electrons must be antisymmetric (change sign) under exchange of the coordinates of any two electrons. Although Eq. (1) is a quantum-mechanical effect, it will be discussed from a classical viewpoint since much insight can be gained in this way. Classically, Eq. (1) can be written as Eq. (2), where θ_{ij} is the angle

$$E_{ij} = -\mathcal{J}_{ij}S_iS_j\cos\theta_{ij} \qquad (2)$$

between the two spins. When $\mathcal{J}_{ij}$ is positive, E_{ij} is minimized for $\vec{S}_i$ and $\vec{S}_j$ parallel; when $\mathcal{J}_{ij}$ is negative, E_{ij} is minimized for $\vec{S}_i$ and $\vec{S}_j$ antiparallel. These two cases are shown schematically in **Fig. 2***a* and *b* as ferromagnetic and antiferromagnetic order, respectively. The magnetic moment vector associated with each spin is proportional to the magnitude of the spin and points along the same line. In the absence of thermal fluctuations, that is, at absolute temperature $T = 0$, the spins of a ferromagnet are all parallel in a single domain so that a large magnetization results (Fig. 2*a*). For an antiferromagnet, again at $T = 0$, there are two sublattices whose spins are oppositely directed so that there is no net magnetization (Fig. 2*b*). As the temperature is raised from near zero, the spins in either case begin to deviate from their original positions until, at the magnetic ordering temperature, a phase transition to the paramagnetic phase takes place (Fig. 2*c*). In the

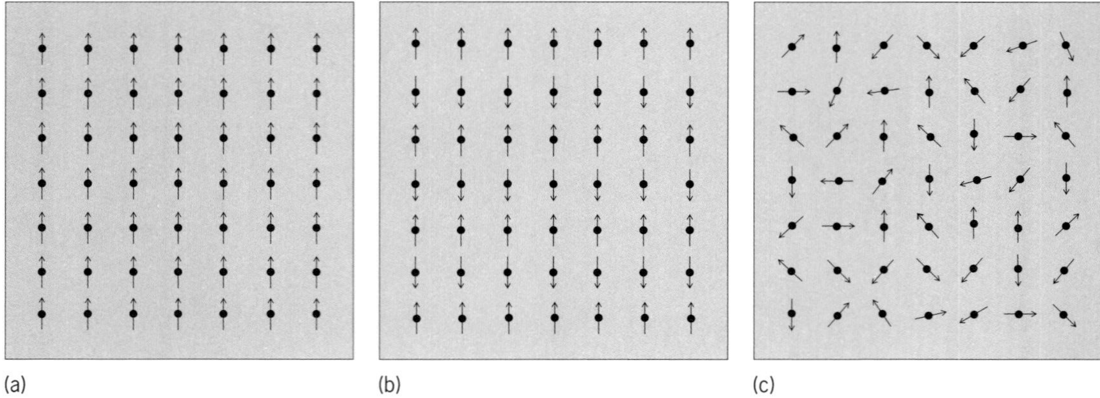

Fig. 2. Ordering of spins in a crystal lattice. Arrows show the direction in which the spin (or the magnetic moment) at a particular lattice site is pointing. (*a*) Ferromagnet. (*b*) Antiferromagnet. (*c*) Paramagnet.

paramagnetic phase, there is a complete lack of long-range order in the spin directions. *See* ANTIFERROMAGNETISM; EXCHANGE INTERACTION.

Magnetism in crystalline alloys. A disordered crystalline alloy is a mixture of two elements in which the atoms of the mixture are found at more or less random positions on a crystal lattice. If the concentration of one element is x (and the other $1 - x$), then some typical examples of disordered binary alloys are $Cu_{1-x}Mn_x$, $Au_{1-x}Fe_x$, and $Ag_{1-x}Pd_x$. If one of the constituents of the alloy has a magnetic moment and therefore spin associated with it, there will be disorder in the atomic positions of the spins, yet it is possible for certain types of magnetic order to exist. In the three examples mentioned above, the manganese (Mn) and iron (Fe) atoms do have localized magnetic moments associated with them.

Spin-glass state. Consider a relatively dilute alloy such as $Cu_{0.95}Mn_{0.05}$. In such a case the conduction electrons will scatter from the manganese impurity atoms and create a disturbance in the net spin density in the vicinity of each manganese atom. This disturbance (net spin up minus spin down) is shown in

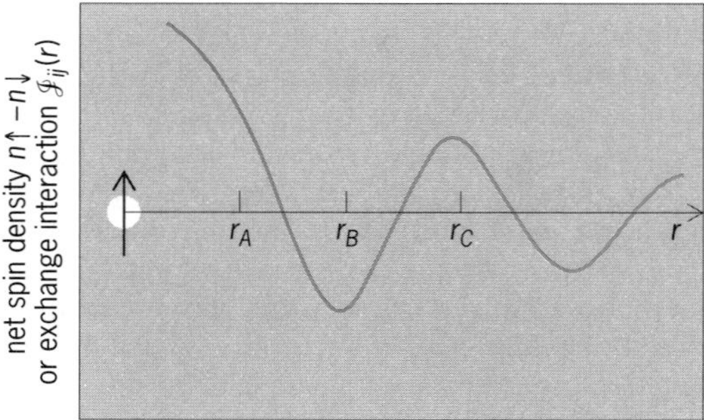

Fig. 3. Net spin density of conduction electrons (number of electrons per unit volume with spin up $n\bar{u}$ minus number with spin down $n\bar{u}$) as function of distance r from magnetic moment. Graph also shows exchange interaction $\mathcal{J}_{ij}(r)$ between magnetic moments i and j as function of distance r between them. Magnetic moment at the origin will couple ferromagnetically to magnetic moments at positions r_A and r_C, but antiferromagnetically to a magnetic moment at r_B.

Fig. 3. If another manganese atom is found in the vicinity of the atom at the origin, it will have a spin that is coupled to the latter according to Eqs. (1) and (2). However, in this case $\mathcal{J}_{ij}$ is a function of the distance r between the two spins and has the behavior shown by Fig. 3. Thus $\mathcal{J}_{ij}(r)$ oscillates in sign and decreases in magnitude as r increases. A magnetic moment at the origin of Fig. 3 will couple ferromagnetically to magnetic moments at positions r_A and r_C, but antiferromagnetically to a magnetic moment at r_B. Since there will be a considerable number of different Mn-Mn interatomic spacings, there will be a mixture of ferromagnetic and antiferromagnetic exchange interactions in such an alloy. This mixture leads to a new magnetic phase called a spin glass. In analogy with the ordinary glass shown in Fig. 1*c*, a spin glass has short-range magnetic order but no long-range magnetic order, even below the spin-glass ordering temperature T_{SG} (**Fig. 4*a***). Above T_{SG}, thermal fluctuations will destroy the ordered spin-glass state, and the magnetic system will become paramagnetic, with no short- or long-range magnetic order (Fig. 4*b*). *See* SPIN GLASS.

Ferromagnetic state. In certain disordered alloys such as $Au_{1-x}Fe_x$, for x values larger than about 0.16, the spin-glass phase is replaced by the ferromagnetic state (Fig. 4*c*). At this iron concentration, called the percolation limit, it is possible for nearest-neighbor ferromagnetic couplings of iron spins to propagate completely throughout the crystal, thus leading to the state in which all iron spins are parallel. This situation depends on $\mathcal{J}_{ij}(r)$ having positive values for nearest neighbors, as depicted schematically in Fig. 3 for $r = r_A$. If the nearest-neighbor magnetic couplings are negative (antiferromagnetic), as is thought to be the case in $Cu_{1-x}Mn_x$, then the ordered state at higher manganese concentrations becomes a random antiferromagnetic state in which, to the extent possible, nearest-neighbor spins are antiparallel.

The question of whether the spin-glass and spin-glass–to–ferromagnet transitions are true or quasi-phase transitions is still a subject of much controversy.

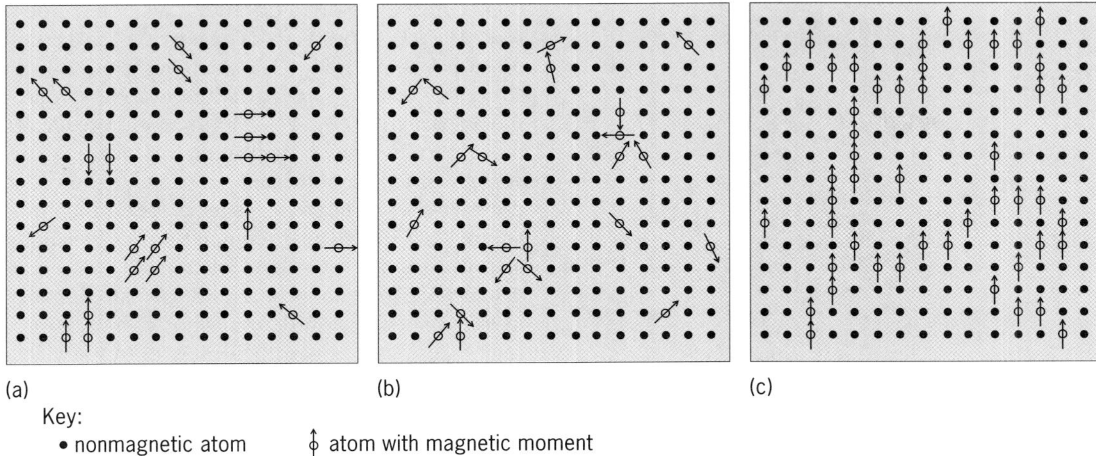

Fig. 4. Arrangement of magnetic moments in crystal lattice of a disordered crystalline alloy. Arrows show the direction in which the magnetic moment at a particular site is pointing. (*a*) Spin-glass state. (*b*) Paramagnetic state. (*c*) Ferromagnetic state.

Speromagnetic and asperomagnetic states. Speromagnetic and asperomagnetic transitions, which occur in substances having amorphous atomic structures, were characterized in the 1970s. In **Fig. 5*a*** a schematic diagram is shown of a glass in which each atom possesses a magnetic moment which is assumed to be coupled to its neighbors by the Heisenberg interaction of Eq. (1). Even though the interatomic spacings are smeared out compared to the case of a crystal, if all the exchange interactions $\mathcal{J}_{ij}$ are positive the state of magnetic order will be as shown in Fig. 5*a*. An example of a ferromagnetic metallic glass is $Gd_{65}Cu_{35}$, in which all the gadolinium (Gd) moments are aligned while the copper (Cu) atoms possess no moments.

A novel state of order can exist in rare-earth glasses such as $Dy_{75}Au_{25}$, $Tb_{75}Ga_{25}$, and $Er_{65}Co_{35}$. In these cases the rare-earth atoms have magnetic moments which are due to both electron spin and orbital angular momentum. The electric fields which exist in such a glass in combination with the spin-orbit interaction make it difficult for the total angular momentum vector $\vec{J}$ and hence the magnetic moment of each rare-earth atom to be aligned in an arbitrary direction. There can be an "easy" axis $\hat{k}_i$ for each $\vec{J}_i$ such that $\vec{J}_i$ tries to align itself parallel or antiparallel to $\hat{k}_i$. A model that describes this situation of the energy of the *i*th ion is given by Eq. (3), where $\hat{k}_i$ is a

$$E_i = -D(\hat{k}_i \cdot \vec{J}_i)^2 \qquad (3)$$

unit vector and *D* the strength of the local anisotropy. The random anistropy model of magnetism in glasses assumes that the amorphous structure leads to completely random anistropy axes $\hat{k}_i$ for each magnetic moment. Then the total energy is given by Eq. (4).

$$E = -\mathcal{J}\sum_{i,j}\vec{J}_i \cdot \vec{J}_j - D\sum_i(\hat{k}_i \cdot \vec{J}_i)^2 \qquad (4)$$

If $\mathcal{J}$ is positive, the first term tends to cause parallel alignment of the $\vec{J}_i$ vectors while the second term causes a scattering of the $\vec{J}_i$ vectors. This leads in the ordered state to either a speromagnetic structure (Fig. 5*b*) or an asymmetric speromagnetic or asperomagnetic structure (Fig. 5*c*). The origin of the term speromagnetism is the Greek word for scattered, *spero*. The speromagnetic state has a more

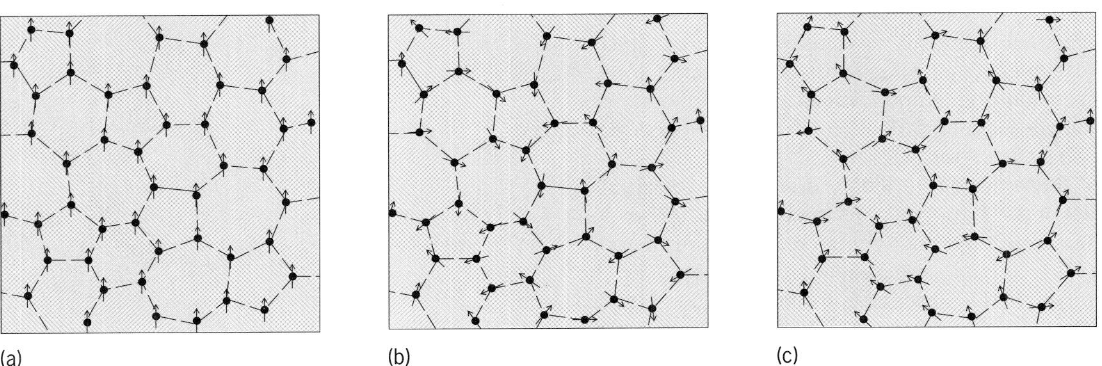

Fig. 5. Arrangement of magnetic moments in (*a*) ferromagnetic glass, (*b*) speromagnetic glass, and (*c*) asperomagnetic glass. Arrows show the direction in which the magnetic moment at a particular site is pointing.

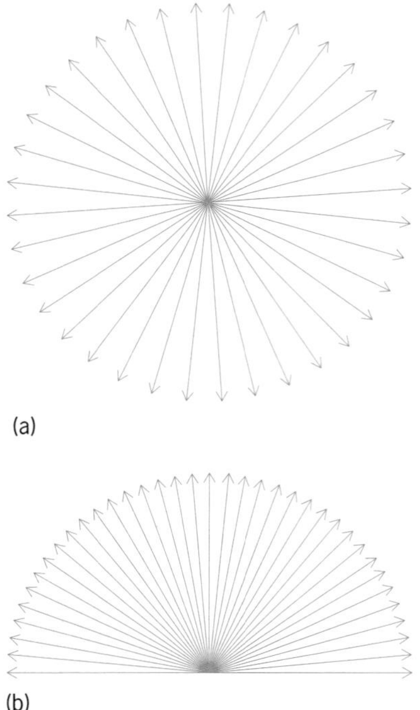

Fig. 6. Distribution of magnetic moments in (*a*) speromagnetic glass and (*b*) asperomagnetic glass.

or less random distribution of moment directions which are frozen in below the magnetic ordering temperature (**Fig. 6*a***). This state of magnetic order has some common features with the spin-glass state, but the origin of the scatter in the frozen moment directions is different. In the asperomagnetic state, the positive exchange interaction is assumed to lead to mostly parallel or nearly parallel near-neighbor moment directions so that the moment distribution would be in the upper hemisphere (Fig. 6*b*). The true ground state of the system, even with all exchange interactions positive, has been the subject of considerable controversy, but research suggests that the ground state is not the asperomagnetic state but rather has some fraction of the moments in the lower hemisphere.

Metallic glasses of the form $Fe_{80}G_{20}$, where G represents a glass-forming element such as boron, phosphorus, or silicon, may have possible uses as soft magnetic materials in transformer cores. Metallic glasses containing rare-earth elements and transition metals may find applications as magnetooptic storage and magnetic bubble computer memory devices. *See* MAGNETOOPTICS.

Order-disorder transition. In an ordered binary alloy AB of the β'-brass type, A occupies the corner position and B the body-centered position of the body-centered cubic structure at absolute zero. The prototype of this structure is β'-brass (β'CuZn; **Fig. 7*a***). The ordered structure is possible if the AB bond is energetically favored over the AA and BB bonds. As the temperature is raised, the thermal vibrations introduce disorder in the structure, the degree of disorder depending on the relative strengths of the various bonds. Above the order-disorder transition temperature, T_{od}, the A and B atoms occupy the lattice sites more or less at random. For example, in the β'-brass structure, above T_{od} the structure is the disordered β-brass structure (Fig. 7*b*). This is an entropy-driven transition where the system minimizes its free energy at a given temperature and pressure. If the order-disorder energy difference is much larger than kT_m, where T_m is the melting temperature and k is Boltzmann's constant, the system will remain ordered until it melts. If this energy difference is small compared to kT_m, the order-disorder transition to a disordered state takes place primarily through diffusion via vacancies. Diffraction techniques such as x-rays can be used to study this transition. *See* DIFFUSION; X-RAY DIFFRACTION.

Structural phase transition. The smallest entity of a perfect crystalline solid that repeats over and over again is known as its primitive unit cell. The structure of a unit cell at a given temperature and pressure is determined by the atomic interactions and the entropy through the minimization of the free energy. A crystal undergoes a commensurate structural phase transition at a certain temperature if its unit cell volume multiplies by an integral number as the temperature is lowered. In a commensurate phase transition, a crystal goes from one periodic structure to another at the transition temperature. The layered compound titanium diseleride ($TiSe_2$) undergoes such a phase transition at $-100°F$ (200 K), with the volume of the unit cell in the low-temperature phase being eight times that in the high-temperature phase. This transition is thought to be driven by charge-density waves (discussed below). A second type of structural phase transition is an incommensurate phase transition, in which a superlattice of periodicity $\{l'\}$ is superimposed on the primary lattice of periodicity $\{l\}$ such that $\{l'\}$ is not contained in $\{l\}$. The superlattice manifests itself as secondary intensity spots around the primary spots in various scattering experiments (x-ray, neutron, electron, and so forth). The incommensurate phase lacks translational periodicity. Paraelectric potassium selenate (K_2SeO_4) undergoes an incommensurate phase transition at 127 K ($-231°F$) that changes to a ferroelectric phase at 97 K ($-285°F$). The ferroelectric phase is commensurate with the paraelectric phase, with the unit cell

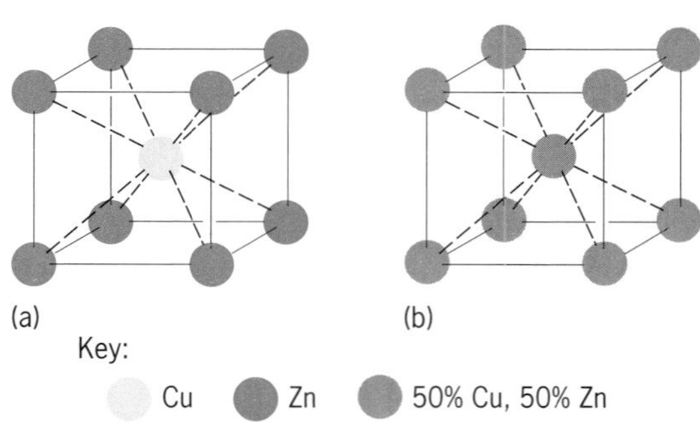

Fig. 7. Structures of (*a*) ordered, or β', brass and (*b*) disordered, or β, brass.

of the former being seven times that of the latter. *See* ELECTRON DIFFRACTION; FERROELECTRICS; NEUTRON DIFFRACTION.

Charge-density waves. These transitions have been observed in low-dimensional conductors, that is, materials that conduct electricity in only one or two dimensions. For example, if a material consists of linear chains of molecules such that the interaction between neighboring molecules on different chains is much smaller than that between neighboring molecules on the same chain, the electrons can move relatively easily along the chain, but motion of electrons from one chain to another is highly unlikely. In this case the material behaves as a quasi one-dimensional conductor. Some examples are $NbSe_3$ and $TaSe_3$ (**Fig. 8***a*). There are also materials whose molecules are arranged in sheets, with the interaction between neighboring molecules on different sheets being much smaller than that between neighboring molecules in the same sheet. In this case the electrons can move relatively easily within the sheet, but motion of electrons between sheets is highly restricted. Such materials behave as quasi two-dimensional conductors. Some examples of two-dimensional conductors are $NbSe_2$ and $TaSe_2$ (Fig. 8*b*).

Below the transition temperature the electronic charge-density is modulated, the periodicity depending upon the topology of the Fermi surface of the electrons. The electronic charge density as a function of position $\vec{r}$ can be represented by Eq. (5), where

$$P(\vec{r}) = P_0(\vec{r})[1 + \alpha \cos(\vec{q}_0 \cdot \vec{r} + \phi)] \quad (5)$$

$P_0(\vec{r})$ is the electronic charge density above the transition (that is, in the normal state), α is the amplitude of the charge-density wave (which is zero above the transition and grows as the temperature is lowered below the transition temperature), $\vec{q}_0$ is the wave vector of the charge-density wave which determines the periodicity, and ϕ is the phase. This modulation

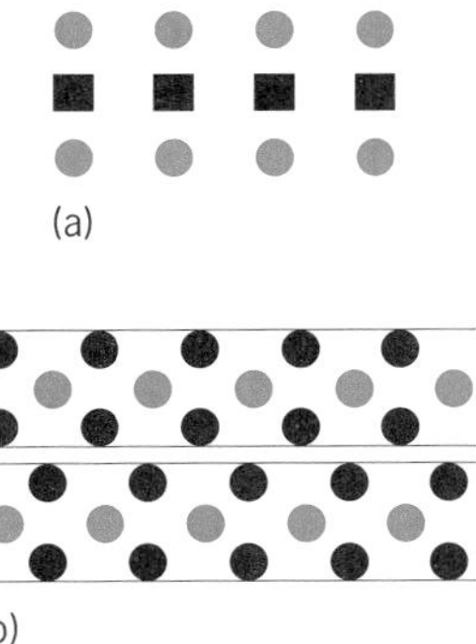

Fig. 8. Schematic representations of low-dimensional conductors. (*a*) One-dimensional conductor. Conducting molecules (circles) are separated by nonconducting molecules (squares). (*b*) Two-dimensional conductor. Conducting atoms (colored circles), for example, niobium, form a sheet. Nonconducting atoms (black circles), for example, sulfur, form sheets above and below the sheet of conducting atoms and thus help to isolate it from the next layer or sheet of conducting atoms.

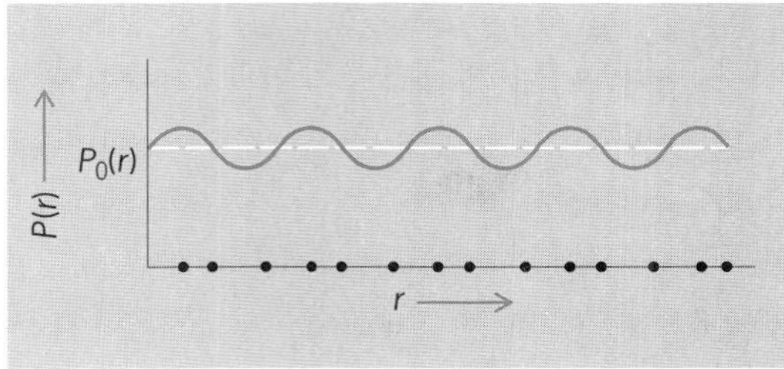

Fig. 9. Charge-density wave. Horizontal straight line shows electronic charge density in absence of wave. Curve shows sinusoidal variation of charge density below transition temperature. Dots at bottom show distorted positions of atoms in lattice.

of the negative electronic charge alone is energetically unfavorable because of the Coulomb interaction. But the modulation of the electronic charge density produces a distortion in the lattice of the ions, such that the attractive Coulomb interaction between electronic charge and ionic charge helps to stabilize the charge-density wave (**Fig. 9**). It is this accompanying lattice distortion that can be observed by x-ray, neutron, and electron diffraction techniques and thus provides evidence for the charge-density-wave transition.

The lattice distortion causes a gap in the electronic energy spectrum, which leads to a reduction in the number of electrons available for conduction and hence causes an increase in the resistivity of the material. The size of the gap grows from zero at the transition temperature to a maximum at the absolute zero of temperature. In some cases the gap is so large that the material becomes an insulator at the onset of the lattice distortion, leading to a metal-insulator transition. *See* BAND THEORY OF SOLIDS; CHARGE-DENSITY WAVE.

Spin-density waves. In a charge-density-wave state below the transition, the modulation in the electronic density has the same phase for electrons of both spins. If the modulation in the density of spin-up electrons is 180° out of phase with the modulation in density of spin-down electrons, then the charge density would remain unmodulated, but there would be a modulation of the spin density. Such a state is called a spin-density-wave state. The electronic density for spin-up (+) and spin-down (−) electrons can then be written in a fashion similar to that for the charge-density-wave case as Eq. (6) [**Fig. 10**]. Spin-density-

$$P_{\pm}(\vec{r}) = {}^1/_2 P_0(\vec{r})[1 \pm \alpha \cos(\vec{q}_0 \cdot \vec{r} + \phi)] \quad (6)$$

wave transitions are much less common than charge-density waves, but are known to exist in metallic chromium below 312 K (102°F). *See* SPIN-DENSITY WAVE.

Metal-insulator transition. The atomic energy levels of the constituents of a solid are broadened into bands due to the overlap of their electron clouds. For semiconductors and insulators, the highest band that is completely full is the valence band. The next band higher in energy than the valence band is called

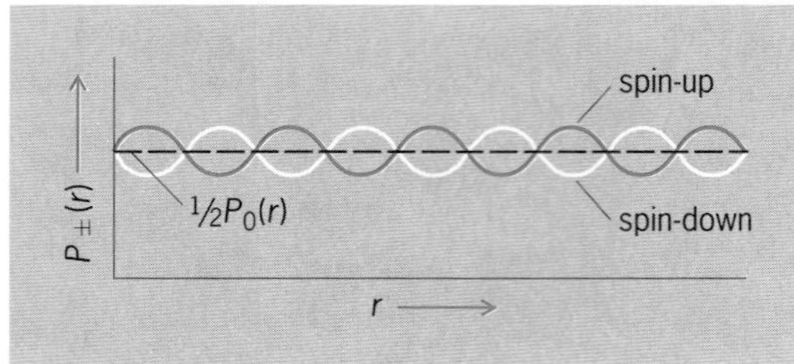

Fig. 10. Spin-density wave. The two curves show the variations of the densities of spin-up and spin-down electrons.

the conduction band. The conduction band is empty in a nonmetal and partly filled in a metal. In response to an applied electric field, the electrons in the conduction band of a metal are accelerated by being excited to a neighboring empty state in the band. Therefore a metal is a good electrical conductor. An insulator with a reasonable band gap (5 eV) between the valence and the conduction bands cannot conduct electricity because the electrons from the valence band cannot be excited to the conduction band with the usual electric fields. In principle, the lattice parameter can be varied to change the band gap and hence induce a metal-insulator transition in a system. This is indeed possible for some systems. For example, iodine becomes metallic under a pressure of 160 kilobars (16 gigapascals). Liquid mercury becomes a semiconductor at high temperature such that its density is below 5 g/cm^3, down from 13.5 g/cm^3 at 68°F (20°C). Many, but not all, metal-insulator transitions can be explained by one-electron band-structure calculations. For the others, it is necessary to go beyond the one-electron model and consider many-body mechanisms. *See* ELECTRICAL CONDUCTIVITY OF METALS; ELECTRICAL RESISTIVITY.

Superfluid transition. Helium gas condenses at 4.2 K (−452°F) and is not known to solidify under atmospheric pressure down to the lowest temperature observed. A pressure of 25 atmospheres (2.5 megapascals) is required to solidify helium. This is due to weak interatomic interactions and the low mass of the helium atom. Thus liquid helium is a quantum liquid in which the energy of zero-point motion is comparable to that of interatomic interactions so that the helium atoms cannot be localized to form a solid. Helium-4 (^{4}He) undergoes a very interesting phase transition at 2.18 K (−455.75°F) from helium I to helium II such that a part of helium II behaves like a normal fluid and a part like superfluid. The superfluid component has zero entropy and zero viscosity, allowing it to flow without resistance through openings of 10^{-2} cm or even smaller. This property is due to the fact that ^{4}He has all of its spins paired, giving it a net spin of zero. Thus ^{4}He is a quantum liquid obeying Bose-Einstein statistics. In Bose-Einstein statistics, there is no restriction on the number of particles that can occupy a given energy level. Thus at 0 K (−459.67°F) all the particles are in the ground state and belong to the superfluid state with zero entropy and zero viscosity. As the temperature is raised, some of the particles occupy the excited states and have finite entropy and viscosity. The number of superfluid atoms N_0 is related to the total number of atoms N by Eq. (7), where T_0 is the transition temperature above which the fluid is normal. If ^{4}He were a

$$\frac{N_0}{N} = 1 - \left(\frac{T}{T_0}\right)^{3/2} \qquad (7)$$

noninteracting Bose-Einstein gas, T_0 would be 3.14 K (−454.02°F), which is in qualitative agreement with the experimental value of 2.14 K (−455.82°F).

The superfluid component of helium II is a very good thermal conductor. This can be used to find T_0 experimentally. Under normal pressure, helium boils at 4.2 K (−452°F) like a typical liquid with many gas bubbles due to poor heat conduction. As the boiling point is lowered by lowering the pressure, the bubbles disappear as soon as the boiling point is the same as the transition temperature. The heat is then conducted away before the gas bubbles can be formed.

The less abundant isotope ^{3}He has a spin of $^1/_2$, and hence ^{3}He liquid obeys Fermi-Dirac statistics. It becomes superfluid below 3 millikelvins by forming Cooper-type pairs of ^{3}He atoms that behave as bosons. This pairing is similar to that occurring in superconductivity, and will not be discussed here. *See* LIQUID HELIUM; SUPERFLUIDITY.

Liquid-crystal transition. An isotropic liquid made up of long organic molecules can undergo a phase change to an anisotropic liquid known as a liquid crystal. The liquid crystal has viscosity similar to that of the liquid but with a long-range order in the orientation of its molecules. Thus a liquid crystal is an intermediate phase between a liquid and a solid. Most of the known liquid crystals have large optical anisotropies so that molecular reorientations produce dramatic changes in optical birefringence, absorption, reflection, scattering, and color. The phase change is determined by a delicate balance of intermolecular forces and changes dramatically because of small outside stimuli such as weak electric fields. These properties are the basis of the display devices used in pocket calculators, wristwatches, clocks, and so forth. Since liquid-crystal displays modify the ambient light instead of emitting light, they use much less power than light-emitting displays. *See* LIQUID CRYSTALS.

D. J. Sellmyer; S. S. Jaswal

Bibliography. B. Bergersen and M. Plischke, *Equilibrium Statistical Physics*, 2d ed., 1994; G. Burns, *Solid State Physics*, 1985; D. Chowdhury, *Spin Glasses and Other Frustrated Systems*, 1986; P. A. Cox, *The Electronic Structure and Chemistry of Solids*, 1987; N. E. Cusack, *The Physics of Structurally Disordered Matter*, 1987; C. Kittel, *Introduction to Solid State Physics*, 7th ed., 1996; K. M. Moorjani and J. M. D. Coey, *Magnetic Glasses*, 1984; N. Mott, *Metal-Insulator Transitions*, 2d ed., 1990; J. M. Yeomans, *Statistical Mechanics of Phase Transitions*, 1992.

Phase velocity

The velocity of propagation of a pure sine wave of infinite extent. In one dimension, for example, the form of the disturbance for such a wave is $y(x,t) = A \sin [2\pi(x/\lambda - t/T)]$. Here x is the position at which the disturbance $y(x,t)$ exists at time t, λ is the wavelength, T is the period which is related to the wave frequency by $T = 1/f$, and A is the disturbance amplitude. The argument of the sine function is called the phase. The phase velocity is the speed with which a point of constant phase can be said to move. Thus $x/\lambda - ft =$ constant, so the phase velocity v_p is given by $dx/dt = v_p = \lambda f$. This is the basic relationship connecting phase velocity, wavelength, and frequency. *See* PHASE (PERIODIC PHENOMENA); SINE WAVE; WAVE MOTION.

For a simple sine wave on a string, one can perceive this phase velocity (provided the wave does not move too quickly) by merely focusing attention on any particular wave crest and observing its apparent motion along the string. A fairly slack, heavy cord has a slow, easily observable phase velocity.

The phase velocity for waves in a medium is determined in part by intrinsic properties of the medium. For all mechanical waves in elastic media, the square of the phase velocity is proportional to the ratio of the appropriate elastic property of the medium to the appropriate inertia property. For example, the square of v_p for transverse waves of small amplitude in a stretched string is $v_p^2 = T/\mu$, where T is the tension in the string and μ is the mass per unit length. The tension is proportional to Young's modulus of elasticity Y for the material from which the string is made, so $v_p^2 \propto Y/\mu$. The phase velocity of electromagnetic waves depends upon the medium as well. In vacuum or (usually to good enough approximation) air, the phase velocity c is given by $c^2 = 1/\epsilon_0\mu_0 \approx 9 \times 10^{16}$ m^2/s^2, where ϵ_0 and μ_0 are respectively the permittivity and permeability of the vacuum. *See* ELECTROMAGNETIC RADIATION; YOUNG'S MODULUS.

Phase velocity may also depend upon the mode of wave propagation. For example, transverse waves in a bar travel with a phase velocity which is different from that for longitudinal waves of the same frequency traveling in the same bar.

Phase velocity will also depend, in general, upon the frequency of the wave. Waves of different frequencies will travel at different speeds, resulting in a phenomenon called dispersion. A beautiful example of this is the dispersion of white light into the colors of the visible spectrum by a prism. There the phase velocity is given by c/n, where n is the index of refraction of the glass from which the prism is made, which depends quite markedly upon the frequency (hence wavelength, hence color) of the light. The equation for determining the disturbance at any place and time, given suitable starting values, is called the wave equation. If this wave equation involves only second-order rates of change of the disturbance with respect to both space coordinates and time, then the phase velocity is frequency-independent. Otherwise, the phase velocity will be frequency-dependent. *See* GROUP VELOCITY; LIGHT; REFRACTION OF WAVES; WAVE EQUATION.

S. A. Williams

Bibliography. D. Giancoli, *Physics for Scientists and Engineers*, 3d ed., 2000; D. Halliday, *Fundamentals of Physics*, 6th ed., 2000; I. G. Main, *Vibrations and Waves in Physics*, 3d ed., 1993.

Phenacetin

One of a general class of medicinals known variously as analgesics, antifebrins, or antipyretics, of which acetanilide is the best known. Phenacetin is an acetyl derivative of *p*-phenetidine. It is made by the reaction of

NaO—⟨benzene ring⟩—NO_2

with C_2H_5Cl to form

C_2H_5O—⟨benzene ring⟩—NO_2

which is reduced to the corresponding amine and acetylated to form phenacetin,

C_2H_5O—⟨benzene ring⟩—$NHCOCH_3$

The reaction is continued as a cyclic process in which phenol, acetic acid, and ethyl chloride are continuously supplied. Phenacetin is less toxic than acetanilide, but it lowers the ability of blood to combine with oxygen. *See* ASPIRIN; HYPOTHERMIA.

Allen L. Hanson

Phenocryst

A relatively large crystal embedded in a finer-grained or glassy igneous rock. The presence of phenocrysts gives the rock a porphyritic texture (see **illus.**).

Granite (quartz monzonite) from the Sierra Nevada of California showing numerous phenocrysts of microcline feldspar in parallel orientation with banded structure of the rock. Hammer is 10 in. (25 cm) long. (*USGS photograph by W. B. Hamilton*)

Phenocrysts are represented most commonly by feldspar, quartz, biotite, hornblende, pyroxene, and olivine. Strictly speaking, phenocrysts crystallize from molten rock material (lava or magma). They commonly represent an earlier and slower stage of crystallization than does the matrix in which they are embedded. Phenocrysts are to be distinguished from certain relatively large crystals (porphyroblasts) which develop late in solid rock as the result of metamorphism or metasomatism. If the origin of a large crystal is in question, the nongenetic term megacryst should be used. *See* CONTACT AUREOLE; IGNEOUS ROCKS; PORPHYROBLAST; PORPHYRY.

Carleton A. Chapman

Phenol

The simplest member of a class of organic compounds possessing a hydroxyl group attached to a benzene ring or to a more complex aromatic ring system. Phenol itself, C_6H_5OH, may also be called hydroxybenzene or carbolic acid. Pure phenol is a colorless solid melting at 42°C (108°F), moderately soluble in water, and weakly acidic (p*K* 9.9).

Phenol has broad biocidal properties, and dilute aqueous solutions have long been used as an antiseptic. At higher concentrations it causes severe skin burns; it is a violent systemic poison. *See* ANTISEPTIC.

Structure and nomenclature. Phenol has the structure shown. Simple substituted phenols, such as the three isomeric chlorophenols, are named as indicated, using the ortho, meta, and para prefixes. In more highly substituted phenols the positions of substitution are indicated by numbers, for example, 2,4-dichlorophenol. Compounds with more than one

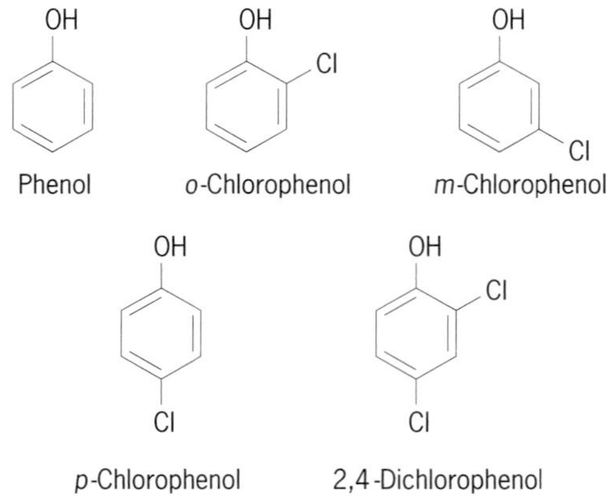

Phenol o-Chlorophenol m-Chlorophenol

p-Chlorophenol 2,4-Dichlorophenol

hydroxyl group per aromatic ring are known as polyhydric phenols, and include catechol, resorcinol, hydroquinone, phloroglucinol, and pyrogallol.

Reactions. The chemistry of phenol can be discussed under two headings.

Reactions of hydroxyl group. Phenols react readily with alkali hydroxides to form salts, reaction (1), and

$$C_6H_5OH + NaOH \rightarrow \underset{\text{Sodium phenoxide}}{C_6H_5ONa} + H_2O \qquad (1)$$

those phenols containing electronegative substituents (for example, trichlorophenol) are even acidic enough to react with bicarbonate.

Like their aliphatic analogs, the alcohols, phenols form esters and ethers. Treatment of phenol with acetic anhydride yields phenyl acetate. The methyl ether of phenol (anisole) can be prepared by treatment of sodium phenoxide with methyl sulfate. Etherification of 2,4-dichlorophenol with sodium chloroacetate, reaction (2), yields 2,4-

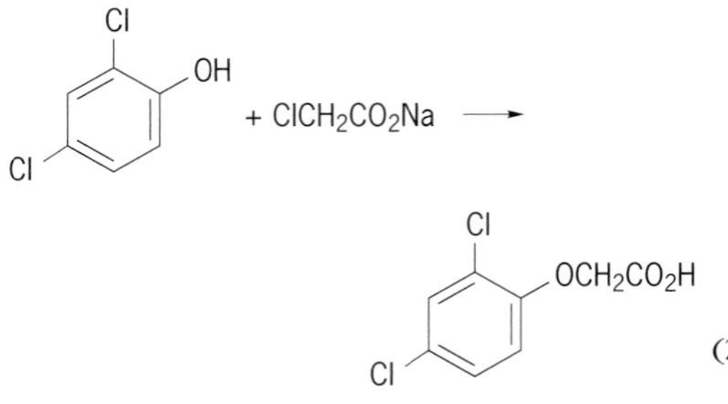

+ $ClCH_2CO_2Na$ → OCH_2CO_2H 2,4-D (2)

dichlorophenoxyacetic acid (2,4-D), which is a powerful herbicide.

Reactions of aromatic ring. Under the conditions for electrophilic substitution reactions (for example, nitration, halogenation, sulfonation, alkylation, and diazo coupling), phenol is one of the most reactive aromatic compounds. Such reactions invariably lead to introduction of the substituent into an ortho or para position, as illustrated by the bromination of phenol, reactions (3). At low temperature in carbon

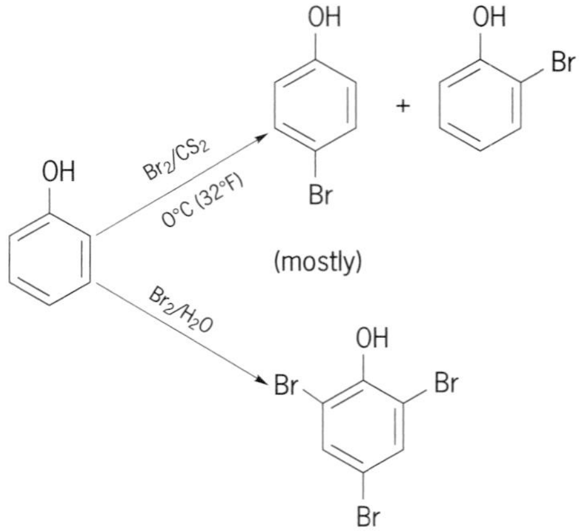

Br_2/CS_2 0°C (32°F) (mostly) (3a)

Br_2/H_2O (3b)

disulfide solution, careful addition of one equivalent of bromine leads to the monobromo products of which the para isomer predominates, reaction (3*a*). Treatment of phenol with excess bromine in water yields 2,4,6-tribromophenol, in which both of the ortho and the para hydrogens have been replaced by bromine, reaction (3*b*). Reaction (3*b*) has been used as a fairly sensitive test for phenol, based upon the rapid precipitation of the insoluble tribromo product.

Two further examples of electrophilic substitution are the reaction with acetone, reaction (4), to form

$$2\,HO-C_6H_4 + CH_3C(=O)CH_3 \rightarrow HO-C_6H_4-C(CH_3)_2-C_6H_4-OH \quad (4)$$

Bis-phenol A

bis-phenol A, an important plastics intermediate, and the carboxylation of sodium phenoxide, reaction (5), to yield salicylic acid and thence aspirin.

$$C_6H_5ONa \xrightarrow[\text{acid}]{CO_2} C_6H_4(OH)CO_2H \rightarrow C_6H_4(OCOCH)CO_2H \quad (5)$$

Salicylic acid; Aspirin

Phenol is particularly sensitive to oxidation. Quinone is formed by chromic acid oxidation of phenol, but most oxidizing agents convert phenol to a complex mixture of products derived from coupling of intermediate phenoxy radicals (oxidative coupling products). Certain 2,4,6-trisubstituted phenols are oxidized to relatively stable free radicals, and this property is used to advantage in the design of phenolic antioxidants. Reduction of phenol to cyclohexanol can be effected by hydrogen in the presence of catalysts such as nickel.

Production. Until World War I phenol was essentially a natural coal tar product. However, synthetic methods replaced extraction from natural sources. There are many possible syntheses of phenol; six commercially significant ones are listed below.

1. The oldest process, that of sulfonation of benzene followed by neutralization of the sulfonic acid thus produced and fusion with caustic soda, was introduced into the United States about 1915. This is shown by reactions (6) and (7).

$$C_6H_6 + H_2SO_4 \rightarrow C_6H_5SO_2OH \xrightarrow{\text{NaOH}} C_6H_5ONa \quad (6)$$

$$C_6H_5ONa \xrightarrow{\text{acid}} C_6H_5OH \quad (7)$$

2. The cumene hydroperoxide process yields the largest production in the United States, reactions (8) and (9). This process grew very rapidly after its

$$C_6H_5-CH(CH_3)_2 + O_2 \rightarrow C_6H_5-C(CH_3)_2-OOH \quad (8)$$

$$C_6H_5-C(CH_3)_2-OOH \xrightarrow{\text{acid}} C_6H_5OH + CH_3-C(=O)-CH_3 \quad (9)$$

introduction in 1955. The coproduct acetone is of considerable value, which undoubtedly contributed to the widespread acceptance of the process.

3. The Raschig process was introduced in 1940 and involves a first-stage chlorination of benzene using an air–hydrochloric acid mixture, as shown by reactions (10) and (11). The by-product HCl is re-

$$C_6H_6 + HCl + {}^1/_2O_2 \rightarrow C_6H_5Cl + H_2O \quad (10)$$

$$C_6H_5Cl + H_2O \rightarrow C_6H_5OH + HCl \quad (11)$$

cycled to the first stage. High temperatures are employed, and high alloy reaction vessels are used to minimize corrosion.

4. The chlorination, or Dow, process was introduced in 1924 and is reasonably straightforward. Starting with chlorobenzene, the process proceeds as in reaction (12). Copper catalysts and high tem-

$$C_6H_5Cl + NaOH \rightarrow NaCl + C_6H_5ONa \xrightarrow{\text{acid}} C_6H_5OH \quad (12)$$

peratures and pressures are employed.

5. A process developed in 1962 starts with toluene and proceeds through oxidation to benzoic acid, as seen in reaction (13). This is followed by further ox-

$$C_6H_5 \rightarrow CH_3 + {}^3/_2O_2 \rightarrow C_6H_5COOH + H_2O \quad (13)$$

idation or decarboxylation in a second step using cupric benzoate as a catalyst, as shown by reaction (14).

$$C_6H_5COOH + {}^1/_2O_2 \rightarrow C_6H_5OH + CO_2 \quad (14)$$

6. The last process starts with cyclohexane, which is oxidized to a mixture of cyclohexanol and cyclohexanone (mixed oil). The next step is hydrogenation to convert all of the mixture to cyclohexanol. This intermediate is then, by classical methods, ring-dehydrogenated to yield phenol and hydrogen, which is recycled, as shown in reactions (15) and (16). Boron compounds are employed as catalysts or mediators.

$$C_6H_{12} + O_2 \rightarrow \text{(mixed oil)} \xrightarrow{\text{hydrogenation}} C_6H_{11}OH \quad (15)$$

$$C_6H_{11}OH \xrightarrow{\text{catalyst}} C_6H_5OH + 3H_2 \quad (16)$$

Uses and derivatives. Phenol is one of the most versatile and important industrial organic chemicals. It is the starting point for many diverse products used in the home and industry. A partial list includes nylon, epoxy resins, surface active agents, synthetic detergents, plasticizers, antioxidants, lube oil additives, phenolic resins (with formaldehyde, furfural, and so on), cyclohexanol, adipic acid, polyurethanes, aspirin, dyes, wood preservatives, herbicides, drugs, fungicides, gasoline additives, inhibitors, explosives, and pesticides.

Naturally occurring phenols. Phenol and the methyl phenols (cresols and xylenols) are obtained from coal tar. Phenol has been found as a minor constituent of various plant materials, including tobacco leaves and pine needles. In the structures shown below, carvacrol, thymol, and vanillin are examples of natural phenols of plant origin. Phenolic rings are also present in such biologically important compounds as the amino acid tyrosine and in estrone, a female sex hormone.

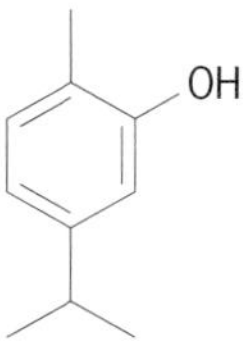

OH OH OH OCH_3 CHO

Carvacrol 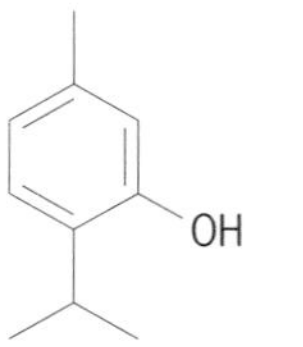 Thymol 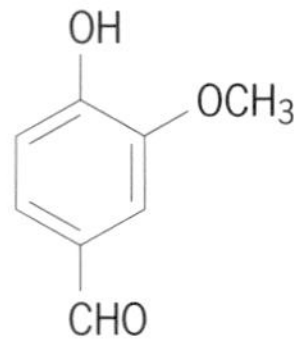 Vanillin

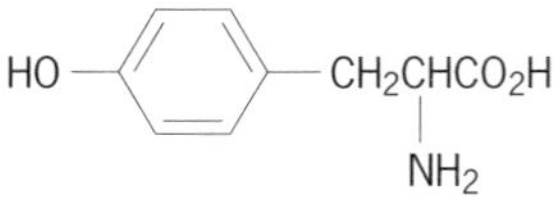

HO–C_6H_4–CH_2CHCO_2H (NH_2)

Tyrosine

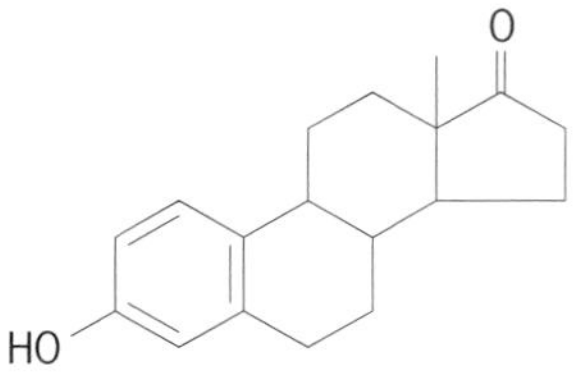

O HO

Estrone

Robert I. Stirton; Martin Stiles

Bibliography. D. H. R. Barton and D. Ollis, *Comprehensive Organic Chemistry*, vol. 4, 1979; S. N. Ege, *Organic Chemistry*, 4th ed., 1998; R. J. Fessenden and J. S. Fessenden, *Organic Chemistry*, 6th ed., 1998.

Phenolic resin

One of the condensation products of phenols or phenolic derivatives with aldehydes such as formaldehyde and furfural. The term phenoplasts is sometimes used to refer to the whole group of products. The phenol-formaldehyde resins, developed commercially between 1905 and 1910, were the first truly synthetic polymers and have found wide usage for electrical insulation, molded objects, shell molds for metals, laminates, adhesives, and many other applications. They are characterized by low cost, dimensional stability, high strength, and resistance to aging. The combination of low cost and good properties is reflected in the fact that phenolic resins are produced in greater volume than any other thermosetting resin.

Phenol is prepared by the hydrolysis of chlorobenzene, by the alkali fusion of sodium benzene sulfonate, by the oxidation of toluene, by the dehydrogenation of cyclohexanol/cyclohexane, or by the decomposition of cumene hydroperoxide, as shown in reaction (1).

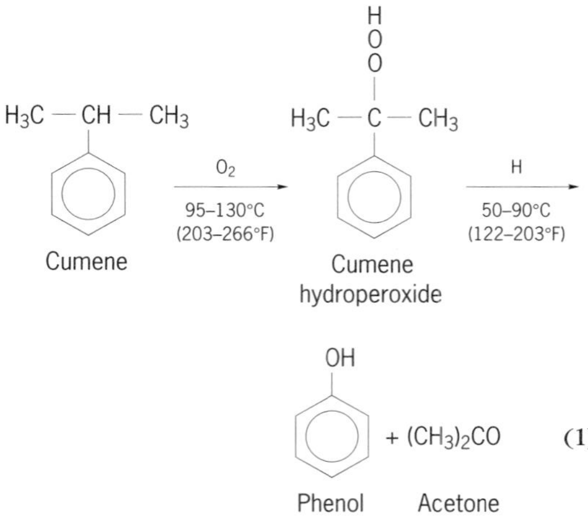

$$\underset{\text{Cumene}}{H_3C-CH(C_6H_5)-CH_3} \xrightarrow[\substack{95\text{–}130°C\\(203\text{–}266°F)}]{O_2} \underset{\text{Cumene hydroperoxide}}{H_3C-C(OOH)(C_6H_5)-CH_3} \xrightarrow[\substack{50\text{–}90°C\\(122\text{–}203°F)}]{H} \underset{\text{Phenol}}{C_6H_5OH} + \underset{\text{Acetone}}{(CH_3)_2CO} \quad (1)$$

Resorcinol, obtained by the alkaline fusion of *m*-benzene disulfonic acid, and *m*-cresol from coal tar are also used.

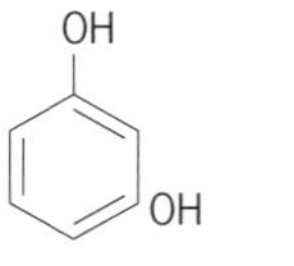

OH OH OH CH_3

m-Dihydroxybenzene or resorcinol *m*-Cresol

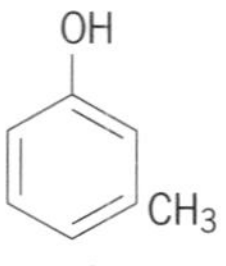

Formaldehyde is produced by the oxidation of methane or methyl alcohol, as in reaction (2),

$$\underset{\text{Methyl alcohol}}{CH_3OH} + O_2 \xrightarrow[\substack{550\text{–}600°C\\(1020\text{–}1110°F)}]{\text{Ag, Fe, or Mo catalyst}} \underset{\text{Formaldehyde}}{HCHO} + H_2O \quad (2)$$

and furfural is obtained by the hydrolysis of oat hulls.

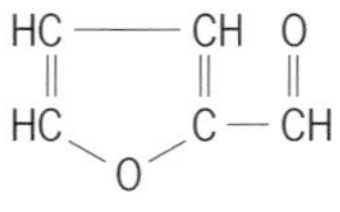

HC—CH O
HC C—CH
O

Furfural

Polymerization. In the presence of an acid or base, phenol and aqueous formaldehyde react to form a solution of phenolic alcohols or methylol derivatives with the methylol groups in the ortho and para positions, reaction (3). This reaction takes place quickly

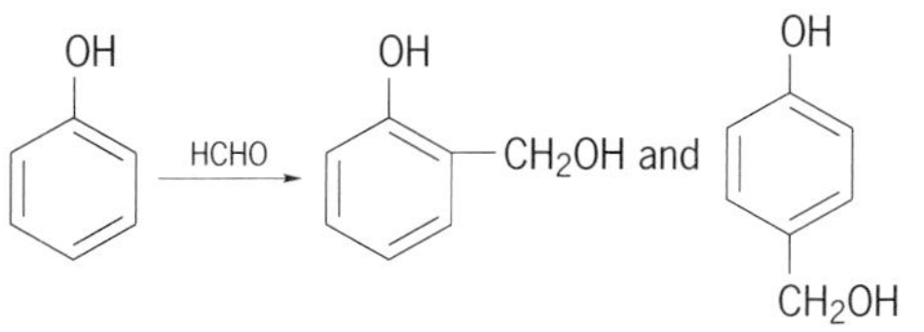

(3)

in a basic medium and slowly in an acidic medium.

The methylol phenols formed initially in a basic medium with formaldehyde in excess condense with each other and with additional formaldehyde to yield an "A-stage" resin or "resole," a brittle resin which is soluble and fusible. The resole resin consists of a mixture of isomers containing free methylol groups, which are available for subsequent cross-linking reactions to form a less-soluble "B-stage" resin. Many structural variations are possible. The structure for a typical resole component is

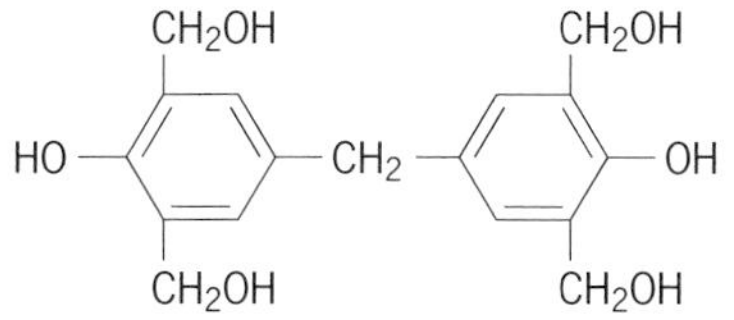

In the presence of acid and less than 0.86 mole of formaldehyde per mole of phenol, the primary alcohols react to yield diphenylmethane polymers called novolacs, which are soluble and fusible and contain an average of 5 or 6 phenol units per molecule. These resins may also be referred to as A-stage resins. Novolacs may also be reacted with epichlorohydrin to yield epoxy polymers. The structure for a typical novolac resin is

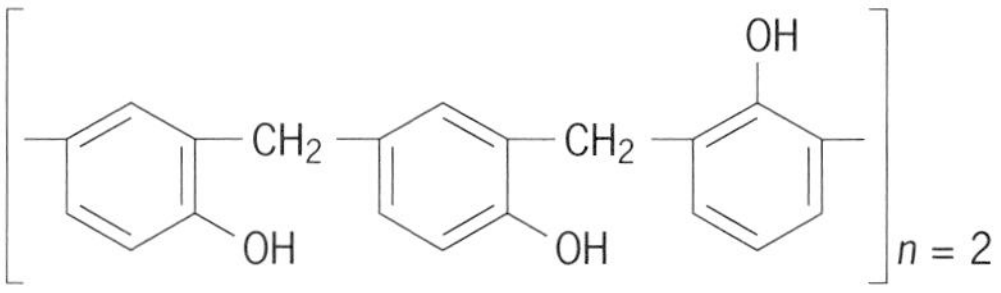

Hardening of all of these is effected by further cross-linking. A resole-type resin is inherently capable of cross-linking itself on heating and is sometimes referred to as a one-stage resin. On the other hand, a novolac has no free methylol groups and must be mixed with an aldehyde to undergo further reaction; hence a novolac is sometimes called a two-stage resin.

In the production of phenolic-resin molding compositions, it is common practice to neutralize, concentrate, and dry the B-stage resin; to mix it with fillers and in, in the case of a novolac, a curing agent or hardener; and finally to compact it into the form of pellets or briquets. Other ingredients may be present also, such as curing accelerators, pigments, lubricants, and plasticizers. The curing agent for a novolac is hexamethylenetetramine, which at the temperature of molding reacts with water to form formaldehyde and ammonia. In the presence of ammonia and the additional formaldehyde, the B-stage resin cures in the mold to yield a highly cross-linked, insoluble, and infusible C-stage product. The structure for a C-stage phenolformaldehyde resin is

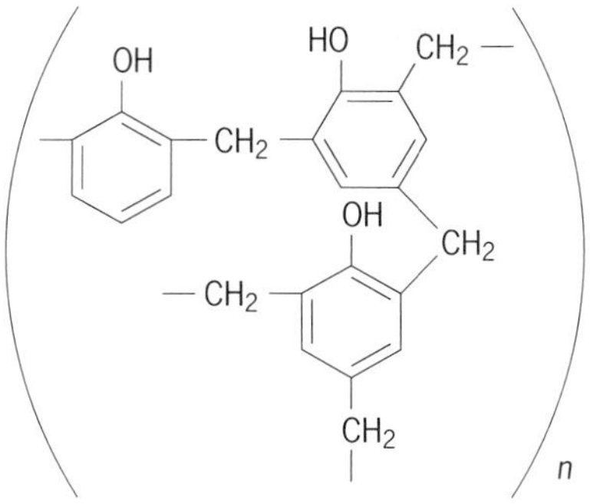

By use of *m*-phenol derivatives, such as resorcinol or *m*-cresol, resins are obtained which cure rapidly at low temperature because the meta substituents activate the ortho and para positions. An ortho or para alkyl phenol which has only two active sites available for reaction (difunctional) can be used, such as *p*-tert-butylphenol

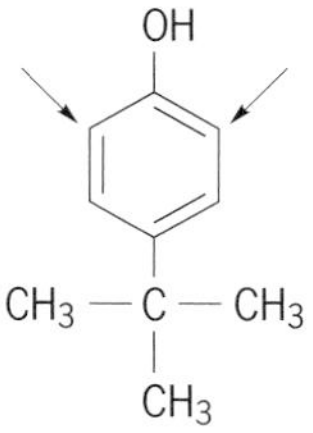

Then oil-soluble, thermoplastic resins are formed instead of the cross-linked materials obtained from the trifunctional phenols just discussed. These products, somewhat more expensive than ordinary phenolic resins, are used in special paint, varnish, and adhesive formulations.

With the use of furfural instead of formaldehyde, the B-stage resin has the unique property of remaining thermosplastic for a relatively long time. The phenol-furfural compositions are useful for molding large complex forms in which extra time is needed for the resin to fill the mold completely.

Fabrication and use. Phenolic resins can be cast from syrupy intermediates or molded from B-stage solid resins. Laminated products can be produced by impregnating fiber, cloth, wood, and other materials with the resin. After heating, laminated sheets can be pressed into any shapes desired. Most of the phenolic plastics can be machined if necessary.

Another important type of phenolic resin product is rigid foam. One type is prepared by incorporating a blowing agent in a curing mixture so that the heat of reaction decomposes the blowing agent. A second type, "syntactic" foam, consists of microscopic hollow spheres of phenolic resin mixed with a curable binder such as an epoxy resin or polyester. Foams of this type have very high strength after curing.

Cured phenolic plastics are rigid, hard, and resistant to chemicals (except strong alkali) and to heat.

Some of the uses for phenolic resins are for making

precisely molded articles, such as telephone parts, for manufacturing strong and durable laminated boards, or for impregnating fabrics, wood, or paper. Phenolic resins are also widely used as adhesives, as the binder for grinding wheels, as thermal insulation panels, as ion-exchange resins, and in paints and varnishes. *See* ADHESIVE; ION EXCHANGE; PHENOL; PLASTICS PROCESSING; POLYMERIZATION; TEXTILE CHEMISTRY.

John A. Manson

Bibliography. J. A. Brydson, *Plastic Materials*, 4th ed., 1982; F. A. Carey and R. J. Sundberg, *Advanced Organic Chemistry*, pt. A: *Structures and Mechanisms*, 4th ed., 2000; A. Knop and L. A. Pilato, *Phenolic Resins*, 2d ed., 1999; S. Schwartz and S. Goodman, *Plastics Materials and Processes*, 1982.

Phenylketonuria

An inborn error of metabolism in which affected individuals lack the liver enzyme phenylalanine hydroxylase (PAH), which is needed to metabolize phenylalanine, an amino acid essential for normal growth and development. If untreated, affected individuals may become severely mentally retarded, become microcephalic, have behavioral problems, develop epilepsy, or show other signs of neurological impairment. Phenylketonuria (PKU) is inherited as an autosomal recessive trait and is found in all ethnic groups but most frequently in individuals of northern European descent. Its incidence is about 1 per 14,000 births in the United States. Classically, persons with phenylketonuria exhibit blood phenylalanine concentrations of 20 mg/deciliter or more (normal concentrations are about 1–2 mg/dl), normal blood tyrosine levels, and excessive phenylalanine metabolites in the urine while on a normal diet. Phenylketonuria variants have blood phenylalanine concentrations of 10–20 mg/dl, but may not have phenylalanine metabolites in their urine unless they have ingested excessive amounts of protein.

Benign hyperphenylalaninemia is the term applied when blood phenylalanine concentrations are 4–10 mg/dl. It causes no clinical symptoms, does not affect intelligence, and requires no treatment. About 1–3% of individuals with phenylketonuria have an atypical form caused by a defect in the synthesis of the enzyme dihydropteridine reductase or the coenzyme tetrahydrobiopterin. This coenzyme is a cofactor that is needed for the conversion of phenylalanine to tyrosine. Infants with deficiencies of tetrahydrobiopterin have high blood levels of phenylalanine and exhibit progressive neurological deterioration when treated with only a low-phenylalanine diet.

The phenylalanine hydroxylase locus has been mapped on the distal tip of the long arm of chromosome 12. The gene has been cloned, and, by using molecular genetic techniques, prenatal identification of phenylketonuria homozygotes and carriers has been achieved in about 90% of affected families. Since phenylalanine hydroxylase is not expressed in amniotic cells, amniocentesis, a common screening procedure, is not useful for the prenatal diagnosis of phenylketonuria. *See* PRENATAL DIAGNOSIS.

Newborn screening programs for phenylketonuria have been successful in identifying most cases within a few weeks of birth. The most widely used screening method is the Guthrie test, which measures the blood phenylalanine level by bacterial inhibition assay using a dried blood spot on filter paper obtained by heel puncture as close as possible to the time of discharge from the hospital nursery. A low phenylalanine diet with restriction of proteins, if initiated early in infancy, can prevent the development of severe intellectual and neurological handicaps that would otherwise occur in virtually all untreated cases. *See* MENTAL RETARDATION.

The success of nationwide screening programs for phenylketonuria created an unexpected problem as women with the disorder reached childbearing age. Retrospective surveys have revealed a high rate of mental retardation, microcephaly, congenital heart defects, and intrauterine growth retardation among offspring of women with untreated maternal phenylketonuria. Therefore, in order to reduce the morbidity associated with the offspring of women with maternal phenylketonuria, a phenylalanine-restricted diet should be followed throughout pregnancy and should probably be initiated prior to pregnancy in order to have maximum effectiveness. *See* HUMAN GENETICS; PROTEIN METABOLISM.

Felix de la Cruz

Bibliography. G. M. Addison et al. (eds.), *Studies in Inherited Diseases: Prenatal and Perinatal Diagnosis*, 1989; F. G. Cunningham, P. C. MacDonald, and N. F. Gant, *Williams Obstetrics*, 21st ed., 2001; N. A. Holtzman et al., Effect of age at loss of dietary control on intellectual performance and behavior of children with phenylketonuria, *N. Engl. J. Med.*, 314:593–598, 1986; R. R. Lenke and H. L. Levy, Maternal phenylketonuria and hyperphenylalaninemia: An international survey of the outcome of untreated and treated pregnancies, *N. Engl. J. Med.*, 303:1202–1208, 1980; A. A. Lidsky, F. Guttler, and S. L. C. Woo, Prenatal diagnosis of classical phenylketonuria by DNA analysis, *Lancet*, 1:549–551, 1985; A. C. Whittle, *Prenatal Diagnosis in Obstetric Practice*, 2d ed., 1995.

Pheromone

A substance that acts as a molecular messenger, transmitting information from one member of a species to another member of the same species. The first pheromone to be characterized chemically was bombykol, an unsaturated, straight-chain alcohol that is secreted in microgram amounts by females of the silkworm moth (*Bombyx mori*) and is capable of attracting male silkworm moths at large distances. The electroantennogram (EAG) technique was developed in connection with this work; the electrical signals that are generated by the pheromone stimulation of an insect's chemoreceptor cells are recorded and analyzed.

A useful distinction has been made between releaser pheromones, which elicit a rapid, behavioral response, and primer pheromones, which elicit a slower, developmental response and may pave the way for a future behavior.

Fungi. Communication via pheromones is common throughout nature, including some eukaryotic microorganisms that exchange vital chemical signals. For example, female gametes of *Allomyces*, an aquatic fungus that reproduces sexually in one part of its life cycle, secrete the sesquiterpenoid alcohol (-)-sirenin, which serves to attract male gametes. Another example is the water mold *Achyla*; the female secretes a steroidal pheromone, antheridiol, which induces the development of male structures. The resultant antheridial branches then secrete a second steroid, dehydro-oogoniol, which induces further development of the female reproductive structures. The later stages in this reproductive process remain to be elucidated chemically. *See* FUNGI.

Cellular slime molds. The cellular slime molds (*Acrasiomycetes*) are soil microorganisms with a particularly interesting life cycle. A mobile, unicellular ameboid form, when the local food supply is exhausted, gathers into large aggregations of up to 10^5 amebas. These unite to form a sorocarp made up of a long, slender stalk that supports a spore-containing fruiting body. A pheromone, originally called acrasin, is responsible for the aggregation. It is cyclic adenosine monophosphate (cAMP), the same compound that serves as a second messenger in many hormonal systems. However, different species of slime mold may utilize different aggregation pheromones.

Algae. In several species of algae, relatively simple hydrocarbons act as sperm attractants. The first natural algal pheromones to be characterized were 1,3,5-octatriene and ectocarpine, from the aquatic species *Fucus serratus* and *Ectocarpus siliculosus*, respectively; these are extremely hydrophobic messenger molecules.

Insects. By far the largest number of characterized pheromones come from insect species. In part this is a consequence of the large number of insect species on Earth (well over a million), and in part it reflects the great interest in this remarkably successful group of animals. In social insects, such as termites and ants, there may be as many as a dozen different types of messages that are used to coordinate the complex activities which must be carried out to maintain a healthy colony. These activities might require specialized pheromones such as trail pheromones (to lead to a food source), alarm pheromones (recruiting soldiers to the site of an enemy attack), or pheromones connected with reproductive behavior. In the honeybee, isoamyl acetate serves as a natural releaser of aggressive behavior. A large number of pheromones have been characterized from important pest species; there has been some elucidation of the biosynthetic pathways whereby these pheromones are produced from fatty acid precursors. The gypsy moth sex attractant, (+)-disparlure, has been characterized; and one of the most complex sex pheromones is that of the American cockroach (*Periplaneta americana*), a pheromone that has been synthesized.

Many female Lepidoptera make use of sex attractants that are most often straight-chain molecules with relatively simple functionalities. An interesting group of nitrogen-containing aphrodisiac pheromones is produced by certain male Lepidoptera. Unlike the females, males produce these compounds in relatively large quantities (around 1–100 micrograms per insect). They are usually applied directly to the female antennae by special organs, present only in the male, during courtship. These pheromones are derived from pyrrolizidine alkaloids, which the insects obtain from plants. Since the alkaloid itself provides protection against predators, such as birds and spiders, the pheromones derived from them may serve to inform a female about the defensive status of a potential mate, thereby serving as a chemical criterion for sexual selection. *See* INSECT PHYSIOLOGY; SOCIAL INSECTS.

Mammals. Much less is known about mammalian pheromones because mammalian behavior is more difficult to study. There are, however, a small number of well-characterized mammalian pheromones from pigs, dogs, hamsters, mice, and marmosets. There is a steroid that is present in the saliva of male pigs which, when sprayed on the face of a sow, induces a rigid stance in the female that facilitates mating. (It has been shown that pigs are effective truffle hunters because truffles excrete this same steroid.) In spite of a great deal of interest in the characterization of possible human pheromones, there have been no definitive chemical and behavioral results.

Research techniques. Pheromone isolation has become much easier since the 1960s as a result of rapidly improving techniques, especially gas chromatography and high-performance liquid chromatography, for the separation of complex mixtures of natural products on a micro scale. The characterization of a pheromone also requires a reliable bioassay. In many cases, several compounds must be present together, and in the correct proportions, in order to convey a message. Once compounds of established biological activity are isolated, their structures may be determined by organic-chemical and spectroscopic techniques. Syntheses, sometimes stereospecific, can provide material for additional biological studies, as well as for confirmation of the proposed structures.

Applications. There is great potential for controlling the behavior of a given species by manipulating its natural chemical signals. For example, pheromones have been used to disrupt the reproduction of certain insect pests. One simple application involves the use of pheromones in baiting traps that can indicate when a particular insect population is growing. This approach can lead to reduced use of pesticides as well as advances in the control of both agricultural pests and disease vectors. *See* CHEMICAL ECOLOGY; CHEMORECEPTION; INSECT CONTROL, BIOLOGICAL.

J. Meinwald

Bibliography. W. C. Agosta, *Chemical Communication: The Language of Pheromones*, 1992; E. S. Albone, *Mammalian Semiochemistry: The Investigation of Chemical Signals Between Mammals*, 1984; B. A. Leonhardt and M. Beroza, *Insect Pheromone Technology: Chemistry and Applications*, 1982; D. W. MacDonald, D. Muller-Schwartz, and S. Natynczok (eds.), *Chemical Signals in Vertebrates*, 1991; G. D. Prestwich and G. J. Blomquist (eds.), *Pheromone Biochemistry*, 1987.

Phlebitis

An inflammation of a vein. Individuals with phlebitis typically experience tenderness, redness, and hardness along the course of the vein. The cause of the inflammation may be related to injury of the vein or infection. The presence of varicose veins and the long-term use of indwelling intravenous catheters or irritating intravenous solutions place individuals at risk of developing phlebitis. In addition, those with certain diseases, including systemic lupus erythematosus, vasculitis, or malignancy, are at increased risk.

The condition is usually diagnosed on the basis of the clinical symptoms described above. If there is uncertainty regarding the diagnosis, a noninvasive ultrasound study can be used to differentiate among inflammation of the lymph system, tissue inflammation, and phlebitis. Two varieties of phlebitis are recognized: phlebothrombosis and thrombophlebitis.

Phlebothrombosis is a condition in which a blood clot develops within an inflamed vein. As the clot enlarges, it may detach and travel to the lung, becoming a pulmonary embolism. Thrombophlebitis begins with an inflammatory reaction in the vein wall. When the lining of the vein is damaged, three reactions influence the development of thrombosis. Initially, damage to the lining results in adherence of white blood cells, coagulation, and a loss of the lining's nonthrombogenic characteristics. Subsequently, the deep lining of the vein is exposed, bringing it into contact with blood and allowing platelets to adhere and aggregate. Finally, the exposed lining and activated platelets result in changes in coagulation, causing more platelets to interact with deep-lining structures. These factors are influenced by the velocity of blood flow in the affected area. *See* EMBOLISM; THROMBOSIS.

Symptomatic thrombophlebitis usually results in a clot which is firmly adherent to the vein wall with a decreased risk of embolizing. Some individuals may develop symptoms suggestive of deep venous thrombosis such as pain and swelling, and should undergo noninvasive ultrasound examination of the deep veins of the leg. This is particularly important in individuals with cancer. There are other factors that can predispose people to phlebitis, including stasis, immobility, tissue destruction, cardiac failure, obesity, varicose veins, and infection.

In the absence of deep venous thrombosis, the goal of treatment of superficial phlebitis is symptomatic relief. Analgesics, warm compresses and elevation of the affected limb may be beneficial. Anticoagulants are not routinely prescribed unless the thrombus is enlarging despite treatment or is in a long leg vein extending to the groin. Late effects of phlebitis include damage to the vein wall and destruction of the venous valves or obliteration of the vein. When the deep veins of the lower extremity are involved, many individuals develop chronic venous insufficiency and its associated morbidity. *See* CIRCULATION; INFLAMMATION.

Lazar J. Greenfield; Mary C. Proctor

Phloem

The principal food-conducting tissue in vascular plants. Its conducting cells are known as sieve elements, but phloem may also include companion cells, parenchyma cells, fibers, sclereids, rays, and certain other cells. As a vascular tissue, phloem is spatially associated with xylem, and the two together form the vascular system. Less is known of phloem than of xylem, partly because of its lesser direct economic importance and partly because the sieve elements function for a short time (usually one season) and then undergo marked structural and functional changes, such as crushing or sloughing off as a result of periderm formation in the case of woody plants. *See* XYLEM.

Sieve elements. Sieve elements differ from phloem parenchyma cells in the structure of their walls and to some extent in the character of their protoplasts. Sieve areas, distinctive structures in sieve element walls, are specialized primary pit fields in which there may be numerous modified plasmodesmata. Plasmodesmata are strands of cytoplasm connecting the protoplasts of two contiguous cells. These strands are often surrounded by callose, a carbohydrate material that appears to form rapidly in plants placed under stress. How the very specialized end walls and associated sieve areas in some sieve elements (sieve-tube members) form is not fully understood, but both callose and cisternae of endoplasmic reticulum are often observed at future pore sites and these might be involved in the selective deposition or removal of cellulosic wall material in these regions. The walls of sieve elements often increase in thickness by the deposition of the so-called nacreous thickening. The protoplast of an immature sieve element is usually indistinguishable from that of a typical parenchyma cell of the same age. In the course of development into a functioning sieve element, however, several distinctive changes appear to occur. Commonly, there is a disintegration or change in appearance of the nucleus. The tonoplast that normally delimits the vacuole from the cytoplasm probably becomes disrupted during preparation of phloem samples for microscopy; thus, mature sieve elements often appear enucleate and without vacuoles. Ribosomes and dictyosomes normally disappear. A plasmalemma (cell membrane), endoplasmic reticulum, mitochondria, and plastids remain, and are generally located next to the wall.

Difficulty in providing an accurate account of the contents of mature, functioning sieve elements lies in the fact that these cells readily show signs of disturbance. This is due to their contents being under a relatively negative water potential (high turgor), because of the high solute (sucrose) concentration. Also, different groups of plants (such as gymnosperms, dicotyledons, and monocotyledons) exhibit different types of sieve elements with contents that are somewhat varied. In some sieve elements the inner lumen appears filled with the contents of the "former" vacuole and some cytoplasmic material. P-protein (formerly called slime) arises as distinct bodies in the cytoplasm of young sieve elements of dicotyledons, but may later disperse throughout the cell. P-protein is the source of the "slime" plugs, accumulations of protein on sieve areas of injured sieve elements. In many species the plastids produce a special kind of starch or proteinaceous material (**Fig. 1***b*). In monocotyledon species, where P-protein is lacking, these plastids or their contents (Fig. 1*a*) might serve to plug the sieve plate pores. Again, it should be emphasized that because of the delicate nature of sieve element contents, many of the structures (nucleus, tonoplast) reported lacking in mature sieve elements are, in fact, destroyed or displaced during tissue processing. Therefore, if one were to reconstruct the pretreatment condition, sieve elements for angiosperms would more than likely contain structures such as those diagrammed in **Fig. 2***a–f*. When a comparison is made between a sieve element that was purposely traumatized (Fig. 1*a*) and one that was not (Fig. 1*b*), it can be seen that sieve element contents look much like those of highly vacuolate parenchyma cells, except that the nucleus might be absent or appear necrotic.

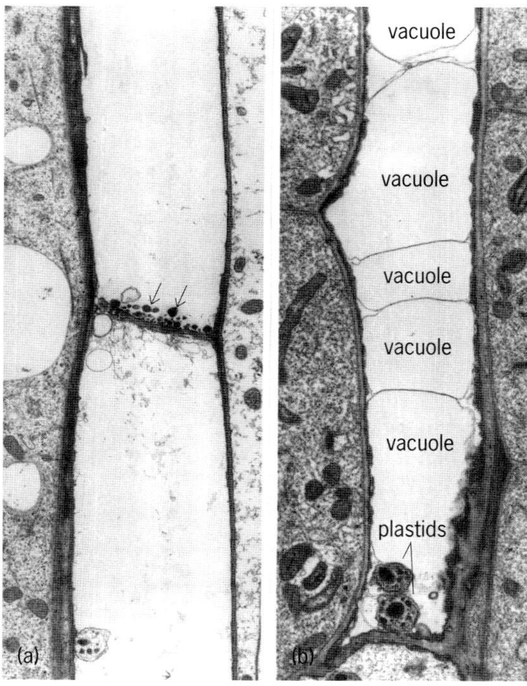

Fig. 1. Electron micrographs showing longisections of sieve elements. (*a*) Sieve elements were traumatized (cut) during pretreatment and their contents (arrows) have surged toward the sieve plate. (*b*) Sieve element with plastids and vacuoles more or less intact as a result of "gentle" pretreatment (no cutting). (*From J. E. Melaragno and M. A. Walsh, Ultrastructural features of developing sieve elements in Lemna minor L.: The protoplast, Amer. J. Bot., 63:1145–1157, 1976*)

Figure 2 shows some features that are somewhat controversial, including the nucleus, tonoplast, P-protein, and sieve plate pores. The nucleus, if retained, appears degenerate or necrotic. The tonoplast persists and may delimit one or more vacuoles in shorter sieve elements (as in Fig. 2*b* and *e*) that are more or less in direct contact with cells providing the source of photoassimilates; or it may become fused with the tonoplasts of other sieve-tube members so as to constitute one continuous vacuole throughout a series of sieve-tube members (as shown in Fig. 2*d* and *f*). The P-protein is shown associated only with sieve-tube members of dicotyledons (Fig. 2*c*, *e*, and *f*) and it persists in the form of discrete bodies that are peripheral in distribution. It should be noted that the P-protein comes in a variety of forms (fibrillar, tubular, and crystalline) and is not restricted to the sieve-tube members but is sometimes found in other phloem parenchyma cells. The pores are relatively unoccluded and are lined by the plasmalemma, and some may be traversed by endoplasmic reticulum and portions of a tonoplast. Sieve-tube members are accompanied by companion cells that have branched cytoplasmic connections in their walls. The connections are shown unbranched in the wall of the sieve-tube members. Plastids, mitochondria, and smooth endoplasmic reticulum are present in sieve elements of both dicotyledons and monocotyledons. Though not distinguished in Fig. 2, the plastids of monocotyledons possess only the unusual crystalline, proteinaceous material, while those of dicotyledons may possess either proteinaceous inclusions or starch.

As a sieve element becomes nonfunctioning, its contents disintegrate. The callose first increases in amount and then disappears along with the connecting strands, leaving empty pores in the wall. The element becomes filled with air and may be crushed later by enlargement of surrounding parenchyma cells.

Sieve cells and sieve-tube members. Typical sieve cells are long elements in which all the sieve areas are of equal specialization, though sieve areas may be more numerous in some walls than in others. In contrast, a sieve-tube member has some sieve areas more specialized than others; that is, the pores, or modified plasmodesmata, are larger in some sieve areas. Parts of the walls containing such sieve areas are called sieve plates. Simple sieve plates have one

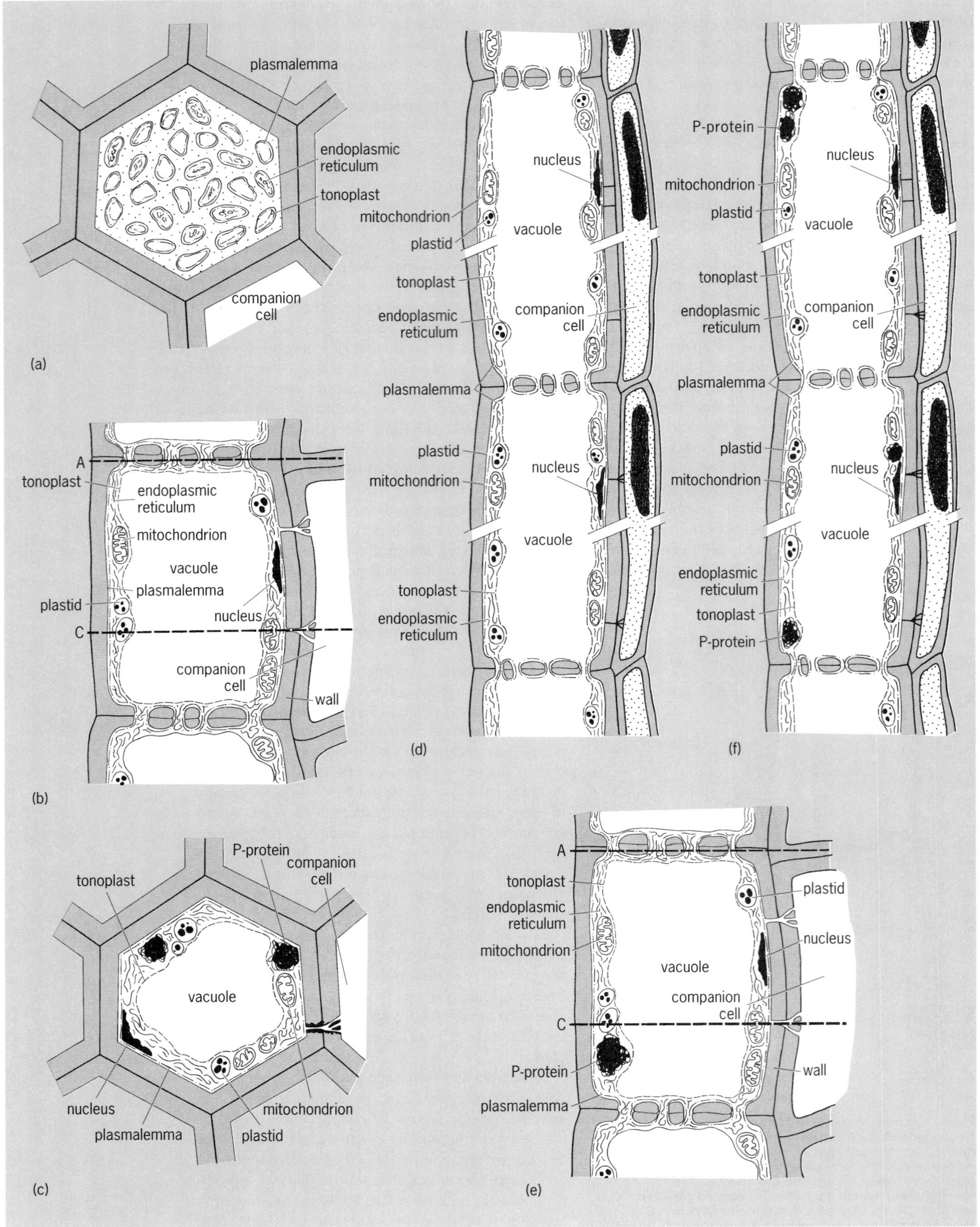

Fig. 2. Mature sieve elements in angiosperms: (*a,b,d*) portions of monocotyledon sieve elements; (*c,e,f*) dicotyledon sieve elements. In *b* and *e* the cell is bisected by lines A and C, referring to views shown in *a* and *c*.

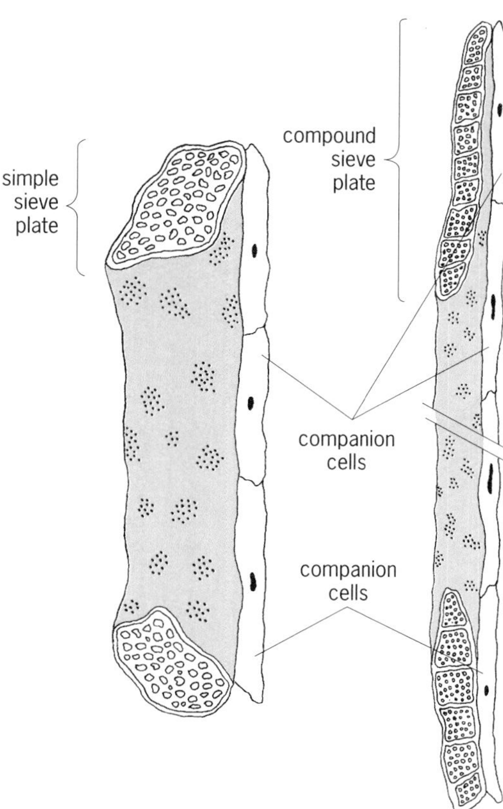

Fig. 3. Sieve-tube members and their sieve plates. Simple sieve plates are shown in the sieve element on the left, and compound sieve plates in the sieve element on the right. Lateral sieve areas are reduced in size and more or less evenly distributed along the lateral walls.

specialized sieve area that generally occurs on a transverse end wall; compound sieve plates have two or more on an oblique end wall (**Fig. 3**). Sieve tubes are composed of an indeterminate number of sieve-tube members arranged end to end. Sieve-tube members are shorter than sieve cells and have become progressively more so with evolutionary change.

Companion cells. Companion cells are specialized parenchyma cells that occur in close ontogenetic and physiologic association with sieve-tube members. They arise from the same meristematic cell that produces the sieve-tube member and vary in size, position, and number, but always retain their nucleus. Some sieve-tube members lack companion cells. The precise functional relationship between these two kinds of cells is unknown, but they become nonfunctioning simultaneously.

Parenchyma cells. Parenchyma cells in the phloem have thin or somewhat thick walls and occur singly or in strands of two or more cells. They store starch, frequently contain tannins or crystals, commonly enlarge as the sieve elements become obliterated, or may be transformed into sclereids or cork cambium cells. Parenchyma cells in secondary phloem may arise from a meristematic cell (phloem initial) that produces only such cells or from one that also eventually produces one or more sieve-tube members and companion cells. Parenchyma cells seem to intergrade with companion cells in the angiosperms.

Fibers. Phloem fibers vary greatly in length (from a fraction of a millimeter in some plants to 20 in. or 50 cm in the ramie plant). The secondary walls are commonly thick and typically have simple pits, but may or may not be lignified. In secondary phloem some fibers do not increase in length beyond the size of their primordia, but others may elongate extensively by apical intrusive growth. In primary phloem immature fibers elongate, sometimes hundreds of times over their original length. The fibers may become septate, are frequently multinucleate, and may intergrade with sclereids. *See* SCLERENCHYMA.

Primary phloem. Primary phloem differentiates from derivatives of the apical meristem. The earliest primary phloem (protophloem) contains sieve elements, with or without companion cells, and parenchyma cells. The sieve elements function for a brief time and then are usually obliterated. The remaining cells may become collenchymatous, as in many leaves, or be transformed into long protophloem fibers, often erroneously called pericyclic fibers. Metaphloem is formed after growth in length of surrounding cells is completed. Sieve elements, companion cells (in angiosperms), and parenchyma cells occur in such phloem, but typical fibers are generally lacking. If secondary phloem is absent, the metaphloem functions throughout the life of the plant. *See* APICAL MERISTEM.

Secondary phloem. Secondary phloem is produced by the same vascular cambium that forms secondary xylem. Such phloem consists of two interpenetrating systems, the vertical or axial and the horizontal or ray (**Fig. 4**). The phloem rays are basically similar to xylem rays, but their component cells differ in typically lacking secondary walls. Moreover, as the girth of the stem or root increases, the older phloem ray cells increase in width and may divide radially. This dilatation does not occur in all phloem rays,

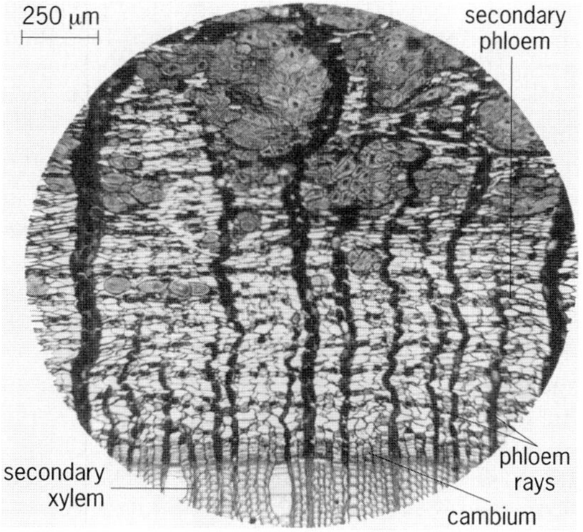

Fig. 4. Photomicrograph of cross section of the secondary phloem of paper birch (*Betula papyrifera*). (*Forest Products Laboratory, USDA*)

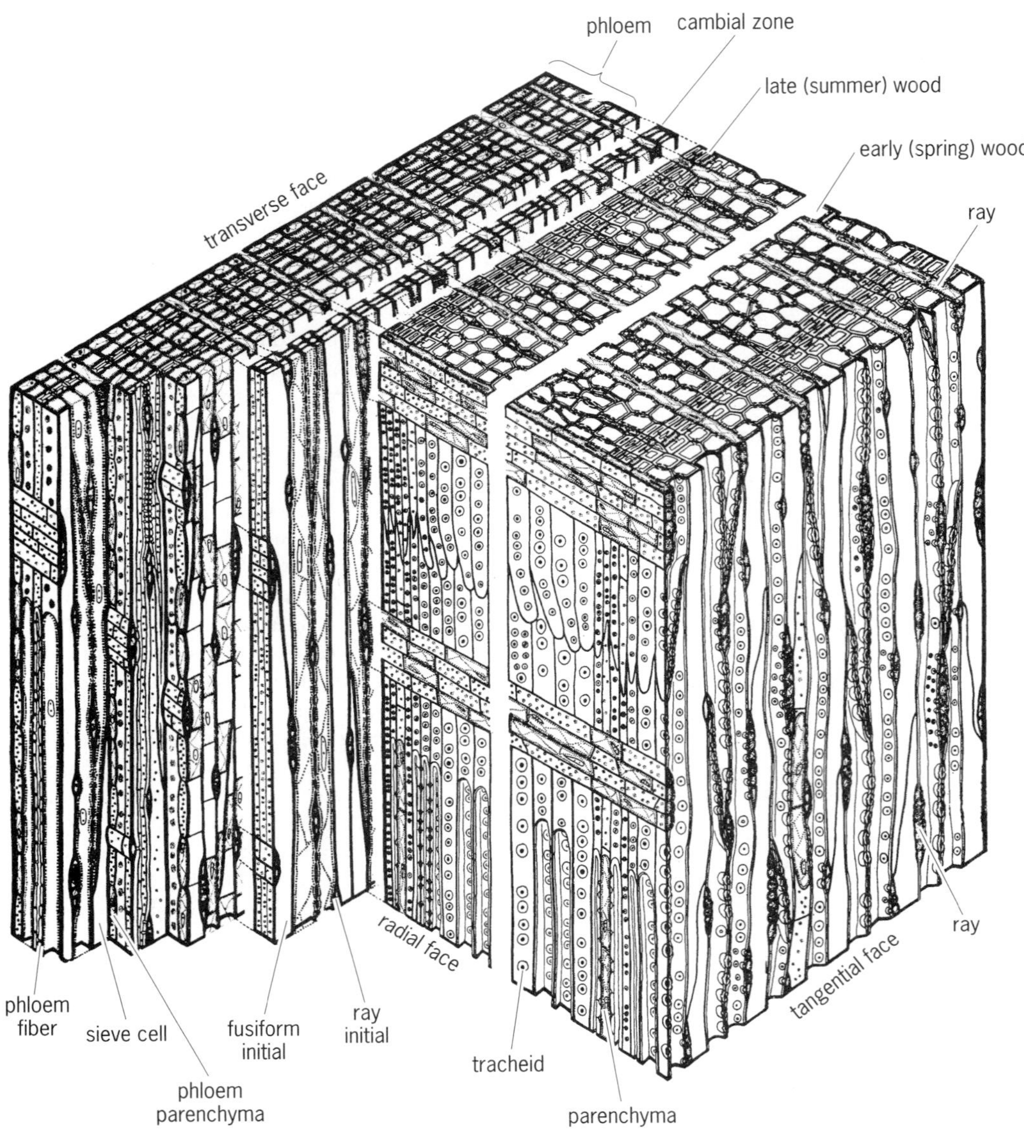

Fig. 5. Block diagram showing the three primary tissues: the secondary xylem, the cambial zone, and the secondary phloem of the conifer (gymnosperm) *Thuja*, or white cedar.

but it is a common feature of secondary phloem and stops only at the time of periderm formation within the ray. The vertical system contains sieve elements, parenchyma, often fibers or sclereids, and infrequently other elements such as laticifers. The fibers may occur singly, in dispersed groups, or in tangential bands.

Conifers and lower vascular plants. The phloem of conifers contains long sieve cells and parenchyma cells and frequently fibers. In secondary phloem of conifers (**Fig. 5**), these cells may be arranged in regularly alternating bands that give an orderly appearance to the phloem as seen in transection. In the ferns and fern allies (lower vascular plants) the phloem is mostly of a primary nature and also consists of sieve cells, parenchyma cells, and fibers. Special structures called refractive spherules occur in the sieve cells of lower vascular plants. Their origin and function are not fully understood, though they are believed to form from the endoplasmic reticulum and Golgi apparatus. Cytochemical studies reveal these to consist of proteinaceous material. *See* PINALES; POLYPODIOPHYTA.

Dicotyledons. The phloem in dicotyledons (**Fig. 6**) has greater diversity of cell structure and of arrangement than that in the conifers. It contains in varying proportions and groupings sieve-tube members, companion cells, parenchyma cells, and often fibers, sclereids, and various other kinds of cells or cell groups, such as secretory cells. The various cells may be arranged in alternating bands or have no regular spatial disposition. The functioning phloem is generally more orderly in appearance than the nonfunctioning. This difference results from partial or total collapse of the older sieve elements and associated companion cells, and frequently from the concurrent enlargement of neighboring parenchyma cells. *See* MAGNOLIOPSIDA.

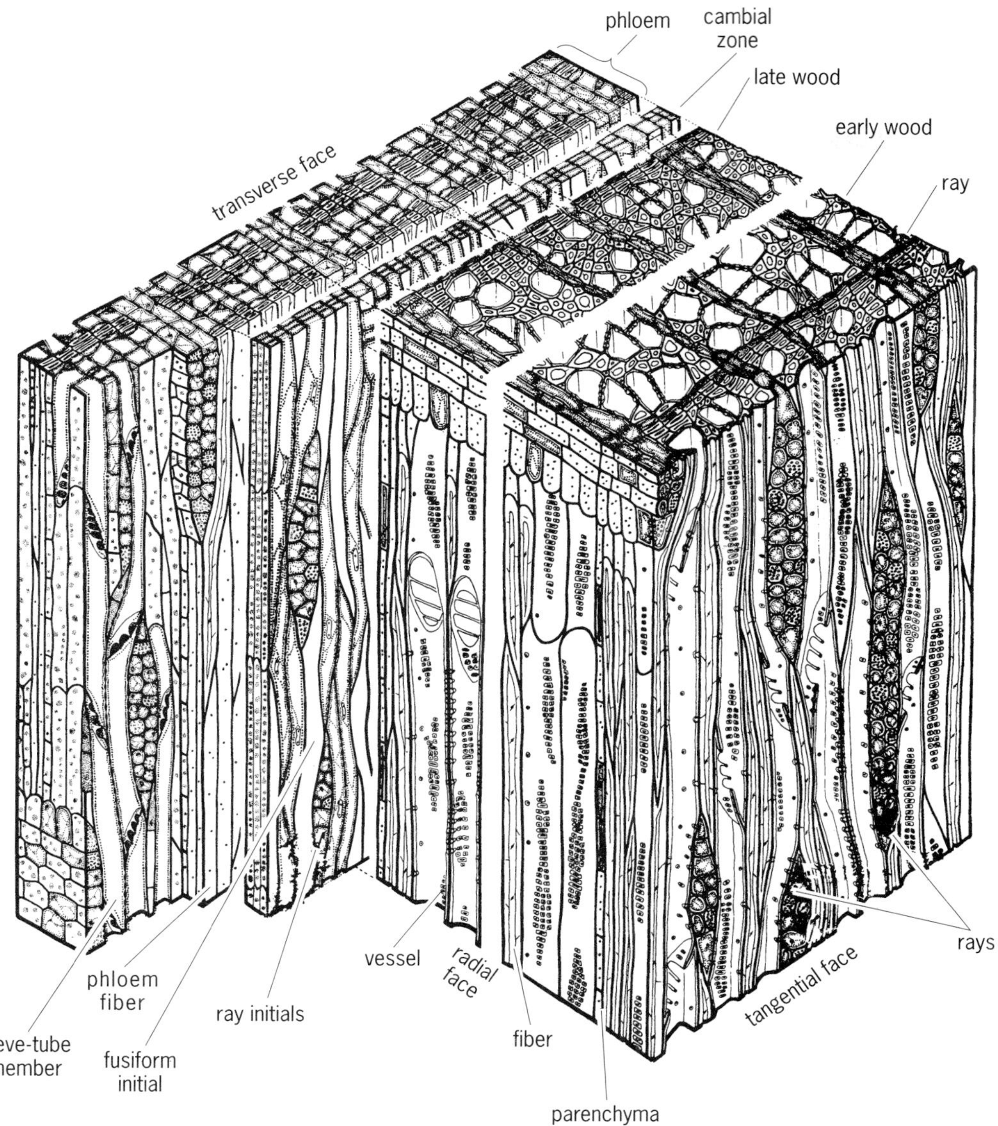

Fig. 6. Block diagram showing the three primary tissues: the secondary xylem, the cambial zone, and the secondary phloem of the dicotyledon (angiosperm) *Liriodendron* (tulip tree).

Moncotyledons. Phloem in monocotyledons is much like that in dicotyledons. However, because no secondary growth is derived from a vascular cambium, there is no secondary phloem in monocotyledons. For this reason much of the phloem, at least in perennial monocotyledons, remains functional for 50–100 years or more, unlike that of some species of woody dicotyledons which might maintain functional phloem for only 5–10 years. In addition, sieve-tube members in monocotyledons appear to lack the P-protein material associated with sieve-tube members of dicotyledons. On the other hand, the sieve-tube members of all monocotyledons thus far investigated possess plastids that contain quasicrystalline, proteinaceous materials. *See* LILIOPSIDA.

Nonvascular plants. Although it may seem a contradiction, there are some nonvascular plants that contain conducting tissues, and cells which may be reminiscent of sieve elements. Examples are the larger mosses (bryophytes) and members of the brown algae (Phaeophyta). Here the special conducting cells are called leptoids and trumpet hyphae, respectively. *See* BRYOPHYTA; PHAEOPHYCEAE; PLANT TRANSPORT OF SOLUTES.

Michael A. Walsh

Bibliography. H. D. Behnke and R. D. Sjolund (eds.), *Sieve Elements*, 1990; J. Cronshaw, W. J. Lucas, and R. T. Giaquinta (eds.), *Phloem Transport*, vol. 1, 1987; A. Fahn, *Plant Anatomy*, 1990.

Phlogopite

A mineral with an ideal composition of $KMg_3(AlSi_3)O_{10}(OH)_2$. Phlogopite belongs to the mica mineral group. It has been occasionally called bronze mica.

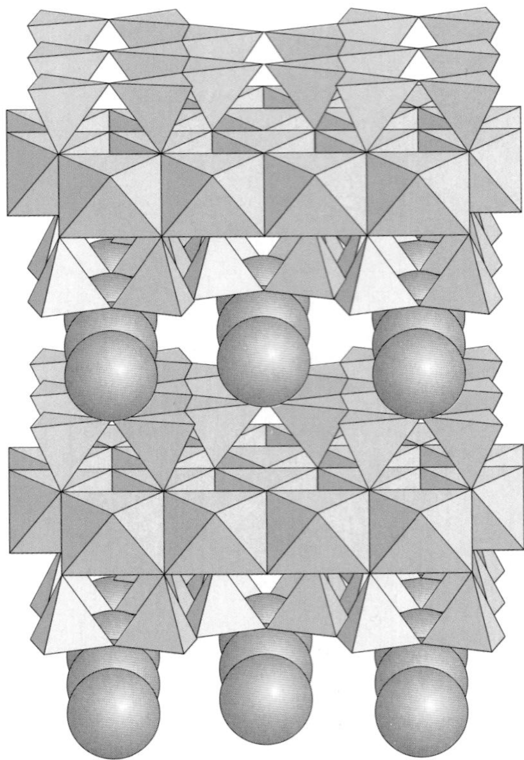

Polyhedral model illustrating the one-layer monoclinic phlogopite atomic structure. The structural unit consists of only one layer; note that both layers shown are identical. The corners of the tetrahedra represent oxygen atoms, whereas octahedral corners represent oxygen or hydroxyl. Potassium ions are depicted as large spheres between and connecting the layers.

Phlogopite is a trioctahedral mica, where all three possible octahedral cation sites are occupied by magnesium (Mg). The magnesium octahedra, $Mg(O,OH)_6$, form a sheet by sharing edges (see **illus.**). As in all micas, tetrahedra are located on either side of the octahedral sheet, which may be occupied by aluminum (Al) or silicon (Si). Adjacent tetrahedra share corners to form a two-dimensional network of sixfold rings, thus producing a tetrahedral sheet. Two opposing tetrahedral sheets and the included octahedral sheet form a 2:1 layer. Potassium (K) ions are located between adjacent tetrahedral sheets in the interlayer region.

Solid solutions between phlogopite and siderophyllite or between phlogopite and lepidomelane are complete, balanced by substitutions of aluminum and silicon, or vacancies in the sites usually occupied by magnesium. Proper usage of the term phlogopite refers to end members, for example, with less than 10% substitutions for magnesium. In the past, the term phlogopite was often used for samples greater than 70% magnesium. Biotite is used to describe the series along or near the hypothetical annite-phlogopite join. Solid solutions between phlogopite and the dioctahedral micas are quite limited. *See* BIOTITE; SOLID SOLUTION.

A perfect basal cleavage occurs because of the layer structure. Specific gravity is 2.86, hardness on the Mohs scale is 2.5–3.0, and luster is vitreous to pearly. Thin sheets are flexible. Color is yellow brown, reddish brown, or green, and thin sheets are transparent. Thermal stability varies greatly with composition, with iron or fluorine substitutions reducing or increasing stability, respectively. Weathering of phlogopite may produce vermiculite. *See* HARDNESS SCALES; VERMICULITE; WEATHERING PROCESSES.

Phlogopite occurs in marbles produced by the metamorphism of siliceous magnesium-rich limestones or dolomites and in ultrabasic rocks, such as peridotites and kimberlites. *See* DOLOMITE; LIMESTONE; PERIDOTITE.

Phlogopite is used chiefly as an insulating material and for fireproofing. It has high dielectric properties and high thermal stability. *See* MICA; SILICATE MINERALS.

Stephen Guggenheim

Bibliography. S. W. Bailey (ed.), *Micas*, Reviews in Mineralogy, vol. 13, 1984.

Phobia

An intense irrational fear that often leads to avoidance of an object or situation. Phobias (or phobic disorders) are common (for example, fear of spiders, or arachnophobia; fear of heights, or acrophobia) and usually begin in childhood or adolescence. Psychiatric nomenclature refers to phobias of specific places, objects, or situations as specific phobias. About one in two individuals report that they are afraid of public speaking; in very severe cases, this is considered a form of social phobia. Social phobias also include other kinds of performance fears (such as playing a musical instrument in front of others; signing a check while observed) and social interactional fears (for example, talking to people in authority; asking someone out for a date; returning items to a store). Individuals who suffer from social phobia often fear a number of social situations.

Types. Although loosely regarded as a fear of open spaces, agoraphobia is actually a phobia that results when people experience panic attacks. Panic attacks are unexpected, paroxysmal episodes of anxiety and accompanying physical sensations (for example, racing heart; shortness of breath) that can occur at any time in susceptible individuals. When individuals have recurrent panic attacks, they often begin to avoid situations where they believe that a panic attack might occur. Agoraphobia develops in approximately two out of three individuals who experience recurrent, spontaneous panic attacks. School phobia is a misnomer, because affected children do not actually fear school. Some fear being away from their parents; that is, they suffer from separation anxiety. Others fear having to perform in front of or interact with their peers or teachers; that is, they suffer from social phobia. Individuals with phobias about dirt and germs may suffer from an anxiety disorder called obsessive-compulsive disorder. Their dirt and germ phobia is a manifestation of an obsession with contamination fears.

Origins. The origin of phobias is varied and incompletely understood. Most individuals with specific phobias have never had anything bad happen to them in the past in relation to the phobia. In a minority of cases, however, some traumatic event occurred that likely led to the phobia. It is probable that some common phobias, such as a fear of snakes or a fear of heights, may actually be instinctual, or inborn. Both social phobia and agoraphobia run in families, suggesting that heredity plays a role. However, it is also possible that some phobias are passed on through learning and modeling, such as when a child sees a parent shy away from encounters with strangers.

Occurrence. Phobias occur in over 10% of the general population. Social phobia may be the most common kind, affecting approximately 7% of individuals. Agoraphobia occurs in approximately 1% of the general population. Most phobias are distressing to the affected person, and in some cases they are even disabling. Some people with social phobia quit school because of the intense distress they experience when they give an oral presentation to the class. Others avoid so many situations where they fear a panic attack that they become housebound.

Reaction. When persons encounter the phobic situation or phobic object, they typically experience a phobic reaction consisting of extreme fearfulness, physical symptoms (such as racing heart, shaking, hot or cold flashes, or nausea), and cognitive symptoms (particularly thoughts such as "I'm going to die" or "I'm going to make a fool of myself"). These usually subside quickly when the individual is removed from the situation. The tremendous relief that escape from the phobic situation provides is believed to reinforce the phobia and to fortify the individual's tendency to avoid the situation in the future.

Treatment. Many phobias can be treated by repeated and prolonged exposure of the individual to whatever is feared. Often, the individual is gradually encouraged to approach the feared object and to successively spend longer periods of time in proximity to it. In some cases, the individual may just have to imagine that he or she is in the situation, but usually it is more effective to have the individual actually encounter it. Cognitive therapy is also used (often in conjunction with exposure therapy) to treat phobias. It involves helping individuals to recognize that their beliefs and thoughts can have a profound effect on their anxiety, that the outcome they fear will not necessarily occur (for example, the likelihood of their plane crashing is very low), and that they have more control over the situation than they realize.

Medications are sometimes used to augment cognitive and exposure therapies. For example, musicians with performance phobias (stage fright) sometimes use beta-adrenergic blocking agents, such as propranolol, prior to the performance. This lowers their heart rate and reduces tremulousness, and leads to reduced anxiety. For agoraphobia and social phobia, certain kinds of antidepressants and anxiolytic medications are often helpful. It is not entirely clear how these medications exert their antiphobic effects, although it is believed that they affect levels of neurotransmitters in regions of the brain that are thought to be important in mediating emotions such as fear. *See* NEUROTIC DISORDERS. Murray B. Stein

Bibliography. American Psychiatric Association, *Diagnostic and Statistical Manual of Mental Disorders*, 4th ed., 1994; R. J. McNally, *Panic Disorder: A Critical Analysis*, 1994; M. B. Stein, *Social Phobia: Clinical and Research Perspectives*, 1995.

Phoenicopteriformes

The flamingos, a small monotypic order of wading birds that includes the family Phoenicopteridae, which has six species found worldwide in tropical marine and fresh waters; one species lives in the high Andes. The flamingos were formally included in the Ciconiiformes and are still placed there by many researchers; others have advocated a close relationship with either the Anseriformes or the Charadriiformes. None of these classifications, however, are strongly supported by available evidence, and so it is best to place these bizarre birds in a unique order. *See* ANSERIFORMES; CHARADRIIFORMES; CICONIIFORMES.

Fossil record. Several Cretaceous fossils, including an Early Cretaceous form, had been labeled flamingos; however, such attributions are most dubious. The earliest definite flamingo fossil is *Juncitarsus*, from the middle Eocene of Wyoming; it possesses several primitive traits for the family. Beginning in the late Oligocene, modern flamingos are found in the fossil records from all areas of the world.

Characteristics. Flamingos are long-legged, long-necked wading birds with a thick bill which is sharply bent downward at the midpoint. The bill is deep, with many transverse lamellae on both the mandible and maxilla, and the tongue is thick and fleshy. During feeding, the head is held upside down with the anterior portion of the bill parallel to and just below the water surface. Water is pumped through the bill by the tongue so that small food particles are trapped between the lamellae. The three anterior toes are webbed, possibly for walking on soft substrates. The long and broad wings enable the bird to fly well. The adult plumage is pink to light red, with black flight feathers. Flamingos are gregarious, often congregative in flocks in excess of 1 million. They breed monogamously in large colonies, and both parents incubate the one or two eggs in a nest formed of a stout pillar of mud 1 ft (0.3 m) high on a mud flat. The downy, gray young leave the nest after hatching to be cared for by both parents, while the older young are kept together in large groups. Most flamingos are found in tropical waters, either in shallow lakes or along the coasts, although one species lives on shallow lakes in the high Andes. *See* AVES. Walter J. Bock

Bibliography. J. Kear and N. Duplaix Hall, *Flamingos*, 1975.

Pholidophoriformes

An extinct actinopterygian group composed of mostly small fusiform fishes of an advanced holostean level that are found in both marine and fresh-water deposits and range from the Middle Triassic to the Lower Cretaceous. They are known from nearly all of the continents but first appeared in Europe, and they are especially common in deposits that were formed along the margins of the ancient Tethys Sea.

Pholidophoriforms are the forerunners of the teleostean stage of development within the actinopterygians. They show many transitional features between the holosteans and the lower teleosts as represented by the Leptolepiformes. For this reason they sometimes are placed in a separate subclass or superorder, the Halecostomi. It is believed that teleostean ancestry can be traced back to the basal members of the most generalized family, the Pholidophoridae. The best-known representative of this family is *Pholidophorus bechei* (see **illus.**), which exhibits holostean features in its enamel-covered ganoid scales, fin rays, and head bones; in the fulcra bordering all the fins with a strong series on the upper caudal lobe; and in the structure of its caudal skeleton. In later pholidophorids there is a tendency for the dermal bones to lose their enamel covering and for the scales to become thin and rounded. Teleostean features are found in the structure of the neurocranium and in the pattern of the sheathing bones on the side of the head. Of particular importance is the shape of the preopercle and its relation to the bones behind the eye, as well as to the underlying hyomandibula and to the opercular series. The skull of *Pholidophorus* is much like that of the contemporaneous *Leptolepis*.

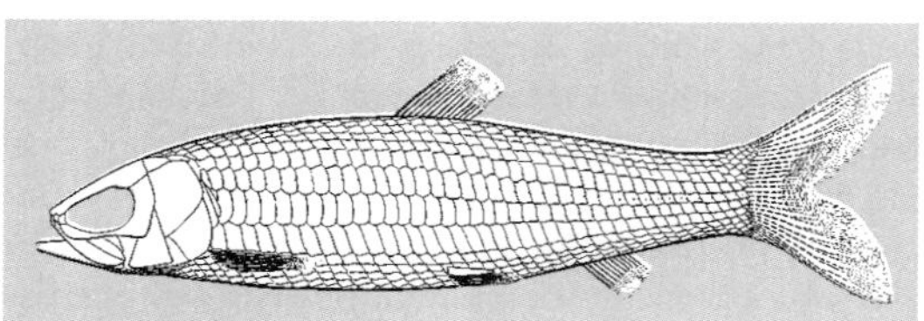

***Pholidophorus bechei*, Lower Jurassic of England; length to 8 in. (20 cm). (*After D. Rayner, The structure of certain Jurassic holostean fishes with special reference to their neurocrania, Phil. Trans. Roy. Soc. London, Ser. B, no. 601, Cambridge University Press, 1948*)**

The axial skeleton is occasionally preserved and shows thin, ringlike vertebral centra. Small ribs and sometimes upper intermuscular bones are present. The caudal skeleton lacks the paired uroneurals found in lower teleosts.

The structure of the mouth with its feeble dentition and the body form suggest that *Pholidophorus* was an open-water, free-swimming plankton feeder. Other pholidophoriforms possessing larger mouths equipped with stronger teeth were probably predacious types. *See* ACTINOPTERYGII; HOLOSTEI; LEPTOLEPIFORMES; TELEOSTEI. Ted M. Cavender

Bibliography. E. Jarvik, *Basic Structure and Evolution of Vertebrates*, 2 vols., 1981; D. V. Obruchev, *Fundamentals of Paleontology*, vol. 11: *Agnatha, Pisces*, 1967; J. Piveteau (ed.), *Traité de Paléontologie*, 1966; A. S. Romer, *Vertebrate Paleontology*, 1966.

Pholidota

An order of mammals comprising the living pangolins, or scaly anteaters, and their poorly known fossil predecessors. All living pangolins are assigned to the genus *Manis*. They are found in Africa south of the Sahara and in southeastern Asia, including certain islands of the East Indies.

Pangolins feed principally on termites and ants. The elongate tubular skull without teeth, long protrusive tongue, small eyes with heavy eyelids, thick skin, strong legs, five-toed feet with large claws, and large tail enable these unique animals to rip open ant nests and termite dens and devour the animals therein. The greatest peculiarity of animals in the genus *Manis* is a covering of all but the undersides of the body by an armor of large imbricating dermal horny scales. Living pangolins are frequently characterized as being animated pine cones. The position and number of hairs in relation to the scales are peculiar to each modern species.

Scattered and rare fossil bones from the Oligocene and Miocene rocks of western Europe have been named *Necromanis*, *Leptomanis*, *Teutomanis*, and *Galliaetatus*, but the validity and complete significance of these names remain doubtful. *Manis* or a closely related genus is found in the Pleistocene of Hungary and southeastern Asia.

Although the phyletic origin of the Pholidota is conjectural, it is believed that the general anatomical similarities of these scaly anteaters to the true anteaters (order Edentata) and to the aardvarks (order Tubulidentata) are the product of convergent evolution between three groups of distantly related eutherian mammals. A common ancestor for anteaters, exclusive of the marsupial anteaters, may be represented by a Late Cretaceous eutherian that is close to or in the order Insectivora. *See* EDENTATA; EUTHERIA; INSECTIVORA; MAMMALIA; TUBULIDENTATA. Donald E. Savage

Bibliography. M. J. Benton, *Vertebrate Paleontology*, 1991; R. L. Carroll, *Vertebrate Paleontology and Evolution*, 1988.

Phonetics

The science that deals with the production, transmission, and perception of spoken language. At each level, phonetics overlaps with some other sciences, such as anatomy, physiology, acoustics, psychology, and linguistics. In each case, phonetics focuses on phenomena relevant to the study of spoken language. The outline presented below follows a speech

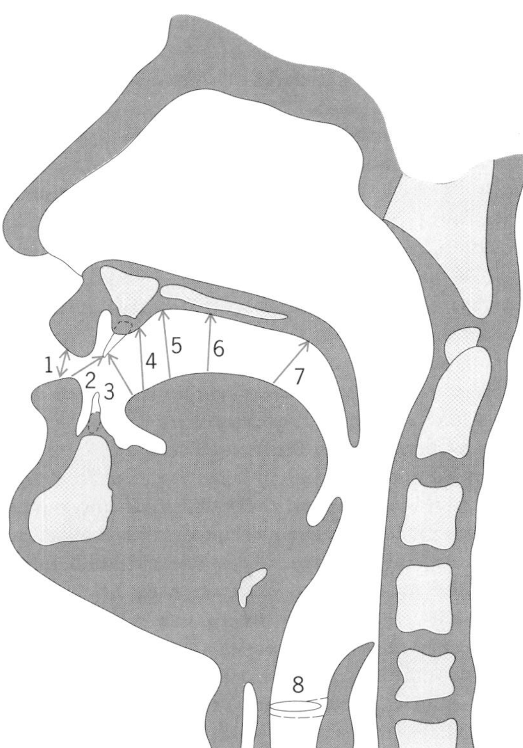

Fig. 1. Sagittal section of the head, showing the vocal tract and identifying points of articulation used in English. 1 = bilabial; 2 = labiodental; 3 = dental; 4 = alveolar; 5 = palatoalveolar; 6 = palatal; 7 = velar; 8 = glottal.

signal from its production by the speaker through the medium of transmission to its perception by the hearer.

Articulation. Speech is normally produced by exhaling air from the lungs through the vocal tract. (Other airstream mechanisms are used in some languages.) The vocal tract extends from the larynx through the pharynx and the oral cavity to the lips. If the velum (soft palate) is not raised, the air also passes through the nasal cavities. The shape and size of the oral cavity can be varied by the movement of active articulators: tongue, lips, and velum. *See* PALATE.

Phoneticians usually describe speech sounds with reference to their point (or place) of articulation (**Fig. 1**) and their manner of articulation. The point of articulation of a sound is the place of maximum constriction within the vocal tract. The great majority of sounds are produced by moving some part of the tongue toward some region on the roof of the mouth. Exceptions are articulations involving lips and those sounds in which the vocal folds serve as articulators (see **table**).

Manners. At most of these points of articulation, sounds can be produced with several manners of articulation. One way to classify manners of articulation refers to the degree of stricture employed in producing the sound. Sounds produced with complete constriction of the vocal tract are stops, or plosives. The term "stop" refers to the fact that the airstream is completely stopped at the point of articulation. The term "plosive" focuses on another aspect of the production of stops: the fact that if a complete closure is formed in the vocal tract, while the lungs continue to produce an egressive airstream, there is a pressure buildup behind the closure until the hold is suddenly released, resulting in a weak explosion.

If the closure is incomplete, but the articulators are brought close enough so that the air passing between them is set into turbulent motion, the resultant sounds are fricatives or spirants. If the articulators are approximated but the constriction remains large enough that air can pass through without friction, the sounds are called approximants—vowellike sounds functioning as consonants. The distinction between vowels and consonants is based primarily on their function: vowels may constitute a syllable nucleus and may be produced in isolation, without any accompanying consonants, while consonants ordinarily constitute syllable margins and normally are not produced in isolation. There are numerous counterexamples, though; in some languages, [r] and [l] sounds can be syllable nuclei, for instance.

Other manners of articulation include nasals, which are produced with a lowered velum, so that the airstream passes through the nasal cavity while an articulation is maintained in the oral cavity; laterals, which are produced in such a way that the airstream passes over one side of the tongue while a constriction is formed by the other side; trills, which are produced by adjusting the tension of the muscles of the articulators and the rate of airflow in

The points of articulation used in English and sample sounds produced at them

Point of articulation	Articulators and movement	Example
Bilabial	Lips	Initial sound of *peel*
Labiodental	Lower lip against upper teeth	Initial sound of *feel*
Dental	Tip of tongue against upper teeth	Initial sound of *thin*
Alveolar	Tip of tongue against the alveolar ridge	Initial sound of *tea*
Palatoalveolar	Blade of the tongue toward a region between the alveolar ridge and the hard palate	Initial sound of *she*
Palatal	Central part of the tongue toward the hard palate	Initial sound of *hue*
Velar	Back of the tongue against the soft palate	Initial sound of *cool*
Glottal	Vocal folds serving as articulators	Occurs in English at the beginning of emphatically produced vowel-initial words

consonants	bilabial	labiodental	dental and alveolar	retroflex	palato-alveolar	alveolo-palatal	palatal	velar	uvular	pharyngeal	glottal
plosive	p b		t d	ʈ ɖ			c ɟ	k g	q ɢ		ʔ
nasal	m	ɱ	n	ɳ			ɲ	ŋ	ɴ		
lateral			l	ɭ			ʎ				
lateral fricative			ɬ ɮ								
rolled			r						ʀ		
flapped			ɾ	ɽ					ʀ		
rolled fricative			ɼ								
fricative	ɸ β	f v	θðlszlɹ	ʂ ʐ	ʃ ʒ	ɕ ʑ	ç j	x ɣ	χ ʁ	ħ ʕ	h ɦ
frictionless continuants and semivowels	w|ɥ	ʋ	ɹ				j (ɥ)	(w) ɣ	ʁ		

(a)

	bilabial	labiodental	dental	alveolar	postalveolar	retroflex	palatal	velar	uvular	pharyngeal	glottal
plosive	p b			t d		ʈ ɖ	c ɟ	k g	q ɢ		ʔ
nasal	m	ɱ		n		ɳ	ɲ	ŋ	ɴ		
trill				r					ʀ		
tap or flap				ɾ		ɽ					
fricative	ɸ β	f v	θ ð	s z	ʃ ʒ	ʐ ʑ	ç j	x ɣ	χ ʁ	ħ ʕ	h ɦ
lateral ficative				ɬ ɮ							
approximant				ɹ		ɻ	j	ɰ			
lateral approximant				l		ɭ	ʎ	ʟ			
implosive	p'			t'		ʈ'	c'	k'	q'		
ejective stop	ɓ ɓ			ƭ ɗ			ƈ ʄ	ƙ ɠ	ʠ ʛ		

(b)

Fig. 2. Consonant chart of the International Phonetic Association. (*a*) Chart used in phonetic transcription before 1989 (*after The Principles of the International Phonetic Association, International Phonetic Association, London, 1949, reprint 1961*). (*b*) Revised chart adopted in 1989; where symbols appear in pairs, the one to the right represents a voiced consonant (*after Report on the 1989 Kiel Convention, J. Int. Phonetic Ass., 19(2):67–80, 1989*).

such a manner that an articulator is set into vibration; taps and flaps, in the production of which an articulator executes a single rapid closure. Most of these sounds can be produced at several points of articulation. Nasals are produced at the same points of articulation as stops (nasals are sometimes called nasal stops); laterals may be dental, alveolar, palatal, or even velar; trills can be produced with the lips, the tip of the tongue, or the uvula; and taps and flaps are produced with the tip of the tongue or the uvula. Symbols for consonant sounds are shown in **Fig. 2**.

Voiced and voiceless sounds. Most of these consonant sounds can be voiced or voiceless; vowels are normally voiced. The terms "voiced" and "voiceless" refer to the presence and the absence of vocal fold vibration. According to the generally accepted myoelastic-aerodynamic theory of phonation, the process consists of the following steps. The vocal folds are approximated and tensed through the action of the laryngeal musculature. The airstream passing through the narrowed opening between the vocal folds (the glottis) increases its velocity, and a negative pressure is generated (the Bernoulli effect) that pulls the folds together. Air pressure from the lungs builds up behind the vocal folds and blows them open. The elasticity of the vocal fold muscles, their inward momentum, and the Bernoulli force cause the vocal folds to return to their original position and close, and thus the cycle is repeated over and over again—the vocal folds are set in vibration, which continues as long as the aerodynamic forces and muscular tension are maintained at appropriate levels. (In certain types of phonation the vocal folds do not close completely as part of the cycle.) Research, primarily utilizing the microbeam technique, has greatly clarified the functions of the individual laryngeal muscles in producing various types of phonation and in controlling the rate of vocal-fold vibration. *See* BERNOULLI'S THEOREM.

Classifying vowels. There are three generally recognized methods for classifying vowels. The traditional method of classification employs the notion "highest point of the tongue" and locates that highest point in a three-dimensional space whose axes are the high-low (vertical) dimension, the front-back (horizontal) dimension, and lip-rounding. While this system has some pedagogical utility, x-ray studies of vowel articulation have shown that the notion "highest point of the tongue" cannot be supported by observation. A modification of this system was developed by Daniel Jones, whose Cardinal Vowel System (**Fig. 3*a***) is based on the following principles. There are two "hinge" vowels that are articulatorily defined: a high front vowel (Cardinal Vowel One) produced with a tongue position as far front as possible and as high as possible without generating friction, and the lowest possible back vowel (Cardinal Vowel Five). Based on these two "hinge" vowels, there are six additional cardinal vowels, for a total of four front vowels and four back vowels, produced so that the "highest point of the tongue" for each of them lies on the extreme outside limits of the articulatory vowel space, and so that they are at equal distances from each other. The cardinal vowels are arbitrarily selected reference points (they do not necessarily coincide with any vowels in a given language), and thus can be used to describe all languages of the world. Of the original eight cardinal vowels, five were unrounded and three rounded; later additions to the system

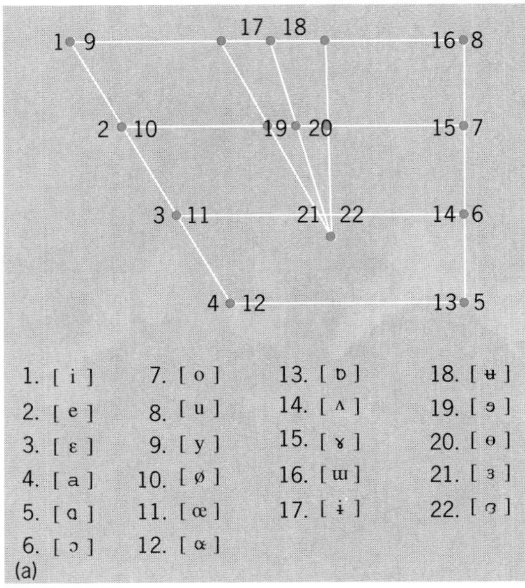

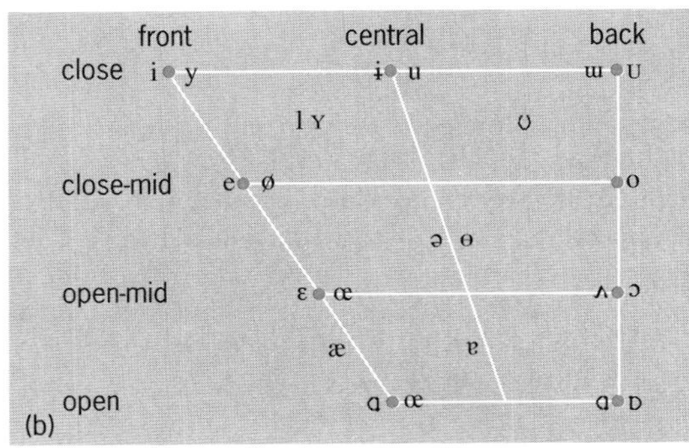

Fig. 3. Charts of vowel sounds. (*a*) Diagram showing the positions of Daniel Jones's Cardinal Vowels. The symbols for the Cardinal Vowels are identified by 1–22. (*b*) Revised chart adopted by the International Phonetic Association in 1989. Where symbols appear in pairs, the one to the right represents a rounded vowel. (*After Report on the 1989 Kiel Convention, J. Int. Phonetic Ass., 9(2):67–80, 1989*)

include the missing rounded/unrounded counterparts of the original (primary) cardinal vowels as well as a series of central vowels.

Attempts have been made to bring the International Phonetic Alphabet more in line with contemporary linguistic thought. One of these attempts tries to incorporate the insights of distinctive features theory into the classification of speech sounds. In phonological theory, the fundamental units of speech are not phonemes (contrastive speech sounds) but distinctive features, which can be used to represent the sounds of all languages of the world. Groups of features tend to be implemented simultaneously to form segments, which have been called bundles of distinctive features. Efforts have been made to develop articulatory as well as acoustic-perceptual correlates of the features. One inventory of features organizes them into two groups depending on whether the vocal tract is relatively open or constricted, corresponding to vocalic and nonvocalic features. The vocalic features are high, low, back, round, nasal, spread glottis, and constricted glottis. The nonvocalic features are sonorant, continuant, coronal, strident, consonantal, anterior, lateral, distributed, and voice. The International Phonetic Association charts in Figs. 2*a* and 3*a* remain a basic tool for interpreting phonetic transcriptions made up to 1989. However, in 1989 the Council of the International Phonetic Association approved a revised version of these charts. The revised consonant chart (Fig. 2*b*) contains columns for the following places of articulation: bilabial, labiodental, dental, alveolar, postalveolar, retroflex, palatal, velar, uvular, pharyngeal, and glottal. The manners of articulation included in the revised consonant chart are plosive, nasal, trill, tap or flap, fricative, lateral fricative, approximant, lateral approximant, ejective stop, and implosive. Modifications of the vowel chart (Fig. 3*b*) are less extensive. The Council also approved various changes in the use of diacritics.

Jones had originally assumed that the equal distance between the vowels meant articulatory distance, but since x-rays showed this not to be true, the definition was changed to refer to auditory equidistance. This definition retains a degree of subjectivity. A relatively more objective way to describe vowels is provided by acoustic analysis.

Acoustic phonetics. This branch of phonetics deals with the manner in which the spoken message is encoded in the sound waves. According to the generally accepted source-filter theory of speech acoustics, sound is generated at a source (which for phonated speech is constituted by the vibrating vocal folds) and passed through the vocal tract. The opening and closing of the vocal folds creates a succession of condensations and rarefactions of air molecules—variations in air pressure—and transforms kinetic energy into acoustic energy. The sound wave that is generated at the glottis can be considered, for practical purposes, a complex periodic wave, and as such it contains energy at frequencies that are multiples of the fundamental frequency (harmonics). *See* HARMONIC (PERIODIC PHENOMENA); SOUND.

The vocal tract acts as a filter, transmitting more energy at those frequencies that correspond to the resonances of the vocal tract than at other frequencies. This modified sound wave is then radiated through the mouth-opening at the lips and propagated through the outside air. Energy concentrations at the resonance frequencies of the vocal tract are referred to as formants (the formants may also be defined as the resonances of the tract).

In principle, the source and filter are independent of each other; consider the fact that the same vowel can be sung at different fundamental frequencies (pitches), and different vowels can be produced at the same pitch. The sound wave can be described by specifying its fundamental frequency, amplitude, and spectrum. The fundamental frequency corresponds to the rate of vibration of the vocal folds. Amplitude is dependent upon the rate of airflow and the efficiency with which muscular energy is converted to acoustic energy in the production of the sound. The

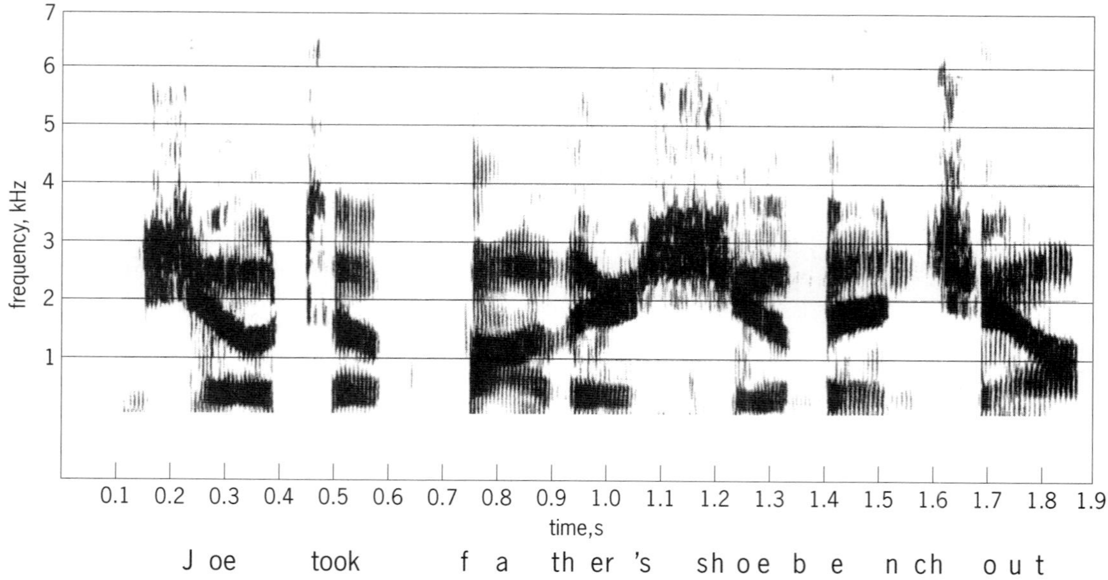

Fig. 4. Broadband spectrogram of the utterance "Joe took father's shoebench out," produced by a male speaker.

spectrum—the distribution of energy in the frequency domain—depends on the size and shape of the vocal tract. In perception, fundamental frequency is correlated with perceived pitch, amplitude with perceived loudness, and spectrum with the phonetic quality of the generated speech sounds. The sound source can also be a nonperiodic one, which is the case with fricatives and the release-bursts of stops. *See* LOUDNESS; PITCH.

Sound spectrograph. While many instruments have been developed for the study of the acoustic structure of spoken language, the sound spectrograph remains a basic tool. The spectrograph is an electronic device that analyzes the acoustic signal into its constituent parts and converts it into a visual display, with time represented on the horizontal axis, frequency on the vertical axis, and intensity conveyed through the darkness of the patterns traced on the spectrogram by a stylus (**Fig. 4**). The selection of analyzing filters makes it possible to concentrate either on the harmonics or on the formant structure of the speech signal. A study of spectrograms of vowels reveals that as a first approximation, the vowels produced with a high tongue position have a first formant with a relatively low frequency, while vowels produced with a low tongue position have a high first formant. Front vowels have relatively high second formants, while back vowels have low second formants. A standard acoustical vowel diagram displays the position of vowels in a two-dimensional acoustic space whose coordinates are the first and second formants (**Fig. 5**). Since pitch perception is nonlinear, it is customary to employ logarithmic scales.

Speech perception. Acoustic phonetics makes it possible to describe what is present in the signal, but such analysis does not explain how the discovered patterns are relevant in speech perception. Various speech synthesizers enable the investigator to produce modifications of the signal and study their perceptual significance. *See* VOICE RESPONSE.

Cues for the perception of the manner of articulation are generally present during the most constricted or central segment of the articulated sound. For the vowels, these cues consist of formant positions and the relative intensities of the formants (for most vowels, the first three formants provide sufficient information for perceptual identification). For consonants, the class of plosives is recognized on the basis of the presence of a gap (corresponding to the closure of the stop) followed by a sudden short burst of energy (corresponding to the explosive release). The class of fricatives is recognized by the presence of noise (aperiodic energy). Nasals are characterized by the presence of additional nasal formants and by the wide bandwidth and relatively weak intensity of the first (oral) formant. Laterals have vowellike formants in the lower part of the spectrum and an

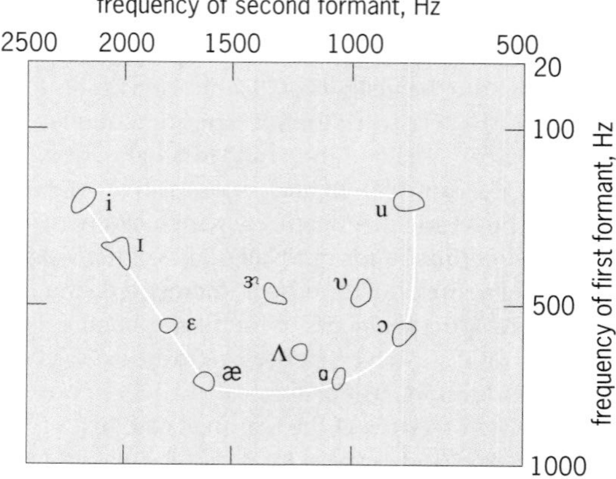

Fig. 5. Acoustical vowel diagram of vowels occurring in the English words *heed, hid, head, had, hod, hawed, hood, who'd, hud, and heard*.

antiresonance in the region where the third formant would be expected.

The point of articulation of a consonant is reflected in the influence of the consonant on an adjacent vowel. Particularly significant perceptually is the frequency position of the second formant at the moment of transition between the consonant and the vowel. Sounds produced at the same point of articulation are characterized by the same frequency region at which the second formant starts its movement toward the target position of the vowel, or toward which it moves in the transition from the vowel toward a postvocalic consonant. This spectral region is called the hub or locus of the consonant. The point of articulation may also be deduced from spectral cues, especially in the case of fricatives and plosive bursts.

Prosodic aspects of language. The subject matter of phonetics is not limited to the production and perception of vowels and consonants; of equal importance are such prosodic and suprasegmental aspects of spoken language as duration, fundamental frequency, and intensity. Suprasegmental features relate to segmental features by constituting an overlaid function of inherent features (features that can be defined with reference to the segment itself). The fundamental frequency of an inherently voiced segment, besides characterizing the segment as voiced, may also serve to signal a tonal or intonational pattern. To be recognizable as a segment, every segment has a certain duration in the time domain; at the same time, that duration may be contrastive (for example, may characterize the segment as being distinctively short rather than long). Every segment also has a certain amount of intensity; whatever the acoustic and physiological correlates of stress, they consist of intensifying phonetic factors already present in a lesser degree. Furthermore, suprasegmental features differ from segmental features by the fact that suprasegmental features are established by comparison of items in sequence, whereas segmental features are identifiable by inspection of the segment itself.

Suprasegmental features are also involved in signaling various boundaries, including the boundaries of morphemes, words, phrases, clauses, and sentences, as well as still-higher-level units such as paragraphs in a longer stretch of speech. For example, modification of the time dimension—use of different degrees of preboundary lengthening—may serve to distinguish between sentence boundaries within a paragraph and those sentence boundaries that simultaneously terminate the paragraph.

As is the case with segmental features, phonetic science is concerned with the production of suprasegmental features, with the way in which the suprasegmental features are manifested in the acoustic representation of the signal, and with the ways in which the encoded information is extracted from the signal through the process of speech perception. This brief sketch has omitted a number of other relevant aspects of phonetics, such as the neurophysiological control of the process of speech production and perception, and the specifically linguistic problems relating to the ways in which the phonological systems of different languages are realized in the phonetic domain. Phonetics also has a number of practical applications in areas such as foreign language teaching, speech therapy, and communications engineering. *See* SPEECH. Ilse Lehiste

Bibliography. J. Clark and C. Yallop, *Introduction to Phonetics and Phonology*, 2d ed., 1995; A. Cutler and D. R. Ladd, *Prosody: Models and Measurements*, 1983; P. Ladefoged, *A Course in Phonetics*, 4th ed., 2000; J. Laver, *Principles of Phonetics*, 1994.

Phonolite

A light-colored, aphanitic (not visibly crystalline) rock of volcanic origin, composed largely of alkali feldspar, feldspathoids (nepheline, leucite, sodalite), and smaller amounts of dark-colored (mafic) minerals (biotite, soda amphibole, and soda pyroxene). Phonolite is chemically the effusive equivalent of nepheline syenite and similar rocks. Rocks in which plagioclase (oligoclase or andesine) exceeds alkali feldspar are rare and may be called feldspathoidal latite. *See* FELDSPATHOID.

Rapid cooling at the surface causes lavas to solidify with very fine-grained textures. Most phonolitic lavas, however, carry abundant large crystals (phenocrysts) when they are erupted, and these are soon frozen into the dense matrix to give a porphyritic texture. Generally very little material congeals as glass. The phenocrysts, many visible to the naked eye, include alkali feldspar, feldspathoids, and mafics. These may be well formed (euhedral) or moderately well formed (subhedral). *See* PHENOCRYST.

Most other features of phonolites can be seen only microscopically. The alkali feldspar is principally soda-rich sanidine and orthoclase. It generally occurs in the rock matrix, but if abundant it may also form as phenocrysts. Plagioclase is not abundant except in nepheline latites where it may form abundant phenocrysts.

Nepheline may occur as euhedral crystals (square or hexagonal), some of which may be phenocrysts. Otherwise it is irregular (anhedral) and interstitial. Nosean, hauyne, and sodalite, as euhedral or partly corroded crystals, may occur as phenocrysts and matrix grains. These twelve-sided (dodecahedral) crystals generally show hexagonal outlines in thin sections of the rock. Eight-sided euhedral crystals of pseudoleucite may occur as phenocrysts in potash-rich rocks. More rounded grains of leucite may form part of the matrix. Leucite is commonly altered to pseudoleucite, but the euhedral outline is retained. Analcite occurs principally as matrix material but in some rocks it is abundant and as large euhedral phenocrysts.

Biotite is not common but may form large strongly resorbed phenocrysts. Amphiboles are usually soda-rich (riebeckite, hastingsite, and arfvedsonite). They may occur as phenocrysts or as interstitial clusters.

They may show resorption or may be replaced by pyroxene. The most important mafic is soda pyroxene. As phenocrysts it is commonly zoned with cores of diopside surrounded by progressively more sodic shells of aegirine-augite and aegirite. Aegirite is the common pyroxene of the rock matrix.

Accessory minerals are varied and include sphene, magnetite, zircon, and apatite.

The structures and textures of phonolite are similar to those of the more common rock trachyte. Fluidal structure, formed by flowage of solidifying lava and expressed by lines or trails of phenocrysts, may be seen without magnification. Under the microscope, flowage is shown by subparallel arrangement of elongate feldspar crystals. *See* TRACHYTE.

Phonolites are rare and highly variable rocks. They occur as volcanic flows and tuffs and as small intrusive bodies (dikes and sills). They are associated with trachytes and a wide variety of feldspathoidal rocks.

The origin of phonolites and related rocks constitutes an interesting problem. There is still a considerable difference of opinion as to how the phonolitic magma (molten material) originates. One theory assumes an origin from basaltic magma by differentiation. Certain early-formed crystals are removed (perhaps by settling), causing the residual magma to approach the composition of phonolite. Another theory supposes these peculiar magmas to form when a more normal rock melt assimilates large quantities of limestone fragments. Volatiles, notably carbon dioxide, are considered by many to play an important role in transferring and concentrating certain constituents (like potassium) in the magma. The great variety of rock types and modes of association strongly suggests that several different mechanisms may operate to form these feldspathoidal rocks. *See* IGNEOUS ROCKS; MAGMA. Carleton A. Chapman

Phonon

A quantum of vibrational energy in a solid or other elastic medium. This vibrational energy can be transported by elastic waves. The energy content of each wave is quantized. For a wave of frequency f, the energy is $(N + {}^1/_2)hf$, where N is an integer and h is Planck's constant. Apart from the zero-point energy, ${}^1/_2hf$, there are N quanta of energy hf. In elastic or lattice waves, these quanta are called phonons. Quantization of energy is not related to the discreteness of the lattice, and also applies to waves in a continuum. *See* QUANTUM MECHANICS; WAVE MOTION.

The concept of phonons closely parallels that of photons, quanta of electromagnetic wave energy. The indirect consequences of quantization were established for phonons just as for photons in the early days of quantum mechanics—for example, the decrease of the specific heat of solids at low temperatures. Direct evidence that the energy of vibrational modes is changed one phonon at a time came much later than that for photons—for example, the photoelectric effect—because phonons exist only within a solid, are subject to strong attenuation and scattering, and have much lower quantum energy than optical or x-ray photons. *See* PHOTOEMISSION; QUANTUM ACOUSTICS; SPECIFIC HEAT OF SOLIDS.

Like photons, phonons can be regarded as particles, each of energy hf and momentum proportional to the wave vector of the elastic or lattice wave. Such a particle can be said to transport energy, thus moving with a velocity equal to the group velocity of the underlying wave. *See* LATTICE VIBRATIONS; PHOTON.

Phonon-electron interactions. Scattering of conduction electrons by lattice vibrations is described by single phonons. An electron absorbs or emits a phonon, changing its own energy by an amount equal to the phonon energy (or its negative, in the case of phonon emission), and changing its own wave vector (which is proportional to its momentum) by an amount equal to the sum of the phonon wave vector (or its negative) and a reciprocal lattice vector, which may equal zero. The change in the electron's energy results from energy conservation, and the change in its wave vector results from constructive interference between the electron wave and the elastic wave. In a discrete lattice, the phonon-electron interaction can be combined with Bragg reflection (involving a nonzero lattice vector), resulting in an Umklapp (flip-over) process. *See* ELECTRON DIFFRACTION; UMKLAPP PROCESS.

At low temperatures, the phonon wave vector is small compared to the nonzero reciprocal lattice vectors, and most processes are said to be normal; that is, the reciprocal lattice vector involved is zero. Thus, the change in the electron wave vector expresses conservation of the combined electron and phonon momentum. The electron-phonon interaction in metals at low temperatures involves scattering of an electron inelastically through a small angle, approximately equal to the ratio of the phonon wave vector to the electron wave vector. Each process is a step in a random walk which gradually changes the direction of the electron. The resulting electrical resistivity varies with temperature in proportion to the product of the square of the temperature and the lattice specific heat (which is itself proportional to the cube of the temperature). The thermal resistivity, however, varies as the lattice specific heat divided by the temperature, because each step changes the electron energy. This makes the thermal resistivity larger than expected from the Wiedemann-Franz law. *See* ELECTRICAL RESISTIVITY; THERMAL CONDUCTION IN SOLIDS; WIEDEMANN-FRANZ LAW.

At low temperatures, when the phonon wave vector is much smaller than the nonzero reciprocal lattice vectors for the important phonons, electron-phonon interactions conserve the combined total momentum of the electron and phonon gases. The net momentum of electrons in an electric current causes net momentum of the phonon gas. This phonon drag sets up phonon heat flow and an addition to the Peltier coefficient. In metals this is of the order of the product of the temperature and the lattice specific heat divided by the electronic charge density. In semiconductors, the specific heat is only that of those phonons which can interact with electrons. At higher temperatures, phonon drag is reduced by other phonon scattering processes and by electron-phonon processes with Bragg scattering (that is, scattering involving a nonzero reciprocal lattice vector). *See* THERMOELECTRICITY.

Other phonon interactions. Magnetic ions and other spin sites change their spin state with the emission or absorption of phonons. At low temperatures, just one phonon is involved, which matches the energy difference of the spin states (a direct process). If all spins have the same level spacing, the phonons are confined to a narrow frequency band, which can sometimes be overpopulated, slowing down the spin-lattice interaction (a phonon bottleneck). *See* PARAMAGNETISM.

Inelastic neutron diffraction also involves single phonons, allowing the phonon frequency to be measured as a function of the phonon wave vector. Phonons form a significant part of thermal excitations in liquid helium, which has also been studied by neutron diffraction. *See* LIQUID HELIUM; NEUTRON DIFFRACTION; SLOW NEUTRON SPECTROSCOPY.

Paul G. Klemens

Bibliography. F. J. Blatt, *Physics of Electronic Conduction in Solids*, 1968; G. Grimvall, *Thermophysical Properties of Materials*, 1986; C. Kittel, *Introduction to Solid State Physics*, 7th ed., 1996.

Phonoreception

The perception of sound by animals through specialized sense organs. A sense of hearing is possessed by animals belonging to two divisions of the animal kingdom: the vertebrates, which form the main subphylum of the phylum Chordata, and the insects, which make up the most important class of the phylum Arthropoda. The sense is mediated by the ear, a specialized organ for the reception of vibratory stimuli. Such an organ is found in all except the most primitive vertebrates, but only in some of the many species of insects. The vertebrate and insect types of ear differ in evolutionary origin and in their modes of operation, but both have attained high levels of performance in the reception and discrimination of sounds. *See* SOUND.

Vertebrates

The vertebrate ear is a part of the labyrinth, located deep in the bone or cartilage of the head, one ear on either side of the brain. A complex assembly of tubes and chambers contains a membranous structure which bears within it a number of sensory endings of different kinds. *See* EAR (VERTEBRATE).

The membranous labyrinth is shown in a generalized schematic form in **Fig. 1**. It is convenient to recognize two divisions, a superior division, which includes the three semicircular canals and the utricle, and an inferior division, which includes the saccule and its appendages, the lagena and the cochlea. The superior division is remarkably uniform in character from the higher fishes upward, but the inferior division shows many variations. The saccule is always present. The lagena is present in all classes except the mammals, although it is missing in occasional species. The cochlea is found in reptiles, birds, and mammals.

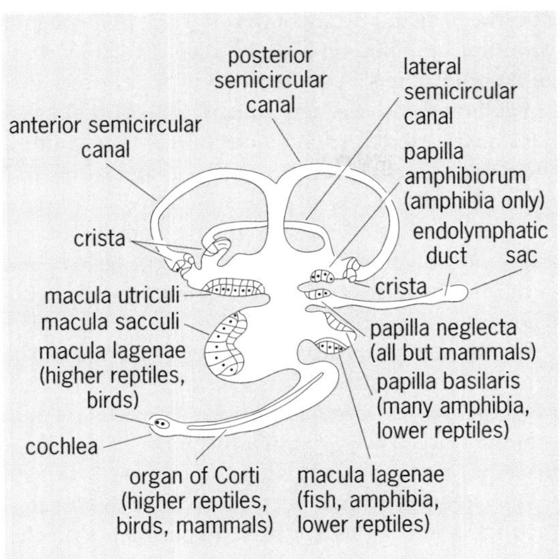

Fig. 1. Generalized sketch of the vertebrate labyrinth. The three cristae, macula ultriculi, and macula sacculi are always present in vertebrates, and the other endings appear as indicated, with a few exceptions.

The sensory endings within these parts of the labyrinth also vary in the vertebrate series. Again there is uniformity for the superior division. There is a crista in each ampulla of the three semicircular canals and a utricular macula. In all but the mammals (with a few individual exceptions), there is a macula neglecta, usually located on the floor of the utricle or close to the junction of utricle and saccule. All forms have a saccular macula. All those with a lagena (in general, all except the mammals) have a lagenar macula. All the amphibians have a papilla amphibiorum, but it is found in no other forms. A basilar papilla appears in certain amphibians, is continued in the reptiles, and then is developed in a more elaborate form as the cochlea of higher reptiles, birds, and mammals.

These endings contain ciliated cells (hair cells) which are supplied by fibers of the eighth cranial (auditory) nerve. In the cristae the cilia of the hair cells are particularly long and are embedded in a gelatinous substance that forms a cap or cupola. In the maculae the cilia are surmounted by a flat plate of gelatinous material in which numerous granules of calcium carbonate (otoliths) are usually embedded. The ciliated cells in the papillae lie on a movable membrane (basilar membrane) and have a membranous covering, the tectorial membrane.

The superior part of the labyrinth generally serves for bodily posture and equilibrium, whereas the saccule and its appendages (lagena and cochlea) serve for hearing. However, there are exceptions to this rule, the most important of which is that in the higher vertebrates, including mammals and probably birds and reptiles, the saccule serves only for equilibrium.

Beginning with the amphibians, which are the earliest vertebrates to spend a considerable portion of their lives on land, there appears a special mechanism, the middle ear, whose function is the transmission of aerial vibrations to the endings of the inner ear. All the vertebrates above the fishes, and certain of the fishes as well, have some type of sound-facilitative mechanism.

Fishes. The maculae of the utricle, saccule, and lagena in the bony fishes have a peculiar form. Instead of numerous calcareous particles there is a single otolith, a large body of distinctive form. The macula neglecta is sometimes lacking.

Few questions have been more actively debated than the ability of fish to hear. Experiments on this question began with G. Parker in 1903, who observed the natural reactions of fish when exposed to a sudden sound, and were carried forward by F. Westerfield and others in 1922 by the introduction of conditioned-response methods. This work culminated in the series of studies by K. von Frisch and his associates, who trained fish to make feeding responses at the sounding of a tone. These experiments proved that fish may be divided into two groups according to hearing ability, those that hear only crudely and those that hear well. The first group includes the great majority of fish species, with what may be called the basic type of labyrinth and lacking any accessory mechanism. The fish that hear well have one of two general types of sound-facilitating structure: either an air vesicle adjacent to some part of the labyrinth or a connection with the swim bladder.

The second of these types of accessory structure is the more common, and is found in a large group of fresh-water fishes known as the Ostariophysi. Between the labyrinth and the anterior part of the swim bladder is a chain of three or four small bones, known from their discoverer, E. H. Weber, as the Weberian ossicles (**Fig. 2**). Weber correctly supposed, when he described this apparatus in 1820, that it serves for the facilitation of sound reception, for it has been demonstrated that the hearing is impaired after removal of the swim bladder and after an interruption of the ossicular chain.

The most extensive study of hearing in fish was carried out by von Frisch and his students on the minnow (*Phoxinus laevis*), one of the Ostariophysi that responds readily to training procedures. This fish was reported to respond to tones over a range from about 32 to 5000–6000 Hz, and in the lower part of its range to be able to discriminate a change of frequency of about 3%. The dwarf catfish [*Ictalurus (Ameiurus) nebulosus*], another ostariophysan, was said to respond to tones as high as 13,000 Hz. Questions have been raised concerning some of the procedures of these experiments, and the tests need to be repeated. More recent studies have not usually indicated upper limits as high as these. Responses only up to 3000 Hz have been obtained in a carefully controlled study of the goldfish (*Carassius auratus*).

In nonostariophysans the observed limits are considerably lower. It was found, in studies of nine species of marine fishes, that the upper limits were around 1000–1200 Hz in eight of them, and there was a limit of 2800 Hz in one, *Holocentrus ascensionis*. Other species have shown even more limited ranges, such as 600 Hz for *Anguilla* and 800 Hz for *Gobius paganellus*.

The mormyrids and labyrinthine fishes, which lack the Weberian ossicles but have an air vesicle adjacent to the labyrinth, have usually shown relatively good hearing, comparable to that of the Ostariophysi.

A number of experiments, mostly on minnows, have dealt with the problem of the particular parts of the fish labyrinth that are concerned with hearing. Removal of the superior portion, which includes the utricle and semicircular canals, does not impair the responses to sound, but seriously affects the posture and swimming ability. After this operation the fish

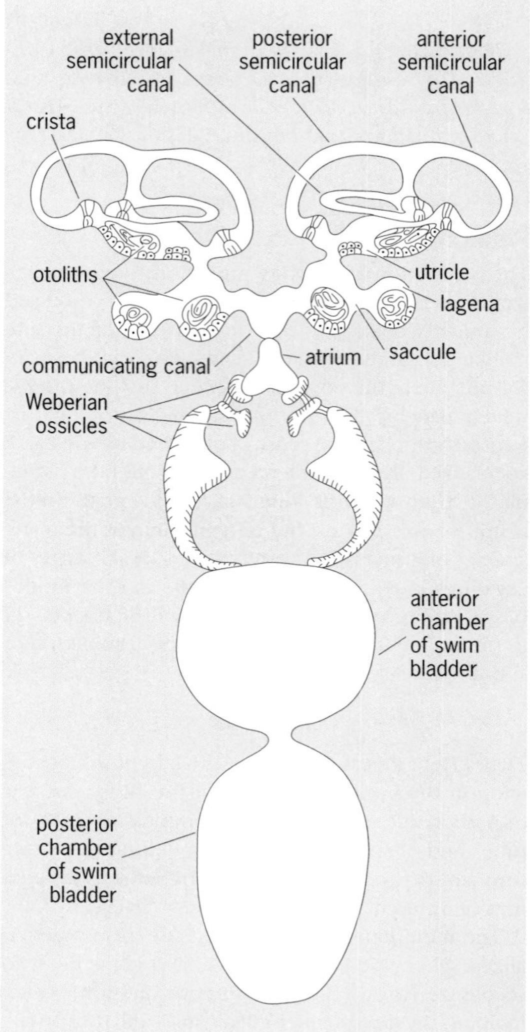

Fig. 2. Diagram (from above) of the two labyrinths of an ostariophysan fish and their connections with the swim bladder through the Weberian ossicles.

may assume an inverted position, and swims erratically. These parts must therefore contain organs of equilibrium. Removal or impairment of either the saccule alone or the lagena alone leaves the fish able to hear, but the removal of both saccule and lagena abolishes the responses to all tones except those of very low frequency which are perceived through skin or lateral-line receptors. Hence, the endings of the saccule and lagena are auditory in function. Exceptional in this respect are the herring and sardine (clupeids), in which an air vesicle is applied to the wall of the utricle; in these the sense of hearing is probably mediated by the utricular macula.

Amphibians. The three orders of amphibians—the Apoda (legless), including wormlike forms such as caecilians; the Urodela (tailed), including mud puppies, newts, and salamanders; and the Anura (tailless), including frogs and toads—all have some type of middle-ear mechanism.

The first two orders include animals whose ears show a great variety of accessory structures, some of which look as though they might function well in sound reception, whereas others seem crude. Of these only the salamander has been studied experimentally. In 1938, S. Ferhat-Akat trained larvae to come for food at the sounding of a tone, and got results for tones up to 244 Hz in one specimen and up to 218 Hz in three others.

Higher amphibians, such as the frog, possess a well-developed middle-ear mechanism, consisting of a disk of cartilage flush with the lateral surface of the head and covered with skin, and a rod of cartilage and bone, called the columella, leading inward from the disk and expanding to form the stapes, which is embedded in an opening (oval window) in the wall of the otic capsule (**Fig. 3**).

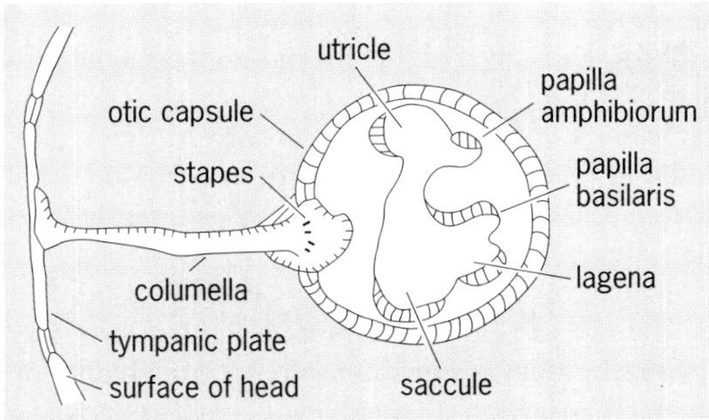

Fig. 3. Simplified diagram of the frog ear.

The active and often loud croaking of frogs in the breeding season has focused attention upon the problem of their hearing. R. Yerkes in 1905 first succeeded in obtaining experimental evidence of their auditory sensitivity by showing that sounds may enhance or inhibit their response to a strong tactual stimulus. Several studies in which the impulses from the eighth cranial nerve were recorded on stimulation with tones showed results only for low frequencies, up to 500–600 Hz, or at most to 1024 Hz. The most extensive study of the electrical responses of the ear by W. Strother in 1959 showed responses in *Rana catesbeiana* over a range from below 100 Hz to about 3500 Hz. The sensitivity was best in two regions, around 500 Hz and around 1500 Hz, and there is reason to believe that these two regions correspond to the actions of the two papillae of the frog, the papilla basilaris and the papilla amphibiorum.

Reptiles. The living reptiles belong to four important groups, represented by snakes, turtles, chameleons and lizards, and crocodiles and alligators.

Many authorities have asserted that snakes are completely deaf, or that their ears are sensitive only to vibrations conducted to the head through the ground. This impression has arisen partly from the fact that snakes do not have any external ear and do not show obvious reactions to sounds. There is no tympanic membrane to receive aerial sound pressures, but its purpose is served by one of the bones of the skull, the quadrate bone, which is loosely attached to the main part of the skull. Although it lies beneath the skin and other tissues of the side of the head, the quadrate bone presents a flat surface for the action of sounds, and communicates them to a thin bony rod (columella) running inward to expand as the stapes in the oval window (**Fig. 4**).

Experiments have shown that electrical potentials are produced in the inner ears of snakes in response to low-frequency sounds, to both the sounds conducted through the substratum and those conducted through the air in the usual way. Hence, it is safe to conclude that snakes have hearing, although only for the lower range of sounds and not as highly sensitive as that of most other animals.

Doubt has often also been expressed about the ability of turtles to hear, but here again the evidence is that they do. They have a well-developed middle ear, including a cartilaginous disk on the side of the head beneath the skin, and a columella leading to a stapes in the oval window of the otic capsule. Two investigators have succeeded in training turtles to make positive reactions to an acoustic signal, although others have failed in this attempt. The electrophysiological method yields positive results. Indeed, the observations show that for low tones,

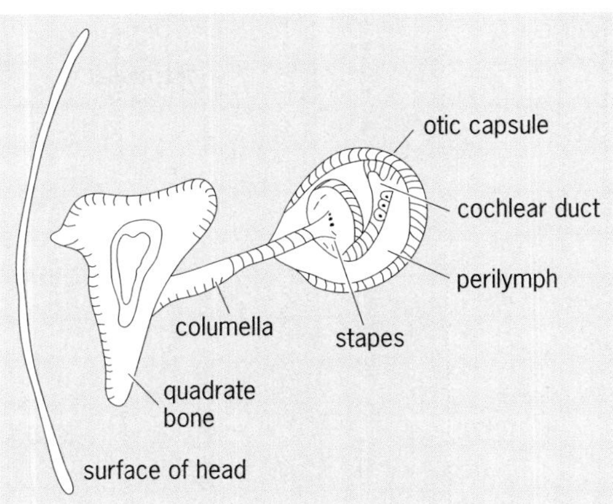

Fig. 4. Diagram of the ear of a snake. The nonauditory parts and endings are not shown.

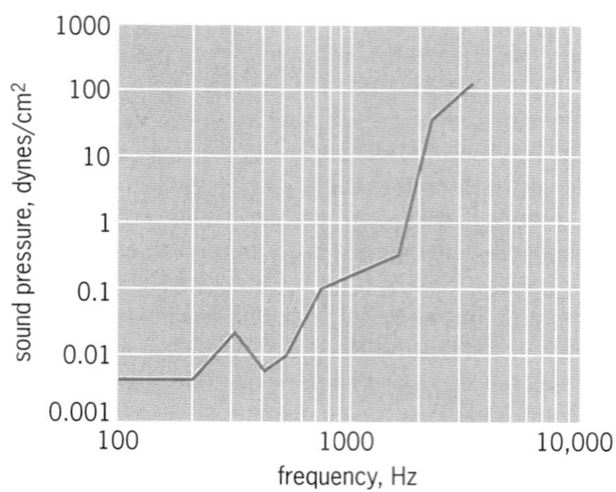

Fig. 5. Auditory sensitivity of a wood turtle (*Clemmys insculpta*), as shown by the potentials produced in its ear by sounds. The curve shows the sound pressure necessary to produce a potential of 0.3 microvolt.

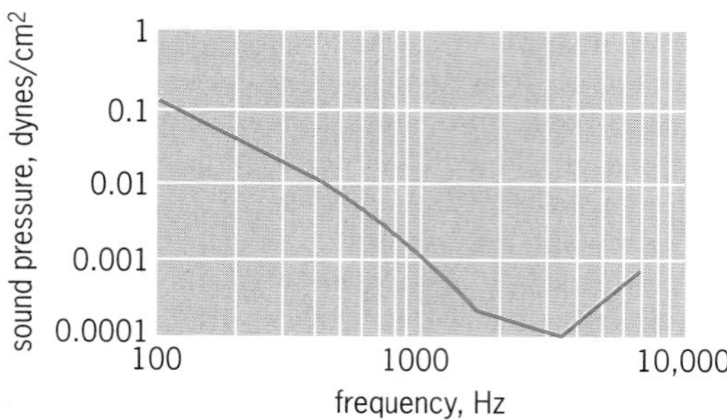

Fig. 6. Graph showing threshold sensitivity in the bullfinch (mean of four birds). (*After J. Schwartzkopff, Uber Sitz und Leistung von Gehöor und Vibrationssinn bei Vögeln, Z. vergl. Physiol., 31:527–608, 1949*)

those of 100–700 Hz, the turtles have excellent sensitivity. The wood turtle (*Clemmys insculpta*) exceeds other species studied (**Fig. 5**).

Structurally the ear of the lizard is superior to that of the turtle. With a few exceptions, there is a membranous drum, an extracolumella, and a columella whose expanded end constitutes a stapes in the oval window. C. Berger was able to train two species of lizards (*Lacerta agilis* and *L. vivipara*) to make feeding movements in response to various sounds, including tones over a range of 69–8200 Hz. Electrical potential studies carried out in a great many species indicate considerable species variations in sensitivity. In general, the geckos, most of which are nocturnal, have the best hearing. The frequency range as represented by the cochlear potentials ranges at least 100–10,000 Hz, though these ears seem especially susceptible to damage by the uppermost frequencies. The structural development of the cochlea varies widely in the different species; thus the iguanids in general have a short and undifferentiated basilar membrane with about 60 hair cells, whereas the gekkonids have a long and tapered basilar membrane with several hundred hair cells (up to 1600 in *Gekko gecko*).

Certain chamelons (*Chamaeleo senegalensis* and *C. quilensis*) have a peculiar auditory structure somewhat similar to that of the snake in that there is no external ear opening or eardrum; but a plate of bone (a part of the pterygoid bone) lying beneath the skin and muscle layers at the side of the head serves as a receptive surface for sounds and transmits the vibrations through the columella to the inner ear.

The crocodiles have an ear representing a distinct advance over other reptiles, and there is a curved cochlea similar to that of birds. Probably they have excellent hearing, although the experimental evidence is scanty as yet. Many general observations indicate that they use sounds in mating activities, and the males are capable of producing loud roars. F. Beach was able to provoke captive animals into roaring and making movements by stimulation with low tones, especially sounds at 57 Hz, but also others at about 300 Hz. E. Wever and J. Vernon found that young caimans gave cochlear potentials in response to sounds over a range of 20–6000 Hz.

Birds. The labyrinth in birds is generally similar to that of the higher reptiles. There is a membranous eardrum, a columella leading inward to the stapes, and a curved cochlea. There are only minor variations in the forms of these structures among the various species.

Most birds have a range of hearing of about 50–20,000 Hz, and an absolute sensitivity that probably approaches that of humans in the medium-high-tone range. The threshold sensitivity of a finch (*Pyrrhula p. minor*) is shown in **Fig. 6**, as determined by J. Schwartzkopff by a training method. In general, the small songbirds are more sensitive than the larger birds, such as chickens and pigeons. The owl is exceptional in having an ear with a particularly large drum membrane and other special features that make for high sensitivity. Thus, an owl in a tree at dusk can hear and locate a mouse rustling in the grass below.

Pitch discrimination in songbirds and parrots is about as keen as that of humans (0.3–0.7%), but in pigeons it is relatively poor (6%).

Mammals. The auditory apparatus attains its highest development in the mammals. The columella of

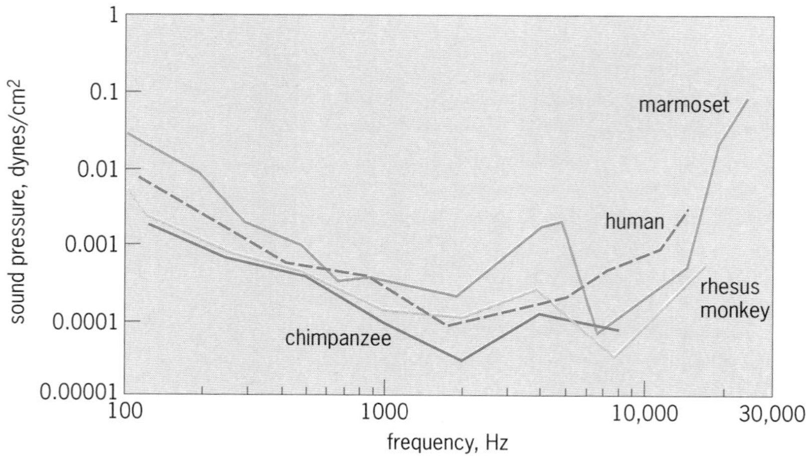

Fig. 7. Auditory thresholds in primates. The curves show the sound pressures that are barely audible for humans, chimpanzees, rhesus monkeys, and marmosets.

lower forms has been replaced by a chain of three ossicles, which connect the tympanic membrane with the inner ear. In the egg-laying mammals, such as the platypus, the cochlea is a curved tube as in crocodiles and birds, but in all other mammals it is a spiral of one to four turns. The great extension of the cochlea and the corresponding multiplication of sensory cells have enhanced the capacities of the mammals to deal with the varieties and complexities of sounds.

Despite intense interest in mammalian hearing, precise information is available on only a few mammals apart from humans. Experimental studies have been carried out on some of the subhuman primates and on a few of the common laboratory animals. Only fragmentary information is available on the many other species of mammals, although it can be assumed from their general behavior, because they seem to respond to much the same range and intensities of sounds that humans do, that their hearing is similar to humans'. *See* HEARING (HUMAN).

Training experiments are easily carried out on the subhuman primates; **Fig. 7** presents threshold curves for the chimpanzee, the rhesus monkey, and a species of marmoset, with the human curve for comparison. It will be noted that these animals have auditory sensitivity similar to man's in the low-tone range, but are superior to man in the high-tone range.

The most extensive studies have been made on the cat. Its sensitivity also is similar to man's over the lower range, but extends far above the human limit to 60,000 Hz or more. Electrical potentials have been recorded from the cat's cochlea to tones as high as 100,000 Hz.

Other animals whose hearing has been studied experimentally are the dog, rat, guinea pig, rabbit, bat, and certain species of mice.

The bats are of special interest, more particularly the small insectivorous species, because they repeatedly produce vocal sounds of high frequency, up to 40,000 Hz or more, as they fly about in search of insect prey. They locate the prey by hearing the echoes of their cries, and they guide themselves in the total darkness of caves by echoes from the walls. The hearing of bats extends far into the high frequencies, up to 100,000 Hz at least. *See* AUDIOMETRY.

Invertebrates

The group of invertebrates which has received the most attention has been the insects. Other arthropods, such as certain crustaceans and spiders, have also been found to be sensitive to sound waves.

The insect ear consists of a superficial membrane of thin chitin with an associated group of sensilla called scolophores. Such an apparatus is shown in simplified form in **Fig. 8**. These ears are found in most species of katydids, crickets, grasshoppers, cicadas, waterboatmen, mosquitoes, and nocturnal and spinner moths. They occur in different places in the body: on the antennae of mosquitoes, on the forelegs of katydids and crickets, on the metathorax of cicadas and waterboatmen, and on the abdomen of grasshoppers. Probably these differently situated organs represent separate evolutionary developments, through the association of a thinned-out region of the body wall with sensilla that are found extensively in the bodies of insects and that by themselves seem to serve for movement perception.

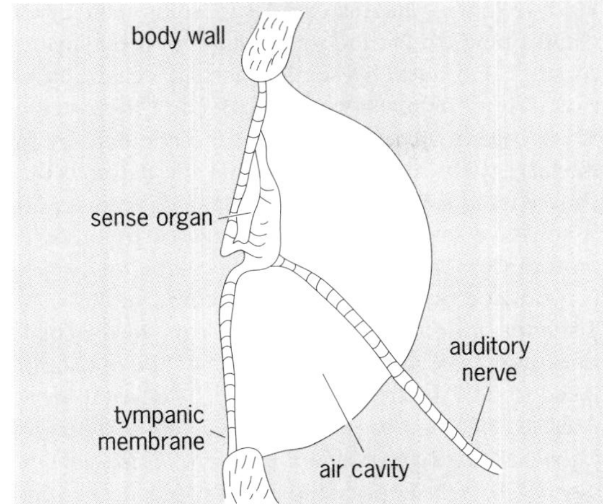

Fig. 8. Diagram of the ear of a grasshopper.

The insects mentioned above are noted for their production of stridulatory sounds made by rubbing the edges of the wings together, or a leg against a wing, or by other means. These sounds are produced by the males and serve for enticing the females in mating.

The sensitivity of insect ears is keenest in the high frequencies. **Figure 9** shows threshold curves obtained on a katydid by observing the potentials produced in the auditory nerve during stimulation with sounds. As will be seen, the sensitivity in this species is greatest in the region of 7000–60,000 Hz. It extends to even higher frequencies, usually as high as 120,000 Hz and sometimes beyond. Other species have distinctly different sensitivity curves, and there is reason to believe that there is a relation to the range of the stridulatory sounds. There is evidence that these sounds are discriminated, for the females of one species respond to the males of their own

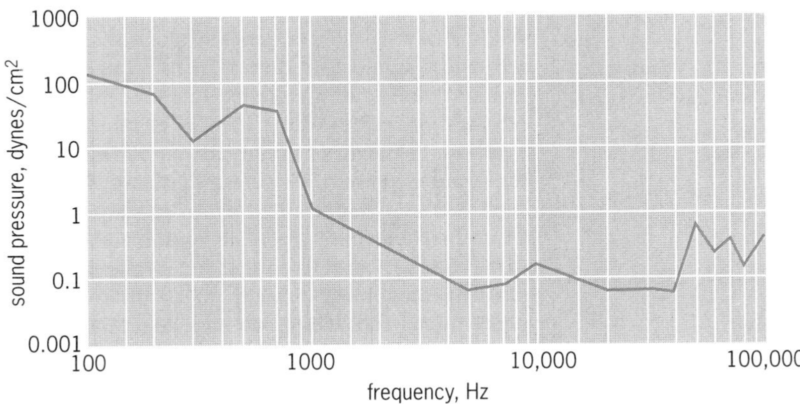

Fig. 9. Auditory thresholds of the katydid (*Conocephalus strictus*). Sensitivity is greatest in the 7000–60,000-Hz region. (*After E. G. Wever and J. A. Vernon, The auditory sensitivity of Orthoptera, Proc. Nat. Acad. Sci., 45:413–419, 1959*)

kind. J. Regen placed a cage of chirping male crickets in a field, and found that females of their species sought them out. This seeking activity ceased, however, when the males were deprived of their stridulating organs or the females were made deaf by removal of their ears. A most striking adaptation is that shown by mosquitoes: The ear of the male mosquito is sensitive only to a narrow range of frequencies around 380 Hz, and this frequency is the one which is produced by the wings of the female in flight. M. Tischner found that when the ear of the male mosquito was made nonfunctional, the mosquito failed to find a mate. Ernest G. Wever

Bibliography. C. R. Pfaltz (ed.), *New Aspects of Cochlear Mechanics and Inner Ear Pathophysiology*, 1990; A. N. Popper and R. R. Fay (eds.), *Comparative Hearing: Mammals*, 1994; A. N. Popper and R. R. Fay (eds.), *Comparative Studies of Hearing in Invertebrates*, 1980.

Phoresy

A relationship between two different species of organisms in which the larger, or host, organism transports a smaller organism, the guest. It is regarded as a type of commensalism in which the relationship is limited to transportation of the guest. The term is credited to P. Lesne following his observations on the biology of a small fly, *Limosina sacra*, which is transported by a scarabeid, one of the dung beetles, into its burrow. These burrows are suitable breeding sites for both animals. *See* POPULATION ECOLOGY.
Charles B. Curtin

Phoronida

A small, relatively homogeneous phylum of animals; in the past they have been grouped with other phyla such as the Annelida, Molluscoidea, and Chordata. Two genera, *Phoronis* and *Phoronopsis*, and about 16 species are recognized at the present time; however, the taxonomy of the group is in need of revision.

Habitat and distribution. Phoronids may occur in vertical tubes placed just below the surface in intertidal or subtidal mud flats, or as feltlike masses of intertwined tubes attached to rocks, pilings, or old logs in shallow water. In both cases the tubes, composed basically of a secreted, parchmentlike material, are encrusted with small particles of sand or shell. A third living habit concerns those phoronids found inside channels, probably self-made, in limestone rock or the shells of dead pelecypod mollusks.

The geographical distribution of phoronids appears to be worldwide in temperate and tropical seas. There are no records of phoronids from the polar regions.

Morphology. The body is more or less elongate, ranging in length from about 1.6 to 8 in. (4 to 20 cm), and bears a crown of tentacles arranged in a double row surrounding the mouth which is usually crescent-shaped (**Fig. 1**). The anus occurs at the level of the mouth and is borne on a papilla immediately outside the double row of tentacles. The digestive tract is therefore U-shaped, the mouth and anus opening close together at one end of the animal. The tentacles rest on a connective tissue base known as the lophophore. The double row of tentacles may form either a slightly indented circle or a complex double spiral. The tentacles vary in number from about 50 to over 300, are ciliated, and create a feeding current which carries food particles to the mouth. Feeding and excretory currents have not been studied in detail. Associated with the mouth is a ciliated flap of tissue known as the epistome. *See* LOPHOPHORE.

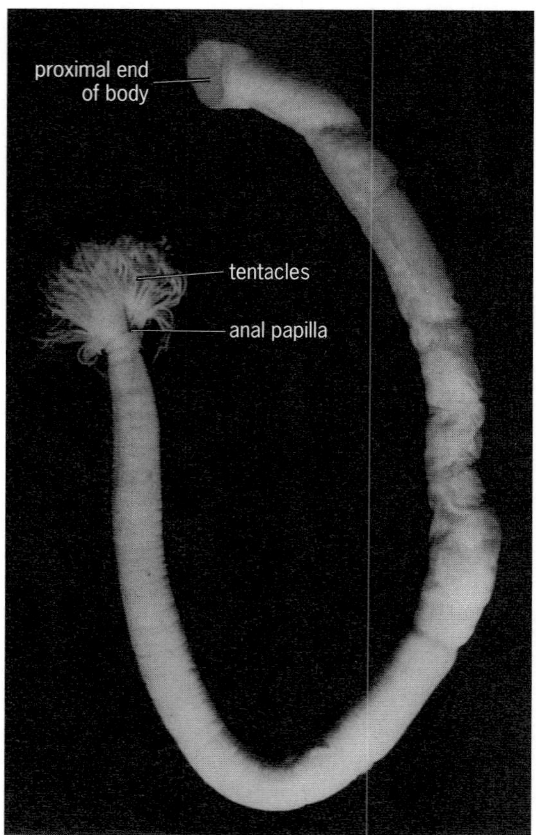

Fig. 1. ***Phoronopsis harmeri*** **removed from its tube. Length is about 8 in. (20 cm).**

The digestive tract consists of an esophagus, stomach, intestine, and rectum. In some species, there is a distinct valve between the esophagus and stomach. The junction of the stomach and intestine occurs at the proximal or aboral extremity of the animal. The food seems to consist chiefly of microscopic phytoplankton. Diatom shells are often found both in the digestive tract and in fecal pellets.

There is a blood vascular system in which elliptical, nucleated corpuscles containing hemoglobin circulate. The vascular system consists basically of two longitudinal vessels, known as the afferent and efferent vessels, which are continuous with one another at the proximal end of the body. Distally, both vessels connect with a pair of semicircular vessels located at the level of the lophophore, immediately below the tentacles. Within each tentacle is a single, blind vessel which branches into two at its base and so connects with both semicircular vessels. In the living animal, corpuscles can be seen pulsating up and down in the tentacular vessels. Associated with the longitudinal blood vessels is a blood sinus surrounding the gut and a large number of blind blood ceca which are particularly numerous in, or may be restricted to, the proximal end of the body. Associated with these ceca is the jellylike fat body, or vasoperitoneal tissue. Found among the large, semifluid cells of this tissue are inclusions of various sorts. Some of these probably consist of guanine or some related form of nitrogenous waste. Others may represent the products of hemoglobin breakdown.

The body cavity is subdivided by a series of longitudinal mesenteries extending from the digestive tract to the body wall. In most phoronids there are four such mesenteries occupying oral, anal, right lateral, and left lateral positions, thus dividing the coelomic cavity into four chambers (**Fig. 2**). There is also a horizontal mesentery near the tentacular end of the body, separating a lophophoral coelom from the four larger and more proximal coelomic cavities.

The two nephridia open on either side of the anal papilla. Each nephridium consists of a duct, coiled once on itself and, usually, a pair of funnels, one opening into each of the oral and anal coelomic cavities. The funnels have extensive folded and ciliated margins.

The body wall consists of an outer layer of epithelial cells, many of them secretory and concerned with the building of the tube, and two layers of muscle. The outermost layer of muscle consists of circular fibers and inside this is a series of bundles of longitudinal fibers. An unpolarized nerve net underlies the external epithelium and continues as a more dense concentration in the form of a ring in the horizontal mesentery. Extending proximally from cell bodies in this ring are one or two giant nerve fibers which taper and disappear at the proximal end of the body. The giant nerve fibers are known in all species except one (*Phoronis ovalis*) and are probably concerned with the rapid retraction of the body into the tube.

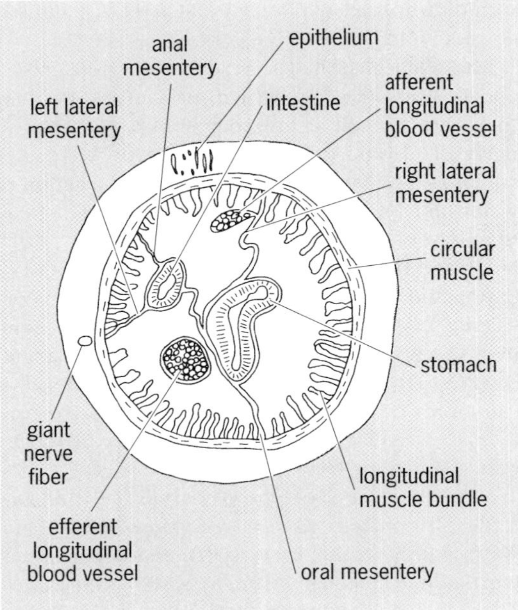

Fig. 2. Cross section through *Phoronopsis*.

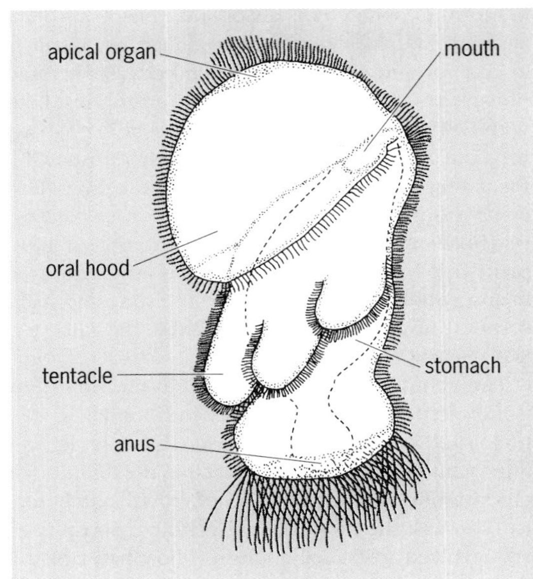

Fig. 3. Actinotroch larva of *Phoronis vancouverensis*.

Reproduction. Reproductive tissue is formed from cells which multiply first on the thin walls of the blood ceca. The phylum includes both dioecious animals and hermaphrodites. As the gonad increases in extent it displaces the vasoperitoneal tissue which shrinks proportionately. When ripe, the gametes are shed into the body cavity and find their way to the nephridia to pass through these organs to the outside. In at least one species (*Phoronis hippocrepia*), the ova are retained in the tentacular crown until the larval stage is reached. All phoronids may reproduce sexually, and in most cases the life history includes the pelagic actinotroch larva (**Fig. 3**). Some species reproduce asexually by transverse fission.

Joan R. Marsden

Bibliography. L. A. Borrandaile et al., *The Invertebrata*, 2d ed., 1935; S. F. Light et al., *Intertidal Invertebrates of the Central California Coast*, 3d ed., 1975; G. E. MacGinitie and N. MacGinitie, *Natural History of Marine Animals*, 1949; S. P. Parker (ed.), *Synopsis and Classification of Living Organisms*, 2 vols., 1982.

Phosphate metabolism

Organic phosphate compounds are present in the structural units of every animal cell, and inorganic phosphate is associated with calcium in bone and teeth. The total phosphorus in the adult human body is about 12 g/kg, with only 1.4 g/kg present in the soft tissues and the remainder in mineralized tissue in the form of apatite crystals. Blood phosphate plays an important role in regulating neutrality, and it is in equilibrium with both bone and cellular organic phosphates. In plasma it is primarily orthophosphate with HPO_4^{2-} and $H_2PO_4^-$ present in a ratio of about 4:1. The blood level is held relatively constant by regulating phosphate excretion by the kidney. This control is primarily mediated by action of parathyroid hormone. Vitamin D enhances the entry of phosphate into bone. Phosphate plays an important role in absorption of sugars from the intestine and reabsorption of glucose from the kidney. *See* PARATHYROID HORMONE; VITAMIN D.

The central role of phosphates in life processes is indicated by their occurrence in ribonucleic acid (RNA) and deoxyribonucleic acid (DNA), which are important in protein synthesis and in the function of chromosomes in the processes of growth and heredity. Through the formation of lecithins, phosphates are involved in fat metabolism. Phosphates play a major role in the conservation and transfer of energy, particularly of the energy produced in the tricarboxylic acid cycle (Krebs cycle), in glycolysis, and in the pentose shunt. They do so by participating in many phosphorylation and transphosphorylation reactions involving sugars and other organic compounds. *See* CARBOHYDRATE METABOLISM; CHROMOSOME; CITRIC ACID CYCLE; LIPID METABOLISM; NUCLEIC ACID.

In phosphorylation reactions, compounds such as phosphocreatine (PC) and adenosine triphosphate (ATP) are formed which are capable of yielding relatively large amounts of free energy (5000–11,000 cal or 21–46 kilojoules per mole) when the phosphate bonds are broken by hydrolysis. ATP and PC have a central role in energy storage and transfer in all tissues. The transfer of phosphate from one organic molecule to another is mediated by enzymes without liberation of inorganic phosphate. *See* ADENOSINE TRIPHOSPHATE (ATP).

Phosphorus-containing coenzyme systems include the pyridine (nicotinamide) and the riboflavin nucleotide systems concerned with oxidation-reduction reactions; coenzyme A, the functional form of pantothenic acid, concerned with transacetylation, acylation, and condensation reactions; the diphosphothiamine system concerned with decarboxylation; and pyridoxal phosphate concerned with transamination. *See* BIOCHEMISTRY; BIOLOGICAL OXIDATION; COENZYME; ENERGY METABOLISM.

Morton K. Schwartz

Bibliography. C. L. Comar and F. Bronner (eds.), *Mineral Metabolism*, vol. 2, pt. A, 1964; L. Stryer, *Biochemistry*, 4th ed., 1995.

Phosphate minerals

Any naturally occurring inorganic salts of phosphoric acid, $H_3[PO_4]$. All known phosphate minerals are orthophosphates since their anionic group is the insular tetrahedral unit $[PO_4]^{3-}$. Mineral salts of arsenic acid, $H_3[AsO_4]$, have a similar crystal chemistry. There are over 150 species of phosphate minerals, and their crystal chemistry is often very complicated. Phosphate mineral paragenesis can be divided into three categories: primary phosphates (crystallized directly from a melt or fluid), secondary phosphates (derived from the primary phosphates by hydrothermal activity), and rock phosphates (derived from the action of water upon buried bone material, skeletons of small organisms, and so forth).

Primary phosphates. These phosphate minerals have crystallized from fluids, usually aqueous, during the late stage of fluid and minor element segregation and concentration. Certain rocks, particularly granite pegmatites, frequently contain phosphate minerals which sometimes occur as enormous crystals. Giant crystals up to 10 ft (3 m) across are known for apatite, $Ca_5(F,Cl,OH)[PO_4]_3$; triphylite-lithiophilite, $Li(Fe,Mn)[PO_4]$; amblygonite, $(Li,Na)Al(F,OH)[PO_4]$; and graftonite, $(Fe,Mn,Ca)_3[PO_4]_2$. Rare-earth phosphates, such as monazite, $(La,Ce)[PO_4]$, and xenotime, $Y[PO_4]$, are known from some pegmatites and are mined for rare-earth oxides. The $[PO_4]^{3-}$ anionic group is often considered a "mineralizer"; that is, it contributes to decreased viscosity of the fluid and promotes the growth of large crystals. Primary phosphates are segregated since they usually do not combine with silicate-rich phases to form phosphosilicates. *See* AMBLYGONITE; APATITE; PEGMATITE.

Primary phosphates also occur with other rock types, notably ultrabasic rocks such as nepheline syenites, jacupirangites, and carbonatites. Enormous quantities of apatite occur with nepheline syenite in the Kola Peninsula, Russia. Carbonatites often contain rare-earth phosphates such as britholite, $(Na,Ce,Ca)_5(OH)[(P,Si)O_4]_3$. Primary phosphates also occur in metamorphosed limestones, usually as apatite. Apatite also occurs with magnetite, $FeFe_2O_4$, segregated from basic rocks such as norites and anorthosites. Rarely, phosphates occur in meteorites, including apatite, whitlockite, β-$Ca_3[PO_4]_2$, and sarcopside, $(Mn,Fe,Ca)_3[PO_4]_2$. *See* CARBONATITE.

Secondary phosphates. A large spectrum of secondary phosphates is known, particularly because they have formed at low temperatures and over a range of pH and pO_2 conditions. Over 50 species are known to have been derived from the action of water on triphylite-lithiophilite. Their crystal chemistry is very complicated; most contain Fe^{2+}, Fe^{3+}, Mn^{2+}, and Mn^{3+}, octahedrally coordinated by $(OH)^-$, (H_2O), and $[PO_4]^{3-}$ ligand groups, and can be considered coordination complexes. The most common species are strengite, $Fe[PO_4](H_2O)_2$; ludlamite, $Fe_3[PO_4]_2(H_2O)_4$; and vivianite, $Fe_3[PO_4]_2(H_2O)_8$; Because they contain transition-metal cations, they are often beautifully colored and are highly prized by mineral fanciers. *See* VIVIANITE.

Members of another group, known as the fibrous ferric phosphates, arise by reaction of phosphatic water with goethite, α-FeO(OH), and occur in certain limonite beds, sometimes in generous quantities. These include rockbridgeite, $Fe^{2+}Fe_4^{3+}(OH)_5[PO_4]_3$, and dufrenite, $Fe^{2+}Fe_5^{3+}(OH)_5[PO_4]_4(H_2O)_2$. There are several other species, but this group of interesting minerals is poorly understood. *See* GOETHITE; LIMONITE.

Rock phosphates. These are phosphates that are derived at very low temperatures by the action of water on buried organic material rich in phosphorus, such as bones, shells, and diatoms. Also included are phosphates obtained by the reaction of phosphatic waters with carbonate materials, such as corals. These waters were enriched in the phosphate anion by previous action upon organic material.

Large beds of phosphatic oolites are known. The mineralogy is poorly understood because of the small grain size. The major mineral is carbonate apatite, but minor amounts of monetite, $CaH[PO_4]$, and brushite, $CaH[PO_4](H_2O)_2$, are also encountered, as well as many gel-like substances which are not strict mineral entities.

Crystals of several hydrated magnesium phosphates have been recovered from bat guanos which were suitably preserved. Common minerals include newberyite, $MgH[PO_4](H_2O)_3$, and struvite, $NH_4Mg[PO_4]\cdot 6H_2O$. Guanos are often marketed for fertilizer because of their high phosphorus content. *See* FERTILIZER.

Phosphate minerals of this category are also known to occur as urinary calculi. Most frequent is apatite. Other species are struvite, newberyite, and whitlockite. Struvite crystals have also been found in canned sardines. *See* MINERAL.

Paul B. Moore

Phospholipid

A lipid that contains one or more phosphate groups. Like fatty acids, phospholipids are amphipathic in nature; that is, each molecule consists of a hydrophilic (water-loving) portion and a hydrophobic (water-hating) portion. Due to the amphipathic nature and insolubility in water, phospholipids are ideal compounds for forming the biological membrane. Phospholipids are present in plasma lipoproteins, and a specialized form of phospholipid called dipalmitoylphosphatidylcholine serves as lung surfactant. High amounts of phospholipids are found in egg yolk, soya bean, and animal tissues such as the brain. Commercially, phospholipids are used to make chocolates, caramels, and many other prepared foods. Phospholipids are important components in cosmetics and are also used as solubilizing and texturizing agents in pharmaceutical preparations. Currently, phospholipids are used to form liposomes for drug delivery. A liposome is a microscopic sphere enclosed by a phospholipid membrane and containing a specific drug for delivery into the body. The phospholipid membrane fuses with the plasma membrane of the target cell, and the drug inside the liposome is emptied into the cell. Some liposomes are preferentially recognized by certain organs in the body, thus facilitating the targeted delivery of drugs. *See* LIPID; LIPOSOMES.

Classes. There are two classes of phospholipids: those that have a glycerol backbone and those that contain sphingosine (synthesized from serine and palmitic acid). Both classes are present in the biological membrane. Phospholipids that contain a glycerol backbone are called phosphoglycerides (or glycerophospholipids), which are the most abundant class of phospholipid found in nature. The simplest form of phosphoglyceride is phosphatidic acid, with the structure below. It contains a glycerol backbone

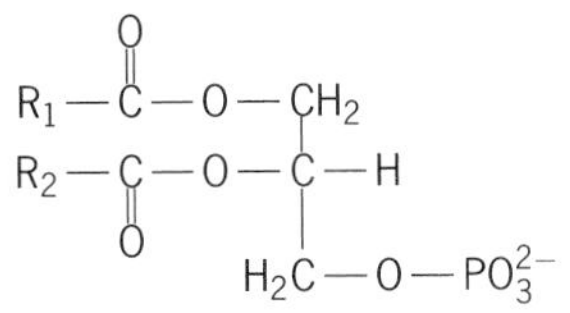

```
        O
        ‖
R1 — C — O — CH2
R2 — C — O — C — H
        ‖         |
        O      H2C — O — PO3^2−
```

with fatty acid esters attached to the sn-1 and sn-2 positions of the glycerol molecule, and a phosphate group attached at the sn-3 position. Other phosphoglycerides share a similar structure, but with appropriate head groups esterified to the phosphate moiety in phosphatidic acid. For example, choline is esterified to the phosphate group in phosphatidic acid to form phosphatidylcholine, and ethanolamine is esterified to the phosphate group to form phosphatidylethanolamine. The most abundant types of naturally occurring phosphoglyceride are phosphatidylcholine (lecithin), phosphatidylethanolamine, phosphatidylserine, phosphatidylinositol, phosphatidylglycerol, and cardiolipin. The structural diversity within each type of phosphoglyceride is due to the variability of the chain length and degree of saturation of the fatty acid ester groups. In general, a saturated fatty acid ester group is attached to the sn-1 position, whereas an unsaturated fatty acid ester group is attached to the sn-2 position.

Sphingomyelin is the major sphingosine-containing phospholipid. Its general structure consists of a fatty acid attached to sphingosine by an amide linkage, producing a ceramide. The alcohol group in the ceramide is esterified with phosphorylcholine, producing sphingomyelin.

Biological roles. The biological membrane is a barrier that separates the inside of a cell from its outside environment. It also serves as a barrier to separate cellular contents into different compartments. A bilayer membrane is formed spontaneously when phospholipids are dispersed in an aqueous solution. In this bilayer structure, phospholipids are arranged in two leaflets with the hydrophobic tails facing each other, and the hydrophilic ends exposed to the aqueous medium. Differences in the head group, the chain length, and the degree of saturation of fatty acids in the hydrophobic end are important factors in determining the shape of the bilayer. For example, in the bilayer the space between the phospholipid molecules and other physical properties are dictated by the chain length and the degree of saturation of the fatty acids. Individual phospholipid molecules are able to move freely in the lateral plane of the bilayer but not in the transverse plane (flip-flop). Small uncharged molecules are able to diffuse through the bilayer structure, but the permeability of larger or charged molecules is restricted. The arrangement of phospholipid molecules into a bilayer in an aqueous medium follows the laws of thermodynamics and represents the structural basis for the formation of all biological membranes. Selected protein molecules, which are responsible for other membrane functions such as channels for molecules (such as polar or large molecules), are embedded in the phospholipid bilayer. *See* CELL MEMBRANES.

For a long time, phospholipids were regarded as merely building blocks for the biological membrane. It was discovered in the mid-1970s, however, that phospholipids participate in the transduction of biological signals across the membrane. For example, when the hormone vasopressin is bound to its receptor on the plasma membrane of a liver cell, the binding sets off a cascade of reactions which result in the activation of the enzyme phospholipase C. The activated phospholipase C catalyzes the breakdown of a special form of phosphatidylinoslitol called phosphatidylinositol bisphosphate, producing diacylglycerol and inositol triphosphate. These two hydrolytic products of phosphatidylinositol bisphosphate are known as second messengers. Second messengers are molecules generated inside the cell in response to the binding of the primary messenger (a hormone) to the receptor. The diacylglycerol that is generated stimulates the activation of protein kinase C, whereas the inositol triphosphate that is generated causes an increase in the intracellular calcium concentration. The activation of protein kinase C and the increase in cellular calcium, initiated by vasopressin binding, lead to the enhanced breakdown of glycogen in the liver cell, thus producing more glucose. The supply of phosphatidylinositol bisphosphate, however, may become exhausted after prolonged hormone stimulation. In order to ensure the continued production of second messengers, the binding of a hormone to its receptor may cause the activation of another type of phospholipase C which is reponsible for the hydrolysis of phosphatidylcholine, a process that produces more diacylglycerol.

A special form of phosphoglyceride, 1-alkyl-2-acetyl-glycero-3-phosphocholine, acts as a very powerful biological mediator. It causes the aggregation and degranulation of blood platelets, and is known as platelet-activating factor (PAF). Other effects of this compound include increasing pulmonary edema, hypersensitivity, acute inflammatory reactions, and anaphylactic shock.

Regulation of metabolism. All organisms have the ability to synthesize phospholipids. In the eukaryotic cell phosphatidylcholine is the major form of phosphoglycerides, whereas in the prokaryotic cell phosphatidylethanolamine predominates. Phosphoglycerides are synthesized either by the condensation of cytidine diphosphate (CDP)-diacylglycerol directly to the head group, or the condensation of 1,2-diacylglycerol to a CDP-bound head group. Since dipalmitoyl-phosphatidylcholine is the major constituent of lung surfactant, failure to produce an adequate amount of this phospholipid by the type II pneumocytes (lung cells) causes the collapse of lung alveoli in newborn babies, a disorder termed respiratory distress syndrome. Phospholipases are responsible for the degradation of phosphoglycerides. These enzymes are found in all tissues and in the pancreatic juice. A number of toxins and venoms have very high phospholipase activity, and several pathogenic bacteria produce phospholipases that dissolve cell membrane and allow the spread of infection. Surprisingly, there are very few inherited diseases associated with the metabolism of phosphoglycerides. Presumably, such genetic defects would be lethal during the early stage of cellular development.

Genetic disorders. Sphingomyelin is synthesized from the condensation of ceramide and CDP-choline. Sphingomyelinase, a lysosomal enzyme, hydrolytically degrades sphingomyelin into phosphorylcholine and ceramide. A genetic disorder caused by a defect in the production of sphingomyelinase, called Niemann-Pick disease, leaves the cell with no or limited or ability to degrade sphingomyelin. In a severe form (type A) of this disease, the liver and spleen are sites of lipid deposits and are therefore tremendously enlarged. The lipid deposits consist primarily of the sphingomyelin that cannot be degraded. Infants with this disease suffer severe mental retardation and death in early childhood. The type A form of this disease is detected predominantly in Ashkenazi Jews, with a carrier frequency of 1 in 100. A number of less severe variants of this disease are known in which minor damage to neural tissues has been detected. In addition, the kidney, spleen, liver, and bone marrow are affected, and the life expectancy of these individuals is reduced to early adulthood. *See* INFANT RESPIRATORY DISTRESS SYNDROME; LIPID METABOLISM.

Patrick Choy

Bibliography. D. E. Vance and J. Vance, *Biochemistry of Lipids, Lipoproteins and Membranes*, Elsevier, 1991.

Phosphorescence

A delayed luminescence, that is, a luminescence that persists after removal of the exciting source. It is sometimes called afterglow.

This original definition is rather imprecise, because the properties of the detector used will determine whether or not there is an observable persistence. There is no generally accepted rigorous definition or uniform usage of the term phosphorescence. In the literature of inorganic luminescent systems, some authors define phosphorescence as delayed luminescence whose persistence time decreases with increasing temperature. According to this usage, luminescence whose persistence time is independent of temperature is called fluorescence regardless of the length of the afterglow; a temperature-independent afterglow of long duration is called simply a slow fluorescence, which implies

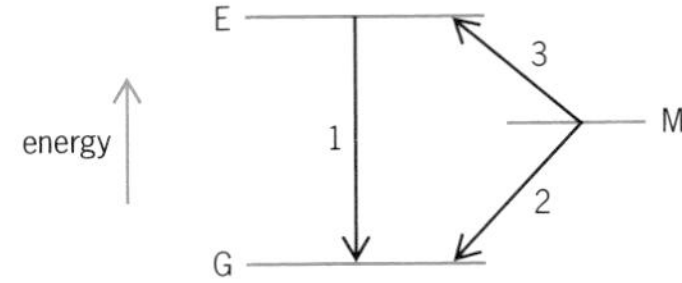

1 = allowed transition
2 = forbidden transition
3 = thermally excited (radiationless) transition

usage

process	1	2	3 followed by 1
inorganic	fluorescence	fluorescence	phosphorescence
organic	fluorescence	phosphorescence	delayed fluorescence

Atomic and molecular transitions involved in luminescence, and corresponding terminology.

that the atomic or molecular transition involved is forbidden to a greater or lesser degree by the spectroscopic selection rules. In nonphotoconductive inorganic systems, phosphorescence arises when some excitation process has placed an atom (or ion or molecule) in a metastable energy state M (from which transitions to the state of lowest energy, or ground state G, are highly improbable or forbidden), and energy from the thermal vibrations of the system subsequently raises the atom to a higher energy state E from which luminescent transitions are highly probable or allowed (see **illus.**). The most common mechanism of phosphorescence in photoconductive inorganic systems, however, occurs when electrons or holes, set free by the excitation process and trapped at lattice defects, are expelled from their traps by the thermal energy in the system and recombine with oppositely charged carriers with the emission of light. In these cases, the level M represents the state of the system with the electron or hole trapped. *See* HOLE STATES IN SOLIDS; SELECTION RULES (PHYSICS).

In the study of organic systems much attention has been given to the G→E→M process, which is the excitation of a molecule from its ground state to an excited state, both usually spectroscopic singlet states, followed by a radiationless transition (called an intersystem crossing) from the singlet excited state to the metastable triplet state M. In the organic literature the term phosphorescence is reserved for the forbidden luminescent transition M → G, while the afterglow corresponding to the M→E→G process is called delayed fluorescence. The spectrum (color) of organic "phosphorescence," defined in this way, is necessarily different from the spectrum of the ordinary fluorescence, because the emitting states (E and M) in the two cases are different and the final (ground) state G is the same. Conversely, the spectra of ordinary fluorescence and delayed fluorescence in organic systems are the same, because the luminescent transition takes place between the same emitting state E and the ground state G in both cases.

The temperature-dependent (M→E→G process) luminescence of a given system can exhibit a wide range of persistence times. At very low temperature where there is insufficient energy available to raise atoms from metastable to emitting states, or to expel electrons from traps, little or no afterglow is observed. At some higher temperature a low-intensity, long-lived afterglow will be observed; at a still higher temperature the afterglow will be brighter but of shorter duration. Finally, at some high temperature where rate of expulsion of atoms from metastable states or rate of expulsion of electrons from traps is very rapid, afterglow can become immeasurably short.

The time dependence of the luminescence intensity (the decay law) can be extremely complex, depending on the number and energies of the metastable states or electron traps involved. Phosphors which give a phosphorescent emission visible to the eye for about half a day at normal temperatures have been synthesized. *See* ABSORPTION OF ELECTROMAGNETIC RADIATION; FLUORESCENCE; LIGHT; LUMINESCENCE. Clifford G. Klick; James H. Schulman

Phosphorus

A chemical element, symbol P, atomic number 15, atomic weight 30.9738. Phosphorus forms the basis of a very large number of compounds, the most important class of which are the phosphates. For every form of life, phosphates play an essential role in all energy-transfer processes such as metabolism, photosynthesis, nerve function, and muscle action. The nucleic acids which among other things make up the hereditary material (the chromosomes) are phosphates, as are a number of coenzymes. Animal skeletons consist of a calcium phosphate.

1	2	3	4	5	6	7	8	9	10	11	12	13	14	15	16	17	18
1 H																	2 He
3 Li	4 Be											5 B	6 C	7 N	8 O	9 F	10 Ne
11 Na	12 Mg											13 Al	14 Si	15 P	16 S	17 Cl	18 Ar
19 K	20 Ca	21 Sc	22 Ti	23 V	24 Cr	25 Mn	26 Fe	27 Co	28 Ni	29 Cu	30 Zn	31 Ga	32 Ge	33 As	34 Se	35 Br	36 Kr
37 Rb	38 Sr	39 Y	40 Zr	41 Nb	42 Mo	43 Tc	44 Ru	45 Rh	46 Pd	47 Ag	48 Cd	49 In	50 Sn	51 Sb	52 Te	53 I	54 Xe
55 Cs	56 Ba	71 Lu	72 Hf	73 Ta	74 W	75 Re	76 Os	77 Ir	78 Pt	79 Au	80 Hg	81 Tl	82 Pb	83 Bi	84 Po	85 At	86 Rn
87 Fr	88 Ra	103 Lr	104 Rf	105 Db	106 Sg	107 Bh	108 Hs	109 Mt	110	111	112	*113*	114	*115*	116	*117*	118

lanthanide series	57 La	58 Ce	59 Pr	60 Nd	61 Pm	62 Sm	63 Eu	64 Gd	65 Tb	66 Dy	67 Ho	68 Er	69 Tm	70 Yb
actinide series	89 Ac	90 Th	91 Pa	92 U	93 Np	94 Pu	95 Am	96 Cm	97 Bk	98 Cf	99 Es	100 Fm	101 Md	102 No

About 90% of the total phosphorus (in all of its chemical forms) used in the United States goes into fertilizers. Other important uses are as builders for detergents, nutrient supplements for animal feeds, water softeners, additives for foods and pharmaceuticals, coating agents for metal-surface treatment, additives in metallurgy, plasticizers, insecticides, and additives for petroleum products. Except for the last four items, these uses involve phosphates. Of the phosphorus compounds utilized in the United States, approximately three-fourths are converted into an

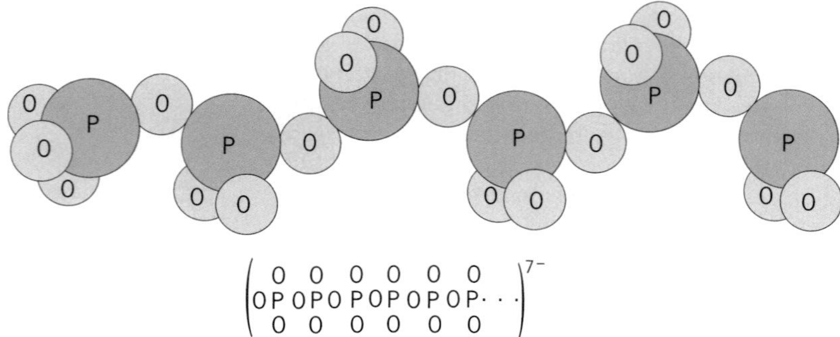

Fig. 1. Long-chain phosphate anion, $(P_nO_{3n+1})^{(n+2)-}$.

impure form of phosphoric acid (wet-process acid) and the remainder into the element (white phosphorus) before further chemical processing into the end products of commerce.

Occurrence and manufacture. Of the nearly 200 different phosphate minerals, only one, fluorapatite, is commercially important. Fluorapatite, $Ca_5F(PO_4)_3$, is mined chiefly from large secondary deposits originating from the bones of dead creatures deposited on the bottom of prehistoric seas and from bird droppings in ancient rookeries. In the United States the major phosphate deposits are in Florida, Tennessee, and the Montana-Idaho region. Other important deposits are found in Morocco, Tunisia, and Russia.

Conversion of phosphate rock (the name given to the common impure form of the mineral apatite) to usable chemicals is accomplished by two major routes: wet acid and elemental phosphorus. In the wet-acid process the phosphate rock is treated with sulfuric acid to obtain a very impure phosphoric acid, plus a precipitate of calcium sulfate. A large body of technology has been developed to achieve easy removal of the calcium sulfate and subsequent concentration and partial purification of the phosphoric acid. Under present-day economic conditions in the United States, the cost of making industrial-grade phosphates via the wet-acid process is about equivalent to the cost of converting the ore to elemental phosphorus and then burning it to give a highly pure phosphoric acid, which is converted into the phosphate. In elemental-phosphorus manufacture phosphate rock, silica, and coke are fed into an electric furnace in which a high-temperature reaction occurs to give the white modification of elemental phosphorus, P_4, a calcium silicate slag, and from iron impurities in the phosphate rock an iron phosphide called ferrophosphorus. Pretreatment of the ore removes most of the fluorine.

Although some elemental phosphorus is used as such in incendiary bombs, in metallurgy, and in the production of organic derivatives and chemicals for matches, most elemental phosphorus is converted to phosphoric acid by reaction with air and water in large burning towers. Elemental phosphorus and phosphoric acid are the starting materials for the synthesis of all other compounds of phosphorus. Phosphoric acid is treated with soda ash, Na_2CO_3, and the resulting orthophosphate composition is then calcined in large rotary converters to make pyro- and tripolyphosphates in very large amounts.

Chemistry. Because of the tremendously large number of compounds based on carbon, descriptive chemistry has been divided into organic, which treats carbon compounds, and inorganic, which deals with the compounds of more than 100 other elements. Work in phosphorus chemistry indicates that there may be as many compounds based on phosphorus as on carbon; the chemistry of phosphorus therefore may become a major branch of chemistry. In organic chemistry it has been customary to group the various chemical compounds based on carbon into families which are called homologous series. This can also be done in the chemistry of phosphorus compounds, even though many phosphorus-based families are incomplete. The best known of the families of compounds based on phosphorus is the group of chain phosphates. Phosphate salts consist of cations, such as sodium, along with chain anions which may have 1–1,000,000 phosphorus atoms per anion. A structural representation of the end of a long-chain, stretched-out phosphate anion is given in **Fig. 1**.

As shown in the figure, the phosphates are based on phosphorus atoms tetrahedrally surrounded by oxygen atoms, with the lowest member of the series being the simple PO_4^{3-} anion (the orthophosphate ion). The family of chain phosphates is based on a row of alternating phosphorus and oxygen atoms in which each phosphorus atom remains in the center of a tetrahedron of four oxygen atoms, as shown in the structural diagram. There is also a closely related family of ring phosphates, a member of which, the trimetaphosphate, is shown in **Fig. 2**.

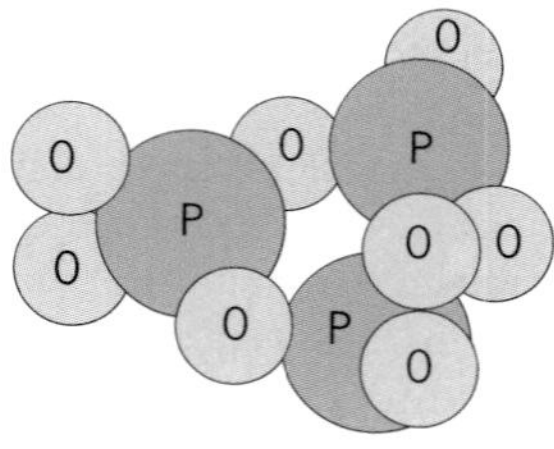

Fig. 2. Ring phosphate anion, $(P_3O_9)^{3-}$.

An interesting structural characteristic of many known phosphorus compounds is the formation of cagelike structures. Such cagelike molecules are exemplified by white phosphorus, P_4, and one of the phosphorus pentoxides, P_4O_{10} (**Figs. 3** and **4**).

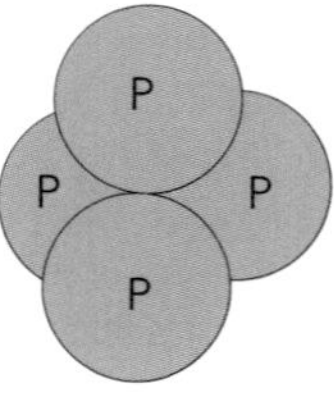

Fig. 3. White phosphorus, P_4.

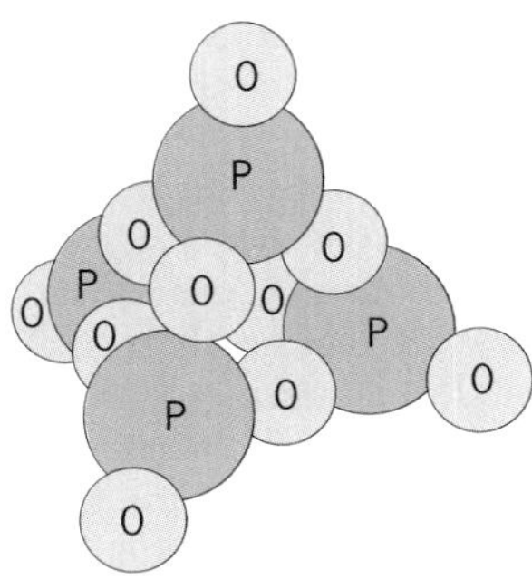

Fig. 4. Phosphorus pentoxide, P_4O_{10}, in vapor state.

Network structures are also common, for example, black phosphorus crystals in which the atoms are bonded together in the form of vast, corrugated planes (**Fig. 5**).

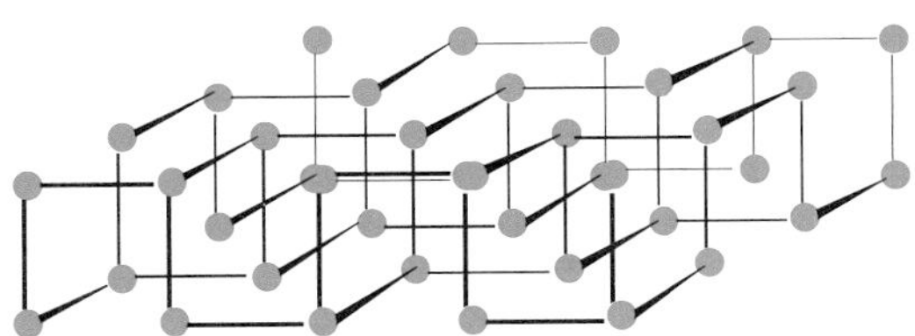

Fig. 5. Black phosphorus, P_n.

In the majority of its compounds, phosphorus is chemically bonded to four neighboring atoms. There is also a large number of compounds in which one of the four neighboring atoms is absent, and in which its place is taken by an unshared pair of electrons. Two typical compounds based on this type of phosphorus are shown in **Figs. 6** and **7**. In addition to the compounds based on quadruply connected phosphorus and those based on triply connected phosphorus, there are also a few compounds in which there are five or six neighboring atoms bonded to the phosphorus. These compounds are very reactive and tend to be unstable because of the use of *d* orbitals in their σ-bond electronic structure. Examples are given in **Figs. 8** and **9**.

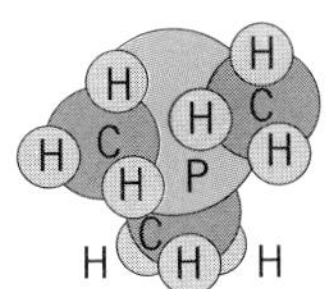

Fig. 6. Trimethyl phosphite, $P(OCH_3)$.

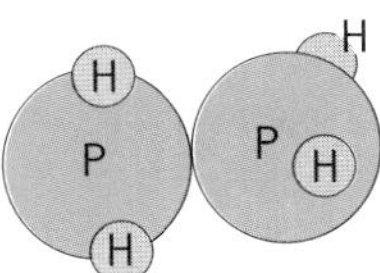

Fig. 7. Biphosphine, P_2H_4.

As for much of inorganic chemistry, structural reorganization plays an important role in the chemistry of phosphorus compounds. Thus, for example, when various mixtures of $POBr_3$ and $POCl_3$ are sealed in a glass tube and allowed to come to equilibrium, the intermediate compounds, $POClBr_2$ and $POCl_2Br$, are formed in various amounts depending on the ratio of the starting materials. The $POBr_3$—$POCl_3$ reorganization involves compounds based on a single phosphorus atom to which is bonded one oxygen and three halogen atoms (chlorine and bromine are halogens). Structural reorganization also occurs between various members of a family or series of compounds. In the polyphosphoryl chloride homologous series, reorganization takes place by exchange of bridging oxygen atoms with chlorine atoms, just as in the $POBr_3$—$POCl_3$ system the exchange is between chlorine and bromine atoms. The various structural units in a polyphosphoryl chloride composition are the monophosphorus compound, $POCl_3$; the end group, $Cl(O)PO_{1/2}$—; the middle group, —$O_{1/2}(Cl)P(O)O_{1/2}$— and the branching group, $OP(O_{1/2}—)_3$, in which the bridging oxygen atoms are shown as $O_{1/2}$, since they are shared between neighboring phosphorus atoms. A typical structure that can be found in a mixture of polyphosphoryl chlorides is shown in **Fig. 10**.

When various ratios of chlorine to oxygen are employed, the distribution of the structural units changes. The ends, middles, and branches do not exist by themselves but must be combined together to form chemical compounds. There is a limit beyond which there is a sufficiently large proportion of branching points that infinite wall-to-wall molecular structures become statistically probable. The presence of such wall-to-wall molecular structures in the mixture of various sized and shaped polyphosphoryl chloride molecules leads to high viscosities and noticeable elastic behavior.

In spite of the fact that families of compounds based on a number of phosphorus atoms are emphasized in this article, the extensive chemical literature before 1950 dealing with phosphorus chemistry was restricted almost entirely to compounds thought to be based on a single phosphorus

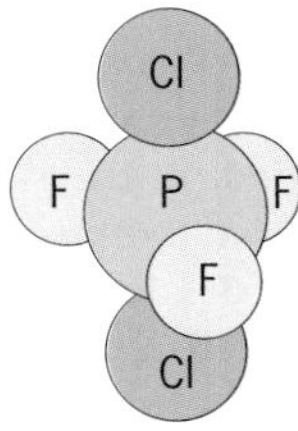

Fig. 8. Phosphorus dichloride trifluoride, PCl_2F_3.

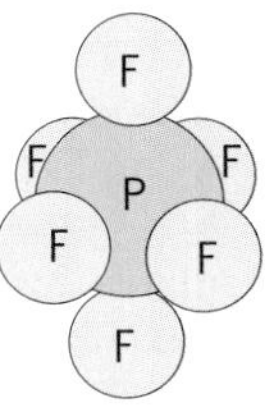

Fig. 9. Hexafluorophosphate anion, PF_6^-.

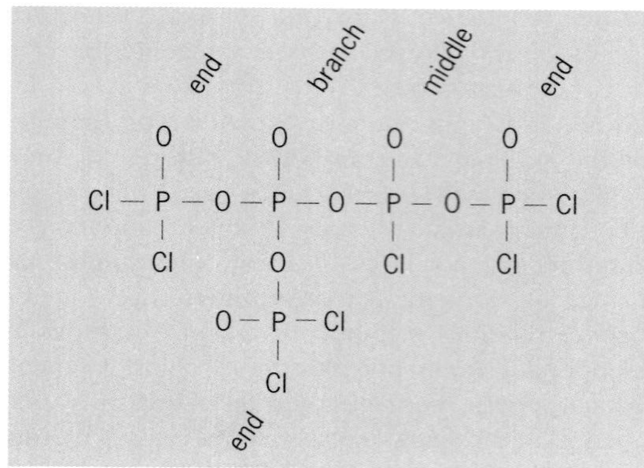

Fig. 10. Isopentaphosphoryl chloride, $P_5O_9Cl_7$.

atom (monophosphorus compounds).

A large number of organic-phosphorus compounds have been prepared. Most of these chemical structures involve three or four neighboring atoms bonded to the phosphorus, but stable structures having two, five, or six neighboring atoms per phosphorus are also known. Some organic-phosphorus compounds exhibit direct bonding between a phosphorus and a carbon atom, while others have C-O-P, C-S-P, or C-N-P linkages as well as various other atomic arrangements. *See* ORGANOPHOSPHORUS COMPOUND.

Principal compounds and uses. Essentially all of the phosphorus used in commerce is in the form of phosphates. The majority of phosphatic fertilizers consist of highly impure monocalcium or dicalcium orthophosphate, $Ca(H_2PO_4)_2$ and $CaHPO_4$. These phosphates are salts of orthophosphoric acid, which is the monophosphorus compound in the phosphate homologous series. Impure dicalcium orthophosphate for fertilizer use is usually called superphosphate, whereas the impure monocalcium phosphate used in this application is called triple superphosphate. *See* FERTILIZER.

Two properties of the family of chain phosphates have led to numerous industrial applications for these compounds. These properties are deflocculation of colloidal particles and formation of soluble complexes with cations. The chain phosphates are strongly adsorbed on the surfaces of inorganic solids and, hence, give these surfaces high negative charges. When finely divided particles bear such high charges, they repel each other and are deflocculated, peptized, or dispersed. An interesting example of this phenomenon is found when a plastic clay-water mass is treated with a chain phosphate. By addition of perhaps a few tenths of 1% of sodium tripolyphosphate to a plastic mass of clay suitably rigid for sculpturing, the clay particles are deflocculated so that the mass liquefies to a consistency similar to that of tomato soup.

The formation of soluble complexes with cations has often been described under the term sequestration, because a complexed ion is sequestered or hidden away in the solution so that it no longer exhibits its normal chemical reactions. The calcium and magnesium of hard water are sequestered by the addition of small (stoichiometric) amounts of chain phosphates so that the water is effectively softened. The complexed calcium will then no longer form precipitates with the carbonate or sulfate in the water to give pipe scale, or with soap anions to give, for example, a ring around the bathtub.

The third member of the family of sodium phosphates, sodium tripolyphosphate, is the major compound used in building synthetic detergents to achieve improved cleaning, primarily by dispersing inorganic soil and softening the water. The average phosphate-containing household detergent produced in the United States for washing clothes consists of about 40% by weight of sodium tripolyphosphate, $Na_5P_3O_{10}$. This compound is used extensively in water softening, as are other members of the homologous series of chain phosphates. The large-volume usage of phosphates in detergent building has led to unwanted growth of algae in inland waters (lakes and rivers) into which the dirty washwaters are discharged. As a result of this fertilizing action, phosphates are considered as water pollutants in those areas where such discharges occur; and in some areas phosphates have been eliminated from detergents by law. For reasonably fast-flowing rivers that discharge directly into the ocean, phosphates are not a problem. *See* DETERGENT; SURFACTANT; WATER SOFTENING.

An interesting water-softening application is found in "threshold treatment" in which tiny traces of a chain phosphate (much less than would be used in sequestering) are used to prevent the formation of pipe scale from hard waters. This application is related to the dispersing action of the phosphates, because traces of phosphate adsorb on the growing surface of the pipe scale as it begins to form, and this inhibits its further growth.

A major pharmaceutical use of phosphates is in toothpastes, in which dicalcium phosphate is the most popular polishing agent. Monocalcium phosphate and sodium acid pyrophosphate, $Na_2H_2P_2O_7$ (the pyrophosphate is the second member of the phosphate family), are employed as leavening agents in cake mixes, refrigerated biscuits, self-rising flour, and baking powder.

Special mixtures based on orthophosphoric acid, H_3PO_4, are used to phosphatize metal surfaces. In this treatment the surfaces become covered with a thin adhering layer of insoluble orthophosphate salts which protect the metal from corrosion and offer an especially adherent base for painting. Automobile bodies, for example, are generally phosphatized before they are painted to prevent rusting in use. Orthophosphate esters find wide use as plasticizers that have flameproofing properties and as gasoline and oil additives.

The phosphorus compound of major biological importance is adenosine triphosphate (ATP), which is an ester of the salt, sodium tripolyphosphate, widely employed in detergents and water-softening compounds. Practically every reaction in metabolism and photosynthesis involves the hydrolysis of this tripolyphosphate to its pyrophosphate derivative, called adenosine diphosphate (ADP). The hydrolysis of chain phosphates occurs through splitting of a P—O—P linkage as indicated in the reaction below.

$$-O-\overset{O}{\underset{O_{-}}{P}}-O-\overset{O}{\underset{O_{-\,-}}{P}}-O + H_2O \longrightarrow$$

$$-O-\overset{O}{\underset{O_{-\,-}}{P}}-O + H-O-\overset{O}{\underset{O_{-\,-}}{P}}-O + H^+$$

In neutral solution at room temperature the rate for this process is extremely slow. However, enzymes increase the rate many thousandfold. The equilibrium between ATP, water, ADP, and the orthophosphate ion is strongly shifted toward the hydrolysis product, ADP and the orthophosphate ion. Because of these facts, organic reactions in biological systems are naturally controlled so that life can exist. A phosphate moiety occurs at every segment of nucleic acid molecules. *See* ADENOSINE DIPHOSPHATE (ADP); ADENOSINE TRIPHOSPHATE (ATP); DEOXYRIBONUCLEIC ACID (DNA). John R. Van Wazer

Bibliography. D. E. Corbridge, *Phosphorus: An Outline of Its Chemistry, Biochemistry, and Technology*, 5th ed., 1995; E. N. Walsh et al. (eds.), *Phosphorus Chemistry: Developments in American Science*, 1992.

Photoacoustic spectroscopy

A technique for measuring small absorption coefficients in gaseous and condensed media, involving the sensing of optical absorption by detection of sound. It is frequently called optoacoustic spectroscopy. Although the technique dates back to 1880 when A. G. Bell used chopped sunlight as the source of radiation, it remained dormant for many years, primarily because of the lack of suitable powerful sources of tunable radiation. However, the usefulness of optoacoustic detection for spectroscopic applications was recognized early in its development, and pollution monitoring instruments (called spectrophones) dedicated for detection of specific gaseous constituents have been used intermittently since Bell's work.

Methods of measuring absorption. During the transmission of optical radiation through a sample (gas, liquid, or solid), the absorption of radiation by the sample can be measured by at least three techniques. The first one is the straightforward detection technique which requires a measurement of the optical radiation level with and without the sample in the optical path. The transmitted power P_{out} and the incident power P_{in} are related through Eq. (1), where

$$P_{out} = P_{in}e^{-\alpha l} \qquad (1)$$

α is the absorption coefficient and l is the length of the absorber. With this technique, the minimum measurable αl is of the order of 10^{-4} unless special precautions have been taken to stabilize the source of radiation.

The second of the techniques is the derivative absorption technique where the frequency of the input radiation is modulated at a low radio frequency or audio frequency, ω_m. The transmitted radiation then contains a time-varying component at ω_m, if the optical path contains absorption which has a frequency-dependent structure. (For structureless absorption, modulated absorption spectroscopy does not provide a signal that can characterize the amount of absorption.) For situations where the absorption has well-defined structure, the modulation absorption spectroscopy can be used to measure αl as small as about 10^{-8} for sufficiently high input powers. The ability to measure the small absorption effects is independent of the input and output power levels for the straightforward measurement technique as long as the noise contributed by the detector is not a factor in determining the signal-to-noise ratio. For the derivative absorption technique, the smallest αl that can be measured varies as $(P_{in})^{-1}$ until the shot noise of the detector begins to be appreciable.

The third technique, optoacoustic detection, is a calorimetric method where no direct detection of optical radiation is carried out but, instead, a measurement is made of the energy, with power P_{abs}, absorbed by the medium from the incident radiation, Eq. (2). Thus the optoacoustic signal, V_{oa}, is given by Eq. (3), where K is the constant describing the con-

$$P_{abs} = P_{in}(1 - e^{-\alpha l}) \qquad (2)$$

$$V_{oa} = K[P_{in}(1 - e^{-\alpha l})] \approx KP_{in}\alpha l \qquad (\text{for } \alpha l \ll 1) \qquad (3)$$

version factor for transforming the absorbed energy into an electrical signal using an appropriate transducer. It has been tacitly assumed that the absorbed energy is lost by nonradiative means rather than by reradiation. The optoacoustic detection scheme implies that the absorbed energy will be converted into acoustic energy for eventual detection.

From Eq. (3), the optoacoustic signal is proportional to the incident power and the absorption-length product αl. Thus, for given sources of noise from the detection transducers, the signal-to-noise ratio improves as the incident energy is increased. Put differently, the smallest amount of absorption that can be measured using the optoacoustic technique varies as $(P_{in})^{-1}$, with no limitation on the level to which P_{in} can be increased for detecting small absorptions. Values of αl as small as 10^{-10} can be measured in the gas phase. The techniques generally used for gases and those used for condensed-phase optoacoustic spectroscopy differ in detail somewhat.

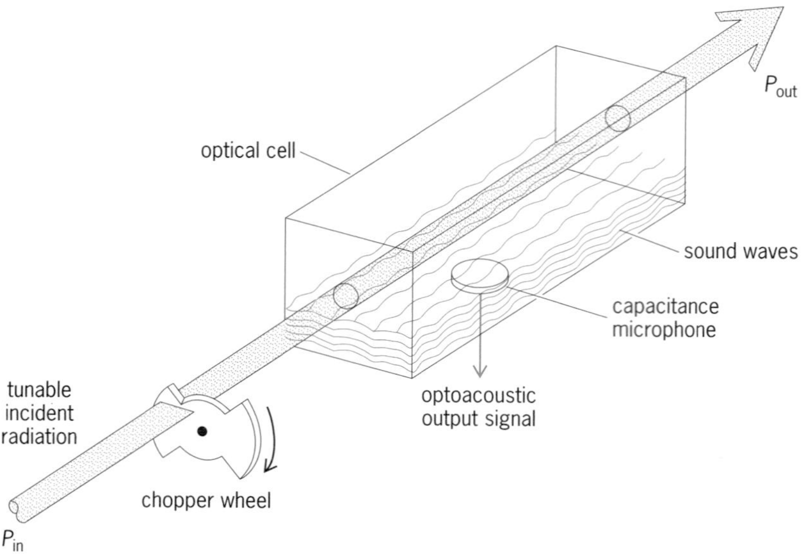

Fig. 1. Optoacoustic cell for gaseous spectroscopy.

Gases. If optical radiation is amplitude-modulated at an audio frequency, the absorption of such radiation by a gaseous medium that has been confined in a cell with appropriate optical windows for the entrance and exit of the radiation and the subsequent nonradiative relaxation of the medium, will cause a periodic variation in the temperature of the column of the irradiated gas (**Fig. 1**). Such a periodic rise and fall in temperature gives rise to a corresponding periodic variation in the gas pressure at the audio frequency. The audio-frequency pressure fluctuations (that is, sound) are efficiently detected using a sensitive gas-phase microphone. The intrinsic noise limitation to the optoacoustic detection scheme arises from the Brownian motion of gas atoms/molecules, and Kreuzer showed that the minimum detectable absorbed power is $P_{min} \approx 3.6 \times 10^{-11}$ W for a 11.9-cm-long (4.7-in) cell. Substituting P_{min} for P_{abs} in Eq. (2), and noting, as in Eq. (3), that $(1 - e^{-\alpha l}) \cong \alpha l$ for $\alpha l \ll 1$, it follows that α_{min} varies as (P_{min}/P_{in}) as indicated above. The usefulness of the optoacoustic detection for measurement of small absorption coefficients became evident with the development of a variety of tunable high-power laser sources which could take advantage of the $(P_{in})^{-1}$ dependence. Using a spin-flip Raman laser tunable in the 5.0- to 5.8-micrometer range, with a power output of approximately 0.1 W, C. K. N. Patel and R. J. Kerl were able to detect α_{min} of approximately 10^{-10} cm^{-1} for a cell length of 10 cm (4 in.). These studies used a miniature optoacoustic cell (**Fig. 2**) with a total gas volume of approximately 3 cm^3 (0.18 in.^{3}). The absorber used was nitric oxide diluted in nitrogen. It is estimated that for a signal-to-noise ratio of approximately 1, and a time constant of 1 s, it is possible to detect a nitric oxide (NO) concentration of approximately 10^7 molecules cm^{-3}, corresponding to a volumetric mixing ratio of approximately 1:10^{12} at atmospheric pressure. *See* LASER.

Fig. 2. Sensitive optoacoustic gaseous spectroscopy cell.

The capability of measuring extremely small absorption coefficients and correspondingly small concentrations of the absorption gases has many applications, including high-resolution spectroscopy of isotopically substituted gases, excited states of molecules and forbidden transitions, and pollution detection. In the last application, both continuously tunable lasers, such as the spin-flip Raman laser and dye lasers, and step tunable infrared lasers, such as the carbon dioxide (CO_2) and carbon monoxide (CO) lasers, have been used as sources of high power radiation. The pollution measurements have demonstrated that the optoacoustic spectroscopy technique in conjunction with tunable lasers can be routinely used for on-line real-time in-place detection of undesirable gaseous constituents at sub-parts-per-billion levels. Specific examples include the measurement of nitric oxide on the ground and in the stratosphere (where nitric oxide plays an important role as a catalytic agent in the stratospheric ozone balance) and measurements of hydrocyanic acid (HCN) in the catalytic reduction of $CO + N_2 + H_2 + \cdots$ over platinum catalysts. These studies point toward expanding use in the future of optoacoustic spectroscopy in pollution detection. *See* LASER; LASER SPECTROSCOPY; STRATOSPHERIC OZONE.

Condensed-phase spectroscopy. A straightforward application of the gas-phase optoacoustic spectroscopy technique to the study of condensed phase (liquid or solid) spectra involves enclosing the condensed phase material within the gas-phase optoacoustic cell (**Fig. 3**). The "photoacoustic" signals generated in the sample due to the absorption of optical radiation are communicated to the gas-phase microphone via coupling through the gas filling the chamber. The inefficiency of such a system is high because of the very poor acoustical match (coupling efficiency approximately 10^{-5}) between the

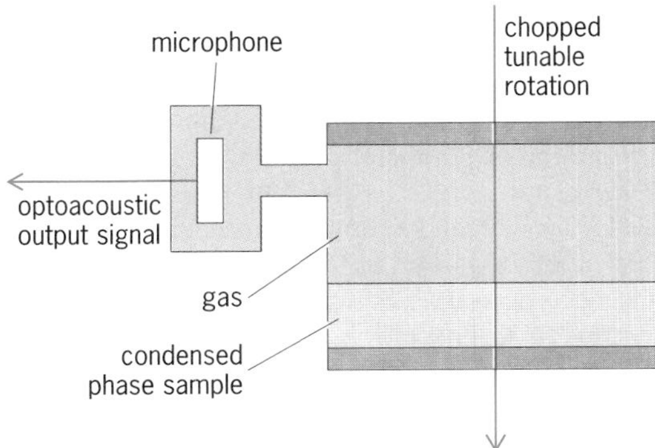

Fig. 3. Arrangement for condensed-phase photoacoustic spectroscopy.

condensed-phase sample and the gas. In reality, because of the large acoustical mismatch, the detection scheme is really "photothermal" rather than "photoacoustic," and this scheme provides a capability of measuring fractional absorption at a level of approximately 10^{-4} when a continuous-wave laser power of approximately 10 W is used. A more severe drawback of the scheme, however, lies in the difficulty of interpretation of the data because of the intimate dependence of the observed optoacoustic signal from the microphone on the chopping frequency, absorption depth, and heat diffusion depth. However, in spite of its shortcomings, the gas-phase microphone technique for condensed-phase optoacoustic spectroscopy has found applications.

A very sensitive calorimetric spectroscopic technique has been developed for the study of weak absorption in liquids and solids. This technique uses a pulsed tunable laser for excitation and a submerged piezoelectric transducer, in the case of a liquid, or a contacted piezoelectric transducer, in the case of a solid, for the detection of the ultrasonic signal generated due to the absorption of the radiation and its subsequent conversion into a transient ultrasonic signal (**Fig. 4**). The major distinction between the above condensed-phase "photoacoustic" spectroscopy technique and the pulsed-source, submerged or contacted piezoelectric transducer technique, is the high coupling efficiency of approximately 0.2 for the ultrasonic signal in the liquid to the submerged transducer, or an efficiency of approximately 0.9 for coupling the ultrasonic wave in a solid to a bonded transducer. Because of this high efficiency, the pulsed-laser, submerged or bonded optoacoustic spectroscopy technique has been shown to be useful for measuring fractional absorptions (that is, values of αl) as small as 10^{-7} when using a laser source with pulse energy of approximately 1 millijoule, pulse duration of approximately 1 microsecond, and a pulse repetition frequency of 10 Hz. There is room for improvement by increasing the laser pulse energy. There is a possibility of the electrostriction effect giving rise to an unwanted background signal, but this signal is not dependent on the light wavelength, and can be minimized by proper choice of experimental parameters.

The pulsed-laser, submerged or bonded piezoelectric transducer technique has a further advantage that time-gating of the ultrasonic signal output can be utilized for the rejection of spurious signals since the sound velocity in condensed media is known and hence the exact arrival time of the real optoacoustic pulse can be calculated. This technique has been used for measurement of very weak overtone spectra of a variety of organic liquids, optical absorption coefficients of water and heavy water in the visible, two-photon absorption spectra of liquids, Raman gain spectra in liquids, absorption of thin liquid films, spectra of solids and powders, and weak overtone spectra of condensed gases at low temperatures. Because of the capability of measuring very small fractional absorptions, the technique is clearly applicable to the area of monitoring water

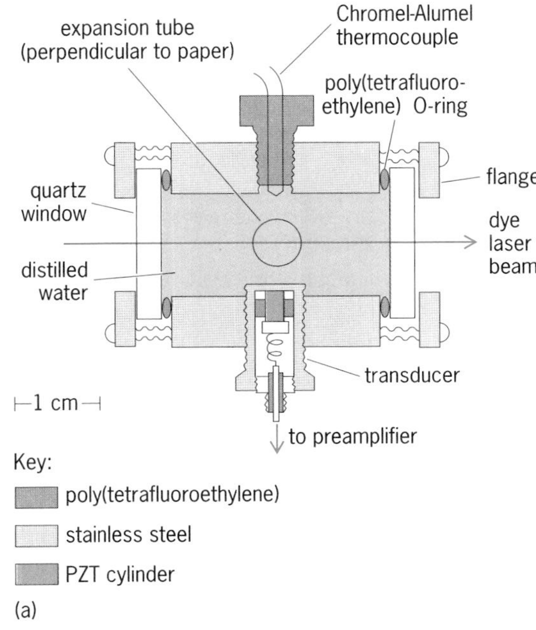

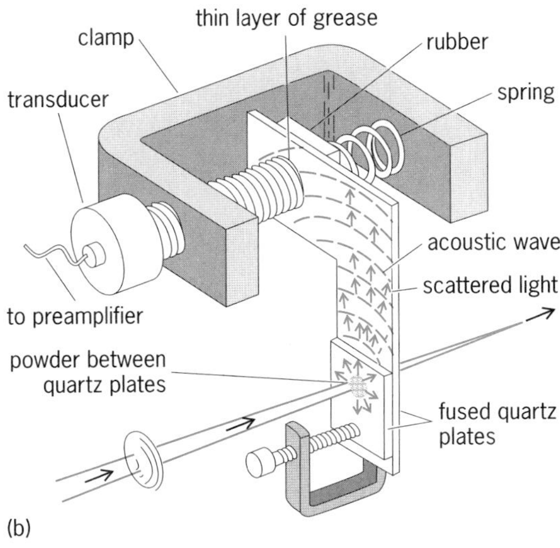

Fig. 4. Arrangement for pulsed-laser (*a*) immersed and (*b*) contacted piezoelectric transducer optoacoustic spectroscopy.

pollution, impurity detection in thin semiconductor wafers, transmission studies of ultrapure glasses (used in optical fibers for optical communications), and so forth. Further, even though in all of the present studies use is made of only the optical radiation, there is no reason to restrict "optoacoustic" spectroscopy to the optical region. By using pulsed x-ray sources, such as the synchrotron light source or pulsed electron beams, the principle described above for a pulsed-light-source, submerged or bonded piezoelectric transducer, gated-detection technique can be extended to x-ray acoustic spectroscopy and electron-loss acoustic spectroscopy. These extensions are likely to have major impact on materials and semiconductor fabrication technology. *See* ABSORPTION OF ELECTROMAGNETIC RADIATION; SYNCHROTRON RADIATION. C. K. N. Patel

Bibliography. P. Hess and J. Pelzl (eds.), *Photoacoustic and Photothermal Phenomena*, 1988; V. S. Letokhov and V. P. Zharov, *Laser Optoacoustic Spectroscopy*, 1986; A. Mandelis (ed.), *Principles and Perspectives of Photoacoustic Spectroscopy*, 1991; M. Rosencwaig, *Photoacoustics and Photoacoustic Spectroscopy*, 1980, reprint 1990.

Photoaffinity labeling

Photoaffinity labeling is a process by which a macromolecule can be labeled at or near its binding or active site. The method can be applied to proteins, nucleic acids, and lipids—all macromolecules—by chemists who need to identify the parts of these molecules that are significant for particular biological functions. Otherwise, x-ray crystallography can frequently determine the complete structure of proteins and sometimes of nucleic acids, although the method does not necessarily highlight the reactive groups in a biomolecule. Nuclear magnetic resonance spectroscopy is increasingly useful in illuminating such structures. *See* NUCLEAR QUADRUPOLE RESONANCE; X-RAY CRYSTALLOGRAPHY.

Many important biological processes involve the formation of a complex between a biopolymer and a chemical reagent; for example, enzymes form complexes with their substrates on the way to catalysis, and antibodies form tight complexes with their antigens. Many biochemical receptors (such as hormone receptors) are complexed with lipid membranes. These complexes cannot usually be crystallized, but information about them can often be obtained by photoaffinity labeling.

In ordinary affinity labeling, a modified substrate that contains a reactive group is chosen. When such a substrate forms a complex with a macromolecule, its reactive substituent can often react with, and so mark, nearby nucleophiles. Although affinity labeling has proved an important method of marking groups essential to enzymatic catalysis, it is grossly inadequate because most of the chemical groups in active sites or binding sites are insufficiently reactive. In many instances, this problem can be solved by photoaffinity labeling.

The idea behind photoaffinity labeling is to place in the binding site a compound that is essentially inert but can be photoactivated at will to yield a highly reactive intermediate. Preferably, this intermediate will react with almost any chemical structure. In the first published example of photoaffinity labeling, a diazoacetyl group was placed on the reactive serine residue of the enzyme chymotrypsin (Chy) [reaction (1); the asterisk indicates an atom of carbon-14].

$$\text{Chy—CH}_2\text{OH} + \text{HC}(=\text{N}_2)\text{—}\overset{*}{\text{C}}(=\text{O})\text{—O—C}_6\text{H}_4\text{—NO}_2 \xrightarrow{\text{pH 6.2}} \text{Chy—CH}_2\text{—O—}\overset{*}{\text{C}}(=\text{O})\text{—CHN}_2 + \text{HO—C}_6\text{H}_4\text{—NO}_2 \quad (1)$$

When the diazo group was photoactivated, it decomposed to yield a carbene [reaction (2)], a derivative of

$$\text{Chy—CH}_2\text{—O—}\overset{*}{\text{C}}(=\text{O})\text{—CHN}_2 \longrightarrow \text{Chy—CH}_2\text{—O—}\overset{*}{\text{C}}(=\text{O})\text{—}\ddot{\text{C}}\text{H} + \text{N}_2 \quad (2)$$

divalent carbon, that will react with almost anything; it will even react with unactivated carbon-hydrogen bonds.

This carbene, and another similar one, marked a number of the amino acids near the active sites of chymotrypsin and trypsin, including a tyrosine residue and the methyl group of an alanine residue. But the process with diazoacetyl chymotrypsin also revealed several of the potential problems with the method. The major reaction product of the carbene proved to be that of internal (Wolff) rearrangement [reaction (3)], rather than an attack on nearby structures.

$$\text{Chy—CH}_2\text{—O—C}(=\text{O})\text{—}\ddot{\text{C}}\text{H} \xrightarrow{\text{Wolff rearrangement}} \text{Chy—CH}_2\text{—O—C(H)=C=O} \quad (3a)$$

$$\text{Chy—CH}_2\text{—O—C(H)=C=O} \xrightarrow{\text{H}_2\text{O}} \text{Chy—CH}_2\text{—O—CH}_2\text{—CO}_2\text{H} \xrightarrow[110^\circ\text{C}]{6N\ \text{HCl}} O\text{-carboxy methylserine} \quad (3b)$$

Another difficulty is nonspecific labeling, that is, labeling of surrounding nucleophiles that occurs when the photoproduct is insufficiently reactive and survives long enough to diffuse out of the pocket where it was formed, only to label amino acids or other structures that are distant from the binding site. Still another problem concerns the occasional transformation of singlet carbenes and nitrenes into the corresponding triplets, which are free-radical structures less reactive than the singlets. Photoaffinity labeling is successful only when the primary photoproducts (the singlet structures) react with their surroundings more rapidly than they undergo intersystem crossing to the relatively unreactive triplets.

Although such problems have also plagued other photoaffinity labeling reagents, they have largely been overcome. Many useful reagents have been prepared, and many successful labeling reactions have been carried out. Computer search of the literature suggests that the method has already been used as many as 5000 times. All 20 of the natural amino acids in proteins have been marked by one reagent or another (see **table**).

The successful reagents include diazirines that (like diazo compounds) yield carbenes on photolysis, arylazides that are photochemically decomposed to nitrenes (highly reactive compounds that contain univalent nitrogen), and ketones that yield free radicals on irradiation. Special structural features (such as trifluromethyl substituents) can enhance the activity of the reagents. The more indiscriminantly reactive the intermediate that results from photolysis, the better; ideally, such intermediates will react with and so identify all the surrounding proteins, lipids, nucleic acids, or other macromolecules. *See* AZIDE.

In order to identify the products of reactions, many of which occur only in low yield, the reagent must be made radioactive. This is usually accomplished with carbon-14 (^{14}C) or tritium (^{3}H), radioactive isotopes of relatively low energy that nevertheless can easily be traced.

Frequently, photoaffinity labeling can be utilized to cross-link two parts of a major assembly of molecules and to reveal its quaternary structure. A good

Examples of photoaffinity labeling reagents

Reagent	Type	Photoproduct	Type
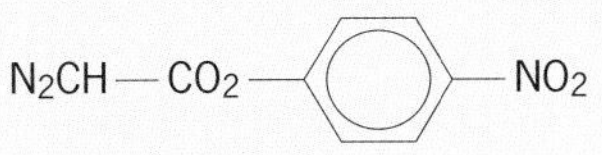 $N_2CH—CO_2—C_6H_4—NO_2$	Diazo	$:CH—CO_2—R$	Carbene
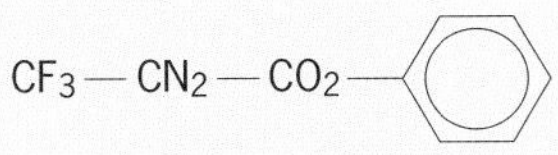 $CF_3—CN_2—CO_2—C_6H_5$	Diazo	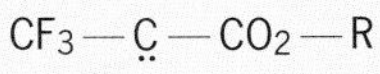 $CF_3—\ddot{C}—CO_2—R$	Carbene
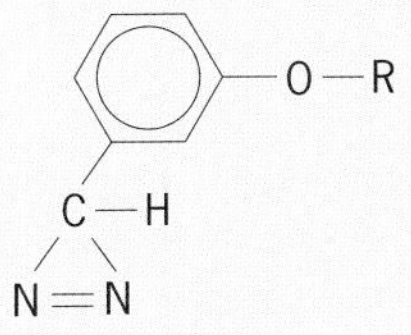 $C_6H_4(—O—R)(—CH{<}N{=}N)$ (diazirine ring: C—H, N=N)	Diazirine	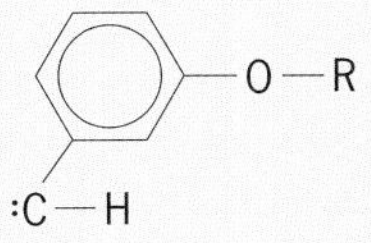 $C_6H_4(—O—R)(—\ddot{C}—H)$	Carbene
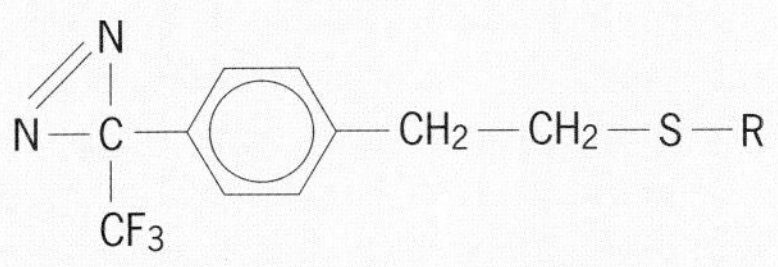 $(N{=}N{>})C(CF_3)—C_6H_4—CH_2—CH_2—S—R$	Diazirine	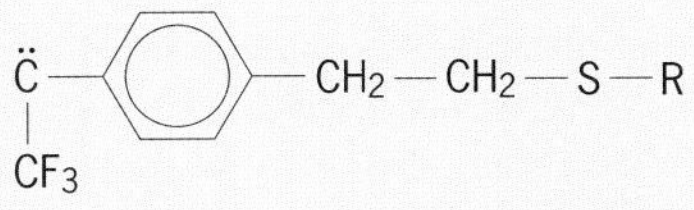 $\ddot{C}(CF_3)—C_6H_4—CH_2—CH_2—S—R$	Carbene
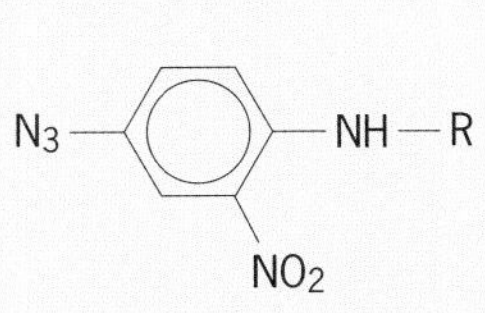 $N_3—C_6H_3(NO_2)—NH—R$	Azide	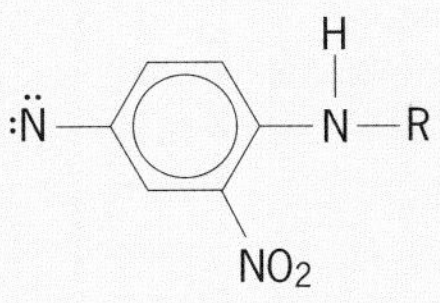 $:\ddot{N}—C_6H_3(NO_2)—N(H)—R$	Nitrene
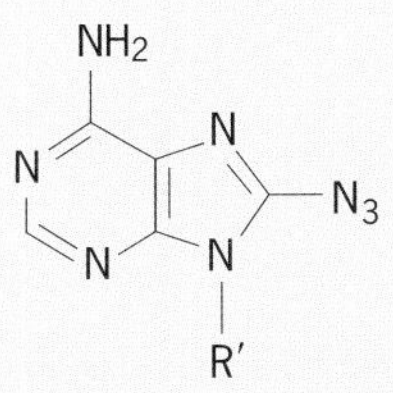 Adenine (NH_2; N-9: R′) with $—N_3$ at C-8	Azide	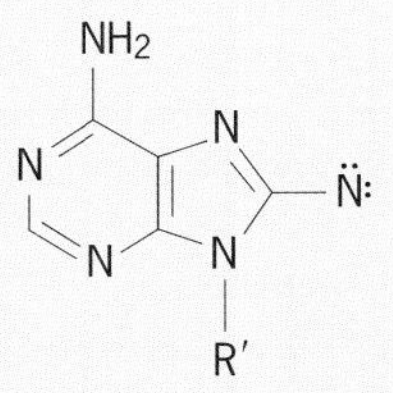 Adenine (NH_2; N-9: R′) with $—\ddot{N}:$ at C-8	Nitrene

example of the utilization of the method is determination of the packing of the protein helices in purple membrane. The relative positions of helices and of the ordering of proteins in membranes can probably be best determined by photoaffinity techniques. F. H. Westheimer

Bibliography. H. Bayley, *Photogenerated Reagents in Biochemistry and Molecular Biology*, Elsevier, 1983; V. Chowdhry and F. H. Westheimer, *Annu. Rev. Biochem.*, 48:293, 1979; F. Kotzyba-Hilbert, I. Kapfer, and M. Goeldner, *Angew. Chem. Int. Ed. Engl.*, 34:1206, 1995; M. S. Platz,, *Photochem. Photobiol.*, 65:193, 1997; J. F. Resek, D. Farrens, and H. G. Khorana, *Proc. Nat. Acad. Sci. USA*, 91:7643, 1994; A. Singh, E. Thornton, and F. H. Westheimer, *J. Biol. Chem.*, 237:3006, 1962.

Photochemistry

The study of chemical reactions of molecules in electronically excited states produced by the absorption of infrared (700–1000 nanometers), visible (400–700 nm), ultraviolet (200–400 nm), or vacuum ultraviolet (100–200 nm) light. Bond making and bond breaking as well as electron transfer and ionization are often observed in both organic and inorganic compounds as a consequence of such excitation. *See* ELECTROMAGNETIC RADIATION.

Electronic absorption. An important generalization sometimes called the first law of photochemistry is that only light that is absorbed can induce chemical change. The absorption of a photon induces an electronic transition in which an electron originally present in a molecular orbital, usually a bonding or nonbonding molecular orbital of the ground state of the absorbing molecule, is promoted to a higher-lying orbital. The excited state produced by absorption of light has a different electronic structure than its ground-state precursor and can reasonably be regarded as an isomeric species with distinct and characteristic chemical and physical properties. *See* PHOTON.

The Born-Oppenheimer approximation allows electronic transitions to be considered as distinct from spins of the electrons and the relative positions of nuclei. The Franck-Condon principle states that because an electronic transition is rapid compared with nuclear motion, the positions of all nuclei remain frozen at the ground-state equilibrium values during an electronic transition. In a Heitler-London plot of the dependence of energy on nuclear configuration (**Fig. 1**), light absorption is thus represented as a straight vertical line: the Franck-Condon principle asserts that a vertical transition (one that takes place without a change in ground-state geometry) is most probable. *See* FRANCK-CONDON PRINCIPLE; RENNER-TELLER EFFECT.

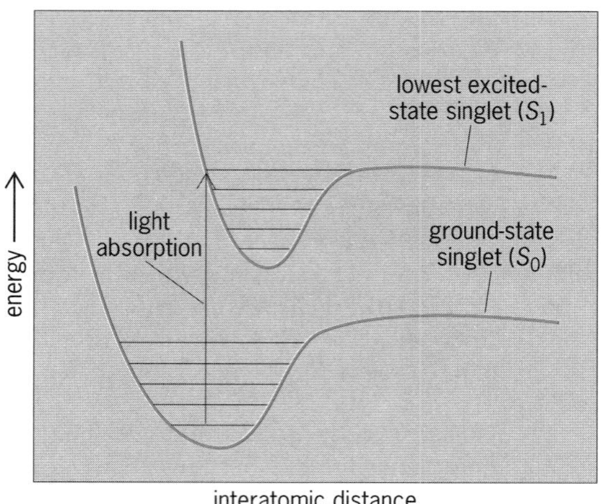

Fig. 1. Representation of a vertical Franck-Condon transition.

Excited state. The probability of producing an excited state upon exposing a ground-state molecule to light is directly proportional to the square of the transition dipole moment, which in turn depends on the electron redistribution attained by the transition. Selection rules are used to define those factors that permit or forbid a specific electron redistribution. The most favorable transitions are those that take place without changes in state symmetry, without spin interconversion, and where excess vibronic energy is minimized. *See* GROUND STATE; SELECTION RULES (PHYSICS).

Most organic molecules exist as ground-state singlets in which all electrons are paired. Because photoexcitation causes the promotion of only a single electron, two singly occupied orbitals are produced upon excitation. If the electronic transition takes place without a spin inversion, these two electrons have opposite spins, and a singlet excited state is produced (**Fig. 2**).

The number of unpaired electrons in a molecule determines its multiplicity: a molecule with no unpaired spins is a singlet; one with one unpaired spin is a doublet; one with two unpaired spin is a triplet; one with three unpaired spin is a quartet, and so forth. If an electronic transition were to take place with a spin inversion, the two singly occupied orbitals would be populated by electrons with parallel spins, producing a triplet excited state. Spin restrictions forbid spin inversion during excitation, and only singlet-singlet electronic transitions are

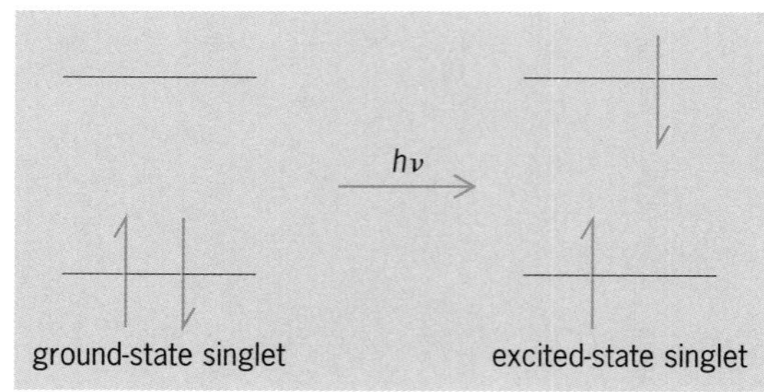

Fig. 2. Spin selection rules dictate that no spin inversion can take place during photoexcitation.

easily observed spectroscopically. After excitation, however, a change in state multiplicity can take place by a process called intersystem crossing. The facility of intersystem crossing is influenced by the magnitude of spin-orbital coupling, which can be enhanced by the presence of a heavy atom (an atom in the third row or below of the periodic table), either bound to the absorbing molecule or present externally as solvent. *See* PERIODIC TABLE; TRIPLET STATE.

For a molecule to be converted to an excited state by photoexcitation, it must have an available state with an energy that exactly matches that of the incident photon, for the energy of each photon is quantized. The energy of a photon is specified by its frequency (ν), from the Bohr frequency condition, as in Eq. (1), where E is the transition energy, h is Planck's

$$\Delta E = h\nu \qquad (1)$$

constant, and ν is the frequency of the radiation. The frequency (ν), often expressed in cm^{-1}, is related inversely to the wavelength (λ) by the expression $\nu = c/\lambda$, where c is the velocity of light. Because energy is directly proportional to frequency, which is inversely proportional to wavelength, higher-energy transitions take place at shorter wavelengths: for example, ultraviolet transitions are of higher energy than those in the visible region of the electromagnetic spectrum. The energy of a given transition can be calculated directly from its wavelength (in nm) by substituting and inserting Avogadro's number (N_A) into the Bohr equation (2). From such a calculation,

$$\begin{aligned}\Delta E &= N_A hc/\lambda \\ &= 1.1963 \times 10^5 \text{ kJ mol}^{-1}/\lambda \text{ (in nm)}\end{aligned} \qquad (2)$$

it can be shown that the energy emitted by a low-pressure mercury arc at 254 nm has an energy corresponding to 471 kJ mol^{-1} (113 kcal mol^{-1}). The lowest energy transition in many molecules is that in which an electron is promoted from the highest occupied molecular orbital (HOMO) to the lowest unoccupied molecular orbital (LUMO). *See* EXCITED STATE; MOLECULAR ORBITAL THEORY.

Transitions. A chromophore is that part of the molecule that accounts for its absorption of light and its photochemical activity. The absorption corresponding to a particular chromophore depends on the type of transition involved in that particular excitation. The promotion of an electron from a π-bonding molecular orbital to a π-antibonding orbital is referred to as a π,π^* (read pi to pi star) transition. Such transitions are frequently encountered in alkenes, alkynes, aromatic molecules, and other unsaturated compounds. Because the spatial overlap of π and π^* orbitals is substantial, such a transition typically has high oscillator strength and a large extinction coefficient (absorbtivity). Promotion of an electron from a nonbonding molecular orbital to a π-antibonding orbital, referred to as an n,π^* transition, involves orbitals that are nearly orthogonal; and it takes place only inefficiently, that is, it has a low oscillator strength and a small extinction coefficient. Such transitions are often encountered in compounds containing carbon-heteroatom or heteroatom-heteroatom double bonds. Because nonbonding molecular orbitals lie at higher energy than bonding ones, n,π^* transitions are of lower energy than the corresponding π,π^* transitions. Both n,π^* and π,π^* transitions are usually found in the ultraviolet region of the electromagnetic spectrum. Transitions involving sigma (σ) bonds (for example $n\sigma^*$ transitions in amines, alcohols, ethers, and alkyl halides and σ,σ^* transitions in alkanes) are usually encountered at the high-energy end of the ultraviolet spectrum or in the vacuum ultraviolet region. *See* CHEMICAL BONDING; ULTRAVIOLET RADIATION.

Electronic transitions in inorganic compounds often involve a metal-based orbital. Promotion of an electron from one metal-based orbital to another is called a *d,d* transition; such transitions are commonly encountered in simple metal complexes. Promotion of an electron from a metal-based orbital to an orbital strongly associated with an attached ligand is referred to as a metal-to-ligand charge transfer (MLCT) transition. Conversely, promotion of an electron from a ligand-based orbital to a metal-based one is called a ligand-to-metal charge transfer (LMCT) transition. The *d,d*, MLCT, and LMCT transitions are usually found in the visible region of the spectrum. *See* COORDINATION COMPLEXES; LIGAND.

Each allowed transition of a compound registers as a band in the absorption spectrum, with the intensity of the transition (measured by its extinction coefficient) being governed by the operative selection rules. The transition intensity of a given absorption is measured by integrating over the whole absorption band. The resulting integrated absorption coefficient is directly proportional to the oscillator strength of the transition. The oscillator strength, a measure of the allowedness of an electric dipole transition compared to that of a free electron oscillating in the three dimensions, is directly related to an experimentally measured value, the extinction coefficient (ϵ). Beer's law is given by Eq. (3), where A is the observed ab-

$$A = \varepsilon bc \qquad (3)$$

sorbance, ε is the extinction coefficient, b is the path length (in centimeters) of the cell used for the measurement, and c is the molar concentration of the absorbing species. This law is used to correlate the observed absorbance with the extinction coefficient and concentration of the absorbing species. *See* ABSORPTION OF ELECTROMAGNETIC RADIATION.

Photophysics. The excited state produced by absorption of a photon is not generally a stable species. After a characteristic lifetime that can vary from femtoseconds (10^{-15} s) to hours, the excited molecule will either relax to its ground-state precursor or

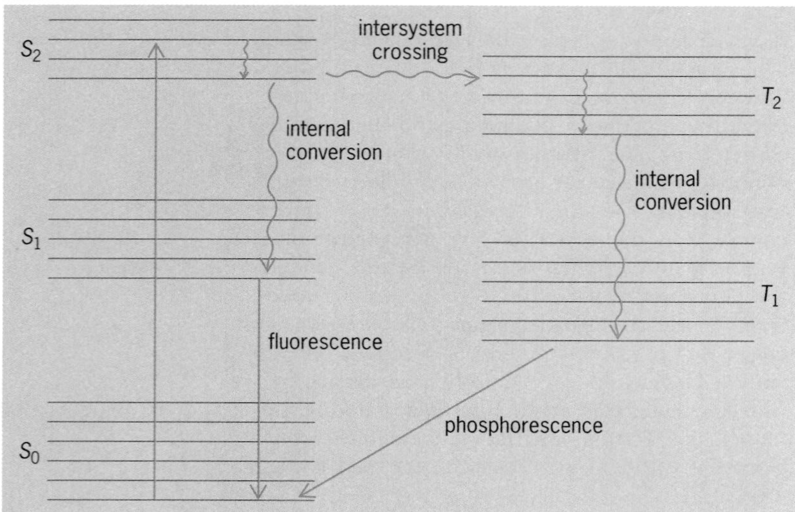

Fig. 3. Jablonski diagram. Solid arrows represent radiative processes; and wavy arrows nonradiative processes. S terms = singlet states; T terms = triplet states.

undergo a chemical transformation. The term photophysics is used to describe nonreactive relaxation processes, which include radiative (taking place with the emission of light) and nonradiative (taking place without the emission of light) pathways.

When absorption of light produces a Franck-Condon excited state that is vibrationally excited, the excess vibrational energy is rapidly dissipated in a process called a vibrational cascade during a period of several femtoseconds to hundreds of picoseconds (10^{12} s), producing the lowest vibrational level of the excited state. When a more highly energetic excited state than the lowest-lying state is produced by the absorption event, relaxation from the upper excited state to a lower one of the same multiplicity also takes place rapidly (typically within a few picoseconds) by a process called internal conversion. Internal conversion is often accomplished isoenergetically, producing a highly vibrationally excited state of the lower-lying excited state, which relaxes very rapidly by a fast vibrational cascade to the lowest vibrational level of the lowest-energy excited state. Internal conversion is so fast, compared with radiative relaxation, that it is generally assumed that radiative processes always occur from the lowest electronic state of a given multiplicity (Kasha's rule) and that the efficiency of the emission is therefore independent of the energy absorbed (Vavilov's rule).

Emission of light associated with a relaxation that does not involve a spin inversion, for example, from an excited singlet state to a ground-state singlet, is called fluorescence. Emission associated with a relaxation in which spin inversion occurs, as from an excited triplet state to a ground-state singlet, is called phosphorescence. Because both fluorescence and phosphorescence result from states produced by fast vibrational relaxation and internal conversion, the emission is observed at a lower energy (longer wavelength) than the absorption that produced the excited state. A transition at a lower energy than a reference wavelength is said to be redshifted; one at higher energy than the reference is said to be blueshifted. In rare cases, thermally activated reverse intersystem crossing from a vibrationally excited triplet state to the excited singlet state can take place; because of the time delay associated with the double intersystem crossing, this emission is called E-type delayed fluorescence. *See* FLUORESCENCE; PHOSPHORESCENCE.

The energies of the lowest singlet and triplet excited states (relative to the ground state) can be obtained from the longest wavelength band of the fluorescence and phosphorescence spectra, respectively. This band is called a 0,0 band to indicate a transition between the lowest vibrational levels of the lowest-lying states. Singlet and triplet energies can also be determined indirectly by measuring quenching efficiencies. The shift between the 0,0 bands for absorption and emission in a single molecule is called its Stokes shift. A small stokes shift is usually observed when the excited state has a geometry similar to the ground state. A Jablonski diagram (**Fig. 3**) is often used to graphically depict the relationship between competing photophysical processes.

The term fluorescence lifetime is used for the time required for the emission of an excited singlet state to decay to half of its initial intensity [typically about 10 nanoseconds (10^{-10} s) for a fluorescent organic molecule]; phosphorescence lifetime, for that of an emissive triplet excited state [typically several microseconds (10^{-6} s) in the absence of a reactive quencher]. The longer triplet lifetime results from the forbiddenness of the phosphorescent emission, which must take place with a concomitant spin flip. The lifetime (τ) is defined as the reciprocal of the rate constant k_r for radiative decay (Eq. 4).

$$\tau = 1/k_r \qquad (4)$$

Emission lifetimes can be measured directly or indirectly. In a direct measurement, the excited state is produced by a laser or electron pulse of very short duration (femtoseconds to nanoseconds), producing an excited state whose emission decay is monitored as a function of time. In an indirect determination, the emission intensity is monitored for solutions containing increasing concentrations of a second chemical species that interacts efficiently (often at a diffusion-controlled rate) with the excited state. A relationship known as the Stern-Volmer equation is used to relate the relative emission intensity to the concentration of the quencher.

Quantum yield, or quantum efficiency, is defined as the number of molecules participating in a given photophysical process or reaction divided by the number of photons absorbed. The quantum yield ranges between zero and one for photoreactions induced by a single photon; values larger than one are indicative of a chain process in which product is formed in a repeating, dark cycle initiated by the photoexcitation. For a photochemical reaction, the number of molecules participating in the reaction is determined spectroscopically or chromatographically as a chemical yield per volume unit per time. The number of photons absorbed is obtained by measuring with a radiometer the light flux per volume unit per time or by employing a chemical actinometer, a known chemical reaction for which the quantum yield is known and accepted as a standard. One of the most frequently used chemical actinometers is the photoreduction of potassium ferrioxalate to ferrous ion (Fe^{2+}), which is determined colorimetrically as its red phenanthroline complex. The quantum yield can also be obtained as the ratio of the rate constant for the desired process to the sum of the rate constants for all competing processes emanating from the same excited state. *See* QUANTUM CHEMISTRY.

Bimolecular excited-state interactions. When an excited state molecule, M*, is produced in the presence of high concentrations of the ground state, M, a complex called an excimer, (MM)*, is produced, namely a dimer that is stable only in an electronically excited state [reaction (5)]. An excimer is stabilized relative

$$M^* + M \rightarrow (MM)^* \tag{5}$$

to an isolated pair because the occupied HOMO interacts strongly with the isoenergetic HOMO of the ground state, inducing an orbital splitting which is energetically favorable. The formation of an excimer is often indicated by a strongly quenched monomer fluorescence coupled with a broad and structureless emission at a wavelength substantially redshifted from that of the monomer. Both singlet and triplet excimers are known, although the existence of the latter are usually inferred from kinetic measurements rather than from direct spectroscopic observation. The binding energy of an excimer is usually less than 85 kJ mol^{-1} (20 kcal mol^{-1}) and typically has a sandwich structure to permit optimal π,π^* interaction.

The bimolecular encounter of two excited triplet states, $^3M^*$, can sometimes lead to a singlet excimer, $^1(MM)^*$, in a process called triplet annihilation [reaction (6)]. If the resulting excimer emits, a time delay

$$^3M^* + {}^3M^* \rightarrow {}^1(MM)^* \tag{6}$$

associated with the triplet lifetime is observed and the resulting fluorescence is called P-type delayed fluorescence.

In parallel to the association of an excimer, the encounter between an excited molecule, M*, and a quencher, Q, can also result in an excited-state complexcalled an exciplex [reaction (7)]. Exciplex for-

$$M^* + Q \rightarrow (MQ)^* \tag{7}$$

mation is most favorable when the energy difference between the LUMOs of M* and Q or the HOMOs of M* and Q are small, namely when the excited state has substantial charge-transfer interaction with the quencher. Like that in excimers, the fluorescence from an exciplex is broad, structureless, and redshifted, and its binding energy can be estimated from a relationship known as the Rehm-Weller equation.

Reaction partners that exhibit strong charge-transfer interactions in the ground state are said to form electron donor-acceptor (EDA) complexes, excitation of which can also yield an emissive exciplex. The presence of a ground-state EDA complex is inferred from the appearance of new redshifted absorption bands or nonadherence to Beer's law upon mixing the electron-rich and electron-poor reagent.

Energy transfer. The process by which an excited state molecule, M*, in an excited singlet or triplet state transfers all or part of its excitation energy to a reaction partner or quencher, Q, is called energy transfer or quenching when the molecule of interest is M [reaction (8)]. This same process is called sen-

$$M^* + Q \rightarrow M + Q^* \tag{8}$$

sitization when the molecule of interest is Q. In the latter case, M is called the sensitizer. Energy transfer permits an exception to the first law of photochemistry in that Q* is produced without having absorbed the incident light.

For energy transfer to take place, an incident wavelength must be chosen so that M is primarily excited, producing an excited state M* whose energy lies above that of Q*. Symmetry selection rules require that all energy transfer events preserve spin multiplicity. Thus, if M* is an excited singlet and Q is a ground-state singlet, M will be produced as a ground-state singlet and Q* as an excited singlet. If M* is an excited triplet and Q is a ground-state singlet, M will be produced as a ground-state singlet and Q* as an excited triplet.

There are several mechanisms by which energy transfer can occur. These include (1) trivial or radiative energy transfer, in which the light emitted in the

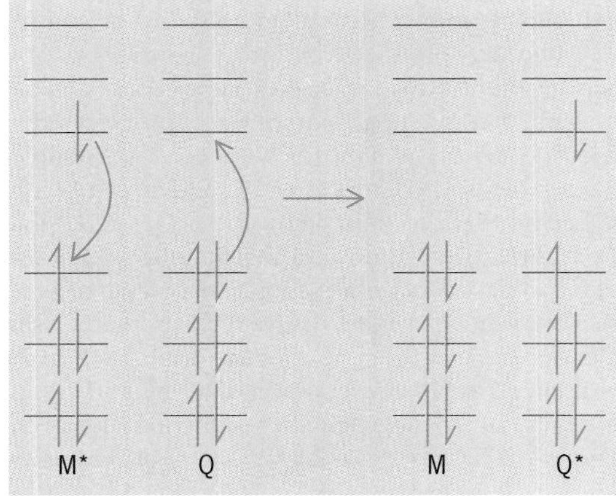

Fig. 4. Dipole-dipole (Förster) energy transfer.

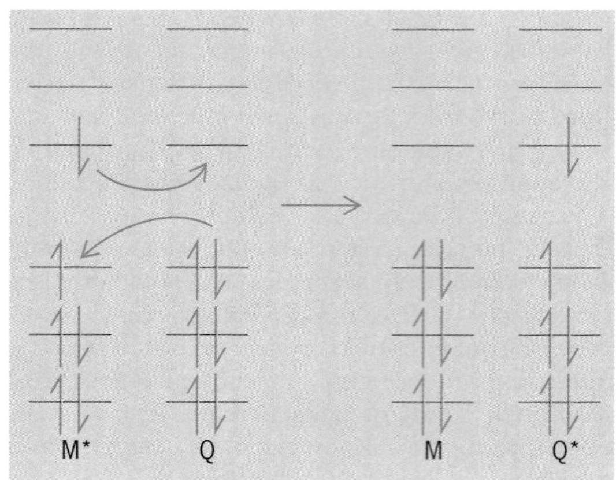

Fig. 5. Electron-exchange (Dexter) energy transfer.

radiative decay of M* is immediately reabsorbed by Q; (2) coulombic or Förster energy transfer, in which dipole-dipole interactions act over long intermolecular distances (**Fig. 4**); and (3) electron-exchange or Dexter energy transfer, in which two electrons are simultaneously exchanged between M and Q during a collision (**Fig. 5**). The first mechanism requires overlap of the absorption spectrum of Q with the emission spectrum of M* but does not require the interacting molecules to be near each other. Its efficiency is governed by those factors and those selection rules that control the oscillator strengths of these two transitions at the common wavelengths. The second mechanism couples M* and Q through a long-range electrostatic interaction. The effect of the oscillating dipole of M* leads to a redistribution of electrons in the HOMOs and LUMOs of the interacting pair, but the electrons of both M* and Q are specifically retained on each molecule (Fig. 4). The magnitude of the dipole-dipole interaction falls off with the sixth power of the distance separating M* and Q. The electron exchange requires a collision in which an electron from the LUMO of M* transfers to the LUMO of Q while an electron from the HOMO of Q simultaneously moves into the vacancy in the HOMO of M* (Fig. 5).

The excited state Q* produced by energy transfer exhibits the same chemical and physical properties as when it is produced by direct light absorption. Sensitization, however, makes it possible to avoid unwanted competing photophysical processes that precede formation of the excited state. For example, higher yields of triplet-derived products can be obtained by a sensitization pathway than are accessible from direct irradiation of a molecule in which the rate of intersystem crossing is low compared with that of radiative or nonradiative relaxation of the singlet state. Therefore, an effective triplet sensitizer will have a small singlet-triplet splitting and a high intersystem crossing yield. Benzophenon [triplet energy = 320 kJ mol^{-1} (76 kcal mol^{-1})] is a commonly used triplet sensitizer.

The triplet energy of a molecule of interest can be determined indirectly by measuring the efficiency of quenching by a series of quenchers with varying, known triplet energies: diffusion-controlled quenching is observed when the triplet energy of the sensitizer is greater than that of the acceptor. The energy transfer can be followed by monitoring the efficiency of reactions derived from the quencher triplet. Some commonly employed triplet quenchers and their sensitized reactions include room temperature solution phase phosphorescence of biacetyl, photosensitized dimerization of 1,3-butadiene, and the geometric isomerization of *cis*-piperylene (*cis*-1,3-pentadiene).

Photoinduced electron transfer. Because photoexcitation promotes an electron to an energetically higher-lying orbital than is occupied in the ground state, both oxidation (removal of the higher-energy electron) and reduction (addition of an electron to the hole created by photoexcitation) are easier in the excited state than in the ground state (**Fig. 6**). That is, the ionization potential (IP), defined as the energy required to remove an electron from the energetically highest-lying orbital to an unbound state, is greater in the ground state than in the excited state; and the electron affinity (ϵ_A), defined at the energy released when an electron is inserted from an unbound state into the energetically lowest-lying molecular orbital, is greater in the excited state than in the ground state. Transfer of an electron from an excited state M* to a quencher Q is called oxidative quenching [reaction (9)], and that from the

$$M^* + Q \rightarrow M^+ \cdot + Q^- \cdot \quad (9)$$

quencher Q to the excited state M* is called reductive quenching [reaction (10). The free-energy change

$$M^* + Q \rightarrow M^- \cdot + Q^+ \cdot \quad (10)$$

associated with these electron transfers can be estimated from free energies of the redox half reactions ($M^*/M^{+/-}$) and $Q/Q^{-/+}$). Here the superscripts indicate one-electron oxidation to produce a cation or one-electron reduction to produce an anion. The free

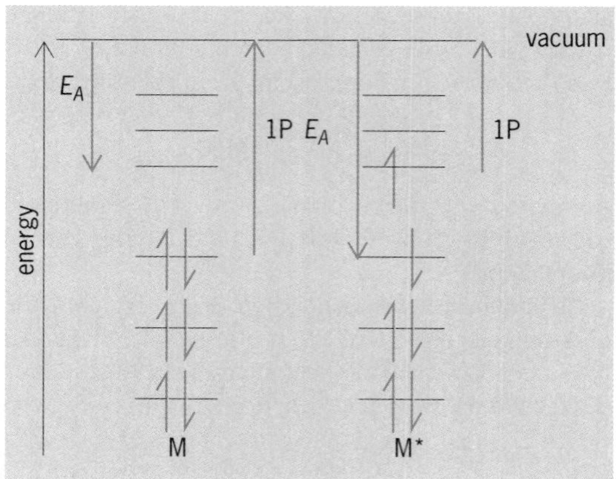

Fig. 6. Diagram showing ionization potentials (IP) and electron affinities (E_A,) of ground (M) and excited states, (M*)

energy of the excited-state redox reaction ($M^*M^{+\cdot/-\cdot}$) differs from its ground-state redox potential by the excited-state energy $\Delta E_{0,0}$, as modified by inclusion of a work term that describes the coulombic energy associated with formation of the two ions ($M^{+\cdot/-\cdot}$ and $Q/Q^{-\cdot/+\cdot}$) held at some fixed distance.

Marcus theory describes the relationship between the energy of activation for a bimolecular electron transfer and the thermodynamic driving force expressed as this free energy. In this theory, the nuclear configuration required for electron transfer is assumed to be sensitive to variations in relative vibrational motions of the interacting molecules (inner-sphere reorganization energy) and in the positions, orientations, and induced polarizations of the surrounding solvent molecules (outer-sphere reorganization energy). Marcus theory predicts that the reaction rate will increase with thermodynamic driving force until the binding energy becomes large and negative, causing the reaction rate to decrease with further increases in exothermicity. This phenomenon is described as the Marcus inverted region. *See* CHEMICAL THERMODYNAMICS.

The radical cation/radical anion pair produced by photoinduced electron transfer exhibits enhanced reactivity compared with that of their neutral precursors. A wide variety of chemical reactions results from these activated species, making photoinduced electron transfer a principal means by which chemical transformations are initiated by light absorption. High yields of chemical products, however, are obtained from these intermediates only if energy-dissipative back electron transfer is suppressed. *See* REACTIVE INTERMEDIATES.

Excited-state reaction surfaces. There are three modes by which a photochemical reaction can proceed: (1) an adiabatic photoreaction, in which all conversions take place completely on the excited-state surface, producing an excited product that relaxes to the observed ground-state product (**Fig.** 7*a*); (2) a diabatic photoreaction, in which the excited-state reactant relaxes directly to the ground state of the product (Fig. 7*b*); and (3) a hot thermal reaction in which the excited-state reactant is converted by internal conversion to a vibrationally excited ground state, which is converted in a thermal transformation to the ground state of the product (Fig. 7*c*). The surface crossing characteristic of an adiabatic photoreaction is facilitated when the energy difference between the two surfaces is minimized: such a minimum is called a reaction funnel, and many photoreactions take place through such funnels.

Photochemical mechanisms. As in all studies of mechanisms of chemical reactions, determining the structure of all products is the first step in the specification of a photochemical reaction. Spectroscopic (nuclear magnetic resonance spectroscopy, electron spin resonance spectroscopy, infrared spectroscopy, mass spectroscopy, x-ray analysis, absorption spectroscopy) and chromatographic (gas, liquid, or thin-layer chromatography) techniques are used to establish product structure and to determine product yields. Monitoring the effect of solvent polarity on reaction rate, the retention or loss of optical activity during the reaction, the positions of isotopic labels, and the success of intermediate trapping experiments can distinguish step-wise chemical reactions (those that proceed through one or more intermediates) from concerted reactions (those that proceed without intermediates). In addition to these mechanistic approaches, the identity and lifetime of the reactive excited state (singlet, triplet, excimer, exciplex, and so forth) and the quantum yields for both product formation and for other competing photophysical processes are required for a full photochemical mechanistic characterization. Time-resolved flash photolysis and pulse radiolysis measurements can, in addition, be sometimes used for direct spectroscopic detection of absorptive or emissive intermediates encountered in a photochemical mechanism, as well as for their kinetic characterization. The addition of specific reactive quenchers or traps, conducting a photoreaction in a low-temperature matrix in which diffusion processes are stopped, and sensitization experiments are effective means for assigning the observed transient absorptions or emissions. The energetics of a well-defined photochemical reaction can be obtained by photoacoustic calorimetry measurements. *See* CHROMATOGRAPHY; MATRIX ISOLATION; PHOTOLYSIS; SPECTROSCOPY.

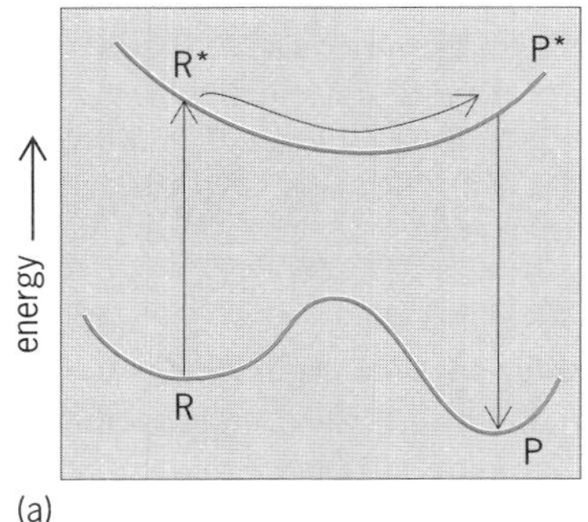

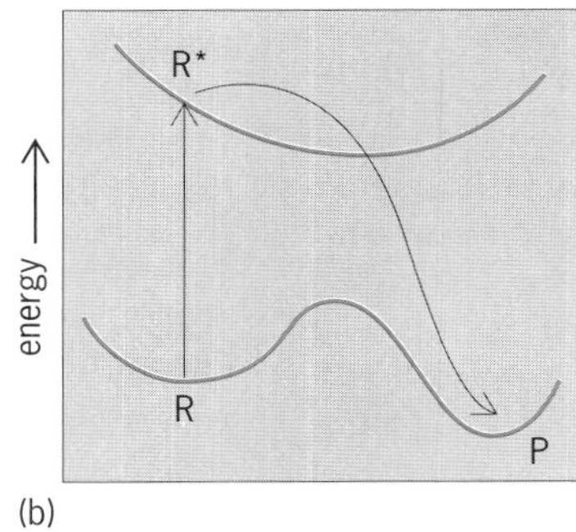

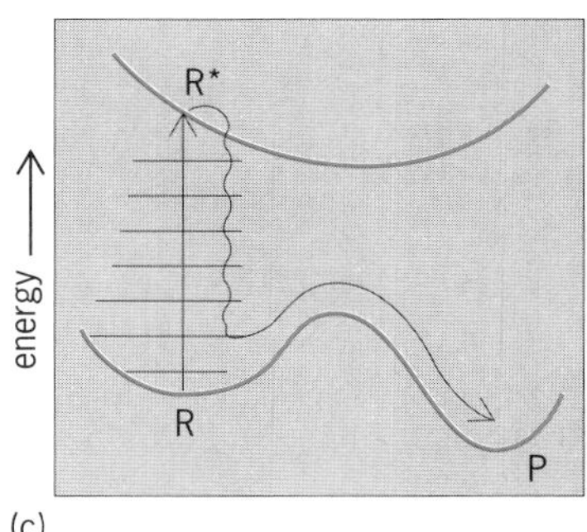

Fig. 7. Modes for photochemical reaction in which reactant R is converted to product P: (*a*) diabatic reaction; (*b*) adiabatic reaction; (*c*) vibrationally excited (hot) ground-state reaction. Excited states are indicated by asterisks.

Concerted photoreactions. There are four common types of concerted photoreactions: (1) cycloaddition reactions, in which the π systems of two separate molecules are joined to produce two new σ bonds in a cyclic product [reaction (11)]; (2) electrocyclic

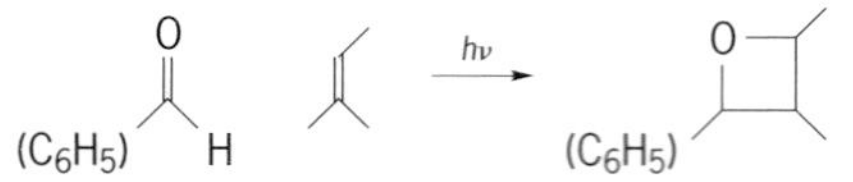

(11)

reactions, in which the ends of a π system within a single molecule are joined by a new σ bond in a cyclic product [reaction (12)]; (3) sigmatropic shifts,

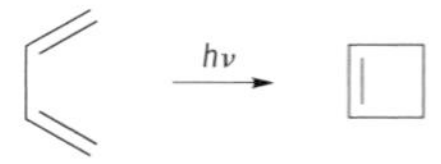

(12)

in which a σ bond migrates along a π system to a new position [reaction (13)]; and (4) cheletropic re-

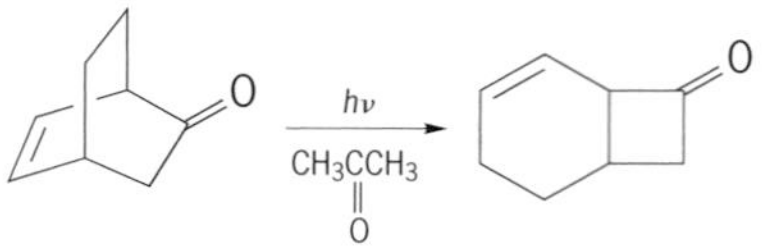

(13)

actions, in which an atom or group of atoms becomes fixed to the ends of a π system [reaction (14)]. The

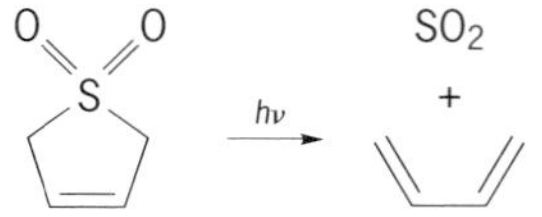

(14)

reverse reactions of each of these types are also concerted and can often be induced by photoexcitation. These reactions are called pericyclic in that they all proceed without intermediates through a cyclic transition state.

The allowedness or forbiddenness of a given pericyclic reaction is dictated by orbital symmetry constraints, expressed in orbital or state correlation diagrams; and the stereochemical sense of allowed pericyclic reactions are defined for molecules containing varying numbers of π electrons by the Woodward-Hoffmann rules. *See* PERICYCLIC REACTION; WOODWARD-HOFFMANN RULE.

Chemiluminescence. Excited states can be produced by thermolysis of highly strained compounds or by highly exothermic electron transfer from a radical anion to a cation. For example, thermolysis of the dioxetane derived from singlet oxygenation of 2,3-dimethylbutane decomposes to two equivalents of acetone, one of which is produced in the excited state [reaction (15)]. The energy released in

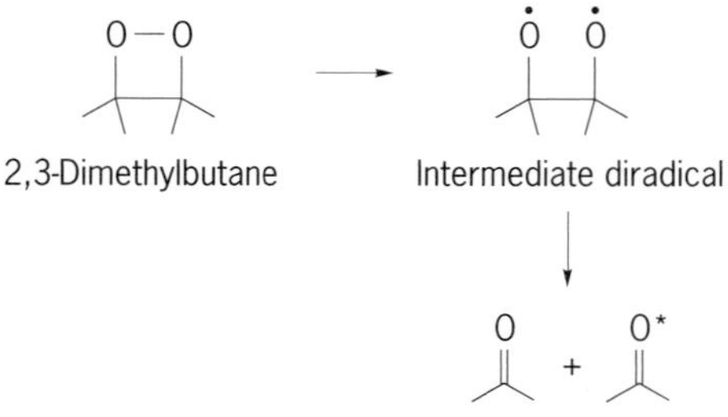

2,3-Dimethylbutane — Intermediate diradical (15)

the cleavage to the intermediate diradical and in the subsequent step to form the two carbonyl groups (—C=O) is substantially more than the triplet energy of acetone. Similarly, back-electron transfer between two electrochemically generated radical ions also releases sufficient energy to produce the luminescent state of the lower-lying singlet. When the radical ions are produced electrochemically, this process is called electrochemiluminescence. *See* CHEMILUMINESCENCE; ELECTROLUMINESCENCE.

Nonhomogeneous environments. Photoreactivity can be substantially affected by the local environment in which an excited state is generated, as has been clearly demonstrated in nature's effect to construct a highly organized reaction center in which the primary events of photosynthesis take place. Since the late 1980s there has been a focus on characterizing altered kinetics and chemical properties that are observed in excited states produced in other than homogeneous solutions in the liquid, gas, or solid state. For example, altered properties are observed when molecules are excited in Langmuir-Blodgett monolayer films, bilayer or multilayer membranes, solid-state crystal lattices, zeolites, normal or reverse micelles, vesicles, colloids, polymers, proteins, enzymes, or antibodies. Similarly, excited-state chemistry is influenced by preadsorption on photoinert surfaces such as silica or alumina or on photoactive ones such as suspended semiconductor particles or metals. For example, environmental detoxification can be accomplished when a pollutant is brought into contact with an excited semiconductor particle. The excited semiconductor often oxidizes the pollutant and reduces oxygen; and in many cases complete breakdown of organic compounds to carbon dioxide, water, and inorganic compounds takes place. Precomplexation with cyclodextrins and other molecular recognition complexants has also been shown to alter chemical reactivity and decay pathways for included molecules. The study of altered photochemical properties induced by noncovalent association with another reagent is called supramolecular photochemistry. *See* CHEMICAL DYNAMICS; INORGANIC PHOTOCHEMISTRY; LASER PHOTOCHEMISTRY; ORGANIC PHOTOCHEMISTRY; PHOTOSYNTHESIS. Marye Anne Fox

Bibliography. M. A. Fox and M. Chanon (ed). *Photoinduced Electron Transfer*, vols. A–D, 1988; A. Gilbert and J. Baggott, *Essentials of Molecular Photochemistry*, 1991; V. Ramamurthy (ed.), Special issue on photochemistry, *Chem. Rev.*, January/February 1993; N. J. Turro, *Modern Molecular Photochemistry*, 1991.

Photoclinometer

A term applied to directional surveying instruments which record photographically the direction and magnitude of well deviations from the vertical. Two instruments of this type are in wide use, the Schlumberger photoclinometer and the Surwell clinograph.

Both instruments record a series of deviation measurements on one trip into and out of the well. From this series of data it is possible to plot quite accurately the course of the well.

In the Schlumberger photoclinometer (**Fig. 1**) the deviation from the vertical is indicated by a small metal ball which rolls in a transparent glass bowl graduated in circular degrees. The direction of the deviation in azimuth is indicated by a magnetic compass. With the instrument suspended by an electrical cable, the positions of the compass and steel ball are photographed on a 35-mm film by operation of electrical controls at the surface which turn on lights in the instrument and snap the camera shutter. After the picture is taken, the film is moved to a new position. Pictures can be taken at a rate of about one per minute. Correlation of the pictures with the depths at which they are taken (known by the length of the suspending cable) yields a measure of the magnitude and direction of deviation of the hole as a function of depth.

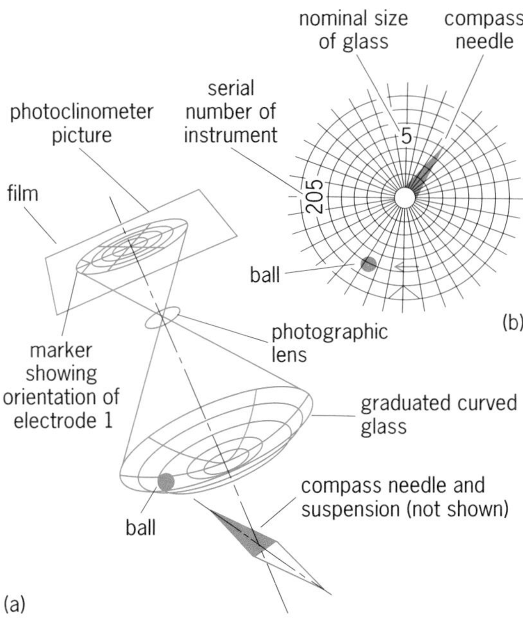

Fig. 1. Schlumberger photoclinometer. (*a*) Principal features. (*b*) Type of record obtained. (*Schlumberger Well Surveying Corp.*)

The Surwell clinograph (**Figs. 2** and **3**) also operates electrically but is powered by batteries contained in the instrument. The deviation from the vertical is indicated by a box level gage and the direction in azimuth by a gyroscopic compass, permitting its use inside steel pipe. This operation is not possible when a magnetic compass is used, unless the pipe is made of special nonmagnetic steel. Since the instrument also contains a watch and a dial thermometer, a simultaneous record of amount and direction of deviation, temperature, and time can be made on 16-mm film. Readings are taken, both descending and ascending, at regular intervals which are preset on the instrument before it is lowered on a wire line into

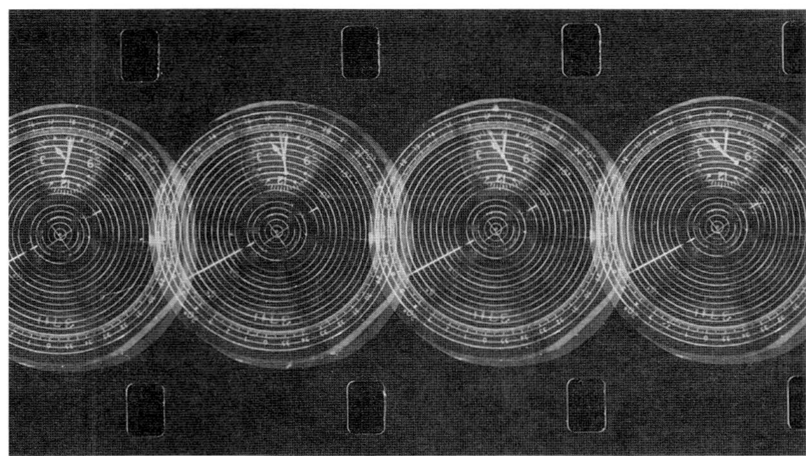

Fig. 2. Motion-picture film records which were made by the Surwell clinograph. (*Sperry-Sun Well Surveying Co.*)

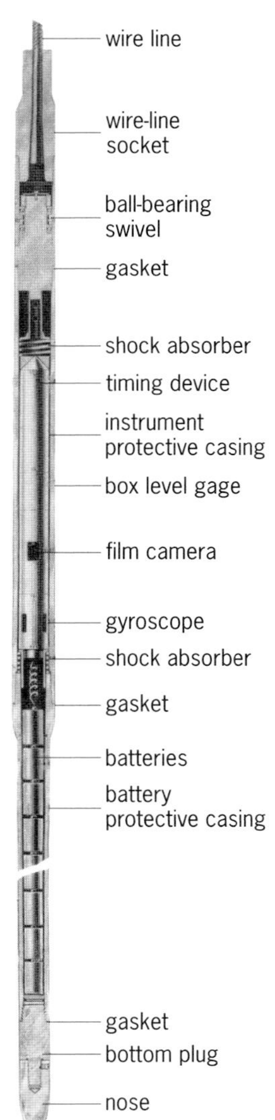

Fig. 3. Vertical section through Surwell clinograph. (*Sperry-Sun Well Surveying Co.*)

the well, thus providing a check on accuracy. Level gages having maximum inclinations of 20, 40, and 55°, respectively, are provided to be used according to the magnitude of deviation. Holbrook G. Botset

Photoconductive cell

A device for detecting electromagnetic radiation (photons) by variation of the electrical conductivity of a substance (a photoconductor) upon absorption of the radiation by this substance. During operation the cell is connected in series with an electrical source and current-sensitive meter, or in series with an electrical source and resistor. Current in the cell, as indicated by the meter, is a measure of the photon intensity, as in the voltage drop across the series resistor.

Photoconductive cells are made from a variety of semiconducting materials in the single-crystal or polycrystalline form. There are elemental types such as germanium, silicon, and tellurium; binaries such as lead sulfide, cadmium sulfide, and indium arsenide; ternaries such as mercury cadmium telluride, lead tin telluride, indium arsenide antimonide, and other combinations of elements. The semiconductor detectors respond to photons that exceed a material-related threshold energy. The bolometer is another important detector and consists of a blackened material having a temperature-sensitive conductivity. This device detects photons over a broad spectrum, utilizing the heating effect of photons. The cells are prepared by growth of the semiconducting materials as nearly pure single crystals, or as polycrystalline films deposited chemically or by evaporation on suitable substrates. *See* BOLOMETER.

Cadmium sulfide cells are used in the visible spectrum for street lighting control and camera exposure meters. Lead sulfide and mercury cadmium telluride, sensitive to infrared radiation, are used for night vision. Infrared photoconductive cells are used for detecting energy loss from buildings and for early detection of breast cancer.

The choice of cell type depends on the application requirements, which include operating temperature, wavelength to be detected, and response time. *See* PHOTOCONDUCTIVITY; PHOTOELECTRIC DEVICES.

Sebastian R. Borrello

Bibliography. N. V. Joshi, *Photoconductivity: Art, Science and Technology*, 1990; W. L. Wolfe and G. J. Zissis (eds.), *The Infrared Handbook*, Environmental Research Institute of Michigan, 1978, reprint 1985.

Photoconductivity

The increase in electrical conductivity caused by the excitation of additional free charge carriers by light of sufficiently high energy in semiconductors and insulators. Effectively a radiation-controlled electrical resistance, a photoconductor can be used for a variety of light- and particle-detection applications, as well as a light-controlled switch. Other major applications in which photoconductivity plays a central role are television cameras (vidicons), normal silver halide emulsion photography, and the very large field of electrophotographic reproduction. The phenomena related to photoconductivity have also played a large part in the understanding of electronic behavior and crystalline imperfections in a variety of different materials. *See* OPTICAL DETECTORS; OPTICAL MODULATORS; PARTICLE DETECTOR; PHOTOCONDUCTIVE CELL; PHOTOGRAPHY; TELEVISION CAMERA TUBE.

Since the electrical conductivity σ of a material is given by the product of the carrier density n, its charge q, and its mobility μ [Eq. (1)], an increase in

$$\sigma = nq\mu \qquad (1)$$

the conductivity $\Delta\sigma$ can be formally due to either an increase in n, Δn, or an increase in μ, $\Delta\mu$. Although cases are found in which both types of effects are observable, photoconductivity ($\Delta\sigma$) in single-crystal materials is due primarily to Δn, with only small effects at low temperatures due to $\Delta\mu$ if photoexcitation decreases the density of charged impurities that scatter charge carriers. In polycrystalline materials, on the other hand, where transport may be limited by potential barriers between the crystalline grains, an increase in mobility $\Delta\mu$ due to photoexcitation effects on these intergrain barriers may dominate the photoconductivity.

The increase in carrier density Δn can be conveniently related to the photoexcitation density f (excitations per unit volume per second) by the simple relation (2), where τ is the lifetime of the pho-

$$\Delta n = f\tau \qquad (2)$$

toexcited carrier, that is, the length of time that this carrier stays free and able to contribute to the conductivity before it loses energy and returns to its initial state via recombination with another carrier of opposite type (that is, electrons with holes, or holes with electrons). In Eq. (2) the photoexcitation term f includes all the processes of optical absorption (excitation across the band gap of the material, excitation from or to imperfection states in the material, generation of excitons that are thermally dissociated to form the free carriers), and the lifetime τ includes all the processes of recombination (free electron with free hole, free electron with trapped hole, free hole with trapped electron). An understanding of the detailed processes of photoconductivity therefore requires a comprehensive understanding of the variation of optical absorption with photon energy, and of the dependence of recombination on imperfection density, capture cross section, photoexcitation intensity, and temperature. *See* ABSORPTION OF ELECTROMAGNETIC RADIATION; BAND THEORY OF SOLIDS; ELECTRON-HOLE RECOMBINATION; EXCITON; HOLE STATES IN SOLIDS; TRAPS IN SOLIDS.

Photosensitivity. Although all insulators and semiconductors may be said to be photoconductive, that is, they show some increase in electrical conductivity when illuminated by light of sufficiently high energy to create free carriers, only a few materials show a large enough change, that is, show a large enough photosensitivity, to be practically useful in applications of photoconductors. There are several ways that the magnitude of the photosensitivity can be defined, depending on the application in mind.

Lifetime-mobility product. Comparison of Eqs. (1) and (2) shows that the basic measure of material photosensitivity is given by the $\tau\mu$ product in the common case where $\Delta\sigma$ results primarily from Δn. The mobility does vary from material to material, but in most practical photoconductors μ has values between 10^2 and 10^4 cm^2/V-s at room temperature, and of course the choice of a particular material is usually dominated by its desirable optical absorption characteristics. The free carrier lifetime τ, on the other hand, can take on a wide range of values from 10^{-9} to 10^{-2} s, depending on the particular density and properties of imperfections present in the material.

Detectivity. One of the major applications for photoconductors has been in the detection of small signals in the infrared portion of the spectrum, where the principal objective is to be able to detect the smallest signal possible with the detecting system. Since in this case the photoconductivity is usually much smaller than the dark conductivity, an ac technique is used in which the light signal is chopped and ac amplification stages are used. The limit to detectability is reached when the light-generated signal is comparable to the electrical noise in the photoconductor. Thus the photosensitivity in this particular case is often defined as a detectivity, which is a normalized radiation power required to give a signal equal to the noise. *See* ELECTRICAL NOISE.

Gain. A third device-oriented definition of photosensitivity is that of photoconductivity gain. The gain is defined as the number of charge carriers that circulate through the circuit involving the photoconductor for each charge carrier generated by the light. The time required for a charge carrier to pass through the photoconductor from one electrode to the other, called the transit time, t_r, is given by Eq. (3), where

$$t_r = \frac{L^2}{\mu V} \tag{3}$$

L is the distance between electrodes and V is the applied voltage. The gain is given then by τ/t_r for each type of possible charge carrier, giving Eq. (4)

$$\text{Gain} = \frac{(\tau_e\mu_e + \tau_b\mu_b)V}{L^2} \tag{4}$$

for the total gain if both electrons and holes contribute. Gains of hundreds or thousands can be readily achieved if the lifetimes are sufficiently long. Gains greater than unity require electrical contacts to the photoconductor that are able to replenish charge carriers that pass out of the opposite contact in order to maintain charge neutrality; such contacts are called ohmic contacts. If nonohmic contacts are used, so that charge carriers cannot be replenished, the maximum gain is simply unity, since only the initially created charge carrier contributes to the current flow. Historically, unity-gain currents of this latter type have been called primary photocurrents, whereas high-gain currents described by Eq. (4) have been called secondary photocurrents.

Spectral response. The variation of photoconductivity with photon energy is called the spectral response of the photoconductor. The curves in the **illustration** typically show a fairly well-defined maximum at a photon energy close to that of the band gap of the material, that is, the minimum energy required to excite an electron from a bond in the material into a higher-lying conduction band where it is free to contribute to the conductivity. This energy ranges from 3.7 eV, in the ultraviolet, for zinc sulfide (ZnS) to 0.2 eV, in the infrared, for cooled lead selenide (PbSe). Photoconductivity associated with excitation across the band gap of the material is called intrinsic photoconductivity. For photon energies smaller than the band gap, the light is not strongly absorbed by the material, and the photoconductivity decreases. For photon energies larger than the band gap, the optical absorption is large, and absorption takes place close to the surface of the material; since the surface has in general more imperfections than the bulk, the carrier lifetime at the surface is generally smaller, and hence the photoconductivity decreases. If the bulk of the material contains a sufficiently high density of imperfections contributing localized levels within the band gap of the material, it is often possible to detect photoconductivity corresponding to optical excitation from an occupied imperfection level to the conduction band or to an unoccupied imperfection level from the valence band of the material. This photoconductivity

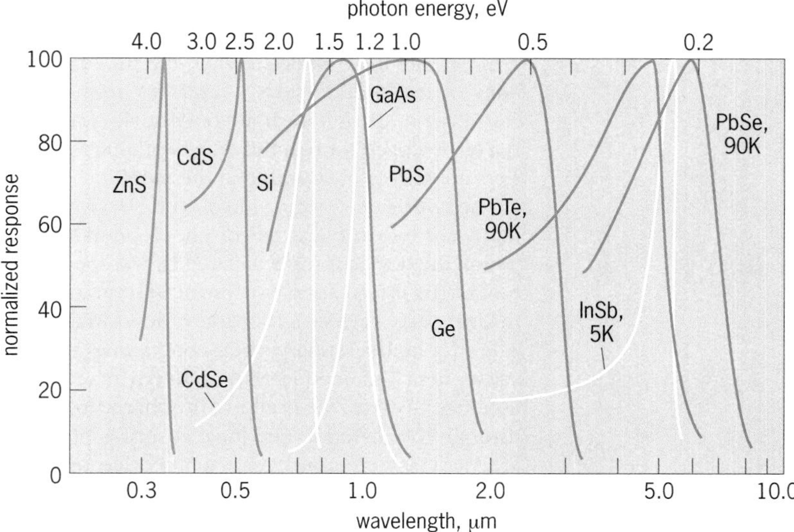

Spectral response of photoconductivity for 10 common photoconducting materials. (*After R. H. Bube, Photoconductivity of Solids, John Wiley and Sons, 1960; reprint, Krieger, 1978*)

occurs for photon energies smaller than the band gap, and is called extrinsic photoconductivity.

Speed of response. A third major characteristic of a photoconductor of practical concern is the rate at which the conductivity changes with changes in photoexcitation intensity. If a steady photoexcitation is turned off at some time, for example, the length of time required for the current to decrease to $1/e$ of its initial value is called the decay time of photoconductivity, t_d. The magnitude of the decay time is determined by the lifetime τ and by the density of carriers trapped in imperfections as a result of the previous photoexcitation, which must now also be released in order to return to the thermal equilibrium situation. If the photoexcitation intensity is high, or the density of imperfections is small, the decay time t_d approaches the lifetime τ as a minimum limiting value. For low light intensities or high imperfection densities, where the density of trapped carriers is much larger than the density of free carriers, the decay of photoconductivity is controlled not by the free carrier recombination rate but by the rate of thermal freeing of trapped carriers, and can be many orders of magnitude larger than the lifetime.

Device forms. Photoconductivity detectors may be made in the form of single-crystal devices or as polycrystalline films, in which the bulk material is homogeneous, or in the form of semiconductor junction diodes. Such junction diodes may be prepared in the form of Schottky barriers, *np* homojunctions, *np* heterojunctions, or more complex *npn* or *pnp* double-junction devices with gain greater than unity. *See* JUNCTION DIODE; MICROWAVE SOLID-STATE DEVICES; SEMICONDUCTOR HETEROSTRUCTURES.

Polycrystalline film photodetectors can be made from a variety of methods involving vacuum evaporation, powder sintering, and chemical solution deposition. The photoconductive behavior can be dominated by quite different effects in different materials systems. For example, the photoconductivity in cadmium sulfide (CdS) films deposited by spray pyrolysis is usually controlled by the modulation of intergrain barriers by photoexcitation, so that $\Delta\mu \gg \Delta n$. On the other hand, standard infrared detecting films of lead sulfide (PbS), deposited from chemical solution, exhibit a photoconductivity for which $\Delta n \gg \Delta\mu$,even though the effects of intergrain barriers are clearly measurable in the mobility.

The television camera vidicon and electrophotography are two applications of photoconductivity in which the device form is dictated by the specific nature of the information processing system involved. In both cases an electrical charge is deposited on one side of a high-resistivity photoconducting material; subsequent illumination of the material increases the conductivity locally and allows the charge to leak off through the material. The local absences of charge are then detected and used to produce or reproduce the original light pattern. The material involved must have special characteristics: it must have a high enough dark resistivity so that the deposited charge does not leak off by ordinary dark conduction, and a high enough photosensitivity so that the charge will leak off as quickly as desired. In the vidicon the charge is deposited by scanning by an electron beam; absence of charge is detected by current flow in a subsequent scanning by the beam. In electrophotography the charge is deposited by a corona discharge; absence or presence of charge is fixed by a subsequent printing process. Richard H. Bube

Bibliography. R. H. Bube, *Photoconductivity of Solids*, 1960, reprint 1978; R. H. Bube, *Photoelectronic Properties of Semiconductors*, 1992; N. V. Joshi, *Photoconductivity: Art, Science and Technology*, 1990; J. Mort and D. M. Pai (eds.), *Photoconductivity and Related Phenomena*, 1976; A. Rose, *Concepts in Photoconductivity and Allied Problems*, 1963, reprint 1978.

Photocopying processes

Processes that use light to generate copies directly from original paper documents. Light is employed to examine an original document and to detect the presence or absence of an image. This examination provides the intelligence required to make each copy. In most cases, light reflected from the original subject directly exposes the medium used to produce an image on a copy. However, in an increasing number of cases, the intelligence-bearing light is converted into an electrical signal, which is later converted back to light to expose the copying medium.

Xerography, the most popular of the photocopying processes, relies on photoconductors and toner powders to create copies. The toner is fused to copy paper by heat or pressure. Microfilming employs silver halide films which must normally be developed and fixed by using wet chemicals. Electrofax, diazo, thermography, and nonmicrofilm forms of silver halide photocopying, although not used as widely as in the past, also have continuing applications.

Various photocopying processes can reproduce black-and-white and color document originals, containing text, line, halftone, and continuous-tone images. In creating copies, users often can elect to reduce or enlarge the size of original document images. In some cases, contrast and exposure controls permit creation of photocopies with image quality superior to original documents.

However, photocopying processes represent just one technology for document duplication. Printing processes, which can produce many copies of a document from one exposure, may be more economical for long-run work. Electronic printers offer another route to paper copies of text and graphic information. However, while such printers yield paper images resembling photocopies, digital input sets them apart from photocopiers which receive their input directly from a paper original. The emergence of dual-personality imaging systems that accept both optical and digital input is making it increasingly difficult to characterize a given piece of equipment as a photocopier or electronic printer. A number of electronic printers equipped with scanner attachments can function as either, and some employ both photocopying and digital imaging processes in generating an individual copy.

This melding of photocopying and printing technologies is expected to continue. Increased future reliance on document storage media such as optical disk plus projected growth of multipurpose local-area networks in offices may hasten the entry of hybrid, multipurpose imaging systems. *See* LOCAL-AREA NETWORKS; OPTICAL RECORDING.

Xerography. There are two classes of xerographic copiers: low-volume units, often called personal or convenience copiers, and high-speed, high-volume units, located in business offices and reproduction centers. Recirculating document feeders, finishers for stapling and binding, sorters, and other peripherals may be added to high-volume units to simplify and speed operations in delivering copy sets. In addition, microprocessor controls enhance their reliability and convenience. Many high-volume units also offer duplexing, the creation of double-sided copies from single-sided originals, in a single machine operation. In addition, some copiers can produce color copies of color originals. *See* MICROPROCESSOR.

The major components of a typical high-volume xerographic copier (**Fig. 1**) include the photoconductor, a primary charger, and systems for exposure, toning, transfer, erasing, and cleaning.

Standard process. The process begins at the primary charger. In this machine, the photoconductor belt is charged to -600 volts. The photoconductor retains the charge on its surface for significant periods of time if it is not exposed to light. Corona charging is the most common charging method. A small-diameter wire is stretched across the photoconductor approximately $^1/_4$ to $^1/_2$ in. (0.64 to 1.3 cm) above its surface. To charge negatively, the wire is raised to a negative potential of several thousand volts, creating a strong electrical field that separates the air molecules into negative and positive ions. The negative ions are repelled from the negatively charged wire to the photoconductor layer which is coated on a conducting surface. The conducting surface is normally held at ground potential.

When the photoconductor belt has moved into the exposure position, the flash lamps illuminate an original document positioned on the platen. With a black-and-white document, black image areas do not reflect light, but white nonimage areas do. The optical system focuses this reflected light on the negatively charged photoconductor, and where light strikes the photoconductor the negative charges are erased. Only areas corresponding to document images retain negative charges.

At the toning station, positively charged toner particles are wiped across the photoconductor surface and are attracted to the negatively charged areas. The greater the amount of negative charge in each area of

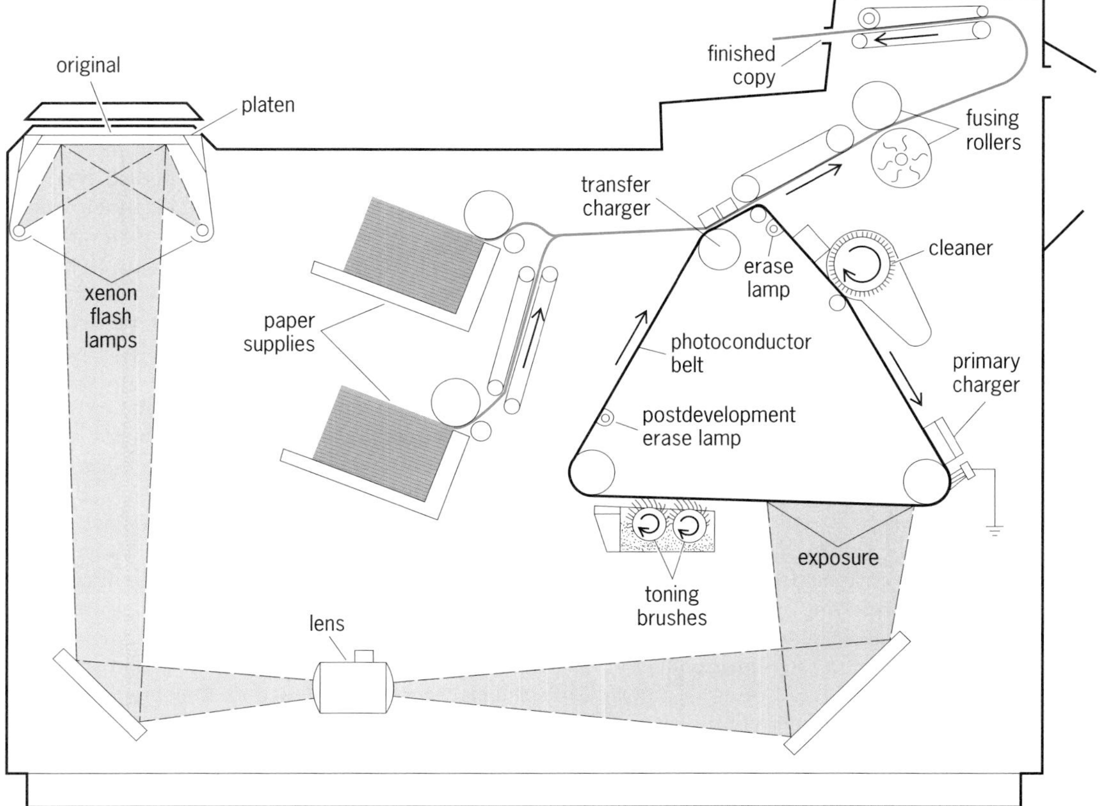

Fig. 1. Diagram of a copier-duplicator. (*Kodak*)

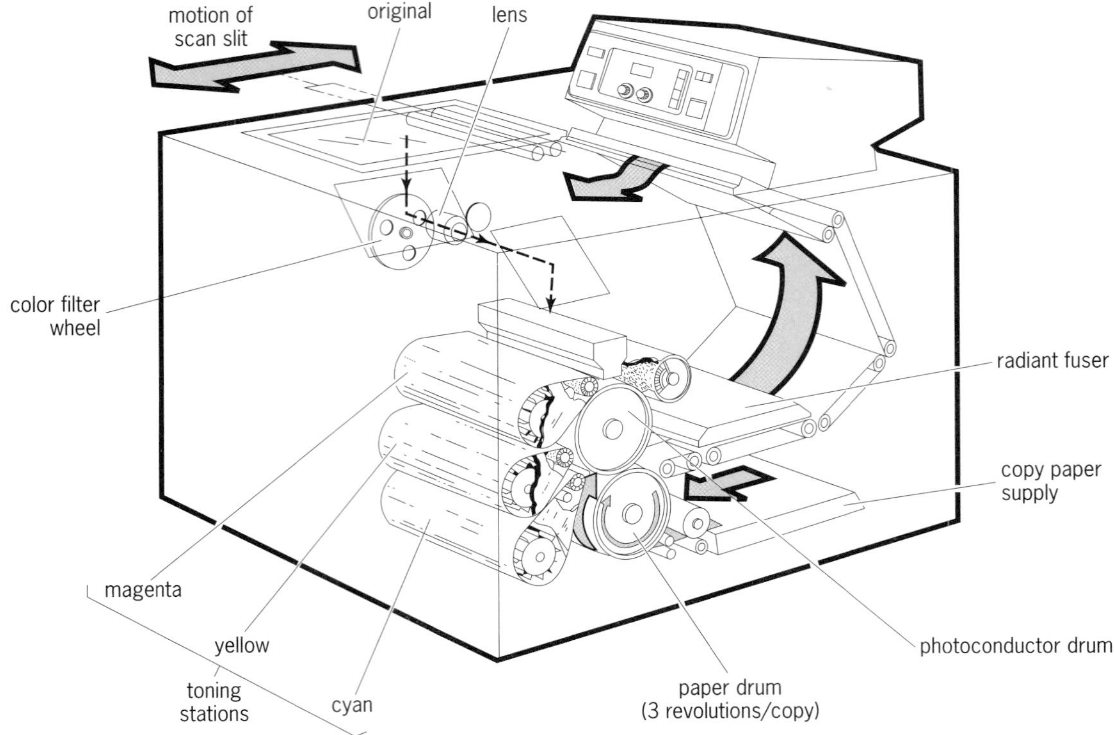

Fig. 2. Diagram of a color copier. (*Xerox Corp.*)

the photoconductor, the greater the amount of toner attracted to it. This principle allows the creation of copies with gradations of gray tones approaching those of the original.

Next the photoconductor's negative field is relaxed by the postdevelopment erase, and the copy paper is brought into contact with the photoconductor. A heavy negative charge is laid down on the back of the paper, by the transfer charger, to attract the positive toner particles to the paper.

The paper then separates from the film and is carried to the hot roller fuser which softens the toner particles and presses them into the paper surface. Meanwhile, the photoconductor is erased, cleaned, and returned to the primary charger, where it is ready to start another cycle.

Process variations. Commercially available machines feature a number of process variations. For instance, some systems use a photoconductor which performs better if it is positively charged. In this case, a negatively charged toner is used. Most organic photoconductors, such as those incorporating zinc oxide and cadmium sulfide, are charged negatively. In contrast, most photoconductors relying on silicon or selenium and its alloys are charged positively.

Some machines transport the photoconductor on a rigid cylinder (**Fig. 2**) instead of using a flexible belt. This simplifies the mechanics of the photoconductor system, but poses optical system complexities. To focus light on a revolving drum, a scanning optical system is needed to project one section of the original at a time onto the appropriate cylinder section. This is done during a continuous exposure by synchronized lamp and scanner movement. Since a belt design can present an extended flat portion, or a virtual frame, for imaging, a simpler optical system can project an entire document on a frame in one brief exposure. A drum must be used for many inorganic photoconductor materials which develop artifacts if they are flexed.

Use of a scanning or stationary optical system is strongly influenced by the choice of photoconductor configuration. Both systems may use a variety of lenses, or lens and mirror configurations, to project the image of the document on the platen onto the photoconductor. Auxiliary lenses of varying focal lengths or zoom lenses are used to project smaller or larger images for copy reductions or enlargements. However, in a scanning system, the scanner's speed must be altered relative to photoconductor rotation speed any time there is a change in copy reduction or enlargement requirements.

Toning systems. Most high-volume machines rely on dry, two-component toning systems. They feature black-pigmented toner particles that form copy images and reusable carrier particles that provide their transport. Chemicals are added to the toner and carrier particles to control the magnitude and polarity of the electrical charging action. Roller-type magnetic brushes move the iron carrier particles and piggybacking toners across the photoconductor surface. Toner-carrier interaction gives toner particles their electrical charge. Toner is added automatically as required to maintain the proper toner-carrier mix.

Many low-volume copiers use monocomponent toning systems. In such systems, very small magnetic particles are incorporated into the toner particles themselves. Rotating magnets in the brushes directly attract the toner particles to wipe them across the photoconductor.

Most monocomponent toners also include agents such as carbon to increase their electrical conductivity. When the brush brings them near a charged image on the photoconductor, the electrical field of the image induces an opposite charge in the conducting toner and the toner is attracted to the image.

Liquid-dispersed toners are used in some machines. Very small toner particles are suspended in an insulating hydrocarbon liquid where they take on an electrical charge. The photoconductor is toned by bringing it into contact with the liquid. Various techniques are used to minimize the amount of liquid which is transferred to the photoconductor with the toner particles. The liquid is usually a highly refined, low-volatility, petroleum distillate. Drying and fixing are normally accomplished by heating the paper.

The heated fusing rollers used with dry toner processes are covered with silicone rubber or Teflon to prevent the toner from sticking to their surfaces. A thin layer of silicone oil often is wicked onto the surface as further insurance against sticking.

Some machines use unheated, hard-surfaced fusing rollers which rely on pressure alone to fix the toner to the paper. The design is simpler, but the resulting copies are glossy and the process can cause changes in the thickness and stiffness of the paper used for copying.

Two types of cleaners are used to remove untransferred toner from photoconductors. Photoconductors with a hard surface can be scraped with a plastic wiper blade. Rotating soft fur brushes are used in other machines. With either technique, the reclaimed toner can be collected in a waste container or recirculated into the toning system.

Color systems. Color xerographic copiers use the same process steps as black-and-white units. However, exposing and toning operations must be repeated three times to produce each of the three overlaid colors on every copy (Fig. 2). Blue, green, and red filters are rotated in front of the lens to make separate exposures for the original document's light reflectance of the primary colors. Then, toner particles of yellow, magenta, and cyan hues are used to develop the resulting photoconductor latent images.

Normally, the three color images are generated in sequence, and are transferred in register onto one sheet of paper. Since this requirement impacts machine speed, color copiers can produce only about one-third as many copies as a corresponding black-and-white copier using the same process velocity. This speed-volume penalty, the extra complexity entailed in color photocopiers, and additional material requirements make color copies considerably more expensive than black-and-white copies.

Microfilming. This is the only black-and-white, silver halide photocopying method used on a large scale to reproduce document originals (**Fig. 3**). John Danzer is credited with making the first microphotograph of a document in 1839, after substituting a form of a microscope lens for his camera's normal lens.

With source microfilming, original paper documents are imaged on very small areas of silver halide film. Fine-grain films and the special lenses in source document microfilmers allow documents to be filmed at reduction ratios of up to 250:1, although 50:1 is the upper limit for most commercial machines. So, for example, a document microfilmer operating at a reduction ratio of 50:1 is capable of imaging more than 25,000 $8^1/_2 \times 11$-in. (22×28-cm)

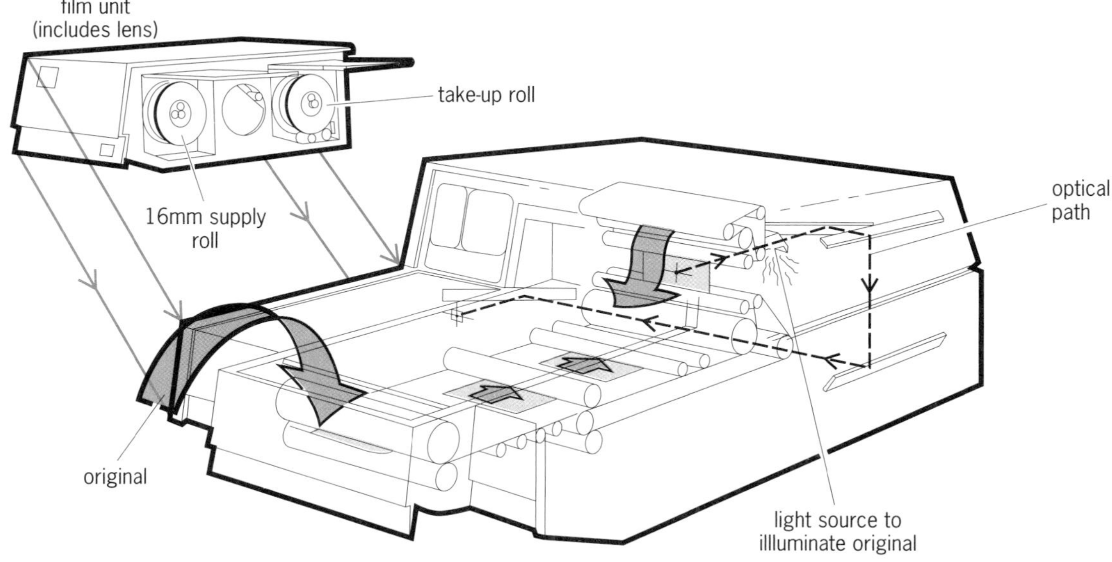

Fig. 3. Diagram of a Kodak Reliant 800 rotary microfilmer.

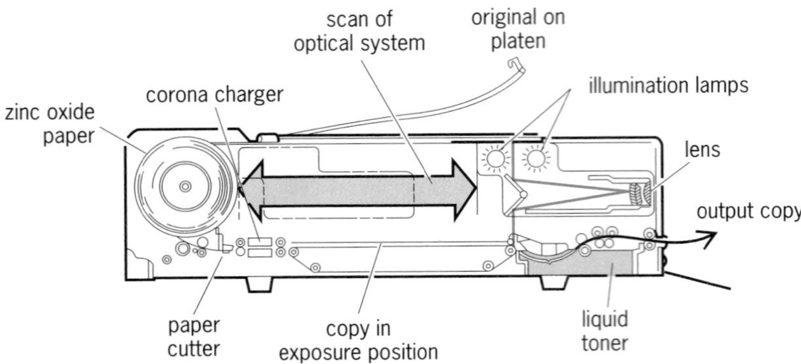

Fig. 4. Diagram of an Electrofax copier.

documents on a single roll of 16-mm microfilm contained in a retrieval magazine.

While some source microfilmers use roll film, others use microfiche, or index-sized film cards (typically 105 × 148 mm or 4.1 × 5.8 in.). If paper documents are discarded after microfilming, users can obtain large file space savings. The microfilm permanence makes this photocopying process popular for archival and document searching purposes.

Since it is impossible to read text reduced to this size on a microfilm frame, projection display units are needed to enlarge the microfilmed images for viewing. When such viewers are combined with printers, they are called microfilm reader-printers. These units may incorporate xerographic or Electrofax printers, or may use printers that employ heat to develop coated papers containing silver behenate. Many microimage terminals not only project microfilmed images for viewing but offer automated computer-assisted retrieval capabilities.

Computer-assisted retrieval systems link microimage terminals to mainframe or minicomputer systems. The addresses of microfilmed documents are stored in computer memory and may be accessed readily from a cathode-ray-tube computer terminal. Image marks placed on the film during microfilming enable the microimage terminals to search for the requested film frames automatically. Some microimage terminals directly linked to computers totally automate such searches. The speed of accessing microfilm photocopies enhances microfilm's value as an inexpensive mass storage medium.

Progress in digital input microfilm technology parallels electronic printing innovation. Computer output microfilmer devices are, in fact, electronic printers which translate digital input into light to create images on film rather than paper. Combination microfilmers/scanners allow images to be stored on optical disk, CD, and/or magnetic disk, in addition to microfilm, and digital workstations allow analog-to-digital microfilm conversion.

Electrofax. The Electrofax process is a simplified xerographic process which eliminates the need for a reusable photoconductor and its associated transfer, cleaning, and erase stations (**Fig. 4**). The photoconductor is an integral part of every sheet of copy paper.

The photoconductive layer on the Electrofax paper is nontoxic zinc oxide dispersed in a flexible resin binder. The paper is negatively charged by a corona charging method. After exposure to light reflected from a document, toner is applied. In early models the toner was dispersed in a liquid and was fixed by evaporating the liquid, often with a heat assist. Later, dry toner particles were applied by a magnetic brush and were heat-fused to the paper.

The need for specially coated papers eventually led to declining use of Electrofax units. The heavy paper tended to mark, did not feel like plain paper, and was more expensive to mail. However, some manufacturers have adapted the Electrofax process to lithographic platemaking, using large process cameras.

Diazo. Contact copying of documents onto chemically treated papers and films is a photocopying process that predates electrophotographic methods by decades, having been developed in the 1920s, and continues to find application niches. For example, engineering firms and other organizations that need to copy large drawings, maps, and other oversize documents continue to use diazo copies or whiteprints.

Diazo copies are produced by placing the original document in direct contact with a diazo paper or film and then exposing it to an ultraviolet light source (**Fig. 5**). The paper is coated with a diazonium compound and an activator. The exposed paper usually is developed by ammonia vapors or anhydrous ammonia gas.

With equipment needs limited to a large light box and a strong light, the process is inexpensive. However, only single-sided documents on a rather thin translucent material can be reproduced, and copy images also are somewhat unstable.

Diazo predecessors include blueprint and brownprint copy methods. The blueprint first appeared in 1842. After this blueprint paper, sensitized with a ferric salt and potassium ferricyanide, is contacted

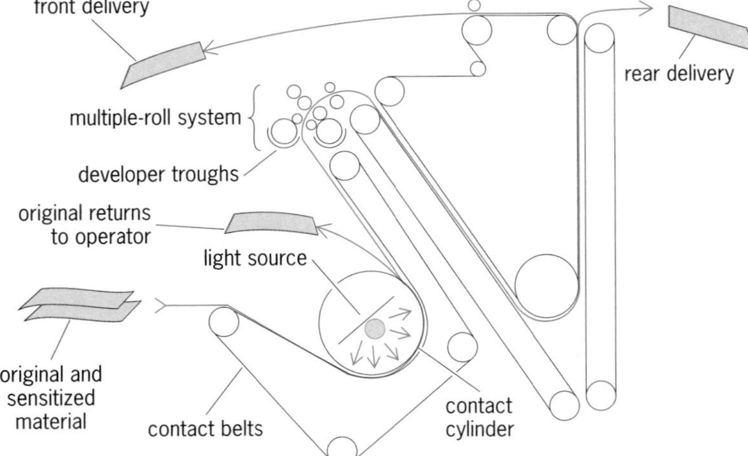

Fig. 5. Flow diagram of whiteprint machine, involving a diazo process. (*Charles Bruning Co.*)

to an original and exposed to a high-intensity light, it is developed in water and dried. Brownprints, intermediate negatives for platemaking, were first used in 1895. Ferric iron and silver salts make the paper light-sensitive. The exposed paper is washed in a hypo solution to remove undeveloped silver.

Thermography. Thermography is a dry, heat-based photocopying process that was introduced in 1949. Updated versions of the process have continuing applications.

The process uses infrared radiation to reproduce an original on a thin, translucent paper coated with a plastic that contains materials that chemically react and darken when subjected to heat. When infrared rays are directed at the original document, they are absorbed by carbon-containing image areas and converted into heat. The heat is then conducted to the treated paper, which turns black in the radiated image areas.

A process limitation is that original document inks must contain carbon or metallic elements, since organic inks are less efficient in converting radiant energy to heat. Also, the thin paper tends to curl and offers limited tone gradation capabilities.

In 1961 the dual-spectrum process was introduced. It combines photographic (light) exposure with thermal (heat) development. With this process, a thin chemically treated intermediate sheet receives a latent image and transfers it by heat to the final copy paper, where a chemical reaction renders it visible. These inexpensive copiers are used by individuals with limited copying needs.

Silver halide. Over the years, many photocopying processes have used the high (super) light sensitivity and versatility of silver halide materials, which produce high-quality archival copies. However, the relatively high cost of silver-based photographic materials, coupled with wet chemistry processing demands, has limited the value of this technology for routine office copying applications. Except for microfilming, black-and-white silver halide photocopying processes are limited to lithographic platemaking. *See* PHOTOGRAPHIC MATERIALS; PRINTING.

Robert I. Edelman; J. Gordon Jarvis; Thomas Destree

Bibliography. P. M. Borsenberger and D. S. Weiss, *Organic Photoreceptors for Xerography*, 1998; R. J. Connors and W. M. Amundsen, *Microfilm: Active and Vital*, 1975; R. M. Schaffert, *Electrophotography*, 2d ed., 1975; E. M. Williams, *The Physics and Technology of Xerographic Processes*, 1992.

Photodegradation

Reduction in the useful properties of materials because of chemical changes resulting from the absorption of light. The chemical changes can include bond scission (especially of the molecular backbone), color formation, cross-linking, and chemical rearrangements. All organic materials can photodegrade, but the process has greatest practical relevance for polymers where scission of the polymer backbone is particularly important. Photodegradations of polymers in the absence of oxygen (photolysis) or using wavelengths shorter (more energetic) than those at the Earth's surface (<280 nanometers) have been studied extensively, but only the more practical situation of polymers exposed to terrestrial sunlight (or its equivalent) in air is discussed in this article.

Rate of degradation. Although all organic polymers can be degraded by light, the rate of degradation varies enormously from polymer to polymer, and is also dependent on the incident wavelengths. Light containing ultraviolet (uv; shorter-wavelength) components is much more destructive than visible light, so that polymers exposed indoors, behind window glass (transmitting >330 nm), will degrade much more slowly than samples exposed outdoors. The sensitivity of various polymers to direct sunlight (wavelength 280 nm) is shown in the **table**. Some aromatic polymers such as aramids absorb light in the visible region (380 nm) and are degraded even by these long wavelengths. In general, most aromatic polymers absorb in the uv regions of terrestrial sunlight (~280–380 nm), whereas all pure, aliphatic polymers do not. Nevertheless, the latter may still photodegrade rapidly in practice when exposed outdoors (weathering conditions). This results from the presence of impurities introduced during polymerization, processing, or storage, impurities which do absorb in the near uv and can trigger the degradation of these "nonabsorbing" polymers.

Mechanisms. The sequence of events following absorption by a light-absorbing group (chromophore) in a polymer is shown in the **illustration**. For many aromatic polymers, such as polyester and the aramids, in which the polymer itself is the chromophore, backbone scission results predominantly from this direct absorption of light energy. For many other polymers, including polyolefins, and poly(vinyl chloride) where only impurities absorb energy from sunlight, scission of a chemical bond by light to give free radicals is followed by reaction of these highly reactive free radicals with atmospheric oxygen. Oxygen can diffuse quite readily through the noncrystalline (amorphous) domains of a polymer and combine with radical sites to promote an oxidative chain reaction. This chain process generates oxidation products but also repeatedly regenerates the reactive radical center. The process may directly cause some backbone scission, but in addition the main oxidation products (hydroperoxide groups) are themselves subsequently destroyed by light to trigger further oxidation and backbone scission (see illus.). Thus a low concentration (1–100 ppm) of light-absorbing impurity groups can trigger extremely long-chain oxidations to produce a host of photolabile products. *See* FREE RADICAL.

Vulnerable materials. Although numerous organic materials will undergo photodegradation, hydrocarbon polymers are particularly vulnerable because their useful properties depend entirely on their high molecular weights, in the tens or hundreds of thousands. Anything that reduces the molecular weight

Weathering lifetimes and absorption limits of common polymers

Polymer	Outdoor lifetime,* years	Cutoff, nm†	Absorption relative to solar limit‡
Polyethylene	0.5–1.0	180	Nonabsorber
Polypropylene	0.2	180	Nonabsorber
Poly(vinyl chloride)	0.5	220	Nonabsorber
Poly(methyl methacrylate)	>20	240	Nonabsorber
Polyamides	3–4	240	Nonabsorber
Polystyrene	0.1§	270	Nonabsorber
(Terrestrial solar limit)		(>280)	
Polycarbonate	0.5	280	Inherent absorber
Polyurethane (MDI-based)	2	280	Inherent absorber
Poly(ethylene terephthalate)	3	310	Inherent absorber
Aramid [poly(paraphenylene terephthalamide)]	0.3	350	Inherent absorber

*Unstabilized, normal impurity levels. Time to 50% loss in tensile, elongation, or impact properties.
†Wavelength at which absorbance of 10-micrometer film reaches 1.0.
‡Solar limit 280–285 μm depending upon location.
§Yellowing.

of polymeric systems will alter the characteristics of these systems and limit their service life. In fact, the scission of as few as one carbon-carbon bond in a thousand in a polymer molecule can completely destroy its useful physical properties. This sensitivity is not observed in lower-molecular-weight substances such as liquid hydrocarbons.

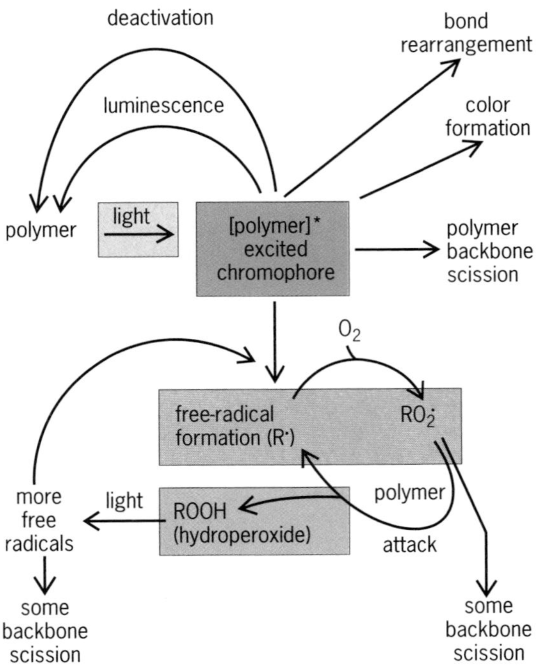

Mechanisms of polymer photodegradation (pathways) and photostabilization (boxes).

Polymers which lack C-H links and which contain bonds of high strength such as C-F in poly(tetrafluoroethylene) are extremely stable to photodegradation. Ironically, among commercially significant polymers, those with chromophores in the repeat unit, poly(ethylene terephthalate), polycarbonate, and such, tend to photodegrade less rapidly than those that formally have no chromophores and should be transparent to terrestrial sunlight, for example, the polyolefins, polystyrene, and poly(vinyl chloride) [see table]. In the former case, although enormous quantities of photons are absorbed during exposure to sunlight, efficient photophysical processes including fluorescence and phosphorescence (see illus.) harmlessly dissipate the absorbed energy without bond scission occurring. Furthermore, even when bond scission does occur, the rigidity of the aromatic polymers aids backreaction to reform the initial bonds before oxygen interception or rearrangement can occur.

The photodegradation of polymers is by no means uniform throughout a molded article or an extruded fiber or film. In the case of a strongly self-absorbing polymer, much of the incident sunlight is absorbed near the "front" surfaces so that fewer of the damaging photons are left to photooxidize the polymers at the back surface. Light absorption in "nonabsorbing" polymers will occur heterogeneously as well because the chromophoric impurities are not located randomly throughout such solids. Furthermore, since neither oxygen nor impurity chromophores are found in the very highly ordered (crystalline) regions of solid polymers, photodegradation occurs in the amorphous zones. Finally, owing to the restrictions to mobility (high viscosity) in solid polymers, the propagation of oxidizing species is restricted to relatively small distances away from the original sites of photon absorption. *See* PHOTON.

Photostabilization. To prevent or reduce the photodegradation of "nonabsorbing" polymers, exclusion of oxygen and other precautions can be taken during synthesis and thermal processing to reduce the number of chromophoric impurities in the finished articles.

A more universal approach to reducing the rates of photodegradation for all types of polymers, however, is the use of low levels of additives. These additives, known as photostabilizers or uv stabilizers, are effective at fractions of a weight percent. These nonpolymeric compounds are normally added to resin pellets prior to fabrication. They can be classified into four categories according to their modes of action (see illus.). (1) Ultraviolet screeners (absorbers or opaque particles) reduce the amount of incident light reaching the chromophores. This is believed to be the most effective practical method of protecting polymers which absorb strongly in the near uv and are damaged by the primary absorption process. (2) Chromophore "quenchers" remove by collision the electronic excitation energy from a chromophore after the absorption of a photon but before an irreversible chemical reaction can occur. (3) Hydroperoxide decomposers react with polymeric hydroperoxide groups, which are both photochemically and thermally unstable, before the relatively weak oxygen-oxygen bond is broken. (4) Radical scavengers react with (deactivate) radicals so that they cannot perpetuate degradative oxidation of the polymer molecules.

Photostabilizers which operate by hydrogen decomposition and radical scavenging are highly effective in extending the lifetime of the "nonabsorbing" polymers which degrade primarily by an oxidative chain reaction. However, the more effective uv stabilizers operate by more than one of the four mechanisms. Thus 2-hydroxybenzophenones operate by both uv screening and radical scavenging, and hindered amine light stabilizers operate by both hydroperoxide decomposition and radical scavenging. Combinations of these two types of additives are particularly effective in prolonging the useful life of polyolefin materials during exposure to sunlight. *See* PHOTOCHEMISTRY; POLYMER; STABILIZER (CHEMISTRY). D. M. Wiles; D. J. Carlsson

Bibliography. D. L. Allara and W. L. Hawkins (ed.), *Stabilization and Degradation of Polymers*, Advances in Chemistry Series, vol. 169, 1978; N. S. Allen (ed.), *Developments in Polymer Photochemistry*, vol. 1, 1980; A. Davis and D. Sims, *Weathering of Polymers*, 1983; N. Grassie (ed.), *Developments in Polymer Degradation*, vol. 7, 1987; J. F. McKeller and N. S. Allen, *Photochemistry of Man-Made Polymers*, 1979; B. Ranby and J. F. Rabek, *Photodegradation, Photo-oxidation and Photostabilization of Polymers: Principles and Applications*, 1975; G. Scott, *Developments in Polymer Stabilization*, vol. 8, 1987.

Photodiode

A semiconductor two-terminal component with electrical characteristics that are light-sensitive. All semiconductor diodes are light-sensitive to some degree, unless enclosed in opaque packages, but only those designed specifically to enhance the light sensitivity are called photodiodes.

Most photodiodes consist of semiconductor *pn* junctions housed in a container designed to collect and focus the ambient light close to the junction. They are normally biased in the reverse, or blocking, direction; the current therefore is quite small in the dark. When they are illuminated, the current is proportional to the amount of light falling on the photodiode. For a discussion of the properties of *pn* junctions *see* JUNCTION DIODE

Photodiodes are used both to detect the presence of light and to measure light intensity. *See* PHOTOELECTRIC DEVICES. W. R. Sittner

Photoelasticity

An experimental technique for the measurement of stresses and strains in material objects by means of the phenomenon of mechanical birefringence. Photoelasticity is especially useful for the study of objects with irregular boundaries and stress concentrations, such as pieces of machinery with notches or curves, structural components with slits or holes, and materials with cracks. The method provides a visual means of observing overall stress characteristics of an object by means of light patterns projected on a screen or photographic film. Regions of stress concentrations can be determined in general by simple observation. However, precise analysis of tension, compression, and shear stresses and strains at any point in an object requires more involved techniques. Photoelasticity is generally used to study objects stressed in two planar directions (biaxial), but with refinements it can be used for objects stressed in three spatial directions (triaxial). *See* NONDESTRUCTIVE EVALUATION.

For biaxial studies, a model geometrically similar to the object to be analyzed is prepared from a sheet of special transparent material and loaded as the object would be loaded.

Use of birefringent phenomenon. Model materials commonly used for photoelasticity are Bakelite, celluloid, gelatin, synthetic resins, glass, and other commercial products that are optically sensitive to stress and strain. The materials must have the optical properties of polarizing light when under stress (optical sensitivity) and of transmitting it on the principal stress planes with velocities dependent on the stresses (birefringence or double refraction). In addition, the material should be clear, elastic, homogeneous, optically isotropic when under no stress or strain, and reasonably free from creep, aging, and edge disturbances. *See* BIREFRINGENCE.

When the stressed model is subjected to monochromatic polarized light in a polariscope, the birefringence of the model causes the light to emerge refracted into two orthogonal planes. Because the velocities of light propagation are different in each direction, there occurs a phase shifting of the light waves. When the waves are recombined with the polariscope, regions of stress where the wave phases cancel appear black, and regions of stress where the wave phases combine appear light. Therefore, in models of complex stress distribution, light and dark fringe patterns (isochromatic fringes) are projected from the model. These fringes are related to the stresses. *See* POLARIZED LIGHT.

When white light is used in place of monochromatic light, the relative retardation of the model causes the fringes to appear in colors of the spectrum. White light is often used for demonstration, and monochromatic light is used for precise measurements.

Polariscope. A basic polariscope used in photoelasticity has a light source (generally monochromatic), a collimating lens, a polarizer, and a quarter-wave plate (**Fig. 1**). This plate is a birefringent material that causes the relative retardation of light to be exactly one-quarter the wavelength of the light. Next in the optical path is a planar model of the object under test and stressed in the direction of the plane. Finally there are a second quarter-wave plate, a polarizer called the analyzer, a focusing lens, and a viewing screen or film. Many variations of this basic transmission-type apparatus are in use. Other lenses may be added and components rearranged. If appropriate mirrors are added, the polariscope converts to a doubling type which is useful for the study of thin models under low stress, as the number of fringes doubles.

A typical isochromatic fringe pattern shows the effect on a flat plate with a central hole, pulled at the upper and lower ends (**Fig. 2**). The congestion of fringes at the boundary of the hole indicates a region of stress concentration, typical of stress behavior at cutouts. To study the exact stress at a given point in the model, the model is gradually loaded (from a condition of no load) and the number of fringe changes (fringe order) at that point is observed. Special equipment is sometimes employed to obtain partial fringe orders and to sharpen vague fringe boundaries. The fringe order is directly related by a calibrated constant to the difference of the principal stresses at that point.

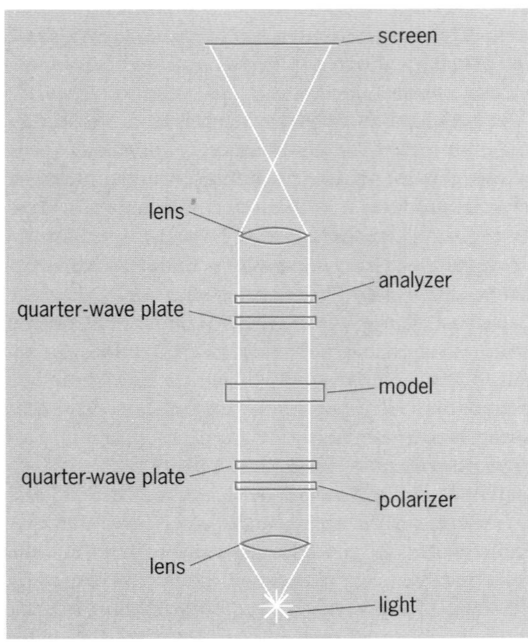

Fig. 1. Basic photoelastic polariscope.

Fig. 2. Isochromatic fringe pattern for plate with hole. (*From M. M. Frocht, Photoelasticity, vol. 2, John Wiley and Sons, 1948*)

Determination of principal stresses. Shear stresses can be related mathematically to the difference of the principal stresses, thereby relating shear stresses directly to fringe order. High shear stresses often cause the material to yield or fail, so that a point of large fringe order indicates a point of potential failure. In many applications of photoelasticity a knowledge of shear stress is all that is needed. This fact makes photoelasticity a simple and direct tool for investigation of complex stress systems. However, if principal stresses and their directions are required, additional experimentation is necessary, as described later. *See* FRINGE (OPTICS).

Isoclinic fringes are a different set of interference patterns made by using white light, removing the quarter-wave plates and rotating the polarizer and analyzer a fixed number of degrees. These fringes represent lines making known angles with the principal planes of stress.

Stress trajectories are lines of principal stress directions over the model, obtained graphically from

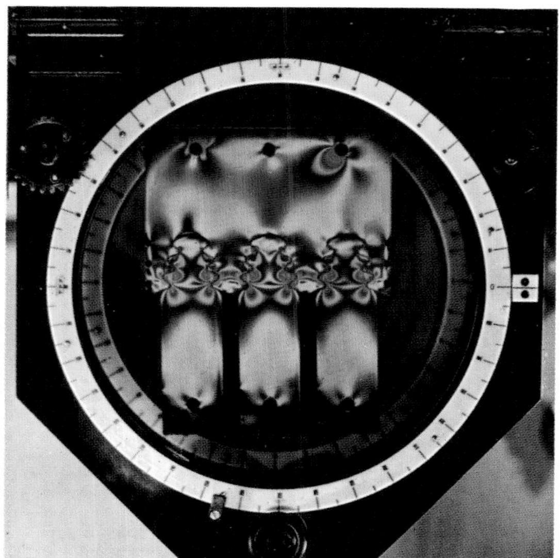

Stress concentrations in plastic models of roots of steam turbine blades under simulated operating conditions. (*Westinghouse Electric Corp.*)

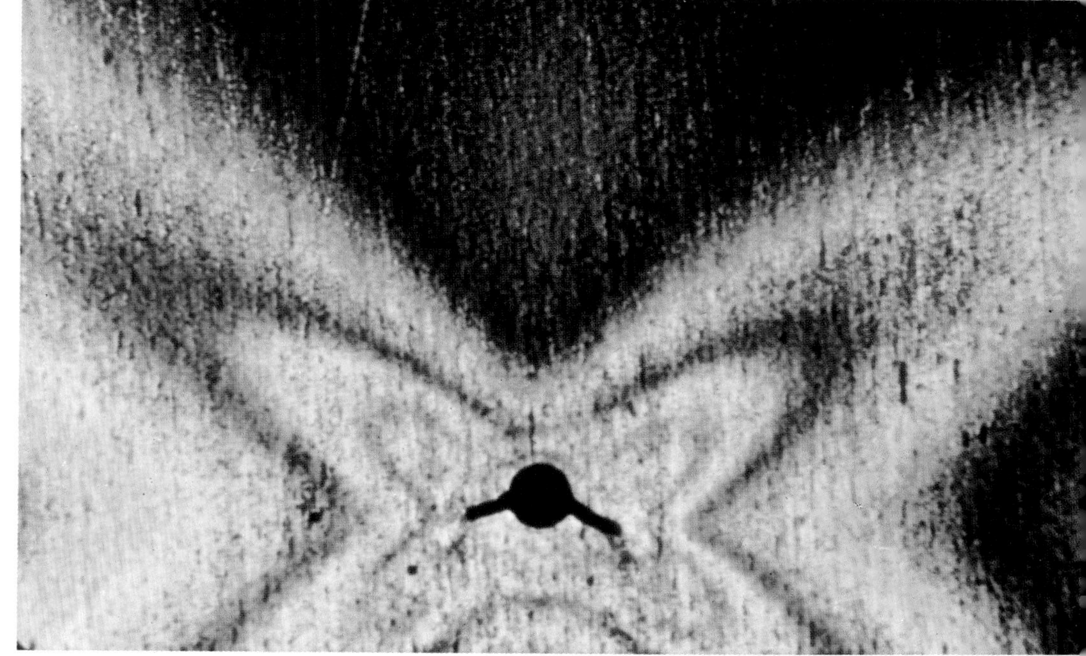

Stress distribution around a circular hole and two saw cuts using photostress technique. Plastic photoelastic coating on actual steel specimen allows study of actual metal part rather than of a plastic model. The proximity of color fringes around the saw cuts indicates high stresses in that region. (*Battelle Memorial Institute*)

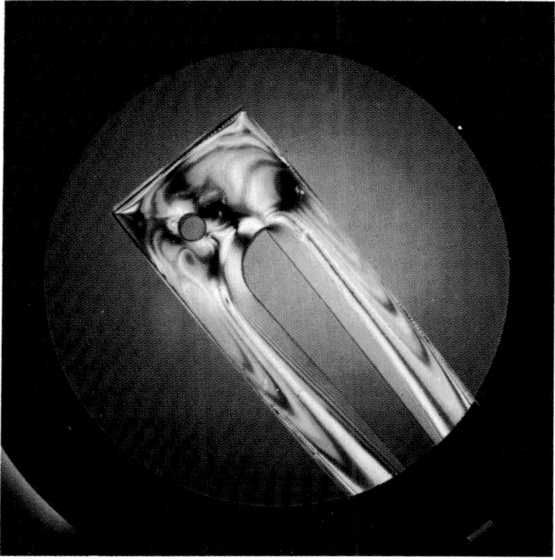

Typical stress patterns in a plastic model viewed under polarized light. (*Bausch & Lomb*)

the isoclinic fringes. Stress trajectories are not lines of constant stress.

The determination of principal stresses requires additional information, which may be obtained in several ways. Principal stresses are determined analytically by differences of the shear stresses based on equilibrium equations. This procedure requires a numerical point by point study of the model, utilizing the shear stresses and stress trajectories. Principal stresses can be found analytically or experimentally by solution of Laplace's equation of elasticity. In principle, this procedure supplies equations pertaining to the sum of the principal stresses at any point in the model. Utilizing equations for the difference of the principal stresses from the isochromatic fringe orders, the stresses may be found by solving the two equations for the two principal stresses. Principal stresses are found experimentally by measuring the thickness of the model under stress. Because thickness changes caused by the Poisson's ratio effect are minute, a sensitive measuring device such as an optical interferometer is needed, although direct-reading thickness gages are sometimes used. The interferometer produces fringe patterns called isopachic fringes. This method essentially provides information regarding the sum of the principal stresses as with Laplace's equation. Another experimental method is to pass polarized light obliquely to the surface of the model. The relative retardations of the light produce interference fringes. These oblique fringe orders can be related to the principal stresses differently from those obtained by isochromatic fringes. Using the information on stresses from the isochromatics in conjunction with the oblique relations, the principal stresses may be obtained. *See* ELASTICITY.

With care, stresses determined by photoelasticity are 98% accurate. With stresses determined, strains may be computed by elastic relations.

Three-dimensional measurements. Three-dimensional photoelasticity is also possible, although the techniques and stress-strain relationships are more involved than for planar objects.

The frozen stress method is well suited for three-dimension studies. Certain optically sensitive materials, such as Bakelite, when annealed in a stressed condition retain the deformation and birefringent characteristics of the initially stressed state when the load is removed. A three-dimensional model may therefore be cut into slices that may then be analyzed individually on somewhat the same principles as are the planar models.

The scattered light technique may also be used for three-dimensional models, such as torsion bars. The scattered light principle is based on the fact that polarized light passing through birefringent materials scatters in a predictable manner, acting as an optical analyzer in a polariscope.

Measurements on actual objects. The reflective polariscope method is essentially a variation of the normal polariscope. A sheet of birefringent material is bonded to the polished surface of the actual object to be studied, with a polarizer and quarter-wave plate interposed between the light and the object. The light (usually white light) passes through the stressed birefringent material and is reflected back through the quarter-wave plate and the polarizer (which now acts as an analyzer). Isochromatic fringes are projected as in a normal polariscope. Dependent on good bonding between the birefringent material and the stressed object, the reflective polariscope has the advantage that it can be used on actual structures under service loads without being reduced to model form. It may be used to find surface stresses and strains on curved or three-dimensional objects. Although photoelasticity is generally limited to elastic behavior, reflective techniques may be used to determine the nonelastic strains (but not stresses) of objects, provided the bonded optical material remains elastic. The technique may also be used with nonheterogeneous materials such as wood and concrete.

Dynamically induced stresses and strains may also be studied by photoelasticity when high-speed motion picture cameras are used to photograph the fringe patterns. Another development in photoelasticity is the use of coherent light to produce holographic interference patterns, instead of the normal fringe patterns. *See* HOLOGRAPHY; STRESS AND STRAIN.

William Zuk

Bibliography. J. W. Dally and W. F. Riley, *Experimental Stress Analysis*, 3d ed., 1991; S. A. Paipetis and G. S. Hollister (eds.), *Photoelasticity in Engineering Practice*, 1985; P. S. Theocaris and E. E. Gdoutos, *Matrix Theory of Photoelasticity*, 1979.

Photoelectric devices

Devices which give an electrical signal in response to visible, infrared, or ultraviolet radiation. They are often used in systems which sense objects or encoded data by a change in transmitted or reflected light. Photoelectric devices which generate a voltage can be used as solar cells to produce useful electric power. The operation of photoelectric devices is based on any of the several photoelectric effects in which the absorption of light quanta liberates electrons in or from the absorbing material. *See* PHOTOVOLTAIC CELL; SOLAR CELL.

Photoconductive devices are photoelectric devices which utilize the photoinduced change in electrical conductivity to provide an electrical signal. Thin-film devices made from cadmium sulfide, cadmium selenide, lead sulfide, or similar materials have been utilized in this application. Single-crystal semiconductors such as indium antimonide or doped germanium are used as photoelectric devices for the infrared spectrum. The operation of these devices requires the application of an external voltage or current bias of relatively low magnitude. *See* PHOTOCONDUCTIVE CELL.

Photoemissive systems have also been used in photoelectric applications. These vacuum-tube devices utilize the photoemission of electrons from a photocathode and collection at an anode. Photoemissive devices require the use of a relatively large bias voltage. *See* PHOTOEMISSION.

Many photoelectric systems now utilize silicon photodiodes or phototransistors. These devices utilize the photovoltaic effect, which generates a voltage due to the photoabsorption of light quanta near a *pn* junction. Modern solid-state integrated-circuit fabrication techniques can be used to create arrays of photodiodes which can be used to read printed information. *See* PHOTODIODE; PHOTOTRANSISTOR.

Photoelectric devices can be used in systems which read coded or printed information on data cards and packages. Similar systems are used to sense and control the movement of objects in perimeter guard systems which sense an intruder by the interruption of a light beam. *See* CHARACTER RECOGNITION.

Most visible and ultraviolet photoelectric devices for use in the infrared must be cooled with the longer-wavelength response devices requiring the most cooling. Richard A. Chapman

Bibliography. R. Bube, *Photoelectron Properties of Semiconductors*, 1992; S. Juds, *Photoelectric Sensors and Controls: Selection and Applications*, 1988.

Photoemission

The ejection of electrons from a solid (or less commonly, a liquid) by incident electromagnetic radiation. Photoemission is also called the external photoelectric effect. The visible and ultraviolet regions of the electromagnetic spectrum are most often involved, although the infrared and x-ray regions are also of interest. For important practical applications of photoemission *see* PHOTOTUBE; TELEVISION CAMERA TUBE.

The salient experimental features of photoemission are the following: (1) There is no detectable time lag between irradiation of an emitter and the ejection of photoelectrons. (2) At a given frequency the number of photoelectrons ejected per second is proportional to the intensity of the incident radiation. (3) The photoelectrons have kinetic energies ranging from zero up to a well-defined maximum, which is proportional to the frequency of the incident radiation and independent of the intensity.

Einstein photoelectric law. These characteristics cannot be explained by J. C. Maxwell's theory of electromagnetic waves. In 1905 Albert Einstein made the clarifying assumption that the radiation had characteristics like those of particles when it delivered energy to electrons in the emitter. In Einstein's approach the light beam behaves like a stream of photons, each of energy $h\nu$, where h is Planck's constant, and ν is the frequency of the photon (**Fig. 1**). The energy required to eject an electron from the emitter has a well-defined minimum value φ called the photoelectric threshold energy. When a photon interacts with an electron, the latter absorbs the entire photon energy. *See* PHOTON.

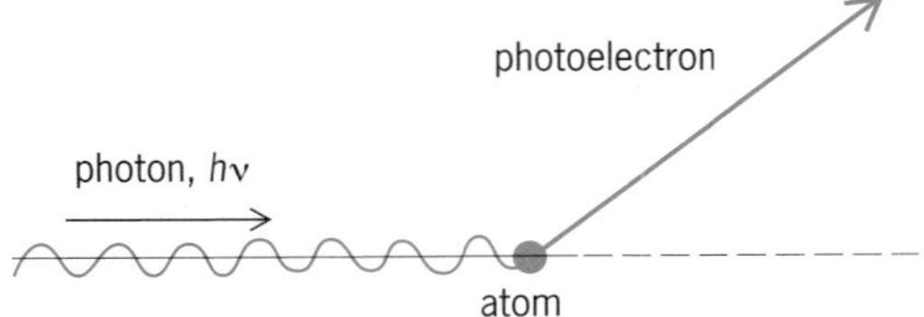

Fig. 1. External photoelectric effect.

For $h\nu$ values below the threshold, photoelectrons are not ejected. Even though the electrons absorb photon energy, they do not receive enough to surmount the potential barrier at the surface, which normally holds the electrons in the solid. The threshold energy φ is associated with a threshold frequency φ/h and a threshold wavelength ch/φ, where c is the velocity of light. For photon energies above φ, the kinetic energies of photoelectrons range from zero up to a maximum value, $E = h\nu - \varphi$. This This is the Einstein photoelectric law, and E is commonly termed the Einstein maximum energy. Careful photoelectric experiments by R. A. Millikan in 1916 fixed h in Einstein's law with considerable precision and furthered its identification with the constant which M. Planck had used in his theory of blackbody radiation. For a discussion of the surface potential barrier *see* SCHOTTKY EFFECT.

Metals. The Einstein law is based only on the photon hypothesis and on the conservation of energy. It does not take into account momentum, which must also be conserved. The incident photon has a momentum $h\nu/c$ which is negligible compared to the change in momentum of the electron when it gains the energy $h\nu$. Thus, it is not possible for a free electron to absorb the entire energy of a photon. In order for this to happen the electron must be bound to another body, which takes up the recoil momentum. *See* COMPTON EFFECT.

Figure 2 shows an energy diagram of the electrons in the metal sodium. There is a potential barrier at the surface, which the electrons must surmount before they can escape. The most easily ejected electrons must acquire 2.3 eV of additional energy from photons in order to do this. This 2.3 eV is the electronic work function, which for a metal is equal to the photoelectric threshold energy. Inside the metal the

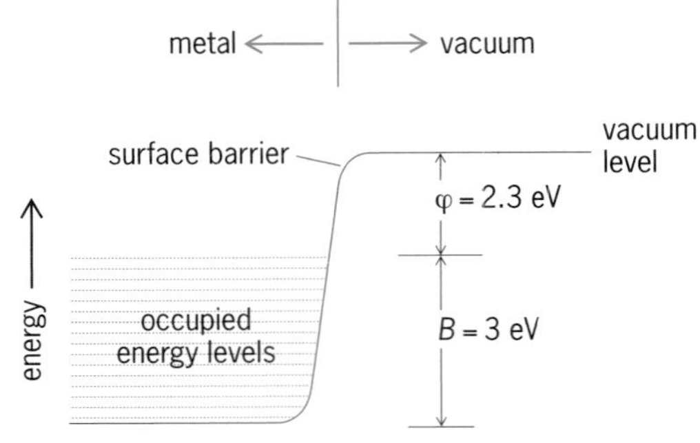

Fig. 2. Energy diagram for electrons in sodium. The photoelectric threshold energy is φ; In a metal φ is equal to the electronic work function. The band of energy levels occupied by almost free electrons has a width B.

electrons occupy a band of energies about 3 eV wide. These electrons are said to be quasi-free. This means that they behave in many ways like a gas of free, noninteracting electrons; nevertheless they move in the periodic potential due to the positive sodium ions, and in this sense they are bound. *See* FREE-ELECTRON THEORY OF METALS.

In this situation two types of photoemission are theoretically possible, the surface effect and the volume effect. In the surface effect, recoil momentum is communicated to the crystal because the electron is coupled to the barrier at the surface during photon absorption. In the volume effect the electron is coupled to the internal periodic potential.

Experimental determination of the relative importance of surface and volume photoeffects in metals is difficult. Experiments by H. Mayer and his collaborators indicate that for potassium the volume effect is predominant for photon energies from the threshold value at 2.1 eV to at least 4 eV. Thus the photoelectric emission increases as the thickness of a potassium film increases. (This would not be true for the surface effect.) Photoelectrons can escape from depths greater than 10^{-6} cm when excited by light in the threshold region.

Thus far the photoelectric threshold has been treated as a sharply defined quantity. This is precisely true for metals only at temperatures near absolute zero. At higher temperatures the upper edge of the band of occupied electron energy states in Fig. 2 is no longer sharp. It becomes diffuse because of thermal agitation. Electrons may be then emitted for photon energies less than the threshold value φ. At ordinary room temperatures, for example, measurable photoemission appears for photon energies as much as 0.2 eV below the threshold. R. Fowler has developed a convenient graphical technique, known as a Fowler plot, for determining the absolute-zero threshold from data taken at higher temperatures on the spectral dependence of the photoelectric yield, which is the number of photoelectrons ejected per incident photon. L. A. DuBridge has developed a similar technique using either the temperature dependence of the photoelectric yield or the distribution of photoelectrons in energy at fixed frequency. These treatments show that the photoelectric yield is approximately proportional to the quantity $(h\nu - \varphi)^2$ when the photon energy $h\nu$ is within about 1 eV of the threshold energy φ. **Figure 3** shows a graph of the spectral dependence of photoelectric yield for some typical emitters. **Figure 4** shows typical energy distributions. Photoelectric yields from metals are of the order of 10^{-3} electron per incident photon when $h\nu - \varphi$ is 1 eV. Photoelectric threshold energies range from 2 eV for cesium to values such as 5 eV for platinum. They vary for different types of crystal faces on the same crystal, and they are exceedingly sensitive to small traces of adsorbed gases.

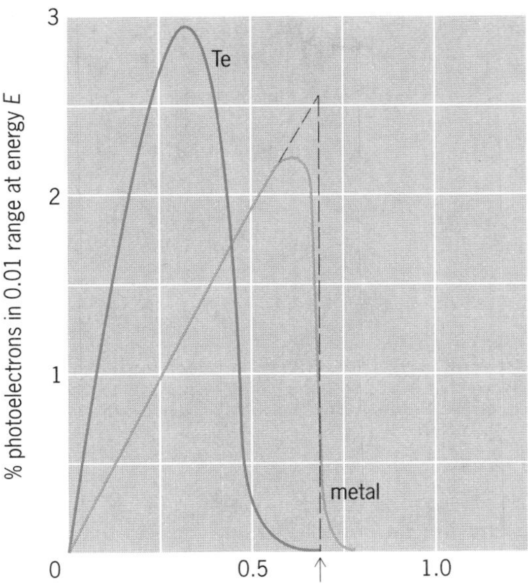

Fig. 4. Energy distributions of photoelectrons from tellurium and a metal having the same work function. The solid curve for the metal shows results for room temperature, and the broken curve shows results for absolute zero. The arrow marks the Einstein maximum energy; photon energy is 5.42 eV.

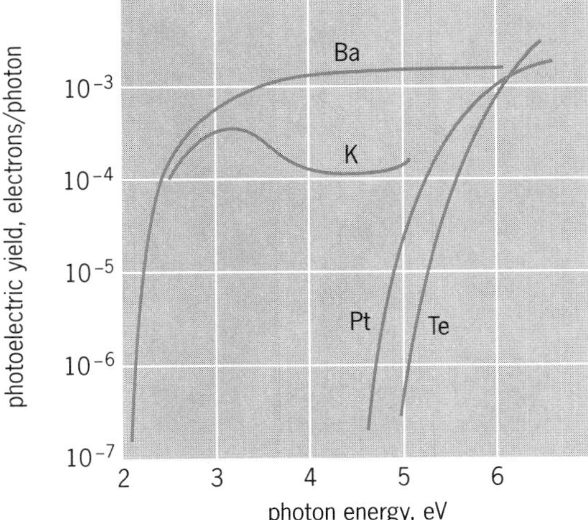

Fig. 3. Spectral distribution of the photoelectric yield from typical samples of barium, Ba; potassium, K; platinum, Pt; and tellurium, Te. Platinum and tellurium have practically the same electronic work function. Note the higher threshold and more steeply rising curve for tellurium, which is a typical elemental semiconductor.

Semiconductors. The photoelectric behavior of semiconductors, such as germanium or tellurium, differs from that of metals. As shown in **Fig. 5**, the electrons in a semiconducting emitter completely occupy a closed band of energies, which lies just below a so-called forbidden energy band. The electrons behave quite differently from those in metals. As a result, the photoelectric threshold energy φ is larger than the electronic work function W. Thus, a semiconductor exhibits a higher photoelectric threshold energy than a metal having the same work function. An example of this is shown for the metal platinum and the semiconductor tellurium in Fig. 3. Both this particular platinum sample (Pt) and the tellurium (Te) have the same electronic work function, about 4.8 eV. The photoelectric threshold of the platinum

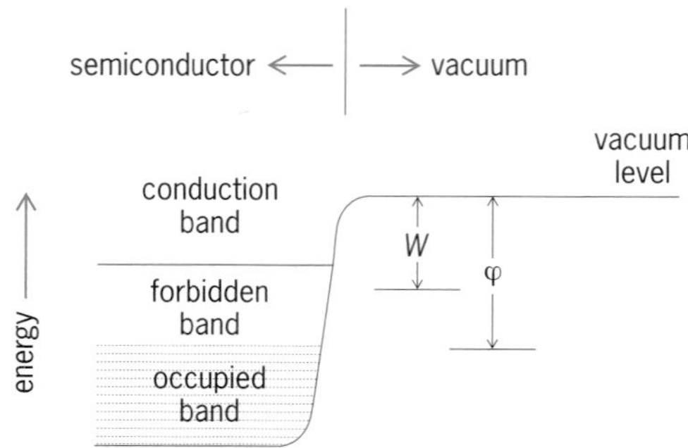

Fig. 5. Energy diagram for electrons in a semiconductor. Occupied band is filled with bound electrons that behave differently from the electrons in metals. As a result, the electronic work function *W* is smaller than the photoelectric threshold energy φ.

is equal to the work function, whereas that for the tellurium is clearly higher. Spectral and energy distributions are shown in Figs. 3 and 4. Clean surfaces of silicon, germanium, and certain semiconducting chemical compounds have been made by cleaving single crystals in ultrahigh vacuum. Both the surface photoelectric effect and the volume effect have been measured. From the measurements of the volume effect, valuable information on the detailed nature of the electron energy bands has been deduced from structure that occurs in photoelectron energy distributions. *See* BAND THEORY OF SOLIDS.

A particularly interesting and important kind of photoemitter is typified by cesium antimonide, Cs_3Sb. This material is a semiconductor having a forbidden energy band about 1.5 eV wide. The photoelectric threshold energy is only slightly higher than this. Electrons excited from the occupied energy band by incident photons cannot assume energies lying in the forbidden band. They must remain in the conduction band shown in Fig. 5. Thus even the slowest ones must retain energies only slightly less than that required for escape. The probability of photoemission is higher than for metals (or for semiconductors that have threshold energies greater than twice the width of the forbidden energy band). Cs_3Sb is sensitive over much of the visible range and can give very high yields, in excess of 0.2 electron per incident photon. It is widely used in practical phototubes. Related compounds can be made with enhanced photoelectric response in the red or ultraviolet regions of the spectrum.

Alkali halides. Three basically different kinds of photoemission are possible for alkali halides: intrinsic, extrinsic, and exciton-induced photoemission.

Intrinsic photoemission. This is characteristic of the ideally pure and perfect crystal. It is thus analogous to the emission already described for metals and semiconductors. It appears only for photon energies higher than the intrinsic threshold. For example, potassium iodide, KI, is an alkali halide having this intrinsic threshold in the far ultraviolet near 7 eV. Apparently the width of the forbidden electron energy band in KI is only about 1 eV less than this. For the same reason that was mentioned for the semiconductor Cs_3Sb, the photoelectric yields are high, in excess of 0.1 electron per incident photon, as shown by section C of the curve in **Fig. 6**.

Extrinsic photoemission. A second kind of emission occurs when a KI crystal contains imperfections in the form of negative ion vacancies (lattice sites from which negative iodine ions are missing). These vacancies can be filled by electrons. Color centers, which absorb visible light, are formed. They may reach concentrations as high as 10^{20} per cubic centimeter. External photoelectrons may be ejected directly from these centers by photons. It is termed an extrinsic process since the light is absorbed by a crystal defect; it is also called direct ionization. The threshold energy for this process is about 2.5 eV. The yields can reach values of the order of 10^{-4} electron per incident photon, as shown in section A of the curve in Fig. 6. The exact value of the yield depends on the concentration of color centers. Most of the incident radiation is lost because it is not intercepted by the centers, which present a limited cross section to the incident photons. *See* COLOR CENTERS.

Exciton-induced photoemission. When color centers are present, another photoelectric process takes place in two stages. Potassium iodide has a sharp optical absorption band, peaking at a photon energy of 5.6 eV. This is the first fundamental or intrinsic optical absorption band. Energy absorbed in this peak does not release free electric charges in the crystal. Rather, it leads to a kind of nonconducting excited state called an exciton state. The exciton can transfer enough energy to color centers to eject photoelectrons from the crystal. This two-stage process is termed

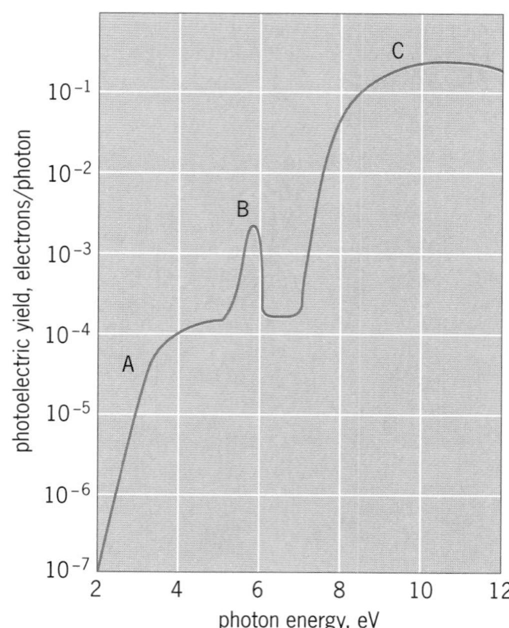

Fig. 6. Spectral distribution of photoelectric yield from potassium iodide containing color centers. Region A of the curve is due to direct ejection of photoelectrons from color centers; the peak B is due to exciton-induced emission; C is due to intrinsic emission.

exciton-induced photoemission. It appears in the peak B on the curve in Fig. 6. It is more efficient than direct ejection of photoelectrons from color centers. The entire crystal is capable of the primary photon absorption, and the energy can be transferred rather efficiently to color centers. Thus the process avoids much of the loss in incident energy that arises from the limited cross section of color centers when they absorb photons directly. *See* EXCITON.

Other compounds. Other ionic crystals, such as barium oxide, behave much like the alkali halides. Direct ejection of photoelectrons from chemical impurities and from energy levels or defects localized at the crystal surface can be important. In addition to these extrinsic processes, exciton-induced emission and intrinsic photoemission both occur.

Compounds such as zinc sulfide behave somewhat like germanium but have higher intrinsic threshold energies, of the order of 7 eV. The photoelectric yields are comparatively low, as for germanium. Extrinsic processes, such as direct ejection of electrons from chemical impurities (or defects), are sometimes detectable but are usually weak.

Certain complex photoemitters are made by letting cesium react with silver oxide to form cesium oxide and silver. They are valuable because they have threshold energies below 1 eV, and thus they are sensitive in the infrared. The photoelectrons appear to be directly ejected either from cesium adsorbed on the oxide surface or from discrete energy levels in the cesium oxide. The yields are about 10^{-3} electron per incident photon. Intrinsic emission from cesium oxide (with yields above 0.01) occurs for photon energies above the intrinsic threshold at about 4 eV. L. Apker

Bibliography. M. Cardona and L. Ley (eds.), *Photoemission in Solids*, 2 vols., 1978, 1979; B. Feuerbacher, B. Fitton, and R. F. Willis, *Photoemission and the Electronic Properties of Solids*, 1978; A. Sommer, *Photoemissive Materials*, 1968, reprint 1980.

Photoferroelectric imaging

The process of storing an image in a ferroelectric material by utilizing either the intrinsic or extrinsic photosensitivity in conjunction with the ferroelectric properties of the material. Specifically, photoferroelectric (PFE) imaging refers to storing photographic images or other optical information in transparent lead-lanthanum-zirconate-titanate (PLZT) ferroelectric ceramics.

Imaging devices. The photoferroelectric imaging device consists simply of a thin (0.1–0.3 mm), flat plate of optically polished PLZT ceramic with transparent electrodes deposited on the two major faces. The image to be stored is made to illuminate one of the electroded faces by using near-ultraviolet illumination in the intrinsic photosensitivity region (corresponding to a band-gap energy of approximately 3.35 eV) of the PLZT. Simultaneously, a voltage pulse is applied across the electrodes to switch the ferroelectric polarization from one stable remanent state to another. Images are stored both as spatial distributions of light-scattering centers in the bulk of the PLZT and as surface deformation strains which form a relief pattern of the image on the exposed surface. The light scattering and surface strains are related to spatial distributions of ferroelectric domain orientations introduced during the image-storage process. These spatial distributions replicate the brightness variations in the image to which the PLZT is exposed.

Stored images may be viewed directly or may be projected by using either transmitted or reflected light. For projection, the image contrast is usually improved by using collimated light and a schlieren optical system. *See* SCHLIEREN PHOTOGRAPHY.

Either total or spatially selective erasure of stored images is accomplished by uniformly illuminating the area to be erased with near-ultraviolet light and simultaneously applying a voltage pulse to switch the ferroelectric polarization back to its initial remanent state, that is, to switch it to the polarization state prior to image storage.

The solid-solution PLZT ceramics can be fabricated with a wide range of compositions and sintering conditions to produce an associated wide range of ferroelectric, dielectric, and electrooptic properties. For example, an important factor in determining the maximum resolution of stored images is the ceramic grain size, which can be controlled to a large extent by the sintering temperature and pressure. Images with resolutions as high as 2500 line pairs per inch (100 line pairs per millimeter) can be stored in plates with 2-micrometer grain size, while the maximum image resolution achieved in 5-micrometer grain size

Fig. 1. Photographic image with high resolution and gray scale stored in a photoferroelectric imaging device.

plates is about 1000 line pairs per inch (40 line pairs per millimeter). *See* CERAMICS; DIELECTRIC MATERIALS; ELECTROOPTICS; SINTERING.

Important potential applications of photoferroelectric imaging devices include temporary image storage with periodic update, projection-type display, spatial-temporal light modulation, and high-density optical information storage. Various types of image processing, including image contrast enhancement, are also offered by the capability of switching from a positive to a negative stored image in discrete steps. Nonvolatile photographic images with gray scale ranges extending from an optical density of about 0.15 to more than 2.0 and with resolutions as high as 1000 line pairs per inch (40 line pairs per millimeter) are routinely stored in PLZT plates with 5-μm grain size (**Fig. 1**). *See* IMAGE PROCESSING; OPTICAL INFORMATION SYSTEMS; OPTICAL MODULATORS.

Intrinsic PFE effect. The intrinsic photoferroelectric effect is characterized by a photoinduced reduction of the coercive voltage V_C (the externally applied voltage required to switch the ferroelectric polarization from a saturation remanent state to an average value of zero) produced by irradiating the PLZT surface with band-gap or higher energy light. The reduction in V_C occurs because the absorbed near-ultraviolet light photoexcites charge carriers into the conduction state, and these carriers are transported, under the influence of the applied field, to the ceramic underlying the absorption region, where they contribute to ferroelectric domain nucleation and reorientation. Retrapped carriers establish space charge fields which, in combination with the photocurrent, provide the mechanisms for the photoinduced reduction of V_C. In any given localized area of the ceramic surface, V_C is reduced by an amount proportional to the exposure energy W_{ex} (product of near-ultraviolet light intensity and image exposure time) in that area. As a result, the stored images faithfully reproduce the gray scale of the input image. *See* BAND THEORY OF SOLIDS.

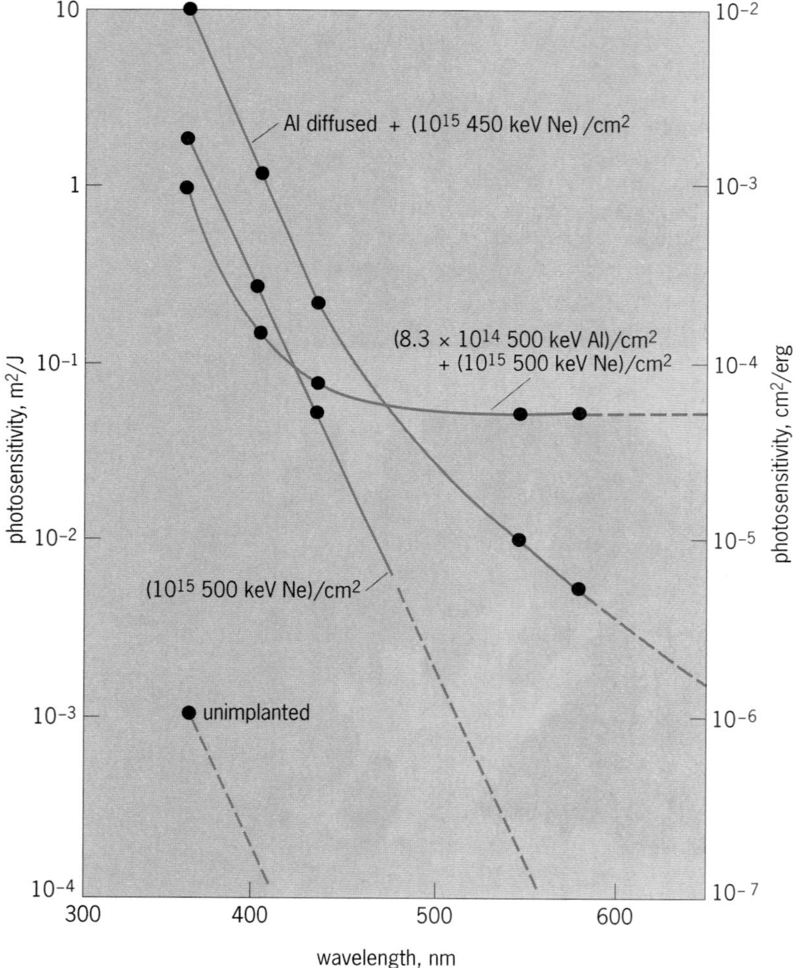

Fig. 2. Wavelength dependence of the photoresponse for aluminum-diffused-plus-neon-implanted PLZT compared to neon-implanted and neon-plus-aluminum-implanted PLZT. The near-ultraviolet response for unimplanted PLZT is also shown.

Photosensitivity enhancement. The exposure energy W_{ex} required to store photographic images in unmodified PLZT is 1-5 × 10^3joules/m^2 (1-5 × 10^6 ergs/cm^2). The relatively high value of W_{ex} (compared to electrophotographic processes or photographic film) tends to limit the scope of practical applications of photoferroelectric imaging devices. Because of the many attractive features of photoferroelectric imaging devices, including erasable and reusable image storage, image contrast enhancement, high-resolution optical information processing, and other optical storage and processing capabilities, various techniques to improve the photosensitivity and to extend the photoresponse from the near ultraviolet throughout the visible spectral region have been explored. Dramatic improvement of the near-ultraviolet photosensitivity, involving reduction of threshold W_{ex} from about 10^3 to 10^{-1} J/m^2 (10^6 to 10^2 ergs/cm^2), has been achieved by implantation of inert ions, such as argon, neon, and helium, and significant photoresponse throughout the visible spectral region has been achieved by co-implantation of chemically active (for example, aluminum or chromium) and inert ions. It has also been found that the combination of thermal diffusion of aluminum followed by neon implantation yields photosensitivity improvement in the near-ultraviolet and blue spectral regions comparable to or better than that obtained by the ion implantation alone. *See* ION IMPLANTATION.

The primary effect of the surface modifications described above is to decrease the dark conductivity of the near-surface region, relative to that of unmodified PLZT, due to implantation-induced defects. These defect states result from atomic displacements produced by energy deposited in nuclear collision processes. In the surface-modified PLZT, a significant fraction of the externally applied voltage can be dropped in the thin (~1 μm), low-conductivity, damaged layer to produce very high drift fields in this near-surface region. Photogenerated charge carriers in this region experience extremely high electric fields and have a high probability of being swept into the underlying ceramic to nucleate and reorient ferroelectric domains. The resulting high photocurrents provide the basis of the ion-implantation enhancement of the photosensitivity.

The photosensitivity S (inverse of W_{ex}) versus wavelength is compared for aluminum-diffused plus neon-implanted and aluminum-plus-neon-implanted PLZT in **Fig. 2**. Also included in Fig. 2 are data for neon-implanted and unimplanted PLZT. Comparison of these data indicates that the major factor contributing to the visible photoresponse is the absorption states associated with aluminum rather than disorder-related defect states produced by the neon implantation. However, the neon implantation is the major contributor to the near-ultraviolet photosensitivity enhancement. This latter conclusion is supported by other published data for PLZT co-implanted with the inert ions argon, neon, and helium which shows that the near-ultraviolet photosensitivity approaches that of aluminum-diffused PLZT, but the visible photoresponse is similar to that of the neon-implanted PLZT.

The dramatic photosensitivity enhancements in both the near-ultraviolet and the visible spectral regions have greatly expanded the variety of possible applications of PLZT as an erasable and reusable photoferroelectric imaging device. *See* ELECTRONIC DISPLAY; FERROELECTRICS; PHOTOCONDUCTIVITY.

Cecil E. Land

Photogrammetry

The practice of obtaining surveys by means of photography. The camera commonly is airborne with its axis vertical, but oblique and horizontal (ground-based) photographs also are applicable. Many industrial and laboratory measurement problems are solved by photogrammetry. Data reduction is accomplished by stereoscopic line-of-sight geometry with use of both analytical and analog methods. *See* AERIAL PHOTOGRAPH; REMOTE SENSING; SURVEYING.

In vertical aerial surveys adjacent photos are overlapped. The two images of the same terrain are then superimposed for three-dimensional viewing by human operators or automated sensors.

In a widely used analog procedure the two photos are placed in the projectors of a stereoplotting instrument. With the aid of visible ground-control points the photos are oriented to the relative positions they had at the instants of exposure. Projections are aimed with the aid of space bars at a plotting table (**Fig. 1**), which can be raised or lowered and moved horizontally. The image is in focus at the center of the disk only when it is at the correct scalar elevation for a given horizontal position. The center of the disk is indicated by a small point of light called the floating dot. To trace a contour, the operator sets the table at its scalar elevation; he moves the table laterally until the elevation's focus is encountered by the floating dot, then lowers the plotting pencil and follows the contour by keeping the image in focus at the floating dot. Planimetry is plotted by moving the floating dot along visible lines (roads or building outlines).

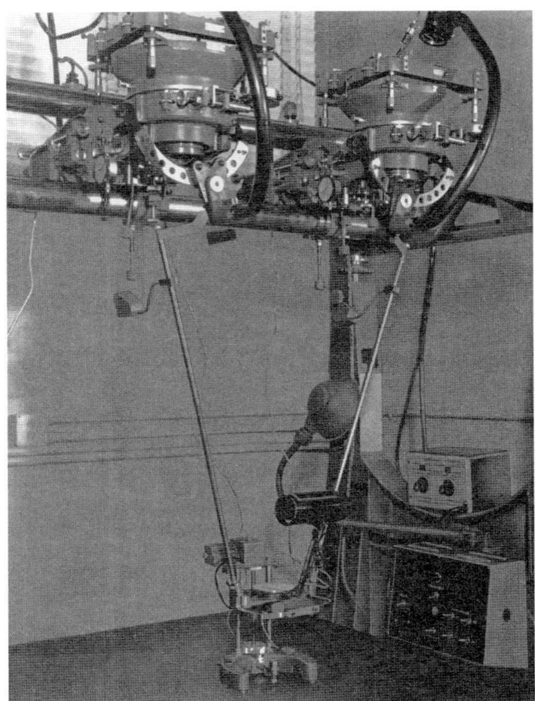

Fig. 1. Stereoplotting instrument, showing projectors and plotting table. (*Kelsh Instrument Co., Inc.*)

Fig. 2. Typical automated stereoplotting system. (*Lockwood, Kessler, and Bartlett, Inc.*)

In a typical automated stereoplotting system (**Fig. 2**) scanning devices substitute for human eyes to sense model-surface slope and thus to control servomechanisms that raise and lower the plotting table (z dimension) and translate it along a succession of closely spaced parallel horizontal dimensions, or ground profiles. Output is available in three forms: digitized x-y-z-coordinate representation of the terrain, contour plotting, and orthophotographs. Each element of the orthophoto represents a vertical view of an element of the stereoscopic image, thereby eliminating the height-displacement distortion characteristic of normal aerials.

Robert H. Dodds

Photographic materials

The light-sensitive recording materials of photography, that is, photographic films, plates, and papers. They consist primarily of a support of plastic sheeting, glass, or paper, respectively, and a thin, light-sensitive layer, commonly called the emulsion, in which the image will be formed and stored. The material will usually embody additional layers to enhance its photographic or physical properties.

Supports. Film support, for many years made mostly of flammable cellulose nitrate, is now exclusively made of slow-burning "safety" materials, usually cellulose triacetate or polyester terephthalate, which are manufactured to provide thin, flexible, transparent, colorless, optically uniform, tear-resistant sheeting. Polyester supports, which offer added advantages of toughness and dimensional stability, are widely used for films intended for technical applications. Film supports usually range in thickness from 0.0025 to 0.009 in. (0.06 to 0.23 mm) and are made in rolls up to 60 in. (1.5 m) wide and 6000 ft (1800 m) long. *See* ESTER; POLYESTER RESINS.

Glass is the predominant substrate for photographic plates, though methacrylate sheet, fused quartz, and other rigid materials are sometimes used. Plate supports are selected for optical clarity and flatness. Thickness, ranging usually from 0.04 to 0.25 in. (1 to 6 mm), is increased with plate size as needed to resist breakage and retain flatness. The edges of some plates are specially ground to facilitate precise registration. *See* GLASS.

Photographic paper is made from bleached wood pulp of high α-cellulose content, free from ground wood and chemical impurities. It is often coated with a suspension of baryta (barium sulfate) in gelatin for improved reflectance and may be calendered for high smoothness. Fluorescent brighteners may be added to increase the appearance of whiteness. Many paper supports are coated on both sides with water-repellent synthetic polymers to preclude wetting of the paper fibers during processing. This treatment hastens drying after processing and provides improved dimensional stability and flatness. *See* PAPER.

Before an emulsion is coated on the support, a "sub" (substratum) is applied to ensure good adhesion of the emulsion layer (**Fig. 1**).

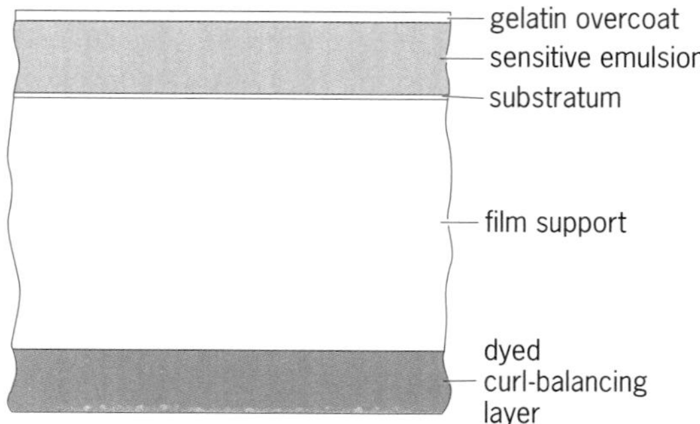

Fig. 1. Schematic cross section of film.

Emulsions. Most emulsions are basically a suspension of silver halide crystals in gelatin. The crystals, ranging in size from 2.0 to less than 0.05 micrometers, are formed by precipitation by mixing a solution of silver nitrate with a solution containing one or more soluble halides in the presence of a protective colloid. The salts used in these emulsions are chlorides, bromides, and iodides. During manufacture, the emulsion is ripened to control crystal size and structure. Chemicals are added in small but significant amounts to control speed, image tone, contrast, spectral sensitivity, keeping qualities, fog, and hardness; to facilitate uniform coating; and, in the case of color films and papers, to participate in the eventual formation of dye instead of metallic silver images upon development. The gelatin, sometimes modified by the addition of synthetic polymers, is more than a simple vehicle for the silver halide crystals. It interacts with the silver halide crystals during manufacture, exposure, and processing and contributes to the stability of the latent image. *See* EMULSION; GELATIN; SILVER.

After being coated on a support, the emulsion is chilled so that it will set, then dried to a specific moisture content. Many films receive more than one light-sensitive coating, with individual layers as thin as 1.0 μm. Overall thickness of the coatings may range from 5 to 25 μm, depending upon the product. Most x-ray films are sensitized on both sides, and some black-and-white films are double-coated on one side. Color films and papers are coated with at least three emulsion layers and sometimes six or more plus filter and barrier layers. A thin, nonsensitized gelatin layer is commonly placed over film emulsions to protect against abrasion during handling. A thicker gelatin layer is coated on the back of most sheet films and some roll films to counteract the tendency to curl, which is caused by the effect of changes in relative humidity on the gelatin emulsion. Certain films are treated to reduce electrification by friction because static discharges can expose the emulsion. The emulsion coatings on photographic papers are generally thinner and more highly hardened than those on film products.

Another class of silver-based emulsions relies on silver-behenate compounds. These materials require roughly 10 times more exposure than silver halide emulsions having comparable image-structure properties (resolving power, granularity); are less versatile in terms of contrast, maximum density, and spectral sensitivity; and are less stable both before exposure and after development. However, they have the distinct advantage of being processed through the application of heat (typically at 240 to 260°F or 116 to 127°C) rather than a sequence of wet chemicals. Hence, products of this type are called Dry Silver films and papers.

Spectral sensitivity. The silver halides (and silver behenates) are normally sensitive only to x-radiation and to ultraviolet, violet, and blue wavelengths, but they can be made sensitive to longer wavelengths by adding special dyes, predominantly polymethines, to the emulsion. The process is known as spectral sensitizing to distinguish it from the chemical sensitizing used to raise the overall or inherent sensitivity of the grains.

Spectrally nonsensitized emulsions are generally termed blue-sensitive or ordinary and exhibit little response beyond 450 nanometers. Emulsions treated with dyes to extend their sensitivity to 500 nm are identified as extended blue, while those with sensitivity extended through the green (550 nm) are termed orthochromatic. Some panchromatic films are sensitized out to 650 nm; others, with extended red sensitivity, respond efficiently out to 700 nm. The upper limit for spectral sensitizing is about 1200 nm, which is in the near-infrared region, but conventional infrared films exhibit little sensitivity beyond 900 nm. Ultraviolet sensitivity is diminished somewhat between 250 and 280 nm and sharply below 250 nm by the strong optical absorption by gelatin. Materials sensitive to much shorter wavelengths can be prepared by several means which result in a high concentration of silver halide crystals at the top surface of the emulsion layer. Films are selected for spectral characteristics to achieve efficient response to various exposing sources and, whenever possible, to provide convenient handling under safelight conditions. Spectral sensitivity is an important consideration in assuring visually satisfying tonal reproduction of colored objects with black-and-white materials or intentionally altering the tonal reproduction to provide greater spectral discrimination than the human eye. The spectral sensitivity of a recording material is altered, in effect, by placing color filters over the camera lens.

Halation. During exposure of a film or plate, part of the radiation entering the emulsion is absorbed; the rest is transmitted into the support. A small fraction of the transmitted light is reflected at the back surface and returned to the emulsion at a point displaced from the incoming radiation, depending upon the angle of incidence and the thickness of the support (**Fig. 2**). This secondary exposure is called halation because a bright point in a scene is recorded as a point surrounded by a halo. Halation can be suppressed to negligible levels by incorporating carbon black or selected dyes in the support, a layer on the back of the support, a layer between the emulsion and the support, or in rare instances, the emulsion. The absorbing layer need not be opaque to be effective since the offending radiation must make two passes through the absorbing layer and is strongly attenuated upon reflection at the back surface. Carbon black is added to some polyester supports to suppress light-piping as well as halation.

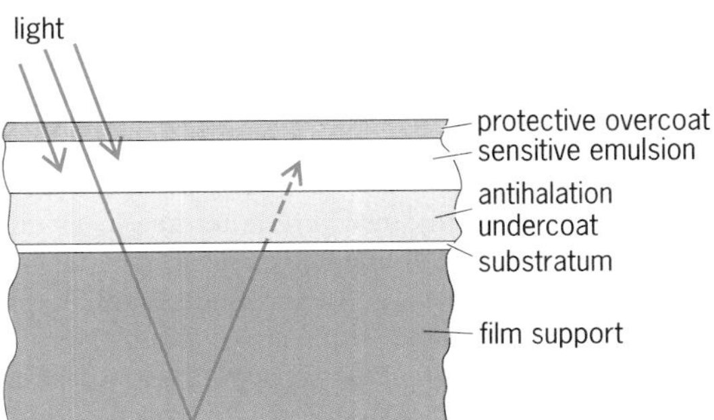

Fig. 2. Suppression of halation by an absorbing medium, shown here as an antihalation undercoat.

Dyes in layers under the emulsion or in curl-balancing layers on the back of some films are bleached or dissolved during processing. The antihalation backings on some plates and films are dissolved or stripped away during processing.

Photographic characteristics. For a discussion of characteristics of sensitive materials, as well as the theory of the photographic process, latent image, sensitometry, and image structure, *see* PHOTOGRAPHY.

All of these characteristics are related to the combination of emulsion, developer, and exposure. The structure of the developed image is of importance in determining the quality of the photograph, especially its ability to reproduce fine detail.

Photographic products. Hundreds of types of films, plates, and papers are available in numerous sizes and configurations for a wide range of applications. Each field of use generally requires special properties. Films for amateur, professional, and technical photography typically range in speed from an exposure index (ISO/ASA) of less than 10 to over 1000, and from very fine grain and high sharpness to rather coarse grain with corresponding loss in definition. They can be obtained in a wide range of contrasts and spectral sensitizations, especially for scientific and technical photography. While most materials are negative-working, a number of products provide positive-working emulsions which result in direct positive reproductions with conventional (negative) processing. Materials of extremely high contrast and density are widely used for graphic reproduction and in the printing industry to obtain high-contrast line and halftone images. Many types of x-ray films are made, most sensitized on both sides and used in combination with fluorescent intensifying screens to minimize patient exposure. Several types of color film are available, designed primarily for pictorial photography.

Photographic papers are offered in both contact and enlarging speeds and with a range of contrasts for producing continuous-tone prints from camera negatives. A second class of enlarging papers provides selective control of contrast through the use of filters during exposure of two-component emulsions. One component is blue-sensitive only and has high contrast, while the other has sensitivity extended to the yellow-green and relatively low contrast. Papers are also distinguished by surface texture and tint of the paper support. Developing agents are incorporated in the emulsions of some photographic papers so that internally controlled development can be accomplished quickly by immersing the paper in an alkaline bath. Color printing papers are made for printing from color negative or positive films.

High-contrast negative and direct-positive papers are used for photographic reproduction of documents, engineering drawings, or line copy. Several papers are designed for oscillographic recording, photographic typesetting, and other specialized applications.

Image-transfer systems. The image resulting from exposure and processing of most photographic materials is retained in the light-sensitive layer or layers. In another and very useful set of products, the image is transferred by chemical means to a receiver sheet which has been placed in contact with the basic photographic material during part or all of the processing step. Careful design of each component in these image-transfer systems generally results in convenient processing and relatively rapid access to the final image. Most widely known are camera-based systems in which continuous-tone silver or dye images are chemically transferred to a receiver sheet or layer from the light-sensitive and sometimes complex image-forming layers. In the original Polaroid-Land process, unexposed silver halide crystals in the emulsion are transferred by a solvent and reduced to form a positive black-and-white image on the receiver in as little as 10 s. Several newer systems form color images through the transfer of dyes to the receiver within 1 to 4 min (**Fig. 3**). These "instant" products are popular for amateur photography but are widely used for professional and technical applications as well, including the photography of cathode-ray-tube images. *See* CAMERA; CATHODE-RAY TUBE.

Other examples of image-transfer systems include materials for producing large color prints or transparencies from either negative or positive color film originals and several high-contrast black-and-white materials used for preparing short-run printing plates. While the image-forming component of most image-transfer systems is usually discarded after being pulled apart from the receiver, it is sometimes retained as a useful film negative. In some instant products intended primarily for amateur photography, all layers are retained as a unit to avoid litter. These integral film structures, including image-forming layers, an image-receiving layer, and chemicals, are handled in sheet form, while two-part structures are offered in both roll (Fig. 3) and sheet configurations.

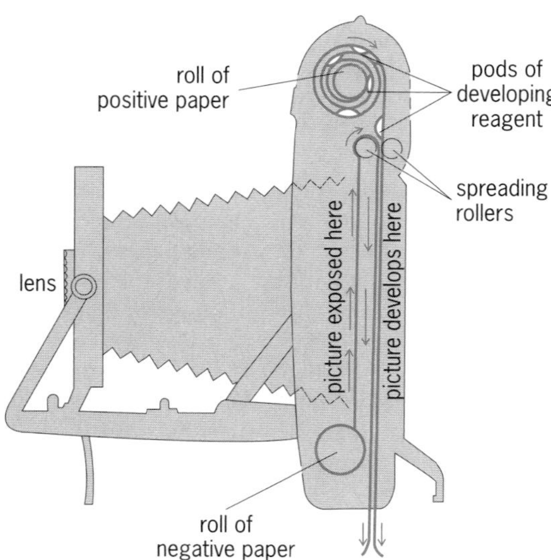

Fig. 3. Cross-sectional view of Polaroid Instant Camera and film. (*Polaroid Corp.*)

Storage of materials. Unexposed films and papers may be damaged by exposure to elevated temperatures and high relative humidities. Protection is provided by packing the materials hermetically at controlled humidities (30–50% relative humidity). Storage at a temperature of 40 to 55°F (4 to 13°C) is sufficient for most materials, but some products require deep-freeze conditions.

The life of processed films and prints depends strongly on processing conditions as well as environmental factors. Thorough fixing and washing are usually a primary requirement. Storage at 30–50% relative humidity and at temperatures not exceeding 70°F (21°C) in carefully selected enclosures is commonly recommended.

Nonsilver processes. Many special-purpose photographic materials use active compounds other than silver salts. In general, they are less sensitive by factors of 100 or more than comparable silver-based products, and they lack the amplification factor associated with the development of silver compounds. Accordingly, they are used mainly in reproduction processes where speed is not essential or in certain recording applications where intense exposures from laser sources are possible. They include inorganic compounds which are photosensitive, especially iron salts and dichromates in colloid layers such as gelatin, glue, albumen, shellac, and polyvinyl alcohol or other synthetic resins which become insoluble on exposure. Other important processes are based on light-sensitive organic compounds, especially diazo compounds; unsaturated compounds such as cinnamic acid derivatives, which are insolubilized by cross-linking on exposure; systems in which polymerization occurs as a result of free-radical formation or exposure; dyes which are formed, bleached, or destroyed directly by exposure to light or electrons; and thermographic systems in which a thermal pattern resulting from intense exposure produces an image by means of melting a composition or initiating a chemical reaction.

Electrophotographic systems rely on the imagewise elimination of a charge by exposure of a uniformly charged photoconductive layer and subsequent development of the remaining charge distribution with charged powders or liquid toners, or by electrolytic deposition of metal on the charge image. Toned images may be fixed by heating so the toner particles are fused either to the photoconductive material or, following physical transfer, to a receiver sheet. Because electrophotographic systems commonly exhibit strong edge effects, they are especially useful in document-copying applications, but continuous-tone systems which rely on careful design of the charging and toning elements have been demonstrated. Many xerographic copiers use coatings of amorphous selenium on rigid supports as the photoconductive layer. The preferred medium in modern copiers of this type consists of an organic photoconductor coated on a flexible support similar to film base. *See* PHOTOCOPYING PROCESSES.

Robert D. Anwyl

Bibliography. N. R. Eldred, *Chemistry for the Graphic Arts*, 3rd ed., 2001; T. Grimm, *Basic Book of Photography*, 4th ed., 1997; T. Grimm, *The Basic Darkroom*, 3d ed., 1999; M. J. Langford, *Advanced Photography*, 6th ed., 1998; M. J. Langford, *Basic Photography*, 6th ed., 1997; S. F. Ray, *Photographic Chemistry and Processing*, 1994; S. F. Ray, *Photographic Data (Technical Pocket Book)*, 1994; S. F. Ray, *Photographic Imaging and Electronic Photography (Technical Pocket Book)*, 1994; L. D. Stroebel (ed.), *Basic Photographic Materials and Processes*, 2d ed., 2000; L. D. Stroebel (ed.), *The Focal Encyclopedia of Photography*, 3d ed., 1996.

Photography

The process of forming stable or permanent visible images directly or indirectly by the action of light or other forms of radiation on sensitive surfaces. Traditionally, photography utilizes the action of light to bring about changes in silver halide crystals. These changes may be invisible, necessitating a development step to reveal the image, or they may cause directly visible darkening (the print-out effect). Most photography is of the first kind, in which development converts exposed silver halide to (nonsensitive) metallic silver.

Since the bright parts of the subject normally lead to formation of the darkest parts of the reproduction, a negative image results. A positive image, in which the relation between light and dark areas corresponds to that of the subject, is obtained when the negative is printed onto a second negative-working material. Positive images may be achieved directly in the original material in two principal ways: (1) In reversal processing, a positive image results when the developed metallic silver is removed chemically and the remaining silver halide is then reexposed and developed. (2) In certain specialized materials, all of the silver halide crystals are rendered developable by chemical means during manufacture. Exposure to light alters the crystals so that development will not proceed. As a result, conventional "negative" development affects only the unexposed crystals and a positive image is formed directly.

The common materials of photography consist of an emulsion of finely dispersed silver halide crystals in gelatin coated in a thin layer (usually less than 25 micrometers) on glass, flexible transparent film, or paper (**Fig. 1**). The halides are chloride, bromide, or bromoiodide, depending upon the intended usage. The most sensitive materials, used for producing camera negatives, consist of silver bromide containing some silver iodide; the slow materials used for printing are usually of silver chloride; materials of intermediate speed are of silver bromide or silver bromide and chloride.

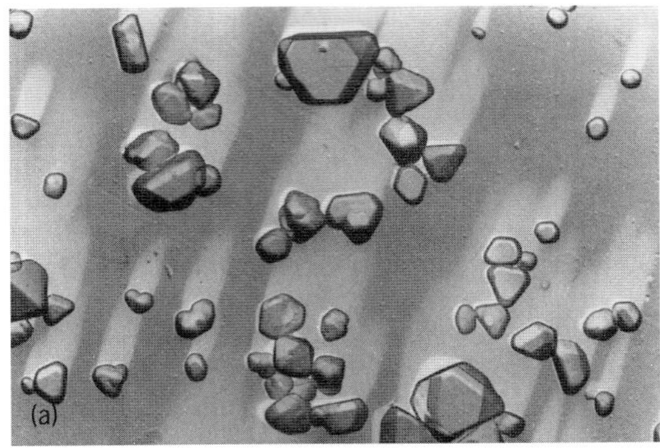

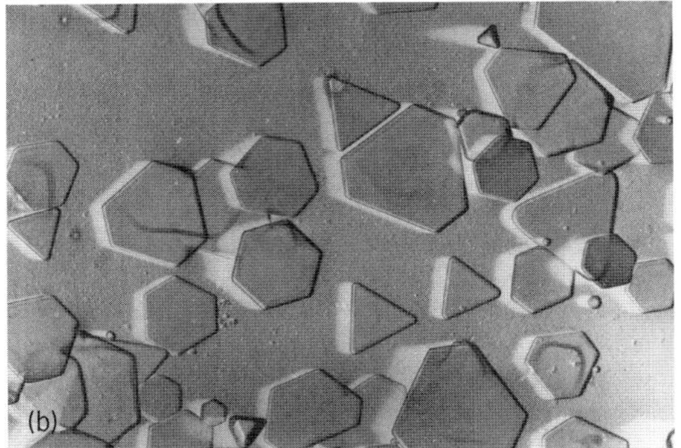

Fig. 1. Silver halide crystals, shown highly magnified. (*a*) Conventional pebblelike crystals. (*b*) Tablet-shaped crystals, used in some modern emulsions, that absorb light more efficiently, resulting in higher film speed.

Following exposure of a sensitized recording material in a camera or other exposing device such as a line-scan recorder, oscillograph, plotter, or spectrograph, the film or plate is developed, fixed in a solution which dissolves the undeveloped silver halide, washed to remove the soluble salts, and dried. Printing from the original, if required, is done by contact or optical projection onto a second emulsion-coated material, and a similar sequence of processing steps is followed.

For about 100 years, the results of practical photography were almost exclusively in black and white or, more precisely, shades of gray. With the introduction of the Kodachrome process and materials in 1935 and of a variety of other systems over the succeeding years, a large and increasing fraction of photography has been done in color. In most color photography, development is basically the same as in black-and-white photography, except that the chemical reactions result in the formation of dye images and subsequent removal of the image-forming silver. Both negative and positive (reversal) color systems are employed. A special class of print materials relies on bleachable dyes instead of conventional silver halide processes. *See* PHOTOGRAPHIC MATERIALS.

Branches and Applications

The several branches of photography may be grouped according to who takes, processes, and uses the photographs. Within the branches, numerous forms or applications of the photographic art may be distinguished by the exposing conditions and the manner in which the photographic materials, methods, and processes have been designed to optimize the results. The divisions are not always distinct, and the subtleties or strengths of one application are often employed to advantage in another.

Amateur photography refers to the taking of photographs by the general public for purposes of recollection, record, or amusement. Historically, the cameras were bulky, film formats were little or no smaller than the prints, and good lighting was required to produce black-and-white photographs of still subjects. Advancements in film technology and sophistication of amateur cameras have enabled the amateur photographer to record color photographs of moving objects on progressively smaller film formats in lightweight pocketable cameras under a wide range of lighting conditions. Processing and printing are provided by commercial photofinishers. Alternatively, the amateur photographer can obtain color prints within minutes by using so-called instant cameras and films which embody their own processing means. Advanced amateurs generally use more costly cameras and a broader range of film materials to produce photographs of esthetic merit for display. They may do their own processing and printing or rely on the assistance of professional finishers.

Photography is practiced on a professional level for portraiture and for various commercial and industrial applications, including the preparation of photographs for advertising, illustration, display, and record-keeping. Press photography is for newspaper and magazine illustrations of topical events and objects. Photography is used at several levels in the graphic arts to convert original photographs or other illustrations into printing plates for high-quality reproduction in quantity. Industrial photography includes the generation and reproduction of engineering drawings, high-speed photography, schlieren photography, metallography, and many other forms of technical photography which can aid in the development, design, and manufacture of various products. Aerial photography is used for military reconnaissance and mapping, civilian mapping, urban and highway planning, and surveys of material resources. Microfilming, which is used to store documents and radiographs in significantly reduced space, represents one of several applications of a more general category, microphotography. Biomedical photography is used to reveal or record biological structures, often of significance in medical research, diagnosis, or treatment. Photography is widely applied to preparing projection slides and other displays for teaching through visual education. Nonamateur uses of photography far exceed the amateur snapshot and home movie fields in volume or commercial significance. *See* PRINTING; SCHLIEREN PHOTOGRAPHY.

Photography is one of the most important tools in scientific and technical fields. It extends the range of vision, allowing records to be made of things or events which are difficult or impossible to see because they are too faint, too brief, too small, or too distant, or associated with radiation to which the eye is insensitive. Technical photographs can be studied at leisure, measured, and stored for reference or security. The acquisition and interpretation of images in scientific and technical photography usually requires direct participation by the scientist or skilled technicians.

Infrared photography. Emulsions made with special sensitizing dyes can respond to radiation at wavelengths up to 1200 nanometers, though the most common infrared films exhibit little sensitivity beyond 900 nm. One specialized color film incorporates a layer sensitive in the 700–900-nm region and is developed to false colors to show infrared-reflecting subjects as bright red. *See* INFRARED RADIATION.

Photographs can thus be made of subjects which radiate in the near-infrared, such as stars, certain lasers and light-emitting diodes, and hot objects with surface temperatures greater than 500°F (260°C). Infrared films are more commonly used to photograph subjects which selectively transmit or reflect near-infrared radiation, especially in a manner different from visible radiation. Infrared photographs taken from long distances or high altitudes usually show improved clarity of detail because atmospheric scatter (haze) is diminished with increasing wavelength and because the contrast of ground objects may be higher as a result of their different reflectances in the near-infrared. Grass and foliage appear white because chlorophyll is transparent in the near-infrared, while water is rendered black because it is an efficient absorber of infrared radiation.

Infrared photography is used in a wide range of scientific and technical disciplines because it permits or greatly enhances the detection of features not obvious to the eye. For example, it has been used for camouflage detection because most green paints absorb infrared more strongly than the foliage they are meant to match, to distinguish between healthy and diseased trees or plants, to detect alterations in documents and paintings, to reveal subcutaneous veins (because the skin is somewhat transparent to infrared radiation), and to record the presence of certain substances which fluoresce in the near-infrared when exposed to visible radiation. While infrared photography can be used to map variations in the surface temperature of hot objects which are not quite incandescent, it cannot be used for detecting heat loss from buildings or other structures whose surface temperatures are usually close to that of the surrounding atmosphere.

One technique for detecting radiation at wavelengths near 10 μm relies on localized transient changes in the sensitivity of conventional photographic materials when exposed briefly to intense radiation from certain infrared lasers, and simultaneously or immediately thereafter to a general low-level flashing with light. Another technique for detecting intense infrared exposures of very short duration with films having no conventional sensitivity at the exposing wavelength is thought to rely on two-photon effects. *See* INFRARED IMAGING DEVICES.

Ultraviolet photography. Two distinct classes of photography rely on ultraviolet radiation. In the first, the recording material is exposed directly with ultraviolet radiation emitted, reflected, or transmitted by the subject; in the other, exposure is made solely with visible radiation resulting from the fluorescence of certain materials when irradiated in the ultraviolet. In the direct case, the wavelength region is usually restricted by the camera lens and filtration to 350–400 nm, which is readily detected with conventional black-and-white films. Ultraviolet photography is accomplished at shorter wavelengths in spectrographs and cameras fitted with ultraviolet-transmitting or reflecting optics, usually with specialized films. In ultraviolet-fluorescence photography, ultraviolet radiation is blocked from the film by filtration over the camera lens and the fluorescing subject is recorded readily with conventional color or panchromatic films. Both forms of ultraviolet photography are used in close-up photography and photomicrography by mineralogists, museums, art galleries, and forensic photographers. *See* ULTRAVIOLET RADIATION.

High-speed photography. Photography at exposure durations shorter than those possible with conventional shutters or at frequencies (frame rates) greater than those achievable with motion picture cameras with intermittent film movements is useful in a wide range of technical applications.

Single-exposure photography. The best conventional between-the-lens shutters rarely yield exposures shorter than 1/500 s. Some focal plane shutters are rated at 1/2000 or 1/4000 s but may take 1/100 s to traverse the film format. Substantially shorter exposures are possible with magnetooptical shutters (using the Faraday effect), with electrooptical shutters (using the Kerr effect), or with pulsed electron image tubes. Alternatively, a capping shutter may be used in combination with various pulsed light sources which provide intense illumination for very short durations, including pulsed xenon arcs (electronic flash), electric arcs, exploding wires, pulsed lasers, and argon flash bombs. Flash durations ranging from 1 millisecond to less than 1 nanosecond are possible. Similarly, high-speed radiographs have been made by discharging a short-duration high-potential electrical pulse through the x-ray tube. *See* FARADAY EFFECT; KERR EFFECT; LASER; STROBOSCOPIC PHOTOGRAPHY.

High-speed serial photographs. The classical foundation for serial frame separation is the motion picture camera. Intermittent movement of the film in such cameras is usually limited to 128 frames/s (standard rates are 16 and 24). For higher rates (up to 10,000 frames/s or more) continuous film movement is combined with optical compensation, as with a rotating plane-parallel glass block, to avoid image smear. Pictures made at these frequencies but projected at normal rates slow down (stretch) the motion according to the ratio of taking and projection rates. Higher rates, up to 10^7 frames/s, have been achieved with a variety of ingenious special-purpose cameras. In some, the sequence of photographs is obtained with a rapidly rotating mirror at the center of an arcuate array of lenses, and a stationary strip of film. In others, the optics are stationary and the film strip is moved at high speed by mounting it around the outside or inside of a rapidly rotating cylinder. To overcome mechanical limitations on the rotation of mirrors or cylindrical film holders at high speeds, image dissection methods have been employed, that is, an image is split into slender sections and rearranged to fill a narrow slit at the film. The image is unscrambled by printing back through the dissecting optics. *See* CINEMATOGRAPHY.

Photomicrography. Recording the greatly enlarged images formed by optical microscopes is indispensable in the life sciences, in metallurgy (where it is known as metallography), and in other fields requiring examination of very small specimens.

Photomicrographic apparatus consists of an optical system in which are aligned an illuminating subsystem, a compound microscope, a simple achromat, and a film holder or transport, all mounted to be free of vibration. Photomicrographs can be made by attaching a simple camera to a complete compound microscope, but a simple achromat, properly designed and located, is superior in this application to a high-quality camera lens of similar focal length. Resolving power is fully as important as magnification, but dependent mainly on the quality of the microscope objective and illumination system rather than the film. Resolving power increases as the exposing wavelength decreases. Because the luminance range of microscopic specimens is generally very low, discrimination of specimen detail is often enhanced spectrally by staining the specimen and inserting suitable filters in the illumination system. Other means for revealing specimen detail rely on controlled illumination, polarization effects, and selecting and processing films to provide higher contrast than is appropriate for general pictorial photography. The practical limitations on resolving power and magnification (approximately 2000×) in optical microscopy can be overcome by using a transmission electron microscope (an electronic analog of the optical microscope), a scanning electron microscope, or an x-ray microscope. *See* ELECTRON MICROSCOPE; LENS (OPTICS); METALLOGRAPHY; OPTICAL MICROSCOPE; RESOLVING POWER (OPTICS); SCANNING ELECTRON MICROSCOPE; X-RAY MICROSCOPE.

Close-up photography. This is the photography of specimens, usually three-dimensional objects, which are little smaller or larger than the film format. By convention, the photography of small objects at magnifications ranging from 1 to 50× is called photomacrography, while the photography of gross specimens at magnifications ranging from 1 to 0.1× is termed close-up photography. Both may be accomplished with conventional camera lenses or with so-called macro lenses, which are specially designed for use at finite object and image distances. Lens apertures must be selected to achieve maximum image sharpness consistent with the depth or thickness of the specimen.

Document copying. Photography is used for reproducing documents of all kinds because (1) it does not introduce errors in the copying process; (2) it provides an inexpensive means for retrieving and disseminating information rapidly; (3) it offers security against loss or disaster because copies can be stored in multiple locations; and (4) it effectively extends the life of perishable records.

Documents are reproduced to essentially the same size as the original or, in some copiers, with modest changes in size. They may also be photographed at reduced size (microfilming) on 16-, 35-, or 105-mm film and printed, as needed, at that size for distribution in microform or with photographic enlargement to provide reproductions at any desired scale. Historically, a variety of silver and nonsilver processes have been used for producing paper copies of business documents at or close to original size. Except for the use of conventional photographic materials in reproducing engineering drawings and other large originals, most copiers now rely on the electrostatic methods (xerography). *See* PHOTOCOPYING PROCESSES.

Microphotography. This is the process of making photographs on a greatly reduced scale, usually from two-dimensional originals. Microfilming is the special technique of copying documents on film, usually in rolls 15, 35, or 105 mm wide. Reduction factors of 10 to 45× are common, though specialized systems employing reductions ranging from 60 to more than 200× have been used. Reductions greater than 60× are customarily made in two steps to overcome physical and optical limitations. In computer output microfilming, the original "document" is displayed on a cathode-ray tube or written on the film with a laser beam. Silver-based microfilms are usually processed to meet archival standards. Microfilm negatives (or prints made on silver, diazo, or vesicular films) may be retained in roll form, cut into strips, assembled into plastic sheaths holding several strips, mounted as individual images in apertures in business machine cards, or printed as an array of images on a sheet of film (microfiche) or, at further reduction, on a film chip (ultrafiche). When retrieved from storage, these microforms are viewed on enlarging readers or readerprinters, the latter incorporating means for making an enlarged paper print quickly from the image in the viewing station. Microfilm readers often include a system for locating selected frames rapidly and automatically by reading indexing information placed alongside the microimage during the recording process.

Microphotography is used also for producing scales, reticles, and, most notably, the sets of masks employed in manufacturing a wide range of microelectronic devices. Reductions from the original artwork by factors of 100 to 1000 are made in two or three steps, depending upon the size of the originals, using materials with progressively higher resolving powers. Historically, artwork was produced manually on large translucent sheets, but high-speed, computer-driven plotters, many using laser beams to expose special films or plates, yield highly complex patterns with high quality at smaller scales. This eliminates one or two photoreduction steps and results in masks of correspondingly higher quality. The plates used in the final step are characterized not only by their high resolving power (greater than 2000 lines/mm) but by their freedom from defects or inclusions in the emulsion layer. As the demands of the microelectronics industry have pushed classical microphotography to fundamental optical limits, it has been necessary to generate patterns at final size by using electron-beam plotters, in some cases directly on resist-coated silicon wafers. *See* COMPUTER GRAPHICS; INTEGRATED CIRCUITS.

Photographic radiometry. The intensity (brightness) and spectral distribution (color) of radiation emitted or reflected by a surface can be measured by photography by comparing the densities resulting from recording a test object with those obtained from photographing a standard source or surface, preferably on adjacent portions of the same film or plate. The method is capable of reasonable precision if the characteristics of the photographic material are accurately known, all of the other photographic conditions are closely or exactly matched, and the standard source is well calibrated. In practice, it is difficult to meet all of these conditions, so photographic radiometry is used mainly when direct radiometric or photometric methods are not practicable. It is even more difficult, though not impossible, when using color films. When photographic radiometry is applied to objects at elevated temperatures which self-radiate in the visible or near-infrared region and the emissivity of the materials is known, the surface temperature of the object may be determined and the process is known as photographic pyrometry. Reference objects at known temperatures must be included in the photograph. *See* PHOTOMETRY; PYROMETER; RADIOMETRY.

Remote sensing. The art of aerial photography, in which photographs of the Earth's surface are made with specialized roll-film cameras carried aloft on balloons, airplanes, and spacecraft, is an important segment of a broader generic technology, remote sensing. The film is often replaced with an electronic sensor, the sensor system may be mounted on an aircraft or spacecraft, and the subject may be the surface of a distant planet instead of Earth. Remote sensing is used to gather military intelligence; to provide most of the information for plotting maps; for evaluating natural resources (minerals, petroleum, soils, crops, water) and natural disasters; and for planning cities, highways, dams, pipelines, and airfields. Aerial photography normally provides higher ground resolution and geometric accuracy than the imagery obtained with electronic sensors, especially when covering small areas, so it continues as the foundation for mapmaking, urban planning, and some other applications. Films designed for aerial photography, both black-and-white and color, have somewhat higher contrast than conventional products because the luminance range of the Earth's surface as seen from altitudes of 5000 ft (1500 m) or more is roughly 100 times lower than that of landscapes photographed horizontally. *See* AERIAL PHOTOGRAPH; PHOTOGRAMMETRY; TOPOGRAPHIC SURVEYING AND MAPPING.

The acquisition of image information with scanning sensors mounted on spacecraft provides an inexpensive means for gathering photographs of large areas of the Earth or the whole Earth at regular intervals (minutes or hours for meteorological satellites, days for Earth resources satellites) or for photographing subjects which cannot be reached with aircraft or approached with spacecraft. Some sensors operate at wavelengths beyond those detected by infrared films. The image information is transmitted to receiving stations on Earth, usually processed electronically to correct for geometric and atmospheric factors, and recorded on a variety of image recorders. Scanning sensors, as well as film cameras, are employed in aerial reconnaissance because they can transmit tactical information to ground stations for evaluation before the aircraft returns to base or is shot down. Synthetic aperture radar, which maps the reflectance of microwaves from the surface of the Earth and other planets, represents another form of remote sensing for both military and commercial purposes in which the information is returned to Earth and reconstructed in photographic form for study. *See* AERIAL PHOTOGRAPH; METEOROLOGICAL SATELLITES; MILITARY SATELLITES; REMOTE SENSING.

Stereoscopic photography. This technique simulates stereoscopic vision. It presents to the two eyes individually two aspects of the subject made from slightly different viewpoints. Relief can be distinguished visually only over moderate distances, although the appearance of relief can be introduced in more distant objects by making photographs at greater distances apart. Stereoscopic photography is possible with a single camera if two separate photographs are made, one after the other, from viewpoints separated by the interocular or other appropriate distance, by displacing the lens of a single camera for the two exposures, or by rotating the object or the camera so as to give a pair of exposures on one film. These methods can be used only with relatively stationary subjects. Most stereoscopic photography is done either by the simultaneous method, in which two photographs are made at the same time with two separate cameras; with a stereoscopic camera, essentially two cameras in one body with matched optical systems and coupled focusing movements; or with single cameras using beam splitters to give two photographs side by side on the film. *See* STEREOSCOPY.

In aerial photography, the stereoscopic effect is achieved by making successive overlapping photographs along the line of flight; in terrestrial photogrammetry, photographs are made from each end of a selected base line; in stereoscopic radiography, two x-ray photographs are made in rapid succession from separate viewpoints.

Stereoscopic photographs are viewed in equipment which presents the right-eye image to the right eye only and the left-eye image to the left eye only. This can be done by separate boxes, each provided with a lens; by open-type viewers with a pair of shielded lenses, sometimes prismatic; by cabinet viewers, with pairs of optics and devices for changing the stereo slides; by grids or lenticular elements permitting each eye to see only its appropriate field; by anaglyphs, in which two images are printed in ink in complementary colors and viewed through spectacles of similar colors such that each filter extinguishes one image; and by the vectograph print, which consists of a reflecting support on which the stereoscopic pairs are placed one over the other in plastic polarizing layers with polarization planes at right angles. Polarizing spectacles are used, the eyepieces arranged so that each eye sees only the appropriate image.

Color stereo prints are made commercially by exposing color film through a fine parallel-slotted grid while swinging the camera in an arc about a principal point in the scene; processing and printing the picture; and then coating it with a plastic and embossing this with parallel lenses, registering closely with the grid pattern of the original photograph.

Nuclear-particle recording. Charged atomic particles give records on photographic emulsions, and the method of recording provides an important adjunct to such detectors as the ionization chamber and Geiger counter. The first studies were done with alpha particles, which produce a track of silver grains in the developed emulsion. Protons were also found to produce tracks, with grains of different spacing. Cosmic rays cause nuclear disintegrations on collision with the atoms in the emulsion; starlike patterns result, consisting of tracks made by particles from the atomic nuclei. Tracks of charged particles can be recorded in special thick emulsions having high silver bromide content, small grains, minimum fog, and appropriate speed or in blocks of emulsions made up of many sheets or pellicles. The grains are made developable by ionization within them caused by impact of the particles. Protons and neutrons may react with atoms of the emulsion, giving rise to charged particles, and so are indirectly recorded. The cloud chamber and the bubble chamber provide means for photographing particle tracks through the use of high-intensity flash. In the cloud chamber the tracks are formed by condensation of moisture from supersaturated vapor on ion trails produced by the particles. In the bubble chamber the medium is a superheated liquid, such as liquid hydrogen, which gives trails of gas bubbles. The chambers are photographed with three cameras whose axes are aligned orthogonally or with a holographic camera to permit accurate determination of track lengths and trajectories. *See* COSMIC RAYS; PARTICLE DETECTOR.

Theory of the Photographic Process

The normal photographic image consists of a large number of small grains or clumps of silver. They are

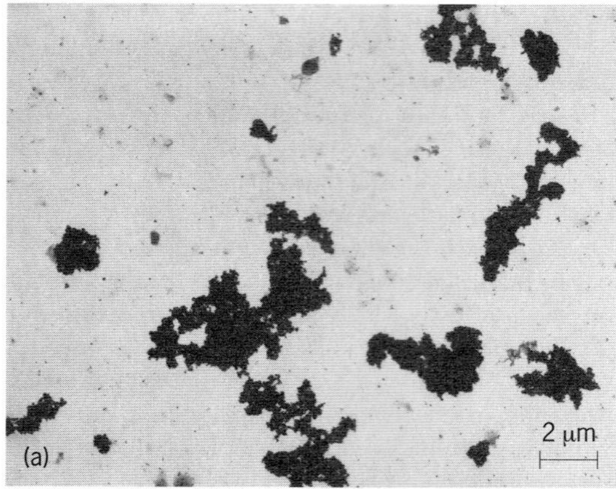

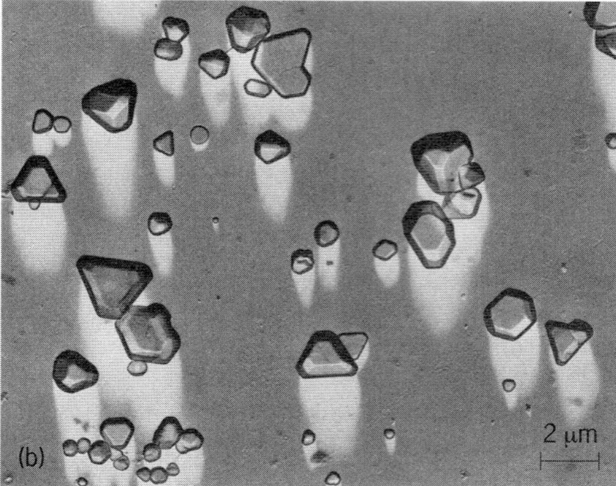

Fig. 2. Photographic silver. (*a*) Developed grains of a photographic emulsion. (*b*) Original silver halide crystals from which the silver grains were developed. Both views are highly magnified. (*Eastman Kodak Co.*)

the end product of exposure and development of the original silver halide crystals in the emulsion and may be two to ten times larger than the crystals, which range in size up to a few micrometers (**Fig. 2**). The shape, size, and size distribution of the crystals are determined by the way the emulsions are made, and are closely related to their photographic properties. In general, high-speed, negative-type emulsions have crystals in a wide range of sizes, while emulsions with low speed and high contrast have smaller crystals that are more uniform in size.

Exposure of the crystals to light causes a normally invisible change called the latent image. The latent image is made visible by development in a chemical reducing solution which converts the exposed crystals to metallic silver, leaving unreduced virtually all of the crystals either unexposed or unaffected by exposure. The darkening which results is determined by the amount of exposure, the type of emulsion, and the extent of development. Density, a physical measure of the darkening, depends on the amount of silver developed in a particular area and the covering power of the silver. With very high exposures, darkening may occur directly, without development, as a result of photolysis. This is known as the print-out effect. Its use is confined mostly to making proof prints in black-and-white portraiture and papers for direct-trace recording instruments.

Latent image. As noted above, the term latent image refers to a change occurring in the individual crystals of photographic emulsions whereby they can be developed upon exposure to light. Early research indicated that development starts at localized aggregates of silver, mainly on the surfaces of the crystals. The existence of these discrete development centers suggested that the latent image is concentrated at specific sites. Subsequent research by numerous investigators relating to the chemical and physical properties of silver halide crystals during irradiation led eventually to a logical and comprehensive explanation for the process of latent image formation, formulated in 1938 by R. W. Gurney and N. F. Mott.

According to the Gurney-Mott theory, absorption of photon energy by a crystal liberates electrons which are able to move freely to sensitivity sites, presumably associated with localized imperfections in the crystal structure and concentrations of sensitizing impurities, where they are trapped and establish a negative charge. This charge attracts nearby positively charged silver ions, which are also mobile. The electrons and ions combine to form metallic silver, and the process continues until sufficient silver has accumulated to form a stable, developable aggregate. The evidence indicates that a minimum of three or four silver atoms is required.

At the exposure levels required to form a latent image, the absorption of one photon by a silver halide crystal can, in principle, form one atom of silver. However, many photons which reach a crystal are not absorbed, and many that are absorbed fail to contribute to the formation of a development center. The overall efficiency of latent image formation is dependent upon the structure of the crystal, the presence of sensitizing dyes and other addenda in the emulsion, temperature, and several other factors, but is essentially independent of grain size.

The efficiency of latent image formation is dependent on the rate of exposure. With low-intensity, long-duration exposure, the first silver atom formed may dissociate before succeeding ions are attracted to the site. Hence, exposing energy is wasted and the emulsion will exhibit low-intensity reciprocity failure. With very intense exposures of short duration, an abundance of sites compete for the available silver atoms with the result that many of the sites fail to grow sufficiently to become developable. In addition, such exposures frequently result in the formation of internal latent images which cannot be reached by conventional surface-active developers.

The Gurney-Mott theory was not universally accepted at first, and there is still uncertainty with regard to some details of the mechanism, but it prompted extensive research and continues to be the best available explanation of latent image formation and behavior. For additional information on the

Gurney-Mott theory *See* PHOTOCONDUCTIVITY; PHOTOLYSIS.

Development. Development is of two kinds, physical and chemical. Both physical and chemical developers contain chemical reducing agents, but a physical developer also contains silver compounds in solution (directly added or derived from the silver halide by a solvent in the developer) and works by depositing silver on the latent image. Physical development as such is little used, although it usually plays some part in chemical development. A chemical developer contains no silver and is basically a source of reducing agents which distinguish between exposed and unexposed silver halide and convert the exposed halide to silver. Developers in general use are compounded from organic reducing compounds, an alkali to give desired activity, sodium sulfite which acts as a preservative, and potassium bromide or other compounds used as antifoggants (fog is the term used to indicate the development of unexposed crystals; it is usually desirable to suppress fog).

Most developing agents used in normal practice are phenols or amines, and a classical rule which still applies, although not exclusively, states that developers must contain at least two hydroxyl groups or two amino groups, or one hydroxyl and one amino group attached ortho or para to each other on a benzene nucleus. Some developing agents do not follow this rule. *See* AMINE; PHENOL.

Alkalies generally used are sodium carbonate, sodium hydroxide, and sodium metaborate. Sulfite in a developer lowers the tendency for oxidation by the air. Oxidation products of developers have an undesirable influence on the course of development and may result in stain.

Developing agents and formulas are selected for use with specific emulsions and purposes. Modern color photography relies mainly on paraphenylenediamine derivatives. So-called fine-grain developers are made to reduce the apparent graininess of negatives. They generally contain the conventional components but are adjusted to low activity and contain a solvent for silver bromide. One fine-grain developer is based on *p*-phenylenediamine, which itself is a silver halide solvent. Some developers are compounded for hardening the gelatin where development occurs, the unhardened areas being washed out to give relief images for photomechanical reproduction and imbibition color printing. In a number of products, the developing agent is incorporated in the emulsion, development being initiated by placing the film or paper in an alkaline solution called an activator. Hardening agents may be included in developers to permit processing at high temperatures.

Monobaths are developers containing agents which dissolve or form insoluble complexes with the undeveloped silver halide during developing.

When film is developed, there is usually a period during which no visible effect appears; after this the density increases rapidly at first and then more slowly, eventually reaching a maximum. In the simplest case, the relation between density and development time is given by Eq. (1), where D is the density

$$D = D_x(1 - e^{-kt}) \qquad (1)$$

attained in time t, D_x is the maximum developable density, and k is a constant called the velocity constant of development.

After processes. These include fixing, washing, drying, reduction, intensification, toning, and bleaching.

Fixing. After the image is developed, the unchanged halide is removed, usually in water solutions of sodium or ammonium thiosulfate (known as hypo and ammonium hypo, respectively). This procedure is called fixing. Prior to fixing, a dilute acid stop bath is often used to neutralize the alkali carried over from the developer. The rate of fixing depends on the concentration of the fixing agent (typically 20–40% for hypo) and its temperature (60–75°F or 16–24°C). Hardeners may be added to the fixer to toughen the gelatin, though they tend to reduce the fixing rate. In so-called stabilization processes, the undeveloped silver salts are converted into more or less stable complexes and retained in the emulsion to minimize the processing time associated with washing. Solutions of organic compounds containing sulfur are often used for this purpose.

Washing and drying. Negatives and prints are washed in water after fixing to remove the soluble silver halide-fixing agent complexes, which would render the images unstable or cause stain. The rate of removal declines exponentially and can be increased by raising temperature and increasing agitation. The rate can be accelerated with neutral salt solutions known as hypo clearing aids, thus reducing washing times. Print permanence can be enhanced with toning solutions (especially selenium toners) too dilute to cause a change in image coloration. After washing, the materials must be dried as uniformly as possible, preferably with dust-free moving warm air, though some papers are often dried on heated metal drums. Drying uniformity is aided by adding wetting agents to the final wash water.

Reduction and intensification. Reduction refers to methods of decreasing the density of images by chemically dissolving part of the silver with oxidizers. According to their composition, oxidizers may remove equal amounts of silver from all densities, remove silver in proportion to the amount of silver present, or remove substantially more silver from the higher densities than from the lower. Intensification refers to methods of increasing the density of an image, usually by deposition of silver, mercury, or other compound, the composition being selected according to the nature of the intensification required.

Toning. The materials and processes for black-and-white photography are normally designed to be neutral in color. The coloration can be modified by special treatments known as toning. This may be done with special developers or additions to conventional developers, but toning solutions used after fixing and washing are more common. These solutions convert the silver image into compounds such as silver sulfide or silver selenide, or precipitate colored metallic salts with the silver image. Dye images can be obtained by various methods.

Bleaching. In reduction and in color processes and other reversal processes where direct positives are formed, the negative silver is dissolved out by an oxidizing solution, such as acid permanganate or dichromate, which converts it to a soluble silver salt. If a soluble halide, ferricyanide, or similar agent is present, the corresponding silver salt is formed and can be fixed out or used as a basis for toning and intensifying processes. Bleach-fixing solutions (blixes) are used to dissolve out silver and fix out silver halide simultaneously.

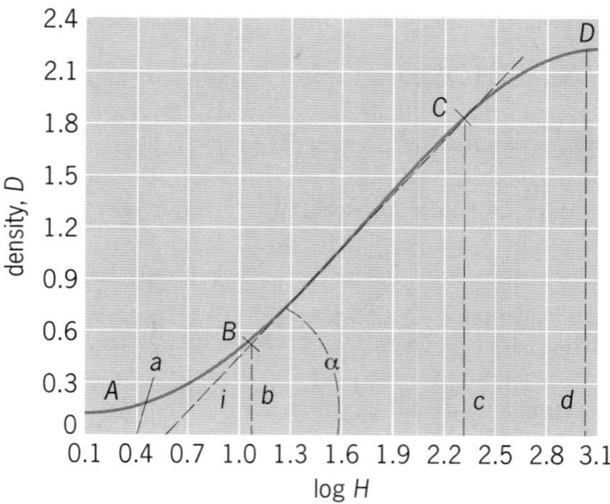

Fig. 3. Hurter and Driffield characteristic curve.

Sensitometry and Image Structure

Sensitometry refers to the measurement of the sensitivity or response to light of photographic materials. Conventional methods of measurement tend to mask the dependence of this response on factors which affect the structure of the photographic image. The simplest method of determining sensitivity is to give a graded series of exposures and to find the exposure required to produce the lowest visible density. Modern sensitometry relies on plotting curves showing the relation between the logarithm of the exposure H and the density of the silver image for specified conditions of exposure, development, and evaluation. Exposure is usually defined as the product of intensity and duration of illumination incident on the material, though it is more correct to define exposure as the time integral of illuminance I ($H = \int I dt$). Density D is defined as the logarithm of the opacity O, which in turn is defined as the reciprocal of transmittance T. If light of intensity I falls on a negative and intensity I' is transmitted, Eqs. (2) hold.

$$T = \frac{I'}{I} \qquad O = \frac{1}{T} = \frac{I}{I'} \qquad D = \log \frac{I}{I'} \qquad (2)$$

Illuminance is the time rate at which light is directed at a unit area of film, exposure is the amount of energy per unit area received by the film, and density is a measure of the extent to which that energy is first converted to a latent image and then, with an infusion of chemical energy from the developer, is amplified to form a stable or permanent silver image. Since density is dependent mainly on the amount of silver per unit area of image, and since the amount of energy required to form a latent image in each crystal is independent of crystal size, it takes more exposure to obtain a specified density with a fine-grain emulsion than a coarse-grain one. However, through advances in emulsion and developer chemistry, modern picture-taking films are roughly 25,000 times more sensitive than the gelatin dry plates introduced in 1880.

Hurter and Driffield curve. The relationship between density and the logarithm of exposure is shown by the Hurter and Driffield characteristic curve, which has three recognizable sections, the largest corresponding usually to the straight line B to C in **Fig. 3**. In this region, density is directly proportional to the logarithm of exposure. In the lower and upper sections of the curve, this proportionality does not apply. These nonlinear regions, AB and CD, are commonly called the toe and shoulder, respectively.

The characteristic curve is used to determine the contrast, exposure latitude, and tone reproduction of a material as well as its sensitivity or speed. It is obtained by using a light source of known intensity and spectral characteristics, a modulator which gives a series of graded exposures of known values, precise conditions for development, and an accurate means of measuring the resulting densities.

Sensitometric exposure. The internationally adopted light source is a tungsten-filament electric lamp operated at a color temperature close to 2850 K combined with a filter to give spectral quality corresponding to standard sensitometric daylight (approximately 5500 K). The light source and exposure modulator are combined to form a sensitometer, which gives a series of exposures increasing stepwise or in a continuous manner. Sensitometers are either intensity-scale or time-scale instruments, depending on whether the steps of exposure are of fixed duration and varying intensity or are made at constant intensity with varying duration. Special sensitometers are used to determine the response of materials to the short exposures encountered in electronic flash photography or in scanning recorders. The exposure may also be intermittent, that is, a series of short exposures adding up to the desired total, though continuous exposures are preferred to avoid intermittency effects. (Below a critical high frequency, the photographic material does not add up separate short exposures arithmetically.) The best sensitometers are of the continuous-exposure, intensity-scale variety, and the exposure time used should approximate that which would prevail in practice. *See* EXPOSURE METER; HEAT RADIATION.

Development. Development is carried out in a standard developer or in a developer specifically recommended for use with the material under test, for the desired time at a prescribed standard temperature, and with agitation that will ensure uniform development and reproducibility. The procedure may be repeated for several development times.

Densitometers. Photographic density is measured with a densitometer. Early densitometers were based on visual photometers in which two beams are taken from a standard light source and brought together in a photometer head. One beam passes through the density to be measured, the other through a device for modulating the intensity so that the two fields can be matched. Visual instruments are seldom used now except as primary reference standards. Most densitometers are photoelectric devices which rely on one or more standard densities which have been calibrated on another instrument.

The simplest photoelectric transmission densitometers are of the deflection type, consisting of a light source, a photosensitive cell, and a microammeter or digital readout, the density being obtained from the relative meter deflections or from readings with and without the sample in place. Other photographic densitometers employ a null system in which the current resulting from the transmission of the test sample is balanced by an equivalent current derived from light transmitted through a calibrated modulator. In another null method, the sample and modulator are placed in one beam and the modulator controlled so that the net transmittance of the combination is constant. Some densitometers are equipped with devices for plotting the characteristic curves automatically. The densitometry of colored images is done through a suitable set of color filters and presents special problems in interpretation.

When light passes through a silver image, some is specularly transmitted and some is scattered. If all the transmitted light is used in densitometry, the result is called the diffuse density. If only the light passing directly through is measured, the density is called specular. The specular density is higher and related to diffuse density by a ratio known as the Callier Q factor. The Q of color films is negligible.

The density of paper prints or other images on reflective supports is measured by reflection. The reflection density is $D_R = \log(1/R)$, where R is the ratio of light reflected by the image to that reflected by an unexposed area of the paper itself or, better, by a reference diffuse light reflector. As with transmission densitometry, the angular relationships between incident and reflected light must be standardized.

Contrast. If extended, the straight-line portion of the characteristic curve BC intersects the exposure axis at the inertia point i on Fig. 3, and the angle α is a measure of contrast. The tangent of α, that is, the slope of the line BC, is known as gamma (γ). As development time or temperature increase, gamma increases, eventually reaching a maximum (**Figs. 4** and **5**). The projection of the straight line BC of Fig. 3 onto the exposure axis at bc is a measure of latitude,

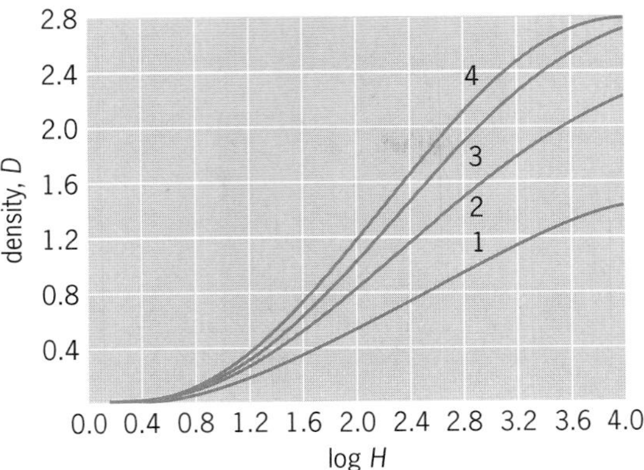

Fig. 4. Characteristic curves for development times increasing in the order 1, 2, 3, 4; horizontal scale is in lux seconds.

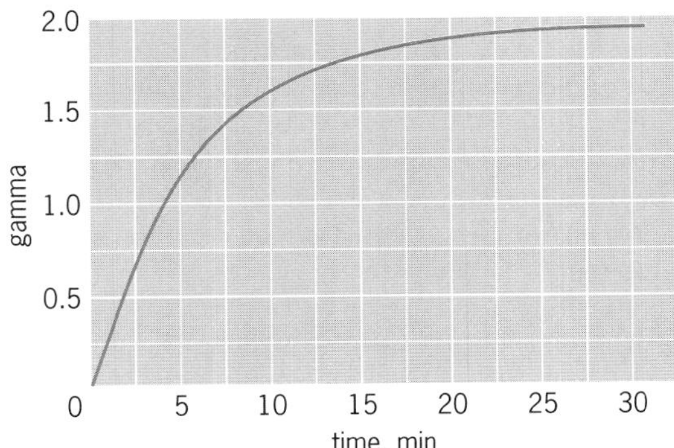

Fig. 5. Gamma versus time of development curve.

and the distance ad represents the total scale, that is, the whole exposure range in which brightness differences in the subject can be recorded.

Another measure of contrast, contrast index, is more useful than gamma in pictorial photography because it is the average gradient over the part of the characteristic curve used in practice. It is determined by measuring the slope of the line joining a point in the toe of the curve and one at a higher density that are separated by a distance related to a specified difference in exposure or density.

Sensitivity. The sensitivity or speed of a photographic material is not an intrinsic or uniquely defined property. However, it is possible to define a speed value which is a numerical expression of the material's sensitivity and is useful for controlling camera settings when the average scene brightness is known. Speed values are usually determined from an equation of the form $S = k/H$, where S is the speed, k is an arbitrary constant, and H is the exposure corresponding to a defined point on the characteristic curve.

Historically, Hurter and Driffield speeds were obtained by dividing 34 by the exposure corresponding to the inertia point, but most modern definitions use much smaller constants and the exposure corresponds to a point in the toe region which is at some arbitrary density above base plus fog or where

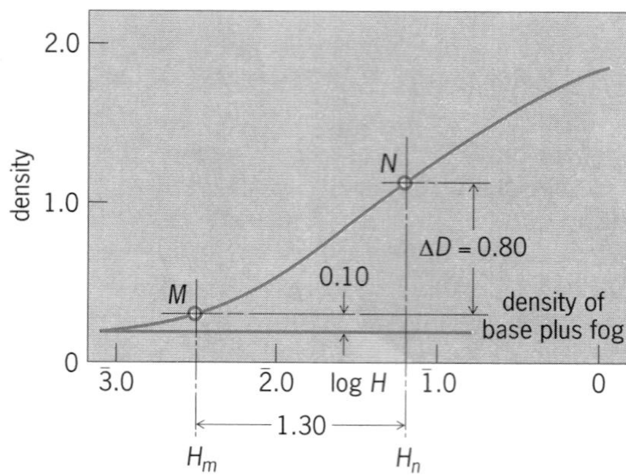

Fig. 6. Determination of the speed of a photographic film according to the 1979 American Standard Method, now adopted internationally. ISO/speed is $0.8/H_M$ when film is processed to achieve gradient defined by points *M* and *N*.

the gradient of the curve has reached some specified value. The goal is always to establish speed values which lead either to correctly exposed transparencies or to negatives that can be printed readily to form pictures of high quality. The constant k usually incorporates a safety factor to assure good exposures even with imprecise exposure calculations.

The present international standard for determining the speeds of black-and-white pictorial films is shown in **Fig. 6**. This standard has been adopted by the American National Standards Institute (ANSI) and the International Organization for Standardization (ISO). Film speeds determined by this method are called ISO speeds and correspond closely to values long known as ASA speeds. Point M is at a density 0.10 above base-plus-fog density. Point N lies 1.3 log H units from M in the direction of increased exposure. Developing time in a standard developer is adjusted so that N also lies at a density 0.8 above the density of M. Then, the exposure H_M (in lux seconds) corresponding to M is the value from which the ISO speed is computed according to $S = 0.8/H_M$. Exposure indexes or speeds for films and papers used in printing and for films and plates used in various technical applications are often determined in the same general manner, but the criteria for required exposure, the processing conditions, and the constant k are usually quite different, reflecting the nature of the application for which the material is intended.

Spectral sensitivity measures the response of photographic emulsions as a function of the exposing wavelength. It is usually determined with a spectrosensitometer, that is, a spectrograph with an optically graded wedge or step tablet over the slit. The contour of the spectrogram for a constant density is related to the spectral response of the material and spectral output of the light source. Ideally, spectral sensitivity curves are adjusted to indicate the response to a source of constant output at all wavelengths. These are known as equal-energy curves.

Reciprocity effects. Defining exposure as the product of illuminance and time implies a simple reciprocal relationship between these factors for any given material and development process. As noted above, the efficiency of latent image formation is, in fact, affected by very long, very short, or intermittent exposures. Variations in the speed and contrast of an emulsion with changes in the rate of exposure are therefore called reciprocity effects. In general, these effects are not dependent upon the exposing wavelength for a single emulsion, but they may vary among the individual emulsions of color materials, thereby affecting color rendition.

Tone reproduction. The relation between the distribution of luminance in a subject and the corresponding distribution of density in a final print is referred to as tone reproduction. It is affected by the sensitometric characteristics of the materials, including their nonlinear aspects, and by subjective factors as well. The exposure and processing of both the negative and print materials are manipulated by skilled photographers to achieve visually appealing results.

Image structure. The granular nature of photographic images manifests itself in a nonhomogeneous grainy appearance which may be visible directly and is always visible under magnification. Granularity, an objective measure of the sensation of graininess, is generally greater in high-speed emulsions and depends also on both the nature of the development and the mean density of the image area being measured. Producing emulsions with increased speed without a corresponding increase in granularity is a constant goal. Graininess in prints increases with increases in magnification, contrast of the print material, and (in black-and-white materials) with mean density of the negative.

The ability of a film to resolve fine detail or to record sharp demarcations in exposure is limited by the turbidity (light-scattering property), grain size, and contrast of the emulsion, and is affected by several processing effects. Resolving power is a measure of the ability to record periodic patterns (bars and spaces of equal width) and is usually expressed as the number of line pairs (bar and space pairs) per millimeter which can just be separated visually when viewed through a microscope. Resolving power values of 100 to 150 line pairs/mm are common but may range from less than 50 to more than 2000.

Sharpness refers to the ability of an emulsion layer to show a sharp line of demarcation between adjacent areas receiving different exposures. Usually such an edge is not a sharp step but graded in density to an extent governed by light scattering (turbidity) and development effects. Subjective assessments of sharpness correlate with a numerical value known as acutance, which is a function of both the rate of change in density and the overall difference in density across a boundary.

A different and generally more useful measure of the ability of a photographic material to reproduce image detail is called the modulation transfer function (MTF). If a sinusoidal distribution of illuminance is applied to the material, the resulting variation in density in the processed image will not, in general, coincide with that predicted by the characteristic curve, again because of localized scattering and development effects. (These effects are excluded in conventional sensitometry.) Modulation transfer is the ratio of the exposure modulation that in an ideal emulsion would have caused the observed density variation to the exposure modulation that actually produced it. This ratio is a function of the spatial frequency of the exposure distribution, expressed in cycles per millimeter. The MTF of a film may be combined with a similar function for a lens to predict the performance of a lens-film combination.

Photographic Apparatus

The camera is the basic exposing instrument of photography. In cameras having shutters within the lens, the entire image is recorded simultaneously; in those with focal plane shutters, exposure is governed by a slit which moves rapidly across the film so that the time required to expose the whole image is necessarily greater than the exposure duration for any line element parallel to the slit. Another important class of exposing instruments includes recorders and plotters in which individual picture elements (spots) are exposed sequentially in some orderly fashion. Other important instrumentation in photography includes means for lighting the subject, devices for handling the photographic materials during processing, equipment for printing by contact or projection, and means for viewing, storing, and retrieving photographs. *See* CAMERA.

Recorders and plotters. The information associated with an image (normally a two-dimensional array of spots of varying density and color) may be acquired in electronic form either directly, as with an electronic camera, or by computation from mathematical descriptions of an object or pattern that has not existed in fact, or indirectly by scanning an original photograph with an electronic sensor. In these cases the information may be retained in electronic or magnetic memory until recalled for viewing. It is often required that the image then be converted to conventional photographic form. Continuous-tone images are usually reconstructed by writing with a single spot which moves along the lines of a raster. Both the number of spots along each line and the number of lines in the height of the image are set by design and may range from several hundred to more than 10,000, depending upon the quality desired. In certain specialized recorders, several independently modulated spots are caused to scan and expose in parallel.

One form of raster imaging recorder consists of a cathode-ray tube or electronic panel display and a camera designed to relay the image to the film or paper, which usually is held stationary. In others, a small spot of light derived from a laser, light-emitting diode, xenon arc, or other intense source is modulated and caused to scan across the photographic material, often with the use of a rotating mirror. In these devices, one of the scanning motions may be achieved by moving the film with respect to the exposing system.

When one is recording maps or engineering drawings, it is often advantageous to use plotters which operate in vector rather than raster mode so that curved lines or straight lines not parallel with the scan lines of a raster will be continuous and time will not be wasted scanning large areas of the photographic materials that are to receive no exposure. Cathode-ray tubes and electron beam recorders (in which the electrons of a specialized cathode-ray tube impinge directly on the film instead of a phosphor screen) are especially useful for vector plotting because there is no mechanical inertia to be overcome when deflecting the electron beam.

The photographic materials used in image recorders and plotters are properly selected with regard for the short exposures which prevail at individual picture elements, the wavelength of the exposing source, the exposure range determined by the modulator, and the application or end use of the recorded image.

Darkroom equipment. Other important instrumentation includes devices for handling photographic materials during processing, printing, viewing, storage, and retrieval. Until processing is virtually completed, most films, plates, and papers must be handled in a darkroom or light-tight container. The darkroom may range from total darkness to subdued illumination from safelights, that is, lamphouses with filters to provide light of a color that will not fog the material in a reasonable time. Other equipment used for manual processing includes sinks with running water and drains; tanks and hangers for processing plates or sheet films vertically; light-tight tanks for processing roll films held in spiral reels; flat trays for processing materials in sheet form individually; thermometers, mixing valves, and other devices for controlling solution temperatures; special containers for washing negatives and prints; dryers, ranging from simple clips for hanging negatives in open air to cabinets with forced warm air and heated drums; clocks and preset timers; printers and enlargers; exposure meters and focusing devices for enlargers; and print trimmers.

Many photographic materials (especially color films and papers) require strict adherence to carefully designed processing cycles. In many technical applications, the quality of the recorded image and the information it represents are critically dependent upon uniform processing conditions. When processing films and prints in commercial quantities, as in the photofinishing, motion picture, and microfilm industries, economic considerations preclude manual darkroom operations. All of these needs are met through the use of various processing machines and automatic or semiautomatic printers. Processing machines commonly use rollers to transport the material through tanks or other stations where developers, fixers, monobaths, activators, bleaches, or other processing chemicals are brought into contact with the emulsion and backing layers by immersion, application of a thin layer, or spraying. Solution temperatures are generally elevated to achieve reduced times. Washing and drying stations complete the processes.

Contact and projection printers. Negatives provide the primary records in most black-and-white photography and a large fraction of color photography. In many applications, the negative is more useful than a print derived from it. However, positive prints from original negatives or positive color transparencies are often required. Such prints are made in contact printers or projection printers (enlargers).

Contact printers are boxes containing lamps, diffusing screens, and a glass top on which a negative is placed. A sheet of printing paper is placed directly over the negative, usually emulsion to emulsion, and a lid with felt or other spongy material on its face is closed to assure good contact through the application of uniform pressure. Alternatively, the negative and print material may be placed in a vacuum printing frame and exposed to a distant point source to retain edge sharpness.

Projection printers, or enlargers, are optical projectors consisting of a lamphouse and light source, a condenser or diffusing sheet to assure uniform illumination, a holder for the negative, a projection lens designed to give optimum quality at relatively low magnifications, a board or easel to hold the sensitized paper, and means for holding these components when aligned and spaced to provide sharp focus at the desired magnification. Color filters are included for color printing. *See* OPTICAL PROJECTION SYSTEMS.

Color negatives are usually analyzed by sophisticated optoelectronic systems to determine the exposures needed for optimum color balance and mean density. In some systems the original is scanned and digitized, the information is processed electronically to achieve effects not possible with conventional printers, and the result is reconstructed as a new negative, from which prints can be made in quantity, or directly as a final print. In either case, the prints are intended to have enhanced esthetic or technical qualities. Robert D. Anwyl

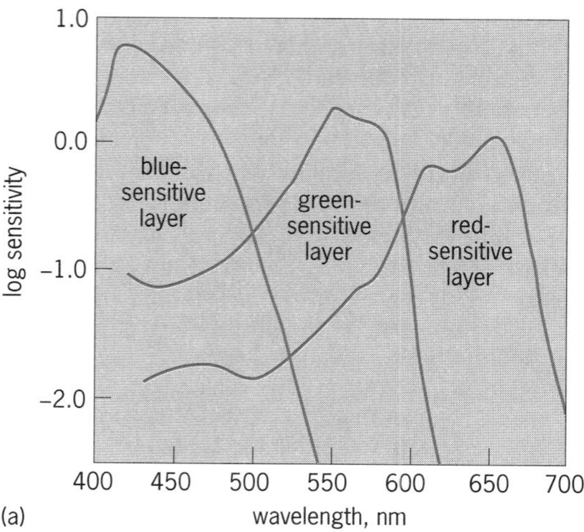

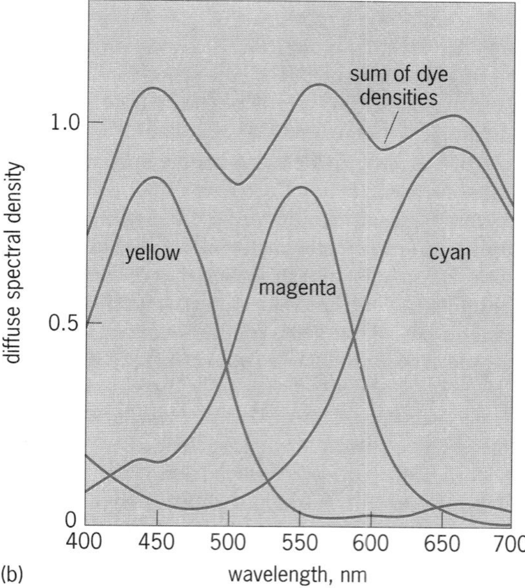

Fig. 7. Subtractive color material. (*a*) Spectral sensitivities of blue-, green-, and red-sensitive emulsions. (*b*) Corresponding spectral densities of dyes that form yellow, magenta, and cyan images. Sum of dye densities, corresponding to uniform exposure at all visible wavelengths, is not flat but yields visually neutral density.

Color Photography

Color photography is the photographic reproduction of natural colors in terms of two or more spectral components of white light. In 1665 Isaac Newton observed that white light could be separated into a spectrum of colors, which then could be reunited to form white light. In 1861 James Clerk Maxwell prepared positive transparencies representing red-, green-, and blue-filter exposures, projected each through the respective "taking" filter, and superimposed the three images to form a full-color image on a screen.

Maxwell's demonstration was the first example of additive color photography. Red, green, and blue are called additive primary colors because they add to form white light. Their complements are the subtractive primary colors, cyan, magenta, and yellow. A cyan, or minus-red, filter absorbs red light and transmits blue and green; a magenta, or minus-green, filter absorbs green and transmits red and blue; and a yellow, or minus-blue, filter absorbs blue and transmits red and green. Subtractive color processes produce color images in terms of cyan, magenta, and yellow dyes or pigments (**Fig. 7**). Two-color subtractive processes (cyan and red-orange), which provide a less complete color gamut, have also been used. *See* COLOR; COLOR FILTER.

Additive color photography. Additive color is most widely represented by color television, which presents its images as arrays of minuscule red, green, and blue elements.

The screen plate system of additive photography uses a fine array of red, green, and blue filters in contact with a single panchromatic emulsion. Reversal processing produces a positive color transparency comprising an array of positive black-and-white image elements in register with the filter elements. Some early screen processes have a hyperfine color stripe screen integral with a silver transfer film.

Lenticular additive photography is based on black-and-white film embossed with lenticules. While the camera lens photographs the scene, the lenticules image onto the film a banded tricolor filter placed over the camera lens. The resulting composite image is reversal-processed and projected through the same tricolor filter to reconstruct the original scene in color.

Subtractive color photography. Many types of subtractive color negatives, color prints, and positive transparencies are made from multilayer films with separate red-, green-, and blue-sensitive emulsion layers.

In chromogenic or color development processes, dye images are formed by reactions between the oxidation product of a developing agent and color couplers contained either in the developing solution or within layers of the film. For positive color transparencies, exposed silver halide grains undergo nonchromogenic development, and the remaining silver halide is later developed with chromogenic developers to form positive dye images. Residual silver is removed by bleaching and fixing.

In the first commercially successful subtractive color transparency film (**Fig. 8**), reversal dye images are formed in successive color development steps, using soluble couplers. Following black-and-white development of exposed grains, the remaining red-sensitive silver halide is exposed to red light, then developed to form the positive cyan image. Next, the blue-sensitive layer is exposed to blue light and the yellow image developed, and finally the green-sensitive emulsion is fogged and the magenta image developed.

Simpler reversal processing is afforded by films with incorporated nondiffusing couplers that produce three dye images in a single color development step.

Color negative films with incorporated couplers are developed with a color developer to form cyan, magenta, and yellow negative images for use in printing positive color prints or transparencies. Colored couplers in color negative films provide automatic masking to compensate for the imperfect absorption characteristics of image dyes. For example, a magenta dye, ideally minus-green, absorbs not only green but also some blue light. The magenta-forming coupler is colored yellow, and in undeveloped areas it absorbs as much blue light as the magenta dye absorbs in developed areas, so that blue exposure through the negative is unaffected by the blue absorption of the magenta dye.

Chromogenic films designed for daylight exposure commonly have an ultraviolet-absorbing dye in the protective overcoat to prevent excessive exposure of the blue-sensitive emulsion. (Ultraviolet exposure of the remaining layers is blocked by the yellow filter layer.) Advances in silver halide, dye, and processing chemistry have resulted in chromogenic films with progressive improvements in speed, color rendition, image quality, and dye stability over early products of this type.

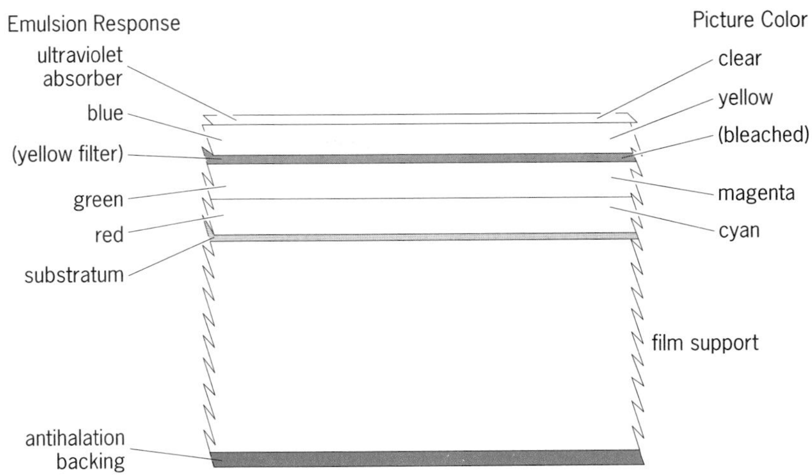

Fig. 8. Schematic cross section of subtractive color transparency film, showing configuration of three emulsion layers with (left) the spectral response of each emulsion and (right) the color of the dye image formed in each layer through chromogenic development. Protective overcoat contains ultraviolet absorbing dye to prevent overexposure of blue-sensitive emulsion with daylight or electronic flash illuminants.

One-step color photography. The introduction in 1963 of a single-structure film and a single-step process marked a significant branching in the history and technology of color photography. The availability of instant color photographs directly from the camera provided a useful alternative to conventional color films and print materials, which require a printing step and separate darkroom processing of each material, particularly in applications where enlargement and multiple prints are not primary requirements.

In the Polacolor process and later one-step, or instant, processes, color prints are produced under ambient conditions shortly after exposure of the film. Processing takes place rapidly, using a reagent—sometimes called an activator—provided as part of the film unit. Important concepts include the design of cameras incorporating processing rollers, the use of viscous reagents sealed in foil pods to provide processes that are outwardly dry, and traps to capture and neutralize excess reagent. The viscous reagent makes possible accurately controlled, uniform distribution. The closed system provides fresh reagent for each picture unit and permits use of more reactive chemicals than are practical in conventional darkroom processes.

Rather than forming dyes during processing, the first one-step process introduced preformed dye developers, compounds that are both image dyes and developers. Development of exposed silver halide crystals causes immobilization of the associated dye developers while dye developers in unexposed regions diffuse to a polymeric image-receiving layer. The positive receiving sheet includes a polymeric

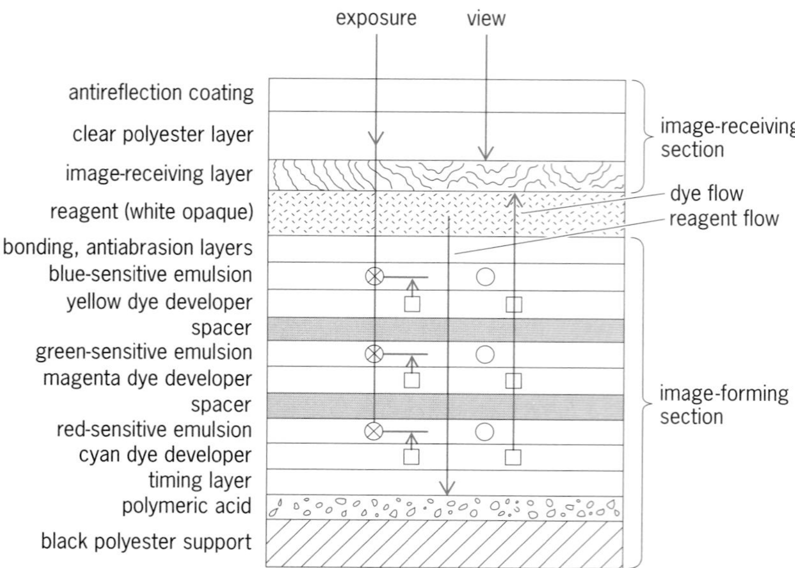

Fig. 9. Schematic cross section of integral one-step film units that provide full-color prints within 2 min after exposure. The dye image diffuses from exposed areas to the image-receiving layer and is viewed against the reflective white pigment of the reagent layer.

acid layer that neutralizes residual alkali. The full-color print is ready for viewing when the negative and positive sheets are stripped apart after 60 s. The process, like most peel-apart systems, is adaptable to large-format photography. A film based on similar principles yields color transparencies suitable for overhead projection.

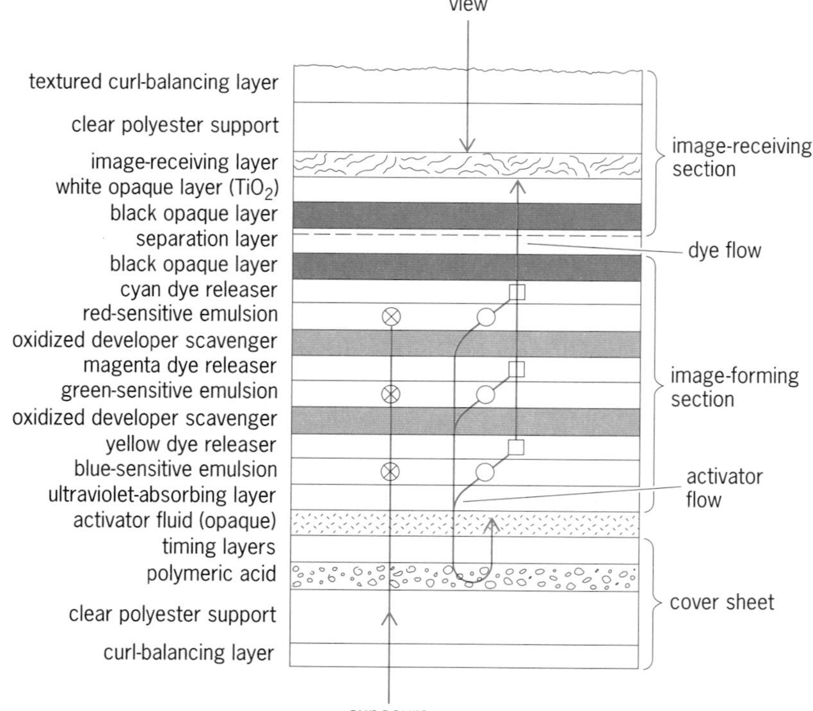

Fig. 10. Schematic cross section of integral one-step film. Dyes released from unexposed areas of image flow to image-receiving layer for viewing against white opaque layer. Image-receiving section may then be separated (at broken line) from image-forming section to obtain a thin print.

The first fully integral one-step color film was also based on dye developers (**Fig. 9**). The film unit comprises two polyester sheets, one transparent and one opaque, with all of the image-forming layers between them. The light-sensitive layers are exposed through the transparent sheet, and the final image is viewed through the same sheet against a reflective white-pigmented layer that is formed by the reagent. Cameras for the system automatically eject the film immediately, driving the film unit between rollers to spread the reagent within the film. Although still light-sensitive when ejected, the emulsion layers are protected from exposure to ambient light by the combined effects of the white pigment and special opacifying dyes included in the reagent. Timing of the process is self-controlled and there is no separation of the sheets. Images on materials which are exposed and viewed from the same side will be laterally reversed unless an odd number of image inversions is introduced in the exposing process. Hence, cameras intended for use with the film incorporate one or three mirrors. The use of three mirrors permits cameras of compact design.

Other instant color films incorporate image-forming and -receiving layers between two polyester sheets, both clear (**Fig. 10**). Direct positive emulsions are selectively exposed through the clear cover sheet and an ultraviolet-absorbing layer. Following exposure, an activator containing potassium hydroxide, a developing agent, and other components is distributed uniformly between the cover sheet and the integral imaging sections as the film unit is ejected from the camera. Carbon black in the activator blocks exposure of the emulsion layers while the activator penetrates them and develops the unexposed silver halide crystals. Oxidized developer resulting from development reacts with dye releasers in an adjacent layer and, in the presence of an alkali, liberates preformed dyes. (Scavenger layers prevent dye release from the wrong layers.) The released dyes migrate through two or more opaque layers to an image-receiving layer. Timing of the process is governed by the neutralizing action of a polymeric acid when the activator penetrates the timing layers.

Because the images formed in the film are viewed through the second clear plastic sheet, they are right-reading. Thus, cameras designed to accept these products need contain no mirrors, though two are incorporated in many such cameras to reduce overall camera size. All layers are retained as an integral unit in the film, but the film (Fig. 10) embodies two extra layers and is partially slit during manufacture so that the picture and border section of the image-receiving section may be separated from the unit after processing is complete. The image-forming section, including the activator pod and trap, is discarded, leaving a thin print. A color slide film produces transparencies instead of thin prints. Following separation from the image-forming section, the transparencies are trimmed and inserted automatically into 2 by 2 in. (5 by 5 cm) slide mounts with a slide-mounting accessory.

Color print processes. Color prints from color negatives are made primarily on chromogenic print materials by using simplified processes similar to those used for the original negatives. Color transparencies are made from such negatives using special print films which are processed like the original negatives. Chromogenic reversal materials, such as specially treated paper and duplicating films, are used to make prints or duplicate transparencies from positive color originals. Alternatively, a color negative intermediate film may be derived from the original positive transparency and used to make multiple prints with negative-working papers.

Color prints and transparencies can be made from original color negatives and transparencies in an image transfer system which uses a single, strongly alkaline activator for processing. As appropriate, a negative- or positive-working film is exposed to the original. After 20-s immersion in the activator, the film is laminated to a reflective (paper) or transparent (plastic) receiver sheet. The two parts are separated minutes later. Other color print films based on release and diffusion of preformed dyes in the presence of alkaline activators include a single-sheet material and a two-sheet graphic arts product.

Color prints and transparencies are also made from original positive transparencies by silver dye bleach processes. The processes are based on the destruction of dyes rather than their formation or release. Following development of silver images in red-, green-, and blue-sensitive emulsion layers, associated subtractive dyes are selectively destroyed by a bleach reaction catalyzed by the negative silver, leaving a positive color image in place. Residual silver salts are removed by fixing.

Dye transfer prints are made by sequential imbibition printing of cyan, magenta, and yellow images from dyed gelatin relief matrices onto gelatin-coated paper or film. Panchromatic matrix films are exposed from color negatives through appropriate color separation filters; blue-sensitive matrix films are printed from black-and-white separation negatives. The images formed by dye transfer and dye bleach exhibit good dye stability.

Color motion pictures. Many of the techniques described above have been used in motion picture processes. Since their introduction in the 1950s, chromogenic negative and positive films have become the most widely used motion picture production materials. Vivian K. Walworth; Robert D. Anwyl

Electronic Devices and Technology

The photographic art has been influenced progressively, especially since 1960, by electronic devices and technology. This influence has affected virtually every element of the photographic process, including exposure control, automatic focusing of cameras, process control, image enhancement through intervention in the printing step, and even acquisition of the original image information.

Early forms of electronic acquisition include the scanning of photographs for transmittal to another location (facsimile) or producing color separations in graphic arts scanners. The scanning of original subjects with television cameras, with point detectors (and oscillating mirrors) mounted in remote-sensing aircraft or spacecraft, or with scanning electron microscopes were forerunners to the charge-coupled arrays and other image-sensing devices that take the place of films or plates in some cameras. The image information may be stored on magnetic media or electronically and presented for initial viewing on a video monitor (cathode-ray tube), but it may also be converted to conventional photographic form by photographing the monitor or with other types of line-scan image recorders. Whenever the image information is in electronic form, it can be manipulated to alter the tonal and chromatic relationships in the resulting image. This manipulation may be introduced to create a print that appears to be more like the original scene than could be achieved without electronic intervention, or it may be used to create images clearly different from the original. The result may be visually arousing in an esthetic sense or reveal image information of scientific or technical significance. At the same time, it is possible to suppress grain noise and enhance edge sharpness. *See* CATHODE-RAY TUBE; CHARGE-COUPLED DEVICES; FACSIMILE; IMAGE PROCESSING; IMAGE TUBE (ASTRONOMY); TELEVISION CAMERA.

The rapid growth in electronic technology and its impact on virtually every other scientific or technical discipline were made possible by the development of integrated circuits, sensor arrays, and other microelectronic devices. These devices, in turn, were made possible by extending the art of microphotography and by using high-resolution photographic plates, photosensitive resists, and other specialized photographic products. Robert D. Anwyl

Digital Photography and Electronic Image Capture

The process of electronic acquisition, the equivalent of taking a photograph, is often referred to as image capture. In the graphic arts industry, the two primary devices for image capture are electronic scanners and digital cameras.

Light intensity is detected in a scanner or digital camera by an electronic photosensor. This is normally a photomultiplier tube (PMT) or charge-coupled device (CCD), although complementary metal-oxide silicon (CMOS) devices are beginning to appear in some systems. *See* PHOTOMULTIPLIER.

When photons strike the sensor, they give up energy. This causes electrons to be emitted, turning the energy of the photons into electrical energy. The number of electrons that are emitted can be measured to determine how many photons struck the capture element, and from this the scanner can generate a value for the intensity of light arriving from the point on the original being analyzed.

Digitization of images. The aim of the digitization stage is to capture all the information from an original that will be needed in the reproduction and convert it into an array of binary numbers that a computer can process. The human visual system actively seeks cues that will give it information about the objects within the visual field, and a reproduction of an image that contains a large amount of detail is almost always preferred to one in which some of the detail has been lost. The more information that the reproduction contains about the original scene—the objects in it, their colors, textures—the more realistic the reproduction appears. *See* IMAGE PROCESSING.

There is, however, a limit to how much information can be used in the reproduction, and acquiring more than this simply makes files (stored binary image data) larger than necessary. Every subsequent file operation will take longer, resulting in a serious loss of productivity. When scanning an original, then, the objective is to capture as much information as possible, while making sure that file sizes are as small as possible. *See* ABSTRACT DATA TYPE; COMPUTER STORAGE TECHNOLOGY.

Electronic scanners. Scanning began to eclipse photographic color separation techniques in graphic arts reproduction during the 1970s, bringing automation to what had been a manual operation. The initial advantage of using a scanner instead of a process camera was that color correction was programmed into the circuitry of the scanner and the laborious retouching of films by masking and etching could be almost eliminated.

The parts of a scanner have functionality similar to the eye and include an optical system to focus light onto the image plane, light-sensitive receptors, and filters that adjust the spectral sensitivity of the receptors and allow three color channels of the image to be recorded separately.

The first color scanners were analog rather than digital, and the image was exposed onto film by the recording unit at the same time as it was being read by the analyze unit. The development of digital scanners, introduced by Crosfield in 1975, created the possibility of digital storage and processing of the image before output to film. Since then, scanner quality and productivity have increased enormously. Preview stations have been added, modular and multistation scanners have been developed to maximize throughput, and workstations for color image assembly and planning have been incorporated.

At the same time that graphic arts scanners evolved into highly efficient and consistent workhorses for quality prepress, businesses began to identify a need for small, simple scanning devices that could be used in the office environment. These flatbed scanners use simple solid-state electronics to record the image instead of the sensitive photomultiplier tubes found in high-end scanners. With their resolution and sensitivity rapidly improving, flatbed scanners soon became of interest to the emerging desktop publishing market. Since then, the quality and price performance of flatbed scanners has reached the point where they are able to challenge specialized commercial drum scanners in the graphic arts market.

The majority of scanners produced today are compact desktop machines. They include flatbed scanners intended mainly for reflection copy, transparency scanners for 35-mm and medium-format transparencies (occasionally taking transparencies up to 4 × 5 in.), and desktop drum scanners similar to the larger commercial systems that accept both reflection copy and transparencies. High-end scanners, designed for trade shops and commercial printers, are normally built as large drum scanners with internal programs for performing color conversions. Some have direct connections to film recorders. High-end scanners are relatively costly, but they generate high-quality output and tend to have higher productivity. *See* COMPUTER PERIPHERAL DEVICES.

Drum scanners. Drum scanners are equipped with one photomultiplier tube (PMT) for each of the red, green, and blue filter signals (with an additional tube for the unsharp masking signal on some systems). The original transparency is mounted on the drum. A beam of light is transmitted from inside the drum, through the transparency, and onto the scanner optics. In the case of reflection copy, light is reflected from the surface of the copy.

By rotating the drum at high speed and moving the analyze head slowly along the drum's axis, the entire surface of the transparency is read in a helical scan.

After being split into red, green, and blue beams by filters, the light strikes the end of the photomultiplier tube. The photomultiplier tube consists of a photocell inside a vacuum tube, with a light-sensitive cathode at one end and an anode at the other (**Fig. 11**). The photocathode emits electrons as it is struck by photons in the light beam. A series of electrodes amplifies the electron stream until it reaches the tube's anode, and the amplification effect allows the PMT to read small changes in lightness with great accuracy over a large density range. The smallest number of photons that can be detected (indicating the darkest areas of the original) is limited by the amount of noise present in the system. The maximum density (degree of opacity) that a PMT can detect is between 3.5 and 4.0, depending on the amount of stray light present. *See* PHOTOTUBE.

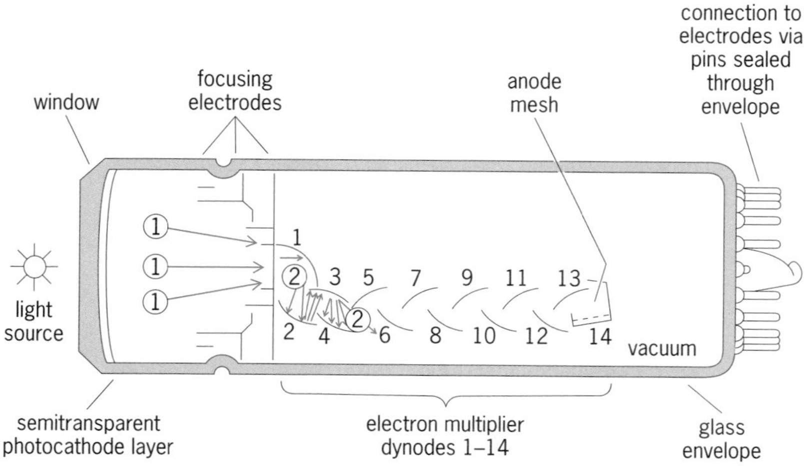

Fig. 11. Photomultiplier tube.

Flatbed scanners. Solid-state sensors (CCD or CMOS) are found in many different kinds of electronic equipment, including camcorders, fax machines, and digital cameras. A CCD consists of an array of tiny electrodes packed closely together on a layer of silicon. When a photon strikes an electrode, it causes an electron to be emitted from the silicon layer. Electrons are tagged with their position and transmitted along pathways to an analog-to-digital converter (ADC), where the voltage for each pixel is measured and given a digital value. The tagging of the electrons to the location of individual pixels is called charge coupling. *See* ANALOG-TO-DIGITAL CONVERTER.

The CCD can detect the presence of a single photon, but the interference caused by stray light, system noise, and crosstalk (interference between the signals produced by adjacent electrodes) is higher than that of the photomultiplier, and thus the ability of CCDs to resolve detail in deep shadow areas tends to be limited. The density range is typically up to 3.0, although a D_{max} of up to 3.5 or even 4.0 is possible. The difference between 3.0 and 3.5 may not seem like very much, but because density is a log scale it implies a threefold increase in sensitivity to detail in dark areas of the original.

The analog-to-digital converter (ADC) turns the continuously varying electron levels into digital values, and the number of bits output by the ADC defines the number of gray levels captured. An 8-bit ADC can output 256 gray levels for each color, but since tonal compression will occur after digitization, the ADC should ideally be able to handle more than 8 bits per color. Some systems employ CCDs that can detect up to 12 bits and then effectively resample down to 8 bits, while others retain all 12 bits for later manipulation.

In a CCD scanner, the data are captured in parallel, and the travel of the scanner head depends on the way the CCD elements are arranged. If the elements are in a row (or linear array), the scanner head makes a single pass over the original for each filter color.

Many scanner CCDs are arranged in a row for each color (a trilinear array), which requires a single pass to scan the image. A two-dimensional array captures the whole image in a single snapshot without the need to move either the sensors or the original. Low-resolution, two-dimensional arrays are used in devices such as camcorders, but they are costly to fabricate in the higher resolutions required in the graphic arts. High-resolution, two-dimensional CCD arrays are found mainly in digital cameras.

When a solid-state sensor is arranged as a two-dimensional array, there is a design choice between a single-shot or three-shot system. In the former, the available elements are divided between the three colors: red, green, and blue. In a three-shot system, the full number of elements on the chip are available for each shot, and the resolution will be higher.

A two-dimensional array can generate considerable amounts of heat during operation, and this can cause noise in the resulting images. In some systems the sensor chip is cooled to minimize this problem.

Fig. 12. High-resolution digital camera backfitted to a conventional large-format camera. (*Phase One*)

Digital cameras. Like conventional cameras, digital cameras come in compact, single-lens reflex, and large-format varieties (**Fig. 12**). Low-resolution compacts are useful for producing classified advertisements and tend to have relatively simple optics, image-sensing electronics, and controlling software. *See* CAMERA.

Digital cameras are often based on existing single-lens reflex camera designs with the addition of CCD backs and storage subsystems. The capture resolution of these cameras is ideal for news photography and other applications with similar quality requirements.

Large-format systems can take the form of digital backs that can be used in conjunction with existing large-format cameras, providing an interchangeable alternative to a film back. Or they can be designed from the start as digital cameras. These large-format systems have adequate resolution for most graphic arts purposes, and in most cases do not require their own storage subsystems since the image is downloaded directly to a host computer. They are most suitable for studio work and are widely used in catalog production, where the volume of product shots means that large savings can be made on film and processing.

When using a digital camera, the level and color temperatures of the lighting will inevitably change from shot to shot. A camera will have a fixed ASA/ISO speed equivalent, and if the brightness of the scene is higher than the camera is designed for, blooming can result where the charge induced in one element spills into adjacent elements. Thomas Destree

Bibliography. A. A. Blaker, *Photography: Art and Technique*, 2d ed., 1988; G. Craven, *How Photography Works*, 1986; G. T. Eaton, *Photographic Chemistry*, 2d ed., 1984; P. Green, *Understanding Digital Color*, 2d ed., 1999; T. Grimm, *Basic Book of Photography*, 4th ed., 1997; T. Grimm, *The Basic Darkroom Book*, 3d ed., 1999; G. M. Haist, *Modern Photographic Processing*, 2 vols., 1979; M. J. Langford, *Advanced Photography*, 6th ed., 1998; M. J. Langford, *Basic Photography*, 6th ed., 1997; R. P. Lovelend, *Photomicrography: A Comprehensive Treatise*, 2 vols., 1970, reprint 1981; *Photographic Imaging and Electronic Photography* (Technical Pocket Book), 1994; S. F. Ray, *Photographic Chemistry and Processing*, 1994; S. F. Ray, *Photographic Data* (Technical Pocket Book), 1994; S. F. Ray, *Photographic Technology and Imaging Science*, 1994; L. D. Streobel (ed.), *Basic Photographic Materials and Processes*, 2d. ed., 2000; L. D. Stroebel (ed.), *The Focal Encyclopedia of Photography*, 3d ed., 1996.

Photoionization

The ejection on one or more electrons from an atom, molecule, or positive ion following the absorption of one or more photons. The process of electron ejection from matter following the absorption of electromagnetic radiation has been under investigation for over a century. The earliest measurements involved the irradiation of metal surfaces by ultraviolet radiation. The theoretical interpretation of this phenomenon, known as the photoelectric effect, played an important role in establishing quantum mechanics. It was shown that, contrary to classical ideas, energy exchanges between radiation and matter are mediated by integral numbers of photons. In the gas phase the photoeffect is called either photoionization (atoms, molecules, and their positive ions) or photodetachment (atomic and molecular negative ions). *See* PHOTOEMISSION.

Photoelectron spectroscopy. Photoionization involves a radiative bound-free transition from an initial state consisting of n photons and an atom, molecule, or ion in a bound state to a final continuum state consisting of a residual ion (or an atom in the case of photodetachment) and m free electrons, that is, reaction (1). In the simplest atomic photoionization

$$nh\nu + X \rightarrow X^{m+} + me^{-} \tag{1}$$

process a single electron is ejected from an atom following the absorption of a single photon. Each mode of fragmentation defines a final-state channel that is characterized by the energy and angular momentum of the outgoing electron as well as the excitation state of the residual ion. Since the photoionization process is endoergic, each channel has a well-defined threshold energy below which the channel is energetically closed. The threshold photon energy for a particular channel is equal to the binding energy of the electron that is to be ejected plus the excitation energy, if any, of the residual ion.

Above threshold, the energy carried off by the outgoing electron represents the balance between the energy supplied by the photon and the binding energy of the electron plus the excitation energy of the residual ion (neglecting the small recoil of the heavy ion). A photoelectron spectrum is characterized by a discrete set of peaks, each peak being associated with a particular state of the residual ion. Information on the excitation state of the ion following photoionization can also be obtained by monitoring the fluorescence emitted in the subsequent radiative decay of the state. One of the earliest applications of photoionization measurements was the investigation of the structure of atoms by determining the binding energies of both outer- and inner-shell electrons by means of photoelectron spectroscopy. *See* ELECTRON SPECTROSCOPY.

Measurements. Photoionization measurements involve the illumination of a target of atoms, molecules, or their ions with a beam of photons from a light source. The binding energies of electrons in atoms and positive ions are sufficiently large that ionization thresholds lie in the ultraviolet or x-ray region of the electromagnetic spectrum. This generally limits the use of lasers as light sources. Traditionally lamps and, subsequently, synchrotron radiation sources have been used in photoionization studies. Lasers, however, are the most frequently used light source for photodetachment studies since the weakly bound electrons in negative ions correspond to thresholds in the infrared and visible. Sources of atoms and molecules are usually static or flowing gas targets, while positive and negative ions are most often produced in the form of beams by using an accelerator. Electromagnetic traps have also been used in photodetachment measurements. The photoionization process can be monitored by studying the absorption of the incident light or by using a more sensitive method based upon detecting one of the products of the breakup, that is, photoelectrons or residual ions. *See* LASER; MOLECULAR BEAMS; NEGATIVE ION; PARTICLE TRAP; SYNCHROTRON RADIATION.

Electron emission pattern. The selection rules for bound-free transitions specify the allowed changes in the quantum numbers characterizing the initial and final states of photoionization. In particular, for an electric dipole-induced process, the orbital angular momentum of the ejected electron is limited to the values $l = l_0 \pm 1$, where l_0 refers to the angular momentum of the bound electron prior to ejection. The outgoing electron carries off this angular momentum, which is manifested in its angular distribution and spin polarization. The partial waves associated with $l = l_0 \pm 1$ interfere as a function of photon energy to determine the shape of the photoelectron emission pattern. It has been demonstrated by using symmetry arguments that the form of the angular distribution for linearly polarized radiation in the electric dipole approximation, assuming an initially randomly oriented target, is given by Eq. (2).

$$f(\theta) = \frac{1 + \beta(3\cos^2\theta - 1)}{2} \tag{2}$$

Here θ is the angle between the photoelectron emission direction and the polarization vector of the radiation, and β represents the asymmetry parameter (a measure of the deviation from isotropy, $\beta = 0$). This parameter, which characterizes the angular distribution, is restricted to the range $-1 \leq \beta \leq 2$. The value of $\beta = 2$ corresponds to the classical dipolar distribution ($\cos^2 \theta$), where the electrons are preferentially ejected in the direction of the polarization vector. Values of $\beta \neq 2$ are the result of interferences between the two partial waves representing the electron. Angular distributions are studied by using angle-resolved photoelectron spectroscopy. *See* ANGULAR CORRELATIONS.

Cross section. The probability that an atom becomes ionized by the absorption of a photon is measured by the photoionization cross section, which is typically 0.1–10 megabarns (1 Mb = 10^{-18} cm^2). To understand the detailed manner in which an atom interacts with radiation requires an investigation of this cross section as a function of photon energy. Above threshold, the partial cross section for a particular channel decreases monotonically with photon energy. New channels open when the photon energy exceeds excited-state threshold values.

A photoionization cross section can also exhibit sharp changes over a narrow range of photon energies. This structure is the result of interference between resonant and direct (nonresonant) pathways to the same final state. The resonant pathway involves the transient photoexcitation of a doubly excited intermediate state which, in turn, rapidly decays by the spontaneous process of autoionization. Resonance parameters such as the energy and width (inversely related to the lifetime against autoionization) can be found by fitting the experimental data to an analytical form called the Fano-Beutler function. *See* RESONANCE (QUANTUM MECHANICS).

The threshold dependence of the cross section is fundamentally different for photoionization and photodetachment. In the final state the ejected electron moves in a strongly attractive and long-range Coulomb field of an ion in the case of photoionization and a weakly attractive and short-range field of a neutral atom or molecule in the case of photodetachment. The enhanced role of correlation in photodetachment is due to the absence of the normally dominant final-state Coulomb interaction. The short-range nature of the final-state interaction in photodetachment causes the cross section in the near vicinity of a threshold to exhibit a dependence on the orbital angular momentum, l, of the outgoing electron. The cross section is zero at threshold and rises to a maximum just beyond threshold. In contrast, because of the long-range final-state interaction, photoionization cross sections are finite at threshold, independent of angular momentum of the outgoing electron.

Use of synchrotron radiation sources. The availability of intense beams of high-energy photons from advanced synchrotron radiation sources facilitates the investigation of photoionization of inner-shell electrons in atoms and inner- and outer-shell electrons in positive ions. Spectroscopic studies of ejected photoelectrons as well as subsequently emitted Auger electrons lead to a better understanding of the final-state dynamics of the photoionization process when an inner-shell vacancy relaxes. *See* AUGER EFFECT.

Resonance ionization spectroscopy. Resonance ionization spectroscopy is a sensitive and selective analytical method based on the photoionization of atoms. Typically it involves two or more resonant photoexcitation steps followed by an ionization step. Similar measurements designed to determine cross sections for photoionization from excited atoms have also employed a sequential photoabsorption scheme. The atom is prepared in an excited state by laser photoexcitation and subsequently photoionized by the absorption of synchrotron radiation. *See* RESONANCE IONIZATION SPECTROSCOPY.

Multiphoton and multielectron processes. There is increasing interest in photoionization processes that go beyond the most probable process of single-photon absorption and single-electron ejection. The high intensity of lasers has made it possible, for example, to study multiphoton absorption processes in which a single electron is ionized by the absorption of two or more photons, each with an energy less than the ionization energy. Interestingly, it has been observed in a process called above-threshold ionization that under certain conditions more photons are absorbed than the minimum number needed to reach the ionization threshold. If the intensity of the laser beam becomes very high, it is possible to observe effects associated with the electron being strongly driven by the oscillating laser field. This quiver motion can modify the energy of the ejected electron.

The availability of intense beams of high-energy photons from lasers and synchrotron sources has facilitated the investigation of ionization processes involving the simultaneous ejection of more than one electron following the absorption of a single photon. Double photodetachment, for example, is particularly interesting near threshold since it involves a three-body final state consisting of two electrons essentially at rest in the field of a singly charged core. The electron correlations that develop in this system are of fundamental interest in atomic structure theory. *See* ATOMIC STRUCTURE AND SPECTRA; LASER SPECTROSCOPY.

David J. Pegg

Bibliography. J. Berkowitz, *Photoabsorption, Photoionization and Photoelectron Spectroscopy*, 1979; R. R. Cordermann and W. C. Lineberger, Negative ion spectroscopy, *Annu. Rev. Phys. Chem.*, 30:347–376, 1979; V. S. Letokhov, *Laser Photoionization Spectroscopy*, 1987; S. T. Manson and A. F. Starace, Photoionization angular distributions, *Rev. Mod. Phys.*, 54:389–405, 1982; J. A. R. Samson, Photoionization of atoms and molecules, *Phys. Rep.*, 28:303–354, 1976; V. Schmidt, Photoionization of atoms using synchrotron radiation, *Rep. Prog. Phys.*, 55:1483–1659, 1992.

Photoluminescence

A luminescence excited in a body by some form of electromagnetic radiation incident on the body. The term photoluminescence is generally limited to cases in which the incident radiation is in the ultraviolet, visible, or infrared regions of the electromagnetic spectrum; luminescences excited by x-rays or gamma rays are generally characterized by special names. The graph of luminous efficiency per unit energy of the exciting light absorbed versus the frequency of the exciting light is called the excitation spectrum. The excitation spectrum is determined by the absorption spectrum of the luminescent body, which it often closely resembles, and by the efficiency with which the absorbed energy is transformed into luminescence.

Photoluminescence may be either a fluorescence or a phosphorescence, or both. Energy can be stored in certain luminescent materials by subjecting them to light or some other exciting agent, and can be released by subsequent illumination of the material with light of certain wavelengths. This type of photoluminescence is called stimulated photoluminescence. In contrast to normal photoluminescence, which is constant in intensity as long as the intensity of the exciting light does not vary, stimulated photoluminescence decreases in intensity as the stored energy is released. *See* FLUORESCENCE; LUMINESCENCE; PHOSPHORESCENCE.

Clifford C Klick; James H. Schulman

Photolysis

The decomposition of matter due to absorption of incident light. When photolysis occurs, it causes definite changes in the chemical composition of the illuminated material. For example, illumination of microcrystals of silver bromide embedded in gelatin results in formation of metallic silver, and is the basis of the photographic process. *See* PHOTOGRAPHIC MATERIALS.

Reactors and features. Numerous metal complexes (for example, oxalate salts), azides, nitrides, and sulfides, and most organometallic compounds undergo decomposition upon illumination, often with concomitant evolution of a gaseous product. In the presence of water and oxygen, illumination of many semiconductors results in their corrosion; for example, an aqueous suspension of cadmium sulfide (CdS) undergoes rapid decomposition upon irradiation with sunlight. Such reactions are especially pronounced if the semiconductor is present in a colloidal form. The outcome of these photochemical processes can often be controlled by addition of adventitious materials. For example, illumination of cadmium sulfide in aqueous solution containing hydrogen sulfide (H_2S) and colloidal platinum results in evolution of hydrogen gas. Here, the semiconductor photosensitizes (or photocatalyzes) decomposition of hydrogen sulfide into its elements, and the reaction can be used to remove sulfides from industrial waste. The same reaction with water used in place of hydrogen sulfide represents the ideal goal of many attempts at the conversion and storage of solar energy. Thus, photolysis of inorganic materials plays an important role in both effluent treatment and energy conversion. *See* COORDINATION COMPLEXES; SEMICONDUCTOR.

Organic materials. The photolysis of organic materials is a particularly rich subject. Many ketones, for example, acetone, abstract hydrogen atoms from adjacent organic matter under illumination. According to the circumstances, the resultant free radicals may be used to initiate polymerization of a suitable monomer or to cause decomposition of a plastic film. Both reactions have important commercial applications, and photoinitiators are commonly used for emulsion paints, inks, polymers, explosives, and fillings for teeth and for development of photodegradable plastics. Photolysis of carbonyl compounds, released into the atmosphere by combustion of fossil fuels, is responsible for the onset of photochemical smog. Many other types of photochemical transformation of organic molecules are known, including isomerization of unsaturated bonds, cleavage of carbon-halogen bonds, olefin addition reactions, halogenation of aromatic species, hydroxylation, and oxygenation processes. Indeed, photochemistry is often used to produce novel pharmaceutical products that are difficult to synthesize by conventional methods. *See* FREE RADICAL; PHOTODEGRADATION; SMOG.

Photolysis causes a photochemical reaction, as distinguished from a thermal reaction. The most important photochemical reaction is green plant photosynthesis. Here, chlorophyll that is present in the leaves absorbs incident sunlight and catalyzes reduction of carbon dioxide to carbohydrate. This is the origin of all fossilized fuels. The reaction also results in oxidation of water to molecular oxygen and, therefore, can be said to be primarily responsible for all life on Earth. There are other photochemical reactions that serve important functions in biology and medicine. One demonstration of the power of photochemistry concerns the treatment of neonatal jaundice. New-born infants suffering from this illness, which is caused by proliferation of bilirubin, are exposed to sunlight. This causes conversion of bilirubin to a form that is more soluble in water and is rapidly excreted from the body. *See* BILIRUBIN; PHOTOSYNTHESIS.

An important feature of all photochemical reactions is that only absorbed light is effective in causing photochemical change. This fact permits photolysis of particular materials simply by optimum selection of the wavelength of incident light. Thus, the ozone layer prevents ultraviolet light from reaching the Earth's surface and thereby acts as a protection against extensive damage that would result from light absorption by biological tissue. One dye in a mixture of similar reagents can be illuminated selectively by using monochromatic light, allowing control of the ensuing chemistry. Furthermore, photonic energy absorbed by one molecule can be transferred, according to certain laws, to similar or different molecules. This process, known as photosensitization, allows transparent materials, such as water, to be photolyzed.

Singlet molecular oxygen. Excitation of an organic molecule results in spontaneous generation of the singlet excited state. In most molecules, this highly unstable excited singlet state may undergo an intersystem-crossing process that results in population of the corresponding (less energetic) excited triplet state, in competition to fluorescence. The excited triplet state, because of spin restriction rules, retains a significantly longer lifetime than is found for the corresponding excited singlet state, and may be formed in high yield.

Almost without exception, these triplet states react quantitatively with molecular oxygen (O_2) present in the system via a triplet energy-transfer process. The resulting product is singlet molecular oxygen. This species is a potent and promiscuous reactant, and it is responsible for widespread damage to both synthetic and natural environments. Indeed, plants and photosynthetic bacteria contain carotenoids to protect the organism against attack by singlet oxygen. The same species is known to be responsible, at least in part, for photodegradation of paint, plastic, fabric, colored paper, and dyed wool. Secondary reactions follow from attack on a substrate by singlet oxygen, resulting in initiation of chain reactions involving free radicals. However, modern technological processes have evolved in which singlet molecular oxygen is used to destroy unwanted organic matter, such as tumors, viruses, and bacteria, in a controlled and specific manner. In photodynamic therapy a dye is injected into a tumor and selectively illuminated with laser light. The resultant singlet oxygen destroys the tumor. Similar methodology can be used to produce photoactive soap powders, bleaches, bactericides, and pest-control reagents. Mosquito larvae, for example, can be destroyed by exposure to sunlight after staining with a photoactive dye. *See* CHAIN REACTION (CHEMISTRY); FLUORESCENCE; TRIPLET STATE.

Flash photolysis. Irradiation of a material with a high-intensity, short-duration pulse of light, delivered from a flash lamp or pulsed laser, can be used to create a high concentration of transient species. This experimental technique, known as flash photolysis, is used to characterize reactive intermediates formed during photolysis. The sample, which may be in the form of a solution, vapor, film, or powder, is irradiated with a single pulse of light of appropriate wavelength in order to generate the initial excited state. After this excitation pulse, a much weaker pulse of light is directed onto the sample in order to record an ultraviolet-visible absorption spectrum of the excited state. By delaying the monitoring pulse with respect to the excitation pulse, a series of spectra can be recorded at different delay times. In this manner the course of reaction can be followed, and important intermediates can be identified. By monitoring formation and deactivation of particular intermediates, kinetic information can be elicited, and the overall mechanism of the photoprocess can be evaluated. In addition to ultraviolet-visible absorption spectra, intermediates may be detected by infrared, Raman, resonance Raman, electron paramagnetic resonance, circular dichroism, magnetic circular dichroism, luminescence, and diffuse reflectance spectroscopic techniques. Time-resolved conductivity, thermal lensing, and optoacoustic calorimetry can be used to gain further information about the photochemistry of the system and to detect heat released into the system during nonradiative decay of transient species. By including compounds possessing certain properties (that is, easily abstractable hydrogen atoms, low oxidation potential, molecular oxygen) in the system, it becomes possible to explore secondary processes and to measure rates of reaction. Transient spectral and kinetic data can be collected on time scales as short as 100 femtoseconds by using readily available laser technology. *See* LASER SPECTROSCOPY.

Bacterial photosynthesis. The most impressive illustration of laser flash photolysis concerns the elucidation of the chemical sequence occurring within bacterial photosynthetic reaction center complexes. Thus, incident sunlight is harvested by arrays of chlorophylls, xanthophylls, and related pigments and transported throughout the array until reaching a trap. Energy migration occurs by way of rapid transfer between closely spaced (1.0–1.5-nanometer) pigment molecules, each transfer taking about 1 picosecond. The trap is a pair of bacteriochlorophyll molecules held in an approximate face-to-face configuration. Upon excitation, an electron is transferred to a bacteriopheophytin molecule, situated some 1.7 nm distant, on a time scale of a few picoseconds. The electron is subsequently transferred to secondary (quinones) acceptors, while secondary (hemes) electron donors serve to reduce the oxidized primary donor. In this manner, charge can be transferred across a membrane and used, via a series of biochemical processes, to produce valuable fuels. The rates and efficiencies of each of the electron transfer steps have been elucidated for various types of photosynthetic bacteria and, with less success, for many of the events occurring in the reaction centers of green plants. *See* PHOTOCHEMISTRY. A. Harriman

Bibliography. V. Balzani and F. Scandola, *Supramolecular Photochemistry*, 1990; J. A. Barltrop and J. D. Coyle, *Principles of Photochemistry*, 1978; G. J. Ferraudi, *Elements of Inorganic Photochemistry*, 1988; K. Kalyanasundaram, *Photochemistry of Polypyridine and Porphyrin Complexes*, 1992; K. Kalyanasundaram and M. Grätzel (eds.), *Photosensitization and Photocatalysis Using Inorganic and Organometallic Compounds*, 1993; N. J. Turro, *Modern Molecular Photochemistry*, 1991.

Photometer

An instrument used for making measurements of light, or electromagnetic radiation, in the visible range. In general, photometers may be divided into two classifications: laboratory photometers, which are usually fixed in position and yield results of high accuracy; and portable photometers, which are used in the field or outside the laboratory and yield results of lower accuracy. Each class may be subdivided into visual (subjective) photometers and photoelectric (objective or physical) photometers. These in turn may be grouped according to function, such as photometers to measure luminous intensity (candelas or candlepower), luminous flux, illumination (illuminance), luminance (photometric brightness), light distribution, light reflectance and transmittance, color, spectral distribution, and visibility. Since visual photometric methods have largely been supplanted commercially by physical methods, they receive only casual attention here, but because of their simplicity, visual methods are still used in educational laboratories to demonstrate photometric principles. *See* ILLUMINANCE; LUMINANCE; LUMINOUS FLUX; LUMINOUS INTENSITY.

Visual photometers. These are luminance comparison devices, and most of them utilize the Lummer-Brodhun sight box or some adaptation of its principles. The sight box is an optical device for viewing simultaneously the two sides of a white diffuse plaster screen illuminated by the light sources that are being compared. The general arrangement of the contrast sight box and the field of view as seen through the eyepiece are shown in **Fig. 1**. The contrast sight box gives greater sensitivity of observation than the equality of luminance prism arrangement and appearance of field of view (**Fig. 2**). If there is a color difference in the two light sources being compared, it may be difficult to make a photometric balance. In this case, methods of heterochromatic photometry, for example, the flicker photometer, may be used.

Flicker photometer. In this photometer a single field of view is alternately illuminated by the light sources to be compared. The rate of alternation is set at such a speed that the color differences disappear; the photometric balance is obtained by moving the sight box back and forth, the criterion of equality of luminance of the two sources being the disappearance of flicker. Other systems of heterochromatic photometry are the cascade method, the compensation of mixture method, and the use of color-matching or color-equalizing optical filters.

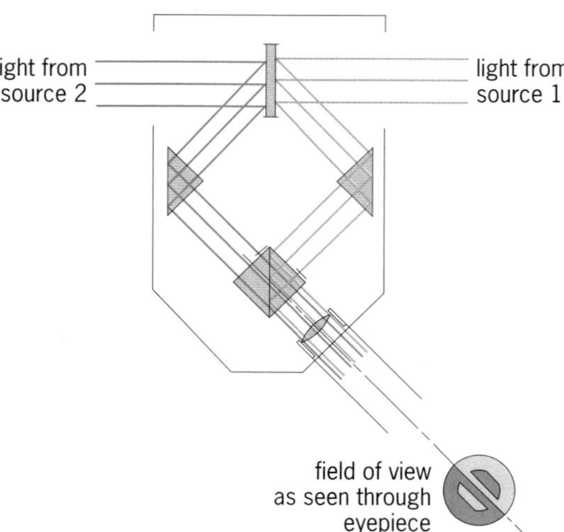

Fig. 1. Lummer-Brodhun contrast sight box.

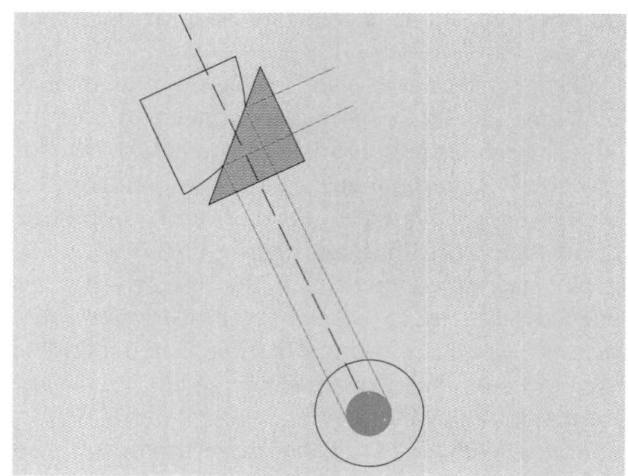

Fig. 2. Diagrammatic sketch of Lummer-Brodhun equality of luminance prism arrangement and appearance of field of view.

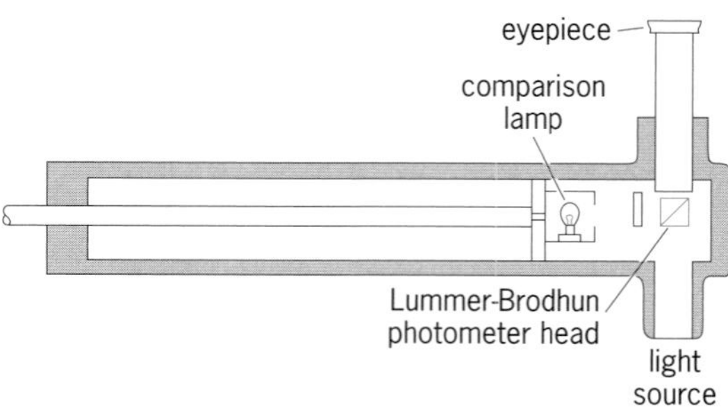

Fig. 3. Essentials of a Macbeth illuminometer.

Illuminometer. An illuminometer is a portable visual photometer, as shown in cross section in **Fig. 3**. The light to be measured enters the instrument and is balanced by a Lummer-Brodhun sight box of the equality of luminance type against a comparison lamp, which is moved along the tube. A control box supplies a calibrated current to the comparison lamp, and calibrated optical filters can be placed in the light paths to correct for color differences in the comparison and measured sources and to extend the range of the instrument.

Luminance meter. This instrument measures photometric brightness and may be of the visual or photoelectric type; it is a major tool for research scientists in visual psychology and illuminating engineering. **Figure 4** shows a photoelectric luminance photometer; a separate unit combines the power supply with the controls and readout meter. The photometer has a telescopic viewing system for imaging the bright surface to be measured on the cathode of a photoemissive tube.

The visual type of illuminometer can also be used to measure the brightness of any surface which is large enough to fill the field of view of the instrument.

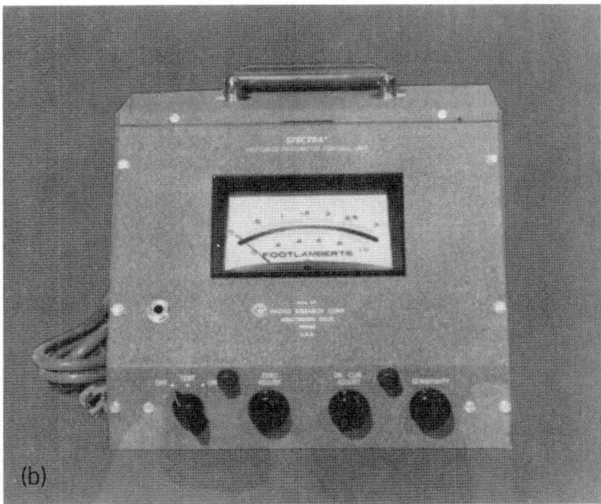

Fig. 4. Spectra Pritchard photoelectric luminance photometer. (*a*) Viewing system. (*b*) Readout meter. (*Photo Research Corp.*)

Photoelectric photometers. Photometers that use barrier-layer (photovoltaic) photocells and photoemissive electronic tubes as light receptors are classified as photoelectric photometers. Since barrier-layer cells generate their own current when illuminated and therefore do not require a power supply for operation, they may be used either in the laboratory or in the portable instruments. The type generally used consists of an iron plate, often circular, coated with a thin layer of selenium, which in turn is covered with a very thin, transparent film of metal such as gold or platinum (**Fig. 5**). A ring of metal is sprayed around the edge of this film; when this metal ring is connected to one terminal of a sensitive microammeter or galvanometer and the other terminal is connected to the iron plate, a current flows through the instrument when light penetrates the metal film to the selenium layer. The photoemissive tube photometers, on the other hand, require an external power supply, and so they are used mainly in laboratory work. Both types of cells should be fitted with optical filters to correct approximately their spectral response to the standard luminosity curve of the eye.

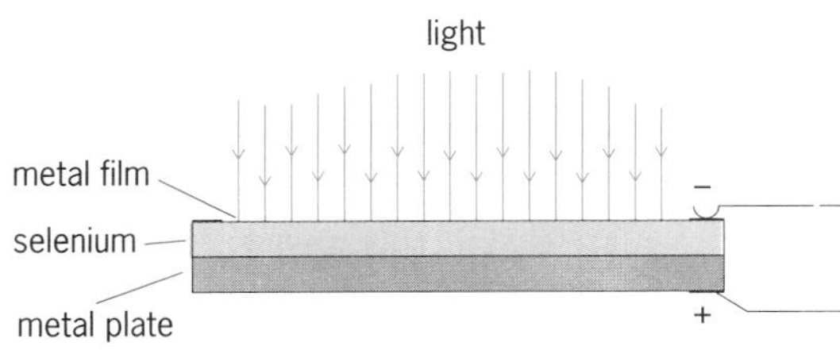

Fig. 5. Barrier-layer photocell.

With the advent of semiconductor electronics, it is possible to construct current amplifiers for use with barrier-layer photocells, since both devices have inherent low-resistance characteristics. The linearity of these cells is also a function of the voltage developed across its terminals. To obtain maximum linearity, the load on the cell should approach zero. Semiconductor operational amplifiers have also been found to fulfill this requirement. Transistor amplifiers can be used with barrier-layer cells to drive digital voltmeters, which have made possible the semiautomation of most observational work in photometry.

Photoelectric illuminometer. The measurement of illumination has been revolutionized since 1931 by the introduction of photoelectric portable photometers employing rugged barrier-layer cells. These instruments eliminate the tedious visual comparison required with previous portable visual illuminometers and have considerably simplified illumination measurements. Barrier-layer cells generate a small electric current of about 6 or 7 microamperes/footcandle when light in the visible range falls upon them. The electric instrument used to measure the cell output should have low resistance (100 ohms or less), making the cell current nearly proportional to the luminous flux falling upon it. The photocell should be corrected to correspond to Lambert's cosine law of incidence. *See* PHOTOMETRY.

In this type of cell the current attains its final value after a short time because of the effects of fatigue and temperature. These effects are minimized by low resistance in the instrument circuit.

Barrier-layer cell photometers are also used extensively in photographic light meters to determine proper lighting and camera diaphragm openings. For such applications the instrument can be calibrated on an arbitrary scale. A simple manual computer is often included to convert the meter reading into the proper diaphragm opening and exposure time for a given type of film. *See* PHOTOGRAPHY.

Photoemissive tube photometer. This photometer utilizes a photoelectric tube whose sensitivity is much lower than that of a barrier-layer cell. This device is more accurate than the barrier-layer cell, and the linearity and stability are sufficient for highest-precision photometric work. Since electronic amplification is required to increase the sensitivity to a practicable level, this type of photometer is not usually used outside the laboratory. In calibration, the dark current which flows when the tube is not illuminated is an important factor and should be subtracted or eliminated by circuitry.

Integrating-sphere photometer. Any of the above photometers can be used with an integrating sphere

Fig. 6. Uhlbricht sphere for measuring luminous flux and efficiency. (*Westinghouse Electric Corp.*)

to measure the total luminous flux of a lamp or luminaire. This is ordinarily made as an Uhlbricht sphere (**Fig. 6**), although other geometrical shapes can be used. The inside surface has a diffusely reflecting white finish which integrates the light from the source. A white opaque screen prevents direct light from the source from falling on a diffusing window, which is illuminated only by diffusely reflected light from the sphere walls. The brightness of this window, which is measured by the photometer outside the sphere, is then directly proportional to the flux emitted by the source if the inside finish of the sphere is perfectly diffusing.

Reflectometer. This instrument serves the dual purpose of a reflectometer and transmissometer and combines integrating spheres and barrier-layer photocells (**Fig. 7**). The absolute reflectance of test surfaces can be determined by measuring the total reflected light when a beam of light strikes the surface. Transmittance can be measured by placing a flat sample of the material in the opening between the two spheres, with the bottom sphere containing the light source for transmittance measurements, and the upper one the light-measuring cells and a collimated beam of light for reflectance measurements.

Light-distribution photometer. One of the most frequently used photometric devices, this measures luminous intensity at various angles from lamps, luminaires, floodlights, searchlights, and the like. Although the most direct method is to tilt the light source so that measurements can be made in various directions, this may be impracticable with sources whose light output is position-sensitive. It is therefore sometimes necessary to use mirror arrangements similar to the one in **Fig. 8**. The source to be measured is placed at a fixed point on the photometric axis (the horizontal line passing through the centers of the source and photocell). The mirror system can then be moved bodily about the axis so that the light emitted by the source in any direction in a vertical plane perpendicular to the photometric axis is reflected to the photocell.

Modern methods of automation have been applied to distribution photometers. They may be controlled by punched cards or tape to give complete programming of the photometer so that no human attention is necessary during the test. The photometric data can be automatically recorded, and curves of luminous intensity and footcandle intensity can be automatically drawn. A modern automated distribution photometer is shown in **Fig. 9**.

Spectrophotometer. Measurements of spectral energy distribution from a light source are made by this device. It measures the energy in small wavelength bands by means of a scanning slit, and the results are presented as a spectral distribution curve.

Visibility meters. These operate on the principle of artificially reducing the visibility of objects to threshold values (borderline of seeing and not seeing) and measuring the amount of that reduction on an appropriate scale. The Luckiesh-Moss visibility meter

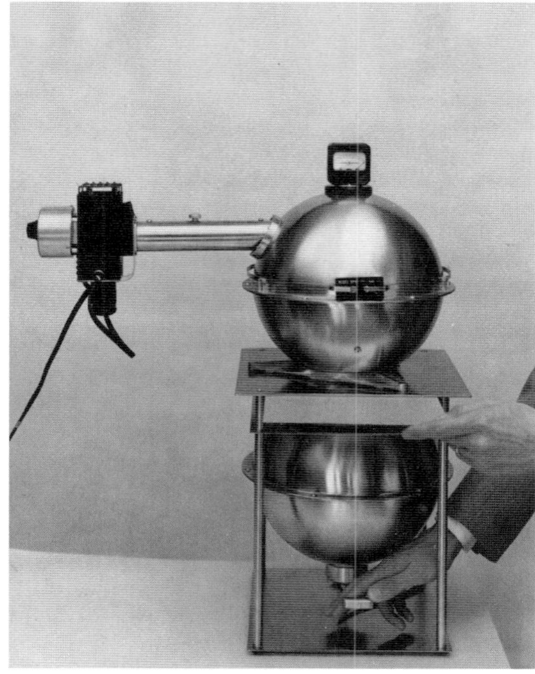

Fig. 7. Dual-purpose reflectometer transmissometer, in which integrating spheres and barrier-layer cell photometers are combined. (*General Electric Co.*)

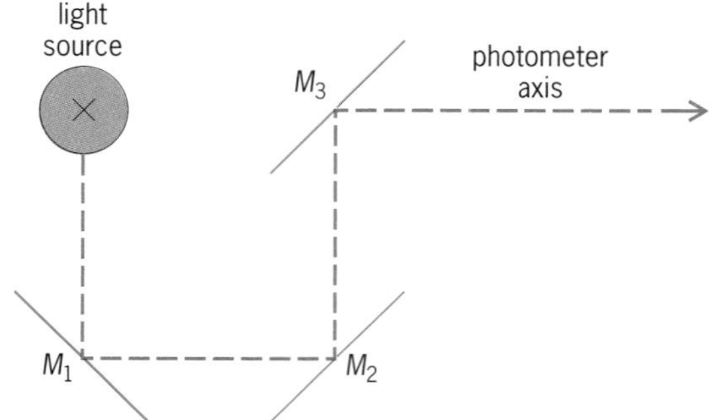

Fig. 8. Mirror system to measure light distribution.

Fig. 9. Modern automated distribution photometer. (*Westinghouse Electric Corp.*)

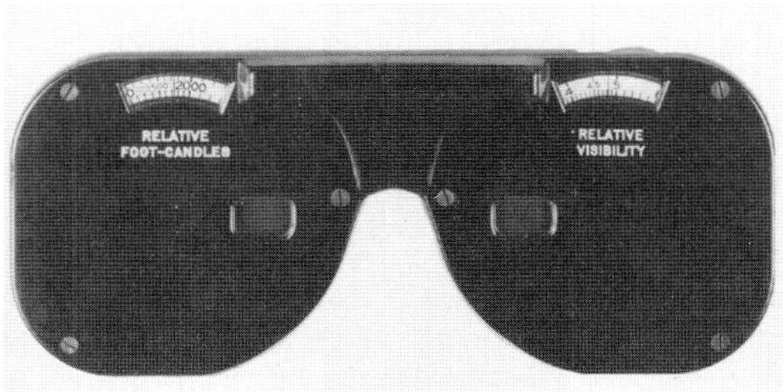

Fig. 10. Luckiesh-Moss visibility meter. (*General Electric Co.*)

is probably the best known and easiest to use. The instrument (**Fig. 10**) consists of two variable-density filters (one for each eye) so adjusted that a visual task seen through them is just barely discernible. Readings are on a scale of relative visibility related to a standard task.

G. A. Horton

Photometry

That branch of science which deals with measurements of light (visible electromagnetic radiation) according to its ability to produce visual sensation. Specifically, photometry deals with the attribute of light that is perceived as intensity, while the related attribute of light that is perceived as color is treated in colorimetry. *See* COLOR; COLORIMETRY.

The purely physical attributes of light such as energy content and spectral distribution are treated in radiometry. Sometimes the word photometry is used to denote measurements that have nothing to do with human vision, but this is a mistake according to modern usage. Such measurements are properly referred to as radiometry, even if they are performed in the visible spectral region. *See* RADIOMETRY.

Relative visibility. The relative visibility of a fixed power level of monochromatic electromagnetic radiation varies with wavelength over the visible spectral region (**Fig. 1**). The relative visibility of radiation also depends upon the illumination level that is being observed. The cone cells in the retina determine the visual response at high levels of illumination, while the rod cells dominate in the dark-adapted eye at very low levels (such as starlight). Cone-controlled vision is called photopic, and rod-controlled vision is called scotopic, while the intermediate region where both rods and cones play a role is called mesopic. *See* VISION.

Originally, photometry was carried out by using the human visual sense as the detector of light. As a result, photometric measurements were subjective. That is, two observers with different relative visibility functions, starting with the same standard and following identical procedures, would assign different values to a test radiation having a different spectral distribution than that of the standard.

In order to put photometric measurements on an objective basis, and to allow convenient electronic detectors to replace the eye in photometric measurements, the Commission Internationale de l'Eclairage (CIE; International Commission on Illumination) has adopted two relative visibility functions as standards. These internationally accepted functions are called the spectral luminous efficiency functions for photopic and scotopic vision, and are denoted by $V(\lambda)$ and $V'(\lambda)$, respectively. *See* LUMINOUS EFFICIENCY.

Thus photopic and scotopic (but not mesopic) photometric quantities have objective definitions, just as do the purely physical quantities. However, there is a difference. The purely physical quantities are defined in terms of physical laws, whereas the photometric quantities are defined by convention. In recognition of this difference the photometric quantities are called psychophysical quantities.

Photometric units. According to the International System of Units, SI, the photometric units are related to the purely physical units through a defined constant called the maximum spectral luminous efficacy. This quantity, which is denoted by K_m, is the number of lumens per watt at the maximum of the $V(\lambda)$ function. K_m is defined in SI to be 683 lm/W for monochromatic radiation whose wavelength is

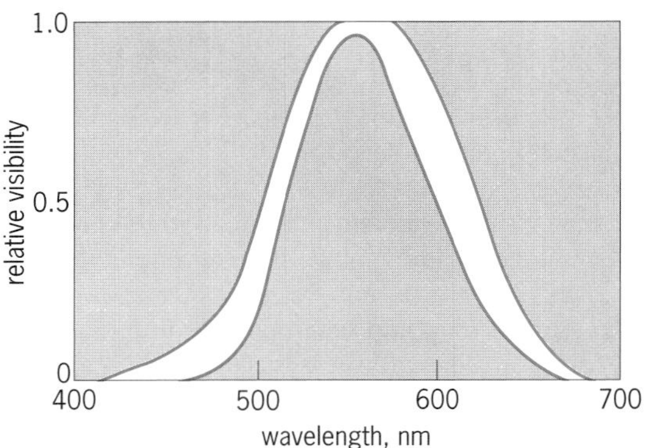

Fig. 1. Range of the relative visibility functions of 125 observers as measured by W. W. Coblentz.

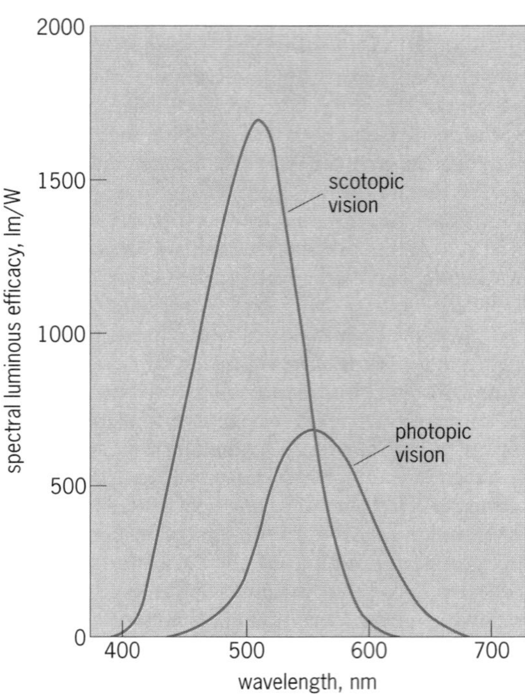

Fig. 2. Internationally accepted standard values for the spectral luminous efficacy of monochromatic radiation for human scotopic (dark-adapted) and photopic (light-adapted) vision.

555 nanometers, and this defines the photometric units with which the photometric quantities are to be measured.

At various times, the photometric units have been defined in terms of the light from different standard sources, such as candles made according to specified procedures, and blackbodies at the freezing point of platinum. According to these definitions, K_m was a derived, rather than defined, quantity. *See* ILLUMINATION; PHYSICAL MEASUREMENT; UNITS OF MEASUREMENT.

Spectral luminous efficacy. The products of K_m and the spectral luminous efficiencies are called the spectral luminous efficacies denoted by $K(\lambda)$. The spectral luminous efficacy functions for photopic and scotopic vision are shown in **Fig. 2**. Currently, there are no internationally accepted spectral luminous efficiency functions for mesopic vision. Thus there are no objectively defined mesopic quantities in use in photometry. However, research suggests the usefulness of linear combinations of the curves shown in Fig. 2, where the relative weight assigned to each curve depends upon the level of illumination. *See* LUMINOUS EFFICACY.

Photometric measurements. Photometric measurements are made by a number of different means. The human eye is still used occasionally, particularly in the mesopic region where no other internationally accepted standard exists. It is also common to make photometric measurements with a detector-optical filter combination in which the filter has been chosen to make the relative spectral response of the combination closely resemble the $V(\lambda)$ function. It is sometimes difficult to get a good enough fit of the spectral response function to the $V(\lambda)$ function over a sufficiently broad spectral region to make accurate measurements on exotic light sources. It is becoming accepted practice to measure the spectral distribution of these sources by conventional spectral radiometric techniques, and to calculate the photometric quantities by multiplying the spectral quantities by $K(\lambda)$ and integrating with respect to wavelength. *See* LIGHT.

Jon Geist

Bibliography. R. McCluney, *Introduction to Radiometry and Photometry*, 1994.

Photomorphogenesis

The regulatory effect of light on plant form, involving growth, development, and differentiation of cells, tissues, and organs. Morphogenic influences of light on plant form are quite different from light effects that nourish the plant through photosynthesis, since the former usually occur at much lower energy levels than are necessary for photosynthesis. While photosynthetic processes are energy-transducing, converting light energy into chemical energy for running the machinery of the plant, light serves only as a trigger in photomorphogenesis, frequently resulting in energy expenditure orders of magnitude larger than the amount required to induce a given response. Photomorphogenic processes determine the nature and direction of a plant's growth and thus play a key role in its ecological adaptations to various environmental changes. *See* PHOTOSYNTHESIS.

Morphogenically active radiation is known to control seed and spore germination, growth and development of stems and leaves, lateral root initiation, opening of the hypocotyl or epicotyl hook in seedlings, differentiation of the epidermis, formation of epidermal hairs, onset of flowering, formation of tracheary elements in the stem, and form changes in the gametophytic phase of ferns, to mention but a few of such known phenomena. Many nonmorphogenic processes in plants are also basically controlled by light independent of photosynthesis. Among these are chloroplast movement, biochemical reactions involved in the synthesis of flavonoids, anthocyanins, chlorophyll, and carotenoids, and leaf movements in certain legumes such as *Albizzia*, *Mimosa*, and *Samanea*. A large number of these processes are affected only by wavelengths of light at the red end of the spectrum. The similarities in the action spectra for these different systems led to the concept that a widely divergent set of developmental and physiological processes might be controlled by a single pigment. This has proved to be the case, the pigment being phytochrome.

Other of these processes are controlled by wavelengths in the near-ultraviolet and blue regions of the spectrum. Again, many of the action spectra are similar, which suggests a single pigment, and the best candidate to date is a flavoprotein. Efforts to isolate, identify, and characterize this blue-sensitive system have lagged far behind similar efforts with phytochrome, and there is no assurance that a flavoprotein is really the photoreceptor in most systems or that the multitude of effects are all mediated by a single pigment.

Pigment characterization. Phytochrome itself is a blue or blue-green pigment consisting of a protein moiety and a light-absorbing group (chromophore). The chromophore, which is responsible for the sensitivity of the pigment to light, is a linear tetrapyrrole of the bilitriene configuration, closely related to the phycobilins, photosynthetic accessory pigments found in certain algae and human bile. Phytochrome exists in two forms which are interconvertible by different regions of the spectrum: P_r, which absorbs maximally in the red (665-nanometer) region of the spectrum, and P_{fr}, which absorbs maximally in the far-red (730 nm). The latter is considered to be the biologically active form. The induction of a particular developmental or physiological response by red light, combined with cancellation of the induction by far-red light given sufficiently soon after red, is generally taken as a diagnostic test for phytochrome action, though such reversibility cannot be shown for every response involving phytochrome.

The conversion of P_r to P_{fr}, and vice versa, does not take place in a single step, but rather through complex sets of intermediates. Although the relative roles of the protein component and the chromophore in phytochrome photoreversibility are not established, available evidence suggests that both transformations involve structural rearrangements of the chromophore accompanied by at least minor conformational changes in the protein moiety.

Phytochrome has been extracted and characterized from dark-growth seedlings of several plants, the bulk of the work being with oat and rye. The protein has a monomer molecular weight of 120,000, and the most common form found in solution is a dimer. Phototransformation to P_{fr} increases the lability of the chromophore to bleaching, and changes the accessibility of the functional groups of various amino acids to externally supplied reagents. In dark-grown seedlings the pigment behaves as a soluble protein; though a small fraction appears to be membrane-bound, the binding partner is not conclusively identified. Formation of P_{fr} is thought to expose an active site that is recognized by some receptor (presumed but not proved to be on a membrane) leading to an alteration in the functional properties of the binding partner. In responses occurring over days or weeks, differential gene activation is almost certainly involved, but rapid responses occurring within seconds or minutes are most likely caused by membrane changes.

The blue-light receptor has not been definitively identified and characterized. A complex sensitive to blue light has been discovered in several fungi, and such a complex has been isolated and partially purified from both fungi and corn. It is associated with a membrane (probably the plasma membrane) and participates in an electron transport chain. Light accelerates the rate of electron transfer from a flavin to a cytochrome. The action spectrum for this process is very much like that for many of the blue-ultraviolet-sensitive physiological and developmental processes, but available evidence proving that this pigment is the photoreceptor is scant. *See* PHYTOCHROME.

Physiological role. Since phytochrome plays a regulatory role in such a variety of physiological and developmental processes, considerable effort has been taken to elucidate the mechanisms of the various responses. It is usually assumed (although far from proved) that the primary action of P_{fr} is the same in all cases; subsequent reaction chains differ widely depending upon the system, the stage of development, or prior environmental conditions.

Seedling development. In order to survive and reproduce, most plants other than fungi need to produce a light-capturing apparatus equipped to carry on photosynthesis efficiently; and this apparatus is the leaf with its chloroplasts. Higher plants have evolved mechanisms to ensure that once a seed has germinated and growth has commenced, little stored energy will be expended on leaf development until the potentially green leaf encounters light conditions adequate for photosynthesis. Rather, the energy is expended in elongation processes which maximize chances that the undeveloped leaves will reach such light conditions. Both phytochrome and the blue-light photoreceptor are key parts of this system for regulating seedling growth and development.

Some seeds, frequently those with only small reserves of food, require light for germination. The requirement for an adequate concentration of P_{fr} before germination progresses ensures that a seed either buried too deeply in soil or heavily shaded by other plants will not germinate and exhaust its stored resources before reaching light sufficient for photosynthesis.

Dicotyledonous seedlings generally have broad expanded leaves, except when grown in darkness. Without P_{fr}, generated by light treatment, the leaves are small, devoid of chlorophyll, and usually folded up and tucked beneath a protective hook of the stem which helps prevent leaf damage as the stem tip pushes up through the soil or litter. With the first light, the hook straightens out to orient the leaves favorably for photosynthesis, and they begin rapid expansion. In beans, rapid elongation of the cells on the inside of the hook causes it to straighten out. Ethylene, a gaseous plant growth regulator, is continually synthesized, escapes from the dark-grown hook, and in some manner regulates growth to maintain the hook. Formation of P_{fr} somehow diminishes ethylene synthesis, and the hook, released from the restraint on growth caused by ethylene, proceeds to open. Meanwhile, in many dicotyledons, the rate of stem elongation begins a dramatic decline. This growth inhibition is mediated not by phytochrome, but rather by a blue-light photoreceptor.

Etiolated dicotyledonous leaves are normally a small fraction of their normal size in light. Not only is enlargement triggered by light, but continued exposure to light is required to maintain growth to normal size. Among monocotyledonous leaves, those of grasses remain tightly rolled in the dark and unfurl upon exposure to light. The tightly rolled leaves can be induced to unroll either by P_{fr} formation or by application of the growth regulator gibberellin. Several workers have shown that isolated etioplasts from dark-grown grass leaves will release biologically active gibberellin within a few minutes upon exposure to red light, with the release prevented by subsequent far red. In this case many of the steps between phytochrome phototransformation and biological response are relatively well known.

To ensure upward growth, seedling stems are especially sensitive in their geotropic curvature response to the stimulus of gravity. To facilitate growth toward a source of light, they are also responsive phototropically. The phototropic response is mediated by light absorbed by the blue-light photoreceptor rather than by phytochrome, but the sensitivity to blue light can be dramatically altered through P_{fr} formation. Thus a delicate balance is achieved between the response designed to raise a plant vertically above its neighbors and one designed to direct it toward light from some particular direction. In addition to the dramatic growth inhibition brought about by blue and ultraviolet light, there is slower but longer-lasting growth inhibition caused by P_{fr}. There are important interactions between this effect of P_{fr} and gibberellins. *See* GIBBERELLIN.

Many roots, if allowed to develop in darkness, are insensitive to gravity. Such roots are free to follow moisture and nutrient gradients as they penetrate the soil. Formation of a minute amount of P_{fr} (probably within the root cap) induces geotropic sensitivity such that the root rapidly turns downward. This response most likely protects a root against breaking through the soil and into the air where it could rapidly become desiccated and killed.

A minimal investment of plant food reserves is spent in the formation of chloroplasts in dark-grown leaves. These chloroplasts (etioplasts) are incapable of photosynthesis because they have no chlorophyll and only a fraction of the needed internal membrane system, and they lack the full complement of enzymes needed to fix carbon dioxide into sugar. Light absorbed both by protochlorophyll and later by chlorophyll itself initiates and sustains development to a mature functioning chloroplast. The light absorbed by protochlorophyll converts it directly to chlorophyll, the principal photosynthetic pigment. The rate of subsequent chloroplast development can be markedly affected by light pretreatments which form small amounts of P_{fr}. The rapid phase of chlorophyll formation and chloroplast development may be advanced several hours by P_{fr}. *See* CHLOROPHYLL.

Continuing vegetative growth. Light-grown plants contain phytochrome; thus the destruction (decay) of the more labile P_{fr} formed in the light does not result in complete disappearance of the pigment. It can be detected spectrophotometrically and immunologically in extracts from green leaf tissue, and its presence is shown by innumerable experiments on light-grown plants. For example, if bean plants are grown under white light, their photosynthetic needs are met. At the beginning of the dark period, plants can be treated with red or far-red light to put phytochrome mainly in either the P_r or P_{fr} form, respectively. The far-red-treated plants have longer stems and petioles. Thus P_{fr} present during the dark period in some manner inhibits elongation. A large portion of this inhibition can be reversed with the growth regulator gibberellin; the manner of interaction is unknown. Many other plants, including woody species such as pines and various broadleaf trees, respond similarly. Supplemental far-red during the light period may have much the same effect as far-red to terminate the light period.

Light passed through green leaves is relatively much richer in far-red than unobstructed sunlight. This fact provides a natural counterpart to the experiments above. Rapid elongation of plants growing in the shade of other plants has a clear survival advantage in aiding them to reach photosynthetically adequate light intensities.

Response to daylength (photoperiodism). In regions with distinct seasons of alternating favorable and unfavorable growing conditions, the precise change in daylength is a more dependable warning of an oncoming seasonal change than are the vagaries of temperature or rainfall. Plants have evolved mechanisms of responding to the shortening days of late summer and fall (short-day plants) or the lengthening days of the spring (long-day plants).

The most obvious physiological response is flower initiation. Flowering is timed to occur soon enough for seed production to be complete prior to the onset of unfavorable conditions. It is also delayed enough to allow adequate vegetative growth to support the formation of fruit and seeds. The protection of growth centers, the meristems of the shoot, is often ensured by photoperiodic induction of dormant buds in trees and shrubs. For example, shortening days cause the formation of abscisic acid, a plant growth inhibitor, in the leaves of birch trees. It is apparently transported to the stem apices, where it stops stem elongation and induces formation of bud scales and formation of a dormant bud. The bud is reactivated only after progression through the cold conditions of winter and exposure to the warmth of spring. *See* ABSCISIC ACID; DORMANCY.

Phytochrome is involved in photoperiodism primarily as a monitor of light. A short-day plant, for instance, needs an uninterrupted dark period of more than some critical number of hours in order to flower at all, to increase the amount of flowering, or to flower at an earlier stage of development. Biochemical processes which occur in leaves during the dark period lead to the formation of a flower-inducing principle which is moved to the meristems in shoot tips. This factor changes the course of development from vegetative to reproductive growth, the formation of flowers. Probably a reduction in flowering inhibitor is also required. If P_{fr} is formed in the leaves by a brief flash of light during the dark period, formation of the floral stimulus is interrupted. If neither of the two parts of the interrupted night is as long as the critical dark period, no flowering ensues. If far-red light treatments immediately follow the interrupting light, thus converting the P_{fr} back to P_r, the short-lived existence of P_{fr} may be insufficient to interrupt synthesis of the flowering stimulus. P_{fr} present for 30 min is sufficient to inhibit flowering in the cocklebur (*Xanthium*); its presence for as little as 30 s is sufficient for the Japanese morning glory (*Pharbitis nil*).

Many plants are capable of differentiating between dark periods differing by only 30 min, and one variety of rice responds very differently to nights varying by only 10 min in length. The nature of this precise measurement of night length is not clear at present. Strong evidence implicates the "biological clock" involved in circadian rhythms, perhaps in combination with phytochrome. If short-day plants are put onto an abnormally long night, 72 h, for example, sensitivity to P_{fr} for inhibition of flowering is not constant, but is maximal about every 24 h with periods of complete insensitivity between. The way in which P_{fr}-produced reactions and the clock are coupled remains unknown.

Long-day plants, by contrast with short-day plants, require a dark period shorter than some critical length in order to flower, and may flower under continuous light. On a normal 24-h daily light and dark cycle, an uninterrupted dark period greater than the critical length inhibits flower induction. If such an inhibitory dark period is interrupted in these plants by P_{fr} formation, the inhibitory effect of the dark period is prevented and the plants flower. Prompt removal of the P_{fr} by far-red light reverses the effect, and the inhibitory state is maintained.

There are several hypotheses aimed at explaining the role of phytochrome in photoperiodism, but these are all conjectural. All that is certain is that P_{fr} indicates daytime in the metabolic circuitry of the leaf. It remains to be seen whether the many diverse phytochrome responses, including the multitude that are photomorphogenic, from the fastest to the slowest, can be traced to a single function of the pigment. *See* PHOTOPERIODISM; PLANT GROWTH.

Winslow R. Briggs

Bibliography. R. E. Kendrick and G. H. Kronenberg (eds.), *Photomorphogenesis in Plants*, 2d ed., 1993; W. Shropshire, Jr. (ed.), *Photomorphogenesis*, 1983; H. Smith and M. G. Holmes (eds.), *Techniques in Photomorphogenesis*, 1984.

Photomultiplier

A very sensitive vacuum-tube detector of light or radiant flux containing a photocathode which converts the light to photoelectrons; one or more secondary-electron-emitting electrodes or dynodes which amplify the number of photoelectrons; and an output electrode or anode which collects the secondary electrons and provides the electrical output signal. It is less frequently known as a multiplier phototube or photomutiplier tube. Because of the very large amplification provided by the secondary-emission mechanism, and the very short time variation associated with the passage of the electrons within the device, the photomultiplier is applied to the detection and measurement of very low light levels, especially if very high speed of response is required.

The first photomultipliers were developed during the late 1930s; applications were generally in astronomy and spectrometry. Development of photomultipliers was stimulated beginning in the late 1940s by their application to scintillation counting. Variations in electron-optical design and the introduction of a variety of photocathode, dynode, and envelope materials led to the use of photomultipliers in a wide variety of applications. Active design programs continue despite the competition from various solid-state detectors. Photomultiplier tubes are available in sizes ranging from $^1/_2$ to 5 in. (13 to 127 mm) in diameter, with photocathodes useful for detecting radiation having wavelengths from 110 nanometers, the deep ultraviolet (using a lithium fluoride window), to 1100 nm, the infrared (with special photocathodes), and with amplification factors ranging from 10^3 to 10^9.

Operation and design. The schematic of a typical photomultiplier in **Fig. 1** illustrates its operation. Light incident on a semitransparent photocathode located inside an evacuated envelope causes photoelectron emission from the opposite side of the photocathode. (Some photomultipliers are designed with the photocathode mounted inside the vacuum envelope to provide for photoemission from the same side of the photocathode on which the light is incident.) The efficiency of the photoemission process is called the quantum efficiency, and is the ratio of emitted photoelectrons to incident photons (light particles). Photoelectrons are directed by an accelerating electric field to the first dynode, where from 3 to 30 secondary electrons are emitted from each incident electron, depending upon the dynode material and the applied voltage. These secondaries are directed to the second dynode, where the process is repeated and so on until the multiplied electrons from the last dynode are collected by the anode.

A typical photomultiplier may have 10 stages of

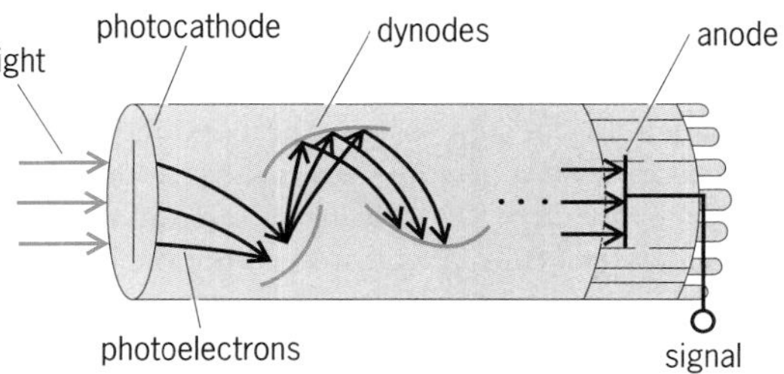

Fig. 1. Schematic of a photomultiplier. (*After R. W. Engstrom, Photomultipliers—then and now, RCA Eng., 24(1):18–26, June–July 1978*)

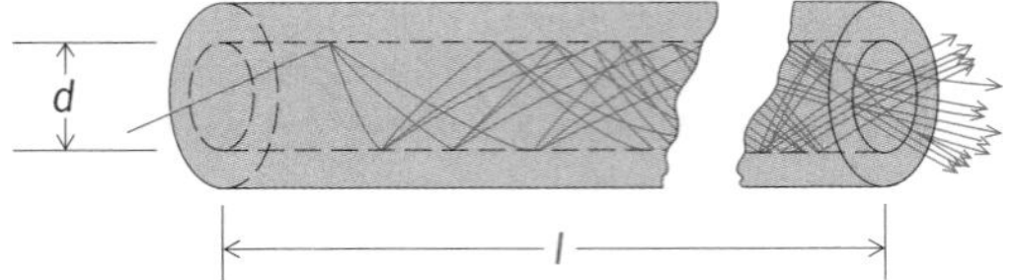

Fig. 2. Continuous-channel multiplier structure. (*After RCA Photomultiplier Manual, Tech. Ser. PT-61, 1970*)

secondary emission and may be operated with an overall applied voltage of 2000 V. In most photomultipliers the focusing of the electron streams is done by electrostatic fields shaped by the design of the electrodes. Some special photomultipliers designed for very high speed utilize crossed electrostatic and magnetic fields which direct the electrons in approximate cycloidal paths between electrodes. Because the transit times between stages are nearly the same, anode pulse rise times as short as 150 picoseconds are achieved.

Another special photomultiplier design is based on the use of microchannel plates. A single-channel multiplier is schematically shown in **Fig. 2**. The channel is coated on the inside with a resistive secondary-emitting layer. Gain is achieved by multiple electron impacts on the inner surface as the electrons are directed down the channel by an applied voltage over the length of the channel. The gain of the channel multiplier depends upon the ratio of its length l to its diameter d; a typical ratio is 50:1. A microchannel plate is formed by combining a large number of the channels in parallel with spacings on the order of 40 micrometers. A very high-speed photomultiplier utilizes a microchannel plate mounted in a closely spaced parallel arrangement between a photocathode and an anode. Time resolutions of less than 100 ps are achieved.

Dynode materials. Common dynode materials used in photomultiplier tubes are cesium antimonide (Cs_3Sb), magnesium oxide, and beryllium oxide. The surface of the metal oxide dynodes is modified by activation with an alkali metal such as cesium which lowers the surface potential barrier, permitting a more efficient escape of the secondary electrons into the vacuum.

Materials characterized as negative-electron affinity (NEA) have been developed and utilized in photomultipliers for both dynodes and photocathodes. A typical NEA secondary-emission material is gallium phosphide (GaP) whose surface has been treated with cesium (GaP:Cs). **Figure 3** shows the energy band model of an NEA material. The surface barrier is reduced by the electropositive cesium layer, and band-bending occurs so that the conduction band may actually lie above the vacuum level as indicated.

When a primary electron impacts an NEA material, secondary electrons are created within the material, and even low-energy electrons escape to the vacuum. The result is a much higher secondary-emission ratio than that achieved in other secondary emitters, especially at higher primary energies when the primary electrons penetrate more deeply into the material. Secondary electrons within an emitter lose energy before reaching the surface, so that in ordinary materials the secondary emission actually decreases with increase in voltage above an energy of perhaps 500 eV. For example. Cs_3Sb reaches a maximum secondary emission of about 8:1 at 500 V. At the same primary voltage GaP:Cs has a secondary emission of 25:1, and secondary emission increases linearly with voltage to a ratio of at least 130:1. *See* SECONDARY EMISSION; SEMICONDUCTOR.

Photocathode materials. The first photomultipliers developed utilized a photocathode of silver oxide activated with cesium (Ag-O-Cs). Although the quantum efficiency of this photocathode is quite low (less than 1% at the wavelength of its maximum response), it is still used because of its near-infrared response out to 1.1 μm.

Most photomultipliers today have photocathodes classed as alkali-antimonides: Cs_3Sb, K_2CsSb, Na_2KSb, Na_2KSb:Cs, Rb-Cs-Sb. These photocathodes have good quantum efficiency in the visible-wavelength range—in some cases exceeding 20%. The NEA type of photocathodes such as GaAs:Cs and InGaAs:Cs have even higher quantum efficiency and response through the visible and into the near infrared. However, they are difficult to fabricate and are more readily damaged by excessive photocurrents. *See* PHOTOTUBE.

Detection limits. Because there is very little noise associated with the amplification process in a photomultiplier, detection limits are primarily determined by the quantum efficiency of the photocathode, by the thermionic or dark emission of electrons from the photocathode, and by the bandwidth of the observation. A typical photomultiplier limited by dark emission is capable of detecting 10^{-12} lumen—the equivalent of the flux on 1 cm^2 from a l-candlepower lamp at a distance of 6 mi (10 km). By selecting a photomultiplier having a photocathode with low dark emission, restricting the effective photocathode area magnetically, and cooling the tube to reduce dark emission further, a background count as low as one or two electrons emitted from the photocathode per second can be achieved. Against this background, with a photocathode having 20% quantum efficiency, the incidence of a flux of only 10 photons per second

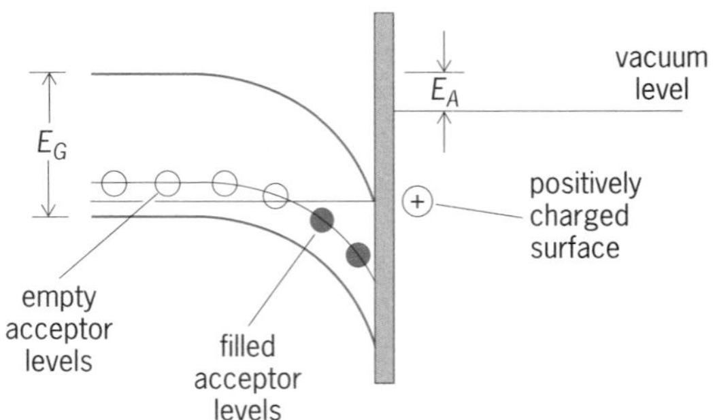

Fig. 3. Semiconductor energy-band model showing negative electron affinity, E_A. Band-gap energy = E_G. (*After RCA Photomultiplier Manual, Tech. Ser. PT-61, 1970*)

can be detected in an observation time of perhaps 10 s. In this technique, referred to as photon counting, each electron originating at the photocathode results in an output pulse which is counted. An evaluation must be made by statistically comparing the cases of source-on and source-off. Such techniques were applied successfully in 1969 in the laser-ranging experiments by using a retroreflector placed on the Moon by the *Apollo 11* astronauts. *See* THERMIONIC EMISSION.

Applications. Ever since their invention, photomultipliers have been found useful in low-level photometry and spectrometry. The many applications of photomulipliers include high-speed inspection of small objects such as fruits, seeds, toys, and other industrial products; pollution monitoring; laser ranging; and process control with transmitted or reflected light to detect flaws in various solid, liquid, or gaseous manufacturing operations. *See* LASER; NONDESTRUCTIVE EVALUATION; PHOTOMETRY; SPECTROSCOPY.

Scintillation counting. The most important applications of photomultipliers are related to scintillation counting. In a scintillation counter, gamma rays produce light flashes in a material such as NaI:Tl. A photomultiplier is optically coupled to the scintillator and provides a count of the flashes and a measure of their magnitude. The magnitude of the scintillation flash is proportional to the energy of the gamma ray, thus enabling the identification of particular isotopes. A whole science of tracer chemistry has developed using this technique and is applied to agriculture, medicine, and industrial problems. *See* GAMMA-RAY DETECTORS; SCINTILLATION COUNTER.

Gamma-ray camera. The gamma-ray camera developed in the late 1950s has proved to be a very valuabel medical tool. In this application, a radioactive isotope combined in a suitable compound is injected into or fed to a patient. Certain compounds and elements concentrate preferentially in particular organs, glands, or tumors: for example ^{131}I concentrates in the thyroid gland. Gamma rays which are then emitted are detected by a large scintillator, and an array of photomulitiplier tubes provides spatial data on the gamma-ray source. The technique is widely used in diagnostic medicine, especially in locating tumors. *See* NUCLEAR MEDICINE; RADIOLOGY.

Computerized tomography. The CT or CAT (computerized axial tomographic) scanner is a similar medical instrument developed during the 1970s. This instrument uses a pencil or fan-beam of x-rays that rotates around the patient. Several hundred photomultiplier-scintillator combinations surround the patient and record density from many different positions. These data are analyzed by a computer, thus providing a cross-section density map of the patient. The CT scanner is particularly useful because of the two-dimensional data provided, although it does not supply the functional information of the gamma-ray camera. Since 1980, silicon photocells have been replacing photomultipliers as the detectors of the scintillations. *See* COMPUTERIZED TOMOGRAPHY.

Positron camera. A similar development is the positron camera. This device uses radioisotopes, such as ^{11}C, ^{13}N, and ^{15}O, that have short half-lives and emit positrons. When a disintegration occurs, the positron is annihilated by an electron, and the result is a pair of oppositely directed gamma rays detected in coincidence with photomultiplier-scintillator detectors on opposite sides of the patient. Functional information is provided and analyzed in a manner similar to that of a CT scanner.

Ralph W. Engstrom

Bibliography. E. L. Dereniak and D. G. Crowe, *Optical Radiation Detectors*, 1984; J. Schanda and I. Ungvari (eds.), *Photon Detectors*, 1987.

Photon

An entity that can be loosely described as a quantum of energy of electromagnetic radiation. According to classical electromagnetic theory, an electromagnetic wave can transfer arbitrarily small amounts of energy to matter. According to the quantum theory of radiation, however, the energy is transferred in discrete amounts. The energy of a photon is the product of Planck's constant and the frequency of the electromagnetic field. In addition to energy, the photon possesses momentum and also possesses angular momentum corresponding to a spin of unity. The interaction of radiation with matter involves the absorption, scattering, and emission of photons. Consequently, the energy interchange is inherently quantized. *See* ANGULAR MOMENTUM; ENERGY; MOMENTUM; SPIN (QUANTUM MECHANICS).

Particlelike behavior. For many purposes, the photon behaves like a particle of zero rest mass moving at the speed of light. Unlike spin-1 particles of finite mass, the photon has only two, rather than three, polarization states. This corresponds to the transverse nature of a classical electromagnetic wave in free space. The particlelike nature of the photon is vividly exhibited by the photoelectric effect, predicted by A. Einstein, in which light is absorbed in a metal, causing electrons to be ejected. An electron absorbs a photon, gaining its energy. In leaving the metal, it loses energy because of interactions with the surface; the energy loss equals the product of the so-called work function of the surface and the charge of the electron. The final kinetic energy of the electron therefore equals the energy of the incident photon minus this energy loss. *See* PHOTOEMISSION.

A second demonstration of the particlelike behavior of photons is provided by the scattering of an x-ray photon from an electron bound in an atom. The electron recoils because of the momentum of the photon, thereby gaining energy. As a result, the frequency, and hence the wavelength of the scattered x-ray, is altered. If the x-ray is scattered through a certain angle, the wavelength is shifted by an amount determined by this scattering angle and the mass of an electron, according to the laws of conservation of energy and momentum. *See* COMPTON EFFECT.

Quantum theory. From a more fundamental view, the photon is the quantum of excitation of a single mode of a radiation field. The dynamical equations for the electric and magnetic energy in such a field are identical to those of a harmonic oscillator. According to quantum theory, the allowed energies of a harmonic oscillator are given by $E = (j + 1/2)hf$, where h is Planck's constant, f is the frequency of the oscillator, and the quantum number $j = 0, 1, 2, \ldots$, describes the state of excitation of the oscillator. This quantum relation was first postulated by M. Planck for the material oscillators in the walls of a thermal enclosure in order to obtain the correct form for the density of radiation in a thermal field, but it was quickly applied by Einstein to describe the state of the radiation field itself. In this picture, j describes the number of photons in the field. *See* HARMONIC OSCILLATOR.

Photon distributions. Different sources of light are characterized by photon distributions with fundamentally different statistical properties. If light from a monochromatic source at high intensity, for instance a laser source, impinges on a photomultiplier, then each signal count corresponds to the absorption of a single photon. If repeated measurements are made, then the fractional root-mean-square (rms) scatter of the readings is exactly as would be expected for independent particles. In the limit of large numbers of absorbed photons, the fluctuations are unimportant, and the photon distribution describes a field with well-defined amplitude and phase, a classical field.

In contrast, for thermal radiation the fractional fluctuation is much larger than for independent particles. The larger fluctuation expresses the fact that photons, having integral spin, obey Bose-Einstein statistics. Such particles tend to clump together, producing large fluctuations. Finally, a radiation field can be considered in which the photon number is precisely determined, called a Fock state. In such a case, the phase of the corresponding electromagnetic field is completely random. Such a state has no classical analog. *See* BOSE-EINSTEIN STATISTICS; SQUEEZED QUANTUM STATES.

Need for quantized fields. Many of the processes by which radiation interacts with matter can actually be described without employing a quantized description of the radiation field or the concept of photons, that is, by treating the radiation field classically. However, the fundamental process of spontaneous emission, in which an excited atom or molecule emits a photon and makes a transition to its ground state, can be described only on the basis of a quantized field. Also, many statistical properties of the radiation field, for instance correlations between photons that are emitted simultaneously in a nonlinear process, are inexplicable in terms of classical fields but have a natural explanation in the quantum theory of radiation and the concept of photons. *See* NONRELATIVISTIC QUANTUM THEORY; QUANTUM ELECTRODYNAMICS; QUANTUM FIELD THEORY; QUANTUM MECHANICS.

Daniel Kleppner

Bibliography. C. Cohen-Tannoudji, J. Dupont-Roc, and G. Grynberg, *Atom-Photon Interactions*, 1992, reprint 1998; R. Loudon, *The Quantum Theory of Light*,, 3d ed., 2000; P. Meystre and M. Sargent III, *Elements of Quantum Optics*, 3d ed., 1999; P. W. Milonni, *The Quantum Vacuum*, 1997.

Photoperiodism

The growth, development, or other responses of organisms to the length of night or day or both. Photoperiodism has been observed in plants and animals, but not in bacteria (prokaryotic organisms), other single-celled organisms, or fungi.

Parameters

A true photoperiodism response is a response to the changing duration of day or night. Responses to the quantity of light energy in photoperiodism are secondary complications. Some species respond to increasing day lengths and decreasing night lengths (for example, by forming flowers or developing larger gonads); this is called a long-day response. Other species may exhibit the same response, or the same species may respond in some different way, to decreasing days and increasing nights; this is a short-day response. Sometimes a response is independent or nearly independent of day length and is said to be day-neutral.

Plant responses. Figure 1 illustrates these phenomena as they apply to the initiation of flowers in plants—probably the most extensively studied photoperiodism response. Flowering can be measured in various ways such as counting the number of flowers on each plant, classifying the size of the buds

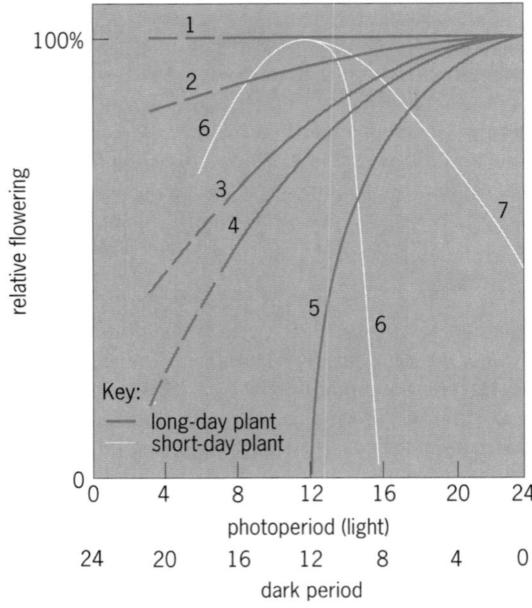

Fig. 1. Flowering (and other) responses to various day lengths. Each line represents a different hypothetical plant.

according to a series of arbitrary stages, or taking the inverse of the number of days until the first flower appears. A truly day-neutral plant flowers about the same amount on all day lengths (curve 1). (With many species, there is little or no flowering when days are unusually short, for example, from 2 to 6 h long.) Curve 2 represents a plant that is slightly but probably insignificantly promoted in its flowering by long days. Plants represented by curves 3 and 4 are quantitatively promoted by long days (to different degrees), although they flower on any day length. A qualitative or absolute long-day plant (5) such as henbane flowers only when days are longer than some minimum, 12 h in the example. A qualitative short-day plant (6) such as cocklebur in this example flowers only when days are shorter than about 15.6 h and nights are longer than 8.3 h. Note that cocklebur also fails to flower if days are shorter than 3–5 h. (In that sense, it is also a long-day plant.) A quantitative short-day plant (7) flowers on any day length but better under short days. Note that different species have different critical day and night lengths, not just the 12 and 15.6 h shown for henbane and cocklebur; yet as the curves show, the long-day and the short-day responses are qualitatively opposite to each other. They can overlap so that both long-day and short-day plants flower on some day lengths (curves 5 and 6).

It is important to emphasize that a true photoperiodism response is to the duration of day or night and not to the total quantity of light received by the organisms. Typically, 8 h at a high light level followed by 16 h of darkness is perceived by the organism as a short day, whereas 8 h at the same high level, plus 8 h of extremely dim light (perhaps a thousandth as bright as the high level), followed by 8 h of darkness, is perceived as a long day.

There are many plant responses to photoperiod (see **table**). These include development of reproductive structures in lower plants (mosses) and in flowering plants; rate of flower and fruit development; stem elongation in many herbaceous species as well as coniferous and deciduous trees (usually a long-day response and possibly the most widespread photoperiodism response in higher plants); autumn leaf drop and formation of winter dormant buds (short days); development of frost hardiness (short days); formation of roots on cuttings; formation of many underground storage organs such as bulbs (onion, long days), tubers (potato, short days), and storage roots (radish, short days); runner development (strawberry, long days); flower formation (strawberry, short days); balance of male to female flowers or flower parts (especially in cucumbers); aging of leaves and other plant parts; and even such obscure responses as the formation of foliar plantlets (such as the minute plants formed on edges of *Bryophyllum* leaves), and the quality and quantity of essential oils (such as those produced by jasmine plants). Note that a single plant, for example, the strawberry, might be a short-day plant for one response and a long-day plant for another response.

Animal responses. There are also many responses to photoperiod in animals (see table), including control of several stages in the life cycle of insects (for example, diapause) and the long-day promotion in birds of molting, development of gonads, deposition of body fat, and migratory behavior. Even feather color may be influenced by photoperiod (as in the ptarmigan). In several mammals the induction of estrus and spermatogenic activity is controlled by photoperiod (sheep, goat, snowshoe hare), as is fur color in certain species (snowshoe hare). Growth of antlers in American elk and deer can be controlled by controlling day length. Increasing day length causes antlers to grow, whereas decreasing day length causes them to fall off. By changing day lengths rapidly, a cycle of antler growth can be completed in as little as 4 months; slow changes can extend the cycle to as long as 2 years. When attempts are made to shorten or extend these limits even more, the cycle slips out of photoperiodic control and reverts to a 10–12 month cycle, apparently controlled by an internal annual "clock." *See* MIGRATORY BEHAVIOR.

Seasonal responses. Response to photoperiod means that a given manifestation will occur at some specific time during the year. Day lengths are calculated as the time from the moment the upper-most part of the Sun (rather than its center) first touches the eastern astronomical horizon, that is, the horizon as it would appear on the open ocean, in the morning until the upper part of the Sun just touches the western astronomical horizon in the evening (**Fig. 2*a***). Thus, the day length at the Equator is slightly longer than 12 h. At the Arctic Circle, 66.5°N latitude, the center of the Sun just touches the astronomical horizon at midnight on the summer solstice, but part of the Sun is visible above the horizon at midnight for a few days before and after the summer solstice, as is evident from the day-length curve for 66.5°N latitude. Because of twilight, the day length perceived by an organism would be somewhat longer than the curves show. In addition, different organisms respond to different light levels during twilight.

Response to long days (shortening nights) normally occurs during the spring, and response to short days (lengthening nights) usually occurs in late summer or autumn. Since day length is accurately determined by the Earth's rotation on its tilted axis as it revolves in its orbit around the Sun, detection of day length and its rate of change (Fig. 2*b*) provides an extremely accurate means of determining the season at a given latitude. Such other environmental factors as temperature and light levels also vary with the seasons but are clearly much less dependable from year to year.

Ecological relationships. All this is clearly important in the distribution of plants and animals. Wild plants, for example, growing in northern latitudes usually flower only when days are long, while plants from more southerly climes respond when days are shorter. The response type may be either long-day or short-day, but the critical day length (the day length that must be exceeded for long-day plants or not

Selected examples of photoperiodism in plants and animals*

Organism	Short/day	Longday	Response promoted by indicated day length
Algae			
Porphyra tenera (Rhodophyta)	X		Production of filamentous phase of spores that gives rise to leafy phase
Scytosiphon lomentaria (Phaeophyta)		X	Production of erect, cylindrical thalli from crustose phase
Bryophytes			
Sphagnum plumulosum	X(?)		Development of reproductive structure
Marchantia polymorpha		X	Development of reproductive structure
Gymnosperms			
Loblolly pine (*Pinus taeda*)		X	Stem growth
Angiosperms (flowering plants)			
Monocots			
Bluestem grass (*Andropogon gerardi*)	X		Flower formation
Wild oats (*Avena sativa*)		X	Flower formation
Onion (*Allium cepa*)		X	Bulb formation
Orchid (*Cattleya trianae*)	X		Flower formation and bud elongation
Yam (*Dioscorea alata*)	X		Tuber development
Dicots			
Soybean (*Glycine max*)	X		Flower formation
Sugarbeet (*Beta vulgaris*)		X	Flower formation
Radish (*Raphanus sativus*)	X		Root thickening
		X	Flower formation
Strawberry (*Fragaria chiloensis*)		X	Runner development
	X		Flower formation
Coneflower (*Rudbeckia* spp.)		X	Stem elongation and flowering
Cucumber (*Cucumis sativus*)	X		Female flowers increased
		X	Male flowers increased
Alfalfa (*Medicago sativa*)	X		Winter hardening
Potato (*Solanum tuberosum*)	X		Tuber development
Dahlia (*Dahlia* spp.)	X		Storage root development
		X	Fibrous root development
Bryophyllum (*Bryophyllum pinnatum*)		X	Development of foliar plantlets
	X		Flower formation
Kalanchoë (*Kalanchoë blossfeldiana*)	X		Dark fixation of CO_2
White mustard (*Sinapis alba*)	X		Flower development
Begonia (*Begonia evansiana*)		X	Formation of aerial tubers
Tulip poplar (*Liriodendron tulipifera*)	X		Cessation of vegetative growth, start of bud dormancy
Insects			
Mite (*Metatetranychus ulmi*)		X	Uninterrupted development
	X		Initiation of diapause
Mulberry silkworm (*Bombyx mori*)		X	Female lays diapause eggs
	X		Female lays developing eggs
Arthropods			
Water flea (*Daphnia pulex*)		X	Hatching of diapaused embryos
Green vetch aphid (*Megoura viciae*)		X	Production of parthenogenetic females instead of sexual females
Pink bollworm of cotton (*Pectinophora gossypiella*)	X		Diapause
Fish			
European minnow (*Phoxinus phoxinus*)		X	Egg maturation
Sickleback (*Gasterosteus aculeatus*)		X	Sexual maturation; male nest-building, spermatogenesis
Reptile			
Lizard (*Xantusia vigilis*)		X	Accelerated gonadal growth
Birds†			
Junco (*Junco hyemalis*)		X	Increase in body weight, marked deposition of fat, gonadal renewal, migratory behavior
Willow ptarmigan (*Logopus logopus*)	X		White winter plumage
Turkey (*Meleogris gallopavo*) and female chicken (*Gallus gallus*)		X X	Acceleration of sexual maturation and stimulation of egg production
White-crowned sparrow (*Zonotrichia leucophrys gambelii*)		X	Testicle development
White-crowned sparrow (*Z. leucophrys gambelii*)	X		Elimination of refractory period so sexual cycle can recur
Mammals			
Sheep (*Ovis* spp.) and goat (*Capra* spp.)	X		Estrus; spermatogenic activity, hair growth
Snowshoe hare (*Lepus americanus*); also ermine (*Mustela erminea cicognani*)		X X	Spring molt, summer pelage, estrus, testicular development
Snowshoe hare (*Lepus americanus*)	X		Autumn molt, winter pelage
Raccoon (*Procyon lotor*)		X	Estrus, male reproductive activity
Red fox (*Vulpes fulva*)	X		Autumn molt, winter pelage spermatogenesis, estrus
Golden hamster (*Mesocricetus aureus*)		X	Testicle development
Horse (*Equus equus*)		X	Estrus and associated hormonal changes in mare

*A given response is sometimes limited to a single genetic strain within a species.
† Rate of gonadal development is proportional to day length in many species of birds—no critical day length.

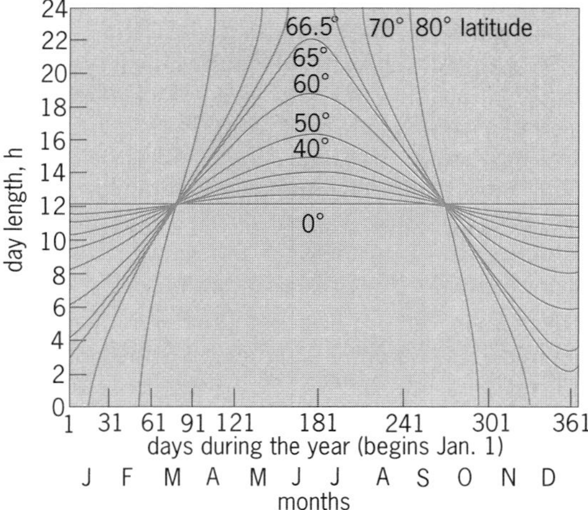

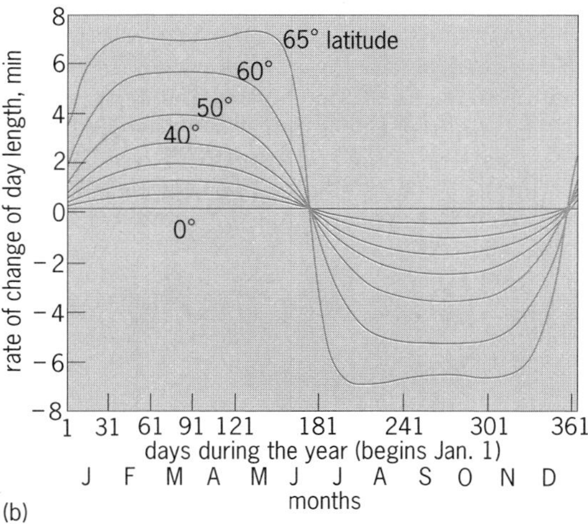

Fig. 2. Relationship of latitude and time of year on the length of day. (*a*) Day length as a function of time of year for various northern latitudes. (*b*) Rate of change of day length during the year. The more rapid rates of change occur during spring and autumn. (*After F. B. Salisbury and C. W. Ross, Plant Physiology, 4th ed., Wadsworth, 1991*)

exceeded for short-day plants) is nearly always shorter for plants in southern locations than for those to the north. This is also true for responses other than flowering. A given species of tree or shrub in Scandinavia, for example, has much longer critical day length for spring shoot elongation when it grows in the far north than when it grows farther south. Plants growing higher on Scandinavian mountains also have longer critical day lengths for stem elongation, meaning that they will begin growth later in the spring.

An examination of the list of responses in both plants and animals confirms these ecological relationships. It is especially evident that flowering of plants and breeding times of animals are often synchronized with the seasons by response to photoperiod. Most important is that the breeding times of members of a population are synchronized with each other by photoperiod. Actual flowering or breeding can occur at almost any time during the year other than coldest winter, but it would not be very successful if members of the same population were not able to breed at the same time. Humans are day-neutral for breeding time and can breed during any season. Indeed, photoperiodism responses in humans have not been clearly demonstrated. Is it possible that the long days provided by artificial lights might account for the fact that humans have become, on average, much taller during the past century or so, responding much as a bean plant would?

Basis for anticipation. Photoperiodism provides for anticipation in the development of organisms. Thus many plants drop their leaves, becoming frost-resistant and dormant, in response to the shortening days of autumn, and insects enter diapause (a dormant condition) in much the same way. Some birds and rodents change the colors of their feathers or fur as winter approaches and then again with the coming of spring, both in response to changing day lengths. Birds also prepare for their migrations by storing body fat and developing restlessness.

Agricultural applications. There are also important applications of photoperiodism in agriculture. Day length, in addition to restricting effects of temperature, moisture, or other factors, may limit the area where a given crop can be grown. Northern cultivars of short-day soybeans, for example, can sometimes be grown successfully only in a belt of latitude about 50 mi (80 km) wide. If they are grown farther north, too much foliage is produced and too few flowers and therefore too few seeds; too far south means an excess of flowers and not enough foliage. Plant breeders who need to have different cultivars flower at the same time to facilitate crossbreeding can often control flowering by controlling day length. This is also done in commercial production of ornamental plants. For example, chrysanthemums, poinsettias, and orchids are caused to flower by shortening days with black curtains over the greenhouse benches, or they are kept vegetative with lights hung above the benches. Crops can also be bred for appropriate critical day lengths, or day-neutral types can be selected so that flowering and fruiting are independent of day length and thus latitude (except as other factors may limit). On the animal side, turkey production is routinely controlled by controlling day length, and chickens lay more eggs on long days, or when the night is interrupted with light.

Discovery. The importance of day length was essentially overlooked until relatively recently, although A. Henfrey had suggested in 1852 that day length might influence plant distribution. The measurement of daily time by plants or animals must have seemed unlikely to nineteenth-century biologists. Nevertheless, Julian Tournois in Paris clearly discovered the short-day flowering responses of hops and hemp, publishing papers in 1910 and 1914. He observed that flowering in these plants occurred when days were shortened, but not when light levels were lowered and days lengthened to give the same quantity of light as with the short days. In Heidelberg, Georg Klebs observed flowering of houseleek plants that were exposed to several days of continuous illumination with electric lights, but he explained this long-day response in terms of nutrition and only suspected that the length of day was involved. Actually, the papers of Tournois and of Klebs had little impact on the science of plant physiology.

W. W. Garner and H. A. Allard discovered photoperiodism also and realized the importance of their finding; they gave the phenomenon its name in 1920. These workers observed, among other things, that Maryland mammoth tobacco flowered in their greenhouses in the winter but not during the summer in the fields. They investigated many of the environmental factors that might differ between summer fields and winter greenhouses, one of these being day length. Their tobacco plants bloomed profusely when subjected to days shorter than about 12 h but remained vegetative when days were longer. They soon tried a number of other species, documenting and naming the short-day, long-day, and day-neutral response types for flowering (**Fig. 3**). The work of Garner and Allard gained much attention, and photoperiodism has been a field of study ever since.

W. Rowan exposed juncos in the middle of winter to artificial light that extended the day, publishing his results in 1925. These birds normally migrate northward only in spring, but when Rowan released them in the middle of winter after their long-day exposure, they immediately flew northward. Since that time the responses listed in the table as well as many others have also been documented in animals.

Complications. Many organisms that are day-neutral in the sense that they will eventually flower or otherwise respond at virtually any day length are nonetheless promoted in their response by some specific day length. Tomato is often cited as a typical day-neutral plant for flowering, but at proper temperatures its flowering is significantly promoted by short days. Plants that require a given day length are said to be qualitative or absolute short-day or long-day plants, while those that would respond anyway but are nonetheless promoted by proper day length are said to be quantitative or facultative in response (Fig. 1). (In the absolute response, the quality, that is, the vegetative or reproductive condition, depends solely on photoperiod.) The majority of plants probably respond quantitatively for flowering; this is also true for stem elongation and many other responses.

Most animal responses are also of the quantitative type, but white-crowned sparrows have remained asexual for up to 2.5 years under short days, and

Fig. 3. Plant responses to photoperiodism. Short-day responses on left and long-day on right for each plant. (*a*) Tomato, a day-neutral plant (110 days). Both plants are flowering, although flowers are difficult to see. Note taller stem under long days, a typical response. (*b*) Cocklebur, a short-day plant for flowering (60 days). (*c*) Japanese morning glory, a short-day plant (35 days). (*d*) Spinach, a long-day plant (35 days). (*e*) Barley, a long-day plant (35 days). (*f*) Radish, a long-day plant (54 days). (*From F. B. Salisbury, The Flowering Process, Pergamon, 1963*)

their capacity to become sexual in response to long days remained unmodified during that time. The reproductive capacity of golden hamsters and sheep can be under complete control of day length for a long period of time, but eventually this control disappears. Hamster testes, for example, shrink to a small size on 12-h days but are full size on 13-h days—a sharp critical day. Placed under short-day regimens, the gonads soon shrink and cease to function, but after 25 to 30 weeks under short days, they begin to grow again, indicating the loss of photoperiodic control.

When, after a given period of time, an animal becomes unresponsive to further photoperiodic treatment of the same type, the animal is said to have entered the refractory period. Often the refractory period can be broken by treatment with an opposite day length. For example, gonadal development in certain sparrows is caused by daily 16-h light periods, but after several months the gonads begin to regress, even on these long days, and further long days will not cause them to develop again. Six weeks of 8-h days (16-h nights) restores the ability of a 16-h day to cause sexual development; that is, short days terminate the refractory period.

A somewhat similar phenomenon can be observed in certain species of plants, although the term refractory is not used by botanists. Some species flower only when short days are followed by long days (as happens in the spring) or when long days are followed by short days (as happens in autumn).

A few plant species (such as sugarcane and purple nutsage) apparently require an intermediate day length; they do not flower if days are either too short or too long. Their counterparts are also known: plants that flower on short or long days, but not on days of intermediate length.

Frequently there is a strong interaction between temperature and day length. Some species (such as Japanese morning glory) have an absolute short-day requirement at high temperatures but are day-neutral at lower temperatures (a few degrees above freezing). Opposite situations are known where plants have absolute or quantitative photoperiod requirements at low temperatures but are day-neutral at higher temperatures. Many plants require or are promoted in their flowering by days to weeks of temperatures just above freezing. Such a low-temperature promotion of flowering is called vernalization. Typically, a vernalization requirement is followed by a long-day requirement or promotion (as with black henbane and many cereals), but sometimes the low-temperature treatment must be followed by short days (as for certain cultivars of chrysanthemum and various grasses). *See* VERNALIZATION.

The complexity of response may be surprisingly intricate. Sweet william (*Silena armeria*) requires long days at 82°F (28°C) to flower but is completely day-neutral at 41 or 90°F (5 or 32°C). It will flower at any temperature when treated with GA_7 (one of the gibberellic acids). *See* GIBBERELLIN.

Species also differ in sensitivity to photoperiod. A common weed, lamb's-quarters (*Chenopodium rubrum*), responds as a minute seedling in a petri dish, and the seed leaves (cotyledons) of Japanese morning glory are highly sensitive to photoperiod. In cocklebur the seed leaves are not sensitive, but the first true leaf becomes highly sensitive when it is half expanded (plants about 30 days old). The bud also becomes competent to flower. Some plants flower only when they are several years old, although it is not known whether they are actually responding to photoperiod.

Frequently, as plants age, they become less dependent on photoperiod for flower induction. Some species show changes in sensitivity to photoperiod within a few days after their seeds sprout; others require many weeks to many months for such changes. This is reminiscent of the eventual loss of photoperiod control over gonadal development in hamsters.

Plants also vary in the number of photoperiodic cycles required for response. Cocklebur, Japanese morning glory, perennial ryegrass, and a few other species respond to a single short day or long day (long day refers to ryegrass). Biloxi soybean requires at least three cycles, while chrysanthemums and many other plants require many more. Indeed, nearly mature flowers will abort from chrysanthemum plants moved from inductive short-day cycles to long days. Such plants as cocklebur that respond to a single cycle become induced; that is, flowering continues even though the conditions that were necessary to cause it are no longer present.

Mechanisms

It has long been the goal of researchers on photoperiodism to understand the plant or animal mechanisms that account for the responses. Light must be detected, the duration of light or darkness must be measured, and this time measurement must be metabolically translated into the observed response: flowering, stem elongation, gonad development, fur color, and so forth. In spite of many years of intensive research, relatively little is known about these mechanisms. Enough is known, however, to suggest that the basic mechanisms differ not only between plants and animals but among different species as well. The roles (synchronization, anticipation, and so on) are similar in all organisms that exhibit photoperiodism, but the mechanisms through which these roles are achieved are apparently quite varied.

Seeing light and dark. In 1938 Karl Hamner and James Bonner investigated whether day or night was more important in photoperiodism. Among other things they interrupted the day with a short period of darkness and the night with a brief interval of light. Darkness during the day had essentially no effect on short-day cocklebur plants, but they completely failed to flower when the long dark period was interrupted by light. Other workers found that long-day plants were promoted in their flowering by a night interruption (**Fig. 4**). Again, the opposite nature of short-day and long-day responses is demonstrated. It was concluded from these experiments that day

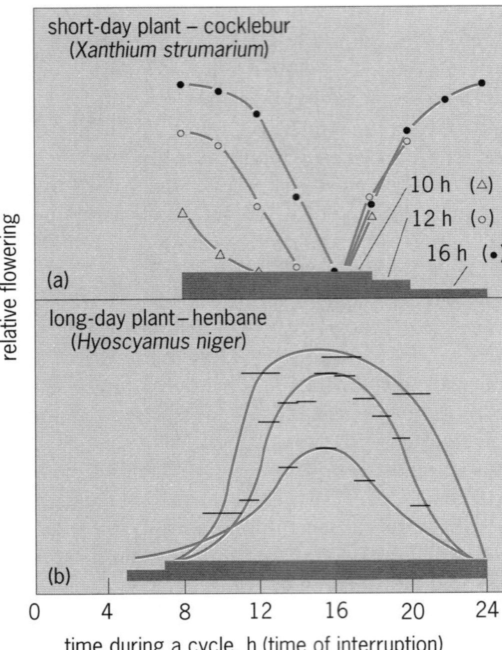

Fig. 4. Effects of a brief interruption of light given to plants during long dark periods. (*a*) Short-day plant inhibited in flowering by light interruptions (most effectively about 8 h after beginning of dark period) (*after F. B. Salisbury* and *J. Bonner, The reactions of the photoinductive dark period, Plant Physiol., 31:141–147, 1956*). (*b*) Long-day plant flowering promoted by a light interruption (1 or 2 h long as indicated by length of horizontal bars). Maximum promotion near middle of dark period (*after H. Claes and A. Lang, Die Blütenbildung von Hyoscyamus niger in 48 Stündigen Licht-Dunkel-Zyklien und in Zyklien mit aufgeteilten Lichtphasen, Z. Naturforsch., 26:56–63, 1947*).

length was of secondary importance to night length, but as noted below, the important role of photoperiod as well as darkness is now recognized. In any event, discovery of the effects of a night interruption led to an unexplored area of research in photoperiodism. This has been pursued mainly with plants.

Light interruption of the dark period. The results of studies on the effective times of light interruption are shown in Fig. 4. Strongest inhibition of flowering in short-day plants comes when the light interruption occurs around the time of the critical night (about 7–9 h for cocklebur plants), but actual effectiveness also depends on the length of the dark period. With short-day cockleburs, the shorter the night, the less the flowering and the longer the time that light inhibits flowering.

The irradiance or brightness level of an effective light interruption depends upon the duration of illumination. Higher levels mean shorter exposure times to produce a given inhibition or promotion—much as brighter light means shorter exposure time in photography. With cocklebur, a few seconds of light from four photo floodlamps held about a meter away are sufficient to saturate the inhibitory process, but complete inhibition can be produced by a naked 7.5-W incandescent bulb hanging all night a little over a meter away from the plants. Thus streetlights sometimes influence the photoperiod responses of nearby plants (for example, leaves might fail to fall in autumn).

Orange-red wavelengths used as a night interruption are by far the most effective part of the spectrum in inhibition of short-day responses and promotion of long-day responses (flowering in most studies), and effects of orange-red light can be completely reversed by subsequent exposure of plants to light of somewhat longer wavelengths, called far-red light. These observations led in the early 1950s to discovery of the phytochrome pigment system, which is apparently the molecular machinery that detects the light effective in photoperiodism of higher plants. (Phytochrome is also present in a few green algae and mosses but not in other algae, in fungi, or in animals.) Phytochrome works in both directions: red reversed by far-red, and far-red by red. Usually, several reversals are possible. Furthermore, many other plant responses can also be influenced in opposite ways by red and far-red light and thus are apparently under the control of the phytochrome pigment system. Red light promotes germination of many wild and a few domesticated seeds (such as lettuce) and formation of pigments in apple skins and other plant parts. Red light also has profound effects on seedlings that have been grown in the dark; for example, it inhibits their stem elongation but promotes leaf expansion and straightening of a hook commonly seen at the tip of the stem in such seedlings. Far-red light will reverse all these effects, which are usually but not always independent of photoperiod. *See* PHYTOCHROME.

Detecting dusk and dawn. In photoperiodism of short-day plants, an optimum response is usually obtained when phytochrome is in the far-red-receptive form during the day and the red-receptive form during the night. Although normal daylight contains a balance of red and far-red wavelengths, the red-receptive form is most sensitive, and so the pigment under normal daylight conditions is driven mostly to the far-red receptive form. At dusk this form is changed metabolically, and the red-receptive form builds up. It is apparently this shift in the form of phytochrome that initiates measurement of the dark period. This is how a plant "sees": when the far-red-sensitive form of the pigment is abundant, the plant "knows" it is in the light; the red-sensitive form (or lack of far-red form) provides a biochemical indication that it is in the dark.

As shown in **Fig. 5**, red light with a brightness indicated by the top of the range for inhibition of dark-period initiation is perceived by the plant as full day; dark measurement is unable to begin at that light level. At the bottom of the range the light is not detected by the plant at all; that is, the plant responds as though it were in total darkness, Thus as far as photoperiodism is concerned, a cocklebur plant responds almost as if it were controlled by a switch. During a period of about 5–11 min during twilight, the plant's mode of response changes from that of full daylight to that of complete darkness. This means that changing light levels during the day, including those caused by clouds, have no effect on the plant's photoperiodism responses.

Detection of moonlight. Figure 5 also shows the range of light levels to which a cocklebur plant is sensitive in the middle of the dark period (such as light from the 7.5-W incandescent bulb mentioned above). The range of sensitivity levels during the middle of a long dark period is about 10 times lower on the logarithmic scale than the range of sensitivity at the beginning of the dark period. Nevertheless, this range is somewhat above the level of light from a full moon. Based on these rather meager and preliminary data, it appears not only that cocklebur plants ignore (as far as photoperiodism goes) changing light levels during the day, but that they are also insensitive to moonlight during the night. A very few experiments with other species in which plants were exposed directly to moonlight suggest that those plants responded to moonlight at a barely detectable level.

Light detection in animals. No phytochrome is found in animals. As a rule, animals are responsive to several wavelengths of light, suggesting that the pigment that must respond to day and night absorbs at several wavelengths. Little work has been done to identify this pigment. A striking discovery is that the response to light can be independent or at least nearly independent of vision. In one experiment the eyes were removed from sparrows. From half of the test animals all the feathers were removed on the top of the head, while the other birds were injected with India ink just below the skin on the top of the head to keep light from reaching the brain. The sparrows were then subjected to extremely dim green light, given for 8 or 16 h each day. Sparrows with the plucked heads responded to the long days as usual with increased gonadal development, but those with the India ink did not. Apparently, the light penetrating the skull of plucked birds was sufficient to induce a photoperiod response in those eyeless birds.

The eyes of mammals must detect the light used in photoperiodism, since blinding several species of rodents held under long days caused the same response as transfer to short days. Even in this case, however, signals from the eyes appeared to be carried by nerve tracts other than the primary optic pathways, indicating that even in mammals vision may be separated from other light effects. As with the cocklebur plants, this could allow the animals' photoperiodism mechanism to "ignore" changing light levels during the day and perhaps at night. It is essential for many animals to be able to see at night, but if they used the same light for both vision and photoperiodism, measurement of day or night might be unreliable. Bypassing the visual system in animal photoperiodism may be analogous to bypassing photosynthesis in plants.

Time measurement. The measurement of time—durations of the day or night—is the very essence of photoperiodism. With the discovery of the reversible nature of phytochrome in higher plants, it was suggested that time measurement might simply be the time required for the pigment to change in the dark from the far-red- to the red-receptive form. If timing were this simple, it would resemble an hourglass, in which the sand runs through after the hourglass has been inverted until it has all run out, just as the phytochrome pigment might change from one form to the other until the change is complete. There are elements of hourglass timing in plants (for example, the longer the dark period, within limits, the more the flowering in various short-day species), and the aphid

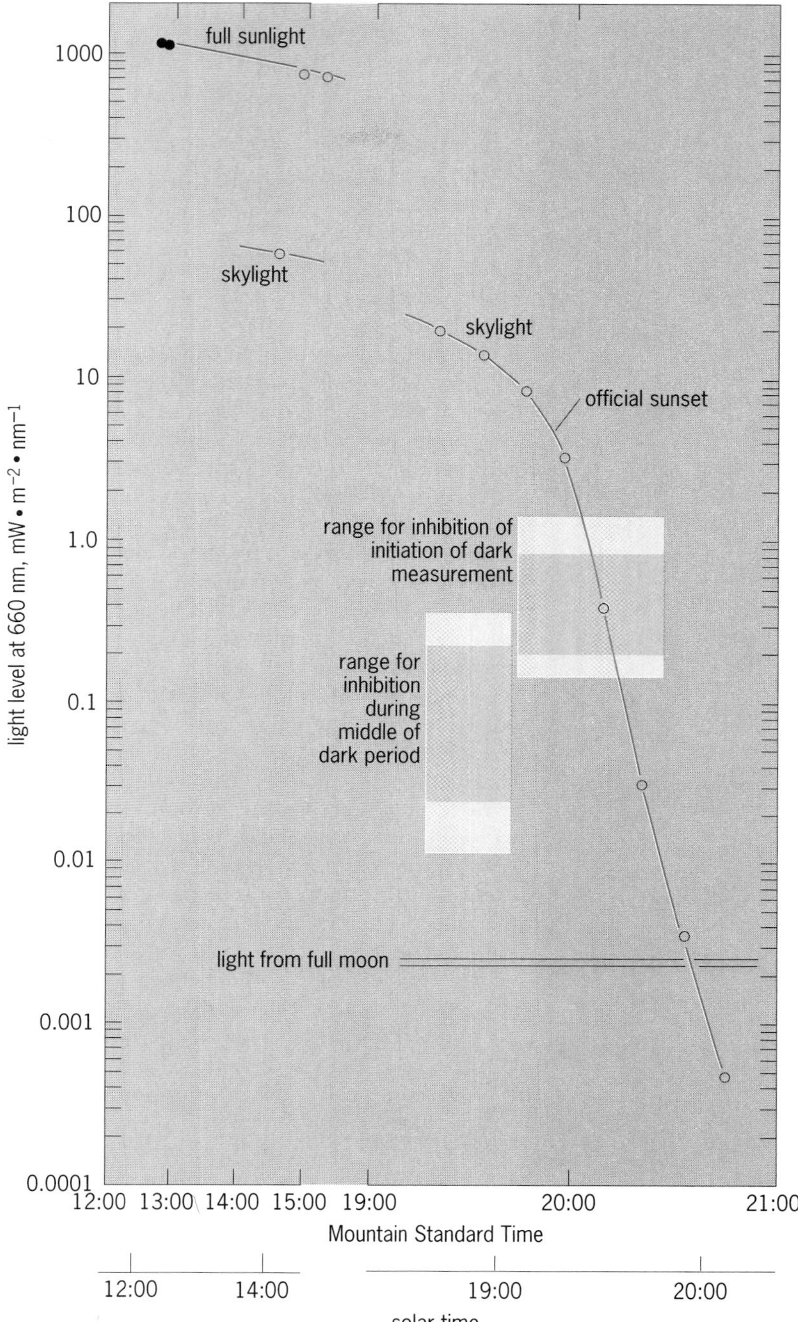

Fig. 5. Light levels at a wavelength of 660 nm (orange-red light) as a function of time on July 26, 1980, at Logan, Utah, and also showing light from a nearly full moon. The shaded areas represent ranges of light levels that inhibit the initiation of measurement of the dark period in cocklebur plants or inhibit flowering when applied for 2 h during the middle of a 16-h dark period. (*After F. B. Salisbury, The twilight effect: Initiating dark measurement in photoperiodism of Xanthium, Plant Physiol., June 1981*)

Megoura seems to be completely controlled by an hourglass timer. At a certain point in this aphid's development, short days lead to production of fertile eggs when males are present; long days indefinitely maintain the virginopara condition (production by virgin females of eggs that develop into aphids). Results of all experiments designed to test the hourglass nature of the clock in this aphid agree with such a clock.

In most cases, however, the clock does not have simple hourglass characteristics. Rather than phytochrome requiring many hours to convert from one form to another, thereby accounting for the critical dark period of a short-day plant, phytochrome conversion appears to be complete in less than 1–3 h. Conversion only initiates a more subtle form of time measurement based on a clock that oscillates from one condition to another, each complete oscillation requiring about 24 h.

The discovery of such a biological clock in living organisms was made in the late 1920s. It was shown that the movement of leaves on a bean plant (from horizontal at noon to vertical at midnight) continued uninterruptedly for several days, even when plants were placed in total darkness and at a constant temperature, and that the time between given points in the cycle (such as the most vertical leaf position) was almost but not exactly 24 h. In the case of bean leaves, it was about 25.4 h. Many other cycles have now been found with similar characteristics in virtually all groups of plants and animals. Daily periods of animal activity under constant conditions provide other good examples. In all cases, the cycling continues, but under constant conditions, where the cycles are "free-running" (not influenced by the usual day/night cycle), the cycle lengths usually differ from 24 h by a few minutes to a few hours. This is strong evidence that the clocks are internal and not driven by some subtle daily change in the environment. Such rhythms are called circadian.

Circadian rhythms usually have period lengths that are remarkably temperature-insensitive, which is also true of time measurement in photoperiodism. Furthermore, the rhythms are normally highly sensitive to light, which may shift the cycle to some extent. Thus, daily rhythms in nature are normally synchronized with the daily cycle as the Sun rises and sets each day. Their circadian nature appears only when they are allowed to manifest themselves under constant conditions of light (or darkness) and temperature, so that their free-running periods can appear.

There is good evidence that the photoperiodism clock has much in common with the circadian clock. **Figure 6**, for example, shows two kinds of supporting evidence. The bottom of the figure shows that soybeans have not only an optimum night length for flowering but also an optimum day length. If nights are long enough, the best flowering occurs when day plus night equals 24 h or multiples thereof; flowering is poor when day and night total about 36 h. The top of the figure shows what happens when light interruptions are given to soybeans at various times during a cycle consisting of 8 h of light plus 64 h of darkness. Light interruptions given around 24 or 48 h after the beginning of the cycle strongly promote flowering, while interruptions given 12–18 h or a little after 36 or 60 h inhibit flowering. Thus, there is a cycling response with period length of about 24 h, just as in the circadian rhythms. Numerous other experiments have demonstrated the similarities between the photoperiodism clock and the clock controlling circadian rhythms. For example, red light can be highly promotive during the day part of an inductive cycle but inhibitory during the night, suggesting that an oscillating timer controls sensitivity to light quality in photoperiodism.

Nevertheless, although the clock controlling circadian leaf movements is similar to the photoperiodism clock, the two can be separated. Thus leaf position bears no dependable relationship to the photoperiodism response of cockleburs, Japanese morning

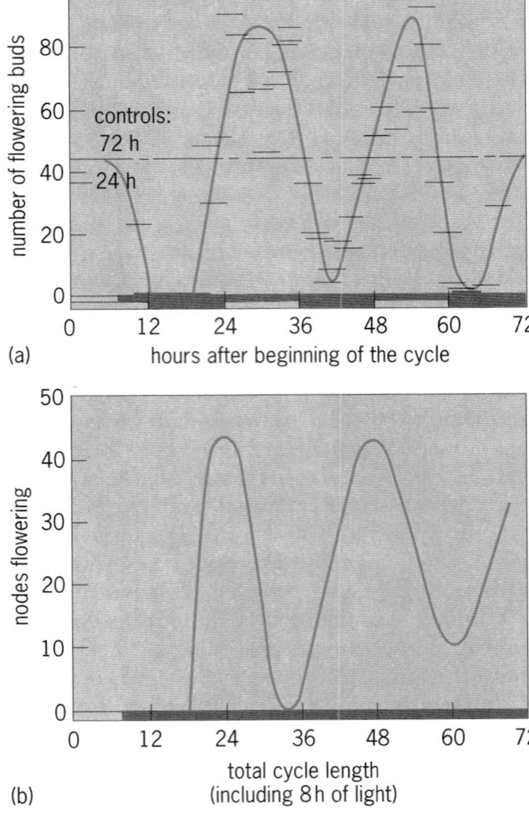

Fig. 6. Examples of a rhythmic response in the flowering process. (*a*) Plants given seven cycles, including 8 h of light and 64 h of darkness (top bar on abscissa). When the long dark period was interrupted with 4 h of light (horizontal bars), flowering was greatly inhibited at some times (12–20, 39, and 64 h) but promoted at other times (24–36 and 48–55 h). If the plant goes through cycles during which light is promotive at one time and inhibitory at another (bottom bar on abscissa), results would be expected. (*b*) Flowering response to 8 h of light plus periods of darkness. Flowering is promoted at total cycles of 24 and 48 h; inhibition occurs at about 34 and 60 h. (*After F. B. Salisbury, The Flowering Process, Pergamon Press, 1963*)

glories, or lamb's-quarters. *See* BIOLOGICAL CLOCKS.

Flowering hormone. The location of the photoperiodic flowering response in plants is the leaf. Exposing the stem and the bud to appropriate day lengths seldom has any effect. Apparently, some substance or other signal is produced in the leaf and moved to the bud where flowers are initiated. If such a substance is not a nutrient, and if it is effective in small quantities, then it would fit the definition of a hormone. M. Chailakhyan named the hypothetical hormone florigen. He also showed that plants that have not been exposed to suitable day lengths can be made to flower by grafting them to plants that have been so exposed. Furthermore, day-neutral sunflower plants can be grafted to a vegetative cocklebur plant under long days, causing the short-day cocklebur to flower. Movement of this hypothetical florigen can be studied by removing the leaves of cocklebur or Japanese morning glory at various times following exposure to a single dark period. If leaves are removed immediately after the dark period, plants remain vegetative, as though the hormone had not had time to move out of the leaf. If leaves are removed several hours later, however, the plants flower nearly as well as when leaves are not removed at all.

So far, attempts to extract florigen have mostly failed. Extracts from flowering plants have occasionally caused nonflowering plants to flower, but they induce only a minimal level of flowering, and the results can seldom be repeated by other investigators and sometimes not even by the investigator making the original report.

There is also evidence for a negative-acting inhibitor of flowering produced by photoperiodism. In a few species (such as short-day strawberries and long-day henbane), removing the leaves results in flowering under noninductive day lengths, as though the leaves normally produce an inhibitor under such day lengths. More than 40 years after his original experiments, Chailakhyan (working with I. A. Frolova and A. Lang) grafted short-day and long-day cultivars of tobacco to day-neutral cultivars and then grew them under various light conditions. When exposed to short days, the short-day graft partners caused the day-neutral plants to flower earlier, and under long days, long-day partners also promoted flowering of day-neutral partners. These results confirm the presence of a positive-acting florigen. But when the short-day graft partners were maintained under long days, flowering in the day-neutral species was slightly or not at all retarded; under short days, the long-day tobacco completely prevented flowering of the day-neutral partner. Thus, the evidence for flowering inhibitors was nearly as impressive in this experiment as evidence for florigen itself.

Various chemicals known to influence plant growth in different ways have been applied to plants, and flowering has then been studied. Although numerous effects have been observed, no clear picture has emerged, with the possible exception of the effects of various gibberellins (plant hormones that cause rapid elongation of plant stems and other responses). These substances will often substitute for long days or for cold treatment (vernalization), causing flowering of long-day plants held under short days or of vernalization-requiring plants never exposed to cold. Gibberellins seldom influence flowering of short-day plants.

Contrasted to the lack of knowledge about plant hormone responses to photoperiodism, there is clear evidence that photoperiodism affects hormones known to affect sexual reproduction in animals, particularly in vertebrates. The Japanese quail, for example, becomes sexually active when transferred from short to long days. Follicle-stimulating hormone (FSH), luteinizing hormone (LH), and testosterone increase in the quail under these conditions. Certain varieties of sheep, on the other hand, breed in the fall and are thus short-day animals for sexual activity. But again FSH and LH increase concurrently with the developing sexual activity. The situation is clearly highly complex in animals as well as plants. For example, in several mammals the pineal body (attached to the third ventricle in the brain) secretes an antigonadotropin that may cause testicular regression under unfavorable light schedules. Melatonin is known to be secreted by the pineal body in the golden hamster and the grasshopper mouse, and administration of this material suppresses testicle growth in both animals. Both are highly sensitive to photoperiod, but melatonin has no effect in two day-neutral species, the common house mouse and the laboratory rat. Thus there seems to be a relation of melatonin to photoperiodism, but since melatonin does not cause testicular regression in hamsters with their pineals removed, it may well act indirectly by causing the pineal to produce an unknown antigonadatropic material. *See* ENDOCRINE MECHANISMS; FLOWER; PLANT GROWTH; PLANT HORMONES; PLANT METABOLISM.

Frank B. Salisbury

Bibliography. S. D. Beck, *Insect Photoperiodism*, 2d ed., 1980; L. N. Edmunds, *Cellular and Molecular Bases of Biological Clocks*, 1987; F. B. Salisbury and C. W. Ross, *Plant Physiology*, 4th ed., 1992; B. M. Sweeney, *Rhythmic Phenomena in Plants*, 2d ed., 1987; A. T. Winfree, *The Timing of Biological Clocks*, 1986.

Photophore gland

A highly modified integumentary gland which arises from an epithelial invagination into the dermis. It becomes cut off from its site of origin and develops into a luminous organ composed of a lens and a light-emitting gland, at the back of which is a pigmented reflector of probable dermal-cell origin. These luminous bodies occur in deep-sea teleosts and elasmobranchs which live in areas of total darkness. *See* CHROMATOPHORE; EPITHELIUM; GLAND.

Olin E. Nelsen

Photoreception

The process of absorption of light energy by plants and animals and its utilization for biologically important purposes. In plants photoreception plays an essential role in photosynthesis and an important role in orientation. Photoreception in animals is the initial process in vision. *See* PHOTOSYNTHESIS; TAXIS.

The photoreceptors of animals are highly specialized cells or cell groups which are light-sensitive because they contain pigments which are unstable in the presence of light of appropriate wavelengths. These light-sensitive receptor pigments absorb radiant energy and subsequently undergo physicochemical changes, which lead to the initiation of nerve impulses that are conducted to the central nervous system.

Morphology of photoreceptors. While their gross structures differ widely, photoreceptors of a variety of animals (protozoa, coelenterates, flatworms, annelids, mollusks, echinoderms, arthropods, and chordates) examined with the high-resolution electron microscope appear to have in common a fine structure featuring the presence of membranous organelles with appreciable surface area.

Vertebrates. The structure of the vertebrate eye, exemplified by the human eye, is illustrated in **Fig. 1**. The retina is inverted so that the light-sensitive (photoreceptor) cells point toward the back of the eye (Fig. 1*b*). The visual pigments are located within the distal segments of the two types of photoreceptors, the rods (Fig. 1*b* and *c*) and the cones (Fig. 1*b*). The distal segment of the rod cell (Fig. 1*c* and *d*) consists of a stack of disks enclosed by a membrane. Electron microscopic investigation of rod distal segments fixed with osmium tetroxide or potassium permanganate or a freeze-drying procedure reveals each disk to be composed of three electron-dense membranes separated by two clear spaces of lower electron density; the outer two membranes are approximately 2 nanometers thick, the middle membrane about 4 nm thick, and the spaces about 3.5 nm thick. These findings, plus the results of investigations of the morphogenesis of rods, indicate that each disk is formed by the juxtaposition of two membranes caused by the invagination of the surface membrane of the distal segment. The disks are spaced 10–20 nm apart, and each distal segment contains many disks oriented perpendicularly to the light path; in the frog the estimated number of disks per rod distal segment is 1000. The disks are usually attached to a cilium which runs the length of the distal segment and connects the distal and proximal segments of the rod cell (Fig. 1*d*). The structure of distal segments of cones is in general similar to that of the rod distal segment, but the disk membranes are somewhat thicker in the cones of some vertebrates, and in some cones the cilium appears to be absent. *See* ELECTRON MICROSCOPE.

Examination of the rod distal segments with polarized light reveals a marked dichroism, suggesting that the membranes contain light-absorbing pigment molecules whose long axes are oriented at right angles to the longitudinal axis of the distal segment and parallel to the transverse plane of the disks, but in this plane there is no evidence of further orientation. It has been demonstrated that if vitamin A deficiency is prolonged to the point where loss of

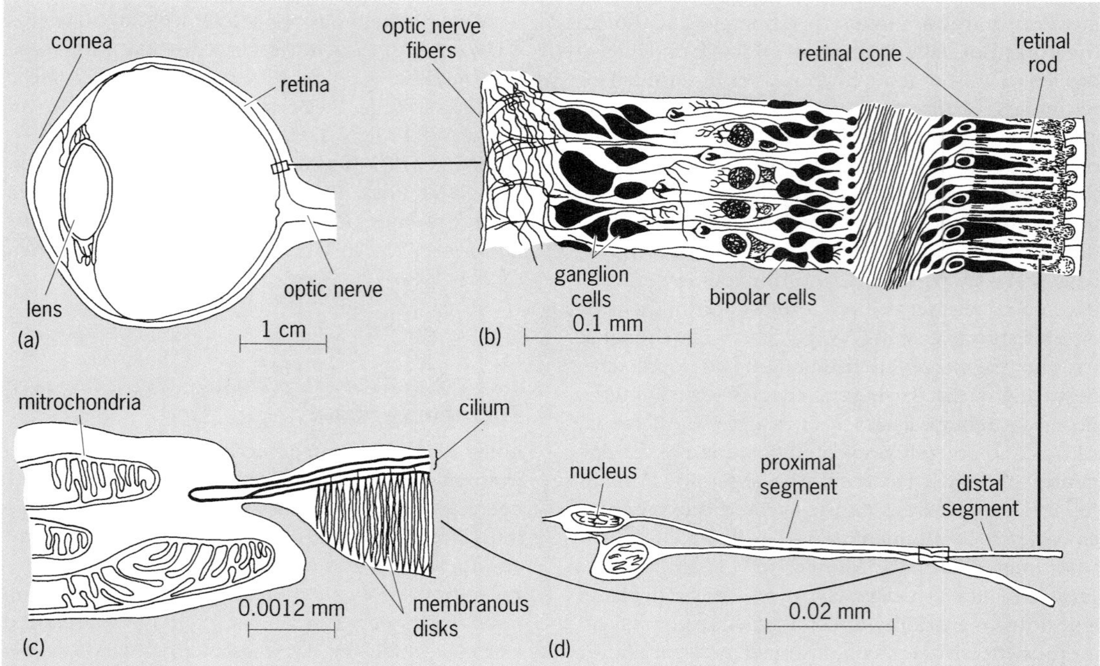

Fig. 1. Diagrammatic representation of gross and fine structure of photoreceptors of a typical vertebrate. (*a*) Section through eye; (*b*) layers of retina; (*c*) retinal rods (*after S. L. Polyak, The Retina, University of Chicago Press, 1941*). (*d*) Rod (*after F. DeRobertis, J. Biophys. Biochem. Cytol., 2(3):319–329, 1956*)

the visual protein opsin occurs, there is widespread degeneration, that is, swelling and dissolution of the disks in distal segments of rods. These and other considerations have led to the view that the disk membranes contain the visual pigment and that the spaces between disk membranes contain a double layer of lipoid molecules. *See* DICHROISM (BIOLOGY); EYE (VERTEBRATE).

Arthropods. The compound eyes and ocelli of arthropods are morphologically different from vertebrate eyes in many respects, but the presence of a membranous organelle situated in the light path is a common characteristic. The lateral eye of the horseshoe crab (*Limulus polyphemus*) illustrates that the compound eye is an aggregate of ommatidia (**Fig. 2***a*), each ommatidium being a packet of photoreceptor cells with its own lens system (Fig. 2*b* and *c*). The structures of particular interest are those parts of the retinula cells in each ommatidium that form the rhabdomeres (Fig. 2*c*), which are the membranous organelles of appreciable surface area. Rhabdomeres of arthropods are composed of microvilli, tiny tubular extensions of the membrane of retinula cells, which in *Limulus* interdigitate with microvilli from the apposed surfaces of the membranes of adjacent retinula cells (Fig. 2*d* and *e*). Collectively the rhabdomeres constitute the rhabdome in the center of which is the dendrite of the eccentric sense cell (Fig. 2*c*), the membrane of which also contributes some microvilli to the rhabdome. (Eccentric cells of the *Limulus* type are not characteristic of arthropod compound eyes; similar cells have been found thus far only in the compound eye of the silkworm moth.) The membranes of the microvilli are approximately 7 nm thick and under high magnification with the electron microscope consist of two electron-dense boundaries separating a space of lower electron density approximately 3.5 nm wide. In those regions where microvilli make contact with each other, the outer boundaries fuse and form five layered membranes approximately 15 nm thick. In all arthropod eyes examined, the rhabdome is situated at or near the center of each ommatidium beneath the crystalline cone and directly in the light path. The microvilli are oriented in regular array with their long axes perpendicular to the long axis of the ommatidium. The diameters of the microvilli range from approximately 40 to 150 nm.

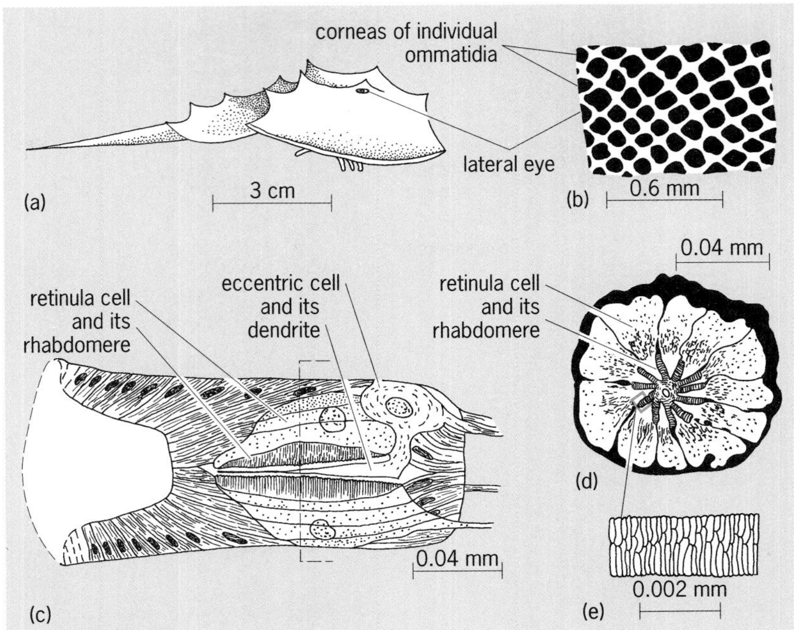

Fig. 2. Gross and fine structure of photoreceptors of the horseshoe crab (*Limulus*). (*a*) The organism. (*b*) The eye. (*c*) Longitudinal section through ommatidium. (*d*) Cross section through ommatidium; (e) microvilli of rhabdomere (*after W. H. Miller, Ann. N.Y. Acad. Sci., 74(2):204–209, 1958*).

Although the basic structural plan of those arthropod eyes which have been examined is similar, differences in the architecture of the rhabdome exist, and in many arthropods and some cephalopods totaling more than 90 species these differences endow the eyes with the ability to discriminate among lights polarized in different planes. In the compound eye of the land crab (*Cardiosoma quanhumi*) this ability is in all probability conferred by the unique arrangement of the microvilli in the rhabdome (**Fig. 3***d*). The microvilli are arranged in plates or layers along the long axis of the ommatidium (Fig. 3*a*). In each plate the microvilli are straight and parallel but are perpendicularly oriented in alternate layers. The microvilli contributing to the alternate layers originate from different retinula cells (Fig. 3*b* and *c*). It has been demonstrated that the electrical response of any one retinula cell within an ommatidium of the *Cardiosoma* eye illuminated by flashes of polarized light rotated through 360° was modulated in amplitude with a period of 180°. Similar electrical observations have been made on the eyes of the crab (*Carcinus maenas*), two species of wolf spiders (*Arctosa variana* and *Lycosa tarentula*), and the lobster (*Homarus vulgaris*). Compound eyes of worker honeybees, which are known to orient their flight by the plane of polarization of light reflected from the sky, possess a rhabdome which is roughly circular in cross section and consists of four quadrants in which the microvilli are parallel to those of the opposite quadrant but perpendicular to the microvilli of the adjacent quadrants.

There is consensus among investigators that visual pigments located within the rhabdomeres are responsible for the absorption of radiant energy, as well as the analysis of polarized light in arthropod eyes. Studies with polarized monochromatic light on the eye of the chalky mutant of the blowfly (*Calliphora erythrocephala*) indicate that the visual pigment is indeed located within the rhabdomeres and that the pigment molecules are oriented with their major axes parallel to the long axes of the microvilli. By virtue of this orientation, photons vibrating in a direction parallel to the microvilli are more readily absorbed.

Mollusca. Molluscan eyes, some of which are as sophisticated in gross structure as the vertebrate eye, contain membranous organelles of varying form. The eye of the squid and octopus contain rhabdomeres

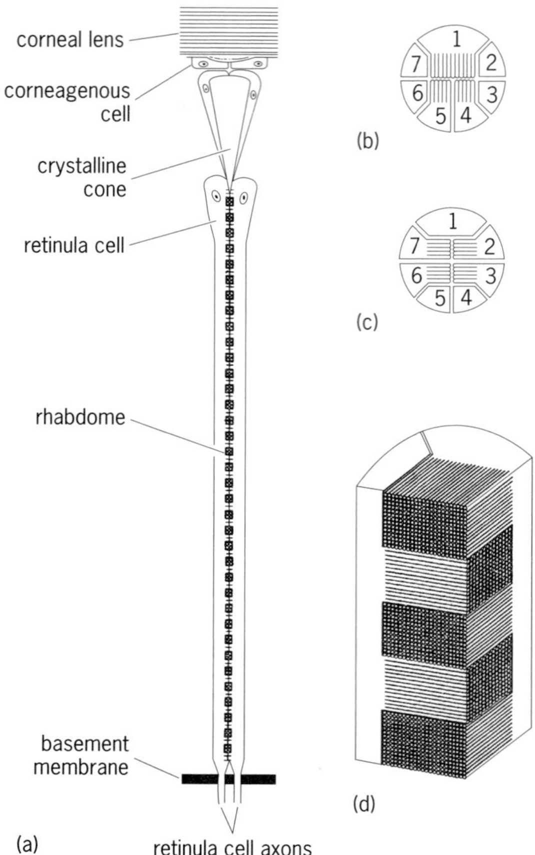

Fig. 3. Schematic diagram of the structure of a typical ommatidium from the crab compound eye. (*a*) Longitudinal section through the optic axis of an ommatidium. (*b, c*) Cross sections through different levels of the ommatidium, showing retinula cells from which the microvilli in the alternating layers originate. (*d*) Three-dimensional representation of fine structure of the rhabdome, showing the alternating layers of the tightly packed microvilli. (*After C. G. Bernhard, ed., Functional Organization of the Compound Eye, Pergamon Press, 1966*)

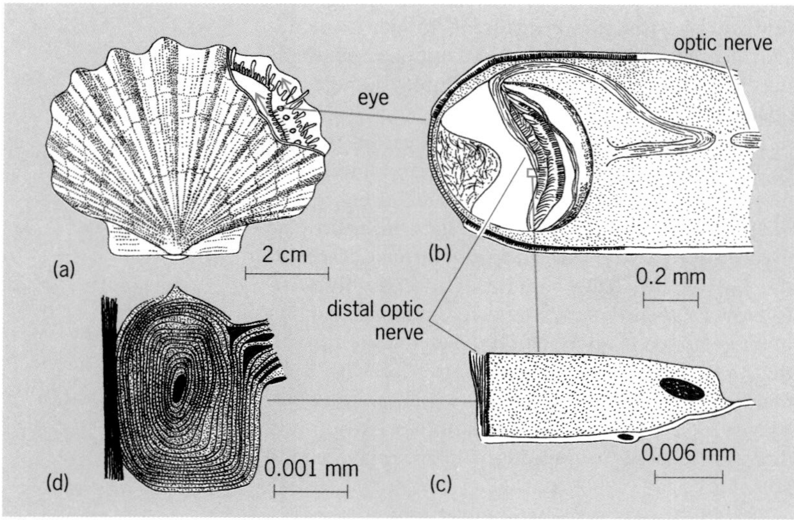

Fig. 4. Gross and fine structure of photoreceptors of the scallop (*Pecten*). (*a*) The organism. (*b*) Longitudinal section through eye; (*c*) distal sense cell (*after W. J. Dakin, Quart. J. Microsc. Sci., 4:49–112, 1910–1911*). (*d*) Appendage (*after W. H. Miller, Ann. N. Y. Acad. Sci., 74(2):204–209, 1958*).

which are combined into rhabdomes and are composed of microtubules. The photoreceptor of *Pecten* (Fig. 4) has a unique membranous organelle, consisting of concentrically arranged light and dark bands which together constitute an oval or spherical body about 1 micrometer in diameter (Fig. 4*d*). The dark bands of this organelle are about 5 nm thick and appear to be continuous with the stalks of cilia which are attached to basal bodies within the cytoplasm of the receptor cell. *See* EYE (INVERTEBRATE).

Chemical behavior of photosensitive pigments. The more than 100 light-sensitive pigments of vertebrate and invertebrate photoreceptors which have been investigated appear to have a similar chemical constitution. Rhodopsin, the visual pigment in the distal segments of rod cells of many vertebrates and in the rhabdomeres of the invertebrates so far studied, consists of a protein conjugated with a carotenoid. The latter has been identified as the aldehyde of vitamin A, called retinal. Two types of retinal exist: retinal$_1$ (see **Fig. 5**), and retinal$_2$, characterized by an additional double bond in the ring between carbon atoms 3 and 4. (Hereafter retinal without a subscript refers to retinal$_1$.) The protein moiety of rhodopsin, called opsin, has not been extensively characterized, and it is widely held that characteristics of opsins from different species vary and that this variance imparts species-specific properties to the rhodopsins. The absorption maximum of free retinal is at 387 nm; when it is combined with the protein opsin (molecular weight about 40,000), the visual pigment rhodopsin is formed and the absorption maximum shifts to higher wavelengths (498 nm in cattle rhodopsin, a visual pigment extensively investigated). The agreement of the absorption spectrum of many rhodopsins with the spectral sensitivity of the activity of photoreceptors from which the pigments were extracted clearly implicates these pigments in light reception.

Ultrastructure and rhodopsin. Evidence already cited suggests that rhodopsin forms an integral and highly organized part of the ultrastructure of vertebrate retinal rod distal segments and of the microstructure of invertebrate membranous organelles. In the frog, rhodopsin contributes about 40% of the weight of the outer segments of rods. Studies with polarized light indicate that rhodopsin molecules are highly oriented in vertebrate rods and in the dipteran rhabdomere; in both, the major axes of the pigment molecules are perpendicular to the long axis of the photoreceptor and, in the dipteran rhabdomere, parallel to the microvilli. Cattle rhodopsin has a phospholipid component which, if totally removed, destroys the stability of the pigment. The possibility exists that this phospholipid fraction represents a part of the structural fabric of the rod distal segment to which the rhodopsin molecule is bound.

Isomers of retinal. The bleaching and resynthesis of rhodopsin depend upon the existence of several isomers of retinal. The structural formulas of the most important pair are illustrated in Fig. 5. Only 11-*cis*-retinal, when incubated with opsin, forms

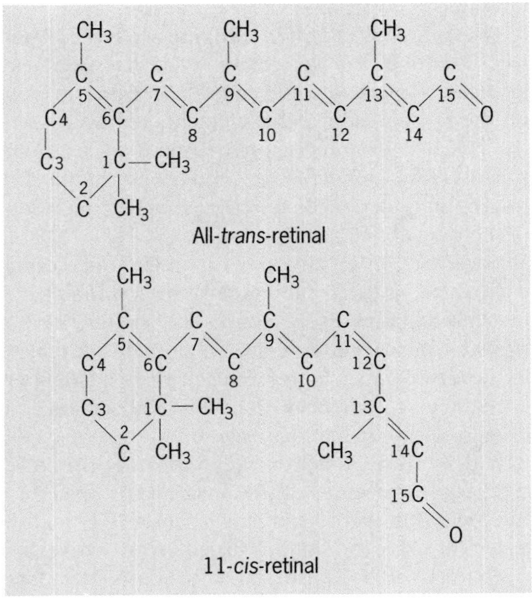

All-*trans*-retinal

11-*cis*-retinal

Fig. 5. Structure of two retinal isomers.

rhodopsin. Illumination of a solution of rhodopsin results eventually in a release of retinal from the opsin in the all-*trans* form. Between these two events a number of intermediate transitory stages occur. The mechanism of bleaching of rhodopsin involves the absorption of a quantum of light energy by the molecule, the isomerization of 11-*cis*-retinal to all-*trans*-retinal, and via intermediate stages the hydrolysis of this incompatible isomer from its site on the protein.

Dark adaptation. After exposure to light, recovery of sensitivity by the receptor cell takes place; this process is referred to as dark adaptation. It is dependent upon an adequate supply of 11-*cis*-retinal and the rate of recombination of opsin with this active retinal isomer. The 11-*cis*-retinal apparently is supplied partly by the action of an enzyme, retinal isomerase, and upon the all-*trans*-retinal released from bleached rhodopsin. In addition, an equilibrium exists between vitamin A and retinal that is catalyzed by alcohol dehydrogenase and diphosphopyridine nucleotide. Opsin traps the 11-*cis*-retinal which is formed to reconstitute the visual pigment.

Other visual pigments. In addition to the rhodopsins, several other visual pigments have been identified in the retinas of vertebrates. Fresh-water fishes, some amphibians, and certain reptiles possess visual pigments which contain retinal_2. Retinal_2, when combined with rod opsin, forms a visual pigment called porphyropsin with an absorption maximum at 522 nm. Cones presumably contain a different opsin which combines with retinal to form a pigment called idopsin. This pigment, which has been isolated from chicken retinas, has an absorption maximum at 562 nm and regenerates more rapidly than rhodopsin. These properties of this cone pigment presumably account for the demonstrable shift in sensitivity to longer wavelengths in going from rod to cone vision in a variety of animals (the Purkinje shift) and the obviously faster recovery of cone sensitivity known to occur during dark adaptation. A fourth pigment, the existence of which is indicated by spectral sensitivity data, has been synthesized by incubating retinal_2 with cone opsin. The resulting pigment is called cyanopsin and has an absorption maximum at 620 nm. Microspectrophotometric measurements of absorption spectra of single cones in the tadpole and chicken suggest that a cone pigment having the absorption characteristics of cyanopsin exists in nature.

The properties of visual pigments are thus determined by the type of retinal, as well as the visual protein. Within the rhodopsin class, for example, the range of absorption maxima is 562–430 nm in the retinal_1 group and 620–510 nm in the retinal_2 group. This considerable range would seem to indicate that the visual proteins make important contributions to the characteristics of the visual pigments.

Considerable evidence indicates that cone pigments exist which have not yet been successfully isolated by extraction. The techniques of reflection photometry and microspectrophotometry of individual cones have produced results which indicate three populations of cones containing pigments with absorption maxima in ranges 445–450, 525–535, and 555–570 nm. The absence of specific cone sensitivities in certain human retinas has been shown to correlate well with the major types of color blindness. *See* COLOR VISION.

Invertebrate visual pigments. The visual pigments of invertebrates have not been as extensively studied as those of vertebrates, but those that are known contain retinal_1 (range in absorption maxima from 420 nm in the honeybee to 520 nm in *Limulus*) and seem to be chemically similar to the vertebrate pigments. There is evidence that the visual pigments in all rhabdomeres of arthropod eyes need not be identical; in the fly, rhabdomeres 1–6 contain a pigment with an absorption maximum at 510 nm, and rhabdome number 7 contains a pigment with a maximum at 470 nm. Photoreceptors of the squid eye contain rhodopsin which, following photoexcitation, exhibits isomerization of 11-*cis*-retinal to the all-*trans* form, but the retinal moiety remains attached to the protein. Photosensitive pigments based upon retinal have now been isolated from the eyes of insects, crustaceans, and arachnids.

Electrical activity of photoreceptors. Despite much information about the structure of photoreceptors and the chemistry of visual pigments, the mechanism which links photochemical events to the initiation of nerve impulses is still obscure.

Vertebrates. It has been known since the late 1800s that the vertebrate retina generates an electrical response upon illumination. The record of this electrical response, called the electroretinogram (ERG), is polyphasic in waveform with an initial cornea-negative component (the *a* wave), a large cornea-positive component (the *b* wave), and an off-response following cessation of illumination. In many cases a third relatively slow cornea-positive response (the *c* wave) is recorded which originates presumably in the pigment epithelium. It is established that the initial limb of the *a* wave and the off-response are of receptor origin and that the *b* wave is contributed by neural and perhaps glial cells located in front of the receptor layer. Nevertheless the ERG is an accurate index of visual sensitivity; it has been shown that human dark adaptation and spectral sensitivity curves determined electroretinographically agree well with those obtained using subjective criteria, and both methods yield results which agree with the absorption spectra of visual pigments measured in place or in extracts. Measurement of the ERG is therefore a useful method for obtaining data in a variety of animals and has often been used in preference to behavioral techniques. The ERG has also been used clinically in the early diagnosis of diseases involving the retina.

By using glass micropipets filled with solutions of potassium chloride, intracellular receptor potentials have been recorded from the proximal (inner) segments of sense cells in the retina of the carp and the salamander (*Necturus maculosus*). The light-initiated intracellular potentials are sustained and of polarity such that the cell interior becomes increasingly negative. That is, the membrane potential increases and the electrical response is a hyperpolarizing one rather than the depolarizing type of response characteristic of nerve and other types of excitable tissues. By histological localization of a dye deposited by the recording micropipet, it has been determined that these receptor potentials were recorded from the proximal segments of the sense cells. In experiments with the carp retina the intracellular potentials were recorded under conditions permitting determination of the spectral sensitivities of single cones. The results of many experiments indicate three populations of cones of decreasing frequency of occurrence and with different spectral sensitivity maxima as follows: 611 ± 23 nm, 462 ± 15 nm, and 529 ± 14 nm. These maxima agree reasonably well with absorption maxima of single cones in the goldfish retina determined by microspectrophotometry, but the red maximum is displaced toward longer wavelength than in the retinas of higher vertebrates.

An electrical response of relatively small magnitude without detectable latency (less than 25 microseconds) has been recorded from both vertebrate and invertebrate photoreceptors in response to brief but intense light flashes. This response, aptly termed the early receptor potential, is mediated by the visual pigments but is resistant to agents such as low temperature, drugs, and removal of Na^+ and H_2O, which depress or abolish the late receptor potential. Similar rapid (also slower) light-evoked potentials have been recorded from the pigment epithelium of the toad eye, from green leaves, and from human skin. It is held that the early response of pigmented systems is the result of a charge displacement, which in the case of rhodopsin is coincident with the initial stages of bleaching, and that the early receptor potential is not directly involved in visual excitation.

The technique of using micropipets to record from single units in the vertebrate retina has been extensively applied, yielding considerable information on the responses of retinal elements removed from the primary photoreceptors by one, two, or more synapses. Such studies have been primarily directed toward an understanding of the neural organization of the retina.

Invertebrates. Invertebrate photoreceptors, especially those of the compound eye of *Limulus*, have been used to study the responses in nerve fibers from primary photoreceptor cells. Experiments in which impulse discharges from a single primary nerve fiber are studied in response to various intensities and durations of illumination delivered to the sense cell have demonstrated that (1) the frequency of impulse discharge up to some maximum is linearly related to the logarithm of the stimulus intensity; (2) within a certain critical duration, usually somewhere between 0.1 and 1.0 s, stimuli in which the product of intensity and duration is constant produce discharges of similar frequency and number; and (3) latency of the impulse discharge decreases as the intensity is increased. In addition, these electrophysiological recording techniques from primary sensory fibers have been used to measure spectral sensitivity, dark adaptation, interaction between ommatidia in the same eye, and the effect of manipulation of the environment in the immediate vicinity of photoreceptors.

The electrical response to illumination of arthropod ommatidia and many other invertebrate photoreceptors has been recorded from single light-sensitive cells by using micropipet electrodes. In the majority of cases the response is a sustained depolarization (that is, the cell interior becomes less negative) of gradually diminishing magnitude which may, depending upon the previous history of illumination and the stimulus parameters, exhibit an initial maximum.

In the retinula cells of *Limulus* and other arthropods spike potentials generally associated with the existence of nerve impulses are rarely recorded. However, responses usually consisting of sustained depolarization on which spike potentials are superimposed have been recorded from the eccentric cell in the lateral eye of *Limulus*. The light-induced depolarization of the eccentric cell is regarded as a generator potential, which initiates the nerve impulses in the axon. Spike potentials of variable but small amplitude have been recorded from retinula cells in the eye of *Limulus* but seem to be reflections of the spike potentials generated by the eccentric cell. The visual significance of the retinula cell in the *Limulus* lateral eye, which exhibits an electrical response to illumination that is independent of the eccentric cell response (except for the spike potentials), is not yet known. In the eyes of other arthropods nerve impulses are presumably initiated in structures behind the eye by virtue of the potential changes which occur in the retinula cells.

Light and electrical changes. It may thus be generally true that energy, absorbed by a photosensitive pigment located in the membranous organelles of photoreceptor cells, acts to produce a sustained electrical change, depolarization in invertebrates and hyperpolarization in vertebrates, by increasing the permeability of the cell membrane, and that this potential change acts as a generator to initiate nerve impulses in the visual pathway. *See* VISION.

Verner J. Wulff

Bibliography. D. J. Cosens and D. Vince-Pru (eds.), *The Biology of Photoreception*, 1984; A. Fein and J. S. Levine, *The Visual System*, 1985; M. G. Holmes (ed.), *Photoreceptor Evolution and Function*, 1997; L. M. Hurvich, *Color Vision: An Introduction*, 1981; K. N. Leibovic, *Science of Vision*, 1990; J. Marvullo, *Color Vision*, 1989; S. P. McKee (ed.), *Optics, Physiology, and Vision*, 1990; H. Steave (ed.), *The Molecular Mechanism of Photoreception*, 1986.

Photorespiration

Light-dependent carbon dioxide release and oxygen uptake in photosynthetic organisms caused by the fixation of oxygen instead of carbon dioxide during photosynthesis. This oxygenation reaction forms phosphoglycolate, which represents carbon lost from the photosynthetic pathway. Phosphoglycolate also inhibits photosynthesis if it is allowed to accumulate in the plant. The reactions of photorespiration break down phosphoglycolate and recover 75% of the carbon to the photosynthetic reaction sequence. The remaining 25% of the carbon is released as carbon dioxide. Photorespiration reduces the rate of photosynthesis in plants in three ways: carbon dioxide is released; energy is diverted from photosynthetic reactions to photorespiratory reactions; and competition between oxygen and carbon dioxide reduces the efficiency of the important photosynthetic enzyme ribulose-bisphosphate (RuBP) carboxylase. There is no known function of the oxygenation reaction; most scientists believe it is an unavoidable side reaction of photosynthesis. *See* PHOTOSYNTHESIS.

Significance. The rate of photosynthesis can be stimulated as much as 50% by reducing photorespiration. Since photosynthesis provides the material necessary for plant growth, photorespiration inhibits plant growth by reducing the net rate of carbon dioxide assimilation (photosynthesis). Plants grow faster and larger under nonphotorespiratory conditions, in either low oxygen or high carbon dioxide atmospheres. Most of the beneficial effects on plant growth achieved by increasing CO_2 may result from the reduced rate of photorespiration. *See* PLANT GROWTH.

Precise measurements of the rate of photorespiration are difficult since carbon dioxide released in photorespiration is subject to refixation in photosynthesis while the measurement is being made. A simple method that has been devised for estimating the rate of photorespiration assumes that the properties of the photosynthetic enzyme ribulose-bisphosphate carboxylase/oxygenase determines the ratio of oxygenation (which initiates photorespiration) to carboxylation (which initiates photosynthesis). This ratio, which is directly proportional to oxygen level and inversely proportional to carbon dioxide level, is expressed by the equation below, where ϕ is the

$$\phi = x\frac{O}{C}$$

ratio of oxygenation to carboxylation at 77°F (25°C), O is the concentration of oxygen (in percent), and C is the concentration of CO_2 (in parts per million). The factor x, which varies directly with temperature, accounts for the specificity of ribulose-bisphosphate carboxylase/oxygenase for carbon dioxide over oxygen and the drop in carbon dioxide level from outside to inside the leaf. Photorespiration is stimulated by high temperature and inhibited by low temperature. For example, the factor x for 77°F (25°C) is 6.7, for 59°F (15°C) the factor is 4.1, and for 95°F (35°C) the factor is 9.4. The effect of oxygen and carbon dioxide on photorespiration (**Fig. 1**) is greater at low carbon dioxide concentration and is less at high carbon dioxide concentration.

Treatments that cause the stomata of leaves to close often cause the carbon dioxide concentration inside leaves to fall, thus stimulating photorespiration. One example of this phenomenon is the effect of water stress (drought). The stomata on leaves close when water is in short supply. This reduces the rate of loss of water from leaves but also stimulates photorespiration by reducing the concentration of carbon dioxide inside the leaf.

Path of carbon. Photorespiration begins when a molecule of ribulose bisphosphate is oxygenated (**Fig. 2**). For each oxygenation event, one molecule of phosphoglycerate, the normal product of photosynthesis, is formed. The other part of the ribulose

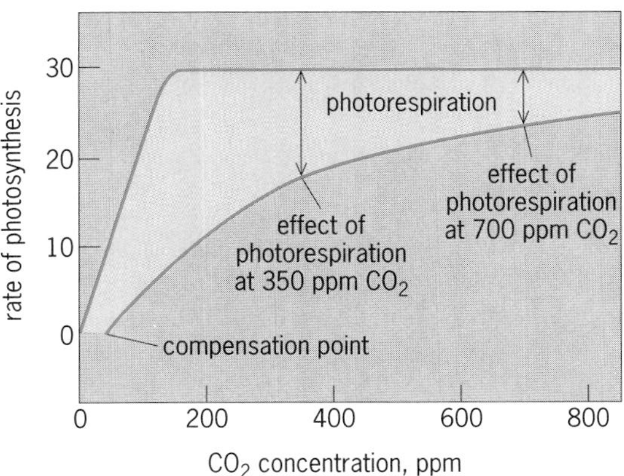

Fig. 1. Relation between photorespiration, photosynthesis, and carbon dioxide concentration. The top curve is the response of photosynthesis of a leaf to carbon dioxide when photorespiration is eliminated by reducing the oxygen concentration to 2%. The bottom curve is the response of photosynthesis to carbon dioxide when photorespiration occurs in the presence of 21% oxygen.

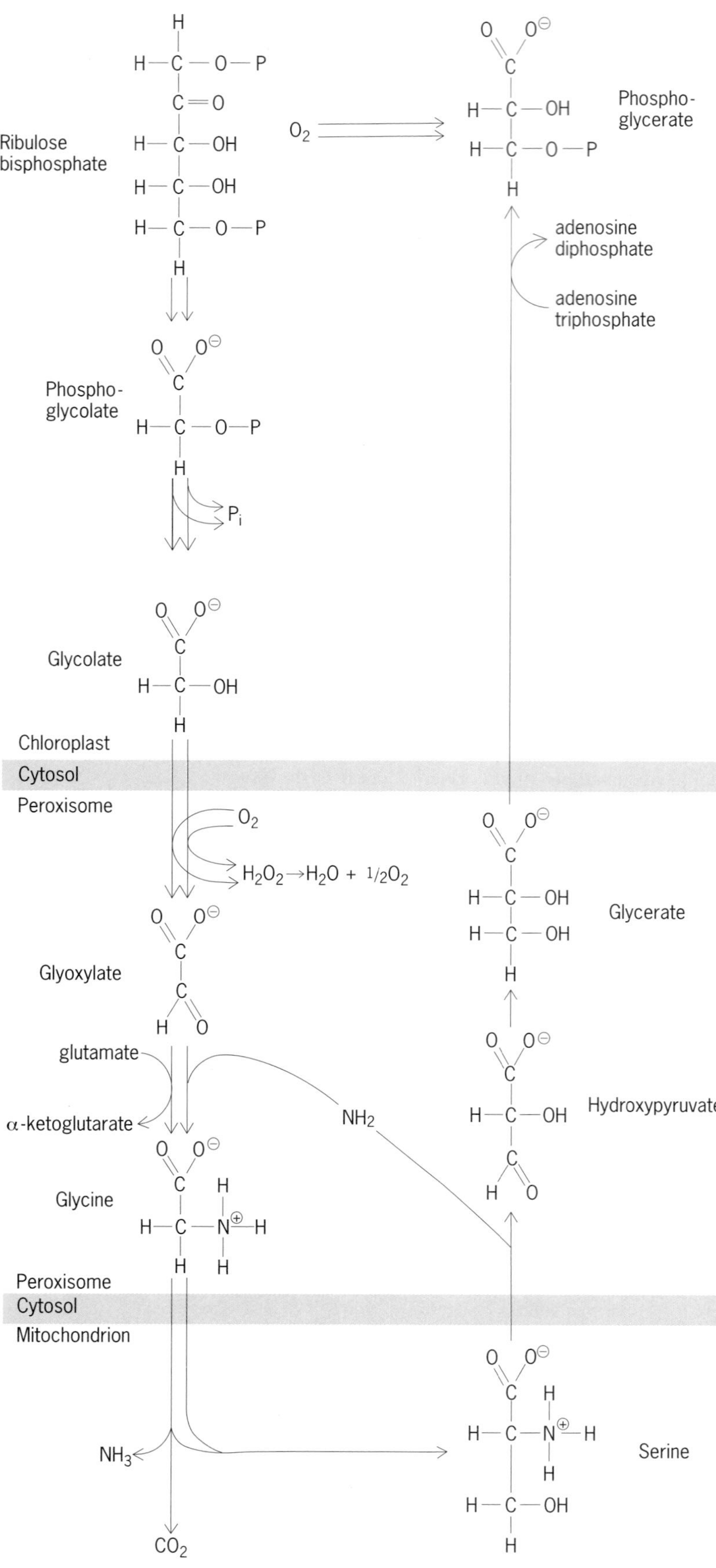

Fig. 2. Path of carbon in photorespiration. Phosphate (P) is normally a 50:50 mixture of HPO_3^- and PO_3^{2-}. Each arrow indicates a molecule following the pathway for every two oxygenations.

bisphosphate molecule forms phosphoglycolate, which is not a normal part of photosynthesis. Phosphoglycolate undergoes a series of reactions in three different compartments of the cell: the chloroplast, the peroxisome, and the mitochondrion. These reactions lead to the release of one carbon dioxide and one ammonia molecule for every two oxygenations. Most of the carbon (75%) is returned to the phosphoglycerate pool. The conversion of phosphoglycolate to phosphoglycerate and the recovery of the ammonia consume energy from the light reactions of photosynthesis, leaving less energy available for carbon dioxide fixation.

The first step in the metabolism of phosphoglycolate is removal of the phosphate group. This is catalyzed by the chloroplast enzyme phosphoglycolate phosphatase. The role of this enzyme in photorespiration was confirmed by discovery of a mutant plant having no phosphoglycolate phosphatase. This plant grows perfectly well under nonphotorespiratory conditions; in normal air, however, which causes photorespiration, it will not grow because it accumulates phosphoglycolate, which inhibits one of the reactions necessary for photosynthesis. Seven other classes of mutants have been found, and in each case the mutant plants can complete their life cycle when grown under nonphotorespiratory conditions but die when grown in normal air. This indicates that there are at least eight steps in plant metabolism that are essential for photorespiration but are not necessary for normal plant growth. *See* ENZYME.

The next step in the path of carbon in photorespiration is the oxidation of glycolate to glyoxylate, which yields hydrogen peroxide as a by-product. The reaction occurs in the peroxisome, a specialized organelle that contains catalase, an enzyme that breaks down the hydrogen peroxide to water and oxygen.

Amino groups are next added to the glyoxylate to make the amino acid glycine. One-half of the amino groups comes from the conversion of serine to hydroxypyruvate, and the other half comes from glutamate. The glycine molecules have two fates. Half of the glycine will be taken apart, releasing a carbon dioxide and ammonia molecule. Because of this release of carbon dioxide, the observed rate of photosynthesis is less than the actual rate of carbon dioxide fixation by one-half of the rate of oxygenation. The remaining carbon is attached to a derivative of folic acid, then transferred to a second glycine molecule to form serine, another amino acid. The serine transfers the amino group back to glyoxylate and becomes hydroxypyruvate, which is next reduced to glycerate, and then phosphorylated to phosphoglycerate. *See* AMINO ACIDS.

Compensation point. During photorespiration, carbon dioxide is released at one-half the rate of oxygenation. When the rate of oxygenation is twice the rate of carboxylation ($\phi = 2$), the carbon dioxide released by photorespiration equals the carbon dioxide taken up by carboxylation, and so the rate of photosynthesis is zero. This is called the compensation point (Fig. 1). The compensation point in most plants at 77°F (25°C) corresponds to a carbon dioxide concentration of about 50 ppm. This differs from the value calculated from the equation primarily because there is no drop in carbon dioxide concentration from outside to inside the leaves at the compensation point, but also because of a small effect of mitochondrial respiration.

Path of nitrogen. Fixed nitrogen is lost as ammonia, a volatile gas, during photorespiration. Most of the ammonia is recovered by the GS-GOGAT pathway, named after the major enzymes involved [glutamine synthetase (GS) and glutaryl:oxoglutarate aminotransferase (GOGAT), also called glutamate synthase (**Fig. 3**)]. Under normal conditions of photosynthesis, 5–20 ammonia molecules from photorespiration are fixed by the GS-GOGAT pathway for every one ammonia fixed for protein synthesis. The GS-GOGAT pathway consumes energy and therefore contributes to the loss of energy attributed to photorespiration.

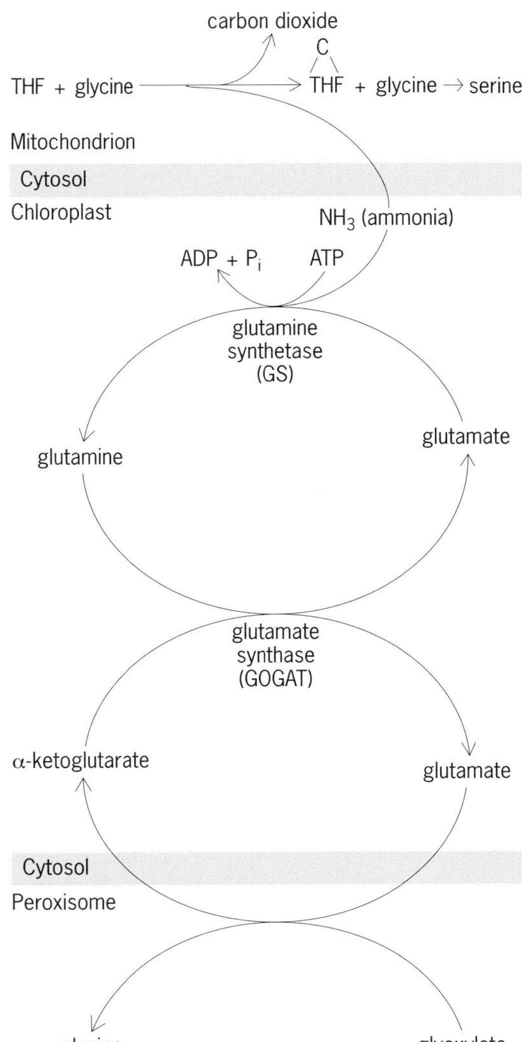

Fig. 3. GS-GOGAT ammonia cycle, which recovers the ammonia lost as a result of glycine metabolism during photorespiration. Tetrahydrofolate (THF) is a compound that can transfer single carbon atoms, in this case from one glycine to another.

Energetics. Three molecules of adenosine triphosphate (ATP) are required for each carboxylation in photosynthesis: one is used for converting ribulose phosphate to ribulose bisphosphate, and one is needed to convert each of the two phosphoglycerate molecules produced by carboxylation to sugars. For oxygenation, 3.5 ATP molecules are required: one for converting ribulose phosphate to ribulose bisphosphate and one for converting the phosphoglycerate to sugar. In addition, three other reactions that require ATP are conversion of glycerate to phosphoglycerate, reduction of phosphoglycerate to sugar, and recovery of ammonia. These three reactions occur once for every two oxygenation events and so are required at one-half of the rate of oxygenation. Therefore, the rate of ATP (energy) use is 3.5 $[2 + (3 \times 0.5)]$ per oxygenation. *See* ADENOSINE TRIPHOSPHATE (ATP).

Effect of carbon dioxide. The increase in the level of carbon dioxide in the atmosphere is reducing the rate of photorespiration, an effect that can be calculated by using the equation. In 1958, when precise carbon dioxide monitoring was begun at Mauna Loa, Hawaii, the carbon dioxide level was 315 ppm, which causes a ratio of oxygenation to carboxylation of $\phi = 0.45$. By 1988, the carbon dioxide concentration had increased to 350 ppm, which reduced the ratio to $\phi = 0.40$. Therefore, the rate of photorespiration has fallen by over 10% in 30 years.

This decline in photorespiration may be responsible for an increase in the yearly oscillation in carbon dioxide concentration in the atmosphere. In the Northern Hemisphere, the carbon dioxide concentration is highest in May, then falls through the summer, reaching a minimum in October. This oscillation averaged 5.5 ppm (May to October decline) in the mid-1960s but averaged over 6 ppm in the mid-1980s. Stimulation of photosynthesis by the increased carbon dioxide concentration is believed to cause this change, and the stimulation of photosynthesis is caused, in large measure, by the inhibition of photorespiration.

Photorespiration in nature. Photorespiration is a process that is wasteful to the plant and may not be essential for plant survival. It is, therefore, surprising that a mutation eliminating photorespiration has not arisen. Many scientists believe that this indicates that oxygenation and the subsequent photorespiration are unavoidable side reactions of photosynthesis.

There are, however, some plants that avoid photorespiration under certain conditions by actively accumulating carbon dioxide inside the cells that have ribulose-bisphosphate carboxylase/oxygenase. Many cacti do this by taking up carbon dioxide at night and then releasing it during the day to allow normal photosynthesis. These plants are said to have crassulacean acid metabolism (CAM). Another group of plants, including corn (*Zea mays*), take up carbon dioxide by a special accumulating mechanism in one part of the leaf, then transport it to another part of the leaf for release and fixation by normal photosynthesis. The compound used to transport the carbon dioxide has four carbon atoms, and so these plants are called C_4 plants. Plants that have no mechanism for accumulating carbon dioxide produce the three-carbon compound phosphoglycerate directly and are therefore called C_3 plants. Most species of plants are C_3 plants.

The C_4 plants, CAM plants, and many algae avoid photorespiration by maintaining a very high concentration of carbon dioxide in the part of the cell containing ribulose-bisphosphate carboxylase/oxygenase, which suppresses photorespiration. The ribulose-bisphosphate carboxylase/oxygenase in carbon dioxide-accumulating plants often exhibits less specificity for carbon dioxide over oxygen than the enzyme of C_3 plants, but photorespiration is still reduced because the carbon dioxide concentration is kept high by the carbon dioxide-accumulating mechanism.

The cost of actively accumulating carbon dioxide in many C_4 plants is about 2 ATP molecules, less than the cost of photorespiration in C_3 plants [2.5 ATP molecules at 77°F (25°C)]. However, as the carbon dioxide level in the atmosphere increases, the cost of photorespiration in C_3 plants will fall, while the cost of carbon dioxide pumping in C_4 plants will remain constant. Hence, C_3 plants benefit more than C_4 plants from the increasing CO_2 concentration. *See* PLANT RESPIRATION.

Thomas D. Sharkey

Bibliography. N. N. Artus, S. C. Somerville, and C. R. Somerville, The biochemistry and cell biology of photorespiration, *CRC Crit. Rev. Plant Sci.*, 4:121–147, 1986; D. W. Husic, H. D. Husic, and N. E. Tolbert, The oxidative photosynthetic carbon cycle or C_2 cycle, *CRC Crit. Rev. Plant Sci.*, 5:45–100, 1987; A. J. Keys et al., Photorespiratory nitrogen cycle, *Nature*, 275:741–743, 1978; W. L. Ogren, Photorespiration: Pathways, regulation, and modification, *Annu. Rev. Plant Physiol.*, 35:415–442, 1984; T. D. Sharkey, Estimating the rate of photorespiration in leaves, *Physiol. Plant*, 73:147–152, 1988.

Photosphere

The apparent, visible surface of the Sun. The photosphere is a gaseous atmospheric layer a few hundred miles deep with a diameter of 864,000 mi (1,391,000 km; usually considered the diameter of the Sun) and an average temperature of approximately 5800 K (10,500°F). Radiation emitted from the photosphere accounts for most of the solar energy flux at the Earth. The solar photosphere provides a natural laboratory for the study of dynamical and magnetic processes in a hot, highly ionized gas.

When studied at high resolution, the photosphere displays a rich structure. Just below it, convective motions in the Sun's gas transport most of the solar energy flux. Convective cells penetrate into the stable photosphere, giving it a granular appearance (see **illus.**) with bright cells (hot rising gas) surrounded by dark intergranular lanes (cool descending gas). A typical granule is approximately 600 mi (1000 km) in diameter. Measurements of horizontal velocity reveal a larger convective pattern, the supergranulation, with a scale of approximately 20,000 mi (30,000 km); the horizontal motion of individual granules reveals intermediate-scale (3000-mi or 5000-km) convective flows.

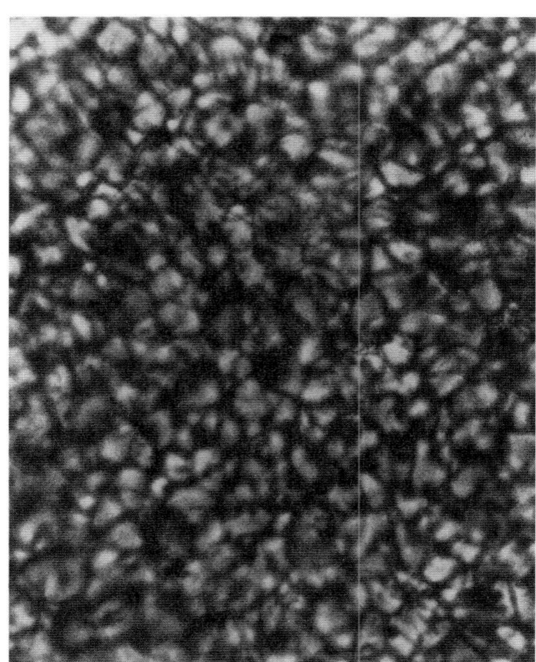

Image of the solar photosphere made in white light near 500 nm, using the 75-cm (30-in.) Vacuum Tower Telescope of the National Solar Observatory at Sacramento Peak, Sunspot, New Mexico. (*National Solar Observatory*)

Magnetic fields also play an important role in shaping photospheric structure. Magnetic flux tubes, smaller than the intergranular lanes, penetrate the photosphere vertically. When many of these tubes are forced together, probably by convective motions deep within the Sun, they suppress the convective motions near the surface, cool the atmosphere, and form dark magnetic pores and sunspots.

Gravitational, acoustic, and magnetic waves are propagated through, and at some frequencies trapped in, the photosphere. The trapped waves, most of which have periods of about 5 min, represent global oscillations of the Sun and provide information about its interior when analyzed by seismological techniques. *See* HELIOSEISMOLOGY; SUN.

Stephen L. Keil

Bibliography. R. J. Bray, R. E. Loughhead, and C. J. Durrant, *The Solar Granulation*, 2d ed., 1984; R. Giovanelli, *Secrets of the Sun*, 1984; R. W. Noyes, *The Sun, Our Star*, 1982; H. Zirin, *Astrophysics of the Sun*, 1988.

Photosynthesis

Literally, synthesis of chemical compounds in light. The term photosynthesis, however, is used almost exclusively to designate one particularly important natural process of this type: the manufacture in light of organic compounds (primarily certain carbohydrates) from inorganic materials by chlorophyll- or bacteriochlorophyll-containing cells. This process requires a supply of energy in the form of light, since its products contain much more chemical energy than its raw materials. This is clearly shown by the liberation of energy in the reverse process, the combustion of organic material with oxygen. *See* CHLOROPHYLL; PLANT RESPIRATION.

In chlorophyll-containing plant cells and in cyanobacteria, photosynthesis involves oxidation of water (H_2O) to oxygen molecules, which are released into the environment. In contrast, bacterial photosynthesis does not involve O_2 evolution—instead of H_2O, other electron donors, such as H_2S, are used. Both types of photosynthesis are discussed below.

The light energy absorbed by the pigments of photosynthesizing cells, especially by the pigment chlorophyll or bacteriochlorophyll, is efficiently converted into stored chemical energy. Together, the two aspects of photosynthesis—the conversion of inorganic into organic matter, and the conversion of light energy into chemical energy—make it the fundamental process of life on Earth: it is the ultimate source of all living matter and of all life energy.

Under favorable external conditions, photosynthesis is a remarkably fast process. For example, with an adequate supply of carbon dioxide and light, a green algal cell will produce as much as 30 times its own volume in oxygen every hour. The rate of photosynthesis can be varied by varying the supply of carbon dioxide (CO_2), the intensity or color of illumination, or the temperature. The rate of photosynthesis depends also on the age, nutrition, and physiological condition of the organism, factors which are much more difficult to define and control precisely.

The total turnover of photosynthesis on Earth has been estimated in two ways: by averaging the yields of organic matter per unit area of field, forest, steppe, and ocean; and by determining the average utilization of incident solar energy by vegetation-covered areas (which is on the order of 1% if the whole solar spectrum is taken into consideration, or 2% if only visible light is considered). Both procedures lead to numbers of the magnitude of 10^{11} tons of carbon transferred annually from the inorganic into the organic state. This corresponds to about 10^{18} kcal (10^{15} kWh) of light energy stored annually. The estimate is rough, mainly because of uncertainty as to the average rate of photosynthesis in the world's oceans. *See* BIOGEOCHEMISTRY.

Plant Photosynthesis

The net overall chemical reaction of plant photosynthesis is shown in Eq. (1), where $\{CH_2O\}$ stands for a carbohydrate (sugar).

$$H_2O + CO_2 + \text{light energy} \xrightarrow[\text{enzymes}]{\text{chlorophyll}} \{CH_2O\} + O_2 \quad (1)$$

The photochemical reaction in photosynthesis belongs to the type known as oxidation-reduction, with CO_2 acting as the oxidant (hydrogen or electron acceptor) and water as the reductant (hydrogen or electron donor). The unique characteristic of this particular oxidation-reduction is that it goes "in the wrong direction" energetically; that is, it converts chemically stable materials into chemically unstable products. Light energy is used to make this "uphill" reaction possible, and a considerable part of the light energy utilized is stored as chemical energy.

Multistage process. Photosynthesis is a complex, multistage process. Its main parts are (1) the primary photochemical process in which light energy absorbed by chlorophyll is converted into chemical energy, in the form of some energy-rich intermediate products; and (2) the enzyme-catalyzed "dark" (that is, not photochemical) reactions by which these intermediates are converted into the final products—carbohydrates and free oxygen. These reactions of photosynthesis can be grouped into three phases (**Fig. 1**). Phase 1 is the transfer of electrons from an unknown intermediate in phase 2 to some intermediate acceptor capable of reducing CO_2. This is the light phase of photosynthesis. Phase 2 is the evolution of oxygen from dehydrogenated water. It requires an oxygen-evolving complex as well as chloride, calcium, and manganous ions and involves several steps. Phase 3 is the reduction of CO_2 by a series of dark reactions. The use of radioactive carbon (carbon-14) as a tracer has given considerable insight into the nature of these reactions.

If the rate of photosynthesis is plotted as a function of light intensity, a curve results which shows first a proportional increase, then a gradual saturation. This saturation may have various causes, one of which is the limitation of CO_2 supply from the outside. Further increase of light intensity becomes of no use when all CO_2 molecules reaching the cell are used up as fast as they arrive. Carbon dioxide concentration can thus act as a limiting factor. (The same principle applies to the effect of increasing CO_2 concentration in weak light when the reaction is light-limited.)

When the supply conditions for CO_2 and light are

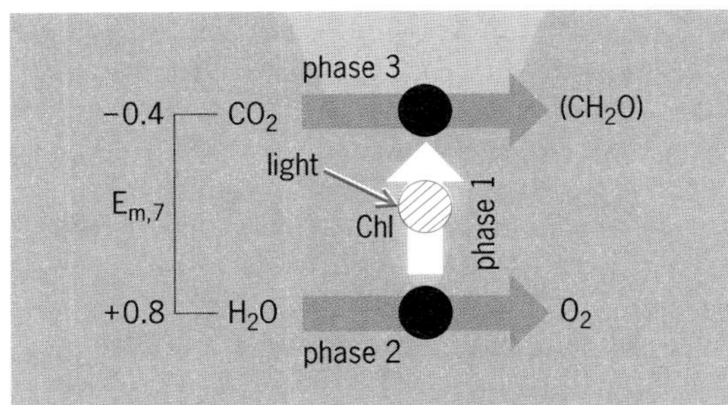

Fig. 1. Schematic illustration of photosynthesis. Phase 1, the light reaction, is the transfer of electrons by light-excited chlorophyll (Chl). Phase 2, oxidation of water, consists of enzymatic reactions converting dehydrogenated water to free oxygen. Phase 3, reduction of carbon dioxide, consists of enzymatic reactions converting carbon dioxide and light-supplied electrons to carbohydrates (CH_2O). $E_{m,7}$ is the oxidation-reduction potential at pH 7.0.

most favorable, the rate of photosynthesis still shows saturation. This is generally attributed to the need for the completion of photosynthesis of at least one (and more likely, several) light-independent enzymatic reactions. An enzyme-catalyzed reaction has a certain maximum rate; the several enzymes involved in photosynthesis impose ceilings on the maximum speed at which photosynthesis as a whole can proceed, with each enzyme functioning as a bottleneck of limited capacity in the reaction path. *See* ENZYME.

Hill reaction. Various observations suggest that the immediate action of light in photosynthesis involves the transfer of electrons from the primary reductant (electron donor), the reaction center chlorophyll molecule, to an electron acceptor, the primary oxidant (Fig. 1, phase 1). This is then followed by electron flow from H_2O to the oxidized reaction center chlorophyll molecule (Fig. 1, phase 2), and from the reduced primary acceptor to a molecule of nicotinamide adenine dinucleotide phosphate ($NADP^+$).

The reaction in phase 1 and phase 2 of Fig. 1 resembles the Hill reaction (named after its discoverer, R. Hill) in which illuminated chlorophyll-bearing organelles (chloroplasts) produce oxygen from water without the concomitant reduction of CO_2 but with the reduction of added, less stable oxidants, such as a quinone, ferricyanide, or 2,6-dichlorophenol indophenol. Since the minimum quantum requirement (number of quanta required to evolve one oxygen molecule) and other kinetic characteristics of the Hill reaction prove to be similar to those of photosynthesis, it has been assumed that in the Hill reaction, the primary photochemical apparatus of photosynthesis is preserved more or less intact. In the Hill reaction, however, the coupling of the primary photochemical process with the enzymatic mechanism which brings about the reduction of CO_2 is easily impaired by the mechanical destruction of the chloroplast's outer membrane.

Quantum process. In photosynthesis, the energy of light quanta is converted into chemical energy. In the conversion of 1 mole of CO_2 and 1 mole of H_2O into 1 mole of carbohydrate group and 1 mole of oxygen, according to Eq. (1), about 112 kcal of total energy or, under natural conditions, about 120 kcal of potential chemical energy (free energy) are stored. Light is absorbed by matter in the form of quanta or photons. *See* ABSORPTION OF ELECTROMAGNETIC RADIATION; PHOTON.

The reduction of one molecule of CO_2 to the carbohydrate level requires the use of four hydrogen atoms as expressed by Eq. (2). A minimum quantum

$$CO_2 + 4H \longrightarrow [CH_4O_2] \longrightarrow [CH_2O] + H_2O \qquad (2)$$

requirement of eight or more would thus permit two

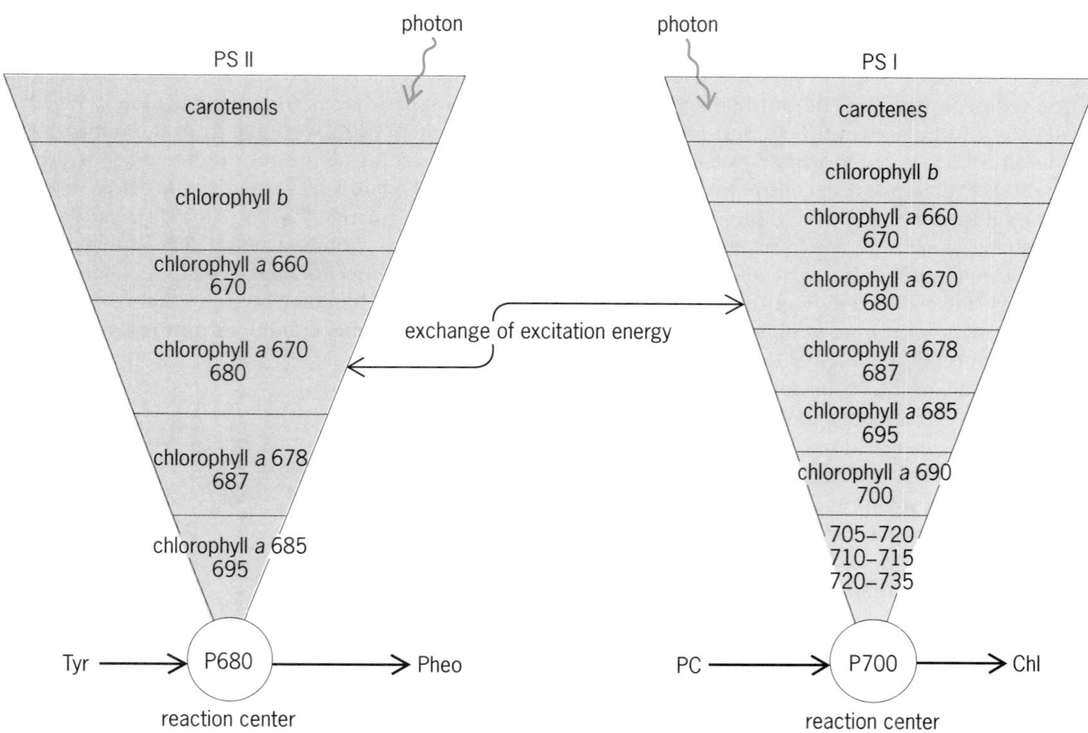

Fig. 2. Working model for the distribution of pigments in the two photosystems of higher plants. Relative abundance is indicated by width of band; chlorophyll *b*, for example, is more abundant in PS II. Chlorophyll *a* is the most important pigment in both systems, but each has a characteristic assortment of spectral forms, distinguished by their maxima in the red end of the spectrum, in nanometers, and fluorescence maxima. In PS I, the primary reaction is the oxidation of P700 (the reaction-center chlorophyll *a* molecule—the primary electron donor) and the reduction of Chl (a chlorophyll *a* molecule acting as the primary electron acceptor, also called A_0); this is followed by the reduction of $P700^+$ by plastocyanin (PC). Photosystem II is similar, but the primary electron acceptor is a pheophytin molecule (Pheo), while the electron donor to $P680^+$ is a tyrosine (TYR) molecule, usually referred to as Z.

quanta to be used for the transfer of each hydrogen atom (or electron) from H_2O to CO_2.

Two-quanta hypothesis. A specific mechanism in which two quanta are used to transfer one hydrogen atom in photosynthesis was suggested by experiments of Robert Emerson. Emerson had discovered that the "maximum quantum yield of photosynthesis" (number of O_2 molecules evolved per absorbed quantum), while constant at the shorter wavelengths of light (red, orange, yellow, green), declines in the far-red above 680 nanometers (the "red drop"). Later, Emerson showed that this low yield could be enhanced if both chlorophyll *a* and *b* are simultaneously excited (only chlorophyll *a* absorbs above 680 nm). This effect, known now as the Emerson enhancement effect, suggested that two pigments must be excited to perform efficient photosynthesis, and thus indicated involvement of two light reactions in photosynthesis, one sensitized by light absorption in chlorophyll *a* and one by absorption in another pigment (for example, chlorophyll *b*). Experiments by others enlarged Emerson's observation by suggesting that plants contained two pigment systems. One (called photosystem I, or PS I, sensitizing reaction I) contains the major part of chlorophyll *a*; the other (called photosystem II, or PS II, sensitizing reaction II) contains some chlorophyll *a* and the major part of chlorophyll *b* or other auxiliary pigments (for example, the red and blue pigments, called phycobilins, in red and blue-green algae, and the brown pigment fucoxanthol in brown algae and diatoms). It appears that efficient photosynthesis requires the absorption of an equal number of quanta in PS I and in PS II; and that within both systems excitation energy undergoes resonance migration from one pigment to another until it ends in special molecules of chlorophyll *a* called the reaction centers; the latter, labeled as (P700 or P680), then enter into the chemical reactions (**Fig. 2**).

Hill and F. Bendall proposed that one of these reactions is the transfer of an electron from some intermediate in the conversion of water to oxygen to cytochrome b_6 (shown later to be a plastoquinone molecule), while the other is the transfer of an electron from cytochrome *f* to an intermediate in the conversion of CO_2 to carbohydrate. The intermediate transfer of hydrogen (or electron) from reduced plastoquinone (plastoquinol) to cytochrome *f* can occur without energy input because the former is a stronger reductant than the latter. Experimental evidence for the existence of two pigment systems and the key role of plastoquinone and cytochrome *f* in this sequence has been provided. However, the specific role of cytochrome b_6 has not been confirmed.

Photosynthesis is conceived of as a set of at least five reactions, two of which are light reactions (PS I and PS II) and three of which are dark reactions (**Fig. 3**). The PS II reaction is the one most closely associated with O_2 evolution. The final result of this set of reactions is the oxidation of water to O_2 and the reduction of a plastoquinone (an oxidation-reduction catalyst). Light absorbed by the major part of the accessory pigments is ultimately transferred to a chlorophyll *a* molecule (P680), which is assumed to be in a favorable position to act as an energy trap (or reaction center). The P stands for pigment and the numerical designation gives the location, in nanometers, of the maximum decrease in the absorption in the red or the near-infrared region of the spectrum when the pigment is illuminated by the bright actinic light. The primary light reaction (Fig. 3) is now shown to be an electron transfer from P680, within 3 picoseconds, to the electron acceptor pheophytin (Pheo), powered by excited P680. The P680 recovers by accepting an electron from a tyrosine residue, Z, within 20–400 nanoseconds, as shown in reactions (3).

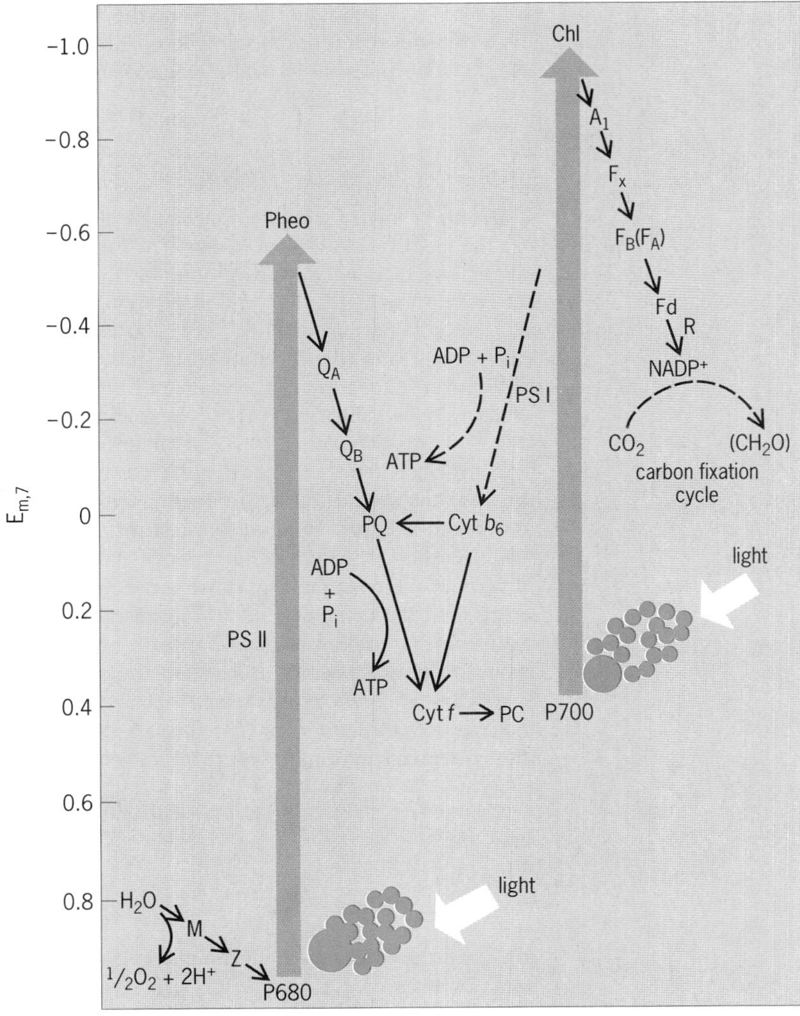

Fig. 3. Schematic representation of electron flow and chemical reactions in the two light steps in photosynthesis. Z (a tyrosine residue) is the electron donor to the oxidized reaction-center chlorophyll *a* P680 of photosystem II (PS II), and M (a manganese-containing protein) is the charge accumulator that leads to O_2 evolution. Pheo (pheophytin) and Chl (chlorophyll *a*) are the primary electron acceptors of PS II and PS I, respectively. Q_A and F_X (an iron-sulfur center) are the stable electron acceptors of PS II and PS I, respectively. Q_B is another bound plastoquinone, while PQ is plastoquinone, Cyt is cytochrome, and PC is plastocyanin. A_1 is an electron acceptor of PS I, F_B and F_A are iron-sulfur centers, Fd is ferredoxin, and R is the Fd–$NADP^+$ reductase. The pigment-containing antenna units I and II involved in the process are indicated by solid circles. $E_{m,7}$ is the oxidation reduction potential at pH 7.0.

$$\text{P680} + h\nu\,\text{II} \longrightarrow \text{P680*} \qquad (3a)$$

$$\text{Pheo} + \text{P680*} \longrightarrow \text{Pheo}^- + \text{P680}^+ \qquad (3b)$$

$$\text{Z} + \text{P680}^+ \longrightarrow \text{Z}^+ + \text{P680} \qquad (3c)$$

The components of reactions (3), that is, P680, pheophytin, and Z, and the next intermediates, Q_A and Q_b (bound plastoquinones), are located on a protein complex consisting of two polypeptides, named D-1 and D-2 (approximately 32,000 daltons molecular mass each). The oxidation product, the strong oxidant Z^+, is utilized to oxidize water and liberate O_2 and protons (Fig. 3).

The PS I reaction is the one most closely associated with the reduction of $NADP^+$. Light absorbed by most of the chlorophyll *a* molecules is ultimately transferred to the reaction center chlorophyll *a* and energy trap of PS I, the P700. The primary light reaction of PS I is oxidation of P700 and the reduction of an electron acceptor chlorophyll *a* (Chl *a*) within a few picoseconds, as shown in reactions (4).

$$\text{P700} + h\nu\,\text{I} \longrightarrow \text{P700*} \qquad (4a)$$

$$\text{Chl}\,a + \text{P700*} \longrightarrow \text{Chl}\,a^- + \text{P700}^+ \qquad (4b)$$

The three dark reactions, mentioned above, are (1) the electron flow from Chl a^- (A_0^-) to $NADP^+$ via an intermediate A_1 (phylloquinone?), the iron-sulfur centers (F_X, F_B, F_A), and ferredoxin (Fd); (2) the electron flow from $Pheo^-$ to P700 via several electron carriers (plastoquinones Q_A, Q_B, and PQ, Rieske iron-sulfur center, cytochrome *f*, and a copper protein plastocyanin); and (3) the electron flow from H_2O to oxidized tyrosine Z^+ via an Mn complex.

O_2 evolution. The mechanism of O_2 evolution is the least known part of the photochemical process. All oxygen liberated in photosynthesis originates in water. Based on the measurements of the amount of O_2 evolved in single brief (10-microsecond) saturating light flashes, it has been suggested that four oxidizing equivalents must accumulate on the O_2-evolving complex before it can oxidize H_2O to O_2. It appears as if an oxygen "clock" exists in which the state (S) of the O_2-evolving complex undergoes the sequential reaction (5) where the subscripts on S

$$S_0 \longrightarrow S_1 \longrightarrow S_2 \longrightarrow S_3 \longrightarrow S_4 \;(\longrightarrow O_2,\ \text{returning to } S_0) \qquad (5)$$

represent the oxidizing equivalents accumulated on the complex.

Manganese (Mn) is required for O_2 evolution. Plants grown in a manganese-deficient medium lose their capacity to evolve O_2. The charge accumulator of the reaction mechanism is a manganese complex; manganese is known to exist in several valence states. In addition to manganese, chlorine and calcium ions also function at the O_2-evolving site, though the mechanism of action is not known. The O_2-evolving complex includes at least four "intrinsic" proteins having molecular weights of 34,000, 32,000, 9000, and 4500 (D-1, D-2, and two subunits of cytochrome *b* 559). In addition, three "extrinsic" polypeptides of molecular weights 33,000, 24,000, and 18,000 are also involved, although cyanobacteria contain only the 33,000 protein. Further research is needed to understand the biochemical mechanism of O_2 evolution.

Photophosphorylation. Chromatophores from photosynthetic bacteria and chloroplasts from green plants, when illuminated in the presence of adenosine diphosphate (ADP) and inorganic phosphate, use light energy to synthesize adenosine triphosphate (ATP); about 10 kcal of converted light energy is stored in each molecule of the high-energy phosphate, ATP. This photophosphorylation could be associated with some energy-releasing step in photosynthesis, such as the electron flow from PS II to PS I. When phosphorylation is associated with noncyclic electron flow from H_2O to $NADP^+$, it is called noncyclic photophosphorylation. *See* ADENOSINE DIPHOSPHATE (ADP); ADENOSINE TRIPHOSPHATE (ATP).

There is also evidence to suggest that light reaction I (PS I) can be reserved, at least in isolated chloroplasts; electrons, instead of going to $NADP^+$, may simply return to an intermediate (such as a cytochrome, plastoquinone, plastocyanin, or P700) and thus close the cycle. It has been suggested that cytochrome b_6 may be an intermediate in this back reaction of PS I. This type of electron flow, mediated by added cofactors and ADP and inorganic phosphate (P_i), leads to the production of ATP and has been termed cyclic phosphorylation. It is not clear if such a cycle exists in the plant.

It has been shown that light produces a high-energy state, and that the actual phosphorylation occurs in the dark. Furthermore, if chloroplasts are first suspended in an acidic medium and then transferred to an alkaline medium in the presence of ADP and P_i, phosphorylation occurs without the need of light. All these experiments have been interpreted in terms of a hypothesis by P. Mitchell, in which light produces a H^+ ion gradient (ΔpH) across the lamellar membranes in the chloroplasts, and the energy dissipation of this H^+ ion gradient via the coupling factor (or) ATP synthase, a protein complex present in the lamellae) leads to phosphorylation. In addition, an electric field ($\Delta\Psi$) generated across the thylakoid membrane as a result of the initial light-induced charge separation can also drive photophosphorylation. Reagents that dissipate the electric field or the H^+ gradient also inhibit phosphorylation. The two together are referred to as proton motive force (**Fig. 4**).

Photosynthetic unit. The concentration of the special chlorophyll *a* molecules (P700 or P680) that engage in the chemical reactions is one in several hundred chlorophyll molecules; energy absorbed by other pigments is effectively transferred to these special molecules (called energy traps or reaction centers). The groups of antenna (bulk) molecules with

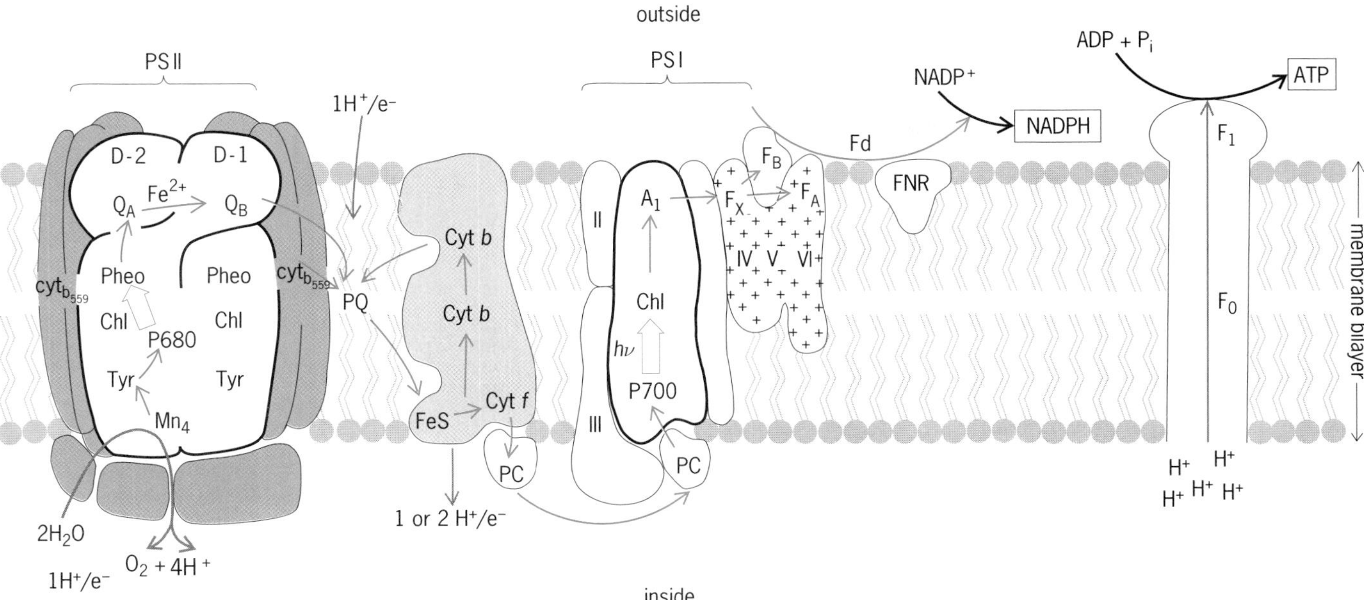

Fig. 4. Special representation of the electron and proton transport mechanisms for plant photosynthesis, located on the inner membrane of a chloroplast. The shapes of the proteins are largely hypothetical. The shaded proteins are the intrinsic and extrinsic polypeptides of the oxygen-evolving complex. The proteins containing P680 and P700 are PS II and PS I reaction center complexes, respectively. Abbreviations as in Fig. 3. FeS = Rieske iron-sulfur center. The iron-sulfur centers F_X, F_B, and F_A are on complexes IV, V, and VI, respectively. Complex I is PS I, and the functions of complexes II and III are unknown. The coupling factor protein has two components: F_0 (embedded in the membrane) and F_1 (extrinsic, and active in ATP synthesis). FNR stands for ferredoxin–$NADP^+$ reductase.

their energy traps are often referred to as photosynthetic units.

Emerson and W. Arnold showed how the light reaction in photosynthesis can be separated from the dark reaction by the use of brief, intense light flashes, separated by intervals of darkness of variable duration. They found that the yield of a single flash was maximum when the interval between the consecutive light flashes was at least 0.04 s at 1°C (33°F). This, then, is the minimum time required for the efficient utilization of the products from the light reactions. Emerson and Arnold further observed that under optimal conditions the maximum yield from a single flash was one O_2 molecule per 2400 chlorophyll molecules present. Since a minimum of eight quanta are required to evolve one O_2 molecule, it can be envisioned that the absorption of eight quanta of light by a group of 2400 chlorophyll molecules results in the evolution of one O_2 molecule. However, it is now known that the two light-reaction mechanisms of photosynthesis require the transfer of four electrons through two light reactions for every molecule of oxygen evolved. Thus there are at least eight photoacts leading to the evolution of one O_2 molecule. Therefore, the ratio of one O_2 per 2400 chlorophylls means one photoact per 300 chlorophyll molecules. By using spinach grown in moderate to high light intensity, it has been shown that there is one active PS II (P680) and one active PS I for a total of about 600 chlorophyll molecules present. If these chlorophyll molecules are equally divided in photosystems I and II, there is one reaction center per 300 chlorophyll molecules in each system. This is the commonly accepted size of one physical unit in higher plants and algae.

Photochemical apparatus. The primary photochemical stage of the photosynthetic process appears to be closely associated with certain structural elements found in plant cells. All algae (except the prokaryotic algae such as green *Prochloron* and the cyanobacteria, the blue-greens), as well as all higher plants, contain pigment-bearing organelles called chloroplasts. In the leaves of the higher land plants, these are usually flat ellipsoids about 5 micrometers in diameter and 2.3 μm in thickness; 10–100 of them may be present in an average cell of leaf parenchyma. Under the electron microscope, all chloroplasts show a layered structure with alternate lighter and darker layers roughly 0.01 μm in thickness. *See* CELL PLASTIDS.

In algae the number and shape of chloroplasts are much more variable; for example, the green unicellular alga *Chlorella* contains only one bell-shaped chloroplast.

Two main types of chloroplasts are known. In lamellar chloroplasts, the layered structure extends more or less uniformly through the whole chloroplast body. In granular chloroplasts, this structure is emphasized in certain cylindrical sections (the grana) and is less pronounced in the area between them (the stroma region). When such granular chloroplasts are permitted to dry out and disintegrate, stacks of disks break off the structure and appear as cylindrical grana in the electron microscope. The photochemical apparatus is less complex in blue-green algae.

The unit of photochemical apparatus in both advanced and primitive plants may be a lamella consisting of two submembranes forming a saclike disk called a thylakoid. The light reactions of photosynthesis are intimately associated with this membrane. The fine structure of thylakoid membranes is far from clear; various-sized particles and surfaces have been observed. It has not been possible to prove the correlation of the fine structure with function except in the case of coupling factor; these particles are found on the top of the thylakoid membrane in the unstacked regions. Their removal stops phosphorylation activities.

Accessory pigments. In addition to chlorophyll *a*, the one pigment present in all (oxygen-evolving) photosynthetically active plants, there are other chlorophylls, such as chlorophyll *b* in the green algae and higher plants. In brown algae, chlorophyll *c* replaces chlorophyll *b*. There are also nonchlorophyllous pigments belonging to two groups: (1) The carotenoids, so called because of similarity to the orange pigment of carrots, are a variable assortment of pigments found in all photosynthetic higher plants and in algae. (2) The phycobilins, or vegetable bile pigments, are chemically related to animal bile pigments. The phycobilins are either red (phycoerythrin) or blue (phycocyanin). Both types are present in special granules called phycobilisomes in red algae (Rhodophyta) and blue-green algae (Cyanophyta or cyanobacteria); the red pigment prevails in red algae and the blue pigment prevails in blue-green algae. Another phycocyanin called allophycocyanin is also present in blue-green algae. *See* CAROTENOID; PHYCOBILIN.

Light absorbed by accessory pigments does contribute to photosynthesis. This is known from measurements of the so-called action spectra of photosynthesis. In such measurements, photosynthesis is excited by monochromatic light, and the production of oxygen per incident quantum of light is measured as a function of wavelength. The observed spectral variations in the yield of photosynthesis can be related to the proportion of light absorbed at each wavelength by the different pigments in the cells. Measurements of this kind have led to the conclusion that quanta absorbed by carotenoids are 50–80% as effective as those absorbed by chlorophyll *a* in contributing energy to photosynthesis. An exception is fucoxanthol, the carotenoid that accounts for the color of brown algae (Phaeophyta) and that supplies light energy to photosynthesis about as effectively as chlorophyll *a*. The red and blue pigments of the Rhodophyta and Cyanophyta are also highly effective. They can be as effective as chlorophyll or somewhat less, depending, among other things, on the physiological status of the algae and the color of the light to which they have become adapted. The primary function of all these pigments is to harvest the light energy and transfer it to reaction-center chlorophyll molecules.

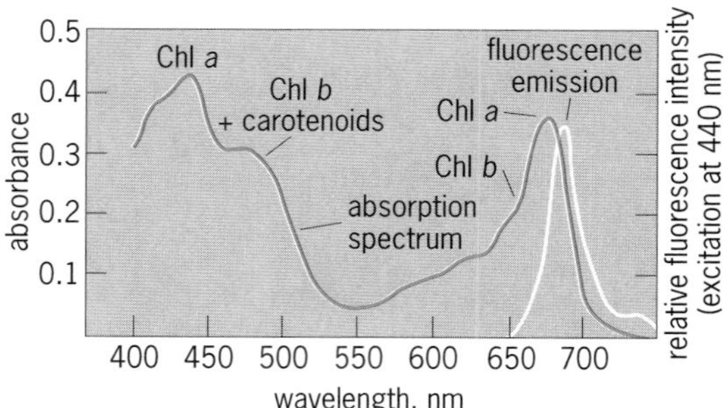

Fig. 5. Absorption spectrum of a corn (*Zea mays*) leaf. Pigments responsible for specific bands are shown. Also shown is the fluorescence emission of chloroplasts from a corn leaf.

Energy transfer between pigment molecules. Chlorophyll *a* in plant cells is weakly fluorescent: this means that some of the light quanta absorbed by it (up to 6%) are reemitted as light (**Fig. 5**). Observations of the action spectrum of chlorophyll *a* fluorescence in different plants have suggested close parallels with the action spectrum of photosynthesis. In other words, fluorescence of chlorophyll *a* in the plant can be excited also by light absorbed by the accessory pigments. Excitation of chlorophyll *a* fluorescence by light quanta absorbed by phycoerythrin requires transfer of the excitation energy from the excited phycoerythrin molecule to a nearby chlorophyll molecule (as in acoustic resonance, where striking one bell causes another nearby bell to ring). Therefore, light quanta absorbed by accessory pigments, such as carotenoids and phycobilins, may contribute to photosynthesis by being transferred to chlorophyll *a*. By this mechanism, red algae, growing relatively deep under the sea where only green light penetrates, can supply the energy of this light to chlorophyll *a*, which has a very weak absorption in the green region of the spectrum.

If excitation energy can be transferred efficiently in the chloroplasts from accessory pigments to chlorophyll *a*, there is a good probability that a simililar transfer occurs also between different chlorophyll *a* molecules themselves. Excitation-energy transfer among chlorophyll *a* molecules or among phycobilin molecules, and excitation-energy transfer from accessory pigments (donor molecules) to chlorophyll *a* (acceptor molecules) or from various short-wavelength forms of chlorophyll *a* to the long-wavelength forms of chlorophyll *a* can be demonstrated. The most widely accepted hypothesis, Förster's hypothesis, is that energy transfer is preceded by thermal relaxation in the donor molecules. The efficiency of energy transfer depends upon three basic factors: orientation of acceptor molecules with respect to the donor molecule; overlap of the fluorescence spectrum of the donor molecule with the absorption spectrum of the acceptor molecule; and the distance between the two molecules. The function of most of the pigments (including most of the chlorophyll *a* molecules) is to act as an antenna, harvest the energy, and transfer to very few (1 in

300) reaction-center molecules (P700 and P680), depending upon the pigment system (Fig. 2). Energy is thus trapped and used for photochemistry. *See* PLANT PIGMENT.

Chemical role of chlorophyll. The question arises as to how does the chlorophyll *a* molecule, ultimately in possession of the absorbed quantum of energy, utilize it for an energy-storing photochemical process, such as the transfer of a hydrogen atom from a reluctant donor, H_2O, to a reluctant acceptor (perhaps $NADP^+$). It has been shown that chlorophyll acts as a typical oxidation-reduction catalyst—that is, by being itself first oxidized and then reduced. Support for this concept is provided by observations of reversible photochemical oxidation and of reversible photochemical reduction of chlorophyll in solution, and the comparison of these data with those in chloroplasts. Studies of changes in the absorption spectrum of photosynthetic cells in light show that a small fraction of a special form of chlorophyll *a* (P700), absorbing maximally at 700 and 430 nm, is in an oxidized state during illumination. This is the reaction center of PS I. The reaction center of PS II (P680) has been suggested also to undergo oxidation-reduction. Furthermore, a chlorophyll *a* molecule appears to be chemically reduced when P700 is oxidized to $P700^+$ in PS I, as noted earlier [see Eq. (4)]. The detailed chemical nature of P700 and P680 remains unknown. There is a good possibility that these components, although similar to chlorophyll *a*, are chemically distinct entities.

Govindjee; R. Govindjee

Carbon Dioxide Fixation

The light-dependent conversion of radiant energy into chemical energy as ATP and reduced nicotinamide adenine dinucleotide phosphate (NADPH) serves as a prelude to the utilization of these compounds for the reductive fixation of CO_2 into organic molecules. Such molecules, broadly designated as photosynthates, are usually but not invariably in the form of carbohydrates such as glucose polymers or sucrose, and form the base for the nutrition of all living things, as well as serving as the starting materials for fuel, fiber, animal feed, oil, and other compounds used by people. Collectively, the biochemical processes by which CO_2 is assimilated into organic molecules are known as the photosynthetic dark reactions, not because they must occur in darkness, but because light—in contrast to the photosynthetic light reactions—is not required.

The route by which CO_2 is assimilated was studied for over a century when it was discovered that photosynthesis leads to the accumulation of sugars and starches. Details of the assimilation of CO_2 were worked out in the 1950s, when the availability of paper chromatographic techniques and $^{14}CO_2$ allowed M. Calvin, A. A. Benson, and J. A. Bassham to develop the outlines of the reductive pentose phosphate cycle, now usually called the C_3 cycle. The C_3 cycle forms the primary, or basic (with other, feeder pathways occurring in some plant types), route for the formation of photosynthate from CO_2.

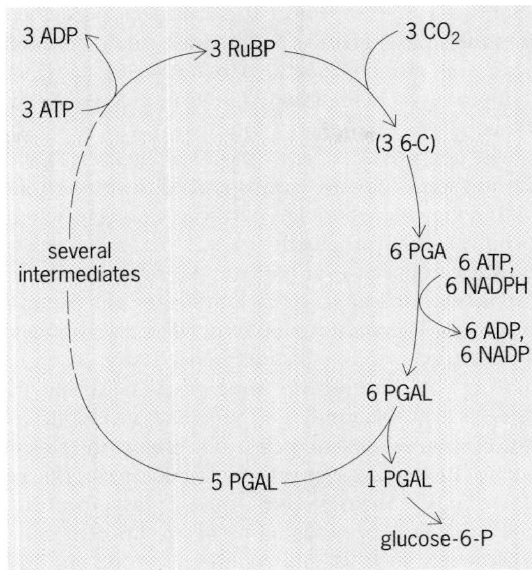

Fig. 6. Schematic outline of the Calvin (C_3) carbon dioxide assimilation cycle.

C_3 photosynthesis. The essential details of C_3 photosynthesis can be seen in **Fig. 6**. Three molecules of CO_2 combine with three molecules of the five-carbon compound ribulose bisphosphate (RuBP) in a reaction catalyzed by RuBP carboxylase to form three molecules of an enzyme-bound six-carbon compound. These are hydrolyzed into six molecules of the three-carbon compound phosphoglyceric acid (PGA), which are phosphorylated by the conversion of six molecules of ATP (releasing ADP for photophosphorylation via the light reactions). The resulting compounds are reduced by the NADPH formed in photosynthetic light reactions to form six molecules of the three-carbon compound phosphoglyceraldehyde (PGAL). One molecule of PGAL is made available for combination with another three-carbon compound, dihydroxyacetone phosphate, which is isomerized from a second PGAL (requiring a second "turn" of the Calvin-cycle wheel) to form a six-carbon sugar. The other five PGAL molecules, through a complex series of enzymatic reactions, are rearranged into three molecules of RuBP, which can again be carboxylated with CO_2 to start the cycle turning again.

It should be noted that the enzyme that incorporates CO_2 into an organic compound, RuBP carboxylase, can comprise up to half of the soluble protein in C_3 chloroplasts, and most likely is the most abundant protein found in nature. RuBP carboxylase has eight large polypeptide subunits and eight small subunits. Interestingly, the small subunit polypeptide is produced (as a larger, precursor form) in the cytoplasm from mRNA encoded in the nucleus. The precursor polypeptide is then transported across the chloroplast membrane (the mature form of this polypeptide cannot be transported in this manner), processed into the shorter, mature polypeptide, and combined with large subunits (encoded in the chloroplast DNA and produced in the stroma) to form the mature enzyme.

The net product of two "turns" of the cycle, a six-carbon sugar (glucose-6-phosphate) is formed either within the chloroplast in a pathway leading to starch (a polymer of many glucose molecules), or externally in the cytoplasm in a pathway leading to sucrose (condensed from two six-carbon sugars, glucose and fructose). This partitioning of newly formed photosynthate leads to two distinct pools; starch is stored in the photosynthesizing "source" leaf cells, and sucrose is available either for immediate metabolic requirements within the cell or for export to "sinks" such as developing reproductive structures, roots, or other leaves. Factors within the photosynthesizing cell, such as energy requirements in different compartments (such as mitochondria, cytoplasm, and chloroplasts) of the cell, along with energy needs of the plant (such as increased "sink" requirements during different developmental stages) and external, environmental factors (such as light intensity and duration) ultimately regulate the partitioning of newly formed photosynthetic product (PGAL) into starch or sucrose.

This profound control of photosynthate partitioning is accomplished primarily through the regulation of the enzymes which convert PGAL to sucrose in the cytoplasm. Under conditions where sink demand is low (and sucrose is not transported through the phloem away from source leaf cells), metabolic effectors accumulate in the cytoplasm which lower the activities of the sucrose-forming enzymes. This results in a condition that reduces PGAL export from the chloroplast, and hence, more PGAL is retained in the chloroplast for starch formation. Also, under conditions which cause low chloroplast PGAL levels (such as low light), PGAL transport out of the chloroplast is restricted, resulting in decreased substrate for sucrose formation, increasing the relative amount of starch production. The energy status of the cell affects sucrose formation (and therefore, photosynthate partitioning) because cytoplasmic uridine triphosphate (used in the formation of sucrose) levels are dependent on ATP generation, and also because PGAL export to the cytoplasm is coupled obligatorily to inorganic phosphate (formed when ATP is metabolized in the cytoplasm) import into the chloroplast.

Photosynthetic induction phenomenon. Carbon dioxide assimilation in plants does not begin immediately upon illumination; there is a lag of several minutes before assimilation attains a rapid rate. This lag is called photosynthetic induction, and is not limited by light reactions because high levels of ATP and NADPH are found almost immediately. Two possible limiting mechanisms have been considered: the buildup of intermediates and the activation of enzymes involved in assimilation reactions.

D. Walker concluded that induction represents the time needed to build up the intermediates of the C_3 cycle, in an autocatalytic manner, from newly formed photosynthate to concentrations sufficient for assimilation to proceed at a rate controlled by the prevailing environment (CO_2 levels, temperature, light intensity, and other factors). The autocatalytic nature of the cycle can be best understood by considering that the net product of one "turn" of the cycle (representing three carboxylations), a PGAL molecule, can be fed back into the cycle, and that the rate of carboxylation during this lag phase is dependent on the level of newly formed RuBP. The level of RuBP would double after five carboxylations. This would increase the RuBP level so that the next five carboxylations would occur in a shorter amount of time, resulting in an exponential increase in the rate of photosynthesis until factors other than intermediate levels become limiting.

Not only is the cycle autocatalytic, but the initial carboxylation catalyst, RuBP carboxylase, as well as glyceraldehyde-3-phosphate dehydrogenase and the enzymes sedoheptulose and fructose biphosphate phosphatase, requires activation. These catalysts are inactivated in the dark and activated in the light. Several conditions are required for activation, including high concentrations of Mg^{2+}, CO_2, and a reductant and high pH. These conditions are facilitated by light-dependent processes but are reversed in darkness. This regulatory mechanism conveniently allows for the synthesis pathway to be "shut off," preventing a futile cycle during the night, when starch reserves are mobilized to meet cell energy requirements via intermediates which, if C_3 cycle enzymes were activated, would be reconverted to starch. It is uncertain whether the rates of light-activated enzyme catalysis are comparable with those of the induction phenomenon, but buildup of intermediates, together with activation and inactivation of catalysts, offers an explanation and working hypothesis for the induction phenomenon.

C_4 photosynthesis. Initially, the C_3 cycle was thought to be the only route for CO_2 assimilation, although it was recognized by plant anatomists that some rapidly growing plants (such as maize, sugarcane, and sorghum) possessed an unusual organization of the photosynthetic tissues in their leaves (Kranz morphology). The researches of H. Kortschak and coworkers in Hawaii and of M. D. Hatch and R. C. Slack in Australia demonstrated that plants having the Kranz anatomy utilized an additional CO_2 assimilation route now known as the C_4-dicarboxylic acid pathway (**Fig. 7**). Carbon dioxide enters a mesophyll cell, where it combines with the three-carbon compound phosphoenolpyruvate (PEP) to form a four-carbon acid, oxaloacetic acid, which is reduced to malic acid or transaminated to aspartic acid. The four-carbon acid moves into bundle sheath cells, where the acid is decarboxylated, the CO_2 assimilated via the C_3 cycle, and the resulting three-carbon compound, pyruvic acid, moves back into the mesophyll cell and is transformed into PEP, which can be carboxylated again. The two cell types, mesophyll

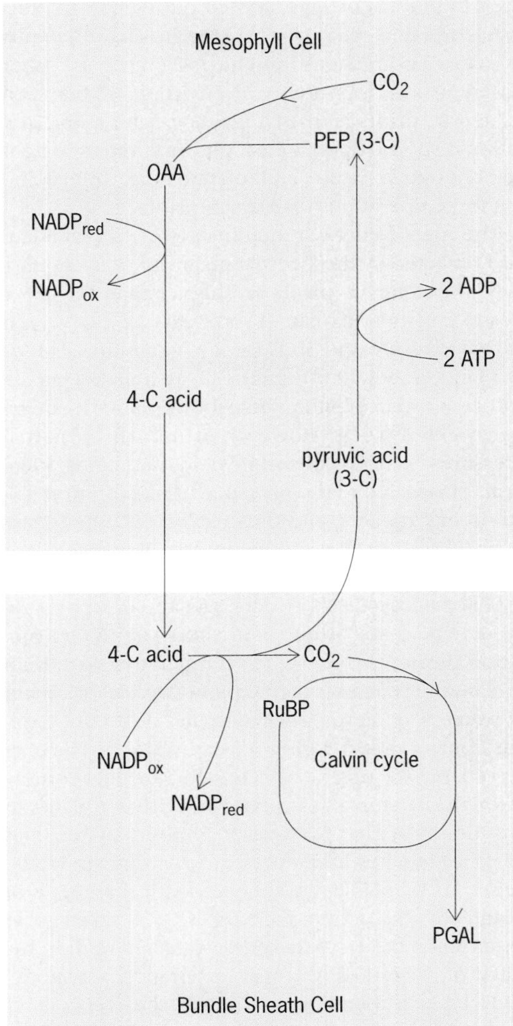

Fig. 7. Schematic outline of the Hatch-Slack (C_4) carbon dioxide assimilation route in two cell types of a NADP-ME-type plant.

and bundle sheath, are not necessarily adjacent (sedges are an example), but in all documented cases of C_4 photosynthesis the organism had two distinct types of green cells. As depicted in **Fig. 8** extensive transport of metabolites must occur between the two cell types in C_4 plants. It is unknown how this directional transport is facilitated, although the presence of plasmodesmata forming a cytoplasmic continuum between the two cell types may be involved.

C_4 metabolism is classified into three types, depending on the decarboxylation reaction used with the four-carbon acid in the bundle sheath cells.

1. NADP-ME type (sorghum),

$$\mathrm{NADP^+ + malic\ acid} \xrightarrow{\text{NADP-malic enzyme}} \mathrm{pyruvic\ acid + CO_2 + NADPH} \quad (6)$$

2. NAD-ME type (*Atriplex* species),

$$\mathrm{NAD^+ + malic\ acid} \xrightarrow{\text{NAD-malic enzyme}} \mathrm{pyruvic\ acid + CO_2 + NADH} \quad (7)$$

3. PCK type (*Panicum* species),

$$\mathrm{Oxaloacetic\ acid + ATP} \xrightarrow{\text{phosphoenol pyruvate carboxykinase}} \mathrm{PEP + CO_2 + ADP} \quad (8)$$

In addition to differing decarboxylation reactions, the particulars of the CO_2 fixation pathway in NAD-ME and PCK plant types differ from those depicted in Fig. 8 with respect to the three-carbon compound transported from bundle sheath to mesophyll cells. With NAD-ME types, the three-carbon compound can be either pyruvic acid or alanine, and in PCK types this compound is PEP. Therefore, the three variations in the C_4 pathway necessarily predicate different energy (ATP and NADPH) usage in the two cell types.

The generation of ATP from ADP, and NADPH from NADP via noncyclic electron flow through photosystem I (PS I) and photosystem II (PS II) is tightly coupled: neither compound can be produced without sufficient substrate for both. Therefore, the different usage of ATP and NADPH in the mesophyll and bundle sheath chloroplasts of the three C_4 plant types (due both to variations in the pathway of carbon flow in the photosynthetic cycle and to variations in partitioning of portions of the pathway between cell types) is supported by variations in the photochemical apparatus which allow for differing ability to produce ATP without concomitant NADPH production. These alternative pathways of ATP production (which result in different ratios of ATP:NADPH produced) are cyclic and pseudocyclic

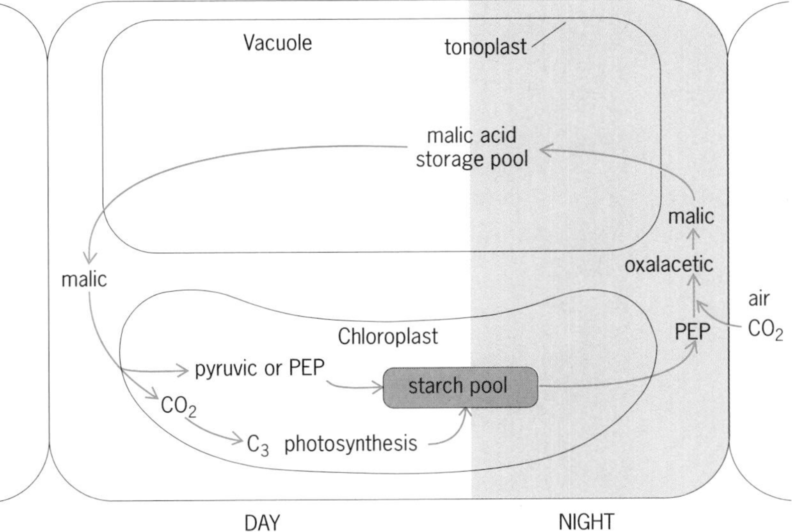

Fig. 8. Scheme for the flow of CO_2 within a single crassulacean acid metabolism cell over a day, showing initial dark CO_2 fixation, malic acid storage in the vacuole at night, followed by decarboxylation and the C_3 cycle the next day.

photophosphorylation, with the cyclic pathway considered the major pathway of uncoupled ATP production in chloroplasts, and the pseudocyclic pathway possibly acting as a "fine-tuning" modulator.

Variations in the photochemical apparatus which indicate enhanced cyclic photophosphorylation capacity (utilizing only PS I) are a high chlorophyll *a*/*b* ratio, low Chl/P700 ratio, and a low PS II reaction. These characteristics are found in bundle sheath chloroplasts of NADP-ME–type plants, indicating that the primary function of the photochemical apparatus in these chloroplasts is the generation of ATP. NADPH is supplied via the decarboxylation of malic acid to support the C_3 cycle activity (PGA conversion to PGAL) in these chloroplasts. Assays of chlorophyll *a*/*b* ratio, Chl/P700 ratio, and PS II activity indicate that NAD-ME mesophyll chloroplasts also have a primary role of cyclic photophosphorylation, while NAD-ME bundle sheath chloroplasts have a primary role of noncyclic electron flow. In PCK-type plants, mesophyll chloroplasts appear to have a photochemical apparatus similar to C_3 chloroplasts, while bundle sheath chloroplasts appear to have a low PS II activity. The enhanced ability of PCK bundle sheath chloroplasts to produce ATP via cyclic photophosphorylation supplies the extra ATP needed to convert pyruvic acid to PEP. These variations in the C_4 pathway and photochemical apparatus among the C_4 plant types demonstrate the close relationship that has evolved between light reactions and the biochemical processes of carbon dioxide assimilation, and show the highly integrated cooperation between the cell types involved.

Functions of C_4 cycle. The concentration of CO_2 in air is about 0.03% by volume, a concentration that does not fully saturate the C_3 cycle when it is operating at capacity. It would be necessary to have about 0.1% CO_2 to saturate the C_3, which can only be achieved under controlled conditions (CO_2-enriched greenhouses or growth chambers). Leaf photosynthesis in C_4 plants, however, is fully saturated at air CO_2 concentrations. Thus, C_4 photosynthesis may be considered to be an evolutionary adaptation to current-day CO_2 levels in air. During the C_4 cycle, CO_2 is continually fed via biochemical reactions in mesophyll chloroplasts to RuBP carboxylase in bundle sheath chloroplasts so that air CO_2 is not rate-limiting. Apparently the spatial compartmentalization of portions of CO_2 assimilation into two cell types not only allows C_4 plants to assimilate air CO_2 rapidly, but also partly explains other physiological characteristics and responses to the external environment of C_4 plants. These include their high efficiency of water use (as water vapor exits through the same stomatal pores through which CO_2 enters the leaf, and since the C_4 plant is more efficient at fixing CO_2 than C_3 plants, more CO plants, more CO_2 is incorporated per unit water lost), their greater efficiency of nitrogen usage (as RuBP carboxylase is produced only in bundle sheath cells in C_4 plants, only 10–35% of the leaf nitrogen is tied up in this enzyme, as opposed to 40–60% in C_3 plants), and their high rates of sugar formation which can facilitate the rapid growth rates seen in such C_4 plants as maize, sugarcane, sorghum, and crabgrass. Other differences in response to the environment between C_3 and C_4 plants is that C_4 plants exhibit a nonsaturating response curve of leaf photosynthesis to light levels found in nature, and tolerate more salinity and higher temperatures than do C_3 plants.

The higher energy requirements of C_4 plants are also reflected by the fact that quantum yields of photosynthesis for C_3 plants are higher than for those possessing the auxiliary C_4 system. At 2% oxygen partial pressure and 30°C (86°F), quantum yield for C_3 plants is about 0.073 mole CO_2 assimilated per absorbed einstein of light, while for C_4 plants the quantum yield is 0.054. However, at normal O_2 partial pressures (21% O_2) quantum yields are almost identical. This is due to the presence of high photorespiration in C_3 plants, and thus represents a net quantum yield rather than a true photosynthetic yield. *See* PHOTORESPIRATION.

CAM photosynthesis. Under arid and desert conditions, where soil water is in short supply, transpiration during the day when temperatures are high and humidity is low may rapidly deplete the plant of water, leading to desiccation and death. By keeping stomata closed during the day, water can be conserved, but the uptake of CO_2, which occurs entirely through the stomata, is prevented. Desert plants in the Crassulaceae, Cactaceae, Euphorbiaceae, and 15 other families have evolved, apparently independently of C_4 plants, an almost identical strategy of assimilating CO_2 by which the CO_2 is taken in at night when the stomata open; water loss is low because of the reduced temperatures and correspondingly higher humidities. Although these succulent plants with thick, fleshy leaves were known since the nineteenth century as being unusual, the biochemical understanding of the process did not occur until the 1960s and 1970s when the details of C_4 photosynthesis were being worked out. First studied in plants of the Crassulaceae, the process has been called crassulacean acid metabolism (CAM).

In contrast to C_4, where two cell types cooperate, the entire process occurs within an individual cell; the separation of C_4 and C_3 is thus temporal rather than spatial. At night, CO_2 combines with PEP through the action of PEP carboxylase, resulting in the formation of oxaloacetic acid and its conversion into malic acid. The PEP is formed from starch or sugar via the glycolytic route of respiration. Thus, there is a daily reciprocal relationship between starch (a storage product of C_3 photosynthesis) and the accumulation of malic acid [the terminal product of nighttime CO_2 assimilation (Fig. 8)].

As in C_4 plants, there may be variations in the decarboxylase which provides the CO_2 for assimilation via the C_3 cycle. In some CAM plants (pineapple) phosphoenol carboxykinase (PCK) is involved, while in others (cactus) the decarboxylase in the NADP-malic enzyme (NADP-ME type) is involved. The role of NAD-ME is uncertain, although it has

Some characteristics of the three major plant groups

Characteristics	C_3	C_4	CAM
Leaf anatomy in cross section	Diffuse distribution of organelles in mesophyll and palisade cells with less chloroplasts in bundle sheath cells if present	Layer of bundle sheath cells around vascular tissue with a high concentration of chloroplasts; layers of mesophyll cells around bundle sheath	Spongy, often lacking palisade cells; mesophyll cells have large vacuoles
Theoretical energy requirement for net CO_2 fixation (CO_2:ATP:NADPH)	1:3:2	1:5:2	1:6.5:2
Carboxylating enzyme	RuBP carboxylase	PEP carboxylase, then RuBP carboxylase	Darkness: PEP carboxylase; light: mainly RuBP carboxylase
CO_2 compensation concentration, ppm CO_2	30–70	0–10	0–5 in dark
Transpiration ratio, g H_2O/g dry weight increase	450–950	250–350	50–55
Maximum net photosynthetic rate, mg CO_2/(dm^2 leaf)(h)	15–40	40–80	1–4
Photosynthesis sensitive to high O_2	Yes	No	Yes
Photorespiration detectable	Yes	Only in bundle sheath	Difficult to detect
Leaf chlorophyll *a*/*b* ratio	2.8 ± 0.4	3.9 ± 0.6	2.5–3
Maximum growth rate, g dry wt/(dm^2 leaf)(day)	0.5–2	4–5	0.015–0.018
Optimum temperature for photosynthesis	15–25° C (59–77° F)	30–40° C (86–104° F)	About 35° C (95° F)

been found in some CAM plants. The **table** summarizes some physiological differences between C_3, C_4, and CAM plants.

Other CO_2 assimilation mechanisms. Both the C_4 cycle and CAM involve the synthesis of oxaloacetic acid, which is also one of the intermediates in the tricarboxylic acid (TCA) cycle of respiration. In the late 1960s a light-driven reversal of the TCA cycle was discovered by B. Buchanan, D. Arnon, and coworkers in Berkeley. This CO_2 fixation cycle, called the reductive carboxylic acid cycle, results in the net synthesis of pyruvic acid via the reversal of the three decarboxylation steps in the TCA cycle (pyruvic to acetyl coenzyme A, isocitric to α-ketoglutaric, and succinyl CoA to succinic acids). The pathway has been detected in the photosynthetic bacteria. *See* CITRIC ACID CYCLE.

In most photosynthetic bacteria, the C_3 cycle is functional despite some differences in detail. The green sulfur bacteria, however, carry out C_3 photosynthesis poorly or not at all. *Chlorobium thiosulfatophilum,* lacking the key enzyme RuBP carboxylase, utilizes a reductive carboxylic acid cycle in which reduced ferredoxin drives the TCA cycle in reverse, resulting in carboxylation reactions much like those of the cycle discovered by Buchanan and Arnon.

Heterocysts of blue-green algae do not have a functional C3 cycle because, in contrast to the normal cells of these algae, the heterocyst cell (implicated in nitrogen fixation) lacks the key enzyme RuBP carboxylase. It has been suggested that CO_2 fixation in heterocysts may be via PEP carboxylase as in C_4 and CAM photosynthesis. Guard cells in C_3 plants which regulate the opening of stomatal pores for gas exchange in leaves also lack RuBP carboxylase and apparently use PEP carboxylase exclusively to fix CO_2.

Martin Gibbs; Gerald A. Berkowitz

Bacterial Photosynthesis

Certain bacteria have the ability to perform photosynthesis. This was first noticed by S. Vinogradsky in 1889 and was later extensively investigated by C. B. Van Niel, who gave a general equation for bacterial photosynthesis. This is shown in reaction (9). Photo-

$$2H_2A + CO_2 + \text{light} \xrightarrow[\text{enzymes}]{\text{bacteriochlorophyll}} \{CH_2O\} + 2A + H_2 \quad (9)$$

synthetic bacteria cannot use water as the hydrogen donor and are incapable of evolving oxygen. The prokaryotic cyanobacteria (also called blue-green algae) are excluded in this discussion of bacterial photosynthesis since their photosynthetic system closely resembles that found in eukaryotic algae and higher plants discussed above. Photosynthetic bacteria can be classified in four major groups.

1. Nonsulfur purple bacteria (Rhodospirillaceae). In these bacteria, H_2A is usually an organic H_2 donor, such as succinate or malate; however, these bacteria can be adapted to use hydrogen gas as the reductant. They require vitamins for their growth and usually grow anaerobically in light, but they can also grow aerobically in the dark by using respiration to utilize organic compounds from the environment. They are thus facultative photoheterotrophs. Examples of this group are *Rhodospirillum*

rubrum and *Rhodobacter sphaeroides*.

2. Sulfur purple bacteria (Chromatiaceae). These cannot grow aerobically, and H_2A is an inorganic sulfur compound, such as hydrogen sulfide, H_2S; the carbon source can be CO_2. These bacteria are called obligate photoautotrophic anaerobes. An example is *Chromatium vinosum*.

3. Green sulfur bacteria (Chlorobiaceae). These bacteria are capable of using the same chemicals as Chromatiaceae but, in addition, use other organic H_2 donors. They may then be called photoautotrophic and photoheterotrophic obligate anaerobes. An example of the green sulfur bacteria is *Chlorobium thiosulfatophilum*.

4. Green sliding bacteria (Chloroflexaceae). These are primarily photoorganotrophic bacteria which can grow under anaerobic conditions in light by photosynthesis or in aerobic conditions in the dark by using respiration to utilize organic compounds from the environment. They are thermophilic bacteria found in hot springs around the world. They also distinguish themselves among the photosynthetic bacteria by possessing mobility. An example is *Chloroflexus aurantiacus*.

Bacteria are capable of photophosphorylation, which is the production of adenosine triphosphate (ATP) from adenosine diphosphate (ADP) and inorganic phosphate (P_i) using light as the primary energy source. Several investigators have suggested that the sole function of the light reaction in bacteria is to make ATP from ADP and P_i. The hydrolysis energy of ATP can then be used to drive the reduction of CO_2 to carbohydrate by H_2A in reaction (9).

Photochemical apparatus. Photosynthetic bacteria do not have specialized organelles such as the chloroplasts of green plants. Electron micrographs of certain photosynthetic bacteria show tiny spherical sacs, with double-layered walls, as a result of invaginations which form stacks of membranes (**Fig. 9***a*). Other photosynthetic bacteria have invaginations which form thylakoids (Fig. 9*b*). These structures, called chromatophores, contain the photosynthetic apparatus and can be easily isolated by mechanical disruption of bacteria followed by differential centrifugation. Isolated chromatophores are the basic preparation for biochemical and biophysical studies of bacterial photosynthesis.

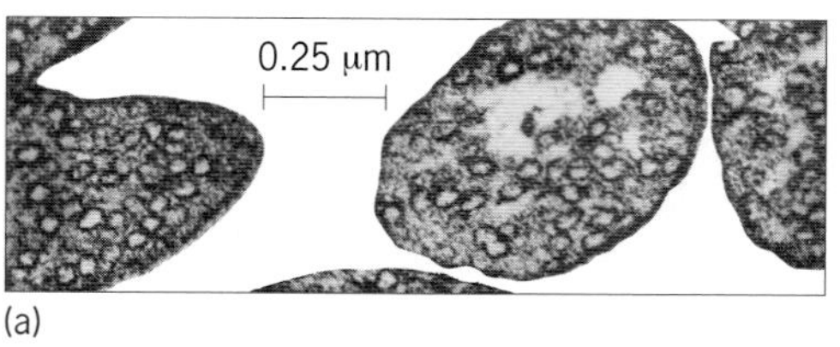

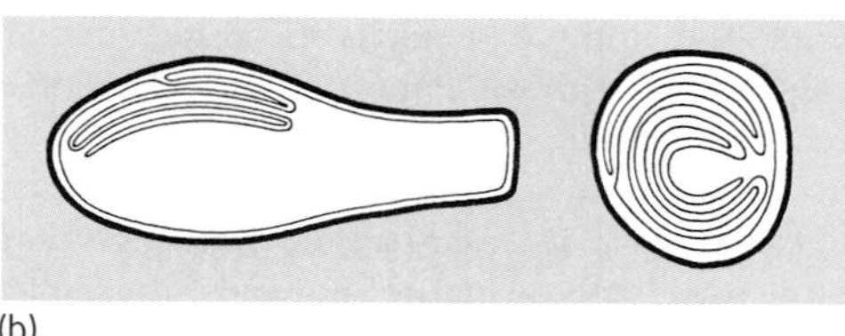

Fig. 9. Photosynthetic bacteria. (*a*) Electron micrograph of *Rhodobacter sphaeroides* with vesiclelike invaginations (*from T. W. Goodwin, ed., Biochemistry of Chloroplasts, vol. 1, Academic Press, 1966*). (*b*) Pictorial representation of a stacked invagination in a photosynthetic bacterium; at left is a longitudinal section and at right is a transverse section (*after R. Whittenbury and A. G. McLee, Archiv für Microbiologie, 59:324–334, 1967*).

Reaction centers. The pigment bacteriochlorophyll (Bchl) is a necessary ingredient for bacterial photosynthesis. There are specialized Bchl molecules in bacteria which engage in the primary chemical reactions of photosynthesis. In addition to these specialized molecules, there are 40–50 Bchl molecules referred to as antenna, whose sole function is to harvest light energy and transfer it to reaction center molecules. This is similar to the photosynthetic unit of plants. Each reaction center contains a special pair (dimer) of Bchl molecules that engage in chemical reactions after they trap the absorbed light energy. They are also called the energy traps of bacterial photosynthesis.

The energy trap in *Rhodobacter sphaeroides* had been identified as P870. Such identification is carried out with a difference (absorption) spectrophotometer. In this instrument a weak monochromatic measuring beam monitors the absorption of the sample; a brief but bright actinic light given at right angles to the measuring beam initiates photosynthesis. When photosynthesis occurs, changes in absorption take place. **Figure 10** shows the absorption spectrum of reaction centers isolated from *R. sphaeroides*. These changes are measured as a function of the wavelength of measuring light. A plot of the change induced in *R. sphaeroides* reaction centers by an actinic light flash, as a function of the wavelength of measuring light, is the difference absorption spectrum (Fig. 10*b*). This spectrum is due largely to the photooxidation of the Bchl dimer, P870.

If P870 is the energy trap, then the following criteria must be met: (1) It must undergo a reduction or oxidation reaction, since this is the essential reaction of photosynthesis. The decrease in absorption at 870 nm (Fig. 10) is an oxidation reaction since chemical oxidants cause a similar change. (2) The quantum yield (number of trap molecules oxidized per absorbed photon) must be very high (close to 1.0). (3) The primary light reaction should occur at very low temperatures, down to 1 K (−460°F or −273°C). (4) The above photochemical reaction should be extremely fast, that is, in the picosecond range.

All the above criteria are fulfilled by P870, and thus it is the reaction center of bacterial photosynthesis in *Rhodobacter sphaeroides*. Among other reaction centers that have been identified and studied extensively are P890 in *Chromatium vinosum*, P960 in *Rhodopseudomonas viridis*, and P840 in *Chlorobium*. Each species of bacteria has only one type of reaction center, unlike plants, which utilize both PS I (P700) and PS II (P680) reaction centers. The reaction centers from plants have been identified

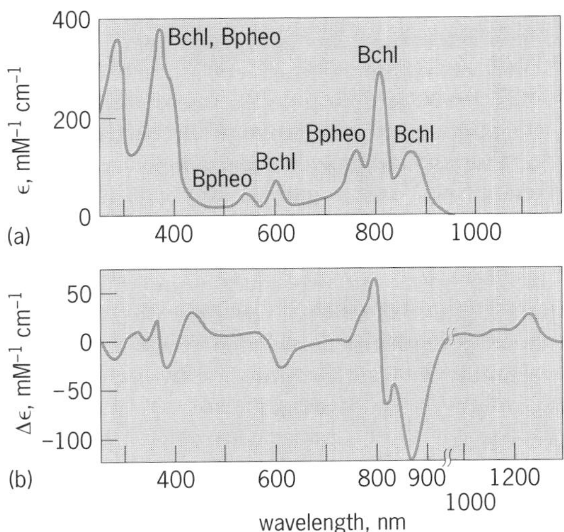

Fig. 10. Plots of (*a*) absorption spectrum and (*b*) the light-induced absorption changes in it, as occurring in reaction centers isolated from carotenoidless mutant R-26 of *Rhodobacter sphaeroides*. In *a*, bands attributed to bacteriochlorophyll and bacteriopheophytin are labeled Bchl and Bpheo, respectively. The ordinate in *a* is the millimolar extinction coefficient; in *b*, it is the differential extinction coefficient. (*After R. K. Clayton, Photosynthesis: Physical Mechanisms and Chemical Patterns, Cambridge University Press, 1980*)

by means similar to those used for bacterial reaction centers. Reaction centers can be isolated as a pure protein, which has served the important function of providing a well-defined system in which primary reactions of photosynthesis can be studied. A milestone in bacterial photosynthesis was reached in the early 1980s by the crystallization of *Rhodopseudomonas viridis* reaction centers. These crystals enabled a 0.3-nm resolution of the molecular structure of the reaction center to be obtained.

Although isolated reaction centers are able to absorb light and convert it to chemical energy, the antenna pigment system in chromatophores (or in whole cells) absorbs most (>90%) of the light. The antenna funnels this energy to the reaction center. Antenna Bchl molecules are bound to protein in a specific manner; this binding and pigment-pigment interactions modify the properties of the pigment and define the absorption maxima and the width of the absorption band. An example is B800 (B represents Bchl, and the number indicates the wavelength of the absorption peak in nanometers) found in *Rhodobacter sphaeroides* (**Fig. 11**).

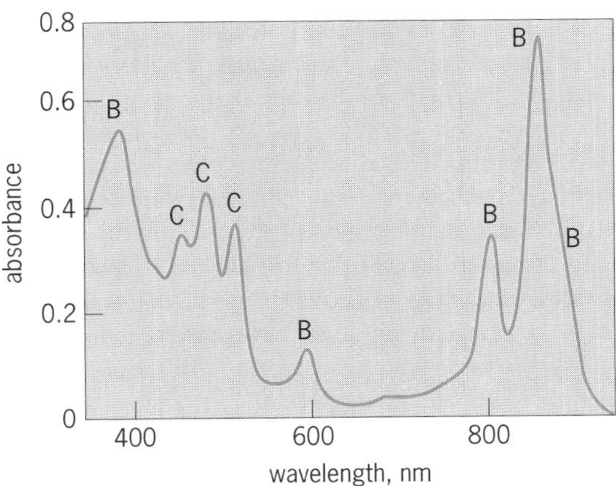

Fig. 11. Absorption spectrum of chromatophores from the bacterium *Rhodobacter sphaeroides*. Absorption bands attributed to Bchl *a* are labeled as B, and those attributed to carotenoids as C. (*After R. K. Clayton, Photosynthesis: Physical Mechanisms and Chemical Patterns, Cambridge University Press, 1981*)

Components of photosynthetic bacteria. These bacteria contain the usual components of living material: proteins, lipids, carbohydrates, deoxyribonucleic acid (DNA), ribonucleic acid (RNA), and various metals. However, the specific components of interest to the electron transport system of bacterial photosynthesis are quinones, pyridine nucleotides, and various iron-containing proteins (cytochromes, ferredoxins, Rieske iron-sulfur centers, and others) in addition to the photosynthetic pigments which capture light energy.

In contrast to plastoquinones found in plants, bacteria contain substituted benzoquinones called ubiquinones (UQ or coenzyme Q) and substituted naphthoquinones called menaquinones (MK or vitamin K_2 which act as electron acceptors. The purple bacteria have a pool of UQ (about 25 UQ per reaction center) which mediates transfer of electrons and protons between protein complexes in the chromatophore membrane. However, MK is found only in some bacteria, usually in a smaller quantity (about 1–2 MK molecules per reaction center) than the more plentiful UQ. Menaquinone's function is probably limited to electron transfer within the reaction center. In contrast to plants which contain $NADP^+$, the major pyridine nucleotide in bacteria is nicotinamide adenine dinucleotide (NAD); it is present in large quantities and seems to be active in photosynthesis. Among the various cytochromes, the *c*-type cytochromes and the *b*-type cytochromes are the important ones for bacterial photosynthesis.

Pigments. Most photosynthetic bacteria contain Bchl *a*, a tetrahydroporphyrin. The chlorophyll of green plants, by contrast, is a dihydroporphyrin. In diethyl ether, Bchl *a* has absorption maxima at 365, 605, and 770 nm. The infrared band of various antenna Bchl *a* has maxima at 800 (B800), 850 (B850), or 890 nm (B890). These antenna absorption bands in the bacterial cell are due to the formation of complexes of Bchl *a* with different proteins.

The reaction center protein (composed of L, M, and H subunits) from *Rhodobacter sphaeroides* binds four Bchl *a* and two bacteriopheophytin (Bph; similar to Bchl but does not contain magnesium). Two of the Bchl form the energy trap P870. Another Bchl and a Bph are involved in the transfer of electrons within the protein. The function for the remaining "voyeur" Bchl and Bph is unknown. The exact locations of these chromophores in the reaction center protein was first established in the crystals of *Rhodopseudomonas viridis* reaction centers. Similar information is now available for *Rhodobacter*

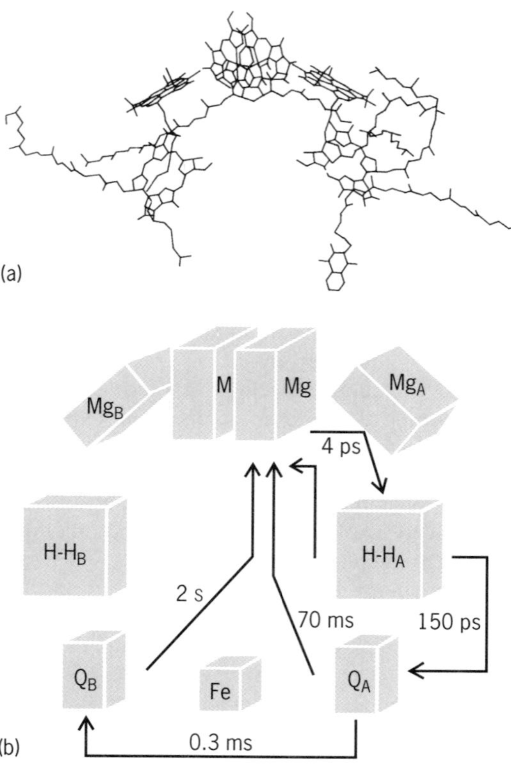

Fig. 12. Reaction centers of bacteria. (*a*) Structure of the reaction center of the purple nonsulfur bacterium *Rhodopseudomonas viridis* as determined by x-ray analysis of the crystalline preparation. Only the components of the electron transport chains are shown (*after J. Deisenhofer et al., X-ray structure analysis of a membrane protein complex: Electron density map at 3 Å resolution and a model of the chromophores of the photosynthetic reaction center from Rhodopseudomonas viridis, J. Mol. Biol., 180:385–398, 1984*). (*b*) A simplified representation of the donor-acceptor complex based on the x-ray data and on spectroscopic data for *Rhodobacter sphaeroides.* The blocks define the aromatic ring systems of bacteriochlorophyll (M-Mg and Mg), bacteriopheophytin (H-H), the quinones (Q), which are ubiquinone and menaquinone, and Fe^{2+}. M-Mg is the primary electron donor, a dimer of bacteriochlorophyll *a* (*Rhodobacter sphaeroides*) or *b* (*Rhodopseudomonas viridis*). Subscripts A and B label the two potential electron transfer pathways, of which only pathway A appears active. The arrows show the various electron transfer reactions with their half-times. Note that Q_B is absent in the crystal of *R. viridis* (*after J. F. Norris and G. Van Brakel, Photosynthesis, in Govindjee, J. Amesz, and D. C. Fork, eds., Light Emission by Plants and Bacteria, Academic Press, 1986*).

sphaeroides reaction centers (**Fig. 12**).

The bacterium *Rhodopseudomonas viridis* utilizes an antenna with an infrared band at 1015 nm. The isolated Bchl from this species has absorption maxima at 368, 582, and 795 nm in diethyl ether, and has been designated Bchl *b*. The reaction center of *R. viridis*, P960, uses Bchl *b* and Bph *b* much in the same way as P870 in other bacteria utilize Bchl *a*.

The green bacterium *Chlorobium* contains a small amount of Bchl *a*, but a large quantity of another type of chlorophyll called chlorobium chlorophyll; the latter exists in two forms. In the cell, the red absorption band is at 725 or 740 nm; the names Bchl *c* and Bchl *d* are assigned, respectively. The Bchl *a* has been shown to be associated with the reaction center, while the Bchl *c* acts as antenna.

The second group of pigments is the carotenoids, which have absorption peaks from 450 to 550 nm. The carotenoids of photosynthetic bacteria are of great variety and include some which are found in green plants, for example, the lycopenes. However, some are typical only of bacteria: γ-carotene, which is found in large quantities in green sulfur bacteria, and spirilloxanthol, which is found mainly in purple bacteria. Carotenoids function to prevent photooxidation and destruction of antenna bacteriochlorophyll. They also function in bacterial photosynthesis by transferring their absorbed energy to bacteriochlorophyll. Similar roles are found for carotenoids in plants and cyanobacteria.

Transfer of excitation energy. Light energy absorbed by the carotenoids is transferred to Bchl with varying efficiency (30–90%), as demonstrated by the method of sensitized fluorescence. When light energy is absorbed by carotenoids, only the fluorescence of bacteriochlorophyll (B875) is observed. By the same method, efficient (almost 100%) energy transfer has been demonstrated from B800 to B850 to B875. The high (almost 1.0) quantum yield of P870 oxidation, when bacteria are excited in the antenna pigments, is a clear demonstration of an extremely efficient excitation energy transfer by antenna pigments and trapping in reaction centers.

The lifetime of the excited state of antenna Bchl in the bacterial cell is of the order of 1–2 nanoseconds. The excitation energy must be channeled from the antenna pigments to the energy traps within this time for efficient photosynthesis to occur. In reaction center preparations, it takes only 3 picoseconds to create a definitively stable charge separation (see below) after the absorption of light. Moreover, the lifetime of the physical state or states preceding P870 oxidation is <3 ps. Thus, it appears that within a few picoseconds of receiving excitation energy, the reaction center has converted the absorbed light energy into chemical energy. Similar reactions occur in plant photosynthesis.

Mechanisms of electron transport. The first act of photosynthesis is the absorption of light by various pigments. As discussed above, light energy absorbed by the carotenoids B800 and B850 is transferred to B875 and finally to the reaction centers, where the primary reaction occurs: the oxidation of the reaction center Bchl dimer leads to bleaching of P870 and reduction of an acceptor (**Fig. 13**). In the current model, P (short for P870 and so on) is oxidized to P^+ and an intermediate I is reduced to I^- within a few picoseconds; I includes a Bchl monomer and a Bph molecule. The reduced I^- transfers the electron to an iron-quinone complex, reducing the primary quinone (Q_A) to a semiquinone within 100–200 ps. For most bacteria Q_A is ubiquinone, though for those containing both menaquinone and ubiquinone the menaquinone functions as Q_A. Although an iron atom is in this complex, and is within 0.5–1.0 nm of the quinone, its presence is not necessary for the reduction of Q_A, nor does the iron undergo redox changes. The function of this nonheme iron in the reaction center is unknown. In plant photosynthesis, PS II contains Q_A, which is a bound plastoquinone; the function of the iron there is also unknown.

The photooxidized donor Bchl dimer, P^+, can be re-reduced by a cytochrome *c* in 1–30 microseconds, thus oxidizing the cytochrome. In *Rhodobacter sphaeroides* and a number of other species, this cytochrome is soluble cyt c_2. In other bacteria (for example, *Rhodopseudomonas viridis*) the cytochrome that donates electrons to P^+ is an integral part of the reaction center. The photochemical reactions and the electron transfers in the reaction center are summarized in reaction (10).

$$\mathrm{PIQ_A} \xrightarrow{h\nu} \mathrm{P^*IQ_A} \xrightarrow{3\ \mathrm{ps}} \mathrm{P^+I^-Q_A} \xrightarrow{200\ \mathrm{ps}}$$

$$\mathrm{P^+IQ_A^-} \xrightarrow[1\text{–}30\ \mu\mathrm{s}]{\mathrm{cyt}\ c \ \to\ \mathrm{cyt}\ c^+} \mathrm{PIQ_A^-} \qquad (10)$$

After this set of reactions, the electron is transferred from Q_A^- to Q_B (a bound UQ), producing $Q_AQ_B^-$. In a subsequent absorption of a photon, the $Q_A^-Q_B^-$ state is created, which is followed by electron transfer from Q_A^- to Q_B^-, forming Q_BH_2 with the uptake of two protons. The bound quinol (Q_BH_2 is replaced by a UQ molecule. This cycle is known as the two-electron gate and is summarized in reaction (11) [omitting the early photochemical steps illustrated in reaction (10)]. The same cycle occurs in photosystem II of plants, except that the electron donor to P^+ is tyrosine Z, and Q_BH_2 is another plastoquinol instead of ubiquinol. The molecular detail is so similar in plants and bacteria that many of the herbicides which act to inhibit PS II electron transfer from Q_A^- to Q_B are also potent inhibitors of electron transfer from Q_A^- to Q_B in photosynthetic bacteria. However, one difference involves a unique role of CO_2/HCO_3^- in reaction (11) in PS II of plants and cyanobacteria, but not in photosynthetic bacteria.

The mechanism of proton uptake in the two-electron gate is not completely known. The first proton does not bind directly to the semiquinone (Q_A^- or Q_B^-), but instead it binds to a protonatable amino acid of the reaction center. The net result from the absorption of two photons is the formation of a ubiquinol in the membrane, the oxidation of two cyt *c*, and the removal of two protons from the cytoplasm of the bacterial cell. In plants and cyanobacteria, the quinol is plastoquinol, and it is water that is ultimately oxidized.

The doubly reduced ubiquinone (QH_2, quinol) through a cyclic pathway serves to re-reduce the oxidized cytochrome (cyt c^+). This cyclic reaction (Fig. 13) is coupled to the production of ATP via the creation of a proton gradient (more accurately a proton motive force) across the membrane. Just as in plants, the proton motive force (which includes two components: a membrane potential, and a proton gradient) is used to drive ATP synthesis. Protons move down the potential gradient through the ATPase to contribute energy to drive the ADP + $P_i \rightarrow$ ATP reaction. This overall mechanism is consistent

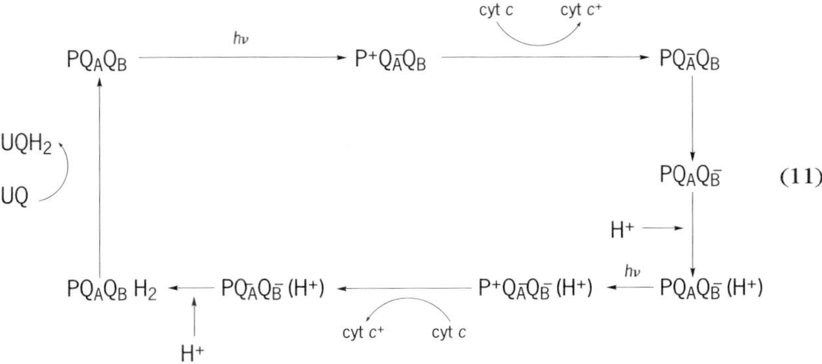

with P. Mitchell's chemiosmotic theory. The quinol produced by the two-electron gate mechanism binds to the cytochrome *b-c* complex (an integral membrane protein) which contains two *b*-cytochromes, a *c*-cytochrome, a Rieske iron-sulfur center, and two quinone binding sites. Plants also contain a similar complex, where cytochrome *b* is replaced by cytochrome b_6, and cytochrome *c* is replaced by cytochrome *f*. The mechanism is strikingly similar, on a molecular level, to that of noncyclic electron transfer from photosystem II to plastocyanin via the plastoquinone pool and the cytochrome *b-f* complex. The path of the electrons and protons through this complex is still a matter of controversy. In all likelihood, it includes a pathway called a Q-cycle by Mitchell; this cycle incorporates two different redox-linked pathways for the electrons. For each quinol oxidized by this complex, two cyt *c* are reduced, two protons are removed from the quinol, an additional two protons are removed from the cytoplasm, and these four protons are released into the intermembrane space. Absorption of two photons leads to the translocation of four protons across the membrane. Data on the bacterial ATPases suggest that $2H^+$ or $3H^+$ are needed to make an ATP (Fig. 13).

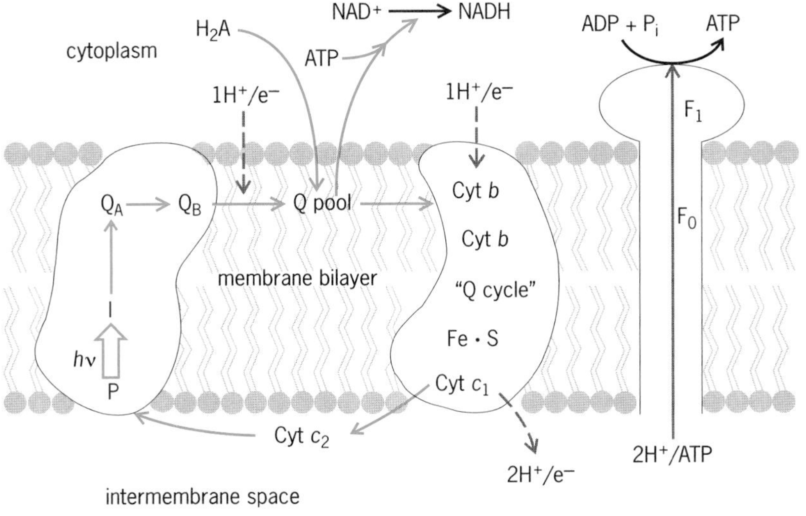

Fig. 13. Electron and proton transport for purple photosynthetic bacteria. For details and explanation of symbols see the text. The shapes of the proteins are largely hypothetical, as in the equivalent diagram for plant photosynthesis (Fig. 4).

The mechanism described here for the generation of ATP from light energy is largely from studies on *Rhodobacter sphaeroides* and is generally valid for other purple photosynthetic bacteria.

The mechanisms for oxidizing the reduced substrate H_2A [reaction (9)] are known in much less detail than those for photophosphorylation. Most substrates feed electrons into the quinone pool, and the resulting quinol can be used by the cytochrome *b-c* complex. An example of this is succinate, which reduces quinone via a succinate dehydrogenase. In bacteria which have a low potential cytochrome *c* bound to the reaction center (such as cyt c_{551} in *Chromatium vinosum*), it has been hypothesized that electrons from some substrates can be fed into the reaction center through this cytochrome. The electrons for the reduction of NAD^+ in purple photosynthetic bacteria are from the quinone pool, but these electrons require additional energy gained perhaps from the hydrolysis of ATP.

Alternatively, especially in some green bacteria, the primary stable acceptor of electrons in the reaction center may not be a quinone but an acceptor with a negative enough oxidation-reduction potential to directly reduce NAD^+. In several green bacteria, this electron acceptor has been shown to be an iron-sulfur (Fe·S) center instead of a quinone. The midpoint redox potential of this Fe·S center is much lower than that for the quinone acceptor in the purple photosynthetic bacteria. This Fe·S center can then directly reduce a ferredoxin, and this can drive the $NAD^+ \rightarrow NADH$ reaction. The reduced ferredoxin may also feed electrons into a cytochrome *b* complex from which a soluble cyt *c* could be reduced, thus allowing cyclic electron transfer to occur. This scheme is very reminiscent of photosystem I driven reactions in plant photosynthesis. However, not all green photosynthetic bacteria follow the above pattern, but, instead, they resemble more the purple photosynthetic bacteria.

The reduced pyridine nucleotide NADH and the ATP made in the light reactions are then utilized to convert carbon sources into carbohydrates. The pathway of carbon involves a reversal of either the Krebs cycle or the Calvin cycle with some modifications. *See* BACTERIAL PHYSIOLOGY AND METABOLISM.

Govindjee; R. J. Shopes

Bibliography. M. Baltschefsky (ed.), *Current Research in Photosynthesis*, 1990; J. Barber (ed.), *Current Topics in Photosynthesis*, vols. 1–8, 1976–1988; C. C. Black, Photosynthetic carbon fixation in relation to CO_2 uptake, *Annu. Rev. Plant Physiol.*, 24:253–286, 1973; R. K. Clayton and W. R. Sistrom (eds.), *The Photosynthetic Bacteria*, 1978; G. Edwards and D. Walker, *C_3, C_4: Mechanisms and Cellular and Environmental Regulation of Photosynthesis*, 1983; M. Gibbs and E. Latzko, *Photosynthetic Carbon Metabolism and Related Processes*, 1979; T. W. Goodwin (ed.), *Plant Pigments,* 1988; Govindjee (ed.), *Bioenergetics of Photosynthesis*, 1975; Govindjee (ed.), *Photosynthesis*, vol. 1: *Energy Conversion by Plants and Bacteria*, vol. 2: *Development, Carbon Metabolism and Plant Productivity*, 1982; Govindjee et al. (eds.), *Molecular Biology of Photosynthesis*, 1989; Govindjee, J. Amesz, and D. C. Fork (eds.), *Light Emission by Plants and Bacteria*, 1986; R. P. F. Gregory, *Biochemistry of Photosynthesis*, 3d ed., 1989; D. D. Hall and K. K. Rao, *Photosynthesis*, 6th ed., 1999; D. W. Lawlor, *Photosynthesis: Molecular, Physiological, and Environmental Processes*, 3d ed., 2001; A. K. Mattoo, J. B. Marder, and M. Edelman, Dynamics of the photosystem II reaction center, *Cell*, 56:241–246, 1989; P. Mohanty and J. A. Govindjee (eds.), *Photosynthesis: Photoreactions to Plant Productivity*, 1993; D. G. Nicholls, *Bioenergetics*, 2d ed., 1997; E. I. Rabinowitch and Govindjee, *Photosynthesis*, 1969; C. Sybesma (ed.), *Advances in Photosynthesis Research*, vols. 1–4, 1984; I. P. Ting and M. Gibbs, *Crassulacean Acid Metabolism*, 1982.

Phototransistor

A semiconductor device with electrical characteristics that are light-sensitive. Phototransistors differ from photodiodes in that the primary photoelectric current is multiplied internally in the device, thus increasing the sensitivity to light. For a discussion of this property *see* TRANSISTOR.

Some types of phototransistors are supplied with a third, or base, lead. This lead enables the phototransistor to be used as a switching, or bistable, device. The application of a small amount of light causes the device to switch from a low current to a high current condition. *See* PHOTOELECTRIC DEVICES.

W. R. Sittner

Phototube

An electron tube comprising a photocathode and an anode mounted within an evacuated glass envelope through which radiant energy is transmitted to the photocathode. A gas phototube contains, in addition, argon or other inert gas which provides amplification of the photoelectric current by partial ionization of the gas. The photocathode emits electrons when it is exposed to ultraviolet, visible, or near-infrared radiation. The anode is operated at a positive potential with respect to the photocathode. *See* ELECTRICAL CONDUCTION IN GASES; ELECTRON TUBE.

Characteristics. A phototube responds to radiation over a limited range of the spectrum that is determined by the photocathode material. Radiant sensitivity, shown in the **illustration** as a function of wavelength, is the photoelectric current emitted per unit of incident monochromatic radiant power. Sensitivity on the short-wavelength side of the curves is limited by the transmittance of the glass envelope. Electron affinity of the photocathode determines the long-wavelength threshold of sensitivity. *See* PHOTOEMISSION.

Average cathode characteristics

Spectral sensitivity characteristic*	Cathode material	Wavelength of maximum response, nm	Peak radiant sensitivity, mA/W	Peak cathode quantum efficiency, %	Luminous sensitivity, μA/lumen†	Remarks
S-1	Cs_2O, Ag	800	2.2	0.3	25	
S-3	Rb_2O, Ag	420	1.8	0.5	6.5	
S-4	Cs_3Sb	400	40	12.4	40	
S-5	Cs_3Sb	340	49	17.8	40	Ultraviolet transmitting window
S-8	Cs_3Bi	365	2.3	0.8	3	
S-10	Bi, Ag, O, Cs	450	20.3	5.6	40	Semitransparent
S-11	Cs_3Sb	440	48	13.5	60	Semitransparent
S-13	Cs_3Sb	440	47	13.2	60	Semitransparent; ultraviolet transmitting window
S-17	Cs_3Sb	490	85	21.4	125	Semitransparent, on reflecting substrate
S-20	(NaKCs)Sb	420	64	18.8	150	Semitransparent

*These characteristics, shown in the illustration, refer to typical phototubes rather than to photocathodes.
†Light source is a tungsten-filament lamp operated at a color temperature of 4700°F (2870 K).

Typical phototube characteristics are summarized in the **table**. Quantum efficiency, or photoelectron yield, is the number of electrons emitted per incident photon. It is tabulated at the wavelength of maximum response. For photometric applications a useful parameter is luminous sensitivity: the photoelectric current per lumen incident from a specified source of light. A source commonly used is a tungsten-filament lamp operated at a color temperature of 4700°F (2870 K). *See* INCANDESCENCE; LUMINOUS FLUX; PHOTON.

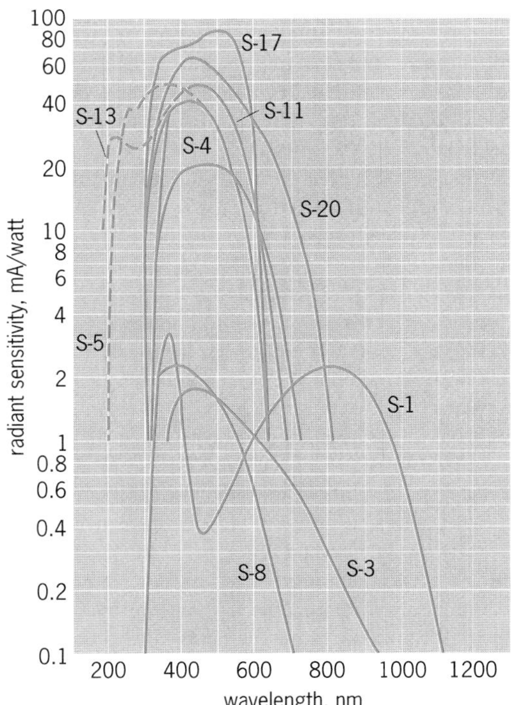

Curves of the average spectral sensitivity characteristics of some typical phototubes.

Photocathodes. Photocathodes are semiconductors which contain one or more of the alkali metals sodium, potassium, rubidium, or cesium chemically combined with bismuth, antimony, or silver oxide. The cathode surface contains a critical excess of the alkali metal which enhances photoelectric emission by decreasing the affinity of the surface for electrons. Negative affinity for electrons is achieved with the gallium arsenide:cesium (GaAs:Cs) and indium gallium arsenide:cesium (InGaAs:Cs) photocathodes used in photomultipliers. Phototubes also emit electrons thermionically at ambient temperatures. This "dark current," observed in the absence of all irradiance, increases almost exponentially with temperature. Thermionic emission from the cesium antimonide (CsSb) photocathode is about 10^{-15} A/cm^2 at 68°F (20°C). *See* PHOTOMULTIPLIER; SEMICONDUCTOR.

Applications. Vacuum phototubes are used as detectors of radiant energy in the spectral range from 200 to 1100 nanometers. Since the photoelectric current is directly proportional to the intensity of the radiation, these tubes are used in radiometers, photometers, and colorimeters. By virtue of their narrow pulse response, vacuum phototubes are also used to measure the intensity of very short pulses of light generated by lasers and visible nuclear radiation. Gas phototubes can be used in light-operated relays and for the reproduction of sound from motion picture film, although their response to intensity-modulated light is limited to frequencies below 15 kHz. Vacuum as well as phototubes have been replaced in many applications by semiconductor photodiodes and photovoltaic cells. *See* COLORIMETRY; LASER; PHOTODIODE; PHOTOMETER; PHOTOVOLTAIC CELL; RADIOMETRY.

James L. Weaver

Bibliography. M. Cardona and L. Ley (eds.), *Photoemission in Solids*, 1978 and 1979; E. L. Dereniak and D. G. Crowe, *Optical Radiation Detectors*, 1984; A. Sommer, *Photoemissive Materials*, 1968, reprint 1980.

Photovoltaic cell

A device that detects or measures electromagnetic radiation by generating a current or a voltage, or both, upon absorption of radiant energy. Specially designed photovoltaic cells are used for power generation, as in solar batteries or solar cells, and for sensitive detection of electromagnetic radiation in radiometry, optical communications, spectroscopy, and other applications. An important advantage of the photovoltaic cell in these particular applications is that no separate bias supply is needed—the device generates a signal (voltage or current) simply by the absorption of radiation.

Most photovoltaic cells consist of a semiconductor *pn* junction or Schottky barrier in which electron-hole pairs produced by absorbed radiation are separated by the internal electric field in the junction to generate a current, a voltage, or both, at the device terminals. The influence of the incident radiation on the current-voltage characteristics of a photovoltaic cell is to add a negative current, equal to the generated photocurrent, to the dark current-voltage characteristic of the diode and thus shift the current-voltage characteristic downward by the magnitude of the photogenerated current, as shown in the **illustration**. Under open-circuit conditions (current $I = 0$) the terminal voltage increases with increasing light intensity (points A), and under short-circuit conditions (voltage $V = 0$) the magnitude of the current increases with increasing light intensity (points B). When the current is negative and the voltage is positive (point C, for example), the photovoltaic cell delivers power to the external circuit. In this case, if the source of radiation is the Sun, the photovoltaic cell is referred to as a solar battery or solar cell. When a photovoltaic cell is used as a photographic exposure meter, it produces a current proportional to the light intensity (points B), which is indicated by a low-impedance galvanometer or microammeter. For use as sensitive detectors of infrared radiation, specially designed photovoltaic cells can be operated with either low-impedance (current) or high-impedance (voltage) amplifiers, although the lowest noise and highest sensitivity are achieved in the current or short-circuit mode. Another mode of operation of a *pn* junction diode as a photodetector involves the application of a reverse bias voltage to the diode. In this case, the photogenerated current is directly proportional to the incident power, and the diode is said to be operated in the photodiode mode rather than the photovoltaic mode. *See* EXPOSURE METER; JUNCTION DIODE; OPTICAL DETECTORS; PHOTODIODE; PHOTOELECTRIC DEVICES; PHOTOVOLTAIC EFFECT; RADIOMETRY; SEMICONDUCTOR; SEMICONDUCTOR DIODE; SOLAR CELL. Gregory E. Stillman

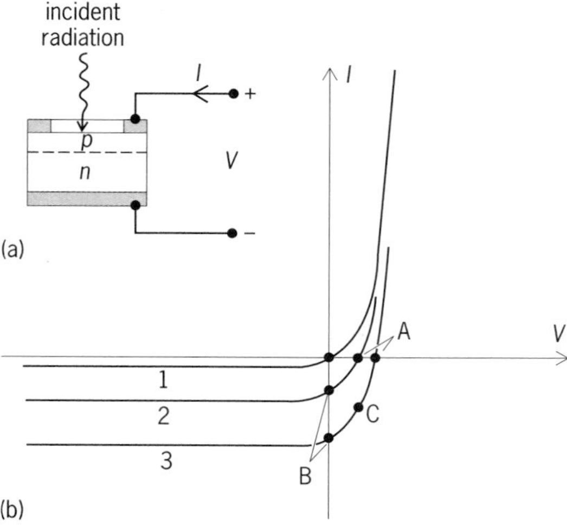

The *pn* junction photovoltaic cell. (*a*) Cross section. (*b*) Current-voltage characteristics. Curve 1 is for no incident radiation or the dark current-voltage characteristic, and curves 2 and 3 for increasing incident radiation.

Bibliography. B. K. Das and S. N. Singh (eds.), *Photovoltaic Materials and Devices*, 1985; H. J. Moller, *Semiconductors for Solar Cells*, 1993; L. D. Partain (ed.), *Solar Cells and Their Applications*, 1995; R. J. Van Overstraeten and R. P. Mertens, *Physics, Technology and Use of Photovoltaics*, 1986; C. M. Wolfe, N. Holonyak, Jr., and G. E. Stillman, *Physical Properties of Semiconductors*, 1989; K. Zweibel, *Harnessing Solar Power: The Photovoltaics Challenge*, 1990.

Photovoltaic effect

The conversion of electromagnetic radiation into electric power through absorption by a semiconducting material. Devices based on this effect serve as power sources in remote terrestrial locations and for satellites and other space applications. Photovoltaic powered calculators and other consumer electronic products are widely available, and solar photovoltaic automobiles and aircraft have been demonstrated.

Principles. The basic requirements for the photovoltaic effect are (1) the absorption of photons through the creation of electron-hole pairs in a semiconductor; (2) the separation of the electron and hole so that their recombination is inhibited and the electric field within the semiconductor is altered; and (3) the collection of the electrons and holes, separately, by each of two current-collecting electrodes so that current can be induced to flow in a circuit external to the semiconductor itself.

There are many approaches to achieving these three requirements simultaneously. A very common approach for separating the electrons from the holes is to use a single-crystal semiconductor, for example, silicon, into which a *pn* junction has been diffused. Silicon is often chosen because its optical band gap permits the absorption of a substantial portion of solar photons via the generation of electron-hole pairs. The fabrication of such a device structure causes a local transfer of negative charges from the *n* layer into the *p* layer, bending the conduction and valence bands in the vicinity of the p-n boundary, and thereby creating a rectifying junction. Electrons generated in the p region can lower their energy by migrating into the n region, which they will do by a random walk process in the electric-field-free region far from the junction, or by drift induced by the electric field in the junction region. Holes created in the n region, conversely, lose energy by migrating into the p region. Thus the presence of such a junction leads to the spontaneous spatial separation of the photogenerated carriers, thereby inducing a voltage difference between current-carrying electrodes connected to the p and n regions. This process will continue until the difference in potential between the two electrodes is large enough to flatten the bands in the vicinity of the junction, canceling out the internal electric field existing there and so eliminating the source of carrier separation. The resulting voltage is termed the open-circuit voltage V_{oc}, and approximates the built-in voltage associated with the pn junction in the dark, a value which cannot exceed the band gap of the semiconductor. *See* HOLE STATES IN SOLIDS; SEMICONDUCTOR; SEMICONDUCTOR DIODE.

In the limit when the device is short-circuited by the external circuit, no such buildup of potential can occur. In this case, one electron flows in the external circuit for each electron or hole which crosses the junction, that is, for each optically generated electron-hole pair which is successfully separated by the junction. The resulting current is termed the short-circuit current J_{sc} and, in most practical photovoltaic devices, approaches numerically the rate at which photons are being absorbed within the device. Losses can arise from the recombination of minority carriers (for example, electrons in the p-type region, holes in the n-type region) with majority carriers. *See* ELECTRON-HOLE RECOMBINATION.

Power generation. For a photovoltaic device to generate power, it is necessary to provide a load in the external circuit which is sufficiently resistive to avoid short-circuiting the device. In this case, the voltage will be reduced compared to the open-circuit voltage because a continuing requirement exists for carrier separation at the junction; thus some band bending and its associated internal field must be retained. A maximum power conversion point in the available current-voltage (I-V) output plane exists for any photovoltaic device operating in a given photon flux. Device efficiency is defined as this maximum power output divided by the incident radiative power. The maximum power is usually expressed as the product of V_{oc}, J_{sc}, and F.F., the fill factor, which is the ratio of the area in the I-V plane defined by the maximum power output divided by the area defined by the product of V_{oc} and J_{sc}.

Alternative materials. While this explanation of the photovoltaic effect has been given in terms of the familiar crystalline silicon pn photovoltaic device, many alternative materials and device structures have been explored since the discovery of the photovoltaic effect in the nineteenth century. In fact, the first all-solid-state photovoltaic device appears to have used a semiconducting layer of glassy selenium illuminated through a semitransparent gold electrode in what would now be considered a metal-semiconductor Schottky barrier configuration.

Amorphous materials tend to have low carrier mobilities and, therefore, short diffusion lengths of minority carriers prior to recombination. Thus, devices utilizing amorphous materials tend to require internal electric field–induced carrier drift in order to achieve efficient carrier separation. Fortunately, the optical absorption of amorphous films can be matched so well to the incident solar radiation that devices as thin as 0.5 micrometer can be utilized, thus permitting the presence of electric fields throughout most of the absorbing layer thickness. *See* AMORPHOUS SOLID.

Multiple-layered devices. Various multiple-layered device configurations based on doped and undoped alloys of amorphous silicon have been developed for photovoltaic devices used in applications ranging from solar watches and calculators to remote power generators. The photovoltaic effect in these devices is particularly intriguing since it is possible to build up so-called tandem devices by stacking one device electrically and optically in series above another. In addition to the increased voltage and concomitant reduction in the required current-carrying capability of electrode grid structures, such devices permit, in principle, an increased efficiency of solar photovoltaic energy conversion. This is achieved by matching the band gap in the upper device with the higher-photon-energy portion of the solar spectrum, thereby achieving a higher voltage, while matching the lower device with the lower-photon-energy portion, thereby utilizing a higher portion of these lower-energy photons than could be efficiently achieved in a single device.

Other configurations and phenomena. The junction in photovoltaic devices, normally configured to lie in the plane normal to the incident radiation, can be contoured to promote significant lateral drift of carriers to point-source contracts on the back surface of the device, where they do not occlude any of the incident radiation. Such device structures, when they also include surface texturing to enhance trapping of incident photons, can increase solar conversion efficiency. Finally, while band bending and internal fields are implicitly combined to create rectifying junctions in a single material (homojunctions), these two charge-separating mechanisms can be utilized independently, in principle, in heterojunctions, which are junctions created between two dissimilar semiconductor materials. Extension of the thin-film deposition techniques used to create tandem amorphous devices appears to have the potential for creating structures, including expitaxial structures, in which the Fermi level and the conduction and valence band energies are independently controllable throughout the device thickness. Such control could permit the consideration of practical photovoltaic device structures which today exist only as theoretical possibilities. *See* SEMICONDUCTOR HETEROSTRUCTURES; SOLAR CELL. John P. de Neufville

Bibliography. B. K. Das and S. N. Singh (eds.), *Photovoltaic Materials and Devicesz*, 1985; A. L. Fahrenbruch and R. H. Bube, *Fundamentals of Solar Cells*, 1983; H. J. Moller, *Semiconductors for Solar Cells*, 1993; L. D. Partain (ed.), *Solar Cells and Their Applications*, 1995; S. M. Sze, *Physics of Semiconducting Devices*, 2d ed., 1981; R. J. Van Overstraeten and R. P. Mertens, *Physics, Technology and Use of Photovoltaics*, 1986; K. Zweibel, *Harnessing Solar Power: The Photovoltaics Challenge*, 1990.

Phreatoicoidea

A suborder of the Isopoda and class Crustacea. The body is subcylindrical, appearing laterally compressed, mainly because of the downward development of the pleura of the pleon (see **illus.**). The first and occasionally the second thoracic segment is fused with the head. Antennules are shorter than the antennae. The eyes may be large, small, or absent. Mouthparts are primitive. The first four pairs of pereiopods are directed forward, while the posterior three pairs are directed backward. The first pair is subchelate. The pleon has six distinct segments, with the last being fused with the telson but marked from it by a suture. Pleopods are broad, foliaceous, and branchial in function, while the uropods are biramous and lateral. The suborder is divided into two families, the Amphisopidae, having both mandibles with a lacinia mobilis, and the Phreatoicidae, in which only the left mandible retains a lacinia mobilis. *See* CRUSTACEA; ISOPODA.

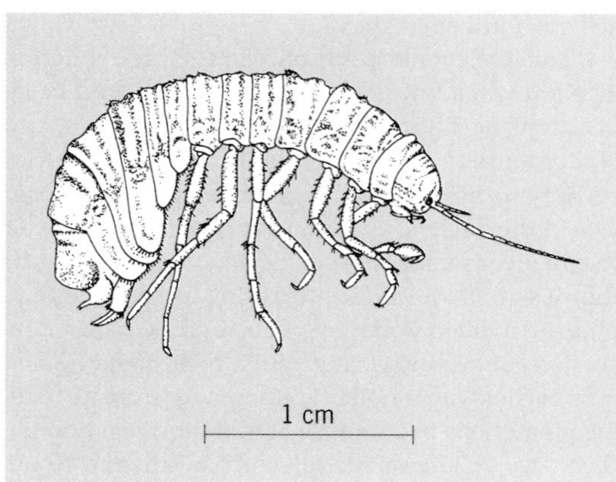

***Onchotelson brevicaudatus*, adult male.**

The suborder is an ancient one and includes a fossil, *Protamphisopus wianamattensis*, from the Triassic beds of New South Wales. Three extant species are recorded from Australia, Tasmania, New Zealand, and South Africa and one that is subterranean from India.

Most species occur in fresh water. Several are blind, subterranean forms and one occurs in hot water from deep artesian bores. A few are semiterrestrial, burrowing forms. Edith M. Sheppard

Bibliography. G. Nicholls, *Pap. Proc. Roy. Soc. Tasmania*, 1943:1–145, 1944:1–157; S. P. Parker (ed.), *Synopsis and Classification of Living Organisms*, 2 vols., 1982; E. Sheppard, *Proc. Zool. Soc. London*, 1927:1, 81–124.

Phrynophiurida

An order of Ophiuroidea in which the vertebrae usually articulate by means of hourglass-shaped surfaces and the arms are able to coil upward or downward in the vertical plane. There is usually a leathery integument, in which calcareous granules or platelets are embedded. Most species are found in deep water, and often the arms are tightly coiled about the branches of black corals, upon which Phynophiurida feed. Of the families, the Gorgonocephalidae often have branched arms, the Asteronychidae have a large disk and slender arms, and the Asteroschematidae have a small disk and stout arms.

The foregoing families share a number of characteristics and are grouped in one suborder, Euryalina. One remaining family, the Ophiomyxidae, differs in having a soft, unprotected integument, like that of *Ophiocanops*, but lacks the peculiar features of the gut and gonads in oegophiurids. For reasons too specialized to discuss here, it appears best not to associate the Ophiomyxidae with the Euryalina, in the Phrynophiurida, placing the family in a distinct suborder, the Ophiomyxina. *See* ECHINODERMATA; OEGOPHIURIDA; OPHIUROIDEA. Howard B. Fell

Bibliography. R. A. Boolootian (ed.), *The Physiology of Echinodermata*, 1966; H. B. Fell, Evidence for the validity of Matsumoto's classification of the Ophiuroidea, *Publ. Seto Mar. Biol. Lab.*, 10:145–152, 1962.

Phycobilin

Any member of a class of intensely colored pigments found in some algae that absorb light for photosynthesis. Phycobilins are open-chain tetrapyrroles structurally related to mammalian bile pigments, and they are unique among photosynthetic pigments in being covalently bound to proteins (phycobiliproteins). In at least two groups of algae, phycobiliproteins are aggregated in a highly ordered protein complex called a phycobilisome.

Occurrence. Phycobilins occur only in three groups of algae: cyanobacteria (blue-green algae), Rhodophyta (red algae), and Cryptophyceae (cryptophytes), and are largely responsible for their distinctive colors, including blue-green, yellow, and red. Five different phycobilins have been identified to date, but the two most common are phycocyanobilin [structure (**1**)], a blue pigment, and phycoerythrobilin (**2**), a red pigment. In the cell,

(1)

(2)

these pigments absorb light maximally in the orange (620-nanometers) and green (550-nm) portion of the visible light spectrum, respectively. A blue-green light (495-nm) absorbing pigment, phycourobilin, is found in some cyanobacteria and red algae. A yellow light (575-nm) absorbing pigment, phycobiliviolin (also called cryptoviolin) is apparently found in all cryptophytes but in only a few cyanobacteria. A fifth phycobilin, which absorbs deep-red light (697 nm), has been identified spectrally in some cryptophytes, but its chemical properties are unknown. *See* CRYPTOPHYCEAE; CYANOPHYCEAE; RHODOPHYCEAE.

Phycobilins are associated with the photosynthetic light-harvesting system in chloroplasts of red algae and cryptophytes and with the photosynthetic membranes of cyanobacteria, which lack chloroplasts. Phycobilins are covalently bound to a water-soluble protein that aggregates on the surface of the photosynthetic membrane. All other photosynthetic pigments (for example, chlorophylls and carotenoids) are bound to photosynthetic membrane proteins by hydrophobic attraction. Phycobiliprotein can constitute a major fraction of an alga. In some cyanobacteria, for example, fresh-water or marine *Synechococcus*, phycobiliproteins can account for more than 50% of the soluble protein and one-quarter of the dry weight of the cell. *See* CELL PLASTIDS.

Phycobiliproteins are classified primarily by their absorption spectrum (**Fig. 1**), which depends on the protein structure and the number and kind of

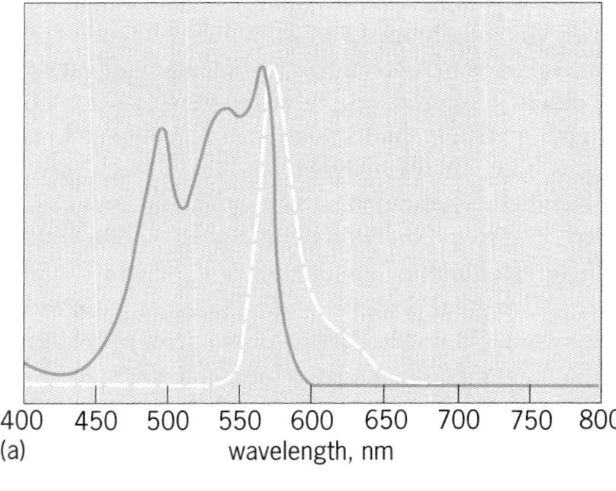

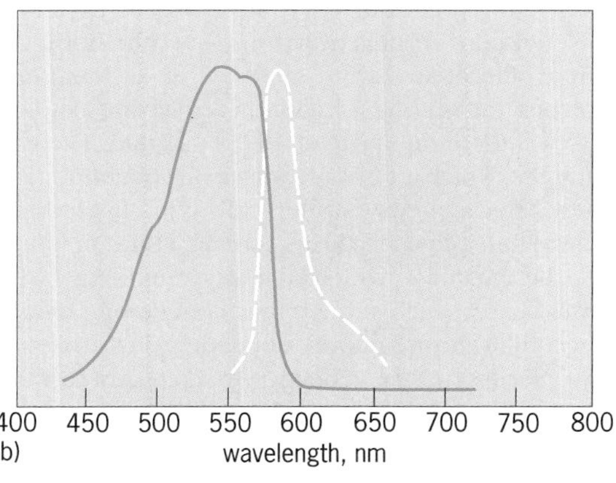

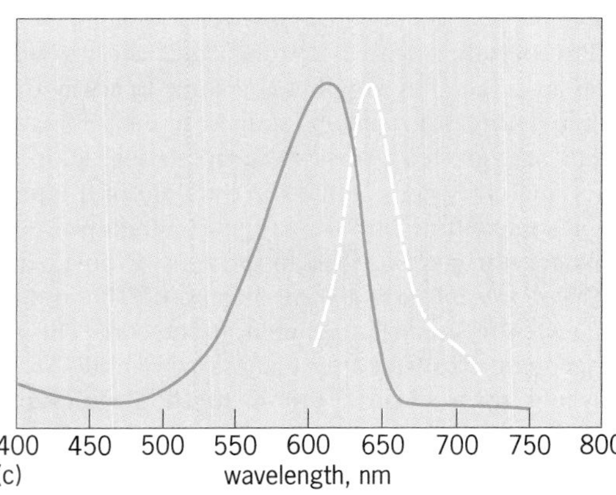

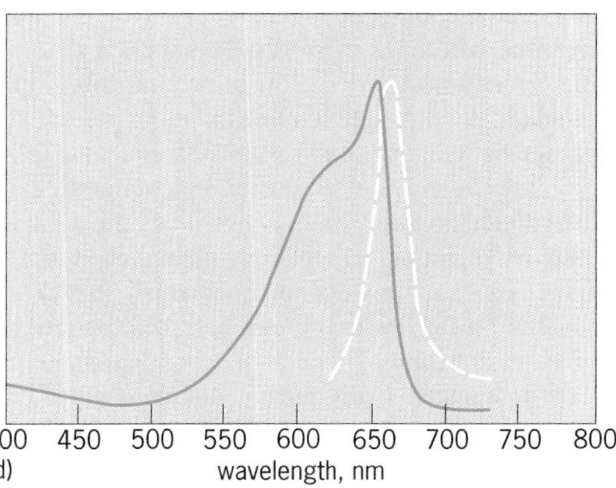

Fig. 1. Absorption spectra (colored lines) and fluorescence emission spectra (white lines) for several isolated phycobiliproteins. (*a*) R-phycoerythrin. (*b*) B-phycoerythrin. (*c*) C-phycocyanin. (*d*) Allophycocyanin. (*After M. N. Kronick, The use of phycobiliproteins as fluorescent labels in immunoassay, J. Immunol. Meth., 92:1–13, 1986*)

Some phycobiliproteins and their characteristics*

Phycobiliprotein	Distribution	Wavelength of absorption peaks in visible light spectrum, nm	Wavelength of maximum fluorescence emission, nm	Presence of phycobilins: α subunit	β subunit	γ subunit
Allophycocyanin	Cyanobacteria, Rhodophyta	650	660	1 phycocyanobilin	1 phycocyanobilin	
Allophycocyanin B	Cyanobacteria, Rhodophyta	671, 618	680	1 phycocyanobilin	1 phycocyanobilin	
C-phycocyanin	Cyanobacteria, Rhodophyta	620	637	1 phycocyanobilin	2 phycocyanobilin	
R-phycocyanin	Rhodophyta	617, 555	636	1 phycocyanobilin	1 phycocyanobilin, 1 phycoerythrobilin	
Phycoerythrocyanin	Cyanobacteria	568, 590	619	1 phycobiliviolin	2 phycocyanobilin	
Phycocyanin 645	Crytophyceae	645, 585	660	1 697†	2 phycocyanobilin 1 phycobiliviolin	
C-phycoerythrin	Cyanobacteria	565, 540	577	2 phycoerythrobilin	4 phycoerythrobilin	
R-phycoerythrin‡	Rhodophyta	568, 545, 498	578	2 phycoerythrobilin	2 phycoerythrobilin, 1 phycourobilin	1 phycoerythrobilin, 3 phycourobilin
B-phycoerythrin	Rhodophyta	545, 563, 498	575	2 phycoerythrobilin	3–4 phycoerythrobilin	2 phycoerythrobilin, 2 phycourobilin
b-Phycoerythrin	Rhodophyta	545, 563	570	2 phycoerythrobilin	4 phycoerythrobilin	
Phycoerythrin 555	Cryptophyceae	545	585	? phycoerythrobilin	? phycoerythrobilin	

*Not included are two known phycoerythrins and phycocyanins from cryptophytes, as well as several phycoerythrins from marine cyanobacteria, which contain unusually high amounts of phycourobilin.
†Unidentified chromophore with peak absorption at 697 nm.
‡Variable phycobilin composition.

phycobilins that are attached. There are three major classes, allophycocyanin (APC), phycocyanin (PC), and phycoerythrin (PE) [see **table**]. All cyanobacteria and red algae contain allophycocyanin and phycocyanin, while some species of these groups also contain phycoerythrin. Cryptophytes possess either phycocyanin or phycoerythrin, but not both. All three biliproteins are composed of at least two polypeptide subunits, α and β, each having a molecular weight in the range of 16,000–22,000. The two subunits, which are always present in equal amounts, form stable aggregates of typically two, three, or six pairs. Some phycoerythrins contain a third polypeptide, designated γ, which is usually found singly with six α-β pairs. Each subunit may contain one to four phycobilin chromophores, depending on the biliprotein. Because of the tendency to aggregate to varying degrees, molecular weights of phycobiliproteins range widely between 35,000 and 240,000.

Function. Phycobilins are photosynthetic accessory pigments that absorb light efficiently in the yellow, green, orange, or red portion of the light spectrum, where chlorophyll *a* only weakly absorbs. Differences among species in phycobiliprotein may be related to the light environment in which the algae grow. For example, cyanobacteria that grow in shallow fresh-water environments generally contain red-light-absorbing phycocyanin as the primary accessory pigment, whereas cyanobacteria from the open ocean also possess phycoerythrin, which absorbs the blue-green and green light that penetrates to greater depths.

Energy transfer. Light energy absorbed by phycobilins is transferred with greater than 90% efficiency to chlorophyll *a*, where it is used for photosynthesis. Phycobiliproteins are associated with phycobilisomes, which are organelles attached to the surface of thylakoid membranes. Phycobilisomes are found in all cyanobacteria and red algae (but not cryptophytes) and contain allophycocyanin, phycocyanin, and sometimes phycoerythrin. The morphology of phycobilisomes is somewhat variable, but the most prevalent form is hemidiscoidal, that is, in the shape of a semicircular disk standing with its straight edge facing the membrane (**Fig. 2**). When this type of phycobilisome is isolated from the cell and viewed with an electron microscope, it is possible to see a core of two or three small disks with usually six radiating rods. The rods resemble stacked disks and vary in number depending on species and growth conditions. The individual disks contain phycobiliprotein, usually in a dimeric, trimeric, or heximeric α-β aggregate. Additional proteins, called linkers, which may have phycobilins attached, are also present and have been shown to be essential for coordinating the assembly of the phycobilisome and determining the aggregation state and spectroscopic properties of the phycobiliproteins. Complete phycobilisomes have a molecular weight ranging 7–15 × 10^6 and contain 300–800 phycobilin chromophores.

Phycoerythrin, phycocyanin, and allophycocyanin are aligned in the phycobilisome in a way that supports directional energy transfer to the photosynthetic reaction center inside the membrane. Allophycocyanin is always confined to the core of the phycobilisome, while phycocyanin and (when present) phycoerythrin are located in the rods. If phycoerythrin is present, it is always located on the end of the rod opposite the attachment to the core. This arrangement allows energy to flow "downhill" from phycoerythrin to phycocyanin to allophycocyanin to chlorophyll *a*. Energy transfer occurs by dipole-dipole interaction between donor and acceptor chromophores and depends on the precise alignment of the phycobilins within the rod and core structures. Light energy absorbed by the phycobilisome

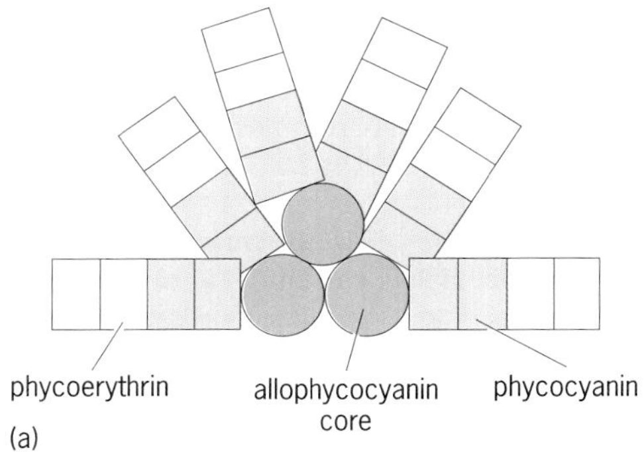

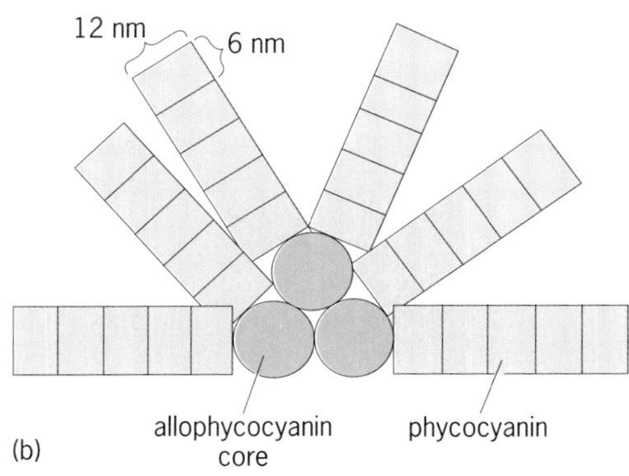

Fig. 2. Structure of a hemidiscoidal phycobilisome of *Tolypothrix tenuis* under different light conditions. (*a*) When illuminated by white light, the phycobilisome contains phycoerythrin, phycocyanin, and allophycocyanin. Energy absorbed by phycoerythrin is transferred to phycocyanin and allophycocyanin. The allophycocyanin core proteins are attached, via a linker protein, to the photosynthetic membrane, which is not shown. (*b*) When illuminated by red light, the phycobilisome undergoes complementary chromatic adaptation, in which phycoerythrin is no longer produced but additional phycocyanin is produced. (*After R. MacColl and D. Guard-Friar, Phycobiliproteins, CRC Press, 1987*)

is passed primarily to photosystem II, the oxygen-evolving complex of the photosynthetic light reactions.

Physiology. Several environmental factors can influence the phycobiliprotein content. Algal cells grown under low light intensity may have up to 20 times more phycobiliprotein than those grown under high light intensity. This response increases the ability of the alga to absorb light when it is in limited supply. Some cyanobacteria and red algae are also influenced by the color of the growth light and exhibit a phenomenon called complementary chromatic adaptation. For example, when the cyanobacterium *Tolypothrix tenuis* is grown under red light, it produces red-light-absorbing phycocyanin as its accessory pigment. But when grown under green light, it produces green-light-absorbing phycoerythrin along with small amounts of phycocyanin (Fig. 2). This response is controlled by an unidentified photoreversible pigment in a manner similar to the action of phytochrome but with absorption maxima near 545 and 645 nm. This process is regulated at the level of deoxyribonucleic acid (DNA) transcription. *See* PHYTOCHROME.

Phycobiliprotein concentration also depends on the availability of nutrients, including nitrogen, carbon dioxide, phosphorus, sulfur, and iron. Nutrient starvation generally causes a loss of phycobiliprotein, with the rod proteins of the phycobilisomes being lost more rapidly than the core proteins. In cyanobacteria, this reduction is due to specific proteolytic degradation of biliprotein present in the cell and repression of synthesis of new biliprotein.

Isolation. Phycobiliproteins are highly water-soluble and readily leach out of broken cells in low-ionic-strength buffer. They can be purified by using ammonium sulfate precipitation and a wide variety of chromatographic techniques. Phycobilins can be cleaved from the apoprotein by refluxing in methanol or other alcohols or by incubating in concentrated acid, but the two techniques yield different chromophore structures. Intact phycobilisomes have been isolated from a number of algae by using a higher-ionic-strength buffer that stabilizes the structure. They are then purified by using sucrose gradient centrifugation.

Chemistry. Phycoerythrobilin and phycocyanobilin are structural isomers but differ in the number of conjugated double bonds. This has a primary influence on the color of light absorbed by the molecules. Phycoerythrobilin, with seven alternating single and double bonds, absorbs light of shorter wavelength than phycocyanobilin, which has nine alternating single and double bonds. Free phycobilins in solution generally exhibit broad absorption peaks and weak fluorescence, whereas phycobilins attached to the apoprotein exhibit narrow absorption peaks and strong fluorescence. Binding of the phycobilin to the apoprotein also affects the wavelength of maximum absorption, reduces the ability of the chromophore to complex metals, and increases the stability of the chromophore.

Both phycocyanobilin and phycoerythrobilin are linked to their respective apoproteins through a sulfur atom of the amino acid cysteine. It is generally accepted that at least one binding site on the chromophore is through ring A of the tetrapyrrole. Phycobilins in some PEs may be doubly linked through rings A and D. The phycobilins are derived from 5-aminolevulinic acid, a precursor of cyclic tetrapyrroles such as heme and chlorophyll. Heme, an iron-containing porphyrin that is also found in hemoglobin, undergoes ring opening to form biliverdin. Biliverdin is then enzymatically isomerized and reduced to form phycocyanobilin.

Evolution. Immunochemical and amino acid sequence data have been used to investigate the evolutionary relationships between the different phycobiliproteins. Immunological reactivity analysis indicates consistent relatedness between allophycocyanin, phycocyanin, and phycoerythrin derived from cyanobacteria and red algae. Further, there is no cross reaction between any of the three types of phycobiliproteins. This evidence supports the idea that eukaryotic red algae evolved from the more primitive prokaryotic cyanobacteria. Amino acid sequences of the three phycobiliproteins, however, do exhibit some degree of homology, suggesting that they evolved from a common biliprotein.

Phycobiliproteins from cryptophytes do not show the same high degree of relatedness either to cyanobacteria or to red algae. Cross reactivity is generally greater with red algae than with cyanobacteria, suggesting a closer evolutionary link between cryptophytes and red algae than between cryptophytes and cyanobacteria. Also, cryptophyte phycoerythrins and phycocyanins exhibit cross reactivity, unlike the phycoerythrins and phycocyanins from the other two algal groups.

Biotechnology. Phycobiliproteins can be used as fluorescent labels of specific biological molecules or cells by covalently binding biliproteins to an antibody or other molecule. This is done by placing a reactive site on the antibody using a heterobifunctional reagent, then reacting with a derivatized sulfhydryl group on the biliprotein to form a sulfur bridge. This procedure yields a specific antibody with high fluorescence yield similar to free phycobiliprotein. Different phycobiliproteins, such as B-phycoerythrin, R-phycoerythrin, and C-phycocyanin, can be used to obtain specific absorption and fluorescence emission characteristics. This technique has become particularly important in studying blood disorders, such as acquired immune deficiency syndrome (AIDS). In the latter application, blood cells are treated with specific fluorescent antibodies and screened for the presence of antigen by using fluorescence microscopy or flow cytometry. *See* ACQUIRED IMMUNE DEFICIENCY SYNDROME (AIDS); IMMUNOFLUORESCENCE; PHOTOSYNTHESIS. Todd M. Kana

Bibliography. A. N. Glazer, Light guides directional energy transfer in a photosynthetic antenna, *J. Biol. Chem.*, 264:1–4, 1989; T. W. Goodwin, *Plant Pigments*, 1988; R. P. F. Gregory, *Biochemistry of Photosynthesis*, 3d ed., 1989; R. MacColl and D. Guard-Friar, *Phycobiliproteins*, 1987.

Phylactolaemata

A class of ectoproct bryozoans. Phylactolaemates have lophophores which are markedly U-shaped (or rarely nearly circular but still kidney-shaped) in basal outline, and relatively short, wide zooecia; these animals dwell only in fresh water. *See* BRYOZOA.

Morphology. Phylactolaemate colonies are either encrusting threadlike networks of relatively isolated zooecia with solid chitinous uncalcified walls, or small to large masses of gelatinous material in which the individual zooids are embedded side by side without definite separating zooecial walls. Stolons are not present. The colonies lack ovicells or monticules, they are not polymorphic, and they are not divisible into distinct endozone and exozone regions.

Quite simple in construction, phylactolaemate zooecia lack diaphragms; the aperture, round in outline and as wide as the zooecium, lacks any operculum-like covering. Just inside the zooecial wall or the gelatinous material lies an epidermis made of tall thick cells; beneath this is found a thin layer of longitudinal and circular muscles; finally, a peritoneum of flat thin cells is the innermost body-wall layer.

The mouth of a phylactolaemate zooid is overhung by a small liplike epistome. Large passageways directly connect the proximal portions of the visceral cavities of adjacent zooids. The lophophore bears moderate to large numbers of tentacles (20–106, where counted). Individual phylactolaemate zooids are always hermaphroditic, as are the individual colonies, consequently.

Life cycle. Each phylactolaemate zygote develops into only one larva, which is sheltered within a soft internal sac (embryo sac) on the body wall. Freely swimming for less than a day, the larva, after attaching to a substrate, develops directly into an ancestrular zooid without degeneration of its larval structures. New zooids produced by asexual budding begin as saclike invaginations on the inner side of the body wall of the parent zooid; rudiments of internal soft-part organs (polypide) appear first, followed immediately by formations of new body wall (cystid) materials. Although never degenerating to form brown bodies, phylactolaemate zooids do produce disk-shaped, cold- and drought-resistant resting bodies (statoblasts), which can germinate or hatch to form new colonies asexually when favorable environmental conditions return.

History and classification. Only a few (about 50) phylactolaemate species exist, all classified in a single order, the Plumatellida (or Plumatellina). Exclusively fresh-water, the phylactolaemates may have evolved relatively recently from ctenostomes, although some workers have suggested that phylactolaemates might be very primitive ectoprocts surviving as evolutionary relics. Possibly appearing as early as the mid-Cretaceous, phylactolaemates were certainly established by late Cenozoic times; they have always been a relatively minor bryozoan group. *See* CTENOSTOMATA. Roger J. Cuffey

Phyllite

A type of metamorphic rock formed during low-grade metamorphism of clay-rich sediments called pelites. Phyllites are very fine grained rocks with a grain size barely visible in a hand specimen. They have a well-developed planar element called cleavage defined by alignment of mica grains and interlayering of quartz-rich and mica-rich domains. Typically, mica grains show the greater alignment, although other mineral components (quartz, carbonate, and feldspars) may show a preferred shape orientation. Where all minerals of a particular type show the same degree of alignment and the fabric is well developed throughout the rock, the fabric is termed a penetrative fabric. Cleavage surfaces in phyllites have a glittery, lustrous sheen due to light reflecting off grains of chlorite and muscovite. The mineralogy of phyllites is dependent on chemical composition; typical minerals in phyllites are chlorite, muscovite, and quartz. Other minerals that may be present in phyllites formed during low-grade metamorphism include chlorotoid, garnet (rarely), sodium-mica, and sulfide minerals. These minerals are typically millimeter in scale and resolvable in a hand specimen by using a 10× hand lens. *See* CHLORITE; MUSCOVITE; QUARTZ.

Phyllites form during intermediate stages of progressive metamorphism of pelites within the greenschist facies. Initial stages of metamorphism transform pelites into slates, which are characterized by very fine grains, scarcely visible with a microscope. These rocks have a strong cleavage along which the rock is easily split. Phyllites are slightly coarser grained, and the rock cleavage may be slightly irregular because of mineral grains protruding across the cleavage planes. Schists are still coarser and are distinguished by mineral grains that are millimeter to centimeter in scale. Cleavage planes in phyllites usually cut across primary sedimentary layers and, where the rocks are folded, cleavage planes are subparallel to the axial surface of folds. *See* FACIES (GEOLOGY); SCHIST; SLATE.

Cleavage formation in phyllites involves the interaction between several different processes, including mechanical rotation of minerals; mechanical rotation of minerals crystallizing during metamorphism; preferred growth of minerals at the time of crystallization; and change in shape of mineral grains during deformation and metamorphism, possibly as a result of mineral dissolution.

Phyllite is found in most regionally metamorphosed terranes in the world, including the Appalachians of eastern North America, the Scottish Highlands, and the Alps. *See* METAMORPHIC ROCKS.

Matthew W. Nyman

Phyllocarida

A subclass of the crustacean class Malacostraca containing the extant order Leptostraca and the fossil order Archiostraca (see **illus.**). The Phyllocarida has a long fossil record, and many early fossil taxa were referred to this subclass. However, studies of presumed phyllocarids from the Burgess Shale have shown that only the archiostracans agree with the definition of the Phyllocarida. The apparent antiquity of the Phyllocarida, together with the foliacious structure of their thoracic appendages, gave rise to the hypothesis that fossil phyllocarids encompassed the ancestral stem of the Eumalacostraca. However, eumalacostracan stocks never show such phyllocarid features as thoracic filter-feeding mechanisms and a brood pouch formed with the setose tips of the thoracic endopods. Thus a consensus has been reached that the Phyllocarida represents an early, and perhaps dead-end, offshoot of the Malacostraca rather than the stem group for eumalacostracan evolution.

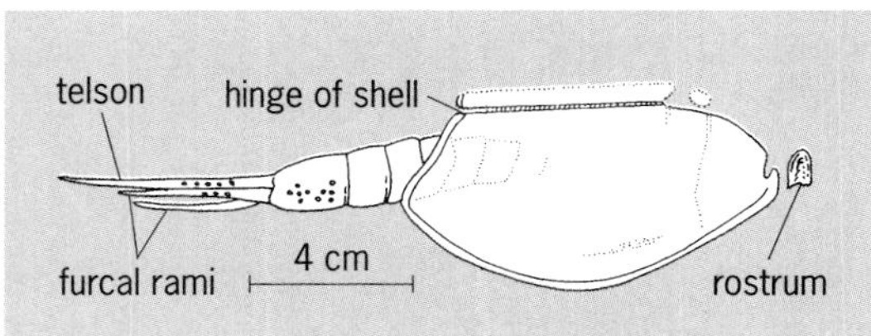

***Ceratiocaris stygia*, a fossil leptostracan of the subclass Phyllocarida. (*After T. R. Jones and H. Woodward, A Monograph of British Palaeozoic Phyllopoda (Phyllocarida Packard), Palaeontographical Society, 1888–1889*)**

Phyllocarids are distinct from other malacostracan crustaceans because of two other characteristics considered to reflect the primitive condition, which strengthen the hypothesis of early separation from the main evolutionary line. The first is the presence of a bivalve carapace. The dorsal carapace hinge seen in archiostracans is absent in leptostracans, but both orders have the characteristic adductor muscle that holds the two sides together. The second is an abdomen consisting of seven fully formed somites and terminating in a telson that bears caudal rami. Only the anterior six somites bear appendages. *See* CRUSTACEA; LEPTOSTRACA; MALACOSTRACA.

Patsy A. McLaughlin

Bibliography. E. Dahl, The subclass Phyllocarida (Crustacea) and the status of some early fossils: A neontologist's view, *Vidensk. Meddr. Dansk Naturh. Foren.*, 145:61–76, 1984; J. K. Lowry (ed.), *Papers from the Conference on the Biology and Evolution of Crustacea*, Austral. Mus. Mem. 18, 1983; F. R. Schram, Crustacean phylogeny, *Crustacean Issues*, vol. 1, 1983.

Phylogeny

The genealogical history of organisms, both living and extinct. Phylogeny represents the historical pattern of relationships among organisms which has resulted from the actions of many different evolutionary processes. Phylogenetic relationships are depicted by branching diagrams called cladograms, or phylogenetic trees. Cladograms show relative affinities of groups of organisms called taxa. Such groups of organisms have some genealogical unity, and are given a taxonomic rank such as species, genera, families, or orders. For example, two species of cats—say, the lion (*Panthera leo*) and the tiger (*Panthera tigris*)—are more closely related to each other than either is to the gray wolf (*Canis latrans*). The family including all cats, Felidae, is more closely related to the family including all dogs, Canidae, than either is to the family that includes giraffes, Giraffidae. The lion and tiger, and the Felidae and Canidae, are called sister taxa because of their close relationship relative to the gray wolf, or to the Giraffidae, respectively.

Cladograms thus depict a hierarchy of relationships among a group of taxa (**Fig. 1***a*). Branch points,

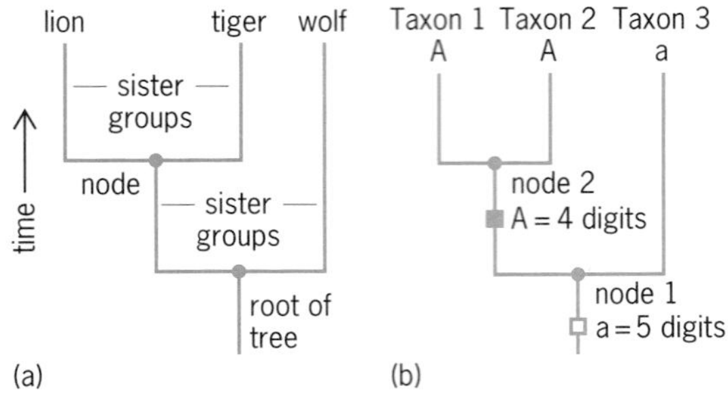

Fig. 1. Phylogenetic trees. (*a*) A tree representing a hierarchical pattern of sister-group relationships (those taxa descended from a common ancestor). (*b*) Relationships are determined by identifying derived characters; in this case the condition of five digits is primitive, whereas the loss of a digit is derived and unites taxa 1 and 2.

or nodes, of a cladogram represent hypothetical common ancestors (not specific real ancestors), and the branches connect descendant sister taxa. If the taxa being considered are species, nodes are taken to signify speciation events. The goal of the science of cladistics, or phylogenetic analysis, is to discover these sister-group (cladistic) relationships and to identify what are termed monophyletic groups—two or more taxa postulated to have a single, common origin.

Methodology. The acceptance of a cladogram depends on the empirical evidence that supports it relative to alternative hypotheses of relationship for those same taxa. Evidence for or against alternative phylogenetic hypotheses comes from the comparative study of the characteristics of those taxa. Similarities and differences are determined by comparison of the anatomical, behavioral, physiological, or molecular [such as deoxyribonucleic acid (DNA) sequences] attributes among the taxa. A statement that two features in two or more taxa are similar and thus constitute a shared character is, in essence, a preliminary hypothesis that they are homologous; that is, the taxa inherited the specific form of the feature from their common ancestor. However, not all similarities are homologs; some are developed independently through convergent or parallel evolution, and although they may be similar in appearance, they had different histories and thus are not really the same feature. In cladistic theory, shared homologous similarities are either primitive (plesiomorphic condition) or derived (apomorphic condition), whereas nonhomologous similarities are termed homoplasies (or sometimes, parallelisms or convergences). This distinction over concepts and terminology is important because only derived characters constitute evidence that groups are actually related.

As evolutionary lineages diversify, some characters will become modified. Examples include the enlargement of forelimbs or the loss of digits on the hand. Thus, during evolution the foot of a mammal might transform from a primitive condition of having five digits to a derived form with only four digits (Fig. 1*b*). Following branching at node 1, the foot in one lineage undergoes an evolutionary modification involving the loss of a digit (expressed as character state A). A subsequent branching event then produced taxa 1 and 2, which inherited that derived character. The lineage leading to taxon 3, however, retained the primitive condition of five digits (character state a). The presence of the shared derived character, A, is called a synapomorphy, and identifies taxa 1 and 2 as being more closely related to each other than either is to taxon 3. Distinguishing between the primitive and derived conditions of a character within a group of taxa (the ingroup) is usually accomplished by comparisons to groups postulated to have more distant relationships (outgroups). Character states that are present in ingroups but not outgroups are postulated to be derived. To determine relationships among the species of cats (family Felidae), comparisons to characters in other carnivore outgroups such as dogs (Canidae) or bears (Ursidae) would help distinguish primitive and derived conditions within cats.

Systematists have developed computer programs that attempt to identify shared derived characters (synapomorphies) and, at the same time, use them to construct the best phylogenetic trees for the available data. Conceptually, the two procedures—determining the polarity of character transformations and finding the best phylogenetic tree—are logically related (**Fig. 2**). For example, three alternative phylogenetic trees are possible for three taxa.

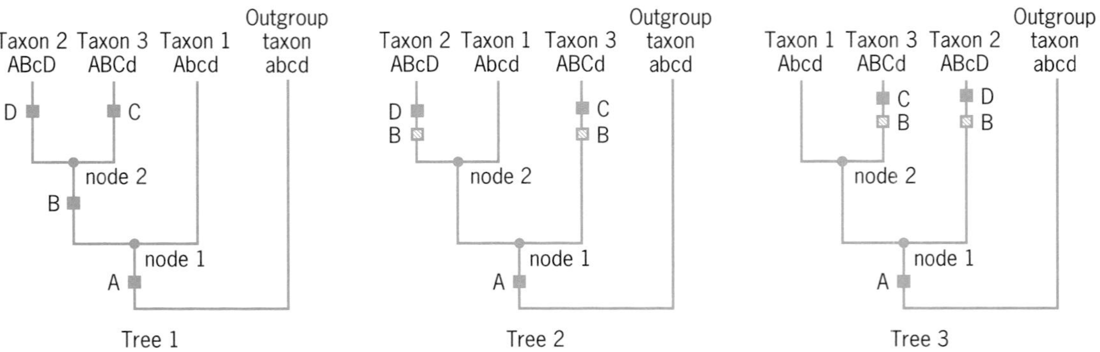

Fig. 2. Alternative trees for three taxa. On tree 1, taxa 2 and 3 are united by a postulated synapomorphy, B. On trees 2 and 3, no taxa are united by a synapomorphy and B evolves in parallel (cross-hatched). Tree 1 assumes fewer character-state transformations for all the data and is thus preferred.

A comparative analysis of the three might reveal that they share four characters, A–D, each of which has two alternative states. Comparisons to one or more outgroup taxa help the systematist interpret the polarity of these character states (in this example, uppercase letters are taken to be the derived conditions). Given the distributions of the character states shown in Fig. 2, postulated character transformations can be assigned to the branches of each alternative tree. In all three trees, the derived condition, A, unites the three ingroup taxa but cannot help choose among the three alternatives for their interrelationships. When the second character is examined, it is seen that the postulated derived condition, B, unites taxa 2 and 3 to the exclusion of taxon 1, which has the primitive condition, a. The derived conditions of the third and fourth characters are each found in one taxon only: C in taxon 3 and D in taxon 2, respectively. Thus, only the transformation from b to B carries phylogenetic information. In addition, tree 1 also better explains the distributions of the different character states because it requires that b transform to B only once, whereas trees 2 and 3 require two independent origins of B from b. Thus, tree 1 is the simplest hypothesis given the available data and is to be preferred. Similarly, computer programs can process data sets having large numbers of taxa and characters, examine all possible character transformations on a set of trees, and then choose that tree which accounts for the fewest number of changes.

The interpretation of the status of a character shared among taxa depends upon an assessment of its history on a given tree. Thus, in tree 1 the derived character B is interpreted as a synapomorphy (it unites taxa 2 and 3); that is, B is inferred to be homologous in the two taxa. Within the framework of threes 2 and 3, however, the derived character B is interpreted as having evolved independently in taxa 2 and 3. In this case, character state B is interpreted as being homoplasious (parallel) and thus not homologous.

Value of phylogenies. Understanding how species and groups of species are interrelated is fundamental to interpreting observations across all of biology. Knowledge of phylogenetic relationships provides the basis for classifying organisms. Ideally, all taxonomic groups—genera, families, orders, and so on—should be monophyletic. Some groups, such as birds and mammals, are monophyletic; that is, phylogenetic analysis suggests they are all more closely related to each other than to other vertebrates. However, other traditional groups, such as reptiles, have been demonstrated to be nonmonophyletic (some so-called reptiles, such as dinosaurs and their relatives, are more closely related to birds than they are to other reptiles such as snakes). A major task of the science of systematics is to search for monophyletic groups. Classifications based on monophyletic groups are termed natural classifications.

Phylogenies are also essential for understanding the distributional history, or biogeography, of organisms. Knowing how organisms are related to one another helps the biogeographer to decipher relationships among areas and to reconstruct the spatial histories of groups and their biotas. *See* BIOGEOGRAPHY.

Biologists use phylogenies to make predictions about the characteristics of organisms that are still poorly studied. If one species produces a particular compound that has useful pharmacological properties, it is probable that its close relatives will share an identical or similar compound. The same is true for other structural, behavioral, or physiological characters.

History of life. Life has diversified on Earth for about 4 billion years. Systematists have described nearly 1.4 million living species, most of which are arthropods (insects, spiders, and their relatives), as well as tens of thousands of fossil species. It is estimated that from 3 million to perhaps as many as 100 million living species remain to be discovered, and the number of unknown fossil species is no doubt equally vast.

Much knowledge about the history of life has been derived from studies of comparative anatomy. Advances in understanding the history life have been made by using information derived from comparisons of DNA sequences, which largely involves diversification within the microbial world. It is often believed that fossils represent the direct ancestors of other fossils or of living organisms and, therefore, that fossils are necessary to reconstruct phylogeny. Because of the vast panoply of taxonomic and structural diversity seen throughout deep evolutionary time, reconstructing its history is a complex undertaking, and many aspects are still not well understood. Many major groups of organisms appear to have diversified in an evolutionary burst about 600 million years ago. Rapid radiations, especially those taking place long ago, pose particular difficulties for reconstructing phylogeny, for often relatively few morphological and molecular characters evolve between branching events that are so close to one another in time. However, there has been an upsurge in interest in reconstructing phylogeny, so knowledge of the history of life should improve substantially. *See* ANIMAL EVOLUTION; ANIMAL SYSTEMATICS; TAXONOMIC CATEGORIES. Joel Cracraft

Bibliography. D. R. Brooks and D. H. McLennan, *Phylogeny, Ecology, and Behavior*, 1991; N. Eldredge and J. Cracraft, *Phylogenetic Patterns and the Evolutionary Process*, 1980; B. Fernholm, K. Bremer, and H. Jornvall (eds.), *The Hierarchy of Life*, 1989; P. L. Forey et al., *Cladistics: A Practical Course in Systematics*, 1992; A. B. Smith, *Systematics and the Fossil Record: Documenting Evolutionary Patterns*, 1994.

Phymosomatoida

An order of regular sea urchins, class Echinoidea, characterized by imperforate tubercles, complex ambulacral compounding in which one or more elements are occluded from the perradial suture, and a stirodont lantern with unfused epiphyses. They comprise two families, one with crenulate and the other with noncrenulate tubercles. They first appeared in the Lower Jurassic and are probably paraphyletic, since they include the ancestors of camarodonts. The two extant genera are each known from a single species: *Glyptocidaris*, from a depth of 30–490 ft (10–150 m) around northern Japan, and *Stomechinus*, a common inhabitant of rocky shores around the Indo-West Pacific. They are epifaunal grazers. *See* ECHINODERMATA; ECHINOIDEA. Andrew Smith

Physical anthropology

The subfield of anthropology that deals with human and nonhuman primate evolution, the biological bases of human behavior, and human biological variability and its significance. Some refer to the field as biological anthropology in order to signal the close links with other biological sciences. The term physical anthropology is largely an American and British invention; in most European and many other countries physical anthropologists are the only anthropologists, while persons who study behavioral aspects of the human condition are known as archeologists, ethnologists, linguists, or prehistorians. In Great Britain, and increasingly in the United States and elsewhere, the term human biology is also used instead of physical anthropology, especially in relation to curricula in the health sciences.

Paleoanthropology. Paleoanthropology is the multidisciplinary study of human evolution as evidenced by fossils, artifacts, and their geological and burial site contexts. Physical anthropologists organize expeditions and direct excavations that lead to discoveries of fossil Hominidae, and then engage in the tedious repair and reconstruction of specimens, their anatomical description, comparison with other specimens, and placement in hominid phylogeny.

Morphological paleoanthropologists must have detailed knowledge of human and other primate anatomy and the principles of taxonomy in order to restore and interpret their discoveries. In addition to traditional anatomical descriptions and measurements, it is requisite that the variations in samples be presented and, when possible, be tested for statistical significance. Because of the fragmentary nature of many fossils, one of the most difficult problems is to decide whether the new discoveries belong with previously described species or represent new underscribed ones. For instance, D. C. Johanson initially announced that several hominid species were represented by a small sample of specimens from the Pliocene Hadar Formation in northern Ethiopia. But after many, more complete specimens were collected and compared, he and T. D. White concluded that only one species, *Australopithecus afarensis*, inhabited the site. P. V. Tobias, on the other hand, concluded that the collection could be accommodated in *A. africanus*, which had been described previously from South Africa. Y. Coppens suggested that there are two species of Hominidae at Hadar, including one of *Homo*.

Similar controversies have raged over the hominid status of *Ramapithecus*, a Eurasian and eastern African Miocene form; whether robust and gracile *Australopithecus* from South Africa are one or more species; and whether *H. habilis* is a distinct species or just another gracile *Australopithecus*.

Proper geophysical dating is also essential for placing fossils in the larger picture of human evolution. If there are no datable minerals associated with a hominid fossil, paleontologists may be able to link associated animal species from new sites with those of geochemically dated sites elsewhere. The known record of the Hominidae extends certainly to 3.5 million years ago (m.y.a), as evidenced by the radiometrically dated footprint trails at Laetoli, northern Tanzania. Before 3.5 m.y.a., however, the record is obscure.

In order to depict the ways of life of human ancestors, carefully mapped contextual information from fossil sites is needed, including plant and animal remains that indicate the climate, habitat and availability of food and water, and the artifacts and archeological features that show how the hominids might have processed food, protected themselves from predators, and sheltered themselves from daily and seasonal vicissitudes.

The hominid fossils may show evidence of violence, cannibalism, or gnawing by animals. The teeth and other cranial structures may indicate crudely the texture of ingested foods, and fossil (coprolites) may provide direct dietary information. The densities, age profiles, and morphology of individuals at hominid sites provide clues to how successful they were in comparison with modern humans and other primates. *See* FOSSIL HUMANS.

Paleoprimatology. Physical anthropologists also look to nonhuman primates for clues to human physical history and status as mammals, and for analogies to the behavior and cognitive abilities of human ancestors. Like paleoanthropologists, paleoprimatologists employ methods of other paleobiologists to collect, describe, and interpret fossil specimens phylogenetically and functionally. Contextual information, including the manner in which specimens were deposited and their possible alteration over time, is vital. Paleoprimatologists usually do not have to contend with archeological materials but are faced with an enormous time span, from the Late Cretaceous period, 75 m.y.a., to the beginning of the Holocene epoch, 10,000 years ago.

Numerous fossils from Oligocene deposits in the Egyptian Fayum show that the earliest ancestors of catarrhine primates (apes, hominids and Old World monkeys) were similar to the highly arboreal New World monkeys in many features. Further, a good case has been made that Old World monkeys do not represent a stage in the evolution of the Hominoidea (apes and hominids) but instead are a later offshoot of the Hominoidea.

Comparative primate morphology. Recent primates, including humans, are the current end products of evolution. Anatomical studies of modern primates are essential for reconstructing the morphologies of fossil forms and for revealing their singularity. Carefully controlled comparative studies on extant primates hold the greatest promise for modeling the functional morphology, physiology, and habitat preferences of extinct forms. Since 1970, advanced techniques and concepts from rehabilitation medicine, orthopedics, orthodontics, radiology, and neurology have been used to establish a truly functional morphology of the primates.

Locomotor anatomy. Electromyography is used to document the activities of specific muscles in nonhuman primates as they engage in a variety of positional behaviors (postures and locomotion). Classic inferences from dissections of cadavers have had to be revised because of electromyographic research. For instance, it has been shown that the bulky upper limb muscles of apes are not active during quiet suspension from overhead supports; instead, special features of their joints are the chief supporting structures. *See* ELECTROMYOGRAPHY.

Cineradiography (x-ray motion pictures) reveals otherwise inaccessible details about the movements of bones and other structures in joints during regular and facultative locomotion. For example, cineradiographic films of bipedal chimpanzees have been used to interpret features of australopithecine (early hominid) limb bones. *See* RADIOGRAPHY.

Force plates record the directional forces of primate feet and hands as they walk over them, which can be compared interspecifically to show, for instance, how closely nonhuman primates approach human patterns of walking. High-speed motion pictures and computer analyses of limb segment positions from films and video tapes are providing details on the ways that primates move on different substrates. Comprehensive gait and kinematic analyses of many living species may serve as a solid basis for modeling hypothetical locomotor patterns of fossil species.

Cranial and dental anatomy. It is no longer common to assign each new fossil primate tooth or bit of skull to a new species, largely because of studies on the crania, jaws, and teeth of living primates have shown that considerable variation is the rule even within a single species.

Electromyography and the use of strain gages to study cranial structures in several primate species have helped to explain the mechanics of chewing and other features related to processing foods. These studies help to identify probable (and to rule out unlikely) feeding habits of fossil forms. They also assist in the selection of animal models for experiments that will contribute to the development of more efficient human orthodontic devices and procedures. *See* BIOMECHANICS.

Scanning electron microscopy, x-rays, serial tomographs, and enlarged photographs reveal details of dental cusp, root, and crest size and shape and abrasive patterns that are correlated with particular diets in living primates and other mammals. This information is valuable for inferring the diets of fossil species.

Neuroanatomy. Although bipedal habit and manipulatory hands set humans apart from other primates, it is the large brain that is usually held to be the great advance of humankind. Hence considerable research is devoted to discerning the differences between human and nonhuman central nervous systems and to documenting the changes in brain size and shape during anthropoid phylogeny.

Like other areas of primatology, neuroanatomy is not the sole province of anthropologists. Indeed only since about 1960 have physical anthropologists centered research careers within it. The steady accumulation of fossil hominid endocasts and crania from which latex endocasts can be made have attracted anthropologists to join other scientists in the quest to elucidate human brain evolution. A theory was developed indicating that the major change from ape to hominid brains was one of internal reorganization of the central nervous system and that this preceded the dramatic increase in gross size. It is also believed by a few workers that speech areas exist on the endocasts of *H. habilis*, a species that dates to 2.0 m.y.a.

Anthropologists are especially keen to learn the extent to which nonhuman primates have cerebral asymmetries that correlate with behavioral traits, whether any of these may provide clues for the evolution of human language, and the extent to which expansion of the brain is a mere consequence of increased body size in the Hominidae. Apes and some monkeys also exhibit differences in size between the left and right hemispheres, but links to handedness and other behaviors remain elusive. Because of their costliness and for sound humanitarian and conservational reasons, the closest primates to humans (chimpanzees, gorillas, and orangutans) are not likely to be used in drastic experiments that could inform questions about the meanings of their asymmetries. The failure of apes to learn to speak, despite intensive efforts to teach them to do so, indicates that they lack the neurological basis for human speech. *See* APES.

Molecular anthropology. The explosion of molecular biology that followed the cracking of the genetic code attracted physical anthropologists who wished to test hypotheses about the propinquity of humans with the apes, and the relationships of other primates to one another and to other creatures. Some have endeavored to find "molecular clocks" that could tell when species diverged from one another. Assuming that genetic changes occur at fairly steady rates and given a few well-dated fossils, the time when the living species may have branched from one another can be calculated. Such a molecular clock was to be used to judge which fossils were part of hominid and other lineages. These studies have been invaluable for excluding monkeys and prosimians from immediate hominid ancestry, and for pinpointing the time interval during which the hominids diverged from apes (12–4 m.y.a.). However, they have inspired a variety of heterodox taxonomic schemes which have African apes or even all apes in the Hominidae. It will probably be some time before compelling paleontological and genetic discoveries will lead to resolution of the controversies that surround these studies. For example, M. Goodman has steadfastly argued that molecular evolution slowed down in hominid lineage, and hence that the clocks of V. Sarich and other scientists are unreliable. Further, little is known about the relationships between molecules and the morphological features that distinguish the apes and humans. *See* PROTEINS, EVOLUTION OF.

Primate behavior and ecology. Although firmly rooted in comparative psychology, behavioral primatology has also become a major section of physical anthropology; anthropologists have contributed mainly through field studies.

An original goal of primate field studies by anthropologists was to find predictable relationships between specific habitats and the patterns of sociality of the primates that inhabit them. This, it was thought, would allow the identification of the social organization of fossil species for which there is information of the paleohabitats. Perhaps it could be learned whether ancestral hominids were monogamous or polygamous and whether one sex dominated the other and served as the stable core in social groups. As commonly occurs in behavioral science, however, the greater the variety of focal studies and the longer the span of observations, the fewer generalizations pertaining to the human career and condition that could be sustained.

Because of their close genetic relationships to humans, chimpanzees have emerged as the most popular model for early hominid sociality. But theorists emphasizing different aspects of their behavior arrive at markedly different models for early hominid behavior. For instance, while one worker has stressed the importance of female choice and maternal behavior in a flexible social system, another has placed the males fully in charge of their monogamous units.

The efforts of primate field behavioralists, and ecologists in general, will have to be directed increasingly toward conservation and gaining appreciation for their subjects by local people and government officials, since habitat destruction is occurring at alarming rates in the tropics. Because primates compete for natural resources and serve so well as models for many human diseases and physiological processes, hard choices have to be made between their conservation and the improvement of human lives. Their utility as models in medical science increases greatly as knowledge of their natural adaptations accrues.

Human variation. The term human variation is rapidly replacing its historical predecessor "race" in anthropology because the latter carries so much negative connotation. Many scientists believe that the concept of race should be abandoned. Instead, researchers should simply record the gene frequencies and biological traits of human populations that are otherwise identified only by their geographic localities. This genotypic and phenotypic information would be interpreted in terms of historical and proximate selective forces in each environment. For example, the Asian sources of paleo-Indian populations that immigrated to the New World, and the deployment of their successors in North America and South America over the past 12,000 years, may be sought by the modern population biological approach. *See* HUMAN GENETICS.

According to the modern perspective, people should no longer be labeled Caucasoid, Mongoloid, Negroid, Australoid, Nordic, Black, Brown, White, and so forth. While this may cause problems in communication as anthropologists attempt to apply their knowledge practically to forensic and medical problems, in the long run humanity will benefit from the exclusion of racial typologies.

Skeletal biology. Though this field became moribund because of its largely descriptive nature and long history of abuse by racial typologists, some creative skeletal biologists have adopted functional approaches from primatological anthropology, and have been discerning the genetic determinants of nonmetric (discrete) and measurable (continuous) traits that would allow them to document the local history of burial populations in an area. For example, in a study of one macaque population, the history of which was well documented, the validity of nonmetric skeletal traits for revealing genetic links within and between populations was established. This permitted others to use cremated and other fragmentary human cemetery remains to establish the history of Native Americans in the Lower Illinois River Valley.

Paleopathology. Human bones and teeth sometimes reflect diseases and mechanical trauma during the lifetime of an individual. Rehydrated soft tissues of mummies may show evidence of parasitic infestations and lesions, such as those from cancer and tuberculosis. Such studies allow reasonable correlations of disease features with demography, ecology, diet, and social factors.

Unfortunately, most diseases cannot be specifically identified from skeletal remains, though improved chemical and microscopic techniques now reduce the number of possible conditions in many cases. The original hope of skeletal biologists that the places or origin and subsequent spread of ancient diseases, such as syphilis, could be documented now seem illusory.

The roots of medical practice itself are found in prehistory via evidence such as set fractures and trephined (surgically perforated) skulls. Whether the wounds were fresh, infected, or healed when the person died can be determined by the state of bones from the injured area. It is also possible to detect caretaking behavior by other group members toward afflicted individuals from the extent of their treatment and evidence of survival despite severe or repeated injuries.

Skin, hair, and eyes. Racial typologies were not built from bones alone. Skin color and hair texture were also fundamental to historical arrangements of humankind into subgroups. In the modern view, however, skin color is chiefly of medical concern, such as for predicting susceptibility to melanomas. Further, the evolutionary problem of how and why light and dark epidermal pigmentation arose is of interest, and has produced hypotheses relating to human physiology and ecology. Dark skin, including that produced by tanning, may prevent hypervitaminosis D from overexposure to sunlight. Light skin may allow sufficient vitamin D to be metabolized by people exposed to low sunlight. There has also been the suggestion that epidermal structures functioned as sexually attractive features in human and prehuman reproductive behavior, but the true extent of such effects is not known.

Fingerprints failed to confirm racial typologies because they are too individualistic. Instead, they have been a boon to forensic and pediatric scientists. *See* FINGERPRINT.

Growth, physique, and aging. The field of growth studies has developed robustly as part of physical anthropology since the 1940s. Standards for the appearance and ossification of bones and for sexual maturation have been established so that congenital, nutritional, and other environmental effects can be detected and often corrected clinically in children and adolescents. It has been established from global nutritional surveys that small adult size is correlated with dietary insufficiency. The field promises to grow further as longitudinal studies are established with more populations and as the nutritional status and general medical condition of people improve. Since 1975, physical anthropologists have begun to apply anthropometric and microscopic techniques to the study of aging in an effort to understand why some people have greater longevity. *See* AGING; NUTRITION.

Nutritional anthropology. Refined chemical assays promise to reveal past diets from hominid skeletal remains. Hence paleonutrition has emerged as a special focus for technical research. Trace elements, such as strontium (Sr), sodium (Na), zinc (Zn), and calcium (Ca), may indicate the nutrients that were incorporated in bones, so that classes of food items can be inferred. For instance, because strontium from plants is taken up by bones much more than by flesh, the extent to which people were herbivorous (high Sr) versus carnivorous (low Sr) can be assessed in burial populations. Unlike strontium, zinc occurs at higher levels in meat than in leafy plants. Thus it can be used to check results from strontium assays, provided that there is no evidence for dietary nuts or mollusks, which also have relatively high levels of zinc. The kind of carbon (via) ^{12}C-^{13}C ratios) in bones may indicate the types of plants that were eaten. This is an important tool for following the spread of maize agriculture in the New World. Protein and vitamin D deficiencies are indicated by flatness of the skull base relative to the skulls of better-nourished people.

Blood group genetics and disease. Many human features such as stature, head shape, epidermal pigmentation, and fingerprints are caused by undetermined multifactorial genetic systems whose phenotypic expression is commonly affected by environmental factors. In contrast, the genetics of blood group systems (such as ABO, MNS, and Rh) are based on single gene loci and are well understood.

Susceptibility to a specific disease has not been definitely linked to a particular human blood type; but there are numerous, usually deadly, single-locus diseases that afflict humans. Cogent arguments have been advanced that the gene for sickle-cell anemia was maintained in human populations because, when paired with a nonsickling gene, producing a heterozygous state, it conferred resistance to malaria. If the gene is paired with its counterpart, the resulting homozygous condition is lethal. The relationship of other genetic polymorphisms with diseases is much less well documented. Diseases are obvious candidates to act as selective agents for human blood groups. Alternatively, the distribution of blood groups in human populations could be the result of random factors, such as genetic drift and recombination. Compelling evidence that would resolve the selection-drift controversy is not available. *See* BLOOD GROUPS; POPULATION GENETICS.

Human adaptability and ecology. Homo sapiens is one of the most versatile species on Earth. The invention and ramification of culture has permitted people to survive on impoverished islands and in climatic extremes of high altitude, deserts, and polar regions. In addition to many clever technological and social conventions, people show physiological and ontogenetic characteristics that enhance their survival and ability to work in harsh environments. The extent to which these features are genetically determined is still largely unknown.

Human ecologists need not focus on exotic people in extreme climates. Urban settings offer challenges to anthropologists who would understand the interaction of human biology and culture in adapting to dietary insufficiency and oversufficiency; new and unconquered diseases; air, noise, and water pollution; and the sheer number and variety of fellow urbanites and their pets. Nutritional factors and different histories of disease, however, make control of comparisons between populations problematic. *See* HUMAN ECOLOGY.

Forensic anthropology. Forensic anthropology is growing quickly, as anthropologists are called as expert witnesses regarding not only classic sorts of criminal evidence (fingerprints, blood types, and skeletal remains) but also grisly exhibits such as bloody footprints and bite marks on murder victims. Forensic anthropology grew out of the need to identify the war dead and individuals who are killed in domestic disasters such as airplane crashes. In criminal cases, forensic anthropologists operate best if contextual evidence is preserved, particularly when the victims were hidden in graves. Here the techniques of field archeology can be as important as those of physical anthropology. *See* ANTHROPOLOGY; ARCHEOLOGY; FORENSIC ANTHROPOLOGY. Russell H. Tuttle

Bibliography. L. Aiello and C. Dean (eds.), *An Introduction to Human Evolutionary Anatomy*, 1990; B. G. Campbell, *Humankind Emerging*, 9th ed., 2001; R. L. Ciochon and R. S. Corruccini, *New Interpretations of Ape and Human Ancestry*, 1983; F. E. Johnston (ed.), 1930–1980 Jubilee Issue, *Amer. J. Phys. Anthropol.*, 56:327–557, 1981; S. Mader, *Human Biology*, 7th ed., 2001; H. Nelson and R. Jurmain, *Introduction to Physical Anthropology*, 8th ed., 1999; P. L. Stein and B. M. Rowe, *Physical Anthropology*, 7th ed., 1999; R. H. Tuttle, D. P. Buxhoeveden, and G. W. Cortright, Anthropology on the move: Progress in experimental studies of nonhuman primate positional behavior, *Yearbook of Physical Anthropology*, vol. 22, pp. 187–214, 1979.

Physical chemistry

The branch of chemistry that deals with the interpretation of chemical phenomena and properties in terms of the underlying physical processes, and with the development of techniques for their investigation. Chemical physics refers to a branch of physical chemistry in which the emphasis is on the interpretation and analysis of the physical properties of individual molecules and bulk systems, instead of their reactions. Theoretical chemistry is another major branch, in which the emphasis is on the calculation of the properties of molecules and systems, and which uses the techniques of quantum mechanics and statistical thermodynamics. For the present purpose it is convenient to regard physical chemistry as dealing with three aspects of matter: its equilibrium properties, the structure of molecules and bulk matter, and matter's ability to change. *See* PHYSICAL ORGANIC CHEMISTRY; QUANTUM MECHANICS.

Chemical thermodynamics. The study of matter in a state of equilibrium constitutes the field of chemical thermodynamics. In particular, chemical thermodynamics provides a technique for discussing the response of a system to a change in the external conditions (such as the change in the boiling and freezing point of a substance when the applied pressure is changed, or a mixture when the composition is modified), and for rationalizing the energy changes that occur in the course of a chemical reaction. The branch of thermodynamics dealing with the latter is called thermochemistry. Chemical thermodynamics also provides a framework for the determination of the maximum work that may be generated by a system undergoing a specified change, and it therefore provides a way of establishing bounds for the efficiencies of a variety of devices, including engines, refrigerators, and electrochemical cells. Thermodynamics is used in chemistry to assess the position of equilibrium of a chemical reaction (that is, how far it will proceed before appearing to stop), and to determine what conditions are necessary in order to optimize the yield of a particular product. The branch of chemical thermodynamics dealing with ionic reactions occurring in the presence of electrodes constitutes the field of equilibrium electrochemistry. *See* THERMOCHEMISTRY.

Chemical thermodynamics is based on the laws of thermodynamics. In chemistry the most important thermodynamic properties are the enthalpy H and the Gibbs energy G (which is also called the free energy). The change in enthalpy may be identified as the heat transferred to a system during a specified change at constant pressure. Although absolute enthalpies cannot be measured, the enthalpies of compounds relative to their elements can be determined, and extensive tables of these quantities have been compiled. From them it is possible to predict the heat available (or required) for a particular reaction. Not the whole of the heat output of a reaction is available to do work (such as mechanical work, or the work to drive other reactions toward a particular desired product). The Gibbs energy expresses the maximum work, other than work of expansion, that a specified process may generate at constant pressure, and as such it may be used to assess whether one reaction may be used to drive another in an unnatural direction. For example, the assessment of the changes in the Gibbs energy that accompany biochemical reactions may be used to discuss the processes that occur in living cells, where the ingestion of food leads ultimately to growth, mechanical work, and nervous activity. *See* ENTHALPY.

The basis of the Gibbs energy is the tendency of energy to attain a condition of greatest dispersal. This tendency is expressed generally in terms of the entropy S of a system: as the dispersal of energy increases, so does the entropy. The concept of entropy may be given a sharp and quantitative definition, and entropies of substances have been determined (through measurements of heat capacity or by making use of spectroscopic data) and tabulated. The crucial feature for the present purpose, though, is that whereas the entropy of the universe increases whenever a spontaneous change occurs, a chemist is normally interested in the changes that occur in the system under investigation. The Gibbs energy focuses attention on changes in the properties of the system itself because, provided the pressure and temperature are constant, G automatically takes into account the changes in entropy of the surroundings. To determine whether a specified change has a natural tendency to occur at constant temperature and pressure, it is necessary only to assess whether that change is accompanied by a decrease in the Gibbs energy of the system. If a change is accompanied by an increase of Gibbs energy, it may still be achieved by coupling the system to another in which a change is occurring such that the overall change in the Gibbs energy of the combined system is negative. Because the Gibbs energy depends on the composition of the system, it is possible to use the Gibbs energy to predict the composition of the system when the system has attained equilibrium (such as when some reaction mixture attains some constant composition).

The crucial equation is $\ln K = \Delta_r G^{\ominus}/RT$, where K is the equilibrium constant of the reaction (a function of the concentration of the components at equilibrium), R the gas constant, T the absolute temperature, and $\Delta_r G^\ominus$, the standard reaction Gibbs energy (the change in the Gibbs energy under certain specified, standard conditions). This equation lies at the heart of chemical thermodynamics and is of great practical importance. *See* CHEMICAL EQUILIBRIUM; CHEMICAL THERMODYNAMICS; ENTROPY; FREE ENERGY.

Equilibrium electrochemistry. In equilibrium electrochemistry, attention is focused on systems in which a chemical reaction may release energy by driving electrons through an external circuit. The potential difference between the two electrodes immersed in the reaction mixture is related to the Gibbs energy for the reaction by $E = -\Delta_r G/F$, where F is the Faraday constant, the magnitude of the charge per mole of electrons (the product of the Avogadro constant and the fundamental charge; F = 96,485 coulombs mol^{-1}). The potential difference under these conditions is denoted E, and is called the electromotive force (emf) of the cell. One object of electrochemistry is to measure the emf of cells, and then to use the tabulated results to discuss the position of equilibria in reactions in aqueous solution (via ΔG) and their response to changes of conditions. Practical applications include chemical analysis, the assessment of the power generation and storage capabilities of electrochemical cells and fuel cells, the discussion of tendencies to corrosion, and the analysis of potential differences across biological membranes (such as those that are responsible for the propagation of nerve impulses). *See* ELECTROCHEMISTRY; ELECTROMOTIVE FORCE (EMF).

Quantum chemistry. The principal role of quantum mechanics in chemistry is in the discussion of atomic and molecular structure and in the interpretation of spectroscopic data. In the branch of physical chemistry known as computational quantum chemistry, interest centers on the numerical solution of the Schrödinger equation in order to obtain the wavefunctions of electrons in molecules and to predict and understand the shapes of molecules. In ab initio calculations the computations are done without any appeal to experimental data, and an attempt is made to predict properties from first principles (that is, from the masses and charges of the electrons and nuclei constituting the molecule). In semiempirical calculations (which are identified by initials such as AM1 and MINDO), some of the computational difficulties are circumvented (but with some loss of reliability) by incorporating experimental data and adjusting parameters so that certain experimental quantities (such as enthalpies of reaction) are reproduced. It is now possible to calculate the shape, electron distribution, and spectroscopic properties of large molecules. One application is referred to as quantum pharmacology, where a pharmacologically active molecule is first screened by calculating and analyzing its electron distribution. To some extent, this type of screening calculation eliminates lengthy and expensive laboratory trials. Computational quantum chemistry is so developed that it is capable of being used to map the changes in the structures of molecules while they are in the course of reaction, when atoms and groups of atoms are being transferred from one molecule to another. *See* COMPUTATIONAL CHEMISTRY; QUANTUM CHEMISTRY; SCHRÖDINGER'S WAVE EQUATION.

Spectroscopy. Spectroscopic techniques are used not only to identify molecules present in a sample but also to determine their shape, size, and electron distribution. The techniques fall into four categories as follows.

Absorption spectroscopy. In absorption spectroscopy, the basic observation is the intensity of radiation absorbed at different frequencies (or wavelengths). When the exciting frequency lies in the microwave region (wavelengths in the vicinity of 1 cm), the absorption is due to the excitation of rotational motion of the molecule. Hence the technique may be used to assess the moment of inertia, and therefore the shape and size of the molecule. When the exciting radiation lies in the infrared region (wavelengths in the region of 1000 nm), the vibrational modes of the molecule are excited, and hence the absorption spectrum gives information about the stiffness of bonds. Furthermore, because different groups of atoms absorb in characteristic regions, infrared spectroscopy is also used to identify groups of atoms present in a molecule, and therefore to identify the molecule (this application is called fingerprinting). Absorption in the visible and ultraviolet region of the spectrum indicates the occurrence of an electronic transition, that is, a shift of electron density from one region of the molecule to another. *See* INFRARED SPECTROSCOPY; MICROWAVE SPECTROSCOPY.

Electronic spectroscopy is used for identification, and provides a basis for the discussion of the processes involved in photochemistry (chemical reactions induced by the absorption of radiation), in fluorescence and phosphorescence, and in laser action. Sometimes the incident radiation is so energetic that its absorption results in the ejection of electrons from the molecule. The analysis of the energies of these electrons leads to information about the energy levels of electrons in the molecules under investigation; these photoelectron spectroscopy techniques are designated uv-PES when ultraviolet radiation expels outer electrons, and X-PES when x-rays are used to eject the more tightly bound electrons (formerly known as ESCA, standing for electron scattering for chemical analysis).

Emission spectroscopy. In emission spectroscopy, the spectrum of frequencies present in electromagnetic radiation emitted from excited atoms is monitored. Apart from its use to identify the composition of mixtures, emission spectroscopy is used to determine the state of molecules immediately after they have reacted, and hence gives valuable information about the processes occurring during reaction.

Raman spectroscopy. In Raman spectroscopy, electromagnetic radiation is scattered from molecules. In the process, some energy may be transferred from the molecule to the radiation, or vice versa, with the result that additional frequencies appear in the radiation emerging from the sample. Rotational vibration Raman spectra give information that complements that obtained from the corresponding absorption techniques. *See* RAMAN EFFECT.

Resonance techniques. These techniques constitute the fourth type of spectroscopy. They depend on excitation frequencies of the molecule being brought into resonance with the surrounding electromagnetic radiation by modifying the external conditions. In nuclear magnetic resonance (NMR), the principal example of this technique, the relevant energy levels are those arising from the different possible orientations of the magnetic moments of the nuclei in the molecules, and resonance occurs at radio frequencies. Nuclear magnetic resonance is a powerful technique for structural analysis and identification. Other resonance techniques include electron spin resonance (ESR or EPR) and Mössbauer spectroscopy. *See* ELECTRON PARAMAGNETIC RESONANCE (EPR) SPECTROSCOPY; ELECTRON SPECTROSCOPY; MOLECULAR STRUCTURE AND SPECTRA; MÖSSBAUER EFFECT; NUCLEAR MAGNETIC RESONANCE (NMR); PHOTOCHEMISTRY; SPECTROSCOPY.

Diffraction techniques. Other major techniques for the investigation of molecular structure are based on diffraction, the interference between waves (of electromagnetic radiation) leading to patterns of constructive and destructive interference. These techniques depend on the observation of the direction through which radiation and particles are scattered when they impinge on a sample. The principal example of these techniques is x-ray diffraction, for through it detailed information may be obtained regarding the arrangement of atoms in crystals and very complex, large, biologically important molecules, such as proteins and deoxyribonucleic acid (DNA). Also, there are diffraction techniques based on electrons and neutrons. *See* DEOXYRIBONUCLEIC ACID (DNA); PROTEIN; X-RAY DIFFRACTION.

Other techniques. Other techniques for investigating structure include the electric and magnetic properties of molecules, in particular, the determination of electric polarizabilities and dipole moments (which give information about electron distribution, and are important for the discussion of the dielectric properties), magnetic properties, and the properties based on optical birefringence, such as optical activity and the Faraday effect. In the case of macromolecules and colloids, important techniques used for structural analysis include x-ray diffraction, osmometry, mass spectrometry, and viscosity measurements. *See* DIPOLE MOMENT; FARADAY EFFECT; MASS SPECTROMETRY; OPTICAL ACTIVITY; OSMOSIS; VISCOSITY.

Statistical thermodynamics. Structural properties and thermodynamic properties are brought together by statistical thermodynamics. This major theoretical procedure gives a way of predicting the thermodynamic properties of substances in terms of the properties of their individual molecules. Statistical thermodynamics represents a grand synthesis of the two major aspects of physical chemistry. It provides an understanding of the bulk, macroscopic properties of molecules and is an important technique for calculating otherwise inaccessible properties from readily available spectroscopic data.

Transport processes. The third major branch of physical chemistry is concerned with physical and chemical change. In particular, it is concerned with the rate of change (whereas thermodynamics is concerned with the possibility of change, and the criteria of spontaneous change). Physical change includes the diffusion of one substance into another, or the migration of ions in an electrolyte solution. The simplest version of the former are the transport properties of gases. These include thermal conductivity and viscosity. They are treated most simply in terms of the kinetic theory of gases, in which a gas is regarded as a swarm of noninteracting points; modern physical chemistry is concerned with the role of intermolecular forces in determining the transport properties. Ion migration gives rise to questions of the mobility of ions in a variety of solvents, and to the description of diffusion processes in general. The application of thermodynamics to change in general constitutes the field of nonequilibrium thermodynamics. *See* DIFFUSION; GAS; TRANSPORT PROCESSES.

Chemical kinetics. The major aspect of chemistry is change, and change is therefore also a major aspect of physical chemistry. Chemical change may be studied at a variety of levels. Empirical chemical kinetics is the study of reactions in order to determine how their rates depend on the concentrations of the participants in the reaction and on the conditions (mainly the temperature). Investigation of the time dependence of reactions (a time dependence that can be observed from the order of minutes to femtoseconds) yields a detailed picture of the sequence of molecular transformations involved in a complex chemical reaction. Each step in the sequence depends on the concentration of the participants and an empirical constant referred to as the rate constant. The aim of theoretical chemistry is to account for the value of the rate constant and its dependence on the temperature. The earliest broadly successful approach led to the Arrhenius law, which is still used as a general guide to the way that the temperature affects the rate of a reaction. The law asserts that the rate constant k varies with temperature as $k = A\ \exp(-E_a/RT)$, where A and E_a are empirical parameters, the latter being known as the activation energy of the reaction. A wide variety of simple reactions conform to this law, and it can be interpreted on the basis that only molecules possessing sufficient energy are able to undergo reaction when they encounter other species; the exponential factor indicates the fraction of collisions that satisfy the energy requirement. *See* CHEMICAL DYNAMICS; SHOCK TUBE; ULTRAFAST MOLECULAR PROCESSES.

Molecular reaction dynamics. Modern approaches to the prediction and explanation of observed reaction rates have been based either on a statistical view of the reaction (the activated complex theory or transition state theory), where the principles of statistical thermodynamics are applied to a system evolving with time; or on a fundamental molecular dynamics approach, where the trajectories of molecules are calculated as they undergo reaction—techniques then being used to convert these trajectories to values of the rate coefficients. The latter approach constitutes the field of molecular reaction dynamics. Experimental techniques have also been developed for observing individual molecular encounters and reactions. These are based on molecular beams, where a diffuse beam of one reactant is directed into the path of another, and the pattern of molecular scattering, including the electronic and vibrational states of the products, is interpreted in terms of the forces acting between the reactants during the reactive encounter. These techniques bring investigations closer to an atomistic and molecular interpretation of chemistry than anything that preceded them, but the relation of trajectories and rate coefficients and the relation of gas-phase events to those in solution remain problems. *See* MOLECULAR BEAMS.

Surface chemistry. Chemical kinetics is so important that it is rich in applications and extensions. An important extension is to the reactions that occur on surfaces; these are the processes involved in heterogeneous catalysis. The study of surface chemistry breaks down into the analysis of the steps that lead to the affixation of the species to the surface (that is, the study of adsorption processes), the determination of the structure of the adsorbed species, and finally the reactions, and escape, of the adsorbed species. The entire subject, including the related processes occurring at liquid surfaces, constitutes the field of surface chemistry. This field has been transformed by the introduction of scanning tunneling microscopy, which gives a detailed portrayal of the surface on an atomic scale. A special application of surface chemistry is to the stability of colloidal suspensions of species in fluids, and another is to the processes that occur at the interface between an electrode and the solution in which it is immersed. Electrode reactions are governed largely by the rate of electron transfer between the ions in the solution and the electrode. The field of dynamical electrochemistry finds important applications in electrochemical power generation and storage, in corrosion, in electrodeposition, and in electrocatalysis. *See* ADSORPTION; COLLOID; ELECTRON MICROSCOPE; HETEROGENEOUS CATALYSIS; SCANNING TUNNELING MICROSCOPE; SURFACE AND INTERFACIAL CHEMISTRY; SURFACE PHYSICS. P. W. Atkins

Bibliography. P. W. Atkins, *Concepts in Physical Chemistry*, 1995; P. W. Atkins, *Physical Chemistry*, 6th ed., 1998; D. A. McQuarrie and J. D. Simon, *Physical Chemistry: A Molecular Approach*, 1997; I. Prigogine and S. A. Rice (eds.), *Advances in Chemical Physics*, ongoing series.

Physical geography

The study of the Earth's surface features and associated processes. Physical geography aims to explain the geographic patterns of climate, vegetation, soils, hydrology, and landforms, and the physical environments that result from their interactions. Physical geography merges with human geography to provide a synthesis of the complex interactions between nature and society.

The roots of physical geography may be traced to classical Greek, Roman, and Arab scholars who provided basic insights into the many facets of mathematical geography, the study pertaining to the configuration, measurement, and motion of the Earth. Observations from the great voyages and explorations in the fifteenth to eighteenth centuries afforded additional, sometimes fanciful, descriptive information about the Earth. However, most explanations of the Earth remained for geographers of the nineteenth century, as in the rigorous and methodical work of Alexander von Humboldt (1769–1859) in South America and the perceptive exploration of the American west by John Wesley Powell (1834–1902).

By the beginning of the twentieth century the basic content of physical geography was well established and comprised a number of areas of specialization. Climatology, the scientific study of climates, concerns the total complex of weather conditions at a given location over an extended time period; it deals not only with average conditions but with extremes and variations. Geomorphology is the interpretive description and explanation of landforms and the fluvial, glacial, coastal, and eolian process that operate on them. The forms, processes, and patterns within the biosphere, including vegetation and animal distributions, are studied as biogeography. With strong ties to fluvial geomorphology, geographic hydrology concerns the scientific study of water from the aspects of distribution, movement, and utilization. Soil geography, with emphasis on the origin, characteristics, classification, and utilization potential of soils, provides an area of specialization with links to land use. Ultimately, the physical geography of a region is understood through an integration of the multiple aspects.

The methods used to derive and present the findings of both specialized and integrative studies changed appreciably during the twentieth century. Initially, emphasis was on the organization of newly acquired information about the Earth's surface. Simple models of landscape evolution (for example, William Morris Davis's geomorphic cycle), procedural classification of climate (such as that formulated by Vladimir Köppen) and soils (for example, Curtis F. Marbut's great soil groups), and regional descriptions were typical areas of research and study. New methods and interpretive systems were introduced around the mid-1950s.

Of considerable importance was the change of methodological focus from form (describing the distribution of phenomena) to studies of processes that created the geographic patterns. This change was aided by the use of deductive scientific enquiry employing mathematical and statistical tools. The advent of computer technology enabled large spatial data sets to be analyzed and physical and stochastic models to be constructed. Significant research on a variety of topics, from the dynamics of climates to quantitative models of stream flow, and from systems science thinking to analysis of biogeochemical cycles, reflect the emphasis on physical geographic process.

Maps have always been important tools of the physical geographer. Maps, used or derived, range in scale from those providing intimate details of hill slope erosion to world maps of vegetation biomass, from explorer maps of landscapes to global wind systems. Modern maps have been greatly enhanced by technological advances, especially the development of remote sensing and geographic information systems (GIS).

Remote sensing, the acquisition of data from a distant location, was for many years based upon aerial photography. The development of nonphotographic imagery provides the physical geographer with data and new analytic techniques. Earth-observing satellites, such as NASA's *Landsat* series and the French *SPOT* series, permit, for example, analysis of sequential spatial changes in remote global regions; maps of global cloud cover and sea surface temperatures are examples. Remotely sensed and other derived data are used to construct a geographic information system, an assemblage of computer software and hardware that enables spatial data to be manipulated and displayed. Typically, the system uses a variety of spatial sets to be overlain in order to produce a map composite.

While some physical geographers still complete research dealing with broad theoretical problems on such subjects as chaos or systems theory, the majority of practitioners deal with more specialized topics. Although physical geographers may study aspects of hydroclimatology or glacial geomorphology, they retain the inherent spatial, process-oriented context associated with the discipline. Underlying this philosophy are a number of basic tenets. First, the linkages between the patterns and processes that operate at different scales, from the microscale to the global level, need to be defined and explained. Second, the careful accumulations of field data and the remotely sensed data need to be reconciled. Third, application of basic research to environmentally sensitive issues is necessary to produce a better understanding of nature-society interrelationships. The integrative perspective of geography is at the basis of such endeavors.

Physical geography is of considerable importance in dealing with complex environmental issues. Applied physical geographers use their expertise to complete research on a variety of topics. Contributions toward understanding the impacts of global warming, flood and drought analysis, coastal erosion, and loss of biodiversity typify the research.

Further considerations of physical geography appear in topical and regional articles. For principal topical discussions *See* AERIAL PHOTOGRAPH; CARTOGRAPHY; CLIMATOLOGY; CONTINENT; EARTH RESOURCE PATTERNS; ECOLOGY; GEOGRAPHIC INFORMATION SYSTEMS; GEOLOGY; GEOMORPHOLOGY; HILL AND MOUNTAIN TERRAIN; HYDROLOGY; MATHEMATICAL GEOGRAPHY; METEOROLOGY; PHYSICAL GEOGRAPHY; PLAINS; TERRAIN AREAS.

For physical geography of the continents and major island areas *see* AFRICA; ANTARCTICA; ARCTIC AND SUBARCTIC ISLANDS; ASIA; AUSTRALIA; EAST INDIES; EUROPE; NEW ZEALAND; NORTH AMERICA; PACIFIC ISLANDS; SOUTH AMERICA; WEST INDIES.

John E. Oliver

Bibliography. M. N. DeMers, *Fundamentals of Geographic Information Systems*, John Wiley, 1997; A. Goudie (ed.), *The Encyclopedic Dictionary of Physical Geography*, 2d ed., Blackwell, 1994; K. J. Gregory, *The Nature of Physical Geography*, Edward Arnold, 1985; A. N. Strahler and A. Strahler, *Introducing Physical Geography*, John Wiley, 1998.

Physical law

A highly ambiguous term that designates four different concepts: (1) objective pattern (or natural regularity), (2) formula purporting to represent an objective pattern, (3) law-based rule (or uniform procedure), and (4) principle concerning any of the preceding. To avoid confusion, these concepts will be called laws of types 1, 2, 3, and 4, respectively.

Four types of laws. For example, Newton's second law of motion, $ma = F$, is a law of type 2. It represents, to a good approximation, the actual behavior (law of type 1) of medium-size particles moving slowly relative to the speed of light. Alternative laws of motion, such as the relativistic and quantum-mechanical ones, are different laws of type 2 representing the same objective pattern or law of type 1 to even better approximations. One of the rules (laws of type 3) associated with Newton's second law of motion is: In order to set in motion a stationary particle, exert a force on it. Another is: In order to stop a moving particle, exert on it a force in the opposite direction. An example of a law of type 4 is: Newton's laws of motion are invariant under a Galileo transformation. *See* GALILEAN TRANSFORMATIONS; NEWTON'S LAWS OF MOTION.

A physical law of type 1, or objective pattern, is a constant relation among two or more properties of a physical entity. In principle, any such pattern can be conceptualized in different ways, that is, as alternative laws of type 2. The history of theoretical physics is to a large extent a sequence of laws of type 2. Every one of these is hoped to constitute a more accurate representation of the corresponding objective pattern or law of type 1, which is assumed to be constant and, in particular, untouched by human efforts to grasp it. Likewise, the history of engineering is to some extent a sequence of laws of type 3, or law-based rules of action, of which there are least two for every law of type 2. As for the laws of type 4, or laws of laws, they are of two kinds: scientific and philosophical. The general covariance principle is of the first kind, whereas the hypothesis that all events are lawful is a philosophical thesis. Unlike the former, whose truth can be checked with pencil and paper, the principle of lawfulness is irrefutable, though extremely fertile, for it encourages the search for pattern. *See* ENGINEERING; THEORETICAL PHYSICS.

Basic and derived laws. Not all formulas are called physical laws. For example, the regularities found by curve fitting are called empirical formulas. In physics a formula is called a law if and only if it meets the following conditions: it is part of a theory (hypothetico-deductive system), and it has been satisfactorily confirmed by measurement or experiment at least within a certain domain (for example, for small mass densities or high field intensities). Thus, the basic assumptions of all the standard physical theories are laws, and so are their logical consequences. In particular, the usual variational principles, such as Hamilton's, are basic laws. However, the equations of motion and field equations entailed by such principles are derived laws (theorems); so are the conservation laws entailed by the equations of motion and field equations. However, the distinction between basic and derived laws is contextual: what is a principle in one theory may be a theorem in another. For example, Newton's second law of motion is a theorem in analytical dynamics, and the first principle of thermodynamics is a theorem of statistical mechanics. *See* CONSERVATION LAWS (PHYSICS); CURVE FITTING; HAMILTON'S PRINCIPLE; PHYSICAL THEORY; SCIENTIFIC METHODS; STATISTICAL MECHANICS; THERMODYNAMIC PRINCIPLES; VARIATIONAL METHODS (PHYSICS).

Mario Bunge

Bibliography. M. Bunge, *Philosophy of Physics*, 1973; M. Bunge, *Scientific Research*, 1967; L. Sklar, *Philosophy of Physics*, 1992.

Physical measurement

Quantitative information on physical conditions, properties, or relations essential for coordination of activities, efficiency of communication, and understanding of the nature of things in science and engineering and in much of everyday life. Time, distance, mass, temperature, force, power, and all other physical quantities (or parameters or variables), as well as the properties of matter, materials, and devices, must be described and measured in terms which have the same meaning for everyone. The measuring device or instrument is calibrated (that is, the functional relationship between its indication and the magnitude of the measured quantity is determined) by direct or indirect comparison with a standard which embodies, possesses, or generates a fixed or reproducible magnitude of the physical quantity which is taken as the unit or some multiple or fraction of the unit.

Any measured quantity may thus be expressed by a number (the magnitude ratio) and the name of the unit, for example, a length of 1.54 meters. The general area of scientific activity relating to standards and units and the accuracy of measurement is called metrology. *See* UNITS OF MEASUREMENT.

Units and Standards of Measurement

From earliest history, nations have had standards for length, volume, and mass. These have differed from country to country and from time to time, so that a large number of units for mass, length, volume, and area came to be in widespread use by the eighteenth century, some by the same name being of different size in different areas.

Metric system. The basic unit of length in the decimal metric system was defined as one ten-millionth of the Earth's polar quadrant (as determined from latitude surveys), and is termed the meter. The basic unit for mass was defined as the mass of a cubic decimeter of water, to be called the kilogram. The English-speaking countries have adopted common definitions for the inch as 2.54 centimeters exactly, or 1 yard = 0.9144 meter, and for the pound as 0.453 592 37 kilogram. However, the United States gallon is only about five-sixths of the imperial gallon, and other differences remain between United States and imperial measures of capacity.

The United States has adopted the Metric Conversion Act, declaring that "the policy of the U.S. shall be to coordinate and plan the increasing use of the metric system in the United States," and established the U.S. Metric Board "to coordinate the voluntary conversion to the metric system." This board has actively promoted metric education, not only in schools but in industry and everyday life. In accordance with the voluntary nature of the programs, the United States may expect many years of mixed units; for example, football fields will probably remain 100 yards in length; track and field events will adapt to international custom. Many technical and some nontechnical publications now give measurement data in both customary and metric units.

However, English units have become almost universal in some worldwide industries—for example,

TABLE 1. SI base and supplementary units

Quantity*	Unit name	Symbol
SI base units		
Length	meter	m
Mass	kilogram	kg
Time	second	s
Electric current	ampere	A
Thermodynamic temperature	kelvin	K
Amount of substance	mole	mol
Luminous intensity	candela	cd
SI supplementary units		
Plane angle	radian	rad
Solid angle	steradian	sr

*Quantity here and in Tables 2, 3, 4, and 7 means a measurable attribute.

TABLE 2. SI derived units with special names

Quantity	Unit name	Symbol	Expression in terms of other units	Expression in terms of SI base units
Frequency	hertz	Hz		s^{-1}
Force	newton	N		$m \cdot kg \cdot s^{-2}$
Pressure, stress	pascal	Pa	N/m^2	$m^{-1} \cdot kg \cdot s^{-2}$
Energy, work, quantity of heat	joule	J	$N \cdot m$	$m^2 \cdot kg \cdot s^{-2}$
Power, radiant flux	watt	W	J/s	$m^2 \cdot kg \cdot s^{-3}$
Quantity of electricity, electric charge	coulomb	C	$A \cdot s$	$s \cdot A$
Electric potential, potential difference, electromotive force	volt	V	W/A	$m^2 \cdot kg \cdot s^{-3} \cdot A^{-1}$
Capacitance	farad	F	C/V	$m^{-2} \cdot kg^{-1} \cdot s^4 \cdot A^2$
Electric resistance	ohm	Ω	V/A	$m^2 \cdot kg \cdot s^{-3} \cdot A^{-2}$
Conductance	siemens	S	A/V	$m^{-2} \cdot kg^{-1} \cdot s^3 \cdot A^2$
Magnetic flux	weber	Wb	$V \cdot s$	$m^2 \cdot kg \cdot s^{-2} \cdot A^{-1}$
Magnetic flux density	tesla	T	Wb/m^2	$kg \cdot s^{-2} \cdot A^{-1}$
Inductance	henry	H	Wb/A	$m^2 \cdot kg \cdot s^{-2} \cdot A^{-2}$
Celsius temperature	degree Celsius	°C		K
Luminous flux	lumen	lm		$cd \cdot sr$*
Illuminance	lux	lx	lm/m^2	$m^{-2} \cdot cd \cdot sr$*
Activity (of a radionuclide)	becquerel	Bq		s^{-1}
Absorbed dose, specific energy imparted, kerma, absorbed dose index	gray	Gy	J/kg	$m^2 \cdot s^{-2}$
Dose equivalent, dose equivalent index	sievert	Sv	J/kg	$m^2 \cdot s^{-2}$

*In this expression the steradian (sr) is treated as a base unit.

dimensions of oil-drilling equipment, or altitude measurement in aviation. Product sizes can of course remain the same if economic considerations make it desirable; they can be expressed easily in metric units. The widespread practice in aviation flight control is to separate aircraft flying in various directions by 500, 1000, or 2000 feet in altitude. Altimeter pointers, making one revolution for each 1000 feet, provide a conveniently readable index. To use a metric separation basis (300 meters, or even 500 meters) would result in reading inconvenience and some loss of flight levels or of flight safety. Also, by international agreement, through the International Civil Aviation Organization, the unit for speed in air navigation (as in marine navigation) is the knot, defined as 1 nautical mile (1 minute of arc of the Earth's surface) per hour. In view of existing practice, the International Committee on Weights and Measures (CIPM) has accepted the nautical mile and the knot with a number of other units to be used temporarily with the International System. Thus it is likely that there will always be exceptions to uniformity, requiring special knowledge of special units for at least some people even as the whole world "goes metric" in principle.

International System of Units (SI). At present the International System of Units (abbreviated SI, from the French Système International d'Unités) is constructed from seven base units for independent quantities plus two supplementary units for plane angle and solid angle (**Table 1**). Units for all other quantities are derived from these nine units. In **Table 2** are listed 19 SI derived units with special names. These units are derived from the base and supplementary units in a coherent manner, which means they are expressed as products and quotients of the nine base and supplementary units without numerical factors. All other SI derived units, such as those in **Tables 3** and **4**, are similarly derived in a coherent manner from the 28 base, supplementary, and special-name SI units. For use with the SI units, there is a set of 16 prefixes (**Table 5**) to form multiples and submultiples of these units. For mass, the prefixes are to be applied to the gram instead of to the SI unit, the kilogram. *See* DIMENSIONAL ANALYSIS.

TABLE 3. Some SI derived units expressed in terms of base unit

Quantity	Name	Symbol
Area	square meter	m^2
Volume	cubic meter	m^3
Speed, velocity	meter per second	m/s
Acceleration	meter per second squared	m/s^2
Wave number	1 per meter	m^{-1}
Density, mass density	kilogram per cubic meter	kg/m^3
Current density	ampere per square meter	A/m^2
Magnetic field strength	ampere per meter	A/m
Concentration (of amount of substance)	mole per cubic meter	mol/m^3
Specific volume	cubic meter per kilogram	m^3/kg
Luminance	candela per square meter	cd/m^2

TABLE 4. Some SI derived units expressed by means of special names

Quantity	Name	Symbol	Expression in terms of SI base units
Dynamic viscosity	pascal second	Pa · s	$m^{-1} \cdot kg \cdot s^{-1}$
Moment of force	newton meter	N · m	$m^2 \cdot kg \cdot s^{-2}$
Surface tension	newton per meter	N/m	$kg \cdot s^{-2}$
Power density, heat flux density, irradiance	watt per square meter	W/m^2	$kg \cdot s^{-3}$
Heat capacity, entropy	joule per kelvin	J/K	$m^2 \cdot kg \cdot s^{-2} \cdot K^{-1}$
Specific heat capacity, specific entropy	joule per kilogram kelvin	J/(kg · K)	$m^2 \cdot s^{-2} \cdot K^{-1}$
Specific energy	joule per kilogram	J/kg	$m^2 \cdot s^{-2}$
Thermal conductivity	watt per meter kelvin	W/(m · K)	$m \cdot kg \cdot s^{-3} \cdot K^{-1}$
Energy density	joule per cubic meter	J/m^3	$m^{-1} \cdot kg \cdot s^{-2}$
Electric field strength	volt per meter	V/m	$m \cdot kg\ s^{-3}\ A^{-1}$
Electric charge density	coulomb per cubic meter	C/m^3	$m^{-3} \cdot s \cdot A$
Electric flux density	coulomb per square meter	C/m^2	$m^{-2} \cdot s \cdot A$
Permittivity	farad per meter	F/m	$m^{-3} \cdot kg^{-1} \cdot s^4 \cdot A^2$
Permeability	henry per meter	H/m	$m \cdot kg \cdot s^{-2} \cdot A^{-2}$
Molar energy	joule per mole	J/mol	$m^2 \cdot kg \cdot s^{-2} \cdot mol^{-1}$
Molar entropy, molar heat capacity	joule per mole kelvin	J/(mol · K)	$m^2 \cdot kg \cdot s^{-2} \cdot K^{-1} \cdot mol^{-1}$
Exposure (x- and γ-rays)	coulomb per kilogram	C/kg	$kg^{-1} \cdot s \cdot A$
Absorbed dose rate	gray per second	Gy/s	$m^2 \cdot s^{-3}$

TABLE 5. SI prefixes

Factor	Prefix	Symbol	Factor	Prefix	Symbol
10^{24}	yotta	Y	10^{-1}	deci	d
10^{21}	zetta	Z	10^{-2}	centi	c
10^{18}	exa	E	10^{-3}	milli	m
10^{15}	peta	P	10^{-6}	micro	μ
10^{12}	tera	T	10^{-9}	nano	n
10^{9}	giga	G	10^{-12}	pico	p
10^{6}	mega	M	10^{-15}	femto	f
10^{3}	kilo	k	10^{-18}	atto	a
10^{2}	hecto	h	10^{-21}	zepto	z
10^{1}	deka	da	10^{-24}	yocto	y

The SI units together with the SI prefixes provide a logical and interconnected framework for measurements in science, industry, and commerce.

Many other physical quantities less commonly used are not included in the SI tables, for example, further time and space derivatives. Likewise, many properties of matter or materials are not listed. Units for all of them can, of course, be expressed as a function of the base units or other derived units.

Natural units. In some cases, quantities are commonly expressed in terms of fundamental constants of nature, and use of these constants or "natural units" is acceptable. *See* FUNDAMENTAL CONSTANTS.

Typical examples of natural units, with their symbols, are:

elementary charge	e
electron mass	m_e
proton mass	m_p
Bohr radius	a_0
electron radius	r_e
Compton wavelength of electron	λ_c
Bohr magneton	μ_B
nuclear magneton	μ_N
speed of light	c
Planck constant	h

Units acceptable for use with SI. Certain units which are not part of the SI are used so widely that it is impractical to abandon them. The units that are accepted for continued use with the International System are listed in **Table 6**. It is likewise necessary to recognize, outside the International System, the following units which are used in specialized fields:

electronvolt	eV
unified atomic mass unit	u
astronomical unit	AU
parsec	pc

Logarithmic measures such as pH, dB (decibel), and Np (neper) are acceptable. *See* ASTRONOMICAL UNIT; ATOMIC MASS UNIT; DECIBEL; ELECTRONVOLT; NEPER; PARSEC; PH.

The units shown with an asterisk in **Table 7** are used in limited fields and have been authorized by the CIPM. It is recommended that the term weight be avoided in technical publications except under circumstances in which its meaning is completely clear. It is also recommended that the terms atomic weight and molecular weight be replaced by relative atomic mass and relative molecular mass in accordance with established international practice. *See* ATOMIC MASS;

TABLE 6. Units in use with the International System

Name	Symbol	Value in SI unit
Minute	min	1 min = 60 s
Hour	h	1 h = 60 min = 3600 s
Day	d	1 d = 24 h = 86,400 s
Degree	°	1° = (π/180) rad
Minute	′	1′ = (1/60)° = (π/10,800) rad
Second	″	1″ = (1/60)′ = (π/648,000) rad
Liter	L*	1 L = 1 dm^3 = 10^{-3} m^3
Metric ton	t	1 t = 10^3 kg
Hectare	ha	1 ha = 10^4 m^2

*An alternate symbol for liter is "l." Since "l" can be easily confused with the numeral 1, the symbol "L" is recommended for United States use.

TABLE 7. Examples of conversion factors from non-SI units to SI*

Quantity	Unit name	Symbol for unit	Definition in SI units
Length	inch	in.	2.54×10^{-2} m
Length	nautical mile*	nmi	1852 m
Length	angstrom	Å	10^{-10} m
Velocity	knot*	kn	(1852/3600) m/s
Cross section	barn*	b	10^{-28} m^2
Acceleration	miles per hour	mph	10^{-2} m/s^2
Mass	pound (avoirdupois)	lb	0.45359237 kg
Force	kilogram-force	kgf	9.80665 N
Pressure	millimeter of mercury at 0°C	mmHg	133.322 Pa†
Pressure	atmosphere	atm	101,325 Pa
Pressure	torr	torr	(101,325/760) Pa
Pressure	bar*	bar	10^5 Pa
Stress	pound-force per square inch	lbf/in.^{2}	6894.757 Pa†
Energy	British thermal unit (International Table)	Btu	1055.056 J†
Energy	kilowatt-hour	kWh	3.6×10^6 J
Energy	calorie (thermochemical)	cal	4.184 J
Activity (of a radionuclide)	curie*	Ci	3.7×10^{10} Bq
Exposure (x- or γ-rays)	roentgen*	R	2.58×10^{-4} C · kg^{-1}
Absorbed dose	rad*	rd	1×10^{-2} Gy
Dose equivalent	rem*	rem	1×10^{-2} Sv

*The Committee for Weights and Measures has sanctioned the temporary use of these units.
†Approximate; all other conversion factors are exact.

MOLECULAR WEIGHT; RELATIVE ATOMIC MASS; RELATIVE MOLECULAR MASS.

The internationally accepted definitions for the seven base units are given below, with brief descriptions of the standards in use for the most precise measurements and calibrations in terms of the units so defined.

Mass. The kilogram (kg) is equal to the mass of the International Prototype Kilogram. The International Prototype is a platinum-iridium cylinder preserved at the International Bureau of Weights and Measures at Sèvres, France.

Prototype no. 20 is kept at the U.S. National Institute of Standards and Technology; equivalent prototypes are kept by other countries. Mass is the only one of the base quantities for which the standard is an arbitrarily defined object. No basic property of matter involving mass can be measured with more precision than is possible in comparing kilogram masses by weighing, about 1 part in 10^8.

The standard prototype kilogram embodies the unit of mass; since masses may be compared by weighing, other mass standards are easily adjusted for equality with the unit, within the uncertainty set by the reproducibility of the weighing equipment and the weighing procedure. Sets of standard masses (or weights) are obtained by adding or subdividing unit masses. Within practical ranges (say, 10^{-9} to 10^4 kilograms) any given mass can be measured by weighing it against combinations of standard masses, with an uncertainty equal to the smallest standard used in the weighing. Most weighing balances have additional means for indicating the mass required to remove the remaining imbalance. *See* BALANCE; MASS; WEIGHT MEASUREMENT. William A. Wildhack

Length. The meter is defined in terms of time and the speed of light: "The meter is the length of the path traveled by light in a vacuum during a time interval of 1/299 792 458 of a second." This definition defines the speed of light to be exactly 299 792 458 m/s and defines the meter in terms of the most accurately known quantity, the second. *See* LIGHT.

The most accurate method of realizing the meter is by means of an interferometrically measured distance by fringe counting in which each vacuum fringe is a half wavelength from the next one. This wavelength, λ, is obtained from the measured frequency, f, using the relation $\lambda = c/f$, where c is the value of the speed of light in vacuum. One meter is exactly $2/\lambda$ fringes in length. In order that diffraction effects be minimized, the use of visible radiation is dictated. To this end, major standards laboratories have measured the frequencies of several lasers stabilized to narrow molecular absorptions in the visible and near-infrared spectral regions. These stabilized lasers now serve as standards of length. Also, the old length standard, the krypton-86 lamp, may still be used, but less accuracy results. **Table 8** lists the stabilized lasers that have been recommended for use in realizing the meter. Blocks, bars, screws, and tapes are calibrated by direct comparison with the radiation of these special lasers by means of optical interferometers. *See* INTERFEROMETRY; WAVELENGTH STANDARDS.

Using the time of flight method of measurement, as

TABLE 8. Stabilized lasers used in realizing the meter

Absorbing molecule	Transition	Component	Wavelength, femtometers
CH_4	ν_3, P(7)	$F_2^{(2)}$	3 392 231 397.0
$^{127}I_2$	17–1, P(62)	O	576 294 760.27
$^{127}I_2$	11–5, R(127)	i	632 991 398.1
$^{127}I_2$	9–2, R(47)	o	611 970 769.8
$^{127}I_2$	43–0, P(13)	a_3	514 673 466.2

the new definition suggests, interplanetary distances (of the order of 10^{11} meters) can be very accurately measured. For example, the distance from Earth to Mars can be measured with an accuracy of a few meters. *See* LENGTH. Donald A. Jennings

Time interval. The second (s) is the duration of 9 192 631 770 periods of the radiation corresponding to the transition between the two hyperfine levels of the ground state of the cesium-133 atom.

Frequency (s^{-1}) is the reciprocal of the period for regularly repetitive events; the unit of frequency has been named the hertz (Hz).

As these definitions imply, a time standard (clock) is the combination of a constant-frequency generator and a period (or cycle) counter.

Oversimplified, the cesium frequency generator involves a beam of cesium atoms (evaporating off a heated bit of cesium metal through collimating holes) passing through a nonhomogeneous magnetic field which separates the atoms as to their energy states, a variable-frequency electromagnetic field stimulating the transition, and a detector of the degree of stimulated transition. An automatic feedback control system varies the frequency of the applied radio-frequency field to achieve maximum effect (resonance), which occurs when the applied frequency is equal to the natural frequency of the transition. In the best equipments the stability and accuracy correspond to an uncertainty of 1 in 10^{12} or even 1 in 10^{13}.

Using refinements of conventional techniques, the cesium-stabilized frequency is divided down to provide other standard frequencies, some of which are made widely available by radio broadcast.

Time intervals are obtained by counting periods of any of the signals of known frequency.

There are other standards besides the cesium beam, among them the hydrogen maser, rubidium clocks, and quartz frequency standards and clocks. Their frequency is controlled by comparison with a cesium standard, either directly, or by means of radio transmissions.

The second was long defined, for physical measurements as well as for civil affairs, as 1/86,400 of the time required for an average complete rotation of the Earth on its axis with respect to the Sun. Because of the slight slowing of the Earth's rotation rate, now averaging about 1 second per year (that is, 3 parts in 10^8) but with erratic and unexplained fluctuations, the universal second thus defined is not a constant. A time scale called Coordinated Universal Time (UTC) recommended by the General Conference of Weights and Measures (CGPM) in 1975 is defined in such a manner that it differs from international atomic time (TAI) by an exact whole number of seconds. This difference is adjusted occasionally by the use of a positive or negative leap second at the end of certain months to keep UTC in agreement with the time defined by the rotation of the Earth with an approximation better than $^9/_{10}$ second. The legal times of most countries are further offset by a whole number of hours (time zones and "summer time"). *See* ATOMIC CLOCK; ATOMIC TIME; DYNAMICAL TIME; EARTH ROTATION AND ORBITAL MOTION; FREQUENCY MEASUREMENT; TIME. William A. Wildhack

Temperature. The kelvin (K), the unit of thermodynamic temperature, is the fraction 1/273.16 of the thermodynamic temperature of the triple point of water.

The unit kelvin and its symbol K should also be used to express an interval or differences of temperature.

Absolute zero is defined as the condition in which all kinetic energy of random motion has been abstracted from the atoms or molecules. While this condition can be approached very closely, it cannot be attained. Its attainment would, in fact, be counter to the laws of thermodynamics, since heat can be abstracted only in an environment of lower temperature. The problems of attaining or measuring extremely low temperatures are both complicated and facilitated by the quantum nature of the energy of internal motions of the atoms or molecules (spins, vibrations, or rotations), as well as the inherent zero-point energy of particles in close proximity to others. *See* ABSOLUTE ZERO; CRYOGENICS; LOW-TEMPERATURE PHYSICS; LOW-TEMPERATURE THERMOMETRY; THERMODYNAMIC PRINCIPLES.

There are a number of physical phenomena that in theory relate thermodynamic temperature to other quantities which can be measured quite accurately, even if subject to many corrections; for example, pressure and volume of gases, specific heats of elements, electrical noise (voltage) in resistors, radiation intensity and spectral distribution, and speed of sound waves in gases. These and others are used in various experiments to determine the numerical values of temperatures below and above the triple point of water, that is, to establish the thermodynamic temperature scale. None of these phenomena, however, provides convenient methods for practical measurements over the entire range of achievable temperatures from 10^{-6} to 10^6 K. *See* TEMPERATURE MEASUREMENT; THERMOMETER.

To provide convenient and adequately accurate means for practical realization and measurement of temperature, the International Temperature Scale is used, based on the assigned values of the temperatures of a number of reproducible equilibrium states (defining fixed points), on standard instruments calibrated at those temperatures, and on vapor-pressure temperature relationships. Interpolation between the fixed-point temperatures is provided by formulas used to establish the relation between indications of the standard instruments and values of International Temperature. An extensive revision, which came into effect in 1990, is called the ITS-90.

The defining fixed points are established by realizing specified equilibrium states of phases of pure substances; vapor pressures of helium-3 and helium-4; the triple point or boiling points (under specified pressures) of hydrogen; triple points of neon, oxygen, argon, mercury, and water; the melting point of gallium; and freezing points of indium, tin, zinc, aluminum, silver, gold, and copper. These fixed points are distributed over the range from 3.0 to 1357.77 K (−270.15 to 1084.62°C). *See* TRIPLE POINT.

The standard instrument used from 3.0 to 24.5561 K (−270.15 to −248.5939°C) is the interpolating constant-volume gas thermometer. *See* GAS THERMOMETRY.

The standard instrument used from 13.8033 to 1234.93 K (−259.3467 to 961.78°C) is the platinum resistance thermometer.

Above 1234.93 K (961.78°C, the freezing point of silver), the International Temperature is defined by Planck's radiation law, with 1234.93 K (961.78°C) as the reference temperature and the value of the second radiation constant, c_2, taken to be 0.014388 meter kelvin. Also, the gold or copper freezing-point temperature may be used as the reference point. *See* HEAT RADIATION.

Above the silver freezing point, the standard techniques and the standard instruments of radiation thermometry are employed, in conjunction with a blackbody at the temperature that is to be measured and at the reference temperature. *See* PYROMETER.

While the temperature of the triple point of water is defined to be exactly 273.16 K, the present uncertainty in independently reproducing it in different laboratories and with different types of cells and apparatus is about 0.0002 K, or about 1 part in 10^6. The (1-standard-deviation) uncertainties in values of the thermodynamic temperatures of the other defining fixed points of the ITS-90 are estimated to be not more than 0.003 K up to the freezing point of indium at 429.7485 K (156.5985°C), and not more than 0.060 K at the freezing point of copper at 1357.77 K (1084.62°C), but the uncertainties increase to about 7 K at the melting point of tungsten (3687 K or 3414°C).

The degree Celsius (°C), earlier known as the degree Centigrade, has the same magnitude as the kelvin. The Celsius scale assigns 0°C to the freezing temperature of water (273.15 K), so that $t/°\text{C} = T/\text{K} - 273.15$.

On the Fahrenheit scale, the freezing point of water is 32°F and the degree Fahrenheit, °F, is 5/9 of a kelvin, so that $t/°\text{F} = (t/°\text{C} \times 1.8) + 32$. *See* TEMPERATURE.

B. W. Mangum

Electric current. The ampere (A) is that constant current which, if maintained in two straight parallel conductors of infinite length and of negligible circular sections, and placed 1 meter apart in a vacuum, would produce between these conductors a force equal to 2×10^{-7} newton per meter of length.

The experimental realization with highest precision is difficult. It is impractical to measure the force in the idealized geometry; coils of many turns are used. The force is measured by a balance in terms of the local acceleration of gravity. The local acceleration of gravity may be calculated for a given latitude, longitude, and elevation by interpolation from geodetic and gravity surveys; measured by local pendulum experiments (seldom better than 1 part in 10^6); or measured by very elaborate time-of-fall experiments using standard atomic frequencies and wavelength (with uncertainties less than 1 in 10^7). *See* CURRENT BALANCE; CURRENT MEASUREMENT.

Constant values of other related electrical quantities—voltage, resistance, capacitance, and inductance—can be maintained more easily than the ampere, some with much better precision. Capacitance of a tubular capacitor of variable length can be calculated quite accurately. The step from capacitance to resistance uses special bridges at various frequencies, and thus the ohm can be determined to about 1 part in 10^7. *See* CAPACITANCE MEASUREMENT; INDUCTANCE MEASUREMENT; RESISTANCE MEASUREMENT.

The known current going through a standard resistor provides a known voltage which can be used to calibrate voltage standards such as the electrochemical cells on which the United States legal volt is based. *See* VOLTAGE MEASUREMENT.

A dc voltage can be maintained with much better precision and stability in terms of a measured frequency by means of the ac Josephson effect. The uncertainties in the ampere and the volt are believed to be about 5 parts in 10^6, but the volt is maintained (thanks to the Josephson apparatus) with a precision and stability better than 5 parts in 10^8. *See* ELECTRICAL UNITS AND STANDARDS; JOSEPHSON EFFECT.

Luminous intensity. The CGPM, in 1979, redefined the base SI unit candela as the luminous intensity, in a given direction, of a source that emits monochromatic radiation of frequency 540×10^{12} hertz and of which the radiant intensity in that direction is 1/683 watt per steradian.

The new definition for the candela is based on monochromatic radiation rather than, as previously, on white light, and provides a definite numerical relationship between the photometric quantities and the watt. It thus links the fields of photometry and radiometry.

The particular frequency, 540×10^{12} hertz (or 555-nanometer wavelength in a vacuum), was chosen because the "spectral luminous efficiency" curves for light- and dark-adapted vision have very closely the same value at this frequency. The CIPM had previously ratified the spectral luminous efficiency curves—or weighting functions—as adopted by the International Commission on Illumination. *See* ILLUMINATION; LIGHT; LUMINOUS EFFICACY; LUMINOUS EFFICIENCY; LUMINOUS INTENSITY; PHOTOMETRY; RADIOMETRY.

Amount of substance. The mole is the amount of substance of a system which contains as many elementary entities as there are atoms in 0.012 kilogram of carbon-12. When the mole is used, the elementary entities must be specified, and may be atoms, molecules, ions, electrons, other particles, or specified groups of such particles. *See* GRAM-MOLECULAR WEIGHT; MOLE (CHEMISTRY).

Uncertainty in Practical Measurements

Although the seven base units, and others derived from them, are thus exactly defined, their practical availability requires the development and refinement of standard devices or apparatus to realize each of them with high precision. Extensive theoretical studies and laboratory experiments are involved in the selection and refinement of operating principles and in design, construction, and operation of these standards. Once compared with the base standard, other subordinate standards and reference instruments, specimen objects, signal sources (generators or modifiers of the quantity), and so on can be used for further calibrations or measurement.

Comparison of standards. As noted above, a kilogram mass standard can be calibrated only through a series of comparisons, starting from the International Prototype. The units for the other five base quantities, and all quantities derived solely from them, are in principle independently realizable; that is, the standard apparatus may be constructed equally well in many laboratories. In practice, however, inevitable minor differences between standards constructed independently, even with equal care, and among the instruments, environments, and operators individually are bound to introduce small discrepancies. Periodic comparison of standards and the resolution of these discrepancies are required for compatibility among domestic standards laboratories, as well as internationally. In the United States, the National Institute of Standards and Technology (NIST; formerly National Bureau of Standards) provides calibration services for industrial, educational, and other governmental standards laboratories and cooperates with them in conducting measurement agreement comparisons. Periodic comparisons of NIST standards with those of other countries are made through the International Bureau of Weights and Measures, international scientific organizations, or direct arrangement.

Frequency and time comparisons within the United States are made between the NIST, the U.S. Naval Observatory, and other users of high-precision frequency standards. The data from worldwide astronomical observations and from standards laboratories in many countries are coordinated by the International Bureau of the Hour, which coordinates differences between the International Atomic Time scale (a running count of atomic seconds) and the Coordinated Universal Time scale and announces the time when the UTC offset should be changed by ± 1 second to keep UTC in phase with the solar year. Most radio broadcast signals are measured on UTC.

Sources of error. In general, any measurement has less than perfect accuracy; there is some error or uncertainty as to the true, or exact, numerical ratio between the magnitude of the measured quantity and the unit. Associated with the measuring instrument itself are the errors made in its calibration, the uncertainty in the constancy or reproducibility of the standard by which it was calibrated, and the possible changes of its response after its calibration. Other uncertainties, somewhat more under the control of the observer making a measurement, include those in taking readings, in correcting for environmental effects, and in allowing for characteristics of the instrument as well as of the system or object undergoing measurement.

Material and structural characteristics. Inherent characteristics of materials and structures cause the phenomena of drift, lag, hysteresis, damping, and resonance in measuring instruments or systems, as well as in the systems undergoing measurement.

These general phenomena may affect the relationship of any quantity or condition (mechanical, electrical, thermal, and so on) to any parameter. In terms of instrument reading and measured quantity they may be defined as follows:

1. Drift is the gradual continued change of instrument reading after a change to a different but constant value of the measured quantity.

2. Lag is the failure of the instrument reading to follow changes in the measured quantity instantly. The time constant of an instrument is the time required for the indication to change by $1/e$ (0.37) of a sudden change in the measured quantity in response to this change.

3. Hysteresis results from lag or drift, and is the difference between readings of the measured quantity for corresponding actual magnitudes of that quantity when the quantity is increasing and when it is decreasing.

4. Damping is the dissipation of energy (electrical, magnetic, or mechanical) caused by a change in the measured quantity. With critical damping, the indication changes to its new value with minimum lag without overshoot, following a sudden change in the measured quantity. With less damping, the indication overshoots the new reading and may oscillate about it with decreasing amplitude; with greater damping, the indication changes to its new value more slowly.

5. Resonance is the condition of enhanced response or oscillation which results when the rapidity of change of the measured quantity is close to the natural rapidity of response of the instrument. *See* CIRCUIT (ELECTRICITY); DAMPING; HYSTERESIS; RESONANCE (ACOUSTICS AND MECHANICS); RESONANCE (ALTERNATING-CURRENT CIRCUITS); TIME CONSTANT.

Change of properties. The change of properties of materials or structures with temperature, pressure, humidity, radiation, vibration, or other environmental conditions may cause further uncertainties in the response of the instrument and in the characterization of the quantity or property being measured.

Instrument-measurand interaction. The interaction of the measuring instrument or process with the quantity being measured (measurand) is often not negligible, and uncertainties remain even after corrections; for example, measuring a voltage usually requires some power, which generally tends to lower the measured voltage. On the atomic scale, Heisenberg's uncertainty principle states that it is impossible to determine simultaneously both the position and the momentum of a particle with the product of the uncertainties less than about 10^{-34} joule second. *See* UNCERTAINTY PRINCIPLE.

Error analysis and reduction. Thus, the analysis of experiments to detect all possible sources of error, the design of experimental procedures to minimize them, and the development of mathematical techniques for estimating their probable magnitudes are all important parts of any measurement in which highest accuracy is essential. The effect of random disturbances and reading errors may be minimized by averaging repeated measurements. Averages with various observers may reduce the effects of systematic reading errors of a given observer.

From the dispersion of repeated observations a statistical estimate of the imprecision (often called precision or repeatability) of the measurement may be obtained. *See* STATISTICS.

The inaccuracy (often called accuracy) of a measurement (or calibration) is the total uncertainty, including both the imprecision of observation and the systematic uncertainties associated with the measuring instrument and with the measuring process.

The reduction of uncertainty in measurement is one of the continuing objectives in science; it is also one of the most fruitful contributions to science. Scientific theories live or die as they are tested by more precise measurements. New theories are evolved to fit phenomena revealed by more precise observations, and new advances in many fields follow an improved measurement capability for any one quantity.

Spectacular reductions in uncertainty—by a factor of 10^6 in 20 years—occurred in the determination of time interval and frequency after the introduction and refinement of the atomic standards. For all quantities, the average improvement has been about a factor of 10 in the same time. While further progress becomes continually more difficult and expensive, there appears to be no fundamental reason why the same average rate of improvement could not be maintained, or exceeded, for at least the rest of the century.

Measurement Techniques

The comparison of quantities as to equality, the counting of units, and the determination of the coincidence of events—all involved in physical measurement—may be done by an observer (usually using visual, aural, or tactile faculties) or by instruments which display or record the results, or apply them to automatic computation or control. *See* INSTRUMENTATION AMPLIFIER.

Direct measurement involves comparison with a standard (such as a meter bar) or measurement by a calibrated instrument (such as a voltmeter). Indirect measurements are those derived from measurements of related quantities; for example, the mass of the electron can be derived from measurements on the bending of the path of electrons of a known velocity in a known magnetic field and from separate measurements of the electron charge.

Measuring instruments or systems may respond to the physical quantity to be measured by generating or modifying another quantity or series of quantities. The final (output) quantity, having a functional relationship to the quantity measured, may be compared with a standard of its own kind, appropriately scaled and labeled to represent units of the quantity being measured. For example, in a voltmeter, the voltage across a resistor induces a current which reacts with a magnetic field to generate a force or torque which deforms a mechanical spring and moves a pointer along a scale. Thus the final quantity generated is a length, the displacement of the pointer, which is compared with a scale graduated in intervals of length (or angle) appropriately proportioned to indicate numerically the applied voltage. The proportionality factor is the product of the successive transformation factors from volts to amperes to force to displacement along the scale. Devices that effect this conversion or transformation of one quantity to another are called transducers. *See* TRANSDUCER.

The hundreds of physical phenomena relating one quantity to another supply a rich reservoir of alternatives for the designer of measurement experiments or measuring instruments. For example, the determination of concentration or composition is widely important in both science and industry. An increasing variety of spectroscopies—radio-frequency, infrared, ultraviolet, x-ray, gamma-ray, and so on—utilize atomic and molecular characteristics for absorption, emission, scattering, or reemission of electromagnetic radiation or nuclear particles to determine not only composition but even molecular and crystalline structure. *See* ACTIVATION ANALYSIS; SPECTROSCOPY.

Many atomic phenomena involve several quantities and may serve as a basis for extending the range or accuracy of measurement of each of them, and others. *See* JOSEPHSON EFFECT; LASER; MASER; MICROWAVE SOLID-STATE DEVICES; MÖSSBAUER EFFECT.

Some techniques for reducing uncertainty, extending range, or providing flexibility and convenience in measurements in general are briefly mentioned below.

1. Many natural phenomena and the fundamental properties of matter, such as the speed of light in a vacuum or the charge of the electron, provide constant values of certain quantities or combinations of quantities which may be used in physical measurement or instrument design. For example, the precession frequency of the spin of the proton in the hydrogen atom, or hydrogen compounds, is proportional to the magnetic field, or to a current generating it, which may be determined from this frequency. It in turn is indicated by varying the frequency of a high-frequency electric field, applied at right angles to the magnetic field, to achieve maximum coupling, through synchronized spins, with a similar passive circuit detecting an induced electric field at right angles to both the magnetic and applied electric fields.

2. The quantity to be measured or one related to it in a known manner may be allowed to modulate or attenuate the value of another quantity—as when the thickness of metal or paper is determined from its attenuation of an x-ray beam, which is converted by any of several types of radiation meters (transducers) to a change in voltage or current and then to a visual indication or record. As is common in spectroscopy, uncertainty due to variation of the source may be reduced by taking the ratio of signals with and without the sample in the beam.

3. The measured quantity itself, or its signal, may

be time-modulated to permit better discrimination between the measured quantity and extraneous effects (noise) in the measuring instrument.

4. In weighing on a balance, known weights are substituted for unknown weights previously balanced by counterweights, so that such uncertainties as knife-edge placement and beam lengths do not affect the comparison.

5. The effect of a quantity to be measured may be nearly offset by a similar known and fixed quantity, and the differences measured by an instrument of lesser range and higher sensitivity.

6. If the standard offsetting quantity can be divided into sufficiently small or continuously variable fractions, it can be adjusted for equality to within the smallest limits detectable. This is the so-called null, or balancing, method of measurement.

7. A number of small and equal magnitudes of the same quantity may be combined so that the cumulative magnitude is appropriate for measurement with available methods.

8. The undesired effects of environmental factors, such as temperature and external magnetic fields, may be compensated for in the instrument system by elements which are responsive to the disturbing factor and interact with the measuring or indicating means to offset such effects.

9. Based on careful design of experiments, observations may be made under a wide variety of conditions, with controlled variation in the factors considered as possible sources of error, permitting statistical estimation of the magnitude of the various errors and appropriate correction for them.

10. In measuring properties of materials, uncertainties may be reduced by comparing measurements on the sample with those made on a sample having closely similar, and accurately known, properties. Hundreds of such reference materials or standard samples are available from government or industrial sources with an indication of the value (and the uncertainty) of the characteristic property, be it composition, purity, size, radioactivity, viscosity, or some other. Measurements on such reference materials obviously provide a calibration check of measuring instruments; if their properties are close to those of the unknowns, the differences may be measured by more sensitive methods or devices.

11. Published values for carefully measured properties of generally available materials may obviate the need for repetitive measurements, and may also provide a basis for checking calibration.

12. Data obtained by the "round-robin" circulation of two test pieces, of somewhat different (and undisclosed) values, for measurement by each of a number of individuals or laboratories, provide a simple and practical method for self—as well as group—evaluation. *See* MAGNETOMETER. William A. Wildhack

Bibliography. J. P. Bentley, *Principles of Measurement Systems*, 3d ed., 1995; *Guidelines for the Use of the International System of Units*, NIST Spec. Publ. 811, 1995; *The International System of Units (SI)*, NBS Spec. Publ. 330, 1991; B. W. Mangum and G. T. Furukawa, *Guidelines for Realizing the International Temperature Scale of 1990 (ITS-90)*, NIST Tech. Note 1265, 1990; *NIST Standard Reference Material Catalog*, NIST Spec. Publ. 260, 1995; B. W. Petley, *The Fundamental Physical Constants and the Frontier of Measurement*, 2d ed., 1988; T. J. Quinn, *Temperature*, 2d ed., 1991; R. S. Sirohi and H. C. Radha Krishna, *Mechanical Measurements*, 3d ed., 1993; P. H. Sydenham, R. Thorne, and D. Hancock, *Introduction to Measurement Science and Engineering*, 1992. See also *Analytical Chemistry*; *Instruments and Measurements* (Russian); *Journal of Physical and Chemical Reference Data*; *Journal of Scientific Instruments* (British); *Metrologia*; *NIST Journal of Research*; *Review of Scientific Instruments*.

Physical optics

The study of the interaction of electromagnetic waves in the optical range with material systems. The optical range of wavelengths may be taken as the range from about 1 nanometer (4×10^{-8} in.) to about 1 millimeter (0.04 in.). More narrowly, physical optics deals with the relationship between the atomic structure of a system and the manner in which the system affects light sent into it. The chief founder of this branch of science was Michael Faraday, who in 1845 provided the first clue to the electromagnetic nature of light by showing the optical properties of glass could be altered by a magnetic field. *See* FARADAY EFFECT.

The explanation of the absorption, reflection, scattering, polarization, and dispersion of light by a material medium in terms of the properties of the atoms and molecules making up the medium is the objective of physical optics. In the course of seeking this objective, physicists have found that optical investigations are powerful methods of determining the structures of atoms and molecules and of larger systems composed thereof. *See* ABSORPTION OF ELECTROMAGNETIC RADIATION; ATOMIC STRUCTURE AND SPECTRA; CRYSTAL OPTICS; DIFFRACTION; DISPERSION (RADIATION); ELECTROMAGNETIC RADIATION; ELECTROOPTICS; FLUORESCENCE; INTERFERENCE OF WAVES; LASER; LIGHT; MAGNETOOPTICS; MOLECULAR STRUCTURE AND SPECTRA; POLARIZED LIGHT; REFLECTION OF ELECTROMAGNETIC RADIATION; REFRACTION OF WAVES; SCATTERING OF ELECTROMAGNETIC RADIATION; SPECTROSCOPY. Richard C. Lord

Bibliography. M. Born and E. Wolf, *Principles of Optics*, 7th ed., 1999; F. A. Jenkins and H. E. White, *Fundamentals of Optics*, 4th ed., 1976; A. Mickelson, *Physical Optics*, 1992.

Physical organic chemistry

A branch of science concerned with the scope and limitations of the various rules, effects, and generalizations in use in organic chemistry by means of physical and mathematical methods. It includes, but is not limited to, the dynamics and energetics of

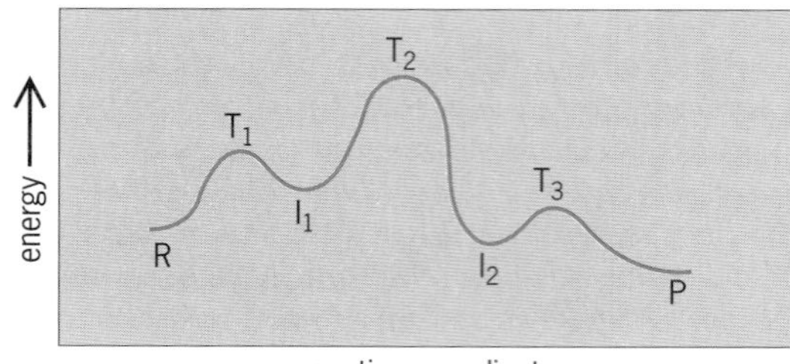

Fig. 1. Energy profile of an organic reaction R → P. T terms indicate transition states and I terms indicate intermediates.

organic chemical transformations, transient intermediates in these reactions, rate comparisons between families of reactions, dynamic stereochemistry, conservation of orbital symmetry, the least-motion principle, the isomer number for a given elemental composition, conformational analysis, nonexistent compounds, aromaticity, tautomerism, strain and steric hindrance, and the double-bond rule. Spectroscopy is the main tool employed, with nuclear magnetic resonance being the most widely used spectroscopic technique. With the advent of modern fast computers, computational chemistry has also become an important tool. *See* NUCLEAR MAGNETIC RESONANCE (NMR); SPECTROSCOPY.

Physical organic chemistry is traditionally distinguished from, yet totally intertwined with, synthetic organic chemistry, which deals with the question of how to obtain desired products from available compounds. This distinction can be illustrated with a diagram (**Fig. 1**) showing how the energy might vary during a chemical reaction in which the reactant R yields a product P (R → P). Whereas the synthetic organic chemist will be interested primarily in the practical problem of how to convert R into P, the physical organic chemist studies the curve or curves connecting R and P as well as the structure and physical properties at all extrema, including R and P. However, the demarcation between synthetic organic chemistry and physical organic chemistry is not sharp. Physical organic chemists have contributed greatly to the understanding of the chemistry of hydrocarbons and their derivatives and have enhanced the repertoire of the synthetic organic chemists. In turn, synthetic organic chemists have made possible the construction of the custom-made, often intricate molecules that physical organic chemists use for their studies. The efforts of both groups, moreover, have made possible the birth of such new fields as molecular biochemistry and computational chemistry. *See* COMPUTATIONAL CHEMISTRY; MOLECULAR BIOLOGY; ORGANIC SYNTHESIS.

Chemical reaction mechanisms. The diagram in Fig. 1 is also useful in discussions of the dynamics of chemical reactions. It is an attempt to portray how the atoms in the reactant molecule R may move in space to their final positions in the product molecule P, and how the potential energy of the system would vary as a function of these positions. A complete correlation would be multidimensional; what is normally shown is a cross section in which the maximum potential energy is in fact a minimum (saddle point). While an essentially infinite number of pathways between R and P can be imagined and followed, the vast majority of the molecules will in practice use the one that makes the least demand on energy to reach the next maximum; this pathway is known as the reaction mechanism. The maxima (T terms in Fig. 1) are known as transition states, and the minima (I terms in Fig. 1) as intermediates.

If a reaction has a single transition state (and, hence, no intermediate), it is known as concerted; alternatively, it is stepwise. A stepwise reaction is simply a succession of concerted steps in which the intermediates are not isolated. Criteria for concertedness have included stereospecificity; inability to detect intermediates either by spectroscopic means or by trapping them with auxiliary compounds to give different products; the rate law; the entropy and volume of activation (temperature and pressure dependence of the rate constant); comparison with reactions thought to be concerted; and isotope, solvent, and substituent effects. *See* ORGANIC REACTION MECHANISM.

In spite of all these criteria, it is exceedingly difficult to prove that a given reaction is concerted, and that there is not some elusive, high-energy intermediate present on the pathway. Indeed, diligent searches have turned up a variety of such species; they include carbanions, carbocations, free radicals and diradicals, radical anions and cations, carbenes, nitrenes, and highly crowded or strained molecules. *See* FREE RADICAL; PHOTOCHEMISTRY; REACTIVE INTERMEDIATES.

Chemical kinetics. The most important route to quantitative information about a reaction is the study of its kinetics. This must begin with an experimental determination of the rate law: the expression that shows how the rate of formation of product $d[P]/dt$ (or loss of reactant: $-d[R]/dt$) depends on the concentration of all species involved in the reaction other than the solvent. The differential might be found to equal $k[R_1]$, $k[R_1]\,[R_2]$, $k[R_1]^2$, and so on; k is known as the rate constant, and the reaction is described as first-order, second-order, and so forth, depending on the total number of concentration terms. One important feature is that the reaction order equals the sum of all molecules that have participated in the formation of the transition state (T_2 in Fig. 1). Thus, for a reaction to be concerted, it is necessary that the order equal the sum of all reactant molecules involved in the stoichiometry; for stepwise reactions, this need not be the case, although it can be (if there is no intermediate following this state). Another important fact is that since at equilibrium the forward and reverse rates are equal, the ratio of the forward rate constant (k_f) to the reverse rate constant (k_r) equals the equilibrium constant K. When the rate law in one direction is established, the one for the reverse reaction is automatically established also, since this constant is defined by the reaction equation. It follows that the forward and reverse paths cannot be different (principle of microscopic reversibility). *See* CATALYSIS; CHEMICAL DYNAMICS.

To establish the rate law, the first step is to consider all the mechanisms that seem likely candidates. For each, a series of differentials is written expressing the change of concentrations with time; this includes the reactants and all intermediates up to the transition state. Since the concentrations of the reactive intermediates are generally not known, it is assumed that the differentials $d[\mathrm{I}]/dt$ equal zero (the steady-state approximation). The expressions are combined into a single differential equation relating the concentration of [R] and time, and this is integrated. The resulting equations are then tested in the laboratory; those that do not fit the experimental data represent mechanisms that are ruled out, and those that do fit correspond to mechanisms that remain as possibilities (there may be several). *See* DIFFERENTIAL EQUATION.

Entry into the thermodynamic quantities (energy, entropy, and volume) of the transition state can then be gained by means of the theory of absolute rates, which has led to Eq. (1), where k represents the rate

$$k = \frac{RT}{Nh}\exp(-\Delta G^{\neq}/RT) \qquad (1)$$

constant, R the gas constant, T the absolute temperature, N the Avogadro constant, h the Planck constant, and $\Delta G^{\neq}$ the difference between the free energies of the transition state and the reactants. Since the value for the entropy (S) is given by $(\partial G/\partial T)p = -S$ and the value for the volume (V) is given by $(\partial G/\partial P)_T = V$, the entropy and volume of the transition state can be derived from measurements of the temperature and pressure dependence of the rate constant, respectively. These quantities give very important insights into the structure of the transition state; thus, volume increases are often an indication of a breaking bond, or of the desolvation of ions. Similarly, solvent effects on the rate constant often reveal substantial polarity changes on the way to the transition state. *See* CHEMICAL THERMODYNAMICS; FREE ENERGY; HIGH-PRESSURE CHEMISTRY; HIGH-TEMPERATURE CHEMISTRY.

Substitution reactions. Substituent effects have played a dominant role in physical organic chemistry. Placing a substituent near the reactive center may have the effect of greatly altering the rate, that is, greatly changing the free energy (stability) of the transition state. Since the effect of a given substituent on the stability of a variety of structures can be studied independently, such a rate effect can provide significant information about the structure of the transition state. These effects are often of a steric nature, particularly if the substituent is placed very close to the reactive center. If the substituent is in conjugation with the reactive center, its influence may be felt electronically. *See* CONJUGATION AND HYPERCONJUGATION; STERIC EFFECT (CHEMISTRY).

In order to separate steric and electronic effects, chemists have resorted to the use of *meta-* and *para-*substituted phenyl groups; these substituents cannot affect the reactive center sterically. Studies of this sort produce the so-called linear free-energy relations, which may be of two types. In one of these, the rate constants of a reaction involving substituted phenyl compounds are plotted logarithmically against the difference in pK_a value of the similarly substituted benzoic acid and benzoic acid itself. The pK_a value is a quantitative measure of acid strength. A straight line is generally obtained. If the slope is positive, the substituent stabilizes or destabilizes the transition state in the same way that it affects the benzoate anion; this generally means that in the transition state the reactive center is developing negative charge. Conversely, if the slope is negative, the transition state involves positive charge, and the larger the slope, the larger the charge. Plots of this type have played an important role in the development of Brönsted theory. *See* ACID AND BASE; PK; SUPERACID.

The other type of linear free-energy relation results from plots of log k of one reaction against that of another. If the reactions are of the same type, a linear plot may be expected; if they are quite different, the correlation will be poor. An example is the plot that results when log k is measured for the reactions of a series of anions with methyl bromide (CH_3Br) to give bromide ion (Br^-) and with difluoramine (NF_2H) to give fluoride ion (F^-). The former is well known to be a simple, direct, and concerted displacement. A linear plot results, suggesting that the fluoride reaction is also a simple displacement; however, the point for hydroxide (OH^-) is far off this line, the reaction with difluoramine being much faster than expected on the basis of the reaction of this ion with methyl bromide. Thus, the hydroxide reactions are not similar. Subsequent investigation revealed that hydroxide reacting with difluoramine acts as a base rather than as a displacing agent, removing a proton (H) to leave an anion (NF_2^-) that then decomposes to give fluoride ion and fluoronitrene (NF·). The latter intermediate can be captured with a variety of trapping agents, and in fact, it has been observed spectroscopically. *See* MATRIX ISOLATION.

Reactivity-selectivity principle. Consideration of the reasons for rate differences in the reaction of a given substrate C—X with a series of anions A^- led to a very useful insight into displacement reactions. The energy profile for the reaction is believed to be essentially the sum of the two bond energy curves of C—X and C—A_i (i = 1, or 2, or 3, etc.; **Fig. 2**). The transition states are presumed to lie near the crossover points of these curves. The energy profiles so obtained suggest that the more stable product is associated with a more stable transition state, and hence it is produced in a faster reaction, while the less stable product is associated with a less stable transition state and thus is produced in a slower reaction. This principle has been generalized to apply to all families of reactions; thus, if a series of organic halide derivatives of comparable stability solvolyzes at greatly different rates, it is believed that the fastest among them produce the most stable intermediate carbocations. The same picture, in fact,

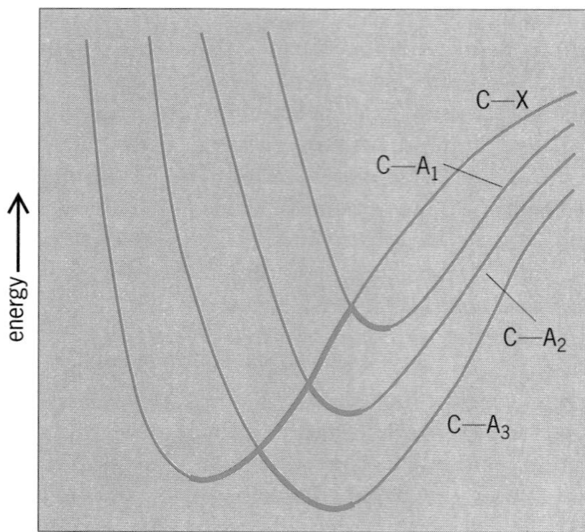

Fig. 2. Profile of the energetics of a series of displacement reactions of substrate C—X with anion A⁻.

suggests another important generalization known as the Hammond postulate (originally stated in more succinct form): the faster a reaction, the "earlier" (more reactantlike) is the transition state, and the slower a reaction, the "later" is the transition state.

The reactivity-selectivity principle is also based on relative reaction rates. The basic assumption is that a highly reactive intermediate, when presented with two substrates or with two sites within a single substrate, will react at the first opportunity and hence be unselective (product ratio near unity); a much more stable species is likely to be more discriminating and lead to larger ratios. While this assumption is basically sound, its sometimes uncritical application to rather dissimilar species or reactions has led to many claims of exceptions that leave its usefulness in doubt.

Stereochemistry. This is also a powerful tool in physical organic chemistry. Experimental work demonstrated that when a chiral compound such as $(-)HCR_1R_2X$ is converted into optically active HCR_1R_2Y by direct displacement with Y, the product obtained has an optical rotation that is opposite to that exhibited by the same material if it is produced in two steps, via initial displacement by A to give intermediate HCR_1R_2A [reaction scheme (2)]. This re-

$$Y + HCR_1R_2X \xrightarrow{\text{direct displacement}} Y{-}CR_1R_2H$$
$$HCR_1R_2X \xrightarrow[-X]{\text{two steps}\; A} A{-}CR_1R_2H \xrightarrow[-A]{Y} HCR_1R_2{-}Y \qquad (2)$$

sult demonstrates that displacement reactions occur with inversion; the reagent approaches in front, and the leaving group departs in the back. *See* OPTICAL ACTIVITY.

Retention of configuration accordingly indicates a two-step reaction (the auxiliary displacing agent may be a solvent molecule or an internal part of the substrate itself). Sometimes racemization is encountered; this may be taken to mean that the substrate suffered initial cleavage to give a trivalent carbon intermediate (for example, a carbocation), which will either be planar or subject to rapid inversion, so that the final capture of the displacing agent can occur at either side with equal likelihood.

Another important area in which configuration of the product is reliably predictable from that of the reactant is the general pericyclic reaction, a type of organic reaction in which a transition can be drawn that possesses a complete cyclic array of parallel *p* orbitals. The stereochemistry is controlled by the tendency to maintain overall symmetry. This principle has overshadowed and in many cases supplanted an earlier one known as the principle of least motion, which held that the nature of the product is largely determined by a need to minimize nuclear displacements. *See* MOLECULAR ORBITAL THEORY; PERICYCLIC REACTION; STEREOCHEMISTRY.

Isomers. The isomer number has perhaps been the most important organizing principle in organic chemistry since its inception. Simply put, it means that for every elemental composition and molecular mass the isomer number can be predicted by writing all possible sequences of the atoms present, obeying the valence numbers of the atoms: four for carbon, three for nitrogen, two for oxygen, one for hydrogen and the halogens, and so forth. On this basis, it is possible to predict that there will be two structural isomers with the composition C_4H_{10} (**1**), two with C_2H_6O (**2**), four with C_4H_9Cl (**3**), and so forth.

$CH_3{-}CH_2{-}CH_2{-}CH_3$ (**1*a***) $\qquad (CH_3)_3CH$ (**1*b***)

$CH_3{-}CH_2{-}O{-}H$ (**2*a***) $\qquad CH_3{-}O{-}CH_3$ (**2*b***)

$CH_3{-}CH_2{-}CH_2{-}CH_2{-}Cl$ (**3*a***) $\qquad (CH_3)_2CH{-}CH_2Cl$ (**3*b***)

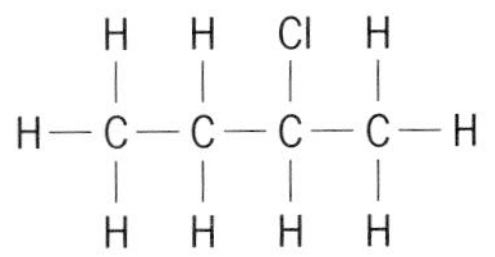

(3c)

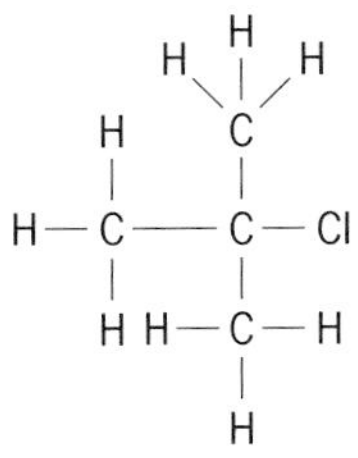

(3d)

Double bonds involving the elements carbon, nitrogen, and oxygen are allowed, as are triple bonds to carbon and nitrogen; thus structure (4) must be included for composition C_2H_4O, and structure (5) for composition C_2H_3N.

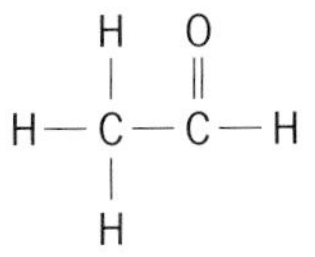

(4)

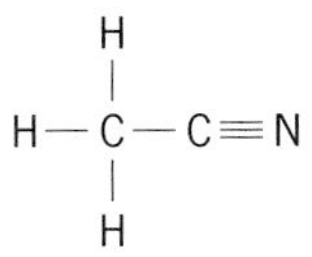

(5)

See CHEMICAL BONDING; ORGANIC CHEMISTRY; VALENCE.

Once reliable atomic weights became available so that elemental compositions could be reported with confidence, this simple rule proved remarkably successful. For example, the isomer rule predicts 35 isomers with the formula C_9H_{20}, and all are known. Furthermore, it is sometimes possible to correlate these structures with the properties of the known compounds. Thus, only one of the two known isomers C_2H_6O reacts with sodium to give hydrogen; since it resembles water in that sense, this must be the isomer with an O-H group.

In view of these successes, organic chemists tried to modify or expand the simple rule when exceptions were encountered; some of these came in the form of extra compounds, and others as missing compounds. Some of the extra compounds required only the revision that anomalous valences sometimes occurred; two for carbon, for instance (C═O), or four for nitrogen (ammonium salts).

The need to specify molecular mass has proved more troublesome: it requires a definition of the concept of a molecule. Such definitions usually refer to covalent bonds as the entities that hold the atoms together, to rule out ionic species such as sodium chloride as candidates. However, there are supramolecular aggregates. Examples are trihydrogen trifluoride (H_3F_3; 6), which remains intact in the gas phase by virtue of the weak and electrostatic hydrogen bonds; diborane (B_2H_6; 7), which exists by virtue of hyperconjugation or no-bond resonance; and charge-transfer complexes (8), in which an electron-

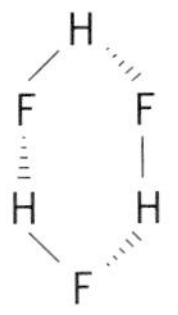

(6)

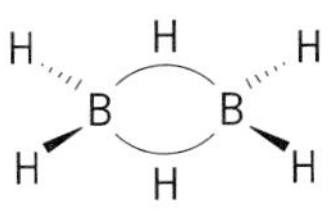

(7)

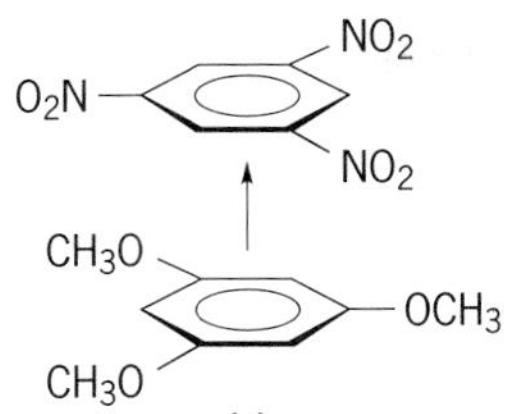

(8)

deficient molecule polarizes and attracts an electron-rich partner. *See* RESONANCE (ACOUSTICS AND MECHANICS).

The synthesis of the so-called catenanes has shown that there are molecules in which component parts are held together by repulsive forces rather than bonds—whether covalent or electrostatic. To mention a specific example: $C_{50}H_{100}$ should, among its isomers, include one in which two macrocyclic $C_{25}H_{50}$ rings are looped together like two links of a chain (9).

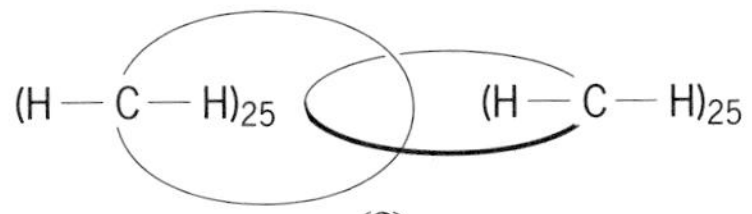

(9)

See MACROCYCLIC COMPOUND; STRUCTURAL CHEMISTRY.

Among the extra compounds, none have affected organic chemistry have drastically than the stereoisomers. It turns out that for all but the simplest compounds a given sequence of the atoms may represent two, more than two, or even many more isomers. An early insight into the real nature of molecules was provided by L. Pasteur, who studies of optical activity of solutions and of crystal structures led him to the conclusion that there must be chiral (and hence, three-dimensional) molecules. The origin of this phenomenon was revealed principally by J. H. van't Hoff, who postulated that the four valences of carbon were directed to the corners of a tetrahedron. *See* MOLECULAR ISOMERISM.

Van't Hoff's work led to correct predictions of far larger numbers of isomers than had been realized before. Further work with stereoisomers revealed that rings containing six carbon atoms or more must be puckered, since in planar form the carbon-carbon bond angles would be 120° or more; the tetrahedral value is slightly less than 110°. *See* CONFORMATIONAL ANALYSIS.

Research has shown that rotation about single bonds if not totally free; the barriers are low so that the interconversion at room temperature is so fast that the isomers (conformers) cannot be isolated. Isolable conformers are possible only in certain especially designed molecules such as the triptycene derivative (10). In this case, rotation about bond a is not free but hindered by the protruding atoms Cl_1

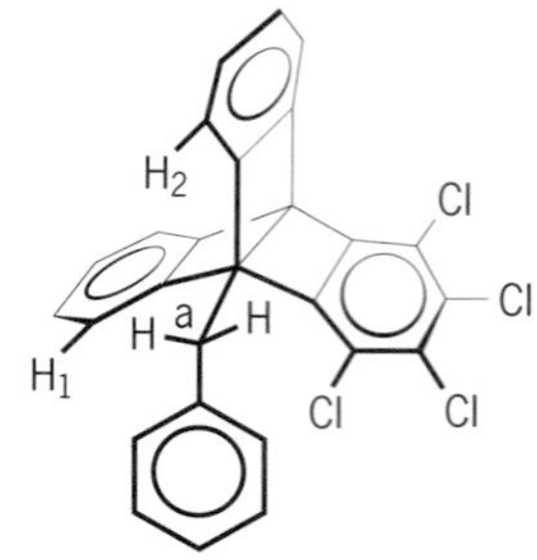

(10)

and H_1. In such instances, the conformations must be counted as isomers.

In small rings, bond-angle deviations from the tetrahedral value cannot be relieved, and high-energy, strained molecules result; nevertheless, many are capable of existence. One of the activities of physical organic chemists has been to search for the limits of strain and crowding that molecules can endure and survive. A few examples of such structures are shown in **Fig. 3***a*; several structures that have so far resisted all efforts at synthesis are shown in Fig. 3*b*.

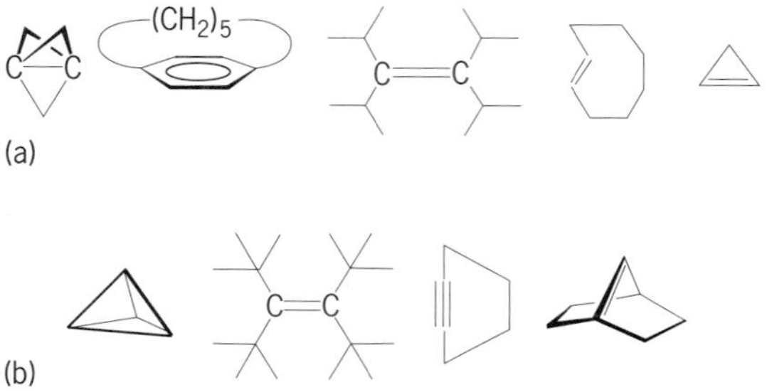

Fig. 3. Examples of high-energy, strained molecules (*a*) Structures that have been synthesized. (*b*) Structures that have not been synthesized.

Many of the nonexistent compounds (allowed by the simple rules of isomer numbers) in fact are transient intermediates in various reactions. Thus, reaction (3) can be shown to produce the strained

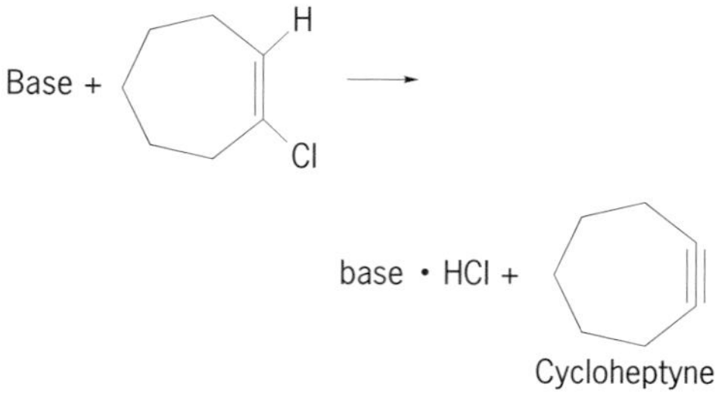

Base + [cycloheptene, H, Cl] → base · HCl + Cycloheptyne (3)

cycloheptyne (bond angles at triply bound carbon are normally 180°), but only if it is conducted at the temperature of dry ice (solidified carbon dioxide); this species can then be detected spectroscopically. As it does not survive warming, the other products observed may be said to form via cyclopheptyne as the intermediate.

There are several additional groups of nonexistent compounds for which other causes have been identified. Thus, enols such as structure (**11**) can usually

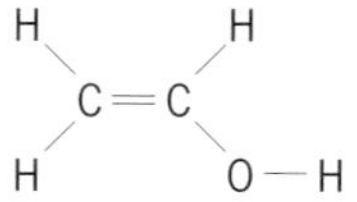

(11)

not be isolated, although they play the role of intermediates in many reactions. The reason for their non existence is not strain or crowding, but simply the fact that the barrier to a more stable isomer is low. The so-called tetrahedral intermediates fall in a similar category, for example, structure (**12**). These

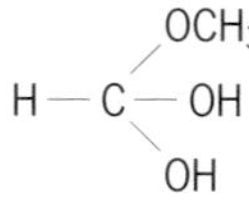

(12)

intermediates invariably eliminate water or alcohol very rapidly and hence cannot be isolated.

There are instances in which neither of two isomers can be isolated, but mixtures of the two can. In other words, the barrier between the two is low, and the equilibrium constant is close to unity. An example is acetoacetic ester, which normally contains about 15% of the enol isomer. Such isomers are known as tautomers. *See* TAUTOMERISM.

William J. LeNoble

Bibliography. F. A. Carey and R. J. Sundber, *Advanced Organic Chemistry, Pt. A: Structure and Mechanisms*, 4th ed., 2000; N. S. Isaacs, *Physical Organic Chemistry*, 2d ed., 1996; W. J. LeNoble, *Highlights of Organic Chemistry*, 1974; J. March, *Advanced Organic Chemistry: Reactions, Mechanisms, and Structures*, 5th ed., 2001.

Physical science

The fields of inquiry to which the general designation science may be appropriately applied are broadly divided into social science and natural science. The latter is further subdivided into biology and physical science. Physical science is generally considered to include astronomy, chemistry, geology, mineralogy, meteorology, and physics. These overlap more or less, as illustrated by astrophysics, chemical physics, physical chemistry, and geophysics. There is overlap, likewise, between the physical and biological sciences, as seen in biochemistry, biophysics, virology, and the close relation between geology and paleontology. The boundaries implied in all such classifications are artificial and consist of regions where one field shades into another. *See* ASTRONOMY; CHEMISTRY; GEOLOGY; METEOROLOGY; MINERALOGY; PHYSICS.

Chemistry and physics differ from astronomy, meteorology, and geology in that they are concerned with the properties of matter and energy encountered upon and within the Earth, the planets, and stars. For this reason, chemistry and physics are not set apart from but rather pervade the other sciences.

To regard the several areas of scientific inquiry as separated by sharp definable boundaries is unrealistic. Cross-fertilization of the different fields has produced some of the most notable advances in science, and an artificial barrier can advantageously be accepted as a challenge by a scientist with an adventurous mind. *See* SCIENCE. Joel H. Hildebrand

Physical theory

A physical theory usually involves the attempt to explain a certain class of physical phenomena by deducing them as necessary consequences of other phenomena regarded as more primitive and less in need of explanation. These more primitive phenomena may at the time the theory is formulated be undiscovered, so that part of the proof of the correctness of the theory consists in demonstrating the existence of the unknown assumed primitive phenomena. A classic example is the kinetic theory of gases, in which the pressure of a gas is explained as arising from the kinetic reactions of colliding molecules, the reality of which was established only later by the discovery of phenomena such as the brownian fluctuations.

The value of a theory depends on both the success with which it coordinates a wide range of presently known facts and its fertility in suggesting places to look for presently unknown phenomena. *See* SCIENTIFIC METHODS. Percy W. Bridgman; Gerald Holton

Physics

Formerly called natural philosophy, physics is concerned with those aspects of nature which can be understood in a fundamental way in terms of elementary principles and laws. In the course of time, various specialized sciences broke away from physics to form autonomous fields of investigation. In this process physics retained its original aim of understanding the structure of the natural world and explaining natural phenomena.

Basic parts. The most basic parts of physics are mechanics and field theory. Mechanics is concerned with the motion of particles or bodies under the action of given forces. The physics of fields is concerned with the origin, nature, and properties of gravitational, electromagnetic, nuclear, and other force fields. Taken together, mechanics and field theory constitute the most fundamental approach to an understanding of natural phenomena which science offers. The ultimate aim is to understand all natural phenomena in these terms. *See* CLASSICAL FIELD THEORY; MECHANICS; QUANTUM FIELD THEORY.

The older, or classical, divisions of physics were based on certain general classes of natural phenomena to which the methods of physics had been found particularly applicable. These consisted of classical mechanics with branches in celestial mechanics, hydrodynamics, and ballistics; heat and thermodynamics; kinetic theory of gases and statistical mechanics; optics; acoustics; and electricity and electromagnetism. These divisions are all still current, but many of them tend more and more to designate branches of applied physics or technology, and less and less inherent divisions in physics itself.

Branches. The divisions, or branches, of modern physics are made in accordance with particular types of structures in nature with which each branch is concerned. Thus particle physics, or high-energy physics, is the most recent branch and is concerned with understanding the properties and behavior of elementary particles, and more particularly of the heavy particles—mesons, baryons, and their antiparticles—which are produced in collisions involving energies in a range measured in billions of electronvolts. The next branch in this classification is nuclear physics, which is concerned with associations of neutrons and protons forming the nuclei of atoms; their structure, properties, and energy states; reactions between nuclei, including scattering processes and radioactivity; and related phenomena, such as the interaction of high-speed nuclear particles with matter. Atomic physics is concerned with the structure and properties of atoms as determined by the electrons outside the nucleus; the states of motion of these electrons, including such topics as energy levels, angular momentum properties, and magnetic moments; and the absorption and emission of radiation by atoms.

Continuing with this classification in ascending complexity there is molecular physics, which is concerned with systems of atoms formed into molecules, the nature of intermolecular forces, chemical binding, vibration and rotation spectra of molecules, and the like. Next in order are solid-state physics; physics of liquids; physics of gases; and plasma physics, which deals with properties of highly ionized atoms forming a mixture of bare nuclei and electrons called an ion plasma.

In this same classification could also be included biophysics, which deals with the application of physical methods and types of explanation to biological systems and structures.

Other more specialized classifications may be made in accordance with particular instruments or techniques, such as x-ray diffraction, neutron diffraction, mass spectrometry, infrared spectroscopy, and seismology. The special field of low-temperature physics is characterized not only by special instruments involved in the production and measurement of low temperatures in the range of liquid helium but also by the phenomena of superconductivity and superfluidity which occur only in this temperature range. Other fields, such as astrophysics and geophysics, are concerned with aspects of other sciences to which physics is applicable.

Mathematical physics is the study of physical phenomena by means of mathematics, and includes the more mathematical parts of all branches of physics, as well as most of the content of statistical mechanics, quantum mechanics, relativity, and field theory.

A distinction is often made between mathematical physics and theoretical physics, in which the latter, although still entirely mathematical in form, is thought of as being more closely related to experimental physics. Neither mathematical nor theoretical physics can really be separated from experimental physics, since a complete understanding of nature can only be obtained by the application of both theory and experiment.

Aim. In every area physics is characterized not so much by its subject-matter content as by the precision and depth of understanding which it seeks. The aim of physics is the construction of a unified theoretical scheme in mathematical terms whose structure and behavior duplicates that of the whole natural world in the most comprehensive manner possible. Where other sciences are content to describe and relate phenomena in terms of restricted concepts peculiar to their own disciplines, physics always seeks to understand the same phenomena as a special manifestation of the underlying uniform structure of nature as a whole. In line with this objective, physics is characterized by accurate instrumentation, precision of measurement, and the expression of its results in mathematical terms.

For the major areas of physics and for additional listings of articles in physics *see* ACOUSTICS; ATOMIC PHYSICS; BIOPHYSICS; CLASSICAL MECHANICS; ELECTRICITY; ELECTROMAGNETISM; HEAT; LOW-TEMPERATURE PHYSICS; MOLECULAR PHYSICS; NUCLEAR PHYSICS; OPTICS; SOLID-STATE PHYSICS; THEORETICAL PHYSICS. William G. Pollard

Physiological acoustics

The study of specific responses that take place in the ear or in the associated central (neural) auditory pathways. The ear is the receptor organ of mechanoacoustic energy, which occurs in the form of sound pressure waves. The responses are elicited by appropriate stimuli of well-defined parameters presented at any level of the auditory system. Such responses may be registered with the aid of various, usually invasive, recording techniques which make use of mechanical, electrical, optical, radiological, or biochemical phenomena, or their combinations. The approach employed by physiological acoustics therefore is purely analytical. This is in contrast to the noninvasive, holistic approach employed by psychoacoustics, which lends itself well to experiments on human subjects. Systematic physiological experiments can be performed only in animals. *See* PSYCHOACOUSTICS.

To bridge the gap between the two fields of inquiry, psychoacoustic techniques are employed in experimental animals that have received appropriate training. Such behavioral studies are combined with physiological methods, for example by setting lesions at selected places in the auditory system and testing various aspects of the animal's hearing function before and after. Postmortem histological and other controls may then be obtained at a time of the experimenter's choosing. Such experiments have shown clearly that the differences in results obtained in humans and in animals are usually in degree, and rarely in principle.

Anatomy

A schematic outline of the anatomy of the central auditory system is given in **Fig. 1**. The fibers of Corti's ganglion cells (or spiral ganglion cells), approximately 25,000 to 30,000 in number, connect peripherally to the hair cells. Centrally, these fibers terminate on a second set of nerve cells in either of the two cochlear nuclei, forming synaptic junctions with them. Fibers of the latter cell continue on a central course, most of them crossing over onto the opposite side of the head, following a general pattern of the nervous system. Additional synaptic junctions, each connecting with the next higher set of neurons, are encountered sequentially: first in the superior olivary complex, where for the first time fibers stemming from both sides are brought together on the same cells; in the nucleus of the lateral lemniscus, an upward running tract; in the inferior colliculus; and in the medial geniculate body. Fibers projecting from here terminate in the auditory cortex, which in humans is almost hidden in a deep furrow, known as the Sylvian fissure.

These various nuclei are much more than mere switching stations, and signal processing is not

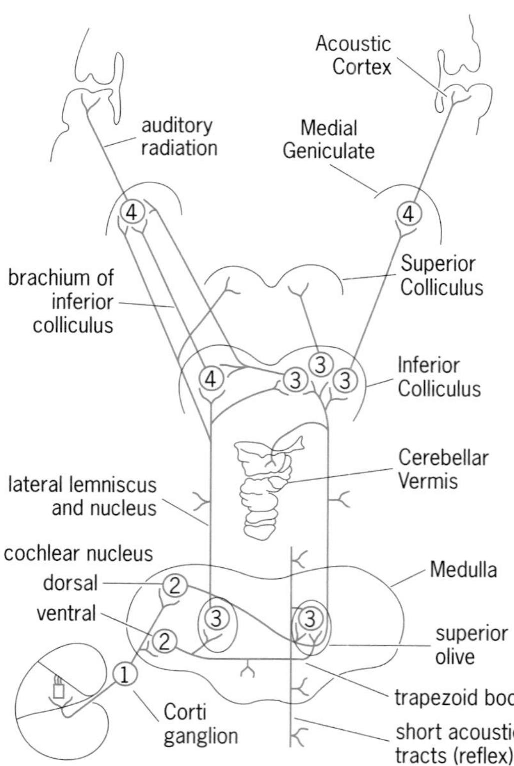

Fig. 1. Schematic outline of the central auditory pathways. Numbers indicate the number of neurons from the cochlea. (*After S. S. Stevens, ed., Handbook of Experimental Physiology, John Wiley and Sons, 1951*)

limited to the cortex. Some specific data handling goes on at each nucleus. What the cortex receives has already been extensively processed at lower levels. The cross-connection between the inferior colliculi on both sides (Fig. 1), for example, obviously aids binaural processing at that level.

Auditory Function

Mechanoacoustic energy received by the ear elicits a number of specific, interlinked events, first in the ear itself, then along the neural pathways, and finally in the auditory cortex. It is only after this sequence of events is completed that a sensation of sound commensurate with the applied signal is experienced by the listener.

Peripheral system. Sound pressure waves originating in the environment enter the external ear and are transmitted through middle and inner ears in the form of mechanoacoustic vibrations.

External ear. The pinna serves to collect sound from the environment in a directional, frequency-dependent manner, reception for most frequencies being maximal for normal incidence. This selectivity aids directional hearing, even of one ear alone. Because of its shape and length, the ear canal resonates at approximately 2700 Hz. The resonance is rather broad (indicating moderate damping of the sound vibration in the ear canal). Hence, the pressure acting on the tympanic membrane is increased over a range of almost two octaves, maximally by about 20 dB at the resonant frequency and for normal incidence. *See* RESONANCE (ACOUSTICS AND MECHANICS).

Middle ear. The middle ear serves to match impedances between the air in the ear canal which has low impedance, and the fluid of the inner ear, which has high impedance. To this end, it employs the principle of a hydraulic lever (in addition to the small mechanical leverage of the ossicular chain): the area of the tympanic membrane is about 20 times larger than that of the stapes footplate. However, this mechanical transformer is optimally efficient over only a limited frequency range from about 1000 to 4000 Hz. One reason for this is that the tympanic membrane is not displaced like a stiff plate; its displacement pattern changes drastically at higher frequencies, at which point it becomes a relatively inefficient driver of the lever system. A second reason is that at frequencies below 1000 Hz the input impedance of the system increases at 12 dB per octave with inverse frequency. It has been shown that the shape of the auditory threshold curve is entirely determined by the middle ear.

In the midfrequency range, the middle ear optimizes the transfer of mechanoacoustic energy into the inner ear, but only at low-input levels. At higher input levels, it acts as a complex, mechanical, nonlinear network, absorbing some of the incident energy and thus to some degree protecting the inner ear against overloading. The middle-ear muscles, the tensor tympani and the stapedius, play an important role in this. By contracting reflexly, they reduce transmission when the input exceeds 70–75 dB SPL. Furthermore, displacements of the tympanic membrane serve to pump air into and out of the mastoid air cells. Hence, these cells serve as a mechanical damping system in exactly the same manner as the holes in the backplate of a condenser microphone. *See* LOUDNESS; MICROPHONE.

The performance of the middle-ear transformer is optimal only when there is no static air-pressure difference across the tympanic membrane. Middle-ear pressure is equalized from time to time via the Eustachian tube, which is briefly forced open by swallowing, yawning, and a number of other activities.

Inner ear. Incident mechanoacoustic energy displaces the stapes footplate, and thus the perilymphatic fluid of the scala vestibuli. Since the stiffness of the basilar membrane varies exponentially over its length (it becomes more compliant toward the helicotrema), the cochlea acts like a hydrodynamic, tapered transmission line (or delay line). The cochlear duct undergoes alternating displacements, which progress in a wavelike manner from base to apex. In this process, the round window at the basal end of the scala tympani acts as a relief valve. The traveling waves form localized displacement maxima at frequency-dependent places; they are located near the base for high frequencies and shift systematically toward the apex with inverse frequency (**Fig. 2**). This distribution of frequency converts frequency information into spatial information; such a frequency distribution is called tonotopic. As the delay-line property demands, there is a cochlear time delay

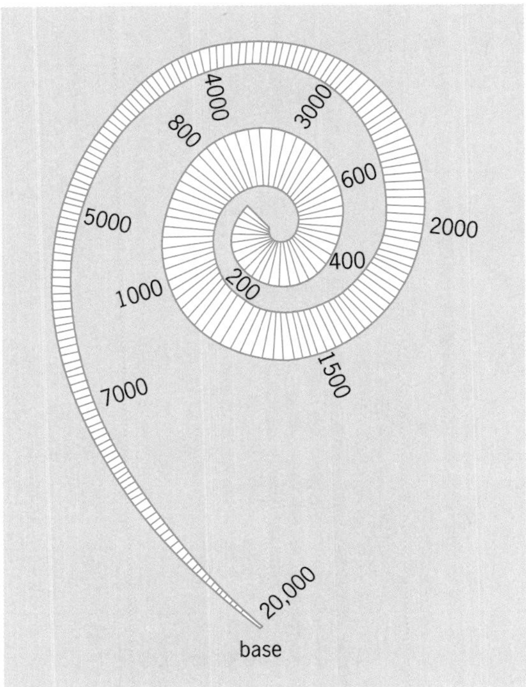

Fig. 2. Distribution of frequency maxima along the basilar membrane. Numbers give frequencies in hertz. Length of basilar membrane = 32 mm, width = 0.04 mm at base, 0.50 mm at apex. (*After O. Stuhlmann, An Introduction to Biophysics, John Wiley and Sons, 1943*)

that maximally (for low frequencies traveling all the way up to the apex) amounts to 4–5 ms. Both the tonotopic distribution and the spatial time delay (systematic with inverse frequency) are preserved throughout the entire auditory system. Seen from an individual location along the basilar membrane, the tuning is very sharp, a bandpath-filter property that is once more maintained throughout the mechanical system and part of the neural system as well. *See* ELECTRIC FILTER.

Shearing motion. The traveling-wave displacements of the cochlear duct, in which the organ of Corti participates, produce tangential shearing motions between the organ proper and the tectorial membrane, to which the sensory hairs are attached. The shearing forces thus generated constitute the ultimate mechano-acoustic input to the hair cells. In this process, a second pressure transformation takes place (at the expense of displacement) and, perhaps, the tuning is also slightly sharpened.

Receptor potentials. At the sensory hair cells, the original mechanoacoustic energy is converted into a chemoelectric form of energy. Two kinds of electrical receptor potentials are generated in each cell being simulated, one ac and the other dc. The ac potential, an analog signal, has the general properties of the applied acoustic stimulus as seen through a bandpass filter, sharply tuned for reasons already discussed. The dc signal, similarly, is mainly evoked by frequencies within the narrow pathband given by the cell's position (**Fig. 3**). The hair cells act as biological amplifiers; that is, their receptor potentials possess more energy than the applied mechanoacoustic signal.

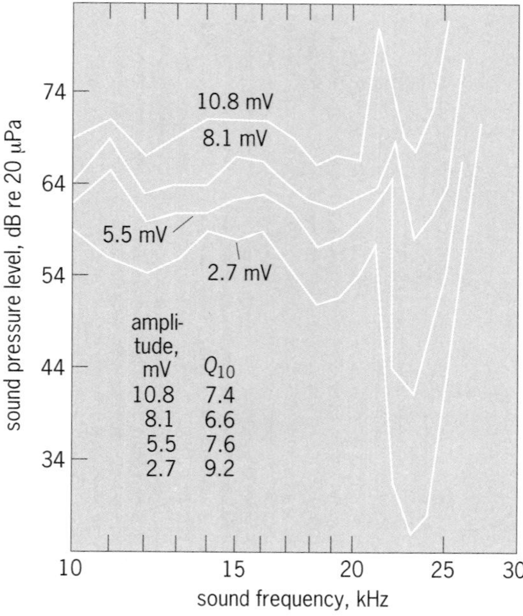

Fig. 3. Isoamplitude curves of the dc receptor potential within one guinea pig inner hair cell of the basal region. Q_{10} is a measure of the width of each curve, equal to the center frequency divided by the bandwidth at 10 dB above threshold. (*After I. J. Russell and P. M. Sellnick, Tuning properties of cochlear hair cells, Nature, 267:858–860, 1977*)

Cochlear microphonics and summating potentials. The ac as well as the dc potentials spill over in the cochlear fluids, where the potentials from many cells are summed; for ac events this occurs according to their local phases, which are determined by the spatial delay of the traveling waves. Both potentials in their extracellular, summed forms can be recovered from the cochlear fluids or from the round-window membrane—they are called cochlear microphonics (ac) and summating potentials (dc) respectively. They are valuable research tools, even though they are rather broadly tuned.

The sensitivity of the ear is extremely high. In a quiet environment, a sound pressure of a mere 20 micropascals is sufficient for threshold stimulation in the most sensitive frequency region, around 3000 Hz. The resulting displacement of the cochlear duct is probably on the order of 100 nanometers. This extreme sensitivity is made possible, in addition to other factors, by the extremely low input of limiting background noise (essentially constituted by brownian motion) at the hair cell. *See* BROWNIAN MOVEMENT.

Hair cell–nerve fiber junction. A sequence of three events leads to the discharge of action potentials in the nerve fiber connected to a given hair cell (1) the receptor potentials of the hair cells; (2) the release of a chemical neurotransmitter, not yet clearly identified, into the synaptic cleft between hair cell and nerve fiber dendrites; and (3) dc generator potentials within the dendrites that, after spatial and temporal integration, finally trigger an action potential. Thus, the original acoustic signal is uniquely encoded as a series of brief electric discharges in a number of nerve fibers. Action potentials are the universal neural signals traveling in all nerve fibers. *See* BIOPOTENTIALS AND IONIC CURRENTS; SYNAPTIC TRANSMISSION.

Neural system. The various synapses of the central auditory pathway (Fig. 1) are much more than mere switching stations—some specific data handling goes on at each of them. What the cortex of the brain receives has already been extensively processed at lower levels, aided by cross-connections between individual cells. For example, the bilateral innervation of superior olivary cells and the cross-linkage between the inferior colliculi on both sides (Fig. 1) aid binaural processing at those relatively low levels, which aids directional hearing, even though ultimate "decisions" are made at the cortical level.

Cochlear nerve fibers. About 90–95% of all cochlear nerve fibers connect exclusively with inner hair cells, each fiber linking up with only one cell, but each cell receiving up to 20 fibers (neural divergence). In contrast, each outer-hair-cell fiber connects to approximately 10 cells in an apparently random sequence, but each cell receives only one branch of one fiber (neural convergence). Since the outer-hair-cell fibers are so few in number, and since the diameter of each is rather small, they are probably rarely encountered by physiologist's electrodes; as a result, their response patterns have not been clearly identified. The following comments therefore apply only to inner-hair-cell fibers, which have been extensively studied.

One reason why the cochlear-nerve-fiber code has only been partly deciphered is that one outer-hair-cell fiber, as mentioned above, connects to more than one cell in random sequence. Furthermore, in each cochlear nerve fiber, there exists a "noise"; that is, random, spontaneous discharges occur that compete with the evoked potentials so that following any discharge, nerve fibers are unable to respond for approximately 1 ms, their so-called refractory period. For these reasons, the evoked firing of all nerve fibers is a highly complicated, probabilistic process. Cochlear nerve fibers increase their firing rate over the spontaneous level when excited by a signal. At higher levels, therefore, the spontaneous firings are reduced in number as the evoked discharges become

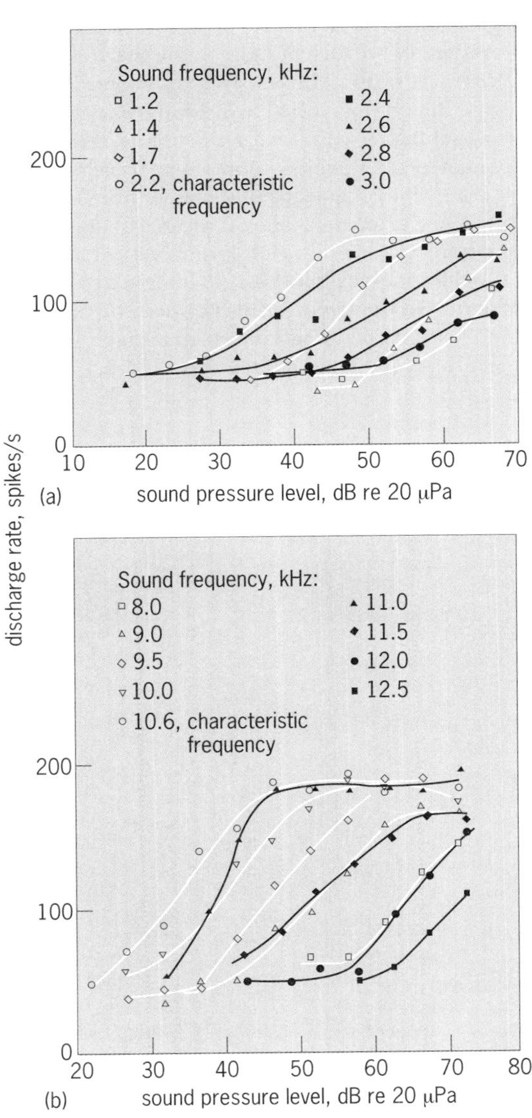

Fig. 4. Intensity/rate functions of two cat cochlear nerve fibers tuned to (*a*) 2.2 kHz and (*b*) 10.6 kHz. (*After W. D. Keidel and W. D. Neff, eds., Handbook of Sensory Physiology, Springer, 1975*)

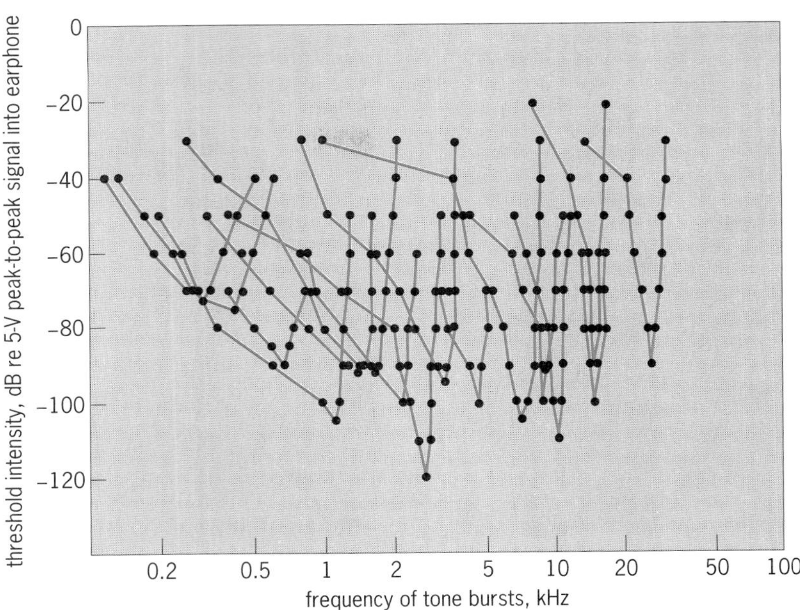

Fig. 5. Tuning curves of a number of nerve fibers in the cat cochlear nerve. (*After N. Y.-S. Kiang, Discharge Patterns of Single Fibers in the Cat's Auditory Nerve, MIT Res. Monogr. 35, MIT Press, 1965*)

dominant. However, the dynamic range of a single fiber is a mere 40 dB, and its linear portion is even smaller (**Fig. 4**). This is in marked contrast to the dynamic range of actual hearing, that is, the differentiation of loudness, which extends over at least 120 dB for frequencies around 3000 Hz.

The tuning of cochlear nerve fibers, based on an increased firing rate, is very sharp. The difference in *Q*-factor between high-frequency and low-frequency fibers shown in **Fig. 5** is more apparent than real. It largely disappears when the data are plotted against a linear frequency scale instead of the conventional logarithmic one. The frequency to which each fiber is most sensitive is known as its best (or characteristic) frequency.

Because of the probabilistic nature of neural responses, most forms of assessment require statistical methods; the histogram is a frequently used tool. **Figure 6** presents some examples of post-stimulus-time (PST) histograms of a number of different fibers tuned to various frequencies. Responses are elicited by repeated and periodically timed signal presentations, usually 5000 in number. The height of each column at each time point represents the frequency of response at that point. Following each brief click signal, there is an extended period of responses, whose frequency of occurrence decays exponentially with time. In high-frequency fibers (best frequencies higher than 3–4 kHz), the firing occurs at random. In low-frequency fibers, however, the firing is periodic, synchronized with the period of the best frequency of the fiber.

The best frequency reflects the location along the cochlea at which the fiber is connected to its hair cell and the frequency of the cochlear traveling wave which has its maximum at that location. Thus, the

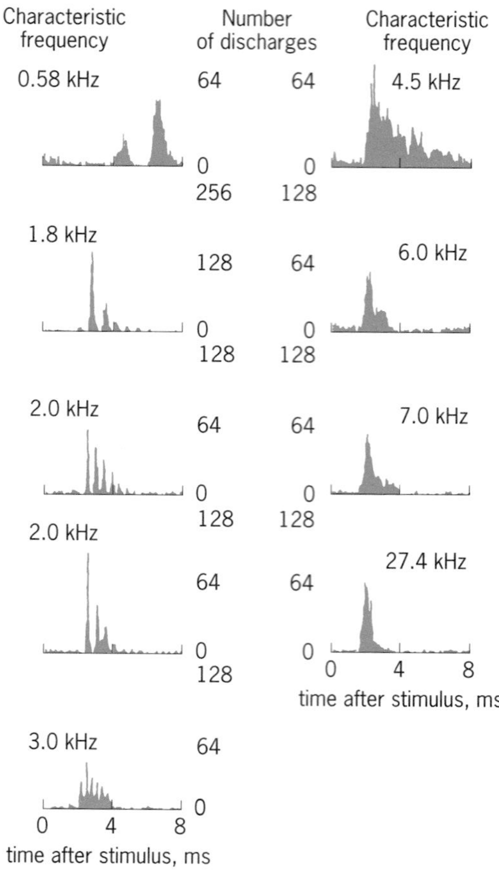

Fig. 6. PST histograms for a number of nerve fibers in a cat with various characteristic frequencies, showing response to click stimulation. Each histogram represents 1 minute of data (600 clicks at click rate of 10 per second). (*After N. Y.-S. Kiang, Discharge Patterns of Single Fibers in the Cat's Auditory Nerve, MIT Res. Monogr. 35, MIT Press, 1965*)

responses of low-frequency fibers, including their exponential time decay, duplicate the mechanical "ringing" occurring locally in each (narrow) cochlear filter. The reason for the two different types of responses is that the firing of high-frequency fibers is triggered by the hair-cell dc receptor potential, with no possibility of time locking, whereas that of low-frequency fibers is triggered by the ac potential, to which responses may become time-locked.

Triggering by the ac potential also allows the firing of low-frequency fibers to be phase-locked to the period of a sine-wave signal. This phase locking can achieve frequency selectivity and sensitivity superior to that based on increased firing rate (Figs. 4 and 5). Signal frequency thus is encoded in two different ways: by phase-locked periodicity, mainly in low-frequency fibers, and by location, mainly in high-frequency fibers.

Intensity is also encoded in a twofold manner: by the firing rate in the best-frequency fiber and by the recruitment of additional fibers tuned to nearby frequencies. These latter fibers start to respond at higher levels, that is, when their tuning curves have become broad enough to include the frequency in question. Integration of such an intensity-staggered rate function over a number of fibers would offer a dynamic range wider than that of a single fiber, but there is no direct evidence for this notion as yet. Although the two codes are seen to overlap partially, there is hardly any confusion on the part of a listener between intensity changes and frequency changes: there is only a very slight pitch change as signals become louder.

Except for the transduction process in the hair cell–nerve fiber junction, there is no evidence for signal processing in cochlear nerve fibers and their associated bipolar ganglion cells. In outer-hair-cell fibers, with their multiple inputs from different cells, there should be some integration, but proof for this assertion is lacking.

Cochlear nuclei. In the cochlear nuclei of the brainstem, there is the first real evidence for neural processing. For example, some cells, aptly named octopus cells, connect across approximately 10 parallel channels. Their tuning curves therefore become considerably broader. In addition to excitation (or no excitation) as found exclusively in cochlear nerve fibers, the outputs from cochlear nucleus cells may show inhibition (that is, a reduction in the rate of spontaneous discharges). Inhibition usually occurs in a frequency region slightly remote from the best frequency, where excitation is invariably found. Furthermore, in addition to the "primarylike" responses of cochlear nerve fibers (Fig. 6), there are three new patterns, descriptively called choppers, pausers, and "on" patterns, although their significance with respect to processing is not understood (**Fig. 7**).

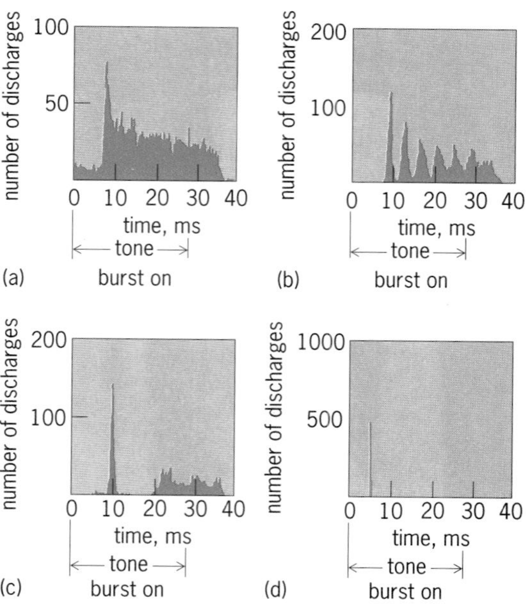

Fig. 7. Response patterns of cells in cat cochlear nuclei, with characteristic frequencies indicated. (*a*) Primarylike response, 3.5 kHz. (*b*) Chopper response, 7.0 kHz. (*c*) Pauser response, 6.6 kHz. (*d*) "On" pattern, 2.1 kHz. (*After R. R. Pfeiffer, Classification of response patterns of spike discharges for units in the cochlear nucleus: Tone burst stimulation, Exp. Brain Res., 1:220–235, 1966*)

Superior olivary complex. Cells in the superior olivary complex (Fig. 1) are the first to receive inputs from both cochleas; in general, those from the same side are inhibitory and those from the opposite side excitatory. Responses from such cells are controlled by differences in binaural inputs, that is, by interaural time or intensity differences. Their actions at this relatively low level presage psychoacoustic phenomena that have to do with binaural hearing, which aids in determining the direction of a given signal source.

Higher centers. The inferior colliculus is a large, conspicuous nucleus. In some lower animals it is even larger, relatively, than in humans. In birds and reptiles, which have a very thin brain cortex, it is the highest nuclear mass of the auditory system. Significantly, even in higher mammals, a number of ascending fibers terminate here. Some auditory tasks, such as amplitude discrimination, are carried out at this level, as has been judged from experiments in which the auditory cortex is surgically ablated. Several acoustic reflexes are controlled in this nucleus, for example the reflex contraction of the middle-ear muscles in response to sounds above 70–75 dB SPL. Also, the inferior colliculus controls the turning of the eyes and of the head toward a sound source, reflexes that aid auditory localization, which is optimal near the midline of the head.

The medial geniculate body belongs to a group of large nuclei, collectively known as the thalamus. All ascending sensory signals (except those of the olfactory receptors) pass through thalamic centers before being distributed to specific cortical areas. How closely the thalamic centers are associated with the cortex is shown by the fact that, when a given cortical area is destroyed, its thalamic nucleus degenerates in a paradoxical, or retrograde, fashion—against the direction of signal flow. The medial geniculate bodies of both sides are interconnected by the commissure of Gudden, another structure that aids in binaural hearing. At this level, there are not only cells that respond to binaural stimulation as such, but also some that register the movement of an auditory target and the direction of such movement.

In all nuclei of the central auditory system, there are well-ordered tonotopic distributions, although multiple representations exist, for instance in the inferior colliculus. Their purpose remains unclear.

Auditory cortex. In contrast to those of the lower levels of the auditory system, cells of the auditory cortex show little, if any, spontaneous activity. A clear-cut tonotopic organization was finally established after considerable experimental difficulties were overcome. Cells of the "primary" auditory area respond directly to acoustic inputs, those of the "secondary" areas only after the primary areas have been paralyzed. The significance of this empirical fact is not yet clear.

With respect to the anatomical and physiological substrates of higher auditory functions, such as

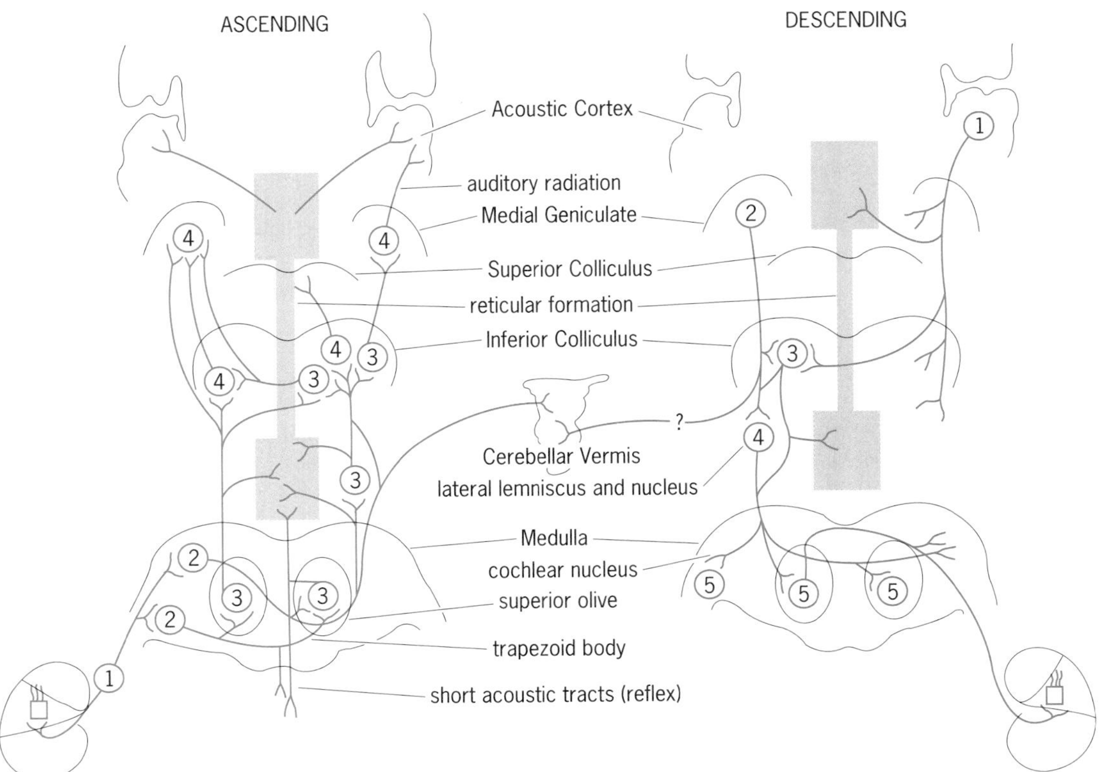

Fig. 8. Ascending (afferent) and descending (efferent) auditory nerve systems. Number labels in the ascending system indicate the number of neurons from the cochlea, while those in the descending system indicate the number of neurons from the auditory cortex. (*After R. Galambos, Some recent experiments on the neurophysiology of hearing, Ann. Oto-Rhino-Laryngol., 65:1053–1059, 1965*)

acoustic memory (both short-term and long-term), auditory associations, and the like, there is little known, except for some serendipitous findings made at the occasion of surgical interventions. Electrical stimulation of specific areas of the temporal lobe surface have given rise to sensations, such as voices known in the past or melodies or ditties long forgotten. However, no systematic studies have been possible on such memory releases, nor is anything known about the underlying mechanisms. The left auditory cortex, to which most of the fibers originating in the right ear project, fulfills mainly intellectual tasks, such as word recognition, whereas the right cortex fulfills emotional tasks, such as the appreciation of music.

Efferent system. Almost duplicating the afferent (ascending) auditory system (Fig. 1) is an efferent (descending) system which is just as elaborate (**Fig. 8**). The final projection of the efferent system to the cochlea is via the olivocochlear bundle of Rasmussen. Specialized nerve endings, different in appearance from their afferent counterparts, link with hair cells or their afferent nerve fibers.

The discovery of the olivocochlear bundle gave the first evidence for the existence of sensory efference, which has since been demonstrated in all other sensory systems. In anesthetized animals, central stimulation of the auditory efferent system was shown to decrease the discharges in cochlear nerve fibers, somewhat akin to the action of a governor controlling an engine; but it also increases the magnitude of the cochlear microphonics. Its function in awake and unrestrained animals has been difficult to demonstrate, but it appears to improve the (physiological) signal-to-noise ratio in the presence of (external) interfering noises, to improve the resistance of the organ to noise-induced damage, and to aid frequency discrimination. *See* CONTROL SYSTEMS; GOVERNOR.

Reticular formation. The reticular formation in the brainstem is the general arousal center of the brain. Sensory inputs of any kind, in stimulating this center, will arouse all other senses, alerting them to the reception of specific inputs of their own so that the whole organism can take appropriate action. When this formation does not function, animals or humans remain in a comatose state, incapable of being aroused by any means. Figure 8, presenting the afferent and efferent systems side by side, shows their connections to the reticular formation. The study of the effects of the auditory system on the reticular formation, and vice versa, has greatly aided understanding of the general function of this formation. *See* RETICULAR FORMATION.

Overview. Understanding of peripheral auditory function is much more advanced than that of its central counterparts. This is because, in contrast to neural functions, peripheral functions can be readily modeled by mechanical and electrical devices, such as mechanical lever systems, electrical transformers, and delay lines, whose working principles are thoroughly understood. No such devices appear to exist that make large-scale use of the operating principles of the brain, for instance of parallel processing in multiple channels. Computers that are often cited as such devices and that indeed come closest in this respect cannot truly duplicate the action of the central nervous system. Existing models of neural functions are mostly phenomenological ones, and the lack of good analytical models makes their study difficult. *See* HEARING (HUMAN); NERVOUS SYSTEM (VERTEBRATE). Juergen Tonndorf

Bibliography. W. L. Gulick, G. A. Gescheider, and R. D. Frisina, *Hearing: Physiological Acoustics, Neural Coding, and Psychoacoustics*, 1989; J. O. Pickles, *An Introduction to the Physiology of Hearing*, 2d ed., 1988; J. Tonndorf (ed.), *Physiological Acoustics*, 1981; G. von Békésy, *Experiments on Hearing*, 1961, reprint 1980.

Physiological action spectra

Representations of the comparative effects of different wavelengths of light on living systems or on the components of living systems. A knowledge of the effects of different wavelengths on living systems helps lead to an understanding of the detailed mechanisms of energy transfer and utilization and to a determination of the essential compounds involved in light action of living systems. The work of action spectroscopy is founded on the firm bases that energy must be absorbed before it is utilized and that each chemical compound has a characteristic absorption spectrum. Therefore, the shape of the action spectrum may lead to the identification of the absorbing molecules. Action spectroscopy has been used extensively to study three classes of compounds: porphyrin-containing proteins, nucleic acid polymers, and plant pigments. *See* ABSORPTION OF ELECTROMAGNETIC RADIATION.

Beneficial effects of light. Proteins which contain porphyrin prosthetic groups, such as hemoglobin and cytochrome, play an important part in the transport of oxygen and in the oxidation-reduction systems of cells. Such prosthetic groups have a high affinity for carbon monoxide (CO), and therefore the respiration of living systems is inhibited by carbon monoxide. Light can remove this carbon monoxide inhibition by causing the dissociation of the porphyrin-CO complex. The extensive studies of O. H. Warburg and collaborators on the action spectra for removal of the carbon monoxide inhibition respiration led to the identification of the respiratory enzyme as a porphyrin-containing protein. The fact that light absorbed not only in the porphyrin ring but also in the protein can cause the dissociation of carbon monoxide indicates that energy may be transferred over distances of about 3 nanometers. *See* CYTOCHROME; HEMOGLOBIN; PORPHYRIN.

Destructive effects of light. The beneficial effect of light mentioned above is to be contrasted with the destructive effect of light in the ultraviolet spectral region. This light can destroy the enzymatic activity

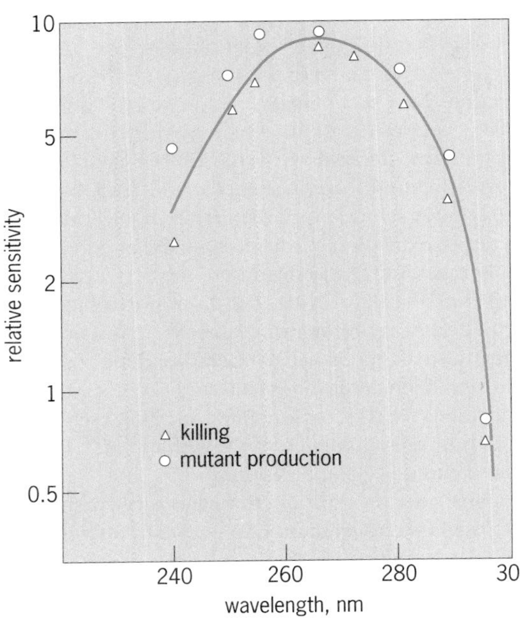

Action spectra for the killing of yeast and the production of mutants in yeast by radiation with ultraviolet light. (*After C. Raut and W. L. Simpson, Arch. Biochem. Biophys., 57(1):218–228, 1955*)

of proteins and can kill bacteria and viruses. Action spectra for the killing and the production of mutants in yeast, shown in the **illustration**, indicate that the most sensitive wavelengths are in the neighborhood of 265 nm. The shape of the action spectrum is very similar to the absorption spectrum of nucleic acids and indicates that nucleic acid polymers are essential components in the duplication of living systems. Some of the effects of ultraviolet light on deoxyribonucleic acid (DNA) are reversed when the system is illuminated by intense visible light. Such reversibility is known as photoreversal or photoreactivation. *See* BACTERIA; NUCLEIC ACID; PROTEIN; ULTRAVIOLET RADIATION (BIOLOGY).

Photosynthesis. Because the Sun is the most important source of energy, it is not surprising that photosynthesis is one of the most thoroughly studied of the fields which concern energy exchanges between light and living systems. The identification of chlorophyll as an essential component of photosynthesis is made by comparing the action spectrum for photosynthesis with the absorption spectrum of chlorophyll. The similarity between the two indicates the role that chlorophyll plays in photosynthesis. On the other hand, the action spectra for photosynthesis in some algae do not resemble the absorption spectrum of chlorophyll. In these cases, light is absorbed in other pigment molecules and the energy is transferred to chlorophyll to be used for photosynthesis. Action spectra have shown the existence of wavelength-dependent effects on such diverse photoperiodic properties of living things as the germination of seeds, the flowering of plants, and the change in coloration of animals with the seasons. All these effects are governed by an elaborate interplay between red and far-red light. *See* PHOTOSYNTHESIS.

Mechanism. The primary process in light action is the absorption of a light quantum by a molecule. The molecule is thus raised to an excited state. The excitation energy may be passed on to other molecules and utilized for chemical reaction, as in photosynthesis, or the extra energy may be used to break bonds and alter chemical structure, as in fluorescent light or ultimately degraded as heat. In the latter case there may be no effect of the absorbed radiation at all.

A particularly interesting feature of excited states is the ability of energy to be transferred over appreciable distances from one molecule to another by a process known as energy transfer. Such processes permit energy absorbed in one molecule to appear in a different type of molecule. The existence of resonance-energy transfers may be inferred from a comparison of the action absorption and the fluorescent spectra of the irradiated material. *See* CHLOROPHYLL; PHOTOPERIODISM; VISION. Richard B. Setlow

Bibliography. D. J. Pizzarello (ed.), *Radiation Biology*, 1982.

Physiological ecology (animal)

Physiological ecology combines the study of physiological processes, the functions of living organisms and their parts, with ecological processes that connect the individual organism with population dynamics and community structure. In the past, physiological ecology focused principally on physiological processes in isolation from each other and without strong quantitative connections to ecological processes. That focus is now rapidly changing into a dynamic quantitative and more integrated discipline. Physiological ecology reaches up to population dynamics and community structure and down to genetic phenomena influencing key endocrine functions affecting behavior, energetics, survival, reproduction, and fitness in diverse environments. *See* POPULATION ECOLOGY.

Methodology and scope. Physiological ecologists focus on whole-animal function and adjustments to ever-changing environments. Short-term behavioral adjustments and longer-term physiological adjustments tend to maximize the fitness of animals, that is, their capacity to survive and reproduce successfully. Physiological ecologists need a breadth of technical competencies in physiology and ecology, in both theory and experiment, and in both laboratory and field. Quantitative and computer skills are essential, and interests that reach beyond these disciplines are needed to forge the quantitative links at higher and lower levels of biological organization.

Among the processes that physiological ecologists study are temperature regulation, energy metabolism and energetics, nutrition, respiratory gas exchange, water and osmotic balance, and responses to

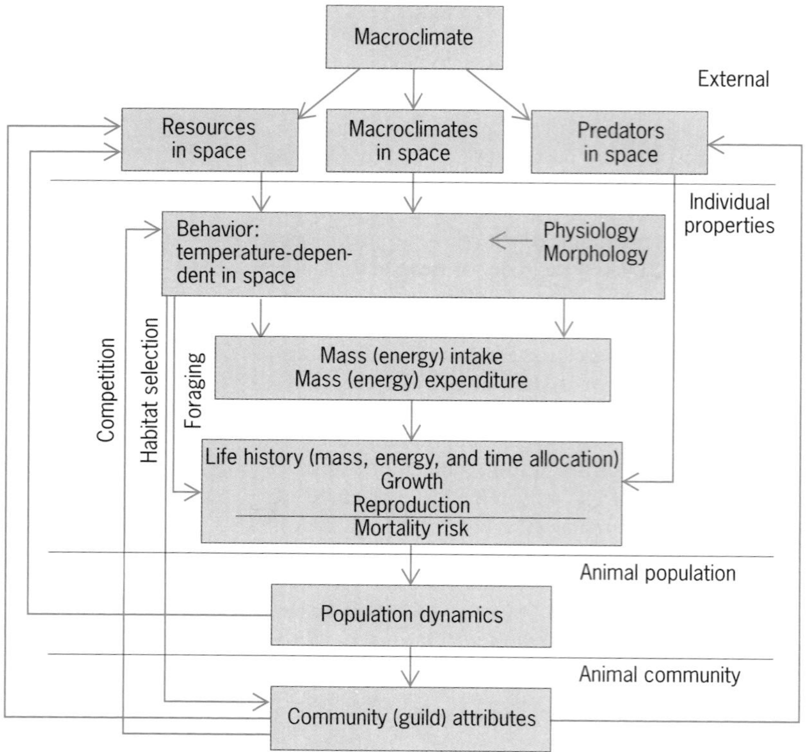

Fig. 1. Relationships between climate, individual animal properties, population dynamics, and community attributes.

environmental stresses. These environmental stresses may include climate variation, nutrition, disease, and toxic exposure. **Figure 1** illustrates the interconnections between climate, microclimates, and individual properties of morphology, physiology, and temperature-dependent behavior. Climate affects animal heat and mass balances, and such changes affect body temperature regulation. Behavioral temperature regulation modifies mass and energy intake and expenditure, and the difference between intake and expenditure provides the discretionary mass for growth and reproduction. Mortality risk (survivorship) also depends on temperature-dependent behavior, which determines daily activity. Activity time constrains the time for foraging and habitat selection, which in turn influence not only mortality risk but also community composition.

Thus, growth, reproduction, and survivorship are the three variables that drive population dynamics. Population dynamics of multiple species strongly affect community attributes, such as species composition and numbers. Competitors in the community may alter temperature-dependent behavior of the species of interest, and thus their energetics and population dynamics. Community attributes may also affect the distribution of predators in space, which affects mortality risk of the species of interest. *See* METABOLISM.

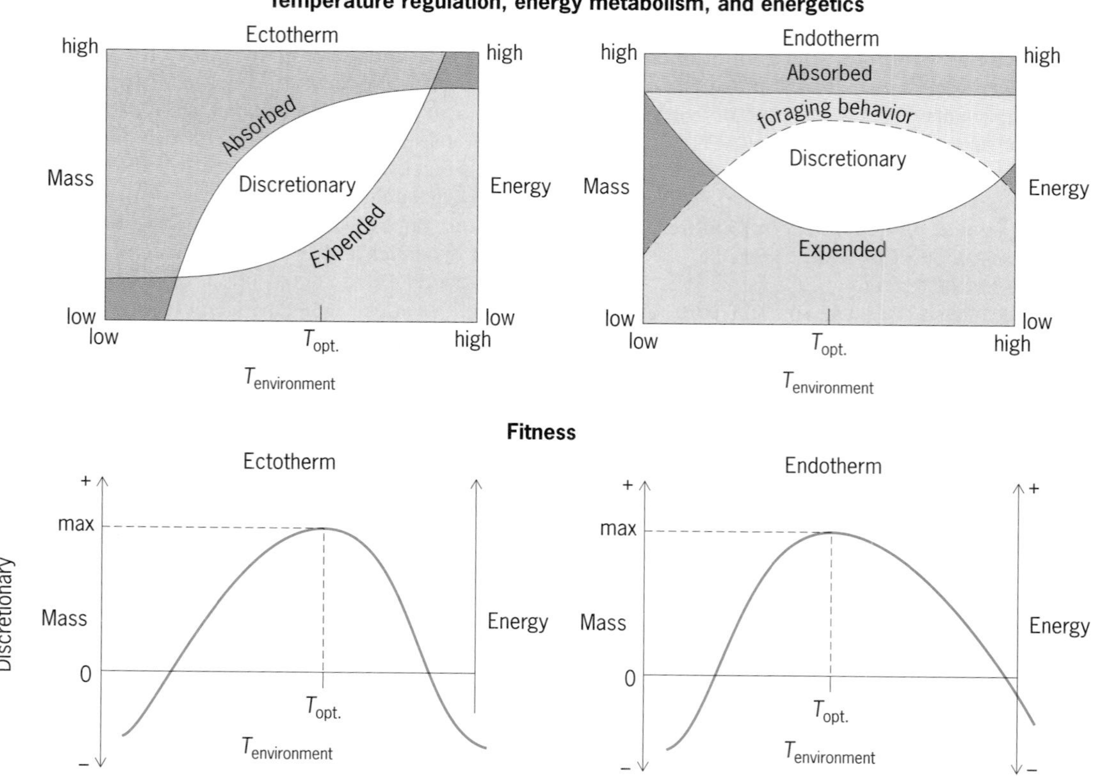

Fig. 2. Relationship between climate, intake and expenditure of mass and energy, and fitness associated with temperature regulation, energy metabolism, and energetics. For endotherms, the maximum potential for absorption is a straight line if body temperature remains constant. Behavior of the animal in cold or heat affects the amount of food ingested and therefore absorbed. The optimal temperature for achieving maximum discretionary mass and energy is different for each different body size. The temperature scale here assumes that air and radiant temperatures are equal.

Temperature regulation, energy metabolism, and energetics. Temperature regulation involves both behavioral and physiological components. The behavioral components typically involve avoiding temperature extremes. **Figure 2*a*** and *b* illustrate the crucial importance of appropriate behavioral temperature regulation and show the energetic consequences to ectotherms (cold-blooded animals) and endotherms (warm-blooded animals) of deviations from optimal temperature. The optimal environmental temperature for each is the temperature that creates the maximum separation between the curves for absorbed mass (chemical energy) and expended mass (chemical energy). A temperature selected that is higher or lower than this optimum reduces the separation, thereby reducing the potential fitness, as illustrated in Fig. 2*c* and *d*.

Endotherms have a maximum potential food absorption that is independent of environmental temperature from a physiological perspective. However, in reality food intake by endotherms is temperature-dependent. This phenomenon was described in domestic animals in the 1950s, but it has only recently begun to be quantified for animals in the wild. The broken curve that is concave downward in Fig. 2*b* represents temperature-dependent foraging behavior. The "absorbed" and "expended" curves show that the fitness of endotherms is as affected by environmental temperature as that of ectotherms, a fact that is not generally appreciated. Thus, temperature regulation, energy metabolism, and energetics are intimately interconnected for both ecto- and endotherms.

Physiological temperature regulation may take many forms. It may involve reduced or increased blood flow to appendages, depending on whether the environment is cold or hot, respectively. This method of modifying an organism's heat balance works best at intermediate and low temperatures. At high temperatures there is little heat transfer advantage to altered blood flow unless the appendage can be placed in an environment where high heat transfer rates are possible. Other mechanisms of physiological temperature regulation involve evaporative cooling by sweating and panting. Since sweating and panting affect water balance and respiratory gas exchange, there are strong constraints on when they may be used effectively. *See* ENERGY METABOLISM.

Respiration. Respiration is an exchange of gases between a living organism and its environment. For all but a few microorganisms, oxygen is absorbed and carbon dioxide is released to the environment. The top half of **Fig. 3** illustrates a respiratory system where gases enter at one location by diffusion, convection, or both. Gases may exit either at the same location or at another part of the body depending on the particular respiratory organ (gill, lung, or tracheae). Because water is added to air entering lungs and tracheae, respiration may affect the animal's overall water balance. Water may be recovered near the exit of these structures if the gas temperatures near the exit are close to external air temperature but below body temperature.

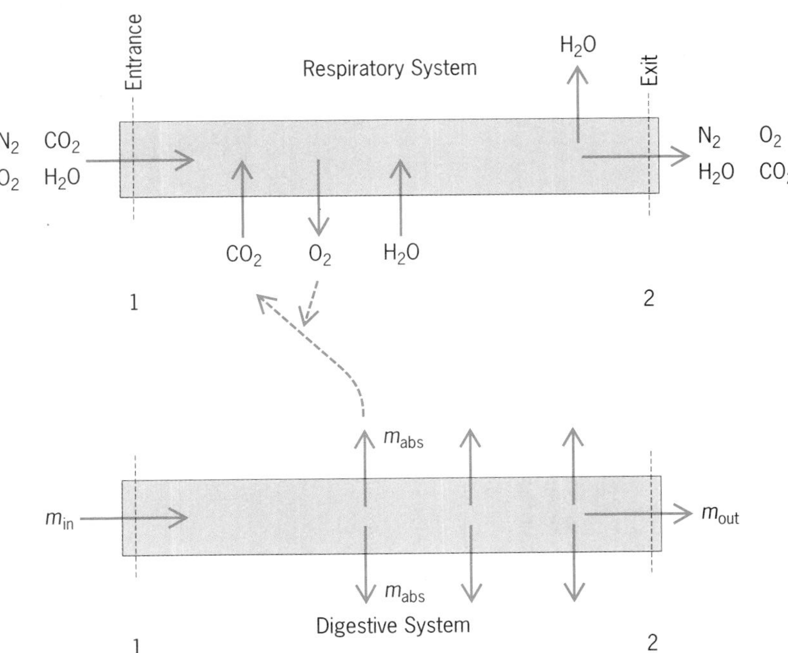

Fig. 3. System diagram models for a respiratory system and a digestive system of an organism. Mass (*m*) flows in at entrance 1 and flows out at exit 2. The type of food in the digestive system affects the ratio of carbon dioxide and oxygen exchanged in the respiratory system and therefore the mass flow and water balance associated with the respiratory system.

The ratio of oxygen absorbed to carbon dioxide released by a respiratory surface is affected by the substrates being oxidized for energy. For example, if carbohydrates are being oxidized, the moles of oxygen consumed are equal to the moles of carbon dioxide released. If protein or lipids are being oxidized, the respiratory quotient, R.Q. (moles carbon dioxide/moles oxygen), may be 0.80 or 0.71, respectively. Thus, as Fig. 3 indicates, the type of food in the digestive system can have an impact on gas transport in the respiratory system. *See* RESPIRATION.

Nutrition. The digestive system processes mass ingested by an organism within a gut or gut cavity. Very primitive organisms that lack a gut absorb nutrients across their cell surfaces, or engulf whole food particles into their cells. Quantitative models of gut function for a variety of organisms have emulated

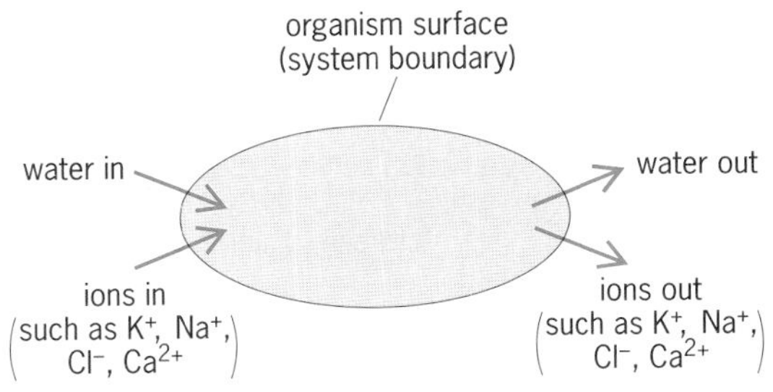

Fig. 4. System diagram for water balance and osmotic balance.

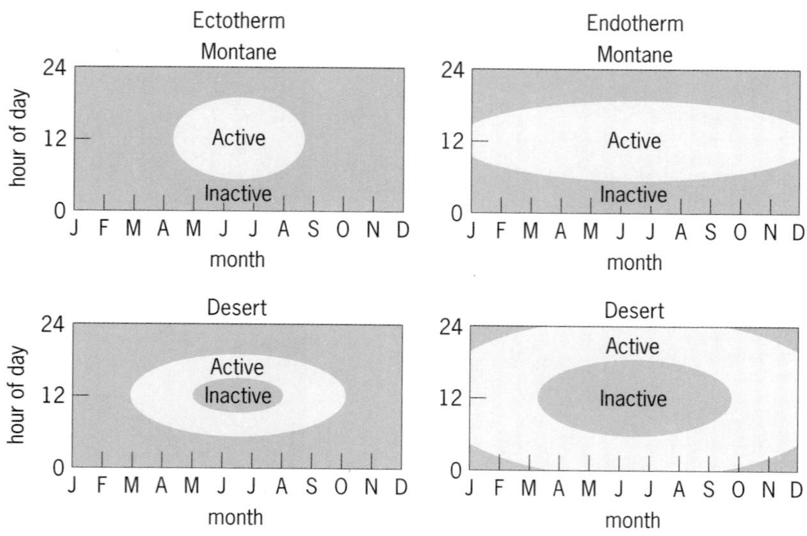

Fig. 5. Temperature-dependent changes in activity time in response to temperature stress. Effects of elevation difference (climate) are represented.

nutrient uptake very well. This is important because quantitative models of physiological processes allow us to understand much better the connections between subsystems in an organism. Good models allow us to explore how an organism's performance may vary as environmental condition changes, as an organism does in natural systems.

The amount of mass absorbed by a digestive system and the chemical energy that mass represents are of crucial importance for fitness (Fig. 2). The efficiency of food absorption by the gut in ectotherms is temperature-dependent. Thus, temperature regulation, energy metabolism, energetics, respiration, and nutrition are all interconnected to help maximize fitness. *See* NUTRITION.

Water and osmotic balance. Mass balances of water, ions, and absorbed nutrients are all involved in the maintenance of osmotic balance in living organisms. Proper osmotic balance is crucial for survival, since the enzymatic reactions of cells and communication between and within cells depends on ionic gradients. As **Fig. 4** indicates, water and ions may flow in across the outer integument, gills, or digestive system surfaces. Water and ions may exit the organism across respiratory surfaces immersed in water, from the kidneys, or from other excretory or digestive organs. Since ion transport under some circumstances may become very expensive energetically, water and ion balance are connected to the energetics of the organism and to the physical environment which modifies those energetics. *See* OSMOREGULATORY MECHANISMS.

Responses to stress. Organisms typically respond to stress in one of two ways. The first and less expensive option is to avoid stress behaviorally, if possible. If that does not work, a physiological response may be required. **Figure 5** illustrates a key behavioral response to thermal stress for ectotherms and endotherms. This particular response, change in activity time, is of central importance to linking individual energetics with population dynamics and community structure. This is because the total amount of time available for activity in a year simultaneously impacts the probability of survival from predation and the time available to gather food and water for growth, reproduction, and fat storage. Survivorship, growth, and reproduction are the key variables that drive population dynamics and community structure. Ectotherms at high elevations have fewer hours of activity than in warmer climates. However, if the climate is too hot at lower elevations, midsummer activity may shift to early morning and late afternoon to avoid the hottest part of the day, also limiting the duration of activity. In contrast, endotherms at high elevations may be active on a year-round basis, but typically significantly restrict their activity in colder months during the daytime. At lower elevations, endotherms may be nocturnal during the hot months and diurnal during winter months. Such changes in activity patterns can have important implications for reducing costs of operation and maximizing fitness.

Changes in activity time directly affect discretionary mass and energy, and survivorship, the currencies associated with fitness. **Figure 6** illustrates how environmental temperatures can modify

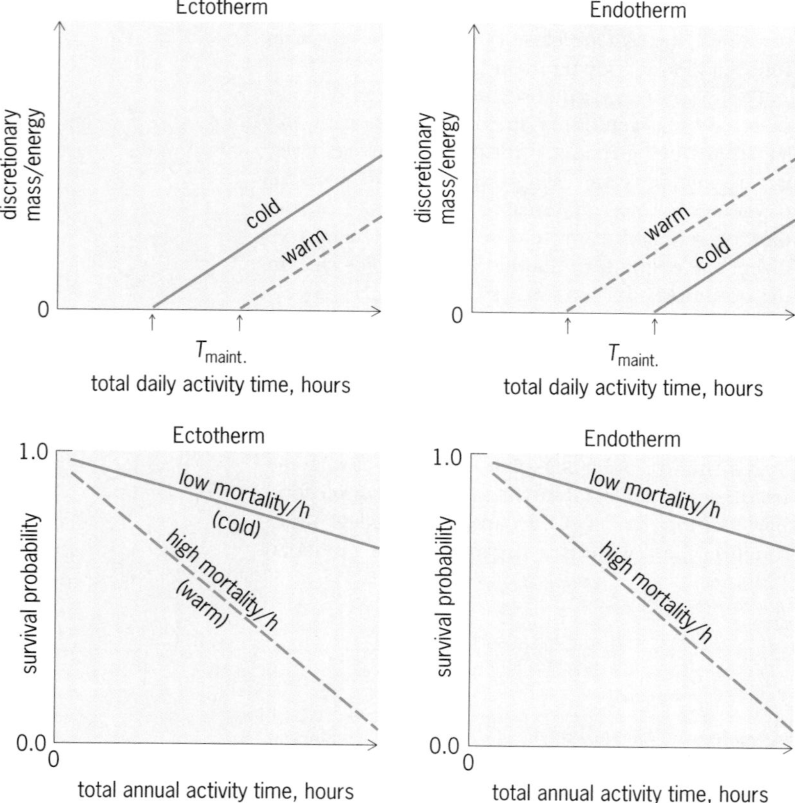

Fig. 6. Activity time connects individual energetics with population level variables of growth and reproduction potential, represented by discretionary mass/energy accumulated during a day. Length of time required to cover maintenance costs before allocating absorbed food to discretionary mass is illustrated for ectotherms and endotherms in different climates. Activity time also affects survival probability, which changes with climate.

discretionary mass and energy because of the different times required in cold versus warm climates to first cover the maintenance costs from absorbed food. Ectotherms have lower metabolic rates at cold temperatures. Therefore the food they absorb and digest may meet maintenance costs sooner than if conditions are warm and metabolic rates higher. The opposite is true for endotherms that are typically maintaining high body temperatures. For them, warmer conditions that are not stressful mean that less time needs to be allocated to obtaining food to meet maintenance costs before they can begin to allocate absorbed food to discretionary mass and energy for growth and reproduction.

Survival probability for both ectotherms and endotherms is also influenced by temperature, as illustrated in the bottom half of Fig. 6. Ectotherms are typically inactive or minimally active under cold conditions, which likely reduces the probability of their being preyed on. In contrast, when warmer temperatures arrive and activity increases, predation may increase, resulting in lower survivorship. For endotherms the influence of temperature on survivorship could be higher or lower with changing temperature, depending upon body temperatures maintained and the types and activity levels of predators present at different temperatures. *See* BEHAVIORAL ECOLOGY; HOMEOSTASIS.

Issues of constraints and trade-offs. Figure 7 illustrates stress responses to climate warming, inadequate nutrition, disease, and toxic stress. For ectotherms, a general warming trend may result in more time being spent at less than optimal higher temperatures for the organism. Because of reduced discretionary energy at a warmer temperature, fitness is reduced. Nutrition stress reduces the amount of available food for absorption relative to abundant food. Fighting disease may cause an animal to spend more time in warmer temperatures. This again reduces discretionary mass and energy, thereby compromising fitness. Many toxins, insecticides, herbicides, and fungicides are immunosuppressants; thus, pesticide exposure may result in a higher probability of contracting disease and increased duration of infection before recovery. Thus, climate warming, disease, and toxic exposure can move an animal to warmer temperatures and reduce discretionary mass, energy, and fitness.

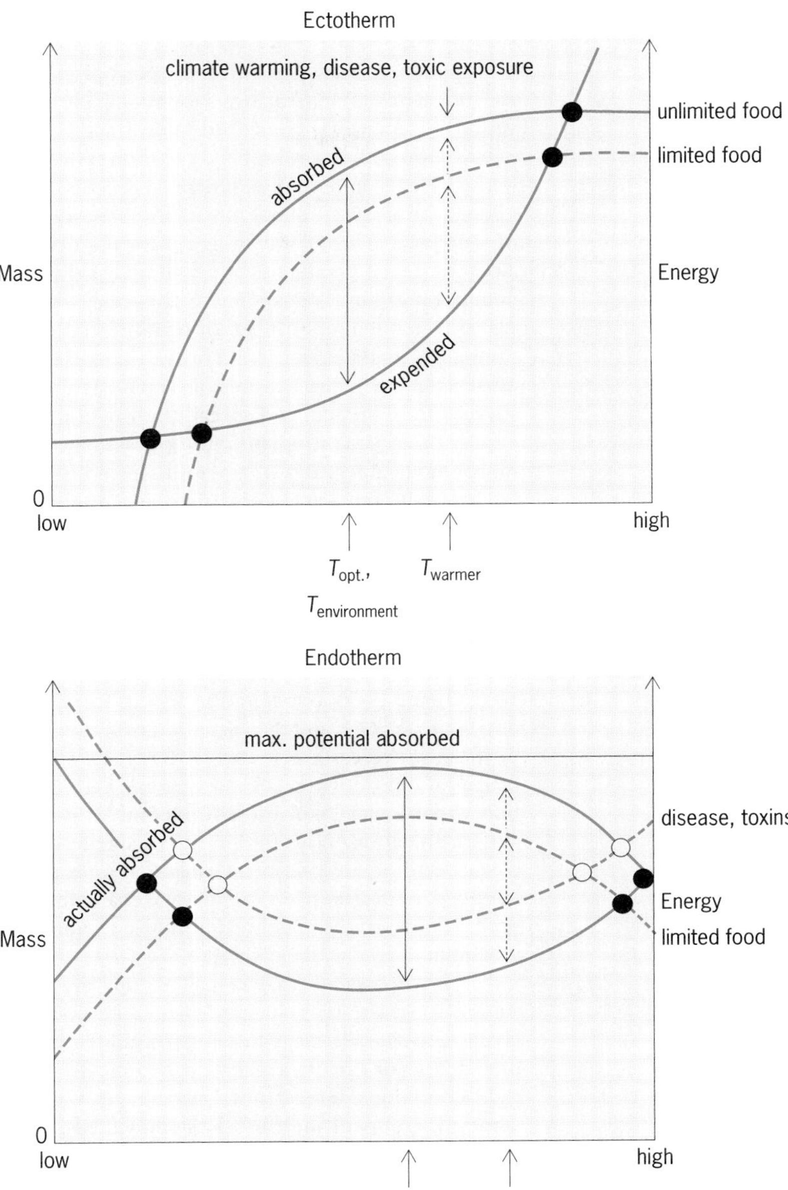

Fig. 7. Trade-offs in responding to environmental stresses of climate, disease, nutrition, and toxic stress. All stresses either affect the temperature that an animal selects in the environment or close the distance between the curves defining absorbed and expended mass. Climate warming which forces an animal to spend more time in a warmer environment decreases discretionary mass because of lower mass intake and higher mass expenditure. Nutritional stress narrows the allowable range of environmental temperatures that represent a positive mass balance for the animal. Disease and toxic exposure also narrow the range of environmental temperatures that represent a positive mass balance; this happens because of decreases in mass intake and increases in mass expenditure.

Endotherms are similarly constrained in their discretionary mass and energy by reduction in nutrition, which decreases absorbed food, and by disease and toxins, which may elevate the costs to maintain a higher body temperature (fever). Sickness also induces reduced activity through suppression of neurological activity. Reduced activity can lead to reduced food intake, which could further constrain discretionary mass and energy. Thus endotherms, like ectotherms, respond in similar ways to climate warming, nutrition, disease, and toxic exposure. Figure 7 illustrates that if these organisms were to try to retain their optimal environmental temperature for maximum discretionary mass and energy, they could compromise their ability to fight disease or toxic exposure. Alternatively, they might spend an inordinate amount of time or resources trying to counter higher average temperatures by evaporation of water (sweating or panting), which could impact water and ion balance and other aspects of organism function.

Warren Porter

Physiological ecology (plant)

The branch of plant science that seeks physiological (mechanistic) explanations for ecological observations. Emphasis is placed on understanding how plants cope with environmental variation at the physiological level, and on the influence of resource limitations on growth, metabolism, and reproduction of individuals within and among plant populations, along environmental gradients, and across different communities and ecosystems. The responses of plants to natural, controlled, or manipulated conditions above and below ground provide a basis for understanding how the features of plants enable their survival, persistence, and spread. Information gathered is often used to identify the physiological and morphological features of a plant that permit adaptation to different sets of environmental conditions. The principal life forms studied are herbs, forbs, grasses, shrubs, trees, and nonvascular plants (such as mosses and lichens). Key life-cycle processes include the acquisition, transport, and loss of mass (CO_2 and H_2O), energy (heat and light), and mineral nutrients; the exchange of gases which influence photosynthesis (CO_2 and H_2O), transpiration (H_2O), and respiration (CO_2 and O_2); and the allocation of fixed energy in the form of carbohydrates (CHO) necessary for life-sustaining activities. Photosynthesis and respiration, in particular, involve a suite of complex physical and chemical reactions within the organelles, cells, tissues, and organs of a plant. Reaction rates are influenced by an array of environmental characteristics such as temperature, and water, nutrient, and light availability.

The environments that plants occupy are often subject to variation or change. The ecophysiological characteristics of these plants must be able to accommodate this or the plants face extinction. Given the right conditions, ample time, and genetic variation among a group of interbreeding individuals, plant populations and species can evolve to accommodate marked ecological change or habitat heterogeneity. If evolutionary changes in physiology or morphology occur on a local (for example, on an isolated mountain top) or regional scale, populations within a single species may diverge in their characteristics. Separate ecological races (ecotypes) arise in response to an identifiable, and often narrow, set of environmental conditions. Ecotypes are genetically distinct and are particularly well suited to the local or regional environment they occupy (locally adapted). Such ecotypes can often increase the geographical range and amplitude of environmental conditions that the species occupies or tolerates. Ecotypes may also occur as a series of populations arrayed over a well-defined environmental gradient called an ecocline. In contrast, if ecotypes are not present, some plant species may still be able to accommodate a wide range of growth conditions through morphological and physiological adjustments, by acclimation to a single factor (such as light) or acclimatization to a complex suite of factors which define the entire habitat. Acclimatization can occur when individuals from several different regions or populations are grown in a common location and adjust, physiologically or morphologically, to this location. Acclimation and acclimatization can therefore be defined as the ability of a single genotype (individual) to express multiple phenotypes (outward appearances) in response to variable growing conditions. Neither requires underlying genetic changes, though some genetic change might occur which could mean that the response seen may itself evolve. Acclimation and acclimatization may also be called phenotypic plasticity. *See* PLANT EVOLUTION.

Studies of metabolic rates in relation to environmental conditions within populations, ecotypes, or species provide a way to measure the tolerance limits expressed at different scales. These data in turn help identify the scales at which different adaptations are expressed, and enhance an understanding of the evolution of physiological processes. Combining observations and measurements from the field with those obtained in laboratory and controlled environment experiments can help identify which conditions may be most influential on plant processes and therefore what may have shaped the physiological responses seen. Laboratory and controlled environment (common garden) experiments also assist in helping identify how much of the variation expressed in a particular metabolic process can be assigned to a particular environmental factor and how much to the plants themselves and the genetic and developmental plasticity they possess.

Methods and measurements. Accurate field measurements are difficult to obtain primarily because of the logistics involved in getting to suitable field locations. The vagaries of uncontrolled environments and their marked influence on rates of metabolism, and the limitations of using the appropriate equipment under unfavorable conditions contribute to the difficulty. Field measurements are, however, essential for providing information about how plants behave under natural conditions. The results obtained from field measurements provide the realistic context against which laboratory and controlled environments experiments can be designed and then compared. One of the most productive and instructive approaches for understanding the ecophysiology of plants combines both field and laboratory observations and experiments.

Solid-state electronics have led to computerized control systems and the miniaturization of key components, so that field instruments and their measurement capabilities have improved dramatically. Portable data acquisition systems which rapidly monitor and record both plant and environmental conditions have also increased the scope of the research and objectives and have extended understanding of how plants function under natural conditions. Satellite imagery and other remotely sensed geographical and botanical information have permitted ways of extending the interpretation of ecophysiological information gathered on individual plants or in well-defined locations. The findings have been extended to the community, ecosystem, and landscape scales. These field techniques have permitted more robust sampling of plants in the field, have added accuracy to the measurements obtained, and have significantly enhanced understanding of plant responses to the environment.

Obtaining information about the precise physiological mechanisms which best explain how plants respond to environmental conditions must often be done by collecting plants from natural environments and then growing them under precisely controlled conditions in the laboratory, growth chamber, or greenhouse. In this way, variables of interest (such as temperature, light, and soils) can be changed and monitored in a well-defined manner while holding all other conditions constant, and then plant responses to those variables can be measured. In addition, particular factors are manipulated, and plant responses in a large number of individuals or in a short span of time can be measured. This permits one to assign and partition the observed variation in a particular physiological behavior into its genetic, developmental, and environmental components. It permits understanding of how plants respond to conditions they may have experienced in the past or might experience in the future, allowing a broader interpretation of observed ecophysiological responses across a wide range of plant species or types. Laboratory analyses may also be essential for understanding how natural environments may have influenced plant function or growth. For example, the analysis of mineral nutrient concentrations of field-collected soil and plant materials may be essential for understanding patterns of plant nutrient acquisition under field conditions. The analysis of tissue characteristics such as the water transport system (xylem), root or leaf characteristics, or storage compounds, performed in the laboratory, are essential for providing detailed information about the ways in which plant form and function permit adaptation to natural conditions. New microscopic, isotopic, molecular, and biochemical methods have significantly expanded knowledge about the underlying basis of the variation seen in the ecophysiological characteristics and responses of plants.

The broad categories of processes studied in the field and in the laboratory under controlled conditions, which define the ecophysiology of a plant, are germination, growth, energy, water, carbon relations, mineral nutrition, and reproduction. These processes are often studied under both optimal and stressful conditions or across the range of conditions that plants experience in the field. Many of the processes studied are clearly influenced by the environmental conditions, but they are also influenced by the growth form, morphology, and life history characteristics of the plant (such as herb, grass, tree, or shrub; woody or herbaceous; evergreen or deciduous; annual or perennial) as well as the shapes and characteristics of the roots, stems, and leaves. *See* PLANTS, LIFE FORMS OF.

Germination. The germination characteristics of plants are properties of the seed, of the seedling, and the factors which influence the transition between the seed and seedling stages. Seeds can possess a range of dormancy mechanisms which inhibit germination, and there are a number of environmental cues which serve to break the dormancy and permit or stimulate the embryonic plant to develop into a seedling and eventually a mature plant. Light, heat (fire), water, smoke, and other factors can influence germination in some species but not in others. Seed size and age (maturity) can also play a role in when and under what conditions seeds germinate. Seedling survival is highly variable and determined by species-specific properties (for example, the tolerance a species has for site conditions) and the variation and severity of growing conditions at the time when seedlings are establishing. Often it is alternating conditions (rather than constant conditions), such as temperature or moisture, which stimulate seeds to germinate, and result in higher germination percentages within a cohort of seeds. Assessing germination success is often estimated by marking seeds in the field or sowing a known number of seeds into plots or growth containers and following their fate over time. Once germination has occurred, there may be physiological and morphological characteristics unique to the seedling or juvenile stage which determine survival and growth rate into the later vegetative stages of development (such as the sapling phase in trees).

Growth. Plant growth in relation to the environment is quantified as changes in dimension (height, length, width, depth), in mass (fresh or dry), or the production of biosynthetic substances such as carbohydrates. Assessments of aboveground plant growth can be made by measuring changes in stem diameters (such as with tree bands which record changes in the diameter of the main trunk) and in the mass, shape, or dimensions of plant canopies (leaves and stems) with hemispherical photographs taken from the ground or digitally recorded images of the canopy from which the leaf area can be estimated. Whole plants or plant parts are harvested or marked and subsequently remeasured at two or more points in time to obtain a measure of growth or growth rate. Growth of belowground parts (rhizomes and roots) can be determined in the field with excavations, with subsamples of soil cores into which roots have grown, or with wedge-shaped Plexiglas boxes or clear tubes embedded in the soil. The boxes can be removed from the soil at different time intervals, and root growth measured along the surface between the box and the soil. Clear boxes can also be used, and root growth determined by tracing roots with colored pens on the box surfaces. The clear tubes can accommodate a small, mobile video camera, which records root presence along set positions of the tube. Large pits can also be excavated, and glass walls installed below the soil surface against which plants are either planted or allowed to grow (called a rhizotron). This permits one to observe the entire soil profile containing roots of large plants and to follow changes in the form and rates of root growth over time. *See* PLANT GROWTH.

Energy relations. The balance between incoming and outgoing (loss) solar radiation determines the temperature of a plant leaf or canopy. The energy relations of a plant therefore require the quantification of both plant features and habitat characteristics, which influence this balance. Radiation absorption, reflection, and transmission of both visible light and infrared light (heat) by the surfaces of a plant and the environment in which it grows from the sun, sky, and ground must be measured, as well as the loss of heat gain by convection, metabolism, and storage, and the loss of heat by evaporation and plant transpiration. Quantification of radiation input and loss is measured as both light and heat. Light is measured with a range of sensors designed to specifically quantify the light used in autotrophic metabolism (photosynthesis) and the light which influences the thermal relations of a plant. Leaf and air temperatures can be measured with noncontact infrared thermometers (sensors) or with fine-wire thermocouples or thermistors, which are in direct contact with surfaces of interest. The amount of energy absorbed by, say, a leaf surface is determined by a balance between energy inputs, features of the leaf (surface texture, size, and color which influence the size and behavior of the boundary layer around the leaf), wind speed and the turbulent heat transfer away from the leaf, the thermal stability of the air surrounding the leaf, and the processes of connective and latent (evaporation and transpiration) heat losses. The temperature of the leaf will affect metabolic processes such as photosynthesis and respiration and leaf-level behaviors such as stomatal opening and closing.

Water relations. The water relations of a plant are determined by its ability to gain access to and absorb water, the status of the plant tissues once water is obtained, and the factors which influence rates of water loss from plant organs such as leaves, stems, and roots. Water absorption and use can be quantified by measuring the changes in plant mass over time by repeatedly weighing potted plants or large blocks of containered soil with plants growing in them (called weighing lysimeters). Water uptake has also been determined by adding a nontoxic chemical (bromide or dye) or isotopic tracers to the water that roots have access to and looking for its appearance in the plant over time and under different conditions. The gradient in the chemical (or water) potential between the soil and the plant root determines if and how quickly water absorption occurs. Water potential is the difference in the free energy content (or chemical potential) between pure water and the water contained inside cells, tissues, or organs. The water potential of pure water is set at a value of zero. Water in the conducting tissues (xylem) of plants or in soils (which contain dissolved solutes, making them solutions) is at a lower water potential than pure water and thus has water potential values that are expressed as negative numbers. Water in live cells or tissues such as leaves or roots also contains dissolved solutes, so these solutions are contained and compartmentalized. This provides a restrictive barrier, and positive pressures build up so that water potentials can have positive values at times called turgor pressure (which prevents leaves from wilting). In actively transpiring plants, water is lost to the atmosphere through minute pores embedded in the leaves called stomata. Stomata can open and close and therefore regulate rates of water loss or transpiration. When stomata are open, water leaving the leaf is replaced at the surface from which it evaporates from the surrounding tissues. As transpiration continues, it creates a gradient toward the leaves from the soil through the roots and stems and into the leaves. Water is pulled up the plant along this gradient, which becomes progressively more negative, from the soil to the atmosphere. Along the gradient there can be resistance to the flow of water which can put the water column under greater tension. If these tensions become too great for the forces which allow water to adhere to itself, bubbles form (because air which forms the bubbles comes out of solution under tension) and the water column can break (become embolized). This happens in roots and stems. Embolized stems or roots can be refilled with water, but if they are not, tissue dysfunction or plant death may result. Instruments called pressure chambers, psychrometers, or hygrometers can measure water potential (the water status) of plants and plant tissues. Other techniques can be used to determine the levels of embolism.

Rates of water loss from leaves can also be quantified with an instrument called a porometer that encloses a leaf in a chamber (cuvette) and determines the transpiration rates by the difference between incoming and outgoing humidity, and the leaf area and flow rates of air through the chamber. Rates of water loss from whole plant canopies can be estimated from measurements of water-flow velocities in the xylem tissues (termed sapflow methods) or from the precise quantification of water vapor loss from entire canopies or stands of plants with instruments placed on a tower that remotely sense changes in the energy status or actual water vapor loss. Water loss can occur from stems, reproductive parts, and roots. Water loss from roots is generally determined either indirectly from changes in the water content of the soil surrounding the roots or by tracing water with added stable or radioactive isotopes from its point of absorption to where it is lost. The amount of water lost from plants is greatest from leaves, though loss of water from roots has been shown to be significant. *See* PLANT-WATER RELATIONS.

Mineral nutrition. All plants require nutrients. These nutrients can come in organic and inorganic (mineral) forms and, depending upon the element, can be required in either large (macro) or very small (micro) quantities. Macronutrients include nitrogen, phosphorus, potassium, sulfur, calcium, magnesium, and iron. Micronutrients include the trace elements manganese, zinc, copper, molybdenum, boron, and cerium. Specific plant groups (such as legumes) may require specific elements (such as phosphorus) in greater amounts because of their importance in a specific metabolic process, such as biological nitrogen fixation; while other plants which inhabit particular substrate types may use or require elements not listed above. Nitrogen is often considered one of the most important elements in limiting plant performance and growth because it is required for so many different physiological processes (including photosynthesis, protein synthesis, and other enzyme-mediated processes). The nutrients required by plants become available by a variety of means, such as geological weathering; soil microbial processes; and biological nitrogen fixation, absorption, and transport through fungal symbionts (called mycorrhizal associations). Changes in soil pH affect the availability of nutrients for plants, as do inputs from anthropogenic sources from fossil fuel combustion (such as nitrogen and sulfur), mining activities (such as heavy metals like copper, magnesium, and iron), or natural disturbances like fire or floods.

Plants and vegetation play a critical role in the cycling of nutrients through the biosphere. The speed and magnitude of ecosystem element cycling is often determined by the interplay between the soils, the climate, the vegetation (form and species), and the microbial community which acts to decompose organic materials in soils and make them available for plant use. Measuring plant nutrient relations in the field is difficult and often requires experimental manipulation such as adding specific fertilizers or stable isotope tracers [for example, ammonium ($^{15}NH_4^+$) or nitrate ($^{15}NO_3^-$)] and following their uptake patterns and fates in the soils, roots, stems, leaves, and reproductive structures (flowers and fruits). Controlled environment experiments are often required to isolate and precisely track plant responses to nutrient additions or deficits and to follow the fate of specific nutrients into the plant and into specific cells, tissues, organs, or constituents. *See* BIOGEOCHEMISTRY; PLANT MINERAL NUTRITION; SYSTEMS ECOLOGY.

Carbon relations. Understanding the carbon relations of plants requires quantification of the processes which influence carbon gain, carbon loss, and carbon allocation to the activities which require the biochemical energy contained in the carbohydrate compounds made in plant photosynthesis. Precise measurements of plant carbon metabolism (photosynthesis and respiration) and the environmental conditions which influence it must be made on field- and laboratory-grown plants. To date, three main photosynthetic pathways have been described, and these show some relationship to the environmental conditions where they are most common. The majority of plants possess the reductive pentose phosphate, or the C_3 pathway (also called the Calvin-Benson cycle; the 3-carbon molecules are the primary product of the biosynthetic reactions). A much smaller set of plants possess the C_4 pathway (or Hatch-Slack cycle), which is biochemically and anatomically more complex. The specialized anatomy (Kranz anatomy) of C_4 plants increases the efficiency of carbon dioxide (CO_2) fixation by virtue of the fact that the initial 4-carbon product (acid) produced in the mesophyll cells is physically transported and sequestered in a ring of cells called the bundle sheath cells. Here it is decarboxylated back to CO_2 and then refixed eventually into carbohydrates in the absence of photorespiration (which competes for the photosynthetic enzymes in C_3 plants). The increased efficiency of CO_2 metabolism often results in very high rates of productivity in C_4 plants; maize (corn) and sugarcane are common crop plant species that possess the C_4 pathway. The third, and least common, photosynthetic pathway, which can often be found in succulent or epiphytic plants, is called crassulacean acid metabolism (CAM; it was first discovered in the succulent plants in the Crassulaceae plant family). Unlike C_3 and C_4 plants which take up CO_2 from the atmosphere through their stomata during the day, CAM plants take up CO_2 at night and close their stomata during the day when temperatures and water loss from plant transpiration are highest. Carbon dioxide taken up at night by CAM plants is fixed into a 4-carbon acid (malic acid); in the daylight, when stomates are closed, the malic acid exits the plant vacuole where it is stored, and enters the chloroplasts where it is decarboxylated and refixed into carbohydrates in a way similar to C_4 plants. Regardless of which of the photosynthetic pathways a plant possesses, all require visible light which drives the light (photo) reactions on the membranes of green plant chloroplasts. The products of these reactions influence the synthesis (light-independent) reactions that convert CO_2 and the carbon compound ribulose bisphosphate into the carbon compounds which eventually are assembled into diverse carbohydrates such as sucrose, glucose, and cellulose (via the C_3 pathway).

At the physiological level, it is known that C_3, C_4, and CAM plants all possess dark respiration (in the mitochondria). In addition, photorespiration occurs, but at rates two to four times higher than dark respiration in C_3 plants. Photorespiration in C_4 plants occurs only in the bundle sheath cells in the presence of high CO_2 concentrations and is therefore very low. Rates of photorespiration in C_3 plants can be very high and can compete for the enzyme ribulose bisphosphate carboxylase-oxygenase (RuBisco), causing overall reductions in the amounts of carbon fixed compared with C_4 plants growing under similar conditions. High O_2 concentrations also cause a marked reduction in photosynthetic rates in C_3 plants because rates of photorespiration are enhanced; this does not occur in C_4 plants. The biochemical differences in plant photosynthetic pathways as well as in the spatial and temporal patterns by which CO_2 is eventually fixed into carbohydrates lead to marked differences in the photosynthetic efficiencies among photosynthetic types. The spatial separation between the mesophyll and bundle sheath in C_4 species and the temporal separation of CO_2 uptake (in the night) and fixation (in the day) in CAM plants commonly leads to greater photosynthetic efficiency compared to C_3 plants when expressed on the carbon income–to–water loss basis.

Differences in the efficiency of photosynthesis (carbon gain per water lost) can be detected by measurements of the actual rates of gas exchange at the leaf level or indirectly by analyzing the ratio of heavy (^{13}C) to light (^{12}C) carbon isotopes that compose the leaf, relative to a known standard; this is termed the carbon isotope ratio ($\delta^{13}C$) and is expressed in units of parts per thousand (‰). The CO_2 assimilated by plants comes in two stable forms, $^{13}CO_2$ and $^{12}CO_2$, and one radioactive form, $^{14}CO_2$. During C_3 photosynthesis, the two stable forms of CO_2 are not taken up equally. Due to differences in the rates of diffusion into the leaf through the stomata and isotopic discrimination against $^{13}CO_2$ by the principal carboxylating enzyme, RuBisco, C_3 plants end up assimilating more $^{12}CO_2$ relative to $^{13}CO_2$ than either C_4 or CAM plants. This means that when the $\delta^{13}C$ of plants with different photosynthetic pathways is measured, C_4 and CAM plants are isotopically heavier (contain more ^{13}C and therefore have higher $\delta^{13}C$ values) than C_3 plants. Because C_3 plants discriminate more than either C_4 or CAM plants, their $\delta^{13}C$ values generally range between -21 and -34‰, while C_4 plants or CAM plants have $\delta^{13}C$ values ranging between -10 and -14‰. CAM plants can have a larger range of $\delta^{13}C$ values than C_4 species depending upon whether they are obligate CAM (-10 to -14‰) or facultative CAM (-15 to -22‰). Negative numbers indicate less ^{13}C present in the plants relative to the source air that they have taken in. Variation in the $\delta^{13}C$ values of plants has been used not only to identify the different photosynthetic pathway types but also to examine photosynthetic efficiency within C_3 species; to trace the diets of animals which feed upon different plants (the $\delta^{13}C$ value of the plant material is transmitted, largely unchanged, to the carbon-containing compounds of the animals that ate it); and to reconstruct changes in the dominant plant species within vast regions (such as grasslands) over long periods of evolutionary time through the analysis of fossil carbon.

The type of photosynthetic pathway that a plant possesses can have a marked influence on its ecology and the environments that it can inhabit. C_3 plants are the most widely distributed on a global basis and can be found in every ecological habitat where plants grow. C_4 plants are commonly found in warmer climatic zones (for example, tropical and desert ecosystems) which can also experience periodic or chronic drought. Some C_4 species do occur in regions with cool or cold winters but hot and often very dry summers; they are not found in the Arctic and alpine regions of the world with short growing seasons and long cold winters. There are also a number of C_4 plants which inhabit aquatic environments where access to CO_2 is restricted. The Earth's most primitive plants possess the C_3 pathway and have appeared most recently in the history of plants on Earth, and C_4 photosynthesis has evolved independently in many plant families that are only distantly related. A few genera of plants contain species with both C_3 and C_4 pathways (such as *Atriplex*) and also C_3-C_4 intermediates (such as *Flaveria*). CAM plants often occur in extremely arid regions or habitats such as deserts where water is often a limiting resource. Many CAM plants are succulents or belong to plant families comprising mostly succulent species; for example, most cacti are CAM plants, as are many members of the pineapple family. What led to the evolution of the C_4 and CAM photosynthetic pathways is still under investigation, but data support the notion that warm temperatures, marked periods of aridity, and global changes in the ambient carbon dioxide may have played a role, particularly in C_4 taxa. The optimum temperature for net photosynthesis in C_3 plants lies between 10 and 25°C (50 and 77°F), while for C_4 and CAM plants it is higher, generally between 30 and 40°C (86 and 105°F).

Several different methods measure rates of photosynthesis and respiration in the field and under controlled laboratory, growth chamber, or greenhouse conditions. They rely on determining the changes in either the CO_2 concentrations passing into and out of a chamber enclosing a leaf with an infrared gas analysis system (based in the laboratory or field portable), or the O_2 which is evolved from leaf material in the presence of nonlimiting CO_2 and light with a high-sensitivity oxygen electrode (in the laboratory only). The infrared gas analysis system allows measurements of both photosynthesis (in the light) and dark respiration (without light) under different conditions in the field and in the laboratory. Such systems allow different environmental variables to be held constant so that precise determinations of the factors which are known to influence rates of photosynthesis can be assessed one at a time. Infrared gas analysis systems are self-contained, computerized, and often automated for controlling conditions and collecting the data. An additional method for determining gross rates of carbon fixation uses radioactively labeled $^{14}CO_2$. Leaves are enclosed in a chamber which contains $^{14}CO_2$, and are allowed to take up the CO_2 for a short period of time. The exposed leaf sample is collected and chemically fixed or frozen and returned to the laboratory, where the determination of how much $^{14}CO_2$ was incorporated into the leaf is done using a liquid scintillation counter. The incorporation of labeled $^{14}CO_2$ into carbon-containing compounds can also be a useful method of determining how and where plants allocate carbon once it is fixed in the process of photosynthesis. Any of these methods can be used to investigate the carbon relation of plants under different environmental conditions, under changing conditions, or in a comparative framework using different species or genotypes (or ecotypes). *See* PHOTOSYNTHESIS; PLANT METABOLISM; PLANT RESPIRATION.

Reproduction. The processes of flowering, fruit production, and seed output are part of plant reproduction. These processes are studied in the field or laboratory by measuring how they change in relation to site or environmental conditions; to an array of physiological variables; and to annual, habitat, and genetic variation within and among plant populations. To understand what influences successful reproduction, these processes must be studied over time and in relation to specific factors such as light, temperature, photoperiod, plant water status, and plant carbon balance, in order to determine which factors are most important in governing flower, fruit, and seed production. This information can be further correlated to the hormonal status of the plants or its pollination ecology to arrive at a picture of what ultimately determines plant persistence or spread over time. Understanding the link between the physiological characteristics of plants, the environmental factors which shape these characteristics, and their influence on plant reproduction is critical for understanding what ecophysiological suites of traits may be favored and thus evolve in plant populations. *See* FLOWER; FRUIT.

Ecophysiology and biophysics. Some of the most promising areas for future research will merge plant physiological ecology with other areas of science. The relationship between plant physiological ecology and biophysics is seen on two scales; the subcellular scale, below the level of whole plants; and the global scale, where whole-plant behavior is related to the impinging environmental conditions. At the subcellular scale, there are a growing number of studies which have used biophysical information about the kinetics of important processes at the level of single electrons to their outcome at the whole-organ (such as leaf) or whole-plant level. At the environmental scale, organismal performance is related to the plant response to the physical environment; and the exchange and feedback of mass and energy between the plant biota and the environment. How temperature, gas exchange, water, and transport processes in the physical world affect or are affected by plants becomes critical to understanding atmosphere-biosphere interactions. This information is used in the formulation of global models of plant production and in predicting how changes in natural resources shape the strength and nature of atmosphere-biosphere interactions. *See* BIOPHYSICS; PLANT PHYSIOLOGY.

Agriculture and conservation. As contemporary agricultural practices begin to incorporate the notion of ecological sustainability, it is increasingly important to understand how the natural vegetation, as a model system, sustains itself in the face of environmental heterogeneity, disturbance, and change. Agriculture is moving away from the intensive cultivation of annual crop plants and toward the use of perennial crops and trees that are sustainable and more subject to natural variation in the environments where they grow, and therefore, an understanding of the physiological and morphological features which permit sustained growth will be necessary. This information will come by borrowing from the knowledge provided by plant physiological ecology. The outcome of more sustainable agricultural practices will be seen in the conservation of the Earth's soils and diversity of species and in the improvement of global water resources. *See* AGRICULTURE.

Ecosystem science. There is an increasing awareness that to fully understand the ecological behavior of large regions on Earth (such as ecosystems), an understanding of the behaviors of the organisms which compose those ecosystems could be critical. What determines the movement of energy, water, and nutrients into, through, and out of a grassland or forest, for example, can best be understood by knowing how the plants which fix, use, and recycle this energy and mass behave. Different plant species possess different chemical, morphological, and physiological features, which determine the speed, magnitude, and efficiency of mass and energy transfer between the atmosphere and the biosphere. In this way, they can influence the processes which occur at the ecosystem scale. The rates of nutrient cycling through an ecosystem, for example, can be dictated by the kinds of plants present; the requirements those plants have for nutrients; and if different species retain, return, or fix particular nutrients, such as nitrogen, within their areas of growth. As such, merging an understanding of the ecophysiological features and behaviors of plants with an investigation of what determines broad-scale ecosystem patterns and processes is yielding new insights.

An acknowledgement of the close relationship between plant physiological ecology and ecosystem science is becoming important in helping scientists to understand not only what may be dictating ecosystem function but how ecosystems will respond to large-scale, human-induced environmental changes now and in the future. A considerable fraction of global change research is focusing on the interface between plant physiological ecology, environmental biophysics, ecosystem science and biogeochemistry, and atmospheric science. Of particular importance is ongoing research focused on understanding the relationship between the Earth's biodiversity, the adaptations and features that the diverse organisms which inhabit the Earth possess, and the effects that these organisms have on ecosystem-level functions as environments continue to change. Loss of biodiversity caused by human-induced habitat destruction (from deforestation, desertification, mining, and pollution) or atmospheric change (from fossil fuel combustion and biomass burning) will mean the loss of species that play key roles in the function, integrity, and stability of ecosystems. For example, the loss of vast areas of tropical trees to timber harvesting and mining practices has changed the climate of the Amazonian basin and decreased the resilience of this ecosystem to additional impacts, either natural or human-induced. Likewise, increasing concentrations of carbon dioxide, soil, and atmospheric pollutants are changing the global carbon, nutrient, and water cycles because of changes in plant species composition or the response of different plant species to these unnatural changes.

The ability to scale up plant responses to the sorts of changes mentioned above to the ecosystem and global levels will therefore be critical to understanding and then mitigating the impacts imposed upon the biosphere in the future. This will require large-scale experiments under both natural and manipulated conditions in the field and under well-controlled conditions; long-term monitoring of plant, soil, water, and atmospheric conditions; and the extension of this information to predict future responses with the use of mathematical models of plant- and ecosystem-level performance in response to environmental change. *See* ECOLOGY; ECOSYSTEM.

Ecophysiology and evolutionary biology. Traditionally, plant physiological ecologists have studied the adaptive features of plants and the environmental characteristics that they surmise have shaped these features. However, these early studies were done with little or no knowledge of the evolutionary origins of the traits under study. Also, because "physiology" is not preserved in the fossil records of plants, elucidating patterns of evolutionary change in physiology is problematic. Today, we can estimate, from indirect analyses of contemporary functional traits, the ancestral physiological states and possible environmental conditions that have led to them using the comparative (for example, comparing features of plants in a group of related species that are known to share a common origin). Understanding the degree of the variation in plant form and function and knowing the evolutionary (phylogenetic) relationships among the taxa being compared is providing new insights into the causes underlying the origins of functional adaptations. Contemporary phylogenetic (evolutionary) methods applied to the study of the plant ecophysiological characteristics are permitting more robust analyses of the nature of character changes that have occurred over evolutionary time. This approach is also providing a way to determine if characters are products of history *or* newly evolved adaptations which have arisen in response to novel environmental challenges. Such an approach is also permitting a deeper understanding of what may have allowed speciation and diversification in some groups of plants. For example, the traits which may have permitted plants to cope with stressful and variable environments when mapped onto the evolutionary tree allows us to evaluate their current and/or past utility. From this, we might also better understand the origin of physiological adaptations, the nature of ecophysiological changes, and the probable trajectory of character and trait evolution. *See* EVOLUTIONARY BIOLOGY. T. E. Dawson

Bibliography. F. A. Bazzaz, *Plants in Changing Environments*, 1996; G. S. Campbell and J. M. Norman, *An Introduction to Environmental Biophysics*, 2d ed., 1998; T. T. Koslowski and S. G. Pallardy, *Physiology of Woody Plants*, 1997; P. J. Kramer and J. S. Boyer, *Water Relations of Plants and Soils*, 1995; W. Larcher, *Physiological Plant Ecology*, 3d ed., 1995; W. H. Schlesinger, *Biogeochemistry: An Analysis of Global Change*, 2d ed., 1995.

Phytamastigophorea

A class of the subphylum Sarcomastigophora, also known as the Phytomastigina. These are the plant flagellates which contain chlorophyll and other pigments, but colorless forms are also included. Grass green is the usually observed color, primarily because the green flagellates are the largest. Those containing an excess of yellow pigments generally are smaller, and fewer species have unusual colors, such as blue or red. Holophytic, saprophytic, and holozoic modes of nutrition occur, and specific chemical components may be demanded by individual species within the group.

Encystment is frequent among phytoflagellates, cyst composition being one method of determining relationships for some colorless species. Reproduction may occur within the cyst or while the organism is active. Gamete formation is largely restricted to Phytomonadina, but life cycles may include an alternation of flagellate with palmella or with ameboid generations. The Phytamastigophorea include 10 orders: Chrysomonadida, Silicoflagellida, Coccolithophora, Heterochlorida, Cryptomonadida, Dinoflagellida, Ebriida, Euglenida, Chloromonadida, and Volvocida. See articles on these groups. *See also* MASTIGOPHORA; PROTOZOA; SARCOMASTIGOPHORA. James B. Lackey

Phytoalexins

Antibiotics produced by plants in response to microorganisms. Plants use physical and chemical barriers as a first line of defense. When these barriers are breached, however, the plant must actively protect itself by employing a variety of strategies. Plant cell walls are strengthened, and special cell layers are produced to block further penetration of the pathogen. These defenses can permanently stop a pathogen when fully implemented, but the pathogen must be slowed to gain time.

The rapid defenses available to plants include phytoalexin accumulation, which takes a few hours, and the hypersensitive reaction, which can occur in minutes. The hypersensitive reaction is the rapid death of plant cells in the immediate vicinity of the pathogen. Death of these cells is thought to create a toxic environment of released plant components that may in themselves interfere with pathogen growth, but more importantly, damaged cells probably release signals to surrounding cells and trigger a more comprehensive defense effort. Thus, phytoalexin accumulation is just one part of an integrated series of plant responses leading from early detection to eventual neutralization of a potentially lethal invading microorganism.

Toxicity. One of the primary features of phytoalexins is their ability to stop the growth of microorganisms. Early workers demonstrated that pieces of live plant tissue could block the growth of fungi nearby. If the tissue was killed, the fungus would grow unchecked. Further experiments showed that extracts of plant tissue that successfully blocked the growth of an invading fungus were equally effective in inhibiting the growth of other microorganisms. The extracts could be added to fungi growing under laboratory conditions, and inhibition of growth and in some cases destruction of the fungi were observed.

The toxicity of phytoalexins is not restricted to fungi. Phytoalexins can also interfere with the growth of other disease-producing organisms (pathogens) important to agriculture, such as bacteria and nematodes. In fact, phytoalexins are even toxic to the plants that produce them. They are carefully localized where they are used, and are produced only when needed. Since they are able to disrupt many basic life processes, it is not surprising that these compounds are able to inhibit a variety of microorganisms.

The method of inhibition by phytoalexins is not completely understood, but membranes appear to be a common target. Processes that require membranes, such as transport of nutrients into and out of cells, and electron transport needed for production of energy, are often affected by phytoalexins.

The effects of phytoalexins are complicated by the ability of some microorganisms to chemically modify or detoxify phytoalexins. As might be expected, organisms that routinely colonize a particular plant are better able to metabolize that plant's phytoalexins than are a randomly selected group of plant pathogens. It is likely that in some cases the ability of a pathogen to degrade its host's phytoalexins determines how successful it will be as a pathogen.

Chemical classification. The tremendous capacity of plants to produce complex chemical compounds is reflected in the structural diversity of phytoalexins (see **illus.**). Each plant species produces one or several phytoalexins, and the types of phytoalexins produced are similar in related species: Members of the potato family (Solanaceae), for example, produce phytoalexins of the terpenoid chemical group; members of the bean family (Leguminoseae), on the other hand, tend to produce phytoalexins which are isoflavonoids.

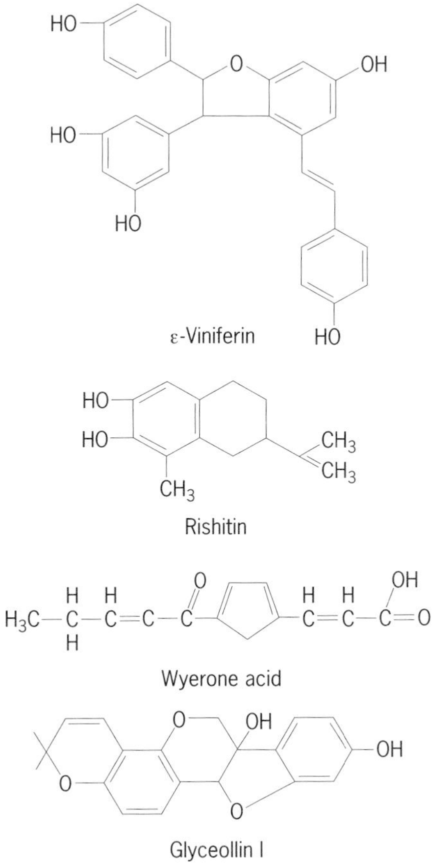

The chemical complexity of phytoalexins led to the use of simple common names for the compounds long before definitive structures had been obtained. The names were derived in many cases from the Latin names of plants in which they were first found. Peas (*Pisum sativum*), for example, produce pisatin as a primary phytoalexin. Other phytoalexins are phaseollin (bean, *Phaseolus*), glyceollin (soybean, *Glycine*), gossypol (cotton, *Gossypium*), orchinol (orchid, *Orchis*), viniferins (grape, *Vitis vinifera*), and ipomeamarone (sweet potato, *Ipomoea*). Of course, the names can also reflect other historical factors as in the case of rishitin (purified from potato variety Rishiri) and wyerone acid (isolated from broad bean at Wye College, United Kingdom).

Role in disease resistance. The diversity, complexity, and toxicity of phytoalexins may provide clues about their function. The diversity of phytoalexins may reflect a plant survival strategy. That is, if a plant produces different phytoalexins from its neighbors, it is less likely to be successfully attacked by pathogens adapted to its neighbor's phytoalexins. Diversity and complexity, therefore, may reflect the benefits of using different deterrents from those found in other plants.

ε-Viniferin

Rishitin

Wyerone acid

Glyceollin I

Structural formulas of four phytoalexins.

Phytoalexins appear to be in the right place at the right time. Healthy plant tissues have no detectable phytoalexins. In those cases in which a pathogen escapes detection and produces disease symptoms, phytoalexins may be produced, but it is too little too late.

Plants that exhibit only minimal damage by a fungal pathogen (those that are resistant) in many cases show a rapid accumulation of phytoalexins in excess of the concentrations needed to inhibit fungal growth. Accumulation is limited to very high concentrations in the cells directly adjacent to the pathogen. The rise in phytoalexins correlates in space and time with cessation of pathogen growth. Moreover, addition of chemicals that block phytoalexin synthesis permits otherwise impotent pathogens to develop and produce disease symptoms. Thus, in at least some host-pathogen interactions, phytoalexins play a critical role in resistance to disease.

Breeding for improved disease resistance has been essential to the success of modern agriculture, yet the function of resistance genes is not understood. Since the same defense mechanisms are triggered by different pathogens in the presence of different resistance genes, it is likely that resistance genes are involved in pathogen recognition. This is an important distinction. Disease resistance genes are not involved in making the biochemical machinery (enzymes) that is needed to produce a defensive response (such as phytoalexin accumulation), but rather they determine whether a particular pathogen will trigger the response. Therefore, it may be more appropriate to call these genes "pathogen recognition" genes.

Pathogen recognition. Plants detect pathogens in a variety of ways, but the most general detection system is based on damage to plant cell walls. Bacteria, fungi, nematodes, and other plant pathogens produce enzymes that degrade plant cell walls. Wall degradation releases nutrients used for pathogen growth and permits further invasion of plant tissues and also releases polysaccharide fragments, which (especially from the pectin portion of walls) stimulate phytoalexin accumulation in neighboring plant cells. Still smaller cell wall fragments are transported to other parts of the plant, making other leaves more resistant to attack. Thus, plants can express both local and systemic resistance.

Pathogens can also be directly detected through the release of components of their own cell walls. Plants produce enzymes that release polysaccharide fragments from fungal cell walls. These polysaccharides, termed elicitors, are produced by many different types of fungal pathogens and are in fact common constituents of fungi in general. Elicitors stimulate phytoalexin accumulation. Thus, the presence of elicitors may be a general signal of invasion that triggers the defensive response of antibiotic production. *See* PLANT PATHOLOGY. Arthur R. Ayers

Bibliography. J. A. Bailey and J. W. Mansfield (eds.), *Phytoalexins*, 1982; R. P. Sharma and D. K. Salunkhe, *Mycotoxins and Phytoalexins*, 1991.

Phytochrome

A pigment that controls most photomorphogenic responses in higher plants. As a result of the dependence of plants on the energy of sunlight and their sessile habit, mechanisms have evolved that allow them to adapt their growth and development to more efficiently seek and capture light and to tailor their life cycle to the climatic seasons. These mechanisms enable the plant to sense not only the presence of light but also its intensity, direction, duration, and spectral quality, which suggests a form of color vision. Plants thus regulate important developmental processes such as seed germination, growth direction, growth rate, chloroplast development, pigmentation, flowering, and senescence, collectively termed photomorphogenesis.

To perceive light signals, plants use several receptor systems that convert light absorbed by specific pigments into chemical or electrical signals to which the plants respond. This signal conversion is called photosensory transduction. The colors of light absorbed by the pigment determine the action spectrum, that is, the colors of light effective in initiating the photoresponse. Pigments used include cryptochrome, a blue light-absorbing pigment; an ultraviolet light-absorbing pigment; and phytochrome, a red/far-red light-absorbing pigment.

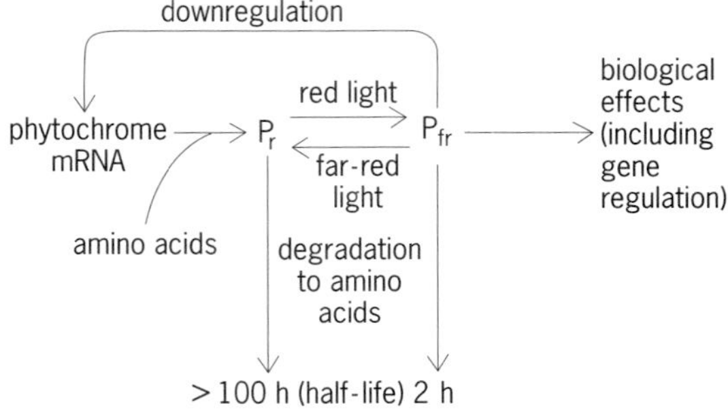

Fig. 1. Phytochrome pathway, showing the photointerconversion between the red-absorbing (P_r) and the far-red-absorbing (P_{fr}) forms, phytochrome degradation, and the control of phytochrome synthesis.

Spectral properties. Phytochrome consists of a compound that absorbs visible light (chromophore) bound to a protein. The chromophore is an open-chain (or linear) tetrapyrrole closely related to the photosynthetic pigments found in the cyanobacteria and similar in structure to the circular tetrapyrroles of chlorophyll and hemoglobin. Phytochrome is one of the most intensely colored pigments found in nature, enabling phytochrome in seeds to sense even the dim light present well beneath the surface of the soil and allowing leaves to perceive moonlight. *See* CHLOROPHYLL; HEMOGLOBIN.

Phytochrome can exist in two stable photointerconvertible forms, P_r or P_{fr}, with only P_{fr} being biologically active. Absorption of red light (near 666 nanometers) by inactive P_r converts it to active P_{fr}, while absorption of far-red light (near 730 nm) by active P_{fr} converts phytochrome back to inactive P_r (**Fig. 1**). Photoconversion of P_r to P_{fr} involves a *cis*-to-*trans* isomerization of one of the double bonds between the four pyrrole rings, a 31° reorientation of the chromophore relative to the protein moiety, and conformational changes within the protein; no additional factors are required for phytochrome phototransformations other than light. As a result, even the purified pigment can be repeatedly photointerconverted between P_r and P_{fr} in the test tube. During the conversion of phytochrome from one form to the other, the molecule passes through a series of intermediates lasting from nanoseconds to milliseconds,

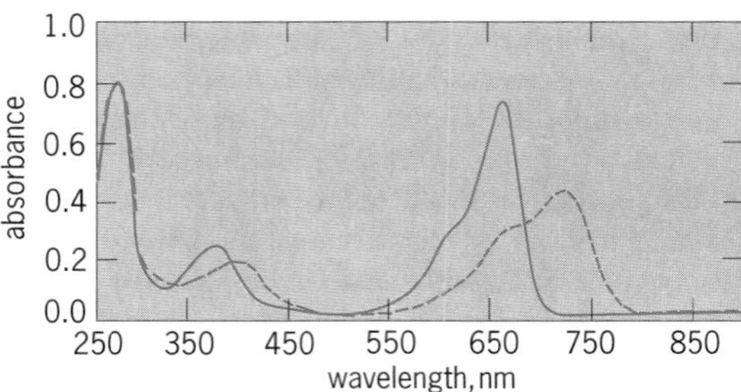

Fig. 2. Absorption spectra of phytochrome in the red-absorbing form (P_r; solid line) and after irradiation with red light producing mainly the far-red-absorbing form (P_{fr}; broken line).

with the final P_r or P_{fr} form appearing within a second after light absorption.

Because the absorption of P_r and P_{fr} overlap in the red region of the spectrum, red light is not able to convert all P_r to P_{fr}. Thus even in saturating red light, a photoequilibrium of 87% P_{fr} and 13% P_r is established (**Fig. 2**). The residual P_r that remains after phytochrome solutions have been irradiated with red light can be observed in the absorbance spectrum as a shoulder at 665 nm. In contrast, because P_r does not absorb light of wavelengths greater than approximately 700 nm, far-red light is able to convert essentially all P_{fr} back to P_r. *See* ABSORPTION OF ELECTROMAGNETIC RADIATION.

Phytochrome protein. The protein moiety of phytochrome ranges in molecular mass from 116 to 127 kilodaltons, depending on the plant species. Genes encoding phytochrome proteins have been characterized from a variety of both monocot and dicot species. In all species studied, phytochrome is synthesized by a small gene family containing several members that encode related but not identical proteins. The complete amino acid sequence of phytochromes within the same plant species and among various plant species ranges in homology from about 60% to greater than 80%, with more closely related species displaying a greater degree of protein homology.

Structural analyses of phytochrome show that it consists of two large domains, an amino-terminal domain of approximately 74kDa that contains the chromophore and a 55-kDa domain responsible for holding the phytochrome dimers together. The spectral properties of the 74-kDa fragment are identical to the intact molecule, indicating that it contains all the domains required for correct chromophore–protein interactions. The amino-terminal region, which is rich in serine or threonine, appears to extend significantly from the surface of the 74-kDa domain, which may explain its susceptibility to proteolytic cleavage.

A model for the quaternary structure of phytochrome (**Fig. 3**) proposes that phytochrome monomers are made of an ellipsoid and a cylinder comprising the 74-kDa and 55-kDa domains, respectively. The two 55-kDa cylinders interact, holding the phytochrome dimer together so that the 74-kDa ellipsoids are oriented in the same plane but 90° from each other. In this way, phytochrome has a cloverleaf shape when observed from above. This asymmetric shape explains why phytochrome behaves in solution as a protein with a molecular size (355 kDa) much larger than its actual mass (250 kDa).

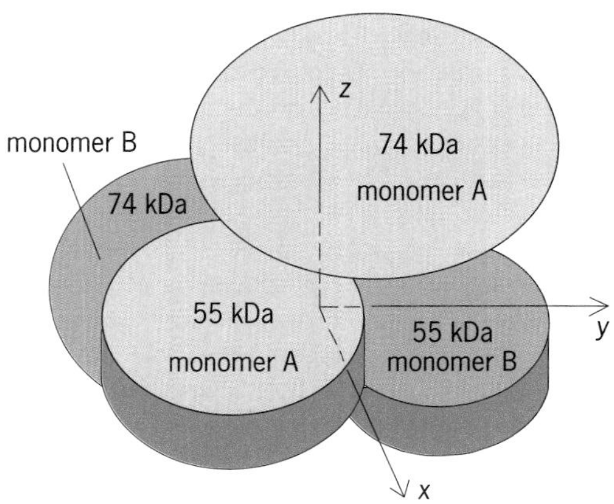

Fig. 3. Molecular model of a phytochrome dimer. The 74-kDa ellipsoids are oriented in the same plane but 90° from each other. (*After S. Tokutomi et al., A model for the dimeric molecular structure of phytochrome based on small-angle x-ray scattering, FEBS Lett., 247:139–142, 1989*)

Biosynthesis and degradation. Plants frequently respond quantitatively to light by detecting the amount of P_{fr} produced. As a result, the amount of P_{fr}—and hence phytochrome—ust be strictly regulated nonphotochemically by precisely controlling both the synthesis and degradation of the pigment. Phytochrome also appears to be posttranslationally modified by phosphorylation and possibly by glycosylation.

In dark-grown (etiolated) seedlings, P_r is continually synthesized. Because its half-life is greater than 4 days, P_r accumulates to high levels within the tissue. However, following transfer of etiolated plants into the light, most phytochrome disappears after photoconversion to P_{fr}, which has a half-life of only 1–2 h. Rapid P_{fr} degradation is the result of the selective recognition of P_{fr} by a proteolytic pathway involving the small protein ubiquitin. Conjugated ubiquitin serves as a recognition signal for a specific protease that degrades only the target protein, releasing ubiquitin in a reusable form.

In addition to enhanced phytochrome degradation following P_{fr} conversion, a feedback mechanism exists whereby P_{fr} inhibits further synthesis of P_r. This inhibition involves not only the repression of continued phytochrome gene transcription into messenger ribonucleic acid (mRNA) but also an enhanced rate of phytochrome mRNA degradation. The repression is highly sensitive to P_{fr}, requiring only 1% of P_r to be converted to P_{fr} to maximally inhibit transcription. Upon the return of irradiated seedlings to the dark, synthesis of phytochrome mRNA slowly resumes.

Etiolated plants thus contain up to 100 times more phytochrome than light-grown plants, allowing a seedling with limited resources to sense even the dimmest light and thus reach the surface of the soil before depleting its energy. Once at the surface, however, detecting light is not as much a problem, and thus much less of the pigment is required to initiate photomorphogenesis.

Occurrence. Phytochrome has been detected in a variety of tissues from all higher plants examined. It also has been found in certain mosses, liverworts, and algae. In etiolated plants, concentrations of phytochrome are especially high in the root cap, crown meristem, and tip of monocot seedlings and the hook and cotyledons of etiolated dicot seedlings. Because of their high phytochrome content, etiolated seedlings are a convenient source from which the protein may be purified. Both the low levels of phytochrome and the presence of chlorophyll have substantially hampered the detection and isolation of the chromoprotein from light-grown tissues, but partially purified preparations of light-grown or "green" phytochrome have been obtained. Green phytochrome appears to differ slightly from that isolated from etiolated seedlings, suggesting that another type of phytochrome is synthesized in light-grown plants.

Subcellular localization. In etiolated seedlings, P_r is found primarily throughout the cytoplasm, with none detected in mitochrondria, chloroplasts, or nuclei. There is no evidence that a significant fraction of phytochrome is associated with membranes of higher plants. Once converted from P_r to P_{fr}, phytochrome changes from a disperse cytoplasmic distribution to one in which the molecules coalesce into many amorphous discrete areas. This sequestration is rapid, occurring within seconds, and requires metabolic energy. In monocots most phytochrome appears to aggregate, whereas in dicots only a portion aggregates while the rest remains dispersed. Following reconversion of P_{fr} to P_r, the pigment slowly reverts to a dispersed form. The function of phytochrome sequestering is unknown, but it may represent molecules that are awaiting degradation.

The failure of localization studies to find phytochrome on membranes or in the nucleus suggests that it does not act by directly altering membrane permeability or gene transcription. However, it is possible that the bulk of phytochrome detected in the cytoplasm is not active or that only a small percentage becomes active following interaction with the appropriate membrane or compartment. The filamentous alga, *Mougeotia*, has a single flat chloroplast that rotates to orient itself perpendicular to light; the movement is phytochrome-mediated. Experiments with microbeams of light demonstrate that the active P_{fr} is stably attached to a structure close to or on the plasma membrane.

Functions. Phytochrome has a variety of functions in plants. Initially, production of P_{fr} is required for many seeds to begin germination. This requirement prevents germination of seeds that are buried too deep in the soil to successfully reach the surface. In etiolated seedlings, phytochrome can measure an increase in light intensity and duration through the increased formation of P_{fr}. Light direction also can be deduced from the asymmetry of P_{fr} levels from one side of the plant to the other. Different phytochrome responses vary in their sensitivity to P_{fr}; some require very low levels of P_{fr} (less than 1% of total phytochrome) to elicit a maximal response, while others require almost all of the pigment to be converted to P_{fr}. Thus, as the seedling grows toward the soil surface, a cascade of photomorphogenic responses are induced, with the more sensitive responses occurring first. This chain of events produces a plant that is mature and photosynthetically competent by the time it finally reaches the surface. Production of P_{fr} also makes the plant aware of gravity, inducing shoots to grow up and roots to grow down into the soil. *See* PLANT MOVEMENTS; SEED.

In light-grown plants, phytochrome allows for the perception of daylight intensity, day length, and spectral quality. Intensity is detected through a measurement of phytochrome shuttling between P_r and P_{fr}; the more intense the light, the more interconversion. This signal initiates changes in chloroplast morphology to allow shaded leaves to capture light more efficiently. If the light is too intense, phytochrome will also elicit the production of pigments to protect plants from photodamage.

Temperate plants use day length to tailor their development, a process called photoperiodism. How the plant measures day length is unknown, but it involves phytochrome and actually measures the length of night. A red light pulse given in the middle of a long night causes plants to believe that they have experienced a short night. Conversely, a far-red light pulse given before a short night makes plants believe that they have experienced a long night. These "night breaks" are widely used commercially to time the flowering of plants. *See* PHOTOPERIODISM.

Finally, phytochrome allows plants to detect the spectral quality of light, a form of color vision, by measuring the ratio of P_r to P_{fr}. When a plant is grown under direct sun, the amounts of red and far-red light are approximately equal, and the ratio of P_r to P_{fr} in the plant is about 1:1. Should the plant become shaded by another plant, the P_r/P_{fr} ratio changes dramatically to 5:1 or greater. This is because the shading plant's chlorophyll absorbs much of the red light needed to produce P_{fr} and absorbs almost none of the far-red light used to produce P_r. For a shade-intolerant plant, this change in P_r/P_{fr} ratio induces the plant to grow taller, allowing it to grow above the canopy.

Mechanisms of action. It is not known how phytochrome elicits the diverse array of photomorphogenic responses, but the regulatory action must result from discrete changes in the molecule following photoconversion of P_r to P_{fr}. These changes must then start a chain of events in the photosensory transduction chain leading to the photomorphogenic response. Many photosensory transduction chains probably begin by responding to P_{fr} or the P_r/P_{fr} ratio and branch off toward discrete end points.

To analyze phytochrome action, several avenues of research have been taken. One has been an attempt

Fig. 4. Morphological effect of oat phytochrome synthesis in transgenic tobacco. (*a*) Transformed plant not expressing oat phytochrome. (*b–d*) Transformed plants expressing oat phytochrome. (*e, f*) Two nontransformed plants.

to assign enzymatic properties to the chromoprotein in the plant. Phytochrome may be a protein kinase capable of adding phosphate groups to other proteins, and given that protein phosphorylation is an important regulator of cell development and metabolism, such an enzymatic activity would be an attractive possibility. *See* ENZYME.

Another avenue has been to identify chemical differences between P_r and P_{fr}. Analyses indicate that in addition to changes in the chromophore, several protein domains change conformation following phototransformation. The most interesting and dramatic changes occur within 6–10 kDa of the amino terminus. This region appears more extended as P_r and becomes more protected as P_{fr}. It has been determined that the amino-terminal region interacts directly with the chromophore, potentially moving to shield the chromophore following photoconversion of P_r to P_{fr}. Other protein domains also undergo conformational changes during phototransformation. They exist within the molecule, and as opposed to the amino terminus, these regions are less exposed as P_r than as P_{fr}

Which changes in conformation during phototransformation are responsible for phytochrome action remains to be determined. An approach to identifying these critical domains has been developed through the use of genetic engineering. It involves the transfer of an oat phytochrome gene into tobacco plants by using the bacterium *Agrobacterium tumefaciens* as the transfer agent. The tobacco plants then express both their own phytochrome and the inserted oat phytochrome, and as a result they have artifically higher levels of the chromoprotein. In light-grown plants, more than nine times more phytochrome is present in the genetically engineered plants, and their growth and development are altered. They exhibit a dwarfed bushy shape and have darker green leaves (**Fig. 4**). These changes indicate that oat phytochrome is morphogenically active in tobacco and can be used as an assay for the functional molecule in the plant. *See* GENETIC ENGINEERING.

Another possible strategy to elucidate phytochrome's mode of action is to trace back through the sensory transduction chain of a particular phytochrome response to identify the initial molecular events. The best-characterized response is phytochrome's control of greening, which involves the development of a photosynthetically competent chloroplast. Chloroplast development requires the synthesis of many proteins encoded by both the nuclear and chloroplast genomes. This synthesis is controlled directly or indirectly by phytochrome and involves transcriptional activation of the corresponding genes. How these genes are activated is unclear, but it requires the synthesis or activation of proteins, commonly called transcription or trans-acting factors, that bind to genes. This binding may facilitate gene transcription. Whereas many phytochrome responses require gene transcription, not all of them do. Phytochrome may function by stimulating a response similar to muscle contraction.

The third possible strategy is the isolation and characterization of mutants in phytochrome responses, including those with reduced levels of chromoprotein. Other mutants show attenuated phytochrome responses but have normal amounts of phytochrome, indicating a defect in the photosensory transduction chain. *See* PHOTOMORPHOGENESIS.

Richard D. Vierstra

Bibliography. R. E. Kendrick (ed.), *Photomorphogenesis in Plants*, 2d ed., 1993; H. McGee (ed.), *A Pigment of the Imagination: USDA and the Discovery of Phytochrome*, 1987; P. B. Moses and N.-H. Chua, Light switches for plant genes, *Sci. Amer.*, 258:88–93, 1988.

Phytoplankton

Mostly autotrophic microscopic algae which inhabit the illuminated surface waters of the sea, estuaries, lakes, and ponds. Many are motile. Some perform diel (diurnal) vertical migrations, others do not. Some nonmotile forms regulate their buoyancy. However, their locomotor abilities are limited, and they are largely transported by horizontal and vertical water motions.

Energy-nutrient cycle. All but a few phytoplankton are autotrophic—they manufacture carbohydrates, proteins, fats, and lipids in the presence of adequate sunlight by using predominantly inorganic compounds. Carbon dioxide, water, inorganic nutrients such as inorganic phosphorus and nitrogen compounds, trace elements, and the Sun's energy are the basic ingredients. This organic matter is used by other trophic levels in the food web for nourishment. The energy in this organic matter ultimately is released as heat in the environment. The inorganic materials associated with this organic matter, on the other hand, are recycled. They are released back into the aquatic environment in inorganic form as a result of metabolic processes at all levels in the food web, and in large part they again become available for reuse by the phytoplankton. Some structural materials, such as the silica in diatoms and silicoflagellates, and carbonates in the scales of certain chrysophytes, are returned to the aquatic environment by chemical dissolution rather than by biological processes. *See* FOOD WEB.

Varieties. A great variety of algae make up the phytoplankton. Diatoms (class Bacillariophyceae) are often conspicuous members of marine, estuarine, and fresh-water plankton. Reproducing mostly asexually by mitosis, they can divide rapidly under favorable conditions and produce blooms in a few days' time. Their external siliceous skeleton, termed a frustule, possesses slits, pores, and internal chambers that render them objects of great morphological complexity and beauty which have long attracted the attention of microscopists. *See* BACILLARIOPHYCEAE.

Dinoflagellates (class Dinophyceae) occur in both marine and fresh-water environments and are important primary producers in marine and estuarine environments. Dinoflagellates possess two flagella; one trails posteriorly and provides forward motion, while the other is positioned more or less transversely and often lies in a groove encircling the cell. This flagellum provides a rotary motion; hence the name dino (whirling) flagellate. In some dinoflagellates the cell wall is thin, while in others it consists of a complicated array of rather thick cellulosic plates, the number and arrangement of which are characters used in identifying genera and species.

The dinoflagellates are often conspicuous members of the marine plankton. Some taxa are bioluminescent. Others are one of the causes of red water, often called red tides, although their occurrence is not related to the tides. Depending on the causative species, red water may be an innocuous discoloration of the water, or if *Gymnodinium breve* is dominant, extensive fish mortalities will be experienced. Several species of *Gonyaulax* occur in bloom proportions in inshore marine waters. They contain a toxin which can accumulate in shellfish feeding upon *Gonyaulax* and which is the ultimate cause of paralytic shellfish poisoning in humans.

Coccolithophorids (class Haptophyceae) are also marine primary producers of some importance. They do not occur in fresh water. This class of algae possesses two anterior flagella. A third flagellumlike structure, the haptonema, is located between the two flagella. Calcium-carbonate-impregnated scales, called coccoliths, occur on the surface of these algae, and are sometimes found in great abundance in recent and ancient marine sediments. *See* COCCOLITHOPHORIDA.

Under certain conditions in subtropical and tropical seas, members of the nitrogen-fixing blue-green algal genus *Trichodesmium* (class Cyanophyceae) can occur in sufficient concentrations to strongly discolor the surface of the sea. Other nitrogen-fixing blue-greens commonly occur in great abundance in eutrophic and hypereutrophic fresh-water lakes and ponds.

Members of still other algal classes occur in marine and estuarine plankton. Their abundance will vary in different environments at different times, and they may even occasionally dominate the standing crop. Much remains to be learned about the identity, physiology, and ecology of the very small (<8 micrometers) flagellates which commonly occur in almost every marine phytoplankton sample. *See* ALGAE.

Communities. Even though marine and fresh-water phytoplankton communities contain a number of algal classes in common (such as Bacillariophyceae, Chrysophyceae, and Dinophyceae), phytoplankton samples from these two environments will appear quite different. These habitats support different genera and species and groups of higher rank in these classes. Furthermore, fresh-water plankton contains algae belonging to additional algal classes either absent or rarely common in open ocean environments. These include the green algae (class Chlorophyceae), the euglenoid flagellates (class Euglenophyceae), and members of the Prasinophyceae. *See* FRESH-WATER ECOSYSTEM; MARINE ECOLOGY.

In estuarine environments in which salinities vary from essentially zero to those of the local inshore ocean salinities, a mixture of organisms characteristic of both fresh-water and marine environments will be encountered. In addition, giving the estuarine plankton a somewhat distinctive appearance, some euryhaline planktonic taxa will be present. *See* ESTUARINE OCEANOGRAPHY.

Samples of phytoplankton in shallow areas of the sea, lakes, and estuaries often contain benthic and epiphytic microalgae. These algae become suspended in the water as a result of strong turbulent mixing so often found in shallow aquatic environments, and are called meroplankton to distinguish them from the holoplankton organisms which are truly planktonic.

While phytoplankton community composition reflects some complicated and as yet rather poorly understood series of biotic and abiotic interaction within ecosystems, the chemical composition of water is recognized as an important factor affecting phytoplankton communities. Many of the differences between the marine, estuarine, and fresh-water phytoplankton communities are associated with changes in salinity, major ion concentrations and ratios, pH, nutrients, and quite possibly trace-element concentrations. *See* ECOSYSTEM.

Eutrophication. Society's common use of lakes as receptacles for wastes, coupled with nutrient-rich runoff from cultivated fertilized land, has had pronounced biological effects not only on the phytoplankton but on other levels in the food web. Domestic sewage and agricultural runoff are quite rich in inorganic phosphorus, which is generally in short supply in most inland bodies of water. The addition of a nutrient which is limiting phytoplankton production, such as phosphorus, to fresh-water environments increases primary production, alters phytoplankton composition, and can lead to dense algal blooms which adversely affect water quality and esthetic and recreational values. The effects of nutrient enrichment, generally known as eutrophication, have been understood by limnologists since the 1920s–1930s, but the American public became aware of the implication only in the 1960s with the widespread use of phosphate detergents and the associated and accelerated deterioration of the quality of lakes and ponds. Advanced techniques of sewage treatment are now available which remove phosphates sufficiently to greatly reduce the impact of sewage effluent upon receiving waters. Important in regulating these cycles are nutrient availability and the presence of a water density gradient, the pycnocline, at some relatively shallow depth (commonly 16.5–165 ft or 5–50 m) below the surface. Above the pycnocline, the location of which is usually well correlated with a zone of rapidly changing water temperature called the thermocline, is a well-mixed region of uniform density and temperature. This layer is called the mixed layer by oceanographers and the epilimnion by limnologists. Within this mixed layer, water motions caused by winds help keep the phytoplankton in suspension within the illuminated surface layer. The pycnocline also greatly reduces the rate of diffusion of nutrients upward into the mixed layer from nutrient-rich deeper waters and thus has a negative effect upon phytoplankton productivity. Any mechanism which introduces nutrients into the mixed layer that are limiting phytoplankton production will enhance primary production. *See* EUTROPHICATION; LIMNOLOGY; WATER POLLUTION.

Seasonal cycles. The phytoplankton in aquatic environments which have not been too drastically affected by human activity exhibit rather regular and predictable seasonal cycles. Coastal upwelling and divergences, zones where deeper water rises to the surface, are examples of naturally occurring phenomena which enrich the mixed layer with needed nutrients and greatly increase phytoplankton production. In the ocean these are the sites of the world's most productive fisheries. *See* UPWELLING.

In marine and fresh waters in temperate to arctic latitudes, the seasonal cycle of phytoplankton is to a large degree influenced by the pycnocline. During winter months a strong pycnocline is generally not present, and any planktonic plants in the water circulate to considerable depths. Even though nutrient concentrations are high during this period of deep vertical mixing, any phytoplankton present do not spend sufficient time in the illuminated surface layers for photosynthesis to exceed respiration. As spring approaches, incident solar radiation leads to the formation of a pycnocline. Phytoplankton then proliferate, giving rise to the annual spring phytoplankton bloom in which diatoms frequently dominate (thus it is often called the spring diatom bloom). The phytoplankton quickly exhaust the nutrients in the newly formed mixed layer, and throughout the summer phytoplankton concentrations generally remain low due to low mixed-layer nutrient concentrations resulting in part from the restricted upward movement of nutrients through the pycnocline. In the fall, incident solar radiation decreases and the pycnocline deepens. Nutrients or phytoplankton living in the pycnocline are incorporated into the deepening mixed layer, and a fall bloom occurs and then wanes as the pycnocline deepens, and finally disappears.

In the tropical oceans and lakes located near sea level, a permanent thermocline exists, and phytoplankton production remains low but may increase if nutrients are injected into the mixed layer as a result of strong wind mixing or other advective processes.

Fossilization. Under favorable conditions, some phytoplankton are incorporated into the sediments and become part of the fossil record. Diatoms, silicoflagellates coccoliths, and the cysts of dinoflagellates are frequently sufficiently abundant and well preserved in lake and ocean sediments to permit paleontologists to reconstruct past environmental conditions and changes in environmental conditions through time. These fossilized remains are also used by stratigraphers to determine the age of sediments. *See* MARINE SEDIMENTS; PALEOECOLOGY; STRATIGRAPHY.

Robert W. Holmes

Bibliography. A. D. Boney, *Phytoplankton*, 2d ed., 1992; G. A. Cole, *Textbook of Limnology*, 4th ed., 1994; C. D. Sandgren (ed.), *Growth and Reproduction Strategies of Freshwater Phytoplankton*, 1992.

Phytotronics

Research using whole plants and conducted under controlled environmental conditions to determine responses to a single or known combination of environmental elements. Originally, the term phytotronics was used to identify research conducted specifically in phytotrons where controlled plant growth units are available for simultaneous use. The name phytotronics is also often applied to any research conducted with whole plants in a controlled environment plant growth chamber or room.

The word phytotronics came into use after Fritz W. Went built the Earhart Plant Research Laboratory at the California Institute of Technology in 1949. This first phytotron consisted of glasshouses and artificially lighted plant growth facilities, each with controlled temperatures and lighting. The first research projects demonstrated the significance of a phytotron in advancing botanical knowledge, and resulted in the building and continuing operation of phytotrons throughout the world.

Scope. A phytotron is a research tool containing a large number of individually controlled environments (**Fig. 1**). These provide the means of studying the effect of each environmental factor such as temperature or light at many levels simultaneously. Thus, phytotronics is usually identified with research that

Fig. 1. Large plant growth chamber with a solid bank of fluorescent incandescent lamps forming the ceiling. The mirrorized walls and the two rows of diffusers with their high volume of airflow provide uniform light, temperature, humidity, and carbon dioxide conditions.

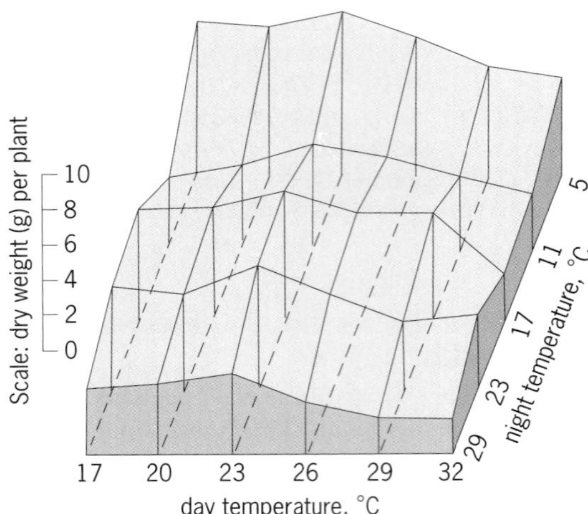

Fig. 2. Results of experiment using five growth chambers and moving the plants to obtain 25 combinations of day-night conditions. Increased total plant dry weight due to a 5°C (41°F) night with all day temperature was unexpected but later confirmed in the New Zealand *Pinus radiata* plantations. °F = (°C × 1. 8) + 32. (*After H. Hellmers and D. A. Rook, Air temperature and growth of radiata pine seedlings, N. Z. J. For. Sci., 3:271–285, 1973*)

involves the use of a range of environmental factors. In most phytotrons, plants are on carts that can be moved easily from one controlled environment to another. This provides the flexibility to greatly increase the number of environmental conditions available over the number of controlled units. For instance, using only five units and redistributing carts daily, 25 combinations of day/night conditions can be obtained (**Fig. 2**). The large number of controlled units in a phytotron also makes it possible to investigate the interaction of the various environmental factors on plants. However, before, or simultaneously with, the study on factor interactions, the effects of the individual factors must be determined if the interaction effects are to be understood. Effects of environmental factors are more often synergistic or antithetic than additive because plant growth is capable of adapting, within limits, to environmental changes.

Phytotronics has greatly extended the precision and scope of botanical research in the laboratory. Phytotronics is used also to complement field research, especially in the area of agriculture and ecology. The building of phytotrons became feasible only after the development of relatively low-cost air-conditioning systems and adequate lighting systems, utilizing high-output fluorescent lamps. Prior to this, whole plant research was conducted under field or slightly modified field or glasshouse conditions. In the field and in ordinary glasshouses the various environmental factors continually fluctuate, making it impossible to determine which factor or combination of factors produces an observed plant response. Phytotrons provide the means of dissecting the environment and studying the effects of the parts by controlling the individual components. Thus, while one factor is held at fixed but different levels in different plant growth units, all other factors are maintained constant or at a known rate of change in all units.

Precision in phytotronic research has increased with improvements in the ability to control the various environmental factors. While temperature and light were the factors studied in most early experiments, improvements in equipment not only provided facilities with greater accuracy of control of these factors, but made it feasible to work with different humidities and with various concentrations of atmospheric gases. Systems are available where the conditions surrounding the roots are controlled separately from the conditions to which the plant top are exposed. As equipment improves, phytotronics will increase in reliability, usefulness, and complexity. One equipment problem is that of lighting, namely, trying to provide artificial light approximating the intensity and wavelength distribution of sunlight received at the Earth's surface in summer.

The growth and development of plants is an expression of their genetic capability as affected by environmental factors. Thus, the potential for phytotronic studies is exceedingly great when one considers the large number of plant species in the world with their genetic variability and the wide range of each environmental factor.

Types of studies. Temperature, probably the most extensively and intensively studied environmental factor in phytotronics, affects the physical process of transpiration as well as all enzymatically controlled physiological processes. Thus, temperature is a dominant environmental factor affecting plant growth and development. Wide ranges of day and night temperature combinations have been used to determine the growth potential of many plant species. Phytotronic studies employing ranges of temperature conditions also have been used to determine the temperature requirements for the development of specific plant parts. These studies include among others the determination of the number of hours at temperatures below a set point, such as 45°F (7°C), required to break bud dormancy in deciduous fruit trees, and the number of heat units, termed degree-hours (temperature × time) required for fruit maturation. Optimum temperature requirements are often different for the various phases of plant development, including seed germination, root, shoot, and leaf growth dry weight, flowering, fruit set, and seed filling. Results from studies of this type are used by plant breeders, glasshouse operators, and agriculturists. The latter two use the information to determine planting and harvesting dates and for selecting plant varieties that give the best product under their expected glasshouse or field temperatures.

While temperature is a basic factor in most phytotronic studies, considerable work has been conducted in which temperature is held at fixed levels, and other environmental factors such as light intensity, light duration (photoperiod), light quality, water stress, humidity, nutrients, carbon dioxide, or air pollutant concentrations are varied. The objective is usually to determine the effects of these environmental factors, either individually or in combination, on a specific phase of plant growth or development.

Results of studies. Examples from a few selected studies serve to illustrate the practical and theoretical values of phytotronics. At least one plant species, tobacco, has been studied to the extent that under controlled environmental conditions plants can now be produced virtually to match field-grown plants at any given age. This requires environmental manipulation which provides stress as well as optimum growth conditions at specific physiological stages of plant development. Thus, plants for experimental work can be provided at any given growth stage throughout the year. In addition, many experiments have been conducted to supply specific information and numbers for use in mathematical plant growth models that have been or are being developed for wild plants as well as agricultural crop species.

Research with temperature as the variable has been conducted on many species. Itchgrass (*Rottboellia exaltata*), a weed introduced into Louisiana and Florida, represents a serious potential weed problem in the warmer moist parts of the United States. Phytotronic studies showed that unless cool-climate ecotypes develop the weed will be less of a problem in the Corn Belt and will probably not be competitive with crops in the northern regions. Some of the results also illustrate the effect of an environmental factor on different constituents of plant growth. Total plant dry weight increased with increased day and night temperatures, but height growth peaked at a cooler night condition, 73°F (23°C). These plants were grown in all 36 temperature conditions simultaneously, thus avoiding any differences that would be introduced due to a time factor. In another study an unusually low night temperature, 41°F (5°C), was found to favor the growth of radiata pine (*Pinus radiata*), which accounts for the more rapid growth of these trees in New Zealand than in their native California. In contrast, growth of young redwood trees (*Sequoia sempervirens*) is severely retarded when night temperature falls below 52°F (11°C). It is probably this factor which determines the northern and eastern limit of its range. Other studies using multiple day/night temperature conditions have also been used to screen plants for introduction to specific areas or conversely to identify areas where exotic plants can be grown.

Phytotronic studies have played key roles in breeding of rice, wheat, and numerous other agricultural crops as well as in matching crops to environmental conditions. For example, the bulb diameter of Creamgold onions was found to be less if the plants grew at temperatures below a 79°F (26°C) day with a 64°F (18°C) night, while the Braeside variety was unaffected, making the latter the better choice for early spring planting.

Many studies have added to or changed basic concepts of what controls plant growth and development. For instance, by alternating weak and intense light on arctic plants it was found that they could be induced to flower, a response previously thought to require darkness. Other studies have shown that enrichment of the atmospheric CO_2 suppresses flowering in some short-day plants under inductive conditions and induces flowering of some long-day plants under noninductive conditions. Such changes in flower induction could be of major ecological importance if the world's atmospheric CO_2 content increases. Also, CO_2-enriched air during the light period or low night temperature can cause a starch buildup to 40–50% of leaf dry weight in some plants. Differences in the physiological response to temperature and day length have been found in races from the northern and southern part of the range of plant species with a wide north-south range.

The identification and confirmation of the acclimation ecotypes concept was obtained by growing arctic and alpine plants simultaneously in the same environmental conditions. Leaf temperature was discovered to control the ratio of chlorophyll *a* to chlorophyll *b*, resulting in size differences of the chlorophyll unit frequently noted in chlorophyll-deficient mutants. Metabolic processes controlling growth, pigments, and secondary plant products are temperature-sensitive and thus have often been studied. Sugar content of sugar beets was found to be markedly increased by cool conditions, a finding now used by growers in timing harvests. The development and aftereffects of water stress have been shown to be affected by temperature and light regimes not only because of the energy aspects per se, but because temperature and light caused pronounced effects on stomatal guard cells and enzymatic systems.

In a study using 18 environmental combinations, three temperatures, three photoperiods, and two light intensities, it was determined that even though light affected CO_2 assimilation it was temperature which regulated the pathway of CO_2 uptake either via organic acids or ribulosediphosphate in crassulacean acid metabolism (CAM) plants. *See* BREEDING (PLANT); PHOTOPERIODISM; PHYSIOLOGICAL ECOLOGY (PLANT); PLANT GROWTH. Henry Hellmers

Bibliography. P. Chouard and N. de Bilderling, *Phytotronics in Agricultural and Horticultural Research*, 1975; P. Chouard, N. de Bilderling, and R. Jacques, *Phytotronic Newsl.*, nos. 1–19, 1971–1979; *Environment Control in Biology*, vols. 1–17, Journal of Japanese Society of Environment Control in Biology, 1962–1979; F. W. Went, *The Experimental Control of Plant Growth*, 1957.

Piciformes

A large order of land birds, second in size only to the Passeriformes, that is found throughout the world, except for the Australian region, and is concentrated in tropical areas. The relationships of the suborder Galbulae to the other piciforms is disputed, with some researchers placing the Galbulae with the Coraciiformes. The evidence for retaining the Galbulae in the Piciformes as well as that for shifting them to the Coraciiformes is inconclusive; here the

Galbulae will be kept with the Piciformes, for which there appears to be somewhat better evidence. The overall affinities of the Piciformes may, in fact, be with other land birds such as the Coraciiformes and Passeriformes. *See* CORACIIFORMES; PASSERIFORMES.

Classifications. The Piciformes is divided into the following three suborders and eight families; the family Picidae is the largest, with 204 species of woodpeckers.

Order Piciformes
  Suborder Galbulae
    Family: Primobucconidae (fossil)
      Galbulidae (jacamars; 17 species)
      Bucconidae (puffbirds; 3 species)
  Suborder Zygodactyli
    Family Zygodactylidae (fossil)
  Suborder Pici
    Family: Capitonidae (barbets; 81 species)
      Ramphastidae (toucans; 33 species)
      Indicatoridae (honeyguides; 16 species)
      Picidae (woodpeckers; 204 species)

Fossil record. The fossil record of the piciforms is quite poor. The earliest fossil barbets and woodpeckers are known from the Miocene, and the earliest honeyguides from the Pliocene. The family Primobucconidae from the Eocene of North America constitutes a small, rather diverse group that require considerable study before their relationships can be determined with any certainty. *Zygodactylus*, from the early Miocene of Bavaria, is a puzzling fossil and is only tentatively included in the Piciformes.

Characteristics. The piciforms are small to medium-sized, hole-nesting land birds. The bill is short to medium-long, straight, and strong, and the wings are of medium length and rounded. The legs are short and strong, with the strong toes arranged in a zygodactylous (yoke) pattern, with two toes forward and two toes back. The tail may have stiffened feathers. The plumage, which varies greatly in hue, is frequently brightly colored and boldly patterned. Piciforms are good fliers and can easily perch and climb, but they walk poorly. Most species feed on insects; toucans, on the other hand, are primarily fruit eaters. Most are solitary, although some barbets and woodpeckers live and breed in social groups. All breed in cavities dug in earthen banks or hollowed out in trees. The eggs are incubated by both sexes, and both parents care for the unfeathered young, which remain in the nests. Except for a few species of woodpeckers, the piciforms are nonmigratory.

Woodpeckers are specialized for climbing on and drilling into trees to excavate nesting cavities and to search for food. Specialized characteristics include the shapes of the bill, some jaw muscles, and the tongue; stiffened tail feathers; and the arrangements of the toes and shape of the claws. Contrary to common belief, the zygodactyl arrangement of the toes is not an adaptation for climbing but for perching. Those specializations vary considerably among the woodpeckers according to their feeding habits.

The honeyguides of Africa and India are nest parasites; they lay their eggs in the nests of other piciforms, usually barbets. They feed commonly at bees' nests, most often consuming the wax combs rather than the insects because they have the digestive enzymes necessary to break down the wax into sugars. At least one species leads mammals, including humans, to the nests; after the nest has been opened to retrieve the honey, the birds feed on any remaining bits.

Economic significance. Although woodpeckers consume insects that damage trees, it is still not clear whether they actually save trees. Several species of woodpeckers have become extinct or seriously endangered because of habitat destruction, and they have become relatively uncommon in Europe because of intensive forest management. A few woodpeckers are considered pests because they drill into wooden poles; more recently they have begun to damage the plastic foam insulation placed on the outside of houses in central Europe. Sapsuckers (*Sphyrapicus*) can kill trees by drilling large numbers of holes in the trunk and major branches for collecting the sap, on which they feed. *See* AVES.

Walter J. Bock

Picornaviridae

A viral family made up of the small (18–30 nanometers) ether-sensitive viruses that lack an envelope and have a ribonucleic acid (RNA) genome. The name is derived from "pico" meaning very small, and RNA for the nucleic acid type. Most of the picornaviruses are stabilized by magnesium chloride against thermal inactivation. The virion is made up of a nucleic acid core surrounded by a capsid of 32 subunits (capsomeres) arranged in the icosahedral form of cubic symmetry. Within the infected cell, virus particles are assembled in the cytoplasm, where they tend to aggregate in crystalline array.

Picornaviruses of human origin include the following genera: *Enterovirus* (polioviruses, 3 types; coxsackieviruses A, 24 types; coxsackieviruses B, 6 types; echoviruses, 33 types) and *Rhinovirus* (more than 70 types). There are also 2 genera of picornaviruses of lower animals: *Aphtovirus* (bovine foot-and-mouth disease) and *Cardiovirus*. Certain plant viruses have characteristics similar to those of the picornaviruses; some RNA-containing bacteriophages also have similar properties. *See* COXSACKIEVIRUS; ECHOVIRUS; ENTEROVIRUS; FOOT-AND-MOUTH DISEASE; POLIOMYELITIS; RHINOVIRUS.

Many picornaviruses cause diseases in humans ranging from severe paralysis to aseptic meningitis, pleurodynia, myocarditis, skin rashes, and common colds. However, subclinical (silent) infection is far more common than clinically manifest disease. Different viruses may produce the same syndrome; on the other hand, the same virus may cause more than a single syndrome.

The host range of the picornaviruses varies greatly

from one type to the next, and even among strains of the same type. They may readily be induced by laboratory manipulation to yield variants which have host ranges and target organs different from those of certain of the naturally occurring "wild" strains; this variation has led to the development of attenuated poliovaccine strains.

The two major subgroups of picornaviruses, the enteroviruses and the rhinoviruses, differ in density of the virus particle; the site commonly inhabited in the human host; acid lability; and cell susceptibility and other conditions for optimal growth in tissue culture. The enteroviruses have a density of 1.34 g/ml, and the molecular weight of their RNA core is about 2×10^6; in contrast, the density of rhinoviruses is 1.38–1.40 g/ml but the RNA core still has a molecular weight of about 2×10^6. The enteroviruses multiply in the alimentary tract and are isolated chiefly from feces; the rhinoviruses are almost always isolated from the nose or throat and seldom if ever from feces. While enteroviruses are stable under acid conditions (pH 3–5), the rhinoviruses are inactivated by 1–3 h at this pH level.

Certain of the enteroviruses (coxsackieviruses of the A group) cannot be grown in tissue cultures, and their cultivation in the laboratory must be carried out in mice. Of those enteroviruses which will grow and produce cytopathic effects in cultured cells, most grow preferentially in primary cultures of human and monkey kidney cells (or in certain continuous human heteroploid cell lines). In contrast, rhinoviruses are more readily isolated in embryonic human kidney or human diploid cell strains than in monkey cells; most strains can be recovered only in cells of human origin. *See* ANIMAL VIRUS; TISSUE CULTURE; VIRUS CLASSIFICATION. Joseph L. Melnick

Bibliography. R. W. Compans et al. (eds.), *Current Topics in Microbiology and Immunology*, vol. 161: *Picornaviruses*, 1990; S. A. Evans (ed.), *Viral Infections of Humans: Epidemiology and Control*, 4th ed., 1997; H. H. Malherbe, *Viral Cytopathology*, 1980; V. M. Zhdanov and D. K. Lvov, *Etiology and Pathogenesis of Viral Infections*, 1989.

Picrite

The term picrite has been used with several different meanings. It is generally considered to include certain medium- to fine-grained igneous rocks composed chiefly of olivine with smaller amounts of pyroxene, hornblende, and plagioclase feldspar (labradorite).

Its feldspar content is slightly higher than that of peridotite and lower than that of gabbro. Certain analcite-bearing types, associated with teschenite, have also been included under the term picrite. A characteristic feature is poikilitic texture in which large pyroxene or hornblende crystals enclose numerous small grains of olivine. *See* GABBRO; PERIDOTITE.

Picrite is rare and is found in small intrusives (sills and dikes). It may also occur in the lower portions of basaltic lava flows where olivine and pyroxene crystals have accumulated under the influence of gravity. *See* IGNEOUS ROCKS. Carleton A. Chapman

Pictorial drawing

A view of an object (actual or imagined) as it would be seen by an observer who looks at the object either in a chosen direction or from a selected point of view. One such view often suffices to give the reader a clear picture of the shape and details of the object. Pictorial sketches often are more readily made and more clearly understood than are front, top, and side views of an object. Pictorial drawings, either sketched freehand or made with drawing instruments, are frequently used by engineers and architects to convey ideas to their assistants and clients. *See* DESCRIPTIVE GEOMETRY; ENGINEERING DRAWING.

In making a pictorial drawing, it is important to select the viewing direction that shows the object and its details to the best advantage. The resultant drawing is orthographic if the viewing rays are considered as parallel, or perspective if the rays are considered as meeting at the eye of the observer. Making perspective drawings with instruments is time-consuming and requires considerable knowledge and skill. There are, however, commercially available devices which make this chore easier and quicker than is the case when conventional instruments are used. Perspective drawings provide the most realistic, and usually the most pleasing, likeness when compared with other types of pictorial views (**Fig. 1**).

Several types of nonperspective pictorial views can be sketched, or drawn with instruments. Although each type has some distortion, all provide a good picture of what the object looks like. They are easier and quicker to make than perspective drawings. The isometric pictorial is especially popular

Fig. 1. Perspective drawing of a residence. (*Home Planners Inc.*)

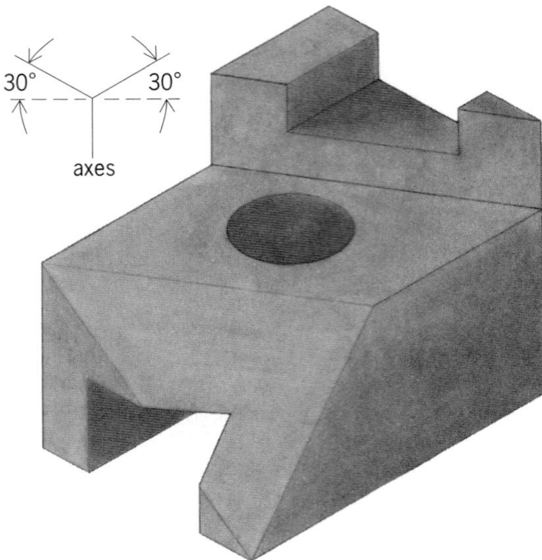

Fig. 2. Isometric drawing; measurements alog each axis are made with the same scale.

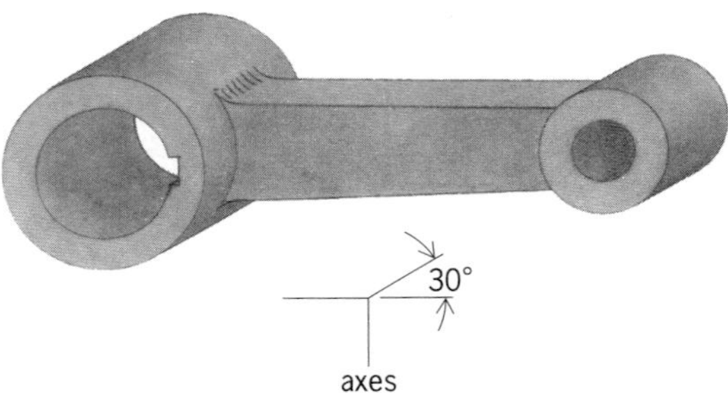

Fig. 3. Oblique pictorial drawing.

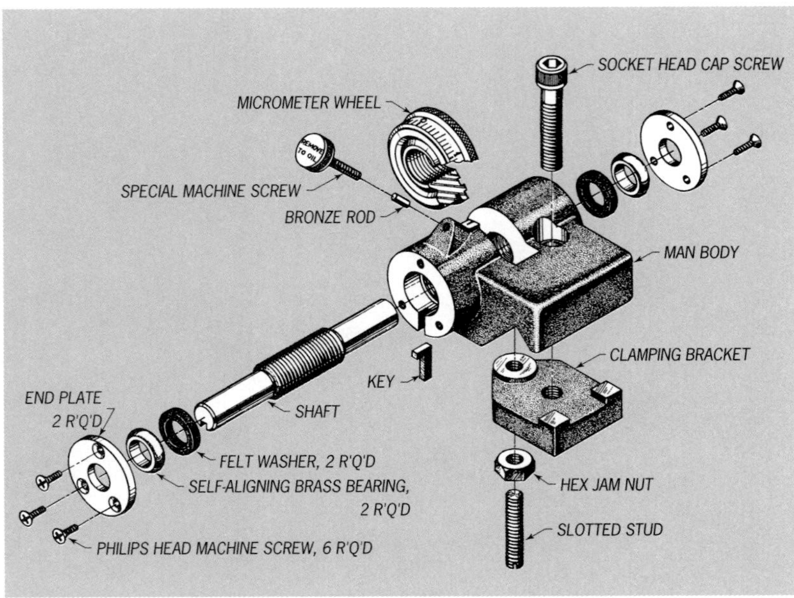

Fig. 4. Exploded-view production illustration; this drawing is isometric.

because of the direction of its axes and the fact that all measurements along these axes are made with one scale (**Fig. 2**). In addition to isometric representation, two other forms—dimetric and trimetric—are sometimes used.

Oblique pictorial drawings, while not true orthographic views, offer a convenient method for drawing circles and other curves in their true shape (**Fig. 3**). In order to reduce the distortion in an oblique drawing, measurements along the receding axis may be foreshortened. When they are halved, the method is called cabinet drawing.

An effective freehand sketch of an object can be made if proper attention is given to viewing direction, proportions, orientation of ellipses, and location of tangent points. Shading is sometimes added to enhance the pictorial drawing.

Shaded exploded-view production illustrations greatly facilitate the learning process in assembly of machines and devices (**Fig. 4**). When this type of illustration is used, the initial assembly of parts into a machine has been found to be three or four times faster than if a conventional assembly drawing is used. Photodrawings can be used to achieve the same visual results. Charles J. Baer

Bibliography. T. E. French et al., *Mechanical Drawing*, 11th ed., 1989; T. E. French, C. J. Vierk, and R. J. Foster, *Engineering Drawing and Graphic Technology*, 14th ed., 1993; C. H. Jensen and J. D. Helsel, *Engineering Drawing and Design*, 5th ed., 1997; D. A. Madsen et al., *Fundamentals of Drafting Technology*, rev. ed., 1994.

Picture tube

A cathode-ray tube used as a television picture tube, also called a kinescope. It might be referred to as a television picture reproducer. Modern television picture tubes use large glass envelopes (**Fig. 1**) that have a light-emitting layer of luminescent material deposited on the inner face. A modulated stream of high-velocity electrons scans this luminescent layer in a series of horizontal lines so that the picture elements (light and dark areas) are recreated.

The number of electrons in the system at any instant of time is varied by electrical pulses corresponding to the signal sent out by the television transmitter. These electrical pulses (picture information) were originally generated by a studio television camera. The home television receiver picks up the signal, and after suitable amplification and detection the picture information is supplied to the picture tube to recreate the original picture.

Construction. This article discusses construction of television picture tubes. For detailed discussion of cathode-ray tube construction *see* CATHODE-RAY TUBE

Glass envelope. A special-composition glass is used in making the envelope to minimize optical defects and to provide electrical insulation for high voltages. It also provides protection against x-radiation and has

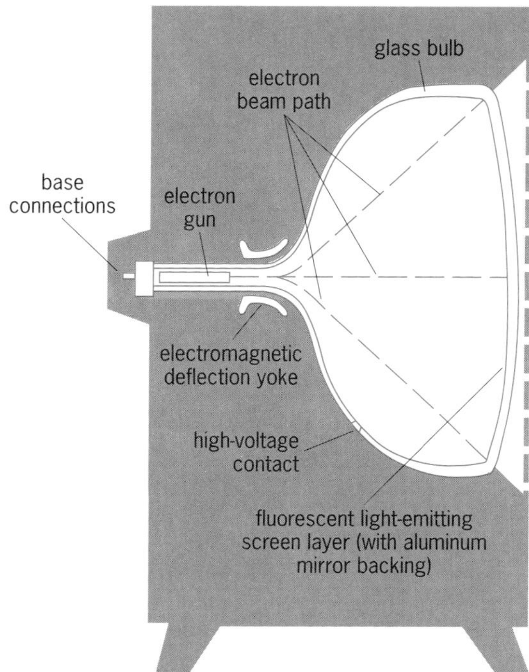

Fig. 1. Black-and-white kinescope.

a light-absorption characteristic that improves picture contrast when it is viewed under bright illumination. In manufacturing the tube, air is evacuated from the glass bulb, and the tube is designed to withstand more than three times the force of atmospheric pressure to provide a safety margin. However, care must be taken in handling evacuated glass bulbs to avoid a dangerous implosion.

Phosphor screen. The luminescent screen is made of a thin layer of phosphors (density of about 3 mg/cm^2). The phosphor materials for monochrome picture tubes are primarily zinc cadmium sulfide (which emits yellow light) and zinc sulfide (which emits blue light). By careful proportioning and mixing of these two phosphors the resultant emanation is a blue-white light. *See* PHOSPHORESCENCE.

The phosphor screen is aluminized by vacuum evaporation from a small molten aluminum pellet. The layer of aluminum, approximately 200 nanometers thick, is deposited on a smooth plastic film placed on top of the luminescent screen. The plastic film is subsequently volatilized and removed in the high-temperature processing of the tube. In the operation of the completed tube, the high-velocity electron beam penetrates the aluminum film, and its energy is transferred primarily to the phosphor screen. Only a small percentage of the electron-beam energy is converted into useful light energy, but this amount is sufficient to produce a brightness of several hundred footlamberts (1 footlambert = 3.4 $candelas/m^2$) in the picture highlights. The reflection of light by the aluminum mirror increases the picture brightness and improves picture contrast by preventing stray light from illuminating the back side of the phosphor screen.

Wall coating. The inside walls of the bulb are coated with graphite to provide electrically conducting surfaces between the screen and the electron gun, and to provide a unipotential field through which the electron beam may travel without being disturbed by stray electrostatic fields. Graphite is a sufficiently poor conductor to minimize eddy-current power absorption from the electromagnetic fields generated by the external deflecting yoke. *See* EDDY CURRENT.

Electron gun. The electron gun produces a stream of high-velocity electrons which are focused to a small spot at the phosphor screen. The electrons are generated by an indirectly heated nickel cathode coated with barium and strontium oxides. A cloud of electrons from the cathode is focused to a crossover point near the cathode by the electric fields between the cathode and the first two grids. This crossover of electrons is then focused electrostatically in a narrow beam by the main lens composed of the next two or more elements of the gun to produce a small spot on the phosphor screen. The voltage on the last element of the gun and on the screen is of the order of 15–25 kV. The intensity of the beam is controlled by the voltage between the cathode and the first grid. This voltage is modulated by the video signal to produce the range of brightness on the screen.

Electromagnetic deflection. The electron beam is deflected electromagnetically to cause it to scan the picture area. This deflection is accomplished by a deflecting yoke, made up of two pairs of shaped coils which fit around the neck of the picture tube (Fig. 1). When pulsating electric currents of proper wave shape and phase are supplied to these coils, they generate magnetic fields which cause the electron beam to bend as it passes through them. By changing the magnitude and direction of the magnetic field, the electron beams are made to scan the screen in a systematic raster of horizontal lines which are sequentially stepped from top to bottom of the screen.

Color picture tube. A color television picture tube differs in several ways from the monochrome picture tube (**Fig. 2**): (1) The light-emitting screen is made up of small elemental (phosphor) areas laid in interlaced arrays, each capable of emitting light in one of the three additive primary colors (red, green, and blue). (2) The electron gun produces three beams, one for each of the primary-color phosphors. (3) A shadow mask assures that each of the three electron beams strikes only the color of phosphor elements intended. Red, green, and blue pictures are superimposed to produce a full range of perceived colors.

Glass envelope. The glass bulb is made in two pieces, the face panel and the funnel-neck region. The separate face panel allows the fabrication of the segmented phosphor screen and the mounting of the shadow mask. The two glass pieces are sealed together by a special frit to provide a strong vacuum-tight seal.

Face panel assembly. The light-emitting colored phosphors on the segmented screen can be either in dot arrays or, now more commonly, in line arrays. Typically, the trios of vertical phosphor lines are spaced

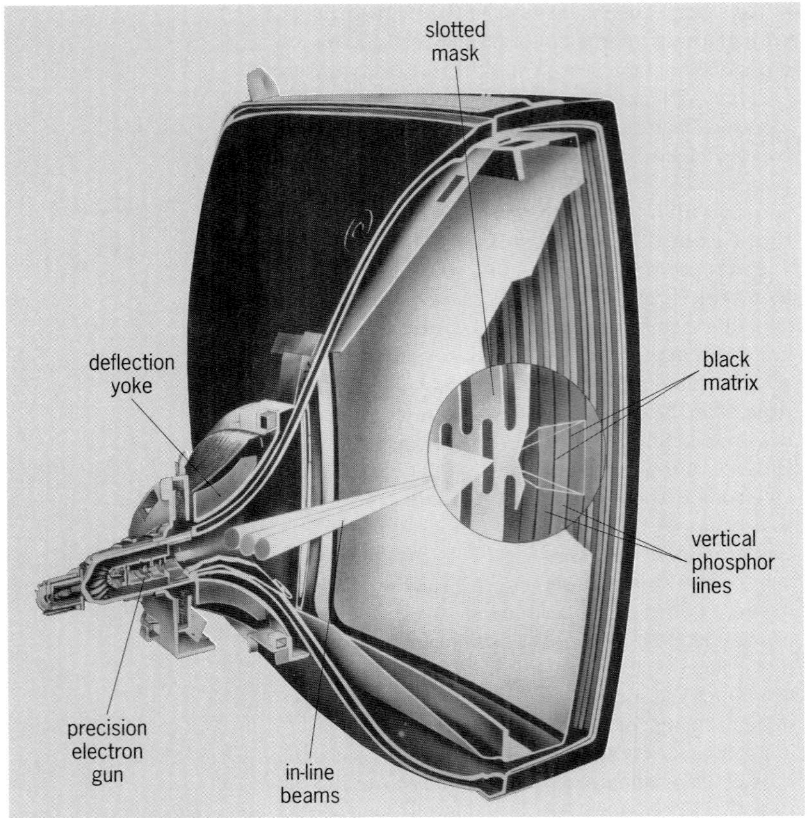

Fig. 2. Color picture tube.

0.6–0.8 mm apart. Most modern tubes use a black matrix screen in which the phosphor lines are separated by opaque black lines. This black matrix reduces reflected light, thereby giving better contrast, and also provides a tolerance for the registration of the electron beam with the phosphor lines.

The shadow mask is made of a thin (0.10–0.17 mm) steel sheet in which elongated slits (one row of slits for each phosphor-line trio) have been photoetched. It is formed to a contour similar to that of the glass panel and is mounted at a precise distance from the glass. The width of the slits and their relative position to the phosphor lines are such that the electron beam from one of the three electron guns can strike only one of the sets of color phosphor lines. The shadow mask "shadows" the beam from the other two sets of phosphor lines.

Electron gun. The electron gun for color is similar to that for monochrome except that there are three guns, usually arranged side by side, or in-line. This triple gun has common structural elements, but uses three independent cathodes with separate beam forming and focusing for each beam.

Deflection yoke. The electromagnetic deflection yoke deflects or bends the beams, as in a monochrome tube, to scan the screen in a television raster. In addition, the yoke's magnetic field is shaped so that the three beams will be deflected in such a way that they land at the same phosphor trio on the screen at the same time. This convergence of the beams produces three images, one in red, one in green, and one in blue, that are superimposed to give a full-color picture. *See* TELEVISION. A. M. Morrell

Bibliography. K. B. Benson and J. Whitaker, *Television Engineering Handbook: Featuring HDTV Systems*, rev. ed., 1992; S. Sherr, *Electronic Displays*, 1993; J. Whitaker, *Electronic Displays: Technology, Design, and Applications*, 1994.

Piezoelectricity

Electricity, or electric polarity, resulting from the application of mechanical pressure on a dielectric crystal. The application of a mechanical stress produces in certain dielectric (electrically nonconducting) crystals an electric polarization (electric dipole moment per cubic meter) which is proportional to this stress. If the crystal is isolated, this polarization manifests itself as a voltage across the crystal, and if the crystal is short-circuited, a flow of charge can be observed during loading. Conversely, application of a voltage between certain faces of the crystal produces a mechanical distortion of the material. This reciprocal relationship is referred to as the piezoelectric effect. The phenomenon of generation of a voltage under mechanical stress is referred to as the direct piezoelectric effect, and the mechanical strain produced in the crystal under electric stress is called the converse piezoelectric effect. *See* POLARIZATION OF DIELECTRICS.

Piezoelectric materials are used extensively in transducers for converting a mechanical strain into an electrical signal. Such devices include microphones, phonograph pickups, vibration-sensing elements, and the like. The converse effect, in which a mechanical output is derived from an electrical signal input, is also widely used in such devices as sonic and ultrasonic transducers, headphones, loudspeakers, and cutting heads for disk recording. Both the direct and converse effects are employed in devices in which the mechanical resonance frequency of the crystal is of importance. Such devices include electric wave filters and frequency-control elements in electronic oscillator circuits. *See* MICROPHONE; ULTRASONICS.

Necessary condition. The necessary condition for the piezoelectric effect is the absence of a center of symmetry in the crystal structure. Of the 32 crystal classes, 21 lack a center of symmetry, and with the exception of one class, all of these are piezoelectric. In the crystal class of lowest symmetry, and type of stress generates an electric polarization, whereas in crystals of higher symmetry, only particular types of stress can produce a pieozoelectric polarization. For a given crystal, the axis of polarization depends upon the type of the stress. There is no crystal class in which the piezoelectric polarization is confined to a single axis. In several crystal classes, however, it is confined to a plane. Hydrostatic pressure produces a piezoelectric polarization in the crystals of those 10 classes that show pyroelectricity in addition to piezoelectricity. The pyroelectric axis is then the axis of polarization. *See* CRYSTALLOGRAPHY; PYROELECTRICITY.

The converse piezoelectric effect is a thermodynamic consequence of the direct piezoelectric effect. When a polarization P is induced in a piezoelectric crystal by an externally applied electric field E, the crystal suffers a small strain S which is proportional to the polarization P. In crystals with a normal dielectric behavior, the polarization P is proportional to the electric field E, and hence the strain is proportional to this field E. Superposed upon the piezoelectric strain S is a much smaller strain which is proportional to P^2 (or E^2). This strain is called the electrostrictive strain. It is present in any dielectric. *See* ELECTROSTRICTION.

Matrix formulation. The relation of the six components T_j of the stress tensor (three compressional components and three shear components) to the three components P_i of the polarization vector can be described by a scheme (matrix) of 18 piezoelectric moduli D_{ij}. The same scheme (d_{ij}) also relates the three components E_i of the electric field to the six components S_j of the strain:

		Compression			Shear		
		S_1 T_1	S_2 T_2	S_3 T_3	S_4 T_4	S_5 T_5	S_6 T_6
E_1	P_1	d_{11}	d_{12}	d_{13}	d_{14}	d_{15}	d_{16}
E_2	P_2	d_{21}	d_{22}	d_{23}	d_{24}	d_{25}	d_{26}
E_3	P_3	d_{31}	d_{32}	d_{33}	d_{34}	d_{35}	d_{36}

The direct effect is obtained by reading this scheme in rows, as in Eq. (1). The converse effect is obtained by reading it in columns, as in Eq. (2).

$$P_i = -\sum_{j=1}^{6} d_{ij}T_j \qquad i = 1, 2, 3 \tag{1}$$

$$S_j = \sum_{i=1}^{3} d_{ij}E_i \qquad j = 1, 2, \ldots, 6 \tag{2}$$

An analogous matrix (e_{ij}) relates the strain to the polarization and the electric field to the stress, as in Eqs. (3).

$$\begin{aligned} P_i &= \sum_{j=1}^{6} e_{ij}S_j \qquad i = 1, 2, 3 \\ T_j &= -\sum_{i=1}^{3} e_{ij}E_j \qquad j = 1, 2, \ldots, 6 \end{aligned} \tag{3}$$

The matrices (d_{ij}) and (e_{ij}) are not independent, but are related by expressions involving the elasticity tensor c^E_{jh} (for constant electric field E), as in Eq. (4).

$$e_{mh} = \sum_{j=1}^{6} d_{mj}c^E_{jh} \qquad \begin{array}{l} m = 1, 2, 3 \\ h = 1, 2, \ldots 6 \end{array} \tag{4}$$

Alternative formulations can be made by introducing the dielectric displacement D or visualizing the simultaneous action of electrical and mechanical stresses. *See* ELASTICITY.

The number of independent matrix elements d_{ij} or e_{ij} depends upon the symmetry elements of the crystal. For the lowest symmetry, all 18 matrix elements are independent, whereas piezoelectric classes of higher symmetry can have as few as one independent element in the matrix (d_{ij}). The matrix takes its simplest form if the natural symmetry axes of the crystal are chosen for the coordinate system.

Electromechanical coupling. The direct piezoelectric effect makes a crystal a generator, and the converse effect makes it a motor. Consequently, a piezoelectric crystal has many properties in common with a motor-generator. For example, the electrical properties, such as the dielectric constant, depend upon the mechanical load; conversely, the mechanical properties, such as the elastic constants, depend upon the electric boundary conditions. The electromechanical coupling factor k can be defined as follows. Suppose electrodes are attached to a piezoelectric crystal and connected to a battery. Then the ratio of the energy stored in mechanical form to the electrical energy delivered by the battery is equal to k^2. In general, k ranges from below 1 to about 30%. In quartz, for example, the coupling is roughly 10%. In ferroelectric crystals, k can approach unity in certain circumstances. *See* FERROELECTRICS.

In quartz, a stress of 1 newton/m applied along the diad axis produces a polarization of about 2×10^{-12} coulomb/m^2 along the same axis. Conversely, an electric field of 10^4 volts/m produces a strain of about 2×10^{-8}. In ferroelectric crystals, such as rochelle salt and potassium dihydrogen phosphate, and in certain antiferroelectrics, such as ammonium dihydrogen phosphate (ADP), these effects can be several orders of magnitude larger.

Molecular theory. Quantitative theories based on the detailed crystal structure are very involved. Qualitatively, however, the piezoelectric effect is readily understood for simple crystal structures. **Figure 1** illustrates this for a particular cubic crystal, zincblende (ZnS). Every Zn ion is positively charged and is located in the center of a regular tetrahedron $ABCD$, the corners of which are the centers of sulfur ions, which are negatively charged. When this system is subjected to a shear stress in the xy plane, the edge AB, for example, is elongated, and the edge CD of the

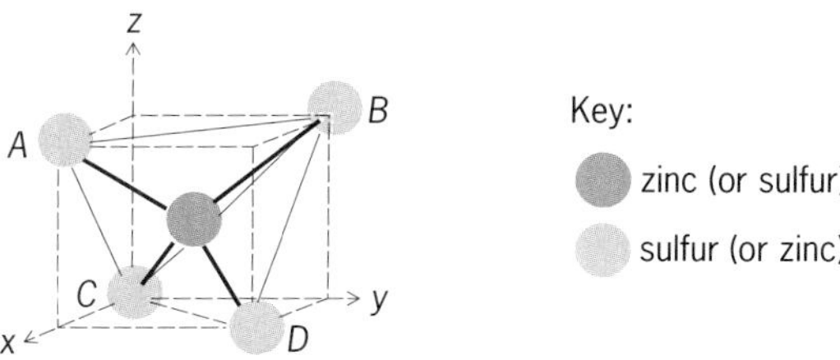

Fig. 1. Tetrahedral structure of zincblende, ZnS. Only part of unit cell is shown. Size of circles has no relation to size of ions.

tetrahedron becomes shorter. Consequently, these edges are no longer equivalent, and the Zn ion will be displaced along the z axis, thus giving rise to an electric dipole moment. The dipole moments arising from different octahedrons sum up because they all have the same orientation with respect to the axes x, y, and z.

Another simple type of piezoelectric structure is encountered in barium titanate as shown in **Fig. 2**. The positive Ti ions are surrounded by an almost regular octahedron of negative oxygen ions. The Ti ions are not in the center of the octahedron, but somewhat displaced along the z axis. This structure already has a dipole moment or spontaneous polarization in the absence of externally applied stresses. It is clear from Fig. 2 that the Ti ion is pushed more off center when the crystal is mechanically compressed in the xy plane or elongated along z. The additional polarization associated with this deformation is the piezoelectric polarization.

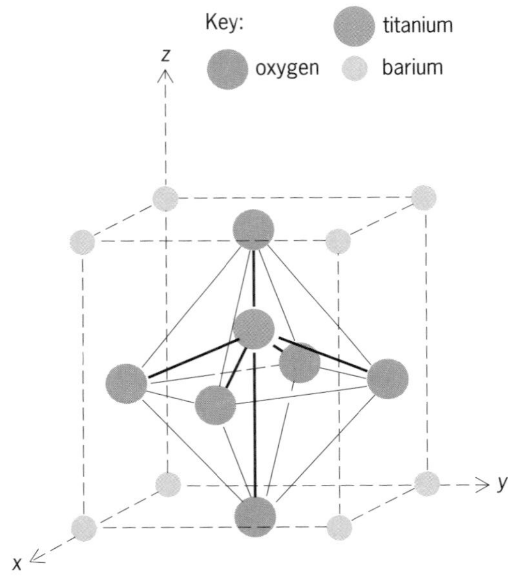

Fig. 2. Unit cell of tetragonal barium titanate, BaTiO3. Deviation from cubic symmetry is exaggerated. Size of the circles has no relation to the size of the ions.

Piezoelectric ceramics. Barium titanate and a few related compounds have the remarkable property that, by means of a sufficiently strong electric field, the direction of the spontaneous polarization can be switched to any one of the x, y, or z axes. This makes it possible to produce polycrystalline samples (ceramics) which are piezoelectric. The electromechanical coupling factors of such ceramics can reach about 50%.

Piezoelectric resonator. The piezoelectric strains that can be induced by a static electric field are very small, except in certain ferroelectrics. Larger strains can be obtained when a piezoelectric crystal is driven by an alternating voltage, the frequency of which is equal to a mechanical resonance frequency of the crystal. The vibrating crystal reacts back on the circuit through the direct piezoelectric effect. In the range of a mechanical resonance, this reaction is equivalent to the response of the network shown in **Fig. 3**, provided that the series resonance frequency of the network is equal to a mechanical resonance frequency of the crystal, as in Eq. (5). An important

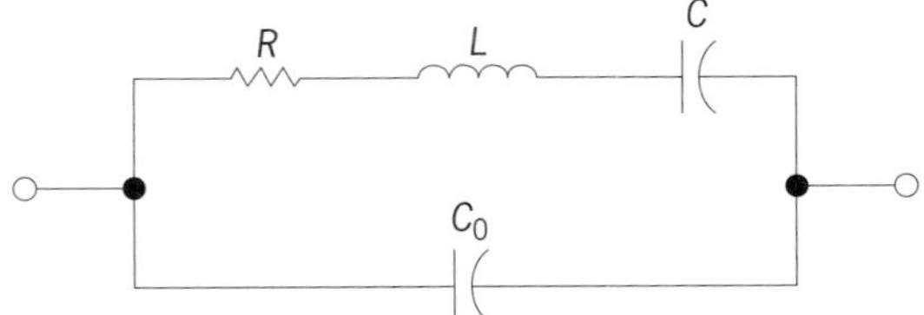

Fig. 3. Network equivalent to a piezoelectric resonator near and at a resonance frequency.

$$f_R = \frac{1}{2\pi\sqrt{LC}} \tag{5}$$

difference between the network of Fig. 3 and the piezoelectric resonator is that the latter has many discrete modes of vibration, whereas the network has only one resonance frequency.

Network elements. The elements L, C, and C_0 of the equivalent network can be calculated from the physical constants of the crystal. Consider, for example, the simple resonator shown in **Fig. 4**. A rectangular crystal bar with the dimensions $l_1 \gg l_2 \gg l_3$ is excited to compressional lengthwise vibrations. The xy faces have adherent electrodes, and the bar is oriented with respect to the natural crystal axes so that an electric field E_3 along z causes a strain S_1 along the bar according to the equation $S_{1(\text{piezoel})} = d_{31}E_3$. A mechanical stress T_1 along the bar causes a strain $S_{1(\text{mech})} = s_{11}^E T_1$, where s_{11}^E is the elastic compliance measured at constant electric field E_3. The resonance frequency for the fundamental lengthwise compressional mode is then given by Eq. (6), where ρ is

$$f_R = \frac{1}{2l_1\sqrt{\rho s_{11}^E}} \quad \text{Hz} \tag{6}$$

the density of the crystal. The parallel capacitance C_0 is the static capacitance of the crystal, as in Eq.

$$C_0 = \frac{8.85\epsilon l_1 l_2}{l_3} \quad \text{picofarads} \tag{7}$$

(7). Here ϵ is the relative dielectric constant along z. For C and L, the analysis yields Eqs. (8) and (9).

$$C = \frac{70.8d_{31}^2 \int_1 \int_2}{\pi^2 S_{11} E \int_3} \quad \text{picofarads} \tag{8}$$

$$L = \frac{\rho(S_{11}^E)^2 \int_1 \int_3}{8d_{31}^2 \int_2} \quad \text{henrys} \tag{9}$$

(All physical constants are in mks units.) For the nth overtone, C_0 and L remain the same, whereas C must be divided by n^2. The losses (damping) represented by the resistance R in Fig. 3 arise, for example, from ultrasonic radiation, friction in the crystal mount, internal friction in the crystal originating in various imperfections, and dielectric relaxation.

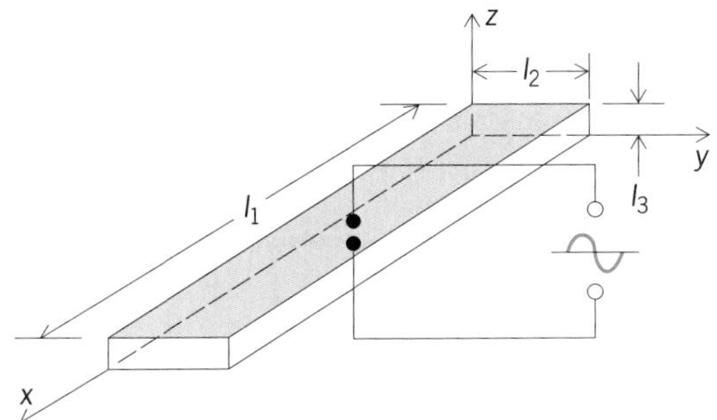

Fig. 4. Simple piezoelectric resonator. A voltage applied to the electrodes shortens or lengthens the bar, thus exciting longitudinal vibrations.

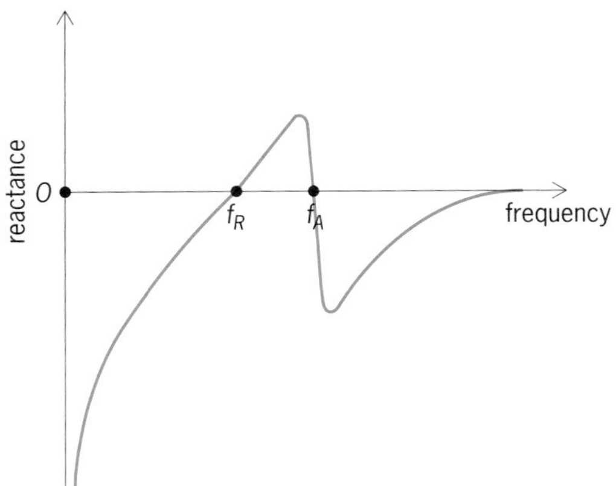

Fig. 5. Diagram showing reactance versus frequency for a piezoelectric resonator.

At the mechanical resonance frequency f_R, the alternating current is maximum and is determined by R. At the antiresonant frequency, given by Eq. (10),

$$f_A = \sqrt{(C_0 + C)/LCC_0} \tag{10}$$

the current is minimum. The difference $\Delta f = f_A - f_R$ increases with increasing electromechanical coupling according to relation (11).

$$\Delta f \approx \frac{4k^2}{\pi^2} \tag{11}$$

The reactance depends upon frequency, as shown in **Fig. 5**. For a typical piezoelectric crystal such as quartz, resonating at about 10^5 Hz, the orders of magnitude given by relations (12) are typical for

$$\begin{aligned} L &\approx 10^2 \text{ henrys} \\ C &\approx 0.02 \text{ picofarad} \\ C_0 &\approx 5 \text{ picofarads} \end{aligned} \tag{12}$$

the elements of the equivalent network. The damping resistance R varies from about 10^2 to 10^4 ohms; that is, the Q factors, given by Eq. (13), are in the

$$Q = \frac{1}{R}\sqrt{\frac{L}{C}} \tag{13}$$

range between 10^6 and 10^4, and the resonances are very sharp. These characteristics cannot be achieved with conventional coils and condensors as circuit elements.

Vibration modes. With piezoelectric resonators of various types, the range from audio frequencies to many megahertz can be covered. The vibration modes frequently used are, in order of increasing frequency: (1) flexural vibrations of bars and plates, (2) longitudinal vibrations of bars and plates, (3) face shear vibrations of plates, and (4) thickness shear vibrations and compressional vibrations of plates. **Figure 6** illustrates some of these modes. The excitation of particular vibration modes can be achieved by proper orientation of the resonator with respect to the natural crystal axes, by proper positioning of the electrodes, and by proper mounting. A simple example is illustrated by **Fig. 7**. A bar is oriented so that an electric field along x causes an expansion or contraction along y. The electrodes are split and cross-connected so that the bar flexes in the yz plane when a voltage is applied. The fundamental flexure mode is easily excited with this arrangement; however, excitation of higher even-numbered flexural modes is also possible. Interesting resonators are possible with

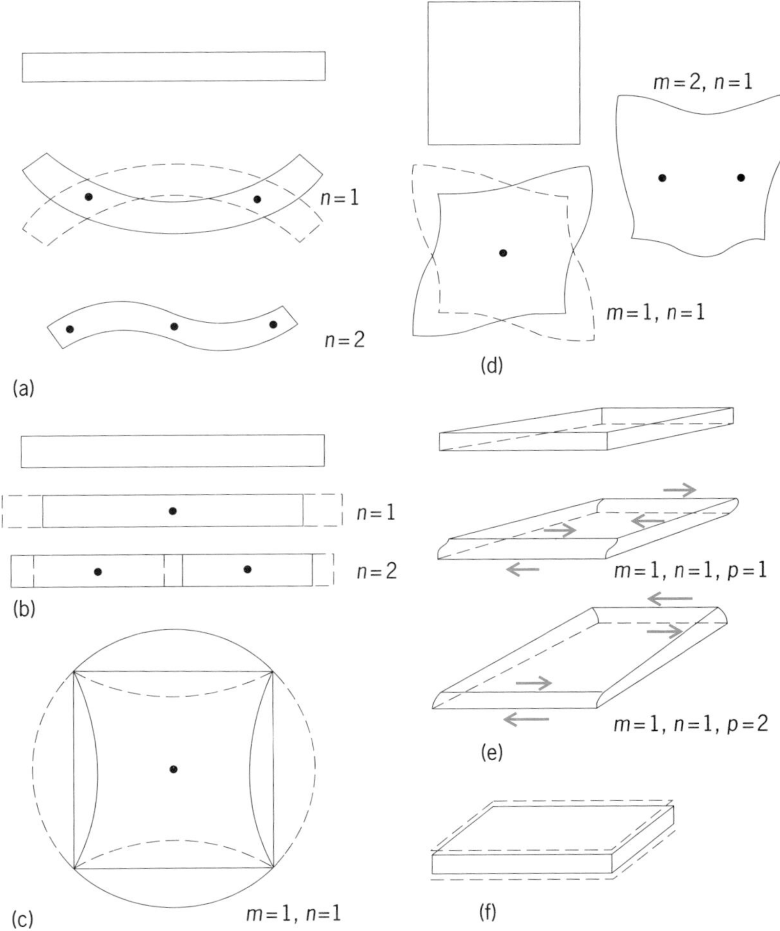

Fig. 6. Diagrammatic representation of examples of vibration modes of bars and plates. (*a*) Flexural vibrations of a bar. (*b*) Longitudinal vibrations of a bar. (*c*) Longitudinal vibration of a plate. (*d*) Face shear vibrations of a plate. (*e*) Thickness shear vibrations of a plate. (*f*) Thickness vibration of a plate.

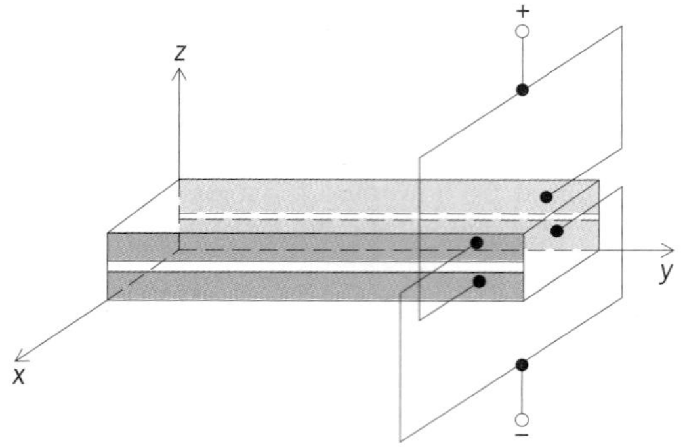

Fig. 7. Excitation of flexure mode by split electrodes.

piezoelectric ceramics (barium titanate type) because different parts of the resonator can be polarized in different directions. *See* VIBRATION.

Common applications. The sharp resonance curve of a piezoelectric resonator makes it useful in the stabilization of the frequency of radio oscillators. Quartz crystals are used almost exclusively in this application. The main advantages of quartz are high Q factor, stability with respect to aging, and the possibility of orienting the resonator with respect to the natural crystal axes so that the temperature coefficient of the resonance frequency vanishes near the operating temperature. **Figure 8** illustrates the orientation of commonly used cuts.

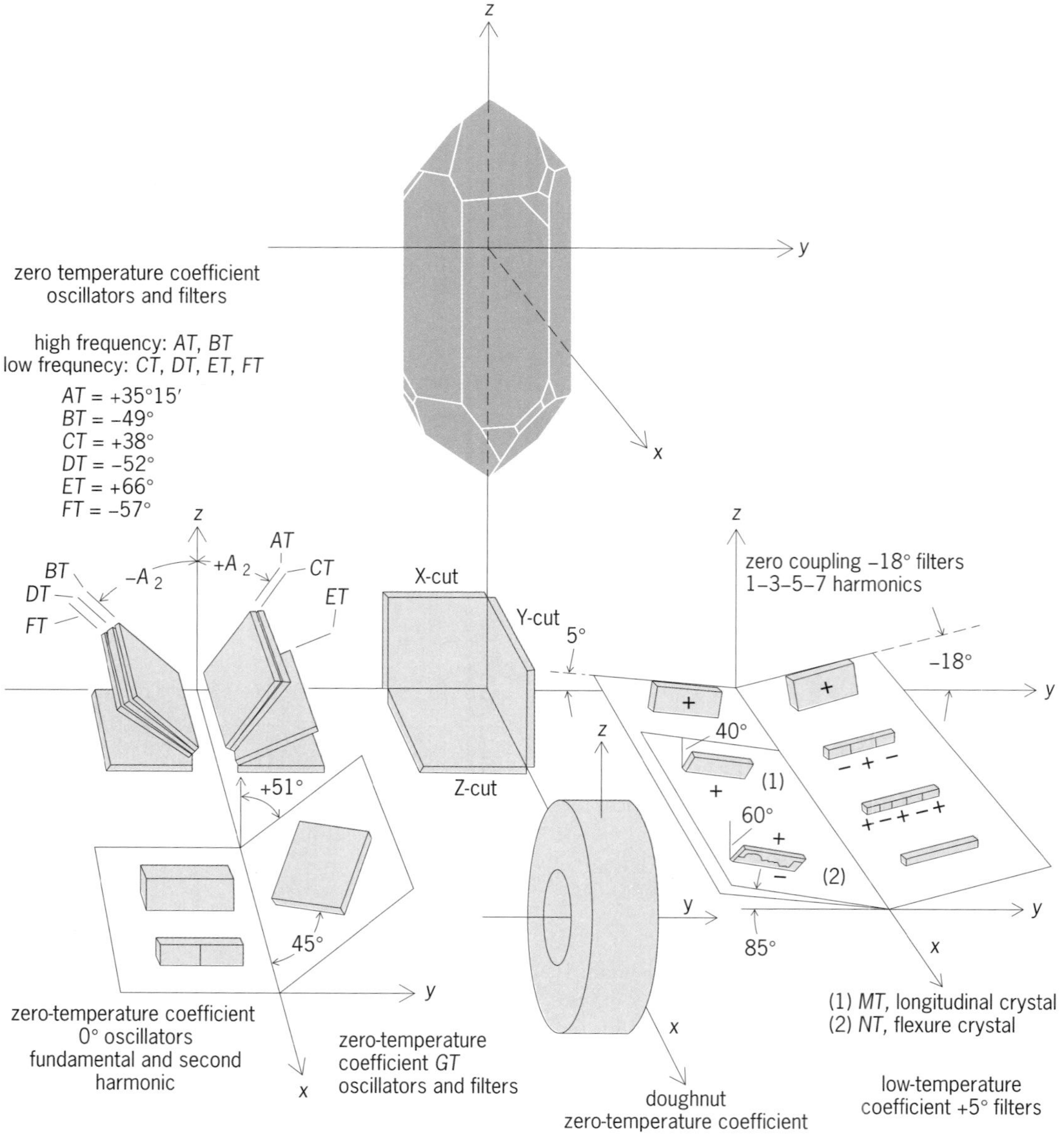

Fig. 8. Orientation with respect to the natural crystal axes of some of the more commonly used special cuts of quartz. (*After W. P. Mason, Piezoelectric Crystals and Their Application to Ultrasonics, Van Nostrand, 1950*)

In vacuum-tube oscillators, the crystal generally is part of the feedback circuit. In the circuit proposed by G. W. Pierce, the conditions for oscillation are not satisfied unless the crystal reactance is positive. Hence, the oscillation frequency is between the resonant and antiresonant frequency of the crystal (Fig. 5). Circuits of this type hold the frequency within a few parts per million. Much greater stability can be achieved with the bridge circuit of L. A. Meacham. Here the oscillation conditions are fulfilled by zero phase shift in the feedback circuit, that is, at the exact series resonance frequency of the crystal. Long-term frequency stability of about 1 part in 10^8 and short-term stability of 1 part in 10^9 can be achieved with such oscillators; for an example *see* QUARTZ CLOCK. For detailed information on the Pierce and Meacham circuits *see* OSCILLATOR.

Selective band-pass filters with low losses can be built by using piezoelectric resonators as circuit elements. With a simple network consisting of resonating crystals only, a passband of twice the difference between resonant and antiresonant frequency can be obtained. For quartz resonators, this passband is about 0.8%. At relatively low operating frequencies, this band is too narrow, and combinations of crystal resonators with coils and condensors are generally used. A synthetic piezoelectric crystal which is often substituted for quartz in this application is ethylene diamine tartrate.

Piezoelectric crystals provide the most convenient means for generation and detection of vibrations in gases, liquids, and solids at frequencies above 10^4 Hz. Quartz, ammonium dihydrogen phosphate, rochelle salt, and barium titanate are frequently used in sonic and ultrasonic transducers. The mechanical impedances of liquids and solids are generally close enough to the mechanical impedance of the piezoelectric crystal so that efficient energy transfer is possible. The intensity of ultrasonic radiation that can be achieved is mainly limited by the mechanical strength of the piezoelectric crystal. The maximum ultrasonic intensity theoretically obtainable in water by means of quartz or ammonium dihydrogen phosphate is of the order 2000 watts/cm^2 and 200 watts/cm^2, respectively. For gases, the mechanical impedance match is so poor that the corresponding values are about 4000 times smaller. However, the mechanical impedance match can be greatly improved by using piezoelectric devices consisting of two differently oriented crystal cuts cemented together in such a way that a voltage applied to the electrodes causes the elements to deform in opposite directions, and a twisting or bending action results. Assemblies of this type (bimorphs) with barium titanate ceramics or rochelle salt are widely used in such devices as microphones, earphones, and phonograph pickup cartridges.

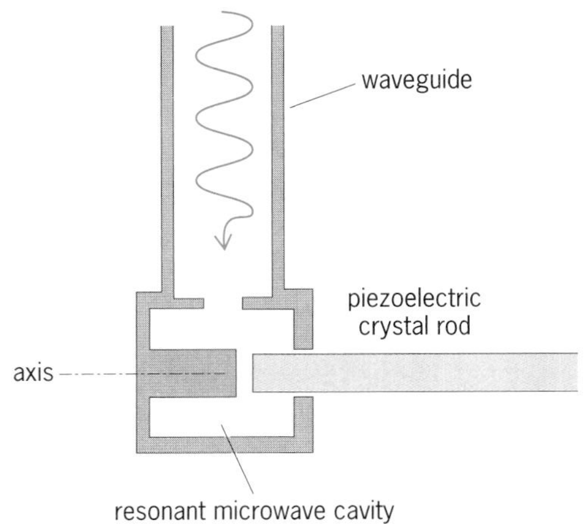

Fig. 9. Diagram showing experimental arrangement for generation of ultrasound at microwave frequencies by means of a piezoelectric crystal.

Ultrasonic waves at microwave frequencies up to 2.4×10^{10} Hz have been generated by means of the piezoelectric effect. The arrangement is shown in **Fig. 9**. The end surface of a piezoelectric crystal rod is exposed to a strong microwave electric field in a resonant reentrant cavity. The ultrasonic waves travel through the rod in a guided wave mode. The attenuation is low only at very low temperatures. H. Granicher

Bibliography. T. Ikeda, *Fundamentals of Piezoelectricity*, 1990; W. P. Mason (ed.), *Physical Acoustics*, vol. 1, pt. A, 1964; J. F. Nye, *Physical Properties of Crystals: Their Representation by Tensors and Matrices*, 1957, paper 1987; C. Rosen, B. Hiremath, and R. Newnham (eds.), *Piezoelectricity*, 1992; G. W. Taylor et al. (eds.), *Piezoelectricity*, 1985; J. Zelenka, *Piezoelectric Resonators and Their Applications*, 1986.

Pigeonite

The name given to the monoclinic pyroxenes of the general formula (Mg,Fe)SiO_3 with some augite in solid solution. Pigeonite bears the same relation to the orthorhombic pyroxenes as augite does to the diopside-hedenbergite series. Pigeonite is the orthorhombic pyroxene equivalent in the volcanic rocks. Most high-temperature metamorphic and igneous orthorhombic pyroxenes were probably originally pigeonite. The small optic angle (2V) distinguishes the mineral from augite and the inclined extinction distinguishes it from the orthorhombic pyroxenes. Pigeonite forms black, brown, or dark-green short stubby crystals with the 87° pyroxene (110) cleavages. The slower cooling rates of the igneous and metamorphic rocks usually permit the augitic materials in solution to exsolve and the remaining monoclinic pyroxene to invert to the orthorhombic form. The original augitic material is evident by the oriented exsolution lamellae in the host orthorhombic pyroxene. The faster cooling rates of the volcanic rocks quenches in the augitic material and thereby preserves the metastable pigeonite at surface temperatures. *See* AUGITE; DIOPSIDE; ENSTATITE; ORTHORHOMBIC PYROXENE; PYROXENE.

George W. De Vore

Pigment (material)

A finely divided material which contributes to optical and other properties of paint, finishes, and coatings. Pigments are insoluble in the coating material, whereas dyes dissolve in and color the coating. Pigments are mechanically mixed with the coating and are deposited when the coating dries. Their physical properties generally are not changed by incorporation in and deposition from the vehicle. Pigments may be classified according to composition (inorganic or organic) or by source (natural or synthetic). However, the most useful classification is by color (white, transparent, or colored) and by function.

White pigments. These pigments are essentially transparent to visible light. Because of the difference in refractive index between the pigment particles and the vehicles, white pigments refract the light from a multitude of surfaces and return a substantial portion in the direction of illumination without significant change in the spectral composition of the light. *See* COLORIMETRY.

The common white pigments are titanium dioxide, derived from titanium ores; white lead, from corrosion of metallic lead; zinc oxide, from burning of zinc metal; and lithopone, a mixture of zinc sulfide and barium sulfate. Pure zinc sulfide and antimony oxide are less commonly used.

Titanium dioxide may be crystallized in the rutile or anatase form, depending on the method of production. It may be further modified by surface treatment to control the rate of chalking and other properties. Rutile titanium dioxide has a higher refractive index than anatase and therefore higher hiding power, but it has a somewhat yellow color. Anatase titanium dioxide provides a purer white.

White lead pigments are the oldest of white pigments and were used extensively to provide excellent hiding power, flexibility, and durability to interior and exterior paints and enamels. Consumer protection rulings have all but removed white lead paints from the market, because leaded paint particles were ingested by children, with toxic effects.

Zinc oxide and lithopone pigments were extensively used in paint formulation, but have been superseded by titanium dioxide. Pure zinc oxide pigment is rarely used. Antimony oxide pigment is used chiefly in certain fire-retardant paints.

Transparent pigments. The refractive indexes of these pigments are very close to the index of the paint vehicle (about 1.54). They are used to provide bulk, control setting, and contribute to the hardness, durability, and abrasion resistance of the paint film. Because they are commonly used to add bulk to other pigments, they are called extenders. Most transparent pigments are natural minerals reduced to pigment particle size. Among the most commonly used transparent pigments are calcium carbonate (ground limestone, whiting, or chalk), magnesium silicate, bentonite clay, silica, or barites (barium sulfate). Transparent pigments often constitute a substantial portion of a protective coating.

Colored pigments. These pigments are available in a wide variety of colors and properties, depending upon the end use. Several hundred have been used, the following being the most common. *See* COLOR.

Red. Iron oxides, often classified by color, include Indian red, Spanish red, Persian Gulf red, and Venetian red, a mixture of iron oxide and calcium sulfate. Other red pigments include cadmium red (cadmium selenide) and organic reds, which are usually coal tar derivatives either precipitated in pigment form (toners) or deposited on a transparent pigment (lakes). Organic reds include toluidines and lithols.

Orange. Chrome orange (basic lead chromate), molybdate orange (lead chromate-molybdate), and various organic toners and lakes are the most common orange pigments.

Brown. Browns are nearly always iron oxides, although certain lakes and toners are used for special purposes.

Yellow. These pigments include natural iron oxides such as ocher or sienna, or synthetic iron oxides, which are stronger and brighter, such as chrome yellow (normal lead chromate) and cadmium yellow (cadmium sulfide), and organic toners and lakes such as Hansa yellow and benzidine yellow.

Green. The most important green pigments are chrome green, a mixture of chrome yellow and Prussian blue; chromium oxide, duller but more permanent; phthalocyanine green, an organic pigment containing copper; and various other organic toners or lakes, often precipitated with phosphotungstic or phosphomolybdic acid.

Blue. The blue pigments include Prussian blue (ferric ferrocyanide, sometimes called milori or Chinese blue, depending upon the shade); ultramarine, an inorganic pigment made by fusing soda, sulfur, and other materials under controlled conditions; phthalocyanine blue, an organic pigment containing copper; and numerous organic toners and lakes.

Purple and violet. These are nearly all organic toners or lakes. Manganese phosphate is a very weak, inorganic purple pigment.

Black. The vast majority of black pigments consist of finely divided carbon-carbon black, lampblack, and bone black—usually obtained by allowing a smoky flame to impinge on a cold surface. Black iron oxide and certain organic pigments are used where special properties are required.

Special pigments. Anticorrosive pigments are used to prevent the formation or spread of rust on iron when the metal is exposed by a break in the coating. The most common are red lead, an oxide of lead, and zinc yellow or zinc chromate, a basic chromate of zinc. Other colored chromates are sometimes used. The color of red leads fades rapidly, and the anticorrosive chromates are usually very weak in tinting strength. Metallic lead is sometimes used for anticorrosive paint.

Metallic pigments are small, usually flat particles of metal, prepared for dispersal in coatings. Aluminum is most commonly used because it leafs and forms a smooth, metallic film. The flakes are sometimes colored. Bronze, copper, lead, nickel, stainless steel, and silver appear occasionally. Zinc dust, or powdered zinc, is used more often because of its excellent adhesion to galvanized iron than because of its appearance.

Luminous pigments radiate visible light when exposed to ultraviolet light. Phosphorescent pigments continue to glow for a period after the exciting light has been removed; these are usually sulfides of zinc and other materials, with small amounts of additives which control the phosphorescent properties. Fluorescent pigments lose luminosity as soon as the exciting light is removed; these pigments may be sulfides, although many organic pigments have this property. *See* LUMINOUS PAINT.

Other specialized pigments include pigments which change color at some predetermined temperature, used to indicate hot areas on motors; pigments which give a pearly appearance; and pigments which conduct electricity for printed circuits.

Coarse materials such as pumice are often added when a nonslippery coating is required. Glass beads give a very high degree of refractivity in the direction of illumination and are often used in center-line paints or for signs where night visibility is required. Intumescent pigments puff up under heat, giving a fire-resistant coating. *See* DYE; PAINT.

C. R. Martinson; C. W. Sisler

Bibliography. R. Lambourne, *Paints and Surface Coatings: Theory and Practice*, 1987; S. Paul, *Surface Coatings*, 1985; H. F. Payne, *Organic Coating Technology*, vol. 2, 1961; D. Stooye (ed.), *Paints, Coatings, and Solvents*, 2d ed., 1998; Z. W. Wicks, N. Frank, and S. P. Pappas, *Organic Coatings Science and Technology: Applications, Properties, and Performance*, 2d ed., 1998; A. D. Wilson, J. W. Nicholson, and W. H. Prosser (eds.), *Surface Coatings*, vols. 1 and 2, 1987, 1988.

Pigmentation

A property of biological materials that imparts coloration. Hence, pigmentation determines the quantity and quality of reflected visible light. The characteristics of light returning from living matter are a function of its chemical and physical properties and, therefore, not only are due to pigments proper but can be of structural origin as well.

The term visible light as used here refers to the spectral sensitivity of the human eye. Such a restricted application of the word pigmentation is useful in the description of colors seen by humans but is not appropriate in a wider biological context. For example, flowers that appear colorless may be "pigmented" to bees because the eyes of these insects are sensitive to the near-ultraviolet and can detect the attenuation of incident solar radiation by ultraviolet-absorbing flower pigments.

Structural colors. These colors may be due to reflection, scattering, or interference. Reflection occurs, for example, at the surface of tiny air pockets in white (or whitish) material such as certain hairs, feathers, butterfly wings, or flower petals. Small structural entities like cellular inclusions (biocolloids), especially purine crystals in cells called iridocytes, scatter preferentially the blue portion of the incoming light. The blue color of weakly pigmented eyes is based on such an effect, while in blue bird feathers the scattering is produced by air spaces of colloidal dimensions. Iridescent colors of many bird feathers or insect wings, finally, are the result of interference at successive layers of biological structures, that is, feather barbules or wing scales, respectively.

Role of pigments. Pigments are essential constituents of the living world. Their contribution to the evolution and maintenance of life, and its manifold expressions, is most evident in the role of chlorophylls and the associated carotenoids of certain bacteria and most plants. These pigments harvest solar light energy for utilization in the photosynthesis of organic material from inorganic precursors. *See* CAROTENOID; CHLOROPHYLL; PHOTOSYNTHESIS.

The outermost structures on the animal skin are pigmented for many reasons, for example, to reduce the animal's visibility against a colored background or to provide optical signals to the other sex or to other species. Conspicuously pigmented flowers attract pollinators, and colored fruits are easily found by animals, which eat them and then disperse the undigested seeds.

The role of pigments in communication depends on the ability of organisms to discriminate between different regions of the solar spectrum. In animals with eyes, this is accomplished by differently colored visual pigments contained in specialized receptor cells. Microorganisms, fungi, and plants also have special pigment systems that permit these organisms to move or grow toward, or away from, light (positive and negative phototaxis and phototropism, respectively). *See* PLANT MOVEMENTS.

Since most organisms are totally dependent on light—at least indirectly—elaborate pigment systems have evolved which tune metabolic and activity patterns to the daily pattern of light and dark, and to the changes in the relative lengths of day and night in the course of a year. The phytochrome of plants and the pigments of the eye or of extraretinal photoreceptor organs of many vertebrates and invertebrates are typical representatives of pigments that correlate biological activity with light-dark cycles (photoperiodism). The pineal gland of higher animals plays a major role in transducing the light signal into regulatory actions of hormones such as melatonin. Simple on-off switches of light-regulated metabolic events, finally, may be provided by colored coenzymes such as flavins and cytochromes. *See* COLOR VISION; CYTOCHROME; PHOTOPERIODISM; PHOTORECEPTION.

In the examples listed above, pigments mediate, in various ways, the beneficial actions of light. Absorbed solar light energy may, however, also have detrimental effects by causing undesirable or even destructive reactions. Pigmentations can, in fact, provide a light-absorbing shield that protects the tissue below from such potentially damaging radiation of the Sun. *See* SKIN.

Chemical classes of pigments. Many pigments, particularly those participating in metabolic reactions, are associated with proteins. Such associations, called holochromes, comprise a colored chromophore bound to an apoprotein. Familiar examples of holochromes are the green pigments of the photosynthetic apparatus, rhodopsin in the retina, hemoglobin of the blood, and the phytochrome of plants which regulates photoperiodic and other light-mediated responses. *See* HEMOGLOBIN; PHYTOCHROME.

All organic biological pigments are polyunsaturated molecules. The main classes are hydrocarbons (polyenes), nitrogen- or oxygen-containing heterocyclic compounds, and substituted or condensed aromatic molecules. Typical representatives of the polyenes are the carotenoids with a long chain of alternating single and double bonds, and the structurally related visual pigment retinal and similar vitamin A derivatives.

Tetrapyrroles are the most common colored nitrogen-containing heterocycles, especially in their cyclic form as porphyrins. Indeed, in this latter form, and complexed with a metal ion such as magnesium or iron, the tetrapyrroles may be the oldest functional pigments in the evolution of the biosphere. As contributors to pigmentation, the chlorophylls (Mg-porphyrins) and the hemes (Fe-porphyrins) are the most important. Metabolic changes, either within the organism or after ingestion of other porphyrin-containing organisms, usually lead to the elimination of the complexed metal ion and often to a conversion of the cyclic arrangement of the four pyrroles to a linear one. Such linear tetrapyrroles are called bilins or bile pigments since they are typical components of the bile, a secretion of the gallbladder. An occlusion of the bile duct leads to an increased level of bile pigments in the blood and causes a yellow pigmentation of the skin (jaundice). Bilins and cyclic degradation products of porphyrins are also found in feces and urine and contribute to the pigmentation of skin, body fluids, and certain organs of many lower and higher animals. Interestingly, bilins (the phycobilins) are also responsible for the unique color of blue-green and red algae, in which they are constitutive pigments of the photosynthetic apparatus. Phytochrome, mentioned earlier, is a bilin as well. *See* BILIRUBIN; JAUNDICE; PHYCOBILIN; PORPHYRIN.

The melanins represent another type of widely distributed nitrogenous pigment. Typical eumelanins are more or less polymeric derivatives of indole formed through oxidative modifications of tyrosine. The simplest representative is the monomeric dopachrome. The lighter-colored phaeomelanins have cysteine residues attached to their indolic constituents and are found in the integuments of some birds and mammals, such as in red and yellow hair. In plants, melanins contribute to the coloration of the sap, and these pigments also occur in blackish or brown fungi and bacteria. But melanins are of greatest significance as skin and hair pigments in invertebrates and vertebrates.

Nitrogen-containing heterocyclic pigments also include the yellow flavins and the pteridines. The latter are best known as the coloring matter of butterfly wings, but they occur in many other insects as well. Pteridines are found also in crustaceans; and in vertebrates such as fishes, amphibians, and birds, pteridines are responsible for the brighter coloration of many irises.

The third class of substances listed above are the derivatives of aromatic hydrocarbons; examples are the quinoid and flavonoid pigments. Brownish quinoid pigments occur in bacteria, plants, fungi, and animals. Since such compounds are, in general, not synthesized by animals, quinoids are most likely to be found in herbivores, notably insects, which ingest the pigment with their food. The red and purple coloring matter of the shells and spines of sea urchins and starfish are echinochromes, a special type of quinoid pigment.

The flavonoids are aromatic molecules comprising an oxygen-containing heterocycle bound to hydroxylated benzene rings. Flavonoid pigments include the flavones and anthocyanins; they have a widespread distribution in the plant kingdom, giving color to flowers, fruits, leaves, and other plant parts. They have also been isolated from animals such as insects. *See* FLAVONOID.

Pigment disposition. The disposition of the pigments is determined by two criteria: their function and their solubility. Photosynthetic and visual pigments are membrane-bound pigment-proteins, while hemoglobin is dissolved in the cytoplasm of blood cells. Flavonoid pigments are typically bound to carbohydrates and are water-soluble; they are usually dissolved in the agueous matrix of cell vacuoles.

Carotenoids are lipid-soluble pigments and, therefore, are often associated with the lipid bilayer of biological membranes, such as chloroplast membranes, the membranes of chloroplast-derived chromoplasts, and cell membranes (for example, bacteria, fungi, and various animal tissues, including nerve). Alternatively, they are dissolved in fat deposits or dispersed as lipid droplets or lipoproteins (as in egg yolk). Conjugated with proteins, the usually yellow or reddish carotenoids can be modified to attain a blue or purplish coloration, as in certain crustaceans. Carotenes of feathers are probably bound to feather-keratin; in the integuments of insects and crustaceans they may be associated with the polysaccharide chitin; and calcium-carbonate-linked carotenes occur in the skeleton of corals.

Animals are unable to synthesize polyenes. Their carotenoid deposits, therefore, are dependent on the amount of dietary polyenes. The best-known examples are the diet-dependent colors of bird feathers and egg yolks and of the skin of infants not exposed to extended periods of sunlight.

In vertebrates, the synthesis of melanins is largely regulated by the neurohormone melanocyte-stimulating-hormone (MSH) and occurs in organelles called melanosomes of special pigment cells referred to as melanocytes or melanophores. Melanocytes are often dendritic, penetrating between the intercellular spaces of tissues, especially in the skin. Melanosomes may be transferred from the melanocytes to other cells, such as to the keratinocytes of the epidermis.

Pteridine-containing cells are called xanthophores when they are yellow and erythrophores when they are red, and contain their pigments in organelles called pterinosomes. Iridophores are structural pigment cells that contain reflecting crystals in so-called reflecting platelets or iridosomes. The common developmental origin of all pigment cells explains their ability to interconvert, and the existence of mosaic cells with different kinds of pigmentary organelles.

Occurrence. The extent and type of pigmentation of organisms are genetically determined, but pigmentation can often be modified within wide limits by environmental factors. Photosynthetic organisms, whose existence depends on the amount of useful light energy absorbed, tend to become more strongly pigmented when their habitat receives little solar radiation. Selective attenuations of portions of the visible spectrum by other organisms causes red and blue-green algae to increase production of those pigments that are capable of absorbing the still-available light colors (chromatic adaptation). Carotenoids, which serve to protect many tissues from light-induced damage, are often (in the bread mold *Neurospora*, for example) synthesized in response to light. Temporary skin darkening observed during certain diseases and, for example, during the menstrual cycle presumably are the result of melanosome dispersing actions of hormones other than melanocyte-stimulating hormone. Ultraviolet-induced tanning is caused by complex sets of events leading to increased melanin production, melanosome formation, and melanocyte proliferation, and may involve a regulation by cellular messengers like prostaglandins. An uncontrolled proliferation of melanocytes with often abnormal melanosomes underlies the formation of melanomas. *See* RADIATION BIOLOGY.

Control. Neural and hormonal control of pigmentation is common in many animals. In humans, the dilation of subcutaneous blood vessels caused by the autonomic nervous system during a moment of embarrassment leads to blushing. Most spectacular, however, are the changes of skin colorations in certain species of reptiles, amphibians, and invertebrates. Sensations received by ocular or extraretinal light receptors can be transmitted to the melanocytes via hormones. The aforementioned melanocyte-stimulating hormone is involved in the regulation of the extent of dispersal of the melanosomes (skin darkening), occasionally in an antagonistic interplay with another hormone, the melanin-concentrating hormone (MCH). Different neurohormones acting in a corresponding fashion are found in invertebrates. The well-analyzed distribution of pigments in the skin of the Florida chameleon *Anolis carolinensis* gives a clue to its various assumed skin pigmentations. Beneath the epidermis occur xanthophores, which reflect yellow light. Below them are located iridophores with reflecting platelets that give rise to scattered blue light. Even farther down are the melanocytes, with long cellular extensions into the upper layers below the epidermis. An upward migration of dispersed melanosomes to the epidermis makes the skin appear brown. When the melanosomes are concentrated in the cell body of the melanocyte, they absorb most of the light that was transmitted by the xanthophore cells and not lost by scattering at iridocytes above them. The radiation returning from the skin of the anole will then contain mainly scattered blue light plus yellow light from the xanthophore. Such light may be all shades from blue to green, depending on the relative abundance of iridocytes and xanthophore cells. *See* CHROMATOPHORE; PROTECTIVE COLORATION. Peter H. Homann

Bibliography. J. T. Bagnara (ed.), *Advances in Pigment Cell Research*, 1988; D. L. Fox (ed.), *Animal Biochromes and Structural Colors*, 1976; T. W. Goodwin (ed.), *Plant Pigments*, 1988.

Pike

Any of about five species of fishes which compose the family Esocidae in the order Clupeiformes, known by a variety of names such as pickerel and muskellunge. These fishes are voracious predators with an elongated beaklike snout and sharp teeth. The head is partly scaled and the body is covered with cycloid scales that have deeply scalloped edges. The body is cylindrical and compressed; thus, these fishes are well adapted for rapid movements as they dart after prey. They prey upon each other, as well as other fishes, amphibians, small aquatic birds and mammals, and rats. All species are edible but are considered second-rate game fishes. *See* SCALE (ZOOLOGY).

Common pike. This species (*Esox lucius*) has a wide distribution and is known as the northern pike in the United States. It is also found in northern and central Europe, as well as in northern Asia. On the

Fig. 1. Two species of fishes that are commonly called pikes. (*a*) Pike (*Esox lucius*) and (*b*) muskellunge (*E. masquinongy*).

Fig. 2. Grass pickerel (*Esox americanus*).

average the fish weigh about 3 lb (1.4 kg; **Fig. 1***a*); however, the record with rod and reel is 46 lb (20.7 kg) and the length over 4 ft (1.2 m). The teeth are directed backward and are constantly breaking and regenerating. When this species spawns near the muskellunge, cross-fertilization of the muskellunge eggs occurs, and the result is the striped tiger muskellunge. The female pike on spawning may release from 10,000 to 20,000 eggs which adhere to aquatic plants, are fertilized by one or more males, and hatch in about 2 weeks. The male matures at about the age of 2, and the female at 4. Large pike may be 20–30 years old and are relatively solitary in their habits.

Muskellunge. This fish (*E. masquinongy*) lives in cooler and deeper waters than the pike (Fig. 1*b*). It is common to the Great Lakes region and upper Mississippi. Its average weight is about 15 lb (7 kg), although the record is 70 lb (32 kg) and length over 5 ft (1.5 m). The muskie is prolific and lays as many as 100,000 eggs, with the fry hatching in about 2 weeks and quickly becoming predacious. Like many fishes at hatching, these fishes have a large yolk sac which provides food for growth during the first weeks. This is absorbed by the muskellunge in about 2 weeks.

Pickerel. The chain pickerel (*E. niger*) is found in the eastern United States, and averages 2.5 ft (0.75 m) in length and 3 lb (1.4 kg) in weight. Like its relatives, pickerel is a ferocious species which eats mice, birds, frogs, and crustaceans. This fish has a preference for clean, quiet weed areas of lakes. The smallest of the pikes is the grass pickerel (*E. americanus*) which reaches a length of 1 ft (30 cm); nevertheless it is as predacious as its relatives (**Fig. 2**). *See* CLUPEIFORMES. Charles B. Curtin

Pile foundation

A special type of foundation that enables a structure to be supported by a layer of soil found at any depth below the ground surface. A pile foundation comprises two basic structural elements, the pile and the pile cap. A pile cap is a structural base, similar to a spread footing, that supports a structural column, wall, or slab, except that it bears on a single pile or group of piles. A pile can be descibed as a structural stilt hammered into the ground. Each pile carries a portion of the pile cap load and transfers it to the soil in the vicinity of the pile tip, located at the bottom of the pile (**Fig. 1**).

The pile and pile cap configuration has provided the basic design solution to the difficult problem of obtaining deep foundation support below areas where poor soil conditions prevail. Poor soil conditions may be difficult to excavate through, and are incapable of supporting structural loads. They are typically characterized by the presence of a soft, compressible layer of clay, high ground-water levels, loosely filled soils, uncontrolled landfills, boulders, abandoned underground structures, and natural bodies of water. By supporting a structure on piles in lieu of spread footings, any adverse soil condition may be virtually bypassed, and adequate foundation support can be obtained at any depth, without the need to perform deep excavation, dewater, and install temporary sheeting and bracing.

While pile foundations offer a simple solution to obtaining deep foundation support through poor soil conditions, they also present major economic concerns. The type of pile, total number and length of piles required, means of pile installation, and size of pile caps are critical factors that may significantly increase foundation costs.

Installation of piles. Piles are commonly driven into the ground by means of an impact (pile) hammer. The pile is hung between a pair of parallel guides or leads suspended vertically from a standard lifting crane. The impact hammer is located above the top of the pile and guided axially by rails incorporated in the leads. To maintain vertical plumbness or to permit driving on a batter, an extendable horizontal member, known as a spotter, is connected from the bottom of the leads to the base of the crane boom.

Most impact hammers are of the steam or diesel type. Steam hammers lift the ram by means of steam pressure and allow the ram to fall by gravity. This is known as a single-acting hammer. If steam pressure is added to the gravitational downward energy, it is called a double-acting hammer. Compressed-air hammers are also used as a substitute for steam.

Diesel hammers lift the ram by explosion of fuel and compressed gas in a chamber between the

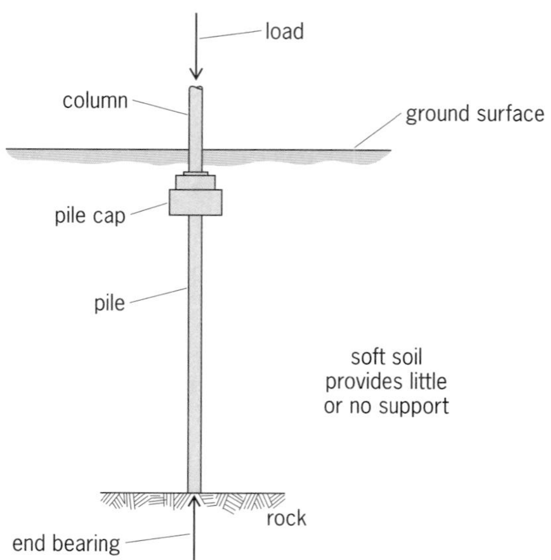

Fig. 1. End-bearing pile.

bottom of the ram and the anvil block. An open-ended hammer is like a single-acting hammer in that it allows the ram to fall by gravity. A closed-end hammer is like a double-acting hammer in that it has a cylinder at the top that allows air to be compressed as the ram rises. The compressed air acts as a spring that limits the upstroke, but returns its stored energy to the ram on the downstroke.

Vibratory hammers are often used when piles are required to be installed to a specific depth, regardless of its load capacity. A vibrating device in the hammer generates a pulsating force in the pile and the surrounding soil, causing the pile to exhibit a controlled up-and-down motion. The weight of the vibratory hammer plus the weight of the pile provide enough downward force to allow the pile tip to penetrate the underlying soil. Vibratory hammers have also proven to be successful pile extractors, provided the upward-pulling capacity of the crane is greater than the weight of the pile hammer.

Determining pile capacity. Piles installed with an impact hammer derive their load capacity from the number of blows per inch of pile penetration. The higher the blow count, the greater the resistance to pile penetration and, therefore, the greater the pile capacity. This oversimplification of the pile-driving phenomenon has resulted in the derivation of various pile-driving formulas that express the relationship between the energy delivered by the hammer to the work done by the pile tip to penetrate a specific distance against soil resistance. Despite the unreliability of pile-driving formulas to accurately determine the load capacity of a driven pile, they have been used for many decades because of their great convenience, and in most cases they are adequate for evaluating low-capacity piles.

A more realistic analysis of the dynamics of pile driving models the driven pile as an elastic bar. A wave equation is used to evaluate the pile's ability to transmit stress waves, imposed by the impact hammer, to the tip of the pile. The capacity of a pile is the net force that develops at the tip of the pile during wave transmission down the length of the pile. The net force at the pile tip must be equivalent to the resistance of the surrounding soil to a downward displacement of the pile.

To verify the actual load capacity of a driven pile, a pile load test is performed. A load is imposed on a driven pile with the use of a jacking frame, a hydraulic jack, and a gage to measure the vertical displacement of the pile under load. The ultimate load capacity of the test pile is equivalent to the maximum load imposed on the pile before failure. Failure of the pile occurs when the anticipated downward displacement of the pile, resulting from elastic shortening of the pile under load, is exceeded. The allowable load assigned to the pile is usually not greater than one-half the load at failure. *See* LOADS, TRANSVERSE.

Development of pile capacity. A driven pile develops its load capacity from the ability of the surrounding soil to develop skin friction along its length and from the ability of the underlying soil to develop point resistance below the pile tip. For these reasons, piles are classified as either friction piles or point-bearing piles, but may actually be a combination of both (**Fig. 2**). The load capacity of a pure friction pile is measured only by the amount of adhesion or friction in the surrounding soil per unit area of pile, multiplied by the actual contact area. The capacity of friction piles is not influenced significantly by the relative difficulty of penetration during driving or by slight differences in the embedded lengths. A pure point-bearing pile, by contrast, depends solely on the adequate contact with or sufficient penetration into a bearing layer to develop its required capacity. Thus, a few more inches of penetration into a firm layer of soil may greatly increase a bearing pile's load capacity.

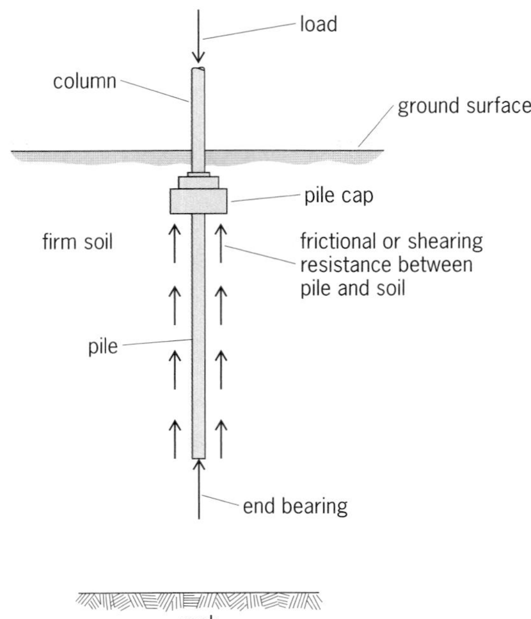

Fig. 2. Friction pile with some support from the end bearing.

In reality, most piles are driven through many layers of soils of different types. This would enable a driven pile to transfer load to the surrounding soil through any combination of skin friction from the pressure of sand grains against the pile, an adhesive bond from the clay particles in close contact with a comparatively rough pile surface, and point bearing from the pile tip embedded into a hard bearing stratum.

Precaution must be taken when designing large pile clusters. A stability analysis must be performed to ensure that the total load capacity of the pile group, which equals the sum of the capacity of each pile, does not exceed the bearing capacity of the underlying soil and the frictional resistance of the surrounding soil.

In addition, soft clays that are undergoing consolidation impose a downward force on a driven pile. This downward force tends to drag the pile down as a

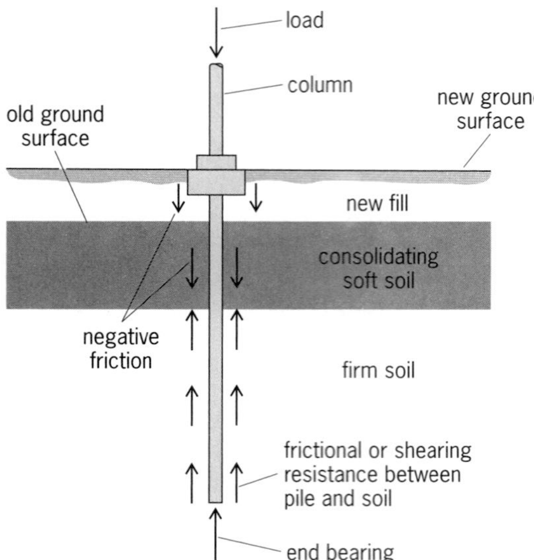

Fig. 3. Pile subjected to negative friction or downdrag. The pile must penetrate a sufficient distance in firm soil to support the external load plus the negative frictional forces.

result of what is known as negative skin friction. This phenonemon imposes additional load on the pile. In effect, negative skin friction reduces the amount of load capacity in the pile available for support of a structure (**Fig. 3**). *See* SOIL MECHANICS.

Types of piles. Piles are available in a variety of sizes, shapes, and materials that enable a particular type of pile foundation to be viable both economically and structurally. For the purpose of identification and discussion, piles are generally classified by their principal materials such as timber, concrete, and steel.

Timber piles. Timber piles, using the trunks of trees with their limbs removed, date as far back as Roman times, and probably still represent the most commonly used type of pile throughout most of the world. Limited by the height of available trees, timber piles 40–60 ft (12–18 m) long are common, with the southern pine being the most common tree type of timber pile used in the United States.

Timber piles are limited to a capacity to no more than 30 tons (27 metric tons) each, because of their inability to be driven into stiff soils of high penetration resistance, without damage. Although damage to the pile tip may be somewhat reduced by using steel point called shoes, the danger of breaking the pile may still exist at the pile butt, the head of the pile that receives the impact from the pile-driving hammer. By limiting the energy rating of the pile hammer and the required number of blows of the hammer, the stress induced in the butt of the pile from the hammer can be significantly reduced.

The unique characteristic of wood is its ability to last indefinitely under water as long as it remains saturated and not allowed to dry. Most timber piles are pressure-treated with creosote to protect the wood above the water table from structural deterioration resulting from dry rot or attack by termites. Treating timber piles with creosote has been known to increase the life of the timber pile in excess of 40 years.

In a marine environment, where timber piles are subject to brackish or salt water and attack by various marine organisms, chemical treatment has not always proven to be an effective means of protection. Alternatively, providing concrete encasement or installing rubber jackets around the timber pile in the area exposed to the marine environment have proven to be a viable means of protection. *See* BORING BIVALVES; SHIPWORM.

Concrete piles. Concrete piles were introduced shortly before 1900. Although a wide variety of concrete piles are available, they are generally divided into two principal categories: cast-in-place and precast piles.

Cast-in-place. Cast-in-place piles entail pouring structural concrete into a hollow shaft of a predetermined size and depth created in the ground. The hollow shaft may be cased or uncased. The casing is usually a thin metal shell or pipe, which acts as a form that is left in the ground. Fundamentally, the casing must be able to resist collapse under lateral pressure from the surrounding soil before it is filled with concrete. Very thin walled shells may not be strong enough to resist damage when driven into the ground and may require the use of a mandrel. A mandrel is a heavy pipe inserted inside the thin shell to provide internal support of the shell and absorb the driving energy from the pile hammer during installation. The mandrel is removed from the thin shell casing prior to casting concrete inside. Very often, pipe casings with a heavier wall thickness are used to eliminate the need for a mandrel, and will in turn contribute to the load capacity of the pile.

Since the casing represents a material cost incorporated into the overall cost of the pile, uncased piles withdraw the casing for reuse as the hole is filled with concrete. Uncased piles are also installed without the use of a casing driven into the ground. This more elaborate method requires drilling a bored hole and filling it with a bentonite slurry to keep the hole from collapsing. Concrete is pumped through a hollow pipe, known as a tremie pipe, that is lowered to the bottom of the hole. As the level of the concrete rises in the bored hole, the tremie pipe is gradually lifted out of the hole to allow the volume of concrete to displace the volume of bentonite slurry. Eventually, all bentonite slurry is replaced by concrete, and the tremie pipe is removed, resulting in a solid column of concrete embedded in the ground. *See* BENTONITE.

The various types of cast-in-place piles differ in the shape, texture, and wall thickness of the casing; the method of casting the concrete in the hole; and the overall pile cost.

Precast. Precast concrete piles are reinforced concrete piles, cast prior to installation in a casting yard. Similar to timber piles, they must be reinforced to resist stresses in the pile due to driving. In addition, precast piles must be reinforced to prevent damage to the pile during handling prior to driving, especially when piles are stored, transported horizontally, and lifted into position. Prestressing of the steel reinforcement in the precast pile tends to reduce tension cracking during handling and driving, and provides an efficient way of withstanding bending stresses.

The primary difficulty in using precast concrete piles is in extending the length of the pile or cutting the excess length off, when there is a variation in the anticipated pile length. However, the advantages of using precast concrete piles are their high load-carrying capacity and their outstanding resistance to deterioration above the water table under ordinary conditions. When exposed to salt water, the precast concrete piles are subject to attack of the steel reinforcement through cracks in the concrete, causing rust to develop and the concrete to spall. Use of high-density concrete and prestressing to minimize tensions cracks serve as the best measures of protection against reinforcing steel corrosion. *See* CONCRETE; PRECAST CONCRETE; REINFORCED CONCRETE.

Steel piles. Steel is the only material used for piling that has a crushing strength comparable to that of hard rock. Available in a wide variety of sizes, steel piles are ideal when conditions call for hard driving, unusually great lengths, or high load capacity. Primarily, steel piles are either steel pipes, usually filled with concrete after being driven, or steel H-sections.

Steel-pipe piles are similar to cased concrete piles, except that the steel-pipe casing is the primary load-carrying component, an therefore it is designed with a heavy wall thickness. When pipe piles are driven open-ended, they must be cleaned out before they are filled with concrete. Often, pipe piles are closed at the bottom with a closure plate or conical point to eliminate the task of cleaning out the pipe and to allow for inspection of the inside of the pile to check for possible damage after driving.

Steel H-sections offer economic advantages not available in other types of piles. Manufactured as a rolled-steel section, the H-section or H-pile can be readily spliced in the field with full-strength welds, and they are available in extremely long lengths. H-piles do not require expensive field fabrication; they can be driven through unconsolidated soils containing timbers, boulders, and other debris with minimum difficulty; and they are designed to carry highly concentrated loads. Protection against marine organisms, termites, dry rot, spalling, and chipping are of no concern when using H-piles. *See* STRUCTURAL STEEL.

Corrosion of steel due to exposure to salt-water spray, wave action, and sand abrasion in waterfront structures is a major concern. The most severe corrosion will occur within the splash zone that extends 2 ft (0.6 m) above and below the water level, where the steel is wetted with a thin film of water saturated with oxygen. By comparison, the area below the water level is oxygen-starved, and the rate of corrosion decreases rapidly with water depth. Similarly, steel piles driven into dry soils or soils that maintain a water table at some depth below the ground surface will not experience significant corrosion, since the amount of oxygen contained below the ground level is negligible.

Cathodic protection is an effective method of preventing corrosion of unprotected steel exposed to seawater. Alternatives involve the use of protective coatings, including flame-sprayed aluminum and a blend of epoxy and polyamide resins. However, any protective coating will reduce corrosion as long as the coating lasts. Maintaining the coating in the tidal region can be expensive and, in many cases, virtually impossible. Another effective means of protecting the steel pile from corrosion, similar to preventing timber piles from deterioration, is by installing a concrete jacket around the pile extending 2 ft (0.6 m) below low water level and 2 ft (0.6 m) above high tide. *See* CORROSION; METAL COATINGS.

Uses. A pile foundation can be used to support any type of structure through any type of adverse soil conditions. Obviously, pile foundations are used to support marine structures and offshore platforms, since they are located over bodies of water. On land, pile foundations are used primarily in locations where poor soil condition exist. Under extreme conditions, all structures including utility lines are supported on piles and pile caps.

Pile foundations have also been used as a means of underpinning. Bracket piles are used to extend the depth of foundation support for shallow footings adjacent to deep excavations. Piles are driven as close as possible to the edge of an existing footing. Pile caps, designed as cantilevered brackets, are installed at the top of the piles that extend beneath the existing footing. The space between the top of the bracket and the underside of the existing footing is wedged and dry-packed to create a positive load transfer from the existing footing to the bracket pile.

Special care must be taken when driving piles near existing structures that are not supported on pile foundations. Because of the displacement of soil that occurs at the pile tip during pile driving, adjacent spread footings are susceptible to settlement. The magnitude of the soil displacement is a function of the shape and cross-sectional area of the pile tip. Closed-end pipe piles, timber piles, and precast concrete piles yield a relatively greater soil displacement than steel H-piles. If the soil displacement directly below the footing can be kept to a minimum, the magnitude of the settlement will be negligible. Experience has shown that the optimum pile to be driven adjacent to a soil bearing structure is the steel H-pile. Open-end pipe piles, despite their relatively small cross-sectional area, are not effective in minimizing soil displacement because they are circular in shape. As the open-end pipe pile is driven into the ground, the soil at the pile tip displaces radially away from the outside face of the pipe and converges radially inward inside the pipe. After driving the pile a certain distance, the converging soil inside the pipe becomes extremely dense and forms an earth plug at the tip of the pile. Unless the earth plug is cleaned out intermittently during the course of driving the pile, the pile will essentially behave as a closed-end pipe pile.

Piles may also be installed by means of jacking. Jacked piles are used as a means of underpinning when job constraints prohibit the use of a pile-driving hammer. Short sections of open-end pipe are set up vertically below an existing footing and jacked down into the ground with a hydraulic jack installed between the top of the pipe and the bottom of the footing. Using the existing structure as a reaction, the hydraulic jack pushes the pipe downward. The next section of pipe is set on top of the previous pipe section and spliced by welding. The advantages of using an open-end pipe for jacking is the ability to remove an obstruction or excavate the soil inside the pipe to reduce the resistance by the soil to jacking. When the jacking operation is completed, the pile is wedged in place to ensure load transfer from the existing footing to the jacked pile.

Steel-sheet piling. The practice of driving structural members into the ground had led to the development of interlocking steel-sheet piling construction. Steel-sheet piles are rolled-steel shapes with interlocking joints along their edges. They are produced in three standard shapes: straight web, arch web, and Z type; and they are available in a graduated series of weights and strengths. The interlocks, which must be sufficiently loose to permit free sliding during installation, are designed to provide strength and watertightness in both longitudinal and transverse directions.

The Z-pile type has the highest bending resistance because of the favorable distribution of steel area away from the axis of bending. For this reason, Z-piles are most economical in the construction of permanent filled bulkheads, land walls, deeper braced cofferdams, and similar wall types that require high beam strength.

The arch web piling is used for similar but generally lighter applications than Z-piling, such as in the construction of low-head cofferdams and shallow bulkheads or walls. The advantages of using arch web piling instead of Z-piling is its relative ease in handling and installation and its adaptability to a wide variety of field conditions.

The straight web piling offers the least amount of bending strength, but the highest value of tension between the interlocks, which is the prime consideration in the construction of large, filled cellular-type structures. Most cellular structures are used as temporary cofferdams to aid in the construction of locks, dams, bridge piers, and land reclamation projects. Cellular cofferdams are also used in various types of permanent marine construction such as deep-water bulkheads, mooring or turning cells, artificial islands, and breakwaters. *See* COASTAL ENGINEERING; COFFERDAM; FOUNDATIONS. Anthony J. Mazzo

Bibliography. R. W. Brown, *Practical Foundation Engineering Handbook*, 2d ed., 2000; R. B. Peck, W. E. Hanson and T. H. Thornburn, *Foundation Engineering*, 2d ed., 1998.

Pilot production

The production of a product, process, or piece of equipment on a simulated factory basis. In mass-production industries where complicated products, processes, or equipment are being developed, a pilot plant often leads to the presentation of a better product to the customer, lower development and manufacturing costs, more efficient factory operations, and earlier introduction of the product. Following the engineering development of a product, process, or complicated piece of equipment and its one-of-a-kind fabrication in the model shop, it becomes desirable and necessary to "prove out" the development on a simulated factory basis.

Facilities. To accomplish this aim, the pilot plant, an intermediate step between development laboratory and production factory, is established. It is provided with personnel and facilities that duplicate as nearly as possible actual manufacturing conditions but are free of the day-to-day necessity of meeting delivery schedules.

This requires personnel who are highly skilled in their field but are not necessarily experts. They must understand the thinking and objectives of development engineers and must also appreciate the day-to-day problems that arise in factory operation with unskilled or semiskilled operators.

Adequate equipment and services should be provided to accommodate the full range of factory requirements. Sometimes, however, it may be impractical to include certain facilities in a pilot plant because of floor space limitations, or because of specialized services required. In such cases, arrangements should be made to perform these pilot-plant functions on regular production equipment in the factory but under pilot plant supervision and financial control.

Objectives. Some of the objectives of a pilot plant operation are as follows.

Quality control. Pilot plant operation demonstrates the ability of processes or equipment to function consistently at the desired quality level under factory operating conditions.

Material usage. To demonstrate economical usage of materials during the pilot operation, the materials used should be truly representative and contain the full range of variability permitted by material purchase specifications.

Process reliability. Pilot operation demonstrates whether the processes involved are practical and realistic and whether operating procedures can be followed in their entirety by the type of personnel used in the factory.

Equipment reliability. To demonstrate the capability of the equipment to produce at speeds and machine efficiencies, which will become standard in the factory, data are recorded during pilot production which show whether the equipment will perform at a satisfactory rate of defectives. Pilot plant experience establishes a list of spare parts which will be initially provided in the factor.

Personnel requirements. Pilot experience establishes the amount and labor grade of production and maintenance personnel required to operate the process or equipment. This is done with the cooperation of industrial engineering and industrial relations departments.

Safety. During pilot runs the safety department has an opportunity to review the process or equipment for compliance with safety requirements.

Factory acceptance. Perhaps the most important objective of a pilot plant is to afford representatives of the factory to which the process or equipment will be transferred an opportunity to witness the operation and to agree that it is ready for the factory. Thus, criticism of the development is minimized and its introduction into regular factory operation accomplished more quickly.

Manufacturing costs. Pilot plant operation provides the opportunity to verify engineering estimates of manufacturing cost and may indicate the necessity for pricing changes or for taking further steps to reduce the cost of the product in order to reach a satisfactory level of profit.

Development costs. The cost of designing and developing processes or equipment and of instructing personnel in their use is often only a fraction of the total cost involved in producing a final efficient process or machine. The debugging or working out of unforeseen problems can consume a great deal of time and add a large amount of expense before the process or machine is ready for factory operation. The pilot plant relieves the factory of the expense of a trial run both in direct cost and in lost production of established products. The close proximity of the pilot operation to engineering and other skilled personnel permits first-hand observation of operation and quicker decisions on required changes. It may also save the expense and time of such personnel traveling to factory locations and the return of equipment from the factory to its point of origin.

The financing of a pilot plant should be on a budgetary basis. Funds for its facilities, personnel, and operation should be provided as part of a manufacturing development budget. The pilot plant should not be expected to finance itself through manufactured products; otherwise management tends to keep the project in the pilot plant longer than necessary in order to create funds for its operation. The pilot plant then loses its value as a development function, and the desire and interest of its personnel to introduce improvements and to tackle new projects become subordinate to that of meeting production schedules. On the other hand, arrangements should be made to credit the operation with a fair value of the product manufactured so that the net amount remaining will represent development cost more accurately. *See* ACTIVITY-BASED COSTING; PRODUCT DESIGN; PRODUCTION ENGINEERING; QUALITY CONTROL.

James E. Woodall

Pilotage

One of four procedures used in navigating an aircraft. The other three are position fixing, homing, and dead reckoning. Pilotage is the procedure of using landmarks, such as cities, towns, rivers, railroads, and prominent highways, to guide an aircraft to a destination. The installation of lights at airports and prominent spots across the country enhanced the ability of the pilot to direct the aircraft. *See* DEAD RECKONING; HOMING.

The introduction of radar brought a new dimension to pilotage because the aviator's ability to conduct pilotage was always seriously hampered by lack of visibility caused by storms, fog, and so forth. Airborne radar operating at microwaves produces very sharp maps of the terrain over which the aircraft is flying. Radar pilotage has even been adapted to missile guidance. Radar map presentation in the missile is compared automatically with a previously constructed map, and the difference is used to furnish missile guidance.

Peter C. Sandretto

Piloting

The form of navigation in which position is determined relative to external reference points, usually fixed points on the Earth. It is the oldest form of navigation. With the development of electronic aids, piloting techniques have been extended far from shore. However, the term piloting is generally associated with nearness of land, where tidal and other currents may be strong, shoals and other underwater obstructions may be in near proximity, and maneuvering room is limited when other vessels are encountered.

Because of these limitations, piloting requires constant vigilance, frequent fixing of position, and a thorough knowledge of the characteristics and locations of landmarks, both natural and artificial, and local regulations. Thus it is not unusual for ships to employ the services of a local expert, called a pilot, to assist in the navigation of the vessel while it enters or leaves port.

Marks. A conspicuous object, structure, or light that serves as an indicator for establishing the position of a craft or otherwise assisting in its safe navigation is called a mark. To be useful, not only must a mark be identified, but its position must be known accurately.

Natural landmarks are prominent features of the landscape. Artificial marks might be prominent buildings or parts of buildings, towers, flagpoles, tanks, and so forth. Additionally, a number of artificial marks have been designed and erected specifically to assist the navigator. These marks are called aids to navigation, and are of several types.

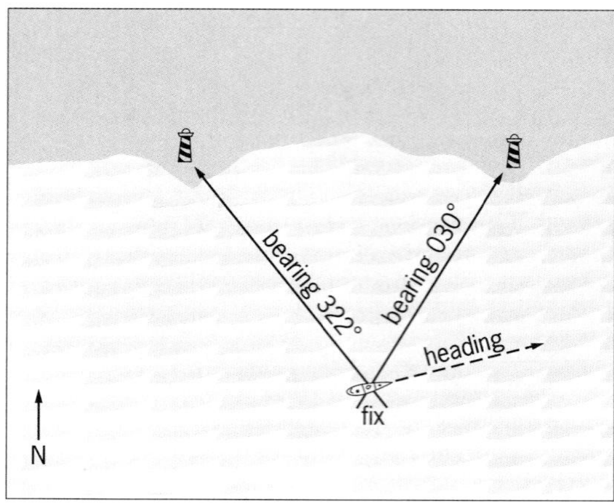

Fig. 1. A fix from two bearings.

Beacons, both lighted and unlighted, and lighthouses are at fixed positions on land or in the water. Buoys, both lighted and unlighted, and lightships float in the water, being moored at desired points. Unlighted aids are called daymarks. These are given distinctive shapes or coloring to assist in identification. A light installed as an aid to navigation is given a distinctive sequence and duration of light and dark periods and of color or colors, by which it can be identified. These characteristics are shown on the charts and given in the light lists available to navigators. *See* BUOY; LIGHTHOUSE.

In addition to visible aids to navigation, bottom topography can be of assistance in locating the position of a vessel. Sound signals transmitted through water or air may be used for navigation. Electronic beacons and positioning systems have been established at a number of places to assist in navigation. *See* ELECTRONIC NAVIGATION SYSTEMS.

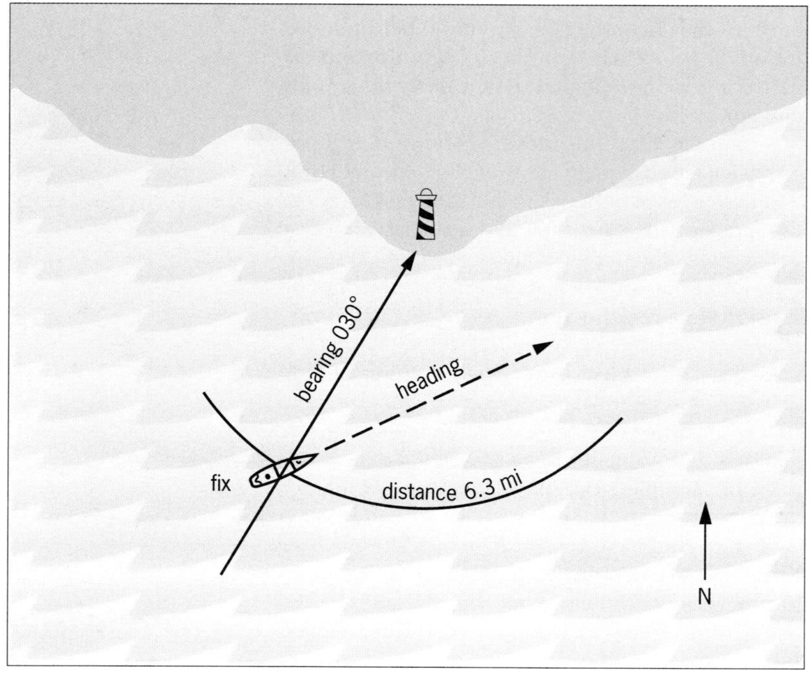

Fig. 2. A fix which is obtained from a bearing and the distance of a single object. 6.3 mi = 10 km.

Measurements. The measurements that are made for piloting purposes are direction, distance, differential distance between two points, and distance to bottom.

Direction. The direction of a landmark is called its bearing, commonly stated as the angular distance from some reference direction, usually true, magnetic, compass, or gyro north. Bearing values are generally given in integral degrees (although half and quarter degrees may be used), using three figures, from 000° at north clockwise through 360°. Relative bearings use the heading of the craft as the reference direction, and may be stated as above, right and left through 180°, or with reference to some part of the craft (as broad on the starboard bow, 10 degrees abaft the port beam, or two points on the port quarter).

Bearings are usually measured (1) by noting when two objects are in range (directly in line); (2) by means of a suitable attachment to a compass or compass repeater; (3) by pelorus, a compasslike instrument without directive properties; or (4) electronically, by radio direction finder, by radar, or by the indication of the receiver-indicator of an electronic system of navigation. *See* DIRECTION-FINDING EQUIPMENT.

Distance. This is now generally measured by radar in nautical miles or yards (meters in countries using the metric system). *See* RADAR.

Differential distance. The difference in distance from the ship to two points is usually measured electronically with the receiver-indicator of a hyperbolic navigation system. The reading is generally in some special unit such as a microsecond or one-hundredth of a lane width. *See* HYPERBOLIC NAVIGATION SYSTEM.

Depth. Depth of water is measured in meters, feet (1 ft = 0.3 m), or fathoms (1 fathom = 6 ft = 1.8 m). Depth measurement is usually made by an echo sounder, a device that emits a sonic or ultrasonic signal downward in the water and measures the time interval until return of an echo from the bottom. *See* ECHO SOUNDER.

Position determination. Traditionally, position by piloting has been determined by lines of position, each indicating a series of possible positions of the craft at time of measurement. A measured bearing provides a straight line of position (actually part of a great circle) passing through the object sighted. A measured distance provides a circular line of position with the object as the center and the distance as the radius. A measured differential distance provides a hyperbolic line of position. A straight or circular line of position is usually plotted directly on the nautical chart. Depth of water generally does not provide a line of position unless the craft crosses a distinctive feature such as the 100-fathom curve. A position, called a fix, is determined by crossing two or more lines of position taken simultaneously or nearly so (**Figs. 1** and **2**).

If there is an appreciable period of time (more than several seconds) between observations, one of the lines of position is adjusted (advanced in the direction of the travel, or retired in the reciprocal direction) a distance equal to the run of the vessel between observations. The resulting position is called a running fix, which is generally somewhat less reliable than a fix because of uncertainty in the speed and direction of travel of the vessel between observations. However, if the objects observed are nearly in the same or reciprocal direction, the passage of time might result in a more favorable crossing angle (90° being ideal), thus more than offsetting the loss of accuracy because of uncertainty in the motion of the vessel. When only one mark is available for observation, there may be no alternative to a running fix.

Position of a vessel can be obtained in other ways than by crossing lines of position plotted on a chart. Certain combinations of successive bearings of the same mark provide immediate indication of position relative to that mark. As an example, bow and beam bearings (the first when the mark is 45° from the direction of travel of the vessel, and the second when the relative bearing, right or left, is 90°) indicate that the distance of the vessel from the mark at the time of the second bearing is equal to the run between bearings, and the direction is perpendicular to the direction of travel. Simultaneous horizontal angles between three objects can be set on a three-arm protractor (called a station pointer by British navigators) and the position determined by fitting the arms of the instrument to the chart symbols representing the objects. The horizontal angles are usually measured by marine sextant. This provides an accurate position sometimes used in hydrographic surveying or in locating the point of letting go an anchor, but seldom for other purposes. A line of soundings plotted on a transparency at the correct distance intervals at the scale of the chart and then matched by trial and error to the soundings shown on the chart may provide a reliable indication of position where the bottom topography has a distinctive pattern. Under favorable conditions, a bottom profile obtained by a recording echo sounder can provide a better position.

A vessel can sometimes be kept in safe water without a fix. In an area where the bottom shoals gradually, a well-chosen danger sounding can give warning in sufficient time to prevent grounding. A danger bearing of some object well ahead can provide an indication of a dangerous situation if the measured bearing exceeds the danger bearing. An off-lying rock or shoal can sometimes be successfully avoided by keeping the measured angle between two objects less (or more) than a preselected horizontal danger angle (**Fig. 3**). Similarly, a vertical danger angle can be established between the top and bottom of a single object.

Electronic charts. Digitization of a nautical chart provides the information needed for display of an electronic facsimile of a published chart. Selective determination of features to be shown makes possible the tailoring of the facsimile to individual requirements. If a radar image is added to the display,

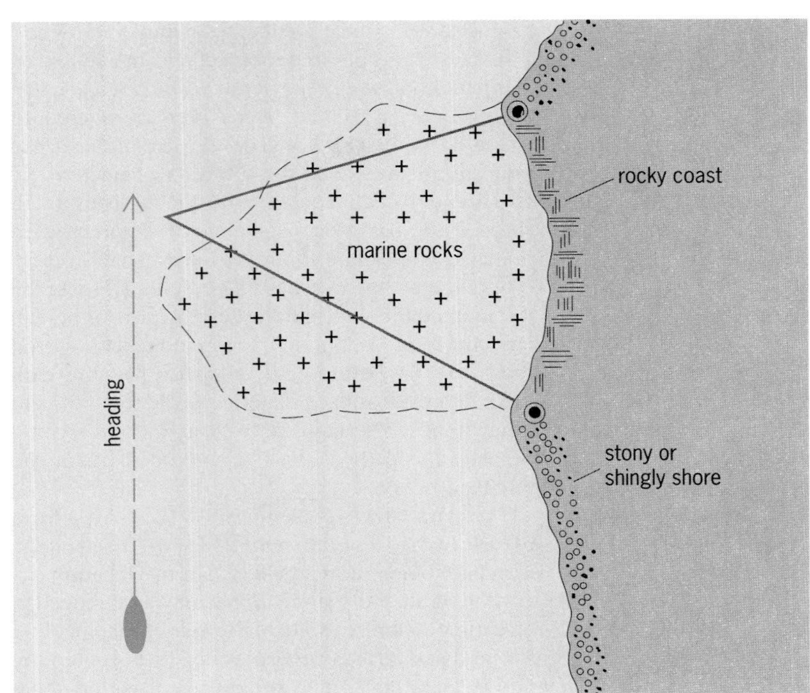

Fig. 3. A horizontal danger angle.

the position of the vessel relative to its surroundings is immediately apparent. A suitable electronic positioning system such as the Global Positioning System (GPS) or differential GPS (DGPS) can be used to display a symbol indicating the position of the vessel, thus providing a check on accuracy of the radar data. Other information can be added as desired. Color can be used to assist in identification of various types of information. The combined display, called an electronic chart display information system (ECDIS), is similar to that of a vessel traffic system (VTS), thus contributing to effective traffic coordination in a harbor. The electronic chart can be corrected by radio, so a vessel entering a harbor can utilize an up-to-date chart. *See* CELESTIAL NAVIGATION; DEAD RECKONING; MARINE NAVIGATION; NAVIGATION; POLAR NAVIGATION; SATELLITE NAVIGATION SYSTEMS.

Alton B Moody

Bibliography. N. Bowditch, *American Practical Navigator*, U.S. Defense Map. Agency Hydrogr. Topogr. Center Publ. 9, vol. 1, 1984, vol. 2, 1981; E. S. Maloney, *Dutton's Navigation and Piloting*, 14th ed., 1985; L. Melton, *Piloting with Electronics*, 1987; H. H. Shufeldt and G. D. Dunlap; *Piloting and Dead Reckoning*, 4th ed., 1999; U.S. Coast Guard Auxiliary Staff, *Coastal Piloting*, 1994.

Piltdown man

The scientifically most successful fraud in the history of anthropology. Between 1908 and 1914 in the English village of Piltdown, parts of a thick human braincase, half of an apelike chinless lower jaw containing two molar teeth, a relatively large lower canine tooth, animal fossils, stone implements, and a 16-in. (41-cm) pointed tool made from a fossilized elephant bone were recovered. Within arguable limits the braincase could be restored as of modern size and shape, with a well-developed forehead; the uniform thickness was the primitive character. The jaw was decidedly apelike, but the surviving teeth showed the flat wear characteristic of premodern humans. The canine tooth was larger than in moderns but smaller than in apes. The glenoid fossa of the jaw joint on the skull had the normal form for human chewing motions, but no corresponding condyle on the lower jaw was found. All parts had the dark color of local fossilized material, as did the tools and other bones. The fauna as well as the tools could be placed as early as the Pliocene or as late as the mid-Pleistocene.

The material was presented in 1912 as a new form of early human, *Eoanthropus dawsoni*. Based on the mandible being more apelike than the recently discovered Heidelberg jaw, *E. dawsoni* was adjudged to be earlier in time. In spite of the incongruity of brain size and jaw form, scientists generally accepted the find at face value. The alternative explanation—that a fossil human skull and ape jaw of similar age had come to rest by chance at this one site—seemed so unlikely that belief in Piltdown man prevailed by default. The subsequent occurrence of two more skull pieces like the first and another worn molar reportedly found near Piltdown overcame the argument of chance deposition.

At that time, radiometric and chemical methods of testing the age or legitimacy of such specimens did not exist, being developed only toward the middle of the twentieth century. In 1953, when the broader picture of fossil humans had made Piltdown incongruous, logic led to the hypothesis of a deliberate hoax. Chemical tests, such as nitrogen loss or fluorine uptake, then showed that none of the human bones could be more than a few centuries old. Also, on re-examination at the same time, signs of forging were evident. All parts, as well as the tools and animal bones, were artificially stained. Teeth in the jaw had been filed unnaturally flat, not worn by chewing; and the isolated canine had also been reduced in size by filing. The jaw itself could be identified as that of an orangutan.

Virtually everyone involved in the discovery and early discussion has been accused of the forgery. It is clear that the forger had anatomical and paleontological knowledge, removing any evidence that might identify the jaw as that of an orangutan. The thickened skull was apparently that of someone with Paget's disease. One scientific effect of the Piltdown find was to lend credibility to other finds of supposedly early large-brained humans. The most important result was to cast doubt on a model for early hominids having a small brain and large, more human teeth. Thus when *Australopithecus* was reported in 1925, it was initially pronounced to be an ape. *See* AUSTRALOPITHECINE; FOSSIL HUMANS.

W. W. Howells

Bibliography. C. Blinderman, *The Piltdown Inquest*, 1986; A. Keith, *The Antiquity of Man*, 2d ed., 1925; R. Millar, *The Piltdown Men*, 1972; J. S. Weiner, *The Piltdown Forgery*, 1955.

Pimento

A type of pepper, *Capsicum annuum*, grown for its thick, sweet-fleshed red fruit. A member of the plant order Polemoniales, pimento is of American origin, and gets its name from the Spanish word designating all sweet peppers. In the United States, however, the term pimento generally refers to the heart-shaped varieties (cultivars) grown in the South for canning and used for stuffing olives and flavoring foods. Perfection is a popular variety. Harvesting begins when the fruits are fully red, usually $2^1/_2$–3 months after planting. Georgia is the only important pimento-producing state. *See* PEPPER; SOLANALES.

H. John Carew

Piña

A fiber, also known as pineapple fiber, obtained from the large leaves of the pineapple plant grown in tropical countries. This natural fiber is white and especially soft and lustrous. In the Philippine Islands, it is woven into piña cloth, which is soft, durable, and resistant to moisture. Piña is also used in making coarse grass cloth and for mats, bags, and clothing. *See* NATURAL FIBER; PINEAPPLE.

M. David Potter

Pinales

An order of the class Pinopsida (=Coniferopsida), of the division Pinophyta (Gymnospermae), with about 50 genera and 600 species still living. All are woody plants, as shrubs or trees, and are often the principal trees of the forests worldwide. Pine, spruce, fir, hemlock, cedar, larch, juniper, cypress, yew, redwood, big tree, kauri, podocarpus, araucaria, and others are all part of this order. The big tree (*Sequoia gigantea*) of California is the largest plant, reaching a height of 330 ft (100 m) and a diameter of 33 ft (10 m), and being over 3500 years old. Leaves are usually needlelike or scalelike, with a few exceptions such as *Agathis* (kauri), *Podocarpus*, *Phyllocladus*, and *Araucaria* (**Fig. 1***a*). Most species are evergreen, bearing their leaves year-round. Modern conifers form some of the most extensive forests in recent times, mainly occurring in the temperate regions or mountainous regions of the subtropics. The families of extant conifers are Araucariaceae, Pinaceae, Taxodiaceae, Cupressaceae, Podocarpaceae, Taxaceae, and Cephalotaxaceae. The phylogenetic relations among these families are practically unknown.

Reproduction. The cones are unisexual (Fig. 1*b* and *c*). In some species, both male and female cones are

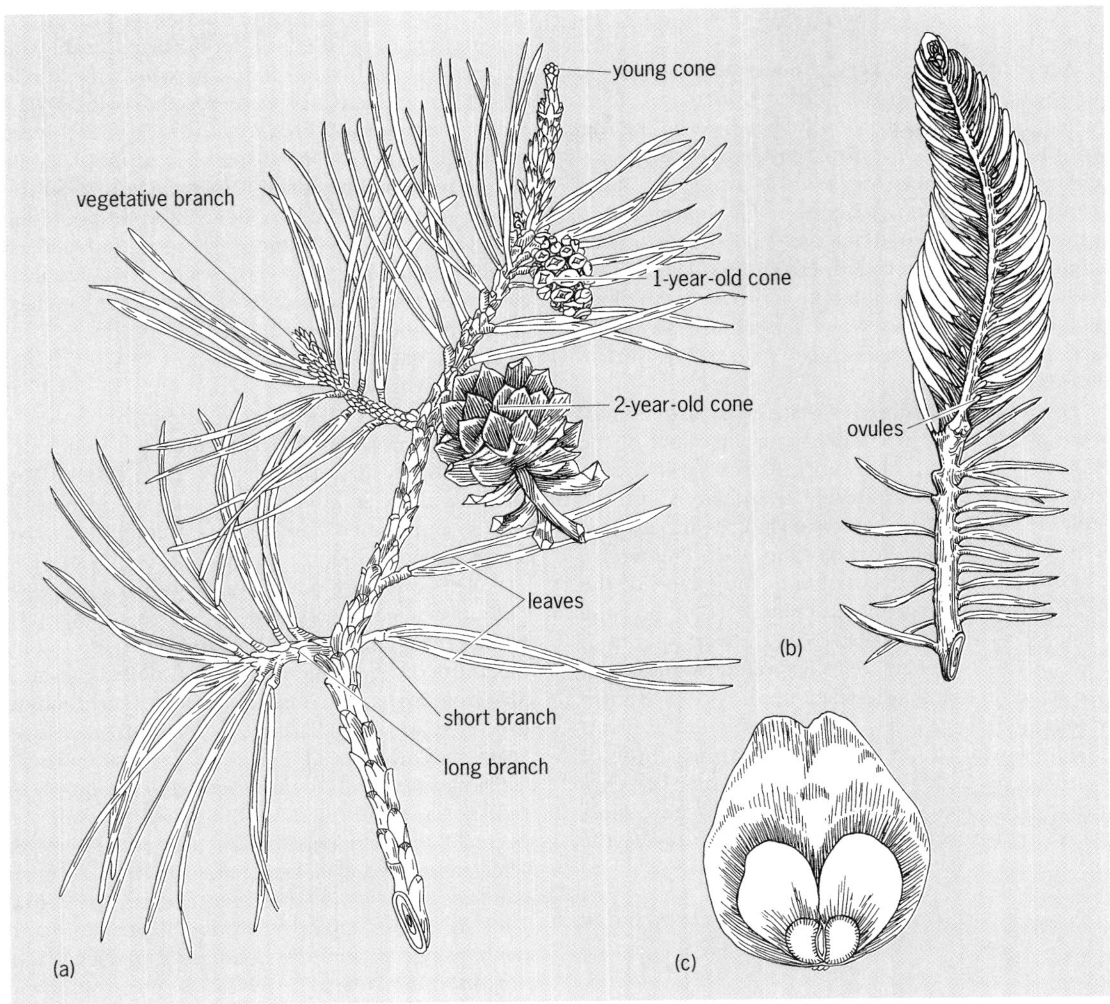

Fig. 1. Leaf, branch, and cone characteristics of conifers. (*a*) Branch of pine with leaves and ovulate cones. (*b*) Ovulate cone of spruce cut longitudinally to show scales with ovules. (*c*) Detail of one scale with two ovules. (*After J. B. Hill et al., Botany, 3d ed., McGraw-Hill, 1960*)

borne on the same plant (monoecious); in others, on separate plants (dioecious). The sperm is nonciliate and nonmotile. In the pine, the pollen grains (microspores) fall from the microsporangia of the staminate (male) cones and are carried by the wind to the pistillate (female) cones, where the microspores filter between the scales of the cone, settling on the megasporangia. At the end of each megasporangium is a tiny pore (micropyle) through which the microspore migrates to the nucellus, the central body of the megasporangium. Here the microspore germinates. Each megasporangium contains a megaspore which also germinates, producing a female gametophyte with two or three archegonia, the egg-producing organs. The microspore produces the pollen tube (male gametophyte) with two sperms. The tube grows at the expense of the nucellus, carrying the sperm to the archegonia, each containing one egg. Here a sperm unites with an egg, forming a zygote which develops into an embryo. Often several eggs are fertilized in the same female gametophyte, so that for a time there are several embryos, but only one matures. Usually, a year intervenes between pollination and fertilization, and another year between fertilization and embryo formation. *See* REPRODUCTION (PLANT).

Economic importance. The conifers are a principal source of lumber and pulp for paper and wood products. Turpentine, tar, resin, and essential oils are some by-products. The group yields little food for humans; some seeds are edible (pine, pinyon, or piñon nuts). *See* PINE NUT; PINE TERPENE. Thomas A. Zanoni

Fossils. Fossils of the earliest conifers consist of leafy shoots from the Middle Pennsylvanian of England. Their vegetative anatomy was similar to that of modern forms, and included uniseriate rays and resin ducts in the pith and leaves. Seed and pollen cones from the Late Pennsylvanian and early Permian of North America and Europe demonstrate that conifers were diverse by that time, although peak conifer diversity occurred during the Mesozoic. The modern families Araucariaceae, Podocarpaceae, Taxodiaceae, and Pinaceae occur in the Triassic; and some modern genera, including *Taxus*, *Torreya*,

Araucaria, and *Sequoia*, were present by the Jurassic.

According to one theory, conifers are a sister group to the extinct Paleozoic order Cordaitales, from which they were previously believed to be descended. The Devonian progymnosperm *Archaeopteris* has been suggested as progenitor of the conifer-cordaite lineage because of its vegetative features, including wood anatomy and simple spirally arranged leaves. A pteridosperm ancestry is also possible, based on the coniferlike reproductive biology of the Callistophytales, which possess saccate pollen and platyspermic ovules. *See* CORDAITALES; PTERIDOSPERMS.

Fossil evidence suggests that the ovulate conifer cone was derived from a structure like that of the Cordaitales, which consists of a long axis with bracts. In each bract axil was a fertile, dwarf shoot with spirally arranged sterile scales and elongate megasporophylls with terminal ovules (**Fig. 2***a*). Progressive flattening of these dwarf shoots and fusion of the scales and megasporophylls culminated in the cone scale of modern conifers. Fertile dwarf shoots of various fossil conifers are interpreted as intermediate stages in this transformation. The Pennsylvanian conifer *Emporia* (Fig. 2*b*) possessed fertile dwarf shoots similar to but more flattened than those of Cordaitales. Early conifers of the family Utrechtiaceae bore ovules laterally on fertile scales. Most early conifers retained sterile scales on the fertile dwarf shoot, although some had only fertile scales.

A more advanced condition occurred in the late Permian family Majonicaceae, in which the fertile dwarf shoot elements were fused basally. Dwarf shoot elements of *Pseudovoltzia* (Fig. 2*c*) were nearly flattened in one plane. The artificial family Voltziaceae includes many Paleozoic and Mesozoic forms, some of which have been interpreted as additional intermediates between Paleozoic and modern conifers. Although some of these are probably ancestors or early representatives of the modern families, others belong to extinct lineages.

By the late Permian, *Ullmania* (Fig. 2*d*) had achieved a flattened, fused cone scale like that of modern conifers. The suggestion is that the cone scale was derived at different times, independently in various lineages. The early Permian families Ferugliocladaceae and Buriadiaceae had simple ovulate cones, and ovules borne singly among vegetative leaves, respectively.

Pollen cones of modern conifers are simple, with pollen sacs borne abaxially on microsporophylls. Fossil pollen cones exhibit greater diversity, and include Paleozoic forms with adaxial pollen sacs and Mesozoic forms with stalked pollen sacs surrounding the microsporophyll base. *See* PINOPHYTA; PINOPSIDA; PLANT KINGDOM. Brian J. Axsmith

Bibliography. C. B. Beck, *Origin and Evolution of Gymnosperms*, 1988; J. A. Clement-Westerhof, Aspects of Permian paleobotany and palynology, 8: The Majonicaceae, a new family of Late Permian conifers, *Rev. Palaeobot. Palynol.*, 52:375–402, 1987; G. Mapes and G. W. Rothwell, Structure and relationships of primitive conifers, *N. Jb. Geol. Paleont. Abh.*, 183:269–287, 1991; C. N. Miller, Mesozoic conifers, *Bot. Rev.*, 43:218–280, 1977.

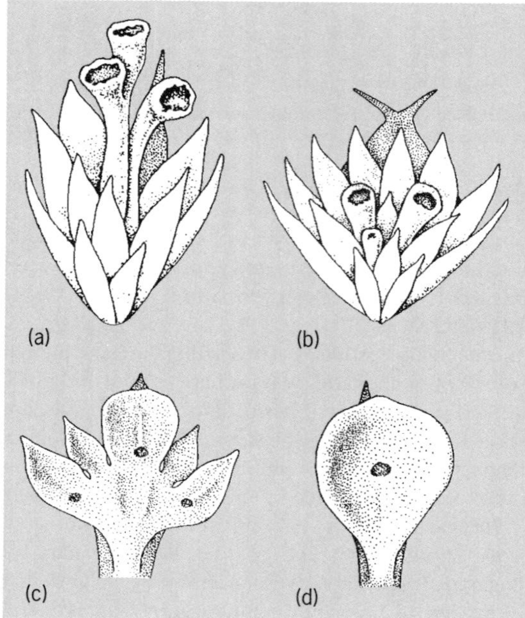

Fig. 2. Fertile dwarf shoots of a cordaite and select fossil conifers representing stages in cone scale evolution. A direct phlogenetic series is not implied. (*a*) In *Cordaites* the ovules were borne terminally on the megasporophylls. (*b*) *Emporia*. (*c*) The dwarf shoot is nearly flattened in one plane with scales fused basally. (*d*) The dwarf shoot of *Ullmania* is completely fused and flattened, forming a true cone scale. (*After G. Mapes and G. W. Rothwell, Structure and relationships of primitive conifers, N. Jb. Geol. Paleont. Abh. 183:269–287, 1991*)

Pinch effect

A name given to manifestations of the magnetic self-attraction of parallel electric currents having the same direction. The effect at modest current levels of a few amperes can usually be neglected, but when current levels approach a million amperes such as occur in electrochemistry, the effect can be damaging and must be taken into account by electrical engineers. Since the late 1940s the pinch effect in a gas discharge has been studied intensively in laboratories throughout the world, since it presents a possible way of achieving the magnetic confinement of a hot plasma (a highly ionized gas) necessary for the successful operation of a thermonuclear or fusion reactor.

Ampère's law. The law of attraction which describes the interaction between parallel electric currents was discovered by A. M. Ampère in 1820 and can be stated as follows: The force of attraction in newtons per meter of length between two thin straight wires r meters apart carrying currents I_1 and I_2 amperes, respectively, is $2 \times 10^{-7} I_1 I_2 / r$. The law applies equally to the attraction between the

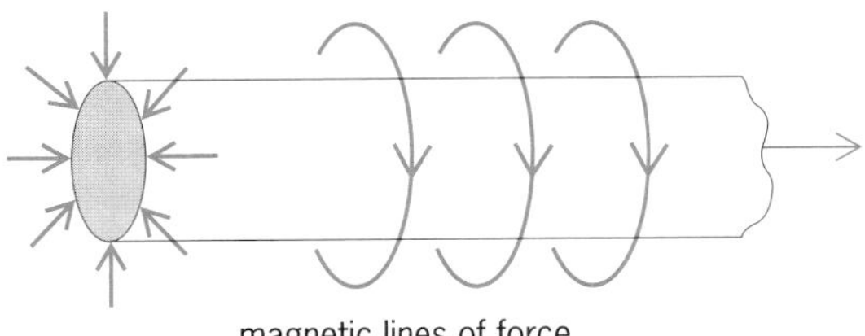

Fig. 1. Pinch pressure on a current-carrying conductor. Arrows at left show direction of pinch pressure.

individual components of a current in a single wire, in which case, for a cylindrical wire of radius r meters carrying a total surface current of I amperes, it manifests itself as an inward pressure on the surface (**Fig. 1**) given by $I^2/2 \times 10^7 \pi r^2$ pascals.

For the electric currents of normal experience, this force is small and passes unnoticed but it is significant that the pressure increases with the square of the current, I^2. For example, at 25,000 A the pressure amounts to about 1 atm (100 kilopascals) for a wire of 1-cm radius, but at 10^6 A the pressure is about 1600 atm or about 12 tons per square inch (160 megapascals).

Manifestations. The pinch effect first showed up practically in certain early types of induction electric furnaces in which large low-frequency alternating currents of the order of 100,000 A were induced at low voltage in a horizontal ring-shaped fused-metal load (**Fig. 2**). At these currents, the pinch pressure can be larger than the hydrostatic pressure exerted by the fused metal, and as indicated above ($I^2/2 \times 10^7 \pi r^2$), the pinch pressure increases as the radius of the conductor decreases. Consequently, once the pinch process starts, the pressure at a narrow neck in the ring of fused metal can squeeze out the fluid metal until the neck pinches off completely, cutting off the current. This led to very uneven heating of the load. The term pinch effect was given to this process by C. Hering in 1907. The technical difficulty was eventually overcome by making the plane of the ring vertical and submerging it deeply below the surface of the fused metal. The force of the pinch effect also manifests itself by the crushing of tubular conductors exposed to large impulsive currents, such as occur in lightning strokes or high-power short circuits, and is used in certain metal-forming techniques.

Thermonuclear applications. One of the conditions for the attainment of a profitable balance between energy expended in heating and energy released in fusion from a thermonuclear reaction in a plasma composed of deuterium and tritium (DT, the most favorable case) is that the temperature shall be not less than about 10 keV (1.16×10^8 K). This enormous temperature can be reached and maintained only if the hot plasma is effectively isolated from the material walls of the container by vacuum. The isolation has the function of preventing cooling by contact with materials at normal temperatures and allowing the plasma to be heated. For the plasma to remain confined under these conditions, its outward pressure must be balanced by the inward pressure of nonmaterial origin, for example, a magnetic field. A profitable energy balance also depends on the density n of the confined plasma and on τ, the time it is confined. The product of $n\tau$ must exceed a certain minimum, which is 10^{14} ions cm^{-3} s for DT. *See* LAWSON CRITERION; MAGNETOHYDRODYNAMICS; PLASMA (PHYSICS); THERMONUCLEAR REACTION.

There are a number of ways in which a magnetic field can be arranged around the plasma to hold it together, and one of these methods is the pinch effect. A fusion reactor using this type of confinement would ideally be a toroidal tube in which the confined plasma would carry a large electric current induced in it by magnetic induction from a transformer core passing through the major axis of the torus. The current would have the double function of ohmically heating the plasma and compressing the plasma toward the center of the tube.

The fundamental equation for the pinch effect in a gas, derived theoretically by W. Bennett in 1934,

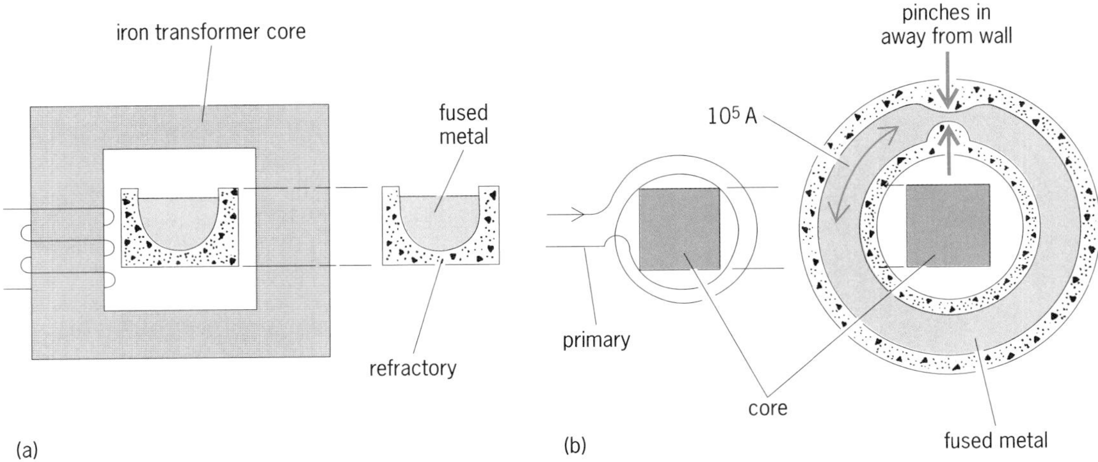

Fig. 2. Early type of ring induction electric furnace. (*a*) Side view. (*b*) Plan view.

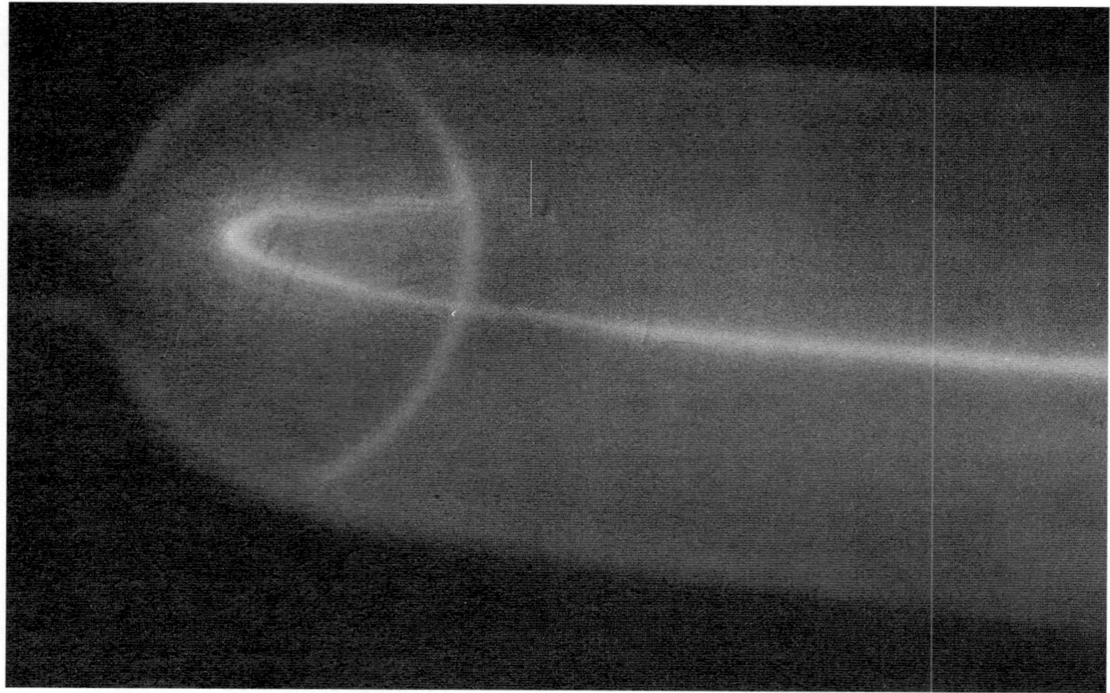

Fig. 3. Xenon pinched discharge in Perhapsatron torus.

gives the current required for the inward pinch pressure to balance the outward gas pressure, as shown in the equation below, where I is the total

$$\frac{I^2}{2 \times 10^7} = Nk(T_i + T_e)$$

current in amperes, N is the number of electrons (also the number of ions) per meter of length of the pinch, $k = 1.4 \times 10^{-23}$ J/K (Boltzmann's constant), and T_i and T_e are the temperatures in kelvins of the ions and electrons, respectively.

Experimental studies. In general, two types of apparatus were used in early studies of the pinch effect: (1) straight discharge tubes of quartz or porcelain, with metal electrodes at each end, intended for short-duration studies, in which the cooling of the plasma by the relatively cold electrodes was slight during the time of the experiment, and (2) toroidal discharge tubes, also composed of quartz or porcelain, in which the pinch was endless and consequently was more effectively confined than in the first type of apparatus, and the current was induced into the discharge by magnetic coupling to a primary winding. In both cases, currents of 50,000–500,000 A were obtained by electric fields of 10–100 V/cm along the pinch. The primary power sources in early experiments used charged capacitors with capacitances of 4–50 microfarads, charged to 10–100 kV.

Instability. Characteristically, as can be shown by high-speed photography, the pinch forms at the inner surface of the discharge tube wall and contracts radially inward, forming an intense line, the pinch, on the axis (**Fig. 3**); the pinch rebounds slightly; the contracted discharge rapidly develops necks and kinks; and in a few microseconds all structure is lost in an apparently turbulent glowing gas which fills the tube. Thus, the pinch turns out to be unstable, and plasma confinement is soon lost by contact with the wall. The cause of the instability is easily seen qualitatively: The pinch confinement can be described as being caused by the magnetic field lines

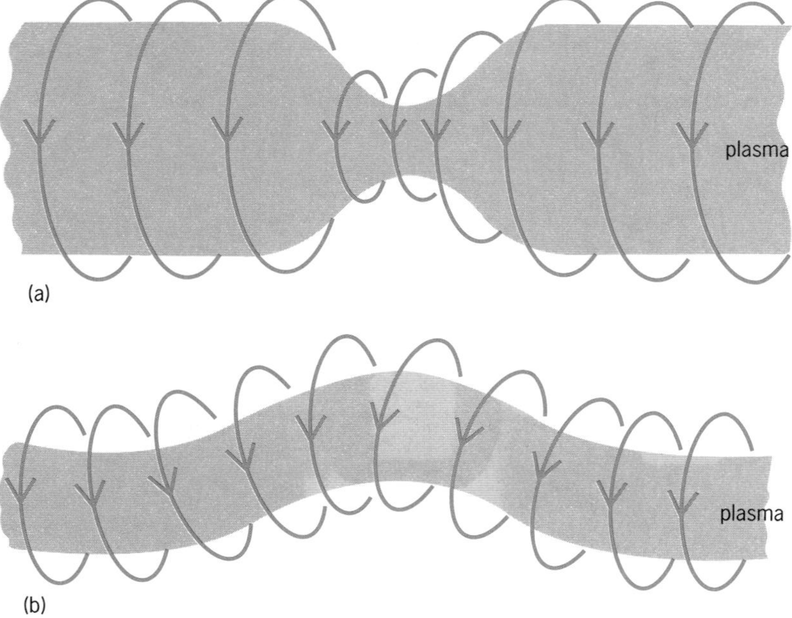

Fig. 4. Instability. (*a*) Sausage type. (*b*) Kink type.

encircling the pinch which are stretched longitudinally but which are in compression transversely (**Fig. 4**). For a uniform cylindrical pinch, the magnetic pinch pressure is everywhere equal to the outward plasma pressure, but at a neck or on the inward side of a kink, the magnetic field lines crowd together, creating a higher magnetic pressure than the outward gas pressure. Consequently, the neck contracts still further, the kink cuts in on the concave side and bulges out on the convex side, and both perturbations grow. The instability has a disastrous effect on τ, limiting it to 10^{-6} s or less in light atom plasmas such as DT.

Neutrons have been produced by deuterium pinches in large numbers. For a time (1952–1953), they were thought to be evidence of thermonuclear reaction, but it has since been shown that they are emitted preferentially in certain directions, and are associated with the instability of the pinch and the violent accelerations that are produced. Such neutrons are then not a product of thermal collisions and are not thermonuclear.

Great efforts have been devoted to overcoming the basic instability of the simple pinch. One such measure was to add an axial magnetic field by means of an external winding around the pinch tube. This might be expected to resist the neck and kink deformation by stiffening the discharge. Also, a conducting wall located close to the discharge tube has the effect of trapping the magnetic field between the pinch and the wall, cushioning and reflecting the moving pinch back to the center.

Studies which involved such modifications were in progress on a worldwide scale during the period 1955–1963, notably including Zeta (Harwell, United Kingdom), Alpha (Leningrad, Soviet Union), and the Perhapsatron (Los Alamos, United States). In general these measures were disappointing, and work on pinches of this type declined.

Reversed-field Z pinch. In the 1970s, encouraged by analysis of Zeta results and improved theoretical understanding of pinch instability, there was renewed interest in the pinch formed by adding another longitudinal field outside the pinch in the opposite direction to that inside the pinch described above. This geometry is known as the reversed-field Z pinch. Confirmation of improved Z-pinch stability was undertaken in ZT-40 (**Fig. 5**; Los Alamos, United States), ETA Beta II (Padua, Italy), HBTX (Culham, United Kingdom), and elsewhere.

High-power pulsed systems. The simple pinch is still of considerable interest because the reciprocal functional relation between n and τ for power production means that there is always the possibility of achieving a net power output no matter how much τ is reduced by instabilities, by the use of very high n. Such plasmas will require extreme high-pulsed electric power to heat and confine them, with perhaps the only limitation being that the resultant output burst of thermonuclear power may destroy the machine. For such pulsed systems, the pinch must always rank highly—it is uniquely the most

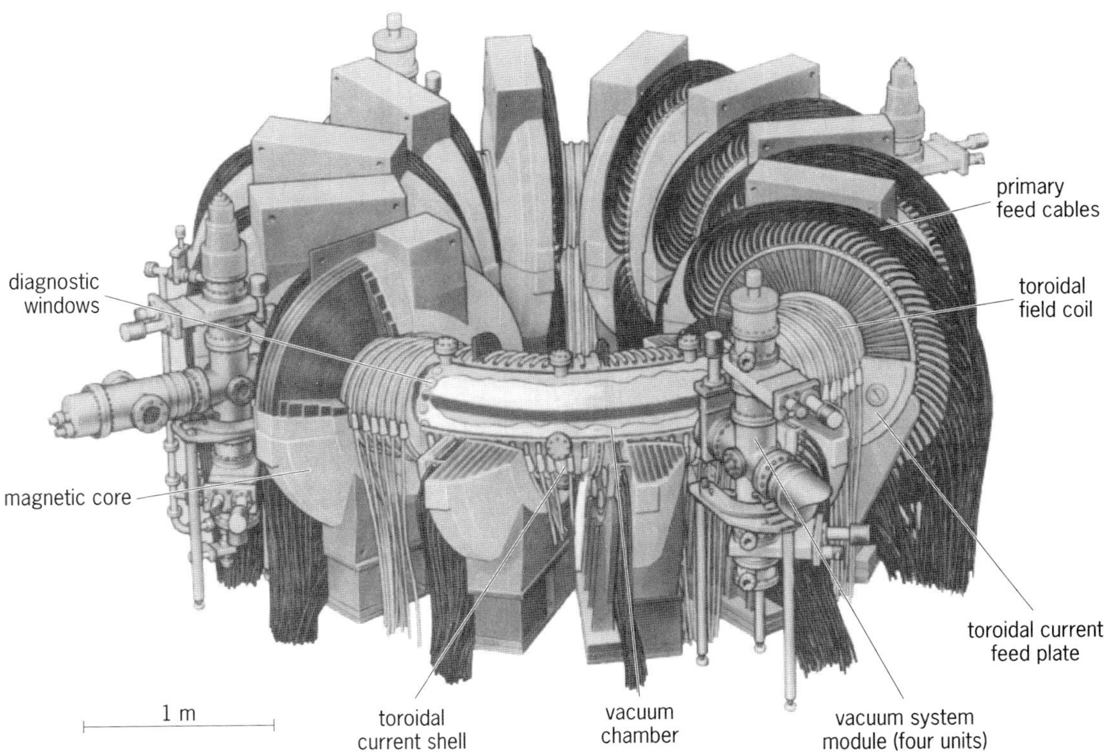

Fig. 5. ZT-40 reversed-field Z-pinch apparatus. 1 m = 3.3 ft. (*Los Alamos National Laboratory*)

efficient magnetic confinement system expressed in terms of magnetic energy expended per unit of plasma energy confined, and furthermore the pressure it confines can exceed the strength of any known material.

Theta pinch. The term theta pinch has come into wide usage to denote an important plasma confinement system which relies on the repulsion of oppositely directed currents and which is thus not in accord with the original definition of the pinch effect (self-attraction of currents in the same direction). Plasma confinement systems based on the original pinch effect are known as Z pinches. **Figure 6** defines the two geometries. The first laboratory thermonuclear reaction was in the small θ-pinch experiment, Scylla, at Los Alamos in 1957.

Tokamak. The stabilized Z pinch became the subject of intense interest when L. Artsimovich of the Soviet Union announced in 1969 the achievement of a 20-ms confinement and an ion temperature of 0.5 keV (5×10^6 K) in the T-3 tokamak experiment. Tokamak is essentially a low-density, slow Z pinch in a torus with a very strong longitudinal field such that the pinch current is below the Kruskal-Shafranov limit. This means that the helical magnetic field lines, resultant from the externally applied field and that of the pinch, do not close, that is, complete one revolution of the minor axis in going around the major axis of the torus once. This is known theoretically to prevent the growth of certain helical distortions of the plasma. The achievements of tokamak experiments are as follows, together with the target values for achieving a positive power balance: $T_i =$ 6 keV for tokamak, target value 10; $n\tau\gamma = 3 \times 10^3$ ions cm^{-3} for tokamak, target value 10^{14}.

The temperatures achieved are higher than the 0.5 keV that could be reached by resistive heating of the plasma by the pinch current, such as was used in the early 1970s. Auxiliary heating is necessary, and the most successful has been the injection of energetic neutral beams.

Construction of several large tokamak installations has been undertaken, including the TFTR (Princeton, United States), JET (European Community at Culham, United Kingdom), JT-60 (Tokyo), and T-15 (Moscow). *See* NUCLEAR FUSION. James A. Phillips

Bibliography. K. Niu, *Nuclear Fusion*, 1989; E. Teller (ed.), *Fusion*, 2 pts., 1981.

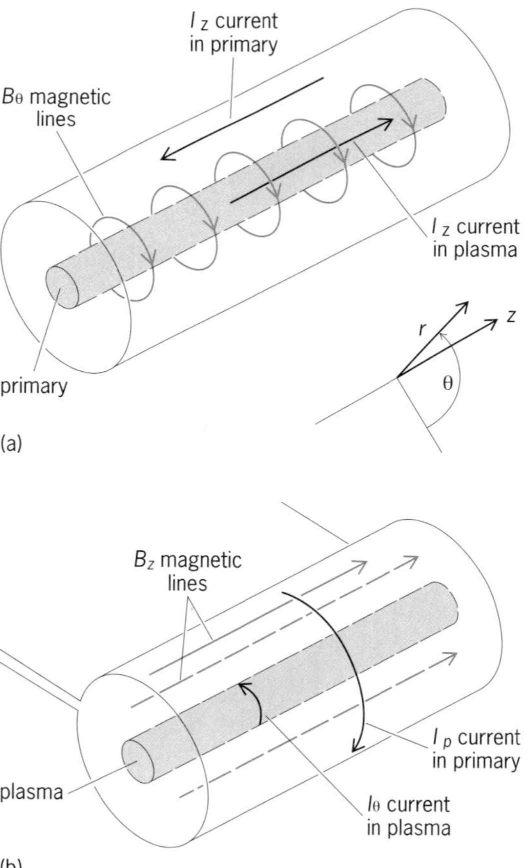

Fig. 6. Two geometries for plasma confinement systems. (*a*) Z pinch. (*b*) θ pinch.

Pine

The genus *Pinus*, of the pine family, characterized by evergreen leaves, usually in tight clusters (fascicles) of two to five, rarely single. There are about 80 known species distributed throughout the Northern Hemisphere. Botanically the leaves are of two kinds: (1) a scalelike form, the primary leaf, which subtends a much shortened and eventually deciduous shoot bearing (2) the secondary leaves or needles. The wood of pines is easily recognized by the numerous resin ducts and by the characteristic resinous odor. The pines may be divided into two classes according to the number of leaves in a cluster. *See* PINALES; SECRETORY STRUCTURES (PLANT).

Soft or white pines. Except for the nut pines, the white pines have five needles in a cluster. Eastern white pine (*P. strobus*), ranging 90–150 ft (27–45 m) in height, is found in the northeastern United States west to the Lake states, adjacent Canada, and the Appalachian Mountain region. Originally it was the most important timber tree of the eastern United States and Canada. The wood of eastern white pine is valuable because it can be easily worked, is light and soft, does not split when nailed, polishes well, and does not warp or swell appreciably. Almost everything—shipmasts, matches, doors, framing, finish, boxes, and crates—has been made from this wood. However, it is restricted to more particular uses because of its increasing scarcity and value.

Of the western white pines, the sugar pine (*P. lambertiana*) is the most important. This is a magnificent Pacific Coast tree attaining a height of about 250 ft (75 m) and having large cones, 10–20 in.

(25–50 cm) long. It is lumbered only in Oregon and California, where it ranks in volume and value with redwood. Other white pines in the West are western white pine (*P. monticola*), a mountain species found almost entirely in Idaho, Montana, and Oregon; limber pine (*P. flexilis*), one of the smaller white pines of the Rockies; and whitebark pine (*P. albicaulis*), also a mountain species with a more northern range. *See* REDWOOD.

The nut pines or piñons are a subgroup of the Southwest with fewer needles, sometimes only one. *See* PINE NUT.

Hard or pitch pines. Red pine (*P. resinosa*), also known as Norway pine, reaches a height of 80 ft (24 m) or more, and is native in the northeastern United States from Maine to Minnesota and adjacent Canada and south along the mountains to West Virginia. The needles are in pairs, and the bark has a red-brown color, hence the name red pine. The wood is fairly soft, but a little harder than that of eastern white pine. The more dense red pine is also stronger and is important commercially for general construction, sash, door, and window frame manufacturing, flooring, boxes, crates, and shipmasts, but it is not durable in contact with the soil. Frequently it is sold in mixture with eastern white pine. About two-thirds of the total volume is in the Northeast and the remainder in the Lake states. Technologically the red pine forms a sort of bridge between the soft white pines and the hard pines.

The hard yellow pines have two or three needles in a cluster. The longleaf pine (*P. palustris*), the loblolly (*P. taeda*), the shortleaf (*P. echinata*), and the slash pine (*P. elliottii*) are the principal coniferous or softwood trees of the southeastern United States and of major economic importance. These four species are the primary producers of lumber and pulpwood in the South. Production of veneer for plywood is also important. The particularly hard and strong wood of the longleaf pine has long made it a preeminent species for structural timbers, flooring, railway car construction, and similar uses. The longleaf and slash pines are the principal trees for the production of turpentine, a distillate obtained from the resin when the trees are tapped. *See* GUM; PINE TERPENE; RESIN.

In the West the hard pines are represented chiefly by the ponderosa pine (*P. ponderosa*), which attains a height of 150–225 ft (45–70 m), and is found in the Rocky Mountain and Pacific Coast regions, including adjacent Canada.

The Austrian pine (*P. nigra*), which has two needles and closely resembles the red pine, has darker bark and is much planted in North America, as is the Scotch pine (*P. sylvestris*), with two shorter, bluish needles (see **illus.**). Other exotics cultivated are mountain pine (*P. mugo*), Japanese black pine (*P. thunbergi*), and Swiss stone pine (*P. cembra*), the last a five-needled species. *See* FOREST AND FORESTRY; TREE.

Arthur H. Graves; Kenneth P. Davis

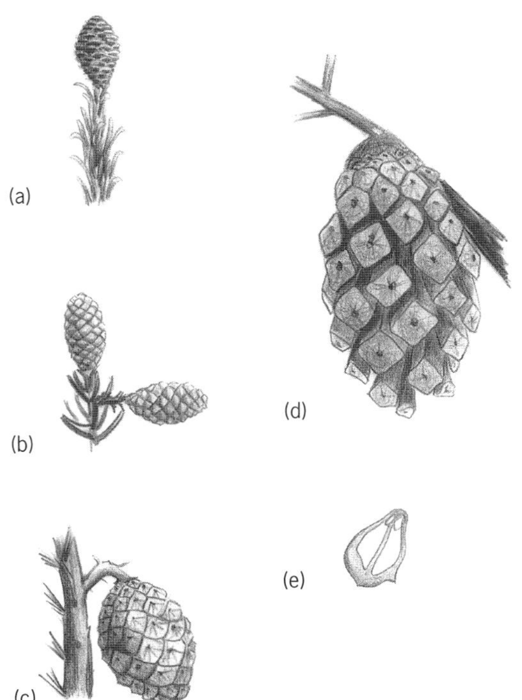

Scotch pine (*Pinus sylvestris*). (*a*) Female (ovulate) cone, unfertilized. (*b*) Fertilized cones. (*c*) Advanced stage in cone development. (*d*) Cone opening to discharge seeds. (*e*) Seed scale showing two winged seeds.

Pine nut

The edible seed of more than a dozen species of evergreen cone-bearing trees in the genus *Pinus*, native to the temperate zone of the Northern Hemisphere. The important nut-producing species are the stone pine (*P. pinea*) of southern Europe; the Swiss stone pine (*P. cembra*), native to the Swiss Alps and eastward through Siberia to Mongolia; and the piñon pine (*P. cembroides* var. *edulis*) of the arid regions of the southwestern United States. The seeds or nuts, variable in size according to species, are borne in cones which take 3–4 years to develop.

The cones of the stone pine are 4–6 in. (10–15 cm) long, each containing a hundred or more nuts $^5/_8$–$^3/_4$ in. (16–19 mm) in length. Cones are picked from the trees before the cone scales separate and later dried in the sun to free the nuts. The nuts are removed from the cone by hand and cracked mechanically to free the kernels which are cleaned and packed for export. Imports to the United States are mostly from Italy and Spain.

The piñon pine (see **illus.**) of Colorado, New Mexico, Arizona, and northern Mexico grows in forests of scattered trees on arid land with only 12–14 in. (30–36 cm) annual rainfall. The trees are dwarf and grow slowly. The relatively small cones open

Piñon pine, *Pinus cembroides* var. *edulis*. (*a*) Unopened cone. (*b*) Opened cone. (*c*) Nuts in shell. (*d*) Branch with needles and old staminate cones.

on the trees, freeing the nuts, which are picked up, mostly by local Indians, and are either used as food or for commerce.

Pine nuts are an important staple food over large areas of the Northern Hemisphere. Only a small part of the nuts produced reaches local or world markets. The raw nuts of some species have a strong turpentine flavor which disappears when they are roasted. *See* PINE; SEED.

Laurence H. MacDaniels

Pine terpene

A major component of the essential oils obtained from various *Pinus* species. The principal terpenes of the oil of southern pines [longleaf pine (*P. palustris*) and slash pine (*P. caribaea*)] are α- and β-pinene. Smaller amounts of moncyclic dienes and 3-carene are also present.

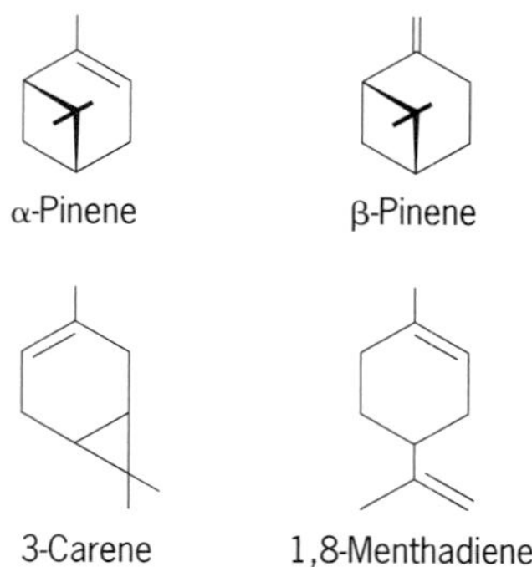

See ESSENTIAL OILS; PINE.

Turpentine. Gum turpentine (gum spirits) is the volatile fraction of the oleoresin that exudes from cuts made in the trunks of live trees. The resin is collected and distilled by a process that yields about 20% turpentine, mainly α- and β-pinene, and 70% rosin; it was the basis of the original naval stores industry.

Wood turpentine is obtained by steam distillation from stumps and other logging residues. The volatile material in this case consists of about 50% turpentine and 30–40% of higher-boiling-point alcohols; the latter fraction is known as pine oil. The bulk of the wood turpentine and pine oil produced by modern industrial processes is a by-product of the sulfate wood-pulping process (sulfate turpentine).

Important uses of turpentine or the purified pinenes derived from turpentine are in terpene resins, as a thinner in paints and varnishes, and as a starting material in the synthesis of other commercially valuable terpenes.

Pine oil. This is a mixture of monoterpene alcohols, mainly α-terpineol, obtained in large amounts mixed with wood turpentine or sulfate turpentine. The term pine oil is also used to designate the essential oil of various species of pine.

Much of the pine oil of commerce is prepared synthetically by acid-catalyzed hydration of α-pinene. This process involves a complex series of reactions that occur via cationic intermediates. A few of these steps are shown in the following reaction scheme.

H^+ $-H^+$ H_2O $-H_2O$ H_2O

α-Pinene   Menthadiene

OH   OH   OH   OH   OH

γ-Terpineol + β-Terpineol   Terpin hydrate   α-Terpineol

The composition of industrial-grade pine oil is

approximately 65% α-terpineol, 20–25% of other monoterpene alcohols, and 10–15% hydrocarbons. Pine oil has surfactant and emulsifying properties and is also a disinfectant. Most of the pine oil manufactured is used in the manufacture of cleansers and textile penetrants. *See* TERPENE; WOOD CHEMICALS.

James A. Moore

Pineal gland

An actively functioning endocrine gland, located in the brain, which secretes melatonin, is strongly regulated by light stimuli, and is an important component of the circadian timing system. The pineal gland or pineal body has the Latin name *epiphysis cerebri*. It is virtually ubiquitous throughout the vertebrate animal kingdom. In nonmammalian vertebrates (lampreys, fishes, amphibians, reptiles, birds), it functions as a photoreceptive third eye and an endocrine organ. In mammals, it serves as an endocrine organ that is regulated by light entering the body via the paired eyes. In all vertebrates, including humans, the pineal gland produces the hormone melatonin, with the highest synthesis and secretion occurring during the night. Despite extensive species variation in anatomy and physiology, the pineal gland generally serves as an essential component of the circadian system which allows animals to internally measure time and coordinate physiological time-keeping with the external environment. *See* BIOLOGICAL CLOCKS; BRAIN.

Anatomy. The pineal gland is an unpaired organ attached by a stalk to the roof of the diencephalon. In frogs and lizards, one component of the pineal complex (the frontal organ or parietal eye) projects upward through the skull to lie under the skin; in all other vertebrates the pineal is located beneath the roof of the skull. No evidence of the pineal has been found in the simplest cyclostomes (myxinoids), and it is lacking in crocodiles, armadillos, and edentates. The size and shape of the pineal varies with species: it is relatively large in cocks, echidnas, rodents, ungulates, and humans. In anamniotes, the pineal has a prominent intrapineal neuronal apparatus transmitting light stimuli to the brain via neural mechanisms. This apparatus is lost in mammals, where endocrine cells predominate.

Across evolution, cells within the pineal gland have progressed from classic photoreceptor cells in the earliest vertebrates, to rudimentary photoreceptors in birds, to classic endocrine cells in mammals. In cyclostomes, two outgrowths project from the dorsal roof of the diencephalon. One, derived from the right side of the brain, is the pineal body; the other outgrowth is the parapineal organ, which is more ventral and derived from the left side of the brain. The pineal gland forms a hollow, knoblike end vesicle directly beneath an area of skin devoid of pigment and lying dorsally between the paired eyes. The dorsal wall of the pineal forms a lens-like structure, while the ventral wall contains sensory cells and ganglion cells with processes passing down the stalk to various nerve centers in the brain. The parapineal of lampreys is essentially similar in structure. Both organs are sensitive to light. In some ganoids and larval teleosts, all tissues overlying the pineal are translucent, and light striking the pineal initiates nerve impulses which are transmitted to the brain. An intracranial parapineal also occurs in teleost fishes. Together, the pineal and parapineal are called the pineal complex. In certain reptiles (*Sphenodon, Sceloporus*, and some other lizards), a parapineal or parietal eye, which is highly specialized and endowed with a distinct lens, occurs along with a glandular pineal body. In snakes, galliform birds, and mammals, the pineal is more thoroughly a glandular structure. *See* PHOTORECEPTION; SENSE ORGAN.

Innervation and biochemistry. The pineal glands of all vertebrate species receive light input by indirect, retinal projections, while fish, amphibian, and bird species also have directly photosensitive pineal organs. In mammals, nerve fibers extend from a variety of sources in the brain to the pineal gland. The best studied of these neural inputs is through the retinohypothalamic tract, which extends from the eyes to the pineal gland in mammals (see **illus.**). Originating in the retina, the majority of the retinohypothalamic fibers project to or around the bilateral suprachiasmatic nuclei in the hypothalamus. These nuclei serve as endogenous oscillators with period lengths close to 24 h. Thus, the suprachiasmatic nuclei function as pacemakers for the circadian system, which regulates daily physiological and behavioral rhythms. From the suprachiasmatic nuclei there are short projections to the paired paraventricular hypothalamic nuclei, and then long descending axons project from these nuclei to synapse on preganglionic sympathetic neurons in the upper thoracic spinal cord. These sympathetic neurons then extend out of the central nervous system to the superior cervical ganglia in the neck region. From there, postganglionic sympathetic axons reenter the cranium and ultimately innervate the pineal gland.

In mammals, information about environmental light and darkness is relayed from the eye to entrain circadian neural activity of the suprachiasmatic nuclei. In turn, the suprachiasmatic nuclei synchronize circadian rhythms in the pineal gland through its sympathetic innervation. One of the best-studied rhythms in the pineal gland is the biosynthesis of the hormone melatonin, with the structure below.

CH_3O–(indole ring, N–H)–$CH_2-CH_2-NH-C(=O)-CH_3$

Melatonin has been found in all vertebrate and nonvertebrate species studied, from single-celled plankton to humans. In vertebrates, pineal endocrine cells take up the amino acid tryptophan from the circulation and convert it to serotonin in two enzymatic steps. The conversion of serotonin to melatonin then takes two additional enzymatic steps which are controlled by adrenergic innervation. When the sympathetic nerves to the pineal release the neurotransmitter norepinephrine, it binds to receptors on the pinealocyte cell-surface membranes. This neural stimulation ultimately regulates melatonin synthesis within the pinealocytes by activating an intracellular second messenger. The second messenger initiates an intracellular cascade of biochemical changes which result in the conversion of serotonin to melatonin. For the most part, melatonin is not stored within the pineal gland. Once synthesized, melatonin is quickly released into the blood. From the circulation, melatonin can be transferred into virtually all of the body fluids such as saliva and cerebrospinal fluid. The primary site for melatonin metabolism is the liver, but some melatonin breakdown occurs in the kidney and brain. In addition to the synthesis of melatonin, pinealocytes have the necessary enzymes for converting tryptophan into a larger family of indole compounds. Furthermore, numerous polypeptides have been localized in the pineal gland. The biological functions of these other pineal indole and peptide constituents are currently unknown.

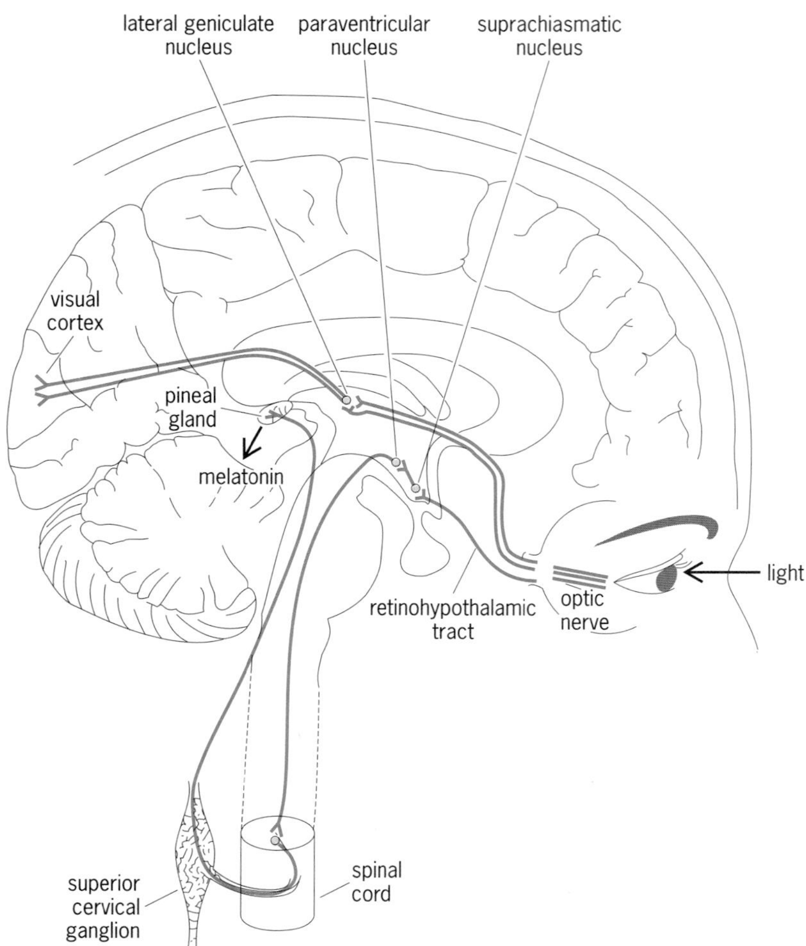

Two anatomically distinct pathways which convey light into the human brain. One pathway extends from the retina to the visual cortex and supports the sensory capacity of vision. The other pathway extends from the retina to the pineal gland for the circadian regulation of melatonin production.

Light regulation of melatonin. In all vertebrate species studied, high levels of melatonin are produced and secreted during the night, while low levels are released during the day. The melatonin circadian rhythm is produced by the endogenous pacemaking activity of the suprachiasmatic nuclei, while the entrainment of this rhythm is coordinated by signals of light and darkness relayed from the eyes. In addition to synchronizing the daily circadian rhythm of pineal melatonin secretion, light has both an acute and longer-term effect on melatonin production. Specifically, unexpected exposure of the eyes to light during the night can induce an acute, rapid decrease in the high nocturnal synthesis and secretion of melatonin. This light-induced melatonin suppression has been documented repeatedly across different vertebrate species and has been used in numerous studies to help determine the ocular, neural, and biochemical physiology of melatonin regulation. In general, the brighter the light stimulus, the greater the suppression of nocturnal melatonin. Different species, however, have widely different sensitivities for light suppression of melatonin.

In addition to light entraining the melatonin rhythm and acutely suppressing nocturnal melatonin, the day length or photoperiod can influence the duration that melatonin production is elevated during the night. This accounts for a seasonal effect of light on the pineal gland. Specifically, in the summer when days are longer and nights are shorter, the duration of increased nocturnal melatonin secretion is shorter than during the winter when nights are longer. This effect of photoperiod length influencing the duration of nighttime melatonin rise has been documented in many species, including humans.

Melatonin regulation of physiology. There is extensive species diversity in the capacity of melatonin to regulate physiology. Numerous species, ranging from insects to mammals, have yearly cycles of activity, morphology, reproduction, or development which are responsive to seasonal changes in day length. Biological regulation relative to annual cycles of daylight is called photoperiodism. Among many species that breed seasonally, melatonin has been shown to be a potent regulator of the reproductive axis in both males and females. Seasonal changes in day length influence melatonin secretion which, in turn, coordinates reproductive physiology in a way that allows for the birth of the young to occur when there is maximum possibility for their survival. Further, the effects of melatonin on the regulation of circadian physiology has been elucidated in many vertebrate species, including humans. In addition, melatonin has been studied in different species for its influence on retinal physiology, sleep, body temperature regulation, immune function, and cardiovascular regulation.

The pineal gland is the primary tissue that secretes melatonin, although the retina and perhaps portions of the intestinal tract have some capacity for melatonin synthesis. Once melatonin is secreted from the pineal gland into the blood, it circulates to target tissues where it binds to melatonin receptors. Melatonin may regulate circadian and reproductive responses by binding to specific membrane-bound receptors in discrete regions in the brain, including the hypothalamic suprachiasmatic nuclei and the pituitary pars tuberalis. A structurally related melatonin receptor also has been localized in the retina and, to a lesser extent, the brain. Outside the nervous system, melatonin binding has been revealed in various peripheral tissues, including the ovaries, gut, and blood vessels. This nonneural binding may be involved in temperature and cardiovascular regulation.

Melatonin has been used as a pharmacological agent for a variety of purposes. It has been used for timing and regulation of reproduction in seasonally breeding livestock such as sheep. In humans, controlled clinical studies have demonstrated that melatonin can be a rapid, mild sleep-inducing compound. In addition, melatonin has a capacity to "reset" the human circadian system when it becomes inappropriately synchronized due to blindness, shift work, or east-west jet travel. Many other claims have been made for the therapeutic and health value of melatonin, but sufficient controlled clinical studies are lacking to support such claims. Although the toxicity of melatonin appears to be low, there are unresolved concerns about the long-term safety of taking it. George C. Brainard

Bibliography. J. Arendt, *Melatonin and the Mammalian Pineal Gland*, Chapman and Hill, London, 1995; S. A. Binkley, *The Pineal: Endocrine and Nonendocrine Function*, Prentice Hall, Englewood Cliffs, NJ, 1988; G. C. Brainard, M. D. Rollag, and J. P. Hanifin, Photic regulation of melatonin in humans: Ocular and neural signal transduction, *J. Biol. Rhythms*, 12(6):537–546, 1997; D. C. Klein, R. Y. Moore, and S. M. Reppert, *Suprachiasmatic Nucleus: The Mind's Clock*, Oxford University Press, Oxford, 1991; R. J. Reiter, Pineal melatonin: Cell biology of its synthesis and of its physiological interactions, *Endocrine Rev.*, 12(2):151–180, 1991; S. M. Reppert, D. R. Weaver, and T. Ebisawa, Cloning and characterization of a mammalian melatonin receptor that mediates reproductive and circadian responses, *Neuron*, 13(5):1177–1185, 1994; L. Wetterberg, *Light and Biological Rhythms in Man*, Pergamon Press, Stockholm, 1993.

Pineapple

A low-growing perennial plant, indigenous to the Americas. The cultivated varieties (cultivars) belong to the species *Ananas sativus* of the plant order Bromeliales.

The edible portion of the pineapple develops from a mass of ovaries on a fleshy flower stock having persistent bracts (see **illus.**). On the cultivated types, the flowers are usually abortive. The leaves are long and swordlike, usually rough-edged, and grow to a height of 2–4 ft (0.6–1.2 m). Commercial plantings

Pineapple (*Ananas sativus*). (*USDA*)

bear fruit at the age of 12–20 months, and may continue to be productive for as much as 8–10 years. Propagation is by suckers or offsets which may be rooted in sand, but are usually set directly in the field where they are to produce. The leafy crowns may also be used as cuttings, but because they are harvested with the fruit, other methods of propagation are more satisfactory. *See* BROMELIALES.

The pineapple, a warm-climate plant, is injured by temperatures below 32°F (0°C). It does best in a dry atmosphere and relatively poor soil, but responds well to fertilizers. The major producing area is Hawaii, where special methods of culture and harvesting have been developed. A paper mulch is used in the production of much of the Hawaiian pineapple crop. The paper is laid in long strips by a special machine which pulls soil over the edges of the strips to hold them in place. The mulch aids in weed control and decreases moisture evaporation from the soil. Because of the tendency to iron chlorosis of plants growing on these soils, spraying with an iron salt is frequently a part of the fertilization program. Careful control of the nitrogen supply in relation to the hours of sunshine, together with the use of hormone sprays under certain conditions, has made possible considerable control over the time of ripening, an important consideration in obtaining maximum year-round use of processing facilities. Pineapples are also grown in the West Indies and other tropical areas, and to a limited extent in southern Florida.

Pineapples are consumed fresh in considerable quantity, but because of distance from markets and the problems of transporting fresh fruit, most of the crop is canned as sliced pineapple or as juice. *See* FRUIT. J. Harold Clarke

Pinnipeds

Carnivorous mammals of the suborder Pinnipedia, which includes 32 species of seals, sea lions, and walrus in 3 families. All species of the order are found along coastal areas, from the Antarctic to the Arctic regions. Although many species have restricted distribution, such as the Baikal seal (*Pusa sibirica*) of Lake Baikal in Russia, the group as a whole has worldwide distribution.

Pinnipeds are less modified both anatomically and behaviorally for marine life than are other marine mammals: Each year they return to land to breed, and they have retained their hindlimbs. They are primarily carnivorous mammals, with fish supplying the basic diet. They also eat crustaceans, mollusks, and, in some instances, sea birds. Their body is covered with a heavy coat of fur, the limbs are modified as flippers, the eyes are large, the external ear is small or lacking, and the tail is absent or very short. Dentition is variable among the families. The intestinal tract is simple, testes are internal in the phocids and external in the otariids, and the males have an os priapi or baculum. *See* COPULATORY ORGAN.

Odobenidae. The single species of this family, the walrus (*Odobenus rosmarus*), is large, growing to 10 ft (3 m) and weighting 3000 lb (1350 kg). It has no external ears but does have a distinct neck region. The upper canines of both sexes are prolonged as tusks which can be used defensively (**Fig. 1***a*). The tusks may reach a length of 3 ft (0.9 m) and weigh 11 lb (5 kg). There are 18 teeth. Walrus populations are about equally distributed in shallow waters of the Atlantic and Pacific oceans in the Arctic polar ice regions. Population densities are continually reduced, threatening the walrus with possible extinction. The walrus feeds on marine invertebrates and fish. The teeth can crush the hardest shells.

Fig. 1. Pinnipeds. (*a*) Walrus (*Odobenus rosmarus*). (*b*) Alaska fur seal (*Callorhinus ursinus*).

Otariidae. This family, the eared seals, includes the sea lions and fur seals (Fig. 1*b*), which are characterized by an external ear. The neck is longer and more clearly defined than that of true seals, the digits lack nails, and there are 34 teeth with a dental formula I 3/2 C 1/1 Pm 4/4 M 1/1. *See* DENTITION.

The California sea lion (*Zalophus californianus*) occurs along the Pacific coast, and is commonly seen in zoos and performing in circuses. It can move along on land since the hindlimbs can be rotated to act as legs. The forelimbs, used for swimming, may have a span of 6 ft (1.8 m) when extended. These animals have been seen swimming at rates over 20 mi/h (9 m/s). An adult male may grow to 7 ft (2 m) long and weigh up to 600 lb (270 kg); the female is smaller. In the wild, males establish harems in the breeding locales with about 40 females. Young are born in the early summer. After the breeding season a migration takes place, involving many immature males as well as adults. This permits more food for the mothers and young left behind in the rookery areas.

The southern sea lion (*Otaria byronia*) is found around the Galapagos Islands and along the South American coast. Stellar's sea lion (*Eumetopias jubatus*) is a large species, the male sometimes reaching a length of 13 ft (4 m) and a weight of about 1300 lb (600 kg), and is found along the Pacific coast. Minor species are the Australian sea lion (*Neophoca cinerea*) and the New Zealand species (*Phocarctos hookeri*).

Phocidae. This is the largest family of pinnipeds, the true seals. It includes the monk seals, elephant seals, common seals, and other less well-known forms. These animals have 30 teeth. The family is unique in that the digits have nails, the soles and palms are covered with hair, and the necks are very short. Most species live in marine habitats; however, the Caspian seal (*Pusa caspica*) lives in brackish water. The only fresh-water seal is the Baikal seal (*P. sibirica*), whose closest relative is the ringed seal (*P. hispida*). These two species perhaps derived from a single species, geographical isolation having caused the branching. It is estimated that there are between 40,000 and 100,000 Baikal seals in Lake Baikal. These animals are hunted for their valuable fur.

One of the best-known species is the Atlantic gray seal (*Halichoreus grypus*), a species fround in the North Atlantic along the coasts of Europe, Iceland, and Greenland. The leopard seal (*Hydrurga leptonyxi*) is found in southern seas around Antarctica, Australia, and South Africa, where they migrate during the postbreeding winter period. These solitary animals are voracious; they prey on penguins and frequently feed on carrion from whalers. They are able to swim rapidly. The southern elephant seal (*Mirounga leonina*) may weigh as much as 4 tons (3.5 metric tons) and reach more than 20 ft (0.6 m) in length. It is found in the southern Pacific, where it breeds on the rocky shores of the islands. The habits of these seals are among the best known of those of the pinnipeds. The cows arrive at the rookeries in early fall to give birth to a single calf. About 2 months later, the bulls fight for possession of the cows to form the harem. Mating occurs in November, after which the adults return to the sea for food. *See* CARNIVORA; MAMMALIA. Charles B. Curtin

Evolution. The major evolutionary lines of extant pinnipeds, and their close fossil allies, are relatively well documented in the fossil record. The earliest fossils showing unmistakable affinities to modern pinnipeds occur in latest Oligocene through early Miocene (approximately 25 million years old) sediments of the northeast Pacific margin.

The nearly complete skeleton of *Enaliarctos mealsi*, which was recovered in central California, represents the fullest available documentation of the evolutionary transition between modern pelagic pinnipeds and their terrestrial carnivoran ancestry. This taxon displays a combination of primitive and advanced skeletal features closely matching that expected for the most recent common ancestor of living pinnipeds. *Enaliarctos* was a capable swimmer; in addition to hands and feet that were modified into flippers, flexure of the vertebral column served as an important propulsive force.

Enaliarctos and several roughly contemporaneous forms were long considered related strictly to walruses and eared seals among living pinnipeds. True seals (Phocidae) were hypothesized to have originated from a terrestrial group separate from that which gave rise to *Enaliarctos*, walruses, and eared seals. The notion that similar environmental demands had shaped myriad pinniped resemblances repeatedly and independently through the process of convergent evolution was central to this dual (diphyletic) view of pinniped origins. More recent anatomical interpretations and biomolecular evidence, however, suggest that pinnipeds stem from a single and exclusive common origin (that is, they are monophyletic) and that *Enaliarctos* represents an outgroup to all modern forms, including true seals (**Fig. 2**).

Other early Pacific fossil pinnipeds include walruses, eared seals, and a distinctive group of extinct forms apparently closely related to true seals. The earliest walruses are approximately 15 million years in age; the morphological hallmark of the group's single modern species, tusks, appear only about 5 million years ago. Walruses previously occupied the midlatitudes; their present Arctic distribution is anomalous in this regard.

Allodesmus and *Desmatophoca*, large-bodied, 22–10-million-year-old Pacific forms, are the best-known representatives of an extinct group formerly considered related to walruses. These are currently viewed as more closely related to true seals than to walruses. Walruses, in turn, share a more recent common ancestor with this extinct group plus true seals, than with the eared seals, as was traditionally thought.

Eared seals (sea lions and fur seals) first appear in the fossil record about 10 million years ago, again in the Pacific. By this point the group had already adopted the specialized mode of forelimb swimming distinguishing modern forms.

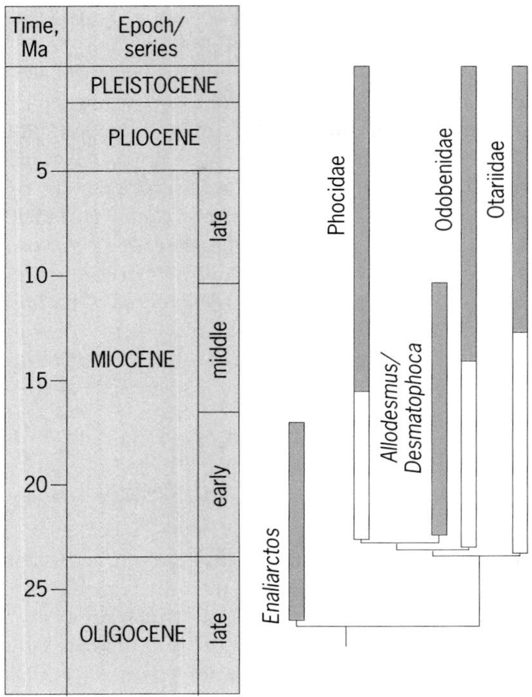

Fig. 2. Evolutionary relationships and stratigraphic ranges of major groups of pinnipeds. Dark bars correspond to known ranges; white bars, to inferred ranges.

In contrast to other major groups of pinnipeds, the early evolutionary diversification of phocids seems to have taken place largely in the Atlantic. Diverse lineages of true seals occur by about 15 million years ago, the earliest record of the group. *See* MIOCENE; OLIGOCENE. Andre Wyss

Bibliography. A. Berta and A. R. Wyss, Pinniped phylogeny, *Proc. San Diego Nat. Hist. Mus.*, 29:33–56, 1993; T. A. Deméré, The family Odobenidae: A phylogenetic analysis of fossil and living taxa, *Proc. San Diego Nat. Hist. Mus.*, 29:99–123, 1993; D. Renouf (ed.), *Behavior of Pinnipeds*, 1991; M. Riedman, *The Pinnipeds: Seals, Sea Lions and Walruses*, 1991.

Pinophyta

One of the two divisions of the seed plants of about 600 to 700 species extant on all continents except Antarctica. The most familiar and common representatives are the evergreen, cone-bearing trees of the Pinales. Because the ovules (young seeds) are exposed directly to the air at the time of pollination, the Pinophyta are commonly known as the gymnosperms, in contrast to the other division of flowering plants, the angiosperms (division Magnoliophyta), which have the ovules enclosed in an ovary. The division Pinophyta consists of three classes: Ginkgoopsida, Cycadopsida, and Pinopsida.

See CYCADOPSIDA; GINKGOOPSIDA; MAGNOLIOPHYTA; PINOPSIDA; PLANT KINGDOM.

Thomas A. Zanoni

Bibliography. C. B. Beck, Gymnosperm phylogeny: A commentary on the views of S. V. Meyen, *Bot. Rev.*, 51:273–294, 1985; S. V. Meyen, Basic features of gymnosperm systematics and phylogeny as evidenced by the fossil record, *Bot. Rev.*, 50:1–111, 1984; C. N. Miller Jr., A critical review of S. V. Meyen's "Basic features of Gymnosperm systematics and phylogeny as evidenced by the fossil record," *Bot. Rev.*, 51:295–318, 1985; G. W. Rothwell, The role of comparative morphology and anatomy in interpreting the systematics of fossil Gymnosperms, *Bot. Rev.*, 51:319–327, 1985.

Pinopsida

The largest and most important class of the division Pinophyta (Gymnospermae), the other classes being Ginkgoopsida and Cycadopsida. There are two orders: the Cordaitanthales, with three extinct families, and the Pinales (Coniferales), with six extinct families and seven families with some extant genera.

The living Pinopsida are woody plants; most are trees with a central axis and excurrent branches. Leaves are simple, alternate, or opposite or in whorls, scalelike or needlelike or rarely planar. The wood lacks vessels and usually has resin canals. Male reproductive structures are aggregated on microsporophylls directly attached to the cone axis. The ovules are borne in compound cones or singly or paired at the end of a stalk (Taxaceae). The main seed plane is tangential to the cone axis (if the seed scale is regarded as a modified dwarf shoot). The embryo has two or more cotyledons. *See* CYCADOPSIDA; GINKGOOPSIDA; PINALES; PLANT KINGDOM.

Thomas A. Zanoni

Pionium

An exotic atom, also called the pi-mu atom, which is similar in structure to the hydrogen atom but with the proton replaced by a pion and the electron replaced by a muon. Pionium is unique among atoms that have been observed in the laboratory in that all of its constituents are unstable particles not found in ordinary matter. The pion, a particle involved in nuclear forces, lives only 2.6×10^{-8} s, and the muon, a particle like the electron except that it is 207 times heavier, lives 2.2×10^{-6} s. The lifetime of this atom is thus determined by the pion lifetime. Due to the large mass of the muon, the atom is about 120 times smaller in radius than is hydrogen. *See* ELEMENTARY PARTICLE; LEPTON; MESON.

Most exotic atoms, such as muonium or positronium, contain only one short-lived particle and are formed by causing that particle to interact with an ordinary atom. This method is impossible for pionium since neither constituent is found in normal material. Instead, pionium is formed during the decay of a certain heavier particle called the neutral kaon. A kaon has many modes of decay, one of which results in the formation of a pion, a muon, and a neutrino ($K_L^0 \rightarrow \pi^+\mu^-\nu$). For this decay, both the muon and pion occasionally have almost exactly the same speed and direction, resulting in the formation of a bound atomic system. Although this process is extremely rare, less than 1 per 10^6 kaon decays, pionium has been observed in the laboratory. This decay of a particle of radius 5×10^{-14} cm (2×10^{-14} in.) into an atom of radius 5×10^{-11} cm (2×10^{-11} in.) is in striking contrast to the usual scheme of things where atoms are broken into smaller constituents. *See* MUONIUM; POSITRONIUM.

Pionium is of interest to physicists for two reasons. First, the rate of formation provides information about the details of kaon decay. The observed formation rate is indeed compatible with present understanding of such decays. Another potentially interesting application of pionium involves atomic spectroscopy. Although pionium is usually formed in the ground state, it is presumably also produced in the excited 2*S* state. A measurement of the Lamb shift, which is the energy difference between the 2*S* and the 2*P* state, would serve to measure the pion radius. *See* ATOMIC STRUCTURE AND SPECTRA; WEAK NUCLEAR INTERACTIONS.

The nomenclature of exotic atoms is not well established, and the name pionium may also refer to the pion-electron atom. Although this atom is probably formed when pions are stopped in gases, there has been no interest in studying it in the laboratory.

Paul A. Souder

Pipe flow

Conveyance of fluids in closed circular ducts. Pipes have been used for thousands of years, but the understanding of the flow details (velocity profile and flow rate) came about during the twentieth century. Equations that accurately predict energy losses have been in use since the midnineteenth century.

Flow in closed conduits is probably the most common way of transporting fluids. Crude oil and its components are moved through pipes in a refinery. Water in the home is transported through tubing. Heated and conditioned air is distributed to all parts of a dwelling in circular or rectangular ducts. *See* PIPELINE.

Flow details. Flow in a closed conduit (circular or otherwise) can be either laminar or turbulent. In laminar flow, the fluid particles move smoothly through the duct in layers called laminae. A fluid particle in one layer stays in that layer. In turbulent flow, flowing fluid particles move tortuously about the cross section, resulting in an effective mixing action. Eddies and vortices are responsible for the mixing, which does not occur in laminar flow. Turbulent flow exists at much higher flow rates than laminar flow.

The criterion for distinguishing between laminar

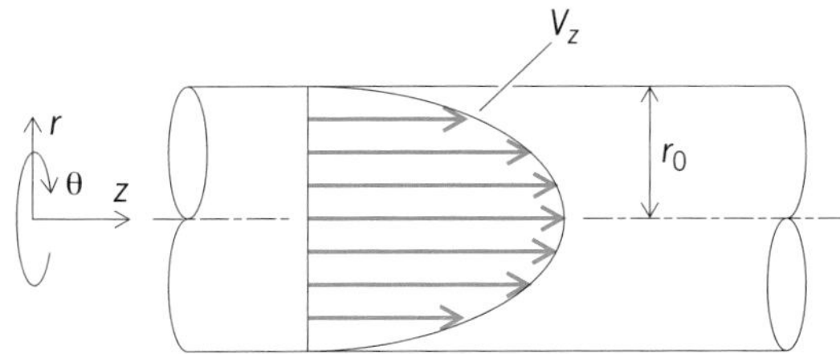

Fig. 1. Laminar velocity profile.

and turbulent flow is this observed mixing action. When injected into a laminar flow in a duct, a dye moves downstream in a threadlike line. When injected into a turbulent flow, a dye disperses quickly. Experiments have shown that laminar flow exists when the dimensionless Reynolds number, Re $= VD/\nu$, is less than 2100; here V is the average velocity, D is the inside diameter of the pipe or tube, and ν is the kinematic viscosity (a property of the fluid). *See* FLUID FLOW; LAMINAR FLOW; REYNOLDS NUMBER; TURBULENT FLOW.

Laminar flow. The axial velocity V_z in laminar flow is given by Eq. (1), in which Δp is the pressure drop

$$V_z = \frac{\Delta p r_0^2}{4\mu L}\left(1 - \frac{r^2}{r_0^2}\right) \tag{1}$$

over the length L, μ is the absolute viscosity (a property of the fluid), r_0 is the inside radius of the duct, and r is the radial coordinate. Equation (1) is of a parabola (**Fig. 1**). The volume flow rate, or discharge, Q, is given by Eq. (2), in which D is the inside diameter of the tube.

$$Q = \frac{\Delta p \pi D^4}{128\mu L} \tag{2}$$

Transition. When the Reynolds number is greater than 2100 but less than 4000, the flow can exhibit laminar, turbulent, or intermittent flow characteristics; the exact transition velocity depends on the nature of the piping system.

Turbulent flow. For very high Reynolds numbers (greater than 4000), instabilities are propagated and form eddies and vortices responsible for the observed mixing action in turbulent flow. Fluid particles move in random paths with large transverse velocity components. The velocity profile in turbulent flow is more uniform than that for laminar flow due to the large transfer of momentum radially across the cross section of the conduit (**Fig. 2**). An equation that gives a fairly accurate representation of the profile is the one-seventh power law, Eq. (3), in which V_z

$$V_z \approx V_{\max}\left(1 - \frac{r}{r_0}\right)^{1/7} \tag{3}$$

is the axial velocity, $V_{\max}$ is the velocity at the pipe centerline, r_0 is the pipe radius, and r is the radial coordinate. The average (turbulent) velocity, V_{avg}, is given by Eq. (4).

$$V_{\text{avg}} \approx \frac{49}{60} V_{\max} \tag{4}$$

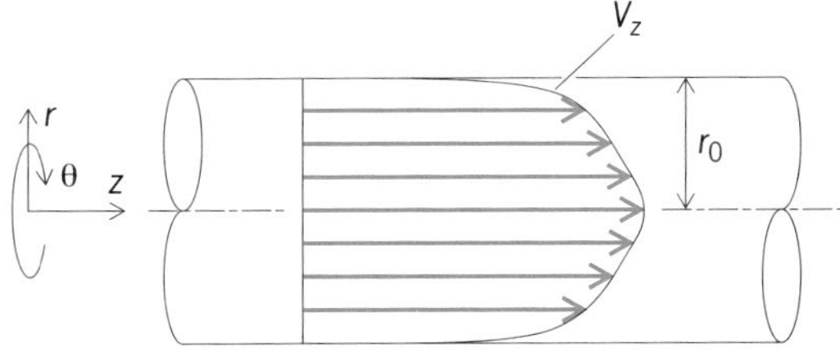

Fig. 2. Turbulent velocity distribution.

Energy loss and friction factor. The energy loss experienced by the fluid is manifested as a pressure drop Δp, which is found in terms of a dimensionless friction factor f as shown in Eq. (5). Here, the

$$\Delta p = \frac{fL}{D}\frac{\rho V^2}{2} \tag{5}$$

density ρ is a property of the fluid, L is the length over which the pressure drop occurs, D is the inside diameter of the pipe or tube, V is the average velocity, and the quantity $\rho V^2/2$ is the kinetic energy of the flow per unit volume. For laminar flow through a circular duct, the friction factor is found in terms of the Reynolds number, as indicated in Eq. (6).

$$f = \frac{64}{\text{Re}} \qquad \left(\begin{array}{c}\text{laminar flow,}\\ \text{circular duct}\end{array}\right) \tag{6}$$

For turbulent flow, the friction factor is dependent upon the wall roughness, the fluid properties, the average velocity, and the pipe diameter; that is, $f = f(V, D, \rho, \mu, \varepsilon)$, in which ε is a measure of the absolute roughness of the conduit wall, having the dimension of length. Values of the roughness ε have been measured for many commercial pipe materials (see **table**). The friction factor for turbulent flow may be obtained from a Moody diagram in which the friction factor f is graphed as a function of the Reynolds number Re, with the relative

Roughness factors for commercial pipe materials

Pipe material	ε, ft (cm)
Cast iron	0.00085 (0.025)
Asphalt-coated	0.0004 (0.012)
Cement-lined	0.000008 (0.00025)
Drawn tubing	0.000005 (0.00015)
Minerals	
Brick sewer; Cement-asbestos; Clays; Concrete	0.001–0.01 (0.03–0.3)
Miscellaneous	
Brass; Copper; Glass; Lead; Plastic; Tin	0.000005 (0.00015)
PVC	Smooth
Steel	
Commercial	0.00015 (0.0046)
Galvanized	0.0002–0.0008 (0.006–0.025)
Wrought iron	0.00015 (0.0046)

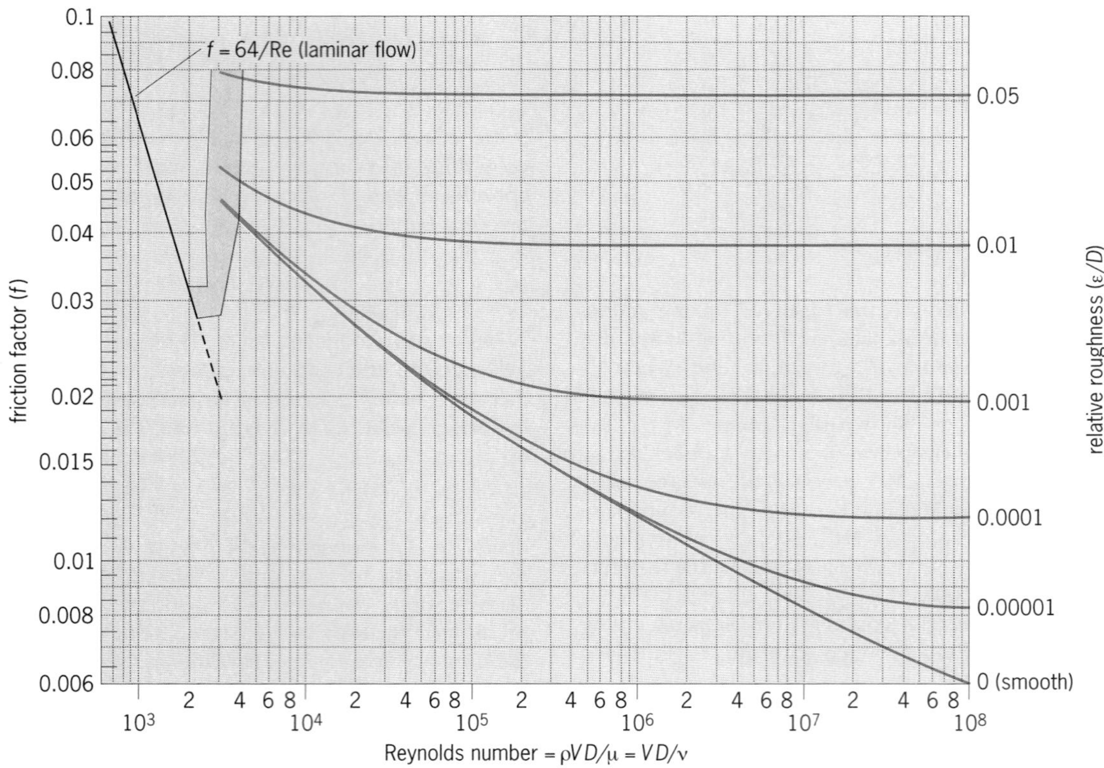

Fig. 3. Moody diagram, showing curves for laminar and turbulent flow.

roughness ε/D as an independent parameter (**Fig. 3**). The Moody diagram is a result of many flow rate and pressure drop measurements made on commercial pipe and tube materials.

A number of equations have been devised to curve-fit the Moody diagram for Reynolds numbers greater than 2100. The older equations require an iterative process when trying to calculate the friction factor f for specified values of the Reynolds number Re and the relative roughness ε/D. More recently published equations overcome this difficulty. An example is the Haaland equation (7), which gives an explicit solution for f in terms of Re and ε/D.

$$f = \left\{-0.782 \ln\left[\frac{6.9}{\text{Re}} + \left(\frac{\varepsilon}{3.7D}\right)^{1.11}\right]\right\}^{-2} \quad \text{(turbulent flow)} \tag{7}$$

For turbulent flow in a smooth pipe, $\varepsilon = 0$, and the expression for the friction factor f reduces to $f = f(\rho VD/\mu)$, in which $\rho VD/\mu$ is the Reynolds number. This relationship is shown as the bottom line on the Moody diagram. For turbulent flow in a rough pipe, the friction factor varies with the Reynolds number and the relative roughness ε/D, represented as all the other lines on the Moody diagram.

As a pipe ages, the wall roughness usually tends to increase linearly with time according to $\varepsilon' = \varepsilon + \alpha t$, in which ε is the roughness of the pipe material when new, ε' is the roughness after a certain time t has elapsed, and α is a constant that must be determined by experiments performed on the specific pipe material–fluid combination.

Calculation methods. Given values of pipe length, diameter, roughness, fluid properties, and volume flow rate, it is possible to find the head loss or the pressure drop using Eq. (5). The Reynolds number and relative roughness are calculated and then used with the Moody diagram (or curve-fit equations) to compute the friction factor. Substituting into Eq. (5) gives the pressure drop.

When the pressure drop is known and the volume flow rate is to be determined, an iterative calculation scheme is required when using the Moody diagram to find the friction factor f. An assumed value for f is obtained from the Moody diagram for the given value of the relative roughness ε/D. It is used to find a trial value for V, which in turn is used to find a Reynolds number. The values of the Reynolds number Re and the relative roughness ε/D can then be used to find a better value for f, and the process is repeated until the calculations converge.

Gases. When a gas or a vapor is the flowing fluid, compressibility effects may have to be considered. For low gas velocities of less than 30 m/s, the calculation methods described in the preceding paragraphs can be used. Otherwise, an equation of state must be used to relate changes in density with changes in pressure and temperature.

Minor losses. Pipes and tubes are joined together with fittings such as elbows, T-joints, and valves. The loss of pressure when a fluid flows through a fitting is called a minor loss, which varies with the kinetic energy of the flow.

William S. Janna

Bibliography. W. S. Janna, *Introduction to Fluid Mechanics*, 3d ed., 1993; R. W. Jeppson, *Analysis of Flow in Pipe Networks*, 1976; J. H. Roberson and C. T. Crowe, *Engineering Fluid Mechanics*, 6th ed., 1996; J. P. Tullis, *Hydraulics of Pipelines*, 1989.

Pipeline

A line of pipe and associated equipment for conveying materials. Pipelines are used for transportation of water, gases, vapors, slurries, sewage, sludge, petroleum, chemicals, foodstuffs, and other process media (**Fig. 1**). They comprise pipe, fittings, flanges, bolting, gaskets, valves, pressure-relieving devices, pumps, specialty items such as strainers, flow nozzles, expansion joints, and supports and hangers.

A tube has a cross section that may or may not be round and has a continuous periphery. A pipe is a tube with a round cross section, and can range in diameter from as little as $^{1}/_{8}$ in. (3 mm) to over 10 ft (3 m).

Wall thickness. Pipe is specified by nominal pipe size (NPS) and schedule or minimum wall thickness. The nominal pipesize identifies the nominal inside diameter for sizes from $^{1}/_{8}$ in. (3 mm) through 12 in. (305 mm) and the actual outside diameters for 14-in. (356-mm) and larger sizes. Along with the nominal pipe size, the pipe wall thickness is stated by schedule number, such as 8-in. (203-mm) nominal pipe size, schedule 40. The schedule number identifies the nominal wall thickness, which in case of 8-in. (203-mm) nominal pipe size, schedule 40, is 0.322 in. (8.2 mm).

When iron pipes were initially standardized, there was only one standard wall thickness for each nominal size of pipe. To satisfy the need for stronger and thicker wall pipes to transport fluids under higher operating pressure, larger uniform thicknesses were designed as extra strong (XS) and double extra strong (XXS), and original thicknesses were retained as standard (STD). These designations have been replaced or supplemented by schedule numbers. Schedule numbers begin with 5 and 5S, followed by 10 and 10S, then progress in increments of 10 through 40, and then in increments of 20 through 160.

Pipe made of special alloys for high-pressure and high-temperature service is very costly and, therefore, is specified by inside diameter and minimum wall thickness rather than nominal pipe size and schedule. Such pipes usually have a nonstandard outside diameter. The special pipe sizes are usually ordered when the material cost savings offset the additional cost of making a special setup for production.

Tube size is generally defined by outside diameter and wall thickness. However, it can be specified by outside diameter and inside diameter, or inside diameter and wall thickness.

Fittings and joints. Fittings, such as elbows, tees, crosses, reducers, returns, flanges, and couplings, are used to join different pipe segments; provide changes in direction or size; install branch lines and instrument connections; and connect pipes with in-line equipment.

Pipe sections, fittings, and in-line equipment are usually joined by welded joints, soldered or brazed joints, flanged joints, threaded joints, bell-and-spigot joints, grooved joints, coupling joints, caulked joints, and other mechanical means. Welded, soldered, or brazed joints are used where maintenance or repair is expected to be minimal. Mechanical joints, such as flanged, threaded, grooved, or coupling joints, offer ease of assembly and disassembly to permit maintenance and repair.

Materials. Metallic pipe and fittings made from cast iron, ductile iron, carbon steel, stainless steel, copper, brass, aluminum, and other metals and alloys are generally used for construction of pipelines or piping systems. Nonmetallic pipes, especially thermoplastic and thermosetting pipes, are excellent for handling corrosive fluids, chemicals, brackish water, and so forth. They are economical but are more susceptible to damage. Vitrified or unglazed clay pipes are used in drainage systems and to convey sewage,

Fig. 1. Pipelines. (*a*) Construction of 300-mi, 24-in. (500-km, 60-cm) pipeline in Colombia that runs over the Andes to the Caribbean coast. (*b*) Construction of 30-in. (75-cm) gas pipeline in Pennsylvania in 1982. (c) Trans-Arabian oil pipeline, built in 1951 to link the Persian Gulf and the Mediterranean Sea. (*Bechtel Group, Inc.*)

industrial waste, and storm water. Reinforced or unreinforced concrete pipe, made from cement, sand, gravel, and water, is used for conveying water, including storm water and cooling water. Because it has good resistance to chemical attack, glass pipe is used for laboratory drainage, chemicals, beverages, food, and pharmaceuticals. Metallic pipes lined with cement-mortar, rubber, epoxy, ceramic, or other materials are used to transport corrosive and abrasive materials. The lining provides protection for the metallic pipe against the corrosive, erosive, and abrasive effects of the flow medium. Also, the linings can be repaired or replaced.

Design. Safety is the primary objective in pipeline design. The codes, standards, or specifications that have jurisdiction over a pipeline or piping system depend on the site of installation, type of service, design considerations, and other requirements. For example, in the United States the water treatment and distribution piping systems in cities and towns are designed and constructed to the standards and specifications published by the American Water Works Association (AWWA), whereas the water-carrying piping systems at a pressure greater than 30 lb/in.^{2} (207 kilopascals) within the boundary of a coal-fired power plant must meet the design requirements of the American Society of Mechanical Engineers Pressure Piping Code. However, the water pipelines for the fire protection systems within the same power plant must be designed in accordance with the requirements of the applicable National Fire Protection Association (NFPA) standards.

Material selection. The materials selected for the pipe, fittings, and in-line equipment must be suitable for the application requirements, including pressure, temperature, and chemical properties of the conveyed fluid, and the environmental conditions. Furthermore, the materials must perform the service safely for the design life of the piping system.

Sizing. Depending upon the required flow rate and the velocity of flow, the cross-sectional area of the pipe is calculated from Eq. (1*a*) in British engineering units and Eq. (1*b*) in metric units. The standard

$$\text{Flow area (ft}^2) = \frac{\text{flow rate (ft}^3/\text{s)}}{\text{mean velocity of flow (ft/s)}} \quad (1a)$$

$$\text{Flow area (m}^2) = \frac{\text{flow rate (m}^3/\text{s)}}{\text{mean velocity of flow (m/s)}} \quad (1b)$$

pipe size that will give the required flow area is selected. **Figure 2** shows the ranges of velocities generally used. However, the actual velocity to be used for a pipeline may differ because of specific design requirements.

The minimum pipe wall thickness required to withstand the sustained fluid pressure can be calculated by Barlow's formula (2), where t_m is the min-

$$t_m = \frac{PD_o}{2S} \quad (2)$$

imum wall thickness in inches; P is the fluid pressure in pounds per square inch; D_o is the outside pipe diameter in inches; and S is the allowable stress for the pipe material at the fluid temperature in pounds per square inch. (Metric or other units may also be used in the equation, so long as t_m and D_o are in the same units, and P and S are also in the same units.)

Most piping codes contain a modified version of Barlow's formula or other guidelines that must be complied with to determine the minimum wall thickness. The calculated minimum wall thickness must be increased to provide for corrosion or erosion; threading or grooving; thinning caused by bending; mechanical strength to prevent damage, collapse, or buckling, of pipe due to external loads; and manufacturing process tolerance, whenever these factors are applicable.

Pressure drop. The flow in a pipe can be laminar, turbulent, or a combination of laminar and turbulent, depending upon the value of a dimensionless number known as the Reynolds number. This quantity can be determined by Eq. (3), where D is the

$$\text{Re} = \frac{Dv\rho}{\mu} \quad (3)$$

internal diameter of the pipe in feet (meters); ρ is the weight density of fluid in pounds per cubic foot (kilograms per cubic meter); v is the mean velocity of flow in feet per second (meters per second); and μ is the absolute viscosity in pound-mass per foot-second or poundal seconds per square foot (kilograms per meter-second or pascal seconds). Usually the flow in a pipe is laminar if the Reynolds number is less than 2000, and turbulent if the Reynolds number is greater than 4000. Flows with Reynolds numbers between 2000 and 4000 are in a transition zone. *See* FLUID FLOW; LAMINAR FLOW; PIPE FLOW; REYNOLDS NUMBER; TURBULENT FLOW.

The pressure drop due to friction can be determined by the Darcy equation (4) for laminar or tur-

$$h_L = \frac{fL}{D}\frac{v^2}{2g} \quad (4)$$

bulent flow of any liquid in a pipe. Here h_L is the pressure drop due to friction, in feet (meters) of liquid; L is the length of pipe, in feet (meters); g is the acceleration of gravity, 32.2 ft/s^2 (9.8 m/s^2); and f is the friction factor, a dimensionless quantity.

The factor fL/D is referred to as the resistance coefficient, and can be determined for straight runs of pipe and various in-line fittings, such as elbows, reducers, and valves. Summation of individual pressure losses through straight runs of pipe, fittings, and other components provides the pressure drop between any two locations in a pipeline due to friction. In calculating pressure losses between two points in a pipeline, the increase or decrease in elevation must be accounted for as loss or gain in feet (meters) of liquid. The entrance and exit pressure losses, as applicable, must be included to determine the total pressure drop.

The friction factor f for laminar flows, that is, when

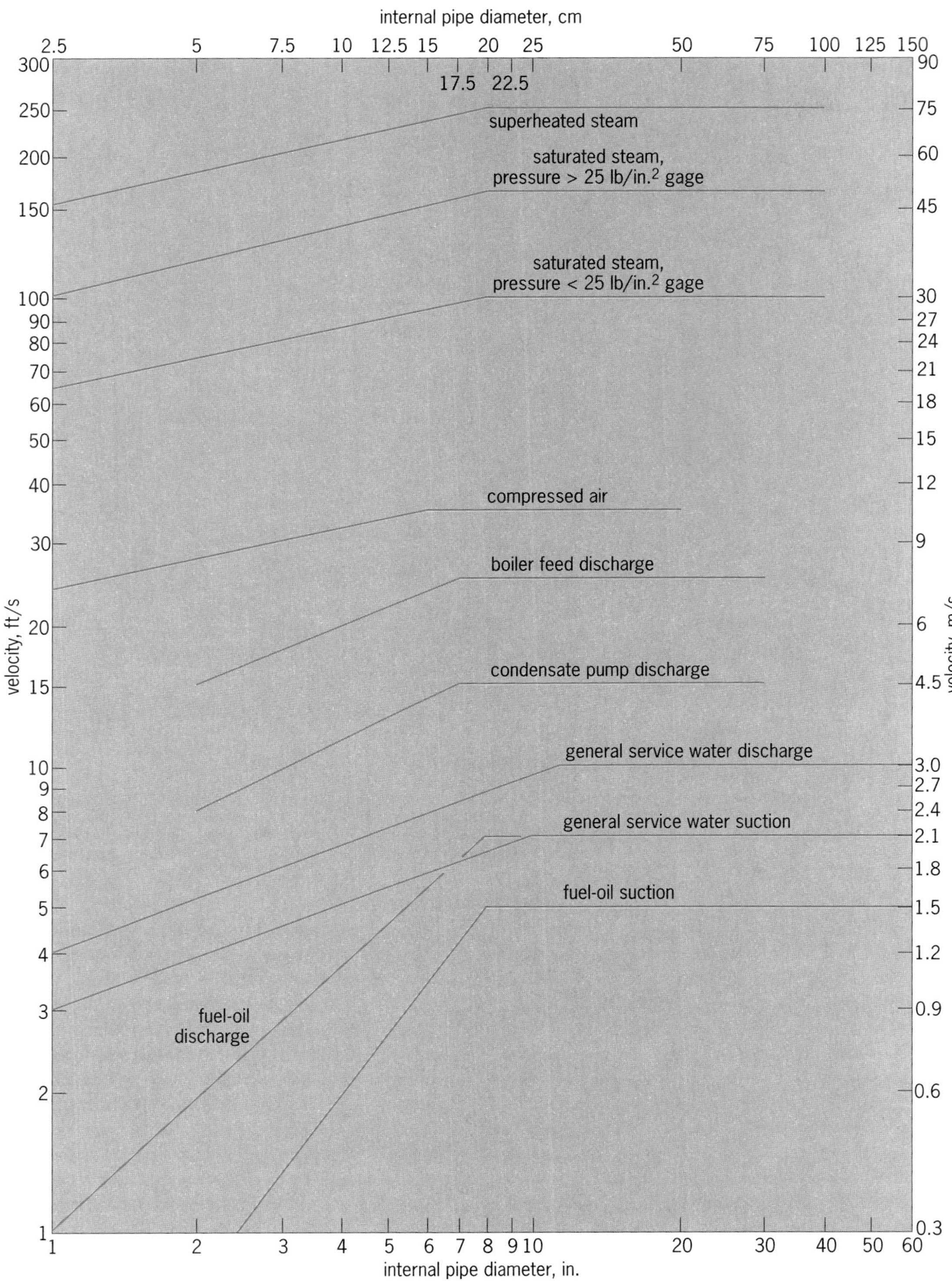

Fig. 2. Velocity of pipe flow as function of internal pipe diameter for various applications. 25 lbf/in.^{2}, gage pressure = 172 kPa above atmospheric pressure.

the Reynolds number is less than 2000, can be calculated from Eq. (5). For turbulent flows, the friction

$$f = \frac{64}{\text{Re}} \tag{5}$$

factor f is dependent upon Reynolds number and the relative roughness of the pipe. **Figure 3** can be used to determine the friction factor from the Reynolds number and the size of the pipe.

Pumping requirements. Pumps are utilized to cause flow against gravity; move liquids from one location to another; overcome pressure losses due to

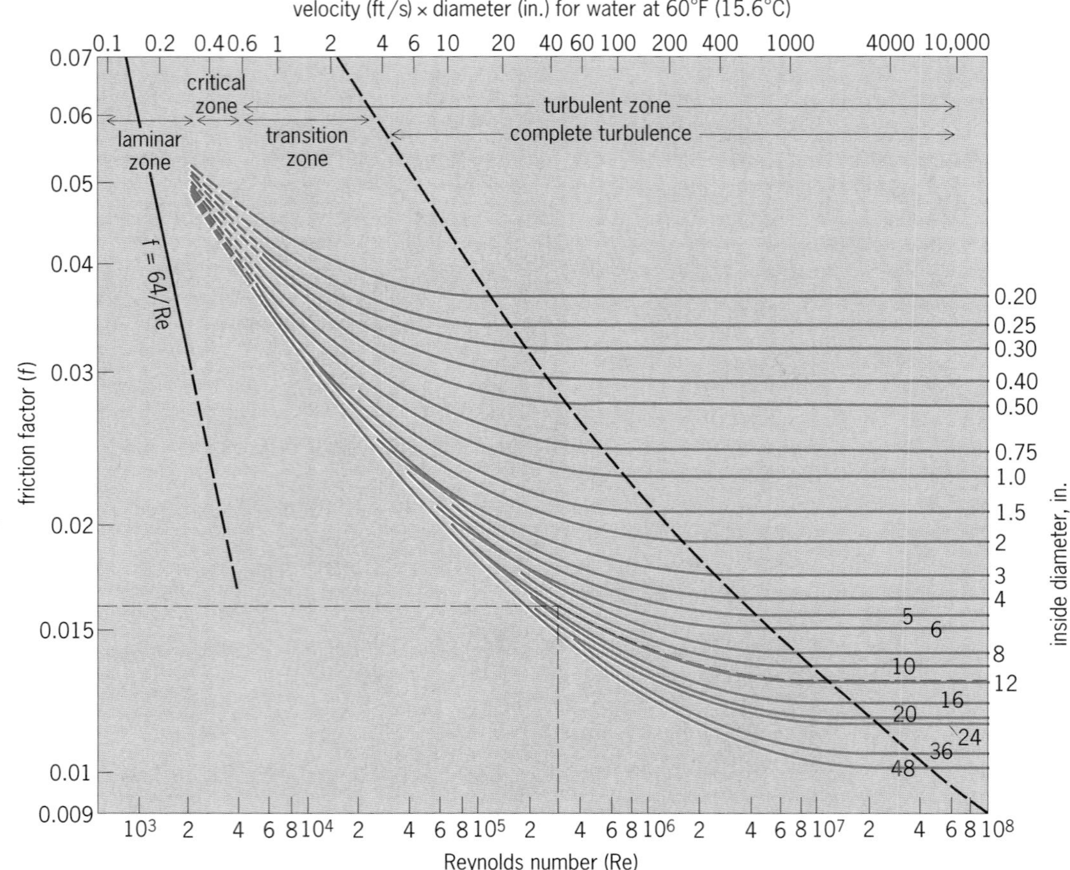

Fig. 3. Friction factor as function of Reynolds number for clean commercial steel pipes with various inside diameters. For example, pipe with inside diameter of 12 in. (305 mm) and flow having Reynolds number of 300,000 has friction factor of 0.016. Upper scale gives product of flow velocity and inside diameter that corresponds to Reynolds number for water at 60°F (15.6°C). 1 in. = 25.4 mm. Velocity in meters per second × diameter in centimeters = 0.77 × velocity in feet per second × diameter in inches.

friction, changes in elevation, and other obstructions; and supply liquid at a desired pressure. The energy imparted by a pump is usually expressed in terms of head, in feet (or meters) of liquid. Many different types of pumps (centrifugal, reciprocating, and rotary) are available to add energy to liquids. Most commonly used is the centrifugal pump. A simple centrifugal pump consists of an impeller in a volute casing with an opening in the center for liquid to enter and a peripheral opening for liquid to exit. The impeller is rotated at high speed by an electric motor or turbine. The impeller vanes impart high velocity to the entering liquid, which is then discharged by the centrifugal force. *See* CENTRIFUGAL PUMP; PUMP.

When the liquid level in the source of supply is below the centerline or impeller eye of the pump, then the pump has to lift the liquid and overcome the friction and entrance losses from the point of supply to the centerline of the pump. This duty of the pump is termed suction lift. When the liquid supply level is above the pump centerline, then the suction head is equal to the static head above the pump centerline minus the suction-line losses.

The total head of a pump must be sufficient to overcome suction lift, friction losses in the discharge pipeline, and pressure losses due to other causes and equipment; and to supply liquid at the required pressure at the point of discharge. The hydraulic power required to develop the total head depends upon the flow rate and the total head. Since no pump is 100% efficient, the brake power of a pump must be greater than the hydraulic power; the brake power is the hydraulic power divided by the pump efficiency. *See* HYDRAULICS.

Other considerations. In designing pipelines so that pressure integrity is maintained, consideration must be given to weight effects, thermal expansion and contraction loads, relief and safety-valve thrust, vibrations, earthquake effects, and impact loads.

Construction. Subassemblies or pipe spools of large piping are generally fabricated in the shop and joined together in proper sequence in the field (Fig. 1*a* and *b*). Smaller piping is usually routed, fabricated, and erected in the field. Quality control is maintained by step-by-step implementation of preapproved fabrication, installation, and examination procedures. Piping joints may require examination by nondestructive methods.

Testing and inspection. After erection, pipelines are cleaned by flushing with water, steam, air, or water mixed with chemical cleaning agents. The pressure integrity of pipeline is then tested by hydrostatic or pneumatic testing. Leaks, if any, are repaired and the pipeline is retested. Mohinder L. Nayyar

Bibliography. *Flow of Fluids*, Crane Tech. Pap. 410, 1988; Ingersoll-Rand Company, *Cameron Hydraulic Data*, 17th ed., 1992; K. K. Kienow (ed.), *Pipeline Design and Installation*, 1990; M. L. Nayyar (ed.), *The Piping Handbook*, 7th ed., 1999; M. P. Sharma and U. S. Rohatgi (eds.), *Multiphase Flow in Wells and Pipelines*, 1992; J. P. Tullis, *Hydraulics of Pipelines*, 1989.

Piperales

A small order of flowering plants (3600 species) in the eumagnoliid group, which is composed of three anomalously woody vines (shrubs) or herbaceous families—the pipeworts (Aristolochiaceae), the black pepper family (Piperaceae), and the lizard's tail family (Saururaceae). The last two families have reduced flowers in dense spikelike flower stems, and the first has medium-sized to enormous flowers that often trap insects for a period before releasing them, covered with pollen.

Black pepper comes from *Piper nigrum* and betel nuts from *P. betle*. Several species of *Aristolochia* have medicinal properties, and some genera in each of these families are commonly grown ornamentals in the temperate zones or house plants, such as *Asarum* (wild ginger), *Peperomia* (pepper elders), and *Houttuynia*. *See* EUMAGNOLIIDS; LAURALES; MAGNOLIALES; MONOCOTYLEDONS. Mark Chase

Pirani gage

A type of instrument used to measure vacuum by utilizing a resistance change due to a temperature change in a filament. This fine-wire filament, one of the four electrical resistances forming a Wheatstone bridge circuit, is exposed to the vacuum to be measured. Electric current heats the wire; the surrounding gas (in the vacuum) conducts heat away from the wire. At a stable vacuum, the wire quickly reaches equilibrium temperature. If the pressure rises, the gas carries away more heat, and the temperature of the wire decreases (see **illus.**). Since the resistance of the filament is a function of temperature, the electrical balance of the Wheatstone bridge is changed. The output meter is usually a microammeter calibrated in torrs or millitorrs (1 torr = 10^2 pascals).

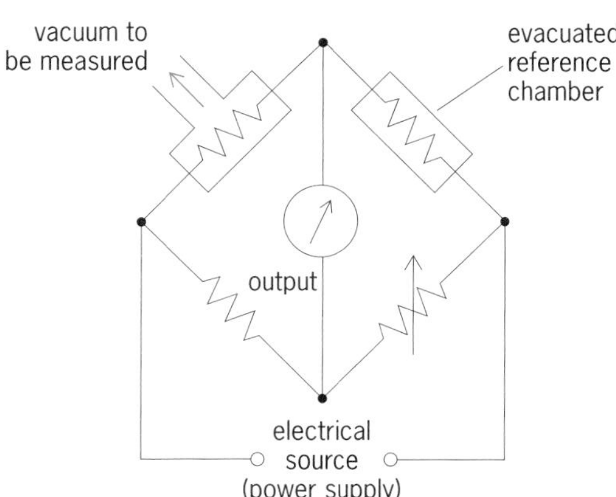

Diagram of a Pirani gage.

The calibration depends upon the thermal conductivity of the gas, and therefore the readings obtained must be adjusted for various gases. Accuracy is of the order of ±5% of scale. Pressure measurement range of this type of gage is usually 1 to 10^{-4} torr. *See* VACUUM MEASUREMENT. Richard Comeau

Bibliography. T. A. Delchar, *Vacuum Physics and Technology*, 1993; J. F. O'Hanlon, *A User's Guide to Vacuum Technology*, 2d ed., 1989; D. Hucknall, *Theory and Practice of Vacuum Technology*, 1989; A. Roth, *Vacuum Technology*, 3d ed., 1990, reprint 1998.

Pisces (constellation)

The Fishes, in astronomy, a zodiacal constellation appearing in the autumn evening sky. Pisces is the twelfth and last sign of the zodiac. It is inconspicuous, having no star brighter than the fourth magnitude. But it is an important constellation because the vernal equinox, which marks the beginning of the astronomical year, is now located in it. Its most distinctive feature is a V-shaped figure, with the fishes' tails toward the point of the V tied together by a ribbon (see **illus.**). The northern fish is poorly defined, but

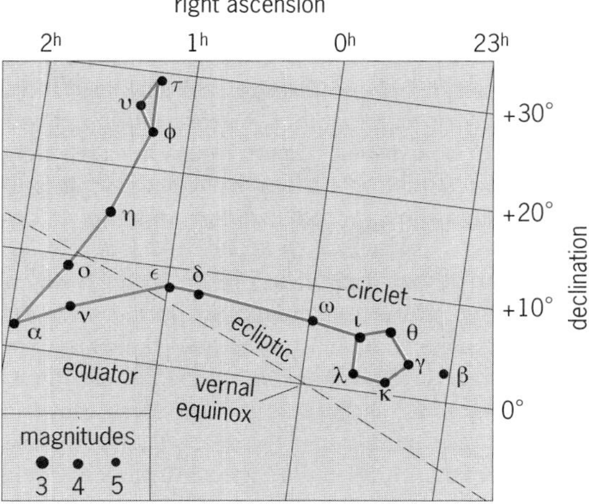

Line pattern in the constellation Pisces. The grid lines represent the coordinates of the sky. The apparent brightness, or magnitude, of the various stars is shown by the sizes of the dots, which are graded by appropriate numbers as indicated.

the western one is marked with a group of stars forming an irregular pentagon, known as the Circlet in Pisces. *See* CONSTELLATION; ZODIAC. Ching-Sung Yu

Pisces (zoology)

A term that embraces all fishes and fishlike vertebrates. In early zoological classifications fishes, like mammals, birds, reptiles, and amphibians, were ranked as a class of the vertebrates. As knowledge of fishes increased, it became apparent that, despite their common possession of gills and fins and their dependence on an aquatic environment, not all fishes were closely related. At least five groups of fishes with modern descendants were already established before the tetrapods appeared. Not only are these groups older, but some are decidedly more divergent structurally than are the four classes of tetrapods. For these reasons several classes of fishes are now recognized. The number of classes varies; one reputable but extreme classification recognizes 11 classes of fishes.

The primary cleavage in vertebrate classification is that separating the jawless fishes, or Agnatha, from those vertebrates with jaws, the Gnathostomata. After recognition of this split, the name Pisces was commonly restricted to the jawed fishes. When these in turn were divided into two or more classes, Pisces was further restricted by some authorities to the bony fishes. Another scheme involves assignment of class names to each of the major constituent groups of jawed fishes, and use of Pisces as a superclass name. In view of the confusion, it seems best to revert to early practice and to employ Pisces as a group name of convenience to embrace all classes of fishlike vertebrates, from jawless fishes to bony fishes. In this sense it has no actual taxonomic status because it cuts across natural classification, dividing the gnathostomes and grouping part of them with the agnaths. *See* GNATHOSTOMATA; JAWLESS VERTEBRATES.

The Pisces include four well-defined groups that merit recognition as classes: the Agnatha or jawless fishes, the most primitive; the Placodermi or armored fishes, known only as Paleozoic fossils; the Chondrichthyes or cartilaginous fishes; and the Osteichthyes or bony fishes. Furture research may demonstrate the need for further division, but this is most likely to involve Paleozoic groups. *See* CHONDRICHTHYES; OSTEICHTHYES; PLACODERMI.

Number of recent species. Present fish classification is not sufficiently precise to permit an accurate tabulation of the number of living species. New kinds are constantly being discovered, others are being synonymized as the result of new research, and the literature is scattered. Nevertheless, estimates by competent ichthyologists are so diverse, ranging from 18,000 to 40,000, that an effort is here made to arrive at a reasonably acceptable approximation. There is indication that most previous estimates are too high. Counts for Recent groups and species include Agnatha, 2 families, about 11 genera, and approximately 45 species; Chondrichthyes, 31 families, some 132 genera, and roughly 575 species; and Osteichthyes, 32 orders, about 357 families, 3570 genera, and about 17,600 species. This last figure broken down gives about 5100 species in 29 orders, nearly 2000 species in the Siluriformes, 3000 species in the Cypriniformes, and 7500 species in the Perciformes. It is to be emphasized that the classification of many groups, especially those in the deep seas and in the tropics, is imperfectly known. Thus, further study will permit refinement of this enumeration of just under 18,300 recognized species of Recent fishes.

Ecology. Fishes live in almost all permanent waters to which they have been able to gain access. In general they have evolved a body conformation and specialized features that adapt them harmoniously to the world about them. Inhabitants of mountain torrents may have peculiar attachment organs; those living in Antarctic waters at a temperature below freezing have made needed physiological adjustments; fishes of the deep sea commonly carry their own light source, and the female anglerfish is assured a mate by the parasitism of the male on her body. In the East Indies some fishes skip with ease over mud flats, and others undertake nightly forays on land. Some fishes mature at extremely small size. A Philippine goby, *Mistichthys*, reaches a length of only 0.5 m (1.25 cm) and is commercially important as a food fish although it takes 31,500 fish to make 1 lb (70,000 fish to make 1 kg); and a Samoan fish, *Schindleria*, attains a weight of only 6 mg. At the other extreme, the whale shark is reputed to reach a length of 60 ft (20 m), and a 38-ft (13 m) individual weighed more than 13 tons (11.7 metric tons).

Adaptive radiation. Because their bodies are supported by water, fishes have been afforded the luxury of diversification in body form not possible for terrestrial animals. A deep, pancake-thin body is not uncommon, and an eellike form has been independently developed in many phyletic lines. Trunkfishes are enclosed in a boxlike casque, and some deep-sea fishes have eyes at the tips of elongate stalks. Long, trailing fins are frequent; sargassum fishes develop appendages that serve as holdfasts and for concealment.

Food habits. Most fishes are more or less carnivorous and predatory, but there is wide diversity in food habits. Many fishes have numerous slender gill rakers with which they strain microorganisms from the water; others have massive teeth and strongly muscled jaws to aid in crushing mollusks or crustaceans. Browsers, nibblers, and grazers employ specially adapted teeth and jaws to scrape vegetation or small attached animals. Some wrasses pluck parasites from larger fishes, and lampreys parasitize other fishes.

Reproductive habits. Reproductive habits are no less varied than feeding behavior. Most fishes are oviparous and scatter their eggs, but nest building and parental care assume a broad spectrum—from a prepared pile of pebbles, through a grassy spherical retreat, to oral incubation or development of a marsupial pouch on the underside of the male pipefish. Viviparity and ovoviviparity have originated along independent lines. Enormously complicated modifications of the anal fin have been evolved to effect insemination of some species in which the young are born alive.

Economics. Fishes play an important role in the lives of most people. Fishing is a way of life in most primitive cultures, and ranks high among recreational activities in highly civilized peoples. Maintenance and care of home aquarium fishes provide an avocation for probably millions of people. An occasional swimmer is killed by sharks; more people die from ciguatera contracted from eating poisonous fish flesh; and venomous fishes take a limited toll in human life and suffering. *See* CHORDATA; VERTEBRATA. Reeve M. Bailey

Pistachio

A tree, *Pistacia vera*, of the Anacardiaceae family. It is native to central Asia and has been grown for its edible nuts throughout recorded history in Iran, Afghanistan, and Turkey and various other countries of the Mediterranean region. Extensive areas in the Sacramento and San Joaquin valleys of California were planted to pistachios in the 1970s, and the first commercial nut crop was harvested in 1977. *See* SAPINDALES.

The pistachio tree, relatively slow-growing, reaches a height and spread of 20–25 ft (6–8 m). It thrives under long, hot summers with low humidity, but needs moderately cold winters to satisfy its chilling requirement. Pistachio is deciduous and has imparipinnate leaves, most often 2-paired. It is dioecious, and both staminate and pistillate inflorescences are panicles that may have 150 or more individual flowers. They lack petals and nectaries and, consequently, are wind-pollinated. One male tree is provided in commercial orchards for every 8–12 female trees. Since the pistachio cannot be propagated from cuttings, seedling rootstocks of *P. atlantica*, *P. terbinthus*, and *P. integerrima* are T-budded with the desired cultivar. These stocks are more resistant to nematodes and other soil-borne organisms than is *P. vera*.

The fruit, a semidry drupe, is borne on 1-year-old wood in clusters similar to grapes and matures in September. The hull (exocarp and mesocarp) at that time slips easily from the shell (endocarp) which has already dehisced, exposing the kernel (see **illus.**).

Hull of the pistachio fruit at maturity separates easily from the dehisced shell that exposes the kernel.

Harvesting in California is done mechanically with tree shakers and catching frames. Mechanical hull removal, washing, and dehydration of the nuts soon after harvesting ensures unblemished, ivorylike shells, eliminating the necessity of dyeing them red, as is commonly done with imported nuts. A yield of 50 lb (22.5 kg) of dried-in-shell nuts per mature tree may be expected in a good cropping year. The tree, however, is noted for its alternate bearing habit, producing a good crop one year followed by little or no crop the next.

Pistachio kernels contain only 5–10% sugars, but their protein and oil content of about 20 and 40%, respectively, make them high in food value. *See* NUT CROP CULTURE. Julian C. Crane

Bibliography. J. C. Crane and B. T. Iwakiri, Morphology and reproduction of pistachio, *Hort. Rev.*, 3:376–393, 1981; R. A. Jaynes (ed.), *Nut Tree Culture in North America*, 1979; J. Maranto and J. C. Crane, *Pistachio Production*, Div. Agr. Sci. Univ. Calif. Leaflet 2279, 1982; W. E. Whitehouse, The pistachio nut: A new crop for the Western United States, *Econ. Bot.*, 11(4):281–321, 1957; J. G. Woodroof, *Tree Nuts*, 2d ed., 1979.

Pitch

The psychological property of sound characterized by highness or lowness. Pitch is one of the two major auditory attributes of simple sounds, the other being loudness.

Simple and complex sounds. A simple sound source, such as a tuning fork, produces an acoustic wave that approximates a perfect sinusoid, and the pitch of a sinusoid wave is almost completely determined by its frequency. Many sounds, however, are complex and contain a number of sinusoidal components. Complex sounds often appear to have a strong pitch, which is the frequency of a sinusoid that appears to match the complex sound. Hence, a tuning fork that vibrates at about 440 Hz will have a pitch very nearly equal to the note A above middle C on the piano. Loudness is determined by the amplitude of the sound vibrations. Loudness and pitch are independent attributes of sound quality. *See* LOUDNESS; TUNING FORK; WAVE (PHYSICS).

Sound sequences. A sequence of different sounds having definite pitches produces a musical tune, making pitch extremely important in music. Practically all musical conventions recognize that doubling the frequency of vibration produces a particular pitch interval, known as an octave. A conventional piano, for example, spans about seven octaves. Western music is based on the 12-tone scale. The frequencies representing the successive tones are separated by roughly equal steps on a logarithmic scale of frequency. The interval between successive notes has a frequency ratio of about $2^{1/12}$ and is called a semitone. Certain musical intervals, since they approximate simple fractions, are given special names. For example, there are seven intervals between C and G; the ratio of the frequencies of vibration between those two notes is nearly 3:2 and is called a perfect fifth. *See* MUSICAL ACOUSTICS; SCALE (MUSIC).

Human hearing. The human auditory system can hear frequencies in the range of 20–20,000 Hz. A hearing test consists of determining individual sensitivity to the frequencies 250, 500, 1000, 2000, 4000, and 8000 Hz. For frequencies between 100 and 4000 Hz, sinusoidal sounds have a clear pitch. Beyond these limits, the pitch of sound is not distinct. Sounds below 100 Hz may be described as rumbles, while those above 4000 Hz may be described as shrill and squeaky. Tunes can be recognized, however, when played in the octave between 6000 and 12,000 Hz. The ability to detect changes in pitch is remarkably acute. The just-detectable change in frequency is about 0.3% for the midfrequency range. Frequency changes are best detected when the sound is loud. Weaker sounds require greater changes in frequency to be detectable. *See* AUDIOMETRY; HEARING (HUMAN).

Mathematical description. Most musical tunes are not produced by simple acoustic sources but by string or reed instruments. Such instruments produce complex sounds—acoustic waveforms consisting of a number of sinusoid vibrations. Many musical sounds are periodic: the pressure waveform repeats itself after some fixed interval of time (period of the wave). According to Fourier's theorem, a periodic signal, such as a periodic sound, is a set of sinusoidal components with frequencies that can be expressed as successive integers. For a sound wave that repeats itself every 1/200 s, the fundamental of the wave is 200 Hz, the reciprocal of the wave's period. Its spectrum may have energy not only at the fundamental frequency but also at the harmonics—successive integral multiples of the fundamental. The pitch of such a periodic sound is generally considered to be equal to the fundamental, or 200 Hz. *See* FOURIER SERIES AND TRANSFORMS.

The missing fundamental. Even though energy at the fundamental as well as many of the lower frequencies might be absent, the pitch of the complex sound is still 200 Hz. This phenomenon is known as the missing fundamental. The fundamental of the human voice is about 100 Hz for males and 200 Hz for females. Since the conventional telephone passes energy only above about 300 Hz, the pitch of a human voice heard on the telephone is an example of the missing fundamental. During the middle of the twentieth century, J. Schouten conducted a number of important experiments on the pitch of the missing fundamental and other such periodic waveforms. His research has stimulated a major reevaluation of how the auditory mechanism is presumed to mediate the perception of pitch. *See* PSYCHOACOUSTICS; SPEECH.

David M. Green

Bibliography. D. M. Green, *Introduction to Hearing*, 1976; W. A. Yost and D. W. Nielsen, *Fundamentals of Hearing: An Introduction*, 4th ed., 2000; W. A. Yost and C. S. Watson (eds.), *Auditory Processing of Complex Sounds*, 1987.

Pitcher plant

Any member of the families Sarraceniaceae and Nepenthaceae. In these insectivorous plants the leaves form deep cups or pitchers in which water collects (**Fig. 1**). Visiting insects, falling into this water, are drowned and digested by the action of enzymes secreted by cells located in the walls of the pitcherlike structure of these plants. The Sarraceniaceae are divided into three genera; *Sarracenia* in eastern North America, *Darlingtonia* in northern California and southern Oregon, and *Heliamphora* endemic

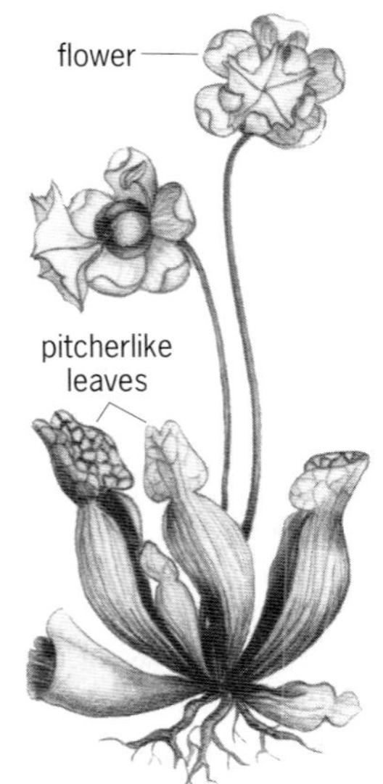

Fig. 1. *Sarracenia purpurea*, a pitcher plant.

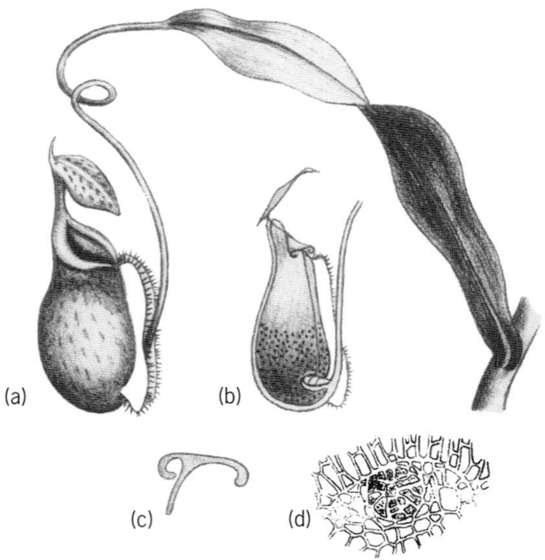

Fig. 2. *Nepenthes*. (*a*) The complete leaf with its pitcher. (*b*) A vertical section through a pitcher. (*c*) A section through the margin of a pitcher. (*d*) Single gland from the lower part of a pitcher

on high mountains in the northern part of South America. The Nepenthes family has only one genus, *Nepenthes*, which occurs in the Old World tropics from China to Australia, chiefly in Borneo. Often these plants climb by tendrils (prolongations of the midrib of the leaf). The end of a tendril may develop into a pitcher, which captures and digests insects (**Fig. 2**). *See* INSECTIVOROUS PLANTS; NEPENTHALES.

Perry D. Strausbaugh/Earl L. Core

Pitchstone

A natural glass with dull or pitchy luster and generally brown, green, or gray color. It is extremely rich in microscopic, embryonic crystal growths (crystallites) which may cause its dull appearance. The water content of pitchstone is high and generally ranges from 4 to 10% by weight. Only a small proportion of this is primary; most is believed to have been absorbed from the surrounding regions after the glass developed. Pitchstone is formed by rapid cooling of molten rock material (lava or magma) and occurs most commonly as small dikes or as marginal portions of larger dikes. *See* IGNEOUS ROCKS; VOLCANIC GLASS.

Carleton A. Chapman

Pith

The central zone of tissue of an axis in which the vascular tissue is arranged as a hollow cylinder. Pith is present in most stems and in some roots (**Fig. 1**). Stems without pith rarely occur in angiosperms but

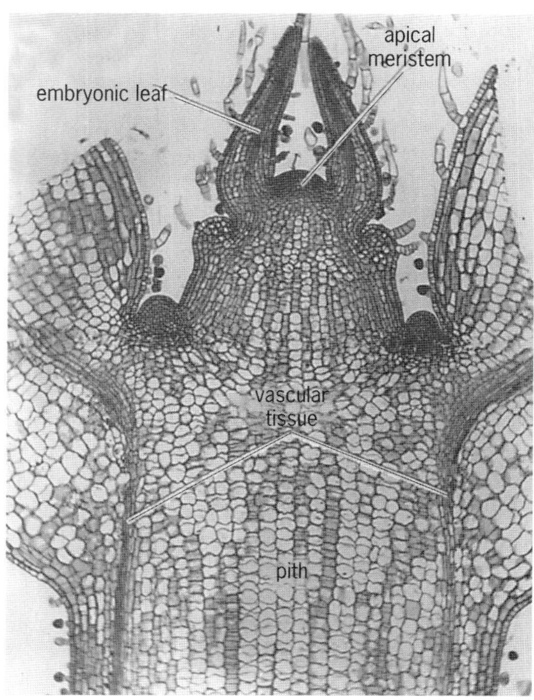

Fig. 1. **Shoot apex of *Coleus blumei*, longitudinal section. (*After R. A. Popham, Laboratory Manual for Plant Anatomy, C. V. Mosby, 1966*)**

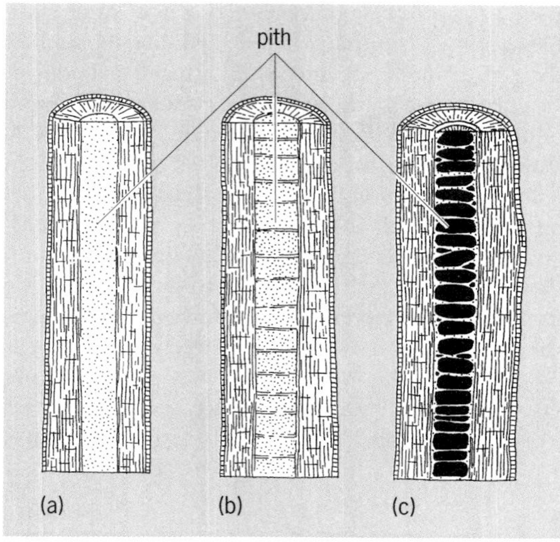

Fig. 2. **Various types of pith in branchlets. (*a*) Continuous. (*b*) Diaphragmed. (c) Chambered. (*After E. L. Core, Plant Taxonomy, Prentice-Hall, 1955*)**

are characteristic of psilopsids, lycopsids, *Sphenophyllum*, and some ferns. Roots of some ferns, many monocotyledons, and some dicotyledons include a pith although most roots have xylem tissue in the center. The pith may be present or absent in the same axis, depending upon size or vigor, the larger segments commonly containing pith.

Pith is composed usually of parenchyma cells often arranged in longitudinal files. This arrangement results from predominantly transverse division of pith mother cells near the apical meristem. The peripheral region consists of small cells with thick walls and remains alive longer than the central region. When the peripheral region is fairly well defined, it is called the medullary sheath or the perimedullary zone. The walls of pith cells may thicken in age and may become hard or remain soft. In some axes the pith may be composed principally of sclereids. In many fern stems, the inner pith is sclerenchymatous, whereas the outer is parenchymatous. In stems of some dicotyledons, plates or nests of sclerenchyma may be interspersed with the parenchyma. Such a pith is called diaphragmed pith (**Fig. 2**). If the parenchyma collapses or is torn during development, the sclerenchyma plates (diaphragms) alternate with hollow zones. Such a pith is said to be chambered. In many stems the entire pith becomes hollow except at the nodes; the nodal diaphragms are sclerenchyma or parenchyma and, in monocotyledons, may contain vascular bundles.

Shape. The shape of the pith in stems of lower vascular plants and in roots of various plants is nearly cylindrical. In stems of higher vascular plants the pith is more or less angled or stellate in cross section. The shape is often characteristic of the plant groups, since it depends on phyllotaxy. In oaks, for example, the pith is five-angled and in alder, three-angled. In stems with cylinders of vascular bundles, the panels of ground tissue between bundles often are called medullary rays or pith rays. In stems in which the vascular bundles occur in a more complex arrangement than a simple cylinder, the limit of the pith is indefinite and when a major cylinder of vascular bundles can be distinguished, the internal bundles are called medullary bundles.

Content. Ergastic materials often are stored in some or all cells of the pith. Secretory cells, or secretory canals, or laticifers may be present. In most stems with considerable secondary growth, the pith dies with the formation of heartwood, although the perimedullary zone may remain alive. In other stems the pith may consist partly or largely of dead cells by the end of the first year. *See* LEAF; PARENCHYMA; PERICYCLE; POLYPODIALES; ROOT (BOTANY); SCLERENCHYMA; SECRETORY STRUCTURES (PLANT); STEM.

H. Weston Blaser

Pitot tube

A device to measure the stagnation pressure due to isentropic deceleration of a flowing fluid. In its original form it was a glass tube bent at 90° and inserted in a stream flow, with its opening pointed upstream (**illus.** *a*). Water rises in the tube a distance, h, above the surface, and if friction losses are negligible, the velocity of the stream, V, is approximately $\sqrt{2gh}$, where g is the acceleration of gravity. However, there is a significant measurement error if the probe is misaligned at an angle α with respect to the stream (illus. *a*). For an open tube, the error is about 5% at $\alpha \approx 10°$.

Shielded probes. The misalignment error of a pitot tube is greatly reduced if the probe is shielded, as in the Kiel-type probe (illus. *b*). The Kiel probe is accurate up to $\alpha \approx 45°$.

Pitot-static probes. In most laboratory flows, there is no free surface, or the fluid is a gas; hence the visual water-rise effect of Pitot's invention is not possible. The modern application is a pitot-static probe, which measures both the stagnation pressure, with a hole in the front, and the static pressure in the moving stream, with holes on the sides. A pressure transducer or manometer records the difference between these two pressures. Pitot-static tubes are generally unshielded and must be carefully aligned with the flow to carry out accurate measurements. *See* BERNOULLI'S THEOREM.

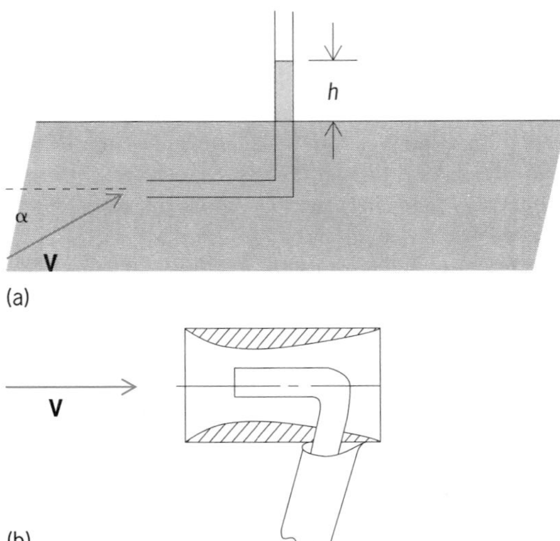

Pitot tubes. (*a*) Original form. Symbols explained in text. (*b*) Modern shielded Kiel probe.

Compressibility effects. When used with gases, estimate of the stream velocity is only valid for a low-speed or nearly incompressible flow, where the stream velocity is less than about 30% of the speed of sound of the fluid. For sea-level air, the stream velocity must be less than 100 m/s (225 mi/h). At higher velocities, estimate of the stream velocity must be replaced with a Bernoulli-type theory, which accounts for gas density and temperature changes. If the gas stream flow is supersonic, or the stream velocity is greater than the speed of sound of the gas, a shock wave forms in front of the probe and the measured static pressure and stagnation pressure are much different from the stream values upstream of the shock. The theory must then be further corrected by complicated supersonic-flow algebraic relations. *See* COMPRESSIBLE FLOW; GAS DYNAMICS; SHOCK WAVE.

A disadvantage of pitot and pitot-static tubes is that they have substantial dynamic resistance to changing conditions and thus cannot accurately measure unsteady, accelerating, or fluctuating flows. Other fast-response devices are used for highly unsteady flows. *See* ANEMOMETER; FLOW MEASUREMENT.

Frank M. White

Bibliography. T. G. Beckwith and R. D. Marangoni, *Mechanical Measurements*, 5th ed., 1993; J. W. Dally, W. F. Riley, and K. G. McConnell, *Instrumentation for Engineering Measurements*, 2d ed., 1993; R. J. Goldstein, *Fluid Mechanics Measurements*, 1983; R. W. Miller, *Flow Measurement Engineering Handbook*, 3d ed., 1996.

Pituitary gland

The most structurally and functionally complex organ of the endocrine system. Through its hormones, the pituitary, also known as the hypophysis, affects every physiological process of the body. All vertebrates, beginning with the cyclostomes, have a pituitary gland with a common basic structure and function. In addition to its endocrine functions, the pituitary may play a role in the immune response.

Developmental embryology. The hypophysis of all vertebrates has two major segments—the neurohypophysis (a neural component) and the adenohypophysis (an epithelial component)—each with a different embryological origin. The neurohypophysis develops from a downward process of the diencephalon (the base of the brain), whereas the adenohypophysis originates as an outpocketing of the primitive buccal epithelium, known as Rathke's pouch. An epithelial stalk connecting the embryonic adenohypophysis with the buccal epithelium commonly disappears during development, leaving the adenohypophysis attached to the neurohypophysis; however, the stalk may persist to form an open ciliated duct in some teleosts or a solid cord in some birds.

The adenohypophysis has three distinct subdivisions: the pars tuberalis, the pars distalis, and the pars intermedia. Development of the pituitary gland is a dynamic phenomenon that involves the formation of Rathke's pouch and its migration to a point of contact with the neural process that will become the neurohypophysis. The pars intermedia forms in the adenohypophysis at its zone of contact with the neurohypophysis. The failure of the pars intermedia to develop in birds, whales, dolphins, elephants, and armadillos can be explained by the failure of the adenohypophysis and neurohypophysis to make contact during embryogenesis. There is no such apparent embryological explanation for the regression of the human pars intermedia after it has formed.

The neurohypophysis comprises the pars nervosa and the infundibulum. The latter consists of the infundibular stalk and the median eminence of the tuber cinereum.

Physiology. The structural intimacy of neurohypophysis and adenohypophysis that is established early during embryogenesis reflects the direct functional interaction between the central nervous system and endocrine system. The extent of this anatomical intimacy varies considerably among the vertebrate classes, from limited contact in cyclostomes to intimate interdigitation in teleosts (**Fig. 1**). Vascular or neuronal pathways, or both, provide the means of exchanging chemical signals, thus enabling centers in the brain to exert control over the synthesis and release of adenohypophysial hormones.

Neurohormones, which are synthesized in specific regions of the brain, are conveyed to the neurohypophysis by way of axonal tracts, where they may be stored in distended axonal endings. Axons may also contact blood vessels and discharge their neurosecretory products into the systemic circulation or into a portal system leading to the adenohypophysis, or they may directly innervate pituitary gland cells.

In most animals, the vascular link is the prime

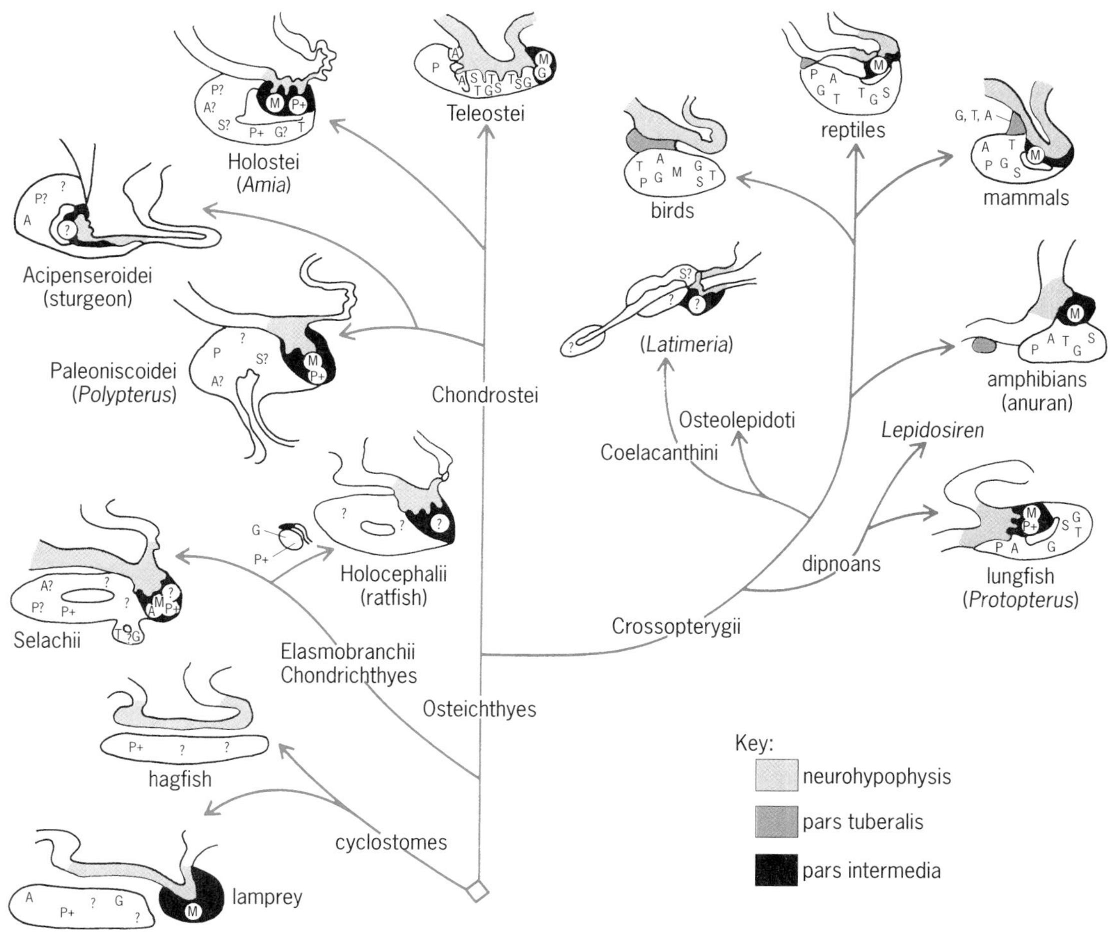

Fig. 1. Phylogenetic tree of generalized vertebrate pituitary glands (midsagittal section). P = prolactin; S = somatotropin; G = gonadotropin; T = thyrotropin; A = adrenocorticotropin; M = melanocyte-stimulating hormone; ? = unidentified function. (*After M. P. Schreibman, Pituitary gland, in P. K. T. Pang and M. P. Schreibman, eds., Vertebrate Endocrinology: Fundamentals and Biomedical Implications, vol. 1, Academic Press, 1986*)

route of information transfer between brain and pituitary gland. A description of this link begins in the tuber cinereum, the portion of the third ventricle floor that extends toward the infundibulum. The lower tuber cinereum, which is known as the median eminence, is well endowed with blood vessels that drain down into the pituitary stalk and ultimately empty into a secondary plexus in the anterior pituitary. The vascular link between the median eminence and the pituitary gland is known as the hypothalamo-hypophysial portal system. The median eminence in humans is vascularized by the paired superior hypophysial arteries. The pituitary gland is believed to have the highest blood flow rate of any organ in the body. However, its blood is received indirectly via the median eminence and the hypothalamo-hypophysial portal system. Most of the blood flow is from the brain to the pituitary gland, with retrograde flow from the adenohypophysis to the hypothalamus, suggesting a two-way communication between nervous and endocrine systems. Although the brain is protected from the chemical substances in the circulatory system by the blood–brain barrier, the median eminence lies outside that protective mechanism and is therefore permeable to intravascular substances.

Histology and cytology. The adenohypophysis is a conglomeration of cell types, specific in appearance and function, that range in organization from masses that are highly segregated by cell type (in teleosts, the bony fishes) to more randomized mixtures of cells (in birds and mammals). Light microscopy with standard cytological stains and electron microscopy have revealed five or six different cell types within the pituitary. This correlates well with the number of hormones thought to originate from the pars distalis. Cytological analysis reveals cells to be identifiable on the basis of physiological function. Immunocytochemistry has facilitated an even more specific identification of pituitary cells and evaluation of their activity.

Comparative morphology. The pituitary gland varies somewhat in complexity and hormonal potential among the vertebrate classes.

Agnatha. Agnathans, the most primitive living vertebrates, are represented by the lamprey and the hagfish. The hagfish pituitary gland appears to be the more primitive of the two. Its neurohypophysis is highly developed, flattened, and saclike, whereas its hypothalamo-hypophysial portal system is absent or markedly reduced but may show species variation.

The adenohypophysis is separated from the neurohypophysis and is not differentiated into zones. Generally, islets of cells are embedded in a poorly vascularized, loose connective tissue that is continuous with the layer separating the adenohypophysis from the neurohypophysis. Several cell types appear to be present in the hagfish. A thyroid gland–stimulating substance may be present, but attempts to demonstrate a factor that stimulates the gonads in the hagfish have not been successful.

Hypothalamic gonadotropin-releasing hormone is delivered to target cells in the adenohypophysis of lampreys by means of simple diffusion between the neurohypophysis and adenohypophysis or by secretion of the hormone into the third ventricle of the brain, where it can reach the pituitary gland. Levels of gonadotropin-releasing hormone are known to fluctuate with reproductive season, suggesting a typical regulatory role on gonad function.

The pars distalis in lampreys is separated from the infundibulum by connective tissue. The cell types display a regional distribution. Adrenocorticotropic hormone, luteinizing hormone, melanocyte-stimulating hormone, and perhaps other hormones are present in the lamprey pituitary gland. *See* JAWLESS VERTEBRATES.

Chondrichthyes. The two extant subclasses of chondrichthyes are the Elasmobranchii (sharks, skates, and rays) and the Holocephali (rat fishes or rabbit fishes and chimaeroids).

In elasmobranch fishes, the neurohypophysis consists of a thin-walled anterior portion—a true median eminence—connected to the pars distalis by a hypothalamo-hypophysial portal system. The adenohypophysis is divided into four regions: an elongated pars distalis (dorsal lobe), with rostral and caudal regions; a ventral lobe attached by a stalk to the caudal pars distalis; and a large pars intermedia that is heavily penetrated by neurohypophysial tissue to form a typical neurointermediate lobe. A pars tuberalis is not present. Follicles and spaces are characteristic of the chondrichthyian pituitary. They range from a highly developed system of vesicles and tubules that communicates with a hollow pars distalis (sharks and dogfish) to a hypophysial cavity that is small or lost entirely (skates and rays). The ventral lobe of the pars distalis is also hollow and often vesicular. Adrenocorticotropic hormone and prolactin are produced, presumably, in the rostral pars distalis; the former may also be found in the pars intermedia. The large cells of the ventral lobe produce thyrotropin (thyrotropes) and gonadotropin (gonadotropes).

The holocephalian neurohypophysis has a pars nervosa and a prominent median eminence that sends many short blood vessels into the rostral and caudal pars distalis, but the entire pars distalis is hollow and is not as clearly divisible into rostral and caudal regions as that of the elasmobranchs (Fig. 1). A ventral lobe is not present. Adult holocephalians have a large compact follicular structure called the Rachendach-hypophyse (pharyngeal lobe). Gonadotropic activity has been identified there in the rabbit fish, making this buccal lobe comparable to the elasmobranch ventral lobe. *See* CHONDRICHTHYES.

Osteichthyes. The subclass Actinopterygii (fish having a bony skeleton and paired ray fins) includes the infraclasses Teleostei, Chondrostei, and Holostei. Only the bony fishes (teleosts) will be considered here.

In those teleosts examined, the adenohypophysis is clearly divided into three regions, the rostral and caudal pars distalis and pars intermedia (Fig. 1).

There is no pars tuberalis. The clarity of this separation is essentially due to a restriction of specific cell types to distinct regions of the gland.

In teleosts, there is a structural intimacy of the neurohypophysis with all three regions of the adenohypophysis, and there is generally no true median eminence or portal system. A unique feature is that the adenohypophysis cells are innervated directly, both neurohypophysial and adenohypophysial cells being in contact with neuronal endings from hypothalamic and extrahypothalamic nuclei. Axons may originate from perikarya as close as the hypothalamic nuclei or from distant extrahypothalamic nuclei.

In general, hormone-producing cells of a particular type are restricted to specific regions of the adenohypophysis. The rostral pars distalis is essentially a single mass of prolactin-producing cells. A narrow band of adrenocorticotropic hormone–producing cells (corticotropes) forms the posterior boundary of the rostral pars distalis. In the caudal pars distalis, cells producing thyroid gland–stimulating hormone (thyrotropes) and somatotropin- or growth hormone–producing cells (somatotropes) are intermingled in islets formed by pervading neurohypophysial tissue. The external boundary of the caudal pars distalis consists of several cell layers of gonadotropes, presumably producing at least two chemically different gonadotropins.

The pars intermedia generally contains two cell types. One is presumed to be the source of melanocyte-stimulating hormone, whereas the other contains glycoprotein and in some species is thought to be a source of a second type of gonadotropin.

The subclass Sarcopterygii consists of the infraclasses Dipnoi and Crossopterygii. The single coelacanth genus, *Latimeria*, and the six species of lungfish have been used to study the evolution of the tetrapod pituitary gland. The neurohypophysis of lungfish (Dipnoi) consists of a median eminence, an infundibular stalk, and a neural lobe. The adenohypophysis is more similar in structure to that of amphibians than to other fishes. Amphibian characteristics of the lungfish pituitary are a well-formed neural lobe, a prominent median eminence, the absence of a saccus vasculosus, and a less obvious regional distribution of cell types. Fishlike characteristics include direct aminergic innervation of the pars distalis cells, interdigitation of the neurohypophysis and pars intermedia, prominent hypophysial cleft between the pars distalis and the pars intermedia, absence of a pars tuberalis, and presence of follicles. The five pars distalis cell types are intermingled and lack the distinct zonation seen in teleosts.

In the infraclass Crossopterygii, *Latimeria chaulumnae* is the only living representative of the Coelacanthiformes. The orthodox features of its pituitary include a typically fishlike neurointermediate lobe ventral to the brain, direct contact between pars distalis and pars intermedia, follicles and tubules containing glycoprotein colloid as in many fishes, a well-developed portal system, and probable median eminence. An exceptional feature is a cylindrical, tubular hypophysial cavity up to 5 in. (12 cm) long that extends from the rostral lobe of a greatly extended, tripartite pars distalis and that contains vascularized masses of adenohypophysial cells (the rostral islets or pars buccalis). These structures are comparable with the Rachendachhypophyse of holocephalians. *See* OSTEICHTHYES.

Amphibians. The amphibian pituitary gland is very much like the hypophysis of the other tetrapods. The adenohypophysis is generally composed of a flat, wide, compact pars distalis. Dorsally it continues with the pars intermedia without an intervening cleft, and anteriorly it is connected to the median eminence by a strand of connective tissue through which the portal vessels pass. The pars tuberalis is seen in amphibians for the first time in vertebrate phylogeny. Specific cell types are scattered, but the pars intermedia forms a distinct and separable band of tissue between the pars distalis and the pars nervosa. One dominant cell type in the pars intermedia predominates and is probably the source of melanocyte-stimulating hormone. *See* PIGMENTATION.

Reptilia. The morphology of the reptilian hypophysis strongly resembles both the amphibian and the avian types. It differs from the amphibian gland in that the pars distalis and the pars intermedia are structurally more intimate, the pars tuberalis may not be present in some groups, and the pars intermedia may be larger than in any other vertebrate. A median eminence is found on the infundibular stalk just anterior to the pars distalis. Portal vessels connect primary and secondary capillary plexuses.

As is typical in tetrapods, the reptilian pars distalis is made up of branching cords of cells that are highly vascularized by intervening sinusoids. Gonadotropes have a scattered distribution. In some snakes and lizards, they are ventrally located in the midline; in other lizards, they are restricted to the rostral lobe. The distribution of hypophysial cell types is similar in reptiles and amphibians. In some reptiles, such as lizards, an epithelial stalk (a remnant of Rathke's pouch) may persist. The pars tuberalis is the most variable component of the gland. Gonadotropes and thyrotropes are present in the pars tuberalis. *See* REPTILIA.

Aves. An epithelial stalk persists in some birds as a connection between the pars distalis and the buccal epithelium. The separation of the adenohypophysis from the neurohypophysis in birds makes the gland less compact than, for example, that of mammals. A well-developed pars tuberalis is located between the median eminence and the pars distalis, although in some species it may lie within the median eminence. The avian median eminence can be regionally differentiated into the "point-to-point" system: a rostral portion supplies blood to the rostral pars distalis, and a caudal area delivers blood to the caudal pars distalis. *See* AVES.

Mammalia. The mammalian pituitary gland shows only minor variations between the various genera (**Fig. 2**). In some mammals, including humans, a

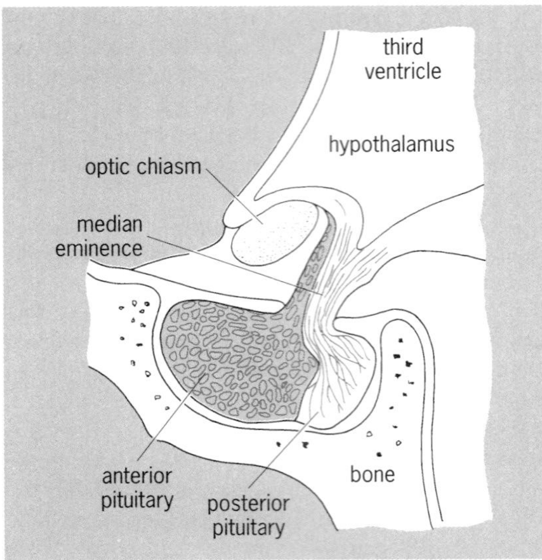

Fig. 2. Relationship of the pituitary gland to the brain and hypothalamus. (*After D. S. Luciano, A. J. Vander, and J. H. Sherman, Human Anatomy and Physiology: Structure and Function, 2d ed., McGraw-Hill, 1983*)

pharyngeal (craniopharyngeal) hypophysis develops from a remnant of Rathke's pouch. Found in the roof of the mouth, it is vascularized and has staining and immunoreactive properties similar to those of the sellar hypophysis.

Another structural variation is the failure of the pars intermedia to form in cetaceans (whales and dolphins), elephants, and armadillos, presumably as a result of a connective tissue barrier that separates the neurohypophysis and the adenohypophysis during development. Although a pars tuberalis is characteristic of the mammalian gland, it may be absent in some animals, such as the sloth and pangolin.

The mammalian adenohypophysis is regulated primarily by its vascular supply. The adult mammalian neurohypophysis is a neurohemal organ that receives aminergic and peptidergic neurons. The median eminence is prominent and forms a primary capillary plexus around much of the infundibular stalk in most species. Most often, portal vessels pass through a well-developed pars tuberalis on their way to the pars distalis. *See* MAMMALIA.

Adenohypophysis hormones. The hormones of the adenohypophysis may be grouped into three categories based on chemical and functional similarities. The first category consists of growth hormone (also known as somatotropin) and prolactin, both of which are large, single, polypeptide chains; the second category consists of the glycoprotein hormones; this family of hormones contains the gonadotropins and thyrotropin. The gonadotropins in many species, including humans, can be segregated into two distinct hormones, follicle-stimulating hormone and luteinizing hormone. In other species only one gonadotropin molecule has been isolated, and in still other species two gonadotropins have been isolated. The third group comprises adrenocortiotropic hormone and melanotropin (MSH; melanocyte-stimulating hormone).

The regulation of the release of pituitary hormones is determined by precise monitoring of circulating hormone levels in the blood and by genetic and environmental factors that manifest their effect through the releasing and release-inhibiting factors of the hypothalamus. *See* ADENOHYPOPHYSIS HORMONE.

Hypothalamic and neuronal involvement. The hypothalamus is located at the base of the brain (the diencephalon) below the thalamus and above the pituitary gland, forming the walls and the lower portion of the third ventricle. It receives major neuronal inputs from the sense organs, hippocampus, thalamus, and lower brainstem structures, including the reticular formation and the spinal cord. Thus, the hypothalamus is designed and anatomically positioned to receive a diversity of messages from external and internal sources that can be transmitted by way of hypothalamic releasing factors to the pituitary gland, where they are translated into endocrine action. *See* NERVOUS SYSTEM (VERTEBRATE).

Neurohypophysis hormones. The neurohypophysis hormones, oxytocin and vasopressin, which are nonapeptides, are synthesized in different neurons of the paraventricular and supraoptic nuclei of the hypothalamus and travel by axonal flow to the terminals in the neurohypophysis for storage and ultimate release into the vascular system. Oxytocin is important in stimulating milk release through its contractile action on muscle elements in the mammary gland. It also stimulates uterine smooth muscle contraction at parturition. Vasopressin affects water retention by its action on certain kidney tubules. Thus, it also affects blood pressure (Fig. 2*a*). *See* LACTATION.

Birds, reptiles, amphibians, and lungfishes have arginine vasotocin and arginine mesotocin as the principal peptides of the posterior pituitary. In birds and reptiles, the nonapeptides primarily originate from the supraoptic and paraventricular nuclei; in poikilothermic (cold-blooded or exothermic) vertebrates, however, the preoptic nucleus is the homolog of the two nuclei found in homeotherms (endotherms). The principal nonapeptide found in various groups of fishes is arginine vasotocin, but a second neural peptide related to oxytocin and sometimes oxytocin itself may be found. Cyclostomes possess only arginine vasotocin.

In several species of birds, both arginine vasotocin and mesotocin are found in separate neurons in the supraoptic and paraventricular nuclei; however, only arginine vasotocin has been localized in the median eminence. *See* NEUROHYPOPHYSIS HORMONE.

Neurohormones. The five neurohormones in vertebrates are growth hormone–releasing hormone, somatostatin, corticotropin-releasing hormone, thyrotropin-releasing hormone, and gonadotropin-releasing hormone. In addition, other peptides found in the nervous system may be released into the median eminence or posterior pituitary and are probably capable of neurohormonal action.

Growth hormone–releasing hormone (GRH)

consists of either 44 (human, ovine, porcine) or 43 (rat) amino acids. In some mammals, including humans, perikarya containing immunoreactive growth hormone–releasing hormone are located in the arcuate (infundibular) and ventromedial nuclei of the hypothalamus.

In mammals, somatostatin, which is also known as somatotropin release–inhibiting hormone (SRIH), contains 14 amino acids or 28 amino acids; both have been found to be identical in all mammals investigated. When released from the median eminence, somatostatin has a neurohormonal function, whereas in the arcuate nucleus it is most likely serving as a neurotransmitter. *See* SOMATOSTATIN.

In sheep, corticotropin-releasing hormone (CRH) is composed of 41 amino acids and affects the release of corticotropin, beta-endorphin, and melanotropin. In each major mammalian group examined, immunoreactive CRH has been localized in many brain areas, the spinal cord, and the gastrointestinal tract, with the greatest concentration in the hypothalamus.

Thyrotropin-releasing hormone (TRH), a tripeptide, controls the release of thyroid gland–stimulating hormone from the pituitary gland. Like somatostatin, thyrotropin-releasing hormone is widely distributed in the brain and in the gastrointestinal tract. Its levels are highest in the median eminence, and the densest collection of cell bodies containing thyrotropin-releasing hormone is in the medial, parvocellular division of the paraventricular nucleus. Studies of its distribution, which have been conducted mainly in mammals, reveal that the distribution of thyrotropin-releasing hormone in frogs and in rats is identical.

Gonadotropin-releasing hormone (GnRH), formerly known as luteinizing hormone–releasing hormone, is a decapeptide (10 amino acids) originally isolated from pig and cattle hypothalami. It regulates the release of luteinizing hormone and follicle-stimulating hormone from the adenohypophysis. There are several structural variants of mammalian gonadotropin-releasing hormone in nonmammalian vertebrates, and four have been sequenced, two from chicken, one from salmon, and one from lamprey. Additionally, several of the variants may exist within a single species, suggesting that there may be different functions for these peptides. Gonadotropin-releasing hormone or GnRH-like peptides have been localized not only in the brain and pituitary gland but also in other areas of the central nervous system and in the gonads, placenta, mammary glands, and pancreas. During evolution, the peptide has been subject to gene duplication and structural change and thus may serve a variety of regulatory functions. In addition to the release of pituitary gonadotropic hormone, gonadotropin-releasing hormone may also function as a neurotransmitter in the central and sympathetic nervous systems, as a paracrine regulator in the gonads and the placenta, and as an autocrine regulator in tumor cells.

The distribution of perikarya containing immunopositive gonadotropin-releasing hormone in the brain has been studied in various mammalian species, particularly the rat. Perikarya that contain immunoreactive gonadotropin-releasing hormone are scattered in a continuum from the olfactory bulb and terminal nerve to the area of the premammillary nucleus. Similarly, immunoreactive gonadotropin-releasing hormone is distributed throughout the brain in nonmammalian vertebrates, although the specifics of this distribution remain controversial. Changes in the distribution of gonadotropin-releasing hormone may be related to the sex, genotype, physiological state, and age of the animals being studied.

In addition to those historically recognized as neurohormones, many of the peptides found in the nervous system are probably released into the median eminence or posterior pituitary and have neurohormonal action.

Neurotransmitters. The better-known neurotransmitters of the central nervous system include the catecholamines (dopamine, epinephrine, and norepinephrine), serotonin, acetylcholine, gamma-amino butyric acid (GABA), histamine, and the opioid peptides (enkephalins, endorphins, dynorphin, neoendorphin, rimorphin, and leumorphin). These substances are distributed widely in the central nervous system and, for most, also in the pituitary gland. As is seen with many of the peptides identified in the nervous system, if a particular amine or neurotransmitter is present in nerve fibers leading to the median eminence, it probably will influence pituitary gland activity via the portal system. Dopamine, serotonin, gamma-amino butyric acid, and acetylcholine are best known for such activity. These neurotransmitters play an important, but poorly understood, role in regulating pituitary function, either directly or by their action on neuropeptide-producing neurons. Understanding the pharmacology of neurotransmitters holds promise for the treatment of basic disorders of the hypothalamic-pituitary axis. *See* ACETYLCHOLINE; ENDOCRINE MECHANISMS; ENDOCRINE SYSTEM (VERTEBRATE); ENDORPHINS; HISTAMINE; HORMONE; NEUROBIOLOGY; NEUROIMMUNOLOGY; PITUITARY GLAND DISORDERS; SEROTONIN.

Martin P. Schreibman

Bibliography. I. Chester-Jones and P. M. Ingleton (eds.), *Fundamentals of Comparative Vertebrate Endocrinology*, 1987; L. J. DeGroot (ed.), *Endocrinology*, 3 vols., 4th ed., 2001; H. Imura (ed.), *The Pituitary Gland*, 2d ed., 1994; E. E. Muller and N. Guiseppe, *Brain Messengers and the Pituitary*, 1988; P. K. T. Pang and M. P. Schreibman (eds.), *Vertebrate Endocrinology*, 4 vols., 1986–1991; W. F. Walker, *Functional Anatomy of the Vertebrates*, 2d ed., 1997.

Pituitary gland disorders

Inborn or acquired abnormalities in the structure or function of the human pituitary gland. Pituitary disorders can stem from any of five disease processes:

tumors and other growths, intrinsic lesions of the anterior lobe, diseases affecting gland function, hypothalamic malfunction, and systemic disease that affects the adenohypophysis.

Tumors and other growths in and near the pituitary may cause failure of hormone secretion or impinge on nearby brain structures. The latter effect can give rise to neurological malfunctions, the most common of which is visual impairment such as narrowing of the visual fields. Intrinsic lesions of the anterior pituitary, usually benign, secrete excessive amounts of a single (rarely multiple) hormone, producing characteristic endocrine syndromes, the most dramatic of which are acromegaly and Cushing's disease. Infections, congenital anomalies, granulomas, vascular disorders, and, rarely, metastatic cancers may induce partial or total failure of one or more pituitary secretions, in which case growth hormone, follicle-stimulating hormone, and luteinizing hormone are the first to fail.

Diseases of the hypothalamus affect pituitary function through the mechanical effects of a mass or through disrupted secretion of the hypophysiotropic peptides (see **table**). Typical manifestations include precocious or delayed puberty; diabetes insipidus; and derangements of sleep, eating, and temperature regulation. The dysfunction may be of congenital, traumatic, inflammatory, or neoplastic origin. However, the most important hypothalamic-pituitary diseases are tertiary hypothyroidism, precocious or delayed puberty, diabetes insipidus, and Kallmann's syndrome, which consists of a deficient sense of smell and lack of sexual development due to inborn failure of the hypothalamus to secrete gonadotropin-releasing hormone. Hypothalamic disorders can cause abnormalities of growth hormone secretion. Whereas most individuals with acromegaly (an excess of growth hormone) have intrinsic pituitary tumors that secrete growth hormone, some have excessive secretion of hypothalamic growth hormone–releasing hormone or insufficient secretion of somatostatin. Pituitary dwarfism, caused by the failure to secrete growth hormone in childhood, is usually due to primary disease of the adenohypophysis, but in some individuals the fundamental dysfunction (opposite to that in acromegaly) lies in the hypothalamus. *See* DIABETES; DWARFISM AND GIGANTISM; SOMATOSTATIN; THYROID GLAND DISORDERS.

The adenohypophysis can be affected subtly by systemic diseases, or more obviously by metastases from breast cancer. Rarely, acute necrosis (death) of the gland occurs with diabetic crises, heart attack, generalized infection, or as a result of using mechanical respirators. Prolonged treatment with large doses of adrenal corticosteroids, as in cases of lupus, asthma, or acute leukemia, can result in the failure of the adenohypophysis to secrete adrenocorticotropic hormone in times of physical stress. A deficiency in adrenocorticotropic hormone is a potentially fatal condition. In addition, manic-depressive psychosis has been linked to an overactivity of the adrenocorticotropic hormone–adrenal cortex axis. *See* ADENOHYPOPHYSIS HORMONE; ADRENAL GLAND DISORDERS.

Anterior lobe malfunction. Malfunctions of the anterior lobe of the pituitary can cause hypopituitarism (pituitary insufficiency) or hyperpituitarism (pituitary excess).

Hypopituitarism. Adult hypopituitarism has many possible causes; more than 50% of cases result from benign or malignant tumors of the gland. Metastatic cancer originating in the breast, colon, or kidney rarely invades and destroys the pituitary. Tumors and

Interrelationships between the hypothalamic hypophysiotropic hormones and the anterior pituitary hormones

Hypothalamic hypophysiotropic hormones			Anterior pituitary hormones			
Hormone	Structure	Function	Hormone	Structure	Cell of origin	Function
Growth hormone releasing hormone	Peptide	Stimulates growth hormone release	Growth hormone	Protein	Somatotrope	Stimulation of bodily growth
Somatostatin	Peptide	Inhibits growth hormone release				
Thyrotropin-releasing hormone	Small peptide	Stimulates thyrotropin release	Thyrotropin*	Glycoprotein	Thyrotrope	Regulation of thyroid gland
		Stimulates prolactin release	Prolactin	Protein	Mammotrope	Postpartum lactation
Dopamine	Amine	Inhibits prolactin release				
Gonadotropin-releasing hormone	Peptide	Stimulates release of follicle-stimulating hormone and luteinizing hormone	Follicle-stimulating hormone	Glycoprotein	Gonadotrope	Stimulation of ovarian follicle; spermatogenesis
			Luteinizing hormone	Glycoprotein	Gonadotrope	Stimulation of corpus luteum, and testosterone secretion by testis
Corticotropin-releasing hormone	Peptide	Stimulates release of adrenocorticotropin, β-endorphin, β-lipotropin, and melanotropin	Adrenocorticotropin	Peptide	Corticotrope	Regulation of adrenal cortex
			β-endorphin†	Small peptide	Corticotrope	Possible effect on stress reaction and memory
			β-lipotropin†	Peptide	Corticotrope	Breakdown of fat
			α, β-, and γ-melanotropins†	Peptide	Corticotrope	Pigmentation of skin

*Also known as thyroid-stimulating hormone.
† The role of these hormones in human physiology and disease is not known.

cysts account for another 15–20%. Still another 10–15% can be traced to granulomas, as in sarcoid, tuberculosis, and syphilis; rare systemic disorders; autoimmune inflammation of the pituitary gland; defects in hypothalamic hypophysiotropic hormone secretion; and injury, including therapeutic x-radiation, surgery, or skull fracture. The most frequent cause of panhypopituitarism, that is, deficiency or absence of all pituitary hormones, is Sheehan's disease, which involves tissue death caused by insufficient blood supply to the gland due to copious blood loss or intravascular clotting and hemorrhage associated with premature separation of the placenta during childbirth.

The clinical picture of adult hypopituitarism depends on the cause. If a pituitary tumor or other mass is present, the first clinical symptoms are usually headache, visual impairment, and visual field defects. Deficiencies of thyrotropin, gonadotropins, and adrenocorticotropin are mild at first; hypothyroidism, hypogonadism, and adrenal insufficiency develop slowly. Still more gradual is the course of Sheehan's disease: full-blown diagnostic features may take 15–20 years to appear. Without treatment, death can follow severe hypoglycemia (low blood sugar) accompanying what would be a minor illness in a normal person. The causes of the hypoglycemia are gross deficiencies of pituitary adrenocorticotropin and growth hormone secretion.

Diagnosis is made by combining computerized tomographic scanning or magnetic resonance imaging with anterior pituitary hormone response tests. Treatment consists of hormone replacement therapy with estrogen or testosterone, thyroxine, and hydrocortisone. Tumors must be destroyed by radiation or surgery.

Chronic deficiencies of single pituitary hormones are usually due to failure of the corresponding specific hypothalamic hypophysiotropic hormone (see table). The most important deficiency for children is the failure to secrete growth hormone, with resulting dwarfism. The usual cause is intrinsic failure of the anterior pituitary somatotropes in the absence of a tumor. Children with hypopituitary dwarfism are small, normally proportioned, and slightly obese. Diagnosis can be established only by documenting subnormal growth over time and demonstrating inadequate secretion of growth hormone during deep sleep. Treatment by frequent injections of synthetic human growth hormone accelerates the rate of linear growth but does not result in full attainment of expected ultimate stature.

Hyperpituitarism. Hyperpituitarism is almost always due to excessive secretion of a single hormone by a tumor of the anterior pituitary. Somatotrope adenomas, which are benign tumors, secrete excessive growth hormone, producing acromegaly or gigantism. The acromegalic person is middle-aged and has enlarged bony prominences of the face, larger-than-normal lips, ears, hands, and feet, a deepened and broadened chest cage, and bowed legs. Diabetes mellitus is the chief metabolic abnormality. Treatment is difficult and usually consists of x-radiation or microsurgical removal of the adenoma. After treatment, plasma growth hormone levels fall slowly.

The most common endocrinologically active tumor (30% of all secretory adenohypophyseal tumors) is the prolactin-secreting lactotrope adenoma. Milk secretion not associated with childbirth is often present in both sexes, and plasma prolactin levels are high. Treatment is controversial, as x-radiation alone is not very effective. The other two choices are removal of the tumor and daily administration over many years of bromocriptine, which attaches to the same cellular receptors as dopamine and thereby inhibits prolactin secretion. Corticotrope adenomas, secreting excessive adrenocorticotropin, occur in Cushing's disease or in those whose adrenal glands have been completely removed. Rare thyrotrope adenomas, which hypersecrete thyrotropin, are found in persons with inadequately treated hypothyroidism. Also, some people, for unknown reasons, lack the normal capacity of thyroid hormone to suppress pituitary secretion of thyrotropin (resistance syndrome). In addition, some have preexisting hyperthyroidism, and so treatment is extremely difficult. The hyperthyroidism can be fairly well controlled by standard endocrine methods, but often the thyrotrope adenoma can be brought under control only by giving a somatostatin analog. The uncommon gonadotrope adenomas secrete excessive follicle-stimulating hormone or, more rarely, luteinizing hormone. Curiously, the tumors produce no endocrine signs, and so they are undetected until they are quite large. Treatment is by microsurgery, x-radiation, or administration of synthetic chemical analogs of gonadotropin-releasing hormone, whose long-term effect is to inhibit the secretion of luteinizing and follicle-stimulating hormones.

These endocrinologically active tumors, together with the hormonally inert tumors of the adenohypophysis, constitute about 10% of all intracranial neoplasms. The hormonally inactive adenomas are noted clinically as neurological or visual problems; the endocrine consequences of pituitary destruction by a mass lesion occur much later. Tumor types are very numerous and can be either congenital or acquired. Diagnosis is made by neurological examination and radiography with computerized tomographic scanning or magnetic resonance imaging. Some individuals show an enlarged sella without pituitary tumor (empty sella syndrome), a developmental or acquired abnormality that can deform the gland. Treatment of pituitary tumors usually involves surgical removal, often combined with radiotherapy.

Posterior lobe malfunction. The two posterior lobe hormones that are secreted by the hypothalamus and stored in the pituitary are oxytocin and vasopressin. Oxytocin is secreted in large amounts at the onset of childbirth. It is not known whether hypersecretion of oxytocin triggers childbirth or is simply a response to it. Women with hypopituitarism and hypothalamic damage or destruction of the neurohypophysis can give birth normally. No known clinical disorders are attributable to abnormal secretion of oxytocin.

Absence or deficiency of vasopressin, the antidiuretic hormone, characterizes diabetes insipidus. Some cases are inherited, sometimes in association with other abnormalities, and are present at birth; some are of unknown cause; and some cases result from trauma, granuloma, tumor, meningitis, encephalitis, or disorders of the blood vessels. Because of the lack of vasopressin, the kidneys cannot excrete urine of normal concentration, and so as much as 3 gal (11 liters) of highly dilute urine are excreted each day. The water loss can lead to vascular collapse, shock, and death. Diagnosis is not simple; the principal disorder that must be ruled out is compulsive, excessive water drinking due to emotional disorder. Intrinsic resistance of the kidney to the action of vasopressin is exceedingly rare. Diagnostic accuracy requires carefully comparing values of plasma vasopressin with the electrolyte content of urine and of plasma under various standardized test conditions. Treatment with desmopressin, a synthetic, long-acting vasopressin analog in the form of a nasal spray, has proved satisfactory.

Vasopressin excess characterizes the syndrome of inappropriate antidiuretic hormone secretion. With this disorder, which has many causes, individuals are lethargic and have altered states of consciousness, seizures, very low serum concentrations of sodium, and variable abnormalities of blood volume. Hypopituitarism can produce the syndrome when deficient pituitary adrenocorticotropin secretion leads to subnormal production of hydrocortisone, a hormone that inhibits the release of vasopressin; hydrocortisone reduces plasma vasopressin levels and returns water balance to normal. *See* ENDOCRINE MECHANISMS; NEUROHYPOPHYSIS HORMONE; PITUITARY GLAND.

Nicholas Christy

Bibliography. P. Belchetz, *Management of Pituitary Disease*, 1984; P. M. Black et al. (eds.), *Secretory Tumors of the Pituitary Gland*, 1984; L. J. DeGroot et al. (eds.), *Endrocrinology*, 4th ed., 3 vols., 2001; G. W. Harris, *Neural Control of the Pituitary Gland*, 1955; H. L. Sheehan and V. K. Summers, The syndrome of hypopituitarism, *Quart. J. Med.*, 42:319, 1949.

pK

The logarithm (to the base 10) of the reciprocal of the equilibrium constant for a specified reaction under specified conditions (for example, solvent and temperature). pK values are often more convenient to tabulate and use than the equilibrium constants themselves. The value of K for the dissociation of the HSO_4^- ion in aqueous solution at 77°F (25°C) is 0.0102 mole/liter. The logarithm is $0.008_6 - 2 = -1.991_4$. pK is therefore $+1.991_4$. The choice of algebraic sign, although arbitrary, results in positive values for most dissociation constants applicable to aqueous solutions. The concept of pK is especially valuable in the study of solutions. *See* CHEMICAL EQUILIBRIUM; IONIC EQUILIBRIUM; PH.

Thomas F. Young

Placentation

The intimate association or fusion of a tissue or organ of the embryonic stage of an animal to its parent for physiological exchange to promote the growth and development of the young. It enables the young, retained within the body or tissues of the mother, to respire, acquire nourishment, and eliminate wastes by bringing the bloodstreams of mother and young into close association but never into direct connection (**Fig. 1**). Placentation characterizes the early development of all mammals except the egg-laying duckbill platypus and spiny anteater. It occurs in some species of all other orders of vertebrates except the birds. In fact, in certain sharks and reptiles it is almost as well developed as in mammals. A few examples are also known among invertebrates (*Peripatus*, certain tunicates, and insects).

Placental modifications. With few exceptions the fetal structures used to establish placental relationships with the mother are modifications of organs present in kindred egg-laying (oviparous) species. In the fishes, sharks and rays, and amphibians these include the gill filaments, tail fin, pericardium, and primitive yolk sac (midgut and adjacent ventral body wall). The essential placental modifications of all

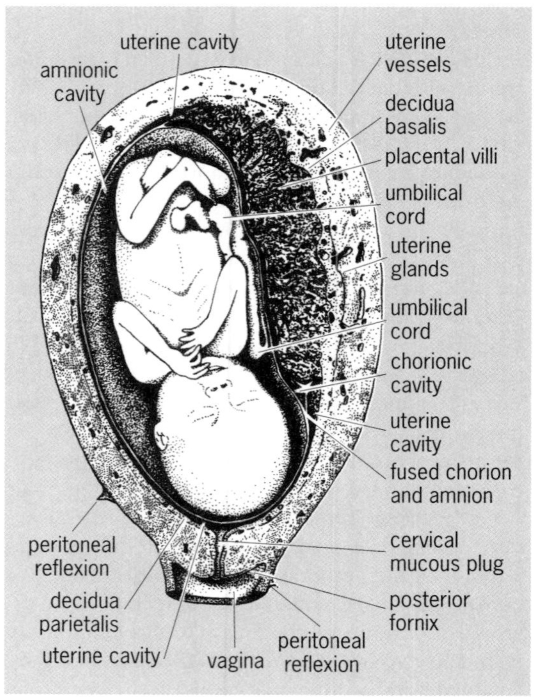

Fig. 1. Pregnant human uterus at $3^1/_2$ months, split sagittally to show fetus, membranes, and placenta.

these are increased surface area and vascularity, and intimate contact with a highly vascularized and often secretory area of the mother. In fishes this highly vascularized area is usually the ovary; in sharks and rays, the uterus; in amphibians, the uterus or the skin. Skin gestation is unique to certain oviparous South American frogs. As the eggs are extruded, they are fertilized by the male and placed on the back (in *Pipa*) or in a skin pouch on the back (in *Notothrema*). Here they become embedded in highly vascular compartments and receive both oxygen and nourishment from the mother's bloodstream. The live-bearing fishes Goodeidae have unique vascular rectal processes, trophotaeniae, which are known in no other group and which establish placental relationships with the ovarian tissue.

In amniotes (reptiles and mammals) the extraembryonic membranes utilized in placentation are specializations of basic membranes found in all oviparous amniotes. In fact, in the beginnings of placentation (portrayed by ovoviviparous lizards and snakes) the large shell-covered and yolk-laden eggs are simply retained within the uterus until hatched. Here apparently the only placental function is respiration; hence there is no obvious modification of the fetal membranes and only an increased and prolonged hypervascularity of the uterine lining.

In a few lizards and snakes and in all marsupial and eutherian mammals, complete placental function is necessary because the eggs are supplied with too little yolk to provide the needs of the embryo for nourishment until birth. Inadequate provision for elimination and storage of nitrogenous waste also occurs in most of these animals. Thus not only must these embryos interchange respiratory gases with the mother's blood, but they must also absorb nutriment and transfer wastes. Full placentation in these forms is provided mainly by anatomical and physiological specialization of three extraembryonic membranes: the chorion, yolk sac, and allantois. The amnion also probably plays a physiological role.

Physiological exchange. Efficient interchange depends on close proximity of large areas of fetal tissues to maternal blood and glandular areas. This is provided in mammals by a remarkable regulatory cooperation between the developing outer layer (trophoblast) of the chorion, together with the vascular yolk sac or allantois or both, and the mother's uterine lining (endometrium). In the typical mammalian placenta, which is always formed by the chorion and the allantoic vessels, the fetal and maternal bloodstreams are as close as a few thousandths of a millimeter from each other (**Fig. 2**). The surface area of the fetal villi which contain the functional fetal capillaries is probably several times larger than the body surface of the female. In humans this ratio is known to be about 8:1. Not only are the two vascular tissues closely approximated, but in many mammals the arrangement of the vessels is such that in the

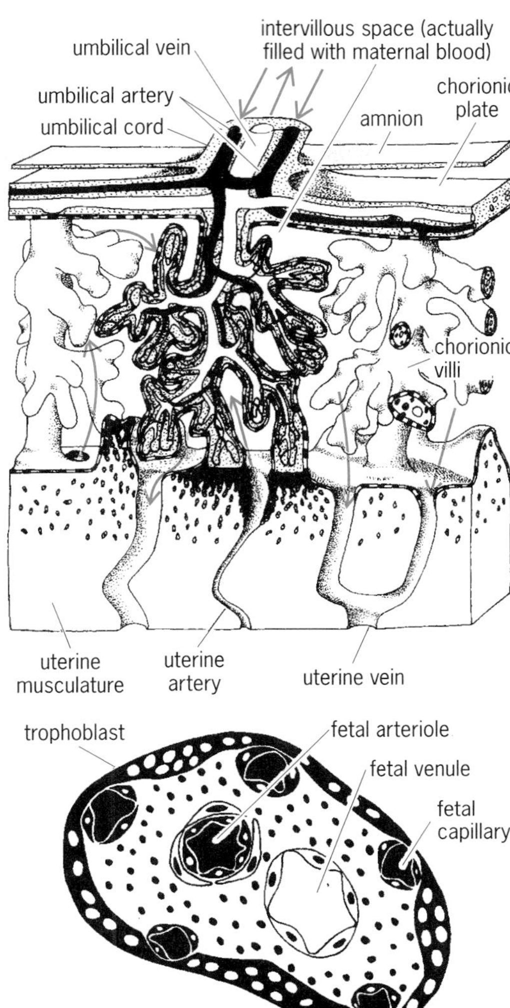

Fig. 2. Block removed from center of human placenta; cut end of branch villus is shown enlarged.

area of interchange the two bloodstreams flow in opposite directions. This counterflow principle could greatly increase the efficiency of interchange, but clear physiological evidence that it actually does so is lacking.

Evidence from electron microscopy shows that the surface of the human villi exposed to maternal blood is covered by multitudes of microvilli (**Fig. 3**). These are minute fingerlike projections of the cytoplasm and its ultrathin covering, the plasmalemma (surface membrane of a cell). Obviously, microvilli increase the surface area manyfold beyond the 8:1 ratio mentioned above. However, the effect of microvilli on physiological interchange between mother and fetus is more complicated than it appears; although microvilli greatly increase the surface area for simple diffusion, they actually reduce it for such processes of absorption as endocytosis, which occurs only on the plasmalemma between the bases of the microvilli. *See* ENDOCYTOSIS.

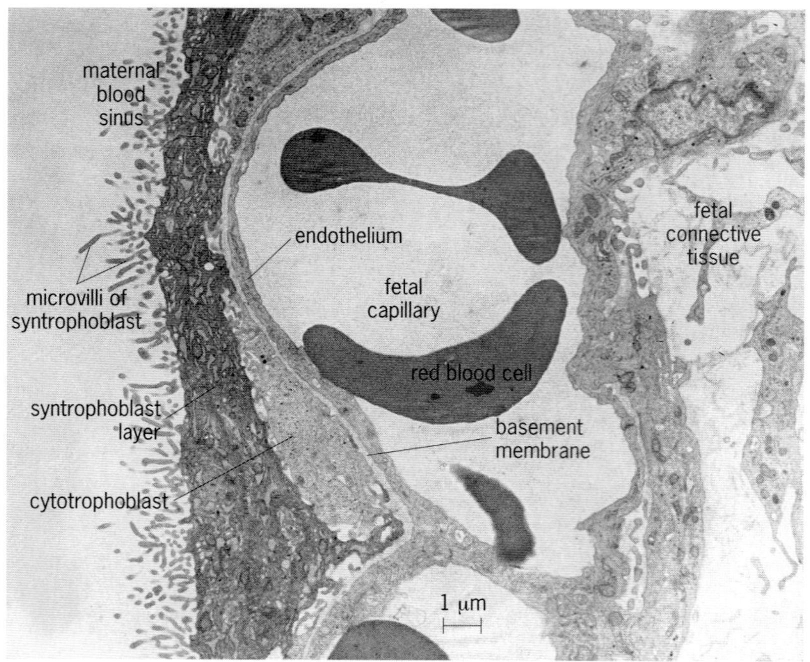

Fig. 3. Low-power electron micrograph of the surface of a section through the placental villus of a macaque monkey at 100 days of pregnancy (normal gestation period 164–168 days). This is very similar to the human placenta of similar stage.

The physiology of interchange through the placental membrane is a fertile field in research. Although simple physical diffusion and osmosis are factors, active membrane transport mechanisms are probably of greater importance. This transport through the placental separation membrane involves work done as the result of energy release within the plasmalemma and adjacent cytoplasm.

Types of placentation. Placentas are classified in various ways, but the most meaningful is based on the identity of the layers making up the separation membrane between the two bloodstreams. On this basis the human placenta is hemochorial; that is, the maternal blood is in direct contact with the chorionic trophoblast. The placenta of the dog is endotheliochorial; that is, the maternal blood is separated from the chorion by the maternal capillary endothelium. Another common condition is the epitheliochorial (**Fig. 4**).

Electron microscopy has shown the existence of ultrathin cellular layers previously missed by light microscopy. For instance, the trophoblastic layer may be single, double, or even triple. It is single in the mature human placenta, double in the rabbit, and triple in the rat. For these, A. C. Enders coined the terms

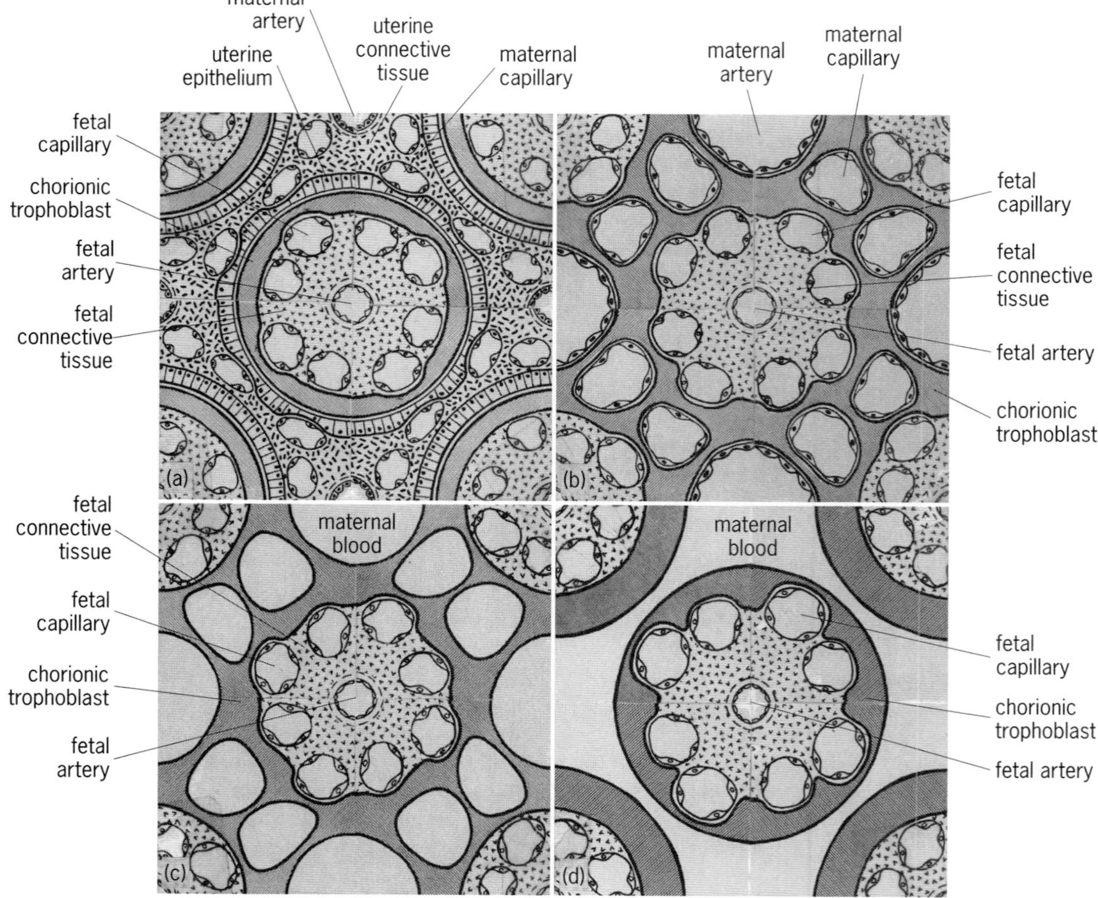

Fig. 4. Schemata of four fundamental types of placentation. (*a*) Epitheliochorial (villous); examples: hoofed animals, whales, lemurs. (*b*) Endotheliochorial (labyrinthine); examples: carnivores, bats, elephants, sloths. (*c*) Hemochorial (labyrinthine); examples: rodents, many insectivores, tarsiers. (*d*) Hemochorial (villous); examples: human, great apes, monkeys.

hemomonochorial, hemodichorial, and hemotrichorial, respectively. *See* FETAL MEMBRANE.

Harland W. Mossman

Bibliography. P. Kaufmann and B. F. King (eds.), *Structural and Functional Organization of the Placenta*, 1982; K. R. Page, *The Physiology of the Human Placenta*, 1993; C. W. Redman, I. L. Sargent, and M. Starkey, *The Human Placenta*, 1993.

Placer mining

The exploitation of placer mineral deposits for their valuable heavy minerals. Placer mineral deposits consist of detrital natural material containing discrete mineral particles. They are formed by chemical and physical weathering of in-place heavy minerals, which are then concentrated through the action of wind or moving water. This concentration can be done through wave and current action in the ocean (beach and offshore placers), glacial action (moraine placers), wind action removing the lighter material (eolian placers), or the action of running water (stream placers). Stream placers are the most important of these deposits because of their common occurrence and their highly efficient concentration mechanisms. Marine placers, primarily beach placers, are the next most economically important, with the potential of offshore placers being the most recent to be recognized and developed. *See* MARINE MINING; ORE AND MINERAL DEPOSITS.

Minerals that are concentrated in placer deposits are a result of differences in specific gravity and, therefore, the economically important deposits are for minerals with high specific gravities (see **table**).

Precious metals, primarily gold and platinum group metals, have been the most important product from placer mines. Their extremely high specific gravity coupled with their low chemical reactivity means that these minerals are efficiently concentrated in a placer environment and can be effectively recovered in a readily usable form. Historically, placer mines were civilization's first sources of gold and platinum and, until recent times, have continued to yield significant production. Although most modern gold is produced from lode, or "hard rock," deposits, the placer deposits of northern Canada, Alaska, and Siberia represent a virtually untapped source of the metal. *See* GOLD; PLATINUM.

Specific gravities of several important placer minerals

Mineral	Specific gravity
Gold	15–19
Platinum	14–19
Diamond	3.5
Cassiterite (SnO_2)	6.8–7.1
Magnetite (Fe_3O_4)	5.2
Rutile (TiO_2)	4.2
Ilmenite ($FeTiO_3$)	4.7
Monazite $(Ce,La,Y,Th)PO_4$	5.0–5.3
Quartz	2.6
Water	1.0

Of more importance than gold are placer diamond deposits. Another important placer mineral is cassiterite, an ore of tin. Additionally, rutile and ilmenite, the principal ores of titanium, are found in commercial quantities only in beach placers. These same types of placers also yield monazite, a source of the rare earths yttrium, lanthanum, cerium, and thorium. *See* CASSITERITE; DIAMOND; ILMENITE; MONAZITE; RUTILE.

Although the importance of placer gold may have decreased over the years, placer mining still is an important component of modern industrial economy.

Methods. Most placer mining operations involve surface mining methods, although underground methods are sometimes used.

Underground mining. Underground mining of placer deposits is limited to some frozen auriferous gravels in the Arctic and the ancient fossilized gold bearing channels typically found in the "mother lode" country of California and the deep leads (placer gravels) of Australia. These types of deposits are typically worked by sinking a shaft to bedrock and then driving a series of drifts and crosscuts along bedrock to form a room-and-pillar mine. *See* UNDERGROUND MINING.

One unique aspect of drift mining is its application to frozen gravels in the Arctic. The temperature of the gravels ranges from 18 and 22°F (−8 and −5°C) with bedrock temperatures between 8 and 14°F (−13 and −10°C). The gravel contains 10% or more of ice by weight and while frozen provides competent support for the mining method. Rather than a conventional drilling and blasting of the frozen gravels (which are very resistant to this), steam points are driven into the faces and the cohesive ice is melted, allowing the loosened gravel to be loaded without fragmentation.

Surface mining. This proceeds in a manner similar to most trip mining operations. First the overburden is removed, and then the placer ground containing the valuable material is processed. This overburden may consist of material similar to the valuable deposit, such as the barren top section of a stream placer deposit where the gold is concentrated near or on bedrock. The character of the overburden may also be distinct from that of the value-containing deposit, for example, the windblown sands covering the diamond-bearing stream channels in South-West Africa. *See* SURFACE MINING.

In general, overburden is removed in a manner similar to the mining of the valuable material, as by mobile power equipment, dredges, and so on. One method unique to placer mining is the use of high-pressure water to remove the overburden. Hydraulicking involves shooting a stream of water against the overburden bank from a large nozzle (2–10 in. or 5–25 cm in diameter) on a device called a hydraulic monitor or hydraulic giant. This method is commonly used when the overburden contains a high percentage of clay or when it is necessary to strip muck from frozen gravels in arctic areas.

Muck consists of fine windblown material (loess), organic material, and water or ice. Hydraulicking has been the only effective method of removing frozen muck. All other methods require that the muck first be thawed. Common practice in Alaska and the Yukon Territory is to thaw with cold water, 50–60°F (10–16°C), pumped into boreholes through $^3/_8$-in.-diameter (10-mm) pipes called sweaters. Cold-water thawing has proven more cost-effective than either steam or hot water. *See* LOESS.

Techniques. For the actual mining of a placer deposit there are basically four different techniques: hand methods, hydraulic mining, use of mobile power equipment, and dredging.

Hand methods. In this technique, once the mainstay of placer operations, the mined material is shoveled into stockpile for hand washing or directly into a rocker box or "long tom" sluice box. This type of placer mining can be traced back to earliest recorded history, and was still a major factor through the Alaskan gold rush at the turn of the century. While hand mining now accounts for only a small portion of total placer production, it is often the most effective method for sampling prospective placer deposits.

Hydraulic mining. As mentioned above, hydraulic giants (monitors) have often been used to remove overburden. In addition, hydraulic pressure can be used to wash away the actual deposit. The resulting slurry is then channeled through sluices and the valuable material is recovered. Due to environmental regulations regarding water quality standards, the use of hydraulic giants in the United States has practically disappeared. They are still employed in the tin mines of Malaysia and Thailand, primarily for overburden removal prior to dredging. Another variation of hydraulic mining, called ground sluicing, uses water under natural pressure to wash the gravel through the sluices. Here, a stream is diverted and used to wash away the gravel bank. If the stream is first dammed and then released all at once, the process is called booming.

Mobile power equipment. The use of diesel powered equipment such as dozers, scrapers, front-end loaders, and hydraulic excavators has caused a revolution in placer mining technology. Not only does this equipment greatly increase the productivity of the miner for a relatively modest cost (compared to a floating dredge), but the mobility and flexibility of the equipment can be used for overburden removal, mining, and reclamation after mining. As a result, most modern placer mining operations will make use of mobile power equipment either in part or in total.

Dredging. Because of the intimate association of placer deposits and water, many of the more valuable deposits are located below the local water table level. If the water inflow is not severe, pumps can be used to dewater the cut and standard surface mining equipment can be used. However, if the water inflow is significant, such as with a modern beach placer or one located under a flowing river, then mining must be done with equipment that can dig below the waterline. This is called dredging.

There are three types of dredging operations used in placer mining: bucketline dredging, hydraulic dredging, and dragline excavation. The bucketline dredge (**Fig. 1**) is a floating excavator which uses a series of buckets attached to an endless rope. The bucketline travels from the floating dredge down to the lower tumbler at the end of the digging ladder, where it digs into the deposit, is pulled back to the surface, and dumps into the concentrating plant on the dredge. After concentration, a stacker conveyor deposits the waste material behind the dredge.

A cut is made by swinging the bucket ladder from side to side at a constant depth. At the end of each cut, the ladder is lowered and a new cut is begun. This continues until bedrock is reached or the maximum cutting depth is attained. Once a series of cuts

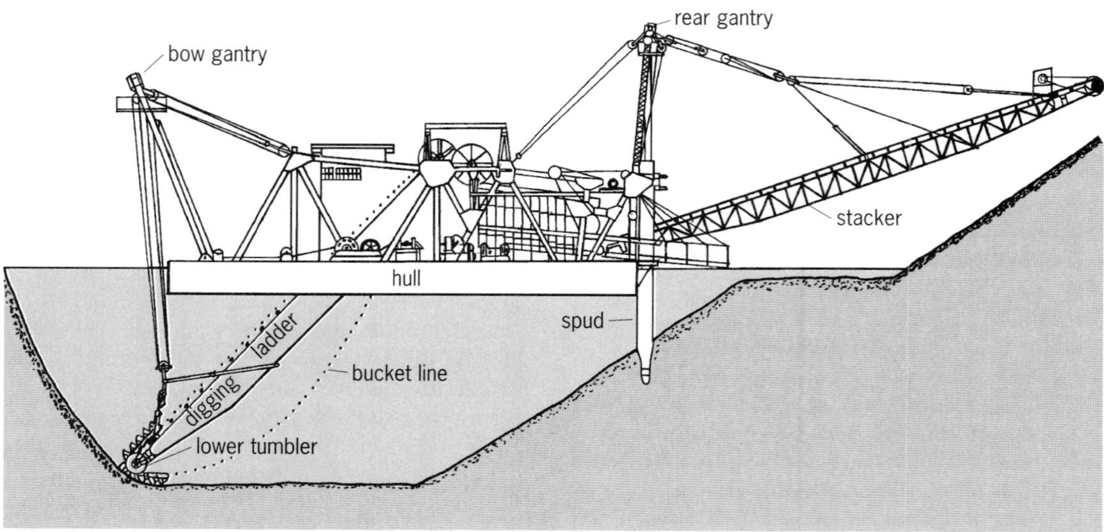

Fig. 1. California bucketline dredge.

Fig. 2. Aerial oblique view upvalley over the site of a placer dredging, in which successive cuts, tailing patterns, and dredge pond are shown. The dredge appears in the left foreground. (*Pacific Aerial Survey, Inc.*)

has been carried to maximum depth, the dredge is moved forward into the newly extended dredge pond and the process is begun again. In this manner, the dredge moves its pond along in front of it, filling in behind with processed gravel (**Fig. 2**).

The size of the buckets ranges 2–20 ft^3 (0.06–0.60 m^3) each. The average speed of the bucketline is usually 20–24 buckets per minute. This implies that the modern large dredge will produce 10,000–15,000 yd^3 (7600–11,500 m^3) per day. A 20-ft^3 (0.6-m^3) bucket dredge, mining tin in Malaysia, produced nearly 900,000 yd^3 (690,000 m^3) in 1 month, digging 150 ft (46 m) below waterline.

Hydraulic dredges are similar to bucketline dredges in that a digging boom is suspended from a floating barge and material is discharged on shore. The principal difference is that, instead of buckets, the excavation is done with a suction pump. A stream of high-pressure water is directed at the digging face near the suction intake to break up the placer material, which is then pumped to the surface. A variation of this process is the cutterhead hydraulic dredge, which uses a rotating cutting head to churn the placer material into suspension. Material mined with hydraulic dredges can be processed onboard the barge, or the slurry can be pumped ashore for processing and disposal.

Hydraulic dredges have found extensive use in offshore and beach placers where sand is the primary material to be moved. Two cutterhead dredges moving sediments and glacial till in Ontario, Canada, each averaged over 2,000,000 yd^3 (1,500,000 m^3) per month for 3 years. On the other end of the scale, small, 2–4-in.-diameter (5–10-cm) suction dredges are often used in rivers and streams for hand operations and prospecting.

The final type of dredging operation uses a dragline or clamshell excavator to dig below water level (**Fig. 3**). Mined material is usually deposited directly into the concentrating plant, which is either crawler-mounted on the shore or barge-mounted offshore. This type of operation is used on smaller deposits where the capital cost of a large dredge cannot be justified.

Concentration. There are two basic principles used to recover valuable minerals from mined placer material: size classification and gravity separation. These apply to all sizes of operations from the smallest hand

Fig. 3. Dragline dredge and pond. (*Bucyrus-Erie Co.*)

mining to the giant bucketline dredges, although in some special cases material adhesion or electromagnetic methods may also be used.

Size classification. This is the screening of the feed to the plant in order to discard oversized material which contains no values. In this manner, the gravity concentrating circuit not only has less material to process but works more efficiently because of the narrowed size range. This technique is effective because valuable placer minerals are discrete grains, separate from the gangue (waste) material and usually fairly small. For example, in a placer gold mining operation where the maximum expected nugget size is $^1/_4$ in. (6 mm), all material larger than 1 in. (25 mm) could be discarded without losing any gold. This would eliminate all of the boulders and cobbles as well as a significant amount of gravel.

Classification is usually done wet either over a series of vibrating screens or through a trommel, a long rotating, inclined cylinder with holes in its surface. In a trommel the material is fed into one end with sufficient water to wash the finer sands out through the holes, while the washed oversize is discharged out the other end, usually onto a stacking conveyor. *See* BULK-HANDLING MACHINES; MECHANICAL CLASSIFICATION; MECHANICAL SEPARATION TECHNIQUES.

Gravity concentration. The principle of gravity separation relies on the difference in specific gravity between the valuable minerals and the gangue.

1. Panning. The most simple gravity concentrator is the gold pan. In North America, these pans are flat-bottomed metal or hard plastic, 2–3 in. (5–7.5 cm) deep and 6–18 in. (15–23 cm) in diameter with sloping sides at a 30–40° angle. In some countries a 12–30-in.-diameter (30–75-cm) conical wooden or metal bowl called a batea is used instead of a flat-bottomed pan. In either case the pan filled with the material is immersed in water, shaken to settle the heavy materials, and swirled to wash away the lighter minerals.

The primary use of panning in modern placer mining is in prospecting and sampling. Some use is made of hand panning to separate gold from black sands (magnetite) where mercury amalgamation is not feasible.

2. Sluice boxes. A step up from hand panning, the sluice box is simply a trough with a series of riffles in the bottom. A mixture of placer material and water flows into one end and is washed down the inclined trough, which can vary in width from 12 to 60 in. (30 to 150 cm) and in length from a few feet to several hundred feet.

The riffles, which can consist of rocks, wooden blocks, dowels, or angle irons, produce eddy currents in the flow, allowing the heavy material to settle out while the lighter material is washed away (**Fig. 4**). Plastic outdoor carpeting (artificial turf) is also sometimes used to produce these eddy currents.

Mercury is often used in sluice boxes processing gold. The liquid mercury (specific gravity = 13.6) settles in behind riffles and is not washed away. When it comes in contact with gold, the gold particles adhere to the mercury, forming an amalgam. During cleanup of the sluice, the amalgam is collected and retorted to separate the gold; the cleaned mercury is recycled.

Undercurrents, often used in conjunction with the primary sluice, are auxiliary sluices which handle specific size fractions of the sluice feed. Screens, gratings, and punch plates within the sluice separate these size fractions and increase the efficiency of the sluicing operation.

Sluices are used to some extent in practically all placer operations. They are efficient, flexible and, because they have no moving parts, very low-cost concentrators.

3. Spinning bowls. Spinning bowls use principles similar to sluices. However, instead of using gravity as the settling force, they use the centrifugal force generated on a rapidly spinning surface. The sized placer material and water is fed into the center of the bowls. The heavy valuable minerals are trapped in baffles similar to riffles, and the lighter material is washed out over the rim. The advantage of the spinners is that a force higher than gravity can be generated, increasing the efficiency of the separation; this

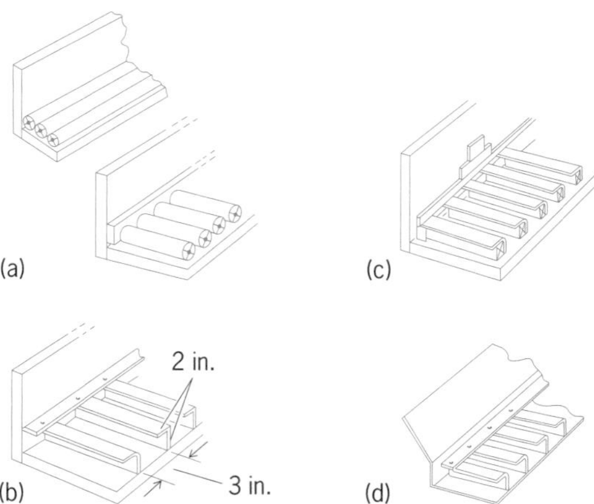

Fig. 4. Types of riffles in partial plan views. (*a*) Pole riffles. (*b*) Hungarian riffle (1 in. = 2.5 cm). (*c*) Oroville Hungarian riffle. (*d*) All-steel sluice. (*After G. J. Young, Elements of Mining, 4th ed., McGraw-Hill, 1946*)

force can be easily adjusted through motor speed to "fine-tune" the recovery.

4. Jigs. Another mechanism used to recover heavy minerals, especially on the tin mining dredges in Malaysia, is the jig. A jig is basically a water-filled box into the top of which the slurried placer minerals are fed. A rapid series of upward pulses are generated in the water at the bottom. The heavy minerals sink through these pulses into a hutch at the bottom, while the lighter material is carried out over the edge.

5. Other methods. These include hand picking, especially in diamond processing; adhesion to specially prepared surfaces, as with the grease tables used to collect diamonds; magnetic separation, to separate magnetite and other iron-containing minerals; and other gravity techniques such as spiral classifiers, vibrating tables, and heavy-media separation.

Environmental concerns. The two major environmental problems associated with placer mining are water pollution and land disturbance. Both the mining and the processing of placer minerals require a great deal of water, and, once used, this water contains large amounts of suspended solids. If the water is allowed to run off into the rivers, these solids can have an adverse impact on the downstream environment. In suspension they can harm aquatic habitats, and when settled out can clog waterways and choke off irrigated crops. In 1884, legislation was passed in California banning hydraulic mining because sediments from the Marysville district were inundating crops in the Sacramento River Delta and causing navigational hazards in the San Francisco Bay.

In the United States, environmental laws, such as the Clean Water Act, set limits on the discharge of suspended solids from a mining operation. To operate within these limits, miners use a series of settling ponds downstream from the mine and recycle as much water as possible. *See* ENVIRONMENTAL ENGINEERING; WATER POLLUTION.

Since most placer mining is surface mining, surface disturbance is necessary, especially where dredging operations create vast piles of cobbles as mining progresses. In some arctic or subarctic areas the ground is very sensitive to disturbance. One method of land reclamation is a mining plan which stockpiles top soil (where possible), recontours the spoil (waste) piles, and then returns the land to useful status. This can be done by revegetation to match the original environment, creation of new recreational or agricultural areas or, in urban areas, creation of new industrial sites. These latter alternatives can be extremely beneficial to the local economy, especially in developing nations. *See* LAND RECLAMATION.

Danny L. Taylor

Bibliography. L. Cope and L. Rice (eds.), *Practical Placer Mining*, 1992; B. A. Kennedy (ed.), *Surface Mining*, 2d ed., 1990; R. S. Lewis and G. B. Clark, *Elements of Mining*, 3d ed., 1964; E. H. Macdonald, *Alluvial Mining: The Geology, Technology and Economics of Placers*, 1983; C. YêKang, *The Combined Use of a Sand Screw, Hydrocyclones, and Gel-Logs to Treat Placer Mine Process Water*, 1988.

Placodermi

A class of fishes (Pisces) known from the Devonian Period, and a few that survived into the base of the Carboniferous. They were true fishes or gnathostomes, and can be distinguished from other fishes by the following characters: the gill chamber extends far under the cranium and is covered laterally by opercula; there is a neck joint between the cranium and the fused anterior vertebrae; often there is also a coaxial joint developed between the dermal bones of the cranial roof and shoulder girdle; the head and shoulder girdle are covered with dermal bones composed typically of cellular bone and superficially of semidentine instead of true dentine;

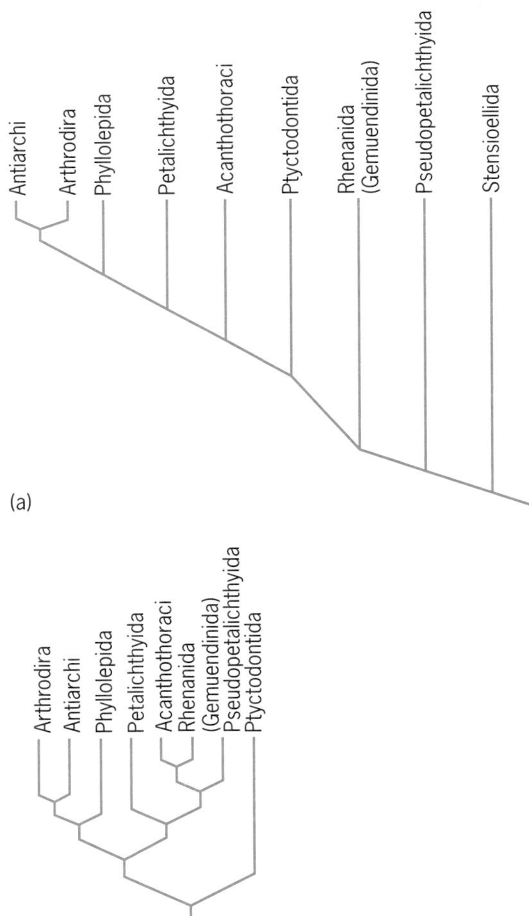

Fig. 1. Two cladograms portraying differing concepts of relationships of orders of placoderms: (*a*) according to Denison (*after R. H. Denison, Placodermi, vol. 2 of O. Kuhn and H. P. Schultz, eds., Handbook of Palaeoichthyology, 1978*); and (*b*) according to Miles and Young (*after R. S. Miles and G. C. Young, Placoderm relationships reconsidered in light of new ptyctodontids from Goga, Western Australia, Linn. Soc. Symp., 4:123–198, 1977*). Note particularly the absence of Stensioellida in *b*, the positions of Ptyctodontida, and alternative associations of some orders. See text for further explanation.

the bones are commonly ornamented with tubercles or ridges; the endoskeleton is cartilage and may be calcified in a globular fashion or perichondrally ossified; the notochord is persistent, and the vertebrae consist only of neural and hemal arches; the tail is diphycercal or slightly heterocercal, and an anal fin is lacking.

Most placoderms were bottom-dwelling fishes, with the head and trunk dorsoventrally depressed; only the Stensioellida and some specialized arthrodires had laterally compressed and deepened bodies, suggesting a more nectic manner of life. Most of them were small or moderate-sized, but a few were, for their time, gigantic, reaching a length of as much as 20 ft (6 m). They were the dominant fishes of the Devonian, and are found in both marine and fresh-water deposits.

There have been many opinions regarding the relationships of the Placodermi. Some have considered them nearest to the Chondrichthyes, but their possession of dermal bones with unique histology as well as other characters makes this unlikely. Others have thought them to be relatives of the Osteichthyes, but this also is questionable. At present, it seems that they may best be considered a separate class of fishes, equivalent to the Chondrichthyes and Teleostomi.

The Placodermi are subdivided into nine orders, each with its own distinct specializations. The evolution and interrelations of these orders are matters of disagreement. According to the hypothesis of R. S. Miles and G. C. Young, primitive placoderms had well-developed head and trunk shields, the trunk shield consisting of a complete ring composed of two median and five paired plates: in different orders the trunk shield was either enlarged or reduced. According to R. H. Denison, the primitive placoderm shoulder girdle was short with few plates, and this was progressively ossified and enlarged in different orders, forming a trunk shield in some. The resultant differences in phylogenetic interpretations are illustrated in the two cladograms in **Fig. 1**. Arranged in order of increasing specialization according to Denison's hypothesis, the placoderm orders are as follows.

Class: Placodermi
- Order: Stensioellida
  - Pseudopetalichthyida
  - Rhenanida
  - Ptyctodontida
  - Anthracothoraci
  - Petalichthyida
  - Phyllolepida
  - Arthrodira
    - Suborder: Actinolepina
      - Wootagoonaspina
      - Phlyctaeniina
      - Brachythoraci
- Order: Antiachi

Stensioellida. This order is based on a *Stensioella heintzi* from the Lower Devonian Hunsrückschiefer of Germany. This is a slender, unarmored fish about 12 in. (30 cm) long (**Fig. 2**), quite different from other members of the class, but shown to be related by the presence of three small bones on its head, by the extension of the gill chamber far under the cranium, and by the articulation between the endocranium and the fused anterior vertebrae. The head is bluntly rounded and mostly covered with small denticles, the orbits are probably lateral, the mouth is ventral, and the jaws carry only small, pointed denticles. The shoulder girdle is short and not expanded into a trunk shield as in most placoderms. The body is gradually tapering and covered with small scales. The pectoral fins are long, slender, and flaplike, and the pelvic fins are small. This may be the most primitive known placoderm, but its relationships cannot be determined with certainty from the single known specimen.

Pseudopetalichthyida. This order is based on two genera, *Pseudopetalichthys* (**Fig. 3**) and *Paraplesiobatis*, from the Lower Devonian Hunsrückschiefer of Germany. They are slender fishes, 4 to 6 in. (10 to 15 cm) long, with prominent, dorsally placed orbits, a ventral mouth, and a skull covered with many small, bony plates and large, lateral gill covers. As

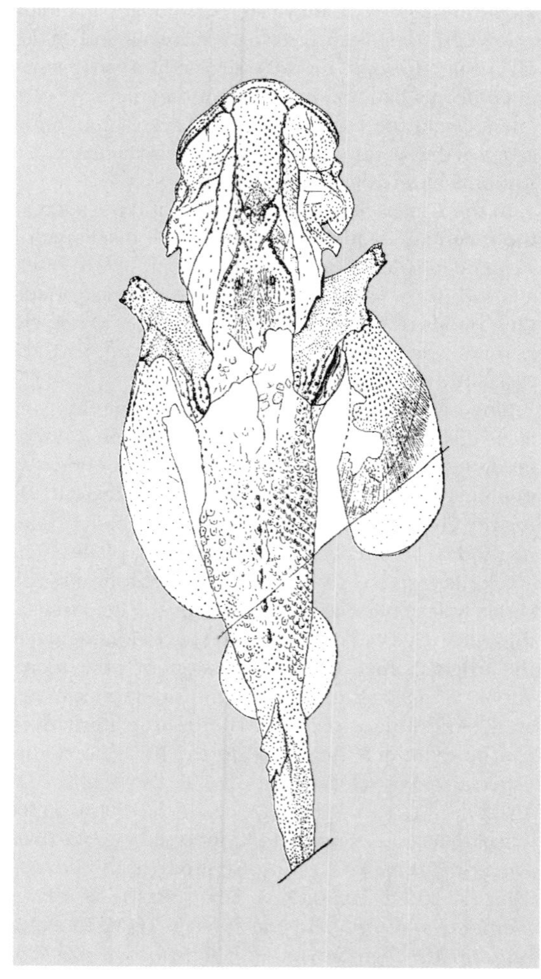

Fig. 2. *Stensioella heintzi* in dorsal view; length about 10 in. (25 cm).

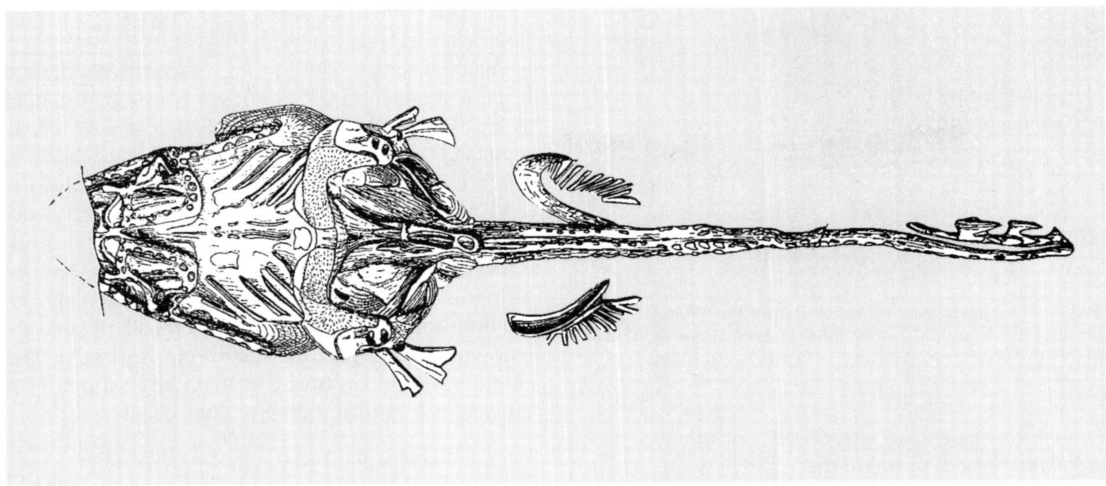

Fig. 3. *Pseudopetalichthys problematica*, ventral view; length about 6 in. (15 cm).

in other placoderms, the gills extend far under the cranium, and there is a well-developed articulation between the cranium and fused anterior vertebrae. The shoulder girdle is short and not expanded into a trunk shield, the pectoral fins are narrow-based, and the pelvic fins are relatively long-based. The trunk and tail may be covered with scales, and there is a series of crested, median dorsal scales. These fishes once were referred to the Stensioellida, but they differ in important respects.

Rhenanida. This order, also known as Gemuendinida, includes the genera *Gemuendina* from the Lower Devonian of Germany, *Asterosteus* from the Middle Devonian of Ohio, and *Jagorina* from the Upper Devonian of Germany. They are flattened, raylike fishes with large, semicircular pectoral fins, dorsal eyes and nostrils, and a terminal mouth (**Fig. 4**). The head shield consists of a few large plates separated by mosaics of small tesserae. The trunk shield is formed of only a single ring of plate, projecting ventrally far under the head.

The flat raylike braincase is not as broad as in arthrodires; it has small lateral processes and a pair of stout posterior processes for articulation with the anterior vertebrae, which are solidly fused into a single structure around the notochord. The jaws and gill covers were highly mobile, and the former could be protruded to open dorsally in front of the head. A stout hyomandibular suspended both the upper and lower jaw cartilages, which are well ossified and free of the braincase and dermal skeleton. There are no tooth plates, although the lips were probably armed with stellate tubercles. Small scales cover the whole body and the pectoral and pelvic fins in *Gemuendina*. The propulsive force in swimming came from the muscular body and not from the winglike pectorals, which were too stiffly armored to undulate as in rays.

The rhenanids are remarkably raylike and, like living skates and rays, were adapted for lying on the bottom of the sea and covering themselves with sand stirred up by the pectoral fins. It has been suggested that rhenanids were directly ancestral to rays, but this cannot be true as they differ from rays in the dorsal position of the nostrils and numerous other anatomical characters. The similarities are due to convergent evolution associated with closely similar modes of life. *See* CHONDRICHTHYES.

Ptyctodontida. These are small fishes from the Devonian and lowest carboniferous of North America, Eurasia, Africa, and Australia. They include both marine and fresh-water species. Usually only the

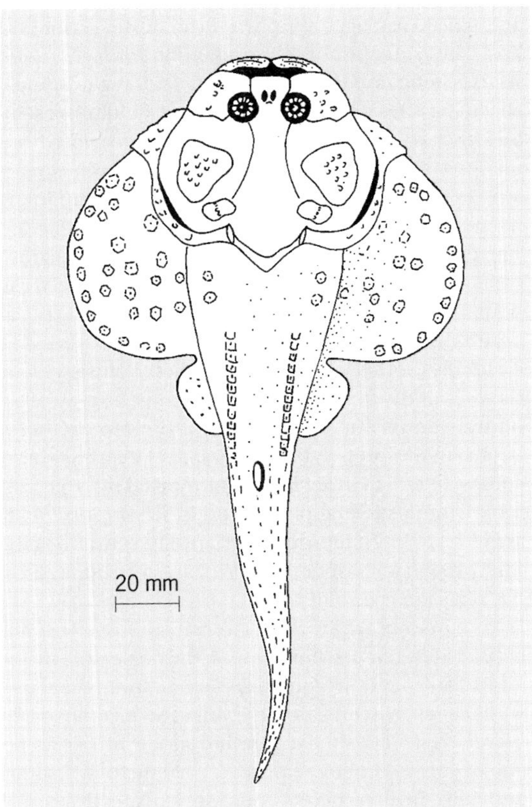

Fig. 4. *Gemuendina stuertzi* from the Lower Devonian of Germany; restoration in dorsal aspect.

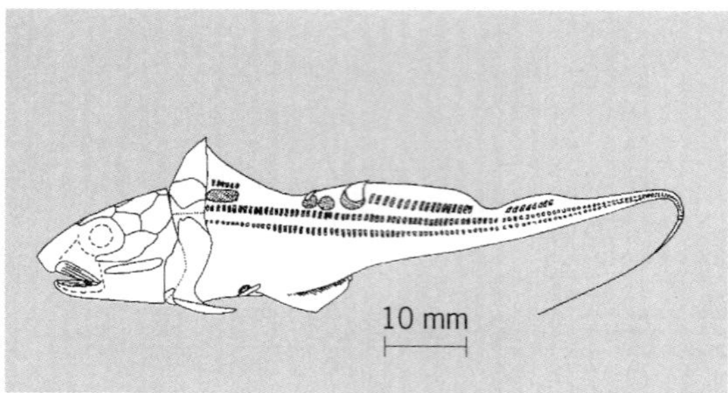

Fig. 5. ***Rhamphodopsis threiplandi*** **from the Middle Devonian of Scotland. Restoration of skeleton with outline of body in lateral aspect.**

stout tooth plates are preserved, but ossified internal skeletons are found in the Scottish genus *Rhamphodopsis* (**Fig. 5**) and the German *Ctenurella*.

Both the head and trunk shields are present, and the joint between them is well differentiated and variable in structure. The head shield has a greatly reduced number of bones which show some similarity to those of petalichthyids in their arrangement and in the central X-anastomosis of pairs of the deeply situated laterosensory lines. The upper jaw cartilages are superficially ossified, large, and firmly attached to the braincase, whereby they gain support. They carry a single pair of large tooth plates which are opposed by a smaller pair of plates in the lower jaw. The tooth plates are variously modified, for shearing in *Rhamphodopsis* and *Ctenurella*, and for crushing with either a large central area (*Ptyctodus*) or separated cups (*Palaeomylus*) of a hard dentinal tissue. Rostral and labial cartilages supported the fleshy snout and lips. The trunk armor is short, and in *Rhamphodopsis* has long median dorsal and lateral fin spines, absent in *Ctenurella*. The vertebral column resembles that of arthrodires, with an unconstricted notochord and separate dorsal and ventral arches with long spines. There are two dorsal fins with an internal skeleton, and a small dorsal lobe in the caudal fin. The pectoral fins are large and mobile with a narrow base. The large pelvic fins are remarkable for their sexual dimorphism associated with internal fertilization, as in Chondrichthyes. In males, claspers bearing a laterally toothed dermal plate project ventrally from the root of the fins; they worked against a spiny plate on the ventral surface. In front of the fins are holocephalanlike prepelvic claspers, also bearing sharp spines. In females the fins are supported by large, immobile endoskeletal plates, and in *Rhamphodopsis* are clothed by numerous, overlapping ventral scales. The prepelvic claspers of males are represented by flat plates in the skin.

Ptyctodonts are of great interest as possible holocephalan ancestors, but because they resemble living rabbitfishes and are quite dissimilar from their fossil relatives, it seems possible that the resemblances merely reflect a similar benthic mode of life as result of convergent evolution. According to R. S. Miles and G. C. Young, ptyctodonts are collateral descendants of all other placoderms, but Denison considers them primitive in retaining a short shoulder girdle. *See* HOLOCEPHALI.

Acanthothoraci. This order, alternatively called Palaeacanthaspidodea, is known from eight genera from the Lower Devonian of Eurasia and arctic Canada. Their relationships to other placoderms is uncertain. The skull is long, rather narrow, and covered with dermal bones and sometimes with a mosaic of small tesserae; it is characterized by the presence of two pairs of paranuchal plates (**Fig. 6**). The orbits are lateral or dorsolateral, and the nostrils dorsal and bounced in front by a premedian plate, unknown in other placoderms except Antiarcha. The shoulder girdle is short and high, and topped by a median dorsal plate that often has a high spine.

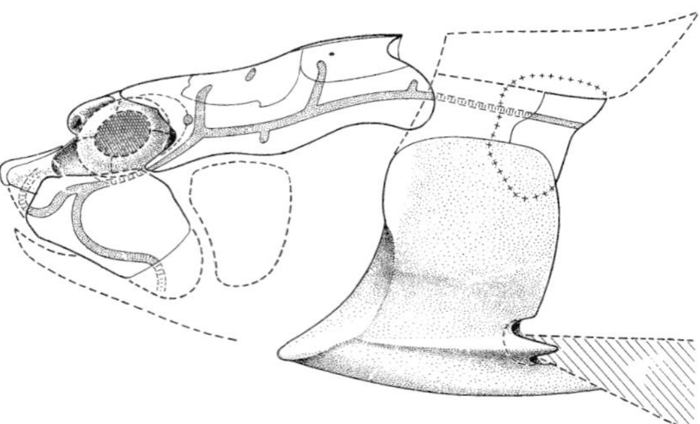

Fig. 6. ***Romundina stellina*****, head and shoulder girdle in lateral view; length about 1.2 in. (3 cm).**

Petalichthyida. This small order of dorsoventrally flattened placoderm fishes ranges from the Lower to the Upper Devonian. They are never common and are an exclusively marine group, seldom exceeding 1.7 ft (0.5 m) in length. Typical genera include *Lunaspis* (**Fig. 7**) from the Lower Devonian of Germany and *Macropetalichthys* from the Middle Devonian of North America and Germany.

The external armor is formed by a head and trunk shield, each composed of large plates. The head is characterized by the dorsal position of the orbits, which are completely enclosed in the shield, and the long occipital region which has a marked effect on the bone pattern in the hind end. The laterosensory lines are deeply sunk in the bones and open to the

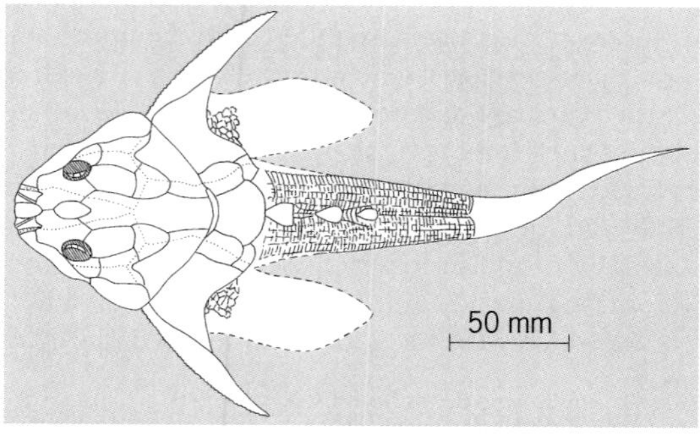

Fig. 7. ***Lunaspis broilii*** **from the Lower Devonian of Germany; restoration in dorsal aspect.**

outside through small pores; further, a pair of canals meet in the center of the shield in a characteristic X-shaped junction. The braincase has a long, narrow occipital region, but is otherwise broad and flat with prominent lateral processes in the postorbital and occipital regions. The jaws and tooth plates are unknown. The trunk armor is short, with long spines in front of the pectoral fins and a poorly differentiated articulation with the head shield. In *Lunaspis* the body posterior to the armor is covered with scales, including three large dorsal ridge scales in place of dorsal fins.

Two genera from the Lower and Middle Devonian of China differ from typical members of the order, the Macropetalichthyidae, in having lateral orbits. They are believed to represent a more primitive family, the Xinanpetalichthyidae.

Although they survived through the Devonian, the evolutionary history of the petalichthyids is poorly known. Evidently they were benthic fishes that occupied a stable adaptive zone. Their adaptations are similar to those of some primitive Arthrodira, but their shorter trunk shield, absence of pectoral fenestrae, and heavily scaled tail suggest that they are more primitive than members of that order.

Phyllolepida. This order is known principally from a single genus *Phyllolepis* from the Upper Devonian. They were fresh-water fishes about 1 ft (0.3 m) long. The genus is widely distributed in Europe, Greenland, and Australia.

The armor is broad and flat; the component plates are easily recognized by their characteristic ornamentation of concentric and transverse ridges. The head shield includes a large central plate, surrounded by smaller plates which are not easily compared to the plates of other placoderms. Shallow grooves which converge on the growth center of the central plate mark the course of the laterosensory lines. There is no sign of the orbits, so unless these fishes were blind the eyes must have lain anterior or lateral to the bony shield (as shown in **Fig. 8**). The trunk shield has fewer plates than in Arthrodira, lacking most of the posterior series of plates of that order.

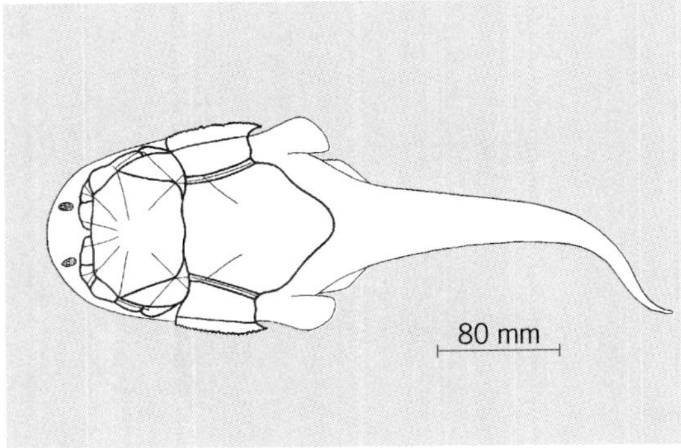

Fig. 8. ***Phyllolepis orvini*** **from the Upper Devonian of Greenland. The restoration of the dermal armor is shown in dorsal aspect. The outline of the front of the head and the body is hypothetical, but shows how the fish might have appeared in life.**

The spinal plates are stout and the joint with the head shield poorly developed. The braincase, tooth plates, and jaws are not known. One specimen from Scotland has elements of the vertebral column. There are separate dorsal and ventral arches, which in life sat astride the notochord.

Phyllolepis was adapted for life on the bottom in streams and floodplains, where it probably fed on organic detritus. Its ancestry is unknown, although a Middle or Upper Devonian genus from Antarctica, *Antarctaspis*, may be a primitive member of the order. New phyllolepid material from Australia has led to the suggestion that phyllolepids may be related to primitive actinolepid arthrodires.

Arthrodira. These are the best-known, most varied, and most successful order of Placodermi, found in fresh-water and marine sediments of Devonian and lowest Carboniferous age in all continents except South America. The pattern of their skull bones is characteristic, the orbits are usually lateral, and the nostrils are anterior. Their jaw suspension is unique: the palatoquadrates attach anteriorly both to the suborbital dermal bones and to the neurocranium; posteriorly they attach to the postsuborbital dermal bones. The hyomandibuli do not aid in jaw suspension, but primitively hinge the gill covers to the endocranium. The dermal jawbones are two pairs of supragnathals and one pair of infragnathals. The dermal shoulder girdle is enlarged to form a trunk shield which typically consists of three median and eight paired bones. The dermal cranio-thoracic joint is primitively of a sliding type, but in most it is ginglymoid, with condyles on the trunk shield fitting in glenoid fossae on the skull. Long, low scapulocoracoids support the pectoral fins, which are primitively narrow-based and articulate through fenestrae in the trunk shield. In more advanced arthrodires, the lateral walls of the trunk shield are reduced so that the pectoral fenestrae can enlarge and open behind, permitting the bases of the pectoral fins to become long-based.

The three suborders of the order Arthrodira as listed in the classification of the Placodermi represent formal taxonomic units; as many as six such units are recognized in alternative classifications. There is no clear evidence that these suborders are monophyletic. Thus, the suborders may be treated informally as evolutionary grades, without an implication that the included species had a common specific base. This way of viewing the relationships of the arthrodires was suggested by Miles in 1969 in his study of their feeding mechanisms and is accepted in the following discussions.

The most primitive known arthrodires are from the Lower Devonian. Along with a number of Middle Devonian forms, they represent the organizational level of the Actinolepina. In the most primitive genera the cranial and thoracic shields are extensive, covering about 60% of the body. The remainder of the body was heavily scaled (**Fig. 9***a*). The intershield joint was not present. The plates tended to slide past one another, in contrast to those of less primitive arthrodires in which a joint was

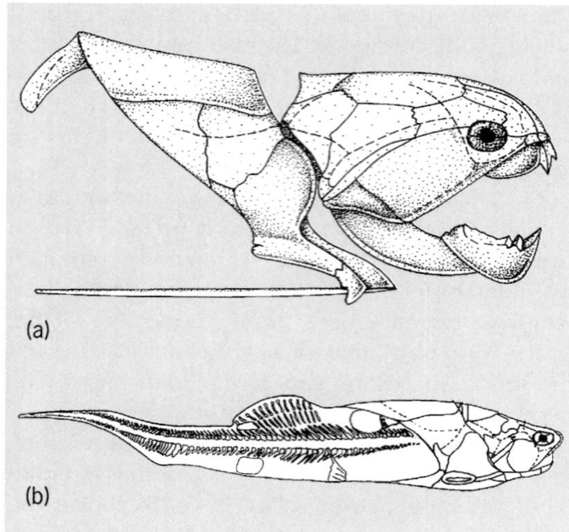

Fig. 9. Representative arthrodires. (*a*) Cranial and thoracic shields of giant Upper Devonian *Dunkelosteus* (*Dinichthys*), a brachythoracic arthrodire showing reduction of thoracic shield, large nuchal gap, and elevated cranio-thoracic joint. Length of originals 8–10 ft (about 3 m) (*after E. H. Colbert, Evolution of the Vertebrates, John Wiley and Sons, 1955*). (*b*) *Coccosteus*, a moderately derived Middle Devonian brachythoracic arthrodire, showing the structure of the postcranial axial skeleton and fins. Note fenestra for pectoral fin located in the basal-lateral portion of the thoracic shield. Original about 13.5 in. (35 cm) in length (*after R. S. Miles and T. S. Westoll, Trans. Roy. Soc. Edinburgh, vol. 67, 1968*).

present. Altogether some 15 genera have been assigned to this group based on specimens from deposits in Europe, North America, eastern Asia, and Australia. Such primitive forms were at the base of a complex, adaptive radiation of arthrodires which climaxed in the Upper Devonian.

Among the probable derivatives of the Actinolepina were the distinctive form *Wuttagoonaspis* of the Middle Devonian of Australia, the sole genus of the suborder Wuttagoonaspina, and members of the Phlyctaeniina. Both groups are similar in structure to the Actinolepina, but a cranio-thoracic joint is present. The spines that lay anterior to the pectoral fins are strongly developed, and the median dorsal plates are elongated. *Dicksonosteus* from the Lower Devonian (Fig. 9*b*) is a fairly representative example. A number of lines of specialized genera, predominantly Middle Devonian in age, arose from Lower Devonian forms. About 20 are assigned to the family Phlyctaenidae and another 10 to the family Holonematidae. In all of them the basic structural plan is preserved, but superimposed upon it are specializations related to various modes of feeding and locomotion, indicating considerable habitat diversification. The Brachythoraci include an extensive array of arthrodires grouped into five superfamilies. Approximately 60 genera of Brachthoraci have been assigned to the listed superfamilies. In addition, some 35 others are not currently placed in any of them, being either too specialized or too poorly known. Together these genera provide the tangible evidence of a massive adaptive radiation of the derived arthrodires.

As the Devonian radiations of the arthrodires were taking place, the cartilaginous fishes (Chondrichthyes) and the bony fishes (Osteichthyes) were undergoing somewhat similar expansions. Sharks in particular paralleled the predaceous arthrodires, but were more active. Members of both groups undoubtedly were direct competitors with the placoderms and probably played a role in what appears to have been a rather rapid placoderm extinction during the Late Devonian and very early lower Carboniferous. Changes in invertebrate animals and in the physical environment also contributed to the demise of placoderms. Just how rapid this extinction actually was and its precise causes remain uncertain, masked by the incompleteness and biases of the fossil record as now known. What is clear is that while the last of the arthrodires died out, the cartilaginous and bony fishes remained successful and continued to expand their roles in the waters of the oceans and the continents. *See* OSTEICHTHYES.

Antiarcha. This is a highly specialized order, characterized by heavy armor consisting of a short head shield and a very long, flat-bottomed trunk shield (**Fig. 10**). Most antiarchs are small or moderate in size, the largest reaching a length of about 48 in. (120 cm). The head shield is so greatly modified that it is difficult to homologize its bones with those of other placoderms. The eyes and nasal openings occupy an orbital fenestra on top of the skull; in front of them is a premedian plate, unknown in most other placoderms. The mouth is ventral. There is an articulation between the dermal bones of the head and trunk shields, but contrary to the condition in Arthrodira, the condyles are on the head shield, and the glenoid fossae on the trunk shield. The trunk shield has both anterior and posterior median

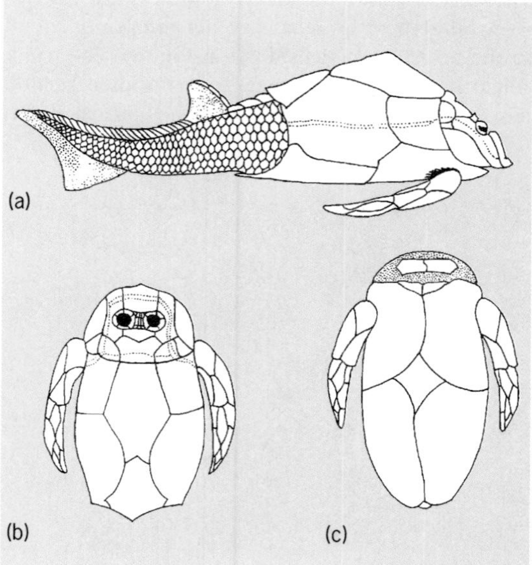

Fig. 10. *Pterichthyodes* (*Ptericthys*), a Middle Devonian antiarch. (*a*) Lateral view showing scale-covered tail; about 15 cm (6 in.) in length. (*b*) Dorsal and (*c*) ventral views of the armor. (*After Traquair*)

dorsal plates. There are peculiar pectoral appendages, completely covered by small plates, usually jointed near midlength, and articulating in axial fossae on the sides of the trunk shield. The body behind the shield is scaled or naked, and has one or two dorsal fins and a heterocercal caudal fin; pelvic fins are possibly represented by "frills" behind the shield of *Bothriolepis*.

Antiarcha were predominantly fresh-water stream and lake dwellers, though a few lived in the sea. They were benthic forms, propelling themselves along the bottom by their pectoral appendages, and surely feeding on the bottom; one genus, *Bothriolepis*, has been shown to have been a mud grubber by the preserved content of its spiral intestine. Antiarcha probably evolved from primitive Arthrodira. The most primitive, the Yunnanolepidae, come from the Lower Devonian of China. The other families, Asterolepidae, Bothriolepidae, and Sinolepidae, are found in the Middle and Upper Devonian and perhaps lowest Carboniferous of North America, Greenland, Eurasia, Australia, and Antarctica.

Robert H. Denison; Everett C. Olson

Bibliography. R. H. Denison, Further considerations of placoderm evolution, *J. Vert. Paleontol.*, 3:69–83, 1983; R. H. Denison, *Placodermi*, vol. 2 of *Handbook of Palaeoichthyology*, edited by O. Kuhn and H. P. Shultz, 1978; B. G. Gardiner, The relationships of the placoderms, *J. Vert. Paleontol.*, 4:379–395, 1984; R. S. Miles and G. C. Young, Placoderm relationships reconsidered in light of new ptyctodontids from Goga, Western Australia, *Linn. Soc. Symp.*, 4:123–198, 1977.

Placodontia

A small but interesting order of marine reptiles of the infraclass Sauropterygia that are known only from deposits of Triassic age of Europe and the Near East (Israel). As the name implies, they are reptiles with grossly specialized dentitions—flat-crowned teeth are located in both the upper and lower jaws and on the palate—that probably functioned as crushing devices for hard-shelled prey (see **illus.**).

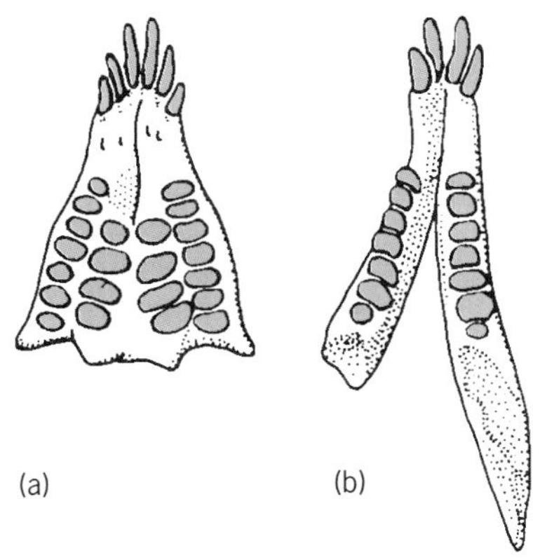

Dentition of *Paraplacodus broilii*, Triassic, Switzerland. (*a*) Upper jaw. (*b*) Lower jaw.

Morphology. The modification of the dentition had its effect on the entire skull, which became massive in relation to the body; and the coronoid region of the mandible, to which most of the masticatory muscles were attached, became greatly extended upward, giving it a superficial similarity to a mammalian jaw. The postcranial skeleton shows a number of aquatic adaptations. The thorax was box-shaped and the digits of hand and foot were probably webbed; the joint surfaces of the limb bones suggest that the animals did little if any walking on land. The advanced placodonts, for example, *Placochelys*, resemble sea turtles in overall appearance and in the fact that their body is encased in an armor of dermal bones similar to that of the dermochelyid sea turtle *Psephophorus*. The more generalized genera lack such an armor and possess a long tail. *See* DENTITION.

Evolutionary trends. In the relatively brief span of Triassic time during which the differentiation of the placodonts is documented, a number of striking evolutionary trends are evident. The most obvious are the acquisition of a dermal armor and a progressive reduction of the number of teeth in the dentition, coupled with an increase in the size (relative to skull size) of the remaining teeth from *Paraplacodus* to *Placodus* to *Cyamodus* to *Placochelys*. Ontogenetic tooth reduction occurs in the genus *Cyamodus*. In *Placochelys* and *Henodus* the anterior teeth have been lost and probably replaced by a horny beak. The entire dentition of *Henodus* consists of one crushing tooth per jaw quadrant.

One early form, *Helveticosaurus*, has primitive, conical, pointed teeth, and is thus not readily identifiable as a placodont; its assignment to this order is based on features of the postcranial skeleton.

Although evolutionary trends are observable through time, primitive and fairly specialized forms such as *Helveticosaurus, Paraplacodus*, and *Cyamodus* occur together in the same beds and these three genera are the earliest placodonts on record. This would indicate that the placodonts had a considerable, though unknown, prehistory and it seems probable that they had a wider geographic distribution than the present record would indicate. *See* ANIMAL EVOLUTION.

Phylogeny. Although *Helveticosaurus* represents a very primitive genus of placodonts, it contributes little toward the question of the phyletic origin of the order. This is partly because of the notable marine adaptations of the postcranial skeleton which tend to obscure the characters that might have linked it with its presumed terrestrial ancestors. In terms of skull morphology the placodonts show similarities to the sauropterygians, but the postcranial skeleton, especially the vertebral column, is so different that a derivation of the placodonts from this order seems quite improbable. *See* DIAPSIDA; REPTILIA; SAUROPTERYGIA.

Rainer Zangerl

Bibliography. M. J. Benton, *Vertebrate Paleontology*, 1991; R. L. Carroll, *Vertebrate Paleontology and Evolution*, 1988.

Plague

An infectious disease of humans and rodents caused by the bacterium *Yersinia pestis*. The sylvatic (wild-animal) form persists today in more than 200 species of rodents throughout the world. At present, contact with wild rodents and their fleas leads to sporadic human disease. The explosive urban epidemics of the Middle Ages, known as the Black Death, resulted when the infection of dense populations of city rats living closely with humans introduced disease from the Near East. The disease then was spread both by rat fleas and by transmission between humans. During these outbreaks, as much as 50% of the European population died; occasionally, entire towns disappeared. *See* INFECTIOUS DISEASE.

In modern times, sylvatic plague has been concentrated in the southwestern United States, southern Russia, India, Indochina, and South Africa. In the United States, ground squirrels, mice, wood rats, prairie dogs, chipmunks, and, rarely, rabbits can be carriers. Rodent infection (and subsequent human disease) occurs most often in the spring and summer; its incidence varies year to year because of animal die-off, but it rarely completely disappears from a region. Feline predators of rodents, such as mountain lions and domestic cats, generally die when infected with *Y. pestis*; however, canines, such as foxes, coyotes, and dogs, often recover and may serve as serologic sentinels to indicate the presence of the disease in wild rodents. Humans can acquire the disease when domestic cats or dogs carrying plague-infected rodents or their fleas return to rural homes.

After infection by *Y. pestis*, fleas develop obstruction of the foregut, causing regurgitation of plague bacilli during the next blood meal. The rat flea, *Xenopsylla cheopsis*, is an especially efficient plague vector, both between rats and from rats to humans. Human (bubonic) plague is transmitted by the bite of an infected flea; after several days, a painful swelling (the bubo) of local lymph nodes occurs. Bacteria can then spread to other organ systems, especially the lung; fever, chills, prostration, and death may occur. Plague pneumonia develops in 10–20% of all bubonic infections.

Pneumonia starts as a secondary lung infection during bubonic/solpticemic disease; it may then be transmitted from human to human by bacterial aerosols that are spread by coughing. It has a rapid course in close quarters and is almost universally fatal if untreated. In some individuals, the skin may develop hemorrhages and necrosis (tissue death), probably the origin of the ancient name, the Black Death. The last primary pneumonic plague outbreak in the United States occurred in 1919, when 13 cases resulting in 12 deaths developed before the disease was recognized and halted by isolation of cases.

Since about 1960, the incidence of sporadic human plague in the western United States has been rising. Although urban rat-related human outbreaks are now rare, they represent a continual threat, which has generally been contained by public health rat-control measures. Spread from wild rodents into urban rats, however, occurred as recently as 1983 in Los Angeles.

Bubonic plaque is suspected when the characteristic painful, swollen glands develop in the groin, armpit, or neck of an individual who has possibly been exposed to wild-animal fleas in an area where the disease is endemic. A culture of blood or of material aspirated from a bubo will grow the organism in 2–5 days. Immediate identification is possible, however, by microscopic evaluation of bubo aspirate stained with fluorescent-tagged antibody. Serologic diagnosis, shown by a rise in antiplague antibody in the blood, takes 2–3 weeks. Antibiotics should be given if plague is suspected or confirmed. Such treatment is very effective if started early. The current overall death rate, approximately 15%, is reduced to less than 5% among patients treated at the onset of symptoms. *See* IMMUNOFLUORESCENCE; MEDICAL BACTERIOLOGY. Darwin L. Palmer

Bibliography. T. Butler, *Plague and Other Yersina Infections*, 1983.

Plains

The relatively smooth sections of the continental surfaces, occupied largely by gentle rather than steep slopes and exhibiting only small local differences in elevation. Because of their smoothness, plains lands, if other conditions are favorable, are especially amenable to many human activities. Thus it is not surprising that the majority of the world's principal agricultural regions, close-meshed transportation networks, and concentrations of population are found on plains. Large parts of the Earth's plains, however, are hindered for human use by dryness, shortness of frost-free season, infertile soils, or poor drainage. Because of the absence of major differences in elevation or exposure or of obstacles to the free movement of air masses, extensive plains usually exhibit broad uniformity or gradual transition of climatic characteristics.

Distribution and varieties. Somewhat more than one-third of the Earth's land area is occupied by plains. With the exception of ice-sheathed Antarctica, each continent contains at least one major expanse of smooth land in addition to numerous smaller areas. The largest plains of North America, South America, and Eurasia lie in the continental interiors, with broad extensions reaching to the Atlantic (and Arctic) Coast. The most extensive plains of Africa occupy much of the Sahara and reach south into the Congo and Kalahari basins. Much of Australia is smooth, with only the eastern margin lacking extensive plains. *See* TERRAIN AREAS.

Surfaces that approach true flatness, while not rare, constitute a minor portion of the world's plains. Most commonly they occur along low-lying coastal margins, the lower sections of major river systems, or the floors of inland basins. Nearly all are the products of extensive deposition by streams or in lakes or shallow seas. The majority of plains, however, are distinctly irregular in surface form, as a result of valley-cutting by streams or of irregular erosion and deposition by continental glaciers.

Plains are sometimes designated by the situations in which they occur. In common speech a coastal plain is any strip of smooth land adjacent to the shoreline, though in geology the term is usually restricted to such a plain that was formerly a part of the shallow sea bottom. An example that fits both definitions is the South Atlantic and Gulf margin of the United States. Intermontane plains lie between mountain ranges, and basin plains are surrounded by higher and rougher land. Upland plains (sometimes loosely termed plateaus) lie at high elevations, or at least well above neighboring surfaces, while lowland plains are those lying near sea level, or distinctly below adjacent lands.

Types of plains are also sometimes designated according to the processes that have produced their distinctive surface features. These differences are discussed below.

Origin. The existence of plains terrain generally indicates for that area a dominance of the erosional and depositional processes over the forces that deform the crust itself. Most of the truly extensive plains, such as those of interior North America or that of northwestern Eurasia, represent areas which have experienced nothing more severe than slow, broad warping of the crust over tens or hundreds of millions of years. Throughout that time the gradational processes have been able to maintain a relatively subdued surface. Certain other areas, including the upland plains of central and south-central Africa and southern Brazil, have suffered moderate general uplift in late geologic time and have not yet been subjected to deep valley cutting.

Many plains of lesser extent, however, have been formed in areas where crustal deformation has been intense. Most of these represent depressed sections of the crust which have been partially filled by smooth-surfaced deposits of debris carried in by streams from the surrounding mountains. Examples are the Central Valley of California, the Po Plain of northern Italy, the plain of Hungary, the Mesopotamian plain, the Tarim Basin of central Asia, and the Indo-Gangetic plain of northern India and Pakistan.

Surface Characteristics

The detailed surface features of plains result mainly from local erosional and depositional activity in relatively recent geologic time. Each of the major gradational agents—running water, glacial ice, and the wind—produces its own characteristic set of features, and any given section of plains terrain is characterized predominantly by features typical of one particular agent.

Features associated with stream erosion. Plains sculptured largely by stream erosion, which are far more widespread than any other class, are normally irregular rather than flat. Integrated valleys and the divides between them are the hallmarks of stream sculpture, and the differences among stream-eroded plains are generally expressible in terms of the size, shape, spacing, and pattern of these features.

The shallow depth of valleys on plains usually indicates that the surface has not been uplifted, in late geologic times, far above the level to which the local streams can erode. Wide valleys commonly indicate weak valley-wall materials, active erosion on the valley sides by surface runoff and soil creep and, in many instances, a long period of development. Resistant materials and permeable upland surfaces, on the other hand, favor narrower valleys and steeper side slopes.

Differences in valley spacing or in the degree to which the plain has been dissected by the development of valley systems are especially striking. Ideally, tributary growth is considered to proceed progressively headward from the major streams into the intervening uplands. For this reason, plains on which tributary valleys are few, major valleys are widely spaced, and broad areas of uncut upland remain (**Fig. 1**) are often termed youthful. If tributary valleys have extended themselves into all parts of the

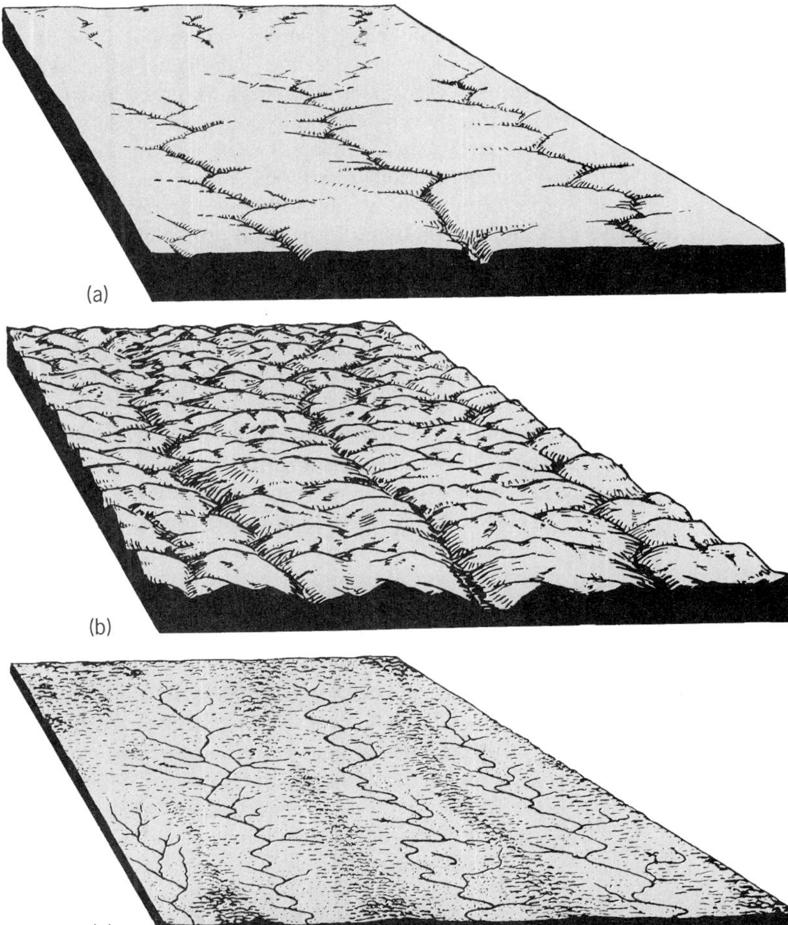

Fig. 1. Ideal stages in the progressive development of a stream-eroded landscape. (*a*) Youth. (*b*) Maturity. (*c*) Old age. (*After G. T. Trewartha, A. H. Robinson, and E. H. Hammond, Elements of Geography, 5th ed., McGraw-Hill, 1967*)

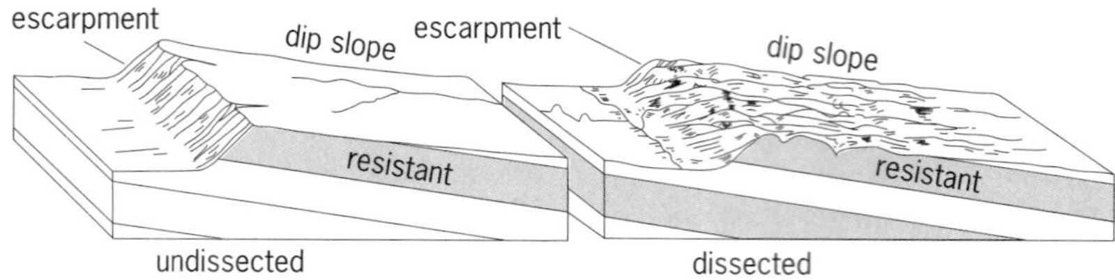

Fig. 2. Surface form and structure of cuestas. Dissected form at right is more typical, especially in humid regions. (*After G. T. Trewartha, A. H. Robinson, and E. H. Hammond, Elements of Geography, 5th ed., McGraw-Hill, 1967*)

surface so that the area is almost wholly occupied by valley-side slopes, the surface is called mature. If valleys have so widened and coalesced that only a few narrow, low divide remnants are left, the surface is said to have reached old age. However, not all stream-eroded surfaces follow the ideal sequences, and many factors other than time affect their development.

Some plains, for example, remain persistently youthful because unusual permeability of the surface material, gentle or inadequate rainfall, excessive flatness, dense vegetation cover, or resistance of the rocks either fail to provide the surface runoff that could carve tributary valleys or else inhibit the erosive capacity of those smaller streams that do form. An example which combines several of these causal factors is provided by the High Plains, which stretch from southwestern Nebraska into the Panhandle of Texas. By way of contrast, some other plains appear to have reached maturity very quickly as a result of the generation of copious surface runoff by heavy rains and an initially rolling surface of low permeability, combined with readily erodible materials. The Dissected Till Plains of northern Missouri and southern Iowa are an example.

Although valleys of great breadth are not uncommon, extensive plains of ideal old age are rare. Valley widening is at best a slow process, and the reduction of low, gentle-sloped, well-vegetated divides requires immense lengths of time without interruption by uplift or environmental changes that initiate a renewal of valley deepening. Some modern scholars suggest that certain wide-valleyed, long-sloped, or gently undulating plains with an unbroken vegetation cover may in fact represent plains on which erosional development has practically ceased. *See* STREAM TRANSPORT AND DEPOSITION.

The pattern of valleys on plains depends chiefly upon the initial slope of surface and the pattern of outcrop of rock materials of contrasting resistance. In the absence of strong contrasts the pattern is usually branching and treelike as in Fig. 1. Where there is great differential resistance to erosion, the unusually resistant rocks form drainage divides or uplands, whereas weak rock belts are soon excavated into broad valleys or lowlands. Where erosional plains bevel across gently warped rock strata of varying resistance, the belts of outcrop of the more resistant strata form strips of higher, rougher country, with an abrupt escarpment on one margin and a more gradual dipslope in the direction toward which the strata are inclined (**Fig. 2**). These features, called cuestas, are common in the Middle West and Gulf Coastal Plain and in western Europe. The wolds and downs of England and the côtes of northeastern France are cuesta ridges.

Most plains that develop in dry climates are characterized predominantly by stream-produced landforms, in spite of infrequent rain. The development of valley systems and erosional features follows the same general rules as it does in humid regions. However, some differences in relative rates and relative significance of weathering and erosion produce distinctive landscape characteristics in arid lands. First, rock decomposition is very slow, so that the surface accumulation of weathered material is normally thin and coarse textured. Second, the sparse vegetation affords to the naked surface little protection against the battering and washing of the occasional torrential rains. As a result, the upper slopes become strongly gullied and often partially stripped of their covering material, leaving much bedrock exposed. Because of the short duration and local nature of the rains and hence the intermittent character of stream flow, however, most of the debris load is dropped in the neighboring basins and valley floors, "drowning" broad areas beneath plains of silt, sand, or gravel. Hence denuded and gullied upper slopes and broad depositional flats in the lowlands are characteristic features of desert plains. *See* DESERT EROSION FEATURES.

Features produced by solution. Features resulting from underground solution characterize several rather extensive areas of plains. The principal features of this class are depressions, or sinks, produced by collapse of caverns underneath. Solution is also an active process in erosion by surface streams, but its effects upon valley form have been little studied.

Significant ground-water solution is largely confined to areas underlain by thick limestones. As subsurface cavities are progressively enlarged by solution, more and more drainage is diverted to subterranean channels. Surface streams become fewer, often disappearing into the ground after a short surface run. Eventually solution cavities near the surface collapse, forming surficial depressions of various sizes. Some are shallow and inconspicuous; others are great steep-walled pits or elongated enclosed valleys. Some of the small depressions contain lakes, because their outlets are plugged with clay. The most extensive areas of solution-featured plains in the United States are in central and northern Florida and in the Panhandle of Texas. In both of these areas shallow sink holes, some of them lake-filled, are numerous. Some of the areas of most active solution work have developed surfaces far too rough to be called plains. This is true of the Mammoth cave area of west-central Kentucky and especially so for the mountainous area of great sinks and solution valleys in the Dalmation Karst of Yugoslavia. *See* KARST TOPOGRAPHY.

Features associated with stream deposition. As a group, alluvial plains (so called from the term alluvium, which refers to any stream-deposited material) are among the smoothest and flattest land surfaces known. Stream-deposited plains fall into three classes: (1) floodplains, which are laid down along the floors of valleys; (2) deltas, formed by deposition at the stream mouths; and (3) alluvial fans, deposited at the foot of mountains or hills.

Floodplains. The flat bottomlands so common to valley floors develop wherever a given segment of a stream system is fed more sediment by its tributaries or upstream reaches than it is able to carry. Most of the excess sediment is deposited in the stream bed, usually in the form of bars. In flood time, some may be strewn across the whole width of the valley floor. Because of continued deposition and choking of the channel, repeated flooding, and the ease with which banks composed of loose alluvium collapse, streams continually shift their channels on floodplains. On silty plains the channel is usually highly sinuous or meandering, while on a coarse sandy or gravelly floodplain the channel is braided, that is, broad, shallow, and intricately subdivided by innumerable sandbars (**Fig. 3**). In either case, many loops or strands of abandoned channels, now mostly dry and partially filled with sediment, scar the floodplain surface. On broad silty floodplains slightly elevated strips, called natural levees, are formed by active deposition immediately adjacent to the channel when the velocity of flow is abruptly checked as the water leaves the swift current of the channel to spread thinly over the plain during floods.

On major floodplains the ground-water table is everywhere close to the surface, and swampy land is common in the abandoned channels and shallow swales. For this reason, any slightly higher features, such as natural levees, are especially sought after for cultivation, town sites, and transportation routes. Though harassed by a high water table, recurrent floods, and shifting channels, floodplains, especially silty ones, are often prized agricultural lands because of the level, easily tilled surface. In some instances the alluvium is also more fertile than the soils of the surrounding uplands.

Here and there along the sides of valleys, benches and terraces of alluvium stand above the level of the present floodplain. These are remnants of earlier floodplains that have been largely removed because of a renewal of downcutting by the streams. These alluvial terraces may be valuable agricultural lands and serve well for town sites and transportation routes because they stand above flood levels. *See* FLOODPLAIN.

Deltas. The surface features of deltas are essentially the same as those just described. As a rule, however, the delta surface is even less well drained than the floodplain surface and at its outer margin may merge with the sea through a broad belt of marshy land. As the delta grows, continual obstruction of the stream mouth produces repeated diversion and bifurcation of the channel, so that the stream commonly discharges through a spreading network of distributaries (**Fig. 4**).

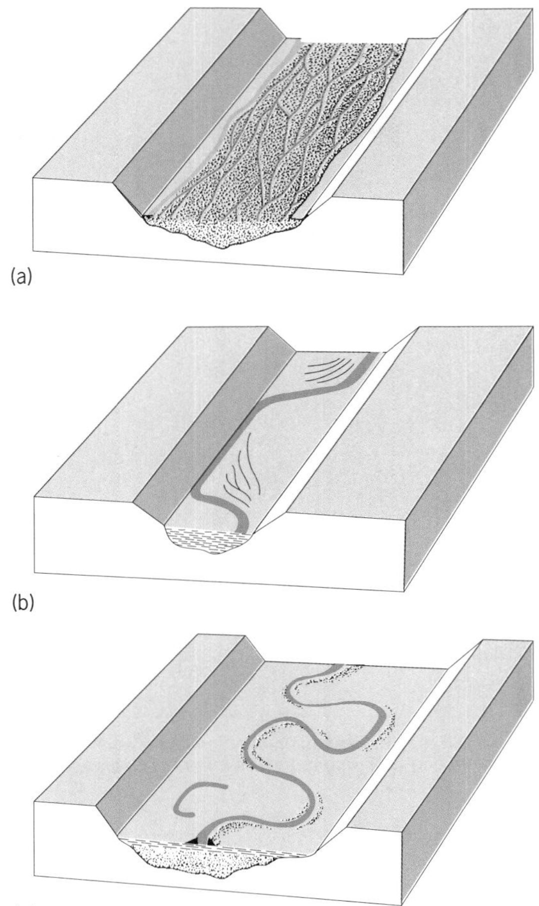

Fig. 3. Types of floodplains. (*a*) With braided channel. (*b*) With sinuous channel but too narrow for free meandering. (c) With freely meandering channel. (*After G. T. Trewartha, A. H. Robinson, and E. H. Hammond, Elements of Geography, 5th ed., McGraw-Hill, 1967*)

Many deltas, like those of the Mississippi, the Nile, the Danube, or the Volga, are immense fan-shaped features that have produced broad coastal bulges by their growth. Others, like those of the Colorado, the Po, or the Tigris-Euphrates, though no less extensive, are less apparent on the map because they have been built in large coastal embayments. Some great rivers have no true deltas because their sediment load has been dropped in some interior settling basin. For example, the Great Lakes remove most of the sediment from the St. Lawrence system, and the Congo deposits most of its load in its broad upland basin.

Like floodplains, deltas are sometimes highly valued as agricultural lands, though they have even worse problems of poor drainage and frequent flooding. The Nile delta and the huge, silty delta plain of the Hwang Ho, in north China, are famous centers of cultivation. Many deltas, of which that of the Mississippi is a good example, are too swampy to permit tillage except along the natural levees. The Netherlands, occupying the combined deltas of the

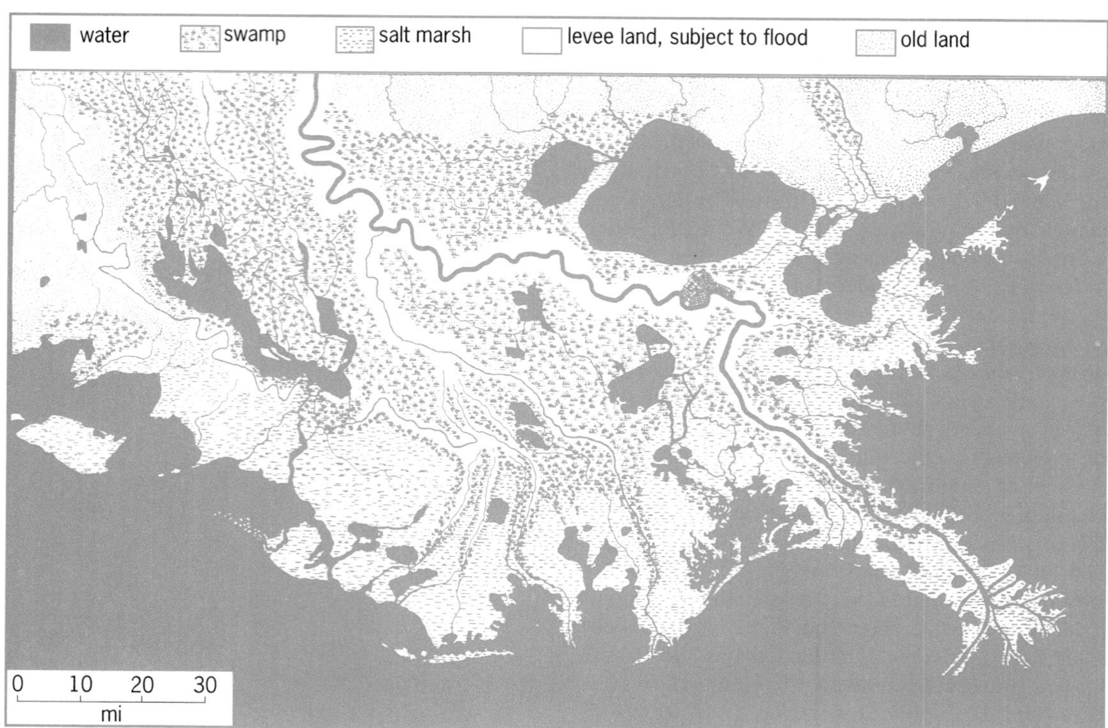

Fig. 4. **Salt marsh, swamp, and natural levee lands in the Mississippi River delta. 1 mi = 1. 6 km. (*After G. T. Trewartha, A. H. Robinson, and E. H. Hammond, Elements of Geography, 5th ed., McGraw-Hill, 1967*)**

Rhine and Maas, stands as an example of what can be done toward reclamation when population pressure is great. *See* DELTA.

Alluvial fans. If a stream emerges upon a gentle plain from a steeply plunging mountain canyon, its velocity is abruptly checked, and it deposits most of its load at the mouth of the canyon. Because of the tendency toward repeated choking and diversion of the channel, the deposit assumes the form of a broad, spreading alluvial fan, essentially similar to a delta, even to the diverging distributary channels (**Fig. 5**). Usually, however, the gradients developed are steeper than those on a delta because sediments are coarser.

Small individual alluvial fans are common features in mountainous country, especially where the climate is dry except for occasional torrential showers. Particularly significant, however, are the rows of alluvial fans that have coalesced to form extensive, gently sloping piedmont alluvial plains at the bases of long, precipitous mountain fronts. The city of Los Angeles is built on such a plain. Still larger ones occupy much of the southern part of the Central Valley of California and stretch eastward from the Andes in northwestern Argentina, Paraguay, and eastern Bolivia.

Because of their smoothness and ease of tillage, alluvial fans, like other alluvial surfaces, are often especially amenable to cultivation. They are particularly significant in drier areas, partly because of the ease with which water may be conducted by gravity from the mountain canyon to any part of the fan, and partly because the thick, porous alluvium itself serves as a reservoir in which ground water is naturally stored.

Lake plains and coastal plains. Closely allied to stream-deposited plains are surfaces that represent former lake bottoms or recently exposed sections of the former shallow sea floor. These nearly featureless plains have been formed by the deposition of sediments carried into the body of water by streams or wave erosion and further distributed and smoothed by the action of waves and currents. Former beach lines and other shore features often provide the only significant breaks in the monotonous flatness. In some places, shallow valleys have been cut by streams since the surface became exposed. The lower parts of lake plains and the outer margins of coastal plains are often poorly drained.

The flat surfaces upon which Detroit, Toledo, Chicago, and Winnipeg stand are all lake plains, as are the famed Bonneville Salt Flats of western Utah. The south Atlantic and Gulf margins of the United States and much of the Arctic fringe of Alaska and Siberia are examples of newly emerged coastal plains. Some of these plains represent valuable agricultural land; others are excessively swampy or sandy. *See* COASTAL LANDFORMS; COASTAL PLAIN.

Features due to continental glaciation. On several occasions during the last $1\text{–}2 \times 10^6$ years (Pleistocene Epoch), immense ice caps, comparable to the one which now covers Antarctica, developed in Canada and Scandinavia and spread over most of northern North America and northwestern Eurasia (**Fig. 6**). The last such ice sheet (called in North America the Wisconsin) reached its maximum extent about 18,000 years ago and did not disappear

Fig. 5. Alluvial fans in the Mojave Desert, southeastern California. (*J. L. Balsley, U.S. Geological Survey*)

until 5000–6000 years ago. *See* GLACIAL EPOCH; GLACIATED TERRAIN.

These great glaciers significantly modified the land surfaces over which they moved. Except near their melting margins, they were able to remove and transport not only the soil but also much weathered, fractured, or weak bedrock. This material was subsequently deposited beneath the ice or at its edge, wherever melting released it. The marks of glacial sculpture are unsystematically distributed swells and depressions; numerous lakes, swamps, and aimless streams (**Fig. 7**); and an irregular mantle of debris (glacial drift), some of which is clearly not derived from the local bedrock upon which it now lies. Such surfaces often contrast strongly with the systematic valley and divide patterns characteristic of stream erosion.

Most of the area that was covered by ice during the Pleistocene is now more or less thickly covered with drift, the thickest cover generally occurring in regions of weak rocks. Characteristic glacial landforms, however, are largely confined to those areas that were occupied by the ice during middle and late Wisconsin times. Earlier glacial features have been modified beyond recognition by surface erosion and soil creep. Hence several large areas, including parts

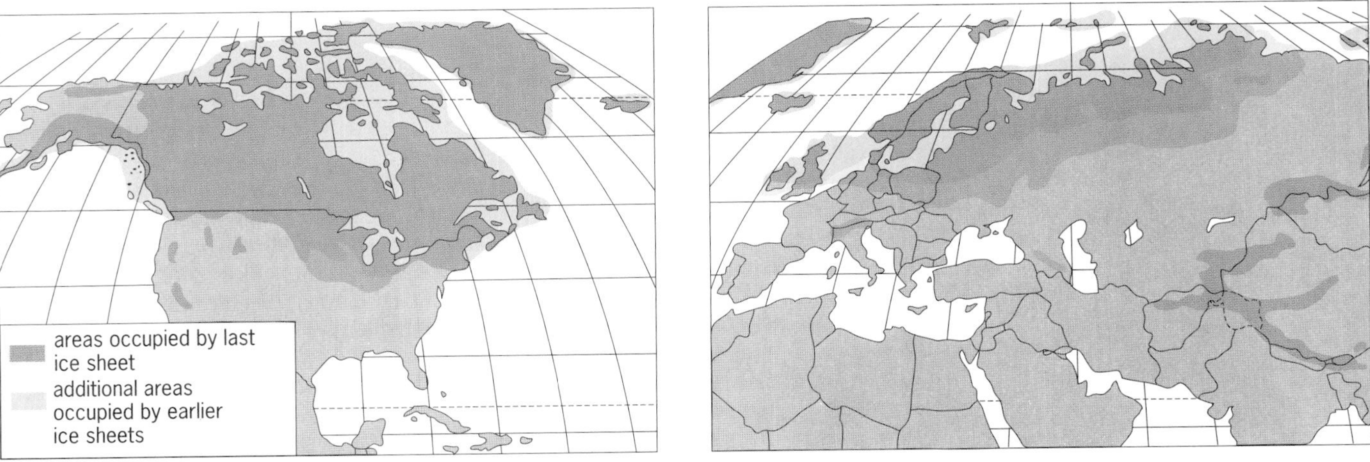

Fig. 6. Extent of former continental glaciers in North America and Eurasia. (*After G. T. Trewartha, A. H. Robinson, and E. H. Hammond, Elements of Geography, 5th ed., McGraw-Hill, 1967*)

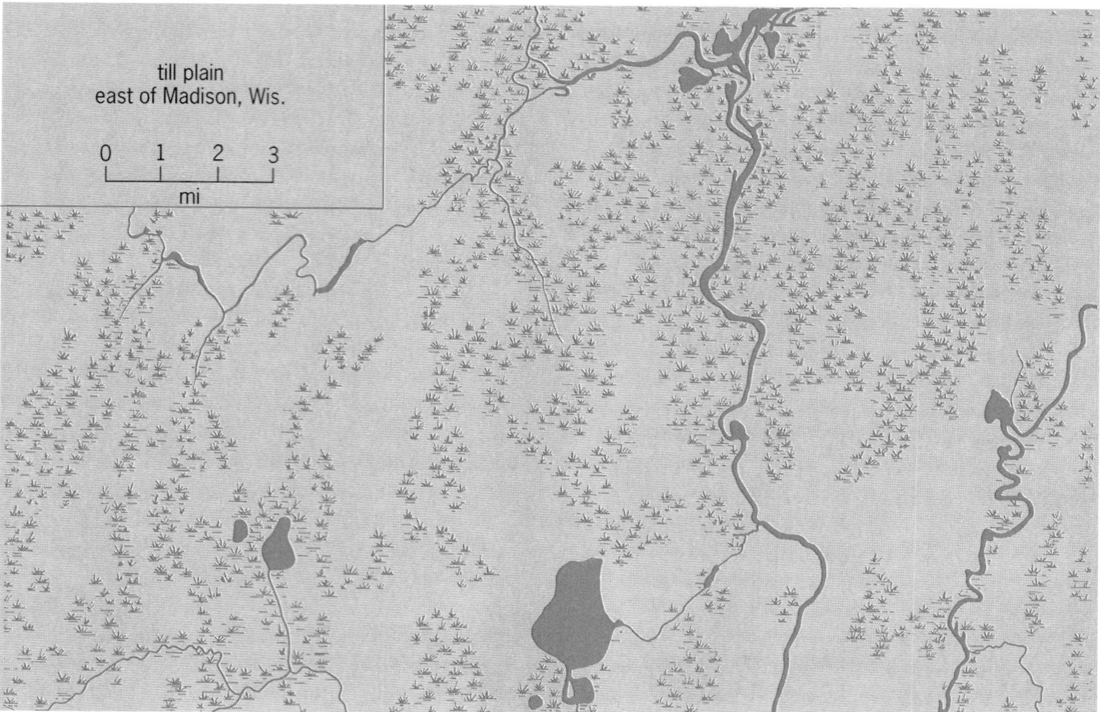

Fig. 7. Diagram of the drainage pattern of a glaciated plain in eastern Wisconsin. 1 mi = 1.6 km. (*After G. T. Trewartha, A. H. Robinson, and E. H. Hammond, Elements of Geography, 5th ed., McGraw-Hill, 1967*)

of the midwestern United States and southern Russia, though drift-mantled, display the surface features characteristic of stream erosion.

Drift-covered plains offer a wide range of potentialities for human use, the critical characteristics being chiefly the adequacy of drainage and the texture of the soils. Excessive roughness is rarely a problem.

Till. Much of the drift represents mixed rock and soil deposited directly by melting beneath the ice sheet or at its edge. This material, called till, is as a rule most thickly deposited in the valleys and thinly over the ridge tops, thus reducing terrain irregularity. The surface of the till sheet itself is usually gently rolling, with poor and unsystematically patterned drainage. Hummocky, often stony ridges, called marginal moraines, mark lines along which the fluctuating edge of the ice remained stationary for a considerable time. In several localities, notably in eastern Wisconsin and western New York, are swarms of smooth, low drift hills called drumlins, all elongated in the direction of ice movement.

The surfaces of stony till plains are usually more irregular than those on silt or clay till. Northeastern Illinois has a remarkably smooth surface developed on silt and clay till, much of it eroded from the Lake Michigan basin. Eastern Wisconsin, northern Michigan, western New York, and southern New England, on the other hand, have more rolling surfaces underlain by till having a high content of stone and sand. In a few areas, especially in southern New England and in the marginal moraines elsewhere, the till is so stony as to impede cultivation.

Outwash. Some of the debris transported by the ice is carried out beyond the glacial margin by streams of meltwater. This material, called outwash, may be deposited as a floodplain (here called a valley train) along a preexisting valley bottom, or it may be broadcast over a preexisting plain. In either case the surface will usually be smooth, with features typical of alluvial plains. Unlike the heterogeneous, unsorted, and unstratified till, outwash material is usually layered and sorted in size. The fine material of silt and clay size is carried out downstream, leaving the coarser sands and gravels to form the outwash deposits. Most of the gravel and sand pits that abound in glaciated areas are developed in outwash plains.

Where outwash was laid down over already deposited till surfaces after the ice had melted back from its maximum extent, the surface of the outwash plain is sometimes pitted and lake-strewn, the depressions having formed as a result of the melting of relict ice masses that were buried by outwash deposition.

Patches and ribbons of outwash are common in glaciated areas, and in a few places, into which unusual quantities of meltwater were funneled, there are very extensive sandy plains. Noteworthy in this respect are the southern Michigan and northern Indiana area and Europe immediately south of the Baltic.

Lacustrine plains. Also present in and around the glaciated areas are numerous plains marking the beds of former lakes that resulted from the blocking of rivers by the glacial ice itself. During the melting of the last ice sheet, while the St. Lawrence and other northward-flowing rivers were still ice-dammed, the

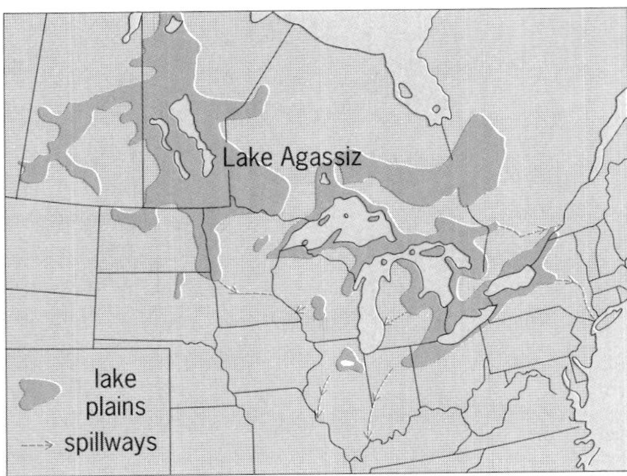

Fig. 8. Plains of former glacial lakes in North America. (*After G. T. Trewartha, A. H. Robinson, and E. H. Hammond, Elements of Geography, 5th ed., McGraw-Hill, 1967*)

Great Lakes basins were much fuller than now and overflowed to the southward. When lake levels were eventually lowered, lacustrine plains were exposed, notably about Chicago, at the western end of Lake Erie, and about Saginaw Bay in Michigan. One of the most featureless plains of North America occupies the former bed of an immense lake (Lake Agassiz) that was present in late glacial time in southern Manitoba, northwestern Minnesota, and eastern North Dakota (**Fig. 8**).

Thin-drift areas. Considerable areas in central and eastern Canada and northern Fennoscandia have only a thin and discontinuous drift cover. These hard-rock regions yielded little debris and, especially in the rougher sections, the small drift supply was thinly strewn in the valleys and depressions. Lakes and swamps occupy drift-blocked valleys and shallow erosional depressions, rapids and waterfalls are common, and broad expanses of scoured bedrock are exposed. Such areas offer little opportunity for agricultural use and usually support only an open, patchy, and often stunted forest growth.

Features reflecting wind action. Features formed by the wind are less widespread and, as a rule, less obtrusive than the forms produced by streams and glaciers. Since the wind can attack only where the surface is almost free of vegetation, its work is strongly evident only in arid regions and in the vicinity of beaches and occasionally exposed river beds in more humid lands. Today plowed fields are also important prey for wind erosion.

The most striking and significant wind-produced features are sand dunes. Where sand is exposed to strong winds, it is moved about for short distances and accumulates in heaps in the general vicinity of its place of origin. Sand dunes assume many forms, from irregular mounds and elongated ridges to crescent-shaped hills and various arrangements of sand waves, depending apparently upon the supply of sand, the nature of the underlying surface, and the strength and directional persistence of the wind.

Nearly all the truly extensive areas of sand dunes are found in the Eastern Hemisphere, especially in the Sahara and in Arabia, central Asia, and the interior of Australia. Most of the dunes have been whipped up from alluvium that has been deposited in desert basins and lowlands.

In several regions of the world, notably in north-central Nebraska and in the central and western Sudan south of the Sahara, extensive areas of dunes have become covered by vegetation and fixed in position since they were formed, suggesting the possibility of climatic change. *See* DUNE.

Other wind-formed features are polished and etched outcrops of bedrock that show the effects of natural sand-blasting; gravel "pavements" resulting from the winnowing out of finer material from mixed alluvium; and shallow blowouts, which are depressions formed by local wind erosion.

The finer silty material moved by the wind is spread as a mantle over broad expanses of country downwind from the place of origin. Though usually thin, this mantle in places reaches a thickness of a few tens of feet, thus somewhat modifying the form of the surface. The extensive deposits of unstratified, buff-colored, lime-rich silty material known as loess are believed to have originated from such wind-laid deposits. Loess is abundant in the central United States, eastern Europe and southern Russia, and in interior northern China. It yields productive soils but is subject to gullying erosion and has the facility of maintaining remarkably steep slopes, so that erosional terrain developed on deep loess is often unusually rough and angular. *See* LOESS.

Interrupted Plains

Plains interrupted by some features of considerable relief occur widely and merit independent treatment. They may be divided into two contrasting groups: (1) tablelands, which are upland plains deeply cut at intervals by steep-sided valleys or broken by escarpments, and (2) plains with hills or mountains, in which the excessive relief is afforded by steep-sided eminences that rise above the plain. Both of these types of surface, with their combination of plain and rough land, suggest histories of development that combined extensive gradation and strong tectonic activity.

Tablelands. These are essentially youthful plains that have been cut to unusual depth by valleys (**Fig. 9**). The plain first had to be brought to a level hundreds or even a few thousands of feet above the level to which streams can erode. In most cases, this elevation was accomplished by the broad uplifting of an erosional or alluvial plain, but in a few instances the plain was built to high level by the deposition of many thick sheets of lava.

It is also necessary that while a few major streams have cut deep canyons, large areas of the upland plain remained largely uncut by tributary development, a condition requiring special circumstances. Such inhibited tributary growth is usually the result of either (1) slight local surface runoff of water because of aridity, extreme flatness of the upland, or highly permeable material; or (2) the presence, at the upland level, of very resistant rock strata that

Fig. 9. View of an ideal tableland: Canyon de Chelly National Monument, northeastern Arizona. (*Spence Air Photos*)

have permitted only the most powerful streams to cut through. Hence tablelands are predominantly a dry-land terrain type. Those that do occur in rainier climates usually indicate the presence of an exceptionally resistant and permeable cap-rock layer.

The cliffs and escarpments that are common features of tableland regions are sometimes fault scarps, produced by the breaking and vertical dislocation of the crust during uplift. More often, however, they are simply steep valley sides that have been worn back long distances from their original positions under the attack of weathering and erosion. Sometimes a once extensive tableland is so encroached upon by the retreat of its bordering escarpments that nothing is left but a small, flat-topped mesa or butte.

Tablelands are the least widespread of the major terrain types, presumably because of the limited circumstances under which they can develop. The most extensive examples occur in the American continents. Especially noteworthy are the Colorado Plateaus, largely in northern Arizona and southern Utah, which represent a complex erosional plain that has been greatly uplifted and then deeply carved by the Colorado River and its major tributaries. Preservation of large sections of the upland plain has been favored by dryness and by the presence of nearly horizontal resistant rock strata. The few major streams that have cut deep canyons are all fed from moister mountainous areas round about.

The Columbia Plateau in eastern Washington is a porous lava plain, cut by the Columbia River and a few tributaries. The northern Great Plains, lying east of the Rocky Mountains from Nebraska northward into Alberta, are an old erosional plain on weak rocks, now crossed by valleys of moderate depth that have been cut by streams issuing from the Rockies. The Cumberland Plateau in Tennessee is maintained by a resident sandstone caprock. The Patagonian Plateau of southern South America is a somewhat similar surface, locally reinforced by extensive lava flows. Parts of the upland of interior Brazil, though in a moist environment, retain a tableland form because of the resistance of thick sandstone beds and, locally, lavas at the upland level.

Though there are significant exceptions, tablelands as a group suffer, in their economic development, from the difficulty of passage through their narrow gorges and across their many escarpments, and in most cases from the dryness of their upland surfaces.

Plains with hills or mountains. These are much more widespread than tablelands and are extensively represented in each continent. Surfaces included under the general heading vary from plains studded with scattered small hills and hill groups to mountain-and-plain country in which high, rugged ranges occupy almost as much space as the plains between them (Fig. 5).

Surfaces of the general type can be produced by two quite different lines of development. (1) They may be mountain, hill, or tablelands that have been brought to the erosional stage of early old age, in which case the isolated hills represent the only remnants of a once extensive highland. (2) They may represent areas in which separated hills or mountains have been constructed by volcanic eruption or by folding or buckling of the crust, the land between them having remained smooth from the outset, or having been smoothed by processes of erosion and deposition since the mountains were formed.

Examples of the first course of development are limited in the United States to small areas in southern New England and in the Appalachian Piedmont just east of the Blue Ridge. Many patches occur in the glacially scoured sections of northern and eastern Canada and similarly in northern Sweden and Finland. Extensive areas are found in the southern regions of Venezuela and Guiana, in the upland of eastern Brazil, and especially on the uplands of western, central, and southern Africa.

In such areas the plains are typical late-stage erosional-depositional surfaces, usually gently undulating. The remnant hills (monadnocks) rise abruptly from the plain, like islands from the sea. Although monadnocks commonly represent outcrops of unusually resistant rocks that have withstood erosion most effectively, many owe their existence solely to their position at the headwaters of the major stream

Fig. 10. Death Valley, California, a basin of interior drainage in the Basin-and-Range province. (*Courtesy of John S. Shelton*)

systems, where they are the last portions of the highland to be reduced.

Surfaces on which spaced mountains have been constructed by crustal deformation or volcanic activity occur extensively in the great cordilleran belts of the continents, and rarely outside of those belts. The largest of all such regions is the Basin-and-Range section of western North America, which extends without interruption from southeastern Oregon through the southwestern United States and northern Mexico to Mexico City. The majority of the rugged mountain ranges of this section are believed to represent blocks of the crust that have been uplifted or uptilted and then strongly eroded. The plains between them are combination erosional-depositional surfaces, some of which have clearly expanded at the expense of the adjacent mountains. As the mountains have been reduced by erosion, there have evolved at their bases smooth, gently sloping plains that are in part erosional pediments, closely akin to old-age plains, and in part piedmont alluvial plains. Many basins have interior drainage; in these the floor is likely to be especially thickly alluviated and to contain a shallow saline lake or alkali deposit in its lowest part (**Fig. 10**).

Other areas of somewhat similar terrain are found in the central Andes, in Turkey and the Middle East, and in Tibet and central Asia. The Tibetan and Andean sections are noteworthy for the extreme elevation (12,000–15,000 ft or 4–5 km) of their basin floors. In general the ranges and basins in the Asiatic areas are developed on a grander scale than those in North America.

It is a curious circumstance that practically all these regions are dry and exhibit the intensely eroded mountain slopes and alluvially drowned basin floors characteristic of that climatic realm. For this reason they are only locally useful to inhabitants, in spite of the large amount of smooth land that they afford. There are, however, many important oases, usually near the bases of the mountains or along the courses of the few streams that have wandered in from moister adjacent regions. Edwin H. Hammond

Bibliography. D. Briggs and P. Smithson, *Fundamentals of Physical Geography*, 1993; D. J. Easterbrook, *Surface Processes and Landforms*, 1992; A. Goudie, *The Changing Earth: Geomorphological Processes and Time*, 1995; A. N. Strahler and A. H. Strahler, *Elements of Physical Geography*, 4th ed., 1990.

Planck's constant

A quantity, designated by the letter h, which occurs repeatedly in those fundamental formulas of physics that describe the microworld, the world of atoms and nuclei and elementary particles. Historically, h occurred first in 1900 in work of Max Planck on the theory of blackbody radiation, but its physical significance was at first not clear. Some physicists would claim that it is still not clear. *See* HEAT RADIATION.

In 1905 Albert Einstein sought to clarify Planck's work, as well as certain paradoxes that had arisen concerning the nature of light, by suggesting that light has a dual nature—that in some way, which he made no attempt to explain, it is both wave and particle. The particle, called a light quantum or photon, is massless, travels always at the speed of light c, and has an energy E which is related to the light wave's vibrational Eq. (1), frequency ν by where the

$$E = h\nu \qquad 6.62608 \times 10^{-34}\ \text{joule} - \text{second} \tag{1}$$

numerical value of h is also given. *See* FUNDAMENTAL CONSTANTS; PHOTON.

If a photon has an energy, it also has a momentum. Maxwell's theory of the electromagnetic field states that if a portion of the field traveling in a given direction has an energy E, its momentum p is given by E/c. Putting this into Eq. (1) one encounters the ratio ν/c, which can also be written in terms of the wavelength λ of the corresponding light wave through the universal relation $\lambda\nu = c$. This leads to Eq. (2).

$$p = \frac{E}{c} = \frac{h\nu}{c} = \frac{h}{\lambda} \tag{2}$$

See MAXWELL'S EQUATIONS.

By the early 1920s, Eqs. (1) and (2) had been tested and verified in several experimental situations, notably those involving the photoelectric effect and the Compton effect. In 1923 Louis de Broglie suggested that these relations govern the behavior of forms of matter other than photons. He proposed experiments in which electrons, previously thought of as particles, would appear as waves with wavelengths given by Eq. (2), and in a few years this too had been verified. *See* COMPTON EFFECT; ELECTRON DIFFRACTION; PHOTOEMISSION.

Other occurrences of h in fundamental physics stem from these early considerations. In Niels Bohr's provisional atomic theory of 1913, the angular momentum of an orbiting electron is an integral multiple of $h/2\pi$, a quantity now denoted by the symbol $\hbar$, and de Broglie was later able to relate this proposal to Eq. (2). Out of these considerations grew the modern theory of quantum mechanics, in which h plays a fundamental role. It occurs also in Heisenberg's principle of uncertainty or indeterminacy, which places limits on the accuracy with which it is possible to know and describe the microworld. *See* ATOMIC STRUCTURE AND SPECTRA; QUANTUM MECHANICS; UNCERTAINTY PRINCIPLE. David Park

Planck's radiation law

A law of physics which gives the spectral energy distribution of the heat radiation emitted from a so-called blackbody at any temperature. Discovered by Max Planck early in the twentieth century, this law laid the foundation for the advent of the quantum theory because it was the first physical law to postulate that electromagnetic energy exists in discrete bundles, or quanta. *See* HEAT RADIATION; QUANTUM MECHANICS. Heinz G. Sell; Peter J. Walsh

Plane curve

The locus of points in the euclidean plane that satisfy some geometric or algebraic definition. Not all sets of points deserve to be called a curve, but the distinction is somewhat arbitrary. For most of this article, a curve is considered to be the locus of a set of points that satisfy an algebraic or transcendental equation in two variables.

In general, a curve can be defined by geometric properties; an equation in cartesian coordinates $h(x,y) = 0$; an equation in polar coordinates $p(r, \phi) = 0$; a pair of parametric equations $x = f(t)$, $y = g(t)$; or an equation involving invariants such as arc length and radius of curvature. For example, the deltoid may conveniently be defined as the locus of points on the circumference of a circle C_1 rolling inside another circle C_2 of three times its radius, as points satisfying the cartesian equation (1), as points satisfying the

$$(x^2 + y^2)^2 - 8ax(x^2 - 3y^2) + 18a^2(x^2 + y^2) = 27a^4 \quad (1)$$

parametric equations (2), or as points satisfying the

$$\begin{aligned} x &= r(2\cos t + \cos 2t) \\ y &= r(2\sin t - \sin 2t) \qquad -\pi \le t \le \pi \end{aligned} \quad (2)$$

implicit equation (3), where s represents arc length

$$9s^2 + \rho^2 = 64r^2 \quad (3)$$

and ρ the radius of curvature. *See* ANALYTIC GEOMETRY.

Properties. The most interesting geometric properties are those preserved by linear transformations, especially translations, rotations, reflections, and magnifications. Useful geometric properties include the number of branches into which the curve is divided; the number and degree of nodes, cusps, isolated points, and flex points; the number of loops; symmetries; branches that go to infinity; and asymptotes.

These terms can be defined informally as follows. A branch is a maximal smooth continuous portion of the curve. A multiple point is a point in the plane that lies on two or more branches; its degree is the number of branches involved. A node is a multiple point where the branches cross. A cusp is a multiple point where the branches meet but do not pass; that is, each of the branches ends at that point. An isolated point is a point of the curve through which no branches pass. Multiple and isolated points are collectively termed singular points. Any point that is not singular is termed ordinary. A flex (or point of inflection) is a point on the curve whose tangent cuts the curve. A smooth closed branch forms a loop. A curve is symmetric about a line L if every line perpendicular to L intersects the curve at equal distances from L on opposite sides of L; that is, portions of the curve form mirror images about L. The curve is symmetric about a point P if every line through P intersects the curve at equal distances from P in opposite directions. An asymptote is a line toward which a branch approaches as it moves to infinity from the origin; the curve and line are said to intersect at infinity.

In addition to these geometric properties, the form of the defining equation is of interest. This can be an algebraic (polynomial in x and y) or transcendental equation. In the former case, quadratic, cubic, and quartic equations are of special interest.

Once a coordinate system has been chosen, and the defining equation is known (in any of the forms, though the parametric form is usually the most useful), various properties of the curve can be defined in terms of the equation. These include the locations of x- and y-intercepts, local maxima and minima, flexes, nodes, and cusps. For any point P on the curve, it is generally possible to specify the equation for the tangent line, radial angle (between the x axis and a line between P and the origin), arc length (from some fixed point on the curve), curvature, and center of curvature. Not all of these features may exist for every point of the curve.

Sketching a curve. It is possible to approximate the curve C represented by Eq. (4) by selecting points

$$h(x, y) = 0 \quad (4)$$

on the x axis at increasing distances from the origin, solving for y for each point, and plotting the resulting (x,y) pairs. Some experimentation is required to decide how close together the points on the x axis must be. Iterative techniques are in general necessary to find approximations to y. If the parametric equations are known or can be found, the process becomes easier, since x and y can be evaluated directly for each value of t to yield the coordinates of the points to be plotted. The process is greatly facilitated by using the defining equation (4) and its partial derivatives to help determine singular points, symmetries, and asymptotes of the curve C.

Derived curves. Once the properties of some curve C are known, they may possibly be used to derive other curves. Auxiliary points, lines, and curves are commonly used.

The evolute of C is the locus of the center of curvature. An involute of C is the locus of a fixed point P on a tangent that rolls along C without slipping. A parallel to C is the locus of points a fixed distance from the points P of C, measured along a line perpendicular to the tangent to C at P. *See* INVOLUTE.

A pedal of C is defined with respect to a point O. For every point P on C, the intersection is found of the tangent at P with the line through O perpendicular to the tangent. These intersections form the pedal of C with respect to O.

To define the inverse of C, a point O and a constant k are chosen. For every point P on C, a line segment L is drawn to O, and a point P' on L is so selected that the distance from O to P, multiplied by the distance from O to P', is equal to k. The locus of such points forms an inverse of C with respect to O.

For strophoids, two points O_1 and O_2 are chosen. For each point P on C, the line L is drawn through O_1 and P, and two points Q_1 and Q_2 are found on L such that the distance from P to Q_1 and the distance from P to Q_2 are both equal to the distance from P to O_2. The locus of such points forms the strophoid of C with respect to the two points.

A roulette of C with respect to a second curve C' and a fixed point P on C is the locus of P as C rolls along C' without slipping.

Other types. Plane curves can be defined in many other ways than algebraic or transcendental equations, for example, solutions of differential or integral equations. Plane curves are formed by many special functions, such as trigonometric, exponential, hyperbolic, gamma, Bessel, elliptic, and probability functions. *See* DIFFERENTIAL EQUATION; INTEGRAL EQUATION; SPECIAL FUNCTIONS.

Limit processes of various kinds define curves that were once termed pathological. An example is the Koch curve. An equilateral triangle is drawn with sides of length l. The middle third of each side is replaced by an outward-pointing equilateral triangle of side $l/3$. The process is continued: the middle third of each line segment is replaced with such an extension. The Koch curve is the set of points formed by taking the limit of the construction. It is continuous everywhere, differentiable nowhere (no tangents exist), and has infinite length. A variation of the procedure can be used to define curves that pass through every point in a closed region of the plane. *See* FRACTALS.

Algebraic and transcendental curves. The **illustration** shows some of the many plane curves of sufficient historical interest to have received names.

The right strophoid (illus. *a*) is the strophoid of a line with respect to two points O and A, where A is the foot of the perpendicular from O to the line. It can also be defined as an inverse of a rectangular hyperbola, or as a pedal of a parabola. It is algebraic of degree 3 (cubic), has a single node at the origin, is symmetric about the x axis, and has asymptote $x = a$. It is expressed by Eq. (5).

$$y^2(a - x) = x^2(a + x) \qquad (5)$$

The trident of Newton, (illus. *b*), also known as the parabola of Descartes, has a discontinuity at $x = 0$ and one flex; the y axis is an asymptote. It is expressed by Eq. (6).

$$xy = ax^3 + bx^2 + cx + d \qquad ad \neq 0 \qquad (6)$$

The cardioid (illus. *c*) is an inverse of a parabola with respect to the focus. It can also be defined as the pedal of a circle with respect to a point on the circumference. The curve is a closed quartic with a cusp and is symmetric about the x axis. It is expressed by Eq. (7).

$$(x^2 + y^2 - 2ax)^2 = 4a^2(x^2 + y^2) \qquad (7)$$

See CARDIOID.

The deltoid (illus. *d*), or tricuspid, has three cusps and three lines of symmetry and forms a closed curve. It is expressed by Eq. (1).

The devil on two sticks (illus. *e*), or devil's curve, is a quartic with two infinite branches, a loop, and a node at the origin. It is symmetric about both axes and the origin. Asymptotes are $x \pm y = 0$. It is expressed by Eq. (8).

$$y^4 - a^2y^2 = x^4 - b^2x^2 \qquad (8)$$

The lemniscate of Bernoulli (illus. *f*), or hyper-

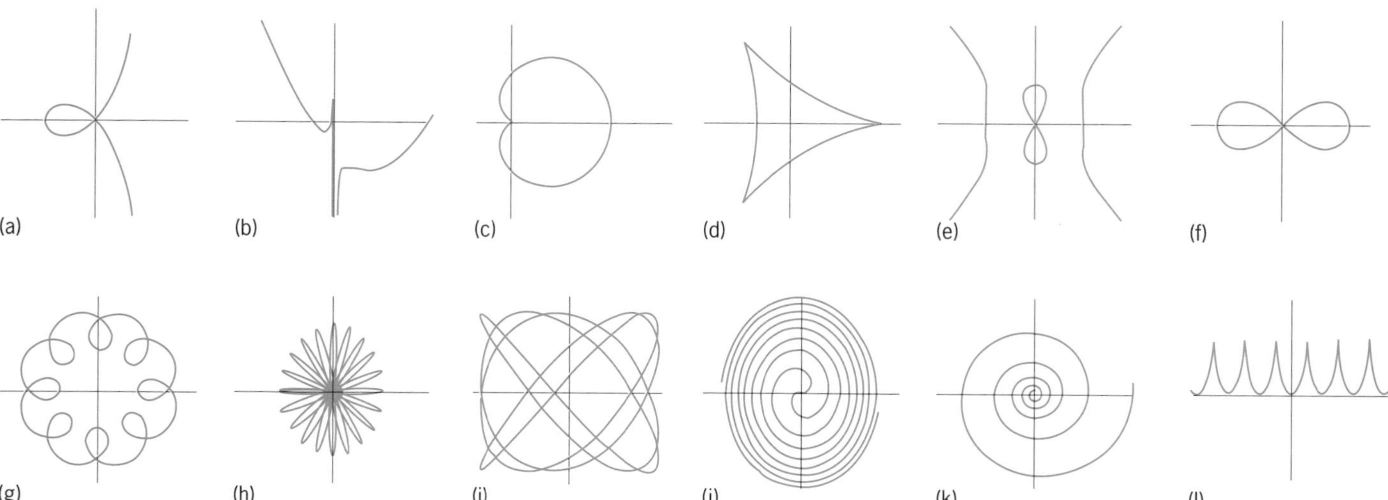

Plane curves. (*a*) Right strophoid. (*b*) Trident of Newton. (*c*) Cardioid. (*d*) Deltoid. (*e*) Devil on two sticks. (*f*) Lemniscate of Bernoulli. (*g*) Epitrochoid. (*h*) Rhodona. (*i*) Bowditch curve. (*j*) Fermat's spiral. (*k*) Logarithmic spiral. (*l*) Cycloid.

bolic lemniscate, is a closed quartic with a single segment and one node (at the origin). There is one loop, and the curve is symmetric with respect to both axes and the origin. It is expressed by Eq. (9).

$$(x^2 + y^2)^2 = a^2(x^2 - y^2) \tag{9}$$

See LEMNISCATE OF BERNOULLI.

The epitrochoid (illus. *g*) is the roulette traced by a point attached to a circle rolling about the outside of another fixed circle. Depending on the ratio of the radii of the circles, the resulting curve may be algebraic or transcendental; there are many loops. The epitrochoid is best defined parametrically by Eqs. (10).

$$\begin{aligned} x &= m \cos t - b \cos \frac{m}{b} t \\ y &= m \sin t - b \sin \frac{m}{b} t \end{aligned} \tag{10}$$

The rhodona (illus. *h*), or rose, is the locus of the polar equation (11). If m is an even integer, there are

$$r = a \cos m\phi \tag{11}$$

$2m$ petals; if m is an odd integer, there are m petals. *See* ROSE CURVE.

The Bowditch curve (illus. *i*), or curve of Lissajous, is defined by Eqs. (12). The curve is algebraic if n

$$\begin{aligned} x &= a \sin (nt + d) \\ y &= b \sin t \end{aligned} \tag{12}$$

is rational; transcendental otherwise. Depending on the parameters, the curve may be closed or not, may have zero to many nodes, and generally forms a variety of complex shapes. *See* LISSAJOUS FIGURES.

Fermat's spiral (illus. *j*) is defined by the polar equation (13) with parameter $m = 2$. It is a special case of the archimedean spirals.

$$r^m = a^m \phi \tag{13}$$

The sinusoidal spirals form a family given by the polar equation (14). Special cases are the logarithmic

$$r^n = a^n \cos n\phi \tag{14}$$

spiral (illus. *k*; $n = 0$), Cayley's sextic ($n = {}^1/_3$), cardioid ($n = {}^1/_2$), circle ($n = 1$), lemniscate of Bernoulli ($n = 2$), parabola ($n = -{}^1/_2$), and line ($n = -1$). *See* PARABOLA.

The cycloid (illus. *l*) is the locus of a point attached to a circle rolling on a line. The shape varies somewhat, depending on whether the point is inside or outside the circle or on the circumference. It is defined by the parametric equations (15).

$$\begin{aligned} x &= at - b \sin t \\ y &= a - b \cos t \end{aligned} \tag{15}$$

See CYCLOID.

J. Dennis Lawrence

Bibliography. E. Brieskorn and H. Knorrer, *Plane Algebraic Curves*, 1986; W. Fulton, *Algebraic Curves*, 1989; J. D. Lawrence, *A Catalog of Special Plane Curves*, 1972; D. H. von Seggern, *CRC Handbook of Mathematical Curves and Surfaces*, 2d ed., 1992; R. J. Walker, *Algebraic Curves*, 1950, reprint 1991.

Plane geometry

The branch of mathematics that deals with geometric figures, that is, collections of points that all lie in the same plane (coplanar). Although the words "point" and "plane" are undefined concepts, for elementary applications the intuitive meanings will serve: a point is a location, and a plane is a flat surface. For similar definitions, together with a discussion of the postulates and axioms (assumed truths) used in plane geometry, *see* EUCLIDEAN GEOMETRY.

Dimensions and measures. There are three spatial dimensions; geometric figures are classified as being zero-, one-, two-, or three-dimensional. Plane geometry deals only with geometric figures having fewer than three dimensions. Three-dimensional geometric figures, called solids, are dealt with in another branch of euclidean geometry. *See* SOLID (GEOMETRY).

A dimension is any measurement associated with a geometric figure that has units of length. A length can be described by giving a number together with a standard unit of length (such as centimeter, foot, or mile), expressing a defined and often-used distance between two points. The measure of a geometric figure is a number multiplied by a power of a length, with the result giving information about the size of the figure. The power of the length will be 1, 2, or 3, depending on whether the figure is one-, two-, or three-dimensional. *See* UNITS OF MEASUREMENT.

A point is the only zero-dimensional geometric figure, because it has no size, only location. A point has no measure.

A curve is a one-dimensional geometric figure, since its measure is the first power of a length. A curve may be thought of as the path followed by a moving point (although, more precisely, a curve is a collection of different points). If the moving point never changes its direction of motion, the curve is a straight line (or merely line). The measure of a curve (its length) is the distance that the point moves. Some curves have finite length (for example, a circle) while others have infinite length (for example, a straight line).

A region is a two-dimensional geometric figure, since its measure (the area) is the second power of a length. A region is bounded if it can be surrounded by a curve of finite length, in which case the region has a finite area. The boundary of a region is the curve such that points on one side of the curve are points of the region and points on the other side are not. For bounded regions these points are, respectively, interior points and exterior points. Points on the boundary are normally considered part of the region. A region's shape often can be described by giving the name of the boundary curve. "Circle" and "square," for example, can be applied to either a curve or to the region enclosed by that curve.

A unit square is a square whose sides have a length of one unit. (Here "unit" can be any unit of length, but usually a standard unit is chosen.) The area enclosed by a unit square is the result of squaring the length of a side, giving 1 unit^2, or one square unit. Any region having finite area has a measure equal to the number of unit squares required to cover the region, counting whole squares and fractional parts of squares. many regions have areas that can be determined by using formulas.

Lines, line segments, and rays. Exactly one line passes through two given points. The part of a line between (and including) two points is called a line segment, with the two defining points being the end points of the segment. That part of a line that lies on one side of a point (together with that point) is called a ray. Individual points of interest in a discussion are commonly named (and labeled in a drawing) with uppercase letters. Names of segments, rays, and lines can be formed by using the names of two points of the figure, together with lines or arrows that overlie the names (see **table**).

Angles. An angle is the geometric figure formed by joining two rays having a common end point. Each ray is a side of the angle; the common end point is the vertex of the angle.

An angle can be named by using the angle symbol, $\angle$, followed by the names of three points: a point on one ray, the vertex, and a point on the other ray. The vertex must be the middle-named point. An angle might be named by using $\angle$ followed by a single symbol, often the uppercase-letter name of the vertex, or perhaps a lowercase Roman or Greek letter, such as alpha (α), theta (θ), or phi (ψ) [see table].

Although an angle is the union (joining) of two rays, at least one angle is formed whenever two lines, segments, or rays intersect. In a drawing, it may be necessary to identify a particular angle by drawing a small arc (part of a circle) whose ends lie on the sides of the angle. An angle identified by an arc is labeled by placing its name somewhere close to the arc. Sometimes arrows are needed at the ends of the arc to avoid possible misinterpretation.

The concept of the measure of an angle may be understood by imagining that one ray of an angle is held fixed but the other ray is hinged at the vertex

Symbols commonly used in geometry

Example	Meaning	Comments
$\overline{AB}$	Line segment having end points A and B	Could be named $\overline{BA}$
AB	Length of $\overline{AB}$	$AB = 5$ means $\overline{AB}$ is 5 units long
$\overleftrightarrow{XY}$	Line containing points X and Y	Could be named $\overrightarrow{YX}$
$\overrightarrow{PQ}$	Ray with end point P and containing Q	
$\angle V$	Angle with vertex V	Use only if no more than two rays have end point V
$\angle RST$	Angle formed by $\overrightarrow{SR}$ and $\overrightarrow{ST}$	Could be named $\angle TSR$
$\angle x$	Angle named x	
$\angle x = 30^\circ$	Angle named x has measure 30°	
$\overset{\frown}{JK}$	Minor arc with end points J and K	
$\overset{\frown}{JLK}$	Major arc that contains point L	
$\odot C$	Circle with center C	
$\triangle XYZ$	Triangle with vertices X, Y, and Z	Could be named $\triangle YZX$, $\triangle XZY$, and so forth
▭ $ABCD$	Rectangle with vertices A, B, C, and D	Adjacent letters must be adjacent vertices
□$PQRS$	Square with vertices P, Q, R, and S	P and R are not adjacent vertices
▱$ABCD$	Parallelogram with vertices A, B, C, and D	
$ABCD\ldots X$	Polygon with vertices $A, B, \ldots, X$	Adjacent letters must be adjacent vertices
$\overline{AB} \perp j$	A segment ($\overline{AB}$) is perpendicular to a line (j)	
$\overrightarrow{ZP} \parallel \overline{TV}$	A ray ($\overrightarrow{ZP}$) is parallel to a segment ($\overline{TV}$)	
$\overline{AB} \cong \overline{XY}$	Two segments ($\overline{AB}$ and $\overline{XY}$) are congruent	Congruent segments have equal lengths
$\triangle ABC \sim XYZ$	Two triangles ($\triangle ABC$ and $\triangle XYZ$) are similar	

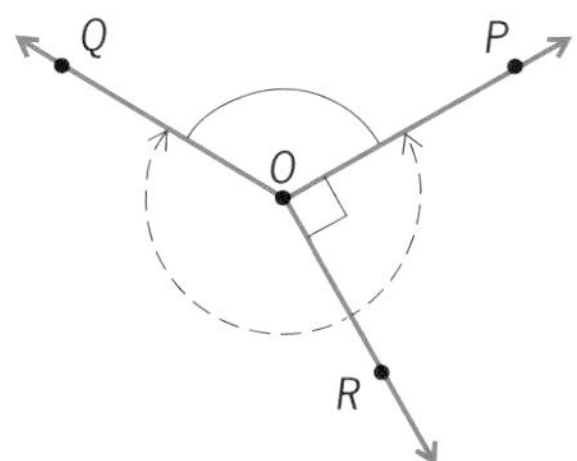

Fig. 1. Three rays with a common end point. Here, ∠POQ (solid arc) = $^1/_3$ revolution (an obtuse angle), and ∠POQ (dotted arc) = $^2/_3$ revolution (a reflex angle). Line segment $\overline{OP}$ is perpendicular to $\overline{OR}$, so ∠POR (square at vertex) = $^1/_4$ revolution = 90° (a right angle).

and allowed to rotate in the plane. A measure of the angle is a number, together with some unit of angular measure, that tells how much the hinged ray would need to be rotated so that it would overlie the fixed ray. If a ray were to be rotated exactly one revolution, it would return to its original position. The measure of an angle can be what fraction of one revolution would enable one side of the angle to become coincident with the other. (The measure of an angle is not a dimension, since it does not have units of length.)

In most applications of geometry that involve angles, the measure of an angle is of such importance that an angle is defined as the two rays together with the measure. In plane geometry, all rotations have positive measure, regardless of the direction of rotation (clockwise or counterclockwise). In some applications it is important which ray is rotated (the initial side) and which is held fixed (the terminal side), and whether the rotation of the initial side is clockwise (a negative rotation) or counterclockwise (a positive rotation). In some applications the measure of the angle is the only property of importance, with the locations of the two sides being immaterial. *See* TRIGONOMETRY.

If the measure of an angle has importance, it may be necessary to indicate the measure with an arc in a drawing (**Fig. 1**). If a rotation of one-third revolution is required to bring a ray, $\overrightarrow{OP}$, into coincidence with a second ray, $\overrightarrow{OQ}$, following a given arc, then the angle *POQ* that is identified with this arc has a measure of $^1/_3$ revolution. However, a second arc can be drawn between the rays $\overrightarrow{OP}$ and $\overrightarrow{OQ}$, which is followed by the ray $\overrightarrow{OP}$ when it is rotated in the opposite direction to bring it into coincidence with $\overrightarrow{OQ}$. The angle *POQ* that is identified with this second arc has a measure of $^2/_3$ revolution. If an angle is drawn without an arc, it normally is assumed that the measure is the smallest rotation that would bring one side into coincidence with the other.

The revolution is a convenient unit of angular measure for many applications. However, a more commonly used unit is the degree (°), which is defined by the equation 360° = 1 revolution. In the modern-day use of calculators and computers, fractions of degrees are most conveniently expressed by using decimal fractions; however, another subdivision of degrees is firmly entrenched in many applications: 1 degree = 60 minutes (1° = 60′), and 1 minute = 60 second (1′ = 60″).

An angle of measure $^1/_4$ revolution, or 90°, is called a right angle. Two lines (or rays or segments) that intersect so as to form a right angle are said to be perpendicular (see table). In a drawing, the fact that two geometric figures are perpendicular can be indicated by placing a small square near the intersection (Fig. 1). Two angles are complementary if their measures have a sum of 90°. An angle of measure $^1/_2$ revolution, or 180°, is called a straight angle. Two angles are supplementary if their measures have a sum of 180°.

Other angles can be classified according to how their measures compare with the measures of a right angle and a straight angle: an acute angle has a measure between 0 and 90°; an obtuse angle has a measure between 90 and 180°; and a reflex angle has a measure between 180 and 360° (Fig. 1).

Angles that share a side but whose arcs do not overlap are adjacent angles. Two nonperpendicular intersecting lines form two acute angles and two obtuse angles. Each acute angle and an obtuse angle form a pair of adjacent supplementary angles. Knowledge of the measure of one of these angles is sufficient to determine the measures of the other three.

If two coplanar lines do not intersect (no matter how far extended), those lines are parallel (see table). A transversal is a straight line that intersects two other lines, which themselves might intersect or be parallel. Parallel lines intersected by a transversal form two sets of angles, with each set containing four angles of equal measure. Certain words describe the relative positions of the arcs of angles in a set (**Fig. 2**). The arcs of vertical angles lie on opposite sides of the same vertex. Corresponding angles have different vertices, but their arcs have the same positions relative to the vertices. Alternate angles have different vertices, and their arcs lie on opposite sides of the transversal. The arcs of interior angles lie

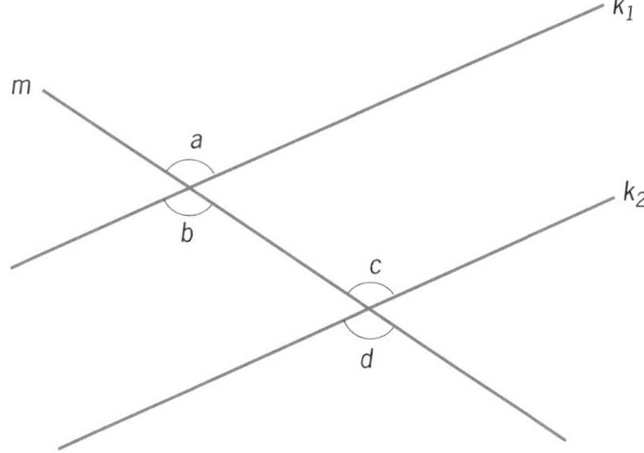

Fig. 2. Relationships between the obtuse angles formed when parallel lines, k_1 and k_2, are intersected by a transversal, *m*. The angles ∠*a* and∠*b* are vertical angles, as are ∠*c* and ∠*d*. The angles ∠*a* and ∠*c* are corresponding angles, as are ∠*b* and ∠*d*. The angles ∠*b* and ∠*c* are alternate interior angles; ∠*a* and ∠*d* are alternate exterior angles. Similar relationships exist between the acute angles formed.

between the parallel lines, while those of exterior angles lie outside the parallel lines.

Polygons. A polygon is the geometric figure formed when line segments are joined end to end so as to enclose a region of the plane. The polygon having the least number of sides (three) is the triangle. Three lines, no two of which are parallel, generally intersect in three points, and the union of the three line segments having those points as end points forms a triangle. (Although a triangle is a one-dimensional geometric figure, sometimes the word means a region enclosed by a triangle.) An angle whose sides are two of those segments and whose arc lies inside the triangle is an interior angle of the triangle; an exterior angle of the triangle is any angle that is adjacent and supplementary to an interior angle (**Fig. 3**). The interior angles of any triangle have measures that sum to 180°. This fact allows the determination of the measures of all angles formed by the intersections of three line if the measures of only two angles with different vertices are known. *See* POLYGON.

Congruent and similar geometric figures. Two geometric figures are congruent (≅) if they have exactly the same shape and size. As examples, any two lines or any two rays are congruent; two segments are congruent only if they have the same length; two angles that have equal measures are congruent.

If two geometric figures are congruent, one figure could be made to overlie the other by a combination of these types of motion: translation (sliding), rotation (twisting), and reflection about a line (flipping over). The parts of the geometric figures that would then coincide are called corresponding parts, where a part of a geometric figure is any set of points associated with that figure.

Two triangles are congruent if any one of the following is satisfied:

1. Each side of one triangle is congruent to a corresponding side of the other triangle.

2. Two sides and the included angle of one triangle are congruent to the corresponding sides and included angle of the other triangle. (An angle is included if the two sides of the triangle are sides of the angle.)

3. Two angles and the included side of one triangle are congruent to the corresponding angles and included side of the other triangle. (A side of a triangle is included if it is a side of each of the two angles.)

Two geometric figures are similar (∼) if they have the same shape but (perhaps) have different sizes. An

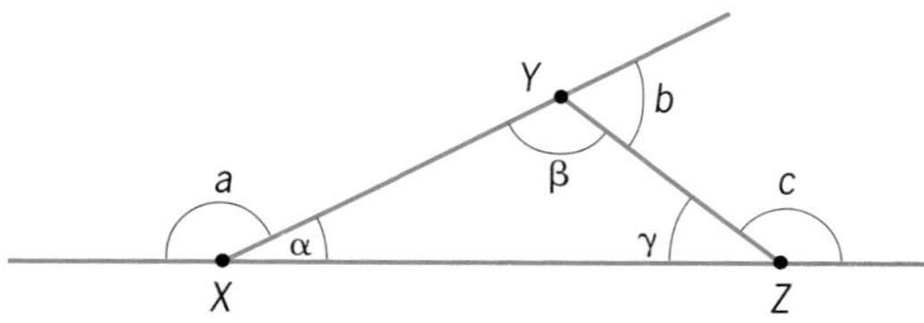

Fig. 3. Triangle *XYZ*, with interior angles $\angle\alpha$, $\angle\beta$, and $\angle\gamma$; and exterior angles $\angle a$, $\angle b$ and $>\angle c$.

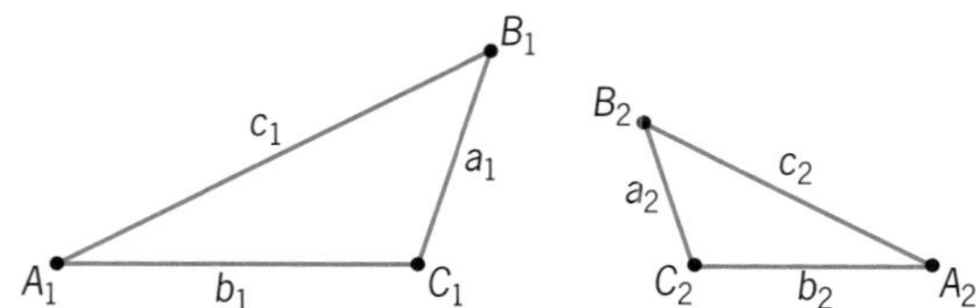

Fig. 4. Two similar triangles, $A_1B_1C_1$ and $A_2B_2C_2$, with the corresponding parts have the same letter names but different subscripts. Thus $\angle A_1 \cong \angle A_2$; $\angle B_1 \cong \angle B_2$; $\angle C_1 \cong \angle C_2$; $a_1/a_2 = b_1/b_2 = c_1/c_2 = 3/2$.

examples, any two line segments are similar; any two squares are similar. If two geometric figures are similar, one figure could be made to overlie the other by a combination of translation, rotation, reflection, and either expanding or shrinking. The parts that would then coincide are corresponding. Two similar geometric figures will have corresponding angles that have equal measures, and will have corresponding dimensions that are proportional.

Most common applications of similar geometric figures involve triangles (**Fig. 4**). If the angles of one triangle have, respectively, measures that equal the measures of the angles of another triangle, those triangles are similar. If all pairs of corresponding sides have lengths that are proportional, those triangles are similar. (Neither of these two facts applies to polygons having more than three sides.)

Circles. A circle is a collection of points in the plane, all of which are the same distance from another point, called the center. The region bounded by a circle sometimes is called a disk, but usually the word "circle" is used, relying on context to distinguish between the curve and the region. Circumference means either the circle that is the boundary curve of a disk or the distance around that circle. A radius of a circle is any line segment that joins the center and a point of a circle. A chord is any line segment whose end points lie on the circle. A diameter is any chord that contains the center. Circles having the same center are concentric.

Any straight line and a given circle have two points, or one point, or on points of intersection. A secant is a line that intersects a circle in two points. A tangent is a line that intersects a circle in only one point, called the point of tangency. A tangent line is perpendicular to the radius that joins the center with the point of tangency (**Fig. 5**).

Associated with a circle are several important dimensions, including the lengths of a radius and a diameter. These lengths often are merely called the radius and the diameter of the circle, usually denoted, respectively, by r and d. These symbols appear in almost all formulas involving the length of the circumference C (called the circumference) or the area A enclosed by a circle (called the area of the circle). Relationships between these variables for any circle are given by Eqs. (1).

$$d = 2r \qquad C = \pi d \qquad A = \pi r^2 \qquad (1)$$

In these formulas, π (the lowercase Greek letter pi)

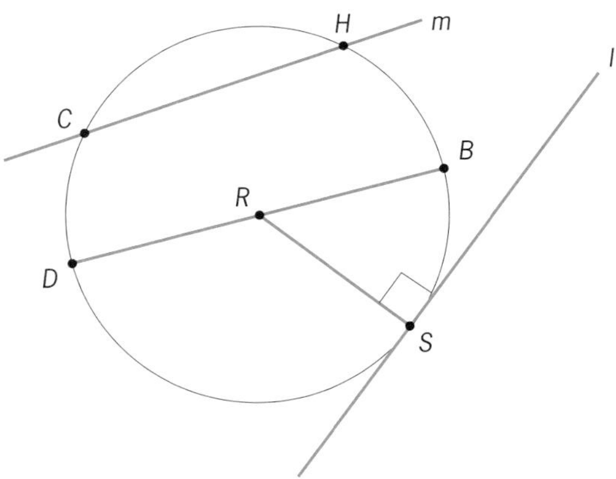

Fig. 5. Circle with center at *R*. Here $\overline{CH}$ and $\overline{DB}$ are chords; $\overline{RS}$, $\overline{DR}$, and $\overline{RB}$ are some radii; $\overline{DB}$ is a diameter; *m* is a secant line; and *l* is a tangent line with point of tangency *S*.

represents the irrational number (value 3.141592. . .) that is usually defined as the ratio of the circumference to the diameter of any circle.

The end points of any chord divide a circle into two arcs: the larger major arc and the smaller minor arc. If the chord is a diameter, the arcs are of equal size and are semicircles. Rays from the center that pass through the end points define two central angles: a reflex angle that corresponds to the major arc, and an obtuse or acute angle that corresponds to the minor arc.

Any angle whose vertex lies on the circle and whose sides both intersect the circle at points other than the vertex is an inscribed angle. The points where the sides of an inscribed angle intersect the circle define a chord, which in turn defines two central angles corresponding to the major and minor arcs. The central angle corresponding to the arc that does not contain the vertex of the inscribed angle has twice the measure of the inscribed angle (**Fig. 6**). Therefore, any inscribed right angle intersects a circle at the end points of a diameter.

A unit circle is a circle having a radius of length 1 unit, and will have a circumference equal to 2π units. Since an arc's central angle is directly proportional to the length of that arc, an angle's measure can be given by the number of units of arc length on the unit circle. Such a unit of angular measure is called a radian, and one revolution equals 2π radians.

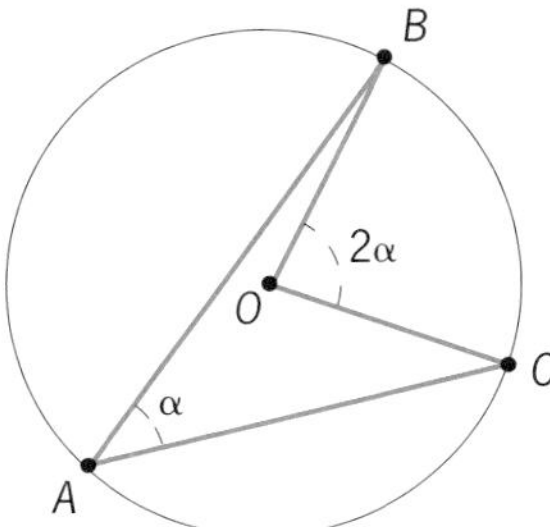

Fig. 6. Circle with inscribed angle and central angles, showing relationship. Here, $\angle BAC = \angle\alpha$ is an inscribed angle, $\angle BOC$ is the central angle of $\widehat{BC}$, and reflex $\angle BOC$ is the central angle of $\widehat{BAC}$. Then $\angle BOC = 2 \cdot \angle BAC$.

All circles are similar, so the length s of a circular arc of a circle of radius r is given by Eq. (2), where θ is

$$s = r\theta \tag{2}$$

the measure of the central angle in radians.

A circular sector is the region bounded by two radii and an arc of a circle. Since the area, A_s, of a circular sector is directly proportional to the measure of the arc's central angle θ, Eq. (3) follows, where θ

$$\frac{\theta}{2\pi} = \frac{A_s}{A} = \frac{A_s}{\pi r^2} \tag{3}$$

is measured in radians, so that the area of the circular sector is given by Eq. (4).

$$A_s = {}^1\!/_2 r^2\theta \tag{4}$$

A circular segment is the region bounded by a circular arc and the chord that joins the end points of that arc. The area, A_g, of a circular segment of a circle having radius r and central angle θ is given by Eq. (5), where θ is measured in radians.

$$A_g = {}^1\!/_2 r^2(\theta - \sin\theta) \tag{5}$$

See CIRCLE. Harry L. Baldwin, Jr.

Bibliography. H. L. Baldwin, Jr., *Essentials of Geometry*, 1993; D. Hilbert and S. Cohn-Vossen, *Geometry and the Imagination*, 1952, reprint 1999; P. G. O'Daffer and S. R. Clemens, *Geometry: An Investigative Approach*, 2d ed., 1992; I. Todhunter, *The Elements of Euclid*, 1891.

Planer

A machine for the shaping of long, flat, or flat contoured surfaces by reciprocating the workpiece under a stationary single-point tool or tools. Usually the workpiece is too large to be handled on a shaper.

Planers are built in two general types, open-side or double-housing. The former is constructed with one upright or housing to support the crossrail and tools. The double-housing type has an upright on either side of the reciprocating table connected by an arch at the top.

Saddles on the crossrail carry the tools which feed across the work. A hinged clapper box, free to tilt, provides tool relief on the return stroke of the table. A variation is the milling planer; it uses a rotary cutter rather than single-point tools. *See* WOODWORKING.

Alan H. Tuttle

Planet

A relatively small, solid celestial body moving in orbit around a star, in particular the Sun.

Planets of the Solar System

Besides Earth, the eight known planets of the solar system are Mercury, Venus, Mars, Jupiter, Saturn, Uranus, Neptune, and Pluto; in addition, over

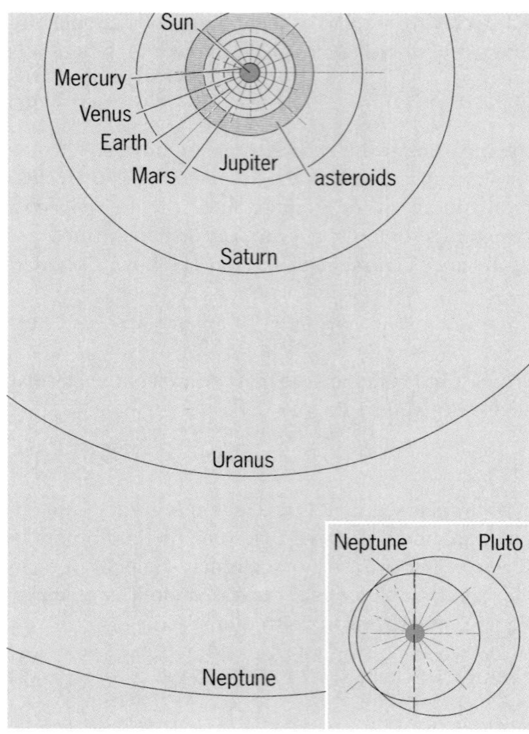

Fig. 1. Plan of the solar system. (*After L. Rudaux and G. de Vaucouleurs, Larousse Encyclopedia of Astronomy, Promethens Press, 1959*)

20,000 minor planets, or asteroids, mostly located between the orbits of Mars and Jupiter, are known (**Fig. 1**). *See* ASTEROID; JUPITER; MARS; MERCURY (PLANET); NEPTUNE; PLUTO; SATURN; URANUS; VENUS.

Classification. There are two basic groups of planets in the solar system: the small, dense, terrestrial planets—Mercury, Venus, Earth, Mars, and Pluto—and the giant or Jovian planets—Jupiter, Saturn, Uranus, and Neptune. With the exception of Pluto, the terrestrial planets are all located within the inner solar system. The low-density Jovian planets extend outward from Jupiter to the remote outer reaches of the solar system. This distribution is not accidental, but is related to the fractionation of rocky, icy, and gaseous materials during the early stages of formation of the solar system. *See* PLANETARY PHYSICS; SOLAR SYSTEM.

Each of the main planets from Earth to Pluto is accompanied by one or more secondary bodies called satellites. Many of the smallest satellites are not observable from Earth, but were discovered during spacecraft flybys. *See* SATELLITE (ASTRONOMY).

The planets may also be divided into inferior planets, Mercury and Venus, located inside Earth's orbit, and superior planets, from Mars to Pluto, circulating outside Earth's orbit.

Kepler's laws. The motions of the planets in their orbits around the Sun are governed by three laws discovered by Johannes Kepler at the beginning of the seventeenth century.

First law: The orbit of a planet is an ellipse, with the Sun at one of its foci.

Second law (the law of areas): As a planet revolves in its orbit, the radius vector (the line from the Sun to the planet) sweeps out equal areas in equal intervals of time.

Third law (the harmonic law): The square of the period of revolution P is proportional to the cube of the orbit's semimajor axis a; that is, for all planets the ratio P^2/a^3 is a constant.

If a is expressed in astronomical units and P in sidereal years, $P^2/a^3 = 1$. One astronomical unit (AU) is the mean distance from Earth to the Sun and is approximately equal to 92.96×10^6 mi (149.6×10^6 km). Otherwise, the constant of the harmonic law is given by Newton's law of gravitation as $G(M + m)/4\pi^2$, where M and m are the masses of the Sun and the planet, and G is the constant of gravitation. *See* ASTRONOMICAL UNIT; EARTH ROTATION AND ORBITAL MOTION; GRAVITATION; YEAR.

Kepler's laws are true only when the mutual perturbations of the motions of the planets by the others are neglected. *See* CELESTIAL MECHANICS; KEPLER'S LAWS.

Planetary configurations. In the course of their motions around the Sun, Earth and other planets occupy a variety of relative positions or configurations (**Fig. 2***a*), the principal of which are designated as

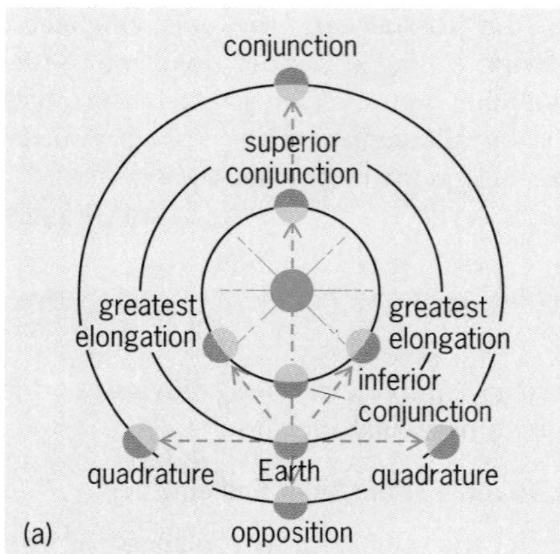

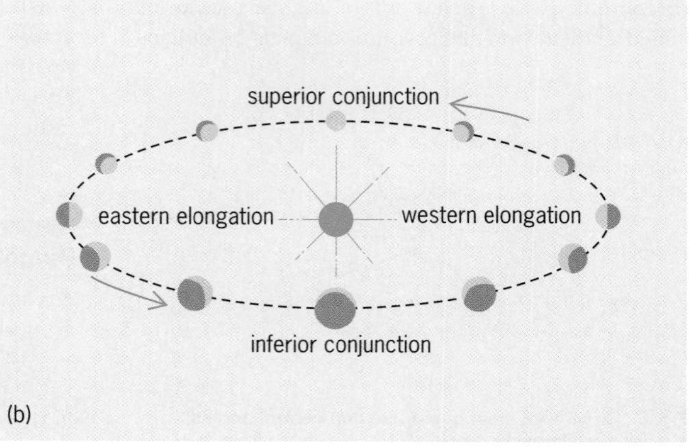

Fig. 2. Planetary configurations and phases. (*a*) Positions of Earth and other planets relative to the Sun. (*b*) Phases of inferior planets (Mercury and Venus). (*After L. Rudaux and G. de Vaucouleurs, Larousse Encyclopedia of Astronomy, Prometheus Press, 1959*)

follows: The inferior planets are in conjunction with the Sun when closest to the Earth-Sun line, either between Earth and the Sun (inferior conjunction) or beyond the Sun (superior conjunction). On rare occasions when the planet is very close to the plane of Earth's orbit at the time of an inferior conjunction, a transit in front of the Sun is observed. *See* TRANSIT (ASTRONOMY).

Between conjunctions, the geocentric angular distance from the planet to the Sun, or the elongation, varies up to a maximum value; the greatest or maximum elongations of Mercury and Venus are 28° and 47°, respectively. The superior planets are not so limited, and their elongations can reach up to 180° when they are in opposition with the Sun; when the elongation is ±90°, they are in quadrature (eastern or western) with the Sun.

The telescopic aspect of the disks of the planets varies according to their configurations, which determine the angle between the directions of illumination and observation, or the phase angle. Between inferior conjunction and greatest elongations, the interior planets show crescent phases, like the Moon between new moon and first or last quarters (Fig. 2*b*); between greatest elongations and superior conjunction, they show a gibbous phase, like the Moon between quarters and full moon. At superior conjunction, they show a circular disk, fully illuminated and seen face on, while during transits, the dark side is profiled against the Sun. The superior planets show their full phase at both conjunction and opposition and a gibbous phase near quadrature, at which time the unilluminated portion of the disk is at a maximum.

Apparent motions. The combinations of the orbital motions of Earth and of any other planet give rise to complicated apparent motions of that planet as seen from Earth. Because the orbits of the main planets are, except for Pluto, only slightly inclined to the plane of the orbit of Earth, the apparent paths of the planets (except Pluto) are restricted to the zodiac, a belt 16° wide centered on the ecliptic. The ecliptic is the path in the sky traced out by the Sun in its apparent annual journey as the Earth revolves around it. Along this path, the apparent motions of the inferior planets with respect to the Sun are alternatively westward, from greatest elongation through inferior conjunction to greatest elongation, then eastward, from greatest elongation through superior conjunction to greatest elongation (Fig. 2). The mean motion of the superior planets is always westward. *See* ASTRONOMICAL COORDINATE SYSTEMS; ECLIPTIC.

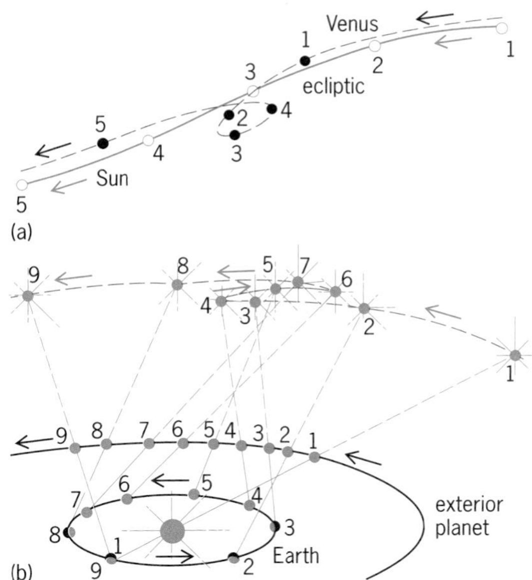

Fig. 3. Apparent motions as observed from Earth (*a*) of an interior planet with respect to the Sun and (*b*) of an exterior planet with respect to the fixed stars. (*After L. Rudaux and G. de Vaucouleurs, Larousse Encyclopedia of Astronomy, Prometheus Press, 1959*)

The apparent motions with respect to the celestial sphere, that is, to the fixed stars, appear for the inferior planets as oscillations back and forth about the position of the Sun steadily moving eastward among the stars. For the superior planets, the apparent motion is generally eastward or direct, but for short periods near the time of opposition it is westward or retrograde (**Fig. 3**). At times when the direction of the apparent motion on the sphere reverses, the planet appears to be stationary.

The mean interval of time between successive returns to the same place with respect to the stars is the sidereal period, which is established by the true motion of revolution of the planet in its orbit around the Sun. The mean interval of time between

TABLE 1. Elements of planetary orbits

Planet	Symbol	Mean distance from Sun (semimajor axis of orbit) AU	10^6 mi	10^6 km	Sidereal period of revolution Years	Days	Synodic period, days	Mean orbital velocity mi/s	km/s	Orbital eccentricity	Orbital inclination, degrees
Mercury	☿	0.387	36.0	57.9	0.241	87.97	115.88	29.75	47.87	0.206	7.00
Venus	♀	0.723	67.2	108.2	0.615	224.70	583.92	21.76	35.02	0.007	3.39
Earth	⊕	1.000	93.0	149.6	1.000	365.24		18.51	29.79	0.017	0.00
Mars	♂	1.524	141.6	227.9	1.881	686.93	779.94	14.99	24.13	0.093	1.85
Jupiter	♃	5.203	483.6	778.3	11.857	4,330.60	398.88	8.12	13.07	0.048	1.30
Saturn	♄	9.555	888.2	1429.4	29.424	10,746.9	378.09	6.01	9.67	0.056	2.49
Uranus	⛢	19.22	1786.	2875.	83.75	30,588.7	369.66	4.24	6.83	0.046	0.77
Neptune	♆	30.11	2799.	4504.	164.72	59,799.9	367.49	3.41	5.48	0.009	1.77
Pluto	♇	39.54	3676.	5916.	248.0	90,589.	366.72	2.95	4.75	0.249	17.14

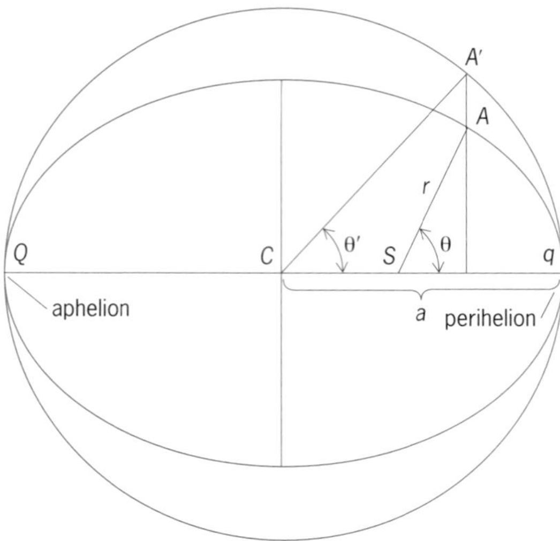

Fig. 4. Elliptic motion of a planet. Symbols explained in text.

successive returns of the same configuration with respect to the Sun (for example, conjunctions or oppositions) is the synodic period, which governs the apparent motion of the planet as seen from Earth (**Table 1**).

Elliptic motion. The motion of a planet having an elliptical orbit of semimajor axis a, with the Sun at the focus S, brings the planet on each revolution to the perihelion q and to the aphelion Q, the points of the orbit respectively nearest to and farthest from S. If C is the center of the ellipse, the semimajor axis is $a = Cq = CQ$. The eccentricity of the ellipse is $e = CS/Cq = CS/a$, from which the perihelion distance is $Sq = a(1 - e)$ and the aphelion distance is $SQ = a(1 + e)$. At any other point, the Sun-planet distance is $r = a(1 - e^2)/(1 + e \cos\theta)$, where θ is the angle $\angle qSA$, termed the true anomaly. At that same point the velocity is $v = [2GM_\odot(1/r - 1/2a)]^{1/2}$, where $M_\odot$ is the mass of the Sun. If A' is the point on the principal circle of radius a whose projection in the ellipse is A (**Fig. 4**), the eccentric anomaly θ' is the angle $\angle qCA'$, so that $r = a(1 - e \cos\theta')$. If the planet is at perihelion at time T and returns to it at time $T + P$, the mean angular velocity (or mean motion) is $n = 2\pi/P$, and at any time t the mean anomaly is $M = n(t - T)$. *See* ELLIPSE.

The relation between the mean and eccentric anomalies, $M = \theta' - e \sin\theta'$, is known as Kepler's equation; its solution gives θ' and, consequently, r at any time t when the orbital elements a, e, n, and T are known. *See* KEPLER'S EQUATION.

Orbital elements. The position of a planet in its orbit and the orientation of the orbit in space are completely defined by seven orbital elements (**Fig. 5**). These are (1) the semimajor axis a, (2) the eccentricity e, (3) the inclination i of the plane of the orbit to the plane of the ecliptic, (4) the longitude ☊ of the ascending node N, (5) the angle ω from the ascending node N to the perihelion q, (6) the sidereal period of revolution P, or the mean (daily) motion $n = 2\pi/P$, and (7) the date of perihelion passage T, or epoch E.

If the plane of a planet's orbit is inclined to the plane of the ecliptic, their intersection NN' is the line of nodes; in its motion, the planet crosses the plane of the ecliptic from south to north at the ascending node N and from north to south at the descending node N'. The longitude of the ascending node is the angle ☊ $= \angle$ ♈SN, measured in the plane of the ecliptic from the vernal equinox ♈. The longitude of perihelion is $\tilde{\omega} =$ ☊ $+ \omega = \angle$ ♈$SN + \angle NSq$, the second angle being measured in the plane of the planet's orbit (Fig. 5). The location of the plane of the orbit in space is defined by i and ☊, the orientation of the ellipse in this plane by ω, its form by e, and its size by a, and the position of the planet on the ellipse by P and T (and by the time t). *See* ORBITAL MOTION.

Determination of orbital elements. Accurate observations of the positions of the planets with respect to background stars or with respect to the celestial coordinates are used to determine the elements of their orbits. In principle, three observations of two coordinates (right ascension and declination) and the laws of elliptic motion are sufficient to determine the six independent elements of a planetary orbit, since by Kepler's third law a^3 is proportional to P^2. In practice, as many observations as possible are combined, and the equations solved by the method of least squares; the elements for a given epoch so obtained are subject to variations and corrections allowing for planetary perturbations. Tables of the motions of the planets for several centuries past and future have been established, from which the yearly ephemerides are extracted in a form convenient for immediate use. *See* EPHEMERIS; LEAST-SQUARES METHOD.

The main characteristics of the planetary orbits are given in Table 1.

Planetary sizes. The apparent diameter of a planet may be determined telescopically by means of a filar micrometer or, preferably, a birefringent or double-image micrometer, or it may be measured on large-scale photographs taken through telescopes or by planetary spacecraft. If the apparent diameter of a planet is d'' when its distance to Earth is Δ, the

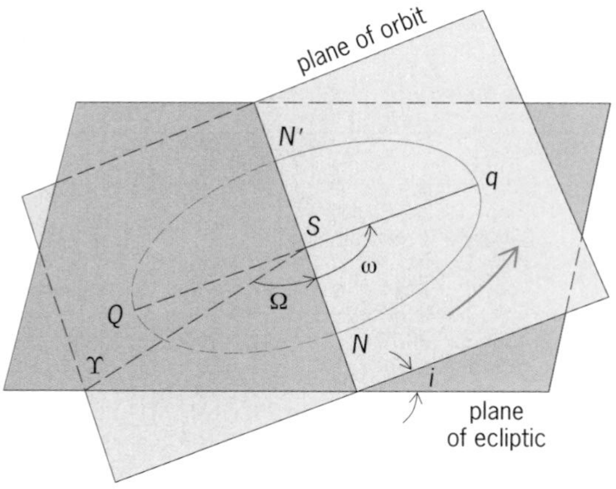

Fig. 5. Orbital elements, which determine the position of a planet in its orbit and the orientation of the orbit in space. Symbols explained in text.

TABLE 2. Physical characteristics of the Sun's planets

Planet	Equatorial radius (r_e) (Earth=1)	Equatorial radius mi	Equatorial radius km	Ellipticity	Volume (Earth=1)	Mass (Earth=1)	Density, g/cm³	Escape velocity mi/s	Escape velocity km/s	Rotation period	Obliquity, degrees[1]
Mercury	0.38	1,515	2,440	0.000	0.055	0.055	5.43	2.5	4.4	58 d 15.5 h	0.1
Venus	0.95	3,761	6,052	0.000	0.854	0.815	5.20	6.5	10.4	243 d 0.5 h	177.4[2]
Earth	1.00	3,963	6,378	0.0034	1.000	1.000	5.52	7.0	11.2	23 h 56 m 23 s	23.45
Mars	0.53	2,110	3,396	0.0069	0.151	0.107	3.34	3.1	5.0	24 h 37 m 23 s	25.19
Jupiter	11.21	44,423	71,492	0.0649	1408.	317.710	1.33	37.0	59.5	9 h 55 m 30 s[3,4]	3.12
Saturn	9.45	37,449	60,268	0.0980	844.	95.162	0.69	22.1	35.5	10 h 39 m 22 s[3,5]	26.73
Uranus	4.01	15,882	25,559	0.0229	64.	14.535	1.32	13.2	21.3	17 h 22.2 m[3,6]	97.86[2]
Neptune	3.88	15,389	24,766	0.017	59.	17.141	1.64	14.6	23.5	16 h 6.6 m[3,7]	29.56
Pluto	0.18	715	1,150	?	0.006	0.002	2.0	0.7	1.1	6 d 9 h 17.6 m	119.6[2]

[1] Obliquity is the tilt of the equator with respect to the orbit plane.
[2] Venus, Uranus, and Pluto are considered to have retrograde rotation.
[3] Internal (System III) rotation period, the rotation period of the planet's core, as deduced from its magnetic field.
[4] Jupiter's equatorial (System I) rotation period is 9 h 50.5 m.
[5] Saturn's equatorial rotation period is 10 h 14.0 m.
[6] Uranus's equatorial rotation period is about 18.0 h.
[7] Neptune's equatorial rotation period is about 18.8 h.

planet's diameter is $D = \Delta \sin d'' = \Delta d''/206{,}265$, where d'' is measured in seconds of arc, and both Δ and D are expressed in astronomical units.

Polar flattening is perceptible only on the planets Jupiter and Saturn. A planet's polar and equatorial radii, r_p and r_e, can be used to establish its mean radius $r = (r_p + r_e)/2$ and its ellipticity $\epsilon = 1 - (r_p/r_e)$. The mean radius may also be expressed in terms of the mean radius of Earth (3963 mi or 6378 km). The relative surface area is then very nearly equal to r^2 and the relative volume to r^3. For the nearer planets with a solid surface (Mercury, Venus, and Mars), the diameter can be determined more precisely by radar, using the time-delay technique. *See* RADAR; RADAR ASTRONOMY.

Masses, gravity and density. The mass of a planet is deduced easily if it has one or more satellites. If a is the mean distance (semimajor axis) of the satellite's orbit and P its period of revolution expressed respectively in astronomical units and sidereal years, the mass m of a planet, expressed as a fraction of the Sun's mass, is given through Newton's law of gravitation by $m = a^3/P^2$. This assumes that the mass of the satellite relative to that of the planet, and of Earth relative to that of the Sun, may be neglected, which is nearly always the case within the accuracy of the data. Since the ratio $m_\oplus/M_\odot$ of the masses of Earth and the Sun can be determined similarly, planetary masses can also be expressed in terms of $m_\oplus$.

If the planet has no satellite (Mercury and Venus), its mass can be derived only from the perturbations that it causes in the motions of the other planets, occasional comets, or passing spacecraft. Since the perturbations are small, the masses so obtained are generally of low accuracy. The most precise values are those derived from the perturbations of the orbits of space probes. *See* PERTURBATION (ASTRONOMY).

Once the mass m and the radius r of a planet are known in terms of Earth's mass and radius, its surface gravity and mean density relative to Earth are given by $g = m/r^2$ and $\rho = m/r^3$, respectively. Multiplication by 981 and 5.552 gives the corresponding values in cgs units (cm/s² and g/cm³, respectively).

From r and m follows also the escape velocity V_1 that permits a projectile (or a molecule) to leave the planet on a parabolic orbit: $V_1 = (2Gm/r)^{1/2}$; this is $\sqrt{2}$ times the velocity of an hypothetical satellite moving in a circular orbit close to the surface of the planet. These elements are listed in **Table 2**. *See* ESCAPE VELOCITY.

Rotation periods. The period of rotation of a planet can be determined by several methods: (1) Direct observation of permanent surface markings (Mars) or of long-duration cloud formations in its atmosphere (Jupiter, Saturn, Uranus, Neptune) is the classical method. (2) The line-of-sight velocity difference between the opposite equatorial limbs, determined either by means of radar sounding (Mercury, Venus) or spectroscopically, can be combined with the diameter to yield a rotation period. (3) Since the giant planets lack solid surfaces, their rotation rates are found by timing the cyclic pattern of radio energy emanating from the planets' magnetospheres, which rotate in synchrony with their deep interiors. (4) When the apparent diameter of the disk is too small for any of these methods (Pluto, asteroids), a determination of the periodicity of the light variations due to the changing presentation of bright and dark regions of the surface provides a fairly accurate value of the rotation period. The rotation periods of all the main planets are now well determined (Table 2).

Planetary radiations. The electromagnetic radiation received from a planet is made up of three main components: the visible reflected sunlight, including some ultraviolet and near-infrared radiation; the thermal radiation due to the planet's heat, including both infrared radiation and ultrashort radio waves; and the nonthermal radio emission due to electrical phenomena, if any, in the planet's atmosphere or in its radiation belts.

Planetary brightness. The apparent brightness of a planet, as measured by visual, photographic, and electronic means, is usually expressed in the stellar magnitude scale; it varies in inverse proportion to the squares of the distances r from the Sun and Δ from Earth. The fraction of the incident light reflected at full phase compared with the fraction that would be reflected under the same conditions by an equivalent perfect diffuse reflecting disk is called the geometric albedo. It is a measure of the backscattering reflectivity of the planet's visible surface. The visual albedos of the planets vary between 5 and 70%. *See* ALBEDO; MAGNITUDE (ASTRONOMY).

Thermal radiation. The thermal radiation from a planet can be measured either with a radiometer at wavelengths of 8–14 micrometers, 17–25 μm, and 30–40 μm (which are partially transmitted by the Earth's atmosphere) or with a radio telescope at wavelengths between 1 mm and 30 cm. In either case, the amount of energy corresponds to that which would be received under the same conditions from a perfect radiator of the same size at a certain temperature T, called the blackbody temperature of the planet. Its relation to the actual temperature depends on the thermal and radiative properties of the atmosphere and surface of the planet. Jupiter radiates nearly twice as much energy as it receives from the Sun. From this it is inferred that the planet has an internal source, perhaps primordial heat continuing to escape long after Jupiter's formation 4.6×10^9 years ago. Saturn and Neptune (but not Uranus) also emit more energy than they receive from the Sun. While gravitational separation of the planet's hydrogen and helium has been suggested as one possible mechanism for this excess energy, no single solution appears to account for the entire energy outflow. *See* HEAT RADIATION; INFRARED ASTRONOMY; RADIO ASTRONOMY; RADIOMETRY.

Nonthermal radiation. Large radio telescopes have recorded nonthermal radio emission at decimeter and decameter wavelengths from Jupiter. The decameter emission takes the form of irregular bursts of noise originating within the planet's atmosphere. Voyager spacecraft revealed that powerful electric currents exist inside the Jovian magnetosphere, particularly one called a flux tube linking higher latitudes on the planet with satellite Io. Since the observed decametric radiation is modulated by the orbital position of Io, this current loop may be responsible for the outbursts.

Planetary atmospheres. The principal constituents of the atmospheres of the terrestrial planets (Pluto, a special case, is discussed below) are carbon dioxide, nitrogen, water, and (on Earth only) oxygen; Mercury has a very tenuous envelope dominated by atoms of sodium and potassium. The atmospheres of the giant planets are composed primarily of hydrogen and helium, with lesser amounts of methane, ammonia, and water. Atmospheric motions are driven by temperature gradients—in general, those existing between the warm equatorial regions and the cooler polar areas. An atmosphere thus tends to redistribute heat over the planetary surface, lessening the temperature extremes found on airless bodies.

On planets having relatively dense atmospheres, heat from the Sun is trapped by the greenhouse effect. That is, visible radiation from the Sun passes readily through the atmosphere to heat the planetary surface, but infrared radiation reemitted from the surface is constrained from escaping back to space by the lower transparency of certain atmospheric gases (especially carbon dioxide and water vapor) to longer wavelengths. For example, although Venus absorbs approximately the same amount of energy from the Sun as does Earth, the greenhouse effect is responsible for heating the surface of Venus to a much higher temperature, approximately 750 K (900° F). *See* VENUS.

Directly through wind erosion or indirectly by carrying water vapor which can precipitate as rain, the atmospheres of the terrestrial planets are a major factor in modifying surface structure and rearranging the distribution of surface materials. Mercury, being virtually airless, exhibits a relatively unmodified surface, very similar in appearance to that of the Moon. Pluto, however, is covered with a methane ice frost and thus maintains a tenuous atmosphere of methane vapor (at least when in the part of its orbit nearest the Sun).

Possible unknown planets. During the nineteenth century, an unexplained irregularity in the motion of Mercury was thought by some investigators to be caused by an unknown planet circulating between the Sun and Mercury, called Vulcan, which was looked for in vain. This irregularity was satisfactorily explained in 1915 by Einstein's general theory of relativity. It is now certain that no intra-Mercurial planet larger than 50 km can exist. The possibility of one or more planets circulating beyond the orbits of Neptune and Pluto has also been discussed, but there is no compelling evidence for the existence of such planets. *See* RELATIVITY. J. Kelly Beatty

Extrasolar Planets

Interest in the possibility that planets might orbit stars originated with the revolutionary proposal by Nicolaus Copernicus in 1543 that the Earth itself is a planet and the Sun is a star. However, it is technically very challenging to detect planets orbiting Sun-like stars. Planets do not generate light; they shine only by the reflected light of the host star. The largest (and easiest to detect) planet in the solar system, Jupiter, 10^9 times fainter than the Sun. Taking a picture of a Jupiter-like planet orbiting another star would be similar to taking a photograph of a firefly near a megawatt searchlight. Although such a photograph is beyond current technology, a number of strategies have been suggested that might be capable of taking direct images of Jupiter-like, or even Earth-like, planets within a few decades.

In the meantime, the presence of extrasolar planets must be deduced by indirect means. The two most developed techniques, astrometry and Doppler spectroscopy, rely on the gravitational perturbations that giant planets impose on their host stars. For example, the Sun and Jupiter jointly orbit a common center of mass, which lies on the line connecting the Sun and Jupiter, just outside the surface of the Sun. A hypothetical alien astronomer could detect the presence of Jupiter either by nothing that the position of the Sun is periodically wobbling about against the background stars (astrometry), or by measuring the periodic velocity variation of the Sun (Doppler spectroscopy) as Jupiter pulls the Sun about their common center of mass. Such measurements would reveal the orbital period and the magnitude of the wobble. From these data, the orbital radius and mass of the unseen Jupiter could be calculated from Kepler's third law of planetary motion and the principle of momentum conservation. *See* ASTROMETRY; DOPPLER EFFECT.

Doppler technique. The Doppler spectroscopy method provided the first definitive detections of extrasolar planets orbiting normal stars. Any wave (sound or light) emitted by a moving object will be shifted as observed by a stationary observer. An object moving toward an observer will emit light or sound that is shifted toward higher frequencies (blueshift), while an object moving away from an observer will emit waves shifted to lower frequencies (redshift). The velocity of the emitting object can be directly deduced from the magnitude of this frequency (or wavelength) shift. Common examples of the Doppler effect include the changing pitch of a train whistle as a train passes a stationary observer, and the so-called radar guns used to measure the speed of cars and baseballs.

Stellar Doppler shifts are measured by collecting a star's light with a telescope and passing this light into a spectrometer. A spectrometer is basically a sophisticated prism that spreads the starlight into its component colors, showing the rainbow of red, orange, yellow, green, blue, and finally violet. A more detailed inspection of a star's spectrum reveals that a number of dark lines interrupt this continuous rainbow (**Fig. 6**). Each of these lines is due to the absorption of light, at a specific wavelength, by atoms in the outer atmosphere (photosphere) of the star. In addition to revealing the chemical composition of stars, these lines serve as wavelength markers. Doppler velocities are calculated based on the measured wavelength shift of these absorption lines. *See* ASTRONOMICAL SPECTROSCOPY; FRAUNHOFER LINES.

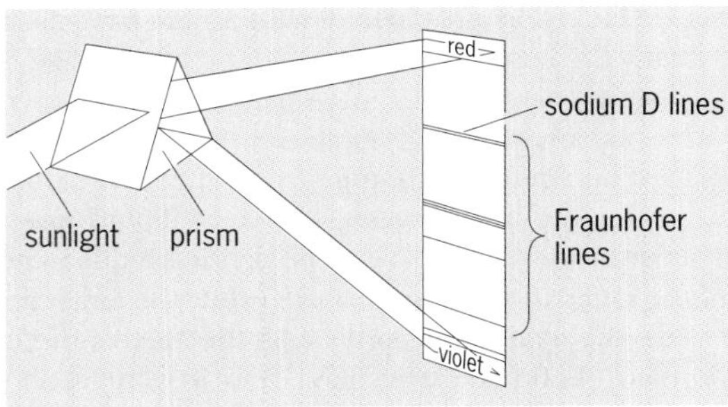

Fig. 6. Fraunhofer lines in the solar spectrum. Passing sunlight (or starlight) through a prism (or spectrometer) reveals the presence of a number of dark lines that interrupt the continuous (rainbow) spectrum. These lines act as wavelength markers, needed to measure Doppler shifts. (*After J. M. Pasachoff, Contemporary Astronomy, 4th ed., Saunders College Publishing, 1989*)

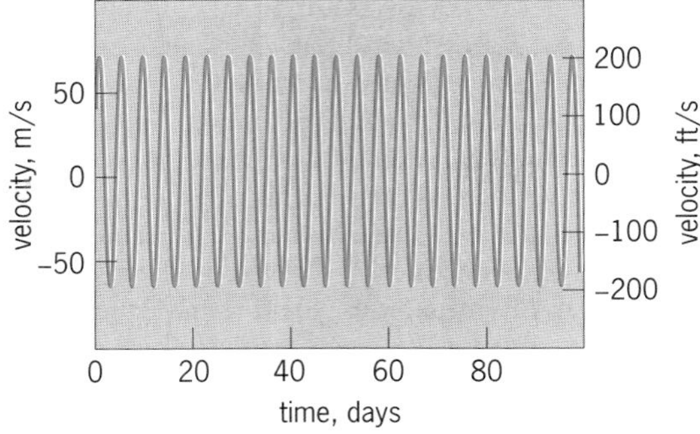

Fig. 7. Doppler velocities for 51 Pegasi from the Lick Observatory Planet Search program. These observations were made between October 1995 and January 1996. The curve is an orbital fit to the data that yields a period of 4.23 days. The semiamplitude of the velocity variations is 187 ft/s (57 m/s), indicating a companion having a mass one-half that of Jupiter at an orbital distance of only 5 $\times$ 10^6 mi (8 $\times$ 10^6 km), 1/20 of the Earth-Sun distance.

51 Pegasi planets. The first confirmed discovery of an extrasolar planet orbiting a normal star was announced in October 1995 by Michel Mayor and Didier Queloz. They had been surveying 140 stars for $1^1/_2$ years with a specialized Doppler-velocity spectrometer, with which they had achieved a precision of 50 ft/s (15 m/s). With this instrument, Mayor and Queloz discovered that the Sun-like star 51 Pegasi periodically changes in its velocity by 187 ft/s (57 m/s) every 4.2 days (**Fig. 7**). This implies that a planet with about one-half of a Jupiter mass orbits the star at a distance of only 5 $\times$ 10^6 mi (8 $\times$ 10^6 km), 20 times closer than the Earth is to the Sun. This result was completely unexpected. Virtually all theoretical predictions of planet formation suggested that giant Jupiter-like planets should form more than 3 astronomical units away from their host stars.

Though this odd planet around 51 Pegasi was theoretically troublesome, observationally it is the easiest type of planet to detect. Its close orbital distance increases the gravitational tug on the host star, thus increasing the Doppler-velocity signal relative to more distant planets. The 4-day period allows many orbits to be followed in just a few months. In contrast, Jupiter induces a meager 43-ft/s (13-m/s) velocity on the Sun, and 12 years are required to follow a single orbit.

51 Pegasi's planet orbits at a distance of only 9 stellar radii from the star. At this close distance, the planet is heated to 2400°F (1300°C). In the solar system, the giant planets (Jupiter and Saturn) are composed primarily of hydrogen and helium gas. Assuming that this is also true of 51 Pegasi's planet, there were initial concerns that the high temperature might give the gas enough kinetic energy to achieve escape velocity, literally boiling the planet away. Numerous calculations have now demonstrated that even at this high temperature hydrogen gas will not escape from the planet. It is also possible that the planet is primarily rocky, like a giant nickel-iron bowling ball, unlike anything seen in the solar system.

Many kinds of planets. After 51 Pegasi's discovery, Geoffrey W. Marcy and R. Paul Butler obtained sufficient computer resources to quickly analyze their observations of 107 Sun-like stars that they had surveyed since 1987. Within a year they had discovered six additional planets. Three of them had short-period orbits like 51 Pegasi while the others had very different characteristics.

The star 70 Virginis was found to have a companion with a minimum mass of 7.4 Jupiters, completing an orbit every 117 days, at a distance of about 0.5 astronomical unit. At this distance, the temperature of the planet is calculated to be 194°F (90°C), or slightly cooler than boiling water. Thus, liquid water, in the form of aerosol droplets, might exist in the atmosphere of this object. Unlike the 51 Pegasi planets and the planets in the solar system, the orbit of this object is not circular, but oval shaped (eccentricity ~0.40).

The large mass and eccentric orbit of this object have led some astronomers to suggest that it is a brown dwarf and not a planet. Brown dwarfs are intermediate objects between stars and planets. They form in a manner similar to stars but differ from stars in that they lack sufficient mass to ignite nuclear fusion in their central cores. Brown dwarfs are usually thought to have masses 20–80 times larger than Jupiter. However, Marcy and Butler's survey of 107 stars did not find any companion objects in this mass range. *See* BROWN DWARF.

The first extrasolar planet with characteristics similar to the planets in the solar system was found orbiting the star 47 Ursae Majoris. This planet has a minimum mass 2.6 times greater than Jupiter, a 3-year orbital period, a relatively circular orbit (eccentricity ~0.13), and an orbital distance of 2 astronomical units. Such a planet in the solar system would share a number of characteristics with Jupiter and look like Jupiter's big brother. R. Paul Butler

Planetary orbits. Doppler surveys for extrasolar planets have been expanded to include about 2000 stars, providing a nearly complete sample of Sun-like stars within 30 parsecs (100 light-years). By May 2001, 67 extrasolar planets had been discovered orbiting normal stars (**Fig. 8**). Of these, 20 orbit at distances less than 0.15 AU from their parent stars (with periods ranging from 3 to 30 days), indicating that at least several percent of all stars have extremely close planetary companions. These 51 Pegasi-like planets are thought to have formed considerably farther from their host stars and then migrated inward when their orbits were destabilized by gravitational interactions with other planets or with material in the circumstellar disk. *See* PROTOSTAR.

Most of the remaining 47 planets that have been discovered orbiting at distances of 0.15 AU or more travel in elliptical orbits like that of the companion of 70 Virginis. Indeed, all but four of these orbits have eccentricities greater than 0.1, whereas Jupiter and the other giant solar system planets have eccentricities of around 0.05 or less (Table 2). This result, like the discovery of the 51 Pegasi-like planets, was unexpected. The nearly circular orbits of solar system planets had led to the expectation that extrasolar planets would travel in circular orbits as well. Planets probably form from disks of gas and dust following circular orbits, and friction within these disks can be expected to circularize the planetary orbits. Several explanations have been proposed for the elliptical orbits: gravitational scattering among giant planets; gravitational perturbations exerted by a companion or passing star, or by the protoplanetary disk; or instabilities in the disk.

The gravitational scattering of giant planets and their resulting elliptical orbits may result in Earth-sized planets being ejected from their planetary systems into interstellar space. For example, Earth and Mars would probably have suffered this fate if Jupiter traveled in an eccentric orbit. In reality, its circular orbit promotes the stability of the circular orbits of the other solar system planets, and Jupiter may also play a protective role by acting as a gravitational sink, swallowing up comets that would otherwise devastate Earth. However, the formation of planetary systems may normally be a chaotic process, and the solar system, in which planets travel in circular, coplanar orbits that are far enough apart to maintain the system's stability, may require special initial conditions. This has implications for the development of advanced life forms, which probably could not exist in the presence of extreme temperature fluctuations that elliptical orbits would entail.

Reliable detection of a planet by the Doppler technique requires data over a time span of at least two orbits, so the total length of time that observations have been carried out places an upper limit on the orbital periods that can be detected. By Kepler's third law, this time span limits the size of detectable planet orbits as well. In May 2001, the longest period detected was 7.5 years, and the greatest orbit radius was 3.5 AU. Some more years of observation are required to detect a planet similar to Jupiter, with a period of 12 years.

Meanwhile, as the time span of observations has extended, multiple-planet systems have been discovered. A three-planet system around the star Upsilon Andromedae was discovered in 1999, and two-planet systems around the five other stars were discovered by May 2001. In two of these systems, the two planets are locked in a resonance with orbital periods in a nearly 2:1 ratio.

Planetary masses. The mass distribution of known extrasolar planets (**Fig. 9**) shows that less massive planets are found much more frequently. The Doppler technique can measure only a minimum mass. This quantity would have to be divided by the sine of the inclination of the planet's orbit (the angle between the axis of the orbit and the Earth's direction) to obtain the true mass, and in general this inclination is unknown. However, since the orbits can be expected to be randomly oriented in space, the statistics of the measured minimum masses should reflect those of the true masses. More massive planets are easier to detect because they exert a greater force on their host stars, so there is a

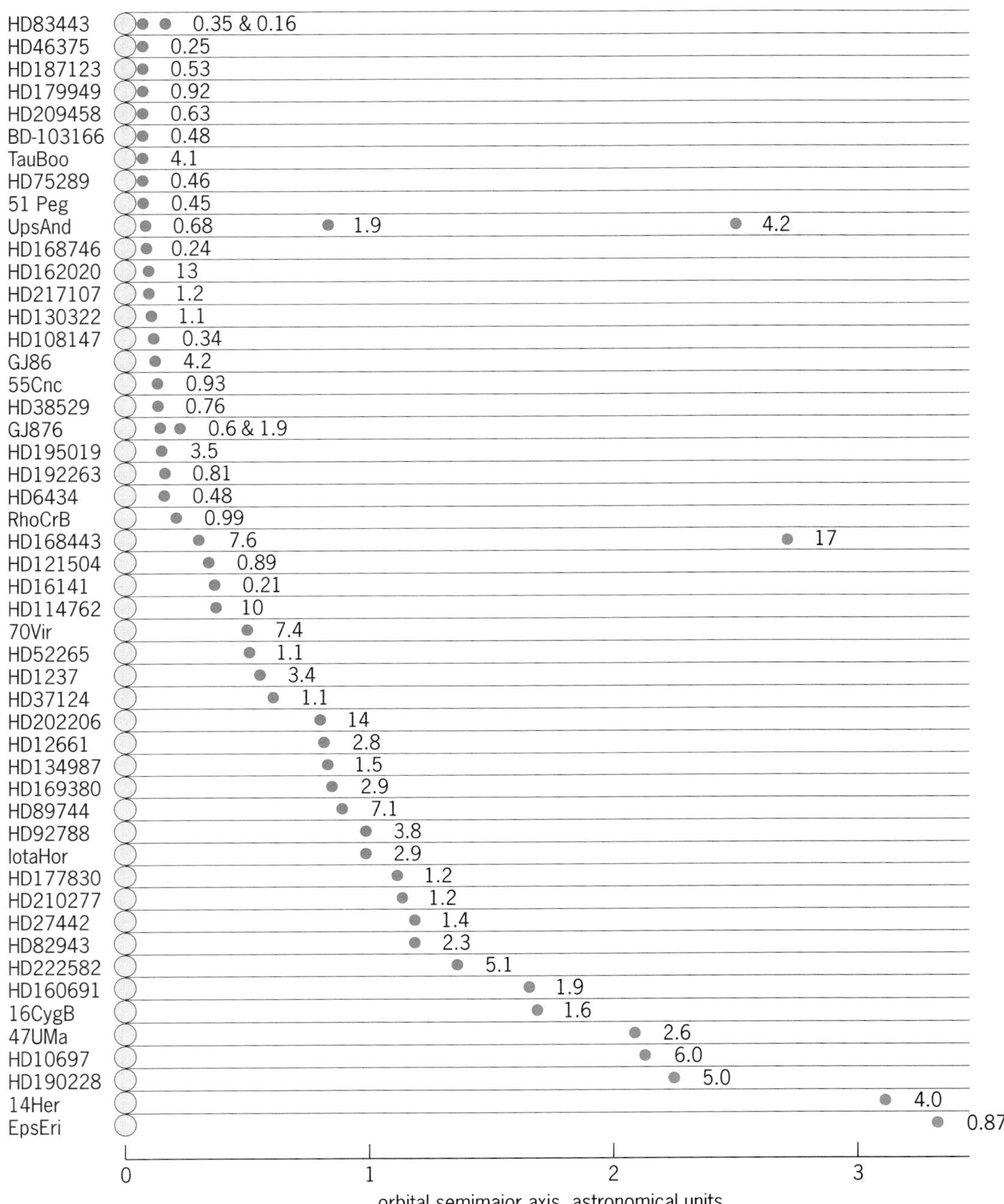

Fig. 8. Masses and orbital radii of the 55 known extrasolar planets orbiting normal stars, as of January 2001. The masses shown are minimum masses and are measured in units of Jupiter's mass, M_J, which is 318 times the Earth's mass. (*After G. W. Marcy, R. P. Butler, et al., http://exoplanets.org/pub.html*)

selection effect favoring the discovery of more massive companions. The greater number of less massive planets detected in spite of this bias implies that many more lower-mass planets should be discovered as detection techniques improve. In 2000, for the first time, four planets were discovered with minimum masses less than that of Saturn, ranging down to about 50 Earth masses (for the outer of two planets orbiting HD 83443). The Doppler technique may eventually be able to detect Neptune-mass planets (17 Earth masses) in close orbits.

The most massive object discovered is the outer of two companions of HD 168443, with a minimum mass of about 17 Jupiters. The status of the small number of more massive objects detected (greater than about 8 Jupiters) is unclear, as their masses are intermediate between those of the other planets and brown dwarfs, and the mechanism of their formation and their true nature are not known.

Transit of an extrasolar planet. Because many of the extrasolar planets orbit extremely close to their host stars, astronomers realized that if one of them were in an approximately edge-on orbit it would periodically pass in front of the parent star, reducing its brightness. The Doppler data make it possible to predict the timing of such an event, known as a transit,

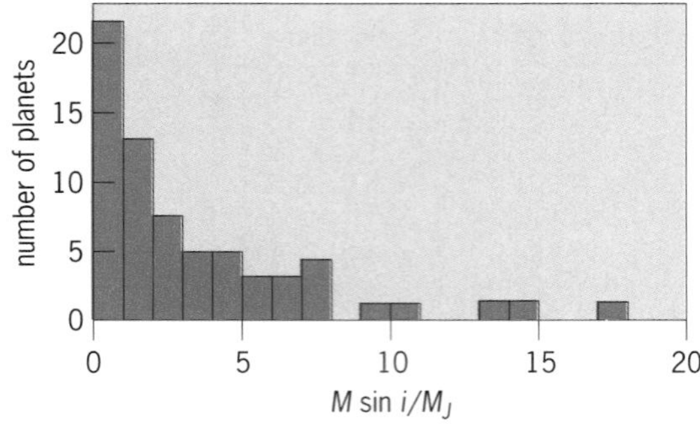

Fig. 9. Histogram of mass distribution of the 67 known extrasolar planets orbiting normal stars, as of May 2001. The masses shown are minimum masses ($M \sin i$, where M is the true mass and i is the orbital inclination), and are measured in units of Jupiter's mass, M. (*After G. W. Marcy, R. P. Butler, et al., http://exoplanets.org/entry.html*)

although not whether it will actually occur. In 1999, the transit of a planet orbiting the star HD 209458 at a distance 0.045 AU was observed as a 2% reduction in the star's brightness at the predicted time. This was the first direct observation of an extrasolar planet, answering objections of previously doubtful astronomers. Moreover, the observation of this and subsequent transits made it possible to calculate that the orbit's axis was inclined 86° (nearly perpendicular) to the Earth's direction. This allowed a calculation of the planet's true mass (0.7 Jupiter mass), and the magnitude of the star's brightness reduction enabled an estimate of the planet's diameter (1.55 times that of Jupiter). This information confirmed that the planet was a gas giant, markedly bloated by the heat of the nearby star, and not a brown dwarf or a huge solid object. These results for one planet increase the probability that most other observed extrasolar planets are of a similar nature. *See* TRANSIT (ASTRONOMY).

Prospects. The Doppler technique will probably continue to provide the majority of extrasolar planet discoveries through the first decade of the twenty-first century. By 2010, several hundred planets will probably have been discovered, and the ongoing surveys should provide the first indications of the fraction of planetary systems that are similar to the solar system, with giant planets in circular orbits of 5 AU and more.

During the decade, optical and near-infrared interferometric systems should come into operation, combining beams of light from several 8–10-m (320–400-in.) telescopes. These systems will measure the wobble in a star's position induced by an orbiting planet. For nearby stars, they should be comparable in planet-finding capability to the Doppler technique, and should complement it. Space-based interferometric missions should be even more accurate, detecting planets as small as 10 Earth masses with orbital periods of up to 5 years. *See* INTERFEROMETRY; TELESCOPE.

Space-borne transit telescopes may be the first observatories to detect Earth-size planets. They would observe crowded star fields to record slight decreases in stellar brightness caused by transits of orbiting planets. Another photometric technique to detect terrestrial planets is based on microlensing. This is a phenomenon in which a foreground star acts as a gravitational lens when it happens to pass in front of a background star, causing the background starlight to brighten and to bend through a ring-shaped region called the Einstein radius. If the foreground star has a planet near the Einstein radius, this planet acts as a secondary lens and superposes a bright, narrow peak on the original light curve. *See* GRAVITATIONAL LENS.

Further in the future are space-borne observatories that would use nulling interferometry with an array of large space telescopes to cancel the glare of the parent stars, to directly image both Jupiter-size and Earth-size planets, and to observe their atmospheric spectra, looking for evidence of life.

Planets around pulsars. Before extrasolar planets were discovered around normal stars, the discovery of planets orbiting a pulsar was announced in 1992 by Alexander Wolszczan and Dale A. Frail. The discovery was based on precise measurements of arrival times of radio pulses from the 6.2-millisecond pulsar PSR B1257+12. Deviations from the predicted times were shown to result from motions of the pulsar due to the gravitational forces of planets with minimum masses of 2.8 and 3.4 Earth masses orbiting the pulsar at respective distances of 0.36 and 0.47 AU. Further observations revealed the presence of a third planet with a mass of only 0.015 Earth mass, 0.19 AU away from the pulsar, as well as perturbations in the motions of the two larger planets resulting from their mutual gravitational interaction. The planets were probably formed after the supernova explosion that resulted in the pulsar, since it is unlikely that they could have survived this explosion. In addition, a Jupiter-sized planet has been detected orbiting at least 10 AU from the millisecond pulsar PSR B1620-26; the pulsar also has a nearby white dwarf companion. *See* PULSAR. Jonathan F. Weil

Bibliography. J. K. Beatty et al. (eds.), *The New Solar System*, 4th ed. Sky Publishing, Cambridge, MA, 1999; D. Fischer, Prowling for planets, *Mercury Mag.*, 29(4):13–17, July-August 2000; W. K. Hartmann, *Moons and Planets*, 4th ed., Wadsworth Belmont, CA, 1998; M. D. Lemonick, *Other Worlds: The Search for Life in the Universe*, Simon and Schuster, 1998; K. Lodders and B. Fegley Jr., *The Planetary Scientist's Companion*, Oxford University Press, New York, 1998; G. Marcy and P. Butler, Hunting planets beyond, *Astronomy*, 28(3):42–47, March 2000; D. Morrison, *The Planetary System*, 3d ed., Addison-Wesley, New York, 1996; J. H. Shirley and R. W. Fairbridge (eds.), *Encyclopedia of Planetary Sciences*, Chapman & Hall, London, 1997; P. R. Weissman et al. (eds.), *Encyclopedia of the Solar System*, Academic Press, San Diego, 1998; J. A. Wood, *The Solar System*, 2d ed., Prentice Hall, Englewood Cliffs, NJ, 1999.

Planetarium

An instrument that projects the stars, Sun, Moon, planets, and other celestial objects upon a large hemispherical dome, showing their motions as viewed from the Earth or space near the Earth. Days and years may be compressed into minutes. There are over 100 major planetariums around the world with domes 50 ft (15 m) or more in diameter; and there are also over 1000 smaller planetariums in communities, schools, and colleges.

The term planetarium originally applied to a mechanical model (also known as an orrery) that depicted the motions of the planets. Today the term

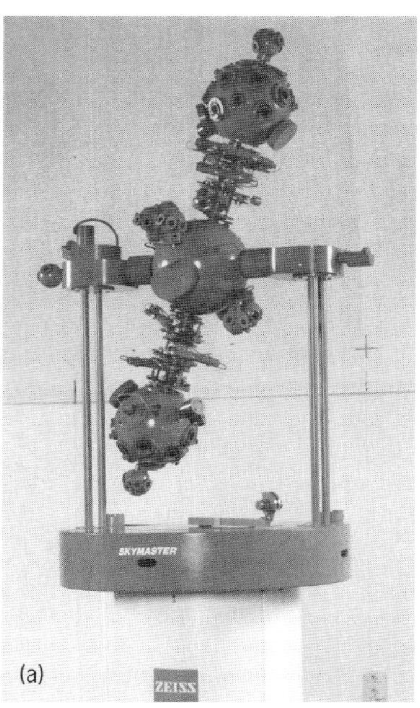

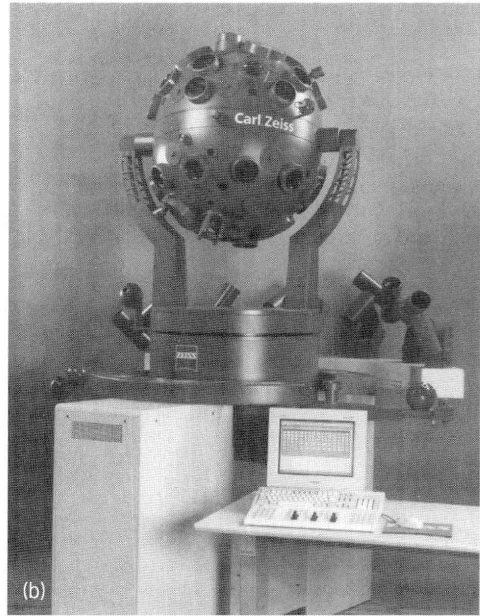

Fig. 1. Planetarium projectors. (*a*) Skymaster. (*b*) Starmaster. (*c*) Universarium M-VIII. (*d*) MS 8. (*e*) Cosmoleap 10. (*f*) Infinium α II. (*g*) GX. (*h*) G 1518. (*i*) GSS Helios. (*Parts a-c from Carl Zeiss Company: d-f from Minolta Corp. g-i from GOTO*)

refers to an optical projector. Most planetariums now have mechanical movements, but planetariums projecting computer-generated displays have also been developed. Additional optical devices and computer controls are common. The term planetarium also refers to the theater or building that houses the projector.

Development. Models of the sky date from early times. A celestial globe was made by Claudius Ptolemy of Alexandria around A.D. 150. The Gottorp Globe, a 10-ft (3-m) globe built in the mid-sixteenth century, is hollow, and up to 10 people may sit inside to see the sky which is painted on the interior. This globe is on display at the Lomonosov Museum, St. Petersburg, Russia. A larger globe, constructed in 1912, and which is on display at the Chicago Academy of Sciences, is 15 ft (4.5 m) in diameter and seats 17 spectators. The metal sphere has 692 holes to depict the stars.

Motions of the Earth, Moon, and planets have also been represented from early times with various mechanical models. About 1682, C. Huygens designed an elaborate model that showed the planets out to Saturn. Around 1712, J. Rowley built the original orrery, in which, by turning a crank, the Moon's motion around the Earth could be seen as well as the Earth's motion around the Sun, thus explaining the lunar phases and the Earth's seasonal relations. *See* EARTH ROTATION AND ORBITAL MOTION; MOON; PLANET.

The invention of the projection planetarium by W. Bauersfeld in 1919 solved the problem of presenting stars and planetary motions in a realistic fashion in a domed theater. The first such planetarium, built by the Carl Zeiss Company of Germany, opened in Munich in 1923.

Types of projectors. Many projectors are patterned after the basic design of the Carl Zeiss Company. Star spheres at each end of the projector show 8900 stars down to magnitude 6.5; 32 lenses (16 located in each globe) are used to project the stars. Cages between the two star spheres contain projectors for the Sun, Moon, and planets. The center part of the machine houses the driving motors. Additional projectors show such effects as variable stars, solar and lunar eclipses, the Milky Way, comets, and various circles and coordinates. Depending on the manufacturer, there may be constellation outlines, clouds, and built-in zoom effects for the planets.

Of the 25 large-model projectors built by Zeiss prior to 1940, several are still in service and have been joined by many new models. Zeiss offers three models for various dome diameters. These are the Skymaster, the Starmaster ZMP, and the Universarium M-VIII for dome diameters of 20–36 ft (6–11 m), 40–60 ft (12–18 m), and 60–80+ ft (18–24+ m) respectively (**Fig. 1***a–c*). The Skymaster, the Starmaster, and the Universarium project 7000, 8900, and over 9000 stars respectively. All are computer-controlled and user-friendly. Both the Starmaster and the Universarium use fiber optics to increase star brilliance. A device also makes the stars twinkle.

Minolta planetariums also come in several models: the MS 6 and the MS 8 for domes ranging 21–30 ft (6.5–9 m), the Cosmoleap 8 and 10 for domes 25–40 ft (7.5–12 m), and the Infinium α II, β II, and γ II for domes 40–88 ft (12–27 m) [Fig. 1*d–f*]. The MS series projects over 3500 stars, the Cosmoleap series from 3900 to 6500 stars, and the Infinium series from 9200 to 23,000 stars. All series are computer-controlled.

GOTO planetariums offer various models, including the GX for domes 29–43 ft (9–13 m) and projecting 6500 stars, the G 1518si for domes 49–59 ft (15–18 m) and projecting 8500 stars, and the GSS Helios for domes 59–88 ft (18–27 m) and projecting up to 25,000 stars (Fig. 1*g–i*). Each of the

Fig. 2. Spitz planetarium. (*a*) System 1024 (*Spitz Space Systems, Inc.*). (*b*) Space Voyager, in a theater that has a 76-ft (23-m) dome tipped at a 257°; angle and seats 344 people. (*R. H. Fleet Space Theater, Balboa Park, San Diego*).

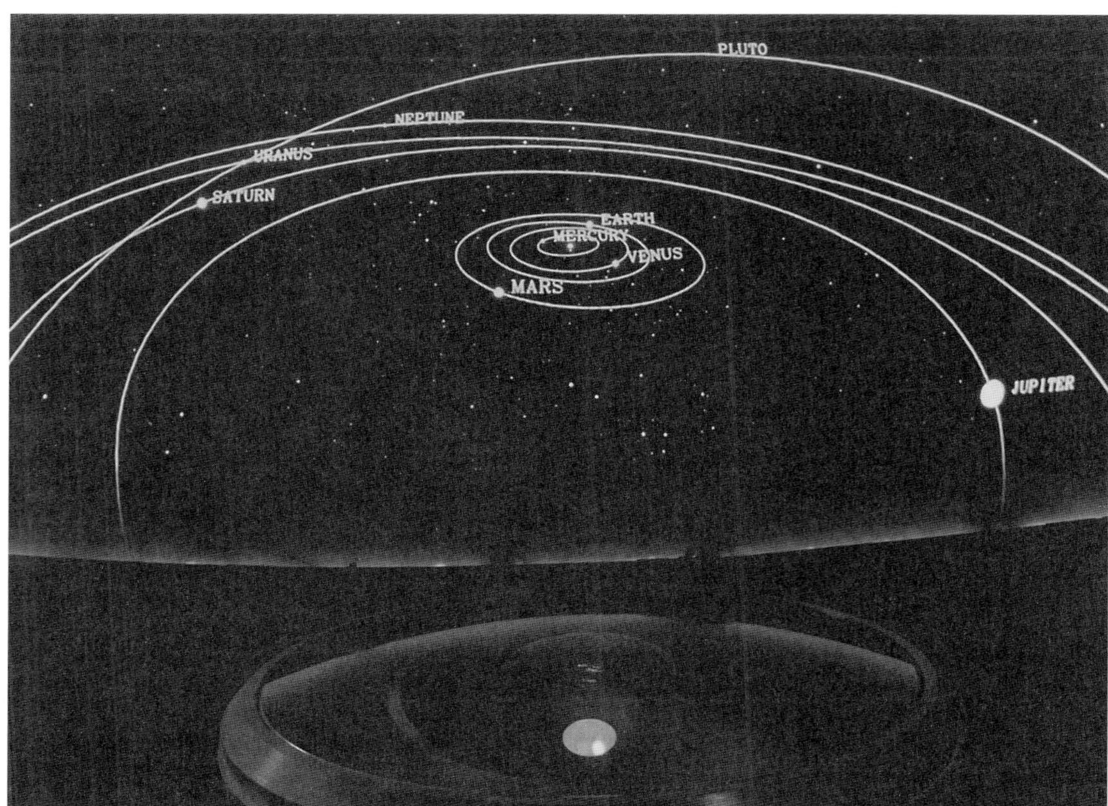

Fig. 3. Digistar II computer-graphics television projection planetarium. The projection lens is in the foreground, with dome projection above. (*Evans & Sutherland*)

planetarium projectors of various manufacturers may be mounted on a hydraulic hoist, permitting it to descend below the floor for parts of the program.

Another design philosophy, begun in the late 1940s when A. Spitz designed a small planetarium for school classrooms and museums, was to manufacture a small and relatively inexpensive projector that would do for schools and small communities what the larger machines had done for the cities. Early models of Spitz Space Systems planetariums featured planetary motions through innovative planet analog mechanisms. One of these models became the most common projector for 24–30-ft (7–9-m) domes, while another was manufactured for dome diameters of 50 ft (15 m). Later models include System 1024, a medium-sized projector designed for the 24–40-ft (7–12-m) dome size. Spitz's Space Voyager, designed for major theaters, has a star-ball which is 4 ft (1.2 m) in diameter (**Fig. 2***a–b*).

The Zeiss Universarium, the Minolta Infinium, the GOTO GSS Helios, and the Spitz Space Voyager are usually installed in tipped domes with all-sky 70-mm motion picture projection systems. Another innovation in planetarium design, based on computer-graphics television projection, is the Evans & Sutherland Digistar II (**Fig. 3**). The projector, which stands only 54 in. (about 1.4 m) high, utilizes a high-resolution cathode-ray tube with a special 160°; wide-angle lens for projection onto domes ranging in diameter 20–80 ft (6–24 m). The planetarium may have a horizontal or tilted dome. Software and documentation include stellar, planetary, and constellation data files. The star and planet positions are fed into the high-intensity cathode-ray tube projector by a computer and thence onto the dome via the wide-angle lens. *See* CATHODE-RAY TUBE; COMPUTER GRAPHICS.

Mechanical aspects. The Digistar II has, essentially, no moving parts, whereas the other projectors described above all feature similar mechanical operation derived from the basic Zeiss design. These mechanical-optical projectors all show the apparent motions of the sky which are, in reality, a reflection of the Earth's motions, or the motion of the observer. These mechanical projectors are geocentric, that is, they show the sky from an Earth-centered viewpoint. For example, daily motion depicts the turning of the sky (rising and setting of the Sun, Moon, planets, and stars) caused by the Earth's rotation on its axis. A variable-speed motor turns the entire projector. Latitude motion shows the changing aspect of the sky as the observer travels north or south over the Earth. Thus as one moves toward the North Pole, the North Star rises higher. This is accomplished by rotation of the entire projector around a horizontal (east-west) axis.

Annual motion uses auxiliary projectors to the star globe to depict the changing monthly phases of the Moon, the eastward motion of the Sun along the ecliptic, as well as the annual (yearly) motion of the planets in the zodiac. More complicated mechanical linkages are required to represent these motions, because of the more complex mathematical ratios of planetary periods. Numerous gears are required to produce the desired accuracy. Some instruments also approximate the variable planetary speeds. To project a planet, for example, Mars, in its correct position in the zodiac, requires not only a gear train to represent the period of Mars but also a gear train to represent the Earth's position (because the planet is observed from the moving Earth). A sliding rod links the Earth drive to the planet drive, which in turn aligns the planet projector in the correct direction in space. The Moon-phase projector not only shows where the Moon is among the stars each hour but also the correct phase. The projector also depicts the regression of the lunar nodes. Some planetariums also have built-in solar and lunar eclipse projectors as well as planet zoom effects. *See* GEAR TRAIN; LINKAGE (MECHANISM).

In addition to the rotation of the Earth upon its axis and its annual journey around the Sun, the Earth has a wobbling motion much like that of a top. The wobbling motion of the Earth's axis, called precession, is caused by the Moon and the Sun pulling on the equatorial bulge of the Earth. The Earth takes 25,800 years for one precession cycle. Planetariums reproduce this effect by rotating the projector around an axis which is inclined 23.5°; to the daily-motion axis, to show how the position of the pole star changes as well as the movement of zodiacal constellations in relation to the equinox. *See* PRECESSION OF EQUINOXES.

The Spitz system 1024 also utilizes mechanical analogs to reproduce the aforementioned motions. It does not follow the basic Zeiss design, however, because of the necessity for miniaturization. The daily, latitude, and precession motions are similar to those of the Zeiss, but the annual motion for the planets is based upon the Tychonic system rather than the Copernican or Keplerian.

For the Space Voyager, mechanical linkages are replaced by computer control. The majority of projectors of the various manufacturers are computer-controlled. The "high-end" projectors that appear as large "starballs" are actually two hemispheres attached to a collar assembly that contains the servo motors. For example, the Zeiss Universarium, the Minolta Infinium, and the GOTO GSS Helios utilize separate zoom projectors to show the Sun, the Moon, and the planets (Fig. 1*c*, *i*, *f*). These projectors are located next to the base of the starball and are controlled by computer rather than mechanical linkages. Several manufacturers offer multiple images for these projectors—for example, comets, asteroids, and solar and lunar eclipses—in addition to the usual planets. These computer-controlled multiaxis projectors show planetary and solar motions not only from the Earth's surface but from any viewpoint in space near the solar system. For example, it is possible to view the planetary motions from Mars.

The Digistar II is also fully computer-controlled with no moving parts; it is not limited to geocentric projection, but may depict space views out to 650 light-years from the Sun (a limitation governed by knowledge of accurate star positions). The effect of a three-dimensional journey through our region of space may be simulated; for example, the viewer can travel toward and then through the stars of Orion (Fig. 3). Charles F. Hagar

Bibliography. R. L. Beck and D. Schrader, *America's Planetariums and Observatories: A Sampling*, 1991; C. F. Hagar, *Planetarium: Window to the Universe*, 1980; H. C. King, *Geared to the Stars*, 1978.

Planetary gear train

An assembly of meshed gears consisting of a central or sun gear, a coaxial internal or ring gear, and one or more intermediate pinions supported on a revolving carrier. Sometimes the term planetary gear train is used broadly as a synonym for epicyclic gear train, or narrowly to indicate that the ring gear is the fixed member. In a simple planetary gear train the pinions mesh simultaneously with the two coaxial gears (see **illus.**). With the central gear fixed, a pinion rotates about it as a planet rotates about its sun, and the gears are named accordingly: the central gear is the sun, and the pinions are the planets.

In operation, input power drives one member of a planetary gear train, the second member is driven to provide the output, and the third member is fixed. If the third memberis not fixed, no power is delivered. This characteristic provides a convenient clutch action. A clutch or brake band positioned about the intermediate member and fixed to the gearbox housing serves to lock or free the third member. The holding device itself does not enter into the power path.

TABLE 1. Speed ratios for simple planetary train

Fixed member	Input member	Output member	Overall ratio*	Range of ratios normally used
Ring	Sun	Carrier	$\frac{N_R}{N_S} + 1$	3:1–12:1
Carrier	Sun	Ring	$\frac{N_R}{N_S}$	2:1–11:1
Sun	Ring	Carrier	$\frac{N_S}{N_R} + 1$	1.2:1–1.7:1

*N_S = number of sun teeth, N_R = number of ring teeth.

TABLE 2. Speed ratios for compound planetary train

Fixed member	Input member	Output member	Overall ratio*	Range of ratios normally used
Ring	Sun	Carrier	$\frac{N_R N_{P_1}}{N_S N_{P_2}} + 1$	6:1–25:1
Carrier	Sun	Ring	$\frac{N_{P_1} N_R}{N_S N_{P_2}}$	5:1–24:1
Sun	Ring	Carrier	$\frac{N_S N_{P_1}}{N_R N_{P_1}} + 1$	1.05:1–2.2:1

*N_S = number of sun teeth; N_{P1} = number of first-reduction planet teeth; N_{P2} = number of second-reduction plan et teeth; N_R = number of ring teeth.

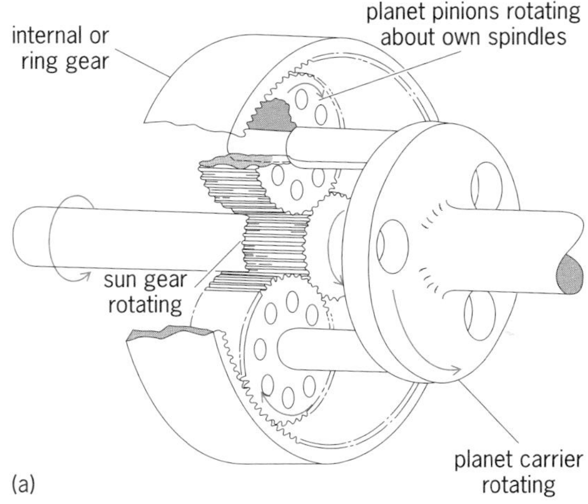

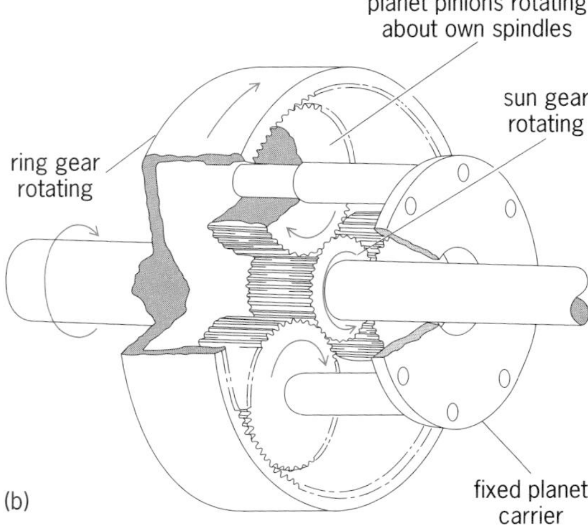

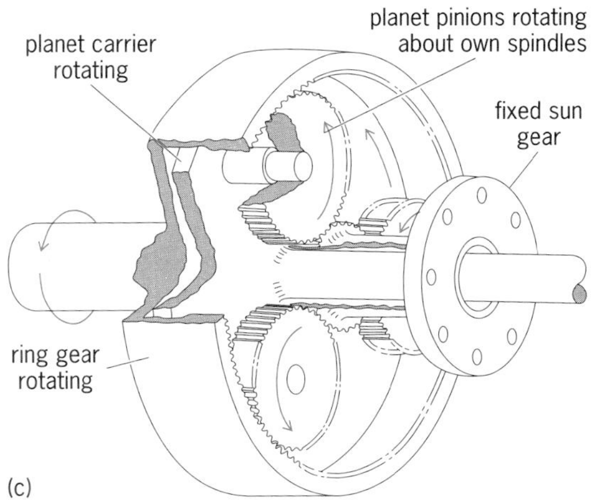

Cutaway isometrics of modes of operation for simple planetary gear train. (*a*) Ring gear locked. (*b*) Planet carrier locked. (*c*) Sun gear locked. (*After D. W. Dudley, ed., Gear Handbook, McGraw-Hill, 1962*)

Any one of these three elements can be fixed: the sun gear, the carrier, or the ring gear. Either of the two remaining elements can be driven and the other one used to deliver the output. There are six possible combinations, although three of these provide velocity ratios that are reciprocals of the other three. The principal ratios for a simple planetary gear train are given in **Table 1**. The ratios are entirely independent of the number of teeth on each planet. However, to assemble the unit, Eq. (1), which does involve the

$$N_R = N_S + 2N_P \tag{1}$$

number of teeth on each planet, must be satisfied; $N_R =$ number of teeth on the ring gear, $N_S =$ number of teeth on the sun gear, and $N_P =$ number of teeth on each planet. Typically more than one planet is used to distribute the load through more than one mesh and to achieve dynamic balance. If the several planets are to be equally spaced, assembly can be made only if Eq. (2) is followed.

$$\frac{N_R + N_S}{\text{Number of planets}} = \text{an integer} \tag{2}$$

Two simple planetary gear sets running on a common sun gear are known as a Simpson gear train. It is widely used in automotive automatic transmissions. In a compound planetary train, two planet gears are attached together on a common shaft. One planet meshes only with the central sun gear, the other only with the ring gear. As in simple planetary trains, there can be several of these planet pairs distributed around the train to distribute the load and achieve balance. **Table 2** shows the ratios obtained for the three principal arrangements.

To assemble the train, the pitch diameters of the gears must agree with Eq. (3), where D_R, D_s, and D_P

$$D_R = D_S + D_{P_1} + D_{P_2} \tag{3}$$

designate ring gear, sun gear, and planet diameters, respectively. If the planets are to be equally spaced about the train, the number of teeth must agree with Eq. (4).

$$\frac{N_{P_1}N_R + N_{P_2}N_S}{\text{Number of planets}} = \text{an integer} \tag{4}$$

See AUTOMOTIVE TRANSMISSION; GEAR TRAIN; RECIPROCATING AIRCRAFT ENGINE.

John R. Zimmerman; Donald L. Anglin

Bibliography. D. W. Dudley, *Handbook of Practical Gear Design*, 1984; J. E. Shigley and C. R. Mishke, *Mechanical Engineering Design*, 5th ed., 1989; J. E. Shigley and J. J. Uicker, *Theory of Machines and Mechanisms*, 2d ed., 1995.

Planetary nebula

A gaseous shell thrown off by a dying star just before the star settles down to become a degenerate white dwarf. Planetary nebulae are among the brightest and best studied nebular objects in the sky, although they are generally a few thousand light-years from Earth. The many shapes of planetary nebulae reflect the processes that occur inside most stars late in their lives. The Sun is likely to eject a planetary nebula in about 5 billion years. *See* SUN; WHITE DWARF STAR.

Planetary nebulae appear small, round, and greenish in color, like the planet Uranus, when seen through small telescopes—hence the origin of the name by astronomers of the nineteenth century. However, planetary nebulae are not planets but large (0.1–10 light-years), expanding (10–100 km/s; 6–60 mi/s), highly symmetrical gaseous clouds of stellar ejecta. The shapes of planetary nebulae vary (see **colorplate**), and are denoted as round, elliptical, or bipolar depending on their outlines. Delicate filaments, knots, and bubbles of material characterize their interiors. These features evolve as stellar winds of increasing speed (from 100 to 1000 km/s; 60 to 600 mi/s) plow into the older, much slower gas ahead of them. Inferences made from images of very young planetary nebulae suggest that their shapes are determined early in the ejection process, and evolve only slowly as the nebulae grow. However, the details of the shaping processes are not yet understood.

Planetary nebulae last for several thousand years before they expand and become too diffuse to be seen as discrete objects. Human lifetimes are far too short for their evolution to be noticed, although their existence is fleeting from a cosmic perspective.

Planetary nebulae emit electromagnetic radiation in many regions of the spectrum, including radio, infrared, visible, and ultraviolet. Complex patterns appear when the spectrum of light from a given nebula is analyzed. From these data the temperature, density, velocity, and chemical nature of the gas are measured. Such studies have shown that the gas ejected from the parent stars of planetary nebulae is enriched in carbon and nitrogen. This knowledge helps in understanding the origin of the solar system. The light output of planetary nebulae can also be used to determine the distances to nearby galaxies and, thence, to estimate the cosmic distance scale. Just as the distance to a car is estimated from the apparent brightness of its headlights, the distance to a galaxy can be derived from the brightnesses of its planetary nebulae. *See* ASTRONOMICAL SPECTROSCOPY; GALAXY, EXTERNAL; HUBBLE CONSTANT.

A planetary nebula becomes visible only as the cool outer layers are ejected, exposing the hotter inner regions, or core, of the star to view. The fuel-starved core continues to cool, much like the filament of an electric oven, for billions of years after ejecting its nebula. Like the old stars from which they are believed to have evolved, planetary nebulae are distributed throughout the Milky Way Galaxy. They show no clear signs of crowding in the spiral arms where dust clouds and the new stars that form from them are found. *See* MILKY WAY GALAXY.

About 2000 planetary nebulae have been cataloged. Ten to twenty times as many are believed to exist in the Milky Way Galaxy; however, they are too faint or obscured by foreground dust clouds to be identified.

Normal stars such as the Sun form planetary nebulae shortly after evolving through the red-giant phase of their lives. At the end of its life, the Sun will swell so much that it will engulf the planets Mercury and Venus. About half of the daytime sky on Earth will be filled with light from its huge ember-red surface, and the Earth's temperature will rise until its atmosphere and eventually its oceans evaporate into space. After a few thousand years, as the last of its available nuclear fuel is depleted, the Sun will eject its remaining outer layers, containing at least one-third of its mass. This gas will sweep past the Earth at 10 km/s (6 mi/s) and quickly exit the solar system. In another few thousand years, only the tiny, hot core of the Sun will still be visible. This remnant of a star will have a temperature of about 100,000°C (180,000°F), and the winds that it will generate will approach speeds of 1000 km/s (600 mi/s). In spite of its tiny size, the solar remnant will be a hundred to a thousand times brighter than the present Sun. Its harsh ultraviolet light will ionize and sterilize the Earth's surface. No life could survive. *See* STAR; STELLAR EVOLUTION.

From stellar death comes future life, however. The winds from planetary nebulae cool as they sweep outward, and dust particles of graphite (carbon), iron, and other types of cosmic dust particles. These carbon-rich dust particles merge with others from the detritus of other dead stars, and become encrusted in water ice. These materials will condense into future generations of stars and planetesimals. Their masses, lifetimes, and large numbers suggest that planetary nebulae are the most prolific source enriching the interstellar medium with carbon and, to a lesser degree, nitrogen. *See* INTERSTELLAR MATTER.

Bruce Balick

Bibliography. L. H. Aller, *Physics of Thermal Gaseous Nebulae*, 1984; H. J. Habing and H. J. G. L. M. Lamers (eds.), Planetary Nebulae, *Int. Astron. Union Symp.*, vol. 180, 1997; J. Kaler, *Astronomy!*, rev. ed., 1997; D. E. Osterbrock, *Astrophysics of Gaseous Nebulae and Active Galactic Nuclei*, 1989; J. Pasachoff, *Astronomy: From the Earth to the Universe*, 5th ed., 1999.

Planetary physics

The study of the structure, composition, and physical and chemical properties of the planets of the solar system, including their atmospheres and their immediate cosmic environment.

Extent of knowledge. While the atmosphere and the oceans have been explored quite thoroughly, probes in the form of deep holes have barely sampled the outermost layer of the Earth's crust. A great deal has been deduced about the interior of the Earth as the result of the application of principles of physics and chemistry and knowledge of the properties of materials. Knowledge of the structure of the deep interior of the Earth principally depends upon studies of the properties of the seismic (earthquake) waves which penetrate through the innermost regions and provide knowledge of conditions there. The Earth's magnetic field also has very deep roots in the interior, and the knowledge obtained from studies of the variations in the magnetic field are supplementary

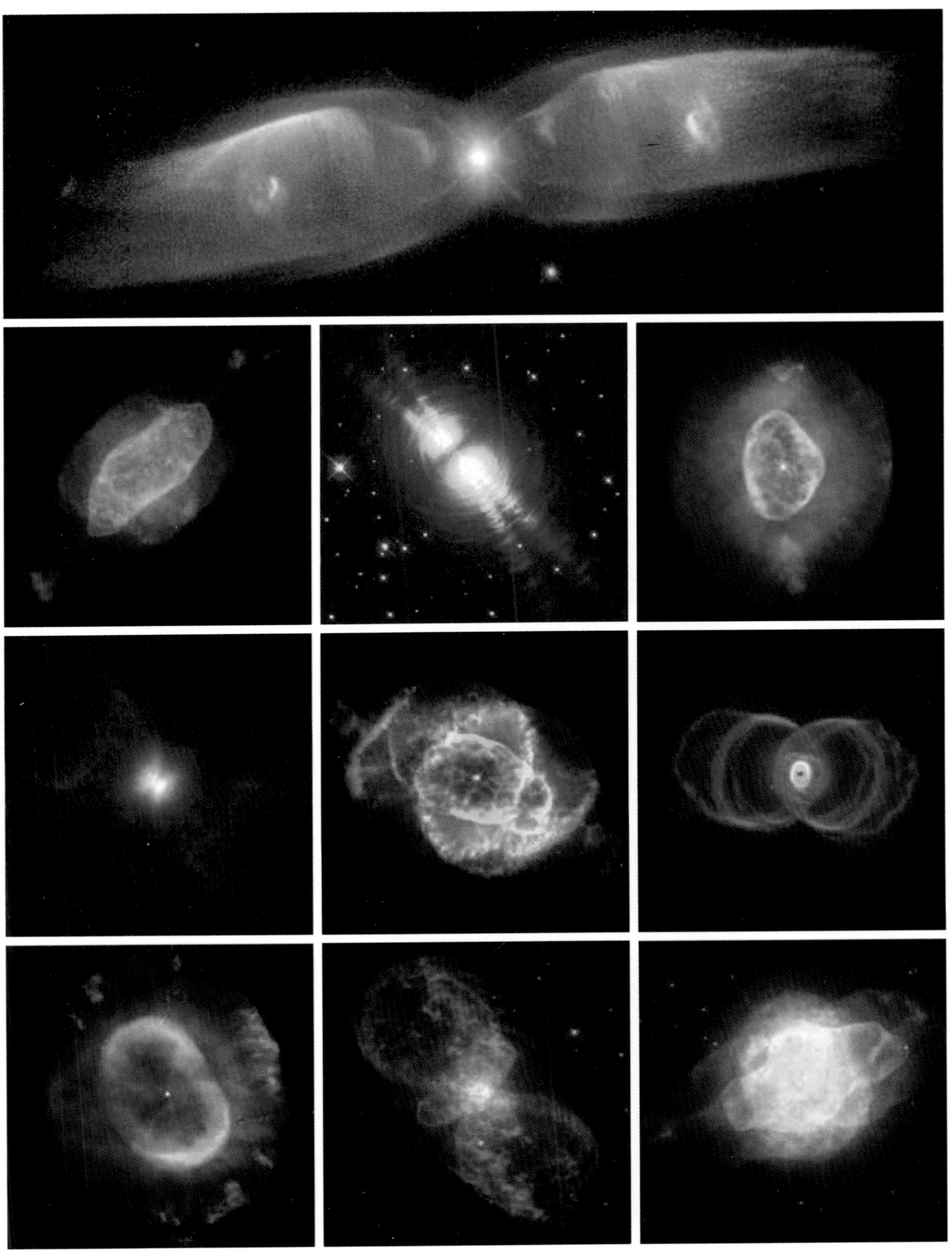

Images of various planetary nebulae taken by the Hubble Space Telescope. The colors are exaggerated in order to highlight the delicate features seen in these objects. (*NASA and the Space Telescope Institute*)

to that obtained from seismic waves. *See* GEOMAGNETISM; HIGH-PRESSURE CHEMISTRY; HIGH-PRESSURE MINERAL SYNTHESIS; HIGH-PRESSURE PHYSICS; SEISMOLOGY.

Knowledge of the other planets in the solar system is much less than that of the Earth. Quite a bit is known about the Earth's Moon as the result of the lunar samples which have been brought back for study in terrestrial laboratories, and of the instruments which astronauts left operating upon the surface of the Moon, which continued to gather a variety of physical information for a period of years following the lunar landings in the Apollo program. A great deal is known about the surface features of Mars, owing to the long series of images taken of the surface by Mars-orbiting spacecraft such as *Mariner 9* and the two *Viking* orbiters. Both the former Soviet Union and the United States have landed spacecraft upon the surface of Mars, but so far these have given relatively little information about the interior of the planet. In the case of Venus, the very thick hazy atmosphere obscures the surface of the planet, and detailed knowledge of the planetary surface depends on increasingly better-resolution radar measurements. Extensive low-resolution radar maps were obtained with terrestrial radar, particularly at Arecibo. Improved resolution was obtained with the radar altimeter on board the *Pioneer* Venus orbiting spacecraft, and the highest-resolution images (about 0. 6 mi or 1 km) have been produced for part of the planet by radar on board a Russian *Venera* spacecraft. Both the former Soviet Union and the United States have landed spacecraft on the planet, but these have been primarily concerned with atmospheric investigations. For Mercury, reasonably good images of one of the hemispheres were taken by a fly-by spacecraft. For the outer solar system, images of Jupiter and its four large Galilean satellites, and of Saturn and its larger satellites and ring system, have been returned by the *Voyager* spacecraft. *See* SPACE PROBE.

Construction of models. Planetary scientists attempt to synthesize their information about the structure and properties of each of the planets by constructing models of them. These models make use of the laws of physics and chemistry and are considered to be successful when they reproduce all of the known measured information about a planet. Sometimes it is possible to fit all of this information using more than one model, indicating a considerable uncertainty about the interior properties of the planet, but nevertheless such model building is a useful exercise, because it tends to limit many of the properties of the planet to certain ranges of values.

The most obvious gross properties of the planet are its mass and its radius. In constructing a model, it is required that the model be in hydrostatic equilibrium. This means that at any interior point in the model, the pressure must be great enough to sustain the weight of the overlying mass of material. Thus, given the mass and the radius, estimating the interior pressures through the principles of hydrostatic equilibrium, and knowing something about the compressibilities of materials, it is generally possible to place constraints upon the interior composition of the planets.

All of the planets rotate to some degree, and some of them spin quite rapidly. The effect of the spin is equivalent to a force pressing outward in the equatorial plane of the planet, which helps to sustain the weight of the material in that plane, producing an equatorial bulge. Knowledge of the equatorial bulge and the rate of spin provides additional information about the distribution of mass in the interior of the planet.

Another property which is difficult to measure, but which is known in some detail for the Earth and very crudely for the Moon, is the rate of heat flow from the interior of the planet. For the Earth and the Moon this heat flow appears to result from radioactive heating of the interior of the body, and the knowledge of this heat flow gives information about the distribution of temperature in the interior. In the case of the Earth and the Moon, this interior heat flow is very small compared to the heat which is received from the Sun. In the case of Jupiter and Saturn, the heat flow from the interior is comparable to the heat received from the Sun, so that it has been possible fairly easily to measure this heat flow for those planets because their temperatures are significantly greater than would be expected on the basis of the heat received from the Sun alone. *See* EARTH, HEAT FLOW IN.

Classes of chemical composition. The planets in the solar system have an extremely wide range of properties. This distribution of characteristics can be understood in part from a knowledge of the more abundant elements in nature and their volatility properties.

Approximately 98% of matter in the Sun, and therefore also presumably in the matter from which the Sun and the solar system were formed, consists of the gases hydrogen and helium. Most of the remaining material consists of carbon, nitrogen, and oxygen, which in the presence of very large amounts of hydrogen tends to form methane, ammonia, and water. These substances are collectively called ices, and they evaporate at relatively low temperatures. Both the light gases hydrogen and helium and the ices are of quite low abundance on the Earth and the other inner planets in the solar system. What comprises the bulk of the material in these planets is the rocky material, constituting only about 3 parts in 1000 of the solar mix of elements, and among the rocks the most abundant elements are magnesium, silicon, iron, aluminum, calcium, and chromium, all of which are present in rocks predominantly in the form of oxides. In the relatively undifferentiated rocks which fall to Earth from outer space, called meteorites, iron appears both in the form of oxide and as the metal. In the surface rocks of the Earth, iron is almost entirely in the form of the oxide, since the metallic iron has predominantly collected near the center of the Earth to form the core. *See* ELEMENTS, COSMIC ABUNDANCE OF; ELEMENTS, GEOCHEMICAL DISTRIBUTION OF; METEORITE.

The differences in the volatilities of these materials, which correlate with the properties of the planetary bodies in the solar system, give information about the properties of the environment in which the planets formed in the solar system. The inner planets, composed predominantly of rocks, evidently formed in a rather hot environment, so that the volatile gases and ices were not condensed and did not collect along with the rocky material, which presumably was condensed. The comets, residing at very large distances from the Sun in the solar system, appear to be mixtures of rocky materials and of the ices. The outer giant planets, Uranus and Neptune, appear to be primarily composed of materials heavier than hydrogen and helium, probably mixtures of rocky and icy materials. The two largest planets in the solar system, Jupiter and Saturn, are much closer in composition to that of the Sun itself, although studies of the interior structures tend to indicate that there is some degree of enrichment in the heavier elements. These differences in composition thus indicate that the tendency to collect hydrogen and helium depends not upon the ability to condense hydrogen and helium into solid form, which would require extremely low temperatures, but rather upon the size of the body which was formed, the larger bodies being more successful in gravitationally capturing the elusive hydrogen and helium. Comets are quite small and are devoid of these materials; Uranus and Neptune are intermediate in mass between the Earth on the one hand and Jupiter and Saturn on the other, and they have been only moderately successful in obtaining hydrogen and helium. However, Jupiter and Saturn were very successful in obtaining these gases. *See* COMET.

Since these compositional classes provide a natural means for dividing the planetary objects within the solar system into separate groups, the structure of the various planetary groups will be discussed in turn.

Giant planets. The giant planets are Jupiter, Saturn, Uranus, and Neptune.

Jupiter. Jupiter, the most massive planet in the solar system, is only about 1/1000 of the mass of the Sun. It comes closest in composition to that of the Sun itself. If the composition of Jupiter truly matches that of the Sun, then it would contain in its total mass the equivalent of about one earth mass of rocky material. However, the best attempts to construct models of the interior of Jupiter indicate the amount of material heavier than hydrogen and helium is significantly in excess of that which would be expected for the solar composition. There are probably something like 10 to 20 earth masses of rock and ice in the interior of Jupiter, which is an enrichment of a factor of three to six over the solar composition if the ice-to-rock ratio in the interior of Jupiter is the solar ratio, which is not known. Even this enhanced amount of material amounts to only a few percent of the total mass of Jupiter. The considerable uncertainty in the amount of heavy-element enrichment in the Jovian interior results from the uncertainties in the extrapolation of the properties of hydrogen and helium to very high pressures and temperatures such as those found in the interior of Jupiter. It is not even clear whether these heavier materials have settled to the center of Jupiter, or whether they are suspended in the atmosphere which is being continually mixed throughout the different interior levels of Jupiter due to convective motions.

One of the interesting properties of hydrogen at higher pressures is its tendency to form a conducting metal, metallic hydrogen. Because hydrogen is a simple substance, the physical calculations that lead to the expected transformation from molecular to metallic hydrogen are reasonably certain, but the precise pressure at which this transformation takes place is still quite uncertain. It appears to be somewhat in excess of 10^6 atm (10^{11} pascals). Most of the mass of Jupiter exists at a pressure considerably in excess of this amount, so that metallic hydrogen is anticipated to form a substantial portion of the interior mass of the planet. *See* JUPITER.

Saturn. Saturn has about only one-third of the mass of Jupiter, but nevertheless it also is predominantly composed of hydrogen and helium, and in this case it is definitely clearer that there are heavier elements in excess of solar composition within the interior of Saturn. Again, it is not known whether these heavy elements maintain the solar composition ratio between the ices and rocky materials, and the precise amount of enrichment is therefore uncertain, depending upon this ratio. However, the total amount of heavy materials in the interior of Saturn is comparable to the excess amount in Jupiter. *See* SATURN.

Heat flow and helium segregation. Attempts have been made to construct evolutionary sequences of models of Jupiter and Saturn which would follow the changes in structure that take place as the planets cool off after their formation. The research has suggested that Jupiter should still be radiating away its interior heat of formation at about the rate which is actually observed as an excess heat flow from the interior, wheras the amount of primordial heat still emerging from Saturn is expected to be much less than is observed. The explanation of this discrepancy may lie in another interesting property expected for a mixture of helium and hydrogen at higher pressures. Below some temperature which is still quite uncertain, it is expected that helium will collect to form small bubbles within the hydrogen; these bubbles, being heavier, will then sink through the hydrogen toward the center of the planet. Not only does this lead to a greater mass concentration toward the center of the planet, but it also releases additional gravitational potential energy, thereby enhancing the heat flow from the interior. It has been suggested that the interior of Jupiter is still sufficiently hot to have prevented this segregation of helium from hydrogen, whereas the interior of Saturn is sufficiently cooler so that a significant amount of such segregation has and is continuing to occur, thus leading to the observed heat outflow from Saturn.

Uranus and Neptune. Uranus and Neptune are quite similar planets, being 14.5 and 17.2 times the mass of the Earth, respectively. Approximately three-quarters of this mass is expected, on the basis of model building, to consist of materials heavier than hydrogen and helium. The precise numbers will depend upon whether these materials are in the solar ratio of ices to rock, which is not known. If one assumes this ratio to be valid, then each of the planets contains approximately four earth masses of rock and approximately twice that much in the form of ices. The remaining hydrogen and helium form a very deep atmosphere. *See* NEPTUNE; URANUS.

Physical composition. Nowhere in the interiors of the giant planets can anything resembling a solid surface be expected. The temperatures in the interiors are very uncertain and can be estimated only as the result of model construction, but they tend to be thousands to tens of thousands of degrees Celsius. The pressures range up to the order of 10^7 atm (10^{12} Pa) and higher. Under these circumstances all materials behave like fluids. There may be a certain amount of compositional stratification, with denser fluids underlying lighter ones.

This issue of stratification is significant in connection with one of the interesting properties of the interior, the transport of gravitational potential energy released in the deep interior to the surface. The thermal conductivity within the interiors of these planets appears to be much too small to do this job efficiently, even in regions of metallic hydrogen. Conduction may be required to transport heat from a layer of one composition to a neighboring layer of different composition. But within a layer of any given composition, the transport of heat appears to require convection. Convection consists of an irregular pattern of overturning motions within a fluid, similar to that which occurs when one boils water within a pot. It has been argued on this basis that the interiors of the giant planets are primarily engaged in convective motions which transport heat outward. *See* HEAT TRANSFER.

Terrestrial planets. The terrestrial planets include Mercury, Venus, Earth, and Mars. The Earth's Moon may also be considered a terrestrial planet.

Earth. The prototype for the terrestrial planets, and the one about which the most is known, is the Earth. The Earth consists of a thin upper crust composed of rocks of relatively low density and low melting points, overlying a much thicker mantle composed predominantly of metallic silicates and oxides, which in turn overlies a substantial core, which is composed of much denser materials, believed predominantly to be iron with other elements, either alloyed or in solution. Most of the core is liquid, but there is a smaller inner core which appears once again to be solid, and which probably has some compositional differences relative to the outer core. *See* EARTH INTERIOR.

On the scale of volatility, the Earth is a very refractory place. Most of the materials in its composition condense at quite high temperatures in a gas of solar composition, usually considerably in excess of 2200°F (1200°C). Under such circumstances, most of the iron is expected to be metallic, and since metallic iron is so much heavier than other typical rocky material, such as magnesium silicates, it is natural for the metallic iron to collect at the center of the planet. The detailed seismic evidence indicates that the core of the Earth is not pure iron, but also has some admixture of lighter elements, probably some combination of oxygen, silicon, and sulfur. Several percent of the core must also be nickel, which has properties very similar to that of iron.

The overlying mantle is composed of the oxides and silicates of the metals which are more abundant in nature. Many phase changes take place as such material is subjected to increasing pressure, and some of the increasing density with depth in the Earth's mantle is due to such phase changes.

Among the many different mineral phases which are present within the Earth, there is a natural sorting process for those minerals which combine a relatively low melting point with low density. Such minerals melt easily and tend to find their way to the surface of the Earth through such cracks or pores as become available. In this way the crust of the Earth is formed predominantly of such materials through tectonic activity.

One of the major revolutions in thinking in the earth sciences has come with the realization that the Earth is a very dynamic place. The position of the Earth's pole has changed dramatically in location with respect to the surface throughout the history of the Earth, and the land masses themselves have drifted about from one part of the surface to another. This continental drift is rendered somewhat easier by the relatively large mass of the Earth and hence the fairly rapid rate with which the temperature increases into the Earth's interior, thereby weakening the materials and allowing them to deform and flow more easily. *See* EARTH; PLATE TECTONICS.

Venus. The next most massive planet within the inner solar system is Venus, which has slightly more than four-fifths of the mass of the Earth. Venus has a very thick atmosphere, and the temperature at its surface is very much higher than is typical of the Earth's surface. The conditions make it very difficult to land spacecraft which can operate for appreciable lengths of time such as would be required to obtain seismic signals from the interior of the planet. On these grounds it can only be conjectured that the interior of the planet is probably much like that of the Earth, with a core, a mantle, and a crust. The *Pioneer* Venus orbiter radar altimeter has found some major structural features on the surface of the planet, suggestive of extensive tectonic activity, but also, to the extent that some of the features are correctly determined to be large craters, indicating that surface weathering processes take place very slowly. The extent to which the crust of Venus is subject to extensive continental drift motions is quite unknown. *See* VENUS.

Mars. The mass of Mars is approximately one-tenth that of the Earth, and hence significant differences in the internal structure are to be expected. There appears to be less of a density contrast between the core of Mars and that of its mantle, suggesting that the amount of lighter material allied with the core—say, possibly sulfur—is increased relative to the Earth. Because the planet is smaller, the temperature increases less rapidly with depth than in the case of the Earth, and hence Mars should have a somewhat more rigid outer mantle and crust than the Earth. There is no indication that large amounts of continental drift have taken place on Mars. On the other hand, tectonic activity has clearly played a large role in the history of Mars, since the planet can be roughly divided into a hemisphere which is of predominantly ancient and heavily cratered terrain, and another hemisphere which is of much younger and less heavily cratered material. The density of craters on the surface of a planet such as Mars, with so little atmosphere that incoming massive bodies are not significantly impeded in striking the planet, is a measure of the relative age of the surface which has been exposed to space. Since the cratering rate apparently fell off rapidly throughout the first few hundred million years of the history of the inner solar system, differences in crater density frequently represent age differences of some few hundred million years back in the heavy cratering epoch. *See* MARS.

Mercury. Mercury has only about half the mass of Mars, but has several distinct planetary characteristics. The mean density of Mercury is very high, indicating that Mercury probably has an abnormally large core predominantly composed of metallic iron. There is much evidence of extensive tectonic activity, although, like Mars, the increase of temperature below the surface of Mercury probably occurs sufficiently slowly so that the crust and upper mantle are relatively rigid, and nothing resembling continental drift has probably taken place. Mercury is a very heavily cratered planet, with the craters of a given size apparently having been produced by smaller projectiles than in the case of Mars. The reason is that at the distance of Mercury from the Sun such infalling projectiles tend to have higher velocities than they do near the orbit of Mars, so that the resulting impacts are more energetic. *See* MERCURY (PLANET).

Moon. Although the Earth's Moon is technically a satellite, it makes sense to describe it as a planetary body, and planetary scientists consider the twin bodies of the Earth and the Moon as interesting examples of the extremes of planetary physics ranging from relatively large bodies to relatively small but still chemically differentiated objects. The Moon has a history which includes extensive episodes of melting and differentiation, much of which can be reconstructed on the basis of the returned lunar samples. The upper layers of the Moon, which is only just over 1% of the mass of the Earth, are quite rigid, and there is no evidence for extensive horizontal motions of the structural units.

The Moon is unique in the solar system in having a relatively low density among the inner planets, and at best a very small core, indicating that the planet is practically devoid of metallic iron. Relative to the Earth, it is also highly depleted in the more volatile elements. This unusual compositional pattern presumably requires an explanation in the mechanisms which resulted in the formation of the Moon, about which there has been much controversy. *See* MOON.

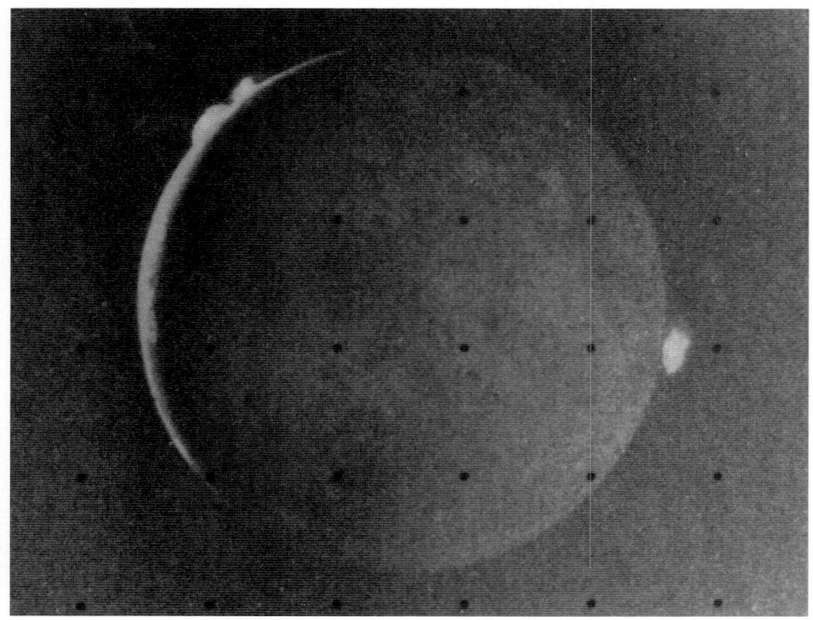

Photograph of Io taken by *Voyager 2* on July 10, 1979, from a distance of 74,000 mi (1,200,000 km). Three volcanic eruption plumes are visible on the limb, all previously seen by *Voyager 1*, 4 months earlier. (*NASA*)

Major satellites. The four Galilean satellites of Jupiter—Io, Europa, Ganymede, and Callisto—have masses which are all roughly comparable to the mass of the Earth's Moon. It is therefore quite clear that they should be considered as planetary bodies in their own right by planetary scientists. The detailed images of these satellites returned by the *Voyager* spacecraft which passed through the Jupiter system revealed them to be very interesting places with many rich, complex, and exotic properties.

The most spectacular of these planetary bodies is undoubtedly Io. This satellite has a surface characterized by large deposits of sulfur and sulfur dioxide, which is in a state of continual change. It appears to have at any time several active volcanos, each of which is likely to be spewing a stream of gas and entrained rocky particles about 60 mi (100 km) or so above the surface (see **illus.**). Such volcanic plumes spread the gases and rocky material from the volcano over a considerable portion of the surrounding terrain. This vigorous tectonic activity is understood to arise from a combination of orbital perturbations of Io by the other Galilean satellites and tidal damping by Jupiter, which results in the dissipation in the interior of Io of very large amounts of heat.

The Galilean satellites appear to represent a composition class which is slightly more volatile-rich than the pure rocky materials characteristic of the inner solar system. In particular, the sulfur content is likely to be considerably higher. The water and carbonaceous contents may also be much higher. If Io ever had much water, it appears to have been lost from the body quite early in its history. It has been suggested that Io has a sufficiently large reservoir of sulfur that this may form an effective fluid layer, or ocean, underlying the solid surface crust.

Europa, Ganymede, and Callisto all appear to have outer crusts composed of water ice. There are a variety of surface markings which indicate a history of cracking, cratering, and in some instances renewal of the icy surfaces. The mean density of Europa is sufficiently high that the planetary body is probably primarily composed of rock. On the other hand, both Ganymede and Callisto have a significantly reduced density, suggesting that quite thick layers of an icy mantle are likely to be part of those planetary bodies. It would not be surprising if the amount of tidal and radioactive heating in the interiors of these bodies was sufficient to maintain a substantial portion of this icy mantle in the form of a liquid brine. Nothing is known about the character of the underlying rocky core.

The Saturnian satellite system contains only one satellite comparable in mass to the Galilean satellites, Titan (the many other satellites are all much smaller in mass). Titan has a significantly higher volatile content than the Galilean satellites. It has an extensive atmosphere (virtually unique in the solar system) largely composed of methane. It is not known what lies at the bottom of this atmosphere, but it has been reasonably speculated that there is a transition layer of heavier hydrocarbons. The satellite has a relatively low density, characteristic of an extensive content of ices, quite likely more than just water ice as in the Galilean satellites. The atmosphere is completely opaque, and hence it is not known whether Titan has surface relief.

Oceans. With the possible exception of the Galilean satellites, the Earth is the only planet in the solar system having oceans. Mars is too cold for substantial bodies of liquid water to exist upon its surface, although there is evidence in its surface features that water once ran through a number of channels for at least short periods of time. Mars also has a substantial amount of ice in its polar caps. The atmosphere of Venus is sufficiently hot that if any liquid water were to be placed upon its surface, it would quickly be evaporated into steam. Venus contains very little water in its atmosphere, giving rise to the question of whether Venus has ever had substantial amounts of water at any time in its past, or whether it has found some mechanisms for getting rid of the bulk of it. Both the Moon and Mercury are very dry.

Within the oceans of the Earth, a complex set of currents and motions takes place. Many of these currents are driven by slight differences of density within the oceans, which in turn arise because of variations of the amount of dissolved salt, or salinity. Some differences in salinity arise from the evaporation of water into the atmosphere from the oceans, followed by rain upon the land and the runoff of salt-depleted water from the rivers. Other differences in salinity arise from the formation of ice from the oceans, which results in a concentration of salt within a liquid phase. *See* SEAWATER.

Large-scale currents are also set up in the oceans as the result of temperature differences within the ocean waters, resulting from the preferential heating of the water in the tropical regions of the Earth. The oceans play an important role in the transport of heat from the Equator toward the poles of the Earth. *See* OCEAN CIRCULATION.

Atmospheres. There are certain general principles which govern the structure and dynamics of planetary atmospheres. In most cases these atmospheres receive their primary heat input from above, resulting from heating due to the Sun. Most atmospheres contain some form of haze or condensed layers in the form of clouds, which results in a reflection of a portion of the incident sunlight back into space where it has not contributed to the deposition of heat within the atmosphere. The remainder of sunlight is either absorbed within the atmosphere or transmitted or scattered downward to the ground where absorption takes place. The heat thus received by the ground must be reradiated into the atmosphere, which will transmit some of it and absorb some of it. The absorbed radiation from the ground will in part be reradiated by the atmosphere toward the ground, adding to the heating effect that has taken place as the result of the original receipt of the corresponding energy. This enhancement of the heating effect is commonly called the greenhouse effect, despite the fact that the mechanism by which a greenhouse keeps warm is somewhat different. *See* GREENHOUSE EFFECT.

The temperature at the surface of the planetary body therefore depends in a complex manner on the properties of the overlying atmosphere, as well as upon the distance of the planet from the Sun. The atmosphere of Venus is very much hotter relative to the Earth than would be expected purely on the basis of the relative distances from the Sun. The difference appears to arise from the extensive operation of the greenhouse effect within the very thick atmosphere of Venus. Russian spacecraft landed upon the surface of Venus have found quite large amounts of illumination by sunlight there, indicating that significant amounts of solar energy do manage to penetrate to the ground of that planet.

The only terrestrial planets with atmospheres are Venus, Earth, and Mars. Both Mars and Venus have atmospheres composed predominantly of carbon dioxide. If all of the carbonate rocks of the Earth had the carbon dioxide extracted from them, the Earth also would have a thick predominantly carbon dioxide atmosphere very similar to that of Venus. Thus the difference between these two planets arises to a large extent from the ability to form carbonate rocks, which is a function of temperature. At the high temperature of the ground surface of Venus, carbonate rocks are broken down, and the carbon dioxide is released to the atmosphere; the thick atmosphere of Venus is therefore the stable state under the circumstances. In the case of the Earth, it appears that water has played an important role in the formation of carbonate rocks from carbon dioxide, and such water has not been present significantly in liquid form on Mars. There is no evidence that Mars has a substantial reservoir of carbon dioxide in the form of carbonate rocks.

The element of next greatest abundance in the atmospheres of Mars and Venus is nitrogen. This happens to be the predominant element in the atmosphere of the Earth. The next most abundant element in the terrestrial atmosphere is oxygen, which is maintained there predominantly as the result of the operation of life upon the surface of the Earth. Any planet with a large content of oxygen in the atmosphere is very likely to be extensively populated by living organisms.

The atmospheres of the terrestrial planets appear to be substantially mixed as a result, in part, of convective processes which transport heat, and in part from winds which are produced by pressure differences and which cause mixing by stirring up the atmosphere. At a sufficiently great height in the atmosphere, mixing is no longer effective, and a gravitational stratification of the components of the atmosphere takes place, with the lighter components of the atmosphere extending to greater heights. At these great altitudes, solar ultraviolet radiation produces an extensive amount of ionization of these atmospheric constituents, producing a plasma layer at the top of the atmosphere which is called the ionosphere. At a sufficiently great height the molecules of the atmosphere are in free ballistic trajectories; this region is called the exosphere. *See* ATMOSPHERE; IONOSPHERE.

The same principles of physics and photochemistry also apply to the atmospheres of giant planets, but the details are considerably different, because the predominant constituents are hydrogen, helium, and methane. At significantly lower levels in the atmosphere, a substantial amount of ammonia appears, and this forms a layer of ammonia clouds, probably contaminated by some amount of hydrogen sulfide which forms a compound with ammonia. At a still lower level in the atmosphere it is expected that water clouds will be present.

In the case of the giant planets, the rapid rotation gives rise to a distinctive banded structure along parallels of latitude within the atmosphere, because Coriolis forces make it difficult for the convective motions within the atmosphere to transport material significantly across parallels of latitude.

Magnetospheres. Some of the planets contain substantial magnetic fields; others do not. Within the inner solar system, the Earth possesses a relatively strong field, Mercury a relatively weak one, and if Venus and Mars contain significant intrinsic fields, they are sufficiently weak that they have not been confirmed. On the other hand, it is known by direct measurement that Jupiter and Saturn have very strong magnetic fields.

The generation of planetary magnetic fields appears to depend upon a combination of planetary rotation with an inner convecting layer having significant electrical conductivity. These conditions appear to be met in the core of the Earth and in the metallic hydrogen mantles of Jupiter and Saturn.

One of the most striking features of planetary magnetospheres is the trapping of energetic particles within them. This gives rise to a great variety of phenomena within the Earth's magnetic field, but in the case of Jupiter the effects are so strong that radiation damage significantly affects the operational lifetime of any space probe inserted into the magnetosphere. *See* MAGNETOSPHERE; PLANET; VAN ALLEN RADIATION.

A. G. W. Cameron

Bibliography. J. K. Beatty, C. C. Petersen, and A. Chaikin (eds.), *The New Solar System*, 4th ed., 1998; J. T. Bergstralh et al., *Uranus*, 1991; T. Gehrels (ed.), *Jupiter*, 1976; T. Gehrels and M. S. Matthews (eds.), *Saturn*, 1984; H. Hunter, T. M. Donahue, and V. I. Moroz (eds.), *Venus*, 1983; H. Jeffreys, *The Earth*, 6th ed., 1976; H. H. Kieffer et al., *Mars*, 1992; D. Morrison and T. Owen, *The Planetary System*, 2d ed., 1993; F. Vilas, C. R. Chapman, and M. S. Mathews (eds.), *Mercury*, 2d ed., 1996.

Plant

An organism that belongs to the Kingdom Plantae (plant kingdom) in biological classification. The term is also loosely used to indicate any organism that is not an animal. In the most commonly accepted modern classification (the five-kingdom system), fungi, bacteria, and algae belong to the kingdoms Fungi, Monera, and Protista, respectively, rather than to the plant kingdom as they did in the outmoded two-kingdom system. The study of plants is called botany. *See* BOTANY; CLASSIFICATION, BIOLOGICAL.

Characteristics. In the five-kingdom system, the Plantae share the characteristics of multicellularity, cellulose cell walls, and photosynthesis using chlorophylls *a* and *b* (except for a few plants that are secondarily heterotrophic). Most plants are also structurally differentiated, usually having organs specialized for anchorage, support, and photosynthesis. Tissue specialization for photosynthetic, conducting, and covering functions is also characteristic. Plants have a sporic (rather than gametic or zygotic) life cycle that involves both sporophytic and gametophytic phases, although the latter is evolutionarily reduced in the majority of species. Reproduction is sexual, but diversification of breeding systems is a prominent feature of many plant groups. *See* REPRODUCTION (PLANT).

Diversity. A conservative estimate of the number of described species of plants is 250,000. There are possibly two or three times that many species as yet undiscovered, primarily in the Southern Hemisphere. The number of known species is, however, large enough that plants are further categorized into nonvascular and vascular groups, and the latter into seedless vascular plants and seed plants. The nonvascular plants include the liverworts, hornworts, and mosses. The vascular plants without seeds are the ground pines, horsetails, ferns, and whisk ferns; seed plants include cycads, ginkgos, conifers, gnetophytes, and flowering plants. Each of these groups constitutes a division in botanical nomenclature, which is equivalent to a phylum in the zoological system.

Function. Plants are essential to the survival of virtually all living things. Through photosynthesis, plants provide all of the food for human and animal consumption. Plants also are the source of such diverse products as building materials, textile fibers, gums, resins, waxes, rubber, perfumes, dyes, and tanning materials. Most of the drugs that are used in modern medicine were originally derived from plants, although many others came from fungi. In the past, medical doctors needed to be as knowledgeable about plants as they were about their patients. Plants also provide habitats for wildlife and birds, offer shelter, and contribute to soil-building processes. Plants of the Carboniferous era provided the energy that is now used as oil, coal, and gas. The importance of plants to the existence of humans makes the preservation and study of plants an essential component of environmental studies. *See* PHOTOSYNTHESIS; PLANT ANATOMY; PLANT KINGDOM; PLANT PHYSIOLOGY; PLANT TAXONOMY. Meredith Lane

Bibliography. H. T. Hartmann et al., *Plant Science: Growth, Development and Utilization of Cultivated Plants*, 3d ed., 2000; R. M. Klein and D. T. Klein, *Fundamentals of Plant Science*, 1987; P. H. Raven, R. F. Evert, and H. Curtis, *Biology of Plants*, 6th ed., 1998.

Plant anatomy

The area of plant science concerned with the internal structure of plants. It deals both with mature structures and with their origin and development.

Plant anatomists dissect the plant and study it from different planes and at various levels of magnification. At highest magnification they examine the smallest units of plant structure, the cells; at intermediate magnification they observe the organized aggregations of these cells, the tissues; and at low magnification they determine the arrangement and interrelations of tissues in plant organs such as root, stem, leaf, and flower. At the level of the cell, anatomy overlaps plant cytology, which deals exclusively with the cell and its contents. Sometimes the name plant histology is applied to the area of plant anatomy directed toward the study of cellular details of tissues. *See* PLANT ORGANS; PLANT TISSUE SYSTEMS.

Katherine Esau

Bibliography. H. A. De Bary, *Comparative Anatomy of the Vegetative Organs of the Phanerogams and Ferns*, trans. by F. O. Bower and D. H. Scott, 1884; A. J. Eames, *An Introduction to Plant Anatomy*, 2d ed., 1947, reprint 1977; A. Fahn, *Plant Anatomy*, 4th ed., 1990; E. M. Gifford, Jr., *Morphology and Evolution of Vascular Plants*, 3d ed., 1989; G. F. Haberlandt, *Physiological Plant Anatomy*, 4th ed., 1928, reprint 1979; B. D. Jackson, *A Glossary of Botanic Terms, with Their Derivation and Accent*, 1986; P. J. Kramer, *Plant and Soil Water Relationships*, 1949; A. C. Leopold et al., *Plant Growth and Development*, rev. ed., 1975.

Plant-animal interactions

The examination of the ecology of interacting plants and animals by using an evolutionary, holistic perspective. For example, the chemistry of defensive compounds of a plant species may have been altered by natural-selection pressures resulting from the long-term impacts of herbivores. Also, the physiology of modern herbivores may be modified from that of thousands of years ago as adaptations for the detoxification or avoidance of plant defensive chemicals have arisen. Other examples of plant-animal interactions include pollination biology, fruit and seed dispersal ecology, herbivory on plants by invertebrates (such as insects) and vertebrates (such as ungulates), ant-plant interactions, the ecology and evolution of carnivorous plants, and community and ecosystem patterns of animal distributions resulting from the availability of plant resources.

The application of the theories based on an understanding of plant-animal interactions provides an understanding of problems in modern agricultural ecosystems. For instance, the problem of limited genetic variation in crop species impacts on attempts to enhance a crop's genetically based resistance to pests. Similarly, plant-animal interactions have practical applications in medicine. For example, a number of plant chemicals, such as digitalin from the foxglove plant, that evolved as herbivore-defensive compounds have useful therapeutic effects on humans. *See* AGROECOSYSTEM.

Types of interactions. The evolutionary consequences of plant-animal interactions vary depending on the effects on each participant. Interaction types range from mutualisms, that is, relationships which are beneficial to both participating species, to antagonisms, in which the interaction benefits only one of the participating species and negatively impacts the other. Interaction types are defined on the basis of whether the impacts of the interaction are beneficial, harmful, or neutral for each interacting species (see **table**). In some cases, these distinctions of types can be unclear because the effects of interactions, whether beneficial, harmful, or neutral, can change in time and space. Amensalisms (that is, interactions that are neutral to one species but harmful to a second species) and neutralisms (that is, interactions that are neutral to both interacting species) are beyond the scope of most studies of plant-animal interaction.

Effects of interaction types for each species*

Interaction	Effect on species A	Effect on species B
Mutualism	+	+
Commensalism	+	0
Antagonism	+	–
Competition	–	–
Amensalism	0	–
Neutralism	0	0

*+ = beneficial, – = harmful, 0 = neutral.

Antagonism. The high frequency of antagonistic interactions in most environments can be attributed to the energy and nutrient needs of organisms. For example, plants are concentrated parcels of such resources for herbivores, as are insects for carnivorous plants. As a consequence of the strong influences that antagonisms have on individual plants, many plants have evolved physical defenses (for example, spines, thorns, or hairs) or chemicals that repel or poison potential herbivores. Some herbivores have counteradapted by evolving resistance to these defenses or mechanisms to avoid them entirely. There are several forms of antagonism, including herbivory, parasitism, and predation.

Mutualism. The numerous examples of mutualisms include many of the relationships between pollinators and flowers or between fruits and fruit-eating birds, mammals, or reptiles. The plant-produced reward (for example, pollen, nectar, or fruit) benefits the pollinator or disperser, while the animal's activities (for example, carrying pollen to another blossom or dispersing seeds) profit the plant. However, not all pollinator and fruit disperser and plant interactions are mutualisms. Some flowers and fruits deceive their pollinators or dispersers and provide no benefit to the animal. The frequency of mutualisms within a community depends on the richness of antagonistic interactions since mutualisms often evolve from antagonisms. Mutualisms themselves can create new resources (nectar, fruits) that promote the development of additional interactions. Thus, mutualisms contribute to the complex organization of communities and biodiversity.

Commensalism. The effects of commensalism are beneficial for one of the two interacting species but are neutral for the other. Examples include the use of trees as nesting places by birds or the temporal succession of mammalian herbivores on the Serengeti Plains of East Africa. In both cases, the relationship of animals to plants influences the animals' distribution and abundance. In the Serengeti, plants indirectly affect a commensalistic relationship involving the zebra, wildebeest, and Thomson's gazelle. The zebra, which migrates first, eats tall, protein-poor grass stems; the wildebeest follows and consumes shorter, more-nutritious grass leaves; while Thomson's gazelle, the last to migrate, feeds on the nutritious regrowth of grasses and herbs. The wildebeest is the commensal of the zebra, and Thomson's gazelle is the commensal of the wildebeest. However, the relationship between these herbivores and their plant foods is antagonistic.

Competition. This is an interaction in which one organism consumes a resource that would have been available to, and might have been consumed by, another. Within the context of plant-animal interactions, competition typically affects outcomes in indirect ways. For example, competition among herbivores for plant-food resources can strongly influence the distribution of competing animals. Variations in adaptations to feeding often lead to segregation of similar species within a habitat and consequently allow coexistence. *See* ECOLOGICAL SUCCESSION. Warren G. Abrahamson

Fossil record. Plants and animals interact in a variety of ways within modern ecosystems. These may range from simple examples of herbivory (animals eating plants) to more complex interactions such as pollination or seed and fruit dispersal. Conversely, animals also rely on plants for food and shelter. For example, various invertebrates ingest algae, which provide a source of photosynthetically produced food in the absence of normally ingested microorganisms. The complex interactions that have existed between these organisms over geologic time not only have resulted in an abundance and diversity of organisms in time and space but have also contributed to many of the evolutionary adaptations found in the biological world.

Paleobiologists have attempted to decipher some of the interrelationships that existed between plants and animals throughout geologic time. This field of study has been slow to emerge because there is a relative absence of detailed information about many plants and animals. In addition, the ecological setting in which the organisms lived in the geologic past is being analyzed in association with the fossils. Thus, as paleobiologists have increased their understanding of certain fossil organisms, it has become possible to consider some of the aspects of the ecosystems in which they lived, and in turn, how various types of organisms interacted.

Herbivory. Perhaps the most widespread interaction between plants and animals is herbivory, in which plants are utilized as food. A common example involves partially chewed leaves or areas along the margin where the lamina has been removed. One reason why there have not been many documented examples of partially chewed leaves is the bias of collecting complete, perfect specimens. In some cases, the outline of the area that has been removed provides insight into the type of organism responsible for the herbivory. Another method for determining the extent of herbivory in the fossil record is by analyzing the plant material that has passed through the digestive gut of the herbivore. For example, in Carboniferous coal balls there are numerous coprolites (fecal pellets) of various sizes and shapes. Some of these contain plant fragments of the same type, while others are made up of the leaves and stems of different types of plants. Still other coprolites contain only pollen grains, suggesting that the herbivores may have been more specific in their nutritional requirements. In addition, the size and shape of the coprolites can be used to determine their source.

The analysis of some coprolites in Carboniferous coal balls suggests that swamps were inhabited by mites, collembolans, millipedes, and cockroaches. Chemical systems that made leaves less palatable for herbivores allowed some of the plants to defend themselves against herbivory. Other plants evolved defense mechanisms in the form of hairs, thick seed coats reinforced with fibrous tissues, and various mechanical methods to protect the more nutritious reproductive organs. Other plants evolved the tree habit in which leaves and reproductive organs are positioned some distance from the forest floor. It is believed that the evolution of flight by certain invertebrates evolved in response to plants repositioning their reproductive organs in the upper canopy of the forest.

The stems of some fossil plants show tissue disruption similar to various types of wounds occurring in plant parts that have been pierced by animal feeding structures. As plants developed defense systems in the form of fibrous layers covering inner, succulent tissues, some animals evolved piercing mouthparts that allowed them to penetrate these thick-walled layers. In some fossil plants, it is also possible to see evidence of wound tissue that has grown over these penetration sites. *See* COAL BALLS; HERBIVORY.

Mimicry. Another example of the interactions between plants and animals that can be determined from the fossil record is mimicry. Certain fossil insects have wings that are morphologically identical to plant leaves, thus providing camouflage from predators as the insect rested on a seed fern frond.

Pollination. The transfer of pollen from the pollen sacs to the receptive stigma in angiosperms, or the seed in gymnosperms, is an example of an ancient interaction between plants and animals. It has been suggested that pollination in some groups initially occurred as a result of indiscriminate foraging behavior by certain animals, and later evolved specifically as a method to effect pollination. The rewards offered by plants can be documented in certain fossils in the form of specialized glands and cavities that secreted nectar, or of expendable pollen as sources of food. The particular structure and organization of flowers, size of floral parts, and nature of the anthers represent sources of information in fossil flowers that can be used to study ancient pollination systems. In some instances, the morphology of the flower can be used to determine the type of insect pollinator. In addition, the size, shape, and organization of fossil pollen grains provide insight into potential pollination vectors. Far less information is available about the role of animals in the dissemination of fossil fruits and seeds. However, the presence of spines and hooked projections on certain types of seeds suggests that they may have been transported by larger animals. In addition, many fossil species bore seeds designed to pass through the digestive guts of animals that ate fleshy reproductive organs such as fruits. *See* FOSSIL; PALEOBOTANY; POLLINATION. Thomas N. Taylor

Bibliography. W. G. Abrahamson (ed.), *Plant-Animal Interactions*, 1989; W. C. Ciepet, Insect pollination: A paleontological perspective, *BioScience*, 29:102–108, 1979; H. F. Howe and L. C. Westley, *Ecological Relationships of Plants and Animals*, 1988; A. C. Scott and T. N. Taylor, Plant/animal interactions during the upper carboniferous, *Bot. Rev.*, 49:259–307, 1983; J. N. Thompson, *Interaction and Coevolution*, 1982; S. L. Wing and B. H. Tiffney, The reciprocal interaction of angiosperm evolution and tetrapod herbivory, *Rev. Paleobot. Palynol.*, 50:179–210, 1987.

Plant cell

The basic unit of structure and function in nearly all plants. Although plant cells are variously modified in structure and function, they have many common features. The most distinctive feature of all plant cells is the rigid cell wall, which is absent in animal cells. The range of specialization and the character of association of plant cells is very wide. In the simplest plant forms a single cell constitutes a whole organism and carries out all the life functions. In just slightly more complex forms, cells are associated structurally, but each cell appears to carry out the fundamental life functions, although certain ones may be specialized for participation in reproductive processes. In the most advanced plants, cells are associated in functionally specialized tissues, and associated tissues make up organs such as the leaves, stem, and root. Although a substantial body of knowledge exists concerning the features of various types of plant cells, there is a great gap in knowledge of how cells become specialized for particular functions and associations.

Plant and animal cells are composed of the same fundamental constituents—nucleic acids, proteins, carbohydrates, lipids, and various inorganic substances—and are organized in the same fundamental manner. A characteristic of their organization is the presence of unit membranes composed of phospholipids and associated proteins and in some instances nucleic acids.

For convenience this article first considers the structure of the cell in the interphase or nondividing stage and then in the dividing stage.

Interphase Cell

For many years it was usual to employ the term protoplasm to refer collectively to the cell components responsible for imparting the characteristics of life. The term has undergone some modification, and all the structures internal to the cell membrane are frequently referred to as protoplasts. However, techniques such as electron microscopy and cytochemistry have made it increasingly possible to define structures and to ascribe particular functions to

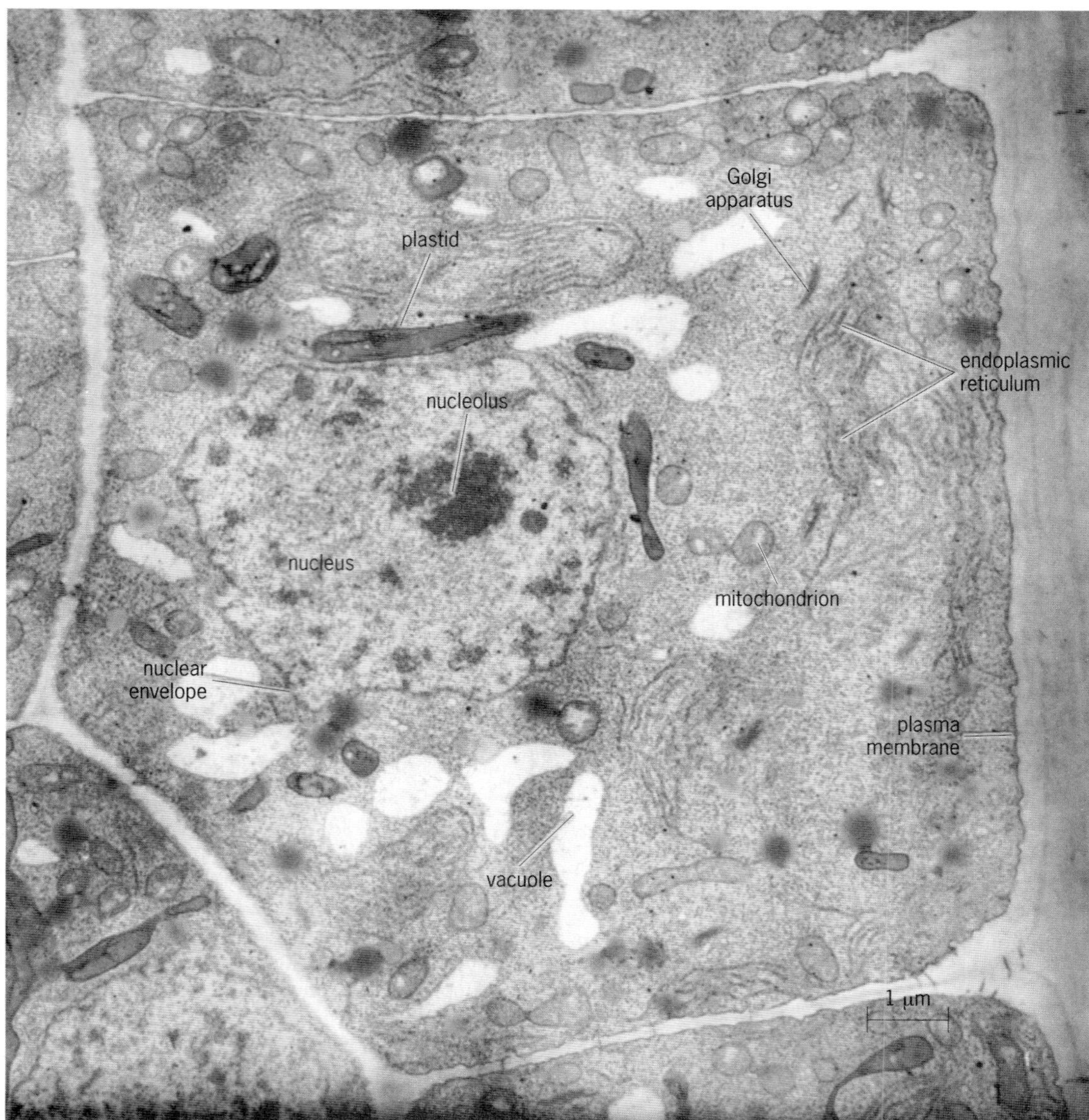

Fig. 1. Spinach (*Spinacia oleracea*) root epidermal cell showing the nucleus and the cytoplasm with inclusions. In plant cells, vacuole size often determines the position of the nucleus. (*Courtesy of M. Dauwalder*)

the two major components of the protoplast: the nucleus and the cytoplasm (**Fig. 1**). Furthermore, it is possible to localize specific activities in given regions or substructures in these components. *See* CYTOCHEMISTRY; ELECTRON MICROSCOPE.

Nucleus. The nucleus of the undifferentiated plant cell is generally centrally located, whereas in mature plant cells it may be displaced to the edge of the cell by the presence of a large vacuole. The nuclear material consists of a ground substance and chromatin, which at the time of division is resolved into chromosomes. Nuclear components of key importance are deoxyribonucleic acid (DNA), ribonucleic acid (RNA), and proteins. Continuing syntheses of DNA, RNA, and proteins, notably histones, are predominant activities in the nucleus, although comparable processes also take place elsewhere in the cell. The singular importance of the relationships of DNA to RNA and proteins lies in the fact that the hereditary characteristics of the cell, encoded in DNA molecules, are transmitted in a complex sequence via RNA to proteins. *See* NUCLEIC ACID.

The same double-helix form characteristic of animal cell DNA is demonstrated for DNA isolated from plant cell nuclei, and the evidence indicates that the coding of proteins follows the same pattern in both cell types. *See* GENETIC CODE.

Different types of RNA are formed in the nucleus, transfer RNA (tRNA), ribosomal RNA (rRNA), and presumably messenger RNA (mRNA), all of which participate in the synthesis of proteins by ribosomes.

One or more nucleoli are found in the nuclei of all plant cells. Chemically, plant nucleoli contain various types of RNA and relatively large amounts of protein, components which presumably play distinctive roles in the transmission of hereditary information to various sites in the cell. Chromatin DNA is structurally intermingled with the components of the nucleoli. The appearance of the nucleolus changes through development, probably in relation to control of cellular activities.

As in animal cells, the nucleus of the plant cell is bounded by an envelope. The envelope is a complex structure, consisting of two membranes with a lumen (called the perinuclear space) between, and with discontinuities (called pores) fairly regularly spaced throughout. These pores are regular in outline and may be bounded by rings, or annuli. They are frequently seen to contain material that stains differ-

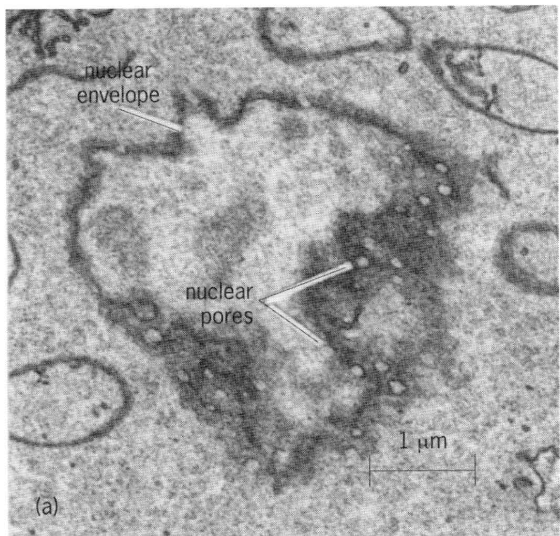

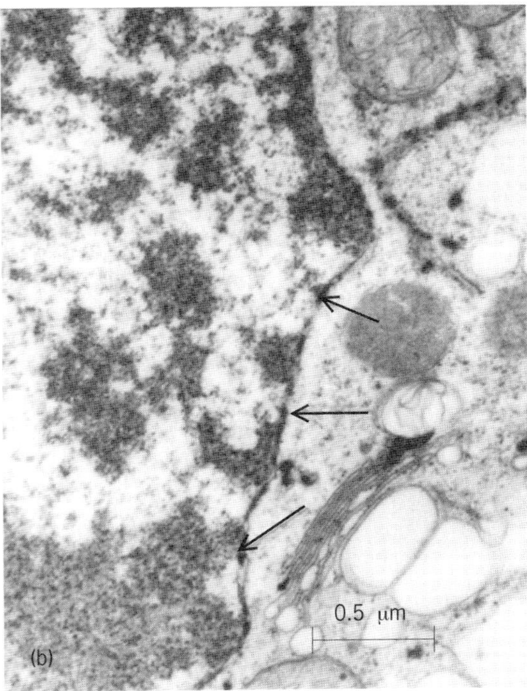

Fig. 2. Cell nucleus. (*a*) Somewhat tangential section (*courtesy of H. H. Mollenhauer*). (*b*) Section showing the nuclear envelope. The dark lead precipitate (arrows) indicates presence of acid phosphatase within the perinuclear space (*courtesy of C. W. Goff*).

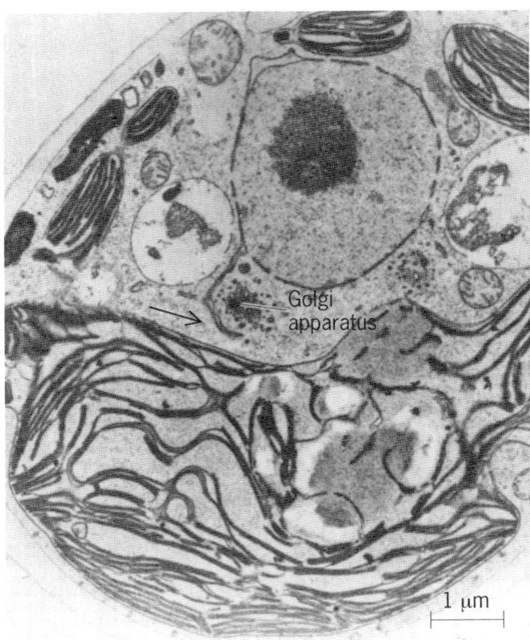

Fig. 3. Section of a *Chlamydomonas* cell showing extension of the nuclear envelope (arrow) to the vicinity of the Golgi apparatus. (*Courtesy of P. L. Walne*)

ently from the adjacent regions, and it has been suggested that interchanges, particularly of large molecules, between nucleus and cytoplasm take place through the pores rather than across the membranes of the envelope (**Fig. 2***a*). The perinuclear space has been relatively little studied, but some cytochemical studies suggest the presence of enzymes (Fig. 2*b*). The nuclear envelope, with a metabolically active lumen, apparently is a complex system for controlling nucleocytoplasmic exchanges.

The inner membrane of the nuclear envelope is frequently in close association with portions of the chromatin. The outer membrane may be studded with ribosomes. In some instances extensions of the nuclear envelope continue outward into the cytoplasm, often in specific association with particular organelles (**Fig. 3**). In instances in which segments of nuclear envelope have particular architectural modifications, and these same modifications are seen elsewhere in the cell, it has been suggested that the membrane may be transferred from the nuclear envelope, perhaps transporting perinuclear material with it. No substantial experimental evidence has been provided for this, but it does seem clear that the nuclear envelope is a dynamic structure from which vesicles may be evolved in certain stages of cellular activity. One example is the frequent appearance of what appear to be vesicles being pinched off from the nuclear envelope and transported to the Golgi apparatus. *See* CELL NUCLEUS.

Cytoplasm. The region of the cell between the nuclear envelope and the plasma membrane, or plasmalemma, bounding the cell is best considered in terms of a matrix throughout which are distributed various organelles and inclusions. Metabolic activities may be generally distributed throughout the matrix, confined to specific regions, or clearly carried out within the organelles. Although more research has been directed to the analysis of the chemical composition and activities of the organelles than of the matrix, it is apparent that in addition to water the matrix contains, at any given time, various minerals and the recognized organic components of the cell.

The organelles, which are increasingly interpreted as compartments in which certain metabolic activities are localized, are bounded by unit membranes. The molecular components (phospholipids and proteins) of the membranes are subject to rapid turnover. The membranes act as sites for the synthesis or breakdown of materials and frequently, as in mitochondria, are structurally highly specialized for these activities. Therefore, far from being simply selective barriers to the movement of materials, the membranes of the plant cell are dynamic structures which play key roles in metabolism.

Ribosomes. Conspicuous among the components of the cytoplasmic matrix are particles, approximately 20 nanometers in diameter, composed of RNA and proteins; these particles are functionally related to similar ones in the nucleus, mitochondria, and plastids. Formerly called ribonucleoprotein bodies but now generally known as ribosomes, these particles function in the relay of the coding system from nucleic acids in the nucleus and as sites of protein synthesis. A characteristic feature of the plant cell is the occurrence in chloroplasts of ribosomes, distinguished by RNA components of different molecular size from those of the cytoplasmic matrix.

The RNA component of the cytoplasmic ribosomes of plant cells can be separated into two (sometimes three) molecular sizes, as can that of the animal cells that have been investigated. When studied by highly specialized techniques, each ribosome is seen to be composed of two distinguishable subunits, but whether this relates to the presence of rRNA molecules of different sizes and what its importance in protein synthesis may be remain subjects for further exploration.

In many instances the cytoplasmic ribosomes are seen in spiral or helical arrangements called polyribosomes or polysomes, in which the ribosomes are interconnected by thin strands of mRNA. The current interpretation is that these arrangements represent ribosomes moving along strands of mRNA which impart information from the DNA and thus code protein synthesis. In all types of cells some of the ribosomes in the cytoplasm appear to be free, while others are attached to the surface of the membranes of the endoplasmic reticulum or to the outer membrane of the nuclear envelope. *See* RIBOSOMES.

Endoplasmic reticulum. The character of the cytoplasmic compartment known as the endoplasmic reticulum has been known only since electron microscopy has been adapted to the study of cells. This compartment may be tubular or lamellar; each form appears to be a profile of two membranes with a lumen between. In plant cells the endoplasmic reticulum is commonly lamellar; but frequently it appears to pass through the plasma membrane and sometimes even the cell wall, and it then becomes tubular. Extensions of the endoplasmic reticulum to and through the cell wall, which provide a form of continuity from one cell to another, form part of the cytoplasmic strands called plasmodesmata.

The endoplasmic reticulum is an architecturally regular structure only in a few types of plant cells. It frequently shows dilatations, sometimes containing crystals or other distinctively staining material. The endoplasmic reticulum is a labile structure, and the manner in which its profiles are associated differ with the stage of develoment and metabolic activity. In certain stages numbers of profiles are seen to be stacked, frequently parallel to the surface of the cell (**Fig. 4**). The profiles may also surround the nucleus or seem to encompass any of several types of organelles. The endoplasmic reticulum may be smooth or rough; that is, the outer surfaces of the membranes may be studded with ribosomes. Smooth endoplasmic reticulum occurs less abundantly in plant cells than in animal cells.

Experiments using animal cells indicate that the endoplasmic reticulum sequesters into its lumen proteins or protein precursors, presumably syntheized at the ribosomes, and that it is the site of synthesis of at least some of the lipid components of the cell. There is no reason to suppose that the same does not obtain in plant cells. Cytochemical analyses have demonstrated a number of enzymes either in the lumen or associated with the membranes of

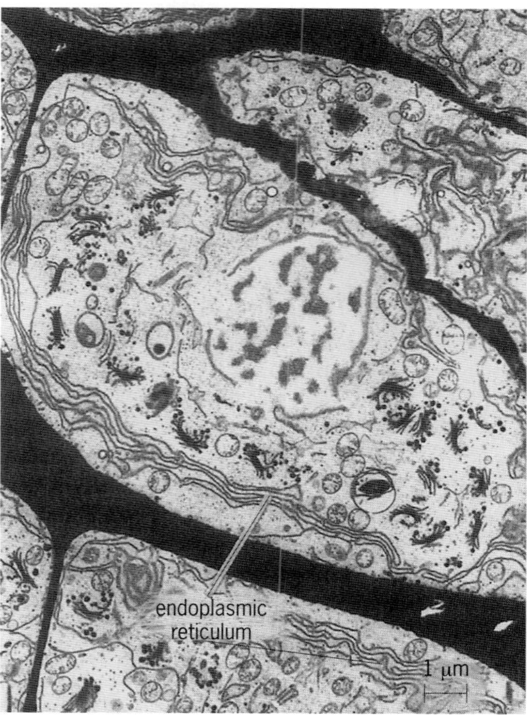

Fig. 4. Portions of maize root cap cells showing profiles of endoplasmic reticulum stacked more or less parallel to the cell surface. (*Courtesy of M. Dauwalder*)

the endoplasmic reticulum. It is known that in animal cells certain substances are transferred from the endoplasmic reticulum to other sites in the cell. For example, in many secreting cells it can be shown by radioautography that proteins or protein precursors move from the endoplasmic reticulum to the Golgi apparatus, where they become part of protein-containing secretory products. The question of how material is transferred from the endoplasmic reticulum to the Golgi apparatus remains unanswered. The most common hypothesis, that the material is transferred in small membrane-bound vesicles, is lent some support by the appearance of blebs in the membrane of the endoplasmic reticulum near the Golgi apparatus and by the occasional appearance of numerous small vesicles between the two structures. Whatever the mechanisms, it seems likely that the patterns of transfer are comparable in both animal and plant cells, though there are obviously differences in physiology. *See* AUTORADIOGRAPHY.

The introduction of toxic substances or other modifications of the metabolic state of the cell sometimes causes extensive proliferation of the endoplasmic reticulum, usually of the rough type. This development has been noted in plant and animal cells and appears to be a compensatory response to deleterious conditions. It may well be that proliferation of the endoplasmic reticulum provides additional reaction surface and more enzymes which increase the capacity of the cell to detoxify substances or otherwise adjust to unfavorable circumstances. *See* ENDOPLASMIC RETICULUM.

Mitochondria. Much of the work involved with isolation and chemical analysis of mitochondria has been done with plant cells. The mitochondria are centers of energy conversion. They function in electron transport and oxidative phosphorylation and act as respiratory control centers of the cell.

Mitochondria are of the same fundamental structural pattern in plant and animal cells. They are bounded by a double-membrane system with the inner membrane projecting into the lumen to form cristae (**Fig. 5**), which may be either tubular or sheetlike, depending upon the type of cell and its activity. In general there is less extensive development of the cristae in the mitochondria of plant cells than in those of animals. This may reflect the fact that plant cells generally have substantially lower respiratory rates. In the few types of plant cells characterized by relatively high respiratory rates, the extent of the cristae more nearly resembles that in animal cells.

Cristae bear stalked structures in ordered arrangements. The integrity of these structures is essential for the conduct of certain metabolic activities involved in the respiratory cycle. In experiments with animal cells mitochondrial activity disappeared when this integrity was destroyed, and there was a recurrence of the activity when the integrity was restored. Inasmuch as both the structure and the indicated activities of mitochondria are similar in plant and animal cells, the indicated relationships presumably hold for plant mitochondria.

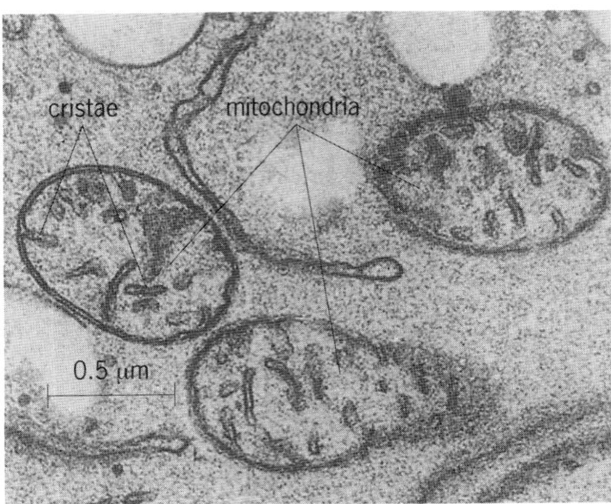

Fig. 5. Three mitochondria showing two bounding membranes and cristae formed by extension of inner membrane into lumen. (*Courtesy of H. H. Mollenhauer*)

That other activities are also involved in the functioning of the mitochondria is suggested by the frequent development of crystalline inclusions (**Fig. 6**) and by the presence of nucleic acids. Mitochondria contain a small portion of the total DNA and RNA complement of the cell. Since ribosomelike particles are also present in the mitochondria, these organelles can probably synthesize some portion of their own proteins. In fact, there is evidence that amino acids are moved into the mitochondria. The presence of nucleic acids within the mitochondria provides the basis for a hypothesis, supported by

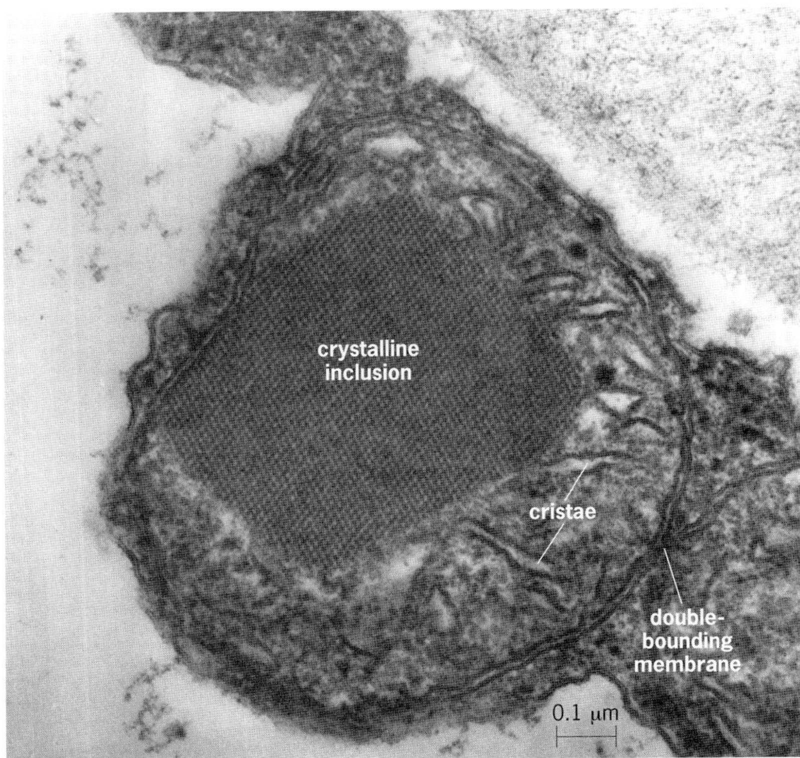

Fig. 6. A highly magnified mitochondrion. Components of crystalline inclusion are in an ordered array. (*Courtesy of H. J. Arnott*)

studies of the biogenesis of mitochondria, that these organelles have at least some degree of autonomy in replication.

In the living cell, mitochondria are in constant motion, frequently in patterns that show the larger number of them shuttling back and forth between one cell site and another, presumably in relation to the transfer of energy. When in motion they tend to appear more elongate than in electron microscope preparations, but both light and electron microscope studies show them to be pleomorphic. *See* CITRIC ACID CYCLE; MITOCHONDRIA.

Golgi apparatus. The Golgi apparatus is a component of all plant cells, with the possible exception of certain fungi and some highly specialized cells such as mature sperm. This organelle has been most studied in relation to secretory activities of glandular cells in animals, and its presence in plant cells suggests that secretory activity is widespread.

In many plant cells the Golgi apparatus clearly functions in secretion, but the ubiquitous occurrence of the organelle suggests that it may have other roles in cellular activity. Although many aspects of its function are still obscure, it is apparent that certain materials are either sequestered into its saccules (**Fig. 7**) or synthesized there, or they are variously combined in the saccules to form complex secretion products. These products are then separated from the saccules as membrane-bound vesicles and transported to and through the plasmalemma. Not all cell secretion takes place, however, via the Golgi apparatus. In many types of secreting plant cells the Golgi apparatus is responsible for packaging and exporting a substantial part of the components of the wall that is built up around the cell. In such cells the direction of movement of the secretory product is well established.

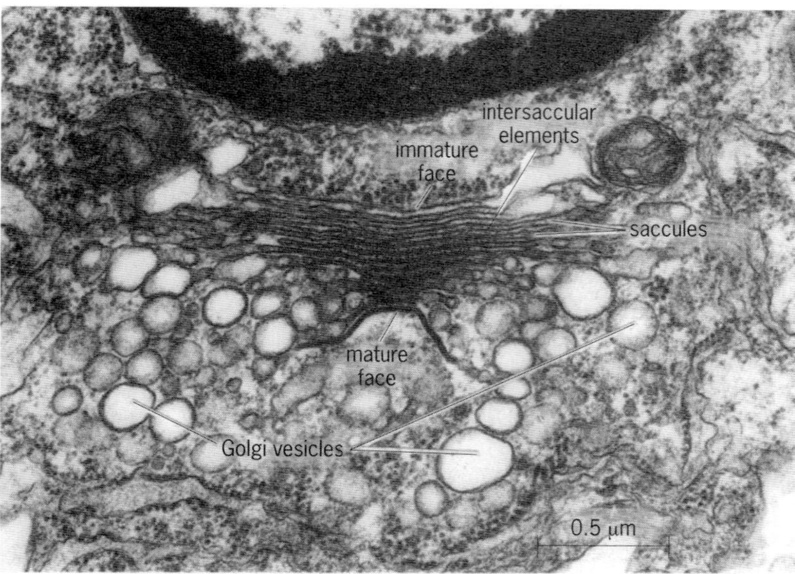

Fig. 7. Portion of a *Nitella* cell showing a Golgi apparatus composed of a stack of saccules, the contents of which stain differentially from the immature face to the mature face. Note also the intersaccular elements and the associated Golgi vesicles produced from the saccules. (*Courtesy of R. Turner*)

The Golgi apparatus is a region characterized by membrane assembly and extension and to which, in animal cells at least, protein or protein precursors are transferred from the endoplasmic reticulum and either elaborated further or combined with carbohydrates. In both animal and plant cells the Golgi apparatus is also the site of synthesis of certain polysaccharides. It also functions in "complexing" activities, which may involve lipid-protein combinations and the addition to the materials assembled there of such elements as iron or sulfur.

The structural and functional similarity of plant- and animal-cell Golgi apparatus supports the concept of a high degree of comparability in their metabolic activities. Differences appear to be related to the presence of more highly specialized organ systems in the animal than in the plant. *See* GOLGI APPARATUS.

Microtubules. As in the case of animal cells, certain of the newer procedures for fixation, notably the use of glutaraldehyde, demonstrate the existence in plant cells of numerous microtubules; these are elongate elements usually with a diameter of 20 nanometers or somewhat more, and presumably composed of polymerized protein.

There is some evidence that there are distinct classes of microtubules distinguished by differences in diameter, location, and orientation. In interphase plant cells the most prominent type is found in conspicuous numbers close to the surface of the cell and in some instances seemingly parallel to it. The number appears to differ considerably from one cell type to another or one stage of activity to another and quite possibly with variations in preparation procedures.

It has been observed that cellulose microfibrils laid down in the wall frequently parallel microtubules inside the cytoplasm. This has led to the speculation that in some manner the orientation of the microtubules controls in part the orientation of the microfibrils. During growth, orientation of the two appears to be modified more or less simultaneously. The spindle fibers of the dividing cell and of the microtubules of the interphase cell are not only structurally similar but both seem to have some part in determining the direction of movement of certain vesicles within the cell. In some instances, as in motile sperm formed by algae and other lower plants, microtubules are arranged in a consistent pattern, which suggests that they may act as a sort of cytoskeleton. In these instances they appear to be related to, if not immediately derived from, the centrioles. Knowledge concerning the exact nature and functional role of the microtubules is extremely limited. However, it has been observed that they are composed of subunits in patterns similar to the basic structures of components of flagella of motile cells. *See* CILIA AND FLAGELLA.

Plastids. Perhaps the most conspicuous and certainly the most studied of the features peculiar to plant cells is the presence of plastids. Among the organisms generally classified as plants, only the bacteria, the blue-green algae, and some fungi lack plastids. The plastids are membrane-bound organelles with an inner membrane system. Chlorophylls and other pigments are associated, generally in noticeably regular molecular arrangements, with the inner membrane system. The extensiveness of the inner membrane varies greatly with the functional state of the plastid. The chlorophylls, or green pigments, have been the most studied of the plant pigments because of their conspicuous role in fundamental energy processes. The other pigments apparently sometimes play a part in energy processes, sometimes screen out particular wavelengths of light, and sometimes are degradation products. *See* CHLOROPHYLL.

The chloroplast was once thought to develop only from a structure termed a proplastid, and it was assumed that it might have a different origin than other types of plastids. It is fairly clear, however, that a membrane-bound structure, representing a compartment in which certain activities are localized, is characteristic of immature plant cells and that from this structure the several different types of plastids may develop, including the chloroplasts. Chloroplast membranes are characterized in most plants as associations of electron-dense disk-shaped grana and interconnecting stroma, which are lighter on electron micrographs (**Fig. 8**). Chloroplasts also contain what appear to be storage lipids.

The chloroplasts are responsible for the green color of many parts of plants and are functional in photosynthesis as the site where light energy is transformed to chemical energy; the chromoplasts are responsible for certain colors in flowers and fruits; in the amyloplasts starch is stored to provide a reservoir of energy for use in the plant's development; and in elaioplasts, the lipoidal material is the energy reservoir. Such colorless plastids as amyloplasts and elaioplasts are frequently termed leucoplasts.

Like the mitochondria, plastids contain DNA and RNA, which provide a basis for protein synthesis within the plastid as well as the possibility of some self-determination in replication. It has been known since the early 1900s that plastids have a genetic system that is to some extent independent of the genetic system of the nucleus; now it is known that these organelles replicate by division. Replicative ability and the presence of a genetic system which (like that of the mitochondria) may differ somewhat from that of the nucleus have given rise to the speculation that plastids may derive from organisms which invaded non-plastid-bearing cells with the result that an enduring symbiotic relationship was established. This idea finds support in that the RNA components of plastid ribosomes more nearly resemble those of bacteria than those of the nuclear, cytoplasmic, or mitochondrial ribosomal particles found in the plant cell. Whatever the origin of the plastids, these structures are important because of the carbohydrate-synthesizing ability of chloroplasts; color-producing function of the chromoplasts in attracting insects and other pollinators; and the energy reservoir for growth and development, as well as the contribution to plant nutritional value, provided by the storage plastids.

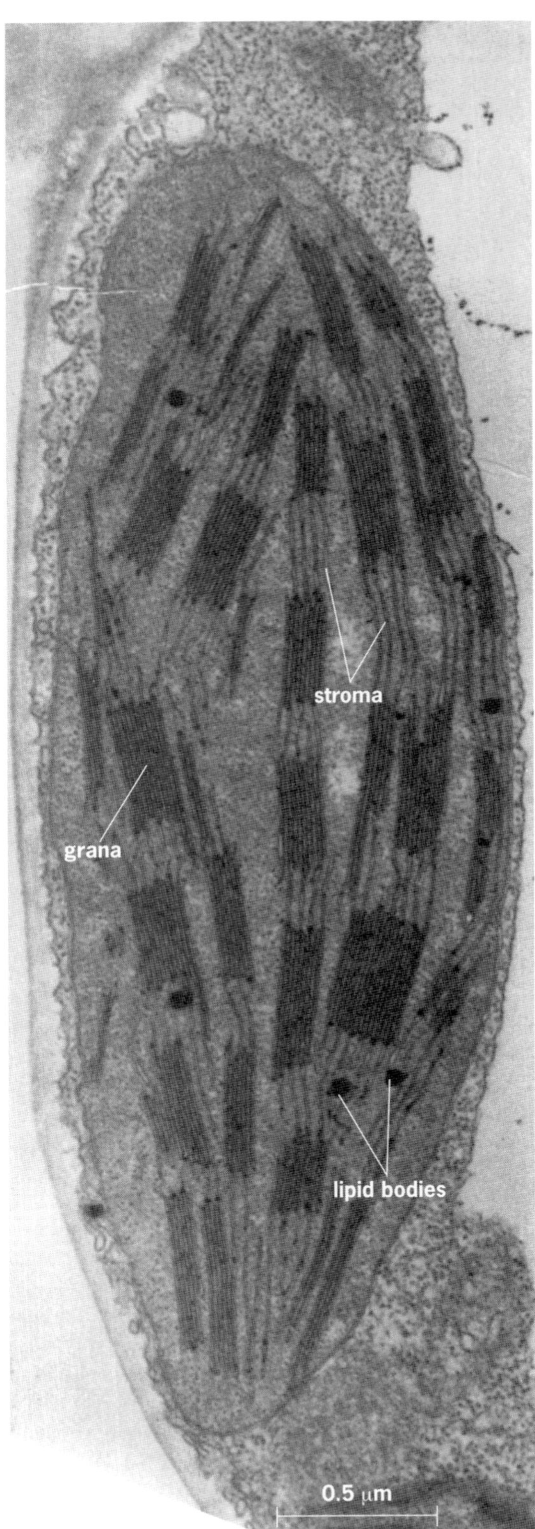

Fig. 8. Electron micrograph of chloroplast of a tomato fruit cell. (*Courtesy of S. W. Rosso*)

Changes in plastid form, distribution, and quite possibly metabolic efficiency have been conspicuous features of plant evolution. At any evolutionary level, however, the presence of plastids, and all that this implies metabolically, provides a sharp distinction between a plant and an animal cell. Plastids impart to plants their unique status in the biotic world. *See* CELL PLASTIDS; PLANT KINGDOM.

Vacuoles. Vacuoles of different types are characteristic of almost all sorts of cells, at least at certain stages of development. Many types of plant cells have large, centrally located vacuoles that make up the greatest part of the total volume of the cell. In meristematic cells vacuoles are generally small and are characterized by contents that stain darkly with certain procedures. The contents of these vacuoles seem to be utilized in the process of development and then are replaced by water. At a certain stage in this process the vacuoles, which are bounded by a membrane (the tonoplast) that seems to differ in thickness from the other membranes of the cell, fuse to form the large central vacuole. The increase in volume of this vacuole is important in the growth of the plant cell.

Since plants have no organ system to eliminate waste products, it has long been supposed that materials no longer usable in the metabolism of the cell are channeled into the vacuoles and there sealed off from the cytoplasm. This may happen, but it is now apparent that there may be several types of vacuoles in plant cells, some of which may function in a more complex fashion. An example is the cells of the roots of the water hyacinth (*Eichhornia*), in which the membrane of the vacuole may become greatly extended, intrude into the vacuole, and delimit the sites at which crystal formation takes place (**Fig. 9**). Another suggestion concerning metabolism within the vacuole is referred to below in lysosomal activity.

The colors of many flowers and in some instances tissues of other plant organs or structures result from the presence of pigments dissolved in the vacuolar fluid. Anthocyanins in the vacuoles produce red-blue colors, depending upon the relative acidity of the fluid in the vacuole. The origin of vacuoles in the plant cell has been inadequately studied. One common suggestion is that vacuoles arise as dilatations of the endoplasmic reticulum. In animal cells vacuoles of different sorts are known to have different origins, with the endoplasmic reticulum, the plasmalemma, and perhaps the Golgi apparatus being involved. It seems likely that various vacuoles in plants also may have different origins. *See* VACUOLE.

Lipid droplets. Another conspicuous feature of many types of plant cells is the presence of large numbers of lipid droplets. These may be what some investigators have called spherosomes. They frequently are abundant in cells of embryos or in root or shoot apices, in which cell division and cell growth activities are building up. These lipids are apparently stored material which may support these activities, and they differ in chemical composition from the lipids of the membranes. Lipid droplets are less numerous in more mature plant cells. These droplets are unique in having a structural boundary that is not the typical unit membrane.

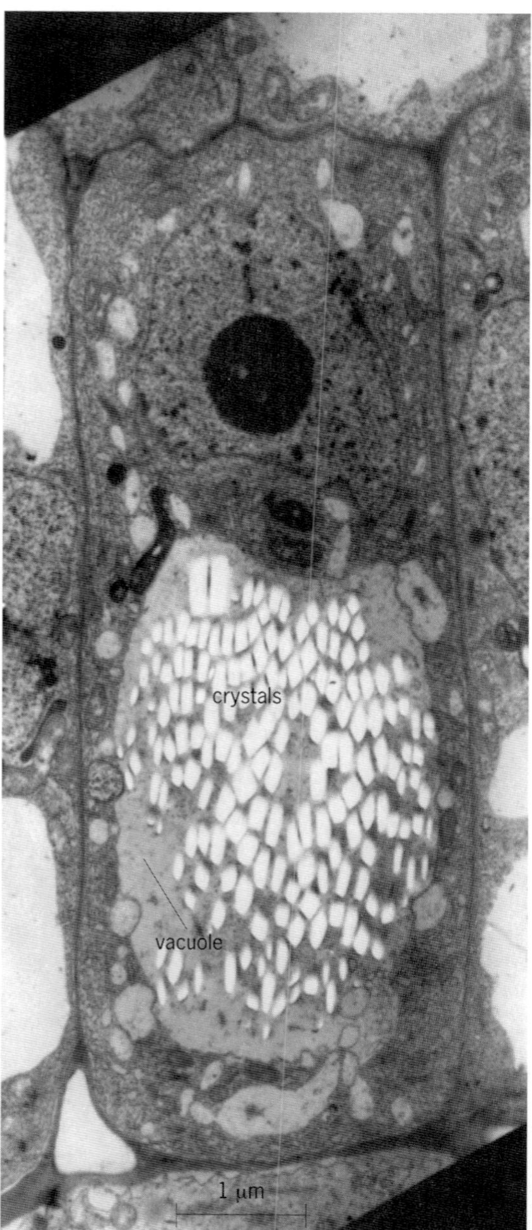

Fig. 9. ***Eichhornia*** **cell showing crystals developing within the vacuole. (*Courtesy of H. J. Arnott*)**

Other organelles and inclusions. Specialized types of plant cells are characterized by other inclusions, some membrane-bound and some crystalline (crystalline inclusions may also occur within the membrane-bound organelles, for example, in the mitochondria). Some types of plant cells have what appear cytologically to be microbodies, though these structures have not been demonstrated to contain the same complement of enzymes as animal cell microbodies. "Multivesiculate" bodies, a catch-all category for bodies showing various types of internal membrane structure, are common in some plant cells, particularly degenerating ones; "dense" bodies, quite possibly representing accumulations of substances in dilatation of the endoplasmic reticulum, are sometimes seen also. Various types of storage bodies, some containing large amounts of protein, for example, are characteristic of the cells of grains and fruits. In addition, plant cells infected with viruses may show numerous bizarre structures that have not yet been interpreted. Normal plant cells may contain other structures at specific stages, such as the centrioles which are characteristic of division stages in some lower plants. Still other structures, such as the pyrenoid and the eyespot of cells of certain algae and the Woronin bodies of certain fungi, distinguish the cells of particular groups of plants.

Lysosomal activity. One apparent difference between higher plant cells and many animal cells is the absence of readily and distinctly definable lysosomes. Many types of animal cells are characterized by the presence of lysosomes: membrane-bound entities which contain a number of hydrolytic enzymes, among which acid phosphatase is generally used as a marker. It seems likely, however, that lysosomal activity occurs in plant cells just as it does in animal cells, because acid phosphatase, and quite possibly the other lysosomal enzymes as well, can sometimes be demonstrated in plant vacuoles, in certain storage bodies, and in other membranous structures, notably the endoplasmic reticulum, in the vicinity of the Golgi apparatus. In plant cells in which digestive activity is a predominant process, lysosomal enzymes are abundant, for example, in cells containing aleurone grains from which protein is released to support early growth of the germinating seedling. These and other storage bodies appear to have high lysosomal activity, though their functions are quite different from those of the digestive vacuoles with lysosomal activity in animal cells. Some lower plant cells, for example, *Euglena*, contain bodies with lysosomal activity that are more nearly comparable to those of the animal cell.

The apparent distinction between plant and animal cells with respect to definable lysosomes may rest in part upon the absence from plants of cells that function in such processes as detoxification and therefore require continuing destruction and renewal of their components, as is the case in liver and white blood cells. *See* LYSOSOME.

Plasmalemma. The cytoplasm is bounded externally by a membrane called the plasma membrane or plasmalemma. Whereas the membranes of the compartments in the cytoplasm separate certain activities from the matrix, the plasmalemma separates the activities of the protoplast from the surrounding environment, whether this is fluid or adjacent cells. The plasmalemma is a typical lipoprotein membrane and is known to be selective with respect to the passage of ions and small molecules into or out of the cytoplasm. While the nature of the exchanges through the plasmalemma is inadequately known, both specific ion exchanges and forms of active transport, energy-requiring exchanges, are involved.

However, neither ion exchange nor active transport explains the passage of all materials through plant cell membranes. These membranes, even those of fairly highly specialized cells in differentiated tissues, often allow the passage of large molecules; such movement into the cytoplasm constitutes endocytosis; movement outward, exocytosis. Exocytosis, such as is involved in some secretion, has been much studied in plant cells; endocytosis has been recognized in only a few instances. The plasma membrane is a very reactive structure. Studies (largely confined to animal cells) have indicated that, like other cellular membranes, there is a rapid turnover of phospholipid and protein molecules in the plasmalemma. Thus the membranous barrier between the cell contents and the environment, once thought to be a more or less passive structure, is now interpreted as a very dynamic one. *See* CELL MEMBRANES; CELL PERMEABILITY.

Motion within cells. Nearly all cells are characterized by several types of motion, ranging from a fundamental motion, so-called Brownian movement, to movements of cellular components along particular pathways to general cyclosis or protoplasmic streaming. All movement in the cell is arrested if the cell is fixed for study under a microscope. Thus the static picture of any cell fails to convey one of its most vital aspects. *See* CELL MOTILITY.

Cell wall. A distinctive feature of the plant cell is the wall surrounding the cell outside the plasma membrane (**Fig. 10**). In mature cells this wall may be variously thickened and sculptured. It is composed largely of cellulose, arranged in microfibrils which have distinctive patterns of orientation and hemicelluloses and pectic substances. The wall is a limiting factor with respect to cell growth beyond a certain stage in development, and it precludes cellular mobility.

Fig. 10. Portion of a plant cell wall showing the fibrillar nature of cellulose. Microfibrils are oriented in a nearly circular arrangement surrounding intercellular connections (a pit field). (*Courtesy of J. A. S. Marshall*)

The cell wall is formed in two stages, and because of this and differences in composition the botanist generally distinguishes between a primary and a secondary wall. The primary wall is laid down at the time of cell division and extended during cell growth. It is composed mostly of pectic substances and hemicelluloses but also contains cellulose and a small amount of protein. The secondary wall is laid down after the cell has ceased to grow and may vary greatly in structure and composition. It has more cellulose than the primary wall and sometimes contains substantial amounts of lignin and other substances. In certain types of cells the secondary wall is uniformly thickened; in others it is thickened in bands or in spiral patterns; and in still others the thickenings are only at the corners. A significant portion of the primary wall is secreted by the Golgi apparatus. The largely cellulosic secondary wall appears, however, to be built from materials which move out of the cytoplasm by some other mechanism.

The walls of plant cells with their loosely arranged microfibril structure constitute intercellular pathways through which materials important in the economy of the cell can be moved from one cell to another or even from the external environment of the plant to the plasmalemma of nonsurface cells. *See* CELL WALLS (PLANT).

Variations. Any generalized consideration of the plant necessarily must omit detailed attention to variations characteristic of particular cell types. These variations are myriad and it is possible to touch on only a few. At maturity some cells, such as those of xylem vessels, lose all their cytoplasmic and nuclear contents. Others, such as the sieve-tube elements of the phloem, at maturity contain cytoplasm but no nucleus. The extent of the endoplasmic reticulum, the number of ribosomes, mitochondria, and plastids, and the Golgi apparatus vary widely with the type and functional state of the cell; as does the degree of vacuolation and the relative thickness and character of the walls. All these variations reflect differences in the functioning and therefore the metabolic activities of the cells. *See* PHLOEM; XYLEM.

Cell growth. The growth of the plant cell invariably involves four key features: the addition of more cytoplasmic material produced by metabolic reactions involving both the absorbed nutrients and the products of photosynthesis; intake of substantial amounts of water, usually greatly expanding the volume of the central vacuole; extension of the plasmalemma; and extension of the primary wall.

The details of accretion during growth are relatively obscure in all organisms, though the specific steps involved in many syntheses are well known. Presumably, most of the additional water enters by osmosis. Other mechanisms may be involved in the movement of water from the cytoplasm into the vacuoles. The plasmalemma apparently increases in extent by intussusception, in which more molecular units from the cytoplasmic pool are inserted between those already in the structure. How the primary wall is extended, though much studied, is still not definitely known. The interrelationship between the extension of the plasmalemma and the growth of the primary cell wall has yet to be elucidated.

The growing plant cell is distinguished by remarkable correlation between the increase in volume of the protoplast, which is significantly related to vacuolation, and the extension of the bounding wall. To what extent wall-growth processes are directly dependent upon influences from the cytoplasm and the nucleus is unknown. *See* PLANT GROWTH.

Differentiation. The range of differentiation of cells in a highly evolved plant is not as great as the comparable range in a given higher animal, and the complexity of the organization of cell types into tissues and organs is not as great. In part, the differences relate to the higher degree of localization of a great range of metabolic processes, including those of the stimulus-response, circulatory, and extracellular digestive systems of higher animals. There is nonetheless a vast number of different sorts of plants and an enormous degree of diversification in cell structure and metabolism and in the various end products of metabolism. *See* PLANT ANATOMY; PLANT MORPHOGENESIS.

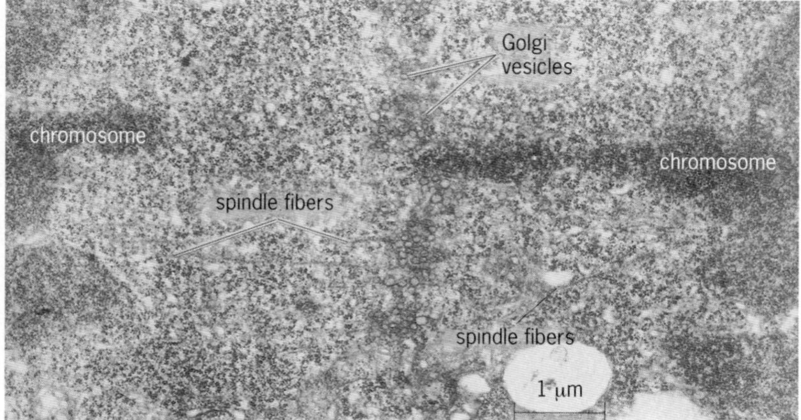

Fig. 11. Dividing cell showing aggregation of small Golgi vesicles in equatorial region and spindle fibers. Chromosomes have moved toward the poles. In right part of cell a chromosome still extends as far as forming plate. (*Courtesy of H. J. Arnott*)

Dividing Cell

Despite innumerable early theories to the contrary, it has long been apparent that plant and animal cells arise only by division of preexisting cells. This observation also pertains to some of the compartments of cells—the nucleus, the mitochondria, plastids, and probably the Golgi apparatus. The manner in which other components, such as the endoplasmic reticulum and vacuoles, behave in cell division is less clear.

Cell division begins with the constitution of the chromatin in the nucleus into the chromosomes. As the chromosomes are resolved, the nucleoli either disappear or are substantially reduced. There follows a breakdown of the nuclear envelope and the development of what is called a mitotic figure. *See* CELL DIVISION; CHROMOSOME.

The mitotic figure includes so-called spindle fibers,

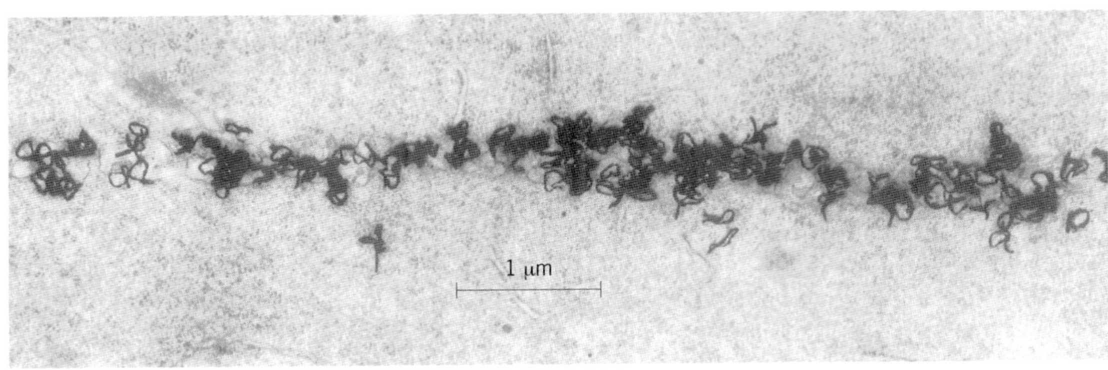

Fig. 12. Radioaudiograph of early stage in cell-plate formation in maize root cell treated with tritiated glucose. Silver grains at plate area result from glucose, transported via Golgi apparatus to the forming wall. (*Courtesy of T. P. Leffingwell*)

some of which are attached to the chromosomes and some of which pass between the chromosomes through the equator of the spindle (**Fig. 11**). The spindle fibers are of the same order of magnitude as, and seem to have a chemical composition comparable to, that of the microtubules. In mitosis the most conspicuous difference between plant and animal cells is the absence of centrioles in all but a few types of lower plant cells. In the animal cell two centrioles, one located at each end of the spindle, appear to represent "anchor points," because the spindle fibers seem to radiate from the centrioles. The spindle figure of plant cells appears to be firmly anchored without any demonstrable centrioles. *See* MITOSIS.

During mitosis the chromosomes are separated toward the poles of the spindle figure. At each pole a new nucleus is organized along with reconstitution of the nuclear envelope and redevelopment of the nucleoli in association with specific chromosome sites, known as nucleolar organizer regions. During mitosis the segmented nuclear envelope appears to surround the spindle figure. The manner in which the nuclear envelope is reconstituted is not known.

The partition of the cytoplasm into two masses, one surrounding each new nucleus, is a vastly different process in most plant cells from that in most animal cells. With few exceptions, this cytoplasmic division (cytokinesis) takes place in plants by a process known as cell-plate formation; a plate-shaped structure consisting of cell-wall material bounded by new plasma membrane appears in the mass of cytoplasm in the center of the cell and extends outward, ultimately connecting with the original walls. The initial stage in cell-plate formation is the movement of small vesicles produced by the Golgi apparatus into the spindle figure of the dividing cell and their clustering around spindle fibers. These small vesicles appear to form a group of islands so aligned as to determine the plane of cell division. Then larger vesicles produced by the Golgi apparatus move into this plane, and both types of vesicles fuse, with some as yet undefined involvement of the endoplasmic reticulum. The Golgi apparatus produces new plasmalemmas and initial wall components. The latter are composed largely

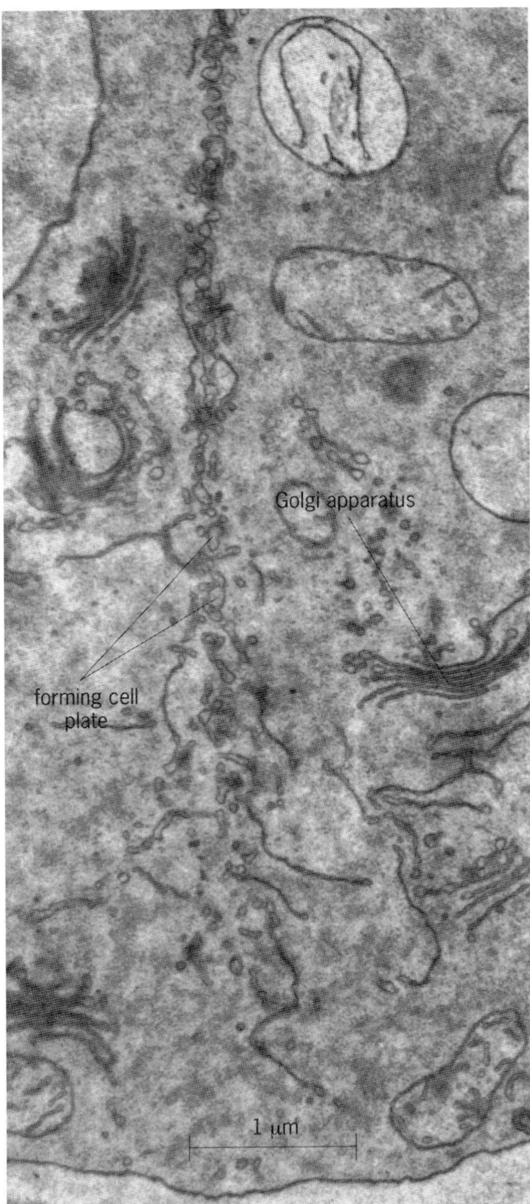

Fig. 13. A stage of cell-plate development showing the close association of the Golgi apparatus with the forming cell plate. (*Courtesy of H. H. Mollenhauer*)

of pectic substances and hemicelluloses. Radioautographic experiments have shown that glucose purposely incorporated into the cell during cell-plate formation collects in the Golgi apparatus, where it becomes part of one or more of these wall substances and from where it moves directly into the new wall (**Fig. 12**). Golgi apparatus are frequently seen clustered around the edges of the cell plate (**Fig. 13**). When the plate is fully extended to the existing walls of the cell, cytokinesis is complete, with the two new cells being separated by both plasma membrane and wall. *See* CYTOKINESIS.

In most cases the mitochondria and plastids, and possibly the Golgi apparatus, appear to increase in number more or less concurrently with the division of the cell. Each is apparently also capable of increasing without cell division, for cell differentiation following the period of cell growth is often characterized by increases in the number of one or more of these organelles.

Segments of endoplasmic reticulum and the various small organelles and inclusions seem to be randomly distributed between the daughter cells, but not enough is known about developmental activities of these components to make any conclusive statements about the pattern of their distribution in cell division. There is substantial evidence that certain classes of compounds, including the so-called plant hormones, exert control over cell division, but the mechanism by which this control is exerted has not been clarified. There is little evidence concerning the means by which certain cells are caused to divide while others do not, and the determination of the plant form. It is clear that cell reproduction involves not only the structural developments treated here but also significant metabolic changes.

W. Gordon Whaley

Plant Protoplasts

Although the plant cell is the basic unit of structure and function in nearly all plants, the term protoplast has been used to describe the organized entity of the living components of the plant cell which carry out active metabolism, biosynthesis, and energy transfer, in contrast to the extracellular, essentially metabolically inactive secondary plant matter. Plasmolysis studies on plant cells established that a functional semipermeable plasma membrane surrounds the protoplast.

For many years plant protoplasts remained largely cytological curiosities of little interest to physiologists, biochemists, and geneticists. However, the introduction of an enzymatic method for the isolation of protoplasts in 1960 stimulated renewed interest. Cell wall–degrading enzymes (cellulases) were employed, and large numbers of protoplasts could be isolated from various plant organs and also from plant tissue cultures. Pectinases were later additionally utilized to facilitate cell separation and wall degradation. A typical preparation of protoplasts isolated enzymatically from leaves of *Coleus* is shown in **Fig. 14**.

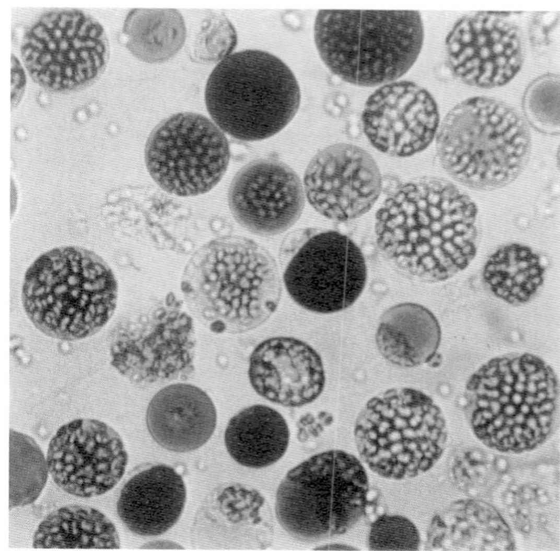

Fig. 14. Protoplasts isolated from *Coleus* leaves.

Experimental value. One of the most useful activities exhibited by protoplasts is that of cell wall regeneration. Studies on cell wall regeneration by protoplasts have provided a wealth of new information, particularly on the early stages of wall synthesis and the early organization of the wall. Protoplasts offer an important biological entity for many physiological and biochemical studies. For instance, chloroplasts isolated from protoplasts possess excellent structural and biochemical integrity, and fractionation of protoplasts has supplied large quantities of mature plant vacuoles. Plant protoplasts have also provided a system for plant virus research via laboratory culture.

Cell division. Once cell wall regeneration has become initiated, protoplasts cultured in suitable media under sterile conditions exhibit mitosis and cytokinesis similar to that normally found in cultured cells undergoing division. The behavior of protoplasts in this respect is extremely variable, ranging from failure even to divide following wall regeneration, to the ready establishment, at high frequency, of cell colonies and whole fertile plant regeneration. This is an area where much work is still required; ignorance of the control mechanisms for cell division in many species, particularly crop species, is delaying the use of protoplasts for crop improvement.

Genetic manipulation. The use of protoplasts for genetic manipulations centers on two major areas: the use of protoplasts in transformation studies, and obtaining new genetic variability by protoplast fusion.

Transformation. Protoplasts provide an ideal single-cell system for these basic studies in plant cell biology. Plasmids can be incubated with protoplast suspensions, and a microbiological approach can be used to follow plasmid interactions. If regeneration of whole plants from transformed protoplasts is possible, then a sexual analysis of somatic transformation will also be possible. It has been demonstrated that isolated *Agrobacterium* plasmids, incubated with *Petunia* protoplasts in the presence of poly-L-ornithine (a known stimulator of virus uptake by protoplasts), are transformed and become capable of growing and dividing in the absence of growth substances, and of synthesizing crown gall tumor-specific opines, such as octopine.

Fusion. Fusion of protoplasts appears to be initiated when the plasma membranes are able to come sufficiently close to one another to enable membrane flow to take place. Membrane fusion requires that the membranes first be brought into apposition at molecular distances of 1 nm or less. Most protoplasts are negatively charged, and the addition of the fusion-inducing agent (or fusogen) probably eliminates this charge. They no longer repel each other, and apposition at molecular distances, with consequent membrane flow, can take place with resultant fusion of the plasma membranes. Calcium ions at high pH (9.5–10.5) and polyethylene glycol are the most commonly used fusogens for protoplasts. Within the *Petunia* genus, sexually incompatible species can be crossed by this process of protoplast fusion and, coupled with the selective culture of somatic hybrid cells, into whole flowering hybrid plants. Horticulturally, this new somatic amphidiploid hybrid between *P. parodii* and *P. parviflora* is of interest, because it is suitable for the introduction of the hanging-basket habit of the sexually isolated *P. parviflora* species into the cultivated *Petunia*. *See* BREEDING (PLANT); CELL (BIOLOGY); SOMATIC CELL GENETICS.

E. C. Cocking

Bibliography. A. Fahn, *Plant Anatomy*, 4th ed., 1990; J. L. Hall, *Plant Cell Structure and Metabolism*, 2d ed., 1986; N. Harris and K. J. Oparka (eds.), *Plant Cell Biology: A Practical Approach*, 1994; B. King (ed.), *Cell Biology*, 1986; F. B. Salisbury and C. W. Ross, *Plant Physiology*, 4th ed., 1992.

Plant communication

Movement of signals or cues, presumably chemical, among individual plants or plant parts. These chemical cues are a consequence of damage to plant tissues and stimulate physiological changes in the undamaged "receiving" plant or tissue. Communication among plants involves induced plant defenses, which are chemicals produced actively by an individual plant in response to attack by insects or disease agents.

Plants produce a wealth of secondary metabolites that do not function in the main, or primary, metabolism of the plant, which includes photosynthesis, nutrient acquisition, and growth. Since many of these chemicals have very specific negative effects on animals or pathogens, ecologists speculate that they may be produced by plants as defenses. Plant chemical defenses either may be present all of the time (constitutive) or may be stimulated in response to attack (induced). Those produced in response to attack by pathogens are called phytoalexins. In order to demonstrate the presence of an induced defense, the chemistry of plant tissues or their suitability to some "enemy" (via a bioassay) must be compared before and after real or simulated attack. Changes found in the chemistry and suitability of control or unattacked plants when nearby plants are damaged imply that some signal or cue has passed from damaged to undamaged plants. Responses in undamaged plants are related to the proximity of a damaged neighbor. *See* PHYTOALEXINS; PLANT METABOLISM.

Nature of the response. Responses by undamaged plants or plant tissues to other plants include altered metabolism, reduced quality for herbivores, and emission of damage-related volatiles that attract predators or parasites of the attacking herbivores. The plant phenolics comprise a large, diverse group of secondary metabolites. Monomers may be toxic to animals or microbes, and polymers (tannins) are thought to bind to enzymes and other proteins, inhibiting growth and other important metabolic functions. *See* PHENOL.

Nature of the cues. The most frequently proposed cue is ethylene, a volatile isoprenoid molecule produced by almost all plants. Ethylene influences a wide range of physiological activities in plants, including senescence, leaf abscission, fruit ripening, reproduction, and phenolic synthesis. It is produced in large quantities by various plant species in response to wounding, freezing, and pathogen infection. Resistance of plants to viruses and other pathogens can be elevated by exposing plant tissues to ethylene, and blocking ethylene production in virus-infected plants can deprive them of acquired immunity. Ethylene regulates the expression of genes for enzymes in the biosynthetic pathways for secondary metabolites. However, there are no direct studies of ethylene as the cue in communication among whole plants. Many other small, physiologically active isoprenoid and terpenoid compounds are produced by plants; studies of damaged tissue have identified dozens of possibilities, but no cues have been confirmed. *See* ETHYLENE.

Methyl jasmonate is the only chemical that has been identified clearly as an external signal effecting communication among plants. This enzymatic product of lipid metabolism is produced via activation of lipoxygenase enzymes in most plant species when tissues are damaged or infected by pathogens. Hence, this volatile compound, a component of many perfumes, may be a widespread, even interspecific, communication device among plants.

Ecological and evolutionary significance. It has not been demonstrated conclusively that plant communication occurs frequently in the field or significantly protects plants from enemies. Presumably, defensive responses cued in advance by nearby plants could increase an individual plant's resistance to subsequent attack. If effective, such a response would reduce the spread or intensity of a widening infestation in a plant community.

Because wind currents dilute and redirect emitted volatiles, only a very large infestation, such as one that affects many plants or entire trees, seems likely to produce signals that are uniform or strong enough to cue whole undamaged plants. It is more likely that cues may contact some tissues and not others, which may explain the broad chemical heterogeneity among plant parts and individuals. The nonuniformity of cue reception also suggests that communication-induced resistance is unlikely to be observed readily in nature.

The evolutionary origins and significance of the communication phenomenon are difficult to decipher. Plants can control the composition of volatiles emitted when tissues are damaged; however, damage-specific emissions are most often directed at attracting protective predators or parasites. Volatiles such as ethylene or methyl jasmonate may be no more than necessary metabolic by-products of responses to attack and may not have evolved originally as signals among plants. There would be little or no advantage accruing to individual plants sending cues to unrelated neighbors unless those neighbors are genetic relatives. Hence, communication would be most useful to and frequent among clonal species or species with small genetic neighborhoods in which an individual is surrounded by relatives, including offspring. Yet, both ethylene and methyl jasmonate appear to influence the metabolism of many plants, regardless of sender or receiving species.

There is obvious advantage accruing to plants that are able to respond to nearby damage in a way that increases their resistance to enemies. This is likely if the nearby damage is caused by some threat to the receiver, which would be true if neighbors are of the same species or share a common enemy. As a result, communication would tend to be more common in single-species stands.

Response to external cues may be more important within individual plants than between them. Although some induced (single individual) responses to attack are known to be cued by internal factors, it is possible that self-cuing could be external as well. Both ethylene and jasmonate are important signals coordinating within-plant responses to damage or infection. Communication among plant individuals thus may be an incidental consequence of contact with a self-cuing mechanism. *See* ALLELOPATHY; PLANT PATHOLOGY; PLANT PHYSIOLOGY.

Jack C. Schultz

Bibliography. R. M. Amasino, *Cellular Communication in Plants*, 1993; D. T. Dennis and D. H. Turpin, *Plant Physiology, Biochemistry and Molecular Biology*, 1991; D. P. Verma, *Molecular Signals in Plant-Microbe Communication*, 1991.

Plant evolution

The process of biological and organic change within the plant kingdom by which the characteristics of plants differ from generation to generation. Understanding of the course of plant evolution is based on several lines of evidence. The main levels (grades) of evolution have long been clear from comparisons among living plants, but the fossil record has been critical in dating evolutionary events and revealing extinct intermediates between modern groups, which are separated from each other by great morphological gaps (evolutionary changes in many characters). Plant evolution has been clarified by cladistic methods for estimating relationships among both living and fossil groups. These methods attempt to reconstruct the branching of evolutionary lines (phylogeny) by using shared evolutionary innovations (for example, presence of a structure not found in other groups) as evidence that particular organisms are descendants of the same ancestral lineage (a monophyletic group, or clade).

Many traditional groups are actually grades rather than clades; these are indicated below by names in quotes. Convergence, where the same new feature evolves in different lines, and reversal, where an older condition reappears, confuse the picture by producing conflicts among characters. These are resolved by using as many characters as possible and applying the principle of parsimony, that is, seeking the tree that requires the fewest total character changes, aided by a computer. Progress has also come from studies of molecular data [deoxyribonucleic acid (DNA) sequences of the same gene in different groups]; bases at any point along the DNA can change like any other kind of character and can be analyzed in the same way.

Most botanists restrict the term plants to land plants, which invaded the land after 90% of Earth history (see **illus.**). There is abundant evidence of photosynthetic life extending back 3.5 billion years to the early Precambrian, in the form of microfossils resembling cyanobacteria (prokaryotic blue-green algae) and limestone reefs (stromatolites) made by these organisms. Larger cells representing eukaryotic "algae" appear in the late Precambrian, followed by macroscopic "algae" and animals just before the Cambrian. *See* ALGAE; EUKARYOTAE; FOSSIL; PROKARYOTAE.

Origin of land plants. Cellular, biochemical, and molecular data place the land plants among the "green algae," specifically the "charophytes," which resemble land plants in their mode of cell division and differentiated male and female gametes (oogamy). Land plants themselves are united by a series of innovations not seen in "charophytes," many of them key adaptations required for life on land. "Charophytes" have a haploid life cycle, in which the zygote undergoes meiosis, and single-celled reproductive structures; but land plants have an alternation of generations, with a haploid, gamete-forming (gametophyte) and diploid, spore-forming (sporophyte) phase. Their reproductive organs (egg-producing archegonia, sperm-producing antheridia, and spore-producing sporangia) have a protective layer of sterile cells. The sporophyte, which develops from the zygote, begins its life inside the archegonium. The spores, produced in fours by meiosis, are air-dispersed, with a resistant outer wall that prevents desiccation. *See* CHAROPHYCEAE; REPRODUCTION (PLANT).

Land plants have been traditionally divided into "bryophytes" and vascular plants (tracheophytes). These differ in the relative role of the sporophyte,

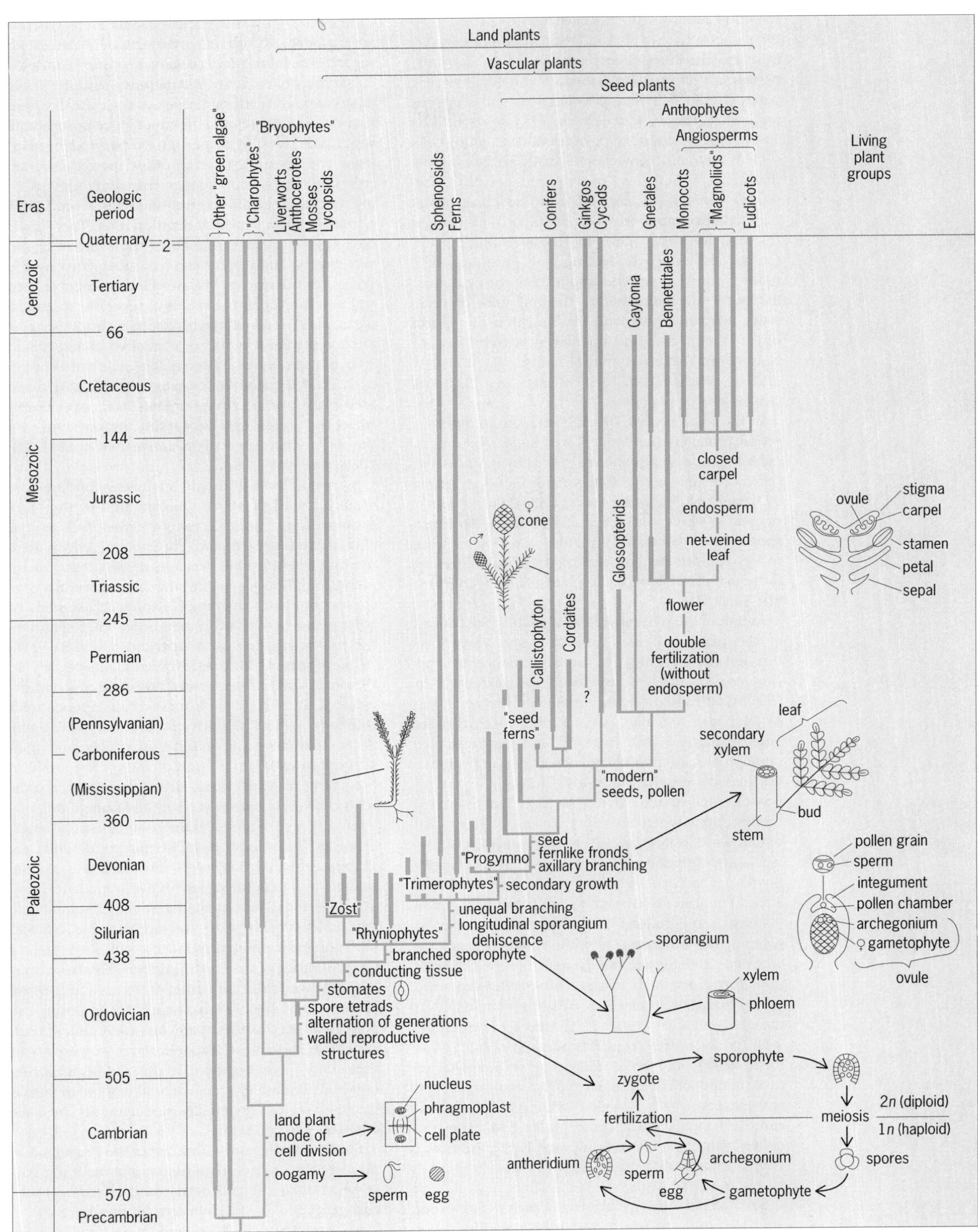

Inferred evolutionary relationships of major groups of plants, including some of their most important evolutionary innovations, and their distribution in geologic time (thick lines: known geologic range). Groups in quotes represent grades (levels of evolution).

which is subordinate and permanently attached to the gametophyte in "bryophytes" but dominant and independent in vascular plants. In vascular plants, tissues are differentiated into an epidermis with a waxy cuticle that retards water loss and stomates for gas exchange, parenchyma for photosynthesis and storage, and water- and nutrient-conducting cells (xylem, phloem). However, cladistic analyses imply that some "bryophytes," namely anthocerotes (hornworts) and mosses, are closer to vascular plants than others (liverworts, or hepatics). Both groups resemble vascular plants in having stomates, and mosses also have conducting cells, although of a more primitive type. This implies that the land-plant life cycle originated before the full suite of vegetative adaptations to land life, and that the sporophyte began small and underwent a trend toward elaboration and tissue specialization. *See* EPIDERMIS (PLANT); PHOTOSYNTHESIS; PRIMARY VASCULAR SYSTEM (PLANT).

In the fossil record, the first recognizable macroscopic remains of land plants are Middle Silurian vascular forms with a branched sporophyte, known as "rhyniophytes." These differed from modern plants in having no leaves or roots, only dichotomously branching stems with terminal sporangia. However, spore tetrads formed by meiosis are known from older beds (Middle Ordovician); these may represent more primitive, bryophytic plants. *See* BRYOPHYTA; RHYNIOPHYTA.

In one of the most spectacular adaptive radiations in the history of life, vascular plants diversified through the Devonian. At the beginning of this period, vegetation was low [perhaps at most 1 ft (30 cm) tall] and probably confined to wet areas, but by the Late Devonian, size had increased in many lines, resulting in large trees and forests with shaded understory habitats. Of the living groups of primitive vascular plants, the lycopsids (club mosses), with pointed, one-veined leaves (microphylls), branched off first, along with the extinct "zosterophyllopsids," which were leafless but resembled lycopsids in having lateral sporangia. A second line, the "trimerophytes," which showed incipient differentiation of a main trunk and side branches, and terminal sporangia with longitudinal dehiscence, gave rise to sphenopsids (horsetails), with simple leaves in a whorled arrangement, and ferns (filicopsids), with pinnately compound leaves (fronds) derived from whole branch systems. This radiation culminated in the coal swamp forests of the Late Carboniferous, with tree lycopsids (Lepidodendrales), sphenopsids (*Calamites*), and ferns (Marattiales). Remains of these plants make up much of the coal of Europe and eastern North America, which were then located on the Equator; these plants had many peculiar anatomical features apparently requiring wet and constant conditions. *See* LYCOPODIALES; MARATTIALES; SPHENOPHYTA.

Seed plants. Perhaps the most significant event after the origin of land plants was evolution of the seed. Primitive seed plants ("gymnosperms") differ from earlier groups in their reproduction, which is heterosporous (producing two sizes of spores), with separate male and female gametophytes packaged inside the pollen grain (microspore), and the ovule (a sporangium with one functional megaspore, surrounded by an integument, which develops into the seed). The transfer of sperm (two per pollen grain) from one sporophyte to another through the air, rather than by swimming between gametophytes living on or in the soil, represents a step toward independence from water for reproduction. This step is comparable to the evolution of the amniote egg in vertebrates, and it must have helped plants invade drier areas than they had previously occupied. In addition, seed plants have new vegetative features, particularly secondary growth, which allows production of a thick trunk made up of secondary xylem (wood) surrounded by secondary phloem and periderm (bark). Together, these innovations have made seed plants the dominant organisms in most terrestrial ecosystems ever since the disappearance of the Carboniferous coal swamps. *See* ECOSYSTEM; PTERIDOSPERMS; SEED.

A major breakthrough in understanding the origin of seed plants was recognition of the "progymnosperms" (for example, *Archaeopteris*) in the Middle and Late Devonian. These plants, which were the first forest-forming trees, had secondary xylem, phloem, and periderm, but they still reproduced by spores, implying that the anatomical advances of seed plants arose before the seed. Like sphenopsids and ferns, they were apparently derived from "trimerophytes." The earliest seed plants of the Late Devonian and Carboniferous, called "seed ferns" because of their frondlike leaves (a convergence with true ferns), show steps in origin of the seed, by fusion of branchlets surrounding the megasporangium into the integument. Origin of the typical mode of branching in seed plants, from buds in the axils of the leaves, occurred at about the same time.

Among seed plants, coniferopsids (fossil cordaites, living conifers, and possibly ginkgos), with fan-shaped to needlelike leaves, have often been considered an independent line of evolution from "progymnosperms." However, cladistic analyses indicate that coniferopsids were derived from "seed ferns" with coniferlike pollen and seeds, similar to the Late Carboniferous genus *Callistophyton*. This event may have involved the elimination of fronds and their replacement by scale leaves, which occur around the buds in all seed plants. Cordaites and conifers both appear in the Late Carboniferous; comparisons of their early members indicate that the seed-bearing cone of conifers is a complex structure, in which the cone scales are actually modified branches. *See* CORDAITALES; PINALES.

Seed plants became dominant in the Permian during a shift to drier climate and extinction of the coal swamp flora in the European-American tropical belt, and glaciation in the Southern Hemisphere Gondwana continents. Early conifers, with small, needle-like leaves, predominated in the tropics; extinct glossopterids, with simple, deciduous, net-veined leaves, inhabited Gondwana. Moderation of climate in the Triassic coincided with the appearance of new seed plant groups, including the living cycads and ginkgos (today reduced to one species, *Ginkgo biloba*) and the extinct Bennettitales (cycadeoids), with cycadlike pinnate leaves but flowerlike reproductive structures, as well as more modern ferns. Many Mesozoic groups show adaptations for protection of seeds against animal predation, while flowers of the Bennettitales constitute the first evidence for attraction of insects for cross-pollination, rather than transport of pollen by wind.

Angiosperms. The last major event in plant evolution was the origin of angiosperms (flowering plants), the seed plant group that dominates the modern flora. Angiosperms show numerous innovations that allow more rapid and efficient reproduction. The flower, typically made up of protective sepals, attractive petals, pollen-producing stamens, and ovule-producing carpels (all considered modified leaves), favors more efficient pollen transfer by insects. The ovules are enclosed in the carpel, so that pollen germinates on the sticky stigma of the carpel rather than in the pollen chamber of the ovule. The carpels (separate or fused) develop into fruits, which often show special adaptations for seed dispersal. Other advances include an extreme reduction of the gametophytes, and double fertilization whereby one sperm fuses with the egg, and the second sperm with two other gametophyte nuclei to produce a triploid, nourishing tissue called the endosperm. Angiosperms also developed improved vegetative features, such as more efficient water-conducting vessels in the wood and leaves with several orders of reticulate venation. These features may have contributed to their present dominance in tropical forests, previously occupied by conifers with scale leaves. *See* RAINFOREST.

The origin of angiosperms has been considered a great mystery of plant evolution, since angiosperms were thought to appear in diverse, modern forms during the Early Cretaceous, with no obvious links to other groups. However, most botanists believed that the most primitive living angiosperms are "magnoliid dicots," based on their "gymnosperm"-like pollen (with one aperture for germination), wood anatomy, and flower structure. Studies of Cretaceous fossil pollen, leaves, and flowers confirm this view by showing a rapid but orderly radiation beginning with "magnoliid"-like and monocotlike types, followed by primitive eudicots (with three pollen apertures), some related to sycamores and lotuses. *See* MAGNOLIOPHYTA.

There is still a gap between angiosperms and other groups, but both morphological and molecular data imply that angiosperms are monophyletic and most closely related to Bennettitales and Gnetales, a seed plant group that also radiated in the Early Cretaceous but later declined to three living genera. Gnetales also experience a type of double fertilization, but without formation of endosperm, an exclusively angiospermous innovation. Since all three groups have flowerlike structures, suggesting that the flower and insect pollination arose before the closed carpel, they have been called anthophytes. The closest relatives of anthophytes are controversial, whether advanced "seed ferns" (*Caytonia,* glossopterids) or coniferopsids. These relationships, plus problematical Triassic pollen grains and macrofossils with a mixture of angiospermlike and more primitive features, suggest that the angiosperm line goes back to the Triassic, although perhaps not as fully developed angiosperms. Within angiosperms, it is believed that "magnoliids" are relatively primitive, monocots and eudicots are derived clades, and wind-pollinated temperate trees such as oaks, birches, and walnuts (amentiferae) are advanced eudicots. However, "magnoliids" include both woody plants (for example, magnolias and laurels) and herbs (for example, waterlilies and peppers), and their flowers range from large, complex, and insect-pollinated to minute, simple, and wind-pollinated. These extremes are present among the earliest Cretaceous angiosperms, and cladistic analyses disagree on which is most primitive.

Although plant extinctions at the end of the Cretaceous have been linked with radiation of deciduous trees and proliferation of fruits dispersed by mammals and birds, they were less dramatic than extinctions in the animal kingdom. During the Early Tertiary, climates were still much milder than today, but mid-Tertiary cooling led to contraction of the tropical belt and expansion of seasonal temperate and arid zones. These changes led to the diversification of herbaceous angiosperms, such as composites and grasses, and the origin of open grassland vegetation, which stimulated the radiation of hoofed mammals, and ultimately the invention of human agriculture. *See* AGRICULTURE; FLOWER; PALEOBOTANY; PLANT KINGDOM. James A. Doyle

Bibliography. C. B. Beck, *Origin and Evolution of Gymnosperms*, 1988; E. M. Friis, W. G. Chaloner, and P. R. Crane, *The Origins of Angiosperms and Their Biological Consequences*, 1987; L. E. Graham, *Origin of Land Plants*, 1993; P. M. Richardson, Origin and relationships of the major plant groups, *Ann. Missouri Bot. Grad.*, 81:403–567, 1994; W. N. Stewart and G. W. Rothwell, *Paleobotany and the Evolution of Plants*, 1993.

Plant geography

The study of the spatial distributions of plants and vegetation and of the environmental relationships which may influence these distributions. Plant geography (or certain aspects of it) is also known as phytogeography, phytochorology, geobotany, geographical botany, or vegetation science.

History. Plant geography arose as a result of the great botanical voyages of the early 1800s, especially those by Alexander von Humboldt, sometimes called the father of plant geography. Early work involved classification of physiognomically and ecologically similar plant types, and so plant geography and plant ecology remained somewhat indistinguishable. Around 1900 the study of vegetation began to split into plant ecology (which focused more on process) and plant geography (including vegetation geography). Two main perspectives had arisen: a historical perspective concerned with migration, dispersal, and the historical development of floras; and an environmental perspective concerned with environmental constraints and ecological relations influencing plant and vegetation distributions.

Floristic plant geography. A flora is the collection of all plant species in an area, or in a period of time, independent of their relative abundances and relationships to one another. The species can be grouped and regrouped into various kinds of floral elements based on some common feature. For example, a genetic element is a group of species with a common evolutionary origin; a migration element has a common route of entry into the territory; a historical element is distinct in terms of some past event; and an ecological element is related to an environmental preference. Aliens, escapes, and very widespread species are given special treatment. An endemic species is restricted to a particular area, which is usually small and of some special interest. The collection of all interacting individuals of a given species, in an area, is called a population.

The idea of area is fundamental to plant geography. An area is the entire region of distribution or occurrence of any species, element, or even an entire flora. The description of areas is the subject of areography, while chorology studies their development. The local distribution within the area as a whole, as that of a swamp shrub, is the topography of that area. Areas are of interest in regard to their general size and shape, the nature of their margin, whether they are continuous or disjunct, and their relationships to other areas. Closely related plants that are mutually exclusive are said to be vicarious (areas containing such plants are also called vicarious). A relict area

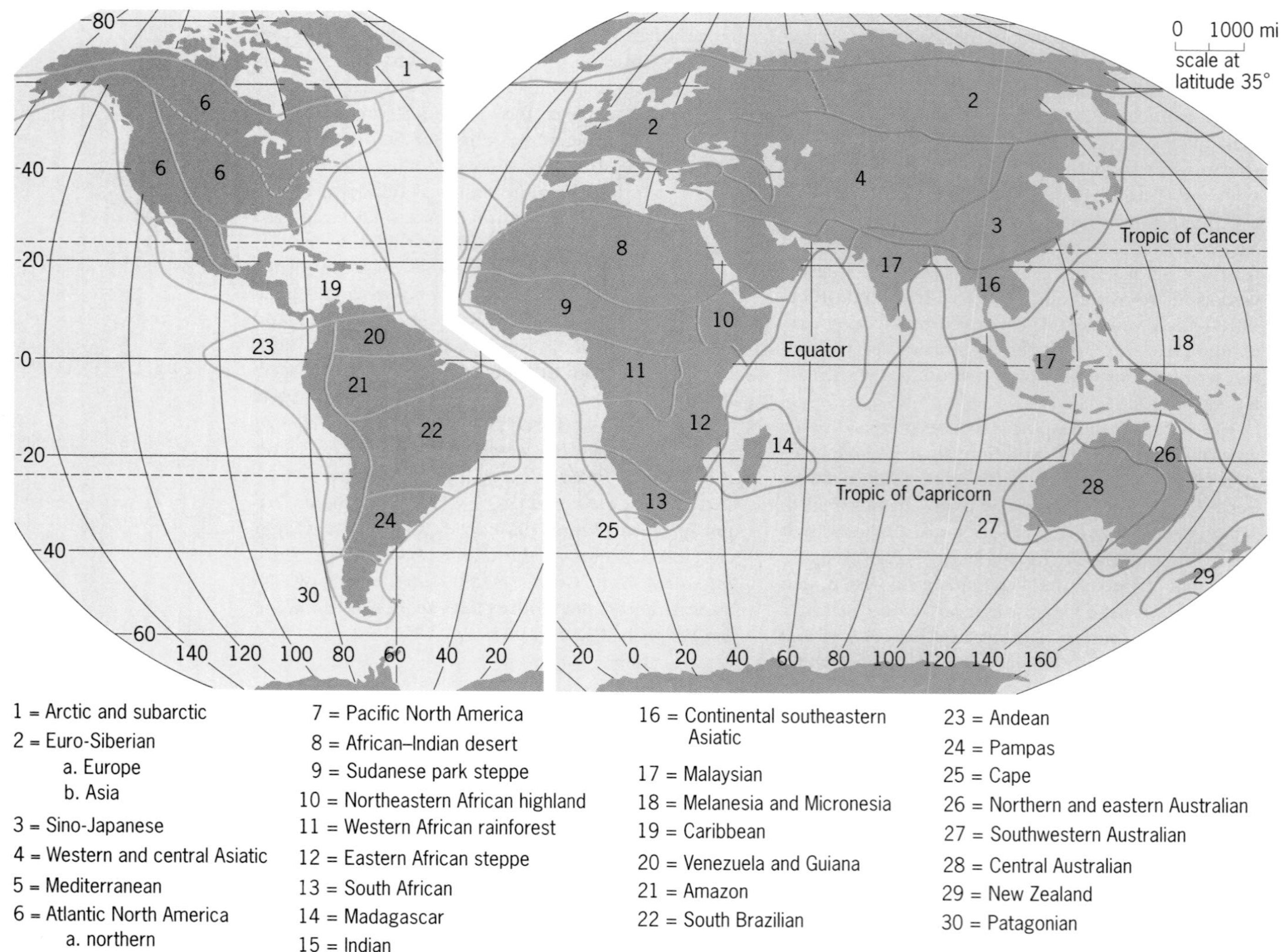

Fig. 1. Floristic regions of world. 1 mi = 1. 6 km. (*After R. Good, Geography of Flowering Plants, 3d ed., Longmans, 1964*)

Wet forest (not a true rain-forest), Olympic National Park, Wash. (*R. Scheich, National Audubon Society*)

(*Below*) *Scutellinia scutellata*, a wooly disk fungus and typical life form of deciduous forests. (*D. Renfro, National Audubon Society*)

(*Upper right*) Jeffrey pine, a symmetrical pine of western North America. (*O. Andrews, National Audubon Society*)

(*Middle right*) Early fall in a Wisconsin forest. (*C. Ott, National Audubon Society*)

(*Lower right*) Organpipe and saguaro cactuses. (*H. L. Parent, National Audubon Society*)

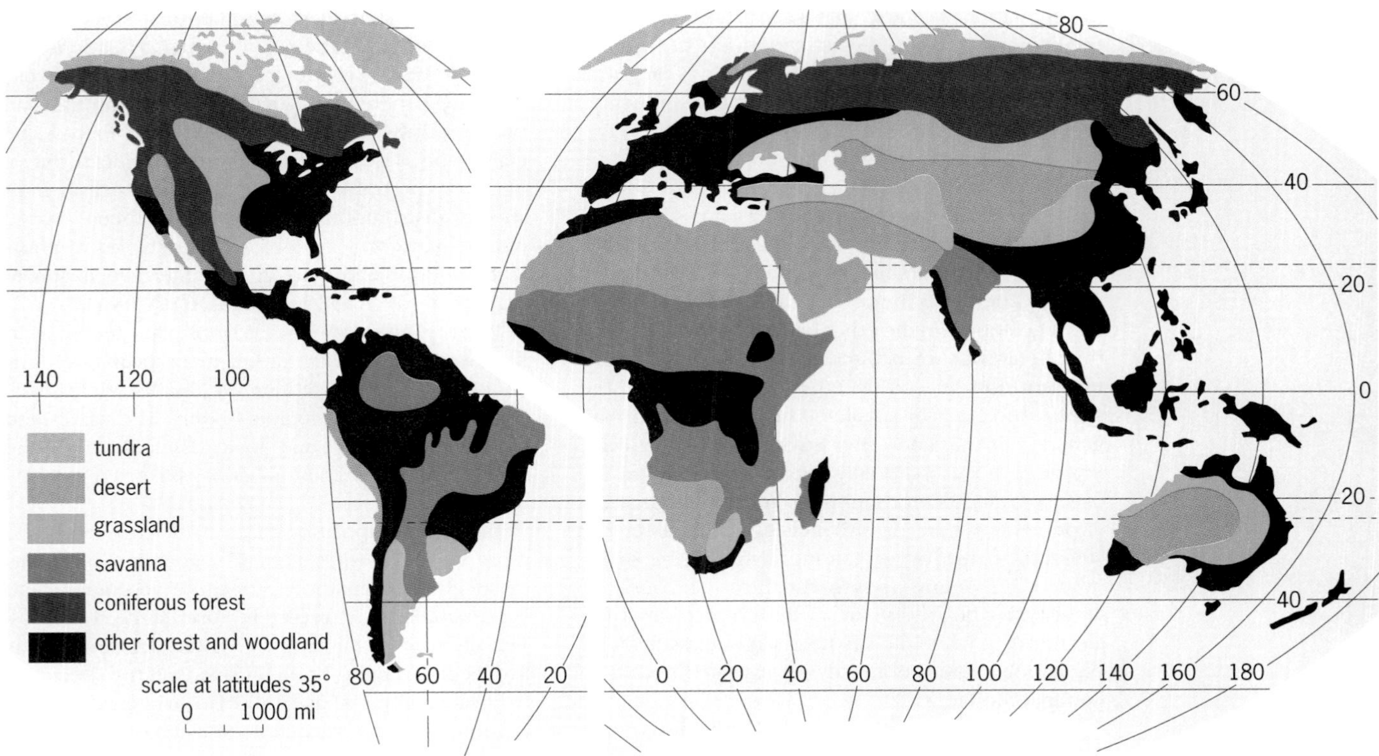

Fig. 2. Map of the world, showing the distribution of physiognomic vegetation types. 1 mi = 1.6 km. (*After R. Good, Geography of Flowering Plants, 3d ed., Longmans, 1964*)

is one surviving from an earlier and more extensive occurrence. On the basis of areas and their floristic relationships, the Earth's surface is divided into floristic regions (**Fig. 1**), each with a distinctive flora.

Floras and their distribution have been interpreted mainly in terms of their history and ecology. Historical factors, in addition to the evolution of the species themselves, include consideration of theories of shifting continental masses, changing sea levels, and orographic and climatic variations in geologic time, as well as theories of island biogeography, all of which have affected migration and perpetuation of floras. The main ecological factors include the immediate and contemporary roles played by climate (temperature, water availability, wind), soil, animals, and humans. *See* ISLAND BIOGEOGRAPHY; PALEOBOTANY; PALEOECOLOGY.

Vegetational plant geography. Vegetation, a term of popular origin, refers to the mosaic of plant life found on the landscape. The vegetation of a region has developed from the numerous elements of the local flora but is shaped also by nonfloristic physiological and environmental influences. Vegetation is an organized whole, at a higher level of integration than the separate species, composed of those species and their populations. Vegetation may possess emergent properties not necessarily found in the species themselves. Sometimes vegetation is very weakly integrated, as pioneer plants of an abandoned field. Sometimes it is highly integrated, as in an undisturbed tropical rainforest. Vegetation provides the main structural and functional framework of ecosystems, which have been actively studied since 1950. *See* ECOSYSTEM.

Plant communities are an important part of vegetation and are studied in plant ecology and various subfields (such as phytosociology, plant sociology, vegetation science, or phytocoenology). No definition of the plant community has gained universal acceptance. In part, this is because of the high degree of independence of the species themselves. Thus, the community is often only a relative social continuity in nature, bounded by a relative discontinuity, as judged by competent botanists. *See* ECOLOGICAL COMMUNITIES.

In looking at vegetation patterns over larger areas, it is the basic physiognomic distinctions between grassland, forest, and desert, with such variants as woodland (open forest), savanna (scattered trees in grassland), and scrubland (dominantly shrubs), which are most often emphasized (**Fig. 2**). These general classes of vegetation structure can be broken down further by reference to leaf types (such as broad or needle) and seasonal habits (such as evergreen or deciduous). Geographic considerations may complete the names of the main vegetation formation types, also called biomes (such as tropical rainforest, boreal coniferous forest, or temperate grasslands). Such natural vegetation regions are most closely related to climatic patterns and secondarily to soil or other environmental factors. *See* ALTITUDINAL VEGETATION ZONES.

Vegetational plant geography has emphasized the mapping of such vegetation regions and the interpretation of these in terms of environmental (ecological) influences. Distinction has been made between potential and actual vegetation, the latter becoming more important due to human influence. *See* VEGETATION AND ECOSYSTEM MAPPING.

Some plant geographers and other biologists are dissatisfied with these more general approaches and point to the effects of ancient human populations, natural disturbances, and the large-herbivore extinctions and climatic shifts of the Pleistocene on the species composition and dynamics of so-called virgin vegetation. On the other hand, it has been shown that the site occurrence and geographic distributions of plant and vegetation types (that is, above the species level) can be predicted surprisingly well from general climatic and other environmental patterns. Problems occur especially when substrates are unusually young or nutrient-poor and in marginal environments where disturbance can emphasize the stochastic nature of vegetation, permitting one or the other of competing types to gain the upper hand. Unlike floristic botany, where evolution provides a single unifying principle for taxonomic classification, vegetation structure and dynamics have no single dominant influence.

Plant growth forms. Basic plant growth forms (such as broad-leaved trees, stem-succulents, or forbs) have long represented convenient groups of species based on obvious similarities. When these forms are interpreted as ecologically significant adaptations to environmental factors, they are generally called life forms and may be interpreted as basic ecological types. These life forms not only may represent the basic building blocks of vegetation but may also provide a convenient way of describing vegetation structure without having to list each individual species. *See* PLANTS, LIFE FORMS OF.

In general, basic plant types may be seen as groups of plant taxa with similar form and ecological requirements, resulting from similar morphological responses to similar environmental conditions. For example, deciduous leaves are generally "softer" and photosynthesize more efficiently in favorable environments. On the other hand, they may lose more water and require more energy and nutrients for their construction (over the plant's life-span) than "harder," longer-lived evergreen leaves. Similarly, larger plants with larger total leaf area may be vulnerable to greater water loss, but may also have more extensive root systems for more effective water uptake. The vegetation of a particular site will be composed of plants with particular combinations of such form characters which permit the plant to function successfully in a particular environment. When similar morphological or physiognomic responses occur in unrelated taxa in similar but widely separated environments, they may be called convergent characteristics (for example, the occurrence of broad-sclerophyll shrubs in the world's five mediterranean-climate regions).

Research. Work in plant geography has involved a variety of approaches. J. Grime recognized three basic plant "strategies" (competitors, stress tolerators, and ruderals), their environmental relationships, and how these may interact. As a result of much work in ecology, there has been an increased emphasis on plant processes and ecophysiology. The focus especially is on plant water and energy budgets throughout the year and how these may influence or limit species distributions. Such work has been carried out, in particular, in nonforest vegetation of grassland, tundra, semidesert, and mediterranean-climate regions. *See* PHYSIOLOGICAL ECOLOGY (PLANT).

The revolutionary discovery of plate tectonics in the twentieth century raised many new questions about past plant migrations and the historical development of taxa and regional floras. The value (and validity) of cladistics and of vicariance theory in the study of plant geographic history has been strongly debated.

As human populations alter or destroy more and more of the world's natural vegetation, problems of species preservation, substitute vegetation, and succession have increased in importance. This is especially true in the tropics, where deforestation is proceeding rapidly. Probably over half the species in tropical rainforests have not yet even been identified. Because nutrients are quickly washed out of tropical rainforest soils, cleared areas can be used for only a few years before they must be abandoned to erosion and much degraded substitute vegetation. Perhaps the greatest current challenge in plant geography is to understand tropical vegetation and succession sufficiently well to design self-sustaining preserves of the great diversity of tropical vegetation. *See* BIOGEOGRAPHY; ECOLOGY; RAINFOREST. Elgene O. Box

Bibliography. E. O. Box, *Macroclimate and Plant Forms: An Introduction to Predictive Modeling in Phytogeography*, 1981; R. Daubenmire, *Plant Geography, with Special Reference to North America*, 1978; J. P. Grime, *Plant Strategies, Vegetation Processes, and Ecosystem Properties*, 2d ed., 2001; J. O. Rieley and S. Page, *Ecology of Plant Communities*, 1990; D. Tilman, *Plant Strategies and the Dynamics and Structure of Plant Communities*, 1988.

Plant growth

An irreversible increase in the size of the plant. As plants, like other organisms, are made up of cells, growth involves an increase in cell numbers by cell division and an increase in cell size. Cell division itself is not growth, as each new cell is exactly half the size of the cell from which it was formed. Only when it grows to the same size as its progenitor has growth been realized. Nonetheless, as each cell has a maximum size, cell division is considered as providing the potential for growth. *See* CELL (BIOLOGY).

Comparison with animals. While growth in plants consists of an increase in both cell number and cell size, animal growth is almost wholly the result of an increase in cell numbers. Another important difference in growth between plants and animals is that animals are determinate in growth and reach a final size before they are mature and start to reproduce. Plants have indeterminate growth and, as long as they live, continue to add new organs and tissues. In animals growth of the different parts of the body is more or less simultaneous; in plants growth is restricted to the growing points or meristems. Therefore, in an animal most body cells attain about the same age, and the individual dies as a unit; but in a plant new cells are produced all the time, and some parts such as leaves and flowers may die, while the main body of the plant persists and continues to grow. The basic processes of cell division are similar in plants and animals, though the presence of a cell wall and vacuole in plant cells means that there are certain important differences. This is particularly true in plant cell enlargement, as plant cells, being restrained in size by a cellulose cell wall, cannot grow without an increase in the wall. Plant cell growth is thus largely a property of the cell wall. *See* ANIMAL GROWTH.

Changes in rate of growth. If the size of a plant (or any organism), plant part, or single plant cell is plotted against time, the resulting line produces an S-shaped, or sigmoid, growth curve (**Fig. 1**). This may have an initial lag phase, and then a phase of steadily increasing growth rate. During this latter phase, termed the log phase, the logarithm of the size plotted against time yields a straight line, since each unit of previous growth gains its own growth increment. The growth rate of an individual stem or root then stabilizes at an approximately constant rate. This is because a growing plant usually has the same number of dividing, growing, and maturing cells in its stem, and always maintains the same proportion of growing cells in its different stages. Thus, the regular growth rate of the whole plant is the integration of thousands of sigmoid growth curves of the individual cells. As the plant matures, the growth rate may slow down or stop completely, to be followed by senescence, or those changes which precede the death of the plant or plant part.

Cell division. Cell division in plants, as in animals, involves the division of the nucleus by mitosis. The various organelles, such as proplastids (precursors of chloroplasts), probably also divide. Following the formation of two daughter nuclei, a new cell wall starts to form between the two nuclei at a right angle to the plane of cell division. The wall starts as a coalescence of vesicles (membrane-bounded spheres) produced by an organelle in the cell termed the Golgi apparatus or dictyosome. These vesicles contain the necessary chemicals for cell wall formation, and as they coalesce the contents polymerize to form the initial layer of the new cell wall, which is termed the middle lamella and composed mainly of calcium pectate. (Calcium is an essential nutrient for plants, and lack of calcium prevents the formation of the middle lamella so that the apex dies.) The membranes of the vesicles become the membranes delimiting the new cell. The new cell wall starts to form across the plane between the two new cells and fuses with the

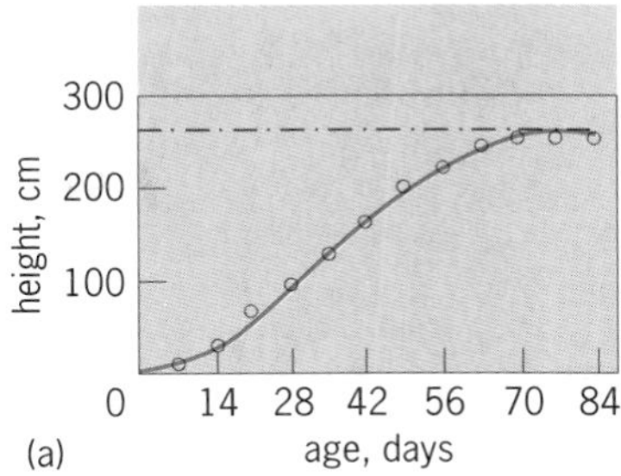

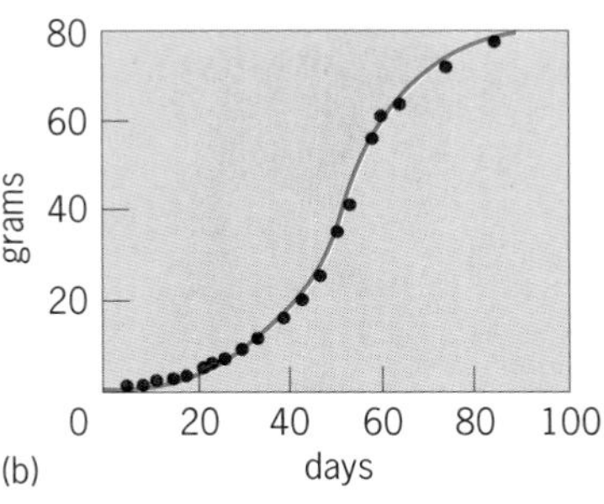

Fig. 1. Graphs showing growth curves. (*a*) Growth in height of sunflower stem; 1 cm = 0.4 in. (*after E. P. Odum, Fundamentals of Ecology, Saunders, 1953*). (*b*) Growth of entire corn plant as measured by increase in dry weight; 1 g = 0.035 oz (*after D. W. Thompson, On growth and Form, Cambridge, 1942*).

old wall at the edges. Further deposition of cell wall polysaccharides into the new wall occurs through continued fusion of Golgi vesicles with the new wall, and perhaps also by direct synthesis on or just below the cell membrane lining the new wall. *See* CELL DIVISION; CELL WALLS (PLANT); MITOSIS; PLANT CELL.

Cell division in plants takes place in discrete zones called meristems (**Fig. 2**). In these meristems the cells divide and grow back to their original size before dividing again. Just beyond this region of cell division a much greater increase in cell size takes place. The principal meristems are at the tip (apex) of the stem and root (or each branch of the stem and root); in the vascular or conducting tissue of the stem and root (the vascular cambium) where the meristem is shaped like a hollow cylinder; and at the exterior surface of woody plants (the cork cambium). Cell division in the stem and root apices occupies only a small zone of about 0.04–0.08 in. (1–2 mm) at the tip of the stem or root of a herbaceous plant. As the cells are continuously produced at the stem and root apices, the daughter cells steadily emerge from the zone of cell division and start to elongate. Elongation itself is also confined to a region just below the tip and may extend for about 1.2–1.6 in. (3–4 cm) below the stem bud and only about 0.2–0.4 in. (0.5–1 cm) behind the root tip in herbaceous plants such as a pea plant (Fig. 2). *See* APICAL MERISTEM; LATERAL MERISTEM.

The stem and root apical meristems produce all the primary (initial) tissues of the stem and root. The cylindrical vascular cambium produces more conducting cells at the time when secondary thickening (the acquisition of a woody nature) begins. The vascular cambium is a sheet of elongated cells which divide to produce xylem or water-conducting cells on the inside, and phloem or sugar-conducting cells on the outside. Unlike the apical meristems whose cell division eventually leads to an increase in length of the stem and root, divisions of the vascular cambium occur when the part of the plant has reached a fixed length, and lead only to an increase in girth, not in length. The final meristematic zone, the cork cambium, is another cylindrical sheet of cells on the outer edge of older stems and roots of woody plants. It produces only new outer cells, which differentiate into the corky layers of the bark so that

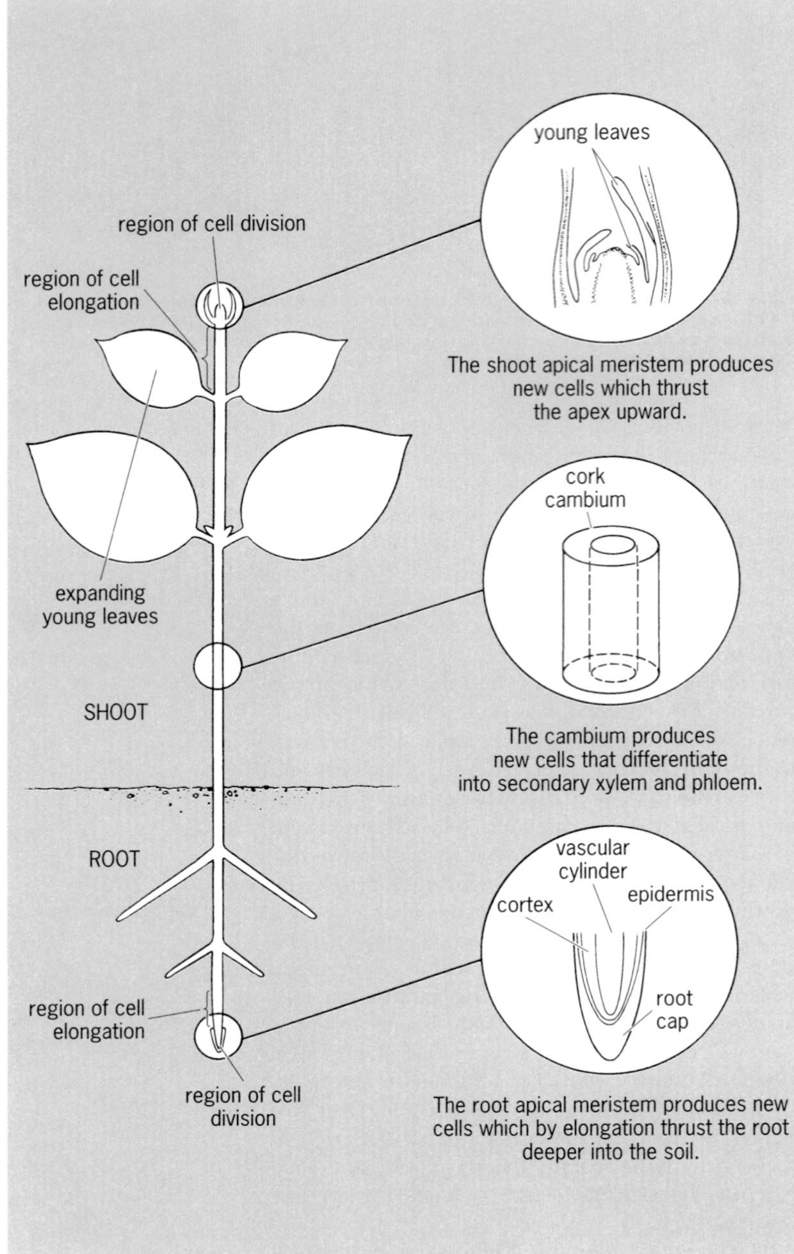

Fig. 2. Schematic drawing of a plant showing the regions of cell division and cell elongation.

new protective layers are produced as the tree increases in circumference. *See* BUD; PERIDERM; ROOT (BOTANY); STEM.

Cell division, followed by enlargement, also takes place in certain areas of the plant not strictly regarded as meristems. Cell division continues for a short time in the newly formed leaves following their production on the stem apex. Fruits and seeds are also active areas of cell division and growth, and will be discussed later.

Cell enlargement. The young meristematic plant cell is approximately isodiametric. It has a thin cell wall composed of cellulose (made of glucose units) and other mixed polysaccharide polymers (containing xylose, glucose, galactose, arabinose, and other minor sugar constituents, as well as some protein). The interior contains dense protoplasmic contents indicative of intense metabolic activity and protein synthesis. Unlike mature plant cells, vacuoles are usually absent or very small. As this cell grows, it does not increase equally in all dimensions, but usually elongates to a final length many times its width and 20–150 times its original size. This cell expansion is primarily due to water uptake. Protein synthesis occurs during cell growth, but at a rate insufficient to keep up with the increase in cell size. As a result, the cytoplasm which originally filled the cell now comes to occupy only the periphery of the cell (**Fig. 3**), and the cell becomes largely filled with a huge central vacuole. This vacuole is bounded by a membrane called the tonoplast, and contains a water solution of salts, sugar, and various secondary metabolites that form a reserve of metabolites and a depository for waste products, as well as a reservoir for compounds involved in the defense of the plant against diseases and predators. The vacuole arises initially as many small vacuoles from either Golgi vesicles or the endoplasmic reticulum (a membrane system throughout the cytoplasm). These then coalesce to form the large central vacuole (**Fig. 4**). Movement of water into the cell increases the size of the vacuole.

While the length of the cell greatly increases, the thickness of the wall remains constant and its density does not change. This is because wall synthesis occurs at the same time as normal elongation. The dry weight of the cell, 80% of which can be accounted for by the wall, increases from the apex to a point where cell expansion ceases, and it is largely due to a synthesis of wall polysaccharides. After cell expansion is completed, there is a slow further increase in

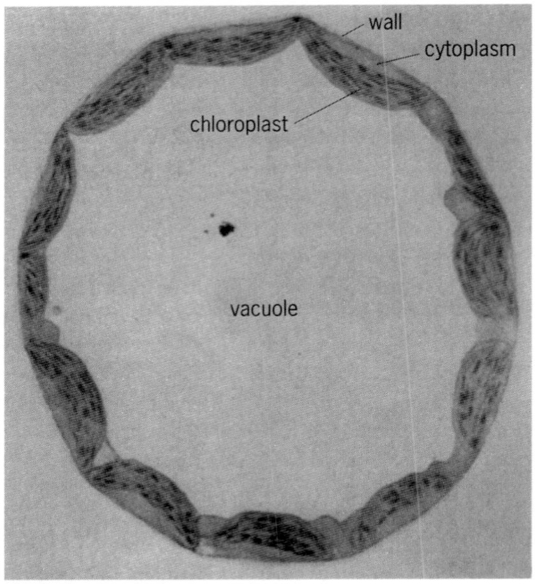

Fig. 3. Electron micrograph of a cell in a leaf. Expansion of the central vacuole is the principal change in cell content during growth. (*From A. W. Galston, P. J. Davies, and R. L. Satter, The Life of the Green Plant, 3d ed., Prentice-Hall, 1980*)

the wall through the deposition of secondary wall material.

The rate of cell extension can be very rapid. In a corn (*Zea*) root, increases of 40% per hour occur, while in the grass *Phleum* 100% per hour has been noted. Such rates occur, however, only in a narrow cell zone, and while the cells are expanding they do so at an average rate of about 20% per hour. The extent to which growth is visible in the whole plant depends on the rate of cell growth and the size of the extending zone. Corn plants are often noted to grow several centimeters per day, and leaf stalks of the water lily *Victoria regia* may increase from 3.6 to 27 in. (9 to 68 cm) in 24 h.

Elongation of a plant cell is dependent on the extension of the cell wall. Because the cell forms a rigid "wooden box" around the cell, there can be no increase in cell size unless the wall stretches. In plant cells there is always a stretching force pushing outward against the cell wall. This force, termed turgor pressure, results when the presence of dissolved solutes in the vacuole causes water uptake by the cell through osmosis. However, despite this pressure inside all cells, only cells in a narrow growing zone elongate. This is because wall extension is restricted and can occur only in the presence of certain growth hormones, most notably auxin and in some cases gibberellin.

During cell elongation the tangential stress on a cell is much greater than the longitudinal stress, yet the cell enlarges predominantly in a longitudinal direction. The reason for this unidirectional enlargement is the cellulose microfibril construction of the cell wall which, when stretchable, permits elongation with little lateral increase.

Factors affecting growth. Plant growth is affected by internal and external factors. The internal controls are all the product of the genetic instructions carried in the plant. These influence the extent and timing of growth and are mediated by signals of various types transmitted within the cell, between cells, or all around the plant. Intercellular communication in plants may take place via hormones (or chemical messengers) or by other forms of communication not well understood. Plants lack the nervous system of animals; yet transmembrane electrical potentials do occur, and action potentials may be transmitted from cell to cell, particularly in plants with rapidly moving parts, such as *Mimosa pudica* (the so-called sensitive plant) or insectivorous plants with actively moving traps, such as *Dionea* (Venus' flytrap). Cells may also be able to detect the presence of neighboring cells through pressure or gradients of various chemicals such as nutrients, though again little is known of the role of such signals. Something must, however, determine that central stem cells develop into pith, while those at the surface become epidermal cells. The genetic makeup of the plant, residing in the genes of the chromosomes in the nucleus, controls both the production of and the response to intercellular signals, as well as regulating development within the cell. Genes do not all operate at once; to do so would only produce chaos. They operate in both a predetermined and externally modulated sequence to control the growth and destiny of the cells, ultimately leading to the formation of the mature organism. *See* GENETICS; PLANT MORPHOGENESIS.

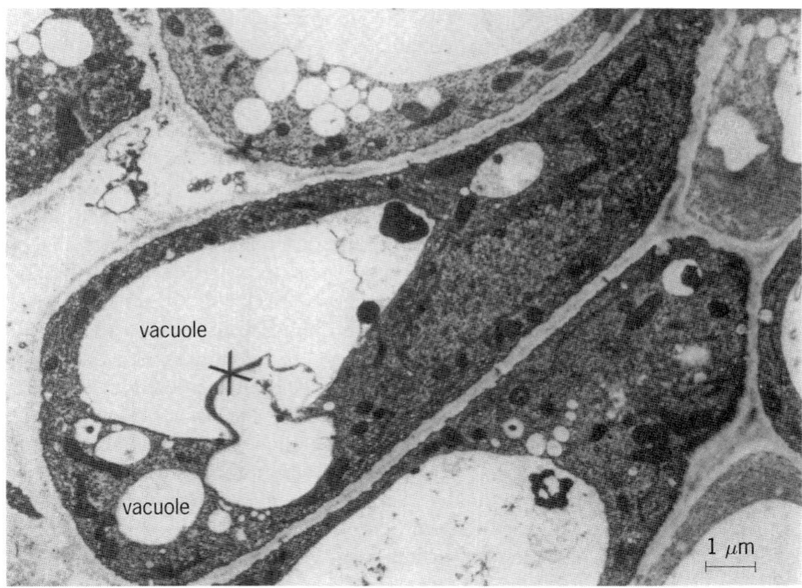

Fig. 4. Young parenchyma cells in culture which have begun to enlarge by development of one or more vacuoles in the cytoplasm. There are many small vacuoles in the upper cell, while vacuoles in the lower cell are coalescing to form a single large vacuole. Note the fusion of two vacuoles at X, and breakdown of the membranes and cytoplasm between them. (*From A. W. Galston, P. J. Davies, and R. L. Satter, The Life of the Green Plant, 3d ed., Prentice-Hall, 1980*)

Internal controls—hormones. The prime means of communication among the different parts of the plant is through the plant hormones. There are several hormones (or groups of hormones), each of which may be produced in a different location, that have a different target tissue and act in a different manner. The groups of hormones that are definitively known are briefly described below. Their role in the integration of growth will be described later. *See* PLANT HORMONES.

Auxin. This hormone is primarily represented in nature by one chemical, indoleacetic acid (IAA). Some other indole derivatives exist in plants (such as indoleacetonitrile in cabbages), but these are generally thought to act as precursors of IAA. Several synthetic derivatives also exist, and these have practical uses in agriculture and horticulture. IAA has numerous roles in plants as a stimulator of cell growth or division or both. *See* AUXIN.

Gibberellin. The gibberellins are a family of isoprenoid compounds with over 50 naturally occurring members which differ slightly in their chemical structure. Each plant species has several gibberellins, though usually no more than 6–10, and they are thought to represent different steps in an interconversion pathway, with some of the gibberellins being biologically active, while some are inactive. Some gibberellins vary in their activity in different species. Gibberellins also enhance cell division and elongation, but in a different way and in different cells from auxin. *See* GIBBERELLIN.

Cytokinins. These hormones are adenine derivatives. There are two main natural compounds, isopentenyl adenine and zeatin (so named because it was discovered in *Zea*, or maize), and both of these also exist in their riboside and riboside phosphate forms in addition to the free base. Cytokinins are primarily active in stimulating cell division. They may also induce cell enlargement but only in leaves, and they also act to delay senescence. *See* CYTOKININS.

Ethylene. Although unusual in that it is a gas, ethelyene is very active in inducing processes normally associated with senescence, particularly fruit ripening, flower fading, and leaf abscission. *See* ETHYLENE.

Abscisic acid. This is a single isoprenoid compound. It has a general growth-inhibiting action and may be involved in the induction of dormancy. It is also synthesized in stomatal guard cells (cells surrounding pores in the leaf epidermis) under conditions of water shortage, when it causes the stomata to close, restricting further loss. *See* ABSCISIC ACID.

External controls—the environment. The external environments of the root and shoot place constraints on the extent to which the internal controls can permit the plant to grow and develop. Prime among these are the water and nutrient supply available in the soil. Because cell expansion is controlled by cell turgor, which depends on water, any deficit in the water supply of the plant reduces cell turgor and limits cell elongation, resulting in a smaller plant. As each cell can expand only for a certain interval of time before its new size is fixed by changes in the cell wall properties, any shortage of water, even if temporary, will result in reduced growth if it exceeds the time period over which any cell expands. Water shortage also reduces cell division, but this is resumed when the water supply returns, provided the shortage is not too prolonged.

It is, in fact, quite common for plants to experience a water deficit during the daytime, because the rate of water loss through transpiration will exceed water uptake on a hot dry day even if there is plenty of water available in the soil. This deficit is, however, made up at night when transpiration ceases, though water uptake continues. Over such a period the cell walls remain stretchable so that as the deficit is made up during the night, and the turgor pressure recovers, there is a spurt of growth to about the same extent as would have occurred in a 24-h period with no water deficit. *See* PLANT-WATER RELATIONS.

Mineral nutrients are needed for the biochemical processes of the plant. When these are in insufficient supply, growth is less vigorous, or in extreme cases it ceases altogether. *See* PLANT MINERAL NUTRITION.

An optimal temperature is needed for plant growth. The actual temperature range depends on the species. Temperate species often grow best at moderate temperatures (for example, 59–68°F or 15–20°C for peas), while tropical species are adapted to higher temperatures nearer 86°F (30°C). In general, metabolic reactions and growth increase with temperature, though high temperature becomes damaging. Most plants grow slowly at low temperatures (32–50°F or 0–10°C), and some tropical plants are damaged or even killed at low but above-freezing temperatures. Many temperate-zone plants can survive subfreezing temperatures if they are acclimated by slowly declining temperatures and short day lengths prior to the freezing period. When actively growing, however, virtually all plants are frost-sensitive. *See* COLD HARDINESS (PLANT).

Light. Light is important in the control of plant growth. It drives the process of photosynthesis which produces the carbohydrates that are needed to osmotically retain water in the cell for growth. These carbohydrates form the basis for all the organic compounds that are produced by the plant and comprise the protoplasm and cell walls. In addition, the carbohydrates are oxidized to produce the energy for metabolism and cell maintenance, both of which are needed for growth. Similarly, carbon dioxide, the raw material for photosynthesis, is needed for plant growth. Under conditions of low light or CO_2, enhancement of growth can be obtained by increasing the levels of either component. *See* PHOTOSYNTHESIS; PLANT RESPIRATION.

Phytochrome. The color of incident light is also important in determining plant growth. While

Fig. 5. Effect of far-red light in promoting stem elongation in *Chenopodium album*. Both plants received an equal amount of white light, but the plant at the right received additional far-red light; therefore less of its phytochrome is in the P_{fr} form. (*From Morgan and Smith, Nature, 262:210–212, 1976*)

chlorophyll, the pigment in photosynthesis, absorbs blue and red light, the plant uses another pigment called phytochrome to detect the light quality of its environment.

Phytochrome is a biliprotein (a protein with a straight-chain tetrapyrrole chromophore). It responds to red (660-nanometer) and far-red (731-nm) light by changing between two forms—a red-absorbing form (P_r) and a far-red-absorbing form (P_{fr}). While natural light does not consist of red and far-red flashes, the ratio of red to far-red in the incident light is changed by the chlorophyll in a leaf canopy. Sunlight is high in red light compared with far-red light and causes a high proportion of the phytochrome to exist in the P_{fr} form. By contrast, the chlorophyll in leaves absorbs the red while allowing the far-red to pass. Thus, under a leafy canopy the ratio of red to far-red is low, and the phytochrome in an understory plant will be predominantly in the P_r form. This change has profound effects on the plant. Many seeds need P_{fr} to germinate and will not do so beneath other plants because of the P_r level. A high P_r also causes etiolation so that the plant becomes tall and spindly (**Fig. 5**). Such enhanced stem growth, however, has adaptive advantage in speeding up the growth of the plant through the leafy canopy above into the sunlight needed for photosynthesis.

The photomorphogenetic effect of light absorbed by phytochrome is also intensity-dependent. The reason for this is thought to be the speed of cycling between the two forms of phytochrome which will increase as the light intensity increases. It is proposed that various intermediates between the two forms are also needed together with a high level of P_{fr} in order for there to be a maximum photomorphogenetic response. The typical plant form is therefore produced only in sunlight or in other high-intensity light that gives a high proportion of phytochrome in the P_{fr} form. The other role for phytochrome is the measurement of photoperiod in the control of numerous developmental phenomena which will be dealt with below. *See* PHOTOMORPHOGENESIS; PHYTOCHROME.

Control of cell division. In the absence of auxin and cytokinin, no cell division takes place. The function of auxin in cell division is unknown. However, there is some idea of the role of cytokinin. In the absence of cytokinins, DNA replication takes place in the nucleus, but the nucleus does not divide. When cytokinins are added, protein synthesis takes place rapidly and division ensues (possibly because proteins are needed for nuclear division). This protein synthesis appears to depend on preformed messenger ribonucleic acid (mRNA), and it has been proposed that cytokinins act by binding to ribosomes (the protein synthesis machinery), thereby allowing the translation of mRNA for cell division proteins.

Control of cell elongation. Under normal turgor pressure the cell wall is moderately elastic, but no permanent deformation takes place in the absence of auxin. If a dicotyledon (such as the pea) stem or grass (such as oat or corn) coleoptile segment is taken and placed in a flowing aerated solution in an apparatus such as that shown in **Fig. 6**, a slow growth will occur in the absence of external auxin. When auxin is added, there is often an immediate, slight, transitory decrease in the growth rate of the stem for unknown reasons. There is a lag period of 10–15 min following the application of the hormone, and then growth increases over a further 10–15 min to reach a new auxin-enhanced rate of growth 10–20 times the original rate.

Auxin does not work in all tissues in all species. In the case of mature grass stems, gibberellin enhances growth while auxin is without effect. In general, however, gibberellin produces hypoelongation of intact stems on a plant but has little effect on stem segments.

Auxin is needed for cell growth and acts to increase the plasticity (permanent stretchability) of the cell wall. Auxin does not, however, act directly on the wall, since it has no effect on dead cells which are mechanically held under tension. Active cellular metabolism is needed for auxin action, and the driving force for wall extension is the turgor pressure. If auxin is added to active stem cells in an osmotic medium which prevents cell elongation, and the stem cells are subsequently killed by freezing and thawing, then the cell walls will stretch when placed

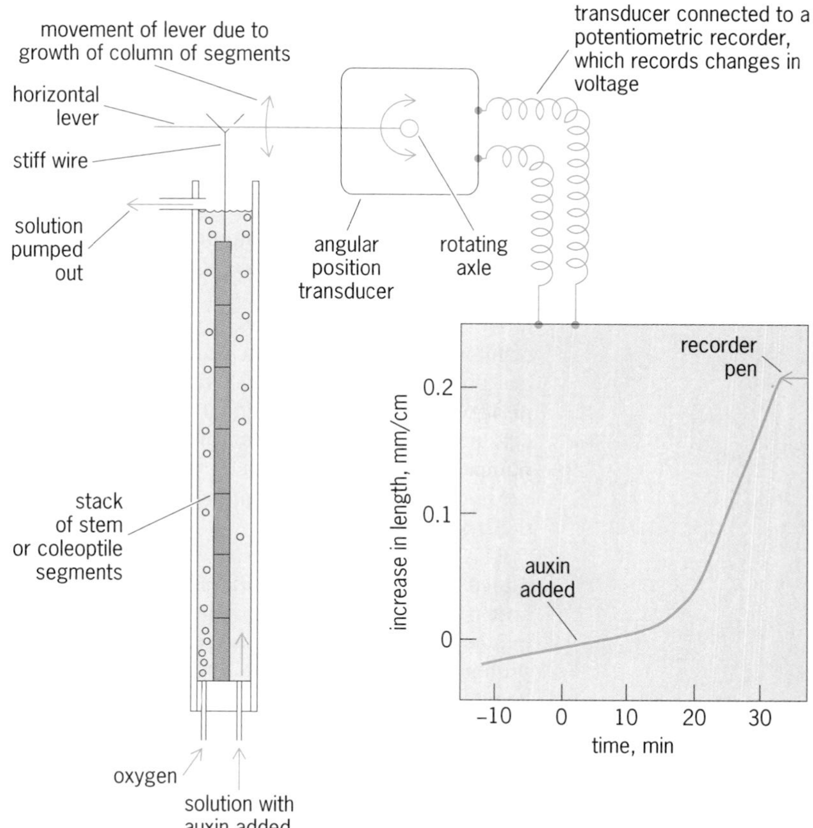

Fig. 6. Apparatus for recording rapid changes in growth, with a typical growth-tracing resulting from the addition of indoleacetic acid to a buffer at about pH 6.5. (*After A. W. Galston, P. J. Davies, and R. L. Satter, The Life of the Green Plant, 3d ed., Prentice-Hall, 1980*)

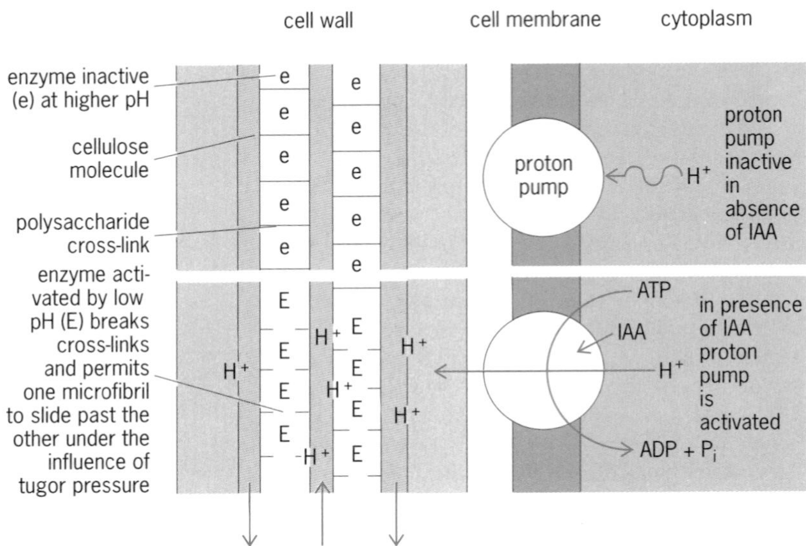

Fig. 7. Proton pump theory to explain the action of indoleacetic acid in inducing wall extensibility and cell elongation. (*After A. W. Galston, P. J. Davies, and R. L. Satter, The Life of the Green Plant, 3d ed., Prentice-Hall, 1980*)

under tension. Thus some change has been brought about in the wall which is subsequently expressed as wall extension. The nature of this change became clear when it was found that the action of auxin, in both living cells and dead cells held under tension, can be mimicked by acid solution of pH 5. In addition, it was found that whan auxin is added to living cells, hydrogen ions (H^+) are pumped out of the cells into the cell walls, causing the walls to become acidic (**Fig. 7**). The increased acidity of the wall medium causes the enzymatic breaking of polysaccharide cross-links in the wall which normally hold the wall rigid. In this state the cellulose microfibrils can slide over one another so that the wall extends (**Fig. 8**). Resynthesis of the cross-links is subsequently needed to fix the new position of the microfibrils, and this is followed by auxin-stimulated cellulose synthesis so that the wall maintains its original thickness. The location of the auxin-promoted proton pump is currently under investigation. The cell membrane is the location of most such proton pumps, but auxin-binding studies seem to designate the endoplasmic reticulum membrane as the site of the auxin action.

RNA and protein synthesis are also needed for continued auxin-induced growth, and auxin enhances both these processes. The proteins needed for auxin-induced growth have been termed growth-limiting proteins, although their nature is as yet unknown. Possible roles for such proteins are as the proton pump or as enzymes needed for the breaking and resynthesis of the wall cross-links.

The action of gibberellin in inducing cell elongation may also involve wall acidification, although the acidification measured so far has been much less than that produced by auxin.

Vegetative growth. The pattern and extent of plant growth is the result of the various internal and external influences on the plant (**Fig. 9**). The roots supply water and nutrients to the shoots while the shoots supply photosynthates. Cytokinins are produced in the shoot tips, where they stimulate cell division, and also in the roots, from which they move to the shoot in the xylem stream. A supply of cytokinins is needed for leaf expansion so that when the root supply is reduced, by root pruning, for example, the resulting leaves are small, thus restricting the water loss. Cytokinins are also needed for the prevention of senescence. When a leaf ages, it can be rejuvenated by applying cytokinin. Since the cytokinins travel in the water stream from the root, it is likely that heavily shaded leaves, which lose less water, become deficient in cytokinins and thus senesce and are shed from the plant.

Auxin is produced primarily in the leaf primordia

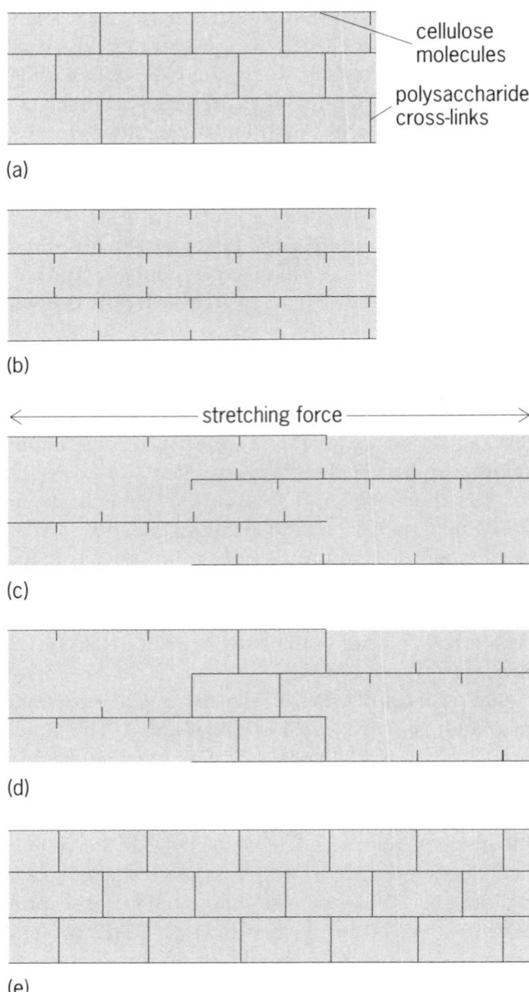

Fig. 8. Model suggesting mechanism of wall elongation under the influence of auxin. (*a*) Original wall structure. (*b*) Cross-links are broken and wall is loosened. (*c*) Under the influence of turgor pressure, cellulose molecules slide past each other, and the wall stretches. (*d*) Cross-links reform to fix the wall in its newly strtched position. (e) More wall material is synthesized. The wall is now elongated but still has its original thickness. The process is then repeated. (*After A. W. Galston, P. J. Davies, and R. L. Satter, The Life of the Green Plant, 3d ed., Prentice-Hall, 1980*)

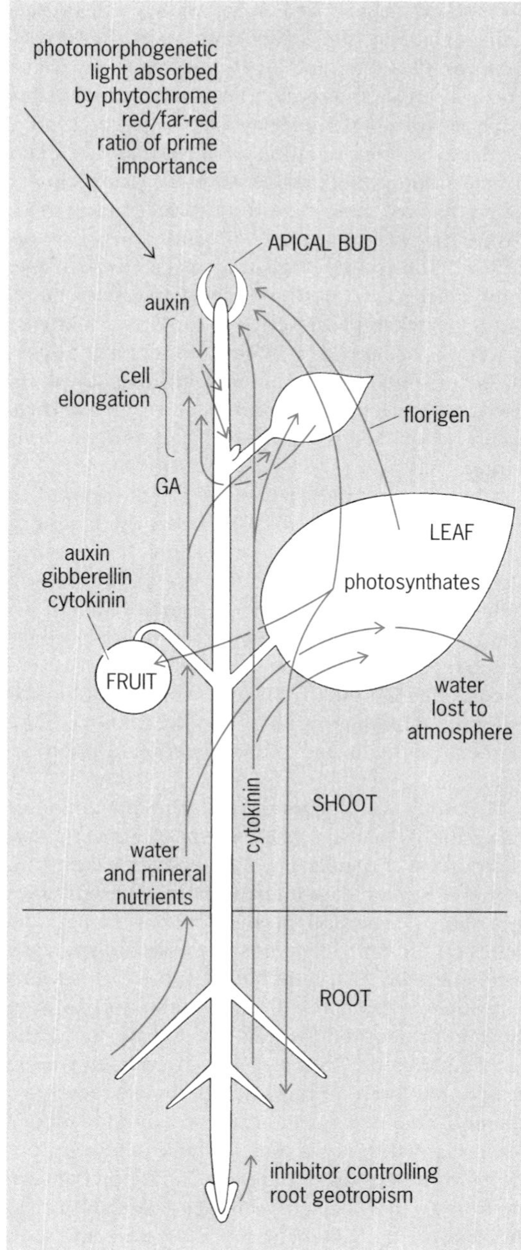

Fig. 9. Schematic drawing of a plant with the interrelationships of some factors involved in growth and development.

and young leaves of the bud. It moves basipetally downward through an active polar (unidirectional) transport mechanism primarily in the young xylem vascular cells and, in so doing, promotes the continued development of xylem cells up toward the growing bud. The auxin that reaches the root is mainly destroyed by an enzyme, IAA-oxidase, though some accumulates in the growing point of the root and the branch root initials where it promotes cell division.

Just below the bud, auxin promotes the elongation of the cells to produce stem growth. Gravity and light cause the transport of auxin toward the downward or dark side of the stem. The resultant higher concentration of auxin causes more rapid cell growth on that side and thus a bending toward the opposite side. Auxin transport may not be involved in tropistic stem movements, but evidence favors the possibility. *See* PLANT MOVEMENTS.

In the root, gravity causes a horizontal root tip to bend and grow downward. This is mediated by a growth inhibitor moving backward from the root cap to the growing zone of the root on the lower side of the horizontal root.

Another effect of auxin in the growing plant is the prevention of growth of lateral buds so that apical dominance exists. The lateral buds near the apical bud remain dominant, while those at a distance from the apical bud start to grow out. It is generally considered that it is auxin produced by the apical bud which inhibits the growth of the lateral buds and that it does so through the production of ethylene. As the distance between the apical and lateral bud increases through the growth of the leader, not only does the level of auxin at the lateral bud decrease, but cytokinins coming from the root oppose the auxin and cause the growth of the lateral branch. *See* APICAL DOMINANCE.

As auxin is transported morphologically downward, it accumulates at the base of stem cutttings. Here it promotes the formation of callus and root initials which then develop into roots so that a rooted cutting results. This can be enhanced by the application of auxin to the cut base of a shoot.

The site of gibberellin production is less definite. It is produced by young leaves and possibly by roots. It is involved primarily in stem growth. Many dwarf plants can be turned into tall plants by application of gibberellin. It also induces the bolting or rapid elongation of flowerstalks of biennial plants. Such plants grow as a rosette of leaves for most of their growth, and then the flowerstalk rapidly elongates just prior to flowering. It is probable that natural tall and bolting plants have higher levels of active gibberellins than dwarf or rosette plants, but because of the number of gibberellins present, each of which has a different biological activity, most of the exact gibberellin levels and changes have yet to be elucidated. In addition, the increase in gibberellins which lead to bolting is possibly of a transitory nature.

Growth of fruits and seeds. Fruits and seeds are rich sources of hormones. Initial hormone production starts upon pollination and is further promoted by ovule fertilization. These hormones promote the growth of both seed and fruit tissue. In some cases, application of the hormone to the deseeded or unfertilized fruit will cause the development of seedless or parthenocarpic fruit. The hormones vary with species: auxin induces fruit development in the strawberry, and gibberellins induce fruit development in grapes; pea seeds contain high levels of gibberellin; and developing corn grains are rich sources of cytokinins. Fruits grow initially by cell division, then by cell enlargement, and finally sometimes by an increase in air spaces. *See* FRUIT.

The growth of a seed starts at fertilization. A small

undifferentiated cell mass is produced from the single-celled zygote. This proceeds to form a small embryo consisting of a stem tip bearing two or more leaf primordia at one end and a root primordium at the other. Either the endosperm or the cotyledons enlarge as a food store. Very specific seed storage proteins are synthesized on endoplasmic reticulum-bound ribosomes, and the proteins are passed into the lumen of the endoplasmic reticulum to form storage-protein bodies. Messenger RNA for the enzymes of germination is also synthesized during the latter part of embryo development. The synthesis of storage proteins and the prevention of the translation of mRNA for the enzymes of germination are brought about by a rise in the level of abscisic acid. *See* REPRODUCTION (PLANT); SEED.

Totipotency. It is possible to grow almost any plant species in tissue culture. This consists of taking a piece of sterile tissue and placing it on a jellylike agar medium containing sucrose, mineral nutrients, vitamins, and hormones. The piece of tissue can be meristematic, but fully differentiated parenchyma cells can also be made to grow. In general, growth occurs only if the medium contains both auxin and cytokinins. However, some tissues, notably tissues isolated from crown galls (whose cells have incorporated a plasmid from an infecting bacterium, though the bacterium may subsequently be absent), are able to synthesize their own hormones and grow vigorously with no further additions of hormones. Most tissues initially develop into an undifferentiated mass of cells, but manipulation of the hormone levels can cause the differentiation of roots and shoots, and ultimately a mass of small plantlets on the callus tissue. These plantlets can be removed and then grown into mature plants. This method is becoming important in the mass propagation of many horticultural species such as orchids and chrysanthemums.

If the callus tissue is transferred to a slowly agitated liquid medium, some single cells may slough off the callus, and in an appropriate hormonal regimen these cells start to develop as if they are fertilized zygotes going through all the stages of embryo development until they develop into plantlets.

The dedifferentiation of mature parenchyma, the subsequent differentiation into whole plants, and the formation of embryos and whole plants from single diploid cells without sexual fusion, show that during differentiation genes are simply turned on and off, rather than any loss of genetic material taking place. Under appropriate nutritional and hormonal signals, the genes can again be switched on and cause a previously mature cell to revert to an embryonic state. The ability of a mature cell to retain all the genetic instructions to build a whole plant is known as totipotency.

Flowering. At a certain time a vegetative plant ceases producing leaves and instead produces flowers. This often occurs at a particular season of the year. The determining factor for this event is day length (or photoperiod). Different species respond to different photoperiods.

Photoperiod classes. Three basic classes can be recognized, namely, short-day plants (such as *Chrysanthemum*, *Pharbitis*, and *Xanthium*), long-day plants (such as cereals and spinach), and day-neutral plants (such as tomato). Short-day plants flower if they receive a shorter duration of light than a certain critical photoperiod, while long-day plants flower if they receive more than the critical photoperiod. Day-neutral plants are not influenced by photoperiod and flower only when they reach a certain age. Some species have more complex day-length requirements, for example, a series of long days followed by a series of short days, which ensures that flowering occurs only in the autumn. Some long-day plants also have a prior cold requirement, termed vernalization, which is normally satisfied the previous winter.

It has been found that when plants respond to photoperiod, they are responding to the length of darkness rather than the length of light. If a short-day plant is given a long night, it flowers, but if that long night is interrupted by a short period of light, flowering is prevented. Experiments on the color of light which regulates flowering in darkness show that red is most effective and that the effect of red is inhibited by far-red. This demonstrates the involvement of the pigment phytochrome in the detection of photoperiod.

This detection occurs through the interaction of either the P_{fr} or the P_r with an endogenous rhythm in the plant. If a short-day plant is given a long dark period of several days interrupted at different times by a short period of light, the effect of the light depends on the time when it is given. If it is given at a time when the rhythm in the plant has, by previous exposures, entered a "light-anticipating" phase, then the light promotes flowering, but if given during the "dark-anticipating" phase of a 24-h cycle, flowering is inhibited, even if the plant has already received a longer dark period than normally required to induce flowering. This shows that phytochrome does not measure the period of darkness simply by converting from P_{fr} to P_r, which occurs in darkness. Instead the state of the phytochrome must coincide with the state of the plant rhythm. In short-day plants, for flowering to occur, the light phase of the rhythm and P_{fr} and the dark phase of the rhythm and P_r must coincide. (Normal white light converts phytochrome predominantly to the P_{fr} form.) In order for long-day plants to flower, the opposite must occur: the rhythm must be disrupted by the conversion of phytochrome to the P_{fr} form during the dark phase of the rhythm. *See* FLOWER; PHOTOPERIODISM.

Flowering hormone. The light signal for flowering is received by the leaves, but it is the stem apex that responds. Exposing even a single leaf to the correct photoperiod can induce flowering. Clearly, then, a signal must travel from the leaf to the apex. Grafting a plant that has been photoinduced to flower to one not so induced can cause the noninduced plant to flower. It has been proposed that a flower-inducing hormone travels from the leaf to the stem apex and there induces changes in the development of the cells such that the floral morphology results. This hypothetical hormone has been named florigen. In some long-day plants gibberellin will induce flowering, but gibberellin is clearly not the universal florigen. It may be that florigen is a mixture of substances.

Juvenility. Some perennial plants will not produce reproductive structures until they are several years old. This is termed juvenility and is a problem to the growers of fruit-tree crops and to tree breeders particularly in the lumber industry. The causes appear to be hormonal. Many conifers have now been brought to a reproductive state while quite young by the application of nonpolar gibberellins.

Dormancy. At certain stages of the life cycle, most perennial plants cease growth and become dormant. Plants may cease growth at any time if the environmental conditions are unfavorable. When dormant, however, a plant will not grow even if the conditions are favorable. Dormancy is normally an adaptation that allows the plant to survive an adverse environmental situation, such as the cold of winter. In temperate zones many trees stop growing after the initial spurt of spring growth. This period is called summer dormancy and is distinguished from winter or true dormancy in that growth can be started again relatively easily by conditions such as defoliation by insects or overfertilization. True dormancy is induced by declining photoperiods and temperatures of autumn. A truly dormant tree can withstand temperatures of tens of degrees below 32°F but will not grow if simply placed in favorable growing conditions. Dormancy is broken by a period of low temperatures (6–12 weeks) just above freezing (about 41°F or 5°C. This has normally occurred by mid to late winter, and on the return of higher temperatures and longer photoperiods in spring, growth will resume.

Seeds may also show dormancy when shed from the plant. This dormancy may be due to immature embryos, in which case it is broken simply by time or by chemical or physical restrictions to germination. Seeds may have impermeable coats that prevent the entry of water and oxygen. Such coats are broken down by abrasion or microbial action over a period of time in the soil. Dormancy may also be imposed by growth inhibitors either in the seed coat or in the embryo itself. This chemical inhibition of growth can be removed or overcome by a period of low temperatures (similar to that required to break bud dormancy), light, or the leaching of the inhibitor by washing. Each of these represents adaptations to different natural conditions. Dormancy thus prevents germination prior to winter, under heavy shade, or during dry conditions in a desert.

The exact nature of the growth inhibitors controlling dormancy is uncertain. Abscisic acid is certainly involved in some species, but the presence of high levels of abscisic acid in vigorously growing shoots shows that it alone cannot be the dormancy inducer. Rather dormancy is brought about by a balance between abscisic acid (and perhaps other growth inhibitors) and growth-promoting hormones, particularly gibberellins and cytokinins. Following a dormancy-breaking cold or light treatment of seeds, gibberellin content often increases, and gibberellins or cytokinins or both increase in woody plants after the breaking of dormancy by a cold period. *See* DORMANCY.

Seed germination. A seed is an embryo plant usually with a food store in the cotyledons or the endosperm tissue. It is formed following sexual reproduction, and serves to disperse the species and, in some cases via a resistant stage, to carry the species through a period of cold or drought. Seeds may or may not be dormant as described above. A non- or postdormant seed will start to grow as soon as it is provided with suitable conditions of water, oxygen, and warmth. There is a rapid initial uptake of water primarily due to the imbibition of colloids. There is then a transitory stage during which no growth takes place. During this time cellular reorganization following the previous dry stage takes place, and the synthesis of proteins begins. The synthesis of proteins often precedes the synthesis of mRNA which codes for the synthesis of proteins. This is because seeds contain genetic instructions stored in previously synthesized mRNA. This information is translated into the enzymes needed during early germination as soon as germination begins. In cotton, abscisic acid has been found to prevent germination by blocking the translation of the information stored in mRNA. Following the initial protein synthesis, RNA synthesis also starts. After the lag period, growth recommences, primarily through the osmotic uptake of water. The embryonic axes (stem and root) may also increase in dry weight at the expense of the food reserves, but an absolute increase in dry weight cannot occur until the plant grows into the light and starts making is own carbohydrates through photosynthesis. Once the plant reaches the light, the stem straightens up and slows down its elongation rate (which is higher in darkness), the leaves expand, and chlorophyll is synthesized. All these changes are mediated through the pigment phytochrome.

Gibberellin has been found to play a specific role in cereal grains in addition to that of breaking dormancy. On imbibition gibberellin is produced by the embryo and diffuses to the aleurone layer—a protein-rich layer just below the surface of the grain. Here it induces the formation of various hydrolytic enzymes, notably amylase, which diffuses into the starchy endosperm and hydrolyzes the starch to soluble sugars for use by the embryo. The gibberellin has been shown to promote the transcription of mRNA specific for amylase.

Senescence, fruit ripening, and flower fading. Senescence refers to those processes which follow the cessation of growth and are a prelude to the death of the whole plant or plant part (such as a leaf, flower, or fruit). As with any other developmental stage in the plant, senescence appears to be under precise hormonal control. Whole plant senescence in annual plants follows flowering and fruiting and is prevented or delayed if the flowers are removed. Because it also occurs in male pollen-producing plants, and in some cases can be prevented by environmental manipulation or hormonal applications in plants with rapidly growing seeds, senescence cannot be caused by the nutrient drain put on the plant by developing seeds. The developing seeds cause senescence either by diverting growth-promoting hormones from the vegetative growing points or by exporting senescence-inducing substances to the rest of the plant. Which mechanism is operating is not yet known. Senescence is prevented by growth hormones which vary with species (for example, gibberellins in peas or auxin and cytokinins in soybean).

The senescence of leaves is associated with a breakdown of chlorophyll and the associated enzymes of photosynthesis. At the same time there is a pronounced increase in protease, the enzyme responsible for protein breakdown. The yellowing of older leaves can often be prevented by cytokinin application.

Flower fading and fruit ripening represent two specialized cases of senescence. Both are promoted by ethylene applications. The fading of orchids or morning glory flowers, or the ripening of numerous fruits, is associated with the pronounced production of ethylene by the flowers or fruits. In certain fruits (such as the apple), this ethylene production combined with a drop in the auxin level triggers a sharp rise in the respiration rate, termed the respiratory climacteric, which is associated with ripening.

Leaf abscission. As a perennial plant grows, new leaves are continuously or seasonally produced. At the same time the older leaves are shed because newer leaves are metabolically more efficient in the production of photosynthates. A total shedding of tender leaves may enable the plant to withstand a cold period or drought. In temperate deciduous trees, leaf abscission is brought about by declining photoperiods and temperatures. Leaf abscission occurs at a zone of specialized cells at the base of the leaf stalk and is also under precise control. Before abscission can take place, an aging of the abscission-zone cells must occur. This is prevented in vigorous leaves by a continual supply of auxin by the leaf blade. When this auxin supply decreases, aging is hastened by ethylene produced by the abscission-zone cells. In the latter stages of abscission, this ethylene promotes the synthesis and release of cellulase by the abscission-zone cells. The cellulase degrades the cellulose of the cell wall of the abscission-zone cells. Thus the connections between the cells are weakened and break, causing the leaf to shed. *See* ABSCISSION; PLANT MORPHOGENESIS. Peter J. Davies

Bibliography. D. E. Fosket, *Plant Growth and Development: A Molecular Approach*, 1994; A. W. Galston, P. J. Davies, and R. L. Satter, *The Life of the Green Plant*, 3d ed., 1980; J. R. Porter and D. W. Lawler (eds.), *Plant Growth: Interactions with Nutrition and Environment*, 1991; T. A. Steeves and I. M. Sussex, *Patterns in Plant Development*, 2d ed., 1989; P. F. Wareing (ed.), *Plant Growth Substances*, 1983.

Plant hormones

Organic compounds other than nutrients that regulate plant development and growth. Plant hormones, which are active in very low concentrations, are produced in certain parts of the plants and are usually transported to other parts where they elicit specific biochemical, physiological, or morphological responses. They are also active in tissues where they are produced. Each plant hormone evokes many different responses. Also, the effects of different hormones overlap and may be stimulatory or inhibitory. The commonly recognized classes of plant hormones are the auxins, gibberellins, cytokinins, abscisic acid, and ethylene. Circumstantial evidence suggests that flower initiation is controlled by hypothetical hormones called florigens, but these substances remain to be identified. A number of natural or synthetic substances such as brassin, morphactin, and other growth regulators not considered to be hormones nevertheless influence plant growth and development. Each hormone performs its specific functions; however, nearly all of the measurable responses of plants to heredity or environment are controlled by interaction between two or more hormones. Such interactions may occur at various levels, including the synthesis of hormones, hormone receptors, and second messengers, as well as at the level of ultimate hormone action. Furthermore, hormonal interactions may be cooperative, antagonistic, or in balance.

The term plant growth regulator is usually used to denote a synthetic plant hormone, but most of the synthetic compounds with structures similar to those of the natural hormones have also been called hormones. For instance, the synthetic cytokinin kinetin is considered a hormone.

Occurrence and function. The occurrence and function of each plant hormone or group of hormones are given below.

Auxins. In most plants, the principal auxin is indole-3-acetic acid, but some plants contain other compounds that cause many of the same responses. Indole-3-acetic acid may also be present as various conjugates such as indoleacetyl aspartate. It is synthesized from the amino acid tryptophan, which is found primarily in leaf primordia, young leaves, and developing seeds. Movement of auxin is from cell to cell, and normal transport in stems and petioles is from young leaves and through other living cells, including phloem parenchyma and parenchyma cells. Auxin movement is polar, always occurring in stems preferentially in a base-seeking direction, and is not a response to gravity.

Auxin causes cell enlargement and stem growth, and in combination with cytokinin it also stimulates cell division. A high auxin-to-cytokinin ratio promotes root formation in tissue culture. Auxin is also responsible for apical dominance, inhibiting lateral bud development, stimulating the initiation of roots, and inducing fruit setting and growth. Exogenous auxin often causes inhibition of root growth, an action partly caused by ethylene, because auxins stimulate many kinds of plant cells to produce ethylene. Auxin also causes differential elongation of cells, which results in organ curvature. When the plant is exposed to light on one side or to gravity when horizontal, the auxin is diverted to the darker side or to the lower side. *See* APICAL DOMINANCE.

Auxin is thought to cause the cell wall to become plastic, permitting osmotic water to enter and causing the cell to swell. It is theorized that cell-wall plasticity is related to the breaking of acid-labile bonds and that indole-3-acetic acid works by facilitating the release of hydrogen ions. *See* AUXIN.

Gibberellins. Gibberellins, or gibberellic acids, are found in angiosperms and gymnosperms and probably also in mosses, ferns, algae, and at least two fungi but apparently not in bacteria. More than 60 gibberellins have been discovered in various fungi and plants. Gibberellins are synthesized in many parts of the plant, especially in actively growing areas such as embryos and meristematic or developing tissues. A number of synthetic compounds inhibit gibberellin synthesis and cause dwarfing. Gibberellins have the unique ability among plant hormones to promote extensive growth in many intact plants. They also promote germination of dormant seeds and growth of dormant buds, acting as a substitute for low temperatures, long days, or red light. In some plants, gibberellin can substitute for the long-day requirement or for an inductive cold period that leads to flowering. Gibberellins also stimulate the mobilization of foods and mineral elements in seed storage cells. *See* GIBBERELLIN.

Abscisic acid. This hormone, present in all vascular plants, has also been found in two fungi as well as bacteria and algae. Abscisic acid is synthesized in leaves and transported from the leaves by way of the phloem. It promotes dormancy and senescence, and is also the agent that mediates stomatal closure under the effect of drought. This hormone appears to behave as much as a promoter as an inhibitor. For instance, it induces storage protein synthesis in seeds and counteracts the effect of gibberellin on amylase synthesis in germinating cereal grains. Various experts interpret differently the importance of abscisic acid in causing abscission of leaves, flowers, and fruit. Some have concluded that it does cause abscission but much less effectively than exogenous ethylene; others make a strong case for an important role of endogenous abscisic acid in causing abscission. *See* ABSCISIC ACID; ABSCISSION.

Ethylene. The gas ethylene is synthesized in all parts of seed plants, particularly in tissues undergoing senescence or ripening. Only a few bacteria reportedly produce ethylene, and no algae are known to synthesize this compound. Although ethylene causes downward curvature (epinasty) of leaves by promoting elongation of cells on the upper side, it usually inhibits elongation of stems and roots, especially in dicots. The induction of flowering in some plants by ethylene is unusual, because the gas inhibits flowering in most species. Ethylene promotes abscission of leaves and fruits and is often used to hasten the ripening process. *See* ETHYLENE.

Cytokinins. These hormones are synthesized in root and actively dividing cells in the shoots of all plants. They also exist in some pathogenic bacteria. Cytokinins promote cell division and organ formation; they delay senescence, increase nutrient transport, and antagonize the inhibiting action of auxin in apical dominance. They also promote chloroplast development and chlorophyll synthesis and increase cell expansion in dicot cotyledons and in leaves. *See* CYTOKININS.

Other substances. A number of substances not considered to be hormones nevertheless influence plant growth and development, some profoundly. A group of steroid derivatives called brassins or brassinosteroids have distinct growth-promoting activity in some plants, especially in stems. The structure of one brassin, brassinolide, is chemically similar to ecdysone, the molting hormone of insects. Triacontanol, a 30-carbon saturated primary alcohol, significantly enhances the growth of maize and rice plants when sprayed on the foliage of seedlings.

Morphactins, a group of synthetic growth regulators, appear to inhibit indole-3-acetic acid transport and seed germination and to antagonize gibberellins. Polyamines, another group of plant-growth regulators, are widespread in all cells. Development is affected in plants whose polyamine content has been genetically altered. Jasmonic acid and its methyl ester, which occur in several plants and in oil of jasmine, inhibit growth of certain plant parts and strongly promote senescence of detached oat leaves.

Isolation and identification. The development and improvement of chromatographic materials and instrumentation, together with immunologic techniques, provide high resolution and reproducibility and enable the isolation of minute amounts of plant hormones. The methods used for extraction and identification depend upon the species of hormones to be analyzed. For auxin, the most commonly used extractants are methanol and to some extent ethanol, although diethyl ether and acetone have also been used. Gibberellins are usually extracted from plant materials by homogenization in methanol or 80% aqueous acetone. Abscisic acid and related compounds are usually extracted with aqueous methanol, but acetone, ethanol, and mixtures of solvents have also been used. Various chromatographic methods can also be applied to separate abscisic acid from other components. Since ethylene is a gas that is released out of the producing tissue, and since ethylene contained in air or other gases can be directly analyzed by gas chromatography, isolation of the gas is not necessary in most cases. Cold methanol, ethanol, or mixtures of solvents have been used to isolate cytokinins, and perchloric acid has been employed for cytokinin nucleotide extraction. After a plant hormone has been extracted, it is separated by various forms of chromatography, including paper, thin-layer, column, and high-performance liquid chromatography. Immunoassay and gas chromatography–mass spectrometry are commonly used to characterize plant hormones. *See* CHROMATOGRAPHY; GAS CHROMATOGRAPHY; IMMUNOASSAY; MASS SPECTROMETRY.

Applications. There are a number of applications of plant hormones in agriculture, horticulture, and biotechnology. Synthetic auxins, such as 2,4-dichlorophenoxyacetic acid, are used as weed killers. Auxins are also used to counteract the effects of hormones that promote the dropping of fruit from trees. Spraying auxins on pear and apple trees is a standard practice to make the trees retain and ripen more of their fruits.

Gibberellins are used extensively to increase the size of seedless grapes: when applied at the appropriate time and with the proper concentration, gibberellins cause fruits to elongate so that they are less tightly packed and less susceptible to fungal infections. Gibberellins are also used by some breweries to increase the rate of malting because they enhance starch digestion. Celery plants, which are valued for the lengths and crispness of their stalks, respond favorably to gibberellin treatment. Gibberellins have also been sprayed on fruits and leaves of navel orange trees to prevent several rind disorders that appear during storage. They are used commercially to increase sugarcane growth and sugar yields.

Cytokinins and auxins are used in plant cell culture, particularly in cultivating genetically engineered plants. The ability of cytokinins to retard senescence also applies to certain cut flowers and fresh vegetables. Ethylene has been used widely in promoting pineapple flowering; flowering occurs more rapidly and mature fruits appear uniformly, so that a one-harvest mechanical operation is possible. Because carbon dioxide in high concentrations inhibits ethylene production, it is often used to prevent overrippening of picked fruits. Ethylene is also used for accelerating fruit ripening. *See* HORMONE; PLANT GROWTH; PLANT PHYSIOLOGY. Chong-maw Chen

Bibliography. P. J. Davies, *Plant Hormones and Their Role in Plant Growth and Development*, 1987; P. Gresshoff, *Plant Biotechnology and Development*, 1992; T. C. Moore, *Biochemistry and Physiology of Plant Hormones*, 2d ed., 1989.

Plant keys

Artificial analytical constructs for identifying plants. The identification, nomenclature, and classification of plants are the domain of plant taxonomy, and one basic responsibility of taxonomists is determining if the plant at hand is identical to a known plant. The dichotomous key provides a shortcut for identifying plants that eliminates searching through numerous descriptions to find one that fits the unknown plant.

A key consists of series of pairs (couplets) of contradictory statements (leads). Each statement of a couplet must "lead" to another couplet or to a plant name (thus the term "lead"). Each couplet provides an either-or proposition wherein the user must accept one lead as including the unknown plant in question and must simultaneously reject the opposing lead. The user then proceeds from acceptable lead to acceptable lead of successive couplets until a name for the unknown plant is obtained. For confirmation, the newly identified plant should then be compared with other known specimens or with detailed descriptions of that species.

Keys are included in monographic or revisionary treatments of groups of plants, most often for a genus or family. Books containing extended keys, coupled with detailed descriptions of each kind of plant (taxon), for a given geographic region are called manuals or floras, though the latter technically refers to a simple listing of names of plants for a given region.

Types of keys. Several types of artificial keys may be constructed. Generally, modern keys are dichotomous, always composed of pairs of opposing statements as described above. Keys that are polychotomous (that is, have three or more optional lead statements simultaneously) are confusing and time-consuming and should be avoided. Most commonly, keys are presented in two forms, the indented or yoked key and the parallel or bracketed key. Examples of each are presented below for the leaves of the six representative trees and shrubs of North America shown in the **illustration**.

Indented or Yoked Key

1. Leaf venation palmate	
2. Leaves once compound	Plant *e*
2′. Leaves simple	
3. Margins lobed; leaves opposite	Plant *a*
3′. Margins entire; leaves alternate	Plant *c*
1′. Leaf venation pinnate	
4. Leaves once compound	Plant *d*
4′. Leaves simple	
5. Margins serrate; stipules leafy	Plant *f*
5′. Margins lobed; stipules absent	Plant *b*

Parallel or bracketed key

1.	Stem bearing two leaves per node (opposite)	2
1′.	Stem bearing one leaf per node (alternate)	3
2.(1.)	Leaves simple	Plant *a*
2′.	Leaves once palmately compound	Plant *e*
3.(1)′.	Leaves once pinnately compound	Plant *d*
3′.	Leaves simple 4	4
4.(3′.)	Venation palmate	Plant *c*
4′.	Venation pinnate	5
5.(4′.)	Leaf margins lobed Plant *b*	Plant *b*
5′.	Leaf margins serrate	Plant *f*

In the indented or yoked key, the two leads of a couplet are arranged in yokes, the first lead separated from the second by all the former's subordinate yokes. Each subordinate yoke, in turn, is indented under its preceding lead. This type of key permits easy visualization of plant groups, each of which is based on similar morphology. However, the indented key consumes large amounts of page space because of progressive indentation, particularly in extended keys where the two leads of the couplet may be separated by several pages. In the parallel key, the two leads of every couplet are adjacent. The parallel key occupies much less space (a distinct advantage for long keys), but it is difficult to visualize relationships among the different groupings of plants.

In dichotomous keys, the couplets usually are consecutively numbered, lettered, or otherwise symbolized for convenience. The two leads of a given couplet either bear the same number or symbol (3 versus 3) or differentiating numbers or symbols (3 versus 3′, la versus lb, or a versus aa). The number or symbol of the previous couplet's lead may be placed parenthetically after the number or symbol in the first lead in a couplet in parallel keys. This enables the user to backtrack through the key, which could prove advantageous after taking a wrong choice or otherwise getting lost in the key. Leads will provide either a name of the plant or a number for the next couplet in the key that should be considered.

A synoptic key, presented in either yoked or parallel key form, uses technical terms to emphasize evolutionary relationships (phylogenetic classification) among groups of plants, usually higher-ranking groups such as divisions, classes, or orders. Synoptic keys are not designed to identify individual plants but to illustrate general trends without concern for minor exceptions. Synoptic keys differ from keys designed specifically for identification of species, in that the latter use obvious, convenient characters regardless of evolutionary importance.

There are also several specialized keys. The pictorialized key includes illustrations to depict character states difficult to visualize as they arise in the key and to present for each plant an overall sketch. The card sort key consists of a set of cards, one for each kind of plant. Every card bears holes along its margins, each hole assigned to a specific character as written on the card face. If a plant possesses a certain character, the hole is cut away to the margin (notched), but if that plant does not possess a character, the hole is left intact. A specific notched hole may refer, for example, to "leaves simple"; if the same hole had been left intact, then it refers to "leaves not simple." A rod inserted through this hole position for the entire deck of cards will retain all cards with "leaves not simple" and cards that fall away from the deck will have notched holes, meaning "leaves simple." Those remaining on the rod are not considered further. The cards that fell away are then reassembled into a deck and the user goes to another character, and another, until the last card to fall away provides for the unknown plant its name, description, and illustration on the obverse face. Also available are computer programs for keys that will automatically generate dichotomous keys from data matrices prepared and coded for them.

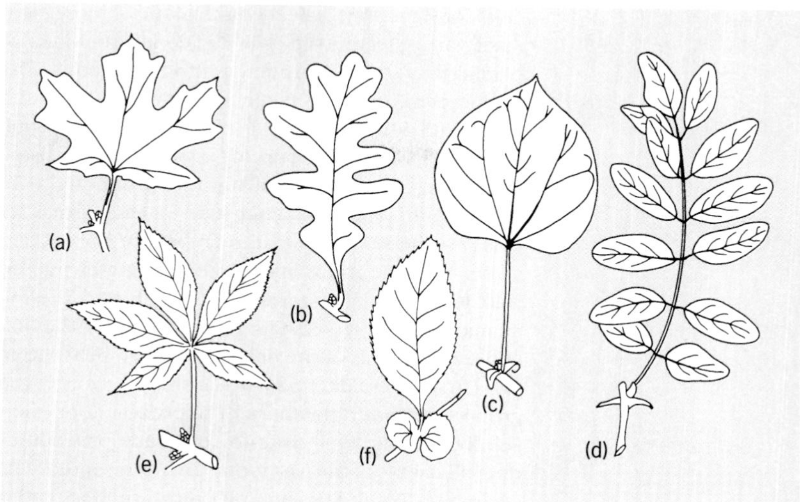

Leaves of six representative trees and shrubs of North America. (*a*) Leaves are simple and opposite; margins are lobed. (*b*) Leaves are simple, margins are lobed, and stipules are absent. (*c*) Leaves are simple and alternate; margins are entire. (*d*) Leaf venation is pinnate; leaves were once compound. (*e*) Leaf venation is palmate; leaves were once compound. (*f*) Leaves are simple, margins are serrate, and stipules are leafy.

Constructing keys. In making artificial keys, simultaneous visualization of differences and similarities among plants is necessary. To accomplish this, a comparison chart (see **table**) may be used. For each plant type considered, character expressions (states) of all major contrasting characters need to be tabulated, such as "simple" or "compound" under the character "leaf type." Then it is possible to quickly ascertain and then indicate simultaneously on the table the differences and likenesses among all plant types.

A comparison chart of character states for leaves of six trees and shrubs of North America shown in the illustration

	Character states (plants *a–f*)					
Character	*a*	*b*	*c*	*d*	*e*	*f*
Leaves per node	**two**	one	one	one	**two**	one
Venation	**palmate**	pinnate	**palmate**	pinnate	**palmate**	pinnate
Leaf type	simple	simple	simple	**compound**	**compound**	simple
Leaf (let) margin	**lobed**	**lobed**	entire	entire	*serrate*	*serrate*
Stipules	**absent**	**absent**	*leafy*	spines	**absent**	*leafy*

In constructing the first couplet, one or more characters are selected that will divide all the kinds of plants (taxa) to be considered into two groups. The character expressions (states) of all members of the first group differ from the character-states of all members of the second group for each character used. Therefore, all members of the first group are alike in having the same character-state for each character used. At the same time, all members of the second group are alike in having the contrasting character-state for each character used. By using these differing character-states to separate the two groups, the first and second leads of the first couplet can be written. (The first lead refers to the first group, preferably the group with fewer members.) The process continues, subdividing the first group by the same procedures until all names of the first group are represented. The second lead of the first couplet is returned to, and is made to lead to the next consecutive number or symbol. No numbers or symbols used for the first group are repeated. The second group continues to be subdivided in like manner until all plant names of the second group are represented. Either the indented (yoked) or parallel key form can be used.

Rules. Certain general guidelines should be followed to construct good keys:

1. Place characters of a given lead in descending order of importance such that the best diagnostic character is first and the most variable last.
2. Make sure that both leads of a couplet are of parallel construction: the same characters in the same sequence and no characters omitted in either of the leads.
3. In opposing leads of each couplet, the character expressions (states) for all characters must be mutually exclusive. In difficult separations, a combination of characters may be necessary, or the same plant may be entered more than once in various portions of the key.
4. The first word of both leads of a couplet should be identical.
5. The first word of consecutive couplets should not be the same.
6. Avoid negative choices if possible. For example, use "leaves compound" versus "leaves simple" but do not use "leaves compound" versus "leaves not compound," or "leaves compound" versus "leaves not as above."
7. Place the name of the plant part before the description. For instance, use "fruits red" but do not use "red fruits."
8. Avoid overlapping variation limits. "Spine 1.5–5.5 cm long" should not be paired with "spine 4.5–8.6 cm long." Avoid vague descriptive parameters, such as "long" or "short," and "leaves narrowly lanceolate" or "leaves broader." Rather, use exact measurements. Avoid redundancy, such as having the same phrase in both leads of one couplet.
9. Use obvious characters where possible. Avoid characters such as chromosome numbers and physiological parameters that are not determinable from dried herbarium specimens. Also, avoid geographic distribution or habitat as a sole character of lead.
10. In dioecious species having separate staminate and pistillate plants, couplets must include characters for both flower types since the plant to be identified will have only one type of flower. Use flower and fruit characters for those plants that do not bear flowers and relatively mature fruits at the same time.
11. Use characters of reproductive plant parts in preference to characters of vegetative plant parts because the reproductive parts are usually less variable, because they have been exposed to influencing environmental factors for a shorter period of time.
12. For each dichotomy, it is most efficient to divide a group of plants into equal or subequal halves; do not splinter off one kind of plant at a time from the main group. An example of a key modified in a historic dichotomous key format is shown below.

Suggestions for the use of keys. Keys are valuable tools but require care in their use. Much material of the unknown plant should be examined to learn of its variability; to work most keys, both reproductive and vegetative materials are required. If possible, diseased, mutant, grazed, juvenile, or otherwise aberrant material should be avoided. Learn the terminology presented in the key. Read, in their entirety, both leads of each couplet before making a decision as to which lead to accept. Remember that the first character of a lead is the most diagnostic. If the user cannot accept either lead of a given couplet, perhaps because the specimen of the unknown plant is incomplete, then accept both leads and continue from each until the unknown becomes identified. If neither lead is workable, then start over and seek out previous errors in judgment. If the unknown does not fit within the key, it may be new to that flora or new to science. *See* PLANT KINGDOM; PLANT TAXONOMY.

Donald J. Pinkava

Bibliography. A. Cronquist, *Evolution and Classification of Flowering Plants*, 2d ed., 1988; M. J. Dallwitz, *User's Guide to the DELTA System: A General System for Coding Taxonomic Descriptions*, 2d ed., 1984; J. E. Forester, *Sort Guide to the Trees of the U.S. and Canada*, 1963; H. D. Harrington, *How to Identify Grasses and Grasslike Plants*, 1977; G. H. M. Lawrence, *Taxonomy of Vascular Plants*, 1951; G. W. Prescott, *How to Know the Aquatic Plants*, Picture Key Nature Series, 2d ed., 1980; A. E. Radford et al., *Vascular Plant Systematics*, 1974.

Plant kingdom

The worldwide array of plant life, including plants that have roots in the soil, plants that live on or within other plants and animals, plants that float on or swim in water, and plants that are carried in the air. Fungi used to be included in the plant kindom because they looked more like plants than animals and did not move about. It is now known that fungi are probably closer to animals in terms of their evolutionary relationships (although not similar enough to be considered animals). Also once included in plants were the "blue-green algae," which are now clearly seen to be bacteria, not plants, although they are photosynthetic (and presumably the group of organisms from which the chloroplasts present in true plants were derived). The advent of modern methods of phylogenetic DNA analysis has allowed such distinctions, but even so, what remains of the plantlike organisms is still remarkably divergent and difficult to classify.

Plants range in size from unicellular algae to giant redwoods. Some plants complete their life cycles in a matter of hours, whereas the bristlecone pines are known to be over 4000 years old. Plants collectively are among the most poorly understood of all forms of life, with even their most basic functions still inadequately known, including how they sense gravity and protect themselves from infection by bacteria, viruses, and fungi (they do not have immune systems like those of vertebrates but nonetheless ward off such attacks). Furthermore, new species are being recorded every year. The application of new approaches to the understanding of this diversity of form and function is yielding major insights, but this increased level of knowledge does not facilitate classification; the stream of new information means that anything suggested today may need to be modified tomorrow.

Classification of plants has relied historically on the use of hierarchy to express how the various named taxa are related. Although the relationships between the taxa are less than certain, there is still information in the levels of the traditional hierarchical scheme. The idea has been that species of plants are grouped into ever larger, inclusive categories to make a formal hierarchy. The white oak (*Quercus alba*), for example, may be classified as follows:

Kindom Plantae
  Subkingdom Embryobionta
    Division Magnoliophyta
      Class Magnoliopsida
        Subclass Rosidae
          Order Fagales
            Family Fagaceae
              Genus *Quercus*
                Species *alba*

Within the land plants, a great deal of progress has been made in sorting out phylogenetic (evolutionary) relationships of extant taxa based on DNA studies, and the system of classification listed below includes these changes. The angiosperms or flowering plants (Division Magnoliophyta) have recently been reclassified based on phylogenetic studies of DNA sequences. Within the angiosperms, several informal names are indicated in parentheses; these names may at some future point be formalized, but for the present they are indicated in lowercase letters because they have not been formally recognized under the Code of Botanical Nomenclature.

It is known that the bryophytes (Division Bryophyta) are not closely related to each other, but which of the three major groups is closest to the other land plants is not yet clear. Among the extant vascular plants, Lycophyta are the sister group to all the rest, with all of the fernlike groups forming a single monophyletic (natural) group, which is reflected here in the classification by putting them all under Polypodiophyta. This group is the sister to the extant seed plants, within which all gymnosperms form a group that is sister to the angiosperms. Therefore, if Division is taken as the highest category within Embryobionta (the embryo-forming plants), then the following scheme would reflect the present state of knowledge of relationships (an asterisk indicates that a group is known only from fossils). See separate articles on names marked by daggers.

Subkingdom Thallobionta (thallophytes)†
  Division Rhodophycota (red algae)
    Class Rhodophyceae†
  Division Chromophycota†
    Class: Chrysophyceae (golden or golden-brown algae)†
      Prymnesiophyceae†
      Xanthophyceae (yellow-green algae)†
      Eustigmatophyceae†
      Bacillariophyceae (diatoms)†
      Dinophyceae (dinoflagellates)†
      Phaeophyceae (brown algae)†
      Raphidophyceae (chloromonads)†
      Cryptophyceae (cryptomonads)†
  Division Euglenophycota (euglenoids)
    Class Euglenophyceae†
  Division Chlorophycota (green algae)†
    Class: Chlorophyceae†
      Charophyceae†
      Prasinophyceae
Subkingdom Embryobionta (embryophytes)†
  Division Rhyniophyta*†
    Class Rhyniopsida†
  Division Bryophyta†
    Class Hepaticopsida (liverworts)†
      Subclass Jungermanniidae†
        Order: Takakiales†
          Calobryales†
          Jungermanniales†
          Metzgeriales†
      Subclass Marchantiidae†
        Order: Sphaerocarpales†
          Monocleales†
          Marchantiales†
    Class: Anthocerotopsida (hornworts)†
      Sphagnopsida (peatmosses)†
      Andreaeopsida (granite mosses)†
      Bryopsida (mosses)†
      Subclass: Archidiidae†
        Bryidae†
        Order: Fissidentales†
          Bryoxiphiales†
          Schistostegales†
          Dicranales†
          Pottiales†
          Grimmiales†
          Seligeriales†
          Encalyptales†

Order: Funariales†
Splachnales†
Bryales†
Mitteniales†
Orthotrichales†
Isobryales†
Hookeriales†
Hypnales†
Subclass: Buxbaumiidae†
Tetraphididae†
Dawsoniidae†
Polytrichidae†
Division Lycophyta†
Class Lycopsida†
Order: Lycopodiales†
Asteroxylales*†
Protolepidodendrales*†
Selaginellales*
Lepidedendrales*†
Isoetales†
Class Zosterophyllopsida*†
Division Polypodiophyta†
Class Polypodopsida†
Order: Equisetales†
Marattiales†
Sphenophyllales*
Pseudoborniales*
Psilotales
Ophioglossales†
Noeggerathiales*
Protopteridales*
Polypodiales†
Class Progymnospermopsida*
Division Pinopsida
Class Ginkgoopsida†
Order: Calamopityales*
Callistophytales*
Peltaspermales*
Ginkgoales†
Leptostrobales*
Caytoniales†
Arberiales*
Pentoxylales*
Class Cycadopsida†
Order: Lagenostomales*
Trigonocarpales*
Cycadales†
Bennettiales*
Class Pinopsida†
Order: Cordaitales*†
Pinales†
Podocarpales
Gnetales†
Division Magnoliophyta (angiosperms, flowering plants)†
Class Magnoliopsida†
unplaced groups: Amborellaceae, Ceratophyllaceae, Chloranthaceae, Nymphaeaceae, etc.
eumagnoliids†
Order: Magnoliales†
Laurales†
Piperales†
Winterales
monocotyledons†
Order: Acorales†
Alismatales†
Asparagales†
Dioscoreales†
Liliales†
Pandanales†
commelinids
Arecales†
Commelinales†
Poales†
Zingiberales†
eudicotyledons†
(basal eudicots)
Order: Ranunculales†
Proteales†
Buxales
Trochodendrales†
(core eudicots)
Order: Berberidopsidales
Gunnerales
Dilleniales†
Santalales†
Caryophyllales†
Saxifragales†
Rosidae†
Order: Vitales
Myrtales†
Geraniales†
Crossosomatales
(eurosid I)
Order: Celastrales†
Cucurbitales
Fabales†
Fagales†
Malpighiales†
Oxalidales†
Rosales†
Zygophyllales†
(eurosid II)
Order: Brassicales†
Malvales†
Sapindales†
Asteridae†
Order: Cornales†
Ericales†
(euasterid I)
Order: Garryales
Gentianales†
Lamiales†
Solanales†
(euasterid II)
Order: Apiales†
Aquifoliales
Asterales†
Dipsacales†

See DEOXYRIBONUCLEIC ACID (DNA); PLANT EVOLUTION; PLANT PHYLOGENY; PLANT TAXONOMY.

Mark W. Chase; Michael F. Fay

Plant metabolism

The complex of physical and chemical events of photosynthesis, respiration, and the synthesis and degradation of organic compounds. Photosynthesis produces the substrates for respiration and the starting organic compounds used as building blocks for subsequent biosyntheses of nucleic acids, amino acids and proteins, carbohydrates and organic acids, lipids, and natural products.

Energy Metabolism

The processes involved in the utilization and generation of energy in plants are photosynthesis and respiration.

Photosynthesis. Photosynthesis, which forms all organic compounds that are used for growth, development, and maintenance processes, has two aspects: primary processes that use light energy to form stable, high-energy compounds [reaction (1)],

$$\text{NADP} + \text{ADP} + \text{P}_i \xrightarrow[\text{light}]{\text{chlorophyll}} \text{ATP} + \text{NADPH} \qquad (1)$$

and carbon metabolism that produces organic compounds from CO_2 by using the energy of reduced nicotinamide adenine dinucleotide phosphate (NADPH) and adenosine triphosphate (ATP) produced by the primary processes [reaction (2)].

$$\text{CO}_2 + \text{H}_2\text{O} + \text{NADPH} + \text{ATP} \rightarrow \text{CH}_2\text{O} + \text{NADP} + \text{ADP} + \text{P}_i \qquad (2)$$

Primary processes. A photosynthetic action spectrum shows major efficiency peaks in the blue and red. Quantum considerations of the chlorophyll molecule and light absorption indicate that blue light acts similarly to red. The electrochemical span for photosynthesis goes from hydrogen to oxygen, a potential difference of about 1.2 eV. Red light of 700 nanometers has 167 kilojoules einstein^{-1}, sufficient to drive the reaction over 1.2 eV (115 kJ) even with an efficiency of 70%.

A photosynthetic unit is a minimum-sized particle that can absorb radiant energy and carry out the photochemical reactions. A unit is visualized to have about 200 chlorophyll molecules, carotenoids, cytochromes, and other components.

According to the modern concept of the photochemistry of photosynthesis, there are two different reaction centers, each composed of a specialized chlorophyll *a*. Associated with these reaction centers are accessory pigments, including chlorophyll *a* and chlorophyll *b*, that act as light-harvesting pigments or antenna pigments transferring energy to the centers. The reaction centers themselves make up only about 1% of the total pigments of the chlorophyll apparatus. The reaction center named photosystem I is a low-redox-potential system functioning in the reduction of reducing agents; the reaction center named photosystem II is a high-redox-potential system capable of oxidizing water.

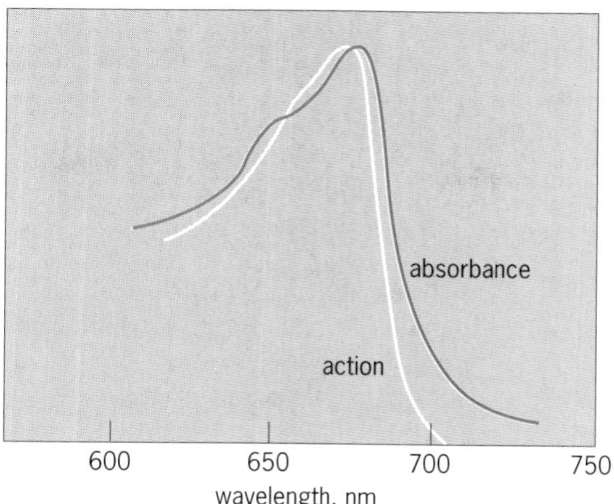

Fig. 1. Absorption spectrum for chlorophyll and action spectrum for photosynthesis, showing the red drop phenomenon in which absorption of red light occurs farther into the red than photosynthesis.

The concept of two centers developed from the fact that the absorption spectrum for chlorophyll *a* extends further into the red than the action spectrum for photosynthesis (**Fig. 1**), and that low levels of red light (<680 nm) given simultaneously with low levels of far-red light (>680 nm) result in more photosynthesis than when the two lights are given separately. **Figure 2** illustrates the present two-light-system hypothesis for photosynthesis.

After light excitation of P700, a special form of chlorophyll *a*, electrons are passed to a primary acceptor. Rapid transfer of electrons to additional carriers ultimately results in reduction of NADP to NADPH [reaction (3)]. Thus light excitation of photosystem

$$\text{PS I} \rightarrow \text{ferredoxin} \xrightarrow[\text{Fd-NADP-reductase}]{\text{NADP}} \text{NADPH} \qquad (3)$$

I produces a strong reductant and a weak oxidant. There is some evidence that the primary acceptor may be a bound ferredoxin, or at least the ferredoxin is close to the primary electron acceptor. The primary acceptor appears to be an iron-sulfur protein. The donor of electrons to the oxidized P700, which come from photosystem II, is probably plastocyanin and cytochrome *f*.

Whereas the data concerning the redox agents of photosystem I are based primarily on electron spin resonance and absorption spectroscopy, data from photosystem II also come from studies of fluorescence. Thus, somewhat more is known about this system, which functions in the oxidation of water and consequent evolution of oxygen [reaction (4)]

$$2\text{H}_2\text{O} \rightarrow 4\text{H}^+ + 4e^- + \text{O}_2 \qquad (4)$$

and the transferring of electrons to photosystem I. The reaction center of photosystem II is named P680.

Coupling of the two centers through the weak reductant of photosystem II and the weak oxidant of photosystem I drives phosphorylation, producing

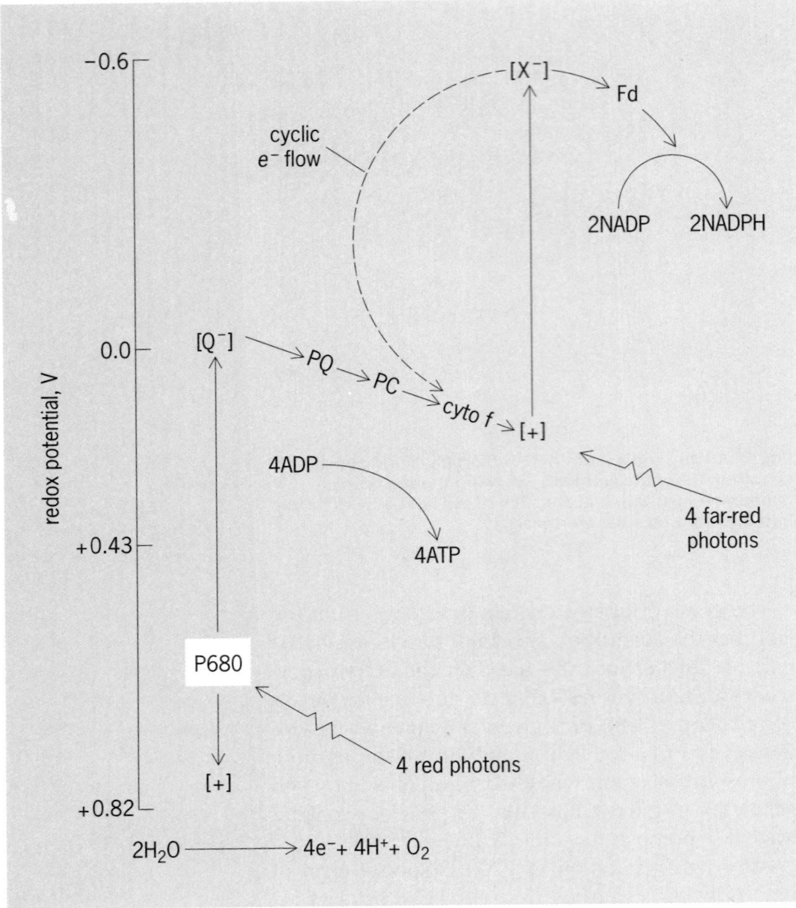

Fig. 2. Z scheme for electron transport in photosynthesis, showing noncyclic electron flow from water in photosystem II to NADP in photosystem I, and cyclic electron flow in which ATP is still formed, but not NADPH. PQ = plastoquinone, PC = plastocyanin, cyto f = cytochrome f, Fd = ferredoxin, Q = quinone.

two ATPs per electron pair. There are two sites for phosphorylation. Phosphorylation is driven by chemiosmotic coupling in which proton gradients are established across chloroplast membranes. The main features of the hypothesis are as follows: there is a vectorial transport of protons across the membrane; electron flow is coupled to proton transport; counterion flux across the membrane prevents positive-ion accumulation; and the proton gradient creates a proton-motive force sufficient to drive phosphorylation.

Electron transport from water to NADP, forming ATP in the process and involving both photosystems, is called noncyclic electron transport. When photosystem II is uncoupled from photosystem I [as by 3(3,4-dichlorophenol)-1,1-dimethylurea, or DCMU], red light will still drive phosphorylation, but there is no NADP reduced. Electrons are cycled from photosystem I to the strong reductant and back through the electron transport system to the photosystem I center. This cyclic electron flow produces ATP but no NADPH.

Under special conditions in the absence of CO_2, chloroplasts consume oxygen and produce ATP but not NADPH. Electron transport is from water to oxygen, a process called pseudocyclic photophosphorylation [reaction (5)].

$$H_2O \rightarrow \text{PS II} \rightarrow \text{PQ} \rightarrow \text{cyt } f/\text{PC} \rightarrow \text{PS I} \rightarrow O_2 \qquad (5)$$

Cyclic photophosphorylation and perhaps pseudocyclic photophosphorylation may produce the extra ATP that is required for cellular activities other than the carbon metabolism of photosynthesis.

The chloroplast envelope is not permeable to ATP and NADPH, yet there is evidence that ATP and NADPH from photosynthesis are used in nonchloroplast metabolism. Glucose and sucrose do not pass readily across the envelope, as do malate and the triose phosphates. The latter are transported by a phosphate exchange reaction. It is visualized that shuttles carry ATP and reducing power across the envelope as shown in **Fig. 3**. *See* ADENOSINE DIPHOSPHATE (ADP); ADENOSINE TRIPHOSPHATE (ATP).

Carbon metabolism. The path of carbon in photosynthesis has been known since the 1950s (**Fig. 4**). Carboxylation catalyzed by ribulose bisphosphate carboxylase produces 2 moles of phosphoglycerate, the first product of photosynthesis. Reduction of the latter forms sugar, glyceraldehyde phosphate. Because the first product is a three-carbon compound, the process is referred to as C_3 photosynthesis.

In the 1960s, a modification of the pathway was discovered in sugarcane, wherein the first detectable product of carboxylation is a four-carbon organic acid, oxalacetate. Because of this, the process is referred to as C_4 photosynthesis. The oxalacetate is rapidly reduced to malate or aminated to form aspartate. The initial carboxylation of P-enolpyruvate (phosphoenolpyruvate, or PEP) carboxylase occurs in the outer mesophyll cells (the C_4 plants have Kranz anatomy). Malate or aspartate is transferred to the inner bundle sheath cells, where decarboxylation forms CO_2 and a three-carbon compound (such as pyruvate). The CO_2 is then assimilated through the ribulose bisphosphate carboxylase reaction. The three-carbon compound is recycled back to the mesophyll cells and converted back to the acceptor molecule, P-enolpyruvate, by a special enzyme called pyruvate inorganic phosphate dikinase. In some C_4 species, amino acids are transported rather than organic acids.

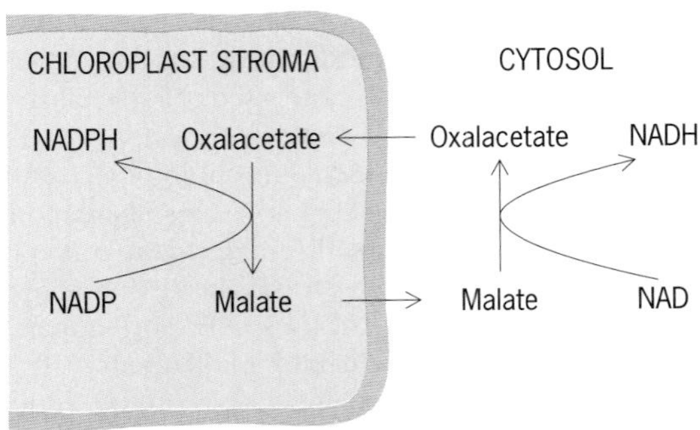

Fig. 3. Malate shuttle hypothesis; a possible mechanism transferring reducing power from NADPH in the chloroplast to NADH in the cytosol. Other possible schemes involve transport of triose phosphates.

The chloroplasts of C_4 species have typical-looking chloroplasts in the mesophyll cells, but they lack ribulose bisphosphate carboxylase. The P-enolpyruvate carboxylase is in the cytosol. Both photosystems are present. Ribulose bisphosphate carboxylase is present in the large lamellar chloroplasts of the bundle sheath cells, but they have a reduced photosystem II.

In C_3 species there are 3 ATP and 2 NADPH required for CO_2 fixation (Fig. 4), whereas in C_4 species additional ATP is required, primarily to regenerate the P-enolpyruvate.

In most plants (C_3 species in particular) there is light-dependent oxygen consumption and CO_2 evolution called photorespiration that occurs in leaf microbodies. The enzyme ribulose bisphosphate carboxylase has oxygenase activity forming phosphoglycerate and glycolate phosphate (**Fig. 5**). The glycolate is oxidized in the microbodies by glycolate oxidase, consuming more oxygen and ultimately producing glycine and serine. Complete oxidation of the latter yields CO_2. In some C_3 plants, as much as 50% of the carbon fixed by photosynthesis can be lost through photorespiration with no conservation of energy, prompting researchers to search for plants with reduced photorespiration. Because C_4 photosynthesis keeps the CO_2 level high in the vicinity of ribulose bisphosphate carboxylase, there is less photorespiration in C_4 species.

Still another modification of photosynthesis occurs in succulent plants, called crassulacean acid metabolism (CAM) because it has been extensively studied in the Crassulaceae. In CAM plants, stomata are closed during the day and open at night. CO_2 entering the plant at night is fixed into malic acid through the P-enolpyruvate carboxylase/malate dehydrogenase couple. Malic acid accumulates in large water-storing vacuoles [100–200 meq kg^{-1} (fresh weight)]. During the subsequent day period, malic acid is decarboxylated, and the CO_2, which is trapped inside the leaf because of closed stomata, is assimilated by the C_3 process. Since CO_2 is taken up through open stomata at night when the evaporative demand is low, water loss is minimized. The transpiration ratio (water loss to carbon gained) is 5 to 10 times less in CAM than in C_3 and C_4. Thus CAM is an adaptation to drought and water-stressed environments.

Some CAM plants shift from C_3 to CAM in response to water stress. Thus CAM is adaptive and flexible, presenting interesting research possibilities. *See* PHOTOSYNTHESIS.

Respiration. Respiration is the oxidative degradation of organic compounds to yield usable energy usually in the form of reduced pyridine nucleotides (NADH and NADPH) and triphosphate nucleotides (such as ATP). It is a process of all living organisms, but in plants photosynthate is the immediate substrate. The process starting with hexose is shown in reaction (6). Complete combustion of hexose yields

$$C_6H_{12}O_6 + 6O_2 \rightarrow 6H_2O + 6CO_2 + \text{energy} \quad (6)$$

about 2900 kJ mol^{-1}. During respiration there is

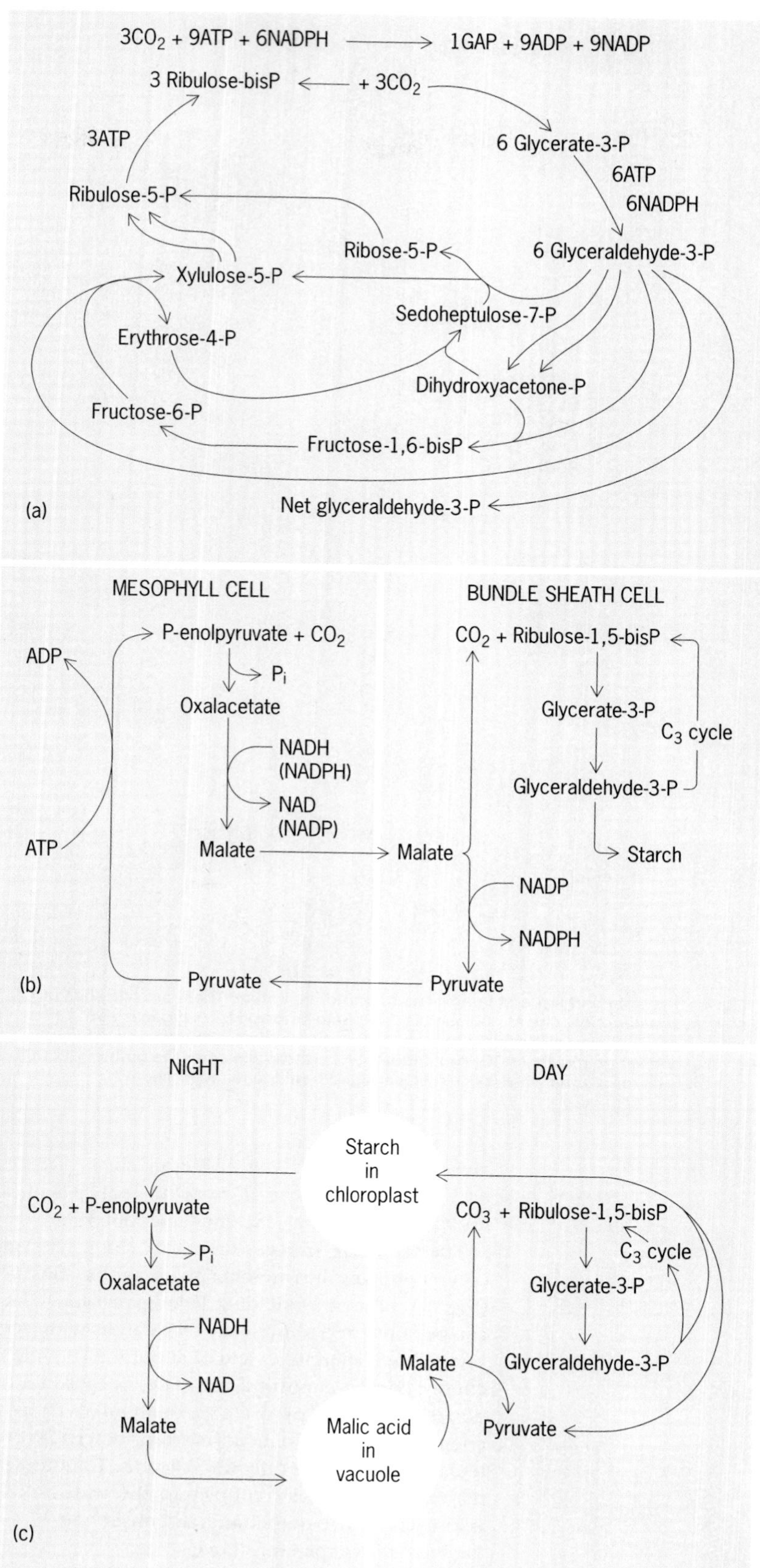

Fig. 4. Photosynthesis reactions. (*a*) C_3 photosynthesis. The reductive pentose cycle of photosynthesis in which 3 moles of CO_2 result in 1 net glyceraldehyde-3-P and the two modifications, C_4 photosynthesis and CAM photosynthesis. (*b*) C_4 photosynthesis. (*c*) CAM photosynthesis. In *b* and *c*, there is an initial carboxylation that forms malate or other four-carbon compound, followed by decarboxylation and subsequent assimilation to form glyceraldehyde-3-P as in the reductive pentose cycle.

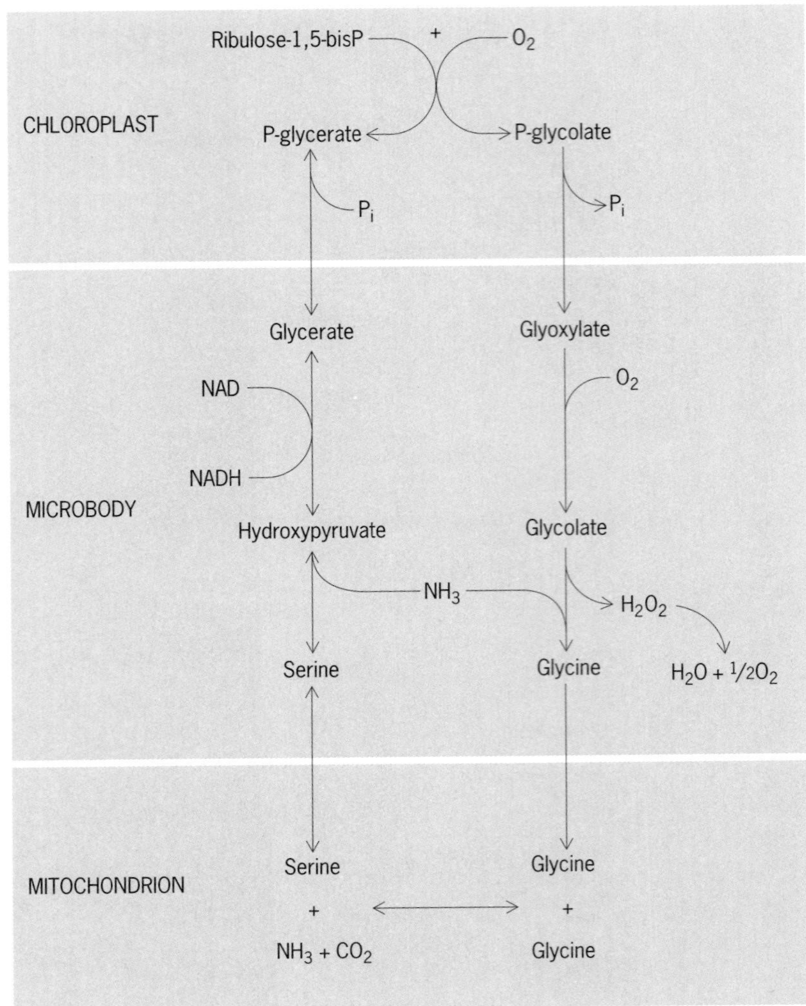

Fig. 5. Glycolate and glycerate pathways of photorespiration. There are two sites of oxygen consumption, one at ribulose bisphosphate carboxylase/oxygenase in chloroplasts and one at glycolate oxidase in the microbodies. Complete oxidation of glycine and serine results in CO_2 production. Glycerate could enter the pathway from the chloroplasts, or alternatively, it could be salvaged from the microbodies.

yielded about 36 ATP or ATP equivalents (free energy of hydrolysis of ATP = 30 kJ mol^{-1}) with an energy conservation efficiency of about 37%.

The oxidative process ordinarily starts with glycolysis utilizing hexose-phosphate as the substrate (**Fig. 6**), ultimately yielding P-enolpyruvate, which can be converted to pyruvate by catalysis with pyruvate kinase. There is a yield of about 6 ATP with an energy conservation of about 31%, because oxidation of hexose to 2 pyruvate yields about 6 ATP. If the complete oxidation of hexose is considered (2900 kJ mol^{-1}), glycolysis is only 6% efficient. To complete respiration, glycolysis couples to the tricarboxylic acid cycle in the mitochondria, conserving 37% of the energy as explained above.

An alternative oxidative pathway for hexose is through the pentose phosphate pathway (Fig. 6), yielding about 24 moles of ATP for an energy conservation of about 25%.

Organic acids are oxidized in the tricarboxylic acid cycle. Pyruvate in the mitochondria is decarboxylated by pyruvate dehydrogenase, a complex reaction involving several steps [reaction (7)]. The

$$\text{Pyruvate} + \text{CoA} + \text{NAD} \rightarrow \text{acetyl CoA} + \text{NADH} + CO_2 \quad (7)$$

acetyl CoA can act as a substrate. Alternatively, P-enolpyruvate still in the cytosol can be carboxylated by P-enolpyruvate carboxylase, forming oxalacetate. Oxalacetate is reduced by malate dehydrogenase, forming malate. In plants, malate is a better substrate for entry into the mitochondria than pyruvate.

Although the complete oxidation of acetate or other organic acid such as malate in the tricarboxylic acid cycle in the mitochondria yields some ATP (see Fig. 6), the reduced compounds flavins adenine dinucleotide ($FADH_2$) and NADH are usually formed. These two compounds feed electrons into a chain of oxidation-reduction electron transport compounds, collectively called the electron transport system, and ultimately to O_2 (**Fig. 7**). The oxidation of $FADH_2$ and NADH by O_2 is sufficient to form 2 or 3 ATP with a yield of about 24 moles ATP from the 2 moles of acetate.

Cyanide, carbon monoxide, azide, and antimycin A will inhibit respiration. In some plant tissues, however, such as aged potato tissue and in the flowers of Araceae, respiration continues in their presence, but only 2 ATP are formed rather than the usual 6 per electron pair transferred to oxygen. This observation indicates that two of the sites are bypassed presumably by an alternative oxidase system (Fig. 7). The excess energy is lost as heat.

Oxidative phosphorylation is driven by chemiosmotic coupling similar to photophosphorylation in chloroplasts.

The respiratory quotient (RQ) of plant tissues is close to 1, indicative of hexose oxidation, but during germination of seeds that store carbohydrate, RQs increase from about 1 to near 1.3, suggesting that the substrate shifts from carbohydrate to organic acids. During germination of seeds that store lipid, the RQ is low, indicative of fatty acid oxidation. *See* PHOTORESPIRATION; PLANT RESPIRATION.

Organic Constituents of Plants

The principal organic compounds of plants are nucleic acids, amino acids, proteins, carbohydrates, organic acids, lipids, and natural products.

Nucleic acids. Nucleic acids, found in virtually all plant cells, are complex polymers with masses of 10^7 daltons or more for native deoxyribonucleic acid (DNA) and 10^6 daltons for some kinds of ribonucleic acids (RNA). DNA contains the genetic code, and RNA functions in protein synthesis and regulation. *See* GENETIC CODE.

Nucleic acids are polymers of substituted purines and pyrimidines. DNA and RNA are composed primarily of adenine, guanine, cytosine, 5-methlylcytosine, and thymine, except that RNA has uracil rather than thymine. Also, small quantities of rare,

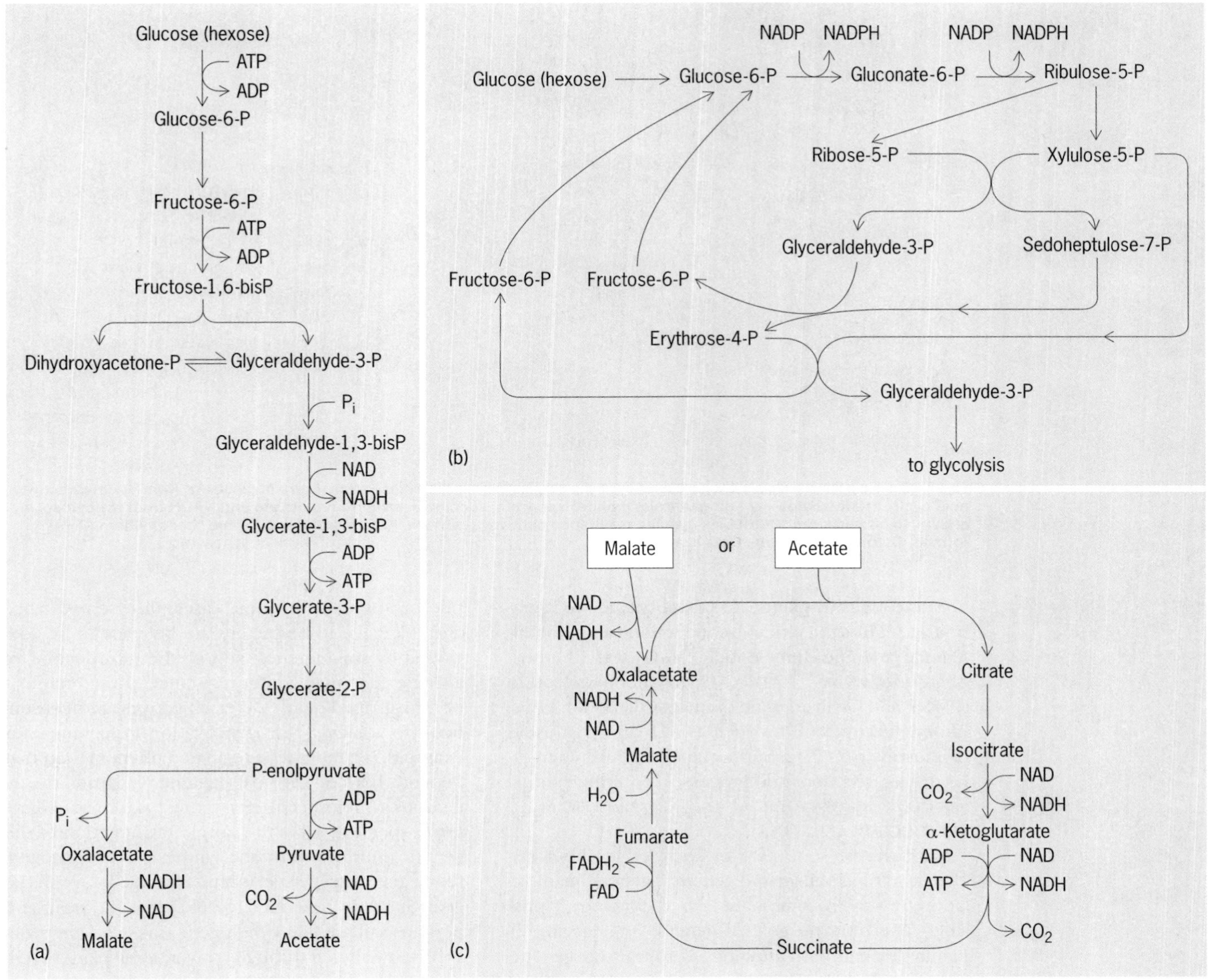

Fig. 6. Glycolysis and the generation of acetate and malate that act as (*a*) tricarboxylic acid cycle substrates, (*b*) the pentose phosphate pathway of hexose oxidation, and (*c*) the tricarboxylic acid cycle.

substituted bases probably function as recognition sites. In the nucleic polymer, the bases exist as nucleotides in which there is a ribose (in RNA) or a deoxyribose (in DNA) bridge between the base and a phosphate. DNA and RNA polymers are formed by polymerization of the deoxyribotides (DNA) or ribotides (RNA) through phosphate diester bonds. Ordinarily, the DNA exists in a double-helical form, whereas the RNA is single-stranded.

The central dogma of cell biology is diagrammed in **Fig. 8**. DNA of the genome (such as in the nucleus) has the information for protein synthesis. This information is transcribed into messenger RNA (mRNA). The mRNA migrates into the cytoplasm and in cooperation with ribosomal RNA and ribosomes is translated and proteins as synthesized. The process is aided by transfer RNAs acting as amino acid carriers. After peptide and protein synthesis, there may be further modification to form active enzymes.

Chloroplasts and mitochondria also have a genome composed of circular DNA not unlike prokaryote DNA. The chloroplast DNA has a mass of about 9×10^7 daltons. Chloroplasts and mitochondria also have RNA and ribosomes with the capacity for protein synthesis, but much of their protein is the result of the nuclear genome. Some organelle membrane proteins are coded for by the organelle genomes. In the special case of the chloroplast, the large subunit of the carboxylating protein, ribulose bisphosphate carboxylase, is coded for by the chloroplast-DNA, while the small subunit is coded for by the nuclear-DNA. *See* CELL PLASTIDS; MITOCHONDRIA.

Polypeptide synthesis systems have been isolated, primarily from wheat germ. Using plant DNA, nucleotides, RNA polymerase enzymes, and other factors, mRNA can be produced in culture. Polypeptides can be prepared with the purified mRNA, amino acids, and proper cofactors.

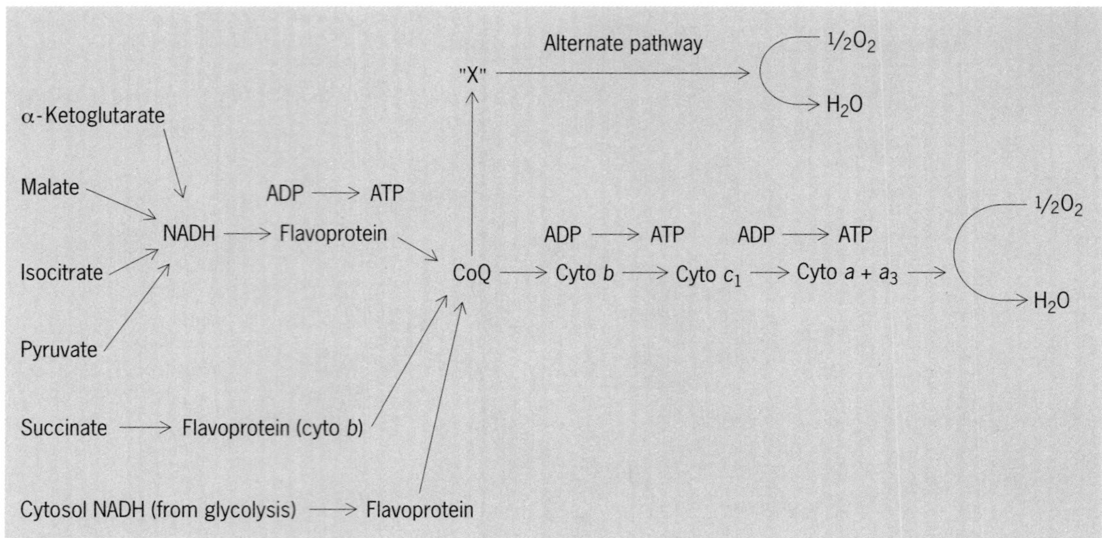

Fig. 7. Scheme for electron transport to oxygen in mitochondria during respiration. Entry of electrons from the tricarboxylic acid cycle intermediates—α-ketoglutarate, malate, isocitrate, succinate—and from pyruvate and NADH from the cytosol are shown. In addition, the alternative, cyanide resistance pathway is shown. Note that there are three "sites" where ATP is formed. Cyto = cytochrome. CoQ = a quinone.

Nucleotides in plants are of importance to metabolism such as the oxidation-reduction factor, nicotinamide adenine dinucleotide (NAD) and its phosphate derivative (NADP), flavin mononucleotide (FMN), and flavin adenine dinucleotide (FAD). Phosphorylated nucleotides such as ATP and guanosine triphosphate (GTP) are important in cellular energetics driving reactions and processes requiring energy. *See* DEOXYRIBONUCLEIC ACID (DNA); NUCLEIC ACID; RIBONUCLEIC ACID (RNA).

With the present state of knowledge about nucleic acid metabolism and genetics, genetic manipulation through recombinant DNA technology is possible. The ultimate goal of genetic engineering of plants through biotechnology is to transfer genetic information in the form of genes from one organism to another in order to alter the metabolism of the host plant. Specific tasks under intensive study using these techniques include the improvement of disease resistance; the development of agricultural plants that are resistant to herbicides; improved cold tolerance; increase in the quality of seed storage proteins for human consumption; increased general stress tolerance to salinity, drought, and heat; introduction of the capacity for nitrogen fixation; and increase in the efficiency of photosynthesis, perhaps by reducing photorespiration. In some cases, disease resistance and tolerance to herbicides are believed to be controlled by single genes, and so the introduction of a single gene may confer resistance of the recipient or transformed plant. Other objectives are more difficult to achieve. Nitrogen fixation in legumes, for example, is a multigene-regulated phenomenon that depends both on the host (legume plant) and the infectious symbiotic nitrogen-fixing bacterium (*Rhizobium* species). Thus the simple introduction of the genetic information for the synthesis of the nitrogen-fixing gene nitrogenase is unlikely to accomplish the goal. Similarly, tolerance to cold, drought, heat, and perhaps salt are likely multigene traits and are probably best handled through conventional plant breeding technology. *See* GENETIC ENGINEERING.

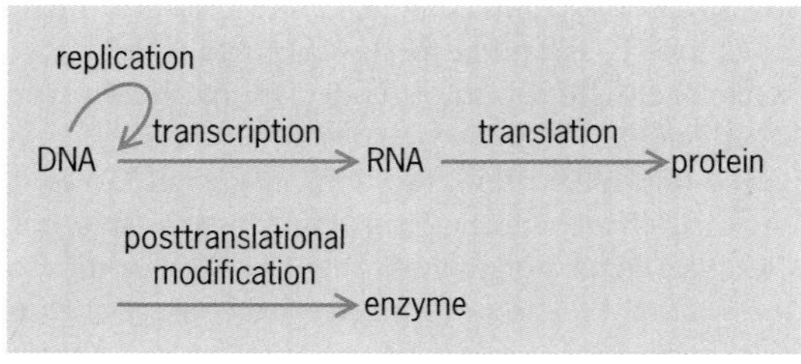

Fig. 8. Diagram of the central dogma of cell biology.

Amino acids and proteins. About 22 amino acids are found in proteins, and probably many others play some role in plant metabolism. Most are of the L configuration, but a few of the nonprotein amino acids are D. Polymerization of amino acids by covalent linkage of an amino group of one and the carboxyl of another (CO—NH) forms peptides. A few small plant peptides such as glutathione (a tripeptide) are found, but they are not common. Larger polypeptides making up proteins are important as storage for nitrogen and carbon, as structural components of membranes and, to a lesser extent, cell walls, and as enzymes that regulate cellular reactions.

Seed storage proteins are important from the standpoint of energy reserves in the developing seedling. They are synthesized during seed development and stored in membrane-bound protein bodies. During germination, the proteins rapidly disappear as the nitrogen and amino acids are mobilized for growth and development of the young seedling. Because of considerations for human nutrition, most is known about storage proteins in cereals and legumes. Almost 70% of the world's protein supply comes from plants, with 35% from seed proteins. Although storage proteins are species-specific (such as zein in corn), they can readily be manipulated through traditional plant breeding technology. For the most part, the storage proteins of cereals are prolamines that are deficient in lysine and frequently also in tryptophan; hence, by themselves, they are poor protein sources from a human nutritional viewpoint. Rice is, however, high in lysine, and the artificial genus *Triticale*, a cross between wheat and rye, is also high in lysine. In legumes, which have substantially more storage protein than cereals, the proteins are mostly of the globulin type, which have adequate lysine but are deficient in sulfur amino acids such as the essential methionine. *See* CEREAL; LEGUME; SEED; TRITICALE.

Two kinds of membrane proteins can be differentiated: peripheral proteins that are easily removed by dilute salt treatment, and integral proteins that are tightly bound. Both membrane proteins play a role in metabolism as enzymes and may function in transport. The cell wall structural protein extensin is rich in hydroxyproline.

Enzyme proteins are of two types: Constitutive proteins that are always present include those regulating respiration, fatty acid synthesis, and protein synthesis. Adaptive enzymes appear during specific periods. As an example, enzymes of the glyoxylate cycle that couple fatty acid oxidation during lipid breakdown to carbohydrate synthesis are present just during germination.

Many membrane proteins evidently function in transport, although there is no direct evidence. Soluble proteins may play a role in the regulation of enzymic proteins; for example, there is an activating protein called activase for ribulose bisphosphate carboxylase. Many proteins are activated by inorganic elements such as calcium: environmental signals such as light bring about membrane transport of calcium, which acts as a secondary messenger binding to proteins and activating them. One such specific protein of plants is calmodulin, which is activated by calcium and in turn regulates metabolism at the level of protein activity.

Enzymes exist in more than one form. Multiple forms of one enzyme are called isozymes and have the same catalytic function but different properties. They are readily separated from each other by electrophoresis, usually on starch or acrylamide gels. Isozymes may function in different pathways that have a common reaction; they may be localized in different cellular compartments, or they may be redundant, acting in conservation during evolution, or simply be the result of heterozygosity. Isozymes may be nonallelic (different gene loci) or allelic (same locus). Most nonallelic isozymes will also exist as allelic isozymes since most gene loci are heterozygous. Also, isozymes may result from posttranslational modifications. The carbohydrate moiety of glycoproteins (such as peroxidases) may differ, resulting in isozymes.

There are three nonallelic isozymes of malate dehydrogenase, each of which occurs in a different compartment (microbodies, mitochondria, and the cytosol), and each functions in a different metabolic pathway (the glyoxylate cycle or photorespiration in microbodies, the tricarboxylic acid cycle in mitochondria, and organic acid synthesis in the cytosol; **Fig. 9**). These nonallelic isozymes are known to exist as allelic isozymes, the result of heterozygosity at

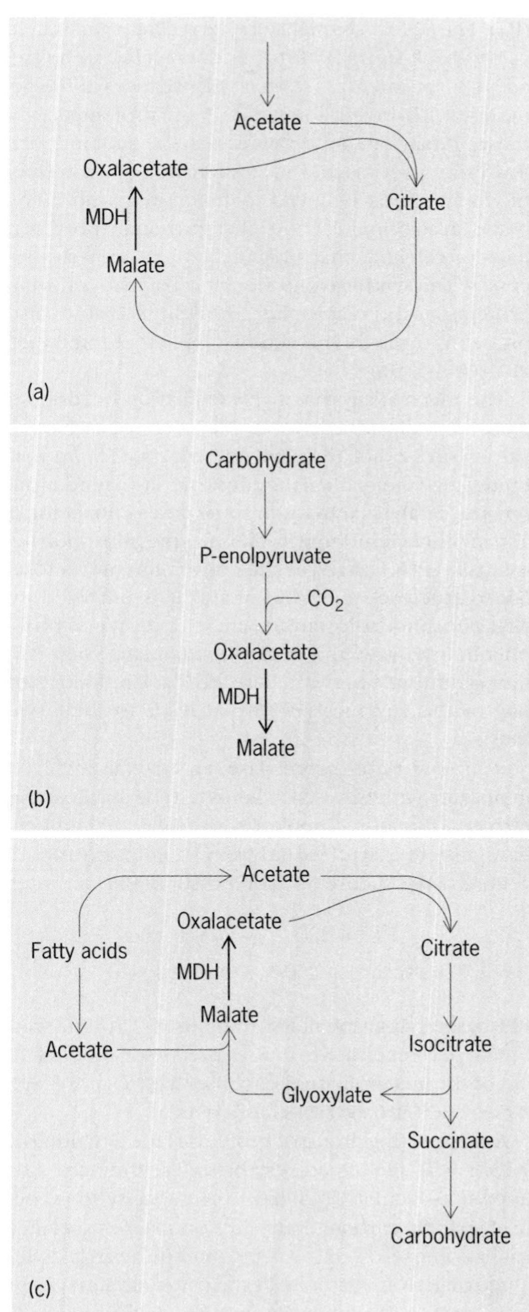

Fig. 9. Function of malate dehydrogenase (MDH) isozymes in different metabolic pathways. The abbreviated metabolic charts show the common steps of the (*a*) tricarboxylic acid cycle in mitochondria, (*b*) dark CO_2 fixation in cytosol, and (*c*) the glyoxylatecycle in microbodies called glyoxysomes that have MDH.

individual gene loci. Isozymes are known for almost all enzymes that are found in plants.

Little is known about the regulation of protein synthesis, but it is assumed that inductive and repression-derepression systems are functional in plants similar to those in animals and bacteria. Light plays an important role in the regulation of protein synthesis through the blue pigment phytochrome. Phytochrome is a protein with an open tetrapyrrole chromophore that exists in two forms, red-absorbing form (P_r) that is converted to a far-red-absorbing form (P_{fr}). The P_{fr} form, which absorbs red light and is converted back to the P_r form, is considered to be the active form and is associated with changes at the genetic level, bringing about mRNA and protein biosynthesis. It plays an important role in the light-induced flowering process and in germination. In addition, phytochrome is believed to function at the membrane level, bringing about changes in transport, perhaps of calcium, that ultimately affect growth processes. The synthesis and regulation of many enzyme proteins of the chloroplast are light-mediated, also probably through the action of phytochrome. *See* PHYTOCHROME.

The mechanism of activity regulation occurs primarily by allosteric effects. Many enzymes have a site or sites other than the active catalytic site that binds small molecules (the allosteric site). Binding alters the catalytic activity, in some cases enhancing it and in others inhibiting it. The enzyme phosphofructokinase (PFK) catalyzes the conversion of fructose-6-P to fructose-1,6-bisphosphate. It is inhibited by ATP, phosphoenolpyruvate, and citrate. When phosphoenolpyruvate and citrate accumulate, their further synthesis is prevented by feedback inhibition of one of the enzymes (PFK) that leads to their synthesis.

P-enolpyruvate carboxylase, an enzyme involved in malate synthesis, is also known to be an allosteric enzyme of plants. Coupled with malate dehydrogenase, malate is synthesized from P-enolpyruvate [reactions (8)]. Glucose-6-P, a precursor of PEP, activates

$$\begin{aligned} \text{PEP} + \text{HCO}_3^- &\rightarrow \text{oxalacetate} + \text{P}_i \\ \text{Oxalacetate} + \text{NADH} &\rightarrow \text{malate} + \text{NAD} \end{aligned} \qquad (8)$$

PEP carboxylase and malate inhibits it. Thus PEP carboxylase is regulated by both a precursor and a product of the metabolic sequence. *See* AMINO ACIDS; ENZYME; PROTEIN; PROTEIN METABOLISM.

An interesting group of proteins is the agglutinins, which will agglutinate erythrocytes; they are also known as lectins. They have the capacity to recognize specific carbohydrate sites on macromolecules such as those of cell walls. Common examples include ricin from castor bean and concanavalin A from jack bean. Although their physiological function is not entirely clear, they may play a role in cell-to-cell recognition. For example, it is believed that they are responsible for the recognition of specific roots by the symbiotic nitrogen-fixing bacteria of the genus *Rhizobium*. Recognition accounts for the legume species' specificity of the rhizobia. They also may be important in protection against plant pathogens by specific binding and agglutination. *See* LECTINS.

Nitrogen. Despite being the most abundant element in the atmosphere, atmospheric nitrogen cannot be utilized directly for plant metabolism. The first step in the incorporation of nitrogen into plant organic matter is nitrogen fixation through symbiotic or free-living nitrogen-fixing prokaryotic organisms. Nitrogenase catalyzes the fixation of atmospheric nitrogen [reaction (9)]. Nitrogenase is a large complex

$$\text{N}_2 + 6e^- + n\text{MgATP} \rightarrow 2\text{NH}_4^+ + n\text{MgADP} + n\text{P}_i \qquad (9)$$

protein containing both molybdenum and iron that is found only in prokaryotes; thus no plant is capable of nitrogen fixation. After NH_4^+ formation, there is a two-step enzymatic reaction incorporating the fixed nitrogen into amino acids. Alternatively, NO_3^- in the soil is available to plants. After uptake, the NO_3^- is reduced to ammonia (NH_3) or ammonium (NH_4^+), depending upon pH. The two-step reaction is catalyzed by nitrate reductase to form nitrite [reaction (10)]. The nitrite is then reduced to ammonia by nitrite reductase [reaction (11)]. Actual assimilation

$$\text{NO}_3^- + 2\text{H}^+ + 2e^- \rightarrow \text{NO}_2^- + \text{H}_2\text{O} \qquad (10)$$

$$\text{NO}_2^+ + 6e^- + 6\text{H}^+ \rightarrow \text{NH}_3 + \text{H}_2\text{O} + \text{OH}^- \qquad (11)$$

of the ammonia takes place by a second two-step reaction [(12) and (13)]. It is catalyzed first by glu-

$$\text{Glutamate} + \text{NH}_3 + \text{ATP} \rightarrow \text{glutamine} + \text{ADP} + \text{P}_i \qquad (12)$$

$$\text{Glutamine} + \alpha\text{-ketoglutarate} \rightarrow 2 \text{ glutamate} \qquad (13)$$

tamine synthetase and then by glutamate synthase, with the sum of the two reactions being glutamate, adenosinediphosphate, and inorganic phosphate [reaction (14)]. Once the nitrogen is an integral com-

$$\alpha\text{-Ketoglutarate} + \text{ATP} + \text{NH}_3 \rightarrow \text{glutamate} + \text{ADP} + \text{P}_i \qquad (14)$$

ponent of glutamate, other nitrogenous organic compounds can be synthesized by catalysis with transaminases and other enzymes of nitrogen metabolism. *See* NITROGEN FIXATION.

Carbohydrates and organic acids. Carbohydrates are among the most abundant biochemicals constituting the initial products of photosynthesis and the primary substrates for respiration. When phosphorylated, they are metabolically reactive. The six-carbon hexose sugars glucose and fructose when phosphorylated are primary substrates for respiration, and the disaccharide sucrose is the primary organic compound transported, although other nonreducing sugars such as raffinose, stachyose, and verbascose are translocated in some species. The simple monosaccharides in plants may have an amino group (the amino sugars), a carboxyl group (the sugar acids), or exist as O-glycosides.

Oligosaccharides have six or fewer monosaccharides linked by an O-glycosidic bond between the aldehyde group of one to a hydroxyl of another. The larger polysaccharides are important from a structural and storage role. The most abundant carbohydrate is cellulose, a major constituent of cell walls. It is a straight-chain polymer of glucose linked through β-(1→4) O-glucosidic bonds and may be 200,000 to 2×10^6 daltons in mass. The common storage polysaccharide of plants, starch, is a mixture of branched (amylopectin) and straight chains (amylose). The linkages are α-(1→4) O-glucosidic bonds in amylose, and amylopectin and the branching of amylopectin is through α-(1→6) O-glucosidic bonds. The fructosans inulin with α-(2→1) linkages and levan with α-(2→6) linkages are other storage polysaccharides.

Their biosynthesis is poorly understood, but sucrose may illustrate a general mechanism. One enzyme that can catalyze sucrose synthesis is sucrose phosphate synthase. Sucrose phosphate is formed from uridine diphosphate glucose and fructose-6-phosphate [reaction (15)], followed by hydrolysis of the sucrose-phosphate [reaction (16)]. The biosyn-

$$\text{UDP-glucose} + \text{fructose-6-P} \rightarrow \text{sucrose-P} + \text{UDP} \qquad (15)$$

$$\text{Sucrose-P} \rightarrow \text{sucrose} + P_i \qquad (16)$$

thesis of other polysaccharides takes place by similar mechanisms in which the monosaccharides as nucleotide derivatives (usually of adenosinediphosphate or uridine diphosphate) are added to an existing saccharide, resulting in chain elongation [reaction (17)]. Breakdown of the polysaccharides is usually by hydrolysis [reaction (18)].

$$n\text{ADP-glucose} + \text{saccharide primer} \rightarrow \text{polysaccharide} + n\text{ADP} \qquad (17)$$

$$\text{Starch} + nH_2O \rightarrow n\text{ maltose} \qquad (18a)$$

$$\text{Maltose} + H_2O \rightarrow 2\text{ glucose} \qquad (18b)$$

In seeds that store starch, germination is accompanied by hydrolysis of starch according to reactions (18) to produce the hexoses glucose and fructose. These hexoses are converted to sucrose according to reactions (15) and (16). The sucrose is transported to growing and developing sites within the young seedling and is used as the primary energy source, being metabolized to CO_2 and water and producing energy in the form of ATP.

Cell wall mannans (polymers of mannose), xylans (polymers of xylose), and mixed polysaccharides with more than one kind of sugar are called hemicelluloses, having properties similar to cellulose. The pectins and pectic acids are common polysaccharides of cell walls and are mixtures of arabans, galactans, and galacturonic acid.

The gums and mucilages are hydrophilic polymers of rhamnose, arabinose, xylose, galactose, glucuronic acid, and galacturonic acid. Because of their large water-holding capacity, they may play a part in plant-water relations or simply be storage carbohydrates.

Organic acids derived from the carbohydrates are substrates in the tricarboxylic acid cycle (such as citrate, malate, succinate, and isocitrate), play a role as counterions, in osmotic balance, and as buffers, and are products of photosynthesis. Plants accumulate specific organic acids. Thus apple has malic acid, citrus has citric acid, *Oxalis* has oxalic acid, *Fumaria* accumulates fumaric acid, and grapes accumulate tartaric acid.

Organic acids are highly compartmented in plant cells. Malic acid occurs in mitochondria as one of the substrates for respiration [reaction (19)], and

$$\text{Fumaric acid} + H_2O \rightarrow \text{malic acid} \qquad (19a)$$

$$\text{Malic acid} + \text{NAD} \rightarrow \text{oxalacetic acid} + \text{NADH} \qquad (19b)$$

there is another pool in the cytosol (and perhaps vacuole) that does not mix with the mitochondrial pool. This malic acid is synthesized by a carboxylation of P-enolpyruvate as mentioned previously. The enzymes catalyzing the common reaction sequences are isozymes. *See* CARBOHYDRATE; CARBOHYDRATE METABOLISM.

Lipids. Plant lipids are a heterogeneous group of compounds that include the neutral lipids (fats and oils), phospholipids, glycolipids, terpenoids, and waxes having the common characteristic of being hydrophobic. They are soluble in organic solvents and slightly soluble in water. They are storage compounds, components of membranes, and protective substances of exposed surfaces such as cutin and suberin.

Neutral lipids yield fatty acids and glycerol upon hydrolysis [reaction (20)]. They can be saponified

$$\begin{array}{l} CH_2OOR \\ | \\ CH\ \ OOR \\ | \\ CH_2OOR \\ \text{Triglyceride} \end{array} + 3H_2O \longrightarrow \begin{array}{l} CH_2OH \\ | \\ CH\ \ OH \\ | \\ CH_2OH \\ \text{Glycerol} \end{array} + \begin{array}{c} 3ROOH \\ \\ \\ \\ \\ \text{Fatty acids} \end{array} \qquad (20)$$

with alkali such as NaOH or KOH to form glycerol and the salts of fatty acids (the soaps) [reaction (21)].

$$\begin{array}{l} CH_2OOR \\ | \\ CH\ \ OOR \\ | \\ CH_2OOR \end{array} + 3KOH \longrightarrow \begin{array}{l} CH_2OH \\ | \\ CH\ \ OH \\ | \\ CH_2OH \end{array} + 3ROOK \qquad (21)$$

Fatty acids are usually even-numbered and include butyric acid (4 carbons), caproic acid (6), caprylic acid (8), capric acid (10), lauric acid (12), myristic acid (14), palmitic acid (16), stearic acid (18), arachidic acid (20), behenic acid (22), lignoceric acid (24), and cerotic acid (26). Palmitic acid is the most common, followed by lauric, myristic, and stearic acids. The unsaturated fatty acids oleic acid, linoleic acid, and linolenic acid—18-carbon fatty acids with one, two, and three double bonds—are also abundant.

The phospholipids of membranes are substitution products of phosphatidic acid (**1**). Free phospha-

$$\begin{array}{l} CH_2\,COOR \\ | \\ CH\ \ COOR \\ | \\ CH_2OPO_3H_2 \end{array}$$

(**1**)

tidic acid is rare in plants, but the phosphatides phosphatidylglycerol, phosphatidylethanolamine, phosphatidylcholine, phosphatidylserine, and phosphatidylinositol derived from phosphatidic acid by bonding to the phosphate are common.

Important constituents of chloroplast membranes are the glycolipids, monogalactosyl-diglyceride and digalactosyl-diglyceride. Unsaturated fatty acids such as linolenic acid are frequent components of the R groups.

The waxes that function primarily in protection differ from the triglycerides in that long-chain alcohols replace the glycerol.

Some plants store waxes rather than triglycerides. Jojoba (*Simmondsia chinensis*) stores a wax ester with an average content of 40 to 42 carbon atoms and is composed of a long-chain monounsaturated alcohol and a long-chain monounsaturated fatty acid. This wax, which is similar in properties to sperm whale oil, is the primary energy source for the germination and initial growth of the jojoba plant. The wax has attracted attention because of its potential use as a lubricant and a nonnutritive oil for human use. It is presently used extensively in cosmetics. *See* JOJOBA; WAX, ANIMAL AND VEGETABLE.

Biosynthesis of the saturated fatty acids begins with acetyl coenzyme A (CoA) adding to a small sulfur-containing carrier protein (ACP-SH; **Fig. 10**). Through the action of a specific enzyme the acetyl group is added to malonyl-S-ACP, forming acetoacetyl-S-ACP (**2**). Reduction of the beta keto group results

$$CH_3 - \overset{\overset{\large O}{\|}}{C} - CH_2 - \overset{\overset{\large O}{\|}}{C} - S - ACP$$

(**2**)

in butyric acid. Subsequent additions of acetyl groups cause chain elongation. Unsaturation of the final product would produce unsaturated fatty acids, but evidence indicates that unsaturation may occur before chain elongation is finished.

Oxidation of fatty acids occurs by a mechanism similar to the biosynthesis but in reverse (Fig. 9). First, the triglyceride is hydrolyzed to glycerol and free fatty acids. The β-oxidation pathway proceeds by formation of the coenzyme A derivative (R-SCoA), followed by oxidation to form a double bond between the alpha and beta carbons. Addition of water and then oxidation to form the beta keto group allows displacement with CoA-SH to form acetyl CoA and a new fatty acid CoA derivative with two less carbons. The process is continued until only acetyl CoA remains. Alpha oxidation removes one carbon at a time rather than two, and probably plays a role in the metabolism of odd-numbered fatty acids.

In seeds that store oil and lipids, such as castor bean and soybean, the mobilization of reserves is much more complicated than in starch-storing seeds and tubers. The triglycerides or wax esters are metabolized as discussed above to produce acetate. The acetate as the acetyl CoA derivative is metabolized in the glyoxylate cycle (Fig. 9*c*) to produce succinate and ultimately carbohydrate by reactions of gluconeogenesis. The latter is essentially a reversal of glycolysis (Fig. 6*a*). Succinate is converted to malate (Fig. 6*a*), and then the synthesis of P-enolpyruvate starts the pathway to hexose. The enzyme responsible for the catalysis is most likely P-enolpyruvate carboxykinase [reaction (22)], effectively overcom-

$$\text{Oxalacetate} + \text{ATP} \rightarrow \text{P-enolpyruvate} + \text{ADP} + CO_2 \qquad (22)$$

ing the energy barrier of the reaction from acetate to

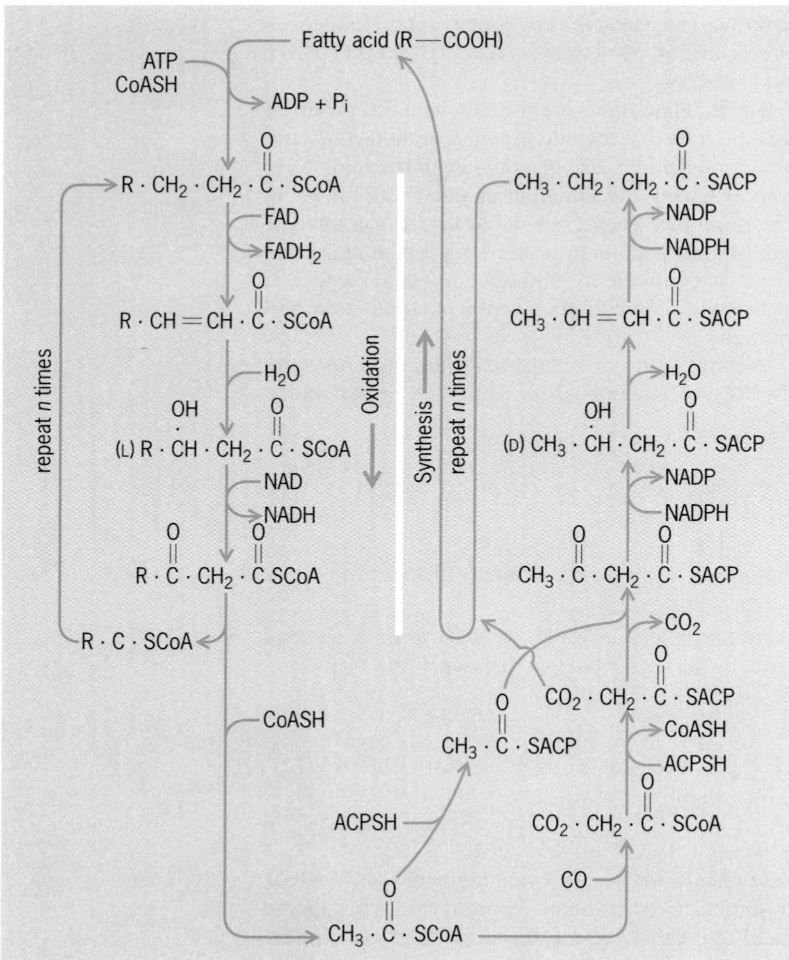

Fig. 10. Metabolic flow chart for fatty acid metabolism showing oxidation (β-oxidation) and biosynthesis. Note that in the synthetic pathway, NADPH is the reducing agent, the D form of the β-hydroxyketo acid is formed, and the active intermediates are derivatives of acyl carrier protein (ACP-SH), whereas in the oxidative direction, NAD is the oxidizing agent, the L form of the β-hydroxyketo acid is formed, and the active intermediates are derivatives of coenzyme A (CoA-SH).

P-enolpyruvate. Subsequent reactions up to fructose-1, 6-bisP proceed with little difficulty. The conversion from fructose-1, 6-bisP to fructose-6-P is catalyzed by a phosphatase [reaction (23)]. Fructose

$$\text{Fructose-1, 6-bisP} + H_2O \rightarrow \text{fructose-6-P} + P_i \qquad (23)$$

is converted to glucose, and then glucose and fructose are converted to sucrose catalyzed by sucrose phosphate synthase [reactions (15) and (16)]. The sucrose is transported to sites of energy requirement. *See* LIPID; LIPID METABOLISM.

Natural products. Natural products are those compounds synthesized by secondary reactions from the primary carbohydrates and amino acids. They include many medicines, narcotics, poisons, stimulants, essential oils, resins, and compounds with teratogenic and carcinogenic properties. Such compounds as opium, quinine, digitoxin, cocaine, caffeine, and rubber are well known. They may be storage or end products. Many are allelochemics that affect the growth, health, or behavior of other organisms such as repellents, growth retardants, or attractants for pollinators and herbivores. *See* SECRETORY STRUCTURES (PLANT).

Natural products incllude phenolics, alkaloids, terpenoids, and porphyrins. Phenolics have an aromatic ring and at least one hydroxyl group such as the cinnamic acids, coumarins, ligini, flavonoids, and tannins. The polyketide phenolics are similar to the above except their biosynthesis is different. Rather than resulting from the shikimate pathway (**Fig. 11**), they are the result of cyclization of linear poly-β-keto chains (**3**). They frequently have hydroxyls in

$$CH_3-\overset{\overset{\displaystyle O}{\|}}{C}-CH_2-\overset{\overset{\displaystyle O}{\|}}{C}-CH_2-\overset{\overset{\displaystyle O}{\|}}{C}-$$

(3)

the meta positions, unlike those from the shikimate pathway. *See* PHENOL.

Alkaloids have basic properties because of a nitrogenous heterocyclic ring or rings. They include piperidine alkaloids, aromatics, betacyanins, isoquinoline and indole alkaloids, purine alkaloids, and aliphatic alkaloids. They are mainly synthesized from the amino acids tryptophan, phenylalanine, glutamate, orinithine, and lysine. *See* ALKALOID.

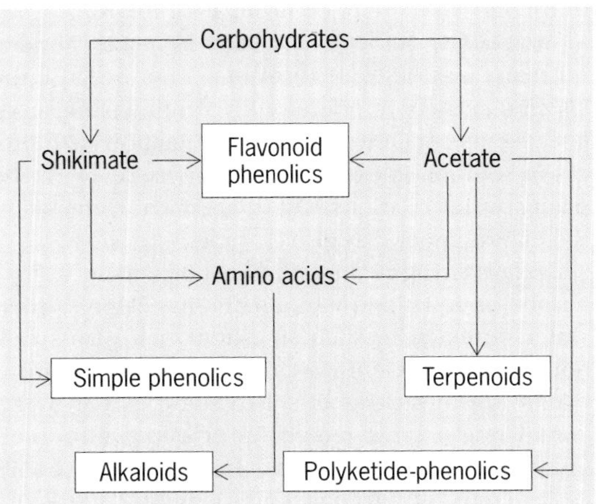

Fig. 11. Abbreviated flow chart for the synthesis of the various natural products found in plants.

Terpenoids are lipids that may be acyclic or cyclic, with a few to several hundred carbon atoms. They include essential oils, steroids, and large polymers such as rubber. They are synthesized by condensation of isoprene (isopentenyl pyrophosphate), usually by a head-to-tail linkage. They are grouped according to number of isoprene units in their backbone: hemiterpenoids with one, monoterpenoids with two, sesquiterpenoids with three, diterpenoids with four, triterpenoids with six, tetraterpenoids with eight, and polyterpenoids with more than eight. *See* RUBBER; TERPENE.

Porphyrins are compounds synthesized from pyrrole usually by cyclization of four pyrroles to form a tetrapyrrole (**4**). Substitutions of the pyrrole rings

(4)

result in the different prophyrins. Magnesium porphyrin is chlorophyll, and the iron porphyrins of plants are the cytochromes. The pyrrole ring is synthesized from δ-aminolevulinic acid that is produced by condensation of succinate and glycine [reaction (24)]. In chlorophyll biosynthesis, however, there is

$$\begin{matrix} CO_2H \\ | \\ CH_2 \\ | \\ CH_2 \\ | \\ CO_2H \end{matrix} \; + \; \begin{matrix} CO_2H \\ | \\ CH_2\,NH_2 \end{matrix} \; \longrightarrow \; \begin{matrix} CO_2H \\ | \\ CH_2 \\ | \\ CH_2 \\ | \\ C{=}O \\ | \\ CH_2\,NH_2 \end{matrix} \; + \; CO_2 \qquad (24)$$

evidence that the δ-aminolevulinic acid comes from glutamate rather than succinate and glycine. *See* CHLOROPHYLL; PLANT CELL; PLANT GROWTH; PLANT PHYSIOLOGY; PORPHYRIN.

Irwin P. Ting

Bibliography. G. Edwards and D. A. Walker, *C_3, C_4: Mechanisms, and Cellular and Environmental Regulations of Photosynthesis*, 1983; D. Grierson and S. N. Covey, *Plant Molecular Biology*, 1988; P. W. Ludden and J. E. Burris, *Nitrogen Fixation and CO_2 Metabolism*, 1985; T. Robinson, *The Organic Constituents of Higher Plants*, 5th ed., 1983; P. K. Stumpf and E. E. Conn, *The Biochemistry of Plants: A Comprehensive Treatise*, 16 vols. 1980–1988; I. P. Ting, *Plant Physiology*, 1982.

Plant mineral nutrition

The relationship between plants and all chemical elements other than carbon, hydrogen, and oxygen in the environment. Plants obtain most of their mineral nutrients by extracting them from solution in the soil or the aquatic environment. Mineral nutrients are so called because most have been derived from the weathering of minerals of the Earth's crust. Nitrogen is exceptional in that little occurs in minerals: the primary source is gaseous nitrogen of the atmosphere.

Mineral Nutrients in Relation to Growth

Some of the mineral nutrients are essential for plant growth; others are toxic, and some absorbed by plants may play no role in metabolism. Many are also essential or toxic for the health and growth of animals using plants as food.

Essential nutrients. In 1699 John Woodward published experiments describing how the growth of spearmint in water increased with the extent to which the water was contaminated by soil. His water culture, or hydroponics, technique has been progressively refined and used extensively in subsequent research aimed at identifying the specific components of his "peculiar terrestrial matter," and six basic facts have been established: (1) plants do not need any of the solid materials in the soil—they cannot even take them up; (2) plants do not need soil microorganisms; (3) plant roots must have a supply of oxygen; (4) all plants require at least 14 mineral nutrients; (5) all of the essential mineral nutrients may be supplied to plants as simple ions of inorganic salts in solution; and (6) all of the essential nutrients must be supplied in adequate but nontoxic quantities. *See* HYDROPONICS.

These facts provide a conceptually simple definition of and test for an essential mineral nutrient. A mineral nutrient is regarded as essential if, in its absence, a plant cannot complete its life cycle. The test for an essential mineral nutrient involves growing plants in water cultures containing a mixture of salts of all the elements thought to be essential: omitting any essential nutrient prevents growth (**Fig. 1**).

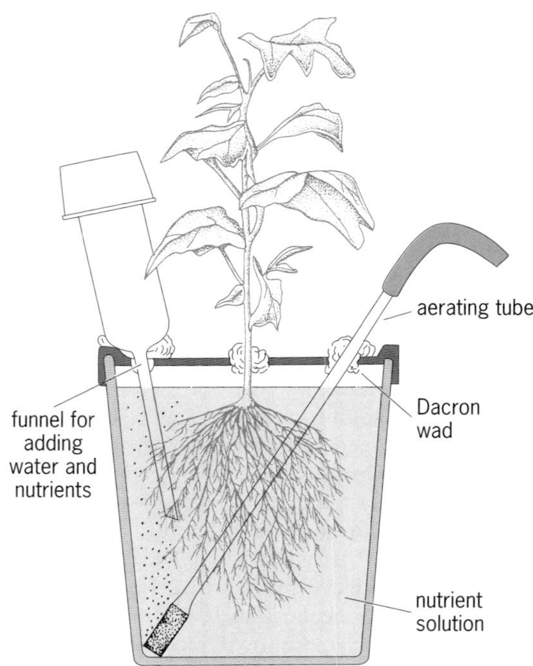

Fig. 1. **Setup for growing plant in solution culture. (*After E. Epstein, Mineral Nutrition of Plants: Principles and Perspectives, John Wiley and Sons, 1972*)**

This test worked well for those nutrients which plants require in relatively large amounts, that is, the essential macronutrients—nitrogen, sulfur, phosphorus, calcium, potassium, and magnesium. It also worked well for iron which, although it is not required in large amounts and hence is regarded as an essential micronutrient or trace element, is rapidly precipitated from aerated solutions of inorganic salts by formation of insoluble ferric hydroxide. As a result, these seven chemical elements were all recognized as essential mineral nutrients before 1900.

However, the delineation of other essential mineral nutrients has proved much more difficult largely because of the extremely small quantities which plants require and the difficulties of removing them from the environment. With the progressive development of better techniques for purifying water and salts, the list of essential nutrients has continually expanded. By 1960, boron, manganese, zinc, copper, molybdenum, and chlorine were added, making 13 chemical elements which are universally recognized as essential mineral nutrients for all plants. Evidence has been accumulated in support of a fourteenth element, nickel, being essential. In addition, sodium and silicon have been shown to be essential for some plants, beneficial to some, and possibly of no benefit to others. Cobalt has also been shown to be essential for the growth of legumes when relying upon atmospheric nitrogen. Claims that two other chemical elements may be essential micronutrients (vanadium and selenium), while more firmly based than claims for the essentiality of yet other mineral nutrients, have still to be firmly established.

Toxicities of nutrients and other elements. Mineral nutrients may be toxic to plants either because the specific nutrient interferes with plant metabolism or because its concentration in combination with others in solution is excessive and interferes with the plant's water relations. Other chemical elements in the environment may also be toxic.

Salt toxicity. High concentrations of salts in soil solutions or aquatic environments may depress their water potential to such an extent that plants cannot obtain sufficient water to germinate or grow. Plants can offset this effect to some extent by accumulating organic molecules or salts within their own cells. Some desert plants growing in saline soils can accumulate salt concentrations of 20–50% dry weight in their leaves without damage. But salt concentrations of only 1–2% can damage the leaves of many species. *See* PLANT-WATER RELATIONS; PLANTS, SALINE ENVIRONMENTS OF.

Specific ion toxicity. A number of elements interfere directly with other aspects of plant metabolism. For example, sodium is thought to become toxic when it reaches concentrations in the cytoplasm that depress enzyme activity or damage the structure of organelles, while the toxicity of selenium is probably due to its interference in metabolism of amino acids and proteins as the result of its partial replacement of sulfur in their structure.

The ions of the heavy metals, cobalt, nickel, chromium, manganese, copper, and zinc are particularly toxic in low concentrations. The toxicities of these ions may result from their multivalent properties, which permit them to interfere with the function of proteins and other large molecules by forming cross linkages through adsorption to negatively charged sites. Plants are particularly sensitive to heavy-metal toxicities when the concentration of calcium in solution is low; increasing calcium increases the plant's tolerance. Aluminum and boron are also toxic at low concentrations. Aluminum is toxic only in acid soils for the reasons discussed below. Boron may be toxic in soils over a wide pH range, and is a serious problem for sensitive crops in regions where irrigation waters contain excessive boron, as in southern California, or where the soils contain unusually high levels of boron, as in parts of southern Australia.

Tolerance to toxic elements also varies with plant species and cultivars. In some cases, tolerance is accompanied by an unusually high accumulation of the normally toxic element in plant tops, where its toxic activity may be inhibited by excretion from the cell to the leaf surface, secretion into cell vacuoles, or combination with specific compounds synthesized by the cell.

Acid soils. All plants grow poorly on very acid soils (pH $\leq$ 3.5); some plants may grow reasonably well on somewhat less acid soils. Several factors may be involved, and their interactions with plant species are complex. High concentrations of hydrogen ions may inhibit and distort root growth, but in soils other factors such as deficiencies of calcium, phosphorus, or molybdenum or toxicities of manganese and especially aluminum are usually more important. Many soil clays contain a high proportion of aluminum within their crystal structure, and in acid soils aluminate ions are released into solution. Aluminate ions are very toxic to root growth, stunting roots by inhibiting cell division. The harmful effects of soil acidity in some areas have been exacerbated by industrial emissions resulting in acid rain and in deposition of substances such as ammonium ions and sulfur dioxide, which increase the acidity on further reaction in the soil. The resulting acidity may release into the soil solution, streams, and lakes aluminate ions from soil clays, with consequent damage to plants and animals in these ecosystems. *See* ACID RAIN.

Nutrients in relation to animal growth. The elemental composition of plants is important to the health and productivity of animals which graze them. With the exception of boron, all elements which are essential for plant growth are also essential for herbivorous mammals. Animals also require sodium, iodine, and selenium and, in the case of ruminant herbivores, cobalt. As a result, animals may suffer deficiencies of any one of this latter group of elements when ingesting plants which are quite healthy but contain low concentrations of these elements.

Animal requirements for mineral nutrients may also differ quantitatively from plants. In addition, nutrients in forage may be rendered unavailable to animals through a variety of factors that prevent their absorption from the gut. It is thus possible for pastures to have a concentration of an element such as magnesium or copper that is adequate for their healthy growth but deficient for the animals grazing them.

Plants and animals differ also in their tolerance of high levels of nutrients, sometimes with deleterious results for grazing animals. For example, the toxicity of high concentrations of selenium in plants to animals grazing them, known as selenosis, was recognized when the puzzling and long-known "alkali disease" and "blind staggers" in grazing livestock in parts of the Great Plains of North America were shown to be symptoms of chronic and acute selenium toxicity. The soils of these areas had unusually high concentrations of selenium and supported several characteristic indigenous species with remarkably high concentrations of selenium that would be toxic to most other plants. Since that discovery, selenosis has been diagnosed in grazing animals on localized seleniferous soils in most other continents. Selenium in agricultural drainage waters has also been blamed for outbreaks of disorders in plants and animals in northern California.

Nutrients in Plant Metabolism

In order to utilize mineral nutrients for metabolism, plants must be able to absorb and distribute the nutrients within the plant body. In addition, plants require each nutrient in specific amounts for one or more unique functions. *See* PLANT METABOLISM.

Absorption of nutrients into cells. When transferred from water to mineral salt solutions, plant cells generally absorb nutrients rapidly. Two distinct phases are evident—an initial, rapid phase which reaches equilibrium within 20 or 30 min, and a slower phase which continues at a constant rate for a longer period (**Fig. 2**).

The first phase results from diffusion into spaces within and between the cell walls (the water free space) and adsorption of ions onto negative and positive charges within the cell wall and on the cytoplasmic surface (the Donnan free space). For cations, adsorption is usually very large owing to their ability to exchange with equivalent quantities of hydrogen ions which readily dissociate from the large amounts of pectic acids in the cell wall. Anions are adsorbed to a much smaller extent, there being few positively charged sites with anions available for exchange;

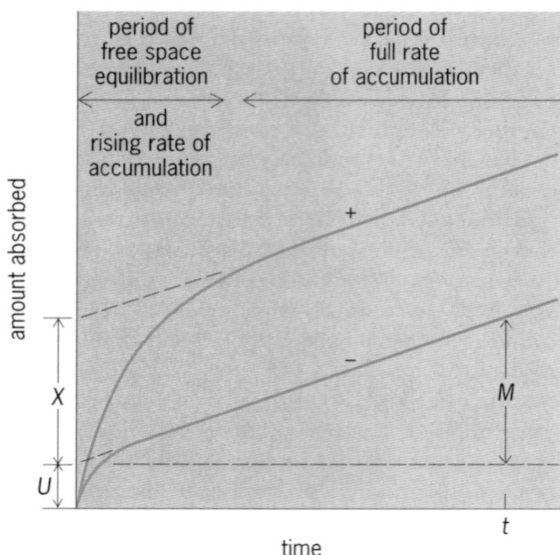

Fig. 2. Diagrammatic representation of sorption of anions and cations by a plant cell, plotted against time. U = amount of ion pairs in the free space; X = amount of cation exchange; M = amount absorbed within cells in time t. (*After G. E. Briggs, A. B. Hope, and R. N. Robertson, Electrolytes and Plant Cells, Blackwell, 1961*)

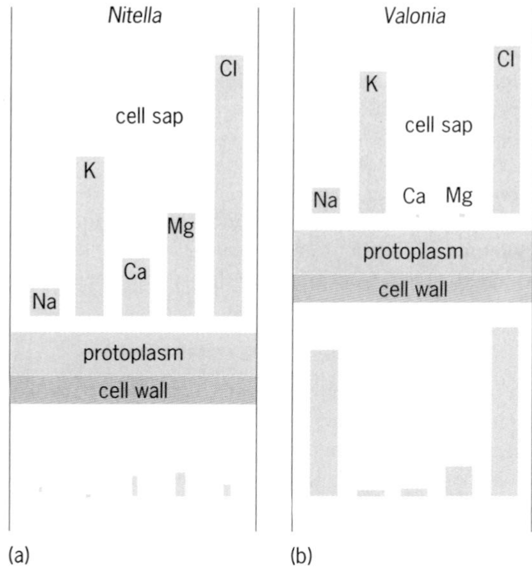

Fig. 3. Differences in absorption of ions by plants growing in (*a*) low-salt (pond-water) and (*b*) high-salt (seawater) environments. Relative concentrations of ions are expressed by the size of each bar. (*After P. J. Kramer, Plant and Soil Water Relationships, McGraw-Hill, 1969*)

proteins probably provide the reactive groups for anion exchange.

The second, slower phase of nutrient absorption results from the movement of nutrients across the outer membrane of the cytoplasm into the interior of the living cell. From the cytoplasm, nutrients may then move into organelles such as chloroplasts or mitochondria or move across the internal cellular membrane into the cell vacuole. The second phase of nutrient absorption is sensitive to temperature, anaerobiosis, and metabolic poisons, indicating that it is under metabolic control whereas the first phase is not. Metabolic absorption into the cell is clearly the significant process for the supply of nutrients for the metabolic activities of the cytoplasm.

The absorption of any one ion may be strongly inhibited or enhanced by other ions. The strongest ion antagonism is shown between nutrient ions which are chemically similar (for example, K^+ versus Rb^+, Ca^{2+} versus Sr^{2+}, Cl^- versus Br^-).

Plant cells can absorb nutrients to much higher concentrations than those of the surrounding solution, such as in the absorption by kelps of iodine to 10,000 times or more its concentration in seawater. Such absorption shows a strong selectivity for some nutrients. Accumulation of nutrients within the vacuoles of plant cells shows similar behavior. Nutrient selectivity varies greatly among plants and with environmental conditions. For example, all the ions of **Fig. 3** reach a higher concentration in the sap of *Nitella* (a fresh-water plant) than in the external solution. In *Valonia* (a seawater plant) the cell sap contains primarily potassium and chloride, whereas the seawater contains chiefly sodium and chloride.

To achieve these levels of selectivity and of concentration gradients, plant membranes must be efficient at discriminating between different nutrients and in keeping apart ions of the same nutrient. Most schemes envisage that membranes achieve these properties through specific sites within the membrane for binding anions and cations. Proteins with capacities to bind specific ions have been identified in animal, microbial, and plant cell membranes.

Differences in concentrations of mineral ions across membranes can result from either passive or active movement of the ions. Ions can accumulate in

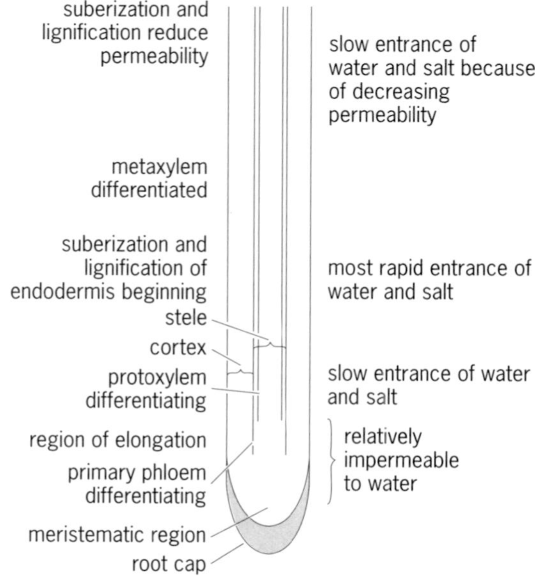

Fig. 4. Diagram of a young root, showing relation between anatomy (left) and regions that absorb (right) water and salt. (*After P. J. Kramer, Plant and Soil Water Relationships, McGraw-Hill, 1969*)

response to electrochemical potentials arising from differential permeability of the membrane or of active absorption of other ions. Ions can also accumulate from their active movement across membranes against electrochemical potentials. *See* ION TRANSPORT.

Energy supply for anion accumulation has been related to electron transfer processes, while that for cation accumulation has been related to mechanisms involving adenosine triphosphate (ATP) in reactions with specific cation-activated ATPase enzymes within the membrane. Alternative proposals envisage that electron transport through the membrane is linked to a separation of charge which establishes a H^+ ion gradient across the membrane.

Absorption of nutrients into roots. Most of the ions which reach plant leaves are absorbed and transported to the xylem in a region of the root where xylem elements are mature and are not enclosed by impermeable outer layers (**Fig. 4**). This region contains a central stele with dead xylem cells surrounded by a continuous ring of living cells (**Fig. 5**). It is surrounded by a cortex consisting of a continuous cylinder of cells, the endodermis, and several layers of loosely arranged, thin-walled parenchyma. Outside the cortex, an epidermis forms a single compact layer of cells. Many of the epidermal cells develop long, fine root hairs which protrude for several millimeters into the medium. A thin skin of some material, possibly pectin, covers the outside of the hairs and the epidermal cells. *See* ROOT (BOTANY).

The cytoplasm of plant cells forms a continuous body (the symplasm) throughout the plant by virtue of interconnecting strands of cytoplasm which pass through pores in the walls of adjacent cells. The walls around individual cells also form a continuous fabric (the apoplast) throughout the plant. Water and nutrients can move readily through the free space of the apoplast except where, as at the endodermis, the cell walls are impregnated with waxes. The endodermis thus presents a barrier to the movement of water and nutrients, allowing passage only after absorption into the cytoplasm. To reach the xylem, nutrients are absorbed into the symplasm external to the endodermis. In the symplasm they can pass through the endodermis and into the parencyma cells surrounding the xylem. Here they are excreted into the dead xylem cells, where their concentrations may exceed those of the outside solution by more than a hundredfold.

Many of the characteristics of ion absorption by roots and excretion into root xylem are similar to those of absorption into individual cells. Thus, ion absorption by roots is metabolically controlled and shows similar patterns of selectivity and ion antagonism. As a result, many factors influence nutrient absorption by roots. Low oxygen and low temperature inhibit it. Low calcium or boron and high acidity or aluminum have marked and variable effects, probably resulting from deleterious effects of each of these environmental factors on membrane structures.

The concentration of the absorbing ion may also be very important (**Fig. 6**). However, the relationship of nutrient concentration to nutrient absorption into the root may vary with the form of the nutrient, the concentration and nature of other ions in solution (**Fig. 7**), the nutrient status of the plant (**Fig. 8**), and plant species. Moreover, the plant root interacts with and responds to its environment by excreting ions and organic substances and by generating reductive activity and enhanced capacities for absorption of nutrients (Fig. 8). The excretion of hydrogen or hydroxyl ions in response to the balance of cation and anion absorption is especially important in control of solution pH in hydroponic systems. Reductive activity at the root surface is also important for the absorption of iron in most or all plants, while excretion of organic compounds with the capacity to form complexes with iron (siderophores) is a unique and valuable property of grass roots. *See* ABSORPTION (BIOLOGY).

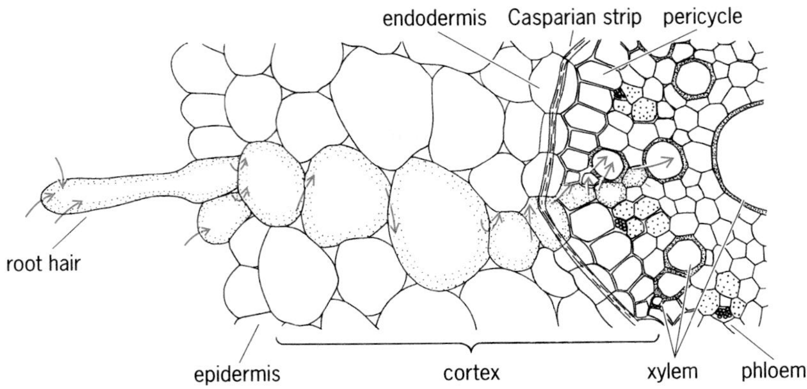

Fig. 5. Part of a wheat root transection in the region of mineral absorption and transport to xylem. (*After K. Esau, Plant Anatomy, John Wiley and Sons, 1965*)

Nutrient distribution. Within the plant body, nutrients move rapidly over long distances by being carried passively along with either water of the transpiration stream or food materials of the translocation stream. Nutrients may move from these streams to

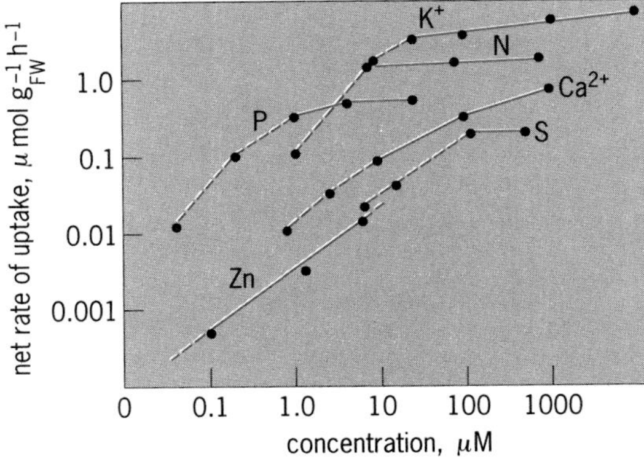

Fig. 6. Relation of the concentration of various nutrients in solution to their rates of absorption by plants. The data are plotted on a log/log basis to show the full range of concentrations used. Broken lines indicate that growth was depressed from nutrient deficiency or toxicity. (*After U. Luttge and M. G. Pitman, eds., Encyclopedia of Plant Physiology, new ser., Springer-Verlag, 1976*)

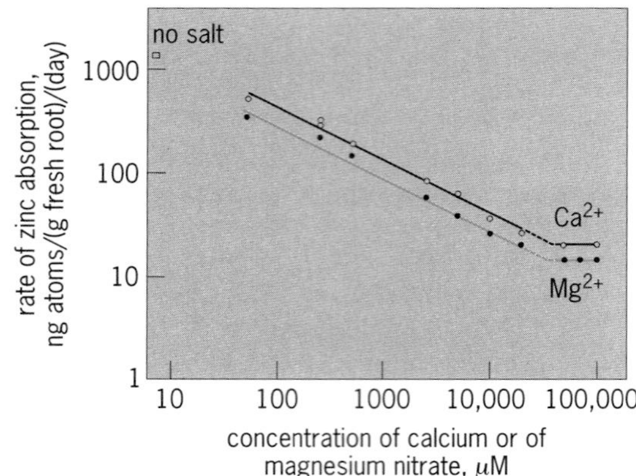

Fig. 7. Ion antagonism in wheat seedlings. As concentrations of calcium and magnesium nitrate increase, absorption of zinc decreases. The data are plotted on a log/log basis to show the full extent of the antagonism and the concentration of antagonizing ions. (*After F. M. Chaudhry and J. F. Loneragan, Zinc absorption by wheat seedlings and the nature of its inhibition by alkaline earth cations, J. Exp. Bot., 23:552, 1972*)

adjacent cells through either the cell walls or the cytoplasm. Appreciable quantities of nutrients may also be lost from leaves by leaching in rain. *See* PLANT TRANSPORT OF SOLUTES.

Xylem transport. Nutrients absorbed into the xylem sap pass to the leaves in the transpiration stream. During passage to the leaves, nutrients move rapidly into and out of cells of the stem. Transpiration rates have little effect on nutrient translocation to plant tops atlow levels of nutrient supply, but large effects at high supply when high transpiration strongly enhances transport to leaves. Within the leaves, nutrients move from the xylem sap into the leaf cells by the same mechanisms which govern their absorption by root cells. *See* XYLEM.

Phloem translocation. Phosphorus, potassium, and nitrogen move readily from leaves in the phloem to roots and other organs. By contrast, little if any calcium, boron, manganese, or iron moves from leaves to other organs. Nutrients like copper, zinc, and sulfur behave in an intermediate manner. The high mobility of nitrogen and phosphorus is associated with their incorporation into organic compounds in the phloem: in the case of phosphorus, phosphorylation of sugars is probably an important mechanism governing the movement of sugars into and out of phloem tissues. The reasons for the failure of elements to move out of leaves are not known—they could arise from precipitation as insoluble waste products outside or inside the phloem, from failure to move across cytoplasmic boundaries into the phloem, or by retention in metabolically active compounds or organelles.

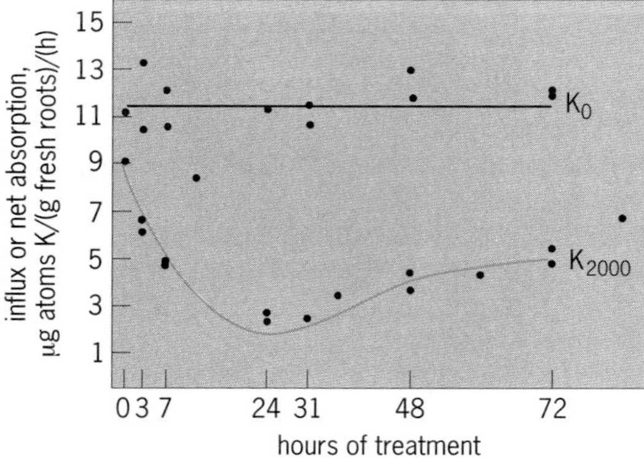

Fig. 8. Amount of potassium in plants influences the absorption of potassium. The rate of potassium absorption is constant in seedlings which receive no previous potassium (K_0). Treatment with potassium (K_{2000}) before the absorbing period quickly depressed subsequent absorption. (*After C. Johansen, D. G. Edwards, and J. F. Loneragan, Potassium fluxes during potassium absorption by intact barley plants of increasing potassium content, Plant Physiol., 45:601, 1970*)

When deficiencies develop, the concentrations of phloem-mobile nutrients generally drop to low values in old leaves but stay high in actively growing and metabolizing organs. By contrast, the concentrations of phloem-immobile nutrients remain high in old leaves but drop sharply in young organs. *See* PHLOEM.

Leaching from leaves. Mineral nutrients may be leached from leaves and stems in rain, fog, or dew. Leaching is favored in plants given an excessively high nutrient supply and from organs at maturity. As much as 80% of both potassium and boron and more than 50% of calcium and manganese have been reported to have leached from leaves.

Requirements for and function of nutrients. Plants are able to absorb nutrients from extremely low concentrations in solution. Figure 6 indicates the concentrations of some nutrients which are required in solution in order to sustain a reasonable growth rate. Plants do not need a continuous supply of those nutrients which are mobile in phloem; an excess supply stored in leaves or available to only part of the root system may be moved and used in any other part of the plant. By contrast, nutrients such as calcium, which are phloem-immobile, must be supplied continuously to the root system. If these nutrients become exhausted from the root environment, the root growth stops and both root and young shoots become deficient, no matter how great an excess is stored in old leaves or is available to other parts of the root system (**Fig. 9**).

The behavior of phloem-immobile nutrients creates difficulties in relating the nutrient content of plant shoots to their actual requirements or the onset of nutrient deficiency (**Fig. 10**). Hence plant composition is not always a definitive guide to the nutrient requirements or nutrient status of plant tissues. Indeed, the concentration of most nutrients can vary over a wide range without affecting growth. In addition, plants may contain appreciable concentrations of mineral elements such as aluminum which do not appear to be needed at all.

By analyzing plants grown under carefully controlled conditions, it is possible to estimate quantitative requirements for nutrients. For leaves, estimates of nutrient requirements have also been made

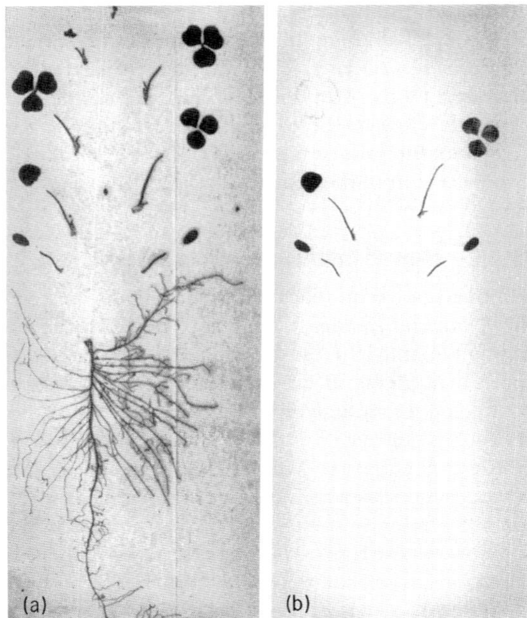

Fig. 9. Failure of nutrients to move from old leaves as shown by the behavior of radioactive calcium. (*a*) Plant grown in nutrient solution with ^{45}Ca and then transferred to solution without calcium. (*b*) Autoradiogram showing that high ^{45}Ca concentrations did not move from old leaves even though the young leaves had severe symptoms of calcium deficiency. (*After J. F. Loneragan and K. Snowball, Calcium requirements of plants, Austral. J. Agr. Res., 20:465, 1969*)

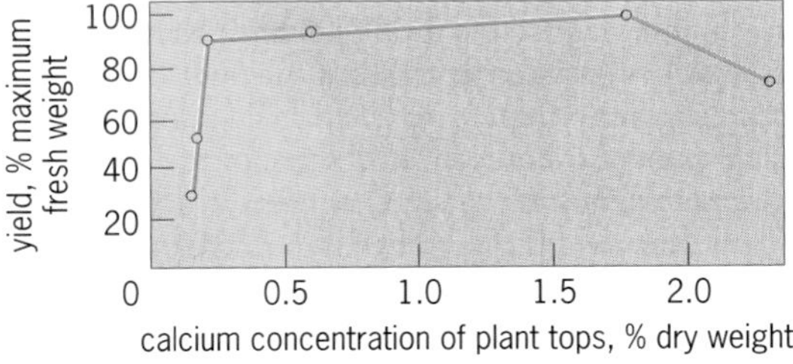

Fig. 10. Relation of yield to calcium concentration in tops of clover plants after growing in solutions with constant concentrations of calcium (open circles). When the plants were grown at a high calcium concentration and then transferred to a very low concentration, calcium deficiency developed in young organs and depressed yield of plant tops by 50% even though they contained 1.0% calcium. (*After J. F. Loneragan, Plant efficiencies in the use of B, Co, Cu, Mn, and Zn, in M. J. Wright, ed., Plant Adaptation to Mineral Stress in Problem Soils, Cornell University Press, 1976*)

by relating their nutrient concentrations to leaf function or growth or shoot growth. These relationships in leaves of defined physiological age, such as the youngest fully developed leaf, have been used as critical values for diagnosing nutrient deficiencies in plants. **Table 1** lists some values for green leaves. In total, leaves require only 4% of their dry weight as essential mineral nutrients, leaving the bulk of their weight as the nonmineral nutrient elements hydrogen, carbon, and oxygen. Since living leaves generally contain more than 90% water, they require for active metabolism less than 0.5% of their weight as essential mineral nutrients.

Historically, the essential mineral nutrients have been treated in two groups—the macronutrients and micronutrients, but the data of Table 1 show that the distinction is somewhat arbitrary. From the point of view of function, the essential mineral nutrients may be more appropriately grouped according to their chemical form and metabolic activities. On these bases, they may be placed into four broad groups, although the activities of some nutrients cannot be confined to a single group.

Group 1. These nutrients occur in plants as reduced atoms covalently bound into the structure of organic matter—namely, nitrogen and sulfur. Both are essential components of proteins, where they provide reactive groups which interact with other nutrients and metabolites. Nitrogen and sulfur are generally absorbed as the oxyanions nitrate and sufate which, apart from small quantities of sulfate in a few organic compounds, must be reduced before being used by the plant.

Group 2. These nutrients are covalently bound to carbon, hydrogen, and oxygen in organic structural compounds and metabolites but, unlike nitrogen and sulfur, in fully oxidized form—namely, phosphorus and possibly boron and silicon. Phosphorus is always present as mono- or polyorthophosphate; in the pH range of cells it functions as a weak acid. It occurs in phospholipids, which are basic components of plant membranes, in the nucleic acids of chromosomes, and in ATP. Phosphorus also occurs as an oxyanion in a large number of intermediary metabolites and is especially prominent in those concerned with transfer of energy.

Boron and silicon may behave in some ways analogous to phosphorus, functioning only in their fully oxidized forms. Their oxyacids are much weaker acids than the two acidic species of the phosphate

TABLE 1. Concentrations of nutrient elements in dry plant leaves at levels considered just adequate*

	Concentration in dry matter	
Element	μg/g or %	Relative number of atoms
Molybdenum	0.1 μg/g	1
Nickel	0.1	2
Copper	3	50
Zinc	20	300
Manganese	20	400
Boron	10	1,000
Iron	100	2,000
Chlorine	100	3,000
Sulfur	0.1%	30,000
Calcium	0.2	50,000
Phosphorus	0.2	60,000
Magnesium	0.2	80,000
Potassium	1.0	250,000
Nitrogen	2.5	1,300,000
Oxygen	45	30,000,000
Carbon	45	40,000,000
Hydrogen	6	60,000,000

*Modified from E. Epstein, *Mineral Nutrition of Plants: Principles and Perspectives*, John Wiley and Sons, 1972.

ion. Consequently their oxyanions are present largely as undissociated acids. However, nothing is known of organic compounds which boron and silicon may form in plants. Boron deficiency has an immediate and marked effect on the function of cell membranes and cell extension, while silicon seems to exist mainly as opal in cells and cell walls where it acts as a structural stiffener of leaves.

Group 3. Nutrients of this group are present largely in ionic form either free in solution or reversibly bound by electrostatic forces to charged sites in the cell—namely, potassium, sodium, magnesium, calcium, manganese, and chloride. They are thought to function by osmotic adjustment of cell activities, as potassium does in the guard cells of stomata. They are also essential for some enzymic reactions where their ions are thought to act by modifying the shape and orientation of enzymes and substrates; each is specifically required for the activity of one or more enzymes, and all of them can activate some enzymes in a general way. Calcium and, to a lesser extent, magnesium ions have the additional role of neutralizing and stabilizing cell structures, such as membranes and walls, probably by forming links between adjacent negative charges through cationic exchange.

In addition, magnesium has a structural role in chlorophyll, and manganese is tightly bound in chloroplasts, where it is required for the photolysis of water; in this activity, manganese behaves more like a member of group 4 than group 3.

Group 4. These nutrients are firmly bound into proteins as metalloproteins: copper, iron, molybdenum, and zinc. Metalloproteins govern a wide range of metabolic reactions, including many involving oxidation and reduction. All except zinc can act as electron carriers by undergoing reversible oxidation-reduction, such as iron in cytochrome enzymes, copper in a wide range of oxidase enzymes, and molybdenum in nitrate reductase. Zinc is also required by some oxidation-reduction enzymes but does not itself change valency. It also activates many other enzymes. Of the many physiological expressions of zinc deficiency, the best known is the disturbance of growth of the shoot apex resulting from a decreased concentration of the growth hormone indole acetic acid.

Mineral Nutrition of Plants in Soils

Although not required for growth, organic matter, solid mineral particles, and soil microorganisms strongly modify the behavior and concentration of mineral nutrients in the soil solution. In addition, root activities strongly modify soil properties at the root-soil boundary.

The concentration of mineral nutrients in the soil solution varies widely from values of less than $10^{-8}M$ for some micronutrients to $10^{-1}M$ for calcium in some soils (**Table 2**). The main forms of some nutrient ions in soil solutions also vary widely.

Nutrient reactions with soil materials. When plants are grown in water culture, the quantities of nutrients in solution are usually sufficient for 2 or 3 weeks' growth. This is generally achieved by having high concentrations of essential nutrients. Soil solutions, which have much lower concentrations of many nutrients (Table 2), can supply the plant's requirements only for very short periods of time. For plants to obtain an adequate supply, they rely on the solid phase of the soil to replenish nutrients rapidly as nutrients are withdrawn from the soil solution. The capacity of a soil to rapidly replenish a nutrient in solution is called buffering capacity, an important property governing nutrient response. Some of the many processes involved are (each is important for those nutrients in parentheses) chelation and complex formation by soil organic matter (Al, Cu, Fe, Ni, Zn);

TABLE 2. Concentrations of ions in common soil and nutrient solutions and dominant forms of nutrients in some soil solutions

	Soil solution		
Nutrient element	Form*	Concentration	Nutrient solutions
Calcium	Ca^{2+}	1–3 mM	1–5 mM
Magnesium	Mg^{2+}	1–4	1–2
Potassium	K^+	0.2–2	2–10
Sodium	Na^+	0.5–20	0–1.4
Chlorine	Cl^-	0.2–30	0–10
Nitrogen	NO_3^-	2–10	3–15
	NH_4^+		1–11
Sulfur	SO_4^{2-}	1–3	1–7
Phosphorus	$H_2PO_4^-$, HPO_4^{2-}	0.05–5 μM	500–2000 μM
Molybdenum	$HMoO_4^-$, MoO_4^{2-}	0.02–0.08	0.03–0.5
Silicon	H_2SiO_4, $HSiO_4^-$	100–1000	0
Boron	H_3BO_3, $H_2BO_3^-$	3–1000	4–50
Cobalt	Co^{2+}, $Co(OH)^+$, Co^{2+} – OM	0.007–0.2	0–0.2
Zinc	Zn^{2+}, $Zn(OH)^+$, Zn^{2+} – OM	0.03–3	0.4–1
Manganese	Mn^{2+}, $Mn(OH)^+$, Mn^{2+} – OM	0.02–68	2–10
Copper	Cu^{2+} – OM	0.01–0.06	0.03–1
Iron	Fe^{3+} – OM	0.4–5	5–100

*OM indicates that the nutrient may be complexed with soluble organic matter: in nutrient solutions, only Fe^{3+} is added as an organic complex, all others being added as simple ions.

cation exchange on organic matter and clay minerals (K, Na, Ca, Mg); specific adsorption of cations on hydrous oxides of iron and aluminum (Co, Cu, Mn, Zn); ligand adsorption of anions on iron and aluminum hydroxide (P, Mo, B, Si, S); mineral decomposition and precipitation (all except N); and synthesis and decomposition of organic matter (N, S, P).

Microorganisms. Soil organisms influence plant nutrition in a variety of ways, including synthesis and degradation of soil organic matter, damage to roots, competition for nutrients, modification of soil properties, conversion of gaseous nitrogen to forms of nitrogen which plants can use, and formation of special structures in association with the root (mycorrhizal roots).

The conversion of atmospheric nitrogen into forms which plants can use for protein synthesis is especially important. A number of free-living soil bacteria can do this, the most notable being the blue-green photosynthetic cyanobacteria and certain free-living bacteria (*Azotobacter*, *Clostridium*). Other nitrogen-fixing bacteria form loose associations with plants, growing on leaf or root surfaces. The most important organisms are the *Rhizobium* bacteria, which form a close symbiotic association in special nodular structures on the roots of most leguminous species and of some other species. Plants having an effective symbiotic relationship with an appropriate *Rhizobium* can convert as much atmospheric nitrogen as they require for growth. After consumption and excretion by animals or death and decay of the plant, the fixed nitrogen is converted to soil organic matter or released to the soil solution by the activities of microorganisms in mineral forms suitable for general plant use. *See* NITROGEN CYCLE; NITROGEN FIXATION; SOIL MICROBIOLOGY.

Mycorrhizal root structures are also important, as they increase the effectiveness of roots in obtaining phosphorus and possibly other nutrients in short supply. *See* MYCORRHIZAE.

Soil–root interface. At the soil–root junction (the rhizosphere), soil properties are changed by plant activities. Such changes may be vital to nutrient absorption since young roots absorb plant nutrients from the soil solution in the rhizosphere. Young roots are coated with a mucilaginous material that extends some distance into the soil. This mucilaginous material and the many other organic materials known to exude from roots can modify the behavior of nutrients in the rhizosphere directly through complexation and cation exchange. The excretion of siderophores by grasses is especially important in making iron available from calcareous soils. Organic materials also act as food for microorganisms, encouraging growth and other activities. *See* RHIZOSPHERE.

Plants also influence the nutrient environment in the rhizosphere by absorbing both water and nutrients. Plants draw water to their roots from some distance away in the soil. Nutrients dissolved in the soil solution move passively with this mass flow of water. At the root surface they may either accumulate or be depleted by rapid absorption (**Fig. 11**). If depleted, they may be replaced by diffusion from the bulk soil. At the same time, any imbalance in absorption of anions and cations results in pH changes at the root surface that may impact on the availability of nutrients from the soil particles.

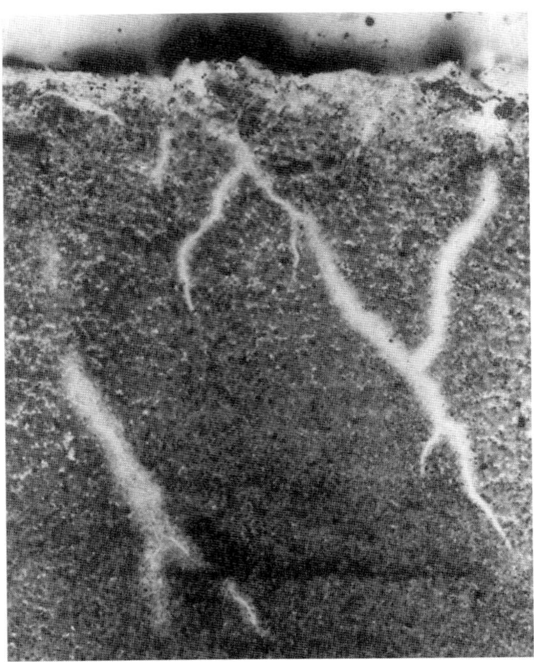

Fig. 11. Autoradiogram of wheat roots growing in soil to which radioactive calcium has been added. The white areas have had calcium removed from them, showing how the root has modified the soil in its environment. (*After H. F. Wilkinson, J. F. Loneragan, and J. P. Quirk, Calcium supply to plant roots, Science, 161:1245, 1968*)

In soils, nutrients can move to roots only over limited distances so that the ability of root systems to explore the soil may become very important for the acquisition of nutrients. Thin roots with fine root hairs present a far larger surface area for interaction with soil nutrients than thick roots, and mycorrhizae extend the area of contact even further. The efficiency of root systems is also enhanced for several nutrients by their capacity to proliferate in zones of high nutrient supply. *See* PLANT GROWTH; PLANT PHYSIOLOGY; SOIL CHEMISTRY. J. F. Loneragan

Bibliography. D. T. Clarkson and J. B. Hanson, The mineral nutrition of higher plants, *Annu. Rev. Plant Physiol.*, 31:239, 1980; H. Marschner, *Mineral Nutrition in Higher Plants*, 2d ed., 1994; K. Mengel and E. A. Kirkby, *Principles of Plant Nutrition*, 4th ed., 1987; M. L. Van Beusichem, *Plant Nutrition: Physiology and Applications*, 1990; A. Wild (ed.), *Russell's Soil Conditions and Plant Growth*, 11th ed., 1988.

Plant morphogenesis

The origin and development of plant form and structure. Morphogenesis may be concerned with the whole plant or a plant part (for example, the shoot, root, leaf, or flower) or with the subcomponents of a structure (such as a tissue or surface hairs).

The inception and development of plant form is an orderly process which is controlled by internal (genetic, biochemical, and physiological) and external (environmental) factors. Techniques used in the study of plant morphogenesis include light microscopy, scanning and transmission electron microscopy, tissue and organ culture, surgical and grafting experiments, chemical (natural and synthetic) treatments and their analyses, and protein and enzyme determinations. Mutants have helped identify specific genes, and the molecular mechanisms by which they interact with other genes, in the morphogenesis of plant organs and tissues. Chimeras have proven valuable in determining the roles of specific tissue layers and the cellular interactions between different layers in the formation of a structure. Tissue cultures have provided insights into the regenerative ability and totipotency of plant cells, and the factors that promote shoot, root, and embryo development. *See* CHIMERA; TISSUE CULTURE.

Basic processes. The development of a plant's form is primarily determined by its genetic constitution. At different stages of development, genes transcribe messages that are expressed through a complex network of processes, leading to the formation of a structure. The development of a structure follows a distinct pattern, but such patterns are not always rigid and can be modified in many cases by external stimuli. However, such stimuli are usually effective only until the particular pattern of development becomes stably fixed or developmentally committed. The following are some of the basic processes involved in plant morphogenesis.

Polarity. The establishment of differences at the two ends of a structure is called polarity. In plants, polar differences can be recognized very early in development, that is, at the zygote stage. In the zygote, cytological differences at the two ends of the cell establish the position of the first cell division, and thus the fate of structures produced from the two newly formed cells. During the development of a plant, polarity is also exhibited in the plant axis (in the shoot and root tips). If a portion of a shoot or root is excised and allowed to regenerate, the end toward the shoot tip always regenerates shoots whereas the opposite end forms roots. Polarity is also evident on the two sides of a plant organ, such as the upper and lower surface of a leaf, sepal, or petal.

Differential growth rates. The diversity in plant form is produced mainly because different parts of the plant grow at different rates. Furthermore, the growth of an individual structure is different in various dimensions. Thus the rate of cell division and cell elongation as well as the orientation of the plane of division and of the axis of cell elongation ultimately establish the form of a structure. Such differential growth rates are, however, not disorganized but rather are very well orchestrated by genetic factors. Although the absolute growth rates of various parts of a plant may be different, their relative growth rates, or the ratio of their growth rates, are always constant. This phenomenon is called allometry (or heterogony), and it supports the concept that there is an interrelationship between the growth of various organs of a plant body. Allometric studies are useful in analyzing different growth patterns in plants. *See* PLANT GROWTH.

Interdependence of plant organs. During development, either the removal of or changes in one part of the plant may drastically affect the morphogenesis of one or more other parts of the plant. This phenomenon is called correlation and is mediated primarily through chemical substances, such as nutrients and hormones. For example, at different stages of development, the various substances produced in the leaves affect the formation of flowers, the falling of leaves, the growth of lateral buds, and the differentiation of vascular tissues in the shoot. The roots produce substances which affect the growth of the shoot and lateral buds and the development of reproductive organs in flowers. Similarly, the terminal shoot apex inhibits the growth of lateral branches, presumably by either supplying or depriving lateral buds of certain growth substances. Thus, the mechanical removal of the shoot apex results in the growth of lateral buds—a practice commonly used by plant growers to produce bushy plants. *See* APICAL DOMINANCE.

Symmetric and asymmetric growth. The developmental pattern of plant structure follows a precise and definite order. In different plant species, however, the pattern may be variable. For example, the arrangement of leaves (phyllotaxy) on the shoot axis is symmetrical; it could be spiral, opposite, or whorled. In the case of spiral arrangement, the production of leaves follows a definite mathematical sequence which may vary in different plant species. Also, the pattern of leaf arrangement may vary at different stages of plant development. The development of plant structures (such as embryos) also follows a definite pattern, although it may vary in different species. *See* LEAF.

In the development of certain structures, however, asymmetric patterns occur. For example, asymmetric cell divisions are involved in the formation of stomata in grasses, and in the production of hairs of many plants. Similarly, asymmetric growth on the two sides of the shoot or a root results in the bending phenomenon (tropism) in plants. *See* PLANT MOVEMENTS.

The normal symmetric pattern of development in plants can be broken down by the invasion of pathogens (such as bacteria or nematodes), resulting in the formation of tumors, galls, and other abnormal forms. Likewise, when cut ends of stems or roots heal, or when isolated segments of plant tissues are cultured, disorganized growth often occurs, producing a mass of cells called a callus. *See* PLANT PATHOLOGY.

Regulation. The various processes involved in the morphogenesis of plants, described above, are affected by a number of internal and external factors, as described below.

Genetic factors. The ultimate factors controlling the form of a plant and its various organs are the genes. In general, several genes interact during the development of a structure, although each gene plays a significant role. Thus, a mutation in a single gene may affect the shape or size of a leaf, flower, or fruit, or the color of flower petals, or the type of hairs produced on stems and leaves. There are at least two classes of genes involved in plant morphogenesis: regulatory genes that control the activity of other genes, and effector genes that are directly involved in a developmental process. The effector genes may affect morphogenesis through a network of processes, including the synthesis and activity of proteins and enzymes, the metabolism of plant growth substances, changes in the cytoskeleton and the rates and planes of cell division, and cell enlargement. *See* GENE ACTION.

The number of sets of chromosomes is also known to affect the size of structures. Polyploids ($4n$, $6n$, or $8n$) are generally larger in size than the diploids ($2n$), which are usually larger than the experimentally induced haploids ($1n$). *See* POLYPLOIDY.

Biochemical factors. The development of plant form is regulated by the synthesis and activity of a large variety of proteins and enzymes that are encoded by genes. During the development of a structure, the occurrence and activity of certain proteins and enzymes, or different forms of the same enzyme (isoenzymes), are known to change at various stages. Also, stage-specific or organ-specific proteins and isoenzymes are known to occur. Mutants with differences in leaf form or in flower morphology exhibit different protein levels or enzymatic activity.

Plant growth substances. Nearly all aspects of plant form and structure are known to be affected by plant growth substances. Five classes of naturally produced plant growth substances, also called plant hormones or growth regulators, are auxins, gibberellins, cytokinins, ethylene, and abscisic acid. These hormones, along with the many synthetic compounds, form a long list of substances which affect plant growth and development.

Unlike animal hormones, each plant hormone is generally involved with a variety of developmental phenomena. For example, auxins affect the movement of plants toward light (phototropism), stimulate fruit growth in the absence of pollination (parthenocarpy), induce the production of roots on the main root and on stem cuttings, promote the differentiation of water-conducting tissue (xylem), and inhibit the growth of lateral buds. Gibberellins stimulate seed germination, stem elongation, and flowering of certain plants, promote the differentiation of male reproductive organs in flowers, and alter the leaf shape of many plants. Cytokinins play a role in bud growth, in fruit development, and in the development of male and female reproductive organs. They also delay the senescence of leaves. Ethylene causes an increase in the diameter of the shoot of young seedlings and promotes upright growth in horizontally growing plants. It also induces the formation of female flowers on male plants and enhances the ripening of fruits. Abscisic acid is generally inhibitory in nature, and is known to induce dormancy in seeds and buds, inhibit stem growth, enhance leaf abscission, and regulate stomatal opening in leaves. In some plants, it also affects leaf form. A number of synthetic growth regulators are used commercially for inducing rooting of stems and leaves, for producing short sturdier plants, and for enhancing fruit ripening. Inhibitors generally tend to retard the growth of shoots and are used by horticulturists in producing short and sturdier plants. *See* ABSCISIC ACID; AUXIN; CYTOKININS; ETHYLENE; GIBBERELLIN.

Each class of plant growth substances is produced in more than one region of the plant and transported to different regions. Some of the hormones are transported primarily in one direction (for example, auxins travel in the basal direction), but others (gibberellins and cytokinins) travel in all directions. Also, during the formation of a structure, the endogenous levels of growth substances change; some are higher at earlier stages, whereas others are more abundant at later stages. This has led to the concept that the formation of a structure involves a critical balance of a number of endogenous growth substances at various stages of development and that a change in this balance generally results in the modification of the form. *See* PLANT HORMONES.

Plant form is also known to be affected by nutritional factors, such as sugars or nitrogen levels. For example, leaf shape can be affected by different concentrations of sucrose, and the sexuality of flowers is related to the nitrogen levels in the soil in some species. Inorganic ions (such as silver and cobalt) have also been known to affect the type of flower produced. *See* PLANT MINERAL NUTRITION.

Environmental factors. Although genes are the ultimate controlling factors, they do not act alone, but interact with the existing environmental factors during plant development. Environmental factors, including light, temperature, moisture, and pressure, affect plant form.

Light has a dramatic effect on plant form. For example, a seedling grown in the dark shows considerable elongation of the shoot with little or no growth of leaves, in contrast to a light-grown seedling in which leaves expand and shoot growth is stunted. The intensity of light also strongly influences the development of a structure; under high light intensity, leaves are usually thicker and more pubescent than under low light. Similarly, the quality of light also affects the morphogenesis of plants. For example, under red light fern gametophytes grow as elongated filamentous structures, but under blue light normal flattened gametophytes are produced. Many of the developmental phenomena, such as seed germination and flowering of plants, are affected by red light. Red light effects are mediated through the pigment phytochrome. *See* PHYTOCHROME.

Temperature has a strong effect on a number of plant developmental processes, such as the growth of shoots and buds, the flowering of plants, and the type of flower produced. Some winter crops, (for example, wheat and rye) require a chilling factor (vernalization) for flowering to occur in the next spring. Also, low temperatures tend to favor the development of female reproductive over male organs in some plants. *See* VERNALIZATION.

The availability of water also drastically affects plant form. Plants grown in dry-arid conditions produce small thick leaves which are often rolled on the upper surface, in contrast to plants grown underwater, which produce thin and highly dissected leaves. Furthermore, in some cases, if land plants with entire leaves are submerged in water, they too develop dissected leaves. *See* PLANT-WATER RELATIONS.

Many of the environmental effects described above are believed to be mediated through a change in the balance of endogenous plant hormones. *See* PHYSIOLOGICAL ECOLOGY (PLANT); PLANT PHYSIOLOGY. V. K. Sawhney

Bibliography. D. E. Fosket, *Plant Growth and Development: A Molecular Approach*, 1994; R. F. Lyndon, *Plant Development: The Cellular Basis*, 1990; T. A. Steeves and I. M. Sussex, *Patterns in Plant Development*, 2d ed., 1989.

Plant movements

The wide range of movements that allow whole plants or plant parts to transport or reorientate themselves in relation to changed surroundings in order to execute specific functions. For example, small free-floating aquatic plants migrate to regions optimal for their activities, and anchored plants move to facilitate spore or seed dispersal. There are two types of plant movement: abiogenic movements, which arise purely from the physical properties of the cells and therefore take place in nonliving tissues or organs; and biogenic movements, which occur in living cells or organs and require an energy input from metabolism.

Abiogenic Movements

Drying or moistening of certain structures causes differential contractions or expansions on the two sides of cells and hence causes movements of curvature. This differential behavior occurs because the composition and physical properties of cell walls vary from place to place in the organ and according to direction in the walls. Such movements are called hygroscopic and are usually associated with seed and spore liberation and dispersal. Examples of such movement occur in the "parachute" hairs of the fruit of dandelion (*Taraxacum officinale*), which are closed when damp but open when the air is dry to induce release from the heads and give buoyancy for wind dispersal. The elaborate teeth (peristome) on the rim of the capsule (spore case) of mosses respond to humidity changes, closing and opening the capsule mouth and allowing the periodic release of the spores. In some grasses (such as *Stipa pennata*), the fruit has a long thin spine (awn) at one end. This is straight when moist, but as it dries the half nearest the fruit twists into a tight coil, while the apical half stays straight and perpendicular to the coil. This acts as a lever when caught by a nearby object, allowing humidity changes to cause the hygroscopic coil to screw the fruit into the ground. In Queen Anne's lace (*Daucus carota*), under moist conditions the whole ripe umbrella-shaped umbel that makes up the infructescence contracts into a kind of ball by the curling-in of the individual rays. Under dry conditions the rays straighten, promoting the release of the small fruitlets. An extreme inverse form of this behavior is found in the rose of Jericho (*Anastatica hierochuntica*), where the whole plant forms a ball when it dries out and then spreads open again in rain to allow raindrops to disperse the seed.

Hygroscopic movements are relatively slow, but there are some which, after tensions increase due to water loss, take place suddenly by explosive release. An example is the "slingshot" fruit dispersal mechanisms of *Geranium* spp., where the five seed-containing parts in each flower are fixed to the base of a long five-partite beak, each part of the beak constituting a spring. When the spring has dried out sufficiently, it detaches the seed container at its base, coils up violently, and projects the seed from its now open container for a considerable distance (**Fig. 1*a***). Similar mechanisms are seen in members of the pea family, where the familiar pods, when dry, suddenly split apart into two valves which coil up, shooting out the seeds (Fig. 1*b*).

Another type of abiogenic movement is due to changes in volume of dead water-containing cells. In the absence of a gas phase, water will adhere to lignocellulose cell walls. As water is lost by evaporation from the surface of these cells, considerable tensions can build up inside, causing them to decrease in volume while remaining full of water. The effect is most commonly seen in some grasses of dry habitats, such as sand dunes, where longitudinal rows of cells on one side of the leaf act as spring hinges, contracting in a dry atmosphere and causing the leaf to roll

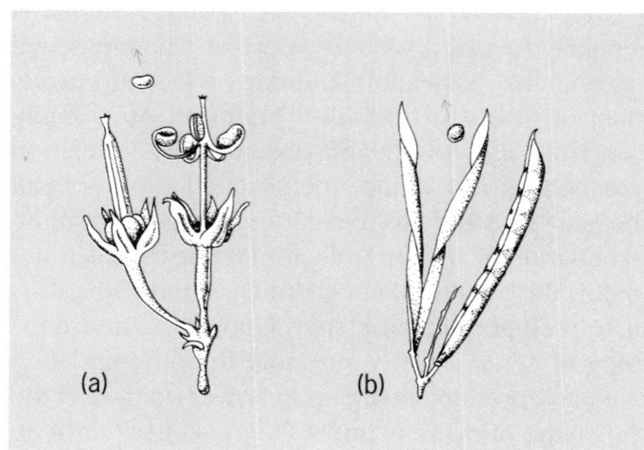

Fig. 1. Hygroscopic movements. (*a*) Fruits of *Geranium dissectum*: the "springs" of the fruit on the right have been released, and one seed is being slung out of the lobe of the capsule. (*b*) Seed pods of *Lathyrus vernus*: the two halves of the pod on the left have split apart, curled up explosively, and ejected one seed. (*After H. Straka, Encyclopedia of Plant Physiology, vol. 17, pt. 2, Springer-Verlag, 1962*)

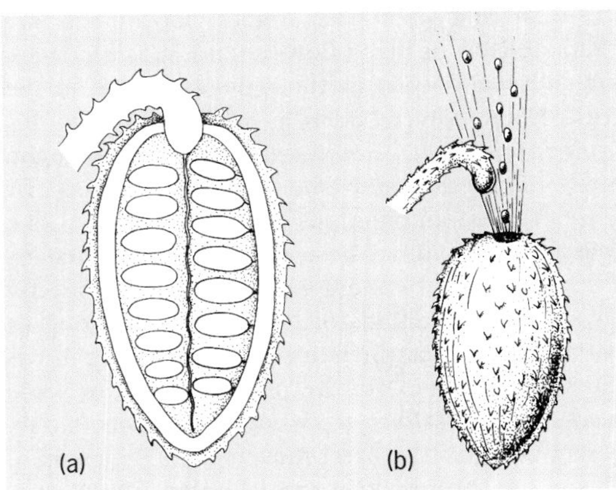

Fig. 2. Fruit of the squirting cucumber (*Ecballium elaterium*). (*a*) Section showing seeds embedded in a pulpy matrix inside a spiny outer covering. (*b*) Fruit in process of squirting as a result of hydrostatic pressure built up in the matrix. (*After H. Straka, Encyclopedia of Plant Physiology, vol. 17, pt. 2, Springer-Verlag, 1962*)

up into a tight cylinder, thus minimizing water loss by transpiration. The most dramatic example is seen in sporangia of many ferns, which are encircled by a ring of special cells, the annulus. These cells are constructed with thick inner and thin outer walls. When water is lost and internal tensions increase, a tangential pull is exerted at a weak point in the ring (the stomium) which consequently ruptures, and the annulus, curling outward, opens up the sporangium, exposing the spores. This opening becomes larger as water loss continues until internal tensions in the annulus cells become so great that the water collapses from the walls, leaving the cells largely filled with water vapor. The outwardly directed tension is thus suddenly released, and the annulus springs back toward its original position, throwing the spores violently out into the air.

Yet other movements depend on changes in the turgidity of cells. Though not strictly abiogenic, since they occur in living cells and thus rely, at least to a small extent, on energy from metabolism, they are included here for convenience. The most important of such movements are those of guard cells which surround the stomatal pores in the epidermis of leaves and other green organs. Stomata allow the exchange of gases, such as carbon dioxide and oxygen in photosynthesis and respiration and the escape of water vapor in transpiration. Guard cells have differentially thickened walls. When the turgidity of these cells rises due to water absorption, their shape alters, causing the pore to widen; when turgidity falls, the cells collapse, closing the pore.

Other movements controlled in a similar way are the opening of grass flowers, which have special hinge cells at the base of the covering scales (glumes). A dramatic example where turgidity is used for seed dispersal is the squirting cucumber (*Ecballium elaterium*). Here increasing turgor, which can rise to 3–4 bars (300–400 kilopascals) inside the fruit as it becomes ripe, blows out the stalk like a champagne cork and with it the liquid seed containing pulp (**Fig. 2**).

A seed-dispersal mechanism, resembling in its action that of the fruit of *Geranium* (Fig. 1*a*), depends on tensions set up not by water loss from the seed-containing parts but by the differential water uptake by separate cell layers causing differential swelling of the affected tissue. An example of this type is found in touch-me-not (*Impatiens noli-tangere*). When the fruit is ripe and fully turgid, the slightest mechanical disturbance causes the detachment of the seed-containing valves at the base; explosive coiling of these valves then ejects the seed up to a distance of a meter (3.3 ft) or so.

Biogenic Movements

There are two types of biogenic movement. One of these is locomotion of the whole organism and is thus confined to small, simply organized units in an aqueous environment. The other involves the change in shape and orientation of whole organs of complex plants, usually in response to specific stimuli. These movements arise either from differential growth on opposite sides of the organ concerned or from local reversible changes in cell volume caused by the loss or gain of water.

Locomotion. In most live plant cells the cytoplasm can move by a streaming process known as cyclosis (**Fig. 3**). Energy for cyclosis is derived from the respiratory metabolism of the cell. The mechanism probably involves contractile proteins very similar to the actomyosin of animal muscles. How these contractile proteins act between the rigid outer cell wall and the fluid cytoplasm to produce this streaming is not yet known. In the slime molds the whole plant body moves across substrata by a similar mechanism. The separating movement of chromosomes during nuclear division, a phenomenon common to plants and animals, is also mediated by fibers constituted of similar contractile proteins.

Cell locomotion is a characteristic of many simple plants and of the gametes of more highly organized ones. Motility in such cells is produced by cilia,

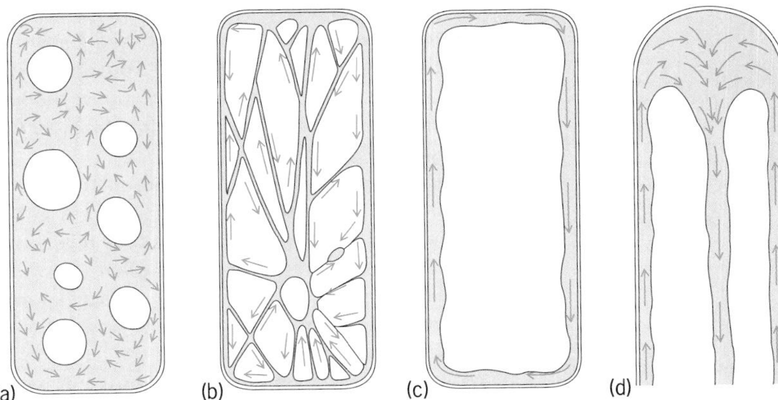

Fig. 3. Common types of protoplasmic streaming (cyclosis) in plant cells: (*a*) agitation; (*b*) circulation; (*c*) rotation; and (*d*) fountain streaming. (*After N. Kamiya, Encyclopedia of Plant Physiology, vol. 17, pt. 2, Springer-Verlag, 1962*)

whiplike protein structures 10–25 micrometers long and about 0.2 μm thick, anchored in the peripheral layers of the cell and projecting into the surrounding medium. There is a complex internal structure of parallel threads (axonemes) running the length of the cilium. There are two axial axonemes and nine in an outer ring, and it is thought that the rhythmic and sequential contraction of the members of this outer ring gives the cilium either a whiplike or a screwlike motion which is responsible for propelling the organism through its aqueous environment. *See* CILIA AND FLAGELLA.

Cell locomotion is usually not random but is directed by some environmental gradient. Thus locomotion may be in response to specific chemicals; this is called chemotaxis. Light gradients induce phototaxis; temperature gradients induce thermotaxis; and gravity induces geotaxis. One or more of these environmental factors may operate to control movement to optimal living conditions. *See* CELL MOTILITY.

Movement of organs. In higher plants, organs may change shape and position in relation to the plant body. When bending or twisting of the organ is evoked spontaneously by some internal stimulus, it is termed autonomous movement. The most common movements, however, are those initiated by external stimuli such as light and the force of gravity. Of these there are two kinds. In nastic movements (nasties), the stimulus usually has no directional qualities (such as a change in temperature), and the movement is therefore not related to the direction from which the stimulus comes. In tropisms, the stimulus has a direction (for instance, gravitational pull), and the plant movement direction is related to it.

Autonomous movements. The most common movement is circumnutation, a slow, circular, sometimes waving movement of the tips of shoots, roots, and tendrils as they grow; one complete cycle usually takes from 1 to 3 h. In horizontal underground stems (rhizomes) there may be an up-and-down hunting movement that maintains the organ at a constant distance from the soil surface. Tendrils may show very extensive swings as they "search" for support. In twining plants this circumnutation is an essential part of the climbing mechanism, where the slenderness of the stem demands rigid external support. The direction of twining is characteristic of the species. It is usually anticlockwise as viewed from above, but a few species twine clockwise, for example, the hop (*Humulus lupulus*) and honeysuckle (*Lonicera periclymenum*).

These movements are due to differential growth, but some may be caused by turgor changes in the cells of special hinge organs and are thus reversible. Such is the case with the telegraph plant (*Desmodium gyrans*), whose leaflets show regular spontaneous gyrations, with cycle times as short as 30 s under warm conditions.

Nastic movements. There are two kinds of nastic movements, due either to differential growth or to differential changes in the turgidity of cells. They can be triggered by a wide variety of external stimuli.

Photonastic (light-dark trigger) movements are characteristic of many flowers and inflorescences, which usually open in the light and close in the dark. Night flowers, such as white campion (*Melandrium album*), close by day and open by night. Many photonastic movements are complicated by a superposed autonomous movement with a 24-h (circadian) rhythm. These rhythms are started by a change in light conditions, but may continue for days under conditions of constant illumination or darkness. A similar phenomenon is the "sleep movements" (nyctinasty) in the pinnate leaves of some plants (such as *Albizzia*), where the leaflets fold together at night. The latter movement is the result of changes in the turgor in special organs (pulvini) at the bases of the leaflets. The turgor changes are produced by the extrusion of potassium ions from cells on one side of the pulvinus, leading to a rise in water potential and associated loss of water from those cells. A special blue pigment, phytochrome, is the receptor for these light responses.

Thermonasty (temperature-change trigger) is seen in the tulip and crocus flowers, which open in a warm room and close again when cooled. Crocus is particularly sensitive, responding to a temperature change as small as 0.2–0.5°C (0.35–1°F). The movement is caused by differential growth responses at

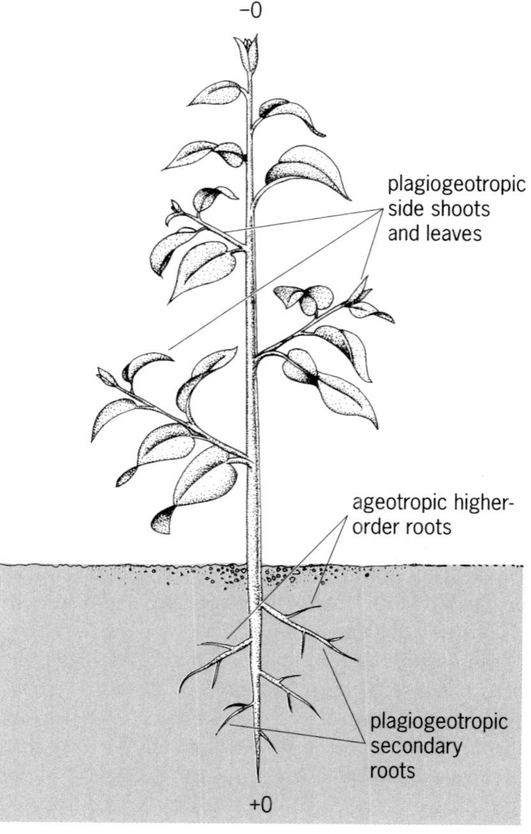

Fig. 4. Geotropic responses of plant organs: −0 = negatively geotropic main shoot; +0 = positively geotropic main root. *(After L. J. Andus, Physiology of Plant Growth and Development, ch. 6, Geotropism, McGraw-Hill, 1969)*

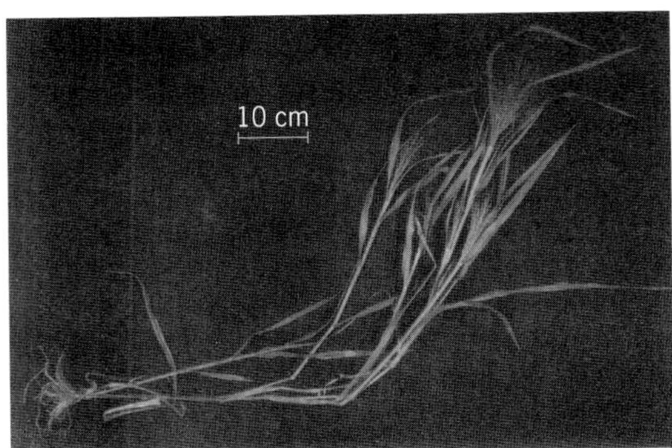

Fig. 5. Knee-forming ability of a witchgrass (*Panicum capillare*) shoot. (*Courtesy of P. B. Kaufman*)

the bases of the tepals (floral leaves).

The most striking nastic movements are seen in the sensitive plant (*Mimosa pudica*). Its multipinnate leaves are very sensitive to touch or slight injury. Leaflets fold together, pinnae collapse downward, and the whole leaf sinks to hang limply. Only one leaflet needs to be touched or damaged for the stimulus to pass throughout the leaf, or even the whole plant, in a rapid sequence of collapse at a rate of 4–30 mm/s. The conducted stimulus is complex, having a rapid electrical action-potential component and a slower hormone-transport component. The turgor mechanism operates on the same principles as those involving pulvini in nyctinasty detailed above. The whole process is reversible, and recovery takes place in 15–20 min. The phenomenon is known as seismonasty.

Epinasty and hyponasty occur in leaves as downward and upward curvatures, respectively. They arise either spontaneously or as the result of an external stimulus, such as exposure to the gas ethylene in the case of epinasty; they are not induced by gravity.

Tropisms. Of these the most universal and important are geotropism (or more properly gravitropism) and phototropism; others include thigmotropism and chemotropism.

1. Gravitropism. In geotropism, the stimulus is gravity. The main axes of most plants grow in the direction of the plumb line with shoots upward (negative gravitropism) and roots downward (positive gravitropism). Lateral organs such as branches, leaves, and underground or surface stems grow normally at a specific angle to the plumb line (plagiogravitropism; **Fig. 4**). Angular displacement of such organs from these orientations induces curvatures to reestablish the normal position. In most organs curvatures are due to differential changes in growth rate in those parts of the organ which are still growing. In grass stems, however, where the curvatures are confined to the nodes, which do not grow in the vertical position, displacement by wind or heavy rain induces growth on the lower side of the node, making a "knee" which erects the stem to its normal upright position (**Fig. 5**).

One part of a plant may influence the gravitropic behavior of another. For example, the branches of many conifers are plagiogravitropic, but if the top of the leading shoot is removed, then, after a lag, neighboring branches will become negatively gravitropic and curve upward, indicating that the main shoot has in some way induced plagiogravitropic behavior on the laterals.

There have been many theories about the mechanism by which a plant senses the force of gravity, but the oldest and still most acceptable identifies the sensors as heavy starch-containing bodies known as amyloplasts, which can be observed to sediment to the lower internal surface of containing cells. These amyloplasts, called statoliths, occur, with one or two doubtful exceptions, in all higher plant organs which respond to gravity, even in species which do not otherwise manufacture storage starch (**Fig. 6**).

Gravitropic curvatures are caused by differing concentrations of hormone on the two sides of the growth zone. In shoots this hormone appears to be the auxin indol-3-ylacetic acid, which is synthesized in the apex of growing shoots and in young expanding leaves; it is then transported to the zone of cell extension, where it controls elongation growth. A higher concentration on the lower side of the growth zone will cause faster growth there and consequently an upward curvature of the shoot. In most shoots this difference in auxin concentration seems to be due either to a downward migration from upper to lower side of the growth zone or to a more active transport from the tip down the lower side. The "knee-forming" nodes of grasses and cereals seem to have a different mechanism. Here the displacement of the stem into a slanting or horizontal position stimulates either the production of auxin or its release from inactive conjugates in the tissues on the underside of the node and, in consequence, its greatly accelerated growth. The sedimented starch statoliths in this lowermost part of the node are presumably in some way involved in this change of auxin status. In roots, where curvature is also due to the induction of differential growth rates, the nature of the

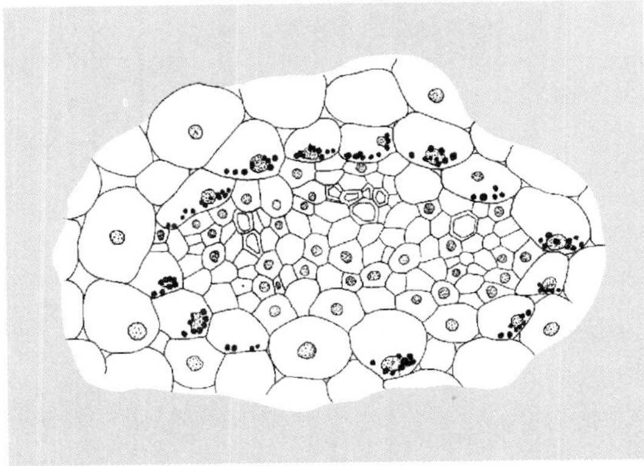

Fig. 6. Transverse section of a vascular strand in the first shoot of an onion (*Allium cepa*) seedling showing sheath of cells containing amyloplasts (small black granules) which have sedimented to the lower side of the cells. (*After H. E. Street and H. Öpik, The Physiology of Flowering Plants, Edward Arnold, 1970*)

hormone responsible is still in doubt since here the lower side grows more slowly than the upper. It is thought that the growth inhibitor abscisic acid or a close chemical relative is involved, although indol-3-ylacetic acid may also participate in some way.

2. Phototropism. In phototropism the stimulus is a light gradient, and unilateral light induces similar curvatures; those toward the source are positively phototropic; those away from the source are negatively phototropic. Main axes of shoots are usually positively phototropic, while the vast majority of roots are insensitive. Leaves usually respond by aligning their surfaces perpendicular to the incident rays and are said to be diaphototropic. Some plants change their responses during development, as can also happen in gravitropism. For example, ivy-leaved toadflax (*Linaria cymballaria*) grows on walls, and the flower stalks turn toward the light to facilitate pollination; when the fruit ripens, they turn toward the wall to bury the fruit in a crevice, ready for germination.

The first shoots (coleoptiles) of grasses and cereals have been used extensively for phototropic studies because of their great sensitivity and simple structure (**Fig. 7**). As for geotropism, differences in auxin concentration between the light and dark sides of the coleoptile seem to be responsible for the curvature; less auxin accumulates on the lighter side. However, in other plants, for example, sunflower (*Helianthus annus*) seedling hypocotyls, auxin seems not to be directly involved; instead more growth inhibitor (possibly abscisic acid) accumulates on the lighter side. The shorter wavelengths of the visible spectrum (that is, blue light) are the effective ones, implying that yellow pigments are concerned with detection of a light gradient across the organ. Whether these pigments are flavins or carotenoids or both remains to be established. The nature of light action in causing hormone concentration differences is also uncertain, but it does not seem to be photodestruction of auxin. A modification of hormone transport down the two sides may be involved.

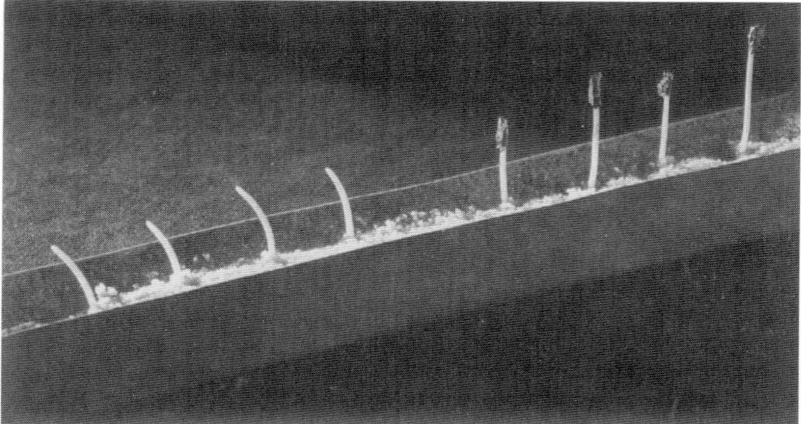

Fig. 7. Phototropism of oat (*Avena sativa*) coleoptile. The four coleoptiles on the left show the growth curvatures induced by light coming from the left. The four on the right have had their tips covered with small caps of aluminum foil. (*After L. J. Andus, Plant Growth Substances, 2d ed., Leonard Hill Ltd., 1959*)

Many fungi also show gravitropic and phototropic responses. However, since growth processes of these organisms differ greatly from those of higher plants, the mechanisms must be very different.

3. Thigmotropism. In thigmotropism (sometimes called haptotropism), the stimulus is touch; it occurs in climbing organs and is responsible for tendrils curling around a support. The curvature is due to a deceleration of growth on the contact side and an acceleration on the other. In many tendrils the response may spread from the contact area, causing the tight coiling of the basal part of the tendril into an elaborate and elastic spring.

4. Chemotropism. Chemotropism is induced by a chemical substance. Examples are the incurling of the stalked digestive glands of the insectivorous plant *Drosera* and incurling of the whole leaf of *Pinguicula* in response to the nitrogenous compounds in the insect prey. A special case of chemotropism concerns response to moisture gradients; for example, under artificial conditions in air, the primary roots of some plants will curve toward and grow along a moist surface. This is called hydrotropism and may be of importance under natural soil conditions in directing roots toward water sources. *See* ABSCISIC ACID; AUXIN; PLANT GROWTH; PLANT HORMONES; PLANT PHYSIOLOGY; ROOT (BOTANY). Leslie J. Audus

Bibliography. W. Haupt and M. E. Feinleib (eds.), *Encyclopedia of Plant Physiology*, new ser., vol. 7: *Physiology of Movements*, 1979; M. B. Wilkins, *Advanced Plant Physiology*, 1988.

Plant nomenclature

The generally accepted system of naming plants. The study of plant variation and kinds leads to classification—the grouping of individual plants, the grouping of the groups, and the organizing of the groups (taxa) into ranks. Nomenclature deals with the naming of the taxa, the goal being to establish one correct scientific name for a given taxon with a given definition.

There are two fundamentally different kinds of plant names, vernacular (common) and scientific. Vernacular names are culturally limited and differ from one society to another. For example, the beech tree is also known as *Buche* (German), *hêtre* (French), *faggio* (Italian), and *buna zoku* (Japanese). The scientific name of the beech, *Fagus*, is culturally neutral and universally recognized by botanists.

For most purposes, plants are considered in terms of three ranks: family, genus, and species. There are ranks above family (such as orders), ranks lower than species (such as subspecies), and even intercalated ranks (such as subgenera). Scientific family names are based on generic names and end in -aceae, such as Fagaceae (the oak family), which includes two major genera of trees, *Fagus* and *Quercus*.

The basic scientific plant name is a binomial, a combination of a Latin capitalized generic name and a Latin lowercase specific epithet, such as *Quercus alba*, the common American white oak. Scientific names may seem awkward, but to the extent that they are universally learned, they are also universally understood. The names also have meaning. *Quercus* is an ancient Roman word for the trees that provided the best and hardest woods, and *alba* means white, presumably a reference to the whitish undersides of the leaves.

History. The binomial system of nomenclature was first widely applied to plants by Carolus Linnaeus in his *Species Plantarum* (1753). Within a generation, binomial names displaced the previously used descriptive polynomials. For example, J. F. Gronovius (1739) called the white oak *Quercus foliis superne latioribus opposite sinuatis, sinubus angulisque obtusis* (oak with leaves broader upward, oppositely sinuate, [and] blunt sinuses and angles). Such polynomials had become increasingly cumbersome as more and more plant species were discovered in the New World during the eighteenth century.

The century following Linnaeus was an authoritarian period, in which plant nomenclature was dominated by botanical scholars such as J. B. P. de Lamarck, K. Willdenow, and their successors. As the numbers of taxa and workers increased, it became obvious that some rules must be established to prevent unnecessary name changes. This led to A. de Candolle's "Laws," ratified in 1867 and considered to be the best guide for botanical nomenclature.

Shortly thereafter, it became evident that the "Laws" were not consistent with past usage, and either the names or the rules had to change. Several individuals and institutions had devised their own rules, which led to chaos. Eventually the first *International Rules of Botanical Nomenclature* was ratified by the botanical community in 1906. This introduced the concept of conservation of generic names in order to maintain well-known names that did not follow its guidelines. The third edition of the Rules (1935) codified the type method for determining the application of scientific names. The fourth edition was the first to be called the *International Code of Botanical Nomenclature* (ICBN), and has been followed by a new edition about every 6 years. Each new edition is produced after an International Botanical Congress, during each of which several hundred proposals to amend the ICBN are considered and voted upon by delegates to the Congress.

Name changes. There are three major reasons that scientific names should be changed: discovery that the type specimen of a name is not what had been assumed, discovery that a hitherto-overlooked name applies, or redefinition of the taxon.

Types control the application of names. Botanists describe their new taxa with at least one specimen in hand. Although the taxon can be redefined (usually as new material becomes available), it can never be redefined so as to exclude all the original (type) material. On occasion, however, the type material of a well-known name has been found to pertain to a completely different species. The ICBN has a formal procedure to maintain usage.

Priority controls when different names apply to the same taxon. Essentially the oldest valid name published after 1753 has rights that newer names do not. Relatively few names are changed because of overlooked earlier names.

Redefinition of taxa causes most name changing. Practically every new collection prompts a redefinition of a taxon. The new collection could establish a link between what had been thought to be separate species and result in two species becoming one. Conversely, careful study might show the opposite—what was thought to be a single, variable species is actually a mixture of two or more species. Taxonomy dictates which taxa are to be recognized and how they are to be defined, and nomenclature provides the names for the taxa. When the scope of the taxon is changed because of new knowledge, the name usually must be changed.

Refinements. Plant hybridization and horticulture lead to special problems that fit uneasily into the Linnaean binomial system. Plants more often hybridize than do animals. The ICBN devotes an appendix to hybrid nomenclature. Hybrid species, called nothospecies, can be named by one of two methods, a hybrid formula (giving the parents) or a binomial with a multiplication sign. *Verbascum lychnitis* L. × *V. nigrum* L. is a hybrid formula. (The "L." indicates that the plants were named by Linnaeus.) It is also correctly known as *Verbascum* × *schiedeanum* Koch. The appendix provides for naming more complex hybrids, but the multiplication sign is a constant feature.

The *International Code of Nomenclature for Cultivated Plants* deals with cultivated plants, generally known as cultivars. Cultivars are usually given fancy names or epithets that can be distinguished from scientific names in that they are capitalized and are not italicized. In addition, the names or epithets are either preceded by cv., the abbreviation of cultivar, or indicated by quotation marks: *Scilla hispanica* cv. Rose Queen and *Scilla hispanica* 'Rose Queen' are both correct for the "Rose Queen" cultivar of *Scilla hispanica*. *See* PLANT KINGDOM; ZOOLOGICAL NOMENCLATURE.

Dan H. Nicolson

Bibliography. C. D. Brickell et al. (eds.), International Code of Nomenclature for Cultivated Plants—1980, *Regnum Vegetabile*, 104:1–32, 1980; W. Greuter et al. (eds.), International Code of Botanical Nomenclature, *Regnum Vegetabile*, 118:1–328, 1988; E. E. Sherff et al., Symposium on Botanical Nomenclature, *Amer. J. Bot.*, 36:1–32, 1949; M. L. Sprague et al., A discussion on the differences in observance between zoological and botanical nomenclature, *Proc. Linn. Soc. Lond.*, 156:126–146, 1944.

Plant organs

Plant parts having rather distinct form, structure, and function. Organs, however, are interrelated through both evolution and development and are similar in many ways.

Roots, stems, and leaves are vegetative, or asexual, plant organs. They do not produce sex cells or play a direct role in sexual reproduction. In many species, nevertheless, these organs or parts of them (cuttings) may produce new plants asexually (vegetative reproduction). Sex organs are formed during the reproductive stage of plant development. In flowering plants, sex cells are produced in certain floral organs. The flower as a whole is sometimes called an organ, although it is more appropriate to consider it an assemblage of organs. *See* REPRODUCTION (PLANT).

Root. The root is usually the underground part of the plant axis. It may consist of a dominant primary seedling root (taproot) with subordinate branch roots, as in carrots and beets; or it may be composed, as in grasses, of numerous branched roots of similar dimensions (fibrous roots). Collectively, all the roots of a plant are known as its root system. Roots anchor the plant, absorb water and mineral salts in solution from the soil, and conduct these to the stem. Organic food and growth substances received from the stem move to the areas of growth and storage in the roots. *See* ROOT (BOTANY).

Stem and leaves. The stem is usually the aerial part of the plant axis and bears leaves. The stem and leaves together constitute the shoot. In some species the major portion of the stem grows horizontally beneath the surface of the soil, and thus is called a rhizome, or underground stem. The stem conducts water and minerals from the roots to all parts of the shoot, and food materials and growth substances from the shoot to the root. The stem may also serve as a storage organ for water and food. Green leaves containing chlorophyll, when exposed to light and air, carry on photosynthesis. As a by-product of this process, oxygen is returned to the atmosphere. Leaves also return large amounts of water vapor to the air through transpiration (evaporation). Some leaflike structures are protective (bud scales), others are fleshy types in which food and water may accumulate. The first leaves on a seed plant are called cotyledons. *See* LEAF; STEM.

Flower, fruit, and seed. The flower is often interpreted as a modified shoot bearing floral organs instead of leaves. These organs are the sepals, petals, stamens, and carpels. The sepals and petals are sterile leaflike appendages. Sepals, like leaves, are commonly green. Petals, containing little or no chlorophyll, are usually white or have some color other than green. The sepals collectively constitute the calyx, the petals the corolla. The calyx and corolla form the perianth or floral envelope. The stamens and carpels are the reproductive floral organs and produce sex cells. A stamen is usually composed of a slender stalk, called the filament, at the tip of which is an anther. The anther is divided into four or fewer pollen sacs in which pollen grains develop. When mature, the pollen sacs open, and the pollen is transferred by wind, water, insects, or man to the tips of the carpels, a process called pollination. The assemblage of stamens is called the androecium, a term implying the male nature of this part of the flower. The carpel assemblage is called the gynoecium to indicate its female nature. Pistil is another term used to designate the female part of the flower. A single carpel may form a pistil, or two or more may be combined into a compound pistil. The pistil generally consists of an enlarged basal part called the ovary. The apex of the ovary usually narrows into a stalk, called the style, that terminates in a sticky surface, the stigma. The ovary contains one or more ovules. The egg cell produced in each ovule becomes fertilized by the sperm brought to the ovule by the pollen tube. The latter is an outgrowth of a pollen grain that became attached to and germinated on the stigma. The pollen tube grows through the style to the ovule where it discharges the sperm. After the egg is fertilized by a sperm, the ovary, sometimes together with other floral parts, develops into a fruit. The ovules become seeds. *See* FLOWER; FRUIT; PLANT PHYSIOLOGY; PLANT TISSUE SYSTEMS; SEED. Katherine Esau

Bibliography. R. M. Klein and D. T. Klein, *Fundamentals of Plant Science*, 1987; J. D. Mauseth, *Plant Anatomy*, 1988.

Plant pathology

The study of disease in plants; it is an integration of many biological disciplines and bridges the basic and applied sciences.

Nature of Plant Disease

An understanding of the dynamic nature of disease in plants combines a knowledge of botanical sciences (for example, anatomy, cytology, genetics, and physiology); of disciplines that relate to causation of disease (for example, bacteriology, mycology, nematology, and virology); of disciplines such as ecology and meteorology which relate environmental factors to the severity of disease; and of those divisions of science that involve the impact of plant diseases upon society (for example, economics and sociology). As a science, plant pathology encompasses the theory and general concepts of the nature and cause of disease, and yet it also involves disease control strategies, with the ultimate goal being reduction of damage to the quantity and quality of food and fiber essential for human existence.

Diagnosis. Diagnosis of disease depends in part upon visible changes in color, form, and structure of the plant (**Fig. 1**). These changes, or symptoms, may involve death of cells and tissues (as in blights, rots); inhibition of plant growth and development (as in chlorosis, stunting); or increase in cell number and size (as in galls, witches'-brooms). Visible structures of the pathogenic agent, or signs, may also be present, as in bacterial ooze or powdery mildew. Identification of the cause of disease is included in diagnosis and becomes a basis of subsequent strategies in control. Many pathogenic agents may be isolated and grown in artificial culture. Such agents may then be used to inoculate healthy plants

Fig. 1. Characteristic appearance of apple scab on leaves and fruit. ***(H. H. Lyon, Cornell University)***

to test pathogenicity. These steps constitute Koch's Postulates, or Rules of Proof of Pathogenicity. Many other procedures are used in diagnosis: light and electron microscopy, biochemical tests (chromatography, electrophoresis), physical tests (determination of buoyant density, ultraviolet fluorescence), serological tests (agar diffusion, enzyme-linked immunosorbent assay or ELISA), and molecular techniques.

Etiology. Etiology is the study of the causes of disease, and symptomology is the study of the symptoms expressed by the plant. The primary agent may be infectious, that is, transmissible from plant to plant (bacterium, fungus, or virus); or it may be noninfectious, that is, nontransmissible (for example, excess of nitrate). Infectious and noninfectious agents are considered pathogens because they cause disease. The term pathogen, however, is usually reserved for disease-inducing organisms, viroids, and viruses. The term abiotic agent is used for nonliving causes of disease, such as excess or deficiency of a nutrient element. A parasite is an organism that derives food from its host. A pathogenic organism also usually derives food from its host, the susceptible plant, but it causes disease.

Epidemiolgy. Epidemiology is a study of how disease develops in populations. A sudden, widespread outbreak of disease is called an epidemic. The term epiphytotic is often used to denote such an outbreak in plants. The behavior and interactions of pathogens, plants, and environmental factors are considered quantitatively by epidemiologists.

Kinds of Plant Diseases

Diseases were first classified on the basis of symptoms. Three major categories of symptoms were recognized long before the causes of disease were known; necroses, destruction of cell protoplasts (rots, spots, wilts); hypoplases, failure in plant development (chlorosis, stunting); and hyperplases, overdevelopment in cell number and size (witches'-brooms, galls). This scheme remains useful for recognition and diagnosis.

When fungi, and then bacteria, nematodes, and viruses, were recognized as causes of disease, it became convenient to classify diseases according to the responsible agent. If the agents were infectious (biotic), the diseases were classified as being "caused by bacteria," "caused by nematodes," or "caused by viruses." To this list were added phanerogams and protozoans, and later mollicutes (mycoplasmas, spiroplasmas), rickettsias, and viroids. In a second group were those diseases caused by such noninfectious (abiotic) agents as air pollutants, inadequate oxygen, and nutrient excesses and deficiencies.

Other classifications of disease have been proposed, such as diseases of specific plant organs, diseases involving physiological processes, and diseases of specific crops or crop groups (for example, field crops, fruit crops, vegetable crops). Disease can be related to seven gross physiological functions of healthy plants: food storage, hydrolysis and utilization of stored food, absorption of water and nutrients by roots, growth and development of meristems, water translocation in the xylem, photosynthesis, and translocation of foods through the phloem. Diseases could then be grouped on the basis of their effect upon these essential functions. This system has much to offer. It directs attention to the disease phenomenon rather than to the pathogens, many different kinds of which could be responsible for a specific disease syndrome, for example, seedling blights. The system allows concentration on the environment as it affects the disease process, rather than on the plant-specific agents. It also forms a logical basis for exploration of possible control measures because it is related to specific stages of growth and development of the plant. Carl W. Boothroyd

Crop Losses to Disease

Diseases of agricultural crops are studied mainly because of the losses they cause; reliable crop loss data are therefore of primary importance in any plant protection program. It is not sufficient to state that diseases cause a loss; the magnitude of the loss must be quantified so that it can be related to the potential economic gain resulting from control measures. Information on crop losses is required by decision makers in all agencies involved in plant protection to allow formulation of sound policies, meaningful allocation of resources, and assignment of priorities. Similarly, the farmer must know the cost of the projected loss before it is decided how much to spend on control.

Assessment. The methodology to assess losses involves two phases. The first phase is an experimental study to develop a disease-loss model that describes the relationship between the amount of disease and the corresponding loss. The second phase involves large-scale surveys conducted to assess the amount of disease in commercial crops. The disease-loss models are then used to translate amount of disease into crop losses. This approach is time-consuming, but some countries, such as the United Kingdom and Canada, have demonstrated the usefulness of generating annual crop loss data for major crops. Crop loss studies in the United Kingdom in 1967 first demonstrated the importance of foliage diseases of barley and the need for fungicides, hitherto considered unnecessary but now applied to more than half of the total acreage of barley.

Such detailed information is generally not available, but there is a great need for reliable information on crop losses. Many countries, including the United States, have been involved in the development of programs on crop losses. The United States is the only country that has attempted to estimate losses due to disease for all crops. However, the reliability ranges from subjective estimates based on consensus of opinion to objectively generated experimental data.

Monocultures. The evolvement of crop monocultures with increasing genetic homogeneity increases the vulnerability of crops to diseases, and this could contribute to higher losses in the future. A leaf-blight epidemic on corn in the United States in 1970 dramatically demonstrated the susceptibility of modern varieties. This single epidemic caused a sudden and unpredicted $1 billion loss and seriously disrupted world food supplies. Substantial losses due to disease on a world basis continue to occur despite the continuous introduction of many resistant varieties and the large annual expenditures for fungicides.

W. Clive James

Symptoms of Plant Diseases

Symptoms are expressions of pathological activity in plants. They are visible manifestations of changes in color, form, and structure: leaves may become spotted, turn yellow, and die (**Fig. 2**); fruits may rot on the plants or in storage (**Fig. 3**); cankers may form on stems (**Fig. 4**); and plants may blight and wilt. Diagnosticians learn how to associate certain symptoms with specific diseases, and they use this knowledge in the identification and control of pathogens responsible for the diseases.

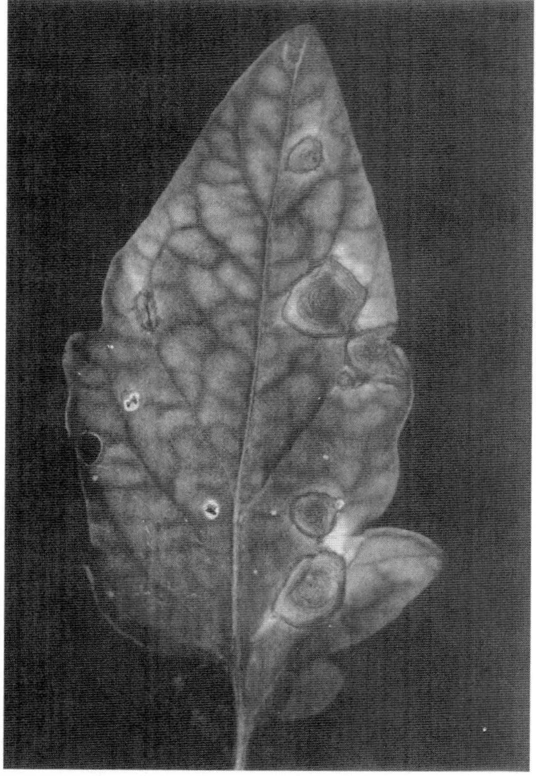

Fig. 2. Early blight of tomato. (*H. H. Lyon, Cornell University*)

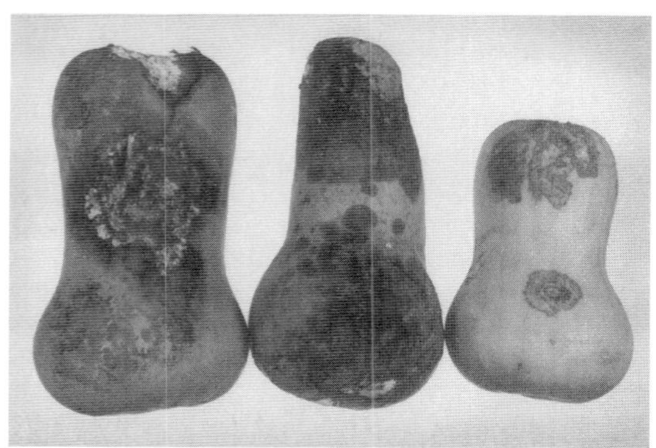

Fig. 3. Black rot of squash. (*H. H. Lyon, Cornell University*)

Fig. 4. White pine blister rust. (*H. H. Lyon, Cornell University*)

Those symptoms that are external and readily visible are considered morphological. Others are internal and primarily histological, for example, vascular discoloration of the xylem of wilting plants (**Fig. 5*a*** and ***b***). Microscopic examination of diseased plants may reveal additional symptoms at the cytological level, such as the formation of tyloses (extrusion of living parenchyma cells of the xylem of wilted tissues into vessel elements; **Fig. 6**).

It is important to make a distinction between the visible expression of the diseased condition in the

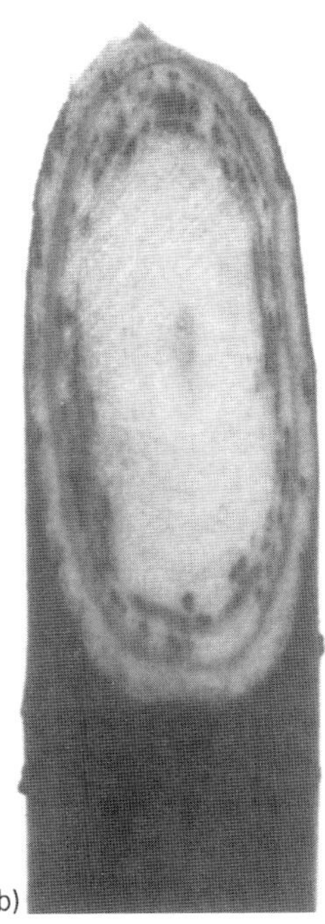

Fig. 5. **Dutch elm disease. (*a*) Wilting of foliage. (*b*) Vascular discoloration in xylem. (*H. H. Lyon, Cornell University*)**

plant, the symptom, and the visible manifestation of the agent which is responsible for that condition, the sign. The sign is the structure of the pathogen, and when present it is most helpful in diagnosis of the disease. For example, large galls may form on the ear of a corn plant. The gall is a symptom of disease and a typical expression of activity by the corn smut fungus. At a later stage of maturity that gall will break and dusty black spores will be exposed. The spore mass is the sign of that pathogen, and in association with the gall the problem is recognizable as corn smut (**Fig. 7**).

All symptoms may be conveniently classified into three major types because of the manner in which pathogens affect plants. Most pathogens produce dead and dying tissues, and the symptoms expressed are categorized as necroses. Early stages of necrosis are evident in such conditions as hydrosis,

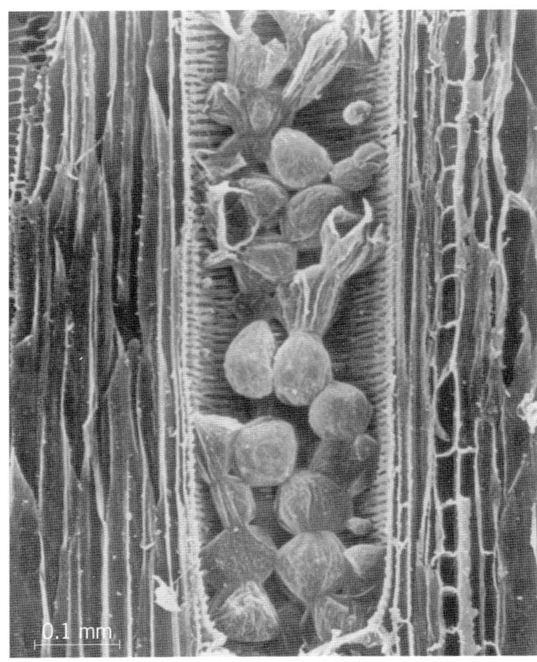

Fig. 6. **Tylose formation in xylem vessel of banana with Fusarium wilt. (*G. E. VanderMolen, University of Rhode Island*)**

Fig. 7. **Gall formation in boil smut of corn. (*H. H. Lyon, Cornell University*)**

Fig. 8. Dwarfing and mosaic in cucumber mosaic; healthy fruit on left. (*H. H. Lyon, Cornell University*)

Fig. 9. Crown gall of tomato. (*H. H. Lyon, Cornell University*)

wilting, and yellowing. As cells and tissues die, the appearance of the plant or plant part is changed, and is recognizable in such common conditions as blight, canker, rot, and spot.

Many pathogens do not cause necrosis, but interfere with cell growth or development. Plants thus affected may eventually become necrotic, but the activity of the pathogen is primarily inhibitory or stimulatory. If there is a decrease in cell number or size, the expressions of pathological activity are classified as hypoplases; if cell number or size is increased, the symptoms are grouped as hyperplases. These activities are very specific and most helpful in diagnosis. In the former group are such symptoms as mosaic, rosetting, and stunting, with obvious reduction in plant color, structure, and size (**Fig. 8**). In the latter group are gall (**Fig. 9**), scab, and witches'-broom, all visible evidence of stimulation of growth and development of plant tissues. *See* CROWN GALL.

C. W. Boothroyd

Causative Agents of Disease

The primary agents of plant disease are fungi, bacteria, viruses and viroids, nematodes, parasitic seed plants, and a variety of noninfectious agents.

Fungi. More plant diseases are caused by fungi than by any other agent. The fungi include a vast variety of living organisms, all lacking chlorophyll. They are microscopic except for some species which at one stage in their life cycle produce visible fruiting bodies containing spores. The somatic phase (thallus) typically is threadlike or filamentous and grows in a branching manner in all directions over or within plant tissues or other substrates. This thallus, known as mycelium, is usually septate, but septa or cross walls are lacking in some groups. Each filament of the fungal mycelium, called a hypha, is tubular and contains protoplasm, one or more nuclei, and most of the organelles, except chloroplasts, found in higher plant cells. Fungi typically reproduce by means of microscopic spores that may be formed by both asexual and sexual processes. The majority of fungi have spores that have one dimension from 5 to 50 micrometers. *See* FUNGI.

The fungi that cause plant disease derive their food from the plant (host) and are called parasites. Those that can live and grow only in association with living plant tissues are obligate parasites. Some fungi obtain their food from dead organic matter and are known as saprobes or saprophytes. Still others can utilize food from either dead organic matter or from living plant cells, and are referred to as either facultative parasites or facultative saprophytes.

Fungi are grouped into classes primarily on the basis of the morphology of their reproductive stages. The classes with organisms that cause plant disease are Plasmodiophoromycetes, Chytridiomycetes, Zygomycetes, Oomycetes, Ascomycetes, Basidiomycetes, and their asexual stages.

During the past three decades, the use of new technologies in cell biology, ultrastructure, and molecular systematics has resulted in a reevaluation of what are fungi. Currently, they are recognized as the Kingdom Fungi with four Phyla: the Chytridiomycota, Zygomycota, Ascomycota, and Basidiomycota. The Form-class Deuteromycetes, which traditionally encompassed asexual states of fungi, has been largely abandoned in recent literature. The Plasmodiophoromycetes have been shown to align closely with Protista and the Oomycetes to the Stramenopila near the yellow-green algae (Chromista).

Plasmodiophoromycetes. The Plasmodiophoromycetes, called the endoparasitic slime molds, are a group of organisms related to protozoans along with slime molds. They do not have the threadlike mycelial phase of most fungi; the thallus is a naked multinucleate mass of protoplasm known as a plasmodium. Resting cells and thin-walled sporangia give rise to biflagellate zoospores. The zoospores penetrate root cells of the host and become ameboid, and the plasmodium develops. Only a few species in this class cause disease in cultivated plants. *Plasmodiophora brassicae*, the agent of club root of crucifers, is the best known.

Chytridiomycetes. Chytrid fungi produce motile cells (zoospores) with a single flagellum. Only a few species are economically important as causative agents (pathogens). A true mycelium is lacking in the species that cause plant disease. The simple thallus is entirely converted into reproductive structures called sporangia. Zoospores develop within the sporangia. One of the best-known disease-causing chytrids is *Synchytrium endobioticum*, the agent of black wart of potato.

Oomycetes. The Oomycetes include species fungi causing destructive plant diseases, most of which are in one order, Peronosporales. Included are the water molds, the downy mildews, and the white blisters (white rusts). *Plasmopara viticola*, the cause of downy mildew of grape, and *Phytophthora infestans*, the cause of late blight of potato and tomato, are examples of Oomycetes. All the plant parasitic species of the class have a filamentous, branched, coenocytic mycelium that grows between or within the plant cells. In some species, haustoria, which are modified hyphal branches, penetrate the host cells. Asexual reproduction occurs by sporangia which, upon germination, produce biflagellate zoospores. In some species of downy mildew fungi the sporangia germinate by a germ tube.

Well-differentiated sex organs, antheridia and oogonia, are formed by these groups. Following fertilization, which is accomplished when a nucleus from the antheridium fuses with a nucleus in the oogonium, a thick-walled oospore is produced. The oospores germinate by germ tubes or by zoospores. The downy mildews derive their name from the abundance of mycelial branches (sporangiophores) bearing sporangia which grow out from the stomates of diseased leaves. This results in visible downy patches on the leaf. The species of *Albugo*, causative agents of white blisters, produce sporangia in compact masses under the host epidermis. A whitish blister results, and when the epidermis is ruptured a powdery white crust is evident.

Zygomycetes. Most of the Zygomycetes are saprobes, living on decaying plant and animal matter. One species, *Rhizopus stolonifer*, is a facultative parasite and can cause a serious rot disease of fleshy plant organs after harvest. Examples are soft rot of sweet potatoes and strawberries. *Rhizopus* has a filamentous mycelium without cross walls (nonseptate). Long aerial branches with spherical black spore bodies (sporangia) develop from the mycelium growing in or on the rotting fleshy roots or fruits. This cottony, whiskerlike growth speckled with black sporangia is the diagnostic sign of *Rhizopus*. The sporangia contain numerous small spores. A thick-walled resting spore called a zygospore develops as a result of sexual reproduction.

Ascomycetes. The Ascomycetes are the sac fungi, so named because the sexual or perfect-state spores are produced in a saclike cell, the ascus. Many such cells (asci), each typically containing eight ascospores, develop in differentiated fruiting bodies called ascocarps. Two kinds of ascocarps recognized for a long time are the perithecium and the apothecium. Perithecia are spherical or flask-shaped, completely closed or opening by a small pore. Apothecia are open and cup- or saucer-shaped. In later studies, two additional types of ascocarps, pseudothecia and ascostromata, have been described. They are similar in appearance to perithecia. Ascocarps may be single or aggregated and are visible signs of many of the plant-infecting Ascomycetes. However, in most instances the ascocarps must be studied microscopically to be certain of their identity. In addition to the ascospore state, most ascomycete fungi have an asexual or conidial spore state. The conidia are produced in tremendous numbers on branches of the septate mycelium or in specialized fruiting bodies.

The Ascomycetes are the largest class of fungal pathogens. Included here are the powdery mildew fungi, which are obligate parasites with a superficial mycelium but with haustoria penetrating the host epidermal cells. They produce enormous numbers of conidia that appear to the unaided eye as a whitish powdery coating on the host plant. The ascocarps have characteristic appendages that are of taxonomic value in separating the genera.

Other examples of Ascomycetes are *Claviceps purpurea*, causing the ergot disease of cereals; *Monilinia fructicola*, the cause of brown rot of stone fruits; *Venturia inaequalis*, the cause of apple scab; *Ceratocystis ulmi*, the cause of the Dutch elm disease; and *Endothia parasitica*, which has nearly eliminated the American chestnut in North America.

Basidiomycetes. Examples of Basidiomycetes that cause plant disease are the smuts, rusts, and the wood-decay fungi of forest trees. All fungi in this class produce basidiospores borne on the outside of a specialized cell or spore-bearing structure called the basidium. Fusion of two nuclei followed by meiosis occurs in the basidium prior to basidiospore formation. Four basidiospores on each basidium are usual. Four or five kinds of spores or reproductive cells may be produced by the rust fungi. These are aeciospores, urediniospores, teliospores, and basidiospores; spermatia (pycniospores), which are male sex cells, are also present. The various kinds of spores are usually produced in large numbers in pustules, sori, or other visible fructifications. The germinating teliospore of the rusts is the basidium. Some rust species have all the spore stages on one host (autoecious), whereas two different kinds of host plants are essential with heteroecious rust fungi. The rust fungal fructifications may be yellow, orange, red, brown, or black, and often powdery but are sometimes gelatinous or waxy. Rust fungi include

Puccinia graminis, the causative agent of black stem rust of cereal grains; *Cronartium ribicola*, the cause of white pine blister rust; and *Gymnosporangium* species, which cause the cedar apple and cedar hawthorn rusts. About 4000 species of rust fungi are known, and all are obligate parasites.

Most of the economically important smut fungi are parasites of corn, sorghums, the cereal grains, and other grasses. The visible signs of these fungi are black, dusty masses of teliospores in host plant tissues. In the more destructive diseases, the spores develop in and destroy the flower parts so there is a total loss of grain yield. *Ustilago* species cause smuts of wheat, barley, corn, and oats. Species of *Tilletia* cause stinking smut or bunt of wheat.

The wood decay- and root rot–causing fungi of forest trees include *Armillaria mellea* and numerous species of *Fomes* and *Polyporus*. All of these fungi produce enormous numbers of basidiospores on gills (lamellae) or in pores or tubes on the underside of large complex fruiting bodies (basidiocarps). The basidiocarps are referred to as mushrooms, conks, or bracket and shelf fungi.

Deuteromycetes. There are several thousand species of fungi that so far as is known reproduce only by asexual spores, called conidia. Lacking a sexual (perfect) state, they have been referred to as imperfect fungi and placed in an artificial Form-class Deuteromycetes. The classification of the Deuteromycetes is based on the morphology of the conidial states and the mode of development of the conidia. Conidia of these species vary greatly in size, shape, and color, and may be one- to many-celled. They are formed singly or clustered on specialized hyphal branches, or in large numbers in specialized fructifications known as pycnidia, acervuli, or sporodochia. Many Deuteromycetes cause serious leaf spots and flower blights, as well as fruit rots, stem cankers, and wilts. Examples include species of *Septoria*, *Gloeosporium*, *Diplodia*, *Alternaria*, and *Verticillium*. C. Wayne Ellett; J. W. Kimbrough

Bacteria. Eleven Gram-negative and seven Gram-positive genera of bacteria are associated with diseases of flowering plants. Species affected by these bacterial diseases include the major cereal, forage, vegetable, and fruit crops. The bacterial pathogens of plants are almost exclusively associated with plants and do not cause serious diseases of humans or animals.

Types of bacterial diseases. Each bacterial species produces distinctive symptoms on the host(s) that it attacks. Typical symptoms caused by bacterial plant pathogens, and the main genera of plant pathogenic bacteria are illustrated in **Fig. 10**.

The bacteria most commonly associated with plant diseases are Gram-negative, non-spore-forming, rod-shaped cells and include the genera *Agrobacterium*, *Acidovorax*, *Burkholderia*, *Erwinia*, *Ralstonia*, *Pantoea*, *Pseudomonas*, *Rhizomonas*, *Xanthomonas*, *Xylella*, and *Xylophilus*. Some of the bacterial species that are Gram-positive are in the genera of *Arthrobacter*, *Bacillus*, *Clavibacter*, *Clostridium*, *Cutrobacterium*, *Rhodococcus*, and *Streptomyces*. Species of *Bacillus* and *Clostridium* produce endospores that are survival mechanisms for the bacteria. Two additional groups of prokaryotes that are commonly associated with plants and are extremely difficult or virtually impossible to culture are the spiroplasmas and phytoplasmas (formerly termed mycoplasmalike organisms or MLOs). Both groups of these prokaryotes lack a true cell wall and are relatively recent members of the Mollicutes. These latter bacterial plant pathogens are associated with yellows-type diseases and various plant growth abnormalities; they inhabit phloem tissue in diseased plants. The spiroplasmas are helical cells that can be cultured on complex, artificial media.

Most of the typical bacterial plant pathogens are motile as a result of the rotation of threadlike flagella attached to the inner and outer membranes. These flagella are involved for movement in water. The flagella are arranged randomly around the cell surface (peritrichous) in *Erwinia* and *Agrobacterium* species. Several other plant-pathogenic bacteria, such as *Xanthomonas*, have a single flagellum (monothrichous) attached to one end (polar). *Pseudomonas* species contain a tuft of flagella at one end (lophotrichous).

Bacterial cells divide by binary fission and produce millions of cells in a short period of time. Thus, a bacterium that multiplies every 30 min will produce more than 10^6 cells in less than 10 h. During periods of optimal conditions for bacterial multiplication after infection of plant tissue, disease can develop extensively and quickly.

Classification. Bacteria cannot be identified to species based on colony characteristics, cell morphology, or staining procedures. Therefore, biochemical, serological, genetic techniques, fatty acid analysis, protein profiles, and sensitivity to phages have been useful to classify species. An equally important test to identify bacteria is to demonstrate pathogenicity on a particular host or hosts.

Means of spread and survival. Many foliar bacterial pathogens are disseminated long distances by splashing rain and by wind-blown mist. When rain events occur, bacterial cells ooze from infected stem or leaf tissue or through lenticels or stomata. During rain events, tiny droplets and aerosols that contain bacteria are produced and carried to healthy plants. Bacteria may also be moved from plant to plant by insects, in irrigation water, and by cultural activities used during crop production (such as pruning). Several bacterial plant pathogens require an insect vector for dispersal. The bacterium that causes Pierce's disease of grapes requires a leafhopper insect to transmit, infect, and cause disease.

Most bacterial plant pathogens survive adverse conditions in the host plant. Those bacterial species that incite diseases of perennial hosts are particularly well adapted to long-term survival. Some bacteria have been reported to survive in the rhizosphere of grasses during the winter months. Few bacterial species are able to survive for long periods in the

Conventional color and infrared color aerial photographs, taken simultaneously with a dual camera, aid research on the distribution and spread of wheat diseases. (*C. J. Johannsen, Laboratory for Agricultural Remote Sensing, Purdue University*)

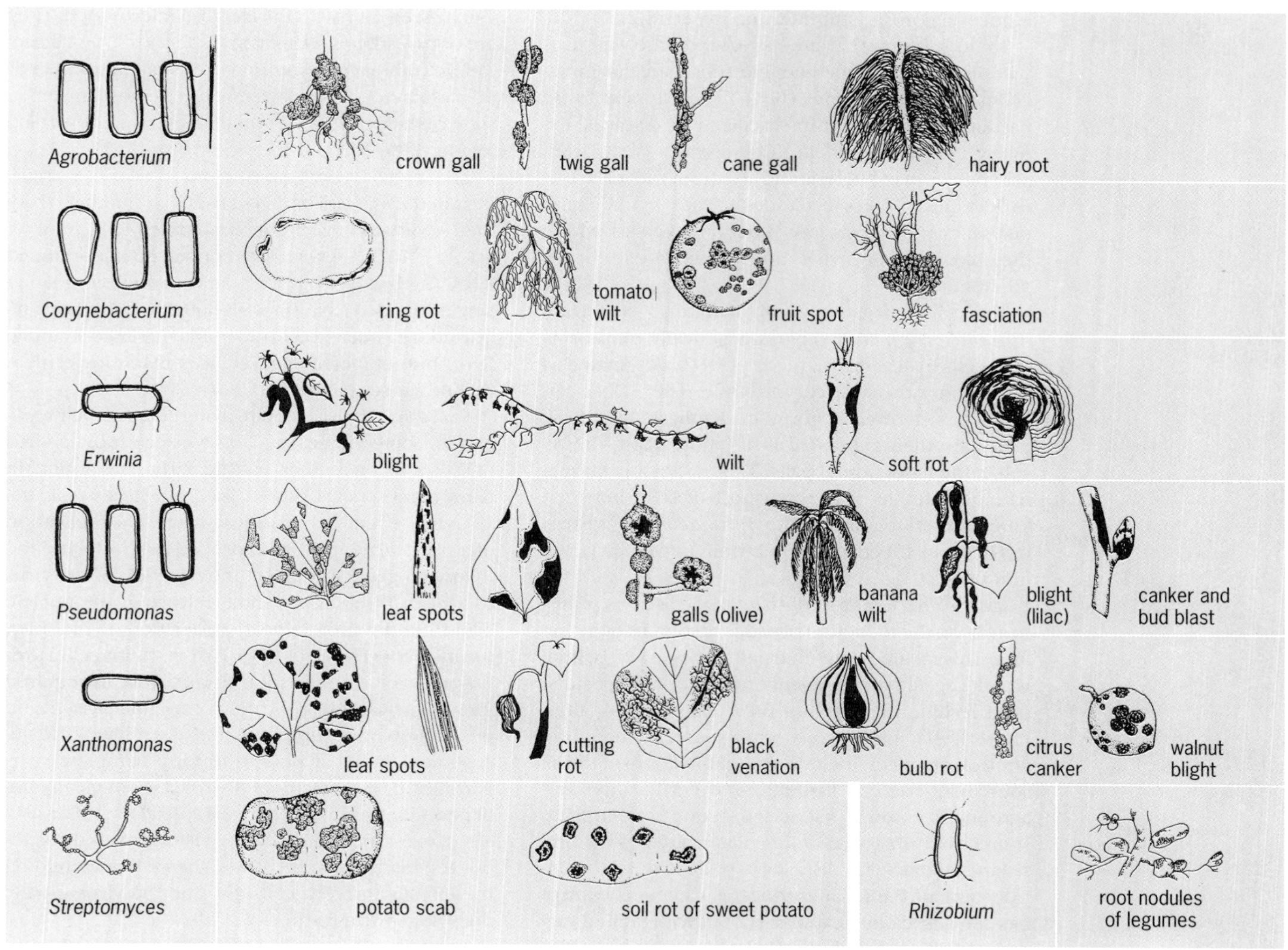

Fig. 10. Genera of bacteria and types of symptoms caused by species in these genera. (*After G. N. Agrios, Plant Pathology, 2d ed., Academic Press, 1978*)

soil free of the host plant; however, the pathogen of bacterial wilt of tomato and potato is reported to survive for many years in fallow soil. Some species survive in the rhizosphere of nonhost plant species. Many bacteria are able to survive on or in seed. Plant pathogenic bacteria are also able to survive in irrigation water.

Pathways of entry. Bacterial plant pathogens are not able to penetrate directly through the leaf or stem; instead they enter through natural openings or wounds. Bacteria can enter through stomata or hydathodes in leaves, and through lenticels in corky tissue. Fresh wounds are excellent portals for bacterial ingress. Chewing insects and nematodes produce wounds that favor entry by bacteria. A leaf miner insect causes large wounds in the leaves of citrus trees which greatly accelerates the spread of bacterial canker. Cultural practices such as pruning and cultivation may create wounds and increase the likelihood of infection by bacteria such as the crown-gall bacterium.

Mechanisms of pathogenesis. The mechanisms associated with pathogenesis are quite variable and are not completely understood. One interesting mechanism has been elucidated recently. Many bacterial pathogens of animals as well as bacterial pathogens of plants have a unique secretion system that is associated with pathogenesis. It is termed the type III secretion system. Gram-negative bacteria use this system to secrete proteins across their inner and outer membranes into the host cell. This enables Gram-negative bacteria to inject pathogenicity-related proteins into the cytosol and nucleus of the eukaryotic host. It is unique among secretion pathways since it delivers virulence proteins into host cells.

Bacterial cells produce many compounds that are involved in disease development. Soft rot bacteria produce enzymes that hydrolyze pectic substances between the cell walls of plants to destroy the integrity of the tissue. This results in a mushy appearance of the host structure. Some bacteria produce phytotoxins that are low-molecular-weight compounds. These phytotoxins directly injure plant cells and influence the course of disease development or symptoms. One toxin, coronatine, is produced by several bacterial plant pathogens. Coronatine induces chlorosis, stunting, and hypertrophy.

Some bacteria that are associated with vascular-wilt diseases produce gumlike substances or extracellular polysaccharides (EPS). These appear to be particularly important in vascular wilt diseases. Extensive wilting occurs in these diseases partly from accumulation of bacteria and also from the release of extracellular polysaccharides. The extracellular polysaccharides pass into functional vessels where they block xylem cavities and accumulate in the pit membranes.

The Gram-negative, soil-inhabiting bacterium *Agrobacterium tumefaciens* genetically transforms host cells by inserting a piece of DNA, designated as T-DNA, into the host chromosome. The T-DNA carries genes involved in hormone synthesis, and these genes are then expressed in the host plant; the result is the induction of galls. This is the only known natural example of the transport of DNA between kingdoms of organisms. The *Agrobacterium* system has been an effective tool to introduce foreign genes into plants.

Control. The control of bacterial diseases relies heavily on the reduction of primary sources of inoculum. This includes bacterial pathogens in seeds, budwood, vegetative propagative material, young plants, crop residue that has not fully decomposed, alternative hosts, and various other sources. Therefore, control practices include the elimination of these sources of bacteria through seed certification and agronomic cultural practices to reduce or eliminate sources of primary inoculum. Plant varieties with resistance to bacterial diseases have developed to control some of the foliar pathogens. Chemical control has been used quite extensively, but with limited success. Antibiotics and copper bactericides have provided some control, but numerous bacterial species have been developed resistance to these compounds. The strategies have also included the use of biological control with microbes antagonistic to the bacterial pathogen or microbes that induce a systemic-acquired resistance (SAR) in the plant. These microbes are applied prior to infection by the bacterial pathogens, and this results in reduced disease. Biological control has been used successfully to control crown gall, incited by *Agrobacterium tumefaciens*, with a nonpathogenic strain of *A. radiobacter*, known as K84. This strain is antagonistic to the causal agent because it produces an inhibitory compound termed a bacteriocin. More recently, biotechnological research has resulted in the cloning of resistance genes and the transfer of these genes to other genotypes to the same plant species or to related plant species. J. B. Jones

Viruses and viroids. Viruses and viroids are the simplest of the various causative agents of plant disease. The essential element of each of these two pathogens is an infective nucleic acid. The nucleic acid of viruses is covered by an exterior shell (coat) of protein, but that of viroids is not. *See* PLANT VIRUSES AND VIROIDS.

Approximately 400 plant viruses and about 10 viroids are known. Because of their small size, they can be seen only with the electron microscope. They are of two shapes, isometric ("spherical") and anisometric (tubular). Isometric viruses range from 25 to 50 nanometers in diameter, whereas most anisometric viruses are of 12–15-nm diameter and of varying length (200–2000 nm).

The nucleic acid of most plant viruses is a single-stranded RNA (mol wt $1.5–4 \times 10^6$); a number of isometric viruses have a double-stranded RNA (mol wt $10–20 \times 10^6$). A few viruses contain double-stranded DNA (mol wt $4–5 \times 10^6$), and several containing single-stranded DNA (mol wt 800,000) have been reported. The nucleic acid of viroids is a single-stranded RNA, but its molecular weight is much lower than that of viruses (about 120,000).

Replication. Viruses require living cells for their replication. Some viruses, such as tobacco mosaic virus (TMV) and cucumber mosaic virus, are found in many plant species; others, such as wheat streak mosaic virus, occur only in a few grasses. The replication of a plant virus appears to proceed according to the following general scheme: introduction of the virus to a plant through a wound, release of the nucleic acid from the protein coat, association of viral RNA (or messenger RNA of DNA viruses) with cellular ribosomes for its translation to the proteins required for virus synthesis, replication of the nucleic acid and production of coat protein, and assembly of the nucleic acid and coat protein to form complete virus particles. The replication of viroids is not clearly understood at present. Cell-to-cell spread of viruses usually occurs, and eventually the virus spreads throughout the plant. In some plants, the cells surrounding the initially infected cells die, and the virus usually does not spread further.

Transmission. Viruses are transmitted from plant to plant in several ways. Grafting scions to infected rootstocks often results in virus infection of the scion. Cuttings taken from infected mother plants will usually be diseased. Contact through leaf rubbing between healthy and diseased plants causes minute wounds through which viruses enter healthy plants. The majority of viruses are transmitted by vectors such as insects, mites, nematodes, and fungi which acquire viruses during feeding upon infected plants. Some viruses are transmitted to succeeding generations by infected seed. Viroids are spread mainly by contact between healthy and diseased plants or by the use of contaminated cutting tools.

Control. The control or prevention of virus diseases involves breeding for resistance, propagation of virus-free plants, use of virus-free seed, practices designed to reduce the spread by vectors, and, in some cases, the deliberate inoculation of plants with mild strains of a virus to protect them from the deleterious effects of severe strains. R. I. Hamilton

Nematodes. Nematodes (roundworms) are one of the most common groups of invertebrate animals and are found in a wide variety of aquatic habitats, both marine and fresh water, from the Antarctic sea bottom to thermal springs. All soils that support plant life contain nematodes living in the water films that surround soil particles. Most nematodes feed primarily on microscopic plants, animals, and bacteria, but a few, such as hookworms, pinworms, and ascarids, are parasites of animals; another relatively small group of nematodes parasitize plants. *See* NEMATA.

Movement is serpentine, and there are no body appendages. Reproduction is by the laying of eggs. Males may or may not be present, and adults of some species are hermaphroditic. All species known have four larval (juvenile) stages and an adult stage separated by molts. The first molt takes place within the egg shell. In some species, adult females become swollen and sedentary. Plant-parasitic nematodes are distinguished by their small size, about 1 mm average length, and mouthparts that are modified to form a hollow stylet which is inserted into plant cells. Food intake of nematodes is by suction provided by a muscular esophagus.

All of the plant parasites are placed into two orders. The order Tylenchida have stylets with basal knobs, a three-parted esophagus (pharynx), and an ornamented integument (cuticle). The order Dorylaimida have a simple tooth, a two-parted esophagus, and a smooth cuticle.

In nature, all nematodes ingest living food, and plant feeders are considered obligate parasites in that they do not grow and reproduce in the absence of a living host plant. Feeding of ectoparasitic species is restricted to the root surface, whereas endoparasites move into plant tissues. Some species become sedentary, others remain migratory during feeding.

Injury to plants. Plant injury is of three general types and is related to feeding habits. Migratory endoparasites destroy tissues as they feed, producing necrotic lesions in the root cortex. Other migratory endoparasites invade leaf tissues and produce extensive brown spots. Sedentary endoparasites do not kill host cells, but induce changes in host tissues, which lead to an elaborate feeding site or gall (**Fig. 11**). A unique feature of these galls is that the host plant cells surrounding the nematode's head become enlarged, with thickened walls, dense cytoplasm, and many enlarged nuclei. These syncytia, or giant cells, are concentrated centers of nutrients that have been diverted from normal metabolism of the host. A few species produce galls on stems or in seeds, and galls are also produced by some ectoparasites. The third general type of symptom is produced by certain migratory ectoparasites, where root tips are devitalized and cease to grow without any associated swelling or necrosis.

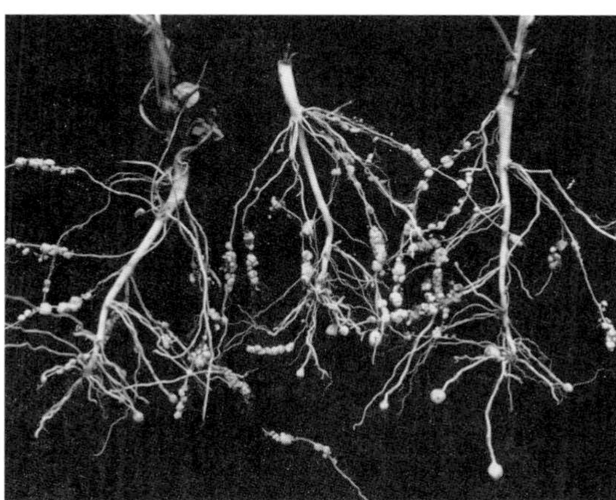

Fig. 11. Root knot nematode galls on roots of potato.

In addition to the plant injury that they cause directly, nematodes are important factors in disease complexes. Lesions and galls provide entrance courts for soil fungi and bacteria, and many diseases caused by soil-borne pathogens are more severe when nematodes are present. Plant varieties resistant to soil-borne fungi sometimes lose this resistance when previously parasitized by nematodes. Important viruses such as grapevine fan-leaf and peach stem pitting are transmitted by nematodes of the order Dorylaimida.

Important species. Root knot nematodes (*Meloidogyne*) are worldwide in distribution with many species that attack most kinds of plants (Fig. 11). The major symptoms are swollen galls on roots. Nematode eggs are produced in jellylike masses on the surface of galls.

Cyst nematodes (*Globodera* and *Heterodera*) are so called because eggs are retained inside the swollen female cuticle, which becomes tanned and leathery. There is relatively little root swelling, and cysts remain attached to roots by their embedded necks.

Lesion nematodes (*Pratylenchus*) and foliar nematodes (*Aphelenchoides*) produce brown necrotic spots on roots and leaves, respectively.

Stem and bulb nematodes (*Ditylenchus*) produce bloated swellings near the soil line. The swollen areas contain large numbers of nematodes which sometimes form resistant desiccated tufts of "narcissus wool." Clovers, alfalfa, and vegetable crops are also important hosts of stem and bulb nematodes.

Seed gall nematodes (*Anguina*) produce blackened "cockles," which are grains that have been largely replaced with thousands of nematodes. Seed gall nematodes remain viable for many years when dried, and the pathogen is transmitted when galls are planted with healthy seed.

Control. Control of plant-parasitic nematodes often is based on selection of nonhosts for crop rotations or nematode-resistant varieties when they are available. Some plant species, such as marigolds, release compounds into the soil that are toxic to nematodes. Some lessening of injury is achieved by irrigation and by fertilization, particularly with potash, since many nematode-induced diseases lead to potassium deficiency. Animal manures, compost, and other organic amendments enhance the buildup of natural enemies of nematodes, particularly the nematode-trapping fungi. Chemical control usually consists of treatment of soils with liquid chemicals that volatilize, producing toxic fumes which move through the soil mass. The high cost of soil fumigation and the requirement for special equipment and preparation restrict this practice to high-cash-value crops. Some of the systemic organophosphate and carbamate insecticides are effectively used, at higher dosages, to control nematodes in living plants.

R. A. Rohde

Seed plants. Many parasitic seed plants (estimated at nearly 3000) attack other higher plants. They occur in at least 15 different families. In some families (for example, the mistletoes Loranthaceae and Viscaceae) all members are parasitic; in others, only a single genus is parasitic in an otherwise autotrophic family (for example, the dodder, genus *Cuscuta*, of the morning glory family Convolvulaceae).

Most parasitic plants are terrestrial; that is, the parasitic connection with the host plant is through the roots. Other parasitic plants grow on the aboveground parts of the host. Some plants are classed as semiparasites because they can live in the soil as independent plants for a time, but are not vigorous or may not flower if they do not become attached to a suitable host.

The nutritional status of parasitic plants ranges from total parasites with no chlorophyll (for example, the broomrapes, Orobanchaceae) to plants that are well supplied with chlorophyll and obtain primarily water and minerals from their hosts (many mistletoes).

Some parasitic plants cause serious damage in agricultural crops. Among the most damaging are the broomrapes and dodder. An Asian plant, the witchweed (*Striga asiatica*), has recently become established in the southeastern United States and is a serious threat to corn and other crops. The U.S. Department of Agriculture has begun an intensive research and control program to combat this damaging weed.

Parasitic plants, particularly the mistletoes, are also serious pests in the forest, orchard, and ornamental trees in many parts of the world. The dwarf mistletoes (genus *Arceuthobium*) are among the most damaging pests of coniferous forests in the western United States (**Fig. 12**); they cause losses of billions of board-feet of wood annually.

Parasitic higher plants of economic importance are being combatted in many ways: by breeding resistant crops, using less-susceptible species, applying chemicals, and by cultural means such as pruning out infected branches or removing diseased plants.

Frank G. Hawksworth

Noninfectious agents of disease. Plants with symptoms caused by noninfectious agents cannot serve as sources of further spread of the same disorder. Such noninfectious agents may be deficiencies or excesses of nutrients, anthropogenic pollutants, or biological effects by organisms external to the affected plants.

Deficiencies or excesses of nutrients. For proper growth, plants need periodically renewed supplies of nitrogen, phosphate, and potash, plus a number of essential minor elements. Nutrient availability may be affected by the acid-base reaction of the soil. *See* FERTILIZER; PLANT MINERAL NUTRITION.

Plants deficient in nitrogen appear chlorotic and develop yellowed older leaves. A phosphate-deficient plant may be darker green than normal with reddened or purpled leaf tips or edges. Plants deficient in potash develop light green leaves punctured with dead spots at the tips and along the margins. Deficiencies of some essential minor elements are associated with a variety of diseases (see **table**).

Fig. 12. Many small plants of a dwarf mistletoe (*Arceuthobium vaginatum*) parasitizing a ponderosa pine branch. This is the most damaging disease of ponderosa pine in many parts of the West.

Naturally occurring excesses of sodium salts are the common cause of alkali injury in many arid regions of the world. A similar problem is caused by the application of deicing salts to highways. Polluting the soil adjacent to the travel lanes, this salt can kill affected plants.

Injury from excessive nitrogen is common on house plants and on seedlings in greenhouses. Plant damage has also been reported from soils excessive in aluminum, arsenic, boron, calcium, chromium, copper, fluoride, iodine, iron, lithium, magnesium, manganese, and nickel. Too much of some of these elements can unbalance the intake or metabolism of other essential nutrients.

Anthropogenic pollutants. The deicing salt, just mentioned, is a common example of an anthropogenic pollutant.

On the farm, plant-damaging pollution may be caused by careless use of pesticides. Although insecticides and fungicides can harm plants if not used properly, mishandled herbicides are by far the most

Examples of plant diseases associated with deficiencies of minor elements

Deficient minor element	Plant disease
Boron	Internal cork of apples
Calcium	Blossom-end rot of tomatoes
Copper	Exanthema of citrus
Iron	Maple chlorosis
Magnesium	Sand drown of tobacco
Manganese	Gray speck of oats
Molybdenum	Whiptail of cauliflower
Zinc	Little leaf of peach

damaging to plants. Herbicide sprays may drift unnoticed onto adjacent fields; herbicide residues may remain in soil to damage a succeeding crop; or herbicide residues may seep into wells used for irrigation. Herbicides may also contaminate other pesticide sprays or fertilizers when spraying or mixing equipment is not properly cleaned.

Another source of pollution on the farm is ethylene generated naturally by ripening fruit. Ethylene, a powerful gaseous plant hormone, affects the growth habits and metabolism of most plants. If apples are left in storage with carrots, for example, the ethylene from the apples will produce a bitter-tasting metabolite in the carrots.

Off the farm, anthropogenic air pollutants are generated by industrial processes, and by any heating or transportation method that uses fossil fuels. The most common air pollutants that damage plants are sulfur oxides and ozone. Sulfur oxides are produced when sulfur-containing fossil fuels are burned or metallic sulfides are refined. Human-generated ozone is produced by sunlight acting on clouds of nitrogen oxides and hydrocarbons that come primarily from automobile exhausts.

There are other plant-damaging air pollutants as well. Gaseous fluorides come from factories that manufacture aluminum, steel, ceramics, and phosphate fertilizers. Ethylene, mentioned before, also comes from waste hydrocarbons emitted by poorly functioning internal combustion engines and oil-burning space heaters. Ethylene is not found in natural gas, but is a component of manufactured gas. This source may again become a problem if there is a return to synthetic fuels. Other air pollutants that have damaged plants are the smog-component peroxyacetyl nitrate (PAN), chlorine escaping from water treatment plants, and cement dust. Methane, the principal component of natural gas, is not itself toxic to plants, but may act as food for methane-consuming bacteria in the soil. By proliferating, these bacteria can kill plant roots indirectly by depleting soil oxygen.

The symptoms of air pollution injury include leaf flecking, spotting, and mottling; red discoloration of leaves and fruits; withering of flower parts; stunting; and death of plant parts on the entire plant.

Some air pollutants, such as sulfur oxides and fluorides, accumulate within the exposed plant. Others, such as ozone and PAN, cannot be detected in the injured plant. The combination of sulfur oxides plus ozone is more toxic to some plants than either pollutant alone.

The severity of the damage depends on the specific pollutant or mixture of pollutants, the pollutant concentration, the length of exposure, the plant cultivar, the condition of the plant when it was exposed, and environmental conditions. Outdoors, air pollution episodes usually occur when temperature inversions confine and concentrate the pollutants. *See* AIR POLLUTION; WATER POLLUTION.

Biological effects by external organisms. The effect of apple ethylene on carrots, mentioned earlier, may be considered an example in this category.

Another group of such disorders may be classed as allelopathic. This term describes the harmful effect on one plant produced by toxic secretions from another plant or plant residue. For example, most plants grow poorly or not at all in the soil over the roots of black walnut trees. The alleged cause is the secretion of a quinonic compound by black walnut roots.

Tobacco frenching is a disease apparently caused by toxic diffusates from soil microorganisms, such as *Bacillus cereus*, that do not colonize tobacco plants. The suspected toxicants are isomers of isoleucine.

Saul Rich

Pathogenic Life Cycles of Agents

Plant pathogens initiate and induce disease in plants in a series of steps that make up the disease cycle and include overwintering of the pathogen, production or liberation of inoculum, dissemination of inoculum, inoculation, penetration, infection, and growth or multiplication of the pathogen leading to production of new inoculum. The new inoculum may cause one or several secondary disease cycles during the growing season, or it may overwinter and cause infection the following growing season. Disease cycles of a typical plant-pathogenic fungus, bacterium, mycoplasma, virus, and nematode are shown in **Figs. 13–17**.

Overwintering. Pathogens overwinter in infected plants (such as trees), in surviving belowground parts of plants, in some other perennial plant, in plant debris, in or on seeds, in some vector such as an insect, or in the soil. Viruses, mycoplasmas, and protozoa overwinter as such inside living cells. Bacteria usually overwinter when they are in contact with living cells. Fungi, parasitic higher plants, and nematodes overwinter, respectively, mainly as spores, seeds, or eggs either on, or usually apart from, the infected plant.

Production and liberation of inoculum. Inoculum, that is, the number of individual pathogens that can be liberated and disseminated to start new infections, may consist of the vegetative body of the pathogen, as in bacteria, viruses, and so on, or as with fungi and parasitic higher plants, it may consist of spores (Fig. 13) and seeds, respectively. The first inoculum of the growth season is called primary inoculum, and causes primary infections. Some pathogens (viruses, mycoplasmas, protozoa, some bacteria) never release inoculum outside the plant, but it is transferred instead with living cells (such as buds, grafts), or within vectors (such as insects; Fig. 15). Fungi, parasitic higher plants, most bacteria, and nematodes produce inoculum outside the host plant, or else it exudes or moves out of the plant, usually in the presence of moisture (Figs. 13, 14, and 17).

Dissemination of inoculum. Inoculum is disseminated with the help of wind (Fig. 13), running or splashing water (Figs. 14 and 17), insects (Fig. 15), other animals, and humans directly or indirectly by contaminated tools. A few pathogens move on their own for a few centimeters (bacteria, fungal zoospores) or up to a meter (nematodes).

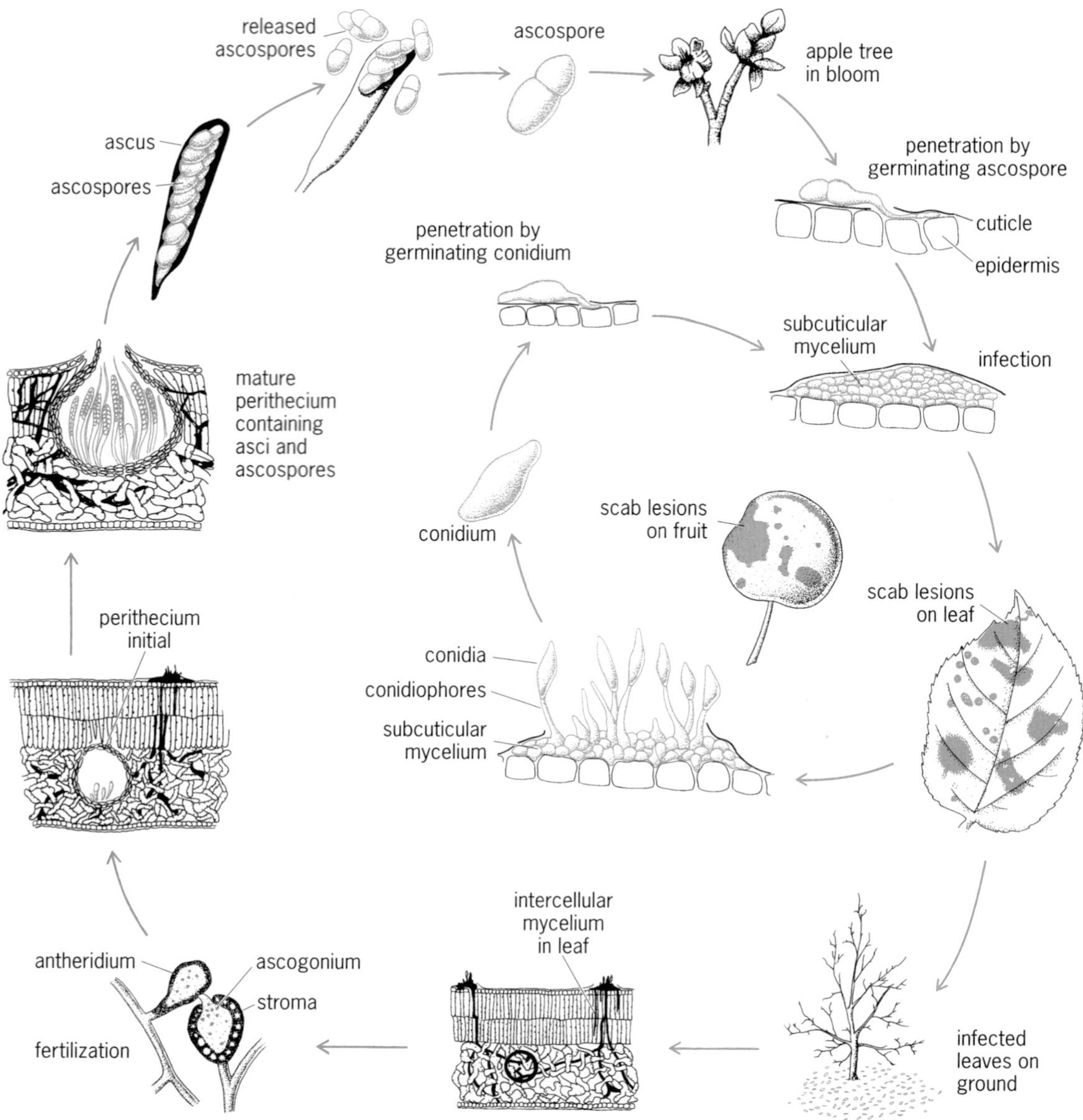

Fig. 13. Development of the disease cycle of apple scab caused by the fungus *Venturia inaequalis*. (*After G. N. Agrios, Plant Pathology, 2d ed., Academic Press, 1978*)

Inoculation. Inoculation is the coming into contact of the pathogen with a susceptible part of the host plant in a way that could result in disease. With some pathogens inoculation is accomplished when the pathogen comes in contact with an intact surface, a wound, or a natural opening of the host plant (Figs. 13, 14, and 17); with others the pathogen must be brought into contact with a wounded living cell (Figs. 15 and 16).

Penetration. Viruses, mycoplasmas, protozoa, and some bacteria are inserted into plants by vectors and thereby accomplish penetration concurrently with inoculation (Figs. 15 and 16). Fungi, most bacteria, parasitic higher plants, and nematodes come into contact with the outer surface of the plant and enter it through wounds (Fig. 14), through natural openings such as stomata and lenticels, or through direct penetration of the cuticle (Fig. 13) and cell walls (Fig. 17). These pathogens penetrate susceptible plants directly by secreting enzymes that break down the plant cell walls and by exerting physical pressure on the plant.

Infection. Pathogens cause infection the moment they begin to obtain nutrients from their host plants. Infection continues either until the plant defends itself successfully and kills, or stops the effects of, the pathogen, or until the plant is finally killed by the pathogen. Infection takes place when suscept and pathogen are compatible with each other. In some instances, initial recognition of such compatibility seems to be made possible by the interaction of complementary molecular configurations of specific pathogen proteins called lectins with specific sugars contained in more complex molecules at the plant surface.

During infection, pathogens produce enzymes that disintegrate plant cells and neutralize the activity of cellular components, toxic substances (toxins) that inactivate plant proteins and destroy the function of plant membranes, and growth regulators that upset the hormonal balance in the plant. The plant also produces enzymes, toxic substances, and growth regulators that contribute to its defense effort against the pathogen.

The reaction between suscept and pathogen is determined by the DNA present in the chromosomes of each. Recent work, however, indicates that some diseases are determined by additional genetic material present in pathogens as relatively small circular strands of DNA called plasmids. Thus, plasmid DNA carries the genes responsible for production of toxins by some pathogens and for inducing uncontrolled cell division and gall formation (malignancy) by others (Fig. 14).

When infection occurs, parts of the plant are destroyed or altered in ways characteristic of the suscept-pathogen combination. These manifestations by the infected plant are the symptoms. The period of time between initiation of infection and appearance of symptoms (incubation period) is characteristic of the particular suscept-pathogen combination and the prevailing temperature; it varies from 2 days to 3 weeks in most diseases and up to 2–3 years in some diseases of trees.

Growth or multiplication of the pathogen. During infection most pathogens are inside the plant and increase only in numbers without increasing much in size; others, especially fungi and parasitic higher plants, increase primarily in size. Plant pathogenic bacteria, mycoplasmas, and protozoa multiply by fission, that is, each individual pathogen splits into two identical pathogens; viruses are replicated by the

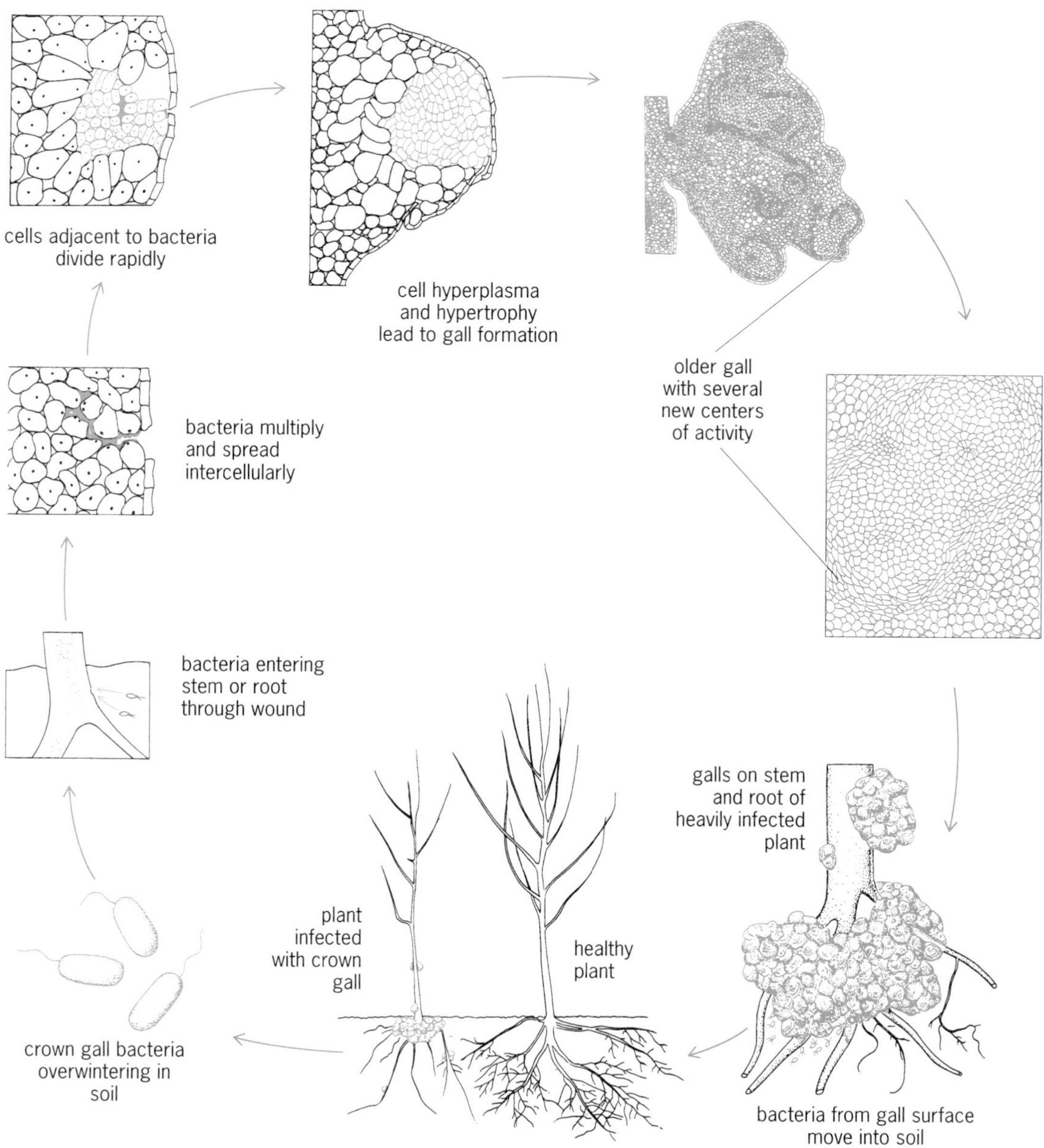

Fig. 14. Development of the disease cycle of crown gall caused by the bacterium *Agrobacterium tumefaciens*. (*After G. N. Agrios, Plant Pathology, 2d ed., Academic Press, 1978*)

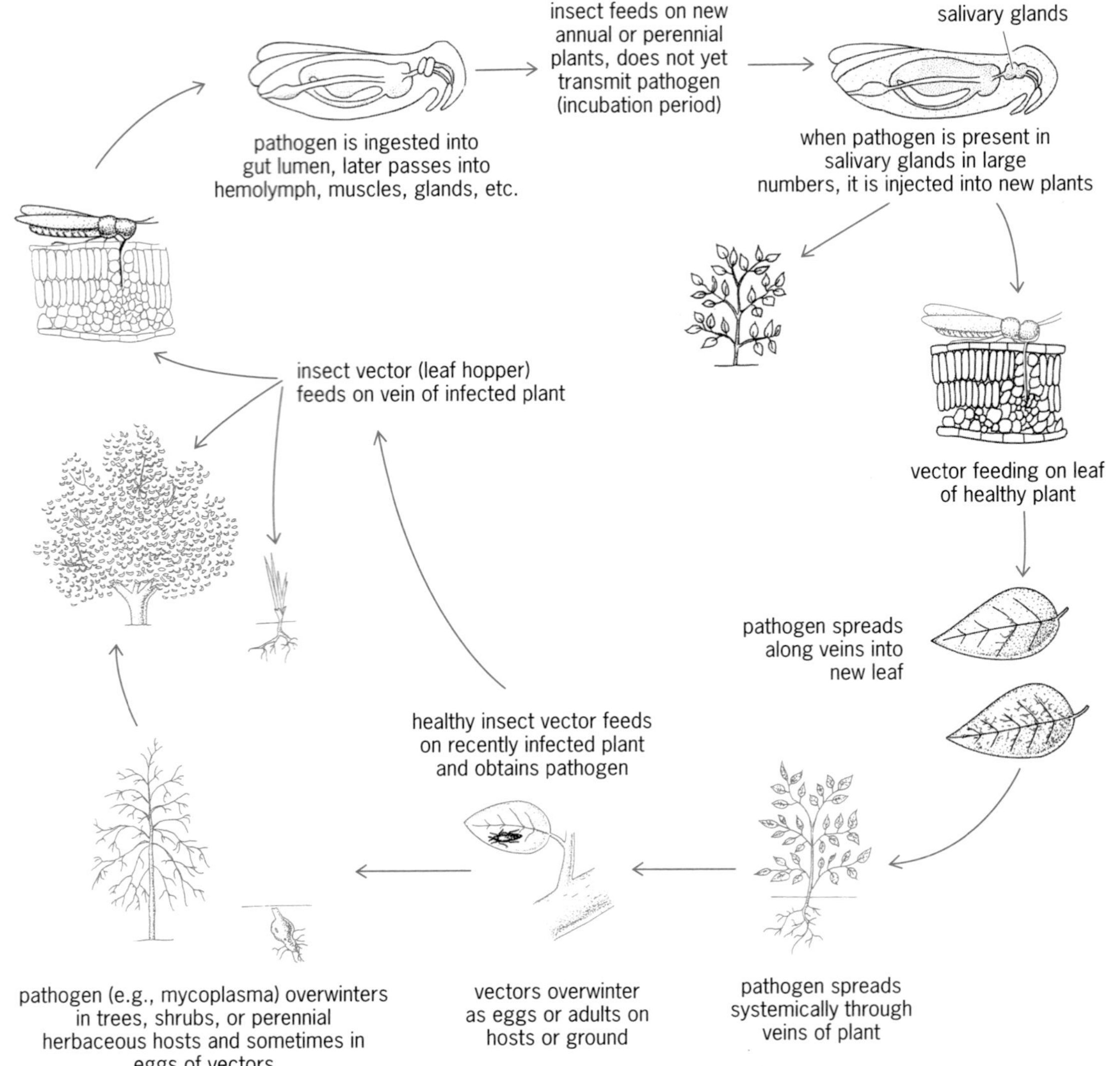

Fig. 15. Development of a typical disease cycle caused by mycoplasma, virus, and so on, transmitted by an insect vector such as a leafhopper. (*After G. N. Agrios, Plant Pathology, 2d ed., Academic Press, 1978*)

millions by the host cell (Fig. 16); and some nematodes lay eggs that hatch into more nematodes inside the plants (Fig. 17). Viruses, mycoplasmas, protozoa, and some bacteria spread as such through the entire plant, particularly in the conductive tissues (phloem or xylem), whereas most bacteria and nematodes spread only into localized areas of plants. Parasitic fungi grow into their host plants as roots grow into soil, and almost all of them produce masses of countless spores at, above, or slightly below the plant surface (Fig. 13). Parasitic higher plants send food-absorbing structures into host plants and multiply by producing seeds outside the hosts.

Thus, in or on every infected plant, thousands, millions, or even billions of new inoculums are produced which, when disseminated to susceptible plants under favorable weather conditions, initiate new (secondary) infections. Secondary infections may be repeated from a few to 20 or more times during each growing season, depending on the pathogen. At the end of the growing season, though, pathogens go into dormancy or produce specialized hardy structures by which they overwinter and thus close the disease cycle. George N. Agrios

Epidemiology of Plant Disease

Epidemiology is the study of the intensification of disease over time and the spread of disease in space. The botanical epidemiologist is concerned with the interrelationships of the host plant (suscept), the pathogen, and the environment, which are the components of the disease triangle. With a thorough knowledge of these components, the outbreak of disease may be forecast in advance, the speed at which the epidemic will intensify may be determined, control measures can be applied at critical periods, and any yield loss to disease can be projected. The maximum amount of disease occurs when the host plant is susceptible, the pathogen is aggressive, and the environment is favorable.

Epidemiologically, there are two main types of diseases: monocyclic, those that have but a single infection cycle (with the rare possibility of a second or even third cycle) per crop season; and polycyclic, those that have many, overlapping, concatenated cycles of infection per crop season. For both epidemiological types, the increase of disease slows as the proportion of disease, y, approaches saturation or 100% ($y = 1.0$) [**Fig. 18**].

Monocyclic diseases. Fungal root rots, bacterial and fungal wilts, damping-off, basal-stem cankers, and other diseases caused by soil-borne pathogens (including nematodes) usually develop over time with a single infection cycle per crop season. Some foliar diseases that have only one infection cycle per season are also considered as monocyclic.

Epidemic development. The progress of monocyclic diseases when plotted against time resembles the inverted J-shaped curve of normal growth. The rate at which monocyclic diseases develop is determined primarily by the amount of pathogenic inoculum at the beginning of the season, and secondarily by the environmental conditions and host resistance.

The curves of monocyclic diseases are usually fit well by the monomolecular model, $y_t = 1 - \exp(-k_M t)$, where y_t is the proportion of disease at time t, and k_M is the monomolecular epidemic rate. Such curves can be linearized with the transformation $Y = \ln[1/(1 - y)]$. The rate parameter, k_M, can be used for comparative purposes.

Control. For many soil-borne pathogens, resistance of the plants to disease is not known or is incomplete. Therefore, the control of monocyclic diseases consists mostly in the reduction of the inoculum of the pathogen before the crop is planted. Sanitation measures such as burning, burying, or removal of crop debris, crop rotation, soil treatment with mulches, fumigation, sterilization, or solarization are frequently used. Additionally, careful management of the soil moisture, pH, and fertility also reduces initial infections. A reduction in initial inoculum has the additional benefit of slowing the monomolecular rate, k_M. Some postharvest diseases develop monocyclically, and treatment with chlorinated water followed by cool temperatures in transit and storage slow the epidemic rate.

Polycyclic diseases. The rusts, mildews, leaf spots, blights, and other foliar diseases develop polycyclically. The pathogens have multiple infection cycles during a single crop season. Even from a single leaf spot or pustule, the propagules of the pathogen are spread by wind, rain, and vectors (including humans) to neighboring or distant plants. After a latent period that follows infection by the propagules, the

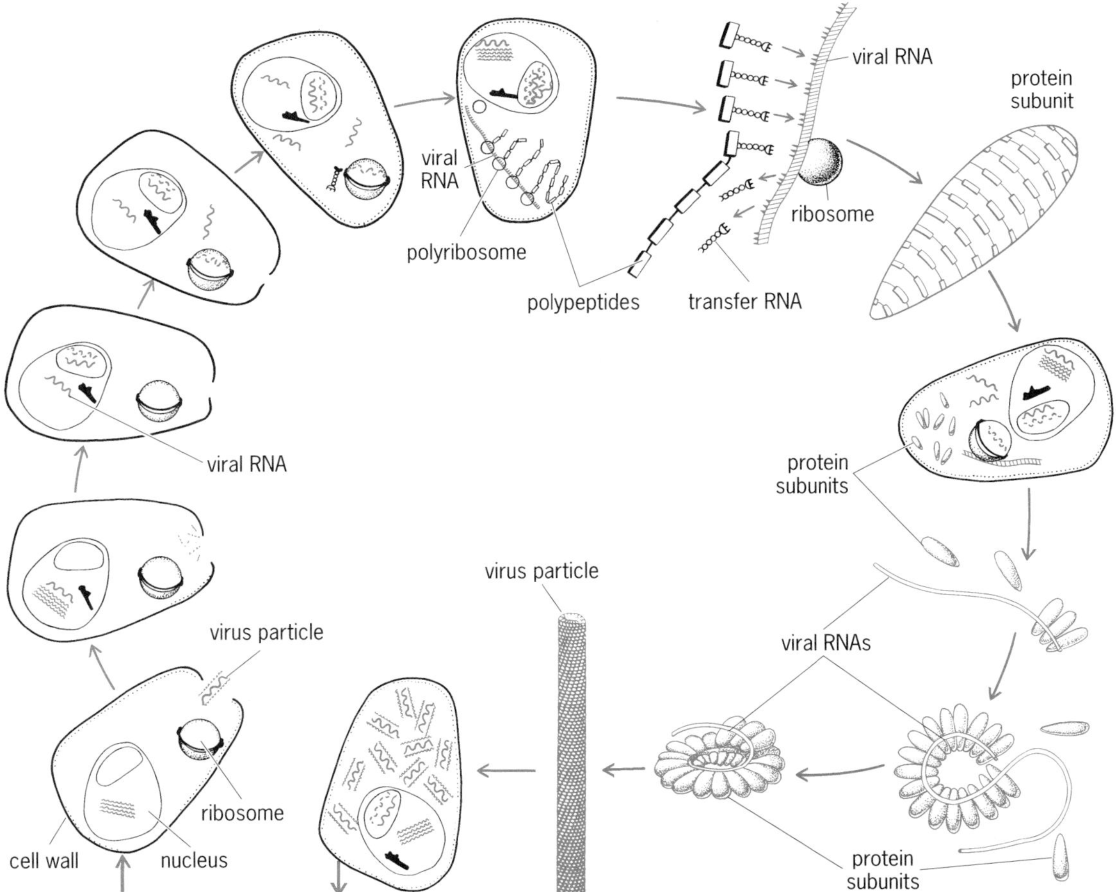

Fig. 16. Sequence of events in the infection of a plant cell by a virus. Two heavy arrows indicate the beginning (left) and the end (right) of the cell infection by the virus. (*After G. N. Agrios, Plant Pathology, 2d ed., Academic Press, 1978*)

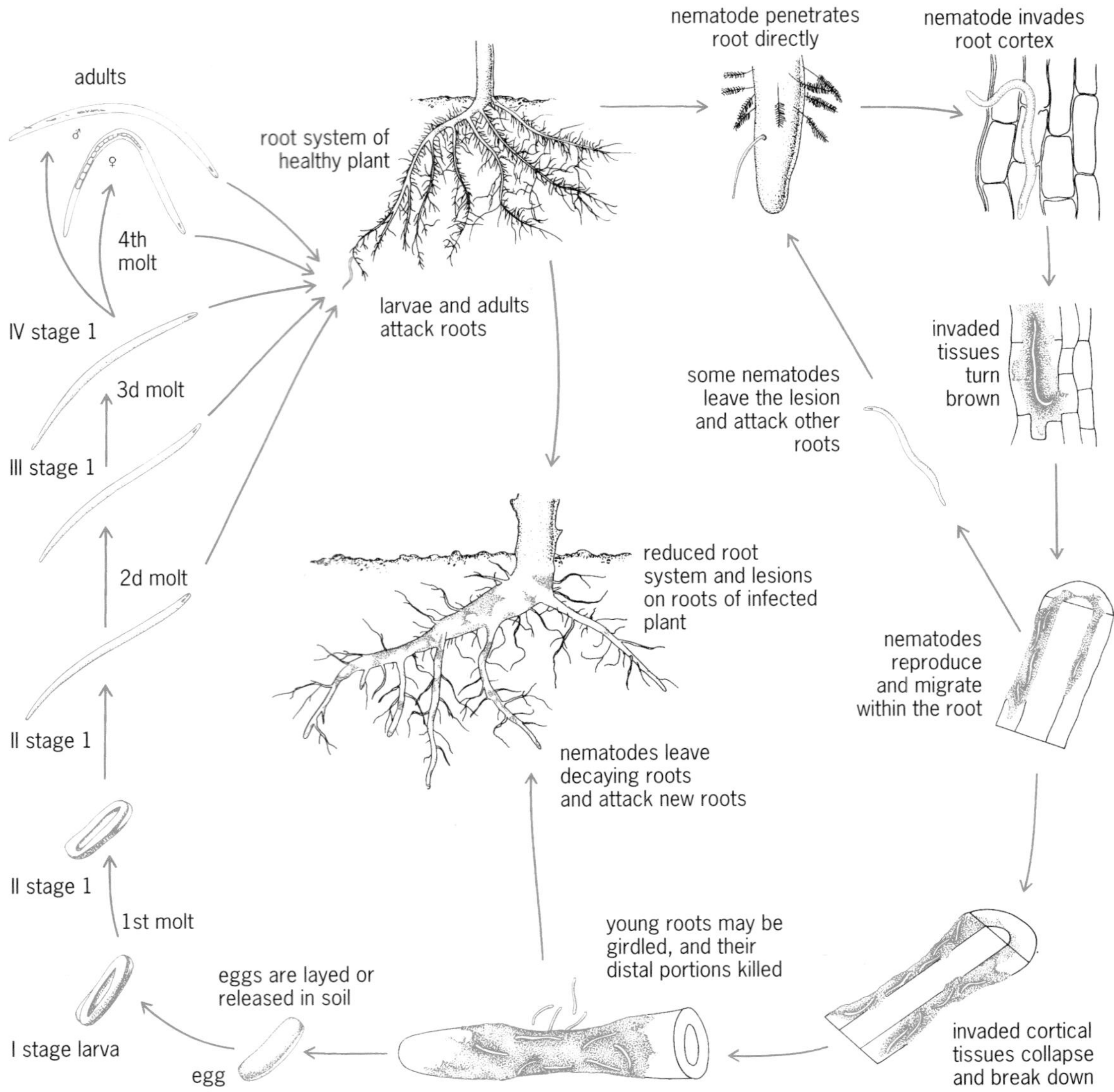

Fig. 17. Development of a typical nematode disease caused by the nematode *Pratylenchus*. (*After G. N. Agrios, Plant Pathology, 2d ed., Academic Press, 1978*)

pathogen produces inoculum, that is, more propagules, on the newly developed lesions and another cycle begins. These early cycles may produce "waves" of newly appearing disease. Inoculum continues to be borne on the old lesions as well, resulting in lesions of many different ages. Under conditions of favorable environment, infection may occur every day. For some plant pathosystems, such as the rusts, the lesions seldom enlarge in size or they expand only slowly. For other plant pathosystems, such as the blights, there is considerable lesion expansion, sometimes 1–25 mm^2 or more per lesion per day. With these latter pathosystems, the area of tissue involved with lesion expansion far exceeds the area occupied by new lesions. A plant pathosystem with short and frequent cycles of infection coupled with rapid lesion expansion (such as potato late blight) is characterized by a rapidly developing and devastating epidemic.

Epidemic development. The source of inoculum to begin polycyclic epidemics may be from infected seed, infested crop refuse, diseased volunteer plants, weeds, or alternate hosts, or it may come as influx of inoculum or as infested vectors from distant sources. From this beginning, the developmental rate of the polycyclic epidemic depends upon the magnitude of the components of the disease traingle. The plotted curve for the progress of polycyclic epidemics is more or less sigmoidal in shape. For convenience to analyze such curves, epidemiologists linearize them. Symmetrical-sigmoidal curves can be linearized with the logistic transformation, $Y = \ln[y/(1-y)]$. Some asymmetrical-sigmoidal curves are linearized with the Gompertz transformation, $Y = -\ln[-\ln(y)]$, or other appropriate transformation. The logistic model equation to describe symmetrical-sigmoidal curves is $y_t = 1/[1 + b(-k_L t)]$, where y is the proportion of disease at time t, k_L is the logistic rate parameter, and b is related to the amount of initial disease (y_0) as $b = 1/(1 - y_0)$. The Gompertz model equation to describe asymmetrical-sigmoidal curves is $y_t = \exp(-B \exp(-k_G t))$, where y_t is as before, k_G is the Gompertz rate parameter, and $B = -\ln(y_0)$.

The slope of the line fitted by linear regression through the transformed values of disease proportions is the epidemic rate. This rate parameter (k_L, k_G, etc.) is used to compare epidemics, for example, to determine the reduction in the rate of disease progress because of host resistance or the application of various measures of control. Epidemics that have occurred under different environmental conditions or in different geographical areas would be expected to have different rates of disease progress.

If the sigmoidal curve of disease progress is effectively linearized by one of the various models, disease at any future time can be predicted with reasonable accuracy. This prediction model is expressed as $Y_2 = Y_1 + k\Delta t$, where $\Delta t = (t_2 - t_1)$, Y_2 and Y_1 are the transformed disease proportions y_2 and y_1 for times t_2 and t_1, respectively. This equation is interpreted as disease y_2 at time t_2 is dependent upon current disease y_1 (at time t_1) plus the disease that will occur at epidemic rate k for the period of time between times t_1 and t_2.

Control. The general predictive equation with transformed values, $Y_t = Y_0 + k\Delta t$, can be used to develop strategies for disease control. A reduction in any of the three factors, y_0, k, or Δt, on the right side of the equation would cause lower disease y_t on the left side of the equation at any future point in time. The three epidemiological principles of disease control therefore are (1) to reduce initial inoculum (and initial disease, y_0); (2) to slow the rate (k) of the epidemic; and (3) to shorten the time (Δt) of exposure of the crop to the pathogen (**Fig. 19**). The control of disease would be maximized by a combination of the three principles.

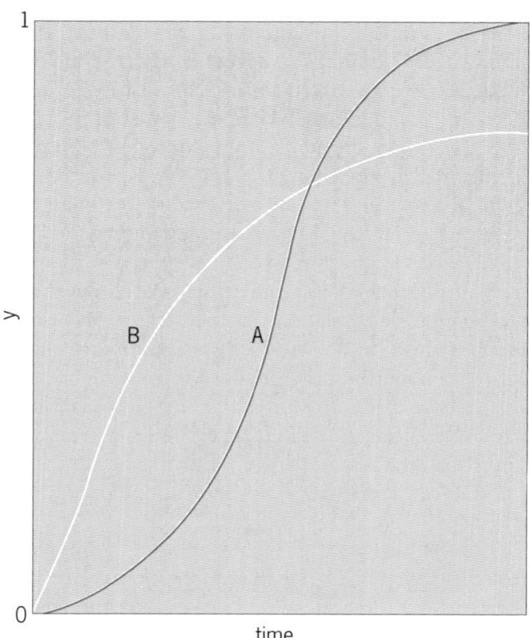

Fig. 18. Increase of proportion of plant disease (*y*) over time for a polycyclic disease (curve A) and a monocyclic disease (curve B).

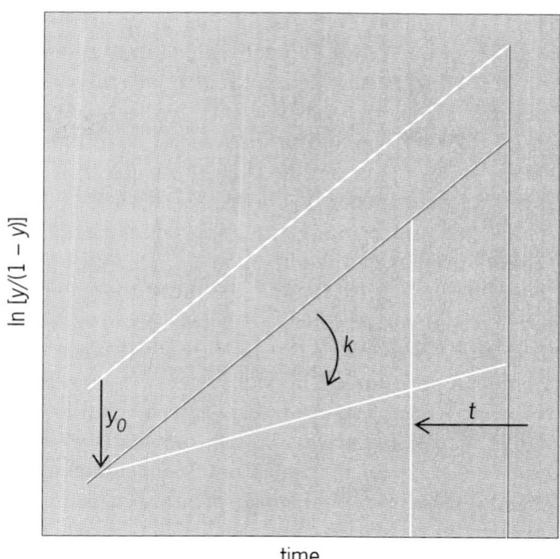

Fig. 19. Diagram of the three epidemiological principles of plant disease control: reduction of initial inoculum (y_0), slowing of epidemic rate (*k*), and shortening time (*t*) of exposure of a crop to the pathogen. Disease (*y*) is expressed as logistic transformation.

The reduction in the amount of inoculum that would be available to start an epidemic is accomplished by procedures usually characterized as sanitation. Some of these effective measures are seed treatment with hot water and chemicals, seed indexing, disease-free transplants, removal of infested crop debris and infected volunteer plants, elimination of alternate hosts, and resistance against specific pathogen races.

The rate k of the developing epidemic can be slowed by removal of diseased tissue (pruning), spraying with fungicidal or bactericidal chemicals, nonspecific host resistance, barrier or trap crops, vector repellants, avoidance of dispersal of the pathogen, and management of the environment (irrigation, nutrition, wind, light, geographic orientation, and relative humidity). A good example of this management is the intentional thinning of leaves on grapevines to obtain better air movement through the vineyard. This practice will reduce the lengthy periods of high relative humidity that are favorable for infection of grape tissue by the pathogens.

The time of exposure to the pathogen can be shortened by planting short-season varieties, planting vigorous seed and transplants, maintaining optimum fertility and moisture, and controlling insect pests.

If the amount of initial inoculum is reduced to a very low level, then to slow the epidemic rate becomes an easier task. All control practices are more effective when the population of the pathogen is low. Once the proportion of visible disease approaches 0.03–0.05 (3–5%), the epidemics of most pathosystems become nearly impossible to control and some yield loss will likely occur. If the latent period of a pathogen is 7 days and the epidemic is proceeding at a conservative epidemic rate, say $k_L = 0.15$, then the 3–5% disease would increase to 8–15% disease in 7 days, solely from the infections that were latent.

Spatial components. In addition to the increase of disease over time, the diseases spread over distance. Frequently, disease begins as one or more distinct foci in the crop area. In time, the individual foci may overlap and their discreteness may be lost. A gradient of disease intensity exists from the focal center to the periphery. The steepness of the gradient is largely dependent upon the dispersal characteristics of the pathogen; that is, rain-splash dispersal of propagules results in a steep gradient (considerable change in disease intensity over a short distance), whereas wind dispersal of propagules results in a flat gradient (**Fig. 20**). The J-shaped curve of disease intensity over distance can usually be linearized by $\log_{10}(y)$ or $\ln[y/(1 - y)]$ versus $\log_{10}(d)$, where d = distance.

The spatial pattern of disease plants received considerable attention in the 1990s. The aggregation (or randomness) of diseased plants is interpreted by the epidemiologist to determine modes of dispersal of inoculum, to determine efficiency and movement of vectors, and to locate initial sources of inoculum. Some of the analytical methods employed are doublet analysis, ordinary runs, spatial autolag correlation, geostatistics, the fitting of disease incidences to various distance or cluster models, dispersion indices, and variance-to-mean ratio.

Disease estimation. The assessment of disease in time and space is accomplished by visual estimation of disease intensity. Plant pathologists use percentage keys, disease rating scales, pictorial keys, remote sensing (color or infrared photography), the digitization of color images of the crop, and the reflected radiation off the crop canopy. Sometimes the spores of fungal pathogens are monitored by volumetric air samplers. The populations of bacterial pathogens are monitored by washing the bacterial cells off leaf surfaces and culturing on selective media. With these techniques, the buildup of pathogenic populations can be quantified and disease-favorable weather periods can be determined. From these estimates of disease and pathogenic propagules, the epidemic can be followed in its entirety.

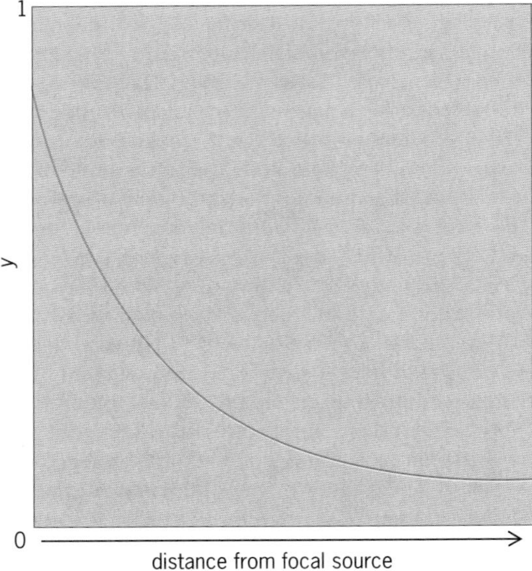

Fig. 20. Decreasing intensity of disease with increasing distance from source. Steepness of the gradient depends on dispersal characteristics of pathogen.

Crop loss. The loss of yield to disease has been the primary reason to study plant pathology. In the past, plant pathologists have made the correlations of crop loss to some measure of disease intensity, for example, to the amount of disease at a certain crop stage, the maximum amount of disease observed, and the area-under-the-disease-progress curve. Sometimes these correlations are excellent for a given crop, but they have had very little predictive value because disease in another season may occur earlier or later, the growth of the crop will be different among seasons, considerable difference in environmental conditions will occur, and so on. A more recent approach is to correlate yield to the duration of the area of healthy leaves (HAD) and the amount of radiation intercepted by this area (HAA). The theory is that yield is determined by this healthy leaf area, rather than that loss is determined by the amount of disease. The HAD/HAA concept has been found to be a generally reliable determinant of crop yield for many plant pathosystems. The photosynthetic competence of leaves is well correlated to leaf color; therefore, crop yield in the future may be predicted from satellite-collected color images of crops.

Epidemic modeling. Models have been developed for many of the important plant pathosystems. These models may be simple in design to forecast likely disease occurrence based on certain weather criteria. Simulation models may be quite complex in which the various components of the pathogen's life cycle (inoculum production, dispersal, catch, infection, latent growth, and lesion expansion) are quantified (**Fig. 21**). Plant growth is carefully determined and mathematically described. By varying the parameters of the model, the experimenter can determine vulnerable places in the cycle where control measures would be most effective. Epidemic progress brought about by differences in environment, plant resistance, or control procedures can be characterized. The simulation exercise could lead to more effective strategies of disease control. The simulation models can also be used to examine epidemiological theories.

Emerging epidemics. Toward the end of the twentieth century, several plant disease epidemics emerged or reemerged to cause concern among the agriculturists of many continents. A new strain of potato late blight appeared that was resistant to the fungicide (metalaxyl) most commonly used for its control. This new race had other important features:

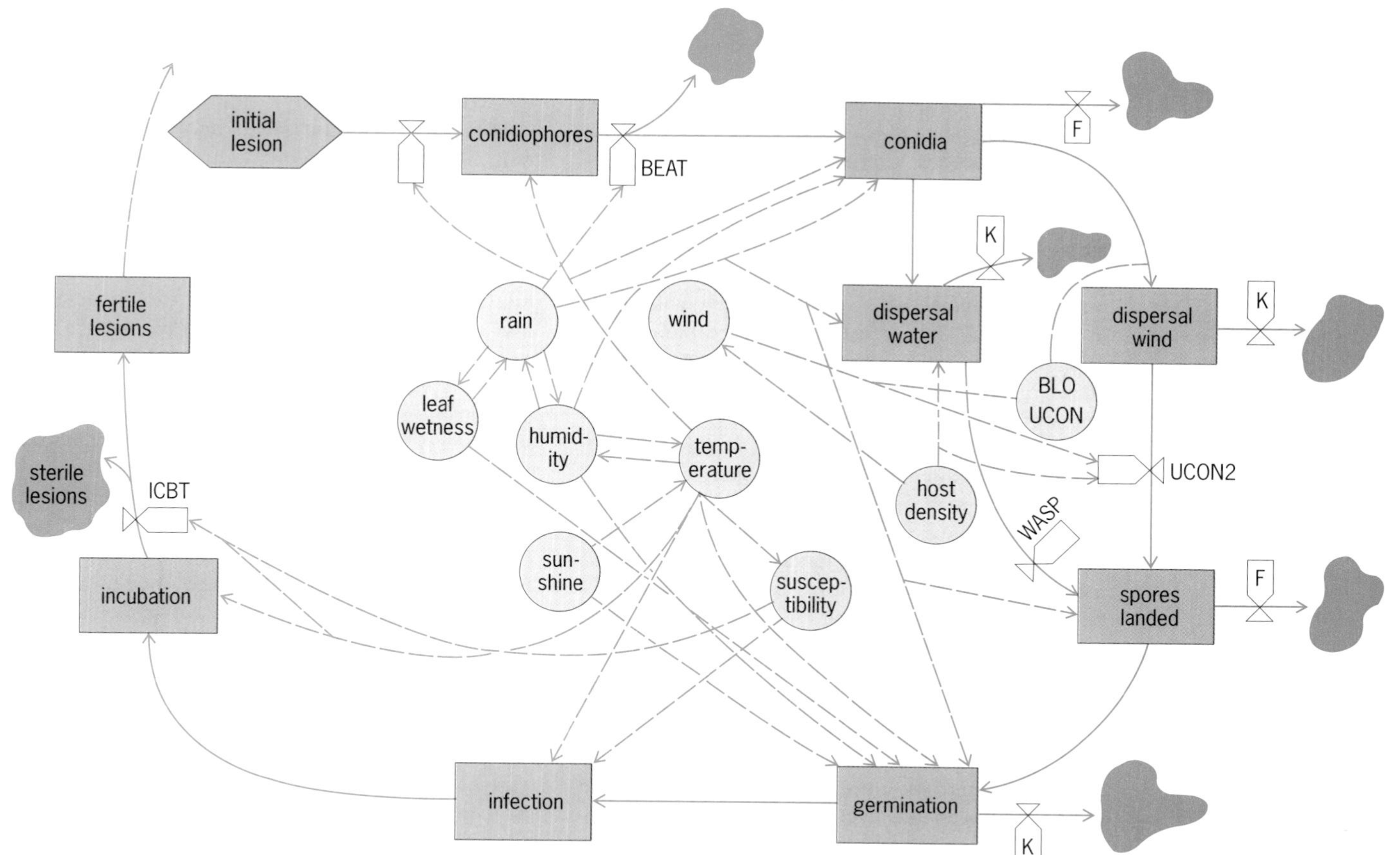

Fig. 21. Flow diagram for a plant disease simulator. The stages in the pathogen's life cycle are quantified with respect to environmental parameters. (*After J. Kranz et al., Z. Pflanzenkr., 80:181–187, 1973*)

with different mating types, there was the opportunity to have sexual recombination which could lead to more new races; with this sexual compatibility, more sexually derived oospores would be available for overseasoning to initiate new epidemics in succeeding crops. Additionally, this new strain was more aggressive pathogenically in that the haulms were attacked severely, resulting in more rapid plant death and more infection of the underground tubers. There was considerable controversy over the origin of this new strain. Nonetheless, it has spread rapidly to attack crops in most of the potato-growing regions of the world.

In the cereal-growing regions of the northern United States, epidemics of a fungus disease, fusarium head-blight of wheat and barley, have become severe, causing considerable losses in crop yield and grain quality. The agronomic change to no or low tillage over the past decades may have been largely responsible for the widespread occurrence of this disease.

In the citrus-growing regions of South America and Florida, epidemics of bacterial canker have been of great concern. In Florida, past epidemics have been controlled by the broad-scale eradication of infected and neighboring trees. However, outbreaks of bacterial canker keep reappearing in commercial groves and in dooryard citrus trees in urban Miami. In South America, canker is so prevalent in Argentina, Uruguay, Paraguay, and some states in Brazil that eradication is not an economical option. The departure from eradication as a control measure may occur in the future for those areas currently under threat.

Another epidemic of current interest is black sigatoka of banana. This disease is more difficult to control than gray sigatoka. Black sigatoka was found in the late 1990s at the edge of the tropical, banana-growing region of Brazil, and its presence is a severe threat to that crop. Another disease, karnel bunt of cereals, has been found occasionally in shipments of grain to the United States. This pathogen is under strict quarantine in the United States, as an introduction of this fungal pathogen into the grain-growing regions may cause considerable losses.

R. D. Berger

Control of Plant Disease

Control of plant disease is defined as the maintenance of disease severity below a certain threshold, which is determined by economic losses. Diseases may be high in incidence but low in severity, or low in incidence but high in severity, and are kept in check by preventing the development of epidemics.

The principles of plant disease control form the basis for preventing epidemics. However, the practicing agriculturist uses three approaches to the control of plant disease: cultural practices, disease resistance, and chemical pesticides.

Cultural practices. Cultural practices affect the environmental requirement of the suscept-pathogen-environment triangle that is necessary for plant disease development. Changes in the atmospheric, edaphic, or biological environment favorable to the plant and unfavorable to the pathogen will result in less disease. Important atmospheric parameters include temperature, humidity, light, and wind; edaphic parameters are physical and chemical properties of soils, soil reaction (pH), temperature, and moisture; and the biological parameters are all of the organisms that affect plant and pathogen survival and growth. The environment that is important is the small volume near the plant surface, termed the microclimate, which may be much different than that measured at a weather station. Cultural practices may be very effective in altering a plant's microclimate, but generally have no effect on the climate of an area.

Crop rotation, crop location, mineral nutrition, cultivation, and irrigation practices are a few cultural practices that influence disease development. Each may exert its influence on a specific phase of the pathogen life cycle. Crop rotation may influence overwintering or survival of pathogens between crops, while crop location may be important for reducing dissemination of pathogen propagules to young plants. Land preparation may affect the production and liberation of inoculum. Mineral nutrition may reduce the amount of colonization by the pathogen once inside the plant, which in turn may reduce the amount of inoculum produced. Proper cultivation may reduce wounding and thus avenues of entrance for pathogens. Proper irrigation may reduce successful germination of fungal spores and penetration of plants by parasites by reducing the wet period of foliage and soils. Successful use of cultural practices in disease control depends upon knowledge of how each practice affects the pathogen or plant. *See* AGRICULTURAL SOIL AND CROP PRACTICES.

Disease resistance. Resistant crop plant varieties have been used extensively for the control of plant diseases. For the farmer, the planting of a resistant cultivar is the easiest and often the most effective means of disease control. Resistant cultivars of many crops have been developed.

Resistant plants occur naturally in populations of crop species, but cannot be detected unless an epidemic occurs. One source of resistant plants is the large population of commercially grown plants that are subjected to natural epidemic. This source is the most desirable because selected plants contain advantageous yield and quality characteristics. The other main source of resistant plants is among natural populations. Progeny of resistant plants multiply more rapidly than progeny of susceptible plants in the wild, if a pathogenic organism is present. Most resistant plants come from this second source. The U.S. Department of Agriculture maintains a seed bank of wild plants of crop species collected from around the world.

Progeny of resistant plants and crosses of resistant plants with high-yield, high-quality susceptible plants must be screened before selections can again be made. For this screening, artificial epidemics are created in glasshouses or in the field, from which surviving plants are again crossed with commercial cultivars. Progeny are screened at each generation, and crosses are continued until a high-yield, high-quality resistant plant is obtained.

Two types of resistant plants may be selected from a population of a crop species. One type is plants with a very high level of resistance, determined by one or a few genes. This type is easily selected among progeny, but is vulnerable to development of more virulent strains of the pathogen. The second type of plants exhibit resistance at a lower level determined by several genes. However, once established in a high-yield, high-quality cultivar, this resistance tends to be durable to the development of more virulent strains of the pathogen.

A high level of resistance acts epidemiologically to reduce the initial inoculum, because it reduces the population of the pathogen that will cause disease to be near zero. The lower level of resistance acts epidemiologically to reduce the rate of disease progress, thus reducing the percentage of disease in each pathogen cycle. *See* BREEDING (PLANT); GENETIC ENGINEERING.

Chemical pesticides. Chemicals that are used to control plant diseases are used either as seed treatments, soil treatments, or protective sprays to plants. Although each use is important for plant disease control, the greatest use of chemicals is, by far, as protectants.

Seed treatment chemicals are used to control pathogenic organisms in and on seeds before they are planted, and around seeds when they are planted in soils. There is no chemical that will control pathogens at all locations, since not all chemicals enter seeds and none is effective against all organisms. Thiram and captan are the most common fungicides for treatment of fungi on seed and around seed in the soil. Sodium hypochlorite is used to control bacteria on seeds. Oxathiins are used on cereal seeds to control the smut fungi that occur in the seeds. Hot water treatments may be used to kill pathogens inside some seeds.

Certain volatile chemicals, such as methyl bromide, used to treat soils for eradication of pathogens have a broad spectrum of activity and kill all plant and animal life in the soil. Other volatile compounds may kill only animal life and thus control plant pathogenic nematodes—a mixture of dichloropropane and dichloropropene is such a nematocide. Many of these volatile compounds are toxic to crop plants, however, and treatment must be made before planting. Enough time must elapse between treatment and planting for fumigants to leave the soil.

Most chemicals applied to plants as protectants prevent germination of spores and invasion by pathogenic organisms since once the pathogen is inside the plant the chemicals are ineffective. Protective chemicals usually have very low solubility in water so they are not washed completely from plants by rains. Nevertheless, several applications must be made to keep new growth covered and to replace chemicals that may be inactivated by weathering processes.

The spectrum of pathogens controlled by protective sprays varies widely. For example, fixed copper compounds have broad activity and control most fungi and bacteria. Maneb controls most fungi, but is ineffective against bacteria. Benomyl controls only a few fungi. Level of control of a specific fungus may also vary. Copper, maneb, and benomyl all control *Cercospora apii*, the pathogen of early blight of celery, but benomyl provides the best control. Thus, specific compounds must sometimes be used for best control.

Chemical pesticides are strictly regulated to ensure against improper use. Pesticides must be registered with the federal government and must contain information on the label for safe use of the product. Chemicals must be tested for safety, and tolerances are set for limits of all pesticides in food products. *See* PESTICIDE. Robert E. Stall

Bibliography. G. N. Agrios, *Plant Pathology*, 3d ed., 1988; G. W. Bruehl, *Soilborne Plant Pathology*, 1986; O. D. Dhingra and J. B. Sinclair, *Basic Plant Pathology Methods*, 2d ed., 1994; C. B. Kenaga, *Principles of Phytopathology*, 2d ed., 1986; G. B. Lucas, C. L. Campbell, and L. T. Lucas, *Introduction to Plant Diseases: Identification and Management*, 2d ed., 1992; O. C. Maloy, *Plant Disease Control: Principles and Practices*, 1993; D. C. Sigee, *Bacterial Plant Pathology: Cell and Molecular Aspects*, 1993; J. A. von Arx, *Plant Pathogenic Fungi*, 1993; B. M. Zuckerman, W. F. Mai, and R. A. Rohde (eds.), *Plant Parasitic Nematodes*, vols. 1 and 2, 1971.

Plant phylogeny

The evolutionary chronicle of plant life on the Earth. Understanding of this history is largely based on knowledge of extant plants, but the fossil record is playing an increasingly important role in refining and illuminating this picture. Study of deoxyribonucleic acid (DNA) sequences has also been revolutionizing this process in recent years. In spite of a remarkably improved situation, controversy over interpretation of the available data can be expected to continue. The molecular data (largely in the form of DNA sequences from several genes) summarized below have been demonstrated to be highly correlated with other information. *See* PHYLOGENY.

Algae. "Algae" was once a taxonomic designation uniting the lower photosynthetic organisms, but ultrastructural and molecular data have uncovered a bewildering diversity of species. Algae are now recognized as 10 divergent lineages on the tree of life that join organisms as distinct as bacteria and eukaryotic protozoans, ciliates, fungi, and embryophytes (including the land plants). In a biochemical context, the term "algae" defines species characterized by chlorophyll *a* photosynthesis (except Embryophyta); some of their descendants are heterotrophic (secondary chloroplast loss). Despite the variety of species it encompasses, the term "algae" also retains phylogenetic relevance.

Cyanophyta (blue-green algae or bacteria) include the prokaryotic algae and have a fossil record dating back 3 billion years. Cyanophytes flourished on the early Earth, their chlorophyll *a* photosynthesis increasing atmospheric oxygen and ultimately favoring eukaryote evolution. Eukaryotes differ fundamentally from prokaryotes by possessing a nuclear membrane and an endomembrane system that probably evolved by internalization of plasma membrane invaginations. Evolution of the eukaryotic cytoskeleton was another key event and enabled new types of movement, including phagotrophy (ability to engulf particles into food vacuoles). Phagotrophy had a profound impact on eukaryote evolution because entire prokaryotes could be ingested, or they could remain metabolically active in the host's food vacuole, leading to the establishment of an endosymbiosis (symbiosis in which one species lives inside another). Substantial evidence indicates that eukaryotic mitochondria and plastids evolved from such primary endosymbiont events. All eukaryotic plastids probably derive from a single primary endosymbiont event between an early eukaryote and a cyanophyte (producing the ancestral phototrophic eukaryote) and, therefore, cyanophytes and plastids share a common ancestor. Hence there is a paradox: algal plastids are closely related, but the host eukaryote lineages are not. The paradox is resolved by the secondary endosymbiont hypothesis, which concludes that six of the nine eukaryotic algal lineages have acquired plastids by engulfing and establishing endosymbioses with phototrophic eukaryotes. "Algae" in a phylogenetic context, therefore, refers to Cyanophyta and the diverse eukaryote lineages that have received their descendent plastids by vertical and horizontal transmission. *See* CHLOROPHYLL; PHOTOSYNTHESIS.

Three eukaryotic algal lineages (Glaucophyta, Chlorophyta, and Rhodophyta) have hypothetically evolved directly from the ancestral phototrophic eukaryote. All have plastids surrounded by two membranes: the inner derived from the cyanophyte plasma membrane; the outer, the plastid envelope, from the eukaryotic food vacuole membrane. Glaucophyta are a division of considerable evolutionary interest because their plastids retain many features of the ancestral cyanophyte. In fact, glaucophytes were considered to be eukaryotes with endosymbiotic cyanophytes living in vacuoles until recent evidence indicated that these structures are true plastids. If there was a single primary endosymbiont event for eukaryotic plastids, Glaucophyta must be related to Rhodophyta (red algae) and Chlorophyta (green algae, here including Charophyceae) that have also evolved from the ancestral phototrophic eukaryote. There is no evidence for such an association; Glaucophyta may have derived plastids from an independent primary endosymbiont event. Furthermore, evidence for a relationship between Rhodophyta and Chlorophyta is weak, and the hypothesis of a single primary endosymbiont event generating eukaryotic plastids remains tentative. Both Rhodophyta and Chlorophyta have additional phylogenetic significance. Rhodophyta completely lack flagella in all life history stages; and this distinction, as well as similarities between rhodophyte chloroplasts and cyanophytes, prompted the proposal that Rhodophyta were the earliest eukaryotes. This view has been rejected by contemporary systematic data. Chlorophyta is the division from which the embryophytes (including the land plants) evolved, and some workers recognize a division, Streptophyta, for embryophytes and their specific green algal relatives. *See* CYANOPHYCEAE; RHODOPHYCEAE.

Eukaryotic lineages derived from secondary endosymbiont events have four (sometimes three) plastid membranes (inner two from the plastid envelope and the third from the plasma membrane of the phototrophic eukaryote, the fourth from the food vacuole membrane of the secondary eukaryote host). Two divisions, Cryptophyta and Chlorarachniophyta, are of paramount phylogenetic importance because of a miniature nucleus (nucleomorph) in the compartment (derived from cytoplasm of the phototrophic endosymbiont) between the inner two and outer two plastid membranes. Cryptophyta are a lineage of flagellate species of unresolved phylogenetic position on the tree of life that nevertheless shares many plastid features with Rhodophyta. Although the host cryptophyte is not related to rhodophytes, the nucleomorph DNA is clearly of red algal nuclear origin, betraying the phylogenetic origin of cryptomonad plastids and supporting the secondary endosymbiont hypothesis. Similarly, Chlorarachniophyta host cells are related to filose amebas, but their plastids are unequivocally of chlorophyte origin. Euglenophyta also have chlorophyte plastids that are considered to be derived by secondary endosymbiosis despite absence of a telltale nucleomorph. The plastids are so similar to the chlorophyte counterpart that euglenoids were included in Chlorophyta by some workers, but recent molecular data indicate that the host is related to a line of eukaryotic flagellates (Kinetoplastida). *See* CRYPTOPHYCEAE.

The final three lineages are loosely the chromophytic algae but, despite plastid similarities, are not close relatives and have probably derived their chloroplasts by independent secondary endosymbiosis events with related phototrophic eukaryotes. Contemporary phycologists lack consensus on a taxonomic label for the descriptively named heterokont chromophytes. This diverse assemblage includes the well-known Bacillariophyceae (diatoms) and Phaeophyceae (brown algae), to name but two of the many classes. These algae are now included in the Heterokonta (Stramenopila or Chromista) with exfungal lineages (such as Oomycota) and certain heterotrophic flagellates (Bicosoecida). These diverse organisms are united by a common flagellar arrangement (heterokont) in which the anterior flagellum has two rows of stiff tripartite hairs, whereas the posterior flagellum is smooth. Molecular data substantiate this feature for uniting this bewildering array of species. *See* BACILLARIOPHYCEAE; CHRYSOPHYCEAE; PHAEOPHYCEAE.

Haptophyta (Prymnesiophyta) is a group of unresolved phylogenetic affinity from the eukaryote crown on the tree of life. Although similar to the heterokont chromophytes in plastid attributes, they differ in flagellar arrangement and have an additional appendage (haptonema) unique to this group. *See* PRYMNESIOPHYCEAE.

Dinophyta also have a characteristic flagellar arrangement (dinokont), as well as plastid and pigment attributes, that distinguish them from other chromophytic algae. Dinoflagellates are close relatives of flagellate Apicomplexa and Ciliata in sharp contrast to earlier views. As an example, the mesokaryote hypothesis argued that dinoflagellates, with permanently condensed chromosomes that lack histones, were intermediate on the evolutionary path from prokaryotes to eukaryotes—a hypothesis since rejected. *See* CHROMOPHYCOTA; DINOPHYCEAE.

It is evident from the previous text that the Algae, a group of species linked phylogenetically through their plastid ancestry, represents a diverse assemblage of species branching from all points on the tree of life. Their importance to understanding eukaryotic evolution cannot be overstated, ensuring them a place of prominence in the continuing realm of phylogenetic investigation. Gary W. Saunders

Embryobionta. Embryobionta, or embryophytes, are largely composed of the land plants that appear to have emerged 475 million years ago. The evidence indicates that land plants have not evolved from different groups of green algae (Chlorophyta) as suggested in the past, but instead share a common ancestor, which was a green alga. Land plants all have adaptations to the terrestrial environment, including an alternation of generations (sporophyte or diploid and gametophyte or haploid) with the sporophyte generation producing haploid spores that are capable of resisting desiccation and dispersing widely, a cuticle covering their outside surfaces, and separate male and female reproductive organs in the gametophyte stage. The life history strategies of land plants fall into two categories that do not reflect their phylogenetic relationships. The mosses, hornworts, and liverworts represent the first type, and they have expanded the haploid generation, upon which the sporophyte is dependent. Several recent analyses of DNA data as well as evidence from mitochondrial DNA structure have demonstrated that the liverworts alone are the remnants of the earliest land plants and that the mosses and hornworts are closer to the vascular plants (tracheophytes). The tracheophytes include a large number of extinct and relatively simple taxa, such as the rhiniophytes and horneophytes known only as Silurian and Devonian fossils. All tracheophytes are of the second category, and they have expanded the sporophyte generation. Among extant tracheophytes, the earliest branching are the lycopods or club mosses (*Lycopodium* and *Selaginella*), and there are still a diversity of other forms, including sphenophytes (horsetails, *Equisetum*) and ferns (a large and diverse group in which the positions of several families still are not clear). Within the tracheophytes, many groups have a much reduced but still free-living gametophyte and a dominant sporophyte. The sporophytes bear the haploid spores in sporangia that may be lateral (lycopods and psilophytes), terminal (horsetails and some ferns), or on the lower surfaces of leaves (most ferns). *See* EMBRYOBIONTA.

All seed plants take the reduction of the gametophyte generation a step further and make it dependent on the sporophyte, typically hiding it within reproductive structures, which are either cones or flowers. The first seed plants originated at least by the Devonian, and they are known to have a great diversity of extinct forms, including the seed ferns. There are two groups of extant seed-bearing plants, gymnosperms and angiosperms. In the gymnosperms, the seeds are not enclosed within tissue derived from the parent plant. There are four distinct groups of extant gymnosperms, often recognized as classes: Cycadopsida, Gnetopsida, Ginkgoopsida, and Pinopsida. For many years, the gnetophytes were considered to be the closest extant relatives of the angiosperms, but the accumulating molecular evidence indicates that they may instead be members of the gymnosperm group, which collectively are the closest relatives of the angiosperms.

The gnetophytes are composed of three families, Ephedraceae (only the genus *Ephedra*, the source of ephedrine), Gnetaceae (*Gnetum*), and Welwitschiaceae (*Welwitschia*, known from southwestern Africa, and one of the most bizarre plants in existence). The cycads, which look like large ferns or palms, are not at all related to either; many are commonly grown ornamentals (sago palm), and all produce large conelike and sometimes leafy reproductive structures with large seeds. *Ginkgo* (the maidenhair tree) is often grown as a street tree. The most important of the gymnosperms, both ecologically and commercially (seeds eaten as food and timber production), are the conifers (Pinopsida), which include the pine (*Pinus*), spruce (*Picea*), fir (*Abies*), yew (*Taxus*), monkey puzzle (*Araucaria*), and cedar (*Cedrus* and *Juniperus*).

The angiosperms (also flowering plants or Magnoliopsida) are the dominant terrestrial plants, although the algae collectively must still be acknowledged as the most important in the maintenance of the Earth's ecological balance (fixation of carbon dioxide and production of oxygen). In angiosperms the seeds are covered by protective tissues derived from the parental plant. The nature of the ancestors of the flowering plants is still the object of a great deal of speculation, and a great deal more research will be required before their evolutionary history can be elucidated. Clearly, there were a large number of extinct lineages of seed plants, but which of these might have been close to the lineage that ultimately led to the angiosperms is highly uncertain. There are no generally accepted angiospermous fossils older than 120 million years, but the lineage is clearly much older based on DNA clocks and other circumstantial lines of evidence such as their current geographic distributions.

Traditionally the angiosperms have been divided into two groups, monocotyledons (monocots) and dicotyledons (dicots), based on the number of seed leaves. However, DNA sequence data have demonstrated that, although there are two groups, these are characterized by fundamentally different pollen organization, such that the monocots share with a group of dicots pollen with one pore whereas the rest of the dicots have pollen with three (or more) pores.

This difference in pollen organization places the monocots within the magnoliids or primitive dicots such as Laurales (avocado, cinnamon, and their relatives), Magnoliales (magnolias), and Piperales (black pepper, peperomias, and pipeworts). In general, magnoliids lack well-organized flowers, such that the number and orientation of parts are highly irregular, and these flowers are pollinated by animals with simple behavorial patterns, such as flies and beetles. The exceptions to this pattern are the more derivative groups of monocots, such as the orchids and gingers, in which the flowers are highly organized and pollinated by behaviorally complex animals (bees, butterflies, and mammals).

Of those groups with uniaperturate pollen, the monocots are by far the most numerous and important ecologically and economically (all the major grains). The most primitive monocots are *Acorus* (Acorales) and alismatids (pondweeds) and aroids (both Alismatales), all of which are predominantly wetland or aquatic plants. The lilioid or petaloid monocots are composed of generally the showiest families and include the lilies (Liliales), orchids and irises (both Asparagales), and yams (Dioscoreales). The most derivative group of monocots includes the palms (Arecales), gingers (Zingiberales), grasses and their relatives such as the sedges and rushes (Poales), and spiderworts, bloodroots, and pickerel weeds (Commelinales).

The triaperturate angiosperms or eudicots are more numerous than the magnoliids, and the details of their phylogenetic relationships, like those of the magnoliids, have only recently been resolved by the collection of large numbers of DNA sequences. The first branching groups include the buttercups and poppies (Ranunculales), lotuses, plane tree, and proteas (all Proteales), boxwoods (Buxaceae), and gunneras (Gunneraceae), which, like the magnoliids, generally lack well-organized flowers. There are four highly derivative eudicot lineages: the relatively small sandalwood and mistletoe families (Santalales), the much larger and more important caryophyllids (including some carnivorous families such as the sundews and tropical pitcher plants, as well as the amaranths, cacti, and carnations), and then the two largest sets of orders, the rosids and asterids. As opposed to the magnoliids and early branching eudicots, all four of these most derived groups have flowers arranged in well-defined whorls of largely definite numbers of parts. The rosids typically have two whorls of stamens, although some have amplified the numbers in each position, with unfused petals. The asterids have a single whorl of stamens fused to the fused petals, although some also have more stamens and may appear to have free petals due to extensive elongation of the petal apices late in floral development. The rosids include the familiar members of the rose family (apples, peaches, strawberries; Rosaceae of Rosales), euphorbs (Euphorbiaceae of Malpighiales), beans (Fabaceae of Fabales), beeches, oaks and birches (several families in Fagales), mustards (Brassicaceae of Brassicales), cotton and hibiscus (Malvaceae of Malvales), and maples and citrus (Sapindaceae and Rutaceae of Sapindales). Asterid orders and families include rhododendrons and ebenies (Ericaceae and Ebenaceae of Ebenales), dogwoods and hydrangeas (Cornaceae and Hydrangeaceae of Cornales), gentians and coffee (Gentianceae and Rubiaceae of Gentianales), mints and snapdragons (Lamiaceae and Scrophulariaceae of Lamiales), daisies and bellflowers (Asteraceae and Campanulaceae of Asterales), and carrots and ivies (Apiaceae and Araliaceae of Apiales).

Mark W. Chase

Bibliography. D. Bhattacharya (ed.), *Origin of Algae and Their Plastids*, Springer-Verlag, 1997; L. E. Graham and L. W. Wilcox, *Algae*, Prentice Hall, 2000.

Plant physiology

That branch of plant sciences that aims to understand how plants live and function. Its ultimate objective is to explain all life processes of plants by a minimal number of comprehensive principles founded in chemistry, physics, and mathematics.

Extent. Organisms are distinguished from the inanimate world by three characteristics. First, they take up materials—water and inorganic and organic chemical compounds—from the environment and process them in various ways, usually producing more complex organic compounds. Second, while they may release part of the absorbed materials, in the original or a modified form, into the environment, they incorporate a major part into their bodies. This incorporation can be in the form of an increase in the size of the existing body and its parts, or of addition of more of the same structures that are already present in the organism (for example, more leaves, shoots, and roots); or it can be in the form of production of new, or at least greatly modified, structures (for example, flowers and fruits, or tubers). The former, quantitative changes are known as growth; the latter, qualitative ones are known as development. Third, organisms use part of the newly acquired substance for reproduction, that is, to produce offspring identical to the parent organisms.

In addition to these three, another somewhat different but no less characteristic and important property of organisms is their continuous response to the changing environment. Plants, except for some relatively simple ones, and for reproductive cells of some complex plants, spend their entire life at a single site and thus cannot escape their environment. Thus, interaction with the environment is particularly essential and elaborate. It endows plants with a considerable measure of adaptability, that is, ability to adjust to adverse environmental conditions such as freezing temperatures in winter or periods of high temperatures and lack of water in summer, and enables them to use environmental signals to adjust their entire life to the given environment. For example, most perennial plants enter a period of dormancy in fall or early winter, and do not resume full activity until the following growing season. Both the establishment of dormancy and its termination occur in response to environmental (climatic) signals and factors, the former in many cases to the shortening days of late summer and fall, the latter to the low temperatures of winter. The response may, however, be delayed until the causative factor is no longer present; thus, while dormancy may already terminate in midwinter, active growth does not start until temperatures increase in spring. Essential developmental processes, like flowering, may similarly be induced or delayed by certain environmental conditions. In some cases it has been found that the effect of an environmental factor extends to the offspring of a plant, even though this offspring has never directly experienced the action of this particular factor. *See* COLD HARDINESS (PLANT); DORMANCY; PHOTOPERIODISM.

Classification, methods, and objectives. Plant physiology seeks to understand all the aspects and manifestations of plant life summarized in the preceding section. In agreement with the major characteristics of organisms, it is usually divided into three major parts: (1) the physiology of nutrition and metabolism, which deals with the uptake, transformation, and release of materials, and also their movement within and between the cells and organs of the plant; (2) the physiology of growth, development, and reproduction, which is concerned with these aspects of plant function; and (3) environmental physiology, which seeks to understand the manifold responses of plants to the environment. The part of environmental physiology which deals with effects of and adaptations to adverse conditions, and which is receiving increasing attention, is stress physiology.

Levels of organization. Plant physiological research is carried out at various levels of organization and by using various methods. The main organizational levels are the molecular or subcellular, the cellular, the organismal or whole-plant, and the population level.

Work at the molecular level is aimed at understanding metabolic processes and their regulation, and also the localization of molecules in particular structures of the cell but with little if any consideration of other processes and other structures of the same cell. It is generally carried out with tissues or organs, the cellular structure of which has been disrupted, and often after separation (fractionation) of the cell contents into their components—the cell structures and organelles and the unstructured or soluble fraction.

Work at the cellular level often deals with the same processes but is concerned with their integration in the cell as a whole. Since most cells are small, work is usually carried out with pieces of tissues. Tissues are selected which consist of as uniform cells as can be found, or with relatively undifferentiated tissues (so-called callus tissues) or cell suspensions.

Research at the organismal level is concerned with the function of the plant as a whole and its different organs, and with the relationships between the latter. Research at the population level, which merges with experimental ecology, deals with physiological phenomena in plant associations which may consist either of one dominant species (like a field of corn) or of numerous diverse species (like a forest). Work at the organismal and to some extent the population level is increasingly carried out in facilities permitting maintenance of controlled environmental conditions (light, temperature, water and nutrient supply, and so on). The purpose is to obtain results that can be quantitatively reproduced; a certain weakness that has to be kept in mind is that natural conditions cannot be exactly duplicated and that results obtained under controlled environments may not be directly applied to the situation in the field. *See* PHYSIOLOGICAL ECOLOGY (PLANT); PHYTOTRONICS.

Methods. The most elementary method of plant physiology is the descriptive method, in which a life process is studied and described under a certain set of conditions, at particular stages or throughout the life of the plant, with the goal of establishing its characteristics. The comparative method consists in the study of a process in plants belonging to different morphological, taxonomical, ecological, or physiological types, but usually kept under identical conditions. The purpose here is to establish variations of a process that are characteristic of different plant types; these variations may permit insights into the nature of the process in question. The correlational method seeks to recognize relationships between the course of a process and variations in the environment or in other processes. Such variations, too, often permit insights into the nature of a process, or the interdependence of two (or more) processes, although it is usually difficult to determine their precise causal relationship from correlational evidence alone. The most sophisticated method, however, is the experimental method in which the process under study is modified by experimental treatments, and in this manner insights are gained not only into its course but also into its causality. Experimental treatments are very diverse; they may be mechanical or surgical, physical (special conditions of light, temperature, and so on), or chemical.

However, the categories of plant physiological research areas, levels of organization, and methods, as enumerated, are made for convenience and are highly artificial. Thus, studies of metabolism often merge with studies of growth and development as the latter depend on metabolic processes. Studies of one and the same process can be carried out first at one and then at another level of organization, and may involve different methods, either simultaneously or in sequence. Attempts at categorizing plant physiological research in too rigid a manner may be counterproductive.

Objectives. Plant physiology strives to be an exact science, which means essentially two things. First, it is a quantitative science which expresses its findings in numbers and in chemical, physical, and mathematical units. Because all organisms and hence also their life processes exhibit natural variation, this quantitative approach is subject to certain limitations. However, these limitations can in turn be defined by statistical methods, and a statistical evaluation of the results is often an essential part of research. Second, plant physiology uses—or should use—the scientific method developed by other exact sciences; that is, the ultimate objective of plant physiology is to explain plant life by as few as possible comprehensive principles which are grounded in the principles of chemistry and physics, and the principles of mathematics as far as these underlie physical and chemical principles. However, physiological phenomena exhibit far more diversity than purely physical and chemical phenomena, and considerably more even than certain other biological phenomena.

Plant physiology may never succeed in reducing its phenomena to a relatively small number of principles. More likely, in addition to principles which are valid for all plants, and some of which may be identical with principles of animal and human physiology, there will be principles valid only for certain kinds of plants, or perhaps certain kinds of life processes of plants.

Relation to other biological disciplines. A life process may exhibit certain characteristic differences in different plant types, and life processes do not proceed in a void but in organs, tissues, cells, and cell organelles—which, in turn, are produced and affected by life processes. For these reasons plant physiology depends upon information from other plant sciences (taxonomy, morphology, anatomy, cytology). The ultrastructure of cells as studied with the electron microscope has gained increasing importance in plant physiological research because, in conjunction with molecular studies, it permits the precise localization of a process within the cell.

Similarly, increasing attention is directed at the localization of processes in different parts of the plant, as well as the distribution of the products of these processes between these different parts. The precise spatial localization of processes within the plant and its cells is called compartmentation; the distribution of products is called partitioning.

Outside the plant sciences, a biological discipline that is acquiring increasing importance in plant physiological work is genetics. The plant can be considered as the product of a continuing, complex interaction between its genotype and the environment. However, for a long time the genotype of plants used in plant physiological experimentation was taken for granted. This situation is, however, changing. Some of the major impulses for this development were provided by two possibilities: generating hybrid plants by fusion of protoplasts, and transferring selected genetic material from one organism to another by the use of plasmids. Both procedures permit genetic engineering, that is, deliberate and to some extent directed manipulations of the genetic information of organisms. *See* GENETIC ENGINEERING.

Relation to other natural sciences. Because of the great importance of environmental conditions for the function of plants, plant physiology may draw on any science that deals with environmental phenomena and problems: meteorology, climatology, soil science, and others. However, the two natural sciences that have a unique position in relation to, and in fact within, plant physiology are physics and chemistry. The areas of life sciences which are based on the direct use of chemical and physical methods are called biochemistry and biophysics, respectively. The share particularly of biochemical work in plant physiology, and the share of biophysical plant research, has increased over the years.

However, while such work is an essential, integral part of plant physiological work, it should be clearly realized that plant physiology is a good deal more than the simple sum of plant biochemistry and plant biophysics. Biochemistry and biophysics are predominantly analytical disciplines which focus primarily on one process and seek to understand it with no concern for any surrounding or parallel processes; to accomplish this, they frequently destroy the integrity of the system. Physiology aims at understanding the function of the whole organism. It strives at preserving the latter's integrity, and to do this it needs more than the principles and methods of biochemistry and biophysics. It is interesting in this context that there has been an increasing trend in biochemistry and biophysics themselves to study chemical and physical processes in the intact organism.

Plant physiology and the world. Plants—apart from some autotrophic microorganisms—are the only organisms capable of converting inorganic (mainly carbon and nitrogen) to organic matter by making use of solar energy. Thus, plants are the direct or indirect source of food for humans and all animals, and are also a source of fiber, fuel, and feedstock. While plants are profoundly dependent on the environment, they greatly affect the environment in turn. The peat bogs which cover wide areas of low-lying land in some parts of the world are products of plant life, and plants can be important factors in protecting land against erosion by wind and water. However, these are only individual and isolated cases of plant activity affecting environment. *See* PHOTOSYNTHESIS.

The most important general role of plants in relation to the environment is their function as a link in major natural cycles. Two examples may suffice. It is estimated that plants remove annually more than 2×10^{11} tons (1.8×10^{11} metric tons) of carbon dioxide from the Earth's atmosphere, transform it into plant matter and other forms of life, and ultimately release it back into the atmosphere. It has also been estimated that one-third of the water that is annually precipitated on the United States passes through plants which take it up from the soil and release it into the air. Thus, disruptions of the plant cover may result in serious changes of the climate and hence of the condition of the land—changes which quite generally are of an unfavorable nature. Many regions of the Earth have become arid because of excessive exploitation and destruction of their original vegetation. Unfortunately, such uncontrolled and excessive exploitation of the natural vegetation is still continuing in many parts of the world. *See* BIOGEOCHEMISTRY; DESERTIFICATION.

Plant physiology has helped greatly in understanding many of these plant-related phenomena, so far particularly in relation to agriculture. For example, widespread and routine use of certain chemicals for preventing premature fruit drop or as selective weed killers is largely based on plant hormone research. There are, however, other important problems of food production which can be solved only with the cooperation of plant physiology. *See* PLANT HORMONES.

Breeding work having to do with crop plants was formerly directed primarily at increases of yield, with little attention to such aspects as the water and fertilizer demands of the plant. The result is that the most productive crop varieties are often wasteful in terms of the amounts of water and of mineral nutrients which they need for optimal performance. Yet water is becoming a scarce commodity in some parts of the world, and the production of mineral fertilizers is a process requiring large amounts of energy, which is also getting scarce as well as expensive. Thus, efforts have been initiated to breed crop varieties for more efficient water and fertilizer use, and similarly varieties which can perform on land that is considered unsuitable for agriculture. Plant physiology can also play an important function in relation to the role of plants in the climate by providing the basis for the development of plants which can be used for reversal of the results of excessive exploitation of natural vegetation. *See* BREEDING (PLANT); FERTILIZER; PLANT GROWTH; PLANT METABOLISM; PLANT RESPIRATION; PLANT-WATER RELATIONS.

Anton Lang

Plant pigment

A substance in a plant that imparts coloration. The photosynthetic pigments are involved in light harvesting and energy transfer in photosynthesis. This group of pigments comprises the tetrapyrroles, which include chlorophylls (chl) and phycobilins, and the carotenoids. The light-absorbing groups of these molecules, the chromophores, contain conjugated double bonds (alternating single and double bonds), which make them effective photoreceptors. Their effectiveness in harvesting light for photosynthesis is shown in **Fig. 1**. The sum of the absorption spectra of the chlorphylls and the carotenoids, evident in the absorption spectrum of a green leaf, is equivalent to the action spectrum of photosynthesis (Fig. 1*a*). *See* CHLOROPHYLL; PHOTOSYNTHESIS; PIGMENTATION.

The second major group comprises the anthocyanins, intensely colored plant pigments responsible for most scarlet, crimson, purple, mauve, and blue colors in higher plants. About 100 different anthocyanins are known. Unlike the chlorophylls and carotenoids, which are lipid-soluble chloroplast pigments, the anthocyanins are water-soluble and are located in the cell vacuole. Chemically, they are a class of flavonoids and are particularly closely related, both structurally and biosynthetically, to the flavonols. Their value to the plant lies in the contrasting colors they provide in flower and fruit, against the green background of the leaf, to attract insects and animals for purposes of pollination and seed dispersal. *See* FLAVONOID.

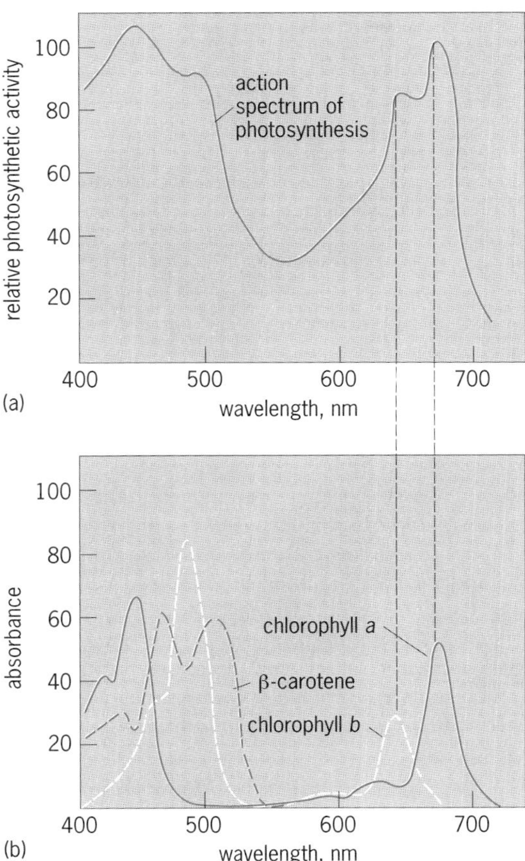

Fig. 1. Photosynthetic spectra of higher plants. (*a*) Action spectrum of photosynthesis showing the wavelengths of visible light that support photosynthesis. Measurements of photosynthetic activity include O_2 evolution or NADPH formation. (*b*) Absorption spectra of chlorophyll *a*, chlorophyll *b*, and β-carotene. (*After J. Darnell, H. Lodish, and D. Baltimore, Molecular Cell Biology, W. H. Freeman, 1986*)

Photosynthetic Pigments

The conversion of light energy to chemical energy is initiated when light-harvesting pigments absorb light of appropriate wavelengths and a π electron in the conjugated double-bond system of each molecule is raised from the ground state to an excited, higher-energy state. The transition is very rapid, and the excited state has a very short lifetime (15 nanoseconds for chl *a*).

When an acceptor molecule is not available to the excited pigment, the excited electron may return to the ground state, with the energy reemitted as a quantum of light. Because of the loss of some energy during the return from the excited to the ground state, the reemitted light, or fluorescence, has less energy and is therefore at a longer wavelength. The difference between the wavelength of maximal absorption and fluorescence is referred to as the Stokes shift and is approximately 8 nanometers for chlorophylls. *See* FLUORESCENCE.

When an acceptor molecule is available, the energy may be transferred from each excited pigment to an adjacent, unexcited one by inductive resonance. This requires proximity (less than 3 nm) and appropriate orientation of the adjacent pigment molecules, as well as an overlap between the fluorescence emission spectrum of the excited molecule and the absorption spectrum of the acceptor molecule. Inductive resonance is random until the energy is transferred from the different antenna pigments to antenna chl *a* in reaction centers and from there to special chl *a* molecules in which the excited state is at a lower energy level. Each of these special molecules is oxidized by expelling a high-energy electron, which reduces an electron acceptor to initiate electron transport. Light energy thus induces the formation of a strong oxidant in the form of chl a^+ and a high-energy electron. The ultimate result of this process is chemical energy in the form of adenosine triphosphate (ATP) and reducing power in the form of reduced nicotinamide adenine dinucleotide phosphate (NADPH) or reduced nicotinamide adenine dinucleotide (NADH).

The range of absorption maxima of photosynthetic pigments in the visible spectrum (see **table**) is due to the different light-harvesting pigments that evolved from the ancestors of higher plants—photosynthetic bacteria, and algae, which lived, as they still do, in aqueous environments. The absorption maxima thus represent the evolution of light-harvesting systems adapted to light available at different depths of a body

Absorption and fluorescence maxima of photosynthetic pigments in living cells*

Pigment	Absorption maxima, nm							Fluorescence emission maxima, nm
	Ultraviolet	Blue	Green	Yellow	Orange	Red	Near-infrared	Red to near-infrared
Chlorophyll *a*		436				676		685
Chlorophyll *b*		480				650		—
Bacteriochlorophyll *a*	375			590		800	850 890	900
Bacteriochlorophyll *b*		400		605			835–850 1015–1035	1020–1050
Bacteriochlorophyll *c*		457				750		770
Bacteriochlorophyll *d*		446				730		771
Allophycocyanin	380					650 654		660–680
Cyanobacterial-phycocyanin	380				620			650
Red algal-phycocyanin	380		553		615			640
Cyanobacterial-phycoerythrin	380			565				578
Red algal-phycoerythrin	380	498	540	565				578
Bangiophycal-phycoerythrin	380	498	545	563				575
Carotenoids		430 →	→	560				
β-Carotene		463, 498						

*The absorption and fluorescence maxima are representative for each group of pigments and may vary slightly among organisms.

of water. Many of these spectra are redshifted when they are compared to the spectra of the free pigments in solution. This is attributed to a polarizing effect on the conjugated double bonds by the protein to which the pigments are bound in the living cell. The specific functions of the various pigments, as well as the different functions of the same pigments, appear to be associated with their noncovalent or covalent binding to specific proteins in or on the thylakoid membrane. The proteins may localize and keep the chromophores in a proper orientation for efficient energy transfer.

Chlorophylls. Chlorophylls are cyclic tetrapyrroles, or porphyrins, in which the four pyrrole nitrogens are coordinated with a magnesium atom. They are embedded in photosynthetic membranes, the thylakoids, in noncovalent association with specific intrinsic membrane proteins. This is believed to occur by ligand formation between the tetrapyrrole magnesium and a histidine nitrogen from the protein. Chlorophylls are an integral part of the photosynthetic apparatus of all higher plants, algae, cyanobacteria, and prochlorophytes. Chlorophyll *a* is found in all oxygen-evolving organisms functioning both as a light-harvesting pigment and as the special energy-trapping reaction-center core pigment (P), either P680 for photosystem II (PS II) or P700 for PS I. It absorbs maximally in the blue and red regions of the visible spectrum (see table). There are several distinct species of chl *a*; certain fluorescent species are due to chl *a* in PS I, while others arise from chl *a* in PS II. The different chl *a* species reflect different chlorophyll-protein complexes. At room temperature, PS II is the only fluorescent photosystem. PS I quenches PS II fluorescence, and the efficiency of energy transfer from PS II to PS I under various conditions is usually measured by following the fluorescence yield, that is, the decrease in fluorescence with time.

Other chlorophylls that harvest and transfer the captured light energy to chl *a* may also be present. They differ from chl *a* primarily in side-group substitutions on the pyrrole rings. The antenna pigments do not fluoresce in the living cell because the energy is efficiently transferred to chl *a*. Chlorophyll *b* is found in higher plants, green algae, euglenoids, and prochlorophytes; brown algae, diatoms and dinoflagellates contain chl *c*; and some red algae contain chl *d*. Cyanobacteria and most red algae contain only chl *a*.

Phycobilins. Linear, open-chain tetrapyrroles (phycobilins) are covalently linked chromophores of phycobiliproteins found on the surface of thylakoid membranes in highly ordered aggregates (phycobilisomes). Phycobilisomes function as antenna complexes, harvesting and transferring light energy to chl *a* of membrane-embedded PS II, in cyanobacteria, red algae, and cyanelles. The phycobiliproteins of cryptomonads have a similar function but contain one of only two phycobiliproteins, phycocyanin or phycoerythrin, which are located in the thylakoid lumen and are not organized into phycobilisomes.

The major phycobiliproteins comprising the phycobilisomes fall into four classes based upon distinctive absorption spectra associated with their chromophores, which are covalently bound via ring A of the phycobilin in a thioether linkage to specific cysteinyl residues in the protein. These include phycoerythrin, which contains phycourobilin chromophores; phycoerythrocyanin, which contains phycocyanobilins and phycobiliviolins; phycocyanin, which contains phycocyanobilins; and allophycocyanin, which also contains phycocyanobilins. The distinct absorption spectra of the different phycobiliproteins is due to interaction of the chromophores with the protein to which they are bound as well as to the way in which they are aggregated to form phycobilisomes. All red algae and cyanobacteria contain allophycocyanin and phycocyanin, and many have, in addition, phycoerythrin, or phycoerythrocyanin. There is a directed energy transfer from phycoerythrin or phycoerythrocyanin to phycocyanin, from phycocyanin to allophycocyanin, and from allophycocyanin to chl *a* in PS II. This is shown in isolated phycobilisomes as fluorescence emission from allophycocyanin and little or no emission from the other excited phycobiliproteins, indicating efficient energy transfer. The table shows that phycobiliproteins exhibit the widest absorption range (490–620 nm), providing those organisms that contain them with great adaptability in harvesting light for photosynthesis. *See* PHYCOBILIN.

Carotenoids. Carotenoids involved in photosynthesis include the hydrocarbon carotenes and their oxygenated derivatives, the xanthophylls. Most have 40-carbon isoprenoid backbones with 11 conjugated double bonds; a substituted cyclohexene is usually found at each end of the molecule. β-Carotene, lutein, violaxanthin, and neoxanthin are the major carotenoids found in higher-plant chloroplasts, approximately 25% of which are carotenes and 45% lutein, the predominant xanthophyll. Carotenoids are found embedded in the membranes of all photosynthetic organisms in noncovalent linkage with specific proteins. They function to protect against photooxidative damage and as light-harvesting pigments, absorbing light in regions of the visible spectrum where chlorophylls absorb little, or not at all. *See* CAROTENOID.

Reaction center pigments. Reaction centers are the smallest multiprotein complexes isolated from thylakoid membranes functional in either PS I or PS II. They usually include small antenna chl *a* proteins. Reaction centers, embedded in the membranes in a highly ordered arrangement, are surrounded by light-harvesting complexes which, though probably not essential, ensure efficient energy transfer. The reaction-center cores are the smallest membrane complexes capable of specific energy transduction. Historically, each reaction center with its light-harvesting chlorophylls (approximately 300) is referred to as a photosynthetic unit. Special chl *a* molecules in the reaction-center cores function as the primary electron donors in the two reaction centers of oxygen-evolving photosynthetic organisms. The special environment created by the association of these chl *a* molecules with certain proteins appears to result in a lower amount of energy being necessary to excite an electron out of the chlorophyll. Shirley Raps

Anthocyanin

Anthocyanins are glycosides, and on acid hydrolysis yield sugars and colored aglycones called anthocyanidins. The most common aglycone is 3,5,7,3′,4′-pentahydroxyflavylium, or cyanidin, so called after the cornflower (*Centaurea cyanus*), from which it was first isolated; the structure is shown below.

The other 15 known anthocyanidins all have the same basic structure as cyanidin and differ only in having fewer or more hydroxyl groups or in being *O*-methylated. The other common anthocyanidins are pelargonidin (structure as in cyanidin without the 3′-hydroxyl), peonidin (3′-methyl ether of cyanidin), delphinidin (extra hydroxyl at 5′), and its 3′-monomethyl and 3′,5′-dimethyl ethers, petunidin and malvidin. The sugar present in anthocyanins may be glucose, galactose, rhamnose, xylose, or arabinose, a combination of two or three of these monosaccharides, or a combination of a monosaccharide with an aromatic hydroxy acid. Sugars are usually attached to the 3-hydroxyl, giving 3-glucosides, or to the 3- and 5-hydroxyl groups, giving 3,5-diglucosides. The presence of these sugars accounts not only for the sap solubility of anthocyanins but also for their natural stability, the free aglycones being rather water-insoluble and unstable.

Anthocyanins are cations, the positive charge being distributed among the various carbons and oxygens of the conjugated ring system. They are normally isolated as chlorides, since in extracting them from plant tissues into alcohol or water, it is necessary to add hydrochloric acid in order to retain the color. The pigments probably occur in plant tissue in association with organic acid anions, and the actual color in the plant may be determined by the acidity of the cell sap as well as by the presence or absence of other constituents in the sap. Anthocyanin-metal complexes have been isolated from the petals of a number of blue-flowered plants, such as the cornflower. These complexes have been shown to contain flavone and polysaccharide as well as metals such as iron and aluminum. The color of the isolated pigments fades above pH 4, and in alkali turns blue, with destruction of the pigment by aerial oxidation. Anthocyanins have characteristic absorption properties in the visible spectrum (**Fig. 2**), and are easily separated and identified by paper or thin-layer chromatography by using aqueous hydrochloric acid as a solvent. Characterization is facilitated by the fact that their mobilities on filter paper are closely correlated with the number of hydroxyl, methoxyl, and sugar groups present. The R_f values assigned to each pigment represent the observed migration of a chromatographic zone relative to the solvent front.

The purpose of anthocyanin pigmentation in leaf tissue is obscure. While permanent leaf coloration due to anthocyanin is not usual, transient anthocyanin synthesis in either young or autumnal

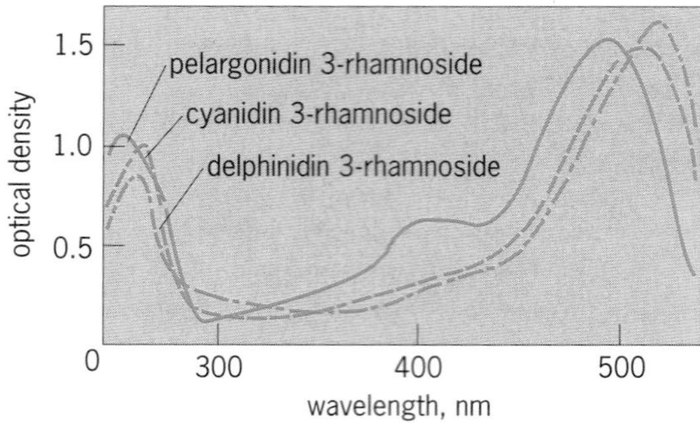

Fig. 2. Absorption spectra of three anthocyanins.

leaves is quite common. Moreover, red coloration often builds up in the leaves of plants suffering from mineral deficiency, environmental stress, disease, or wounding. There is little doubt that under such circumstances anthocyanin synthesis is closely connected with carbohydrate metabolism. The concentration of anthocyanin in certain species of *Eucalyptus* has been found to increase with the altitude at which the tree grows, and in this case pigment production may well be a protective device against ultraviolet radiation. Jeffrey B. Harborne

Bibliography. T. W. Goodwin (ed.), *Plant Pigments*, 1988; R. P. F. Gregory, *Biochemistry of Photosynthesis*, 3d ed., 1989.

Plant propagation

The deliberate, directed reproduction of plants using plant cells, tissues, or organs.

Traditional method. Asexual propagation, also called vegetative propagation, is accomplished by taking cuttings, by grafting or budding, by layering, by division of plants, or by separation of specialized structures such as tubers, rhizomes, or bulbs. This method of propagation is used in agriculture, in scientific research, and in professional and recreational gardening. It has a number of advantages over seed propagation: it retains the genetic constitution of the plant type almost completely; it is faster than seed propagation; it may allow elimination of the nonfruiting, juvenile phase of the plant's life; it preserves unique, especially productive, or esthetically desirable plant forms; and it allows plants with roots well adapted for growth on poor soils to be combined with tops that produce superior fruits, nuts, or other products.

Cuttings. Propagation by cuttings involves removing branches, leaves, roots, or bud-bearing stems from a donor plant and causing each cutting to regenerate the other parts requisite to a new, complete plant. The plant part chosen for making cuttings varies with the plant species, because different species retain regenerative capacities to differing extents in their various parts. In propagating coleus plants, for example, branch cuttings are removed from the plant, caused to root, and then placed in soil. Coleus leaf cuttings also root, but they form only a bigger leaf, not a whole plant. Leaf cuttings are, however, used for propagating African violets. One or more buds are regenerated on the petiole (stalk) of the African violet leaf when the petiole is embedded in sand or vermiculite. The bud forms new leaves and the cutting roots, thus establishing a new plant. In horticultural practice, root formation on cuttings is often increased and accelerated by the application of auxin, a root-inducing chemical that mimics the plant's own root-forming hormone. *See* AUXIN; VERMICULITE.

In propagation by cuttings, periodic misting with water helps prevent desiccation. Shading and special disease-preventing precautions may also be necessary during this vulnerable period before wounds have healed and roots have been regenerated. In scientific research, the use of cuttings allows all test plants to be derived from one plant; thus a clone of plants is established. Such a procedure greatly diminishes biological variation, a source of random experimental error. However, if an infectious disease organism infests the clone, it may cause a spontaneous change in test responses.

Grafting. Grafting, another method of vegetative propagation, follows a highly developed set of horticultural practices in which two different plants are combined into one plant that has some of the desirable traits of both. All grafts consist of at least two parts—the scion, which is the shoot-bearing or shoot-producing top part, and the stock, which is the root-bearing bottom part. In this way, a rootstock that tolerates unfavorable soil conditions or resists soil-borne disease organisms can be combined with a scion that produces pest-resistant fruit to yield a superior plant.

Grafting offers an opportunity to study the effects of one part of a plant on the functioning of its other parts. For example, evidence for a flowering hormone is based heavily on grafting experiments in which stock and scion are taken from different species or from different pretreatment environments. Evidence indicates that a hormone, florigen, almost universal in plants, is produced in leaves and is translocated to the shoot apex, where it evokes flower initiation.

In all grafting, the objective is to establish a firm junction between scion and stock. This junction must be traversed by new and fully functional vascular tissues before the graft is successful. Except at the growing tips of plants, vascular tissues are produced by a layer of rapidly dividing cells between xylem and phloem called the cambium, which is found in both stems and roots. The aim in grafting, to join the cambium of the scion to the cambium of the stock, can be accomplished by any of several grafting techniques (see **illus.**). In most grafts, wound tissue (callus) is formed by each of the graft partners at their juxtaposed surfaces. There the callus cells of one partner are made to adhere to those of the other, as in the whip graft (illus. *a*). Through cell divisions on both sides, the cambium of both stock and scion is

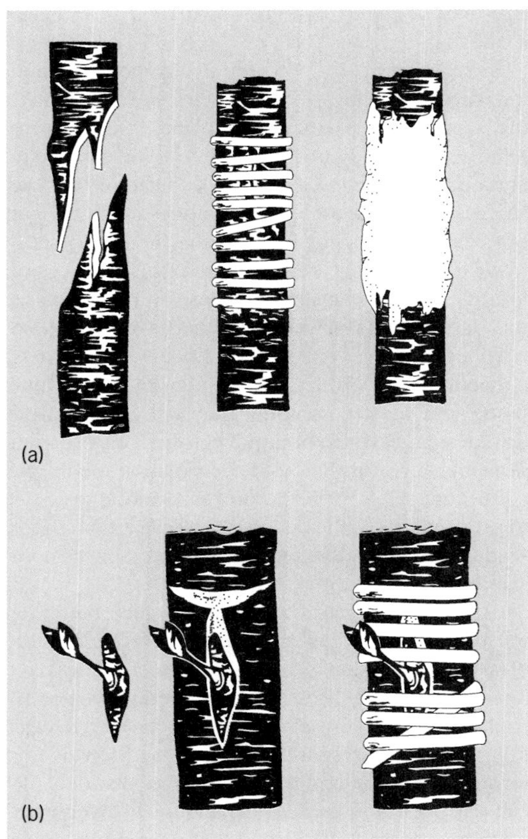

Two grafting techniques. (*a*) Whip graft, with paraffin to seal in moisture. (*b*) T-budding.

extended into the callus. Both xylem and phloem are produced by this new cambium, thus establishing vascular continuity across the graft union. *See* PLANT TISSUE SYSTEMS.

Budding, or bud grafting, is a specific grafting technique of which T-budding or shield budding, the most common form, can be used successfully when bark can be easily removed from the wood (illus. *b*). Vertical and horizontal cuts are made through the bark of the stock plant in the shape of a T. The two cuts, each about $1^3/_4$ in. (4.5 cm) long, are made through the bark down to the underlying wood. The bud or scion is removed from a budstick by cutting out a shield- or tongue-shaped piece of wood and bark. The cut starts about $^1/_3$-$^1/_2$ in. (0.9- 1.3 cm) below the bud and widens and deepens as it is continued upward. About $^1/_3$ in. (0.9 cm) above the bud, a horizontal cut is made so as to remove the bud and its underlying bark and wood as one piece. (If the wood can be removed from the bud-bearing piece without tearing the bark, it is done before inserting the bud into its grafted position.) The bark flaps at both sides of the T are spread open slightly, and the shield-shaped piece, with its original orientation maintained, is inserted into the T. The graft, but not the bud, is wrapped tightly with rubber budding strips or with strips of polyethylene or a paraffin-coated film to seal in water. When the graft union has completed its development, the original shoot on the stock above the grafted bud is cut off. T-budding is used by nurseries to propagate roses, most fruit-tree species, and many ornamental shrubs.

The extent to which successful grafts can be made is limited. Different varieties of the same species can almost always be grafted successfully, but different species in the same genus are usually not graft-compatible. In some cases, a graft between species succeeds in one direction only. For example, a Marianna plum scion can be grafted on a peach stock, but the reciprocal graft is not successful. Species from different genera are seldom compatible.

The physiological basis for graft failure or graft incompatibility is not fully understood. According to one theory, metabolites produced by one partner are toxic to tissues of the other or are converted to toxins by enzymes present in the other. The clearest example is that of pear grafted on quince. Quince produces cyanogenic glycoside that is capable of releasing poisonous cyanide when acted on by the enzyme glycosidase, which is produced in tissues of the pear. The cyanide kills pear cells which otherwise would have been involved in producing a graft union. Another theory supposes the existence of sophisticated cellular mechanisms for recognizing foreign cells that are similar to the immune system reactions in animals.

Layering. Layering is used to propagate plants by causing roots to develop on stems, a phenomenon that occurs rarely in nature. This is accomplished by burying a portion of the stem, which is still attached to the plant, in order to keep it dark, warm, and moist. To further hasten root formation, the bark of the portion being buried may be cut or ringed.

In tip layering, only the stem tip is buried. Tip layering is used to propagate trailing blackberries and black raspberries, and duplicates the natural propagation methods of those plants. In simple layering, one portion of the stem, still attached to the plant, is buried. Filberts, for example, are routinely propagated by simple layering. In compound layering, more than one portion of the same stem is buried. Muscadine grapes are propagated by this method.

With air layering, a few inches of a stem are girdled or cut and then enclosed in sphagnum moss to provoke root formation. The moss is wrapped in transparent, nonporous film and is kept slightly damp until the roots form. The rooted piece of stem, including an axillary bud or a large leafy shoot at its upper end, is cut off and planted in its own pot. Air layering is used to propagate large plants such as *Dieffenbachia, Ficus*, pecan, and *Magnolia*. Stool layering involves pruning a plant so that it develops several branches near its base. The undersides of those branches are buried in soil until they form roots. The rooted branches are then cut off and set out as individuals. Plants that are propagated by stool layering include currants, gooseberries, and clonal rootstocks of pear and apple. *See* AGRICULTURAL SCIENCE (PLANT); BREEDING (PLANT); REPRODUCTION (PLANT).

C. E. LaMotte

Culture technology. Tissue cultures and protoplast cultures are among the techniques that have been investigated for plant propagation; the success of a specific technique depends on a number of factors. Practical applications of such methods include the clonal propagation of desirable phenotypes and the commercial production of virus-free plants.

Tissue culture. Plant tissue cultures are initiated by excising tissue containing nucleated cells and placing it on an enriched sterile culture medium. To prevent growth of undesirable microorganisms, the plant tissue itself must first be surface-sterilized, usually by immersing it in a dilute bleach solution and then rinsing repeatedly with sterile water. Another approach is to surface-sterilize seeds, germinate them on sterile medium solidified with agar, and later remove axenic tissue from the seedlings. The response of a plant tissue to a culture medium depends on a number of factors: plant species, source of tissue, chronological age and physiological state of the tissue, ingredients of the culture medium, and physical culturing conditions, such as temperature, photoperiod, and aeration.

For unknown reasons, different plant species often exhibit unique requirements for successful growth in culture; it has been suggested that variation in products of secondary metabolism may explain observed variation in the responses. Within a single plant, different source tissues often show wide variation in response to a given culture condition, which may be due to different endogenous concentrations of either growth-promoting or growth-inhibiting substances.

Though most tissue culture media have similar compositions, recipes suitable for a given species and type of tissue must be determined by trial and error. Standard media include appropriate amounts of macronutrient and micronutrient elements, vitamins, a carbon source (usually glucose or sucrose), and plant hormones (usually an auxin and a cytokinin). In addition, many popular media contain complex additives such as coconut milk, yeast extract, and casein hydrolysate.

Identical plant tissues may respond quite differently when cultured on media with different compositions. Plant hormones have the strongest influence of all ingredients on the response of a given plant tissue. In general, relatively high concentrations of auxins stimulate callus proliferation and possibly root development, while relatively high concentrations of cytokinins induce development of shoots. In some plant species, somatic embryos that exhibit patterns of development similar to those exhibited by sexual embryos can be induced to form from callus tissue that is first cultured on a medium containing a high auxin concentration, then subcultured to a medium with little or no auxin. *See* PLANT HORMONES.

Differences in the physical environment used for culturing plant tissues can also profoundly affect their growth and development. For example, callus cultures grown in the light usually produce chlorophyll, while those grown in the dark do not. Light is usually required for shoot formation from callus tissue.

Protoplast culture. Though technically more demanding, successful culture of plant protoplasts involves the same basic principles as plant tissue culture. Empirical methods are used to determine detailed techniques for individual species; such factors as plant species, tissue source, age, culture medium, and physical culture conditions have to be considered.

Plant protoplasts are most often isolated by using fungal enzymes (usually a cellulase and a pectinase) that degrade the cell walls. Protoplasts have been isolated from leaves, roots, stems, petals, hypocotyls, cotyledons, coleoptiles, cell suspensions, and callus tissue, but most commonly they are isolated from leaf mesophyll tissue or rapidly growing cell suspensions. It is critical that both the isolation media and culture media contain the proper osmotic pressure to prevent lysis of the delicate protoplasts; mannitol, sorbitol, glucose, and sucrose are common osmotic stabilizers. *See* PLANT CELL.

Isolated protoplasts are injured cells, and they must replace their cell wall before they are capable of sustained division. Trial-and-error techniques have not yet led to the discovery of procedures whereby protoplasts of many plant species can be successfully induced to resynthesize a cell wall, divide and form callus tissue, and then regenerate plantlets. Because recalcitrant species can be found in diverse taxonomic orders, it is assumed that epigenetic blocks (such as failing to recognize hormonal signals) or physiological blocks (such as not receiving requisite environmental signals) to cell division will eventually be discovered and overcome.

Applications of culture technology. Perhaps the most obvious use of plant tissue culture is in clonal propagation of desirable phenotypes. For example, elite hybrid *Cymbidium* orchids are rapidly propagated in tissue culture by hormonal stimulation of shoot meristem production from excised protocorms. In this way, changes that might cause loss of desirable flower traits during sexual reproduction can be avoided. However, tissue culture procedures have been shown to enhance mutation frequency, and true genetic clones are not always produced.

Meristem culture is also used in the commercial production of virus-free plants, including chrysanthemums, potatoes, and dahlias. Many viruses move through the plant's vascular tissue, and because vascular connections to the apical meristem are not continuous, excision and culture of apical meristems provides a means for eradicating viruses.

Embryos that form when two sexually incompatible species are crossed are often arrested in their development because of incompatibility between the developing hybrid embryo tissue and the maternal tissue of the ovule. However, if these embryos are excised from the ovule and cultured in a nutrient medium, they sometimes develop into mature hybrid plants that may exhibit desirable agronomic traits.

Protoplasts from two sexually incompatible species have been fused to form somatic hybrids.

Hybrid plants have been regenerated from some somatic hybrid protoplasts. Commercially valuable hybrids have yet to be produced by this means, but the technique has considerable potential.

Both tissue and protoplast cultures have been explored as means for introducing limited genetic variation into agronomically useful cultivars. When individual mesophyll cells are isolated from a leaf and cultured as either protoplasts or tissue cultures, they often exhibit somatic variation. (Cytological studies show that chromosomal rearrangements and duplications are common in cultured plant cells.) This somaclonal variation may be useful in selecting cells that tolerate phytotoxins, herbicides, and environmental stresses, and the plants regenerated from these cells may exhibit the same desirable phenotypes.

Large-scale culture of plant cells for production of pharmaceuticals and agrochemicals has resulted in limited success. Though many patent applications have been filed, the technologies required for cost-effective cultivation of plant cells in large vats have not yet been developed.

Cryopreservation of plant cells has been proposed as an alternative to seed stores for preservation of genetic resources. The cells are stabilized by immersion in cryoprotectants such as dimethyl sulfoxide and glycerol, slowly frozen to −321°F (−196°C), and then stored in liquid nitrogen. They can be thawed in warm water, poured into a semisolid medium, and once again grown as a cell suspension in liquid within a week or two. This approach is especially relevant for storage of vegetatively propagated species and species whose seeds are short-lived. Both shoot meristem cultures and cell suspension cultures have been advocated as appropriate candidates for cryopreservation. The approach has been most successful when applied to cell suspension cultures. Cryopreservation works for years, but it is not known if it will be suitable for storage over decades. *See* CRYOBIOLOGY.

Several elaborate procedures have been devised to introduce foreign deoxyribonucleic acid (DNA) into plant cells. One widely applicable method involves cloning desirable genes into an *Agrobacterium*-derived plasmid vector, introducing this vector into *Agrobacterium*, and then allowing the *Agrobacterium* to infect an excised piece of plant tissue. As part of its normal infection process, the bacterium inserts the cloned genes into plant cells for integration into the plant's chromosomes. The transformed plant cells can be selected to form callus tissue from which transformed plants can be regenerated, allowing direct introduction of desirable genetic traits. *See* GENETIC ENGINEERING.

Finally, plant protoplasts have provided unique opportunities to study both the plasma membrane and relatively undamaged intracellular organelles. Purified plasma membranes, tonoplasts, and intact vacuoles are routinely isolated from protoplasts for study of their transport properties. *See* TISSUE CULTURE.

Karen Grady Ford

Bibliography. L. C. Fowke and F. Constabel (eds.), *Plant Protoplasts*, 1985; H. T. Hudson, D. E. Kester, and F. Davies, *Plant Propagation: Principles and Practices*, 5th ed., 1990; N. F. Jensen, *Plant Breeding Methodology*, 1988; N. C. Stoskoph, D. T. Tomes, and B. R. Christie, *Plant Breeding: Theory and Practice*, 1993; I. K. Vasil (ed.), *Cell Culture and Somatic Cell Genetics of Plants*, 3 vols., 1984–1986; I. K. Vasil and T. A. Thorpe (eds.), *Plant Cell and Tissue Culture*, 1994.

Plant respiration

A biochemical process whereby specific substrates are oxidized with a subsequent release of carbon dioxide, CO_2. There is usually conservation of energy accompanying the oxidation which is coupled to the synthesis of energy-rich compounds, such as adenosine triphosphate (ATP), whose free energy is then used to drive otherwise unfavorable reactions that are essential for physiological processes such as growth. Respiration is carried out by specific proteins, called enzymes, and it is necessary for the synthesis of essential metabolites, including carbohydrates, amino acids, and fatty acids, and for the transport of minerals and other solutes between cells. Thus respiration is an essential characteristic of life itself in plants as well as in other organisms.

Since respiration involves release of CO_2, it appears superficially to be the reverse of photosynthesis, the fixation of CO_2 by green plant tissues in light. However, these two processes are very different biochemically. Moreover, aerobic respiration usually occurs in an organized system of enzymes located in the cytoplasmic organelles, called mitochondria, while the cellular site of photosynthesis is the much larger chlorophyll-bearing organelle, the chloroplast. Some photosynthetic tissues have been recognized as having a special system of respiration, known as photorespiration, that functions only in light. Most of this discussion, however, is concerned with the easily measured oxygen, O_2, uptake or CO_2 release in various plant organs or the "dark" respiration in green plant tissues.

Factors affecting respiration. Although other methods of measurement are available, most of the recorded rates of respiration have been obtained by manometric determination of O_2 uptake or CO_2 evolution, or by measurement of the rate of loss of dry weight. Examples of respiratory rates in different plants and plant organs are presented in **Table 1**. Like most enzymatic processes, respiration increases about twofold for every 18°F (10°C) rise in temperature between 32 and 95°F (0 and 35°C). *See* ENZYME; MANOMETER.

Respiratory quotient. The ratio of the CO_2 evolved to the O_2 taken up is called the respiratory quotient (RQ), and its value is frequently close to 1.0. Since sucrose and starch are the most common forms of carbohydrate storage in plants, an RQ of 1.0 would be expected if the hexose derived from these

TABLE 1. Rates of respiration of plant species and organs*

Species	Plant part	Temperature, °F (°C)	Respiratory rate, microliters O_2/(100 mg fresh wt)(h)	Respiratory quotient, CO_2/O_2
Carrot	Intact root	50 (10)	1.4	1.1
		75.2 (24)	3.0	1.1
Potato	Tuber	50 (10)	0.2	0.9
		75.2 (24)	0.6	1.0
Maize	Seedling	77 (25)	127	1.0
Castor bean	Seedling	86 (30)	133	0.4
	Germinating seed	—	—	0.7
Tobacco	Leaf segment	77 (25)	33	1.3
Maize	Intact leaf	78.8 (26)	68	1.0
Lemon	Intact fruit	32 (0)	0.13	1.2
		69.8 (21)	1.1	1.0
		100.4 (38)	2.9	1.4
Tomato	Intact fruit	75.2 (24)	2.2	1.1

*After P. L. Altman and D. S. Dittmer (eds.), *Metabolism*, Federation of American Societies of Experimental Biology, 1968.

substances were the primary respiratory substrate and if it were completely oxidized, as in reaction (1).

$$C_6H_{12}O_6 + 6O_2 \rightarrow 6CO_2 + 6H_2O + 674,000 \quad (1)$$
$$\text{Hexose} \qquad RQ = 1.0$$

When plants show an RQ close to unity, this suggests that carbohydrate is respirated, as must be the case in starchy plant parts such as carrot root and potato tuber (Table 1). However, it is conceivable that hexose could be respired and result in a different RQ if some of the CO_2 evolved were consumed again, if the hexose were not completely oxidized, or if the diffusion of O_2 into the tissue were hindered.

When the respiratory substrate is a fatty acid, such as stearic acid, its complete oxidation would produce an RQ less than 1.0, as shown in reaction (2).

$$CH_3(CH_2)_{16}COOH + 26O_6 \rightarrow 18CO_2 + 18H_2O \quad (2)$$
$$\text{Stearic acid} \qquad RQ = 0.69$$

Thus respiration of a tissue rich in lipid material, such as a castor bean seedling, produces an RQ below 1.0, indicating that fatty acids are the substrates of respiration by this tissue.

The complete oxidation of organic acids, such as citric acid or malic acid, would create an RQ greater than 1.0, as shown in reaction (3). As shown in

$$2\,HO-\overset{\displaystyle CH_2-COOH}{\underset{\displaystyle CH_2-COOH}{\overset{|}{\underset{|}{C}}}}-COOH + 9O_2 \longrightarrow 12CO_2 + 8H_2O \quad (3)$$
$$RQ = 1.33$$
$$\text{Citric acid}$$

Table 1, tobacco leaf and lemon fruit, known to contain high concentrations of organic acids, have an RQ above 1.0 as might be expected.

Respiratory losses. Respiration rates in air of actively growing tissues, such as seedlings and leaves, are rather rapid. The dry weight of most plant tissues represents about 10% of the fresh weight, and since 22.4 microliters of CO_2 are equivalent to 0.044 mg, it follows that leaf and seedling tissue (Table 1) lose about 1% dry weight per hour because of respiration. In leaves in bright light the rate of net photosynthesis is 10–20 times the dark respiration, and some CO_2 evolution usually takes place in light. Thus respiration must play a large part in determining the net yield of crop plants.

Variations in respiration. The rate of respiration decreases in leaves with age. As is shown in **Table 2**, leaves of shade-tolerant species grow relatively more slowly than shade-intolerant species, even when both are grown in full sunlight. Dark respiration in a number of instances was also consistently less in shade-tolerant plants, suggesting that this characteristic enables these species to survive the adverse light conditions common in a dense forest.

Another example of a changing respiration rate is frequently encountered during postharvest storage of many fruits. Typical examples of avocado, banana, pear, and apple are presented in **Fig. 1**. The rate of respiration increases to a maximum value, known as

TABLE 2. Effect of shade-tolerance on growth and dark respiration of leaves of species grown in full sunlight*

Species	Relative leaf growth rate, mg/(g)(h)	Dark respiration of leaf disks at 25°C (77°F), mg loss dry wt/(g dry wt)(h)
Shade-tolerant trees		
Sugar maple	0.99	1.8
Red oak	1.7	—
Shade-intolerant trees		
Black birch	4.8	3.6
Smooth sumac	3.0	—
Crop plants		
Tomato	12.6	—
Tobacco	31.4	9.0
Kentucky bluegrass	13.2	—
Sunflower	6.5	3.5
Arable weeds		
Common sorrel	10.2	7.3
Common portulaca	12.4	—
Pigweed	—	5.4

*After J. P. Grime, *Nature*, 208:161, 1965.

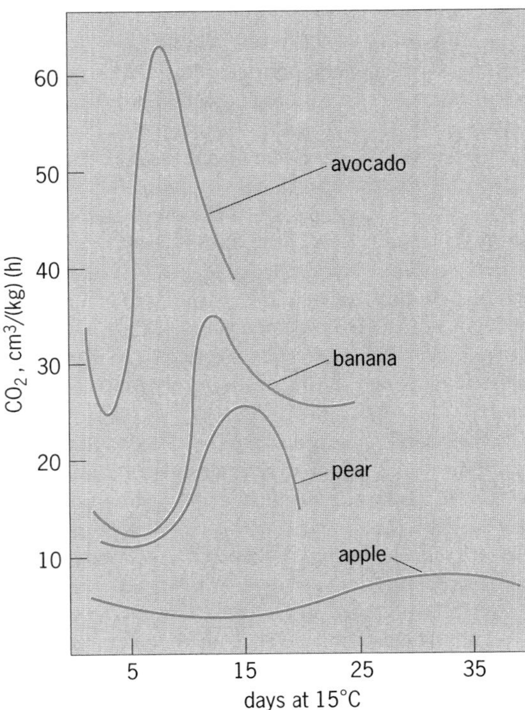

Fig. 1. The course of the climacteric rise in fruit respiration during the postharvest period. 15°C = 59°F. (*After J. B. Biale, Annu. Rev. Plant Physiol., 1:183, 1950*)

the climacteric rise, followed by a decrease in respiration. The intensity and the length of time needed to obtain the maximal rate varies considerably with the tissue. The metabolic reasons for this increased capacity for respiration are still not clearly understood, although it is known that an increase in many synthetic reactions accompanies the increased respiration.

The ambient oxygen concentration affects the respiration rate of plant tissues, but this is difficult to assess accurately because plant tissues are frequently bulky and hence offer a large diffusive resistance to the incoming O_2. Nevertheless, leaves show maximal rates of CO_2 evolution in the dark with 2–3% O_2 in the atmosphere (as compared with 20% O_2 present in normal air), and even fruit tissues generally show half-maximal rates of O_2 uptake at about 6% O_2.

Mechanism of respiration and fermentation. Overall aerobic respiration is the end result of a sequence of many biochemical reactions that ultimately lead to O_2 uptake and CO_2 evolution. In the absence of O_2, as may occur in bulky plant tissues such as the potato tuber and carrot root and in submerged plants such as germinating rice seedlings, the breakdown of hexose does not go to completion. The end products are either lactic acid or ethanol, which are produced by anaerobic glycolysis or fermentation.

Anaerobic glycolysis or fermentation. The sequence of reactions of this pathway is shown in **Fig. 2**. Although the complete aerobic combustion of 1 mole of glucose results in the release of 674,000 cal (2.82 megajoules), only about 57,000 cal (239 kilojoules) per mole are produced when 2 moles of lactic acid are synthesized during anaerobic glycolysis. Beginning with starch, 1 mole of ATP is consumed (at the phosphohexokinase-catalyzed reaction) while 4 moles of ATP are synthesized, for a net gain of three

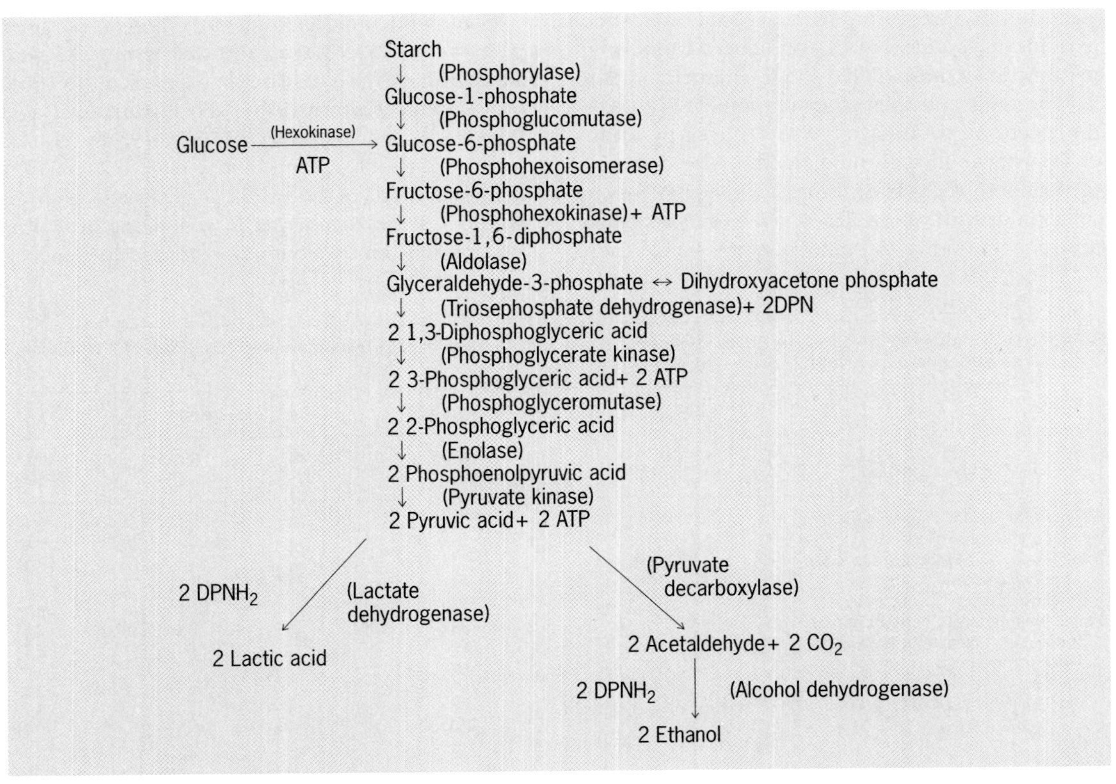

Fig. 2. Reaction sequence for anaerobic glycolysis. The soluble enzymes are shown in parentheses.

energy-rich phosphate bonds. Since each ATP contains about 8000 cal (33 kJ) per mole, the gain of 24,000 cal (100 kJ) represents about 42% of the potentially available energy conserved as chemical energy. When starting with the fermentation of glucose, an extra ATP is consumed in the hexokinase-catalyzed step, so that the net gain is then two energy-rich bonds. The energy yield by fermentation is very much smaller than in aerobic respiration, as discussed later.

The enzymes associated with anaerobic glycolysis have been isolated from many plant tissues, but more often ethanol and not lactic acid is the final product. An example is the rice plant, which can germinate underwater, and in the absence of O_2 produces CO_2 and ethanol in equivalent amounts.

Krebs cycle. In aerobic tissues, pyruvic acid produced during glycolysis is completely oxidized with the accompanying synthesis of much more ATP. Pyruvic acid oxidation takes place in the mitochondria by means of a cyclic sequence of reactions which begins when the first product of pyruvate oxidation, acetyl coenzyme A, reacts with oxaloacetic acid to produce citric acid. Oxaloacetic acid is eventually regenerated. Thus the cycle (also known as the citric acid cycle), which accounts for 5 moles of O_2 uptake and 6 moles of CO_2 per mole of hexose, can be repeated. The oxidative steps of the cycle are summarized in **Table 3**. *See* CITRIC ACID CYCLE.

Reduced pyridine nucleotides [diphosphopyridine nucleotide (DPN), also known as nicotinamide adenine dinucleotide (NAD); and triphosphopyridine nucleotide (TPN), also known as nicotinamide adenine dinucleotide phosphate (NADP)] are produced in four of the reactions of the Krebs cycle. The electrons are transported from reduced diphosphopyridine nucleotide ($DPNH_2$) or reduced triphosphopyridine nucleotide ($TPNH_2$) in the mitochondria to flavoproteins and then to cytochromes. Eventually the electrons are transferred to O_2 itself by means of the enzyme cytochrome oxidase. The oxidation of each pair of electrons from $DPNH_2$ or $TPNH_2$ results in the synthesis of 3 moles of ATP by oxidative phosphorylation in the mitochondria; hence these reactions have a P/O ratio of 3.0. An additional phosphorylation, not connected with the oxidation of reduced pyridine nucleotides, takes place during the oxidation of α-ketoglutarate so that a P/O ratio of 4.0 is obtained in this step of the cycle. The oxidation of succinate, however, begins at the flavoprotein level and results in a P/O ratio of only 2.0. Thus the Krebs cycle accounts for 5 of the 6 moles of O_2 consumed during glucose oxidation (1 mole arises from the triosephosphate dehydrogenase reaction of glycolysis), 6 moles of CO_2, and 30 moles of ATP synthesized. In terms of conservation of chemical energy, the Krebs cycle is therefore about 12 times more efficient than anaerobic glycolysis per mole of glucose oxidized.

Oxidation of the various substrates of the Krebs cycle has been shown in mitochondria isolated from a variety of plant tissues, and P/O ratios similar to those described have been obtained. Some of the tissues investigated include pea and spinach leaves, bean seedlings, and avocados and apples.

Electron transport in mitochondria. Under aerobic conditions electrons are transported from NADH or other substrates to oxygen by a pathway found in mitochondria that ultimately reacts with cytochrome oxidase (cytochrome a_3), as shown in **Fig. 3**; this oxidase is sensitive to cyanide as well as being inhibited by carbon monoxide in a reaction that is reversed by visible light. The pathway is also strongly inhibited by antimycin A. Cytochrome oxidase has not been isolated in a soluble form, but its presence has been demonstrated in isolated mitochondria from a large number of species, and it is functionally the most important of the plant oxidases. *See* BIOLOGICAL OXIDATION; CYTOCHROME; MITOCHONDRIA.

Plant mitochondria frequently possess another pathway of electron transport that terminates with an uncharacterized oxidase that is resistant to 1 mM cyanide and is sensitive to salicylhydroxamic acid (SHAM), as shown in Fig. 3. The alternate oxidase is the main route of respiratory electron transport in mitochondria of the spadix of *Arum maculatum*, where it serves a function in generating heat. It is also present and accounts for appreciable quanti-

TABLE 3. Summary of reactions in Krebs cycle concerned with O_2 uptake, CO_2 evolution, and formation of ATP during complete oxidation of pyruvate

Oxidative reaction	Reduced product	Per mole of glucose: Atoms O_2 uptake	Moles CO_2	Moles ATP	Ratio P/O
Pyruvate ⟶ acetyl CoA + CO_2 (pyruvate dehydrogenase)	$DPNH_2$	2	2	6	3
Isocitrate ⟶ α-ketoglutarate + CO_2 (isocitrate dehydrogenase)	$TPNH_2$	2	2	6	3
α-Ketoglutarate ⟶ succinate + CO_2 (α-ketoglutarate dehydrogenase)	$DPNH_2$	2	2	8	4
Succinate ⟶ fumarate (succinate dehydrogenase)	Flavin	2		4	2
Malate ⟶ oxaloacetate (malate dehydrogenase)	$DPNH_2$	2		6	3
NET		10	6	30	

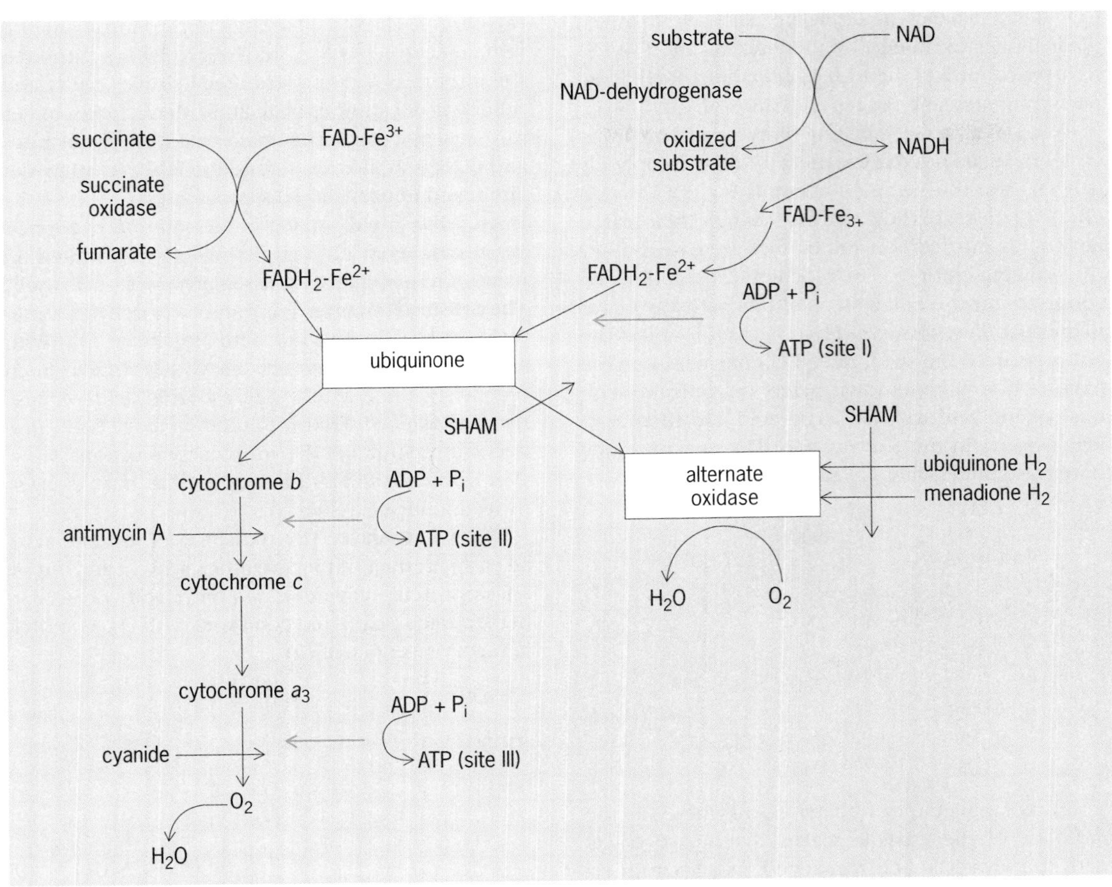

Fig. 3. Scheme of electron transport in mitochondria showing the normal (cyanide-sensitive) pathway of respiration with three phosphorylation sites from NAD-linked dehydrogenases and two phosphorylation sites from succinate. The alternate oxidase (SHAM-sensitive) pathway has only one phosphorylation site.

ties of respiration in many other plant tissues including leaves. Since this pathway produces only about one-third as much ATP for each pair of electrons transported to oxygen, understanding the possible function and regulation of this pathway would appear to be of great importance in plant metabolism.

Alternate pathways of respiration. In addition to anaerobic glycolysis and the Krebs cycle, there are two other sequences of biochemical reactions related to respiration that are important in plant tissues: (1) The pentose phosphate pathway permits an alternate mechanism for converting hexose phosphate to pyruvate, and (2) in germinating fatty seeds the reactions of the Krebs cycle are modified so that acetyl coenzyme A is converted to succinic acid and then to hexose by a pathway called the glyoxylate cycle.

Pentose phosphate pathway. Most plant tissues have abundant glucose-6-phosphate dehydrogenase, an enzyme that catalyzes the reactionshown in (4).

$$\text{Glucose-phosphate} + \text{TPN} \rightarrow \text{6-phosphogluconate} + \text{TPNH}_2 \quad (4)$$

A second enzyme, 6-phosphogluconate dehydrogenase, carries out an oxidative decarboxylation at C-1 of phosphogluconate to produce pentose phosphate according to reaction (5).

$$\text{6-phosphogluconate} + \text{TPN} \rightarrow \text{Ribolose-5-phosphate} + CO_2 + \text{TPNH}_2 \quad (5)$$

Additional enzymes are present in many plant tissues, for example, pea and spinach leaves, that convert the pentose into hexose and triose. The latter may then end up as pyruvic acid by the usual reactions of anaerobic glycolysis. The net result of the pentose phosphate pathway is thus to provide an alternate mechanism of converting hexose to triose. However, the CO_2 released this way is derived from the C-1 of hexose, while CO_2 evolved during anaerobic glycolysis comes first from the C-3 and C-4 of hexose. Then during the operation of the Krebs cycle CO_2 would arise next from C-2 and C-5, and finally from C-1 and C-6 of hexose.

Hence if tissues are supplied with radioactive ^{14}C-glucose labeled either in the C-1 or C-6 position, the relative rates at which the ^{14}C is found in respiratory CO_2 from these two kinds of molecules provides a rough estimate of the relative contributions of the glycolytic and pentose phosphate pathways. If the ratio of the $^{14}CO_2$ from C-6/C-1 equals 1.0, only glycolysis is assumed to occur; but if this ratio is less than 1.0, it suggests that a portion of the respiratory

CO_2 derived from the alternate pathway. In many plant tissues examined by this method the C-6/C-1 ratio was often less than 1.0, especially in older aerial parts; meristematic tissues had ratios of 1.0.

Glyoxylate cycle. Germinating fatty seeds, such as caster bean, do not oxidize the fatty acids completely to CO_2, and this helps to account for the low RQ (Table 1). Instead, they convert much of their stored fatty acids into carbohydrates that are essential for the growing embryo. The biochemical reactions responsible for the initial steps of this pathway occur in microbodies, known as glyoxysomes, found in the endosperm of the seed. Acetyl coenzyme A derived from fatty-acid breakdown reacts enzymically with oxaloacetic acid to yield citric acid and then isocitric acid, as in the Krebs cycle. The enzyme isocitrate lyase then catalyzes reaction (6).

$$\underset{\text{Isocitric acid}}{\begin{array}{c}COOH\\|\\HOCH\\|\\HC-COOH\\|\\CH_2\\|\\COOH\end{array}} \longrightarrow \underset{\text{Succinic acid}}{\begin{array}{c}COOH\\|\\CH_2\\|\\CH_2\\|\\COOH\end{array}} + \underset{\text{Glyoxylic acid}}{\begin{array}{c}CHO\\|\\COOH\end{array}} \qquad (6)$$

Another molecule of acetyl coenzyme A in the presence of the enzyme malate synthetase reacts with the glyoxylic acid produced, as shown in reaction (7). Malic acid is readily oxidized (with DPN)

$$\underset{\text{Glyoxylic acid}}{\begin{array}{c}CHO\\|\\COOH\end{array}} + \underset{\text{Acetyl CoA}}{\begin{array}{c}CH_3\\|\\C=O\\|\\S-CoA\end{array}} \longrightarrow \underset{\text{Malic acid}}{\begin{array}{c}COOH\\|\\HOCH\\|\\CH_2\\|\\COOH\end{array}} + CoA-SH \qquad (7)$$

to oxaloacetic acid, which is needed to regenerate the cyclic process. Hence the net effect of the glyoxylate cycle is shown in reaction (8).

$$2\ \text{Acetyl CoA} + \text{DPN} \rightarrow \text{succinic acid} + 2\ \text{CoA} + \text{DPNH}_2 \qquad (8)$$

The succinic acid thus produced from acetyl coenzyme A is converted by several consecutive reactions into phosphoenolpyruvic acid, and this substance then produces carbohydrate by a reversal of the glycolytic sequence.

Copper-containing oxidases. Plant cells frequently become discolored after injury; this "browning" is caused by the oxidation of phenolic compounds catalyzed by the phenol oxidases. These enzymes contain copper as the prosthetic group and are inhibited by the same inhibitors as cytochrome oxidase, but inhibition by carbon monoxide is not reversed by light. During oxidation of the phenolic substrate to the corresponding quinone, the cupric form of the enzyme is reduced to the cuprous form, and the reduced copper is then reoxidized by O_2. Although this enzyme is widely distributed in leaves, fruits, and tubers, it does not appear likely that it participates normally as a carrier between respiratory substrates and O_2; it may, however, function in the synthesis of lignin and in providing disease resistance.

Ascorbic acid oxidase is another copper-containing enzyme that is widely distributed in plants, frequently being associated with cell walls. The substrate, vitamin C, is reduced to dehydroascorbic acid by the enzyme with uptake of O_2. With enzymes isolated from plants, systems have been reconstructed whereby the ascorbic acid–oxidase reaction can be linked with oxidation of $TPNH_2$ produced by respiratory dehydrogenase reactions. However, such a role for this oxidase has not yet been proved in intact tissues.

Glycolate oxidase. The oxidation of glycolic acid, an early product of photosynthesis, is carried out by a flavoprotein enzyme that is universally present in higher green plants and combines with O_2, as shown in reaction sequence (9).

$$\underset{\text{Glycolic acid}}{\begin{array}{c}CH_2OH\\|\\{}^*COOH\end{array}} + O_2 \longrightarrow \underset{\text{Glyoxylic acid}}{\begin{array}{c}CHO\\|\\{}^*COOH\end{array}} + H_2O_2 \longrightarrow \underset{\text{Formic acid}}{HCOOH} + {}^*CO_2 \qquad (9)$$

The glycolic acid is first oxidized to glyoxylic acid and hydrogen peroxide, and then the hydrogen peroxide oxidizes the glyoxylic acid to formic acid and CO_2. The CO_2 is thus derived from the carboxyl-carbon atom of glycolic acid. Evidence indicates that the glycolate oxidase reaction is the primary source of respiratory CO_2 by many photosynthetic tissues in light during the process known as photorespiration. *See* PHOTORESPIRATION; PHOTOSYNTHESIS; PLANT GROWTH; PLANT METABOLISM. Israel Zelitch

Bibliography. H. Beevers, *Respiratory Metabolism in Plants*, 1961; A. L. Moore and R. B. Beechley (eds.), *Plant Mitochondria: Structural, Functional, and Physiological Aspects*, 1987; J. M. Palmer, The organization and regulation of electron transport in plant mitochondria, *Annu. Rev. Plant Physiol.*, 27:133–157, 1976; F. B. Salisbury and C. W. Ross, *Plant Physiology*, 4th ed., 1992; I. Zelitch, *Photosynthesis, Photorespiration, and Plant Productivity*, 1971.

Plant taxonomy

The area of study focusing on the development of a classification system, or taxonomy, for plants based on their evolutionary relationships (phylogeny). The assumption is that if classification reflects phylogeny, reference to the classification will help researchers focus their work in a more accurate manner. This principle is illustrated by the example of a mother and daughter, with a great deal being known about the genetic traits of the mother. The genetic traits exhibited by the daughter can be largely predicted in this case, whereas for two people of unknown relationships no conjecture is possible regarding their genetic similarity. The task is to make phylogeny reconstruction as accurate as possible. How this phylogenetic system of classification is to be constructed has been subject to varying opinions over time. The basic unit of classification is generally accepted to be the species, but how a species should be recognized has been intensely debated. A historical perspective is useful in explaining the nature of this controversy. *See* PLANT KINGDOM; PLANT PHYLOGENY.

History. The earliest classifications of plants were those of the Greek philosophers such as Aristotle (384–322 B.C.) and Theophrastus (372–287 B.C.). The latter is often called the father of botany largely because he listed the names of over 500 species, some of which are still used as scientific names today. All of Theophrastus' plants were arranged into three categories: herbs, trees, and shrubs. This was a practical solution but one that was not based on an estimate of how closely related the species were. In the next 1600 years, as in most areas of science, little progress occurred in plant taxonomy. It was not until the fifteenth century that there was renewed interest in botany, much of which was propelled by the medical use of plants. Up to and including that time, medicines were taken directly from botanical sources (this is largely true even today). The Doctrine of Signatures stated that the appropriate medical use of plants was indicated by their shape or color; for example, plants such as the hepatic, of which the leaves were shaped like a liver, and the orchid, of which the underground tubers were shaped like testicles, were supposed to have been intended by God to treat liver disease and male infertility.

Historically if a plant was not of medical use, it was likely not classified at all. This was changed by Carolus Linnaeus, a Swedish botanist who in 1753 published his *Species Plantarum*, a classification of all plants known to Europeans at that time. Linnaeus's system was based on the arrangement and numbers of parts in flowers, and was intended to be used strictly for identification (a system now referred to as an artificial classification as opposed to a natural classification, based on how closely related the species are).

Binomial nomenclature. In *Species Plantarum*, Linnaeus made popular a system of binomial nomenclature developed by the French botanist Gaspard Bauhin (1560–1624). In this system, which is still in use, each species has a two-part name, the first being the genus and the second being the species epithet. For example, *Rosa alba* (italicized because it is Latin) is the scientific name of one species of rose; the genus is *Rosa* and the species epithet is *alba*, meaning white (it is not a requirement that scientific names be similar to common names or have real meaning, although such relevance is often the case). The genus name *Rosa* is shared by all species of roses, reflecting that they are thought to be more closely related to each other than to species in any other group.

Our ideas about classification have changed drastically since the time of Linnaeus, but we still use his system of binomial nomenclature today. The rules of botanical nomenclature (and those of zoology as well, although they are not identical) are part of an internationally accepted Code that is revised (minimally) at an international congress every 5 years. *See* PLANT NOMENCLATURE.

Natural classification system. In addition to his system of binomial nomenclature, Linnaeus introduced the Sexual System of classification, based on the number of male and female flower parts. The system was not widely used even in his day, with many of his French contemporaries preferring to use a system based more on assessments of how the plants were related (a more "natural" system). Their concept of relatedness was drastically different from that used today, due to the fact that Charles Darwin's theory of evolution (published in *Origin of Species*, 1859) came much later, but the eighteenth-century French botanists made major improvements in how plants were described and how the various categories above the level of genus were to be defined, and set us on a course to produce the natural taxonomic systems that took on its modern meaning after the time of Darwin. Today, we understand that the best classification system is one that reflects the patterns of the evolutionary processes that produced these plants. *See* PLANT EVOLUTION.

Modern taxonomic categories. Use of common names in science and horticulture is not practical because in each language and country common names will vary; for example, "bluebell" means something different in northern and southern North America as well as in England and Scotland. Scientific names are internationally agreed upon so that a consistent taxonomic name is used everywhere for a given organism. In addition to genus and species, plants are classified by belonging to a family (Rosaceae, the rose family, for example); related families are grouped into orders, and these are typically grouped into a number of yet higher and more encompassing categories. In general, higher categories are composed of many members of lower types—for example, a family may contain 350 genera, but some may be composed of a single genus with perhaps a single species if that species is distantly related to all others. An example is the genus of the sycamore or plane tree, *Platanus*, which is the only genus in its family, Platanaceae.

Many botanists use a number of intermediate categories between the level of genus and family, such as subfamilies, tribes, and subtribes, as well as some between species and genus, such as subgenera and sections, but none of these categories is formally mandated. They are useful nonetheless to reflect intermediate levels of relatedness, particularly in large families (composed of several hundreds or even thousands of species). Below the level of species, some botanists use the concept of subspecies (which is generally taken to mean a geographically distinct form of a species) and variety (which is often a genetic form or genotype, for example a white-flowered form of a typically blue-flowered species, or a form that is ecologically distinct).

Phylogenetic classification. In the twentieth century, several approaches to constructing a natural system of classification were used. For many years, there were several competing systems that were generally similar in overall ideas of how plants were related but that differed in minor, though important, ways. Circumscription of many families has been highly consistent since the French botanists of the seventeenth century, but the grouping of families into orders has been much more variable. For example, the important agronomic family comprising the legumes (Fabaceae: beans, peas, lentils, and the like) has been presented as most closely related to several different families, such as Connaraceae, Polygalaceae, Rosaceae, and Saxifragaceae, depending on the author. The reason for these differing ideas was reliance on one or a few characteristics as the most important indicators of relationships. The different characters, such as pollen and floral morphology (shape and structure), growth form (tree, shrub, herb), and chemistry (whether alkaloids or steroids were present, for example), did not appear to reflect the same patterns of relatedness, and authors were likely to reach different conclusions based on their opinions of which characteristics were more important. *See* PLANT ANATOMY; PLANT PHYSIOLOGY.

The basic idea that plant classification should reflect evolutionary (genetic) relationships has been well accepted for some time, but the degree to which this could be assessed by the various means available (by studying morphology, chemistry, or chromosome numbers, for example) differed. It has only recently become possible to assess genetic patterns of relatedness directly by analyzing DNA sequences. Although most botanists have supported the idea that classification should summarize and synthesize all available information, this was rarely attempted, even though in the latter half of the twentieth century computerized methods became available to construct such syntheses. Part of the problem was the incompleteness of the information base available; some families of plants were much better studied than others (often due to the number of economically important species in a family), and completing these surveys was time-consuming and expensive. As a result, attempts to perform computerized analyses of large groups of plants, such as the flowering plants (angiosperms), were rare, and the missing data raised questions about the validity of these studies. The result was that plant taxonomy continued to be practiced as it had in the late nineteenth century, and the plethora of contradictory classifications frustrated scientists who wished to consider the evolutionary implications of their research.

DNA technology. In the 1990s, DNA technology became much more efficient and less costly, resulting in a dramatic upsurge in the availability of DNA sequence data for various genes from each of the three genetic compartments present in plants (nuclear, mitochondrial, and plastid or chloroplast). In 1998 a number of botanists collectively proposed the first DNA-based classification of a major group of organisms, the angiosperms or flowering plants. For the first time, a classification was directly founded on assessments of the degree of relatedness made with objective, computerized methods of phylogeny reconstruction. Other data, such as chemistry and morphology, were also incorporated into these analyses, but by far the largest percentage of information came from DNA sequences—that is, relatedness was determined mostly on the basis of similarities in plants' genetic codes. The advantages of such a classification were immediately obvious: (1) it was not based on intuition about which category of information best reflected natural relationships; (2) it ended competition between systems based on differing emphases; (3) the analysis could be repeated by other researchers using either the same or different data (other genes or categories of information); and (4) it could be updated as new data emerged, particularly from studies of how chromosomes are organized and how morphology and other traits are determined by the genes that code for them. *See* DEOXYRIBONUCLEIC ACID (DNA); GENETIC MAPPING.

Cladistics. At the same time that DNA data became more widely available as the basis for establishing a classification, a more explicit methodology for turning the results of a phylogenetic analysis into a formal classification became popular. This methodology, called cladistics, allowed a large number of botanists to share ideas of how the various taxonomic categories could be better defined. Although there remain a number of dissenting opinions about some minor matters of classification, it is now impossible for scientists to propose alternative ideas based solely on opinion. *See* TAXONOMY.

Mark W. Chase; Michael F. Fay

Bibliography. D. Bhattacharya (ed.), *Origin of Algae and Their Plastids*, Springer-Verlag, 1997; L. E. Graham and L. W. Wilcox, *Algae*, Prentice Hall, 2000.

Plant tissue systems

Most plants are composed of coherent masses of cells called tissues. Large units of tissues having some features in common are called tissue systems. In actual usage, however, the terms tissue and tissue system are not strictly separated. A given tissue or a combination of tissues may be continuous throughout the plant or large parts of it.

Although classification of tissue systems may be based on structure or function, the two aspects usually are combined. Plant tissues are primary or secondary in origin. The primary arise from apical meristems, the perennially embryonic tissues at the tips of roots and shoots (**Fig. 1**). The primary tissues include the surface layer, or epidermis; the primary vascular tissues, xylem and phloem, which conduct

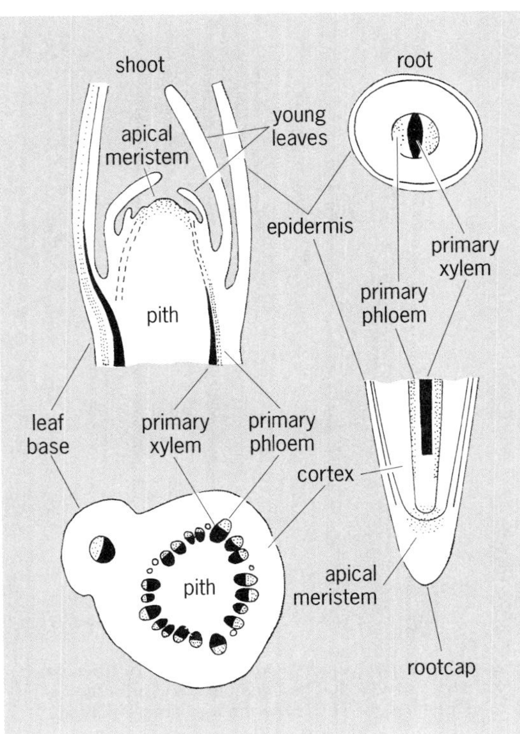

Fig. 1. Primary plant tissue systems.

water and food, respectively; and the ground tissues. The ground tissues are parenchyma (chiefly concerned with manufacture and storage of food) and collenchyma and sclerenchyma (the two supporting tissues). In the stem and root, the vascular tissues and some associated ground tissue are often treated as a unit, the stele. Ground tissue may be present in the center of the stele (pith) and on its periphery (pericycle). The ground tissue system enclosing the stele on the outside is the cortex. It may have a hypodermis peripherally and an endodermis next to the stele.

The secondary tissues (**Fig. 2**) arise from lateral meristems, and their formation is mainly responsible for the growth in thickness of stems and roots. They comprise secondary vascular tissues and the protective tissue called periderm. Secondary growth may build up a massive core of wood, but the outer tissue system, the bark, remains relatively thin because its outer or older part becomes compressed and, in many species, is continuously sloughed off.

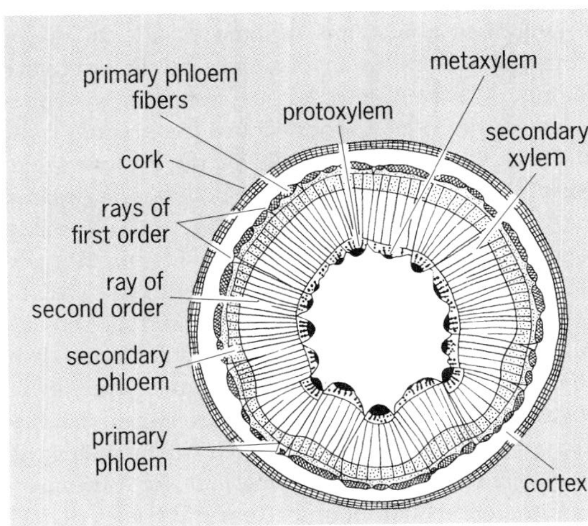

Fig. 2. Cross section of *Prunus* stem, showing secondary tissues, which arise from lateral meristems.

The production of flowers instead of vegetative shoots results from physiological and morphological changes in the apical meristem, which then becomes the flower meristem. The latter, however, produces tissue systems fundamentally similar to those in the vegetative body of the plant. See separate articles for detailed discussion of the various plant tissue systems. Katherine Esau

Plant transport of solutes

The movement of organic and inorganic compounds through plant vascular tissues. Transport can take place over considerable distances; in tree species transport distances are often 100–300 ft (30–100 m).

This long-distance transport is necessary for survival in higher land plants in which specialized organs of uptake or synthesis are separated by a considerable distance from the organs of utilization. Diffusion is not rapid enough to account for the amount of material moved over such long distances. Rather, transport depends on a flowing stream of liquid in vascular tissues (phloem and xylem) that are highly developed structurally.

The movement of organic solutes occurs mainly in the phloem, where it is also known as translocation and where the direction of transport is from places of production, such as mature leaves, to places of utilization or storage, such as the shoot apex or developing storage roots. Organic materials translocated in the phloem include the direct products of photosynthesis (sugars) as well as compounds derived from them (nitrogenous compounds and plant hormones, for example). Some movement of organic solutes does occur in the xylem of certain species. Inorganic solutes or mineral elements, however, generally move with water in the xylem from sites of uptake in the roots to sites where water is lost from the plant, primarily the leaves. Some redistribution of the ions throughout the plant may then occur in the phloem.

Organic Solutes

Organic compounds are found in both the phloem and xylem of higher plants; the relative importance of these tissues as conducting channels in organic transport depends on both the nature of the organic compound transported and the site of origin of the compound in the plant.

Importance of phloem. The phloem is the predominant tissue for the translocation of organic solutes, particularly sugars. This has been established convincingly by data from ringing (girdling) experiments

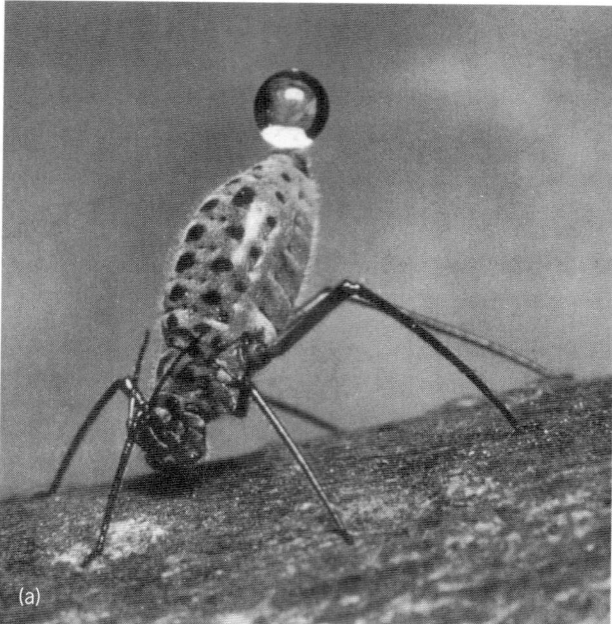

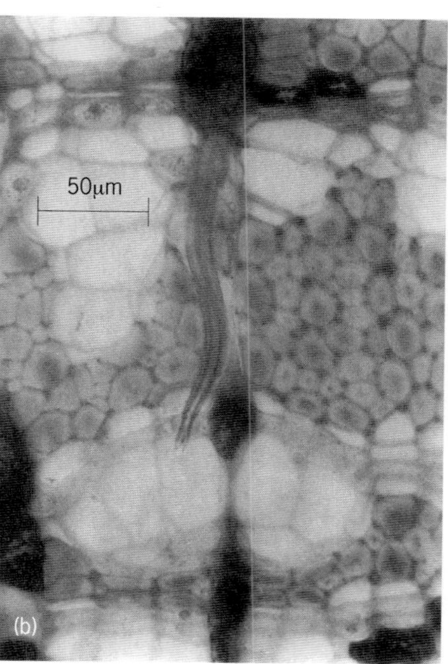

Fig. 1. Phloem-feeding aphid. (*a*) Droplet of sieve-tube sap being released by the aphid *Longistigma caryae.* (*b*) Transverse section through the bark of *Tilia americana,* showing the tip of an aphid stylet bundle that had been exuding prior to sectioning. Note that the tips of the stylets are inside a single sieve-tube member. (*M. H. Zimmermann, Harvard Forest, Harvard University*)

in which the continuity of the phloem is interrupted by removal of a narrow ring of bark external to the xylem. While the bark on the side of the girdle nearest the leaves swells somewhat and remains healthy, the bark on the side of the girdle which is separated from the leaves dries up and eventually dies due to a lack of organic nutrients. More elaborate experiments using radioactive tracers also support the conclusion that the sieve elements of the phloem are the major transport pathway for the conduction of sugars and other organic nutrients synthesized in the leaves. *See* PHLOEM.

The nature of the organic substances translocated in the phloem has been determined primarily by chromatographic analyses of sieve-element exudates collected from incisions in the bark or from severed stylets of phloem-feeding aphids (**Fig. 1**). Although the phloem sap obtained by incision or aphid stylet is probably not identical to the contents of the undisturbed sieve element, it does give data consistent with radioactive tracer experiments and thus appears to provide a valid picture of the compounds in transit.

In most plant species, sugars form the bulk of the substances moved in the phloem, with sieve-element sap containing 10–25% sugar by weight. Sucrose is the dominant sugar in the sap of most species analyzed. However, some species translocate nonreducing oligosaccharides or sugar alcohols, or both, as well as sucrose. The oligosaccharides belong to the raffinose family and include raffinose, stachyose, and verbascose. These sugars form a series consisting of sucrose (a disaccharide) with one, two, or three galactose residues joined to form the tri-, tetra-, and pentasaccharides. The transported sugar alcohols are generally mannitol or sorbitol. Athough the hexoses glucose and fructose are frequently abundant throughout the remainder of the plant, analysis of sap from many different species reveals a consistent absence of hexose sugars in the sieve elements. Other organic substances are found in the sieve-element sap in addition to sugars, but these are generally much less concentrated than the carbohydrates. Included are nitrogenous substances, particularly amino acids and amides, as well as plant hormones, vitamins, and sugar phosphates. *See* CARBOHYDRATE; OLIGOSACCHARIDE.

Role of xylem. Xylem sap may also contain organic substances, including nitrogenous compounds, carbohydrates, and plant hormones, especially cytokinins and gibberellins. In some species, the xylem constitutes an important channel for transport to the shoots of organic nitrogenous compounds synthesized in the roots. Carbohydrates transported from the leaves via the phloem provide the carbon "skeletons" for these compounds; the necessary inorganic nitrogen either is absorbed from the soil or is made available to the plant by nitrogen-fixing microorganisms. *See* NITROGEN FIXATION; PLANT HORMONES.

In other species, however, organic nitrogenous solutes are more concentrated in the phloem sap than in the xylem sap, and xylem nitrogen is primarily in the form of nitrate. In these species, nitrate that has arrived in the leaves via the xylem stream is then incorporated into organic compounds. Redistribution of nitrogenous compounds from the leaves occurs mainly in the phloem in all species. During senescence (yellowing and progessive deterioration) of leaves, large amounts of nitrogen- and phosphorus-containing organic compounds are exported from the leaves in the phloem tissue.

During translocation through the phloem of the stem of a plant, a considerable quantity of sugar may move into lateral tissues, such as the cortex, phloem parenchyma, or xylem parenchyma cells, and be stored as starch available for later remobilization. In spite of the proximity of the phloem and of carbohydrate reserves, only negligible quantities of sugar appear in sap from the xylem elements during most of the growing season; the average maximal concentration seldom exceeds 0.05%. During the dormant season, however, when the flow of water in the xylem elements practically ceases, sugars often accumulate in readily detectable amounts. As an extreme example, in sugar maple (*Acer saccharum*) a concentration as high as 8% has been reported in the spring before the leaves emerge. Even in this species, however, the phloem is still the most important channel for transport of sugars. *See* XYLEM.

Source–sink relationships in phloem. In general, organic compounds are translocated in the phloem from areas called sources, which are regions of synthesis or mobilization, to areas of utilization called sinks, such as young leaves, root tips, shoot tips, and developing fruits. Mature photosynthesizing leaves are the major sources; materials are transported from a given source leaf (S) both acropetally and basipetally, that is, in the direction of the shoot tips and root tips, respectively (**Fig. 2**). The distribution pattern from a leaf also depends on the pattern of vascular connections with other organs (**Fig. 3**). *See* PHOTOSYNTHESIS.

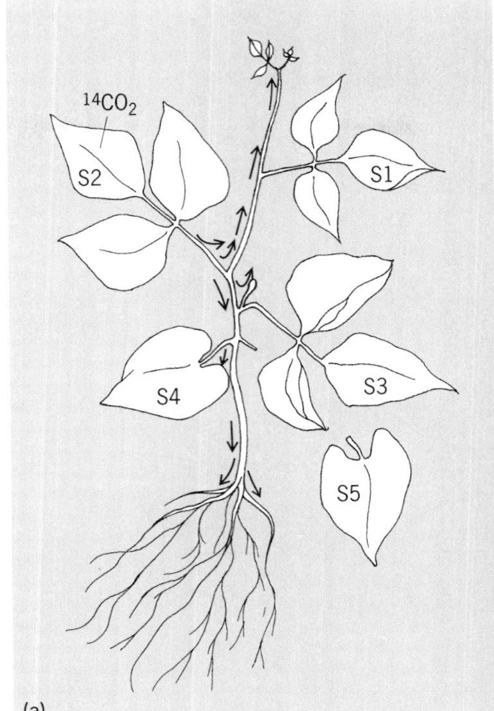

Fig. 2. Movement of ^{14}C sucrose through a bean plant. Sources are indicated by S1–S5. $^{14}CO_2$ was supplied to the youngest fully mature leave (S2) of a bean plant for 2 h in the light. Radioactive sucrose was synthesized from the $^{14}CO_2$ by the leaf and translocated throughout the plant. The plant was then dismembered, dried, mounted on a piece of paper, and placed next to x-ray film for a few days. (*a*) Mounted bean plant. The arrows indicate movement from the treated leaf S2 to the various sinks. (*b*) Developed and printed x-ray film, with radioactivity revealed (white). (*D. R. Geiger, University of Dayton*)

Early in its development a leaf is dependent on the import of a significant quantity of organic material via the phloem. Not until the leaf is about half expanded does it begin to export the products of photosynthesis. Export starts in the tip region of the leaf and extends baseward until the entire leaf is exporting. For a time the blade is both importing and exporting organic material. A mature leaf may export nearly half of the carbohydrate produced during photosynthesis; much of the rest will be accumulated as starch, which will contribute to translocated materials during darkness. Only small amounts are used daily for the metabolic needs of the leaf cells. *See* LEAF.

The process by which sugars enter the conducting cells of the phloem in source regions is known as phloem loading. Loading requires metabolic energy, since the sugar concentrations are higher in the sieve elements than in the surrounding cells of the source. The mechanism of loading has been most widely studied in sugarbeet and corn leaves. In these species, sucrose exits the photosynthesizing cells of the leaf and enters the cell-wall space (apoplast) in the immediate vicinity of the smallest veins of the leaf. Sucrose enters the sieve elements of the phloem from the apoplast by first binding to a specific carrier protein located in the cell membrane of the sieve elements and associated parenchyma cells. The carrier transfers the sucrose across the membrane to the interior of the conducting cells and releases it to the long-distance transport stream. In sugarbeet and corn the carrier does not directly utilize metabolic energy; rather, energy is expended to establish conditions in which the carrier can operate. Unloading is the process by which sugars leave the phloem in sink regions. Pathways of unloading and the site of energy utilization during unloading differ depending on the species, organ, and developmental stage studied. For example, the process of unloading appears to differ in growing tissues, such as roots and young leaves, and in storage tissues, such as sugarcane stems.

Allocation and partitioning of organic solutes. Research on the transport of organic solutes emphasizes the study of two processes: allocation and partitioning. Allocation refers to the fate of photosynthetic products in both source and sink leaves. For example, in source leaves some of the sugar formed during photosynthesis is used to synthesize storage materials, such as starch, for use by the plant during the night period, and some is used to synthesize transport sugars, including sucrose, for translocation out of the leaf. Partitioning refers to the relative distribution of transport sugars among the various sink tissues of the plant. Both of these

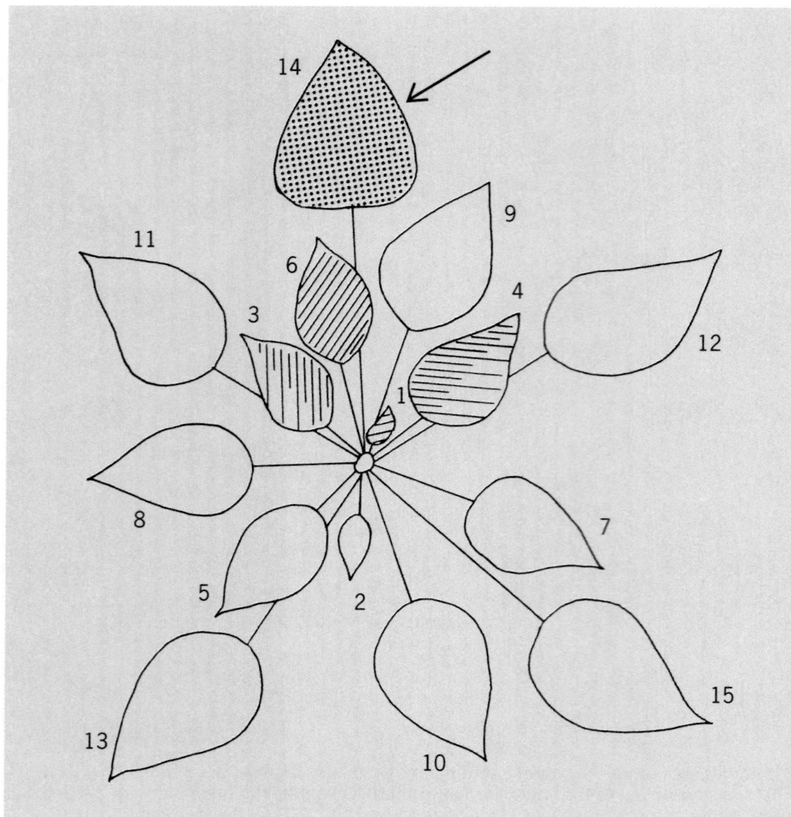

Fig. 3. Distribution of radioactivity in the leaves of a beet plant 1 week after administration of $^{14}CO_2$ for 4 h to a single leaf, indicated by arrow. Hatching indicates the approximate intensity of radioactivity from autoradiographs of leaves. Most of the ^{14}C is translocated to young leaves on the same side of the plant as the labeled source leaf. The leaves are numbered to show relative age, with 1 being the youngest.

processes are important because they determine the proportion of photosynthetic product that will reach edible portions of the plant and because they are a factor in improved crop productivity.

Allocation. Carbon dioxide that is converted into organic compounds (sugars) during photosynthesis is said to be fixed during the procedure. The mechanisms that determine the fate of this fixed carbon throughout the plant constitute a complex and finely tuned system. One example of such control in source leaves will be examined here. Rates of sucrose and starch synthesis must be coordinated in source leaves, and part of the allocation between translocation and storage occurs via a transport protein, called the phosphate translocator, located in the membrane of the chloroplast. The process of sucrose synthesis in the cytoplasm of a source cell results in the release of inorganic phosphate in the cytoplasm. The phosphate translocator exchanges phosphate in the cytoplasm for one of the early products of photosynthesis from the chloroplast. In this way, carbon destined for export from the leaf exits the chloroplast, and phosphate required during photosynthesis enters the chloroplast. Whenever the demand for sucrose by other parts of the plant is high, the rate of sucrose synthesis in the source tends to be high, and more phosphate is released and exchanged for carbon from the chloroplast. This sequence of events serves to divert fixed carbon away from the synthesis of starch when carbon is required in other parts of the plant. Other regulatory processes ensure that the rate of starch synthesis is maintained and that all the available fixed carbon is not drained away from the chloroplast. In sink leaves, imported carbon compounds either are stored in various forms or are utilized for metabolic processes, depending on the type of sink involved.

Partitioning. Source and sink regions interact within a single plant to control rates of export and to regulate the distribution of photosynthate between various sinks. Increasing the photosynthetic production of sugars in a source leaf, by increasing the light intensity or carbon dioxide level around the leaf, increases the rate of sugar export by the source. Increasing the temperature of a given sink region increases the proportion of photosynthate arriving at that sink from the source areas. Experiments which alter the balance between source and sink regions demonstrate the interactions between the two regions. Removing or shading all the source leaves except one on a plant usually results in an immediate redistribution of the assimilate exported from the remaining active source, with no change occurring in export rate. In these short-term experiments, the sink tissues appear to compete for available sugars, with the strongest sinks diverting more of the translocation stream to themselves. Increased sink strength may depend in part on the ability of the stronger sink tissues to lower the turgor pressure in the translocating cells to a greater extent. Over the long term (on the order of 2 days or more), removing or shading all the source leaves except one leads to compensatory increases in both photosynthesis and export by the remaining active source. Long-term source–sink interactions are probably mediated by plant hormones, such as cytokinins, and by nutrients, such as potassium, phosphate, and sucrose.

Mechanism of phloem translocation. The mechanism of phloem translocation is not known with certainty. Proposed mechanisms fall into two classes: one stresses the role of the conducting tissues in generating the moving force, and the other views the regions of supply and utilization as the source of this force. In the former group are mechanisms that depend on cytoplasmic streaming, electroosmosis, and activated diffusion in the sieve elements. The second group of theories, which has received more general acceptance in spite of a number of admitted limitations, includes a variety of mass-flow mechanisms based on a theory developed principally by E. Münch about 1930. Theories of translocation must account for the following important observations: polarity, bidirectional movement, velocity, energy requirement, turgor pressure, and phloem structure.

Polarity. Organic solutes move in the phloem from sources to metabolically active sinks, where growth or accumulation is occurring. The direction of translocation is determined more by physiological states of source and sink areas than by structural relations.

Bidirectional movement. Translocation in the conducting tissues may occur in opposite directions. Solutes may move both up and down a stem or petiole simultaneously, even in the same phloem bundle, but it has not been shown that this bidirectional movement can occur within a single sieve element.

Velocity. Translocation is rapid and conveys large quantities of solutes. Velocity measurements with radioactively labeled compounds and by growth rate studies reveal that the velocity of translocation is about 0.4–0.8 in./min (1–2 cm/min), with velocities over 7.9 in./min (20 cm/min) reported for a rapid component of translocation. The mass-transport rate for sugarbeet has been measured at 9 mg sucrose/h out of a 15.5-in.^{2} (100-cm^2) leaf. In the sausage tree (*Kigelia africana*) as much as 1.1 oz/day (32 g/day) of organic material must pass through the slender stem to account for the dry-weight gain of the cluster of four fruits. These rates are many orders of magnitude greater than could be explained by diffusion.

Energy requirement. The high rates of directed mass transfer necessitate a considerable expenditure of energy both to move the stream of solutes and to maintain the structural integrity of the conducting tissue. Living phloem cells are necessary for translocation of organic solutes in sieve elements. Killing the cells in a localized zone of the stem by heating is as effective a barrier to transport of phloem-limited solutes as is ringing. Treatments such as cooling lower the metabolism of a tissue and are used to evaluate the role of energy in transport in the treated region. In cold-tolerant plants, translocation continues at an undiminished rate through a petiole held at 32°F (0°C) for many hours, indicating that energy expenditure in the conducting tissue is primarily required for maintenance of structure of the phloem rather than for moving the translocate stream. On the other hand, cooling source or sink regions slows translocation markedly. *See* COLD HARDINESS (PLANT).

Turgor pressure. Sieve-element contents are under positive turgor pressure. Pressures have been measured in the phloem of tree species up to 2.0 megapascals (20 atm), and in herbaceous species up to 1.0 MPa (10 atm).

Phloem structure. The sieve elements of flowering plants (angiosperms) lack a nucleus, have a specialized protoplast, and are interconnected by areas called sieve plates (**Fig. 4**). Because of their high degree of specialization, sieve elements are not expected to be a major source of energy for longitudinal transport. However, the phloem parenchyma cells associated with terminal regions of the phloem (companion cells) are large and have dense cytoplasm with numerous mitochondria. These cells appear to be well suited to controlling the concentration of solutes in the sieve elements and to supplying energy for loading and perhaps unloading as well. While evidence for the presence of cytoplasmic strands passing through sieve plates and traversing sieve elements in the past led to interest in theories that provide a role for such strands, many modern studies have shown the sieve plates to be essentially devoid of cytoplasmic material (Fig. 4).

Mass-flow theory. Although none of the theories proposed to explain phloem translocation are without limitations, and much work in evaluating them remains, the mass-flow theory, as modified by several investigators, has received wide acceptance. According to this theory, sieve elements accumulate solutes by phloem loading which requires metabolic energy. Companion cells associated with the phloem endings in source regions also accumulate solutes by phloem loading and deliver them to the sieve elements. At these sites the sugar concentration in the sieve elements is high, and this causes the entry of water by osmosis from surrounding cells and from the xylem in particular; the resulting hydrostatic pressure causes water and dissolved solutes to flow along under the force of turgor pressure. At points of utilization or storage, solutes are removed from the sieve elements, probably by a process requiring metabolic activity. The resultant lowering of solute concentration causes a lowering of the turgor pressure, also aiding translocation stream movement (**Fig. 5**). Information that supports this theory includes the requirement for metabolic energy in sources and sinks but not in path tissues to drive translocation, and evidence for open pores in sieve plates which thus provide a low-resistance pathway for solution flow. However, the mass-flow theory is not tenable for all species; for example, the sieve areas in gymnosperm sieve cells appear to be obstructed.

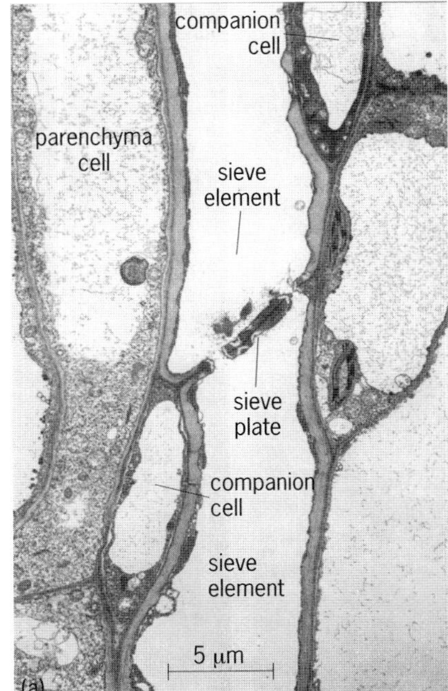

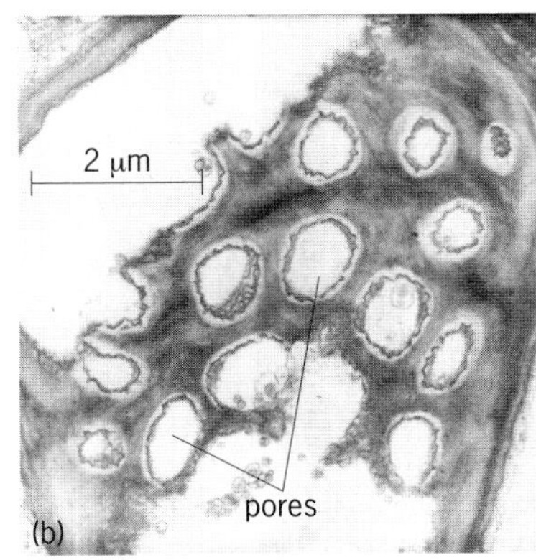

Fig. 4. Phloem in the stem of squash, as seen in electron micrographs. (*a*) Longitudinal view of parts of two mature sieve elements and a sieve plate (*from R. F. Evert, W. Eschrich, and S. E. Eichhorn, P-protein distribution in mature sieve elements of Cucurbita maxima, Planta, 103:193–210, 1968*). (*b*) Face view of a simple sieve plate (one sieve area per plate) between two mature sieve elements (*courtesy of R. F. Evert*).

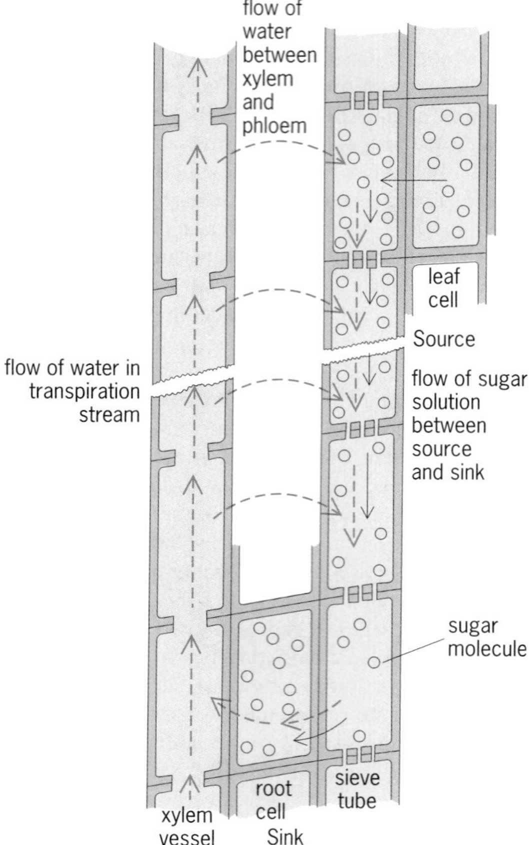

Fig. 5. Diagram of mass-flow theory; water flow is represented by colored arrows and sugar flow by black arrows. (*After P. H. Raven, R. F. Evert, and S. E. Eichhorn, Biology of Plants, 4th ed., Worth Publishers, 1986*)

Inorganic Solutes

Inorganic solutes or mineral elements are found in both the phloem and xylem of higher plants. The initial distribution of minerals following uptake by the roots occurs within the vessels of the xylem and involves a general distribution to all transpiring surfaces. Transpiration is the loss of water from plants in the form of vapor, principally from the leaves, and is essentially an evaporation process. A more specific redistribution of some elements may then occur within the phloem.

Importance of phloem and xylem. In the ringing experiments mentioned above, the bark containing the phloem can be removed from a shoot with no immediate effects on the movement of water and dissolved mineral elements from the roots to the shoots. From these experiments and from experiments with radioactive tracers (**Fig. 6**), it can be concluded that water and dissolved solutes move upward primarily in the tissues of the xylem.

Xylem sap can be collected for analysis in two ways. In herbaceous plants, xylem sap exudes from the root system of detopped plants, driven by a phenomenon called root pressure, and can be easily collected. Sap can also be extracted from shoots and twigs by suction. Sap collected by either method contains relatively low concentrations of dissolved solutes (0.05–0.40% by weight) compared to phloem sap. Most of the solutes in xylem sap are inorganic cations and anions, with potassium, calcium, magnesium, and sodium as the major cations, and with phosphate, chloride, sulfate, and sometimes nitrate as the major anions.

Analysis of phloem sap and studies of export of radioactively labeled ions from leaves indicate that some mineral ions move in the phloem as well. Potassium, magnesium, and phosphate are the major ions moving extensively in the phloem. Other elements such as sulfate and nitrate appear to move in the phloem primarily in organic forms, while still others, calcium being one example, move very little in phloem tissues. The low concentration of calcium in the phloem results primarily from its inability to enter the sieve elements, and is further maintained by the low solubility of calcium phosphate salts. One consequence of the differing conduction of elements in the phloem is that the symptoms of mineral deficiencies can differ according to the phloem conduction of the elements involved. Sink tissues, including young leaves, can withdraw nutrients from older leaves via the phloem. Thus, when a plant is deficient in an element which can be transported in the phloem, the symptoms appear earlier and are most

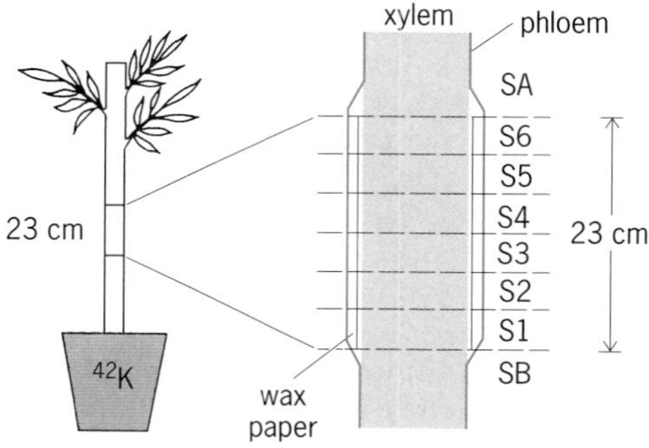

stripped branch (xylem and phloem separated by wax paper)

		ppm ^{42}K in phloem	ppm ^{42}K in xylem
above strip	SA	53	47
stripped section	S6	11.6	119
	S5	0.9	122
	S4	0.7	112
	S3	0.3	98
	S2	0.3	108
	S1	20	113
below strip	SB	84	58

Fig. 6. Demonstration, using radioactive potassium supplied to the roots, that xylem is the channel for the upward movement of water and minerals. Wax paper is inserted between the xylem and phloem to prevent lateral transport of the isotope. The relative amounts of radioactive potassium in each of the segments (S) is shown at the bottom. (*After P. H. Raven, R. F. Evert, and S. E. Eichhorn, Biology of Plants, 4th ed., Worth Publishers, 1986*)

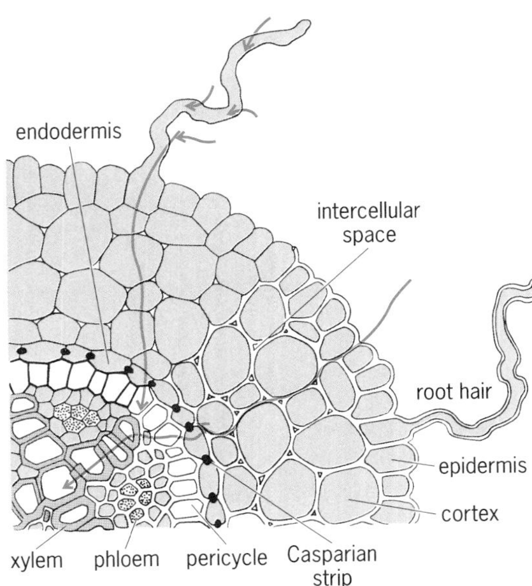

Fig. 7. Cross section of a root, showing the pathway of uptake of water and minerals. Note that water and ions may be taken up into the cytoplasm of the root cells even before they reach the barrier of the Casparian strip (upper part of figure). (*After P. H. Raven, R. F. Evert, and S. E. Eichhorn, Biology of Plants, 4th ed., Worth Publishers, 1986*)

pronounced in the older leaves. On the other hand, the deficiency symptoms of an ion which is not transported in the phloem initially affect young, actively growing tissues.

Entry into xylem. The absorption of inorganic nutrients by plants involves several sequential steps: movement of ions through the soil to root surfaces (by diffusion and mass flow); lateral movement across the roots through cell walls and root cells; release to the xylem elements; and movement to the aerial portions of the plant in the xylem stream (**Fig.** 7). The initial movement of mineral ions into the root occurs primarily by diffusion through cell walls external to cell membranes. However, in young roots at least, there exists a barrier to free diffusion of ions across roots. This barrier, called the Casparian strip, consists of an impregnation of the cell wall of the endodermis of the root with suberin, which renders the cell wall impermeable. When they reach this strip, the inorganic solutes cannot continue to cross the root through cell walls but must enter the cytoplasm of the living cells. The solutes then pass from cell to cell until they reach the vascular tissues, where they are released from the cells into the xylem elements (which are dead at maturity).

Although movement of ions within the cell walls and xylem elements is essentially a passive process, the passage across the cell membrane into the cytoplasm of the root cells is a selective transport process, requiring metabolic energy. Another active transport step may be involved when ions leave the cytoplasm of the xylem parenchyma cells and enter the xylem vessels. Thus the uptake process in young roots may show both selectivity and a capacity to accumulate ions because of these membrane transport steps. Although it is generally recognized that most salt absorption occurs in the youngest region of the root (immediately behind the root tip), significant amounts of water and salt do enter through older roots in woody species. For the most part these roots are rendered impermeable by the processes of secondary growth; however, it appears that water and solutes can enter through lenticels, fissures between plates of bark, and openings caused by the death of small branch roots. This nonselective movement occurs, as in the soil, both by diffusion and by mass flow. As a result of this nonselective component of ion uptake, practically all dissolved ions of the root environment are found in varying quantities in the aerial portions of plants.

Mechanism of xylem transport. Once in the xylem elements, inorganic solutes move with water to the transpiring surfaces. The model for the ascent of sap in the xylem which is correct according to all present evidence is called the cohesion hypothesis. According to this hypothesis, water is lost in the leaves by evaporation from cell-wall surfaces; water vapor then diffuses into the atmosphere by way of small pores between two specialized cells (guard cells). The guard cells and the pore are collectively called a stomate. This loss of water from the leaf causes movement of water out of the xylem in the leaf to the surfaces where evaporation is occurring. Water has a high internal cohesive force, especially in small tubes with wettable walls. In addition, the xylem elements and the cell walls provide a continuous water-filled system in the plant. Thus the loss of water from the xylem elements in the leaves causes a tension or negative pressure in the xylem sap. This tension is transmitted all the way down the stem to the roots, so that a flow of water occurs up the plant from the roots and eventually from the soil. The velocity of this sap flow in tree species ranges from 3 to approximately 165 ft/h (1 to 50 m/h), depending on the diameter of the xylem vessels.

The movement of inorganic solutes within the transpiration stream might suggest that all mineral elements move upward at the same rate. However, some ions appear to move more slowly than others due to adsorption on the walls of the xylem elements and accumulation by living cells bordering the xylem elements. Some have suggested that sodium, for example, might be prevented from reaching toxic levels in leaves by uptake in upper parts of roots and lower parts of shoots. In addition, certain ions are accumulated more in the roots of some species and are thus prevented from reaching the xylem stream.

The rate of transpiration may affect the rate of ion uptake. Rapid water movement carries ions in the soil solution to the root surfaces and reduces the salt concentration in the root xylem, facilitating release of ions into the xylem sap.

In the absence of transpiration, the active uptake of ions by root cells may cause an osmotic influx of water into the roots, building up a positive pressure in the xylem elements and providing an alternate driving force for water and ion movement. This phenomenon, called root pressure, only occurs under special conditions and in some species. *See* PLANT MINERAL NUTRITION; PLANT-WATER RELATIONS. S. Sovonick-Dunford

Bibliography. J. W. Einset, *Plant Growth and Development*, 1988; T. J. Flowers and A. Yeo, *Solute Transport in Plants*, 1992; O. Nelson, *Plant Transposable Elements*, 1988; M. A. Schuler and R. E. Zelinski, *Methods of Plant Molecular Biology*, 1988; K. R. Stern, *Introductory Plant Biology*, 7th ed., 1997.

Plant viruses and viroids

Plant viruses are pathogens which are composed mainly of a nucleic acid (genome) normally surrounded by a protein shell (coat); they replicate only in compatible cells, usually with the induction of symptoms in the affected plant. Viroids are among the smallest infections agents known. Their circular, single-stranded ribonucleic acid (RNA) molecule is less than one-tenth the size of the smallest viruses.

Viruses. Viruses can be seen only with an electron microscope (**Figs. 1** and **2**). Isometric (spherical) viruses range from 25 to 50 nanometers in diameter, whereas most anisometric (tubular) viruses are 12 to 25 nm in diameter and of various lengths (200–2000 nm), depending on the virus. The coat of a few viruses [for example, the large and complex rhabdoviruses such as lettuce necrotic yellows virus (LNYV) or the tomato spotted wilt virus] is covered by a membrane which is derived from its host.

Over 800 plant viruses have been recognized and characterized. The genomes of most of them, such as the tobacco mosaic virus (TMV) or the potato virus Y, are infective single-stranded RNAs; some RNA viruses (for instance, the wound tumor virus) have double-stranded RNA genomes. Cauliflower mosaic virus and bean golden mosaic virus are examples of viruses having double-stranded and single-stranded deoxyribonucleic acid (DNA). The genome of many plant viruses is a single polynucleotide and is contained in a single particle, whereas the genomes of brome mosaic and some other viruses are segmented and distributed between several particles. There are also several low-molecular-weight RNAs (satellite RNAs) which depend on helper viruses for their replication. *See* DEOXYRIBONUCLEIC ACID (DNA); RIBONUCLEIC ACID (RNA).

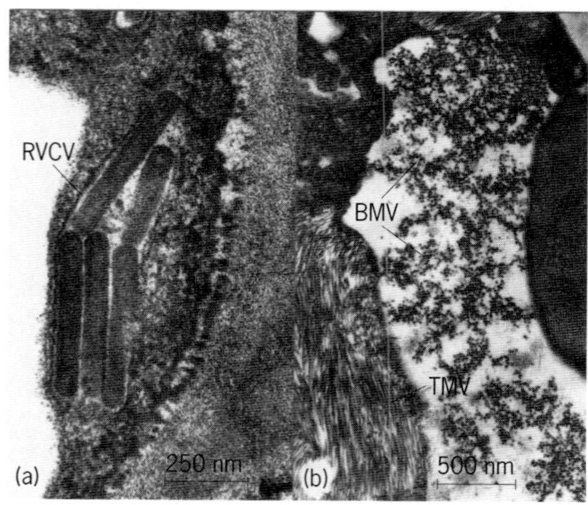

Fig. 2. Infected leaf cells. (*a*) Cell of raspberry showing particles of raspberry vein chlorosis virus (RVCV) (*courtesy of R. Stace-Smith*). (*b*) Leaf cell of barley infected with brome mosaic virus (BMV) and tobacco mosaic virus (TMV) (*from R. I. Hamilton and C. Nichols, The influence of bromegrass mosaic virus on the replication of tobacco mosaic virus in Hordeum vulgare, Phytopathology, 7:484, 1977*).

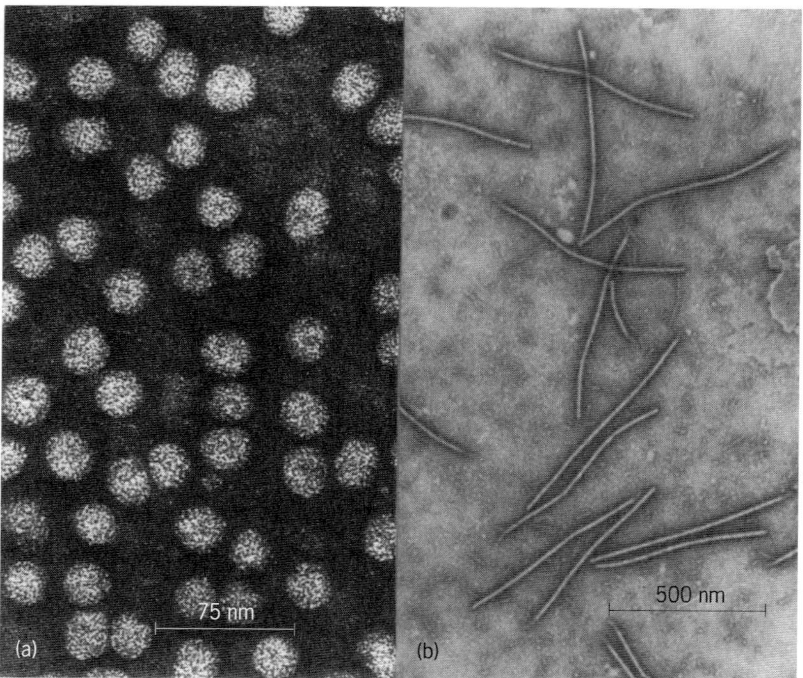

Fig. 1. Representative plant viruses in purified virus preparation obtained from infected leaves. (*a*) Tobacco streak virus (isometric). (*b*) Pea seed-borne mosaic virus (anisometric).

Purified viruses suitable for characterization can be obtained from infected tissues by ultracentrifugation and other procedures. The diversity of particle size, chemical composition of nucleic acids and proteins, and other characteristics has resulted in the classification of plant viruses into 42 genera and 11 families, with more groups being formed as knowledge of the properties of viruses increases.

Replication. The natural hosts of plant viruses are widely distributed throughout the higher-plant kingdom, including some algae and fungi. Some viruses (TMV and cucumber mosaic virus) are capable of infecting over a hundred species in many families, whereas others, such as wheat streak mosaic virus, are restricted to a few species in the grass family. The replication of single-standard RNA viruses involves release of the virus genome from the coat protein; the association of the RNA with the ribosomes of the cell; translation of the genetic information of the RNA into specific proteins, including subunits of the coat protein and possibly viral RNA-synthesizing enzymes (replicases); transmission by vectors and diseases induction; synthesis of noninfective RNA using parental RNA as the template; and assembly of the protein subunits and viral RNA to form complete virus particles.

In other RNA viruses such as lettuce necrotic yellows virus, an enzyme which is contained in the virus must first make a complementary (infective) copy of the RNA; this is then translated into enzymes and coat protein subunits. The replication of double-stranded RNA viruses such as wound tumor virus is basically similar to that of lettuce necrotic yellows virus.

With double-stranded DNA viruses such as cauliflower mosaic virus, viral DNA is uncoated in a newly infected cell and transported to the nucleus, where it associates with histones to form a closed circular minichromosome. Two major RNA species (35S and 19S) are transcribed from the minichromosome by a host-encoded enzyme and are translated in the cytoplasm to produce virus-associated proteins. The 35S RNA serves as the template for a viral enzyme which transcribes it to viral DNA, which is then encapsidated to form virus particles. For example, the single-stranded DNA geminiviruses are usually found in nuclei of infected plants. Their circular genomes are encapsidated by coat protein arranged to form a couplet of particles. Synthesis of progeny DNA occurs in nuclei via a double-stranded DNA replicative structure associated with host-encoded DNA synthesizing enzymes.

Symptoms. Symptoms are the result of an alteration in cellular metabolism and are most obvious in newly developing tissues. Infected leaves often show a mosaic of light green-yellow (chlorotic) areas and areas of normal green color (**Fig. 3*a***); virus-induced chlorosis is associated with the development of abnormal chloroplasts. Infection of plants with two or more viruses is common, especially in perennial plants or those propagated by cuttings; specific diseases (for example, severe streak of tomato and rugose mosaic of potato) are caused by dual infection with tomato mosaic virus and potato virus X, and potato virus X and potato virus Y, respectively. In some plants, depending on the virus, the initial infection does not spread because cells surrounding the infected cells die, resulting in the formation of necrotic lesions (Fig. 3*b*). Such plants are termed hypersensitive. The size and shape of leaves (Fig. 3*c*) and fruit (Fig. 3*d*) may be adversely affected, and in some instances plants may be killed. Not all virus infections produce distinctive symptoms; potato infected with some North American strains of potato viruses S and X are symptomless.

Fig. 3. Symptoms induced by virus infection. (*a*) Mosaic in leaves of raspberry infected with raspberry vein-banding mosaic disease induced by rubus yellow net virus in association with black raspberry necrosis virus. (*b*) Necrotic lesions in tobacco leaves induced by rub inoculation of cherry leaf roll virus. (*c*) Deformation of cherry leaves caused by infection with cherry raspberry leaf virus. (*d*) Necrotic rings on tomato fruit induced by infection with tomato mosaic virus. (*Courtesy of R. Stace-Smith*)

Transmission. Some viruses, especially those infecting trees, can only be transmitted experimentally by grafting. The continued propagation of infected plants by cuttings also transmits viruses. Contact transmission occurs through wounds caused by leaves of infected plants rubbing those of adjacent healthy plants or by the handling of diseased and healthy plants.

The most common mode of transmission for many viruses is by means of vectors, mainly insects (predominantly aphids and leafhoppers), and to a lesser extent mites, soil-inhabiting fungi, and nematodes which acquire viruses by feeding on infected plants. Viruses transmitted by one class of vector are rarely transmitted by another, and there is often considerable specificity between strains of a virus and their vectors. Wound tumor virus infects its leafhopper vectors so that they transmit the virus for life, while the aphid vector of potato virus Y, which carries the virus on its mouthparts, loses the ability to transmit the virus within an hour or two unless it again feeds on a diseased plant.

Some viruses (such as barley stripe mosaic virus and soybean mosaic virus) are transmitted to succeeding generations mainly by embryos in seeds produced by infected plants; over 200 viruses are transmitted in this way. Some are also vectored by insects or nematodes. Infected pollen can transmit virus to healthy flowers, resulting in infected seed and, in some cases (ringspot in cherry and bushy dwarf disease in raspberry), infection of the seed-bearing plant.

Control of virus diseases. Virus diseases are difficult to control because of the complex interaction of virus strains, vectors, overwintering sources such as weeds, and the crop to be protected. Breeding for resistance utilizing genes for hypersensitivity to a particular virus may be useful because the virus is usually limited to the few cells bordering the point of infection. Some success is being obtained in breeding plants that are not suitable hosts for aphid development, thus reducing the probability that they will be infected by aphid-borne viruses. Plants genetically engineered to express specific viral proteins such as the coat protein are often highly resistant to infection by related viruses. *See* GENETIC ENGINEERING.

In vegetatively propagated plants (such as the potato or fruit trees), important virus diseases are being controlled by tissue culture of plants from virus-free stem tips selected from infected plants. Seed-transmitted viruses are controlled by the use of virus-free seed. Crop rotation and chemical treatment of fields aid in the control of soil-borne viruses, primarily by reducing the population of their soil-borne fungal and nematode vectors. Inoculation of young tomato and citrus plants with mild strains of tobacco mosaic virus and citrus tristeza virus, respectively, often protects such plants from damaging effects of severe strains.

Viroids. Only about 30 viroids are known, but they cause very serious diseases in such diverse plants as chrysanthemum, citrus, coconut, and potato. They can also be isolated from plants that do not exhibit symptoms. Viroids are mainly transmitted by vegetative propagation, but some, such as potato spindle tuber viroid, are transmitted by seed or by contact between infected and healthy plants. Tomato planta macho viroid is efficiently transmitted by aphids. *See* PLANT PATHOLOGY; VIROIDS; VIRUS.

Richard I. Hamilton

Bibliography. T. Hohn and J. Schell, *Plant DNA Infections Agents*, 1987; K. Maramorosch (ed.), *Viroids and Satellites: Molecular Parasites at the Frontier of Life*, 1991; R. E. F. Matthews, *Fundamentals of Plant Virology*, 1997; P. Pirone and J. G. Shaw (eds.), *Viral Genes and Plant Pathogenesis*, 1990; A. Bruce Voyles, *The Biology of Viruses*, 1993; T. Wilson, A. Michael, and J. W. Davies (eds.), *Genetic Engineering with Plant Viruses*, 1992.

Plant-water relations

Water is the most abundant constituent of all physiologically active plant cells. Leaves, for example, have water contents which lie mostly within a range of 55–85% of their fresh weight. Other relatively succulent parts of plants contain approximately the same proportion of water, and even such largely nonliving tissues as wood may be 30–60% water on a fresh-weight basis. The smallest water contents in living parts of plants occur mostly in dormant structures, such as mature seeds and spores. The great bulk of the water in any plant constitutes a unit system. This water is not in a static condition. Rather it is part of a hydrodynamic system, which in terrestrial plants involves absorption of water from the soil, its translocation throughout the plant, and its loss to the environment, principally in the process known as transpiration.

Cellular water relations. The typical mature, vacuolate plant cell constitutes a tiny osmotic system, and this idea is central to any concept of cellular water dynamics. Although the cell walls of most living plant cells are quite freely permeable to water and solutes, the cytoplasmic layer that lines the cell wall is more permeable to some substances than to others. This property of differential permeability appears to reside principally in the layer of cytoplasm adjacent to the cell wall (plasma membrane, or plasmalemma) and in the layer in contact with the vacuole (vacuolar membrane, or tonoplast). This cytoplasmic system of membranes is usually relatively permeable to water, to dissolved gases, and to certain dissolved organic components. It is often much less permeable to sugars and mineral salts. The permeability of the cytoplasmic membranes is quite variable, however, and under certain metabolic conditions solutes that ordinarily penetrate through these membranes slowly or not at all may pass into or out of cells rapidly. *See* CELL (BIOLOGY); PLANT CELL.

Osmotic and turgor pressures. If a plant cell in a flaccid condition—one in which the cell sap exerts no pressure against the encompassing cytoplasm and cell wall—is immersed in pure water, inward osmosis of water into the cell sap ensues. Osmosis may be defined as the movement of solvent molecules, which in living organisms are always water, across a membrane that is more permeable to the solvent than to the solutes dissolved in it. The driving force in osmosis is the difference in free energies of the water on the two sides of the membrane. Pure water, as a result of its intrinsic properties, possesses free energy. Since there is no certain way of measuring the free energy of water, it is arbitrarily given a value of zero when exposed only to atmospheric pressure. The free energy of water is influenced by temperature, but this discussion will be restricted to isothermal systems. This free energy has been designated water potential and may be expressed in either energy or pressure units. It is more meaningful to use pressure units in considerations of the water potential of plants.

Inward osmosis of water takes place under the conditions specified above because the water potential of the cell sap is less than that of the surrounding pure water by the amount of its osmotic pressure. Introduction of solutes always lowers the potential of water by the amount of the resulting osmotic pressure. If the osmotic pressure of the cell sap is 15 atm (1.5 megapascals), then the water potential of the cell sap is 15 atm (1.5 MPa) less than that of pure water at the same temperature and under the same pressure.

The gain of water by the cell as a result of inward osmosis results in the exertion of a turgor pressure against the protoplasm, which in turn is transmitted to the cell wall. This pressure also prevails throughout the mass of solution within the cell. If the cell wall is elastic, some expansion in the volume of the cell occurs as a result of this pressure, although in many kinds of cells this is relatively small.

Because of the solutes invariably present, the cell sap possesses an osmotic pressure. The osmotic pressures of most plant cell saps lie within the range of 5–40 atm (0.5–4 MPa), although values as high as 200 atm (20 MPa) have been found in some halophytes (plants that can tolerate high-solute media).

The osmotic pressures of the cells of a given plant tissue vary considerably with environmental conditions and intrinsic metabolic activities. More or less regular daily or seasonal variations occur in the magnitude of cell sap osmotic pressures in the cells of many tissues. It is the osmotic pressure of the cell sap, coupled with the differential permeability of the cytoplasmic membranes and the relative inelasticity of the cell walls, which permits the development of the more or less turgid condition characteristic of most plant cells.

With continued osmosis of water into the cell, its turgor pressure gradually increases until it is equal to the final osmotic pressure of the sap. Subjection of the water in the cell sap to pressure increases its water potential by the amount of the imposed pressure. In the example given above, disregarding the usually small amount of sap dilution as a result of cell expansion, the water potential of the cell sap is reduced 15 atm (1.5 MPa) because of the presence of solutes (the osmotic pressure is the index of this lowering of water potential) and raised 15 atm (1.5 MPa) as a result of turgor pressure when maximum turgor is reached. Hence, when a dynamic equilibrium is attained, the water potential of the cell sap is zero and thus is equal to that of the surrounding water, a necessary condition if equilibrium is to be achieved.

If the same cell in a flaccid condition is immersed in a solution with an osmotic pressure of 6 atm (0.6 MPa), inward osmosis of water occurs, but does not continue as long as when the cell is immersed in pure water. Disregarding sap dilution, a dynamic equilibrium will be attained under these circumstances when the turgor pressure of the cell sap has reached 9 atm (0.9 MPa), because at this point the water potential in the cell sap and the water potential in the surrounding solution will be equal. Since the water potential of the cell sap was originally diminished 15 atm (1.5 MPa) because of the presence of solutes and then raised 9 atm (0.9 MPa) because of turgor pressure, the net value of the water potential in the cell at equilibrium is −6 atm (−0.6 MPa). This is also the value of the water potential in the surrounding solution, as indexed by its osmotic pressure. At dynamic equilibrium the number of water molecules entering the cell and the number leaving per unit of time will be equal. *See* OSMOREGULATORY MECHANISMS.

Water potential. As the examples just given indicate, the effective physical quantity controlling the direction of osmotic movement of water from cell to cell in plants or between a cell and an external solution is the water potential (WP) of the water. In most plant cells this quantity is equal to the osmotic pressure (OP) of the water less the turgor pressure (TP) to which it is subjected. Since the osmotic pressure is the index of the amount by which the water potential of a solution is less than that of pure water under the same conditions, it must be treated as a negative quantity. Turgor pressures usually have a positive value, but if water passes into a state of tension, they are negative. Examples of the occurrence of water under tension in plants will be discussed later in this article. In an unconfined solution the water potential is equal to the osmotic pressure, since there is no turgor pressure. In a fully turgid plant cell the water potential is zero, since the turgor pressure is equal to the osmotic pressure. In a fully flaccid plant cell the turgor pressure is zero and the water potential is equal to the osmotic pressure. The interrelationships among these quantities may be expressed: WP = (−OP) + TP.

Since the osmotic pressure must be treated as a negative quantity and since turgor pressures rarely exceed osmotic pressures, the water potentials in plant cells are negative in value. The only well-known exception to this is root pressure, discussed later, during which small positive water potentials can be generated in the water-conductive tissues.

The relationships expressed in the equation are illustrated graphically in **Fig. 1**, in which allowance has been made for the effect of the volume changes which are characteristic of some kinds of cells with shifts in turgor pressure. The conditions which would prevail in the cell if the cell sap passed into a state of tension (negative pressure) are indicated by the dotted extension of curves to the left.

Cell-to-cell movement of water in plants always occurs from the cell of greater (less negative) water potential to the cell of lesser (more negative) water potential. Such movement of water in plant tissues apparently often occurs along water potential gradients in which the water potential of each cell in a

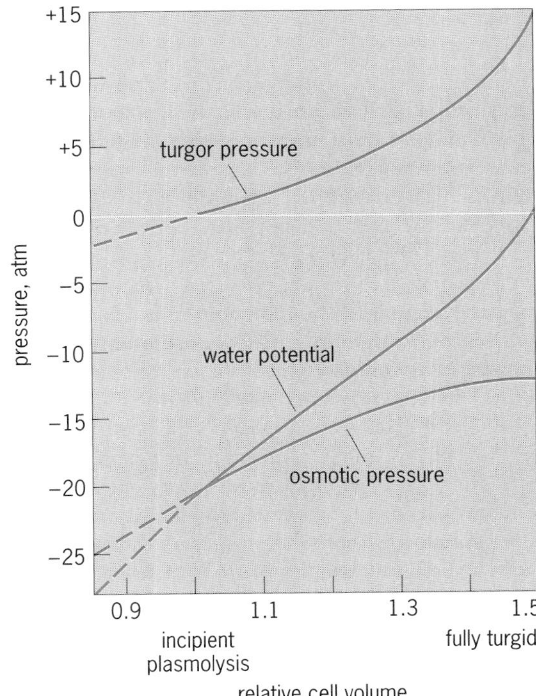

Fig. 1. Curves showing interrelationships among osmotic pressures, turgor pressures, water potentials, and volumes of a plant cell. 1 atm = 10^2 kPa.

series is less (more negative) than that of the preceding one.

Plasmolysis. If a turgid or partially turgid plant cell is immersed in a solution with a greater osmotic pressure than the cell sap, a gradual shrinkage in the volume of the cell ensues; the amount of shrinkage depends upon the kind of cell and its initial degree of turgidity. When the lower limit of cell wall elasticity is reached and there is continued loss of water from the cell sap, the protoplasmic layer begins to recede from the inner surface of the cell wall. Retreat of the protoplasm from the cell wall often continues until it has shrunk toward the center of the cell, the space between the protoplasm and the cell wall becoming occupied by the bathing solution. This phenomenon is called plasmolysis. If a cell is immersed in a solution with an osmotic pressure which just slightly exceeds that of the cell sap, withdrawal of the protoplasm from the cell wall would be just barely initiated. The stage of plasmolysis shown in Fig. 1 is called incipient plasmolysis, and it is the basis for one of the methods which is commonly used in measuring the osmotic pressure of plant cells.

Imbibition. In some kinds of plant cells movement of water occurs principally by the process of imbibition rather than osmosis. The swelling of dry seeds when immersed in water is a familiar example of this process. Imbibition occurs because of the more negative water potential in the imbibant as compared with the water potential in some contiguous part of the system. An equilibrium is reached only when the water potential in the two parts of the system has attained the same value. The water potential in a dry seed is extremely low, being equal in value, but with a negative sign, to its imbibition pressure. The water potential of pure water is zero; hence movement of water occurs into the seed. Even if the seeds are immersed in a solution of considerable osmotic pressure, which in an unconfined solution is an index of its negative water potential, imbibition occurs. However, if the osmotic pressure of the solution is high enough (of the order of 1000 atm or 10 MPa), the seed will not gain water from the solution and may even lose a little to the solution. In other words, if the water potential of the solution is negative enough, imbibition does not occur.

However, a more negative water potential in the imbibant, as compared with the surrounding or adjacent medium is not the only condition which must be fulfilled if imbibition is to occur. Many kinds of seeds swell readily if immersed in water but not when immersed in ether or other organic solvents. On the other hand, rubber does not imbibe water but does imbibe measurable quantities of ether and other organic liquids. Certain specific attraction forces between the molecules of the imbibant and the imbibed liquid are therefore also a requisite for the occurrence of imbibition.

In an imbibitional system the imbibition pressure IP of the imbibant is the analog of the osmotic pressure in an osmotic system; hence in such a system water potential is related to the imbibition pressure and turgor pressure: $WP = (-IP) + TP$. The imbibition pressure may be regarded as an index of the reduction in water potential in an imbibant insofar as it results from attractions between the molecules of the imbibant and water molecules. For an unconfined imbibant immersed in water, the negative water potential initially equals the imbibition pressure, since there is no turgor pressure. The more nearly saturated such an imbibant becomes, the smaller its imbibition pressure and the less negative its water potentials. A fully saturated imbibant has a zero imbibition pressure and a zero water potential.

Stomatal mechanism. Various gases diffuse into and out of physiologically active plants. Those gases of greatest physiological significance are carbon dioxide, which diffuses into the plant during photosynthesis and is lost from the plant in respiration; oxygen, which diffuses in during respiration and is lost during photosynthesis; and water vapor, which is lost in the process of transpiration. The great bulk of the gaseous exchanges between a plant and its environment occurs through tiny pores in the epidermis called stomates. Although stomates occur on many aerial parts of plants, they are most characteristic of, and occur in greatest abundance in, leaves. *See* EPIDERMIS (PLANT); LEAF.

Each stomate or stoma (plural, stomates or stomata) consists of a minute elliptical pore surrounded by two distinctively shaped epidermal cells called guard cells. Stomates are sometimes open and sometimes closed; when closed, all gaseous exchanges between the plant and its environment are greatly retarded. The size of a fully open stomate differs greatly from one species of plant to another. Among the largest known are those of the wandering Jew (*Zebrina pendula*), whose axial dimensions average 31 by 12 micrometers. In most species the stomates are much smaller, but all of them afford portals of egress or ingress which are enormous relative to the size of the gas molecules that diffuse through them. The number of stomates per square centimeter of leaf surface ranges from a few thousand in some species to over a hundred thousand in others. In many species of plants stomates are present in both the upper and lower epidermises, usually being more abundant in the lower. In many species, especially of woody plants, they are present only in the lower epidermis. In floating-leaved aquatic species stomates occur only in the upper epidermis.

Rates of transpiration (loss of water vapor) from leaves of the expanded type often are 50% or even more of the rate of evaporation from a free water surface of equal area under the same environmental conditions. Loss of water vapor from leaves may occur at such relatively high rates despite the fact that the aggregate area of the fully open stomates is only 1–3% of the leaf area. Much more significant is the fact that the rate of diffusion of carbon dioxide, essential in photosynthesis, into the leaves through the stomates is much greater than through the equivalent area of an efficient carbon dioxide–absorbing surface.

Although some mass flow of gases undoubtedly occurs through the stomates under certain conditions, most movement of gases into or out of a leaf takes place by diffusion through the stomates. Diffusion is the physical process whereby molecules of a gas move from a region of their greater diffusion pressure to the region of their lesser diffusion pressure as a result of their own kinetic activity. Molecules of liquids and solids (to a limited extent), molecules and ions of solutes, and colloidal particles also diffuse whenever the appropriate circumstances prevail.

Diffusion of gases through small pores follows certain principles which account for the high diffusion capacity of the stomates. In the diffusion of a gas through a small pore, an overwhelming proportion of the molecules escape over the rim of the pore relative to those escaping through its center. Hence, diffusion rates through small apertures vary as their perimeter rather than as their area. The less the area of a pore, therefore, the greater is its diffusive capacity relative to its area. Therefore, a gas may diffuse nearly as rapidly through a septum pierced with a number of small orifices, whose aggregate area represents only a small proportion of the septum area, as through an open surface equal in area to the septum. The high diffusive capacity of the stomates can be accounted for in terms of these principles. Since diffusion of gases through stomates is proportional to the perimeter of the pore rather than to its area, the diffusion rate through a partially open stomate is almost as great as that through a fully open stomate.

In general, stomates are open in the daytime and closed at night, although there are many exceptions to this statement. The mechanism whereby stomates open in the light and close in the dark seems to be principally an osmotic one, although other factors are probably involved. Upon the advent of illumination, the hydrogen ion concentration of the guard cells decreases. This favors the action of the enzyme phosphorylase, which in the presence of phosphates causes transformation of insoluble starch into the soluble compound glucose-1-phosphate. The resulting increase in the solute concentration of the guard cells causes an increase in their osmotic pressure, and hence also in the negativity of their water potential. Osmotic movement of water takes place from contiguous epidermal cells, in which there is little daily variation in osmotic pressure, into the guard cells. The resulting increase in turgor pressure of the guard cells causes them to open. With the advent of darkness or of a relatively low light intensity, the reverse train of processes is apparently set in operation, leading ultimately to stomatal closure.

Light of low intensity is, generally speaking, less effective than stronger illumination in inducing stomatal opening. Hence stomates often do not open as wide on cloudy as on clear days, and often do not remain open for as much of the daylight period. A deficiency of water within the plant also induces partial to complete closure of the stomates. During periods of drought, therefore, stomates remain shut continuously or, at most, are open for only short periods each day, regardless of the light intensity to which the plant is exposed. Opening of the stomates does not occur in most species at temperatures approaching freezing. Hence in cold or even cool weather stomates often remain closed even when other environmental conditions are favorable to their opening. Nocturnal opening occurs at times in some species, but the conditions which induce this pattern of stomatal reaction are not clearly understood.

Transpiration Process

The term transpiration is used to designate the process whereby water vapor is lost from plants. Although basically an evaporation process, transpiration is complicated by other physical and physiological conditions prevailing in the plant. Whereas loss of water vapor can occur from any part of the plant which is exposed to the atmosphere, the great bulk of all transpiration occurs from the leaves. There are two kinds of foliar transpiration: (1) stomatal transpiration, in which water vapor loss occurs through the stomates, and (2) cuticular transpiration, which occurs directly from the outside surface of epidermal walls through the cuticle. In most species 90% or more of all foliar transpiration is of the stomatal type.

Stomatal transpiration. The dynamics of stomatal transpiration is considerably more complex than that of cuticular transpiration. In the leaves of most kinds of plants the mesophyll cells do not fit together tightly, and the intercellular spaces between them are occupied by air. A veritable labyrinth of air-filled spaces is thus present within a leaf, bounded by the water-saturated walls of the mesophyll cells. Water evaporates readily from these wet cell walls into the intercellular spaces. If the stomates are closed, the only effect of such evaporation is to saturate the intercellular spaces with water vapor. If the stomates are open, however, diffusion of water vapor usually occurs through them into the surrounding atmosphere. Such diffusion always occurs unless the atmosphere has a vapor pressure equal to or greater than that within the intercellular spaces, a situation which seldom prevails during the daylight hours of clear days. The two physical processes of evaporation and diffusion of water vapor are both integral steps in stomatal transpiration. Physiological control of this component of transpiration is exerted through the opening and closing of the stomates, previously described.

Environmental factors. Light is one of the major factors influencing the rate of transpiration because of its controlling effect on the opening and closing of stomates. Since stomatal transpiration is largely restricted to the daylight hours, daytime rates of transpiration are usually many times greater than night rates, which largely or entirely represent cuticular transpiration. Since leaves in direct sunlight usually have temperatures from one to several degrees higher than that of the surrounding atmosphere, light also has a secondary accelerating effect on transpiration through its influence on leaf temperatures. Increase in leaf temperatures results in an increase in the pressure of the water vapor molecules within the leaf.

The rate of diffusion of water vapor through open stomates depends upon the steepness of the vapor pressure gradient between the intercellular spaces and the outside atmosphere. When the vapor pressure in that part of the intercellular spaces just below the stomatal pores is high relative to that of the atmosphere, diffusion of water vapor out of the leaf occurs rapidly; when it is low, water vapor diffusion occurs much more slowly.

Temperature has a marked effect upon rates of transpiration, principally because of its differential effect upon the vapor pressure of the intercellular spaces and atmosphere. Although leaf temperatures do not exactly parallel atmospheric temperatures, increase in atmospheric temperature in general results in a rise in leaf temperature and vice versa. On a warm, clear day such as would be typified by many summer days in temperate latitudes and with an adequate soil water supply, increase in temperature results in an increase in the vapor pressure of the intercellular spaces. Such a rise in vapor pressure occurs because the vapor pressure corresponding to a saturated condition of an atmosphere increases with rise in temperature, and the extensive evaporating surfaces of the cell walls bounding the intercellular spaces make it possible for the intercellular spaces to be maintained in an approximately saturated condition most of the time. An increase in temperature ordinarily has little or no effect on the vapor pressure of the atmosphere, and this is especially true of warm, bright days, when transpiration rates are the highest. Hence, as the temperature rises, the vapor pressure of the intercellular spaces increases relative to that of the external atmosphere, the vapor pressure gradient through the stomates is steepened, and the rate of outward diffusion of water vapor increases.

Wind velocity is another factor which influences the rate of transpiration. Generally speaking, a gentle breeze is relatively much more effective in increasing transpiration rates than are winds of greater velocity. In quiet air, localized zones of relatively high atmospheric vapor pressure may build up in the vicinity of transpiring leaves. Such zones retard transpiration unless there is sufficient air movement to prevent the accumulation of water vapor molecules. The bending, twisting, and fluttering of leaf blades and the swaying of stalks and branches in a wind also contribute to increasing the rate of transpiration.

Soil water conditions exert a major influence on the rate of transpiration. Whenever soil conditions are such that the rate of absorption of water is retarded, there is a corresponding diminution in the rate of transpiration.

Daily periodicity of transpiration. The rate of every major plant process, including transpiration, is measurably and often markedly influenced by the environmental conditions to which the plant is exposed. Many of the environmental factors exhibit more or less regular daily periodicities, which vary somewhat, of course, with the prevailing climatic conditions. This is especially true of the factors of light and temperature. Many plant processes, including transpiration, therefore exhibit daily periodicities in rate that are correlated with the daily periodicities of one or more environmental factors. *See* PHOTOPERIODISM.

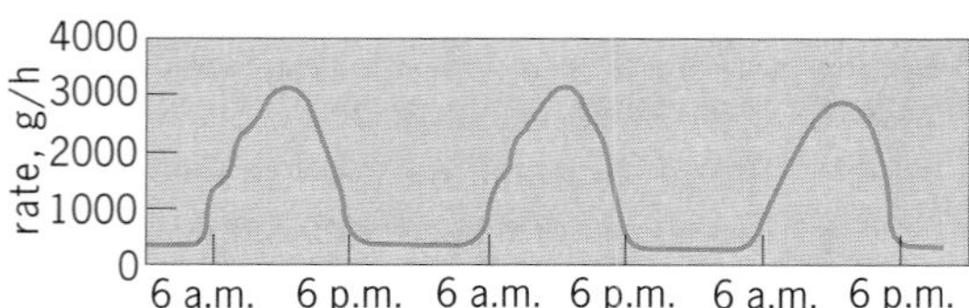

Fig. 2. Daily periodicity of transpiration of alfalfa on three successive clear, warm days with adequate soil water. Rate of transpiration expressed as grams per hour per 6-ft^2 (0.17-m^2) plot of alfalfa. 1 g = 0.035 oz. (*After M. D. Thomas and G. R. Hill, Plant Physiol., 12:285–307, 1937*)

The daily periodicity of transpiration in alfalfa, as exhibited on three clear, warm days with adequate soil water available, is illustrated in **Fig. 2**. A similar daily periodicity of transpiration is exhibited under comparable environmental conditions by many other species. During the hours of darkness, the transpiration rate is relatively low, and in most species water vapor loss during this period may be regarded as entirely cuticular or nearly so. The transpiration rate shows a steady rise during the morning hours, culminating in a peak rate which is attained in the early hours of the afternoon. The increase in transpiration rate during the forepart of the day results from gradual opening of the stomates, beginning with the advent of light, followed by a steady increase in the steepness of the vapor pressure gradient through the stomates, which occurs as a result of increasing atmospheric temperature during the morning and earlier afternoon hours.

In most plants, if transpiration is occurring rapidly, the rate of absorption of water does not keep pace with the rate at which water vapor is lost from the leaves. In other words, the plant is gradually being depleted of water during the daylight hours. In time the resulting decrease in the water content of the leaf cells results in a reduced vapor pressure within the intercellular spaces, and a diminution in the rate of transpiration begins. Stomates also start to close as a result of the diminished leaf water content, and their closure is accelerated during the later part of the afternoon by the waning light intensity. By nightfall complete closure of virtually all stomates has taken place, and water vapor loss during the hours of darkness is again restricted largely or entirely to the relatively low rate of cuticular transpiration. It is noteworthy that the peak rate of transpiration occurs during the early afternoon hours, correlating more closely with the daily temperature periodicity than with the daily periodicity of light intensity.

Under environmental conditions differing considerably from those postulated in the preceding discussion, patterns of transpiration periodicity may show a considerable variance from the one described. On cloudy days, for example, stomates generally open less completely than on clear days, and a curve for daily transpiration periodicity presents a greatly flattened appearance as compared with those shown in Fig. 2. A cool temperature, even in a range somewhat above freezing, greatly diminishes and may even cause cessation of stomatal transpiration, resulting in a pronounced modification in the daily march of transpiration periodicity. A deficient soil water supply is probably the most common cause of departures from the pattern of transpiration described above. A reduction in soil water content below the field capacity (optimum water availability) results not only in a general flattening of the transpiration periodicity curve, but frequently also in appearance of the peak of the curve somewhat earlier in the day. Since, even in temperate zone regions, drought periods of greater or less severity are of common occurrence during the summer months and in many habitats are the rule rather than the exception, transpiration periodicity curves of this flattened and skewed-peak type are undoubtedly of frequent occurrence.

Magnitude of transpiration. Transpiration of broad-leaved species of plants growing under temperate zone conditions may range up to about 0.0102 oz per square inch (5 g per square decimeter) of leaf area per hour. Sufficient quantities of water are often lost in transpiration by vegetation-covered areas of the Earth's surface to have important effects not only on soil water relations, but also on meteorological conditions. The quantities of water lost per acre by crops, grasslands, or forest are therefore a matter of basic interest. An acre of corn (maize), for example, transpires water equivalent to 15 in. (38 cm) of rainfall during a usual growing season. Transpiration of deciduous, largely oak, forest in the southern Appalachian Mountains has been estimated as equivalent to 17–22 in. (43–56 cm) of rainfall per year. Marked variations occur in such values from year to year, however, depending upon prevailing climatic conditions.

Significance of transpiration. Viewpoints regarding the significance of transpiration have ranged between the two extremes of considering it a process that is an unavoidable evil or a physiological necessity. Neither of these extreme views appears to be tenable. Some of the incidental effects of transpiration appear to be advantageous to the plant, but none of them is indispensable for its survival or even for its adequate physiological operation. Likewise, while some of the incidental effects of transpiration appear to be detrimental to the plant, none of them is so in such a critical fashion that survival of plants, considered in the aggregate, is endangered.

Transpiration is a necessary consequence of the relation of water to the anatomy of the plant, and especially to the anatomy of the leaves. Terrestrial green plants are dependent upon atmospheric carbon dioxide for their survival. In terrestrial vascular plants the principal carbon dioxide-absorbing surfaces are the moist mesophyll cells walls which bound the intercellular spaces in leaves. Ingress of carbon dioxide into these spaces occurs mostly by diffusion through open stomates. When the stomates are open, outward diffusion of water vapor unavoidably occurs, and such stomatal transpiration accounts for most of the water vapor loss from plants. Although transpiration is thus, in effect, an incidental phenomenon, it frequently has marked indirect effects on other physiological processes which occur in the plant because of its effects on the internal water relations of the plant.

Water Translocation

In terrestrial rooted plants practically all of the water which enters a plant is absorbed from the soil by the roots. The water thus absorbed is translocated to all parts of the plant. In the tallest trees (specimens of the coast redwood, *Sequoia sempervirens*) the distance from the tips of the deepest roots to the tips of the topmost branches is nearly 400 ft (122 m), and water must be elevated for this distance through such trees. Although few plants are as tall as such redwoods, the same mechanisms of water movement are believed to operate in all vascular species. The mechanism of the "ascent of sap" (all translocated water contains at least traces of solutes) in plants, especially tall trees, was one of the first processes to excite the interest of plant physiologists.

Water-conductive tissues. The upward movement of water in plants occurs in the xylem, which, in the larger roots, trunks, and branches of trees and shrubs, is identical with the wood. In the trunks or larger branches of most kinds of trees, however, sap movement is restricted to a few of the outermost annual layers of wood. This explains why hollow trees, in which the central core of older wood has disintegrated, can remain alive for many years. The xylem of any plant is a unit and continuous system throughout the plant. Small strands of this tissue extend almost to the tip of every root. Other strands, the larger of which constitute important parts of the veins, ramify to all parts of each leaf. In angiosperms most translocation of water occurs through the xylem vessels, which are nonliving, elongated, tubelike structures. The vessels are formed by the end-to-end coalescence of many much smaller cells, death of these cells ensuing at about the same time that coalescence occurs. In trees the diameters of such vessels range from about 20 to 400 μm, and they may extend for many feet with no more interruption than an occasional incomplete cross wall. In gymnosperms no vessels are present, and movement of water occurs solely through spindle-shaped xylem cells called tracheids. Vertically contiguous tracheids always overlap along their tapering portions, resulting in a densely packed type of woody tissue. Individual tracheids may be as much as 0.2 in. (5 mm) in length. Like the vessels, they are nonliving while functional in the translocation of water. Small, more or less rounded, thin areas occur in the walls of vessels and tracheids that are contiguous with the walls of other tracheids, vessels, or cells. Structurally, three main types of such pits are recognized, but all of them appear to facilitate the passage of water from one xylem element to another. *See* XYLEM.

Root pressure. The exudation of xylem sap from the stump of a cutoff herbaceous plant is a commonly observed phenomenon. Sap exudation ("bleeding") from the cut ends of stems or from incisions into the wood also occurs in certain woody plants, such as birch, currant, and grape, especially in the spring. A single vigorous grapevine often loses a liter or more of sap per day through the cut ends of stems after spring pruning. This exudation of sap from the xylem tissue results from a pressure originating in the roots, called root pressure. A related phenomenon is that of guttation. This term refers to the exudation of drops of water from the tips or margins of leaves, which occurs in many species of herbaceous plants as well as in some woody species. Like sap exudation from cut stems, this phenomenon is observed most frequently in the spring, and especially during early morning hours. The water exuded in guttation is not pure, but contains traces of sugar and other solutes. Guttation occurs from special structures called hydathodes, which are similar in structure to but larger than stomates. In most species water loss by guttation is negligible in comparison with the water lost as vapor in transpiration. Like xylem sap exudation, guttation results from root pressure.

Root pressure is generally considered to be one of the mechanisms of upward transport of water in plants. While it is undoubtedly true that root pressure does account for some upward movement of water in certain species of plants at some seasons, various considerations indicate that it can be only a secondary mechanism of water transport. Among these considerations are the following: (1) There are many species in which the phenomenon of root pressure has not been observed. (2) The magnitude of measured root pressures seldom exceeds 2 atm (200 kPa), which could not activate a rise of water for more than about 60 ft, and many trees are much taller than this. (3) Known rates of xylem sap flow under the influence of root pressure are usually inadequate to compensate for known rates of transpiration. (4) Root pressures are usually operative in woody plants only during the early spring; during the summer months, when transpiration rates, and hence rates of xylem sap transport, are greatest, root pressures are negligible or nonexistent. *See* ROOT (BOTANY).

Water cohesion and ascent of sap. Although invariably in motion, as a result of their kinetic energy, water molecules are also strongly attracted to each other. In masses of liquid water the existence of such intermolecular attractions is not obvious, but when water is confined in long tubes of small diameter, the reality of the mutual attractions among water molecules can be demonstrated. If the water at the top of such a tube is subjected to a pull, the resulting stress will be transmitted, because of the mutual attraction (cohesion) among water molecules, all the way down the column of water. Furthermore, because of the attraction between the water molecules and the wall of the tube (adhesion), subjecting the water column to a stress does not result in pulling it away from the wall.

The observations just mentioned have been made the basis of a widely entertained theory of the mechanism of water transport in plants, first clearly enunciated by H. H. Dixon in 1914. According to this theory, upward translocation of water (actually a very dilute sap) is engendered by an increase in the negativity of water potential in the cells of apical organs of plants. Such increases in the negativity of water potentials occur most commonly in the mesophyll cells of leaves as a result of transpiration.

Evaporation of water from the walls of the mesophyll cells abutting on the intercellular spaces reduces their turgor pressure and hence increases the negativity of the water potential in such cells. Consequent cell-to-cell movements of water cause the water potentials even of those mesophyll cells which are not directly exposed to the intercellular spaces to become more negative. The resulting decrease in the water potential of those cells directly in contact with the xylem elements in the veinlets of the leaf induces movement of water from the vessel or tracheids into those cells. Whenever transpiration is occurring at appreciable rates, water does not enter the lower ends of the xylem conduits in the roots as rapidly as it passes out of the vessels or tracheids into adjacent leaf cells at the upper ends of the water-conductive system; therefore the water in the xylem ducts is stretched into taut threads, that is, it passes into a state of tension. Each column of water behaves like a tiny stretched wire. The tension is transmitted along the entire length of the water columns to their terminations just back of the root tips. Subjection of the water in the xylem ducts to a tension (in effect, a negative pressure) increases the negativity of its water potential. The water potential in the xylem conduits is also subject to some lowering because of the presence of solutes, but xylem sap is usually so dilute that this osmotic effect is rarely more than a minor one. As a result of the increased negativity of the water potential in the xylem elements, movement of water is induced from adjacent root cells into those elements in the absorbing regions of roots.

The tension engendered in the water columns can be sustained by them because of the cohesion between the water molecules, acting in conjunction with the adhesion of the boundary layers of water molecules to the walls of the xylem ducts. The existence of water under tension in vessels has been verified in a number of species of plants by direct microscopic examination. There is some evidence that, under conditions of marked internal water deficiency, the tensions generated in the water columns are proliferated into the mesophyll cells of leaves and cells in the absorbing regions of roots. Conservative calculations indicate that a cohesion value of 30–50 atm (3–5 MPa) would be adequate to permit translocation of water to the top of the tallest known trees. However, under conditions of internal water deficiency, tensions considerably in excess of 50 atm (5 MPa) are probably engendered in the water columns of many plants, especially woody species. *See* PLANT TRANSPORT OF SOLUTES.

Water Absorption

This process will be discussed only from the standpoint of terrestrial, rooted plants. Consideration of the absorption of water by plants necessitates an understanding of the physical status of the water in soils as it exists under various conditions.

Soil water conditions. Even in the tightest of soils, the particles never fit together perfectly and a certain amount of space exists among them. This pore space of a soil ranges from about 30% of the soil volume in sandy soils to about 50% of the soil volume in heavy clay soils. In desiccated soils the pore space is occupied entirely by air, in saturated soils it is occupied entirely by water, but in moist, well-drained soils it is usually occupied partly by air and partly by water. In a soil in which a water table is located not too far below the surface, considerable quantities of water may rise into its upper layers by capillarity and become available to plants. In arid regions, however, there ordinarily is no water table. Even in many humid regions the water table is continuously or intermittently too far below the soil surface to be an appreciable source of water for most plants. In all soils lacking a water table or in which the water table is at a considerable depth, the only water available to plants is that which comes as natural precipitation or which is provided by artificial irrigation. If water falls on, or is applied to, a dry soil which is homogeneous to a considerable depth, it will become rapidly distributed to a depth which will depend on the quantity of water supplied per unit area and on the specific properties of that soil. After several days further deepening of the moist layer of soil extending downward from the surface virtually ceases, because within such a time interval capillary movement of water in a downward direction has become extremely slow or nonexistent. The boundary line between the moist layer of soil above and the drier zone below will be a distinct one. In this condition of field equilibrium the water content of the upper moist soil layer is, in homogeneous soils, essentially uniform throughout the layer. *See* SOIL.

The water content of a soil in this equilibrium condition is called the field capacity. Field capacities range from about 5% in coarse sandy soils to about 45% of the dry weight in clay soils. The moisture equivalent of a soil, often measured in the laboratory, is usually very close in value to the field capacity of the same soil. It is defined as the water content of a soil which is retained against a force 1000 times gravity as measured in a centrifuge. A soil at its field capacity is relatively moist, but is also well aerated. Soil water contents at or near the field capacity are the most favorable for growth of most kinds of plants.

A considerable proportion of the water in any soil is unavailable in the growth of plants. The permanent wilting percentage is the generally accepted index of this fraction of the soil water. This quantity is measured by allowing a plant to develop with its roots in soil enclosed in a waterproof pot until the plant passes into a state of permanent wilting. The water content of the soil when the plant just passes into this condition is the permanent wilting percentage. The range of permanent wilting percentages is from 2 to 3% of the dry weight in coarse sandy soils to about 20% in heavy clay loams. About the same value is obtained for the permanent wilting percentage of a given soil, regardless of the kind of test plant used.

The potential of soil water is always negative and has two major components. One is the osmotic pressure of the soil solution, which in most kinds of soils is only a fraction of 1 atm (0.1 MPa), although saline and alkali soils are marked exceptions. The other component is the attractive forces between the soil particles and water molecules, which may attain a very considerable magnitude, especially in dry soils. In moist soils, those at the field capacity or higher soil water content, the former of these two components is principally responsible for the negative soil water potential. In drier soils the latter component is almost solely responsible for the negative water potential. In the majority of soils the water potential is only slightly less than zero at saturation, not less than −1 atm (−0.1 MPa) at field capacity, and in the vicinity of −15 atm (−1.5 MPa) at the permanent wilting percentage. With further reduction in the water content of a soil below its permanent wilting percentage, its water potential increases rapidly in negativity at an accelerating rate. Almost no water can be absorbed by plants from soils with such extremely negative water potentials; for this reason the permanent wilting percentage is the index of soil water unavailable to plant growth (**Fig. 3**).

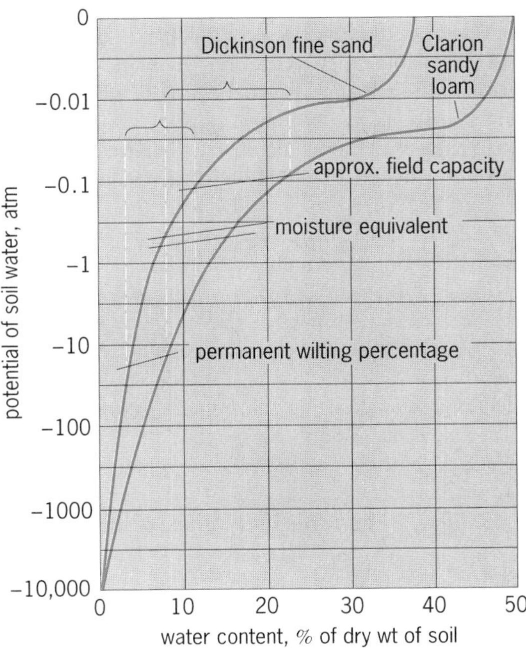

Fig. 3. Relation between soil water potentials and soil water contents of two soils over the entire range of soil water contents. Horizontal braces show the range for each soil over which water is readily available to its plants. Vertical scale is logarithmic. 1 atm = 10^3 kPa. (*After M. B. Russell, Amer. Soil Sci. Proc., 4:51–54, 1939*)

Root growth and water absorption. The successively smaller branches of the root system of any plant terminate ultimately in the root tips, of which there may be thousands and often millions on a single plant. As generally employed, the term root tip refers to the region extending back from the apex of the root for a distance of at least several centimeters. The terminal zone of a root tip is the root cap. Just back of this are the regions in which cell division and cell elongation occur and in which all growth in length of roots takes place. Just back of these regions, in the majority of species, is the zone of root hairs. Each root hair is a projection from the epidermal cell of which it is an integral part. A single root tip may bear thousands of root hairs, ranging in length from a few millimeters up to about a centimeter. In most species the root hairs are short-lived structures, but new ones are constantly developing just back of the growing region of the root as it elongates. Most absorption of water occurs in the root tip regions, and especially in the root hair zone. Older portions of most roots become covered with cutinized or suberized layers through which only very limited quantities of water can pass.

Whenever the potential of the water in the root hairs and other peripheral cells of a root tip is less than that of the soil water, movement of water takes place into the root cells. If the soil water content exceeds the field capacity, water may move by capillarity toward the region of absorption from portions of the soil not immediately contiguous with the root tips, and the supply of readily absorbable water is maintained in this way. Elongation of the roots, although slower in most species in relatively wet soils, also helps maintain contact between the root tips and untapped areas of soil water.

Many plants much of the time grow in soils with a water content in the range between the field capacity and the permanent wilting percentage. In this range of soil water contents, capillary movement of water through the soil is extremely slow or nonexistent, and an adequate supply of water cannot be maintained to rapidly absorbing root tips by this means. In such soils maintenance of contact between the root tips and available soil water is assured only by continued elongation of the roots through the soil. Mature root systems of many plants terminate in millions of root tips, each of which may be visualized as slowly growing through the soil, absorbing water from around or between the soil particles with which it comes in contact. The aggregate increase in the length of the root system of a rye plant averages 3.1 mi/day (5 km/day). Calculations indicate that the daily aggregate root elongation of this plant is adequate to permit absorption of a sufficient quantity of water from soils at the field capacity to compensate for daily transpirational water loss.

Mechanisms of water absorption. As previously indicated, tension, generated in the water columns of a plant, most commonly as an indirect result of transpiration, is transmitted to the ultimate terminations of the xylem ducts in the root tips. The water potential in the water columns is diminished by the amount of the tension generated, which is equivalent to a negative pressure. As soon as the water potential in the water columns becomes less than that in contiguous cells in the root tip, water moves from those cells into the xylem. This activates further cell-to-cell movement of water in a lateral direction across the root and presumably in the establishment of gradients of water potentials, diminishing progressively in magnitude from the epidermal cells, including the root hairs, to the root xylem. Whenever the water potential in the peripheral root cells is less than that of the soil water, movement of water from the soil into the root cells occurs. There is some evidence that, under conditions of marked internal water stress, the tension generated in the xylem ducts will be propagated across the root to the peripheral cells. If this occurs, water potentials of greater negativity could develop in peripheral root cells than would otherwise be possible. The absorption mechanism would operate in fundamentally the same way whether or not the water in the root cells passed into a state of tension. The process just described, often called passive absorption, accounts for most of the absorption of water by terrestrial plants.

The phenomenon of root pressure, previously described as the basis for xylem sap exudation from cuts or wounds and for guttation, represents another mechanism of the absorption of water. This mechanism is localized in the roots and is often called active absorption. Water absorption of this type only occurs when the rate of transpiration is low and the soil is relatively moist. Although the xylem sap is a relatively dilute solution, its osmotic pressure is usually great enough to engender a more negative water potential than usually exists in the soil water when the soil is relatively moist. A gradient of water potentials can thus be established, increasing in negativity across the epidermis, cortex, and other root tissues, along which the water can move laterally from the soil to the xylem. There is evidence, however, that a respiration mechanism, as well as an osmotic one, may be involved in the correlated phenomena of active absorption, root pressure, and guttation.

Environment and absorption. Any factor which influences the rate of transpiration also influences the rate of absorption of water by plants and vice versa. Climatic conditions may therefore indirectly affect rates of water absorption, and soil conditions indirectly affect transpiration. Low soil temperatures, even in a range considerably above freezing, retard the rate of absorption of water by many species. The rate of water absorption by sunflower plants, for example, decreases rapidly as the soil temperature drops below 55°F (13°C).

Within limits, the greater the supply of available soil water, the greater is the possible rate of water absorption. High soil water contents, especially those approaching saturation, result in decreased water absorption rates in many species because of the accompanying deficient soil aeration. In the atmosphere of such soils the oxygen concentration is lower and the carbon dioxide concentration is higher than in the atmosphere proper. In general, the deficiency of oxygen in such soils appears to be a more significant factor in causing retarded rates of water absorption than the excess of carbon dioxide. This retarding effect on water absorption rate is correlated with a retarding effect on root respiration rate. *See* PLANT RESPIRATION.

Likewise, if the soil solution attains any considerable concentration of solutes, water absorption by the roots is retarded. In most soils the concentration of the soil solution is so low that it is a negligible factor in affecting rates of water absorption. In saline or alkali soils, however, the concentration of the soil solution may become equivalent to many atmospheres, and only a few species of plants are able to survive when rooted in such soils.

Wilting. Daily variations in the water content of plants, more marked in some organs than in others, are of frequent occurrence. The familiar phenomenon of wilting, exhibited by the leaves and sometimes other organs of plants, particularly herbaceous species, is direct visual evidence of this fact. In hot, bright weather the leaves of many species of plants often wilt during the afternoon, only to regain their turgidity during the night hours, even if no additional water is provided by rainfall or irrigation. This type of wilting reaction is referred to as a temporary, or transient, wilting and clearly results from a rate of transpiration in excess of the rate of water absorption during the daylight hours. As a result, the total volume of water in the plant shrinks, although not equally in all organs or tissues. In general, diminution in water content is greatest in the leaf cells, and wilting is induced whenever the turgor pressure of the leaf cells is reduced sufficiently.

Even on days when visible wilting is not discernible, incipient wilting is of frequent occurrence. Incipient wilting corresponds to only a partial loss of turgor by the leaf cells and does not result in visible drooping, folding, or rolling of the leaves. Leaves entering into the condition of transient wilting always pass first through the stage of incipient wilting. Occurrence of this invisible first stage of wilting is almost universal on bright, warm days on which environmental conditions are not severe enough to induce the more advanced stage of transient wilting.

Confirmation of the inferred cause of transient wilting has been furnished by investigations of the comparative daily periodicities of transpiration and absorption of water (**Fig. 4**). As illustrated in this figure, there is a distinct lag in the rate of absorption of water as compared with the rate of transpiration. During the daylight hours the tissues of the plant are being progressively depleted of water, whereas the store of water within the plant is being steadily replenished during the night hours. The lag in the rate of absorption behind the rate of transpiration during the daylight hours appears to result largely from the relatively high resistance of the living root cells to the passage of water across them.

Both incipient and transient wilting should be distinguished from the more drastic stage of permanent wilting. This stage of wilting is attained only when there is an actual deficiency of water in the soil, and a plant will not recover from permanent wilting unless the water content of the soil in which it is rooted increases. In a soil which is gradually drying out, transient wilting slowly grades over into permanent wilting. Each successive night recovery of the plant from temporary wilting takes longer and is less complete, until finally even the slightest recovery fails to take place during the night.

Although the stomates are generally closed during permanent wilting, cuticular transpiration continues. Plants in a state of permanent wilting continue to absorb water, but at a slow rate. Restoration of turgidity is not possible, however, because the rate of transpiration even from a wilted plant exceeds the rate of absorption of water from a soil at the permanent wilting percentage or lower water content. During permanent wilting, therefore, there is a slow but steady diminution in the total volume of water within the plant, and a gradual intensification of the stress in the hydrodynamic system. Tensions in the water columns of permanently wilted plants are relatively high and have been estimated to attain values of 200 atm (20 MPa) in some trees, although this is probably an extreme figure.

As previously mentioned, the permanent wilting percentage, an important index of soil water conditions, is defined as the soil water content when a plant just enters the condition of permanent wilting. The sunflower is the most commonly used test plant in making determinations of the permanent wilting percentage of a soil. Permanent wilting of the basal pair of leaves, judged to have occurred when they fail to recover if placed in a saturated atmosphere overnight, is taken as the critical point in the measurement. The range of soil water contents between the first permanent wilting of the basal leaves of sunflower plants and the permanent wilting of all the leaves is called the wilting range. The water content of the soil at the time all the sunflower leaves have become wilted is termed the ultimate wilting point. In general, the wilting range is narrower in coarse-textured soils than in fine-textured soils, and may be 10–30% of the soil water content between the field capacity and the ultimate wilting

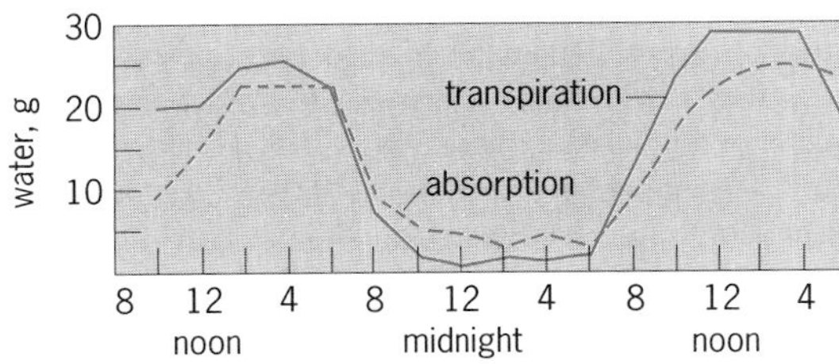

Fig. 4. Comparative daily periodicities of transpiration of water and absorption of water in the loblolly pine (*Pinus taeda*). 1 g = 0.035 oz. (*After P. J. Kramer, Plant and Soil Water Relationships, McGraw-Hill, 1937*)

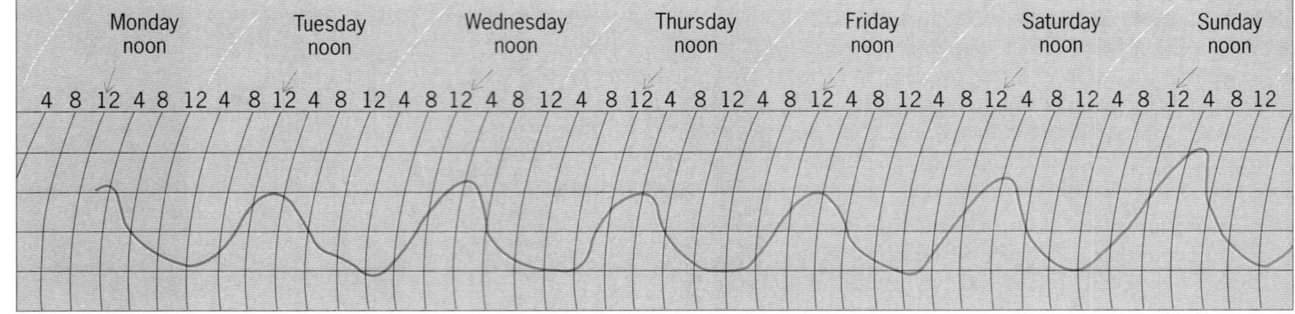

Fig. 5. Daily variations in diameter of lemons. (*After E. T. Bartholomew, Amer. J. Bot., 13:102–117, 1926*)

point. Although plants cannot grow while the soil in which they are rooted is in the wilting range, many kinds of plants can survive for considerable periods under such conditions. This is especially true of many shrubby species indigenous to semidesert areas.

Internal redistributions of water. For convenience, the processes of transpiration, translocation of water, and absorption of water are often discussed separately, although there is a close interrelationship among these three processes. The hydrodynamic system of a plant is essentially a unit in its operation, and changes in the status of the water in one part of a plant are bound to have effects on its status in other parts of the plant.

Whenever a plant is saturated, or nearly so, with water, differences in water potential from one organ or tissue to another are minimal. But whenever the rate of absorption of water lags behind the rate of transpiration, an internal water deficit develops in the hydrodynamic system of the plant, which in turn favors the establishment of marked differences in water potential from one part of the plant to another. Under such conditions redistribution of some of the water present from some tissues or organs of a plant to others generally occurs.

Internal movements of water from fruits to leaves and vice versa seem to be of common occurrence. Mature lemon fruits, while still attached to the tree, exhibit a daily cycle of expansion and contraction (**Fig. 5**). The lemon fruits begin to contract in volume early in the morning and continue to do so until late afternoon. Since transpirational loss from a lemon fruit is negligible, it is obvious that during this part of the day, corresponding to the period of high transpiration rates from the leaves, water is moving out of the fruits into other parts of the tree. Most of this movement probably occurs into leaves. During the daylight hours, the water potentials of the leaf cells presumably decrease until they soon are lower than those of the fruit cells, thus initiating movement of water from the fruits to the leaves. During the late afternoon and night hours, the volume of the fruits gradually increases, indicating that water is now moving back into the fruits. During this period, transpirational water loss from the leaves is small, leaf water contents increase, and the water potential of the leaf cells becomes less negative. Less of the absorbed water is translocated to the leaves than during the daylight hours, and more can move into the fruits, despite the relatively high water potential of the fruit cells. Marked daily variations take place in the diameters of lemon fruits, even under environmental conditions which result in no observable wilting of the leaves.

In growing cotton bolls, however, as long as enlargement is continuing, increase in diameter continues steadily both day and night and even during periods when the leaves are severely wilted. Movement of water is obviously occurring into the growing bolls without interruption during this period. Once the bolls cease enlarging, however, reversible daily changes occur in their diameter, similar in pattern to those which take place in mature lemon fruits (**Fig. 6**). Similarly, it has been shown that in a tomato plant the topmost node, within which growth in length occurs, continues to elongate at approximately the same rate both day and night. The stem below the first node, however, shrinks measurably in length during the daytime and elongates equally at night, undoubtedly as a result of reversible changes in the turgidity of the stem cells. The growing cells in the terminal node of the tomato stem continue to obtain water during the daylight hours while the rest of the stem is losing water, and some water utilized in their growth probably comes from the cells of the lower nodes.

In general, as the last two examples illustrate,

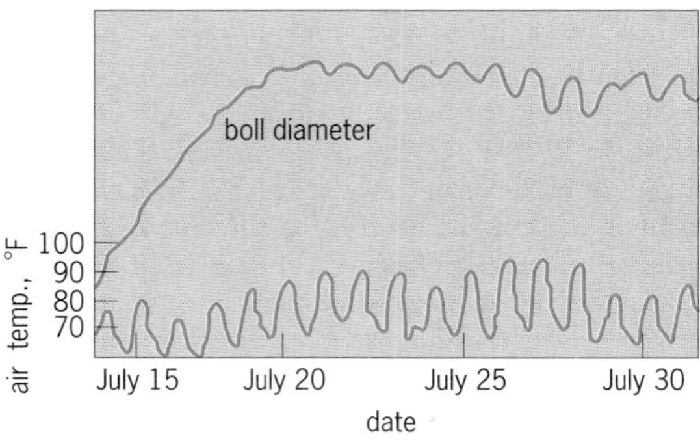

Fig. 6. Daily variations in diameter of a cotton boll (still growing for first 5 days). °C = (°F − 32)/1.8. (*After D. B. Anderson and T. Kerr, Plant Physiol., 18:261–269, 1943*)

actively meristematic regions such as growing stems and root tips and enlarging fruits, under conditions of internal water deficiency, apparently develop more negative water potentials than other tissues. Hence water often continues to move toward such regions even when an internal water deficiency of considerable magnitude has developed within the plant. However, under conditions of drastic internal water deficit, approaching or corresponding to a state of permanent wilting, growth of all meristems is greatly retarded or inhibited. *See* APICAL MERISTEM.

Drought resistance. The term drought refers, in general, to periods during which the soil contains little or no water which is available to plants. In relatively humid climates such periods are infrequent and seldom of long duration except in certain local habitats. The more arid a climate, in general, the more frequent is the occurrence of droughts and the longer their duration. Most species of plants can survive short dry periods without serious injury, but a prolonged period of soil water deficiency is highly injurious or lethal to all except those species of plants that have a well-developed capacity for drought resistance.

Most species which grow in semidesert regions, such as those of the southwestern United States and adjacent Mexico, or in locally dry habitats must be drought resistant in one sense of the word or another. Annual species which grow in many arid regions during short rainy spells are exceptions to this statement. Such species complete their brief life cycle from seed to seed during a period when soil water is available, and they can survive in arid regions because they evade rather than endure drought.

Succulents, such as cacti, are found in most semidesert regions, and are also often indigenous to locally dry habitats such as sand dunes and beaches in humid climate regions. Succulents are able to survive long dry periods because of the relatively large quantities of water which accumulate, in some species in fleshy stems and in other species in fleshy leaves, during the occasional periods when soil water is available. Many succulents can live for months on such stored water.

Those plants which are drought resistant in the truest sense are those whose cells can tolerate a marked reduction in water content for extended periods of time without injury. Many shrubby species of semidesert regions have this property. Certain structural features undoubtedly aid in the survival of such plants for long periods of arid habitats. Many xerophytes (plants that can endure periods of drought) have extensive root systems in proportion to their tops; such a structural characteristic aids in maintaining a supply of water to the aerial portions of the plant longer than would otherwise be possible. Other drought-resistant species are characterized by having diminutive leaves; the transpiring surface of the plant may thus be small relative to the absorptive capacity of the roots. In still other species the leaves abscise (fall) with the advent of the dry season, thus greatly reducing transpiration in the period of greatest internal stress in the hydrodynamic system.

Despite any structural features which may help maintain their internal water supply, shrubby plants of semiarid regions regularly undergo a gradual depletion in the store of water within them and a gradual intensification of the stress prevailing in the internal hydrodynamic system over dry periods often lasting for months. Only drought-resistant species can endure this condition, which is in essence a state of permanent wilting, for long periods of time without injury. A fundamental factor in the drought resistance of plants therefore appears to be a capacity of the cells to endure a substantial reduction in their water content without suffering injury. This capacity probably is based in part on structural features of the cells of such species and in part on the distinctive physiological properties of their protoplasm. *See* ECOLOGY; PLANT MINERAL NUTRITION.

Bernard S. Meyer

Bibliography. J. Kolek (ed.), *Physiology of the Plant Root System*, 1992; T. T. Kozlowski and A. J. Riker (eds.), *Flooding and Plant Growth*, 1984; T. T. Kozlowski and A. J. Riker (eds.), *Deficits and Plant Growth*, 1983; P. J. Kramer (ed.), *Water Relations of Plants*, 1983; E. Weiss, *Plants Tolerant to Arid and Semiarid Conditions*, 1988.

Plantaginales

An order of flowering plants, division Magnoliophyta (Angiospermae), in the subclass Asteridae of the class Magnoliopsida (dicotyledons). The order consists of only the family Plantaginaceae, with about 250 species. Within its subclass the order is marked by its small, chiefly wind-pollinated flowers that have a persistent regular scarious corolla. The perianth and stamens of the flowers are attached directly to the receptacle (hypogynous) and there are typically four petals that are joined at the base to form a tube. The plants are herbs or seldom half-shrubs with mostly basal, alternate leaves. The common plantain (*Plantago major*) is a familiar lawn weed of this order. *See* ASTERIDAE; MAGNOLIOPHYTA; MAGNOLIOPSIDA; PLANT KINGDOM.

Arthur Cronquist; T. M. Barkley

Plants, life forms of

A term for the vegetative (morphological) form of the plant body. A related term is growth form but a theoretical distinction is often made: life form is thought by some to represent a basic genetic adaptation to environment, whereas growth form carries with it no connotation of adaptation and is a more general term applicable to structural differences.

Life-form systems are based on differences in gross morphological features, and the categories bear no necessary relationship to reproductive structures, which form the basis for taxonomic classification. Features used in establishing life-form classes include deciduous versus evergreen leaves, broad versus needle leaves, size of leaves, degree of protection

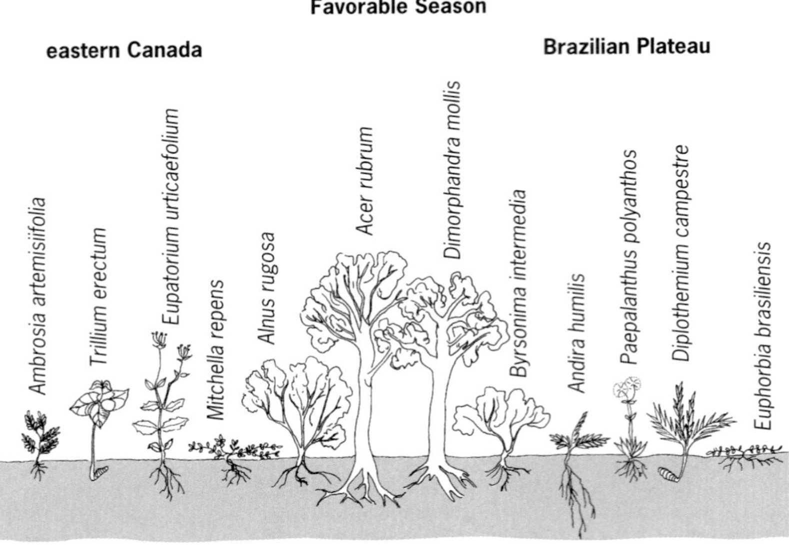

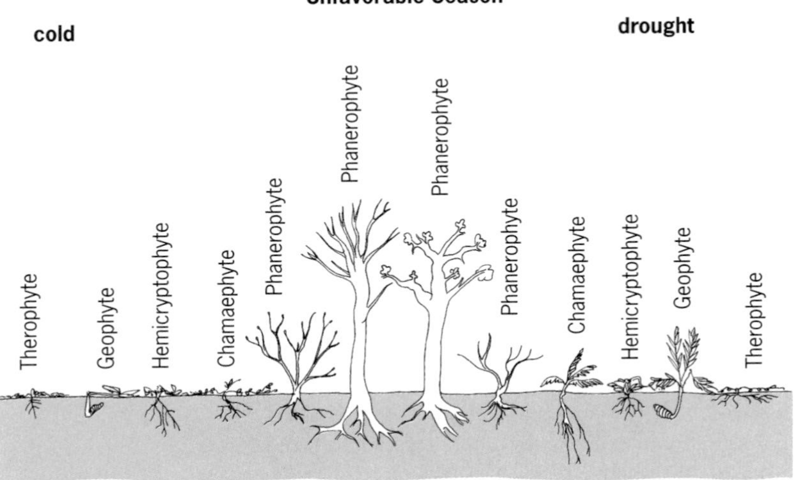

Life forms of plants according to C. Raunkiaer. (*After P. Dansereau, Biogeography: An Ecological Perspective, Ronald Press, 1957*)

afforded the perennating tissue, succulence, and duration of life cycle (annual, biennial, or perennial). Thus the garden bean (family Leguminosae) and tomato (Solanaceae) belong to the same life form because each is an annual, finishing its entire life cycle in 1 year, while black locust (Leguminosae) and black walnut (Juglandaceae) are perennial trees with compound, deciduous leaves.

Climate and adaptation factors. There is a clear correlation between life forms and climates. For example, broad-leaved evergreen trees clearly dominate in the hot humid tropics, whereas broad-leaved deciduous trees prevail in temperature climates with cold winters and warm summers, and succulent cacti dominate American deserts. Although cacti are virtually absent from African deserts, members of the family Euphorbiaceae have evolved similar succulent life forms. Such adaptations are genetic, having arisen by natural selection. However, since there are no life forms confined only to a specific climate and since it is virtually impossible to prove that a given morphological feature represents an adaptation with survival value, some investigators are content to use life forms only as descriptive tools to portray the form of vegetation in different climates.

Raunkiaer system. Many life-form systems have been developed. Early systems which incorporated many different morphological features were difficult to use because of this inherent complexity. The most successful and widely used system is that of C. Raunkiaer, proposed in 1905; it succeeded where others failed because it was homogeneous and used only a few obvious morphological features representing important adaptations.

Reasoning that it was the perennating buds (the tips of shoots which renew growth after a dormant season, either of cold or drought) which permit a plant to survive in a specific climate, Raunkiaer's classes were based on the degree of protection afforded the bud and the position of the bud relative to the soil surface (see **illus.**). They applied to autotrophic, vascular, self-supporting plants. Raunkiaer's classificatory system is:

- Phanerophytes: bud-bearing shoots in the air, predominantly woody trees and shrubs; subclasses based on height and on presence or absence of bud scales
- Chamaephytes: bud within 10 in. (25 cm) of the surface, mostly prostrate or creeping shrubs
- Hemicryptophytes: buds at the soil surface, protected by scales, snow, and litter
- Cryptophytes: buds underneath the soil surface or under water
- Therophytes: annuals, the seed representing the only perennating tissue

Subclasses were established in several categories, and Raunkiaer later incorporated leaf-size classes into the system.

By determining the life forms of a sample of 1000 species from the world's floras, Raunkiaer showed a correlation between the percentage of species in each life-form class present in an area and the climate of the area. The results (see **table**) were expressed as a normal spectrum, and floras of other areas were then compared to this. Raunkiaer concluded that there were four main phytoclimates: phanerophyte-dominated flora of the hot humid tropics, hemicryptophyte-dominated flora in moist to humid temperate areas, therophyte-dominated flora in arid areas, and a chamaephyte-dominated flora of high latitudes and altitudes.

Subsequent studies modified Raunkiaer's views. (1) Phanerophytes dominate, to the virtual exclusion of other life forms, in true tropical rainforest floras, whereas other life forms become proportionately more important in tropical climates with a dry season, as in parts of India. (2) Therophytes are most abundant in arid climates and are prominent in temperate areas with an extended dry season, such as regions with Mediterranean climate (for example, Crete). (3) Other temperate floras have a predominance of hemicryptophytes with the percentage of phanerophytes decreasing from summer-green

Examples of life-form spectra for floras of different climates

Climate and vegetation	Area	Life form*				
		Ph	Ch	H	Cr	Th
Normal spectrum	World	46	9	26	6	13
Tropical rainforest	Queensland, Australia	96	2	0	2	0
	Brazil	95	1	3	1	0
Subtropical rainforest (monsoon)	India	63	17	2	5	10
Hot desert	Central Sahara	9	13	15	5	56
Mediterranean	Crete	9	13	27	12	38
Steppe (grassland)	Colorado, United States	0	19	58	8	15
Cool temperate (deciduous forest)	Connecticut, United States	15	2	49	22	12
Arctic tundra	Spitsbergen, Norway	1	22	60	15	2

*Ph = phanerophyte; Ch = chamaephyte; H = hemicryptophyte; Cr = geophyte (cryptophyte); and Th = therophyte.

deciduous forest to grassland. (4) Arctic and alpine tundra are characterized by a flora which is often more than three-quarters chamaephytes and hemicryptophytes, the percentage of chamaephytes increasing with latitude and altitude.

Most life forms are present in every climate, suggesting that life form makes a limited contribution to adaptability. Determination of the life-form composition of a flora is not as meaningful as determination of the quantitative importance of a life form in vegetation within a climatic area. However, differences in evolutionary and land-use history may give rise to floras with quite different spectra, even though there is climatic similarity. Despite these problems, the Raunkiaer system remains widely used for vegetation description and for suggesting correlations between life forms, microclimate, and forest site index. *See* PLANT GEOGRAPHY.

Mapping systems. There has been interest in developing systems which describe important morphologic features of plants and which permit mapping and diagramming vegetation. Descriptive systems incorporate essential structural features of plants, such as stem architecture and height; deciduousness; leaf texture, shape, and size; and mechanisms for dispersal. These systems are important in mapping vegetation because structural features generally provide the best criteria for recognition of major vegetation units. *See* ALTITUDINAL VEGETATION ZONES; VEGETATION AND ECOSYSTEM MAPPING. Arthur W. Cooper

Bibliography. H. C. Bold, *Morphology of Plants and Fungi*, 5th ed., 1990; S. A. Cain, Life forms and phytoclimate, *Bot. Rev.*, 16(1):1–32, 1950; P. Dansereau, *A Universal System for Recording Vegetation*, 1958; R. Daubenmire, *Plant Communities*, 1968; C. Raunkiaer, *The Life Forms of Plants and Statistical Geography*, 1934.

Plants, saline environments of

All plants tolerate saline environments to some extent in that they require salts in order to grow. However, some plants have evolved which flourish where the salt concentrations (notably of sodium chloride) are 400 mol m^{-3} and beyond. These plants are known as halophytes. The majority of plant species are unable to tolerate one-quarter of this salt concentration and are called glycophytes.

Saline environments. Much of this discussion is directed toward the adaptations shown by terrestrial plants to salinity. However, it must be remembered that the greater proportion of the surface of the Earth is covered not by land but by water and that this water contains a considerable concentration of dissolved salts (**Table 1**). Although many different ions are present in seawater, the point to note here is the high concentration of sodium and chloride ions (about 500 mol m^{-3}). These ions, in fact, predominate in most saline environments, since not only can the sea affect low-lying coastal regions but it may have had a historical influence on land no longer close to an ocean. Other ions which are commonly found, particularly in saline ground waters, are potassium, calcium, and magnesium, together with sulfates, carbonates, and borates. Two important categories of soil resulting from the presence of these salts in combinations are sodic soils, dominated by sodium carbonate (pH may be greater than 10), and acid sulfate soils, dominated by sulfates (pH may be as low as 3). Saline soils are dominated by sodium chloride itself. All the salt-affected soils tend to occur in lower latitudes, where evapotranspiration rates are high.

Apart from deserts and oceans, the other most extensive naturally saline areas are coastal marshes which are periodically inundated by seawater; these

TABLE 1. Chief cations and anions in ocean seawater with a salinity of 355‰*

Cations and anions	Concentration, mol m^{-3}
Sodium	483
Magnesium	55
Calcium	10
Potassium	10
Strontium	0.1
Chlorine	558
Sulfate	29
Bromine	0.8
Borate	0.4

*Salinity is the weight of total salt per kilogram of water after drying at close to 500°C (932°F).

form mangrove swamps in tropical climates. *See* BOG; SALT MARSH; SWAMP, MARSH, AND BOG.

Anthropogenically formed saline soils have become more common in areas where irrigation is practiced. The added water often contains moderate quantities of ions, and these are left in the soil when the water evaporates, unless they can be leached through the soil profile by a controlled excess of water. However, even this can exacerbate the situation if it leads to an increase in the height of the water table and then to capillary rise of soil solution, eventually leaving the salts in the soil profile. This problem has led to losses of otherwise good agricultural land since, once saline, the soil becomes relatively impermeable to water as the structure breaks down, and this condition is difficult to treat. *See* SALINE EVAPORITES.

It is difficult to determine what proportion of the world's surface is actually saline. The oceans cover about 71% of the Earth, but as far as the land is concerned, the estimated area depends upon the criteria used to define a saline soil. The salt concentration at which any soil is said to be saline is rather arbitrary, since it is subject to wide fluctuation depending upon rainfall and evaporation. However, a generally accepted view would be that if the soil solution contains 0.5% dissolved salts, the soil would be classed as saline. So far only a small proportion of the world has been mapped in detail, and estimates of the extent of salt-affected land vary widely from 1.3×10^6 to 3.7×10^6 mi^2 (3.4×10^6 to 9.5×10^6 km^2). An area of 3.0×10^6 mi^2 (7.7×10^6 km^2) would be equivalent to that of the contiguous United States. Of particular concern are statistics indicating that about 13% of cultivated lands and 30–50% of irrigated lands are salt-affected.

Plants. There are many plant species adapted to live in saline environments. All those that normally grow in seawater are clearly halophytes. However, it is sometimes more difficult to classify terrestrial plants based on observations of their natural habitat, since the concentration of salts varies so much with the weather. It is possible to be more precise if the plants are grown in solution culture under controlled conditions of temperature, humidity, and light intensity. Under these circumstances, those which complete their life cycle in the presence of about 1% sodium chloride (say, 170 mol m^{-3}) are undoubtedly halophytes. Plants that will not survive in a concentration of 100 mol m^{-3} sodium chloride would be classed as glycophytes (**Fig. 1**); in between these concentrations is a gray area occupied by what might be called salt-tolerant glycophytes.

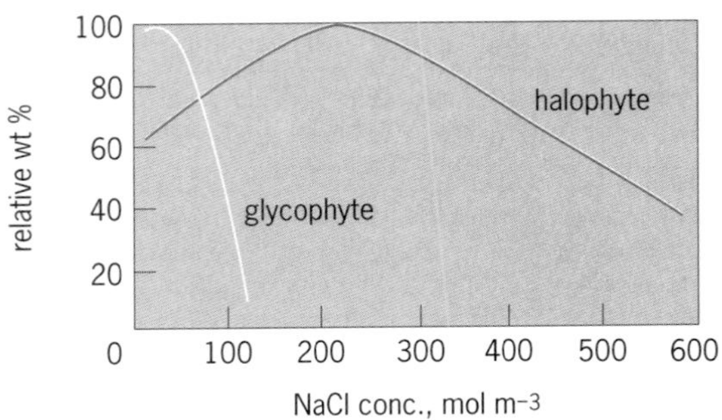

Fig. 1. Relationship between growth and sodium chloride concentration for halophytes and glycophytes. The diagram assumes that the plants are growing in a culture solution which accounts for about 10–20 mol m^{-3} of ions.

Marine flora. The flora of the oceans is dominated by algae, with few flowering plants adapted to this environment. The angiosperms and benthic algae are restricted to the ocean fringes, which make up about 2% of the total area. The remainder of the deep water is home to the phytoplankton, mainly diatoms, dinoflagellates, and silicoflagellates. The algae of the coastal regions are predominantly brown (Phaeophyta) or red (Rhodophyta), although the green algae (Chlorophyta) are well represented (**Table 2**). The Cyanophyta and the Euglenophyta are largely confined to fresh water, although salt-tolerant genera are known. The distribution of the various algal species depends on the temperature of the water and the degree of exposure to air which the algal thallus can withstand. This gives rise to the well-known zonation of algae seen on the seashore. *See* ALGAE.

Apart from the algae, the marine flora is restricted to a few angiosperms; bryophytes, pteridophytes, and gymmosperms are completely absent. As far as the angiosperms are concerned, there are no marine dicotyledonous plants and only one order of monocotyledons, the Helobiae, with some 12 genera and about 49 species (Table 2). Most of these sea grasses, as they are known, grow in warmer waters, and since they cannot withstand long exposure to air, they occur at or about the level of the lowest tides. As far as is known, they appear to have evolved from terrestrial salt-tolerant plants.

TABLE 2. Some examples of common halophytes from marine and terrestrial habitats

	Marine Halophytes		
	Algae		
Chlorophyta	Phaeophyta		Rhodophyta
Acetabularia	*Alaria*		*Bostrychia*
Caulerpa	*Ascophyllum*		*Ceramium*
Chaetomorpha	*Chorda*		*Chordaris*
Codium	*Chordaria*		*Corallina*
Enteromorpha	*Fucus*		*Corollopsis*
Halimeda	*Halidrys*		*Porphyra*
Ulva	*Laminaria*		*Rhodymenia*
	Pelvetia		
	Sea grasses		
Cymodocea	Halodale	Ruppia	Zostera
	Terrestrial Halophytes		
Chenopodiaceae	Gramineae		Mangrove
Atriplex	*Aeluropus*		*Avicennia*
Arthrocnenum	*Phragmites*		*Aegiceras*
Beta	*Puccinellia*		*Aegialtis*
Halimione	*Spartina*		*Rhizophora*
Halogeton			*Sonneratia*
Salicornia			
Suaeda			

Terrestrial flora. The distinction between terrestrial and aquatic plants is largely a matter of the degree to which the plant is predominantly in air or in water. For example, salt marsh plants are terrestrial but must withstand submergence, while algae of the upper shore are mostly submerged but must withstand periodic exposure to the atmosphere.

The terrestrial halophytes are nearly all flowering plants, and there are few, if any, salt-tolerant mosses, liverworts, or even gymnosperms. Among the ferns, there are just two families with salt-tolerant genera, the Pteridaceae and the Ophioglossaceae. Among the flowering plants, salt tolerance occurs in many different families and appears to have been polyphyletic in origin. Salt-tolerant species are to be found in about one-third of the families of flowering plants. However, about half of the approximately 500 genera know to contain halophytic species belong to just 20 families. Among the monocotyledonous plants, grasses dominate as halophytes, with 109 species, followed by salt-tolerant sedges, with 83 species. Within the dicotyledonous plants, the Chenopodiaceae is the preeminent family for salt tolerance with 44 of its 100 genera containing salt-tolerant species (there are 312 such species); examples are *Atriplex*, *Salicornia*, and *Suaeda* (Table 2). Other dicotyledonous families with significant numbers of halophytes are the Asteraceae (53 species) and the Aizoaceae (48 species).

It is a notable feature of terrestrial saline environments, however, that many of the genera have worldwide distribution patterns, even though individual species and particular dominants vary with the mean temperature. For example, arctic salt marshes commonly have large populations of *Puccinellia phryganoides,* whereas *P. maritima* is often dominant in temperate marshes. *See* MAGNOLIOPHYTA.

Tropical salt marshes, the so-called mangrove swamps, differ from those of temperate zones in being dominated by trees and shrubs rather than by herbs. There are at least 12 genera from eight different families (Table 2), and besides their salt tolerance, these trees are remarkable for their root systems, which are adapted to provide stability in soft mud and facilitate gas exchange in what is often an anaerobic substrate (**Fig. 2**). Apart from coastal areas, saline environments are also frequent inland, especially in the hotter and drier regions of the world. In such environments, plants must be adapted not only to tolerate the high concentrations of salts in the soil but also to be drought-resistant. Many of the adaptations which have evolved to combat the former stress are advantageous in surviving the latter (and vice versa). *See* MANGROVE.

Crops. Crops, like other plants, differ in the degree to which they will tolerate salt in the soil. Few, if any, reach the level of tolerance seen in halophytes. The growth of most crops is thus reduced significantly by the presence of salt in the soil, leading to serious reductions in yields. How much salt different crops will tolerate is frequently categorized by the effect of salinity on the relative yield, that is, the yield obtained in a saline soil compared with a control, nonsaline soil. Salinity is commonly quantified by the electrical conductivity of a saturated soil extract. One procedure is to determine the conductivity that reduces the yield to half of that in the nonsaline soil. Among the most tolerant crops are barley (*Hordeum vulgare*), cotton (*Gossypium hirsutum*), and sugarbeet (*Beta vulgaris*); at the other end of the scale are onion (*Allium cepa*), rice (*Oryza sativa*), and beans (*Phaseolus vulgaris*). A more elaborate approach to quantifying tolerance is to determine the maximum salinity (again characterized by the electrical conductivity of a saturated extract of the soil) that has no discernible effect on the yield (the threshold) and the rate of loss of yield at higher salinities.

Within crop species, there is also information about the relative tolerance of individual varieties. For example, although rice is generally rather sensitive to salt, the older variety Pokkali is much more salt-tolerant than modern varieties released by the International Rice Research Institute. Similar information is available for many crop species, but rather little is known about the reasons for the differences in tolerance. The relative tolerance of varieties is often known to farmers or has been discovered by screening programs aimed at evaluating differences. Differential tolerance has rarely been developed by plant breeders, in spite of considerable effort in the case of some crops. It seems likely that breeders have not been succesful because they have treated salt

Fig. 2. The mangrove *Avicennia marina* in a river estuary at Dangar Island, Australia. Aerial roots are growing out of the mud.

tolerance as a single genetic character such as plant height or flower color. Where it exists, salinity tolerance requires a complex of physiological characteristics. For rice, these characteristics include vigorous growth producing a large plant, the ability to exclude sodium (and chloride) ions from the shoots, the ability to maintain a lower salt concentration in younger as opposed to older leaves, and the ability to tolerate accumulated ions. In wheat, the balance between the uptake of sodium (one of the ions dominating saline soils and not required by the plant) and potassium (a chemically similar ion that is an essential plant nutrient) is such a physiological attribute. The regulation of this characteristic appears to be under the control of a gene or genes on the long arm of a particular chromosome (chromosome 4D), which is provided to the wheat hybrid by the species *Aegilops squarrosa*. However, much less is known about the genetics of most of the characteristics involved in tolerance, which accounts for the lack of success of planned breeding for tolerance within crops.

Adaptations. Many terrestrial halophytes, having small leaves which are often succulent, look somewhat different from glycophytes (**Fig. 3**). In many species the leaves bear glands which secrete salt. These visible differences are the external manifestations of more fundamental differences in physiology.

Water potential. Perhaps the basic problem faced by all halophytes is to retain water within the plant in the face of large external osmotic forces. The concept of water potential has proved to be useful in understanding water movement. The water potential is a function of the free energy of water, and it determines the direction in which water moves; just as for electricity, water flows from high to low potential. Water potential may be expressed in either energy or pressure units, but pressure units are usually used for plants. The only complicating feature of the concept is that the potential of pure free water—which is the high end of the scale—is set at zero so that water moves from less negative to more negative potentials. Where plants are growing in fresh water, the water potential around their roots is close to zero, as would be the potential of a soil at field capacity. Most plants grow at about zero to −15 bars (−1.5 megapascals), and at the lower potential would be close to, or at, the stage of permanent wilting—that is, they would be unable to take up water from such a low water potential, and water might even leave the plant and move into the soil. The water potential of normal seawater is about −23 bars (−2.3 MPa), so that halophytes must be adapted to take up and retain water from much lower potentials than glycophytes can normally cope with. Analysis shows that they achieve this by accumulating salts (**Fig. 4**) and thereby lowering the potential of the water in their cells, so that a positive gradient is maintained between plant and environment. However, this answer itself poses a further problem, that of living with a high salt content. *See* PLANT-WATER RELATIONS.

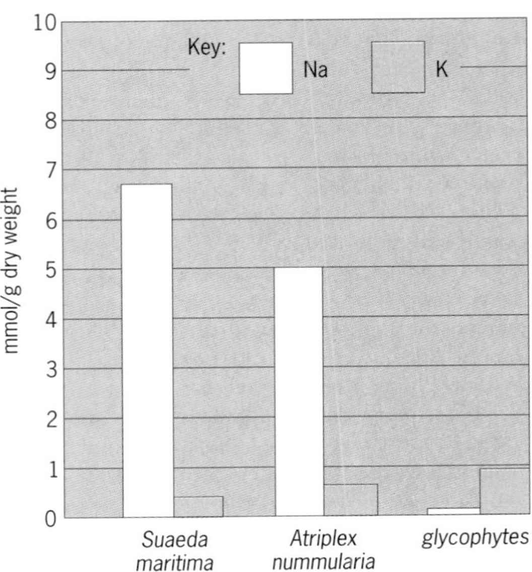

Fig. 4. Sodium and potassium contents in the leaves of two halophyte species growing in solution culture containing about 300 mol m^{-3} Na^+ and 6 mol m^{-3} K^+. Mean data are given for glycophytes growing in the presence of 1 mol m^{-3} Na^+ and 5 mol m^{-3} K^+.

Ion content and distribution. The ion content of a halophyte growing in a saline environment may be some five times greater than that of a glycophyte in its natural surroundings. This may mean that the concentration of sodium ions is between 500 and 1000 mol m^{-3} when expressed on the basis of the plant water content. The ability to live with such a high ion content is not unique to plants; a group of bacteria live in extremely saline environments and may have ion contents which approach 5 kmol m^{-3}. These organisms, of which *Halobacterium* and *Sarcina*

Fig. 3. *Suaeda maritima*, the annual sea blite, a halophytic member of the Chenopodiaceae growing at a coastal site in northwest Spain.

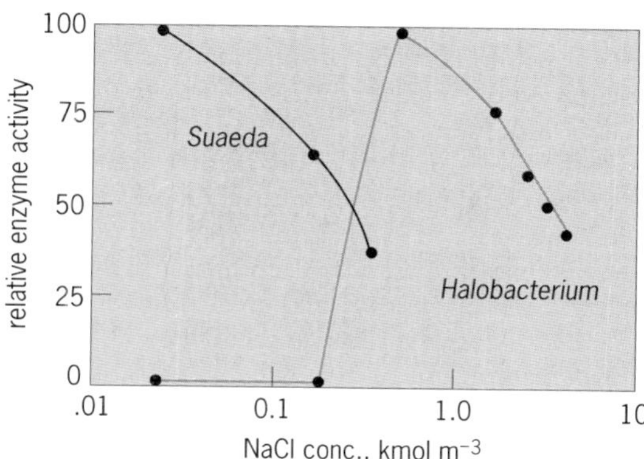

Fig. 5. Effect of sodium chloride on activity of the enzyme malate dehydrogenase from a halophilic bacterium (*Halobacterium*) and a halophyte (*Suaeda*), measured in the laboratory.

have been particularly well investigated, are collectively known as halophilic bacteria. They possess unique proteins that actually require high concentrations of salts to carry out their enzyme activities. It is possible, then, for enzymes to function at high ionic strengths. However, when the activity of enzymes which had been extracted from halophytes was tested for salt tolerance, it was found that they did not require high salt concentrations, and indeed, under these conditions, the activity of many was drastically reduced (**Fig. 5**). Halophytes thus differ from the halophilic bacteria in that their metabolism does not seem to function at high salt concentrations. *See* HALOPHILISM (MICROBIOLOGY).

Plant cells are, of course, more complex than those of bacteria, and in particular they contain a large central vacuole. Consequently it is possible for a plant to sequester ions within its vacuole and to maintain the cytoplasm at a relatively lower ion content. However, it is very difficult to gain direct evidence of the intracellular localization of ions, so that formerly researchers had to rely on indirect evidence such as the sensitivity of enzymes to high salt concentrations from which to infer the subcellular distribution of ions. However, advances in x-ray microanalysis have allowed the measurement of the subcellular distribution of ions within halophyte cells, and these confirm the earlier conclusions that the bulk of the ions which bring about osmotic adjustment are localized within the vacuole, leaving the cytoplasm at an ion concentration which is not very different from that of a glycophyte cell (**Table 3**). This leads then to the question of how the water potential of the cytoplasm is, in turn, adjusted.

Compatible solutes. Analysis of the leaves of halophytes has shown that they contain rather high concentrations of a number of organic compounds. Of these, perhaps the most common are the amino acid proline and the quaternary ammonium compound glycine betaine. Other important organic solutes found in salt-tolerant higher plants are polyols such as mannitol and sorbitol and the cyclitol, pinitol. Individual species of halophytes contain different compounds or combinations of compounds, and it has been postulated that these organic solutes are responsible for the osmotic adjustment of the cytoplasm. Two lines of evidence have given rise to this hypothesis. First, the quantity of solute present increases enough with increasing external salt concentration to bring about the osmotic adjustment of the relatively small cytoplasmic compartment. Second, an important feature of these compounds is that they do not inhibit enzyme activity even at high (1.0 kmol m^{-3}) concentrations: because of this they have become known as compatible solutes. Although limited, experimental evidence supports the view that concentrations of glycine betaine in the cells of halophytes are higher in the cytoplasm than in the vacuoles.

Terrestrial halophyte. The picture which emerges, then, of a terrestrial halophyte is a plant which must have a low water potential in order to maintain the water content of its cells. It achieves this by accumulating large quantities of ions, particularly sodium and often chloride, in its vacuoles and adjusting the cytoplasmic water potential by means of a suitable compatible solute. Remarkably, however, evidence suggests that in spite of very high osmotic pressures in the cell sap, turgor pressure is generally less than 4 bars (0.4 MPa) in *Suaeda maritima*.

Although ions are freely available in the media in which halophytes grow, compatible solutes must be synthesized and consequently may be some drain on the energy resources of the plant. Restriction of such solutes to the cytoplasm rather than requiring their presence in the whole cell volume clearly lessens the demand, although the energy costs of salt tolerance are still poorly understood.

Ion accumulation must be carefully regulated so that all halophytes restrict, in some way, the entry of ions into the transpiration stream and hence their accumulation in the shoot. Xylem ion concentrations are frequently less then 10% of the external concentration. This accumulation must be carefully matched with the growth rate if the overall ion concentration in the leaf cells is to be maintained approximately constant.

Some halophytes have a secondary means of controlling the ion content of their shoots in that they can remove excess salts through the operation of specialized glands (**Fig. 6**) which are normally situated on the leaves. These glands, which generally

TABLE 3. Ion concentrations in the mesophyll cells of a mature leaf of the halophyte *Suaeda maritima*

Cell compartment	% of cell volume	Concentration, mol m^{-3} Na	K	Cl
Cytoplasm	11	136	21	71
Chloroplasts	1.5	208	44	196
Vacuole	82	494	20	352
Cell wall	5.5	554	40	394

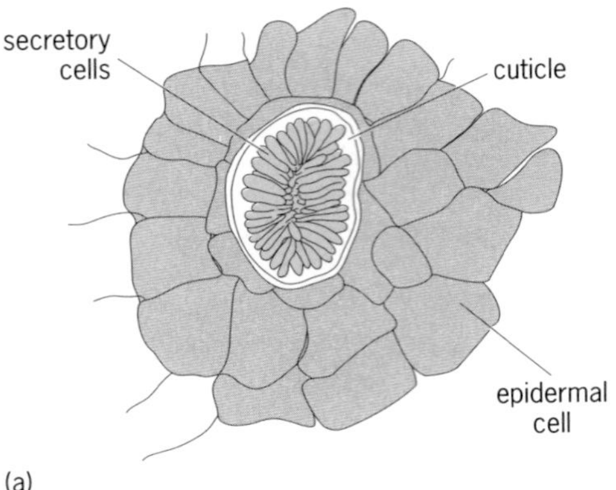

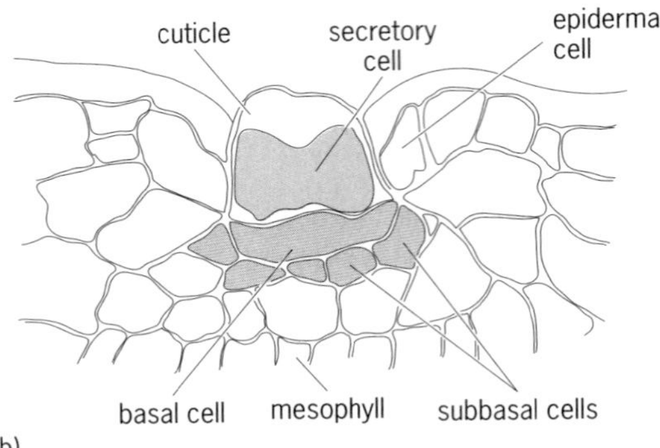

Fig. 6. Salt gland from the mangrove swamp species *Aegiceras corniculatum*. (*a*) Secretory cells surrounded by cuticle and epidermal cells (*after W. P. Anderson, ed., Ion Transport in Plants, Academic Press, 1973*). (*b*) Section through the gland.

have between 2 and about 40 cells, depending upon plant species, contain highly active cells with dense cytoplasmic contents and can excrete ions from the symplast to the exterior of the plant. *See* SECRETORY STRUCTURES (PLANT).

Marine halophyte. In general, marine halophytes appear to possess many of the adaptations seen in their terrestrial counterparts. Of course, they do not have to face the problems of transpiration, but many must be adapted to low light intensities and to the often violent movement of the sea. Many live in estuaries or other regions where the salinity of the water can vary considerably over relatively short periods of time. Such species utilize compatible solutes, often sugar alcohols, to adjust their water potential, and they can both synthesize and break down the compound they use in relation to the external salinity.

Uses of halophytes. Halophytes are obviously most valuable in areas of high salinity, where other plants will not grow. They are used to provide cover on otherwise infertile soils during land reclamation, to provide grazing, and to ornament saline gardens. The seed of the chenopod, *Salicornia europaea*, might prove a valuable source of edible oil. Halophytes are also a most important research material, since it is only by investigating the ways in which they are adapted to their environment that humans can hope to understand the phenomenon of salt tolerance and develop more resistant crop species, such as appears to be possible with barley and tomato, to alleviate the problems of secondary salinization. *See* PLANT PHYSIOLOGY.

T. J. Flowers

Bibliography. P. Adam, *Saltmarsh Ecology*, 1990; J. H. Cherry (ed.), *Environmental Stress in Plants: Biochemical and Physiological Mechanisms*, 1989; T. J. Flowers, M. A. Hajibagheri, and N. J. W. Clipson, Halophytes, *Quart. Rev. Biol.*, 61:313–336, 1986; H. G. Jones, T. J. Flowers, and M. B. Jones (eds.), *Plants under Stress*, 1989; E. V. Maas, Salt tolerance of plants, *Appl. Agr. Res.*, 1:12–26, 1986; D. Pasternak, Salt tolerance and crop production: A comprehensive approach, *Annu. Rev. Phytopath.*, 25:271–291, 1987; A. R. Yeo et al., Screening of rice (*Oryza sativa L.*) genotypes for physiological characters contributing to salinity resistance, and their relationship to overall performance, *Theor. Appl. Genet.*, 79:337–384, 1990.